THE HISTORY OF CARTOGRAPHY

VOLUME TWO, BOOK TWO

THE HISTORY OF CARTOGRAPHY

1
Cartography in Prehistoric, Ancient, and Medieval Europe and the Mediterranean

2.1
Cartography in the Traditional Islamic and South Asian Societies

2.2
Cartography in the Traditional East and Southeast Asian Societies

2.3
Cartography in the Traditional African, American, Arctic, Australian, and Pacific Societies

3
Cartography in the European Renaissance

4
Cartography in the European Enlightenment

5
Cartography in the Nineteenth Century

6
Cartography in the Twentieth Century

THE HISTORY OF CARTOGRAPHY

VOLUME TWO, BOOK TWO

Cartography in the Traditional East and Southeast Asian Societies

Edited by
J. B. HARLEY
and
DAVID WOODWARD

Associate Editor
JOSEPH E. SCHWARTZBERG
Assistant Editor
CORDELL D. K. YEE

THE UNIVERSITY OF CHICAGO PRESS • CHICAGO & LONDON

J. B. Harley was professor of geography at the University of Wisconsin—Milwaukee.
David Woodward is professor of geography at the University of Wisconsin—Madison.

The University of Chicago Press, Chicago 60637
The University of Chicago Press, Ltd., London

Printed in the United States of America

03 02 01 00 99 98 97 96 95 94 5 4 3 2 1

ISBN 0-226-31637-8 (v. 2, bk. 2)

♾ The paper used in this publication meets the minimum requirements of the American National Standard for Information Sciences—Permanence of Paper for Printed Library Materials, ANSI Z39.48-1984

Editorial work on *The History of Cartography* is supported in part by grants from the Division of Research Programs of the National Endowment for the Humanities and the Geography and Regional Science Program of the National Science Foundation, independent federal agencies. For a complete list of foundations, organizations, and individuals who supported the editorial work, see pages v and vi.

The costs of publishing this book have been defrayed in part by three awards:

A publication grant from the National Endowment for the Humanities, an independent federal agency.

The 1992 Hiromi Arisawa Memorial Award from the Books on Japan Fund with respect to *Peasant Uprising in Japan* published by the University of Chicago Press. The award is financed by the Japan Foundation from generous donations contributed by Japanese individuals and companies.

A publication grant from the Chiang Ching-kuo Foundation for International Scholarly Exchange.

Library of Congress Cataloging-in-Publication data will be found on the last page of this book

Financial Support

Federal Agencies

Division of Research Programs
of the National Endowment for the Humanities

Geography and Regional Science Program
of the National Science Foundation

Foundations and Institutions

Gaylord and Dorothy Donnelley Foundation
The Henry Luce Foundation, Inc.
The Japan Foundation
The Johnson Foundation
The Andrew W. Mellon Foundation
The National Geographic Society
Rand McNally
The Luther I. Replogle Foundation
The Hermon Dunlap Smith Center for the History of Cartography, The Newberry Library

Organizations

California Map Society
The Chicago Map Society
Geography and Map Division of the Special Libraries Association
Map and Geography Round Table, American Library Association
Map Society of British Columbia
The Mercator Society, New York Public Library
Michigan Map Society
The New York Map Society
North East Map Organization
The Rocky Mountain Map Society
Washington Map Society

Founders

Roger S. and Julie Baskes
Arthur Holzheimer
Mr. and Mrs. Kenneth Nebenzahl
Mr. and Mrs. Roderick Webster

Benefactors

Arthur L. Kelly
Bernard Lisker
Glen McLaughlin
Chuck and Lia Palmer
George Parker
David M. Rumsey

Patrons

Julie A. Anderson
Richard B. Arkway
Frederick and Howard Baron
Stephen D. and Nancy Brink
Clive A. Burden
Rand Burnette
JoAnn and Richard Casten
Mr. and Mrs. James R. Donnelley
Ralph Ehrenberg
Johan W. Eliot
Clifton F. Ferguson
Richard and Dorothy Fitch
Gerald F. Fitzgerald
Joseph H. Fitzgerald
William B. Ginsberg
John M. Gubbins
Warren Heckrotte
Francis H. Heller
Francis Herbert
Robert A. Highbarger
John S. Josey
Jay I. Kislak
Edward D. Kleinbard
Dee Longenbaugh
Scott A. Loomer
Douglas W. Marshall
Martayan Lan, Inc.
George F. McCleary, Jr.
Barbara Backus McCorkle
Donald L. McGuirk, Jr.
Braham Norwick
Harold L. Osher
William Sherman Reese
Jack L. Ringer
Arthur H. Robinson
Joseph E. Schwartzberg
Richard H. Sigel
Stephen Stares
Thomas and Ahngsana Suarez
Antiques of the Orient Pte. Ltd, Michael J. Sweet
Iain C. Taylor and North by West

Additional support from

Daniel M. Amato
W. Graham Arader III
Patricia and Stanley K. Arnett, II
James Axtell
Stanley Balzekas
Gwendolyn R. Barckley
Thomas R. Beall
Yasha Beresiner
Aníbal A. Biglieri
Stephen A. Bromberg
Michael Burack
Charles A. Burroughs
Fred A. Cazel, Jr.
Barbara Mae Christy
Sherry Coatney and David Niemi
Tim Coss
Cray Research Foundation
Gerald Danzer
Richard Dittman
Michael J. Dubin
Elizabeth F. Dunlap
Oliver C. Dunn
Clinton R. Edwards
C. Eide
Herbert F. Ellis
Edward B. Espenshade, Jr.
Lucy A. Fellowes
Bruce Fetter
Norman Fiering
Robert L. Fisher
John Fondersmith
Theodore N. Foss
John Frye
Edward Garcia
Robert Graebner
Mark A. Green
Eugene M. Grossman
Peter J. Guthorn
Susanne A. Haffner
J. Scott Hamilton
John B. Henderson
Bangbo Hu
Alice C. Hudson
Murray Hudson
IBM
Kit S. Kapp
Elton R. Kerr
Anne and Lawrence Knowles
Josef W. Konvitz
Steven Kosakowski
G. Malcolm Lewis
Janice and Chingliang Liang
Catharine McClellan
Michael McGuire
Allen H. Meyer
Jack and Carmen Miller
Mr. and Mrs. Paul J. Mode, Jr.
John T. Monckton
Mark Monmonier
Gene Moser
Curtis J. Musselman
Mr. and Mrs. Jerome J. Nerenberg
Alfred W. Newman
Judy Olson
Theodore W. Palmer
Richard P. Palmieri
Douglas T. Peck
Mary Pedley
Edward F. Penico
Carla Rahn and W. D. Phillips, Jr.
Miklos Pinther
Francesco Prontera
Jean M. Ray
Dennis Reinhartz
Charles D. Reynolds
John R. Ribeiro
Walter W. Ristow
Steve Ritchie
Pierre L. Sales
Stephen E. Schalk
Don Schnabel
Cherie Semans
David Charles Sheldon
Robert B. Shilkret
John D. Shugrue
Lawrence Slaughter
Thomas R. Smith
John P. Snyder
Margaret Sowers
Bruce N. Spring
Mr. and Mrs. Martin Steinmann
Richard J. A. Talbert
G. Thomas Tanselle
Norman and Elizabeth Thrower
Richard M. Ugland
Richard Umansky
Carol Urness
Diane D. Vasica
Leonard Vis
Rainer Vollmar
Stephen J. Walsh
Daniel Gilbert Watters
Ann H. Wells
James A. Welu
Louis Werner
Scott D. Westrem
Joan Winearls
Eric W. Wolf
John Wolter
Alberta and Clifford Wood
Jennifer Woodward
Jeanne and Stephen Young
Rick Ray Zellmer and Erica Schmidt
His Royal Highness The Duke of Cornwall

Contents

Illustrations
with Tables and Appendixes

Color Plates

Figures

TABLES

APPENDIXES

Preface

Before introducing this third book in the *History of Cartography*, I must acknowledge the crucial role of Brian Harley in its origin and development. His perception, knowledge, wisdom, and dynamism were so vital to the History of Cartography Project, and these qualities so influential, that the remaining volumes will become his lasting memorial.[1]

The conception of the History of Cartography Project can be pinpointed to a specific time and place. In 1975 Brian proposed writing a four-volume "Mapping of North America."[2] I had come to know him well since his Kenneth Nebenzahl, Jr., Lectures on Mapping the American Revolutionary War, and late in May 1977 I visited him in Devon, where he taught at the University of Exeter. We were walking on a country footpath leading up to Highweek Church, near Brian's home in Newton Abbot, when his American project again came up in conversation. Instead of his four-volume history of North American cartography, I suggested, could we not jointly edit a four-volume worldwide "History of Cartography"? By the next month we were exchanging outlines and general ideas, and that summer we wrote a preliminary letter describing the project to Allen Fitchen, then humanities editor at the University of Chicago Press. The idea was enthusiastically received.

We were interrupted by other activities, and immediate plans for the multivolume *History* were put on hold. It was not until 1980 that the first grant proposal to the National Endowment for the Humanities was written; the project officially began in August 1981. Our original intent was to include the cartography of all non-Western societies in volume 1. That volume was to include not only discussions of the maps of the prehistoric, ancient, and medieval West and the traditional cartographies of Asia, but also those of the indigenous societies of Africa, the Americas, the Arctic, Australia, and the Pacific islands. The utter impracticality of this view was soon apparent as the richness and diversity of the non-Western mapping traditions emerged. By 1982 it became obvious that we should devote a separate volume (volume 2) to the traditional cartographies of Islamic and Asian societies. Even then we assumed that a single "Asian" volume would suffice.[3] As the amount of material coming within our definition of cartography continued to expand, however, it became necessary to divide volume 2 into two books. We were faced with covering a wider spectrum of cultural and historical artifacts, a task that demanded enormously expanded editorial efforts and stretched the patience of our authors, staff, and publisher.

Our definition of "map" proved to be at the same time liberating and overwhelming. In the Preface to volume 1 we had suggested that "maps are graphic representations that facilitate a spatial understanding of things, concepts, conditions, processes, or events in the human world."[4] This definition explicitly focused on both the form and the function of maps, for we believed these twin aspects were intimately linked. Realizing from the outset that these volumes could not embrace the whole history of geographic perception, thought, and writing, we strongly believed that this approach to maps as material culture should be stressed for the light it shed on those broader issues of human spatial and environmental understanding and communication. We did not see "representations" as replications or mirrors, but rather viewed them as

1. See also the detailed obituaries, including David Woodward, "John Brian Harley, 1932–1991," *Special Libraries Association, Geography and Map Division Bulletin* 167 (1992): 50–52; idem, "Brian Harley, 1932–1991," *Map Collector* 58 (1992): 40; idem, "J. B. Harley: A Tribute," *Imago Mundi* 44 (1992): 120–25; William Ravenhill, "John Brian Harley," *Transactions of the Institute of British Geographers*, n.s., 17 (1992): 120–25; Matthew H. Edney, "John Brian Harley (1932–1991): Questioning Maps, Questioning Cartography, Questioning Cartographers," *Cartography and Geographic Information Systems* 19 (1992): 175–78; Peter J. Taylor, "Politics in Maps, Maps in Politics: A Tribute to Brian Harley," *Political Geography* 11 (1992): 127–29; "John Brian Harley, 1932–1991," *Cartographica* 28, no. 4 (1991): 92–93.

2. Correspondence, J. B. Harley to David Woodward, 4 February 1975.

3. On the appropriateness of this emphasis we were led astray by the standard general histories of cartography. For example, it is revealing to compare the length of the present volume with the number of pages Leo Bagrow devoted to the same areas: six and a half pages to Chinese cartography, half a page each to Central Asia and Korea and three to Japan, and a third of a page to Southeast Asia. See Leo Bagrow, *History of Cartography*, rev. and enl. R. A. Skelton, trans. D. L. Paisey (Cambridge: Harvard University Press; London: C. A. Watts, 1964; reprinted and enlarged, Chicago: Precedent, 1985), 197–208.

4. J. B. Harley and David Woodward, eds., *The History of Cartography* (Chicago: University of Chicago Press, 1987–), 1:xvi.

highly conventionalized human constructs. By using the term "graphic," it was certainly not our intent to limit our attention to the traditional two-dimensional mapping media or to scaled and "plan-view" representations, realizing that, for example, the recording of a ritual or a dance could be intended to convey spatial information. Our use of the phrase "graphic representations" was thus intended to embrace any visual representation whose spatial structure conformed to that of the referent. The second part of the definition—to "facilitate a spatial understanding"—focused on the function of these representations in human societies.

Based on such reasoning, we thus sought to include "maps" that had previously been ignored or marginalized because they did not fit the Western model of scaled orthogonal representations of the physical world. For *Cartography in the Traditional Islamic and South Asian Societies*, the most striking effect of this decision was the inclusion of representations of cosmographical worlds no less "real" to those societies than were or are the physical worlds of Western culture. It is our preoccupation with our view of reality that inhibits Westerners' understanding of "Eastern" cartographies on their own terms. More fundamental still, this focus implies that the Western paradigm of mapping—now in universal use in topographic mapping—is appropriate to all other modes of mapping. What our treatment of the history of Asian cartography may reveal, however, is that the very success of this Western cartographic paradigm may have impoverished and dehumanized modern cartography. As Cordell Yee persuasively argues in this volume and elsewhere, pictorial representations on traditional Chinese maps had a more central role:

> The mapmaker saw art—poetry, calligraphy, and painting—as essential to the task. To such a practitioner, a map is a fusion of image and text, of the denotative and the expressive, of the useful and the beautiful. In the twentieth century, modern mathematical cartography displaced traditional techniques and put an end to this idea of maps. Whether this was progress remains an open question.[5]

As a consequence of broadening the horizons of the *History* beyond maps of the physical world to include those of the metaphysical world, it became necessary to seek out and enlist authors not previously involved in the traditional history of cartography or even geography, thus establishing an enormously enriching dialogue with scholars in other disciplines such as history, anthropology, philosophy, art history, and literary criticism. Our conceptual, editorial, managerial, and diplomatic tasks were also concomitantly increased.

In this book we have continued to try to create a framework that is coherent both geographically and historically in order to bring out the cultural interplay. As with volume 2.1, the basis of the framework for volume 2.2 is geographical. East Asia (defined by Nathan Sivin and Gari Ledyard in their introduction as China, Japan, Korea, and Vietnam) is treated separately from Southeast Asia. Vietnam, although normally thought of as falling within Southeast Asia, is here grouped with East Asia, with which it was closely associated in cartographic tradition. Another ad hoc category designed for this book is Greater Tibet, which fits in neither East nor Southeast Asia. Such shoehorning of complex and diverse cartographies into apparently arbitrary categories has not always resulted in a comfortable fit. But pragmatic considerations had to prevail as we sought to divide the increasingly voluminous material into books of manageable size.

As with the previous two books in the series, we have been frustrated by the paucity of surviving maps from the early period (pre-1500). Our interpretation of Chinese cartography, for example, has to rest on a handful of artifacts gleaned largely from archaeological investigations. Knowledge of Tang cartography is based almost exclusively on literary allusions. Even our acquaintance with much of the more recent Song cartography is based on Ming and Qing copies. To confuse matters further, Chinese historians often date maps by their content, not by when the artifact was made: thus a map may be described as "Song dynasty" when its content dates from Song, even though the object may have been made in Qing. Moreover, it is often difficult to determine the faithfulness of later copies and to disentangle the influence of the later culture on the copy. On principle, wherever possible, we have sought to select illustrations closest to the date of the original, striving always to illustrate the artifact in such a way as to retain the clues to its age and origin.

For later periods of Chinese history, we were frustrated by the sheer quantity of material. Since the literature tends to focus on individual maps or particular exhibitions, and since systematic surveys of collections worldwide are generally lacking, it is hard to get a clear idea of what is available. Thus we have tended to rely on well-known institutions and have obviously overlooked many artifacts.

In romanizing East Asian alphabets, we have tried to follow commonly accepted practices. For Chinese, we chose pinyin over Wade-Giles because the former is becoming more widely used, even in scholarly books on pre-1949 China. It is also the romanization system generally recommended by *The Chicago Manual of Style*. For Japanese, we have used Kenkyusha's *New Japanese-English Dictionary*, following *The Chicago Manual of*

5. Cordell D. K. Yee, "A Cartography of Introspection: Chinese Maps as Other Than European," *Asian Art* 5, no. 4 (1992): 29–47, esp. 46.

Style. For Korean, the standard McCune-Reischauer system has been used. In the footnotes and Bibliographical Index, for works written in Chinese, Japanese, Korean, and Vietnamese the authors are listed with surnames first. In the footnotes, works written by East Asian authors in Western languages follow each author's own common usage (the surname may be first or last).

Although the Wylie system of orthography is widely used among Tibetologists, it provides little guidance to nonspecialists in respect to pronunciation. We have followed the most common spellings employed in the texts we have cited and have sometimes provided the Wylie spellings in parentheses. Our spellings, however, provide at best a rough approximation of the proper Tibetan pronunciation.

For the numerous vernacular languages of Southeast Asia, systems of orthography vary widely. Again we have generally followed the usage of the authors cited. For related names and terms that may be rendered in either Sanskrit or Pali, our preference is for the former. Where Pali forms were more appropriate, we have noted their use. For the forms of personal names in Tibet, Southeast Asia, and Mongolia, we have followed the most common spellings employed in the texts cited. Place-names in the volume in general follow *Webster's New Geographical Dictionary* (with the notable exception that we have rendered Chinese place-names in pinyin).

Ambiguities often arise in the transliteration of Chinese, Japanese, and Korean when using any romanization system, and to clarify these as far as possible we have provided the corresponding characters of these languages in the Bibliographical Index (for names of authors, editors, titles) and in the General Index (for items such as map titles, terminology, and people). For reasons of economy, characters are seldom given in the text or footnotes. In footnotes we include the original form of pagination or other referencing for ancient texts; modern editions of these works are also cited wherever possible. For all languages, titles of works (and their translations) and personal names are usually given in full only on the first use; a shortened form is used thereafter.

The historical framework is rather more complex. The cultures described in this book use various chronological systems, and sometimes we have had to maintain these to preserve each culture's sense of historical time. But since these volumes are primarily designed for the Western reader, and since it is generally helpful to relate the chronologies of the East to the more familiar ones of Europe, particularly in the periods of greatest interaction between them, most dates are given as B.C./A.D.

In thanking the many people who have contributed chapters to this book, I speak also on behalf of Brian Harley and the entire staff of the History of Cartography Project. We owe our specialist authors an enormous debt for their patience over the past decade (often in spite of extremely difficult personal circumstances) as the scope and focus of the book changed dramatically. It is indeed a privilege to name them here: Catherine Delano Smith, John B. Henderson, Guntram Herb, Gari Ledyard, Kazuhiko Miyajima, Joseph E. Schwartzberg, Nathan Sivin, F. Richard Stephenson, Kazutaka Unno, John K. Whitmore, and Cordell D. K. Yee. They define dedication in scholarship. Only they know the extent of expansion, rewriting, and recasting that resulted from editorial efforts and the demands of the Press's four readers. I hope they now share our pride in the result. Our authors have also generously helped us in our search for additional authors, and they have given many critical readings of several chapters. Of the authors who have been with us from the beginning, Tony Campbell, Catherine Delano Smith, G. Malcolm Lewis, and Joseph E. Schwartzberg have continued to offer sage advice and to be steadfast friends of the project.

Cordell Yee joined the project as assistant editor in 1988. His contribution was pivotal: not only did he write the chapters on Chinese terrestrial cartography in a refreshingly revisionist mode, but he offered support and counsel at every stage of the book's compilation. Without his input, the Chinese section of the book would have been severely impoverished. Another staff member whose contributions are more hidden but are nevertheless crucial is Kevin Kaufman, research specialist, who wrote entire sections, tables, and appendixes. A special debt is owed to the two full-time staff members of the project who have been with us through a bewildering series of personal and professional vicissitudes: Jude Leimer and Susan MacKerer. Jude Leimer has been managing editor since 1982 and has provided the editorial and managerial continuity so crucial to a project of this type. She has controlled the daily operations of this work—in liaison with the University of Chicago Press and with authors, advisers, and editors—with such determination and strength of personality that I can only describe her role as indispensable. Likewise, Susan MacKerer has become essential as administrator of finances and records, ensuring our compliance with complex regulations of both the university and our funding agencies. She has also fulfilled a crucial function in identifying possible sources of funding for the project and keeping in contact with our growing list of donors. Particularly important in shaping this book, however, was her experience in the Department of East Asian Languages and Literature. She knew Cordell Yee and actively recruited him for the Project; she also suggested that we apply to the Henry Luce Foundation, Inc., which provided major funding for East and Southeast Asian research.

Brian Harley's Milwaukee office of the Project and the

Office for Map History of the American Geographical Society Collection closed the year following his death. Nevertheless, the support of Ellen Hanlon and Mark Warhus needs thankful acknowledgment. They provided much logistical support in the preparation of this book during the years since 1986, when Brian was at the University of Wisconsin—Milwaukee. They and Brian's family—Karen, Claire, and Sarah—were a very close presence during the difficult weeks around Christmas 1991.

At the early stages of the book's planning, several people helped us define our approach and lay the groundwork. In particular, we would like to acknowledge the contribution of Mei-ling Hsu, who introduced us to many Chinese colleagues and laid out for us the extent of the research material for the early period. Cao Wanru, Shannon McCune (whose death saddened us), Niu Zhongxun, Kazutaka Unno, and Yi Chan (Chan Lee) all made valuable contributions to initial planning. Simon Potter acted as an invaluable liaison for our Japanese authors, editing large sections, arranging translations, tracking down illustrations, and providing general on-site support for what were to become extremely complex chapters. For checking bibliographical references and quotations at various stages of the book, we are indebted to Chiu-chang Chou, Pin-chia Feng, Jooyoun Hahn, Ingrid Hsieh-Yee, Bangbo Hu, and Chu-ming Luk, as well as many translators: Yuki Ishiguro, Kiyo Sakamoto, Daniel Samos, Atsushi Taira, Agatha Tang, Qingling Wang, and Kenneth White. Tsai-fa Cheng helped with Chinese philological matters relating to the six cartographic principles of Pei Xiu.

Our illustrations editors were first Guntram Herb and then Christina Dando. Both have doggedly pursued libraries and archives in the remotest corners of the world through mail, courier (both commercial and personal), fax, and telephone. Owing to their efforts, we have been able to select the highest-quality illustrations in often difficult circumstances. For special help with finding or obtaining illustrations, in addition to our authors, we thank Cao Wanru, the Harvard Yenching Library, and Simon Potter. Line drawings and reference maps were skillfully prepared by the University of Wisconsin Cartographic Laboratory in the Department of Geography at Madison led by Associate Director Onno Brouwer and his manager Daniel H. Maher.

In addition to its role in helping to define the scope and methods of the history of cartography, the *History* is intended as a basic work of reference. This entails continued and close attention to bibliographical accuracy. Paula Rebert and her successor Barbara Whalen have exactingly tracked down and checked arcane references and identified quotations in sometimes obscure publications in many languages. They have been helped by the excellent library facilities on our campus and the efficient Inter-Library Loan department at Memorial Library headed by Judith Tuohy. Other essential library and office help has been provided by Karen Beidel and Charles Dean.

Home institutions play a major, if sometimes unacknowledged, role in the stability of long-term projects such as ours. It is indeed a pleasure to acknowledge the Departments of Geography at the University of Wisconsin—Madison and Milwaukee and their respective graduate schools for their support of the Project.

None of our work would be possible without the financial support of the many funding agencies, foundations, and individuals listed on the financial support page (pp. v and vi). We continue to be especially grateful to the National Endowment for the Humanities and the National Science Foundation for their faith in, as well as their generosity to, the *History*. From the private sector, we acknowledge the initial support of the Andrew W. Mellon Foundation, the National Geographic Society, the Gaylord and Dorothy Donnelley Foundation, the Hermon Dunlap Smith Center for the History of Cartography at the Newberry Library, and the Luther I. Replogle Foundation.

Special mention must be made of several foundations whose support was particularly associated with the research on East and Southeast Asia. The Johnson Foundation sponsored a conference at the Wingspread Conference Center, Racine, Wisconsin, in November 1982 that laid the groundwork for the decision to create a separate volume for Asian cartography. The Henry Luce Foundation, Inc., provided two generous grants to the project, and the Japan Foundation helped fund our chapters on the traditional cartography of Japan.

As the project grows, we are depending more and more on identifying private sources of funding. Richard Arkway, Art Kelly, Jack Monckton, and Kenneth Nebenzahl have been particularly helpful in advising us in this regard. Several antiquarian map dealers printed a call for funding in their catalogs, which expanded our funding base considerably: Richard B. Arkway, James E. Hess (The Antiquarian Map & Book Room), John T. Monckton, Jonathan Potter, Thomas and Ahngsana Suarez, Michael Sweet (Antiques of the Orient Pte., Ltd.), Martin Torodash (Overlee Farm Books), and the firm of Martayan Lan and Augustyn.

As in volume 2, book 1, we are delighted to have the opportunity to thank several people at the University of Chicago Press. Penelope Kaiserlian, associate director, has continued to be one of our staunchest supporters and has appeared miraculously and almost ubiquitously at historical and geographical conventions to promote the book. Alice Bennett, our copyeditor par excellence since volume 1, has improved the consistency and accuracy of the text. Robert Williams, designer, has proved again that his initial design for the books in the series meets the challenge of innumerable plates, illustrations, tables, and appendixes.

To the foregoing corporate acknowledgments a personal note must be added. I am only too painfully aware that the editorial "we" must now change to "I." I cannot begin to say what Brian might have said, but I know he would have wanted to acknowledge the help of very many friends. They know who they are and what their role in this book has been. For my part also, personal debts are growing too rapidly to specify. Some, however, are exceptional. I thank my parents Max and Kathleen in England for all their love and support; they have continued to follow our progress with great anticipation as each volume has appeared. My wife Ros and son Justin in Madison and my daughter Jenny in New York have now learned to live with and enjoy the challenges provided by "the Project."

To these people and everyone else—named or unnamed—who had a hand in this book, I extend my warmest thanks.

David Woodward

1 · Prehistoric Cartography in Asia

CATHERINE DELANO SMITH

INTRODUCTION

It has been established that in prehistory humans had both the mental capacity and the communicative and graphic skills to make maps.[1] We can also take it for granted that whatever maps have survived are most likely to be found in the rock and mobiliary art of the period. For Asia as for the western parts of the Old World, prehistoric maps are to be sought in paintings, engravings, and pecked or "bruised"[2] figures executed on the walls of caves and rock shelters, on cliffs, rock outcrops, and large boulders. Some examples may also be found on pottery and in the decoration of metalware and slate, bone, or wood artifacts.

A search of the archaeological literature for reports and illustrations of Asian prehistoric art reveals a wealth of depictions of space, or objects or events in space, in rock art and mobiliary art. Given the enormous task of checking the archaeological literature of the entire continent, not to mention the attendant problems of the accessibility and language of that literature, my aims here are modest: to draw attention to a few examples of Asian rock art in which prehistoric cartographic thinking is manifested and to offer a framework for future studies. I hope also to broaden the vista on the conceptual history of a "mapping impulse" by showing that it is not only in Europe, the Middle East, and North Africa that the origins of mapmaking can be traced far back into prehistory.[3]

THE MAPPING IMPULSE IN PREHISTORIC ART

Pictures and maps are conceptually linked, being two forms on a continuum of graphic communication.[4] Inasmuch as pictures and maps are not usually seen as interchangeable, the one merely a substitute for the other, some sort of distinction has to be made and a line drawn between a picture of a place and a map of a place. That line, however, has meaning only in the context of the thinking and intentions of the original artist. It is the message that gives the graphic image its form. The obvious difficulty in the case of prehistoric art is in recovering the original function and decoding a message that had a meaning in prehistoric times but is now wholly unfamiliar. The result tends to be a distinction between picture and map that some may find unacceptably arbitrary. The line has to be drawn, however, and my aim here is to suggest a more rational basis for such classifications than has been characteristic of either the antiquarian literature on prehistoric maps or, indeed, traditional attitudes to the origins of mapmaking in the history of cartography. An underlying premise throughout this chapter is that the focus should be narrowed to those images that lend themselves to cartographic analysis. Others, made up of no more than a line or two, may have been just as much part of prehistoric spatial communication; given their incomplete survival or fragmentary form, however, no rational arguments can be applied to them and they have to be set aside.

One problem is to distinguish between pictures and picture maps in prehistoric art. Picture maps, as I have already suggested, have a mixture of pictorial perspectives, some elements being shown in profile and some in plan.[5] The former are usually anthropomorphic or animal figures, sometimes landscape features such as buildings. The plan element is usually an enclosure represented by a single or double line. Indeed, such an enclosure is often the crucial iconic indicator of a map in prehistoric art, since it portrays space—an area in which features (e.g., houses) may be distributed or events (e.g., trapping of

1. Catherine Delano Smith, "Prehistoric Maps and the History of Cartography: An Introduction," idem, "Cartography in the Prehistoric Period in the Old World: Europe, the Middle East, and North Africa," and G. Malcolm Lewis, "The Origins of Cartography," all in *The History of Cartography*, ed. J. B. Harley and David Woodward (Chicago: University of Chicago Press, 1987–), 1:45–49, 54–101, and 50–53, respectively.

2. Reported mainly from India. The color of the rock surface is changed by hammering (bruising), but there is no perceptible roughness: Douglas Hamilton Gordon, *The Pre-historic Background of Indian Culture* (Bombay: N. M. Tripathi, 1958), 114.

3. Delano Smith, "Cartography in the Prehistoric Period" (note 1). The topic is briefly introduced for South Asia in volume 2, book 1 of this *History*.

4. W. J. Thomas Mitchell, *Iconology: Image, Text, Ideology* (Chicago: University of Chicago Press, 1986), 9–10.

5. Delano Smith, "Cartography in the Prehistoric Period," 62 (note 1).

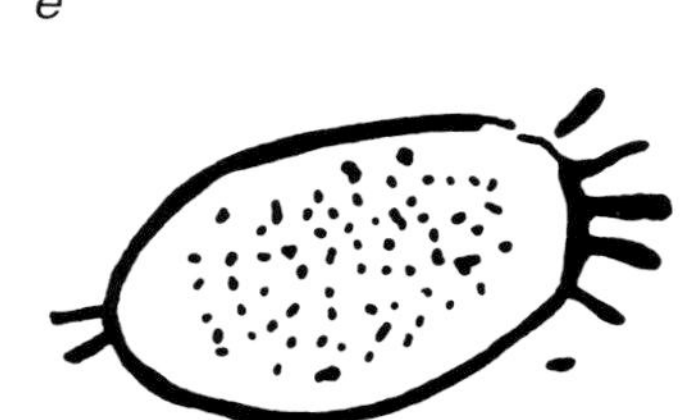

FIG. 1.1. EARLY REPRESENTATION OF ENCLOSURES IN PLAN. The idea of depicting an enclosure in plan was widespread in Asia in prehistoric times. These examples are drawn from the archaeological literature.

(*a*) A rock painting from Mirzapur described as people dancing within an enclosure.
After Rai Sahib Manoranjan Ghosh, *Rock-Paintings and Other Antiquities of Prehistoric and Later Times*, Memoirs of the Archaeological Survey of India, no. 24 (Calcutta: Government of India, Central Publication Branch, 1932; reprinted Patna: I. B. Corporation, 1982), 18 and pl. XXIa (fig. 2).
(*b*) In this rock painting, dating from the Mesolithic (ca. 8000–2500 B.C.), found at Bhimbetka and said to represent a child burial and mourning family, the open circle represents the outline of the tomb.
After Vishnu S. Wakankar, "Bhimbetka—The Prehistoric Paradise," *Prachya Pratibha* 3, no. 2 (July 1975): 7–29, fig. 7, no. 9.
(*c* and *d*) From historical Syunik, Armenia.
After Grigor Hovhannesi Karakhanyan and Pavel Geworgi Safyan, *Syownik'i zhayrhapatkernerě* (Yerevan, 1970), figs. 176.2 and 237.2.
(*e*) Among the figures of horsemen and animals engraved on rock 90 in Script Valley (Bichigtin-Am), Bayan Khongor Province, Mongolia, is a subcircular feature very similar to some features in Franco-Cantabrian Paleolithic and Mesolithic cave art that were interpreted as "game enclosures."
After N. Ser-Odzhav, *Bayanligiyn Khadny Zurag*, ed. D. Dorj (Ulan Bator, 1987), 118, fig. 89.
(*f*) Petroglyph from Yin Shan, China, captioned as showing a "dance, presumably inside a hut."
After Chen Zhao Fu, *Cina: L'arte rupestre preistorica*, Italian trans. Giuliana Aldi Pompili (Milan: Jaca Books, 1988), 178 and fig. 25.

animals) may take place. Thus, in a picture map the focus of interest may be the event itself rather than the place where it is happening. Nevertheless, such representations embody the essential concepts of a map: distributions are shown, and the idea of plan representation is manifested. One must bear in mind that spatial distributions are represented in prehistoric art according to what we recognize as topology, the key principle of which is contiguity: what is next to something is shown next to it, while considerations of direction, distance, and shape are of little or no importance.[6] A prehistoric map may have been perfectly "accurate" on these terms though it appears "distorted" to the modern eye. This is particularly true of prehistoric plan maps.

Asian prehistoric art includes a number of rock paintings or engravings that reflect these cartographic considerations. For example, in one of the paintings in the rock shelters of Mirzapur (Uttar Pradesh, India) a rectangle with a double outline including a scalloped outer border encloses a group of stick figures (fig. 1.1*a*). The painting is said to represent "four persons drawn in a pale red colour . . . dancing within an enclosure."[7] For those who would argue that the enclosing lines might be nothing more than an abstract frame provided by the artist, let me stress that frames or borders in any form are almost never found in European rock art. There is no evidence that the situation is different in Asia. Moreover, such

6. Delano Smith, "Cartography in the Prehistoric Period," 67–68 (note 1).

7. Rai Sahib Manoranjan Ghosh, *Rock-Paintings and Other Antiquities of Prehistoric and Later Times*, Memoirs of the Archaeological Survey of India, no. 24 (Calcutta: Government of India, Central Publication Branch, 1932; reprinted Patna: I. B. Corporation, 1982), 18 and pl. XXIa (fig. 2).

pictures tend to make sense only if interpreted as part of an assemblage, despite the lack of evidence that the contiguity we see today was intended by the original artist. Another example is found among the Mesolithic paintings in a Bhimbetka rock shelter (Vindhya Range, Madhya Pradesh, India). This painting has been described as representing a "child burial and mourning family" within a hut or some sort of enclosed space (fig. 1.1*b*).[8] Yet another example comes from paintings in the village of Kollur (Tamil Nadu) in southern India. Said to date from the end of the first millennium B.C., the design is described as showing animals and "a few lines [which] suggest a place of defence, possibly something like a primitive fort."[9] Similar depictions have been reported from Armenia. One subcircular enclosure, from historical Syunik (near modern Sisian), contains two stick figures (fig. 1.1*c*).[10] Another irregular shape, drawn with double lines, can be seen as an empty enclosure remarkably similar to that described on the Rajum Hani' stone (from Jordan) as an animal pen (fig. 1.1*d*).[11] From Mongolia comes a report of a rock-carved outline similar to those of Paleolithic Europe interpreted by Henri Breuil as hut or game enclosures (fig. 1.1*e*),[12] from China we have an outline with four figures inside what may be a hut (fig. 1.1*f*), and from India come two examples of figures in a hut with a tiger-skin rug on the floor (fig. 1.2*a*, *b*). In all these examples, huts and buildings are represented in plan, as if seen from above. In view of this, they can be considered part of the story of the early history of mapmaking. There are also, however, images in prehistoric rock art in which all the components are in profile, as in figure 1.2*c*. These we classify as pictures rather than as picture maps or antecedent picture maps.

Two outstanding assemblages in Asian prehistoric rock art merit close attention, particularly for the way they illustrate the problems of identifying a prehistoric map. Both show what purport to be groups of buildings depicted in profile, but whereas one example readily fits our criteria for a cartographic representation of place, by the same token the other remains in our view essentially a picture of a place, rather than a map of that place.

The first are the Boyar petroglyphs (Minusinsk, on the Yenisei River). There are two main groups, known respectively as the Major and Minor Boyar petroglyphs.[13] It is the former that has received the most attention in the literature. They are said to date from the first millennium B.C. In style, the markings belong to the art of the stock-raising Khakass tribesmen who inhabited the surrounding steppes during the Bronze Age. They were first discovered by archaeologists in 1904. The pecked figures decorate outcrops of red Devonian sandstone at the top of a steep hill in such a way as to form a continuous frieze. Visible from afar, they present a view that has been described as "a majestic sight" giving the place "an air of unique originality."[14]

The frieze is crowded with figures of animals (deer and cattle), humans engaged in various activities, and domestic utensils (pails or bowls like those still used for koumiss) (fig. 1.3). About sixteen individual buildings are discernible, all shown in profile. There are two types—what seem to be log houses with sloping roofs, and others that reflect the conical timber-framed tents of the local yurts. The whole scene is thought to portray a village on the occasion of some ceremony or feast. Apart from the huts, no landscape features are depicted. Nothing is shown in plan. Moreover, there is nothing to confirm that the assemblage was created, or intended, as a single composition. On the contrary, although we are invited to regard it not as a "mere collection of drawings but [as] a single whole composed of well-developed individual compositions," the suggested link is simply the more or less orderly arrangement of the buildings. Moreover, it is admitted that the creation of the various figures and groups occurred over a period, albeit rather short.[15] It is thus difficult to see the Major Boyar petroglyph as a picture map rather than a picture, however attractive, of a

8. Vishnu S. Wakankar, "Bhimbetka—The Prehistoric Paradise," *Prachya Pratibha* 3, no. 2 (July 1975): 7–29, esp. fig. 7, no. 9; reprinted (but without some illustration and appendix material) in *Indische Felsbilder von der Arbeitsgemeinschaft der Ge-Fe-Bi* (Graz: Gesellschaft für Vergleichende Felsbildforschung, 1978), 72–93, esp. 93 and fig. 7, no. 9. Some semicircular lines have been interpreted as indicating a snare or trap, as in the case of one Mesolithic painting from Putli Karar in Madhya Pradesh, India, described as a scene in which a deer is heading "towards snare traps placed on poles": Erwin Neumayer, *Prehistoric Indian Rock Paintings* (Delhi: Oxford University Press, 1983), fig. 26d.

9. N. S. Ramaswami, "Prehistoric Rock Paintings Discovered in Tamil Nadu," *Indian News*, 6 February 1984, 7. I owe this reference to Joseph E. Schwartzberg.

10. Grigor Hovhannesi Karakhanyan and Pavel Geworgi Safyan, *Syownik'i zhayrhapatkernerě* (Rock carvings of Syunik), in Armenian with Russian and English summaries (Yerevan, 1970), fig. 176.2. For other examples of Armenian rock art, see A. A. Martirosyan and A. R. Israelyan, *Naskal'nye izobrazheniya Gegamskikh gor* (The rock-carved pictures of the Gegamskiy Khrebet) (Yerevan, 1971), text in Armenian and Russian, with English summary, 54–66.

11. Karakhanyan and Safyan, *Syownik'i zhayrhapatkernerě*, fig. 237.2 (note 10). For the Rajum Hani' stone, see Delano Smith, "Cartography in the Prehistoric Period," 61 and fig. 4.3 (note 1).

12. N. Ser-Odzhav, *Bayanligiyn Khadny Zurag* (Rock drawings of Bayan-Lig), ed. D. Dorj (Ulan Bator, 1987), 118 (fig. 89). I am grateful to Paul Dimond of the Far Eastern Department of the Foreign and Commonwealth Office, London, and to the British Embassy, Ulan Bator, for their kind help in obtaining this work. For examples from Breuil and others in a European context, see Delano Smith, "Cartography in the Prehistoric Period," esp. 68–69 and fig. 4.10 (note 1).

13. M. A. Devlet, *Bol'shaya Boyarskaya pisanitsa / Rock Engravings in the Middle Yenisei Basin* (Moscow: Nauka, 1976), in Russian and English.

14. Devlet, *Rock Engravings*, 14 (note 13).

15. Devlet, *Rock Engravings*, 15 (note 13).

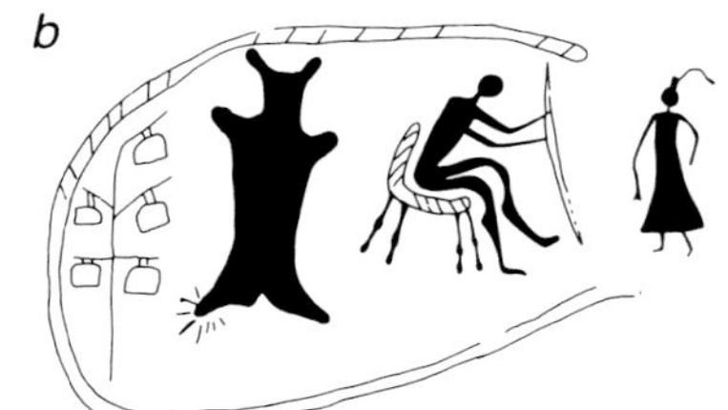

FIG. 1.2. DEPICTION OF ENCLOSURES AND BOUNDARY LINES IN INDIAN ROCK ART. Huts and compounds may be portrayed viewed from above (in plan, as in examples *a* and *b*), or from ground level (in profile or perspective, as in example *c*). The encircling line in the painting shown in *a* has been described as indicating a shelter or boundary and the scene in *c* as taking place in "a tent-like structure.... The double line in the foreground could be a stone arrangement to hold the vertical posts of the structure"; Erwin Neumayer, *Prehistoric Indian Rock Paintings* (Delhi: Oxford University Press, 1983), 89. Examples *a* and *b* are from Satkunda (Madhya Pradesh). Although they were probably painted early in the historical period, their style is derived from that of the Chalcolithic (last millennium B.C.). Example *c* dates from the Mesolithic and comes from Lakhajoar.
Sizes of the originals: 22 × 35 cm; 22 × 34 cm; 24 × 38 cm. After Neumayer, *Rock Paintings*, 136 (figs. 128 and 129) and 89 (fig. 44).

real or imagined prehistoric settlement "at the moment of a traditional calendar feast."[16] In the absence of any enclosing line, it is impossible to decide whether the drawing was originally intended to represent a number of buildings scattered over an area (as if in three dimensions) or whether the huts were just drawn wherever there was suitable space on the cliff face.

Picture Maps

In contrast, a cliff painting from Cangyuan (Yunnan Province, southwestern China) can be shown to fit our definition of a picture map.[17] It dates from the last millennium B.C. At first glance, it is not dissimilar to the Boyar petroglyphs (fig. 1.4). As in the Boyar petroglyphs, a large number of animals (dogs and pigs), humans, and buildings are shown, all in profile. Here the buildings appear to be huts, ten or more in number, each supported by the stilts or piles characteristic of traditional local vernacular architecture. A single line forms a rough ellipse, while several other more or less straight lines converge on each other or on the ellipse. All the huts are within the enclosure, and their disposition leaves little room to doubt that the ellipse represents the village boundary or fence, to which lead paths or tracks along which animals and people are walking in the direction of the village compound.[18] Particularly interesting is the way each perimeter hut has been arranged in relation to the village fence, its supporting piles neatly meeting the boundary line, so that the huts on the far side of the enclosure had to be drawn upside down according to the dictates of topology.[19] Thus, in the Cangyuan cliff painting—unlike the Boyar petroglyph—spatial relationships are clearly demonstrated. Moreover, the composition does not seem to be a case of accidental superimposition, since the neat meeting of the critical lines implies that they were executed in relation to each other. Rather more confidently, then, we can call the Cangyuan painting a (picture) map.

Plan Maps

Identifying objects and landscape portrayed from above is always difficult and is particularly so in rock art, where

16. Devlet, *Rock Engravings*, 18 (note 13).

17. Wang Ningsheng, *Yunnan Cangyuan bihua di faxian yu yanjiu* (The rock paintings of Cangyuan County, Yunnan: Their discovery and research) (Beijing: Wenwu Chubanshe, 1985), illustration on 35, description on 33–34 (English abstract only). Chen Zhao Fu, *Cina: L'arte rupestre preistorica*, Italian trans. Giuliana Aldi Pompili (Milan: Jaca Books, 1988), illustrates the map as a line drawing (102–3), although there are photographs of the cliff in general (pl. 14) and of other painted groups (pls. 64–69).

18. Wang, *Yunnan Cangyuan bihua*, 33–34 (note 17); Chen Zhao Fu, "Ancient Rock Art in China," *Bollettino del Centro Camuno di Studi Preistorici* 23 (1986): 91–98, esp. 97; Yunnan Sheng Lishi Yanjiusuo Diaochazu (Investigative team of the Yunnan Historical Research Institute), "Yunnan Cangyuan yahua" (Cliff paintings of Cangyuan, Yunnan), *Wenwu*, 1966, no. 2:7–16, 38, and for an English summary, Richard C. Rudolph, ed., *Chinese Archaeological Abstracts*, Monumenta Archaeologica, vol. 6 (Los Angeles: Institute of Archaeology, University of California, 1978), 556 (with Lin Sheng given as author).

19. Similar topological depictions can be found on European maps and in topographical paintings in the seventeenth century. See, for example, P. D. A. Harvey, *The History of Topographical Maps: Symbols, Pictures and Surveys* (London: Thames and Hudson, 1980), 59 (fig. 29), 96–97 (figs. 53 and 54); the road maps of John Ogilby or Emmanuel Bowen, where the hill signs are inverted to indicate "up" or "down" gradients from the traveler's point of view; and other examples from sixteenth-century Holland in Cornelis Koeman, "Die Darstellungsmethoden von Bauten auf alten Karten," *Wolfenbütteler Forschungen* 7 (1980): 147–92.

FIG. 1.3. VILLAGE SCENE FROM THE LATE BRONZE AGE. Petroglyphs from the Boyar Ridge, Minusinsk, form a frieze carved into a south-facing rock face. The only landscape features depicted are the buildings (rectangular log huts and more rounded or conical yurts), all shown in profile. The section here (the right two-thirds of the Major Boyar petroglyph) forms a continuous frieze. Despite the impression of a spatial dimension, it is difficult to be sure that a perspective view was intended and that the figures forming the top row of the frieze as it appears on the cliff face are supposed to be in the distance, a problem that would be resolved were there a surrounding line suggesting an enclosure. Natural features are absent. Moreover, the individual figures are so disconnected that there must be uncertainty about whether the assemblage is a palimpsest or a composition.
Size of the entire original: ca. 1.5 × 9.8 m. After M. A. Devlet, *Bol'shaya Boyarskaya pisanitsa/Rock Engravings in the Middle Yenisei Basin* (Moscow: Nauka, 1976), fig. VI.

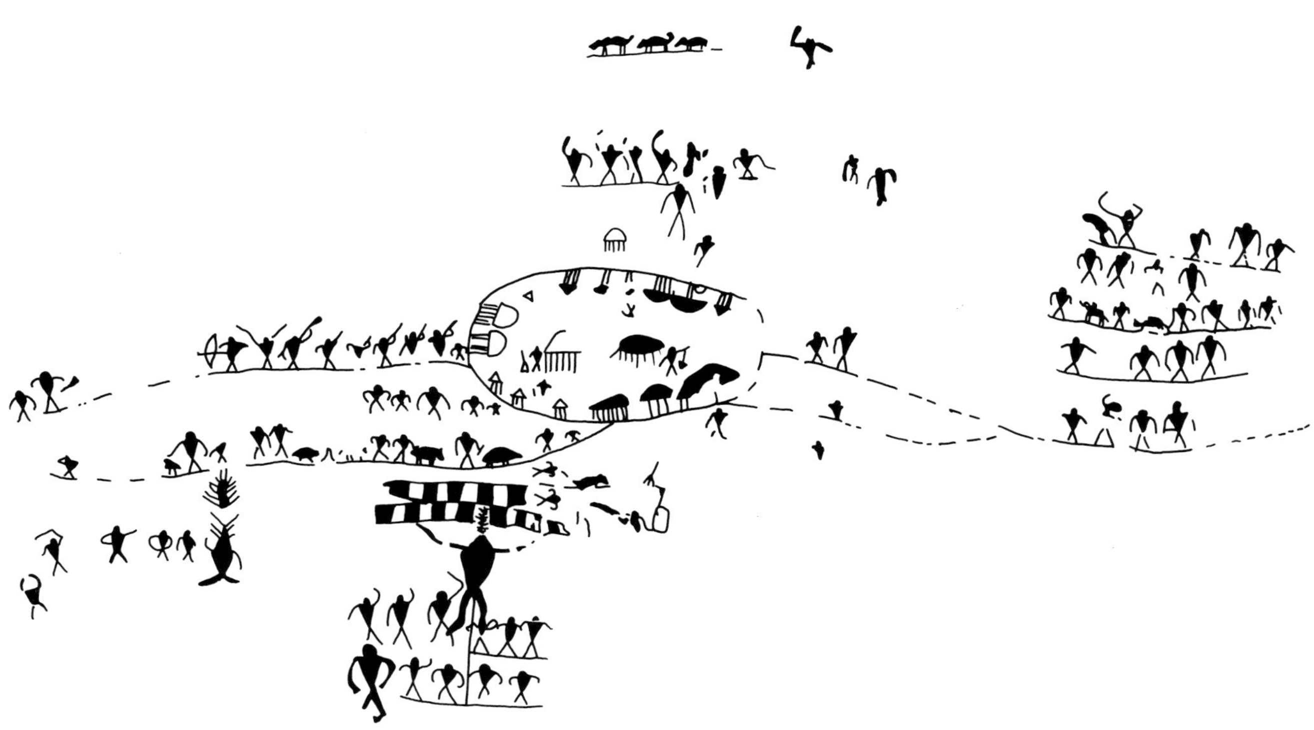

FIG. 1.4. PICTURE MAP OF A LATE PREHISTORIC VILLAGE. This picture map, at Cangyuan, Yunnan Province, China, is painted in red on a cliff face. The huts are supported on piles, and the proximity of most to the village fence or boundary is made clear according to the rules of topology. Paths lead to the village. One large hut, or perhaps two, occupies the central open space. Nine other major groups of pictographs were discovered in the area in 1965.
Size of the original: ca. 175 × 310 cm. After Wang Ningsheng, *Yunnan Cangyuan bihua di faxian yu yanjiu* (Beijing: Wenwu Chubanshe, 1985), 35.

all external evidence is lacking.[20] Plan maps from the historical period are usually recognized from their context, from their titles or other writing, by the correspondence of their images with other cartographic artifacts, or by the familiarity of the drawn outline. Nothing like this can exist for the prehistoric period. Interpretation of a prehistoric image as representing an object or a place as seen from above has to rest primarily on intrinsic visual characteristics. Only then can circumstantial evidence be taken into consideration together with ethnographic analogy.

In an earlier attempt to reduce the arbitrariness of those intuitive interpretations of plan maps that have swelled the antiquarian literature, I suggested that specific criteria, derived from a systematic analysis of the visual aspects of a modern large-scale topographical map, can serve as a model against which to judge the "cartographicness" of a prehistoric, or other wholly "silent," image.[21] I proposed various diagnostics, but three can be singled out as the key ones: the degree to which an assemblage of individual motifs, or signs, can be regarded as having been intended as a single composition, however many separate technical operations might have been needed in its production; the relevance of the individual motifs or signs (they should include landscape features); and their relative frequency (bearing in mind that a single sign, e.g., for a hut, does not on its own constitute a map). With these criteria in mind, I searched the Asian literature for potential map examples. Although nothing as large or complex as the petroglyph assemblages of Bedolina or Giadighe in the Italian Alps[22] has yet been reported from Asia, smaller groups of figures, some strikingly similar to those of Mont Bégo (France), are to be found in the upper Yenisei River valley, the Altai Mountains, and in Mongolia.

The petroglyphs of Mugur-Sargol are found on rocks and cliffs in the valley of the upper Yenisei. Several subjects are portrayed. One of the most common is faces, often with horns or antennae, thought to represent the painted or masked faces of local shamans.[23] There are also geometric figures in which various combinations of rectangular or subrectangular outlines and internal markings have been systematically arranged to create four different signs: solid outlines (squares or rectangles); compartmentalized outlines (generally squares); stippled outlines; and empty outlines. Each of the petroglyphs in question comprises usually one solid shape or a subdivided shape and one or more stippled or empty shapes (fig. 1.5). They have been interpreted by archaeologists as representations—in plan—of the local herders' yurts and stockyards (fig. 1.6). Devlet, for instance, comments on the variety of Mongolian-type yurts found among the Tuva, noting that "in plan such dwellings look like houses in the petroglyphs of Mugur-Sargol," and that similar "hut and enclosures" or "hut and yards" petroglyphs can be found in the Altai Mountains.[24] Many were never completed, and others have since been damaged by erosion and are now too fragmentary to be deciphered with any confidence.

The Mongolian rock paintings differ from the petroglyphs of the upper Yenisei in content as well as technique. The paintings are found in the Transbaikal region as well as in Mongolia proper. They show a single rectangular outline within which there may be either an irregular scatter or just a few rows of dots (figs. 1.7 and 1.8). Also within the rectangle, or sometimes beside it, are one or two anthropomorphic figures. Sometimes there is also the upper part of a bird, with outstretched wings. The composition is thought to represent, in plan, the graves of local tribesmen. These stone-built burial places show up clearly in the surrounding grasslands, each grave outlined by stone slabs set on edge and covered with a scatter of pebbles or small stones (fig. 1.9). The graves are thought to date from any time since the Bronze Age.[25] If some appear comparatively new, this is explained as the effect of wind erosion, which has removed, or prevented, any covering of soil.[26] Featured in rock art, such "grave plans" seem to be rich in religious symbolism.

20. Various attempts to portray subjects in different planes of vision have been reported from Asian rock art. See, for instance, Douglas Hamilton Gordon, "The Rock Engravings of Kupgallu Hill, Bellary, Madras," *Man* 51 (1951): 117–19, esp. 118, where fig. 1a is said to show a woman as viewed from above while the man beside her is shown in profile. Ya. A. Sher, *Petroglify Sredney i Tsentral'noy Azii* (Petroglyphs of Middle and Central Asia) (Moscow: Nauka, 1980), 202–5, attempts an analysis of the relative frequency of plan and profile representation in a number of pictures of animals harnessed to vehicles such as plows or wagons, the plan representations emerging in the minority (18–25 percent): the "harnessed animals are also shown in profile but in such a way as if they are placed on a flat surface with backs and legs mixed up; these we shall call in plan" (p. 202).

21. Delano Smith, "Cartography in the Prehistoric Period," 61–62 (note 1).

22. Delano Smith, "Cartography in the Prehistoric Period," 78–79 and figs. 4.28, 4.29 (note 1).

23. M. A. Devlet, *Petroglify Mugur-Sargola* (Petroglyphs of Mugur-Sargol) (Moscow: Nauka, 1980), e.g., 226 and 229; and idem, *Petroglify Ulug-Khema* (Petroglyphs of Ulug-Khem) (Moscow: Nauka, 1976), 10–25, figs. 5, 6, 7, and 13.

24. For instance, Devlet, *Petroglify Mugur-Sargola*, 234; see also Devlet, *Petroglify Ulug-Khema*, 27 (both note 23).

25. E. A. Novgorodova, *Alte Kunst der Mongolei*, trans. Lisa Schirmer (Leipzig: E. A. Seemann, 1980), pl. 62, "Plattengrab vor der Freilegung," and idem, *Mir petroglifov Mongolii* (The world of Mongolian petroglyphs) (Moscow: Nauka, 1984), 93 (fig. 34).

26. Folke Bergman, "Travels and Archaeological Field-work in Mongolia and Sinkiang—A Diary of the Years 1927–1934," in *History of the Expedition in Asia, 1927–1935*, 4 vols., by Sven Anders Hedin (Stockholm: [Göteborg, Elanders Boktryckeri Aktiebolag], 1943–45), 4:1–192, esp. 4–6.

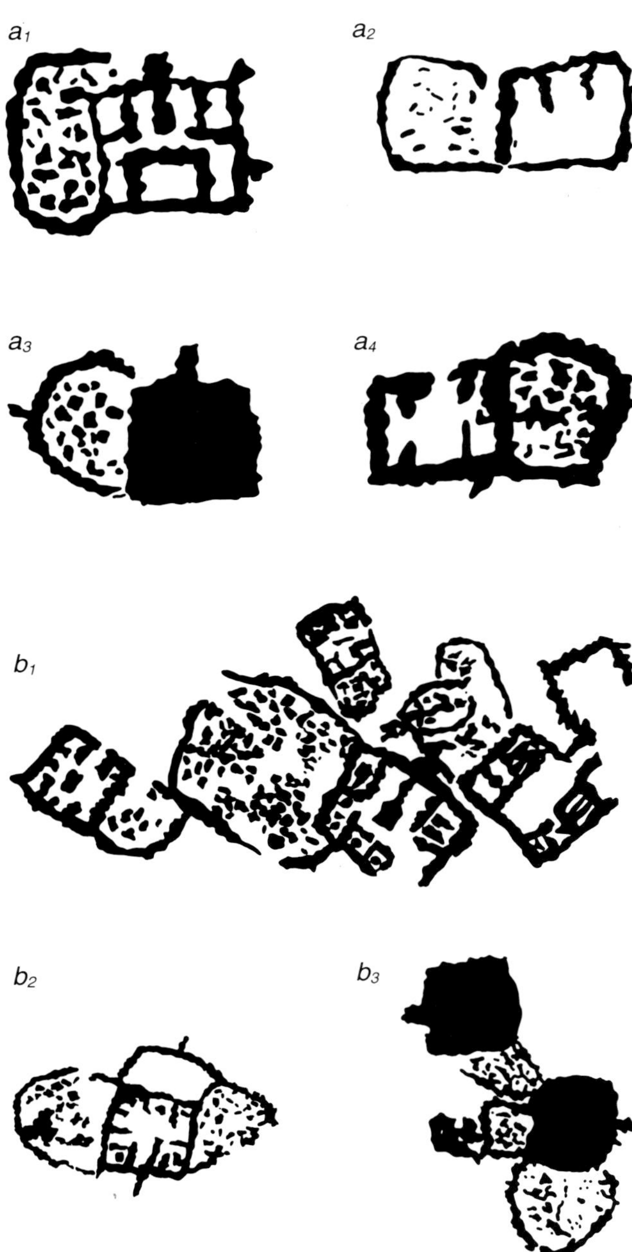

FIG. 1.5. "HUT AND YARDS" DEPICTED IN PLAN. These petroglyphs come from Mugur-Sargol. The solid and compartmentalized shapes have been interpreted as representing the winter yurts typical of the middle Yenisei region, and the stippled outlines as the adjacent stockyards or enclosures. Examples in *a* are relatively simple combinations of hut signs, suggesting an individual hut or yurt with stockyard as shown in figure 1.6, and enclosure signs, whereas those reproduced in *b* are much more complex arrangements (to judge from the interconnecting lines) and could thus be representing a group or village of yurts.
After M. A. Devlet, *Petroglify Ulug-Khema* (Moscow: Nauka, 1976), 26–27 (parts of figs. 16 and 17); idem, *Petroglify Mugur-Sargola* (Moscow: Nauka, 1980), 234 (fig. 17.2).

FIG. 1.6. PICTURE OF A YURT. This home of herdsmen is seen from higher up the mountainside. It is this familiar perspective that inspired local rock artists to represent the homestead in plan. Note the rectangular chimney holes of the yurt, the surrounding stockyards and pens, and the way the closely grazed turf stands out in contrast to the rough grass beyond, the whole making a distinctive landscape complex effectively captured in the rock art. This example comes from the upper Yenesei.
From M. A. Devlet, *Petroglify Mugur-Sargola* (Moscow: Nauka, 1980), 235.

Okladnikov suggested, in accordance with traditional beliefs, that the dots represent the souls of those buried there, that the role of the bird (probably the sacred eagle) is protective, and that the human figures, usually depicted with joined hands, act as intercessors.[27] Some archaeologists have gone further and, like Novgorodova, iden-

27. A. P. Okladnikov, *Der Hirsch mit dem goldenen Geweih: Vorgeschichtliche Felsbilder sibiriens* (Wiesbaden: F. A. Brockhaus, 1972), 148; this is a translation of Okladnikov's *Olen' zolotye roga* (Deer with the golden antlers) (Leningrad, 1964). I am grateful to Guntram Herb for help with the translation of parts of *Der Hirsch*. See also A. P. Okladnikov, *Ancient Population of Siberia and Its Cultures* (Cambridge: Peabody Museum, 1959), 48; idem, "The Petroglyphs of Siberia," *Scientific American* 221, no. 2 (1969): 78–82, esp. 78–79; D. Dorzh, "Rock 'Art Galleries' of Mongolia," *Canada Mongolia Review* 1, no. 2 (1975): 49–55, esp. 50; and Esther Jacobson, "Siberian Roots of the Scythian Stag Image," *Journal of Asian History* 17 (1983): 68–120, esp. 100.

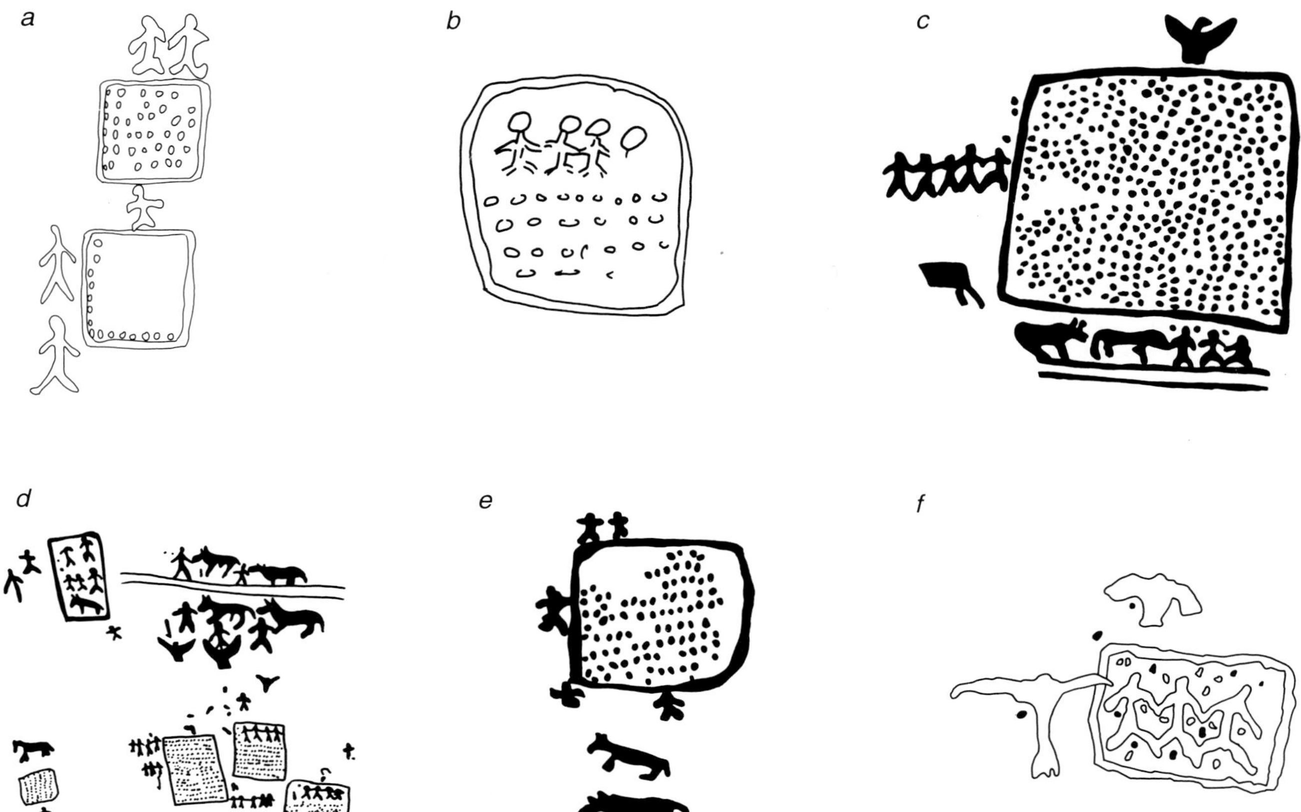

FIG. 1.7. GRAVE PLANS FROM SEVERAL SITES IN MONGOLIA. These rock figures are found in the same parts of Mongolia characterized by a certain type of traditional burial that can still be seen on the grass steppes and that they seem to match (see fig. 1.9). It has been suggested that the rectangular outline with an internal pattern of dots represents these stone-edged graves and the pebbles scattered over their surface. The anthropomorphic figures might represent those buried there or their relatives who, hands joined, intercede for the departed. The bird is widely used in Asia to represent the human soul, and the eagle is associated with some Siberian cults.
(*a*) From Hövsgöl-Nuur, Mongolia. After A. P. Okladnikov, *Petroglify Mongolii* (Leningrad: Nauka, 1981), 88 (fig. 2).
(*b*) From Dood-Chulgan, Mongolia. After E. A. Novgorodova, *Alte Kunst der Mongolei*, trans. Lisa Schirmer (Leipzig: E. A. Seemann, 1980), pl. 72.
(*c*) From Ich-Tengerin-Am, northern Mongolia. After A. P. Okladnikov and V. D. Zaporozhskaya, *Petroglify Zabaykal'ya*, 2 vols. (Leningrad: Nauka, 1969–70), 2:238 (fig. 67.2).
(*d*) Ocher paintings from Gachurt, Mongolia, said to represent the earthly world and netherworld. After E. A. Novgorodova, *Mir petroglifov Mongolii* (Moscow: Nauka, 1984), 93 (fig. 34).
(*e*) Ocher paintings from the end of the first millennium B.C. at Gachurt, Mongolia. After Novgorodova, *Mir petroglifov Mongolii*, 92 (fig. 33).
(*f*) After Okladnikov and Zaporozhskaya, *Petroglify Zabaykal'ya*, cover illustration of volume 1.

tified these paintings as Late Bronze Age or Early Iron Age representations of the "world of the living and world of the dead."[28]

Celestial Maps

When it comes to searching for early examples of celestial maps—maps of all or part of the visible sky or even of a particular constellation—Asia presents something of a paradox. On the one hand, this is a continent that saw some of the earliest developments in astronomy, as well as in the complex of interrelated ideas and beliefs that Berthelot has called "astrobiology" (use of the stars to determine the agricultural seasons).[29] On the other hand, neither rock art nor the history of cartography literature seems to have explored the contribution of prehistoric astronomy and celestial mapping.[30]

As in historical tribal societies, in prehistoric societies

28. Novgorodova, *Alte Kunst der Mongolei*, 113 (note 25).

29. René Berthelot, *La pensée de l'Asie et l'astrobiologie* (Paris: Payot, 1949). Nontropical Asian peoples on the whole determined the seasons by the sun rather than the stars.

30. To judge from available literature, where little is said about the cup marks and cup-and-ring marks that are so common in Europe and that are the most open to interpretation as maps of stars. They are mentioned in Chen, *Cina*, 181 (note 17).

FIG. 1.8. GRAVE PLANS FROM ICH-TENGERIN-AM, MONGOLIA. Described erroneously by Okladnikov (1972) as Bronze Age rectangular courts or yards (*hose*), these markings from Ich-Tengerin-Am are consistent with all others from Mongolia recognized elsewhere by him, as by other archaeologists, as representations of graves. (See A. P. Okladnikov, *Der Hirsch mit dem goldenen Geweih: Vorgeschichtliche Felsbilder sibiriens* [Wiesbaden: F. A. Brockhaus, 1972], 148 [fig. 41]. See also his earlier descriptions of them as grave plans in Okladnikov and Zaporozhskaya, *Petroglify Zabaykal'ya*, 2:54, drawing 4, and Okladnikov, *Ancient Population of Siberia and Its Cultures* [Cambridge: Peabody Museum, 1959], 48.)
From E. A. Novgorodova, *Alte Kunst der Mongolei*, trans. Lisa Schirmer (Leipzig: E. A. Seemann, 1980), fig. 72.

celestial observation would have been closely related to some fundamental aspects of life. In areas devoid of obvious landmarks (such as the sandy deserts of the Gobi, the snowy wastes of the high plateaus, or out at sea), stars would have been used in wayfinding.[31] In regions of seasonally undifferentiated climate (such as the humid tropics of Southeast Asia), the appearance and disappearance of certain constellations, notably the Pleiades, are still used to mark the agricultural calendar.[32] Through association of ideas, these stars may have come to signify fertility.[33] Finally, in all parts of Asia as elsewhere in the world, astronomy is closely linked with cosmology, especially through the need for celestial observations to fix ground points for the earthly counterparts of a society's cosmological ideas. Yet examples of such interpretations in prehistoric art are lacking. There may be general discussion of, for example, the way the orientation and layout of settlements and the location and placing of ceremonial monuments and other microcosmological symbols may depend on the sighting and position of certain stars or the relevance of the notion of the earth as

31. For a discussion of the need for maps in wayfinding in primitive societies, see Delano Smith, "Cartography in the Prehistoric Period," 59 and references therein (note 1).

32. I. C. Glover, B. Bronson, and D. T. Bayard, "Comment on 'Megaliths' in South East Asia," in *Early South East Asia: Essays in Archaeology, History and Historical Geography*, ed. R. B. Smith and W. Watson (New York: Oxford University Press, 1979), 253–54, refer to the Tetum speakers of Portuguese Timor, who use the Pleiades to determine the start of each stage of rice cultivation, agreeing that "the need for astronomical calculations is present in parts of South East Asia, despite the low latitudes and the relative lack of seasonal climatic variation" (254).

33. A. H. Christie, "The Megalithic Problem in South East Asia," in *Early South East Asia: Essays in Archaeology, History and Historical Geography*, ed. R. B. Smith and W. Watson (New York: Oxford University Press, 1979), 242–52.

FIG. 1.9. TRADITIONAL BURIAL PLACE FOUND IN MONGOLIA. Stone slabs set upright or on edge define the tomb, usually containing more than one skeleton. Smaller stones litter the grass covering. The force of the wind tends to prevent the accumulation of soil over the burial places with the result that they may sometimes have been as commonplace in the landscape as the dwellings of living herdsmen.
From E. A. Novgorodova, *Alte Kunst der Mongolei*, trans. Lisa Schirmer (Leipzig: E. A. Seemann, 1980), fig. 62.

a microcosm rather than as part of the macrocosm.[34] Taking the rather special case of the design of capital cities on cosmic models, Wheatley thinks that by the time the first Chinese urban centers were taking form (second millennium B.C.), their layout would have been reflecting beliefs that could even then have been "as ancient as man himself . . . that had taken their rise coevally with the human mind, and [that] had become so inextricably interwoven with the pattern of human thought that they were not consciously recognized as beliefs at all."[35] And Needham has suggested that one reason astronomy came to be a science of cardinal importance in China so early was that observation of the stars had arisen "naturally out of that cosmic 'religion,' that sense of the unity . . . of the universe."[36] But other than pointing to evidence that equinoxes and solstices were being determined as early as the Shang dynasty (sixteenth to eleventh century B.C.) and that solar eclipses were recorded by the end of the third millennium B.C. at the earliest,[37] there is a general silence on the prehistoric manifestation of such activity in celestial mapmaking. It would be surprising, however, if this "very long and continuous tradition of celestial cartography"[38] were not to be found somewhere in the prehistoric record. We have to await the evidence.

34. A substantial literature is available on the relation between cosmological belief and the layout of towns throughout the historical period. For China, see Paul Wheatley, *The Pivot of the Four Quarters: A Preliminary Enquiry into the Origins and Character of the Ancient Chinese City* (Chicago: Aldine, 1971), esp. chap. 5, "The Ancient Chinese City as a Cosmo-magical Symbol"; for the Greco-Roman world see, for a start, Joseph Rykwert, *The Idea of a Town: The Anthropology of Urban Form in Rome, Italy and the Ancient World* (London: Faber and Faber, 1976); for landscape in general, in Renaissance Europe, see Denis E. Cosgrove, *Social Formation and Symbolic Landscape* (London and Sydney: Croom Helm, 1984), and also Douglas Fraser, *Village Planning in the Primitive World* (New York: George Braziller, 1968).

35. Wheatley, *Pivot*, 416 (note 34).

36. Joseph Needham, *Science and Civilisation in China* (Cambridge: Cambridge University Press, 1954–), vol. 3, with Wang Ling, *Mathematics and the Sciences of the Heavens and the Earth* (1959), 171.

37. Needham, *Science and Civilisation*, 3:284 and 409 (note 36).

38. Needham, *Science and Civilisation*, 3:265 (note 36).

FIG. 1.10. CELESTIAL DIAGRAM FROM THE *JIN SHI SUO*. Despite the very early documented development of astronomy in China, few examples of constellations or groups of constellations have so far been reported from prehistoric art. Yet the idea of joining dots by lines to represent asterisms, as in this tomb painting featuring the Great Bear, is known to go back at least to the Han period (206 B.C.–A.D. 220).
From Feng Yunpeng and Feng Yunyuan, *Jin shi suo* (Collection of carvings, reliefs, and inscriptions, 1821); modern edition in 2 vols., Guoxue Jiben Congshu, vols. 157 and 158 (Taipei: Taiwan Shangwu Yenshuguan, 1968), 2:164–65.

Meanwhile, there are a few pointers on what to look for. The lines on certain pieces of Chinese Neolithic pottery have been interpreted as celestial symbols and the circles as sun and moon signs.[39] In Han tomb decoration, at the start of the historical period in China, it was already conventional to represent groups of stars by "ball-and-link" patterns—dots or circles connected by lines (fig. 1.10).[40] Armenia and adjacent parts of Central Asia appear to have been a particular focus of prehistoric and early historical astronomical activity. Recently discovered petroglyphs in the mountains of Armenia are said to contain representations of "various constellations" and "ingenious calendars."[41] Here too are found geometric symbols of "the sun, the moon, the lightning, the stars, as well as whole complexes of symbols, which bear in themselves the concept of the stellar system," said to be "directly related with myths and legends originating from the cult of the starry sky, the moon and the luminaries." Cosmological myths are given anthropomorphic or animal forms, each associated with a star. The sun is shown as a "radiant wheel," often connected with the idea of a chariot, drawn by bulls. Many of these rock markings

39. T. I. Kashina, "Semantika ornamentatsii neoliticheskoy keramiki Kitaya" (Semantics of ornamentation of China's Neolithic pottery), in *U istokov tvorchestva* (At the sources of art) (Novosibirsk: "Nauka," 1978), 183–202, cited in Ildikó Ecsedy, "Far Eastern Sources on the History of the Steppe Region," *Bulletin de l'Ecole Française d'Extrême-Orient* 69 (1981): 263–76, esp. 271 n. 18, who warns against seeing meanings in the signs without first studying the relevant cult or culture.

40. Needham, *Science and Civilisation*, 3:276–82 and figs. 90 (p. 241) and 102 (in pl. XXVI) (note 36).

41. Miroslav Kšica, *Umění staré Eurasie: Skalní obrazy v SSSR* (The art of ancient Eurasia: Rock pictures in the Soviet Union) (Brno: Dům Umění, [1974]), 71 (summaries in Russian, German, English, and French).

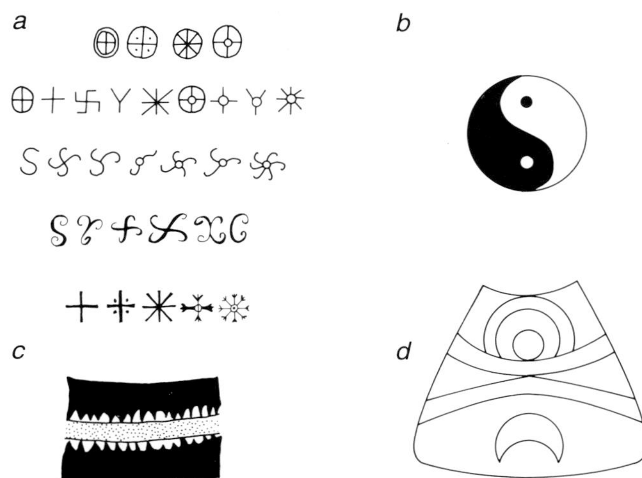

FIG. 1.11. MAPPING THE COSMOS. The nature of the cosmos, the relation between this world and the next, and the means of access from this world to the next were as important preoccupations of prehistoric peoples as of those living in historical times. More common than the occasional map are signs associated with these preoccupations. Four groups of these are shown here:
(*a*) An array of what are widely known as "sun and fire" symbols.
After F. Fawcett, "Notes on the Rock Carvings in the Edakal Cave, Wynaad," *Indian Antiquary* 30 (1901): 409–21, esp. 413; he found many examples from this array among the prehistoric rock paintings in Indian caves and rock shelters.
(*b*) The Chinese "yang and yin" sign encapsulates fundamental life principles also personified in sun gods and moon and earth goddesses and signified by the signs in *a*.
(*c*) The color red is widely associated with death and the afterlife. In the "death pattern" on Chinese mortuary pottery, the central band of dots is red and indicates the area forbidden to the living; the triangles mark their territory, the earthly world.
After Johan Gunnar Andersson, *Children of the Yellow Earth: Studies in Prehistoric China*, trans. from the Swedish by E. Classen (New York: Macmillan, 1934), fig. 137.
(*d*) Rock-cut design from Japan, possibly "descriptive of [a] cosmic myth."
After Neil Gordon Munro, *Prehistoric Japan* (Yokohama, 1911), 192.

are also said to be found as decoration on pottery from the third millennium B.C. and later to have been incorporated into Urati hieroglyphs (800–600 B.C.).[42] Manuscripts taken in the Middle Ages from early pagan sanctuaries confirm the meaning of each celestial sign. A prehistoric astronomical observatory has been excavated at Metsamor (Armenia) dating back to perhaps the third millennium B.C., its axes coinciding with those of modern observatories and its azimuth oriented to the star Sirius.[43] Star signs—"octagonal stars contained in a trapezium"—are carved on stone structures in the highest part of the observatory. Yet despite the usually close association of astronomical and cosmological observance, when Okladnikov uncovered an arrangement of stone slabs placed like rays around a circular platform dating from the Bronze Age, he associated this with the practice of a sun cult rather than with astronomy.[44]

Cosmological Maps

As already noted, the dividing line between celestial and cosmological beliefs in early societies is a fine one. Both ethnographic and archaeological literature testify to the cosmological and eschatological beliefs of various Asian societies in historical and prehistoric times. If there are any lingering doubts about the meaning, or meanings, of the various signs, the archaeologist has recourse to the ethnographic literature, for the use of these signs tends to have remained an important aspect of traditional art in many regions of Asia. Maps made up of these signs are found throughout the region and from all times in the historical period.

Two broad groups of cosmological signs can be distinguished. In one group can be put the general signs of a cult, such as sun signs. In the other are the signs that relate to a specific aspect of that cult, such as the question of access from this world to the next. The former can include a formidable variety of cosmological signs. For Asia, Fawcett reproduced no fewer than thirty-two types of "sun and fire" symbols, characteristic of Asian art in general, many of which he had found among prehistoric rock paintings in the Edakal cave (Wynaad, India), still the focus of an annual religious pilgrimage (fig. 1.11*a*, *b*).[45] The signs ranged from the familiar quartered circles, crosses, swastikas, and stars to curving Y shapes, S shapes, and variously rayed circles, and they were packed so densely that Fawcett commented, "It takes a protracted and close study to make anything of them."[46] For the historian of cartography, however, it is the second group of signs that holds the greatest potential interest, since they imply a spatial relationship between this world, the underworld, and the next world or heaven, as well as the journey between the worlds. The tree sign, for instance,

42. Martirosyan and Israelyan, *Naskal'nye izobrazheniya Gegamskikh gor*, 58 (note 10).

43. A. A. Martirosyan, "Sémantique des dessins rupestres des Monts de Guégam (Arménie)" (Moscow, 1971), 8; limited circulation, unpublished papers omitted from the proceedings: *Actes du VIIIe Congrès International des Sciences Préhistoriques et Protohistoriques*, 3 vols. (Belgrade, 1971); David Marshall Lang, *Armenia: Cradle of Civilization*, 3d corrected ed. (London: George Allen and Unwin, 1980), 263–64; and "Astronomical Notes from Prague," *Sky and Telescope*, November 1967, 297, where E. S. Parsamian presents these findings.

44. Okladnikov, *Ancient Population*, 24 (note 27).

45. F. Fawcett, "Notes on the Rock Carvings in the Edakal Cave, Wynaad," *Indian Antiquary* 30 (1901): 409–21, esp. 413.

46. Fawcett, "Edakal Cave," 413 (note 45).

may stand for the Sacred Tree, an *axis mundi* equivalent to the Mount Meru of Hindu, Jain, or Buddhist cosmology. Or standing for the Tree of Life, it may represent the highest degree of unity, in which case it signifies the idea of upper world and netherworld combined.[47] Even more spatially explicit is the "death pattern" decoration found on some prehistoric mortuary pottery in China (fig. 1.11*c*). Here a red (or violet) band between two lines of zigzags is said to demarcate the zone reserved for the dead and forbidden to the living.[48] Likewise, the labyrinth design, representing the journey between the world of the living and that of the dead, encompasses the idea of a map between the two worlds.[49] Labyrinth signs are one of the most common motifs in prehistoric art in Asia as elsewhere. They feature prominently in, for example, the caves and rock shelters of Bhopal District, central India, the richest zone of prehistoric rock art in the subcontinent, where 90 percent of the region's rock shelters are concentrated into the Vindhya, Mahadeo, and Kaimur sandstone hills.[50] Another cosmological motif, found especially in eastern Siberia and southeastern Asia, on the one hand, and in the Finnish-Baltic region, on the other, is the ship, the means of transporting the soul to the next world. Shown without oars, the ship is seen as propelled by divine force alone.[51] Depicted in red paint on river cliffs at Shishkino (upper Lena River, Siberia), "a whole row of boats [in which human figures are sitting, arms upraised] is seen floating in a line on the sacred river into the world of the dead."[52] Modern Ngaju Dayaks in Borneo have explained their *tiwah* (feast of the dead) ceremonies, in which the spirit Tempon Telon acts as a Bornean Charon and corresponds to the prehistoric ship sign.[53] Other ship signs are found in the decoration of the bowl or the tympanum of Bronze Age kettle drums of Borneo and other parts of Indonesia. Occasionally the coffin itself takes the form of a boat.[54] All these signs signify essentially similar beliefs. All, like the maps on coffin bases from ancient Egypt,[55] testify to a universal preoccupation, the question of life after death and the means of access to the next world.

On their own, however, signs do not constitute maps. At most, they are a pointer to the cosmological structures of the people who made them. In a cosmological map, such signs would be used to indicate the location of the various parts of the cosmos. Inevitably, though, little can be understood of the graphic forms of a society's cosmological beliefs without insights from ethnography. Thus, the curved lines decorating a panel of a Japanese prehistoric jar would not have been described as a possible illustration of the origin of the world without some knowledge of Ainu mythology (fig. 1.11*d*).[56] Similarly, "generally speaking, where we find a bird, it symbolizes heaven or the upper world, whereas a fish or a sea serpent represents the water or the nether world."[57] The Mongolian grave plans, with their eagles suggesting, if not actually confirming, the association of the petroglyphs with the traditional beliefs of the region, have already been discussed. From ethnography we also learn that local cosmic beliefs may embody class distinctions. Among the Dayaks of Borneo, formerly only the upper classes went to a sky world on death,[58] the lower classes being destined for the netherworld.

---

47. For example, Abraham Nicolaas Jan Thomassen à Thuessink van der Hoop, *Indonesische siermotieven* ([Batavia]: Koninklijk Bataviaasch Genootschap van Kunsten en Wetenschappen, 1949), 274–75 (pl. CXXIX), text in Dutch, Malay, and English.

48. Johan Gunnar Andersson, *Children of the Yellow Earth: Studies in Prehistoric China*, trans. from the Swedish by E. Classen (New York: Macmillan, 1934), 315 and fig. 137. See also Hanna Rydh, "On Symbolism in Mortuary Ceramics," *Bulletin of the Museum of Far Eastern Antiquities* 1 (1929): 71–120 and plates. The use of red in these contexts is itself symbolic. Red is taboo for the living, devoted to the mortuary culture, symbolizing blood as the strongest carrier of life, offered to the deceased on their departure for the dreaded journey into the Land of the Dead (Andersson, *Children*, 69). Kšica, *Umění staré Eurasie*, 71 (note 41), also comments on the use of red in cave painting and on red ocher sprinkled over human skeletons in prehistoric burials in the Urals.

49. Delano Smith, "Cartography in the Prehistoric Period," 87–88 and footnotes (note 1). R. K. Sharma and Rahman Ali, *Archaeology of Bhopal Region* (Delhi: Agam Kala Prakashan, 1980), pl. 15, shows a labyrinth pattern surrounded by triangular decorations. Vishnu S. Wakankar, "Painted Rock Shelters of India," *IPEK: Jahrbuch für Prähistorische und Ethnographische Kunst* 21 (1964–65): 78–83, esp. pl. 59, fig. 4, illustrates a labyrinth with seven bulls' heads outside.

50. Yashodhar Mathpal, *Prehistoric Rock Paintings of Bhimbetka, Central India* (New Delhi: Abhinav Publications, 1984), 14.

51. Horace Geoffrey Quaritch Wales, *Prehistory and Religion in South-east Asia* (London: Bernard Quaritch, 1957), 51; Engkos A. Kosasih, "Rock Art in Indonesia," in *Rock Art and Prehistory: Papers Presented to Symposium G of the AURA Congress, Darwin, 1988*, ed. Paul Bahn and Andrée Rosenfeld (Oxford: Oxbow Books, 1991), 65–77, esp. the summary of distribution in the last sentence; and A. L. Siikala, "Finnish Rock Art, Animal Ceremonialism and Shamanic Worldview," in *Shamanism in Eurasia*, 2 vols., ed. Mihály Hoppál (Göttingen: Edition Herodot, 1984), 1:67–84.

52. Okladnikov, *Ancient Population*, 43 (note 27).

53. Victor Goloubew, "L'Age du Bronze au Tonkin et dans le Nord-Annam," *Bulletin de l'Ecole Française d'Extrême-Orient* 29 (1929): 1–46, esp. 36–37. Martirosyan and Israelyan, *Naskal'nye izobrazheniya Gegamskikh gor*, 59 (note 10), mention boat representations in Armenian rock art.

54. Goloubew, "L'Age du Bronze au Tonkin," 36–37, and pl. XXIX(C) (note 53).

55. A. F. Shore, "Egyptian Cartography," in *The History of Cartography*, ed. J. B. Harley and David Woodward (Chicago: University of Chicago Press, 1987–), 1:117–29, esp. 120 and pl. 2.

56. Neil Gordon Munro, *Prehistoric Japan* (Yokohama, 1911), 285 and fig. 85 (on p. 180). The myth concerned is an Ainu belief with analogies in Japanese and Russian folklore. It describes how the world with its surrounding sea is supported on the back of a fish.

57. Thomassen à Thuessink van der Hoop, *Indonesische siermotieven*, 40 (note 47), for just one example.

58. Wales, *Prehistory and Religion*, 92 (note 51). See also chapter 17 below, esp. figs. 17.1 and 17.2.

FIG. 1.12. A POSSIBLE COSMOGRAPHICAL SYMBOL. This design is incised on a pottery vessel discovered at Lingyanghe, Shandong Province. The vessel dates from 2900–2400 B.C. The round element at the top of the design is generally taken to be the sun. The crescent-shaped element has been taken to be the moon or a cloud. The bottom element has been interpreted as representing a fire or a mountain. Depending on how one combines these separate interpretations, one can read the design in several ways, for example, as showing the union of the celestial and terrestrial, or as representing a sunrise, or as signifying heat. Size of the vessel: 62 cm high; 29.5 cm diameter. By permission of the Shandong Provincial Museum.

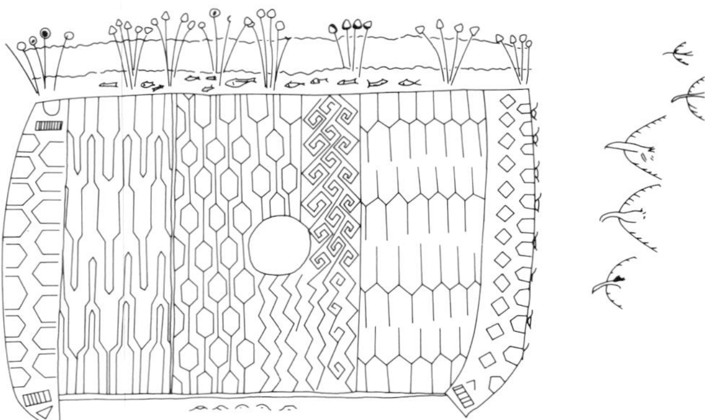

FIG. 1.13. MESOLITHIC COSMOLOGICAL MAP FROM INDIA.
Size of the original: 44 × 75 cm. After Erwin Neumayer, *Prehistoric Indian Rock Paintings* (Delhi: Oxford University Press, 1983), 68 (fig. 26e).

Only one example of a cosmological map of prehistoric date is included in the present list (appendix 1.1, no. 2). In other cases, even where ethnographic evidence has been used to interpret the decoration of a prehistoric artifact as a cosmological representation, too few details are given for us to identify the artifact or include the figure in question (fig. 1.12).[59] A single rock painting from India has been expounded by archaeologists as portraying the Mesolithic cosmos (fig. 1.13). The painting comes from a rock shelter at Jaora, on the Malwa Plateau (Madhya Pradesh). The three parts of the cosmos are represented. The top band of the design is easily recognized as portraying, in naturalistic style, a watery environment. A shoal of fish (associated with the netherworld) swims among clumps of reeds or water plants. Also in naturalistic style are the five birds in flight outside the main design, said to represent the air, or the upper world. In the rest of the picture, the signs are more stylized. In the right-hand and lower borders are water birds. An empty circle or disk, presumably the sun, marks the center of the composition. From it, rays—zigzag lines—run to the lower border. The four intricately patterned bands filling the rest of the panel have been suggested as representing the earth. As a whole, the picture "can be understood as a symbolic depiction of the Mesolithic cosmos."[60]

Problems of Interpretation

Appendix 1.1 contains twenty-two examples of prehistoric maps drawn from the rock art of Asia. It includes one picture map, twenty plan maps (mostly from Mugur-Sargol), and one cosmological map. Compared with the fifty prehistoric maps itemized for Europe alone,[61] the present list may seem short for so vast a continent as Asia. It is based on available literature, and the archaeological interpretations therein, and one of its most obvious limitations reflects the dearth or unavailability of that literature. For instance, although reports of rock

59. Wales, *Prehistory and Religion*, 69 (note 51), describes a type of design found on drum membranes that shows the drum represents a microcosm: "A transverse line separates the celestial region, with its sun and moon, from the earth below, and sometimes again the underworld beneath." He states his intention of applying this interpretation to the Bronze Age art of the Dongson but does not specify particular prehistoric examples that I can list here.

60. Neumayer, *Indian Rock Paintings*, 14 and fig. 26e and caption (note 8).

61. Delano Smith, "Cartography in the Prehistoric Period," 93–96 (app. 4.1) (note 1).

art have existed from the seventeenth century onward for parts of Central Asia, Siberia, and India—and from the fourth century for at least one region of China—the serious study of rock art in most Asian countries has gathered momentum only in the past two decades. This is generally true even for India, where the initial discovery in the nineteenth century of painted rock shelters and the realization that the art was the product of indigenous people in prehistoric times predated by a dozen years the better-known discovery in Spain of the Altamira cave paintings.[62] Prevailing Eurocentric and colonial views, however, may also be held to account for the neglect until recently of Indian prehistoric art.[63] Other factors affecting the availability of literature for the present survey are the difficulty of identifying and obtaining relevant secondary sources and language problems. Yet another hazard is dating Asian rock art and matching Asian prehistoric chronology to European dates. Specific problems of discovering maps in prehistoric art have already been reviewed elsewhere.[64]

Notwithstanding such difficulties, two salient general facts readily emerge. In the first place, Asia contains an amazing wealth of prehistoric rock art. Little of this has as yet received full study, and undoubtedly even more remains to be discovered to complete the distribution map (fig. 1.14). Second, Asian rock art is similar in form, subject matter, stylistic range, occurrence, and archaeological context to that already encountered in Europe and other parts of the Western Old World. This adds weight to the prevailing opinion among archaeologists that we are dealing here with a record of some of the most fundamental thoughts, anxieties, and perceptions of humankind. Cosmological ideas, the recording of parts of the earthly world and of the skies, were as universal in prehistoric times as throughout history. Something of them is to be glimpsed in the petroglyphic and pictographic record. One way to approach the problem of recovering those prehistoric messages is through a better understanding of the archaeological and ethnographic context of the medium, Asian prehistoric rock art in general.

The prehistoric period in Asia tends to be described in terms of the chronology of European scholars, the main divisions defined by way of life and economy. For most of Asia, absolute dates are in short supply or disputed, especially for the early part of the period. Another complicating factor is that prehistoric ways of life survived well into historical times, and Epipaleolithic life-styles were still to be found in parts of the continent at the start of this century. Usually, however, the end of the prehistoric era and the dawn of historical times was marked by the arrival of linear scripts. In one or two regions, such as Elam, an intervening "protohistoric" period was characterized by pictograph or cuneiform writing (see fig. 1.15). Thus, nonliterate and literate cultures coexisted in Asia throughout most of the historical period. It was primarily through religious and, later, commercial expansion that literacy was conveyed to other parts of the continent. For instance, Hinduism and Buddhism took Indian scripts east and south into peninsular Southeast Asia and the Malay Archipelago, while Confucianism took Chinese characters to Vietnam, Korea, and ultimately (but not before the sixth century A.D.) to Japan, and Islamic traders took the Arabic script to South Asia and the coastal zones of Southeast Asia.

In this way, Asian rock art reflects a fascinating but sometimes bewildering mixture of prehistoric and historical economies and life-styles and nonliterate and literate cultures. Prehistoric economies and nonliterate cultures have lingered longest in the deepest interiors, most distant plains, and peripheral regions. For the historian of cartography these contrasts have two main consequences. One disadvantage is the difficulty of distinguishing the prehistoric rock art from that of similarly nonliterate but historical cultures.[65] The other is advantageous; it leads to the availability of a rich ethnographic record, illuminating the role of rock art in prehistoric life and shedding light on the possible meanings of the various motifs and symbols it contains.

The map in figure 1.14 shows the main areas of rock art in Asia. Such a continental overview is useful to help redress the tendency to report archaeological discoveries in terms of national units, resulting in an inappropriately fragmentary picture. Many of the apparently empty portions of the map are best regarded as areas of ignorance, awaiting fieldwork or the dissemination of its reporting, rather than as areas entirely devoid of rock art. As might be expected, it is the center of the continent that is least known—the Himalayas and the Tibetan Plateau. To the north of this, rock art is found virtually everywhere in Central Asia and Siberia. Some of it has been known to Westerners since the seventeenth century, though new discoveries are regularly being made. As is so often the case, especially with prehistoric rock art, the petroglyphs and pictographs are found in local concentrations in high, seemingly inaccessible, scenically dramatic localities. Over 100,000 petroglyphs, for instance, first discovered in 1982–83 at Saimaly-Tash (in the Fergana mountains), all lie above 3,200 meters and within a pass that has been

62. Mathpal, *Rock Paintings of Bhimbetka*, 12 (note 50).

63. Neumayer, *Indian Rock Paintings*, 1 (note 8).

64. Delano Smith, "Cartography in the Prehistoric Period," esp. 55–63 (note 1).

65. For a recent comment on the technical aspects of rock art dating, see Ronald I. Dorn, Margaret Nobbs, and Tom A. Cahill, "Cation-Ratio Dating of Rock-Engravings from the Olary Province of Arid South Australia," *Antiquity* 62 (1988): 681–89.

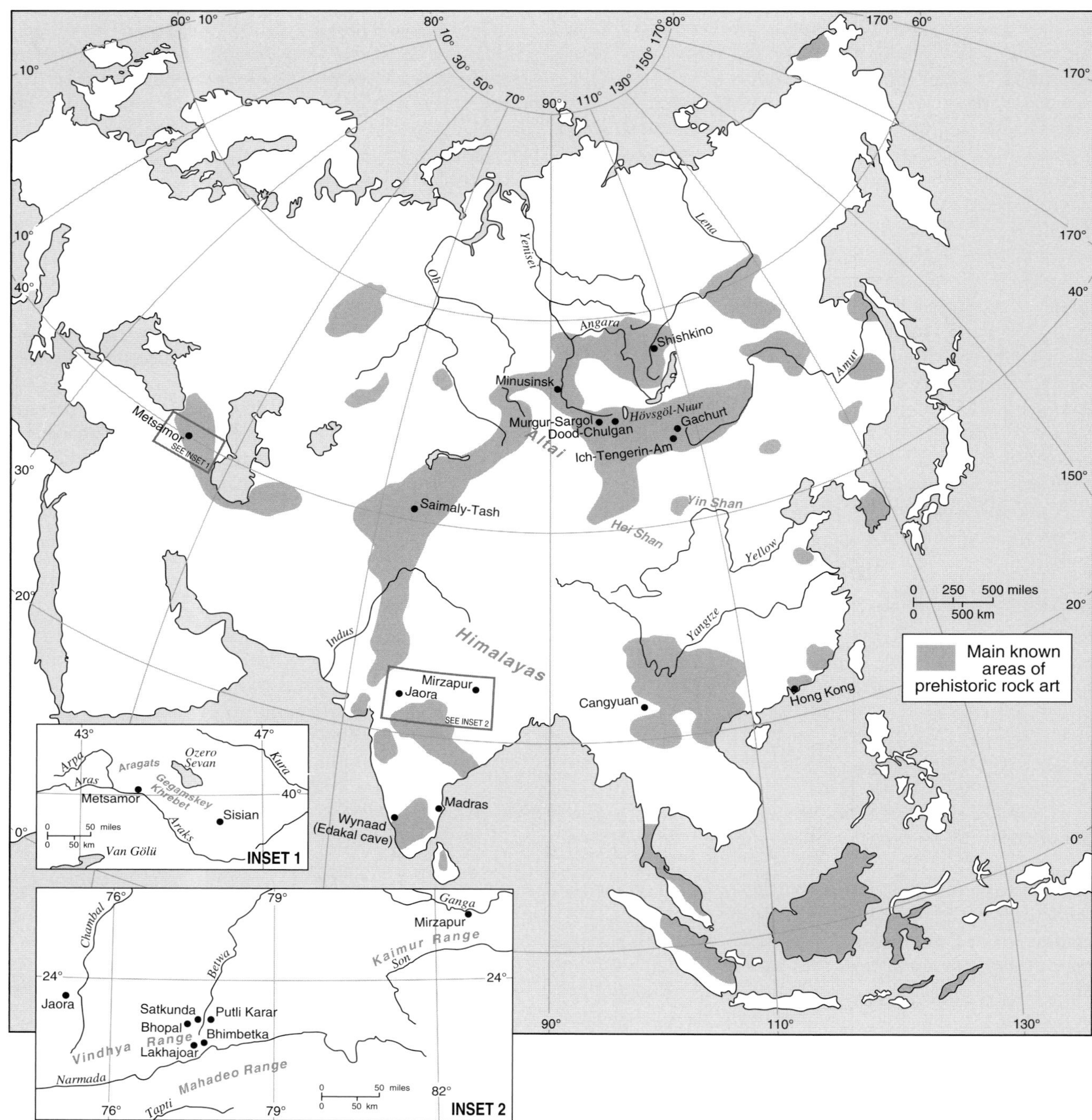

FIG. 1.14. REFERENCE MAP FOR THE STUDY OF ROCK ART IN ASIA. This map shows many locations of examples listed in appendix 1.1 and places mentioned in the text.

described as a "stupendous" place, a typical cult site, surrounded by impassable mountains on all sides but the north.[66]

South of the Himalayas, prehistoric rock art has a similarly widespread distribution over the Indian subcontinent. Here the study of paintings in caves and rock shel-

66. Grégoire Frumkin, *Archaeology in Soviet Central Asia* (Leiden: E. J. Brill, 1970), 45–46; see also the connection made by Kšica between such high-altitude concentrations (" 'closest' to the sun," as at Saimaly-Tash) and the preponderance of astral signs in some of these areas (e.g., Armenia) (*Umění staré Eurasie*, 72 and 71 respectively [note 41]). Most are petroglyphs carved into hard rock such as granite, slate, sandstone, limestone, even basalt.

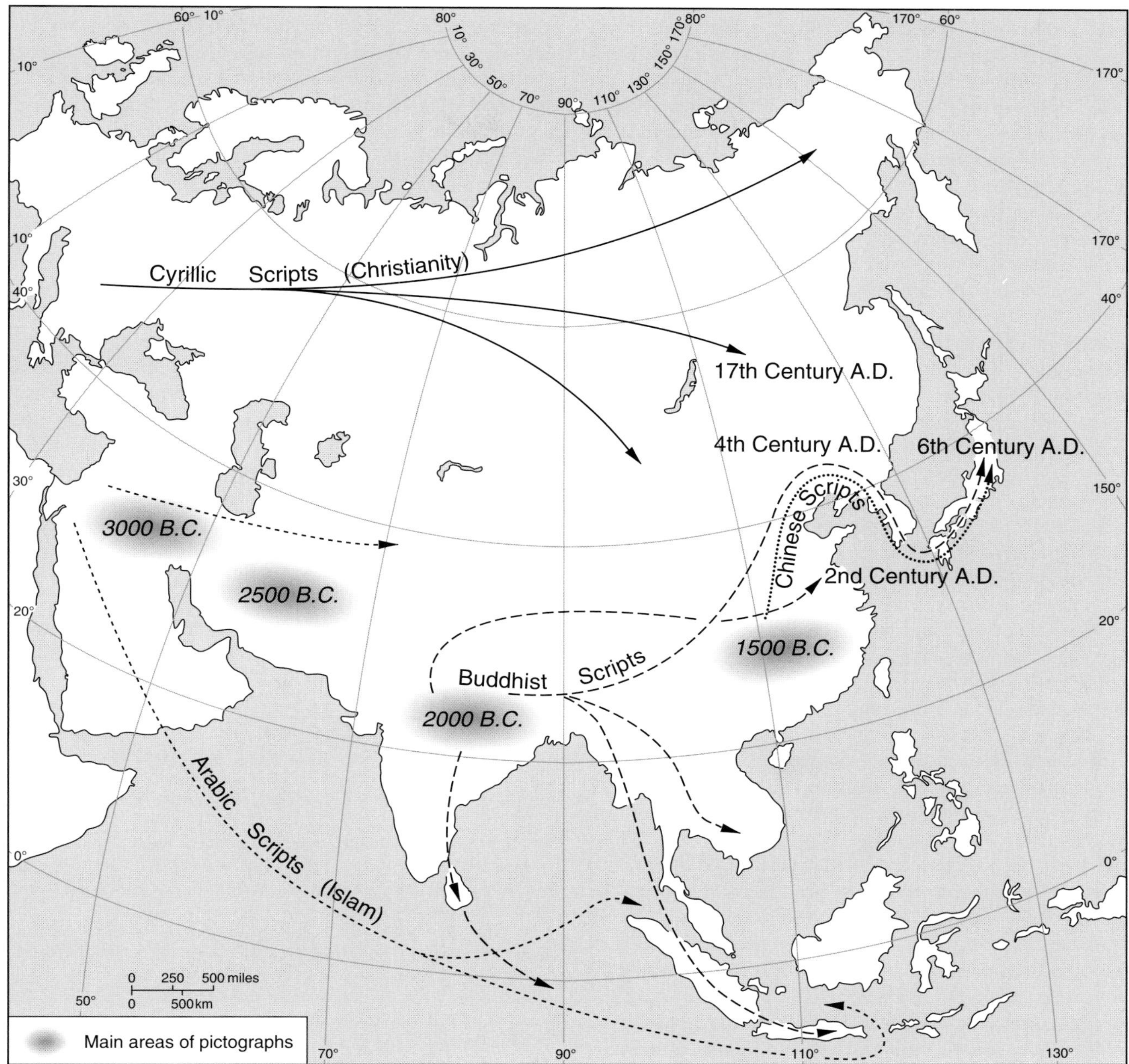

FIG. 1.15. REFERENCE MAP OF THE SPREAD OF LITERACY AT THE END OF THE PREHISTORIC PERIOD. Based on David Diringer, *Writing* (New York: Frederick A. Praeger, 1962), 162.

ters started in the second half of the nineteenth century, after 1867, the year Carlleyle made the link between paintings on the cave walls and stratified stone chippings that could be dated to the Mesolithic.[67] Further work on rock paintings (including that by Cockburn in Mirzapur) was followed by the first literature dealing with the petroglyphs of Kerala (southwestern India).[68] By the 1930s, Gordon was attempting to establish a rock art chronology for India. One of his conclusions, however, was that the paintings were of no "considerable antiq-

67. See Neumayer, *Indian Rock Paintings*, 1–4 (note 8), for a summary. Archibald Carlleyle's account remained unpublished until Vincent A. Smith wrote "Pygmy Flints," *Indian Antiquary*, July 1906, 185–95; Neumayer quotes some of Carlleyle's words (1–2).

68. John Cockburn, "On the Recent Existence of *Rhinoceros indicus* in the North Western Provinces, and a Description of a Tracing of an Archaic Rock Painting from Mirzapore Representing the Hunting of This Animal," *Journal of the Asiatic Society of Bengal* 52, pt. 2 (1883): 56–64, and idem, "Cave Drawings in the Kaimūr Range, North-West Provinces," *Journal of the Royal Asiatic Society of Great Britain and Ireland*, 1899, 89–97; Fawcett, "Edakal Cave," 409–21 (note 45).

uity" (compared with those of Spain, which date from the Upper Paleolithic), thus reinforcing traditional archaeological opinion, which held that indigenous art could not have made any contribution to Indian culture.[69] Only with the start of Wakankar's work on Indian rock art in 1957 have the indigenous prehistoric antecedents of Indian culture been taken seriously.[70]

Southeast Asia is a geographically fragmented and comparatively little known area. Not more than thirty rock art sites have been reported, scattered among the islands and peninsular parts of Southeast Asia. Moreover, despite efforts to establish dates, the various local styles "have proved so diverse that the problems of age, origin and meaning of the paintings have in most cases remained unsolved."[71] Despite the proximity of India, and closeness of cultural contact with it from time to time during the prehistoric period, the rock art of neighboring Burma and Indochina contains none of the dynamic painted scenes characteristic of the subcontinent. Cave paintings in the Shan highlands of Burma show only wild cattle, stags, and human hands in outline, apart from the equally ubiquitous "sun symbol."[72] Some rock art is reported from Hong Kong, but none of the curvilinear patterns on the rocks there, said to date mostly from the Bronze Age, can be regarded as a map, least of all—from the sound of it—the meandering single line of the so-called pirate map, evidently a popular interpretation.[73]

China and Mongolia, however, are a different matter. Cliff paintings of horselike figures at Yin Shan (Inner Mongolia) were being written about as early as the fourth century A.D., although the reference attracted attention only recently.[74] Since then, over a thousand rock paintings have been discovered in that area alone. Under the Chinese Republic, this type of art has been seen as the product of "minority ethnic nationalities" and of folk practices, as opposed to that of the literate peoples of the early urbanized and culturally advanced heartlands.[75] In some areas, such as Yin Shan, the pictographs are the accumulation of activity in both prehistoric and historical millennia; in others, such as Hei Shan (Gansu Province), they are thought to be wholly prehistoric in origin. Already the recently intensified study of rock art in China has resulted in the discovery of over thirty-six major zones or sites of prehistoric paintings and petroglyphs.[76]

In western Asia, the distribution of prehistoric rock art merges with those areas previously described under the headings Middle East and European Russia.[77] For instance, the Caucasus is now known to be rich in rock art sites, though few were known before 1967. In the Armenian Caucasus, one twelve-month visit by a group of archaeologists resulted in the discovery of over 100,000 rock carved or painted figures, all said to be "remarkably alike, if not identical" in style, within a straight-line distance of two hundred kilometers from the Aragats to the mountains around Sisian.[78] As far as the cartographic material is concerned, the main theme in this region seems to be celestial or cosmological. There are said to be numerous representations of the sun, individual constellations, and astronomical calendars.[79]

It is important to keep the cartographic content of all this prehistoric art in perspective. Figures usually described in the literature as "abstract" or "geometric" that may on closer inspection be interpretable as maps of one sort or another constitute a very small part of the entire rock art corpus.[80] Far and away the most common subjects are mammals (wild and domesticated), birds, and fish, followed by human or anthropomorphic figures and parts of the human figure (hands, faces, or masks). Weapons and tools are the third most important subject group.

The function or purpose of the art as reflected in the subjects portrayed may also reflect environmental or cultural restrictions. For instance, if plows and plowing scenes are found less frequently in some areas than in others (or in Asia in general compared with Europe), this may be because in much of Asia there was little or no plowing until comparatively recently. Even in the previous century, in Siberia for example, cultivation was restricted to parts of the Yenisei and Angara valleys. Elsewhere, nomadic hunters and gatherers peopled the

69. Gordon, *Pre-historic Background*, 98 (note 2). See also M. E. Gordon and Douglas Hamilton Gordon, "The Artistic Sequence of the Rock Paintings of the Mahadeo Hills," *Science and Culture* 5 (1939–40): 322–27 and 387–92; and Neumayer, *Indian Rock Paintings*, 3 (note 8).

70. For a list of Vishnu S. Wakankar's papers see the bibliography in Neumayer, *Indian Rock Paintings*, 46 (note 8). See also Robert R. R. Brooks and Vishnu S. Wakankar, *Stone Age Painting in India* (New Haven: Yale University Press, 1976).

71. Heinrich Kusch, "Rock Art Discoveries in Southeast Asia: A Historical Summary," *Bollettino del Centro Camuno di Studi Preistorici* 23 (1986): 99–108, quotation on 99.

72. Kusch, "Rock Art Discoveries," 106 (note 71).

73. William Meacham, *Rock Carvings in Hong Kong* (Hong Kong: Christian Study Centre on Chinese Religion and Culture, 1976), 33. Some of the petroglyphs have been "known for centuries" although, unusually, no legends or ancestral links seem to be associated with them.

74. Li Daoyuan (d. 527), *Shui Jing zhu* (River classic commentary), a book written about the sixth century A.D. The reference is cited by Chen, "Ancient Rock Art," 91 (note 18), and idem, *Cina*, 35 (note 17).

75. Chen, "Ancient Rock Art," 92–93 (note 18).

76. Chen, "Ancient Rock Art," map on 94 (note 18).

77. Delano Smith, "Cartography in the Prehistoric Period," 70–73 (note 1).

78. Martirosyan and Israelyan, *Naskal'nye izobrazheniya Gegamskikh gor*, 58 (note 10).

79. Kšica, *Umění staré Eurasie*, 71 (note 41).

80. There is always the problem of bias through the reporting of the most aesthetically pleasing pictures (notably animals and scenes of human activity).

steppes, deserts, and mountains of regions from Mongolia and the Transbaikal to Kazakhstan, Turkmenistan, and Afghanistan much as they would have done throughout the prehistoric period. In India and China, where farming arrived very early (fifth millennium B.C., and possibly earlier), it was confined, until the first millennium B.C., mainly to the valleys associated with the great Asian civilizations. Away from the Indus or the Huanghe (Yellow River), in the mountains of Kashmir and Nepal as in the hills of central and southern India, hunting provided the economic basis of life and the social framework until the introduction of herding of domesticated stock.

Given such long persistence of prehistoric ways of life in many parts of Asia, there is much for the historian of cartography to learn from ethnographers as well as from archaeologists. The possible significance of those dot-filled outlines painted on rocks in the grassy steppes of Mongolia as "grave plans"[81] could not have been guessed without reference to the surviving tombs and traditional burial practices and their links with prehistory. The use of constellations by certain rice-growing tribes of Southeast Asia to mark the agricultural seasons was investigated at the beginning of this century.[82] In India—in Bengal, Bihar, among the Gonds of Mandla, the Gallas of the south, the Pardhas, the Rathvas, and the Saoras, for instance—surviving practices of ritualistic wall painting have likewise been witnessed by anthropologists and ethnographers.[83] These researchers report that pictures relating to agricultural fertility sometimes include scenes of agricultural activity similar, we find, to those depicted on the rocks.[84] We learn too that the paintings are made in response to a particular crisis; that the painting is seen as integral to the remedy; that, although a shaman may in due course be involved, the drawing may be produced by any member of the tribe having sufficient skill and knowledge.[85] We are told, too, how the Rathvan painting of creation myths "is a multidimensional activity" in which music, dance, and trance are all essential parts of the productive ritual; how a consecrated area is set aside for the painting (a conclusion archaeologists usually reach only by deduction in the case of prehistoric rock art); and of the careful and deliberate way the figures are created, the tribal artists observing practices that may go back not just centuries but millennia.[86] Elsewhere in Asia, such as those regions of damp tropical climate where rock paintings are poorly preserved (Southeast Asia in general and Indonesia in particular), the focus shifts to the symbolism of motifs used in weaving and carving, where, as in Indonesian designs, cosmological symbolism is a recurrent theme.[87]

Rarely if ever does the word "map" appear in the archaeological or ethnographic literature on Asian art. Instead, a rock painting will be described as "portraying" or "showing" a village with its boundary line or fence.[88] The tympanum of an ancient drum or a textile will likewise be described in terms of "symbolizing" the upper world and netherworld of the cosmos.[89] Arguably, all that such semantic reticence highlights are the cartographic preconceptions of the observer. These limit the definition of a "map" to wayfinding devices or to depictions of an area according to mathematical coordinates. The observer's mind tends to remain closed to an entire range of well-documented experience, wholly relevant to the history of cartography though scarcely touched on in the traditional literature. The stress in recent history of cartography literature is to see the social use of maps and to be ready to abandon the "notion of mapping as plotting of resemblance" for certain periods.[90] Like medieval societies, prehistoric societies were sacred, not profane. For each individual in such archaic societies, the world was "fraught with messages."[91] Some of these messages, painted or carved by prehistoric people in Asia as elsewhere, have to do with place: the skies, the homesteads and fields, the graves of this world, and the unknown configurations of the next. We need a liberal as well as an informed approach to the surviving evidence of these prehistoric messages and to the archaeological and ethnographic contexts that illuminate them. Only

81. Hedin, *Expedition in Asia*, 1:109 (note 26). The graves are described as "walled squares of stones within which had been placed smaller stones," the largest measuring eight by four meters. On the significance of the symbols, see above, pp. 6–7 and note 27.

82. Charles Hose, "Various Methods of Computing the Time for Planting among the Races of Borneo," *Journal of the Straits Branch of the Royal Asiatic Society*, no. 42 (1905): 1–5.

83. Verrier Elwin, *The Tribal Art of Middle India: A Personal Record* (Bombay: Geoffrey Cumberlege, Oxford University Press, 1951), 183–214; D. H. Koppar, *Tribal Art of Dangs* (Baroda: Department of Museums, 1971); and Jyotindra Jain, *Painted Myths of Creation: Art and Ritual of an Indian Tribe* (New Delhi: Lalit Kala Akademi, 1984).

84. Elwin, *Tribal Art of Middle India*, 191–92 (note 83).

85. Koppar, *Tribal Art of Dangs*, 117 (note 83), also makes it clear that Dang paintings are in effect votive offerings, dedicated only to the god or gods to be propitiated. He points out that "in all these drawings there is not one single theme but a combination of several themes."

86. Jain, *Painted Myths*, ix–xii (note 83).

87. Thomassen à Thuessink van der Hoop, *Indonesische siermotieven*, 13 (note 47).

88. Wang's words are, "something people can now clearly recognize as a picture of a village" (*Yunnan Cangyuan bihua*, 33 [note 17]). I am grateful to Cordell Yee for this translation of Wang.

89. Wales, *Prehistory and Religion*, 69 (note 51); Thomassen à Thuessink van der Hoop, *Indonesische siermotieven*, 274–75 (pl. CXXIX) (note 47).

90. Jonathan J. G. Alexander, review of volume 1 of *The History of Cartography*, "Mapping the Medieval World," *Journal of Historical Geography* 16 (1990): 230–33, esp. 232.

91. Mircea Eliade, *The Sacred and the Profane: The Nature of Religion*, trans. Willard R. Trask (New York: Harcourt, Brace and World, 1959), 146.

then will the relevance of prehistoric art to the history of cartography be properly understood.

Conclusion

In searching for examples of prehistoric maps, the historian of cartography is looking beyond the mere analysis of pictorial forms, seeking the origins of concepts used in sometimes very different ways and circumstances throughout history. Also sought is an understanding of what these images express. The conclusion among students of rock art is that rock art expressed fundamental human anxieties and preoccupations. Okladnikov, one of the most eminent and experienced archaeologists in Siberian prehistory, drew attention to the high regard in which traditional Yakuts and Tungus held the cliff drawings of their region, believing they constituted "a form of writing, filled with profound and significant content."[92] He records how during the previous century Vitashevskiy was told that

> on the Olekma, upstream from a point where the river Nyukzha empties into it, was depicted the whole *sirkaartata;* that is, a map of the whole earth, the whole universe. It included the phases of the moon, from two days old to full, the sun, and the Great Bear *(Arangas Sulus).* These drawings, in their opinion, were made by the *khaya-ichchite* himself—the presiding spirit of the place—and the images depicted appear and disappear from time to time.[93]

We accept that, notwithstanding their specialized form, maps are no less social documents than other forms of art and text. The diversity, even unfamiliarity, of early forms of maps is no surprise. No more than the rock art in which they are preserved, itself "far from a drab collection of drawings dealing with the same, identically treated subjects," are the earliest maps from widely distributed territories likely to reflect identical practices or cultural predilections.[94] Allowing for cultural diversity, however, the study of maps in prehistoric art in Asia, as in Europe, does underline that the expression of spatial relationships is one of the great traditions of human existence. It also demonstrates that tribal societies have had a role as carriers of fundamental cartographic concepts. Established histories of maps and mapmaking have tended to ignore prehistoric cartography and to belittle the map products of nonliterate historical times.[95] One reason for this neglect has been an excessive focus on the spatial aspects of the cartographic image at the expense of its equally important temporal, personal, contextual, and evaluative components.[96] Yet images, like their constituent signs, should be studied, as Geertz reminds us, not only as a means of communication but above all as a means of thought.[97] Faced with an unfamiliar context and still less immediately recognizable images such as those of prehistoric rock and mobiliary art, it is admittedly not always easy to see these drawings as significant symbols and as vehicles of thought about terrestrial, celestial, or cosmological space.

This chapter, as a summary of accessible knowledge, is designed to point out further research directions and to structure future dialogue. The starting points are encouraging. For in Asia there is the rich bonus of the ethnographic literature and the illumination this offers of the cartographic images found in the prehistoric rock art record. There is also the promise of discoveries yet to come. But even with the present state of knowledge, it is clear that in Asia, as in Europe and other parts of the world, the history of the human mapping impulse starts extremely early.

---

92. A. P. Okladnikov, *Yakutia before Its Incorporation into the Russian State*, ed. Henry N. Michael (Montreal: McGill-Queen's University Press, 1970), 212, referring to V. Vitashevskiy, ed., "Izobrazheniya na skalkh po r. Olekme" (Drawings on the cliffs along the Olekma River), *Izvestiya Vostochno-Sibirskago Otdela Imperatorskago Russkago Geograficheskago Obshchestva* (East Siberian department of the Imperial Russian Geographical Society's News) 28, no. 4 (1897).

93. Okladnikov, *Yakutia*, 212 (note 92), referring to N. B. Kyakshto, "Pisanitsa Shaman-Kamnya" (The cliff drawings of Shaman-Kamnya), *Soobshcheniya Gosudarstvennoy Akademii Istorii Materialnoy Kul'tury* (GAIMK: Report of the State Academy for the History of Material Culture), July 1931, 29–30. Okladnikov identified the cliff drawings as those of the "Shaman-Kamnya" (stone) on which are depicted "animals, hunting scenes, the sun, moon, and stars, the sun being given features of the human face" (450 n. 22).

94. A. P. Okladnikov and A. I. Martynov, *Sokrovishcha tomskikh pisanits* (Treasures of the Tomsk petroglyphs) (Moscow, 1972), 252.

95. As summarized by Delano Smith, "Prehistoric Maps," 45–49 (note 1).

96. Joseph Michael Powell, *Mirrors of the New World: Images and Image-Makers in the Settlement Process* (Folkestone, Eng.: Dawson; Hamden, Conn.: Archon Books, 1977), 18, makes this point, developing the ideas of Kenneth Ewart Boulding, *The Image* (Ann Arbor: University of Michigan Press, 1956).

97. Clifford Geertz, *Local Knowledge: Further Essays in Interpretive Anthropology* (New York: Basic Books, 1983), 120.

APPENDIX 1.1 LIST OF PREHISTORIC MAPS

This appendix enumerates, with locations and citations to the relevant literature, the sites and artifacts in which cartographic representations have been identified. References included in this list are only those where a cartographic interpretation has been suggested or commented on; the general literature is cited in the text. The identification of maps included here has been derived from a variety of disciplines, and in some cases their interpretation may be still regarded as controversial. This seems an appropriate juncture, however, to set out this corpus as a basis for future discussion and elaboration.

Number of Map, Province, State, or County	Commune and/or Locality (italics indicate the usual name in the literature); Description; Nature of Site; Type of Markings; Map Type; Date	Measurements	Reference; Observations; Figure Number in Text (if illustrated)
1 Yunnan Province, China	*Cangyuan*; cliff face; picture map; first millennium B.C.	1.8 × 3.2 m	Wang, *Yunnan Cangyuan*, 35; Chen, *Cina*, 102–3; figure 1.4
2 Madhya Pradesh, Bhopal District, India	*Jaora*; rock shelter; painting; cosmological map; Mesolithic (8000–2500 B.C.)		Neumayer, *Rock Paintings*, 14 and fig. 26e; figure 1.13
3 Hövsgöl Province, Mongolia	*Dood-Chulgan*; rock; painting; plan map ("grave plan"); prehistoric		Novgorodova, *Alte Kunst der Mongolei*, pl. 72; figure 1.7*b*
4 Hövsgöl Province, Mongolia	*Dood-Chulgan*; rock; painting; plan map ("grave plan"); prehistoric		Novgorodova, *Alte Kunst der Mongolei*, pl. 71
5 Hövsgöl Province, Mongolia	*Dood-Chulgan*; rock; painting; plan map ("grave plan"); prehistoric		Novgorodova, *Alte Kunst der Mongolei*, 104
6 Hövsgöl Province, Mongolia	*Hövsgöl-Nuur*; rock; painting; plan map ("grave plan"); prehistoric	approx. 70 × 35 cm	Okladnikov, *Petroglify Mongolii*, 88 (fig. 2); figure 1.7*a*
7 Töv Province, Mongolia	*Gachurt*; rock; painting; plan map ("grave plan"); Bronze Age		Novgorodova, *Mir petroglifov Mongolii*, 92 (fig. 33); figure 1.7*e*
8 Töv Province, Mongolia	*Gachurt*; rock; painting; plan map ("grave plan"); prehistoric		Novgorodova, *Mir petroglifov Mongolii*, 93 (fig. 34); figure 1.7*d*
9 Töv Province, Mongolia	*Ich-Tengerin-Am*; rock; painting; plan map ("grave plan"); prehistoric	approx. 50 × 55 cm	Okladnikov and Zaporozhskaya , *Petroglify Zabaykal'ya*, 2:238 (fig. 67.2); figure 1.7*c*
10 Töv Province, Mongolia	*Ich-Tengerin-Am*; rock; painting; plan map ("grave plan"); prehistoric		Okladnikov, *Der Hirsch*, 148 (fig. 41); figure 1.8
11 Unknown	Rock; painting; plan map ("grave plan"); prehistoric		Okladnikov and Zaporozhskaya, *Petroglify Zabaykal'ya*, cover illustration of vol. 1; figure 1.7*f*
12 Tuva Republic	*Mugur-Sargol*; stone 198; rock; petroglyph; plan map ("hut and yards"); prehistoric	approx. 10 × 25 cm	Devlet, *Petroglify Ulug-Khema*, 52; Devlet, *Petroglify Mugur-Sargola*, 143; figure 1.5*a*(1)
13 Tuva Republic	*Mugur-Sargol*; stone 283; rock; petroglyph; plan map ("hut and yards"); prehistoric	approx. 10 × 25 cm	Devlet, *Petroglify Ulug-Khema*, 74; Devlet, *Petroglify Mugur-Sargola*, 205; figure 1.5*a*(4)
14 Tuva Republic	*Mugur-Sargol*; stone 198; petroglyph; plan map ("hut and yards"); prehistoric		Devlet, *Petroglify Ulug-Khema*, 52; Devlet, *Petroglify Mugur-Sargola*, 143; figure 1.5*b*(3)

APPENDIX 1.1 (*continued*)

Number of Map, Province, State, or County	Commune and/or Locality (italics indicate the usual name in the literature); Description; Nature of Site; Type of Markings; Map Type; Date	Measurements	Reference; Observations; Figure Number in Text (if illustrated)
15 Tuva Republic	*Mugur-Sargol*; stone 257; petroglyph; plan map ("hut and yards"); prehistoric	40 × 55 cm	Devlet, *Petroglify Ulug-Khema*, 65; Devlet, *Petroglify Mugur-Sargola*, 195; it is illustrated with an attached line that could represent a path aligned with trees, posts, or other features
16 Tuva Republic	*Mugur-Sargol*; stone 257; petroglyph; plan map ("hut and yards"); prehistoric	38 × 30 cm	Devlet, *Petroglify Ulug-Khema*, 65; Devlet, *Petroglify Mugur-Sargola*, 195
17 Tuva Republic	*Mugur-Sargol*; stone 283; petroglyph; plan map ("hut and yards"); prehistoric	20 × 36 cm	Devlet, *Petroglify Ulug-Khema*, 75; Devlet, *Petroglify Mugur-Sargola*, 205; figure 1.5*a*(2)
18 Tuva Republic	*Mugur-Sargol*; stone 283; petroglyph; plan map ("hut and yards"); prehistoric	15 × 30 cm	Devlet, *Petroglify Ulug-Khema*, 75; Devlet, *Petroglify Mugur-Sargola*, 205
19 Tuva Republic	*Mugur-Sargol*; stone 283; rock; petroglyph; plan map ("hut and yards"); prehistoric		Devlet, *Petroglify Ulug-Khema*, 73; Devlet, *Petroglify Mugur-Sargola*, 205
20 Tuva Republic	*Mugur-Sargol*; stone 283; petroglyph; plan map ("hut and yards"); prehistoric; a comparatively large group		Devlet, *Petroglify Ulug-Khema*, 74; Devlet, *Petroglify Mugur-Sargola*, 205; figure 1.5*b*(1)
21 Tuva Republic	*Mugur-Sargol*; stone 283; petroglyph; plan map ("hut and yards"); prehistoric	65 × 90 cm	Devlet, *Petroglify Ulug-Khema*, 74; Devlet, *Petroglify Mugur-Sargola*, 205; figure 1.5*b*(2)
22 Tuva Republic	*Mugur-Sargol*; rock; petroglyph; plan map ("hut and yards"); prehistoric	approx. 20 × 30 cm	Devlet, *Petroglify Ulug-Khema*, in fig. 16; Devlet, *Petroglify Mugur-Sargola*, 234 (fig. 17.2); figure 1.5*a*(3)

References: Chen Zhao Fu, *Cina: L'arte rupestre preistorica*, Italian trans. Giuliana Aldi Pompili (Milan: Jaca Books, 1988); M. A. Devlet, *Petroglify Murgur-Sargola* (Moscow: Nauka, 1980); idem, *Petroglify Ulug-Khema* (Moscow: Nauka, 1976); Erwin Neumayer, *Prehistoric Indian Rock Paintings* (Delhi: Oxford University Press, 1983); E. A. Novgorodova, *Alte Kunst der Mongolei*, trans. Lisa Schirmer (Leipzig: E. A. Seemann, 1980); idem, *Mir petroglifov Mongolii* (Moscow: Nauka, 1984); A. P. Okladnikov, *Der Hirsch mit dem goldenen Geweih: Vorgeschichtliche Felsbilder sibiriens* (Wiesbaden: F. A. Brockhaus, 1972); idem, *Petroglify Mongolii* (Leningrad: Nauka, 1981); A. P. Okladnikov and V. D. Zaporozhskaya, *Petroglify Zabaykal'ya*, 2 vols. (Leningrad: Nauka, 1969–70); Wang Ningsheng, *Yunnan Cangyuan bihua di faxian yu yanjiu* (The rock paintings of Cangyuan County, Yunnan: Their discovery and research) (Beijing: Wenwu Chubanshe, 1985).

2 · Introduction to East Asian Cartography

NATHAN SIVIN AND GARI LEDYARD

SCOPE

"Asia" is a peculiarly European notion. It is an eccentric one by, say, Indian or Chinese standards. Asians seldom find the idea of Asia meaningful unless they have adopted Western categories of thought, and with them European ways of lumping things together. The word describes a nonentity. Asia is not a continent; it is divided by topographic obstacles almost impassable before modern times; it embraces a rich assortment of societies, cultures, and states; and through history it has juxtaposed some of the world's richest and poorest, as it does today.

Nor has any word resembling "East Asia" expressed a sense of unity among the countries that have drawn on Chinese institutions, ideologies, and techniques. The term did not exist in the vocabularies of any of its peoples until it was introduced from the West. We find no counterpart of "Europe" or "Western civilization" reflecting a shared consciousness, part of everyone's identity, that does not stop at frontiers. Except for Buddhist pilgrims, itinerant merchants, and occasional diplomats, East Asians, especially the ruling classes, stayed at home.

Whether the idea of East Asia has any value for the history of cartography depends on how one defines it. As a general term for Asia east of the great mountainous barriers, it has no significance beyond geography. Some of the cultures of that region adapted Chinese institutions and creeds, and others responded to the influence of India as borne by Hinduism and Buddhism. The border between these two zones shifted through history. Central Vietnam remained part of the Indianized Champa state until the 1470s, and the Mekong basin belonged to the Khmer world until the mid-eighteenth century. "East Asia" has of course been defined and redefined to fit various geopolitical ambitions. Those who pushed the Greater East Asia Co-Prosperity Sphere, that watchword of Japanese territorial ambitions in 1940–41, would have been delighted to include India.

"East Asia" is useful mainly as a cultural label. We mean by it the parts of Asia that were governed by a system based more or less on the bureaucratic hereditary monarchy of China. The connection was more than political. In these countries the elite before this century generally were educated in the Chinese classics and literature, wrote in classical Chinese, and were exposed to Neo-Confucianism. The last began as a quest for private self-cultivation leading to conscientious public lives, based on the teachings of Zhu Xi (1130–1200) and others. It kept that emphasis in independent lineages, but these influenced only a minority of intellectuals. The upper crust generally accepted and enforced state-sponsored orthodoxies based on rigid authoritarian interpretations of Zhu's philosophy, beginning in China early in the fourteenth century. In the fifteenth century in Korea and Vietnam, and in the seventeenth century in Japan, the state began promoting such orthodoxies. The illiterate or barely literate majority of the population were indirectly affected by the shared "Confucian" ideology of their rulers. Unlike the peoples of Southeast Asia, they directly experienced Buddhism in its Mahayana version, as recorded in Chinese-language canons that literate priests and monks studied.

These "East Asian" characteristics unite China, Japan, Korea, and Vietnam, the last with qualifications. Although these influences did not strongly affect the southernmost parts of its modern territory until about five hundred years ago, the northern part entered the Chinese cultural sphere in the second century B.C. Through history the north was repeatedly conquered and repeatedly fought free of Chinese control. Its elite even when independent wrote in the Chinese language, and continued to do so for some time after all of Vietnam became a French colony in 1884.

To sum up, despite its geographic and political shortcomings, "East Asia" is unobjectionable as a cultural designation. That suits the needs of this volume. Some maps are geographic, but cartography is culture.

As the authors make clear, mapmaking and map use in the four countries have been locally diverse, but before the twentieth century they were as marked by commonalities as what we find in Europe (which is, after all, a smaller part of the world). But those commonalities do not extend to Southeast Asia—Thailand, Cambodia, Laos, Burma, and Malaya—and so our remarks do not extend to it either. As for the Mongol and Manchu peoples of northeast Asia and the Tibetans, on the whole their elites depended on China for neither language nor forms of governance. When Mongols and Manchus conquered part or all of their rich agricultural neighbor, however, their leaders quickly grasped both language and bureaucracy. That does not make them essentially East Asian in culture, but this volume is an obvious place for reports on Greater Tibet and (more briefly) Mongolia.

The Variety of East Asia

To assume that adaptations of Chinese culture made East Asia uniform would leave us unable to account for the distinct visual worlds that the maps in this volume reveal. Vietnam, Korea, and Japan had their own cultures long before they came under Chinese cultural influence. Their material cultures, from food and housing to ceramics and metal-working; their archaic ruling structures; their religious traditions, both ancestral rituals and popular forms of reverence for nature and gods; the vernaculars in which they spoke, thought, and remembered: all of these differed fundamentally.

The people of these countries were not passive recipients of Chinese influence. They welcomed it at times, at other times rejected it,[1] and to a remarkable degree decided for themselves what suited their own circumstances. Buddhism and then Neo-Confucianism indeed became a cultural cement, but despite their dependence on common canons they varied locally in important ways.[2] Chinese popular religion had some influence on folkways elsewhere, but its specialized outgrowth Daoism (which borrowed extensively from esoteric Tibetan religion and other traditions of Buddhism) had practically none.[3]

Another example of discriminate appropriation is classical Chinese itself. Koreans, Japanese, and Vietnamese made it a language of learning much as Europeans did Latin, but what they recorded would not have been understandable in many respects to Chinese contemporaries. In any case few Chinese were curious about foreign writings. They also used the written characters to transcribe their own vernaculars, eventually abbreviating and stylizing them to form the Japanese *kana* and Vietnamese *chữ-nôm* scripts. The alphabet that has replaced Chinese graphs in Korea is an essentially independent invention, and modern Vietnamese still use the spelling that the French imposed on them.

These adaptations are scarcely surprising. It is natural to think of China as a homogeneous cultural and political unit, but it is too large for that, even today. Just as dynastic change was often barely felt in distant provinces, the diversity of which we have just given examples is mirrored within the country. We can, for instance, find consistent local variations in the writing of classical Chinese, and recorded dialects used graphs (some of them not found in so-called Mandarin) in idiosyncratic ways. Just as histories of all Europe are not a staple genre, it is becoming clear that China is too large for much of the generalization customary among historians.[4] The linguistically diverse peoples of the isolated southern provinces also chose and adapted from the culture of the political and commercial centers.

Thus we see that East Asia did not begin as culturally Chinese, and as every reader knows, it did not end that way. After discussing the ancient Chinese terminology for maps, we will return to the cartographic multiplicity that these civilizational differences bred. (For a generalized time line of East Asian history, see table 2.1.)

Terms

The Chinese written language was, until about the third century B.C., largely built on the idea that (proper names aside) one word should be expressed by one graph. This led to magnificent concision, but since the number of graphs remained limited, it meant that a given graph might stand for many related ideas. Ambiguity was later avoided by combining graphs to make compounds; but in the first phases of writing, one avoided it by making sure the context was specific. Chinese writers did that with great skill. The problem with reading the early classics is not generally that their diction is unclear, but that scholars today know too little about the resonances of individual

1. On this point see, for example, Masayoshi Sugimoto and David L. Swain, *Science and Culture in Traditional Japan: A.D. 600–1854* (Cambridge: MIT Press, 1978). This book organizes its narrative around alternating periods of "cultural waves," first from China and then from Europe, and periods in which contacts with foreign countries were cut off as new ideas and usages were assimilated.

2. It is well known that this is true of Buddhism. There has been much less comparative study of Confucianism, but see in particular William Theodore de Bary and Irene Bloom, eds., *Principle and Practicality: Essays in Neo-Confucianism and Practical Learning* (New York: Columbia University Press, 1979).

3. Except, remarkably, among peoples on the non-Mahayana southern periphery of China. This topic has only recently come under active scrutiny by anthropologists. See Michel Strickmann, "The Tao among the Yao: Taoism and the Sinification of South China," in *Rekishi ni okeru minshū to bunka: Sakai Tadao Sensei koki shukuga kinen ronshu* (Peoples and cultures in Asiatic history: Collected essays in honor of Professor Tadao Sakai on his seventieth birthday) (Tokyo: Kokusho Kankōkai, 1982), 23–30. The peoples studied in northeast Thailand and Laos are non-Han but originated within the historical borders of China. We refer to the Daoist religious movements, not to the early philosophic classics that became part of the common literary heritage first of East Asia and then of the world.

4. The most widely used set of smaller units are the physiographic "macroregions" of G. William Skinner. His most eloquent argument for a history based on them is his "The Structure of Chinese History," *Journal of Asian Studies* 44 (1985): 271–92.

Table 2.1 Generalized Time Line of East Asian History

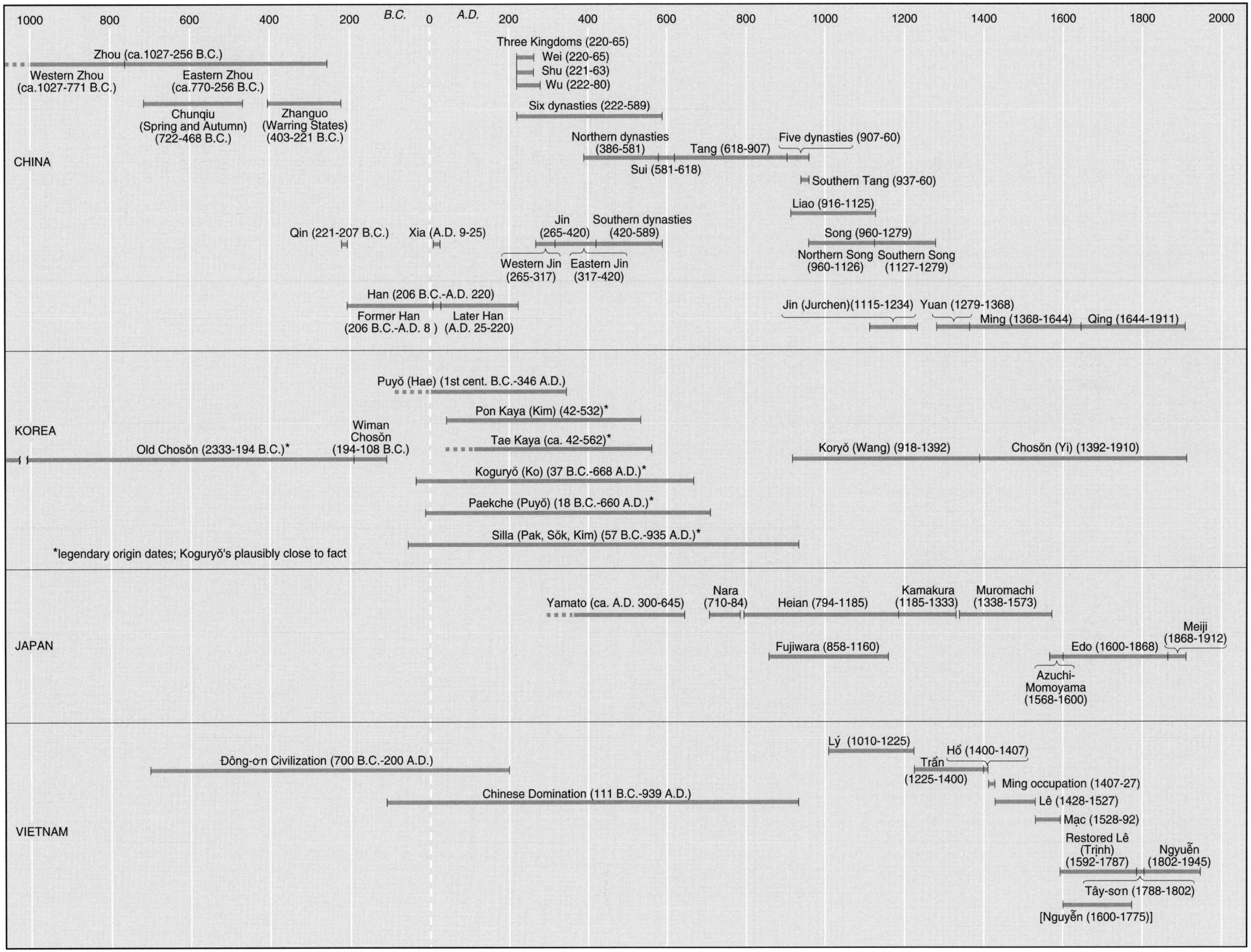

1000 800 600 400 200 B.C. 0 A.D. 200 400 600 800 1000 1200 1400 1600 1800 2000

CHINA

Zhou (ca.1027-256 B.C.)
Western Zhou (ca.1027-771 B.C.)
Eastern Zhou (ca.770-256 B.C.)
Chunqiu (Spring and Autumn) (722-468 B.C.)
Zhanguo (Warring States) (403-221 B.C.)
Qin (221-207 B.C.)
Xia (A.D. 9-25)
Han (206 B.C.-A.D. 220)
Former Han (206 B.C.-A.D. 8)
Later Han (A.D. 25-220)
Three Kingdoms (220-65)
Wei (220-65)
Shu (221-63)
Wu (222-80)
Six dynasties (222-589)
Jin (265-420)
Western Jin (265-317)
Eastern Jin (317-420)
Southern dynasties (420-589)
Northern dynasties (386-581)
Sui (581-618)
Tang (618-907)
Five dynasties (907-60)
Southern Tang (937-60)
Liao (916-1125)
Song (960-1279)
Northern Song (960-1126)
Southern Song (1127-1279)
Jin (Jurchen)(1115-1234)
Yuan (1279-1368)
Ming (1368-1644)
Qing (1644-1911)

KOREA

Old Chosŏn (2333-194 B.C.)*
Wiman Chosŏn (194-108 B.C.)
Puyŏ (Hae) (1st cent. B.C.-346 A.D.)
Pon Kaya (Kim) (42-532)*
Tae Kaya (ca. 42-562)*
Koguryŏ (Ko) (37 B.C.-668 A.D.)*
Paekche (Puyŏ) (18 B.C.-660 A.D.)*
Silla (Pak, Sŏk, Kim) (57 B.C.-935 A.D.)*
Koryŏ (Wang) (918-1392)
Chosŏn (Yi) (1392-1910)

*legendary origin dates; Koguryŏ's plausibly close to fact

JAPAN

Yamato (ca. A.D. 300-645)
Nara (710-84)
Heian (794-1185)
Kamakura (1185-1333)
Muromachi (1338-1573)
Fujiwara (858-1160)
Azuchi-Momoyama (1568-1600)
Edo (1600-1868)
Meiji (1868-1912)

VIETNAM

Đông-ơn Civilization (700 B.C.-200 A.D.)
Chinese Domination (111 B.C.-939 A.D.)
Lý (1010-1225)
Trần (1225-1400)
Hồ (1400-1407)
Ming occupation (1407-27)
Lê (1428-1527)
Mạc (1528-92)
Restored Lê (Trịnh) (1592-1787)
Tây-sơn (1788-1802)
Ngyuễn (1802-1945)
[Nguyễn (1600-1775)]

words to interpret them as readers did when they were written. In all four societies literacy (defined by contemporary standards) was rare, and the elite tended to share great provinces of allusion and symbol.

Nevertheless, in reading ancient documents one often comes across a word that, regardless of context, refers to what modern readers consider two or more quite different things. It would be foolish to assume that the language of the time was incapable of resolving the ambiguity. As we can see in one instance after another, the linguistic ambiguity is there because the ancients' convictions about what ought to be kept separate do not happen to agree with ours. We can learn from this equivocality if we are attentive to their opinions about the fitness of things without being distracted by our own. The etymology of "map" is a case in point.

Tu is the word used consistently in archaic writing to designate maps, but it never referred to maps alone.[5] One can draw no conclusions about etymology from the form of the graph. Compilers of dictionaries classify it by the "enclosure" radical, shown in the modern form at the far left of figure 2.1 as the box around the outside. This system of radicals was applied very late, probably first in *Shuowen jiezi* (Explanation of writing and explication of graphs), the great etymological dictionary compiled ca. 100. Radicals often have nothing to do with the original meanings of graphs. The "enclosure" box often occurs as a part of graphs whose meaning has to do with the enclosure of space, but one cannot conclude from this observation that the character represents a map of some sort. In its forms in early bronze inscriptions, of which figure 2.1 shows examples, the design inside the box varies too greatly to invite speculation about what it portrays. The character as a whole is certainly not a simple pictograph. If it is an ideograph (what early lexicographers called *huiyi*), present knowledge is not adequate to decipher its origin.

Nor for that matter was the knowledge of early etymologists adequate to reliably explain graphs that had been in use for nearly two millennia; their explanations are sometimes far-fetched. For instance, the *Shuowen jiezi* defines *tu* as "difficulty in planning" (*huaji nan ye*), a meaning that does not occur in earlier texts. The dictionary analyzes it into an enclosure and the enclosed part. The latter, it avers, means "difficult." But it is easily seen that, of the three early bronze forms in the figure, only the second even remotely resembles what the dictionary says is the original form. This definition is useful only as an indication that scholars in the second century A.D. connected its origins with planning (the Han meaning of the compound *huaji*). It is interesting, but not significant, that they did not connect this origin with mapping.[6]

Tu is unusual in lacking both homophones with clearly

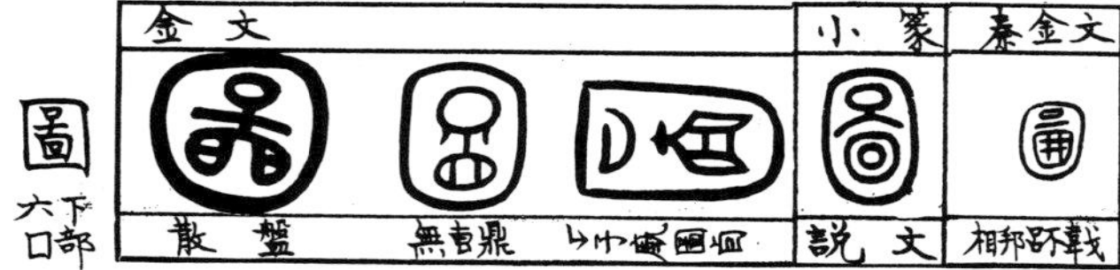

FIG. 2.1. EARLY FORMS OF THE GRAPH *TU*. The small graph at the far left is the modern form as written with a brush. The graph at far right is the "small seal" version associated with the dictionary *Shuowen jiezi*. The three graphs in the middle are from Zhou dynasty bronze inscriptions. The first is from the ninth-century basin, the *San pan*, discussed below. Reproduced from Zhang Xuan (Chang Hsüan), *Zhongwen changyong sanqian zixing yishi/The Etymologies of 3000 Chinese Characters in Common Usage* (Hong Kong: Hong Kong University Press, 1968), 171.

related meanings and other characters that share its graphic form. Ingenious etymological speculations have accumulated over the centuries, but the means to choose between them are so far missing.

Tu as it occurs in writings before 300 B.C. has a great many meanings. It can refer to pictures, diagrams, charts, and tables. As a verb it can refer to planning, anticipating, giving thought to, or dealing with something. In many documents, as Cordell Yee shows below, it is impossible to tell which sense is meant, because we do not share the knowledge of context that ancient authors expected of their readers. Some instances are clear enough. One of the earliest occurs in an inscription of 359 graphs on a large bronze basin that was probably cast in the mid-ninth century B.C. It commemorates the settling of a boundary dispute between San and a state so obscure

5. The classic reconstruction of early phonetics, Bernhard Karlgren, *Grammata Serica Recensa* (Stockholm, 1957), reprinted from the *Bulletin of the Museum of Far Eastern Antiquities* 29 (1957): 1–332, esp. 37, item 64a–c, proposes the Old Chinese reading **d'o* (roughly 700 B.C.). More recent authorities posit an assortment of alternatives: *dag* (Li Fanggui), *dɔ* (E. G. Pulleyblank), *da* (Axel Schuessler), *d/la* (William Baxter). By the second century A.D. the word's pronunciation in northern China was perhaps closer to **d'ân̆*. Its Early Middle Chinese pronunciation (ca. A.D. 600) was *dwo*. See Edwin G. Pulleyblank, *Lexicon of Reconstructed Pronunciation in Early Middle Chinese, Late Middle Chinese, and Early Mandarin* (Vancouver: UBC Press, 1991), 311; Axel Schuessler, *A Dictionary of Early Zhou Chinese* (Honolulu: University of Hawaii Press, 1987), 615–17; and William H. Baxter, *A Handbook of Old Chinese Phonology* (Berlin: Mouton de Gruyter, 1992), 649.

6. Xu Shen, comp., *Shuowen jiezi*, 12:13a, in *Shuowen jiezi yuezhu* (*Shuowen jiezi* with a simple commentary), ed. Zhang Shunhui, 3 vols. (completed 1971; Loyang: Zhongzhou Shuhua She, 1983). Karlgren (*Grammata Serica Recensa*, item 847a–d [note 5]) uncharacteristically speculates that *hua* depicts a hand drawing a map, but even if this were true, *huaji* clearly refers to planning. It is possible that *nan* is a kind of wordplay (paranomasia), meant to be taken not literally but by the way its sound (something like **t'nân* in the second century A.D.) echoes that of **d'ân̆*, probably the contemporary pronunciation of *tu*.

that it has never found its way into the geographical dictionaries. After officials set up boundary posts and exchanged solemn pledges never to encroach on each other's territories, a *tu* was executed. Whether this was precisely a map is impossible to say, but the context makes it reasonably certain that it was some sort of diagram rather than a written record.[7]

It is interesting to compare the meanings of *tu* with those of its counterparts elsewhere. Late Greek *chartes*, from which "cartography" derives, refers to a sheet of papyrus, and late Latin *mappa* means "a cloth"; both are words for the material on which a map is drawn. In the main premodern Islamic languages we find a number of words derived from roots for "form," "draw," or "paint," at least as broad in scope as *tu*. The word for map in most Indian languages, derived from an Arabic word, may mean not only a picture but a general description or report. With respect to *tu*, China would thus be intermediate in concreteness between Europe and India.[8] It shares with modern European languages the overlap of sense between "map" and "plan."

In the fourth and third centuries B.C., nouns began more often to take the form of two-character compounds, which restrict the possibilities of ambiguity at the cost of added information. A number of terms that mean "map" and nothing else appear beginning at this time. The most prevalent, *ditu*, combines *tu* with *di*, the common graph for land or place.[9]

The appearance of compound words does not mean we are no longer frustrated by passages that may or may not be about maps. The problem, as the authors of this section explain, is a matter not of language but of cultural practice. Graphical maps and informational texts in China form a unit that it is not always possible to dissociate, an important point to which we will return.

Contents

Chapters 3–14 on East Asia use the broad definition of "map" characteristic of this *History*. They include separate chapters on geographic cartography, with attention to the cosmological and religious significance of maps, in China, Korea, Japan, and Vietnam, and on celestial cartography in China, Japan, and Korea. The chapters on geographical maps in China present a strikingly new interpretation of their general character. Its potential utility for the history of cartography is so great that it calls for full statement and documentation. To understand in what sense this argument is a new departure, it is necessary to pause briefly over the historiography of Chinese mapmaking. But we should emphasize that the rest of the volume is amply innovative in ways that reflect the new scope and emphases of the series.

Historiography

Western historical studies in East Asian cartography began just after the turn of the century with an essay by Chavannes.[10] This excellent contextual study of what were then the two oldest extant Chinese maps was published in Hanoi and then ignored for decades by European compilers of histories. Chavannes, probably the greatest Sinologist of his era and certainly the broadest, rather than using the maps primarily to illuminate Song culture, concentrated on the question of accuracy in mapping. So did his successors.

In China the "evidential research" scholars who, beginning in the seventeenth century, undertook a massive critical study of the classical heritage were curious about old maps as well as every other relic of antiquity.[11] Modern studies began with two papers in a geological journal in 1911.[12] A book-length survey by Wang Yong (1958)

7. The *San pan* is one of the most celebrated bronzes in the old Palace Museum, now in Taipei. A rubbing is clearly reproduced in *Chinese Cultural Art Treasures: National Palace Museum Illustrated Handbook*, 3d ed. (Taipei: National Palace Museum, 1967), diagram 10. Because two characters in the sentence after the one we summarize are unreadable, the syntax is somewhat uncertain. For translated examples of the earliest meanings, "to plan," "planned," "a plan or map," see Schuessler, *Early Zhou Chinese*, 615–17 (note 5). This book adduces only a few bronze inscriptions, however, and those that document *tu* are not among them. For additional classical meanings see Karlgren, *Grammata Serica Recensa* (note 5).

8. See the preface and Ahmet T. Karamustafa, "Introduction to Islamic Maps," in *The History of Cartography*, ed. J. B. Harley and David Woodward (Chicago: University of Chicago Press, 1987–), 1:xv–xxi, esp. xvi–xvii and nn. 7 and 13, and vol. 2.1 (1992), 3–11, esp. 7–8.

9. A chapter with this title occurs in the eclectic *Guanzi* ([Book of] Master Guan), which was compiled in the first century B.C. from materials written as early as the fifth century. The chapter seems to be a fragment of a lost military manual and probably belongs to a late stratum. It is translated in W. Allyn Rickett, trans., *Guanzi: Political, Economic, and Philosophical Essays from Early China* (Princeton: Princeton University Press, 1985–), 1:387–91. Some commentators give *ditu* an alternative sense of terrain (a rare but not unique usage). This rather than "maps" may be the meaning of the title. The subject of the chapter is in fact terrain and a number of other concerns of the tactician. Although Rickett does not query the translation "maps," see his remarks on 389.

10. Edouard Chavannes, "Les deux plus anciens spécimens de la cartographie chinoise," *Bulletin de l'Ecole Française d'Extrême-Orient* 3 (1903): 214–47. Earlier writings on the topic, such as William Huttman, "On Chinese and European Maps of China," *Journal of the Royal Geographical Society* 14 (1844): 117–27, are much less adequately informed.

11. For instance, Hu Wei's 1697 reconstruction of a map in the *Yu gong*; see his *Yu gong zhuizhi* (Using an awl to gauge the depths of the *Yu gong*), in *Huang Qing jing jie* (Explications of the classics from the Imperial Qing, compiled 1825–29), ed. Ruan Juan and Yan Jie, 27:53b.

12. Zhang Yi, "Zhongguo gudai ditu zhi bijiao" (A comparison of ancient Chinese maps), *Dixue Zazhi* 2, no. 5 (1911): 1–8, and Tao

largely defined the field but has been partly superseded by a new general history.[13]

There are now a few specialists, of which the best known, in addition to the authors of these chapters, are Cao Wanru in China, Mei-ling Hsu in the United States, Funakoshi Akio in Japan, and Yi Ch'an (Chan Lee) in Korea. They share with their predecessors the viewpoint of the cartographer or geographer; that is, a concern with ancient mapmakers as predecessors of modern mapmaking technique (and, particularly in the People's Republic of China, science). Scholars in East Asia for more than two generations have concentrated on finding the documents, setting them in order, and describing them to a readership of cartographers and others. They have studied them as unique objects whose documentary and social matrices are only incidentally relevant. This largely antiquarian and philological effort has accelerated with the general growth of institutions in the history of science and with the remarkable frequency of important archaeological discoveries since the 1950s.

The positivist view of ancient cartography as a gestating technology is still prevalent in China, Japan, and Korea.[14] The philological work is generally high in quality and continues to add important sources to the accessible record, but evaluations tend to stress documents and "achievements." This is particularly true in the People's Republic of China, where the official view of history makes science an unproblematically progressive force, and where the imperatives of nationalism prod historians to find Chinese technical priorities. The result of this scientism has naturally been an emphasis on geographic information, accuracy of scale, and elaboration of map signs. There has been little attention to the socioeconomic, aesthetic, and moral dimensions. There has been none at all to the abstract cosmic diagrams of which John Henderson shows the historic importance in chapter 8, or to the maps of visionary space that were important in Buddhist and Daoist practice (see chapter 11 for Japanese Buddhism, and chapter 15 for Tibetan Buddhism).

The most important inquiry in the West, with not a little influence in East Asia, has been that of Joseph Needham. His 1959 essay of sixty-odd pages, "Quantitative Cartography in East and West," was (as the title indicates) no exception to the positivist and progressivist trend, and its "East" was China. But it was pathbreaking in several respects. He embedded it in a general reconnaissance of science and civilization. It focused on well-informed comparisons with European mapmaking. Needham's catholic view of science led him to make religious cosmology, Eastern and Western, an integral part of the inquiry, even though he did not examine it in depth. His curiosity about concrete routes of transmission led him to explore the role of the Islamic world as an intermediary between East and West.[15]

Cordell Yee shows below the fundamental flaws in Needham's demonstration that map grids evolved steadily from the Han up to the point when Chinese practice was replaced by European methods in the seventeenth century. Needham's essay was nevertheless a carefully articulated and fully documented argument, meant to encourage and ease studies that would test and improve upon it. In that it succeeded. Because of its erudition, it has not been superseded in any important respect until the survey in this volume, which not only corrects the interpretive errors of this pioneering work but replaces its problematic with a much more commodious one.

The Means and Ends of Cartography

This volume presents the challenge of several humanists to views that see mapmaking exclusively as science and technology. As Yee puts it, "To achieve literacy in traditional Chinese cartography, one needs grounding in the history of science and technology, art, literature, government, economics, religion, and philosophy—in short, the polymath range of the mapmakers" (p. 228 below). This is clearly not a view that excludes or diminishes the technical dimension of mapmaking. Yee sees it as one dimension of a larger picture, none of which will make sense without a rounded view.

Maps assuredly have been more or less accurate spatial representations that could guide exploration. But as Yee remarks below, the fact that Chinese cartography shared "the aesthetic principles of painting and poetry" could

Maoli, "Zhongguo dituxue faming zhi yuanshi ji gailiang jinbu zhi cixu" (The origins of cartographic invention and steps toward reform and progress in China), *Dixue Zazhi* 2 (1911): no. 11, 1–9, and no. 13, 1–9.

13. Wang Yong, *Zhongguo ditu shi gang* (Brief history of Chinese cartography) (Beijing: Sanlian Shudian, 1958). This is a revision of two chapters in Wang Yong's *Zhongguo dilixue shi* (History of geography in China) (1938; reprinted Taipei: Shangwu Yinshuguan, 1974). Four chapters are translated in Donald J. Marion, "Partial Translation of *Chung-kuo ti-t'u shih kang* by Wang Yung: A Study of Early Chinese Cartography with Added Notes, an Introduction and a Bibliography" (M.A. thesis, Graduate Library School, University of Chicago, 1971). The recent history is Lu Liangzhi, *Zhongguo dituxue shi* (History of Chinese cartography) (Beijing: Cehui Chubanshe, 1984). The most complete bibliography of essays in the field is in Yan Dunjie, *Zhongguo gudai kejishi lunwen suoyin 1900–1982* (Index of essays on the history of ancient Chinese science and technology, 1900–1982) (Nanjing: Jiangsu Kexue Jishu Chubanshe, 1986), 127–32 and 907, 79 items to 1982. For Western publications and publications on cartography elsewhere in East Asia one must use more general bibliographies.

14. We do not know what historical research and publication are under way in Vietnam.

15. Joseph Needham, *Science and Civilisation in China* (Cambridge: Cambridge University Press, 1954–), vol. 3, with Wang Ling, *Mathematics and the Sciences of the Heavens and the Earth* (1959), 525–90.

provide as adequate a theme for its history as could mensuration (p. 164). An impartial survey of maps' place in history cannot privilege one role or another. It is equally worthy of reflection that they were used for education, for aesthetic appreciation, to express emotional states, to represent power, to settle disputes, to symbolize submission or subordination, and to promise immortality. Yee is not merely suggesting that all these functions be studied as part of one picture; he and to varying extents the other authors show by example how it can be done.

In considering integrally the uses of mapmaking, Yee reveals again and again the wishful thinking to which the sentimental view of cartography as a technical march of progress has led. He does not brush aside, as his predecessors have done, these important facts: that too few maps survive from before the eleventh century to permit generalizations about practice in the first millennium A.D., much less earlier; that it was usual even for careful planimetric maps to show certain features such as mountains and buildings in elevation; that a consistent scale over a whole map is the exception; that the grid found on Chinese maps is not based on a coordinate system but is primarily an aid for estimating distances between points; and that even so grids were still not the norm in officially sponsored nineteenth-century maps. To the contrary, he treats these characteristics as valuable evidence for situating cartography in the culture that created it. He thus corrects once and for all the common fallacy that cultures that reward literary, artistic, and bureaucratic pursuits cannot encourage technological achievement.

No one will deny, we are sure, that before we compare the achievements of two cultures, both ought to be understood in their own contexts. It remains true that this is practically never done, since it demands of specialists an effort at polymathy of precisely the sort Yee has described. But his example and those of the other authors will no doubt encourage others.

John Henderson's discussion of cosmological diagrams in chapter 8 is also characteristically innovative. He shares with Yee a strong sense of development and change over the two-thousand-year life span of imperial China. His contribution provides a model for discussing nongeographical maps. His examination of philosophical cosmology furnishes an interesting contrast with the visionary luxuriance of Indian cosmographic mapping as described by Joseph E. Schwartzberg in chapter 15 of book 1 of this volume. In this book Kazutaka Unno and Schwartzberg reveal the Japanese and Tibetan Buddhist visions of the spiritual macrocosm. F. Richard Stephenson and Kazuhiko Miyajima, both historians of astronomy, provide a great deal of reliable information about Chinese, Japanese, and Korean star maps. The survey of Korean cartography by Gari Ledyard is attentive to the cultural and social particularities that have shaped cartography, and to interactions with China and the West. John K. Whitmore has provided the little information about mapping in Vietnam that is available in present political circumstances.

Text and Map

Another portentous conclusion that emerges from this volume is that in the Chinese tradition, and to a large extent throughout East Asia, the map is not always the appropriate unit of study for the history of cartography. This conclusion was natural as a result of taking seriously the characteristics of East Asian maps—unavoidable, that is, for scholars free of the parochiality that has led many historians of science to take as their task explaining Chinese inferiority. Yee and his colleagues make a solid case that what appear to be limitations of East Asian maps arise from the fact that they are not used alone but are read in conjunction with text. Regularly it turns out that a map contains little quantitative information. In some cases, of course, the makers were incapable of providing it, but in others they intended their map merely to give a quick impression of spatial relationships alongside a text that itemizes distances and directions in great detail. A map mainly meant to complement verbal description, as was common even in late gazetteers, has no need for a scale.

Anyone familiar with East Asian art is aware that painting and calligraphy, depiction and writing, share methods, materials, languages of gesture, and aesthetics. It is obvious, once Yee points it out, that there is no firm line, even that of convention, between painting and map. Cartography was not the province of specialists. Any magistrate was expected to draw a presentable map when one was needed, because he was trained to write beautifully and to paint. His staff would probably do the legwork for a survey if the map were to be based on a new one rather than on old records, and he might have an artist on his staff, but there is no evidence that even the central government had a special cartographic staff except when special projects made it necessary to organize one.

It was obviously not easy for a literatus to meet the technical demands of mapmaking without a good bit of experience. But if gentlemen valued the fusion of description and subjective experience when they saw it in landscape painting (or in the calligraphy of a landscape poem), it is not surprising that they should be willing to combine depiction of mountains and buildings in elevation, as we experience them, with an overall view from above. This is another instance of the commonsense demand for multiple viewpoints that naturally limited the role of convergent perspective in the Chinese visual arts. The chapters in this book make all of this clear. They teach us how important it is not just to look at the map, but to analyze the interaction between the graphic depiction and the legend on the map, and between the map and the textual description that accompanies it—or, more often, that the map accompanies.

We also learn that the shaping force of China on East Asian cartography was far from total. Both local practice and competing influences from elsewhere made the maps of each country distinct.

Although in China, as we have remarked, grids were meant primarily for estimating distances, the first grids in Korea (1791) established a countrywide standard. National maps had already been evolving for two generations toward fixed scale and orientation, a uniform grid for the whole country, and the more or less total banishment of text. The two maps developed on this basis in the nineteenth century used a grid primarily as coordinates. Their maker, Kim Chŏngho, who had no technical peer among makers of national maps in China, numbered his ranks and files to help users find places. Local maps made by others employed the same coordinate system.

There are many such examples of ideas that originated in one civilization but were used creatively in another. The science of siting (or "geomancy"), which studies the flow of vital substance (*qi*) through the contours of the land in order to find dynamically balanced sites for buildings and tombs, was Chinese in origin, but only Koreans used it as a structure for national cartography. Again, Islamic materials that the Chinese had found useful only for a map of the "Great Ming" made it possible for the Koreans to produce a genuine world map, the *Kangnido*.

Unno's contribution makes clear the profound cartographic difference between Japan's Buddhist and Shinto cultures on the one hand and the Chinese and Korean combination of bureaucracy and "Confucian" orthodoxy on the other. In addition, from the seventeenth century on, merchants in the evolving Japanese urban milieu commercialized mapmaking. The travelers who filled the roads could choose from dozens of competing route maps, all in the colorful Edo style. In China and Korea, mapmaking was seldom tied to exploration, but the Bakufu's agents struggled through voyage after voyage to establish the outlines of Hokkaidō and Sakhalin. There is no doubt that Vietnamese cartography, once we know enough about it, will turn out to be equally distinctive.

Another point previously obscured by wishful thinking is the extremely limited impact of European cartographic methods and conceptions in China. In the early seventeenth-century Jesuit missionaries were able, in the exceptional circumstances of the Manchu occupation, to demonstrate the superiority of their eclipse prediction technique and in short order take over the Directorate of Astronomy. They succeeded because they and their hosts had the same uncomplicated view of what constituted a better prediction.

There was no such agreement about what constituted the best map. Chinese mapmakers did not see their task as projecting a spherical earth on a flat surface using rigorously geometric procedures. They did not see the earth partitioned by lines of longitude and latitude. They were familiar with these concepts long before the missionaries arrived, for their mathematical astronomy used ecliptic as well as equatorial coordinates; but they had no reason to project them on the earth. Their essentially numerical approach to astronomical prediction did not oblige them to decide whether the earth was flat, discoidal, or spherical. It is not precisely that mapmakers were convinced the earth was flat: that question did not arise in connection with their work. They simply acted as if they were transferring points from a very large flat surface to a smaller one.

Chapter 7 takes a fresh look at Chinese maps from the seventeenth century on and shows how little they were affected by Western innovation. That is true even though the maps Europeans produced for the 1718 national atlas, based on the Beijing prime meridian, represented the state of the art. Yee points out that because the eighteenth-century Jesuit maps rigorously used a standardized scale, they did not need a text that listed distances. They were among the first in China that were in principle independent of text. Shen Kuo in the eleventh century had claimed hyperbolically that his map of China could if necessary be reconstructed from his text; here were maps from which a geography handbook could be read.

But the Jesuit atlas did not change provincial and local practices; the graticule and scale were not widely adopted, even in late official maps, and no standard ever emerged for local government practices. The modernizers who spread information about the rest of the world were not cartographically up to date except in the maps they copied from foreign publications. Their aim was to improve policy, not mapmaking. Mixed pictorial modes remained common up to the twentieth century, and the popularity of religious and magical maps was not threatened. This is understandable given the discrepancy at the end of the imperial era in European and Chinese values, especially those that affected the weight given purely technical criteria.

European cartographic methods also had little impact in Korea but a great deal in Japan. Despite the Tokugawa regime's strict policy forbidding contact with foreign countries, Western maps continued to be imported and copied, and their sale flourished. Japanese were avid for marine charts, adding new data to the Portuguese originals. They were used for navigation to Southeast Asia and even became cultural icons, awarded to graduating navigators as a sort of diploma.

These differences in response and adaptation open up a broad set of questions about cultural interaction. Inquiries in this direction may yield something better than the trendy notions of "development" and "technology transfer," which assume that only suicidal irrationality keeps the "underdeveloped" or "less developed" from following the American path to the promised land.

Implications

It may seem paradoxical that this broader view of mapmaking as more than scientific brings the history of cartography into the mainstream of the history of science. The latter field has in recent years moved decisively away from the narrow focus on technical concepts and activity that was the norm fifty years ago. No longer obsessed by the myth of an unending march of progress, historians of science are now much better at explaining how values pervade theories, what scientific practice shares with the rest of human activity, and why the benefits and dangers of applying new knowledge are inseparable.

Equally stimulating in this audacious new look at old maps are its implications for other domains of the history of cartography. When Yee observes that Chinese maps are redolent of "power, duty, and emotion," what comes to mind is their lack of uniqueness in this respect. This redolence is true of all maps, geographic, cosmological, and religious; they differ in how it is true, and to what extent.

The studies in this book suggest that scrutinizing every dimension and connection of maps in a way sensitive to the unity of precision and aspiration is likely to yield a more adequate understanding of cartography in every place and time, including our own. There is need for more descriptions of individual maps as elegant and perceptive as these. There remain wonders of the quantitative imagination to be found and admired. But in demonstrating so persuasively that the antiquarian and technical approaches alone tempt us into blind alleys, and that a more attentive reading of all the evidence can keep us out of them, the contributors show how much can be gained from the broadened definition of cartography on which this *History* is based.

Cartography in China

3 · Reinterpreting Traditional Chinese Geographical Maps

CORDELL D. K. YEE

My interest in this chapter and the following four is traditional Chinese geographic mapping—that is, Chinese mapping of the earth before its Westernization in the late nineteenth and early twentieth centuries. One of the first lessons one learns when studying this subject is that the traditional periodization used in scholarship is unsatisfactory. The traditional scheme takes the rise and fall of China's ruling houses as constituting distinct periods (see table 2.1). Such a scheme may have been useful for organizing material dealing with political and institutional history, and as will be seen in a later chapter, cartography was intimately connected to that history. But cartographic developments do not neatly parallel changes in politics. Historians of cartography in the past, however, have tried to tie cartography to dynastic changes in ways I have found misleading. For example, Wang Yong, a pioneer in the study of traditional Chinese maps, once claimed that Tang (618–907) cartography was superior to that of the Song dynasty (960–1279), even though virtually no cartographic artifacts from the Tang are extant.[1] Other writers have made similar claims: that the Yuan (1279–1368) and Ming dynasties (1368–1644) represent the high point of Chinese cartography, that scientific cartography in China began in the third century or as early as the Former Han (206 B.C.–A.D. 8)—depending on one's sources. Given that we have only a handful of artifacts that warrant cartographic interest for the period from the first through the tenth century, these claims seem overconfident. That such claims are made at all suggests there are at least two major premises underlying them: that history is best seen as a march of progress through a generally ascending series of high points; and as a corollary, that cartographic history is best seen as a movement toward increased mathematization or quantification—toward cartography in its modern manifestation.

In this book I am questioning these premises, and one of the results of this questioning is the organization of the material. For the most part I have chosen to arrange the material thematically. I believe this is the best way to carry out searching examinations of the questions raised by the maps and other sources—inquiries that might be limited if we were forced to keep dynastic time frames in mind. These extended inquiries are obtained at some cost, however. With a thematic approach one risks losing a clear sense of chronology, and one sacrifices the power of narrative to maintain a sense of direction. By keeping the focus on ideas or themes, one also risks losing sight of the maps themselves. Detailed descriptions of artifacts can disrupt the flow of an argument or at least make it harder to follow, and so in the thematic chapters that follow this one, artifacts are dealt with in only as much detail as is necessary to support the arguments presented.

The loss of chronology and detail would be regrettable, especially when at least part of the audience for this book—collectors and cartobibliographers, for example—could reasonably be expected to take an interest in such matters. One of the aims of this opening chapter, therefore, is to discuss artifacts in greater detail than elsewhere in this section and to give some sense of their chronology.

We should remember that, in an important sense, the history of Chinese cartography has yet to be written. There are large gaps in the artifactual record. Between the Later Han (25–220) and the end of the ninth century, for example, there are almost no maps. For the Ming and Qing periods (1644–1911) one faces the opposite problem: a superabundance of maps. The primary source material on the cartography of these periods outbulks that of all previous dynasties combined. In addition to thousands of maps in the imperial archives and thousands of gazetteers, there are collections of memorials and other documentary material that need to be examined for evidence of map use within the Ming and Qing administrations. Thus there are abundant opportunities for further research (see appendix 3.1).

I would like to acknowledge the help of Kevin Kaufman and the coeditors of the *History of Cartography* in the preparation of this chapter.

1. Wang Yong, *Zhongguo dilixue shi* (History of geography in China) (1938; reprinted Taipei: Shangwu Yinshuguan, 1974), 70, 74. Chavannes made similar claims, saying that advances in cartography were made during the Tang. See Edouard Chavannes, "Les deux plus anciens spécimens de la cartographie chinoise," *Bulletin de l'Ecole Française d'Extrême Orient* 3 (1903): 214–47, esp. 244.

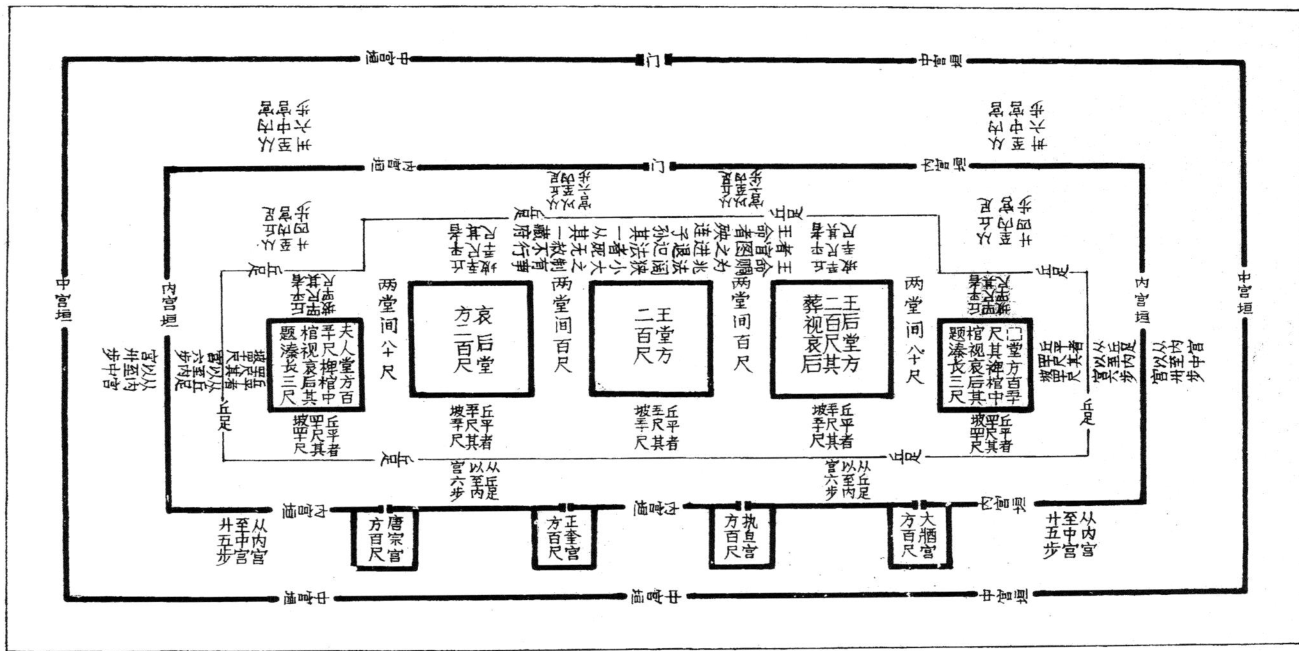

FIG. 3.1. *ZHAOYU TU* ENGRAVED ON BRONZE. Notes on the map give dimensions of the mausoleum in terms of *chi* (feet) and *bu* (paces). The *chi* is thought to have been equal to twenty-two or twenty-five centimeters, while estimated values for the *bu* range between five and seven *chi*. Below is a reconstruction of the *zhaoyu tu* with modern Chinese graphs. The original bronze plate bears graphs in their archaic form. Size of the original: 48 × 94 cm (ca. 1 cm thick). Photograph courtesy of China Pictorial Publications, Beijing. Reconstruction from Cao Wanru et al., eds., *Zhongguo gudai ditu ji* (Beijing: Wenwu Chubanshe, 1990–), vol. 1, fig. 3.

In this chapter, artifacts through the fourteenth century are well represented, since they are relatively few. For later periods the coverage is somewhat more selective, since maps from those periods, especially the Qing, are discussed more fully later in this section as part of an examination of the Westernization of Chinese cartography.

The descriptions in this chapter are not provided solely for the sake of presenting artifacts or for the sake of chronology. Those purposes are subordinate to another one: to introduce some of the themes and issues explored in depth in the chapters that follow. In doing so, I will refer to the artifacts and related documents and show how they raise questions of central concern to historians of cartography—questions that have been largely overlooked.

Chinese Mapping: A Mathematical Tradition?

The previous scholarship on traditional Chinese cartography is characterized by a remarkable unity of approach. It has generally tried to interpret Chinese cartography as a mathematical or quantitative tradition—characterized by attention to scale, conventional abstract signs, and practical function, such as planning, administrative, and military. Mapping so conceived is mathematical in at least two respects. First, it involves quantification and the reduction of topographical features to signs that aid the presentation of quantitative information. Second, it serves purposes that often involve the application of mathematics. If scale, abstract signs, and practical function characterize Chinese mapping, then it might be said to constitute a rational discipline, or science, of cartography. The current literature suggests that a science of cartography developed in China at least by the Former Han and continued through the early Qing, when Chinese cartography became Westernized. The case for a continuous mathematical tradition of Chinese mapping once rested largely on a few textual sources and a few maps, notably the 1136 *Yu ji tu* (Map of the tracks of Yu [legendary emperor famous for flood control]), which Chavannes said was "the result of a long scientific evolution."[2] According to recent scholarship, that case has been strengthened by recent archaeological discoveries. According to the mathematical or quantitative interpretation of traditional Chinese cartography, these artifacts record an inevitably unsuccessful, but nonetheless essentially scientific, attempt to draw maps to scale. The claims of proponents of this interpretation will be submitted to critical analysis. But first I will present the artifactual and textual evidence for this interpretation.

The earliest artifact lacks a title, but Chinese researchers refer to it as the *zhaoyu tu*, which, because of the ambiguity in the graph *tu*, can be rendered as mausoleum map or plan. It was discovered in 1978 in a tomb unearthed in Pingshan Xian (County), Hebei Province.[3] The tomb is that of King Cuo of the Zhongshan kingdom, a small state of the Zhanguo (Warring States) period (403–221 B.C.). He was buried about 310 B.C., so the *zhaoyu tu* dates from at least the fourth century B.C.[4]

The *tu* is engraved on a bronze plate (fig. 3.1). It is oriented with south at the top and is believed to represent a walled area of about 191 by 414 meters.[5] It depicts in plan five sacrifice halls, four smaller buildings, an inner and outer wall, and a baseline marking the foot of the grave mound. These features are represented with lines of gold and silver inlay. The five sacrifice halls were intended to cover the tombs of King Cuo, his two queens, and two other members of the royal family. At the Pingshan site, two tombs have been unearthed, those of King Cuo and Queen Ai. The other three tombs were never built, evidently because several years after the king's interment, the Zhongshan kingdom fell to another state.[6] All this suggests that the *zhaoyu tu* was a plan for construction rather than a map depicting actual structures.

The *tu* is annotated, and the notes include a transcription of a decree issued by the Zhongshan king. This decree was probably responsible for the *tu*'s survival, because it provided for the preservation of the plan: "one [copy] to accompany [the burial] and one to be stored in the archives."[7] The notes on the *tu* also name the objects represented in it and provide linear measurements for the dimensions of the buildings and the distances between them.

The next group of artifacts in the current canon comprises seven maps, drawn in ink on four wooden boards

2. Chavannes, "Les deux plus anciens spécimens," 236 (note 1).

3. The discovery was first reported by the Hebeisheng Wenwu Guanlichu (Hebei Province Cultural Relic Agency), "Hebeisheng Pingshan xian Zhanguo shiqi Zhongshanguo muzang fajue jianbao" (Excavation of the tombs of the Zhongshan kingdom of the Zhanguo period at Pingshan County, Hebei Province), *Wenwu*, 1979, no. 1:1–31.

4. Fu Xinian, "Zhanguo Zhongshan wang Cuo mu chutu di 'zhaoyu tu' ji qi lingyuan guizhi di yanjiu" (A study of the mausoleum map unearthed from the tomb of King Cuo of the Zhanguo period's Zhongshan kingdom and the planning of the mausoleum), *Kaogu Xuebao*, 1980, no. 1:97–118, esp. 97.

5. Yang Hongxun, "Zhanguo Zhongshan wang ling ji zhaoyu tu yanjiu" (A study of the mausoleum of the king of Zhanguo period's Zhongshan kingdom and the mausoleum map), *Kaogu Xuebao*, 1980, no. 1:119–38, esp. 127–29.

6. Liu Laicheng and Li Xiaodong, "Shi tan Zhanguo shiqi Zhongshanguo lishishang di jige wenti" (Tentative discussion of certain problems in the history of the Zhanguo period's Zhongshan kingdom), *Wenwu*, 1979, no. 1:32–36, esp. 33.

7. Hebeisheng Wenwu Guanlichu, "Zhongshanguo muzang fajue jianbao," 5 (note 3).

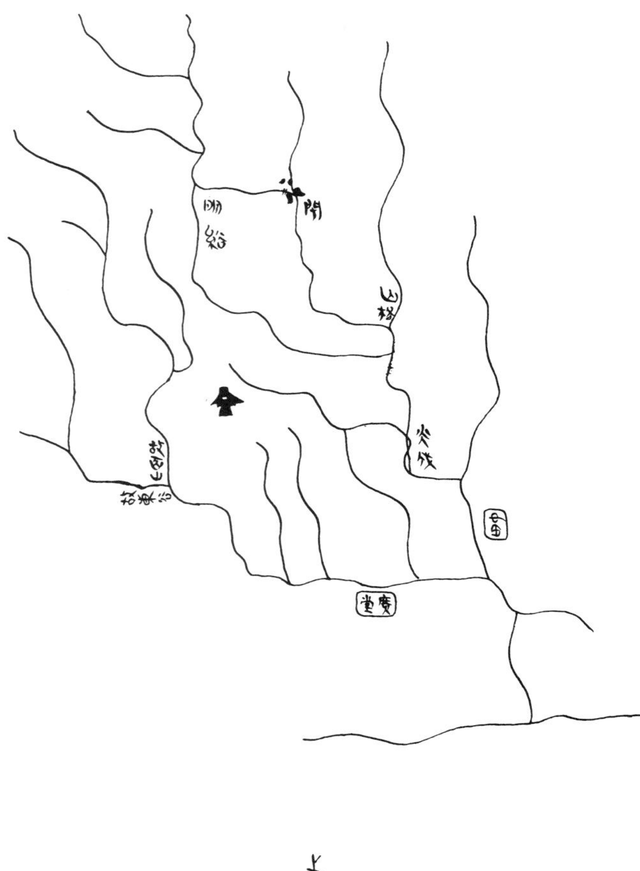

FIG. 3.2. FANGMATAN MAP, VERSO OF BOARD 1. The features represented on the maps found at Fangmatan are primarily linear: mountains, rivers and streams, and roads (see the reconstruction on the right). Some place-names also appear on the maps.

Size of the original: 26.7 × 18.1 cm. By permission of Wenwu Chubanshe. Reconstruction from Cao Wanru et al., eds., *Zhongguo gudai ditu ji* (Beijing: Wenwu Chubanshe, 1990–), vol. 1, fig. 13.

(figs. 3.2 to 3.4). Six of the maps are on both sides of three boards. The maps were discovered in 1986 at Fangmatan forestry station, Tianshui, Gansu Province. They had been buried in a tomb dating from perhaps 239 B.C. and belonging to an officer in the Qin army whose personal name was Dan.[8] The boards on which the maps were drawn are all about one centimeter thick and range in width from 26.5 to 26.8 centimeters and in height from 15 to 18.1 centimeters. The maps all depict parts of the same region, an old administrative district called Gui Xian, which researchers have identified as the Wei River valley and its tributaries, cutting through a section of the Qinling Mountains. This region, which includes the Fangmatan grave site, held strategic importance as a defensible pass through which transportation was funneled from the west into the heart of the Qin state. The maps represent rivers and tributaries by black lines. Gullies, passes, transportation checkpoints, and stands of various kinds of trees, including pine, fir, cedar, and orange, are identified

8. The maps were first described in He Shuangquan, "Tianshui Fangmatan Qin mu chutu ditu chutan" (Preliminary study of the maps excavated from the Qin tomb at Fangmatan in Tianshui), *Wenwu*, 1989, no. 2:12–22.

The date used here follows He Shuangquan's analysis of records on the deceased found in the tomb. The records, on eight bamboo slips, are transcribed in He Shuangquan, "Tianshui Fangmatan Qin jian zongshu" (Comprehensive account of the Qin bamboo slips from Fangmatan in Tianshui), *Wenwu*, 1989, no. 2:23–31, esp. 28–29. The records note that Dan, who served as a military officer and participated in a northern campaign, injured someone in the face and afterward killed himself. He was buried outside the city and came back to life three years later. The records mention years but do not identify the reigning monarch. They suggest, however, that the ruler reigned for at least ten years. It is possible to date the tomb earlier than 239 B.C. as proposed by He Shuangquan. Zhang Xiugui, for example, would push the date of the tomb back to around 300 B.C. by identifying a Qin ruler who also ruled for at least ten years and launched northern expeditions. See Zhang Xiugui, "Tianshui 'Fangmatan ditu' di huizhi niandai" (Date of the maps from Fangmatan, Tianshui), *Fudan Xuebao*, 1991, no. 1:44–48.

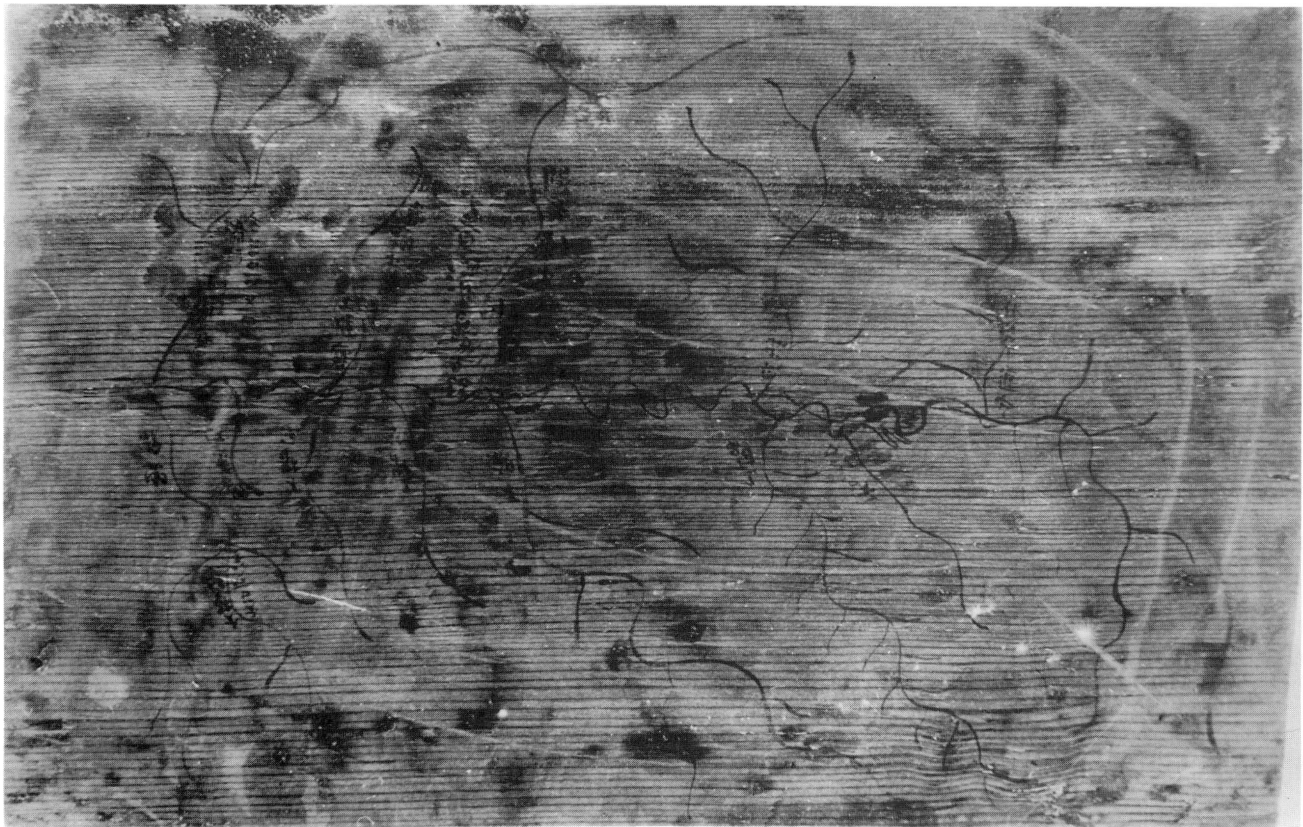

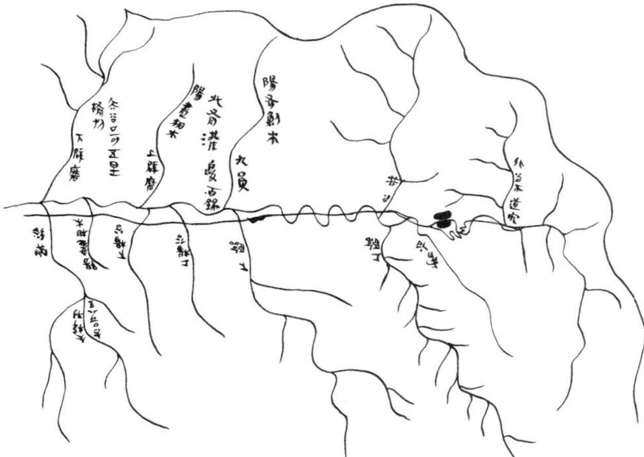

FIG. 3.3. FANGMATAN MAP, RECTO OF BOARD 3. (Reconstruction below.)
Size of the original: 18.1 × 26.5 cm. By permission of Wenwu Chubanshe. Reconstruction from Cao Wanru et al., eds., *Zhongguo gudai ditu ji* (Beijing: Wenwu Chubanshe, 1990–), vol. 1, fig. 5.

by labels. The names of settlements are enclosed in squares. There is some overlap between the maps in geographic coverage, and there are discrepancies in the location of certain features.[9] Like the *zhaoyu tu*, the wooden maps bear annotations giving distances, but they do not indicate the points these distances are meant to connect. No directions are marked on the maps. One map has a label indicating which side is the top, and this has been found to correspond with a northern orientation. The other maps, however, are oriented in different directions.

One other discovery from Fangmatan is worth mentioning. In another tomb dating from 179–141 B.C. (Former Han) a fragment of what some believe to be a map was found inside a coffin on the chest of the deceased. The fragment has been described as a piece of yellow paper (fig. 3.5). Black lines are used to depict mountains, rivers, and roads. The fragment is too small to permit a positive identification of the area depicted,

9. There are conflicting interpretations regarding the relationship among the map images. He Shuangquan thinks that six of the maps can be combined to form a composite image of the district. Cao Wanru has argued that one of the images, one that He places at the center of the composite image, is a general map of the district, while the others are detail maps of that area. See He, "Tianshui Fangmatan Qin mu chutu ditu chutan," 14, 16 (note 8); and Cao Wanru, "Youguan Tianshui Fangmatan Qin mu chutu ditu di jige wenti" (Several problems concerning the maps excavated from the Qin tomb at Fangmatan in Tianshui), *Wenwu*, 1989, no. 12:78–85, esp. 80, and idem, "Ancient Maps Unearthed from Qin Tomb of Fangmatan and Han Tomb of Mawangdui: A Comparative Research," *Journal of Chinese Geography* 3, no. 2 (1992): 39–50.

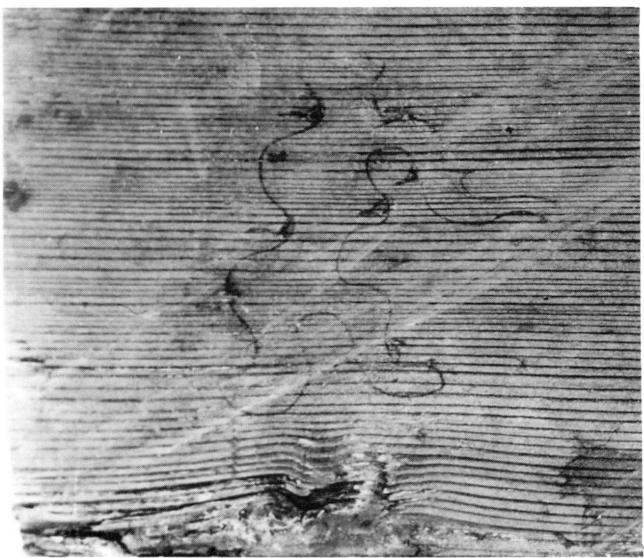

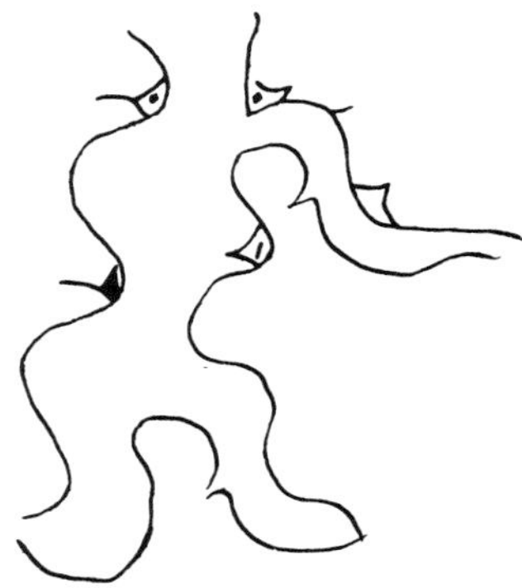

FIG. 3.4. FANGMATAN MAP, DETAIL OF THE VERSO OF BOARD 3. This detail is about one-fourth of the entire board (reconstruction on the right).

Size of the entire original: 18.1 × 26.5 cm. By permission of Wenwu Chubanshe. Reconstruction from Cao Wanru et al., eds., *Zhongguo gudai ditu ji* (Beijing: Wenwu Chubanshe, 1990–), vol. 1, fig. 9.

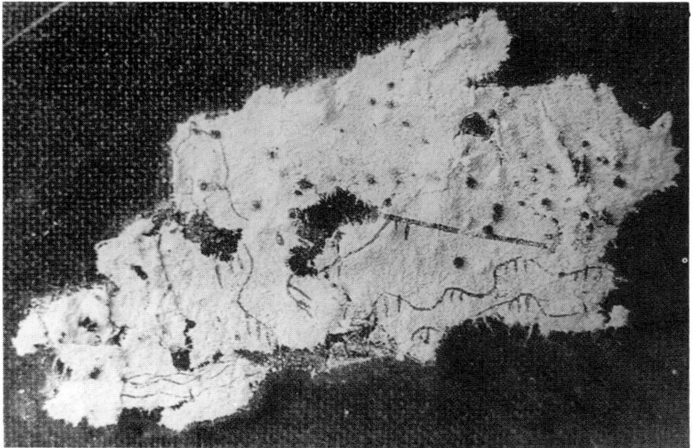

FIG. 3.5. MAP FRAGMENT ON PAPER. Discovered at Fangmatan, this fragment dates from between 179 and 141 B.C. Size of the original: 2.6 × 5.6 cm. By permission of Wenwu Chubanshe.

One of the three maps was in tatters when it was found; its condition has made interpretation difficult (fig. 3.6). The upper half of the map contains an irregular closed curve filled in with oblique lines, a yellow line, squares, and rectangles. What these features represent is unclear but it may well have been the Wei River valley region shown in the earlier Fangmatan maps.[10]

The Fangmatan map fragment is roughly contemporaneous with three silk maps found in 1973 in a tomb at Mawangdui on the outskirts of Changsha in Hunan Province. All three of these maps depict portions of the Changsha state, whose territory during the early Han included present-day Hunan and adjacent portions of Guangdong and Guangxi. The person buried in the tomb was evidently a high-ranking official in the state.[11] His burial took place in 168 B.C.; thus the maps must have been drawn somewhat earlier.

10. The interpretation of this artifact has been subject to debate. When it was produced and whether it is indeed a map have not been established conclusively. The description here is based on that given by Gansusheng Wenwu Kaogu Yanjiusuo and Tianshui Beidaoqu Wenhuaguan (Institute of Archaeology, Gansu Province, and Cultural Center of Beidao District of Tianshui), "Gansu Tianshui Fangmatan Zhanguo Qin Han muqun di fajue" (Excavation of the tombs from the Qin state of the Warring States period and from the Han dynasty), *Wenwu*, 1989, no. 2:1–11, esp. 9. Chen Qi-xin and Li Xing Guo have challenged the dating of this artifact and its identification as a map, arguing that it could have fallen into the coffin sometime after the coffin had decayed. They further maintain that the black lines on the paper could have been stains from the black paint on the coffin. See Chen Qi-xin and Li Xing Guo, "The Unearthed Paperlike Objects Are Not Paper Produced before Tsai-Lun's Invention," *Yearbook of Paper History* 8 (1990): 7–22. In the same journal Wang Ju Hua also challenges the dating of the paper artifact. He does not, however, question whether it is a map. See Wang Ju Hua, "The Inventor of Paper Technology—Ts'ai Lun," *Yearbook of Paper History* 8 (1990): 156–63. It is worth noting that the authors of the last two articles have not based their arguments on firsthand inspection of the artifact in question.

11. Tomb 3 at Mawangdui, in which the maps were found, is believed to be that of a son of Li Cang, marquis of Dai, chancellor to the prince of Changsha. Li Cang was buried in Mawangdui tomb 2 in 186 B.C. Tomb 1 was that of Li Cang's wife, who was buried not long after 168 B.C. Li's son may have served in the military as a general.

because of gaps in the map. The lower part of the map shows a city with an outer and inner wall (fig. 3.7).[12]

The other two maps found at Mawangdui are in better condition. They have been restored, and detailed accounts of them have been published.[13] Both are oriented with south at the top. One represents the southern part of the Changsha state and is referred to as a topographic map, because it emphasizes mountainous areas and the courses of rivers (figs. 3.8 and 3.9). The names of the mountains are not given, but rivers, as well as county seats, are identified. The map was drawn with vegetable colors. The third map (figs. 3.10 and 3.11) is thought to represent a portion of the terrain shown on the topographic map—more specifically, part of the southernmost portion. This area was of military significance since it bordered on Nanyue, a reluctant tributary state to the Han. Because it shows the locations of army installations and headquarters, the third map is thought to have had military applications. It is notable for its use of color: military-related features, roads, and some settlements are shown in red; rivers and streams in light blue-green; and other features and lettering in black. The map also bears annotation. For some settlements, distances from other settlements and the number of households are given.

All the maps described so far have been valued for their "modern" appearance. Their mode of presentation seems to be planimetric, and the manner of depiction tends toward conventionalization, for instance, in the representation of settlements, mountains, and trees. In

FIG. 3.6. HAN SILK MAP UNEARTHED AT MAWANGDUI. This map was unearthed in 1973 from tomb 3 at Mawangdui. It was drawn on the same piece of silk as an illustration showing breathing exercises and calisthenics. The tattered condition of the map has made interpretation difficult. The lower half of the map seems to represent a walled city (see fig. 3.7).
Size of the original: 48 × 48 cm. Photograph courtesy of Cao Wanru, Institute for the History of Natural Sciences, Academia Sinica, Beijing.

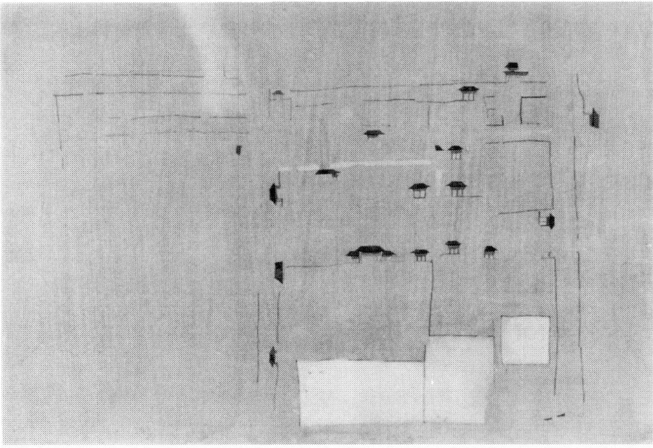

FIG. 3.7. DETAIL OF THE HAN SILK MAP. A copy of the lower center portion of the map in figure 3.6. The outer wall of the city, as drawn on the map, measures 18.8 by 20 centimeters.
Photograph courtesy of Cao Wanru, Institute for the History of Natural Sciences, Academia Sinica, Beijing.

12. Information on this map is taken from Han Zhongmin's description in Cao Wanru et al., eds., *Zhongguo gudai ditu ji* (Atlas of ancient Chinese maps) (Beijing: Wenwu Chubanshe, 1990–), 1:18. Han conjectures that the map as a whole represents the mausoleum and the city of the marquis of Dai, Li Cang. Cao Wanru, however, believes that the map depicts cities and towns of the southern Changsha state; see Cao Wanru, "Maps 2,000 Years Ago and Ancient Cartographical Rules," in *Ancient China's Technology and Science*, comp. Institute of the History of Natural Sciences, Chinese Academy of Sciences (Beijing: Foreign Languages Press, 1983), 250–57, esp. 251.

13. Information on the Han silk maps is drawn largely from the following studies: Mawangdui Han Mu Boshu Zhengli Xiaozu (Study Group on the Han Silk Manuscripts from Mawangdui), "Changsha Mawangdui sanhao Han mu chutu ditu di zhengli" (Restoration of the maps excavated from Han tomb 3 at Mawangdui, Changsha), *Wenwu*, 1975, no. 2:35–42; idem, "Mawangdui sanhao Han mu chutu zhujun tu zhengli jianbao" (Preliminary report on the restoration of the military map excavated from Han tomb 3 at Mawangdui), *Wenwu*, 1976, no. 1:18–23; Tan Qixiang, "Erqian yibaiduo nian qian di yifu ditu" (A map from more than 2,100 years ago), *Wenwu*, 1975, no. 2:43–48; and Zhan Libo, "Mawangdui Han mu chutu di shoubei tu tantao" (Investigation of the garrison map excavated from the Han tomb at Mawangdui), *Wenwu*, 1976, no. 1:24–27. English-language treatments of these maps have been written by A. Gutkind Bulling, "Ancient Chinese Maps: Two Maps Discovered in a Han Dynasty Tomb from the Second Century B.C.," *Expedition* 20, no. 2 (1978): 16–25; Mei-ling Hsu, "The Han Maps and Early Chinese Cartography," *Annals of the Association of American Geographers* 68 (1978): 45–60; and Kuei-sheng Chang, "The Han Maps: New Light on Cartography in Classical China," *Imago Mundi* 31 (1979): 9–17.

FIG. 3.8. TOPOGRAPHIC MAP FROM MAWANGDUI. Both this map and those shown in figures 3.6 and 3.10 date from the Han dynasty and were found in a lacquer box. They had been folded up, and by the time they were unearthed they had disintegrated around the folds. The folded sections had also become stuck together, making restoration difficult. This topographic map consists of thirty-two pieces. Following the orientation of the graphs, south is at the top.
Size of the original: 96 × 96 cm. By permission of Wenwu Chubanshe.

addition, the artifacts have been interpreted as examples of scale mapping. There is some textual support for this. An astronomico-mathematical work, the *Zhoubi suan jing* (Arithmetical classic of the Zhou gnomon, ca. 200 B.C.), describes the effect of altering scale on the size of a map: "Whenever a scale of one *fen* to a thousand *li* was used, one drew a square map of eight *chi* and one *cun*. In present usage, one draws a square map of [half the size, or] four *chi* and five *fen*. A *fen* [in this case] is equal to two thousand *li*."[14] Modern scholars have

14. *Zhoubi suan jing*, *Siku quanshu* edition, A3.2a.

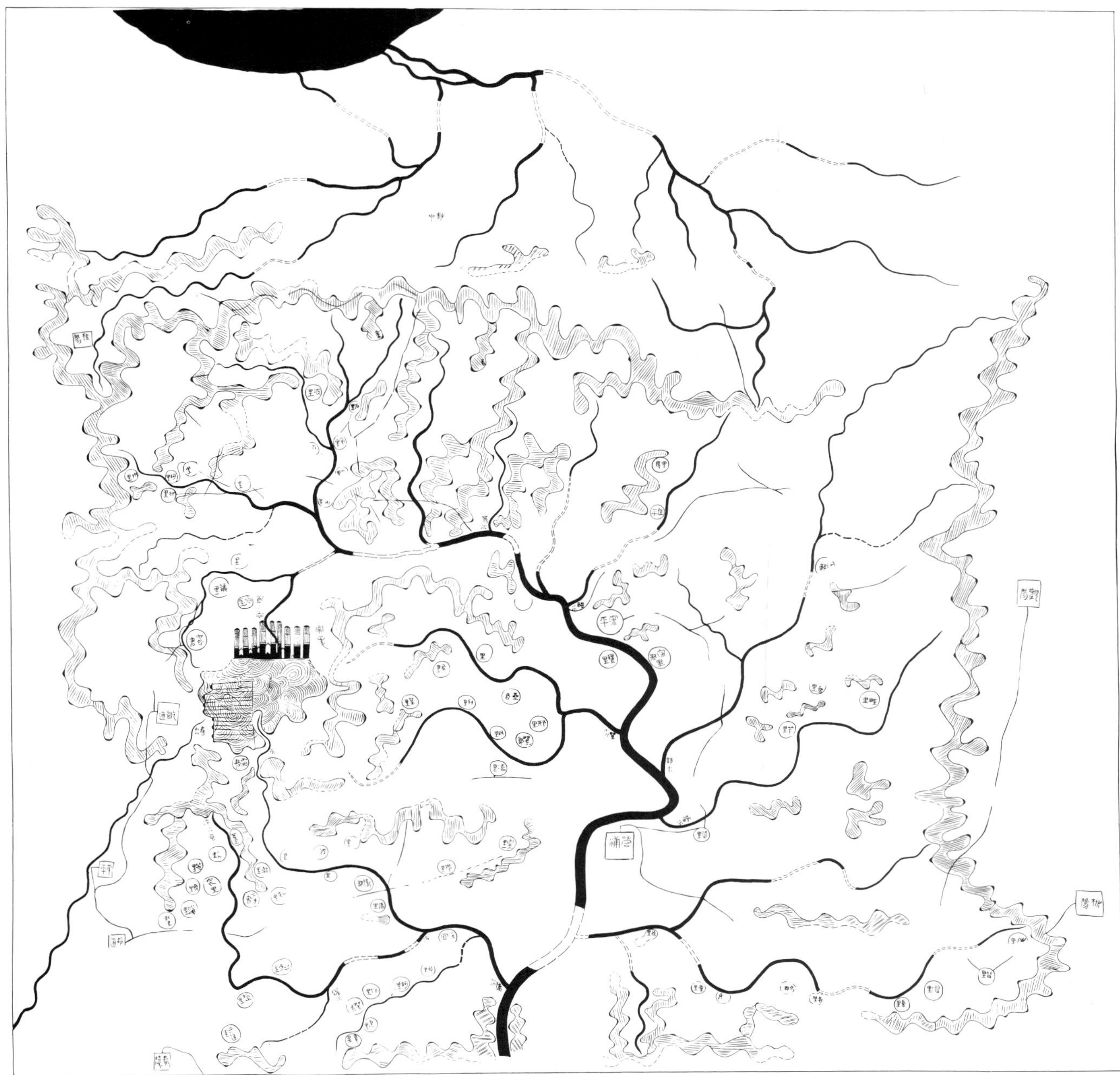

FIG. 3.9. RECONSTRUCTED TOPOGRAPHIC MAP FROM MAWANGDUI. The map is thought to represent an area mainly lying between 110° and 112°30′E and between 23° and 26°N. From *Gu ditu lunwenji* (Essays on ancient maps) (Beijing: Wenwu Chubanshe, 1977).

tended to interpret the pre-Han and Han artifacts in light of such references to scale and have taken pains to establish scale mapping as a general practice during the Han and even earlier. From dimensions marked on the early artifacts and from comparisons of the early maps with modern maps, researchers in China have calculated the scales of the early map images. It has been determined that the scale for the area within the baseline of the *zhaoyu tu* is about 1:500; outside it, this scale is not maintained. The scale of the Fangmatan maps has been determined to be about 1:300,000. As for the Mawangdui maps, the scale has been found to vary between 1:150,000 and 1:200,000 in the central portion of the topographic map and between 1:80,000 and 1:100,000 in the central

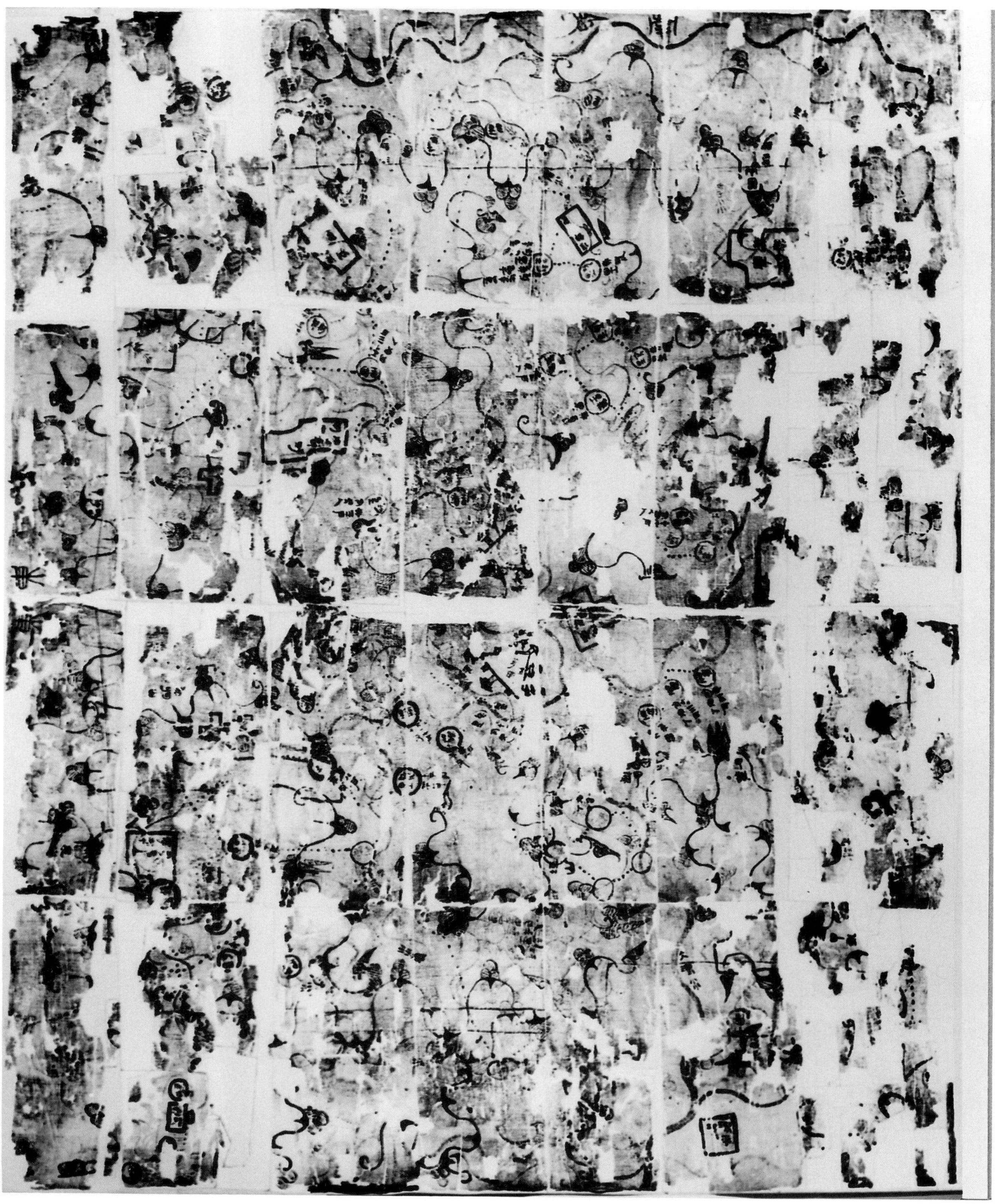

FIG. 3.10. GARRISON MAP FROM MAWANGDUI. This map consists of twenty-eight pieces. South is at the top and labeled. Size of the original: 98 × 78 cm. By permission of Wenwu Chubanshe.

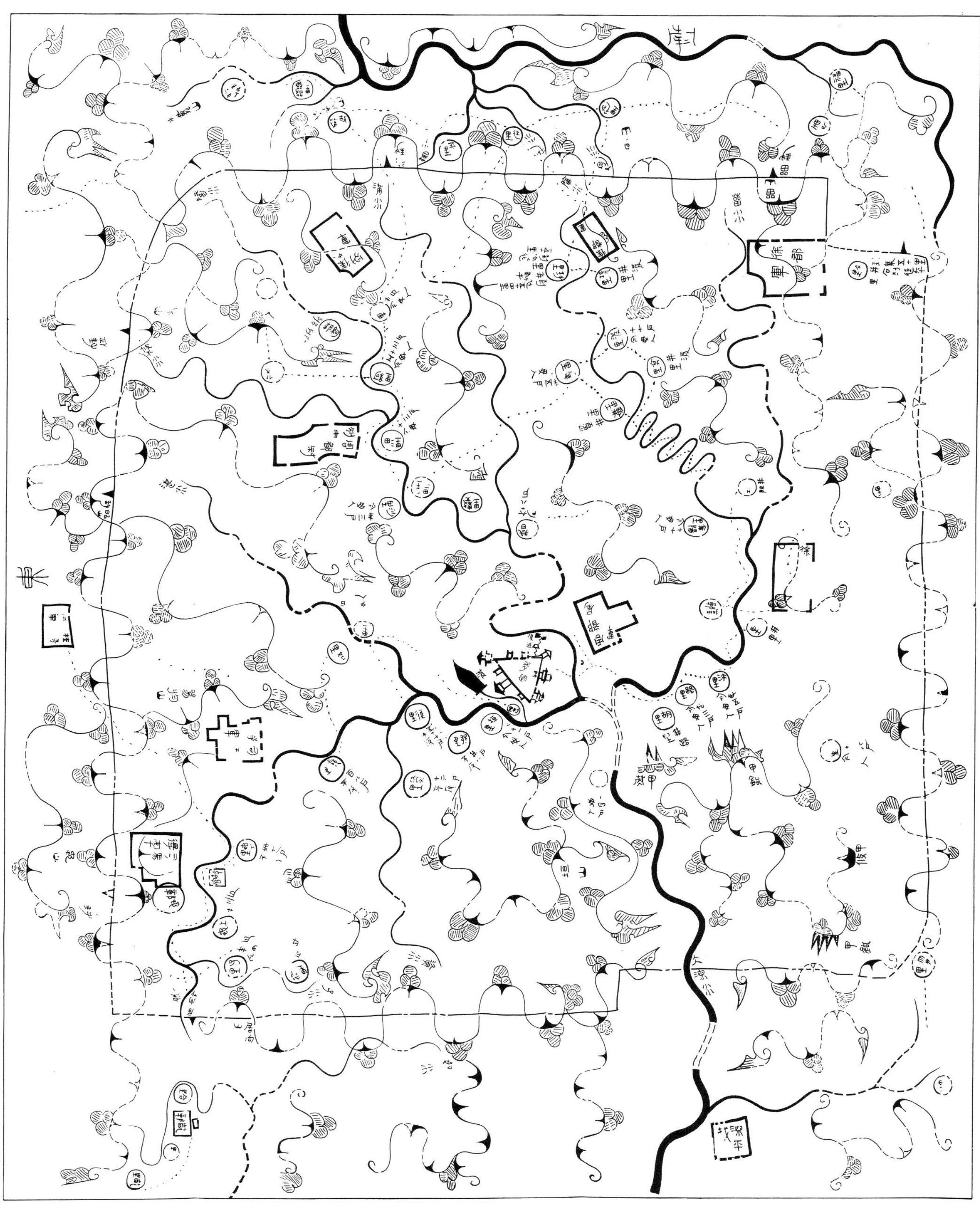

FIG. 3.11. RECONSTRUCTED GARRISON MAP FROM MAWANGDUI. From *Gu ditu lunwenji* (Essays on ancient maps) (Beijing: Wenwu Chubanshe, 1977).

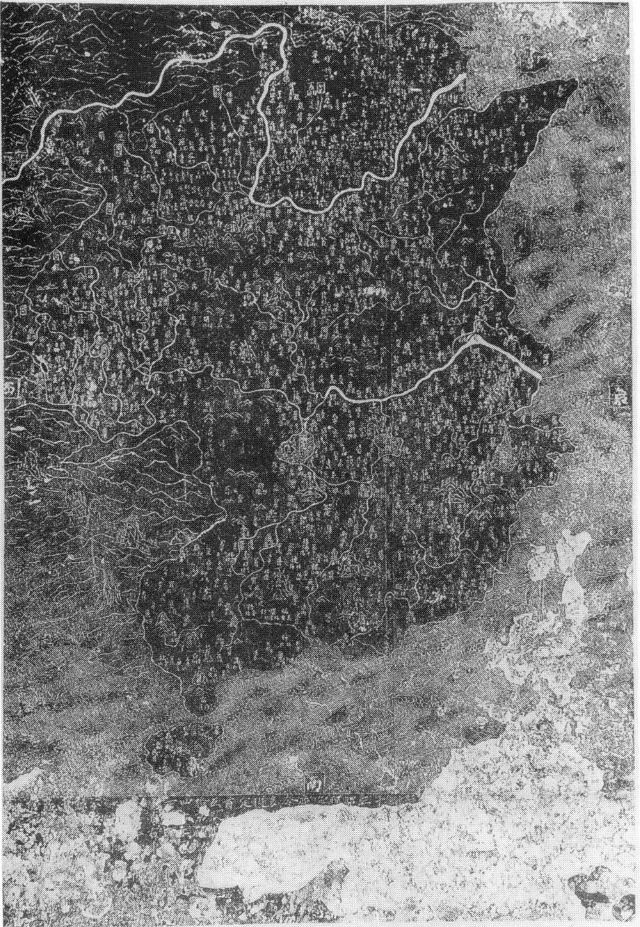

FIG. 3.12. *JIU YU SHOULING TU*. Engraved on stone in 1121. Size of the original: 130 × 100 cm. Photograph courtesy of Cao Wanru, Institute for the History of Natural Sciences, Academia Sinica, Beijing.

portion of the military map. This is a "remarkably small" scale error, in the words of one researcher.[15]

The next group of artifacts dates from about twelve centuries later than those found at Mawangdui—from the Song dynasty. The early artifacts described above were in the main local maps. Although textual sources contemporaneous with and later than those artifacts attest to maps of the empire, no such artifacts survive from the Tang or before. The earliest examples are from the Song, and these suggest that mapping had progressed to the point that imperial maps of "high" quality could be produced.[16] The task of compiling maps of the entire empire must have been time consuming, if the first surviving artifacts are any indication. They are engraved in stone, the medium of commemoration, suggesting that the makers thought the maps were important enough to merit long-term preservation in the manner of classic texts, which were often engraved in stone as well. The stone maps were set up in schoolyards, studied by students, and frequently copied.

The earliest extant example of such a comprehensive map is the *Jiu yu shouling tu* (Map of the prefectures and counties of the nine districts [the empire]). It was engraved on stone in 1121 in Rongzhou (in present-day Sichuan Province) and erected in the courtyard of the prefectural school there. After being lost for several centuries, the map was rediscovered by archaeologists in 1964.

The *Jiu yu shouling tu* names more than 1,400 administrative units (fig. 3.12). Its scale is said to be 1:1,900,000.[17] Coastal features such as the Shandong peninsula, Hangzhou Bay, the Leizhou peninsula, and Hainan Island are clearly recognizable. Seas and lakes are depicted using waves on which are scattered sailing vessels. Mountains are drawn pictorially, and tree symbols indicate that slopes are forested. The course of the Huanghe, or Yellow River, however, was outdated at the time the map was engraved. In 1121 the river ran northward before emptying into the ocean near Tianjin. The *Jiu yu shouling tu* shows the river flowing eastward, emptying near the present-day border of Hebei and Shandong provinces, which suggests that it was based on an earlier map made when the Yellow River actually followed the eastward course. This direction of flow occurred twice: from 1069 to 1081 and from 1094 to 1099.[18]

Another map of the empire, the *Hua yi tu* (Map of Chinese and foreign lands), was engraved on stone in 1136 (fig. 3.13), but information on the map suggests that the image was perhaps compiled during 1117–25. The *Hua yi tu* provides about five hundred place-names and identifies about thirteen rivers and tributaries, four lakes, and ten mountain ranges. Information on foreign lands is provided in notes. The map lacks an expressed scale, and its image is regarded as deficient in certain respects: it locates the sources of the Changjiang (Yangtze River) and Yellow River in the wrong places, and its depiction of the coastline does not adequately represent the Liaodong and Shandong peninsulas.

Such deficiencies do not appear on a map carved on the opposite side of the same stele as the *Hua yi tu*. This

15. Hsu, "Han Maps," 49 (note 13).

16. The Tang mapmaker Jia Dan (730–805) is credited with producing imperial maps of high quality, but none of his maps survive. The *Hua yi tu* (Map of Chinese and foreign lands), engraved on stone in 1136 and described below, may have been based on a map by Jia Dan, but how closely the stone map conforms to its possible ancestor is impossible to determine.

17. Chen Feiya et al., eds., *Zhongguo gudai dilixue shi* (History of ancient Chinese geography) (Beijing: Kexue Chubanshe, 1984), 306.

18. Zheng Xihuang, "*Jiu yu shouling tu* yanjiu" (A study of the *Jiu yu shouling tu*), in *Zhongguo gudai ditu ji*, ed. Cao Wanru et al. (Beijing: Wenwu Chubanshe, 1990–), 1:35–40, esp. 35.

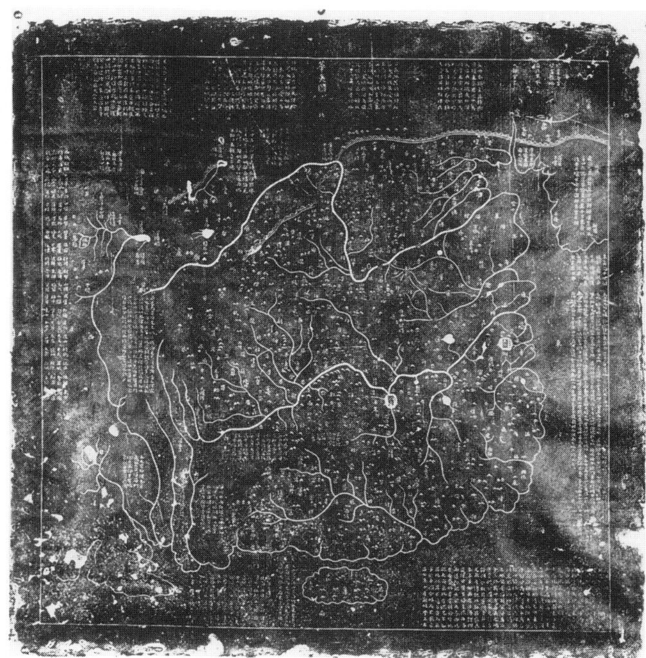

FIG. 3.13. *HUA YI TU.* Engraved on stone in 1136, it may have been based on a map by the Tang scholar Jia Dan. On the left is the stele and on the right is a rubbing of the map. On the other side of this stele is the 1136 *Yu ji tu* (Map of the tracks of Yu), figure 3.14, but it is oriented in the opposite direction so that when one map is right side up (in this case, fig. 3.14), the other is inverted. Thus it would have been difficult to display both of these maps on the stele at the same time. The stele may have been intended not for display but as a means of reproduction. The image could be transferred by "rubbing"—placing a sheet of paper over the engraved surface and applying ink to the paper so as to produce a white-on-black reproduction. An inscription on the lower right side of the map (when the map is right side up) says: "The venerable Jia Wei [Jia Dan] of the Tang dynasty listed several hundred countries. Now we [or I] have selected the well-known ones and recorded them here."
Size of the original: 79 × 79 cm. Photographs courtesy of Cao Wanru, Institute for the History of Natural Sciences, Academia Sinica, Beijing.

is the *Yu ji tu* (Map of the tracks of Yu), also engraved in 1136 (fig. 3.14). Internal evidence suggests that the map image was originally drawn somewhat earlier: for example, there are no names of prefectures and counties established after 1100. Until the discovery of the Mawangdui maps, the *Yu ji tu*, measuring about eighty centimeters square, was one of the most celebrated examples of the precocious development of Chinese cartography. The reasons are not hard to fathom. The map's representation of China's coastline is remarkably close to twentieth-century representations, and the map marks the first known appearance of the Chinese cartographic grid, consisting of a latticework of squares of equal size superimposed on a map image. The squares serve to indicate scale: each grid increment, or side of a square, represents a fixed ground distance. A note engraved on the *Yu ji tu* states that each side of a square represents one hundred *li*, and on that basis the map scale is assumed to be about 1:4,500,000.[19]

A second version of the *Yu ji tu*, at Zhenjiang, Jiangsu (fig. 3.15),[20] was carved in stone in 1142, commissioned by the director of the local prefectural school. It is similar to the 1136 version in many ways: in its use of the square grid and an expressed scale of one side of a square to one hundred *li*, in the dimensions of the image, in its depiction of mountains and rivers, and in the names of the prefectures and counties. There are some differences, however. The 1142 map image does not distinguish between main streams and tributaries. An inscription on the 1142 version corroborates the suggested date of 1100 for the original compilation: "Carved in the first month of the third year of the Yuanfu reign period [1100] according to the model copy at Chang'an."

It might seem that in comparing the *Jiu yu shouling tu* and *Hua yi tu* with the *Yu ji tu*, one sees in the last increasing technical competence in the production of map images. One sign is the use of the grid as a scaling device, suggesting the importance of scale imaging to

19. Cao et al., *Zhongguo gudai ditu ji*, 21 (note 12).

20. There are reports of a third stone *Yu ji tu*, now lost, that was once kept at Jishan Xian, Shanxi. The map is said to have had a square grid with a scale of "one [side of a] square to one hundred *li*." See Lu Liangzhi, *Zhongguo dituxue shi* (History of Chinese cartography) (Beijing: Cehui Chubanshe, 1984), 156.

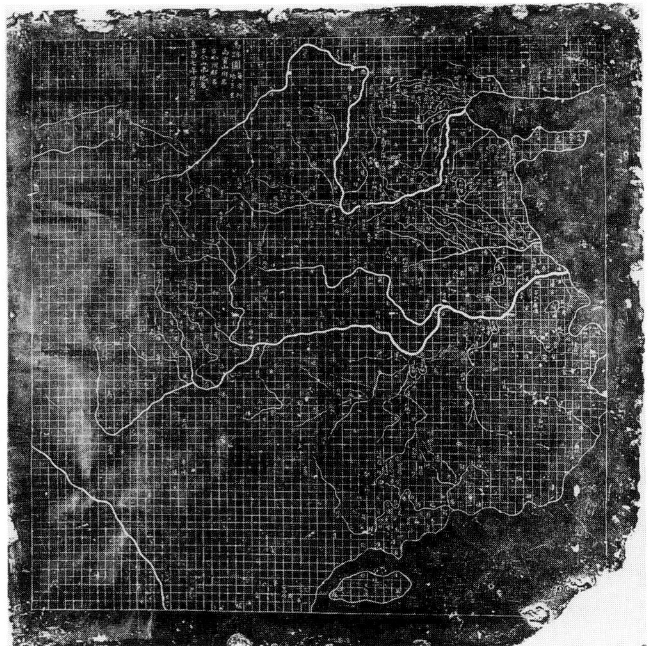

FIG. 3.14. *YU JI TU*, 1136. (See also fig. 3.13.) The stele is on the left and a rubbing is on the right.
Size of the original: 80 × 79 cm. Photographs courtesy of Cao Wanru, Institute for the History of Natural Sciences, Academia Sinica, Beijing.

Chinese mapmakers. The grid certainly contributes to the "modern look" of the *Yu ji tu*, because it resembles the graticule used on modern maps. As a consequence, the *Yu ji tu* is often described as indicating the advanced level of mathematical cartography in traditional China: "Anyone who compares this map with the contemporary productions of European religious cosmography . . . cannot but be amazed at the extent to which Chinese geography was at that time ahead of the West."[21]

The accuracy of the *Yu ji tu*, especially in its representation of the rivers and coastline, is remarkable, and even more so given that so little else like it from earlier times has survived. Precisely how its maker achieved such accuracy is unknown. At most one can speculate that the *Yu ji tu* was based on information from gazetteers and other geographical writings, and from earlier imperial and regional maps that, though no longer extant, are often mentioned in textual sources. In addition, there is no disputing that by the time the *Yu ji tu* was produced, the Chinese had already established the foundations for a mathematical cartography of the type described by proponents of the quantitative approach. It is clear from textual sources that the instrumentation and surveying techniques for performing the direct and indirect measurements required to produce a map like the *Yu ji tu* had been developed well before the twelfth century (see pp. 115–16).

For proponents of the quantitative approach, the artifacts described so far provide a context for certain surviving texts on mapmaking, some written by mapmakers, others by imperial historians. In line with quantitative interpretations of the artifacts, these texts are interpreted as advocating scale mapping and the employment of mathematical techniques, as will be outlined here. One of the seminal texts on cartography is a statement of mapmaking principles by Pei Xiu (223–71). That statement stresses the importance of measurement and scale for achieving fidelity to geographic actualities (see pp. 110–13). Some, including Joseph Needham, have interpreted Pei's statement as advocating the use of the cartographic grid, but there is no evidence for this. Before the discovery of the pre-Han and Han maps, Pei Xiu was thought to have initiated a mathematical tradition. If current interpretations of the *zhaoyu tu* and the Fangmatan and Mawangdui maps are accepted, however, Pei's principles of measurement were understood well before his time, so that his statement represents a culmination of

21. Joseph Needham, *Science and Civilisation in China* (Cambridge: Cambridge University Press, 1954–), vol. 3, with Wang Ling, *Mathematics and the Sciences of the Heavens and the Earth* (1959), 547.

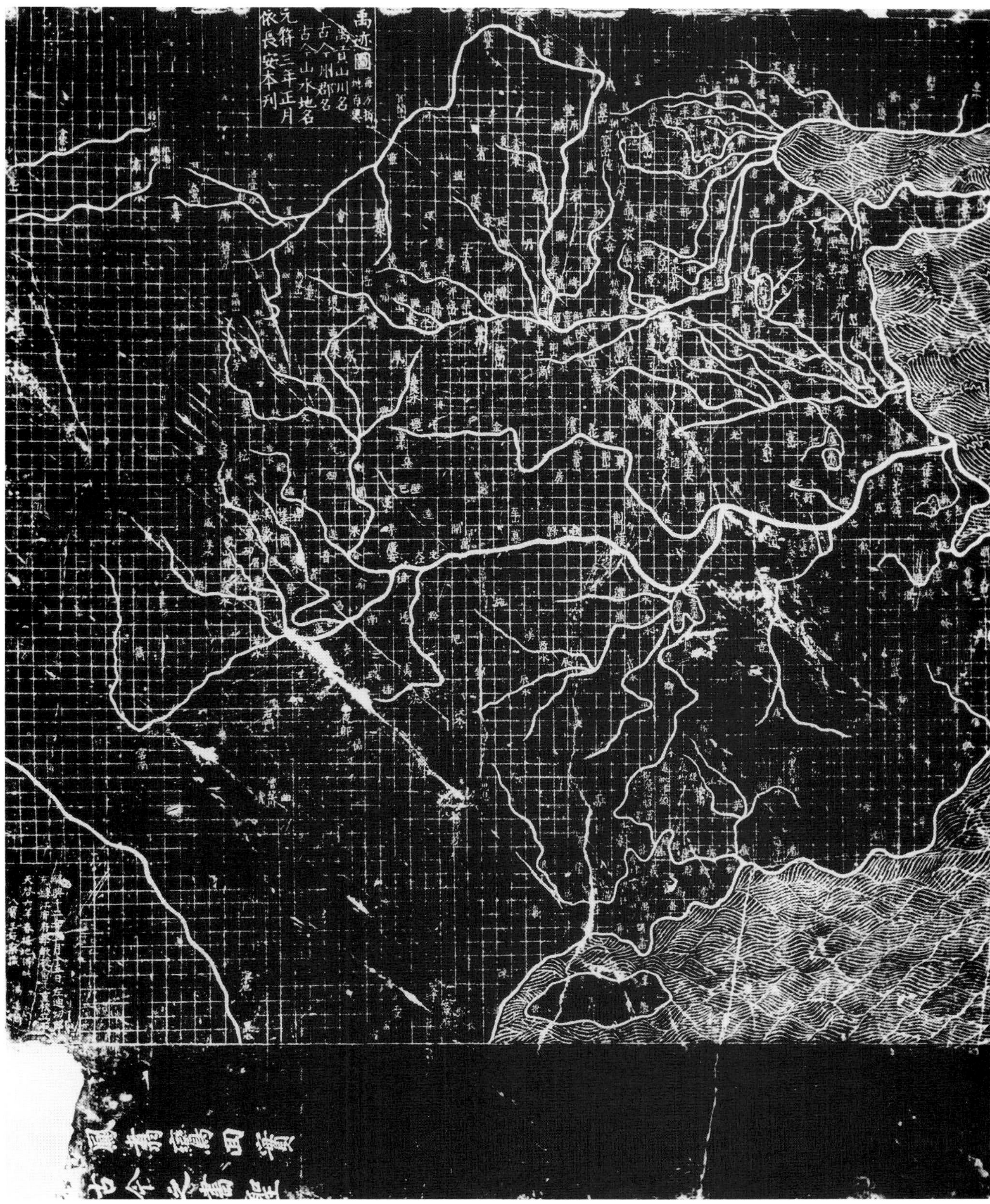

FIG. 3.15. RUBBING OF THE *YU JI TU*, 1142.
Size of the original: 83 × 79 cm. Photograph courtesy of Cao Wanru, Institute for the History of Natural Sciences, Academia Sinica, Beijing.

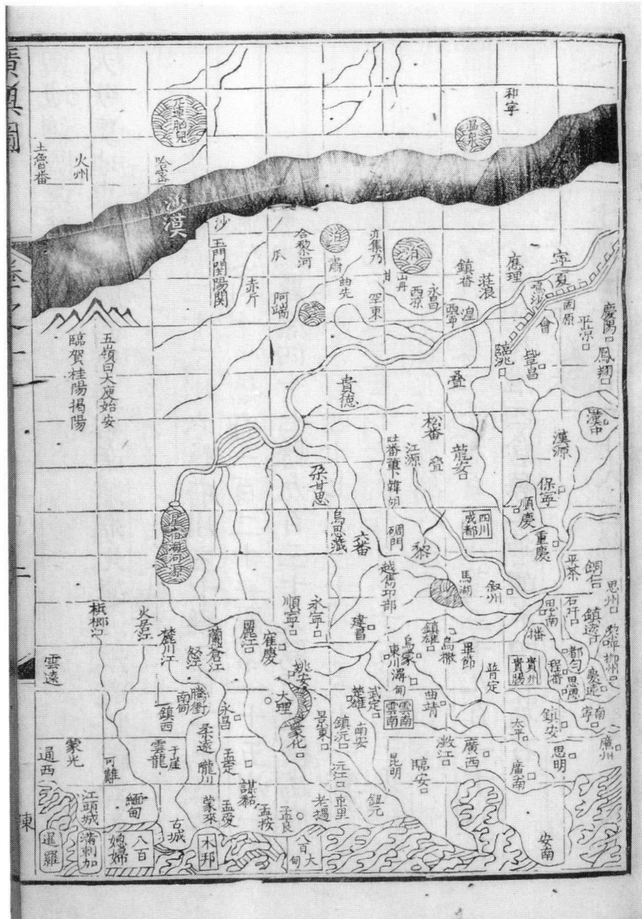

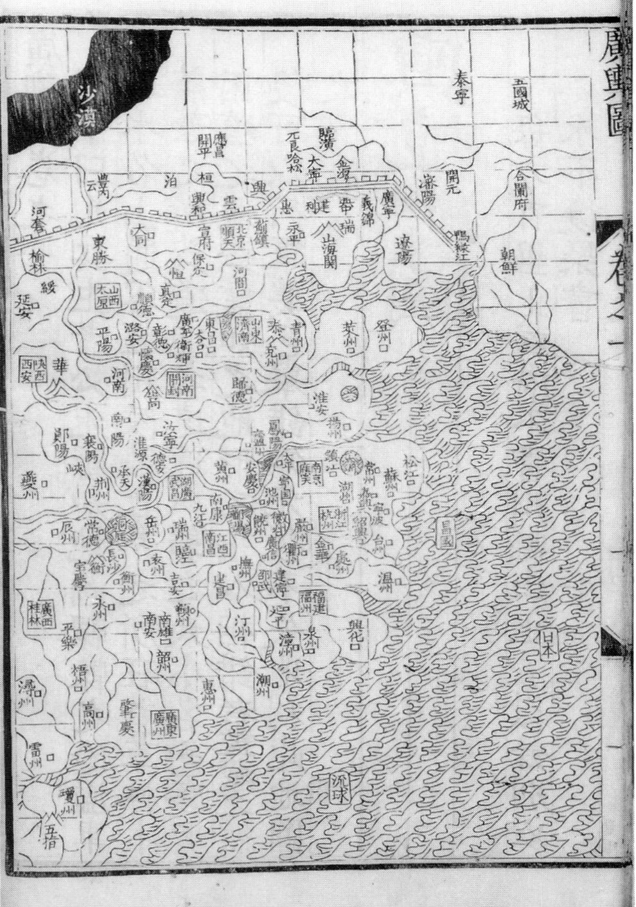

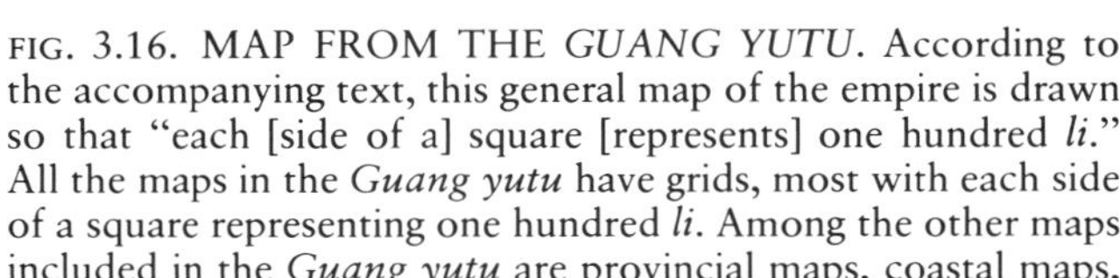

FIG. 3.16. MAP FROM THE *GUANG YUTU*. According to the accompanying text, this general map of the empire is drawn so that "each [side of a] square [represents] one hundred *li*." All the maps in the *Guang yutu* have grids, most with each side of a square representing one hundred *li*. Among the other maps included in the *Guang yutu* are provincial maps, coastal maps, hydrological maps, and maps of Korea, Vietnam, and Japan. This illustration is taken from the *Guang yutu* dated 1799, which is an exact copy of the 1579 edition.
Size of the original: 28.5 × 41 cm. By permission of the British Library, London (15261.e.2), 1b–2a.

tradition, not an innovation. To judge from the present artifactual record, that tradition did not include a cartographic grid.

The next important figure for the exponents of the mathematical interpretation is Jia Dan (730–805), who cites Pei Xiu's attention to measurement as a model. In one historical account, Jia Dan is described as having ordered a scale map to be made: "He ordered an artisan to paint the *Hainei Hua yi tu* [Map of Chinese and foreign lands within the seas] on a scroll. It was three *zhang* wide and three *zhang* and three *chi* high. Its scale was one *cun* to one hundred *li*."[22] Since he was familiar with Pei's principles and admired them, Jia Dan has been regarded by some as a user of the square grid. Shen Kuo (1031–95) can be added to the list of mathematical cartographers. He uses terminology similar to Pei Xiu's in a list of his own mapmaking methods. In addition, at least one scholar suspects that Shen Kuo was responsible for the *Yu ji tu*, thereby linking Shen more closely to a "grid" tradition.[23]

That tradition is said to extend to the sixteenth century with the work of Luo Hongxian (1504–64): grids appear on the maps in his *Guang yutu* (Enlarged terrestrial atlas,

22. Liu Xu et al., *Jiu Tang shu* (Old history of the Tang, compiled 940–45), chap. 138; see the edition in 16 vols. (Beijing: Zhonghua Shuju, 1975), 12:3786.

23. Cao Wanru, "Lun Shen Kuo zai dituxue fangmian di gongxian" (On Shen Kuo's contributions to cartography), *Keji Shi Wenji* 3 (1980): 81–84. Cao's argument is based on Shen Kuo's presence near Chang'an (present-day Xi'an) in the years 1080–82, a period coinciding roughly with the time when the information reflected on the *Yu ji tu* was compiled. She reasons that since Shen Kuo compiled a map of the empire and had high standards of mapmaking that are reflected on the stone map, he could have been the author of the original image.

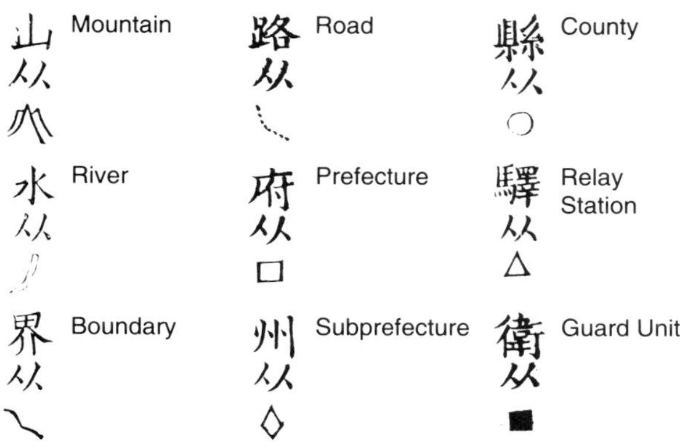

FIG. 3.17. SIGNS FROM THE LEGEND ON THE *GUANG YUTU*. In the preface to the *Guang yutu*, Luo Hongxian lists signs for mountains, rivers, boundaries, roads, prefectures, subprefectures, counties, cities, postal relay stations, and military units of varying size.
Compiled from Luo Hongxian, *Guang yutu*, 6th ed. (1579; reprinted Taipei: Xuehai Chubanshe, 1969), preface, 3a–b.

ca. 1555) (fig. 3.16). Luo's work was based on the *Yutu* (Terrestrial map, 1320) by the Yuan mapmaker Zhu Siben (1273–1337), and in the preface to his atlas Luo says that Zhu used grids on his map. Furthermore, Zhu's own preface, preserved in Luo's work, says that among the maps he consulted was a *Yu ji tu* engraved on stone in present-day Hubei.[24] If this map had a grid, it seems that Luo and Zhu were consciously following a tradition of grid mapping.

Luo's atlas has also drawn attention because it is the earliest known use of a map legend on a Chinese map (fig. 3.17). In his preface, Luo writes: "The names and forms of mountains and streams, and cities and towns cross each other so that they cannot all be written down. Instead, to save words twenty-four [signs] are used to avoid confusion and supplement [the main map image]."[25] Luo's list is not comprehensive, however, since it does not include the signs he used for the Great Wall, deserts, or lakes.

Luo's atlas was once valued for its use of abstract signs. It was taken as an indication that mapmakers were moving away from pictorialism. Wang Yong wrote that before Luo Hongxian, "drawing was often the work of painters. Maps were comparatively pictorialized: [the depiction of] things such as mountains and streams, cities and passes often approached the manner of realistic painting. Simple signs were not greatly used. In contrast, Luo Hongxian commonly and consistently used signs."[26] Above I said that the *Guang yutu* had been valued for its use of signs. It is still valued, but its use of signs appears less than unique in light of the Mawangdui maps.

The Mawangdui maps are adduced as evidence that mapmaking had already broken away from pictorialism. They have no legends, but they are regarded as sophisticated in their use of signs. On the "topographic" map, for example, signs have been identified for streams, mountains, prefectures, subprefectures, and roads.[27] The symbols for the rivers have often been remarked as widening as one moves downstream, which is taken as an indication that the mapmaker was trying to represent an increase in the volume of water. Some have stated that the representational methods used on this map, particularly those for rivers and mountains, are equal to those on maps from the Qing and even the twentieth century: this seems to include the *Yu ji tu* and *Guang yutu*.[28] The representation of Jiuyi Shan (Nine Beguiling Mountains) has received much attention (figs. 3.18 and 3.19). It consists of nine bars of varying length, with three grades of shading within them. The bars have been taken to represent the heights of the nine peaks, in which case the topographic map is also regarded as being in part a contour map. Kuei-sheng Chang, however, goes further, saying that the shading of the bars is "undoubtedly intended to convey other basic geographic information, possibly rainfall, temperature and/or cloudiness."[29] Under this interpretation, then, the "topographic" map becomes in part an early climatic map. The garrison map, it has been claimed, shows similar sophistication in the use of signs, particularly in representing mountains with wavy lines. One researcher believes that it expresses "the primitive but basic notion of contouring, the outline of the symbol being a *contour* which describes both the shape and size of the 'mountain' as the cartographer visualized it."[30]

If the proponents of a quantitative tradition of Chinese are correct, scale mapping has had a history of roughly seventeen centuries based on textual sources, or if pre-Han and Han artifacts are accepted as evidence, of about twenty-two centuries. This interpretation of traditional Chinese mapping practice is apparently buttressed by contextual considerations. As Needham and researchers writing in his wake have been fond of pointing out, the mathematical and mensurational foundation for a math-

24. See Zhu Siben's original preface to the *Yutu*, in Luo Hongxian, *Guang yutu*, 6th ed. (1579; reprinted Taipei: Xuehai Chubanshe, 1969), 1a–b, esp. 1a.

25. Luo, *Guang yutu*, preface, 3a (note 24).

26. Wang Yong, *Zhongguo ditu shi gang* (Brief history of Chinese cartography) (Beijing: Sanlian Shudian, 1958), 68–69.

27. Hsu, "Han Maps," 51 (note 13).

28. Tan, "Erqian yibaiduo nian qian di yifu ditu," 44–45, 47–48; Chang, "Han Maps: New Light," 14 (note 13).

29. Chang, "Han Maps: New Light," 10 (note 13). There is, however, no independent evidence for this interpretation, and Chang offers no support for it.

30. Hsu, "Han Maps," 55 (note 13).

FIG. 3.18. SIGN FOR JIUYI SHAN (NINE BEGUILING MOUNTAINS). Shown in the photograph is the sign for Jiuyi Shan that appears on the topographic map found at Mawangdui (fig. 3.8). A number of utilitarian functions have been proposed for its shape and shading. What seems to be a shadow image of part of the sign appears in the representation of an adjacent lake. This suggests that the sign may have had aesthetic value: the mapmaker may have been trying to show the reflection of the mountains in the lake.
Size of the detail: ca. 24 × 15 cm. By permission of Wenwu Chubanshe.

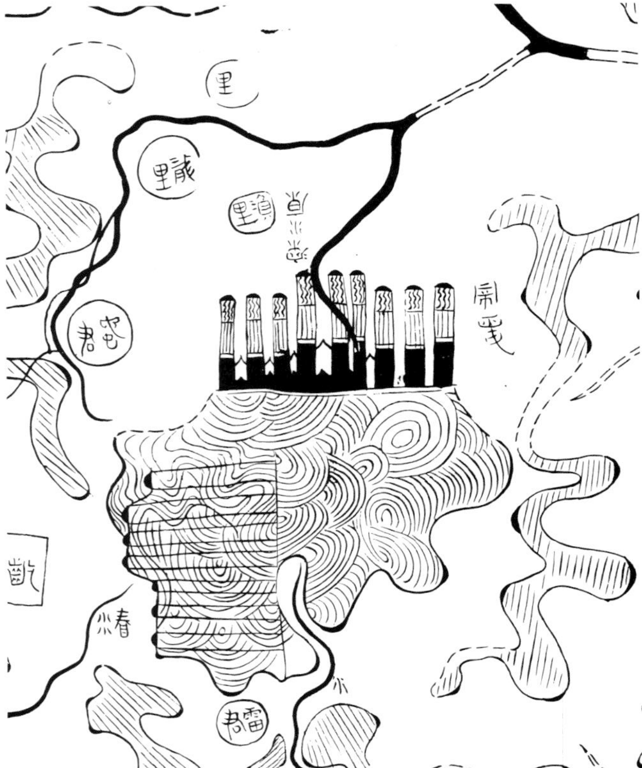

FIG. 3.19. RECONSTRUCTION OF THE SIGN FOR JIUYI SHAN. The shadow image shows up more clearly here (compare fig. 3.18).
From *Gu ditu lunwenji* (Essays on ancient maps) (Beijing: Wenwu Chubanshe, 1977).

ematical cartography was laid even before Pei Xiu. Such a foundation existed at least by the Han, if the Mawangdui maps are admitted as evidence: "The detail and accuracy of these maps could not have been achieved without substantial field surveying using relatively sophisticated techniques."[31] Such mensurational techniques were developed not only for cartography but also for related endeavors such as navigation and astronomy, and the evidence for these techniques often comes from those other fields.

The greatest achievement in navigation is associated with Zheng He (1371–1433), who from 1405 to 1433 led seven maritime expeditions as far west as the coast of eastern Africa and perhaps as far south as Kerguelen Island in the Indian Ocean. A navigational chart contemporaneous with Zheng He's voyages has been preserved in the *Wubei zhi* (Treatise on military preparations), compiled about 1621 by Mao Yuanyi (1594–ca. 1641). Mao provides no information on the source or maker of the chart, but some have traced it to Zheng He's voyages.[32]

31. Hsu, "Han Maps," 55 (note 13). See also Yang Wenheng, "Shilun Changsha Mawangdui sanhao Han muzhong chutu ditu di shuli jichu" (On the mathematical foundation of the maps excavated from Han tomb 3 at Mawangdui), *Keji Shi Wenji* 3 (1980): 85–92, esp. 86. Even granting that the instruments and geometry necessary for surveying had been developed by the Han, in speaking of early Chinese maps one must be wary of applying the word "survey" in its modern sense: gathering information about a well-defined area by the use of techniques applied consistently throughout. Much of the information reflected on the Mawangdui maps, for example, could have been obtained from direct observation, but it might have been gathered in patchwork fashion, in bits and pieces, not systematically, from several sources: direct measurements, existing records, even local lore about travel time. All of this information could be combined, compared, and adjusted to create a map giving approximate locations with the same accuracy as the Mawangdui maps. There are European examples of this.

32. Mills, however, traces the map to Mao Kun (1512–1601), Mao Yuanyi's grandfather, who was a member of an admiral's staff. Mills believes the map was the work of a cartographer and his assistants who

The chart depicts a voyage from Nanjing to the island of Hormuz to ports on the east African coast. It is thought to have been originally a strip measuring about 20.5 by 560 centimeters, with the image on it divided into forty sheets when included in the *Wubei zhi*. The strip format has been found to result in angular, linear, and directional distortions. The orientation of the chart, for example, changes frequently (fig. 3.20), and the map scale has been found to vary from section to section according to the amount of detail represented. Sea routes are depicted with broken lines, and sailing instructions are given in notes on the chart. Some scholars have checked the instructions against modern charts and found that, in contrast to the map image, the instructions are accurate for the most part with regard to distance and direction.[33] The chart is thus taken as evidence of sophisticated navigational techniques based on astronomical observations and the use of the magnetic compass.

Astronomical techniques also reached a high level of development. During the Tang and Yuan, for example, the Chinese have been credited with what are interpreted as geodetic surveys. The Tang survey involved observations from a series of stations forming a chain about 3,800 kilometers long. The survey determined, among other things, that a north-south distance of about 351 *li* corresponded to a change of one degree in the altitude of the Pole Star. This information, some have speculated, was used to produce large-area maps such as the *Yu ji tu*.[34] The Yuan "survey of the four seas" (*sihai ceyan*) was more extensive than the Tang survey, involving twenty-seven observation stations within an area covering about 5,000 kilometers from north to south and about 2,700 kilometers from east to west.[35] Measurements of the length of the sun's shadow at noon were taken, often with gnomons as tall as twelve meters (fig. 3.21). The determinations and measurements from these surveys were intended for the correction of calendars, but they had potential cartographic applications. From the length of the sun's shadow, for example, it is possible to compute latitude.

Because of their cartographic applications, potential or actual, the meridian "surveys" and navigation have been discussed in recent histories of Chinese cartography as indicative of its mathematical foundations. The only difficulty here is that the cartographic application of astronomical techniques has yet to be demonstrated, and the earliest evidence for the cartographic application of seafaring techniques exists in the *Wubei zhi*, which dates from the seventeenth century. Thus evidence of mensuration in fields often associated with cartography still does not give much support to the idea of a continuing tradition of measured mapping.

The Use and Abuse of Cartographic History: Flaws in the Quantitative Approach

The quantitative interpretation is tenable if one insists on linking mathematics and cartography, so that only those images that bear signs of measurement are considered cartographic and those associated with text are beneath cartographic consideration. Uniformity of scale, I agree, is important for achieving a certain kind of fidelity to geographic reality, one emphasizing the representation of linear distances. As I stated above, the pre-Han and Han artifacts can be interpreted as evidence that the idea of scale was understood. But there are some problems with that interpretation. For example, the deteriorated condition of some of the artifacts when excavated makes determination of exact scale difficult: there are gaps in the artifacts, and scale determinations can vary depending

compiled the chart from earlier ones. Others have conjectured that the map had origins in the Islamic world. See Ma Huan, *Ying-yai Sheng-lan: "The Overall Survey of the Ocean's Shores" [1433]*, ed. and trans. J. V. G. Mills (Cambridge: Cambridge University Press, 1970), 239–41.

33. Mills, for example, remarks that in some cases the details on the chart are "astonishingly accurate," but he also finds that the instructions for the main voyage from Ceylon to Hormuz are "defective" and that the instructions for sailing along the east African coast are "rudimentary and inadequate." See Mills, *Ying-yai Sheng-lan*, 248 (note 32). Hsü Yü-hu finds the distance and bearing information to be "surprisingly" accurate for routes along the Southeast Asian coast. See Hsü Yü-hu (Xu Yuhu), *Mingdai Zheng He hanghai tu zhi yanjiu* (Study of Zheng He's nautical chart from the Ming period) (Taipei: Xuesheng Shuju, 1976), 7. The difference in opinions may be related to disagreement as to what distance corresponded to the Chinese *geng*, or watch, equal to 2.4 hours. Distance on the chart is given in terms of *geng*, and one's calculation of that distance will vary depending on how fast one believes a Chinese sailing vessel could travel. Estimates of sailing speed range from about twelve to twenty nautical miles per *geng*. A useful collection of studies on the chart is Zhongguo Hanghai Shi Yanjiuhui (Research Association for the History of Chinese Navigation), ed., *Zheng He yanjiu ziliao xuanbian* (Selected research materials on Zheng He) (Beijing: Renmin Jiaotong Chubanshe, 1985).

34. Arthur Beer et al., "An 8th-Century Meridian Line: I-Hsing's Chain of Gnomons and the Pre-history of the Metric System," *Vistas in Astronomy* 4 (1961): 3–28, esp. 16.

35. This "survey" is described in Song Lian et al., *Yuan shi* (History of the Yuan, compiled 1369–70), chap. 48; see the edition in 15 vols. (Beijing: Zhonghua Shuju, 1976), 4:1000–1001. The same history also records an attempt by Yelü Chucai (1190–1244) to correlate celestial measurements and terrestrial distance between Beijing and Samarkand. Yelü, an astrologer who served the Mongol leader Chinggis Khan, noticed a two- or three-hour time lag between celestial occurrences predicted in the official Chinese calendar, which had been made for use in Beijing, and their occurrence in Samarkand. To overcome this, Yelü compiled a new calendar for Samarkand, using an adjustment factor to calculate the time difference of astronomical phenomena observed in regions east or west of there. This factor, however, did not find cartographic application, though it could have been used to determine longitude in the Western sense of a geographical coordinate. See *Yuan shi*, chap. 52 (4:1119–20).

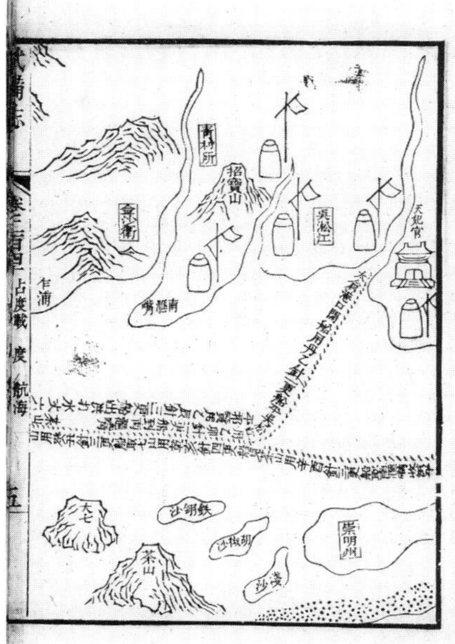

FIG. 3.20. NAUTICAL CHART FROM THE *WUBEI ZHI*. Shown here are the first six pages of the nautical chart, which cover the route from the Nanjing area to just south of the Huangpujiang (Whangpoo River). The chart bears no scale indications, but notes give compass directions and sailing times for routes between various places. Orientation changes as one

on how one fills in those gaps. In the case of the Mawangdui maps, it is important to note that scale consistency decreases as one moves away from the center. It is not sufficient to consider absolute variation in scale across a whole map, as several researchers have done; one must also take into account the distribution of that variation. Some researchers on early Chinese maps have tried to establish uniformity of scale by hiding its variability, by giving the scale as a range. It also will not suffice to argue that the center is of main interest, and therefore that it is drawn to scale or exhibits a narrower range of scale variation. For defense purposes, which the garrison map is thought to have served, the periphery might very well have been of main interest. There is not enough evidence, in either case, to judge. We should also remember as we consider the scales given for the early maps that none of the artifacts in question bears an expressed scale, and that the "rounding off" of the ratio may imply more systematizing than actually existed.[36]

In addition, emphasis on scale and accuracy has tended to divert attention from other questions about traditional Chinese mapping. For example, more attention could be paid to methods of graphic production. It seems that much of the technological basis for map production had been laid by the end of the Han dynasty. Maps continued to be produced with brush and ink on silk and paper, and to be engraved on various media, up to the twentieth century. This continuity is one reason for not dividing up Chinese cartographic history into dynasties. As far as methods of graphic production are concerned, there was one technological innovation: the development of woodblock printing during the eighth century, which made it easier to reproduce and thereby to disseminate maps. Thus, after the tenth century we have many more maps extant than for earlier periods.

One should also note that the same methods of production were employed in the graphic arts in general, so that it is possible to conjoin cartography and other visual arts. The failure to explore this connection has led to the devaluation of a large body of artifacts that fall within our definition: "graphic representations that facilitate a spatial understanding of things, concepts, conditions, processes, or events in the human world."[37] If one views Chinese artifacts with this definition in mind, one gets a

36. Han Zhongmin has suggested that too much emphasis has been placed on ascertaining the accuracy of the Mawangdui maps. He proposes a different reconstruction of the garrison map, one that would result in a map of ninety-six by ninety-six centimeters instead of ninety-eight by seventy-eight centimeters. See Han Zhongmin, "Guanyu Mawangdui boshu gu ditu di zhengli yu yanjiu" (Concerning the restoration and study of the ancient silk maps from Mawangdui), in *Zhongguo gudai ditu ji*, ed. Cao Wanru et al. (Beijing: Wenwu Chubanshe, 1990–), 1:12–17. Curiously, no illustration of Han's proposed reconstruction appears. Han does imply that the map would no longer be as accurate as is thought: his reconstruction apparently would alter the degree of correspondence between the map and the area it supposedly depicts.

37. Preface to *The History of Cartography*, ed. J. B. Harley and David Woodward (Chicago: University of Chicago Press, 1987–), 1:xvi.

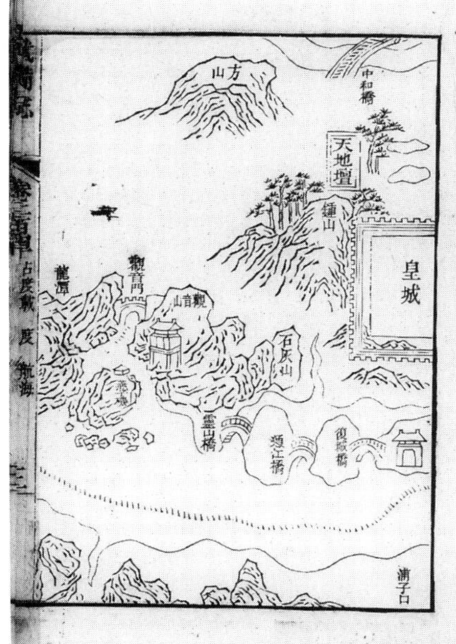

moves across the map. On the first two pages, starting from the right, south-southeast is at the top. On the third, fourth, and fifth pages, south is generally at the top. On the sixth page, west is at the top. Size of each page: ca. 14.5 × 10 cm. Courtesy of the Asian Division, Library of Congress, Washington, D.C. (F701.M32.1).

different impression of Chinese mapmaking—that scale maps tend to be isolated examples.[38] One might object that looking at artifacts with a broader conception of maps is merely another instance of arbitrariness, of imposing a different preconception on Chinese artifacts. A broader conception, the following chapters show, is sanctioned by the context: the artifactual record itself demands a broader understanding of the map. A focus on scale mapping is insufficient to account for the broad range of production and the variety of representational practices. Furthermore, what are taken to be scale maps often have much in common with artifacts not usually regarded as cartographic. The modes of representation on the Mawangdui maps, for example, appear less striking when one sees similar curves representing mountains on a map of a burial site dating from the tenth century and bearing no scale indications (fig. 3.22).

Since the artifacts singled out for attention under the quantitative interpretation are viewed as instances of scale mapping, they are regarded as "scientific" and as serving secular, utilitarian purposes. Thus they are described as topographic military, economic, or administrative maps. Such descriptions, however, tend to overlook the contexts in which the artifacts were found—for example, tombs of members of the ruling class. The placing of maps in tombs seems to confirm that the functions of maps from at least the Han went beyond the secular (see pp. 77–80). It is clear from literary sources, cited in detail in the following chapters, that maps did serve utilitarian purposes, such as administration and military planning. But textual and artifactual evidence makes it clear that, for the literary elite, maps also served religious functions. They could depict the arrangement of implements used in rites. They could be used to locate auspicious sites for buildings. They could serve as talismans to ward off demons or as representations of power to secure passage to the otherworld. They could record astrological information and thus help interpret heavenly signs.

All this suggests that to understand early Chinese maps, we need to study them in the context of the beliefs and values of the ruling elite, not "abuse" them by imposing modern conceptions of cartography. As is described elsewhere, maps in Chinese culture were used not only for representing distances, but also for demonstrating power, for education, and for aesthetic appreciation. A conception of Chinese cartography as a rational, mathematical discipline for understanding space has led to a failure to

38. The range of cartographic styles in traditional China is suggested by the artifacts collected in *Zhongguo gudai ditu ji*, ed. Cao et al. (note 12). Many of the maps illustrated in this atlas have no expressed scale and are highly pictorial. The maps selected for extended analyses, however, are generally those that researchers are able to evaluate in terms of accuracy and scale—namely, those that support the notion of progress within a quantitative tradition. The atlas is useful for collecting a large number of Yuan and pre-Yuan maps, but it also represents a missed opportunity to examine the Chinese map tradition whole. Here I attempt to correct the imbalance.

FIG. 3.21. YUAN OBSERVATIONAL TOWER. This observational tower in Dengfeng Xian (in present-day Henan Province) once contained a gnomon about twelve meters tall. Extending horizontally on the ground from the north side of the tower is a graduated scale more than thirty-six meters long, which was used to measure the sun's shadow.

From Zhongguo Shehui Kexueyuan Kaogu Yanjiusuo (Archaeological Research Institute, Chinese Academy of Social Science), *Zhongguo gudai tianwen wenwu tuji* (Album of ancient Chinese astronomical relics) (Beijing: Wenwu Chubanshe, 1980), 14.

consider the full range of cartographic functions. It also leads to an anomaly: if the all-important goal was scale mapping, the preservation of such disparate images as the *Hua yi tu* and *Yu ji tu* on the same stone stele seems somewhat curious.

One researcher attempts to explain this by identifying two "parallel" traditions of Chinese cartography, one mathematical or "analytic" and the other "descriptive."[39] The former is concerned with measurement and is therefore a "science"; the latter is concerned with "information" and less concerned with accuracy. There is, however, no evidence that Chinese mapmakers saw themselves as working in two traditions distinguished by their attention to measurement, and the identification of two traditions still does not explain why parallel traditions would meet on the same stone. In the history of cartography, there are, to be sure, many examples of maps from different traditions appearing side by side. In Renaissance atlases, for example, Ptolemaic and "modern" world maps appear together. But in such instances the two types of maps are both predicated on the importance of mathematical representation to cartography. This does not seem to be the case with the two twelfth-century stone maps at Xi'an. The claim advanced by advocates of the quantitative approach is that these maps represent qualitatively different ways of mapmaking, one more advanced, more scientific than the other. Thus, one might argue that the maps on the stone stele are products of a transitional period.

The problem with that interpretation is that ten centuries seems rather long for a transition, and that the transition, as we will see below, does not seem to have been fully realized. The evidence does not support the view that Chinese maps were perceived within their cultural contexts as representing different traditions. Maps like the *Hua yi tu*, in short, were not regarded as inferior to maps like the *Yu ji tu*.

The preservation of both maps on the same stone sug-

39. See Hsu, "Han Maps," 56–59 (note 13).

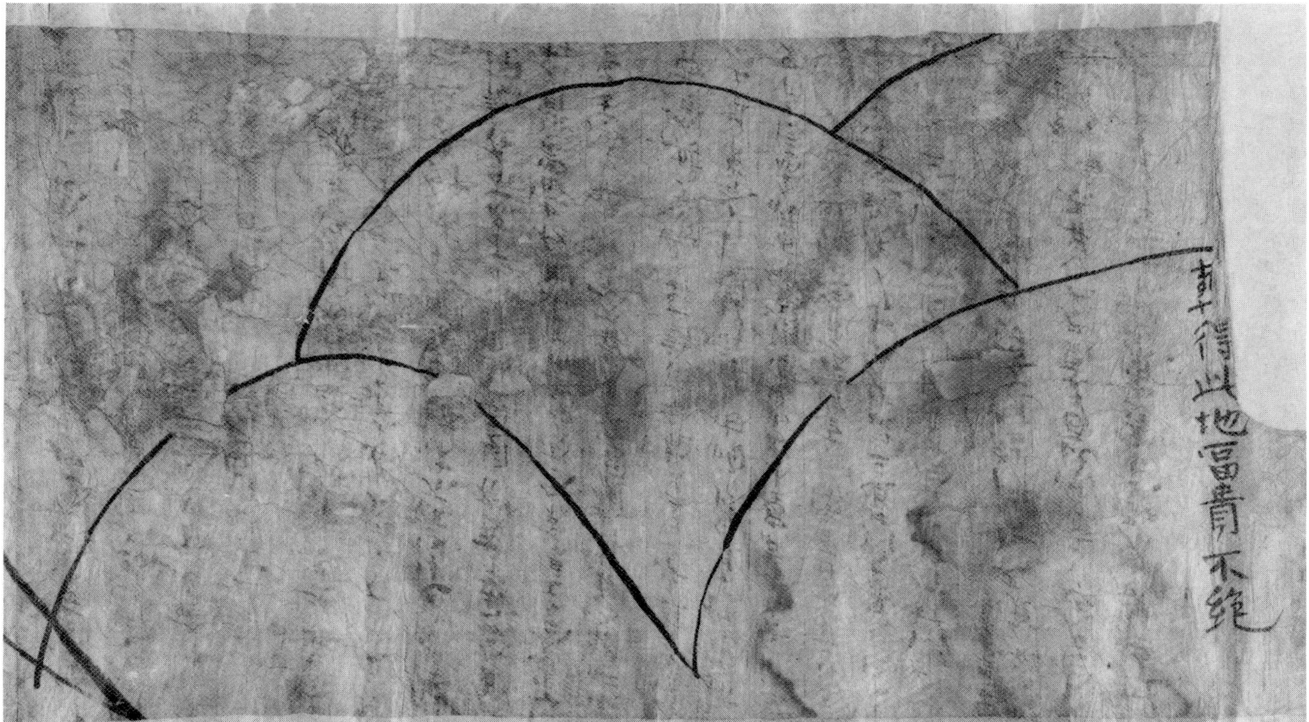

FIG. 3.22. MAP OF AN AUSPICIOUS SITE FOR A FAMILY TOMB (DETAIL). This map, discovered at Dunhuang in the Cave of the Thousand Buddhas, was drawn perhaps in the tenth century. A note on the map says: "Anyone buried here will have good fortune."

Size of the entire original: ca. 25 × 229 cm; this detail: 24 × 39.5 cm. By permission of the British Library, London (Stein no. 3877 [Dunhuang MS. 6971]).

gests that considerations other than scale and mathematical correspondence to geographic actuality were considered important for mapmaking. The artifactual record provides ample evidence to undermine the idea that the *Yu ji tu* established a foundation for a climax of mathematical cartography during the Yuan and Ming, or that there was a general advance in measured mapping from the twelfth through the seventeenth century.[40] The maps in the *Lidai dili zhizhang tu* (Easy-to-use maps of geography through the dynasties, 1098–1100, supplemented 1162), for instance, are roughly contemporaneous with the *Yu ji tu* but do not adhere as closely to geographic actuality. China's outlines are depicted as being more squarish than they really are, and the Shandong peninsula almost disappears (figs. 3.23 and 3.24). Furthermore, islands representing other countries are nestled in an encircling ocean.

From a quantitative point of view, the image does not necessarily improve with later maps, for example, those in the *Da Ming yitong zhi* (Comprehensive gazetteer of the Great Ming, 1461). According to the gazetteer's preface composed by the emperor Yingzong (r. 1436–49, 1457–64), the work of compiling geographic information for the gazetteer began when the emperor Chengzu (r. 1403–24) ordered scholars to gather maps and documents from the empire's prefectures and towns and to combine them into one work.[41] This project was not completed, however, until the emperor Yingzong resumed the project, which was completed in 1461.

The main body of the comprehensive gazetteer opens with a preface to a "comprehensive map" of the empire and contiguous lands (fig. 3.25). The compilers of the gazetteer state that they provided this map at the beginning of the work so that one could "turn to the map and view the entire world, and the extent of the territory would be clearly in view as if one were looking at one's own palm."[42] The gazetteer was based on more than thirty years of information gathering, but the quality of

40. Exponents of the view that traditional Chinese cartography reached its peak during the Yuan and Ming include Joseph Needham and, more recently, Lu Liangzhi. See Needham, *Science and Civilisation in China*, 3:551–56 (note 21); and Lu, *Zhongguo dituxue shi*, 99 (note 20).

41. See the emperor's preface to Li Xian et al., *Da Ming yitong zhi*, 1b; see the edition in 10 vols. (Taipei: Tailian Guofeng Chubanshe, 1977), 1:2.

42. Li et al., *Da Ming yitong zhi*, preface (by compilers), 2b (1:56) (note 41).

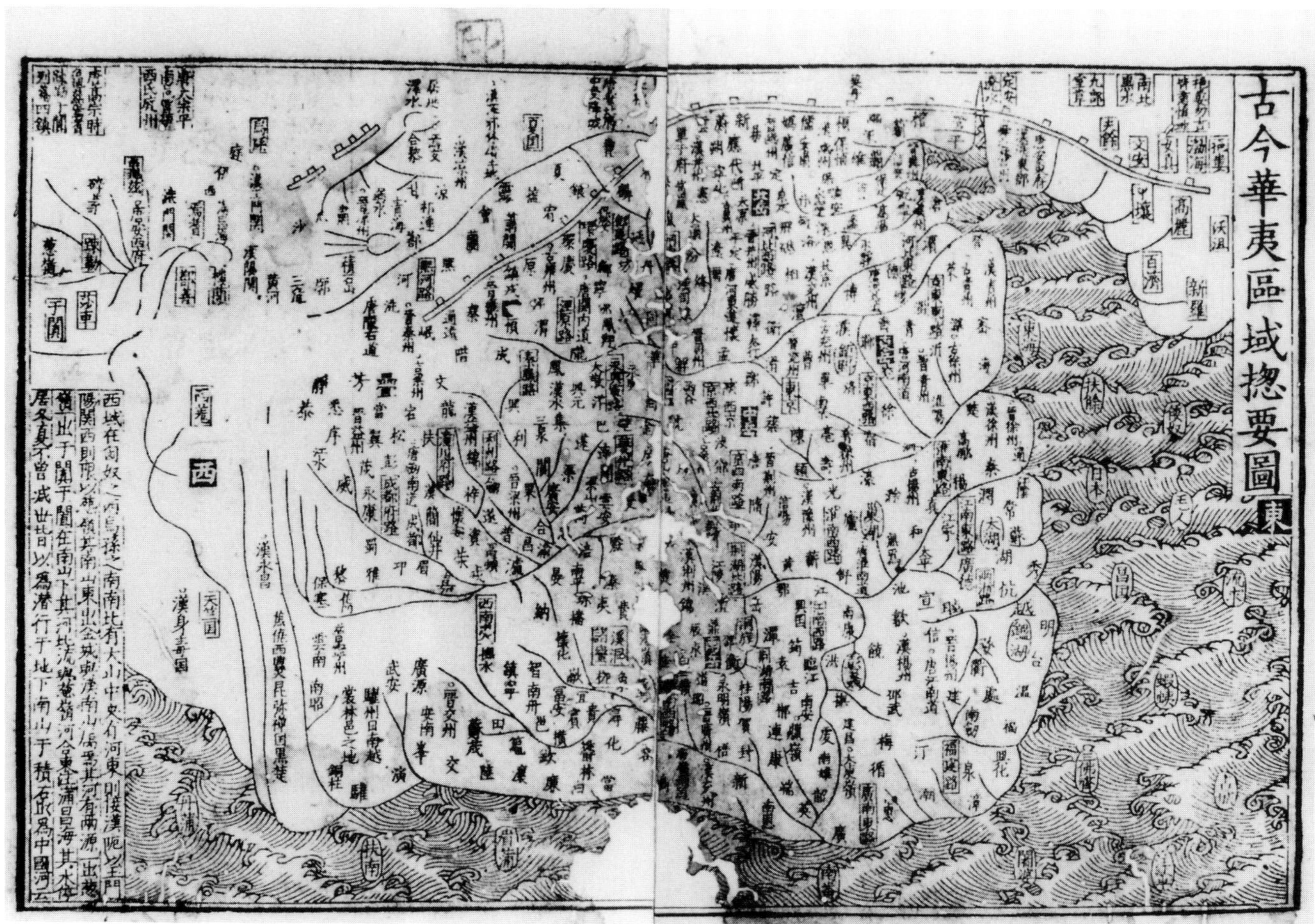

FIG. 3.23. "GUJIN HUA YI QUYU ZONGYAO TU" (GENERAL MAP OF THE ANCIENT AND PRESENT TERRITORIES OF CHINA AND FOREIGN COUNTRIES). This map is from the *Lidai dili zhizhang tu*, a collection of forty-four historical maps of dynastic territories from the time of Ku, legendary male ancestor of the Shang, to the Song dynasty. Each map is accompanied by explanatory notes. The maps shown here and in figure 3.24 are from a Southern Song edition.
Size of the original: unknown. Photograph courtesy of Cao Wanru, Institute for the History of Natural Sciences, Academia Sinica, Beijing.

the geographic information reflected in the general map does not quite fulfill the compilers' intentions. According to the criteria of the quantitative approach, the map represents a step backward from the two stone *Yu ji tu*, which predate the gazetteer by more than three hundred years. The Ming map is less detailed than the Song map, particularly in the depiction of the courses of rivers and the configuration of the coastline. Unlike the Song map, the Ming map lacks a grid and shows Yunnan and even what is now Tibet as having a coastline. The Ming map also makes greater use of pictorial elements, notably in the representation of mountains, which are merely named on the Song map.

The high degree of generalization in the Ming map may be at least partly explained by its extensive textual supplementation. The Song map has very little textual accompaniment; the cartographic image is dominant. In contrast, despite the importance attached to visual representation in the emperor's order, maps actually constitute only a small portion of the *Da Ming yitong zhi*. The work contains fifteen other maps besides the general map discussed above, so that maps occupy a mere thirteen leaves out of more than 2,800. These fifteen maps represent subdivisions of the territory depicted on the general map—the two capital districts and thirteen provinces—and introduce textual descriptions of the geographic subdivisions.[43]

43. The text itself subdivides these divisions further and is organized into a series of treatises, most devoted to a single prefecture (*fu*) or tributary state. The treatises are further subdivided into sections on topics such as local history, geography, and customs, as well as famous officials, itinerant scholars, exemplary women, and Daoist and Buddhist clergy. Such categories of information became standard in later gazetteers.

FIG. 3.24. "TANG YIXING SHAN HE LIANGJIE TU" (MAP OF THE TWO BOUNDARIES FORMED BY MOUNTAINS AND RIVERS ACCORDING TO YIXING OF THE TANG). The map illustrates Yixing's idea that the mountains and rivers of China formed two natural boundaries that defended against foreign invaders. The map shows two mountain ranges running from west to east, converging at Taihua Shan and then separating again. One is labeled "northern defense," and the other "southern defense." From the *Lidai dili zhizhang tu* (see fig. 3.23).
Size of the original: unknown. Photograph courtesy of Cao Wanru, Institute for the History of Natural Sciences, Academia Sinica, Beijing.

The ratio of image to text in traditional Chinese geographic works varies, but it is not uncommon to find a disproportionate amount of text. One example is the *Guang yutu* mentioned earlier, which consists of a book made up mostly of textual accompaniment to maps: about one hundred pages of maps and about three hundred pages of notes. Another is the *Gujin xingsheng zhi tu* (Map of advantageous terrain past and present).[44] This map was printed from a woodblock in 1555, making it contemporaneous with Luo Hongxian's *Guang yutu*. Unlike the *Guang yutu*, however, the *Gujin xingsheng zhi tu* lacks an expressed scale as well as the square grid (plate 1). One might claim that this difference is related to function.[45] The *Gujin xingsheng zhi tu* is heavily annotated with information regarding changes in toponyms and administrative status, and this feature combined with its title might suggest that it was meant to aid historical scholarship, not to provide a sense of distances. The problem with this line of argument is that the *Guang yutu* also provides historical information about the places it describes. It too could serve as a reference for historical scholarship and was itself based on such scholarship.

The grid on the general map of the empire in the *Guang yutu* raises questions of how seriously the grid was to be taken as a gauge of distance. Each square on the map is supposed to represent 100 *li*, or about 55 kilometers. Thus the distance from Xi'an to the coast directly east is given as about 600 *li*, or about 330 kilometers. This is about one-third the actual distance. The squares on the

44. This map is described in Ren Jincheng, "Xibanya cang Ming ke *Gujin xingsheng zhi tu*" (The *Gujin xingsheng zhi tu* printed during the Ming and preserved in Spain), *Wenxian* 17 (1983): 213–21.

45. As Ren has done in "Xibanya cang Ming ke *Gujin xingsheng zhi tu*," 214 (note 44).

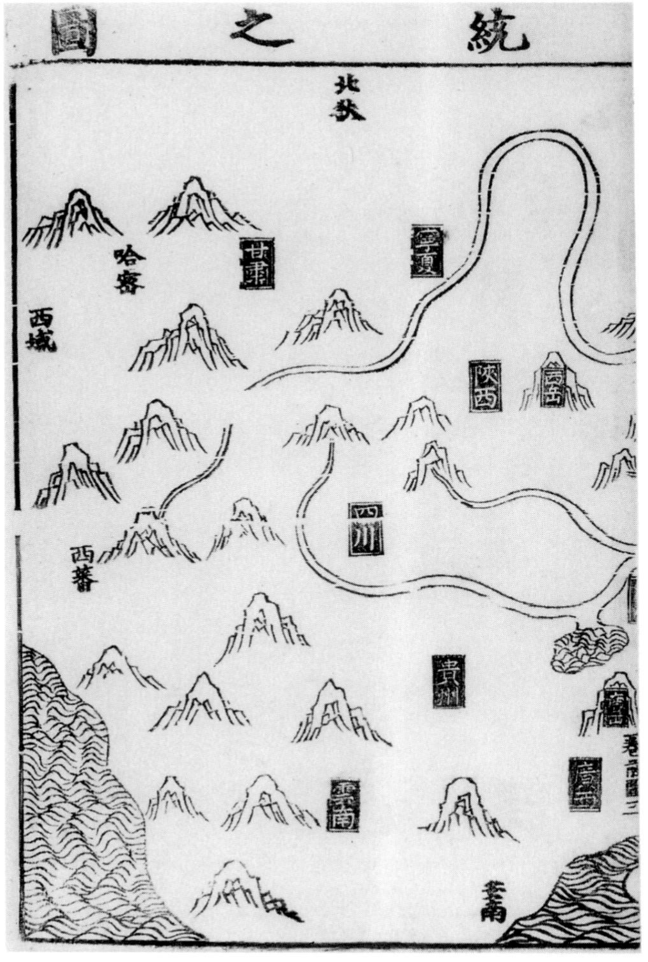

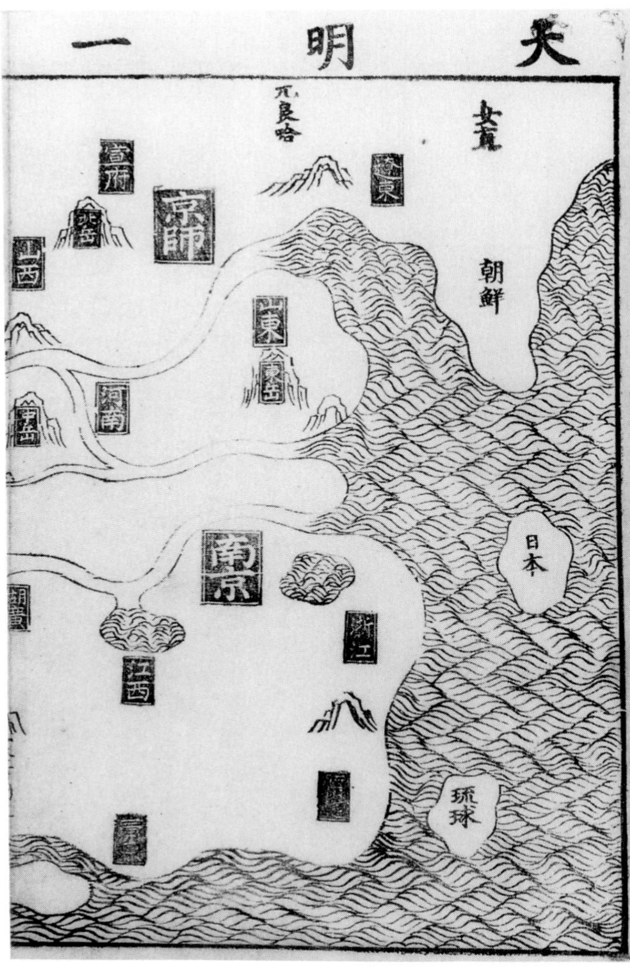

FIG. 3.25. MAP OF CHINA IN THE *DA MING YITONG ZHI*.
Size of each page: 26 × 18 cm. From Li Xian et al., *Da Ming yitong zhi* (completed 1461), prefatory chapter, unpaginated. Reproduced courtesy of the Harvard-Yenching Library, Harvard University, Cambridge.

Yu ji tu (dated 1136) also represent 100 *li*, but the corresponding distance on the stone map is about 2,500 *li*, which is about one-third longer than the actual distance. Thus grid use on the *Guang yutu* and the maps based on it may be suspect as an indicator of the advanced state of scale mapping. In fact, not all mapmakers fully appreciated the utility of the square grid. In one edition of the *Guang yutu*, Fuchs has pointed out, some squares of the grid are elongated into rectangles, apparently in an attempt to fill wider pages.[46] Such elongated grids would still be of use as a scaling device, but uniformity of scale along the horizontal and vertical axes of the map would be lost.

The converse of what I said above about grids also needs to be heeded: the lack of scale or a grid on a map does not necessarily indicate a lack of geographic understanding or even a lack of utility. A case in point is the maps that often accompanied reports describing water conservancy projects. These were produced by regional and local officials and drawn using brush and ink. Like most traditional Chinese maps, they typically lack the square grid and scalar indications. This does not mean measurements were not performed. Quantitative information such as distances and dimensions was often written on paper strips, which were then pasted to the map (fig. 3.26). To appreciate the practical value of the map, one has to appreciate the complementary relationship between image and text, which might appear on the map itself or accompany it on separate sheets, as in a book.

46. Walter Fuchs, *The "Mongol Atlas" of China by Chu Ssu-pen and the Kuang-yü-t'u*, Monumenta Serica Monograph 8 (Beijing: Fu Jen [Furen] University, 1946), 21.

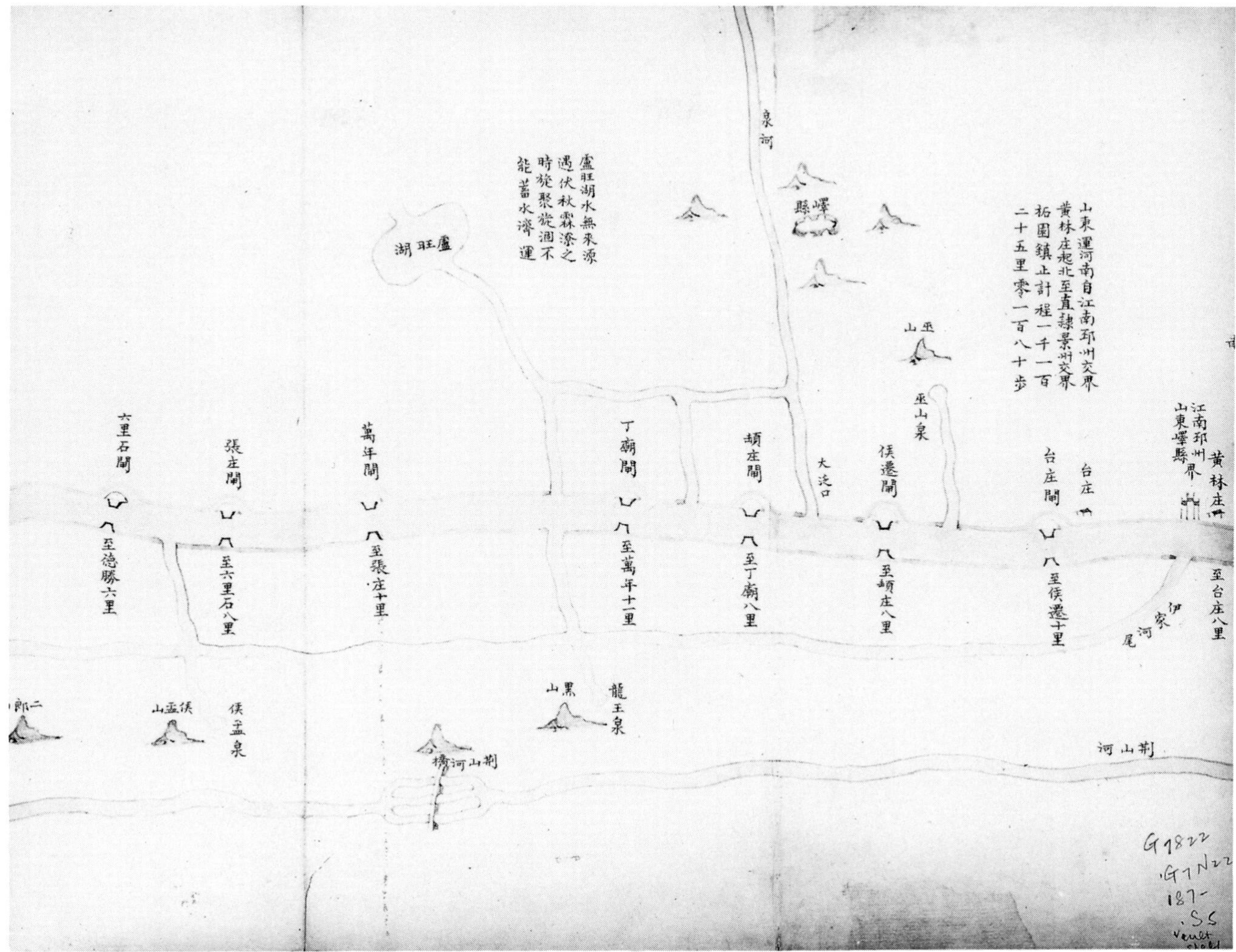

FIG. 3.26. WORD AND IMAGE ON A QING HYDROLOGICAL MAP (DETAIL). This map dating perhaps from the 1870s depicts the water control system on the Grand Canal in Shandong Province. It was drawn using watercolor and ink, and labels indicating distances were pasted on.
Size of the entire original: 23 × 382 cm. Courtesy of the Geography and Map Division, Library of Congress, Washington, D.C. (G7822.G7N22 187-.S5 Vault Shelf).

In studying Chinese maps, we may also have to adjust our norms of utility. Two nineteenth-century examples can help clarify what I mean. The first is a hydrological map depicting the Yangtze River and associated waterways in Hubei Province (plate 2). The second is a map of the city of Hangzhou (fig. 3.27). The hydrological map is largely planimetric and shows signs of measurement: notes give distances between certain landmarks. The city map is also generally planimetric. So far there is nothing in these maps to undermine a quantitative interpretation, that is, not until one considers a curious feature common to both maps. In certain places on these maps—in representing hills and buildings, for example—the mapmakers depart from a planimetric mode of representation and represent things pictorially in a manner highly suggestive of painting. These examples cannot be dismissed as aberrant, for pictorial representation in hydrological mapping goes back at least as far as the Yuan (fig. 3.28). Such examples also cannot be dismissed as the expression of nonutilitarian or decorative intentions. Even the nautical chart in the *Wubei zhi* represents coastal areas in a manner similar to landscape painting,[47] as do other maps used

47. Researchers bound to the quantitative interpretation have taken pains to explain away this feature. Niu Zhongxun, for example, says that the map distinguishes between smooth and rough islets, so that it shows a greater concern for the scientific than the artistic. See Niu Zhongxun, " 'Zheng He hanghai tu' di chubu yanjiu" (Preliminary study of Zheng He's nautical chart), in vol. 1 of *Zheng He xia Xiyang lunwenji* (Collected essays on Zheng He's expedition to the Western Ocean), ed. Zhongguo Hanghai Shi Yanjiuhui (Research Association for the History of Chinese Navigation) (Beijing: Renmin Jiaotong Chubanshe, 1985), 238–48, esp. 243.

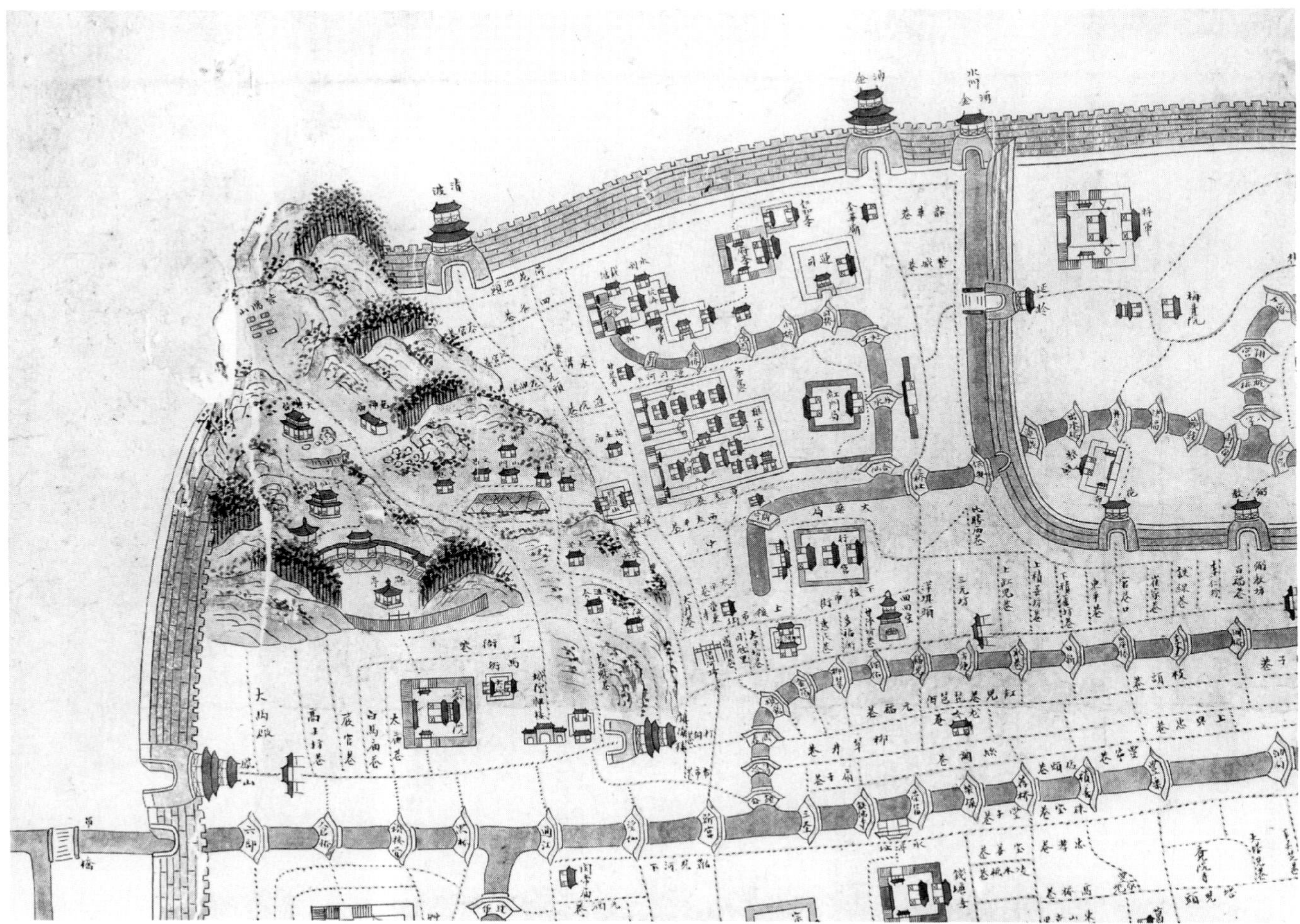

FIG. 3.27. DETAIL FROM THE *HANGZHOU CHENG TU* (MAP OF HANGZHOU). This map of Hangzhou dates perhaps from the nineteenth century. North is to the right. Streets are drawn in plan, while buildings, temple grounds, and gates are represented pictorially. Shown here is the southwest corner of the city, where the landscape is dominated by hills.
Size of the entire original: 63 × 94 cm. Courtesy of the Geography and Map Division, Library of Congress, Washington, D.C. (G7824.H2A5 18--.H3 Vault Shelf).

for navigation (plate 3). In addition, coastal defense maps from as late as the nineteenth century are often drawn in a pictorial manner (plate 4 and fig. 3.29). What we seek to understand in the following chapters is why. It is also curious to find these examples of pictorialism so late in the imperial period, when European methods supposedly supplanted the traditional methods.

Conversely, a planimetric mode of representation does not always indicate practical function. A case in point is a map of Jingjiang Fu Cheng (Prefectural City) carved on the stone cliff of Yingwu Shan (a mountain north of Guilin) in 1271–72 (fig. 3.30). Construction of the city began in 1258 to provide a defensive bulwark against Mongol forces. The city walls were completed before 1270, and the map was engraved shortly afterward. The city fell to the Mongols in 1277. The engraved surface measures about 340 by 300 centimeters. It shows some of the main streets of the city, along with moats, city walls, watchtowers, and gates. Because of the prominence given to features of military importance, some have interpreted the map as serving defensive purposes.[48] But this interpretation is not convincing, since the map was carved outside the walls of the city even as the Mongols were advancing. In addition, qualities often associated with "military" maps—flexibility, portability, and secrecy—are not among those of the map of Jingjiang Prefecture. Its engraving on stone suggests a commemorative function similar to that of the *Yu ji tu*.

If being engraved on stone is an indication of value,

48. See Chen et al., *Zhongguo gudai dilixue shi*, 309 (note 17); and Lu, *Zhongguo dituxue shi*, 152 (note 20).

FIG. 3.28. MAP OF THE SOURCE OF THE YELLOW RIVER. This map was based on information in the *Heyuan zhi* (Treatise on the source of the Yellow River, 1315) by the geographer Pan Angxiao (fl. thirteenth century). South is at the top. Size of the original: 13.5 × 17.4 cm. From Tao Zongyi, comp., *Nancun chuogeng lu* (Notes taken by Nancun [Tao Zongyi] while at rest from plowing, 1366), *Sibu congkan* edition, 22.3a–b.

moreover, then it is clear that planimetric maps like the *Yu ji tu* and *Hua yi tu* were not the only ones considered worthy of long-term preservation. Two maps of the sacred mountain Hua Shan (Mount Hua) in Shaanxi Province—one from the Ming and one from the Qing—illustrate this point (figs. 3.31 and 3.32). Both make use of pictorial representation, and both have no expressed scale—characteristics not valued by proponents of a quantitative tradition. But both show clear signs of intent to convey knowledge of place: on both maps, place-names are given. In addition, pictorial representation was not necessarily perceived as detracting from the maps' utility. An inscription on the Ming map says the map was made as an aid to the visitor to the mountain. Bearing this out, the map depicts climbing paths, steps cut into the sides of cliffs, and bridges over gaps.

Pictorial maps, of course, did not serve only practical purposes. On the Qing map of Hua Shan, a number of "supernatural" features are emphasized; a palm print of a giant on the east peak, an image of a transcendent being on the west peak, and between them a waterfall interrupted by a cavern. The Qing map differs from the Ming map in its depiction of man-made objects: as Munakata has pointed out, it deemphasizes them through the use of thin lines so they are overwhelmed by the natural landscape.[49] The intended result seems to be a mixture of awe and humility.

The artifactual record suggests that scale mapping was not the primary concern of Chinese mapmakers, although

49. Kiyohiko Munakata, *Sacred Mountains in Chinese Art* (Urbana: University of Illinois Press, 1991), 57.

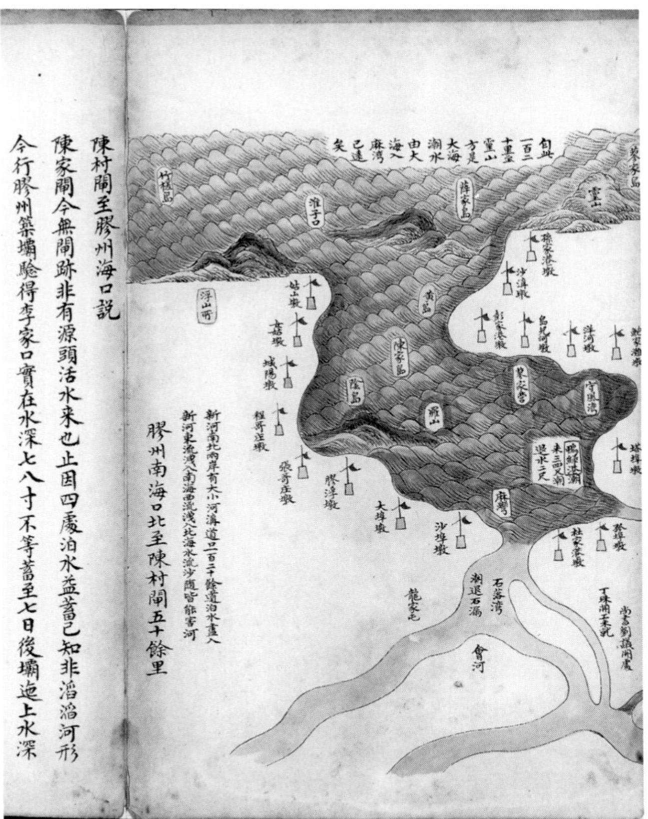

陳村閘至膠州海口說
陳家閘今無閘跡非有源頭活水来也止因四處泊水益蓄已知非滔滔河形
今行膠州築壩驗得李家口實在水深七八寸不等蓄至七日後壩進上水深
膠州南海口北至陳村閘五十餘里

FIG. 3.29. DETAIL FROM THE *WANLI HAIFANG TU* (TEN THOUSAND *LI* MAP OF MARITIME DEFENSES, 1705?). This map represents the Chinese coast from Hainan Island to the Liaodong peninsula. It bears notes on defense establishments and makes some policy recommendations. Garrisons and military bases are indicated by flagpoles. Coastal fortresses and prefectural cities are represented by pictorial signs, as are mountains. North is to the left.
Size of the entire original: 30 × 274 cm. Courtesy of the Geography and Map Division, Library of Congress, Washington, D.C. (G7821.R4 1705.W3 Vault Shelf).

they certainly understood its principles. What links the maps claimed for a mathematical tradition with the vast body of nonmathematical artifacts is the close relation between word and image. The *zhaoyu tu*, the Fangmatan wooden maps, the Mawangdui silk maps, and the Ming navigational chart, we should recall, all lack scale indications. Distances are given in notes on the maps themselves. With later maps, such as the *Guang yutu*, the textual accompaniment could fill up chapters of books. In addition, as is documented in subsequent chapters, the figures described above as advocates of quantitative cartography all recognized a complementary relationship between image and text. Previously this characteristic of Chinese cartography has been regarded as a hindrance to the development of an independent science of cartography in China. Wang Yong, for example, labels maps that rely on textual accompaniment, employ pictorial representation, and lack grids as "immature and backward."[50]

The same attitude toward text characterizes recent research on the nautical chart in the *Wubei zhi*. One researcher states that it is the first nautical chart that can be used without lengthy textual explanation, a characteristic that anticipates modern charts.[51] This assertion cannot be maintained when one considers that without the lengthy notes on the map itself, the map would be of little use for navigational purposes. The orientation and scale of the image vary across the map, and the notes make one aware of the changes. The textualism seen on the nautical chart, as I argue in a later chapter, was seen as essential to mapmaking, and far from indicating backwardness, it was a sign of scholarly rigor.[52] Given the relative paucity of cartographic images drawn to scale in the existing corpus, one can surmise that textual sources provided much of the information for maps like the *Yu ji tu*. Thus attempts to isolate such maps from the nonmathematical artifacts by claiming that makers of unmeasured maps were not interested in the method of presentation seem forced.[53] Mapmakers could express their concern for measurement in different ways, and they had available different methods of presentation for different purposes.

50. Wang, *Zhongguo ditu shi gang*, 50 (note 26). Elsewhere Wang makes a distinction between maps accompanied by text and "pure maps" (*Zhongguo dilixue shi*, 74 [note 1]). Recently, however, there have been some signs that this disparaging attitude toward the relation between cartographic image and text may change. At the conclusion of a study of the Mawangdui maps, Han Zhongmin mentions that for the most part the achievements of Chinese surveying were not integrated into traditional Chinese mapmaking. He attributes this to the use of textual descriptions and to the influence of Chinese painting. Unfortunately, he does not develop these ideas beyond posing the question whether the influence of painting on mapmaking is a reflection of China's long "feudal" history. See Han, "Guanyu Mawangdui boshu gu ditu di zhengli yu yanjiu," 16–17 (note 36). Han's reticence is understandable, since he is a scholar holding a decidedly unorthodox opinion in a political system not very tolerant of heterodoxy. Another scholar who has foreshadowed my work here is Tan Qixiang. In his preface to the *Zhongguo gudai ditu ji* (note 12), Tan devotes several paragraphs to the connection between textual scholarship and Chinese geography. He points out the importance of maps to the documentary culture of traditional China. On this see pp. 73–77.

51. Zhu Jianqiu, "'Zheng He hanghai tu' zai woguo haitu fazhanshizhong di diwei he zuoyong" (The place and role of Zheng He's nautical chart in the history of the development of our country's nautical charts), in *Zheng He xia Xiyang lunwenji*, 1:229–37, esp. 231 (note 47).

52. The close relation between text and image is even more crucial to an understanding of Chinese religious mapping. Without consulting accompanying text, it is very easy to misread Daoist maps.

53. Hsu, "Han Maps," 59 (note 13).

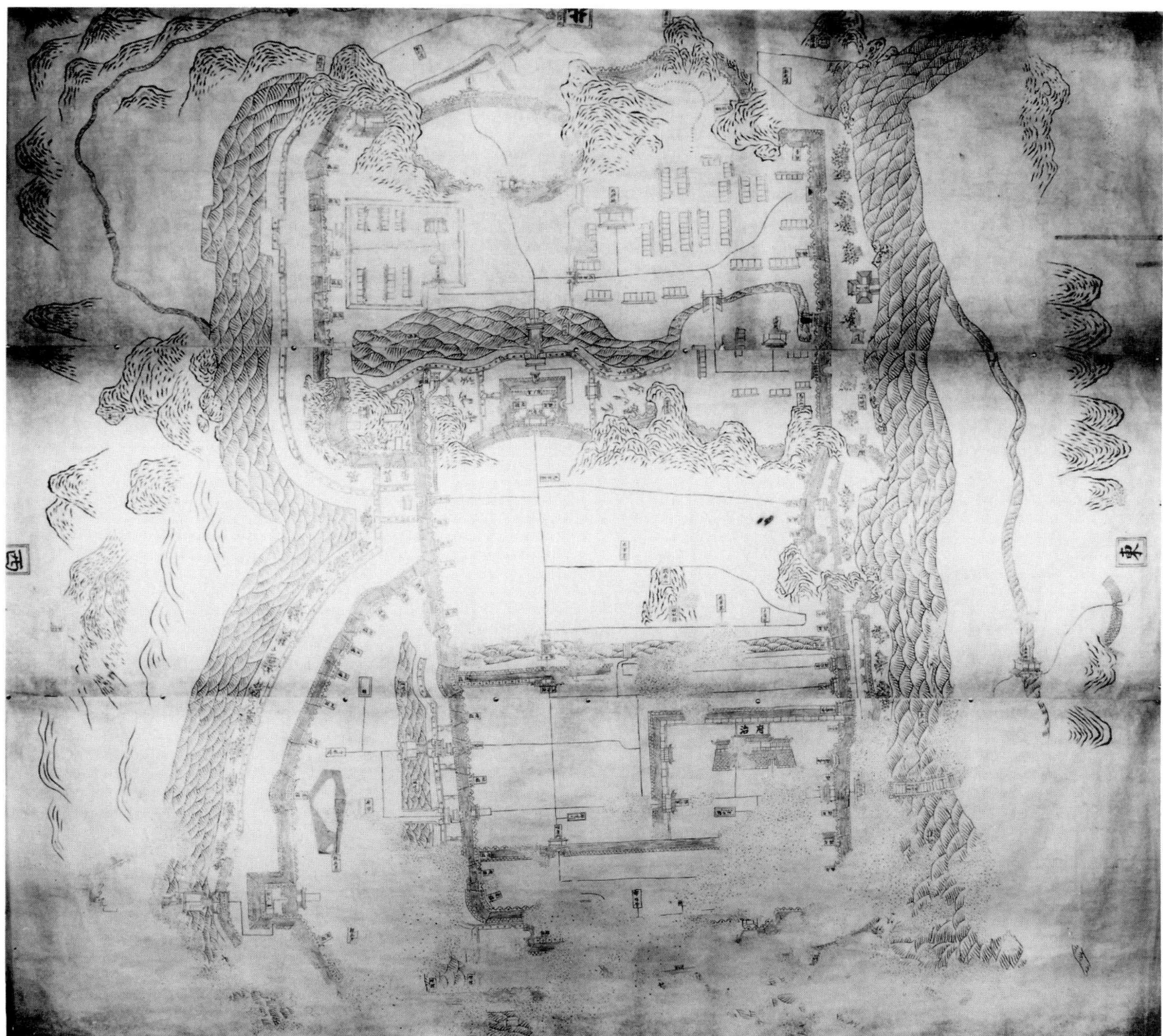

FIG. 3.30. MAP OF JINGJIANG PREFECTURAL CITY, 1271–72. A copy of the map carved on a cliff of Yingwu Shan, north of Guilin in Guangxi Zhuang Autonomous Region. North is at the top. Not shown on the copy is the upper part of the map, which contains annotations describing the construction of the city, listing its dimensions, recording the cost of labor and material, and identifying those in charge of the project. Size of the entire original: 300 × 340 cm. Photograph courtesy of Cao Wanru, Institute for the History of Natural Sciences, Academia Sinica, Beijing.

Toward a Revision of the Chinese Map Tradition

I therefore suggest that a quantitative interpretation of traditional Chinese cartography is inadequate for understanding what constitutes a map in Chinese culture. A Western model of scientific cartography has been pursued too rigorously in previous work on Chinese maps; its relevance even for Western cartography has come to be questioned.[54] The challenge in the following pages is to articulate a coherent alternative to that interpretation of Chinese cartography. In questioning it, I by no means imply that traditional Chinese cartography was deficient

54. See, for example, Stephen Toulmin, *Cosmopolis: The Hidden Agenda of Modernity* (New York: Macmillan, 1990), and David N. Livingstone, "Science, Magic and Religion: A Contextual Reassessment of Geography in the Sixteenth and Seventeenth Centuries," *History of Science* 26 (1988): 269–94.

FIG. 3.31. RUBBING OF A STONE MAP OF HUA SHAN (*TAIHUA SHAN TU*), 1585. The map was made as an aid to visitors to the sacred mountain. The upper portion contains a prayer to the god of the mountain. The prayer was written by the first emperor of the Ming dynasty after he dreamed he was taken to the top of the mountain.
Size of the original: 113 × 60 cm. By permission of the Field Museum of Natural History, Chicago (244848).

or backward. What I mean to convey is that it needs to be understood on its own terms—terms that do not necessarily identify knowledge with number. Just what those terms are and how they affect the canon of traditional Chinese cartography will be discussed in the following chapters.

Those chapters attempt to construct an alternative history by establishing contexts for traditional Chinese geographic mapping. This has been done both by examining the maps themselves and by studying contemporaneous texts where available. The alternative history I am proposing is developed in four overlapping stages. The first deals with the political context of Chinese mapping. The second examines the roles of measurement and textual scholarship. The third explores the relation of traditional Chinese mapping to the arts. The fourth corrects the view that Chinese cartography fused with European cartography in the late imperial period. That view assumes that Chinese cartography was backward and prepared to make use of superior European techniques. In other words, that view of Chinese cartographic history rests on the mathematical interpretation I have begun to question here. My approach can be described as historicist in the sense that I have tried to reconstruct the purposes, functions, and contexts of traditional Chinese mapping practice. A possible objection to this approach is that it entails relativism, the notion that the validity or rightness of ideas is relative to a tradition and therefore that universal standards cannot be formulated. Historical reconstruction, however, need not lead to relativism; it may yield a kind of pluralism—the possibility that excellence in cartography can be measured by social, aesthetic, and even religious criteria as well as scientific ones, and that Europe did not hold a monopoly on all these forms of excellence throughout the history of mapmaking. This pluralism implies that modern cartography has something to learn about diversity of excellence from many traditions, not that of Europe alone.

We are just beginning to understand how to read traditional Chinese maps. According to the corrective view advanced here, Chinese conceptions of mapping differed enough from European ones to make assimilation difficult up till the end of the nineteenth century. It is not that Chinese mapping was nonmathematical, it was more than mathematical. Traditional Chinese conceptions of knowledge differ from those that have been applied to traditional Chinese maps in previous accounts. Traditional maps were products of scholarly enterprises, and under Chinese conceptions, they possessed intellectual value. Under those conceptions, a "good" cartographic image did not necessarily tell how far it was from one point to another. But it might, as is detailed in the following chapters, tell us about such things as power, duty, and emotion.[55]

FIG. 3.32. RUBBING OF A STONE MAP OF HUA SHAN (*TAIHUA QUAN TU* [COMPLETE MAP OF TAIHUA]), 1700. The map was made to commemorate a Qing official's climb up the mountain. The palm print on the east (left) peak of the mountain is larger than it appears on the Ming map (fig. 3.31).
Size of the original: 135 × 69 cm. By permission of the Field Museum of Natural History, Chicago (116470).

55. Modern high-precision maps often involve such things. A map labeling Tibet "Xizang," for example, says something different from one that labels it "Tibet." But insofar as such maps express political ideology, they are often regarded as departing from the scientific enterprise of cartography.

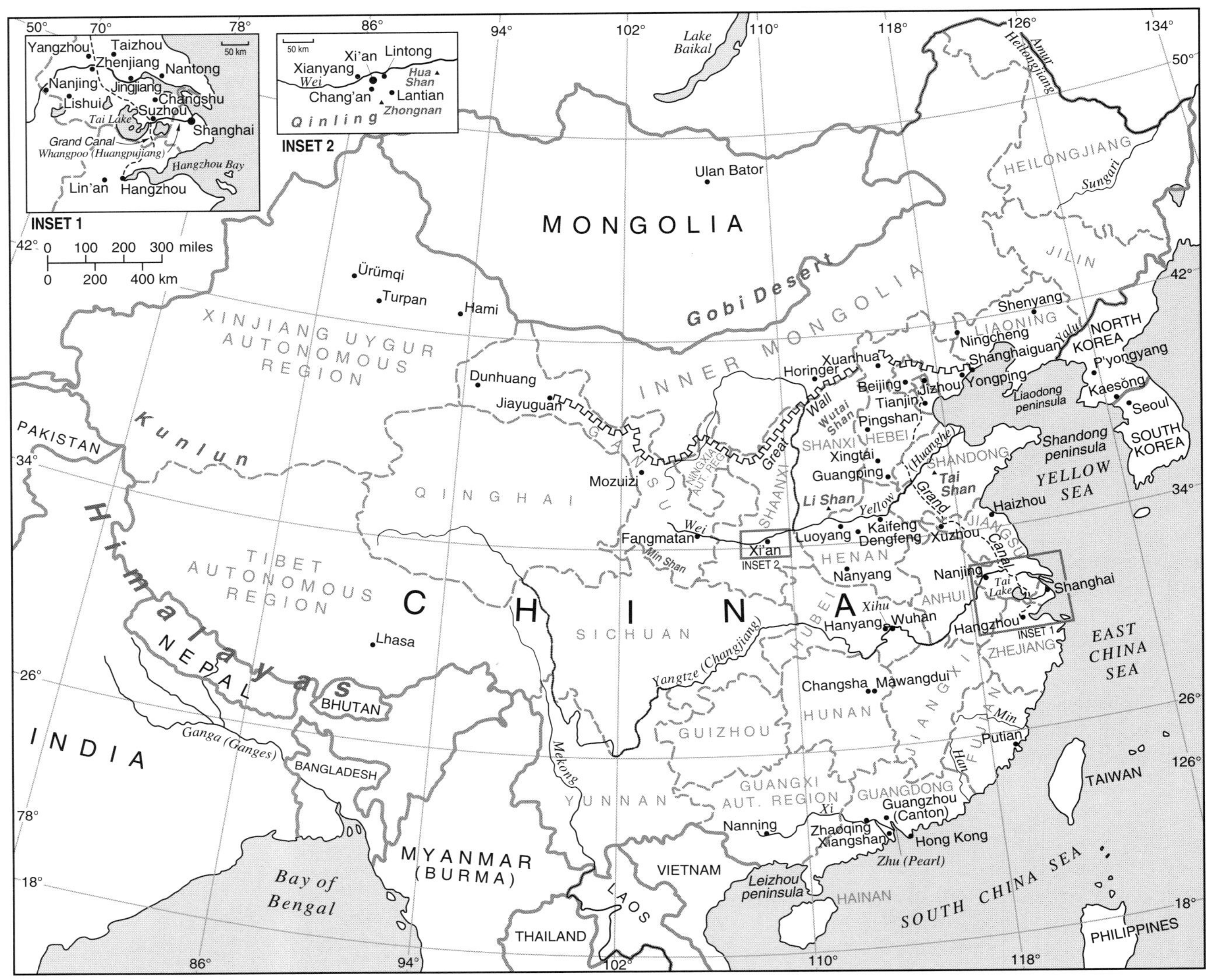

FIG. 3.33. REFERENCE MAP FOR THE STUDY OF CHINESE CARTOGRAPHY. This map shows the location of places in China mentioned in chapters 3–9.

Appendix 3.1 Chronological List of Selected Maps, Fourth Century B.C. through the Yuan Dynasty

Name of Map, and Author (if known)[a]	Date	Dimension (cm)	Medium	Location	Figure Number in Text
Zhaoyu tu (mausoleum map or plan)	Produced between 323 and 315 B.C.	48 × 94	Bronze plate inlaid with gold and silver	Hebei Provincial Institute of Cultural Relics	Fig. 3.1
Seven maps found at Fangmatan[b]	Produced ca. 239 B.C.	26.7 × 18.1; 15 × 26.6; 18.1 × 26.5; 16.9 × 26.8	Ink on wood (four boards)	Gansu Provincial Institute of Archaeology and Cultural Relics	Figs. 3.2 to 3.4
Map fragment found at Fangmatan	Drawn ca. 179–141 B.C.	2.6 × 5.6	Ink on paper	Gansu Provincial Institute of Archaeology and Cultural Relics	Fig. 3.5
Topographic map found at Mawangdui	Drawn before 168 B.C.	96 × 96	Ink on silk	Hunan Provincial Museum	Fig. 3.8
Garrison map found at Mawangdui	Drawn ca. 181 B.C.	98 × 78	Ink on silk	Hunan Provincial Museum	Fig. 3.10
City or mausoleum map found at Mawangdui	Drawn before 168 B.C.	48 × 48	Ink on silk	Hunan Provincial Museum	Fig. 3.6
Map of Ningcheng	Former Han	120 × 318	Tomb mural	Horinger County, Inner Mongolia	Fig. 6.13
Map of Fanyang	Former Han	94 × 80	Tomb mural	Horinger County, Inner Mongolia	Fig. 6.12
Manor map	Former Han	191 × 300	Tomb mural	Horinger County, Inner Mongolia	Fig. 6.16
Yu gong diyu tu, Pei Xiu	Third century	Unknown	Not extant		
Fangzhang tu, Pei Xiu	Third century	ca. 3 × 3 m	Not extant		
Map of Wutai Shan	Tenth century	4.6 × 13 m	Cave mural	Dunhuang, Gansu Province	Fig. 6.18
Shouling tu, Shen Kuo	Eleventh century	Unknown	Not extant		
Jiu yu shouling tu	1121	130 × 100	Engraved on stone tablet	Sichuan Provincial Museum	Fig. 3.12
Hua yi tu	1136	79 × 79	Engraved on stone	Shaanxi Provincial Museum	Fig. 3.13
Yu ji tu	1136	80 × 79	Engraved on stone	Shaanxi Provincial Museum	Fig. 3.14

[a]These are the pre-Ming maps most often referred to. Illustrations of all the extant maps in this appendix, as well as those of numerous other maps, appear in Cao Wanru et al., eds., *Zhongguo gudai ditu ji* (Atlas of ancient Chinese maps) (Beijing: Wenwu Chubanshe, 1990–), vol. 1, *Zhanguo-Yuan* (Warring States to the Yuan dynasty).

[b]For the location of place-names in this chapter and chapters 4–9, see fig. 3.33.

APPENDIX 3.1 (*continued*)

Name of Map, and Author (if known)[a]	Date	Dimension (cm)	Medium	Location	Figure Number in Text
Yu ji tu	1142	83 × 79	Engraved on stone	Zhenjiang Provincial Museum	Fig. 3.15
Dili tu	1247	101 × 179	Engraved on stone	Suzhou Stone Tablets Museum	Fig. 4.11 (copy)
Pingjiang tu	1229	279 × 138	Engraved on stone	Suzhou Stone Tablets Museum	Fig. 6.6
Map of Jingjiang Fu Cheng	Probably 1272	340 × 300	Engraved on cliff	Yingwu Shan, north of Guilin, Guangxi Zhuang Autonomous Region	Fig. 3.30
Yutu, Zhu Siben	1320	ca. 2.3 × 2.3 m	No longer extant		

4 · Chinese Maps in Political Culture

CORDELL D. K. YEE

Much of the history of Chinese cartography is bound up with that of Chinese political culture. Political culture, as used here, refers to the institutions and practices of the ruler and the class of scholar-officials that arose to assist rulers in their duties. This social stratum was the literary elite, those who "worked with their minds."[1] It was, as Balazs has put it, "numerically infinitesimal."[2] At the end of the twelfth century, for example, the imperial bureaucracy consisted of about 42,000 officials, drawn from a scholarly pool of about 200,000, itself comprising less than one-fifth of 1 percent of an estimated total population of 123 million. During the Qing dynasty (1644–1911), the population grew from 200 million to more than 300 million, and the scholarly pool increased to about 2 million. Prospects for a would-be official, however, seem to have worsened. The number of civil appointments within the bureaucracy totaled only about 20,000, with several thousand other posts obtainable by purchase. In other words, there was roughly one official per 10,000 population.[3] Despite their small numbers, the scholar-officials were "omnipotent by reason of their strength, influence, position, and prestige, held all the power and owned the largest amount of land. . . . Their social role was at one and the same time that of architect, engineer, teacher, administrator, and ruler."[4]

The connection of maps with this sphere of Chinese society goes back to textual sources dating from the Eastern Zhou (ca. 770–256 B.C.), one of the periods of disunity that punctuate Chinese history. It is traditionally divided into two parts, the Chunqiu (Spring and Autumn, 722–468 B.C.) and Zhanguo (Warring States, 403–221 B.C.), named after the titles of two ancient histories of the period.[5] The fifth through third centuries B.C. in general were marked by intellectual ferment. Political advisers representing "a hundred schools of thought" roamed from state to state offering rulers competing advice on statecraft, the central problem of which was, as classically defined by Confucius (551–479 B.C.), how to restore the harmony last attained during the Western Zhou dynasty (ca. 1027–771 B.C.). The intellectual climate is described thus by Ban Gu (A.D. 32–92) in the *Han shu* (History of the Former Han): "[The various masters] all arose when the way of the kings had weakened, the feudal lords ruled by strength, and the lords and rulers of the age differed in their likes and dislikes. Thus the theories of the nine schools arose like wasps [that is, were as numerous as wasps because of the various predilections of rulers]."[6]

The textual record, supported by artifacts, shows that cartography was implicated in the Eastern Zhou's general intellectual orientation toward statecraft—a connection that continues in later periods. According to one Chinese classic, the *Zuozhuan* (Zuo's tradition [of interpreting the *Chunqiu*]), "The great affairs of state lie in ritual and warfare."[7] Not surprisingly, most of the earliest possible

1. The idea of government by an intellectual elite goes back as far as the philosopher Mencius (372–289 B.C.), who said: "Some work with their minds. Some work with their physical strength. Those who work with their minds rule others. Those who work with their physical strength are ruled by others" (*Mengzi*, 3A.4; see *Mengzi yinde* [Concordance to Mencius], Harvard-Yenching Sinological Index Series, suppl. 17 [1941; reprinted Taipei: Chengwen Chubanshe, 1966], 20).

2. Etienne Balazs, *Chinese Civilization and Bureaucracy: Variations on a Theme*, trans. H. M. Wright, ed. Arthur F. Wright (New Haven: Yale University Press, 1964), 16.

3. A convenient compilation of population data for China from the Han through the mid-twentieth century is John D. Durand, "The Population Statistics of China, A.D. 2–1953," *Population Studies* 13 (1960): 209–56. A more detailed compilation of population statistics from the Han through the Qing is Liang Fangzhong, *Zhongguo lidai hukou, tiandi, tianfu tongji* (Population, field acreage, and land tax statistics for China through the dynasties) (Shanghai: Renmin Chubanshe, 1980). For population statistics from the Ming dynasty onward, one may also consult Ping-ti Ho, *Studies on the Population of China, 1368–1953* (Cambridge: Harvard University Press, 1959). Information on the composition of the literate elite can be found in Ping-ti Ho, *The Ladder of Success in Imperial China: Aspects of Social Mobility, 1368–1911* (New York: Columbia University Press, 1962); and John W. Chaffee, *The Thorny Gates of Learning in Sung China: A Social History of Examinations* (Cambridge: Cambridge University Press, 1985).

4. Balazs, *Chinese Civilization and Bureaucracy*, 16 (note 2).

5. The *Chunqiu* chronicles political events from 722 to 480 B.C., and the *Zhanguo ce* (Intrigues of the Warring States) is a collection of historical narratives purporting to describe the period 403–221 B.C.

6. Ban Gu, *Han shu* (compiled first century A.D.), chap. 30; see the edition in 12 vols. (Beijing: Zhonghua Shuju, 1962), 6:1746. There is a partial translation of the *Han shu* into English: Homer H. Dubs, trans., *The History of the Former Han Dynasty*, 3 vols. (Baltimore: Waverly Press, 1938–55).

7. *Zuozhuan* (ca. 300 B.C.), *Cheng* 13, in *Chunqiu jingzhuan yinde*

references to cartography occur in the context of those two activities.

Maps, Ritual, and Warfare

One such reference perhaps occurs in the *Shu jing* (Book of documents). In the chapter entitled "Luo gao" (Announcement concerning the capital Luo), which may predate Confucius, the duke of Zhou describes how the site of the eastern capital Luoyi (near today's Luoyang) was chosen:

> I prognosticated about the region of the Li River north of the He; I then prognosticated about the region east of the Jian River, and west of the Chan River; but it was the region of Luo that was ordered [by the oracle]. Again I prognosticated about the region east of the Chan River; but again it was the region of Luo that was ordered. I have sent a messenger to come [to the king] and to bring a *tu* [chart or map] and to present the oracles.[8]

The text provides no further information about the map or chart that was sent to the king; it is unclear whether the chart was a talisman representing unseen forces like those that survive in the Daozang (Daoist canon) or a representation of physical features. But this much can be ventured: it was associated with hieratic activity as a function of the state.[9]

A Han commentator on the *Shi jing* (Book of odes), Zheng Xuan (127–200), regards one poem in the "Zhou song" (Hymns of Zhou) as describing the Zhou kings' use of maps for divination. Translated according to Zheng's reading, the poem, which may date from the Western Zhou, would read:

> Oh, august are those Zhou!
> They ascended the high mountains,
> They followed the mountain's long narrow ridges,
> Truly they followed the chart [of the mountains
> and rivers] and combined the rivers [in their
> sacrifices];
> All under the vast heaven,
> To [the wishes of] all those [lands] they responded;
> That was the [heavenly] appointment of the Zhou.[10]

Zheng Xuan comments that the chart or map he believes the poem refers to is being used to "order" a sacrifice conducted on a mountain.[11] It is doubtful, however, that the poem actually does refer to a chart or map—it does so only if one accepts Zheng's reading of a character that has several meanings other than "chart" or "map."[12] Zheng may be reading the poem in light of sacrificial practices of his own time, not necessarily explaining Western Zhou sacrificial practices, but perhaps providing an indication of the use of maps in sacrifices during the Han (206 B.C.–A.D. 220).

As the preceding examples suggest, the evidence for the ritual uses of maps before the Han dynasty may be

(Concordance to the *Chunqiu* and its commentaries), 4 vols. (1937; reprinted Taipei: Chengwen Chubanshe, 1966), 1:234. Some speculate that the *Zuozhuan* was originally a history unrelated to the *Chunqiu* and that its passages were subsequently reorganized under the appropriate sections in the *Chunqiu*.

8. The Chinese text used here is that included in "The *Book of Documents*," ed. and trans. Bernhard Karlgren, *Bulletin of the Museum of Far Eastern Antiquities* 22 (1950): 1–81. Karlgren's translation (p. 51) has been modified slightly.

9. Here as elsewhere, one encounters the problem of the ambiguous application of the Chinese graph *tu*. In texts from the Zhanguo period, the graphs *di* (地 earth, land) and *tu* (圖 drawing, illustration) appear as a compound for the first time—*ditu*, denoting "geographic maps." The graph *tu* as a word by itself can also denote "map." Its interpretation, however, presents some problems, since its semantic range is not restricted to "map," and often context does not allow one to determine whether a text is referring to maps or illustrations. As a verb, *tu* can mean "to anticipate" or "to hope"; it can also mean "to scheme" or "to plan"—often in the negative sense of "plot against." The Han etymological dictionary *Shuowen jiezi* (Explanation of writing and explication of graphs) assigns *tu* the meaning "onerous to plan." See Xu Shen, comp., *Shuowen jiezi* (compiled ca. 100), s.v. *tu*, in *Shuowen jiezi gulin* (Collected glosses to the *Shuowen jiezi*), 12 vols., ed. Ding Fubao (Taipei: Shangwu Yinshuguan, 1959), 5:2722b. As a noun, the graph *tu* can be used in a similar sense to mean a "scheme," "plot," or "plan." This meaning may have been extended to include visual representations of plans, and thus *tu* may have acquired its sense of "drawing" or "chart." It had clearly acquired this sense by the early Han. The *Erya* (Progress toward correctness), a dictionary believed to have been compiled during the Qin or early Han, uses the graph *tu* in this sense. *Erya*, 2.91; see *Erya yinde* (Index to the *Erya*) (1941; reprinted Taipei: Chengwen Chubanshe, 1966), 7. The *Erya* also lists "plot" or "plan" as a synonym for *tu* (1a.12 [p. 1]).

10. *Shi jing*, ode 296. The Chinese text is that in *The Book of Odes*, ed. and trans. Bernhard Karlgren (Stockholm: Museum of Far Eastern Antiquities, 1950; reprinted 1974), 253. Karlgren's translation of this ode has been revised here, in large part to reflect Zheng Xuan's reading.

11. *Mao shi Zheng jian* (Zheng [Xuan]'s commentary on Mao [Heng]'s version of the *Shi* [*jing*], second century), *Sibu beiyao* edition, 19.17b.

12. Bernhard Karlgren explains why Zheng Xuan's reading should be rejected in "Glosses on the *Ta Ya* and Sung Odes," *Bulletin of the Museum of Far Eastern Antiquities* 18 (1946): 1–198; reprinted in Bernhard Karlgren, *Glosses on the Book of Odes* (Stockholm: Museum of Far Eastern Antiquities, 1964), esp. 172. There also does not seem to be any archaeological evidence to support Mao Heng's reading. Archaeologists have turned up examples of bones and tortoise shells—with verbal inscriptions—used in divination, but so far no examples of maps. The engravers of the inscriptions carved "straight or sinuous boundary lines" on the bones and shells, thereby "forming a series of compartments or 'spheres of influence' " to show where one inscription ends and another begins. See David N. Keightley, *Sources of Shang History: The Oracle-Bone Inscriptions of Bronze Age China* (Berkeley and Los Angeles: University of California Press, 1978), 53–54. Shang divination practices are also discussed in Kwang-chih Chang, *Shang Civilization* (New Haven: Yale University Press, 1980), 31–42, 202–3. The inscriptions on the oracle bones are not without geographic significance. Keightley has used them to map the Shang state. See David N. Keightley, "The Late Shang State: When, Where, and What?" in *The Origins of Chinese Civilization*, ed. David N. Keightley (Berkeley and Los Angeles: University of California Press, 1983), 523–64, esp. 532–39.

equivocal, but there is little doubt that maps representing geographic knowledge were being produced and used regularly in military affairs. As might be expected during a period characterized by warfare, a number of Zhanguo texts discuss the military value of maps and geographic knowledge. The *Sunzi* (Master Sun['s art of war]), a military treatise believed to date from about the fourth century B.C., contains a chapter titled "Topography" (*dixing*), which maintains that topographic knowledge is often the key to victory or defeat:

> Topography is an aid in warfare. To evaluate the enemy, calculate the chances of victory, and analyze the difficulties of the terrain and the distances involved are the way of the superior general. Those who know this [the difficulties of the terrain and the distances] and engage in war will surely prevail; those who do not know this and engage in war will surely be defeated.[13]

The *Sunzi* makes no mention of maps, and their value can only be inferred from the text: knowledge of distances could have been represented in maps. The military application of maps, however, is more explicit in the *Guanzi* ([Book of] Master Guan), sections of which date from the third century B.C.: "In military affairs, it is the duty of military commanders to examine maps, consult with the court astronomer, estimate accumulated stores, organize the brave warriors, acquire a broad knowledge of the realm, and determine strategy."[14] The value of maps is also discussed in a chapter entitled "Maps" (*ditu*): maps are essential for planning troop movements, avoiding potential obstacles, and gaining advantage from the terrain.

> All military commanders must first examine and come to know maps. They must know thoroughly the location of winding mountain passes; streams that may inundate their chariots; famous mountains; passable valleys; arterial rivers; highlands and hills; the places where grasses, trees, and rushes grow; the distances of roads; the size of city and suburban walls; famous cities and deserted ones; and barren and cultivated lands. They should completely store up [in their minds] the ways in and out of and the contrasts in the terrain; afterward they can move their troops and raid towns. In the disposition [of troops] they will know what lies ahead and behind, and will not fail to take advantage of the terrain. This is the constant value of maps.[15]

The role of maps in military affairs evidently was not limited to securing victory. They were also used as a token of defeat or surrender. The *Han Feizi*, a philosophical text dating from the third century B.C., describes this manner of map use in the course of refuting the argument that, as a matter of survival, a small state must serve a larger one:

> To serve a great power always requires substantial concessions, wherefore one must offer up one's map and submit, and put the state seal in pawn for military aid. If the map is offered up, the territory will be cut up; if the state seal is handed over, prestige will be diminished. When the territory is cut up, the state will be cut up; when prestige is diminished, the government will fall into chaos.[16]

The suggestion here is that maps are vital to state security: to offer a map to another state is to render one's own state vulnerable to attack and dismemberment; giving up a map of one's country is tantamount to giving up one's country.

This principle is illustrated by Jing Ke's assassination attempt on the king of Qin. One account of this attempted assassination occurs in the *Zhanguo ce* (Intrigues of the Warring States). The prince of Yan, a small state threatened by the large state of Qin, commissions Jing Ke to kill the king of Qin. In order to win audience with the king, Jing Ke arrives in Qin with the head of one of Yan's generals and a map of Dukang, a fertile region of Yan. The king of Qin interprets these offerings as tokens of respect and fear and is pleased to greet Jing Ke inside his palace. Jing Ke presents the map to the king, who unrolls it, exposing a poisoned dagger. Jing then seizes the king's sleeve and, gripping the dagger, thrusts at him. The king evades the thrust, and a chase ensues. Jing Ke hurls the dagger but misses his target, and he then is beheaded by the king's attendants. Qin's response to the assassination attempt is to attack Yan, and eventually Qin succeeds in forming an empire, comprising itself and its six former rival states.[17]

Political Culture and Documentary Scholarship

One of the major accomplishments of the Qin dynasty

13. *Sunzi*, *Sibu beiyao* edition, 10.10b–11a.

14. *Guanzi*, *Sibu beiyao* edition, 2.6b. The translation is essentially that in W. Allyn Rickett, trans., *Guanzi: Political, Economic, and Philosophical Essays from Early China* (Princeton: Princeton University Press, 1985–), 1:389 n.

15. *Guanzi*, 10.7a–b (in a modification of the translation by Rickett, *Guanzi*, 1:389–90 [note 14]).

16. *Han Feizi*, 49.14, in *Han Feizi suoyin* (Concordance to *Han Feizi*), ed. Zhou Zhongling et al. (Beijing: Zhonghua Shuju, 1982), 858. The translation modifies that of W. K. Liao, trans., *The Complete Works of Han Fei Tzŭ*, 2 vols. (London: Arthur Probsthain, 1939–59), 2:292.

17. *Zhanguo ce*, *Sibu beiyao* edition, 31.3b–8a. A translation of this account can be found in J. I. Crump, Jr., trans., *Chan-kuo Ts'e* (Oxford: Clarendon Press, 1970), 553–61. Another account appears in Sima Qian, *Shi ji* (Records of the grand historian, completed ca. 91 B.C.), chap. 86; see the edition in 10 vols. (Beijing: Zhonghua Shuju, 1959), 8:2526–38.

(221–207 B.C.) was the establishment of a centralized bureaucracy, setting a pattern for subsequent dynasties. One of the distinguishing features of the bureaucratic state was its emphasis on documentation, which played a key role in maintaining communication and control over a large territory. Maps were part of this documentary system of administration, whose philosophical foundations were laid during the Zhanguo period.

One of the earliest possible references to administrative maps occurs in the *Lun yu* (Analects [of Confucius]), a text believed to have been compiled during the fifth or fourth century B.C.: "If he [Confucius] saw a person dressed as a mourner, he would salute in respect; if he saw a person bearing official documents [*ban*], he would salute in respect."[18] The text of the *Lun yu* gives no details about what these documents were, but Zheng Xuan, one of the earliest commentators on this text, says they consisted of "charts and documents" (*tuji*).[19] Whether these included maps is unknown. What is clear from Confucius's salute is his reverence for documents—an attitude shared by other political thinkers and reflected in the documentary scholarship that was to develop within the bureaucracy.

The *Zhanguo ce*, dating from perhaps the third century B.C., attests to the political use of maps of large areas like those mentioned in the *Zhou li*. It tells how the scholar-politician Su Qin persuaded the king of Zhao to join forces with other states against the Qin by saying: "I have examined a map [*tu*] of the empire, and according to it, the territory of the princes is five times larger than that of the Qin. . . . If the six states were to join forces, head west and attack Qin, Qin would be smashed."[20] During the same period, the political philosopher Han Feizi (d. 233 B.C.) suggests that maps are essential to administration: "The laws are codified in maps [or charts] and books, kept in government offices, and promulgated among the people."[21]

The *Xunzi*, attributed to the Confucian philosopher Xun Qing (ca. 300–230 B.C.), indicates that the value placed on maps and other documents exceeded their administrative utility:

> [High officials] preserve the laws and regulations, the weights and measures, the maps [or charts] and books. They do not know their significance, but take care to preserve them, not daring to decrease or increase them. They hand them down from father to son for the use of kings and dukes. Thus, although the Three Dynasties have fallen, their administrative practices and laws still survive.[22]

Maps and other archival materials are revered, according to this account, because their preservation helps ensure the continuity of institutions. Another late Zhanguo work, the *Guo yu* (Discourses of the states), contains a brief statement that adds another dimension to the cultural value of maps: "If you open the teachings of former kings and examine their documents, maps [*tu*], punishments and laws, then when observing their successes and failures, you will understand everything."[23] The maps or illustrations are useful not only for spatial understanding but for moral understanding: they function in part as guides to ethical behavior. The text does not say how maps fulfill this function, but records from later periods attest to their educational usefulness (see below, pp. 86–87).

The interest in preserving cultural inheritance is also reflected in the practices that developed later in imperial China. Important documents, including maps, were often carved in stone, and documents of each dynasty were incorporated into official histories, which included treatises on such topics as geography and government organization. The government's interest in preserving cultural inheritance, as well as controlling public opinion, gave it a stake in collating and editing literary and philosophical works. If government service was a proper concern of the scholar, as Confucian philosophy stipulated, then literary scholarship was also a proper concern of the government.[24] This association of scholarship and politics perhaps helps to account for the continuity of government practice after the Qin and Han.

Maps in Han Political Culture

The Qin recognized the importance of documents, in accordance with the ideas expressed in classical texts. The Qin government, for example, collected maps (*tu*) and documents from the six states it conquered. The *Shi ji* (Records of the grand historian) refers to this collection in its biography of Xiao He (d. 193 B.C.), an intimate and adviser of Liu Bang (256–195 B.C.) or, as he came to be

18. *Lun yu*, 10:18, in *Lun yu zhengyi* (Orthodox interpretation of the *Lun yu*), ed. Liu Baonan (1791–1855), *Sibu beiyao* edition, 13.12b.

19. *Lun yu zhengyi*, 13.12b (note 18).

20. *Zhanguo ce*, 19.2b (note 17). For another translation of this passage, see Crump, *Chan-kuo Ts'e*, 290 (note 17).

21. *Han Feizi*, 38.19, in *Han Feizi suoyin*, 835. The translation modifies that of Liao, *Han Fei Tzŭ*, 2:188 (both in note 16).

22. Xun Qing, *Xunzi*, chap. 4, in *Xunzi yinde* (A concordance to *Xunzi*) (1950; reprinted Taipei: Chengwen Chubanshe, 1966), 10. The translation here modifies that of Dubs, see *The Works of Hsüntze*, trans. Homer H. Dubs (London: Arthur Probsthain, 1928), 57.

23. *Guo yu*, *Sibu beiyao* edition, 3.7a.

24. A concise historical survey of the relation between scholarship and government in China can be found in R. Kent Guy, *The Emperor's Four Treasuries: Scholars and the State in the Late Ch'ien-lung Era* (Cambridge: Council on East Asian Studies, Harvard University, 1987), 10–37. One particularly detailed study of the scholar-state relationship is David McMullen, *State and Scholars in T'ang China* (Cambridge: Cambridge University Press, 1988).

known, Gaozu, first emperor of the Qin's successor, the Former Han (206 B.C.–A.D. 8). When Liu Bang took Xianyang, the capital of the Qin dynasty, Xiao He sought out the ordinances, maps, and documents in the Qin chancellery before Liu's generals burned down the city. The maps and documents proved especially useful to Liu Bang. With them he was fully informed about "the empire's strategic passes, its household population, the vulnerable points along its frontiers, and the hardships suffered by the people."[25] The Qin maps and documents apparently survived into the first century, for the *Han shu*, composed during that time, refers to them in its *dilizhi* (geographical treatise).[26]

Under the Han, the Qin-style bureaucratic government continued. Details about the actual uses of maps within the government are sparse, but an idealized description of a bureaucratic government, the *Zhou li* (Ritual forms of Zhou), mentions a wide range of administrative uses for maps. The *Zhou li* is first mentioned in Han texts and purports to describe the Western Zhou's government institutions: a structure of six major organs, each headed by a principal minister with sixty subordinates. Since there is no evidence beyond the *Zhou li* itself that the Western Zhou ever developed such an elaborate government apparatus, the *Zhou li* may represent a Han attempt to justify its practices by claiming ancient precedent. It may thus provide an indication of the range of map use conceived during the Han period.

The maps (*tu*) mentioned in the *Zhou li* are used by officials in a variety of contexts. According to the *Zhou li*, maps have some economic uses. A kind of resource map is associated with the *gongren* (mining superintendent), who "presides over areas with gold, jade, tin, and precious stones, and issues strict prohibitions to preserve them." When there is need to extract them, the *Zhou li* adds, the *gongren* "inspects maps of their locations and gives them [to miners]."[27] Terrain maps are used by the *sixian* (director of defense works) "in order to be fully informed about the obstacles presented by mountains, forests, streams, and marshes [in the nine regions] and to open roads through them."[28]

Maps referred to in the *Zhou li* are also used within the bureaucracy for auditing and accounting purposes. The *sikuai* (accountant) uses the information in maps and other documents to review the administration and manage the accounts of government officers.[29] Another official, the *sishu* (manager of writings), is said to use land maps (*tudi zhi tu*) "in order to be fully informed about incoming and outgoing wealth."[30] Other officials use several types of boundary maps. A kind of cadastral map seems to be used by the *xiao situ* (vice minister of education) to settle land disputes among the people.[31] Boundary maps of narrower scope seem to be the responsibility of the *zhongren* (grave maker), who "distinguishes the tombs in [royal burial grounds] and makes maps of them."[32] The verb "distinguish" suggests that the *zhongren* indicated where one tomb site began and another ended, perhaps by drawing boundaries. Another kind of boundary map appears to be used by the *suiren* (supervisor of exterior districts) "to demarcate fields and wilderness."[33] There are also maps of larger scope, presumably depicting political subdivisions. One class of official, the *xingfang shi* (supervisor of territories), establishes the territory of the feudal states and regulates the boundaries of their fiefdoms.[34] Maps of the empire are used by the *zhifang shi* (overseer of feudatories) to help control imperial territory as well as to distinguish the populations, finances, agricultural products, and livestock of various administrative subdivisions and various tribes.[35] The *da situ* (grand minister of education) seems to have a similar function: he "presides over the maps of the territory of the feudal states and [the registration] of the number of their people in order to help the king secure the feudal states"; by means of maps he kept informed of each region's territorial extent, geographic features, and natural resources.[36] The *da situ*'s interest in geographic and demographic information was shared by another official, the *xiaozai* (junior steward), who is said to use population registers and maps to administer villages.[37]

---

25. Sima Qian, *Shi ji*, chap. 53 (6:2014) (note 17). Another account of Xiao He's deed appears in the *Han shu*, chap. 39 (7:2006) (note 6). The phraseology of the *Han shu*'s account closely parallels that of the *Shi ji*.

26. *Han shu*, chaps. 28A and 28B (6:1586, 1622) (note 6). By the Western Jin (265–317), the Qin maps were lost. Pei Xiu (223–71) reports them as no longer in the imperial archives (see Fang Xuanling et al., *Jin shu* [History of the Jin, compiled 646–48], chap. 35; see the modern edition in 10 vols. [Beijing: Zhonghua Shuju, 1974], 4:1039).

27. *Zhou li*, *Sibu congkan* edition, 4.37a. The *Zhou li* has been translated into French by Edouard Biot: *Le Tcheou-li; ou, Rites des Tcheou*, 3 vols. (1851; reprinted Taipei: Chengwen Chubanshe, 1969).

28. *Zhou li*, 7.26a (note 27).

29. *Zhou li*, 2.19b (note 27). According to the commentator Zheng Xuan, the maps referred to here are "representations of the land" (*tudi xingxiang*). Neither the text of the *Zhou li* nor the accompanying commentary supplies further details about these maps.

30. *Zhou li*, 2.20a (note 27).

31. *Zhou li*, 3.24b–25a (note 27). The *Zhou li* also refers to *dantu* (vermilion charts), on which minor contracts, those between individuals, were written. Zheng Xuan speculates that these contracts bore illustrations and representations (*tuxiang*). See *Zhou li*, 9.27b–28a.

32. *Zhou li*, 5.45b (note 27). A surviving burial district map (*zhaoyu tu*) from the Zhanguo period is discussed and illustrated on pp. 36–37. That map differs from the maps of burial grounds described in the *Zhou li* in that it was used as a plan for construction, not as a diagram of an already existing site.

33. *Zhou li*, 4.23b (note 27).

34. *Zhou li*, 8.30b (note 27).

35. *Zhou li*, 8.24b (note 27).

36. *Zhou li*, 3.10b–11a (note 27).

37. *Zhou li*, 1.21b (note 27).

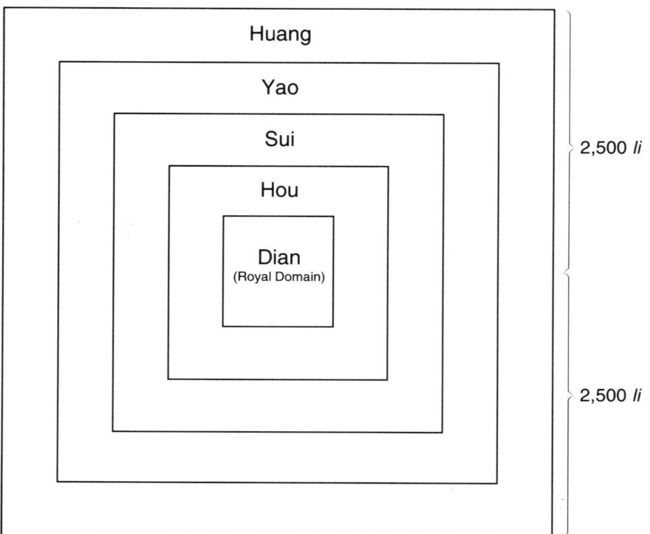

FIG. 4.1. DIAGRAM OF THE *WU FU*, OR FIVE DEPENDENCIES. These are political subdivisions described in the "Yu gong."
After Bernhard Karlgren, "Glosses on the *Book of Documents*," *Bulletin of the Museum of Far Eastern Antiquities* 20 (1948): 39–315, esp. 159.

The preceding examples suggest the importance of maps, within the governmental system described in the *Zhou li*, for political control. Beyond that, there is tacit recognition that a ruler should be well versed in geography. Two officials were responsible for keeping the king informed on geography. Both closely attended the king on royal inspection tours, presumably to point out things of geographical and historical interest. One of the two officials, the *tuxun* (royal scout), explained maps, and the second, the *songxun* (travel guide), was charged with explaining *fangzhi* (local records).[38]

This last idea, the importance of geography for rulers, finds classical precedent in the "Yu gong" (Tribute of Yu), a chapter in the *Shu jing*, dating from at least the late Zhanguo period.[39] In the course of narrating the accomplishments of the legendary emperor Yu, reputed to have founded the Xia dynasty, the "Yu gong" purports to describe geographic features that serve as the basis for dividing territory into *jiu zhou* (nine regions or provinces). The Yan region, for example, supposedly between the Ji and Yellow rivers, is characterized by rich, black soil, luxuriant grass, and tall trees. The "Yu gong" also describes a second scheme of geographic subdivision, *wu fu* (five dependencies), based not on natural features of the land, but on political status:

> In the central kingdom he [the emperor] conferred lands and clan-names. . . . Five hundred *li* [in each direction from the capital] are the *dian fu* [royal domain proper]. . . . Five hundred *li* [in each direction outside the *dian fu*] are the *hou fu* [princes' zone]. . . . Five hundred *li* [in each direction outside the *hou fu*] are the *sui fu* [pacification zone]. . . . Five hundred *li* [outside the *sui fu*] are the *yao fu* [zone of compacts]. . . . Five hundred *li* [in each direction outside the *yao fu*] are the *huang fu* [wild zone].[40]

The translation given above follows an interpretation that yields a scheme of five concentric squares, the largest with sides of five thousand *li* (fig. 4.1).[41] During the Han dynasty, this geometrical regularity proved attractive to interpreters of the "Yu gong," as the *jiu zhou* were reconceptualized as a series of nine concentric squares (see fig. 8.2 and p. 207). The scheme was no longer based on physical or natural features.

During the Han, the interest manifested by the "Yu gong" in geographic inventory and political subdivision continues. The *Han shu*, much of which was compiled under official auspices, contains a geographic treatise, or *dilizhi*. This *dilizhi*, like those of later dynastic histories, is organized by administrative subunits. Most of the treatise is devoted to the geography, history, and demography of various *jun* (commanderies, equivalent to the prefectures of later periods) and *guo* (regions subordinate to the emperor). The treatise contains no maps, but Qin maps, as suggested above, and the "Yu gong" were among its sources.

Maps other than those of the Qin dynasty are mentioned in textual sources about the Han—for example, the *yudi tu* (maps of the empire).[42] None of these comprehensive maps is extant, and textual sources give few details about their content and physical format.[43] We do know, however, that these maps, as in the past, had their

38. *Zhou li*, 4.34b–35a (note 27). The contents of these local records are unspecified; no titles of such works from this period have survived. Titles of *fangzhi* ascribed to the Han dynasty have been preserved, but nearly all surviving works date from after the Tang. *Fangzhi* (local gazetteers) of later periods will be covered in more detail below.

39. Gu Jiegang dates the "Yu gong" from the late years of the reign of Duke Zhuang of Lu (r. 693–662 B.C.)—from an oral communication to Herrlee Glessner Creel, cited in Creel's *Studies in Early Chinese Culture, First Series* (Baltimore: Waverly Press, 1937), 99 n. 2. Elsewhere Gu dates the "Yu gong" from the Zhanguo period. See Gu Jiegang et al., eds., *Gushi bian* (Essays on ancient history), 7 vols. (1926–41; reprinted Hong Kong: Taiping Shuju, 1962), 1:206–10.

40. The Chinese text used here is that included in Karlgren, "The *Book of Documents*" (note 8). Karlgren's translation (p. 18) has been modified slightly.

41. See Bernhard Karlgren, "Glosses on the *Book of Documents*," *Bulletin of the Museum of Far Eastern Antiquities* 20 (1948): 39–315, esp. 159. The "Yu gong" itself contains no illustrations; its idealized symmetries seem to have rendered them otiose.

42. According to Sima Zhen, a Tang commentator on the *Shi ji*, the term *yudi tu* derives from the ancient idea of heaven as a chariot cover and earth as a chariot chassis (*Shi ji*, chap. 60 [6:2110] [note 17]).

43. Han *yudi tu* survived into the Liuchao (Six Dynasties) period. A Liuchao commentator on the *Han shu* used a Han *yudi tu* as a reference. See *Han shu*, chap. 6 (1:189 n) (note 6).

administrative and military uses. In his commentary on the *Zhou li*, Zheng Xuan says that Han *yudi tu* resemble the territorial maps used by the *da situ*.[44] Guangwudi (the Emperor Guangwu) (5 B.C.–A.D. 57) brought a *yudi tu* with him during his campaign to reestablish the Han dynasty. At one point he unrolled it and assessed his prospects to one of his generals: "The empire's commanderies and kingdoms are like this; I have just taken a small part. How could you have said before that my worries about taking the whole empire were unfounded?"[45] Besides the emperor, feudal princes apparently had their own copies of *yudi tu*, with which they could make military plans. Liu An (d. 122 B.C.), the prince of Huainan, is said to have used a *yudi tu* to position his troops.[46] The maps also had ritual uses. Presentation of *yudi tu* to the emperor was apparently part of the ceremonies associated with enfeoffment.[47]

A number of sources besides the Qin maps would have been available for the compilation of the Han *yudi tu*. The *yudi tu* might have used information contained in maps received in tribute from foreign peoples. In A.D. 46 the southern branch of the Xiongnu, nomads who often raided northern China, dispatched a map to the Han government before seeking annexation.[48] The makers of *yudi tu* may also have relied on maps submitted by administrative units and subordinate states. All the extant maps from the Han seem to have been produced by local administrations, and among the information conveyed by these maps are topographical features, such as mountains and streams, and the locations of cities, settlements, and military outposts. Maps, population and land registers, and financial accounts of local administrations were presented annually to the central government. Ban Gu describes a practice of map submission in his "Dongdu fu" (Eastern capital rhapsody): "The Son of Heaven receives maps and registers from the four seas."[49] These documents were reviewed by the counselor-in-chief (*chengxiang*) in preparing the state budget. They were also used by the *yushi zhongcheng* (palace aide to the censor-in-chief) to review local administration.

The examples discussed above might suggest that the Han government was merely a passive recipient of geographic information, but it did seek out such information for maps in an active manner. One of the responsibilities of the *yushi zhongcheng* was to "preside over" maps and documents (*tuji*).[50] Survey teams were also dispatched to gather information. Under the emperor Wu (140–87 B.C.), Han envoys traced the Yellow River (Huanghe) to its source, in a mountainous area rich in precious stones. After the envoys returned, the emperor "consulted old maps and documents (*tushu*) and gave the name 'Kunlun' to the mountains from which the river flows."[51] Some geographic information also came from military expeditions. The general Li Ling (d. 74 B.C.) once left Juyan, in present-day Gansu Province, with five thousand foot soldiers and went north for thirty days, making maps of the terrain he passed.[52] Another military official, Li Xun (first century A.D.), while subjugating "northern barbarians" (*bei di*), mapped the "mountains and streams, and villages and fields" that he passed. The finished product consisted of more than one hundred sections and greatly pleased the emperor.[53]

The association of maps and political culture, well established by the Qin and Han as shown above, had ritualistic implications. The equation of territory with political power seems to have led to attaching political power to maps themselves, particularly in matters pertaining to the afterlife. This would help to explain why all Han maps discovered so far have been found in tombs. Maps seem to have been buried in the tombs of local officials as symbols of their past temporal power and to ease their passage to the otherworld. The practice of map

44. See *Zhou li*, 3.10b (note 27).

45. Fan Ye, *Hou Han shu* (History of the Later Han, compiled fifth century A.D.), chap. 16; see the edition in 12 vols. (Beijing: Zhonghua Shuju, 1965), 3:600.

46. *Han shu*, chap. 44 (7:2149); and for another example of a feudal lord possessing a *yudi tu*, see *Han shu*, chap. 53 (8:2417) (note 6).

47. *Shi ji*, chap. 60 (6:2110) (note 17); *Hou Han shu*, chap. 1B (1:65) (note 45). The *Dongguan Han ji* (Han records from the eastern tower, compiled ca. first to second century) by Liu Zhen mentions that the emperor Ming of the Han (r. 57–75) used a *yudi tu* to assess the size of his son's fiefdom (*Sibu beiyao* edition, 2.4a).

48. *Hou Han shu*, chap. 89 (10:2942) (note 45). Another possible reference to maps received from "barbarian" peoples occurs in the *Han shu*. In 35 B.C. the emperor received a slain "barbarian" (*hu*) leader's "charts and documents" (*tushu*), and these were shown to "worthy ladies," a group of wives ranking just below the empress (*Han shu*, chap. 9 [1:295] [note 6]). According to some commentators, these documents described the configuration of the leader's "lands, mountains, and streams."

49. In Xiao Tong, comp., *Wen xuan* (Literary selections, completed ca. 526–31), ed. Hu Kejia (1809; reprinted Kyōto: Chūbun Shuppansha, 1971), 1.25b. The translation is essentially that in David R. Knechtges, trans. and annotator, *Wen Xuan; or, Selections of Refined Literature* (Princeton: Princeton University Press, 1982–), 1:165.

50. *Han shu*, chap. 19A (3:725) (note 6). The text does not not specify whether this official supervised map production or collection or both, but at least it provides a clear indication that the administrative value of maps was recognized during the Han. The "usurper" Wang Mang (45 B.C.–A.D. 23), who reigned between the Former and Later Han, consulted those who "understood geography and maps and documents" when reforming government institutions and revising geographical subdivisions (*Han shu*, chap. 99B [12:4129]) (note 6). In doing this, Wang Mang maintained that he was following precedents established in classic texts like the "Yu gong" and *Zhou li*. For a translation of this episode, see Dubs, *History of the Former Han*, 3:319–23 (note 6).

51. *Shi ji*, chap. 123 (10:3173) (note 17). Another account of this expedition appears in the *Han shu*, chap. 61 (9:2696) (note 6).

52. *Han shu*, chap. 54 (8:2451) (note 6). The text provides no further details about these maps.

53. *Hou Han shu*, chap. 51 (6:1683) (note 45). As is often the case, the text supplies no further details about these maps.

FIG. 4.2. CLAY MODEL OF FIELDS FOUND IN A HAN TOMB.
Height of the figures: ca. 13–15 cm. From Joseph Needham, *Science and Civilisation in China* (Cambridge: Cambridge University Press, 1954–), vol. 6, *Biology and Biological Technology*, pt. 2, *Agriculture*, by Francesca Bray (1984), fig. 27.

FIG. 4.3. CLAY MODEL OF A POND FOUND IN A HAN TOMB.
Size of the original: 9 × 28 × 28 cm. By permission of Wenwu Chubanshe.

burial may be related to another burial practice—that of placing in tombs pieces of clay wrapped in cloth (*bu tu*) and clay models of fields and ponds (figs. 4.2 and 4.3).[54] These artifacts are believed to have symbolized landholdings and would presumably help the tombs' occupants win a place of respect in the spirit world.

Such appears to be the motivation behind the model commissioned by Qin Shihuang (r. 221–210 B.C., founding emperor of the Qin) for his mausoleum, in present-day Lintong Xian (County), Shaanxi. Sima Qian (ca. 145–ca. 85 B.C.) describes the model and the process of its construction in the *Shi ji*:

> When Shihuang first came to power, he hollowed out and arranged Li Shan (Mount Li). After he unified the empire, conscripts from all over the empire were sent there—more than 700,000. They tunneled through to three subterranean springs, poured in copper, and made the outer coffin. Palaces, a hundred offices, wondrous contrivances, and rare exotica were sent to fill the tomb. Artisans were ordered to make mechanical crossbows and bolts, so that anyone who tunneled near would be shot. They used mercury to make the hundred streams, the [Yangtze] Jiang, the [Yellow] River and the great sea; devices made the mercury circulate from one course to the other. Above, they prepared the patterns of heaven; below, they prepared the configurations of the earth. They used seal oil to make lamps, which, it was reckoned, would burn forever without going out. . . . Trees and grass were planted to make the tumulus resemble a mountain.[55]

The mausoleum has yet to be excavated, so the *Shi ji*'s account of the three-dimensional map cannot be confirmed. Even if the tomb is excavated, it may still be hard to judge the veracity of the *Shi ji*'s account, since the mechanical crossbows did not prove to be an effective deterrent—plunderers during the late Qin or early Han period may have taken much of the tomb's contents.[56] There is, however, evidence that mercury was used in some way in the mausoleum. Chang Yong and Li Tong have analyzed the mercury content of the soil in the area and found a pocket of unusually high mercury concen-

54. On these burial practices, see Wang Zhongshu, *Han Civilization*, trans. Kwang-chih Chang et al. (New Haven: Yale University Press, 1982), 207–8.

55. *Shi ji*, chap. 6 (1:265) (note 17). Another description of the tomb, not as elaborate as that in the *Shi ji*, appears in the *Han shu*, chap. 51 (8:2328) (note 6). In addition to the three-dimensional model, sacrificial victims were also placed in the tomb. The *Shi ji*'s account says: "Ershi [the second emperor] said: 'It is not proper to send away those of my father's ladies who had no sons.' Thus all of them were ordered to follow the first emperor into death; those put to death were a multitude. When the coffin had been set down, some said that since the laborers and artisans made the devices and knew all about the stored treasures, the great value of the treasures would leak out. Thus after the burial and the collection of treasures, the middle gate was shut, and the outer gate closed, to immure all the laborers, artisans, and stockpilers; no one ever came out again."

56. According to one account, in thirty days of plundering about the end of the Qin, 300,000 people were unable to remove all the tomb's treasures. See Li Daoyuan, *Shui jing zhu* (River classic commentary, ca. sixth century), chap. 19, in *Shui jing zhu jiao* (Corrected *Shui jing zhu*), ed. Wang Guowei (Shanghai: Renmin Chubanshe, 1984), 621. The tomb is said to have been plundered once again during the Liuchao period. See Li Xueqin, *Eastern Zhou and Qin Civilizations*, trans. Kwang-chih Chang (New Haven: Yale University Press, 1985), 254.

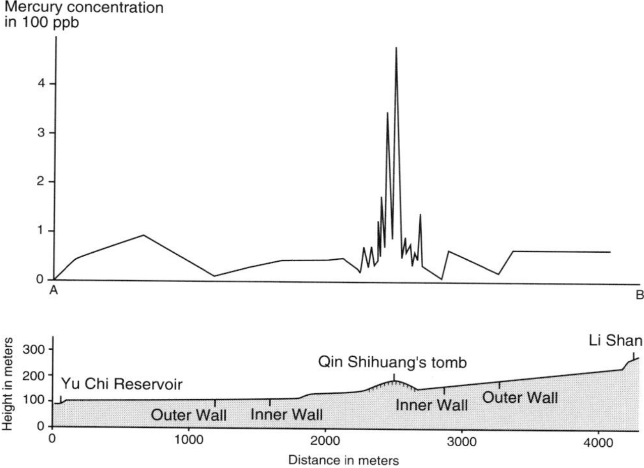

FIG. 4.4. GRAPH OF THE CHANGES IN MERCURY CONCENTRATION IN THE SOIL AT THE SITE OF QIN SHIHUANG'S TOMB. The levels of mercury directly above the tomb are about four times the levels in the surrounding area. After Chang Yong and Li Tong, "Qin Shihuang lingzhong maicang gong di chubu yanjiu" (Preliminary study of the mercury interred in Qin Shihuang's tomb), *Kaogu*, 1983, no. 7:659–63 and 671, esp. 663.

tration just above the emperor's tomb (fig. 4.4). They conclude that this verifies the *Shi ji*'s description of the mercury-filled streams and sea.[57]

Some idea of the dimensions of the model empire can be gathered from the mausoleum's external dimensions. The perimeter of the tomb mound is about four hundred meters square, and at its apex the mound stands about forty-three meters high (fig. 4.5).[58] The three-dimensional map of the emperor's realm and the heavens above seem to have had little utility other than to symbolize Qin Shihuang's temporal power as the Son of Heaven. Such trappings of power might have been useful in establishing the emperor in the spirit world: with his body placed amid the cosmographic model, the emperor, in death as

57. Chang Yong and Li Tong, "Qin Shihuang lingzhong maicang gong di chubu yanjiu" (Preliminary study of the mercury interred in Qin Shihuang's tomb), *Kaogu*, 1983, no. 7:659–63 and 671.

58. See Li, *Eastern Zhou and Qin*, 251–54 (note 56); Maxwell K. Hearn, "The Terracotta Army of the First Emperor of Qin (221–206 B.C.)," in *The Great Bronze Age of China: An Exhibition from the People's Republic of China*, ed. Wen Fong (New York: Metropolitan Museum of Art, 1980), 353–68, esp. 357; and Robert L. Thorp, "An Archaeological Reconstruction of the Lishan Necropolis," in *The Great Bronze Age of China: A Symposium*, ed. George Kuwayama (Los Angeles: Los Angeles County Museum of Art, 1983), 72–83.

FIG. 4.5. VIEW OF THE SITE OF QIN SHIHUANG'S TOMB. Photograph courtesy of the William A. Dando family.

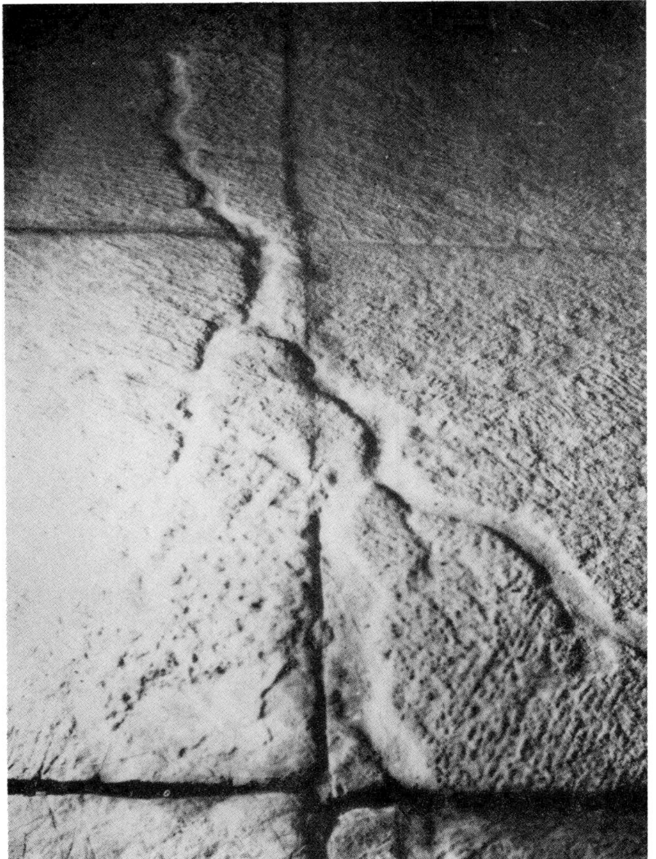

FIG. 4.6. RELIEF MODEL FOUND IN A TOMB FROM THE SOUTHERN TANG DYNASTY. The grooves on the floor of the emperor Li Bian's tomb (*left* is the western part of the floor, *right* is the eastern part) represent rivers. Size of the floor: 6.03 × 5.9 m. From Nanjing Bowuyuan (Nanjing Museum), *Nan Tang erling fajue baogao* (Report on the excavation of two Southern Tang mausoleums), ed. Zeng Zhaoyue (Beijing: Wenwu Chubanshe, 1957), 34 (nos. 3 and 4).

in life, would still serve—albeit symbolically—as the intermediary between heaven and earth.[59]

The Continuity of Qin and Han Practices

The description of Qin Shihuang's tomb seems to have influenced the interior design of the tombs of later emperors: celestial paintings are found in the tombs of the Han, Tang (618–907), Liao (916–1125), and Song (960–1279) dynasties, and a three-dimensional model of the cosmos has been found in a tomb of the Southern Tang (937–60).

The occupant of this last tomb was the Southern Tang's first emperor, Li Bian (r. 937–43). His short reign allowed little time, about six years, for the construction of his tomb. In contrast, Qin Shihuang's tomb took nearly forty years to complete.[60] The relatively short construction period, coupled with the Southern Tang's limited financial resources and, because of its reduced territorial

59. The emperor's trappings of power would have been recognized by the spirit-world bureaucracy, attested to by artifacts dating from the first half of the second century B.C. The Han dynasty's earthly bureaucracy was based on the Qin model, but it is not clear whether the Han's spirit-world bureaucracy had a similar Qin antecedent. Han literature, however, does recount Qin Shihuang's belief in the existence of immortal beings. The literature on Chinese burial practices and conceptions of the afterlife is fairly extensive. On Chinese burial practices, see Robert L. Thorp, "Burial Practices of Bronze Age China," in *The Great Bronze Age of China: An Exhibition from the People's Republic of China*, ed. Wen Fong (New York: Metropolitan Museum of Art, 1980), 51–64; and idem, "The Qin and Han Imperial Tombs and the Development of Mortuary Architecture," in *The Quest for Eternity: Chinese Ceramic Sculptures from the People's Republic of China*, ed. Susan L. Caroselli (Los Angeles: Los Angeles County Museum of Art, 1987), 17–37. On Chinese conceptions of the afterlife, see Albert E. Dien, "Chinese Beliefs in the Afterworld," also in *Quest for Eternity*, 1–15; Michael A. N. Loewe, *Ways to Paradise: The Chinese Quest for Immortality* (London: George Allen and Unwin, 1979); Ying-shih Yü, "Life and Immortality in the Mind of Han China," *Harvard Journal of Asiatic Studies* 25 (1964–65): 80–122; and idem, " 'O Soul, Come Back!' A Study in the Changing Conceptions of the Soul and Afterlife in Pre-Buddhist China," *Harvard Journal of Asiatic Studies* 47 (1987): 363–95.

60. Before the Song dynasty, construction of an emperor's tomb

holdings, its relatively small labor pool, perhaps account for the relative spareness of the model when compared with the sumptuousness ascribed to the model in Qin Shihuang's tomb. The Southern Tang model, as it survives, may also be incomplete, since the tomb had been looted before its excavation in 1950–51.[61]

The tomb is at the southern foot of Gao Shan (Mount Gao), near Nanjing, and occupies a mound about 5 meters high and 30 meters in diameter. It consists of three main chambers—front, middle, and rear—each with side chambers serving as repositories for burial objects. The rectangular rear chamber measures 5.9 meters wide and 6.03 meters long, making it the largest of the three main chambers. Its size corresponds to the importance of its contents: it housed the emperor's coffin and the three-dimensional model.

The coffin was placed on a brick platform extending from the middle of the chamber's rear wall. On the floor of the chamber are carved two meandering and branching grooves, one at each side of the bed, presumably representing the Yellow and Yangtze rivers, supposedly modeled in Qin Shihuang's mausoleum (fig. 4.6). There is no evidence that mercury flowed through the model rivers, as it was said to have done in Qin Shihuang's tomb, or that the model included representations of mountains, as the *Shi ji*'s description of Qin Shihuang's tomb implies, or that there were buildings like those reportedly made for Qin Shihuang's tomb. But similar to the description of Qin Shihuang's tomb, the ceiling of Li Bian's chamber has a painting of the heavens. In addition to constellations, the painting has a red sun on its eastern panel, a full moon on its western panel, and pole stars on the northern and southern panels (fig. 4.7). As in the case of Qin Shihuang's tomb, the model in Li Bian's tomb seems to have little practical utility. Along with clay figures presumably representing the emperor's retinue, it seems to have been intended to ease the emperor's transition to the spirit world—on the platform his body would have symbolically occupied an intermediate position between heaven and earth.

In addition to funerary uses, other patterns of governmental map use established during the Qin and Han periods seem to have continued in later dynasties. In the Qing dynasty, for example, charts showing the arrangement of altars and ritual objects accompany texts describing the sacrifices carried out by the emperor as intermediary between heaven and earth (fig. 4.8). As for the administrative uses of maps, the elevation of the *Zhou li* to canonical status during the Tang dynasty seems to have made it almost the last word. Later writers probably saw no need to describe in detail what was already covered in such an authoritative text. Consistent with this, treatises on government from the Tang onward say little that is original about maps. During the nineteenth century,

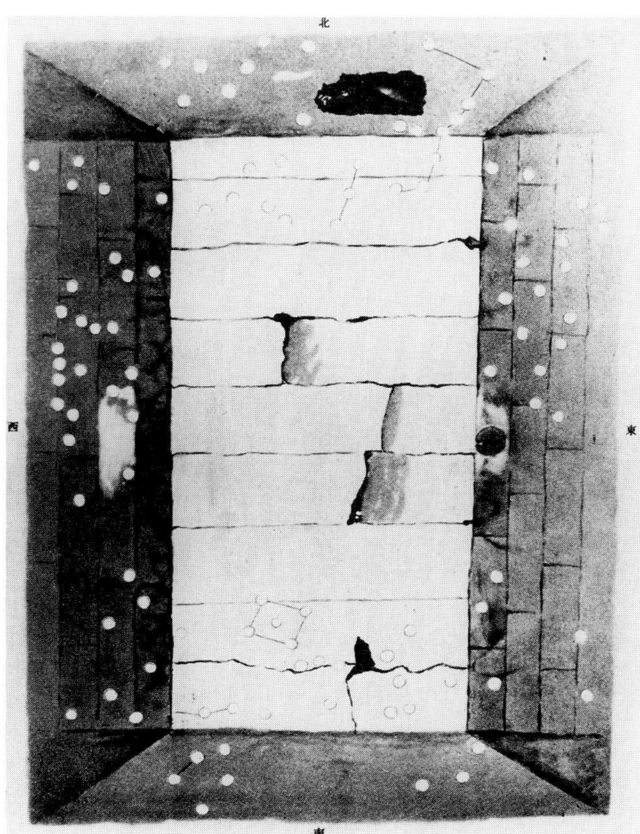

FIG. 4.7. STAR MAP PAINTED ON THE CEILING OF THE TOMB OF LI BIAN, AN EMPEROR OF THE SOUTHERN TANG. North is at the top.
Size of the ceiling: 6.03 × 5.9 m. From Zhongguo Shehui Kexueyuan Kaogu Yanjiusuo (Archaeological Research Institute, Chinese Academy of Social Science [Academia Sinica]), *Zhongguo gudai tianwen wenwu tuji* (Album of ancient Chinese astronomical relics) (Beijing: Wenwu Chubanshe, 1980), 74.

for example, compilers of gazetteers, compendiums that typically contain maps, referred to the *Zhou li* as the definitive statement on the value and use of maps in government. The preface of one eighteenth-century gazetteer begins: "Why do we make large gazetteers? They were

began in the year of his accession. During the Song, this system changed: construction of an emperor's tomb began after his death, and it was stipulated that an emperor had to be buried within seven months of his death. This meant there was little time for the elaborate preparations of the kind attributed to Qin Shihuang's tomb. For more information on Chinese imperial burial practices, see Robert L. Thorp, *Son of Heaven: Imperial Arts of China* (Seattle: Son of Heaven Press, 1988).

61. Further information about the excavation of this tomb may be found in Zeng Zhaoyue et al., eds., *Report on the Excavation of Two Southern T'ang Mausoleums: A Summary in English* (Beijing: Cultural Objects Press, 1957); and Zeng Zhaoyue et al., eds., *Nan Tang er ling fajue baogao* (Report on the excavation of two Southern Tang tombs) (Beijing: Wenwu Chubanshe, 1957).

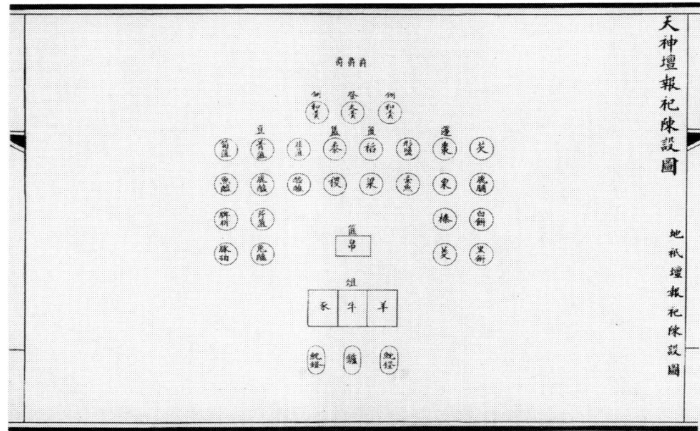

FIG. 4.8. CHART SHOWING THE ARRANGEMENT OF RITUAL OBJECTS AND OFFERINGS USED IN IMPERIAL SACRIFICES THANKING THE HEAVENLY SPIRITS. The placement of sacrificial animals—pig, cow, and sheep—is indicated in the three boxes near the bottom of the chart. Other offerings include silk, indicated in the single box, and various grains, among the circles.
Size of the original: 20 × 29 cm. From *Qinding Da Qing huidian* (Imperially commissioned collected statutes of the Great Qing), 75 vols. (1899), *tu* 13.12b–13a. Reproduced courtesy of the Harvard-Yenching Library, Harvard University, Cambridge.

first created in the *Zhou guan* [Institutes of Zhou, i.e., *Zhou li*]."[62]

There is some warrant for this assertion of continuity. The titles of the officials concerned with maps may have differed depending on the period, and there may have been differences in administrative organization, but the historical record does not suggest any significant alteration of the general description of map use given for the Han. As in the *Zhou li*, maps after the Han served a variety of government purposes. They were used in public works projects, especially in water conservancy. The record for the Qing dynasty is particularly full in this regard: memorials to the emperor describing repairs on levees and canals were often accompanied by maps (see pp. 101–2). Also in accordance with the *Zhou li*, maps continued to help resolve boundary disputes.

As before, maps were also valued for their utility in warfare. The cartographer Pei Xiu (223–71), for example, attributes the military success of the Western Jin (265–317) in part to accurate maps of rival states.[63] In recognition of the military value of maps, *yudi tu* continued to be a responsibility of the ministry of war through the Qing dynasty. During the Song dynasty, for example, the *bingbu shangshu* (minister of war) "used maps of the empire, prefectures and counties to be fully informed about its territory." Serving under the minister of war was the *zhifang langzhong* (director of the bureau of operations), who with his assistant "presided over the empire's maps and documents [*tuji*] in order to be fully informed about the extent of regional territory and the distances of routes to prefectures and outposts."[64] Maps were also used by commanders in the field for planning strategy and as accompaniments of reports to the throne. Commanders would sometimes go to great lengths to obtain maps of enemy territory. In a campaign against Yao tribesmen in 1552, Mao Kun (1512–1601) sent spies into enemy areas and had them draw maps in invisible ink. Using these maps, Mao Kun was able to construct a three-dimensional model of the area and plan attacks on seventeen tribal strongholds. They were all taken in a single day.[65]

As in the Han, maps later played a part in foreign relations. Tributary states often presented maps to the Chinese government as signs of subservience. In 648, for example, the general Wang Xuance defeated the kingdom of Kāmarūpa, which then presented the Chinese emperor with "wondrous things and curious objects and a map."[66] In 1721 the emperor reported to his grand secretaries that a Russian emissary had presented a map of his country to the court. The map was significant to the emperor because it confirmed an ancient account—once thought to be "wanton fiction"—of an impassable sea of ice to the far north, where "rats" the size of elephants (mammoths) once roamed.[67]

62. "Jiu xu" (Old preface), in *Xiangshan xian zhi* (Gazetteer of Xiangshan County [in modern Guangdong Province]) (1750; reprinted Taipei: Taiwan Xuesheng Shuju, 1968), 1a. Gazetteers are described more fully later in this chapter.

63. *Jin shu*, chap. 35 (4:1040) (note 26). Pei Xiu may have come to appreciate the strategic importance of maps while helping to plan Sima Zhao's successful campaign against his rival Zhuge Dan (d. 258). At the time of this campaign, Sima Zhao, enfeoffed the prince of Jin in 254, was the most powerful leader of the Wei dynasty (220–65). See *Jin shu*, chap. 35 (4:1038) (note 26).

64. Tuotuo et al., *Song shi* (History of the Song, 1346), chap. 163; see the edition in 40 vols. (Beijing: Zhonghua Shuju, 1977), 12:3855–56.

65. See Chaoying Fang and Else Glahn, "Mao K'un," in *Dictionary of Ming Biography, 1368–1644*, 2 vols., ed. Luther Carrington Goodrich and Chaoying Fang (New York: Columbia University Press, 1976), 2:1042–47, esp. 1043–44. The use of three-dimensional maps for military planning goes back to at least A.D. 32, when the general Ma Yuan (14 B.C.–A.D. 49) made a three-dimensional map in front of the emperor and used it to explain military strategy: "He collected rice to make hills and valleys, pointed out topographical features, indicated the routes to be followed by massed armies, and discriminated the bends and turns [in the routes] so that the situation was clearly understood" (*Hou Han shu*, chap. 24 [3:834] [note 45]).

66. Ouyang Xiu et al., *Xin Tang shu* (New history of the Tang, compiled 1032?–60), chap. 221A; see the edition in 20 vols. (Beijing: Zhonghua Shuju, 1975), 20:6238. Another account of this military campaign appears in Liu Xu et al., *Jiu Tang shu* (Old history of the Tang, compiled 940–45), chap. 198; see the edition in 16 vols. (Beijing: Zhonghua Shuju, 1975), 16:5308. According to both accounts, Wang Xuance was aided by about one thousand Tibetan soldiers and seven thousand Nepalese cavalry.

67. *Da Qing Shengzu Ren (Kangxi) huangdi shilu* (Veritable records

As the last example suggests, maps received in tribute were useful for revising geographic knowledge. The central government, however, often sought geographic information on foreign lands in a more active manner. During the Tang, for example, the court of state ceremonial (*honglu si*) was responsible for the accommodation and entertainment of foreign guests. The office also served as a source of geographic information for the bureau of operations: "Whenever foreign guests arrived, the court for diplomatic relations inquired about their country's mountains and streams and its customs and terrain. It then made a map and presented it to the emperor, a copy of which was sent to the bureau of operations."[68] To supplement the information provided by foreign guests, Chinese embassy missions were expected to gather geographic information, which would have military value in case hostilities broke out. This information included distances between major places, topographic descriptions, locations of fortifications, and maps.

When the central government was weak, maps were exchanged with adjoining states as part of treaty arrangements. In one case the central government was so weak that it found itself sending maps to an invading force—precisely the act of submission described in pre-Han texts like the *Han Feizi*. The Song emperor in 1126 concluded a treaty with a Jurchen (Ruzhen) prince-commander in which he agreed to pay indemnity and cede territory. In recognition of the territorial concessions, new borders were drawn, as described in the Song government's oath letter: "Now it has already been agreed and determined that north of the counties and garrison towns administered by the whole southern area comprising Zhongshan, Taiyuan, and Hejian prefectures, prefectural troops will draw a boundary. There are other maps showing the situation before the Jin [Jurchens] arrived and established border outposts."[69] Maps of the three prefectures were delivered to the Jurchens along with the oath letter.

Presenting maps to foreigners must have been humiliating to the Song government, which had previously been apprehensive about the mapmaking and map taking of foreign tribute missions to China. One memorial, written in 1089, warned that Korean envoys might draw maps of Chinese mountains and rivers and pass this geographic information to the Khitans (Qidans).[70] In his *Mengxi bitan* (Brush talks from Dream Brook) Shen Kuo (1031–95) recorded the activities of a Korean tribute mission:

> During the Xining reign period [1068–77], Gaoli [Korea] sent tribute. In every province and country through which [the mission] passed, they asked for maps; and in each case these were made and presented to them. Mountains and rivers, thoroughfares and roads, the difficulty of the terrain—there was nothing that was not completely recorded. When they arrived at Yangzhou, they reconnoitered the prefecture and obtained maps. At this time the minister Chen Xiugong [Chen Shengzhi, 1011–79] was prefect of Yangzhou [in present-day Jiangsu Province], and he deceived the emissaries. He asked to see all the maps of the Liang Zhe circuits [in present-day Jiangsu] that they had been supplied with, so that he could copy their patterns and remake them for presentation. When the maps came, he gathered them together and burned them, and wrote a report on the whole affair.[71]

This suspicion against foreigners who made maps of China surfaced again in the Qing, when China was governed by a ruling house with foreign origins. In 1805, by which time the Manchu court had been Sinicized, the emperor's ministers reported that letters written in a European language and a map had been intercepted. After an investigation, an Italian named Adeodato confessed that he had sent the letters and map to the pope:

> I am an Italian and the Superior of the [Xitang, or Western church]. These places on this map show the regions in which Chinese converts worship according to our religion. Because the regulations of our various orders are so different, there are always disputes when new missionaries of different orders arrive in [Beijing]. Therefore we have used signs to distinguish the various orders. . . . I desire to send this map to the Pope to

---

of Shengzu, emperor Ren [Kangxi], of the Great Qing, compiled ca. 1739) (1937; reprinted Taipei: Hualian Chubanshe, 1964), 191.19a–20b. For a translation of this record, see Lo-shu Fu, comp., trans., and annotator, *A Documentary Chronicle of Sino-Western Relations (1644–1820)*, 2 vols. (Tucson: University of Arizona Press, 1966), 1:133.

68. *Xin Tang shu*, chap. 46 (4:1198) (note 66).

69. *Da Jin diaofa lu* (Record of the Great Jin's consolation [of the people] and punishment [of the guilty], compiled ca. twelfth century), *Baibu congshu jicheng* edition, 1.25a. The Jurchens, a tribal people from China's northeast frontier, conquered northern China during the twelfth century and established a Chinese-style dynasty known as the Jin (1115–1234).

70. Herbert Franke, "Sung Embassies: Some General Observations," in *China among Equals: The Middle Kingdom and Its Neighbors, 10th–14th Centuries*, ed. Morris Rossabi (Berkeley and Los Angeles: University of California Press, 1983), 116–48, esp. 139. The Khitans, a tribal people from China's northeast frontier, established a Chinese-style dynasty in northern China in the tenth century. Their state was known as the Liao (916–1125).

71. Shen Kuo, *Mengxi bitan* (ca. 1088), chap. 13, par. 234, in *Xin jiaozheng Mengxi bitan* (Newly edited *Mengxi bitan*), ed. Hu Daojing (1957; reprinted Hong Kong: Zhonghua Shuju, 1975), 144. The same paragraph number can be used to locate this passage in *Mengxi bitan jiaozheng* (*Mengxi bitan* edited), 2 vols., ed. Hu Daojing, rev. ed. (1960; reprinted Taipei: Shijie Shuju, 1961). For another translation of this passage, see Joseph Needham, *Science and Civilisation in China* (Cambridge: Cambridge University Press, 1954–), vol. 3, with Wang Ling, *Mathematics and the Sciences of the Heavens and the Earth* (1959), 549–50. For an extensive account of Shen's life and work, see Nathan Sivin, "Shen Kua," in *Dictionary of Scientific Biography*, 16 vols., ed. Charles Coulston Gillispie (New York: Charles Scribner's Sons, 1970–80), 12:369–93.

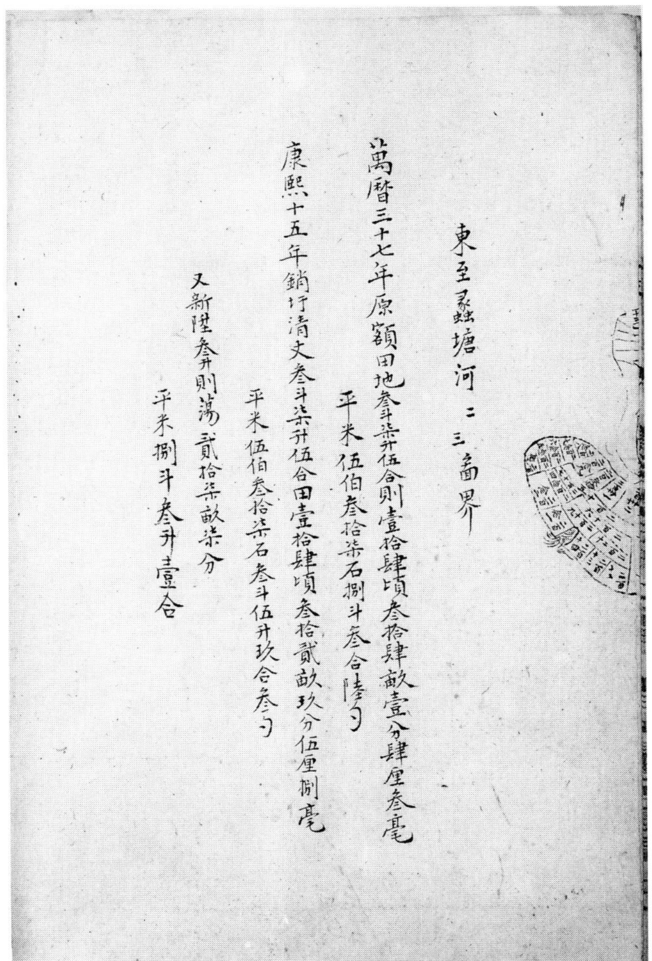

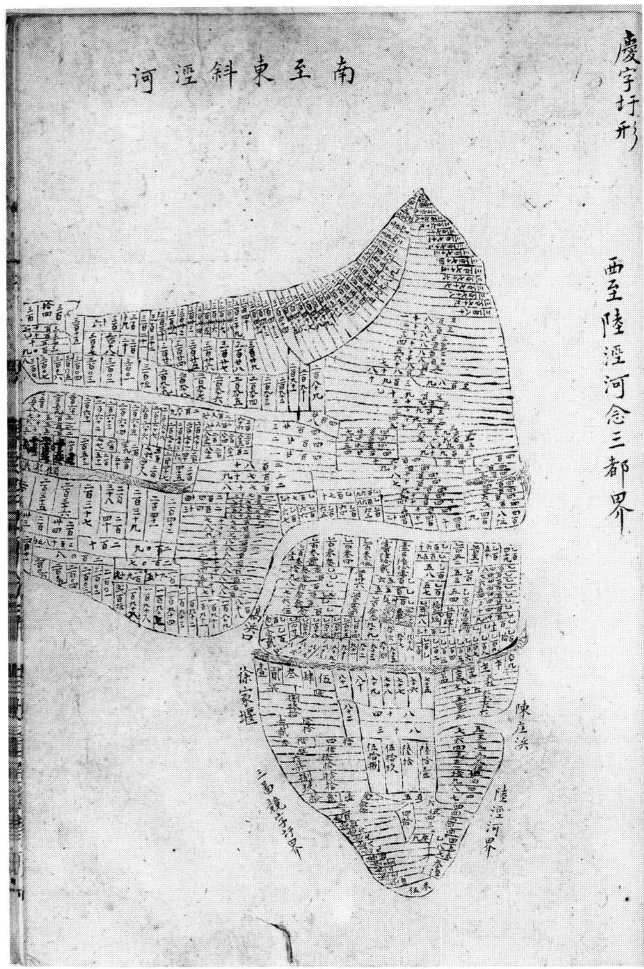

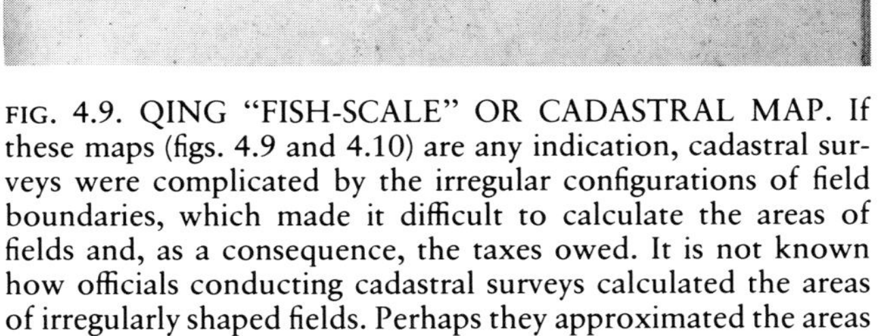

FIG. 4.9. QING "FISH-SCALE" OR CADASTRAL MAP. If these maps (figs. 4.9 and 4.10) are any indication, cadastral surveys were complicated by the irregular configurations of field boundaries, which made it difficult to calculate the areas of fields and, as a consequence, the taxes owed. It is not known how officials conducting cadastral surveys calculated the areas of irregularly shaped fields. Perhaps they approximated the areas by abstracting more regular geometric shapes, such as squares and triangles, from the actual configurations. This depiction is of fields in Yuanhe Xian (County), in modern Jiangsu Province, from a fish-scale map register.
Size of each folio: 37 × 23 cm. By permission of the Institute of Oriental Culture, University of Tokyo.

> let him know in which regions there are Chinese converts, and to which order they belong, so that in the future, he may send missionaries of the accepted order to a particular church and thus avoid controversy.[72]

The Chinese ministers asked Adeodato how he obtained the map, and he replied: "This was an old map in the [Xitang]. After I became the Superior, I found it in a wastepaper basket."[73] This explanation left the Chinese officials dissatisfied: If Adeodato had picked up the map by accident, they asked, how could he understand its symbols? According to the Chinese, Adeodato's response to further questioning was unclear, and the official conclusion was that Adeodato "may be concealing something."[74]

From a legal point of view, Adeodato's map was especially incriminating, since by imperial decree only Europeans were allowed to practice Christianity. Furthermore, Europeans were to have no social contacts with the Chinese populace, and Chinese citizens were forbidden to practice the heretical religion from the West. Adeodato's map provided evidence that breaches of imperial decree were widespread, and as a result Catholicism was officially denounced, Catholic publications were col-

72. Fu, *Documentary Chronicle*, 1:351 (note 67).
73. Fu, *Documentary Chronicle*, 1:351 (note 67).
74. Fu, *Documentary Chronicle*, 1:352 (note 67).

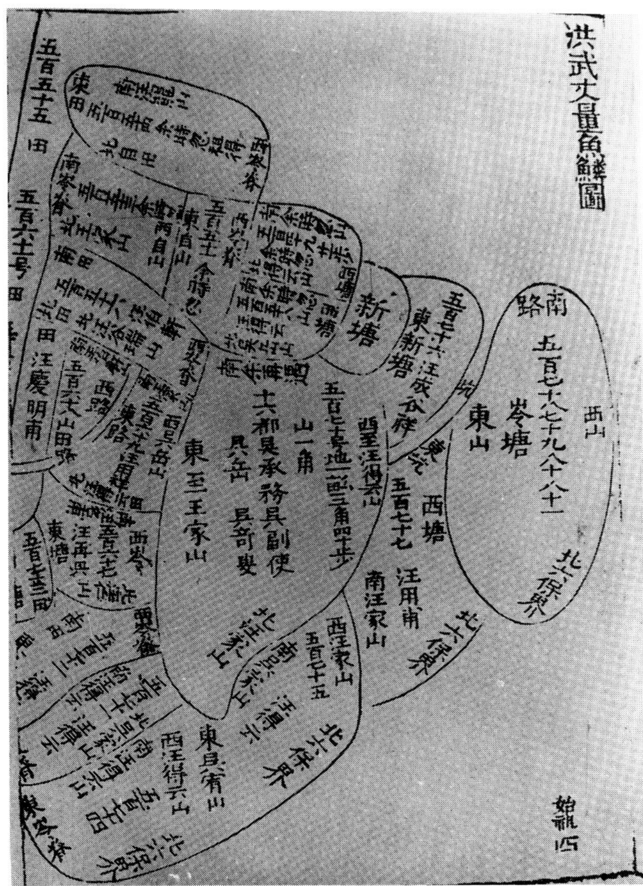

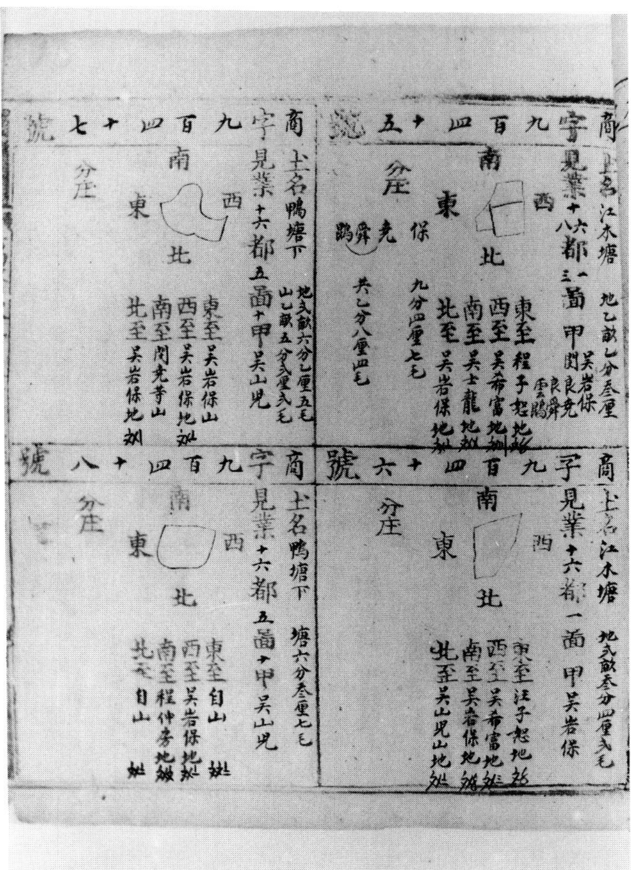

FIG. 4.10. MING "FISH-SCALE" OR CADASTRAL MAPS. During the Ming dynasty, fish-scale map registers contained general maps of contiguous landholdings (*left*, apparently a Ming copy of a fish-scale map) and tables containing maps of individual fields (*right*, from a Ming fish-scale map register). The tables, keyed to a general map, also list such information as registration numbers and areal measurements and describe the eastern, western, southern, and northern boundaries of the fields. The maps within the tables do not seem to be drawn to consistent scale. According to the annotations, the plot in the upper left is supposedly somewhat larger than that in the lower right. But according to the images, the plot in the lower right is larger.

The map on the left is a copy of the "Hongwu zhangliang yulin tu" (Fish-scale map of measurements from the Hongwu reign period [1368–98]) preserved in *Wu shi xianying zhi* (Record of the Wu family's ancestral graves, 1635). The map on the right is in *Ming Wanli jiu nian yulin tuce* (Fish-scale map register of the ninth year of the Ming dynasty's Wanli reign period [1573–1620], 1581), 1:1.

Size of the originals: left, not known; right, 39 × 29 cm. Photograph courtesy of the Museum of Chinese History, Beijing.

lected and burned, and practicing Christians were persecuted and barred from office.[75]

Behind the manifest anti-Christian message of the government's actions, there was also a geographic one: geographic information about the Chinese empire and especially maps were the preserve of the Chinese government. In accord with what the *Zhou li* implied about the importance of geography for rulers, ruling houses after the Han collected geographic information on their territorial holdings. The practice of map submission by regional and local governments continued, though the periodicity of such submissions varied. During the Tang dynasty, prefectural maps were submitted to the bureau of military affairs every two years; during the Song, the schedule was changed to every five years.[76] *Zhuanyun shi* (fiscal commissioners) during the Song dynasty were required to submit maps of their jurisdictions (known as *lu*, circuits) to the central government at ten-year intervals, while prefectural governments presented maps in each intercalary year, or about once every three years.[77]

At times the central government commissioned its own local surveys. In 1387, for example, the imperial government ordered a cadastral survey of Zhejiang and southern

75. See Fu, *Documentary Chronicle*, 1:352–58 (note 67).

76. *Tang huiyao* (Important records of the Tang, completed 961), chap. 59; see the edition in 3 vols. (Taipei: Shijie Shuju, 1963), 2:1032.

77. See *Song shi*, chap. 441 (37:13041) (note 64).

FIG. 4.11. *DILI TU* (GEOGRAPHIC MAP). This copy of a map by Huang Shang was engraved on stone by Wang Zhiyuan in 1247 and erected at the Suzhou prefectural school.
Size of the original: 179 × 101 cm. Municipal Museum (Stone tablets), Suzhou. Photograph courtesy of Cao Wanru, Institute for the History of Natural Sciences, Academia Sinica, Beijing.

Jiangsu: survey teams "measured fields all around, registered them using numbers, recorded the names of the owners and the dimensions of the fields, and compiled them into registers, shaped like fish scales and called fish-scale map registers." In addition to boundaries, these fish-scale maps (*yulin tu*) represented such land features as elevation and soil type.[78] Such map registers were in use as early as the Song dynasty and continued to be used through the Qing dynasty (figs. 4.9 and 4.10). The government's interest in the cadastral survey and resulting map registers was twofold: the registers could be used to help resolve legal disputes over property rights, and they also served as an aid for tax accounting; with them, government officials could estimate the revenue base from land taxes and keep track of individual taxpayers. The survey commissioned in 1387, for example, was ordered as a result of widespread tax evasion: the wealthy landowners in Zhejiang and Jiangsu were registering their lands under the names of relatives, neighbors, and even servants. Rectifying the situation required more than twenty years: the fish-scale registers for the area were presented to the throne in 1398.[79]

Although maps were useful, as the preceding examples illustrate, in fulfilling ritualistic, military, and administrative purposes, their value extended beyond those areas. As early texts like the *Xunzi* and *Zhou li* suggest, geographic knowledge was part of the necessary equipment of the ruler, and maps could be used to transmit cultural values. A geographic map titled *Dili tu* (Geographic map), made for this purpose perhaps in 1193, was copied on stone in 1247. The original map does not survive, but the stone copy (fig. 4.11) was preserved in a Confucian temple in Suzhou, along with stone copies of three other documents made at the same time. According to the inscription on the *Dili tu*, all four documents were presented to the prince of Jia by his tutor.[80] The other three documents consisted of a synoptic table of the history of China, a celestial chart, and a plan of Suzhou, an important city of the Southern Song (1127–1279). According to the *Song shi* (History of the Song), the documents were made by the prince's tutor Huang Shang for the purpose of educating the future ruler. The program of instruction proved timely, if one accepts Chavannes's 1193 date for the documents, for the prince rose to the throne as the emperor Ningzong (r. 1194–1224) the following year.[81]

78. Zhang Tingyu et al., *Ming shi* (History of the Ming, 1739), chap. 77; see the edition in 28 vols. (Beijing: Zhonghua Shuju, 1974), 7:1881–82.

79. For further discussion of cadastral surveys in China, see Ho, *Population of China*, 101–35 (note 3). For further discussion of fish-scale maps, see Niida Noboru, "Shina no tochi daichō 'gorinsetsu' no shiteki kenkyū" (Historical study of Chinese land register "fish-scale" maps), *Tōhō Gakuhō* (Tokyo) 6 (1936): 157–204; and Chao Kang (Zhao Gang), "Ming-Qing diji yanjiu" (Study of Ming and Qing land records), *Zhongyang Yanjiuyuan Jindaishi Yanjiusuo Jikan* 9 (1980): 37–59.

80. A transcription and French translation of the inscription and the notes on the *Dili tu* appear in Edouard Chavannes, "L'instruction d'un futur empereur de Chine en l'an 1193," *Mémoires concernant l'Asie Orientale* 1 (1913): 19–64. In the inscription, the engraver of the stone copy identifies himself as Wang Zhiyuan; nothing is known about him other than what he says in the inscription. He says that he found the *Dili tu* and three other documents in what is now Sichuan, where the prince of Jia once resided. He took them to Suzhou, where he engraved them in stone to ensure their transmission to future generations.

81. *Song shi*, chap. 393 (34:12,000) (note 64); and Chavannes, "L'instruction" (note 80).

The didactic function of the celestial chart and geographic map is implied by the *Song shi*'s summary account of Huang's poems to the prince, advising him on how to apply astronomical and geographic knowledge: "[Huang] wished that the prince would observe phenomena and realize that advancement in learning is like the ceaseless motion of the heavens, and that he would unroll maps and consider that half the territory of his ancestors had fallen into foreign hands and had yet to be recovered."[82] During the Southern Song, much of China was under the control of the Jin, and the *Dili tu* is drawn so as to remind its user of his mission. The map shows the locations of the capitals of the Northern Song (960–1126) and Liao dynasties, both of which had fallen to the Jin—thus the map's user could see how much territory was lost and had to be recovered. If the representation of lost capitals was not enough, notes on the *Dili tu* refer to the example of another emperor, Guangwudi of the Han, who recovered the empire from a usurper. In an incident already cited, Guangwudi was reminded of the difficulty of recovering the empire when he looked at a map. Though the forces of the Han emperor were numerically weak, he was able to prevail, the *Dili tu*'s notes imply, because of his superior virtue. The parallel would not have been lost on the future Southern Song ruler, who would also suffer from a numerical disadvantage. The application of the parallel would also have been clear: the future ruler could compensate for lack of military strength by self-cultivation, by preparing himself to rule through moral example.

The usefulness and pervasiveness of maps at all levels of government could reasonably be expected to have contributed to the development of specialists in mapmaking. No such class of specialists seems to have appeared, however—at least not until the end of the Qing dynasty. It seems that map use was so widespread that all officials were expected to have or acquire some skill in producing them. In general, however, there was little sense of professionalism; maps were seldom carefully drawn; practices were inconsistent; and there were no generally shared standards.[83] As with mapping, there does not seem to have been an official class of specialists in surveying; knowledge of mensuration seems to have been expected of all officials. The Kangxi emperor himself (1654–1722) showed officials how to take measurements for a general survey of China and led armies on topographic surveys: "We personally marshalled the six armies and went to Ningxia, where we stayed for twenty days. We extensively viewed land configurations, roaming about and inspecting the forms of mountains."[84]

The official class, Balazs has stated, was "firmly against any form of specialization."[85] This generalization, though not so true for the lower levels of government, especially during the Song and after, holds for the highest levels of the bureaucracy.[86] There, in a hierarchy of subjects that may seem opposed to modern intellectual values, literary learning was more valued than technical knowledge. Those who passed the mathematics examination, for example, were relegated to low-level posts; the highest posts were reserved for those who passed the literary examination. Knowledge of the written word was the way to official success, and this fact of life helped reinforce the reverence for documents expressed in the classics.

Occasionally, however, the government recognized the mapmaking skills of particular individuals and commissioned them to produce maps. The scholar-official Shen Kuo, for instance, received no formal schooling and seems to have been trained in mapmaking on the job. He served in a series of local posts in which he designed and supervised land reclamation schemes requiring surveys. Several of Shen's posts in the central government—for example, *jiaoshu lang* (editor in the palace library), *sitian jian* (director of astronomy), *chafang shi* (investigation commissioner)—gave him access to documents and instruments useful to mapmaking, though his chief responsibilities often lay in military and fiscal affairs. In 1075, while in the service of the central government, he made an inspection tour of the Khitan frontier. During this tour he carefully recorded topographic features—mountains, rivers, and roads—and represented them on a wax topographic map. He presented a wooden copy of the wax map to the emperor, who called his ministers together to view it and then ordered all frontier prefectures to make wooden maps to be stored in the inner treasury.[87] Although Shen Kuo was not officially classified as a mapmaker, his mapmaking talents, as the example of his three-dimensional model shows, were recognized: he was commissioned in 1076 to compile an atlas of all Chinese territory.

---

82. *Song shi*, chap. 393 (34:12,001) (note 64).

83. The central government made an attempt to standardize cartographic practices in the late Qing, but it allowed deviations from them when provincial governments reported that the announced standards would delay the submission of maps. See p. 195.

84. Aixinjueluo (Aisingioro) Xuanye (Kangxi emperor), *Kangxi di yuzhi wenji* (Collected commentaries of the Kangxi emperor) (1733; reprinted Taipei: Taiwan Xuesheng Shuju, 1966), 32.2a. See also Jonathan D. Spence, *Emperor of China: Self-Portrait of K'ang-hsi* (1974; reprinted New York: Vintage–Random House, 1975), 73.

85. Balazs, *Chinese Civilization and Bureaucracy*, 17 (note 2).

86. The generalization, however, can be qualified. First, the large number of polymaths in the Northern Song contrasts with the narrower attitude from the Southern Song, one that tended to discourage nonliterary pursuits by scholars. Second, in the Northern Song financial experts could reach the highest levels of government.

87. Shen, *Xin jiaozheng Mengxi bitan*, chap. 25 (256); see also Sivin, "Shen Kua," 380 (both in note 71).

Astrology and Celestial Mapping in Political Culture

One aspect of mapping did fall under the purview of specialists—celestial mapping. The need for accurate calendars to regulate agriculture and the precisely timed ritual functions of the court, such as seasonal sacrifices, contributed to the establishment of an office devoted to astrology. The *Zhou li* mentions at least two offices concerned with celestial phenomena, the *fengxiang shi* (royal astronomer) and the *baozhang shi* (royal astrologer). The *fengxiang shi* was concerned with keeping time in accordance with a number of celestial phenomena and calculating the positions of those phenomena.[88] The officer's calculations of those positions, according to Zheng Xuan, were equivalent to a calendar.[89] The *baozhang shi*'s duties illustrate the inseparability of astronomy from astrology: he was responsible for "recording the changes and movements of the planets, sun and moon, so as to observe movements in the world and to distinguish good and bad fortune." The connection between celestial and earthly phenomena was also reflected in administrative divisions. The *baozhang shi* divides the territory of the nine regions into areas dependent on particular stars, and all investitures have "distinct stars by which one can prognosticate their misfortune or good fortune."[90]

During the Han, the head of the astrological office was called the *taishi ling* (grand astrologer). To a large extent, his duties paralleled those of the astrologers mentioned in the *Zhou li*: he was responsible for constructing the annual calendar, identifying auspicious and inauspicious days for rituals, and recording portents and omens.[91] These remained the essential responsibilities of the court astrologers in later dynasties, though their exact titles varied. The astrologer's purpose in recording celestial and terrestrial anomalies went beyond the needs of agriculture. The goal was to correlate them with political events. This is made clear by Sima Qian, an astrologer-historian of the first century B.C.: "Looking up, [rulers] contemplated the signs in the heavens and, looking down, observed their counterparts on earth." The ancients, Sima Qian says, prognosticated on the basis of celestial omens and anomalies in order to "conform to the exigencies of the time."[92] The *Han shu* makes the point that administrative mistakes on earth are reflected in anomalies in the heavens: "Thus the enlightened lord beholds them [celestial omens] and awakens, governing himself and rectifying his affairs; if he ponders his faults, then misfortune will be eliminated and good fortune will result."[93] Understanding celestial phenomena was regarded as a means of legitimizing political power, a way of preserving the mandate of heaven. Unexplained phenomena, those that could not be predicted, were cause for political apprehension, as Sivin has remarked:

> An astronomical system was a complete set of mathematical techniques for calculating an ephemerides which provides both positions and dates of characteristic phenomena for the sun, moon, and planets. Once a system was officially adopted, it became part of the Emperor's ritual paraphernalia . . . because the ability to predict moved celestial events from the realm of the ominous to that of the rhythmic and intelligible. The Emperor was thus enabled to know Nature's [Dao] so that his social order might be kept concordant with it. Failure of the official system to predict was necessarily a sign of moral imperfection, a warning that the monarch's virtue was not adequate to keep him in touch with the celestial rhythms.[94]

The correlation of natural anomalies with political happenings required collecting data on both. Thus the duties of court astrologers included compiling historical records. In some cases, astrologers noted that celestial disturbances coincided with terrestrial phenomena, as shown by these records from the Tang:

> In the fifth month of the eighteenth year [of the Zhenguan reign period] [644], a streaming star [meteor] fell out of the Eastern Wall [in Pegasus], with a sound like thunder. The prognostication said: "Those with a sound like thunder are signs of anger [on earth? in heaven?]."[95]

> In the tenth month of the fourth year [of the Yonghui reign period] [653] a woman from Muzhou [in present-day Zhejiang], Chen Shuozhen, revolted; the prefect of Wuzhou, Cui Yixuan, executed her; a star fell on the rebel camp.[96]

> In the second month of the second year [of the Jinglong reign period, 708] a heavenly dog [explosive meteor] fell in the southwest, with a sound like thunder; the wild pheasants all shrieked.[97]

The astrologers did not correlate every heavenly anomaly with a particular earthly event. More often, the astro-

---

88. *Zhou li*, 6.44b (note 27).

89. *Zhou li*, 6.44b (note 27).

90. *Zhou li*, 6.45a–b (note 27).

91. *Hou Han shu*, "Baiguan zhi" (Treatise on the hundred offices) (12:3572) (note 45). The grand astrologer was also expected to possess expertise in documentation: he supervised examinations for candidates for offices responsible for handling documents: *shangshu* (imperial secretary) and *shishu lingshi* (calligraphy clerk). See Hans Bielenstein, *The Bureaucracy of Han Times* (Cambridge: Cambridge University Press, 1980), 19.

92. *Shi ji*, chap. 27 (4:1342–43) (note 17).

93. *Han shu*, chap. 26 (5:1273) (note 6).

94. Nathan Sivin, *Cosmos and Computation in Early Chinese Mathematical Astronomy* (Leiden: E. J. Brill, 1969), 7.

95. *Xin Tang shu*, chap. 32 (3:842) (note 66).

96. *Xin Tang shu*, chap. 32 (3:842) (note 66).

97. *Jiu Tang shu*, chap. 36 (4:1321) (note 66).

logical records note only that a certain disturbance—eclipse, meteor, comet, red pneuma—occurred at a certain time, as in this example from the Song:

> On the *dingmao* [fourth] day of the second month of the first year [1049] of the Huangyou reign period, a comet emerged from the void; at dawn it appeared in the eastern region, and it pointed to the southwest. It passed through Purple Tenuity [a constellation consisting largely of the stars of Draco] to Harvester [a *xiu* or lunar lodge]. In all it lasted 114 days before vanishing.[98]

The sheer quantity of such observations has fostered the impression that Chinese astrology was largely empirical, that Chinese astrologers rarely engaged in theory. Theory in a nontechnical sense, however, did underlie the practice of data collection—a theory that correspondences existed between heaven and earth. If enough data were collected, then the links between natural anomalies and earthly events might be induced.[99]

Chinese astrologers worked out at least one system of correspondences between sectors of the heavens and particular Chinese cities and provinces. This system appears in the explanatory text to the stone copy of the celestial map presented to the prince of Jia (p. 547, fig. 13.21). According to this text, phenomena in certain regions of the heavens affect particular political subunits: "Whenever the sun and the moon mutually eclipse or an unusual phenomenon of the stars or planets occurs, the fortune or misfortune of the corresponding regions may be predicted by knowing in each case the given correlation."[100]

The Proliferation of Geographic Records

The desire to correlate heaven and earth suggests another reason besides administration and defense for the Chinese government's interest in compiling geographic information, such as could be used in maps. Geography was considered a subdivision of history, and geographical treatises, as well as the dynastic histories of which they were a part, were products of the astrological bureau and its institutional descendants: Chinese histories typically include treatises on astrology (*tianwen zhi*) and geography. These treatises preserved the empirical data for inferring relationships between celestial and terrestrial phenomena. Maps, as mentioned above, supplied some of the data for those treatises. Not coincidentally, some of those known for geographic mapping were astrologers. Zhang Heng (78–139), for example, presented a topographic map (*dixing tu*) in the year 116, while holding the office of grand astrologer.

Because the imperial state claimed that its foundations were those of the celestial order, it is easy to understand the proliferation of geographic works after the Han.[101] The collection of data on terrestrial phenomena was not restricted to the imperial historians. Much of the heavens could be viewed from a single observatory; this was not the case with the earth. For information on terrestrial occurrences not directly observable by the astrologer-historian, the central government had to rely on data submitted from various localities, often in the form of *fangzhi* (local gazetteers), a term already encountered in the *Zhou li*. These gazetteers, compiled under central or local government auspices, are compendiums of information pertaining to a particular region, typically an administrative subunit: provinces, prefectures, counties. Typically, gazetteers included a section of maps. The central government collected gazetteers from its administrative subunits and used them to compile comprehensive gazetteers of the entire empire. These also often contained maps.

Only a few examples of these works predating the Song survive, but we know the titles of several hundred compiled before the Song. The bibliographic section in the *Sui shu* (History of the Sui, compiled in the seventh century) lists nearly 140 geographic works, including gazetteers and maps. The compilers of the *Sui shu* suggest that the gazetteers in particular follow a precedent established by the "Yu gong" and the *dilizhi* of the *Han shu*.

The *Sui shu* provides no details about the content of the gazetteers it lists, but it does mention that in some cases compiling such works required the use of documents submitted by local administrations. About 610 the emperor ordered all the prefectural governments to "arrange their [records concerning] customs and products and their maps, and submit them to the secretariat."

98. *Song shi*, chap. 56 (4:1227) (note 64).

99. For an extended discussion of the relation between Chinese astrology and documentary scholarship, see Shigeru Nakayama, *Academic and Scientific Traditions in China, Japan, and the West*, trans. Jerry Dusenbury (Tokyo: University of Tokyo Press, 1984)—a very suggestive work.

100. This translation is that of W. Carl Rufus and Hsing-chih Tien, *The Soochow Astronomical Chart* (Ann Arbor: University of Michigan Press, 1945), 7. According to Rufus and Tien, the correlation consisted of four elements: "first, a branch [one of the twelve places toward which the Dipper's seven stars point during the twelve months] with its compass direction; second, the position in the sky designated by an asterism; third, a kingdom; and fourth, a region" (7 n). A table illustrating the scheme appears on pp. 12–13.

101. A bibliography of more than two thousand of these works through the Yuan has been compiled by Zhang Guogan in *Zhongguo gu fangzhi kao* (Study of ancient local gazetteers in China) (Beijing: Zhonghua Shuju, 1962); only about fifty of them survive. Zhang's list may be complemented by *Zhongguo difangzhi lianhe mulu* (Union catalog of Chinese gazetteers), ed. Zhongguo Kexueyuan Beijing Tianwentai (Beijing Observatory, Chinese Academy of Sciences) (Beijing: Zhonghua Shuju, 1985), which lists more than eight thousand extant gazetteers, dating from the Song through the Republican period.

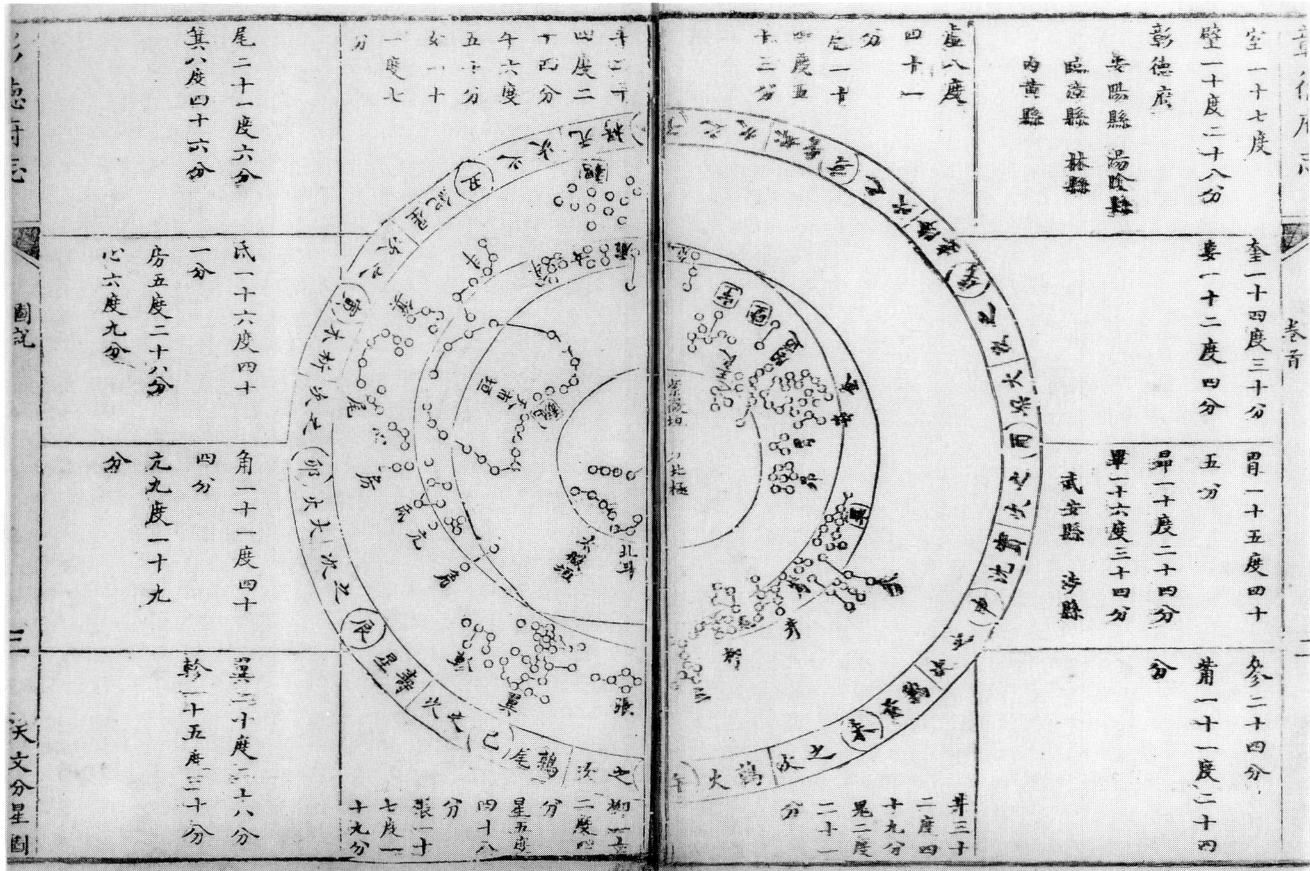

FIG. 4.12. STAR MAP FROM A LOCAL GAZETTEER. Size of the original: 20 × 28 cm. From the *Zhangde fu zhi* (Gazetteer of Zhangde Prefecture [in modern Henan Province], 1787), *tushuo* (illustrated explanation), 2b–3a. Reproduced courtesy of the Harvard-Yenching Library, Harvard University, Cambridge.

These records were used to compile three compendiums of information on various administrative subdivisions.[102] All three are now lost, so it is uncertain whether they contained maps.

The titles of other geographic works often include the compound *tujing* (illustrated classic) or *tuzhi* (illustrated record), suggesting the inclusion of illustrations and perhaps maps. None of the existing editions of gazetteers before the Song, however, contain maps. The preface to the *Yuanhe junxian tuzhi* (Illustrated record of the prefectures and counties of the Yuanhe reign period [806–20]), a Tang gazetteer, states that a *tu* headed each section devoted to a garrison town (*zhen*).[103] But since the present text of the work contains no illustrations, it cannot be determined whether the *tu* mentioned in the preface were what would now be called maps.

Unlike the early gazetteers mentioned above, those from the Song dynasty onward usually contain maps. In addition, they typically contain sections devoted to topics like geography, history, biography, and customs and mores. This information not only was of use in compiling comprehensive gazetteers, but also was useful to officials sent by the central government to serve in regional and local posts. By the Qing dynasty, gazetteers were regarded as essential to local administration, and thousands of them were compiled in that dynasty alone. The editors of an eighteenth-century gazetteer state: "That prefectures have gazetteers is a constant principle"—thus the surprise of one newly appointed official upon discovering that his subordinates had never heard of a gazetteer.[104]

Although the administrative usefulness of these gazetteers eventually became primary, their compilers did

102. Wei Zheng et al., *Sui shu* (compiled 629–56), chap. 33; see the edition in 6 vols. (Beijing: Zhonghua Shuju, 1973), 4:982–88.

103. Li Jifu, *Yuanhe junxian tuzhi* (written 814), *Baibu congshu jicheng* edition, preface, 2b.

104. See *Zhengding fu zhi* (Gazetteer of Zhengding Prefecture [in modern Hebei Province]) (1762; reprinted Taipei: Taiwan Xuesheng Shuju, 1968), *shou* (prefatory material) 7a; and "Jiu zhi xuba" (Preface to the old gazetteer), *Tongzhou zhi* (Gazetteer of Tongzhou [in modern Hebei Province, near Beijing]) (1879; reprinted Taipei: Taiwan Xuesheng Shuju, 1968), 1a–b.

not lose sight of their original purpose: as an aid to astrological prediction. Compilers of Qing gazetteers often included sections on astrology; sections on field allocation, which identifies the celestial region to which a territory corresponds; and sometimes even star maps, presumably to help correlate the terrestrial and celestial (fig. 4.12 and p. 198, fig. 7.22). Sometimes the central government had an immediate need for the kinds of information contained in gazetteers. Their compilation, however, usually took several years. Faster means of reporting the information needed to correlate earthly and celestial events were therefore developed. During the Qing, for example, the data usually included in gazetteers were regularly reported to the throne in the form of palace memorials (*zouzhe*). In these memorials provincial officials were required to submit information on natural disasters, such as droughts and floods, and on the agricultural economy, including such matters as grain prices, harvests, and weather conditions. Such information was sought because it might yield clues about the will of heaven: "We consider that if our administration falters below," the Kangxi emperor said in an edict of 1689, "calamities will come in reponse from above."[105]

Gazetteer Maps

The maps in local gazetteers seem to have served primarily as complements to the verbal descriptions. Although literary sources attest to the use of maps with indications of scale during the Liuchao (Six Dynasties) period (222–589), gazetteer maps before the Qing dynasty rarely if ever bear such indications, and if they were based on actual measurements, they do not seem intended for presenting quantitative information.[106] One reason may have been that such information was redundant, since distances and directions were given verbally in the geographic sections. The verbal accounts of particular areas were often more detailed than the maps of those areas. In this passage from a Southern Song gazetteer, bearing and distance are provided with respect to Lishui Xian (county in present-day Jiangsu Province): "The Baima bridge is 40 *li* southeast of the county seat. The Meitang bridge is 120 *li* southeast of the county seat. The Dengbu bridge is 120 *li* southeast of the county seat."[107] The map depicting the area described in this passage (fig. 4.13) does not indicate the locations of these bridges, suggesting that it was intended to complement, not substitute for, the verbal description. The idea that map and text were complementary persisted till the end of the nineteenth century. According to the statement of principles in a gazetteer dating from 1894, "Narration of events without maps is not clear; and maps without explanation are not intelligible."[108]

The limitations of gazetteer maps as sources of information are recognized in the *Zhouxian tigang* (Essentials of prefectural and county [government]), by Chen Xiang (1017–80). In a section titled "Xiang hua ditu" (Drawing detailed maps), Chen warns against relying solely on maps in *tujing*, from which one can "get only a rough, general understanding" of an area.[109] For a thorough understanding, one must have new maps. After assuming a local post, a new official "must order that maps be drawn in detail to record the areal extent of districts and security groups, the dwelling places of the populace, the lengths of roads, and the number of mountains, forests, and fields."[110] Maps submitted by various districts are then combined into one large map, which is placed in a corner of the office. Chen's account does not make clear whether the newly arrived official combines the maps himself or whether his subordinates are delegated that task. Nor does it specify whether the various maps are merely fastened together or whether a large map is drawn based on them. What is clear is that this large map aids immensely in the conduct of government business: with it the official has "the people, land configuration, mountains, forests, streams, and marshes all in view; and whenever there are legal disputes, taxes and levies, droughts, and arrests, everything can be seen at a glance."[111] None of these composite maps seems to have survived, and as a consequence we do not know how such a variety of information was represented.

105. *Shengzu Renhuangdi shengxun* (Sacred instructions of Shengzu, the emperor Ren [Kangxi], presented 1732), in *Da Qing shichao shengxun* (Sacred instructions of the ten reigns of the Great Qing) (Taipei: Wenhai Chubanshe, 1965), 3.2a. For a study of the Qing memorial system, see Silas H. L. Wu, *Communication and Imperial Control in China: Evolution of the Palace Memorial System, 1693–1735* (Cambridge: Harvard University Press, 1970).

106. For an example of the problems involved in reading distances from gazetteer maps, see the caption to figure 4.13.

107. Zhou Yinghe, comp., *Jingding Jiankang zhi* (Gazetteer of Jiankang [Nanjing] of the Jingding reign period [1260–64], compiled ca. 1261), *Siku quanshu* edition, 16.44b.

108. *Guangping fu zhi* (Gazetteer of Guangping Prefecture [in modern Hebei Province]) (1894; reprinted Taipei: Taiwan Xuesheng Shuju, 1968), "Fanli" (Principles), 1a. This quotation is remarkably close to one from the medieval European mapmaker Paolino Veneto: "There is needed moreover a twofold map, [composed] of painting and writing. Nor wilt thou deem one sufficient without the other, because painting without writing indicates regions or nations unclearly, [and] writing without the aid of painting truly does not mark the boundaries of the provinces of a region in their various parts sufficiently [clearly] for them to be descried almost at a glance." Quoted in Juergen Schulz, "Jacopo de' Barbari's View of Venice: Map Making, City Views, and Moralized Geography before the Year 1500," *Art Bulletin* 60 (1978): 425–74, esp. 452.

109. Chen Xiang, *Zhouxian tigang* (eleventh century), *Baibu congshu jicheng* edition, 2.16b.

110. Chen, *Zhouxian tigang*, 2.16b (note 109).

111. Chen, *Zhouxian tigang*, 2.16b (note 109).

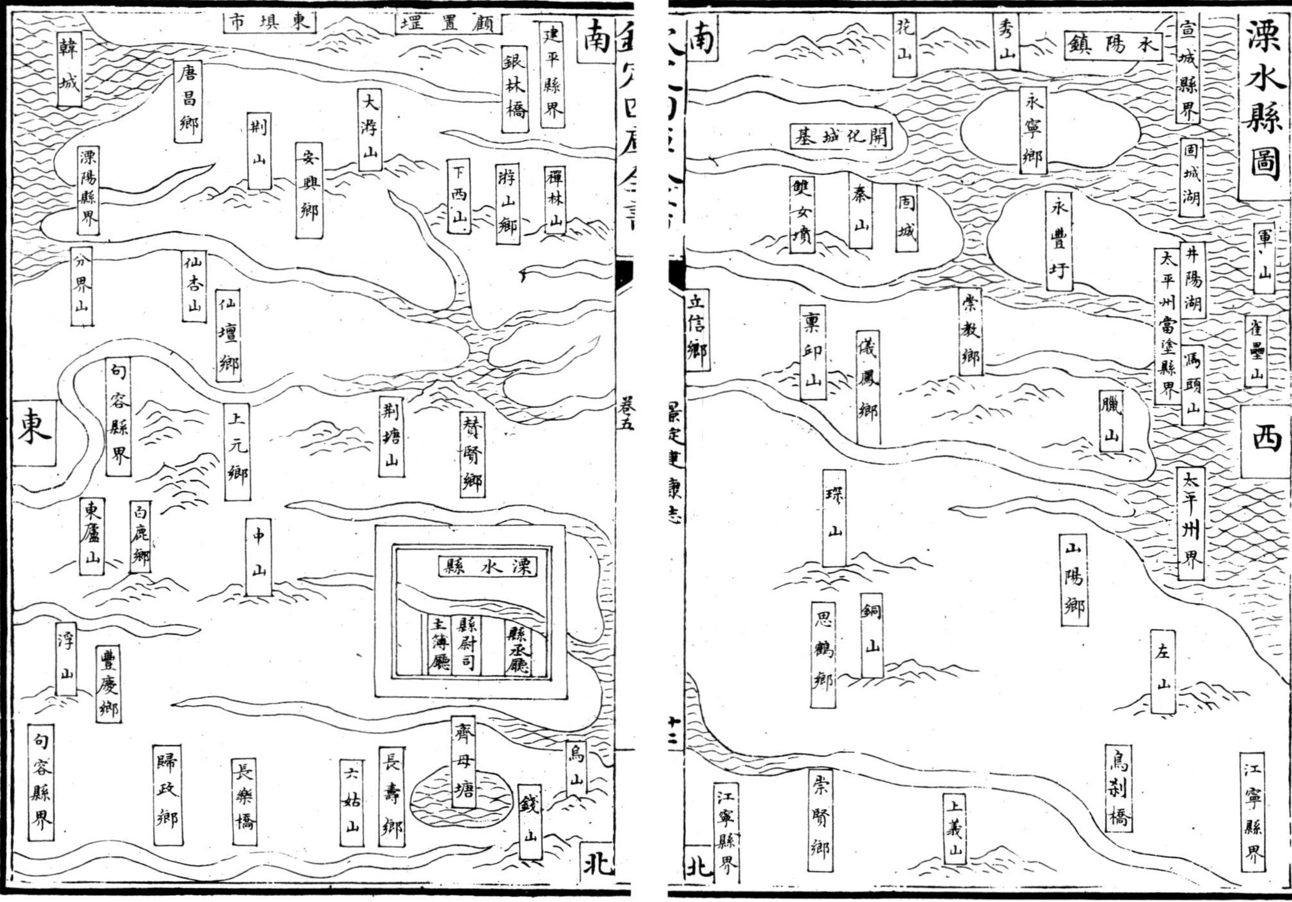

FIG. 4.13. MAP OF THE AREA AROUND LISHUI XIAN. This map comes from a Qing edition of a Song gazetteer. South is at the top. According to the gazetteer, Zhong Shan (Mount Zhong) and Donglu Shan (Mount Donglu) are both fifteen *li* southeast of Lishui Xian (represented by the large squarish box). On the map, however, Zhong Shan (to the immediate left of Lishui Xian) appears to lie almost midway between Lishui Xian and Donglu Shan (the second mountain to the left of Zhong Shan).
Size of each image: ca. 13.5 × 9 cm. From Zhou Yinghe, comp., *Jingding Jiankang zhi* (Gazetteer of Jiankang [Nanjing] of the Jingding reign period [1260–64], compiled ca. 1261), *Siku quanshu* edition, 5.12a–b.

The limitations that Chen Xiang perceived in gazetteer maps during the Song were not eliminated in later dynasties. Partly as a result of these limitations, the Qing court took an interest in the cartographic methods introduced by Jesuit missionaries. A survey of the entire country was undertaken with Jesuit involvement from 1708 to 1717. In 1719 the emperor reported to his ministers that the maps based on that survey had been completed. The ministers responded with a critique of previous mapping efforts: "Formerly territorial maps and geographic records were always copied from beginning to end from other versions, transmitting hearsay and guesswork. Although there are complete works, in the end these are hardly reliable. Sometimes they do not distinguish between the main body and branches of mountains and streams; sometimes the positions of counties and districts vary from their actual locations. From antiquity till the present, there has not been a definitive work."[112]

Maps, Scholarship, and Cultural Continuity

Whatever their shortcomings, however, maps still served as valuable sources and tools for scholarship. During the Qing dynasty, historical cartography became an important branch of textual scholarship—thus the appearance of maps depicting, for example, the geography of the Han dynasty and the changing course of the Yellow River. The state sponsored much of this scholarship, since the aim of textual scholarship was to recover the way of

112. *Shengzu Ren (Kangxi) huangdi shilu*, 283.10b (note 67).

the ancients as reflected in classic texts and histories, which had supposedly been misinterpreted by scholars of preceding periods. If the ideas of the ancient sages could be understood, the Qing textualists believed, the administration of the empire could be improved.[113]

Much of the scholarship of the Qing centered on two earlier geographic works. One was the "Yu gong," described previously, studies of which often included cartographic reconstructions of the classic text's topographic descriptions (fig. 4.14). The other work was the *Shui jing zhu* (River classic commentary), compiled by the scholar-official Li Daoyuan (d. 527). This work is, as its title suggests, based on an earlier text known as the *Shui jing* (River classic), thought to be written by Sang Qin and perhaps dating from about the middle of the third century.[114] The *Shui jing* no longer exists as an independent text. In Li Daoyuan's work, text and commentary have been run together, and much textual scholarship has focused on separating the two. The original text is believed to have indicated the origin and course of 137 rivers.[115] In contrast, Li Daoyuan's work expands on this base to provide geographic and historical information on more than 1,200 rivers and the areas they pass.

The title of Li's work is thus slightly misleading, since it is far more than a commentary. It is virtually an independent work of scholarship, drawing occasionally on Li Daoyuan's own observations and usually on a variety of textual sources—among them the "Yu gong," the *dilizhi* of the *Han shu*, and gazetteers. In this the appeal of Li's work to Qing textualist scholars becomes clear. In it they found a precedent for their own methods, at the same time focusing attention on a text that epitomizes the multidimensional role of maps in Chinese political culture.

Like the Qing scholars, Li Daoyuan sees his work as a kind of historical criticism. He recognizes the value of previous attempts to compile geographic information, but he sees deficiencies as well. Part of his motivation in expanding the *Shui jing* is to correct the defects in classic texts. In the preface to his commentary, Li writes: "In the past Yu the Great made a record comprehensive [in scope] but incomplete [in detail] of mountains and streams; the records in the *dilizhi* are simple and incomplete; the annals and administrative regions of the *Shang shu* [Book of the Shang, or *Shu jing* (Book of documents)] are sketchy; as for what the rhapsodies on capitals describe, that genre does not convey meaning; although the *Shui jing* roughly describes the courses of rivers, it still leaves out their tributaries."

Accordingly, *Shui jing zhu* is organized around systems of rivers and waterways. In his preface Li Daoyuan appears to take for granted the economic importance of these waterways for agriculture, transportation, and communication. Instead he stresses cosmological principles to justify his attention to water: "The *Yi* [*jing*] says that

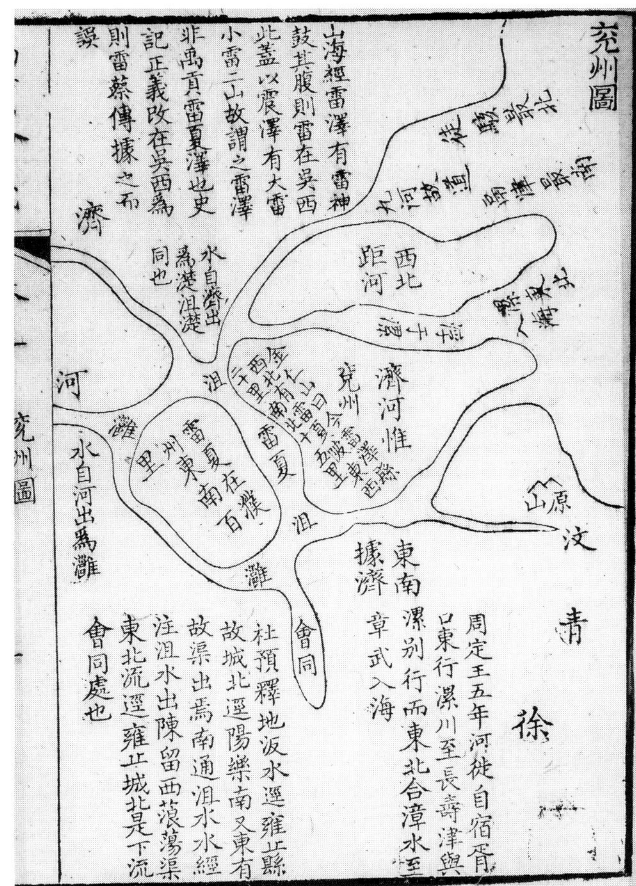

FIG. 4.14. MAP OF YAN ZHOU. This is one of the nine regions described in the "Yu gong," as reconstructed by a Qing textual scholar. In this instance, text seems to dominate map image.
Size of the original: 19.5 × 10.5 cm. From Xu Wenjing, *Yu gong huijian* (Collected commentaries on the "Yu gong," 1753), *tu* 10a. Reproduced courtesy of the Harvard-Yenching Library, Harvard University, Cambridge.

heaven, as the one, gave rise to water; thus *qi* [primal energy] originates in the North and is the progenitor of things. The *Xuanzhong ji* [Records within the occult] says: 'The most plentiful thing under heaven is water; it buoys up the heavens and supports the earth; high and low there is nothing it does not reach; among the ten thousand things there is nothing it does not enrich.' " Li implies that his work, with its focus on water, will help

113. This is to state the ideal of the Qing textualists. In practice, the political efficacy of textual scholarship was not always obvious.

114. The basis of this dating is the *Sui shu*, which says that Guo Pu (276–324) also wrote a commentary to the *Shui jing*. This commentary has been lost, so it is not entirely certain that Guo Pu and Li Daoyuan were annotating the same work.

115. See Wu Ze's preface to the *Shui jing zhu jiao*, 1–2 (note 56).

further understanding of the workings of the cosmos—water is a primordial substance and a vital element.

The *Shui jing zhu* itself contains no maps, but it is almost inconceivable that a work of such scope—ranging over China and the regions considered to be its vassals and tributaries—could be compiled without reference to them. Li Daoyuan in fact says that he used maps. In the case of the Ru River in Henan, he found that textual sources gave divergent accounts of its source. To resolve the confusion, he consulted maps of mountains and streams as well as gazetteers, but to no avail, and then ordered that its origin be searched out.[116] In some instances Li Daoyuan provides map titles, but he gives hardly any descriptions of the maps themselves.[117] One exception to this is a description of the *Hetu* (Yellow River chart): "The chart records the [Yangtze] Jiang, [Yellow] River, small mountains and streams, and the field allocation [system of correspondence between celestial "fields," or lunar lodges, and terrestrial regions] of the provinces and border areas."[118]

For geographic information Li had maps of varying scope: comprehensive maps like one entitled *Yu gong tu* (Map of the Tribute of Yu), regional maps like the *Jing zhou tu fuji* (Supplementary notes to the map of the Jing region), and mountain maps such as the *Kai shan tu* (Map of the Kai Mountains). These maps are no longer extant, but they were evidently annotated. Li Daoyuan, for example, quotes from the *Kai shan tu*'s description of Qi Shan (Mount Qi, in present-day Hunan): "The streams around it wind and twist; the mountain itself is high and steep."[119] The information Li Daoyuan obtains from maps, however, is not limited to topographic features. In one case, what seems to be a divination map or chart, the *Ruiying tu* (Map [or chart] of auspicious signs), is cited as a source of ornithological information: it attests to "three-legged crows, red crows, and white crows."[120] A comprehensive map titled *Kuodi tu* (Map of included territories) is cited for religious or mythological information. Li Daoyuan records this map's description of the river god Fengyi: "Fengyi always rides a cloud chariot, harnessing two dragons."[121] In its citation of maps for a wide range of information, the *Shui jing zhu* illustrates what was said earlier about their importance in preserving cultural inheritance: Li Daoyuan is using maps to do just that. His aims go beyond accuracy of topography and hydrology.

Li Daoyuan's stress on continuity was shared by Qing textual scholars—at least initially, for later it seems that textual research was done for its own sake. In this essay of generalization, I have attempted to avoid the latter pitfall and maintain the former emphasis on continuity. That emphasis should not be construed as meaning that Chinese political culture was monolithic and unchanging. Chinese politics were marked by tension between ruler and bureaucrats and conflicts between factions of bureaucrats and between levels of government. With a few exceptions, most notably during the Qing, the evidence on cartography does not generally reflect those internal divisions. Thus the ruling elite has been treated here as a more or less homogeneous group. The cartographic record does reflect, however, that maps could be adapted to new uses as political circumstances changed. During the Ming dynasty, for example, as China's sea power and contact with seafaring peoples grew, military map use expanded to include not only territorial but also maritime mapping. Those foreign contacts, especially during the Qing dynasty, also led to some changes in cartographic techniques and standards among the intellectual and political elite. But through the successive dynastic changes, the institutions and practices of that stratum of society do show a high degree of constancy. As far as maps are concerned, that social stratum can be characterized by the complex of attitudes reflected in the kinds of information Li Daoyuan gleans from maps—a complex in which the work of administration insinuates itself into cosmology, geography, and history. The patterns of map use within Chinese political culture may have shown remarkable consistency, but the manifestations of those patterns were subject to variation.

In a sense, map use among the Chinese elite illustrates the doctrine of *yiben wanshu*, "one root but ten thousand manifestations": "The ten thousand are one, and the one is realized in the ten thousand manifestations. The ten thousand and the one each have their proper place."[122] Neo-Confucians, most notably Zhu Xi (1130–1200), adhered to this doctrine as a guiding principle for their investigation of things (*gewu*). Of course, Neo-Confucian thinkers never offered this doctrine of *yiben wanshu* as the basis of methodology in the history of cartography. But perhaps this chapter may be taken as evidence that they spoke more truly than they knew.

At the same time, one should remember that the conclusions reached here apply only to that elite—and even then only provisionally—and cannot be safely extended to Chinese culture as a whole. Map use among the general populace is a topic that has barely been explored, even

116. *Shui jing zhu jiao*, chap. 21 (663) (note 56).

117. A listing of the maps and other sources cited by Li Daoyuan appears among the prefatory material of the *Shui jing zhu jiao* (note 56).

118. *Shui jing zhu jiao*, chap. 1 (5) (note 56). The subject of *Hetu* is discussed at greater length below in chapter 8 on Chinese cosmography.

119. *Shui jing zhu jiao*, chap. 20 (646) (note 56).

120. *Shui jing zhu jiao*, chap. 13 (431) (note 56).

121. *Shui jing zhu jiao*, chap. 1 (5) (note 56).

122. Zhou Dunyi, *Tong shu* (Comprehensive treatise [on the *Yi jing*, or Book of changes], ca. 1055), *Sibu beiyao* edition, 5a.

though the possible sources of information are voluminous. A considerable number of potential sources regarding map use within Chinese political culture, especially at the regional and local levels, also remain to be tapped. In some ways, then, this essay of generalization may be premature. Its methods will undoubtedly strike some as old-fashioned, focusing on incident rather than on institutional structures. More sophisticated studies, however, await more detailed knowledge of the inner workings of the bureaucracy in relation to map production.

5 · Taking the World's Measure: Chinese Maps between Observation and Text

CORDELL D. K. YEE

Historical scholarship has tended to measure the achievement of premodern Chinese cartography in terms of its application of numerical technique. Treated in this manner, Chinese cartography seems to parallel the course of European cartography. The modern form of the latter has often been regarded as the product of its alliance with number, so much so that modern maps are commonly viewed as scientific and value-free.[1] In the history of Chinese cartography, there is ample evidence of an alliance between number and map. There are, for example, attempts to formulate a coherent "code" specifying the methods to be used in translating empirical reality into quantitative terms. The quantitative techniques involved in mapmaking do not seem to have developed in response to cartographic needs per se; they were imported from established disciplines such as mathematical astronomy and water conservancy.[2] It is these codes and techniques that historians of Chinese cartography adduce to support their claims for a scientific tradition. The codes, however, do not seem to have been institutionalized or widely accepted. Part of the reason seems to have been a belief that the function of maps extended beyond the representation of nature knowledge to the transmission of cultural values and the preservation of political power. Contrary to the impression given in current accounts, in Chinese cartography there was no general tendency to eliminate human value from maps or to minimize its presence. As a result, cartography in China encompassed not only numerical techniques, but also what would now be considered humanistic concerns. The map allied itself with number and text. The two were not opposed—both were associated with value and power.

THE GOVERNMENT INTEREST IN MEASUREMENT

In *Oriental Despotism*, Wittfogel argues that the development of the Chinese bureaucracy could be linked to what he termed a "hydraulic" economy, one in which agriculture was supported by extensive irrigation projects requiring direction by an agromanagerial elite. The success of this economy depended on the making of accurate calendars to regulate agricultural activities, and this task also fell to the managerial elite. To the calendar measuring time, one could add the map for measuring and controlling space.

Wittfogel's analysis of the Chinese economy has been disputed, since the basic pattern of Chinese government, including the use of astrology, was created before rulers undertook extensive irrigation projects. Historians of Chinese science have also taken issue with him, since he proposes that hydraulic societies stifle creativity and are characterized by stagnation. According to Wittfogel, a centralized, bureaucratic political structure tended to "paralyze the search for scientific truth and social

1. One recent manifestation of this alliance is the use of digital computers for mapmaking. The literature on cartography is replete with references to its scientific character. In *Cartographical Innovations*, for example, maps are said to be scientific documents and works of art. The same work, however, suggests that this dual characterization does not apply to modern maps: "The 19th century mapmaker under his new name of cartographer belonged to the age of scientific communication in which cartography was establishing itself both as a science and as an industry." See Helen M. Wallis and Arthur H. Robinson, eds., *Cartographical Innovations: An International Handbook of Mapping Terms to 1900* (Tring, Hertfordshire: Map Collector Publications in association with the International Cartographic Association, 1987), XI and XVIII. John Keates, while reviewing a volume of essays exploring the relation between cartography and art, speaks of a "proper scientific and mathematical foundation for maps." See Keates's review of *Art and Cartography: Six Historical Essays*, ed. David Woodward (Chicago: University of Chicago Press, 1987), in *Cartographic Journal* 25 (1988): 179–80, esp. 179. For an exposition of cartography in terms of communication or information theory, see Arthur H. Robinson and Barbara Bartz Petchenik, *The Nature of Maps: Essays toward Understanding Maps and Mapping* (Chicago: University of Chicago Press, 1976). The authors of this work assert that "although cartography is often dubbed 'an art and science,' it is important to understand that it is also an exercise in engineering" (p. 108). This characterization again suggests the alliance of cartography and number—mapmaking as a kind of applied mathematics.

2. Here I use Stephen Toulmin's two-tiered definition of an intellectual discipline as having a well-defined subject matter and an explanatory (or procedural) ideal. See Toulmin's *Human Understanding* (Princeton: Princeton University Press, 1972–), vol. 1. In China, technical fields were defined according to subject matter and procedural ideal, though with epistemological bases different from those of European disciplines. See Nathan Sivin, "Science and Medicine in Imperial China—The State of the Field," *Journal of Asian Studies* 47 (1988): 41–90, esp. 43–44.

improvement."[3] Scholarship on Chinese science, however, has shown that Chinese civilization was capable of considerable innovation. Further weakening Wittfogel's interpretation of Chinese culture is the lack of evidence that the calendar served more than ritual purposes.

While admitting that some of Wittfogel's general conclusions do not withstand scrutiny, we can still recognize the aptness of at least one of his specific points: the masters of hydraulic civilization were "singularly well-equipped to lay the foundations for two major and interrelated sciences: astronomy and mathematics."[4] In large part, the mensurational and computational techniques that were applied to cartography do seem to have developed in response to bureaucratic or "managerial" needs, especially in the areas of astrology, water conservancy, and cadastral survey. Measurement, of course, was useful in other areas of government activity—for example, city planning and road construction. But given the agricultural basis of the Chinese economy and the importance of ritual to the ruling elite, a focus on astrology, water conservancy, and land measurement is perhaps justified.

The political interests involved in astrology and water conservancy went beyond those Wittfogel suggests. The accurate prediction of celestial phenomena was linked to the fortunes of the state: according to the correlative cosmology that characterized much traditional Chinese thought, heavenly anomalies signified political malfeasance that could lead to loss of political authority (see pp. 88–89 and 209–10). Water conservancy was linked to the fortunes of the political establishment in a more literal way: canals served as a major means of transporting to the capital grain and other commodities collected as taxes, in addition to supporting the agriculture that generated those revenues. The *Shi ji* (Records of the grand historian) contains the following account of the uses of canals: "All the canals could be used for the passage of boats; if there was surplus water, then they were used for irrigation, so that the people could benefit from them. Wherever the canals passed, they drew water forth from them, using it to irrigate the ditches in their fields. There were tens and hundreds of thousands of these ditches—rather, an incalculable number of them."[5]

In state service, those involved in astrology and water conservancy employed a number of mensurational techniques that had cartographic applications. By the Han dynasty (206 B.C.–A.D. 220), properties of right triangles and circles and their applications in astronomy, especially for determining distances between celestial objects and the earth and calculating the periodicity of certain celestial phenomena, were already understood. One text, the *Huainanzi* ([Book of the] Master of Huainan), traces the origins of units of measurement to celestial phenomena: "The *cun* arises from the millet seed. The millet seed arises from the sun. The sun arises from form. Form arises from shadow. This is the root of measurement."[6] The relationships among sun, shadow, and measurement are more explicit in the *Huainanzi*'s description of the use of gnomons to determine the height of the sun (illustrated in fig. 5.1):

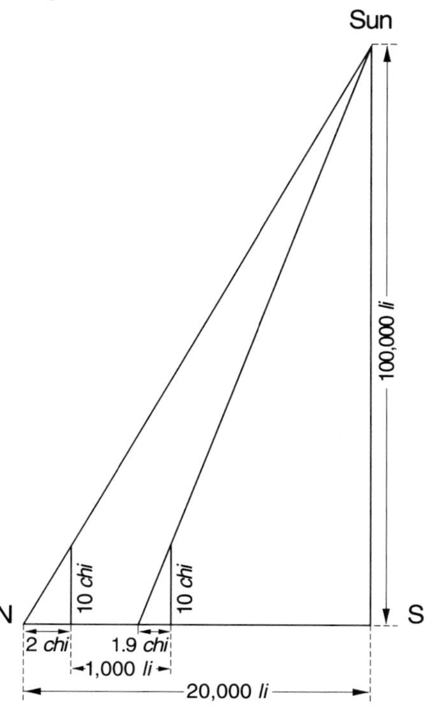

FIG. 5.1. CALCULATING THE HEIGHT OF THE SUN, ACCORDING TO THE *HUAINANZI* ([BOOK OF THE] MASTER OF HUAINAN).
After A. C. Graham, *Later Mohist Logic, Ethics and Science* (Hong Kong: Chinese University Press; London: School of Oriental and African Studies, University of London, 1978), 370.

> If you wish to know the height of the sky, plant gnomons 1 *zhang* [= 10 *chi*] tall directly north and south of one another, 1,000 *li* apart. Measure their shadows on the same day. If that of the northern gnomon is 2 *chi* long, and that of the southern gnomon 1 *chi* and 9 *cun* [10 *cun* = 1 *chi*], then every thousand *li* farther south the shadow shortens by a *cun*, and 20,000 *li* southward there is no shadow, since it is directly under the sun. Where the shadow is 2 *chi* and one obtains a height of 1 *zhang*, the height is five times the distance

3. Karl A. Wittfogel, *Oriental Despotism: A Comparative Study of Total Power* (New Haven: Yale University Press, 1957), 9.

4. Wittfogel, *Oriental Despotism*, 29 (note 3).

5. Sima Qian, *Shi ji* (completed ca. 91 B.C.), chap. 29; see the edition in 10 vols. (Beijing: Zhonghua Shuju, 1959), 4:1407.

6. *Huainanzi* (ca. 120 B.C.), attributed to Liu An (d. 122 B.C.), in *Huainanzi zhu* (Commentary on the *Huainanzi*, third century), ed. Gao You, chap. 9; see the modern edition (Taipei: Shijie Shuju, 1962), 141.

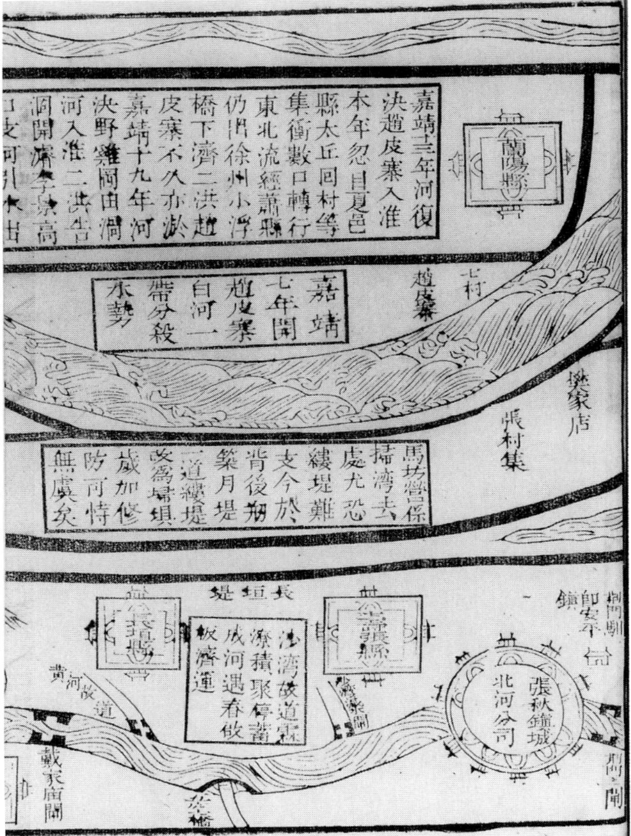

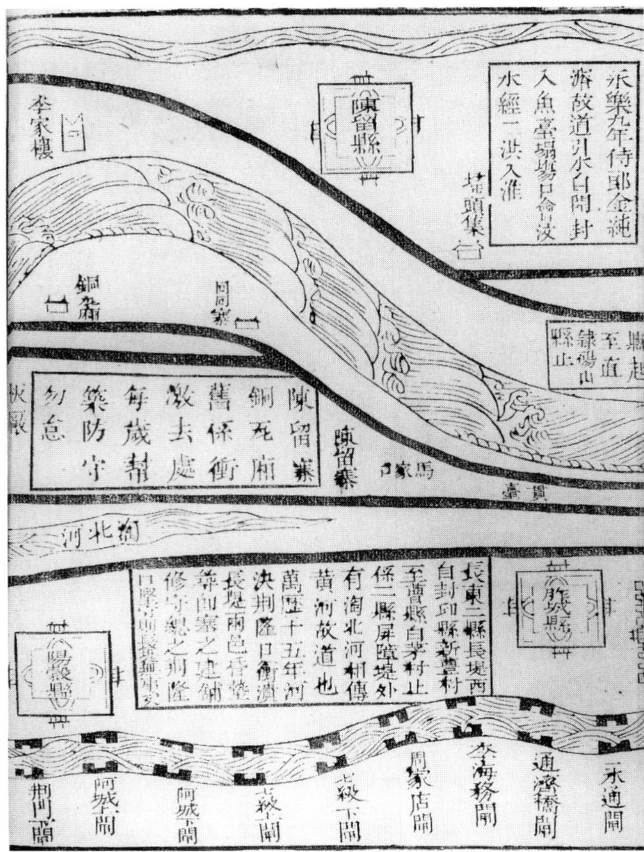

FIG. 5.2. PLANIMETRIC HYDROLOGIC MAP. Portion of a map of the Huanghe, or Yellow River (compare this with fig. 5.3 and plate 5). Size of each page: 19.5 × 14 cm. From Pan Jixun, *Hefang yilan* (General view of river control, 1590), 1748 edition, chap. 1, unpaginated. Reproduced courtesy of the Harvard-Yenching Library, Harvard University, Cambridge.

> southward. If you take the number of *li* from this point to the place under the sun [20,000 *li*] and multiply it by five, you get 100,000 *li*: this is the height of the sky.[7]

The properties of similar triangles used for indirect measurement in this example were also applied in cartography, as we will see below.

From textual sources, it appears that water conservancy projects, as well as land allocation schemes, were stimuli for techniques of direct measurement. During the Han, water conservancy was believed to have had a long association with mensuration: in the *Shi ji*, the emperor Yu is described as holding a plumb line, a carpenter's square (*ju*), and a pair of compasses (*gui*) while surveying for his work in flood control.[8] The *Zhou li* (Ritual forms of Zhou) purports to describe a Zhou dynasty (ca. 1027–256 B.C.) bureaucracy that had already standardized the depths and widths of canals—an action that presupposes the existence of mensurational techniques. Since at least the late Zhanguo (Warring States) period (403–221 B.C.), mensuration itself was regarded as linked to the well-being of the state. According to an account in the *Zuozhuan* (Zuo's tradition [of interpreting the *Chunqiu*, or Spring and autumn annals]), a text dating from about 300 B.C., mensuration served as an instrument of fiscal policy:

> Wei Yan was made marshal of Chu, and Zimu [the chief minister] commissioned him to regulate the gov-

7. *Huainanzi zhu*, chap. 3 (54) (note 6). The ancient Greeks also noticed that gnomon shadows shorten as one proceeds southward toward the tropic. Eratosthenes (ca. 275–194 B.C.) attributed this decrease to the curvature of the earth and used it to calculate the earth's circumference (see Germaine Aujac and the editors, "The Growth of an Empirical Cartography in Hellenistic Greece," in *The History of Cartography*, ed. J. B. Harley and David Woodward [Chicago: University of Chicago Press, 1987–], 1:148–60, esp. 154–56). In contrast, the *Huainanzi* ascribes the decrease in shadow length to the decrease in angle as one gets closer to the place directly under the sun. The calculation of the sun's altitude presupposes that the earth's surface is essentially level. The *Huainanzi* does give dimensions for the earth: "28,000 *li* east to west and 26,000 *li* north to south" (chap. 4 [56]), but how these dimensions were derived is not stated. Chinese notions of the shape of the earth are discussed later in this chapter.

8. *Shi ji*, chap. 2 (1:51) (note 5).

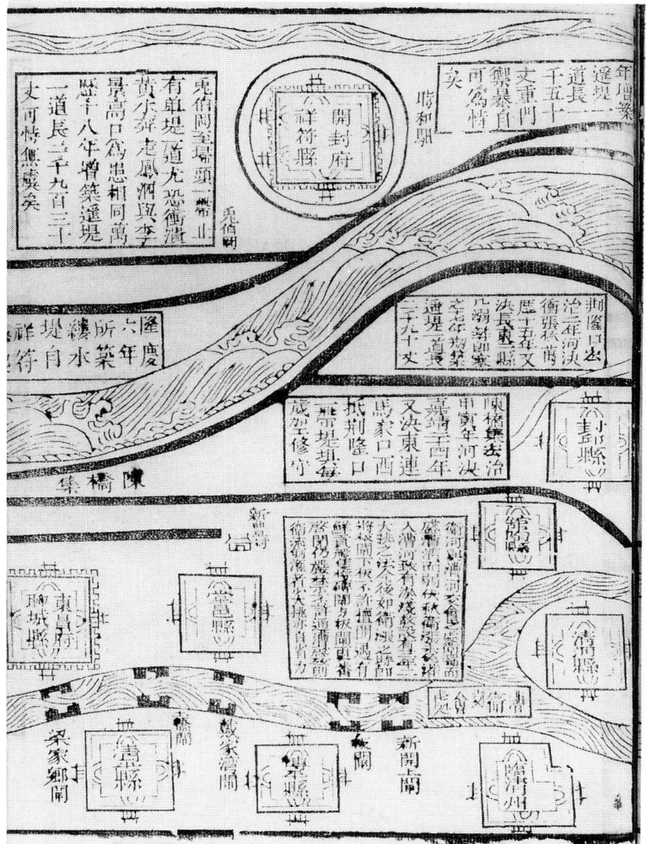

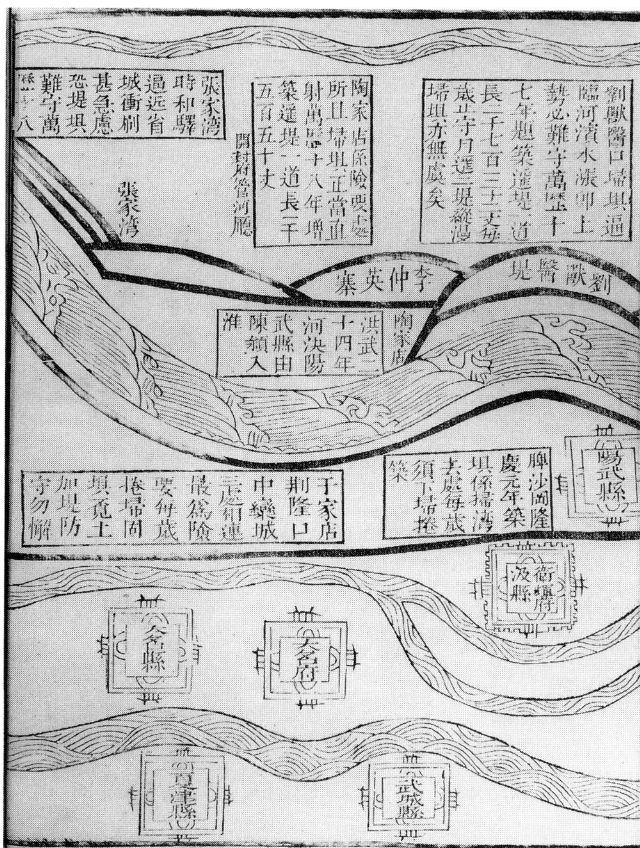

> ernment levies and count the number of cuirasses and weapons each fief had to contribute. . . . Wei Yan made a register of arable lands, measured the mountains and forests, calculated the area of marshlands, distinguished the highlands and the downs, listed salt tracts, enumerated the borders of flooded areas, measured the area of diked reservoirs, regularized balks to divide the plains between embankments, assigned low wetlands for pasturage, divided fertile land into units shaped like the graph *jing* [well; that is, into grids]. He determined the levies due, fixing the number of chariots and horses to be contributed, and assessing the numbers of chariot drivers, foot soldiers, and armored soldiers with shields that had to be mustered. He presented the results to Zimu. They were all proper.[9]

As this last example suggests, in China as elsewhere, government support for numerical techniques had as its primary purpose not the representation of nature, or reality, for its own sake, but the perpetuation of political power. The understanding of nature was secondary to the practical matter of maintaining the government. Land, for example, was measured as a means of calculating taxes owed. Information on ownership, extent of holdings, and boundaries was recorded in land registers. Those that survive from the Tang (618–907) consist entirely of text, but from the Song (960–1279) onward such registers were often accompanied by maps. The maps in the registers gave a sense of the configuration of the fields measured, while annotations and accompanying text provided the measurements and other information.

This dual form of record keeping could also be used to help prevent tax evasion, as has been mentioned elsewhere (p. 86). The Qing scholar Gu Yanwu (1613–82) transcribes this account of the situation under the Ming: "Local governments kept records of the number of male

9. *Zuozhuan* (ca. 300 B.C.), *Xiang* 25, in *Chunqiu jingzhuan yinde* (Concordance to the *Chunqiu* [Spring and autumn annals] and its commentaries), 4 vols. (1937; reprinted Taipei: Chengwen Chubanshe, 1966), 1:307. The translation here is based on that of James Legge in *The Chinese Classics*, 5 vols. (1893–95 editions; reprinted Hong Kong: Hong Kong University Press, 1960), vol. 5, *The Ch'un Ts'ew with the Tso Chuen*, 517.

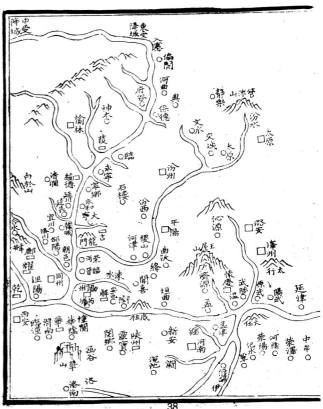

FIG. 5.3. PORTION OF A MAP OF THE YELLOW RIVER. The rivers are represented planimetrically and the mountains in a more pictorial form. Size of each page: ca. 14 × 11 cm. From Jin Fu, *Zhihe fanglüe* (Summary of river-control methods), ed. Cui Yingjie (1767; photo-reprinted Taipei: Guangwen Shuju, 1969), 38–39.

members and the property of each household. Each village kept maps of fields that were close together. [The imperial government] could check the maps to see whether fields were desolate or cultivated, and there was no covering up of the owner of each field."[10] Under the Ming, four sets of population records were made by local officials, one each for county, prefectural, and provincial governments. The fourth copy was sent to the imperial government and had a yellow cover to signify this, thus the generic term for these registers: *huangce*, or yellow books. The registers were often accompanied by cadastral maps, which depicted the boundaries of contiguous landholdings in a manner that resembled the scales of a fish. These maps were thus known as *yulin tu*, or fish-scale maps (see fig. 4.10 above).[11]

For presenting the results of cadastral surveys, both forms of representation, the textual and the cartographic, were regarded as essential. Gu Yanwu records this statement: "One can see that maps and registers complement each other, and neither is dispensable. Maps link people to the land, so they are the foundation of taxation. Registers list the farmland of each household, so that [the

10. *Zhenjiang fu zhi* (Gazetteer of Zhenjiang Prefecture [in present-day Jiangsu Province], date not given but presumably the 1596 edition), quoted in Gu Yanwu, *Tianxia junguo libing shu* (Treatise on the advantages and disadvantages of the commanderies and states of the empire, preface written in 1662) (1811; reprinted Taipei: Taiwan Shangwu Yinshuguan, 1981), 7.80a. As Gu recognizes, however, the system seldom worked to perfection. Local officials and clerks were susceptible to corruption and could be bribed to falsify records. In many instances the central government had to intervene: "In the twelfth year of the Hongwu reign period [1379], there was an investigation of the farmland all over the empire. [The government] sent students from the imperial academy to measure the fields, make maps, and number them" (*Zhenjiang fu zhi*, quoted in Gu, *Tianxia junguo libing shu*, 7.80a).

11. Sometimes such maps were included in the yellow books, and sometimes they made up registers of their own, *yulin ce* or *yulin tuce* (fish-scale registers or fish-scale map registers).

Previous scholarship has tended to use the fish-scale maps as a means of gauging the efficiency or inefficiency of the imperial tax collection process, while slighting their cartographic interest. As a result, there are no catalogs of fish-scale maps, though visitors to China have noted their existence in various libraries and museums. See Frederic Wakeman, Jr., ed., *Ming and Qing Historical Studies in the People's Republic of China* (Berkeley: Institute of East Asian Studies, University of California, Berkeley, Center for Chinese Studies, 1980). A useful survey of land tenure systems in Ming China is Chao Kang and Ch'en Chung-i (Zhao Gang and Chen Zhongyi), *Zhongguo tudi zhidu shi* (History of Chinese

FIG. 5.4. WATER CONSERVANCY WORKERS PERFORMING MEASUREMENTS FOR CUTTING A CANAL. This is a detail from a Qing illustration.
Size of the full page: 29 × 28 cm. From Linqing, *Hongxue yinyuan tuji* (Illustrated record of my life experiences—[traces of] a goose [treading on] snow, 1847), vol. 11, unpaginated. Reproduced courtesy of the Harvard-Yenching Library, Harvard University, Cambridge.

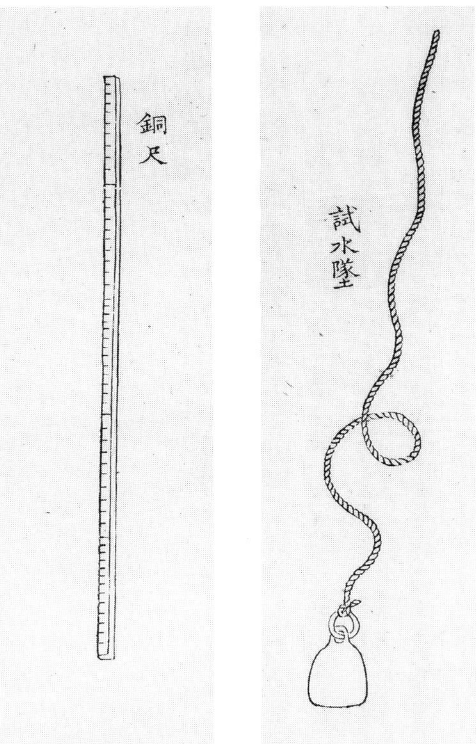

FIG. 5.5. EXAMPLES OF MEASURING TOOLS USED IN WATER CONSERVANCY. On the left is a graduated rod and on the right a plumb line.
Length of the originals: both ca. 16 cm. From Linqing, *Hegong qiju tushuo* (Illustrated explanation of the tools used in river works, 1836), 1.6a and 1.4a. Reproduced courtesy of the Harvard-Yenching Library, Harvard University, Cambridge.

government] can collect taxes and assign civic duties."[12] The dual system of record keeping described here was not unique to cadastral surveys. Similar systems in which text complemented map were also employed in other areas of government concern—for example, water conservancy.

Water Conservancy and Cartography

The state interest in water conservancy projects was accompanied by an interest in specialized hydrological maps representing rivers, canals, and their immediate surroundings. One might expect that mensuration in water conservancy projects would be transferred to maps—that water conservancy projects fostered the production of maps and plans reflective of careful measurement. In this case, however, history does not conform to expectations. Like the surviving corpus of geographic maps, hydrological maps manifest a wide range of representational modes (see figs. 5.2 and 5.3 and plate 5). Some of them are almost purely planimetric, making them suitable for the presentation of quantitative information, except that in some cases such maps lack scalar indications. More often, however, hydrological maps are eclectic in their manner of representation, some showing a mix of planimetric and pictorial modes and others almost completely pictorial. Like geographic maps, hydrological maps were often not intended to stand alone—they were usually accompanied by texts. In many cases, maps or diagrams were drawn to accompany memorials to the throne reporting on water conservancy projects.[13]

---

land-tenure systems) (Taipei: Lianjing Chuban Shiye Gongsi, 1982). On the yellow book system, one may consult Wei Qingyuan, *Mingdai huangce zhidu* (Yellow book system of the Ming period) (Beijing: Zhonghua Shuju, 1961).

12. *Zhenjiang fu zhi*, quoted in Gu, *Tianxia junguo libing shu*, 7.80b–81a (note 10).

13. The practice of appending maps or diagrams (*tu*) to memorials pertaining to water conservancy goes back at least to the Song. The *Song shi* (History of the Song, 1346) preserves at least one memorial stating that it is accompanied by a map or illustration (Tuotuo et al., *Song shi*, chap. 94; see the edition in 40 vols. [Beijing: Zhonghua Shuju, 1977], 7:2332). The map, however, is not reproduced. In addition, at least 200 memorials from the Qianlong reign period (1736–95), out of the roughly 34,000 reproduced by the National Palace Museum in

These memorials reported on such matters as the condition of embankments, the progress of construction and repair work, the expenses incurred for particular projects, and changes in flood conditions. In some cases officials seem to have carried out volumetric calculations to estimate the acreage that certain canals would be able to irrigate.[14] More often, they performed and recorded linear measurements of rivers, canals, and embankments (fig. 5.4). For these measurements, the officials could have used the graduated rods and plumb lines illustrated in a Qing text on hydraulics (fig. 5.5).

The attention officials paid to measurement is evident in these excerpts from memorials of the Qianlong period (1736–95):

> The east embankment of the Feng River should be subdivided again and raised up 2 or 3 *chi* in order to avoid overflowing. Moreover, within the 20 *li* at the southern embankment where the floodwaters collect in the lower reaches, reinforcement should be made to block river sediments.[15]

> Together we measured below the city of Xuzhou [in present-day Jiangsu Province]. The surface of the river was 20 or 30 *zhang* wide and the water 5 or 6 *chi* deep. We also measured the original flooding at Sunjiaji. The surface of the river was 203 *zhang* wide where two dams had already been built in the north and south. These works were 87 *zhang* long. Where dams were as yet unbuilt, the surface of the water was 116 *zhang*, and the water's depth varied from 3 or 4 *chi* to 1 *zhang* and 8 or 9 *chi*.[16]

> From the lakeshore to the base of the dam, there was originally a diversion canal. It measured 35 *li* in length. Beyond the dam, the diversion canal reached to the cliffs of the Yellow River. We measured the course to be 14 *li*. This time, though the lake's water was overflowing for 13 *li* around, we measured the water at lakeside to be 1 *zhang* and 8 or 9 *chi* deep. The water at the head of the diversion canal was only 2 or 3 *chi* deep.[17]

The memorials these passages were taken from all state that a map was appended for the emperor's reference. The Qianlong emperor seems to have attached great importance to the maps presented with memorials. He implicitly criticized officials for not submitting maps with memorials reporting on water conservancy projects, and in some cases he asked for explanation of things that were unclear on the maps submitted.[18] The emperor's orders for the execution of water conservancy work were often based on maps. One official, for example, memorialized that the emperor had used a map to direct work on the expansion of a waterway and thus averted danger.[19]

The maps that accompanied the memorials quoted above apparently have not been preserved, so we do not know what modes of representation were employed in them. What evidence there is suggests that pictorial representation was used to some extent. In the memorials, the verb most often used to describe their manufacture is *hui*, meaning "to paint," especially using color. In fact, one official describes the color symbolism employed on maps presented with memorials: "Within the dikes, [a map submitted previously] used a deep green. Within deep waters, it used a deep blue. For hidden sandbars, it used black ink. Each color was painted in a clear and distinct manner. The map presented this time uses only light colors to give a simple sketch and does not distinguish between shallow and deep water. It thus cannot be understood at a glance, and that being so, your highness ordered that maps presented subsequently by this governor should—according to the former style—be painted so as to distinguish the colors."[20] This memorial dates from 1778, well after the court had been introduced to the European style of cartography. The discrepancy between the two styles of painting described in the memorial suggests a lack of standardized practice within the bureaucracy. In addition, it seems that pictorial technique for the easy recognition of geographic features of interest was just as important as the quantitative information presented in the memorials. This is corroborated by maps now separated from the memorials they once illustrated: these maps are characterized by the use of color and pictorial elements (plates 6 and 7).[21]

The same pictorial representation can be found in maps included in hydrological treatises, where the division of labor is similar to that proposed for map and memorial. In hydrological treatises from the Qing (1644–1911), quantitative information, as well as historical facts, is found in verbal text, and maps provide a sense of the

Taiwan, say that they are presented together with maps or diagrams. These memorials appear in *Gongzhongdang Qianlong chao zouzhe* (Palace memorials from the Qianlong reign period in the palace archives), 69 vols. (Taipei: Guoli Gugong Bowuyuan, 1982–88).

14. See, for example, *Qianlong chao zouzhe*, 1:385–387 (note 13).

15. *Qianlong chao zouzhe*, 2:109 (note 13).

16. *Qianlong chao zouzhe*, 15:656 (note 13).

17. *Qianlong chao zouzhe*, 48:617 (note 13).

18. See, for example, *Qianlong chao zouzhe*, 37:863, 45:662, 6:10 (note 13). We do not have the emperor's exact words, but the officials' responses to them note that the emperor commented that maps were lacking in previous memorials. The officials also promise to attach maps to future memorials on water conservancy projects.

19. *Qianlong chao zouzhe*, 27:156 (note 13).

20. *Qianlong chao zouzhe*, 43:552 (note 13).

21. Under the Qing palace memorial system, memorials and maps were preserved separately, thus the difficulty in matching maps with their corresponding memorials. For further discussion of the relation between map and picture, see chap. 6 below, esp. pp. 139–53.

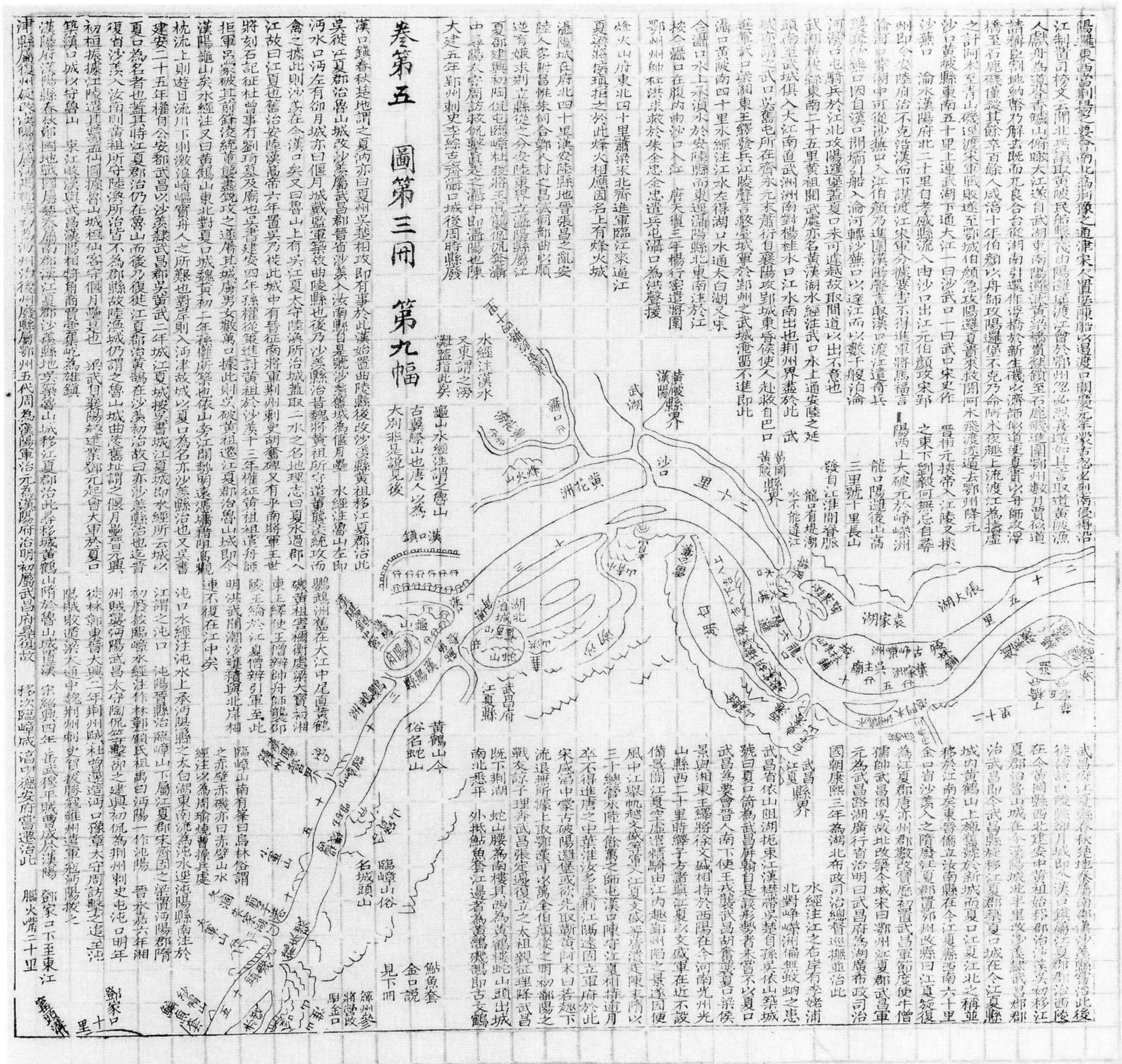

FIG. 5.6. TEXT AND IMAGE ON A QING MAP OF THE CHANGJIANG, OR YANGTZE RIVER. The text on the map gives historical and geographic information, referring to previous scholarship. Size of the original: 26.5 × 27 cm. From Ma Zhenglin, *Changjiang tushuo* (Illustrated account of the Yangtze, 1871), chap. 5, map 9. Reproduced courtesy of the Harvard-Yenching Library, Harvard University, Cambridge.

physical appearance of the area of interest.[22] In some cases the cartographic image is almost overwhelmed by text (fig. 5.6). More typical is a division of map and text, as in the *Xingshui jinjian* (Golden mirror of flowing waters), a hydrological work dating from 1725. It opens with a section of maps predominantly pictorial in their mode of presentation (fig. 5.7). The work is predominantly text, however, and the author's reverence for text is made clear in the narrative. The main descriptions are taken from classic texts such as the "Yu gong" (Tribute

22. A similar situation obtains in the use of maps in legal cases in Europe since the late medieval period. The map does not serve as the final authority. Even in present-day international boundary legislation, lawyers prefer to verbalize, not map.

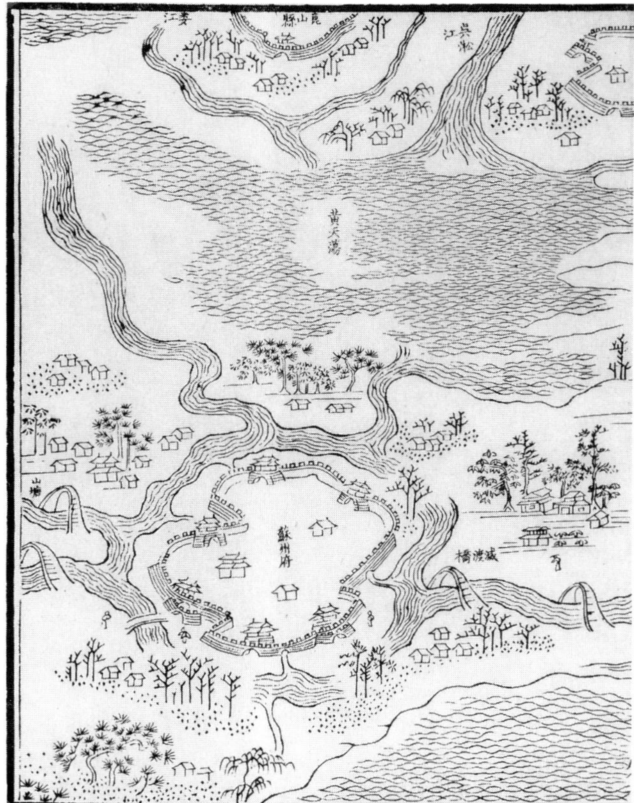

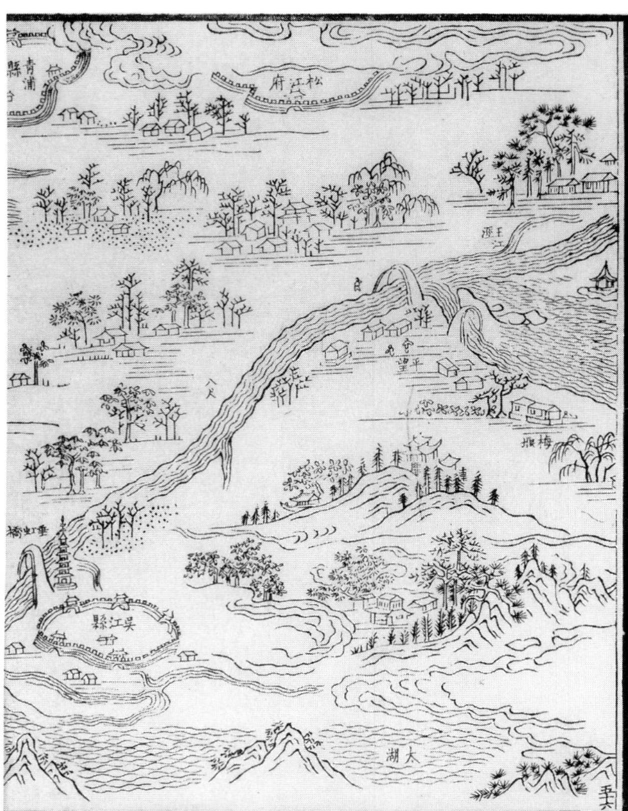

FIG. 5.7. SECTION OF A QING MAP OF THE YELLOW RIVER.
Size of each page: 18 × 13.5 cm. From Fu Zehong, *Xingshui jinjian* (Golden mirror of flowing waters, 1725), 1.55a–56b. Reproduced courtesy of the Harvard-Yenching Library, Harvard University, Cambridge.

of Yu), the *Shui jing* (River classic), and the commentary on the *Shui jing*. The author of the *Xingshui jinjian* intrudes only to supplement, correct, or update information given in a prior text.

The *Xingshui jinjian* was compiled at roughly the same time as the Jesuit atlas of China, which was based primarily not on textual research, but on an actual survey of much of the empire. The maps in the atlas are uniformly planimetric in presentation, bear expressed scales, and indicate latitude and longitude. Looking at the atlas alone, one might be tempted to conclude that Chinese cartography had joined European cartography and left text behind. This view of Chinese cartography is untenable. As the preponderance of text in the *Xingshui jinjian* suggests, one also finds a strong interest in text among Chinese mapmakers even during the Qing.

Evidentiary Scholarship and Cartography

The late Ming (1368–1644) and early Qing dynasties were marked by what Sivin, Elman, and others have described as a revolution in Chinese intellectual discourse, the rise of *kaozheng* (evidential research).[23] At the heart of evidential scholarship was a faith in textual research as a ground for knowledge—primarily knowledge of what was meant in canonical texts. Textual scholarship was a means of recovering the past, a pursuit of prime importance, since the way of the ancients was considered the

23. See, for example, Nathan Sivin, "Wang Hsi-shan," in *Dictionary of Scientific Biography*, ed. Charles Coulston Gillispie, 16 vols. (New York: Charles Scribner's Sons, 1970–80), 14:159–68, esp. 160–61; Benjamin A. Elman, *From Philosophy to Philology: Intellectual and Social Aspects of Change in Late Imperial China* (Cambridge: Council on East Asian Studies, Harvard University, 1984); Liang Ch'i-ch'ao (Liang Qichao), *Intellectual Trends in the Ch'ing Period*, trans. with introduction and notes by Immanuel C. Y. Hsü (Cambridge: Harvard University Press, 1959) (translation of *Qingdai xueshu gailun* [Survey of Qing scholarship, 1921]); and R. Kent Guy, *The Emperor's Four Treasuries: Scholars and the State in the Late Ch'ien-lung Era* (Cambridge: Council on East Asian Studies, Harvard University, 1987), 39–49. For an analysis of the *kaozheng* movement as the logical outcome of preceding intellectual trends, not a departure from them, see Yu Yingshi (Ying-shih Yü), "Qingdai sixiang shi di yige xin jieshi" (New interpretation of Qing intellectual history), in his *Lishi yu sixiang* (History and thought) (Taipei: Lianjing Chuban Gongsi, 1976), 121–56.

pattern for the present. One of the main factors in the rise of this type of scholarship was dissatisfaction with Song and Ming Neo-Confucian discourse, which its detractors claimed imparted to the classics a metaphysical significance inconceivable to their original authors. Also contributing to the development of evidential research was contact with the West. Confronted with European mathematical and scientific achievements, Chinese scholars reexamined their past in an attempt to rediscover their own traditions of astronomy and mathematics. There were those who claimed that Western mathematics and astronomy were actually of Chinese origin—thus ancient Chinese mathematics and astronomy were part of the lost past that needed to be reconstructed.[24]

Given the variety of the kinds of texts to be used in reconstructing the past, it was almost inevitable that some academic specialization occurred—scholars often focused their energies on texts from specific fields, such as mathematical astronomy, mathematics, and history. Regardless of specialty, the evidential scholars used a common method, one based on the examination and comparison of textual sources.

In efforts by *kaozheng* scholars to recover the past, geography had its place—historical geography was a means of reconstructing the past. In this research, maps were used as documentary sources and could also serve as a means of presenting research results, for example, in studies of ancient places. One such study was undertaken by the scholar Xu Song (1781–1848), in the *Tang liang jing chengfang kao* (Study of the walled cities and wards of the two Tang capitals). In the preface to his study Xu explains his motivation: "I delight in reading the *Jiu Tang shu* [Old history of the Tang] and Tang fiction. Whenever they mention palaces and gardens with their crooks and bends and alleys and lanes that fork and branch out, I reach for the *Chang'an zhi* [Gazetteer of Chang'an, by Song Minqiu (1019–79)] to verify them. What I always get are errors, and with the eastern capital the mistakes are similar."[25] Part of the rationale behind

24. See John B. Henderson, "Ch'ing Scholars' Views of Western Astronomy," *Harvard Journal of Asiatic Studies* 46 (1986): 121–48, esp. 138–43.

25. Xu Song, *Tang liang jing chengfang kao* (1848); *Baibu congshu jicheng* edition, preface, 1a.

Xu's study was to correct errors in previous descriptions of the Tang capitals. As might be expected of a work of *kaozheng* scholarship, Xu's method of verification involves textual research, the collation and comparison of a variety of documentary sources: histories, gazetteers, encyclopedias, maps, and inscriptions.

In presenting his research results, Xu makes use of maps, though the great bulk of his work consists of verbal description. In using maps, Xu says he is following ancient precedent: "In antiquity, those engaged in scholarship had maps on the left and history on the right. Maps must be interrelated with history."[26] In accord with the tenets of *kaozheng* scholarship, Xu conceives of maps as tools for historical reconstruction. The intimate connection Xu sees between map (or chart) and history may be a residuum of the earlier Qing reaction against the abstract geography of those Neo-Confucian intellectuals who were more interested in geometric regularity and metaphysical correspondences than in empirical reality and who saw no problem with constructing charts of the nine regions (*jiu zhou*) with no reference to geographic reality. In opposition to this practice, Xu affirms that maps should have a basis in historical actualities. Xu is interested in historical reconstruction not for its own sake, but as a means of clarifying ancient precedent. The purpose of his work, he says, is to serve as "an aid in singing praises of the poetic compositions by Tang exemplars."[27] By Xu's time, Tang poetry had long been regarded as the high point in Chinese literary history, and Xu intends his study as a work of literary scholarship—it is supposed to further one's appreciation of Tang literature.

Much of Tang literature, poetry and fiction, is set in one of the two capitals, Chang'an or Luoyang, which attracted literati seeking careers in government. From the descriptions in Xu's study, one can infer that his study was meant to be used to reestablish the connection between Tang literary works and historical actuality. True to what he says in his preface, Xu provides, for example, a plan and narrative description of the palace city, once used as the residence of the emperor and his court. The plan (fig. 5.8) has no scale indications and seems to serve mainly to indicate the relative positions of various landmarks. More detailed information including measurements is provided in the text, a typical passage from which reads: "The palace wall is 4 *li* from east to west and 2 *li* and 270 *bu* from south to north. In perimeter it is 13 *li* and 180 *bu*. Its height is 3 *zhang* and 5 *chi*."[28] Such descriptions, drawing from dynastic histories, gazetteers, encyclopedias, anecdotes, inscriptions, maps, and other sources, account for almost all of Xu's study.

The maps and chorographic information are supposed to be in the service of literary criticism, but rarely does Xu cite the Tang poetic and fictional works his study is supposed to illumine. Occasionally he quotes a few lines from a poem written at a particular place, and sometimes he merely notes that a certain poet wrote a poem about a certain place. Beyond that, Xu does not discuss how the maps and chorographic information he provides increase one's appreciation for literature. For this reason, Xu's study might seem to fall short in comparison with modern works of literary scholarship, which tend to quote extensively from and to closely analyze the literary texts under study. Xu, however, may be excused for forgoing such quotations and close readings. Readers of Chinese poetry in imperial times expected that poems set in a city would incorporate accurate descriptions of that city. Thus a major aim of literary criticism was to determine whether a poem's description of place was accurate, for example, verifying what was actually visible from a given location. It is therefore not surprising that Xu does not quote a great deal of poetry. To depend on the poetry would have defeated the book's critical enterprise.

The usefulness of cartography as textual criticism, though important for historical reconstruction, was also recognized in projects that aimed at further understanding of the present. The compilation of the imperially sponsored *Da Qing yitong zhi* (Comprehensive gazetteer of the Great Qing realm, completed 1746) involved a number of textual scholars, who collected and compared documents containing geographic information. The gazetteer includes maps, but the bulk of the work consists of verbal descriptions. The maps in the gazetteer's last revised edition, completed in 1820 and printed in 1842, have no grids or meridians and parallels and no expressed scale. It seems that even after Matteo Ricci's world map and after the Jesuit survey of China, a considerable group of Chinese intellectuals still remained unconvinced of the advantages of Western European cartography. Maps, it appears, were useful in providing a quick grasp of the spatial relationships between the landmarks represented, but for detailed information about distance and direction, the preferred means of presentation remained verbal description.

This lack of attention to cartographic representation by Chinese scholars was pointed out in a work roughly contemporaneous with the last revised edition of the *Da Qing yitong zhi*. This work was the *Yinghuan zhilüe* (Short account of the maritime circuit), a world geography completed in 1848 by Xu Jiyu (1795–1873). Xu, who served as governor of Fujian Province, was writing in the aftermath of the Opium War, which demonstrated the empire's vulnerability to foreigners it had regarded as inferior. In order to deal with the new challenge from without and to accurately assess their place in the world,

26. Xu, *Tang liang jing chengfang kao*, preface, 1a (note 25).

27. Xu, *Tang liang jing chengfang kao*, preface, 1a (note 25).

28. Xu, *Tang liang jing chengfang kao*, 1.1a–b (note 25).

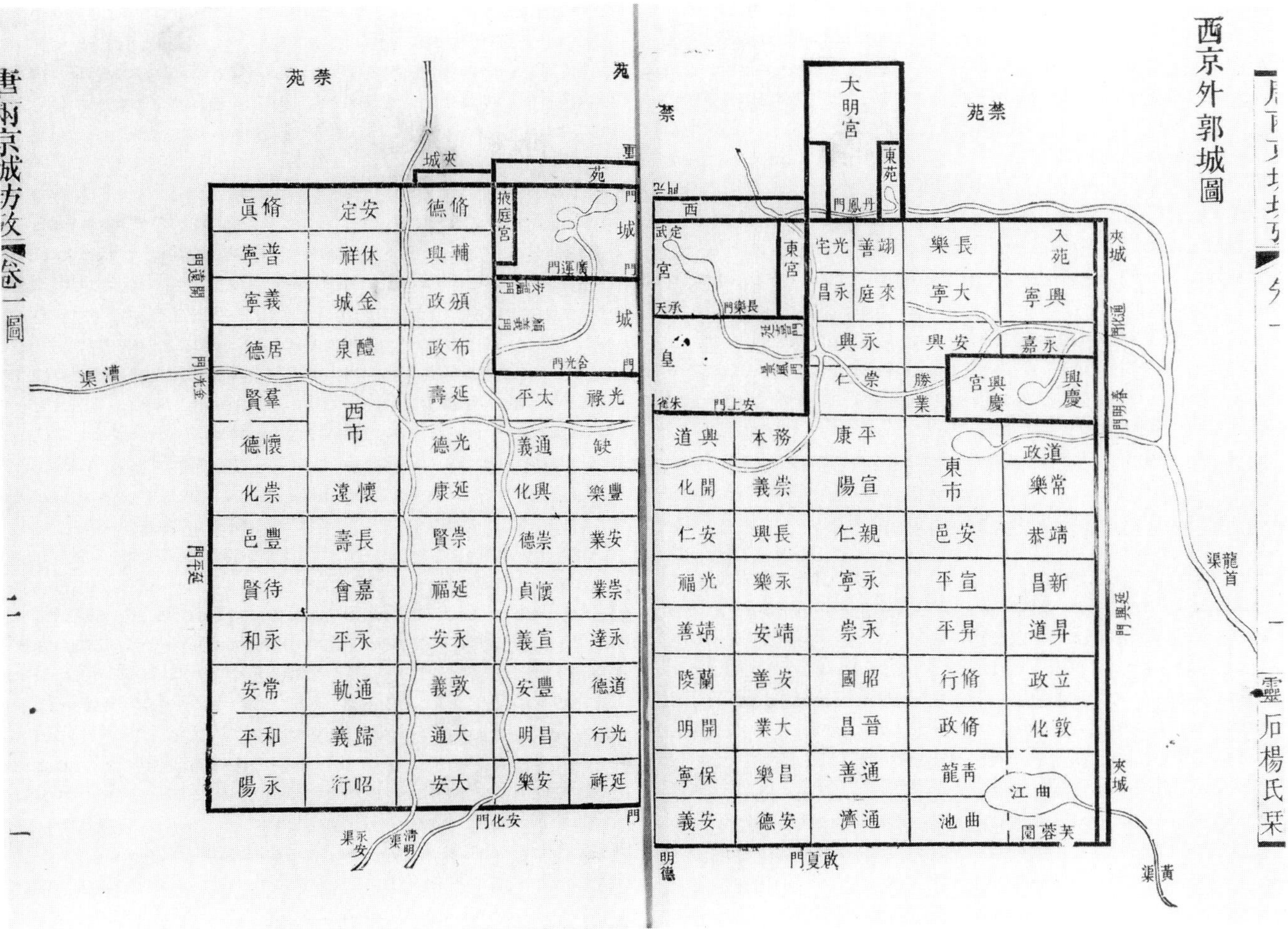

FIG. 5.8. MAP OF CHANG'AN DURING THE TANG DYNASTY.
Size of the original: 26.5 × 29 cm. From Xu Song, *Tang liang jing chengfang kao* (Study of the walled cities and wards of the two Tang capitals, 1848), *tu* 1b–2a. Reproduced courtesy of the Harvard-Yenching Library, Harvard University, Cambridge.

the Chinese needed reliable information about their competitors. This is precisely what Xu Jiyu set out to provide.

In the *Yinghuan zhilüe* Xu countered at least one scholarly tendency of his time—what has been called an inward turn in scholarship during the Qing. The emphasis of *kaozheng* scholarship, as I explained above, was Chinese texts and their role in the reconstruction of the Chinese past. Xu did not reject *kaozheng*, but he extended its application to foreign sources. For his study he did use Chinese sources, such as the dynastic histories, but he "repeatedly sought and collected various kinds" of Western sources. When the accounts in these sources diverged, he followed the most recent, and whenever he met a Westerner, he "would at once open a volume so that person would verify it regarding the topography and present conditions of the various countries beyond our territory."[29] Xu goes on to say: "Whenever I received a book or there was some new information, I at once revised the draft or added to it, so that it was all changed dozens of times."[30]

Although most of his study consists of verbal description, as was typical of Chinese geographic studies, Xu stresses the centrality of maps to his work: "This book takes maps as its leading principle."[31] Chinese maps, however, Xu implies, cannot compare with those of Westerners: "Westerners are good at traveling afar. Their sails and masts encircle the four seas. Wherever they land, they at once pull out pens and draw maps. Therefore their maps are the only ones that are reliable." Xu's dissatisfaction with native cartography stems not only from comparison with foreign models but from what he per-

29. Xu Jiyu, *Yinghuan zhilüe* (completed 1848) (1850; reprinted Taipei: Jinghua Shuju, 1968), preface, 8a. The type of volume opened is not specified.

30. Xu, *Yinghuan zhilüe*, preface, 8b (note 29).

31. Xu, *Yinghuan zhilüe*, "Fanli" (principles), 1a (note 29).

ceives as lack of correspondence to observed reality: "Geography without maps is unclear, but maps not based on observation are incomplete in detail. The world has a form; one cannot simply imagine its extensions and contractions."[32] The challenge to the schemes of abstract geography propounded by Song metaphysicians is clear from the empiricism implied by Xu's emphasis on observation as a basis for mapping. Given Xu's views regarding the merits of Western maps, it is not surprising that Western models served as the basis of all but one of the forty-two maps in the *Yinghuan zhilüe*: "The maps here have been copied in outline from original maps in original books of Westerners. The courses and arteries of rivers are as fine as hair; mountain ranges and cities, large and small, are exhaustively presented."[33]

Xu offered his work ostensibly as a geographic treatise, but his intent went beyond presenting information. The book's political implications were recognized by Xu's colleagues. Those who contributed prefaces to the work praised it for correcting the falsehoods in past Chinese accounts of foreign countries and said it would help the country deal more effectively with foreigners. The ideological import of Xu's work is apparent from the beginning—with its world map. The body of the *Yinghuan zhilüe* begins with a map of the Eastern Hemisphere, in which China appears in the upper right quadrant as a country smaller than Africa (fig. 5.9): thus Xu's early readers saw that China had to adjust its foreign policy.

More than two centuries after the Jesuits had helped to propagate the notion of a round earth among the Chinese, Xu still believed he must educate his readers, members of the literate elite, about the true shape of things: "The shape of the earth is like a ball; on the basis of the period of one revolution, it is divided into warp and woof, and lines are drawn vertically and horizontally across it. Each revolution equals 360 degrees; each degree is equal to 250 Chinese *li*. The sea covers more than six-tenths [of the earth], and land less than four-tenths."[34] Later Xu explains that the earth's landmasses, according to Western divisions, consist of four continents: "Of the four continents, Asia is the largest, and China is located in its southeast portion."[35] Xu seems to have recognized, as Drake has suggested, that the worldview he was offering would meet resistance from those committed to the idea of China's political centrality and geographic importance.[36] Thus he points out that the extent of the Qing empire has never been matched in the past and that China is master among the countries of Asia. Apparently as a concession to those who believed in Chinese preeminence, he also follows his general description of the world with a map of the Qing empire accompanied with a brief geographic description. He will not deal with the empire in detail, because "it is not fitting to speak of it in a history of foreign places."[37] By excluding China from his account of the world, Xu gained a rhetorical advantage: he could, as Drake has observed, avoid explicit comparisons of China with powerful Western nations.[38] The necessity for Xu's rhetoric of silence becomes clear further on in his work: the major power in the world, Xu makes clear, is England, not China.[39]

The praise for England contains implicit criticism of the Chinese government, which had just been humbled by the British. Xu Jiyu's work represents a contribution to a political debate about China's future: the description of England, for example, implies that China might do well to adopt Western technology if it is to match the military power of countries seeking to open China to more trade, and in his choice of maps itself, Xu makes an assertion regarding the state of cartography in China in comparison with that of the West. In copying Western-made maps, however, Xu Jiyu does not simply advocate a "scientific" cartography, one intended to present information dispassionately. Maps, as Xu uses them, constitute a form of political expression—they function as texts with rhetorical intent. This is not surprising in a setting in which—as in England during the nineteenth century—statecraft, not science, was regarded as the highest calling of the intellectual. The best preparation for that calling was not technical, but literary.

The textualism characterizing much Qing cartography is hardly an anomaly, but it has usually been overlooked in favor of the Manchu dynasty's adoption of European techniques. It is, however, a product of the same institutional setting as are the mathematical and mensurational techniques mentioned above. The main practitioners of those techniques were not primarily scientists in any modern sense of the word, but traditional scholars serving political aims. This is not to minimize the importance of quantitative techniques in cartographic theory and practice, but merely to point out an aspect of Chinese cartography often overlooked, one related to the emphasis placed in political life on the written word as a medium of communication.

32. Xu, *Yinghuan zhilüe*, preface, 8a (note 29).

33. Xu, *Yinghuan zhilüe*, "Fanli," 1a (note 29).

34. Xu, *Yinghuan zhilüe*, 1.4a (note 29). On Chinese conceptions of the shape of the world, see below, pp. 117–24.

35. Xu, *Yinghuan zhilüe*, 1.5b (note 29).

36. Fred W. Drake, *China Charts the World: Hsu Chi-yü and His Geography of 1848* (Cambridge: East Asian Research Center, Harvard University, 1975), 58–59.

37. Xu, *Yinghuan zhilüe*, 1.11a (note 29).

38. Drake, *China Charts the World*, 68 (note 36).

39. Among the attractions of England for Xu Jiyu are its naval power and the character of its people: "Their plans are clever and precise, their actions resolute, and their spirit brave and doughty. This places them at the top of the several European countries" (Xu, *Yinghuan zhilüe*, 7.49a [note 29]).

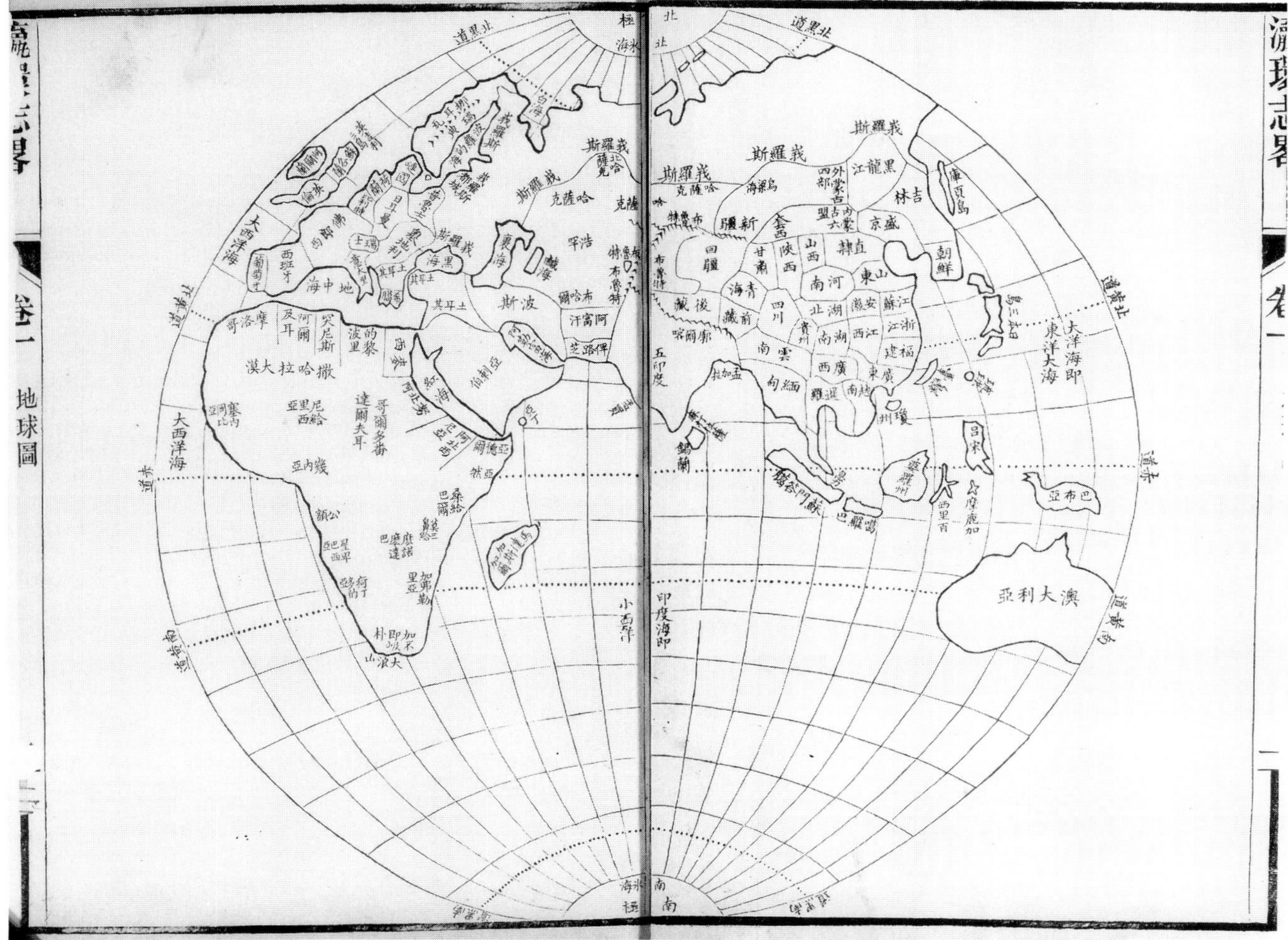

FIG. 5.9. NINETEENTH-CENTURY MAP OF THE EASTERN HEMISPHERE.
Size of the original: 28 × 36 cm. From Xu Jiyu, *Yinghuan zhilüe* (Short account of the maritime circuit, completed 1848, printed 1850), 1.1b–2a. Reproduced courtesy of the Harvard-Yenching Library, Harvard University, Cambridge.

Maps, Measurement, and Text

As a result of that emphasis, traditional Chinese cartography is marked by the coexistence of two major tendencies. These two tendencies can be termed mensurational (or more broadly, observational) and textualist. The latter strand can be seen as having two aspects: first, a reliance on texts as sources of information in the compiling of maps and, second, a reliance on text to complement the presentation of information in maps. The mensurational and textualist tendencies of Chinese cartography are apparent early in its history.

During the Han dynasty at least some intellectuals were conceiving mapmaking in terms of mathematical proportion. They recognized that ratios, or scales, could govern the representation of actual distances on a map. In a passage from the *Han shu* (History of the Former Han), Liu An discusses the idea of map scale in connection with military strategy: "When one uses a map to inspect hills, streams, and strategic passes [of the kingdom of Min-Yue, in present-day Fujian], distances are no more than several inches, but are really several hundred or thousand *li*. Obstructions, narrow passes, forests, and woods cannot all be recorded. When you look at a map, [the route] looks easy, but to traverse it is extremely difficult."[40] Liu An was noted for his patronage of scholarship, but the extent of his involvement with mapmaking is hard to ascertain. Scale was often considered a means of achieving correspondence with observed reality, but here Liu An recognizes that it can lead to distortion of that reality: the choice of scale may limit the number of features a map can represent. The compendium attributed to Liu

40. Ban Gu, *Han shu* (compiled first century A.D.), chap. 64A; see the edition in 12 vols. (Beijing: Zhonghua Shuju, 1962), 9:2778.

An, the *Huainanzi*, illustrates the limitations of maps in a different way. The book's sections on cosmology and geography give an account of world geography but include no maps, relying solely on verbal description. That description poses cartographic difficulties. The *Huainanzi*, Major has shown, gives such divergent information, particularly about the center of the world, that it is impossible to draw a map fully consistent with the text.[41]

Number and Text in Pei Xiu's Cartography

The problem of correspondence between map and reality was also taken up by Pei Xiu (223–71). While serving as minister of works (*sikong*) under the Western Jin (265–317), Pei Xiu compiled the *Yu gong diyu tu* (Regional maps for the "Yu gong" [Tribute of Yu]), now lost. A portion of the preface to that work, however, is preserved in the *Jin shu* (History of the Jin), compiled during the Tang. In the preface, Pei Xiu echoes Liu An's dissatisfaction with Han maps: "None gives a complete record of the famous mountains and great rivers. Although they roughly depict forms, this is not done carefully, and one cannot rely on them."[42] Pei Xiu's major criticism of Han maps is their lack of correspondence to observed reality: they record things that do not exist, incompletely record things that do, and are not drawn with precision. In discussions of Pei's proposed correctives to this state of affairs, historians of Chinese cartography have focused on Pei's six principles of mapmaking: proportional measure (*fenlü*), standard or regulated view (*zhunwang*), road measurement (*daoli*), leveling (or lowering) of heights (*gaoxia*), determination of diagonal distance (*fangxie*), and straightening of curves (*yuzhi*).[43] These are commonly interpreted as involving various types of direct and indirect measurements.

The first principle, proportional measure, seems to involve map scale. It is "a means by which the units of measurement [in the map] are determined" and preserves "the actualities of distance." If a map is made without a proportional measure, "it will not distinguish between what is far and near." The second principle, standard or regulated view, has to do with directional orientation: it deals with the positions of points on the map and their relation to one another. In other words, it preserves "the actualities of relative positions."[44] If the principle is correctly applied, "the curved and straight and the far and near can conceal nothing of their form."[45] The third principle, road measurement, is "a means of determining distance from a point of origin" along a given route: road measurement preserves "the actualities of paths and roads."

The other three principles—leveling of heights, deter-

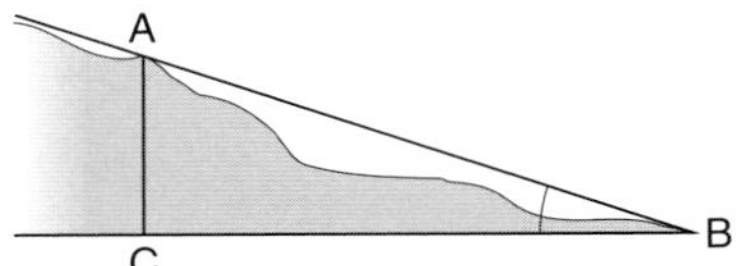

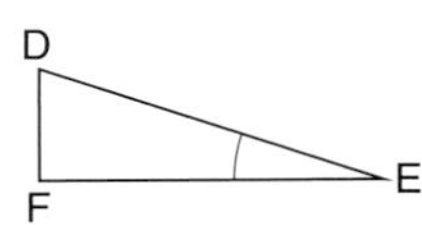

FIG. 5.10. INTERPRETATION OF PEI XIU'S METHOD OF LEVELING HEIGHTS. Three of Pei Xiu's six principles seem to involve the application of geometric principles as a means of indirect measurement. The use of right-triangle geometry seems to be called for by the fourth and fifth principles, *gaoxia* (lowering or leveling of heights) and *fangxie* (determination of diagonal distance). A situation involving the leveling of heights may be pictured as in this figure. Distance AB along an elevation can be measured directly, but the actual horizontal distance between points A and B is distance BC. Distance BC may be calculated by creating a right triangle similar to triangle ABC, for example, DEF, with angle FED equal to angle CBA. Since distance AB and all the dimensions of triangle DEF are known, the distance BC can be derived by applying the principle that corresponding sides of similar right triangles are proportional. This interpretation of *gaoxia* has some contextual support. At least one Han dynasty text, the *Jiuzhang suanshu* (Nine chapters on mathematical art), contains problems that require the use of similar right triangles.

41. See John S. Major, "The Five Phases, Magic Squares, and Schematic Cosmography," in *Explorations in Early Chinese Cosmology*, ed. Henry Rosemont, Jr. (Chico, Calif.: Scholars Press, 1984), 133–66, esp. 133–37. A more detailed account of Chinese cosmography appears in chapter 8 below.

42. Fang Xuanling et al., *Jin shu* (compiled 646–48), chap. 35; see the edition in 10 vols. (Beijing: Zhonghua Shuju, 1974), 4:1039. Some modern scholars would dispute Pei Xiu's characterization of Han maps, especially based on the silk maps discovered at Mawangdui. In Pei's favor, however, the scale of the Mawangdui maps varies and detail decreases as one moves from the center toward the edges. For further discussion of the Mawangdui maps and other Han specimens, see pp. 40–46 and 147–51.

43. See *Jin shu*, chap. 35 (4:1040) (note 42). In this exposition of Pei Xiu's six principles, all quotations from Pei's preface are from this source unless otherwise noted. The translations of Pei's terms are tentative, since it is sometimes unclear exactly what he intended. None of Pei's own cartographic work, which could be expected to illustrate his principles, survives. For another translation, see Joseph Needham, *Science and Civilisation in China* (Cambridge: Cambridge University Press, 1954–), vol. 3, with Wang Ling, *Mathematics and the Sciences of the Heavens and the Earth* (1959), 539–40.

44. Here the translation follows the excerpt from Pei Xiu's preface quoted in the *Yiwen leiju* (Classified compendium of arts and letters), comp. Ouyang Xun (557–641), chap. 6; see the edition in 2 vols. (Beijing: Zhonghua Shuju, 1965), 1:101. Ouyang Xun is a contemporary of the compiler of the *Jin shu*, Fang Xuanling (576–648), but his version of Pei Xiu's preface has some phrases not included in the *Jin shu*.

45. Interpretations of the principle of *zhunwang* vary. It has been interpreted as involving the use of the magnetic compass to determine direction, but this is unlikely since the use of the compass for wayfinding is unattested for Pei Xiu's time. The principle has also been interpreted as involving the use of a cartographic square grid. This interpretation is also unlikely, for reasons given later in this chapter when the grid is discussed. At that time I propose a third interpretation of the principle.

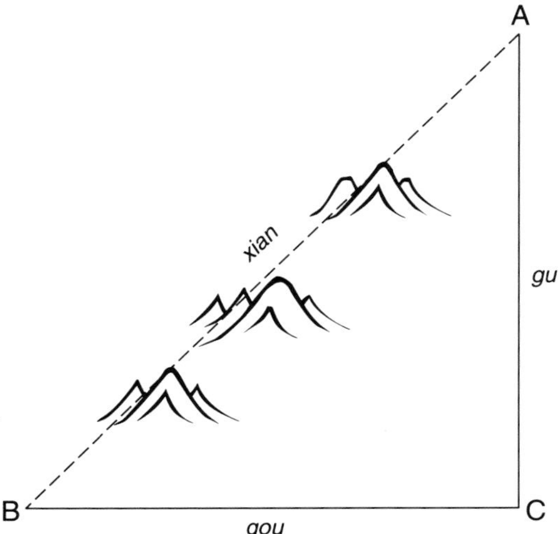

FIG. 5.11. INTERPRETATION OF PEI XIU'S METHOD OF DETERMINING DIAGONAL DISTANCE. Pei's fifth principle seems to require knowledge of the *gougu* (*gou* = base, *gu* = altitude) theorem, the equivalent of the Pythagorean theorem. A problem whose solution requires the *gougu* theorem appears in the *Zhoubi suan jing* (Arithmetical classic of the Zhou gnomon), which attained its present form no earlier than 200 B.C. As expressed by Zhao Junqing, the Han commentator on the *Zhoubi suan jing*, the *gougu* theorem may be used to derive the length of the hypotenuse: "Multiply base [*gou*] and height [*gu*] by their own values, and add these results to get the square of the hypotenuse [*xian*]. Divide this by its square root. This is the hypotenuse." This appears to be the calculation required to determine diagonal distance. Application of the principle seems to involve envisioning two points separated by inaccessible terrain as opposite vertices of a square (*fang*) and then using the *gougu* theorem to determine the length of the diagonal (*xie*) between them. In other words, distance AB across a mountain range can derived by taking the square root of the sum of the squares of distances AC and CB.

mination of diagonal distance, and straightening of curves—are explained even more concisely. Instead of dealing with them individually, Pei Xiu explains them collectively: "The last three principles are applied according to the nature of the terrain, so that differences between hills and plains are taken into account." These three rules seem to deal with the problem of converting actual ground distances, which may be curved in both horizontal and vertical dimensions, to straight-line distances and depicting them on a flat map. Pei Xiu does not provide any information on the way these conversions are to be performed, but he does list the benefits of following them: "The actualities of distance measure are preserved by the leveling of heights, the determination of diagonal distances, and the straightening of curves. Although there are obstacles of steep mountains and vast seas, distant places in inaccessible and strange lands, routes that climb and descend and twist and turn, and difficult curves and slopes, all these can be taken into

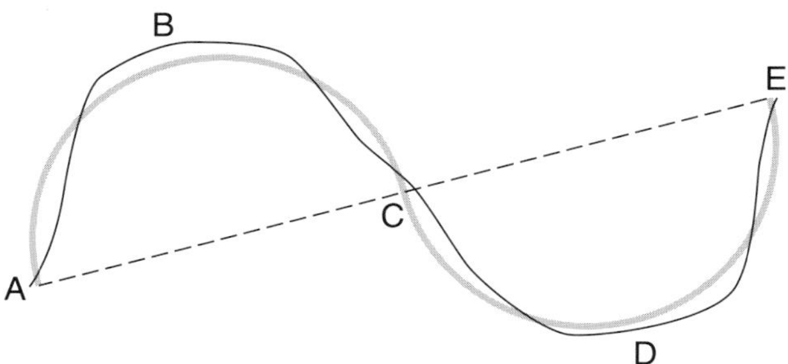

FIG. 5.12. INTERPRETATION OF PEI XIU'S METHOD OF STRAIGHTENING CURVES. In some cases, the application of Pei Xiu's sixth principle, straightening of curves (*yuzhi*), seems to have required knowledge of the properties of circles. Points A and E are joined by a winding road, ABCDE, which has been measured. A cartographer wishing to know the straight-line distance between A and E would be able to derive an approximation by regarding the curve described by ABC and the curve described by CDE as semicircles and segments AC and CE as the diameters of those semicircles. By doubling the lengths of curves ABC and CDE and dividing each of the resulting values by the ratio of circumference to diameter (*yuanzhou lü*), the equivalent of pi, one could derive rough values for AC and CE. By the early Han dynasty, the ratio of circumference to diameter had been approximated as 3, and about Pei Xiu's time, Liu Hui (fl. late third century) calculated the ratio of circumference to diameter as 157 to 50, or about 3.14. For more on the history of Chinese mathematics, see Jean-Claude Martzloff, *Histoire des mathématiques chinoises* (Paris: Masson, 1988).

account and determined." From this passage one can infer that the method of leveling heights seems to be a way of adjusting distances on a map for variations in elevation, "routes that climb and descend." The determination of diagonal distance seems to be a way of calculating the straight-line distance between points separated by natural obstacles such as mountains and seas. Finally, the method of straightening curves seems to be a means of adjusting distances for routes that "twist and turn." What survives of Pei's preface does not specify how one is to realize these principles. If Pei Xiu deliberately refrained from providing those details, he may have believed that the means of applying his rules were common knowledge. In any case, literary sources, predating Pei Xiu, do provide enough information about mensurational and mathematical methods to enable one to reconstruct how he might have applied his principles of indirect measurement (see figs. 5.10 to 5.12).

Pei Xiu stresses that the six principles form a coherent whole. Failure to observe one vitiates the quality of the map, no matter how scrupulously a cartographer has adhered to the others:

> If a map has a standard or regulated view but no road measurement, then in extending [the map] to lands cut

> off by mountains and seas, one will be unable to make connections. If a map shows road distance without leveling heights, determining diagonal distance, and straightening curves, then the figures for distance shown on paths and roads will contradict the facts about what is far and near, and the accuracy of the standard view will be lost. Thus, one must verify a map using these six principles in association.

The six principles constitute the earliest surviving statement in China about the application of mensurational techniques to cartography, and as such they are worthy of attention. But accounts of Pei Xiu's practice point toward another means of verification in cartography. The *Jin shu*, for example, provides this account of the circumstances that led Pei to produce the *Yu gong diyu tu*: "He found that the names of mountains, streams and places in the 'Yu gong' had changed many times since antiquity. Later expositors of those names sometimes forced interpretations, which gradually became obscure and beclouded. Thus Pei examined ancient writings, rejecting what was doubtful, and annotating and listing as far as possible now nameless places that formerly had names. Then he produced the *Yu gong diyu tu* in eighteen sections."[46] In this account of Pei's practice, the six principles are conspicuous by their absence. His work on the *Yu gong diyu tu* does not seem to have involved direct or indirect measurements at all. While at first sight this might seem inconsistent with Pei's six "scientific" principles, it is in fact quite consistent with the interplay of map and text that characterizes premodern Chinese cartography.

Indeed, as Pei's preface also implies, one of his major sources of information was text. In his preface he identifies the types of information he sought from textual sources, one of which is the "Yu gong": "I have studied the mountains and seas, rivers and streams, and the high and low marshlands in the 'Yu gong'; the nine regions of antiquity; the sixteen provinces, the prefectures and principalities, and the counties and cities of the present; border areas and outlying districts; the names of places where ancient kingdoms concluded treaties; navigable waters and roads and paths. Then I produced this map in eighteen sections."[47] Exactly how all this information was reflected in this map is unclear, since it does not survive.

Information concerning another lost map attributed Pei Xiu, the *fangzhang tu* (one-*zhang*-square map) (*zhang* = roughly 3 m), is similarly sketchy. The art historian Zhang Yanyuan (ca. 815 to after 875) lists this as the *dixing fangzhang tu* (topographical one-*zhang*-square map) and classifies it under the category "rare paintings and precious drawings from antiquity."[48] No details other than its authorship are provided, so its mode of representing topography is unknown. The notes of another writer, Yu Shinan (558–638), supply more information. The map was a reduction of an "old great map of the world." Consisting of eighty bolts of silk (each 40 *chi* long, or about 12.2 m), the old map was considered too unwieldy. In accordance with Pei Xiu's first principle, the new map, the *fangzhang tu*, was drawn to scale: "one *fen* equal to ten *li* [1:1,800,000], one *cun* [equal to ten *fen*] equal to one hundred *li*."[49] In showing the locations of "famous mountains and cities," it seems to have measured up to Pei's insistence on correspondence to observed reality. It could substitute for firsthand observation: "Rulers, without descending from their halls, could comprehend the four quarters of the world."[50] Yu Shinan's notes on the *fangzhang tu* do not explain how Pei Xiu made it, whether he merely redrew the old map, accepting its details, or reconfirmed them, relying on textual sources. The latter course, however, would seem to conform more closely to his practice with the *Yu gong diyu tu*.

The textualist aspect of Pei's cartographic practice might be expected when one remembers the value placed on text in the institutional structure in which he worked. That value extended beyond administrative utility or the need to exercise control. Chinese political thinkers early on recognized the temptations to corruption and abuses of power within the bureaucratic ranks, especially among those officials charged with revenue collection. As a counterweight to those temptations, some political thinkers advocated the cultivation of values in literary texts: "A great man's virtue is vast; his writings will be brilliant. A lesser man's virtue may shine; but his writings will be flawed. If an official is esteemed, his writings will be abundant; if virtue is high, literary compositions will multiply."[51]

The connection between literary talent and integrity seems to have been instrumental in advancing Pei Xiu's career. Pei entered government service on the strength of a general's recommendation. According to the *Jin shu*, Pei was singled out not for his mathematical ability, but for his moral virtue and broad learning: "There is not a piece of literature with which he is not familiar."[52] Within

46. *Jin shu*, chap. 35 (4:1039) (note 42).

47. *Jin shu*, chap. 35 (4:1040) (note 42).

48. Zhang Yanyuan, *Lidai minghua ji* (Record of famous painters through the dynasties, completed 847), chap. 3; see the modern edition (Beijing: Renmin Meishu Chubanshe, 1963), 76.

49. Yu Shinan, *Beitang shuchao* (Transcriptions from the Northern Hall, compiled ca. 630) (Taipei: Yiwen Yinshuguan, [1968?]), 96.6a.

50. Yu, *Beitang shuchao*, 96.6a (note 49).

51. Wang Chong, *Lun heng* (Balanced discussions, ca. 82–83), chap. 28; see the modern edition (Shanghai: Renmin Chubanshe, 1974), 431.

52. *Jin shu*, chap. 35 (4:1038) (note 42).

the government, Pei was recognized not only for his cartography but also for his literary ability. Before serving as minister of works, he rose to the post of prefect of the masters of writing (*shangshu ling*). In this capacity he acted as an imperial secretary, handling documents addressed to the emperor and composing his replies and orders.

In light of his considerable literary background, Pei's reliance on textual scholarship in his cartographic practice might be predicted. But focus on the six principles has tended to overshadow the textualism implicit in his practice. This is understandable, since Pei's explanation of the six principles, taken in isolation, does not so much as allude to textual research. But that explanation occurs immediately after his description of the role of textual scholarship as a means of verification in his own work. Furthermore, in the preface Pei implies that his critique of Han maps that introduces his explanation arose from archival research. One should also remember that the preface, as it survives in the *Jin shu*, was selected by compilers who frame it in the context of Pei's textual scholarship. Narrative and quotation need to be considered together, not separately.

As a consequence, to characterize Pei Xiu as originating or even inheriting a "scientific" or quantitative tradition of cartography, as many have done, tells only half the story.[53] There is no denying that quantitative methods constitute an important part of his theory—they are crucial to his aim of achieving correspondence between map and reality. But in practice Pei seems to have relied heavily on textual sources to achieve that correspondence, so that one also cannot deny the significance of textual research, an aspect of "humanistic" learning. Textual sources might provide quantitative information, but their value went beyond that, preserving information about changes in place-names and administrative units. They were, as Qing textual scholars were to recognize, a means of recovering the past. Pei Xiu's legacy to Chinese cartography thus might be described as a blending of the "two cultures" or, in other words, a combination of empiricism and textualism. Both the evidence of the senses and the authority of textual sources served as grounds for knowledge.

Text and Measurement in Later Cartography

The textualist tendencies focused on here are not isolated phenomena but constitute an important element of traditional Chinese cartography. Cartographic writings after Pei Xiu are consistent in their insistence on the importance of textual research and the complementarity of map and text as well as the need for measurement. One indication of this attitude appears in the writing of Jia Dan (730–805), regarded as the foremost mapmaker of the Tang dynasty. Jia recognizes the value of Pei's principles, describing them as a "new conception" of mapmaking, which they can seem to be when they are considered in isolation from developments in collateral fields and evidence from other sources. He also makes explicit what is at best only implicit in Pei's practice, stating that maps require verbal supplementation. It is impossible for representation in a map to approach completeness: "With the various prefectures and armies, one must discuss distances in *li* and numbers of heads; with the various mountains and rivers, one must talk of heads and tails and sources and reaches. On a map, one cannot completely draw these things; for reliability, one must depend on notes."[54] A map, even if drawn in accordance with Pei Xiu's six principles, Jia Dan implies, has its limits. By themselves they cannot, to borrow a phrase from Pei, "preserve actualities." Although one could understand much by looking at a map, one could understand even more through verbal texts. Pei Xiu's practice had implied the importance of texts as sources of information. Jia Dan goes a step further by advocating that texts supplement or accompany maps for a fuller geographic understanding.[55] Apparently others shared Jia's belief, as suggested by the sheer quantity of atlases and gazetteers in which verbal geographic descriptions greatly outbulk maps and illustrations.

An exception to this rule is the stone *Yu ji tu* (Map of the tracks of Yu), dating from 1136. It is drawn with a square grid, bears clear indications of scale, and contains a minimum of verbal annotation. The absence of

53. See, for example, Edouard Chavannes, "Les deux plus anciens spécimens de la cartographie chinoise," *Bulletin de l'Ecole Française d'Extrême Orient* 3 (1903): 214–47, esp. 241; W. E. Soothill, "The Two Oldest Maps of China Extant," *Geographical Journal* 69 (1927): 532–55, esp. 534; Needham, *Science and Civilisation*, 3:538–41 (note 43); and Chen Cheng-siang (Chen Zhengxiang), "The Historical Development of Cartography in China," *Progress in Human Geography* 2 (1978): 101–20. Chen makes the extreme assertion that Pei Xiu's six principles constitute "a perfect discourse on the cartographic art as we know it today." He also downgrades the importance of Claudius Ptolemy because his writings on mapmaking are "mainly about the problem of map projection and can hardly be termed cartography" (104). There is ample evidence to undermine Chen's statements about the modernity of Pei's cartography, as we will see later in this chapter.

54. Liu Xu et al., *Jiu Tang shu* (Old history of the Tang, compiled 940–45), chap. 138; see the edition in 16 vols. (Beijing: Zhonghua Shuju, 1975), 12:3784. None of Jia's maps survive, so the extensiveness of his annotation is unknown. A map perhaps based on one of his works, the *Hua yi tu* (Map of Chinese and foreign lands, engraved 1136), is bordered with lengthy notes (see fig. 3.13 above).

55. The practice of supplementing map with text does not violate Pei's principles, but since none of Pei Xiu's maps survive, we do not know whether they were accompanied by narrative descriptions. Written accounts of his maps are silent on this matter.

verbal description, however, is the exception in traditional Chinese cartography—a fact belied by much scholarship on Chinese cartography.

For example, contrary to accounts that place Shen Kuo (1031–95) firmly in a mathematical tradition, a belief in the complementarity of map and text seems to underlie Shen's cartography. None of the maps he made are extant, but written accounts of them give no hint that they were heavily annotated or accompanied by extensive verbal descriptions. In one instance, Shen says that if his maps were to be lost in the future, they could be reconstructed from information in his writings.[56] This statement can be interpreted in two ways. In support of those who see Shen Kuo primarily as a practitioner of mathematical cartography, this remark indicates that he saw his maps and verbal descriptions as two separate entities. In support of a more textualist view, however, Shen also sees map and text as interchangeable.

Shen's own writings on mapmaking also suggest a combination of empiricism and textualism. In one document Shen says that to make the *Shouling tu* (Map of prefectures and counties), he applied "six principles" (*liu ti*)—the same term Pei Xiu used to name his cartographic rules.[57] This might seem to support a quantitative interpretation of Shen's cartography, but in his *Mengxi bitan* (Brush talks from Dream Brook), he refers to "seven methods" (*qi fa*): "I once made the *Shouling tu*, although I used a scale [*fenlü*] of two *cun* to one hundred *li* [1:900,000], I also adopted a regulated view [*zhunwang*] and mutual inclusions [*hurong*], and indirect verification [*pangyan*], lowering of heights [*gaoxia*], diagonal distance [*fangxie*], and straightening of curves [*yuzhi*]—[in all] seven methods, in order to obtain distances as the bird flies."[58] Shen Kuo does not mention Pei Xiu here, but five of the six technical terms used here—*fenlü*, *zhunwang*, *gaoxia*, *fangxie*, and *yuzhi*—are identical to those used in Pei's statement of six principles. Shen does not explain the meaning of these terms or their application, so perhaps one can assume their equivalence with Pei Xiu's usage. Shen does not explain the remaining two terms, *hurong* and *pangyan*, either. As far as I have been able to determine, the two terms do not recur in later discussions of cartographic method. The parallelism in the morphology of the two terms—both can be read as adjective-noun or adverb-verb compounds—suggests that they are meant to be paired.[59] In the document where Shen speaks of "six principles," he emphasizes the textual research, not the measurements, he performed to prepare the *Shouling tu*: "As an official I consulted documents daily to verify information."[60] Thus the two terms in question may refer to textual criticism. In this context the term *hurong*, or "mutual inclusions," might refer to a process of comparing material common to contemporaneous documents, including maps, as a means of verifying locations and distances, and the term *pangyan*, "collateral" or indirect verification, might refer to the use of later documents to verify information.[61] As we saw earlier, this was the type of research Pei Xiu undertook to correct old maps. Shen's use of the terms *hurong* and *pangyan*, according to the interpretation proposed here, thus represents an attempt to formalize what could only be inferred from accounts of Pei's practice. In addition, under this interpretation, *hurong* and *pangyan* seem to be precursors of the distinction late Ming and Qing dynasty textual critics made between *benzheng* (internal evidence) and *pangzheng* (collateral or external evidence).[62]

The textual emphasis of his account of the *Shouling tu* might give the impression that observation and measurement had only small parts in his practice. Elsewhere, however, such as in an account of his topographic models, Shen does say that he used notes taken while actually traversing the terrain (p. 87). Like Pei Xiu and

56. See Shen Kuo, *Mengxi bitan* (Brush talks from Dream Brook, ca. 1088), chap. 3 suppl., par. 575, in *Xin jiaozheng Mengxi bitan* (Newly edited *Mengxi bitan*), ed. Hu Daojing (1957; reprinted Hong Kong: Zhonghua Shuju, 1975), 322 (the same paragraph number, 575, can be used to locate this passage in *Mengxi bitan jiaozheng* [*Mengxi bitan* edited], 2 vols., ed. Hu Daojing, rev. ed. [1960; reprinted Taipei: Shijie Shuju, 1961]). A European analogy to what Shen describes here is Claudius Ptolemy's *Geography*. Although it contains no maps, it is possible to "reconstruct" them from Ptolemy's coordinates. See O. A. W. Dilke, "The Culmination of Greek Cartography in Ptolemy," in *The History of Cartography*, ed. J. B. Harley and David Woodward (Chicago: University of Chicago Press, 1987–), 1:177–200, esp. 189–90.

57. Shen Kuo, *Changxing ji* (Collected works of [the viscount of] Changxing), chap. 16, in *Shen shi san xiansheng wen ji* (Collected works of the three masters of the Shen clan, compiled 1718), *Sibu congkan* edition, 4.27a.

58. Shen, *Xin jiaozheng Mengxi bitan*, chap. 3 suppl., par. 575 (322) (note 56).

59. For the linguistic analyses here, I am indebted to Tsai Fa Cheng, a specialist in Chinese historical linguistics at the University of Wisconsin–Madison.

60. Shen, *Changxing ji*, chap. 16 (4.27b) (note 57).

61. Another interpretation has been suggested by Cao Wanru in "Lun Shen Kuo zai dituxue fangmian di gongxian" (On Shen Kuo's contributions to cartography), *Keji Shi Wenji* 3 (1980): 81–84, esp. 83. She argues that the passage should be read as listing only six methods, and that the graph for "seven" (*qi*) is a typographical error—a misprint, Cao suggests, for *zhi*, a particle indicating a partitive genitive. There is, however, no bibliographic evidence for such an emendation, and it is possible to read the passage as listing seven methods. Elsewhere Shen does mention applying six embodiments (*liu ti*), but this does not necessarily contradict what he says here. Shen does seem to distinguish "methods" (*fa*) from "embodiments" (*ti*). See Shen's *Changxing ji*, chap. 16 (4.27a) (note 57).

62. The terms *benzheng* and *pangzheng* are usually credited to the philologist Chen Di (1541–1617). To reconstruct the phonological system reflected in the *Shi jing* (Book of odes, ca. twelfth to seventh century B.C.), Chen drew on evidence from the classic itself (*benzheng*) and then on evidence from texts of the same age or slightly later (*pangzheng*).

Jia Dan before him, Shen Kuo seems to regard cartography as an endeavor combining empirical and textual methods.

The extent to which he applied mensurational techniques in his cartography is unclear, but Shen was obviously well versed in such methods. As I have noted elsewhere, Pei Xiu had available to him the mathematics and instruments needed to apply the six principles with some rigor, but it is unknown whether he actually used them. In contrast, our knowledge concerning Shen's application of mensurational techniques is more certain. Shen often describes instruments and techniques applicable to cartographic problems, and in certain instances he may have been able to apply quantitative methods more rigorously than Pei.

This seems particularly true in two areas: direction readings and linear measurements. Before the Song dynasty, there is no clear evidence of the existence of compasses of magnetized iron as distinct from lodestone compasses, nor is there evidence that a dial or card with indications of direction was ever attached to a lodestone.[63] Even without such a dial, rough determinations of direction could be made by suspending a lodestone to indicate south and then using one's arms to approximate the angular deviation from the southerly direction. During the Song, a steel industry developed, allowing the manufacture of alloys able to retain magnetization for long periods, and it is possible that compasses made with a needle-shaped piece of metal—important for its ability to provide precise readings from an attached direction card—were in use. Shen Kuo describes the magnetization of a needle in the *Mengxi bitan*: "Technicians rub the point of a needle with a lodestone; then it can point south. But it always inclines slightly to the east, not pointing directly south."[64] In the same passage Shen also speaks of needles that point north—a situation that results from needles being magnetized at different poles of the lodestone. According to Shen, magnetized needles could be suspended in several ways. They could, for example, be floated on water or balanced on a thin object like one's fingernail or the rim of a bowl. In the first instance, the needle would tend to shake; in the second, it would be likely to fall. The best way to suspend the needle, Shen says, is by a thread: "one takes a single cocoon fiber from new silk and attaches it to the middle of the needle with a piece of wax the size of a mustard seed; then when one hangs it in a windless place, the needle will always point south."[65]

The evidence for the use of a dial or card during the Song is less certain than that for the use of a magnetized needle. Again, Shen Kuo supplies key information. He describes his *Shouling tu* as distinguishing twenty-four directions, each with its own name: "Even if the map is lost in later generations, one may obtain my writings so as to arrange the prefectures and counties according to the twenty-four directions and reconstruct the map quickly without the slightest discrepancy."[66] The *Shouling tu* has been lost, and the writings the map was based on also seem to have disappeared, so it is not clear how Shen Kuo determined direction. His distinguishing twenty-four directions, however, suggests the existence of some means of taking such directional bearings—for example, a magnetized needle with a dial or card marked with those compass points. Shen Kuo, however, does not say that he ever used such a device in his cartography—his description of the magnetized needle occurs in the context of divination. In addition, no such dial or card for navigational or direction-finding purposes survives from the Song, and their use during that period can only be inferred from texts like Shen Kuo's. Examples of maritime compasses with dials indicating twenty-four directions date from the Ming and Qing dynasties.[67]

For linear measurements, Shen Kuo could have used a variety of instruments. Like Pei Xiu, he would have

63. The earliest reference to the "south pointer" probably occurs in the *Han Feizi*, a work dating from the third century B.C. This text states merely that these devices were used to avoid getting lost and to determine direction. No information as to their manufacture or appearance is given. See *Han Feizi*, 6.5, in *Han Feizi suoyin* (Concordance to *Han Feizi*), ed. Zhou Zhongling et al. (Beijing: Zhonghua Shuju, 1982), 737. Another indicator of direction, the south-pointing carriage, was available during Pei Xiu's time, but its cartographic application has not been established. The *Jin shu* contains this description of the device: "The *sinan che*, also called the *zhinan che*, is drawn by four horses. Its lower part was made like a tower with three levels; on its four corners four golden dragons held a plumed canopy with their mouths; and wood was carved into the shape of a transcendent being. It wore feathered garments and stood on top of the carriage. Even if the carriage turned as it moved, the figure's hand always pointed south" (chap. 25 [3:755] [note 42]). This seventh-century text provides no details on how the carriage worked. For a reconstruction of Song versions of the carriage, see André Wegener Sleeswyk, "Reconstruction of the South-Pointing Chariots of the Northern Sung Dynasty: Escapement and Differential Gearing in 11th Century China," *Chinese Science* 2 (1977): 4–36.

64. Shen, *Xin jiaozheng Mengxi bitan*, chap. 24, par. 437 (240) (note 56).

65. Shen, *Xin jiaozheng Mengxi bitan*, chap. 24, par. 437 (240) (note 56).

66. Shen, *Xin jiaozheng Mengxi bitan*, chap. 3 suppl., par. 575 (322) (note 56). The twenty-four directions were named after the twelve earthly branches, eight of the ten heavenly stems, and four of the eight trigrams.

67. Divination boards dating from as early as the Han have been described as "geomantic" compasses. These have bands marked with the names of the twenty-eight *xiu* (lunar lodges) and the twenty-four directions. Describing these boards as compasses, however, is misleading, since they were made for fortune-telling. Their use in navigation or direction finding has not been established. See Michael A. N. Loewe, *Ways to Paradise: The Chinese Quest for Immortality* (London: George Allen and Unwin, 1979), 75–80; and Marc Kalinowski, "Les instruments astro-calendériques des Han et la méthode *liu ren*," *Bulletin de l'Ecole Française d'Extrême-Orient* 72 (1983): 311–419.

been able to use the water level, graduated rod, plumb lines, and a sighting instrument. References to these instruments occur in texts dating from at least the Han dynasty. The forms these instruments took during Pei Xiu's time, however, are a matter of conjecture, for no descriptions of them before the Tang and no illustrations from before the Ming have survived. For Shen Kuo's time, our knowledge of these instruments is more detailed. They are described in a Tang text, and these descriptions are repeated in a Song text, suggesting that in form they had changed little if at all. According to these texts, the water level consisted of a block of wood mounted on a stand with a pivot. Carved into the block of wood were three compartments of the same size, one in the middle and one at each end. The compartments were filled with water, which was able to flow freely between them by a connecting groove. Inside each compartment was a wooden float that rose and sank depending on the level of the water. On top of each float were toothlike sights: "When the three compartments are filled with water, the wooden floats rise. When the three sets of teeth are level, when viewed with one eye, this can be taken as a universal standard."[68]

The water level could be used with a graduated rod and sighting board to determine distance and height. The graduated pole was two *zhang* long and was marked with two thousand divisions. The sighting board is described as being shaped like a square fan and as having a square opening in the center.[69] For distance determination, the sighting board could be used with the water level and graduated rod (figs. 5.13 to 5.15).

The rod and a plumb line could also be used to calculate distance—for example, the width of a stream can be calculated when one is standing on one of its banks (as illustrated in figs. 5.16 and 5.17). For determination of height, the graduated rod could be planted vertically on ground lower than the water level (fig. 5.18). One could then sight along the floats in the level and read the corresponding height on the rod. Elevation could be obtained by subtracting the height of the level from the reading taken from the rod.[70]

From this discussion, one can conclude that by the Song dynasty, if not before, the Chinese possessed the means to produce maps based on direct and indirect measurements. Such mensurational techniques are often used to justify claims of a distinct quantitative tradition of Chinese cartography, one relying heavily on measurement. Such a tradition is identifiable if one ignores the textualism of Chinese cartography, a tendency that coexisted with and in some cases outweighed the observational and mathematical. For example, a conflict between observation and text seems to have arisen regarding the shape of the earth. As described above, the mensurational techniques employed in cartography emphasized straight-

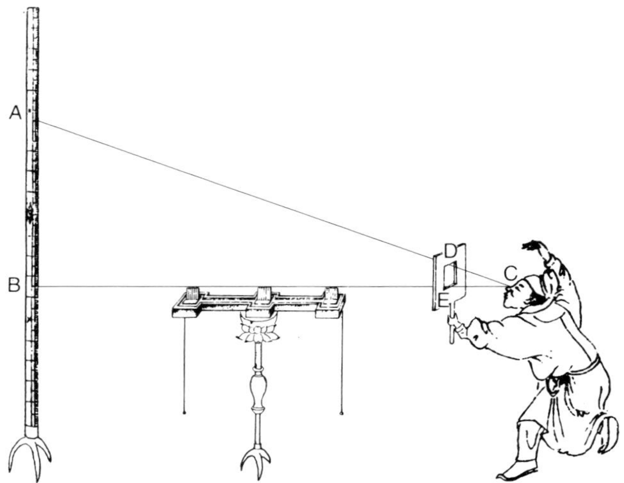

FIG. 5.13. CALCULATION OF DISTANCE USING A SIGHTING BOARD, WATER LEVEL, AND GRADUATED ROD. This method involves the use of similar right triangles, here triangles ABC and DEC. Length AB would be the section of the rod seen through the opening (length DE) of the sighting board. The water level would be used to determine the line BC, which corresponds to the distance being measured. Since AB, DE, and CE are known, it is a simple matter to calculate BC by applying the principle that the corresponding sides of similar triangles are proportional. (This representation is assembled from elements in figs. 5.14 and 5.15.)

68. Li Quan, *Taibai yin jing* (Secret classic of the Grand White [star; i.e., Venus, planet of war], 759) (also known as *Shenji zhidi Taibai yin jing* [Secret classic of the Grand White on wondrous contrivances for subduing an enemy]), *Baibu congshu jicheng* edition, 4.6a. See also Zeng Gongliang, *Wujing zongyao* (Conspectus of essential military techniques, 1044), *Siku quanshu* edition, *qianji* (part 1) 11.3a. The use of three floats when two would have been sufficient might seem curious. The third float may have been added as a means of determining when the board became warped: if the water level was made so that the line connecting the floats was perpendicular to the grain, the third float would reveal when the board was out of line. For an illustration of the water level, see figure 5.14.

69. Li, *Taibai yin jing*, 4.6a (note 68).

70. Shen Kuo's work in waterworks resulted in the invention of another means of determining height. In 1072 he was sent to inspect a land reclamation project in the drainage basin of the Bian River and to make a topographic survey as a preliminary step to dredging and deepening the Bian Canal, a waterway important for the transport of grain tribute to the capital of Kaifeng. In carrying out the survey, Shen measured the difference in elevation between the canal's upper and lower reaches, more than 840 *li* apart. The problem of measuring differences in altitude across a long distance was compounded by the flatness of the terrain. In these circumstances measurements obtained with water level, sighting tube, and graduated rod would not be reliable, since angular variation would be hard to discern. As Shen Kuo recognized, "small errors would be unavoidable." Those small errors, especially in angular measurements, would have led to large ones. To avoid such errors, Shen constructed weirs layer by layer in a temporary channel parallel to the canal. "When the water was level [with one weir], [the section] above it would gradually become shallow and dry. Then another weir would be built to mesh like teeth [with the one below, the result being] like a series of steps. I measured the level of the water

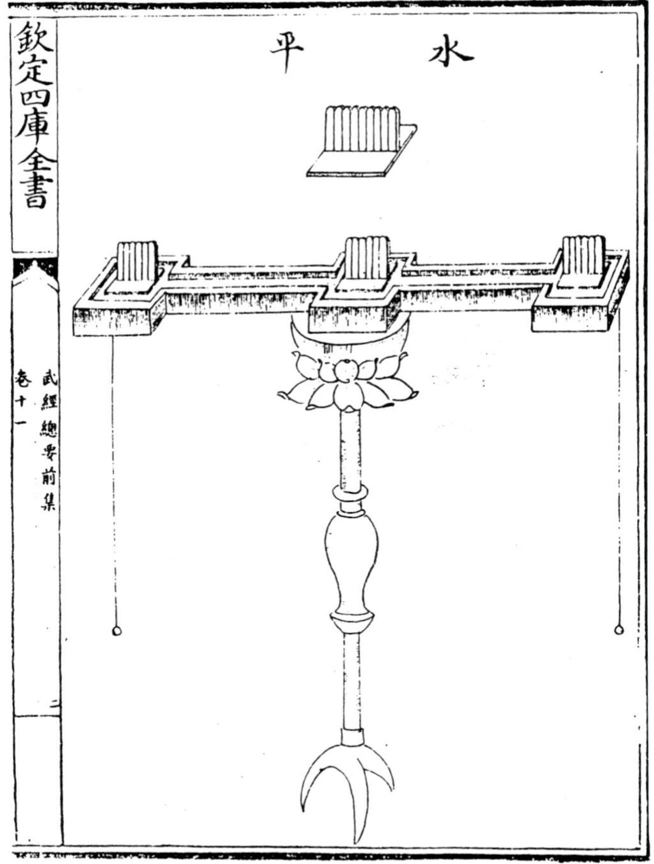

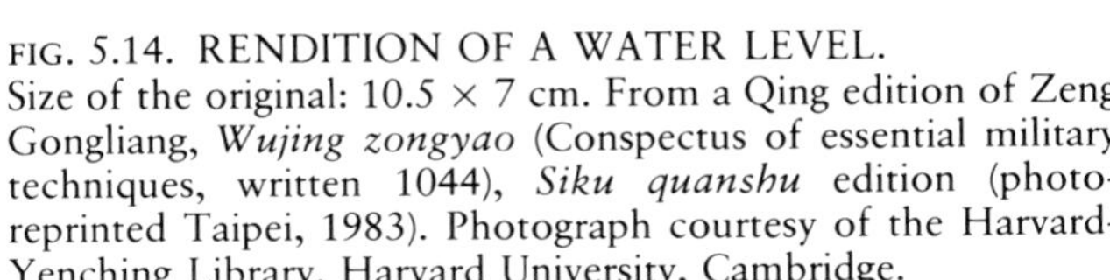

FIG. 5.14. RENDITION OF A WATER LEVEL.
Size of the original: 10.5 × 7 cm. From a Qing edition of Zeng Gongliang, *Wujing zongyao* (Conspectus of essential military techniques, written 1044), *Siku quanshu* edition (photo-reprinted Taipei, 1983). Photograph courtesy of the Harvard-Yenching Library, Harvard University, Cambridge.

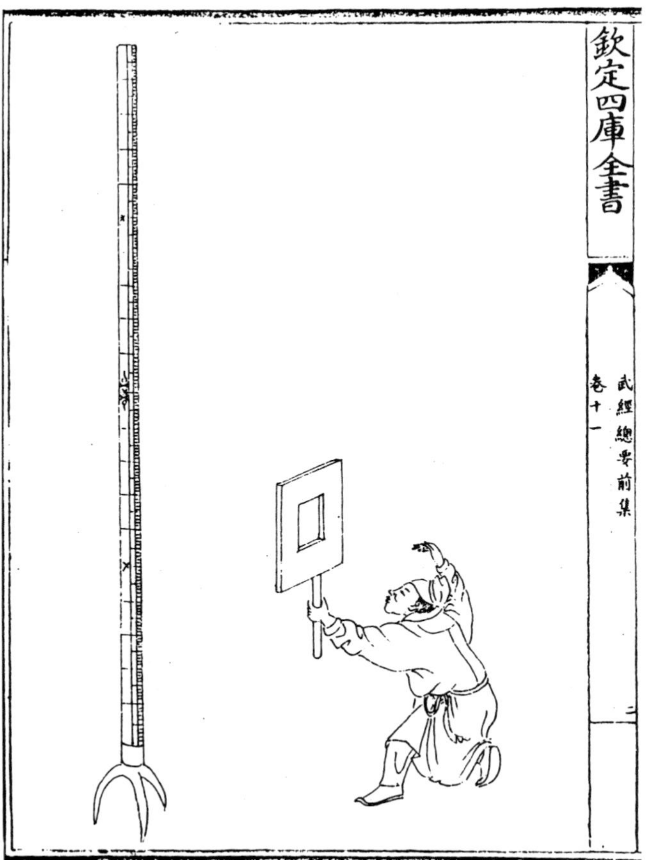

FIG. 5.15. RENDITION OF SIGHTING BOARD.
Size of the original: 10.5 × 7 cm. From a Qing edition of Zeng Gongliang, *Wujing zongyao* (Conspectus of essential military techniques, written 1044), *Siku quanshu* edition (photo-reprinted Taipei, 1983). Photograph courtesy of the Harvard-Yenching Library, Harvard University, Cambridge.

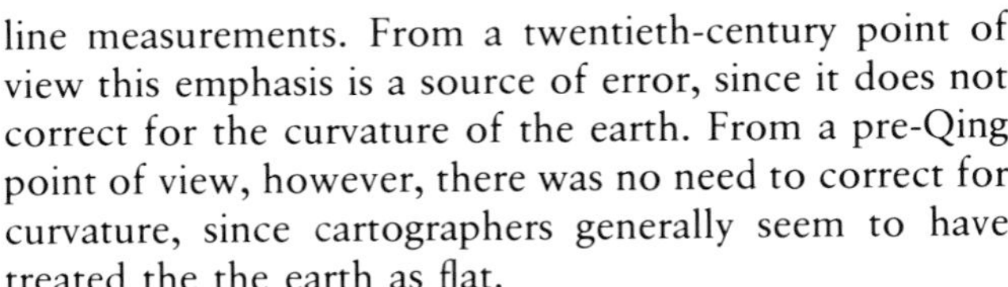

line measurements. From a twentieth-century point of view this emphasis is a source of error, since it does not correct for the curvature of the earth. From a pre-Qing point of view, however, there was no need to correct for curvature, since cartographers generally seem to have treated the the earth as flat.

The Shape of the World: Observation versus Text

This last statement seems to be contradicted by some recent scholarship in the history of Chinese science, according to which the Chinese knew the world was spherical as early as the Han dynasty.[71] The argument for this view rests on a round-earth interpretation of the *huntian* (enveloping heaven) theory, a cosmology developed during the Han. The evidence for this interpretation is equivocal, however, and this ambiguity would have

above and below each weir, and totaled the heights between them. Thus I obtained the actual height of the terrain" (*Xin jiaozheng Mengxi bitan*, chap. 25, par. 457 [250] [note 56]).

Another means of determining height may have been a graduated crossbow sight and trigger assembly. The only reference to this instrument comes from Shen Kuo, who says he saw it unearthed at Haizhou (in present-day Jiangsu Province): "The original idea behind it was that by sighting one's eye along the shaft of an arrow, one could sight a mountain and measure its height in degrees. One could then obtain its height by using the *gougu* method ["base-height" method, equivalent to the Pythagorean theorem; see fig. 5.11] of the mathematicians" (*Xin jiaozheng Mengxi bitan*, chap. 19, par. 331 [194] [note 56]). It is unknown whether the two methods of determining elevation described here were actually used for cartography.

71. See, for example, Cheng Yen-tsu, "Cosmological Theories in Ancient China," *Scientia Sinica* 19 (1976): 291–309, esp. 294–97. In Wallis and Robinson's book the Chinese are credited, apparently based on descriptions of armillary spheres, with making terrestrial globes as early as 260 (*Cartographical Innovations*, 25 [note 1]). Needham, however, is more circumspect, stating that if the Chinese believed in a spherical earth, that belief was not expressed cartographically, at least

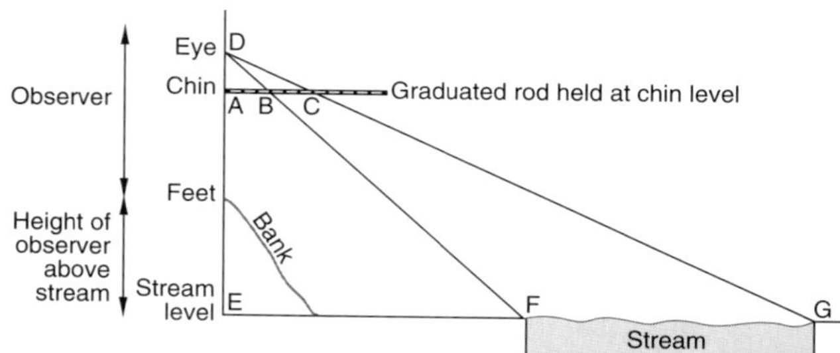

FIG. 5.16. CALCULATING THE WIDTH OF A STREAM WHILE STANDING ON ONE OF ITS BANKS. The length DE, the sum of the eye height of the observer and the height of the bank above the stream, can be determined with a plumb line (perhaps suspended from a rod to obtain the height of the bank). To find the width of the stream (FG), one holds a graduated rod parallel to the ground at chin level and sights and marks two points: the point on the rod (C) that coincides with the far edge of the stream, and the point on the rod (B) that coincides with the near edge of the stream. The length FG can be determined by using two sets of similar triangles: DAC and DEG, and DBC and DFG. The ratio of FG to BC is equal to the ratio of DC to DG, and the ratio of DC to DG in turn is equal to the ratio of DA to DE. The ratio of DA (the distance from observer's eye to chin) to DE is known, as is the distance BC on the rod. Thus the distance FG is easily derived.
After Ulrich Libbrecht, *Chinese Mathematics in the Thirteenth Century: The Shu-shu Chiu-chang of Ch'in Chiu-shao* (Cambridge: MIT Press, 1973), 131.

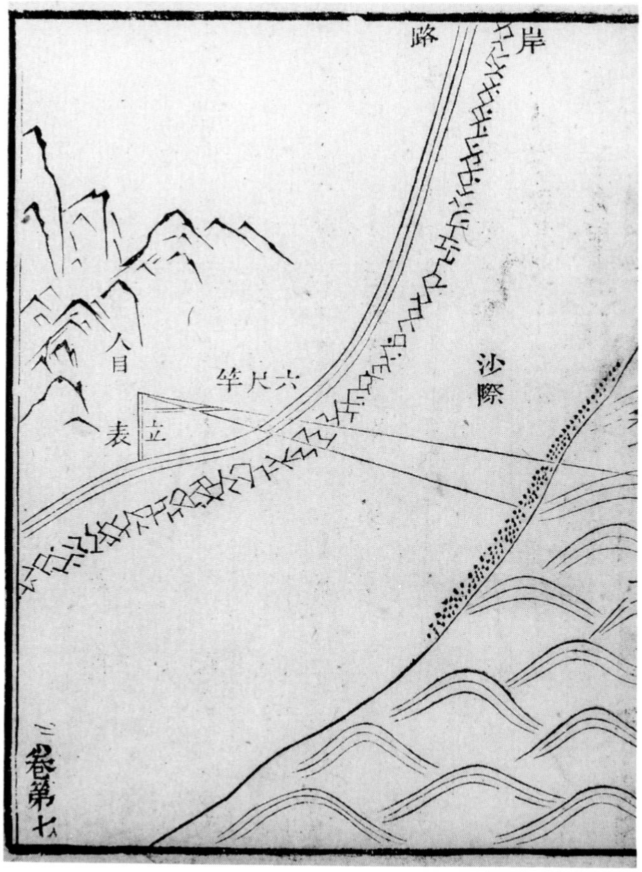

FIG. 5.17. WOODBLOCK ILLUSTRATION SHOWING THE CALCULATION OF THE WIDTH OF A STREAM.
Size of the original: 17.5 × 25 cm. From Qin Jiushao, *Shushu*

allowed mapmakers to hold a view of the earth as flat—a conception perhaps sanctioned by textual sources.

One of the earliest expositions of the *huntian* theory is attributed to the polymath Zhang Heng (78–139): "The enveloping heavens are like a chicken egg, and the celestial form is round like a crossbow pellet; the earth is like the egg yolk, alone occupying the center."[72] The egg metaphor might be interpreted as describing a spherical earth—a reading encouraged by the misconception that the *huntian* theory displaced the *gaitian* (covering heaven) theory, which seems to imply that the earth is flat. This conception of the earth is often traced to the *Zhoubi suan jing* (Arithmetical classic of the Zhou gnomon), which dates perhaps from the early Han: "Turning a carpenter's square creates a circle; combining carpenter's squares forms a square. Squareness belongs to the earth; circularity belongs to heaven. Heaven is round; the earth is square."[73] This is the cosmology that is believed to be symbolized on divination boards and cosmic mirrors dating from the Han (figs. 5.19 and 5.20).

The *huntian* theory is certainly not inconsistent with a conception of the earth as spherical, and it is difficult to imagine that Chinese astronomers who applied the theory in their calendrical calculations would have failed to appreciate the empirical evidence for curvature of the earth's surface. Zhang Heng, for example, states that the

not until the arrival of the Jesuits; see Needham, *Science and Civilisation*, 3:437–38, 498–99 (note 43). Tang Ruchuan makes a strong case that in the *huntian* cosmology the earth was conceived as hemispherical or, more likely, discoidal. See Tang's "Zhang Heng deng huntianjia di tian yuan di ping shuo" (On the theory of Zhang Heng and other uranosphere school cosmologists that the sky is spherical and the earth flat), *Kexue Shi Jikan*, 1962, no. 4:47–58. I am indebted to Nathan Sivin for the information on this citation.

72. Zhang Heng, "Hunyi tu zhu" (Commentary on a diagram of the armillary sphere, ca. 117), in Sun Wenqing, *Zhang Heng nianpu* (Chronological biography of Zhang Heng), rev. ed. (Shanghai: Shangwu Yinshuguan, 1956), 72–75, esp. 72. A clear account of Chinese cosmological theories can be found in Shigeru Nakayama, *A History of Japanese Astronomy: Chinese Background and Western Impact* (Cambridge: Harvard University Press, 1969), 24–43.

73. *Zhoubi suan jing* (ca. 200 B.C.), *Siku quanshu* edition, 1A.15b. See also *Huainanzi zhu*, chap. 3 (44) (note 6), which says, "Heaven is round; earth is square."

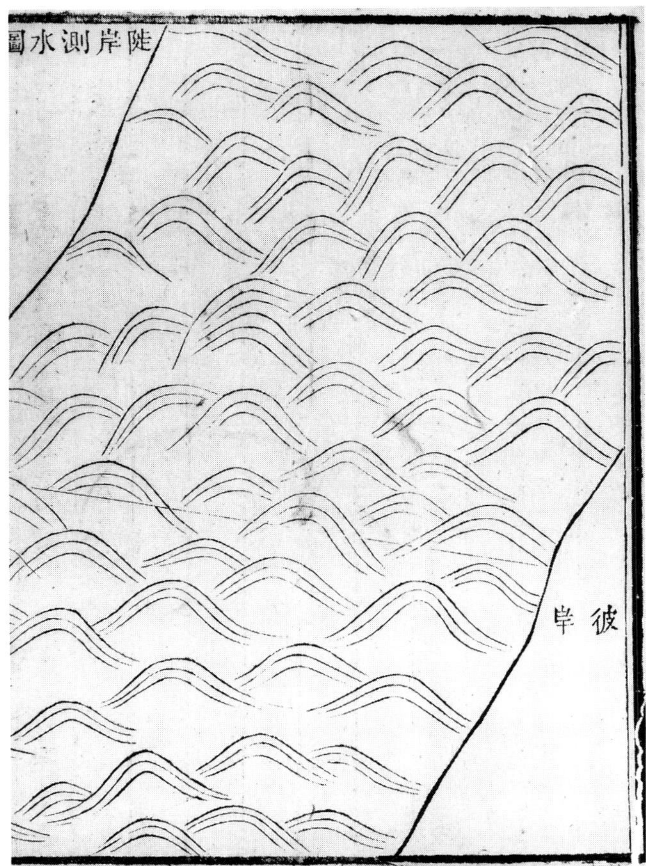

jiuzhang (Mathematical treatise in nine chapters, 1247), ed. Song Jingchang, 1841 edition, 7.17a–b. Reproduced courtesy of the Harvard-Yenching Library, Harvard University, Cambridge.

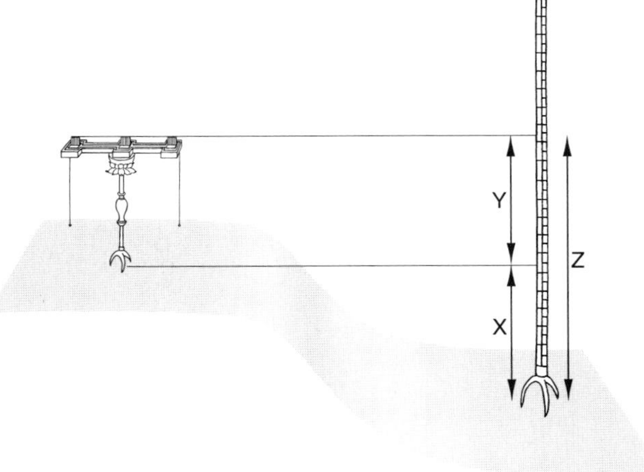

FIG. 5.18. DETERMINATION OF HEIGHT USING A WATER LEVEL AND GRADUATED ROD. To determine height X, read height Z sighting with a water level, then subtract the height of the water level (Y).

moon reflects the sun's light, and that lunar eclipses are caused by the earth's shadow.[74] The Ming encyclopedia *Sancai tuhui* (Illustrated compendium of the three powers [heaven, earth, man], completed 1607) explains lunar eclipses with an illustration showing the earth between sun and moon (fig. 5.21).[75] The resulting shadow, as anyone who has witnessed a lunar eclipse can attest, is at least curved, if not round. From this one might conclude that the earth is flat and round like a disk, but Chinese cosmologists do not seem to have entertained this possibility. If the idea of roundness was considered, the earth's entire surface was regarded as curved. The Song philosopher Zhu Xi (1130–1200), for example, describes the shape of the heavens and earth in terms suggesting curved surfaces: "The form of heaven and earth is like someone putting two bowls together, holding water within. If they are constantly turned with the hands, the water remains inside without spilling. But if the hands are stopped, the water leaks out."[76]

It is thus plausible to suppose that Chinese astronomers had reason to believe the earth was round by the time Matteo Ricci (1552–1610) made his maps depicting a spherical earth.[77] Legends on the cartographic works

74. Zhang Heng, "Lingxian" (Spiritual constitution of the universe, ca. 118), quoted in Sun, *Zhang Heng nianpu*, 79 (note 72).

75. But it also illustrates solar eclipses with the earth between the sun and moon.

76. Zhu Xi, *Zhuzi yulei* (Classified conversations of Master Zhu, 1270), comp. Li Jingde (1473; reprinted Taipei: Zhengzhong Shuju, 1962), 1.6a–b.

77. At least one representation of a round earth was introduced into China before Ricci's. During the Yuan dynasty, Chinese astronomers may have seen a representation of a terrestrial globe, one of seven "representations [simulacra or images] of instruments [*yixiang*]" made in 1267 by a Persian astronomer who later served as director of the Islamic observatory in Dadu (Beijing). It is unclear whether these representations took the form of drawings or models. A description of the representations appears in Song Lian et al., *Yuan shi* (History of the Yuan, compiled 1369–70), chap. 48; see the edition in 15 vols. (Beijing: Zhonghua Shuju, 1976), 4:998–99. The terrestrial globe, which the *Yuan shi* explains as a kind of "geographic record" (*dilizhi*), does not seem to have influenced Chinese cartography or, some have maintained, Chinese astronomy. Yabuuchi comments: "The terrestrial globe that was kept in the Islamic observatory did not interest the Chinese astronomers, who had other ideas about the form of the earth." See Kiyoshi Yabuuchi (Kiyosi Yabuuti), "The Influence of Islamic Astronomy in China," in *From Deferent to Equant: A Volume of Studies in the History of Science in the Ancient and Medieval Near East in Honor of E. S. Kennedy*, ed. David A. King and George Saliba (New York: New York Academy of Sciences, 1987), 547–59, esp. 549. See also Christopher Cullen, "A Chinese Eratosthenes of the Flat Earth: A Study of a Fragment of Cosmology in *Huai Nan Tzu*," *Bulletin of the School of Oriental and African Studies* 39 (1976): 106–27. Cullen asserts that for the Chinese the earth was "at all times flat, although perhaps bulging up slightly," and that Chinese ideas on the shape of the earth remained unchanged from "early times until the first contacts with modern science through the medium of Jesuit missionaries in the seventeenth century" (107). Evidence from sources apparently not considered by Cullen, however, suggests a more complex history of Chinese ideas on the shape of the earth—a history I have tried to summarize here.

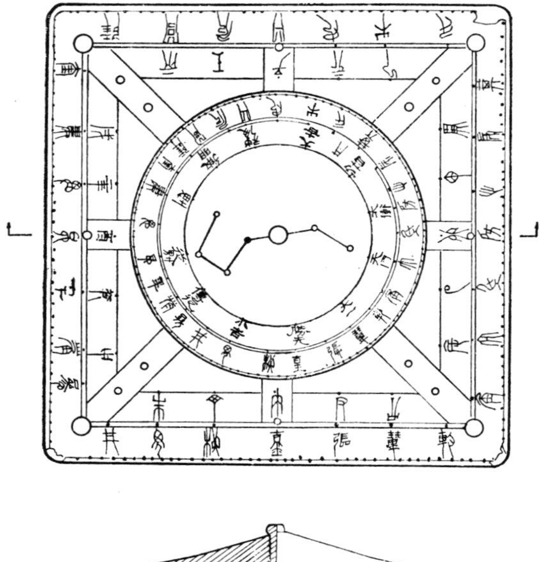

FIG. 5.19. HAN DIVINATION BOARD. The board perhaps symbolizes a round heaven and square earth. The original was found in tomb 62, Mozuizi, Gansu.
Size of the original: 9 × 9 cm (diameter of inner circle, ca. 6 cm). From Gansusheng Bowuguan (Gansu Provincial Museum), "Wuwei Mozuizi sanzuo Hanmu fajue jianbao" (Brief report on the excavations of the three Han tombs at Mozuizi in Wuwei County), *Wenwu*, 1972, no. 12:9–21, esp. 15.

made in China by Ricci and other Jesuits suggest that they were made to contradict a flat-earth conception. But precisely because they held that conception, Chinese intellectuals had trouble accepting Ricci's maps.[78] This seems curious, since Ricci arrived in China centuries after the *huntian* theory had supposedly prevailed over the *gaitian* theory.

The contradiction here is only apparent. As I pointed out above, the *huntian* theory is consistent with a round-earth conception, but that does not mean it is inconsistent with a flat-earth conception. Beyond Zhang Heng's egg yolk image, which is primarily a metaphor for position, the *huntian* theory suggests nothing about the earth's shape, and it is possible to realize the cosmic egg metaphor with an image of a flat earth. Two Ming dynasty encyclopedias, for example, illustrate the *huntian* theory by depicting heaven as round and the earth as flat (figs. 5.22 and 5.23). In one of his writings Zhang Heng, regarded as an adherent of the *huntian* theory, describes three imperial "ritual buildings." One of them uses imagery consistent with the *gaitian* theory: "One has two stories and a double roof, / Eight windows and nine rooms. / It is round like Heaven, square like Earth."[79] This passage appears in a rhapsody (*fu*) contained in the *Wen xuan* (Literary selections), a sixth-century anthology revered by the educated elite from the Tang onward.[80] The majority of that elite did not consist of astronomers, and as I mentioned above, they valued literary training above technical knowledge. Among that group, Zhang Heng would have been proof that despite what the *huntian* theory said, the *gaitian* theory that heaven is round and earth square—a conception expressed in works like the *Huainanzi* and *Zhoubi suan jing*—was still tenable.

Unno has argued that this description is meant to be read metaphorically, that roundness and squareness do not literally refer to shape but suggest metaphysical properties. For support Unno cites Zhao Junqing, who says that roundness and squareness do not refer to the actual shapes of heaven and earth.[81] Zhao, however, was writing

78. See Helen Wallis, "The Influence of Father Ricci on Far Eastern Cartography," *Imago Mundi* 19 (1965): 38–45; and *China in the Sixteenth Century: The Journals of Matthew Ricci, 1583–1610*, trans. Louis J. Gallagher from the Latin version of Nicolas Trigault (New York: Random House, 1953), 325. Trigault's book, originally published in 1615, is based in large part on Ricci's account of China and his mission. Matteo Ricci's own account can be found in *Storia dell'introduzione del Cristianesimo in Cina*, 3 vols., ed. Pasquale M. d'Elia, Fonti Ricciane: Documenti Originali concernenti Matteo Ricci e la Storia delle Prime Relazioni tra l'Europa e la Cina (1579–1615) (Rome: Libreria dello Stato, 1942–49). For a general discussion of Chinese resistance to Western science during the Ming and Qing, see George H. C. Wong, "China's Opposition to Western Science during Late Ming and Early Ch'ing," *Isis* 54 (1963): 29–49. One Chinese intellectual who denounced the round-earth conception promulgated by the Europeans was Wang Fuzhi (1619–92); see p. 225. Ricci's maps also incurred criticism because of their depiction of five major continents—Europe, Africa, America, Asia, and Magellanica—a world situation that reduced the geographic importance of China. The compilers of the *Ming shi* state that the explanation of world geography in terms of five continents is "nonsensical and hard to verify." They do admit, however, that the presence of visitors to China from places named on the map is proof that such places exist. See Zhang Tingyu et al., *Ming shi* (History of the Ming, 1739), chap. 326; see the modern edition in 28 vols. (Beijing: Zhonghua Shuju, 1974), 28:8459. There were, of course, some Chinese intellectuals sympathetic to Western learning. For more discussion of the Chinese response to Ricci's maps, see pp. 174–76.

79. Zhang Heng, "Er jing fu" (Two metropolises rhapsody, ca. 107), in Xiao Tong, comp., *Wen xuan* (Literary selections, completed ca. 526–31), ed. Hu Kejia (1809; reprinted Kyōto: Chūbun Shuppansha, 1971), 3.11b. The translation is that of David R. Knechtges, trans. and annotator, *Wen xuan; or, Selections of Refined Literature* (Princeton: Princeton University Press, 1982–), 1:263. In his "Lingxian," Zhang also describes the earth as being "flat and at rest" (*Zhang Heng nianpu*, 77 [note 72]).

80. Through at least the Song dynasty, the *Wen xuan* was studied by candidates for civil service degrees. See Knechtges, *Refined Literature*, 1:54–55 (introduction) (note 79); and David McMullen, *State and Scholars in T'ang China* (Cambridge: Cambridge University Press, 1988), 223–25.

81. Kazutaka Unno makes these points in two articles: "Japan before the Introduction of the Global Theory of the Earth: In Search of a Japanese Image of the Earth," *Memoirs of the Research Department of the Toyo Bunko* 38 (1980): 39–69; and "Kodai Chūgokujin no chiriteki sekaikan" (The ancient Chinese people's geographical conception of the world), *Tōhō Shūkyō* 42 (1973): 35–51.

FIG. 5.20. HAN COSMIC MIRROR. The mirror is believed to represent a square earth and round heaven. Diameter of the original: 14.3 cm. Courtesy of the Freer Gallery of Art, Smithsonian Institution, Washington, D.C.

perhaps during the third century, a time when the *huntian* theory seems to have been well established, and thus may have been attempting to reconcile astronomical knowledge with textual authority. Although he does not do so, Unno might also have cited Zhang Huang, compiler of the Ming encyclopedia *Tushu bian* (Compilation of illustrations and writings), who says that roundness refers to the motion of the heavens, and squareness to the stillness of the earth.[82] Zhang Huang's reading may be atypical, however, since he was acquainted with Matteo Ricci and may be reacting to Western conceptions of the world. In the *Tushu bian* he includes a copy of Ricci's map of the world, and he defends the spherical conception of the earth. Zhang's interpretation of the "heaven is round, earth is square" formula may have been motivated by a desire to harmonize traditional conceptions with Western ideas.

82. Zhang Huang, comp., *Tushu bian* (compiled 1562–77) (1613; reprinted Taipei: Chengwen Chubanshe, 1970), 29.34b.

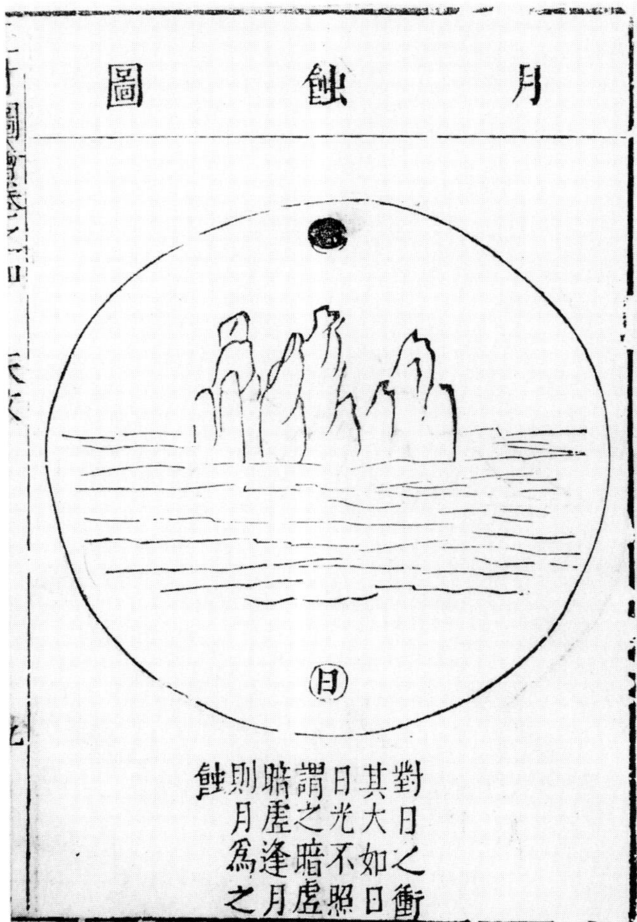

FIG. 5.21. ILLUSTRATION EXPLAINING LUNAR ECLIPSES. Size of the original: 21 × 14 cm. From Wang Qi, comp., *Sancai tuhui* (completed 1607, printed 1609), *tianwen* [astronomy], 4.9a. Reproduced courtesy of the Harvard-Yenching Library, Harvard University, Cambridge.

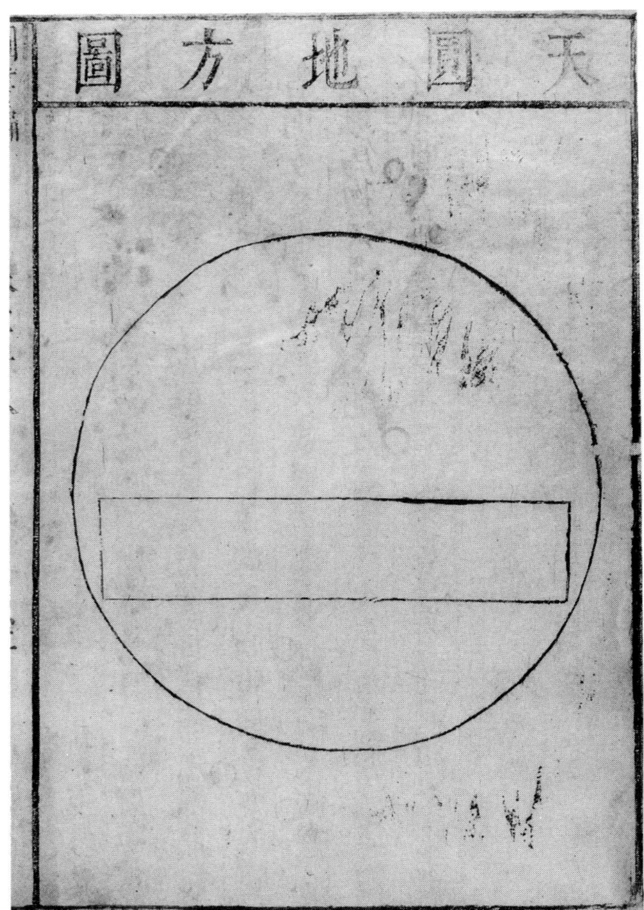

FIG. 5.22. ZHANG HUANG'S DEPICTION OF THE HEAVENS AS ROUND AND THE EARTH AS SQUARE. Size of the original: 22.5 × 15 cm. From Zhang Huang, comp., *Tushu bian* (1613), 28.2a. Reproduced courtesy of the Harvard-Yenching Library, Harvard University, Cambridge.

The persistence of the flat-earth conception despite astronomical evidence to the contrary serves as a reminder that not all astronomical innovations with potential applications in cartography were realized in cartographic practice. This fact supports Sivin's suggestion that there was no "systematic connection between all the sciences in the minds of the people who did them."[83] Elsewhere Sivin points to an explanation for this: traditions of learning were regarded as consisting of lineages of texts. This conception of learning did not preclude innovation, but it did place some restrictions of how much innovation was permissible: "Those who worked in the sciences accepted and carried on their forebears' sense of the enterprise, often modifying it in the process. In certain periods innovation was recognized and prized in astronomy and medicine as in many other aspects of Chinese life, but the new became more acceptable to the extent that precedents showed it was not wholly new."[84] To apply this conception of multilineality to the case at hand, the *huntian* and *gaitian* theories had different textual lineages and tended to be used in different contexts. In addition, they were associated with instruments having different purposes. The *huntian* theory represented the spatial orientation of the armillary sphere, used for the demonstration and observation of celestial motions, and the *gaitian* theory represented that of the gnomon, used to measure the length of the sun's shadow. In these circumstances, the two representations were not necessarily juxtaposed for comparison.

83. Nathan Sivin, "Why the Scientific Revolution Did Not Take Place in China—Or Didn't It?" *Chinese Science* 5 (1982): 45–66, esp. 48.

84. Sivin, "Science and Medicine," 43 (note 2). The idea of textual lineages also applies to Western scientific traditions, but perhaps, as Sivin suggests, to a lesser extent than in China.

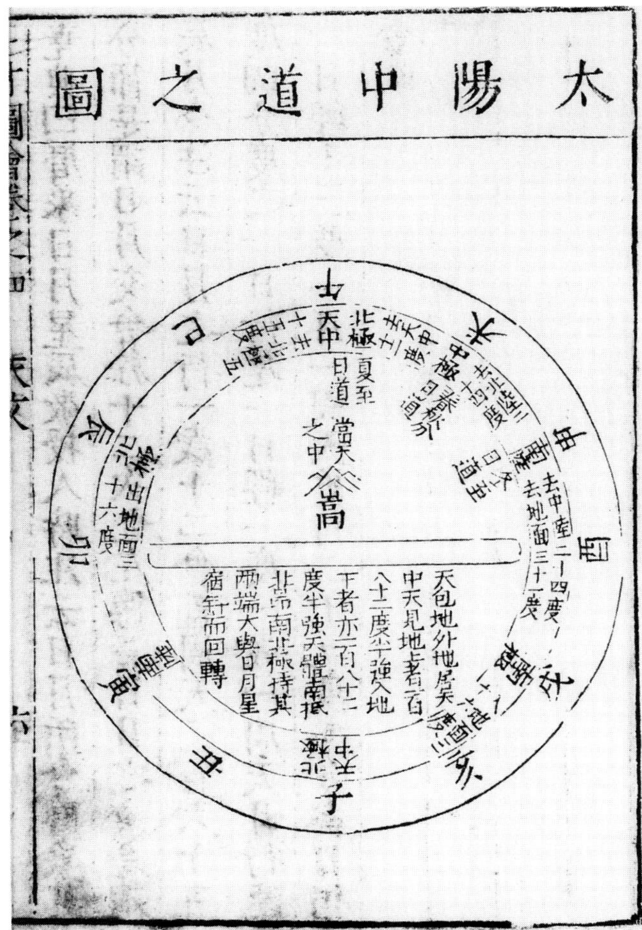

FIG. 5.23. WANG QI'S DEPICTION OF THE HEAVENS AS ROUND AND THE EARTH AS SQUARE.
Size of the original: 21 × 14 cm. From Wang Qi, comp., *Sancai tuhui* (completed 1607, printed 1609), *tianwen* [astronomy], 4.6a. Reproduced courtesy of the Harvard-Yenching Library, Harvard University, Cambridge.

The multilineal conception of learning can perhaps explain why other astronomical innovations did not realize their cartographic potential. For instance, the astronomical survey performed by Yixing (682–727) and Nangong Yue during the Tang dynasty calculated that a change of one degree in celestial latitude (determined by observing a difference of one degree in the altitude of the Pole Star) corresponded to a change of about 351 *li* in terrestrial distance. This ratio makes sense only if one assumes a terrestrial meridian that traces a circle. The ratio would then allow one to determine ground distances indirectly using astronomical measurements and thus seems to have cartographic significance. The ratio determined by Yixing and Nangong Yue was helpful to court astronomers, who could use it to calibrate gnomon shadow lengths, which were used to determine reference dates, such as the summer solstice, important in the construction of calendars.

Outside astronomical circles, however, as Beer and others have suggested, the implications of the ratio would have made it unacceptable to adherents of the *gaitian* theory.[85] After describing the astronomical survey, the compilers of the *Jiu Tang shu* (Old history of the Tang, compiled 940–45) note this discrepancy between the two schools: "Adherents of the *hun* and *gai* schools have not yet been able to reconcile their theories."[86]

Similarly, the results of the "survey of the four seas" (that is, the world), directed by the Yuan astronomer Guo Shoujing (1231–1316), would not have been applied by cartographers adhering to a flat-earth conception. The survey established twenty-seven observation stations at which the height of the North Star in degrees was determined. The results of the survey are recorded in the *Yuan shi* (History of the Yuan), compiled shortly after the fall of the Yuan dynasty. For seven observation sites, including the capital, the *Yuan shi* records, in addition to the height of the North Star in degrees, the gnomon shadow length at the summer solstice and the length of day and night at the summer solstice.[87] This information would have been of use in constructing calendars, one of the main duties of court astronomers. In addition, the correlation of celestial positions with terrestrial locations had potential cartographic significance, since it would have permitted indirect measurement of ground distances based on celestial observations. As in the case of the Tang astronomical survey, however, the cartographic potential does not seem to have been realized.

The correlation of celestial degree measurements with ground distance is possible only on the assumption that the earth's surface parallels the celestial sphere or, in other words, that the earth is round. Those responsible for mapmaking within the political establishment, however, generally seem to have adhered to the flat-earth conception. This idea was sanctioned by a number of ancient texts and, furthermore, was implied by the indirect methods of measuring ground distances described in mathematical texts.

This conclusion, at any rate, is supported by the history of Chinese cartographic theory and practice after the

85. Arthur Beer et al., "An 8th-Century Meridian Line: I-Hsing's Chain of Gnomons and the Pre-history of the Metric System," *Vistas in Astronomy* 4 (1961): 3–28, esp. 25. With regard to the results of the astronomical survey, the point of contention between the adherents of the two theories would not have been whether the observed height of heavenly bodies varied with distance from a given reference point, but the rate at which the height varied. Under the *gaitian* conception, equal increments of arc measurement would not correspond to equal increments of ground distance.

86. *Jiu Tang shu*, chap. 35 (4:1307) (note 54).

87. *Yuan shi*, chap. 48 (4:1000) (note 77).

Tang. Shen Kuo's statement of methods says nothing about correcting for curvature of the earth's surface. A rough gauge of the strength of the flat-earth conception is the Suzhou astronomical chart (fig. 13.21 below). Carved in stone in 1247, it purports to be a copy of a star map prepared for the education of a royal prince. According to the text accompanying the chart, the "essentials" of astrology include a conception of the earth as flat and static: "The body of Heaven is round and the body of Earth is square. The round is in motion and the square is at rest."[88]

Patterns in the use of the cartographic grid also do not support the notion that Chinese cartographers accepted a round-earth conception to any appreciable degree. Grids seem to have been used primarily on small-scale maps representing large areas, such as whole provinces or all of China. Yet these maps are, of course, the ones on which distortion from the earth's curvature would be most apparent. On large-scale maps representing smaller areas such as counties or prefectures, grids are rare. The use of the grid to support a textualist interpretation of Chinese cartographic history might appear to run counter to the grid's standard interpretation as a high point in Chinese mensurational cartography. As we will see, however, the grid, though highly suggestive of quantification, is not incompatible with textualism.

The Cartographic Grid

The first known use of the square grid occurs on the *Yu ji tu* (Map of the tracks of Yu), carved in stone in 1136.[89] A note on the map says, "[A side of] each square converts to one hundred *li*." The grid bears a superficial resemblance to the system of latitude and longitude that developed in Europe, but unlike the graticule, the Chinese grid is not a fixed coordinate system. In contrast to the graticule, which represents a mathematical structure underlying the two-dimensional projection of the terrestrial sphere, the square grid seems to have been superimposed arbitrarily on a given area of interest.

The graticule, as used in Europe, is based on ideas summarized by Claudius Ptolemy (ca. 90–168), the Alexandrian geographer and mathematical astronomer. Under Ptolemy's conception of mapping, space is composed of points, each occupying a specific location that can be determined mathematically on an x-y axis. The graticule provides a framework for organizing and locating specific points, thus permitting projection from a spherical to a plane surface. Geographers of the Renaissance constructed maps on this principle. A rectangular graticule was laid out on a blank plane surface as a way to delimit and organize space and determine its relative scale. Plotted on this base were a series of points in the form of geographical coordinates of latitude and longitude corresponding to features of the earth's surface. The rest of the map—linear and areal features—was filled in relative to these points.[90]

In contrast to European practice, map space was not treated analytically in China; points were located not by coordinates, but solely by distance and direction. In consequence, the grid served a function different from that of the graticule. From the indications of scale on grid maps, often expressed as a certain number of *li* per (side of a) square, one purpose of the grid was to help a map reader calculate distance and area. This contrasts with the function of the graticule, which has been used primarily as a means of locating position and relating the area mapped to the globe of which it is a part.[91] Underlying this difference are divergent views of the earth's shape: the square grid presupposes a generally flat surface. The graticule developed as a means of transferring points on a spherical surface to a plane surface, thus entailing some manipulation of the increments by the cartographer and an inevitable distortion of the image itself.

The origins of the grid are unknown. Several analogous schemes for spatial division or organization, however, predate its first appearance. One is the graph for "field," *tian* 田, which, according to a Han etymological dictionary, represents the configuration of paths in grain fields.[92] Land division by such crisscrossing paths, carried out over a large area, would result in a pattern similar to a grid. So too would the "well-field" system of land allocation, so called because it would create units resembling the graph for "well," *jing* 井 (fig. 8.1 below).[93] Literary

88. As translated in W. Carl Rufus and Hsing-chih Tien, *The Soochow Astronomical Chart* (Ann Arbor: University of Michigan Press, 1945), 2.

89. Contrary to the impression given by Needham (*Science and Civilisation*, 3:543 [note 43]), there is no evidence that square grids appeared on maps before the Song. The claim that Jia Dan used a grid seems to be based on a misreading of the *Jiu Tang shu*. It says that Jia drew a map with "a scale of one *cun* to one hundred *li*," not "one [side of a] square (*fang*) to one hundred *li*." See *Jiu Tang shu*, chap. 138 (12:3786) (note 54).

90. For more on Ptolemy, see Dilke, "Culmination of Greek Cartography" (note 56).

91. Thus Helen Wallis's description of the grid as a "cartographic reference system" is misleading, since it gives the impression that the grid was a coordinate system. See Helen Wallis, "Chinese Maps and Globes in the British Library and the Phillips Collection," in *Chinese Studies: Papers Presented at a Colloquium at the School of Oriental and African Studies, University of London, 24–26 August 1987*, ed. Frances Wood (London: British Library, 1988), 88–96, esp. 88.

92. Xu Shen, comp., *Shuowen jiezi* (Explanation of writing and explication of graphs, compiled ca. 100), s.v. *tian*, in *Shuowen jiezi gulin* (Collected glosses to the *Shuowen jiezi*), 12 vols., ed. Ding Fubao (Taipei: Shangwu Yinshuguan, 1959), 9:6183b.

93. The well-field system is just one manifestation of the nonary square pattern that was influential in Chinese cosmology. See pp. 205–16. As John Henderson points out below, the square pattern also manifested itself in city planning. For an example of this, see figure 8.6.

accounts of land division systems like the well-field system have often been dismissed as idealizations, but there is evidence that some system must have been practiced. Leeming has studied maps and aerial photographs of Chinese terrain and found that an ancient gridlike pattern of land division is often discernible.[94] The cartographic grid may in part represent an attempt to mimic this feature of the landscape.

Another possible source of the grid is Zhang Heng, who is credited with the *Suan wang lun* (Discourse on net calculations), now lost. The title is said to derive from the method of calculations described in the work: according to one commentary, Zhang Heng "cast a net around heaven and earth and calculated on the basis of it."[95] Joseph Needham has speculated that the net was a "rectangular grid system," the celestial coordinates corresponding to the *xiu* (lunar lodges) of traditional Chinese astrology. This speculation seems unlikely, since one cannot speak of a rectangular coordinate system when right ascension was measured in degrees and declination was recorded in various linear units. As for the terrestrial coordinates, Needham runs into the problem commonly encountered in the study of early Chinese cartography—insufficient detail in existing texts.[96] Needham speculates that Zhang Heng's terrestrial system consisted of rectangular grids.[97] Clues about Zhang's methods could normally be expected to come from his practice, but none of his maps are extant. He made a topographical map (*dixing tu*) that apparently survived until the Tang dynasty, since it is recorded in a catalog of Han paintings compiled in the ninth century.[98] Unfortunately, the catalog's compiler provides no information about the map's content or appearance.

Scholars since the seventeenth century have credited Pei Xiu with developing the square grid.[99] But unless one follows Needham, who interprets the second principle, *zhunwang*, as describing a rectangular grid system,[100] there is no evidence that Pei Xiu used a grid in the way it is used on, for example, the *Yu ji tu*. What the principle of *zhunwang* may entail is the use of a single reference point (A in fig. 5.24*a*) for measurements of distance and direction. The advantages become most apparent in small-scale mapping. If multiple reference points are used (A, B, and C in fig. 5.24*b*), cumulative errors in distance and direction measurements occur. A single reference point, however, allows backsighting as a means of checking one's bearings. Thus one can "regulate" one's directional sightings in the manner Pei Xiu implies: "Regulated sighting is a means of rectifying the configuration of this place and that [or in other words, relative position]. . . . If a map has proportional measure [or scale] but lacks regulated sighting, then, though it may be correct in one corner, it will fail in other places."[101] As interpreted here, applying *zhunwang* minimizes cumulative errors in directional measurement.[102]

The grid, though probably not employed by Pei Xiu, was used by later cartographers to facilitate the production of maps, particularly in plotting direction and distance in accordance with Pei's first two principles. In his preface to the *Guang yutu* (Enlarged terrestrial atlas, printed about 1555), Luo Hongxian (1504–64) says that his predecessor Zhu Siben (1273–1337) used the grid in this manner. The faithfulness of Zhu's maps to reality, Luo says, can be attributed to his use of the grid: "His maps employ the method of drawing squares for measuring distance, and as a result, their depiction of reality is reliable. Thus, whether one divides or combines them, east and west match each other without incurring any discrepancy."[103]

The use of the grid, together with scale indications, perhaps warrants the conclusion that maps like the *Yu ji tu* relied at least in part on direct and indirect measure-

94. Frank Leeming, "Official Landscapes in Traditional China," *Journal of the Economic and Social History of the Orient* 23 (1980): 153–204.

95. Quoted in Fan Ye, *Hou Han shu* (History of the Later Han, compiled fifth century A.D.), chap. 59; see the edition in 12 vols. (Beijing: Zhonghua Shuju, 1965), 7:1898 n.

96. Needham, *Science and Civilisation*, 3:537–38 (note 43). Needham's speculation about the net corresponding to the *xiu* seems to have little to support it, unless it is derived from a rectangular projection of the sky. For more on the *xiu*, see the chapters below on celestial mapping.

97. Needham, *Science and Civilisation*, 3:541 (note 43).

98. Zhang, *Lidai minghua ji*, chap. 3 (76) (note 48).

99. See Wang Yong, *Zhongguo dilixue shi* (History of geography in China) (1938; reprinted Taipei: Tiawan Shangwu Yinshuguan, 1974), 57–59. In this work Wang expresses some doubts about Pei Xiu's knowledge of the square grid system, but in his later work he changes his mind. See Wang Yong, *Zhongguo ditu shi gang* (Brief history of Chinese cartography) (Beijing: Sanlian Shudian, 1958), 20. In addition, Needham, *Science and Civilisation*, 3:539–41 (note 43), and Chen, "Historical Development of Cartography," 103–4 (note 53), attribute the cartographic grid to Pei Xiu.

100. Needham, *Science and Civilisation*, 3:539–41 (note 43). Needham's evidence for interpreting *wang* as implying verticality is not convincing. The Han dictionary *Shuowen jiezi* defines *wang* as "going into exile and gazing at [one's homeland] in the distance" (*Shuowen jiezi gulin*, ed. Ding Fubao, 9:5717b [note 92]). The graph *zhun* can mean "regulated" or "standard," thus the translation of *zhunwang* adopted here—regulated viewing or sighting.

101. *Jin shu*, chap. 35 (4:1040) (note 42).

102. This interpretation of *zhunwang* was proposed to me by Tsai Fa Cheng. Gari Ledyard proposes a different interpretation of *zhunwang* as "balanced sighting" from two or more reference points (personal correspondence).

103. Luo Hongxian, *Guang yutu* (ca. 1555), 6th ed. (1579; reprinted Taipei: Xuehai Chubanshe, 1969), preface, 2a. The quotation points to another function of the grid: to provide register marks that ensure accuracy when two sections are joined along their edges to form a single large map.

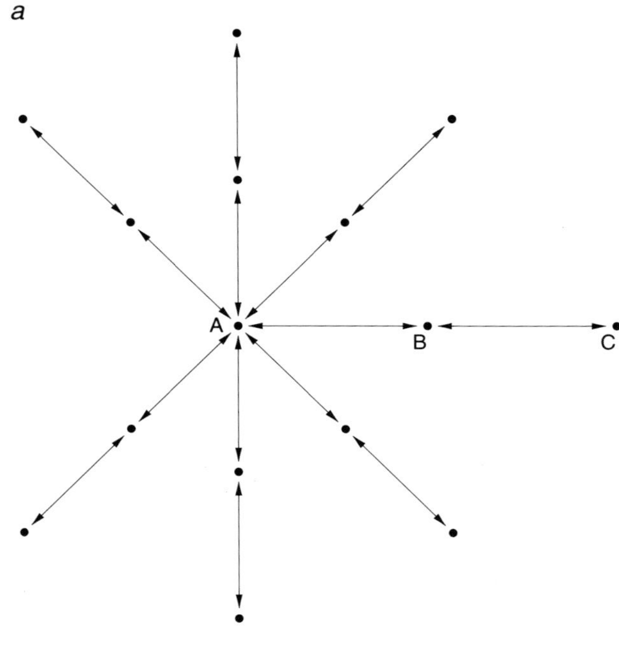

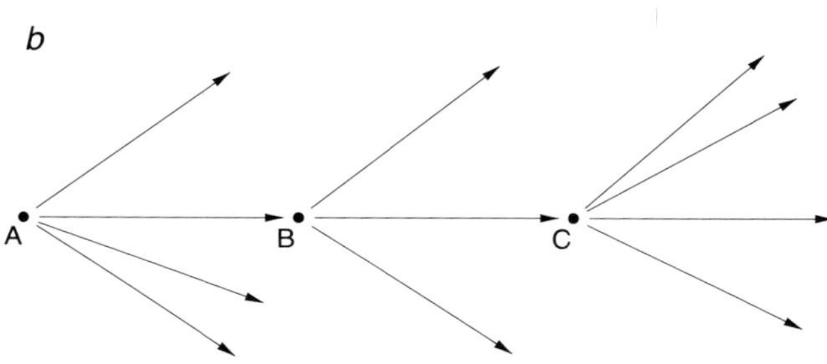

FIG. 5.24. INTERPRETATION OF *ZHUNWANG*. (*a*) The use of a single control point (or "regulated sighting") from which to observe all features (A) permits one to backsight and minimize errors in directional readings. (*b*) The use of multiple reference points for directional measurements results in an accumulation of errors as one moves from point to point (A to B to C) and takes one's bearings on local features.

ments of ground distances.[104] As I stated above, however, grid use does not necessarily imply an attenuation of the textualism discussed earlier. This is particularly true of large-area maps, in which case a mapmaker would often be unwilling or unable to survey all the distances involved and would have to rely on secondary sources. In preparing his map of China, the *Yutu* (Terrestrial map), Zhu Siben relied on earlier maps (including the *Yu ji tu*), literary sources, and personal observations, not necessarily on survey measurements. Zhu, though eclectic in his sources, cautions against using them uncritically. Some foreign countries were omitted from the *Yutu* because Zhu lacked reliable information: "As for the various foreign lands southeast of the South Sea and northwest of the great desert [Mongolia], although they at times send tribute to the court, their distance precludes investigation. Those who speak of them are unable to be specific. Those who are specific cannot be trusted."[105] Similarly, Luo Hongxian consulted literary sources when expanding Zhu's work. This willingness to rely on literary sources, though treating them critically, is perfectly consistent with Chinese cartographic theory and practice from Pei Xiu through Shen Kuo.

The state of Luo's atlas suggests a reason for this. It seems to have been conceived as a treatise: most of the work consists not of maps, but of verbal description. In fact, bibliographic treatises have traditionally classified maps in a manner that emphasizes their character as text. Maps and geographic works are classed together under the rubric "history" (*shi*)—in short, they were considered a form of narrative literature. Thus, during the Qing, maps, as we saw above, were important tools of historical research, both as source material and as a means of presenting results. The evidentiary approach Luo adopts in his cartography, based on careful attention to documentary sources and verification, might be considered as foreshadowing later *kaozheng* practices.

The textualist orientation persisted until at least the end of the nineteenth century when the central government attempted to Westernize Chinese mapmaking practices. The installation of Western techniques stressing measurement and mathematical projections did not displace text-based verification. According to the *huidianguan* (bureau of institutional studies), the work of compiling an atlas from maps submitted by provincial governments was complicated by divergent methods of mapmaking—this despite the promulgation of cartographic standards: "When [new maps] were compared with old accounts, there were still discrepancies and similarities. These were all checked by measuring distances and by widely searching through records in an effort to resolve ambiguities and reach a single conclusion."[106] In a departure from the European ideal, text together with measurement serves as the final authority.

CONCLUSION

In light of the material presented here, to subject Chinese

104. As, for example, P. D. A. Harvey has concluded in *The History of Topographical Maps: Symbols, Pictures and Surveys* (London: Thames and Hudson, 1980), 133–36.

105. Zhu Siben, preface to the *Yutu*, in Luo, *Guang yutu*, 1b (note 103).

106. *Qinding Da Qing huidian* (Imperially commissioned collected statutes of the Great Qing), 24 vols. (1899; reprinted Taipei: Zhongwen Shuju, 1963), 2:1022.

cartographic history to the kind of treatment once accorded early Western maps—as precursors of "modern scientific" cartography based on mathematical techniques—is to give undue weight to fragments of texts isolated from their total contexts. Such a treatment introduces a modern bias toward mathematics as the foundation of scientific knowledge into a context where the bias tended in the opposite direction. The persistence of a textualist attitude among cartographers in premodern China does not lessen the achievement of premodern Chinese cartography, but it shows it in a different light, suggesting the need for a reorientation in the way historians of Chinese cartography have regarded their subject.

The interplay between observation and textualism traced here is proposed as an explanatory model for some of the features observed in Chinese maps—features not always taken into account in previous accounts of Chinese cartography: the lack of consistency in applying the cartographic grid, the complementarity of text and map in Chinese atlases, and the heavy reliance on textual methods of verification even after mensurational techniques had developed to a high degree. The textualist tendency in Chinese cartography suggests that a fruitful approach to the study of Chinese maps might involve interpretive principles commonly invoked in literary study. More specifically, such research might involve what Schleiermacher called "subjective reconstruction": "One seeks to understand the writer intimately . . . to the point that one transforms oneself into the other." The point is "not to understand an ancient text in view of modern thinking, but to rediscover the original relationship between the writer and his audience."[107] Traditional Chinese cartography thus demands to be conceived in broader terms than it has been in previous modern studies. In other words, what is needed is a better match between interpreter and original audience. The Chinese intellectuals who made and read maps held *boxue* (broad learning), not specialization, as an educational ideal and seem to have regarded cartography in those terms.[108] This view of the subject is not unique. Late in the twentieth century, a textualist tendency can be seen in more than a few geographers who work in the academy, a setting in which what one has seen can seem less important than what one cites. Such is the reverence for the published word.

107. Friedrich D. E. Schleiermacher, "*The Hermeneutics*: Outline of the 1819 Lectures," trans. Jan Wojcik and Roland Haas, *New Literary History* 10 (1978): 1–16, esp. quotations on 14 and 6.

108. Thus it was with considerable trepidation that I, a specialist in literary studies, undertook this essay at all, since the research involved would make my incompetencies painfully apparent. I owe considerable debts to those who helped me compensate for those weaknesses by allowing me to stand on their shoulders: David Woodward, Kevin Kaufman, Brian Harley, Jude Leimer, Tsai Fa Cheng, and Nathan Sivin.

6 · Chinese Cartography among the Arts: Objectivity, Subjectivity, Representation

CORDELL D. K. YEE

> Those able to evoke a mood that persists beyond the rendition of physical objects completely understand the art of writing.
>
> —Liu Xie (ca. 465–522)
> *Wenxin diaolong*, chap. 46

> "World" does not refer only to scenes and objects; joy, anger, sorrow, and happiness are also a world within the human heart. Thus, those poems that can describe true scenes and objects and true emotions and feelings are said to have a world; otherwise, they are said to lack a world.
>
> —Wang Guowei (1877–1927)
> *Renjian cihua*, sec. 1

Cartography in China did not emerge as a representational practice fully independent from the visual and literary arts until late in the imperial period, under the influence of Western examples. One would expect it to have done so earlier only if one took Western cartography as the standard or norm. Although some would have it otherwise, cartography in China followed a different course. What contributed to its distinctiveness was an idea of representation that went beyond the duplication of physical forms, that stressed its multifaceted function. Neglect of this multidimensional concept of representation has resulted in a distorted view of Chinese cartographic history—as the development and refinement of a practice based on measurement and mathematical techniques. Chinese maps produced before the twentieth century, however, can be said to fall neatly into such a quantitative tradition only if one takes a too narrow view of what qualifies as a map. There is little doubt that maps were important for representing and understanding the physical world. What is at issue is the mode of that representation and understanding. Geometric and mathematical fidelity to observed reality was not an overarching aim: maps were often placed in contexts where they complemented verbal representations of geographic knowledge. As a means of storing geographic information, both verbal and graphic modes of representation were held to have their uses. What is now called cartography thus had its place in a unified conception of the arts.

To extend an argument begun elsewhere (pp. 91 and 101–4), the distinction between word and visual image, so strong in the Western tradition, is not nearly as sharp in China. This difference in traditions can be traced to differing conceptions of representation and its purposes. As a result of this critical difference, historians of Chinese cartography may need to adjust their definitions of what constitutes cartographic representation. The usual oppositions between visual and verbal, cartographic and pictorial, mimetic and symbolic representation may not apply.

THE RELATION BETWEEN ART AND REALITY

An example suggesting that such a change is necessary occurs in the *Zuozhuan* (Zuo's tradition [of interpreting the *Chunqiu*, or Spring and autumn annals]) and other early texts. The *Zuozhuan*, which perhaps dates from about 300 B.C., states that during the legendary Xia period, visual representations of information useful to travelers were molded into nine cast-bronze caldrons, the *jiu ding*. The historical existence of these caldrons is by no means certain, especially since there is no evidence of Bronze Age culture during the putative period of the Xia.[1] This situation, however, has not stopped previous scholars from speculating on the cartographic nature of the representations on them.[2] The *Zuozhuan*, written centuries after the caldrons were supposedly made, says that "representations" of objects from the "nine regions"

1. The *Zuozhuan*'s account may be supplemented by one in the *Shi ji* (Records of the grand historian, completed ca. 91 B.C.) by Sima Qian. According to Sima Qian's account, the nine caldrons were cast during the reign of Yu and handed down to the Chou rulers. When the Zhou's virtue declined and the Song state's altars were destroyed, the caldrons "sank into the waters and were seen no more." Before the time of Yu, the Huang Di (Yellow Lord) made "three precious caldrons, representing heaven, earth, and man." Sima Qian, *Shi ji*, chap. 28; see the modern edition in 10 vols. (Beijing: Zhonghua Shuju, 1959), 4:1392. The translated excerpts here are based on those in Burton Watson, trans., *Records of the Grand Historian of China*, 2 vols. (New York: Columbia University Press, 1961), 2:49.

2. See, for example, Wang Yong, *Zhongguo ditu shi gang* (Brief history of Chinese cartography) (Beijing: Sanlian Shudian, 1958), 1; and idem, *Zhongguo dilixue shi* (History of geography in China) (1938; reprinted Taipei: Shangwu Yinshuguan, 1974), 16–19.

FIG. 6.1. PAINTED POTSHERDS FROM THE YANGSHAO CULTURE. The radiating pattern on these sherds has been interpreted as a representation of the sun.
Size of the original: approximately 8.5 × 13 cm (at the widest points). Zhengzhou Municipal Museum. From Zhongguo Shehui Kexueyuan Kaogu Yanjiusuo, *Zhongguo gudai tianwen wenwu tuji* (Album of ancient Chinese astronomical relics) (Beijing: Wenwu Chubanshe, 1980), 3.

appeared on them. According to the *Zuozhuan*, the caldrons, in addition to securing the favors of heaven, were partly intended to represent knowledge of the nine regions, knowledge that would be useful to travelers. The knowledge imparted by the representations on the caldrons supposedly had magical-religious efficacy. Armed with this knowledge, travelers could ward off evils.[3] The *Zuozhuan*, however, gives no details about the representations themselves. It says nothing of the objects or information represented, so we cannot determine whether the caldrons can be considered cartographic artifacts.[4]

Some art historians question whether designs with the degree of formal resemblance attributed to the nine caldrons appeared before the *Zuozhuan* was composed. According to Loehr, art with concrete representations developed in late Bronze Age China, that is, during the late Chunqiu (722–468 B.C.) or during the Zhanguo (Warring States) period (403–221 B.C.), about the time the *Zuozhuan* was composed.[5] A fully representational art does not develop until the Han (206 B.C.–A.D. 220). Before that time any representational elements—to which early cartography may perhaps be linked—coexisted with the ornamental. In general, Loehr says, art had few references to reality. By implication, artifacts purporting to promote spatial understanding of geographic reality would have been unlikely. Art dating from before the Shang (ca. sixteenth to eleventh century B.C.) is generally limited to geometric patterns—for example, spirals, zigzags, and interlocked Ts. The same lack of references to reality characterized Shang and Zhou (ca. 1027–256 B.C.) art:

> Seen in the entire perspective of Chinese art history, most of the decorative designs on Shang and [Zhou] bronzes may be characterized as purely ornamental. They were typical creations of a phase that preceded representational art, the art concerned with, and dependent on, reality. The shift toward representation occurred during the Han period. The moment it occurred, the art of the ornament—heretofore the art of greatest consequence—took second place and began to stagnate. It is a noteworthy fact that at the same time the day of the bronze vessel, too, was over.[6]

3. *Zuozhuan* (ca. 300 B.C.), *Xuan* 3, in *Chunqiu jingzhuan yinde* (Concordance to the *Chunqiu* [Spring and autumn annals] and its commentaries), 4 vols. (1937; reprinted Taipei: Chengwen Chubanshe, 1966), 1:182.

4. According to the *Shi ji*, the caldrons were used to "boil offerings to Shangdi [Lord on High] and other spirits" (chap. 28 [4:1392]; translation based on Watson, *Grand Historian*, 2:49 [both in note 1]).

5. Max Loehr, *Ritual Vessels of Bronze Age China* (New York: Asia Society, 1968), 12; and idem, "The Fate of the Ornament in Chinese Art," *Archives of Asian Art* 21 (1967–68): 8–19.

6. Loehr, *Ritual Vessels*, 12 (note 5).

FIG. 6.2. BRONZE *HU* VESSEL. The surface of the vessel is decorated with Chinese graphs and realistic and abstract designs. Size of the original: 30.7 × 17–22.5 cm. Xinyang District Cultural Relics Administration. From Robert L. Thorp, *Son of Heaven: Imperial Arts of China*, exhibition catalog (Seattle: Son of Heaven Press, 1988), 53.

There are, to be sure, referential elements in pre-Han art, as Loehr recognizes. Certain designs on Neolithic pottery, for example, have been interpreted as early forms of numerals. One symbol in particular may be a representation of the sun (fig. 6.1), and another may be an early cosmographical diagram (fig. 1.12 above).[7] In addition, certain zoomorphs in Zhou bronzes have been interpreted as representing actual animals (fig. 6.2). Such examples, however, seem sparse in relation to the existing body of artifacts, and they may turn out to be exceptions that prove Loehr's rule.[8] At any rate, ancient conventions of representation are not well understood—the ritual bronzes were intended for use solely by initiates, who would presumably be versed in any conventions of representation—and before dismissing what seem to be purely decorative designs as nonrepresentational, one should remember that one may be dealing with a cotradition or coexisting practice of abstract representation.

A case in point is the ancient divination manual *Yi jing* (Book of changes), parts of which may date from the Western Zhou (ca. 1027–771 B.C.). The manual consists of oracle texts ("judgments" or "verbalizations") that are "attached" to a system of symbols. The symbols, made up of various combinations of unbroken (yang) and broken (yin) lines, fall into two groups: eight trigrams, made up of three lines, and sixty-four hexagrams, each composed of two trigrams. A commentary on the *Yi jing*, the "Xici zhuan" (Commentary on the attached verbalizations), dating from about the third century B.C., explains the hexagrams as representations of natural phenomena: "The holy sages instituted the hexagrams, so that phenomena might be perceived therein."[9] In the course of the *Yi jing*'s development, each of the hexagrams was given a name, regarded as encapsulating its significance. One hexagram, Qian (heaven), made up of six unbroken lines, embodies "the strong, light, active creative power, whose symbol is heaven," and the hexagram Kun (earth), made up of six broken lines, embodies the "dark, receptive, maternal element" symbolized by the earth.[10] From this one could infer that, by the Han, these two hexagrams represent heavenly and earthly phenomena, and the "Xici zhuan" does suggest that the *Yi jing* provides access to astronomical and geographical knowledge: "The *Yi* contains the measure of heaven and earth; there-

7. The representation of the sun appears on pottery dating from about 3000 B.C., and the "cosmographical" symbol appears on pottery dating from about 2900–2400 B.C. See Zhongguo Shehui Kexueyuan Kaogu Yanjiusuo (Archaeological Research Institute, Chinese Academy of Social Sciences), *Zhongguo gudai tianwen wenwu tuji* (Album of ancient Chinese astronomical relics) (Beijing: Wenwu Chubanshe, 1980), 3, 17, and 113; and Kwang-chih Chang, *The Archaeology of Ancient China*, 4th ed. (New Haven: Yale University Press, 1986), 167–69 and 172. On the origins of Chinese graphs, see Cheung Kwong-yue, "Recent Archaeological Evidence relating to the Origin of Chinese Characters," trans. Noel Barnard, in *The Origins of Chinese Civilization*, ed. David N. Keightley (Berkeley and Los Angeles: University of California Press, 1983), 323–91.

8. For further discussion of representation in Zhou art, see Wen Fong, "The Study of Chinese Bronze Age Arts: Methods and Approaches," 20–34, esp. 29–33, and Ma Chengyuan, "The Splendor of Ancient Chinese Bronzes," 1–19, esp. 6–10, both in *The Great Bronze Age of China: An Exhibition from the People's Republic of China*, ed. Wen Fong (New York: Metropolitan Museum of Art, 1980).

9. "Xici zhuan," A.2, in *Zhou yi yinde* (Concordance to the *Zhou yi* [*Yi jing*]) (1935; reprinted Taipei: Chengwen Chubanshe, 1966), 39. The translation is that in Richard Wilhelm and Cary F. Baynes, trans., *The I Ching or Book of Changes*, 3d ed. (Princeton: Princeton University Press, 1967), 287.

10. Hellmut Wilhelm, *Change: Eight Lectures on the "I Ching,"* trans. Cary F. Baynes (1960; reprinted New York: Harper and Row, 1964), 50, 59. It is not clear whether the names are as early as the hexagrams themselves, which appear in rudimentary form during the Shang, without the Zhou divination texts.

fore it enables us to comprehend the way of heaven and earth. Looking upward, we use it to comprehend the signs in the heavens; looking down, we use it to examine the patterns of the earth."[11] Of course the "Xici zhuan," compiled centuries after the oldest stratum of the *Yi jing*, does not necessarily reflect the original intentions behind the hexagrams. But the possibility that it is transmitting a much older tradition, though seldom accepted, cannot be completely discounted. The rough dates of the "Xici zhuan" fall slightly after the period Loehr identifies as marking the beginning of concrete representationalism in art—the fourth century B.C. The example of the "Xici zhuan" at least opens the possibility that at the time art with clear references to reality was being produced, conventions of interpreting abstract designs with no resemblance to the empirically observed objects as referring to reality had already been established.

These two broad categories—formal resemblance as a means of representing empirical reality and its counterpart, a "supraempiricism" that sought to go beyond physical appearance—were seen not as opposed, but as complementary. For this reason I have tried to avoid terms such as "realism" and "symbolism." In European aesthetic discourse these terms have been applied to tendencies often regarded as countercurrents. To emphasize the complementarity of these two tendencies in the Chinese context, I have applied the term "representation" to both. Both are means of depicting what is conceived as being in the world. The two modes of representation described here do differ, however, in the kinds of things they seek to depict, and I use terms such as "concrete" and "abstract," "objective" and "subjective," and "material" and "spiritual" in an attempt to render that distinction.

The urge to go beyond material form is not usually associated with maps, but this aesthetic tendency does manifest itself in Chinese cartography. This is not to say that formal resemblance had no part in the Chinese art of mapmaking. It is clear from the artifactual and textual record that such an idea of representation had been formulated by the Qin period (221–207 B.C.).[12] Qin Shihuang's tomb, for example, reportedly contains replicas of mountains, streams, and buildings. During the Han, artisans were producing bronze censers clearly recognizable as mountains (fig. 6.3), and soon after, if not during, the Han, literary artists were valuing graphic representations for their correspondence to geographic reality.

FIG. 6.3. BRONZE INCENSE BURNER—RELIEF MODEL OF A MOUNTAIN FROM THE HAN DYNASTY.
Size of the original: height 18.4 cm and diameter of dish 20.3 cm. By courtesy of the Board of Trustees of the Victoria and Albert Museum, London (M. 275–1910).

Literature, Maps, and Representation of the Material World

The literature of the Han dynasty was marked by the appearance of a form known as the *fu* (rhapsody), characterized by the exhaustiveness of its descriptions, particularly of geographic subjects. In light of this geographic emphasis, it is hardly surprising that a number of works refer to mensuration, surveying, and maps. Fidelity to the material world accounts for a large part of the significance of these references, but their importance, as we will see below, extends beyond that aim. An example of the descriptive meticulousness of the Han rhapsody occurs in the literary writings of Zhang Heng (78–139), also known for his work in cartography and astronomy. His "Er jing fu" (Two metropolises rhapsody, ca. 107), bears perhaps a twofold cartographic significance. First is its use of measurement and surveying, often closely associated with cartography, as political metaphor. In a description of how the Han emperor Gaozu established his capital at Chang'an, measurement and surveying are suggestive of the dynasty's initial political order:

> He [Gaozu] measured the diameter and
> circumference,
> Reckoned the length and breadth.
> They built the city walls and moat,

11. "Xici zhuan," A.3, in *Zhou yi yinde*, 40; the translation is based on that of Wilhelm and Baynes, *I Ching*, 293–94 (note 9). The compound translated as "patterns of the earth," *dili*, means the morphology of the land and in later texts is often interpreted as meaning "geography."

12. For more citations of the evidence for this assertion, see Michael Sullivan, *The Birth of Landscape Painting in China* (Berkeley and Los Angeles: University of California Press, 1962), 4–6, 10–15.

Constructed the outer enclosures.
He adopted various patterns from the capitals of
the eight directions,
And never considered following the measures of the
ancient past.[13]

It is doubtful that Zhang Heng's description of the founding emperor, written nearly three centuries after the event, has much basis in literal fact. Zhang is taking advantage of the association, commonplace in much pre-Han and Han political discourse, between measurement and the ideal role of a ruler. As it says in the *Huainanzi* ([Book of the] Master of Huainan), a philosophical text compiled about 120 B.C., "Law is the rule and measure of the empire, and the level and plumb line of the ruler."[14] In the application of law, for example, the ideal ruler is like a suspended plumb line, showing no bias. Zhang uses the trope to suggest that the planning for the new capital reflected a restoration of political order by the founding emperor—order especially welcome after the excesses of the Qin. The emperor's planning followed the example of the ancient sage-kings:

In the past, when a former king planned the city,
He thoroughly examined the nine provinces,
And there was no site that went unsurveyed.
The jade tablet measured the shadow;
It was neither too short nor too long.
They sought a place where all winds and rains
converged,
And then established the Royal City.
They examined curvature, direction, and
topography.[15]

But though the emperor had the right idea in his planning, his work, Zhang suggests, did not amount to a complete renewal of the way of the ancient sage-kings. For his capital, the emperor Gaozu selected a site different from, and hence less favorable than, that chosen by the ancients, and as a result the reigns of succeeding rulers were marked by increasing disorder, culminating in the usurpation of power by Wang Mang (45 B.C.–A.D. 23). When the Han returned to power, it established its capital at Luoyang, the same site chosen by the ancients, and as a result, Zhang implies, political order continued up to his time.

Zhang Heng's rhapsody also contains possible evidence for the political as well as literary use of maps. According to Zhang's description, the founding emperor consulted unspecified "patterns" (*cai*) of other capitals when planning his own. Whether the patterns abstracted from those other capitals took the form of verbal descriptions or city plans is unclear, but city plans or maps predating Zhang Heng have been discovered in tombs (see below, p. 147).

Besides a possible description of map use, Zhang's rhapsody contains phraseology suggesting that it could have been composed with reference to a map. In parts of it, Zhang might be described as constructing a "verbal map." His *fu*'s directional orientation is pronounced, as the account moves left and right, then south and north:

The first capital of the Han house
Lay on the banks of the Wei River. . . .
To the left, there are
The double defiles of the Yao and Han,
The barrier of Taolin,
Connected by the Two Hua peaks. . . .
To the right, there is
The gap of Longdi,
Which partitions China from the barbarian
lands. . . .
At its southern front, there are
Zhongnan and Taiyi,
Twisting upward tall and stately . . .
At its northern rear, there are
High hills and level plains . . .
In the distance, there are
Nine Peaks and Sweet Springs.[16]

There is no external evidence that Zhang Heng used maps to compose his rhapsody, but at least one writer, inspired by Zhang's *fu*, did do so—Zuo Si (ca. 250–ca. 305). In his preface to the "San du fu" (Three capitals rhapsody), Zuo Si describes the preparatory research for his *fu*: "When I first thought of writing the 'Three Capitals' in imitation of the 'Two Metropolises,' for the mountains and streams, cities and towns, I consulted maps. Birds and animals, plants and trees, I have verified in gazetteers."[17] The aim of his research is to enhance the credibility of his *fu*: "One who praises an object considers first and foremost its true nature, and one who honors a deed should base himself on the facts. Without the truth and facts, what can the reader believe?"[18] Zuo Si

13. Xiao Tong, comp., *Wen xuan* (Literary selections, completed ca. 526–31), ed. Hu Kejia (1809; reprinted Kyōto: Chūbun Shuppansha, 1971), 2.5a. The translation is that in David R. Knechtges, trans. and annotator, *Wen Xuan; or, Selections of Refined Literature* (Princeton: Princeton University Press, 1982–), 1:187. The "Er jing fu" was completed about 107. See Sun Wenqing, *Zhang Heng nianpu* (Chronological biography of Zhang Heng), rev. ed. (Shanghai: Shangwu Yinshuguan, 1956), 48.

14. *Huainanzi* (ca. 120 B.C.) attributed to Liu An (d. 122 B.C.), in *Huainanzi zhu* (Commentary to *Huainanzi*, third century), ed. Gao You, chap. 9; see the modern edition (Taipei: Shijie Shuju, 1962), 140.

15. *Wen xuan*, 3.6a–b; the translation is that of Knechtges, *Refined Literature*, 1:249 (note 13).

16. *Wen xuan*, 2.2b–3b; the translation is that of Knechtges, *Refined Literature*, 1:183, 185 (note 13). That the south precedes the north in Zhang's description suggests that south was at the top of the map, if there was a map.

17. *Wen xuan*, 4.13a–b; the translation is that of Knechtges, *Refined Literature*, 1:339 (note 13).

18. *Wen xuan*, 4.13b; the translation is that of Knechtges, *Refined Literature*, 1:339 (note 13).

regards maps as reliable representations of geographic knowledge, so much so that he stakes the veracity of his literary work on them. Their authority is equal to that of firsthand observation. We may recall that it was a prime concern of Pei Xiu (223–71) to make maps that conformed to observed reality, and Zuo Si happens to be roughly contemporaneous with Pei.

A similar belief in the verisimilitude of maps is perhaps expressed by Tao Qian (365–427) in two lines from the first of his series of poems on reading the *Shanhai jing* (Classic of mountains and seas). In these lines, Tao describes the *Shanhai jing*, a geographic work, as containing illustrations of some kind: "I glance over the pictures [maps?] of mountains and seas. / In the space of a nod, I completely [comprehend] the universe."[19] The graph translated as "pictures," *tu*, can also mean "maps," but the context of the poem does not make clear whether maps are specified. What is clear is that Tao Qian takes for granted the verisimilitude of the illustrations, their correspondence to reality. In other words, they have mimetic (in a sense closer to Aristotelian as opposed to Platonic usage) and consequently cognitive value. Although much of the material in the *Shanhai jing* is considered mythical, Tao clearly does not treat it as such. He regards the illustrations in the book as providing knowledge of the universe by their ability to convey reality.

From the text of the *Shanhai jing*, it is not difficult to divine why Tao might have been impressed by the work. The *Shanhai jing*, the present text of which was first edited by Liu Xin (ca. 50 B.C.–A.D. 23), describes the hydrography, minerals, fauna, and flora of various mountain systems. The geographic information contained in the text may have served as glosses to illustrations, or perhaps maps.[20] These illustrations do not survive; whatever illustrations extant editions may contain are renditions of artists after the Song (960–1279). Nevertheless, as scholars since the Song have observed, some passages in the book read much like the descriptive notes found on a map, or like captions pointing to items in an illustration.[21] The following excerpt—besides exemplifying the kind of mineralogical, botanical, and zoological information to be found in the *Shanhai jing*—reads as if one were moving eastward across a map or illustration drawn on a scroll:

> [The head of the Que mountain range] overlooks the western sea. It produces much cinnamon and much gold and jade. There is a plant there whose shape resembles scallions and that has blue blossoms. Its name is *zhuyu*, and if one eats it, one will not starve. . . . And 300 *li* to the east is the mountain called Tangting, which has many *yan*-fruit trees, many white gibbons, much crystal, and much gold. Another 380 *li* to the east is the mountain called Yuanyi. In it there are many strange beasts, and in its waters are many strange fish. . . . Another 370 *li* to the east is the mountain called Niuyang.[22]

Tao Qian, however, in his poetic sequence, is not responding specifically to the text, but is commenting on accompanying illustrations. The poet's response to the illustrations implies a shared interest in verisimilitude between illustrator (perhaps mapmaker) and poet. Such a common interest would not have been entirely coincidental. During the Liuchao (Six Dynasties) period (222–589), verisimilitude, *xingsi*, was a technical term in Chinese aesthetics and literary criticism, particularly in discussions of landscape poetry. The prevalence of geographic topics in literature—in rhapsodies and poems—may have prompted the literary theorist Liu Xie (ca. 465–522) to link the land with poetic creation: "Mountain forests and river banks are actually the deep repository of literary thought."[23] Liu Xie also asserts that verbal art can reproduce the effects of visual art:

> Poets respond to things, endlessly associating ideas. They flow among the ten thousand things, completely submerged in visual and aural sensations. They depict the atmosphere and paint the appearance of things, following their changing aspects. . . . Recently literature has been valued for verisimilitude [*xingsi*]. Writers perceive the true form of landscape, and penetrate the appearance of grass and plants. . . . Thus this technique of apt expression for form can be likened to the ink used for imprinting seals, for an impression made reproduces the seal to the smallest detail with-

19. Tao Qian, "Du *Shanhai jing*" (Reading the *Shanhai jing*, ca. 400), in *Tao Yuanming juan* (Collected materials on Tao Yuanming [Tao Qian]), 2 vols. (Beijing: Zhonghua Shuju, 1962), 2:286–87.

20. Wang Yong speculates that the book was known by at least two other titles, both of which imply the inclusion of illustrations: *Shanhai tu* (Maps/illustrations of mountains and seas) and *Shanhai jing tu* (Illustrated classic of mountains and seas). See Wang's *Zhongguo ditu shi gang*, 1 (note 2). There is little evidence to support Wang's surmises. The first known reference to the *Shanhai jing* occurs in a Han work, the *Shi ji*, which makes no mention of maps, illustrations, or even alternative titles. It calls attention to the "strange things" described in the *Shanhai jing*. See *Shi ji*, chap. 123 (10:3179) (note 1).

21. See Zhang Xincheng, *Weishu tongkao* (Comprehensive study of forged books) (1939; reprinted Taipei: Hongye Shuju, 1975), 575–76.

22. *Shanhai jing* (Classic of mountains and seas, compiled ca. second century B.C.), chap. 1, in *Shanhai jing jiaozhu* (Edited and annotated *Shanhai jing*), ed. Yuan Ke (Shanghai: Shanghai Guji Chubanshe, 1980), 1–3.

23. Liu Xie, *Wenxin diaolong* (The literary mind and the carving of dragons, ca. 500), chap. 46, in *Wenxin diaolong yizhu* (*Wenxin diaolong* interpreted and annotated), ed. Lu Kanru and Mou Shijin, 2 vols. (Jinan: Qilu Shushe, 1981–82), 2:345. A complete English translation of the *Wenxin diaolong* is *The Literary Mind and the Carving of Dragons*, trans. Vincent Yu-chung Shih (Hong Kong: Chinese University of Hong Kong, 1983).

> out further carving and cutting. Thus we can perceive appearances by looking at words, or know the season from the graphs.[24]

Liu Xie's comments affirm the importance of language as a means of representing physical reality: language, Liu suggests, functions not just as a means of preserving utterances, but also as a way of seeing. After all, the Chinese written language, with its pictographic elements, constitutes the basis of calligraphy, traditionally the most highly regarded of the visual arts. A painter's training, in fact, traditionally began with writing. "Learning painting," says a Song dynasty theoretician of landscape art, "is no different from learning calligraphy."[25] Calligraphy and painting both emphasized brushwork, and by practicing calligraphy a painter strengthened hand-eye coordination and gained a sense of arrangement and proportion. In a sense, painting was an extension of language. The interconnections among poetry, calligraphy, and painting were felt to be so strong that the three arts became known as the "three perfections" (*san jue*).

The Dual Function of Representation in Literature

Part of the reason for this strong association among the arts may have been a shared physical medium—the scroll—on which all three arts came to share the same space. Another, and perhaps more important, reason for the unified view of the arts is that language and painting were regarded as having similar representative powers—a notion that contrasts with modern ideas about the primacy of visual modes in representing reality.[26] At the root of this Chinese belief in shared modes of representation is not a simple assertion that linguistic art is like painting, in the manner of the Horatian tag *ut pictura poesis* (as is painting, so is poetry).[27] The idea of representation, or mimesis, in the arts, as it developed during the Han and later, does not end with fidelity to external or empirical reality. Thus, representational value is imparted to designs as seemingly abstract as the trigrams and hexagrams in the *Yi jing*. The artist Wang Wei (415–43), for example, refers approvingly to a statement by the poet-official Yan Yanzhi (384–456) that "a painting should not be merely for the exercise of craft, but when perfected should correspond in form to the images in the *Yi* [*jing*]."[28] That is, a painting would not necessarily look like a hexagram, but it would represent configurations of reality underlying the material world. Similarly, in the case of Zhang Heng's rhapsody, the physical descriptions suggest an abstract political principle. Physical descriptions are a means to another end, that of penetrating appearances to reach underlying realities, an aim attainable through language as well as through the visual arts.

Insofar as the arts strove for formal resemblance, literary artists could learn much from the cartographer or illustrator, as examples cited previously suggest. The attainment of formal resemblance alone did not satisfy Chinese art theorists, and this, as we will see below, had some effect on cartographic representation. The relationship between the literary and visual arts was not one-sided. From literature, the visual artist might come to appreciate another aspect of representation—that representation also has a subjective element.

"Poetry expresses intention," says the *Shu jing* (Book of documents); objective representation, though considered an important aspect of literary art, does not completely account for its value. In the "Wen fu" (Rhapsody on literature), the literary critic Lu Ji (261–303) makes this point clear, using imagery associated with mensurational science:

> [The writer] encages heaven and earth in form; he grinds the ten thousand things against the tip of his brush. . . . Forms have a thousand variations; things do not have a single measure. Various and fleeting, forms are hard to delineate. Words display talent and show one's skill. . . . In the midst of being and nonbeing, one struggles; facing the shallow and the deep, one does not yield. Although one leaves the square and discards the compass, one hopes to probe forms and exhaust appearances.[29]

To paraphrase Lu Ji, even with instruments useful for objective measurement, the square and compass, the forms of the phenomenal world are difficult to capture: there is "no single measure." In trying to capture reality, a writer must do more than faithfully render the objective

24. Liu, *Wenxin diaolong yizhu*, chap. 46 (2:341–45) (note 23). For a detailed discussion of verisimilitude and Chinese poetry, see Kang-i Sun Chang, "Description of Landscape in Early Six Dynasties Poetry," in *The Vitality of the Lyric Voice: Shih Poetry from the Late Han to the T'ang*, ed. Shuen-fu Lin and Stephen Owen (Princeton: Princeton University Press, 1986), 105–29.

25. Guo Xi, *Linquan gaozhi ji* (Lofty aims in forests and springs, eleventh century), in *Hualun congkan* (Collection of treatises on painting), ed. Yu Anlan (Haiyan), 2 vols. (Beijing: Renmin Meishu Chubanshe, 1962), 1:16–31, esp. 18.

26. Manifestations of the visual emphasis in modern culture, especially in the United States, include the importance of television and the decline in the status of the spoken (and written) word. On the primacy of vision in modern culture, see Walter J. Ong, *Orality and Literacy: The Technologizing of the Word* (London: Methuen, 1982).

27. For an account of the history of this tag in the European tradition, see Jean H. Hagstrum, *The Sister Arts: The Tradition of Literary Pictorialism and English Poetry from Dryden to Gray* (Chicago: University of Chicago Press, 1958).

28. Zhang Yanyuan, *Lidai minghua ji* (Record of famous painters through the dynasties, completed 847), chap. 6; see the modern edition (Beijing: Renmin Meishu Chubanshe, 1963), 132.

29. Lu Ji, "Wen fu" (third century), in *Wen xuan*, 17.3b–4b (note 13).

appearance of things: a writer must give up the square and compass. A literary work also represents something of its author. This idea is elaborated by Liu Xie, who, as we saw above, also asserted the power of linguistic art to render objects in the physical world: "The writer is moved emotionally and produces words. Readers open the literary work and enter the writer's sensibility; they follow the waves to trace the source, so that even though it is secluded, it will be manifest. An age may be remote so that we cannot see its face, but reading its literature, we can always see its heart."[30] Representation in literature thus fuses the rendering of physical reality with that of subjective experience. This differs from traditional Western art criticism, which tends to polarize, not unite, objectivity and subjectivity, preferring to limit representation to the former and to regard the rendering of subjective experience as a separate phenomenon—expression.

Painting and Representation

Representation in Chinese painting ultimately serves a dual function similar to that of literature. To be sure, formal resemblance, as I noted above, was an important aim in the visual arts. But as an artistic standard, it was often paired with another: *qiyun*, or "breath-resonance." The use of this compound dates back to at least the sixth century when the critic Xie He (fl. ca. 500–535?) used it to refer to personal and representational vitality, or "breath," and harmoniousness, "resonance," of execution.[31] By striving for *qiyun*, a painter could achieve the purpose of painting, "to express the spirit through form," in the words of Gu Kaizhi, a fourth-century painter.[32] The "breath" that painters tried to express involved two senses of innerness: the innerness of the objects represented and the innerness of the painter. In aesthetic theory the two were inseparable: recognition of the innerness, the underlying essence, of external phenomena required the active perception of an artist. In this way painting could be linked more closely with calligraphy. As one art theoretician expresses it, the two arts "arise in thoughts and feelings and are transferred to silk and paper": "Painting is like calligraphy . . . 'words are the sounds of the mind; calligraphy is the delineation of the mind'; when the sound and delineation take form, whether one is a superior man or an ignoble one is revealed."[33] In theory, then, the visual arts could not only express the artist's emotions, but also reveal the artist's moral quality—both were bound up with the notion of vitality, breath-resonance in the personal sense. An emphasis on subjectivity in representation would naturally lead to deviation from formal resemblance.

In practice, however, the complementarity of the verisimilitude and breath-resonance—or, if one prefers, the dialectic of objectivity and subjectivity—does not express itself so unequivocally. The history of Chinese artistic practice in painting can be viewed as a series of pendulum swings between formal likeness and expression or, in terms taken from Chinese art criticism, between the artisanly and the scholarly. Here artisanly implies professionalism, being supported by and having to satisfy someone else—thus an outward orientation. Scholarly implies amateurism, in its best sense: art performed not as one's livelihood or main pursuit, but at leisure, to satisfy only oneself—thus an inward orientation. At any one time, one tendency might be dominant, but the other would not be entirely forgotten.[34] From the Tang (618–907) to the Northern Song (960–1126), formal resemblance seems to have been dominant, prompting a number of critics of the period to remind artists of the necessity for expressiveness in art—reminders that, as will be seen below, were heeded by mapmakers. The Tang dynasty art historian Zhang Yanyuan, for example, remarked that paintings of his day lacked breath-resonance because artists focused too narrowly on formal resemblance. He advised painters to strive for breath-resonance, for "if a painter sought for only breath-resonance in a painting, formal likeness would be achieved."[35] In the Northern Song, Su Shi (1037–1101), who achieved a lasting reputation in poetry, calligraphy, and painting, helped to initiate a reaction against the emphasis on formal verisimilitude, a reaction that became dominant in succeeding periods as painters increasingly explored the possibilities of self-expression with a less rigid adherence to the observed world. In Su's view, the attainment of formal likeness was hardly the primary end of art: "To discuss painting in terms of formal resemblance / Is a view similar to a child's."[36] According to Su, artists should sub-

30. Liu, *Wenxin diaolong yizhu*, chap. 48 (2:390) (note 23).

31. Xie He, *Gu huapin lu* (Classification of ancient painters, ca. sixth century), ed. Wang Bomin (Beijing: Renmin Meishu Chubanshe, 1962), 1.

32. See Zhang, *Lidai minghua ji*, chap. 4 (118) (note 28).

33. Guo Ruoxu, *Tuhua jianwen zhi* (Record of experiences in painting, 1074), seventeenth-century edition, chap. 1; reprinted in *Songren huaxue lunzhu* (Song treatises on painting), ed. Yang Jialuo (Taipei: Shijie Shuju, 1962), 30–31.

34. A number of scholars have proposed oscillating models for Chinese art history. See, for example, Max Loehr, "Some Fundamental Issues in the History of Chinese Painting," *Journal of Asian Studies* 23 (1964): 185–93; and Wen C. Fong, "Archaism as a 'Primitive' Style," in *Artists and Traditions: Uses of the Past in Chinese Culture*, ed. Christian F. Murck (Princeton: Art Museum, Princeton University, 1976), 89–109. For a discussion of the problem of periodization in the arts, see Maureen Robertson, "Periodization in the Arts and Patterns of Change in Traditional Chinese Literary History," in *Theories of the Arts in China*, ed. Susan Bush and Christian Murck (Princeton: Princeton University Press, 1983), 3–26.

35. Zhang, *Lidai minghua ji*, chap. 1 (3) (note 28).

36. Su Shi, *Dongpo shi jizhu* (Poetry of Dongpo [Su Shi] with collected annotations, compiled twelfth century), ed. Wang Shipeng, *Siku quanshu* edition, 27.22b.

ordinate the attempt to render appearance to the attempt to go beyond appearance: "The artisans of the world may be able to completely render its forms, but as far as its principles are concerned, unless one is a superior person with outstanding talent, one cannot discern them."[37]

Su Shi's sentiments are echoed by one of his contemporaries, Guo Ruoxu (fl. ca. 1075), whose *Tuhua jianwen zhi* (Record of experiences in painting, 1074) was accepted as an authoritative treatise as early as the twelfth century. Most of the precious works of the past, Guo observes, were painted by those of noble character who imparted to their works their "elevated and fine emotions." The lesson to be learned is this: "In general, a painting must encompass breath-resonance in order to be called a treasure of the age."[38] In short, painting, like poetry, is an "imprint of the mind" (*xinyin*). The similarity between the powers of the two arts was not lost on Song dynasty artists. Interartistic comparisons were commonplace. Bush has generalized aptly: "Chinese painting always strove for the status of poetry. . . . Despite changing evaluations of painting's function, most Chinese art critics remained consistent in their low estimate of formal likeness. For them painting like poetry could fuse mood and scene, joining the subjective and objective worlds."[39] Thus Su Shi states, "Poetry and painting are at root the same: / Effortless craft and clear freshness." Because of their common nature, it is possible, as Su Shi suggests elsewhere, to conceive of poems as "paintings without forms" and paintings as "unspoken poems."[40] One of Su's acquaintances, the landscape master Guo Xi (ca. 1010–ca. 1090), suggested that the comparison of poetry and painting was conventional wisdom: "Just as people previously have said, poems are paintings without form; paintings are poems with form."[41] He also discerns a human element in landscape painting, anthropomorphizing one of its major subjects—mountains: "A mountain has streams as blood vessels, grass and trees as hair, mist and clouds as its countenance."[42] By focusing merely on reproduction of form, a painter risks sacrificing the vitality beneath external appearances.

The breath-resonance seen here as uniting painting and poetry also joins art and science. The term *qi*, "breath" or "pneuma," a constituent of the compound *qiyun*, pervades traditional Chinese scientific thought. The link between art and science can also be seen in Guo Xi's passage on mountains, quoted above, where he draws on an idea common in traditional Chinese medical texts: that of body as microcosm. Understanding for artist and scientist involved seeing the resonance or correspondence between human being and nature. Human understanding could lead to natural understanding, and vice versa. Artist and scientist both dealt with vital processes.[43]

Thus Shen Kuo (1031–95), noted for his scientific writing, also demonstrates an understanding of aesthetic theory. Here he expresses the belief that pictorial representation must do more than reproduce external forms:

> The wonders of calligraphy and painting should be sought through spiritual understanding; it is difficult to seek them through material forms. Many of those who look at paintings in this day and age can pick out the flaws in formal likeness, arrangement, and color, and no more. It is rare to see those able to discern the abstruse principles and hidden structures.

According to Shen, a painter could give up adherence to material reality and still communicate something true. For example, a painting in his own home, Shen writes in the *Mengxi bitan* (Brush talks from Dream Brook, ca. 1088), shows a plantain tree in a snowy landscape. The scene is improbable but is one that, in Shen's words, "creates pattern, penetrates the spirit, and is so different

37. Su Shi, *Jingjin Dongpo wenji shilüe* (Arranged and presented prose collection of Dongpo [Su Shi] with brief commentary, presented 1191), ed. Lang Ye, *Sibu congkan* edition, 54.9a–b. Su Shi's use of the word *li*, or "principle," in this and other passages has prompted discussions of Song dynasty aesthetics in terms of Neo-Confucian thought, in which *li* functioned as a basic metaphysical term. The devaluation of formal likeness, expressed by Su Shi and others, however, is not necessarily a consequence of a Neo-Confucian outlook. It is possible to detect ideas and terminology from diverse sources in the same aesthetic writings. Chinese art criticism, Pauline Yu has noted, is highly syncretistic, so that one can find support for just about any philosophical position in a particular critical work (see her "Formal Distinctions in Chinese Literary Theory," in *Theories of the Arts in China*, ed. Susan Bush and Christian Murck [Princeton: Princeton University Press, 1983], 27–53, esp. 27). A problem with this syncretist view is that in the Song and after, it is hard to locate texts outside explicitly religious writing that can be identified as "purely" Buddhist or "purely" Daoist. Thus it may be more reasonable to regard the conventional views of the literate elite as inseparably combining ideas from a number of traditions for a thousand years before the Song. It has been, however, a standard scholarly exercise to identify particular influences or allusions in art. Examples of studies exploring particular philosophical contexts of Chinese art are James F. Cahill, "Confucian Elements in the Theory of Painting," in *The Confucian Persuasion*, ed. Arthur F. Wright (Stanford: Stanford University Press, 1960), 115–40; Richard Mather, "The Landscape Buddhism of the Fifth-Century Poet Hsieh Ling-yün," *Journal of Asian Studies* 18 (1958): 67–79; and Lothar Ledderose, "Some Taoist Elements in the Calligraphy of the Six Dynasties," *T'oung Pao* 70 (1984): 246–78.

38. Guo, *Tuhua jianwen zhi*, chap. 1 (30) (note 33).

39. See Susan Bush, *The Chinese Literati on Painting: Su Shih (1037–1101) to Tung Ch'i-ch'ang (1555–1636)* (Cambridge: Harvard University Press, 1971), 23.

40. Su, *Dongpo shi jizhu*, 27.22b (note 36); Bush, *Chinese Literati on Painting*, 25, 188 (Chinese text) (note 39). Su Shi's comparison resembles that of Simonides of Ceos, who said that poetry is a speaking picture, and painting silent poetry.

41. Guo, *Linquan gaozhi ji*, 1:24 (note 25).

42. Guo, *Linquan gaozhi ji*, 1:22 (note 25).

43. For a detailed treatment of the terminology and ideas of Chinese science, particularly those of traditional Chinese medicine, see Manfred Porkert, *The Theoretical Foundations of Chinese Medicine: Systems of Correspondence* (Cambridge: MIT Press, 1974).

that it attains divine meaning—this is difficult to discuss with common people."[44]

Shen Kuo's writings on art are rarely discussed in relation to his ideas on cartography. In histories of cartography, Shen is usually seen as exemplifying a quantitative tradition. But the evidence is not so univocal.[45] Shen's own writings and other sources suggest that the distinction between map and painting may not have been as clearly demarcated as previous historians have suggested—that what is now considered to be cartography participated in the same "economy" of representation as poetry and painting. Because of this participation, one would expect cartographic artifacts to fall within a representational complex accommodating both rigid adherence and no adherence to material forms.

The Artistic Economy: Common Technologies of Production

At a more basic level than theory of representation, the interrelatedness of calligraphy, painting, and cartography would have been reinforced by their shared technology of production and common physical media: for example, silk, paper, wood, and stone. Silk was often used as a medium because of its elegance, resilience, and light weight, which made it preferable to wood; wood was used for maps as early as the late Zhanguo period. Despite silk's desirability as a base material for artistic works, its expensive production process and social value precluded using it to satisfy ordinary demands. An alternative, inexpensive medium, however, was available from the second or first century B.C.[46] Like silk, paper was held in esteem—even sent to the imperial capital as tribute—but it was less expensive to produce than silk.

On silk, paper, and even wood, maps—like paintings and calligraphic works—were drawn with brush and ink. The Han silk maps discovered at Mawangdui, for example, seem to have been drawn with brush and pigments, though those who have studied these maps firsthand have said little about how they were made. The Tang mapmaker Jia Dan (730–805) is also known to have painted maps, but virtually nothing is known of his drafting methods.[47]

Stone was used as a medium because of its permanence.[48] Engraving a map on stone was a three-stage process, beginning with the drawing of a rough draft, normally on perishable material such as paper, followed by the transfer of the image or text to the stone surface. This image took the form of a provisional outline to guide the hand of the stonecutter. Creating the outline involved two simultaneous processes: incising thin, rectilinear guidelines with a chisel, and incising the image itself. Distinctions were usually made between sections for image or decoration and sections for text. Finally, the carving of the image itself took place: a rounded or V-shaped chisel was used to broaden and rectify the initial cuts.

Paper could be pressed on the engraved image, then tamped with a soft, ink-soaked cloth to produce a reproduction. This aspect of the artifacts engraved in stone allies them technically with those printed from woodblocks, in use in China from at least the eighth century. Both printing and engraving required that an image be transferred from paper to a prepared surface and then cut out with chisels and other cutting tools.[49] Stone surfaces were typically engraved in intaglio, whereas woodblocks were typically, though not always, carved in relief. As a result, their corresponding images would be quite distinct, the woodblock producing a black mirror image on a white background and the stone engraving producing a white-on-black image oriented in the same way as the original.

Although the techniques for reproducing images from

---

44. Shen Kuo, *Mengxi bitan* (ca. 1088), chap. 17, par. 280, in *Xin jiaozheng Mengxi bitan* (Newly edited *Mengxi bitan*), ed. Hu Daojing (1957; reprinted Hong Kong: Zhonghua Shuju, 1975), 169. The same paragraph number (280) can be used to locate this passage in *Mengxi bitan jiaozheng* (*Mengxi bitan* edited), 2 vols., ed. Hu Daojing, rev. ed. (1960; reprinted Taipei: Shijie Shuju, 1961).

45. The idea that Shen exemplified a quantitative tradition is only partly true for his scientific work in areas other than cartography. For example, Shen was aware of the limits of systematic observation and computation for attaining an understanding of the cosmos. See Nathan Sivin, "On the Limits of Empirical Knowledge in the Traditional Chinese Sciences," in *Time, Science, and Society in China and the West*, Study of Time, vol. 5, ed. J. T. Fraser, N. Lawrence, and F. C. Haber (Amherst: University of Massachusetts Press, 1986), 151–69, esp. 159–61.

46. Joseph Needham, *Science and Civilisation in China* (Cambridge: Cambridge University Press, 1954–), vol. 5, pt. 1, by Tsien Tsuen-hsuin, *Chemistry and Chemical Technology: Paper and Printing* (1985), 1–2, 38–40.

47. According to his biography in the *Jiu Tang shu*, Jia Dan painted numerous maps. In one instance, because of the press of official duties, he engaged an artisan to paint a map for him. See Liu Xu et al., *Jiu Tang shu* (Old history of the Tang, compiled 940–45), chap. 138; see the edition in 16 vols. (Beijing: Zhonghua Shuju, 1975), 12:3784 and 3786.

48. Unlike stone, other documentary media such as paper, silk, bamboo, and wood were vulnerable to fire and water.

49. Nicolas Trigault's 1615 published version of Matteo Ricci's diaries includes a description of the transfer of image from paper to woodblock: "The text is written in ink, with a brush made of very fine hair, on a sheet of paper which is inverted and pasted on a wooden tablet. When the paper has become thoroughly dry, its surface is scraped off quickly and with great skill, until nothing but a fine tissue bearing the characters remains on the wooden tablet. Then, with a steel graver, the workman cuts away the surface following the outlines of the characters until these alone stand out in low relief." See *China in the Sixteenth Century: The Journals of Matthew Ricci, 1583–1610*, trans. Louis J. Gallagher from the Latin version of Nicolas Trigault (New York: Random House, 1953), 20–21.

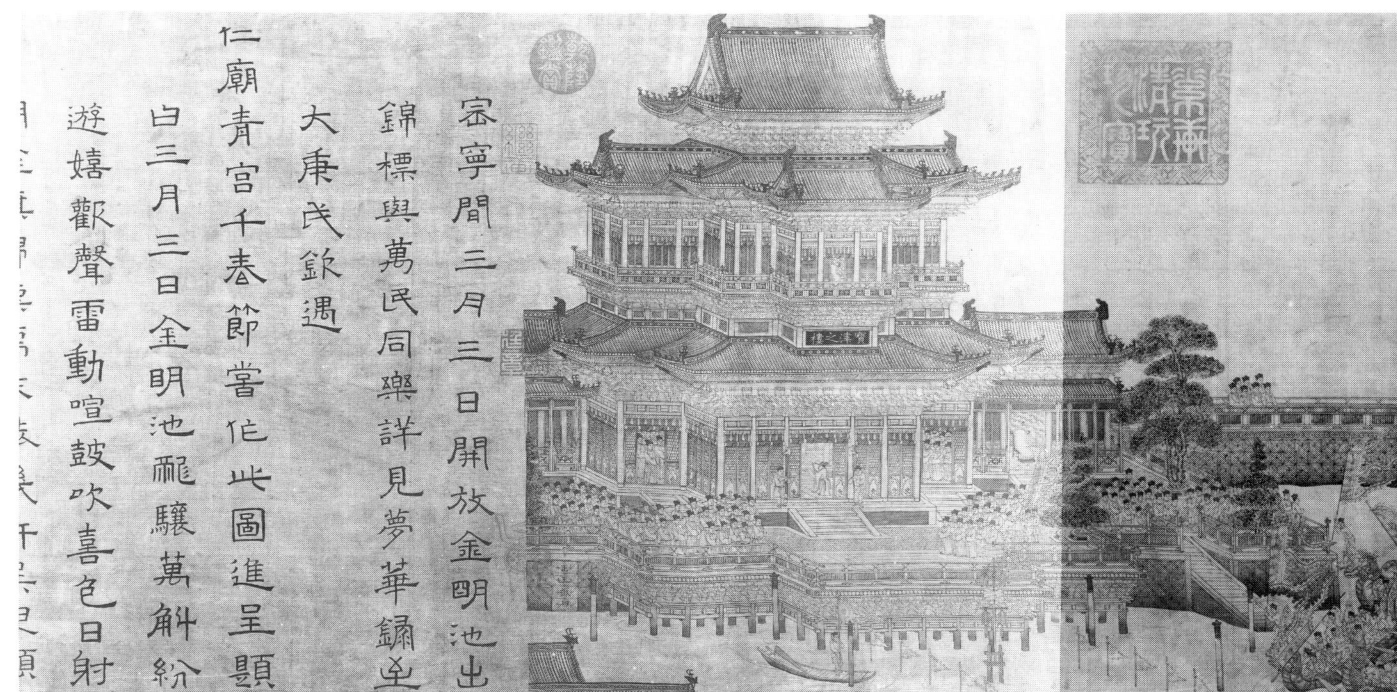

FIG. 6.4. *JIEHUA*, OR RULED-LINE PAINTING. This is a detail from the hand scroll *Longchi jingdu* (Dragon boat regatta on Longchi, 1323), ink on silk, by Wang Zhenpeng. Size of the entire original: 30.2 × 243.8 cm. By permission of the National Palace Museum, Taipei.

stone and from wood are similar, stone and woodblock maps seem to have assumed distinct functions. Tsien describes the distinction using an apt spatial metaphor: documents on perishable materials were intended primarily for "horizontal" communication among contemporaries, and those on more durable materials were for "vertical" communication across generations.[50] Stone stele maps implied permanence, an expression of cultural heritage to be passed down through the ages: "Bronze and stone are engraved on," a Song dynasty writer notes, "because one wants to transmit what is lasting."[51] Thus the difficult and time-consuming task of engraving on a durable material like stone was reserved for only the most authoritative maps. This attitude may offer at least a partial explanation of why the Song stone maps were more carefully drawn than printed gazetteer maps made centuries later. In contrast, the comparatively simpler process of woodblock engraving was intended foremost for duplication. It seems to have implied immediacy: wood was inexpensive; it was softer than stone and thus easier to carve; a woodcut could be carved quickly, as Jesuit missionaries witnessed in the late Ming (1368–1644).[52] Making corrections was a simple matter of recutting or inserting plugs. Stone engravings could also be used for duplication, but their bulk and weight made them impractical for reproduction in large quantities. In addition, "rubbing" stone was much slower and trickier than printing from an inked block.

The aesthetic qualities of woodblock maps compared with those on stone often seem to reflect their distinct communicative roles. Woodblock maps generally lack the attention to arrangement and design, the pictorial detail, and the fine line work often lavished on stone images. One reason for this was that woodblock maps were most often part of gazetteers, where they served primarily as illustrative accompaniments to a text. Their incorporation into text, furthermore, determined their format. Whereas stones and silk scrolls could be cut to a desired size, woodblock maps were restricted to the dimensions of the printed page. Limitations of space may account for some of the "faults" observed in gazetteer maps.

50. Tsuen-hsuin Tsien, *Written on Bamboo and Silk: The Beginnings of Chinese Books and Inscriptions* (Chicago: University of Chicago Press, 1962), 179.

51. Zhao Yanwei, *Yunlu manchao* (Random jottings at Yunlu, 1206), *Siku quanshu* edition, 4.17a.

52. See *China in the Sixteenth Century*, 20–21 (note 49).

Cartography and the Visual Arts: Conceptual and Stylistic Connections

Beyond their shared media, there is theoretical justification for the linkage between cartography and visual art, since both maps and paintings present informational content in the form of visual representations. In other words, map content, like pictorial content, often takes the form of analogues meant to reproduce certain characteristics of the objects represented—characteristics that are apprehended visually. Because of this shared mode of representation, historians of art and cartography have previously speculated about connections between cartography and the visual arts but have felt hampered by a lack of map artifacts.[53] In recent years artifacts for the Han have been discovered, but even without artifacts the evidence for connections between maps and painting is compelling. Like Liu Xie's conception of literature, the idea of painting also has links to geography—it is perhaps connected etymologically to land configuration. According to the Han dictionary *Shuowen jiezi* (Explanation of writing and explication of graphs), the graph *hua* (to paint, painting) means "boundary" and "represents the four boundaries of a field [*tian*]."[54] The *Shuowen* may not be a reliable guide to etymology, but it at least serves as a useful indicator of Han opinion and misconceptions about it. In the case of painting, the *Shuowen*'s opinion seems to have been influential. Since at least the Song dynasty, landscapes have constituted the most highly valued genre of painting, and maps by Zhang Heng and Pei Xiu appear in Zhang Yanyuan's catalog of paintings.[55] Zhang Yanyuan provides another indication that Zhang Heng enjoyed some repute as a painter: he records an anecdote in which Zhang Heng paints a strange beast with his toes.[56]

53. See, for example, Sullivan, *Birth of Landscape Painting*, 35–37 (note 12); Wang, *Zhongguo ditu shi gang*, 25–28 (note 2); and Alexander C. Soper, "Early Chinese Landscape Painting," *Art Bulletin* 23 (1941): 141–64, esp. 149. Soper states that it is unlikely that Han mapmaking "had any important effect on the development of a landscape art," but he seems to leave open the possibility of influence in the opposite direction.

54. Xu Shen, comp., *Shuowen jiezi* (Explanation of writing and explication of graphs, compiled ca. 100), s.v. *hua*, in *Shuowen jiezi gulin* (Collected glosses to the *Shuowen jiezi*), 12 vols., ed. Ding Fubao (Taipei: Shangwu Yinshuguan, 1959), 3:1275a.

55. See Zhang, *Lidai minghua ji*, chap. 3 (76) (note 28).

56. Zhang, *Lidai minghua ji*, chap. 4 (102) (note 28). For a translation of this anecdote, see William Reynolds Beal Acker, trans. and annotator, *Some T'ang and Pre-T'ang Texts on Chinese Painting*, 2 vols. (Leiden: E. J. Brill, 1954–74), 2.1:12–13.

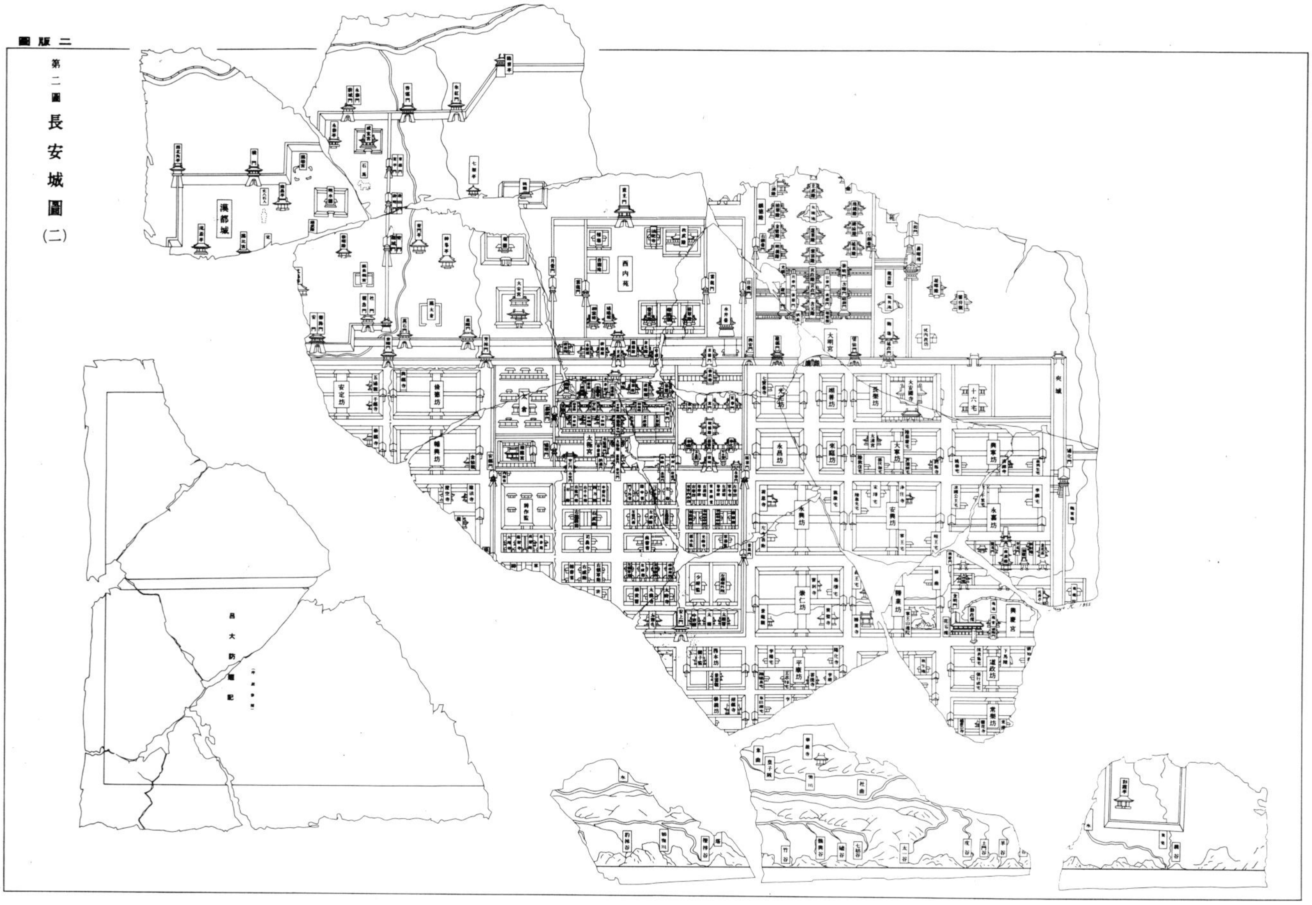

圖版二

第二圖 長安城圖 (二)

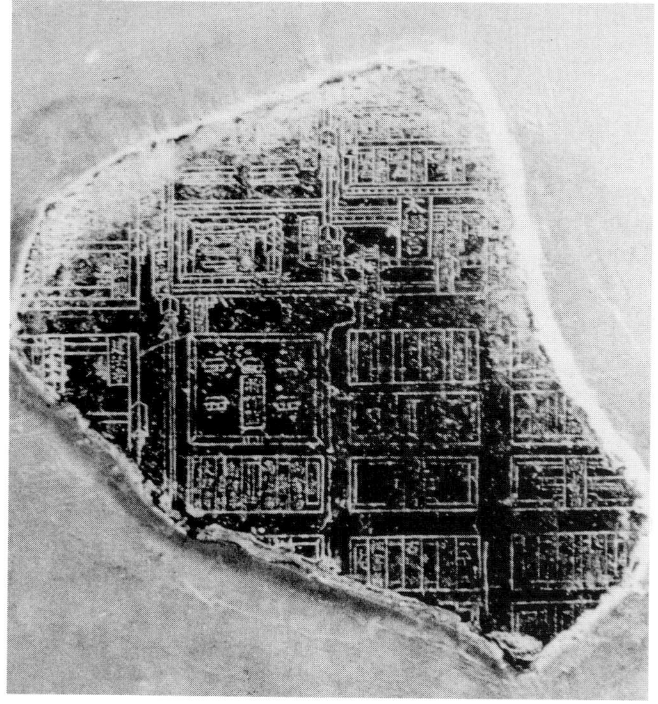

FIG. 6.5. DETAIL OF A MAP OF CHANG'AN, 1080. This detail of a lithographic copy of a rubbing of a stone map of Chang'an, by Lü Dafang (from a modern reproduction of the copy), appears to use ruled lines. To the left is a fragment of the stone map.
Size of the entire original: estimated to have been 200 × 136 cm. Kyōto University Humanities Research Institute. From Hiraoka Takeo, *Chōan to Rakuyō: Chizu* (Chang'an and Luoyang: Maps), T'ang Civilization Reference Series, no. 7 (Kyōto: Jinbunkagaku Kenkyūsho, Kyōto University, 1956), map 2. Photograph of the fragment courtesy of Cao Wanru, Institute for the History of Natural Science, Academia Sinica, Beijing.

Another anecdote, besides illustrating the military significance attached to maps, suggests that mapmaking called for the skills of a visual artist:

> Lady Zhao in the court of the lord of Wu was the younger sister of the prime minister Zhao Da. She was skilled in painting; her technique had no peer. Between her fingers, using colored silk, she could weave brocades with clouds and dragons. The large ones were a full *chi* in size; the small ones were one *cun* square. In the palace she was called the matchless weaver. Sun Quan [r. 228–48, first emperor of Wu] often lamented that the states of Wei and Shu had not been con-

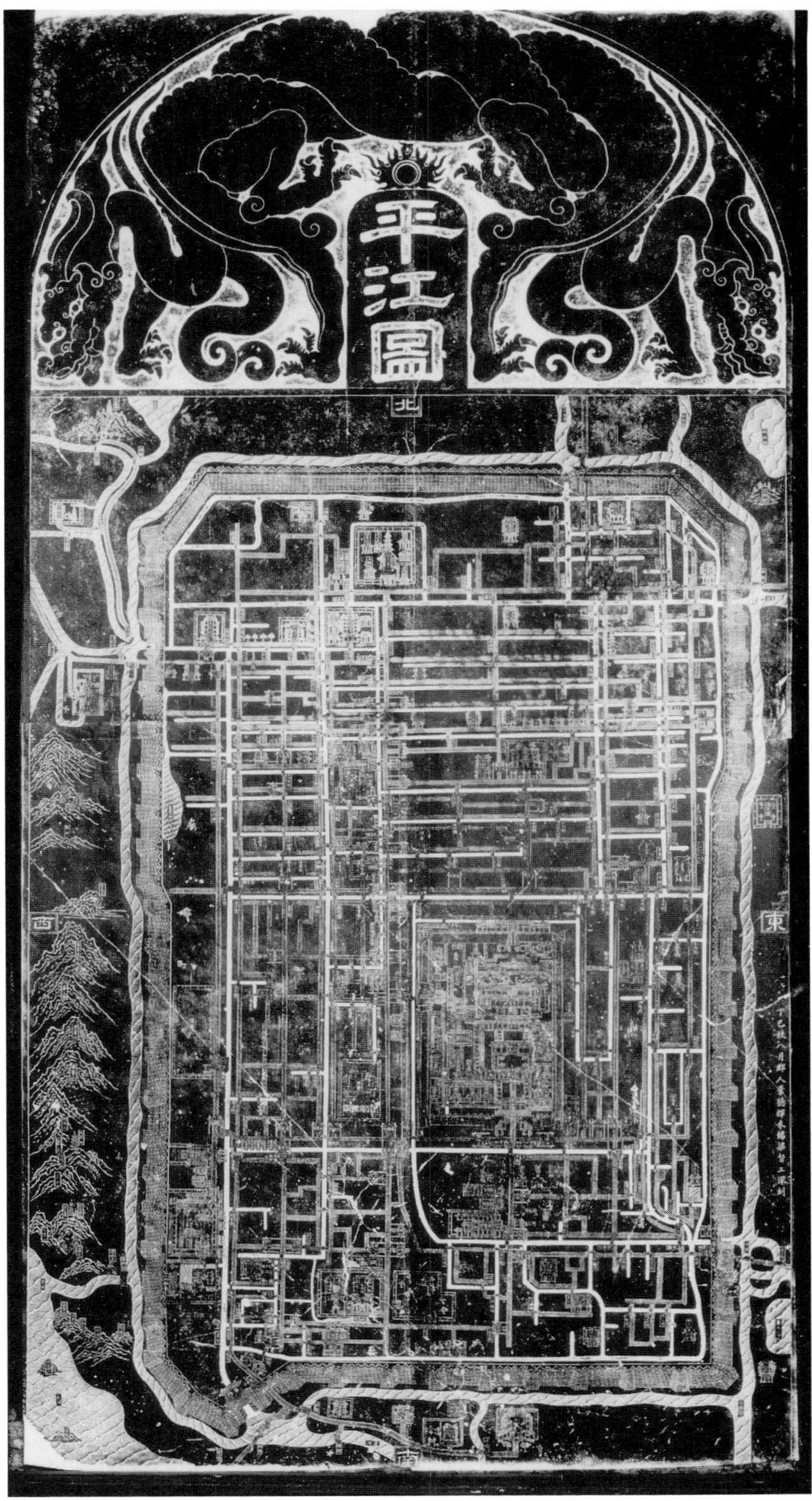

FIG. 6.6. THE *PINGJIANG TU* (MAP OF PINGJIANG PREFECTURE [PRESENT-DAY SUZHOU, JIANGSU PROVINCE]). This is a rubbing of the *Pingjiang tu*, carved on stone in 1229 and seeming to use ruled lines. The map depicts more than 640 features of the landscape, both human and natural. Human features include temples, administrative and military organizations, workshops, bridges, and roads. Natural features include hills and mounds, rivers, lakes, marshes, and streams. Directions are marked on the map; north is at the top. Compare figures 6.5, 6.7, and 6.8.

Size of the original: 279 × 138 cm. Photograph courtesy of Cao Wanru, Institute for the History of Natural Science, Academia Sinica, Beijing.

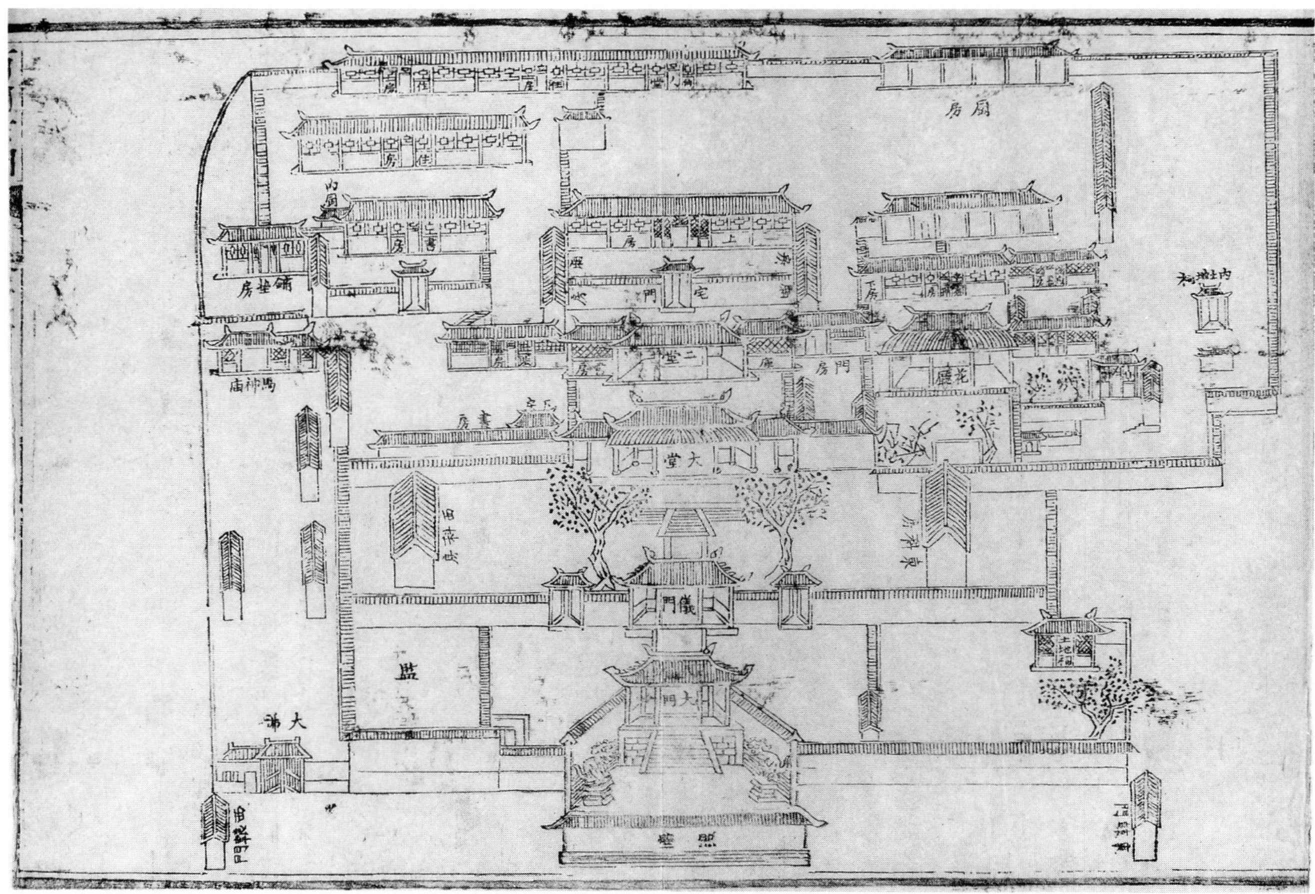

FIG. 6.7. WOODBLOCK MAP OF THE SEAT OF LOCAL GOVERNMENT AT JIZHOU IN PRESENT-DAY HEBEI PROVINCE. Besides its use of what appear to be ruled lines, this example is notable for its variable perspective, used on city maps since the Han dynasty. Compare figures 6.5, 6.6, and 6.8. Size of the original: 19.5 × 29 cm. From *Jizhou zhi* (Gazetteer of Jizhou, 1831), 1.25b–26a. Reproduced courtesy of the Harvard-Yenching Library, Harvard University, Cambridge.

> quered. To facilitate troop movements, he was thinking of getting someone skilled in painting to make maps of their mountains, streams, topography, and military positions. Zhao Da then presented his younger sister, and Sun Quan asked her to draw the rivers and lakes of the nine regions and their mountains. The Lady said: "It is extremely easy for pigments to fade; they cannot be preserved for long. But I am able to embroider a map." She made the various principalities on a square piece of silk, and drew the five peaks, the [Yellow] River and the sea, and cities, as well as the disposition of troops. When the map was finished and presented to the king of Wu [Sun Quan], people called her the "wonder of needlework."[57]

One style of painting seems particularly close to the drafting style often associated with mapmaking: *jiehua* (ruled-line painting), the only category of painting that involved tools other than the brush. It was used for detailed and scaled depiction of objects, especially architectural subjects (see fig. 6.4). Practitioners of this style of painting were skilled not only with drafting tools such as the calibrated ruler, compass, and square, but also with surveying instruments such as the water level and plumb line and with building calculations (*mujing suanfa*).[58]

Although the use of rules and compasses in drawing dates back to antiquity, the category *jiehua* and its terminology seem to have developed in the Song. In fact, one source credits Guo Zhongshu, a painter of the Northern Song, with mastering and defining the measured drawing or "boundary" style. In the beginning, *jiehua* seems to have been somewhat of a pejorative term among

57. Wang Jia, *Shiyi ji* (Gleanings, fourth century), *Baibu congshu jicheng* edition, 8.2a–b. A less detailed account of Lady Zhao's map appears in Zhang, *Lidai minghua ji*, chap. 4 (105–6) (note 28).

58. More detailed discussions of *jiehua* can be found in Robert J. Maeda, "*Chieh-hua*: Ruled-Line Painting in China," *Ars Orientalis* 10 (1975): 123–41; and Joseph Needham, *Science and Civilisation in China* (Cambridge: Cambridge University Press, 1954–), vol. 4, pt. 3, with Wang Ling and Lu Gwei-djen, *Physics and Physical Technology: Civil Engineering and Nautics* (1971), 104–7.

the scholarly elite; it derived from an artisanal craft with tools borrowed from the carpenter's trade—tools that Lu Ji, one will recall, had urged artists to discard. *Jiehua*, like many maps, seems to have occupied a zone between fine art and practical craft, and a number of maps seem to have been produced with the aid of straightedges: for example, city plans and gazetteer maps (figs. 6.5 to 6.8).

The links between painting and cartography seem to have gone beyond technique. There is also some evidence of a conjunction between painting theory and cartographic theory. Pei Xiu's emphasis on correspondence, as I mentioned previously, parallels an emphasis on verisimilitude in aesthetics. Xie He, for example, formulated six principles of painting, two of which are "correspondence to objects, that is, resemblance to forms," and "division and planning, that is, placement and arrangement."[59] Xie He does not elaborate on these principles, but correspondence and arrangement are concerns that Pei Xiu shares. Even before Xie He, Zong Bing (375–443) had thought about the idea of scale in relation to landscape painting: "Now, if one spreads thin silk to capture the distant scene [of the Kunlun Mountains], the form of Kunlun's Lang Peak can be encompassed in a square *cun*. A vertical stroke of 3 *cun* will equal a height of thousands of *ren* [equal to 80 *cun* or 8 *chi*], and a horizontal stretch of several *chi* will form a distance of 100 *li* [equal to 1,800 *chi*]."[60]

59. Xie, *Gu huapin lu*, 1 (note 31). For a translation of Xie's six principles, see Susan Bush and Hsio-yen Shih, comps. and eds., *Early Chinese Texts on Painting* (Cambridge: Harvard University Press, 1985), 40.

60. Zong Bing, *Hua shanshui xu* (Preface to painting landscape, ca. fifth century), in Zhang, *Lidai minghua ji*, chap. 6 (131) (note 28).

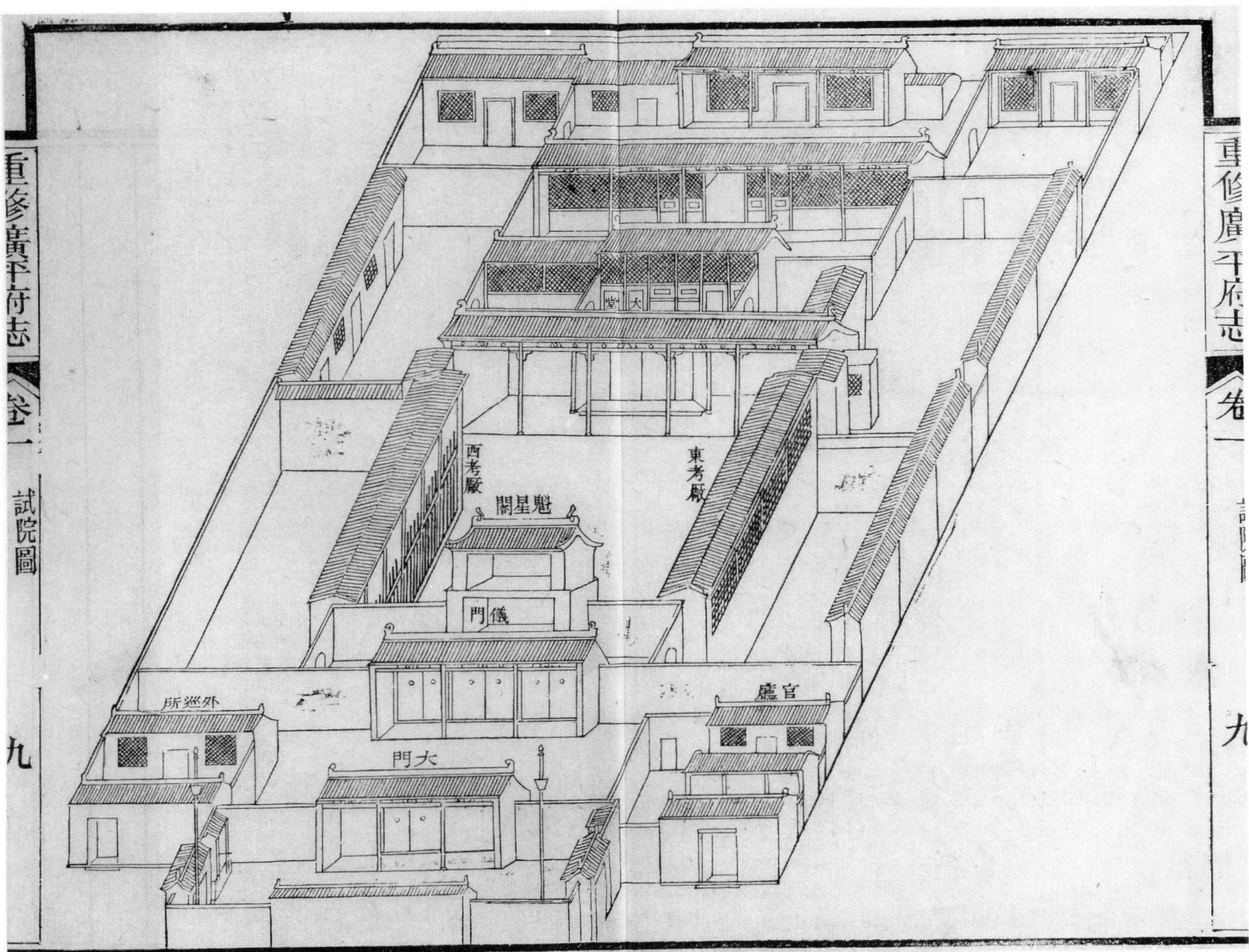

FIG. 6.8. WOODBLOCK ILLUSTRATION OR MAP, 1894. This image is of the examination hall in Guangping Prefecture in present-day Hebei Province (compare figs. 6.5 to 6.7).
Size of the original: 23 × 27 cm. From *Guangping fu zhi* (Gazetteer of Guangping Prefecture, 1894), chap. 1, map 9. Reproduced courtesy of the Harvard-Yenching Library, Harvard University, Cambridge.

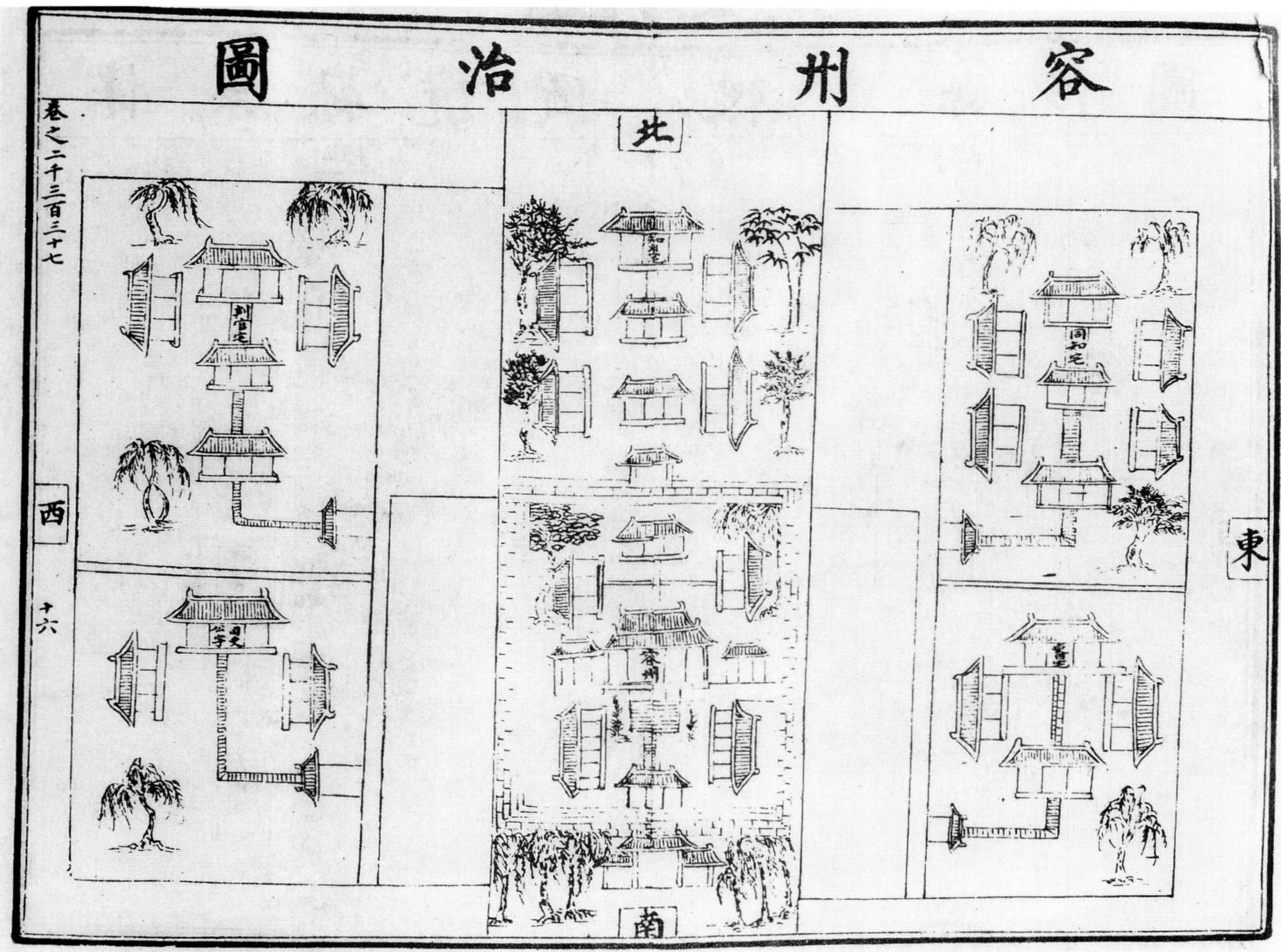

FIG. 6.9. WOODBLOCK MAP OF THE SEAT OF LOCAL GOVERNMENT AT RONGZHOU IN PRESENT-DAY GUANGXI PROVINCE. Here and in figure 6.10 one must change the orientation of the map to view the objects correctly. Size of the original: 15 × 19.5 cm. From *Yongle dadian* (Grand encyclopedia of the Yongle reign period [1403–24], presented 1409), photo-reprint edition (Taipei, 1962), chap. 2337.16ab. Reproduced courtesy of the Harvard-Yenching Library, Harvard University, Cambridge.

In Chinese painting, however, pictorial scale tends to dominate natural scale; that is, the size of an object depicted was determined by needs of design and not rules of geometric perspective. Foreground features might be diminished to avoid obstruction and overemphasis, while distant objects might be enlarged to act as counterpoint to the middle distance and foreground. The practice of variable scale obviously conflicts with Pei Xiu's insistence on consistent map scale to "preserve actualities." Many mapmakers, however, apparently chose to adhere to the pictorial technique of variable scale well into the Ming and Qing (1644–1911), as suggested by an artifactual record containing an abundance of maps that do not maintain a consistent scale. This and other examples of the conjunction between painting and mapmaking perhaps help explain the ambiguity of the graph *tu* in its application to drawings and pictures or to charts and maps.

A shared conception of space also strengthens the bond between mapmaking and painting. To state some broad generalizations, the experience of space was dynamic and fluid, intimately related to one's experience of time. Space, emptiness, was regarded almost as an entity in itself—as such it was boundless and unlimited. Objects could be measured and defined; space could not be so fixed, since it changed with vantage point and time. As a result, no abstract geometrical system governed space, and points within it were not definable or delimitable in any absolute terms. This conception of space, consonant with what Needham has described as an organismic and

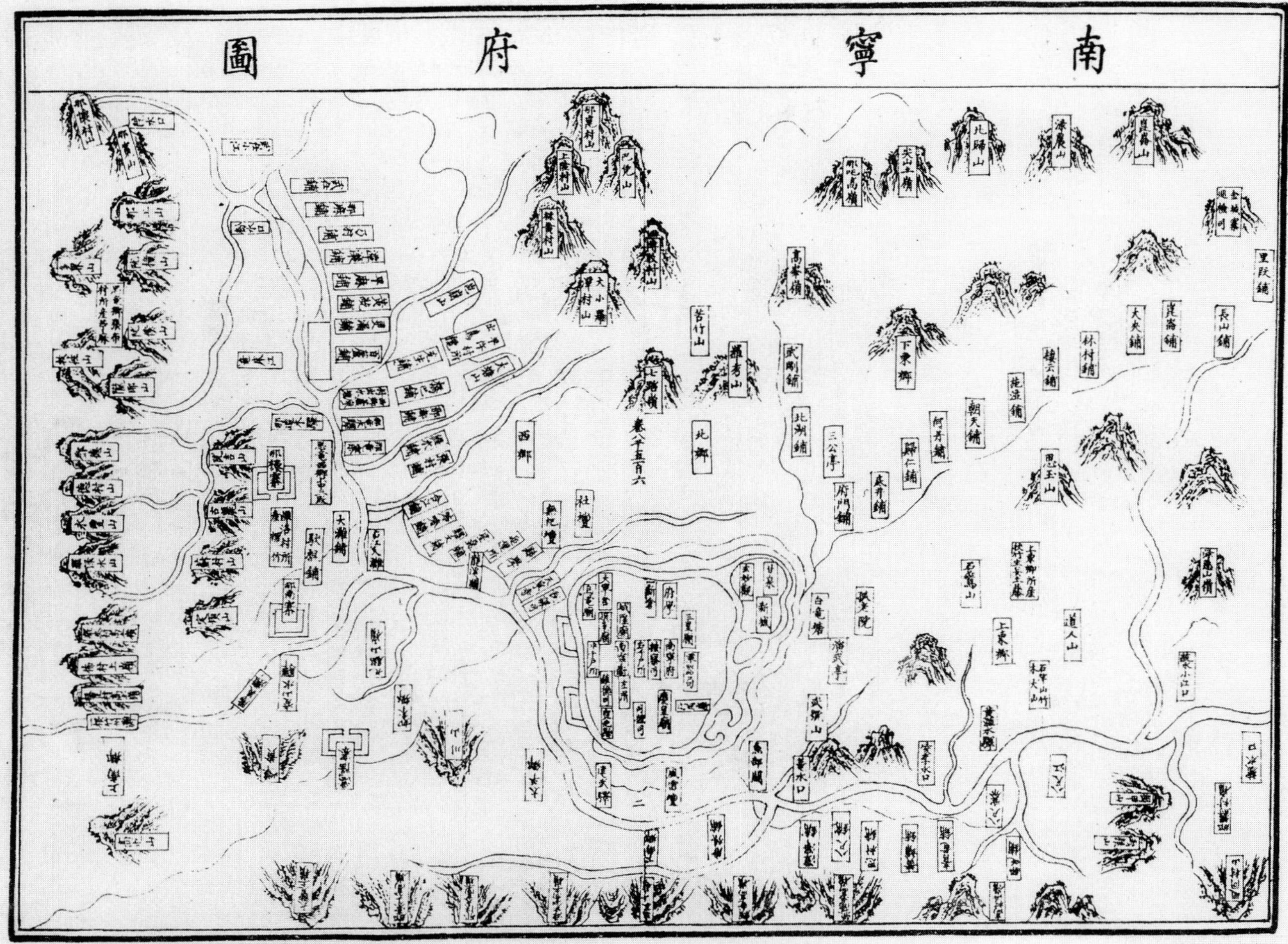

FIG. 6.10. WOODBLOCK MAP OF NANNING PREFECTURE IN PRESENT-DAY GUANGXI PROVINCE. Compare figures 6.9 and 6.11.
Size of the original: 14.3 × 19 cm. From *Yongle dadian* (presented 1409), photo-reprint edition (Taipei, 1962), chap. 8506.2ab. Reproduced courtesy of the Harvard-Yenching Library, Harvard University, Cambridge.

processual worldview, differs from the one that has prevailed among European cartographers and artists since about the fifteenth century—a conception that defines space abstractly, as an entity that is bounded, static, and therefore organizable and measurable. Under this conception, which accords with the scientific tendency toward what Pepper has called "discrete mechanism," space can be viewed as particulate rather than as a continuum and can be represented as a mathematical equation. It is conceived as being made up of a coordinate system of points, each of which has its own discrete identity and can be treated objectively from a single vantage point.[61]

The difference in the two treatments of space leads to differences in the graphic treatment of perspective, the projection of three-dimensional space on a plane surface. In European Renaissance art, depth was represented as a continuously receding ground plane directed toward a vanishing point on the horizon, with a correlative diminution in the height of verticals. The geometry essential to the European artist's use of convergent perspective, however, was generally unknown to, or at least unused by, traditional Chinese painters. The problem of perspective, or *yuanjin* (far-near), in the graphic arts was resolved by different conventions.

61. On the Chinese organismic worldview, see Joseph Needham, *Science and Civilisation in China* (Cambridge: Cambridge University Press, 1954–), vol. 2, with Wang Ling, *History of Scientific Thought* (1956); on discrete mechanism, see Stephen C. Pepper, *World Hypotheses: A Study in Evidence* (Berkeley and Los Angeles: University of California Press, 1942); on the mathematization of space in Europe, see Samuel Y. Edgerton, Jr., *The Renaissance Rediscovery of Linear Perspective* (1975; reprinted New York: Harper and Row, 1976).

FIG. 6.11. RUBBING OF A STONE MAP, *TAI SHAN QUAN TU* (COMPLETE MAP OF TAI SHAN [MOUNT TAI]), PROBABLY QING DYNASTY. Tai Shan, in central Shandong Province, is one of the five sacred mountains of China. This map was apparently meant as a guide for pilgrims to the mountain: it shows the route from the temple of the god of Tai Shan at the foot to the shrines at the summit and identifies places and buildings. The mountain and buildings are presented in elevation, while temple grounds and walls are presented in plan, which creates a multiplicity of ground planes. In addition to this variable perspective, the map makes use of variable scale: the temple complex at the foot of the mountain is disproportionately large, apparently to permit the buildings within it to be represented.
Size of the rubbing: 110 × 62 cm. By permission of the Field Museum of Natural History, Chicago (235581).

One was the use of variable viewpoint (see figs. 6.9 to 6.11 for examples of variable perspective in maps). The standpoint of the observer, instead of being fixed, is movable without restriction or else multiple.[62] Each portion of the composition is drawn with its own viewpoint perpendicular to it at some distance. This convention was useful for composing a spatially dynamic sequence of scenes on, for example, scrolls, one of the traditional media for Chinese painting and maps. Unlike paintings in frames, scroll paintings are often too long to be viewed all at once. A moving viewpoint seems well suited for a medium in which the image passes before the observer section by section as it is unrolled. On sheet maps of restricted length, this technique could be adapted to create multiple ground planes: one might have to imagine oneself rotating, instead of moving laterally, in order to view the depicted objects correctly.

There were, however, exceptions to the general application of variable perspective. Shen Kuo recognized the merits of certain experiments that pointed toward convergent perspective. He says that painters who employed a stationary viewpoint were able to diminish heights and distances accurately but attached too much importance to this diminution. Shen advocates the use of a large viewing area in which the artist combines the vantage points of multiple observers:

> In general, the method of landscape painting is to take a large view of the small, as when a person observes an artificial mountain. If it were the same as the method for [viewing] a real mountain, looking up from below, one would see only a single layer of the mountain; how can one see layer on layer, or its valleys and gorges and other details? Similarly in the case of dwellings and houses, one would not see what was happening in their courtyards and in the lanes behind them.[63]

Elsewhere the idea of multiple viewpoint is described in terms of three depths or perspectives: *gao yuan* (high "distance" or perspective), *shen yuan* (deep distance or perspective), and *ping yuan* (level distance or perspective). The three perspectives are also explained in a Qing dynasty manual of painting:

> Mountains have three perspectives: looking at a peak from below is called *gao yuan*; looking from a mountain in front to one in back is called *shen yuan*; and looking from the near to the far is called *ping yuan*. The effect of high perspective is a precipitous view; the effect of deep perspective is the repetition of layers [or planes]; the effect of level perspective is of an expanse. These principles govern the overall composition. If it is deep but without perspective, it will seem

62. This mode of presentation can also be observed in pre-Renaissance European art and in the art of other cultures. See, for example, Edgerton, *Renaissance Rediscovery*, 7–10 (note 61).

63. Shen, *Xin jiaozheng Mengxi bitan*, chap. 17, par. 283 (170) (note 44).

> shallow. If it is level but without perspective, it will seem too close. If it is high but without perspective, it will seem low.[64]

Distance is represented by height so that objects behind or beyond another object are placed above it on a two-dimensional surface. The resulting composition is a series of plane ground surfaces or terraces, each with its own vanishing point.[65]

Maps as Paintings/Paintings as Maps

The parallels between maps and paintings in the use of perspective seem too striking to be a case of independent development—a suspicion supported by textual sources already cited. To judge from Zhang Yanyuan's lists of paintings, maps were considered a genre of painting during the Tang. Existing artifacts suggest that this view of maps as painting may have been prevalent well before then. This is not to say that purely planimetric representation was unknown. The mausoleum map (*zhaoyu tu*) and wooden maps dating from the Zhanguo period, stone maps from the Song, and some manuscript and printed maps from later periods do seem to employ an essentially planimetric mode. But Chinese maps characteristically present a mixed mode of representation. For example, a tomb mural from the Former Han dynasty (206 B.C.–A.D. 8) discovered at Horinger (Holingol), Inner Mongolia, presents a city plan of Fanyang (present-day Chuwang village in Henan Province) (fig. 6.12). All the structures depicted are presumably on the same horizontal plane, as suggested by the representation of the surrounding walls in plan. The depiction of buildings in elevation, however, creates the impression that there are several horizontal planes. The human figures also seem to be exaggerated in size; they are as large as buildings. The plan may perhaps be taken as an illustration of a technique of variable scale.

A more complex example (fig. 6.13) comes from the same tomb as the Fanyang city plan. This is a city plan of Ningcheng. All the objects represented in the plan are on the same ground plane, but some are depicted obliquely, some in plan, and some in elevation, so that they seem to occupy different ground planes. In the upper left quarter of the plan, the facade of a building is presented in elevation. The scene inside the building is also depicted, and the artist recognizes that a different perspective would allow a fuller representation of that interior scene. Thus, in accordance with what Shen Kuo would write centuries later on variable perspective, the interior of that same building is presented obliquely so that everything inside can be seen.

The use of variable perspective is not restricted to artifacts from the tomb at Horinger. It also appears on two

FIG. 6.12. CITY PLAN OF FANYANG FROM THE HAN DYNASTY. This is a copy of the Fanyang city plan discovered in a Han tomb at Horinger, Inner Mongolia. According to Jan Fontein, comparison of photographs of the original murals at Horinger with photographs of the modern copies shows that the copies are extremely faithful. The copies are important because the originals have begun to deteriorate. See Jan Fontein and Wu Tung, *Han and T'ang Murals Discovered in Tombs in the People's Republic of China and Copied by Contemporary Chinese Painters* (Boston: Museum of Fine Arts, 1976).
Size of the original: 94 × 80 cm. Photograph courtesy of Cao Wanru, Institute for the History of Natural Science, Academia Sinica, Beijing.

of the silk maps discovered in one of the Han tombs at Mawangdui in 1973. One depicts buildings in a manner that forces a viewer to rotate in order to maintain proper orientation (see fig. 3.7 above). The same style of rendering can be seen at the center of the other map in a representation of a military headquarters (plate 8). This map has been described as a garrison map because of the centrality of the headquarters and the map's depiction of various military installations. The scholars responsible for the initial reports on which almost all subsequent work

64. Wang Gai, *Jiezi yuan huapu* (Mustard seed garden manual of painting, 1679) (Taipei: Wenguan Tushu Gongsi, 1967), *chuji* (pt. 1) 19.

65. For more detailed treatment of perspective in Chinese painting, see Benjamin March, "Linear Perspective in Chinese Painting," *Eastern Art* 3 (1931): 113–39; and George Rowley, *Principles of Chinese Painting* (Princeton: Princeton University Press, 1959).

FIG. 6.13. MAP OF NINGCHENG FROM THE HAN DYNASTY. This is a copy of the map of Ningcheng discovered in a Han tomb at Horinger, Inner Mongolia (see also fig. 6.12). Size of the original: 129 × 159 cm. By permission of Wenwu Chubanshe, Beijing.

on this map has been based have argued that this map was drawn to scale and would therefore have been useful in planning strategy. The map bears no scale indications, but based on comparison with a modern map of the area depicted, it was claimed that the silk map was drawn to a fairly consistent scale. This conclusion, however, seems somewhat tenuous: scale in the most consistent portion of the map was found to vary from 1:80,000 to 1:100,000, with greater variance in other portions.[66] A variation of 25 percent may not seem large, but it is not difficult to imagine military situations in which it would be significant—for example, in directing reinforcements to particular locations. Rather than demonstrating that Han mapmakers already practiced Pei Xiu's first principle of consistent scale, the map seems to exemplify the practice of variable scale. This and other evidence to be adduced below suggests that the map may have more in common with painting than with the quantitative cartography advocated by Pei Xiu.

One curious feature of this map lies in its treatment of curved lines, which represent mountains and streams in the same manner that curved lines represent hills, waves, and clouds in late Zhou art. The regularity of the curvature is also striking. If Pei Xiu had seen a map like this, it probably would have raised his ire for its lack of correspondence to observed reality. A tendency toward idealization of landscape, however, has long been recognized as a characteristic of Chinese art. In addition, the pointed and trefoil motifs on the mountain con-

66. Mawangdui Han Mu Boshu Zhengli Xiaozu (Study Group on the Han Silk Manuscripts from Mawangdui), "Mawangdui sanhao Han mu chutu zhujun tu zhengli jianbao" (Preliminary restoration report on the military map found in Han tomb 3 at Mawangdui), *Wenwu*, 1976, no. 1:18–23.

FIG. 6.14. LACQUERED WINE CUP. Cups of this type were discovered in the Han tombs at Mawangdui.
Size of the original: 4.5 × 17.3 × 17.8 cm. By permission of the Hunan Provincial Museum.

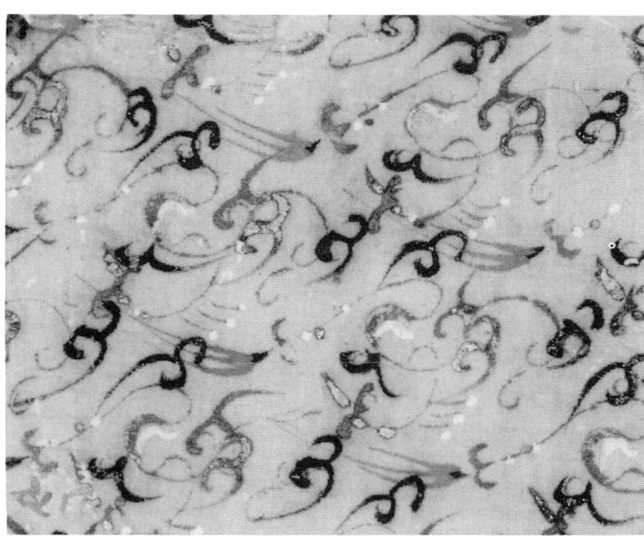

FIG. 6.15. SILK GAUZE FROM THE HAN DYNASTY. This is a small part of a piece of fabric discovered in a Han tomb at Mawangdui. The designs in five colors were painted onto the fabric.
By permission of the Hunan Provincial Museum (no. 340–32).

FIG. 6.16. MANOR MAP FROM THE HAN DYNASTY. This is a copy of a manor map mural discovered in the Han tomb at Horinger, Inner Mongolia (see also fig. 6.12)
Size of the original: ca. 191 × 300 cm. By permission of Wenwu Chubanshe, Beijing.

FIG. 6.17. PAINTING OF WANGCHUAN. Section of a late Ming copy of the *Wangchuan tu* (Picture [or map?] of Wangchuan), by Wang Wei. Size of the entire original: 30 × 480.7 cm. By permission of the Seattle Art Museum (Eugene Fuller Memorial Collection 47.142). See also plate 9.

tours—the latter perhaps deriving from the graph for mountain—seem to lend an abstract quality to the map. This conjecture is corroborated by the appearance of similar motifs on artifacts from the Han tombs at Mawangdui. These motifs are painted on lacquerware and painted and printed on textiles (figs. 6.14 and 6.15). Considered with the regularity of the curves and the lack of scalar indication, the designs seem to point toward a less practical application for the map than has been previously claimed, especially if one were to use it to plan troop movements as some scholars have suggested.

Because of its affinities with visual art, the garrison map ought to be viewed in relation to the other examples of tomb cartography discussed previously. In its mode of representation, the Mawangdui map shares some characteristics with the ones found at Horinger. One Horinger map of a manor (fig. 6.16) depicts roads and mountains with not nearly as much symmetry as the one from Mawangdui. But the variable perspective used on the manor map resembles the mode of representing objects seen on the garrison map: the manor map offers frontal views of some objects and overhead views of other objects lying on the same plane as those depicted frontally. Similarly, the garrison map gives overhead views of streams and military installations but depicts mountains and buildings in elevation.

The garrison map's burial with an official also allies it with the Horinger mural maps. The maps at Horinger, drawn on the inside walls of a tomb, would serve no practical purpose in this world; as symbols of the tomb occupant's official standing and personal wealth, however, they would have helped the deceased gain standing in the afterworld. The entombment of the garrison map from Mawangdui seems to indicate that its function was conceived at least in part as ceremonial, that it was a ritual object, part of the funerary rites. This use would explain the presence of decorative motifs shared with other artifacts buried in the same tomb. Seen in this light, the map appears to have been drawn not to help military planners, but to dignify the official's passage to the otherworld by serving as a representation of his former military authority. This conclusion is also consistent with the message to the otherworld found with the artifacts in the official's tomb. The message was written on a wood tablet by a member of the deceased's household staff: "On *wuchen* [the twenty-fourth day], second month, the twelfth year [of Wendi's reign, 168 B.C.], Household Assistant Fen to Gentleman in Charge of the Dead: a list of mortuary objects is herewith forwarded to you. Upon receiving this document, please memorialize without delay to the Lord of the Dead."[67] The message implies

67. Hunansheng Bowuguan and Zhongguo Kexueyuan Kaogu Yanjiusuo (Hunan Provincial Museum and Institute of Archaeology, Aca-

that the deceased is being entrusted to an otherworldly bureaucracy, and the artifacts are apparently expected to help sustain him there.

Although the conclusions reached here question the validity of earlier conclusions about the garrison map, they do not necessarily undercut the argument that during the Han, the Chinese produced maps based on direct and indirect measurements of distances. The garrison map does correspond closely enough to observed geographic reality for identification of the area depicted. As the regularity of the mountain contours suggests, the mapmaker probably did not rely on survey data. But it is conceivable that the map was based on an archetype relying on such data, perhaps a map considered too valuable in this world to be buried. This would explain why some investigators have insisted that the garrison map reflects a quantitative tradition while overlooking its obvious connections with artistic and political practices. The counterarguments offered here, however, do have their difficulties. There is no evidence, for example, that the map was newly drawn for the burial.

In any case, though, the garrison map exemplifies the difficulty of separating cartographic from artistic traditions. That difficulty is not confined to early artifacts, in which case one might argue that the two traditions would become easier to separate as they "matured." To this possible counterargument, a number of examples can be offered in rebuttal. One of the more notable is a scroll painting by the Tang poet and painter Wang Wei (701–61). The original has been lost, and the painting is known only through painted or printed copies of later date (fig. 6.17 and plate 9). The painting depicted the artist's estate, and copies show that it was executed with a variable perspective or moving focus often associated with the scroll form. The moving focus is accentuated by the spatial organization of the painting: the points of interest are arranged so that space flows completely around them. They are enclosed in what have been described as "space cells."[68] In addition, the ground plane is tilted, but mountains and trees are shown frontally. Also, each point of interest on the painting is labeled.

Wang Wei's painting has generated some debate over its proper classification—landscape painting or map. In

demia Sinica), "Changsha Mawangdui er, sanhao Han mu fajue jianbao" (Preliminary excavation report on Han tombs 2 and 3 at Mawangdui, Changsha), *Wenwu*, 1974, no. 7:39–48, 63, esp. 43. The translation is a modification of that of Ying-shih Yü, "New Evidence on the Early Chinese Conception of Afterlife—A Review Article," *Journal of Asian Studies* 41 (1981): 81–85, esp. 82. The inventory of funerary objects found in the tomb in question consists of 410 wooden strips and boards. According to published reports, the list corresponds fairly well with the objects found, but I have been unable to determine whether the maps are named in the inventory.

68. Sherman E. Lee, *Chinese Landscape Painting* (Cleveland: Cleveland Museum of Art, 1954), 19.

FIG. 6.18. TWO DETAILS FROM A TENTH-CENTURY PAINTING OF WUTAI SHAN. From the mural of Wutai Shan in the Cave of the Thousand Buddhas at Dunhuang. Size of the entire mural: 4.6 × 13 m. By permission of Wenwu Chubanshe, Beijing.

favor of those who would call it a landscape, the painting lacks the high degree of abstraction usually associated—at least in the twentieth century—with maps. Other than the labels, it makes no use of conventional symbols to represent geographic features but tries to approximate the actual appearance of those features. On the other hand, it certainly has elements that suggest cartographic intent: it represents an actual place; it seems intended to promote spatial understanding of that place, as suggested by the labels; and it does manifest some abstraction in its spatial organization into squarish enclosures. The painting's combination of pictorial and cartographic elements makes it suitable for classification under Harvey's term "picture map."[69]

The same can be said of the mural of Wutai Shan (Wutai Mountains, in present-day Shanxi Province) in the Cave of the Thousand Buddhas at Dunhuang (fig. 6.18). Measuring about 4.6 meters high and 13 meters wide, the mural was painted between 980 and 995. As in Wang Wei's painting, the pictorial space of the mural is organized into space cells, though in this case the cells are connected by paths and roads so that a viewer "is not forced to leap barriers in order to pass from place to place but is able to follow easily the same paths taken by travellers in the painting."[70] And like the artifacts discussed above, the Wutai Shan mural is marked by variable perspective: the objects depicted occupy several ground planes. The cartographic character of the mural derives in part from its use of cartouches identifying objects on the ground as well as in the heavens, and in part from the abstraction evident in the pictorial representations: the representations of mountains, buildings, and towns show some uniformity in appearance, implying at least some conventionalization.[71]

Even after the introduction into China of European cartography and its planimetric and abstract modes of representation, Chinese maps still make use of variable perspective and pictorial representation. Both were regarded as proper to cartography. The *Sancai tuhui* (Illustrated compendium of the three powers [heaven, earth, man]), an encyclopedia completed in 1607, classifies landscapes, grid maps, and maps combining pictorial and planimetric modes of representation together as *dili tu*, "geographic illustrations" or maps (fig. 6.19). The encyclopedia is a compilation of earlier as well as contemporaneous materials, so its contents may not be representative of seventeenth-century cartography, but a similar situation involving mixed modes of representation holds for contemporary and even later maps. A prefectural map from an eighteenth-century atlas (plate 10), for

69. See P. D. A. Harvey, *The History of Topographical Maps: Symbols, Pictures and Surveys* (London: Thames and Hudson, 1980). On the basis of Wang Wei's painting, Berthold Laufer concludes that Tang dynasty masters of landscape painting "received a strong impetus for their work from the high development of contemporaneous map-making." A work like Wang Wei's, Laufer goes on to say, is "not intended to represent any landscape, but it is the topography of the Wang river cherished and minutely investigated by the artist." Painters had to practice methods similar to those of the cartographers. See Berthold Laufer, "The Wang Ch'uan T'u, a Landscape of Wang Wei," *Ostasiatische Zeitschrift* 1, no. 1 (1912): 28–55, esp. 53–54.

70. Ernesta Marchand, "The Panorama of Wu-t'ai Shan as an Example of Tenth Century Cartography," *Oriental Art*, n.s., 22 (1976): 158–73, esp. 159.

71. Marchand, "Panorama of Wu-t'ai Shan," 159, 169–70 (note 70).

example, bears some striking stylistic resemblances to Wang Wei's painting. The prefectural map, colored in hues of green and blue, shows a similar spatial organization with an emphasis on cell-like enclosures. It makes use of variable perspective: the ground plane is tilted, yet mountains are depicted in elevation. As in Wang Wei's painting, points of interest are labeled, consistent with the map's function as an aid to understanding spatial relation between objects. The relatively late date of this map suggests that in manner of presentation cartography, even after the introduction of European models, had still not fully distinguished itself from other forms of visual representation.

This conclusion is corroborated by examples from geographic works dating from the sixteenth through the nineteenth century, in which pictorial and planimetric representations are all labeled *tu*. For example, reproduced in figures 6.20 to 6.22 are sequences of *tu* from regional and local gazetteers. If one were to remove the place-names, some of these *tu* would be indistinguishable from what would unequivocally be called landscape pictures. This is especially true in cases where human figures, animals, and boats—elements not usually associated with cartographic representation—make up part of the map content. Compilers of gazetteers saw both planimetric and pictorial representation as essential to their task: grid maps are the "warp," and maps of mountains, streams, and villages are the "woof."[72] In Qing dynasty gazetteers, one influence of representational practices imported from Europe does manifest itself in the greater use of linear or convergent perspective in maps. These techniques did not displace the variable perspective seen in earlier paintings and maps but were juxtaposed to traditional spatial representation (see figs. 6.23 and 6.24).[73] Like grids of traditional Chinese cartography and graticules of European mapmaking, convergent perspective and variable perspective often appear on the same map.

Toward a Redefinition of the Map

The influence on Chinese cartography of European models of both cartography and the visual arts provides yet another indication of the close relation of painting and cartography in Chinese culture. Further evidence appears in an eighteenth-century anthology of poems inscribed on paintings, commissioned by the Kangxi emperor (1654–1722), who also commissioned Jesuit missionaries to undertake a survey of his empire and to draw new maps using European techniques. Despite the introduction of European map styles into the Kangxi emperor's court, the compilers of the anthology express a conception of maps that seems uninfluenced by European ideas. Among the classifications of illustrations used in this collection are *dili* (geographic) and *shanshui* (landscape). According to the statement of principles that introduces the collection, a geographic picture or *tu* is one that gives an overall view of all the mountains, seas, and topographic configurations. A landscape is a picture in which one creates a scene out of nothing "to portray one's impressions" or a picture in which one "embellishes [reality] as one wishes without identifying names so as to make any mountain or river whatever."[74] What makes a geographic map a map, according to this statement, is its depiction of an actual place as opposed to one that is imaginary or even unnamed. A landscape painting seems to allow for more self-expression and greater use of the imagination than a geographic map, although it is unclear how much embellishment would disqualify a picture from being classified as a map.

Unfortunately, the collection of poetry does not contain reproductions of the pictures on which the poems are inscribed so that the distinction between map and landscape might be clearly illustrated. As the definition stands, any painting of an actual place, such as the one by Wang Wei, in which the artist strove for fidelity without embellishment and identified the geographic feature depicted, would be considered a *dili tu* or map. Thus, in the gazetteer maps of the Qing period one often sees both pictorial and planimetric modes of depiction, sometimes within the same representation.

Beneath the term *dili* lies an additional explanation for traditional Chinese cartographic practice. Besides "geography," the term can also refer to the traditional Chinese science of "siting," used to evaluate the configurations of *qi* underlying proposed sites for buildings and tombs. The importance of *qi* in siting hints at another connection

72. *Yuzhou zhi* (Gazetteer of Yuzhou [in present-day Henan Province]) (1835; reprinted Taipei: Taiwan Xuesheng Shuju, 1968), 1.1a–b.

73. As I mentioned above, convergent perspective may have been used, but infrequently, before the arrival of the Europeans. The principles of convergent perspective do not seem to have been widely known. Jesuits are known to have given lectures on the principles of perspective, and several Chinese writers marveled at the realism of European pictorial art. On the influence of European art on Chinese painting, see James Cahill, *The Compelling Image: Nature and Style in Seventeenth-Century Chinese Painting* (Cambridge: Harvard University Press, 1982), 70–105; and Harrie Vanderstappen, "Chinese Art and the Jesuits in Peking," in *East Meets West: The Jesuits in China, 1582–1773*, ed. Charles E. Ronan and Bonnie B. C. Oh (Chicago: Loyola University Press, 1988), 103–26.

74. Chen Bangyan et al., comps., *Yuding lidai tihua shi lei* (Imperially commissioned, classified [collection] of poems inscribed on paintings through the dynasties, commissioned 1707), *Siku quanshu* edition, "Fanli" (principles), 1b. This collection is also discussed by Kazutaka Unno in "Maps as Picture: The Old Chinese Views of Maps" (paper presented at the Thirteenth International Conference on the History of Cartography, Amsterdam and The Hague, 26 June to 1 July 1989). Unno shares the view expressed here that cartographic representation in traditional China was generally not conceived as separate from other forms of visual representation.

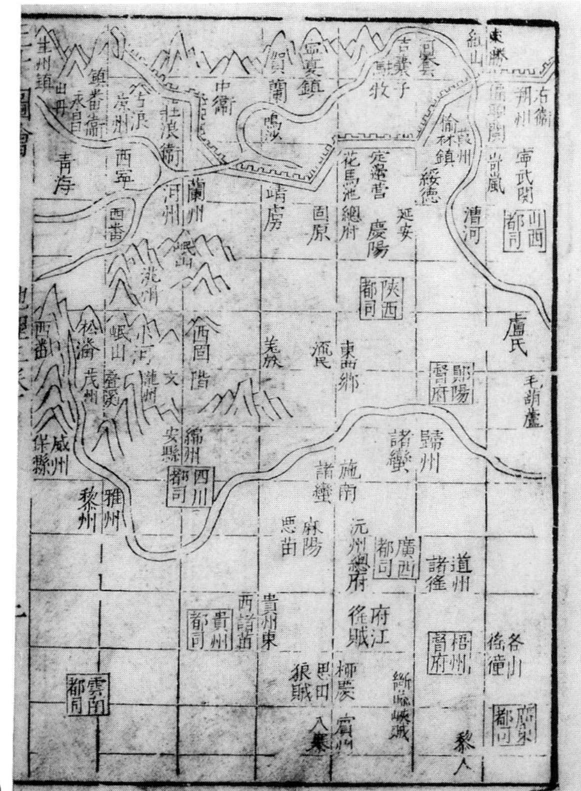

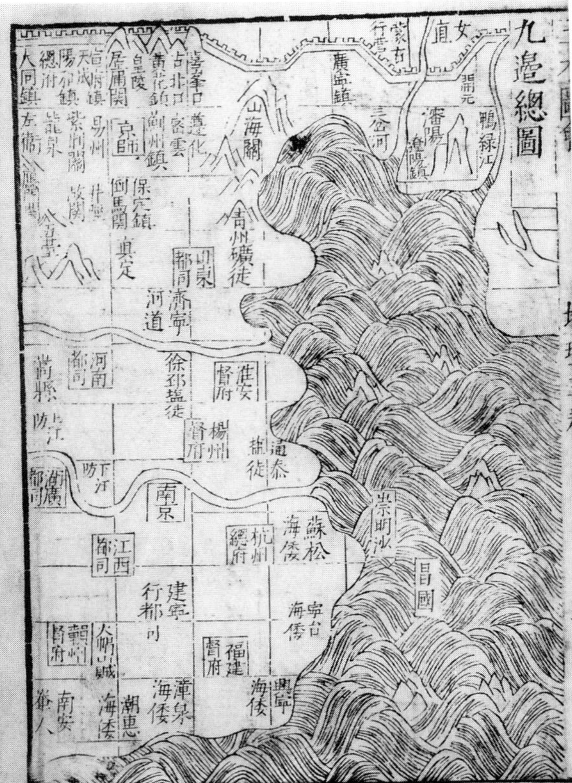

(*a*)

FIG. 6.19. THREE MAPS FROM A MING ENCYCLOPEDIA. Selection of *dili tu* (geographic pictures or maps) from the *Sancai tuhui*. Note the range of representational modes used in these woodblock maps—planimetric, pictorial, and a mixture of both.
(*a*) "Jiubian zongtu" (Comprehensive map of the nine frontiers).
(*b*) Map of Xihu (West Lake), in Hangzhou.
(*c*) Map of Yanglongkeng (in present-day Guizhou Province).
Size of each page: 21 × 14 cm. All from Wang Qi, comp., *Sancai tuhui* (completed 1607, printed 1609), chapters on *dili* (geography): (*a*) 3.1b–2a, (*b*) 9.20ab, (*c*) 12.39a. Reproduced courtesy of the Harvard-Yenching Library, Harvard University, Cambridge.

between traditional Chinese art and science, and the links between siting and the graphic arts are not difficult to discern. Landscape paintings and maps often depicted ideal siting configurations (fig. 6.25). Among practitioners of siting, straight lines were regarded as signs of malign influences, of a lack of *qi*, and traditional Chinese painters tended to hold straight lines in low esteem for the same lack of vitality. Mapmakers, too, show a similar interest in vitality with their predilection for asymmetry and irregularity, qualities best expressed pictorially rather than planimetrically. Even on grid maps, one frequently finds that the uniformity imposed by the squares is broken up by pictorial elements.[75]

The interest in the vitality underlying material forms serves as a reminder that maps fell within a representational complex, that adherence to the physical world took no precedence over the subjective. As discussed above, in Chinese theories of representation, objective presentation does not preclude subjectivity.

75. In one instance, an emperor is said to have altered a map to eliminate straight lines. According to Yue Ke (b. 1183), in 968 the first emperor of the Song was planning to refurbish the capital (present-day Kaifeng) and enlarge the foundations of the city wall. The wall "wound and curved as if earthworms wriggled within it." The secretary-general gathered artisans and produced a plan that "seized upon the square and straight": all four sides had gates, and the wards and marketplaces were arranged in gridlike fashion. The emperor inspected the plan and grew angry, apparently revolted by its straight lines: "He personally took up a brush to smudge them out and ordered that a large circle be made on a sheet of paper. It wound and turned, went up and down and slanted. [The emperor's] side note said: 'Repair the wall according to this.'" See Yue Ke, *Ting shi* (History [recorded] on a stand [by the bed], ca. 1210), chap. 1; see the modern edition (Beijing: Zhonghua Shuju, 1981), 8–9. In the episode just cited, although Yue Ke does not explicitly say so, the emperor seems to be acting in accordance with well-known siting principles. The best introduction to the science of siting, also known by the names *feng shui* (wind and water) and *kanyu* (cover and support, that is, heaven and earth), is Steven J. Bennett, "Patterns of the Sky and Earth: A Chinese Science of Applied Cosmology," *Chinese Science* 3 (1978): 1–26. Siting is often erroneously called "geomancy," which actually refers to an unrelated practice. For further discussion of siting within this volume, see pp. 216–22 and 276–79.

(*b*)

(*c*)

(*a*)

(*b*)

FIG. 6.20. FOUR MAPS OF ZHEJIANG PROVINCE. Compare figures 6.21 and 6.22.
Size of each page: ca. 19.5 × 14 cm. From *Zhejiang tongzhi* (Comprehensive gazetteer of Zhejiang, 1736), from an 1884 edition: (*a*) Chuzhu fu, 1.28b–29a, (*b*) Yuhuan Shan, 1.30b–31a, (*c*) Tianmu Shan, 1.32b–33a, (*d*) Wulin Shan, 1.34b–35a. Reproduced courtesy of the Harvard-Yenching Library, Harvard University, Cambridge.

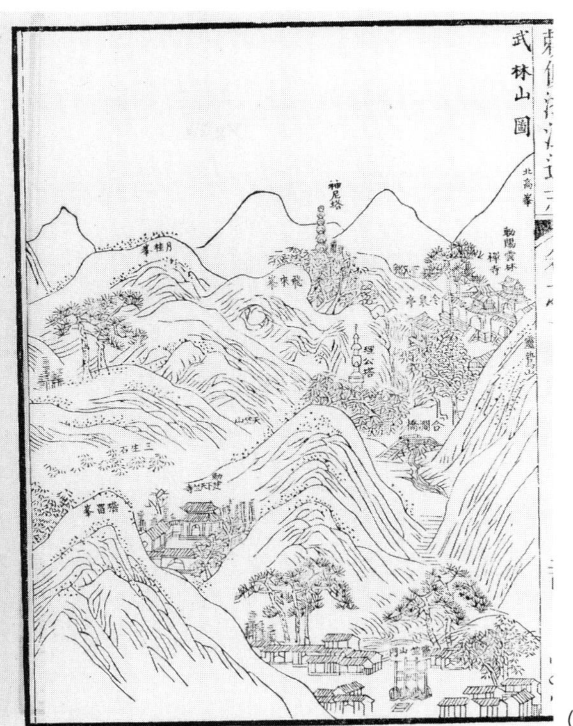

(*c*)

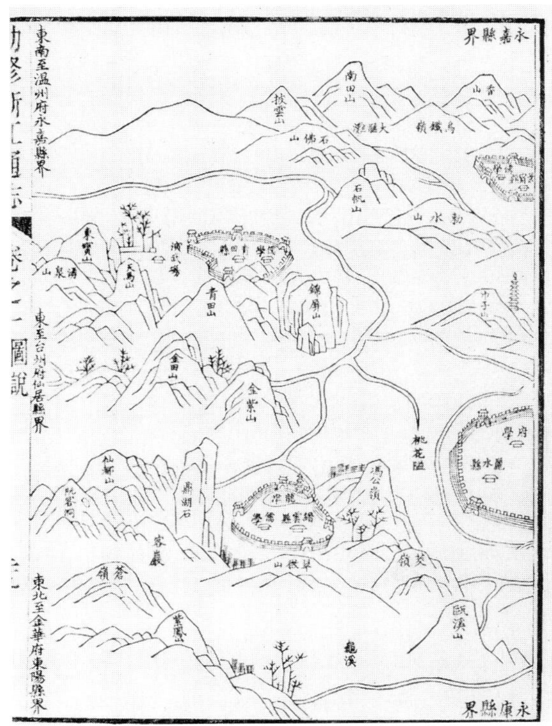

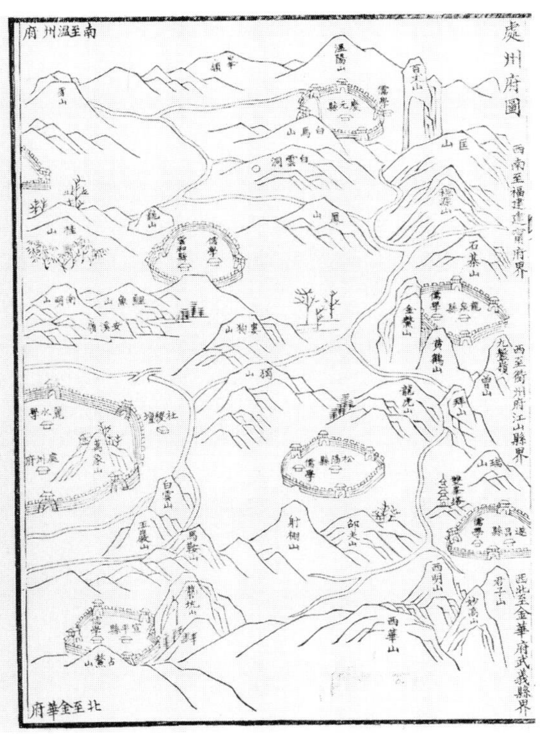

(*d*)

FIG. 6.21. THREE MAPS OF GANSU PROVINCE. Size of each original: 22 × 34 cm. From *Gansu tongzhi* (Comprehensive gazetteer of Gansu, 1736): (*a*) 1.35b–36a, (*b*) 1.36b–37a, (*c*) 1.37b–38a. Reproduced courtesy of the Harvard-Yenching Library, Harvard University, Cambridge.

Representation of physical appearance was a means of understanding underlying realities, the innerness of both object and artist. This method is recognized in a Qing hydrological work: "Those who draw territorial maps must discern details, and later they can abstract their ordering principles. They must also take into account the complete configuration [of the land], and later they can grasp its essential structure."[76] A map, like a painting, is not just a record, but is a product of the mapmaker's intuitive sense of underlying form—mapmaking involves abstraction of external details into something internal, a "mindscape." A cartographic image can thus represent not only physical appearances, but also the mapmaker's memories and reflections. The collection of poems inscribed on paintings referred to above provides ample evidence that maps were read in this way, not only as a means of gaining knowledge of the physical world but as a means of enriching one's subjective world or emotional experience.

As is well known, Chinese landscape paintings have inspired much poetry; in fact, after the Song, a landscape was often not considered complete unless a poem—by the artist or by another—was inscribed on it.[77] The painting and poem would thus exemplify the "three perfections," the entire artifact constitituting a unity of linguistic, calligraphic, and visual art. Perhaps less known is that maps also inspired poetry—poems inscribed on the maps themselves and serving to bring out subjective elements submerged in the attempt to render the physical world. This adds another dimension to the relationship between map and text discussed in other chapters.

These poems suggest that maps were viewed not only for intellectual understanding, as a means of increasing one's understanding of spatial relationships, but also for emotional experience—for example, as a means of renewing one's sense of place. "Space is transformed into place as it acquires definition and meaning," Tuan has noted; place "is a concretion of value," and values carry emo-

76. Wang Niansun, *He yuan jilüe* (Short accounts of the sources of the Yellow River, commissioned and printed 1782) (reprinted Taipei: Guangwen Shuju, 1969), 1.6a.

77. See Michael Sullivan, *The Three Perfections: Chinese Painting, Poetry, and Calligraphy* (London: Thames and Hudson, 1974); and Shen C. Y. Fu et al., *Traces of the Brush: Studies in Chinese Calligraphy* (New Haven: Yale University Press, 1977), 179–80.

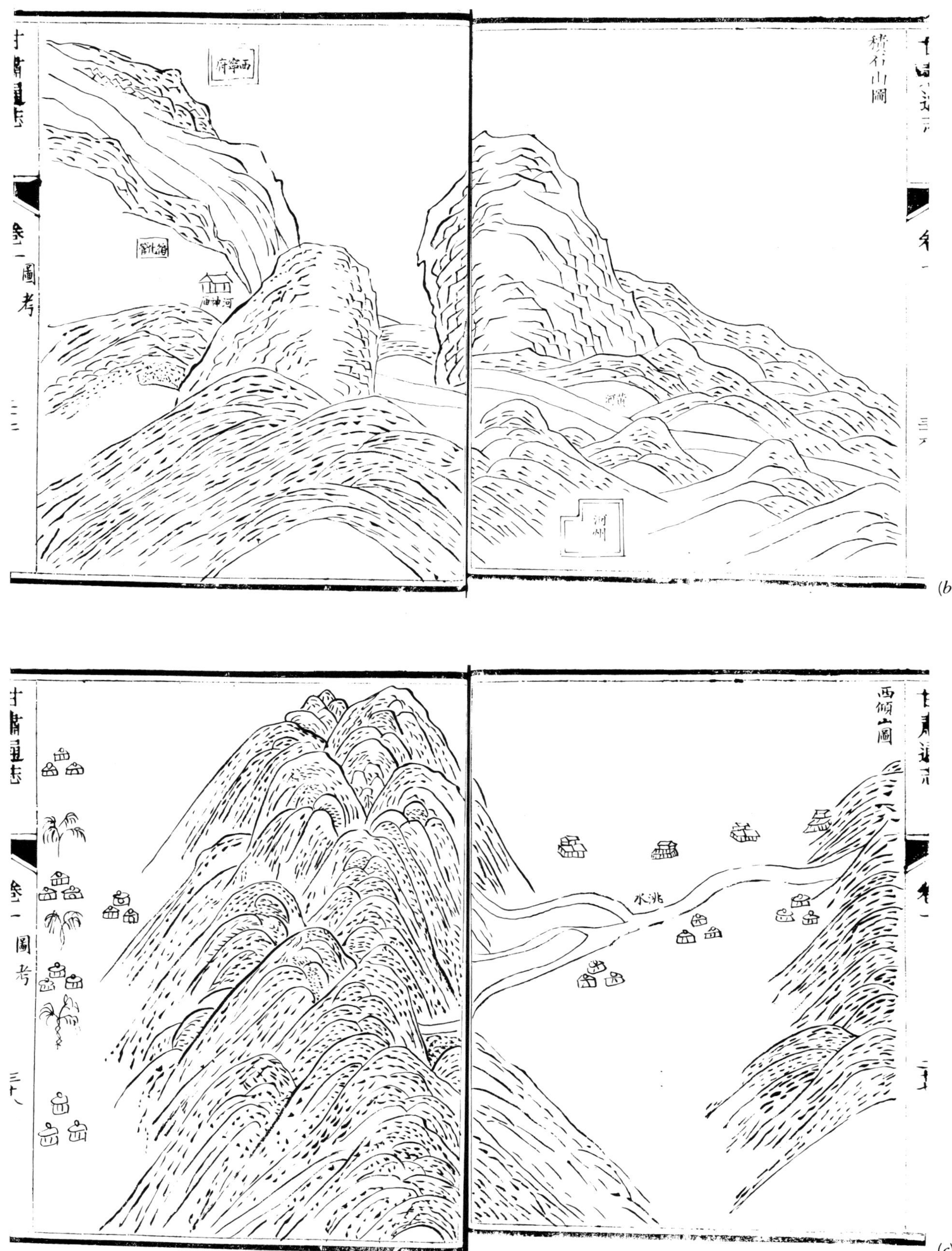

(b)

(c)

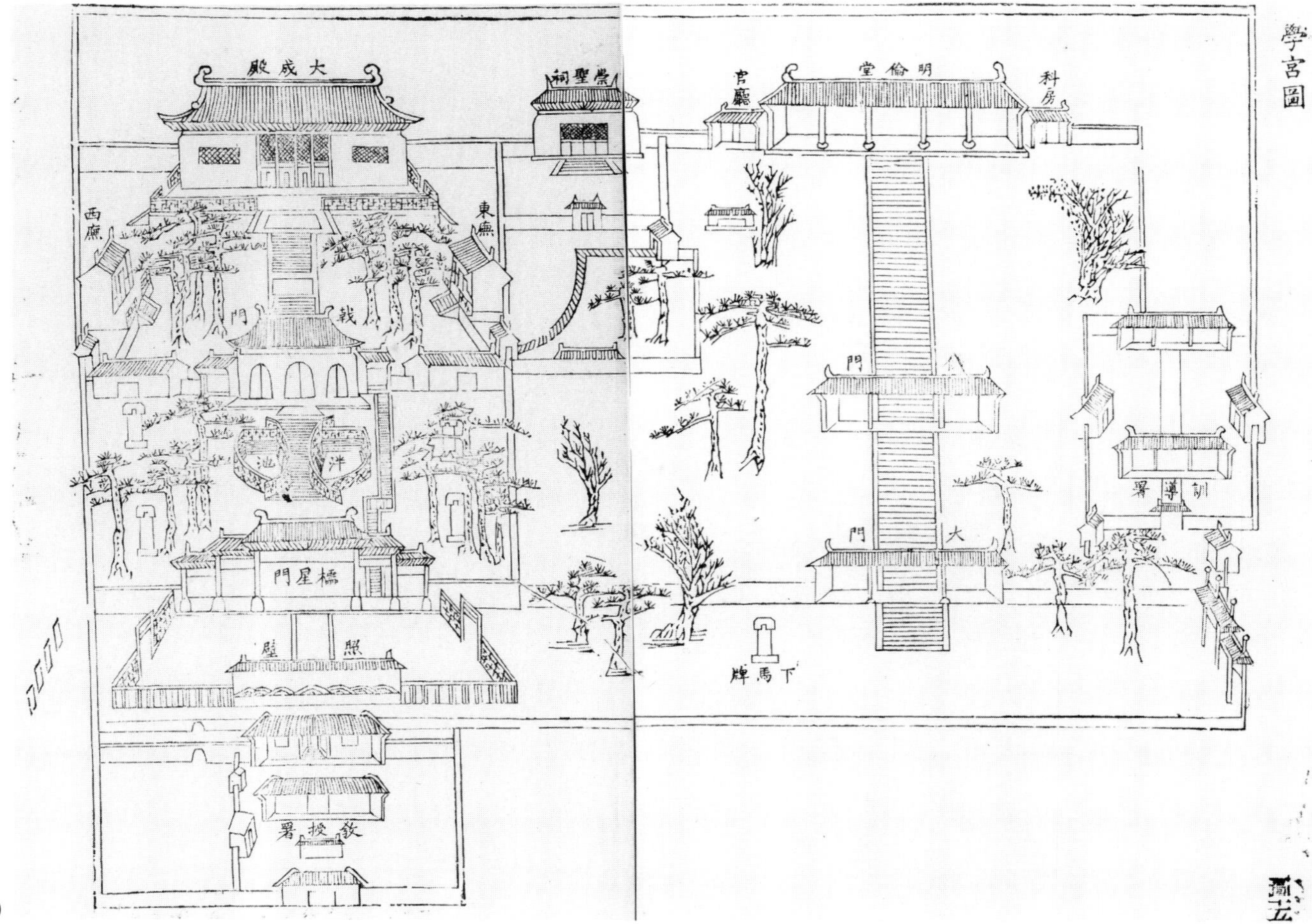

FIG. 6.22. THREE MAPS OF YONGPING PREFECTURE. Size of each original: 26.5 × 32 cm. From *Yongping fu zhi* (Gazetteer of Yongping Prefecture [in modern Hebei Province], 1879): (*a*) *tu* 15ab, (*b*) *tu* 16ab, (*c*) *tu* 17ab. Reproduced courtesy of the Harvard-Yenching Library, Harvard University, Cambridge.

tional weight.[78] This is illustrated by the poetic response of Ding Henian (1335–1424) to a map titled *Changjiang wanli tu* (Ten thousand *li* map of the Changjiang [Yangtze River]), which apparently does not survive. Maps with the same name or similar names do survive, and from these we may gain a sense of what might have been depicted on the map that elicited Ding's response (details from two examples of maps/landscapes of the Yangtze River are figs. 6.26 and 6.27). Although the map artifact is no longer extant, the text of Ding's poem is preserved in the anthology of poems inscribed on paintings and appears in the section on maps (*dili tu*). The description of the lost artifact as a map thus merely follows the classification of the compilers of the anthology. In Ding's case the map is a pretext for a search for origins, a place of familiarity:

> The Changjiang extends thousands and thousands of *li*;
> Where is my old home?
> Suddenly through the clear sky I see stream and trees,
> And vaguely make out Hanyang.[79]

Although the poem is written in response to a map, the text of the poem itself gives no indication that the speaker is looking at a map. Rather, the poet writes as if he were

78. Yi-fu Tuan, *Space and Place: The Perspective of Experience* (Minneapolis: University of Minnesota Press, 1977), 136, 12.

79. In Chen et al., *Yuding lidai tihua shi lei*, 6.8a (note 74). Hanyang is a county in Hubei Province, north of the Yangtze, near its junction with the Han River. In the first line of the poem, Ding engages in some wordplay, describing the Changjiang in terms that justify the literal meaning of its name, "long river."

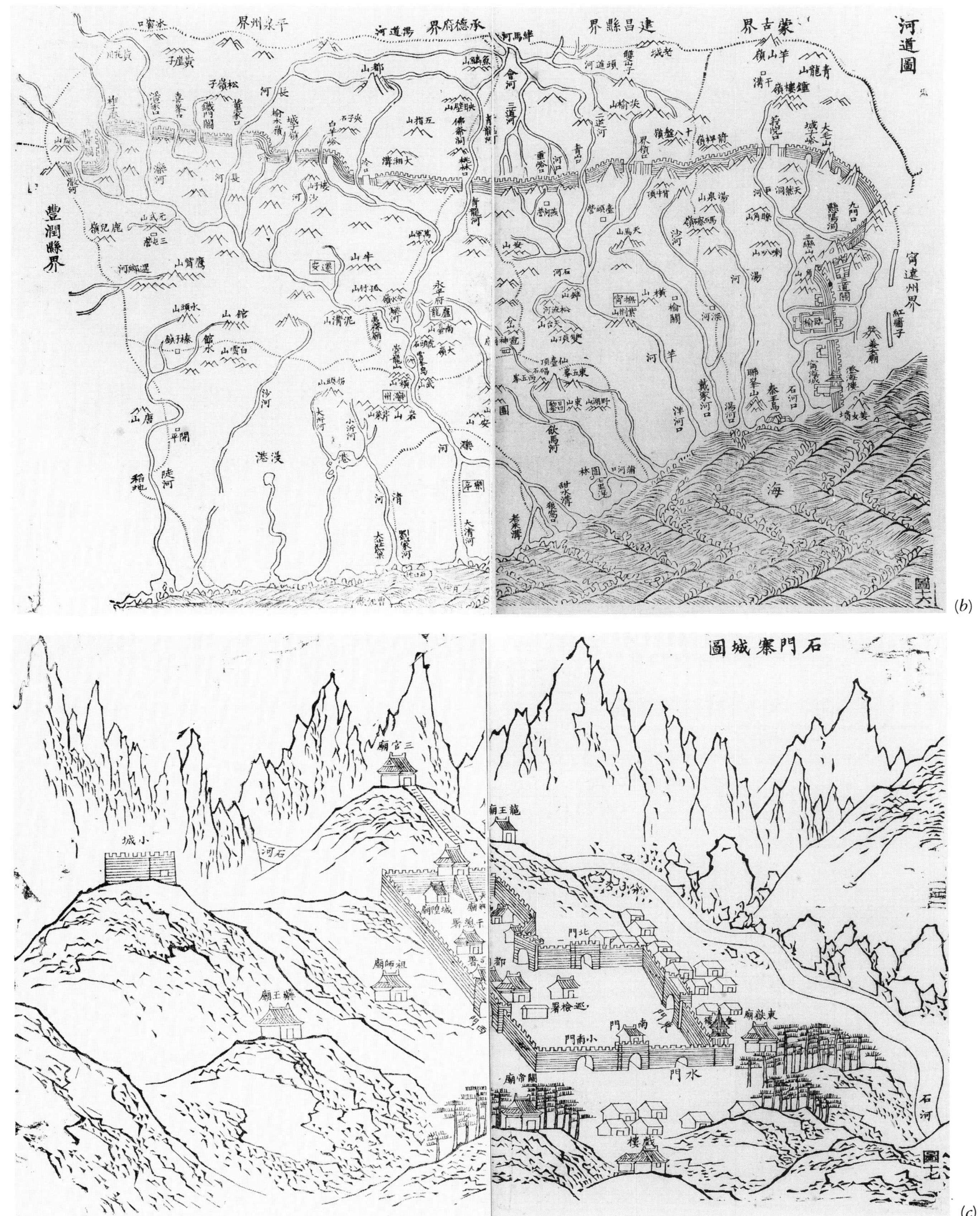

河道圖
蒙古界
建昌縣界
承德府界
平泉州界
寧遠州界
豐潤縣界
海
圖十六
石門寨城圖
三官廟
龍王廟
小城
石河
城隍廟
祖師廟
北門
南門
小南門
水門
東嶽廟
關帝廟
戲樓
巡檢署
圖十七

(b)

(c)

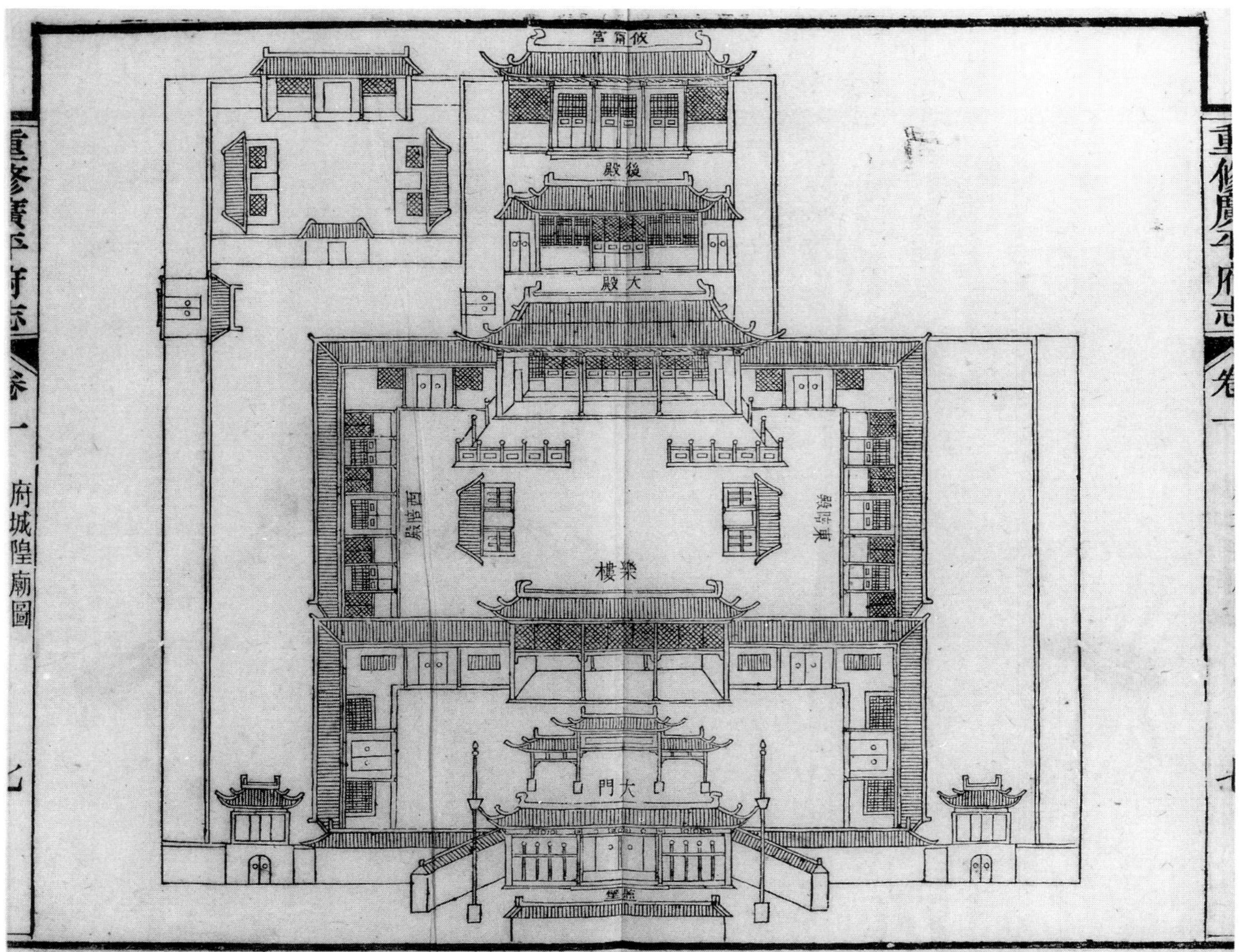

FIG. 6.23. MAP OF THE TEMPLE OF THE GUANGPING PREFECTURE'S GUARDIAN DEITY. See also figure 6.24. Both are examples of convergent perspective in woodblock maps. Size of the original: 23 × 27 cm. From *Guangping fu zhi* (Gazetteer of Guangping Prefecture [in modern Hebei Province], 1894), chap. 1, map 7. Reproduced courtesy of the Harvard-Yenching Library, Harvard University, Cambridge.

viewing an actual scene, not a representation. In effect, the map serves as a substitute for reality, implying a high degree of formal likeness. But in accordance with Chinese aesthetic theory, the physical world and the psychological become fused. Physical descriptions are intertwined with acts of perception: the poet is actively looking ("Where is my old home?"), not just seeing, and the poem closes with an act of recognition—a rediscovery of the familiar in Hanyang.

A similar blending of objective and subjective experience takes place in these lines by the painter-poet Yang Ji (ca. 1334–ca. 1383), which, like Ding Henian's, were inscribed on a map entitled *Changjiang wanli tu*, a work that also does not seem to have survived. Like Ding Henian's poem, Yang Ji's appears in the anthology of poems inscribed on paintings, appearing under the classification of maps. It too gives no hint that the speaker is looking at a map, and the process of viewing the map becomes the occasion of emotional release. The movement of this poem differs from Ding's, however, in that its emotional scope expands as the poem progresses—from the poet's personal situation to more generalized experience of those traveling on the Yangtze, perhaps figures depicted on the map:

> My home lies farther west of the Min Mountains,
> Directly facing where the Min River issues forth.
> In Sanba, under a clear spring sky, the snow first thaws;
> And through a hundred bends and a thousand turns, it flows east.
> The Jiang flows east for ten thousand *li*;

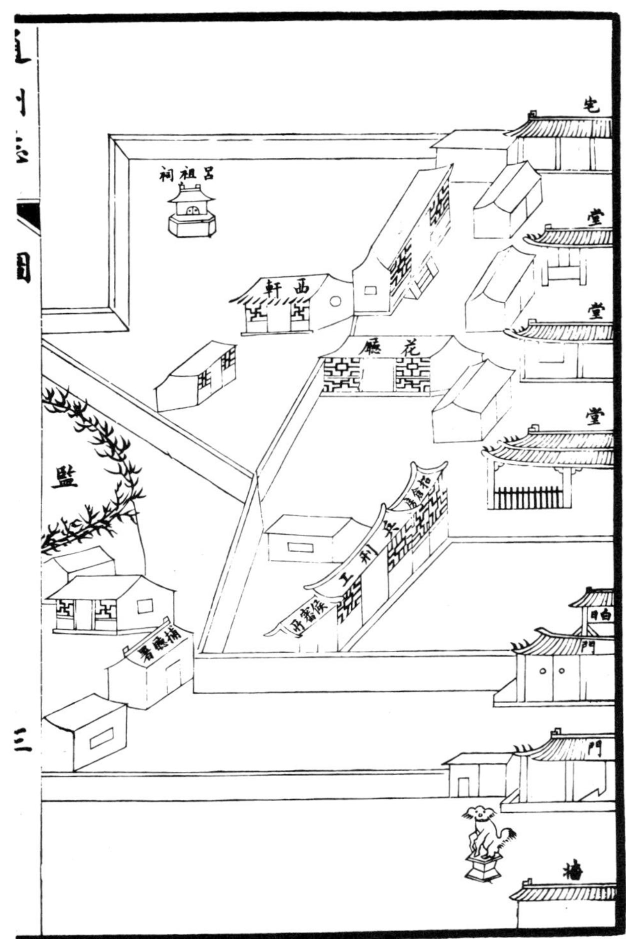

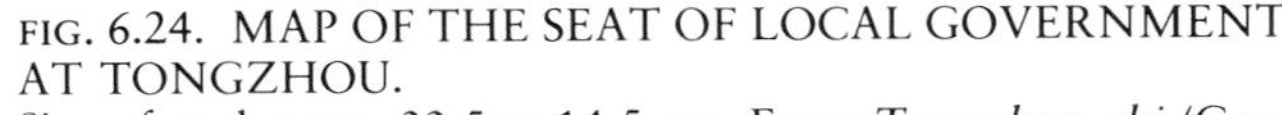

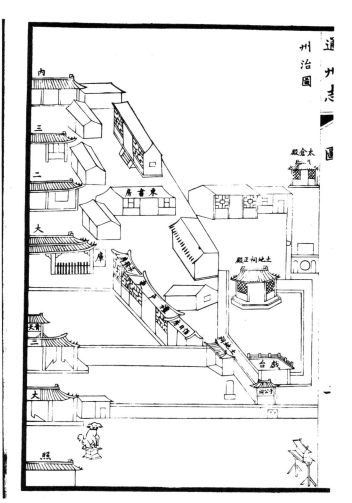

FIG. 6.24. MAP OF THE SEAT OF LOCAL GOVERNMENT AT TONGZHOU.
Size of each page: 22.5 × 14.5 cm. From *Tongzhou zhi* (Gazetteer of Tongzhou [in modern Hebei Province, near Beijing], 1879), *tu* 2b–3a. Reproduced courtesy of the Harvard-Yenching Library, Harvard University, Cambridge.

People now drift and float toward other places.
The mist, waves, and green grass at times draw out
  their sorrow;
The sounds of wind, rain, and gibbon are about to
  break their hearts.[80]

As the preceding examples suggest, traditional Chinese mapmakers worked in an environment where they themselves or map readers might inscribe poems on maps, as they would be expected to do on paintings. To complete the parallel between cartography and painting—to extend the argument for a unity of map and landscape—a cartographic work was often not considered complete until a poem was inscribed on it. The cartographic artifact at its best was a fusion of graphic and verbal art, and like a poetic work, it was meant to be appreciated on at least two levels. The inscribed poems themselves demonstrate that at least viewers read mapmakers' intentions in those terms and responded accordingly. In other words, cartographic forms were meant not only to reproduce but to express. Painters of landscape paintings had similar expectations, and poems inscribed on landscapes are often difficult to distinguish from those inscribed on maps. For example, here is a poem inscribed by the Neo-Confucian philosopher Zhu Xi (1130–1200) on a landscape by Mi Youren (1072–1151). As with the poems inscribed on maps, the poet treats the representation as actuality and animates the scene by inserting a human agent:

80. In Chen et al., *Yuding lidai tihua shi lei*, 6.8b–9a (note 74). The Min Mountains and Min River are in Sichuan. It is possible to translate the last three lines as referring to the speaker in the poem, so that the poem remains personal throughout. In this case the graph *ren*, translated here as "people," would be interpreted as referring to the speaker ("a person").

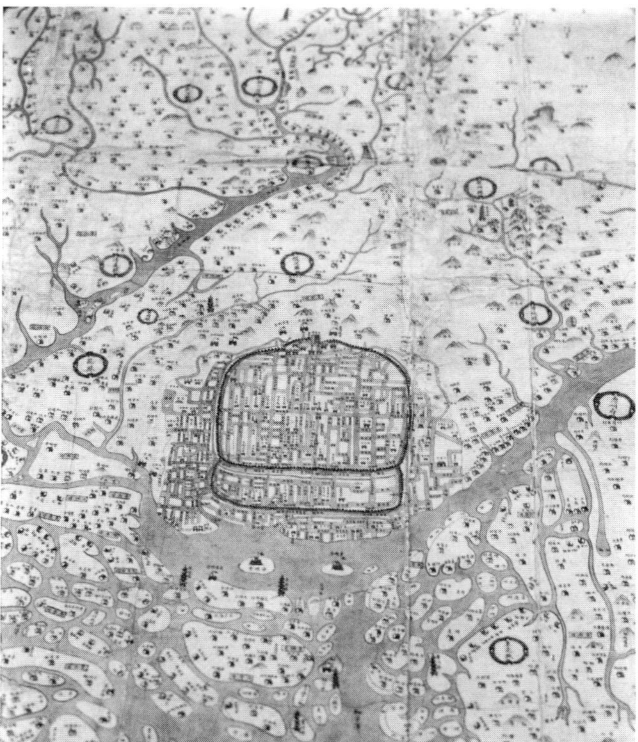

FIG. 6.25. QING MAP OF GUANGDONG. Detail of the *Guangdongsheng quantu* (Complete map of Guangdong Province), drawn in the eighteenth century. The estuary of the Pearl River has been displaced so that it is due south of the city of Guangzhou (Canton) instead of to its southeast, and its two banks are symmetrically aligned in relation to the north-south axis passing through the center of the city. The deviations from actuality are intended to situate the city in a favorable position according to the principles of the science of siting.
Size of the entire original: 151 × 277 cm. Courtesy of the Geography and Map Division, Library of Congress, Washington, D.C. (G7823.K8A5 17--.K8 Vault).

> The mountains of Chu are densely arrayed;
> Trees shed, autumn clouds rise.
> Toward dawn, as soon as I ascend the terrace—
> Behold, the vast Jiang, daily going a thousand *li*.[81]

Combining Fact and Value

On the basis of the evidence considered here, it seems that Chinese cartography often shared the aesthetic principles of painting and poetry. This conclusion is hardly surprising, since cartographers, painters, and poets generally occupied the same social stratum, as part of the intellectual elite, and came from similar educational backgrounds. In some cases the same person might be involved in cartography, painting, and poetry. In such a person's mind, it seems likely that the principles governing those pursuits would have blended together and not formed three disparate sets. The cultural environment was such that self-expression was as important as reproduction of external reality, if not more important. It might therefore be possible to write a history of Chinese cartography showing its assimilation into the fine arts. This calls for an adjustment in the sorts of artifacts selected for attention, artifacts chosen not only for signs of quantification and mensuration, but for expressive and aesthetic value. It has been customary, however, to regard maps with a seemingly utilitarian aim—planimetric grid maps presenting quantitative information and ideally based on direct and indirect measurements—as the defining achievements of Chinese cartography.[82] According to this view, pictorialism and expressionism in cartography are something to be outgrown, just as they were in Europe. Enough has been said in this and other chapters to weaken that view of the Chinese cartographic tradition, which privileges the objective and quantitative over the subjective and qualititive—in short, fact over value. But the split between fact and value was never as strong in China as it has become in the European tradition.

To elevate objectivity over subjectivity, presentation over expression, is to introduce a hierarchy foreign to Chinese aesthetics, an economy of representation that seems to have involved cartography. Maps were made not only to be studied, not only for reference, but also to be enjoyed, and the record shows that mapmakers had users capable of appreciating their works intellectually and aesthetically. If one goal of the history of cartography is to reconstruct the reception of maps by particular users in particular cultures, accounts of Chinese cartography must take these theories of representation into account. When viewing a Chinese map, one may have to consider not only landscape—the representation of visible earth forms—but also what might be called "inscape": "But as air, melody, is what strikes me most of all in music and design in painting, so design, pattern or what I am in the habit of calling 'inscape' is what I above all aim at in poetry. Now it is the virtue of design, pattern, or inscape to be distinctive and it is the vice of distinctiveness to

81. "Ti Mi Yuanhui hua" (Inscribed on a painting by Mi Yuanhui [Mi Youren]), in Chen et al., *Yuding lidai tihua shi lei*, 11.11a (note 74). This poem appears under the *shanshui*, or landscape, classification. Chu is a region comprising much of the Yangtze basin: it encompasses Hubei Province and parts of other present-day provinces. The Jiang is the Changjiang, or Yangtze River.

82. For example, see Edouard Chavannes, "Les deux plus anciens spécimens de la cartographie chinoise," *Bulletin de l'Ecole Française d'Extrême Orient* 3 (1903): 214–47; Joseph Needham, *Science and Civilisation in China* (Cambridge: Cambridge University Press, 1954–), vol. 3, with Wang Ling, *Mathematics and the Sciences of the Heavens and the Earth* (1959); Chen Feiya et al., eds., *Zhongguo gudai dilixue shi* (History of ancient Chinese geography) (Beijing: Kexue Chubanshe, 1984); Lu Liangzhi, *Zhongguo dituxue shi* (History of Chinese cartography) (Beijing: Cehui Chubanshe, 1984).

FIG. 6.26. SECTION OF A THIRTEENTH-CENTURY COPY (?) OF THE *CHANGJIANG WANLI TU*. The *Changjiang wanli tu* (Ten thousand *li* map of the Changjiang [Yangtze River]) was made by Juran (fl. ca. 960–80), a Buddhist priest. The painting presents an oblique view of both banks of the Yangtze and provides place-names.
Size of the entire original: 43.7 × 1,654 cm. Courtesy of the Freer Gallery of Art, Smithsonian Institution, Washington, D.C. (acc. no. 11.168).

become queer."[83] In the Chinese context inscape applies to the representation of the innernesses of earth and mind, which resonate with each other and with the outer patterns of heaven (*tianwen*). Inscape, or a similar idea, may help to account for the discrepancies between image and geographic reality observed in many Chinese maps. Some maps, for example, those of the Daozang (Daoist canon), seem to take Su Shi's injunctions against formal resemblance to their limit, forgoing geographic reality altogether and purporting to represent configurations of spiritual energy or forms beyond the phenomenal world (see figs. 6.28 and 6.29).

The explanatory power of the economy of representation described here, however, still needs to be established more firmly. This requires deeper and more extensive study of the interaction of the arts in China, so that their relation to mapmaking can be more precisely defined. The research presented here surveys only a small portion of the available materials. My purpose has been merely to demonstrate the usefulness of a humanistic approach to Chinese maps, an approach in which the claims of value are recognized, in the hope of stimulating further investigation. The examples discussed above illustrate the value of the visual arts for the study of maps: in its treatment of space and manner of representation, mapmaking seems to have shared a number of conventions with painting and—by virtue of the close relationship between painting and poetry—with literature. Knowledge of those conventions can often shed light on how a particular map was meant to be "read," and disregard for them may lead to misreading or partial understanding, as in the case of the Han garrison map. In many instances one may have to consider the possibility that departures from formal likeness were intentional, that mapmakers held expressive aims paramount.

The conjoining of map and art history that I have argued for here is crucial in another, perhaps more fundamental, respect. Scholarship in Chinese art and cartography often involves the study of purported copies of works made centuries earlier. In many cases such copies are all that is available, but they should be handled with more caution than they have been previously, especially

83. Gerard Manley Hopkins to Robert Bridges, 15 February 1879, in *Gerard Manley Hopkins*, ed. Catherine Phillips (Oxford: Oxford University Press, 1986), 234–35, esp. 235.

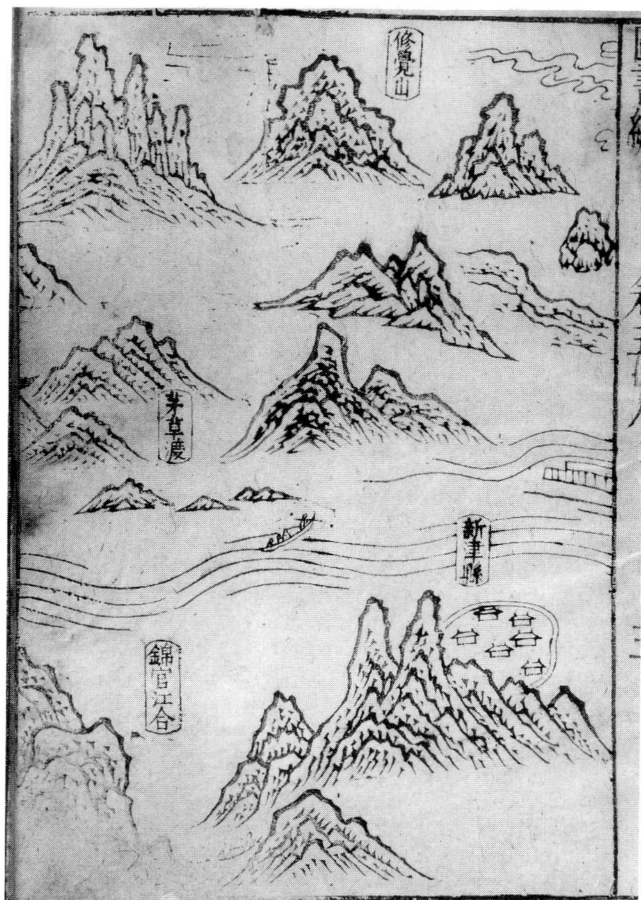

FIG. 6.27. SECTION OF THE "WANLI CHANGJIANG TU" (TEN THOUSAND *LI* MAP OF THE YANGTZE RIVER). As in figure 6.26, this map shows both banks of the river and gives place-names.

Size of each page: 22.5 × 14.5 cm. From Zhang Huang, comp., *Tushu bian* (Compilation of illustrations and writings, compiled 1562–77, printed 1613), 58.2a–3b. Reproduced courtesy of the Harvard-Yenching Library, Harvard University, Cambridge.

since the validity of conceptions about the sequence of styles and modes of representation depends on the establishment of authenticity. What Loehr has said about the history of Chinese art holds also for cartography:

> The authenticity question results in a true paradox: (1) Without knowledge of styles, we cannot judge the authenticity of individual works, and (2) without convictions about authenticity, we cannot form concepts of style. . . . The case for or against authenticity, therefore, rests on convictions, and for the convincingness of a work, style is the foremost criterion. A judgment about authenticity, therefore, remains itself arbitrary and subjective unless it is fully compatible with the criteria of style.[84]

Historians of Chinese cartography need to be sensitive to Loehr's concerns. Some judgments about the appearance and accuracy of maps supposedly made during the Song, for example, have been based on copies dating from the Ming and Qing. The fidelity of the Qing copies to the Song originals cannot be taken for granted. True, copying was taken seriously: Xie He identifies copying, "the transmission of models," as the sixth principle of painting.[85] But copying was not a purely mechanical affair. Manual copying of past masterpieces of painting and calligraphy was a crucial element in defining one's own artistic style, and therefore a copy was supposed not merely to reproduce, but also to introduce something of the copyist. In the case of woodblock printing, there is evidence that fidelity of copies to the physical appearance of original woodcuts was not quite complete, as one might expect from the place of formal resemblance in Chinese aesthetics. Maps in Song and Ming editions of the same work could differ appreciably in appearance (see figs. 6.30 and 6.31).[86]

84. Loehr, "Some Fundamental Issues," 187–88 (note 34).

85. Xie, *Gu huapin lu*, 1 (note 31).

86. There are also differences in the cartographic images on the two

FIG. 6.28. MAP OF TAI SHAN (MOUNT TAI) FROM THE DAOIST CANON. Titled "Dongyue zhenxing tu" (Map of the true form of the sacred peak of the east), it represents the mountain's caverns. Maps like this were meant to harness the spiritual power of the mountain for the benefit of the user: they could repel evil spirits and help one attain immortality.
Size of the page: 11 × 10.5 cm. From *Lingbao wuliang duren shangjing dafa* (Great rituals of the supreme scripture on the infinite salvation of Lingbao [numinous treasure]), 21.16a, in *Zhengtong Daozang* (Daoist canon of Zhengtong reign period [1436–49]), 1,120 volumes, vol. 89. From Commercial Press reprint, 1923–26, reproduced courtesy of the Harvard-Yenching Library, Harvard University, Cambridge.

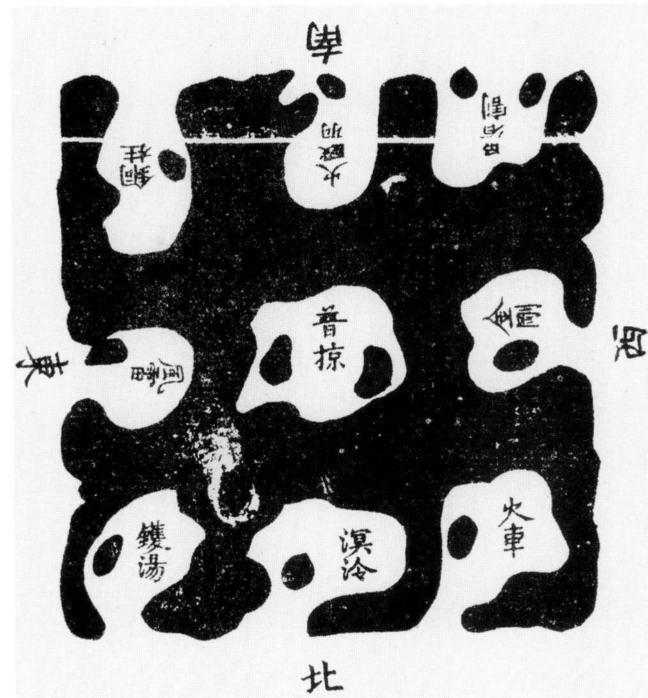

FIG. 6.29. MAP OF THE UNDERWORLD FROM THE DAOIST CANON. The "Jiu yu deng tu" (Map of the lanterns of the nine purgatories) represents the nine chambers where souls of the dead undergo various punishments, such as flogging, boiling, and immersion. The chambers are laid out following the nonary square pattern typical of many cosmographic representations (see pp. 204–5). The map was supposed to be used to help win the release of underworld prisoners.
Size of the page: 11.5 × 11 cm. From Lu Shizhong, *Wushang xuanyuan santian yutang dafa* (Great rituals of the jade hall from the three heavens of the supreme, sublime, and primal, compiled 1158), 14.5b, in *Zhengtong Daozang* (Daoist canon of the Zhengtong reign period [1436–49]), 1,120 volumes, vol. 101. From Commercial Press reprint, 1923–26, reproduced courtesy of the Harvard-Yenching Library, Harvard University, Cambridge.

Before we can draw conclusions about the cartography of a period from later copies, we need to know more about the attitudes of the block cutters themselves and those who drew the paper images that were transferred to the blocks. To what extent were the artists allowed to deviate from the physical appearance of the original images, and to what extent were the block cutters allowed to deviate from the transferred image? On these questions, scholars have yet to reach even tentative conclusions. But answers to such questions are crucial if historians of Chinese cartography are to arrive at even a provisional understanding of the chronology of their own subject. Solutions seem to require the expertise of both historians of science and historians of the arts, since the questions involve technology, production processes, artistic style, and aesthetic theory. In short, what is needed is a merger of fact and value.

The stress here has admittedly fallen on value, but that should not be misconstrued as a denial of the claims of fact. Both are equally important aspects of the cultural realities we seek to understand, and one should not be ignored at the expense of the other. In the field of the history of cartography, it is time to heal what might be called a dissociation of sensibility. This chapter is intended as a first step toward this reunification, and as a first step, it reasserts the value of value—partly as a matter of analytical convenience, but mainly to direct

surviving versions of the *Yu ji tu* (Map of the tracks of Yu) carved in stone. (See figs. 3.14 and 3.15.)

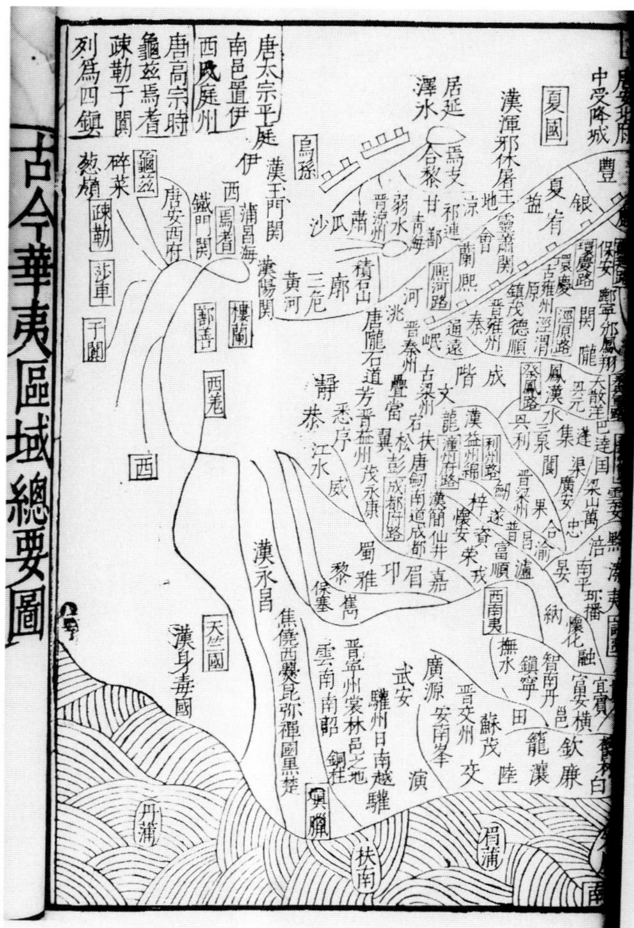

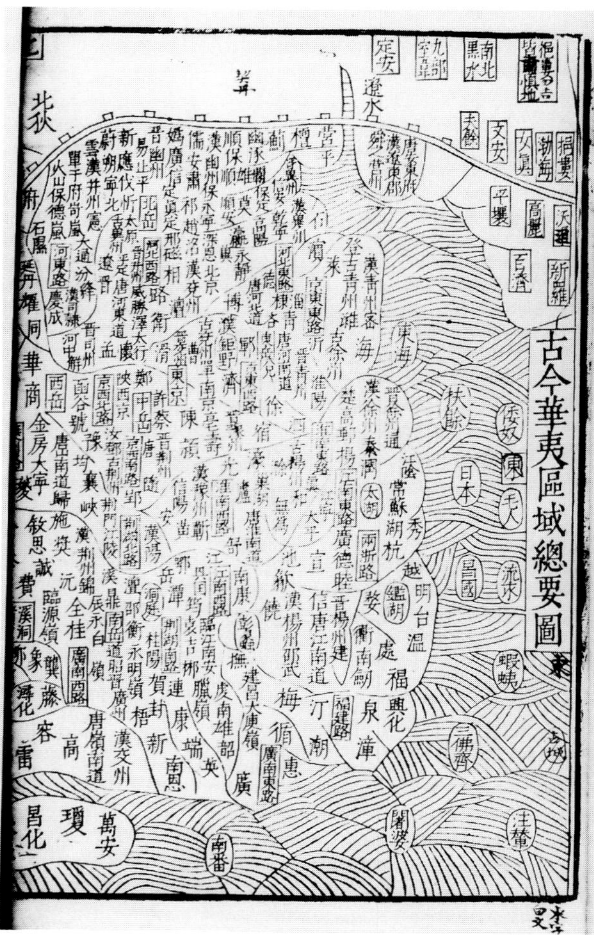

FIG. 6.30. "GUJIN HUA YI QUYU ZONGYAO TU" (GENERAL MAP OF THE ANCIENT AND PRESENT TERRITORIES OF CHINA AND FOREIGN COUNTRIES). This map, as well as figure 6.31—both from a Ming edition of the *Lidai dili zhizhang tu* (Easy-to-use maps of geography through the dynasties, 1098–1100, supplemented 1162)—should be compared with those from a Southern Song edition (figs. 3.23 and 3.24 above). There are differences in the density of the labels, in the forms of certain graphs, and in the depictions of the coastlines.
Size of the original: unknown. Photograph courtesy of Cao Wanru, Institute for the History of Natural Science, Academia Sinica, Beijing.

attention to an area that tends to be slighted in much research into Chinese cartography.[87]

87. Though there is little movement among historians of Chinese cartography toward art history, in Chinese art history there are indications of movement in the opposite direction. Kiyohiko Munakata, for example, consults traditional Chinese maps in order to understand Chinese attitudes toward sacred mountains, a traditional subject of landscapes. See Kiyohiko Munakata, *Sacred Mountains in Chinese Art* (Urbana: University of Illinois Press, 1991).

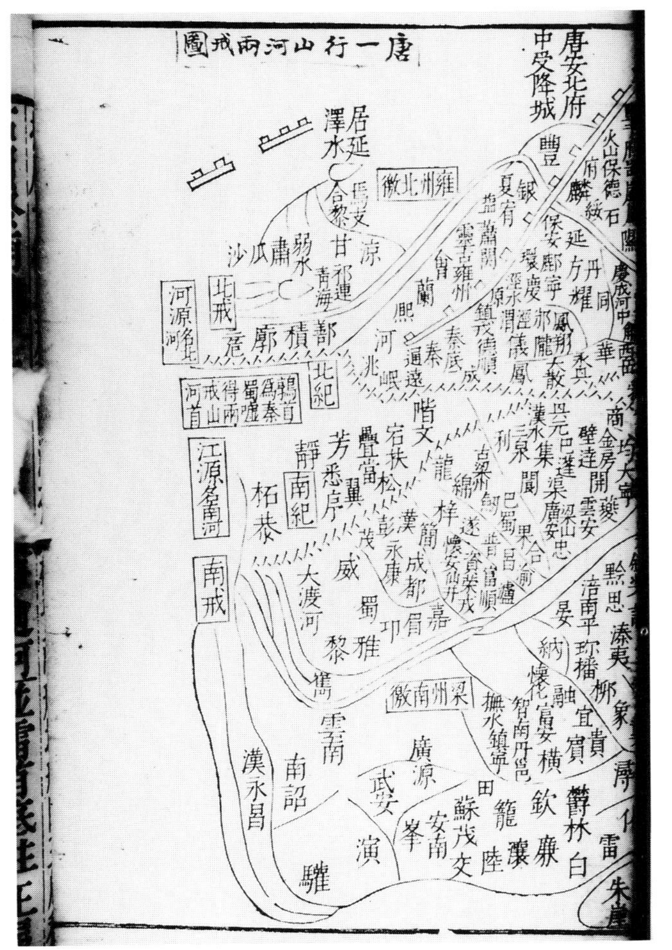

FIG. 6.31. "TANG YIXING SHAN HE LIANGJIE TU" (MAP OF THE TWO BOUNDARIES FORMED BY MOUNTAINS AND RIVERS ACCORDING TO YIXING OF THE TANG). See figure 6.30.

Size of the original: unknown. Photograph courtesy of Cao Wanru, Institute for the History of Natural Science, Academia Sinica, Beijing.

7 · Traditional Chinese Cartography and the Myth of Westernization

CORDELL D. K. YEE

In the preceding chapter I stated that Chinese cartography did not dissociate itself from the visual arts until the nineteenth century. That statement seems to be contradicted by accounts of Chinese mapping in the late Ming (1368–1644) and Qing (1644–1911) dynasties, according to which Chinese cartography assimilated techniques imported from Europe and became a "science" in the Western sense of the word. What this new science of Chinese cartography involved was a conception of the earth as spherical and the use of a coordinate system for locating points on the earth's surface. This entailed the use of mathematical techniques for projecting points on the earth's spherical surface to a plane mapping surface. To judge from previous accounts of Chinese cartography, European cartography so displaced traditional Chinese practices that they disappeared or at least are not worth mentioning. The accounts of late Ming and Qing cartography in works by Wang, Needham, Lu, and others focus on the Jesuit mapping of China.[1] For these historians, small-scale mapping is the measure of all cartography, and so other aspects of cartographic culture are overlooked. Representatives of the earlier tradition are barely mentioned. Accounts like these foster the impression that in the eighteenth century, Chinese and European cartography became indistinguishable.

When European cartography was first introduced into China in the late sixteenth century, the major difference between European and Chinese cartography was that traditional Chinese mapmakers treated the earth as flat. According to previous accounts, that treatment changed after a different world model and Ptolemaic cartographic techniques were brought to China by Jesuit missionaries. Here I examine the Chinese responses to those works, insofar as they were relevant to Chinese cartography. Perhaps lack of response is a better way to describe the situation. For most of the period under discussion, from the late sixteenth century to the beginning of the twentieth, Chinese cartographic practice bears few traces of European influence. The conversion of Chinese cartography to the Ptolemaic system was not as swift or complete as previous accounts have made it seem.

THE INTRODUCTION OF EUROPEAN CARTOGRAPHY

It was not the primary aim of the Jesuits to train the Chinese in European science and technology. In fact, to most Jesuits, even to consider this aim was controversial. The missionaries Alessandro Valignani (1539–1606) and Michele Ruggieri (1543–1607), however, were perceptive enough to see that Sinicization was the only way to secure a foothold in China. Matteo Ricci (1552–1610) followed their line of reasoning, though not without great pressure from his superiors against such a policy. Ricci believed that the way to win Chinese converts to Christianity was through indirect means, rather than by a direct challenge to Chinese values and beliefs. He attempted to win over the intellectual elite by recourse to the scientific achievements of European culture, in mathematics, astronomy, and cartography. Once Chinese intellectuals appreciated the advantages of European science and technology, according to Ricci's line of thought, they might be induced to convert to Christianity. Members of the elite were targeted for attention because the Jesuits saw them as a way to the imperial court. If the emperor could be converted, the rest of the empire would follow. For the Jesuits, then, maps were part of what Jacques Gernet has described as an "enterprise of seduction."[2]

Though scientific aims were not of paramount concern to the Jesuits, their cartographic works had the potential

1. See Wang Yong, *Zhongguo dilixue shi* (History of geography in China) (1938; reprinted Taipei: Shangwu Yinshuguan, 1974); idem, *Zhongguo ditu shi gang* (Brief history of Chinese cartography) (Beijing: Sanlian Shudian, 1958); Joseph Needham, *Science and Civilisation in China* (Cambridge: Cambridge University Press, 1954–), vol. 3, with Wang Ling, *Mathematics and the Sciences of the Heavens and the Earth* (1959); Chen Cheng-siang (Chen Zhengxiang), *Zhongguo dituxue shi* (History of Chinese cartography) (Hong Kong: Shangwu Yinshuguan, 1979); Chen Feiya et al., eds., *Zhongguo gudai dilixue shi* (History of ancient Chinese geography) (Beijing: Kexue Chubanshe, 1984); Lu Liangzhi, *Zhongguo dituxue shi* (History of Chinese cartography) (Beijing: Cehui Chubanshe, 1984).

2. Jacques Gernet, *China and the Christian Impact: A Conflict of Cultures*, trans. Janet Lloyd (Cambridge: Cambridge University Press, 1985), 15.

to effect revolutionary changes in Chinese mapmaking practices. Before 1583, the year Ricci and Ruggieri established a mission at Zhaoqing Prefecture (in present-day Guangdong Province), Chinese mapmakers had used a square grid system as an aid for plotting distance and direction. Through Ricci's maps, however, some mapmakers were introduced to the Ptolemaic system of organizing cartographic space.

On the wall of the mission room at Zhaoqing, according to Ricci's journal, "there was a cosmographical chart of the universe, done with European lettering. The more learned among the Chinese admired it very much and, when they were told that it was both a view and a description of the entire world, they became greatly interested in seeing the same thing done in Chinese."[3] At the request of the prefect of Zhaoqing, Ricci made the map "speak Chinese," drawing a new version: "The new chart was made on a larger scale than the original, so as to give more room for the Chinese written characters which are somewhat larger than our own."[4] Ricci seems to have recognized that traditional Chinese maps were characterized by textual supplementation, since it is recorded that "new annotations were also added, more in keeping with the Chinese genius."[5] The new map was printed in 1584 as the *Yudi shanhai quantu* (Complete geographic map of the mountains and streams). No examples of this edition are extant, but a rendition of it survives in the *Tushu bian* (Compilation of illustrations and writings), compiled by Zhang Huang (1527–1608), who met Ricci in 1595.

If Zhang Huang's rendition (fig. 7.1) can serve as an indication, Ricci's map was based on a map of the Ortelius type, oriented with north at the top. It defined geographic space by a graticule of longitude and latitude lines. The equator is distinguished from other parallels by its thickness; there is no clear indication of a prime meridian, though the map's center lies somewhere in the Pacific Ocean. The Ming empire is erroneously represented as comprising two large islands and part of the larger Asian continent. This image, however, was improved in later editions, in the sense that the whole empire was depicted as occupying part of the Asian continent (see, for example, fig. 7.2).

Ricci's map, if accepted as true, called for some alterations in Chinese worldviews. The conception of the world as round went against a school of cosmographic thought according to which the world was square and had a generally flat surface. According to Ricci's journal, the Chinese "could not comprehend the demonstrations proving that the earth is a globe, made up of land and water, and that a globe of its very nature has neither beginning nor end."[6] Thus, much Chinese mapmaking had proceeded on the assumption that details on one plane, the earth, were being transferred to another plane, the surface on which the mapmaker would plot them. There would be no real distortion: the transfer process was a matter of reducing the real world to fit on a page or scroll. There was no need to devise mathematical formulas to compensate for the earth's curvature; attention to distance and direction were all that were necessary to locate physical features correctly. The proportions could be scaled down with the square grid.

The graticule on Ricci's map would not have been entirely incomprehensible to Chinese intellectuals. It had an analogue in the twenty-eight *xiu* (lunar lodges) of Chinese astronomy, used to determine positions of celestial phenomena. It was this symmetry with the Chinese celestial system that attracted the attention of some Chinese intellectuals and helped convince them of the map's truth. Feng Yingjing (1555–1606), for example, wrote in a preface to the fourth edition of Ricci's world map (1603): "It fixes the surface of the terrestrial sphere by means of the degrees of the celestial sphere; by means of the trajectory of the sun it differentiates the cold zones from the warm zones; it fixes the confines (of the world) by means of the Five Great Continents, and adds to them the specialities of products and the singularities of customs of peoples. What a nice thing!"[7]

Perhaps more difficult to accept was the representation of China as one small country among many in a world covered mostly by water, not land. The great distances that were depicted on the world map and that the Jesuits claimed to have traveled were interpreted as attempts to deceive. One Ming official wrote: "In past years he [Alphonse Vagnoni (1566–1640), Jesuit missionary] and his colleagues claimed to practice Catholicism, but actually deceived the king of Lüsong [Philippines], seized his land, and changed its name to Daxiyang [which Ricci identified as his homeland]. Thus Min [present-day Fujian] and Ao [present-day Guangdong] are close to a shrewd, savage country. How could it [Daxiyang] be 80,000 *li* away?"[8] Another contemporary writer feared

---

3. *China in the Sixteenth Century: The Journals of Matthew Ricci, 1583–1610*, trans. Louis J. Gallagher from the Latin version of Nicolas Trigault (New York: Random House, 1953), 165–66. Trigault, a Jesuit missionary in China, translated and added to Ricci's own account of his activities. Trigault's translation was originally published in 1615.

4. *China in the Sixteenth Century*, 166 (note 3).

5. *China in the Sixteenth Century*, 166 (note 3).

6. *China in the Sixteenth Century*, 167 (note 3).

7. The text and translation of this preface may be found in Pasquale M. d'Elia, "Recent Discoveries and New Studies (1938–1960) of the World Map in Chinese of Father Matteo Ricci SJ," *Monumenta Serica* 20 (1961): 82–164, esp. 129.

8. Shen Que (d. 1624), appointed in 1615 as vice minister of rites in Nanjing, quoted in Zhang Weihua, *Ming shi Ouzhou si guo zhuan zhushi* (Commentary on the chapters on four European countries in the *History of the Ming*) (1934; reprinted Shanghai: Shanghai Guji Chubanshe, 1982), 131.

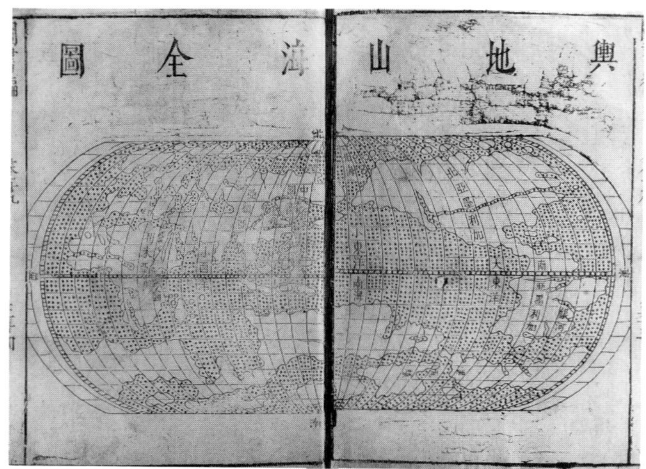

FIG. 7.1. "YUDI SHANHAI QUANTU" (COMPLETE GEOGRAPHIC MAP OF THE MOUNTAINS AND SEAS). A Chinese rendition of the first edition of Matteo Ricci's world map (1584).
Size of each page: 23 × 14.5 cm. From Zhang Huang, comp., *Tushu bian* (1613), 29.33b–34a. Reproduced courtesy of the Harvard-Yenching Library, Harvard University, Cambridge.

that the same tactic used in the Philippines might be applied to China: "When these foreigners claim to have come from 90,000 *li*, the distance claimed is meant to induce us to believe that they have no ulterior motives, so that we will not worry about their swallowing us up."[9]

One feature often remarked on Ricci's world maps is the placement of China near the center. Trigault's version of Ricci's diary asserts that this was a concession to the traditional Chinese belief that China was the center of the world: the Chinese "firmly believe that their empire is right in the middle of it," and they dislike "the idea of our geographies pushing their China into one corner of the Orient."[10] According to the projection on Ortelius's map copied by Ricci, the 180° meridian lay near Asia, somewhere in the Pacific Ocean. Ricci adjusted the image so that this meridian rather than the 0° meridian passed through the center of the map. Thus China also appears near the center.

9. Su Jiyu, a late Ming writer, quoted in Zhang, *Ming shi Ouzhou si guo zhuan zhushi*, 131 (note 8).

10. *China in the Sixteenth Century*, 167 (note 3).

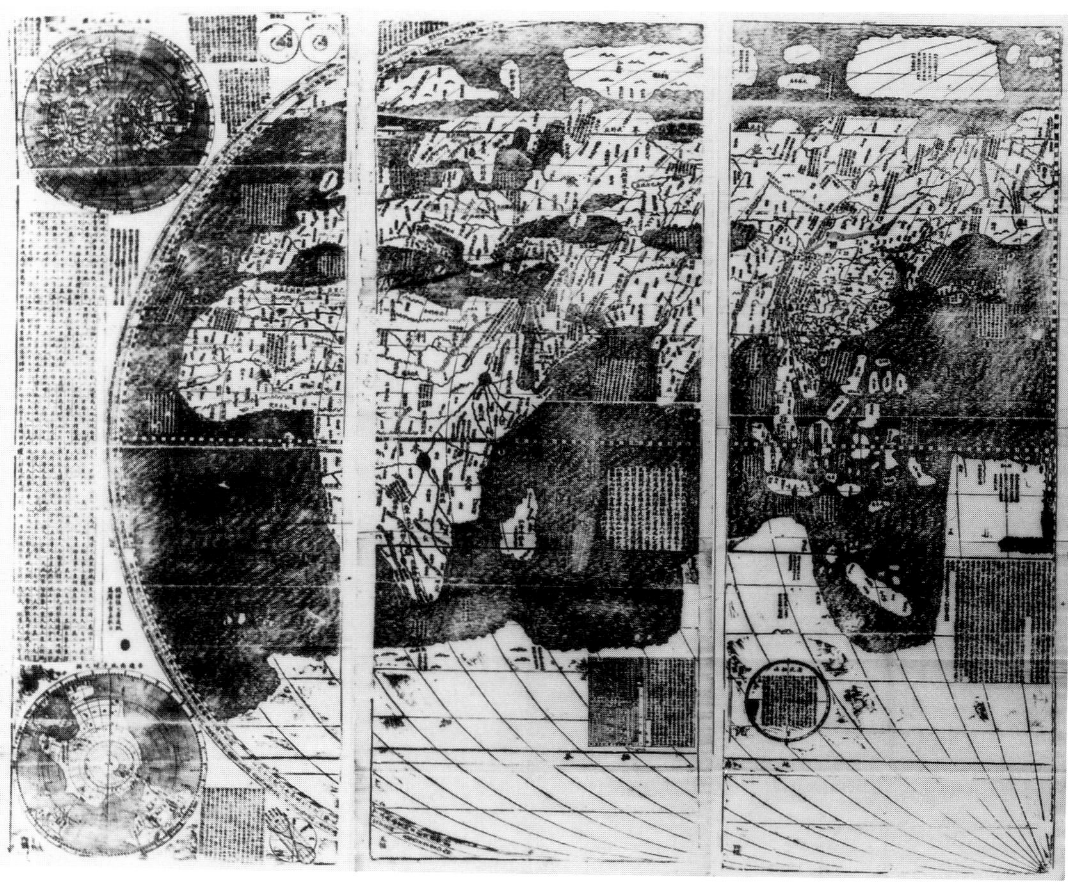

FIG. 7.2. THIRD EDITION OF MATTEO RICCI'S WORLD MAP, 1602.

The idea that this change was made to conform with Chinese conceptions of geographic actuality deserves some scrutiny. The often-quoted Wei Jun (late Ming period), who criticized Ricci for not placing China at the center of the world, does not necessarily reflect prevailing opinion at all levels of Chinese society.[11] The Chinese "Middle Kingdom complex" was somewhat more complicated. For the bureaucratic elite before the seventeenth century, the center of the world may have been near Luoyang at Dengfeng, the site of the great gnomon. But outside those circles there was room for divergent conceptions. One tradition locates the center of the world at the Kunlun Shan (Mount Kunlun), not in China but somewhere to the west. Further evidence appears on Chinese Buddhist maps depicting Jambūdvīpa, in Buddhist cosmology the island continent containing India and surrounding territories. The continent was traditionally represented as an inverted triangle on which China would be a small country to the northeast, not at the center. Chinese Buddhists changed that image, enlarging China, but did not place China at the center (figs. 7.3 and 7.4).

The name "Zhongguo" is often translated as "Middle Kingdom," a name that seems to imply that the Chinese believed their empire was the geographic center of the world. But in its original, quite literal usage, "Zhongguo" referred to the northern Chinese states that were considered the kernel of the Zhou kingdom. It maintained this secondary meaning until the end of imperial China and connoted cultural or political primacy. "Zhongguo" thus does not necessarily imply geographic centrality, but connotes cultural or political centrality.

As a consequence there are several explanations for

11. See, for example, Kenneth Ch'en (Chen Guansheng), "Matteo Ricci's Contribution to, and Influence on, Geographical Knowledge in China," *Journal of the American Oriental Society* 59 (1939): 325–59, esp. 348. Wei Jun is quoted as saying: "China should be in the center of the world, which we can prove by the single fact that we can see the North Star resting at the zenith of the heaven at midnight. How can China be treated like a small unimportant country, and placed slightly to the north as in this map?" Ch'en's article was published in Chinese as "Li Madou dui Zhongguo dilixue zhi gongxian ji qi yingxiang" (Matteo Ricci's contributions to and influence on Chinese geography), *Yu Gong Banyuekan* 5, nos. 3–4 (1936): 51–72.

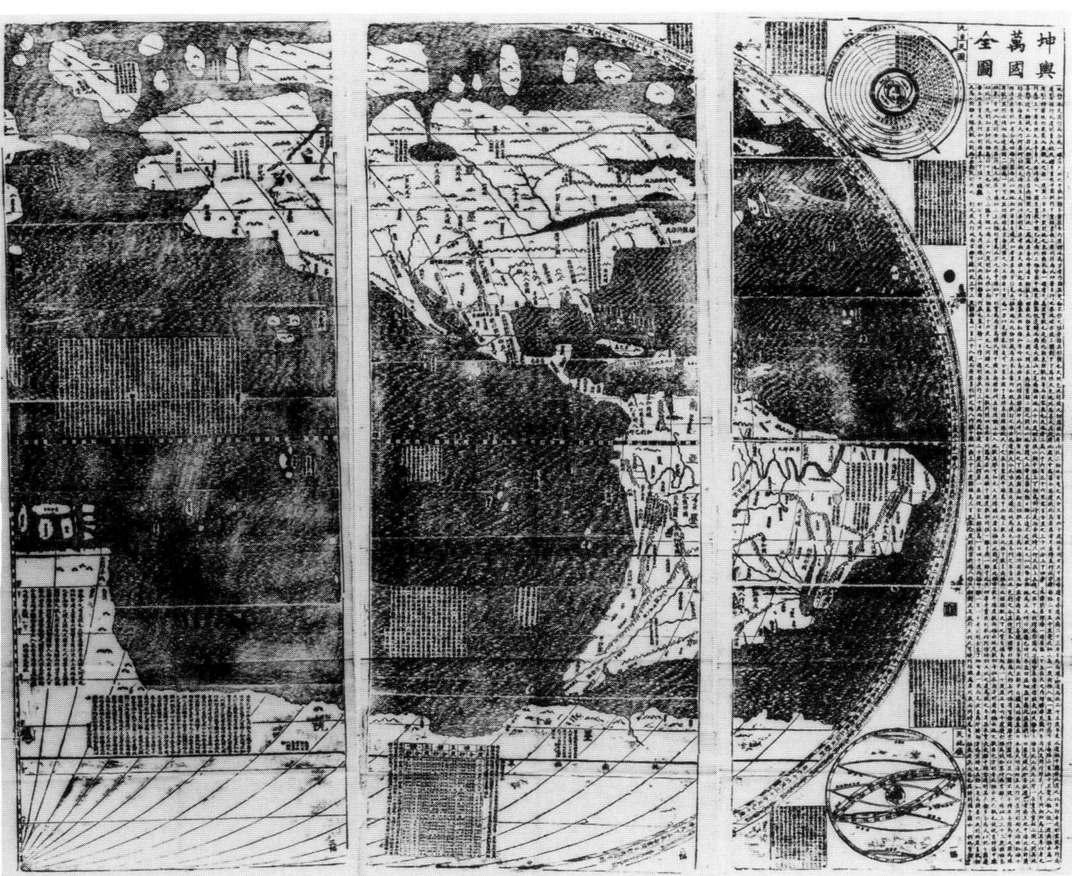

Size of the original: 168.7 × 380.2 cm. Photograph courtesy of Sotheby's, London.

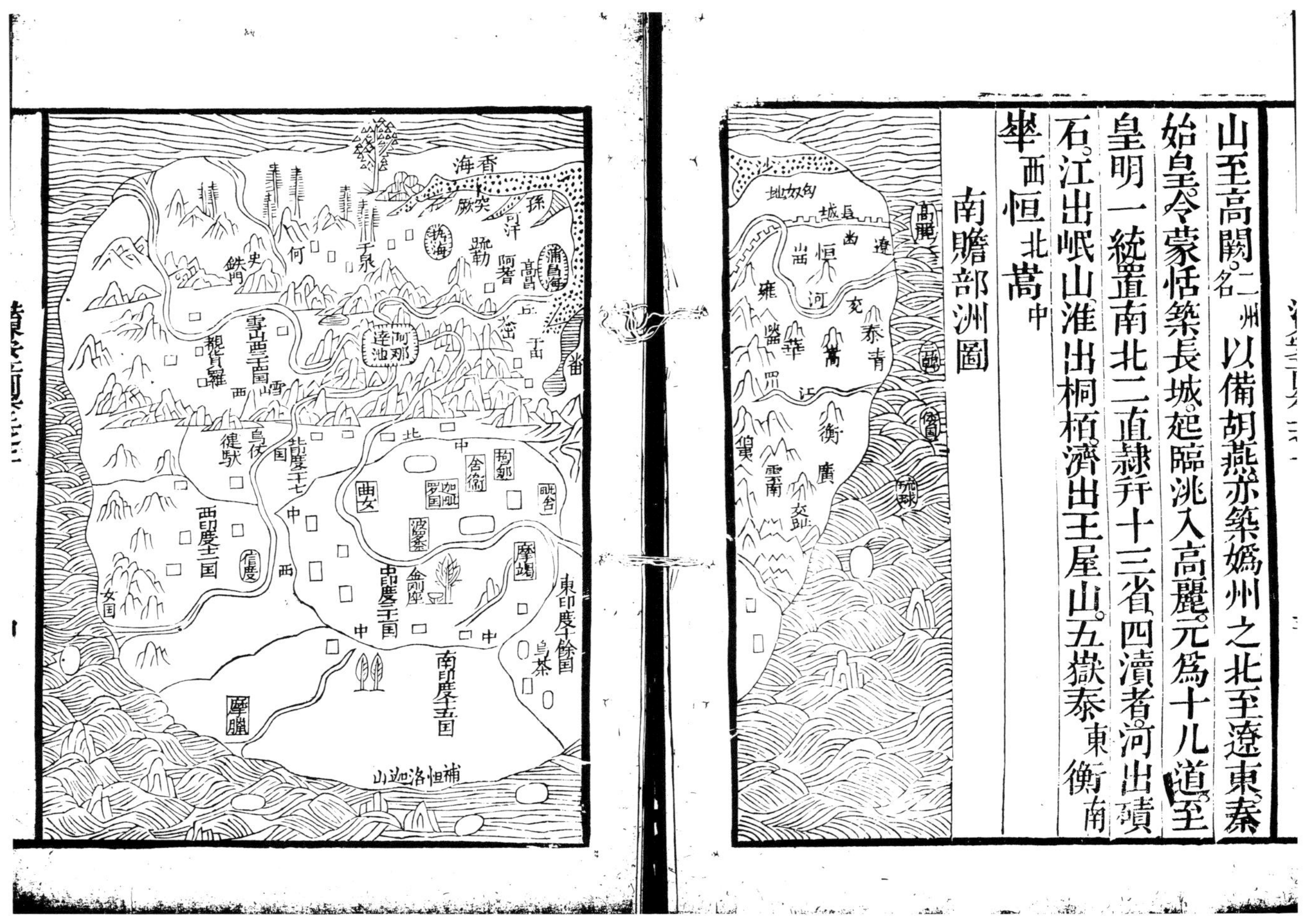

FIG. 7.3. CHINESE BUDDHIST MAP OF JAMBŪDVĪPA. Size of the images: 20 × 14.2 cm and 20 × 5.1 cm. From Renchao, *Fajie anli tu* (Maps of the configuration of Dharmadhātu [physical universe], 1607), 1824 ed., A.3b–4a.

the Sinocentrism of traditional Chinese maps of "all under heaven," or the world. They include a notion of geographic centrality, but perhaps more important, a belief that China was the center of culture, the standard for civilization: peoples desiring cultural attainment gravitated toward China. Although it has been said that "the Chinese are of the opinion that only China among the nations is deserving of admiration,"[12] Ricci's placement of China near the center of his maps could merely reflect the interest his map readers would have had in it.

Previous accounts of the Chinese reception of Ricci's maps have fostered the impression that they enjoyed widespread acceptance in China. For example, Trigault's version of Ricci's journal states that Ricci's map, "frequently revised and refined and often reprinted, found its way into the courts of the Governor and of the Viceroy, where it was greatly admired, and finally into the palace of the King, on his own request."[13] Kenneth Ch'en states: "It appears that no matter where Ricci went he was asked to make maps for the local officials."[14] Ricci himself records that on seeing his world map, one official immediately ordered it to be engraved and gave copies as presents to friends.[15] In another instance, a magistrate had a Ricci map inscribed in stone and distributed rubbings of it to friends. In addition, two of Ricci's closest Chinese friends, Feng Yingjing and Li Zhizao (d. 1630), had copies of Ricci's maps printed from woodblocks.

12. *China in the Sixteenth Century*, 167 (note 3).

13. See *China in the Sixteenth Century*, 168 (note 3).

14. Ch'en, "Matteo Ricci's Contribution," 343 (note 11). For a similar account of the reception of Ricci's maps, see John F. Baddeley, "Father Matteo Ricci's Chinese World-Maps," *Geographical Journal* 50 (1917): 254–70.

15. Matteo Ricci, *Storia dell'introduzione del Cristianesimo in Cina*, 3 vols., ed. Pasquale M. d'Elia, Fonti Ricciane: Documenti Originali concernenti Matteo Ricci e la Storia delle Prime Relazioni tra l'Europa e la Cina (1579–1615) (Rome: Libreria dello Stato, 1942–49), 1:211–12.

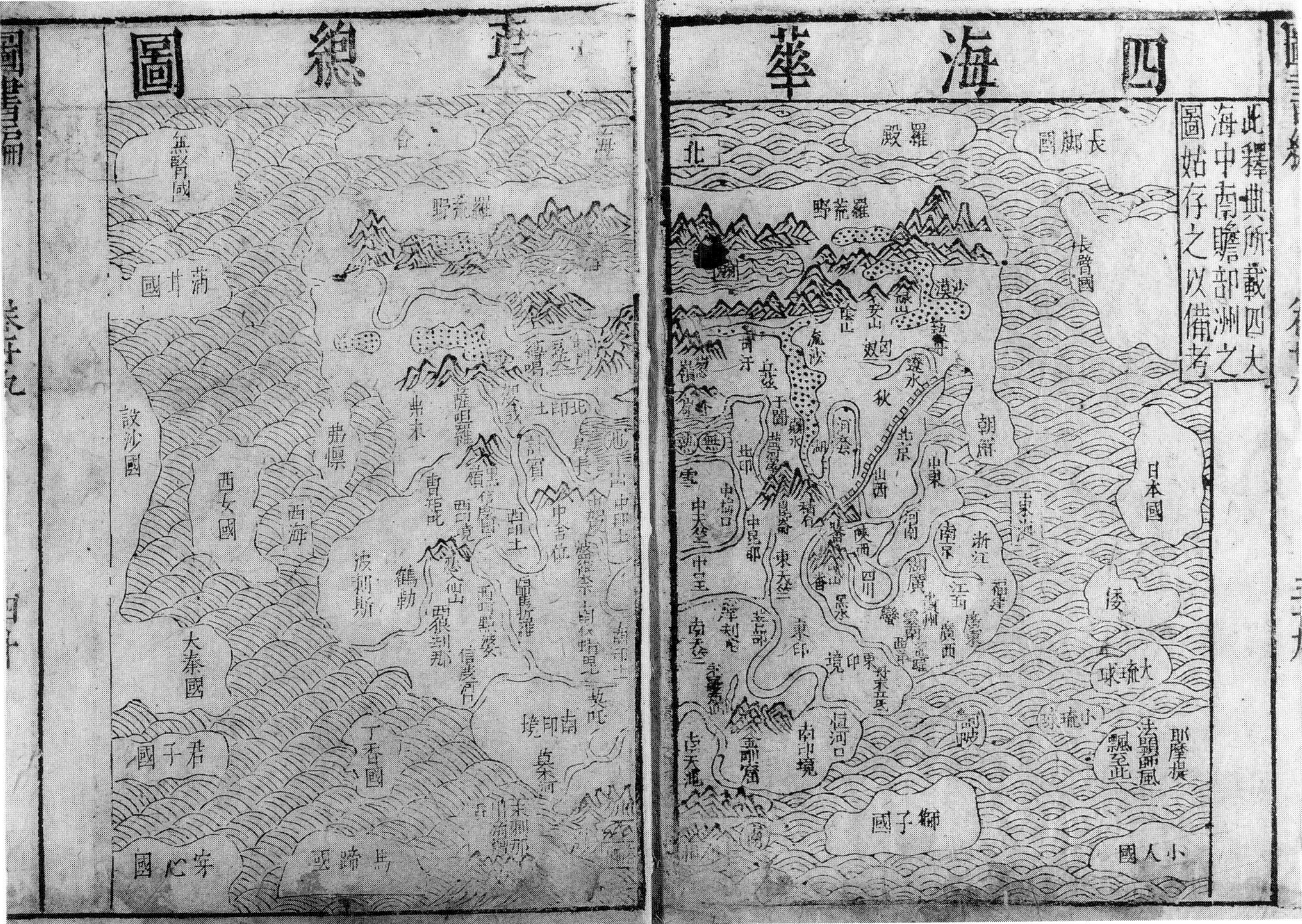

FIG. 7.4. "SIHAI HUA YI ZONGTU" (GENERAL MAP OF CHINESE AND FOREIGN TERRITORY WITHIN THE FOUR SEAS). Another example of a Chinese Buddhist map of Jambūdvīpa (see also fig. 7.3). Size of each page: 22.5 × 14.5 cm. From Zhang Huang, comp., *Tushu bian* (1613), 29.39b–40a. Reproduced courtesy of the Harvard-Yenching Library, Harvard University, Cambridge.

According to Ch'en, several thousand copies of Li's edition were circulated. In 1608 a eunuch showed the emperor a copy of one of Ricci's maps, and the emperor was so impressed that he demanded twelve more copies.[16]

A few scholars thought highly enough of Ricci's maps to include copies of them in their own works. Besides Zhang Huang, already mentioned above, Wang Qi, in the *Sancai tuhui* (Illustrated compendium of the three powers [heaven, earth, man], completed in 1607), reproduces a copy of the second edition of Ricci's map (1600) without the parallels and with only a few place-names (fig. 7.5). Ricci's map of the two hemispheres printed in 1601 was reproduced in Cheng Boer et al., *Fangyu shenglüe* (Compendium of geography, 1612) (fig. 7.6).[17] The maps are accompanied by a list of countries and places with their latitudes and longitudes, and with annotations. In the process of copying, a number of discrepancies in place-names and notes were introduced.

The images in the *Fangyu shenglüe* in turn served as the basis for the world map included in the later *Yutu beikao* (Complete study of maps, compiled ca. 1630) by Pan Guangzu. The *Yutu beikao* begins with a section of maps. The first two are the maps showing the Eastern and Western hemipheres copied from the *Fangyu shenglüe*. Most of the remaining twenty-three maps are based on the maps in the *Guang yutu* (Enlarged terrestrial atlas, ca. 1555) by Luo Hongxian (1504–64) but do not include the square cartographic grid. Pan Guangzu seems to have made no attempt to reinterpret native materials according

16. Ricci, *Introduzione del Cristianesimo*, 2:472–74 (note 15).

17. For a study of the *Fangyu shenglüe*, see Chen Guansheng (Kenneth Ch'en), "*Fangyu shenglüe* zhong geguo dufen biao zhi jiaoding" (Edited table of geographic coordinates for various countries in the *Fangyu shenglüe*), *Yu Gong Banyuekan* 5, nos. 3–4 (1936): 165–94.

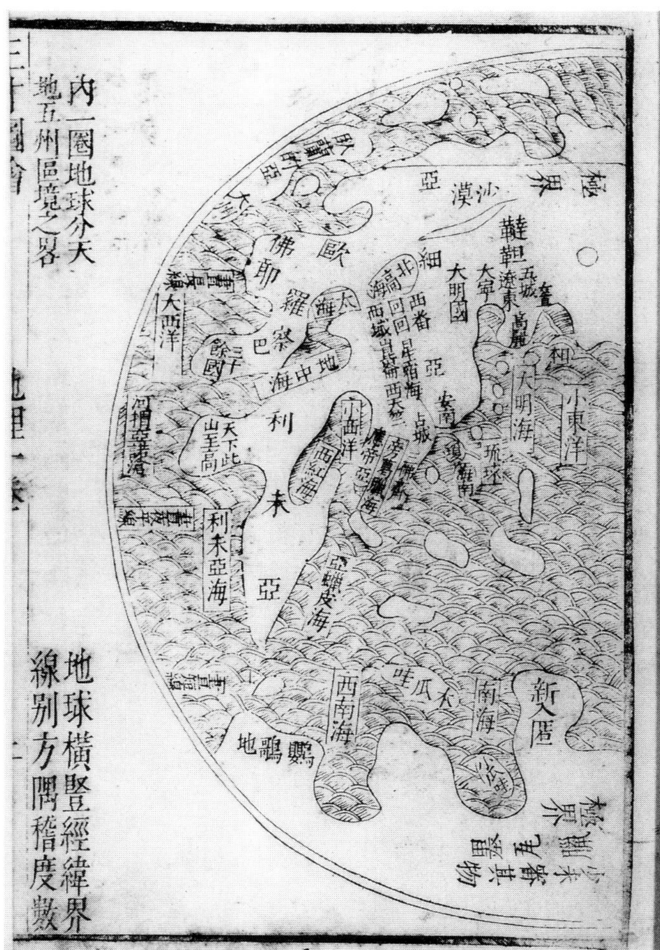

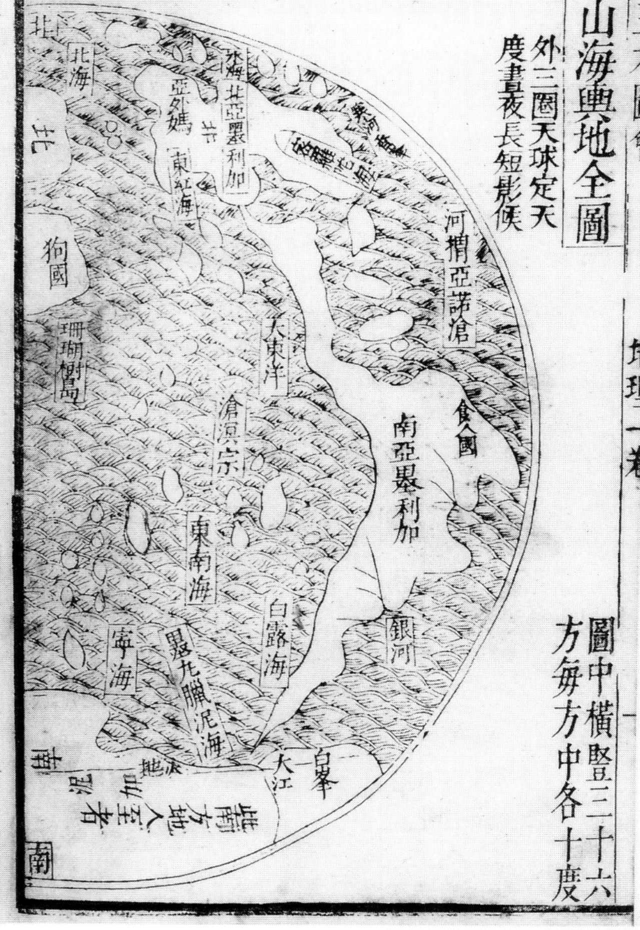

FIG. 7.5. "SHANHAI YUDI QUANTU" (COMPLETE GEOGRAPHIC MAP OF THE MOUNTAINS AND SEAS). A Chinese rendition of the second edition of Matteo Ricci's world map (1602). Size of each page: 21 × 14 cm. From Wang Qi, comp., *Sancai tuhui* (completed 1607, printed 1609), *dili* 1.1b–2a. Reproduced courtesy of the Harvard-Yenching Library, Harvard University, Cambridge.

to European techniques, so it seems doubtful that he really understood them.[18]

Since Ricci's maps had a fairly wide circulation in original editions and reproductions in Chinese works, one of the major questions for historians of cartography has been why they made no lasting impression on Chinese mapmaking. Ch'en offers four reasons: Chinese complacency—a belief that they had nothing to learn from the West; the connection of the world maps with the Catholic religion, which a few years after Ricci's death in 1610 became the object of persecution; the undeveloped state of Chinese science; and careless reproduction by Chinese copyists.[19]

Certainly the information about China reflected in indigenous maps was more reliable than the image presented in the first edition of Ricci's world map. China, as represented in Zhang's rendition of Ricci's map, consists of a section joined to the Asian landmass and two contiguous islands—a representation reflecting the incompleteness of European knowledge of China. Using Chinese maps and geographic works, including the *Guang yutu*, Ricci was able to provide Europeans with more reliable geographic information on China, and the fruits of his research showed in later editions of his world map, as the representation of China changed to conform more closely to that in Chinese maps.

Ricci did perform some measurements of his own to adapt Chinese information to representation using projection. He determined latitude and longitude for a number of places in China. But errors were introduced into

18. See Wang Yong, *Zhongguo dili tuji congkao* (Collected studies on Chinese geographic maps and documents, 1st ed. 1947), rev. ed. (Shanghai: Shangwu Yinshuguan, 1956), 20–21. The *Yutu beikao* was reported to be at the Library of Congress, but it has not been located there. See also p. 409 on the *Huiji yutu beikao juanshu*.

19. See Ch'en, "Matteo Ricci's Contribution," 357–59 (note 11).

his maps because he thought incorrectly that each degree consisted of 250 *li*, when the correct figure was about 194 *li*. In any case, Ricci had the means to teach European techniques of projection, and his Chinese friends and admirers certainly would have had the opportunity to learn those techniques.

But apart from the reproductions of Ricci's maps mentioned above, no Chinese geographic maps from the Ming show evidence of the use of the graticule or any analogous coordinate system. The Chinese reproductions themselves, as Ch'en has pointed out, betray an incomplete understanding of Ricci's maps.[20] Besides omitting the graticule in some cases, they mislabel certain countries, misinterpret Ricci's notes as place-names, and fail to consider geographic extent when locating countries by coordinates. In the *Fangyu shenglüe*, for example, the coordinates for France are given as 45° north latitude and 5° longitude (the Fortunate Isles being 0°).

Historians of cartography have thus seemed to be answering a nonquestion. It is doubtful whether Ricci's maps had much cartographic effect at all. Wide circulation is not necessarily a measure of influence, and exposure does not always mean adoption. It therefore seems that Ricci's influence is more properly spoken of in a European context, in which changing representations of China are directly attributable to his maps.

European Cartography and Qing Mapping

If Chinese cartography had still not joined with European cartography during the Ming, some might argue that it did so during the Qing—again under foreign influence. During the Qing dynasty, Chinese culture was exposed to foreign influence primarily from two sources: conquest by the Manchus and contact with Europeans—first through Jesuit missionary efforts (continued from the late Ming) and then through mercantile expansion by European trading nations. Some changes were wrought by these foreign contacts, especially by the mercantile expansion, but the effect on cartography was hardly profound. To demonstrate this claim, let us examine various levels of mapping during the Qing. Cartographic history during the Qing can be viewed in general terms as a double-layered phenomenon: mapping at the top of the political hierarchy somewhat influenced by foreign cartographic practice, and an indigenous layer below that was resistant to foreign influence until late in the nineteenth century.

Comprehensive Surveys of the Empire

The Manchus were outnumbered about fifty to one by the conquered population, and they ultimately controlled an empire roughly twenty times the size of their original power base. The Manchus had studied their Chinese history and were well aware that foreign dynasties had tended to be short-lived. They were determined not to repeat the mistakes made by their predecessors: a loss of tribal military prowess and factionalism between an indigenous bureaucracy and a foreign aristocracy. The system that evolved under the Qing has been described as a Manchu-Chinese dyarchy, intended to minimize the distinctions between conqueror and conquered.[21] The Manchu rulers also adopted the Chinese language, promoted Chinese culture, and commissioned projects to preserve Chinese cultural artifacts. By such projects, they sought not only to win favor among Chinese intellectuals, but also to control public opinion. By overseeing editing projects, for example, they could ensure that texts expressing antiforeign or anti-Manchu sentiments were censored.

For the expansion and maintenance of political control, reliable geographic information was essential. Whether the Manchus had their own cartographic tradition is unknown. After their conquest of China, the Manchus relied on geographic information from Chinese sources, as they had before the conquest, and also obtained information from comprehensive, empirewide surveys carried out by the Jesuits.

The Jesuit surveys and the resulting maps are probably the most familiar aspects of Qing cartography. Less well known is the comprehensive survey attempted at the beginning of the Qing by the Manchu government. Unlike the Jesuit surveys, this survey did not have as its end the production of maps; it was a tool for consolidating political authority. As a means of securing popular support, the early Manchu government proclaimed that taxes would be "collected entirely according to the original quota recorded in the accounts of the former dynasty."[22] The Ming tax and land records, however, were hopelessly outdated, the most recent being at least twenty-five years old. On 9 June 1646, Dorgon, the prince regent, decided to rectify this situation. He ordered his grand secretary to find out how much land was being cultivated throughout the empire and to audit the tax collection procedures of local governments. As a result,

20. Ch'en, "Matteo Ricci's Contribution," 347 (note 11).

21. This system has been described in Frederic Wakeman, Jr., *The Great Enterprise: The Manchu Reconstruction of Imperial Order in Seventeenth-Century China*, 2 vols. (Berkeley and Los Angeles: University of California Press, 1985); and Robert B. Oxnam, *Ruling from Horseback: Manchu Politics in the Oboi Regency, 1661–1669* (Chicago: University of Chicago Press, 1975).

22. *Da Qing Shizu Zhang (Shunzhi) huangdi shilu* (Veritable records of Shizu, emperor Zhang [Shunzhi], of the Great Qing, compiled ca. 1672) (1937; reprinted Taipei: Hualian Chubanshe, 1964), 17.16b; cited hereafter as *Shizu shilu*. The translation is that of Wakeman, *Great Enterprise*, 1:463 (note 21).

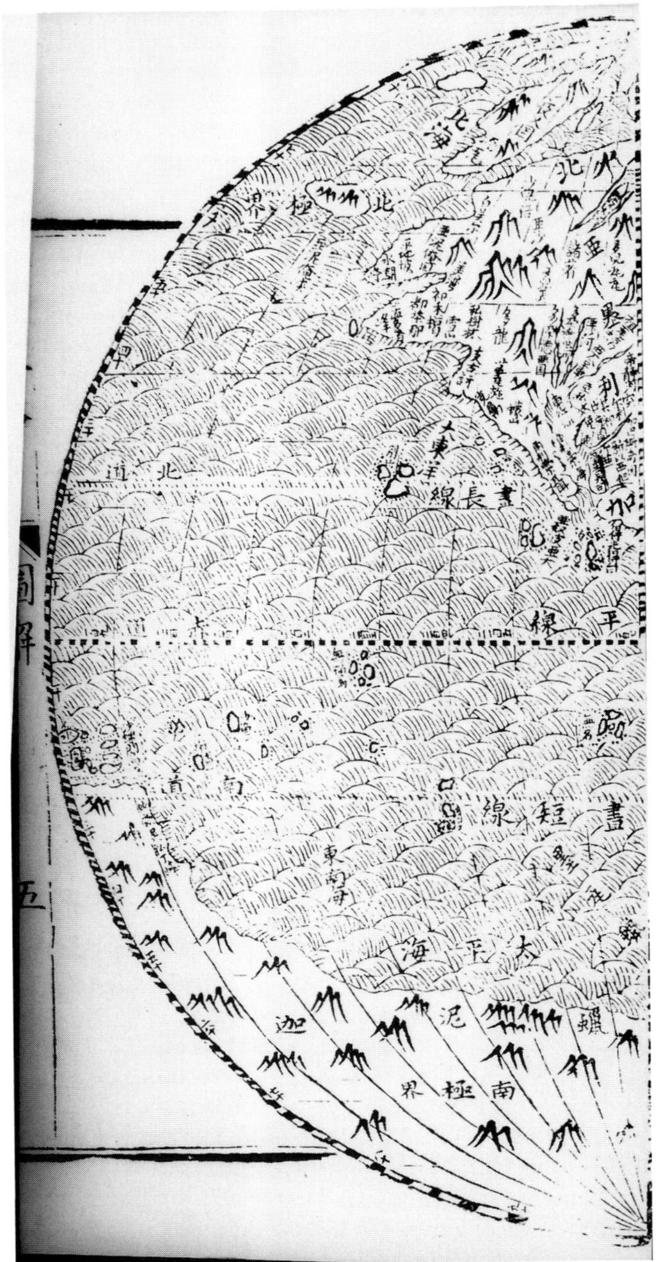

FIG. 7.6. CHINESE RENDITION OF THE TWO HEMISPHERES. After a map of the two hemispheres drawn by Matteo Ricci in 1601. Left, Western Hemisphere; right, Eastern Hemisphere.

Size of the originals: unknown. From the *Fangyu shenglüe* by Cheng Boer et al. Photographs courtesy of Cao Wanru, Institute for the History of Natural Sciences, Academia Sinica, Beijing.

the first empirewide cadastral survey since the 1580s was ordered: "Investigate thoroughly and examine in detail. Decide on comprehensive registers of taxes and service. Present to us for our personal review. Promulgate them throughout the empire."[23] According to Wakeman, however, the determination of taxes and service "really amounted to a reevaluation of quotas rather than a thorough national land survey."[24]

23. *Shizu shilu*, 25.24b (note 22). The translation is based on that in Wakeman, *Great Enterprise*, 1:464 (note 21).

24. Wakeman, *Great Enterprise*, 1:464 n. 119 (note 21).

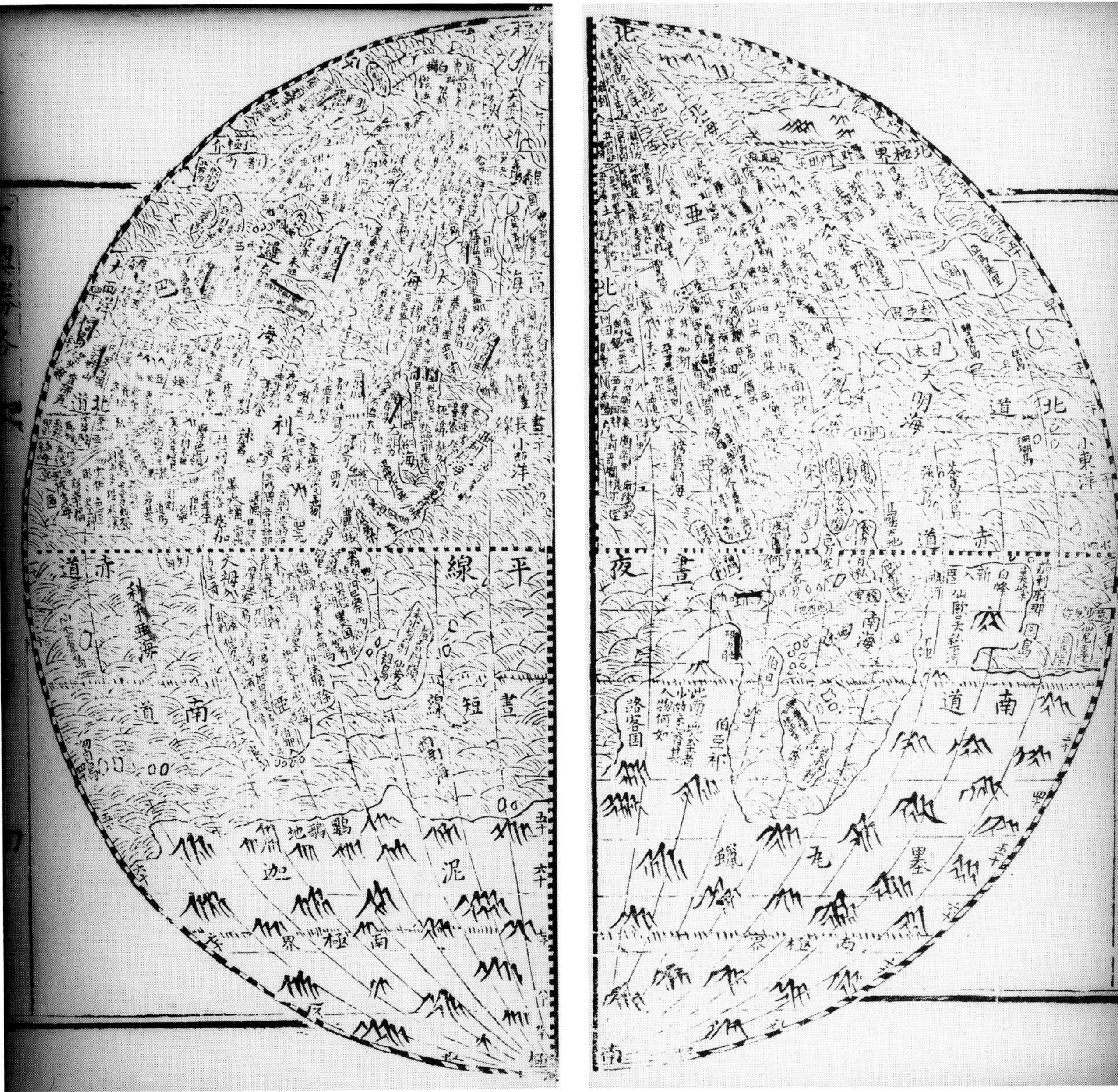

The Manchu rulers' experience with Ming records helped make them receptive to the Jesuits' offer to make better maps of China than those available at that time. When the Jesuits suggested to the Kangxi emperor in 1698 that they should undertake a survey of the empire, both sides knew that the Jesuits had already demonstrated the superior predictive power of their astronomical methods in comparison with traditional Chinese, as well as Islamic, methods. On 29 July 1644, Johann Adam Schall von Bell (1592–1666), a Jesuit missionary, petitioned the throne, offering to rebuild astronomical instruments destroyed by bandits before the Manchus took power and presenting "the prediction of a solar eclipse on the first day of the eighth month of this year [1 September 1644], calculated according to new Western methods": "In some provinces the eclipse will appear earlier, in others later. The various data are listed here for examination. I humbly beg a decree to the [board of rites] to test the measurements publicly at the proper time."[25] Schall von Bell's request was granted in an edict that said in part: "For many years the old calendar has been inaccurate whereas the new methods from the West have often been accurate. We knew this."[26] The test confirmed what the imperial court had already known: "As for the hour, minute, and second, the position, and other details regarding the start of the eclipse, the total eclipse, and the sun's recovery, only the new methods from the West coincided point for point. The Datong [the official Ming method] and the Islamic methods were both erroneous as to the time."[27] On 19 October 1644 the Western calendar was officially adopted, and on 31 October Schall von Bell was named director of the imperial board of astronomy.

During the Kangxi period (1662–1722), the Jesuits were given an opportunity to demonstrate the virtues of their cartographic techniques. They accompanied the emperor on northern expeditions, and they had taught him how to take astronomical measurements and to measure elevations and distances. The Kangxi emperor had a deep interest in mathematics, and he was also interested in learning geography: "Our territory is complicated, broad and vast, extending ten thousand *li*. . . . Climatic conditions vary, and the people's customs differ. These have not been compiled. How is one to know them completely? We observe that writers on geography have been fairly numerous since the Han dynasty. But their accounts vary in their amount of detail, and reports produced then and now differ. We therefore order that a bureau be set up to collect all kinds of documents, verify the gazetteers, and compile a book."[28] This book was to be titled *Da Qing yitong zhi* (Comprehensive gazetteer of the Great Qing realm, completed 1746), and its editors were enjoined to report on strategic passes, mountains, streams, customs, and personages as well as to draw maps.

The lack of uniform practices of representation among Chinese cartographers, as is described below, hampered the production of a comprehensive geographic record such as the emperor envisioned. In 1698 the Jesuit missionary Dominique Parrenin (1665–1759) examined various provincial maps and found errors in the location of prefectures, counties, and cities. He memorialized the emperor and recommended a survey of the empire. The emperor responded by asking Joachim Bouvet (1656–1730) to return to France and recruit more missionaries to come to China. Bouvet went back to France and returned with more than ten Jesuits trained in astronomy, mathematics, geography, and surveying. The emperor put them to the test. About 1705, for example, the emperor commissioned them to survey and map the region of Tianjin, in part to determine whether flooding in the area could be prevented and in part to judge the exactitude of European cartographic methods.[29] The Jesuits completed the map and presented it to the emperor within seventy days; he was satisfied with the results.

In 1707 the emperor commissioned the Jesuits to survey the area around the capital of Beijing and to compare their results with the information on old maps. A new map was completed in six months and presented to the emperor, who inspected it and pronounced it superior to previous efforts. In 1708 he sent Jesuits out to survey and determine the position of the Great Wall. According to the Jesuit missionary Antoine Gaubil (1689–1759), "Those who are interested in the geography of China will perhaps be very pleased to know: first that it is Fr. Parrenin who found the means to nurture in the Kangxi emperor the desire to see a map of the Great Wall; second that the prince was so pleased with the map of the wall made by Frs. Bouvet, Régis, and Jartoux that he resolved to have made the map of all of his vast states in China and Tartary."[30] Gaubil wrote this statement in 1728 and does not specify when Parrenin proposed that a map of the Great Wall be made. Foss seems to identify this proposal with the one for a comprehensive survey,[31] but Gaubil's language does not suggest a comprehensive survey. Parrenin was in China when the survey of the Great Wall was commissioned, and it seems more likely that Gaubil was referring to that survey.

The survey of the Great Wall began nearly a decade of surveying that culminated in the publication of the first Jesuit atlas of China. The emperor apparently saw the political advantages of measured maps: they would improve communication and aid in military planning. The Great Wall itself was vital to both government concerns

25. *Shizu shilu*, 5.24a (note 22).

26. *Shizu shilu*, 5.24a (note 22).

27. *Shizu shilu*, 7.1b (note 22).

28. *Da Qing Shengzu Ren (Kangxi) huangdi shilu* (Veritable records of Shengzu, emperor Ren [Kangxi], of the Great Qing, compiled ca. 1739) (1937; reprinted Taipei: Hualian Chubanshe, 1964), 126.15b–16a.

29. This survey is described in Jean Baptiste Du Halde, ed., *Lettres édifiantes et curieuses, écrites des missions étrangères par quelques missionnaires de la Compagnie de Jésus*, 27 vols. (Paris: Nicolas le Clerc, 1707–49), 10:413–15, reproducing a letter written in 1705 by Jean-François Gerbillon (1604–1707).

30. Antoine Gaubil (1689–1759), *Correspondance de Pékin, 1722–1759* (Geneva: Librairie Droz, 1970), 214. The translation is based on that in Theodore N. Foss, "A Western Interpretation of China: Jesuit Cartography," in *East Meets West: The Jesuits in China, 1582–1773*, ed. Charles E. Ronan and Bonnie B. C. Oh (Chicago: Loyola University Press, 1988), 209–51, esp. 223–24.

31. Foss, "Western Interpretation of China," 223 (note 30).

and thus an understandable choice. The task of measuring the wall fell to Bouvet, Jean-Baptiste Régis (1664–1738), and Pierre Jartoux (1669–1720). On 4 June 1708 they left Beijing and in four days reached Shanhaiguan, where the wall meets the sea. They then followed the wall westward, keeping track of direction with compasses, measuring distance with cords, and determining latitude from the height of the sun. After two months Bouvet was forced to return to Beijing because of illness, but Régis and Jartoux kept on. On 10 January 1709 they returned to Beijing with a map about five meters long, depicting gates, forts, rivers, hills, and mounds. The emperor was pleased with the map and directed that the surveying continue to cover the rest of the empire. Gaubil provides this account of the Jesuits' surveying methods:

> These Fathers requested a quadrant of two feet two inches in radius; they often took care to check it, and they constantly found that it represented elevations too great by a minute. They had large compasses, many other instruments, a pendulum and other things for the execution of the emperor's orders. With cords divided precisely, they accurately measured the way from Peking. . . . On this road they often took by observation the height of the meridian of the sun; they observed at every moment the rhumb and took care to observe the variation and declination of the peak. . . . In all these vast regions, the Fathers . . . have observed the height of the pole, observed the rhumbs . . ."[32]

The survey included tributary states such as Korea, but the Jesuits sometimes encountered difficulties in surveying such areas. In the case of Korea, any measurements they obtained were evidently gotten through subterfuge. Matteo Ripa (1682–1745), a secular priest in Beijing, wrote that the Koreans were "extremely jealous of strangers" and denied entrance to the Europeans:

> This part of the business was consequently executed by a mandarin, purposely instructed by the Jesuits, and then sent thither by the Emperor, under pretext of an embassy: even then they watched every movement of the mandarin so closely, that he could not take a step without being observed by the guards, who never left him, and wrote down all he said or did. Thus, being unable to measure the longitude with a line, he could only calculate the miles by the hour. This ambassador, with whom I was intimately acquainted, informed me that he had only succeeded in taking the sun's altitude by making them believe that the instrument he used was a sun-dial, and that he stopped to look at it in order to ascertain the time.[33]

Ripa's account gives the impression that the Jesuit map of Korea (fig. 7.7) was based on a survey, but this seems to be true only of the northern portion. Ripa's account needs to be supplemented by Régis's statement, reported by Jean Baptiste Du Halde (1674–1743), that a map received by a "Tartar lord" (an envoy) from the Koreans served for the most part as the basis of the Jesuit map of Korea (see pp. 299–305).

The Jesuits' survey of the empire was completed in 1717, and an atlas was presented to the emperor the following year. It was titled *Huangyu quanlan tu* (Map of a complete view of imperial territory), perhaps in recognition of the emperor's desire to be able to view all parts of the empire at a glance.[34] The emperor was pleased with the results, saying that "the mountain ranges and waterways were all in accord with the 'Yu gong' [Tribute of Yu]."[35] The maps in the atlas used a trapezoidal projection; depicted the Qing empire, including Mongolia and Manchuria, east of Hami; and were drawn to a scale of 1:400,000 to 1:500,000. The meridian running through Beijing was adopted as the prime meridian, in part to avoid errors in longitude that would be introduced by adopting a European prime meridian.[36]

The Kangxi Jesuit atlas, as it came to be known, had a complicated publication history. The earliest edition was printed in China with woodblocks and consisted of twenty-eight maps. In 1719 a manuscript version with thirty-two maps was produced. This version was divided into forty-four copperplates engraved by Matteo Ripa, who produced an atlas drawn to a scale of 1:1,400,000.[37]

32. Gaubil, *Correspondance de Pékin*, 214 (note 30). The translation is based on Foss, "Western Interpretation of China," 227–28 (note 30).

33. Matteo Ripa, *Memoirs of Father Ripa, during Thirteen Years' Residence at the Court of Peking in the Service of the Emperor of China*, trans. and ed. Fortunato Prandi (London: John Murray, 1846), 65.

34. Joseph-Anne-Marie de Moyriac de Mailla, *Histoire générale de la Chine ou annales de cet empire*, 13 vols. (Paris: Grosier, 1777–85), 11:314.

35. *Qing shi gao jiaozhu* (Edited and annotated draft history of the Qing, original draft completed 1927), 15 vols. (Taipei: Guoshiguan, 1986–), chap. 290 (11.8773–74). See also *Qing shi* (History of the Qing), 8 vols. (Taipei: Guofang Yanjiuyuan, 1961), chap. 284 (5:4010).

36. Jean Baptiste Du Halde, *Description géographique, historique, chronologique, politique, et physique de l'empire de la Chine et de la Tartarie chinoise*, 4 vols. (Paris: Lemercier, 1735), 1:xxxvi. An English translation of this work was published as *A Description of the Empire of China and Chinese-Tartary, Together with the Kingdoms of Korea, and Tibet*, 2 vols. (London: Edward Cave, 1738–41).

37. Work on the plates began in 1718, perhaps with the first edition as a rough guide. Copies of the copperplate edition survive in England (King George III's Topographical Collection, British Library, London) and Italy (Istituto Universitario Orientale di Napoli). See Foss, "Western Interpretation of China," 234 and 249 n. 93 (note 30), and Helen Wallis, "Chinese Maps and Globes in the British Library and the Phillips Collection," in *Chinese Studies: Papers Presented at a Colloquium at the School of Oriental and African Studies, University of London, 24–26 August 1987*, ed. Frances Wood (London: British Library, 1988), 88–96, esp. 93.

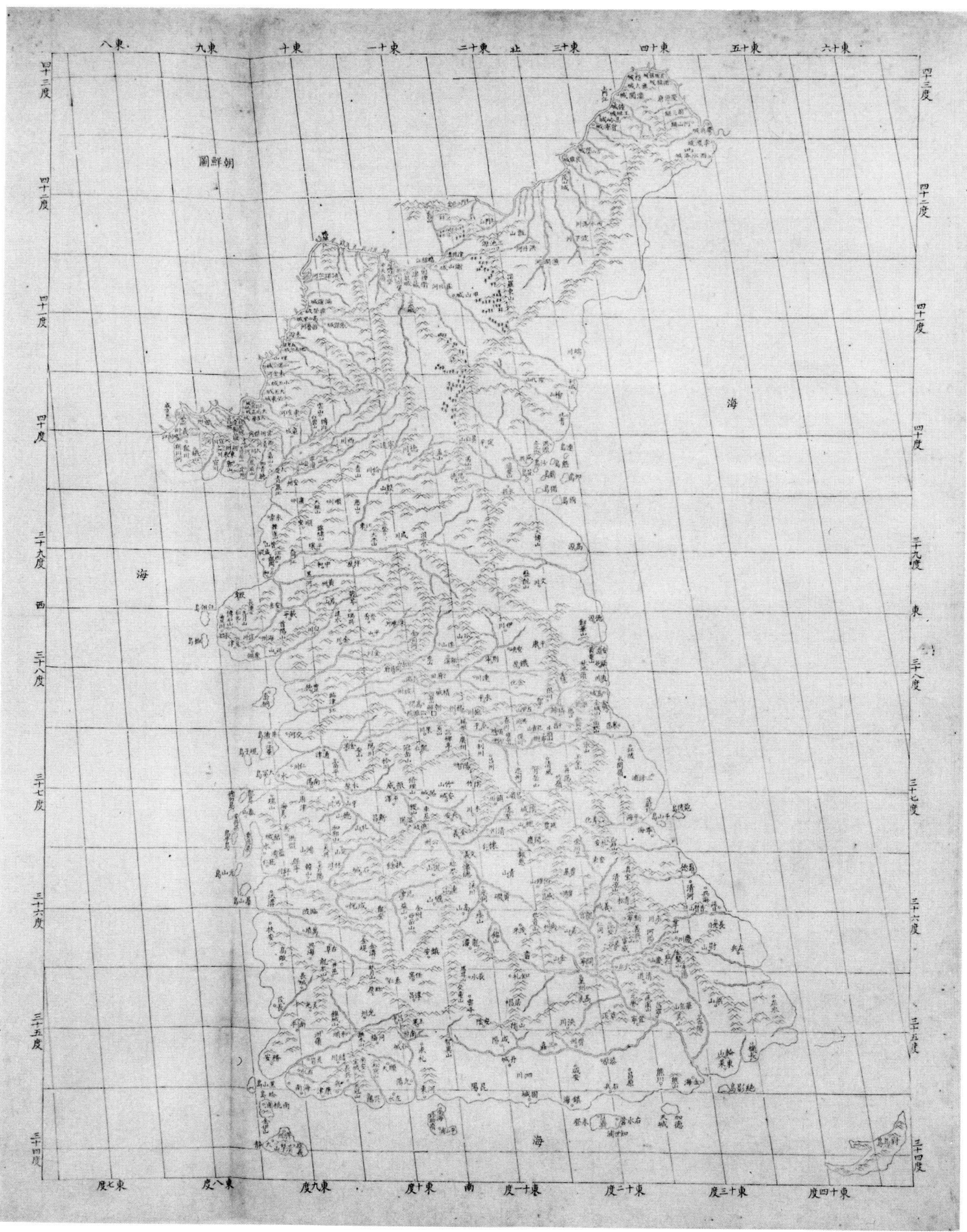

FIG. 7.7. MAP OF KOREA FROM THE *HUANGYU QUANLAN TU*. This map is from the 1721 edition of the atlas. The map is close to modern representations of northern Korea down to about the thirty-ninth parallel, but below that the image suffers in comparison. Seoul, for example, is placed too far from the west coast, and the Han River flows to the southwest instead of the northwest.
Size of the original: 58 × 43 cm. By permission of the British Library, London (Maps C.11.d.15).

reason members of the Chinese elite, who had already seen Jesuits displace native scholars from the astronomical bureau, regarded the project with suspicion. The emperor saw the initial stages of the survey as a kind of contest between cartographic traditions. In 1710, after the Jesuits presented a map of Beizhili, the provincial seat of the imperial government, the emperor examined the map himself and saw that areas he was familiar with and had previously ordered Manchus to measure had been "justly exhibited." He then "signify'd to the Missionaries that he wou'd answer for the Accuracy of it; and that if the rest proved as good, their Performance wou'd satisfy him, and be out of the reach of Criticism."[42] Without European techniques, the atlas could not have been made. The Jesuits' use of native materials was made easier by the emperor's standardization of the units of linear measurement. In 1704 the emperor stipulated that two hundred *li* should correspond to one degree of longitude, basing his decision on geodetic measurements performed by the Jesuit Antoine Thomas (1644–1709). This allowed the Jesuits to convert distance information provided by the Chinese to the European coordinate system. The standardization of map scale in the Jesuit atlas, furthermore, allowed the maps in the atlas to stand independent of text. This was another departure from Chinese tradition, which, as Ricci recognized, tended to view image and text as integral to the cartographic enterprise. Perhaps the divorce of European post-Renaissance science, and cartography in particular, from textual scholarship made it difficult for Chinese intellectuals to accept European maps or recognize them as useful, so that there was less demand for them than previous historians have assumed.

Undoubtedly, the Jesuits made use of Chinese scholarship in compiling the Kangxi atlas. But this was not merely an instance of transmission of Chinese knowledge. The Jesuits' reliance on native materials was necessitated by the scope of the project, covering more area than any previous survey. The survey was directed by about a dozen Jesuit missionaries divided into teams responsible for specific areas. The Jesuits wished to complete the project as quickly as possible, and to determine directly the position of every point deemed worthy of cartographic representation would have been too time consuming. According to Du Halde, the Jesuits determined the latitude and longitude of more than six hundred locations.[43] A "method of triangles" was used to calculate the distances between cities, checked where possible by observation of eclipses.[44] The Jesuits' use of Chinese materials was thus not a matter of uncritical acceptance. The Jesuit atlas may have drawn heavily from Chinese sources, especially for place-names, linear features such as rivers, and areal features such as mountains, but the cartographic theory underlying it was European. Locations were determined according to a coordinate system based on a conception of the world different from that implied by the Chinese cartographic grid. The survey techniques needed to obtain these measurements were unknown to Chinese mapmakers, despite the use of superficially analogous techniques by Chinese astronomers (see pp. 123–24). Moreover, the technology and techniques used to determine position were all of European origin: the quadrant and tables of declination for latitudinal measurement; and for longitudinal measurement, timepieces and telescopes for the observation of the moons of Jupiter or the earth's moon. For these reasons, the Jesuit mapping of China is perhaps better treated as an example of European mapping adjusting to new cultural circumstances and drawing on information already available in China.[45]

SUPPLEMENTAL SURVEYS FOR THE IMPERIAL ATLAS

The Jesuit surveys, though extensive, did not cover the entire empire. To supplement them, the central government commissioned regional surveys, so as not omit any of its territory from the imperial atlas. In the case of Tibet, for example, a descriptive survey of the topography of Tibet was completed in 1711 and a map drawn, but because the map lacked lines of latitude and longitude, it was difficult to incorporate it into the Jesuit atlas. Thus the map was not used. The Kangxi emperor then commissioned a measured survey that was carried out by a mathematician from the imperial board of astronomy. The results of this survey were compiled in a map submitted to Jesuit scholars in 1717 for review. They found a number of mistakes: for example, the city of Lhasa was shown as lying at about 30.5° north latitude, when its actual position is about 29.4°. Thus a team of surveyors was dispatched to recheck certain of the latitudinal and longitudinal measurements. A complete resurvey was not ordered because of a wish not to offend the official trained at the imperial board of astronomy and, perhaps more important, because military conflict with competitors for Tibet made such a survey potentially dangerous. For reasons that are unclear, the position of Lhasa is still

42. Du Halde, *Description of the Empire*, 1:viii (note 36).

43. Du Halde, *Description of the Empire*, 1:viii (note 36). De Mailla lists about 630 points for which latitude and longitude were determined; see *Histoire générale*, 12:179–96 (note 34). According to Matteo Ripa, latitude was determined with "mathematical instruments" and longitude with "long chains" (*Memoirs*, 65 [note 33]).

44. Du Halde, *Description of the Empire*, 1:x (note 36). The "method of triangles" was perhaps triangulation as developed by Gemma Frisius (1508–55) in 1533.

45. Other useful treatments of Jesuit surveys are Foss, "Western Interpretation of China" (note 30); Fuchs, *Der Jesuiten-Atlas* (note 39); and Walter Fuchs, "Materialien zur Kartographie der Mandju-Zeit," *Monumenta Serica* 1 (1936): 386–427.

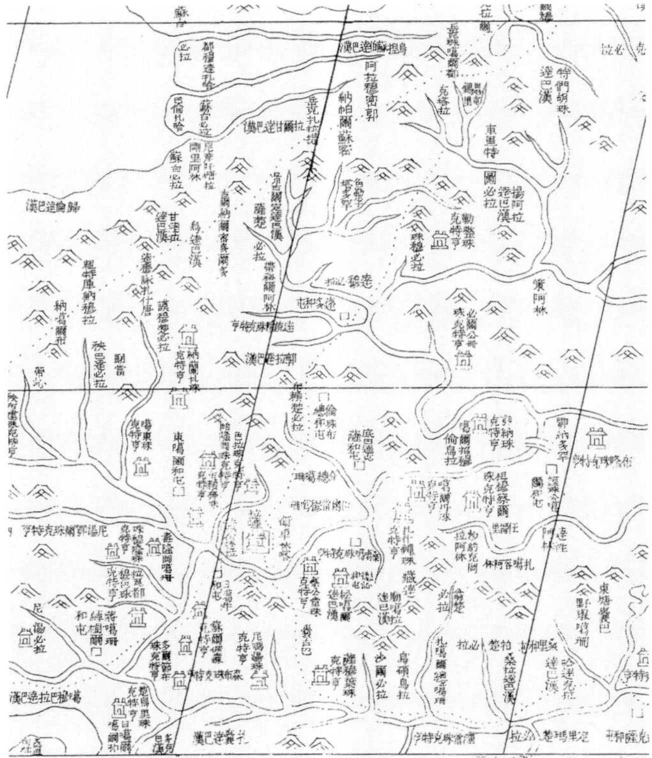

FIG. 7.10. DETAIL SHOWING LHASA FROM THE QIANLONG REVISION OF THE JESUIT ATLAS, 1760. The first Qianlong revision of the Kangxi Jesuit atlas updated information in its predecessor. One of the areas of interest to the imperial government was present-day Tibet. Although the government sent new survey teams to the area, Lhasa (represented by a square just east of the 26°W meridian) still appears above the thirtieth parallel, as in the earlier version.
Size of the image: 27.7 × 47 cm. From *Qingdai yitong ditu* (Comprehensive map of the Qing period, printed 1760) (Taipei: Guofang Yanjiuyuan and Zhonghua Dadian Bianyinhui, 1966), 149–50. This is a photo-reprint of the first edition of the Qianlong Jesuit atlas, which was printed using copperplates.

incorrect on the maps included in the 1721 edition of the Jesuit atlas. Not until the 1750s was a comprehensive survey undertaken. The results were apparently submitted too late to be incorporated into the first Qianlong revision of the Kangxi atlas, printed with copperplates in 1760: Lhasa still appears at the wrong latitude (fig. 7.10).[46]

As in the case of Tibet, the surveying and mapping of what is now Xinjiang was hampered by conflict with the Dzungars. The conflict began late in the seventeenth century when the Dzungars, in an attempt to establish a Central Asian empire, threatened eastern Mongols who were under Qing protection. Not until 1755 did the Qing government believe the Xinjiang area secure enough to dispatch a survey team. The team included the Jesuits Felix da Rocha (1713–81) and Joseph d'Espinha (1722–88), who were commissioned to perform the measurements. The survey took four years to complete, and in 1769 Michel Benoist (1715–74) was commissioned to produce an atlas based on the supplemental surveys and the earlier Jesuit atlas. The Qianlong emperor was already familiar with Benoist's cartography: in 1764 the emperor had ordered him to copy a European world map for display in the throne room of the imperial palace.[47] Benoist's atlas, consisting of maps on 104 sheets, was completed within a year and printed with woodblocks. This work was titled the *Qianlong neifu yutu* (Terrestrial atlas of the inner prefectures of the Qianlong period [1736–95]). A copperplate version was published in 1775. The maps are drawn to a scale of 1:500,000, and toponyms appear in Chinese. Each map is divided into five-degree increments of latitude, so that the atlas has a total of thirteen bands, each five degrees wide—hence its alternative name, *Qianlong shisan pai tu* (Map in thirteen bands of the Qianlong period). As with the Kangxi Jesuit atlas, Needham claims the Benoist atlas as an achievement of Chinese cartography: "Once again China was ahead of all other countries in the world in map-making."[48] But as before, it is hard to see how the claim can be justified, since the technology underlying the atlas was European and applied by Europeans.[49]

Gauging the Extent of Western Influence

Chinese practitioners of cartography at the provincial and local levels seem on the whole to have been untouched by the cartographic innovations introduced at court. Contact with the Jesuits was limited mainly to the imperial court, and after the dissolution of the Society of Jesus in China in 1773, opportunities for Chinese intellectuals to interact with foreign scholars became even more limited for a time. The openness of the Manchu court to

46. See Foss, "Western Interpretation of China," 235–36 (note 30); Fuchs, *Der Jesuiten-Atlas*, 73 (note 39); and Lu, *Zhongguo dituxue shi*, 186–87 (note 1).

47. De Mailla, *Histoire générale*, 11:580 (note 34). The emperor knew of Benoist's cartography from a world map prepared in 1760. It was particularly remarkable because the text on it included the first adequate discussion of Copernicus's doctrine available in China. It is described by Nathan Sivin, "Copernicus in China," in *Colloquia Copernicana 2: Etudes sur l'audience de la théorie héliocentrique*, Studia Copernicana 6 (Warsaw: Zakład Narodowy im. Ossolińskich, 1973), 63–122, esp. 92–103.

48. Needham, *Science and Civilisation*, 3:586 (note 1).

49. In the case of Matteo Ripa, engraver of the plates for the first copperplate edition of the Kangxi atlas, the technology may not have been applied particularly well. Ripa himself wrote that he had had only a single lesson in the technique of engraving with aquafortis (nitric acid), and that his attempts to use the process were not entirely satisfactory: "Thus owing to the inefficiency of the aquafortis, the lines were very shallow, which, added to the badness of the ink, caused the prints to be of the worst possible description." See Ripa, *Memoirs*, 71 (note 33).

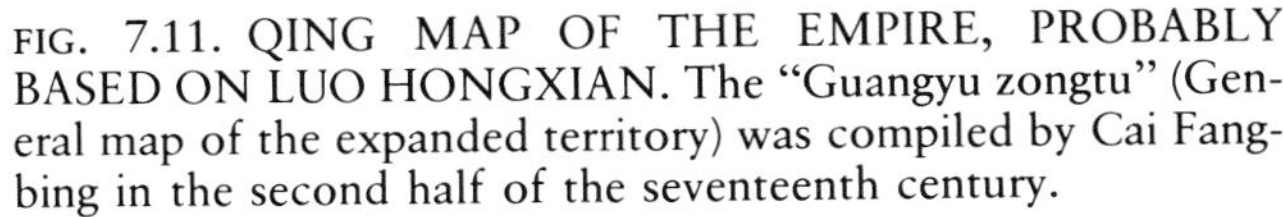

FIG. 7.11. QING MAP OF THE EMPIRE, PROBABLY BASED ON LUO HONGXIAN. The "Guangyu zongtu" (General map of the expanded territory) was compiled by Cai Fangbing in the second half of the seventeenth century.
Size of each page: 22.9 × 13.7 cm. By permission of the Herzog August Bibliothek, Wolfenbüttel (Cod. Guelf. 148 Blankenb. Vb.).

foreign ideas began to wane in the latter part of Qianlong emperor's reign as the emphasis in scholarship turned increasingly inward—to the preservation of Chinese culture.

Although the Jesuit atlases produced during the Kangxi and Qianlong periods underwent various printings, it is not clear how many Chinese had access to them. According to the *Qing shi gao*, the Kangxi atlas was stored in the office of the palace treasury, where it would have been under direct palace control.[50] The cartographic record suggests that the influence of Western cartography was generally not felt among Chinese mapmakers. As a result, traditional Chinese cartographic practices continued unabated.

One atlas of the entire empire produced during the Kangxi period can serve as a gauge of the strength of the Chinese cartographic tradition after Ricci introduced Ptolemaic methods. This is the *Zengding Guangyu ji quantu* (Enlarged and revised complete atlas of the record of the expanded territory), compiled by Cai Fangbing. This work contains a general map of the Qing empire and fifteen provincial maps, all printed from woodblocks. The title bears some similarity to that of the *Guang yutu* (Enlarged terrestrial atlas, ca. 1555) of Luo Hongxian (1504–64), and it is possible that Cai used Luo's maps, or maps deriving from Luo's, as a source. This is suggested by the similarities between the general maps in the two atlases. Cai's is titled "Guangyu zongtu" (General

50. *Qing shi gao jiaozhu*, chap. 290 (11:8774) (note 35).

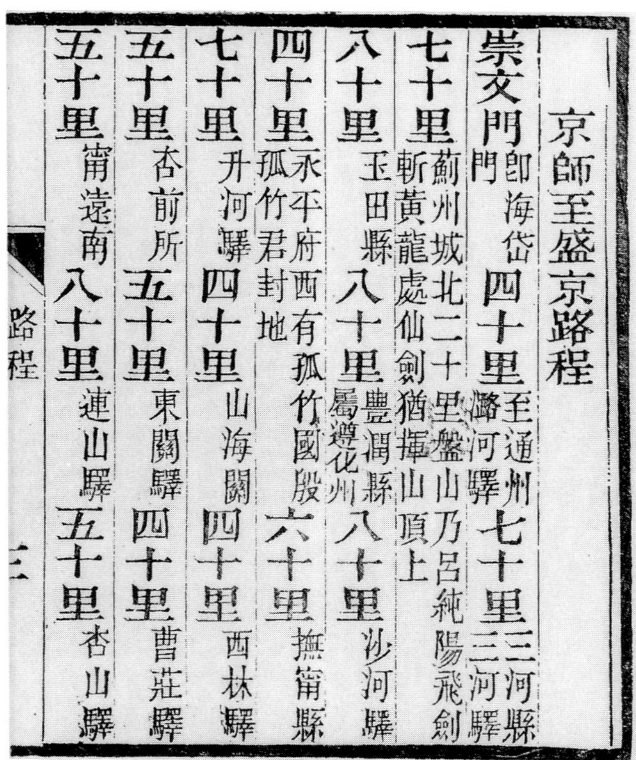
京師至盛京路程
崇文門即海岱門四十里至通州潞河驛七十里三河縣三河驛
七十里薊州城北二十里盤山乃呂純陽飛劍斬黃龍處仙劍猶揮山頂上
八十里玉田縣八十里豐潤縣屬遵化州八十里沙河驛
四十里永平府西有孤竹國殷孤竹君封地六十里撫甯縣
七十里卄河驛四十里山海關四十里西林驛
五十里杏前所五十里東關驛四十里曹莊驛
五十里甯遠南八十里連山驛五十里杏山驛
路程 三

FIG. 7.12. PAGE FROM A ROUTE BOOK. The route from Beijing to Shenyang (Mukden, in present-day Liaoning) as it appears in a guide to Beijing.
Size of the original: unknown. From Yang Jingting, *Chaoshi congzai* (Collected notes for going to market, 1883), 1886 edition. Reproduced courtesy of the Harvard-Yenching Library, Harvard University, Cambridge. This work was originally titled *Dumen jilüe* (Abridged notes on the capital, 1864).

map of the expanded territory) (fig. 7.11), and Luo's is titled "Yudi zongtu" (General map of the empire). Despite the differences in name, the geographic coverage of the two maps corresponds closely. To the west they both reach Turfan, and to the east they reach Korea. Northward they both reach into what is now Mongolia, past the Gobi Desert. Along their southern edges, some differences are easy to detect, one of the most obvious being that Cai's map shows southwestern China as being landlocked and not bounded by ocean as in Luo's map. Cai's map is somewhat more generalized than Luo's: it omits some of the signs used for cities and leaves out the Great Wall. Another difference between the two maps is the absence of the cartographic grid on Cai's map—an omission that brings up the next point to be made: that Chinese cartographers did not believe the square grid was essential to their craft. This implies further that measurement itself, one of the main functions of the grid, was not regarded as essential to a map.

The persistence of this attitude toward measurement calls into question the belief that the arrival of the Jesuits initiated a new period in Chinese cartography. Under this interpretation, expressed by Mills and others, unmeasured mapping is an aberration in Chinese cartography from the late sixteenth century onward: according to Mills, from 1584 to 1842 "Jesuit influence was predominant," and after 1842 "scientific principles gradually triumphed" as the "opening of China effected a major revolution in Chinese cartography." Because of the predominance of Jesuit influence, Mills says, traditional maps can be seen as "fanciful misrepresentations."[51] This interpretation, however, does not seem tenable. European cartographic techniques did not begin to supersede traditional Chinese practice until late in the nineteenth century.

Much of the evidence for this last statement comes from Chinese gazetteers (*fangzhi*), compendiums of information on particular administrative units. During the Qing, more than five thousand gazetteers are known to have been produced. These generally consisted of a section of maps or illustrations (*tu*) followed by sections devoted to topics such as local history, geography, administration, water conservancy, and literature. The central government used gazetteers submitted by local and provincial administrations to compile comprehensive gazetteers of the empire.

Most of the maps in gazetteers were printed with woodblocks. In general they are inconsistent in their use of the square grid and often lack any scalar indications whatever, despite the central government's interest in careful surveys. Grids are rare on maps in gazetteers below the provincial level of administration. They tend to appear on maps published late in the Qing, in the nineteenth century—more than a century after the Kangxi Jesuit atlas. Gazetteer maps during the Qing, as in previous periods, often use pictorial representation, making them ill suited for presenting quantitative information. Generally, small-scale maps tended toward the use of abstract signs, and pictorial elements tended to dominate large-scale maps. The varied representational modes might pose a problem for a map reader attempting to use such maps to derive distances, but they were not intended for that purpose: quantitative information was often given in verbal descriptions of the areas depicted on the maps. This is especially true of route books produced during the period: maps in these books provide a sense of spatial relationship and relative position, while accompanying text provides distances along specified routes (fig. 7.12).[52]

51. See, for example, J. V. Mills, "Chinese Coastal Maps," *Imago Mundi* 11 (1954): 151–68, esp. 152.

52. On these route books, see Timothy Brook, *Geographical Sources*

These generalizations regarding the use of grids, scale indications, and mode of representation also hold true for maps produced under the auspices of the central government, where Jesuit influence was strongest. The *Da Qing yitong zhi*, for example, underwent at least two revisions after it was first completed in 1746. Although it was completed well after the Kangxi Jesuit atlas, its cartography reflects little European influence, confirming the belief that access to the atlas was limited. In the revised version published in 1842, for example, the maps are generally planimetric in mode of representation but do not have grids or expressed scales. As a result it is often difficult to relate the provincial and prefectural maps to the general map of the empire (see figs. 7.13 and 7.14).

The mapping practices of compilers of provincial and local gazetteers were more varied. The maps in the *Shaanxi tongzhi* (Comprehensive gazetteer of Shaanxi Province, 1735), for example, show a mixture of representational modes (figs. 7.15 and 7.16). None of the maps bears a cartographic grid or an expressed scale. Maps of large areas are generally planimetric with some pictorial elements, particularly in the representation of cities and mountains. Maps of smaller areas, such as mountain regions and river basins, are generally pictorial. Some maps, however, balance planimetric representation and pictorial elements. The use of pictorial signs rather than the abstract signs used in the *Guang yutu* is further evidence that cartographic representation was still not clearly distinguished from pictorial.

The mixture of representational modes is perhaps even more striking in maps in gazetteers of prefectures and smaller administrative units. In these compilations, compounds of buildings and even individual buildings are the subjects of maps, and pictorial representation is more heavily employed than in national and provincial gazetteers. I will not belabor this point here, when the maps can speak for themselves (see figs. 7.17 to 7.19 for examples of maps from Qing local gazetteers showing a mixture of planimetric and pictorial representation).

The differences in the use of pictorialism between maps in provincial gazetteers and those in prefectural or county gazetteers makes it tempting to draw some conclusions about different conventions for various administrative subunits. It may be that the decrease in pictorial elements as one moves up through the administrative levels is a response to practical demands. The larger the area to be mapped onto a single leaf, the more difficult it is to represent the area pictorially.

Regional maps, atlases, and other types of maps were not only printed with woodblocks, but also drawn using brush and ink on sheets of paper and on scrolls, which often afforded more cartographic space than the printed page. The exact size of the corpus of manuscript maps from the Qing is unknown, but such maps are known to number in the thousands.[53] Provincial and other medium-scale maps drawn on those media make greater use of pictorial representation than their counterparts in gazetteers. As an example, one can take the map of the Great Wall found in 1952 by Leo Bagrow in the Lateran Museum (fig. 7.20).[54] It was drawn on a scroll, probably between 1680 and 1700, and represents the stretch of the Great Wall extending from Jiayuguan in Gansu Province to Shanhaiguan on the coast at the border of Manchuria, a distance of about 1,700 kilometers. The map presents a mixture of representational modes: the wall itself is rendered pictorially in elevation, as are mountains and villages of non-Chinese tribes; the Yellow River and garrisons within the wall are depicted from overhead. Meijer finds that the scale of the map varies from section to section: "It seems that the map does not at all aim at a faithful representation of the length of the wall."[55] This, as Meijer points out, does not pose a problem for a map reader wishing to obtain distances. Distances between places are given in notes on the map, as well as the strength of the garrisons within the wall and the locations of barbarian tribes within and beyond the wall. The same combination of text and image appears on another map of the Great Wall, made in the first half of the eighteenth century (plate 11).

On other manuscript maps from the Qing, pictorialism is even more evident: the result is often almost indistinguishable from a landscape painting (see esp. p. 153 and figs. 6.20–6.22). This is true even for map types in which careful measurements and scale drawing might seem useful: defense maps and water conservancy maps. The *Huanghe tu* (Map of the Yellow River), dating perhaps from the mid-nineteenth century, can be taken as representative (fig. 7.21 and plate 12). This map was drawn on a scroll and depicts the lower course of the Yellow River in Jiangsu Province before 1853. Some distances are indicated in annotations, but there is no expressed scale. A square grid drawn with a pencil appears to have been added after the map image was drawn with ink and color.[56] The Yellow River and its tributaries are repre-

of Ming-Qing History (Ann Arbor: Center for Chinese Studies, University of Michigan, 1988), 3–25.

53. Just to give an idea of the vastness of the corpus, the card catalog of imperial household maps in the Ming-Qing archives in Beijing alone is reported to fill almost ten foot-long drawers. See Frederic Wakeman, Jr., ed., *Ming and Qing Historical Studies in the People's Republic of China* (Berkeley: Institute of East Asian Studies, University of California, Berkeley, Center for Chinese Studies, 1980), 50.

54. Information concerning this map is drawn from M. J. Meijer, "A Map of the Great Wall of China," *Imago Mundi* 13 (1956): 110–15.

55. Meijer, "Map of the Great Wall," 110 (note 54).

56. This map may be among the first graphic productions in China to exhibit the use of pencil. The modern method of making pencils

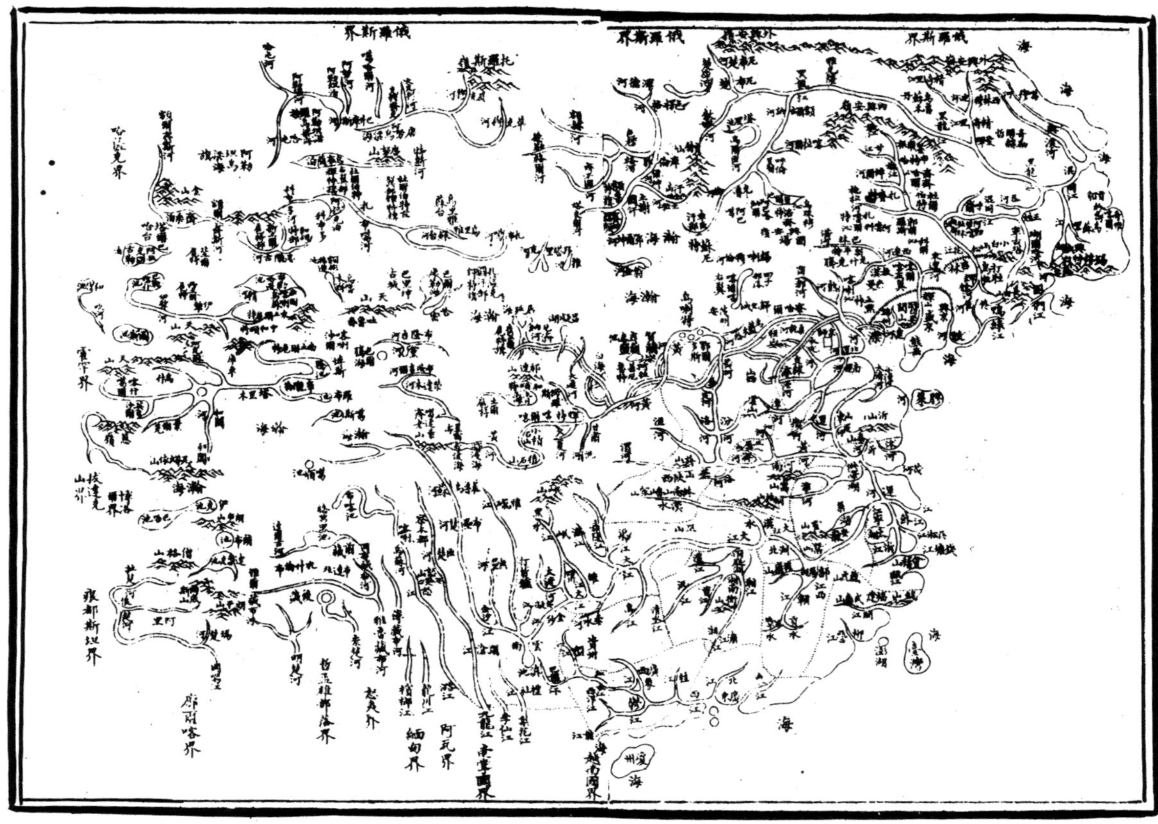

FIG. 7.13. MAP OF THE EMPIRE FROM A QING COMPREHENSIVE GAZETTEER.
Size of the image: 10.5 × 14.5 cm. From *Da Qing yitong zhi* (Comprehensive gazetteer of the Great Qing realm, last revision completed 1820; printed 1842); edition in 11 vols. (Taipei: Taiwan Shangwu Yinshuguan, 1967), 1:8. Also known as *Jiaqing chongxiu yitong zhi* (Revised comprehensive atlas of the Jiaqing reign period [1796–1820]).

sented planimetrically, while cities and mountains are drawn pictorially. The cities are drawn from a bird's-eye view, and mountains are drawn in elevation. The variable perspective these features produce is, as I stated before, characteristic of Chinese painting and early Chinese maps.

The heavy use of pictorial representation should not be taken as an indication that Chinese mapmakers were incapable of producing measured maps. Provincial surveys carried out by Chinese mapmakers did result in maps and atlases. In 1684 an imperial decree commissioned an atlas of Guangdong Province, after various officials had reported that existing accounts of the province were inadequate. In the latter half of that year, each administrative division of the province was inspected in order to gather information on the names and locations of mountains and streams, natural and artificial boundaries, historic and scenic locations, and the distances between them. The atlas was completed in 1685 and consisted of ninety-seven maps. The maps do not have grids, but distance information is given in the accompanying text. Errors in earlier accounts are also noted.[57] Evidently not enough errors were corrected, for in about 1739 the Qianlong emperor ordered a survey of Guangdong Province that resulted in a map printed from a woodblock (plate 13). The map shows administrative divisions and the distances between various places in the province. Cities, mountains, and historical remains as well as trees are represented pictorially. The author of the map states that he compiled it from eighty-eight maps that he made of different parts of the province.

The indigenous maps described here constitute an almost hidden history of late imperial Chinese carto-

with graphite and wood was not invented until 1795. When pencils were introduced into China is not known, but the opening of the treaty ports in 1842 increased the opportunities for importing them.

57. See Arthur Hummel, "Atlases of Kwangtung Province," in *Annual Report of the Librarian of Congress for the Fiscal Year Ended June 30, 1938* (Washington, D.C.: United States Government Printing Office, 1939), 229–31, esp. 230. Unfortunately, the staff at the Library of Congress has been unable to locate this acquisition reported by Hummel.

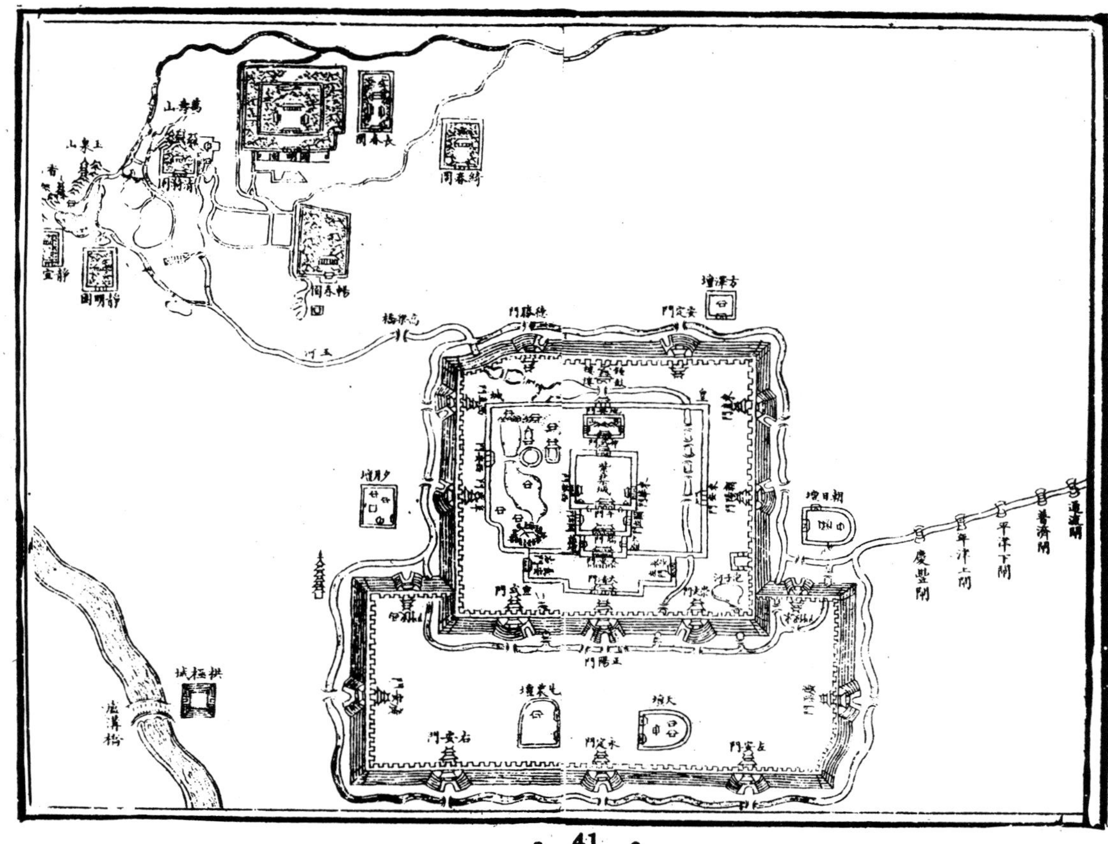

FIG. 7.14. MAP OF THE IMPERIAL CAPITAL FROM A QING COMPREHENSIVE GAZETTEER.
Size of the image: 10.8 × 13.8 cm. From *Da Qing yitong zhi* (1842); edition in 11 vols. (Taipei: Taiwan Shangwu Yinshuguan, 1967), 1:41.

graphy.[58] The sampling presented here (and in the previous chapter) suggests that European science exerted minimal influence on Chinese cartographic practice even late in the Qing. In addition, it is hard to speak of Chinese cartography as becoming more scientific in the Western sense when maps still had religious and magical functions. Notes on a map of Wutai Shan (Wutai Mountain, in present-day Shanxi Province), printed in 1846, advise one to study the map and Buddhist doctrine so that one's "troubles will disappear" and one "will be reincarnated in a blessed place" (plate 14). Maps also served astrological purposes, as they had in the past. An example of this appears in the *Henan tongzhi* (Comprehensive gazetteer of Henan, 1882). Besides geographic maps, it contains maps depicting star groups associated with various localities (fig. 7.22). This was in accordance with the *fenye*, or field-allocation system, by which events in specific sectors of the heavens were correlated with events in specific terrestrial regions (see pp. 208–10). Such star maps suggest the persistence of traditional astronomy in provincial administrations even after the adoption of European astronomy and calendrical science at the imperial court.

Late Qing Manifestations of European Influence

The exceptions to lack of foreign influence on traditional Chinese cartography generally occur after 1842, the year of China's defeat by the British in the first Opium War. The defeat helped initiate a new order in Chinese foreign relations. Many in the Qing establishment still believed in China's cultural superiority and viewed trade concessions to Western countries as a means of appeasing for-

58. One exception to the tendency to overlook such maps is Sen-dou Chang, "Manuscript Maps in Late Imperial China," *Canadian Cartographer* 11 (1974): 1–14. Chang also provides evidence that measured mapping was not the norm among mapmakers in late imperial China, and that political and aesthetic concerns were just as important as measurement in shaping a cartographic image.

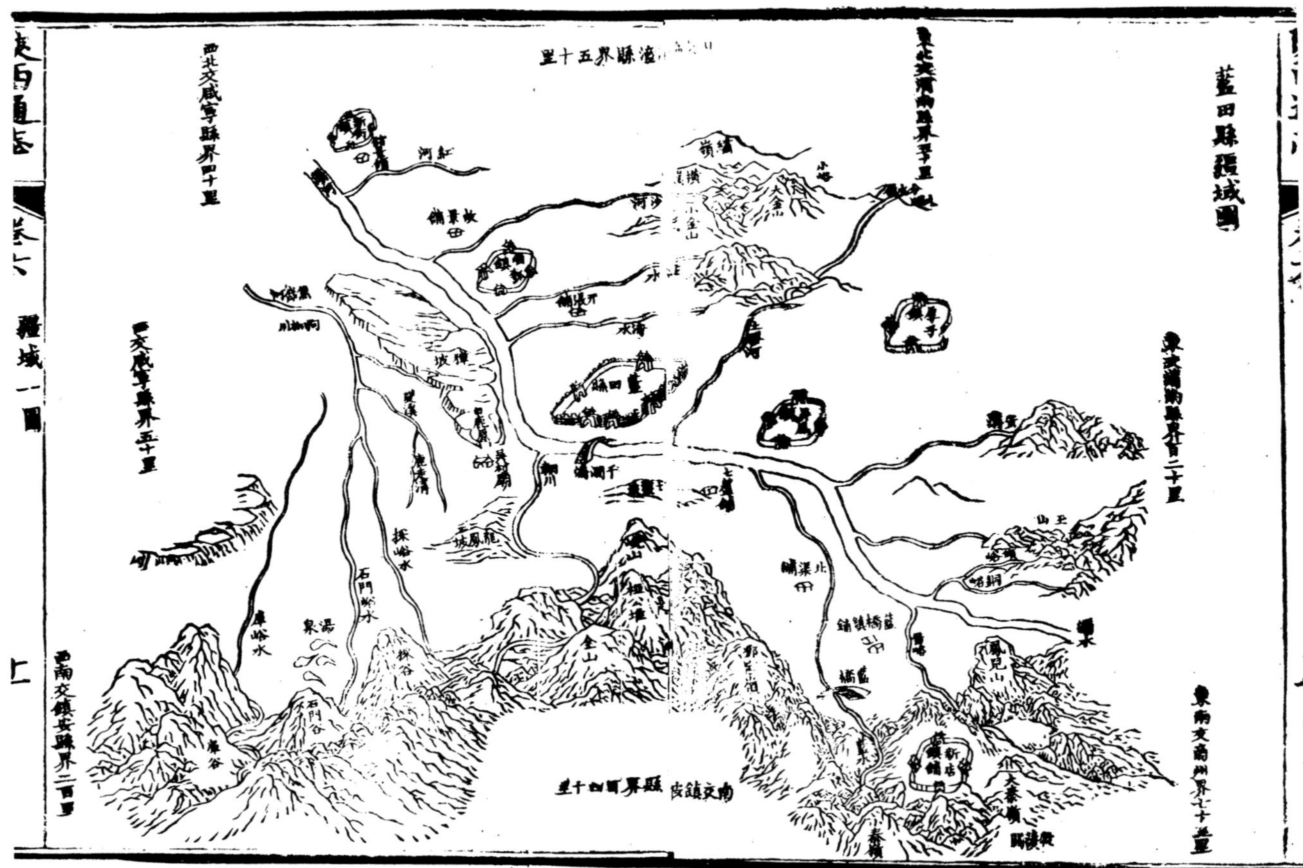

FIG. 7.15. MAP OF LANTIAN XIAN FROM A QING GAZETTEER OF SHAANXI PROVINCE. Size of the image: 9.8 × 13.8 cm. From *Shaanxi tongzhi* (1735; reprinted Taipei: Huawen Shuju, 1969), 6.10b–11a.

eigners. But some Chinese scholars came to realize that China was not the preeminent country on earth—it could no longer treat all other countries as tributary states. To deal with foreigners, China would need more reliable information about the rest of the world. One of the most notable works that were written to meet this need was the *Haiguo tuzhi* (Illustrated record of maritime kingdoms, 1844, 3d ed. 1852) by the scholar-official Wei Yuan (1794–1856). The treatise is important in the political history of the Qing, since it was the first Chinese work to "make a realistic geopolitical assessment of the worldwide dimensions of Western expansion and of its implications for Asian trade and politics."[59] Before Wei Yuan, Qing foreign policy had been directed toward Central Asia, not the maritime regions. Wei presented his challenge to this foreign policy in a traditional format. Like most gazetteers of the Ming and Qing, Wei Yuan's treatise consists of a combination of map and text. For the information presented in his verbal accounts, Wei draws on traditional Chinese as well as European sources. He acknowledges the *Sizhou zhi* (Record of four continents) as an important source. The compilation of the *Sizhou zhi* was directed by Lin Zexu (1785–1850) while he was imperial commissioner in Guangzhou (Canton) in 1839. The work consists of translations of Western-language materials about the West and Sino-Western affairs, and it was intended to shed light on the nature of European objectives in Asia. This was an aim shared by Wei Yuan, who wrote in his first preface to the *Haiguo tuzhi*:

> When defending against an enemy, whether or not one knows the disposition [or configuration, *xing*] of one's enemy makes the difference between success and failure. Similarly, in dealing with one's enemy, whether or not one knows their customs also makes the difference between success and failure. In the past those who controlled the foreigners always inspected the disposition [*xing*] of their enemy and became as familiar with that as with their own desks and chairs.[60]

59. Jane Kate Leonard, *Wei Yuan and China's Rediscovery of the Maritime World* (Cambridge: Council on East Asian Studies, Harvard University, 1984), 2.

60. Wei Yuan, "*Haiguo tuzhi* yuan xu" (Original preface to the *Haiguo tuzhi*), in *Zengguang Haiguo tuzhi* (Expanded *Haiguo tuzhi*), 2a; see the edition in 5 vols. (1852; reprinted Taipei: Guiting Chubanshe,

FIG. 7.16. MAP OF ZHONGNAN SHAN (ZHONGNAN MOUNTAINS). Size of the image: 10 × 14.2 cm. From *Shaanxi tongzhi* (1735; reprinted Taipei: Huawen Shuju, 1969), 8.4b–5a.

Wei saw maps as essential for making foreign countries more accessible to Chinese readers. From a cartographic viewpoint, Wei's work is a blend of Chinese and European practice. The historical maps of China at the beginning of his work are heavily annotated and lack grids as well as graticules. In drafting style, they are practically indistinguishable from those of traditional gazetteers (fig. 7.23): they do not have expressed scales and seem to be based on sketches. The maps of the countries of the world, however, draw on European mapping techniques (fig. 7.24). Unlike the Qing compilers of the *Ming shi* (History of the Ming, 1739; Zhang Tingyu et al.), who a century before had regarded Western geographic accounts with suspicion, Wei Yuan accepts them. Lu Liangzhi has argued that Wei Yuan shows an understanding of the strengths and weaknesses of various projections. For African countries near the equator, Lu says that Wei chose to use the Sanson-Flamsteed projection, which reduces angular distortion at latitudes near the equator. For countries above 45° latitude, he used the Bonne projection, which reduces angular distortion at the upper latitudes. In maps with navigational significance, such as those of Australia and surrounding waters, Wei chose the Mercator projection, on which the constant compass course between two points is a straight line.[61] If this interpretation is correct, Wei's work appears to be much more complex than many European atlases of the time. But the choice of projections may have been made for Wei Yuan by the sources available to him. He may have been merely copying European maps. In one of his prefaces he says that he consulted European sources and that he is "providing" maps, not creating them.[62] He does not claim authorship of the maps, and his book provides no evidence that he was aware that different methods of projection existed. Thus it is hard to credit Wei Yuan with understanding projection.

Wei's treatise, however, was not meant primarily as an exercise in cartography. Its purpose was to promote Westernization. Wei advocated adopting European technol-

1978), 1:7. This is the third edition of the *Haiguo tuzhi*, originally published in 1844.

61. See Lu, *Zhongguo dituxue shi*, 203 (note 1).

62. Wei Yuan, "*Haiguo tuzhi* hou xu" (Postface to the *Haiguo tuzhi*), 3a, in *Zengguang Haiguo tuzhi*, 1:9 (note 60).

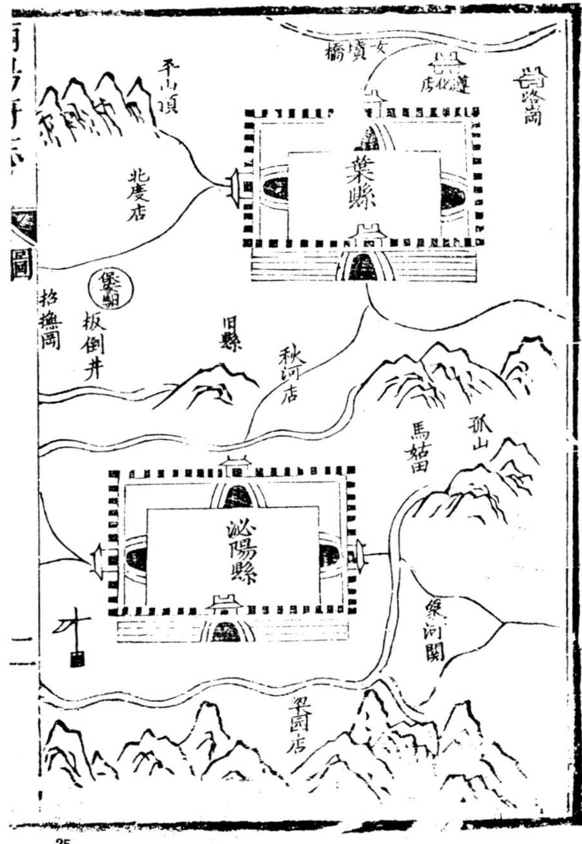

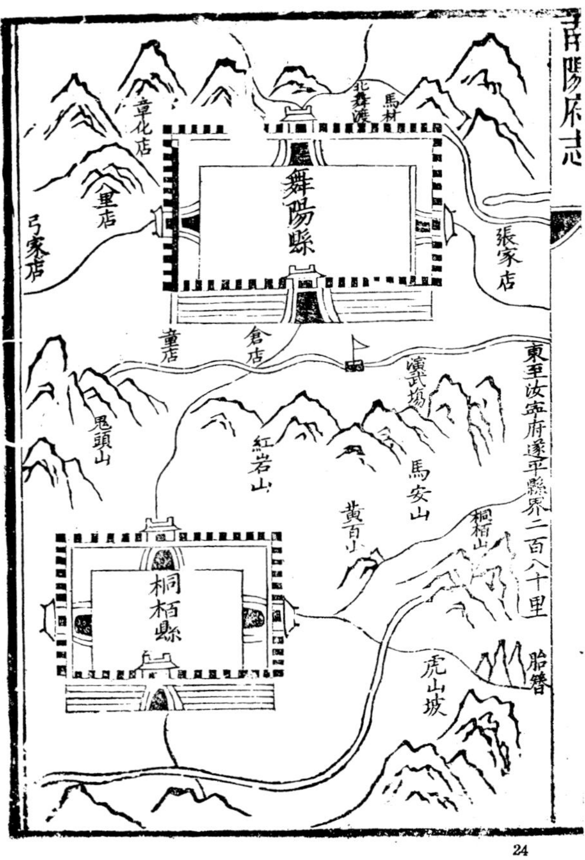

FIG. 7.17. MAP FROM THE *NANYANG FU ZHI*. Size of each page: ca. 16.5 × 10.5 cm. From *Nanyang fu zhi* (Gazetteer of Nanyang Prefecture [in present-day Henan Province]) (1694; reprinted Taipei: Taiwan Xuesheng Shuju, 1968), *tu* 1b–2a.

ogy, particularly in the manufacture of arms and naval vessels, and encouraged the study of European technology. These and other reforms that he put forth as a means of dealing with maritime "barbarians" struck a responsive chord in China and abroad. His work was, for example, translated into Japanese and influenced Japanese views on Westernization.

The need to realize at least some of Wei's proposals became apparent to a number of Chinese intellectuals in the second half of the nineteenth century. During that period China experienced a series of domestic rebellions, among them the Taiping Rebellion, a movement inspired by Christian ideology to expel the Manchus from China. China also suffered setbacks in border disputes—setbacks attributable to a lack of solid geographic information.[63]

As a consequence of a growing sense of national weakness, reformers urging Westernization gained strength in the central government, and they did succeed in promoting the development of industry and communications. Chinese intellectuals also saw mapping practice as needing improvement. The deficiencies of Qing predecessors are admitted by the compilers of a gazetteer dating from 1879: "As far as maps are concerned, the Kangxi edition of this gazetteer is too sketchy, and the distances and locations of mountains, streams, and cities and their outskirts are topsy-turvy and confused. We rechecked the area following those maps, from beginning to end, and there was not one correct place. This is probably a result of clerks and laborers being delegated the task, while [those in charge] did not personally pass through the area

63. The growth of Manchu power coincided with the expansion of the Russian empire across what is now Siberia and into the Amur (Heilongjiang) basin, where tension arose between the two empires. In this confrontation the Qing empire ended up with a loss of territory attributable to errors in surveying and mapping. On the Qing border dispute with the Russian empire, see Joseph Sebes, *The Jesuits and the Sino-Russian Treaty of Nerchinsk (1689): The Diary of Thomas Pereira, S.J.* (Rome: Institutum Historicum S.I., 1961); and John Robert Victor Prescott, *Map of Mainland Asia by Treaty* (Carlton, Victoria: Melbourne University Press, 1975). The Qing also lost territory in border disputes with the Koreans as a result of faulty geographic information. For a detailed study, see Zhang Cunwu, "Qingdai Zhong-Han bianwu wenti tanyuan" (An inquiry into the Sino-Korean border question during the Qing dynasty), *Zhongyang Yanjiuyuan Jindaishi Yanjiusuo Jikan* 2 (1971): 463–503.

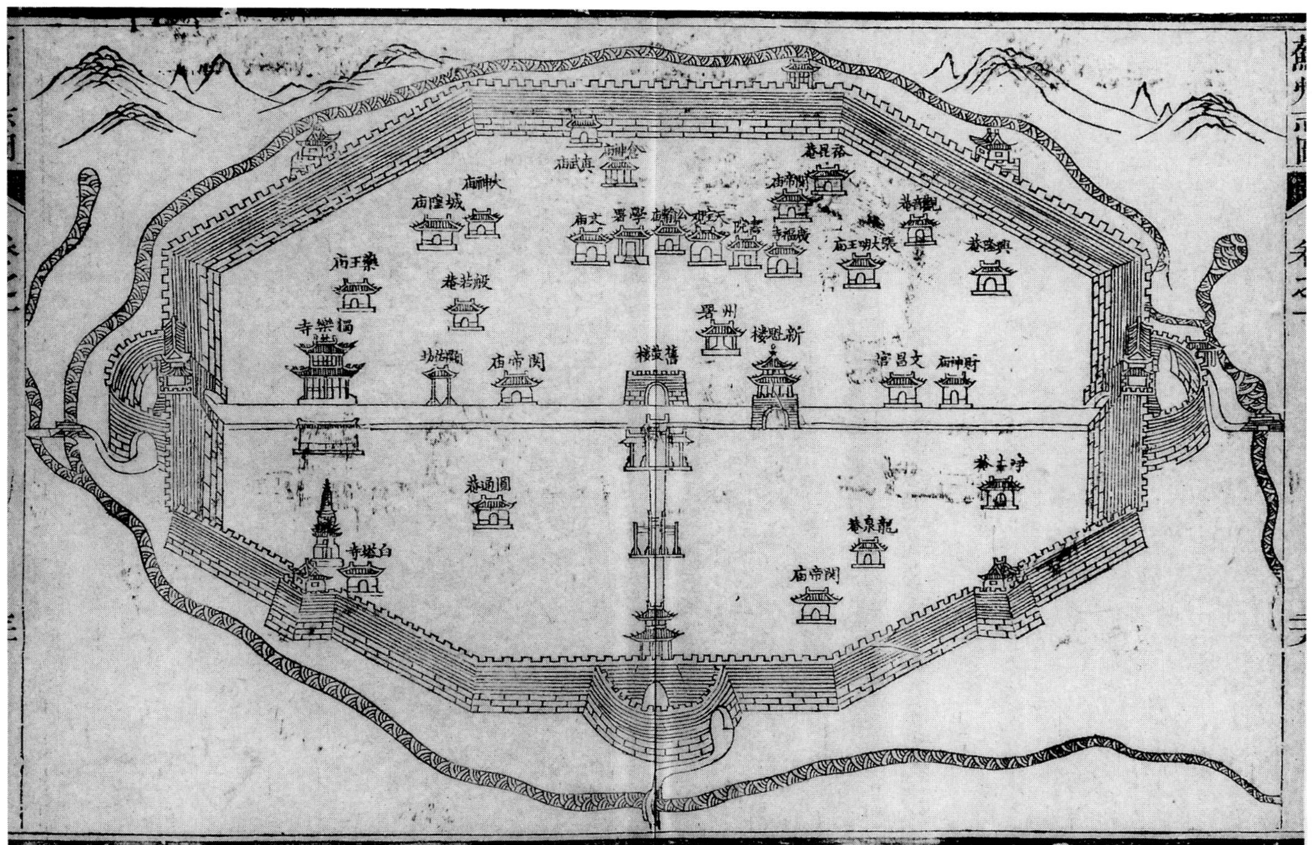

FIG. 7.18. MAP FROM THE *JIZHOU ZHI*.
Size of the original: 19.5 × 29 cm. From *Jizhou zhi* (Gazetteer of Jizhou [in present-day Hebei Province], 1831), 1.29b–30a. Reproduced courtesy of the Harvard-Yenching Library, Harvard University, Cambridge.

and conduct a detailed examination."[64] As a corrective, the compilers of the 1879 gazetteer based their information on trips personally undertaken or delegated to scholars who took notes on the places they passed.

Such care in the recording of information, however, seems to have been exceptional among compilers of local gazetteers even during the late Qing. In the 1890s the central government felt compelled to try to standardize the cartographic practices of regional and local administrations. During that time, the *huidianguan* (bureau of institutional studies) was compiling a new atlas of the empire. Provincial governments were asked to submit maps based on measurements of distances and of latitude and longitude and to submit provincial maps using conic projections. The attempt at standardization failed in large part because of a shortage of scholar-officials knowledgeable in surveying techniques. In 1892 a provincial governor memorialized the throne, lamenting that there was no way to adhere to the new standards: "In each subprefecture and county, scholars familiar with the territory are extremely few. In addition, there are no surveying or drawing instruments. For this reason, things are confused and there is no way to proceed."[65]

It is not known how many Chinese were competent enough to apply European techniques on their own. Those who produced Western-style maps in the late Qing tended to base them on the Jesuit atlases compiled during the reigns of the Kangxi and Qianlong emperors.[66] Sometimes they imitated European maps without much understanding. A manuscript map of the Eastern Hemisphere drawn in 1790, for example, lacks the lines of latitude and longitude (plate 15). Instead, the outer portion of the map is labeled with the twenty-four points of the Chinese compass. The mapmaker seems to have erroneously believed that constant compass bearings on a

64. *Yongping fu zhi* (Gazetteer of Yongping Prefecture [in present-day Hebei Province]) (1879; reprinted Taipei: Taiwan Xuesheng Shuju, 1968), "Fanli" (Principles), 2a–b.

65. Zhang Zhidong (1837–1909), *Zhang Wenxiang gong quanji* (Complete works of the honorable Zhang Wenxiang [Zhidong], 1928) (1937; reprinted Taipei: Wenhai Chubanshe, 1970), 31.12b–13a.

66. An example of such a compilation is the *Da Qing yitong yutu* (Comprehensive geographic map of the Great Qing, 1863), which is discussed below.

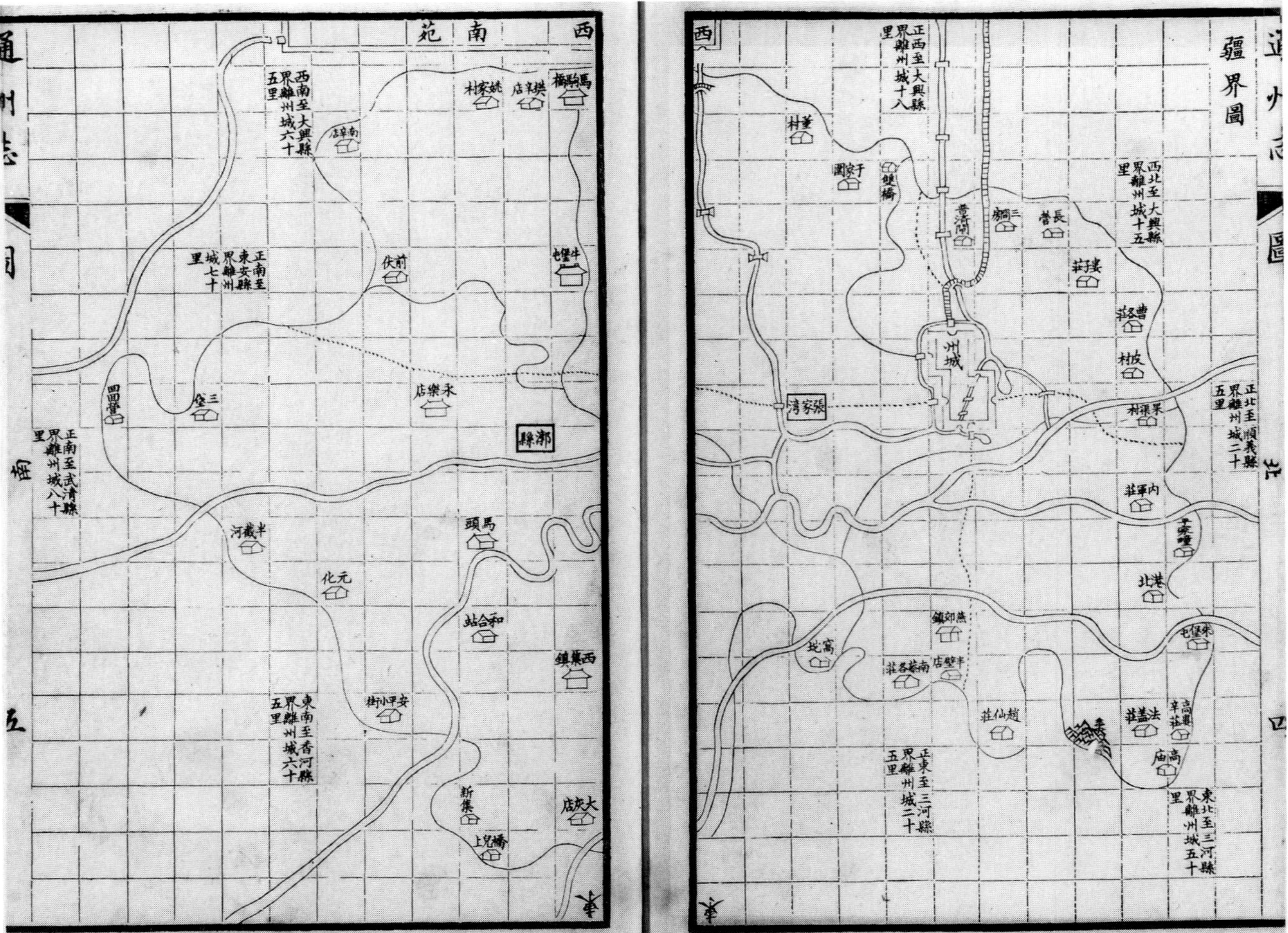

FIG. 7.19. MAP FROM THE *TONGZHOU ZHI*. Size of each page: 22.5 × 14.5 cm. From *Tongzhou zhi* (Gazetteer of Tongzhou [in present-day Hebei Province, near Beijing]) (1879), *tu* 4b–5a. Reproduced courtesy of the Harvard-Yenching Library, Harvard University, Cambridge.

spherical earth lay on a straight line.

In some quarters Pei Xiu (223–71), whose six principles seem to have presupposed a flat earth, was still upheld as the model of cartographic practice. The compilers of a gazetteer dating from 1894 say that old gazetteer maps are crude in their technique, since they often do not use grids: "Now we follow the methods of Pei Xiu of the Jin dynasty."[67]

It seems that human nature may also have been a factor in hindering efforts at cartographic reform, if government attempts to rectify cadastral survey practices are any indication. The central government did try to establish a system of tax accounting based on registers and cadastral maps. But the use of maps and registers was insufficient to maintain fiscal soundness. During the eighteenth century, in the Jiangnan region (Jiangsu, Jiangxi, and Anhui), for example, large landowners were able to circumvent the tax collection despite the use of maps and registers. They would, as Zelin has found, often "divide their holdings into several tens or even several hundreds of different household registrations, each claiming a minute amount of land":

> The name of the household heads for each of these *hu* [households] would be falsified, using the names of dead ancestors, people who had migrated from the area, temple names, and so on. Ownership of the land was rendered difficult to trace, and the small amount of arrears from each *hu* discouraged investigation. Even if they were found out, their familiarity with the

67. *Guangping fu zhi* (Gazetteer of Guangping Prefecture [in present-day Hebei Province]) (1894; reprinted Taipei: Taiwan Xuesheng Shuju, 1968), "Fanli" (Principles), 1a–b. For more on Pei Xiu, see esp. pp. 110–13.

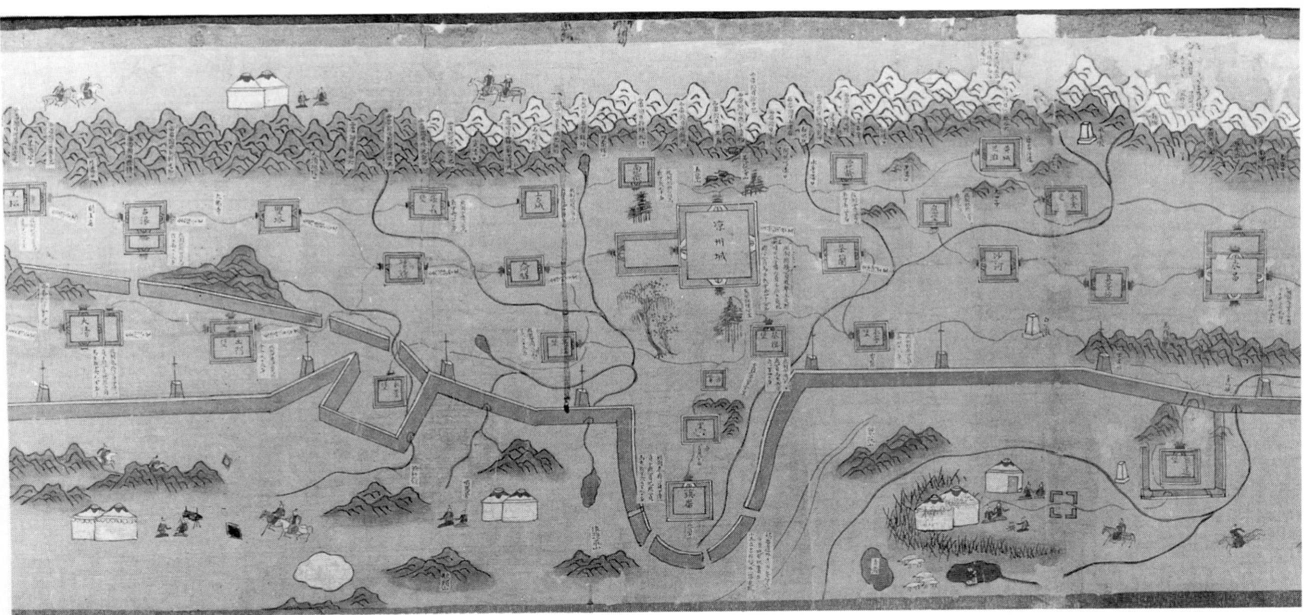

FIG. 7.20. LATE SEVENTEENTH-CENTURY MAP OF THE GREAT WALL (DETAIL). Size of the entire original: 22.5 × 755 cm. By permission of the Monumenti Musei e Gallerie Pontificie, Vatican City.

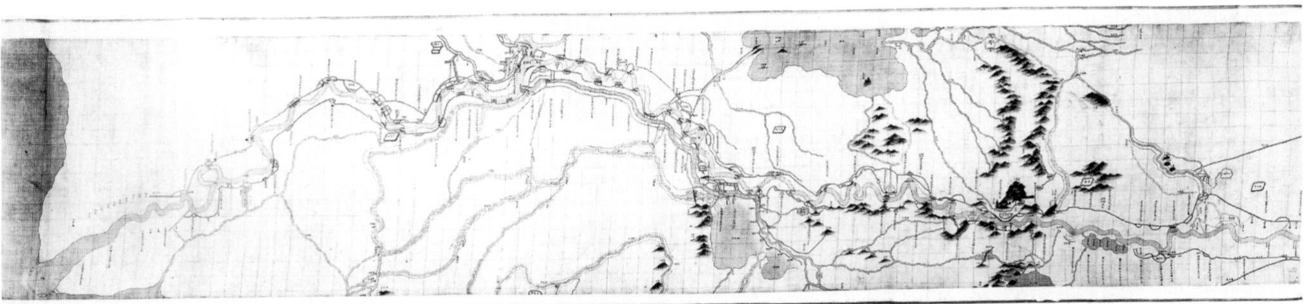

FIG. 7.21. NINETEENTH-CENTURY MAP OF THE YELLOW RIVER. Size of the original: 38 × 182 cm. Courtesy of the Geography and Map Division, Library of Congress, Washington, D.C. (G7822.Y4A5 18--.H9 Vault Shelf).

> malpractices of officials and clerks and their own influence in the community allowed the rural elite to blackmail the regular bureaucracy into granting them almost total immunity from taxation.[68]

Officials hoped that the use of maps and registers would simplify tax collection procedures. But they underestimated the resourcefulness of tax evaders and the venality of local bureaucrats, who would often alter, hide, or even destroy maps and registers so that the amount of taxes in arrears would be nearly impossible to determine.

The lack of consistent practices among provincial governments posed problems to the mapmakers at the *huidianguan*. Because maps submitted by provincial governments were drawn according to different standards, the bureau complained, it was difficult to combine these maps into a comprehensive map of the entire empire without incurring error. After rechecking measurements and consulting textual sources, the *huidianguan* redrew the maps submitted, adopting the following practices as a standard:

> The bureau now . . . uses the method of conical projection for the complete imperial map, without a grid. For the general maps of provinces, a grid with increments representing one hundred *li* is used; and for the individual maps of each prefecture, a grid with increments representing fifty *li* is used. Both of these lack longitude and latitude. The general maps include only the most important information, such as famous mountains and rivers, places where local governments

68. Madeleine Zelin, *The Magistrate's Tael: Rationalizing Fiscal Reform in Eighteenth-Century Ch'ing China* (Berkeley and Los Angeles: University of California Press, 1984), 245.

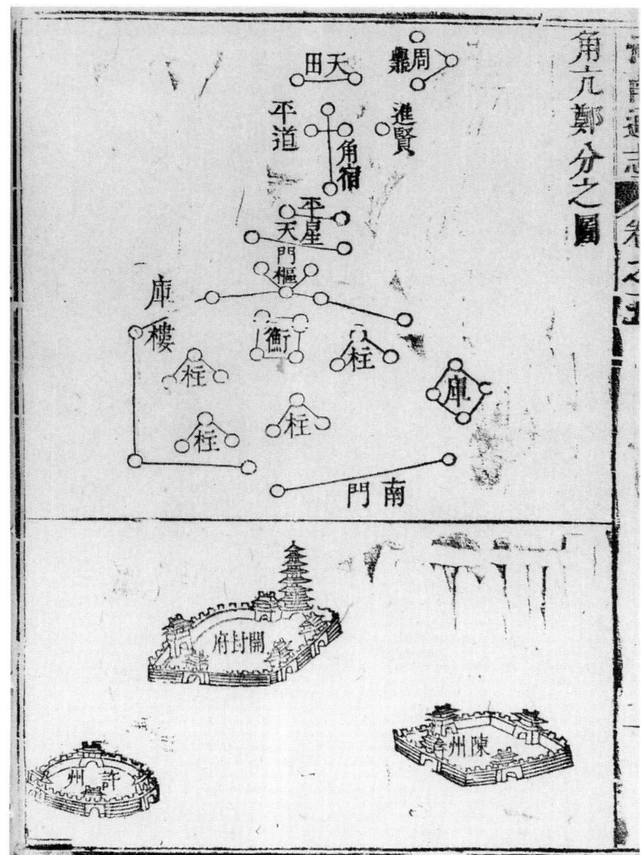

FIG. 7.22. ASTROLOGICAL DRAWING FROM A QING GAZETTEER OF HENAN PROVINCE. Chart of the celestial field over Kaifeng Prefecture.
Size of the original: 22.5 × 16 cm. From *Henan tongzhi* (1882; rev. ed. 1869), 5.1b. Reproduced courtesy of the Harvard-Yenching Library, Harvard University, Cambridge.

> are located, and public and private telegraph lines. Prefectural maps, on the other hand, attempt to be as detailed as possible, including all the mountains, rivers, villages, towns, courier stations, mountain passes, seaports, and islands. As a rule, the general maps should observe the degrees [of latitude and longitude]; and the regional maps should observe the *li* [of distance], so that the general and regional complement each other and present all perspectives. The general provincial maps should be concise, and the prefectural maps greatly detailed.[69]

The bureau's standards draw much from European cartography in their advocacy of the graticule and the tendency toward planimetric representation. Pictorial elements were minimized with the bureau's adoption of a set of standard signs for administrative units and topographic features (see fig. 7.25). In accordance with its announced standards, the *huidianguan* published a map of the empire (fig. 7.26). Not all the signs it adopted, however, were employed—for example, the sign for telegraph lines. In its standards, the bureau also made some concessions to the continued reverence for traditional Chinese cartographic practices, particularly in the continued use of the square grid (fig. 7.27). This was not a new development. Some Chinese mapmakers even before this time often combined grid with projection, demonstrating an imperfect grasp of the principles behind the idea of projection. The two systems are incompatible—a grid of squares each representing equal increments of distance cannot simply be superimposed on a projection from sphere to plane, in which degree increments do not necessarily translate into equal increments of distance.

The use of grid and graticule may itself have been a result of one tendency within the Chinese reform movement—that toward combining Chinese and European learning. This tendency was embodied in the slogan *Zhongxue wei ti, Xixue wei yong* (Chinese learning as the principle, Western learning as the application). The impulse to fuse the two cultures is evident in the statement of principles for the *Da Qing yitong yutu* (Comprehensive geographic map of the Great Qing, 1863). The mapmakers think highly both of old Chinese methods and of the methods represented by the Kangxi atlas and attempt a synthesis of the two. They see the Kangxi atlas as falling within the Chinese tradition of mapping, not as a departure from it.[70] The compilers of the atlas intend that it should represent "distances as a bird flies," that is, as straight lines—therefore the use of the square grid. They are also aware of the Kangxi emperor's linking Chinese units of distance to a degree of longitudinal arc—"two hundred *li* are equal to one degree of longitude [at the equator]"—thus the use of projection.[71] The Chinese method, however, seems to form the *ti* (the body or principle) of the atlas, since the square grid is printed in solid black lines. The horizontal lines correspond to parallels of latitude, but the meridians are indicated with broken lines.

The atlas consists of more than one hundred leaves, which can be assembled to form four sheet maps: two small maps consisting of a few leaves, one of Vietnam and one of Taiwan; and two large maps, one of mainland China with Hainan and one of Asia extending east to

69. *Qinding Da Qing huidian* (Imperially commissioned, collected statutes of the Great Qing), 24 vols. (1899; reprinted Taipei: Zhongwen Shuju, 1963), "Fanli" (Principles), 2:1024–25.

70. This response to European science was common among Qing intellectuals, who often sought precedents for European ideas in traditional texts.

71. Hu Linyi et al., comps., *Da Qing yitong yutu* (1863; Shanghai: Shanghai Shuju, 1896), "Fanli" (Principles), 7b–8a. This work is also known as the *Huangchao Zhongwai yitong yutu* (Comprehensive geographic map of China and foreign countries of the present dynasty).

FIG. 7.23. HISTORICAL MAP BY WEI YUAN. Map of the Western Region during the Han dynasty.
Size of each page: 21 × 12.5 cm. From Wei Yuan, *Zengguang Haiguo tuzhi* (Expanded Illustrated record of maritime kingdoms), from an edition of 1847, 3.7a–b. Reproduced courtesy of the Harvard-Yenching Library, Harvard University, Cambridge.

west from the Pacific Ocean to the Caspian Sea and north to south from the Arctic Ocean to Indochina and India. In the 1896 edition of the atlas, the leaves making up the comprehensive map of East Asia are printed so that each leaf covers four degrees of latitude (or about eight hundred *li*) and each grid increment along the vertical axis corresponds to one-half degree or one hundred *li*. Each square is thus supposed to represent ten thousand square *li*. This would be the case if a flat surface were being mapped, but as the meridians drawn on the maps remind a map reader, the map is actually a projection from a spherical surface to a plane—in this case trapezoidal projection, since the Kangxi atlas was used as a base. In the process of projection some distortion occurs. For example, north and south bearings do not always correspond to the vertical axis of the grid. As one moves away from the prime meridian running through Beijing, the meridians become increasingly oblique, so that the north-south axis is no longer perpendicular to the east-west axis (see fig. 7.28). In addition, with the trapezoidal projection, the scale of one grid increment to one hundred *li* would be true only along the prime meridian and one or possibly two base parallels. For the remainder of the map, the determination of direct distances between points would not be a simple matter of counting squares.

The production of hybrid maps combining grid and graticule in the late nineteenth century casts doubt on Needham's assertion that during the Qing dynasty Chinese cartography had become part of "world" cartography, or one with European cartography. The evidence presented here suggests a different conclusion—that traditional Chinese cartography continued to flourish under Manchu rule, despite the employment of foreign cartographers for imperial mapping projects. The mere fact of transmission does not always translate into reception, and the case of European cartography in China illustrates this point.

Chinese cartography during the late imperial period was much more complex than allowed by those who have focused on the Jesuit surveys. How complex is something that is just now beginning to be appreciated. Contrary to many accounts of Chinese cartography, European influence did not mean the end of the indigenous tradition. It was, the evidence examined so far suggests, actually dominant for much of the late imperial period.[72]

72. Furthermore, the textual scholarship underlying much traditional Chinese cartography also flourished. See esp. pp. 92–95.

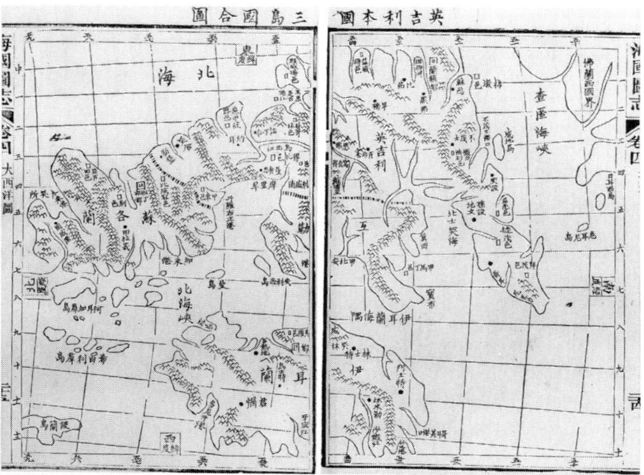

FIG. 7.24. MAP OF THE BRITISH ISLES BY WEI YUAN. The prime meridian falls to the east of London. East is at the top.
Size of each page: 20.5 × 14 cm. From Wei Yuan, *Zengguang Haiguo tuzhi* (edition of 1876), 4.24b–25a. Reproduced courtesy of the Harvard-Yenching Library, Harvard University, Cambridge.

The history of European cartography in China was not a case of a dominant culture imposing its science on a weaker recipient. The response of Chinese cartography to the European model was similar to that of Chinese astronomy: mapmakers, like astronomers, as Sivin says, were "on the whole members of the old educated élite, imbued with its values. Their first impulse was to supplement and strengthen the indigenous science, not to discard it, and their loyalty remained with their ancestral world view."[73]

73. Sivin, "Copernicus in China," 64 (note 47).

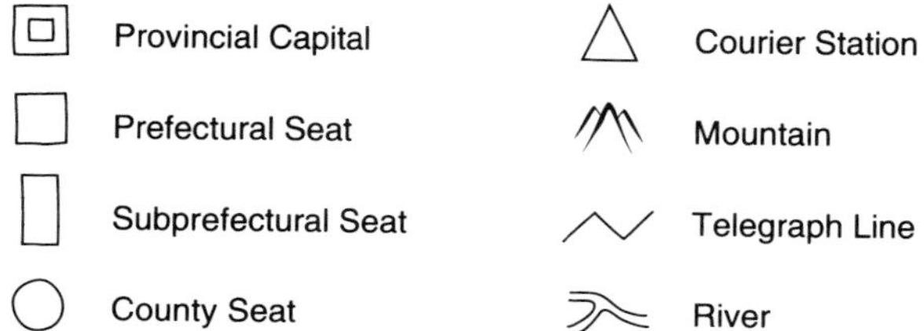

FIG. 7.25. STANDARDIZATION OF MAP CONVENTIONS DURING THE LATE QING. Standard symbols adopted by the *huidianguan* (bureau of institutional studies).

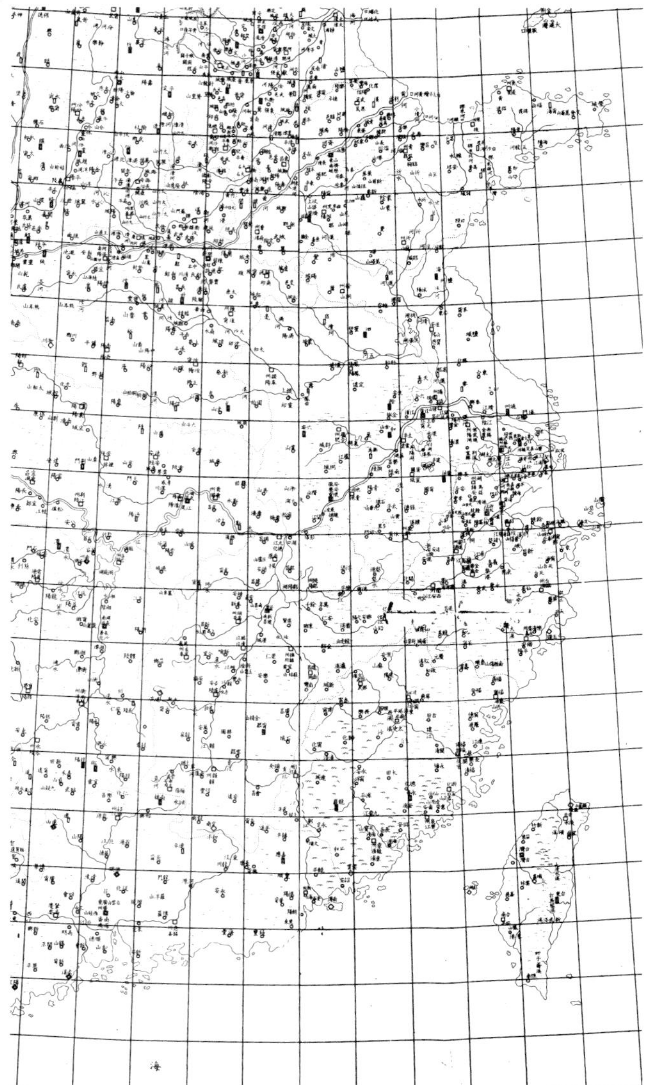

FIG. 7.26. DETAIL FROM "HUANGYU QUANTU" (COMPLETE MAP OF THE EMPIRE, 1899). This map of the empire was drawn using a conical projection. The prime meridian passes through Beijing. The entire map represents an area extending 47° east and 47° west of Beijing (from west of present-day China to about the Kamchatka peninsula) and extending between 18° and 61° north latitude (from Hainan Island to about midway through Siberia). The portion reproduced here shows the eastern part of China.
Size of the entire original: 114.9 × 185.2 cm. From the *Qinding Da Qing huidian* (Imperially commissioned, collected statutes of the Great Qing), 24 vols. (Beijing: Huidianguan, 1899).

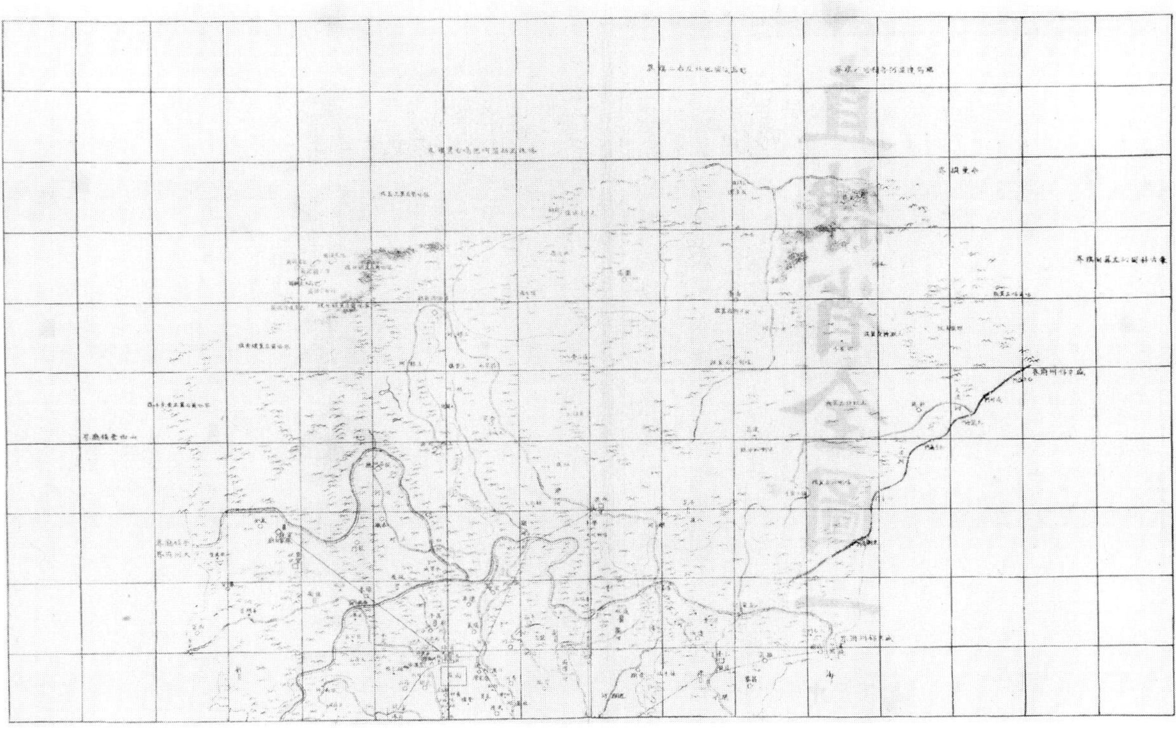

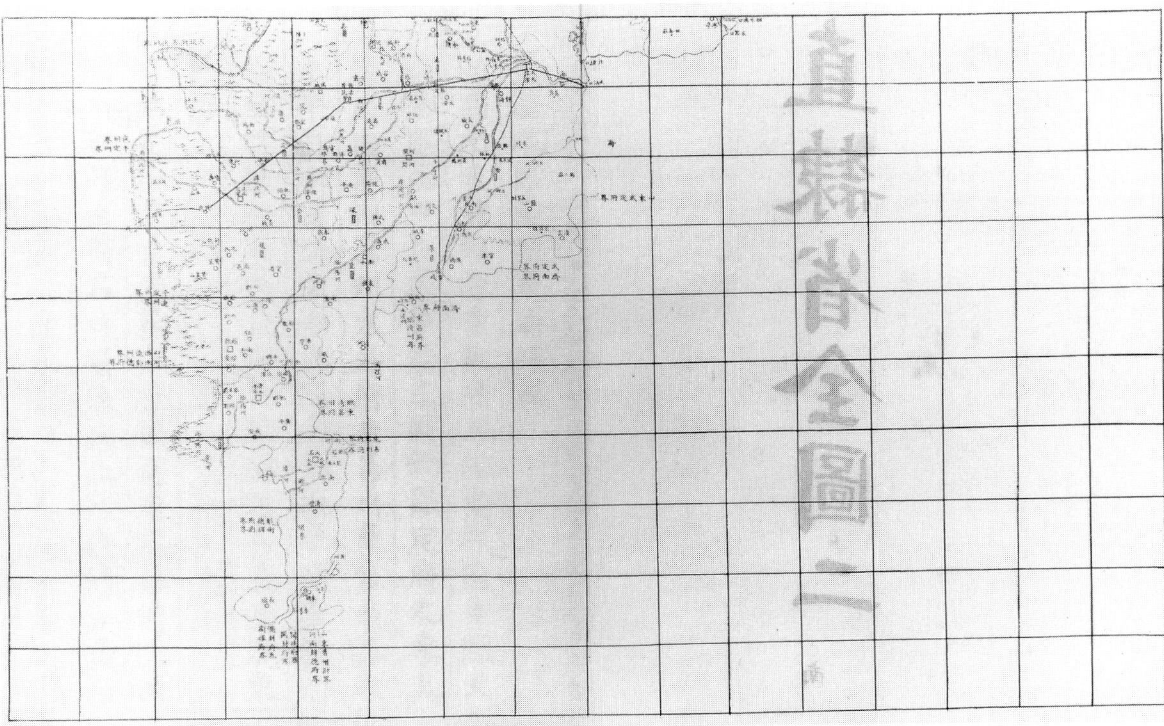

FIG. 7.27. MAPS OF ZHILI, PROVINCIAL SEAT OF THE IMPERIAL GOVERNMENT. These maps present two parts of Zhili. The grid was intended in part to help line up the images on the two maps, but because of discrepancies in the sizes of the squares, the two images do not match precisely. Above, northern Zhili; below, southern Zhili.
Size of the images: 18.9 × 30.4 cm and 18.6 × 29.9 cm. From the *Qinding Da Qing huidian*, 24 vols. (Beijing: Huidianguan, 1899).

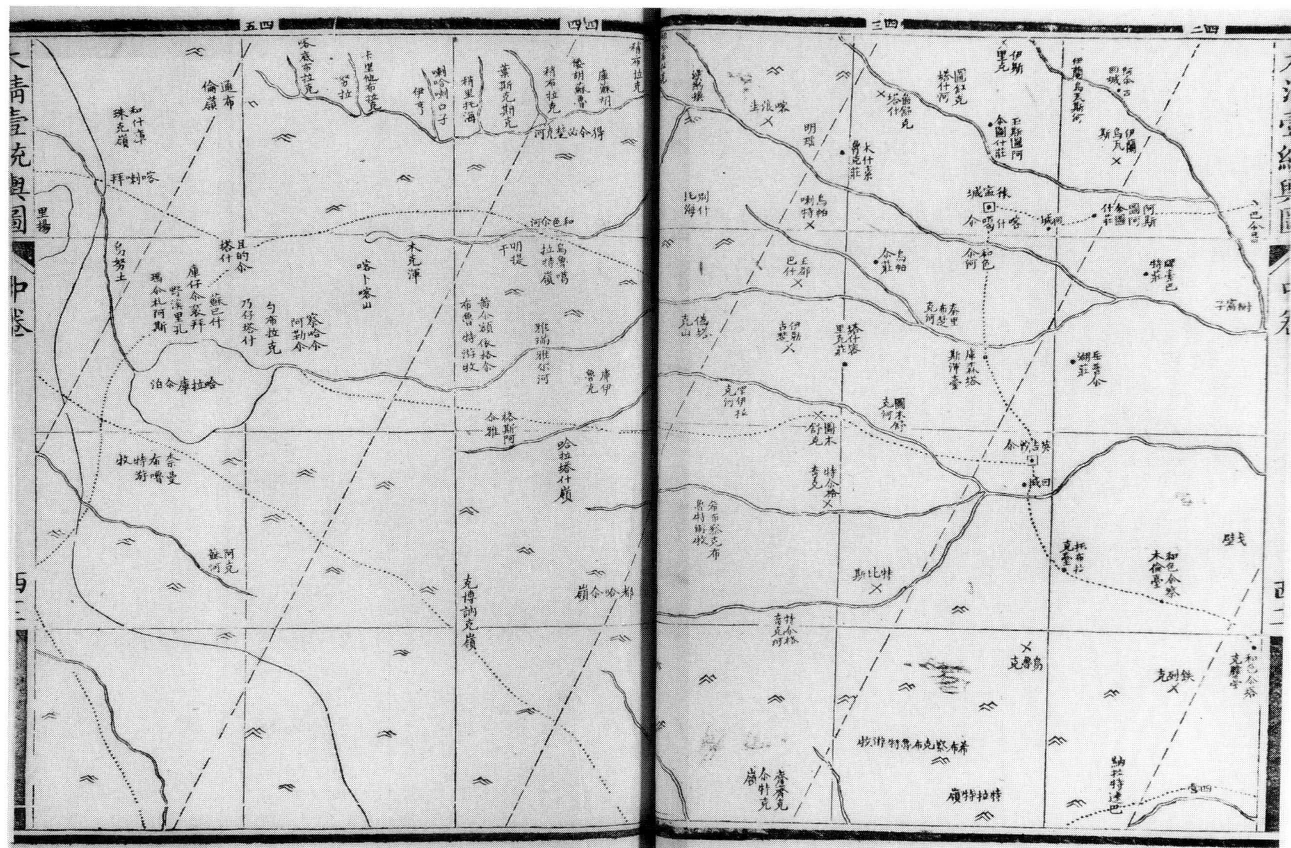

FIG. 7.28. MAP COMBINING GRID AND GRATICULE. The meridians become more oblique as one moves away from the prime meridian running through Beijing. Although the map employs a projection, the grid gives the impression that one can measure direct distances between points along a straight line. This section of the map depicts part of the Kashgar region. Size of each page: 22.5 × 18 cm. From Hu Linyi et al., comps., *Da Qing yitong yutu* (1863), chap. *zhong*, *xi* 11b–12a. Reproduced courtesy of the Harvard-Yenching Library, Harvard University, Cambridge.

8 · Chinese Cosmographical Thought: The High Intellectual Tradition

John B. Henderson

Chinese cosmographical thought of premodern times was not as concerned as its counterparts in Western civilizations with the overall shape of the world or structure of the cosmos. There is no pre-seventeenth-century Chinese equivalent of the medieval European *mappaemundi* or of Western representations of the earth showing its various cosmographical divisions or climatic zones. The widely held conception that China comprised "all under heaven" (*tianxia*), as well as the geographical isolation of Chinese civilization, may have contributed to Chinese cosmographers' lack of interest in outlining, either realistically or schematically, the form of the world as a whole. Whatever the explanation, traditional Chinese cosmographical charts generally represent structures in such microcosmic dimensions as the architectural, the urban, and the agrarian rather than depicting the shape of the earth or the system of the world. Like many Western cosmographical diagrams, however, these Chinese charts were generally based on the supposition that there exist correspondences or correlations, which may be graphically expressed, between various orders of existence or realms of the universe, such as those of heaven, earth, and humanity. But in Chinese cosmographical thought these correspondences were drawn less often between macrocosm and microcosm as among various orders of mundane reality.

In comparison with most of their counterparts in Western and Middle Eastern cosmographical traditions, Chinese cosmographical charts were also generally more subordinate to and less independent of textual descriptions. Although graphic representations of cosmographical schemata were common in China from the Song era (960–1279), those diagrams were often printed principally to illustrate or substantiate a cosmographical conception that received a more authoritative or precise formulation in an accompanying text. For Chinese scholars a picture (or graph) was *not* worth a thousand words, however useful it might be as a visual aid. The relationships between graph and text may well have differed in earlier periods, as the lore surrounding the composition and articulation of the canonical *Yi jing* (Book of changes) implies. But so few of such diagrams have survived from the pre-Song or preprinting era in China that it is difficult to establish just what these relationships were. Hence this chapter must rely largely on verbal descriptions for its account of Chinese cosmographical thought in its formative phase in the Han era (206 B.C.–A.D. 220), though graphic reconstructions dating from the Song and later will be included where they are helpful and relevant.

A glance at these reconstructions reveals that most of them were patterned on the same model, a square divided into nine equal squares resembling the form of a simple magic square or three-by-three grid. Cosmographers in traditional China applied this plan to the conceptualization and arrangement of such diverse kinds of space as astronomical, political, agrarian, urban, and architectural. Just as Sir Thomas Browne, according to Coleridge, saw "Quincunxes [or lozenges] in Heaven above, Quincunxes in Earth below, & Quincunxes in the water beneath the Earth,"[1] so Chinese cosmographers of the Han and later eras regarded the simple nonary square as the basis of proper order in practically every realm of space. This form was at least as important and ubiquitous in premodern Chinese cosmography as the circle was in Greek, medieval European, and Islamic cosmography. One of my principal aims in this chapter is to outline the development, articulation, and later criticism of this dominant cosmographical conception, which made Chinese cosmographical thought, at least that of the high literate tradition, remarkably uniform and well integrated. In so doing, I will focus on two particularly important eras in the history of geometric and nonary cosmography in China. These include the formative phase of the Han period and the seventeenth century, which saw the criticism and decline of established cosmographical conceptions and the emergence of new ones.

I thank Nathan Sivin of the University of Pennsylvania for his valuable comments and constructive criticism on drafts of this chapter.

1. Samuel Taylor Coleridge's letter to Sara Hutchinson, on two and one-half flyleaves of a volume containing Browne's *Vulgar Errors, Religio Medici, Hydriotaphia*, and *Garden of Cyrus* (London, 1658), now in the Berg Collection, New York Public Library; see *Coleridge on the Seventeenth Century*, ed. Roberta Florence Brinkley (Durham: Duke University Press, 1955). The reference is to Browne's *The Garden of Cyrus; or, The Quincunciall, Lozenge, or Network Plantations of the Ancients, Artificially, Naturally, Mystically Considered.*

In composing this outline of the formation and decline of the perennial Chinese cosmography, I found it necessary to consult sources on such diverse topics as microcosmic architectural structures, classical city plans, and ideal agrarian orders, for Chinese cosmographical conceptions were articulated most extensively and illustrated most graphically in these areas. Little of the modern scholarship on these subjects or schemata is devoted to their cosmographical aspects, however, focusing instead on their political, economic, or ritual significance and treating them in isolation from congruent cosmographical conceptions. In short, cosmography is not a well-developed or even a very distinct field of study among Sinologists. Hence there is at this juncture little point in attempting a systematic historiographical or bibliographical overview of the secondary literature on Chinese cosmography.

Foundations of Geometric and Nonary Cosmography

Antecedents of the perennial Chinese cosmography may be traced back to high antiquity. Rectilinear forms, especially squares, figure prominently in Chinese art and artifacts from the dawn of Chinese history. Indeed, the designs on Chinese Neolithic pots dating from as early as 5000 B.C. often consist of "parallel bands or lozenges containing concentric squares, crosses, or diamonds."[2] The art of the earliest Chinese civilization, the Shang, which flourished in the latter half of the second millennium B.C., is also marked by "the imposition of abstract, balanced, geometric patterns over entire surfaces."[3] Even Shang domiciles, palaces, temples, and tombs, according to Chang, "were invariably square or oblong, governed in orientation by the four cardinal directions and dominated in design by a persistent attempt at symmetry." Shang Chinese, moreover, may well have conceived of the form of the political cosmos as square, or at least have seen the Shang realm as oriented toward the four directions, with the "countries beyond the kingdom . . . grouped into four directional classes."[4]

By the Han era, these rectilinear orientations and divisions had been developed into a systematic cosmography in which most realms of space were supposed to be ordered by forms that were not only symmetrical or rectilinear but patterned on the nonary square or three-by-three grid. How this came to be is unknown, though Cammann speculates that these forms "all seem to have been conceived in a determined effort to apply to the greater world the plan of the simple magic square of three."[5] In any case, the invention of this nine-square formation was customarily attributed to one of the legendary sage-kings of high antiquity, either Yu the Great or Fu Xi, who was said to have "differentiated the nine palaces,"[6] the nonary square. But Fu Xi was believed to have first observed such a pattern on the shell of a turtle emerging from the Luo River. So even though the nine-palace formation, the graphic basis of the perennial Chinese cosmography, was credited to an ancient sage-king, its pattern was supposedly taken from the natural world. It was immanent in the structure of heaven and earth, or at least marked on the shell of a remarkable turtle.

As one of the great civilizing inventions attributed to an ancient sage, the nonary square or three-by-three grid, like other inventions such as agriculture, written language, and herbal medicine, was meant to be applied for the development and improvement of human culture and society. It was not, in other words, simply an aspect of the traditional Chinese worldview; it had policy implications and practical uses. Its judicious application in the agrarian realm, in the form of the "well-field system," for example, was supposed to secure the livelihood of the cultivator, the revenue of the state, and the peace and prosperity of the realm as a whole. The construction

2. Michael Sullivan, *The Arts of China*, 3d ed. (Berkeley and Los Angeles: University of California Press, 1984), 7.

3. David N. Keightley, "The Religious Commitment: Shang Theology and the Genesis of Chinese Political Culture," *History of Religions* 17 (February–May 1978): 211–25, quotation on 221.

4. Kwang-chih Chang, *The Archaeology of Ancient China*, rev. and enl. (New Haven: Yale University Press, 1977), 291.

5. Schuyler Cammann, "The Magic Square of Three in Old Chinese Philosophy and Religion," *History of Religions* 1 (summer 1961): 37–79, quotation on 44. Cammann as well as other historians traces the cosmographical applications of this magic-square form in China to a rather shadowy classical thinker, Zou Yan (305–240 B.C.?), whose original works have not survived. According to the account preserved in the *Shi ji* (Records of the grand historian, completed ca. 91 B.C.) of Sima Qian, Zou Yan held that China, which itself consisted of nine regions, comprised the central eighty-first portion of the world and the central ninth of the middle continent of the world. Small encircling seas separated the nine continents from one another, and a large encircling sea marked the outer rim of the earth where it met the dome of heaven. "Mengzi Xunqing liejuan" (Collected biographies of Mencius and Xunqing), in Sima Qian, *Shi ji*, 74.1b; see *Xinjiao Shi ji sanjia zhu* (Newly collated *Shi ji* with three principal commentaries), 5 vols. (Taipei: Shijie Shuju, 1972), 4:2344. The *Shi ji* is the first in the series of twenty-five official histories, and the most celebrated Chinese historical work.

6. Chen Menglei, Jiang Tingxi, et al., comps., *Gujin tushu jicheng* (Complete collection of books and illustrations, past and present, completed 1726, printed 1728), cat. 2, sec. 1, 43.47a; see the edition in 79 vols. (reprinted Taipei: Dingwen Shuju, 1977), 7:449. This imperially sponsored encyclopedia/anthology is the most comprehensive reference work published in Chinese history. For brief descriptions of this excellent but underutilized source, see Ssu-yü Teng and Knight Biggerstaff, comps., *An Annotated Bibliography of Selected Chinese Reference Works*, 3d ed., Harvard-Yenching Institute Studies 2 (Cambridge: Harvard University Press, 1971), 95–96; and Richard J. Smith, *China's Cultural Heritage: The Ch'ing Dynasty, 1644–1912* (Boulder, Colo.: Westview Press, 1983), 3.

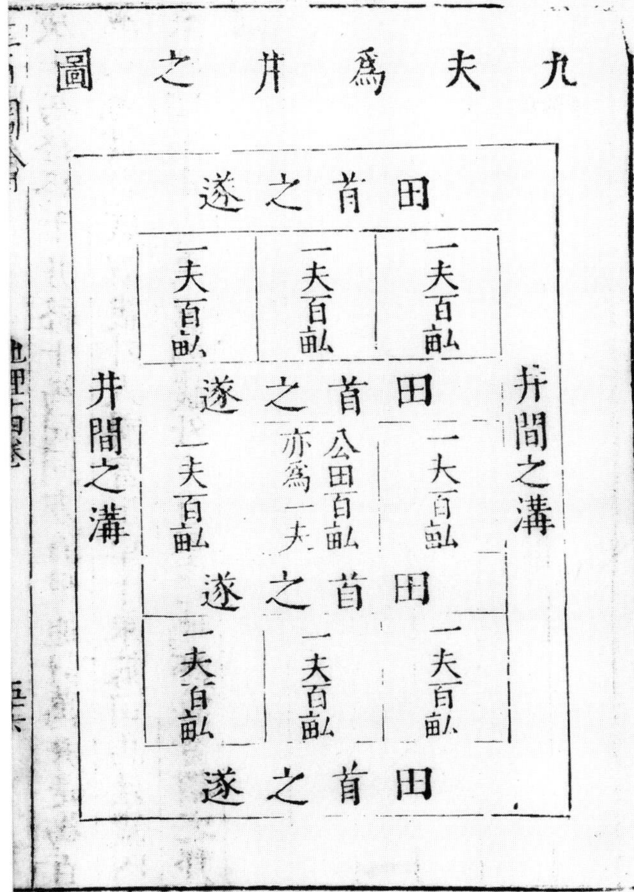

FIG. 8.1. DIAGRAM OF THE NINE UNITS THAT FORM A WELL. This diagram, which appears in a seventeenth-century encyclopedia, illustrates the established "nine-palace" version of the well field. The "public field" occupies the central square, composing one-ninth of the total area. Each of the other eight squares is occupied and cultivated privately by an individual family. Although political reformers often called for the application of this checkerboard arrangement to the division of land, it was seldom implemented in Chinese history.
Size of the original: 21 × 14 cm. From Wang Qi, comp., *Sancai tuhui* (Illustrated compendium of the three powers [heaven, earth, and man], completed 1607, printed 1609), *dili* 14.56a. Reproduced courtesy of the Harvard-Yenching Library, Harvard University, Cambridge.

of an architectural version of the nine-palace formation, moreover, was considered essential for the correct performance of important imperial rituals that harmonized cosmic cycles with human activities.

Thus Chinese reformers, ritualists, and rulers of the Han and later eras applied the nine-palace formation to the ordering of various types of space, ranging from the agrarian to the architectural, with an almost relentless consistency. But in so doing, they often found it necessary to reformulate or reshape earlier schemata designed for ordering space to fit the mold of the nonary square.

Schematic Arrangements of Various Types of Space

The most famous and influential of these schemata was the well-field system, so called because its ideal shape resembles the Chinese character for "well" (井 *jing*), a rough facsimile of the three-by-three grid or nine-palace formation (fig. 8.1). As interpreted by writers in the Confucian tradition, beginning with the great classical philosopher Mencius (372–289 B.C.), this checkerboard schema was supposed to serve as the basis for the mensuration and allotment of agricultural fields. Each family in a group of eight families was to receive one of the nine equal allotments in the arrangement, while the ninth or central unit was to be cultivated in common for the public good. This well-field schema was not, however, simply a system of land mensuration. Mencius spoke of it as the basis of good government.[7] Later Confucian writers came to regard it as "the fundamental institution of the antique social system," which "ensured a rational distribution of land and labor, and removed all cause for strife between rich and poor."[8] Hence Confucian reformers throughout Chinese history repeatedly called for its reinstitution, frequently arguing that a utopian state could be established on such a basis, and even insisting that the three-by-three grid form be applied uniformly in the division of agricultural space. The well-field system served as a vehicle for political and social reform in other East Asian countries besides China, especially in Japan during the epoch-making Taika Reform of the seventh century. Even today the checkerboard pattern of the well-field schema can still be detected throughout a wide area of Japan.[9]

The earliest classical Chinese references to the well-field schema do not, however, give a very precise account of its geometric form. This has led some modern scholars to speculate that it was not originally a mensural scheme at all, much less a geometric formation, but rather a sort

7. *Mengzi yinde* (Concordance to Mencius), Harvard-Yenching Sinological Index Series, suppl. 17 (1941; reprinted Taipei: Chengwen Chubanshe, 1966), 3A, 3.13. For a modern translation of this passage see D. C. Lau, trans., *Mencius* (Harmondsworth, Eng.: Penguin Books, 1970), 99. The *Mengzi*, together with the *Lun yu* (Analects), the *Zhongyong* (Doctrine of the mean), and the *Daxue* (Great learning), compose the Confucian *Sishu* (Four books) canonized by the Neo-Confucian philosophers of the Song era.

8. John W. Dardess, *Confucianism and Autocracy: Professional Elites in the Founding of the Ming Dynasty* (Berkeley and Los Angeles: University of California Press, 1983), 37.

9. John Whitney Hall, *Japan: From Prehistory to Modern Times* (New York: Delacorte Press, 1970), 54.

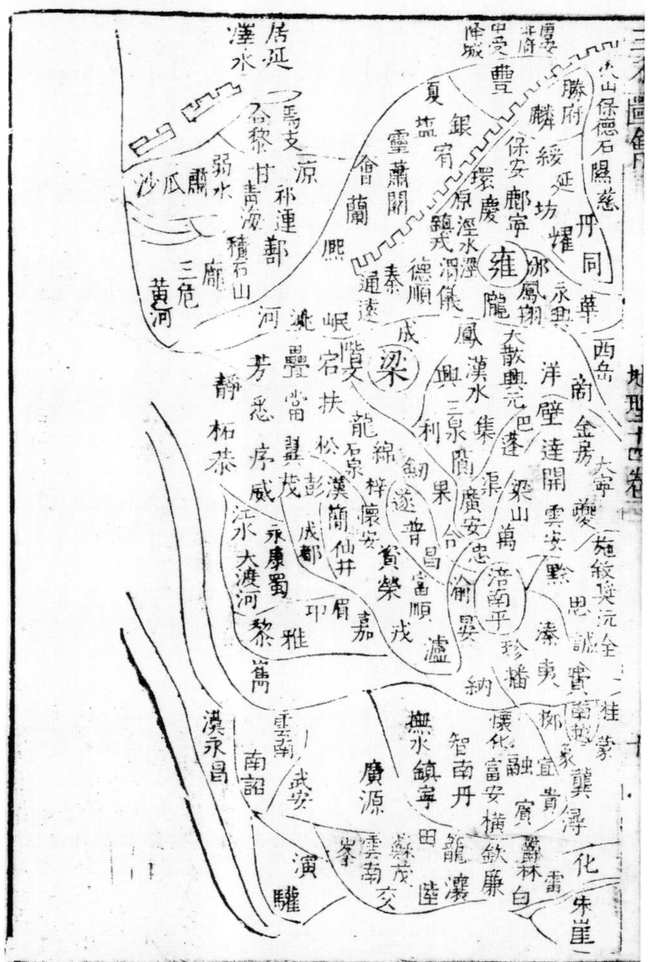

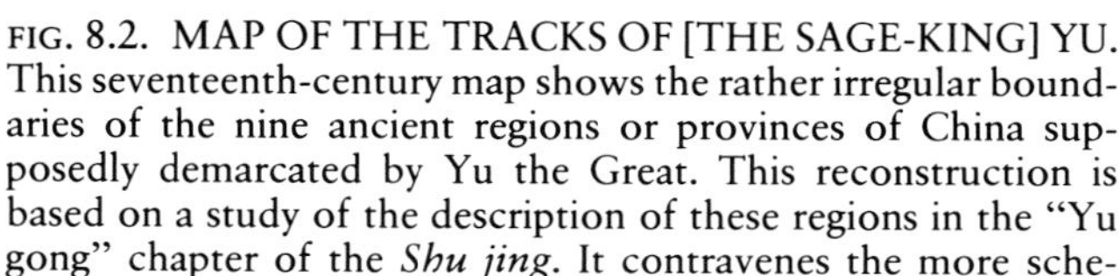

FIG. 8.2. MAP OF THE TRACKS OF [THE SAGE-KING] YU. This seventeenth-century map shows the rather irregular boundaries of the nine ancient regions or provinces of China supposedly demarcated by Yu the Great. This reconstruction is based on a study of the description of these regions in the "Yu gong" chapter of the *Shu jing*. It contravenes the more schematic, geometrized versions of the *jiu zhou* favored by cosmographers of the Han era.
Size of each page: 21 × 14 cm. From Wang Qi, comp., *Sancai tuhui, dili* 14.10ab. Reproduced courtesy of the Harvard-Yenching Library, Harvard University, Cambridge.

of manorial system or perhaps a formula for allocating the produce of the land between the sedentary and swidden members of a community.[10]

Whatever the social or economic origins of the well-field arrangement, by Han times it was conceived primarily as a mensural schema cast in the form of the three-by-three grid or nonary square.[11] Indeed, so closely associated was the well-field system with this nine-palace formation that the great twelfth-century Neo-Confucian philosopher Zhu Xi (1130–1200) posited that all other applications of this grid form, such as the *mingtang* or "luminous hall," a sort of cosmological temple or architectural microcosm, arose from the well field.[12] Needham suggests that the well-field grid may even have helped inspire the idea of a coordinate system in Chinese car-

10. Cho-yun Hsu, *Ancient China in Transition: An Analysis of Social Mobility, 722–222 B.C.* (Stanford: Stanford University Press, 1965), 112; Wolfram Eberhard, *Conquerors and Rulers: Social Forces in Medieval China*, 2d rev. ed. (Leiden: E. J. Brill, 1965), 34–36.

11. See, for example, the commentary under the fifteenth year of Duke Xuan in Fan Ning (339–401), comp., *Chunqiu Guliang zhuan* (Spring and autumn [annals] with the Guliang commentary, fourth century), 7.8b; see the modern edition (Taipei: Xinxing Shuju, 1975), 92. The *Chunqiu* is one of the five Confucian classics (*Wujing*), and the only one supposedly composed by Confucius himself.

12. Zhu Xi in *Li ji jishuo* (Collected explanations of the Record of rituals), annotated Chen Hao (1261–1341), 3.35b–36a, in *Sishu wujing Song-Yuan ren zhu* (Song and Yuan commentaries on the four books and five classics), 3 vols. (Beijing: Zhongguo Shudian, 1984), 2:83–84. The *Li ji* (Record of rituals, ca. first century B.C.), another of the five Confucian classics, is actually an anthology of prose writings of the most diverse sort and provenance, compiled by redactors of the Han era.

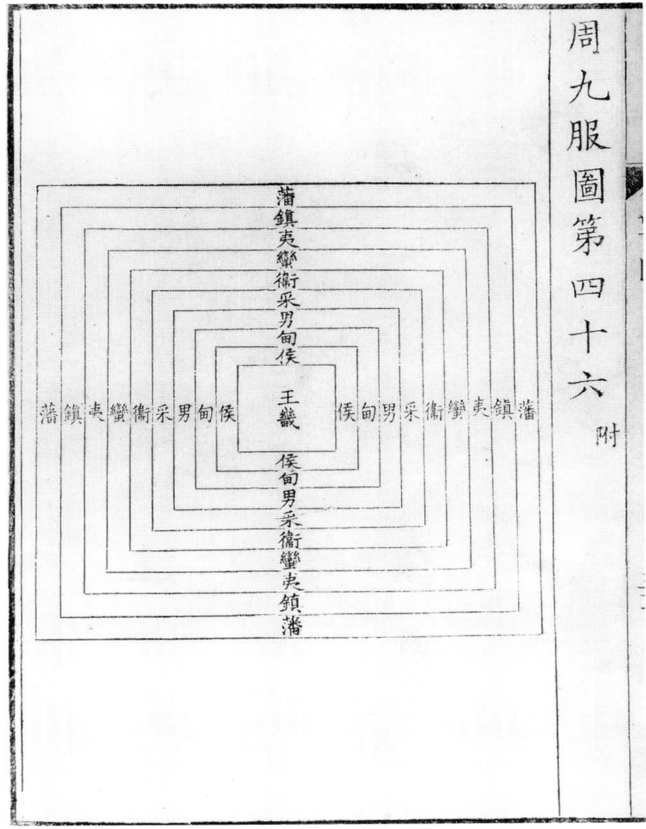

FIG. 8.3. DIAGRAM OF THE NINE DOMAINS OF THE ZHOU. This diagram depicts the type of geometrized political geography devised by Han cosmographers. All the domains in this schema are centered on the royal capital, which occupies the central square. According to this schema, the degree of barbarism increases with the square of the distance from the center, with the lands of the various types of tributaries being closer to the royal domain and the areas inhabited by various grades of barbarians at a greater distance. Like the diagram represented in figure 8.2, this is a seventeenth-century reconstruction based on a reading of a classical text, in this case the *Zhou li* (Ritual forms of Zhou, second century B.C.).
Size of the original: 19.5 × 14.5 cm. From Hu Wei (1633–1714), *Yu gong zhuizhi* (Using an awl to gauge the depths of the *Yu gong*, written 1694–97), from an edition of 1705, *tu* 52b. Reproduced courtesy of the Harvard-Yenching Library, Harvard University, Cambridge.

tography, though he cites no evidence to support this interesting speculation.[13]

Cosmographers of the Han era applied the nonary square to the realm of political geography as well as to the mensuration of agrarian space. In doing so, they found it necessary to resort to more obvious contortions of earlier classical schemata than with the well-field pattern. The primary geographical arrangement outlined in the Confucian canon, the *jiu zhou*—the nine regions or provinces of the realm—is composed of units that are irregularly shaped and spaced. The boundaries of these regions, as described in the "Yu gong" (Tribute of Yu) chapter of the canonical *Shu jing* (Book of documents, dating from at least the late Zhanguo period [403–221 B.C.]), are generally marked by such sinuous and meandering physical features as mountains and rivers (fig. 8.2). However, later accounts of the *jiu zhou*, especially those dating from the third and second centuries B.C., became progressively more schematic and even approach the geometric form of the nonary square. The descriptions of the *jiu zhou* in the early Han compendium the *Huainanzi* ([Book of the] Master of Huainan, written ca. 120 B.C.) attributed to Liu An, for example, locate the nine regions "simply in terms of the eight cardinal directions plus the center."[14] The "Wangzhi" (Royal institutions) chapter of the canonical *Li ji* (Record of rituals, ca. first century B.C.), which also dates from the Han era, remarks that "within the four seas, there are nine regions, each a thousand *li* squared."[15] By conflating the nine regions with the nine domains (*jiu fu*), a series of concentric squares centered on the royal capital as described in another canonical text compiled in the Han era, the *Zhou li* (Ritual forms of Zhou), the later Han commentator Zheng Xuan (127–200) completed the geometrization of the classical *jiu zhou* (fig. 8.3). Thus the influential Tang-era commentator Kong Yingda (574–648) wrote that the nine regions were bounded by straight lines, not by natural physical features.[16]

As with the grid form of the well-field arrangement, this geometrized political geography was not simply an aspect of the traditional worldview or solely a matter for cosmographical speculation. Even as late as the nineteenth century, political reformers in China and Japan proposed to establish squared administrative districts as a prelude to a reign of virtue, or at least the improvement of government.[17] These later reformers, however, were not so much attempting to reimpose a particular cartographic plan, that of the nine regions, as proceeding on the assumption that the geometrization of political districts was an important aspect of serious political reform.

13. Joseph Needham, *Science and Civilisation in China* (Cambridge: Cambridge University Press, 1954–), vol. 3, with Wang Ling, *Mathematics and the Sciences of the Heavens and the Earth* (1959), 541.

14. John S. Major, "The Five Phases, Magic Squares, and Schematic Cosmography," in *Explorations in Early Chinese Cosmology*, ed. Henry Rosemont, Jr. (Chico, Calif.: Scholars Press, 1984), 133–66, quotation on 137.

15. See *Li ji jishuo*, 3.3b, in *Sishu wujing*, 2:67 (note 12).

16. Part 2 of Kong Yingda, "Yu gong wufu tu shuo" (Explanation of the diagrams of the five domains of the Tribute of Yu) in Zhang Huang, comp., *Tushu bian* (Compilation of illustrations and writings, compiled 1562–77), 86.2b, in *Gujin tushu jicheng*, 7:1091 (note 6).

17. See, for example, Hall, *Japan*, 276 (note 9); and William Theodore de Bary, Wing-tsit Chan, and Burton Watson, comps., *Sources of Chinese Tradition* (New York: Columbia University Press, 1960), 728.

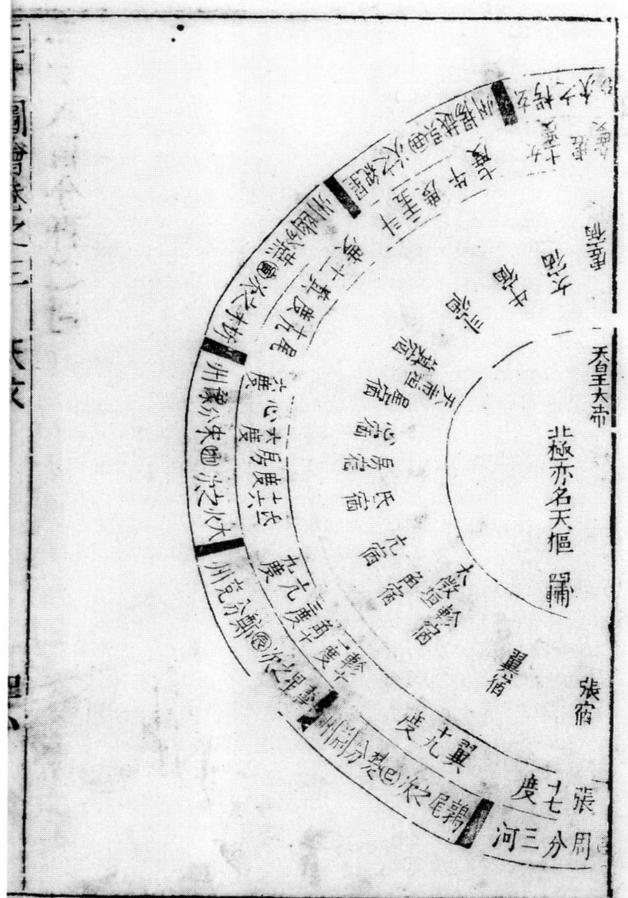

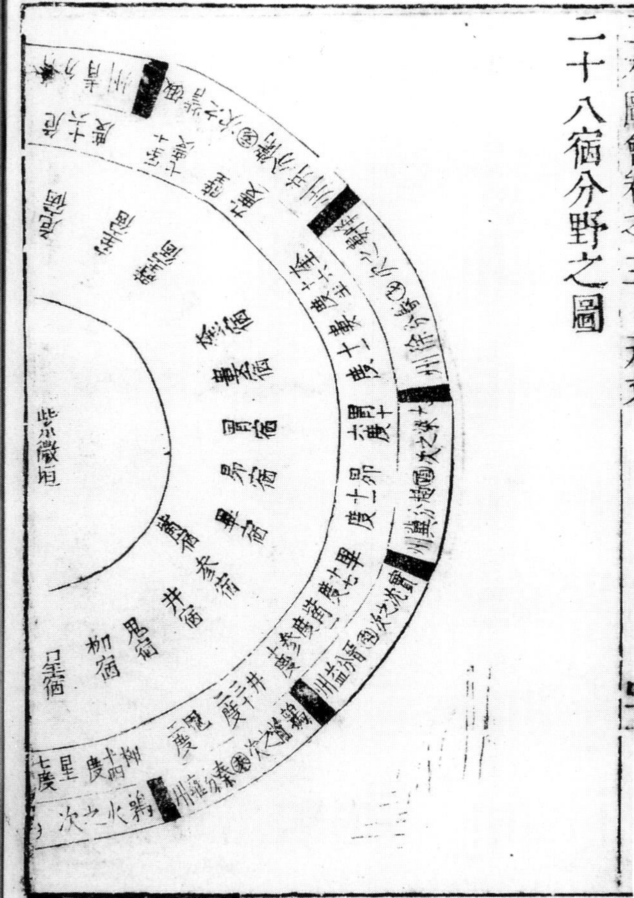

FIG. 8.4. DIAGRAM OF THE FIELD ALLOCATION OF THE TWENTY-EIGHT LUNAR LODGES. This chart shows the correspondences between these sectors of the lunar path through the heavens and the twelve regions or provinces of China. Other versions of the *fenye* correlated the nine terrestrial regions with their celestial counterparts. The main function of this and other *fenye* charts is astrological: an unusual occurrence in a particular heavenly field might portend disaster for an important personage in the corresponding earthly region.
Size of each page: 21 × 14 cm. From Wang Qi, comp., *Sancai tuhui, tianwen* 3.47b–48a. Reproduced courtesy of the Harvard-Yenching Library, Harvard University, Cambridge.

The nine-region schema provided a basis for a third cosmographical conception, the *fenye* or "field-allocation" system, which correlated sectors of the heavens with territories of the earth. To each of the nine geographical regions of China there was "allocated" a corresponding celestial "field." This *fenye* arrangement is a good example of a correlative system, based on the idea that correspondences may be drawn between separate orders of existence or realms of the cosmos.

The origins of the *fenye* system, like those of the well-field arrangement, are obscure. It may well have originated as a cartographic device for tracing the annual progress of the sun, moon, and planets through the heavens.[18] Marking off the band of the ecliptic into a number of sectors would have enabled terrestrial observers to track the movements of these celestial lights in a rough way. If such was indeed the earliest form of the *fenye*, then it would initially have had little to do with nonary cosmography. On the other hand, Wang Yong, a modern historian of Chinese geography, speculates that the *fenye* first arose from the *jiu zhou* or nine-region conception.[19]

Whatever its origins, by Han times the *fenye* had developed into an astrological system pairing earthly regions or states with corresponding heavenly fields or asterisms. The late-classical compendium the *Lüshi Chunqiu* (Master Lü's Spring and autumn [annals]), associated the nine

18. Hashimoto Masukichi, *Shina kōdai rekiho shi kenkyū* (Studies on the history of ancient Chinese calendrical astronomy) (Tokyo: Tōyō Bunko, 1943), 515–17.

19. Wang Yong, *Zhongguo dilixue shi* (History of geography in China) (1938; reprinted Taipei: Shangwu Yinshuguan, 1974), 12.

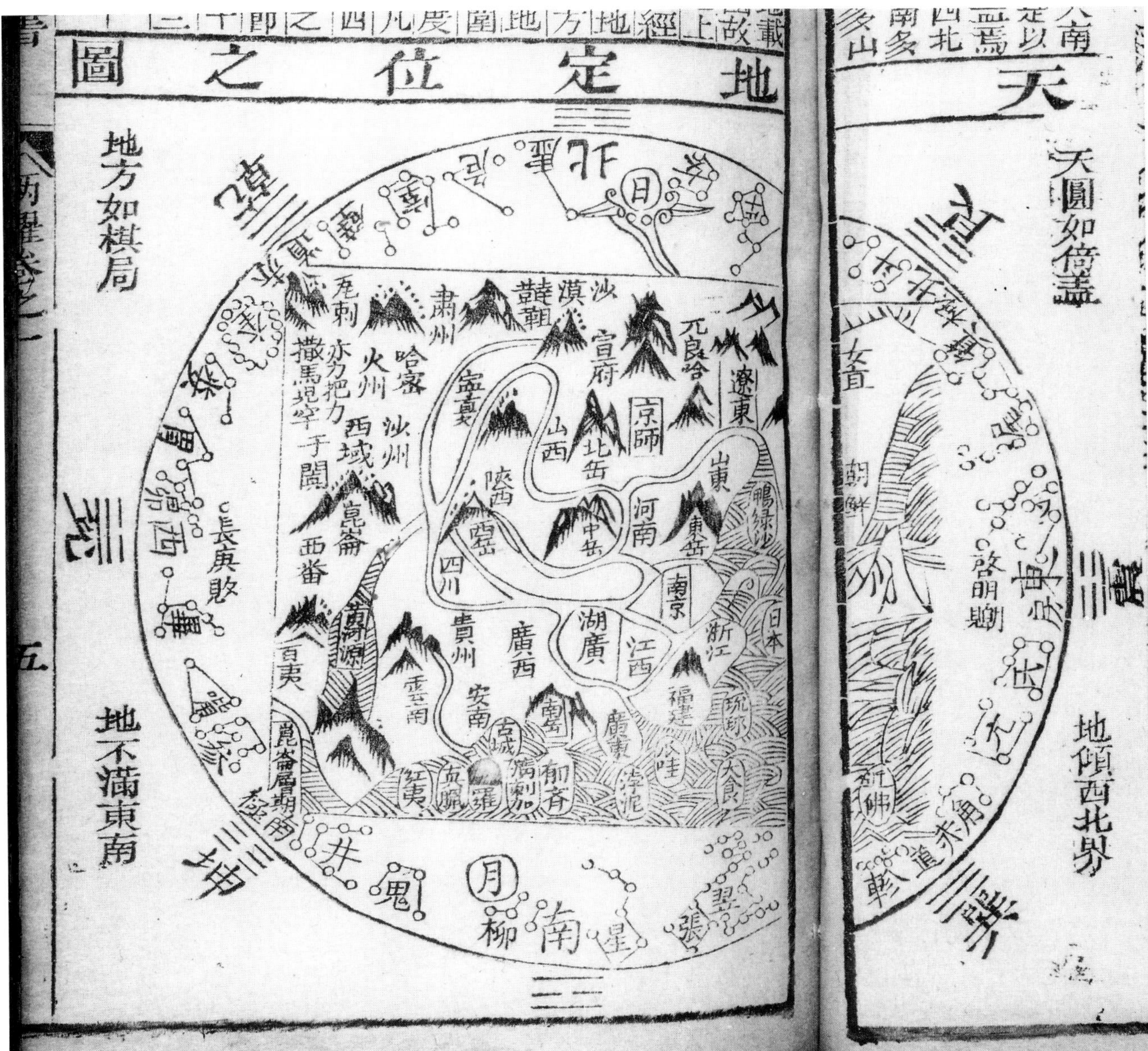

FIG. 8.5. A *FENYE*-INFLUENCED DIAGRAM OF THE "ROUNDED HEAVENS" CIRCUMSCRIBING THE "SQUARED EARTH." This map shows the more modern provinces of China with surrounding tributaries and barbarian regions within the central square, and the twenty-eight lunar lodges, divided into four groupings of seven each, on the peripheral circle. This chart, however, does not specify which lunar lodges correspond to which terrestrial regions. But it does illustrate another important Chinese cosmographical idea, that "the earth is inclined toward the northwest." Thus "the Southeast is mostly watery and the Northwest mostly mountainous." Size of the original: ca. 15.5 cm. From *Xinzeng xiangji beiyao tongshu* (Newly amplified almanac of auspicious images, 1721), 1.4b–5a. By permission of the Oriental and India Office Collections, British Library, London (15257 a 24).

classical regions with nine celestial fields.[20] Later Han texts like the *Huainanzi* correlated earthly territories with other astronomical schemata such as the twenty-eight lunar lodges, the twelve stations of the Jupiter cycle, and the nine paths of the moon (figs. 8.4 and 8.5). An untoward event in any one of these celestial sectors boded ill for the political authorities in the corresponding terrestrial region. A simple example of a *fenye* prognostication or verification, drawn from the "Wuxing zhi"

20. *Lüshi Chunqiu* (ca. third century B.C.), commissioned by Lü Buwei, *Sibu beiyao* edition, 13.1a–b.

(Treatise on the five phases) of the official Han history *Han shu*, is as follows:

> In the seventh month of the third year of the reign of Gaodi [204 B.C.], there was a comet in the *Da jiao* [Great Horn] asterism that disappeared after a little more than ten days. Liu Xiang [77–6 B.C.] suggests that at that time Xiang Yu was the king of Chu and the hegemon of the feudal lords, but that [his rival] the Han had already pacified the three Qin [states] and was separated from Xiang Yu only by the district of Yingyang. Thus the hearts of all in the realm were shifting to the Han, and Chu was about to disintegrate. So the comet purged the royal throne [in the heavens].[21]

The political application of the *fenye* astrological schema required that the boundaries of the corresponding celestial and terrestrial fields be measured with precision. Thus, Chen Zhuo in the third century listed the dimensions of such fields by number of degrees.[22]

Like the well-field and nine-region schemata, the field-allocation system was not confined to the ethereal realms of cosmological speculation. It had practical political uses as well. In fact, most imperial Chinese dynasties from Han times employed official astrologers charged with making political prognostications based on their observations of celestial abnormalities.[23] When such prognostications were taken seriously by those in positions of authority, changes in the boundaries of corresponding celestial and terrestrial sectors could influence the conduct of government. Thus it is not surprising that leading cosmographers, astronomers, and even historians of the Tang (618–907) and Song eras, particularly Li Chunfeng (602–70), Yixing (682–727), and Ouyang Xiu (1007–70), were seriously concerned with proposals to reform the boundaries of the terrestrial divisions in the *fenye* system, as well as to adjust the celestial divisions to new astronomical measurements.[24] Such proposals are not known to have resulted in new land surveys and maps, however. Their purpose was, after all, to bring the ancient *fenye* conceptions into accord with *existing* terrestrial boundaries.

Classical Chinese cosmographical conceptions left a more permanent, palpable mark on urban configurations—the shape of Chinese cities—than on any other arrangement of space. Some of the great capital cities of imperial China, including Beijing, were built or reformed to accord with cosmographical strictures, particularly orientation of the city walls and gates to the four points of the compass and construction on a rectilinear plan very nearly resembling a square. Whereas cosmographers in several other Eurasian civilizations generally only speculated on the form of the ideal city, some of their Chinese counterparts actually saw reasonable facsimiles of their ideal urban designs constructed. Chinese cosmographical models also influenced urban planning in other East Asian countries, especially Korea and Japan. The two imperial capitals of premodern Japan, Nara and Kyōto, were modeled on the rectilinear grid form of the great capital of Tang-era China, Chang'an.

However, cosmographical conceptions apparently had little influence on the general shape of Chinese cities until at least the end of the Han era. According to Wright, early Chinese cities seem to have been irregular and asymmetrical in form. This was true even of the first great imperial capital in Chinese history, Chang'an in the Former Han era (206 B.C.–A.D. 8).[25]

A canonical text that probably dates from this era, the "Kaogong ji" (Artificers' record) chapter of the *Zhou li*, however, established the basic urban plan that greatly influenced the design of later imperial capitals. In addition to prescribing cardinal axiality and orientation, this text explains that "the builders should build the capital city as a square of nine square *li*. On each side of the city walls there are three gates. Within the city there are nine meridional and nine latitudinal streets, each of the former being as wide as nine chariot-tracks."[26] This classical plan of the ideal city thus incorporates nonary numerology, as well as a squared perimeter, into its design, though it does not prescribe that the city be patterned after the three-by-three grid. Wheatley, however, has suggested that the nine streets in this text originally designated nine

21. "Wuxing zhi" in Ban Gu, *Han shu* (History of the Former Han, compiled first century A.D.), 27C.26b–27a, in *Xinjiao Han shu jizhu* (Newly collated *Han shu* with collected commentaries), 5 vols. (Taipei: Shijie Shuju, 1973), 2:1516. The *Han shu* is the second in the series of twenty-five official histories, all but the first covering one particular dynasty, which in the case of the *Han shu* is the Former Han era (206 B.C.–A.D. 8).

22. Ho Peng-yoke, trans. and annotator, *The Astronomical Chapters of the Chin Shu* (Paris: Mouton, 1966), 113.

23. See, for example, Ho Peng-yoke, "The Astronomical Bureau in Ming China," *Journal of Asian History* 3 (1969): 137–57, esp. 144.

24. Part 2 of "Tianwen zhi" (Treatise on astrology) in Liu Xu et al., *Jiu Tang shu* (Old history of the Tang, compiled 940–45), *juan* 36, in *Lidai tianwen lüli deng zhi huibian* (Collected treatises on astrology, astronomy, and harmonics in the standard histories), 9 vols. (Beijing: Zhonghua Shuju, 1976), 3:673; part 1 of "Tianwen zhi" (Treatise on astrology) in Ouyang Xiu et al., *Xin Tang shu* (New history of the Tang, compiled 1032?–60), *juan* 31, in *Lidai tianwen*, 3:718–19, 722. The *Jiu Tang shu* and the *Xin Tang shu* are the official histories covering the Tang era (618–907).

25. Arthur F. Wright, "The Cosmology of the Chinese City," in *The City in Late Imperial China*, ed. George William Skinner (Stanford: Stanford University Press, 1977), 33–73, esp. 42–44.

26. Zheng Xuan, *Zhou li Zhengzhu* (Ritual forms of Zhou with Zheng's commentary, second century), *Sibu beiyao* edition, 41.14b. I have followed the translation of this passage as given in Laurence J. C. Ma, "Peking as a Cosmic City," in *Proceedings of the 30th International Congress of Human Sciences in Asia and North Africa: China 2*, ed. Graciela de la Lama (Mexico City: El Colegio de México, 1982), 141–64, quotation on 144.

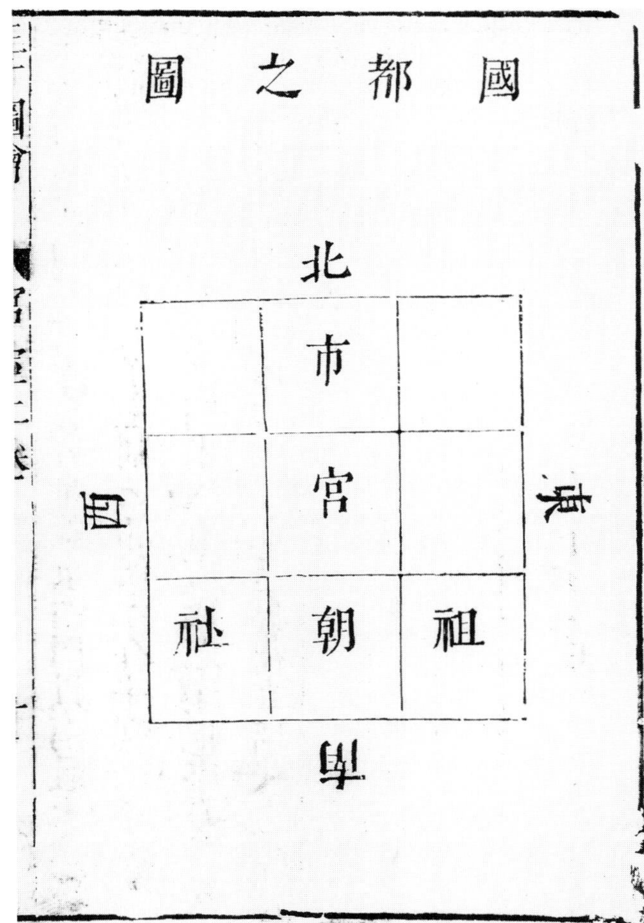

FIG. 8.6. DIAGRAM OF THE NATIONAL CAPITAL. This shows the form of the capital city modeled on the nine-palace plan and oriented to the four quarters, with the imperial palace occupying the central square. The description of the ideal capital in the canonical *Zhou li* (Ritual forms of Zhou, second century B.C.), however, refers to nine streets rather than nine squares or sectors of the city.
Size of the original: 21 × 14 cm. From Wang Qi, comp., *Sancai tuhui, gongshi* 2.11a. Reproduced courtesy of the Harvard-Yenching Library, Harvard University, Cambridge.

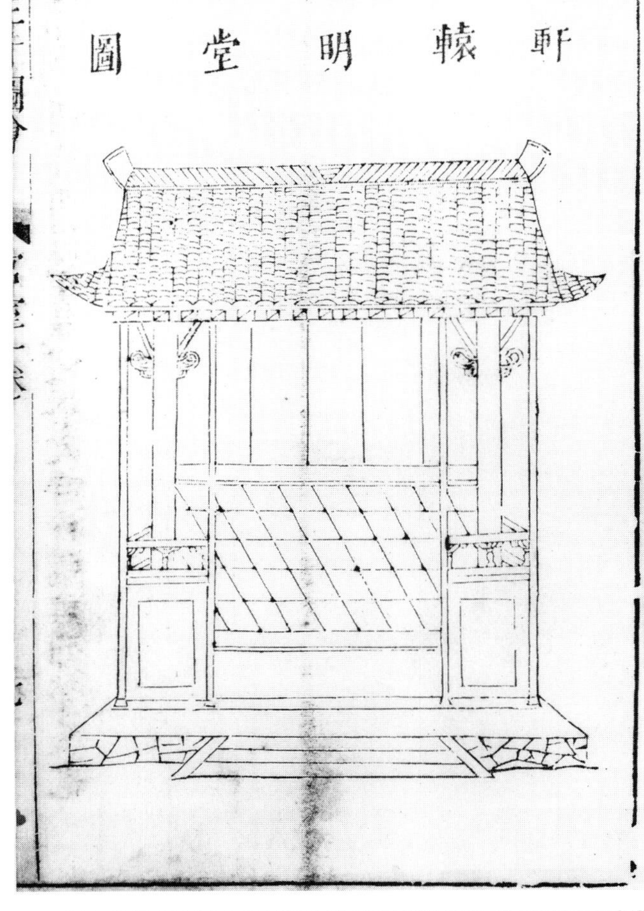

FIG. 8.7. DIAGRAM OF XUAN YUAN'S (OR THE YELLOW EMPEROR'S) LUMINOUS HALL. This seventeenth-century diagram purports to show the earlier, more pristine and less schematic version of the *mingtang*. Conceptions of the *mingtang* favored by Han cosmographers were, however, based on nonary cosmography, as illustrated below in figure 8.8.
Size of the original: 21 × 14 cm. From Wang Qi, comp., *Sancai tuhui, gongshi* 1.7a. Reproduced courtesy of the Harvard-Yenching Library, Harvard University, Cambridge.

squares. In that case, he speculates, "the ideal-type city should have originally comprised a regular nonary layout of eight sectors, pivoted about a central unit consisting of one-ninth of the total area."[27] In other words, the nine-palace formation may at one time have been applied to the plan of the ideal city as well as to the well-field and nine-region configurations (fig. 8.6).

The classical city plan was, however, actually realized in the designs of some of the great imperial capitals of Chinese history, such as Chang'an in Tang times and Beijing in the Ming (1368–1644) and Qing (1644–1911) eras, though such obstacles as existing settlements, topographical irregularities, and geomantic requirements did necessitate some adjustments. This plan, like its counterparts in several other Eurasian civilizations, was designed to illustrate and establish the centrality of the ruler, the Son of Heaven who sat at the "pivot of the four quarters" and mediated between heaven and earth. Hence emperors throughout Chinese history often concerned themselves with the geometric layout and numerological symbolism of the capital and its cultic sites and structures, just as

27. Paul Wheatley, *The Pivot of the Four Quarters: A Preliminary Enquiry into the Origins and Character of the Ancient Chinese City* (Chicago: Aldine, 1971), 414.

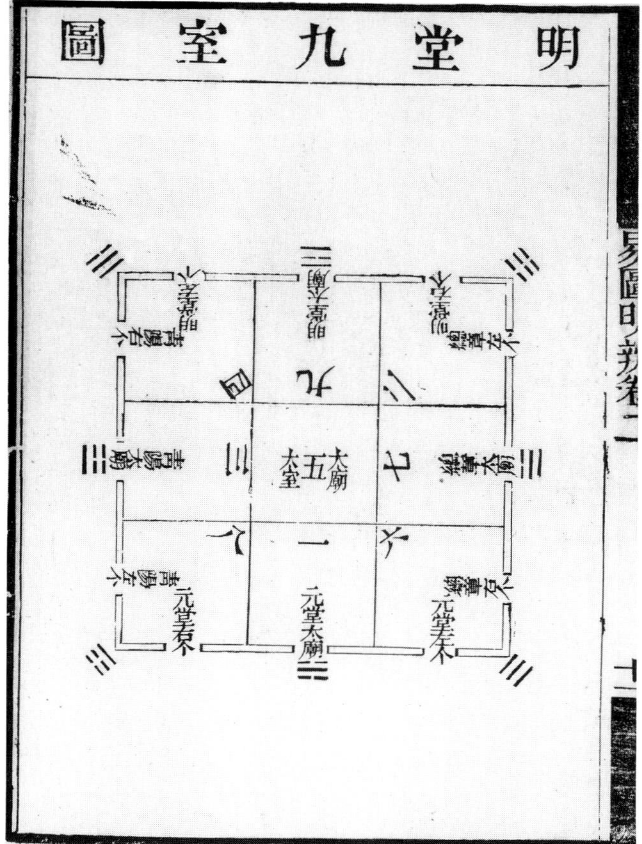

FIG. 8.8. DIAGRAM OF THE NINE CHAMBERS OF THE LUMINOUS HALL. This diagram outlines the standard plan of the *mingtang*, correlating the chambers with magic-square numbers and the eight trigrams associated with the *Yi jing* (Book of changes). Only the central chamber, the "Grand Temple" or "Grand Chamber," lacks a corresponding trigram. This seventeenth-century reconstruction by a famous classical scholar is based on his study of Han cosmographical texts, since no diagrams of the plan of the *mingtang* survived from this earlier period.
Size of the original: 20 × 13.5 cm. From Hu Wei, *Yi tu mingbian* (A clarifying critique of the diagrams [associated with the Book of] changes, 1706), from an edition of 1843, 2.11b. Reproduced courtesy of the Harvard-Yenching Library, Harvard University, Cambridge.

they did with agrarian reform along the lines of the well-field schema.[28]

Perhaps the most important and famous cultic site or structure was the *mingtang* or "luminous hall," a sort of cosmological temple designed for the performance of imperial rites, as well as "an architectural symbol of virtuous government" (fig. 8.7).[29] Unlike the other cosmographical schemata that came to be ruled by the nine-palace formation, particularly the well-field system and the nine regions, there is no clear or detailed description of the *mingtang* in any pre-Han text. The references to this institution in earlier classical sources, such as the *Mengzi* and the *Zuozhuan* (Zuo's tradition [of interpreting the *Chunqiu*], ca. 300 B.C.), are both brief and cryptic, giving no account of its form and little even of its functions. These early sources do not associate the *mingtang* with nonary cosmography, or indeed with cosmography of any sort.

The dearth of detailed descriptions of the *mingtang* in earlier classical texts left a clear field for Han-era scholars to recreate the structure along the lines of their own cosmographical conceptions. There was little to limit the free play of Han cosmographical fancies regarding the *mingtang*, as there was with the nine-region conception, for which an inconveniently detailed classical description existed. Hence, postclassical students of the *mingtang* devised more than twenty different versions of this architectural microcosm.[30] However, the most popular postclassical reconstructions of the *mingtang* were those based on quinary and nonary cosmography and numerology, particularly the latter (fig. 8.8).

Inasmuch as architectural structures are generally more amenable to the expression and elaboration of cosmographical ideas than are agricultural fields, political divisions, and even city plans, these reconstructed *mingtang* extend such conceptions to a greater length than any of the nonary schemata discussed thus far. Cai Yong (133–92), a later Han expositor of the nonary interpretation of the *mingtang*, described the structure as a complete architectural microcosm. Its nine rooms, for example, represented the nine regions, its twelve palaces resonated with the twelve hours of the day, its twenty-eight pillars symbolized the twenty-eight lunar lodges, and its eight inner passages were paired with the eight trigrams of the *Yi jing*. Its base was square, the shape of the earth, its round roof resembled the heavens, and the whole structure was surrounded by water that represented the four seas.[31]

The complex form of the *mingtang* is not architecturally feasible, as Maspero has pointed out.[32] Yet impe-

28. For an example of such imperial concern, see Jeffrey F. Meyer, *Peking as a Sacred City* (Taipei: Chinese Association for Folklore, 1976), 109.

29. Howard J. Wechsler, *Offerings of Jade and Silk: Ritual and Symbol in the Legitimation of the T'ang Dynasty* (New Haven: Yale University Press, 1985), 195.

30. Wang Meng'ou, "Gu mingtang tu kao" (An investigation of the plan of the ancient luminous hall), in Li Yuegang et al., *Sanli yanjiu lunji* (A collection of articles on the three ritual classics) (Taipei: Liming Wenhua Shiye, 1981), 289–300, esp. 295.

31. "Mingtang Yueling lun" (A discussion of the luminous hall [in connection with the] Monthly ordinances) in Cai Yong, *Cai Zhonglang ji* (Collected writings of Cai Zhonglang [Yong], second century), *Sibu beiyao* edition, 10.6a–b.

32. Henri Maspero, "Le Ming-t'ang et la crise religieuse chinoise avant les Han," *Mélanges Chinois et Bouddhiques* 9 (1948–51): 1–70, esp. 66–67.

rial sovereigns throughout Chinese history, beginning with the emperor Wu (r. 140–87 B.C.) of the Former Han era, commissioned the construction of *mingtang*. Some of these, such as that built by the usurper Wang Mang (45 B.C.–A.D. 23) in A.D. 4 and that reconstructed by the Later Han emperor Guangwu (r. 25–57), were apparently modeled as closely as possible on the nonary plan outlined in the Han compendium the *Da Dai Li ji* (The Elder Dai's record of rites). There is archaeological as well as literary evidence that such architectural microcosms were constructed as early as Han times, for an elaborate building, said to be the ruins of Wang Mang's *mingtang*, recently excavated just south of the Han-era capital of Chang'an, was divided into nine rooms situated on a square base and surrounded by a circular watercourse.[33]

Why did some of the most renowned, energetic, and practical imperial sovereigns in Chinese history concern themselves with the construction of a cosmographically and numerologically proportioned architectural microcosm? One of the main reasons was that it provided the appropriate setting for the Son of Heaven's ritual ordering of space and time, especially of the seasons of the year. By moving through the *mingtang* at the appropriate pace, the ruler helped to ensure that the seasons would follow their proper order.

But for this ritual tour of the *mingtang* to be efficacious, the dimensions and parts of this architectural microcosm had to correspond with the larger patterns and rhythms of the cosmos. These included such basic markers and divisions of space and time as the nine regions of the realm, the twelve months of the year, the twenty-four solar periods, and the twenty-eight lunar lodges, and such configurations as the square base (corresponding to earth) and the round roof (corresponding to heaven). Hence debates over the proper form and dimensions of the *mingtang*, which often arose at court in response to an imperial commission to construct such a structure, were not merely antiquarian exercises.[34] For inasmuch as the *mingtang* was supposed to be constructed as an architectural microcosm, the subject of dispute was ultimately the shape and proportions of the cosmos in general, not simply the optimum measurements of a building. Hence it is not surprising that some of the most important figures in Chinese intellectual history, such as Dong Zhongshu (ca. 179–ca. 104 B.C.), Zhu Xi (1130–1200), Wang Fuzhi (1619–92), and Kang Youwei (1858–1927), wrote on the subject of the *mingtang*, though they did not include in their extant works charts or graphs of its proper plan.

The cosmographical pattern or model that inspired the most extensive debate and speculation in traditional China was not, however, the *mingtang*, but the *Luo shu* (Luo River writing) and its companion, the *Hetu* (Yellow River chart). The importance of these schemata in Chinese thought of premodern times is illustrated by the large section devoted to them in the great Chinese encyclopedia the *Gujin tushu jicheng* (Complete collection of books and illustrations, past and present, 1726), in which materials on the *Hetu* and *Luo shu* occupy more space than that given to some of the individual Confucian classics. Indeed, these figures were the pivots of cosmological discourse in the Neo-Confucianism of the Song era, the most influential intellectual movement in the past thousand years of Chinese history. Moreover, later scholars' exposé of the heterodox origins of commonly accepted versions of the *Hetu* and *Luo shu* was one of the most important episodes in the intellectual history of the Qing period.

Unlike the cosmographical conceptions discussed above, neither the *Luo shu* nor its complement, the *Hetu*, was used directly as a model for ordering any type of physical space—astronomical, geographical, agrarian, urban, or architectural. But the *Luo shu* in particular was widely regarded as the original source of nonary cosmography and even of the nine-palace formation. Later commentators on the *Hetu* and *Luo shu*, moreover, interpreted these figures as the paradigms not just of nonary cosmograms, but of all *tu*, a Chinese character that might designate almost any form of graphic representation including charts, diagrams, maps, and illustrations in general. These *tu*, supposedly devised or revealed by the sage-kings of high antiquity, were among the primary sources of order in the world, as important in this regard as the books of the Confucian canon.

As with the *mingtang*, there are no clear or detailed descriptions of either the *Hetu* or *Luo shu* in classical Chinese literature of the pre-Han era, a situation that promoted their cosmological metamorphosis. Lacking precise classical accounts, Han cosmologists could give free rein to their imaginations in reconstituting and reinterpreting these figures. On the other hand, they could not find much support for their speculative reconstructions in classical literature. Of all the extant texts that may be reasonably dated from the pre-Han era, none associates the *Hetu* or *Luo shu* with any particular geometric configuration or numerological system. These early references generally limit themselves to relating that these figures emerged from sacred rivers and that they were good auspices.[35]

Postclassical interpretations of these references, especially those devised in the Han era, often identified the

33. Wright, "Cosmology of the Chinese City," 51 (note 25).

34. For a brief account of one such debate, which arose in the seventh century, see Wechsler, *Offerings*, 207 (note 29).

35. Bernhard Karlgren, "Legends and Cults in Ancient China," *Bulletin of the Museum of Far Eastern Antiquities*, no. 18 (1946): 199–365, esp. 273.

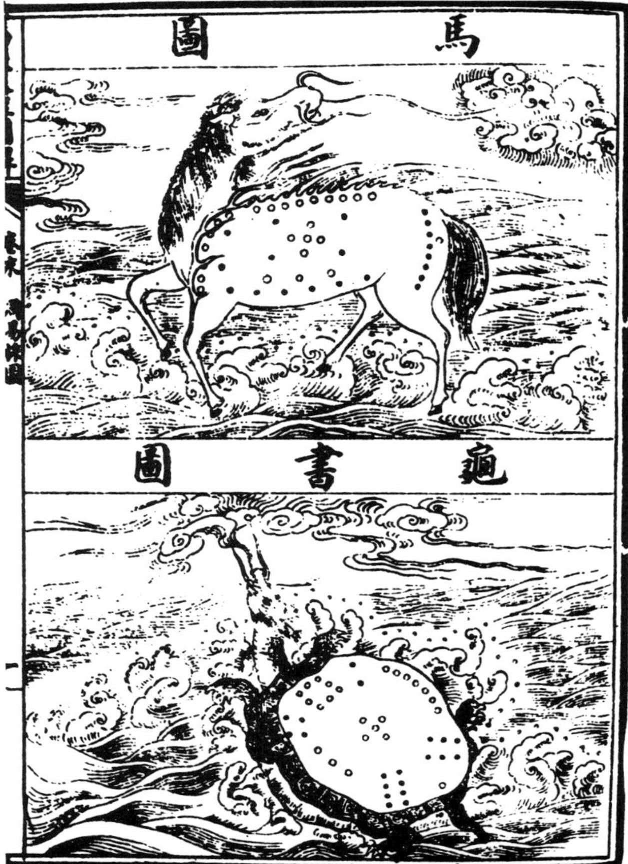

FIG. 8.9. ILLUSTRATION OF A LEGENDARY ACCOUNT OF THE ORIGINS OF THE YELLOW RIVER CHART (*HETU*) AND LUO RIVER WRITING (*LUO SHU*). This diagram is included in a sixteenth-century commentary on the *Yi jing*. The *Hetu* was supposed to have been imprinted on the side of a horse wading in the Yellow River, and the *Luo shu* on the shell of a turtle emerging from the Luo River.
Size of the image: unknown. From Lai Zhide, *Yi jing Laizhu tujie* (Explanations of the diagrams [associated with] the Book of changes with Lai's commentary, 1598) (Taipei: Guangtian Chubanshe, 1975), 1a.

Hetu with the eight trigrams of the *Yi jing* and the *Luo shu* with the "nine categories" or "nine sections" (*jiuchou*) of the "Hongfan" (Great plan) chapter of the *Shu jing* (fig. 8.9).[36] Thus the "Wuxing zhi" chapter of the *Han shu* remarks that the Han scholar Liu Xin (ca. 50 B.C.–A.D. 23),

> held that [the legendary sage-king] Fu Xi continued [the work of] heaven as king, [and thus] received the Yellow River chart; [what he] drew on its pattern was the eight trigrams [of the *Yi jing*]. [The legendary sage-king] Yu regulated the floodwaters and thus was favored with the Luo River writing; [what he] set out on its model was the "Great Plan" [chapter of the *Shu jing*].[37]

Later commentators went on to identify the *Hetu* with roundness and the order of the heavens and the *Luo shu* with squareness and the mensuration of the earth.[38]

The association of the *Luo shu* with the nine-palace formation as well as with nonary numerology and geographical order may also have arisen as early as the Han era.[39] Postclassical commentators on the *Luo shu* even identified this figure with the simple magic square of three. On the other hand, the standard diagram of the *Luo shu*, that favored particularly by the Neo-Confucian cosmologists of the Song era, both dispenses with the squared outline and represents each of the magic-square numbers by figures that resemble knotted cords rather than by numerals (fig. 8.10). Since no graphic representation of the *Luo shu* survives from earlier than the Song, it is difficult to assess the relative antiquity and authority of these different forms of the *Luo shu*. Nor is it easy to reconstruct any genealogy of charts typed as "*Luo shu*," though different versions did exist.

In sum, few of the cosmographical conceptions outlined above—the well-field, nine-region, field-allocation, classical city plan, *mingtang*, and *Luo shu* schemata—could be fit so clearly and perfectly into the nine-palace template as to inspire universal approbation. Indeed, debates on the correct form of some of these cosmograms frequently arose in the postclassical era of Chinese history. Nevertheless, the pervasive influence of this nonary grid form on the cosmographical theory and practice of Chinese civilization is remarkable. Nor was this influence confined to the schemata just named. It may be traced in such cosmographical artifacts as the bronze "TLV" mirrors dating from the Han era, which depict a

36. "Wuxing zhi" in *Xinjiao Han shu jizhu*, 27A.1a–b (2:1315) (note 21); Kong Anguo (fl. ca. 100 B.C.), *Shang shu Kongzhuan* (Documents of antiquity [Book of documents] with Kong's commentary), *Sibu beiyao* edition, 7.lb; Li Daoping, ed., *Zhou yi jijie zuanshu* (Collected commentaries and annotations on the Zhou change [Book of changes]) (Taipei, 1967), 8.793. The *Yi jing*, probably the most complex and controversial of the five Confucian classics, was originally a manual for divination and a collection of omens and anecdotes to which were later appended philosophical and cosmological commentaries ascribed to Confucius.

37. "Wuxing zhi" in *Xinjiao Han shu jizhu*, 27A.1a–b (2:1315) (note 21).

38. Zhu Xi, *Yixue qimeng* (Primer on the study of the [Book of] changes, 1186), 1.3a–b; see the modern edition (Taipei: Guangxue She Yinshuguan, 1975), 1.5–6.

39. Cammann, "Magic Square of Three," 43–44, 61–64 (note 5). However, the Qing cosmological critic, Zhang Huiyan (1761–1802), dated the association of the *Luo shu* with the nine-palace form from a later period, the Eastern Jin era (317–420). See Zhang's *Yi tu tiaobian* (Systematic critique of the diagrams associated with the [Book of] changes, ca. 1800), 3b, in *Zhang Huiyan yixue shishu* (Ten books by Zhang Huiyan on the study of the [Book of] changes), 2 vols. (Taipei: Guangwen Shuju, 1970), 2:958.

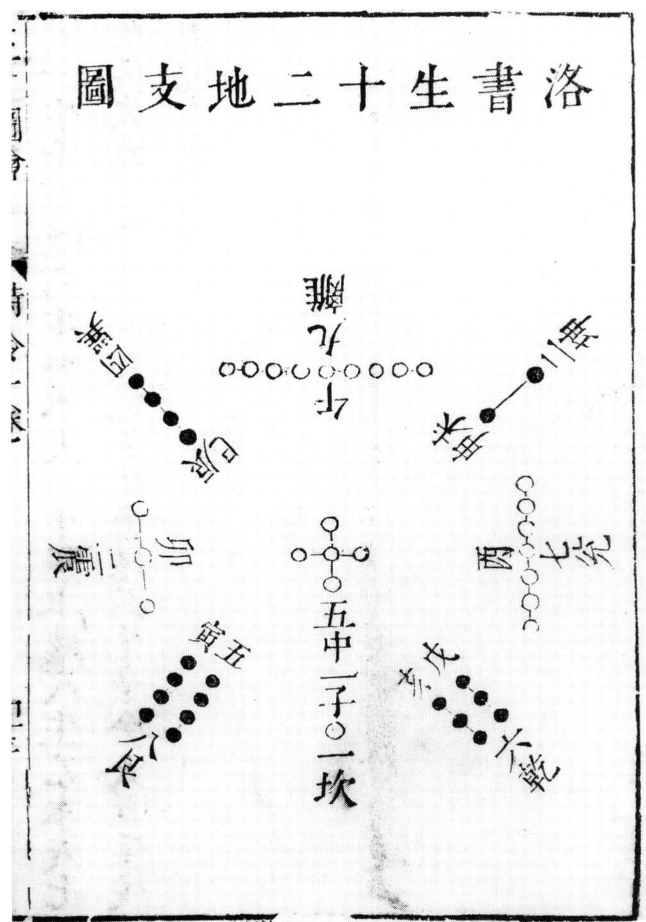

FIG. 8.10. DIAGRAM OF THE *LUO SHU* GIVING BIRTH TO THE TWELVE TERRESTRIAL BRANCHES. This diagram correlates the twelve terrestrial branches, a series of chronograms, with magic-square numbers and with the eight trigrams associated with the *Yi jing*. The numbers are presented not as numerals but in the form of knotted cords. Although the squared outline of the nine-palace form does not appear here, the general configuration of the chart is nonary. Diagrams like this were supposed to have been transmitted secretly from antiquity, but no extant graphic representation of the *Luo shu* dates from before Song times (960–1279). The version shown here is that favored by Song Neo-Confucian cosmologists.
Size of the original: 21 × 14 cm. From Wang Qi, comp., *Sancai tuhui, shiling* 1.43a. Reproduced courtesy of the Harvard-Yenching Library, Harvard University, Cambridge.

rounded heaven encompassing a squared earth. According to Cammann, even nine divisions, representing the nine ancient regions of China, can be projected on the face of some of these cosmic mirrors.[40] Similar patterns appear in the checkered formation of the Han diviner's board as well as the gaming boards used for various types of "chess."

Nonary cosmography in premodern China was not, moreover, the exclusive preserve of orthodox thinkers associated with the Confucian or classical traditions. The nine bureaus of the Daoist hell were sometimes represented in the form of a nonary square, as were some Buddhist mandalas.[41] The Daoist altar used in rites of cosmic recreation also incorporated the nine-palace arrangement along with other cosmological proportions and symbols.[42]

At a more popular level, the influence of the schematic cosmography sketched above may be traced even in Chinese domestic architecture. As Wang remarks, "The typical home in rural Taiwan is . . . constructed, as nearly as possible, to conform to an ideal of perfect symmetry."[43] The second-floor plans of many typical Chinese houses, moreover, approximate the form of the nine-palace model.[44]

Although literary evidence concerning popular cosmographical thought in China is sparse, there is thus some reason to suspect that common people in late traditional China were more concerned and familiar with the cosmographical conceptions articulated by the cultured elite than were their counterparts in medieval Western and Middle Eastern civilizations. Such conceptions were widely circulated through the ubiquitous Chinese almanac (*lishu*), probably the most popular book in late traditional China.[45] They might also have been purveyed through the medium of diviners, many of whom used variations and applications of the nine-palace formation,

40. Schuyler Cammann, "The 'TLV' Pattern on Cosmic Mirrors of the Han Dynasty," *Journal of the American Oriental Society* 68 (1948): 159–67.

41. For a good illustration of a nine-checkered mandala, see Minoru Kiyota, *Shingon Buddhism: Theory and Practice* (Los Angeles: Buddhist Books International, 1978), 93–94.

42. Kristofer M. Schipper and Wang Hsiu-huei, "Progressive and Regressive Time Cycles in Taoist Ritual," in *Time, Science, and Society in China and the West*, Study of Time, vol. 5, ed. J. T. Fraser, N. Lawrence, and F. C. Haber (Amherst: University of Massachusetts Press, 1986), 185–205, esp. 192 and 201; graphic illustrations of this cosmic altar appear on 190 and 191. Other examples of nonary cosmography and numerology in Shangqing Daoism, including the "nine palaces in the head," are given in Livia Kohn, *Early Chinese Mysticism: Philosophy and Soteriology in the Taoist Tradition* (Princeton: Princeton University Press, 1992), 110.

43. Wang Sung-hsing, "Taiwanese Architecture and the Supernatural," in *Religion and Ritual in Chinese Society*, ed. Arthur P. Wolf (Stanford: Stanford University Press, 1974), 183–92, quotation on 183.

44. A sketch of the second floor of a "typical Chinese house" appears in Stephen Skinner, *The Living Earth Manual of Feng-Shui: Chinese Geomancy* (London: Routledge and Kegan Paul, 1982), 110. A similar diagram may be found in Francis L. K. Hsu, *Under the Ancestors' Shadow: Kinship, Personality, and Social Mobility in Village China* (1948; reprinted Garden City, N.Y.: Anchor Books, 1967), 30.

45. On the wide distribution of the *lishu*, see Ch'ing K'un Yang, *Religion in Chinese Society: A Study of Contemporary Social Functions of Religion and Some of Their Historical Factors* (Berkeley and Los Angeles: University of California Press, 1961), 17.

particularly the *Luo shu*, in practicing their art (see, for example, fig. 8.16 below).

Geomancy and its Relation to Cosmography

There did exist countervailing cultural traditions in premodern China that opposed some of the basic principles of geometric cosmography. The most prominent of these was geomancy, or *feng shui* (literally, wind and water), an art concerned primarily with siting of such structures as buildings and graves in auspicious locations so as to benefit the living and pacify the dead. This was accomplished by locating these structures in places where they would harmonize with and draw upon the flow of the energetic pneuma (*qi*) that circulated through such features of the terrain as mountains and streams. Hence the art of *feng shui* is perhaps better characterized by the English term "siting" than by the word "geomancy," though the latter term has become so well established as to make abandoning it rather difficult.[46]

The most widely known and practiced application of geomancy is the identification of proper burial sites for deceased parents and other ancestors.[47] Geomantic prescriptions for the location of graves are, moreover, more specific and theoretical than for other types of sites. But the Chinese art of siting has been applied in a very wide range of spatial dimensions, "from the smallest of spaces—say, a bedroom or even the location of a chair—to the largest, cosmic dimension."[48] In late imperial China, geomancers were frequently called upon to site and orient residences, villages, and even capital cities.[49] They also advised the government on such enterprises as the construction of public buildings, the conduct of military operations, and the management of public works.[50] Even the boundaries of Chinese administrative districts, ranging from villages and townships to whole provinces, frequently follow the irregular and meandering lines prescribed by *feng shui*.[51] Among the numerous contemporary examples of the public use of geomancy is that by the famous architect I. M. Pei in his design for the Fragrant Hill Hotel near Beijing.[52]

Although geomancy in its higher reaches was a rather complex art practiced by skilled specialists, the knowledge of general geomantic rules of thumb was surprisingly widespread in late traditional Chinese society. As Smith remarks, "Virtually everyone in China recognized its basic symbolic repertoire and knew intuitively a good geomantic location on first glance."[53] Even the least educated Chinese, according to the nineteenth-century observer J. J. M. de Groot, "show an astounding amount of knowledge of Fung-shui."[54] Indeed, this art provided some of the basic categories and concepts through which the Chinese man in the rice paddy comprehended the physical environment. It helped bring the cosmology of the high culture down to earth.

As with several of the Chinese arts and sciences, the origins of geomancy have been traced by its practitioners to hoary antiquity, to the same legendary sage-kings who are credited with the delineation of the *Hetu* and *Luo shu* and the demarcation of the nine regions (*jiu zhou*). More historically, there is some archaeological and literary evidence that the establishment of settlements as early as the Shang dynasty of the second millennium B.C. was preceded by careful examination of local topographical conditions, perhaps auguring the existence of a protogeomantic art in that remote era.[55] The cosmographical foundations of siting were established in texts of the classical era, especially the *Yi jing* and the *Guanzi* ([Book of] Master Guan). During the early Han period, *feng shui* apparently was recognized and practiced as an independent art. A Han-era bibliography contains notices of treatises that seem to have been concerned with geomancy, though none of these are now extant.[56] The post-Han migrations of Chinese to the south, with its variegated natural landscapes and exotic aboriginal cultures, evidently stimulated the development of this art, which was codified in the *Zang shu* (Burial book) ascribed to Guo Pu (276–324) and the *Huangdi zhai jing* (The Yellow Emperor's site classic) credited to Wang Wei (415–43).[57]

46. For a brief discussion of why the term "geomancy" may not apply very well to the Chinese art of *feng shui*, see Steven J. Bennett, "Patterns of the Sky and Earth: A Chinese Science of Applied Cosmology," *Chinese Science* 3 (1978): 1–26, esp. 1–2. In premodern China this art was designated by many names besides *feng shui*. Some of these are listed in Richard J. Smith, *Fortune-Tellers and Philosophers: Divination in Traditional Chinese Society* (Boulder, Colo.: Westview Press, 1991), 315 n. 1.

47. Smith, *Fortune-Tellers and Philosophers*, 151 (note 46).

48. Sarah Rossbach, *Feng Shui: The Chinese Art of Placement* (New York: E. P. Dutton, 1983), 2.

49. Sang Hae Lee, "Feng-Shui: Its Context and Meaning" (Ph.D. diss., Cornell University, 1986), 21.

50. Smith, *Fortune-Tellers and Philosophers*, 157–58 (note 46).

51. Lee, "Feng-Shui," 188 (note 49).

52. Lee, "Feng-Shui," 134 n. 158 (note 49).

53. Smith, *Fortune-Tellers and Philosophers*, 171 (note 46).

54. J. J. M. de Groot, *The Religious System of China: Its Ancient Forms, Evolution, History and Present Aspect*, 6 vols. (Leiden: E. J. Brill, 1892–1910), vol. 3, bk. 1, *Disposal of the Dead*, pt. 3, *The Grave* (reprinted Taipei: Chengwen Chubanshe, 1972), 939.

55. Lee, "Feng-Shui," 49 (note 49).

56. Stephan D. R. Feuchtwang, *An Anthropological Analysis of Chinese Geomancy* (Vientiane, Laos: Editions Vithagna, 1974), 16–17. For a list of these titles, as well as other evidence for the early development of *feng shui* in China, see Joseph Needham, *Science and Civilisation in China* (Cambridge: Cambridge University Press, 1954–), vol. 2, with Wang Ling, *History of Scientific Thought* (1956), 359–63, and de Groot, *Religious System*, 3:994–96 (note 54).

57. Steven Bennett questions the attribution of the *Huangdi zhai jing* to Wang Wei; see Bennett, "Patterns," 5 (note 46).

In the Song era, two major schools of geomancy crystallized, one concentrating on cosmology and the other on the forms of the landscape.[58] The first of these, the "Compass" or "Directions and Positions" school, was supposedly developed by Wang Ji (fl. 1030–50) and was especially popular in the southern coastal province of Fujian.[59] Hence it is also sometimes referred to as the Fujian school. This school drew on the metaphysical speculations of the Song Neo-Confucian philosophers and used such cosmological ideas as yin-yang and the five phases (*wuxing*) to analyze sites.[60] In determining the spatial and temporal dimensions of a site, adepts of this Directions and Positions school relied heavily on the geomantic compass (*luopan* or *luojing*), a complicated instrument whose dial might have as many as thirty-eight concentric rings centered on a "celestial pool" that housed a magnetic needle (fig. 8.11). Inasmuch as these rings incorporate "nearly all the Chinese symbols which are used in dealing with time and space," ranging from the eight trigrams to the twenty-eight lunar lodges, it amounted to "an outline of the universe according to traditional Chinese natural philosophy."[61] Like the structures inspired by the classical nonary square, such as the classical city plan and the *mingtang*, it was a concrete model of the cosmos. The compass was applied to the art of siting by aligning the disk with the magnetic needle and then determining the coordinates of the site in question by all the relevant measures of space, time, and cosmic change indicated on the compass dial.[62] Siting by this method left wide latitude for creative interpretation by professional practitioners of the art.

The other major geomantic school, the "Form" or "Forms and Configurations" school, associated with the southern inland province of Jiangxi, supposedly developed from the teachings of Yang Yunsong (fl. 874–88). In contrast with the rather analytical Compass school, with its "abstruse cosmology and elaborate calculations," the Form school was much more popular and down to earth, "by virtue of its emphasis on physical forms and configurations."[63] Its practitioners relied less on the cosmological schemata encompassed by the geomantic compass and more on on-site inspection of the landforms around proposed sites for buildings and tombs, particularly the shapes of mountains and the directions of watercourses, in order to determine the patterns of cosmic energy flow.[64] The approach of this school has been characterized as more "intuitive" and "mystical" than that of the "analytical" Compass school.[65] Indeed, some authorities of this school "describe the experience of finding the right location in terms reminiscent of Chan-Buddhist enlightenment."[66] Nevertheless, the adherents of the Form school did not rely on intuitive insight or visual scanning alone, but made use of schematic topographical maps that showed how the currents and con-

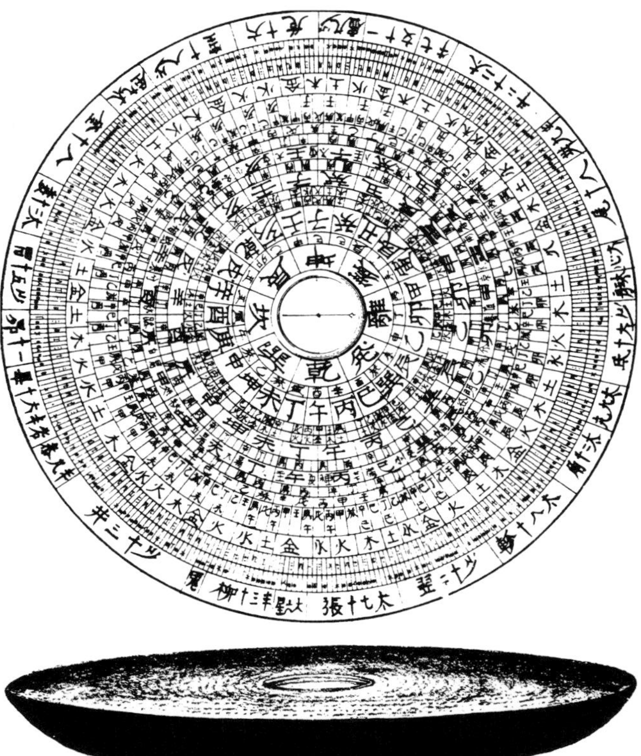

FIG. 8.11. ILLUSTRATION OF A GEOMANTIC COMPASS. This diagram depicts a nineteenth-century geomancer's compass. The innermost circle is called the "celestial pool" and is divided in half by the magnetic needle, betokening the division of the Supreme Ultimate into yin and yang. The first ring contains the eight trigrams of the *Yi jing*, the second has characters representing the twenty-four directions, and so on to the outermost ring, which designates the twenty-eight lunar lodges. The compass included practically all of the systems and series the Chinese used to measure or represent elements of space, time, and cosmic change.
Size of the image: unknown. From J. J. M. de Groot, *The Religious System of China: Its Ancient Forms, Evolution, History and Present Aspect*, 6 vols. (Leiden: Brill, 1892–1910), vol. 3, bk. 1, *Disposal of the Dead*, pt. 3, *The Grave* (reprinted Taipei: Chengwen Chubanshe, 1972), 959.

58. Feuchtwang, *Anthropological Analysis*, 17 (note 56).

59. Lee, "Feng-Shui," 128 and 159 (note 49).

60. Lee, "Feng-Shui," 158–59 (note 49), and Bennett, "Patterns," 3 (note 46).

61. Jeffrey F. Meyer, "*Feng-Shui* of the Chinese City," *History of Religions* 18 (November 1978): 138–55, quotation on 149; Feuchtwang, *Anthropological Analysis*, quotation on 96 (note 56).

62. Bennett, "Patterns," 3 (note 46).

63. Smith, *Fortune-Tellers and Philosophers*, 139 and 138 (note 46).

64. Bennett, "Patterns," 3 (note 46); Skinner, *Living Earth Manual*, 8–9 (note 44).

65. Bennett, "Patterns," 3 (note 46).

66. Smith, *Fortune-Tellers and Philosophers*, 139 (note 46).

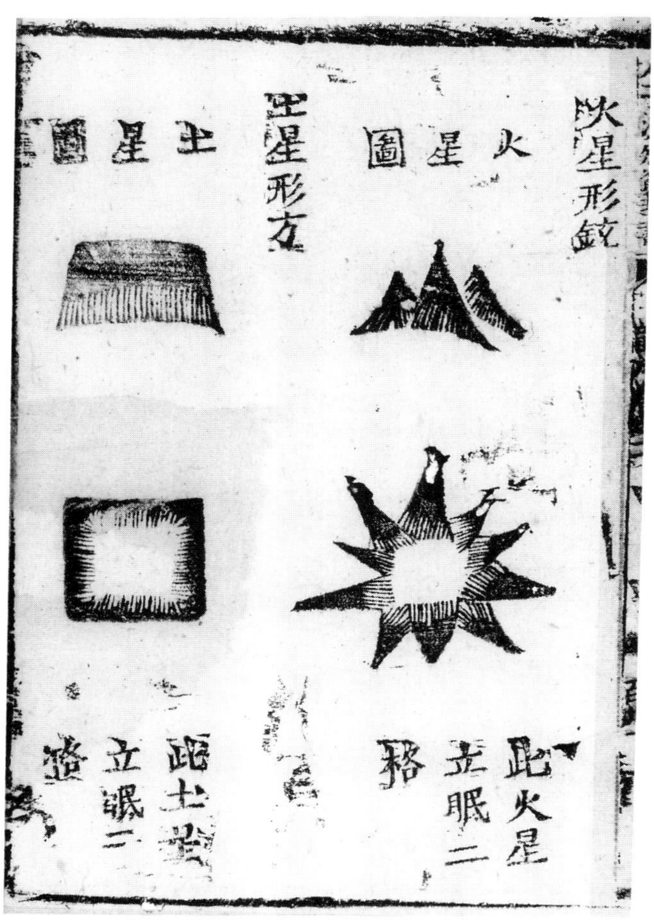

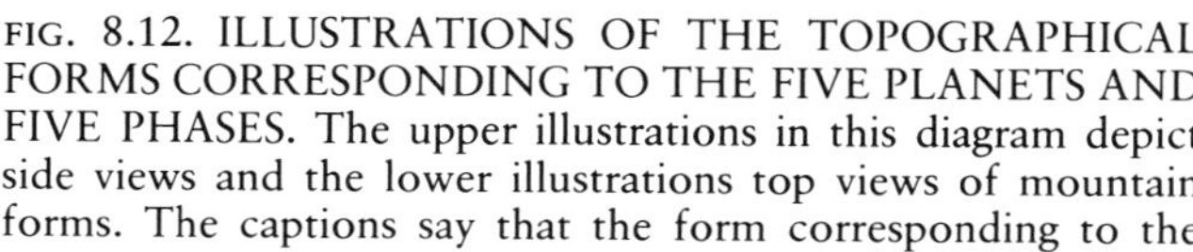

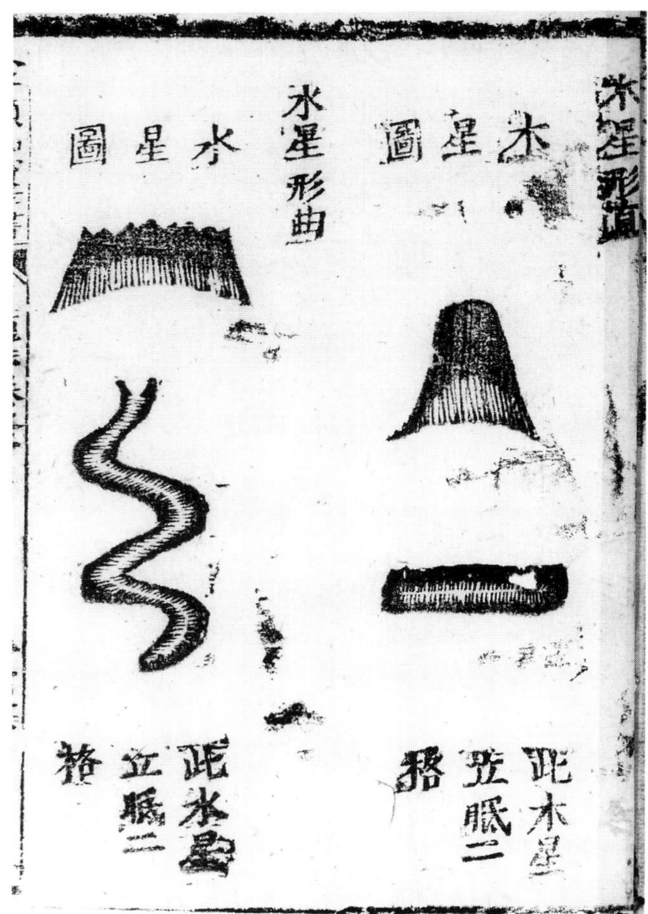

FIG. 8.12. ILLUSTRATIONS OF THE TOPOGRAPHICAL FORMS CORRESPONDING TO THE FIVE PLANETS AND FIVE PHASES. The upper illustrations in this diagram depict side views and the lower illustrations top views of mountain forms. The captions say that the form corresponding to the metal star (Venus) is round, to the wood star (Jupiter) is straight, to the water star (Mercury) is curvy, to the fire star (Mars) is pointed, and to the Earth star (Saturn) is square.
Size of each page: 21 × 14 cm. From Xu Shanji and Xu Shanshu,

centrations of the energetic *qi* or pneuma that pulsed through the ridges and watercourses of the earth could be traced in order to identify auspicious locations. They also cataloged various landforms by a rich metaphorical vocabulary that likened the physical features and relationships of the landscapes to animals (especially the mythical dragon), parts of the human body, celestial bodies, manufactured objects, and even the characters of the Chinese written language.[67] For example, geomancers correlated each of the five planets (and correlatively the five phases of cosmic change) with particular landforms, specifically the shapes of mountains, on the theory that celestial bodies must have their terrestrial counterparts. They correlated gently rounded forms with metal and the planet Venus, elongated trunklike forms with wood and the planet Jupiter, softly undulating forms with water and the planet Mercury, sharply pointed forms with fire and the planet Mars, and squared forms with Earth and the planet Saturn (fig. 8.12).[68]

Although theoretically distinguished from one another, the Compass and Form schools were often combined in practice, in their literature, and even in their cosmographical orientations. Geomancers of all persuasions, for example, were generally in accord regarding the characteristics of the ideal *feng shui* spot, sometimes called the "dragon's lair." Such a location has been characterized by Freedman as "one which nestles in the embrace of hills standing to its rear and on its flanks; it is then like an armchair, comfortable and protecting" (figs. 8.13 and 8.14).[69] In the front of the ideal site should run a watercourse, which must flow neither too swiftly, lest the

67. Smith, *Fortune-Tellers and Philosophers*, 139 (note 46).

68. Smith, *Fortune-Tellers and Philosophers*, 144 (note 46), and Lee, "Feng-Shui," 136–37 (note 49).

69. Maurice Freedman, *Chinese Lineage and Society: Fukien and Kwang-tung* (London: Athlone Press, 1971), 122.

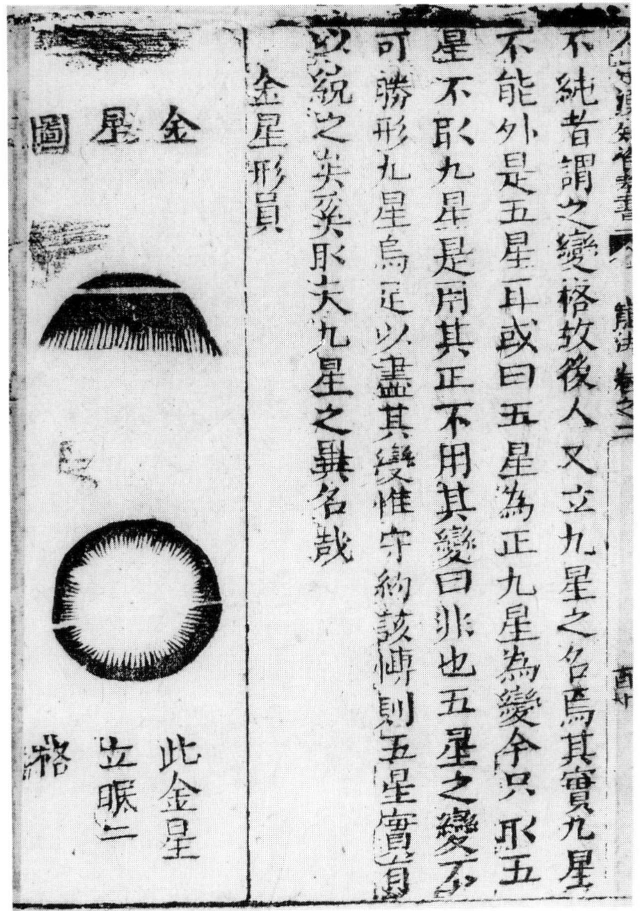
不純者謂之變格故後人又立九星之名焉其實九星
不能外是五星耳或曰五星爲正九星爲變今只取五
星不取九星是用其正不用其變曰非也五星之變不
可勝形九星烏足以盡其變惟守約該博則五星實足
以統之矣奚取夫九星之異名哉
金星形員

金星圖

此金星立眠二格

Dili renzi xuzhi (Everything that geomancers should know, first printed 1559), 1583 edition, 11.120b–121a. Reproduced courtesy of the Harvard-Yenching Library, Harvard University, Cambridge.

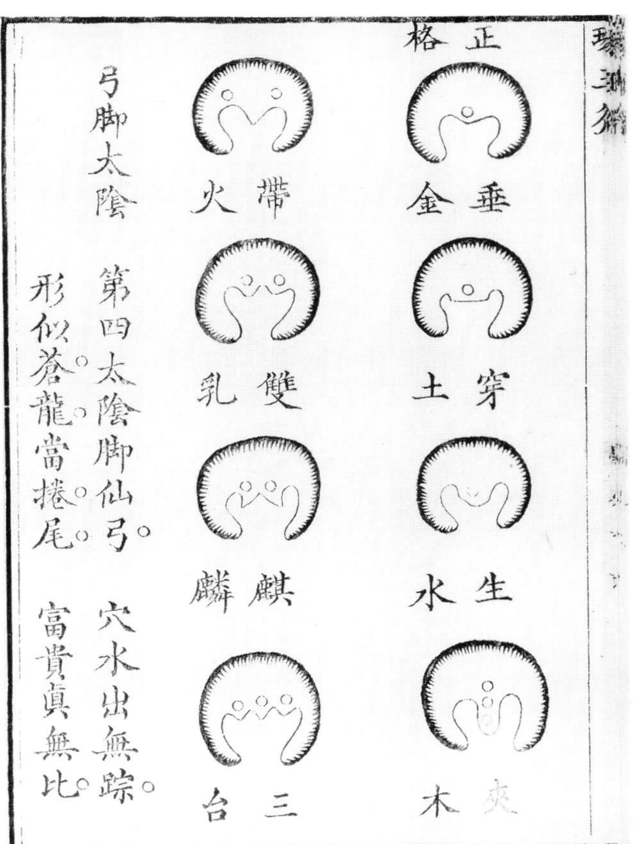

FIG. 8.13. ILLUSTRATIONS OF THE IDEAL *FENG SHUI* SPOT, OR THE DRAGON'S LAIR CONFIGURATION. These drawings show how the site of the grave is protected on three sides by ridges and is open to the south. Compare these rather schematic illustrations with the more complex diagram in figure 8.14.
From Xu Zhimo, *Dili chuoyu fuluantou gekuo* (A summation of songs on the pattern of the earth, chiseled jade, and pared hilltops, 1828), 2:7. Reproduced courtesy of the Harvard-Yenching Library, Harvard University, Cambridge.

good influences be dispersed, nor too sluggishly, lest the flow of cosmic energy stagnate. The wind stirring around the site must be correspondingly moderate. Cosmologically, the *feng shui* spot is the meeting place of the heavenly *qi* and the earthly *qi*, where the yang and yin, or male and female influences, are most closely intertwined and most intensely concentrated.[70] The configuration of this ideal site has been compared to a womb, both "the place of fertility" and "the place all creatures return to at last."[71] From such a place, the spirits of the ancestors were well situated to favorably influence the fortunes of their living descendants.

Whatever one might think of the efficacy of this form of applied cosmography, the aesthetic and hygienic attractions of the *feng shui* spot, situated on a south-facing slope receiving the sun and with an unrestricted view that takes in a gently flowing stream, may be appreciated by all. As Feuchtwang puts it, "Everyone takes unsymbolic pleasure in sunlight, airiness and a good view."[72] Thus it is not surprising that the concept and configuration of the ideal *feng shui* spot were extended to cover dwellings of the living as well as of the dead. This principle is illustrated even in the architecture of rural Taiwanese houses, which "are often shaped like a lair, with two projecting wings."[73]

Although *feng shui* was practiced and appreciated throughout China by the late imperial period, it was "most pervasive and deeply entrenched" in the area of

70. Lee, "Feng-Shui," 189 (note 49); Smith, *Fortune-Tellers and Philosophers*, 143 (note 46).
71. Lee, "Feng-Shui," 191–92 (note 49).
72. Feuchtwang, *Anthropological Analysis*, 117 (note 56).
73. Bennett, "Patterns," 13 (note 46).

FIG. 8.14. ILLUSTRATION OF A NINTH-CENTURY WORK ON GEOMANCY. This chart, which illustrates the *Shier zhang fa* (Twelve staff patterns) attributed to Yang Yunsong (fl. 874–88), exhibits some of the characteristics of an ideal geomantic site according to the "coiled up" pattern. The site, where a grave is situated, is protected, particularly from the north and west, by sinuous ranges of hills. It is also nourished by small streams that combine to flow through an opening to the south, the region from which beneficent influences radiate. Joseph Needham remarks that "the relation of this kind of drawing to physiographic map-making . . . is evident." For Needham's commentary on this particular chart, see his *Science and Civilisation in China* (Cambridge: Cambridge University Press, 1954–), vol. 2, *History of Scientific Thought* (1956), pl. XVIII (fig. 45) caption following p. 360.
Size of the image: 3 × 2.5 cm. From Chen Menglei, Jiang Tingxi, et al., comps., *Gujin tushu jicheng* (Complete collection of books and illustrations, past and present, completed 1726, printed 1728), 79 vols. (Shanghai: Zhonghua Shuju, 1934), 475.38.

the country south of the Yangtze, with its sinuous mountains and undulating watercourses.[74] Mountains are of particular importance in geomancy and, indeed, "receive more attention in feng-shui manuals than any other natural phenomena." Such manuals also give considerable attention to the bends, confluences, and branches of watercourses.[75] To the geomancer, both mountains and rivers literally animated the landscape. They constituted the dragon's veins, through which pulses *qi*, the energetic pneuma that was also conceived as the blood of the dragon whose winding form and features could be seen in all interesting topographical formations.[76] In fact, the whole of China has been depicted cartographically as "a tripartite dragon system with subsidiary trunks and branches" (fig. 8.15).[77]

On the other hand, flat landscapes and relatively straight watercourses, such as those of the North China Plain, were both uninviting and unpromising to *feng shui* practitioners. In fact straight lines, one of the chief building blocks of classical geometric cosmography (which was based in the northern China homeland of Chinese civilization), were "anathema" in *feng shui*.[78] Geomancers looked askance at such lines, which might be straight ridges, watercourses, thoroughfares, housetops, or even telegraph wires, because they easily conducted a sort of evil miasma or noxious *qi* that encouraged the movement of demons. Straight watercourses, moreover, were often suspected of too quickly draining off benign influences from the site in question. Hence geomancers much preferred that meandering, undulating, and even tortuous lines inform the areas around the structures for which they were employed to find the most auspicious locations.[79] They regarded flat landscapes, the kind that were ideal for the situation of the well-field schema and the classical city plan, as "old, tired, [and] worn-down."[80]

Although Confucian scholars in the late imperial period frequently criticized *feng shui* on both ethical and cosmological grounds, the art did influence and interact with aspects of the high intellectual and cultural tradition in premodern East Asia. Geomancers' preference for sinuous ridges and undulating watercourses may well have affected the shapes of such features in Chinese landscape paintings, one of the noblest arts of China. Both geomancers and landscape painters, moreover, conceived of the landscape as the body of the dragon through whose veins, mountain ridges and watercourses, pulsed the vital *qi*. Even the *feng shui* spot, the dragon's lair, frequently appears in landscape paintings, where "dwellings and temples are often ensconced in three-sided enclosures of earth, rock or foliage."[81]

74. Smith, *Fortune-Tellers and Philosophers*, 149 (note 46).

75. Feuchtwang, *Anthropological Analysis*, 121 and 129 (note 56).

76. Feuchtwang, *Anthropological Analysis*, 141 (note 56).

77. Bennett, "Patterns," 13 (note 46).

78. Skinner, *Living Earth Manual*, 25 (note 44). Joseph Needham remarks that in *feng shui* "there was in general a strong preference for tortuous and winding roads, walls and structures, . . . and a strong objection to straight lines and geometrical layouts." See Needham, *Science and Civilisation*, 2:361 (note 56).

79. Skinner, *Living Earth Manual*, 25 (note 44). On this point Jack Potter remarks that "most roads in the New Territories [of Hong Kong] have a serpentine quality that is due more to *feng shui* requirements than to bad engineering." See Jack M. Potter, "Wind, Water, Bones and Souls: The Religious World of the Cantonese Peasant," *Journal of Oriental Studies* 8 (1970): 139–53, quotation on 143.

80. Skinner, *Living Earth Manual*, 37 (note 44). According to Sarah Rossbach, "The worst possible location for a building is featureless ground—a flat plain with no ch'i-formed undulations"; Rossbach, *Feng Shui*, 59 (note 48).

81. Bennett, "Patterns," 13 (note 46).

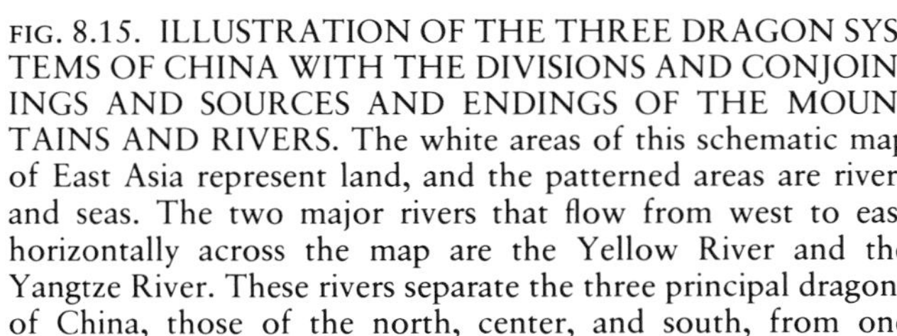

FIG. 8.15. ILLUSTRATION OF THE THREE DRAGON SYSTEMS OF CHINA WITH THE DIVISIONS AND CONJOININGS AND SOURCES AND ENDINGS OF THE MOUNTAINS AND RIVERS. The white areas of this schematic map of East Asia represent land, and the patterned areas are rivers and seas. The two major rivers that flow from west to east horizontally across the map are the Yellow River and the Yangtze River. These rivers separate the three principal dragons of China, those of the north, center, and south, from one another. The characters written inside the squares are the names of China's provinces. The archaic and extremely schematic character of this map is illustrated in that it depicts Korea as an island off the eastern coast of China, due north of Japan. From Xu Zhimo, *Dili chuoyu fuluantou gekuo* (A summation of songs on the pattern of the earth, chiseled jade, and pared hilltops, 1828), 1:6. Reproduced courtesy of the Harvard-Yenching Library, Harvard University, Cambridge.

Both landscape painting and geomancy may be related to the development of cartography in China, the former in that its early forms may have been "the first steps of cartography," and the latter in that it "stimulated cartographic skills."[82] In premodern Korea, the influence of *feng shui* on cartography was apparently even more pronounced. It contributed to the emergence of "a uniquely Korean map style" in which, for example, "mountains are represented not individually but as part of a system of ridgeline" to emphasize the long-distance flow of the terrestial *qi* or pneumatic energy. This feature, Jeon asserts, is absent from most Chinese maps.[83]

In China, geomancy impinged on the high cosmographical tradition by coming into conflict with geometric cosmography, particularly in city planning. As Meyer has noted, the failure of a number of great imperial cities in premodern China to conform precisely to the classical city plan, particularly their departures from rectilinearity, arose more from the requirements of siting than from physical or political obstacles.[84] Moreover, where the local landforms were not propitious for siting a capital city with its imperial palace, they were in some cases modified to suit geomantic requirements. For example,

82. Feuchtwang, *Anthropological Analysis*, 140 (note 56). See also Hong Key Yoon, "The Expression of Landforms in Chinese Geomantic Maps," *Cartographic Journal* 29 (1992): 12–15.

83. Sang-woon Jeon, *Science and Technology in Korea: Traditional Instruments and Techniques* (Cambridge: MIT Press, 1974), 279–80.

84. Meyer, "*Feng-Shui*" (note 61). On such geomantic adjustments in urban designs, see also Skinner, *Living Earth Manual*, 25 (note 44).

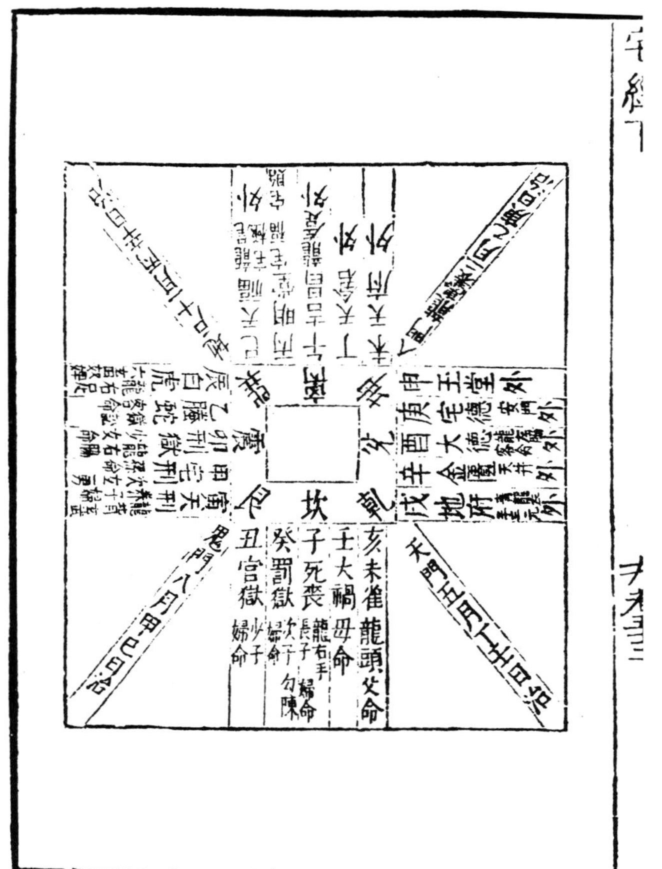

FIG. 8.16. "FATE POSITION DIAGRAM." Even in geomancy the geometric and nonary cosmography outlined above had its place, as exemplified in this "fate position diagram" that appears as an illustration in a Qing edition of the geomantic classic *Huangdi zhai jing* (The Yellow Emperor's site classic). As explained by Bennett, "The inner square contains the trigrams [of the *Yi jing*], which in this case are used to orient the site on the cardinal points. The siting expert then consults the diagram to determine which part of the site will affect certain members of the family" (Bennett, "Patterns of the Sky and Earth: A Chinese Science of Applied Cosmology," *Chinese Science* 3 [1978]: 25). On the use of this *Luo shu* type of diagram in *feng shui*, see also Stephen Skinner, *The Living Earth Manual of Feng-Shui: Chinese Geomancy* (London: Routledge and Kegan Paul, 1982), 47–48, 61–67.
Size of the original: 16.5 × 11.5 cm. From Wang Wei, *Huangdi zhai jing* (Yimen guang du [Extensive documents from the Great Gate], vol. 24), B.1b. Reproduced courtesy of the Harvard-Yenching Library, Harvard University, Cambridge.

in Ming times a huge artificial mound known as "Prospect Hill" was built at the northern end of the Forbidden City to protect it from the noxious influences emanating from that direction. Thus could unhealthy land be treated by geomancers, who have in fact been called "doctors of the earth."[85]

Finally, geomantic orientations may have influenced the criticisms that Confucian scholars in the late traditional era directed against the established geometric and nonary interpretations of the classical cosmograms outlined above. At least there was some congruence between the preferences of geomancers and of these later critics for naturally irregular lines and boundaries (fig. 8.16).

Later Modifications and Criticisms of Traditional Cosmographical Schemata

Debates and disagreements regarding the configurations and practicalities of some of the cosmographical schemata sketched above ranged throughout premodern Chinese history. For example, agrarian reformers who proposed to implement the well-field system frequently encountered spirited opposition from opponents who argued that the plan was unworkable.[86] As early as Song times, such famous scholars as Zheng Qiao (1108–66) and Zhu Xi noted the disjunctions between the irregular boundaries of the classical nine regions outlined in the "Yu gong" chapter of the *Shu jing* and the geometrized reconstitution of these regions formulated by Han commentators.[87] As I noted above, several famous writers of Tang and Song times proposed redesigning the terrestrial components of the field-allocation (*fenye*) system that were out of accord with contemporary political boundaries. Debates over the proper plan and proportions of the *mingtang* were major episodes in the court politics of Han and Tang times, though the participants in these debates are not known to have used charts to support their arguments. And the question of the form and significance of the enigmatic *Hetu* and *Luo shu* was a chief point of cosmological controversy from Han times to the beginning of the modern era. Thus the particular geometric and nonary conceptions and patterns sketched above by no means monopolized the field of cosmographical thought in traditional China. There is no question here of a uniform, monolithic worldview from which departure was inconceivable or dissent unthinkable.

Nevertheless, few major writers before the seventeenth century challenged the basic principles and fundamental

85. Smith, *Fortune-Tellers and Philosophers*, 131 (note 46). Smith adds that geomancers "often employed medical metaphors to explain their ideas, and in fact, a great many specialized in traditional Chinese medicine."

86. See, for example, the introduction to the survey on the land tax in Ma Duanlin (fl. 1273), *Wenxian tongkao* (General study of literary remains, completed ca. 1280) (Taipei: Xinxing Shuju, 1962), 1.2a–4a; English translation in *Sources of Chinese Tradition*, 501–3 (note 17).

87. "Dili lüe" (Monograph on geography) in Zheng Qiao, *Tongzhi* (Comprehensive treatises, ca. 1150), 40.86b; see the modern edition (Taipei: Shijie Shuju, 1970), A.218, 224; Zhu in *Li ji jishuo*, 3.4a, in *Sishu wujing*, 2:67 (note 12).

presuppositions of geometric or nonary cosmography. In some cases, such as with the debates over the proper plan and proportions of the *mingtang*, the opponents of the nine-palace pattern supported models that were just as much in accord with geometric cosmography as was the nonary form. There was little to choose between the nonary and quinary models on this point. Further, the objections that were raised to the implementation of such plans as the well-field system focused more on the social and economic dislocations that applying them might cause than on their lack of accord with the physiography of the land. In sum, even critics of the cosmographical schemata outlined above generally operated in the same mental universe as the Han cosmographers who devised most of these models.

From this perspective, the seventeenth century marks an epoch in the history of Chinese cosmographical conceptions, at least on the level of high intellectual history. For Chinese savants in this era, like their contemporaries in Europe, challenged the cosmological geometry associated with the dominant intellectual traditions of their respective cultures. Bacon condemned the medieval Scholastics for having promoted "the fiction that all celestial bodies move in perfect circles";[88] and Galileo observed "that the surface of the moon is not smooth, uniform, and precisely spherical as a great number of philosophers believe it (and the other heavenly bodies) to be."[89] So Chinese cosmographical critics of the same era blurred the edges of the three-by-three grid form, the nonary square, arguing that this plan did not fit physiographical contours, political or ritual requirements, or classical descriptions.

Seventeenth-century criticisms of the well-field system, for example, focused on the disjunctions between the ideal geometry of this plan and the physiographical irregularities of the surface of the earth. Thus Lu Longqi (1630–93) argued that this model could be applied only in those limited areas where the topography was suitable.[90] And Lu Shiyi (1611–72), though advocating an eventual restoration of the well-field arrangement, practically despaired that the manifold irregularities of either the natural or the artificial aspect of the landscape could ever be made to conform to this geometric template.[91] In sum, just as the circle, according to Bacon and Galileo, did not fit the order of the heavens, so the nonary square, according to Lu Longqi and Lu Shiyi, did not match the contours of the earth.

Seventeenth-century Chinese commentators also noted disjunctions between geometric versions of the nine regions and the natural boundaries of China's ancient provinces as described in the canonical account of these areas contained in the *Shu jing*. Yan Ruoju (1636–1704), the most renowned early Qing authority on this classic, maintained that the legendary sage-king Yu the Great "fixed the frontiers of the nine regions by mountains and rivers."[92] Yan and his contemporaries, especially Hu Wei (1633–1714), apparently reached this conclusion through a close study of the "Yu gong" chapter of the *Shu jing*. Their geographical interests and conceptions, in other words, were generally focused on classical texts. But it is also possible that an acquaintance with the more realistic terrestrial maps of their day led them to doubt the geographical value of schematic cosmography.

Some seventeenth-century criticisms of the astrological *fenye* or field-allocation system developed a line of argument broached by scholars as early as the Tang era, that the boundaries of the terrestrial regions in the schema should be reformed to bring them into accord with natural physiographical divisions.[93] Other commentators emphasized the disjunctions and incommensurability between these regions and the celestial "fields" to which they supposedly corresponded. Fang Yizhi (1611–71), for example, remarked that though the southerly Yangzhou region comprised about half of China's land area, its heavenly "field" included only three of the twenty-eight lunar lodges.[94] Finally, several early and mid-Qing critics of the *fenye* system objected particularly to the precision with which its expositors had attempted to calculate the bounds of the corresponding celestial and terrestrial fields in the schema. Thus Huang Zongxi (1610–95) argued that the *fenye* was originally intended only as a rough outline and was not geared to meet the exacting demands of astrological prediction.[95] Lu Longqi contended that the *fenye* should not be regarded as a system at all, that it had little basis apart from a rather incomplete and haphazard historical record of correspondences between

88. Francis Bacon, *The New Organon*, bk. 1, in *The New Organon and Related Writings*, ed. Fulton H. Anderson (New York: Liberal Arts Press, 1960), 50.

89. Galileo, *The Starry Messenger*, in *Discoveries and Opinions of Galileo*, trans. and annotated Stillman Drake (Garden City, N.Y.: Doubleday, 1957), 21–58, quotation on 31.

90. Lu Longqi, *Sanyutang riji* (Diary from Three Fish Hall, written 1659–92) (Taipei: Shangwu Yinshuguan, 1965), B.113.

91. Lu Shiyi, *Sibian lu jiyao* (Epitome of the record of thinking and sifting, published 1707–13), comp. Zhang Boxing, 19.2a–b, in *Kunzhi ji deng sanzhong* (The record of hard study [of Lo Qinshun] with two other works of the same sort) (Taipei: Guangxue She Yinshuguan, 1975), 191–96.

92. Yan Ruoju, *Shang shu guwen shuzheng* (Inquiry into the authenticity of the ancient text version of the documents, first printed 1745) (Tianjin: Wushi Kanben, 1796), 6B.79b; *Siku quanshu* edition, 6B.99b.

93. Yan, *Shang shu guwen shuzheng*, 6A.6b, 10a–b; *Siku quanshu* edition, 6A.8a, 13a–14a (note 92).

94. Fang Yizhi, *Tongya* (Comprehensive refinement, completed 1636, printed 1666), *Siku quanshu* edition, 11.23a.

95. Huang Zongxi, *Poxie lun* (Discussion confuting heterodoxy, seventeenth century), 5a, in *Lizhou yizhu huikan* (A collection of [Huang] Lizhou's extant works, 1910) (Taipei: Long Yan Chubanshe, 1969), vol. 2.

celestial appearances and terrestrial events. In any case, "it has nothing to do with boundaries."[96]

Unlike Qing commentaries on most of the geometric and nonary cosmograms outlined above,[97] the Qing critique of the *Hetu* and *Luo shu* was a major episode in the intellectual history of the era. This critique was an important aspect of the Qing repudiation of the cosmology and metaphysics associated with the dominant Lixue (Principle) school of Song Neo-Confucianism. The discrediting of the standard versions of these figures probably did more to subvert cosmology and cosmography in the traditional mode than did any other affair in the intellectual history of late traditional China.

As early as the fourteenth century, Confucian scholars pointed out the nonclassical provenance and heterodox associations of the interpretations of the *Hetu* and *Luo shu* accepted by the major Song Neo-Confucian cosmologists. They particularly questioned the identification of the *Luo shu* with the nine-palace formation.[98] Seventeenth-century scholars such as Mao Qiling (1623–1716) meticulously exposed how such figures as the nonary *Luo shu* had been constituted through the arbitrary conflation of unrelated passages in various classical texts.[99]

But sixteenth- and seventeenth-century criticisms of the cosmographical *Luo shu* were not confined to such textual arguments. Gui Youguang (1507–71), for example, contended that *any* precise graphical or numerological representation of the *Hetu* or *Luo shu*, not just the nine-palace formation, distorted their true essence and significance. For the *Hetu* and *Luo shu* were really ubiquitous under heaven. They could not be satisfactorily delimited or depicted by a geometric figure,[100] just as the ineffable *Dao* could not be adequately represented through words or signs.

The famous seventeenth-century scholar Huang Zongxi also attempted to distinguish the early classical *Hetu* and *Luo shu* from later cosmographical reconstructions and elaborations. But instead of affirming the cosmic ubiquity of these figures as had Gui Youguang, Huang argued for their mundaneness and specificity. According to Huang, the *Hetu* and *Luo shu* originally had no cosmological purport or significance. They were really only topographical maps and economic geographies used in high antiquity for administrative and tax purposes.[101] Huang Zongyan (1616–86) similarly contended that in ancient times "the *Hetu* and *Luo shu* were geographical records that recorded the contours of mountains and rivers and the rates of land taxes."[102] These seventeenth-century commentators thus reduced the *Hetu* and *Luo shu*, which Han and Song Neo-Confucians had interpreted as the sources of cosmic geometry and numerology, to the level of practical administrative tools and devices. In so doing, they radically subverted one of the primary bases of traditional cosmographical speculation.

These scholars did not, however, attempt to reconstruct the ancient geographical maps and records that, they claimed, were designated by the terms *Hetu* and *Luo shu*. Nor did they explain precisely how these charts were supposed to have functioned in the government of high antiquity. Their rejection of cosmographical interpretations of the *Hetu* and *Luo shu* did not, in other words, lead directly to the composition of more realistic geographical representations or records.

On the other hand, some of the principal seventeenth-century critics of traditional cosmographical conceptions, particularly Huang Zongxi, Gu Yanwu (1613–82), Hu Wei, and Yan Ruoju, contributed significantly to the revival of geographical studies in that era. Such scholars helped to establish geography as "a precise field of evidential inquiry" and "an important element of concrete studies" (*shi xue*) with applications in such areas as land reclamation and hydraulic works, as Elman has noted.[103] A possible link between these scholars' cosmographical criticism and their empirical geographical studies may be found in their use and conceptualization of *tu* (a Chinese character that may mean anything from "chart" to "schema") not as cosmological diagrams but as realistic maps. The seventeenth century thus marked a change in the orientation toward space within the tradition of Confucian scholarship, from a concern with cosmological geometry to an emphasis on empirical geography.

96. Lu, *Sanyutang riji*, B.46–47 (note 90).

97. I have not included accounts of Qing criticisms of the classical city plan and the *mingtang*, which in any case are only marginally relevant to the history of cartography.

98. Pi Xirui (1850–1908), *Jingxue lishi* (History of classical studies, printed 1907), annotated Zhou Datong (Taipei: Yiwen Yinshuguan, 1966), 291; idem, *Jingxue tonglun* (Comprehensive discussions of classical studies, printed 1907), 4 vols. (Taipei: He-Lo Tushu Chubanshe, 1974), 1:28; "Hetu Luo shu shuo" (Explanation of the Yellow River chart and Luo River writing) in Song Lian (1310–81), *Song Wenxian gong quanji* (Complete collected writings of Song Wenxian, 1810), *Sibu beiyao* edition, 36.3a.

99. Mao Qiling, *Shang shu guangting lu* (Record of a broad understanding of the documents, seventeenth century), *Siku quanshu* edition, 3.8b–10b.

100. "Yi tu lun" (Discussion of the diagrams associated with the [Book of] changes) in Gui Youguang, *Zhenchuan xiansheng ji* (Collected writings of [Gui] Zhenchuan), *Sibu beiyao* edition, 1.1b, 3a–b.

101. Huang Zongxi, *Yixue xiangshu lun* (Discussion of the images and numerologies associated with the study of the [Book of] changes, written ca. 1661) (Taipei: Guangwen Shuju, 1974), 1.14–15.

102. Huang Zongyan, *Yixue bianhuo* (Exposé of delusions regarding the study of the [Book of] changes, seventeenth century), 5b–6a, in *ce* 132 of *Zhaodai congshu* (A collection of books from our illustrious era, 1876), ed. Zhang Chao.

103. Benjamin A. Elman, "Geographical Research in the Ming-Ch'ing Period," *Monumenta Serica* 35 (1981–83): 1–18, esp. 15.

Countercosmography and Anticosmography in Qing Thought

The seventeenth-century cosmographical reformation outlined above was confined largely to the sphere of high intellectual history. Throughout the Qing era, the premises and principles of geometric cosmography continued to inform popular culture and even imperial ritual. But the seventeenth-century cosmological critics apparently left their mark on fields as diverse as landscape painting, moral philosophy, and astronomical and geographical thought. At least there is a certain congruence between the sort of cosmos these critics imagined and that conceived by several prominent Qing painters, philosophers, astronomers, and geographers.

Seventeenth-century landscape painters, in particular, might well have sympathized with the remark of the mid-Qing cosmological critic Zhang Huiyan (1761–1802) that "the ways of heaven and earth are invariably uneven and irregular."[104] Modern historians of Chinese landscape painting have noted the "almost painful distortions" and "tortured forms" in late Ming and early Qing landscape art.[105] In contrast, the work of one of the great Song-era masters, Li Cheng (919–67), has been aptly characterized as animated by a "conviction of a coherence and order underlying surface appearances in nature, the same conviction that inspired Song philosphers to erect the vast and orderly structure of the Neo-Confucian cosmology."[106] A famous painting, *Xishan xing lü* (Travelers among streams and mountains), by Fan Kuan (active ca. 990–1030), a near contemporary of Li's, likewise presents "a cosmological vision as grandiose and all-encompassing" as that found in the cosmographical diagrams devised by the great Song Neo-Confucian cosmologist Shao Yong (1011–77).[107] Such grand visions of cosmic order and coherence were apparently no longer possible, or at least were not very convincing, in Qing times. Inasmuch as Chinese landscape paintings were often conceived as cosmograms, schematic representations of heaven and earth, this is surely a significant development in the history of Chinese cosmographical thought, not just in the aesthetics of landscape painting.

A departure from the values of order, regularity, and symmetry that informed Neo-Confucian cosmology and traditional cosmography is also evident in the moral philosophy of the Qing era. Whereas Song Neo-Confucians regarded imbalances and asymmetries in the constitution of the cosmic *qi*, the energetic pneuma, as the cosmological source of evil in the world,[108] some Qing philosophers accepted and even celebrated such irregularities. Yan Yuan (1636–1704), for example, remarked that deviation from the mean was potentially as moral a position as balance and centrality. For "although open and obstructed, straight and bent all have their differences, the pneuma and pattern that fills the universe is the same everywhere."[109] Hence proper moral cultivation might begin from any point on the moral map, not just the mean position. Yan, indeed, included several such maps in his *Cunxing bian* (Treatise on preserving the nature) as a guide to show how those with different endowments or propensities might improve their moral stance.[110] Although these "maps" are not geographical in any sense, they do illustrate how graphic representations or *tu* came to be applied in moral philosophy as well as in the sciences of the heavens and the earth.

The seventeenth-century recognition of imbalance, asymmetry, and irregularity in the world and a preference for natural rather than geometric lines of division appear in the astronomical and geographical thought of the period, as well as in its landscape art and moral philosophy. These inclinations even influenced the Chinese reception of the European cosmography and astronomy introduced into late Ming China by Jesuit missionaries. Wang Fuzhi, for example, rejected the theory of the spherical earth propounded by the Jesuits, arguing that "since it is in some places level, in others steep, in some places recessed and in others convex, then wherein lies its sphericity? . . . Thus from the earth's inclines, irregularities, heights, depths, and vastness, it is clear that it has no definite form."[111]

---

104. Zhang, *Yi tu tiaobian*, 28a (2:1007) (note 39).

105. Michael Sullivan, *A Short History of Chinese Art* (Berkeley and Los Angeles: University of California Press, 1967), 226; James Cahill, "Style as Idea in Ming-Ch'ing Painting," in *The Mozartian Historian: Essays on the Works of Joseph R. Levenson*, ed. Maurice Meisner and Rhoads Murphey (Berkeley and Los Angeles: University of California Press, 1976), 137–56, esp. 149.

106. James Cahill, *Chinese Painting* (Geneva: Editions d'Art Albert Skira, 1960; reprinted New York: Rizzoli International Publications, 1977), 32.

107. Michael D. Freeman, "From Adept to Worthy: The Philosophical Career of Shao Yong," *Journal of the American Oriental Society* 102 (1982): 477–91, quotation on 484.

108. See, for example, Zhang Zai (1020–77), *Zhangzi zhengmeng zhu* (Master Zhang's correcting youthful ignorance with commentary) (Taipei: Shijie Shuju, 1967), 3.92–93; and Cheng Hao (1032–85), *Henan Chengshi yishu* (Surviving works of the Chengs of Henan, 1168), ed. Zhu Xi, in *Er Cheng quanshu* (Complete works of the two Cheng [brothers], collected 1323), *Sibu beiyao* edition, 2A.1b.

109. Yan Yuan, *Cunxing bian* (completed 1669, first printed 1705), 2.29–30, in *Sicun bian* (Treatises on the four preservations) (Taipei: Shijie Shuju, 1966).

110. For graphic illustration of Yan Yuan's moral maps, his seven "diagrams of human nature" (*xing tu*), see Yan, *Cunxing bian*, 2.24–33 (note 109). Some of these diagrams, with a more detailed explanation of their contents, may also be found in John B. Henderson, *The Development and Decline of Chinese Cosmology* (New York: Columbia University Press, 1984), 236.

111. Wang Fuzhi, *Siwen lu, waipian* (Record of intellectual inquiry, outer chapters, seventeenth century), 63, in *Lizhou Chuanshan wushu* (Five books by [Huang] Lizhou and [Wang] Chuanshan) (Taipei: Shijie Shuju, 1974).

Seventeenth-century scholars and astronomers criticized European celestial cartography and calendrics on similar grounds, arguing that Western astronomers had overlooked natural irregularities and anomalies in their quest for geometric order and mathematical precision. Wang Xichan (1628–82), for example, contended that the Western division of the celestial sphere into 360 degrees was an unnaturally neat schema whose only merit was that it aided computation.[112] Like other prominent scholar-astronomers of his era, Wang preferred to measure the great circles of the sphere by 365¼ degrees (corresponding to the number of days in a tropical year) as Chinese astronomers had done since the Han era. The units in this cartographical schema corresponded to a natural measure of celestial time and space, the sun's daily movement on the ecliptic, even though they added up to a rather irregular and unwieldy sum. Abandoning this natural measure for an artifically uniform schema, said Wang Fuzhi, would be as senseless as arbitrarily rounding off the lengths of Chinese dynasties in order to produce historical units of uniform duration.[113]

Both Wang Xichan and Mei Wending (1633–1721) raised similar arguments against the Western solar calendar introduced by Jesuit missionaries into seventeenth-century China. They objected particularly to the constitution of the months in the Western calendar, pointing out that they did not correspond to lunations or any other natural cycle but were merely conventional divisions of the solar year.[114] The Chinese lunisolar calendar, these scholar-astronomers admitted, had major drawbacks, particularly the problem of intercalation. But the native calendar at least made it possible for cartographic and calendrical standards to conform to natural divisions and cycles, in this case those of the sun and the moon.

Some Ming and Qing scholars posed yet more radical criticisms of both traditional Chinese and contemporary Western cosmographical standards. They did not stop with the contention that demarcations of space and time should follow naturally irregular contours and cycles. Instead, they went on to devise a sort of anticosmography or anticartography that looked askance at the formulation or imposition of clear, definite boundaries of any sort—spatial, temporal, or cosmographical. Wang Tingxiang (1474–1544), for example, objected to a Han-era cosmologist's division of the year into two finely distinguished units of yin and yang months (corresponding to the cold and hot seasons, respectively) on the grounds that these two pneumatic powers were inextricably intermingled: "There was never a case of pure yin without yang or pure yang without yin."[115] Lü Kun (1536–1618) asserted that the famous circular diagram of the yin and yang that depicts them as separated by an S-shaped curve also distorts the face of reality, for the interactions between the yin and yang are too subtle to be portrayed graphically.[116] Even heaven and earth, said Fang Yizhi, mutually interpenetrate in such a way as to preclude drawing a sharp line of distinction between the two.[117]

Thus the cosmographical ideas of late Ming and Qing scholars contravened those that dominated earlier cosmographical thought in China. Yet there are some Chinese antecedents for these critics' departures from the schematic cosmographical conceptions of their orthodox predecessors. One of these might be found in Chinese geomancy, with its preference for naturally undulating and meandering lines, as noted above. Philosophers associated with Daoism, who celebrated the anomalous, the irregular, and even the freakish aspects of nature, might also have struck a responsive chord with seventeenth-century cosmological critics, some of whom wrote commentaries on the Daoist classics. Those who did so, or were otherwise familiar with heterodox literature, might well have appreciated the following passage from an important syncretic text of the fourth century A.D., the *Baopuzi* ([Book of] the master who embraces simplicity):

> Life and death, beginning and end, are indeed the great laws of the universe. Yet the similarities and differences of things are not uniform. Some are this way and some are that. Tens of thousands of varieties are in constant change and transformation, strange and without any definite pattern. Whether things are this way or that, and whether they are regular or irregular in their essential and subsidiary aspects, cannot be reduced to uniformity.[118]

But unlike Daoist philosophers, Qing scholars sometimes attempted to apply such principles and precepts practically to the sciences of the heavens and the earth, particularly astronomy and geography. Whereas Browne saw quincunxes everywhere, "in Heaven above . . . in

112. "Zi xu" (author's preface) of Wang Xichan, *Xiaoan xinfa* ([Wang] Xiaoan's new system, 1663), in *Xiaoan yishu* (Surviving works of [Wang] Xiaoan, comp. ca. 1682), 1.2b, in *ce* 31 of *Muxi xuan congshu* (A collection of books from Muci Studio), comp. Li Shengduo, 40 vols. (1883–91).

113. Wang, *Siwen lu, waipian*, 53 (note 111).

114. Wang, *Xiaoan xinfa*, 1.2a–3a (note 112); Mei Wending, *Lixue yiwen* (Queries on astronomical studies, presented 1702), pt. 2, in *Meishi congshu jiyao* (Epitome of Mei's collected works, printed 1771), ed. Mei Gucheng (d. 1763) (Taipei: Yiwen Yinshuguan, 1971), 47.1b.

115. Wang Tingxiang, *Jiacang ji* (Writings for the family repository, published ca. 1636–37), in *Wang Tingxiang zhexue xuanji* (Selected philosophical works of Wang Tingxiang, 1965) (Taipei: He-Lo Tushu Chubanshe, 1974), 167.

116. Lü Kun, *Shenyin yu* (Groaning words, 1593) (Taipei: He-Lo Tushu Chubanshe, 1974), 1.53.

117. Fang Yizhi, *Dongxi jun* (The adjustment of things, 1653) (Beijing: Zhonghua Shuju, 1962), 95–96.

118. Ge Hong (283–343), *Baopuzi*, *Sibu beiyao* edition, pt. 1 of 8, 2.2a. I have followed the translation of this passage given in *Sources of Chinese Tradition*, 299 (note 17).

earth below," and Han-era cosmographers applied the nonary square, the nine-palace formation, to the ordering of practically every realm of space, some Ming and Qing scholars regarded anomaly (*cha*) and irregularity (*buqi*) as constitutive of the structure of space and time.[119] Anomalies appeared in the heavens in such forms as the precession of the equinoxes, the "annual difference" (*suicha*) between the lengths of the sidereal and tropical years, the inequalities of solar and lunar motion, and the retrograde motion of the planets. On the earth, irregularities took the form of topographical peculiarities and the historical changes in landforms and landscapes that so disturbed seventeenth-century scholars like Gu Yanwu.[120] Such anomalous movements were not simply complicating factors for which adequate adjustment could be made in calculation. Rather, they were woven into the fabric of the cosmos. They made it unlikely that any cosmographical, astronomical, or geographical model, not just the nonary grid, could be serviceable for very long at a time.[121]

Conclusion

Despite the cosmological criticisms expressed by seventeenth-century scholars, vestiges of geometric cosmography and even of the nonary grid have survived in modern Chinese culture. Even the centerpiece of modern-day Beijing might well be regarded as modeled on the nonary square, or at least on the Greek cross. At the center of the capital is the largest public square in the world. To the north stand the monumental Gate of Heavenly Peace (Tiananmen) and the old imperial palace, and due south loom the equally massive Front Gate (Qianmen) and the Mao Zedong Memorial Hall. Symmetrically disposed to the east and west of the square are the Museum of Chinese History and the Great Hall of the People. The total effect of this complex is overpowering in its monumental symmetry.

Inasmuch as most of the structures in this array, including the large public square that stands at the center, were established or erected under Communist rule, their disposition might well be taken as evidence that China's contemporary leaders, like their imperial predecessors, regard the proper ordering of space as an effective means of political control. Indeed, before the advent of such modern-day mass media as newspapers, radio, and television, the ordering of space may well have been the most widely employed instrument of political and cultural propaganda in many civilizations, especially those of South and East Asia. Insofar as the traditional cosmographical orientations outlined in this chapter persist among the folk in such cultures, cosmographical models may still be used to accomplish political goals like enhancing the legitimacy of the ruling powers who have created a model of heaven on the surface of the earth, an order not to be marred by such asymmetrical monstrosities as a goddess of democracy.

119. For illustrations of this viewpoint, which was anticipated by earlier Chinese astronomers, see Wang Tingxiang, *Shenyan* (Prudent words, 1533), in *Wang Tingxiang zhexue xuanji*, 56 (note 115); and Jiang Yong (1681–1762), *Shu xue* (Mathematical studies, ca. 1750) (Shanghai: Shangwu Yinshuguan, 1936), 1.26, 28–29.

120. For Gu's comments on how the physical geography of China, particularly of the once prosperous Northwest, had changed since antiquity, see his essay on the "Jiu zhou" (nine regions) in *Yuanchaoben Rizhi lu* (The original manuscript version of the Record of daily knowledge, 1958) (Taipei: Minglun Chubanshe, 1970), 23.626–27.

121. There is a superficial resemblance between Qing scholars' anticosmography and contemporary chaos theory, which poses "a geometry of the pitted, pocked, and broken up, the twisted, tangled, and intertwined." For just as Qing scholars elevated anomalies and irregularities to the status of ultimate principles, so chaos theorists look upon "pits and tangles" as "more than blemishes distorting the classic shapes of Euclidean geometry. They are often the keys to the essence of a thing." James Gleick, *Chaos: Making a New Science* (New York: Viking, 1987), 94.

9 · Concluding Remarks: Foundations for a Future History of Chinese Mapping

CORDELL D. K. YEE

The maps studied here were generally products of a highly educated elite. The members of this elite possessed the means to produce maps like the *Yu ji tu* (Map of the tracks of Yu), which appear to have a mathematical and mensurational foundation, make consistent use of signs, and stand independent of text. More often than not, however, members of the elite chose not to make such maps. They regarded other interests as important in the production of cartographic images: aesthetic, religious, and political.

Of course, one can also see those interests at work in Western cartography, especially before the fifteenth century. But late in that tradition, after the Renaissance, they tend to become marginal to a central drive toward the mathematizing of landscape. In traditional Chinese cartography, that drive did not suppress other impulses. Chinese mapmaking was resistant to the idea that space should be homogenized to aid quantification. Chinese mapmakers remained acutely aware of locality. Part of this difference is a question of viewpoint: the mapmakers studied here did not make their living by cartography, but were part of an elite oriented toward civil service. The elite worked in localities; as administrators, its members had to be sensitive to local peculiarities. Also contributing to this sense of particularity of place was a strong interest in textuality. The map image worked in tandem with language to communicate something about space and place. As a result, study of place meant more than carrying out a survey: it involved studying the texts associated with that place. At least one more implication of this close relation between image and text is worth remarking. As more than a few titles of the works dealt with in the previous chapters illustrate, the classical Chinese language is highly metaphorical, and so are traditional Chinese maps. The term "metaphorical" is meant to be contrasted with "literal": what was communicated by means of maps did not have to be numerical, measurable, or even directly perceivable. A traditional Chinese map might give a mathematical interpretation of the earth or the heavens, but it might also serve as an instrument of political persuasion, give form to emotional states, or even afford access to transcendent beings. In such ways, maps functioned as a form of rhetoric.

In light of the close relation between text and image, this intersection between the verbal arts and cartography seems more than accidental.[1] The range of information conveyed through maps cut across modern disciplinary boundaries; the split between the sciences and the humanities did not apply. Thus, to achieve literacy in traditional Chinese cartography, one needs grounding in the history of science and technology, art, literature, government, economics, religion, and philosophy—in short, the polymath range of the mapmakers. We have tried to see traditional Chinese mapping practices from the inside by adopting an interdisciplinary approach to Chinese maps. We have found that in a cartographic practice so enmeshed in textuality, measurement itself became a metaphor. It involved seeing something in terms of something else: for example, a wall in terms of "paces" (*bu*) or other arbitrary units of measure. As a way of seeing, measurement was not necessarily regarded as the standard of truth.

For this reason, we have taken exception to an approach that views Chinese cartographic history as tending toward scale mapping, as following a course similar to that of Western cartography, as anticipating much of what makes modern cartography "modern." To be sure, as I suggested in a previous chapter, many of the elements needed to formulate a "modern" style of cartography involving projection from sphere to plane were probably available to Chinese intellectuals by the Yuan dynasty (1279–1368): a spherical conception of the earth and a means of locating points on the earth's surface through astronomical measurements. These elements, however, did not find expression in Chinese mapmaking until Jesuit attempts to Westernize Chinese cartography.

The emphasis in much current research on finding antecedents for modern conceptions and practices has obscured the strength of traditional Chinese cartography. Contributing to this narrow perspective is the tendency

1. Somewhat ironically, the connection is hinted at in the modern Chinese term for scale: *bilichi*. The graph *bi* is also the term for metaphor or simile in traditional Chinese literary criticism. Scale can thus be understood as a yardstick or rule (*chi*) for interpreting the metaphorical relation between the standards (*li*) of map and reality.

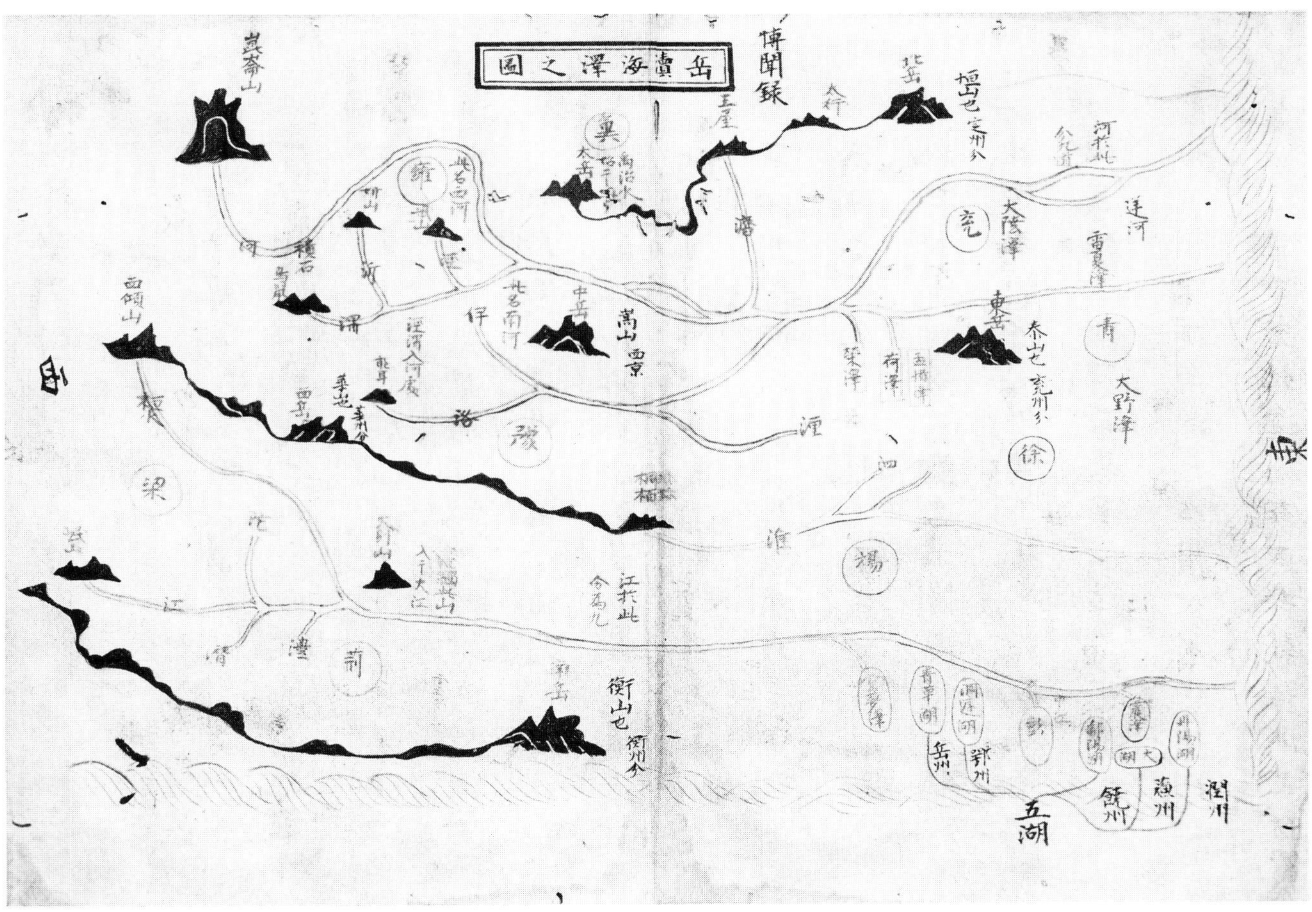

FIG. 9.1. JAPANESE COPY OF A MAP FROM A SEVENTH-CENTURY CHINESE TEXT. This is a Japanese rendition of a map from the *Wuxing dayi* (Main principles of the five phases [or elements: wood, fire, earth, metal, water], ca. 600) by Xiao Ji (d. 614). The map and the text were copied by a Japanese Buddhist priest in the fourteenth century. The map is the only known representation of the original: extant Chinese editions no longer contain it. Titled "Yue du hai ze zhi tu" (Map of mountains, streams, seas, and marshes), it is a general map of China. It is bounded on two sides by oceans and gives the locations of sacred mountains and rivers in black ink. Other place-names are written in red ink. The nine names that are enclosed in circles are the nine regions described in the "Yu gong" (Tribute of Yu). North is at the top.
Size of the original: 33.3 × 46 cm. By permission of Taiichi Takemoto, Gamagori Aichi, Japan.

of research on cross-cultural transmission of ideas to focus on the Westernization of non-European cultures. What has been overlooked—and this is true even of researchers working in China—is the influence of Chinese practices on non-Chinese cultures. Cartographic works, as the following chapters on Korea, Japan, and Southeast Asia document, provide much evidence of this. Many Chinese maps are known through copies made by foreign hands (fig. 9.1).[2] Techniques and instruments used in cartography and graphic production also found their way from China to Japan, Korea, and Southeast Asia. In saying this, I am not suggesting that Chinese cartography so overwhelmed the indigenous practices that they were assimilated into it. That would repeat the mistake once made in the study of the relationship between European and Chinese cartography. The cultures in and around China's sphere of influence did develop their own distinctive map styles and genres, and they responded to Western cartography in different ways. In other cultures, for example, the relation between text and image does not seem to have been quite as strong. Thus European mapping seems to have been adopted more quickly in

2. For discussion of a Korean copy of a Chinese map different from the one illustrated here, see Marcel Destombes, "Wang P'an, Liang Chou et Matteo Ricci: Essai sur la cartographie chinoise de 1593 à 1603," in *Actes du Troisième Colloque International de Sinologie: Appréciation par l'Europe de la tradition chinoise à partir du dix-septième siècle* (Paris: Belles Lettres, 1983), 47–65; and Kazutaka Unno, "Concerning a MS Map of China in the Bibliothèque Nationale, Paris, Introduced to the World by Monsieur M. Destombes," *Memoirs of the Research Department of the Toyo Bunko (the Oriental Library)* 35 (1977): 205–17.

Japan than in China, and less quickly in Korea. Even though cultures in those places developed their own cartographic traditions, there is no denying the power and influence of Chinese mapping practice. It is hard to reconcile this evidence with judgments of "backwardness" and failures of accuracy.

The story told in the preceding chapters is still incomplete. The aim there has been to be more suggestive than definitive, since the body of traditional Chinese maps is too large to be dealt with justly even in two books of this size. There are also a number of unresolved issues and topics that require further research.

One regrettable omission from our account of Chinese cartography is the influence of Buddhism. We have presented a few artifacts associated with Buddhism, a map of a grave site, a mural of Wutai Shan, and two maps of Jambūdvīpa, but we have interpreted them in the larger context of Chinese cartography as a whole. A systematic account of Buddhist influence has not been attempted.

This omission requires a few words of explanation. For the crucial period from the first century through the ninth century, when Buddhism spread throughout China and won official patronage under some rulers, there are no cartographic artifacts. For later periods there are a few artifacts, such as the ones already illustrated, but these provide little evidence of a distinctive Buddhist cartography in China. Examples of Buddhist maps based on Chinese originals have been found in Korea and Japan, and these are discussed in the chapters on those cultures. Modern editions of the Chinese Tripiṭaka (*Da zang jing*) contain a few ritual diagrams showing the placement of objects and some architectural drawings, but again these seem to provide little evidence of a practice separate from those described previously.[3] These are, however, initial impressions that should be tested in a more rigorous treatment.

A similarly regrettable omission is Daoist cartography. Daoism was organized into a number of sects that produced a corpus of sacred texts known as the Daozang (Daoist canon). The canon contains numerous examples of labyrinthine diagrams, nonary charts, star maps, floor plans, and pictorial maps. Study of these maps has just begun, and initial findings tend to reinforce some of the conclusions reached here. The maps in the Daoist canon served a variety of ritual and religious purposes: for example, as aids to communicate with the spiritual world, or as part of the accoutrements for rituals, worn as hats or even ingested. They appear to have been made by an elite—religious adepts, often members of the aristocracy. As members of the upper stratum of society, these scholars placed importance on text, and often their maps cannot be understood independent of text. Unfortunately, time has not permitted a thorough account of Daoist cartography here.[4]

The chapter on cosmographic mapping has pointed to the religious underpinnings of traditional Chinese cartography. Studies of Buddhist and Daoist cartography would deepen and broaden our understanding of those underpinnings. They would also help counterbalance the secularist bias in much recent work in the history of Chinese cartography.

Basic cartobibliographic problems still need to be resolved. One is the development of a suitable vocabulary for describing Chinese maps. This is lacking even from traditional Chinese bibliographic practice, which simply lists titles of maps along with names of authors (if known). Current terminology is also not quite adequate. For example, the "single-sheet" or "loose-sheet" map trade of commercial publishers did not develop in China as in Europe, so many means of categorizing or dating or comparing maps useful to cartobibliographers of Western maps are unavailable to those working with Chinese maps. It is often meaningless to speak of "editions" or "states," since many maps were hand copied. Until someone has determined systematically how much fidelity was required for one map to be called a copy of another, any hypothesis about chronological sequence must be to some extent arbitrary. A clear chronological sequence would provide a firmer basis for discussing stylistic progression and the development of map genres. The practices of copyists in Japan and Korea also need to be studied in order to to determine the usefulness of purported copies to the study of Chinese cartography. It is important to know what, if anything, a fourteenth-century Japanese rendition of a seventh-century Chinese map (see fig. 9.1) allows one to infer about the Chinese original. Such basic research would permit the development of connoisseurship of Chinese maps—something needed since artistry was central, not peripheral, to traditional Chinese mapmaking.

The artifactual record raises at least one more question. If maps were important to the endeavor to understand the cosmos and to preserve tradition, the question arises why so few maps before the Song dynasty (960–1279) have survived. One possible answer is that practically no book illustrations from before the Song have survived, and the attrition of books before the invention of printing was enormous. But if reproducing maps was difficult before the Song, one might reason, maps would be that much more valuable and thus worthy of preservation. This reasoning holds true to a certain extent. The military

3. For Buddhist cartography as it developed elsewhere in Asia, see esp. pp. 254–56, 371–76, 619–38, 714–40, 777–84.

4. But see Judith M. Boltz, "Cartography in the Taoist Canon," *Asia Major*, forthcoming. Boltz has also published a book that reproduces some examples of Daoist maps: *A Survey of Taoist Literature: Tenth to Seventeenth Centuries* (Berkeley, Calif.: Institute of East Asian Studies, 1987).

and administrative usefulness of maps was recognized, and they were collected by the central government and stored in archives. Warfare and the attendant destruction of archives can then be blamed for some loss of maps, as well as other documents. Another cause of the destruction of original maps, however, may be found in the very institutions that encouraged their production. The practices of historical scholarship, the very process of preserving the past, may have also contributed to the loss of original maps. Original documentary materials do not seem to have possessed intrinsic value for their originality; what mattered most was their content. To write a history was primarily to edit and select from original documents rather than to construct a narrative. Once the information from the source material was incorporated into the official account, the original seems to have been considered dispensable. Maps were part of this documentary culture. Once the informational content of a map was incorporated into a description in a geographic treatise of an official dynastic history, for example, it was likely that the map would be discarded. Its visual appearance, as modern scholars have demonstrated, would have been reproducible from the verbal description.[5] Historians of Chinese mapping thus confront the somewhat paradoxical situation of being both reliant on and hindered by the practices of traditional Chinese historiography: political institutions fostered both the production and destruction of the artifacts essential to the historian's work.

A similar problem persists in modern accounts of traditional maps. More interest is shown in the informational content of the map than in its physical description. If a map cannot be studied for content, it almost disappears from the record. Examples of this are the third map found at Mawangdui and the fragment found at Fangmatan, which apparently have not been given much scholarly attention since the reports of their discovery. Though fragmentary, they could shed light on how maps were produced. Our knowledge of map production is based on inferences from other branches of the graphic arts, such as book printing, painting, and calligraphy. We still do not know if Chinese scholar-officials received any training specifically in mapmaking: How did one go about making fish-scale maps, for instance? What was considered important in their manufacture? Answers to such questions seem to require more attention to the craft of mapmaking, and again the allied graphic arts may provide more insight into that craft. Painting, calligraphy, and mapmaking, for example, are all related by their reliance on line. But whether and how the expressiveness of line in painting and calligraphy is transferred to mapmaking is still unknown.

As I suggested above, the range of unanswered questions about traditional Chinese maps is enormous. That fact alone is enough to suggest that the study of traditional Chinese cartography involves far more than comparing traditional Chinese maps with modern works and calculating how small the "errors" are. If our work has succeeded in any way, it is in showing that such a "modernist" approach cannot take the measure of traditional Chinese cartography.

5. For an example of modern reconstructions of Chinese maps from textual sources, see D. D. Leslie and K. H. J. Gardiner, "Chinese Knowledge of Western Asia During the Han," *T'oung Pao* 58 (1982): 254–308.

Cartography in Korea, Japan, and Vietnam

10 · Cartography in Korea

GARI LEDYARD

INTRODUCTION

Koreans have been making and using maps for more than fifteen centuries. Since most of their country's borders were naturally determined by the sea, they had a general concept of Korea's outline at an early date, and their deep consciousness of *samch'ŏlli kangsan* (three thousand *li* of mountains and rivers) gave their mapmakers a general idea of what went within that outline.[1] Underlying these imprints on the national psyche were a strong tradition of administrative and cultural geography and a nationally conceived theory of geomantic analysis. All these factors contributed to the production of interesting maps. While naturally emphasizing their own country, Korean cartographers also showed an enduring interest in the shape of their neighbors' lands and territories; and looking beyond these to the greater world, they produced several carefully studied world maps as well as more traditional cosmographies. Just as Korea's culture freely absorbed many of the features and institutions of Chinese civilization yet retained a strong individual Korean identity, so too Korea's mapmakers, applying general cartographic norms developed in China, adapted these norms to their own circumstances and created maps of both utility and beauty.

That much said, by East Asian standards the antiquity of Korea's surviving cartographic artifacts is not great. As in other countries, time, war, and carelessness have taken a heavy toll on all written artifacts, but especially on paintings and maps. The oldest Korean map to survive today is an important world map dated 1402 (known in three copies, of which the earliest was made around 1470). But even that date is early in terms of the surviving cartographic corpus taken as a whole, which dates mostly from the sixteenth to the nineteenth century. For maps before 1402, we must rely on written records and reasonable inferences that can be based on the general trends of East Asian and Korean cultural history. An inquiry along these lines will show that whereas mapmaking before 1402 emphasized the nation and its local districts, a twelfth-century scholar had already produced a map of the world along Buddhist lines, and a fourteenth-century man had compiled a historical map of Korea and China. The description of the latter is conceived in terms very similar to those evident in the 1402 world map and provides an appropriate link from the unseen to the visible corpus.

It seems convenient to organize Korean maps into four broad categories, proceeding from the more general world and national maps to the more particular regional and local ones. Although this scheme will involve a few chronological discontinuities, the existing corpus is such that most of the more interesting world maps appear relatively early, whereas the great majority of local and topical maps come from the later centuries.

The category of world maps is very heterogeneous, including a few genuine maps of the world, a great variety of East Asian regional maps, and the numerous prints and copies of the quasi-cosmographical *ch'ŏnhado*. Korean scholars sometimes use this term, which can be broadly translated "world maps," for this whole group, but in this chapter it will be reserved for the popular and generally recent maps, often with the terms *ch'ŏnha* or *ch'ŏnhado* in their titles, that present the Sinocentric world—China, Korea, and their proximate East Asian neighbors—surrounded by peripheral rings of exotic or mythical lands and peoples. The origin and development of the *ch'ŏnhado* presents many problems on which scholars still have their differences, but there is no disagreement on the great vogue these maps enjoyed during the eighteenth and nineteenth centuries. They easily account for most of the world map category. Although in terms of the development of maps they are late and perhaps better explained in terms of folklore than of science, still they had a real place in Korean life and have their own absorbing story to tell. The cartographically more significant world maps and East Asian regional maps, though less numerous, have infinitely more variety than the *ch'ŏnhado*, and they generally come earlier. Explaining this seemingly reversed typological development will be one of my major tasks.

1. The phrase *samch'ŏlli kangsan* has long been a part of Korean folklore. The significance of "mountains and streams" in Korean national geomantic theory is evident as early as the tenth century; see the discussion beginning on p. 276 below.

Then there were maps of Korea alone, understandably a large and varied category. The oldest cartographic depiction of the country to survive is the representation of Korea on the 1402 world map, although we have a number of written references to earlier national maps, including one interesting description of a map of Korea said to have been made in the twelfth century or earlier. During the fifteenth century there was an abundance of geographic research, but unfortunately none of the many maps known to have been produced in that period seem to have survived to modern times. However, a map completed in 1463 by Chŏng Ch'ŏk had great influence and is believed to have been taken as a model by later mapmakers, so that we have a reasonably good idea of how the peninsular outline was conceived as well as of the cartographic detail involving rivers and mountains, place-names, and other features. During the early eighteenth century the mapmaker Chŏng Sanggi and his family achieved a genuine revolution in cartographic technique, producing a dramatically improved understanding of the nation's borders, both the long coastlines and the much harder to grasp northern frontier. These techniques were refined and perfected by the nineteenth-century master Kim Chŏngho, who was both a mapmaker and a publisher and popularizer. Although he was familiar with Western mapmaking techniques and made use of geodetic coordinates in his work, the visual appearance of his late traditional maps stayed completely within the evolutionary lines of native cartographic practice. Korea's shift to the styles and methods of Western cartography occurred only toward the end of the nineteenth century, as the nation struggled to come to terms with a new Western world order led (as far as Korea was concerned) by Japan, which was much more threatening than reassuring.

Provincial maps were popularized in the late fifteenth century as part of an important compendium of administrative geography, and they achieved high levels of quality in the eighteenth century, when Chŏng Sanggi made maps of all the provinces on a unified scale, so that they could be used as separate maps or combined to make a single national map. Reforms introduced in 1791 promoted extensive local surveys and were a key impetus both to the mapping of counties and towns and to the compilation of local histories. But whereas national and provincial maps came to achieve a certain level of standardization and cartographic professionalism, town and county maps were made by a great variety of local hands, some very skilled, others quite crude. As we shall see, the background of these country mapmakers was more in painting and drawing than in cartography, and the results are evident in hundreds of local maps that might also pass for bird's-eye-view landscapes, a style that is also well documented for China (see above, chap. 6, esp. pp. 135–37, 144–47).

The last of the four major categories of Korean maps is the so-called defense map, or *kwanbangdo* in the traditional term. These range from long scrolls representing frontiers thousands of *li* long and reaching far beyond Korea to maps of local mountain fortresses. The variety is very great. Many were mounted on screens that probably stood in the offices of defense officials in Seoul or provincial governors; others were in more portable scrolls or folios that were an essential part of the equipment of frontier commanders and military officers. One very interesting variety of defense map was oriented to coastal defense and navigation. The evident purpose of maps in this category was principally to clarify terrain and communications from a military perspective, while cartographic scale, so important in the later national and provincial maps, is decidedly a lower priority. The considerable skill and painterly talent evident in these maps shows that they were mostly made and used in the central government or high military commands, where the resources for maintaining staff artists and mappers were readily available.

The Present State of Korean Cartographic Research

As I have already shown, the corpus of existing Korean maps goes back nearly six hundred years. But the organization of this corpus into a field of study has mainly occurred in the past forty years. Most of the extant artifacts of this long tradition are in Korean collections. The Library of Congress in Washington, D. C., has a substantial Korean map collection, which while mainly devoted to modern holdings also has some significant premodern items, including many of the maps collected by Shannon McCune during his years in Korea. There are some important individual maps in Japanese and other foreign map collections. Aside from an unnamed collection (or collections) of Korean maps in the northern Democratic People's Republic of Korea,[2] the chief holdings are in the southern Republic of Korea. Of these the most important are those of the National Central Library (Kungnip Chung'ang Tosŏgwan) in Seoul and the Seoul National University Library (Sŏul Taehakkyo Tosŏgwan). The latter contains, in addition to its general cartographic collection, the important Kyujanggak (Royal Library), founded in 1776 and built around the books and writings

2. See Mok Yŏngman, *Chido iyagi* (Map conversations) (P'yŏngyang: Kunjung Munhwa Ch'ulp'ansa, 1965). The illustrative material, presumably from northern collections, is of very limited use because of poor reproductions and paper.

assembled by the research librarians of King Chŏngjo (r. 1776–1800) and his immediate successors. The Kyujanggak has many excellent maps and is notable for its nearly complete collection of district (*ŭp*) maps, although most of these come from the very late years of the tradition.

Most other Korean universities also have cartographic collections. Although they cannot all be mentioned here, two of particular distinction are Koryŏ University Library (Koryŏ Taehakkyo Tosŏgwan) in Seoul and Soongsil (Sungsil) University Library in Seoul. The latter collection was assembled by Kim Yangsŏn, a lifetime student of maps who made many scholarly contributions to the field.[3] It contains some particularly important Western maps in the Sino-Jesuit tradition, including one of the few known copies of Matteo Ricci's world map of 1603. This collection is also distinguished for the relatively good dating of the maps in its catalog. Most Korean university collections have catalogs abounding in entries of the category "author unknown, date unknown." Of course this reflects the reality that in the great majority of cases Korean maps have no indication of either mapmaker or date. Still, many of these maps could be dated approximately with a little research and professional judgment; that they have not been limits the value of the catalogs as research tools.

The understanding of Korean maps would be very much less developed were it not for the efforts of Yi Ch'an (Chan Lee), for many years professor of geography at Seoul National University. In addition to a number of specialist articles, he has compiled a comprehensive, large-format album of Korean cartography, *Han'guk ko chido* (Old Korean maps).[4] It contains nearly 120 large reproductions, including 17 in color and many others on double- or triple-sized foldout pages. The tiger's share (Korea has no lions) of the illustrations come from the superb National Central Library (Kungnip Chung'ang Tosŏgwan) collection. In the back of the book is an excellent introduction to Korean maps, with a helpful English synopsis, a bibliography, and a list of the principal holdings of eight important map collections (including those named above). This work has virtually defined the field for the present generation of scholarship and has been indispensable in the preparation of this chapter. I have also benefited greatly from the shorter but conceptually different monograph by Pang Tong'in, *Han'guk ŭi chido* (Korean maps).[5] Pang takes a more developmental approach and provides a very useful summary of Korean cartographic methods. Unfortunately the small format of his book ruled out satisfactory illustrations. Apart from Yi's book, relatively little adequately reproduced illustrative material is easily available.

The research literature to date shows a pronounced emphasis on the bibliographical approach and publication by library organizations and specialists. This is fundamental, and more work along these lines is needed. But with few exceptions, Korean cartographic research so far has not been much concerned with the general cartographic context for Korean maps, especially in reference to Chinese and Japanese map studies. On the other hand, much remains to be done to connect Korean maps to the social, economic, intellectual, and art-historical trends of Korean history. Until we have better understanding along these lines, the full significance of many Korean maps will escape us.

The present essay commences with a summary of historical references to Korean maps and mapping before the fifteenth century. This leads to a discussion of the world map of 1402, the earliest Korean map to survive, and other world maps including those deriving from Western sources of the seventeenth century or later and medieval and late cosmographies based on East Asian traditions. These maps were all based on international literary or cartographic sources of one kind or another and were not the result of Korean explorations or surveys. I then investigate the cultural and technical foundations of Korean national cartography as it developed from the early fifteenth to the late nineteenth century, considering first maps of Korea as a whole, then maps of its provinces and localities, and finishing with the important genre of defense maps that flourished from the seventeenth century on. The conclusion considers the relation of Korean cartography to Korean social, cultural, and historical forces and ends with observations on the circumstances in which traditional cartography, which for several hundred years had strongly resisted the influence of Western cartography, expired virtually overnight when the latter reentered the Korean peninsula driven by the demonic force of Japanese imperialism.

Korean Maps before the Fifteenth Century

As we have seen, Korean cartography has roots in a distant past far predating "the world map of 1402," whose sophistication is in itself evidence of a long mapping his-

3. For the principal contributions of Kim Yangsŏn, whose pen name was Maesan, see *Maesan kukhak san'go* (Selected writings in Korean studies by Maesan) (Seoul: Sungjŏn Taehakkyo Pangmulgwan, 1972).

4. Yi Ch'an, *Han'guk ko chido*, map commentaries by Che Honggyu (Seoul: Han'guk Tosŏgwanhak Yŏn'guhoe, 1977). Note should also be made here of the same author's *Han'guk ŭi ko chido* (Old maps of Korea) (Seoul: Pŏm'usa, 1991). This is a completely new and different work, with over 250 illustrations in large format, mostly in color, and a greatly expanded text. Unfortunately it became available too late to be used in writing this chapter.

5. Pang Tong'in, *Han'guk ŭi chido* (Seoul: Sejong Taewang Kinyŏm Saŏphoe, 1976).

tory. In the course of surveying this preartifactual development, it will be convenient to give, along the way, a sketch of the broader trends in Korean history.

The earliest Korean state to be mentioned in historical sources is Ko Chosŏn (Old Chosŏn), so named to distinguish it from the later Chosŏn dynasty. The origins of this state are not known, but it was certainly in existence by the fourth century B.C. Although it conducted trade and war with Zhao and Yan, two northeastern Chinese states of the Zhanguo (Warring States) period (403–221 B.C.), and shared a border with the latter, it was a fully independent entity based on a local cultural tradition. Ko Chosŏn, whose territory was limited to the eastern Liaodong area and the northwestern part of modern Korea, was conquered by the armies of Han China in 108 B.C. and suffered the partition of its lands into four Chinese commanderies (*jun*), two of which lasted down to the early fourth century. During this time, most of the ancestors of the Koreans lived beyond the Chinese pale, principally the Koguryŏ and Puyŏ peoples in what is now the Dongbei area (formerly Manchuria) of China and the Mahan, Chinhan, and Pyŏnhan peoples in the southern half of the Korean peninsula. Chinese occupation never reached into the southern areas; for all practical purposes it was limited to the region of the modern provinces of North and South P'yŏng'an and North and South Hwanghae. However, the Han military authorities had trade and diplomatic relations with many of the southern peoples, and under the short Wei dynasty (220–65) these ties reached to Japan as well.

By the first century A.D. there was an effective Koguryŏ kingdom, usually beyond Chinese control; and by the third and fourth centuries the southern peoples had organized the states of Paekche, Silla, and Kaya (Korean historiography claims earlier legendary dates for all of these states). Kaya was absorbed into Silla in the sixth century, and Koguryŏ, Paekche, and Silla (the Three Kingdoms) coexisted in relations of alternating alliance or hostility until the year 668. For most of this time, China was divided into the Northern and Southern dynasties, with the north controlled mainly by non-Chinese regimes, so that the three Korean kingdoms suffered little Chinese pressure and were able to develop their own highly individual political and cultural institutions. There was Chinese cultural influence, but it was highly indigenized. Beginning with the unification of China in 589, however, the Korean states began to feel the pressure of Chinese expansionism, and each of them developed either military (Koguryŏ) or diplomatic (Silla and Paekche) strategies to resist it. Silla proved the most adroit, enlisting Tang Chinese help in destroying Paekche and Koguryŏ, thus emerging in 668 as the state called "Unified Silla" (668–935) by modern historians. During the years of the unification struggles (roughly 598–668), there was considerable Chinese institutional influence in all the kingdoms, but especially in Paekche and Silla. One price of the Tang-Silla alliance was the loss of all of Koguryŏ's Manchurian territory, and even some of its peninsular lands south of the Yalu River, to Tang China. From that time, in spite of a Koguryŏ element that survived in partnership with other Manchurian peoples in the state of Bohai (pronounced Parhae in Korean), Manchuria lay outside the limits of Korea's military power if not its political aspirations. By the middle of the eighth century Tang itself lost its position in the northeast, and China would not again establish its presence there until more than five hundred years had passed; for most of the intervening period the Khitans (Qidans), Jurchens, and Mongols dominated the region. In 668, the Korean frontier met that of its northern neighbors in the neighborhood of the modern city of P'yŏngyang, and only over long centuries regained all the land south of the Yalu-Tumen River line (figs. 10.1 and 10.24).

The earliest evidence of Korean maps comes from the Koguryŏ kingdom. A drawing that seems to be a town plan has been found painted on the wall of a tomb near Sunch'ŏn, about fifty kilometers north of P'yŏngyang in northern Korea. The map is labeled *Yodong sŏng* (Liaodong city) and shows, according to those who have seen it, walls, streets, buildings, and a river and mountains.[6] Evidently the occupants of this tomb, though far from Liaodong itself, wished to be associated with it in their eternal rest. During most of the fifth, sixth, and early seventh centuries, Liaodong was within Koguryŏ territory, and this particular tomb probably dates from the earlier part of that period. As with similar contemporary drawings known from China, and many more in both China and Korea almost down to modern times, this map partakes of some of the features of a painting; for instance, not only is the location of gates indicated, but the gates themselves are depicted.

The first literary evidence of mapmaking also comes from Koguryŏ. In 628, during a diplomatic interlude in its long struggle with the Sui and Tang dynasties, Koguryŏ presented a map of its territory, entitled *Pongyŏk to* (Map of the infeudated region), to the Tang court.[7] The title

6. Yi Chinhŭi, "Kaihō go Chōsen kōkogaku no hatten: Kōkuri hekiga kofun no kenkyū" (The development of postwar Korean archaeology: Studies of Koguryŏ wall-painted tombs), *Kōkogaku Zasshi* 45, no. 3 (1959): 43–64, esp. 51–53; the illustration on 52 is too poor to permit interpretation.

7. Liu Xu et al., *Jiu Tang shu* (Old history of the Tang, compiled 940–45), 199A.5321; see the edition in 16 vols. (Beijing: Zhonghua Shuju, 1975); Ouyang Xiu et al., *Xin Tang shu* (New history of the Tang, compiled 1032?–60), 220.6187; see the edition in 20 vols. (Beijing: Zhonghua Shuju, 1975). The report in *Samguk sagi* is derived wholly from these earlier Chinese notices and unfortunately contains no Korean perspective on this map; Kim Pusik (1075–1151), comp., *Samguk sagi*

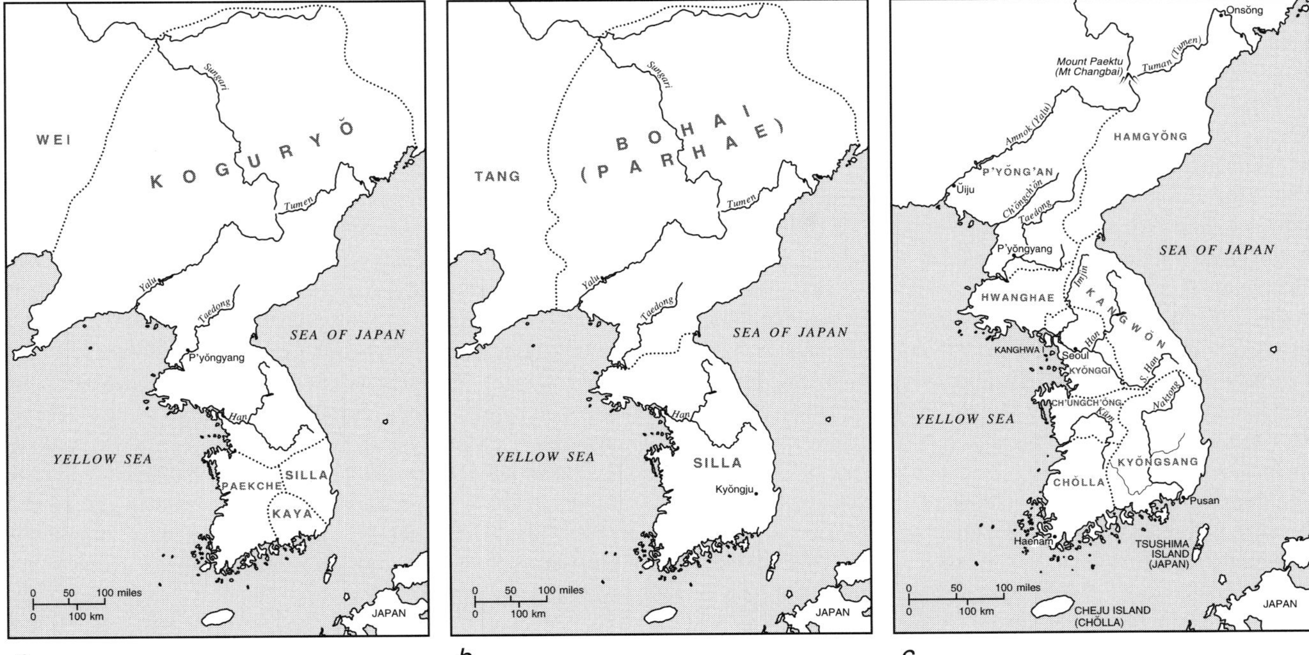

FIG. 10.1. (*a*) THE KOREAN KINGDOMS IN THE SIXTH CENTURY. Kaya was absorbed by Silla in 562. Paekche, the most active of the Korean states in cultural relations with China and Japan, was conquered by a Silla-Tang alliance in 660. Koguryŏ fell to the same combination in 668; its territory went largely to Tang, but by 712 it was the home of Bohai (Korean, Parhae), a state dominated by proto-Jurchen peoples and Koguryŏ military elements.
(*b*) The kingdoms of Silla and Bohai from the late seventh to the early tenth century. Silla was absorbed by Koryŏ in 935; Bohai was overcome by the Khitans (Liao) in 927. Korean peoples had played a leading role in the Bohai state, but with its fall they permanently lost their position in Manchuria.
(*c*) Korea during the Chosŏn dynasty (1392–1910), showing provincial and district names mentioned in the text. For details of the evolution of the northern frontier, see figure 10.24.

neatly sums up the tributary rhetoric of the era but gives no hint of any cartographic detail. Such a map was probably a part of the tributary ritual at that time. In spite of a long (though by no means continuous) tributary association with Chinese dynasties, however, this is the only known case of the ritual presentation of a Korean map.

The kingdom of Paekche, in the early seventh century just before the unification wars, is known to have used "maps and registers" (*tojŏk*) in its local administration.[8] Although we find no reference to maps in Silla's written remains, which are not abundant, Silla in the seventh century could not have been cartographically behind Paekche; its later regional administrative system, which carried out extensive cadastral surveys on the Tang model, could not have functioned without maps. Nor is it likely that the extensive wars of the unification era could have been prosecuted without maps.

The state of Silla during its unified era (668–935) passed through two distinct phases in its political and cultural development. During the first phase, which actually had begun several decades before unification, there was a heavy importation of Chinese institutions and learning. Although so thoroughly adapted to local circumstances that a modern historian of China would hardly recognize them, these institutions brought about Chinese-style results: a centralized administration under which the regional powers and traditional aristocracy were weakened while the royal center in Sŏrabŏl (modern Kyŏngju) was both represented and protected by a strong bureaucracy. During this period, the end of which may be marked by King Hyegong's assassination in 780, Silla reached its cultural high point, represented by the famous monastery Pulguksa, the exquisite Buddhist grotto Sŏkkuram, and the astronomical observation tower the Ch'ŏmsŏngdae, all of which survive today. During the second phase the old aristocracy and regional forces came back to prominence, while central authority was reduced to a shadow. Chinese-style institutions and culture fell back before a wave of nativism. In this atmosphere

(History of the Three Kingdoms, 1145), 20.13b; see the edition in 9 vols. (Kyŏngju, 1512; reprinted Seoul, 1931). For the role of maps in tributary ritual, see above, pp. 72–73.

8. Iryŏn (1206–89), *Samguk yusa* (Legends of the Three Kingdoms) (Kyŏngju, 1512; reprinted Seoul, 1932), 2.25a; for a modern text, see *Samguk yusa*, ed. and trans. Yi Pyŏngdo (Seoul: Tongguk Munhwasa, 1956), 72.

coastal magnates carved out independent positions for themselves in international trade and manipulated the central government, while secession movements evoking the old kingdoms of Paekche and Koguryŏ arose in the areas of their former strength. In 918 the latter, adopting the abbreviated form Koryŏ for its name (from which the name Korea ultimately comes), founded a new dynasty, and from its position of strength it bided its time until both Silla and Later Paekche submitted (935–36).

During the long Koryŏ dynasty (918–1392), much of Korea's culture was elaborated and defined. The Koryŏ kingdom began during a period of chaos and fragmentation in China (the Ten Kingdoms in the south, 902–78, the Five Dynasties in the north, 907–60), and therefore launched its enterprise totally free of Chinese manipulation or meddling. This fundamental fact accounts for the generally independent character of the entire period, in spite of on-again, off-again tributary relationships with Northern Song (960–1126; there were no relations with Southern Song, 1127–1279) and only grudging or forced relations with the various non-Chinese peoples, the Khitans (Chinese, Liao), Jurchens (Jin), and Mongols (Yuan). These latter peoples variously invaded, threatened, or occupied Koryŏ but never interrupted its dynastic continuity or took over its internal administration, quite in contrast to their practice in China or the parts of it they controlled, where the Chinese dynasty was replaced by their own and the bureaucracy was headed by members of the conquering elite. Beginning in 950, the Koryŏ kings put a heavy emphasis on Chinese-style institutions. These left a durable imprint on the organs of the central bureaucracy (which have a Tang, not a Song, look to them), on the social structure (in the form of a Confucian patrilineal ritual and descent system), and on the literature (mostly in classical Chinese and expressed in Chinese genres). The Confucian influence became particularly strong during the period of Mongol occupation of both China and Korea, when the currents of Neo-Confucianism, elaborated in Song during the eleventh and twelfth centuries, flowed into Korea and nurtured patterns of culture and thought that to some degree still remain part of Korean life. On the whole, however, Koryŏ counted itself a Buddhist kingdom, inheriting and strengthening the strongly indigenized Buddhism of the Silla period. It was this Buddhist dimension that reinforced Koryŏ's independence and nativism, sometimes much to the annoyance of the elite Confucians, who characteristically favored more Chinese ways. But except at the very end of the Koryŏ period, in the late fourteenth century, the general result of this mix was a Buddhist-Confucian eclecticism.

Although no authentic Koryŏ maps are known to survive, there is no doubt that Koryŏ had a very respectable cartographic tradition. The level of state organization and local administration dictated a need for maps. Indeed, from 1275 until nearly the end of the dynasty, the department of revenue was known as the directorate of "registers and maps" (*p'ando sa*).[9] Another factor arguing for a sophisticated development of cartography was the near mania in Koryŏ times for geomantic analysis on the national as well as the local level—a phenomenon I will return to (see below, pp. 276–79). It is hardly conceivable that the numerous professional specialists known to have operated in this field over the Koryŏ centuries could have done so without good maps. Finally, references to maps in Koryŏ historical sources are not uncommon. We see, for instance, the twelfth-century scholar Yun P'o composing a Buddhist-inspired map of the Five Indias;[10] we note a Mongol envoy asking for a map of the Koryŏ kingdom in 1281;[11] and we observe Koryŏ officials, in the course of their expulsion of the Mongols in 1356, relying on their maps to reassert national control over a sector of northeastern frontier territory north of Ch'ŏllyŏng (Iron Pass), which had been under direct Yuan rule.[12]

In addition, Koryŏ had a cartographic curiosity that is probably unique in the world: it had a unit of money whose shape resembled the outline of the Korean peninsula. A notice of 1101 reads: "In this year the silver vase [*ŭnbyŏng*] was put into use as [a unit of] exchange. As for the design, it was made with one *kŭn* of silver and resembled the territorial outline of this country."[13] The precise weight of a *kŭn* during the Koryŏ period is unknown, but it would have been in the vicinity of half a kilogram or more—hardly a coin that the average person would often see; indeed, not a coin at all. "Silver vases" seem to have been used mainly in large financial transactions and as ceremonial gifts or rewards. They went out of circulation about the middle of the fourteenth century. (For a possible evocation in the fifteenth century, see below, pp. 295–96.)

Koryŏ's cultural relations with China also show maps

9. Chŏng Inji (1396–1478) et al., comps. and eds., *Koryŏ sa* (History of Koryŏ) (Seoul, 1451; reprinted in 3 vols., Seoul: Yŏnhŭi University Press, 1955), 76.16a.

10. See the memorial inscription on stone "Yun P'o Myoji," erected in Kaesŏng in 1154, and collected in Chōsen Sōtokufu (Government-General in Korea), ed., *Chōsen kinseki sōran* (A comprehensive survey of ancient Korean inscriptions), 2 vols. (Seoul: Chōsen Sōtokufu, 1919), 1:369–71. See also below, pp. 255–56.

11. *Koryŏ sa*, 29.3b (note 9).

12. *Koryŏ sa*, 111.32b (note 9). During its Mongol period, most of Koryŏ's territory had been under direct Koryŏ administration with general Mongol oversight. But one area, corresponding roughly to modern Hamgyŏng Province, had been annexed outright by the Mongols and administered by the Yuan court and its Korean collaborators. This was the territory in question here. Later Ming Chinese claims to this area precipitated the crisis that led to the internal overthrow of the Koryŏ dynasty.

13. *Koryŏ sa*, 33.11a (note 9). The *kŭn* was standardized at six hundred grams in 1902.

figuring in the exchange. Diplomats on their trips to the Song capital were avid buyers of Chinese written materials of all kinds, and these certainly included maps. We have already seen, in an earlier chapter, an instance in which Korean map purchases in China were treated by Song officials as cartographic espionage.[14] Not too many years after that affair, Song authorities heard rumors of long-lost Chinese bibliographical treasures that had survived in Korea, and they requested that the Koryŏ court send copies of any Chinese editions in Korean collections that it deemed rare or unusual. Among the items sent in response, in 1091, were two geographical works that may have contained maps, the *Yudi zhi* (Monograph on the [Imperial] territory, sixth century) in thirty scrolls by Gu Yewang, and the *Guadi zhi* (Inclusive geographical monograph, 638) in five hundred scrolls by Xiao Deyan and Gu Yin.[15] (These works are now lost both in China and in Korea.) Such notices suggest a broader cartographic flow, the details of which can no longer be known.

Going in the opposite direction, there are representations of the Korean peninsula on a few of the older Chinese maps still surviving, and it is likely that some of these images derive from Korean maps that found their way into China one way or another. The oldest such case seems to be on the *Hua yi tu* (Map of Chinese and foreign lands, see above, fig. 3.13). Some believe, apparently because of the similarity of the titles, that this was based on the famous but long lost *Hainei Hua yi tu* (Map of Chinese and foreign lands within the seas), compiled in 801 under the direction of Jia Dan (730–805). The *Hua yi tu* was engraved on stone in 1136 and is now in the Shaanxi Provincial Museum in Xi'an. Although the representation of the peninsula is primitive and cut off on the eastern side by the margin, the mouths of the rivers on the northwest coast are passably close to reality. The depiction of Korea on a slightly earlier map, the *Gujin Hua yi quyu zongyao tu* (General map of the ancient and present territories of China and foreign countries), has a mere suggestion of the peninsula, and its maker probably had no cartographic image of Korea in front of him (see above, figs. 3.23 and 6.30; it was first published in the period 1098–1100 and is available from a woodblock edition of 1162).

Another Chinese representation of Korea is that in the *Guang yutu* (Enlarged terrestrial atlas, ca. 1555) of Luo Hongxian (1504–64) (fig. 10.2). The problem is how to date it, and this is tied to the difficulty of knowing to what degree Luo followed or departed from his model, the now lost *Yutu* (Terrestrial map, 1320) of Zhu Siben (1273–1337). Although Luo gave this piece the title "Chaoxian tu" (Map of Chosŏn), Korean cartographic specialists consider his rendering of Korea to be ultimately based on a Koryŏ map, apparently because of the Zhu Siben connection and because its peninsular outline does not look like that on any known later map.[16] For our purposes it is enough to know that Zhu Siben included a map of Korea in his atlas. Zhu flourished in a period of broad cultural exchange between Korea and China, occasioned by the Mongol requirement that Korean princes reside in Beijing until the death or abdication of their predecessors on the throne in Korea. These princes presided over miniature courts of their own, attended by dozens of Korean officials and advisers whose residence in China sometimes lasted for decades. It is likely that in this period maps of Korea were easy to come by in Beijing.

We can get a reasonably clear idea of the information available on at least one map of Koryŏ from a preface that was probably written in 1402, just a decade after the dynasty's fall. The official Yi Ch'ŏm, in the introduction to his abridged version of a history of the three former Korean kingdoms of Koguryŏ, Paekche, and Silla, described at some length a scroll-mounted map of Koryŏ that he had come across.

> Such a map of Koryŏ could only have appeared after the unification, but we do not know from whose hand it has come. You can observe the line of mountains winding down from Whitehead to the Iron Pass, where the Maple Peaks suddenly spring up. Then [the line] splits to become both the Greater and Lesser Whites, forming Bamboo Pass, Cockstand, Three Rivers Pass, and Surging Sun Mountain. The central highland stretches down to Cloudrest, from which point neither the earth features nor the map scroll go any farther south, into the sea; rather, the pure and pristine matter here mingles and accumulates, which is why the mountains are so high and steep. Indeed, no other mountains can be as great as these. West of this mountainous spine we have the Sal, Pae, Pyŏngnan, Imjin, Han, and Ungjin rivers, all making their courses westward to the sea; but east of it there is only the Kaya River, flowing to the south. Primal matter here flows and there solidifies, and the mountains and rivers form their separate zones. The different climatic areas and

14. See p. 83, above. The shrill complaints of Chinese officials over what they regarded as cartographic espionage were motivated by the belief that the Koreans were really acting for the Khitans (Qidan), the Song dynasty's perennial enemy in the northeast. For excellent background on these matters, see Michael C. Rogers, "Factionalism and Koryŏ Policy under the Northern Sung," *Journal of the American Oriental Society* 79 (1959): 16–25, and idem, "Sung- Koryø Relations: Some Inhibiting Factors," *Oriens* 11 (1958): 194–202.

15. *Koryŏ sa*, 10.23b (note 9). The list of books sent on this occasion goes on for several pages and includes 124 titles in some 4,800 scrolls.

16. For the map, see Luo Hongxian (1504–64), *Guang yutu*, 2.82b–83a; 6th ed. (1579; reprinted Taipei: Xuehai Chubanshe, 1969), 379–81. Notes on Korea and an anachronistic table of Korean provincial organization (Chosŏn provinces and many Koryŏ districts) follow, 83b–85a (382–85).

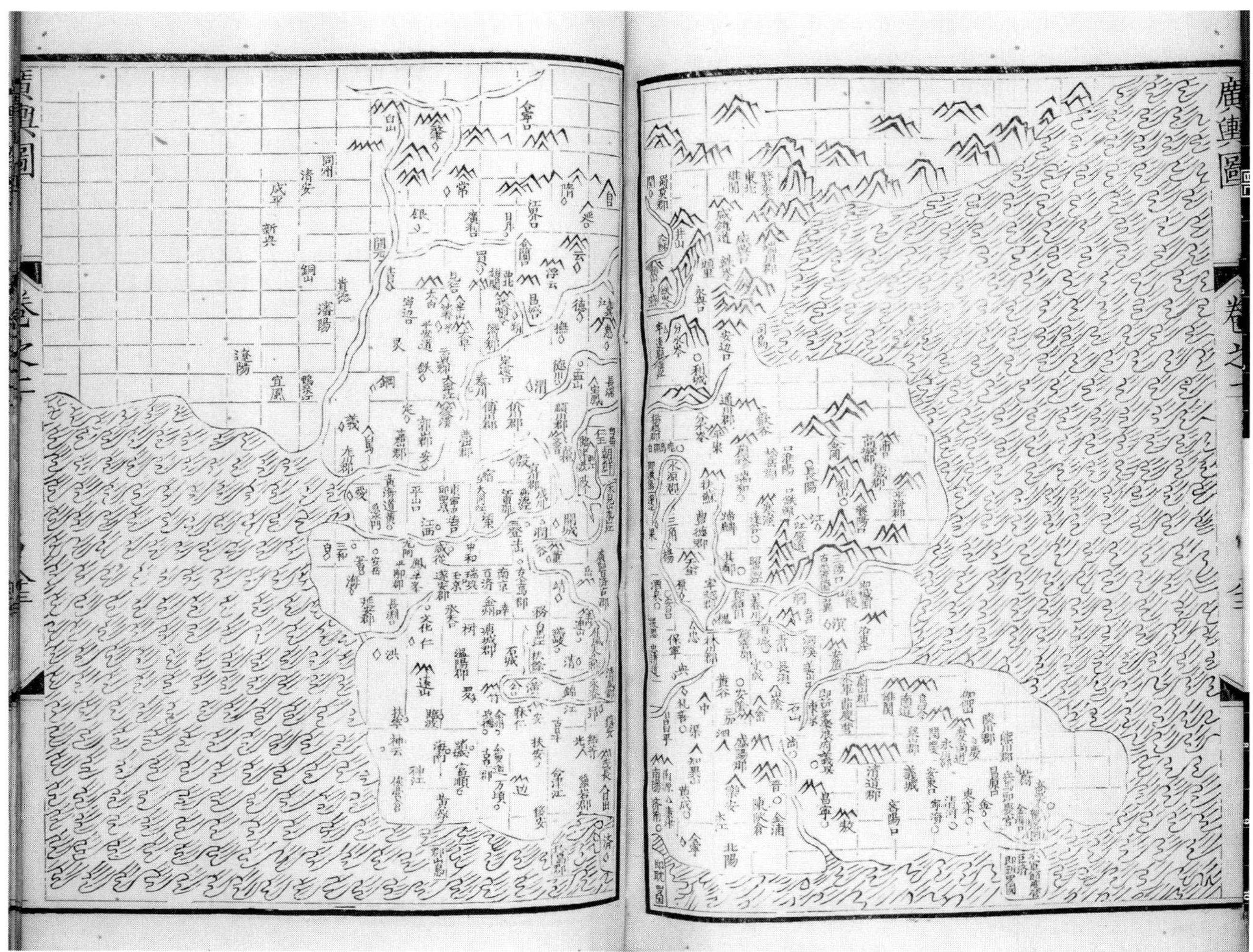

FIG. 10.2. "CHAOXIAN TU" FROM LUO HONGXIAN'S *GUANG YUTU*. Since Luo presumably took this map over from the earlier *Yutu* by Zhu Siben, it is believed to be based on a fourteenth-century Koryŏ map provided with Zhu's distinctive grid. However, the name Chosŏn, the Chosŏn dynasty provincial organization in Luo's accompanying tables, and the indication of the Chosŏn rather than the Koryŏ capital are not features that could have come from Koryŏ or Zhu Siben. Given such problems, and the bizarre distortions of the map itself, this map defies intelligible historical placement.
Size of the original: 28.5 × 39.5 cm. By permission of the British Library, London (15261.e.2).

> the borders between one county and another can be seen just by unrolling the map.[17]

I shall comment later on some of the geomantic thought evident in this passage (below, pp. 278–79); for now it is enough to note that mountain ranges and river courses were prominently indicated, and that both natural climatic zones (*p'unggi chi kuyŏk*; literally, wind and air zones) and administrative boundaries could be seen. I am not aware of any later Korean map that depicts climatic zones, and I suspect that in this very impressionistic description Yi Ch'ŏm meant only to suggest that such zones were implicit in the areas marked off by mountain ranges. But the important point is that he conceived of a map

17. Yi Ch'ŏm, "Samgukto husŏ" (Postface to a map of the Three Kingdoms), in *Tongmun sŏn* (Korean literary anthology), ed. Sŏ Kŏjŏng et al. (1478; reprinted Seoul: Kyŏnghŭi Ch'ulp'ansa, 1977), 92.12b–14a. By "unification" in the quotation, Yi refers to Koryŏ's unification of the Later Three Kingdoms, 935–36. He identifies the occasion of his seeing the Koryŏ map as a tour of duty in the "new capital" (Seoul) in 1396, but the preface was probably written after he was assigned by King T'aejong to a committee for the preparation of an edition of the *Samguk sagi* in 1402; see *T'aejong sillok* (Annals of King T'aejong, r. 1400–1418), 3.31a (see bibliographic note in next paragraph). His review of Korean maps was stimulated by his desire to include maps of the three kingdoms in this new edition of the *Samguk sagi*. But it is also significant that work on the 1402 world map was going on at precisely this time and that Kwŏn Kŭn, who wrote the preface for that map,

as a means to convey such information. Many of the names of mountains and ranges used here are no longer current (although all are identifiable), so it seemed just as well to translate them and give some of the flavor of this man's interaction with the map.[18] But Whitehead (or Mount Paektu), Maple Peaks (P'ung'ak, a name for the Kŭmgang san, or Diamond Mountains, in the autumn), Iron Pass (Ch'ŏllyŏng), and the Greater and Lesser Whites (the T'aebaek and Sobaek mountain ranges) are all key features of Korean physical geography, and we shall come back to them again.

Late in the Koryŏ period, a man named Na Hŭngyu (fl. 1315–76) is said to have compiled a historical map of Korea and China, in which he "set forth the traces of the rise and fall of emperors and kings and of the sundering and joining of territories and regions since the beginning of time."[19] Na had a reputation for learning and a highly interesting career. According to his biography, in addition to making maps he operated a tutoring service for candidates who failed examinations, supervised palace building projects, presided over the production of sculptural decorations such as carved dragons, served as a royal poet and court jester for King Kongmin (r. 1351–74), and in 1375, at an advanced age, went as ambassador to Japan, only to be thrown in jail for a time by his suspicious hosts, who had not seen a Korean ambassador in many years. Na said of his map that "gentlemen of breadth and refinement who love antiquity can look at it and grasp all of heaven and earth in their imaginations."[20] Na's more fanciful claims for his map must be categorized as rhetorical flights. Some of them, such as its showing all territorial changes "since the beginning of time," probably referred to the use of textual notes on the map, in the manner of the later work by Kim Suhong (below, pp. 267–68). Others, such as showing "the rise and fall of emperors and kings," may have been accomplished by lists on the map's margins, as on the world map of 1402. Thus his work, which we can only imagine, serves as a good link to that map, bearing the formal title *Honil kangni yŏktae kukto chi to*, to which we now turn.

World Maps and East Asia Regional Maps

The founder of the Chosŏn dynasty (1392–1910) was Yi Sŏnggye, a military man and native of the northeastern frontier who had risen to fame for his resistance to the Japanese marauders that had plagued Korea throughout the fourteenth century. These were veritable armies, sometimes two or three thousand strong, whose coastal raids often penetrated far into the interior. No town anywhere in the southern provinces was safe from them. Yi's successive victories over this menace throughout the 1380s had brought him a national following. He came to power in 1389 in a spectacular military coup. In 1388 the Ming dynasty (1368–1644), which had just ousted the Mongols from the Liaodong area, demanded that the Koryŏ rulers turn over the lands northeast of the Ch'ŏllyŏng that had been administered directly by the vanquished Yuan dynasty—the same lands we have seen Koryŏ repossessing with its maps in 1356. Koryŏ refused and ordered Yi Sŏnggye to attack Ming forces in Liaodong; but Yi, thinking this policy foolish and ill advised, took over the government instead. The Ming forces, given the change, did not push their demands, and the northeastern territories stayed with Koryŏ. Three years later Yi took the throne himself, bringing the Koryŏ dynasty to an end after almost five centuries of rule.

This was no mere dynastic change. Into power with Yi Sŏnggye came a movement of Neo-Confucian reform that within a generation remade Korea into a completely different kind of kingdom. With the dispossession of the old Koryŏ aristocracy and the disestablishment of Buddhism as a state-protected religion, the reformers launched a political program that proclaimed Confucian priorities in social policy, educational reconstruction, and cultural development. Thousands of monks were laicized and a multitude of slaves manumitted, all to reinvigorate the revenue-producing peasantry, on lands often confiscated from monasteries. A small but dynamic corps of Confucian ideologists rewrote the legal codes, redesigned government institutions and the civil service, and in countless other ways turned the Neo-Confucian intellectual revolution of the Chinese philosopher Zhu Xi

was on the same *Samguk sagi* committee. Yi Ch'ŏm's "primal matter" (*wŏn'gi*) refers to the original physical matter from which the earth, in his Neo-Confucian belief, had been formed. He evidently saw it as gaseous or liquid matter that gradually congealed into solid mountains.

Concerning the structure of the *sillok* (royal annals) collection and its editions: the twenty-seven kings of the Chosŏn dynasty (1392–1910) each have official *sillok*, compiled upon their deaths from contemporary court records. The annals of the first twenty-five kings are gathered in Kuksa P'yŏnch'an Wiwŏnhoe (National History Compilation Committee of the Republic of Korea), ed., *Chosŏn wangjo sillok* (Royal annals of Chosŏn), 48 vols. (Seoul: Kuksa P'yŏnch'an Wiwŏnhoe, 1955–58). The annals of a given king will be cited by that king's posthumous name, followed by the word *sillok* (e.g., *T'aejong sillok*, *Sejong sillok*) with the original *kwŏn* and page numbers, which are photolithographically reproduced along with the text.

The modern volume and page numbers provided by the editors of the *Chosŏn wangjo sillok* are omitted in the citations in this chapter. Not only are they prohibitively general, since one page of the *Chosŏn wangjo sillok* contains four pages of the original *sillok*, but they detract from the integrity of the original *sillok*, which were separately compiled by different people in different circumstances over the centuries.

18. The river names too included many old names: the Sal (modern Ch'ŏngch'ŏn), Pae (Taedong), Pyŏngnan (Yesŏng), Ungjin (Kŏm), and Kaya (Naktong).

19. *Koryŏ sa*, 114.27a–b (note 9).

20. *Koryŏ sa*, 114.27a–b (note 9).

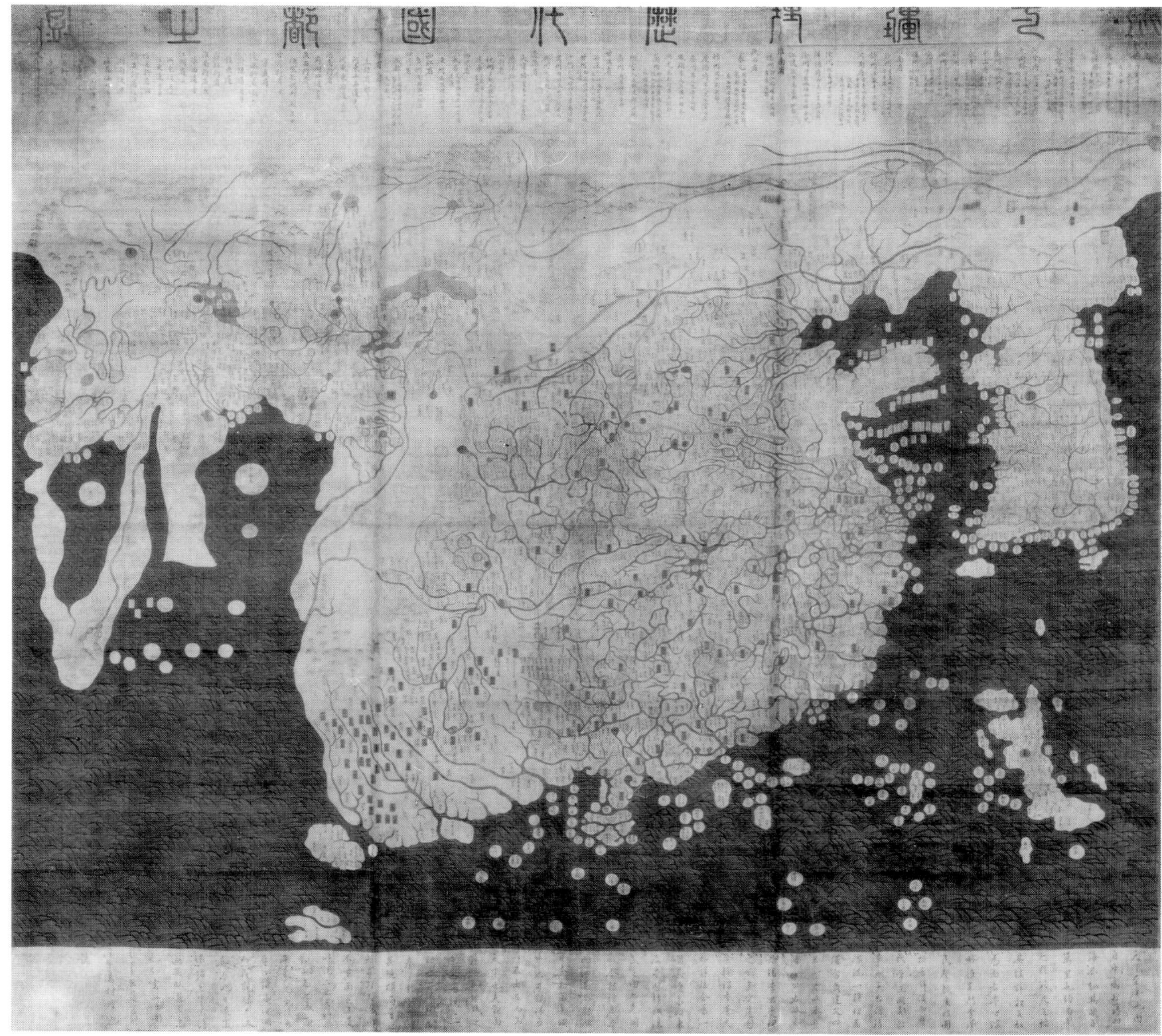

FIG. 10.3. *HONIL KANGNI YŎKTAE KUKTO CHI TO* (1402), BY YI HOE AND KWŎN KŬN, FROM A COPY OF CA. 1470. The *Kangnido*, as it is commonly called, is the earliest known map of the world from the East Asian cartographic tradition and the oldest surviving Korean map. Based on fourteenth-century Chinese maps, one of which had among its sources an unknown Islamic map, the *Kangnido* has clear delineations of Africa and the Arabian Peninsula and a recognizable outline of Europe, but India is submerged in the general Chinese continent. The greatly magnified Korean part (see detail, plate 17), along with Manchurian detail and the Japanese islands, was based on sources available in Korea. Japan, based on a Japanese map brought to Korea in 1402, is oddly placed in the South China Sea. Japan is oriented with west at the top, but the outline itself compares favorably with that of contemporary Japanese maps.
Size of the original: 164 × 171.8 cm. By permission of Ryūkoku University Library, Kyōto, Japan.

(1130–1200) into state orthodoxy. These men, who saw their regime as having the classical "mandate of heaven," had a keen awareness that they were effecting millennial change. Their vision became concrete in a 518-year rule, which apart from China's chronologically problematical Zhou dynasty (ca. 1027–256 B.C.) and Japan's very different monarchical institution, is the longest dynastic duration in East Asian history.

THE WORLD MAP OF 1402

It is no accident that among the early cultural projects

of this new regime we should find a map of the world and a map of the skies—heaven and earth themselves redefined and proclaimed within the cadre of Korea's cultural revolution to demonstrate the new dynasty's cosmic legitimacy. Nor is it mere coincidence that the official guiding both of these projects, Kwŏn Kŭn, was one of the key Confucian scholars among the reformers.[21] The star map, which purports to be a revision of an ancient Koguryŏ map, was engraved on stone in 1395; it is discussed elsewhere in this book (see pp. 560–68). Here I limit myself to consideration of the world map.

The *Honil kangni yŏktae kukto chi to* (Map of integrated lands and regions of historical countries and capitals), hereafter referred to as the *Kangnido*,[22] was completed in 1402. It easily predates any world map known from either China or Japan and is therefore the oldest world map surviving in the East Asian cartographic tradition, and the only one before the Ricci world maps of the late fifteenth and early sixteenth centuries. Although the *Kangnido* is no longer preserved in Korea itself, there are three versions in Japan, of which the oldest is that held by the Ryūkoku University Library (Kyōto), which is dated based on internal evidence to about 1470 (fig. 10.3). The principal distinguishing characteristics of the Ryūkoku copy are its generally excellent condition and its preservation of the original Kwŏn Kŭn preface. Painted on silk and still preserving its colors well, it is a very large map, nearly square at 164 by 171 centimeters. It was first brought to scholarly notice by the Japanese historical geographer Ogawa Takuji in 1928.[23]

The place to begin discussion of this very unusual map is with its preface, the crucial part of which is translated here from the text on the Ryūkoku copy, with reference to the closely similar version in Kwŏn Kŭn's collected works, the *Yangch'on chip*.

> The world is very wide. We do not know how many tens of millions of *li* there are from China in the center to the four seas at the outer limits, but in compressing and mapping it on a folio sheet several feet in size, it is indeed difficult to achieve precision; that is why [the results of] the mapmakers have generally been either too diffuse or too abbreviated. But the *Shengjiao guangbei tu* [Map of the vast reach of (civilization's) resounding teaching] of Li Zemin of Wumen is both detailed and comprehensive, while for the succession of emperors and kings and of countries and capitals across time, the *Hunyi jiangli tu* [Map of integrated regions and terrains] by the Tiantai monk Qingjun is thorough and complete. In the fourth year of the Jianwen era [1402], Left Minister Kim [Sahyŏng] of Sangju and Right Minister Yi [Mu] of Tanyang, during moments of rest from their governing duties, made a comparative study of these maps and ordered Yi Hoe, an orderly, to carefully collate them and then combine them into a single map. Insofar as the area east of the Liao River and our own country's territory were concerned, Zemin's map had many gaps and omissions, so Yi Hoe supplemented and expanded the map of our country and added a map of Japan, making it a new map entirely, nicely organized and well worth admiration. One can indeed know the world without going out of his door! By looking at maps one can know terrestrial distances and get help in the work of government. The care and concern expended on this map by our two gentlemen can be grasped just by the greatness of its scale and dimension.[24]

Both Kim Sahyŏng (1341–1407) and Yi Mu (d. 1409) held high offices during the formative years of the Chosŏn dynasty, although Yi Mu fell afoul of King T'aejong (r. 1400–18) and was later executed for his alleged role in a political plot. Both went to China on diplomatic business during their careers, and it is believed that Kim's trip, completed in the summer of 1399, was the occasion for obtaining the Chinese maps mentioned by Kwŏn Kŭn.[25] Both Kim and Yi probably had administrative experience with maps, since they had reported to King T'aejong on the progress of the land surveys of the northern frontier area in the spring of 1402, just a few months before the world map was made.[26] As high ministers, however, they probably had little time for actual cartographic work. Kwŏn's own role was probably important, even though he insists that he only stood in the background and "enjoyably watched the making of the map."[27] But he was being modest and tactful, since he was younger in age and junior in rank to the two ministers. The real cartographer, even though Kwŏn minimizes his role, was Yi Hoe, whose entire career was in rather low-ranking but often special positions. We will come back to him when I discuss Korean national maps.

21. Even though the two projects were seven years apart, the prefaces for both appear next to each other in Kwŏn's collected works, *Yangch'on chip* (Collected writings of Kwŏn Kŭn) (Chinju, 1674; reprinted Seoul: Chōsen Sōtokufu, 1937 [Chōsen Shiryō Sōkan (Korean historical sources series), no. 13]), 22.1a–2b.

22. This is the title on the Ryūkoku University copy of the map; the short form *Kangnido* is standard in the literature. The title indicated in Kwŏn Kŭn's preface is *Yŏktae chewang honil kangnido* (Map of historical emperors and kings and of integrated borders and terrains); *Yangch'on chip*, 22.2a (note 21).

23. Ogawa Takuji, *Shina rekishi chiri kenkyū* (Studies in Chinese historical geography), 2 vols. (Tokyo: Kobundō Shobō, 1928–29), 1:59–62.

24. The translation is from the text transcribed from the map by Ogawa, *Shina rekishi chiri kenkyū*, 1:60 (note 23); see also Aoyama Sadao, "Gendai no chizu ni tsuite" (On maps of the Yuan dynasty), *Tōhō Gakuhō* (Tokyo) 8 (1938): 103–52, esp. 110–11. These texts differ very little from that in the *Yangch'on chip*, 22.2a–b (note 21).

25. *Chŏngjong sillok* (Annals of King Chŏngjong, r. 1398–1400), 1.17a (note 17). Yi Mu's trip took place in 1407, after the map was finished.

26. *T'aejong sillok*, 4.10b–11a (note 17).

27. *Yangch'on chip*, 22.2b (note 21).

FIG. 10.4. DETAIL OF EUROPE FROM THE *KANGNIDO*. Either the original cartographers or the copyist neglected to provide wave patterns for the Black and Mediterranean seas and the Persian Gulf.
Size of the detail: ca. 48 × 37 cm. By permission of Ryūkoku University Library, Kyōto, Japan.

Judging by Kwŏn's description of the monk Qingjun's *Hunyi jiangli tu*, it was probably an ordinary historical map of China, compiled in the late fourteenth century. Qingjun (1328–92) was a close adviser to the Hongwu emperor (r. 1368–98),[28] who was the founder of the Ming dynasty and himself an erstwhile monk. Apart from its use as a source for the *Kangnido*, nothing is known of Qingjun's map. Its chief contribution to the *Kangnido* is believed to have been the Chinese historical dimension—the indication of the areas and capitals of the earlier dynasties, which was accomplished by a combination of textual notes and cartographic devices. Other than that, the main feature of the *Hunyi jiangli tu* that stuck with the Korean map was probably its name, which reads *Honil kangnido* in Sino-Korean.

The international dimension of the *Kangnido* unquestionably came from Li Zemin's *Shengjiao guangbei tu*. Li is mentioned by the Ming cartographer Luo Hongxian as a contemporary and possibly as an associate of Zhu Siben.[29] Aoyama's careful study of the Chinese place-names on the *Kangnido* shows them in general accord with those on Zhu's map, as preserved in Luo's *Guang yutu*, but with variants that would indicate place-name changes made in 1328–29. This suggests that the *Kangnido*'s source map was made about 1330. Since Zhu explicitly excluded most non-Chinese areas from his map,[30] Aoyama and others have reasoned that Li Zemin must have found his cartographic sources for these areas elsewhere, the only plausible source being Islamic maps, which made their appearance in China under Mongol rule.[31] Luo Hongxian's probable use of the *Guangbei tu* is deduced from his maps of the southeast and southwest maritime regions, and it could well be from the *Guangbei tu* that the *Da Ming hunyi tu* (Integrated map of the Great Ming) derives. But for the missing or incomplete detail in the eastern areas of Manchuria, Korea, and Japan, the latter map, now in the Palace Museum in Beijing, bears a very close resemblance to the *Kangnido*.[32]

Takahashi Tadashi has shown that the *Kangnido*'s Chinese transcriptions of place-names in Southwest Asia, Africa, and Europe come from Persianized Arabic originals. Although some of Takahashi's matches do not command credence in early modern Chinese phonological terms, he generally makes a convincing case. One of the more interesting correspondences is the name placed by the mountains near the Ptolemaic twin lakes that are the source of the Nile. Though it is not on the Ryūkoku copy of the *Kangnido*, the Tenri copy shows the Chinese transcription Zhebulu Hama, which Takahashi identifies

28. Aoyama, "Gendai no chizu ni tsuite," 122–23 (note 24).

29. See Luo Hongxian's preface to the *Jiubian tu* (Map of the nine frontiers), partly quoted in Aoyama, "Gendai no chizu ni tsuite," 123 (note 24).

30. Zhu's preface to his lost *Yutu*, preserved in Luo's *Guang yutu*, quoted in Aoyama, "Gendai no chizu ni tsuite," 105 (note 24). The exclusion, in Zhu's own words, was "the areas southeast of the overflowing seas and northwest of the sandy wastes, and all the bordering tribes and strange territories."

31. See Joseph Needham, *Science and Civilisation in China* (Cambridge: Cambridge University Press, 1954–), vol. 3, with Wang Ling, *Mathematics and the Sciences of the Heavens and the Earth* (1959), 551–56.

32. On this map see the illustration and description of Walter Fuchs, "Pekin no Mindai sekaizu ni tsuite" (On the Ming-period world map in Beijing), *Chirigakushi Kenkyū* 2 (1962): 3–4, with 2 pls.; reprinted in *Chirigakushi kenkyū* (Researches in the history of geography), 2 vols., ed. Chirigakushi Kenkyūkai (Society for Research in Historical Geography) (Kyōto: Rinsen Shoten, 1979), 2:3–4 and pls. 1–2.

with Persianized Arabic Djebel al-Qamar (Mountains of the Moon).[33] All in all, there are about thirty-five names indicated on or near the African continent, most of them in the Mediterranean area.

The European part of the map, which is said to contain some one hundred names, has not yet been the object of an individual study (fig. 10.4). The Mediterranean is clearly recognizable, as are the Iberian and Italian peninsulas and the Adriatic, but until the place-names can be read and interpreted it will be impossible to come to any firm understanding of its sources.[34]

Kwŏn Kŭn observed in his preface that the *Guangbei tu* gave only sketchy treatment to the area east of the Liao River and of Korea. His language suggests that some image of Korea, however deficient, was on the original *Guangbei tu* and that this was supplemented and expanded by Yi Hoe. Yi is known to have produced a map of Korea, called the *P'altodo* (Map of the Eight Provinces),[35] and it was probably a version of this that appears on the *Kangnido*. In any case, this version is the oldest Korean map of Korea to survive. I will discuss it more fully when we consider Korean national maps.

The last major element of the map to be supplied, as far as the Koreans were concerned, was Japan. At this particular time, Korea's relations with the Japanese were very difficult owing to the continuing problem of Japanese marauders, who were beyond the ability of the Ashikaga shogunate to control. Diplomatic initiatives were in progress, and coastal defenses and strategies were undergoing constant development. All of this was backed by a general Korean effort to improve the government's knowledge of Japan, and this involved maps in particular. Pak Tonji, a military man and diplomatic specialist in Japanese affairs, made at least two trips to Japan, one in 1398–99, the other in 1401–2, and the second visit resulted in a map. A later report quoted his statement that in 1402 he had been given a map by the "Bishū no kami, Minamoto Mitsusuke," which "was very detailed and complete. The entire land area was on it, all but the islands of Iki and Tsushima, so I added them and doubled the scale." In 1420, this report states, he formally presented this map to the board of rites, the branch of the Chosŏn government that handled foreign affairs.[36]

It is generally assumed by Korean cartographic specialists that this map was the basis for the representation of Japan on the *Kangnido*. Compared with other maps of Japan in this period, this outline is unusually good: the positioning of Kyūshū with respect to Honshū is quite accurate, and the bend north of the Kantō area is indicated better than on many of the Gyōki-style maps then current. Except for the joining of Shikoku to Honshū, the three main islands make a very decent appearance. But this splendid effort seems to be vitiated by orienting the Japanese islands with west at the top. Furthermore, the whole ensemble is positioned far to the south, so that the first impression a modern observer gets is that the Philippines, not Japan, are under view. A probable explanation is that the mapmakers had run out of space on the right (east) edge of the *Kangnido* and so had to place Japan in the open sea to the south. On the other hand, Chinese maps had long shown Japan off China's southern coast, and the *Kangnido*'s treatment may well reflect this (see below, pp. 272–73). As for the orientation with west at the top, it is possible that this was copied from the map Pak Tonji received from Minamoto Mitsusuke. Indeed, the earliest known map of Japan (805) has this orientation.[37] Interestingly, the Korean makers of the Tenri and Honmyōji copies of the *Kangnido* corrected the Japanese orientation to the north even while substituting more conventional Gyōki-style outlines.

The overall disposition and bulk of the different components of the *Kangnido* at first make an odd appearance. On the one hand, there is nothing formulaic or mandated about its structure, such as a T-O scheme, or the wheel

33. Takahashi Tadashi, "Tōzen seru chūsei isurāmu seikaizu" (Eastward diffusion of Islamic world maps in the medieval era), *Ryūkoku Daigaku Ronshū* 374 (1963): 86–94. Takahashi cites a number of features that are on the Tenri but not the Ryūkoku map, mainly in the African part.

34. Takahashi, "Tōzen seru chūsei isurāmu seikaizu," 89 n. 9 (note 33), cites four Chinese transcriptions from the European part of the map and matches them with names from al-Idrīsī's maps. Without knowing where on the map these names are, however, it is hard to evaluate them. The one hundred names from the European part still await a thorough study by the appropriate specialists. See below, p. 266 and fig. 10.13, for discussion of the *Kangnido*'s Mediterranean area in another context.

35. This may have been the same map as the "map of this country" presented by the State Council to King T'aejong on 6 June 1402 (*T'aejong sillok*, 3.27a [note 17]). This date coincides with the period when Yi Hoe would have been working on the *Kangnido*, which must have been completed by the eighth lunar month of 1402—solar 19 August to 16 September—the date of Kwŏn's preface. Yi Hoe's death date is unknown; the last mention of him I have seen is during May–June 1409, when he was appointed to a supernumerary post in the censorate (*T'aejong sillok*, 17.35a). It is only many years later, in 1482, that Yi Hoe's authorship of the *P'altodo* finds documentary confirmation in a list of maps that the official Yang Sŏngji was seeking to have restricted to official use; see *Sŏngjong sillok* (Annals of King Sŏngjong, r. 1470–94), 138.10b (note 17).

36. These events of 1402 and 1420 are reported retrospectively in 1438; see *Sejong sillok* (Annals of King Sejong, r. 1418–50), 80.21a–b (note 17). For Pak's 1398–99 mission to Japan, which lasted more than seventeen months, see *Chŏngjong sillok*, 1.13a–b (note 17). The Bishū no kami (governor of Bishū) Minamoto Mitsusuke is not otherwise identified. Iki and Tsushima were well-known pirate bases of special interest to the Koreans. On the question of scale, see below, p. 284.

37. See the *Yochi zu* (Land map), pp. 370 and 459. The original map is lost; only a mid-seventeenth-century copy survives. This circumstance suggests the possibility that such a map was available for Pak Tonji in the early fifteenth century. See the illustration in Akioka Takejirō, *Nihon chizu shi* (History of maps of Japan) (Tokyo: Kawade Shobō, 1955), pl. 1.

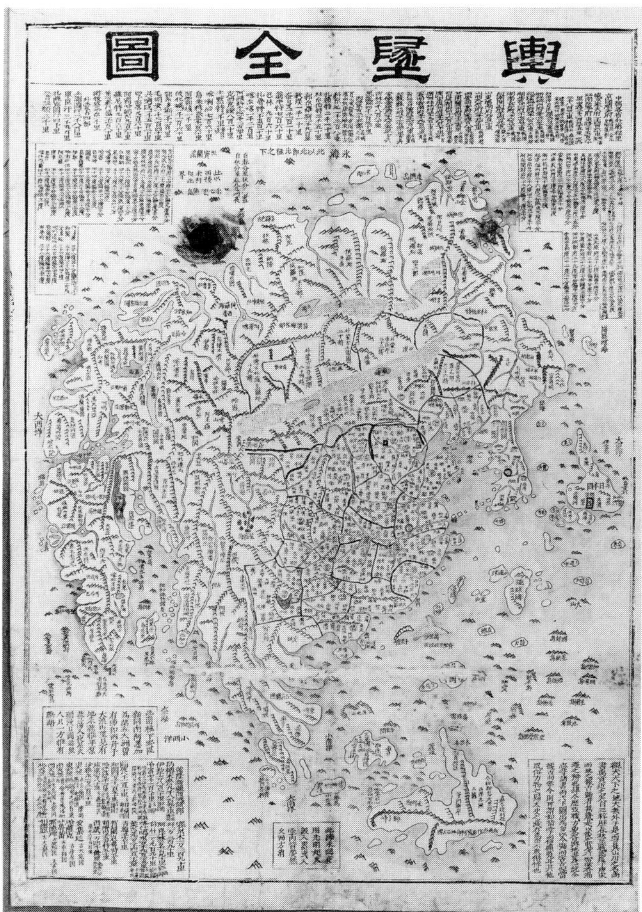

FIG. 10.5. *YŎJI CHŎNDO*, A PARTLY HAND-COLORED WOODCUT OF CA. 1775. This world map, although visibly influenced by seventeenth-century Sino-Jesuit maps then available in Korea, yet evokes the coverage and outline of the *Kangnido* of 1402.
Size of the original: 86.3 × 59.5 cm. By permission of Yi Ch'an, Seoul.

arrangement of the quasi-cosmographic *ch'ŏnhado*, to be discussed shortly. The attempt here was to study the best maps available in China, Korea, and Japan and put together a comprehensive, indeed integrated (*honil*), map that included every known part of the world, truly a breathtaking objective by the cartographic standards of any nation at that time. The result is inevitably strange to our eyes. China and India, like a monstrous cell that has not yet divided, make up a dominating mass that overfills the center of the map. To the west the Arabian Peninsula, with a clearly delineated Persian Gulf, and the African continent, with its tip correctly pointing south (not east, as on many early European maps), hang thinly but with assurance, as if they belong exactly where they are. At the top of Africa the Mediterranean supports a less securely grasped Europe, and the entire north fades into mountains and clouds. On the eastern side of the map a relatively massive Korea, easily occupying as much space as the whole African continent (which, to be sure, is unduly small), identifies itself as a very important place, while Japan, as if randomly flipped off the fingers into the ocean, floats uncertainly in the South China Sea. The relative size and disposition of the three major East Asian countries reflects a plausible Korean view of the world in the early fifteenth century: Korea projecting itself as a major East Asian state, refurbishing its traditional view of China as the major center of civilization and playing its eternal game of keeping Japan as far away as possible. On the other hand, Koreans were telling themselves that theirs was not just an East Asian country but part of the larger world. Their ambition and ability to map that world would validate their position in it.

To say this is to begin to answer the question, What was this map for? A map whose composition was guided by the nation's top educator and Confucian ideologist, and presided over by two ministers of state, was surely destined for a prominent, central place in the capital. It was probably displayed on a screen or a wall in some important palace building frequented by the king and senior officials. But a good understanding of its function is hampered because we know nothing of its history after its completion. The Ryūkoku *Kangnido*, judging by Korean place-name indications, is a copy reflecting place-name changes made about 1470.[38] If its source map was the 1402 *Kangnido*, then this is the last that is heard of this original.

We know little about how the *Kangnido* came to Japan, but the copies probably arrived there independently on three separate occasions. Both the Ryūkoku and Honmyōji copies were evidently part of the loot from Hideyoshi's invasion of Korea (1592–98). The Ryūkoku map was reportedly given by Hideyoshi to the Honganji, an important Buddhist temple in Kyōto. This institution ultimately was divided into two branches, east and west, and the latter (Nishi Honganji) is today associated with Ryūkoku University, which explains the map's present location.[39] The Honmyōji copy, a paper scroll titled

38. Aoyama, "Gendai no chizu ni tsuite," 143–45 (note 24).

39. See Aoyama, "Gendai no chizu ni tsuite," 110 (note 24); and Takahashi, "Tōzen seru chūsei isurāmu sekaizu," 85 and 89 n. 1 (note 33). Takahashi examined an unpublished catalog of the Honganji's books and manuscripts compiled during the 1840s and 1850s and found an item titled *Rekidai teikyō narabini sengi no zu* (The capitals of historical emperors, together with a usurpatious map). The *rekidai* (Korean *yŏktae*) evokes the Korean title of the map. The "usurpatious" probably reflects Japanese umbrage either at Japan's being made part of a world map that listed only foreign "emperors and kings" or at Japan's incorrect orientation and position on the map, both of which could have been seen as detracting from the dignity of the Japanese imperial institution. Such nationalist attitudes were very strong in some Japanese scholarly circles in the mid-nineteenth century, when the Honganji's catalog was being compiled.

Daiminkoku chizu (Map of the Great Ming), was given to that institution by Katō Kiyomasa, its major patron and one of the senior Japanese commanders on the Korean expedition.[40] Nothing is reported concerning the provenance of the Tenri copy, a silk scroll with no title (fig. 10.12 below), but according to a study by Unno it is a "sister map" to the Honmyōji scroll. His persuasive analysis of the place-names indicates that both maps were copied in Korea about 1568, from a version already cartographically distant from the Ryūkoku copy.[41]

This information permits the conclusion that the *Kangnido* was probably often copied in Korea during the fifteenth and sixteenth centuries. There is an arguable possibility that its fortunes intersected with those of the *ch'ŏnhado* in the sixteenth or seventeenth century (see below), and other evidence could extend its existence down to the eighteenth century. The very interesting *Yŏji chŏndo* (Complete terrestrial map) dated about 1775, while clearly influenced by a Sino-Jesuit world map, also shows a strong structural similarity to the *Kangnido*, as its owner, Yi Ch'an, has pointed out (fig. 10.5).[42] Thus Japan is righted and put in its proper place, the respective masses of Korea, China, and Africa are brought into more accurate relation, and England and Scandinavia emerge from Europe. But the map as a whole, and particularly its treatment of India and Africa, strongly evokes the *Kangnido*. This is good evidence that the *Kangnido* tradition was not broken by the Hideyoshi wars but stayed alive in Korea for two more centuries.

AN EIGHTEENTH-CENTURY TERRESTRIAL GLOBE

If the *Kangnido* of 1402 bears witness to a Korean contact with the Islamic cartographic tradition as filtered through Yuan China, a terrestrial globe of the eighteenth century represents a Korean encounter with the new Western cartography as it came into Korea from Ming and Qing China. It is the earliest known Korean application of European cartographic knowledge.

Korea's first news of the West arrived in 1521, when an envoy returning from Beijing reported that a people named the Folangji had conquered a place called Manla and then tried to get permission to trade in Canton.[43] This was of course an echo of the Portuguese conquest of Malacca in 1511. The Folangji were the Feringhi, or Franks, since the time of the Crusades a general term in Islamic lands and waters (including Malacca) for western Europeans of Roman Catholic faith. Although Portuguese missionaries and traders were well established in both China and Japan by the middle of the sixteenth century, they never had any direct contact with Korea. Nor did the Spanish or the Dutch ever establish relations with Korea before the twentieth century.

Yet the West had a significant impact on Korea in both the intellectual and religious spheres, and it was all done, so to speak, by diplomatic pouch. From the time Matteo Ricci (1552–1610) arrived in Beijing, Korean diplomats regularly returned from the Chinese capital with news and books relating to the Jesuits. Thus Ricci's world map of 1602, the *Kunyu wanguo quantu* (Complete terrestrial map of all countries), was brought back to Korea in 1603. Ricci's 1603 edition, titled *Liangyi xuanlan tu* (Map of the heavens and the earth as seen from obscurity), was acquired in 1604. It is held today by the Soongsil University Museum and is one of only a few copies of this edition now in existence.[44]

In 1631 the envoy Chŏng Tuwŏn returned with "three or four hundred ounces [silver] worth" of European books, maps, and manufactures, including the famous *Zhifang waiji* (Unofficial accounts of foreign countries, 1623) by Giulio Aleni (1582–1649), together with a separate five-sheet set of its maps entitled *Wanguo quantu* (Complete maps of all countries); books by Ricci and others on astronomy and mathematics; a telescope with an instruction manual; star maps of both the Northern and Southern hemispheres; a European cannon with an instruction manual; an alarm clock; and many other items. Moreover, Chŏng Tuwŏn continued to maintain contact with the Jesuits through correspondence.[45]

In 1645 the Korean crown prince Sohyŏn, released by the Manchus after nine years of detention in Shenyang as a hostage, spent two months in Beijing before returning home. He is said to have developed a close acquaintance with the Jesuit Johann Adam Schall von Bell (1592–1666), who gave him, in addition to religious writings and articles, books on astronomy and mathematics and a globe.[46]

---

40. See Akioka, *Nihon chizu shi*, 80–81 (illustration) (note 37).

41. Unno Kazutaka, "Tenri toshokan shozō DaiMin kokuzu ni tsuite" (On the 'Map of Ming' held by the Tenri University Library), *Ōsaka Gakugei Daigaku Kiyō* 6 (1958): 60–67, with 2 pls. See below, p. 289, note 166, for further discussion of the Honmyōji map.

42. Yi, *Han'guk ko chido*, 41 (note 4). There is another copy of this map in the Soongsil University Museum (Seoul).

43. *Chungjong sillok* (Annals of King Chungjong, r. 1506–44), 41.11b–12a (note 17).

44. Kim, *Maesan kukhak san'go*, 227–29 (note 3). On 197–213 Kim provides a complete transcription of all of the original prefaces and geographic notes found on this very rare copy of the *Liangyi xuanlan tu*. The Chinese title of this map is highly allusive. *Liangyi* (the twin instrumentalities) refers to yin and yang, earth and heaven, etc.; *xuanlan* is a Daoist term connoting the seeing and understanding of things that are impenetrable.

45. Yi Nŭnghwa, *Chosŏn kidokkyo kŭp oegyo sa* (History of Korean Christianity and foreign relations) (Seoul: Chosŏn Kidokkyo Changmun Sa, 1928), 3–4; Kim, *Maesan kukhak san'go*, 232–33 (note 3). The value in modern terms of three to four hundred ounces (silver) is beyond certain estimation. But it was a large amount of money, and the cannon and the telescope probably represented most of it.

46. Kim, *Maesan kukhak san'go*, 245–46 (note 3); Yamaguchi Masayuki, "Shōken seishi to Tō Jakubō" (Prince Sohyŏn and Tang Ruowang

FIG. 10.6. THE KORYŎ UNIVERSITY *SŎN'GI OKHYŎNG* (DEMONSTRATIONAL ARMILLARY SPHERE). Believed by Needham and others to be the *Sŏn'gi okhyŏng* described in Korean records as having been made in 1669 and repaired and copied during the eighteenth century, but argued here to be of the second half of the eighteenth century. The armillary ring assembly is shown. In addition to a fixed terrestrial component, consisting of the outer horizontal ring with intersecting meridian (double) and equator (single) rings, there were clock-driven rotating solar, sidereal, and moon-path components, although the solar component is now missing. (For the terrestrial globe, see figs. 10.8 and 10.9.)
Diameter of the outer horizontal ring: 41.3 cm. Koryŏ University Museum, Seoul. Photograph courtesy of Gari Ledyard.

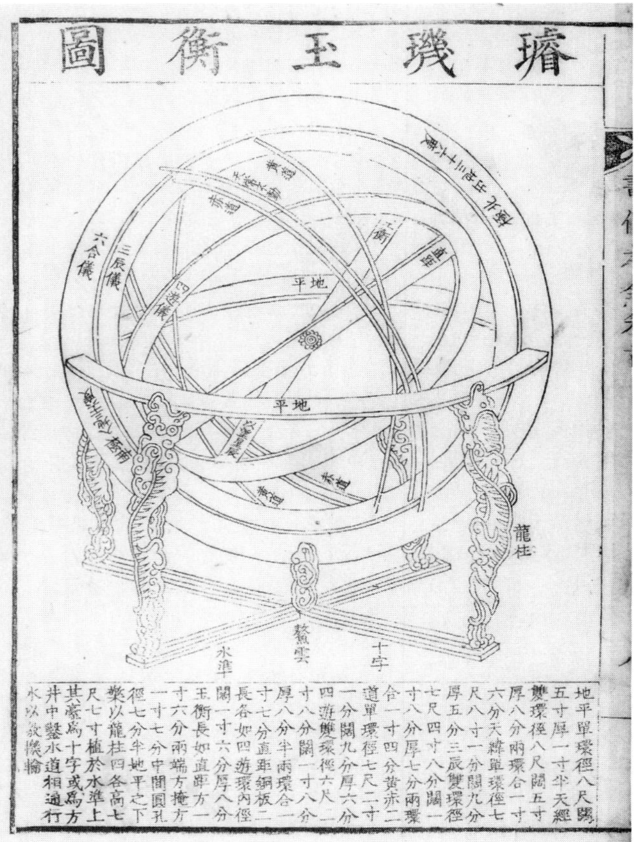

FIG. 10.7. THE *SŎN'GI OKHYŎNG* AS DEPICTED IN A 1620 KOREAN EDITION OF THE *SHUZHUAN DAQUAN*. The *Shuzhuan daquan* (Complete commentaries on the Book of History), a Ming work, was consulted by the makers of the *Sŏn'gi okhyŏng* in 1669. Note the absence of any earth model. Size of the original: ca. 24.5 × 18 cm. By permission of the British Library, London (MS. 15215.e.10, fol. 15v).

Thus there was an abundant background for Korean cartographic responses to the West. Most of these took the form of copies of various Western maps, which continued to be imported as they were acquired either through gift or purchase by Korean official travelers to Beijing. Thus, in 1708 King Sukchong (r. 1674–1720) directed that a copy be made of a map titled *Kunyu tu* (Terrestrial map), said to have been made by Schall von Bell. Judging from its title and cartographic details, this map, of flattened spherical projection, was a reedition of one of Ricci's maps. Two Korean copies resulting from this project are known, each in the form of an eight-panel screen with the 1708 preface of Ch'oe Sŏkchŏng (1646–1715), then Sukchong's chief minister.[47] We will return to this text shortly.

[Adam Schall]), *Seikyū Gakusō* 5 (1931): 101–17, esp. 105 and 113. Yamaguchi's Japanese translation uses the term for "celestial globe," but the text he translates from, a French version of a Latin original, says simply *sphère*.

47. Kim, *Maesan kukhak san'go*, 229–30 (note 3). One of the copies, once in Pongsŏn monastery, southeast of Seoul, was lost in the Korean War; Kim gives the text of Ch'oe's preface that was on that screen. Another copy is illustrated in Chōsen Sōtokufu ([Japanese] Government-General in Korea), *Chōsen shi* (History of Korea), six series comprising thirty-seven vols. (Seoul: Chōsen Sōtokufu, 1932–37), ser. 5, vol. 6, pl. 8, where it is stated to be held by Keijō Imperial University. The present whereabouts of this copy is uncertain; it cannot be found in a map list of that institution's successor, Seoul National University; see Sŏul Kungnip Taehakkyo Tosŏgwan (Seoul National University Library), comp., *Han'guk ko chido haeje* (Bibliographical notices of old Korean maps) (Seoul: Seoul National University, 1971). Although this map, as pictured in the 1930s, was in bad condition, most of Ch'oe's

FIG. 10.8. THE TERRESTRIAL GLOBE IN THE *SŎN'GI OKHYŎNG*. The globe was originally stationary and is so today, but evidence in the machinery (including the missing solar component) suggests that an attempt was once made to effect a diurnal rotation. Meridians are indicated at ten-degree intervals. The view here shows Africa (Liweiya, a transcription of "Libya," but miswritten "Limoya"). The Cape of Good Hope is called Talang Shan, "Big Wave Mountain," Antarctica is Yingwu Di, "Place of Parrots."
Diameter of the globe: ca. 9 cm. Koryŏ University Museum, Seoul. Photograph courtesy of Gari Ledyard.

A much more interesting response, because it represents a creative Korean *application* of Western cartography, is the Western-derived terrestrial globe mounted in a clock-driven armillary sphere of distinctly East Asian inspiration. This instrument, called the *Sŏn'gi okhyŏng* (a kind of armillary sphere), is preserved in the museum of Koryŏ University in Seoul. According to the studies of Joseph Needham and his collaborators, it combines a mid-seventeenth-century armillary assembly with a weight-driven clock mechanism based on a Japanese design, or perhaps imported as an entire movement. When it functioned, the clock not only announced the time with ringing bells and a visual display, but also indicated the movements of the sun and the moon in their regular cycles by means of moving pegs on the appropriate armillary rings (figs. 10.6 and 10.7). At the very center of the circling rings is set planet Earth, its polar axis mounted at an angle of 37°41′ to correspond to the latitude of Seoul. The globe is made of wood and covered

preface is legible in the *Chōsen shi* illustration. Both of these copies include seventeenth-century Western-style ships and sea monsters floating in the ocean spaces, which were not on Ricci's original editions.

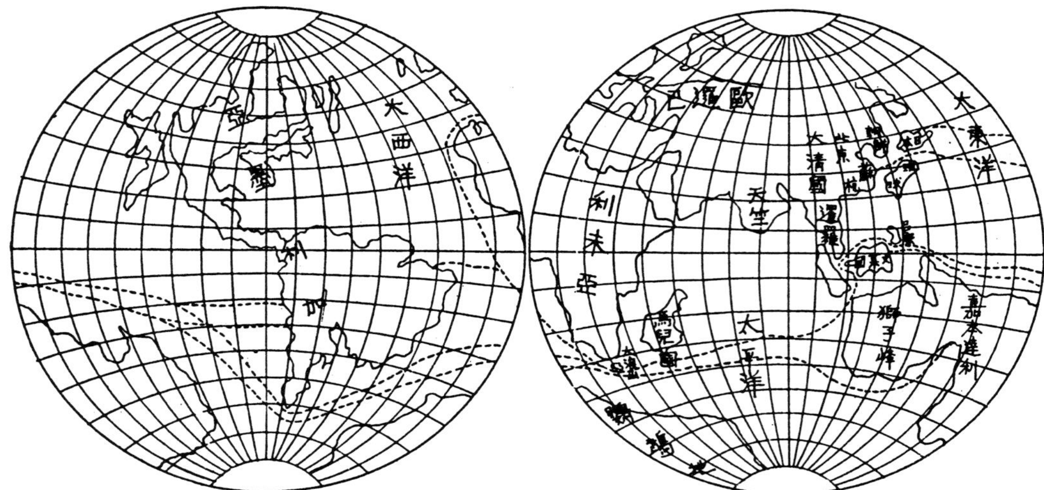

FIG. 10.9. THE SURFACE OF THE TERRESTRIAL GLOBE IN THE *SŎN'GI OKHYŎNG*, DRAWN IN TWO PLANISPHERES.

From W. Carl Rufus and Won-chul Lee, "Marking Time in Korea," *Popular Astronomy* 44 (1936): 252–57, esp. 257.

with a very fine skin of oil-painted paper (fig. 10.8). On the surface is a very accurate rendering, according to standards achieved by the late seventeenth century, of the earth's lands and seas, including Europe, Africa, the Americas, a continent looking like Antarctica (called Yingwu Di, or Place of Parrots—an echo of penguins?), and Australia, labeled Jiabendaliya (Carpentaria) (fig. 10.9). The globe was not originally intended to rotate diurnally, but there are indications that someone may once have tried to alter the mechanism to include this feature.[48]

Beyond the rough indication given above, the date of this terrestrial globe is not easy to determine, largely because of ambiguities in the evidence for the date of the remarkable instrument of which it is a part. The Korean historian of science Sang-woon Jeon has maintained that this instrument is the *Sŏn'gi okhyŏng* known to have been built by Song Iyŏng (fl. 1661–69) in 1669, although the identification is more asserted than argued.[49] Needham and his co-workers have followed Jeon in this conclusion. Although they point out problems in the interpretation of the written sources concerning the 1669 machine and note several occasions when it was repaired, rebuilt, or copied, they conclude that the *Sŏn'gi okhyŏng* in the Koryŏ University Museum is at least the linear descendant of the 1669 device and do not hesitate to refer to it as Song Iyŏng's armillary clock.[50]

Most of the documentation on the 1669 instrument making refers to the water-driven armillary sphere made by Yi Minch'ŏl (fl. 1669–1713).[51] It is not at all clear that this instrument featured a terrestrial globe within its rings. A recollection of this device written in 1687 gives the following details: "The sun and moon each had their rings, and in the centre there was not an alidade, but instead a terrestrial plane [*chip'yŏng*] made of paper with painted mountains and seas."[52] It would stretch the Sino-Korean term *chip'yŏng* (Chinese, *diping*) beyond semantic possibility to interpret it as a globe, for which by this time there was a well-established term, *chigu* (Chinese *diqiu*). Indeed, the more usual meaning of *chip'yŏng* in connection with armillary spheres is the horizontal and flat terrestrial horizon ring to which the main nest of rings is affixed.

48. Needham, *Science and Civilisation in China*, 3:339–82 (note 31), and Joseph Needham et al., *The Hall of Heavenly Records: Korean Astronomical Instruments and Clocks, 1380–1780* (Cambridge: Cambridge University Press, 1986), 115–52.

49. Sang-woon Jeon (Chŏn Sang'un), *Science and Technology in Korea: Traditional Instruments and Techniques* (Cambridge: MIT Press, 1974), 68–72.

50. Needham et al., *Hall of Heavenly Records*, 106–14 (note 48). The authors say that the 1669 description "of Song Iyŏng's armillary clock would have remained merely a tantalising glimpse . . . were it not for the survival, against all odds, of that clock down to the present day" (p. 114).

51. Needham et al., *Hall of Heavenly Records*, 104–11 (note 48), citing and translating extensively from the account of the 1669 work by Kim Sŏkchu (1634–84) included in Hongmun Kwan (Royal Library), comp., *Chŭngbo Munhŏn pigo* (Documentary reference encyclopedia, expanded and supplemented) (Seoul: Empire of Korea, 1908), 250 *kwŏn* in 50 fascicles (cited hereafter as *Munhŏn pigo*), 3.2a–3b, and the description by King Yŏngjo (r. 1724–76) of a copy of the 1669 water-driven instrument made by An Chungt'ae in 1704 and refurbished by him in 1732, in *Munhŏn pigo*, 3.6a–7b. See also *Sukchong sillok* (Annals of King Sukchong, r. 1674–1720), 17.35a–b (note 17).

52. *Munhŏn pigo*, 3.3a (note 51). The translation follows Needham et al., *Hall of Heavenly Records*, 107 (note 48), but rephrases their last clause, "but instead an earth-model of paper with mountains and seas drawn upon it to represent the surface of the earth [*chip'yŏng*]."

The very brief 1669 description of Song Iyŏng's clock-driven armillary sphere, which supposedly is the grandfather of the Koryŏ University Museum instrument, if not the thing itself, says nothing about a terrestrial globe, but simply notes that its armillary assembly is the same as that of Yi Minch'ŏl's water-driven device except that the movement is powered by a weight-driven clock of Western type. Thus it would have had nothing substantially different from the paper "terrestrial plane" already described. King Yŏngjo's description of the 1704/1732 copy of Yi Minch'ŏl's instrument says, "From the south-polar pivot there projected an iron rod bending toward the earth center [*chisim*, Chinese *dixin*, literally earth heart]. It was in the shape of a claw or fork that held a map of mountains and rivers [*sanhaedo*, Chinese *shanhai tu*]."[53]

If these contemporary texts give no support for a terrestrial globe in the 1669 instrument, the Koryŏ University Museum globe itself offers internal evidence that virtually rules out a 1669 date: the name "Carpentaria" (Chinese Jiabendaliya, Korean Kabondallia). This name, now belonging to the great gulf west of the Cape York peninsula of Australia, appeared on some mid-seventeenth-century European maps as a regional name for the northeastern coast of the continent. The earliest appearance is on a Dutch map of 1648; it was commonly seen throughout the 1650s and 1660s.[54]

It would seem to have been impossible that the Koreans, who had to wait for their European knowledge to be filtered through the Sino-Jesuit medium, could have received knowledge of Carpentaria by 1669. The earliest East Asian cartographic appearance of the name is on the *Kunyu quantu* (Complete map of the earth), published in 1674 by Ferdinand Verbiest (1623–88). But the earliest evidence of this map in Korea comes only from 1721.[55] In 1708, as we have seen, the Koreans were still copying the older Ricci maps, which were too early to reflect the name Carpentaria or the identity of Australia as a separate continent.

The evidence thus points to a general eighteenth-century date for the terrestrial globe in the Koryŏ University's *Sŏn'gi okhyŏng*. Although it is possible that its association with that instrument resulted from an eighteenth-century reconstruction or repair of Song Iyŏng's 1669 *Sŏn'gi okhyŏng*, it is equally possible that the globe is evidence against the identification of the Koryŏ University piece as Song Iyŏng's armillary clock or its direct descendant, and indeed this is my belief.[56]

However and whenever this Western global map was mounted at the center of the armillary assembly of a traditional yet innovative *sŏn'gi okhyŏng*, the combination represented a remarkable marriage of Western and East Asian emblems and concepts. As far as I have been able to determine, it is the only known instance of such a combination. It must have been conceived by a mind deeply knowledgeable in East Asian scientific tradition and at the same time appreciative of and open to Western ideas.[57]

53. *Munhŏn pigo*, 3.3a and 7a (note 51); translation altered from Needham et al., *Hall of Heavenly Records*, 108 (note 48). The representation of the earth by the incorporation of a flat or square earth model within the rings of an armillary sphere goes back at least to the third century in China, as Needham has shown; see *Science and Civilisation in China*, 3:350, 383–86 (note 31).

54. *Nova totius terrarum orbis tabula*, by Joan Blaue (1648); see Günter Schilder, *Australia Unveiled: The Share of the Dutch Navigators in the Discovery of Australia*, trans. Olaf Richter (Amsterdam: Theatrum Orbis Terrarum, 1976), 370–71 (map 64). Of the last fourteen maps on Schilder's list (maps 75 to 88), dated from 1652 to 1666, most give Australia the general name Nova Hollandia (or the inverted form), but eight of them also add Carpentaria as a regional name. Thus this name was gaining currency at the very time the Koreans were building their armillary sphere.

55. Kim, *Maesan kukhak san'go*, 234 (note 3). Kim cites the autograph copy of *Yŏnhaengnok* (Record of a journey to Beijing) of Yu T'akki, an official on the Korean embassy of 1721–22. The manuscript is in the Soongsil University Library, Seoul.

56. Other indications also argue against this identification. From the 1669 notice until modern times, no other documentary reference to Song Iyŏng's armillary clock is known, although references to its water-driven mate are relatively abundant. Given that a clock mechanism is more likely than a water drive to go awry, it seems improbable that such a clock could have reached the twentieth century in as good condition as the Koryŏ University instrument. Furthermore, if the information I have cited is true, that the armillary assembly of Song's piece was generally the same as Yi Minch'ŏl's, we have to consider that the 1704/1732 copy of the latter device, which on King Yŏngjo's testimony (*Munhŏn pigo*, 3.7a [note 51]) had a circumference of 12 [Zhou] feet, or about 239 centimeters, was considerably larger than the Koryŏ University instrument, which with a reported diameter of 41.3 centimeters for the largest ring (Needham et al., *Hall of Heavenly Records*, 134 [note 48]) would have a circumference of only 129.7 centimeters. Finally, while the 1669 pieces featured a visual jack-work display to announce the hours, the Koryŏ University piece has a rotating set of twelve medallions for this function. Admittedly, there are many unknowns in our chain of knowledge that could invalidate these observations, but what we know weakens more than strengthens the case for the identity or close relationship of the two armillary instruments.

57. Hong Taeyong (1731–83) was such a man. A genuine polymath—Confucian scholar, traveler to China, mathematician, musician, and expert in military thought, who was knowledgeable about the West and a successful district magistrate—he built and operated a weight-driven armillary clock in his personal study, one of the rare known instances of the private execution of such a project. The instrument took three years to build and was completed before Hong's trip to China in 1765–66. The description of it by his friend Kim Ian (1722–91) is too involved to explicate here, but one phrase is of particular interest: "Its instrumentalities were two, its rings ten." See Hong Taeyong, *Tamhŏn sŏ* (Writings of Tamhŏn, or Hong Taeyong) (Seoul: Sin Chosŏnsa, 1939; reprinted Seoul: Kyŏng'in Munhwasa, 1969), appendix, 8b. The term *liangyi* (twin instrumentalities), which we have already observed in the title of Matteo Ricci's world map of 1603, is a classical phrase connoting "heaven and earth." Kim Ian's similar phrase hints at the inclusion of an earth model within the rings, which themselves would have represented heaven. The Koryŏ University instrument also has ten rings, and its size corresponds to that of Hong's machine, which

The kind of thinking such a mind would have been up against can be gauged by the remarks of Ch'oe Sŏkchŏng in his preface to the 1708 copy of the Ricci map mentioned above. In presenting this flattened spherical projection of the world to his compatriots, Ch'oe took pains to remind them that although the Westerners saw the world as round, the view considered orthodox in Korea was that "the way [*dao*] of the earth emphasizes quiescence, and its inherent character [*de*] is squareness." He went on, "The Western theories are far reaching but devious and boastful; they stray into the unattested and uncanonical. But their learning has been transmitted and received, and we should not be disrespectful or too quick to criticize it; rather, we should preserve it to broaden our knowledge of things foreign."[58]

Part of the reason for such ambivalence was to protect oneself from the political attacks of Korean Confucian zealots. But behind that concern was the existence of an ideological dimension to the Western cultural traffic, for besides science and mathematics, of course, Catholicism itself issued from Beijing's rectories. The same Jesuit who handed out maps and copies of Euclid in Chinese translation also dispensed holy pictures and scapulars. Korean enthusiasm for the science and gadgets was balanced by official apprehension over religious activity of any kind. Neither the Jesuits nor the Koreans ever forgot the double dimension of this cultural traffic. In 1759 Father August von Hallerstein (1703–74), then director of the Qing government's bureau of mathematics, elucidated the point in a letter to his brother in Laybach (now Ljubljana): "We have not come here to promote and correct astronomical tables. But since astronomy is necessary to protect and preserve, if it can, the interests of religion, we will do all that we can to that end."[59]

By the end of the eighteenth century the Jesuits' greatest hope and the Koreans' greatest fear had come to pass: the baptism in Beijing of a Korean diplomatic traveler had engendered in Korea a major Catholic movement that, in the end, not only resisted suppression but in doing so generated ninety-four Korean martyr-saints, far more than have so far been canonized from all other East Asian countries combined. In this kind of atmosphere, there was an understandable hesitation on the part of many officials to have any contact with things Western. But it helped that Korea's cultural tradition provided a large native niche for maps, and the government itself had in the past sponsored the study and copying of Western maps. The *sirhak* (practical learning) scholars of the eighteenth century took an active interest in Western maps and included them in the map albums made for their own studies. In 1834 the scholar Ch'oe Han'gi (1803–77, see below) enlisted the help of his friend, the mapmaker and publisher Kim Chŏngho (fl. 1834–64), in reprinting a Western-style hemispheric map of the world that had been issued in China. This map, entitled *Chigu chŏnhudo* (Map of the front and back [hemispheres] of the globe), had many new names of English rather than Latin origin, and it sent to anyone alert enough to notice a cartographic signal that English mercantilism was replacing Latin Catholicism in speaking for the West in China.[60] By 1860, when British and French marines occupied Beijing and forced the Chinese to accept a new international order, that transfer was complete, and Koreans expected their own capital to be next. Perhaps to refamiliarize Koreans with the West, an unknown printer in that year made the last known Western cartographic import from China: an 1856 Canton reprint of Verbiest's map, which he then reprinted in Seoul.[61] He would have been more up to date had he reprinted Kim Chŏngho's map of 1834.

KOREA AND MAPS IN THE BUDDHIST TRADITION

In importing the famous maps of Ricci and Verbiest and copying them on any number of luxuriously painted screens, the Koreans were responding not so much to new cartographic science as to a new vision of the world. I have already emphasized that Western maps had little impact on Korean cartography in general, and we will see even stronger confirmation of this when we consider Korean national maps in another section. Yet the Western view of the world, with its globes, hemispheres, and graticules, was undeniably intriguing to Koreans. A Korean viewer of these maps, whether knowing much about geography or not, could see that they had to be taken seriously. So far as historical records reveal, no one ever said

in the impression of Kim Ian took up space "enough to seat a man." Although there is no specific mention of a terrestial globe in the device, Hong Taeyong was the first Korean to champion the idea of the diurnal rotation of the earth (reported by his friend Pak Chiwŏn [1737–1805]; see *Tamhŏn sŏ*, appendix, 1a), and the already mentioned signs that someone may once have tried to adapt the Koryŏ University instrument to include such a movement are at least interesting. In my opinion, Hong Taeyong is more plausible than Song Iyŏng as the maker of that instrument.

58. Translated from the preface of Ch'oe Sŏkchŏng, as written on the map illustrated in *Chōsen shi*, ser. 5, vol. 6, pl. 8 (note 47).

59. Aloys Pfister, *Notices biographiques et bibliographiques sur les Jésuites de l'ancienne mission de Chine, 1552–1773*, 2 vols. (Shanghai: Mission Press, 1932–34), 2:754. Father Hallerstein received frequent visits from Koreans during the 1750s and 1760s.

60. Kim, *Maesan kukhak san'go*, 235–246 (note 3). For Ch'oe Han'gi's hemispheric map, see Yi Kyugyŏng (pen name, Oju), *Oju yŏnmun changjŏn san'go* (Oju's extended essays and lengthy notes, undated manuscript of ca. 1840s), 38.180ab–ba; modern edition 2 vols. (Seoul: Tongguk Munhwasa, 1959). Yi identifies the map reprinted by Ch'oe and Kim Chŏngho as the *Wanguo jingwei diqiu tu* (Graticuled global map of all countries) by Zhuang Tingping. Judging by details in Zhuang's preface to the maps, which Yi quotes (but which Ch'oe Han'gi omitted), this map was first produced sometime after the famous Macartney mission to China in 1793.

61. Yi, *Han'guk ko chido*, pl. 4 and p. 29 (note 4).

"there is no continent here" or "that island is out of position." These maps commanded respect.

Still, the world thus mapped, with all its strange shapes, unpronounceable transliterations, and unimaginable distances, could not have presented much for Koreans to relate to. Their concept of the world, deeply implanted by school primers and reinforced through a literate life, would have conditioned them to a very different kind of map, one that related to the classics, the histories, and the geographical concepts of China and Korea. This is why the more or less primitive but incomparably more relevant *ch'ŏnhado* (map of all under heaven), which circulated in great numbers from the seventeenth century on, was in the end more appealing to Koreans. Before dealing with this genre, however, it is necessary to briefly consider an earlier kind of world map, a Buddhist variety, that may have enjoyed Korean esteem in an earlier age. We shall have to investigate this cartographic genre in any case, since some believe it played a role in the origin of the *ch'ŏnhado*.

Buddhist world maps, as they are known today, are essentially a Japanese phenomenon of the period lasting from the middle of the fourteenth to the middle of the eighteenth century. Unno sketches this Japanese development later in this volume.[62] The *Gotenjiku* (Five Indias) genre of maps was inspired by the travels of the Tang monk Xuanzhuang (602–64), as recorded in his famous *Da Tang xiyu ji* (Record of a journey to the western regions of Great Tang). The *Saiiki zu* (Map of the western regions, 1736), formerly at the Hōshō-in Temple, is the clearest Japanese representative of this tradition (this map was destroyed in World War II). It goes back through a lost copy to an earlier lost map that was in Tō Temple in Kyōto and is accompanied by a text saying that the storied monk Kūkai (774–835) brought it back to Japan upon the completion of his studies in China, even implying that it was the original drawn by Xuanzhuang himself. Nobody takes this note literally, but Nakamura believes that it was still a map of Kūkai's time.[63] Unno calls the story a "legend . . . [that] may have been concocted as part of the biography of Kūkai" (see below, p. 374). Nakamura, Unno, and Muroga all assume that the original of the *Gotenjiku* map was Chinese, but after years of searching no one has found even a mention of such a missing link, let alone the item itself.[64] Given the Chinese concept of the "Middle Kingdom" (see pp. 172–73), it is not particularly surprising that a cartographic genre that had only a tiny place for China on the eastern edge of the continent should have failed to gain favor with the Middle Kingdom's inhabitants.

The fact is that the oldest documentary record of a *Gotenjiku* map, and the first known use of such a term, comes from Korea, where the pronunciation of *Gotenjiku* is *Och'ŏnch'uk*. The text is a memorial inscription on stone, erected in Kaesŏng in 1154 for Yun P'o, an aged official who died in that year. In its account of his life, we read: "He also presented an *Och'ŏnch'ukkuk to* [Map of the lands of the Five Indias] based on the *Xiyu ji* of Dharma Preceptor Xuanzhuang of Tang. His Highness admired it and gave him seven spools of Yan thread."[65] The map has long since been lost, and there are no known copies in Korea. Nor is there much sign of any other explicitly Buddhist cartography remaining in Korea.[66]

In spite of this apparent dead end, the Koryŏ official Yun P'o, in the present state of our knowledge, seems to have the most solid claim to be the originator of this map genre, which ultimately found patronage only in Japan. Unno and Muroga have indeed considered Yun's map as a possible origin for the Japanese maps, rightly pointing out that two constant factors in the latter are the concept of Five Indias (*Gotenjiku*) and the character of a Xuanzhuang itinerary, both of which were explicitly present in Yun's *Och'ŏnch'ukkuk to*. But they point to the lack of any indication of Korea on the earliest *Goten-*

62. Chapter 11, esp. pp. 371–76. I have also relied on the thorough study by Muroga Nobuo and Unno Kazutaka, "Nihon ni okonowareta Bukkyō kei sekaizu ni tsuite" (On Buddhist world maps in Japan), *Chirigakushi Kenkyū* 1 (1957): 67–141; reprinted in *Chirigakushi kenkyū*, 1:67–141 (note 32), and their related article, "The Buddhist World Map in Japan and Its Contact with European Maps," *Imago Mundi* 16 (1962): 49–69.

63. Nakamura Hiroshi, "Chōsen ni tsutawaru furuki Shina sekai chizu (Mappemondes antiques chinoises conservées chez les Coréens)," *Chōsen Gakuhō* 39–40 (1966): 1–73, esp. 43–44 (Western pagination). The article, in spite of its Japanese title, is in French except for the opening and concluding parts. It is a correction and expansion of Nakamura's "Old Chinese World Maps Preserved by the Koreans," *Imago Mundi* 4 (1947): 3–22.

64. Muroga and Unno, "Nihon ni okonawareta Bukkyō kei sekaizu ni tsuite," 78–79 (note 62). On 92–108 they discuss the Buddhist-inspired maps in the *Fozu tongji* (History of Buddhism in the [true Tiantai] lineage, written and published during the years 1265–71) and the well-known Renchao map of 1607 (fig. 7.3), but even the earlier work is more than a century after the date of the first Korean reference to *Xiyu ji*-related maps. Compare Muroga and Unno, "Buddhist World Map," 50 (note 62). Nakamura's search will be noted below in the discussion of the sources for the *ch'ŏnhado*.

65. *Chōsen kinseki sōran*, 1:369–71 (note 10). *Koryŏ sa*, 18.3a (note 9), notes his death on 13 June 1154 but gives no other details of his career. "Yan thread" is probably silk thread from northeastern China, possibly intended for a luxury copy of the map. Yan is the classical name for the Beijing area.

66. The only reference I have seen to any Buddhist cartography in Korea is not to a conventional map but to a schematic plan of Xuanzhuang's itinerary, showing his route beginning in Dunhuang and ending in Sri Lanka. This itinerary plan, executed in 1652, was seen by Nakamura Hiroshi in the T'ongdosa, a Buddhist monastery in Yangsan county northwest of Pusan. Korean historians of cartography do not seem to have mentioned this work, so I have no information on whether it still survives. See Nakamura, "Chōsen ni tsutawaru furuki Shina sekai chizu," 55–56 (note 63).

jiku maps as a reason to doubt the connection.[67] However, the memorial inscription gives no grounds for suspecting that Korea was on Yun's map. It was certainly never visited by Xuanzhuang, so why should Yun have included it? It is true that the evolution of this tradition in Japan was away from a mapped itinerary in India and Central Asia and toward a genuine world map: thus Japan makes an early appearance, China grows and grows, and Korea itself ultimately ends up depicted, even with indication of all eight of its provinces. But this was not the early situation, and in the only record we have, Yun's map made no claim to be anything but a map of Xuanzhuang's travels.

But Unno asks the right question. Whoever drew the first map of the great monk's itinerary must have had a base map of some kind: What was that map? Five Indias, as a geographical concept, relates specifically to the Indian subcontinent of Xuanzhuang's day. But Buddhist literature made abundant mention of an earlier concept of India and its world as being the southernmost of the four huge continents thought to make up the terrestrial part of the earth's surface. (The other three were considered uninhabited; therefore all inhabited lands had to fit somehow into the southern Jambūdvīpa.) The Sino-Japanese transcription of this name, Nan (south) sembushū, shows up frequently on the *Gotenjiku* maps.[68] There must have been some tradition of Jambūdvīpa maps that would have provided a foundation on which Yun P'o—or whoever initiated the Five Indias map tradition—traced Xuanzhuang's itinerary.

The case for Yun P'o as the originator of this tradition is of course weakened by the absence of any Korean map that can be traced to him. But nobody has a stronger documentary claim to be the founder, and he should be given more consideration than he has so far received from those who are now the chief custodians of the *Gotenjiku* legacy.[69] That both China and Korea now have so few remaining traces of Buddhist cartography is surely a function of the persecution or neglect of Buddhism during the centuries of the intellectual and social ascendancy of Neo-Confucianism in those countries. In Korea in particular, as we have seen, Confucian-inspired legislation against Buddhism promoted an atmosphere in which donors and patrons would direct their support to other purposes—usually Confucian schools. Japan's Buddhism was institutionally more secure, with patronage strong throughout society, and such Confucian pressures as there were came weakly and only very late. Thus what was once a broader East Asian Buddhist map tradition came to be confined to Japan.

THE POPULAR "WHEEL MAPS" OF THE *CH'ŎNHADO* TRADITION

Whatever their merits, the *Kangnido* of Kwŏn Kŭn and Yi Hoe and the *Och'ŏnch'ukkuk to* of Yun P'o did not survive in their native land. But the *ch'ŏnhado* (map of all under heaven), which made a silent and utterly unnoticed debut at a date we cannot pin down even to the century, steadily gained admirers—and therefore adapters and printers—and by the end of the nineteenth century existed in copies beyond count and could already be found in museums around the world. Scientifically it was naive in the extreme compared with the other world maps we have discussed, but it appealed to all classes of Koreans and to many foreigners.

What was the source of this appeal? For foreigners it was the exotic and the curious. Somehow it was a great Korean souvenir. Nobody could recall seeing anything like it in China or Japan, and it seemed to capture the country for them. For the Koreans, the appeal was of course more complicated, and much harder to explain. On the one hand, the earth of the *ch'ŏnhado* was flat, and in spite of some astute and already venerable Korean writings to the contrary, most Koreans in traditional times either believed the earth was flat or wanted to. Then too, China was in the middle, and that was where the Middle Kingdom belonged. Korea, being a peninsula, always stood out on the map and was close to China, the classical center of civilization. Japan was always shown smaller than Korea, and that too seemed right to most Koreans. Beyond the immediate East Asian region, the geography

67. Muroga and Unno, "Nihon ni okonowareta Bukkyō kei sekaizu ni tsuite," 78–79 and 90 n. 12, and idem, "Buddhist World Map," 50–51 (both note 62). In the former article the authors say, "In general, most maps drawn in Korea are almost exact copies of Chinese maps," so that it is more reasonable to see Yun's map as based on a Chinese map. But as we have seen, no maps of any kind survive from the Koryŏ period, so there can be no basis whatever for this kind of generalization, quite apart from the fact that no earlier Chinese map of the Five Indias is known either. In the latter article the authors even spell Yun P'o's name "Yin-pu," as if he were Chinese—an indication of how hard it has been to get any recognition of the Korean dimension of this question.

68. *Sembu* and the occasionally seen *embu* go back to Middle Chinese transcriptions of the Sanskrit *jambū*. *Shū* is the Japanese form of the Chinese *zhou* (large island or landmass), which was not a transcription but a translation of Sanskrit *dvīpa*. In the form *embudai*, the *-dai* represents the first syllable of *dvīpa*.

69. Until the past decade there were about a dozen Buddhist religious paintings in Japan that had always been called Chinese and attributed to the Song or Yuan dynasty. But a conference and exhibition mounted by the Yamato Bunkakan in Nara in 1978 demonstrated that almost all of these were of Korean origin, dated to the Koryŏ period. See Kikutake Jun'ichi and Yoshida Hiroshi, eds., *Kōrai butsuga* (Korean Buddhist paintings of the Koryŏ dynasty), exhibition catalog (Nara: Yamato Bunkakan, 1978). Perhaps a parallel reexamination of "Chinese" Buddhist cartography is in order.

FIG. 10.10. UNDATED *CH'ŎNHA CHEGUKTO* (MAP OF THE COUNTRIES OF THE WORLD). Maurice Courant, its early owner, dated this map after 1712, but the criterion for this is unclear. Such a dating would apply to all but a very small number of surviving *ch'ŏnhado*. This version is similar to the *Ch'ŏnha ch'ongdo* (General map of the world) in the British Library and shares with it the peculiarity of showing more than fifty countries not found on ordinary versions of the *ch'ŏnhado*, most of them historical countries of Central Asia in contact with China during the Han or Tang dynasty. Although this map has most of the usual fictional countries found on the *ch'ŏnhado*, the extra historical countries give it a more "real" appearance.
Size of the original: unknown. Present whereabouts unknown. From Maurice Courant, *Bibliographie coréenne*, 3 vols. (Paris: Ernest Leroux, 1894–96), vol. 2, pl. 10 (facing 480).

became less real and more exotic—"the land of the tree eaters," "the land of the hairy people," "the land of righteousness and harmony," and so forth. But as we shall see, these countries, though strange and unbelievable, had a long life in literature; and though no credible person had ever seen them, they were not unfamiliar. The Chinese geographical tradition, long since internalized by Koreans and their "civilized" neighbors, had early on claimed this foreign world by naming it, and the user of the map knew the names. The *ch'ŏnhado*, then, both related to and validated the cultural rearing of its users.

Another important feature of the *ch'ŏnhado* was its context. Although occasionally these maps were found on screens or in some other individual format, the usual mode of publication was as the first map in an atlas. After one looked at the world, one could turn the page and

FIG. 10.11. *CH'ŎNHADO* (MAP OF THE WORLD) WITH GRATICULE. Woodcut, probably late nineteenth century. An example of a late "degenerate" *ch'ŏnhado*, in which an enterprising publisher has attempted to make the map look more modern, and perhaps "scientific," by superimposing a spherical graticule over the flat surface of the world. This version is also unusual for eliminating the trees, although the written legends for them remain.
Size of the original: unknown. Courtesy of the Geography and Map Division, Library of Congress, Washington, D.C. (G2330 .Y651 176-? Vault).

peruse more detailed maps of the countries that were important because of their proximity, whose people many Koreans had met and talked with and who shared with Koreans the classical Chinese language, no matter what their vernacular: such were China, Japan, and the Ryūkyūs. These were followed by a general map of Korea and individual maps of its eight provinces. Many such albums also had tables indicating the locations of postal and military stations, figures on population and annual rice production by province, and other notes on history, scenic sites, and such. The albums thus went from the general to the particular, acquainting the user with the world, then proceeding through the neighboring countries to Korea itself, then to the provincial maps and their indication of every one of the country's 328 county seats.

Koreans viewing Western maps of the world could relate to that world only with difficulty, and given the history of the West's relations with East Asia in the nineteenth century, later users might even have found themselves filled with apprehension and fear. This was quite different from the security and familiarity they could find in the *ch'ŏnhado*, which in part flowed from the concept of *ch'ŏnha* itself. This term (Chinese *tianxia*) very much connoted a Chinese world, in which China was at the center, with China's Confucian ethical system the accepted moral basis for civilized life. Even if China's writ did not run everywhere "under heaven," as the term implied, in theory it ought to have. Korea could be and was remarkably aloof from this imperial China, but until the last years of the nineteenth century it never considered itself aloof from its civilization.

Although *ch'ŏnhado* show infinite variety in style and occasionally differ from one another in details of place-names or orthography, the more salient fact is their invariable structural regularity (see plate 16 and fig. 10.10). Most of the countries shown are imaginary, but the roster of them is virtually identical from map to map, and their respective positions on the map are relatively fixed. Thus, while *ch'ŏnhado* appear to be a whimsical assortment of countries and features, in fact nothing is left to the imagination. From the oldest known examples (perhaps from the sixteenth century) to almost the end of the tradition, the content and structure of these maps, somewhat in the manner of a mandala, changed very little. Some late examples show the attempts of enterprising publishers to be "modern" by irrelevantly imposing curved graticules of Western inspiration (fig. 10.11) or by trying to rearrange a Western map of the world in imitation of *ch'ŏnhado* style.[70] However, these degenerate forms simply show that when they were made the day of the *ch'ŏnhado* was coming to its end.[71]

The structure of the *ch'ŏnhado* is simple (see plate 16). A main continent occupies the center of the circular map; it is surrounded by an enclosing sea ring, which itself is surrounded by an outer land ring. Beyond this outer land ring is what appears to be another sea ring, but this area bears no name and is not home to any island or place (with the single exception of the map in the British Library). In the north, on the outer land ring, is a "thousand-*li* coil tree," adjacent to a pond one thousand *li* in circumference. On the east and west sides of the outer land ring are a pair of trees, one marking the place where the sun and moon rise, the other the place where they set. These trees are sometimes shown on the land ring itself, as the northern tree always is, and in one or two examples they are on islands in the sea beyond; but usually they appear as peninsular features solidly joined to the outer shore of the land ring. From the general annular structure of the surrounding sea and land rings comes the term "wheel-map," popularized by Yi Ch'an.

The main continent contains China and Korea, a number of other historically known countries, some well-known mountains and rivers, and a small number of fictional countries and a fictional mountain, for a total of thirty-two place-names. The inner sea ring contains fifty-seven names of island countries, including Japan and the Ryūkyūs, but also Cambodia and Siam likewise treated as islands. All other names in the inner sea ring are fictional. The outer land ring shows fifty-five place-names, all of them fictional countries, peoples, mountains, ponds, or trees.[72] Neither the borders of countries nor the outlines of the islands are drawn; just the names are indicated, usually in cartouches. The names of ponds and lakes are given in circles or ovals; those of mountains are written underneath a mountain symbol.

There is a modest literature on the *ch'ŏnhado*, but it has so far not resolved the principal questions raised by the genre: When and where did it originate? Why was such a seemingly primitive map so popular so late, to the prejudice of more accurate and generally available world maps of Western origin? To the latter question I suggested an answer at the beginning of this section, and I will come back to it again in summing up. The matter of origins is more difficult.

Copies of the *ch'ŏnhado*, as all investigators have accurately lamented, never bear any original date or maker's name. Many *ch'ŏnhado* atlases, including one I own, feature a preface by a man named Yŏ On, self-styled as Kŭmho San'in, "Indigent of Kŭmho," dated according to the sexagenary cycle in the year *kiyu*. This might be 1849 or any other year at sixty-year intervals forward or backward. Kim Yangsŏn assigns 1789, while Nakamura argues

70. For the graticules, see Shannon McCune, "The Chonha Do—A Korean World Map," *Journal of Modern Korean Studies* 4 (1990): 1–8; for both aberrations, see Yi Ch'an, "Han'guk ŭi ko segye chido" (Old Korean world maps), *Han'guk Hakpo* 2 (1976): 47–66 with 9 plates, see esp. pls. 5–6.

71. I suspect that this judgment might also apply to the much-noticed *ch'ŏnhado* in the British Library, which was the object of a study by Henri Cordier, *Description d'un atlas sino-coréen manuscrit du British Museum*, Recueil de voyages et de documents pour servir à l'histoire de la géographie depuis le XIIIe jusqu'à la fin du XVIe siècle, section cartographique (Paris: Ernest Leroux, 1896), 6–12; and to the map published in Maurice Courant, *Bibliographie coréenne*, 3 vols. (Paris: Ernest Leroux, 1894–96), vol. 2, pl. 10 (facing 480) (item 2187). Whereas other *ch'ŏnhado* have only a general indication of "various countries of the western regions" (meaning Central Asia) and "twelve countries of neighboring barbarians," without any individual names, the Cordier and Courant maps and a few others add approximately fifty other names, mostly from the *Han shu* (History of the Former Han, compiled first century A.D. by Ban Gu), but some from the *Shanhai jing*. Cordier says that the British Library manuscript is of the eighteenth century, but this was just his guess, since he was convinced that the genre was very ancient. Courant indicates no date. Nakamura Hiroshi, who owned two copies of this variety, said that "for the most part they are of the late period." See Nakamura, "Chōsen ni tsutawaru furuki Shina sekai chizu," 67 (note 63), where there is also a list of the added countries. To my way of thinking, the very fact that these copies depart from the general norm argues for their lateness. Four out of five known copies of this version are owned by foreign individuals or institutions.

72. See the analytical list of names in Nakamura, "Chōsen ni tsutawaru furuki Shina seikai chizu," 62–68 (note 63). There are also numbered lists in Homer B. Hulbert, "An Ancient Map of the World," *Bulletin of the American Geographical Society of New York* 36 (1904): 600–605, reprinted in *Acta Cartographica* 13 (1972): 172–78; Yi Ik Seup (Yi Iksŭp), "A Map of the World," *Korean Repository* 1 (1892): 336–41, esp. 339–40; Yi Ch'an, "Han'guk ŭi ko segye chido," pls. 5–6 (note 70); and idem, *Han'guk ko chido*, 191–92 (note 4). Unfortunately, all of these enumerations are different. I follow Nakamura's systematic list, which is based on the structure of the map.

TABLE 10.1 Sources of *Ch'ŏnhado* Place-Names

Ch'ŏnhado Division	Total Number of Place-Names	*Shanhai jing*	Classics	Histories	Daoist Works	Unknown
Continent	32	14	3	14		1
Inner sea ring	56	40		5	7	4
Outer land ring	55	49	2		2	2
Total	143	103	5	19	9	7

Note: This table is based generally on the tables in Nakamura Hiroshi, "Chōsen ni tsutawaru furuki Shina seikai chizu (Mappemondes antiques chinoises conservées chez les Coréens)," *Chōsen Gakuhō* 39–40 (1966): 1–73, esp. 62–68, but categories and some data have been revised in accordance with my own research. Since some names appear in more than one source category, the assignments here are sometimes arbitrary, but in general the category has been chosen that best fits the pattern of the unambiguous data.

for 1849.[73] This is the only fixed date that can be associated with the *ch'ŏnhado*, but Yŏ On, whoever he was, was not the maker of the map, which certainly existed before either 1789 or 1849. He was merely a publisher or an editor and characterized himself only as a lover of maps.

An early modern commentator, Yi Ik Seup (Yi Iksŭp), thought the *ch'ŏnhado* an ageless cultural artifact, calling it "the authorized Chosénese map of the world from time immemorial."[74] Kim Yangsŏn moved in the direction of a more useful date when he said that the map goes back to late Koryŏ or early Chosŏn times (fourteenth and fifteenth centuries), but he could suggest as support for this only some similar-sounding map titles, such as *ch'ŏnhado*, *ch'ŏnha ch'ongdo*, and *ch'ŏnha chido*, found in fifteenth- and sixteenth-century records. All these names, and others besides, have been found attached to the *ch'ŏnhado*, but as Kim himself concedes, the fifteenth- and sixteenth-century sources he cites show it is far more likely that they then referred to Ming maps of China, which often were so named.[75] Nakamura, in an especially valuable study that exploits the largest comparative corpus of *ch'ŏnhado* prints and manuscripts in the literature, many of them in his personal collection, concluded that the map in its present form probably did not go back earlier than the sixteenth century.[76] This remains the best consensus view, nobody having made a convincing case for an earlier or a later date.

But this distinctive world map did not spring up out of whole cloth in the sixteenth century. It had much earlier sources, and these are best approached through a study of the place-names. Nakamura made a good start in this area, but many problems remain to be solved. Among them, the filiation of existing manuscript and printed versions has not been satisfactorily organized, variant and corrupt names remain to be sorted out, and a number of names are still unidentified as to source. Above all, why were these names chosen and not others, and what is the basis for their virtually fixed locations on the map?

Nakamura listed 143 names as the basic *ch'ŏnhado* toponymic corpus, classified them according to their positions on the map, and sought out their textual sources, which are summed up in table 10.1. As can be seen, the *Shanhai jing* (Classic of mountains and seas) dominates the *ch'ŏnhado* as a whole, accounting for over 72 percent of its place-names. But the map's central continent has relatively few names from the *Shanhai jing* and a relatively strong representation from the classics and histories. Of the fourteen *Shanhai jing* names, half mark real places and might have been assigned to the classics or histories. None of the names from Daoist literature, all of them fabulous, appear on the central continent. In all, only eight of the continental names are fictional. Thus the core continent of the *ch'ŏnhado* consists primarily of the real and the familiar. It is only when we get to the inner sea ring and the outer land ring that the character of the map turns fabulous.

The *Shanhai jing* contains very ancient Chinese geographical lore that probably achieved its status in a distinct book during the Former Han dynasty (206 B.C.–A.D. 8) but with much later material interwoven with the text. It registers a very early Chinese mental organization of the world, which itself includes fallout from an even earlier mythical tradition that was already in an advanced state of fragmentation when the work was compiled. The eighteen chapters of the received text can be divided into five principal groups:

73. Kim, *Maesan kukhak san'go*, 218 (note 3); Nakamura, "Chōsen ni tsutawaru furuki Shina sekai chizu," 29 (note 63). Kim does not say what led him to 1789, but Nakamura quotes a specific dating formula from a printed copy owned by him that points unambiguously to 1849. In spite of this evidence, Nakamura believes, based on the pseudonym given for Yŏ On, that this person was not Yŏ On at all but a sixteenth-century personage named Im Hyŏngsu (1504–47). It does not seem to have occurred to Nakamura that Im's life span does not include a *kiyu* year (the only proximate *kiyu* years are 1489 and 1549).

74. Yi, "Map of the World," 336 (note 72).

75. Kim, *Maesan kukhak san'go*, 216–26 (note 3).

76. Nakamura, "Chōsen ni tsutawaru furuki Shina sekai chizu," 49 (note 63).

TABLE 10.2 Distribution of *Ch'ŏnhado* Place-Names in the *Shanhai jing*

Ch'ŏnhado Division	(1) *Shanjing* (mountains)	(2) *Haiwai* (beyond the seas)				(3) *Hainei* (within the seas)				(4) *Dahuang* (great wasteland)				(5) *Hainei* (within the seas)
		N	E	S	W	N	E	S	W	N	E	S	W	
CONTINENT (14)	2					1	1	2	3					5
SEA RING (40)														
North (7)		5	2											
East (9)	1		6			1	1							
South (12)	1			9	1							1		
West (12)		5		1	6									
LAND RING (49)														
North (13)										11			2	
East (10)										1	9			
South (18)											4	14		
West (8)											1		7	
TOTAL (103)	4	10	8	10	7	2	2	2	3	12	14	15	9	5

Note: This table is based on the tables of Nakamura Hiroshi, "Chōsen ni tsutawaru furuki Shina sekai chizu (Mappemondes antiques chinoises conservées chez les Coréens)," *Chōsen Gakuhō* 39–40 (1966): 1–73, esp. 62–68, with the same qualifications as noted for table 10.1.

1. the mountain courses (*shanjing*) in the four directions and in the center (chaps. 1–5);
2. the areas in the four directions beyond the seas (*haiwai*, chaps. 6–9);
3. the areas in the four directions within the seas (*hainei*, chaps. 10–13);
4. the areas in the four directions of the great wasteland (*dahuang*, chaps. 14–17); and
5. the area within the seas (*hainei*), a chapter of disparate and often redundant matter (chap. 18).[77]

Of these, groups 1–3 are generally regarded as the earliest textually, while groups 4 and 5 are considered to have been added after the book had taken form. As can be seen in table 10.2, it is groups 2 and 4 that have provided most of the *Shanhai jing*'s contribution to the Korean *ch'ŏnhado*.

There is a clear pattern to the distribution of names. Those *Shanhai jing* names that appear on the *ch'ŏnhado*'s central continent (not a large number) are all from group 1, 3, or 5, that is, the mountain and *hainei* (within the seas) sections, and half of these represent real geographical features or places, as already noted. (The term *hainei* itself, in general literature, is a synonym for the known—i.e., Chinese—civilized world.) The situation is very different for the inner sea ring, most of whose names come from the *haiwai* group (2), and the outer land ring, toponymically supplied exclusively by the *dahuang* group (4). In these areas the mythical and the fanciful hold sway.

Another point made strongly by table 10.2 is the general congruence of directional relationships between the *Shanhai jing* and the *ch'ŏnhado*. Thus a place-name associated with north on the land and sea rings of the Korean map will generally be found in the northern chapters of groups 2 and 4 in the *Shanhai jing*, and likewise for the other directions. This tendency is even stronger than it appears because of perceptible shifts on the *ch'ŏnhado*. In breaking down the land and sea rings by direction, Nakamura's analysis has been kept in table 10.2, because he seems to have divided the rings into directional sectors at the most natural points. But in tracking the names in these corner positions to their respective chapters in the *Shanhai jing*, we find some assigned to a direction ninety degrees off. Thus Nakamura's number 73, Yibi Guo (Land of the One-Armed), is in his southern sea ring on the *ch'ŏnhado* but in the eastern sector of the *haiwai* group, while number 100, Buzhou Shan (Mount Defective), is in his northern land ring on the *ch'ŏnhado* but in the western *dahuang* in the *Shanhai jing*. If these names in corners are corrected for this shift, then fully thirteen that appear to be directionally incongruent between map and text will become congruent, greatly strengthening an already strong tendency in the data. These shifts show regular patterns: those in the inner sea ring move counterclockwise from their *Shanhai jing* source, while those on the outer land ring shift clockwise from the same.[78]

77. I use the excellent modern edition of the *Shanhai jing* edited by Yuan Ke, *Shanhai jing jiaozhu* (Edited and annotated *Shanhai jing*) (Shanghai: Shanghai Guji Chubanshe, 1980), which includes the standard Guo Pu (third century A.D.) and Hao Yixing (1804) commentaries as well as many helpful notes by Yuan himself.

78. In the inner sea ring, Nakamura's no. 73 goes from west in the *Shanhai jing*'s *haiwai* section to south on the *ch'ŏnhado*, nos. 86–88 go from north to west, and nos. 33–34 go from east to north—all counterclockwise shifts. On the outer land ring, nos. 100–101 go from west in the *Shanhai jing*'s *dahuang* section to north on the *ch'ŏnhado*, no. 103 goes from north to east, and nos. 118–21 go from east to

The only conclusion to be drawn from these facts is that the maker or makers of the *ch'ŏnhado* used the *Shanhai jing* in elaborating the basic structure of the map. Other names were added from the histories (the *Han shu* [History of the Former Han, compiled first century A.D.] in particular), from Daoist works (especially in the rings),[79] or from common knowledge (Japan, the Ryūkyūs, Siam), but these were mainly seasoning for a *Shanhai jing* stew. The fact that these other sources were freely used and, more important, that only a sampling of the many available *Shanhai jing* place-names were actually selected for inclusion shows that it was not the intention of the makers of the *ch'ŏnhado* to actually make a map of the *Shanhai jing*. It must rather have been their purpose to make a world map, for which they used this ancient Chinese mine of fabulous geographical lore only as their principal source.

The question arises, then, whether the makers had at their disposal an earlier map of the *Shanhai jing*. The great Qing commentator Hao Yixing (1747–1825) believed there had once been actual *Shanhai jing* maps that "must have contained mountains, rivers, roads, and stages," but that these were already long lost by the time of Guo Pu's commentary in the third century.[80] The *tu* (graphic depictions including maps, illustrations, and charts) mentioned by Guo seem to have been only pictures of the strange people and freakish beasts that inhabited the *Shanhai jing*'s universe. Scholars generally believe that the occasional mention of *tu* in ancient and medieval bibliographic citations in the *Shanhai jing* are all of this genre. The fact is that in the earliest notices of this book in the Former Han, it was already thought of more as a bestiary than as a work of geography.[81] Whether or not there were once maps of the *Shanhai jing*, it appears that no later Chinese cartophile ever attempted to fill the void. The only map that comes close to this function is the *ch'ŏnhado*.

The inner sea ring and the outer land ring are theoretical depictions of the *haiwai* and *dahuang* sections of the *Shanhai jing*. No textual foundation for this ring structure is evident in the *Shanhai jing* itself, nor is there any text that would rule it out. It is one possible theoretical construct. But what of the inner continent, the more or less real core of the *ch'ŏnhado*? What was its cartographic model?

Nakamura believed that it derived from some Chinese map in the Buddhist tradition. It had to come from China, since, in his deeply held view, Koreans had historically not created their own culture but had "slavishly" borrowed it from China.[82] And it had to be a Buddhist map because there was no other tradition of world maps in China. (He believed that Xuanzhuang himself had created the Five Indias genre in the seventh century.) His study of *ch'ŏnhado* place-names showed that none of them came from works compiled later than the eleventh century, so he examined various likely Chinese possibilities between then and the sixteenth century, by which time the *ch'ŏnhado* must have been fixed in its currently known form. At first he thought he saw some signs of the *ch'ŏnhado* in the *Sihai Hua yi zongtu* (General map of Chinese and foreign territory within the four seas; illustrated above, fig. 7.4), which appeared in the famous *Tushu bian* (Compilation of illustrations and writings, completed in 1577), compiled by Zhang Huang (1527–1608). Zhang had attributed this "map of southern Jambūdvīpa within the great sea" to some unnamed Buddhist source. But Nakamura could find no cartographic antecedents for this map earlier than the thirteenth century. Moreover, since typologically it seemed to be only a hybrid between a Buddhist antecedent and the supposed Chinese ancestor of the *ch'ŏnhado*, the latter had to be earlier in date and in any case had to go back to at least the eleventh century. He then explored a Sino-Tibetan map, which had been brought to Japan in the ninth century from Xuanzhuang's own monastery. This showed a greater geographic extent than the Xuanzhuang itinerary

south—all clockwise shifts. I believe these shifts originated not from the *ch'ŏnhado*'s original compiler(s) but from Nakamura's perceptions as to where the directional sectors on the rings were to be marked. He took the understandable course of marking the borders more or less in the northeast, southeast, southwest, and northwest corners, but the original compilers seem to have (implicitly) marked them more flexibly.

79. For instance, the Five Marchmounts (*wuyue*) of the Daoist *xianjing*, or Transcendent spaces, i.e., Mounts Guangsang (east), Li'nong (west), Changli (south), Guangye (north), and Kunlun (center), are shown in a diagram titled *Yuzhong xianjing* (Transcendent spaces in the Ecumene) in the Daoist work *Shangqing lingbao dafa* (Great Praxis of the supernal magic treasure), 10.14a (in the *Zhengtong Daozang* [Daoist canon of the Zhengtong reign period (1436–49)], 1120 fascicles [Shanghai: Shangwu Yinshugwan, 1923–26], fasc. 945). Mount Kunlun is on the central continent of the *ch'ŏnhado*, while the rest are all found in its *neiyang* (inner sea). Aside from the center mountain, none of these mountains are found in the *Shanhai jing*. I do not therefore call the *ch'ŏnhado* a Daoist map; but given these correspondences, the case for doing so would be much better than the case for calling it a Buddhist map.

80. *Shanhai jing jiaozhu*, appendix, 484 (note 77).

81. See Liu Xiu's original presentation notice (late first century B.C.) in *Shanhai jing jiaozhu*, appendix, 477–78 (note 77). Compare Needham, *Science and Civilisation in China*, 3:504–7 (note 31). So pervasive is this view that Needham, writing here in the geography section of his great work, laments (507) that no one has ever made a biological study of the monsters! What we have always needed, and still need, is a study that will clarify the *geographical* foundations of the *Shanhai jing*.

82. Speaking of the origins of the *ch'ŏnhado*, Nakamura writes: "Such material will come, it is to be hoped, from Chinese sources rather than Korean, for this mappemonde is purely Chinese. It bears no trace of anything specially Korean, which is understandable when we consider that the sciences and the arts of Korea were almost always slavishly modelled upon those of China" (Nakamura, "Old Chinese World Maps," 13 [note 63]). The same statement, in French, is made in Nakamura, "Chōsen ni tsutawaru furuki Shina sekai chizu," 36 (note 63).

and therefore had better qualifications as a world map. In addition it was diagrammatic, indicating place-names in relative position in blocked cartouches, just as in the *ch'ŏnhado*. This convinced him that there had been a Chinese prototype of the *ch'ŏnhado* in existence already in the seventh century, which, though later lost in China, would have survived in Korea, where by the sixteenth century it would have achieved its final form.[83]

In spite of all his labors, Nakamura still did not prove that this supposed Chinese prototype was a Buddhist map. It would only have served as one antecedent for the Sino-Tibetan map, which was Buddhist. But his two-sentence argument even for this is totally unconvincing. This alleged Buddhist connection has nonetheless been widely accepted and is now virtually fixed in the Western literature on the *ch'ŏnhado*. Indeed, it has grown. One writer says that the *ch'ŏnhado* "shows how the facts of geography were pressed into the a priori format of a Buddhist diagrammatic view of the cosmos." Another speculates that the trees found at the northern, eastern, and western extremities of the *ch'ŏnhado* may be "Buddhist symbols." Others assert that the *ch'ŏnhado* usually appears as a map in a "Buddhist atlas."[84]

Japanese and Korean scholars have been more reluctant to go along with Nakamura.[85] Indeed, beginning with his sadly biased initial preconception that Koreans are incapable of any cultural originality and following to the end of his argument, one finds more assumptions than evidence. The extra points added by others are also dubious: the atlases that typically contained the *ch'ŏnhado* are utterly without any feature that could be called Buddhist. The famous trees come from the *Shanhai jing*, along with most of the other *ch'ŏnhado* names;[86] of course the *Shanhai jing* itself has no connection with Buddhism. The *ch'ŏnhado* does have two names that are not in the *Shanhai jing* and may have some Buddhist echo: Mount Tiantai, famous for its monasteries and the great syncretic sect that bears its name (but which could just as well have a Daoist resonance), and possibly the name Jiabi (anciently Ka Pi), which is otherwise unexplained but might be an abbreviation for Kapilavastu, kingdom of the Sakyas and birthplace of the historical Buddha. Both of these names were common in general literature and have been counted as historical in table 10.1. No other *ch'ŏnhado* names have any Buddhist associations.

Cartographically, the *ch'ŏnhado* feature that might lead some to see a Buddhist connection is the central mountain, Kunlun, and the four rivers that flow from it. But there are many differences between it and Mount Sumeru on the Five Indias and Jambūdvīpa maps. Kunlun has its own pre-Buddhist, ancient Chinese origin in the "Yu gong" (Tribute of Yu) chapter of the *Shu jing* and in a large body of autochthonous Chinese myth, as reflected in the *Shanhai jing* among many other sources.

The conclusion has to be that the argument for a Buddhist origin for the *ch'ŏnhado* is without merit. Indeed, the sixteenth century would have been one of the least likely periods in Korean history for the ascendency of any Buddhist artifact as an icon of popular culture, much less literary culture. That period saw the extension of Confucian belief and practice to all classes of Korean society through a network of approximately 325 state-supported schools, perhaps another 200 private Confucian academies (*sŏwŏn*), and thousands of informal village schools (*toksŏdang* or *kŭlppang*) where young children sang out their Confucius and Mencius. Buddhist temples were banished from the cities and towns, and the monks, deprived of any significant patronage, managed a bare survival in mountain monasteries. This was not a setting in which Korean society was likely to find fascination with Five Indias, Jambūdvīpa, or any of their putative cartographic offspring.

The Korean cartographic historian Kim Yangsŏn has found a quite different origin for the *ch'ŏnhado*, explaining it as a mapping of the world according to the theories of the ancient Chinese naturalist philosopher Zou Yan (third century B.C.). In support he cites a book called *Hwanyŏng chi* (Description of the Ocean World), written by Wi Paekkyu (1727–98), who is well known as a geographer and naturalist. Wi attributed the following state-

83. Nakamura, "Chōsen ni tsutawaru furuki Shina sekai chizu," 36–56 (note 63).

84. A. L. Mackay, "Kim Su-hong and the Korean Cartographic Tradition," *Imago Mundi* 27 (1975): 27–38, esp. 31; McCune, "Chonha Do" (note 70); Norman J. W. Thrower and Young Il Kim (Kim Yŏng'il), "Dong-Kook-Yu-Ji-Do: A Recently Discovered Manuscript of a Map of Korea," *Imago Mundi* 21 (1967): 30–49, esp. 32. Thrower and Kim refer to the *ch'ŏnhado* by one of its alternative titles, *Sahae ch'ongdo* (General map of the four seas), which I note here with standard romanization and my own translation.

85. For instance, Muroga and Unno, "Buddhist World Map," 51 n. 7 and 57 n. 16 (note 62); Yi, "Han'guk ŭi ko segye chido," 57–58 (note 70).

86. *Shanhai jing jiaozhu*, 9.260, 14.354, 16.394, 17.423 (note 77). There are some variants in the names, but there can be no doubt that all of the *ch'ŏnhado*'s trees come from the *Shanhai jing*. Interesting material on the northern tree that cannot be found in the received text has been found quoted in standard Han works by editor Yuan Ke. According to this, the tree in the north had branches in a 3,000 *li* "coil" (*pan*) and was the residence of two gods who controlled the world's ghosts. The lack of any connection with Buddhism is patent. Mackay suggests that the trees are "perhaps the Cosmic Trees of the Shamans of north-east Asia." This is more plausible than the Buddhist explanation but still not likely. Chinese shamanism connects to the Southeast Asian variety rather than to that of northern and northeastern Asia. His interpretation of the northern tree's name as "axle-tree" is incorrect. Chinese commentaries make it clear that *pan* means "coil" in this name. In spite of the shamanist angle, Mackay generally advocates a Buddhist explanation of the *ch'ŏnhado*. Mackay, "Kim Su-hong," 31–33 and caption to fig. 5 (note 84).

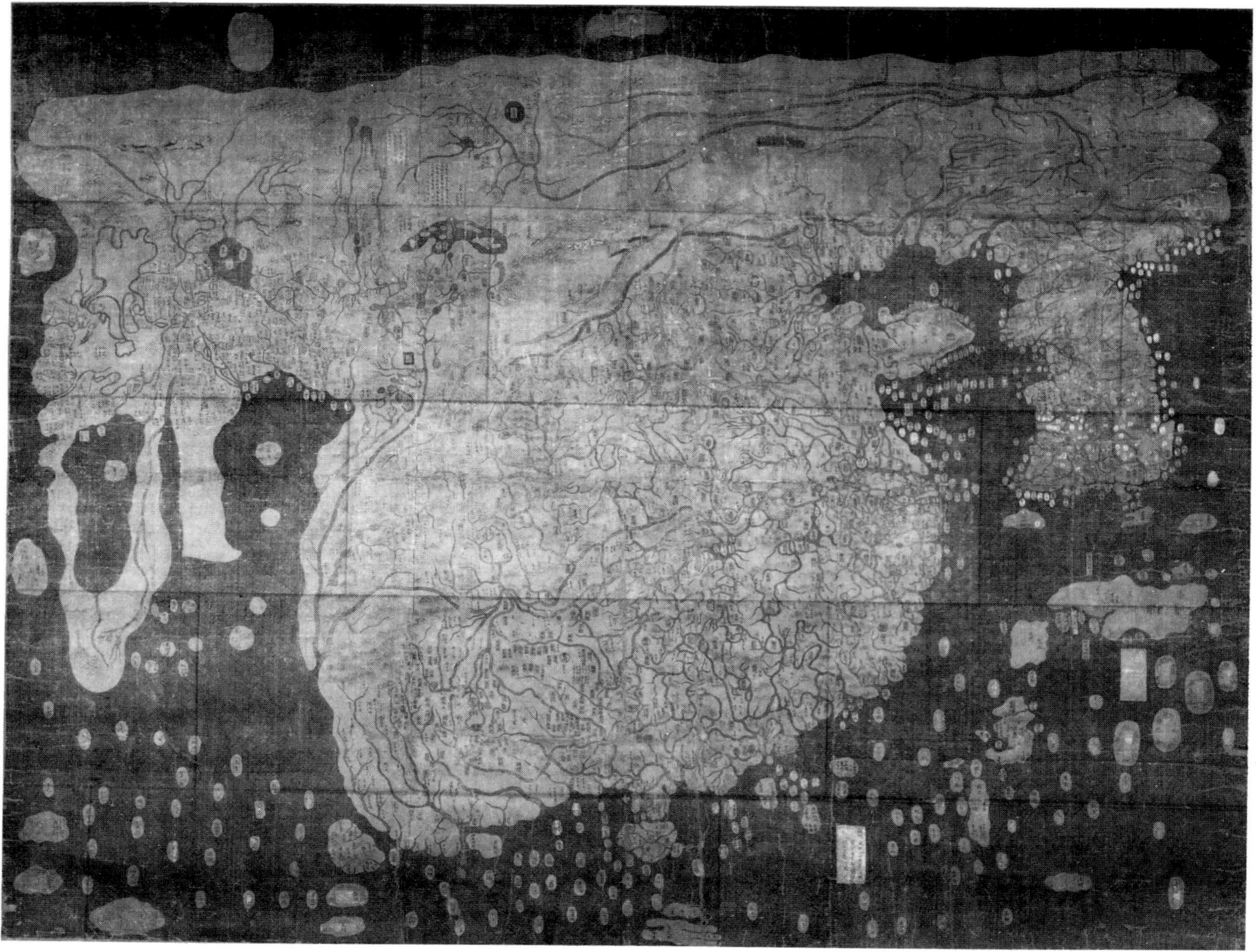

FIG. 10.12. UNTITLED MAP, CALLED "*CHUGOKU ZENZU*" (COMPLETE MAP OF CHINA). It is in fact a copy of, or in the tradition of, the *Honil kangni yŏktae kukto chi to* (or the *Kangnido*, see fig. 10.3). This map, thought of as a "sister map" to the *Kangnido* example in the Honmyōji (in Kumamoto), has been determined by Kazutaka Unno to have been copied in Korea ca. 1568. It differs from the two other examples of the *Kangnido* in that the continent is shown completely surrounded by water.
Size of the original: 135.5 × 174 cm. By permission of Tenri Central Library, Tenri, Japan.

ment to Zou Yan: "As for the sea around China in the four directions, I call it 'the tiny sea' [*bihai*]. Beyond that sea there is a great continent that rings it, and beyond the great continent there is a vast ocean sea circling around. It is only here [that one comes to] the edge of the earth."[87] This description is an excellent evocation of the configuration of the *ch'ŏnhado*. The problem is that this statement cannot be located among the fragments of Zou Yan's long-lost writings that are quoted in authentically ancient works.[88] His biography in Sima Qian's *Shi ji* (Records of the grand historian, completed ca. 91 B.C.) has a similar though still quite different concept of the world, in which China is one mere *zhou* (island or isle-land, signifying continent) out of eighty-one that exist under heaven. As Sima Qian explains Zou's idea, there are nine China-sized *zhou*, "[each] ringed by 'tiny seas' and mutually impenetrable by man or beast, but within a single sector, what [Zou Yan] calls one

87. From Wi Paekkyu, *Hwanyŏng chi*, as quoted in Kim, *Maesan kukhak san'go*, 217 (note 3). I have been unable to find a copy of Wi's original book.

88. None of Zou Yan's writings survive intact. The quotation cannot be found in Ma Guohan's authoritative compendium of quoted fragments of ancient lost writings, Ma Guohan, comp., *Yuhan shanfang ji yishu* (Fragments of lost writings collected in the Jade Box Mountain Studio, compiled 1853), bk. 77 (Zou Yan).

FIG. 10.13. COMPARISON OF CONTINENTAL OUTLINES BETWEEN THE *KANGNIDO* AND THE *CH'ŎNHADO*. At the upper left, the outline of the Tenri *Kangnido*; at the upper right, the outline of a typical *ch'ŏnhado*. The sequence at bottom proposes a development by which the outline of the *ch'ŏnhado*'s inner continent might have evolved. A key element in this hypothetical development is the Arabian Peninsula, which with the Red Sea and the Arabian Sea forms a peninsula between the two rivers on the *ch'ŏnhado*.

[great] *zhou*. There are nine of these, with a vast ocean sea ringing their outer reaches where heaven and earth meet."[89]

There is enough in this account to suggest some typological connection with our puzzling Korean map and perhaps to justify Kim Yangsŏn's term, "the Zou-Yan-style *ch'ŏnhado*." But one wonders if Wi Paekkyu, trying to make sense of Sima Qian's well-known but unclear summary, didn't interpret (and then quote) it while looking at a *ch'ŏnhado*, which was in its heyday during his lifetime.

On the other hand, the toponymic relationship of the map to the *Shanhai jing* is beyond doubt, and its inner sea and outer land rings are filled in a systematic way with *Shanhai jing* names and features. They can only be theoretical projections of the *Shanhai jing*'s geographical divisions, perhaps with some distant influence from Zou Yan's geographic thought.

The outline of the *ch'ŏnhado*'s inner continent, however, is not theoretical, even though it may be in part imaginary. The Korean peninsula, the Yellow and Yangtze rivers, perhaps the bend of the southeastern coastline suggesting the Gulf of Tonkin, the Kunlun Mountains, and other features show a certain basis in cartographic experience and provide enough material to encourage a morphological analysis of the inner continent as a whole. The very awkwardness and asymmetry of its outline suggests something other than a theoretical or imaginary model.

I suggest that the *Kangnido* of 1402, being a documented Korean map of the world and known to have been copied in Korea as late as about 1568, is a plausible candidate for that model. Especially interesting is the Tenri copy because, unlike the other two surviving exam-

89. Sima Qian, *Shi ji*, 74.2344; see the edition in 10 vols. (Beijing: Zhonghua Shuju, 1959). This text is obscure, and Sima Qian, who was not fond of Zou Yan and his doings, certainly brought no clarity to his concepts. For another translation, see Joseph Needham, *Science and Civilisation in China* (Cambridge: Cambridge University Press, 1954–), vol. 2, with Wang Ling, *History of Scientific Thought* (1956), 236.

ples of that map, it presents a single landmass completely surrounded by water (fig. 10.12). Even Africa is shown clearly as a peninsula hanging from Europe, not as an independent continent. An outline of the Tenri map is juxtaposed with that of a typical *ch'ŏnhado* in figure 10.13, together with an inset showing hypothetical transitional stages from one to the other. Two principal clues to such a transition are found on the *ch'ŏnhado*. The first is the triangular peninsula on the inner part of the western half of the continent, just below Mount Kunlun. It is formed between the long Hei Shui (Black River), which flows south into the sea, and the short Yang Shui (Welling River), which flows into the Black. Both are fictional rivers stated in the *Shanhai jing* to rise on Mount Kunlun.[90] It is important to emphasize that this interior peninsular configuration is unique in the overall layout of the continent, being the only case of an inland feature outlined by water and the only case showing a tributary to a river. The second clue is in the large body of water just to the northwest of this peninsula. The most frequently seen name for this feature is Sule (anciently pronounced *su-lek*), but there are variations, and it is likely that it originally had no name at all.[91] This body of water is also unique on the *ch'ŏnhado*, being the inner continent's only inland sea. Other than these two features, the continent is nothing but nine mountains, four large rivers, the Korean peninsula, and the curving coastline.

If we now look at the western part of the *Kangnido* landmass and imagine Africa merged into the main bulk of the continent in such a way that the Arabian Sea and the western Indian Ocean become a long, south-flowing river, as postulated in figure 10.13, the *ch'ŏnhado*'s triangular peninsula emerges as the remains of the Arabian Peninsula and the large inland sea as the residue of the Mediterranean and Black seas. The *Kangnido*'s Arabian Sea and the western Indian Ocean have become on the *ch'ŏnhado* the Black River, and the Red Sea has become the Welling River. To this revised continent it is necessary only to introduce some rounding and contraction (particularly of Korea) and to insert the Yellow, Yangtze, and Chi Shui (Red River) in order to arrive at the *ch'ŏnhado*'s basic outline.

The suggested evolution from the *Kangnido*'s outer continental coastline is concededly arbitrary, dictated by a "known" target outline. It is the inner evolution, that involving the *Kangnido*'s Mediterranean/Black, Red, and Arabian seas and the Arabian Peninsula into their *ch'ŏnhado* counterparts, that is morphologically more significant. This inner evolution is not arbitrary but is morphologically meaningful for its creation of two forms that do not otherwise occur on the *ch'ŏnhado*—an inland sea and a river with a tributary.

Although the judges of the court of plausibility must render their verdict on this explanation of the shape of the *ch'ŏnhado*'s inner continent, it can be argued that every significant landmass on the *Kangnido* can be found in some way folded into its flexible shores. In a way that might seem perverse given the "scientific" content of the *Kangnido* and the "primitive" character of the *ch'ŏnhado*, the latter has achieved a much better balance in the continental landmass; by critically placed rivers it has divided the continent into intelligible parts; and by gentle curves it has suggested the Indian subcontinent and the contour of Vietnam. Historically, the timing is right for the emergence of the *ch'ŏnhado*, which cannot be documented before the seventeenth century but has been reasonably inferred to have arisen sometime in the sixteenth. The Tenri copy of the *Kangnido*, as we have seen, contains Korean place-name changes (vis-à-vis the Ryūkoku copy) that according to Unno's research are datable to about 1568.

If this hypothesis is permitted, the known evolution of the *Kangnido*'s landmass into a single water-surrounded continent may have created the primary cartographic condition for the emergence of the *ch'ŏnhado*'s inner continent. This geographical form, once achieved, would have been supplemented by some clever spirit engrossed with the fantasies of the *Shanhai jing*. Finding the Chinese-character transcriptions of the Arabic names on the African and European parts of the map quite unintelligible, the maker would have ignored them and adapted the forms of Africa and Europe themselves to the requirements of historical imagination. In this way Europe became a narrow strip of land north of the inland sea, and Africa turned into the *xiyu* (western regions) of the Chinese histories, ending up filled with the names of Central Asian "barbarians." On this explanation, the *Kangnido* did not disappear but rather evolved into the *ch'ŏnhado*.

Those who favor the glory of scientific progress over the comforts of culture might consider such a development degenerate. But it does not take much thought to realize that the original *Kangnido*, however great and unique a cartographic achievement, did not and could not relate to the traditional geographic culture of Korea or of East Asia as it existed at the beginning of the fifteenth century.[92] It was a sport, produced by Kwŏn Kŭn

90. *Shanhai jing jiaozhu*, 2.48 and 11.297 (note 77).

91. Sule is the name of a historical Central Asian kingdom (identified with Kashgar) found in Han and Tang historical writings. It is not connected in any way with a sea or lake. Some versions of the map have the lake nameless with the name Sule adjacent to it, along with the names of other ancient Central Asian kingdoms. I suspect that this was the correct appearance. A few versions identify this body of water as a river (while clearly showing it as a lake or sea) named Ruo (weak), Ni (drowning), or Niao (urine) depending on the reading or graphic variant.

92. This point is also made by Pang, *Han'guk ŭi chido*, 76 (note 5).

and Yi Hoe before its time, that could have had only limited cultural meaning to all but a few of those who beheld it. But if my guess is correct, its continental structure, aided by some precocious armchair plate tectonics, found new life through indigenization—as part of the *ch'ŏnhado*.

KOREAN MAPS OF CHINA AND JAPAN

Although up to now the *ch'ŏnhado* has been discussed as a map of "all under heaven," its subject matter was always somewhat more ambiguous given the age-old use of that term to connote the Chinese world specifically, the area where the prestige, if not the actual authority of the emperor of China, held sway. During the Ming (1368–1644) particularly, Koreans commonly referred to maps of China as *ch'ŏnhado* or some close variation. Maps of China were particularly popular after the beginning of the Ming because China was once more Chinese after nearly two and half centuries during which part or all of its territory had been under the rule of alien dynasties. The Mongols were the first non-Chinese people in history to rule not just a part of China but all of it, and for a protracted time. With this situation put to an end, one could once more use classical expressions like *ch'ŏnha/tianxia* without embarrassment.

I have already referred to Na Hŭngyu's lost map of China, probably made sometime in the 1370s, and to the Chinese maps that Koreans imported early in the fifteenth century and incorporated into the *Kangnido*. Although there were definitely Korean-made maps of China—we know of one drawn up in the Hongmun Kwan (Royal Library) in 1469 in addition to Na's and the *Kangnido*[93]—in general most Korean maps of China during the fifteenth and sixteenth centuries were probably made in China and imported.[94] In ordinary cases, Korean-made ones would necessarily have had the character of copies rather than original works.

The conquest of the Ming dynasty by the Manchus (1644) provided a direct stimulus for Koreans to produce their own maps of China. Bitter Korean experiences at the hands of the Manchus, among them invasions in 1627 and 1636–37, a forced severance of relations with the Ming rulers, a humiliating oath of allegiance extracted from the king, and the Manchus' taking of dozens of hostages from the royal family and other leading Korean families, resulted in a nationwide anti-Manchu hostility. At the same time, the role of Ming military forces in aiding Korea against the Japanese invasions of the 1590s had nurtured strong pro-Ming sentiments among Korean literati. One expression of this was the popular (though unofficial) Korean use throughout the Qing centuries of a Ming year title (*nianhao*) in dating: the use of the title Chongzhen, formally restricted to the period 1628–44, was extended by the Koreans down to the late nineteenth century. Another one was a fondness for maps of China showing the Ming capitals and provincial organization rather than those of the Qing.

The Ming dynasty had two capitals (Nanjing and Beijing) and thirteen provinces, for a total of fifteen major units of regional administration. The Qing system corresponded generally with the Ming except that there was only one capital (Beijing), while the old southern capital region and Huguang Province were split into two provinces each and an additional province (Gansu) was added in the west. These were relatively minor changes, but of course everything lay in the symbolism: the changes were those of an illegitimate dynasty as far as the Koreans were concerned. And since Korea insisted on its own standard of legitimacy, it had to produce its own maps of China. The practice must have been widespread, because every published Korean map of China that I have seen from traditional times shows the Ming organization.[95]

An elaborate example of this kind of map is the Kim Suhong woodcut of 1666, a rare case of a Korean map with printed indication of both author and date (fig. 10.14). Titled *Ch'ŏnha kogŭm taech'ong pyŏllamdo* (Comprehensive and synoptic map of the ancient and modern world), it not only gives the Ming organization of the country but displays a cultural panorama of Chinese civilization reaching back to antiquity. Published reproductions of this very interesting work are not legible, but one scholar who has studied it asserts that its legends note "antiquities, scenic wonders, old monasteries, famous men, and filial sons" in appropriate places all over the map.[96] In an earlier chapter, Cordell Yee emphasized the close association of map and text as a perennial feature of East Asian cartography (above, chap. 5), and Kim Suhong's map is a splendid example of the phenomenon in Korea. It might be described as a text on civilization in the shape of a map of China. Although Kim's map appears to have some material of his own, it strongly evokes earlier Chinese maps, notably Liang Zhou's *Qian-*

93. *Yejong sillok* (Annals of King Yejong, r. 1468–69), 6.15a–b (note 17).

94. Two maps of China mentioned in a list of maps discussed in 1482 are not identified by author, as are all the Korean maps in the same list, so presumably they were imported; *Sŏngjong sillok*, 138.10b (note 17). A map mentioned in 1511 is impossible to identify as either Chinese- or Korean-made; *Chungjong sillok*, 14.21a (note 17). Efforts or successes in importing maps of the Ming dynasty are reported during the years 1536–38; *Chungjong sillok*, 81.51b–52a, 84.31b, and 89.9b.

95. For example, Yi, *Han'guk ko chido*, 40–45 and pl. 5 (p. 18) (note 4). The latter bears a date corresponding to 1747.

96. Kim, *Maesan kukhak san'go*, 223–25 (note 3). Kim's copy, now in the Soongsil University Museum, appears to be a manuscript version. Mackay has described two manuscript copies owned by him in "Kim Su-hong" (note 84). A woodblock print of this map, owned by Yi Ch'an, is illustrated in his *Han'guk ko chido*, 40 (note 4).

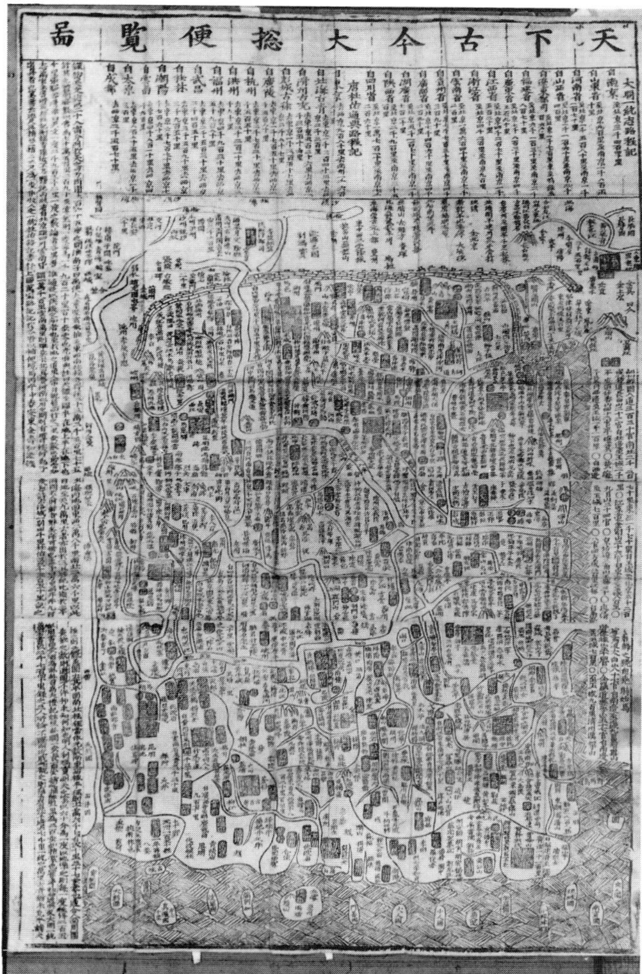

FIG. 10.14. *CH'ŎNHA KOGŬM TAECH'ONG PYŎLLAMDO*, BY KIM SUHONG, 1666. This copy appears to be the only surviving example of a woodcut edition; three hand-copied versions are known. Typically for late Korean maps of China, this map continues to show the Ming dynasty provincial system well after it had been replaced by that of the Qing dynasty. The style, with its dense textual material, imitates a Ming genre that featured references to well-known historical individuals, events, and scenic sites.
Size of the original: 142.8 × 89.5 cm. By permission of Yi Ch'an, Seoul.

kun wanguo quantu gujin renwu shiji (Complete map of the myriad countries of heaven and earth, accounts and reminders of famous people of antiquity and the present) of 1593.[97]

Kim Suhong (1602–81) had a curious and disastrous public career. His official obituary in the Chosŏn dynasty annals is not only critical but slanderous, saying that he was "eccentric and foolish, shunned by all," and that he was "an experienced pervert, contemptibly wicked . . . everyone loathed him."[98] He apparently earned this opprobrium by opposing the standard-bearers of his own party and siding with the opposition in rancorous court debates in 1659 and 1674. This behavior was all the more unusual because of his membership in the Andong Kim lineage, one of the backbone forces of the conservative, anti-Manchu western faction (a local reference with no connection to Western culture or science), led by the famous Song Siyŏl (1607–89), which was then in power. Among many items on Song's agenda was the continued use of the Ming dynasty era title "Chongzhen" for dating purposes. In order to irk Song, Kim Suhong is said to have used the then-current Qing title "Kangxi." In the eyes of his enemies this constituted unfiliality, the most heinous of crimes to a Confucian, because Kim's grandfather, Kim Sangyong, had been a glorious anti-Manchu martyr. On his last stand at Kanghwa fortress in 1637, he had mounted some bags of gunpowder and blown himself up rather than submit to the Manchus. A Western writer admiringly argues that in using the Qing year title Kim had "acknowledged change" and was thus siding with the "progressives" against the "reactionaries."[99] But even allowing for the obviously biased opinion against him, there is little if any evidence for such a view, which in any case fails to consider the highly charged ideological atmosphere. A cartographic jury would have to deliver a contrary verdict: twenty-two years after the establishment of the Manchus in Beijing, Kim's map shows the Ming provincial system without a hint of the existence of the Qing dynasty.

As with Kim's map, so it was with virtually all other Korean maps of China down to the mid-nineteenth century. From elaborate, large-sized sheet maps down to the map of China generally found in atlases, it was the Ming provinces and capitals that were presented to Korean readers. Possibly some of the atlases associated with the *sirhak* (practical learning) school were exceptions to the general phenomenon, but if so they were few. Apart from the Ming organization, which included written indication of the distance of each provincial center from Beijing, a typical atlas map of China emphasized bold and abstract delineation of major features: the Great Wall, the Yellow and Yangtze rivers, the important mountains of the classical and literary tradition, and the principal towns of the empire (fig. 10.15). It was basically in the same carto-

97. This map was once in the Phillips Collection in London; it is illustrated in Sotheby's catalog, *The Library of Philip Robinson*, pt. 2, *The Chinese Collection* (day of sale, 22 November 1988), 76–77 (no. 85), and less satisfactorily in Howard Nelson, "Maps from Old Cathay," *Geographical Magazine* 47 (1975): 702–11, esp. 708.

98. *Sukchong sillok*, 12.12b (note 17). The date of the obituary corresponds to 4 October 1681. According to data in this notice, Kim would have been born in 1602, not 1601 as found in several published references.

99. Mackay, "Kim Su-hong," 27 (note 84).

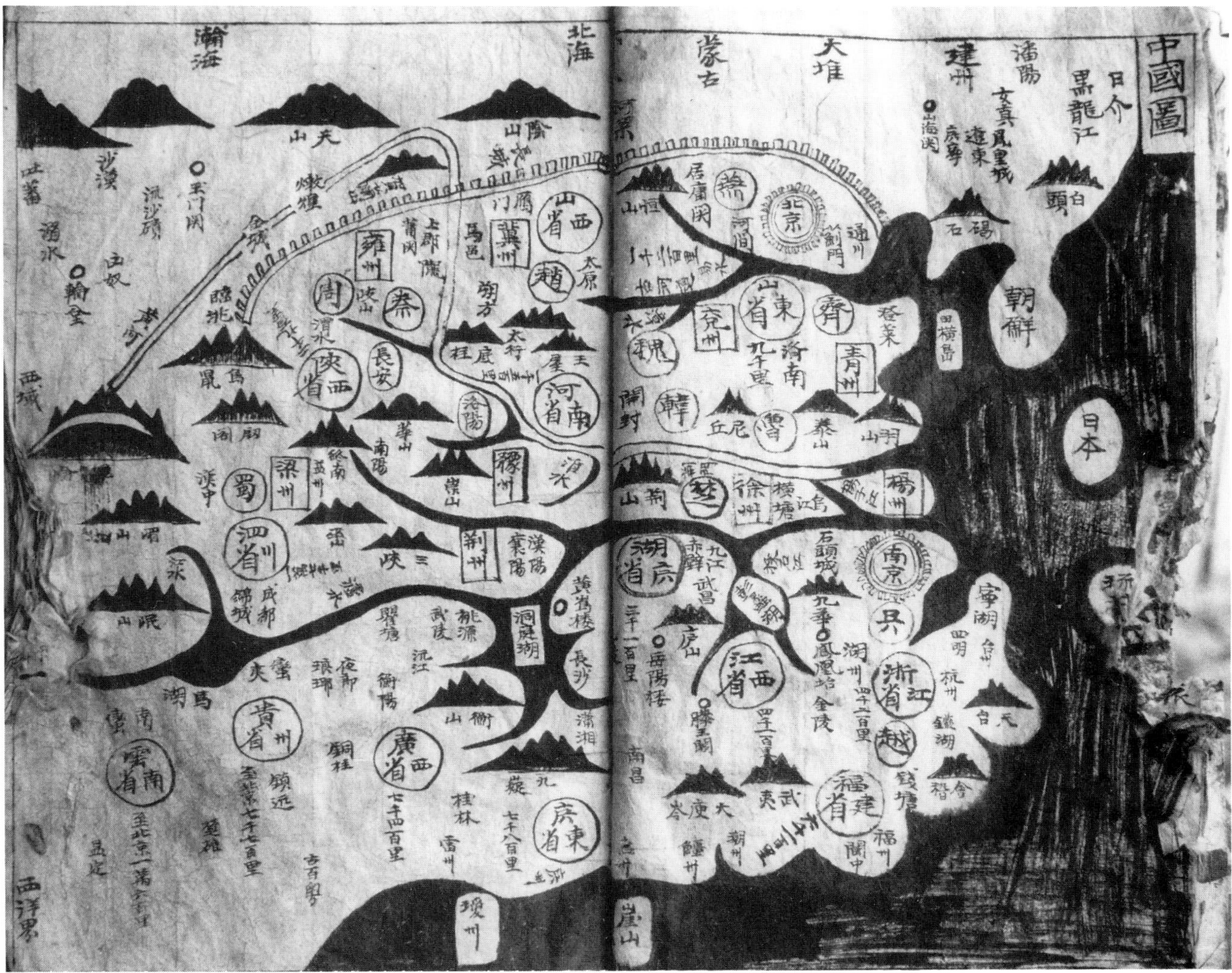

FIG. 10.15. "CHUNGGUKTO" (MAP OF CHINA). From a hand-copied and well-thumbed atlas, with a preface that can be dated 1849. Still indicating, in large circles, the Ming provincial arrangement (discontinued in 1644), this map also shows, with rectangles, the classical "Nine Isle-lands" (*jiu zhou*) marked out by the legendary Yu when he drained the floods; the major states of the Spring and Autumn period, in small circles; principal districts and mountains; and a few historical sites. No attempt is made to reflect the real China of the nineteenth century; rather, it is the China of the Ming dynasty and the classics that is offered.
Size of the original: 32.4 × 26.8 cm. Author's personal collection.

graphic style as the *chŏnhado* and could be considered an enlarged detail from it.

Korea's other major neighbor, Japan, was also a traditional object of cartographic concern on the part of Korea's statesmen and mapmakers, as we have seen in detailing the construction of the world map of 1402. By the standards of most maps of Japan during the Chosŏn dynasty, that effort had produced a superior outline of the Japanese archipelago, even if it was poorly positioned and oriented on the world map itself. Even by comparison with the surviving fourteenth- and fifteenth-century Japanese maps of Japan its outline was unusually good. Except that the island of Shikoku was grafted onto Honshū, with the resultant loss of the Inland Sea, it might have been counted among the best. But of course this map was an adaptation based on a Japanese map brought to Korea in 1402 by Pak Tonji, and at root it represented one particular stream of Japanese cartography.

As is noted below (p. 370 and note 100), the oldest known printed map of Japan was a Korean product published in Sin Sukchu's *Haedong cheguk ki* (Chronicle of the countries in the Eastern Sea) of 1471.[100] This was not

100. Sin Sukchu, *Haedong cheguk ki* (Seoul, 1471; reprinted Seoul: Chōsen Sōtokufu, 1933 [Chōsen Shiryō Sōkan, no. 2]), front section. Although some material was added to this book in later printings, there is no evidence that anything in the original part was ever changed. The

FIG. 10.16. "HAEDONG CHEGUK CH'ONGDO." From the *Haedong cheguk ki* (Chronicle of the countries of the Eastern Sea), by Sin Sukchu, 1471. This is the first of a set of six maps relating to Japan and the Ryūkyūs, the earliest maps of Japan to be printed anywhere. This map establishes the overall spatial relationships within the Japanese archipelago, between it and the Ryūkyūs, and between both of them and Korea. For its time, this map was far ahead of Chinese and even other Korean maps, which often placed Japan at the latitude of the Yangtze and even farther south. The two islands of Iki and Tsushima, in the straits between Korea (upper left corner) and Japan, are grossly exaggerated.
Size of each page: 17.6 × 12.3 cm. From a 1933 photolithographic reprint of a movable-type edition of the *Haedong cheguk ki*, ca. 1506. Photograph courtesy of the Harvard-Yenching Library, Harvard University, Cambridge, Massachusetts.

simply a map of Japan but a collection of maps showing both Japan and the Ryūkyūs and some enlarged cuts of the two islands Iki and Tsushima. Sin Sukchu (1417–75), who in his earlier career had participated in many of the technical research projects of King Sejong (r. 1418–50), served as secretary on an important embassy to Japan in 1443. It was this mission that marked the end of Japanese piracy on Korean shores and laid the foundation for diplomatic exchanges that lasted until the Hideyoshi invasions. Later, as a longtime president of the board of rites, he supervised all of Korea's foreign relations for over a decade. The *Haedong cheguk ki* was a compendium of Japanese and Ryūkyū history, geography, and customs, together with a digest of events, precedents, and protocol related to Korea's relations with the two countries. Sin's lifetime concern with Japanese affairs led him to take the Japan-Korean relationship very seriously. It is said that he petitioned King Sŏngjong (r. 1470–94) from his deathbed never to permit a break in peaceful relations with Japan.

The map collection in the *Haedong cheguk ki* was based on various maps maintained at the board of rites since the early part of the fifteenth century. Pak Tonji's

book proper was printed with movable type, but the map section was produced by woodblock printing.

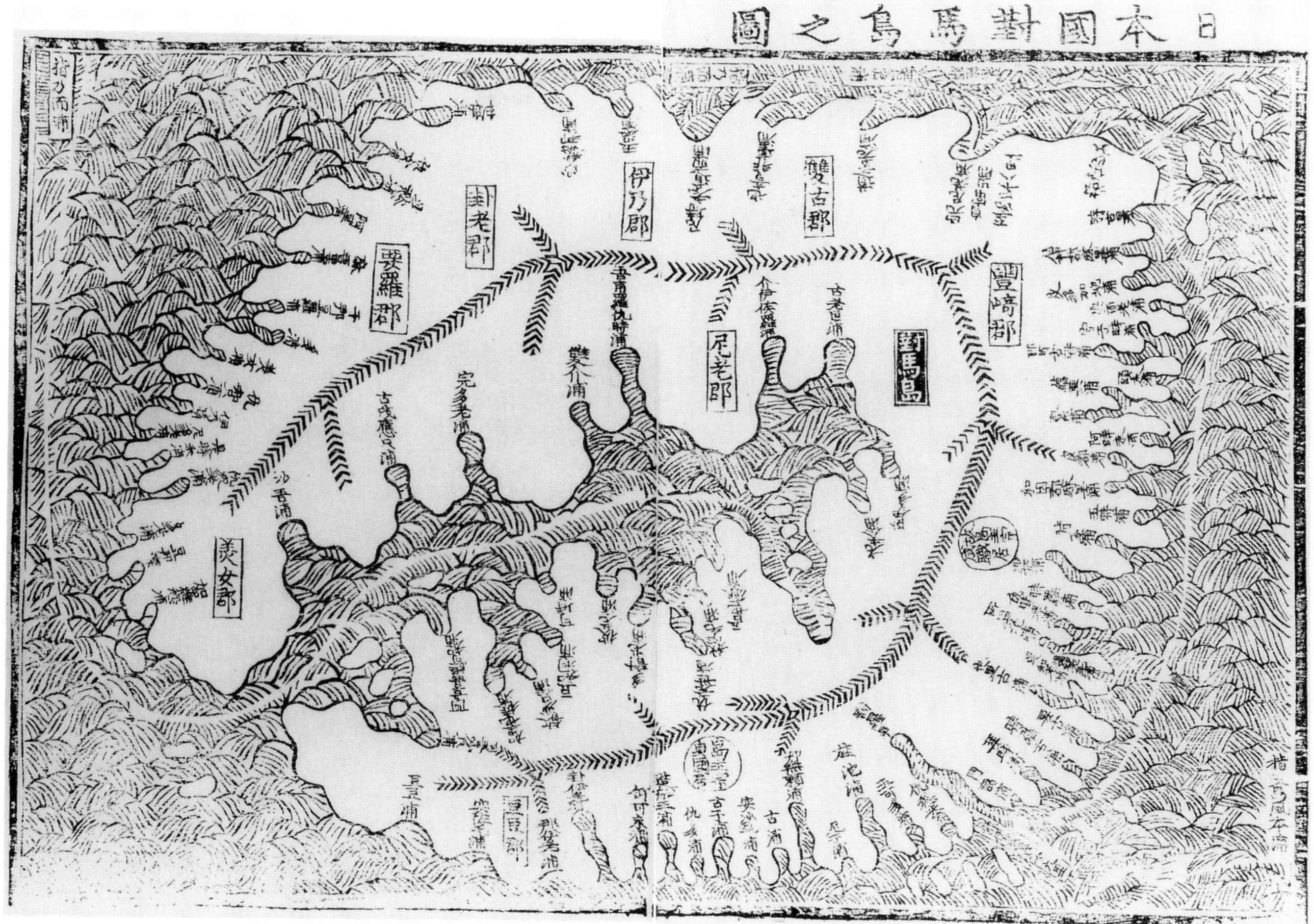

FIG. 10.17. "ILBON'GUK TAEMADO CHI TO." From the *Haedong cheguk ki*, by Sin Sukchu, 1471. Tsushima Island, with its own daimyo, was closer to Korea than to mainland Japan, and over the centuries it played a major intermediary role in economic and political relations between the two countries. The horseshoe shape given to the island, though a distortion, permitted the full indication of virtually every inlet and bay of its coast on a single square page. It also reflected the strategic accessibility of the island through the large inner bay (known today as Asō Bay), which opens to the west. Korean naval forces carried out a major strike against Japanese pirate bases in the bay in 1419. The white line, which was originally colored red, indicates important routes to and from Korea. Where the line crosses the land there was a portage (Japanese *funakoshi*); at the time of the Russo-Japanese War, the Japanese navy removed this isthmus, thus turning Tsushima into two islands. See also figure 10.18.
Size of each page: 17.6 × 12.3 cm. From a 1933 photolithographic reprint of a movable-type edition of the *Haedong cheguk ki*, ca. 1506. Photograph courtesy of the Harvard-Yenching Library, Harvard University, Cambridge, Massachusetts.

map was one of the earlier samples in this collection. Another map of Japan and one of the Ryūkyūs were added in 1453, supplied by the Japanese monk Tōan, who served as an ambassador from the king of the Ryūkyūs (then independent of Japan) in that year.[101] Available copies of the *Haedong cheguk ki* have only black-and-white woodcuts, but early copies appear to have been hand colored. Sin Sukchu's note on format and conventions, just after the table of contents and just before the maps, says: "In the maps, yellow lines are circuit borders, ink lines are district borders, and red lines are roads. . . . The routes use Japanese *li*, one of which equals ten *li* in our country."

The "Haedong cheguk ch'ongdo" (General map of the countries in the Eastern Sea) shows the (then) three main islands of Japan, the Ryūkyūs, the islands of Iki and Tsushima (both greatly exaggerated in size) in the strait between Korea and Kyūshū, and the southeastern corner

101. *Tanjong sillok* (Annals of King Tanjong, r. 1452–55), 7.2b (note 17). Tōan's maps are also noted in the 1482 map list, *Sŏngjong sillok*, 138.10a–b (note 17), and see appendix 10.1.

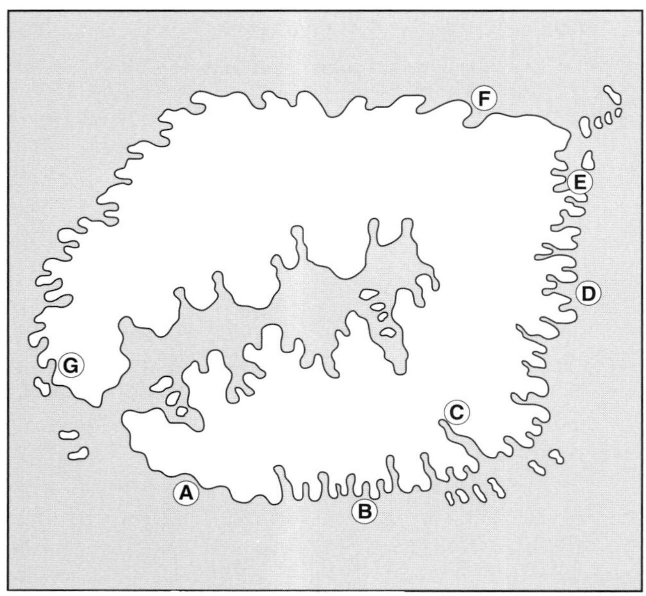

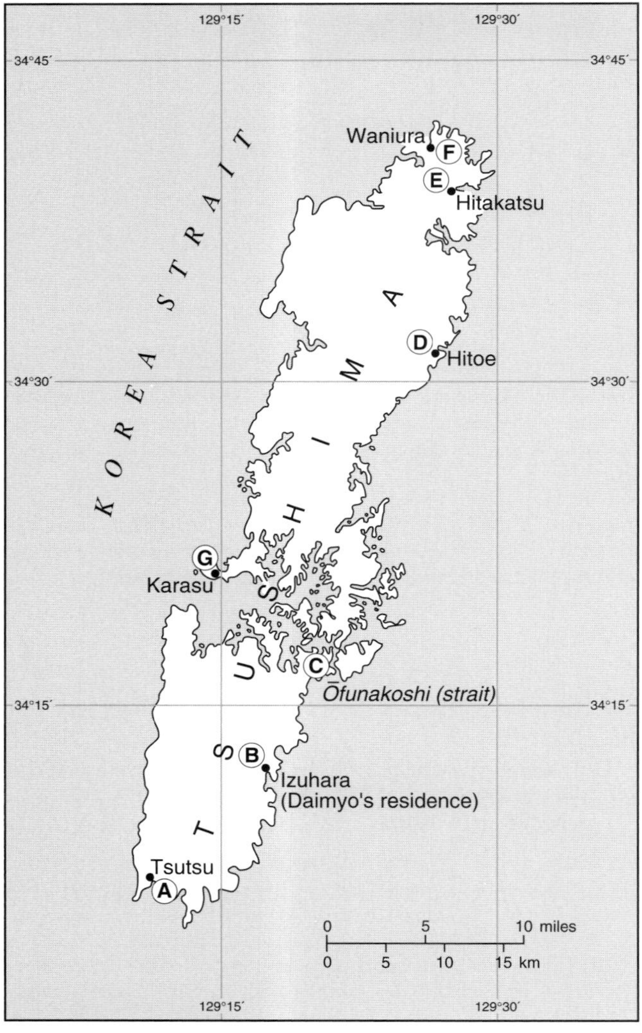

FIG. 10.18. TSUSHIMA ISLAND. Above is the outline of the 1471 map with Roman letters marking some place-names. Below those places are shown on a modern map of Tsushima.

of the Korean peninsula, indicating the three ports where Japanese traders were permitted (fig. 10.16). This general map, showing the relative disposition of all these areas on a single page devoted to them alone, appears to be unique in either Korea or Japan. Following this are two maps, one showing Honshū and Shikoku, the other Kyūshū.

Also of special interest in this mini-atlas is the single-page "Ilbon'guk Taemado chi to" (Map of Japan's Tsushima Island) (fig. 10.17). This island was of special importance to Korea, both as a neighbor and as a regular stop for envoys and other travelers between the two countries. The daimyo of Tsushima was the formal contracting party in relations between the Korean court and Japan. He monopolized the Japanese side of the Korea-Japan trade and was de facto Japan's hereditary agent for Korean affairs. In diplomatic form, Korea considered Tsushima its tributary, but Sin's title for this map, with its "Japan's Tsushima," makes it unambiguously clear that this did not mean it was considered Korean territory. The horseshoe shape given to Tsushima seems to have begun as a convention to get the whole island, which is about seventy-two kilometers long from north to south but only fifteen kilometers wide at its widest point, on a single square page (fig. 10.18). Several writers have expressed puzzlement that Korean maps consistently show Tsushima as one island rather than as two. However the traditional cartographers were completely correct. The Japanese navy, about the turn of the twentieth century, removed the natural isthmus between the upper and lower islands for strategic reasons (a move vindicated by the Japanese victory over the Russian fleet in the battle of Tsushima in 1905). Thus Tsushima's history as two islands has been relatively short.

A persistent problem in representing Japan, shared by cartographers in both China and Korea, was its general location in the eastern sea. The earliest Chinese itinerary to Japan, resulting from an embassy of A.D. 238 (reported in the *Sanguo zhi* [History of the Three Kingdoms] of 297),[102] left the impression, caused by either confusion or deception, that Japan stretched far to the south, with its center roughly at the same latitude as the mouth of the Yangtze and with a climate similar to that of Hainan Island (center at 19°N). The consequences of this early but false impression remained evident on Chinese and Korean maps down to late traditional times. Ming maps

102. See Gari Ledyard, "Yamatai," in *Kodansha Encyclopedia of Japan*, 9 vols. (Tokyo: Kodansha International, 1983), 8:305–7.

routinely show Japan as a small island off the central-southern Chinese coast. I have already noted an even more southerly lay on the Korean *Kangnido* of 1402, and although I attributed it to a paucity of space on the east side of the map, it cannot be ruled out that this was one more reflection of the old problem. Atlases current in Korea in the nineteenth century still show Japan generally south rather than east of Korea. Simple attention to Sin Sukchu's written information would have avoided this error. In the preface to the *Haedong cheguk ki*, he had given a much more accurate indication of Japan's extent and location: "Its territory starts from north of [the mouth of] the Heilong [Amur] River and reaches to the south of our Cheju Island, touching the Ryūkyū Islands; it is extremely long."[103] This would have put Japan's northern tip about at the north end of Sakhalin Island, far beyond its northernmost reach then or ever, but at least it was an error in the needed direction. Since the Japanese themselves are not known to have defined the nation's northern limit in such terms as early as the fifteenth century, Sin's statement probably came from intelligence arising from relations with Manchurian peoples on Korea's northeastern frontier, which were in a very active state during his lifetime.

The maps in the *Haedong cheguk ki* represented the high point in Korean maps of Japan. The book was reprinted in both full and abridged versions at various times down to 1629. In the early eighteenth century it was functionally replaced by another book that served as an official guide to both Chinese and Japanese relations, but that book did not contain maps. Japanese maps in the popular atlases were primitive by the earlier standards. Perhaps this cartographic decline reflects a corresponding psychological deterioration in Korea's relations with Japan following the Hideyoshi wars.

The Foundations of Korean Cartography

The world and regional maps examined so far involved a Korean development or reworking of maps and source materials that for the most part had come into Korea from China, or in the case of the Islamic and Western influences, through China. This cartographic activity did not rest on the Koreans' own direct observations of geographical phenomena or involve their own application of mensuration or graphic projection techniques. Rather, it was a reprocessing—interesting and innovative to be sure—of the maps of others. In making the earlier maps of their own country, which no longer survive but for which I have reviewed the written evidence, they would have had to deal with the more fundamental levels of cartography, in which observed geophysical reality, and not just someone else's maps or data, had to be translated to the dimensions of a piece of paper. It is to these fundamental levels—general geographical knowledge, geomantic conceptions, mensuration techniques, and the problems of projection and scale—that we now turn.

Geographical Studies

Administrative geography has a long and distinguished tradition in Korea. In the main, it followed Chinese models of great antiquity. Common to both countries was a consistent and durable model of bureaucratic centralism, wherein the central government directly administered the nation down to the local level. There was no principle of local autonomy within the state. The emperor or king appointed provincial governors and district magistrates. On lower levels the people chose their own leaders by consensus of the local gentry, but the centrally appointed magistrate had the power, even if it was not often invoked, to replace such leaders. Such a system provided a uniform and systematic structure to local administration and was conducive to centrally compiled directories of regional and local government in which were registered data on population, landholdings, natural resources, and other information affecting revenue and governance. These government files periodically provided the material for the monographic descriptions of the system that over the centuries appeared in the dynastic histories.

The historical aggregate of such sources reflects the administrative history of Korean local communities. Since the central government constantly tinkered with the system in response to various fiscal or political conditions, there were frequent changes to record. A given district may at one time have had its own centrally appointed magistrate but at another time been subordinated to a neighboring district, often incurring a name change in the process. Korean districts were ranked according to their political, economic, and military importance, and these rankings too could change. A rebellion in a given district, for instance, could result in a district's rank being diminished or lost altogether by subordination to another district, usually with unfavorable tax consequences and the loss of some privileges for its residents, while a district that achieved some nationally notable success would be correspondingly promoted. The typical treatise of administrative geography (*chiri chi*) usually began with a chronological account of such changes over long periods.

The earliest surviving treatise on Korean geography is found in the *Samguk sagi* (History of the Three Kingdoms), compiled in 1145 by Kim Pusik.[104] This was an official history, ordered by the king and utilizing govern-

103. Sin, *Haedong cheguk ki*, preface, 1b (note 100).
104. Kim, *Samguk sagi*, chaps. 34–37 (note 7).

ment as well as general sources. For his monograph on geography, Kim relied on an earlier compilation, possibly written as early as the eighth century, whose information reflected the organization of localities as it was during the time of Silla's King Kyŏngdŏk (r. 742–65). Understandably, it has fuller and more reliable data on Silla than it does on the partly contemporary kingdoms of Koguryŏ and Paekche. It is especially valuable for its preservation of Korea's oldest known place-names, which frequently were vernacular forms. Most of these were replaced by the Sinicized forms preferred by Kyŏngdŏk and his age. The official history of the Koryŏ period, the *Koryŏ sa*, compiled on royal order by a team headed by Chŏng Inji (1396–1478) and published in 1451, has a similar treatise on the changes of the Koryŏ period, written principally by Yang Sŏngji (1415–82),[105] who as we will see was one of the major cartographers of the fifteenth century. Because of these two works, it is generally possible to know the historical outlines of most of Korea's districts back to the Three Kingdoms period (approximately fourth century to 668).

As would be expected, the Chosŏn dynasty made many changes in local administration after it came to power in 1392. These were already numerous when King Sejong ordered a comprehensive survey of the nation's provinces and districts in 1424.[106] We are fortunate to have good detail on this survey because of the preservation of the uniform questionnaire sent to every governor and magistrate in the country in that year. The Seoul authorities asked for a great variety of data from each local official. They wanted to know the complete administrative history of his district and its present subordinate areas, its borders and population, distances to neighboring districts, physiographic features, and details on all kinds of economic, social, and religious matters. Thus Seoul had a broad range of data, systematically collected over a uniform reporting period from every one of the 334 districts in the country. The raw responses to this questionnaire survive for a single province, Kyŏngsang on the southeastern coast, which with sixty-six districts was the largest and richest.[107] Of key importance for cartography were the precise data on distances from district to district, collected in a way that could provide a mapmaker with a high degree of redundancy and verification. These materials were supposedly consolidated in the *Sinch'an chiri chi* (Newly compiled geographical monograph), presented to the throne in 1432.[108] This compilation was supposedly incorporated in the geographical monograph in the appendixes to Sejong's official annals (*Sejong sillok*); however, that monograph lacks any reference to many of the categories in the original 1424 questionnaire.[109]

These survey data were updated and supplemented for government files in 1469, when additional data were sought from each district concerning an extensive list of economic and military categories. Of particular interest to mapmakers was the information sought on shipping routes and distances to Seoul, and on the postal network.[110] Much of this information, particularly the military data, was for government use only and was never systematically published. But it would have been available to those responsible for the various cartographic projects that went forward during this period.

In the following decade a much more thorough project was launched that resulted in 1481 in the *Tongguk yŏji sŭngnam* (Complete conspectus of the territory of the Eastern Country [Korea]),[111] a comprehensive geographical reference work along the lines of the *Da Ming yitong zhi* (Comprehensive gazetteer of the Great Ming, 1461), by Li Xian et al. Refined and updated over the following half-century, the *Sŭngnam* is known today through the final edition of 1531.[112] After that date the Chosŏn dynasty did not again publish a survey of the nation's administrative geography until 1770, when a new summation appeared in the encyclopedia *Munhŏn pigo* (Documentary reference encyclopedia). The final edition of this work, *Chŭngbo Munhŏn pigo* (Documentary reference encyclopedia, expanded and supplemented),

105. *Koryŏ sa*, chaps. 56–58 (note 9).

106. *Sejong sillok*, 26.25a (note 17).

107. *Kyŏngsangdo chiri chi* (Administrative geography of Kyŏngsang Province, draft dated in twelfth lunar month of 1425 [January 1426]), and *Kyŏngsangdo sokch'an chiri chi* (Administrative geography of Kyŏngsang Province continued, draft dated 1469) (both reprinted Seoul: Chōsen Sōtokufu Chūsuin, 1938).

108. *Sejong sillok*, 55.7b (note 17).

109. *Sejong sillok chiri chi*, in *Sejong sillok*, chaps. 148–55 (note 17).

110. *Kyŏngsangdo chiri chi* and *Kyŏngsangdo sokch'an chiri chi*, *samok* (specifications), 1–3 (note 107).

111. According to the original preface to the *Tongguk yŏji sŭngnam*, dated during the spring of 1481, the compilation began in 1478. Its chief compilers were listed as No Sasin, Kang Hŭimaeng, and Sŏ Kŏjŏng, but in 1482 Yang Sŏngji—probably justly—claimed it as his work (see *Sŏngjong sillok*, 138.9b [note 17]). Beyond its function as a reference source for administrative geography, the *Sŭngnam* had the character of a literary anthology, since poems and literary essays of particular relevance to the scenery, culture, and history of Korea's districts were liberally scattered throughout the text. In this it followed the model of the *Fangyu shenglan* (Complete conspectus of the [imperial] territory), a Song geographical work whose title is echoed in the Korean work's title (*sŭngnam* is equivalent to the Chinese *shenglan*). See the prefatory matter in Yi Haeng et al., comps., *Sinjŭng Tongguk yŏji sŭngnam* (New and expanded *Tongguk yŏji sŭngnam*) (Seoul, 1531; reprinted Seoul: Tongguk Munhwasa, 1958). This work is cited simply as *Sŭngnam* in later notes.

112. Parts of the original edition and of another edition that appeared during the reign of Prince Yŏnsan (r. 1495–1506) survive in various rare book collections, but the edition of 1531 was thoroughly revised and expanded, being longer than the original by five chapters (*kwŏn*). A comparison of the *Sŭngnam* with the *Da Ming yitong zhi* shows many points of structural and cartographic similarity, and the Ming work was specifically cited in the *Sŭngnam*'s original preface.

appeared in 1908.[113] Although the *Munhŏn pigo* was in some respects more up to date than the *Sŭngnam*, its geographical section was not as complete or as well produced as the latter and never replaced it in popularity or prestige. The *Munhŏn pigo* was a vast reference encyclopedia covering many subjects and fields and was not widely available until 1908, whereas the *Sŭngnam* was a more conveniently produced work of geographical reference only. Another factor affecting usefulness was the inclusion of provincial maps in the *Sŭngnam*, in contrast to the lack of maps in the *Munhŏn pigo*. I will discuss these maps later in this chapter.

In addition to these officially compiled works, the Chosŏn period saw a number of privately written works of great distinction. The *sirhak* scholars of the seventeenth and eighteenth centuries were active in geographical studies and were especially concerned to reconstruct or refine knowledge of the nation's historical territory and boundaries. But for a connection with cartography, no private work was of higher quality than the *Taedong chiji* (Administrative geography of the Great East [Korea]) of Kim Chŏngho.[114] Kim is unquestionably Korea's greatest cartographer, and this work, which was not quite complete at his death, shows that behind his maps there was a broad and deep knowledge of his country. His coverage of each district was not as thorough historically or culturally as that of the *Sŭngnam*, but he provided a greater variety of quantitative data, especially on distances and locations. Aside from the usual administrative history and physiographic data, he gives the names and locations of *myŏn* (subdistrict units) and of such sites as granaries, pastures, bridges, and signal towers. Nor are schools and shrines ignored. Each site is precisely located with respect to the district seat. Each provincial section concludes with detailed sets of tables showing interdistrict distances, and for each district the area under cultivation, the number of households, the number of individuals, and the number of individuals owing military service (i.e., commoners) are provided. There are succinct summaries of the postal and military signal networks, and provincial numerical totals for everything from people to bridges, dikes, and shrines. Especially important are Kim's data, unfortunately incomplete, on geodetic coordinates for several dozen towns and cities.[115] In many of these categories Kim was the first to assemble systematic information nationwide. His rich data base is broadly reflected on his maps.

Behind the official compiling activity were the needs of the bureaucracy, on both national and local levels. Even many of Kim Chŏngho's data must have come from government information, either published or unpublished. But the seventeenth and eighteenth centuries brought significant economic development and diversification to Korean life, and information that had formerly been assembled for the benefit of officials was now required by merchants and travelers of all kinds and was published in handbooks and maps that circulated among the general population. Figure 10.19, which is from a late eighteenth- or early nineteenth-century album of provincial maps, shows a triangular table giving interdistrict distances for all the towns of Hamgyŏng Province.

Other private geographical writers, such as Yi Chunghwan (1690–1753), took a more humanistic approach to the land. His *T'aengni chi* (On selecting a village), probably written in the 1730s or 1740s, is a remarkable investigation on a nationwide scale of physical terrain, climate, soil quality, economic conditions, interesting scenery, and even human behavior and character in the different parts of the country, all with the purpose of finding a place where a gentleman can live.[116] His assessments of his country's regions and localities, some deeply admiring and others devastatingly critical, show him to have been an astute, sometimes opinionated, but always interesting critic of land and landscape. Yi's commentary is marked by an especially deep familiarity with Korea's mountains and rivers and the ways they organize the overall structure of the land. In his sensitivity to the relative balance of mountain ranges and plains, of rivers and watersheds, he shows himself to be one of the most articulate links in Korea's long geomantic tradition.[117]

---

113. The *Munhŏn pigo* (note 51) has a total of twenty-seven chapters (*kwŏn* 13–39) devoted to geography. These are organized on historical principles and have great reference value. Major headings are: territorial extent and borders, administrative history by districts, mountains and rivers, roads, defense fortifications, naval defense, maritime routes, and a few miscellaneous sections. In spite of its late date, the information in this work is solidly traditional in both content and institutional classification.

114. Kim Chŏngho, *Taedong chiji* (manuscript dated 1864) (Seoul: Hanyang Taehakkyo Kukhak Yŏn'guwŏn, 1974). The original manuscript is not completely integrated, and toward the end of the work there are significant differences in format from one chapter to another. The problems include some missing tables and two whole missing chapters. Kim Chŏngho is believed to have died in 1864 and clearly had not completed editorial work on the text. Although this 1974 photolithographic edition, ed. Yi Pyŏngdo, is of great importance, the work deserves a more systematic modern editing to make it fully useful as a reference.

115. Kim includes these coordinates together with a set of coordinates for China's provincial capitals and other miscellaneous information on Korean-Chinese itineraries; Kim, *Taedong chiji*, chap. 28 (note 114).

116. Yi Chunghwan, *T'aengni chi* (manuscript, no date), reprint ed. Ch'oe Namsŏn (Seoul: Chosŏn Kwangmunhoe, 1912). Korean writers also cite this work under the title *Tongguk sansurok* (Record of Korean mountains and rivers), a rare manuscript in the Kyujanggak Collection in the Seoul National University Library.

117. See Ch'oe Ch'angjo, "Chosŏn hugi sirhakchadŭl ŭi p'ungsu sasang" (The geomantic thought of "practical learning" scholars in the late Chosŏn dynasty), *Han'guk Munhwa* 11 (1990): 469–504.

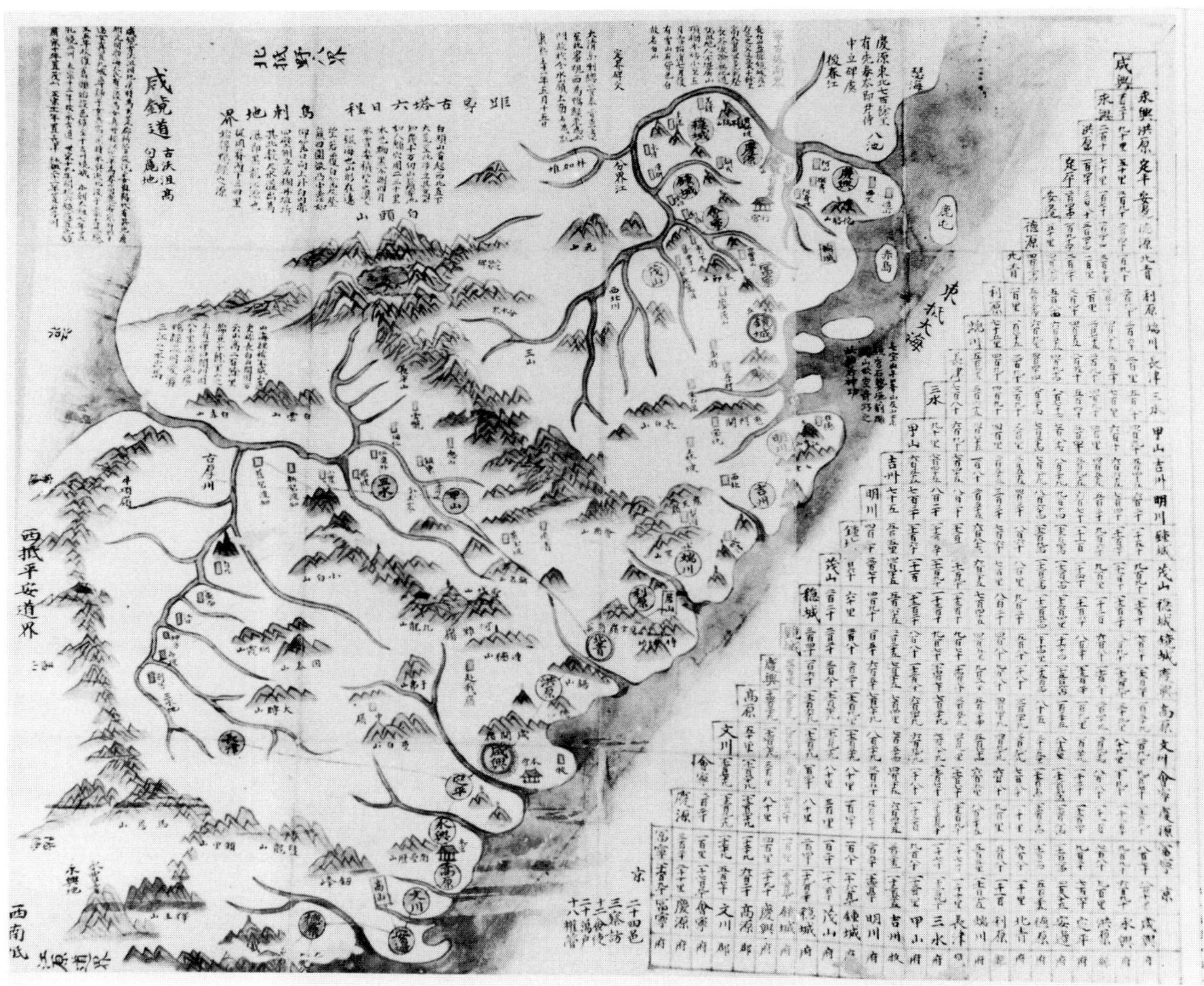

FIG. 10.19. "HAMGYŎNG TO" (HAMGYŎNG PROVINCE). From a late eighteenth- or early nineteenth-century album of national and provincial maps entitled *Tong'yŏ chido* (Map of the Eastern Territory [Korea]). The indication of Changjin as a district dates the execution of this map after 1787. The upper margin has the text of the 1712 boundary inscription and notes on the neighboring Manchu areas of Ula and Ninguta, Mount Paektu, and the history of the province. The modern-looking table of intraprovincial distances is a special feature of this map.
Size of the original: 48.6 × 57.5 cm. By permission of the National Central Library, Seoul (cat. no. Kojo 61-48).

KOREAN GEOMANCY: THE SHAPES AND FORCES OF THE LAND

The geomantic approach to land and life coexisted with the bureaucratic concern for the more practical approaches to geography. It was a more intuitive way of seeing the land, combining various emotional and religious elements in landscape perception with an often insightful and shrewd analysis of the physical features of the earth. Its roots in Korea extended deep into the past. Geomancy, as generally conceived, was a system of beliefs and theories about the land applied to the auspicious siting of graves and dwellings.[118] Even today most communities in Korea do not lack for experts in such knowledge, but in earlier times the general belief in such lore created a great demand for geomantic professionals

118. Two works of great value as introductions to East Asian geomantic theory and practice are Sophie Clément, Pierre Clément, and Shin Yong Hak, *Architecture du paysage en Asie orientale* (Paris: Ecole Nationale Supérieure des Beaux Arts, 1982); and Steven J. Bennett, "Patterns of the Sky and Earth: A Chinese Science of Applied Cosmology," *Chinese Science* 3 (1978): 1–26. Bennett urges the abandonment of the term "geomancy" in favor of "topographical siting" or simply "siting," arguing that the Western practice of geomancy, such

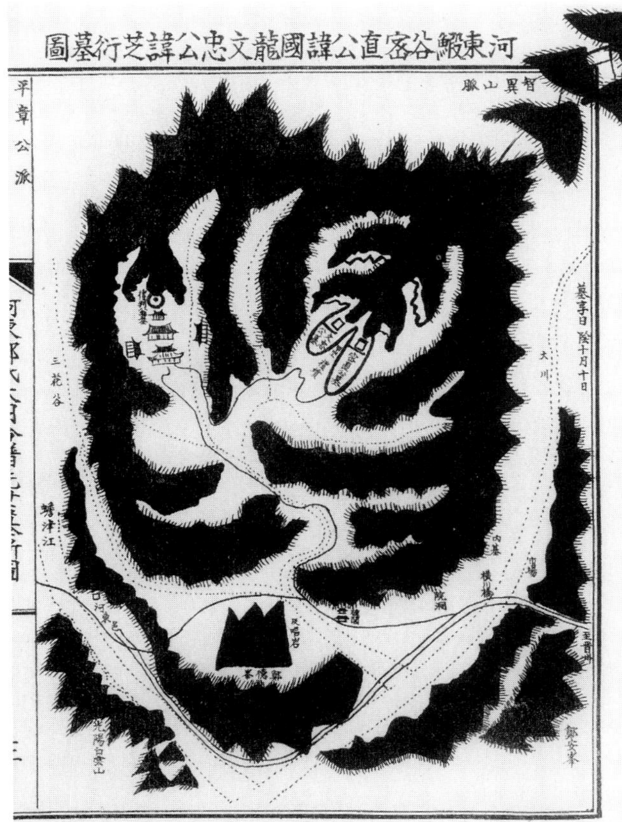

FIG. 10.20. MAP OF A GRAVE SITE. This is a typical woodblock print, of the kind commonly found in genealogies and family histories, of a grave site and surrounding terrain, the most basic type of shapes-and-forces (*hyŏngse*) map. The mountainous ground is in sharp black, rivers and streams are shown as dotted lines doubled to indicate width. At left, a grave and shrine buildings; at right, two grave mounds with the directional phrase "aligned east by northeast." At upper right, a chain of peaks going off the map with the legend "Chirisan artery," graphically emphasizing the arriving forces from Mount Chiri, the dominant mountain of the southern peninsula. The graves belong to two principal ancestors of the Hadong Chŏng lineage, of which the famous eighteenth-century cartographer Chŏng Sanggi is also a member.
Size of the original: 31 × 19 cm. From *Hadong Chŏngssi taedongbo* (Comprehensive genealogy of the Hadong Chŏng lineage), 1:3; published in 1960 in Seoul, copying and updating countless earlier editions. Photograph courtesy of the Harvard-Yenching Library, Harvard University, Cambridge.

and consultants. These experts in *p'ungsu* (wind and water, Chinese *feng shui*), as they were popularly known, or *kam'yŏga* (specialists in celestial and terrestrial forces, Chinese *kanyujia*), as they referred to themselves, produced over time a sizable body of technical knowledge. Their geomantic siting practices generated some specialized mapping conventions that are still commonly seen on maps of burial places (fig. 10.20).

Yet if the geomantic arts had been applied only to the siting of graves and dwellings they would probably be of limited interest for the present chapter. In earlier times geomancy played a broad role on the Korean national stage as well, and its terms and conventions were a significant part of the general political and cultural discourse. Buddhist monasteries and shrines, which during the Koryŏ period were generally patronized by the state, were often sited with the purpose of "supplementing" (*pibo*) terrain perceived to be geomantically defective, a matter of political as well as popular interest. More important, geomancy was indispensable to the selection or evaluation of sites for capitals, whether on the national, regional, or local level. Since the claims of any given area to be a national or branch capital were seen to involve the spiritual relation of that region to the country's entire body politic, geomantic analysis had to be applied within a national framework. The geophysical and geopsychic conclusions the specialists reached thus left a deep impression on Korean national consciousness and identity.

Many elements of Korean geomantic practice have roots in commonsense insights and closeness to the land that go back to the beginnings of communal life. But in the organized and systematized form that swept Korea beginning in the mid-ninth century, geomancy was an import from China.[119] Its theory had undergone much refinement in Chan (Korean Sŏn, Japanese Zen) Buddhist centers in southern China during the late Six Dynasties and Tang periods (317–907). Geomancy was not in itself a matter of Buddhist belief, but rather was an age-old body of Chinese lore long predating the arrival of Buddhism, which the Chan monks applied in seeking the ideal

as it is, has nothing to do with the East Asian practices in question. The argument is correct; the proposed solution is not. "Siting" comes nowhere near describing the Korean dimensions of the geomantic theory and practice that will be treated throughout the rest of this chapter. Precisely because geomancy in the West is of such insignificant consequence (Bennett, p. 1, implicitly ridicules its simple-minded childishness) and is a word most Westerners will never hear in their lives, it is an ideal vessel for refilling with old and new content from East Asian experience. (We have done this with great success for the word "dragon," much to the enhancement of the Western reputation of that much maligned species.) Also in favor of "geomancy" is its easy grammatical convertibility into adjective and adverb; "siting" is an extremely awkward term to use in actual writing, although it is still very useful in general geomantic discussion. Of these two works, that by Clément et al., in spite of its poor apparatus and romanization, is more relevant to Korean concerns and makes good use of Korean source material, much of which it unfairly hides behind an overuse of Chinese romanization in the discussion. It makes particularly good use of Korean maps.

119. See Ch'oe Pyŏnghŏn, "Tosŏn ŭi saeng'ae wa Namal Yŏch'o ŭi p'ungsu chiri sŏl" (Tosŏn's career and geomantic theory in late Silla and early Koryŏ), *Han'guk sa Yŏn'gu* 11 (1975): 102–46; and Michael C. Rogers, "*P'yŏnnyŏn T'ongnok*: The Foundation Legend of the Koryŏ State," *Korean Studies* 4 (1982–83): 3–72, esp. 26–30.

mountain sites for their monasteries. There were a number of schools, but the one that became important in Korea was the *xingshi* (shapes-and-forces, Korean *hyŏngse*) school associated with the late ninth-century Chinese master Yang Yunsong (fl. 874–88). It regarded mountains and streams as conduits for distributing the inner energies of the earth, which, in accordance with physical and environmental conditions and perceptions of yin and yang forces, could be either beneficent or threatening. The object was to find sites where the beneficent influences were dominant by analyzing all the factors. Such a place was often called a *mingtang* (bright hall, Korean *myŏngdang*). Much of the technical vocabulary of such analysis evoked medical lore. Thus mountain chains were arteries (*me* or, in the frequently seen Chinese colloquial reading, *mai*; Korean *maek*), and a geomancer was a kind of earth physician who by taking the pulse of the hills could determine the health of the land.[120]

Such ideas found a ripe area for application in Korea, where it is hard to stand anywhere and not see mountains on the horizon, and usually closer. During the ninth and tenth centuries the mountain paths must have been thronged with monks and geomancers searching out the secrets of the nation's arterial system. At an early date a picture of Korea's montane network had evolved in which Mount Paektu (Whitehead, 2,744 m), a majestic volcanic peak with a spectacular crater lake (Ch'ŏnji, the Pond of Heaven), the source of the Yalu, Tumen, and Sungari (Songhua) rivers, turned out to be also a spiritual power plant that sent potent legitimating forces coursing through Korea's veins. Geomantic theory was one of the chief factors legitimizing the Koryŏ dynasty, whose capital Songdo (modern Kaesŏng) was home to Mount Song'ak, supposedly the terminus of one of Paektu's most efficacious conduits. The monk Tosŏn (827–98) is said to have inspected Mount Song'ak in 875 and confirmed the Paektu connection, predicting that a future king would be born in Songdo because of this.[121] Two years later there duly occurred the birth of Wang Kŏn, founder of the Koryŏ dynasty. After this auspicious beginning and for the remainder of the dynasty, geomancy was routinely involved with political matters of all kinds, especially with schemes to move the capital or to accord or deny some special favor to Koryŏ's branch capitals in P'yŏngyang, Hanyang (now Seoul), and Kyŏngju, on occasion contributing to serious political instability.[122] The influence of geomancy on the national stage continued to be felt in the Chosŏn dynasty, especially in its early years when the new capital at Seoul was established.

During the Koryŏ period it clearly had a major impact on cartographic development. Tosŏn's own epiphany as a geomancer had its setting in a curious map story. Having retired to a hermitage on Mount Chiri, he was one day visited by a stranger who said that he had something important to tell him and would do so at a certain time on the banks of a river by the southern sea. He then disappeared. Tosŏn went down the mountain to the appointed place and once more encountered the mysterious man, who molded the sand on the beach into a model to demonstrate "the concordant and refractory forces of the mountains and streams." As Tosŏn studied the sand map the man vanished again, never to return.[123] The story was later remembered in the name of the nearby settlement, which was called Sado Ch'on (Sand Map Village).[124] Thus Tosŏn's geomantic thought, which not only dominated his age but for the rest of the Koryŏ dynasty inspired imitators to claim his mantle for whatever scheme they wished to pursue, was born in a map in the sand. Unfortunately, none of Tosŏn's writings now survive except in remarks or dicta attributed to him by others.

Although there were indubitable excesses associated with geomancy during the Koryŏ years, the craze did stimulate a positive interest in geophysical exploration, contributing to a growing knowledge of the dimensions of the country and its various regions. Such explorations could hardly have proceeded without extensive mapping. The heightened consciousness of the river and mountain

120. See the useful discussion of this point in Bennett, "Patterns," 6–7, and especially Clément, Clément, and Shin, *Architecture du paysage*, 77–79, 85–87 (note 118). On the translation of the Chinese term *xingshi* (Korean *hyŏngse*), Bennett gives "forms and configurations" (p. 2), Clément et al. speak of "l'Ecole des formes," "la force des formes," and "les aspects des configurations" (p. 85). It is ill advised to take the two syllables *xing* and *shi* as being in possessive relation; a coordinate relation is overwhelmingly shown by the texts. Of all these suggestions, the definition "force" for *shi/se* most effectively suggests the *qi* or energy that surges through those arteries and is the most appropriate for the geomantic contexts under discussion here. I adopt the hyphenated form "shapes-and-forces" to emphasize the tandem character of this unitary concept.

121. "Koryŏ segye" (Koryŏ genealogy), prolegomena to the *Koryŏ sa*, 7b (note 9). See also Rogers, "Foundation Legend," 10–11, 47–50 (note 119); and Yi Pyŏngdo, *Koryŏ sidae ŭi yŏn'gu* (Study of the Koryŏ period) (Seoul: Ŭryu Munhwasa, 1954), 3–61.

122. Agitation in the first half of the twelfth century by the monk Myoch'ŏng to move the capital from Songdo (Kaesŏng) to P'yŏngyang, on the ground that the former was geomantically depleted, led to a civil war and seriously threatened Korea's stability at a time when the nation was under pressure from the Jurchen peoples of Manchuria. See Michael C. Rogers, "The Regularization of Koryŏ-Chin Relations (1116–1131)," *Central Asiatic Journal* 6 (1961): 51–84, esp. 68, and Yi, *Koryŏ sidae ŭi yŏn'gu*, 174–233 (note 121).

123. Memorial inscription for Tosŏn (posthumously National Preceptor Sŏn'gak) by Ch'oe Yuch'ŏng (1095–1174), in *Tongmun sŏn*, 117.20b (note 17); see Rogers, "Foundation Legend," 30–31 (note 119).

124. *Sŭngnam*, 39.9b, 40.27a (note 111). By the time of this fifteenth- or sixteenth-century notice, the name of the village, in the district of Kurye in Chŏlla Province, had been corrupted from Sado (Sand Map) to Sadŭng (Sand Stairs).

arteries also promoted their careful depiction on national maps. I have already drawn attention to Yi Ch'ŏm's prose description of a Koryŏ map he had studied during the 1390s. In this essay, virtually his entire cartographic attention was taken up with mountain ranges and drainage basins. His prose pulsed with such remarks as: "Primal matter here flows and there solidifies, and the mountains and rivers form their separate zones" and the "pristine matter here mingles and accumulates, which is why the mountains are so high and steep."[125] Yi Ch'ŏm's terminology, true to his own age, was Neo-Confucian, but his application of it melded well with Koryŏ geomantic sensibility. Not only did a geomantic perspective emerge from the clear delineation of the mountain ranges and watersheds on the map, but Yi's own perceptions as a cartographic critic were influenced by that perspective.

MAPPING AND MENSURATION IN EARLY CHOSŎN

The general Korean word for map is *chido* (Chinese *ditu*), meaning "earth chart" or "land picture." The second syllable (pronounced *to* in initial positions or when following *-l* or voiceless consonants, *-do* between vowels and after nasals) can mean "map" by itself when attached to other names or terms, such as *chŏndo* (complete map), or *ŭpto* (district map). The semantic range of *to/-do* is very broad, covering just about any kind of pictorial representation and extending to schemes, plots, and other kinds of mental pictures. Schematic representations of philosophical systems, moral concepts, and other such formulations also fell under the rubric of the *to*.[126]

The *hyŏngsedo* (shapes-and-forces map), as we can guess from the term, was particularly associated with geomancy. It is frequently mentioned in the records of the early Chosŏn period, first in connection with the surveys conducted in preparation for moving the capital to Hanyang, and later when King Sejong and his son King Sejo (r. 1455–68) were pushing projects to map the country and its districts. Thus in 1393 we find references to shapes-and-forces maps of Hanyang (Seoul) and its principal building sites,[127] and also for certain cult facilities that were to be sited in the district of Chindong in Chŏlla Province.[128] None of these maps survive, but most likely they were terrain charts showing the shapes-and-forces of mountains and streams and provided with compass readings to indicate the orientation of principal features. Two late traditional examples of this kind of map are shown in figures 10.21 and 10.22.

I have already noted King Sejong's geographical survey of 1424–25, which, in seeking detailed figures on the distances from the seat of each district to its borders in all directions, seems to have envisioned a broad data base for precise maps on both the national and local levels. This is nothing less than we would expect of Sejong, who is justly famous for the scope and quality of his many scientific projects. But Sejong certainly included geomancy within his concept of science. Soon after launching the survey, he recalled the government's records on geomantically positioned monasteries and shrines,[129] as well as documents relating to the "shapes-and-forces of mountains and streams" in the nation's various localities, and ordered this material to be moved from a branch archive in Ch'ungju to the Spring and Autumn Hall (*ch'unch'ugwan*, or office of history) in Seoul, which was handling the geographical survey.[130]

The instructions to governors and magistrates had not mentioned maps, nor were any maps mentioned when the office of history presented its report on the survey, the *Sinch'an chiri chi*, in 1432. But once Sejong had absorbed its results, he inspected the maps on file in the board of war and found them wanting. In 1434 he issued another order to localities for the collection of five kinds of information: the disposition and orientation of all government buildings, the "incoming arteries" of mountains and streams, route distances, distances to district boundaries (once again), and finally, "full and detailed maps."[131] In 1436 he chose Chŏng Ch'ŏk (1390–1475), then a middle-level official in the board of rites, to conduct on-site surveys in the three northern provinces.[132] Just as there

---

125. See the translation of Yi Ch'ŏm's complete remarks from "Samgukto husŏ" above, pp. 241–42 and note 17.

126. We see such an extended usage in the name *Ch'ip'yŏngdo* (Map of order and peace). It was not an earth map but a schematic arrangement of Confucian concepts, each in a box or circle, that were important in the cultivation and activity of a ruler (for example, rectitude of mind, self-cultivation, acceptance of criticism, delegation of authority). It was presented in 1454 to the boy-king Tanjong (r. 1452–55) by Yang Sŏngji (see *Tanjong sillok*, 10.24b–25a [note 17]), who just a few months earlier had been put in charge of a vast project to compile national, provincial, and local maps. Jeon erroneously discusses this "map" in connection with Yang's administrative cartography (*Science and Technology in Korea*, 295 [note 49]). Although Yang was an experienced cartographer, in this instance he was acting more in the tradition of Neo-Confucian philosophers, who had a great fondness for such diagrams. Yang's moral map does not survive, although its outlines could be reconstructed from the notice in the *Tanjong sillok*. Such diagrams are maps in the broad definition accepted in the volumes of the present work: as graphic representations that facilitate spatial understanding of things, concepts, conditions, processes, or events in the human world (preface to volume 1 of *The History of Cartography*)—a formulation that fits the Sino-Korean term *to/-do* remarkably well.

127. *T'aejo sillok* (Annals of King T'aejo, r. 1392–98), 3.3b (note 17).

128. *T'aejo sillok*, 3.1b (note 17).

129. During the Koryŏ period monasteries and shrines were often sited with a view to either making up perceived geomantic deficiencies or adding to geomantic strengths. This was called *pibo* (compensation and reinforcement).

130. *Sejong sillok*, 28.22b (note 17).

131. *Sejong sillok*, 64.30a (note 17).

132. *Sejong sillok*, 71.9a (note 17).

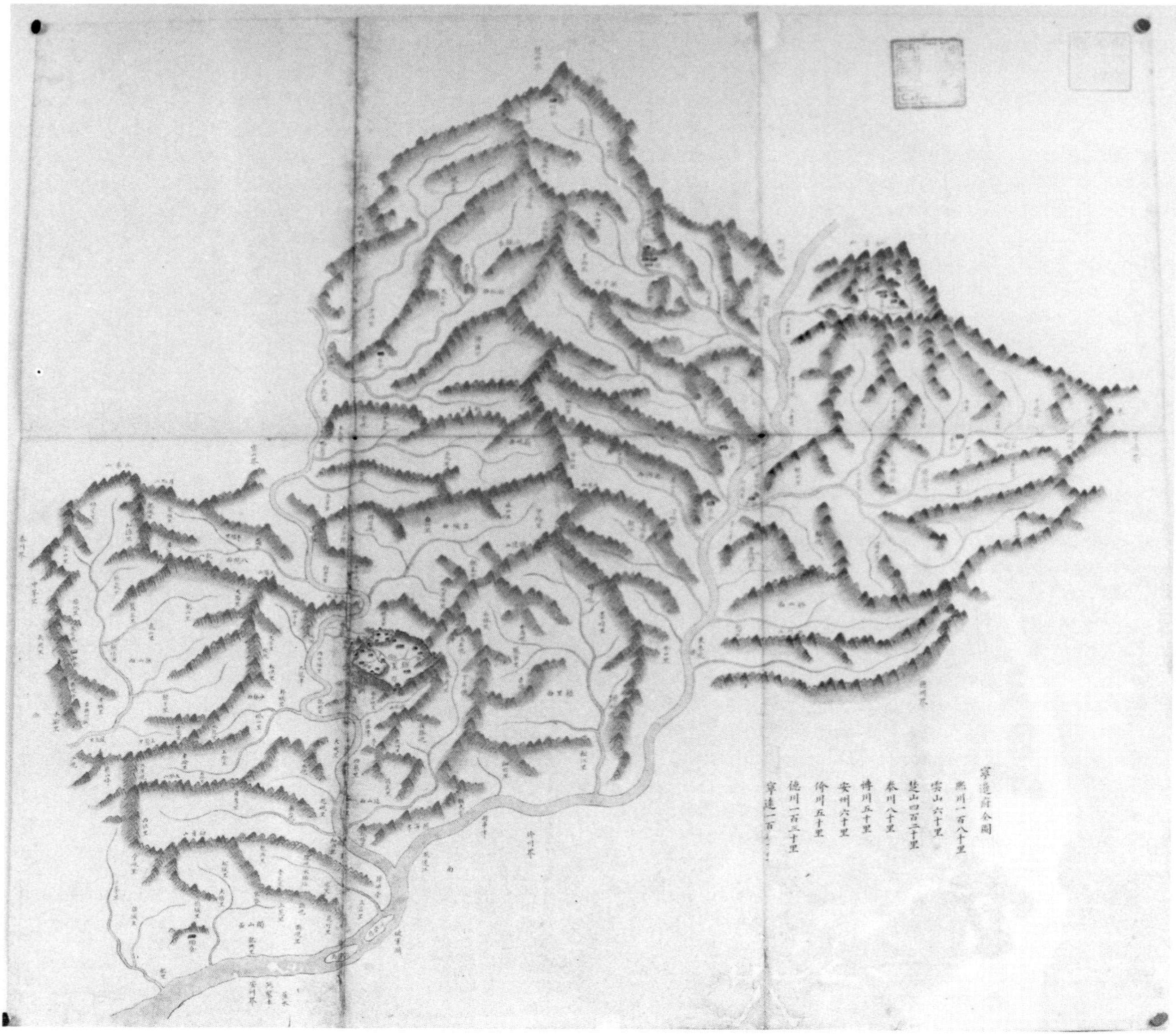

FIG. 10.21. *YŎNGBYŎNBU CHŎNDO* (COMPLETE MAP OF YŎNGBYŎN PREFECTURE). This map, with a strong shapes-and-forces (*hyŏngse*) emphasis, provides a detailed picture of the lay of Yŏngbyŏn's mountains and rivers. Monasteries, shrines, schools, granaries, and post stations are also indicated. The renowned Ch'ŏrong (Iron Jar) mountain fortress, with its precipitous cliffs on all four sides, dominates the lower left, while Mount Myohyang with its serrated peaks stands out at the upper right. Undated, but probably of the late eighteenth or early nineteenth century.
Size of the original: 69.5 × 75 cm. By permission of the National Central Library, Seoul (cat. no. Ko 2702-18).

was a clear concern for geomantic factors in the 1434 instructions, so Chŏng's appointment also involved shapes-and-forces studies, to be followed by *tohwa* (illustrations). The term *tohwa* is most commonly seen in the sense of paintings, but here it should be interpreted as referring to maps. We are uninformed on the outcome of these events, but Chŏng Ch'ŏk later emerged as a cartographer of importance. Although no map by him is known to survive, historians of cartography have given his name to one type of national map of Korea that was current from the fifteenth to the seventeenth century. We see him again in map survey activity in 1454, when Prince Suyang (later King Sejo) led a group to the top of a mountain overlooking Seoul to organize a new map of the city. The prince himself did the sketching, but the group included a master cartographer (Yang Sŏngji), a painter,

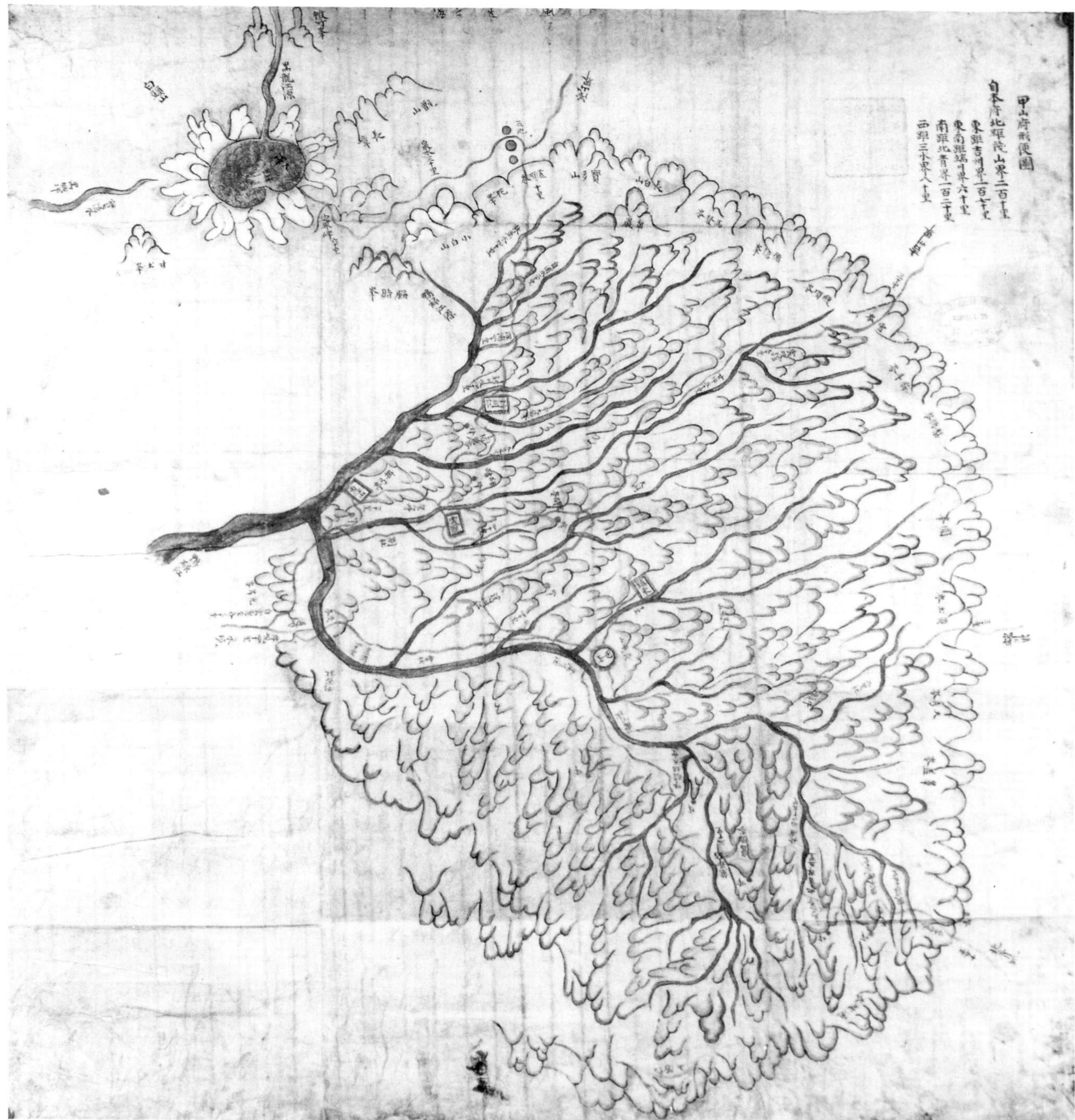

FIG. 10.22. *KAPSANBU HYŎNGP'YŎNDO* (SITUATIONAL MAP OF KAPSAN PREFECTURE). The district of Kapsan administered the vast area to the south of Mount Paektu, shown with its Pond of Heaven (Ch'ŏnji) at the upper left. The district town of Kapsan is indicated by the circle in the center. The shapes-and-forces treatment is drawn in a highly unusual style, giving the mountains and rivers an almost biological appearance. Undated, but probably of the late eighteenth or early nineteenth century.

Size of the original: unknown. By permission of the National Central Library, Seoul (cat. no. Kojo 61-51).

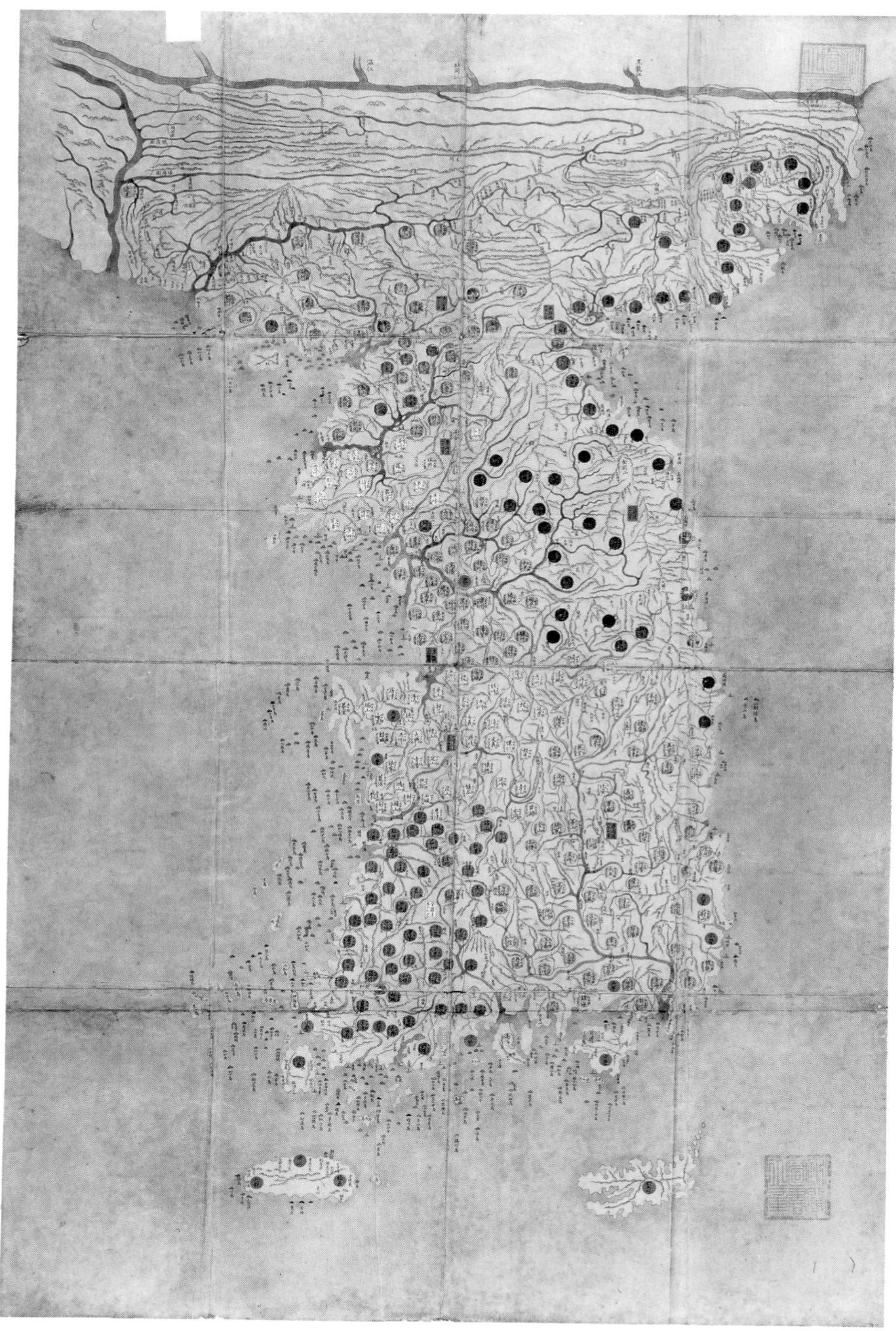

FIG. 10.23. UNTITLED MAP OF KOREA. This map, known as the Naikaku map and in Japan given the title *Chōsenkoku ezu* (Map of Korea), is believed to be a copy of, or in the tradition of, the *Tongguk chido* (Map of the Eastern Country [Korea]) by Chŏng Ch'ŏk and Yang Sŏngji, presented to King Sejo in 1463. Its compressed northern frontier is characteristic of the Chŏng Ch'ŏk style; in other respects it has a high degree of accuracy for its time. Chŏng Ch'ŏk, a known shapes-and-forces specialist, provided a detailed depiction of the nation's mountain ranges (colored green) and river systems (blue), but the map also gives generous indication of district seats, coastal inlets, roads, and islands. The cartouches marking district seats are distinctively colored according to the provinces they are in. Size of the original: In excess of 151.5 × 90.9 cm. By permission of the Cabinet Library (Naikaku Bunko), Tokyo.

a geomancer, a mathematician, and Chŏng, identified as knowledgeable in "shapes-and-forces of mountains and streams."[133]

Sejong's concern for accurate maps, though evident in the sources, seems to have led to only a small cartographic production. A list of maps from a Yang Sŏngji report of 1482 mentions only two maps from Sejong's reign, and both of them were connected with Chŏng Ch'ŏk.[134] This apparent lapse of cartographic interest may have been related to the failing eyesight that afflicted the great king during the last decade of his life, or to his growing concern in those years for projects related to language and literature. In his last six years he invented the Korean alphabet and saw through to completion several major publishing projects involving the vernacular language, both epochal events in Korean cultural history.

In any case, his son Sejo picked up on his mapmaking projects and pushed them with considerable energy. In 1453, while still Prince Suyang but already de facto king in the minority of his nephew King Tanjong (r. 1452–55), whom he later ousted and had put to death, he appointed Yang Sŏngji to oversee a project to produce a general map of Korea and separate maps of each of the eight provinces and 330-odd districts.[135] Compilation of a new geographical monograph was added to his responsibilities in 1455.[136] Later that year Yang presented maps of three frontier districts that had been newly established during the preceding forty years, but that were proving expensive to administer and defend against the nearby Jurchen tribesmen.[137] Partly as a result of Yang's work and recommendations, they were all discontinued as administrative units by Sejo.

In 1456 Yang reported to Sejo that he had been working on mapping steadily for three years. He had investigated everything: mountains and rivers, strategic places, roads and distances, and all the affairs of districts and military garrisons throughout the country. But the great bulk of Yang's report was concerned with cultural and religious matters—on just what it was that made Korea Korea. He saw it as a combination of geography and culture. He comprehensively listed all the mountains and rivers that deserved the special religious attention of the state, in terms of a hierarchy of mountain and river deities to which the state offered, or should offer, special sacrifices. He then reviewed the heroes and great kings of the past, and what was due them, and praised the enlightened Confucian culture that then prevailed in the land. It is a remarkable report. The cartographic substance filled barely a paragraph at the end of many pages devoted to religious and cultural matters.[138] This emphasis serves to remind us of the spiritual power of the land, theoretically clarified, perhaps, by the determinations of the geomancers, but reflected long before the dominance of their ideas in the ancient worship of mountains and rivers. These cults remained active until the end of the Chosŏn dynasty, both on the popular level and through the patronage and officiation of the state and its civil servants.

Thus geomantic conceptions of the landscape, which had become deeply rooted in the national psyche during the long Koryŏ centuries, showed no sign of weakening during the early decades of Chosŏn. This is documented not only in the written sources just reviewed, but in the very few maps of Korea that either date from the fifteenth century or are believed to reflect its style (plate 17 and fig. 10.23). The cartographic approach that might follow from such a shapes-and-forces consciousness is suggested in the following remarks of Sophie Clément and her collaborators concerning the nature of "points where energy is focused or concentrated" on the landscape:

> It is essential that [these points], far from corresponding to a system of coordinates such as longitude and latitude and justified through definition in a universal exterior space, instead be located in relation to the internal structure of the organic system created by the [energy] networks. Territory is a living material, in movement, and not a static and inert mass of elevations and rocks. Thus there can be no territorial segregation of any point of ground. All points are situated in correspondence with the veins and arteries represented by the mountains and streams, elements of nature that themselves transmit life.[139]

In this kind of approach, it would be more important to show the mountains and rivers that transmit the energy, with particular emphasis on the nodal points of the distribution system, than to present the precise geometric relation between the cities and towns that are sited with reference to that system. Such an approach is surely reflected on the so-called Naikaku map (fig. 10.23), thought to be a copy, or in the tradition, of the map of Korea made by Sejong's chief cartographer, Chŏng

133. *Tanjong sillok*, 11.3a (note 17).

134. The two mapping projects were a *p'altodo* (map of the Eight Provinces) and a *yanggye taedo sodo* (large and small maps of the two frontier zones); see *Sŏngjong sillok*, 138.10b (note 17). Chŏng's map of the frontier zones was actually presented in 1451, after Sejong's death, and the *p'altodo* was probably the same as the *Tongguk chido* (Map of the Eastern Country [Korea])—which does not appear on Yang's list—that Chŏng and Yang jointly presented to Sejo in 1463; see *Munjong sillok* (Annals of King Munjong, r. 1450–52), 7.47a, and *Sejo sillok* (Annals of King Sejo, r. 1455–68), 31.25b (note 17). Thus even the projects mentioned for Sejong's reign seem to have been completed only later.

135. *Tanjong sillok*, 8.21b (note 17).

136. *Sejo sillok*, 2.7a (note 17).

137. *Sejo sillok*, 2.39b–41a (note 17). The maps were accompanied by extensive recommendations by Yang on northern defense matters.

138. *Sejo sillok*, 3.24b–32a (note 17).

139. Translated from Clément, Clément, and Shin, *Architecture du paysage*, 216 (note 118). For a similar appreciation of the dynamic nature of the geomantically seen landscape, see David J. Nemeth, "A Cross-Cultural Cosmographic Interpretation of Some Korean Geomancy Maps," in *Introducing Cultural and Social Cartography*, comp. and ed. Robert A. Rundstrom, Monograph 44, *Cartographica* 30, no. 1 (1993): 85–97.

Ch'ŏk. The mountain and river system is meticulously indicated, but no scale is noted and no interdistrict distances are given, even though the government had complete data on these at its disposal.

But this is not to say that spatial relationships were completely ignored. The Naikaku map is a political map and had many features of importance to Seoul administrators, not the least of them the distance of every district seat from Seoul. Moreover, the overall shape of the country was very well grasped: apart from the misshapen northern frontier and the general curve of the east coast (both durable problems to which I shall return), the general dimensions of Korea compare very well with those of modern maps.

Clearly there was a consciousness of scale. The problem is that we have no concrete evidence on how this consciousness was formulated or applied. The earliest indication of scale on a Korean map and the earliest discussion of observed distance and its representation on maps are both due to the eighteenth-century cartographer Chŏng Sanggi (1678–1752). Korean records reflect no knowledge of the scale maps or writings of Zhu Siben or of his sixteenth-century editor Luo Hongxian, nor any awareness of the famous stone-engraved grid map of Song times (above, esp. pp. 47–48). Even the classical six principles of the third-century Chinese cartographic theorist Pei Xiu (223–71) (above, pp. 110–13), which implicitly reflect a consciousness of scale, do not appear to have become a topic of discussion among Korean mapmakers before the nineteenth century (see below, p. 344).

A vague sense of scale seems evident Kwŏn Kŭn's preface to the *Kangnido*, where he remarked on "the greatness of its *kyumo* [scale] and *kungnyang* [dimension]" (above, p. 245). The term *kyumo* (Chinese *guimo*), consisting of two elements meaning respectively "draw a circle" or "compass," and "shape" or "model," classically connotes largeness of vision or planning and has not normally had any reference to cartographic scale in the sense of a ratio between observed distance and its representation on a map. Kwŏn seems to have simply been referring to the impressive size of the map.

We should also note the remark attributed to Pak Tonji concerning the map he brought back from Japan in 1402: "The entire land area [of Japan] was on it, all but the islands of Iki and Tsushima, so I added them and doubled the scale" (above, p. 247). His term *chungmo* means literally to increase, or double, the size of the model or form. Commenting on this map in 1438, King Sejong, finding the Japanese islands too small for his perhaps already failing eyesight, ordered that a copy be made by the process of *kaemo*, "changing (or correcting) the model."[140] Both remarks clearly suggest an increase in scale. The 1402 version of Pak's map is supposedly the basis for the representation of Japan on the *Kangnido* (figs. 10.3 and 10.12), but the latter's odd placement of Japan, and its gross magnification of Korea, create distortions so severe (although of quite different character on the two copies illustrated here) that neither version of the *Kangnido* can give us a clue to what Pak was trying to say. On the other hand, if Sin Sukchu's map of Japan (fig. 10.16), has any connection with Pak's—which is entirely possible—it may provide some help. On this map, the size of Iki and Tsushima is grossly exaggerated with respect to the three main Japanese islands, so much so that Pak's meaning may have been that he added Iki and Tsushima and doubled their size only, leaving Japan relatively smaller. In this interpretation, speculative to be sure, the concept of scale would have been applied selectively to only a part of the map, leading not to more accurate representation of the spatial relationships but to their distortion.

These remarks of Kwŏn Kŭn, Pak Tonji, and King Sejong all concerned foreign areas and ultimately foreign maps. Precisely because the Koreans had foreign maps but lacked any sense of their relative scales, their incorporation of them into the *Kangnido* created distortions in the relative sizes of China, Korea, and Japan (just to confine ourselves to the East Asian portion of the map). Thus the interpretation of their terms—*kyumo*, *chungmo*, *kaemo*—while perhaps showing instinctive awareness of the concept of scale, cannot help us much in determining what they knew about true cartographic scale, that is, the ratio between empirically grasped real distances and their proportional reduction on a map. To deal with that question we have to come back to their maps of their own country, where actual distance measurement was feasible. Therefore I turn now to the standards and techniques of mensuration that developed in tandem with the survey activity of the fifteenth century.

In spite of useful summaries by Sang-woon Jeon (Chŏn Sang'un) and Pang Tong'in,[141] there is still considerable uncertainty concerning the actual design, function, and application of the mensuration standards and devices developed in the mid-fifteenth century by King Sejong and King Sejo. We are much better informed about Sejong's astronomical instruments, primarily because these were the object of a summary essay in his official annals, whereas land survey methods did not receive comparable research attention and no organized summation was made of the work that was done. I will confine myself here to what is known of the efforts to refine linear measure and measuring instruments, including rulers, ropes, odometers, and triangulation devices. I will also review

140. *Sejong sillok*, 80.21a–b (note 17).

141. Jeon, *Science and Technology in Korea*, 294–96 (note 49), and Pang, *Han'guk ŭi chido*, 87–90 (note 5).

TABLE 10.3 Traditional Korean Linear Measure

10 *ri*	= 1 *p'un*	
10 *p'un*	= 1 *ch'on*	"inch"
10 *ch'on*	= 1 *ch'ŏk*	"foot"
6 *ch'ŏk*	= 1 *po*	"pace"
10 *ch'ŏk*	= 1 *chang*	"fathom"
2,160 *ch'ŏk*	= 360 *po* = 1 *li*	"mile"

Source: Ch'oe Hang et al., *Kyŏngguk taejŏn* (Great codex of state administration, compiled 1469, revised 1476) (Seoul: Chōsen Sōtokufu, 1934), 6.1b–2a; Kim Chaero et al., *Sok taejŏn* (Supplement to the Codex, compiled 1746) (Seoul: Chōsen Sōtokufu, 1935), 6.1a. *Ri* and *li*, normally both romanized *i*, are for reasons of clarity given irregular spellings here.

the few scant details on the measuring of polar altitude and its possible application to cartography.

Korea adopted China's decimal foot ruler (*chi*, Korean *ch'ŏk*) at a very early date, and long experience with this linear measure was officially codified in the statutes of the Chosŏn dynasty, the *Kyŏngguk taejŏn* (Great codex of state administration), formally adopted in 1469.[142] By an apparent oversight, the *Kyŏngguk taejŏn* failed to specify the conversion ratio of the foot to long-distance units such as the *li*. However, the ratio of 2,160 *ch'ŏk* (360 *po*) per *li* was specified in the supplement to the *Kyŏngguk taejŏn* published in 1746,[143] and it is implicit in the distance figures given in the *Sŭngnam*, as we shall see. Table 10.3 provides a conspectus of these units. Sino-Korean usage stated the name of each decimal place. For example, the length of an object might be given as 4 *ch'ŏk*, 7 *ch'on*, 2 *p'un*, and 6 *ri*. However, the decimal character of the system permits the reduction of this statement to a simple 4.726 *ch'ŏk* or 47.26 *ch'on*, for example. The difficulty comes in the fact that there were at least five customary *ch'ŏk* rulers current during the early Chosŏn dynasty, each applied for its own specified purposes. Worse still, standards for these varied from place to place and time to time, and the conversion ratios therefore could not be dependably stated.

Sejong's researchers and earlier government specialists, apparently for ideological reasons, wanted to use the *chuch'ŏk* (Zhou [dynasty] foot) as a general standard for institutional purposes. It appears that this measure had been advocated by the Chinese Neo-Confucian reformers of the eleventh and twelfth centuries, who wanted to restore the linear standard of Confucius's own time. A notice in Sejong's annals in 1437 gives a brief history of the question: Sima Kuang (1019–80) is said to have possessed a rubbing of a Zhou ruler from an ancient stone inscription, and this ruler reportedly had been reproduced in the *Jiali*, a popular compendium of Neo-Confucian household ritual. However, frequent reprinting of that work in different formats had reduced this attempt at definition to chaos. In 1393, Korean officials tried to re-establish a definition by adopting as a standard the length of ancestral tablets (*sinju*, Chinese *shenzhu*) from the household temples of several apparently strict Confucian ritualists. Evidently the length of these had some direct connection with the Zhou foot.[144] An official length was determined, a model ruler was made, and this measure was formally adopted in 1437, at the time of the completion of the instruments for Sejong's new astronomical observatory. The Zhou ruler was used in producing all its horological and observational instruments, and in addition it was prescribed as the standard for "all ancestral tablets in the household shrines of the gentry, as well as for route distances in *li*, and for pacing rules at the archery range."[145]

It remained to define the ratios between these and the other current rulers, and for this purpose the *hwangjong* (yellow bell, Chinese *huangzhong*) foot was made the integrating standard. *Hwangjong* was the name of the first note of the classical musical scale, the standard tuning note for musical instruments. This unit had been defined in 1425 as a fundamental step in Sejong's project to reconstruct classical music and manufacture a standard set of orchestral instruments. The *hwangjong* pitch pipe was made the foundation for all measurement—length, volume, and weight.[146] The ratios between the *hwang-*

142. Ch'oe Hang et al., *Kyŏngguk taejŏn* (compiled 1469; revised 1476) (Seoul: Chōsen Sōtokufu, 1934), 6.1b–2a.

143. Kim Chaero (1682–1759) et al., *Sok taejŏn* (Supplement to the Codex, compiled 1746) (Seoul: Chōsen Sōtokufu, 1935), 6.1a.

144. I have been unable to confirm this detail. Modern editions of the *Jiali* (Household rituals) have no drawing of the Zhou ruler. The edition deemed the best by the Qianlong emperor's book reviewers in 1781, and reprinted in vol. 142 of the *Wenyuange Siku quanshu* (Manuscript copy of the *Siku quanshu* in the Wenyuange Library) (Taipei: Shangwu Yinshuguan, 1983), 4.24a, does not mention the Zhou ruler in its specifications for ancestral tablets. Of course the *Jiali* has gone through numberless editions, and its text history, beginning with its authorship (the traditional attribution to Zhu Xi has long been rejected), is probably beyond recovery. But the alleged connection of the Zhou ruler with Neo-Confucian ritual would be worth some serious research given its great importance for post-Song East Asian mensuration.

145. *Sejong sillok*, 77.11a–b (note 17). The *Jiali*, 4.24a (note 144), specifies the length of an ancestral tablet as 1 *chi* (foot) 2 *cun* (inches). If the standard is taken to be the Zhou ruler, then five-sixths of the length of a precisely made tablet would equal one Zhou foot.

146. According to the editors of the modern translation and annotation of the *Kyŏngguk taejŏn*, in 1425 King Sejong's chief musicologist, Pak Yŏn, collected medium-sized millet grains in the district of Haeju, selected one hundred average grains, then lined them up side by side to fix the length of the *hwangjong* foot. The *hwangjong* pitch pipe was defined as a pipe nine *hwangjong* inches long with a circumference of 0.9 inch. The water that would fill that pipe also fixed the standards for volume and weight measures. See Han Ugŭn et al., eds., *Yŏkchu Kyŏngguk taejŏn: Chusŏk pyon* (The annotated *Kyŏngguk taejŏn*: Notes and commentary, translated [from Chinese into Korean] and annotated) (Seoul: Han'guk Chŏngsin Munhwa Yŏng'guwŏn, 1986), 751. See also *Munhŏn pigo*, 91.2b (note 51).

TABLE 10.4 Korean Linear Measurement Standards Current in the Fifteenth Century

Unit	Application	Ratio	Metric Equivalent (provisional)
Hwangjong ch'ŏk	Standard for musical pitch-pipe length	1.000	32.85 cm
Chuch'ŏk	Zhou foot; general linear measurement	0.606	19.91 cm
Choryegi ch'ŏk	Ritual foot; ritual implements	0.823	27.04 cm
Yŏngjo ch'ŏk	Architectural foot	0.899	29.54 cm
P'obaek ch'ŏk	Textile foot; standard bolt width	1.348	44.29 cm

Source: Ch'oe Hang et al., *Kyŏngguk taejŏn* (Great codex of state administration, compiled 1469; revised 1476) (Seoul: Chōsen Sōtokufu, 1934), 6.1b–2a. The metric equivalents are based on the figure of 19.91 centimeters for the Zhou foot, as determined by Wu Chengluo, *Zhongguo duliangheng shi* (History of Chinese weights and measures) (Shanghai: Shangwu Yinshuguan, 1937), 64–66 (table 15). For the likely applicability of this value to Korea, see note 149. These values should be considered provisional pending further research.

jong foot and other foot measures then in use were determined next, probably during the late 1430s and early 1440s. Cast bronze rulers were sent to all district magistrates during the 1440s,[147] and the official ratios were stated in the statutes in 1469 (table 10.4).

The conversion of these various rulers to metric units remains problematic. Jeon measured the *choryegi ch'ŏk* (ritual implement foot) printed in the Sejong annals and found it to be 28.9 centimeters, and on this basis he derived the length of the *hwangjong* foot as 35.1 centimeters and that of the Zhou foot as 21.27 centimeters.[148] However, studies from the Chinese side have produced values of 19.91 and 24.525 centimeters for the same measure. Since the details of Jeon's ritual foot are still unclear, I have provisionally accepted Wu Chengluo's value of 19.91 centimeters, which also passes a test I devised for its applicability to Korean distance data. Provisional metric equivalents of the Korean measures, based on this value, are given in tables 10.4 and 10.5.[149]

147. The bronze standard rulers from the Sejong era were lost over the years, mostly during the Japanese invasions of 1592–98. But in 1740 the court learned of a bronze textile ruler (*p'obaek ch'ŏk*) in the office of the magistrate of Samch'ŏk bearing an office of standards inscription with a date in Sejong's reign corresponding to the period 18 December 1446 to 16 January 1447. Compared with this ruler, the textile rulers then current were found to be anywhere from a half to a full inch (*ch'on*) shorter. Since the original ratios to the other standard rulers were known, King Yŏngjo (r. 1724–76) ordered a new set of bronze standard rulers based on the Samch'ŏk ruler to be cast and distributed; see *Munhŏn pigo*, 91.3b–4a (note 51).

148. Jeon, *Science and Technology in Korea*, 134 (note 49).

149. Jeon, *Science and Technology in Korea*, 131–34 (note 49). Jeon does not give any details concerning his measurement of the *choryegi ch'ŏk* in the *Sejong sillok*. His illustration shows the ruler as depicted in the 1956 photolithographic reprint of the second woodblock edition of 1604. His figure of 28.9 centimeters cannot have been obtained from the 1956 edition (which by actual measurement yields a length of about 18.1 cm). But did this figure come from the 1604 edition or from the original printed edition of 1472 (Seoul National University Library, Kyujanggak #12722)? One hopes it was from the latter, but Jeon does not say.

In addition to Jeon's derived value of 21.27 centimeters for the Zhou foot, there is another that is derivable from table 15 of Wu Chengluo, *Zhongguo duliangheng shi* (History of Chinese weights and measures) (Shanghai: Shangwu Yinshuguan, 1937), 64–66. Wu calculates the ratio of the Zhou foot to the Chinese *shichi* (market foot, 33.3 cm) at 0.5973 and gives a value of 19.91 centimeters for the Zhou foot. Finally, Needham et al., *Hall of Heavenly Records*, 90 (note 48), assume a Zhou foot value of 24.525 centimeters based on the physical measurement of a bronze *liangtian chi* (celestial foot) of Ming times, itself reportedly in agreement with an earlier Chinese "iron ruler" of the sixth century. However, a reading of Yi Shitong, "Liangtian chi kao" (Study of a sky-measuring scale), *Wenwu*, 1978, no. 2:10–17, which Needham and associates cite as their authority, finds no discussion of the Zhou foot and no mention of it as being equivalent to the *liangtian* foot.

In order to decide between these alternatives, I devised a test based on fifteenth-century Korean distance data. I randomly chose from the *Sŭngnam* (note 111) twenty distances in *li* between points that could with reasonable precision be found on a good modern map of Korea (USAF Operational Navigational Chart ONC G-10, 1:1,000,000, Aeronautical Chart and Information Center, United States Air Force, St. Louis, Missouri, 1964) and that were in areas of relatively flat terrain in different parts of the country. I then used an opisometer to measure the same distances in kilometers on this modern map and translated these figures into *li* using the three standards of Wu, Jeon, and Needham. The aggregate of all twenty measured distances was 614 kilometers, corresponding to 1,423 stated *li*, or an average of 2.318 *li* per kilometer. Against this, Wu's figure (2.325 *li*/km) yielded 1,428 *li*, or just 0.3 percent over the stated *li* total; Jeon's figure (2.17 *li*/km) yielded 1,334 *li* (−6.3 percent); and Needham's figure (1.89 *li*/km) yielded 1,160 *li* (−18.5 percent). Needham's figure, if not already suspect, is thus safely eliminated, at least in connection with the Korean use of the Zhou foot. As for the choice between Wu and Jeon, this experiment clearly favors Wu. This experiment assumes that the roads between the forty points of measurement were more or less the same over the centuries (flat terrain in relatively undeveloped areas was chosen). It also assumes that the distances in the *Sŭngnam* were correct; even if some were erroneous, twenty different distances in different parts of the country ought to average out reasonably well. Finally, an opisometer moving over a 1:1,000,000 scale map will necessarily smooth out many twists and turns on the road and will therefore tend to produce a kilometer total on the minus side of the actual figure. A correction made on this basis will move *li* totals calculated from Jeon closer to the stated *li* totals while moving those calculated from Wu away from them. Still, according to my test, Wu's value (determined without awareness of the Zhou foot in Korea, thus removing any circularity in the argument) compels respect. Until the uncertainty about the length of the ritual foot, which Jeon used in calculating his Zhou foot length, is resolved, Wu's equivalent of 19.91 centimeters for the Zhou foot seems the more reliable figure, and it will be applied in the rest of this chapter. There are, of course, plenty of other approaches to this problem. Much more research must be done before historical Korean weights and measures are brought under control.

TABLE 10.5 Zhou Foot Units in the Measure of Route Distances

Unit Equivalents				Metric Equivalents (provisional)		
1 *ch'ŏk*				19.91 cm		
6 *ch'ŏk*	= 1 *po* ("pace")			119.46 cm	1.19 m	
2,160 *ch'ŏk*	= 360 *po*	= 1 *li*			430.05 m	0.43 km
	10,800 *po*	= 30 *li*	= 1 *sik*			12.90 km

Note: This table is derived from Kim Chaero et al., *Sok taejŏn* (Supplement to the Codex, compiled 1746) (Seoul: Chōsen Sōtokufu, 1935), 6.1a. Metric equivalents should be considered provisional; see note to table 10.4.

The officials of the Chosŏn dynasty apparently inherited distance data from their Koryŏ predecessors. There is an example of the application of such data in 1402. In codifying the laws of banishment and defining degrees of distance from the capital, legal specialists noted that Kyŏngwŏn in the far northeast (a place of first-degree banishment) was 1,680 *li* from the capital and that Tongnae in the far southeast (second-degree) was 1,230 *li* away.[150] This suggests that systematic distance data were available at that time.

The results of a questionnaire in 1469, which among many other things asked for the distance of district seats from Seoul, are available for Kyŏngsang Province,[151] while comparable data for the whole country can be found in the *Sŭngnam* (1481; final edition 1531). The latter data are clearly and quite consistently given according to the Zhou foot, but it is interesting to see that the 1469 figures are lower across the board. Aggregate 1469 data from the fifty-five (out of sixty-six) Kyŏngsang districts that have coherent figures for distance from Seoul show a total of 31,260 *li*, which compares with a total 39,711 *li* for the same districts in the *Sŭngnam*. The 1469 figure comes to 78.7 percent of that for the *Sŭngnam*.[152] The *Sŭngnam* unit of measure was clearly smaller than the unit used for the 1469 measurements (and therefore it took more of them to measure the same distances), but not by a proportion that matches the ratio of the Zhou foot to any of the other standards in table 10.4. The only conclusion is that linear measurement units were still not actually standardized by 1469, even though official ratios were established by statute in that year, and that the Zhou foot, though prescribed for distance measurement in 1437, was still not being uniformly applied a generation later. Moreover, if we take the figures for the single district of Tongnae—725 *li*, 96 *po* (or 725.27 *li*) from the 1469 results, and 962 from the *Sŭngnam*—and compare these with the distance of 1,230 *li* given for Tongnae in the banishment discussion of 1402, it seems there was a great variety of customary practice that had to be overcome in the effort to reach national norms.

We have very little information on how measurement was actually carried out. References to surveys of Seoul taken in 1393 speak of crews with rope measures,[153] and given the precision of the 1425 reports (with distance measurements stated to the *po*), it is likely that some such method was applied in measuring long distances as well. But Sejong was clearly looking for something better. In 1441 a new device, called the *kiri koch'a* (*li*-counting drummobile) made its appearance on an extended royal excursion to a hot spring in Ch'ungch'ŏng Province: "On this trip, His Highness for the first time rode in his small chaise and used the *li*-counting drummobile. At each *li* passed by the entourage, a wooden man would automatically beat a drum."[154] Later that same year, we see reference to the use of a "newly made pace-counting measuring stick [*posuch'ŏk*]" to measure distances in P'yŏngan Province. The board of war wanted markers set up or trees planted every thirty *li* along all roads.[155]

It is unclear how extensively the drummobile was used. As far as I can tell, Sejong's trip to the hot spring occasions the only reference to it.[156] I suspect that the *li*-

150. *T'aejong sillok*, 4.9b–10a (note 17).

151. *Kyŏngsangdo chiri chi* and *Kyŏngsangdo sokch'an chiri chi* (note 107).

152. The 1469 data in this discussion are based on figures found in the *Kyŏngsangdo sokch'an chiri chi* (see note 107) under the respective headings for the sixty-six counties of Kyŏngsang Province. In four cases the data are missing or incomplete. In seven cases the 1469 figures are significantly larger than those in the *Sŭngnam*, suggesting different routes for the Seoul measurement or some other problem with the data. These figures are not included in the sample. The *Sŭngnam* data are from the corresponding headings in the *Sŭngnam*, chaps. 22–32 (note 111).

153. *T'aejo sillok*, 3.3b (note 17).

154. *Sejong sillok*, 92.18b (note 17). Pang, in common with several other writers, says that the wooden man beat the drum every ten *li* (*Han'guk ŭi chido*, 88 [note 5]). Apart from the dubious utility of such a large unit of measure, the original reference explicitly says "one *li*."

155. *Sejong sillok*, 93.26a–b (note 17). Pang cites this and several other instances as examples of the use of the drummobile for measuring and survey work, but in none of the cases he cites is the drummobile actually mentioned in the sources (*Han'guk ŭi chido*, 88 [note 5]).

156. As far as we can infer from the sources, the drummobile was a piece of equipment for Sejong's personal use. In this it bears some similarity to its identically named Chinese model, the *jili guche*. Along with the famous *sinan che* or *zhinan che* (south-pointing carriage), the drummobile was part of the imperial equippage of the Western Jin dynasty (265–317) and was treated as a symbol of Chinese imperial prestige. Lost to the barbarians in 317, both of these devices were

counting drummobile was found impractical, if indeed it was ever intended for real survey work. The pace-counting measuring stick would have been much simpler, more accurate, and easier to use. We would expect a distance measure made in 1441 to have been calibrated according to the Zhou foot, yet the 1469 Kyŏngsang measurements were clearly based on a longer, perhaps more customary standard. All we can say with certainty is that sometime before or during the early recensions of the *Sŭngnam* in 1481, either new measurements were made or the old ones were recalculated according to the Zhou foot.

We are only slightly better informed about the triangulation device, called the *kyuhyŏng* (sighting balance), invented by King Sejo in 1467. In the spring of that year, Sejo gave several parties for some friends and technically inclined officials, introducing them to his new gadget, which was also referred to as an *inji ŭi* (ground-marking instrument).[157] The editor of the annals defined this as "a thing for measuring ground distances."[158] A test was arranged out in the country, in the vicinity of Sejong's tomb, but no information is given on what was actually done or what the results were.[159] A description by one of these surveyors, Yi Yuk, quoted in the eighteenth-century encyclopedia *Munhŏn pigo*, leaves no doubt that it was a triangulation instrument of some kind but is frustratingly lean on details.[160] The *kyuhyŏng* was used later that year in making a map of Seoul but seems to go unmentioned after that.[161] In spite of all the uncertainties concerning the drummobile and the sighting balance, the information we do have shows that Koreans were working to make distance measurement more accurate during the middle years of the fifteenth century, and that sufficient precision had been achieved to result in the internally consistent distance figures recorded in the *Sŭngnam* in 1481.

It remains to note efforts to measure the country's length from north to south by means of polar altitude. In 1437 the designers of Sejong's observatory measured the polar altitude of Seoul at "38 degrees [plus] a little."[162] This assumed a celestial circumference of 365.25 degrees. Correction to a 360-degree circle would yield a reading of 37.45°+ as the latitude of Seoul, not unreasonably far off from the modern figure of 37°35′44″N (37.595 degrees). The *Munhŏn pigo* reports that during Sejong's reign he sent three mathematicians to measure the polar altitude at Mount Mani on Kanghwa Island (approximately the same latitude as Seoul), Mount Paektu (symbolically Korea's farthest point north), and Mount Halla on Cheju Island (close to the nation's southernmost point). Unfortunately, as the *Munhŏn pigo*'s editors also noted, no results were recorded.[163] But the purpose of such a measurement was most likely to calculate the total length of the nation north and south from the capital. And since distance figures (accurate or not) already existed for all these points, it was perhaps the intention to perform some kind of check on measured land distances, or perhaps to independently establish a length of the nation for cartographic purposes. Whatever the reason, Sejong might better have sent his team to the town of Onsŏng, which his military officials, following extensive campaigns against the Jurchens, had established within the northern bend of the Tumen River in 1440. Onsŏng was at that time, and still remains, the northernmost town in Korea (42°57′21″), almost a whole degree of latitude farther north than the Ch'ŏnji (Pond of Heaven) in the crater of Mount Paektu (42°00′00″). Korean perception at the time may well have been that Paektu was the farthest point north. We will return to this possibility when we consider the widely noted "flatness" of Korea's northern border on the so-called Chŏng Ch'ŏk maps that show Paektu and Onsŏng at about the same latitude, and on both the national and Hamgyŏng provincial maps in the *Sŭngnam*, where Onsŏng is latitudinally south of Paektu.

Apart from Sejong's interesting polar altitude project, we have no other information on geodetic measurement in Korea until after the introduction of Western survey methods. These first appeared in 1713, when a Manchu envoy arrived in Seoul and sent a Jesuit-trained Chinese surveyor to the center of the city to take a measurement. I shall return to this very curious affair in a later section.

The Shape of Korea

A few months before his death in 1482, Yang Sŏngji reviewed a large number of official compilations that had been edited during his lifetime—many of them in whole or in part by himself—and made proposals concerning their printing and distribution within the government. Prominent among his concerns were maps. He listed twenty important maps, or sets of maps, that were then

recovered, to the Jin emperors' great relief, by successful Chinese campaigns in 409 and 417. See Fang Xuanling et al., *Jin shu* (History of the Jin, compiled 646–48), 25.756 and 25.764; see the edition in 10 vols. (Beijing: Zhonghua Shuju, 1974). Given the treatment in both the Chinese and Korean cases, it remains to be demonstrated that the *li*-counting drummobile found significant practical application as a distance-measuring instrument.

157. *Sejo sillok*, 41.20b–21a and 41.21b (note 17).

158. *Sejo sillok*, 41.12b (note 17).

159. *Sejo sillok*, 41.22a (note 17).

160. *Munhŏn pigo*, 2.32a–b (note 51).

161. *Sejo sillok*, 44.9b (note 17).

162. *Sejong sillok*, 77.9b (note 17). See Jeon, *Science and Technology in Korea*, 102–4 (note 49), and Needham et al., *Hall of Heavenly Records*, 108–9 (note 48).

163. *Munhŏn pigo*, 2.10a–b (note 51). The editors cite the *Kwansanggam ilgi* (Royal observatory daybook) and identify the three mathematicians by name, but they are not otherwise traceable.

in his custody or within his knowledge. It was a mixed list that included a map of Korea from Koryŏ times and a few imports from China and Japan, but that mostly related to cartographic works by himself and other Korean contemporaries. His list is presented in appendix 10.1. It represented the result of the projects ordered by King Sejo, which I have already noted, and thus contained maps on the national, provincial, and district levels in addition to defense maps and a map showing coastal shipping routes.

Two of the listed maps of Korea may survive today in copies. The Japanese and Ryūkyū maps are probably reflected in the *Haedong cheguk ki* (mentioned above). Possibly some echo of the Ming maps might be detected in later Korean or Chinese copies, but apart from this meager residue, the maps on Yang's inventory seem to have left no trace either in other written records or in any identifiable cartographic artifact. This situation is probably at least partly the result of policies favored by Yang himself. He was no champion of the freedom of information. In the same memorial of which this list is a part, he advocated severe restrictions on most government publications. Even the *Koryŏ sa*, the standard history of the Koryŏ dynasty, was "not to be lightly shown to people" lest the details of military campaigns centuries in the past become too well known. As for maps, "they should be put in government offices, but they must not be dispersed among the people."[164] Given such attitudes, most of Yang's maps undoubtedly had very little circulation.

There is one prominent omission on Yang's list. One expects to find the *Kangnido* of 1402, which was almost certainly copied in Seoul during Yang's heyday as a cartographer,[165] but unless item 15, the *TaeMyŏng ch'ŏnhado* (Map of the Great Ming world), is the *Kangnido* in disguise, there is no mention of it.[166] It is the Korea portion of the *Kangnido*, drawn originally by Yi Hoe with Kwŏn Kŭn's guidance, that is the most likely correspondent to item 2, Yi Hoe's *P'altodo* (Map of the Eight Provinces). To be sure, the *Kangnido*'s Korea does not show provincial divisions, but the basic outline of the peninsula and other details are likely to be related.

The first feature of Korea as shown on the *Kangnido* to attract attention is the prominent indication of mountain ranges, not only the main north-south trunk along the eastern coast, but the direct "arteries" extending to Kaesŏng and Seoul, demonstrating their clear connection with that trunk (plate 17). The trunk range does not connect directly with Mount Paektu (which is not specially emphasized and is out of place to the southeast from where it should be), but there is a color variation at the spot where the line is broken, suggesting possible damage or incomplete mending at some point. Such indication of mountain ranges is not found anywhere else on the *Kangnido*, but it well reflects Korean fascination with the "arterial" system in Korea itself, already noted.

The rivers of the peninsula are decently represented, especially in the central and southern regions; the Han and Naktong systems are remarkably well done. Most of the administrative districts of the country are also shown, with provincial capitals and some key military centers in square cartouches. Seoul is emphasized with a large crenellated circle.[167] The indication of harbors and inlets, in oval cartouches, gives an appreciation of the grain transport and coastal defense organization, which was under severe pressure from Japanese pirates at the time this map was made.

The feature that draws the most criticism from Korean scholars is the general outline of the country, which, though it very well approximates the real coastline in the central and southern parts of the peninsula, is marked by a radical shortening and flattening of the northern border area.[168] At the time the *Kangnido* was made in 1402, most of the Yalu-Tumen frontier was still occupied by Jurchen people (the later Manchus) and had not yet been reached by Korean conquest. Yi Hoe and Kwŏn Kŭn very likely followed Koryŏ cartographic models that were already out of date.

Since the flatness of the northern border plagues Korean cartography until the eighteenth century, it will be useful to sketch out the history of the Yalu-Tumen frontier during the Koryŏ and Chosŏn periods. Silla's frontier was marked by a line that stretched from P'yŏngyang in the west to the area just north of modern Wŏnsan on the east coast (for this discussion, see figs. 10.1 and 10.24). In the early years of the Koryŏ period, the new dynasty pushed the frontier north to the mouth of the Yalu River, settling all of the lands in the Ch'ŏngch'ŏn-Taedong basins and successfully holding this territory

164. *Sŏngjong sillok*, 138.10b (note 17).

165. The Ryūkoku copy of the *Kangnido* indicates three Yalu frontier districts (Yŏyŏn, Much'ang, and Unul) as old (*ko*). They were abolished in 1455. On the other hand, it shows two districts, Yewŏn and Such'ŏn, that were discontinued in 1459 and 1466, respectively. The copy therefore seems to have been made by 1470. See Aoyama, "Gendai no chizu ni tsuite," 111–12, 143, and 149 n. 12 (note 24).

166. An argument can be made for this identification. The *Da Ming hunyi tu* (Integrated map of the Great Ming) in the Beijing Museum—which but for the Manchurian, Korean, and Japanese areas strongly resembles the *Kangnido*—and the *Daiminkoku chizu* (Map of the Great Ming), a copy of the *Kangnido* in the Honmyō Temple in Japan, both have "Great Ming" in their titles. See above, at notes 32 and 40. The appearance of the common term *ch'ŏnhado* in the title of item 15 is at least interesting, given my earlier suggestion that the later *ch'ŏnhado* may be an evolutionary sport of the *Kangnido* (pp. 265–67).

167. Needham, *Science and Civilisation in China*, 3:555 (note 31), mistakenly identifies the circle as representing P'yŏngyang.

168. Yi, *Han'guk ko chido*, 198b (note 4); Kim, *Maesan kukhak san'go*, 260–62 (note 3).

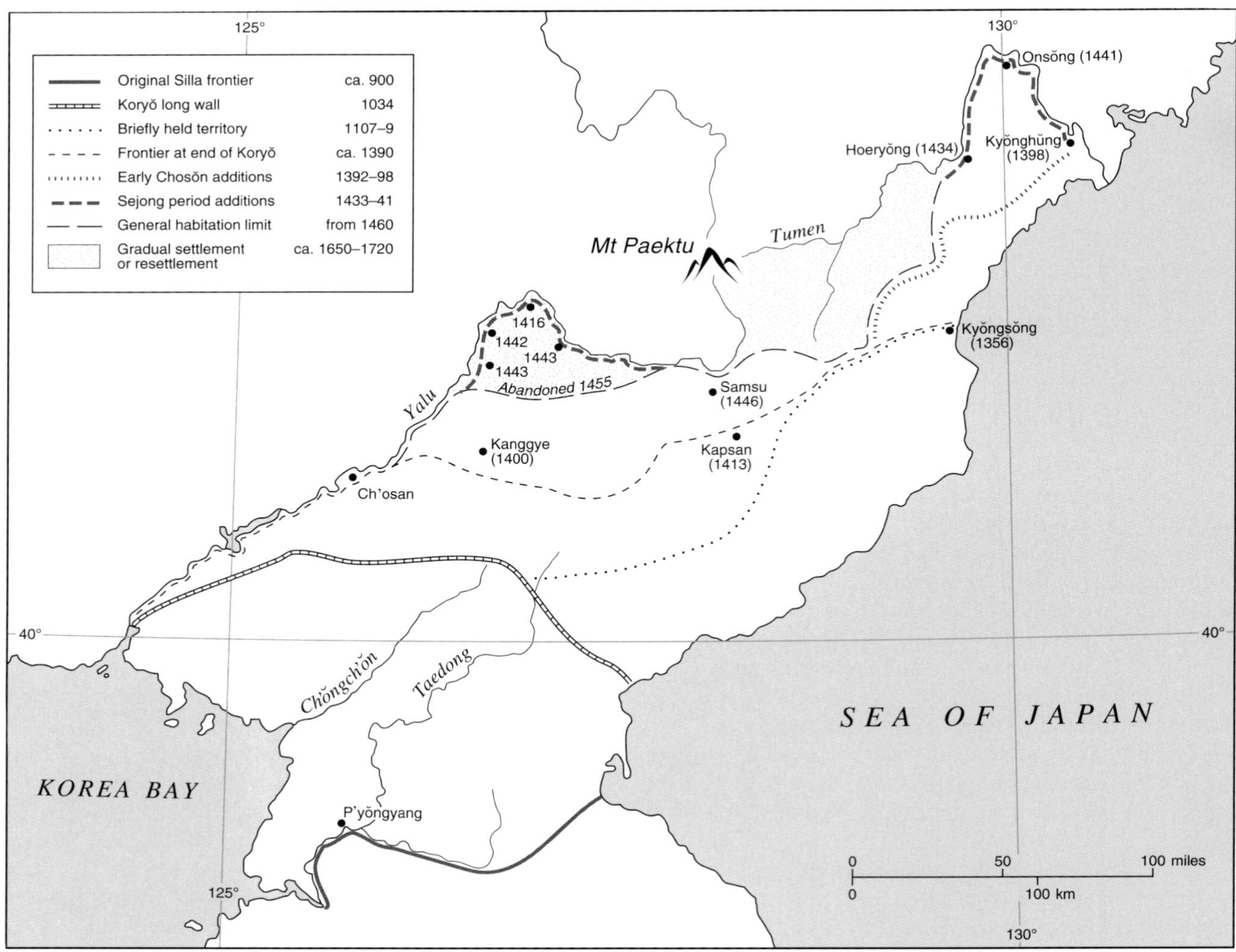

FIG. 10.24. KOREA: NORTHERN BORDER CHANGES FROM THE TENTH TO THE FOURTEENTH CENTURY.

against repeated incursions by the Khitan and Jurchen peoples. This area is clearly marked by the Koryŏ changsŏng (Koryŏ long wall), completed in 1034. In 1107, campaigns led by Yun Kwan (d. 1111) drove the Jurchens from an extensive area on the east coast as far north as Kyŏngsŏng (approx. 41°35′N). Nine towns were established and fortified. The Jurchens took back this area two years later, however, and in the thirteenth century it passed into the hands of the Mongols. Koryŏ did not control it again until 1356, when King Kongmin reclaimed it—on the basis of maps, as we have seen above—in the course of expelling the Mongols. By the end of the Koryŏ dynasty in 1392, Koryŏ administration had reached a line extending from a point east of Ch'osan on the Yalu River on a slightly northeasterly course to Kyŏngsŏng on the east coast. Yi Sŏnggye, the founder of the Chosŏn dynasty, then extended this coastal strip to Kyŏnghŭng (formerly Kongju, approximately 42°36′N) on the lower course of the Tumen River. By 1441, during Sejong's reign, all the lands within the far northern bend of the Tumen had been added. This was the farthest point north reached by any Korean state since Koguryŏ times, and the territory has been held ever since. In the north-central frontier zone Sejong also seized and organized the territory within the northern bend of the Yalu, but it was extremely rough terrain and proved too difficult to hold. Sejo abandoned it in 1455. This land, together with another large and wild region south of the Tumen River in the Paektu area, thereafter remained generally beyond Korean administration. It was not until the seventeenth and early eighteenth centuries that all these uninhabited areas were permanently settled and organized. The frontier from the crater lake and summit of Mount Paektu to the headwaters of the Tumen was fixed by formal agreement with the Qing dynasty in 1712. This was the final step in establishing Korea's present borders.

It can be seen from this discussion and the map in figure 10.24 that, even though most of the cis-Tumen lands had still not been reached by Korean administration when the *Kangnido* was made in 1402, the northern border already slanted visibly northeastward to Kyŏnghŭng, where some of Yi Sŏnggye's ancestors (who had been in Mongol service) were buried. Yi Hoe and Kwŏn Kŭn put Kyŏnghŭng, and therefore the area of the mouth of the Tumen, on the *Kangnido*'s Korea section but considerably underestimated how far north it was. Undoubtedly, in 1402 the geography of this region was still poorly understood. (The Ryūkoku copy has the names of other major towns within the northern Tumen bend, but since they were not established until the 1430s and 1440s, they must have been added at the time this copy was made, probably by 1470.) More problematic than the Tumen frontier is the border on the northwestern corner of the peninsula. Yi and Kwŏn have the Yalu flowing due westward into the Yellow Sea on virtually a horizontal line. The southwesterly course of the lower Yalu was surely well known during Koryŏ times, and it is hard to understand how the 1402 cartographers could have arrived at this treatment. The fact is, however, that the "flattened" northern frontier remained a feature of Korean maps for nearly three centuries after the appearance of the *Kangnido*.

Maps of Korea after the *Kangnido* are usually classified by Korean scholars in four groups. These are (1) the Chŏng Ch'ŏk–style maps, prevalent from the mid-fifteenth to the mid-seventeenth century and identified principally by their flat northern frontier; (2) the *Sŭngnam*-style maps in imitation of the maps drawn for the *Tongguk yŏji sŭngnam*, commonly included in the *ch'ŏnhado* albums and popular until the end of the nineteenth century; (3) the Chŏng Sanggi–style maps, with the first relatively realistic depiction of the northern frontier and the first use of an indicated scale, which probably began to appear during the first half of the eighteenth century and were common until the mid-nineteenth; and (4) Kim Chŏngho's maps, a refinement and completion of the Chŏng Sanggi maps and indeed the grand summation of Korean national cartography, produced from 1834 to 1864 but displaced by cartography of Western inspiration from the late nineteenth century on.

MAPS IN THE CHŎNG CH'ŎK STYLE

We have already encountered Chŏng Ch'ŏk several times in the preceding pages. Although he was King Sejong's principal cartographer and his name is nowadays memorialized in the label for a particular cartographic style, the longish account of his career in his obituary in the *Sŏngjong sillok* (Annals of King Sŏngjong) says not a word about any work with maps.[169] This is typical of Korean historiography of the Chosŏn period, which did not consider scientific or technical achievements during one's government service as an ornament worth mention. Such specialization went against the Confucian scholarly ideal. Chŏng Ch'ŏk's forebears, clerks in local administration, had been relatively humble, but in the early years of the Chosŏn period a clever person of such circumstances might find opportunities. Chŏng passed the civil service examination in 1414 and thereafter rose to the upper-middle reaches of the bureaucracy. He became a favorite companion of Sejong and participated in many of his cultural and scientific projects.

The major map known to have been associated with Chŏng Ch'ŏk is the *Tongguk chido* (Map of the Eastern Country [Korea]), which he and Yang Sŏngji presented to the throne in 1463.[170] This has not survived, but a map of Korea now held by the Cabinet Library (Naikaku Bunko) in Tokyo, bearing the Japanese title *Chōsenkoku ezu* (Map of Korea), is thought to be a copy of it or in its tradition (fig. 10.23). Aoyama Sadao's study of the Naikaku map identified features that suggested a mid-fifteenth-century compilation date,[171] and on this basis it is considered to be the prototypical representative of what has come to be called the Chŏng Ch'ŏk style in peninsular cartography. In any case, it is certainly in the older stratum of Chosŏn period maps.

The Naikaku map represents a considerable advance from the standard of the Korean portion of the *Kangnido*. The rivers, marked clearly with thick blue lines, show considerable accuracy of detail. The arterial system of the mountains, as I have noted, is much more developed than on the *Kangnido*. From the latter's indications, more or less confined to the main east coast spine and a few major offshoots only sketchily indicated, the Naikaku map advances to a thorough depiction of the total montane network of the country. A shapes-and-forces emphasis, textually associated with Chŏng Ch'ŏk in 1436 and 1454 (mentioned above), is evident on this map. One's eye can easily follow a mountainous line from virtually any part of the country directly to Mount Paektu itself. The treatment is extended into Manchuria, a considerable portion of which is included, although it is highly compressed.

The names of the district magistracies in each province are shown in circular cartouches, color-coded according

169. *Sŏngjong sillok*, 58.1b–2a (note 17). The date of Chŏng's death, and of this notice, corresponds to 1 September 1475.

170. *Sejo sillok*, 31.25b (note 17).

171. Aoyama Sadao, "Richō ni okeru nisan no Chōsen zenzu ni tsuite" (On several Yi [Chosŏn] dynasty maps of Korea), *Tōhō Gakuhō* (Tokyo) 9 (1939): 143–71. A fine color reproduction of a Korean copy held by the National History Compilation Committee (Kuksa P'yŏnch'an Wiwŏnhoe) in Seoul may be found in Yi, *Han'guk ko chido*, pl. 10 (p. 24) (note 4).

to province. The concentration of similarly colored cartouches in each province has the advantage of distinguishing the provinces, which are not indicated by borders, from each other. Aoyama lists the colors (his plate is not in color) as follows:[172]

Kyŏnggi	deep yellow
Ch'ungch'ŏng	light yellow
Chŏlla	red
Kyŏngsang	pink
Kangwŏn	green
Hamgyŏng	blue
P'yŏng'an	gray
Hwanghae	white

This system, with occasional variations, is a common though not universal feature of colored maps down to Kim Chŏngho's time. The assignment of colors is generally not fortuitous but responds to the classical Chinese *wuxing* (five phases) system, according to which the center is yellow, east is green (blue), south is red, west is white, and north is black. This provincial color system may also be reflected in an interesting Japanese usage. Hideyoshi's forces in Korea in the 1590s referred to the provinces by color in both military communications and ordinary conversation. The Japanese monk Keinen, who followed his campaigning lord around the southern part of the peninsula during the battles of 1597–98 and left a fascinating diary, refers to Chŏlla as Akakuni (Red Country), Ch'ungch'ŏng as Aokuni (Blue Country), and Kyŏngsang as Shirakuni (White Country).[173] Apart from Chŏlla and red, his names do not match the scheme above, yet the suspicion lurks that this usage reflects Japanese warriors' reliance on Korean maps for their campaigns.[174] Such exploitation of intelligence materials would only have been expected.

Other noteworthy features of the Naikaku map are the indication of principal military centers, as well as the names of hundreds of islands and coastal inlets. Principal roads are shown as red lines, and the distance of each district center from Seoul is marked both in *li* and in number of travel days.[175] Although Korea is depicted on this map in generally realistic proportions, there is no indication of scale, a development that had to wait until the Chŏng Sanggi era. Yi Ch'an estimates the de facto scale to be approximately 1:800,000.[176]

The problematic northern frontier shows a clear improvement over the depiction on the *Kangnido*, but whereas in the west the lower Yalu flows in a moderate southwesterly direction, the upper reaches are about as flat as on the earlier map. On the eastern side the northern frontier rises gently toward the northeast, but not to a realistic degree. The overall impression is still of a flat, compressed north, a feeling that is reinforced by the grossly horizontal aspect given to Manchuria's mountains and drainage basins. One would think that, given the extensive cartographic exploration and measurement that unfolded during the reigns of Sejong and Sejo, this situation would have been corrected by the 1460s. Perhaps the apparent desire to show the general lay of Manchuria claimed the space at the top of the map, leading to the flattening of Korea itself. But this would not explain the flatness on other and later Chŏng Ch'ŏk–style maps, which by this criterion show only regression. Note, for instance, the *Chosŏn'guk p'alto t'onghapto* (Consolidated map of the Eight Provinces of Chosŏn), otherwise an attractive map with an evident Korean folk-art touch, but on which the northern borderline is not merely flat but even slopes to the southeast (plate 18).[177]

The fault surely cannnot be attributed to a lack of distance data for the northern parts of the country, at least not in connection with the post-fifteenth-century examples of the Chŏng Ch'ŏk genre. The *Sŭngnam* recorded a distance of 1,186 *li* from Seoul to the northwestern frontier town of Ŭiju on the Yalu but 2,101 *li* to the town of Onsŏng within the northern bend of the Tumen.[178] Even an approximate recognition of the relative difference in these distances from the capital would have largely solved the border depiction problem.

That the frontier stayed cartographically flat in spite of such data, easily available from the early decades of the sixteenth century on, is not easy to explain, but several factors may have been at work. One might be security considerations. We have already seen Yang Sŏngji so obsessed with military secrecy that he was unwilling to let any maps circulate outside the government. In this context, it is possible that the border outline was left deliberately undeveloped in order to mislead any Chinese or Jurchen military leaders who might come into possession of Korean maps. The fifteenth century was a very active period in the Jurchen area, with the establishment and consolidation of anti-Jurchen commanderies by the Ming authorities throughout southern and eastern Man-

172. Aoyama, "Richō ni okeru nisan no Chōsen zenzu ni tsuite," 157 (note 171).

173. The text of Keinen's diary, *Chōsen nichinichi ki* (Record of days in Korea), is reprinted in *Chōsen Gakuhō* 35 (1965): 55–167. References to provincial color names are on 68, 75–76, and 150 n. 25.

174. The Korean copy of the Naikaku map (note 171) seems to make the Kyŏngsang cartouches white, which would correspond with Japanese military usage, but it is possible that an original pink coloring does not show up in the reproduction. On the same copy Ch'ungch'ŏng seems to be orange, which would be in reasonable harmony with a yellow Kyŏnggi Province to the north and a red Chŏlla to the south.

175. Distance and travel days are not marked on the Korean copy.

176. Yi, *Han'guk ko chido*, 206b (note 4).

177. Other Chŏng Ch'ŏk–style maps with flat or sagging northern borders are illustrated in Yi, *Han'guk ko chido*, 64–65, 70, and 79 (note 4).

178. *Sŭngnam*, 50.38a, 53.1a (note 111).

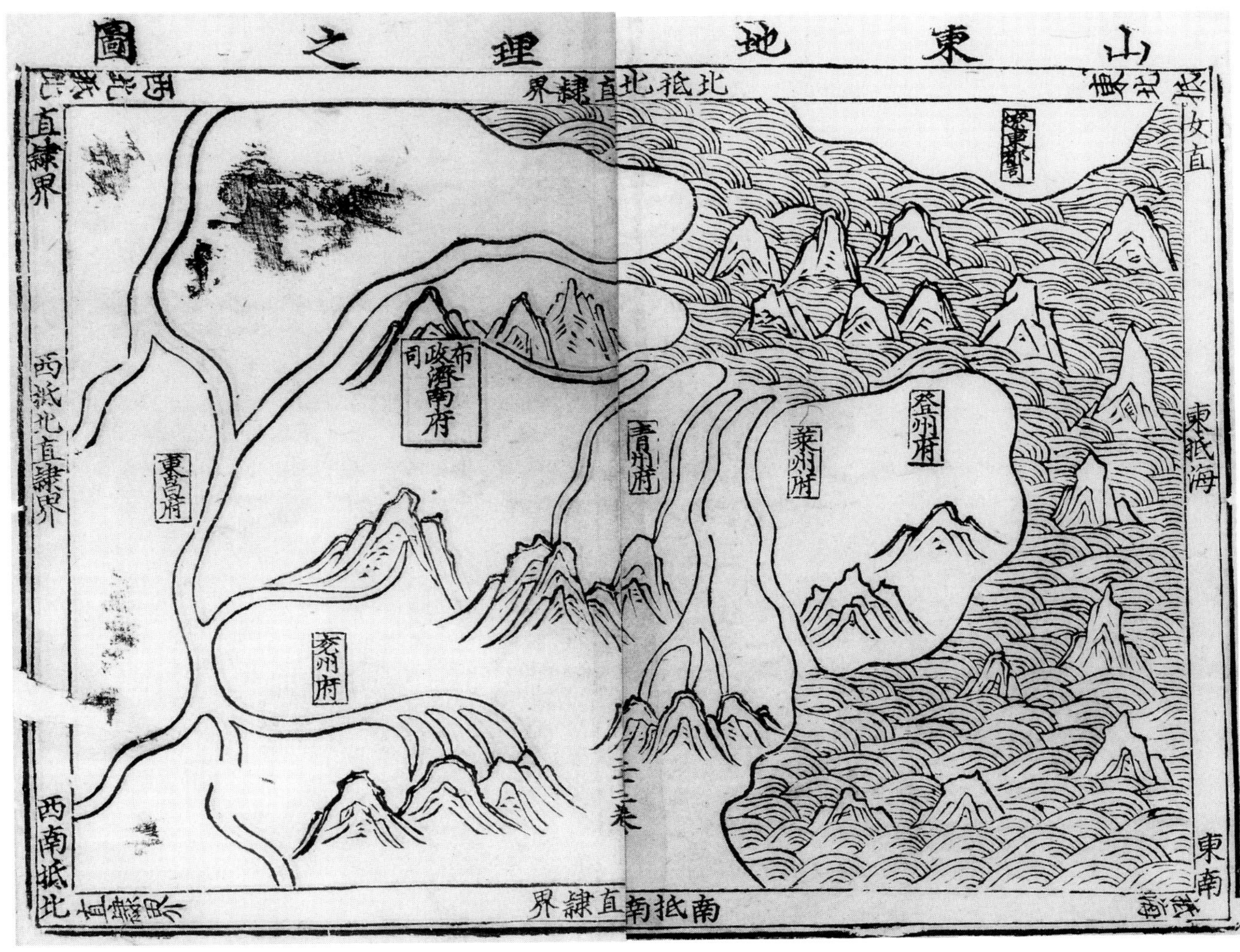

FIG. 10.25. "SHANDONG DILI ZHI TU" (MAP OF SHANDONG'S GEOGRAPHY). From the *Da Ming yitong zhi* (Comprehensive gazetteer of the Great Ming, 1461), frontispiece to *juan* 22. In this Ming Chinese map, only the six prefectures (*fu*) of Shandong Province, and none of its districts, are indicated; mountains and rivers are unnamed. Aside from these features there is only the coastline. This spare, reticent map style had a noticeable influence on the maps made for the Korean equivalent work, the *Tongguk yŏji sŭngnam* of 1481. Still, as can be seen in figures 10.26 and 10.27, the Korean maps were filled out more and drawn more interestingly.
Size of the original: 20 × 27 cm. From a 1505 edition, reproduced courtesy of the Harvard-Yenching Library, Harvard University, Cambridge.

churia and a steady effort by Korea to solidify its own frontier generally along the line of the Yalu and Tumen rivers. Both policies were at the expense of the Jurchens and inevitably also involved a quiet but obvious Sino-Korean competition. Yang Sŏngji stood at the meeting point of cartography and national security. In his later years, his clearly expressed security paranoia had probably overcome his cartographic judgment. But he was not alone in having such concerns.

MAPS IN THE *SŬNGNAM* STYLE

An approach to this question of the border may possibly lie in the maps that were compiled in connection with the *Sŭngnam*, which went through its first compilation stage in Yang's last years. The earlier recensions have not survived, but the maps of an edition probably printed in 1499 are identical with those of the final 1531 edition, showing that their form was already fixed at an early stage in the compilation process. Although the text of the *Sŭngnam* is justly famous for its thoroughness and reliabilty, the maps at first glance seem to be a disappointment. It is known that King Sŏngjong's instructions in 1485 to those preparing a printed edition asked for strict adherence to the style and format of the *Da Ming yitong zhi*, the Great Ming dynasty administrative geography of

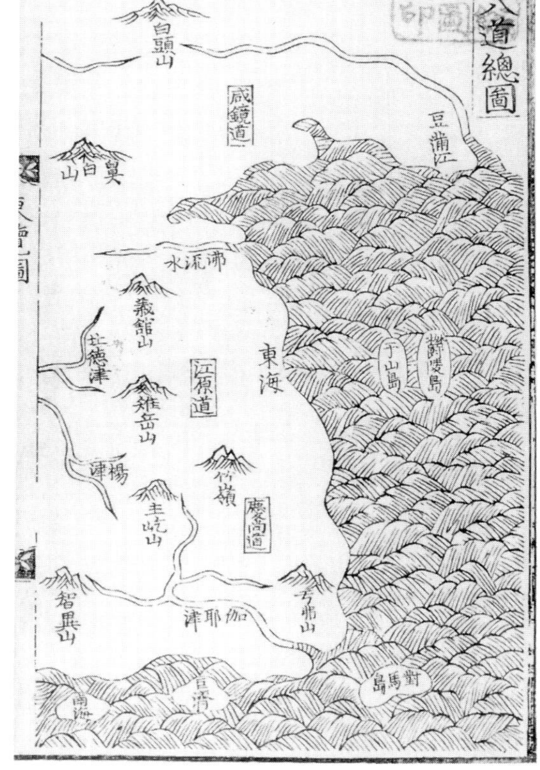

FIG. 10.26. "P'ALTO CH'ONGDO." Frontispiece to *kwŏn* 1 of *Sinjŭng Tongguk yŏji sŭngnam* (1531; first edition 1481). The designers of this map, which accompanies an extensive and highly detailed description of Korea some two thousand pages in length, left the details to the text and concentrated on simplicity. The rivers and mountains shown on it were all objects of major religious sacrifices conducted by the state; the capital and the names of the eight provinces (in thin cartouches) and of some major islands are the only other indications on the map. The courses of the Yalu and Tumen rivers are drastically flattened and the northern borders thereby distorted. This was partly dictated by the rectangular shape of the woodblock used in printing, but national security was also almost certainly a factor.
Size of the original: 27 × 17.1 cm. By permission of Kyujanggak Archives, Seoul National University.

1461.[179] A comparison of that 1461 edition with the available editions of the *Sŭngnam* shows that these orders were followed.[180] The cartographic style of the *Da Ming yitong zhi* may be described as one of exaggerated and highly stylized simplicity. Coastlines are generalized into the broadest of curves; rivers are uniformly shown as straightened, broad strips; mountains are merely suggested with a decorative peak here and there, with no concern for ranges or geophysical patterns. On the provincial maps only major administrative divisions (*fu*) are indicated, with the names in dark rectangular cartouches (fig. 10.25). It would be hard to imagine greater cartographic reticence. The *Sŭngnam* mapmakers in fact found Sŏngjong's mandated model too vacuous, and they not only improved upon it but executed it in a manner more pleasing to the eye. The resulting style then achieved enduring popularity in Korea, especially through the *ch'ŏnhado* albums that probably began to be produced during the sixteenth century.[181]

The "P'alto ch'ongdo" (General map of the Eight Provinces), which headed the first of the *Sŭngnam*'s 155 chapters (*kwŏn*), is illustrated in figure 10.26. The abbreviation of the northeastern area, by comparison with earlier cartographic depictions, is striking in its degree. A distortion not only is registered but seems to have been purposely exaggerated. The people who drew these maps surely knew the true state of affairs, because better maps of Korea were in the very library they worked in. But clearly they had certain purposes in mind, and they also

179. Kim Chongjik's 1486 colophon to the *Sŭngnam*, printed at the end of the 1531 edition (note 111).

180. I have inspected a rare copy of the 1461 edition of the *Da Ming yitong zhi* in the C. V. Starr East Asian Library, Columbia University, New York.

181. Examples of later album maps in the *Sŭngnam* style are illustrated in Yi, *Han'guk ko chido*, 63, 76, and 80 (note 4).

worked under some limitations. Foremost among the latter was the woodblock mode of reproduction. Woodblocks for printing cannot be of unlimited size, and given the inherent properties of grained wood, there is an encouragement to the woodcarver to straighten out the meanderings of rivers or the intricacies of a coastline. Beyond this, it was a cultural habit among woodblock technicians to fill up the block. Given these conventions and the size of the blocks used for the *Sŭngnam* (21.3 × 34.1 cm),[182] the Korean peninsula ended up vertically compressed and horizontally broadened.

But this defect has to be seen in relation to what the compilers were trying to achieve in these maps, which were intended as a sketch reference for the detailed and abundant text. The general map, the "P'alto ch'ongdo," was amply supplemented by eight provincial maps placed in the appropriate sections of the book. Because Sŏngjong intended to give the *Sŭngnam* a broad distribution,[183] militarily useful information had to be strictly absent. This is probably the reason for the undeveloped treatment of the northern borders on both the general map and the maps of the relevant provinces.

The items chosen for inclusion on the "P'alto ch'ongdo" were limited but carefully chosen. They included the obvious features, such as the capital, the eight provinces (marked by cartouches), and principal islands (including the important neighboring Japanese island of Tsushima). Every other name on the map represents a mountain or river, and all but one of these were the object of sacrifices conducted by government officials. (The single exception is Mount Paektu, which in spite of its major importance for Korea was at that time beyond the line of Korean administration.) Five major rivers designated as *tok* (drains) and five prominent mountains classed as *ak* (marchmounts) were officially honored with class 2 sacrifices (*chungsa*), while another group of *myŏngsan* (famous mountains) and *taech'ŏn* (great streams) were celebrated with class 3 sacrifices (*sosa*). The categories and the times for the sacrifices are codified in the *Kyŏngguk taejŏn*.[184] I have already noted the emphasis Yang Sŏngji accorded to geographical features of special religious importance.[185] The reflection of this concern on the *Sŭngnam*'s otherwise sparsely filled general map serves to warn us that in evaluating such maps we must always consider the cultural priorities of the mapmakers. A sixteenth-century reader opening the *Sŭngnam* and looking at this map heading the first volume might well have thought, "This is our country, and these are the mountains and rivers that protect it and make it great." That the map was not magnificent in every cartographic respect was less important to that reader than it might be to modern Korean critics or the world's cartographic scholars generally.

It is also useful to consider popular ideas about the national outline when evaluating the overall shape of the peninsula on this map. Modern Koreans, with modern maps in mind, see the peninsula as shaped like a rabbit. It is imagined sitting up and facing to the left (west), its hind feet and rump on the southern coast, its back along the east coast, its front paws jutting out into the West (Yellow) Sea, its head and nose outlined by the northwestern corner at the Yalu River, and its ears reaching up to the Tumen River bend in the northeast. A Korean woman of education and wit, shown the *Sŭngnam*'s general map, expressed to me her astonishment that her ancestors had no better conception of a rabbit. Doubtless the rabbit sat up only in the time of Chŏng Sanggi, to whom we shall shortly turn.

But it is worthwhile to ask what was the popular conception of the peninsula's shape during the first century of the Chosŏn dynasty. I have found no literature on this subject, but we do know something of the popular idea of the peninsula during the Koryŏ period. I have already noted (above, p. 240) that from the twelfth to the fourteenth century the government minted a special unit of money in silver called the *ŭnbyŏng* (silver vase), whose shape "resembled the territorial outline of this country." The image of a vase was reflected in the popular name for this specie, which was *hwalgu* (broadmouth).[186] Although none of these coins seem to have survived, one may imagine that the main body of the peninsula was the body of the vase, the narrowest width (at about 39°20′N) was its neck, and the flatly flared northeast and northwest corners (in those days before the northern expansion) were its broad lips. In pondering the problematic northern border on the *Sŭngnam*'s general map, with the northeast and northwest corners extending (or drooping) out to the left and right, perhaps we can iden-

182. This is the measurement Yi Ch'an gives (*Han'guk ko chido*, 61 [note 4]) for the woodcut "P'alto ch'ongdo" in the rare edition of the *Sŭngnam* in the National Central Library in Seoul (Kwi 228, 60-3).

183. "King Sŏngjong . . . ordered his Confucianists to make revisions in accordance with the model of the *Da Ming yitong zhi* and to print it for distribution throughout the country, [so that] from the secret archives above, all the way down to private collections below, there would be none that lacked it." From Im Sahong's 1499 colophon to the *Sŭngnam*, reprinted at the end of the 1531 edition (note 111).

184. Pang, *Hanguk ŭi chido*, 97–100 (note 5). See also *Kyŏngguk taejŏn*, 3.34a–b (note 142); and Han et al., *Yŏkchu Kyŏngguk taejŏn: Chusŏk pyon*, 412–13 and 417 (note 146).

185. Yang Sŏngji's memorial listing mountains and rivers meriting national sacrifices (*Sŏngjong sillok*, 3.25b–26a [note 17]) presents quite a different group of names from those represented in the sources in note 170 and on the "P'alto ch'ongdo." In fact, he cited many mountains and rivers now found on the latter for abolition of sacrifices. If there was a political battle over this, it was one that he lost. Perhaps this is one of the reasons he is not listed as one of the compilers of the *Sŭngnam* even though his research and drafts must have been of fundamental importance in its compilation (see note 111).

186. *Koryŏ sa*, 33.lla (note 9).

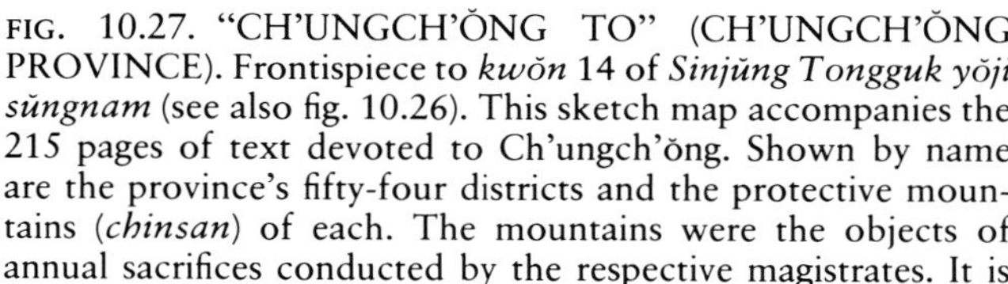
FIG. 10.27. "CH'UNGCH'ŎNG TO" (CH'UNGCH'ŎNG PROVINCE). Frontispiece to *kwŏn* 14 of *Sinjŭng Tongguk yŏji sŭngnam* (see also fig. 10.26). This sketch map accompanies the 215 pages of text devoted to Ch'ungch'ŏng. Shown by name are the province's fifty-four districts and the protective mountains (*chinsan*) of each. The mountains were the objects of annual sacrifices conducted by the respective magistrates. It is interesting to compare this map with the map of Shandong Province—on the China coast just across the Yellow Sea from Ch'ungch'ŏng—in the comparable Ming dynasty geographical reference work, figure 10.25.
Size of the original: 27.3 × 17.7 cm. By permission of Kyujanggak Archives, Seoul National University.

tify the lip at the top of the vase. This is admittedly speculative. But I suspect that if the *Sŭngnam*'s map had been significantly out of accord with some popular conception of the peninsula's shape, whatever it was, it would not have remained in vogue for two more centuries.

Although there are many references to separate maps of provinces in the fifteenth-century literature, the eight that are attached to the relevant sections of the *Sŭngnam* are the earliest that remain available today. The map of Ch'ungch'ŏng Province (on the southwest coast of the peninsula) is shown in figure 10.27. Although in essence they do not go beyond an application of the cartographic style of the *Da Ming yitong zhi*, they are demonstrably superior to the latter, as a scrutiny of the Shandong map (fig. 10.25 above) will make clear.[187] Instead of just major districts, the *Sŭngnam*'s maps indicate every local magistracy, together with its *chinsan* (protective mountain). Each district town had such a mountain, to which the local magistrate made regular sacrifices. Again we see religious factors playing an important role.

Although the clustered mountain symbols add a decorative and interesting element to the map, they do not show the geomantic and arterial connections that have been a major element of Korean maps up to this point. Omission of the arterial network is a major departure in the history of Korean cartography, though not a permanent one. Maps in the *Sŭngnam* style characteristically avoid any demonstration of this tenacious Korean con-

187. Frontispiece to *juan* 22 of *Da Ming yitong zhi* (note 180).

TABLE 10.6 Literary References to Korean Maps between 1484 and 1756

Category	1484–1550	1551–1600	1601–50	1651–1700	1701–56	Total
China, *ch'ŏnha*[a]	3	1	0	0	0	4
National	1	0	0	0	0	1
Provincial	2	0	0	0	0	2
District	5	6	0	1	0	12
Northern frontier	4	0	0	0	6	10
Other defense[b]	3	0	0	6	3	12
Islands	1	0	0	0	3	4
Shapes-and-forces[c]	2	0	0	0	0	2
Cultural[d]	3	0	0	0	1	4
Total	24	7	0	7	13	51

Source: Pang Tong'in, *Han'guk ŭi chido* (Seoul: Sejong Taewang Kinyŏm Saŏphoe, 1976), 191–93. The content and chronological breakdown are mine. Pang's list has forty-two entries with references to fifty-one maps. Military censorship, cultural disruption, and loss of records were heavy during the years 1592–1637 because of the Japanese and Manchu invasions, probably accounting in part for low numbers during this period.

[a]Two of the four items are *ch'ŏnha* maps. In the sixteenth century, *ch'ŏnha* often refers to Ming China, but world maps of the *ch'ŏnhado* type are not to be excluded.

[b]This category includes defense maps of several towns on route from north to Seoul, the important royal refuge fortresses at Namhan and on Kanghwa Island, and several maps bearing on provincial military organization. Two maps of the shapes-and-forces type that relate to key military areas are counted here.

[c]See note b.

[d]This category includes maps of a palace layout, a monastery, a royal tomb, and a scenic mountain.

cern, and to a notable extent the aversion is also reflected on the later Chŏng Ch'ŏk–style maps. Shapes-and-forces thought was such a conspicuous feature of Korean cartography for most of the fifteenth century that this sudden turning away toward the end of that period must have been the result of a conscious policy, or perhaps reflected an intellectual trend among the Korean upper class. Neo-Confucianism, temperamentally wary of the political ramifications of such popular beliefs, had undergone an intense development among the ruling elite in the first hundred years of the Chosŏn dynasty. It probably served to inhibit geomantic theorizing, not so much (not at all, really) the personal level of auspicious siting for dwellings and graves as shapes-and-forces analyses on the national level. But such a deeply running current was not likely to be entirely eliminated in Korea. It strongly reemerged in many of the Chŏng Sanggi–style maps, as it did in the geographical thought of Yi Chunghwan's *T'aengni chi*, which was contemporaneous with them. With Kim Chŏngho in the nineteenth century it was to rise to an epochal, and final, level of cartographic expression. On the other hand, to the very end no maps in the *Sŭngnam* style have any taint of it. It is a curious but evident dichotomy.

The *Sŭngnam*'s maps, intended on the one hand as symbolic, almost iconographic statements of official georeligious relationships and on the other as sketch references to a rich and detailed geographic text, served the purposes of its editors perfectly well. But their influence on national and provincial maps, especially through the albums that began to be popular in the sixteenth century, overwhelmed development that might have been expected in the Chŏng Ch'ŏk style. The great promise evident in the Naikaku map went unrealized in the national maps that followed, up until the eighteenth century. Security concerns are almost certainly responsible for this situation. The Japanese invasions of the 1590s and a long-term series of Manchu pressures (including invasions in 1627 and 1636–37) that continued throughout the seventeenth century combined to suffuse the Korean world with an atmosphere of international wariness. In such times the *Sŭngnam*-style maps, which satisfied general needs but withheld or distorted the details, dominated general circulation, while most other maps were probably restricted to limited circulation within the government and the governing class.

It does not, however, follow from such circumstances that cartographic development was interrupted across the board. Military security concerns were probably also a stimulant to cartographic activity. If it was necessary to keep accurate maps out of the hands of potential or actual enemies, it was just as important to provide them for officials who bore the responsibility for national defense. The number of surviving maps solidly connected to the years between the early editions of the *Sŭngnam* and the emergence of the Chŏng Sanggi maps is not large, but Pang Tong'in has compiled an interesting list of maps that are mentioned in written sources (mostly *sillok* [royal annals] of the various reigns).[188] Pang's list is summed up in table 10.6. It should not be taken as representative of

188. Pang, *Han'guk ŭi chido*, 191–93 (note 5).

all Korean cartographic activity, but it does mention items that came to the attention of the throne and therefore speaks for many other maps now unknown or unrecognized. The preponderance of defense and local maps, together with the low profile of national and provincial maps, probably is a decent representation of what government cartographers were doing during this long period.

INTERLUDE: KOREA, THE JESUITS, AND MAP DIPLOMACY

During the seventeenth century three developments helped transform Korean culture and nurture new trends in cartography. The first of these was the Manchus' overthrow of the Ming dynasty (1644) and their eventual conquest of all of China by 1684. The fall of the Ming, plus the earlier and continued rough handling of Korea by the Manchus, led to abrasive changes in the tributary relationship and to bitter anti-Manchu hostility, and it indirectly encouraged the formulation of a more independent Korean self-consciousness. The result was a burst of cultural creativity on a score of fronts, and a new focus on the defense of the northern frontier and the still unincorporated cis-Tumen territory in the vicinity of Mount Paektu. These developments changed the maps.

Second, new trends in scholarship encouraged a fresh interest in science and pragmatic research. In this movement, which later received the name *sirhak* (practical learning), geography among other disciplines came to be newly fashionable among many scholars.[189] This considerably broadened concern for the accuracy of maps.

Finally, as already described, the principal monuments of early Sino-Jesuit cartography continued to find their way into Korea. Apart from the eighteenth-century globe mounted in the armillary clock—an exotic sport within the history of Korean cartography—Western cartographic styles were not imitated in Korea. But indirectly they provided models of precision and realism and suggested the advantages of a firm geodetic foundation. Although Sejong had flirted with polar altitude determination from different parts of the country, we have no indication of any success in this project or of any cartographic application. But during the early eighteenth century, Korean specialists became acquainted with Western observational and mensural instruments and soon acquired their own. Thus the potential was created for a leap ahead in the quality of cartography.

The rise of the Manchus resulted at first in a favorable development for Korea. About the year 1600, the Jurchens (who only in 1636 began calling themselves Manchus) of the Paektu-Tumen region largely abandoned the Korean frontier area to join in the campaigns of their new paramount leader Nurhaci, leaving their old haunts more open to Korean penetration, which soon began. To meet the growing challenge posed by Russian expansion, the Manchus reorganized the defenses of the Manchurian homeland in the 1650s and 1660s, turning the present Jilin area into an important strategic territory. A Manchu survey team inspected the region of Changbaishan (the Chinese name for Paektusan) in the summer of 1677. By 1679, responding to the Kangxi emperor's expressed curiosity over the hydrography of the southern slope of this great mountain, the Manchus had made or acquired maps of the territory, including virtually the whole of the Korean side of the border from one side of the peninsula to the other. In that year they visited the Korean commander in the north and requested information on "present installations, maps, and 'floating iron' [compass] bearings in the area of Changbaishan and Paektusan." Obligingly, they allowed him to copy their own map. Quickly communicated to Seoul, the news and the map aroused predictable consternation.[190]

Korean officials were shocked and alarmed when they saw the detail of the Manchu maps and realized the full extent of the Kangxi emperor's interest in the Paektusan area. Doubtless this incident stimulated a stronger Korean presence on the southern bank of the upper Tumen, which a few officials had been advocating for some time.[191] The important new district town of Musan was established there in 1684.[192] In 1685 a Manchu survey team, on its own side of the Yalu gathering materials for what would eventually become the standard administrative geography of the Qing dynasty (the *Da Qing yitong zhi* [Comprehensive gazetteer of the Great Qing realm, completed in 1746), was fired upon by Korean hunters. This incident, settled in 1694, led to King Sukchong's being made to pay a hefty indemnity and meet Qing demands for stricter controls on Korean frontier dwellers. In 1699 Korean envoys were ordered by Manchu authorities to execute a map of "Korea's eight provinces,"

189. This should not be taken to mean, as many writers claim, that in devoting themselves to "progressive and modern" studies scholars rejected the supposedly old-fashioned views of "conservative" Confucianism. On the contrary, new insights often led them to a revitalized Confucianism, as in the case of Chŏng Yagyong (1762–1836), whom many consider among the greatest of the *sirhak* scholars.

190. *Sukchong sillok*, 8.56a–b (note 17). For further sources and discussion, see the study by Zhang Cunwu (Chang Ts'un-wu), "Qingdai Zhong-Han bianwu wenti tanyuan" (An inquiry into the Sino-Korean border question during the Qing dynasty), *Zhongyang Yanjiuyuan Jindaishi Yanjiusuo Jikan* 2 (1971): 463–503, esp. 473–75.

191. A memorial advocating such, submitted in 1674 by Hamgyŏng governor Nam Kuman, is quoted in the *Munhŏn pigo*, 18.23a–24b (note 51). See also *Hyŏnjong sillok* (Annals of King Hyŏnjong, r. 1659–74), 21.54a–b (note 17).

192. *Munhŏn pigo*, 18.20a–b (note 51).

with route and distance data.[193] The canny diplomats were able to avoid compliance on that occasion. But if Qing suzerainty meant anything, it was the right to have a map of a tributary territory. Given the determination of the Kangxi emperor to have this map, such pressures could only grow.

Kangxi's project to map his empire took on new energy when the Jesuits joined the effort in 1709. Before the year was over they had mapped Manchuria and the borders of Korea, and in the second half of 1710 they returned to map the Amur region. By 1716, after many interruptions caused by illness, death, and reorganization, they and their Chinese and Manchu assistants had completed the mapping of the entire Chinese empire plus Tibet and Korea. Printed in Chinese editions of 1717 and 1719, and in a definitive revised version in 1721, these maps, in sumptuous versions edited by the cartographer Jean Baptiste Bourguignon d'Anville, were published in Paris in 1735 in Du Halde's famous description of China.[194] The Jesuit draft of the map of Korea and d'Anville's version of it are shown in figures 7.7 (above) and 10.28.

The Jesuits themselves had not been permitted into Korea by the Kangxi emperor. Tributary protocol forbade it, and Kangxi, who had an astute understanding of Korean sensitivities, knew they would never have been allowed in anyway. Therefore the map of Korea had to be elaborated from one that Kangxi's agent—the "Tartar lord" of Western accounts—procured in Korea. An explanation of the Jesuit map of Korea by Father Jean-Baptiste Régis (1664–1738), who with Fathers Pierre Jartoux (1669–1720) and Ehrenberg Xavier Fridelli (1673–1743) had done the Manchurian and Korean frontier work in 1709 and 1710, is given by Du Halde and may be summed up as follows.

The "Tartar lord" was accompanied by a Chinese mathematician and a surveying team trained by the Jesuits. A Jesuit (possibly Régis, but he says only "nous") went with them to Fenghuang, the traditional Chinese gateway to Korea, and stayed there until the mission was completed. The "lord" and his team made observations and took rope measurements from Fenghuang to the Korean entry point at Ŭiju on the Yalu River, and from there to Seoul. The mathematician measured the latitude at Seoul at 37°38′20″N. This figure, by comparison with the measured latitude of Fenghuang, "assured us of the length [of Korea] from the north to its center" and provided a basis for generating equivalents for Korean-supplied distance measurements and therefore for deducing the distance from Seoul to the southern coast. While in Korea the team was under constant surveillance, but the "Tartar lord" was given a copy of a map of Korea held in the royal palace. Editing this map in accordance with the collected data and his earlier measurements on the Manchu side of the frontier, Régis produced the map that came out in the Kangxi atlas and that, in d'Anville's version, was eventually published in Paris.[195]

D'Anville himself had a moderately high estimation of the Korean map and considered it no great disadvantage that the Jesuits had been unable to make their own direct survey. "On the contrary," he wrote, "if any map should pass for correct, it ought to be this one, since it was drawn originally by Korean geographers on the very orders of the king, and the original was kept in his palace. It is from this original that the map given here was derived. It is likely that the missionaries, in examining and determining the frontiers of the kingdom in the north, found no notable differences between their own observations and the borders marked on this map, since [if they had] they would not have failed to mention it. This circumstance alone seems to answer for its precision."[196]

This is a generous appraisal of Korean cartography, particularly as it comes from one as eminent as d'Anville, and it naturally arouses curiosity about this Korean prototype. Because of the relatively accurate northern frontier on the Jesuit map, and because the Chŏng Sanggi-style maps are the only Korean ones of that era to have this feature, scholars have wanted to make a Chŏng-style map the candidate for this role. But as all have acknowledged, since these maps are not known to have surfaced in Korea until long after the d'Anville map had been published in Paris (and at even a greater interval from the actual Jesuit work in the 1710s), this theory cannot stand.[197] D'Anville himself implied that the frontier must

193. Zhang, "Qingdai Zhong-Han bianwu wenti tanyuan," 474 (note 190).

194. See Theodore N. Foss, "A Western Interpretation of China: Jesuit Cartography," in *East Meets West: The Jesuits in China, 1582–1773*, ed. Charles E. Ronan and Bonnie B. C. Oh (Chicago: Loyola University Press, 1988), 209–51, esp. 224–40.

195. Jean Baptiste Du Halde (1674–1743), *Description géographique, historique, chronologique, politique, et physique de l'empire de la Chine et de la Tartarie chinoise*, 4 vols. (Paris: P. G. Lemercier, 1735), 4:424–25. Other brief but interesting notes on the adventures of the "map embassy" in Korea are given by Father Matteo Ripa (1682–1745). He had some misunderstandings about Korea, but his reports of constant surveillance and note taking by Korean officials can easily be corroborated in Korean sources, as we shall see. He identifies the person who took the measurements—an acquaintance of his—as "the ambassador," but the mathematician was assuredly not the ambassador. See Matteo Ripa, *Memoirs of Father Ripa, during Thirteen Years' Residence at the Court of Peking in the Service of the Emperor of China*, selected and translated from Italian by Fortunato Prandi (New York: Wiley and Putnam, 1846), 77.

196. Jean Baptiste Bourguignon d'Anville, *Nouvel atlas de la Chine, de la Tartarie chinoise et du Thibet* (The Hague: H. Scheurleer, 1737), introductory note (my translation).

197. See Shannon McCune, "Some Korean Maps," *Transactions of the Korean Branch of the Royal Asiatic Society* 50 (1975): 70–102, esp.

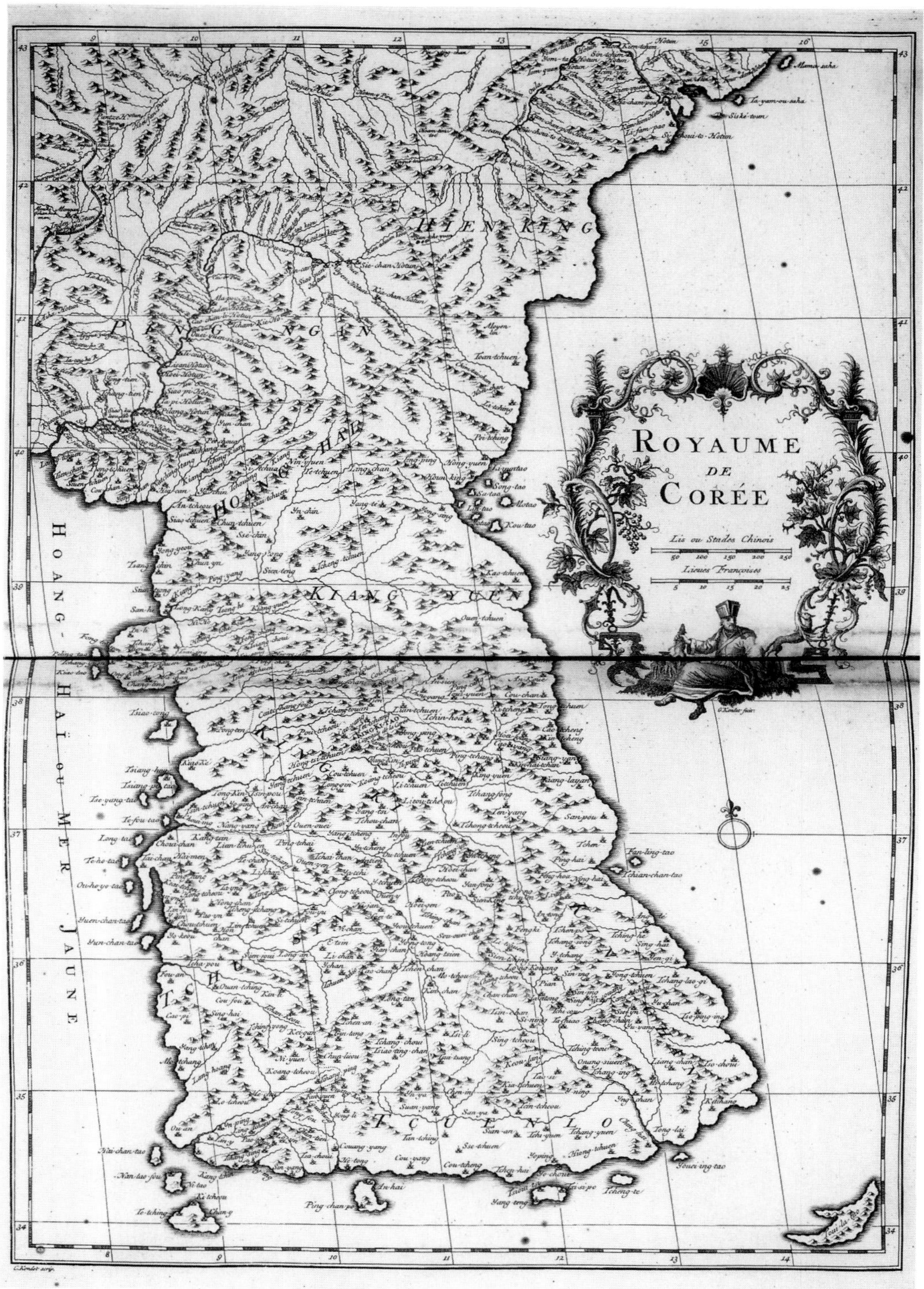

FIG. 10.28. *ROYAUME DE CORÉE.* This engraving of the Jesuit map as edited by d'Anville was first published in Du Halde's *Description . . . de la Chine et de la Tartarie chinoise* in 1735, and again in d'Anville's own *Nouvel atlas de la Chine* in 1737. Compare the earlier Jesuit version done in Beijing, figure 7.7 above. D'Anville's editing consisted essentially of changing the Jesuits' double lines for rivers to a single line, designing more attractive symbols, and adding shading on the coasts. This image of Korea prevailed in Western atlases until the mid-nineteenth century.

Size of the original: 53 × 37 cm. Courtesy of the Ayer Collection, the Newberry Library, Chicago (Ayer 135 A6 1737, map no. 31).

have been more or less the same on both maps or else the Jesuits would have said something. Actually, both views are mistaken: d'Anville was just speculating without real knowledge of Korean cartography and in particular of the frontier question, and the modern scholars failed to take into account that the Jesuits had their own data concerning the northern frontier and would not have had to rely on a Korean map for this feature.

Nevertheless, the problem of a Korean prototype is important, and moreover is one that it is possible to investigate in Korean historical records. Although we will not be able to make an absolute identification of the Korean map by title, we can suggest the type of map involved and a possible candidate. At the same time, we can throw some light on the circumstances in which it was obtained.

The "Tartar lord" was Mukedeng, called the commander of Daxing Ula in Chinese sources and *Hulie zongguan* (commander of the hunt) in Korean sources. More important than his titles was that he was a trusted assistant and personal troubleshooter for the Kangxi emperor. He did not acquire the Korean map until 1713, but he had been involved in Korean affairs for the emperor since 1710. In that year a Korean ginseng smuggler murdered some Chinese merchants on the Manchu side of the Yalu, and Mukedeng convened a court of regional officials in Fenghuang to try the case. Mukedeng demanded that, to make an example of him before the Korean frontier population, the culprit be executed publicly in his Korean hometown, which was Wiwŏn on the middle course of the Yalu. Kangxi, using this affair to pry Korea's tight lid open a bit, instructed Mukedeng himself to oversee the execution, and while in Korea, to follow the Yalu to its headwaters on Mount Paektu, find the source of the Tumen, and then return to report his results. In 1711 the execution was duly carried out. The Korean authorities, arguing that there was no precedent for his proposed oversight of the execution, and protesting that they were quite capable of witnessing it themselves, refused to admit Mukedeng at Ŭiju. But an agent of the Kangxi emperor was not to be deterred by such tactics. Mukedeng went up the river on the Manchu side, crossed over to Wiwŏn on the appointed day, presided over the beheading, and then announced that he was going to Paektusan. Korean officials had no choice but to accompany him—or try to, because this hearty woodsman and warrior proceeded to set an exhausting pace. But the trackless wastes, the heavy rains, and the rampaging Yalu were more than this ill-planned expedition could overcome, and even Mukedeng was forced to give up. He announced that he would return the following year.

In 1712 a better-organized effort with more experienced Korean escorting officials succeeded in reaching the goal. They viewed the magnificent Pond of Heaven in Mount Paektu's ancient crater, identified a source for the Tumen River on the eastern slope of the mountain, and erected a stele with an inscription to mark the border between Korean and Manchu territory for the first time. Not long afterward, an anonymous Korean mapmaker executed a splendid depiction of the area (fig. 10.29).

But this was not enough for Kangxi. In 1713, Mukedeng was sent to Korea once more. To keep Koreans off balance, it was announced that he was bringing an imperial edict bestowing largess upon Korea for the "ascendant peace in the empire" (*tianxia shengping*). This put the matter in a special diplomatic category that compelled reception of the embassy. But no sooner had the mission entered Korea than surveying teams started running up and down the roads with their measuring lines while mathematicians took polar altitude measurements and "floated the iron" to ascertain compass bearings. Long before the mission reached Seoul, informal word had already come that Mukedeng was demanding geographic and cartographic information. When he actually arrived (19 July 1713), he communicated his needs directly to the king.[198]

Mukedeng seems to have made three specific requests. He wanted, first, maps and other information concerning hydrography south of Mount Paektu; second, a complete set of distance data for Korean towns; and third, a general map of the country. The negotiations got off to a bad start. Mukedeng said: "We have never been certain about the river veins and mountain arteries south of Mount Paektu, and we would like to see maps of your honorable country. This is the emperor's command." The king consulted with his advisers on the spot and then replied: "This region is wild and cut off, and we have never mapped it."[199] Such an implausible and ill-advised denial of the existence of maps probably flew in the face of Mukedeng's direct knowledge, and in any case it was embarrassing. For as one of his aides later exclaimed, "How is it conceivable to have a country but not have maps?"[200]

94–102; Pang, *Han'guk ŭi chido*, 217–19 (note 5); Kim, *Maesan kukhak san'go*, 276–80 (note 3). McCune, misled by Régis's failure to distinguish between activities of 1710 and 1713, erroneously puts the appearance of the "Tartar lord" in Korea in 1710.

198. All this activity is summed up, with abundant citation of Korean source materials, in Zhang, "Qingdai Zhong-Han bianwu wenti tanyuan," 475–484 (note 190). For the Korean version of Mukedeng's title, see *Sukchong sillok*, 53.39a (note 17).

199. *Sukchong sillok*, 54.4b (note 17).

200. *Sŭngjŏngwŏn ilgi* (Daily record of the royal secretariat) (transcription and reprint, Seoul: National History Compilation Committee, 1961–), 25:963bb–64aa (39/6/1 = 22 July 1713). Since the original archives were not paginated, page numbers here follow the modern edition. The codes aa and ab identify the top two quarters of the page; likewise ba and bb for the bottom two. Date (Korean reign year/month/lunar day) speeds reference; Western equivalents for lunar dates are given. In the classical rhetoric of Chinese statecraft, to "have the country" (*you guo*) means to rule it.

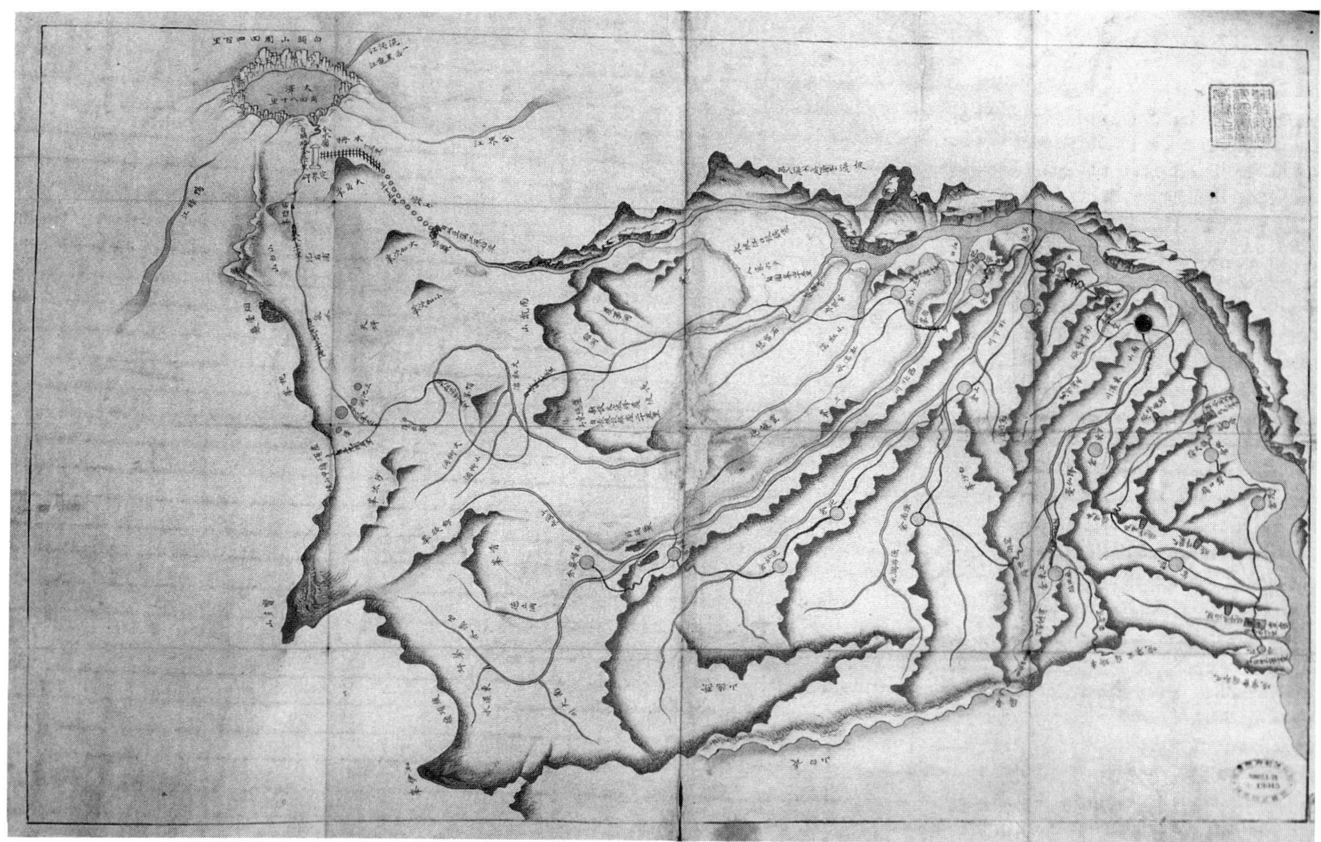

FIG. 10.29. *PUKKWAN CHANGP'A CHIDO* (MAP OF THE CHANGP'A, THE NORTHERN FRONTIER). This map, probably dating from the mid-eighteenth century, shows the region of Mount Paektu and the boundary settled in the border agreement of 1712. Mount Changp'a (Long Slope), left center, which dominated the sources of the Tumen, was the center of a defensive signal-tower network under the command of military officials in Musan (dark circle, right). At the upper left are the Pond of Heaven (Ch'ŏnji) and the stele erected in 1712 on the crest of the watershed between east and west. A wooden fence leads fifteen *li* to a line of cairns, which extends for thirty-five *li* to the place where "the Tumen River starts to flow." The present border, negotiated in 1887 and again in 1909, is to the north of this line.
Size of the original: 71.5 × 109.7 cm. By permission of the National Central Library, Seoul (cat. no. Kojo 61-59).

In the days that followed a response was debated while aides worked to explain away the king's gaffe, assuring the visitors that he had meant to say only there were no *copies* of maps. In the meantime, Mukedeng produced his own maps to emphasize that he knew what he was talking about and could not be deceived. Korean officials now anxiously began looking for a map that would satisfy Mukedeng without giving away too much. A resourceful minister informed the king that "although the nation's defense maps are too detailed, we have just acquired a map that is neither too precise nor too general, with many mistakes on the Paektu rivers. Let us show him this." He proposed comparing it with Mukedeng's own map to show him *his* errors.[201] As far as can be judged from the available sources, this is the way the Paektu river issue was handled. Mukedeng had this map copied; his cartographers redrafted their own map based on it; and a copy was then made and given to the Koreans. "We are taking one copy with us, and leaving one with you, so that our name and fame can spread to this place."[202] Whatever else it achieved, this must at least have put to an end Kangxi's false puzzle of Paektu's southern drainage. Essentially there was none. All waters move or quickly turn to the east or west. It is the mountain "artery" that moves south, not the water "veins."

The question of distance data was no less sensitive, though the Koreans might have given a sigh of relief that the Manchus apparently had not gained possession of a copy of the *Sŭngnam*, with its exhaustive data for every district in the country. On this question too they pleaded that they did not have detailed records, and this time they

201. *Sukchong sillok*, 54.5a (note 17).
202. *Sŭngjŏngwŏn ilgi*, 25:967bb (39/6/6 = 27 July 1713) (note 200).

appear to have convinced Mukedeng. They did give him distances from Seoul to the various extremities of the country.[203]

As for the general map of Korea that went north with Mukedeng, we know only that he requested a *badao ditu* (a map [or *the* map?] of the Eight Provinces, Korean *p'alto chido*). It is impossible to know whether this involved a precise title or a generic term, and in any case it is not certain whether the words are his or belong to Chief Minister Yi Yu (1645–1721), who reported the request.[204] But that Yi Yu used the term *p'alto* suggests the map he gave to Mukedeng had that term in its title.

Of course, the phrase "Eight Provinces" is extremely common in map titles. But if it is to have any precise reference, the most likely would be to the general map from the *Sŭngnam*—the "P'alto ch'ongdo"—or an elaboration of it that combined the sketch maps of each of the eight provinces into a single map. Such a *Sŭngnam*-style map, with this title, dating to the first half of the seventeenth century, is owned by the Korean collector Che Honggyu (fig. 10.30).[205] This map could have provided the range and also the *im*precise information found on the Jesuit map, which has the names of many individual features but often gives them erroneous placement. As already emphasized, it would be no problem at all if the Korean prototype had had flat northern frontiers, because the Jesuits already had their own excellent data for this feature. But interestingly enough, this copy is unique among *Sŭngnam*-style maps in having a modest degree of northeastward lift to its frontier.

The details on the Jesuit map of 1717 and on d'Anville's edition of it are particularly poor for the south, which is understandable since Chinese and Manchu diplomats and officials were never permitted to observe this large region, where by far the greater part of Korea's population and wealth was to be found. Given the general Korean inclination to deceive the Chinese and other foreigners about their country (and the map episode only represents a pervasive habit, nourished by foreign invasions over the centuries), it is not beyond possibility that the map given to Mukedeng could even have been doctored in spots. Particularly interesting is the lay of the capital, the Han River, and the strategic offshore islands of Kanghwa and Kyodong. The original Jesuit map (fig. 7.7) shows Seoul ("Chaoxian," or Chosŏn) too far south and in the center rather than on the west side of the peninsula. The Han River flows from Seoul to the Yellow Sea in a west-southwest direction rather than following its true northwesterly course, and Kanghwa and Kyodong are egregiously misplaced, the latter even misnamed. This must have caught someone's eye in Beijing, because the d'Anville map shows some correction for the islands (including the names), although the errors affecting Seoul and the Han remain. This distortion, if such it is, is significant, since Kanghwa was the site of a highly fortified royal refuge for use in emergencies. Its location, direction, and distance from the capital would have been considered critical information in the hands of an invader, and a motive to distort can be reasonably suspected.

The general shape of the Jesuit map's southern coast is not all that different from that of the map owned by Che Honggyu. The big difference is that the Jesuit map shows the southeastern corner of the peninsula curiously stretched to the right, although this feature is less exaggerated on the d'Anville version. Yet the longitude it indicates for this part of the coastline, approximately 13°50′ east of Beijing (about 129°30′E Greenwich), is about the same as the longitude of the farthest north bend of the Tumen River, for which the Jesuits would have had their own measurements. This alignment is virtually the same as that on modern maps. The Jesuits, who used Dutch data for their map of Japan, probably had longitude data on the southeastern Korean coast from the same sources. Understanding that the southeastern corner had to line up longitudinally with the north bend of the Tumen, they may have "pulled" it to the right. Apart from this freakish pull, the outline of the southern coast and the positions of Cheju and Tsushima islands (indeed, even the shape of Tsushima) are similar to those of the map owned by Che.[206]

In suggesting this provisional hypothesis on the Korean prototype, I must concede some problems. The Bay of Wŏnsan on the upper east coast, although exaggerated on the Che map, is in the direction of accuracy, but there is only a faint hint of it on the Jesuit map. This might suggest a *Sŭngnam*-style map of slightly different proportions as a better candidate. From the Korean point of view, a *Sŭngnam*-style map of some variety would probably have been the ideal answer to what to give Mukedeng. From the beginning, the likely intent of its design had been to conceal, not reveal, specific details of the country's topography, and that was their purpose too.

Although the Koreans' instinct was to resist requests for maps and map data, they had their own requests to make of Mukedong's survey team, and they tried hard

203. *Sŭngjŏngwŏn ilgi*, 25:965aa (39/6/2 = 23 July 1713) (note 200); *Sukchong sillok*, 54.5a–b (note 17).

204. *Sukchong sillok*, 54.8a (note 17).

205. Yi, *Han'guk ko chido*, 62 (note 4). Based on place-names and paper quality, Yi dates it to the reigns of Kwanghae (1608–23) and Injong (1623–49).

206. An alternative but less likely way to explain the eastward pull might be to suggest influence from the map of Korea in Luo's *Guang yutu* (see fig. 10.2 and note 16), which has a similar distortion (among many others). This map is also like the Jesuit draft in calling Korea's capital "Chosŏn" and placing it in the center rather than on the west side of the peninsula. The Jesuits are known to have referred to Luo Hongxian's work. In the case of the China maps this was justified, but if they relied on his map of Chosŏn, they were ill advised.

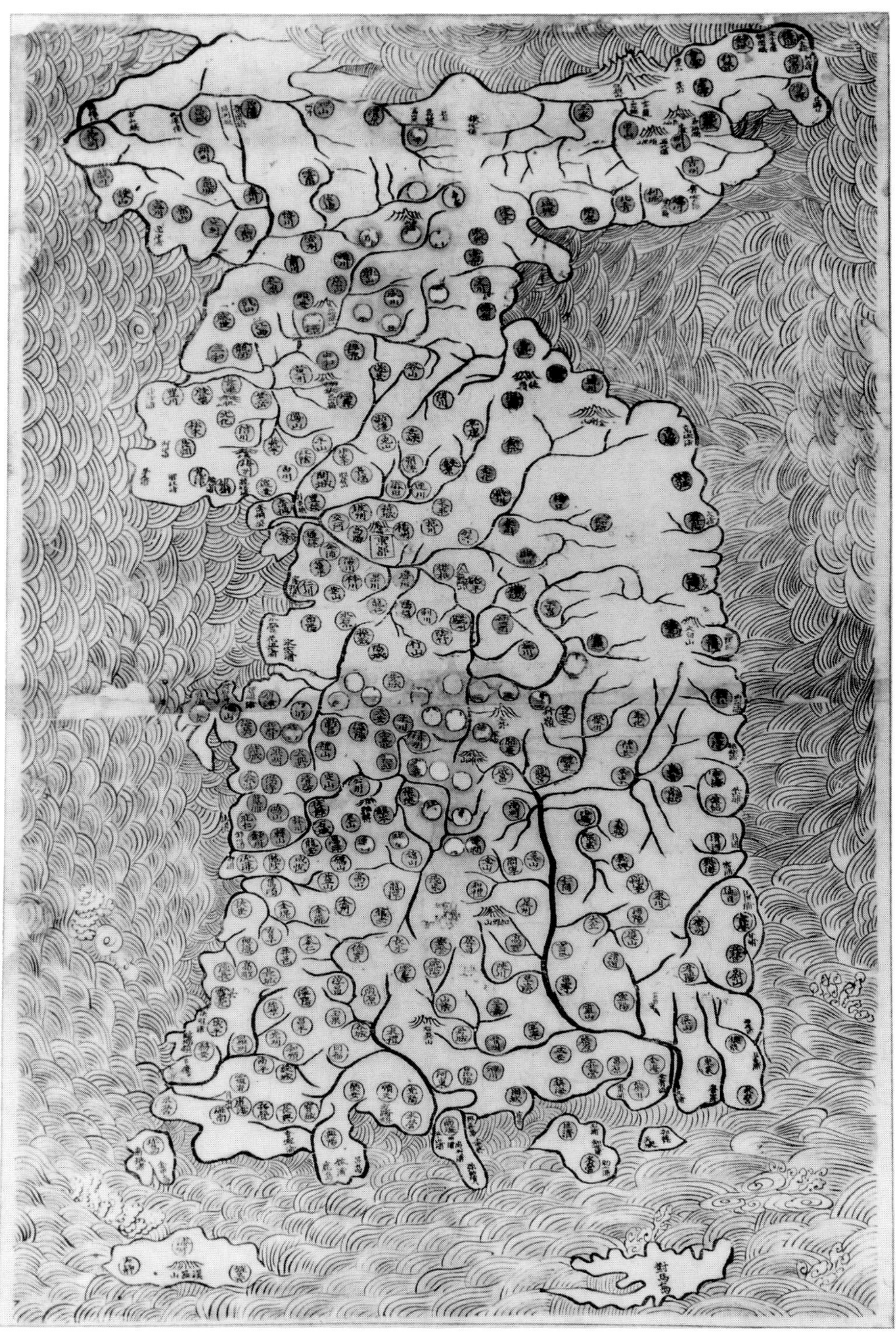

FIG. 10.30. *P'ALTO CH'ONGDO* (GENERAL MAP OF THE EIGHT PROVINCES), FIRST HALF OF THE SEVENTEENTH CENTURY. This woodblock printed map of Korea, dated from place-name indications and paper analysis to the first half of the seventeenth century, is essentially a version of the general map of the same title in the *Tongguk yŏji sŭngnam* with the place-name details added from the *Sŭngnam*'s provincial maps. But it is significantly different in drawing style and in the treatment of the northern frontier, which is more in the Chŏng Ch'ŏk style. A map of such a type, although probably with a less pronounced Bay of Wŏnsan, was detailed enough, and at the same time imprecise enough, to have served as a model for the Jesuit map, which may also have made reference to non-Korean sources.
Size of the original: 106 × 68 cm. By permission of Yi Ch'an, Seoul.

to learn as much as they could about its doings. The king's advisers missed no opportunity to acquire technical information concerning the mission's survey instruments and methods. It was decided to have some young technicians from the *kwansangdae* (royal observatory)—posing as ordinary citizens—strike up an acquaintance with the Chinese technicians and get as much instruction as possible. Interpreters too were mobilized to socialize with the technicians and learn as much as they could.[207]

There was great curiosity about the chief of the surveying team, a Chinese named He Guozhu, who was identified as a mathematician in the Qing government's *qintianjian* (bureau of mathematics). One of Sukchong's ministers, while in Beijing on diplomatic business, had once visited the bureau ("No one there but Westerners," he said) and met a man he thought might be one of He's relatives.[208] This would probably have been He Guozong, Guozhu's older brother. Both were Jesuit trainees and rich in experience on Kangxi's mapping projects.[209] It was He Guozhu who took the much-reported measurement of Seoul's latitude. Notices on the 1713 visit make no mention of it, but it is found in the *Munhŏn pigo* and other sources.[210]

An officer of the observatory, Hŏ Wŏn, in the course of an earlier mission to Beijing, had purchased "mathematics materials, supplementary textbooks and calculating devices," but he needed instruction in order to use them properly. This was mentioned in one of the conferences on the Mukedeng problem, and Hŏ may have used this opportunity to make contact with the surveyors. In any case, he returned to Beijing two years later (1715), visited the bureau of mathematics, and made some further purchases, including "books on eclipses" and "six different kinds of observational and calculating devices."[211]

When Mukedeng finally departed from Seoul (29 July 1713), Korean authorities were doubtless relieved to be finally free of his demands. But there had almost certainly been some compensation for their pains in the information harvested from observations of the survey team and likely contacts with it. The six devices that were acquired in 1715 probably included up-to-date instruments for taking polar altitude and determining latitude. Precise information on eclipses would have been essential for planning observations to determine longitude. And the chance to observe the working methods of probably the most advanced mapmaking operation in the world at that time—Kangxi's project led by the Jesuits in China—should have brought great benefits to Korean mapmakers. Still, as we shall see, it is hard to find proof that these benefits were seized.

MAPS IN THE CHŎNG SANGGI STYLE

We are now in a position to consider the Chŏng Sanggi maps. Unlike his mapmaking predecessors Yi Hoe, Chŏng Ch'ŏk, and Yang Sŏngji, the eighteenth-century master Chŏng Sanggi (1678–1752) did not compete in the civil examinations or serve in the bureaucracy. By all accounts—which are not that extensive—he lived in retirement and devoted himself to the pursuits of a *sirhak* scholar. His writings included studies in political economy, defense, military tactics, medicine, agronomy, and machinery, but today he is famous mainly for his maps, which were the first in Korea to feature the use of scale and the first to reveal, at long last, something close to the actual shape of the country. His descendants, down to at least a great-grandson, seem to have continued his cartographic interests, but aside from his son Chŏng Hangnyŏng, who had a successful official career under King Yŏngjo (r. 1724–76), little is known of them.[212]

It was apparently not until after Chŏng Sanggi's death that his maps came to the attention of the government. During the late 1750s the government was considering whether it ought to recompile the *Sŭngnam*, and the

207. General discussions among the Korean officials on how to approach the Chinese technicians are found in *Sŭngjŏngwŏn ilgi*, 25:948aa–50aa (39/5i/15 = 7 July 1713) and 961ab–ba (39/5i/26 = 18 July 1713) (note 200). In this and later references, "5i" indicates the intercalary fifth month.

208. *Sŭngjŏngwŏn ilgi*, 25:950aa (39/5i/15 = 7 July 1713) (note 200).

209. He Guozhu was probably a younger brother or cousin of the brothers He Guozong (d. 1766) and He Guodong; all were mathematicians and close associates of the Jesuits in the Qing government's bureau of mathematics (*qintianjian*). See Arthur W. Hummel, ed., *Eminent Chinese of the Ch'ing Period, 1644–1912*, 2 vols. (Washington, D.C.: United States Government Printing Office, 1943–44), 1:285–86 and 330. He Guozhu was listed as a mathematical specialist on the compilation team of the *Lixiang kaocheng*, the great compendium of Western mathematics and calendrical science completed in 1723: See He Guozong and Mei Gucheng, comps., *Lixiang kaocheng* (Summation of measurements and observations, 1723); modern reprint in *Siku quanshu zhenben* (Rare editions from the Siku Manuscript Library), 2,400 vols. (Taipei: Taiwan Shangwu Yinshuguan, 1971), 4th series, vols. 151–54, prolegomena roster of compilers, 2b. Korean sources give He Guozhu's title as *wuguan sili*, literally "five officials chief mathematician." The structure of this phrase is puzzling. Kim, in *Taedong chiji*, 28.561bb (note 114), garbled it into "five hundred mathematicians"; Pang, *Han'guk ŭi chido*, 172 (note 5), renders it "five mathematics officials." Although the exact significance of *wuguan* eludes me, it is certain that it can appear in the title of a single individual, including some officials in the bureau of mathematics itself; see *Da Qing huidian* (Qing administrative code) (edition of 1732), 86.5a.

210. *Munhŏn pigo*, 2.10a (note 51); Kim, *Taedong chiji*, 28.561bb (note 114).

211. *Sŭngjŏngwŏn ilgi*, 25:959ba–b (note 200). The nature of his earlier purchases is revealed in the notice of the second visit, in 1715; *Sukchong sillok*, 56.3a–b (note 17).

212. Yi, *Han'guk ko chido*, 207–8, 226–27 (note 4); Yi Hongjik, comp., *Kuksa taesajŏn* (Encyclopedia of Korean history), 4th ed. (Seoul: Samyŏng Ch'ulp'ansa, 1984), 1353.

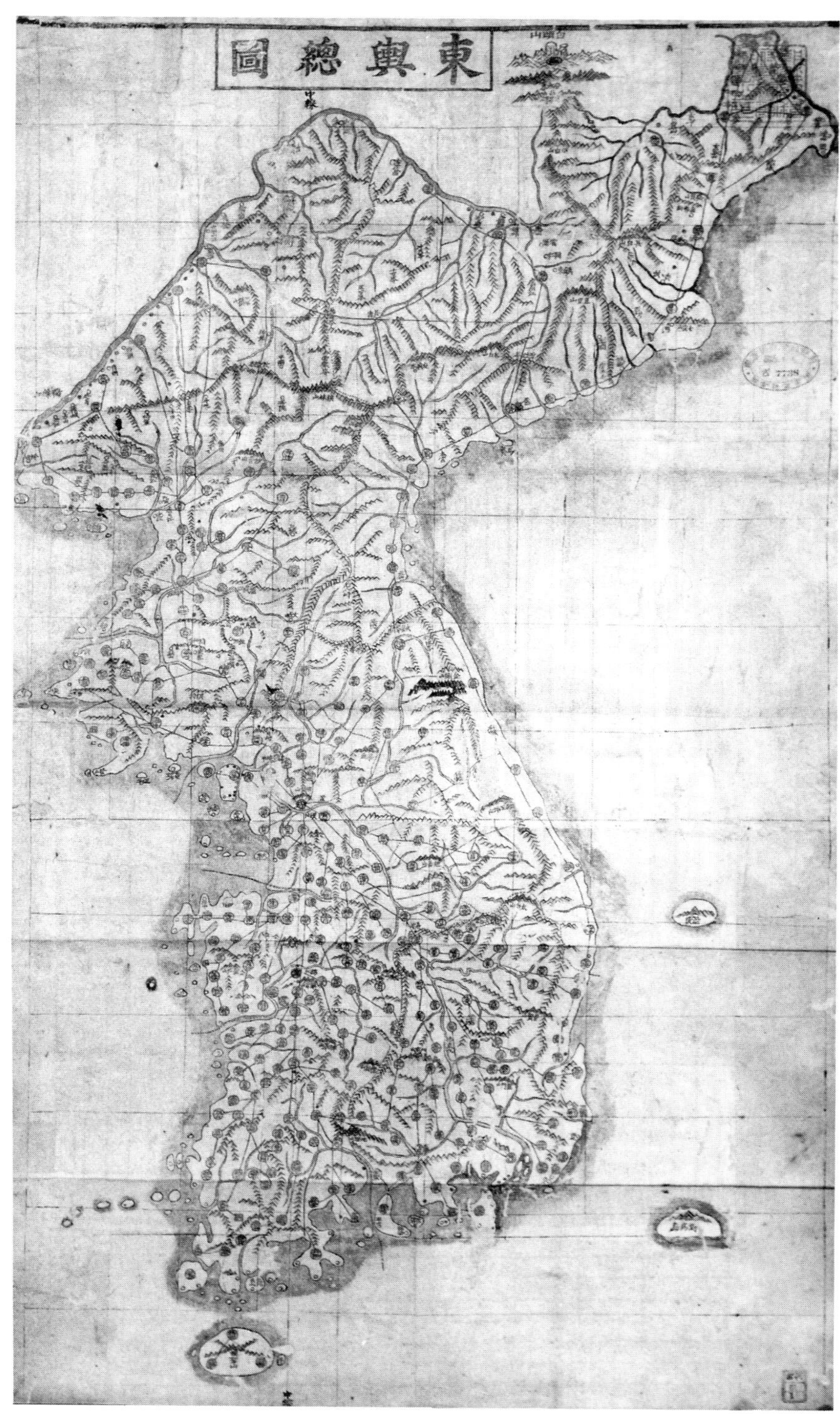

FIG. 10.31. *TONG'YŎ CH'ONGDO* (GENERAL MAP OF THE EASTERN TERRITORY). A single-sheet woodblock print; judging from the nonindication of the town of Changjin as a district, it dates from before 1787. Although provincial borders are not shown, the provinces are distinguished by distinctive colors around the names of their respective districts, as on the earlier Chŏng Ch'ŏk–style maps. Chŏng Sanggi was the first to use a strict scale (although none is indicated on this copy) and the first to draw the nation's outline and borders in a relatively accurate manner. His principal problem was the Yalu River, which bends too far to the north, considerably inflating the area of P'yŏng'an Province. First reported in written sources in 1757, after his death, maps in this style dominated the period 1750–1860.

Size of the original: 98 × 57 cm. By permission of the National Central Library, Seoul (cat. no. Kojo 61-16).

discussion apparently stimulated some presentations of maps to the throne. On such an occasion in 1757, King Yŏngjo came to hear of some maps in the household of Chŏng Hangnyŏng, who then held a minor post in the bureaucracy. They were duly brought in for royal inspection and found to be very detailed in their depiction of mountains, rivers, and roads. They also featured the *paengni ch'ŏk* (hundred-*li* "foot"), a scale that is said by the perhaps overenthusiastic chronicler to have virtually eliminated error in map measurements. "In all my seventy years I've never heard of such a thing as a hundred-*li* foot," Yŏngjo exclaimed in approval, and he sent the map, titled *Tongguk tae chido* (Great map of the Eastern Country), off to the royal library for copying. Provincial maps, received through the same intermediary a few days later, were also reproduced, with copies going to the royal library and to the state defense council (*pibyŏnsa*).[213]

It is possible that Chŏng Sanggi's map was already known among *sirhak* scholars and other map fanciers by the time it came to the king's notice, but this would probably be difficult to prove. No copies of Chŏng's maps bear his name; few are dated, and those that are are late. His provincial maps—the core of his work—were never printed, and all known examples are hand-drawn copies. Maps in his style are known under many different titles, no single one having prevailed as a standard. Given these facts, no one has been able to pinpoint with any certainty the date when Chŏng's maps would have become known. Shannon McCune has suggested 1730 as an approximate date for their first appearance.[214] This accords with the general development of cartographic method and Chŏng's own maturity as a mapmaker. He would have been fifty-two at that time. It is a reasonable guess.

Maps in the Chŏng Sanggi style come both as single maps of all of Korea (fig. 10.31) and as separate maps of the eight provinces (fig. 10.32). There is complete, and intended, convertibility between the two. Chŏng designed his work so that all the provincial maps were of identical scale. Individual parts could be joined together to make a national map or, if desired, any regional combination of provinces. Hand copyists have preferred to duplicate the provincial maps, which afford versatility of purpose and are conveniently bound together in albums, while woodblock publishers probably found the national map a better seller. Modern scholars tend to refer to the single map of the nation as the *Tongguk chido* (Map of the Eastern Country) and to albums of the provincial parts as *P'alto chido* (Maps of the Eight Provinces). Unfortunately, the great variety of known titles on both types still remains.

No copy of the *Tongguk chido* certain to have come from Chŏng Sanggi's hand has been identified. Yi Ch'an has suggested that the most likely candidate is an album titled *P'alto chido*, held by the National Central Library.[215] That copy bears an unsigned and undated preface by a man who says he made the maps, and it may be taken to be the words of Chŏng Sanggi. This preface often appears in album copies on the first page, invariably devoted to a map of one of the halves of Hamgyŏng Province.[216]

On many copies, among them that shown in figure 10.32, the preface is followed by a drawing of what I will call the scale-foot, a small ruler line that, applied to the map, represents 100 *li* (or in tortuous terrain, 120 or 130 *li*; see below). Writers using English often render *ch'ŏk* as "foot"—as I did in the King Yŏngjo anecdote above—but such a translation should be avoided as a general rule lest a reader think that one Korean foot (about 19.9 cm; see table 10.4) is equivalent on the map to 100 *li*. Maps in the Chŏng Sanggi style are nowhere near that large. As measured on the provincial maps owned by the late Yi Pyŏngdo, Chŏng's scale-foot is 8.2 centimeters in actual length and yields a scale of about 1:420,000.[217] A translation of the preface follows:

213. Pang, *Han'guk ŭi chido*, 161 (note 5). See also *Yŏngjo sillok* (Annals of King Yŏngjo, r. 1724–76), 90.8b–9a (note 17). Circumstances, including the same intermediary on both occasions of the map inspections, make it almost certain that these were all Chŏng Sanggi maps. Further government copying took place in 1770. In that year, according to the family genealogy, a team of official copiers came to the Chŏng house, which would then have been headed by his son Hangnyŏng or his grandson Wŏllim (1731–1800), and set up a temporary office (*togam*) for copying a general map of Korea entitled *Taedong yŏjido* (Territorial map of the Great East)—the first time anything like that had ever occurred in Korea, according to the claim of the genealogist. See *Hadong Chŏngssi taedongbo* (Comprehensive genealogy of the Hadong Chŏng lineage) (Seoul: Hadong Chŏngssi Family Association, 1960), 2.57b.

214. McCune, "Some Korean Maps," 101 (note 197).

215. Completely illustrated in Yi, *Han'guk ko chido*, 96–103 (note 4).

216. See illustrations of the first page of the *Chwahae yŏdo* (Terrestrial map of [the country in] the Left [Eastern] Sea) and of the *Tongguk chido* in Yi, *Han'guk ko chido*, 112 and 114 (note 4). An important copy held by the library of the University of California, Los Angeles, also has this preface, although with a number of variants. The UCLA map has been illustrated and described in the article by Thrower and Kim, "Dong-Kook-Yu-Ji-Do" (note 84). This copy is reported to bear a cyclical date, *kich'uk*. On the ground that the name Inch'ŏn is found on the map, and in the belief that that name dates only from 1882, they pick the next available possibility and equate *kich'uk* with 1889. This assumption is baseless; the district has borne the name Inch'ŏn since King T'aejong decreed it in 1413. Barring some other factor that would rule it out, *kich'uk* could just as well be equated with 1829 or 1769, although based on style and appearance 1829 seems the better choice. Apart from the nonstandard romanization, the repeated erroneous rendering of the cartographer's name as Chung Sang-*ik*, and the statement that he completed his map in 1786 (he died in 1752), there are many other errors and misperceptions in this article, which should be used with caution.

217. Yi, *Han'guk ko chido*, 207b (note 4). Three maps from Yi Pyŏngdo's copy are illustrated in Thrower and Kim, "Dong-Kook-Yu-Ji-Do," 44–46 (note 84), who, however, give the de facto scale as 1:400,000.

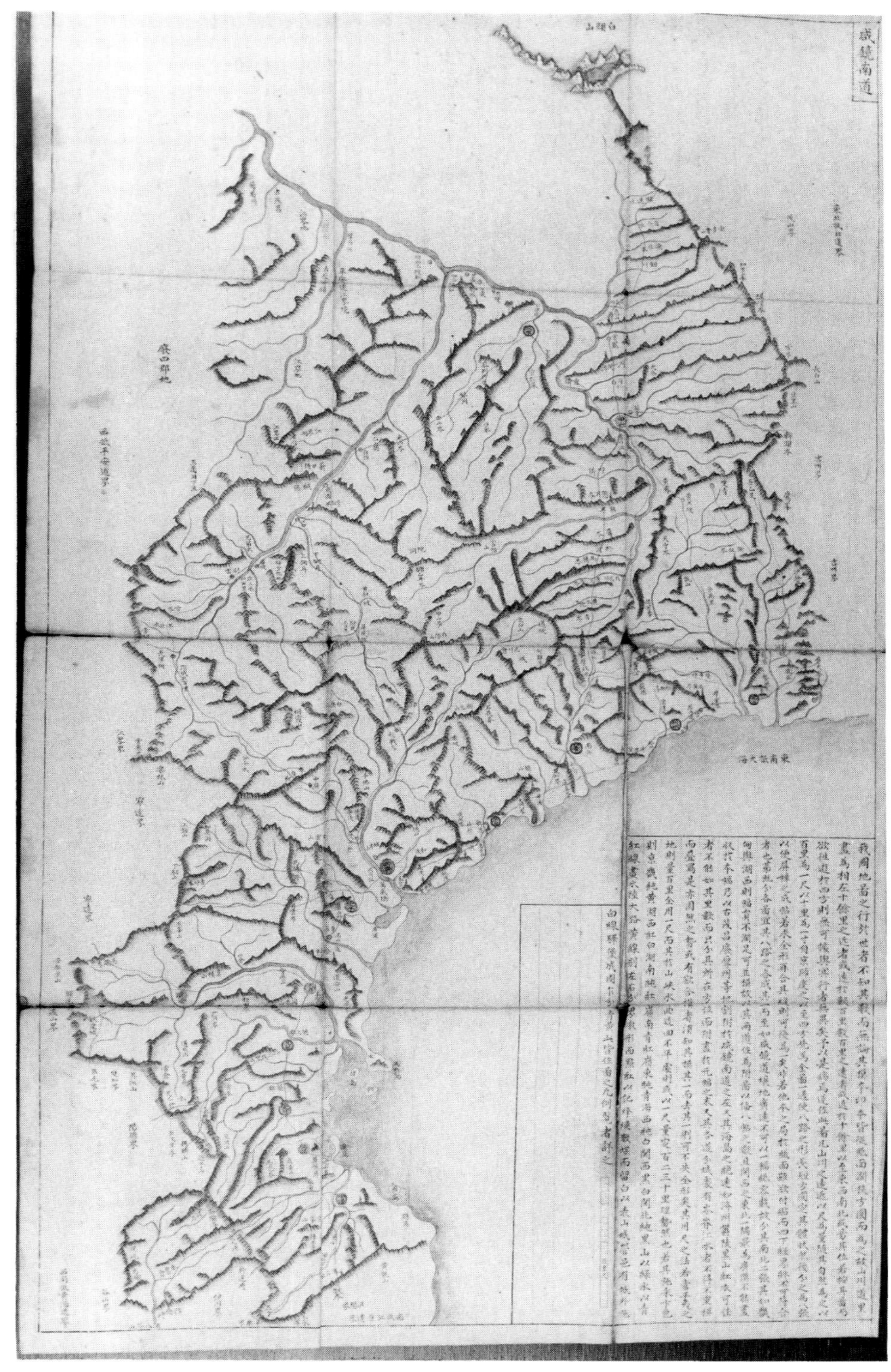

FIG. 10.32. "HAMGYŎNG NAMDO" (SOUTHERN HAMGYŎNG PROVINCE). This sheet, from an album titled *Tongguk chido* (Map of the Eastern Country), is a carefully executed copy of a Chŏng Sanggi–style provincial map. The shapes-and-forces (*hyŏngse*) character of Chŏng's cartography is here strongly emphasized, providing an excellent rendering of the watersheds and drainage of the southern part of Hamgyŏng. District towns and roads, and some post stations and military outposts, are also shown. In the margin at lower right is Chŏng's preface (translated, p. 309), which ends with a drawing of the "hundred-*li* scale-foot" explained in his text. This copy does not show Changjin (just below the upper middle fold at the far left) as a district town, and therefore it was probably executed before 1787.

Size of the original: unknown. By permission of the National Central Library, Seoul (cat. no. Rare 122 Kojo 61–12).

I do not know how many maps of our country are circulating in the world, but whether they are hand copied or printed they are all made to conform to the different dimensions or shapes of paper. For this reason mountains, rivers, and route distances are all in error with respect to each other. A place as close as ten *li* or so might be several hundred *li* away, while one several hundred *li* away might be as close as ten *li*, resulting sometimes in shifts of alignment in north, south, east, and west. If one should wish to take a trip anywhere with such a map, he would not be able to depend on it at all; it would be no different from traveling in the dark.

Mindful of such defects, I have made [my own] maps. Regardless of whether mountains and streams are high or low or distances near or far, I have measured with the scale-foot and made the map in accordance with things as they really are. I make one scale-foot equal one hundred *li*, and one scale-inch equal ten *li*. Spanning from Seoul outward in all directions, I first made a complete map as a unity, so as to fix the relationship of all the land features in the eight provinces regardless of size or curvature. Then I divided [this map] into eight pages convenient for separate and discrete placing in an album. If one seeks the complete picture, he can rematch the borders [of the eight separate maps] and make them into one again. It is not like other maps, which suffer from limitations of paper shape. With those, even if one wanted to paste sheets together at their four borders, he would never be able to match them up.

Since the map is divided into eight sheets, it would be ideal if each of the eight provinces made up a single sheet. However, Hamgyŏng Province is so vast and spread out that it cannot be made to fit on a single piece of paper, so I have divided it into two sheets, north and south. But in a case like Kijŏn and Hosŏ [Kyŏnggi and Ch'ungch'ŏng] provinces,[218] the areas are not large and it suffices to place them together, so I took both provinces and made a combined map, thereby keeping to an eight-page album. In addition, the northeast corner of Kwansŏ [P'yŏng'an] Province extends to such a distance that it cannot all be gathered onto its own sheet, so I took the areas of Former Much'ang and Discontinued Huju, etc.,[219] and added them on to the left side of Hamgyŏng Province. Then too, islands in the sea, such as Cheju, Ullŭng, Hŭksan, Hong'ŭi, and Kaga, are so far out that one cannot go by the number of *li*. I have just assigned them to their direction and drawn them in at the edge of the relevant sheet. When it comes to mountain ranges and rivers at the point where two provincial territories meet, one has no alternative but to copy them twice, making a duplication. This is unavoidable, but if one should wish to make a combined copy, he should understand that if he copies [such features] from the one sheet but not from the other he will not err in the overall picture.

As for the method of applying the scale-foot, in level and regular areas it is appropriate to have 100 *li* correspond to one scale-foot. But in irregular areas such as mountainous terrain and meandering rivers, 120 or 130 *li* may correspond to one scale-foot.

On the assignment of distinctive coloring, Kyŏnggi is solid yellow, Hosŏ white with red, Honam [Chŏlla] solid red, Yŏngnam [Kyŏngsang] blue with red, Yŏngdong [Kangwŏn] solid blue, Haesŏ [Hwanghae] solid white, Kwansŏ white with black, and Hamgyŏng solid black.[220] Green is used for mountains, blue for rivers. Red lines indicate major routes on land or sea; yellow lines show borders between the left and right divisions [of provinces].[221] A platform-shaped symbol with dotted red marks a fire signal; a crenellation with solid white indicates a mountain fortress [*sansŏng*]. If garrisons or district towns are walled, a white line is drawn outside; if post stations or command posts are within enclosures, they are distinguished by blue and yellow, respectively.

These are the conventions and rules used in making this map. Readers should take careful note of them.[222]

With the Chŏng Sanggi–style maps, the shapes-and-forces emphasis of earlier days returned strongly to Korean cartography. Every river flows between two mountain ranges; every mountain range gives birth to rivers on either side. This is only common sense, yet the consistent and clear representation of the aggregate of such facts dramatically reveals the overall character of the landscape in images that common sense cannot easily generate. The dramatic differences in terrain between the mountainous northern areas and the broader river plains of the south, with all that they convey for a sense of the varying economic, demographic, and scenic character of the different regions of the country, are made instantly clear to the eye. It would be easy to give too much emphasis to such conventions, thus bringing on information overload and shattering clarity. It would be even easier to underuse or ignore them and abdicate the responsi-

218. Here and in the lines following the writer uses the popular nicknames for the provinces.

219. Areas abandoned by Sejo in 1455; see above, p. 290.

220. This is slightly different from the color-code mentioned above, though it comes from the same basic Five Phases scheme. Bear in mind that what is being colored is not the map itself, but the cartouches surrounding the names of the various districts—with different colors depending on what provinces they are in. Chŏng is not implying, where two colors are used, that they are blended but means they are used in combination to outline and/or tint the cartouche.

221. Except for Hamgyŏng, the provinces were each divided into two sectors to reflect military commands. These were referred to as the "left province" (east) and the "right province" (west).

222. The translation follows the texts illustrated in the plates in Yi, *Han'guk ko chido*, 96 and 114 (note 4). The first is attached to an album called *P'alto chido*, the second to an album titled *Tongguk chido*. There are only minor variations between the two.

bility to inform. But Chŏng Sanggi and after him Kim Chŏngho managed a clear and pragmatic deployment of this cartographic technique, in balance with the provision of other important information—names, roads, boundaries, and so forth—that truly made physical features intelligible and revealed the shapes-and-forces that governed the life of their community. Indeed, through their maps, many users probably saw shapes-and-forces thought less in terms of the mystic conduits to Mount Paektu than as an aid to physiographic and political-economic understanding.

The elaborate working out of such principles was probably the main stimulus to the discovery of Korea's true shape. With the strict emphasis of shapes-and-forces thought on the direction of the river "veins" and mountain "arteries," and with Chŏng Sanggi's rigorous application of scale, the road was finally open to the successful understanding of the complicated Yalu and Tumen river systems. And when that was done, the actual lines of the northern frontier were necessarily revealed. Koreans of earlier days, surrounded by their "three thousand *li* of mountains and rivers" had instinctively understood shapes-and-forces principles but lacked the methods to relate accurate measurements to precise positions. In the eighteenth century, reliable distance data that had long been available were linked to the cartographic application of true scale. In spite of a few problems that remained to the end on the comprehension of the Yalu's upper-middle course, the Chŏng Sanggi understanding of the northern border was a net improvement over that of Régis and his companions.[223]

That much said, the Chŏng Sanggi–style maps still left room for improvement. His Korea was a bit too long in relation to its width, and the Yalu River strayed too far north to its major bend and then flowed on too southerly a course to its mouth, resulting in an inflated size for the northern part of P'yŏng'an Province. It would be easier to address such problems if we had some understanding of how Chŏng Sanggi actually worked. The technology to make accurate geodetic measurements probably existed in the Korea of his day. Could it have filtered down to him? Even though such a development was wholly plausible in terms of the sociology of knowledge—there was a high degree of social overlap between the scholarly and the bureaucratic worlds—in fact we have no indication that it did. More important, we cannot even be certain that the government itself effectively used the instruments and knowledge its officials continually brought back from Beijing.

The earliest evidence of programmatic government concern with geodetic coordinates comes in 1791. King Chŏngjo had ordered the director of the government's observatory (the *kwansanggam*) to make new calculations on time differentials in the eight provinces so that more precise information on sunrise and sunset and on the beginning points of the twenty-four *chŏlgi* (solar seasons, Chinese *jieqi*)[224] could be incorporated into the calendars supplied to provincial and district officials. He considered it backward that the calendars gave only one set of times for these events for the whole country when he knew that they varied with locality. The director of the observatory, Sŏ Hosu, submitted a report with the requested information, together with interesting remarks and data on geodetic coordinates, and the king urged incorporation of the relevant details in the calendar for the following year. Although objections of an unknown character and origin were raised against this plan and the whole idea was dropped, fortunately the report itself has survived in King Chŏngjo's annals.[225]

Sŏ Hosu began with a general exposition of latitude and longitude. He observed that one degree of longitude represented a time difference of four minutes, and he reported that Seoul was forty-two minutes ahead of Beijing. In terms of distance, he said, each degree of latitude or longitude was roughly equal to two hundred *li*. "Going by straight lines on maps," he continued, "we have used the hundred-*li* scale-foot [*paengni ch'ŏk*] to determine [the following coordinates]."[226] He then gave latitudes

223. The difference in dates prompts the question whether Chŏng might have been influenced by the Jesuit maps. I have found in Korean records no indication that Korean officials on business in Beijing ever bought or saw copies of the very rare 1721 edition of the Jesuit maps. But if they had heard of them, they would have made every effort to buy them. Export of such material at that time was forbidden by law, but Koreans often found ways to get such things home anyway. The Jesuit maps of China were published in pieces in Chen Menglei, Jiang Tingxi, et al., *Gujin tushu jicheng* (Complete collection of books and illustrations, past and present, completed in 1726), which was quickly available in Korea, but this encyclopedia's section on Korea, though voluminous, lacks any map. In any case, Chŏng's treatment of the frontier was more accurate than that of Régis, and it unquestionably came from his own application of scale.

224. These twenty-four periods, which were structured around the equinoxes and solstices, provided farmers and nature watchers with those solar constants that do not jibe with the lunar calendar. A number of popular holidays and festive occasions were also tied to this system. For a complete list of the solar periods, see Jeon, *Science and Technology in Korea*, 90 (note 49).

225. *Chŏngjo sillok* (Annals of King Chŏngjo, r. 1776–1800), 33.36b–37a (note 17). The longitudes, but not the latitudes, are also found in *Munhŏn pigo*, 1.10a–b (note 51). The project ordered by Chŏngjo seems to have been closely modeled on a Chinese discussion of analagous time differences in the onset of the solar seasons in the various Chinese provinces; see *Lixiang kaocheng*, pt. 2, 1.14b–15b (note 209), where Seoul's longitude of 10°30′E and its time differential of +42 minutes, both relative to Beijing, are also documented. This is the earliest record of Seoul's longitude and probably derives from the observations of He Guozhu.

226. *Chŏngjo sillok*, 33.36b–37a (note 17). The Korean use of maps to plot geodetic coordinates also seems to have been suggested by the *Lixiang kaocheng*. A footnote to the Chinese work says: "For the

TABLE 10.7 Korean Geodetic Coordinates Reported in 1791

Site[a]	Latitude[b]			Longitude[c]			
	Reported	Modern	Error	Reported	= Adjusted	Modern	Error
Onsŏng	44°44′N	42°55′N	+109′	2°58′E	= 129°57′E	130°00′E	−3′
Hamhŭng	40°57′N	39°55′N	+62′	1°00′E	= 127°59′E	127°32′E	+27′
P'yŏngyang	39°33′N	39°01′N	+32′	1°15′W	= 125°44′E	125°45′E	−1′
Haeju	38°18′N	38°03′N	+15′	1°24′W	= 125°35′E	125°42′E	−7′
Seoul	37°39′N	37°33′N	+6′	0°00′	= 126°59′E	126°59′E	0
Wŏnju	37°06′N	37°21′N	−15′	1°03′E	= 128°02′E	127°57′E	+5′
Kongju	36°06′N	36°27′N	−21′	0°09′W	= 126°50′E	127°07′E	−17′
Taegu	35°21′N	35°52′N	−31′	1°39′E	= 128°38′E	128°36′E	+2′
Chŏnju	35°15′N	35°49′N	−34′	0°09′W	= 126°50′E	127°09′E	−19′
Haenam	34°15′N	34°34′N	−19′	0°28′W	= 126°31′E	126°36′E	−5′

[a]The eight entries Hamhŭng through Chŏnju are the capitals of the eight provinces; coordinates for all but Kyŏnggi Province are given in *Chŏngjo sillok* (Annals of King Chŏngjo, r. 1776–1800), 33.37a, item dated Chŏngjo 15/10/11 [6 November 1791]. The longitude figures are also given in *Chŭngbo Munhŏn pigo* (Documentary reference encyclopedia, expanded and supplemented) (Seoul: Empire of Korea, 1908), 1.10a–b. Kim Chŏngho, *Taedong chiji* (Administrative geography of the Great East, 1864) (Seoul: Hanyang Taehakkyo Kukhak Yŏn'guwŏn, 1974), 28.561bb–562aa, gives the full coordinates of the provincial capitals, including those for Kyŏnggi, and adds those of Onsŏng and Haenam, respectively the farthest northeast and southwest district seats on the mainland. The modern coordinates were taken from Kwŏn Sangno, *Han'guk chimyŏng yŏnhyŏk ko* (A study of historical changes in Korean place-names) (Seoul: Tongguk Munhwa Sa, 1961). Both traditional and modern data take as their reference point the courtyard in front of the given provincial or district capital building.

[b]The error is the difference in minutes between the reported 1791 latitude in the "reported" column and the modern official latitude in the "modern" column.

[c]The 1791 figures represent longitude east or west of the Seoul meridian. The figures in the "adjusted" column are derived from Seoul's official modern longitude, 126°59′E (Greenwich). The error is the difference in minutes between the longitude in the 1791 "adjusted" column and the official longitude in the "modern" column.

and longitudes for seven of the eight provinces, as measured at their governors' compounds.

Referring to the same report, Kim Chŏngho made a similar statement, saying that the observatory staff, "using the maps held by the state defense council [*pibyŏnsa*], determined the north polar elevation and the degrees east or west of Seoul for the compounds of the provincial governors." He went on to provide a set of equivalents: 200 *li* = 1°, 10 *li* = 3′ (or 180″), 1 *li* (or 2,160 *po*) = 18″, and 120 *po* = 1″.[227] Evidently these were Kim's basic formulas for relating long, medium, and short distance measurements to geodetic space. The conclusion seems unavoidable: the geodetic coordinates of 1791 were determined by distances plotted "in straight lines" (*chikto*) on maps. We know that the state defense council had Chŏng Sanggi–style maps, and the use of Chŏng's term "hundred-*li* scale-foot" is another clue making it likely that these were the maps used in plotting the 1791 coordinates.

In spite of this association of the Chŏng Sanggi maps with geodetic coordinates, we have no evidence that he had used them in making his maps, and the maps themselves do not have graticules. Some examples, such as the undated woodblock edition shown in figure 10.31, have grids drawn in on an already printed copy. In this case, although a vertical line going through Seoul is labeled *chungsŏn* (center line), these lines have nothing to do with meridians or latitudes, and the intersections show no relation to the coordinates of 1791 or those seen in Kim Chŏngho's work.

The figures of Sŏ Hosu's report are given in table 10.7, supplemented by data from Kim Chŏngho. Kim cites the 1791 report and gives identical figures, but he adds coordinates for the missing province, Kyŏnggi, with its governor's compound in Seoul. He also provides coordinates for Onsŏng, in the extreme northeast, and Haenam, in the extreme southwest, which permits the determination of the total length of the Korean mainland. In another part of his vast compendium, Kim gives additional coordinates for nineteen district seats in northern P'yŏng'an Province (see below). It is unclear whether the original 1791 source had a full list of coordinates for the whole country or whether Kim had obtained the supplemental figures elsewhere, possibly even from his own calculations.

longitudes of each province, we in all cases relied on the degrees established on the maps. We have now taken observations of the gnomon shadow to find the times for the solar seasons, and have verified these by time differentials [in the observation] of lunar eclipses. All agree with the data of the maps"; *Lixiang kaocheng*, pt. 2, 1.15b (note 209). I have found no record that in the Korean case these important confirmatory steps were taken.

227. Kim, *Taedong chiji*, 28.561bb–563bb (note 114). A copyist's error yielding "20 *po*" has been emended to "120 *po*" to meet the obvious requirements of arithmetic.

Kim Chŏngho says that Seoul's latitude, 37°39′, was measured by He Guozhu on Chongga (Bell Street, the modern Chongno, which runs east and west), in 1713.[228] The modern official latitude, measured a few blocks south in front of the city hall, is 37°33′29″. If Kim's information is correct then He Guozhu's measurement was off by nearly six minutes, but wherever the fault lies, Kim began with that error. That would have been entirely bearable. The problem, which can be seen clearly in table 10.7, is that the error keeps growing the farther north the data reach, until at Onsŏng it is nearly two degrees. Going south, it is more than half a degree at Chŏnju. (Haenam, much farther south, has a much smaller error, but its reported latitude is suspect since it goes against the clear pattern of the data, which is for the error to grow incrementally in proportion to the distance from Seoul.) This pattern in itself confirms that official distance data centered on Seoul are the foundation of the coordinates of 1791 and those of Kim Chŏngho.

Given Kim's latitudes for Seoul and the district town of Onsŏng, the geodetic distance between these two latitudes would have been 7°05′, which at the ratio of 200 *li* per degree would be equivalent to 1,417 *li*, or 609.3 kilometers (see table 10.5). The actual modern geodetic distance between the same points is 5°24′, which at the metric standard of 111 kilometers per degree of latitude comes to 600 kilometers. In kilometers Kim's difference is only 1.6 percent greater than the modern figure, which shows that his distance figures for Onsŏng were more or less on the mark. But in geodetic distance, Kim's 7°05′ is 31 percent greater than the modern figure of 5°24′.

The cause of this disparity lies in the Zhou foot, which at 19.91 centimeters is short in relation to the other Chinese and Korean linear standards. The formula 200 *li* = 1° came into use in China during the Sino-Jesuit era in the seventeenth and eighteenth centuries, and it was put on record in Chinese textbooks on astronomy.[229] Going by Qing dynasty official definitions of the *li* and of the *chi* (foot) and the standard metric ratio, the 200-*li* formula yields the length of 115.2 kilometers per degree of latitude.[230] This is 4.1 kilometers too long, but the error is considerably less than that resulting from the Korean figure of 86 kilometers, which is short by 25.1 kilometers. Korea, with shorter *li*, used more of them, so that when the 200-*li* formula was applied to determine geodetic coordinates from distances plotted on a scale map, there was an inflation in degrees of latitude. It was not that the distances were wildly wrong in their own terms, but that relative to the Chinese *li* they reached higher numbers and, when divided by 200, created more degrees of latitude. This then, is the cause of all but 1.6 percent of the error indicated in table 10.7.

The longitudes in table 10.7 were stated in degrees east or west of the Seoul meridian, and Seoul was itself fixed, presumably by He Guozhu, at 10°30′ east of Beijing. There is no discernible pattern in the deviations from the modern longitudes, and the swings are extreme—for example, only a −3′ error for Onsŏng, at 2°58′ east of Seoul, but a +27′ error for Hamhŭng, reported at 1° east but actually much closer to the Seoul meridian. This reflects the positions of Hamhŭng and its nearby coastal areas on the Chŏng Sanggi maps, which are noticeably farther east than they should be (Hamhŭng is at a northeast bearing of 19° from Seoul compared with an 11° bearing on modern maps). Evidently the rugged road to Hamhŭng made east-west measurement especially difficult.

Kim Chŏngho gives nineteen sets of coordinates for towns in the general area of northern P'yŏng'an Province. These are presented in table 10.8. In this case his longitudes are based on the P'yŏngyang meridian. It will be seen that they all show a negative error with respect to modern figures. Although the data are quite variable, if the table had been arranged from west to east rather than from north to south, one would notice overall a heavier error for the towns farther west and a somewhat lighter one for those farther east. It might be possible to explain this by a westward exaggeration in the northwest area of the Chŏng Sanggi–style maps used to determine the coordinates. This is actually the case on two of the four Chŏng Sanggi maps illustrated in Yi Ch'an's album.[231] Although this is speculative, the explanation at least has to be sought on a map, since we know that the coordinates were determined with reference to one.

228. Kim, *Taedong chiji*, 28.561bb (note 114). This figure was rounded off from 37°39′15″, which Kim cited from the *Lixiang kaocheng*, pt. 2, 1.16b (note 209). We have seen that Régis himself gave the latitude as 37°38′20″.

229. Wu, *Zhongguo duliangheng shi*, 271–72 (note 149).

230. Wu, *Zhongguo duliangheng shi*, 271–72 (note 149). In stating the equivalence 200 *li* = 1°, the *Da Qing huidian* specifies a *li* of 1,800 *chi* (360 *bu* at 5 *chi* each), using the Qing standard *yingzao chi* (construction foot). At the metric equivalent of 32 centimeters for this measure, the *li* has a length of 576 meters. Kim also referred to the *Huidian*, using the term *zongshu chi* (vertical millet foot), which in the Qing official definition was equal to the *yingzao chi* (Kim, *Taedong chiji*, 28.562ba [note 114]). Kim's text erroneously specifies six *chi* per *bu* instead of five. Since he made no calculations of Chinese *li*, this does not affect the present discussion, but it is of interest as an example of the endless obstacles to conversions between the measurement standards of one East Asian country and another.

231. On modern maps using a Lambert conformal conic projection, the bearing from Seoul to the northwest border town of Ŭiju is about N36°W. This is coincidentally very close to the bearing that could be deduced on two Chŏng Sanggi–style maps, the *Tongyŏ ch'ongdo* and the *Haedong yŏjido* (Yi, *Han'guk ko chido*, 68 and 77 [note 4]). But two others, the *Taedong chido* and the *Chwahae yŏdo* (75 and 78), indicate bearings from Seoul of about 39½° and 40°, respectively. Such a westward shift would contribute substantially to the kind of longitudinal error evident in table 10.8.

TABLE 10.8 Northern P'yŏng'an Province Geodetic Coordinates from Kim Chŏngho

Site	Latitude			Longitude[a]			
	Reported	Modern	Error	Reported	= Adjusted	Modern	Error
Kanggye	42°36′N	40°58′N	+98′	48′E	= 126°32′E	126°36′E	−4′
Wiwŏn	42°41′N	40°53′N	+108′	05′E	= 125°49′E	126°04′E	−15′
Ch'osan	42°25′N	40°50′N	+95′	15′W	= 125°29′E	125°48′E	−19′
Pyŏktong	42°02′N	40°37′N	+85′	39′W	= 125°05′E	125°26′E	−21′
Ch'angsŏng	41°31′N	40°30′N	+61′	1°08′W	= 124°36′E	125°03′E	−27′
Sakchu	41°19′N	40°23′N	+56′	1°12′W	= 124°32′E	125°03′E	−31′
Ŭiju	41°04′N	40°12′N	+52′	1°42′W	= 124°02′E	124°32′E	−30′
Hŭich'ŏn	41°19′N	40°10′N	+69′	24′E	= 126°08′E	126°17′E	−9′
Kusŏng	40°57′N	39°59′N	+58′	48′W	= 124°56′E	125°15′E	−19′
Unsan[b]	41°01′N	39°58′N	+63′	06′W	= 125°38′E	125°48′E	−10′
Yongch'ŏn	40°52′N	39°56′N	+56′	1°29′W	= 124°15′E	124°22′E	−7′
T'aech'ŏn	40°39′N	39°55′N	+44′	29′W	= 125°15′E	125°24′E	−9′
Yongbyon	40°42′N	39°49′N	+53′	01′?	= 125°44′E	125°49′E	−5′
Sŏnch'ŏn	40°35′N	39°48′N	+47′	1°05′W	= 124°39′E	124°55′E	−16′
Ch'ŏlsan	40°45′N	39°46′N	+59′	1°19′W	= 124°25′E	124°40′E	−15′
Pakch'ŏn	40°39′N	39°44′N	+55′	17′W	= 125°27′E	125°35′E	−8′
Kasan[c]	40°33′N	39°43′N	+50′	24′W	= 125°20′E	125°34′E	−14′
Chŏngju	40°33′N	39°42′N	+51′	41′W	= 125°03′E	125°13′E	−10′
Kwaksan[d]	40°35′N	39°41′N	+54′	50′W	= 124°54′E	125°05′E	−11′
P'yŏngyang	39°33′N	39°01′N	+32′	0°00′	= 125°44′E	125°45′E	−1′

Source: Kim Chŏngho, *Taedong chiji* (Administrative geography of the Great East, 1864) (Seoul: Hanyang Taehakkyo Kukhak Yŏn'guwŏn, 1974), *kwŏn* 23, under respective headings. Modern coordinates, same source as in table 10.7. Error is calculated in same fashion as in table 10.7.

[a]The figures in the "reported" column represent longitudes east or west of the P'yŏngyang meridian, as given by Kim Chŏngho. The figures in the "adjusted" column are based on the Greenwich-adjusted longitude of P'yŏngyang, taken from table 10.7 as 1°15′ west of Seoul, itself said by Kim to be 10°30′ east of Beijing (which is 116°23′ east of Greenwich). This derived figure and the modern longitude of P'yŏngyang are virtually identical.

[b]The source is in error for the modern longitude; this figure is taken from a map.

[c and d]These are no longer district towns; coordinates are taken from a map.

The same set of coordinates throws light on another problem with the Chŏng Sanggi maps: the oversized P'yŏng'an Province and the northward exaggeration of the major bend of the Yalu River. Translating the geodetic distances, via the Korean *li* of that time, into kilometers in the same manner as was done earlier for the Seoul-Onsŏng calculation, we can determine the distance between Kim's latitudes for P'yŏngyang and Kanggye, the most northerly town, at 262.3 kilometers. On the other hand, the calculation of the same distance based on modern coordinates and the value of 111 kilometers per degree yields a figure of 217 kilometers, for an eighteenth/nineteenth-century Korean error of +21 percent, while the error in degrees is +56 percent. Thus, in this case, a considerable part of the overall error was apparently due to overlong Korean distance measurements. In using such exaggerated distance figures, Chŏng Sanggi's strict scale method resulted in a considerable distortion of the Yalu border and of the northern and northwestern parts of P'yŏng'an Province in general.

It would be wrong to conclude the discussion of the Chŏng Sanggi maps with too much emphasis on these errors. Overall, Chŏng's application of a consistent scale produced more benefits than problems for a correct grasp of the shape of Korea. His maps were the basis for numerous copies made by both the government and the general public, and they dominated the century between the 1750s and the 1860s.

To sum up, although the potential for scientific determination of geodetic coordinates was present in Korea, such data do not appear to have been a foundation for Chŏng's scale maps. Rather, long after his death, it was his maps that were used to plot the coordinates that made their appearance in 1791 and in the work of Kim Chŏngho. Considering the method used, it is amazing that the coordinates are as good as they are. Such success as they can claim must be credited to Chŏng Sanggi.

THE MAPS OF KIM CHŎNGHO

Kim Chŏngho lived in a very different world than had Chŏng Sanggi. Whereas in the eighteenth century Korea enjoyed relative stability and remained insulated from most international pressures, beginning about 1800 it suffered every imaginable affliction from both within and without. By 1860 even much of the ruling class was impoverished, the government was depleted both materially and morally, the countryside was seething and would soon rebel, and everyone, whether powerful or weak, saw the country as exposed and vulnerable before the Western powers that had humbled China and Japan. In 1864 Kojong, an eleven-year-old boy, came to the throne, and in 1866 his domineering but charismatic father, known as the Taewŏn'gun, launched a bloody persecution of Korea's small but fervent Catholic community, killing many thousands of believers and most of the surviving French missionaries who, replenished from time to time, had been working underground in the country for thirty years. The French punitive assault on Kanghwa Island that soon followed was only one of a number of Western-related incidents in that year. That Korea more than held its own in these encounters helped for a time to mobilize the country behind the Taewŏn'gun but had little long-run effect on the course of events that resulted, in 1910, in Korea's colonization by Japan.

Although we know much about Kim Chŏngho as a geographer and cartographer, we are almost wholly ignorant of the details and circumstances of his personal life. We do not know when or where he was born, or to whom; nor do we have any knowledge of when, where, or how he died. Because of his posthumous fame, many legends were invented or grew around him, and some have gained official sanction through memorialization in government-compiled school textbooks. Thus millions of Koreans today picture the driven mapmaker, oblivious of home and hearth, walking the length and breadth of the land on surveying expeditions, climbing Mount Paektu several times, living in poverty outside Seoul's city walls, and enlisting his daughter's help with the woodblock carving. In the end, according to the legend, he patriotically presented his maps to the government only to be arrested for disseminating national security information. His woodblocks were confiscated and destroyed, and he was left to die miserably in jail.[232]

Like most legends, this one may have some factual core, but it has so far eluded those who have sought it out. As Yi Pyŏngdo has remarked, it is not likely that the government ever took action against Kim's maps. Such an initiative would have left some trace in public and private writings, and too many copies of his maps, and even some of his woodblocks, survive (fig. 10.43, below). "He may have been arrested and jailed for a connection with Catholicism," Yi says, "but it is scarcely conceivable that such things happened because of a connection with maps."[233] Sources on Korean Catholic history make no mention of Kim Chŏngho, but in an age of anti-Western and anti-Catholic paranoia such as the 1860s, it would have been possible for someone with knowledge of things Western to be singled out by some ambitious policeman or zealot for persecution, especially since the ambiguous term *sŏhak* (Western studies) applied alike to Catholicism and Western science, and Kim certainly had some acquaintance with the latter. On the other hand, it is equally possible that because of the poverty and humble status of his family, Kim Chŏngho's death simply went unnoticed in what were hard times for everybody. Whatever the circumstances, he disappeared without a trace.

An important documented fact in Kim's life is that he was a woodblock publisher by trade. We do not know whether he was led to cartography in pursuit of his printing business or to a woodblock printing career in order to market his maps, but there is a clear relation between the two, and it is doubtful whether someone who was not a block carver could have developed his mature cartographic style. When Kim emerges from the mists of his unknown background in 1834, it is as a printer. In a collaboration, which I have already noted, at the request of the *sirhak* scholar Ch'oe Han'gi, he carved the blocks for a Korean edition of a hemispheric map in a Western style that had been published in China sometime after 1793.[234] Although Ch'oe's map, the *Chigu chŏnhudo* (Map of the front and back [hemispheres] of the globe), had some interest for its up-to-date, graticuled hemispheres and an English rather than a Sino-Jesuit background, it was basically a piece of exotica that had nothing to do with Kim's own cartography.

But in the same year 1834, he also completed a general atlas of Korea titled, to evoke one of Korea's age-old poetic sobriquets, *Ch'ŏnggudo* (Map of the Blue Hills [Korea]).[235] In this instance Ch'oe Han'gi returned a favor and contributed a preface. In the total pattern of Kim's career the *Ch'ŏnggudo* can be regarded as a preparatory stage for the culminating work of the early 1860s. But it

232. Yi Pyŏngdo sums up what can be known of Kim's life in "*Ch'ŏnggudo* haeje" (Biographical note to the *Ch'ŏnggudo*), in Kim Chŏngho, *Ch'ŏnggudo* (Map of the Blue Hills, 1834), 2 vols., ed. Yi Pyŏngdo (Seoul: Minjok Munwa Ch'ujinhoe, 1971), 1:6–9; and in "*Taedong chiji* haeje" (Bibliographical note to the *Taedong chiji*), in Kim, *Taedong chiji*, 641–48 (note 114). He attributes most of the details I have cited here to oral tradition. See also Pang, *Han'guk ŭi chido*, 189–90 (note 5). For a typical reflection of this tradition, see *Kugŏ 5-2* (Fifth-grade Korean reader) (Seoul: Ministry of Education, 1987), 76–83.

233. Yi Pyŏngdo, "*Ch'ŏnggudo* haeje," 1:8 (note 232). Compare Yi, "*Taedong chiji* haeje," 643–44 (note 232).

234. Yi Kyugyŏng, *Oju yŏnmun changjŏn san'go*, 38.180ab–ba (note 60). Yi, a nineteenth-century *sirhak* scholar and a contemporary, identifies Kim Chŏngho as the woodblock publisher for Ch'oe's map.

235. The country of the "Blue Hills" (Chinese Qingqiu) is mentioned several times in the *Shanhai jing* in areas later identified by Korean tradition as in Korea. *Shanhai jing jiaozhu*, 1.6, 9.256, and 14.347 (note 77).

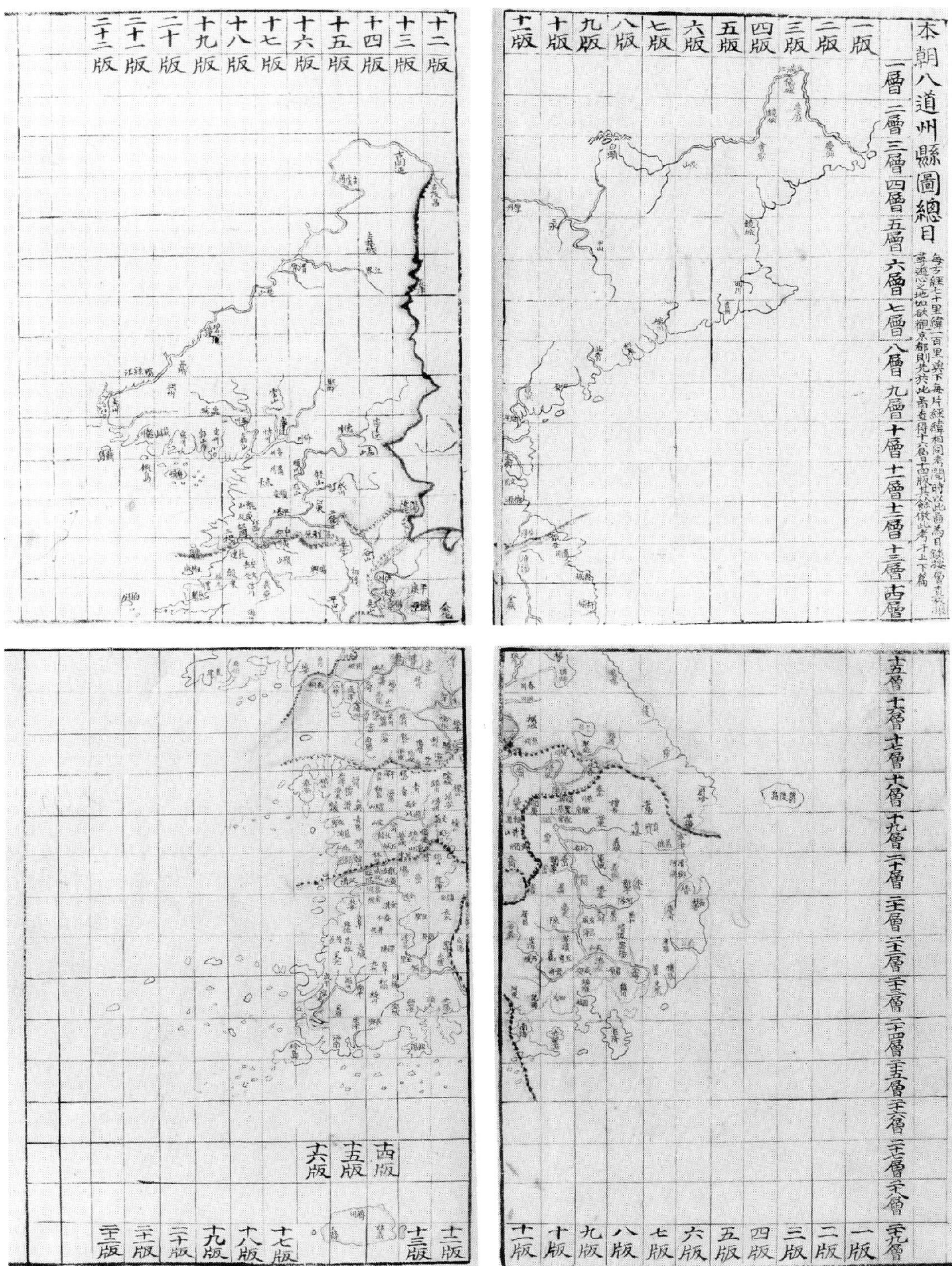

FIG. 10.33. "PONJO P'ALTO CHUHYŎNDO CH'ONGMOK" (GENERAL INDEX TO DISTRICT MAPS IN THE EIGHT PROVINCES OF THE NATION). This is the index grid by Kim Chŏngho for the *Ch'ŏnggudo*, 1834. The top half is in the first volume, the bottom half in the second volume, so that they can be laid out together. Along the right are indicated the twenty-nine horizontal ranks from top to bottom; at the top and bottom are indicated the twenty-two vertical files. Even-numbered ranks are arrayed in one volume, odd-numbered ones in the other, permitting the use of both volumes together to double the contiguous area. Each rectangle of the grid represents an area 70 by 100 *li* (30 × 43 km). The note under the title includes directions for use: "Refer to the rank and seek out the file to find the place you have in mind."
Size of each page: 27.5 × 20 cm. By permission of Kyujanggak Archives, Seoul National University.

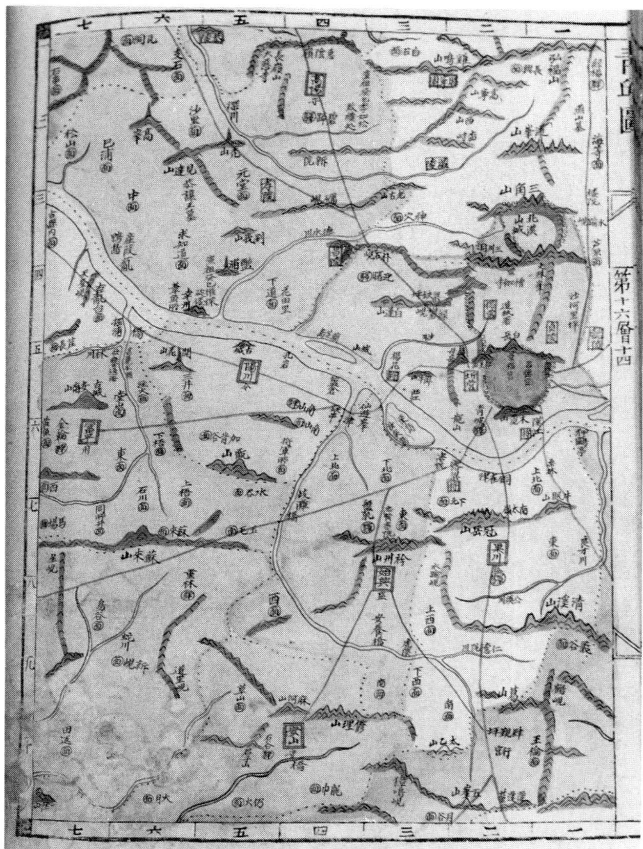

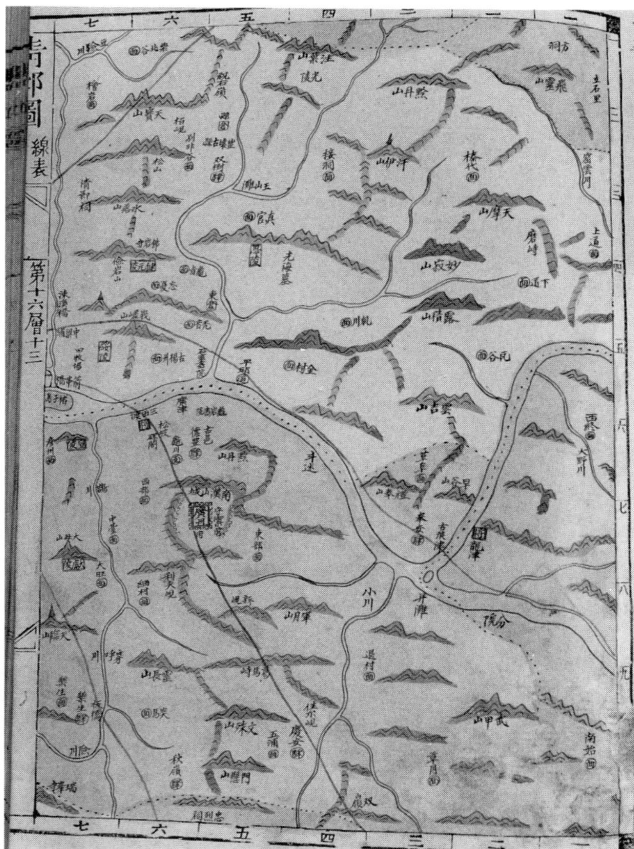

FIG. 10.34. AREA OF SEOUL FROM THE *CH'ŎNGGUDO*. All known copies of this map by Kim Chŏngho are hand drawn; this copy is of unknown date but probably from the mid-nineteenth century. It was made on a printed form ("line guide," or *sŏnp'yo*) with scale units of ten *li* impressed on the margins. In the center woodblock identifying strip are the title of the map and the printed numbers of the rank and files. This section, from rank 16, files 13 and 14 (see reference grid in fig. 10.33), shows the Seoul area. Dotted lines are district boundaries; district names are in square cartouches; names for subdistricts (*myŏn*) and post stations (*yŏk*) are followed by the appropriate characters in circles. Principal mountains are linked with a "sawtooth" pattern to show, in a relatively understated manner, the shapes-and-forces dispositions (compare fig. 10.35).
Size of each page: 35.2 × 23.2 cm. By permission of the National Central Library, Seoul (cat. no. Kwi 239 Ko 61-80).

may also be seen on its own terms as the successful completion of what can only have been a long and arduous labor in assembling, relating, and refining cartographic source materials of many kinds.

The *Ch'ŏnggudo* covers the total territory of Korea, not in a national map or a set of provincial maps, but rather in grid rectangles, assembled in two large albums supplied with a general grid reference map that serves as an index to the pages (fig. 10.33).[236] Each page contains one rectangle of the grid, covering an area seventy *li* east and west and one hundred *li* north and south. Korea is organized into twenty-nine numbered ranks (*ch'ŭng*) covering the country horizontally from top to bottom, and twenty-two numbered files (*p'an*) arrayed vertically from east to west. To find a desired area—for example, Kwangju—one consults the index map and locates Kwangju in rank 22 at file 14. The most unusual feature—indeed, it may be unique—is that the ranks are assigned to alternate albums, the odd numbers to album A, the even numbers to album B, so that it is possible to extend the eye north or south of the focal spot by opening the other album to the corresponding rank and file and placing it above or below the first album. Thus, having found Kwangju at 22-14 in album B, one may place album A above it and open to 21-14, or below it and open to 23-14, to inspect the adjacent areas to the north or south. For

236. On the *Ch'ŏnggudo* in general, see Yi, *Han'guk ko chido*, 86–95 (illustrations) and 208–10 (note 4), and Pang, *Han'guk ŭi chido*, 167–80 (note 5). A manuscript copy of the *Ch'ŏnggudo* has been photolithographically reproduced in its entirety in Kim, *Ch'ŏnggudo* (note 232). The northern half of the grid map is in volume 1, 2–3 (main pagination), and the southern half in volume 2, 2–3, so that both parts may be consulted together.

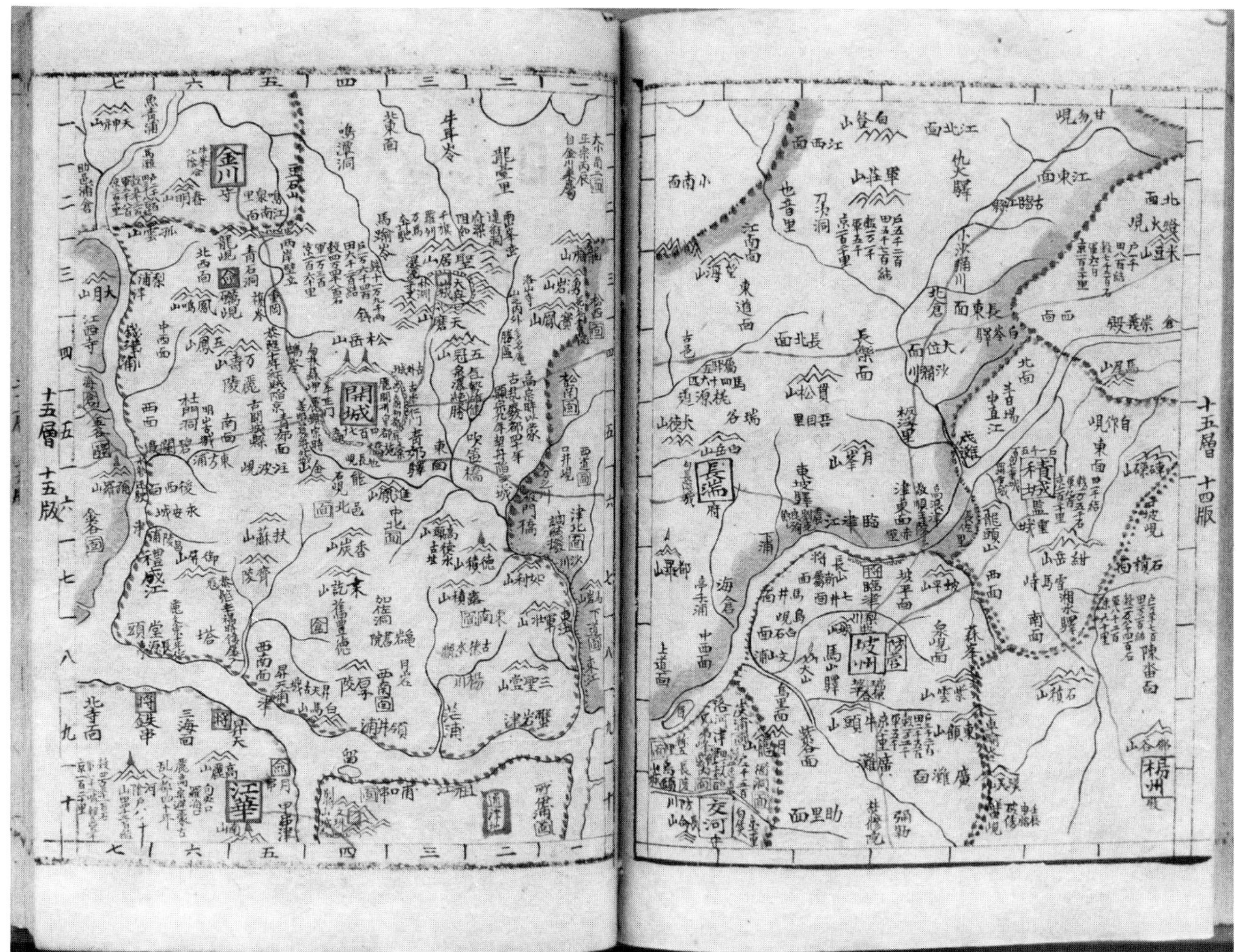

FIG. 10.35. ANOTHER VERSION OF THE *CH'ŎNGGUDO*, SHOWING THE AREA NORTHWEST OF SEOUL. From another hand copy of unknown date, also done on a preprinted form with margin scales. The rank and file numbers are indicated on the right and left margins. In this version, done with less care and skill than the one shown in figure 10.34 and hurt further by undue emphasis on the district borders, there is a richer presence of notes on features and places, and also demographic and economic data that are completely missing on the other copy. Kim Chŏngho argued strongly for such information: "Data on population, arable land, grain production, and military manpower might seem to have nothing to do with maps, but for gauging a district's relative fertility and strength nothing is better. In truth, it is the major factor in political geography." This version also differs in its treatment of mountains, showing only principal peaks and omitting ranges, avoiding the shapes-and-forces treatment in figure 10.34.
Size of each page: 27.5 × 20 cm. By permission of Kyujanggak Archives, Seoul National University (cat. no. Ko 4709-21A).

eastward and westward extensions, one need only turn the pages left or right.

While the grid has a theoretical total of 638 rectangles of 7,000 square *li* each, about half correspond to open sea and are not found in the albums. A total of 313 rectangles contain at least some land, if only part of an island, and each of these has a page. Each page is bordered top and bottom with the 70-*li* scale, and at the sides with the 100-*li* scale, so that distances can be gauged quickly, with a mere glance at the edge of the page. One does not have to make oneself little rulers. The scale has been calculated at 1:160,000.[237]

It goes without saying that at such a large scale many features are found on the *Ch'ŏnggudo* that are not generally seen on earlier Korean maps. For the first time, districts are presented with their full boundaries, and *myŏn* (subdistricts) make their appearance. Sites of

237. Yi, *Han'guk ko chido*, 208 (note 4). Pang, *Han'guk ŭi chido*, 169 (note 5), giving dimensions different from those given by Yi Ch'an, calculates a scale of 1:133,333.

former or defunct district seats are shown, as are post-station networks with indication of the number of stations and horses. We see Buddhist temples, Confucian shrines and schools, military bases and fortifications, and granaries. There are historical notes ("In Sin U's time [1380s] Sim Tŏkpu fought here with the Japanese and was defeated," but in another place, "Sin Nak smashed the Japanese here in 1592") and occasional comments on the weather ("extremely cold area") or physical features ("the true source of the Han River," "twin peaks thousands of fathoms high"). For every district, four items of econodemographic data are written in, followed by the distance from Seoul. Thus, for Changsu in Chŏlla province: "Households, 3,700; arable, 2,700 *kyŏl*; grain production, 12,600 *sŏk*; military reserves, 3,200; 650 *li* from Seoul." The figures are for the year 1828, with tens and digits lopped off.

The *Ch'ŏnggudo* was not printed, although the surviving manuscript versions are copied on woodblock-printed form sheets, with the 70- and 100-*li* scales impressed on the borders and the rank number indicated in the center strip (fig. 10.34). Kim—or someone else—evidently produced these printed and numbered sets of *sŏnp'yo* (line guides) for sale to professional copyists or do-it-yourself enthusiasts. His introductory notes on general procedures (*pŏmnye*) even include instructions on how to copy a map. With such a vast map, with so much detailed information, it is inevitable that the known copies should often be very different from each other. Some are punctiliously executed, others less carefully done (fig. 10.35).

In spite of its obvious achievement, the technical foundations of the *Chŏnggudo* seem not yet to have been investigated. But there is plenty of material in the introductory matter to the maps and in external cartographic evidence to provide an understanding of its background and structure.

The preface by Ch'oe Han'gi sums up the basics of longitude and latitude, recapitulating in balanced parallel clichés what by 1834 had become general knowledge among Chinese and Korean geographers. His notes make the usual statements about polar altitude, observations of eclipses, 200 *li* per degree of latitude, and so forth. He then says that in 1791 King Chŏngjo ordered each of his major and minor districts (then totaling 334) to make surveys with reference to a *sŏnp'yo* of "154 longitudes [*kyŏngdo*] and over 280 latitudes [*wisŏn*]." The details of this project, apparently unknown to the official annals, are unfortunately sacrificed to Ch'oe's greater concern for his mannered and very obscure parallel prose.

Kim Chŏngho, in his general procedures introduction, says that Chŏngjo ordered the districts to map their lands, and that "from this time, there were line guides with vertical and horizontal lines." Kim's wording could be rendered as "longitude and latitude," but in contrast to the impression left by Ch'oe, Kim is clearly referring to a grid and not to geodetic graticules. In fact, he uses the word *sŏnp'yo*, "line guide," as the name for his printed form sheets.

As we have seen, Kim covered all of Korea with twenty-nine horizontal ranks, each with a height representing 100 *li* north and south. Korea's southernmost piece of earth, the tiny islet of Mara, about 30 *li* south of Cheju Island, falls at about the ninetieth *li* of his twenty-ninth rank. At 100 *li* per vertical rectangle, this in effect was undrawn line 289. Thus Ch'oe Han'gi's "over 280 latitudes" neatly fits Kim Chŏngho's grid.[238] The "longitudes" match even more exactly: Ch'oe's 154, divided by Kim's 22 files, produces 7, precisely agreeing with Kim's 70 *li* of east-west distance per horizontal rectangle. Evidently, as part of the 1791 researches I have already discussed, a basic map of Korea was prepared, overlaid with a grid of 154 vertical and 280 horizontal lines, and sent in parts to each district magistrate as *sŏnp'yo* for local maps. In this way the resulting maps, all drawn locally and by many different hands, would share the same format and scale. The necessary conclusion is that the *Ch'ŏnggudo*'s 70-by-100-*li* grid rectangle was designed to accommodate the data and maps from King Chŏngjo's 1791 project.

Some scholars, seeing the 1791 maps in the context of the general geodetic theory outlined by Ch'oe Han'gi's preface, have concluded that geodetic coordinates were determined by astronomical observation as a preliminary foundation for this project.[239] I am unaware of any evidence or record of actual geodetic observations. The coordinates that are on record were plotted from distances on scale maps. If a set of coordinates existed for Korea's district seats, one would expect to find them in Kim Chŏngho's own geographical reference, the *Taedong chiji*, which in many other respects has a close relation to the *Ch'ŏnggudo*. But apart from the coordinates already discussed, none are found in that work.

Several sets of district maps exist that appear to have a connection with the 1791 mapping project mentioned by both Ch'oe and Kim. They feature a standard form with numbered grid lines that fits the description of the

238. The conventions of literary parallelism discourage blatant pairings, and since in the matching word of the previous line Ch'oe had used a numeral, it would have been bad style to use one here. So instead of saying, for instance, "289," he wrote "280-plus."

239. Yi Pyŏngdo, "*Ch'ŏnggudo* haeje," 1:9 (note 232), and Pang, *Han'guk ŭi chido*, 169–72 (note 5). Pang concludes that in 1791 "we can say that features, directions, and locations could be corrected through the composition of maps according to astronomical observations." Yi Ch'an, *Han'guk ko chido*, 209–10 (note 4), is properly cautious, simply referring to coordinates' being measured in 1791, without saying anything about astronomical observations.

sŏnp'yo. Such district maps are found in two separate collections, but their grid systems are identical, and even the line numbers are consistent with each other. Obviously these district maps have a common source, and it can only have been the 1791 project (figs. 10.36 to 10.38).[240] In laying the numbered coordinates found on these maps—some of which have been published by Yi Ch'an—over Kim Chŏngho's *Ch'ŏnggudo* reference map, one quickly sees that if they were proportionately extended all of Korea would have about 77 vertical lines and about 143 horizontal lines. These figures come out to exactly half of Ch'oe Han'gi's 154 longitudes and 280-plus latitudes. This, plus a check of distance relations on the maps themselves, shows that each square on this district map grid had sides of 20 *li*. If its line numbers are doubled, they will be congruent with the coordinates of the *Ch'ŏnggudo*. Evidently mapmakers and copyists

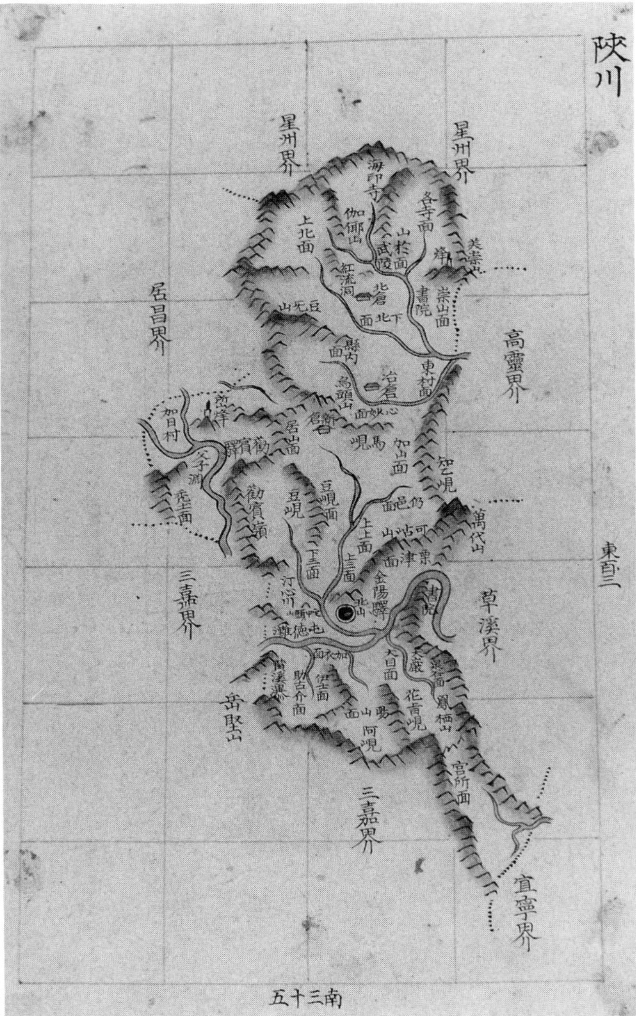

FIG. 10.36. "HAPCH'ŎN" (MAP OF HAPCH'ŎN). This map and the one in figure 10.37 are from the album *Tongguk chido* (dated by Yi Ch'an in Chŏng Sanggi's lifetime, 1678–1752, but the maps pictured here must be of the late eighteenth or early nineteenth century). These maps appear to be related to a 1791 project ordered by King Chŏngjo to map districts according to an overall national grid. Note the line numbers at right and bottom on both maps. The district seat of Hapch'ŏn is near the intersection of "East 103" and "South 35"; that of Kyŏngsan (fig. 10.37) is near "East 101" and "South 28." The grid squares have twenty-*li* sides, while the *Ch'ŏnggudo*'s margin scales have ten-*li* intervals. Using Kyŏngsan as an example, doubling its numbers produces near congruity with the *Ch'ŏnggudo* coordinates 20.2 (twenty *li* from the top of rank 21) and 56 (near left edge of file 8).
Size of the original: 53 × 35.5 cm. By permission of the National Central Library, Seoul (cat. no. Kwi 677 Sunggye ko 2702-22).

240. The albums in question are the *Tongguk chido* (Maps of the Eastern Country) and *Haedong yŏjido* (Terrestrial maps of [the country] east of the sea [Korea]), illustrated in Yi, *Han'guk ko chido*, 141 and 150–51, respectively (note 4). I find I must disagree with Yi Ch'an, and apparently with the bibliographers of the National Central Library, who regard both of these albums as autograph copies of local and provincial maps by Chŏng Sanggi. Apart from their conformity with the written references to the 1791 map project by Ch'oe Han'gi and Kim Chŏngho, and especially the relation of their line numbers to the grid of the *Ch'ŏnggudo*, there is no evidence of the Korean use of grids during Chŏng Sanggi's lifetime (1687–1752). There may be some confusion here between a bar scale, which Chŏng definitely used, and a grid. A grid implies a bar scale, but a bar scale does not necessarily imply a grid.

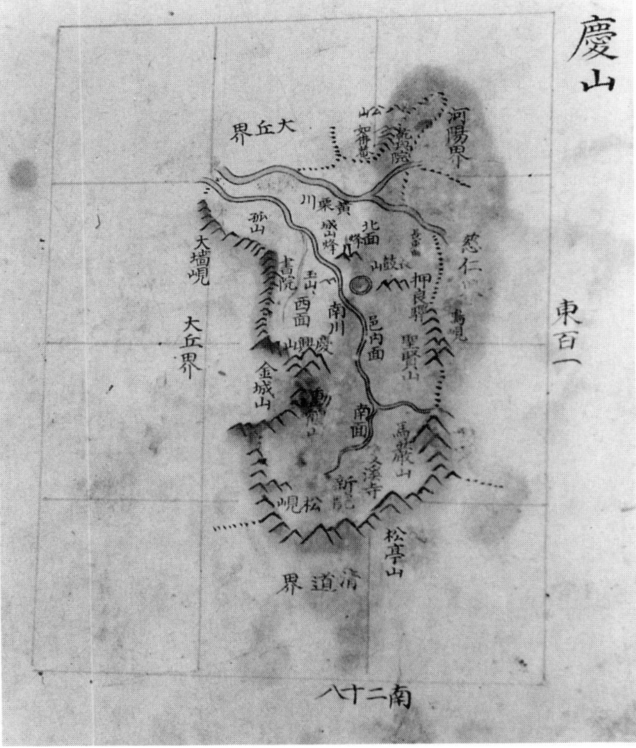

FIG. 10.37. "KYŎNGSAN" (MAP OF KYŎNGSAN). See figure 10.36.
Size of the original: 53 × 35.5 cm. By permission of the National Central Library, Seoul (cat. no. Kwi 677 Sunggye ko 2702-22).

found a grid of 154 by 280+ needlessly fine, to the point that it actually impeded visual comprehension of the map. In fact, Ch'oe Han'gi made exactly that complaint about the 1791 project in his preface to the *Ch'ŏnggudo*.

In his general procedures introduction, Kim has special praise for the district maps of Chŏng Ch'ŏlcho, Hwang Yŏp, and Yun Yŏng,[241] but he criticizes those of nameless others for imprecision and for using nonstandard grids. "Therefore," he says, "I have here used the overall national grid map [*taebok chŏndo*] to fix [my] ranks and files, and arranged them in a booklet, so that both faults can be avoided while information from written descrip-

FIG. 10.38. "CHECH'ŎN, CH'ŎNGP'UNG, TANYANG" (MAP OF CHECH'ŎN, CH'ŎNGP'UNG, AND TANYANG). From the album *Haedong yŏjido*, with dating and dating revision the same as for figures 10.36 and 10.37. This map, although from a different collection and copyist, reflects the same coordinate system as that of the *Tongguk chido*, illustrated in figures 10.36 and 10.37. The vertical line marked "South 35" in that illustration is the same as "South 35" in this one; transferred to the *Ch'ŏnggudo*, this line falls near undrawn line 68 on the left side of Kim Chŏngho's file 10. Exact congruence would have been at line 70 (2 × 35), but since Kim has refined the distance figures, especially in the eastern half of the country, his grid lines often differ slightly from those of the district maps. The *Haedong yŏjido* has 146 maps of districts or district clusters like this one.
Size of the original: 34 × 22.3 cm. By permission of the National Central Library, Seoul (cat. no. Ko 2107-36).

241. Of these three mapmakers, I have found only Chŏng Ch'ŏlcho mentioned in contemporary sources. Evidently he was known for careful brushwork: in 1781, as a lower-ranking censorate official, he was asked to make a copy of the royal portrait; see *Chŏngjo sillok*, 12.28a–b (note 17).

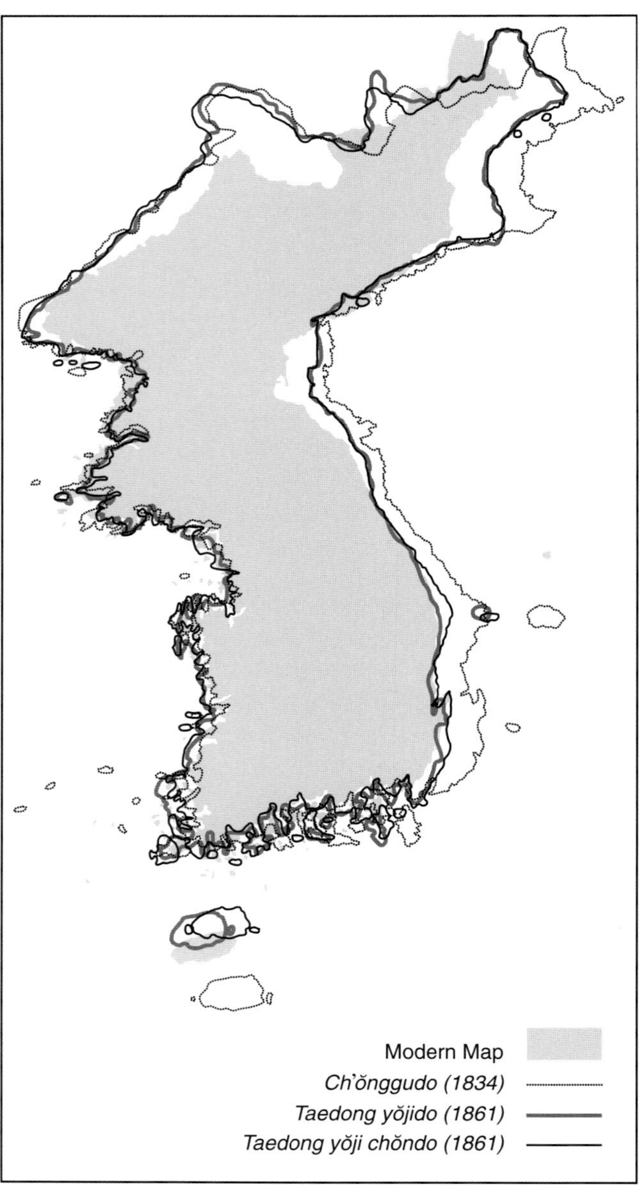

FIG. 10.39. COMPARISON OF PENINSULAR OUTLINES OF KIM CHŎNGHO'S *CH'ŎNGGUDO*, *TAEDONG YŎJIDO*, AND *TAEDONG YŎJI CHŎNDO* WITH THOSE OF A MODERN MAP. In his *Taedong yŏjido* and *Taedong yŏji chŏndo*, Kim Chŏngho considerably corrected the overextension eastward of the east coast of the *Ch'ŏnggudo*, but lacking accurate, astronomically derived longitude data, he was never able to achieve complete success by modern standards. Note the relative undercorrection in the north and the slight overcorrection in the south.
Based on Yi Ch'an, *Han'guk ko chido* (Seoul: Han'guk Tosŏgwanhak Yŏn'guhoe, 1977), 211, but recompiled with the *Taedong yŏji chŏndo* for further comparison.

TABLE 10.9 Relative Length of Korea on Maps by Chŏng Sanggi and Kim Chŏngho

Latitudinal Difference	*Tong'yŏ ch'ongdo*	*Ch'ŏnggudo*	*Taedong yŏjido*	*Taedong yŏji chŏndo*	*ONC* F-9, G-10
Onsŏng–Seoul	15.0 cm	1,495 *li*	1,479 *li*	24.0 cm	607 km
Seoul–Haenam	9.5 cm	853 *li*	863 *li*	14.5 cm	363 km
Total units	24.5 cm	2,348 *li*	2,342 *li*	38.5 cm	969 km
North–south ratio	1.578	1.752	1.713	1.655	1.672

Source: My calculations. The figures show the distance between the latitude of Seoul and the extreme northern and southern latitudes (not including islands) of Onsŏng and Haenam districts, repectively, and comprise the total length of the Korean mainland. The *Tong'yŏ ch'ongdo* (fig. 10.31) and the *Taedong yŏji chŏndo* (fig. 10.47) were measured from large photographs in Yi Ch'an, *Han'guk ko chido* (Old Korean maps) (Seoul: Han'guk Tosŏgwanhak Yŏn'guhoe, 1977). The *li* figures for the *Ch'ŏnggudo* and the *Taedong yŏjido* were counted out using the respective grids of those maps. Modern distance figures were measured on Operational Navigational Charts sheets F-9 and G-10 (St. Louis, Mo.: Aeronautical Chart and Information Center, U.S. Air Force, 1966). The ratios show that Kim Chŏngho must have constantly worked to refine his distance figures, in the end coming remarkably close to modern proportions.

tions and previously compiled maps can be researched with reference to them."[242]

Thus Kim used existing maps related to the 1791 project, and probably others that he must have collected or received elsewhere in the course of his work. His "general procedures" in places reads like a list of instructions to collaborators, and perhaps in another form it filled such a role.

In determining the general shape of the peninsula, Kim started with a preexisting national grid map (*taebok chŏndo*), which must have been compiled from maps of the Chŏng Sanggi style as edited by the 1791 project. The relative disposition of the *Ch'ŏnggudo*'s district seats is more refined and accurate than on the Chŏng Sanggi maps, and the line of flow of the Yalu River is closer to the facts, giving P'yŏng'an Province overall better proportions. But either Kim or the 1791 project team left the major upper bend of the Yalu too far north, and in correcting its general direction had to introduce another bend that made the original distortion even worse. And on the southern part of the east coast there is a peculiar protuberance that is not on the Chŏng Sanggi maps and is hard to explain as anything other than an uncharacteristic error. Finally, as on Chŏng's maps, the east coast is too far from the west coast. Although this problem was considerably rectified on the *Taedong yŏjido* (Territorial map of the Great East [Korea]) in 1861, even there the coastal error in the bend of the Bay of Wŏnsan persisted (fig. 10.39 and table 10.9). Apart from these problems, Kim's outline is generally more accurate than Chŏng's.

Once the districts were generally positioned, the cartographic work focused on them. Kim gives a step-by-step account of how he proceeded at this level, along with a generalized *chido sik* (demonstration map) of an average district with its various typical features by way of illustration (fig. 10.40). For each district he would sketch out a draft map on a separate sheet, first placing the seat with respect to the four margins and laying out the mountains and rivers in general. Then, with the district seat as center point, he would mark twelve directional positions on the margins for reference, using the same characters used to name the hours of the day. Next he would draw concentric circles at ten-*li* intervals, radiating from the seat. Following his tables of distance figures from the seat to the borders, he would draw in the district boundaries with reference to the concentric circles. The next step was to refine the relationship of the rivers and mountains. With data on river length from point to point—say from source to confluence with another river—he would use the concentric circles to fix its course. When the rivers and streams were in, and the general watershed areas therefore known, he would sketch in the principal mountains of the shapes-and-forces network, connecting them with a sawtooth pattern, leaving wide spaces for the plains and open areas. Although he does not mention roads in this section, it was probably at this point that they were drawn in. (In published illustrations of the *Ch'ŏnggudo*, they often show poorly because of color problems.) Finally he would fill in the names of the subdistricts, post stations, garrisons, granaries, reservoirs, shrines, schools, and such, and enter outside the district boundaries the names of the various neighboring districts. In the next stage of the work, when the district maps were all drafted, he would reconcile them with each other and enter them on the general map, removing all the concentric circles and reference directions.

The basic data base for this work was eventually included in Kim's *Taedong chiji* (1864), which had extensive tables of distance figures and demographic data (fig. 10.41) for each district and long lists of administrative, military, and cultural buildings or sites, with distances from the district seat. Kim mined the obvious sources, such as the *Sŭngnam* and the *Munhŏn pigo*, but he also used more up-to-date government data, for instance, the statistics of the state defense council for 1828. He prob-

242. Kim, *Ch'ŏnggudo*, 1:3–4 (second introductory pagination) (note 232).

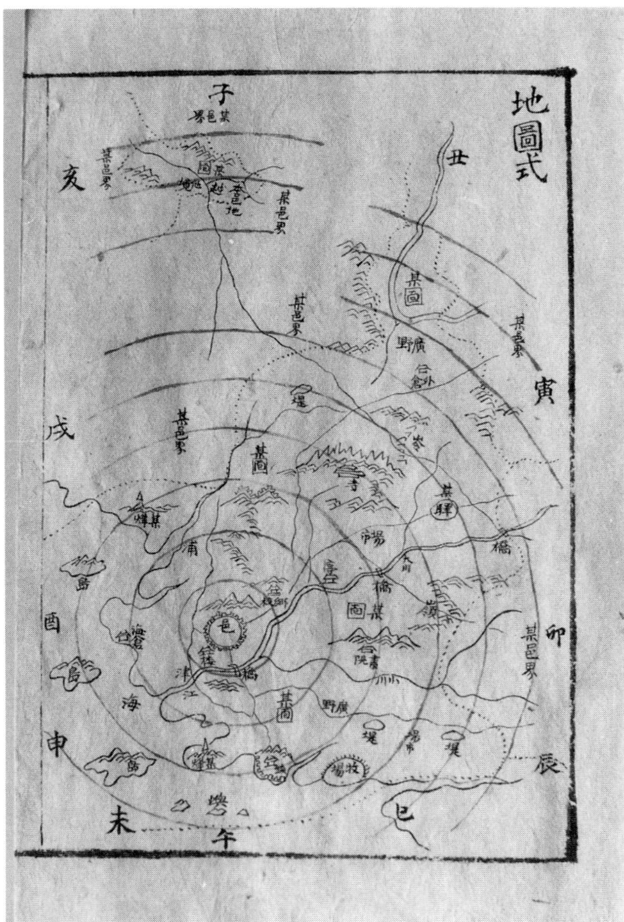

FIG. 10.40. *CHIDO SIK* (DEMONSTRATION MAP OF A DISTRICT). From Kim Chŏngho's general procedures (*pŏmnye*) introduction to his *Ch'ŏnggudo*. Kim drew this as a demonstration to accompany his discussion of the proper procedures for drawing a district map. It also shows the scope of his district coverage. The concentric circles, radiating at ten-*li* intervals from the district seat, and the twelve directional indicators entered boldly at the margins are used in the draft stage as guides for the accurate placement of features and localities; they are removed when the map is transferred to the general map. Size of the original: 27.5 × 20 cm. By permission of Kyujanggak Archives, Seoul National University.

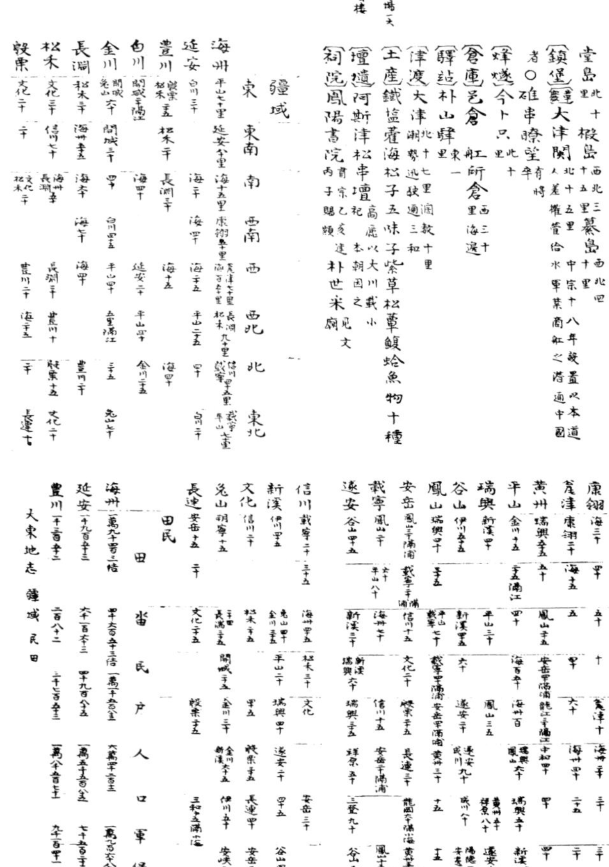
疆域
東 東南 南 西南 西 西北 北 東北
海州 延安 豐川 白川 金川 長淵 松禾 殷栗

大東地志 疆域 民田

FIG. 10.41. A SAMPLE OF KIM CHŎNGHO'S DATA. Distance and demographic data from Kim's *Taedong chiji*, which he used in compiling the *Ch'ŏnggudo* and *Taedong yŏjido*. The two pages shown are from the section on Hwanghae Province (*kwŏn* 18, 390b–391a). At upper left and on most of the lower page is a table of interdistrict distances for Hwanghae Province. Along the top are the names of the districts; along the right, eight directional labels for the eight bars of distance data under each district name, starting at east, then southeast, etc., and around to northeast. The distances are to the borders of neighboring districts or to the sea. Thus the first four bars for the first district, Haeju, read: East, P'yŏngsan 70 *li*; Southeast, Yŏn'an 80; South, the sea 15; Southwest, Kangnyŏng 50; etc. Kim rounds figures off to 0 or 5. On the lower left is the beginning of a table by district of demographic and economic data. Figures are given for arable area, paddyland area, households, population, and military reserves. Kim cites an 1828 government source for these statistics. Scholars believe the manuscript of the *Taedong chiji* was brought to the state now known in 1864, but Kim had begun assembling the data long before; they are already widely reflected on the *Ch'ŏnggudo* of 1834.
From the reprint of the *Taedong chiji*, ed. Yi Pyŏngdo (Seoul: Hanyang Taehakkyo Kukhak Yŏn'guwŏn, 1974), 390–91.

ably relied on connections with government officials and influential friends, such as Ch'oe Han'gi, for this kind of information. Local histories (*ŭpchi*) were consulted where available. Not all of this material was suitable for cartographic expression, but Kim, at least in his early career, saw maps and written descriptions as interdependent and equally necessary.

A particular point of interest in the *Ch'ŏnggudo* is the treatment of mountain ranges, on which we find mixed signals. In one place in his introduction he observes that mountains and watershed areas will naturally be found between the sources for rivers and streams, and he advises that only principal mountains need be entered on the map, writing: "It is not necessary to connect the peaks and join the ridges; it makes a mess of the sheet and can cause errors. Where there is a famous mountain, just draw three or four peaks."[243] This suggests the treatment seen in figure 10.35, where individual mountains appear more or less in the *Sŭngnam* style. However, we have also seen him counseling the connection of the major mountains with a sawtooth pattern, which seems to have been the advice followed by the copyist in figure 10.34. It is possible that this second approach—that is, the classical shapes-and-forces emphasis—was more popular with map buyers, because in the end Kim went in this direction, producing in his 1861 map the grand culmination of shapes-and-forces cartography.

If the period before 1834 is a blank concerning our knowledge of Kim Chŏngho, so are the twenty-seven years after 1834. The diligent research of scholars has not uncovered a single syllable by him or about him from that time until 1861, when the first edition of his *Taedong yŏjido* appeared under his own imprint (figs. 10.43 to 10.46). Since we resume our scrutiny of his career with a printed map of the same general type as the *Ch'ŏnggudo*, we might approach this twenty-seven-year gap by asking why Kim Chŏngho, a printer, apparently never printed his 1834 opus.

Woodblock printing, like any other product, responds to market conditions, and the strongest factor in Kim's not printing his *Ch'ŏnggudo* was likely the economic one. Not all woods are ideal for block carving; the wood must be hard enough to provide a durable, even surface through repeated inkings, yet soft enough for reasonably efficient carving. Nor are acceptable woods always available in sizes sufficient to accommodate an average double page. The block-printed form used to copy the two hand-produced *Ch'ŏnggudo* rectangles illustrated in figure 10.34 measured a bit over thirty-five by twenty-three centimeters; the actual block, including edges and margins, would have been larger, requiring a board no less than twenty-five or twenty-six centimeters wide. Extra cutting, transport, and finishing costs would have added to an already high price for the wood itself. Next came the cost of carving, an uncommon skill requiring special training and long experience. All together, in the case of the *Ch'ŏnggudo*, one would have to calculate materials, tools, labor, damage and wastage, and overhead and storage costs for 162 blocks, the number necessary to provide for the *Ch'ŏnggudo*'s 324 grid rectangles (because some ranks have an odd number of files, eleven of the blocks were only half used, for a total of 313 rectangles). When it was all done, such a printing operation had two strong advantages: with adequate storage and reasonable maintenance, the blocks would have a relatively long income-generating life, and once a correct text or image was achieved, error was permanently eliminated. The disadvantages were high production costs and long-term storage and maintenance costs.

For such an investment to make sense, there would have had to be a heavy initial demand and a steady market for the middle and long term. But Kim's map was very large and specialized and would have been very expensive. Add to that the bad economy and hard times in his day, and it is not likely that buyers were lined up outside his door. Even in relatively prosperous times, printed editions of books were expensive enough in Korea to justify hand copying by individual readers or suppliers. The number of such manuscript copies surviving today is very large relative to printed copies. Many books of great merit were never printed; very few of the classics of the *sirhak* school of scholarship, for example, were printed, and even today some major works are available only in photolithographic reprints of an original manuscript or hand copy. Even if Kim Chŏngho had seen a heavy demand, he would have realized that a printed edition could only end up as the source of an even greater number of hand copies, on which he would realize no profit at all.

One might also ask if Kim was satisfied with the map he had produced. We have already detected in him some ambivalence over whether to add a shapes-and-forces emphasis to the *Ch'ŏnggudo*. Other differences from one copy to another show a variety of tastes in style and coverage among users or copyists (whether to include demographic data, for instance), and some of this hesitation may have been original with Kim himself. Inevitably there were errors, and it would have been expected that Kim would discover them himself or have them pointed out to him. Finally, a cartographer of Kim Chŏngho's intelligence and imagination could hardly go twenty-seven years without having some new ideas about maps. Whatever the factors that led to his decision, the *Ch'ŏnggudo* was never printed, and in 1861 he put out an entirely new map.

In 1861, times were very different than in 1834. As of 1834, no Westerners had penetrated Korea, and the Catholics seemed under control. But by 1861 Korea was in a state of alarm over the West. The previous year, British and French troops had occupied Beijing, burned down the imperial summer palace, and compelled the Chinese to accede to Western demands on trade and diplomatic practice as well as to accept open missionary activity. Many Koreans thought Korea was next, and they considered the French missionaries operating underground in Korea since 1836 a fifth column. Amid their alarm and uncertainties, they began to urgently address military and defense concerns.

Kim Chŏngho's preface to the *Taedong yŏjido* reflects

243. Kim, *Ch'ŏnggudo*, 1:7 (second introductory pagination) (note 232).

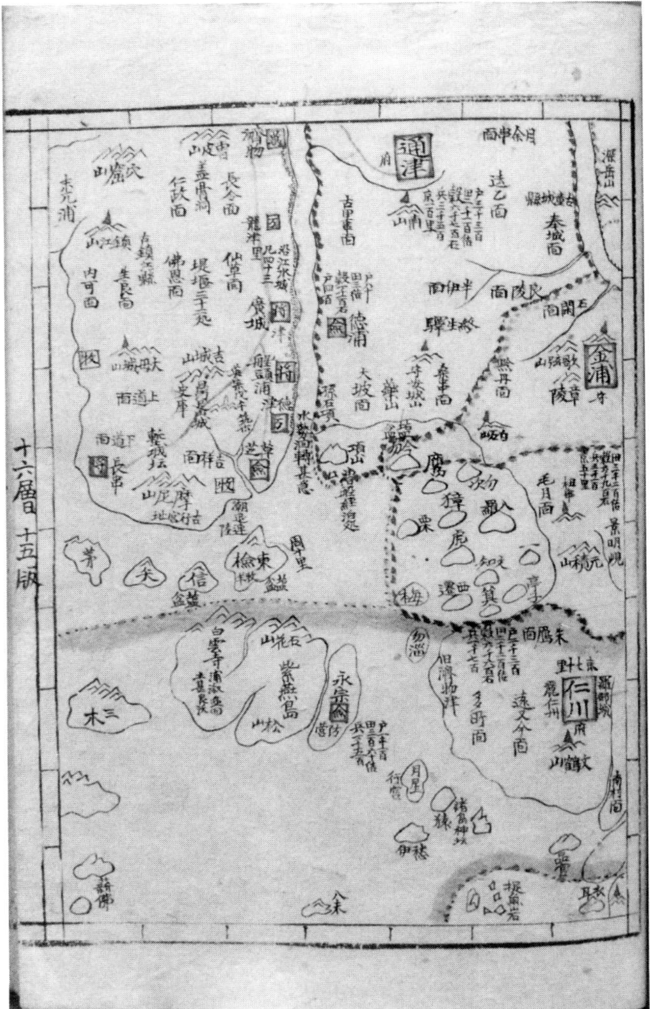

FIG. 10.42. INCH'ŎN AND SOUTHERN KANGHWA ISLAND FROM THE *CH'ŎNGGUDO*, BY KIM CHŎNGHO, 1834. This is rectangle 16–15. Compare the *Taedong yŏjido* treatment of the same area in figure 10.43. At lower right is the district seat of Inch'ŏn (name in rectangle); the modern city centers on "Old Chemul Crossing" (commonly Chemulp'o) on the coast about twenty *li* to the west (each numbered unit on the borders equals ten *li*). The southern part of Kanghwa Island is at upper left; it was the site of several fortified royal refuges for times of emergency: note the fortifications along the channel, designed to protect the island from forces on the mainland. Squares indicate military bases; the enclosed character shows the rank of the commander. Flame-shaped black triangles mark signal stations. The district at the far upper right is Kimp'o, site of the modern international airport.
Size of the original: 27.5 × 20 cm. By permission of Kyujanggak Archives, Seoul National University.

this concern. About a third of it is devoted to a long quotation from the classical Chinese military strategist Sunzi emphasizing deep familiarity with the terrain before the battle takes place. In times of crisis, such preparation is necessary to deal with violent events; in times of peace, the same knowledge is applied in the administration of the state and the people. During the eighteenth and nineteenth centuries, many *sirhak* scholars—Chŏng Sanggi among them—had specialized in military and defense studies, but with no sense of urgency. Between the lines of Kim's preface, one senses that in 1861 the time for study was over and the moment for actual preparation was at hand. Whether such an atmosphere also created a demand for a map like Kim's, we can only speculate. But beyond economic factors, one should not discount patriotism or national need, and perhaps even special financial help from men of means who might have wanted the map published. If there is anything to the legend that Kim was thrown in jail for compromising national security, at least he himself had suggested his own concern with that issue in putting his map before the public.

The *Taedong yŏjido* was not just a corrected and improved edition of the *Ch'ŏnggudo* but a new map.[244] Evidently Kim had continued his research and travels, refined his distance measurements, and redrawn the entire country both generally and in detail. He devised a new method for indicating scale and introduced a new system of symbols. He drastically changed his drafting style and his treatment of mountains and of islands. He abandoned textual notes and econodemographic data, eliminated subdistricts, and considerably developed his coverage of roads. Finally, he completely redesigned his format, abandoning the alternating albums in favor of long folded strips that were more compact and permitted even greater flexibility of display. Let us examine some of these changes.

The shape of the country was improved, through a substantial realignment of the east coast. For us it is easy to recognize this, simply by comparing the new east coast with that on a modern map (fig. 10.39). But there were no modern maps for him to see. He must have achieved this by a general refinement and shortening of east-west measurements across the mountainous eastern spine of the country. Failing that, the only thing that could explain his more accurate placement of the coastline would be an astronomically based set of longitude data. Although such progress may have occurred and is not to be ruled out, the fact is that we have no concrete knowledge of such observations or measurements, either by Kim or by others in Korea. Kim's new alignment of the coast fell short of what was needed, especially in the north, but it was a decided improvement over that of the *Ch'ŏnggudo* and was the most accurate ever determined before modern times.

244. See general treatments of the *Taedong yŏjido* in Yi, *Han'guk ko chido*, 210–11 (note 4), and Pang, *Han'guk ŭi chido*, 180–89 (note 5).

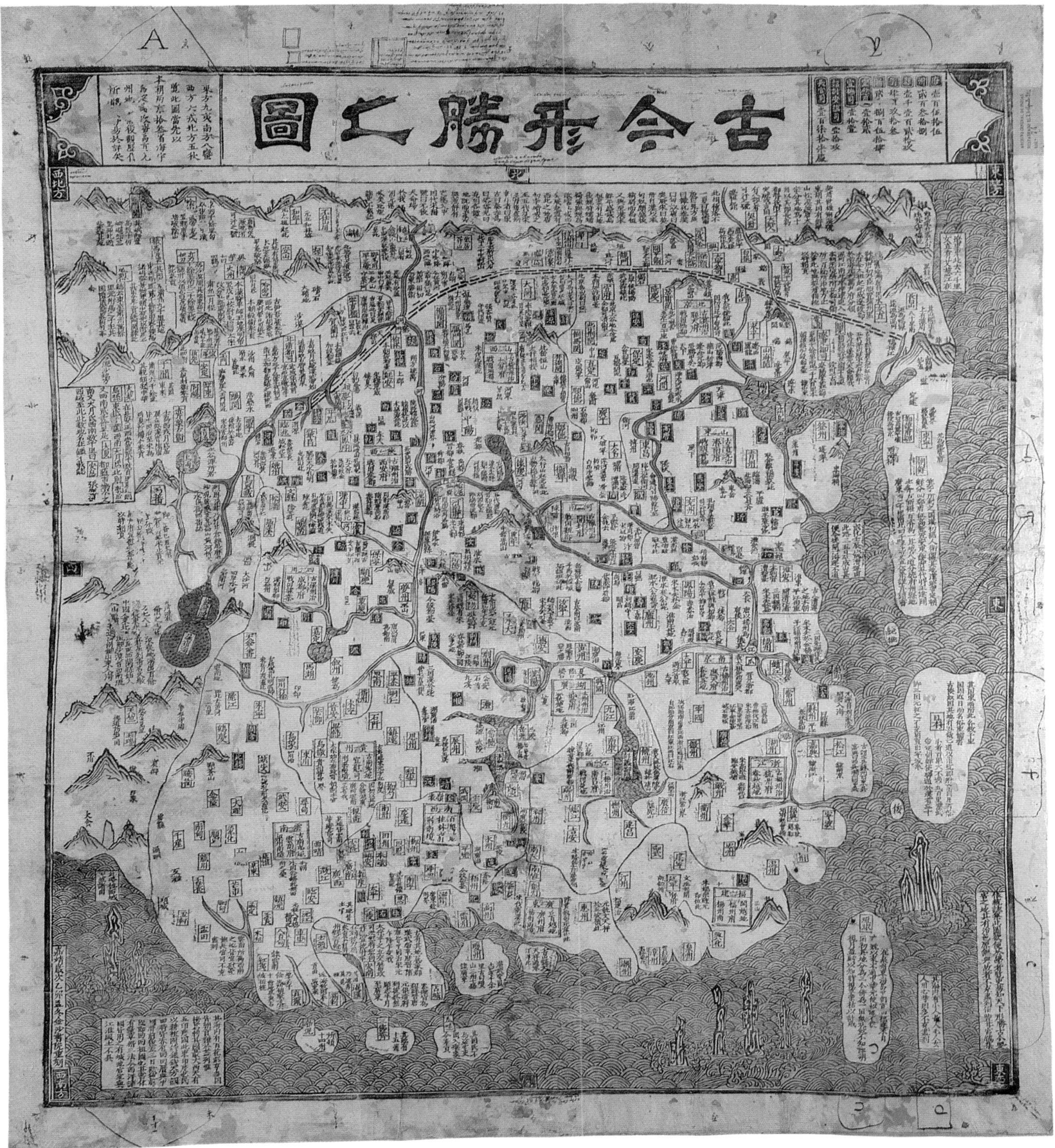

PLATE 1. *GUJIN XINGSHENG ZHI TU* (MAP OF ADVANTAGEOUS TERRAIN PAST AND PRESENT, 1555). (See p. 59.) This map represents territory from Samarkand in Central Asia to Japan and from present-day Mongolia to Java and Sumatra in Southeast Asia. The map was printed with black ink from a woodblock and colored by hand afterward. The Huanghe (Yellow River) is yellow, and the Changjiang (Yangtze River) is blue. Mountains are represented pictorially, as is the Great Wall. Much of the map is covered by notes describing changes in place-names and administrative status. This map was sent to Spain in 1874 by the Spanish viceroy in the Philippines.
Size of the original: 115 × 100 cm. By permission of Archivo General de Indias, Seville.

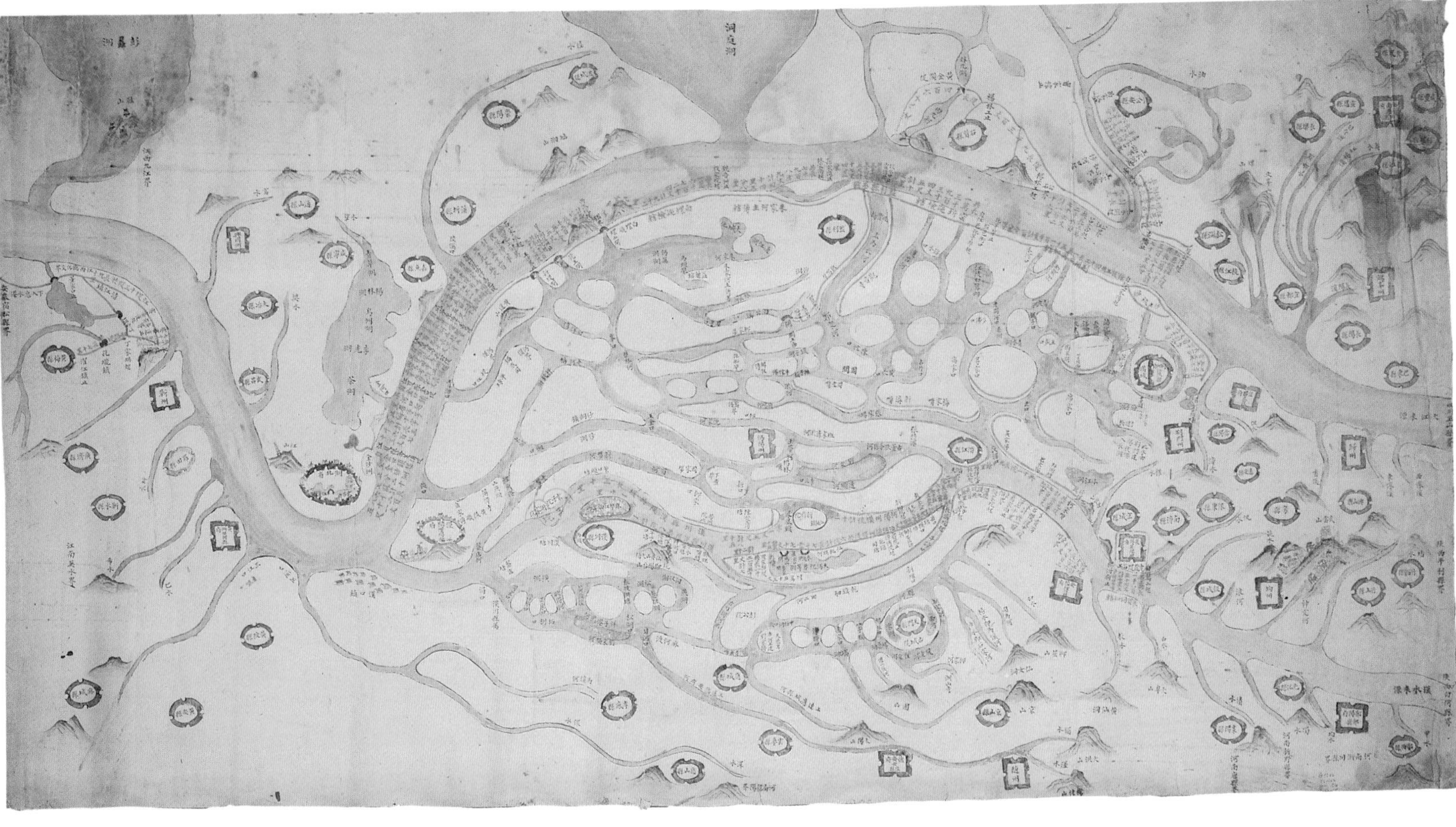

PLATE 2. *CHANGJIANG TU* (MAP OF THE YANGTZE RIVER). (See p. 61.) This map shows flood control works on the Yangtze and Han rivers in Hubei Province. It is oriented with north at the bottom. Size of the original: 74 × 140 cm. Courtesy of the Geography and Map Division, Library of Congress, Washington, D.C. (G7822.Y3N22 18--.C4 Vault).

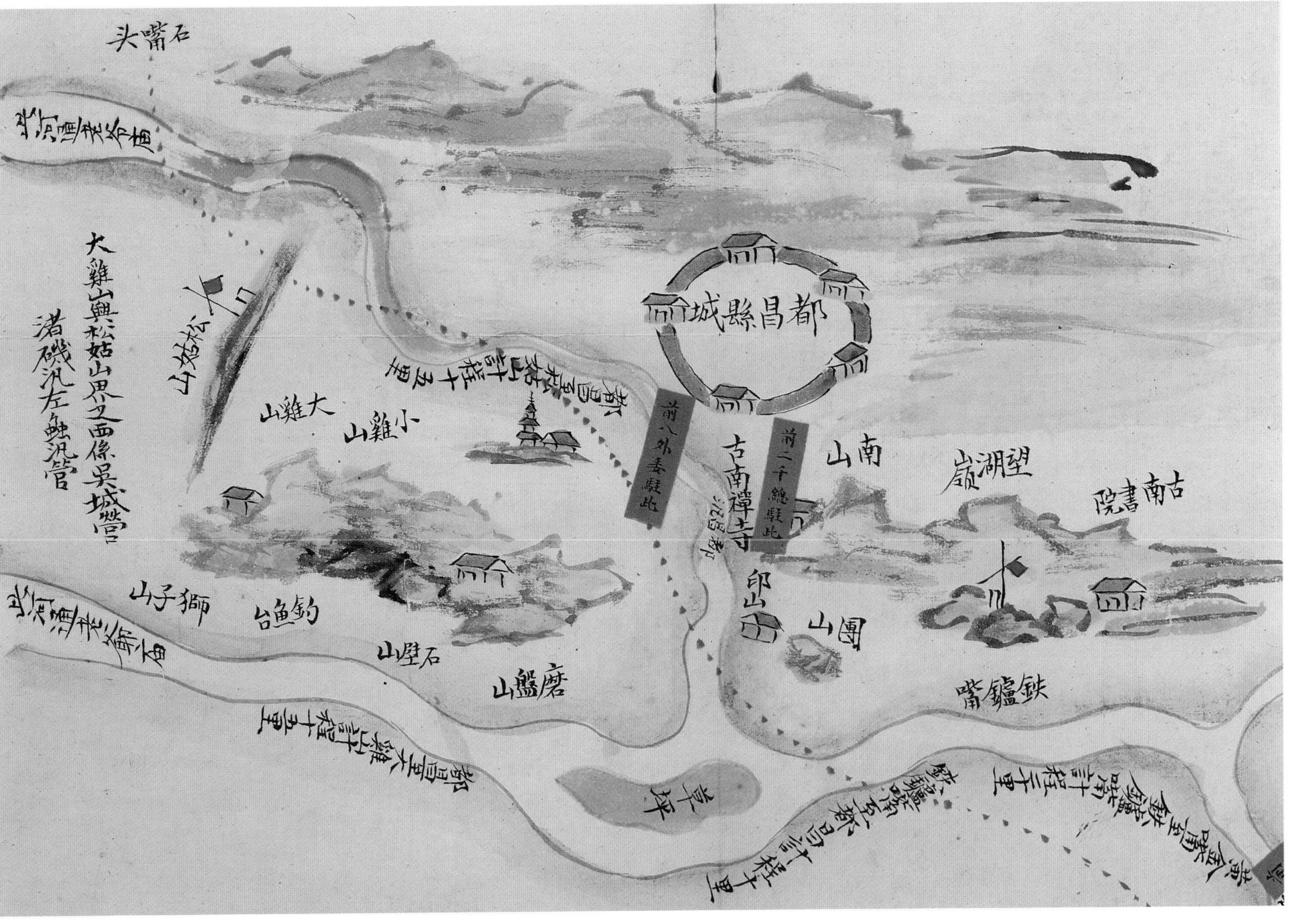

PLATE 3. WATER POLICE MAP, 1850. (See pp. 61–62.) This map represents the jurisdiction of the water police station at Tongzhou near the mouth of the Yangtze River. The red strips on the map give patrol instructions, distance information, and information about tributaries. Size of the original: 63.5 × 96 cm. By permission of the Staatsbibliothek zu Berlin—Preussischer Kulturbesitz, Kartenabteilung (E 1155).

PLATE 4. DETAIL OF A CHINESE COASTAL MAP ON A SCROLL. (See p. 62.) This map represents the Chinese coast from Korea to Annam. According to the accompanying text, the map was intended as an aid to surveillance of the coast for the protection of commerce. Strategic points and harbors are labeled. Mountains, walls, bridges, and temples are represented pictorially. Counties (*xian*) are represented as red squares. Size of the entire original: 30 × 900 cm. By permission of the Staatsbibliothek zu Berlin—Preussischer Kulturbesitz, Kartenabteilung (E 530).

PLATE 5. SECTION OF A MAP OF THE GRAND CANAL LINKING BEIJING WITH HANGZHOU, EIGHTEENTH CENTURY. (See p. 101.) The thousand-mile length of the canal is represented on a scroll. While the waterway is drawn in plan, mountains are presented in elevation and cities and towns from a bird's-eye view. Compare this map with figures 5.2 and 5.3. Size of this detail: ca. 49 × 47 cm. By permission of the British Library, London (MS. Or. 2362).

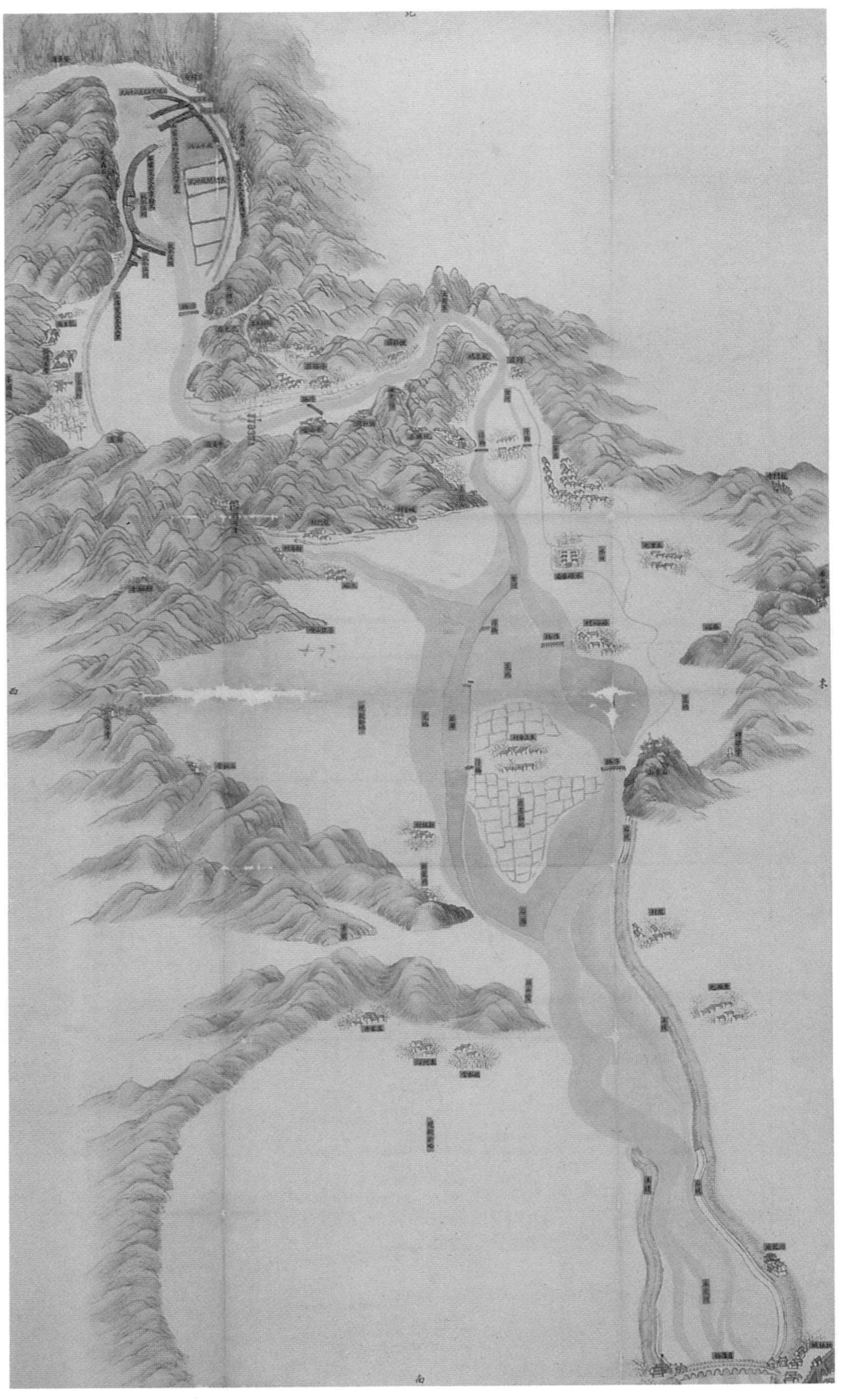

PLATE 6. QING MAP OF THE YONGDING RIVER. (See p. 102.) This map, presented to the emperor with a memorial, depicts conservancy works along the Yongding River (in present-day Hebei Province). It is oriented with north at the top. Size of the original: ca. 95 × 55 cm. By permission of the National Palace Museum, Taipei, Taiwan.

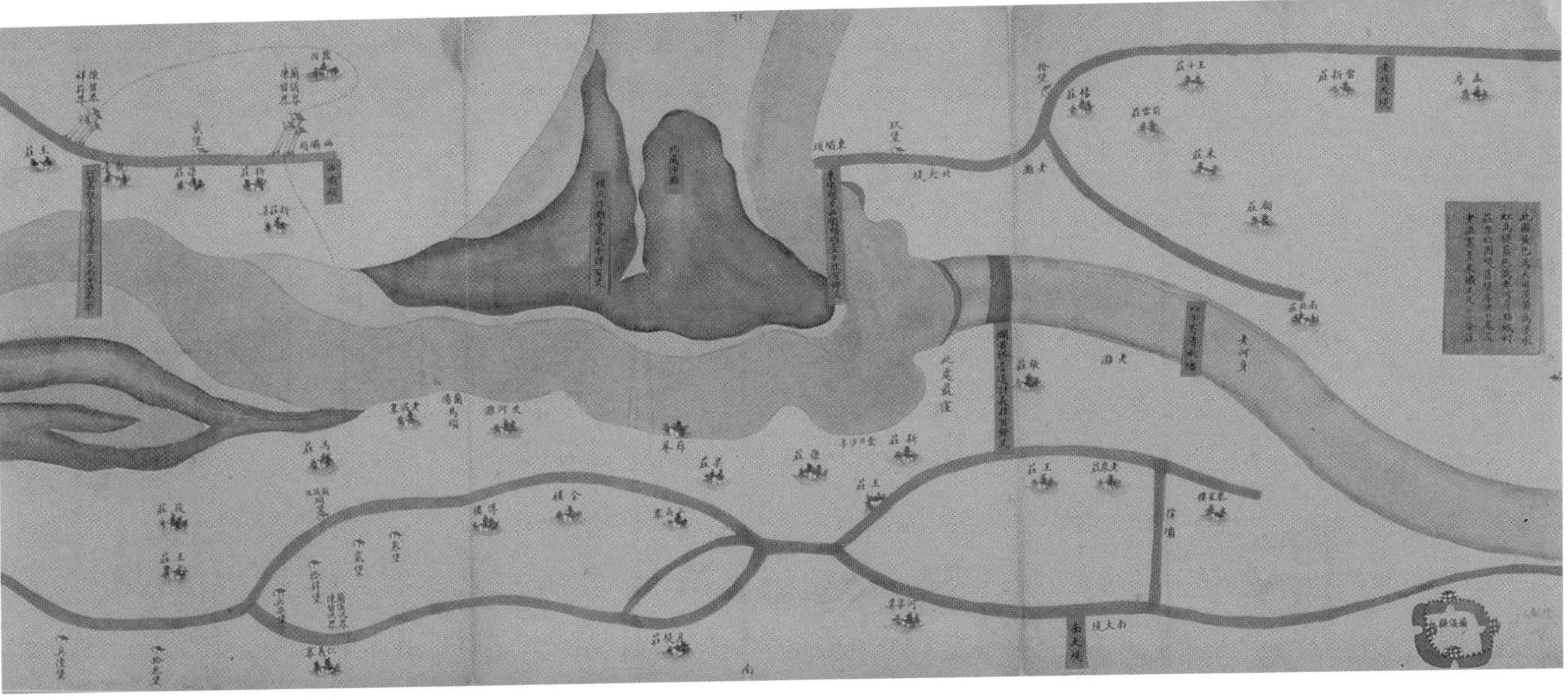

PLATE 7. QING MAP OF THE YELLOW RIVER. (See p. 102.) This map depicts the conservancy works along the Yellow River at Lanyi Xian (in present-day Henan Province). It was presented to the emperor with a memorial. Size of the original: ca. 30 × 76 cm. By permission of the National Palace Museum, Taipei, Taiwan.

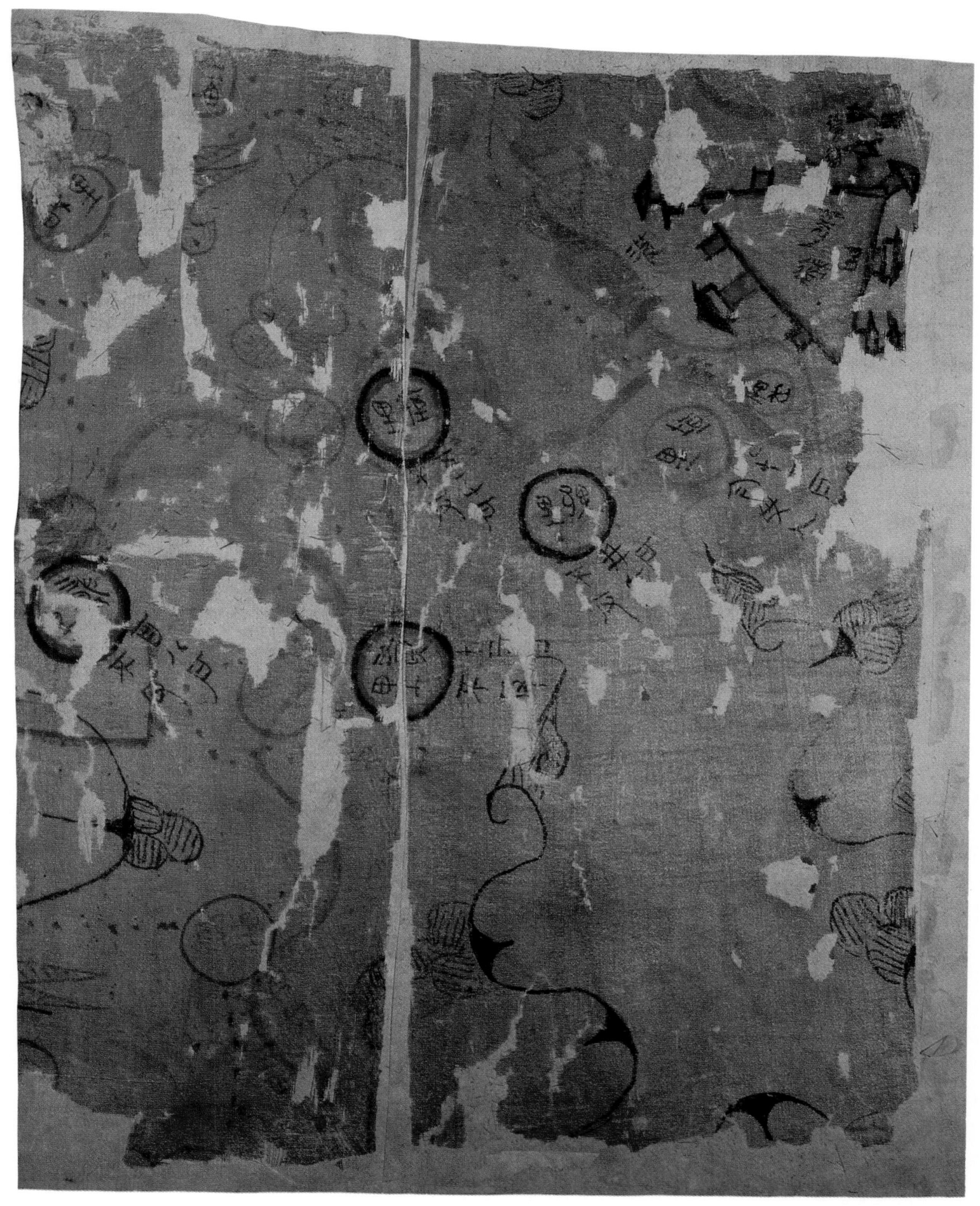

PLATE 8. DETAIL OF A SILK MAP FROM THE HAN DYNASTY. (See pp. 147–48.) Detail of the garrison map drawn on silk discovered in Han tomb 3 at Mawangdui. See also figure 3.10 above. Size of this detail: ca. 25 × 19 cm. By permission of Wenwu Chubanshe, Beijing.

PLATE 9. DETAIL FROM THE PAINTING OF WANG-CHUAN. (See p. 151 and fig. 6.17.) Size of the detail: unknown. By permission of the Seattle Art Museum (Eugene Fuller Memorial Collection 47.142).

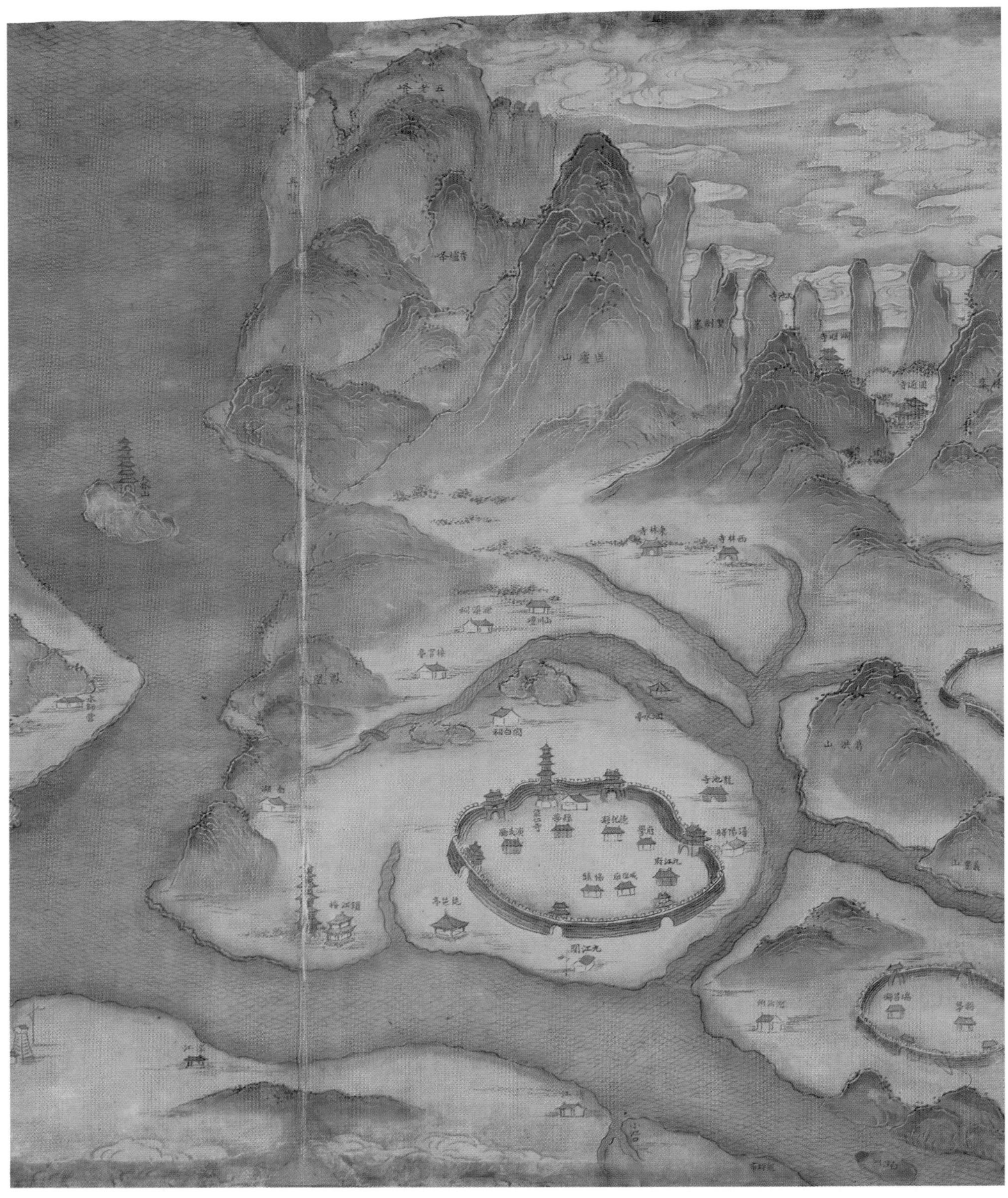

PLATE 10. DETAIL FROM A PREFECTURAL MAP FROM AN EIGHTEENTH-CENTURY MANUSCRIPT ATLAS OF JIANGXI PROVINCE. (See pp. 152–53.) Size of the entire original: ca. 40 × 53 cm; this detail: ca. 35 × 27 cm. By permission of the British Library, London (Add. MS. 16356).

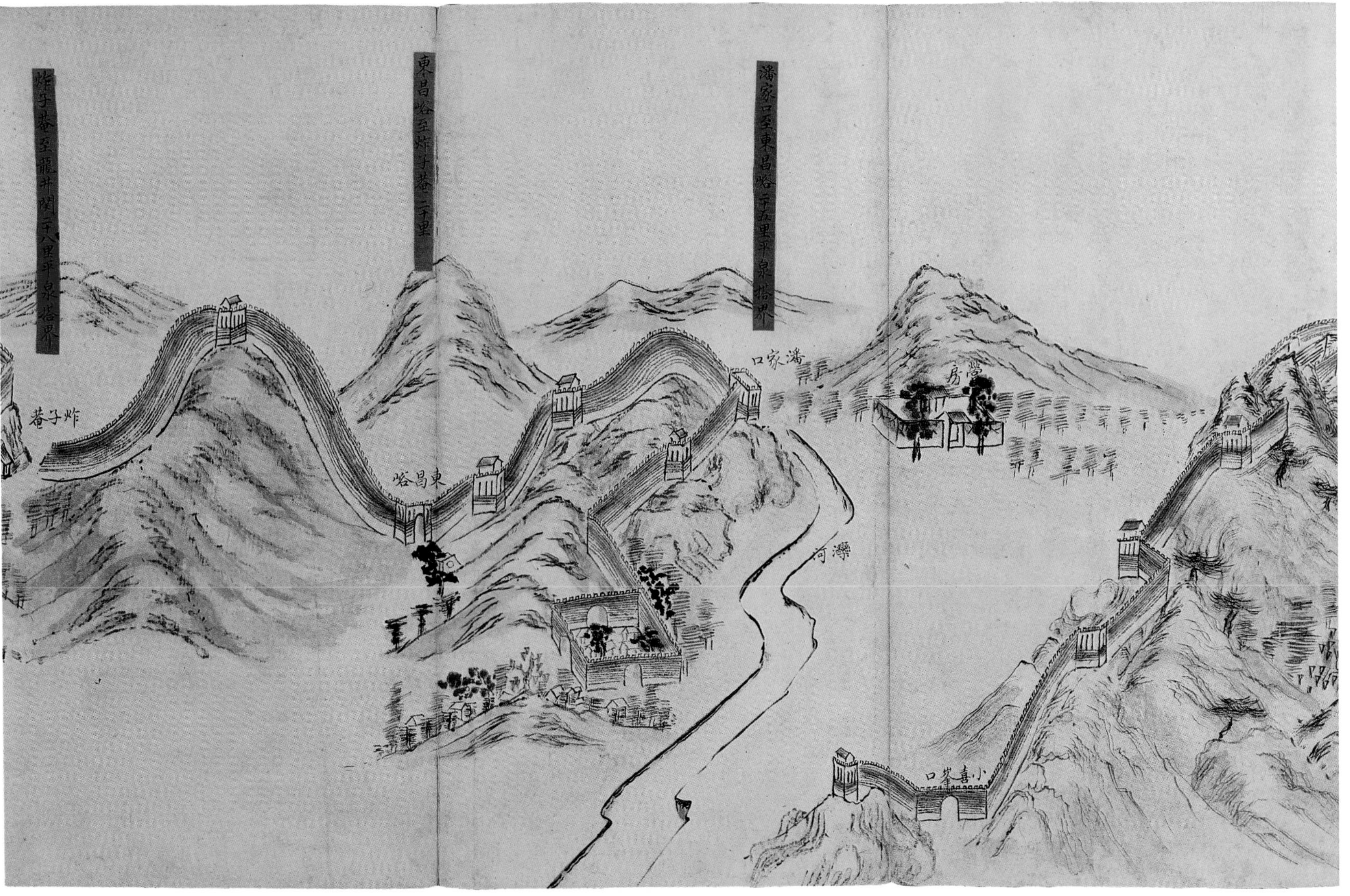

PLATE 11. DETAIL FROM AN EARLY EIGHTEENTH-CENTURY MAP OF THE GREAT WALL. (See p. 189.) The entire map, drawn in the late Ming, represents the section of the Great Wall extending from Shanhaiguan to Luowenyu, a span of about six hundred kilometers. Distance between guard stations on the wall is given on red strips pasted to the map. Size of the entire original: 32 × 600 cm. By permission of the Staatsbibliothek zu Berlin—Preussischer Kulturbesitz, Kartenabteilung (19 271).

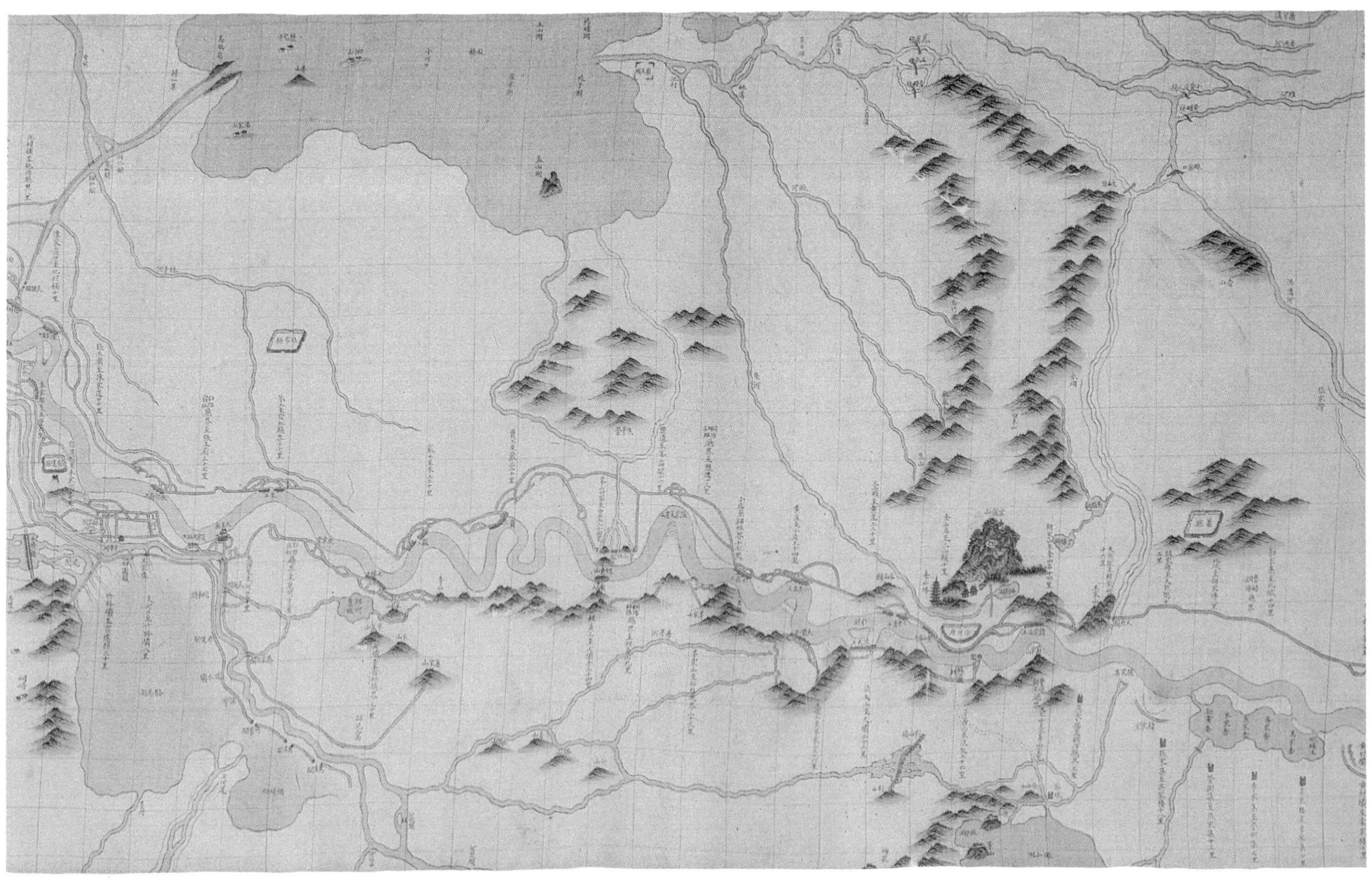

PLATE 12. DETAIL OF A NINETEENTH-CENTURY MAP OF THE YELLOW RIVER. (See pp. 189–90 and fig. 7.21.) Size of the detail: ca. 38 × 64 cm. Courtesy of the Geography and Map Division, Library of Congress, Washington, D.C. (G7822.Y4A5 18--.H9 Vault Shelf).

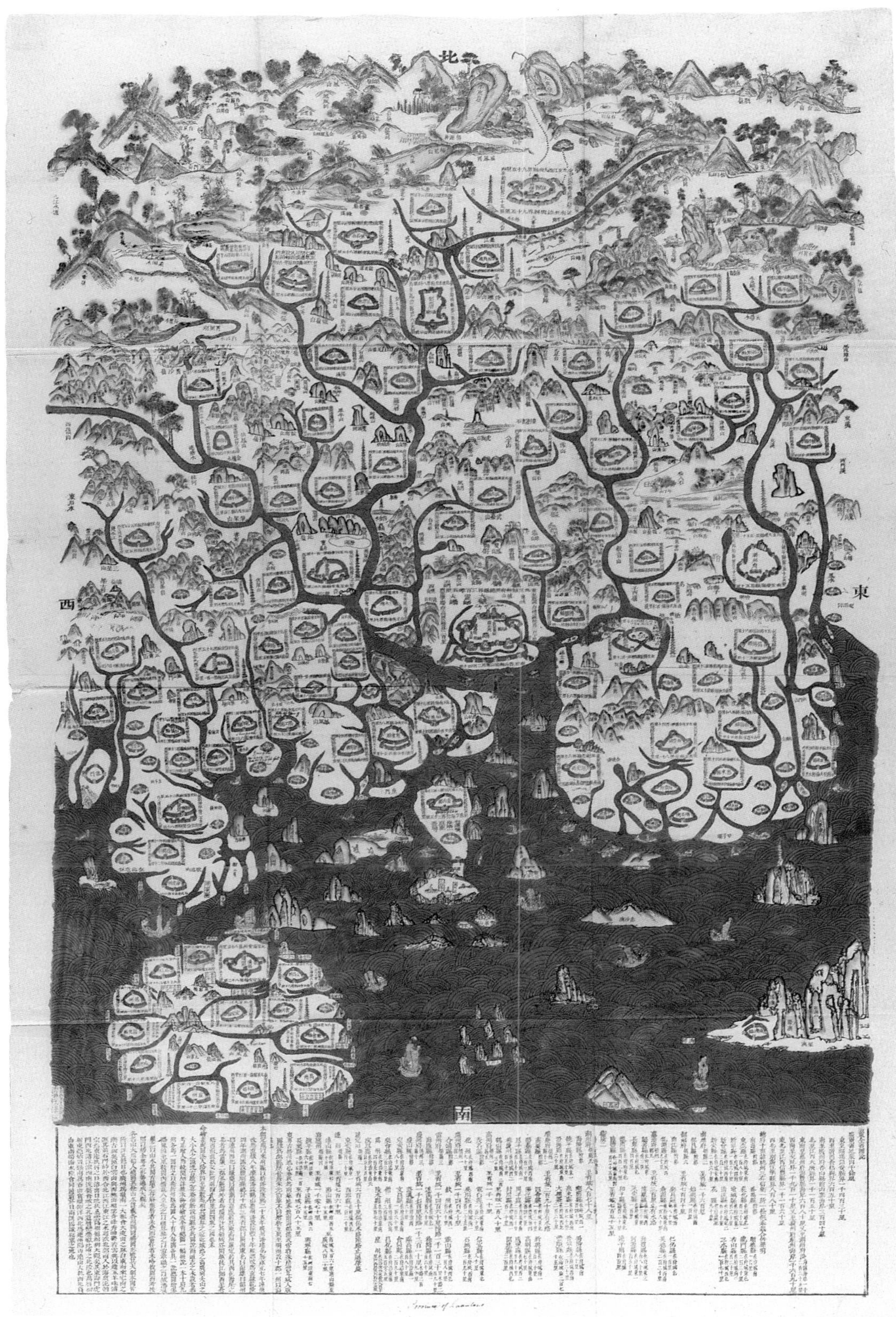

PLATE 13. MAP OF GUANGDONG PROVINCE, CA. 1739. Size of the original: 163.5 × 103 cm. By permission of the owner. Photograph courtesy of Sotheby's, London. (See p. 190.)

PLATE 14. PART OF A LATE QING MAP OF WUTAI SHAN. (See p. 191.) The map presents a panoramic view of Wutai Shan (in present-day Shanxi Province). The dark lines on the map were produced by a woodblock; the rest of the linen surface was hand colored. The map bears inscriptions in Chinese, Manchu, and Mongolian. It was made in 1846. The portion shown here is slightly less than one-quarter of the map. A nearly identical map is described and illustrated in Harry Halén, *Mirrors of the Void: Buddhist Art in the National Museum of Finland* (Helsinki: Museovirasto, 1987), 142–59.

Size of this detail: ca. 59 × 64 cm. Courtesy of the Geography and Map Division, Library of Congress, Washington, D.C. (G7822.W8A3 1846.W8 Vault).

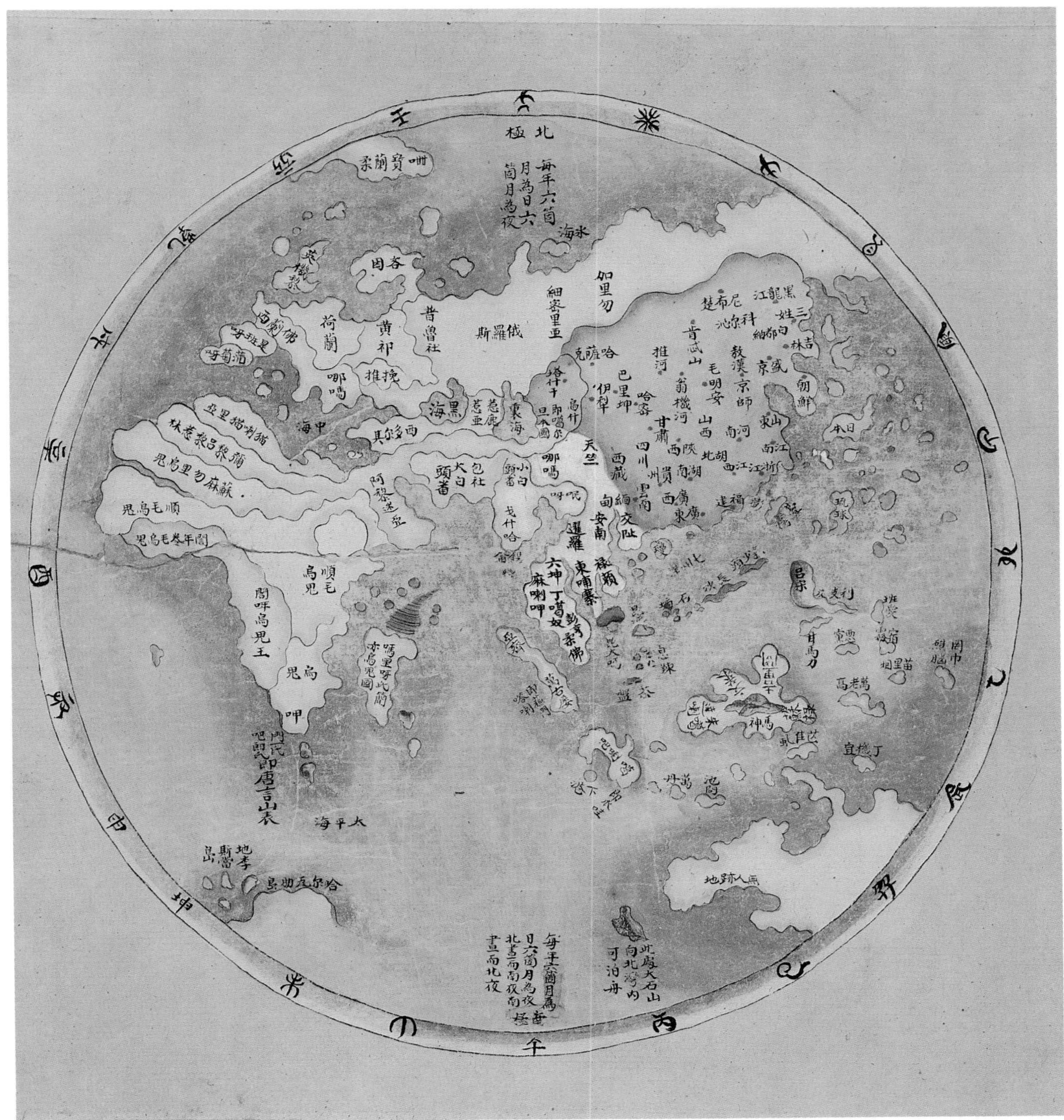

PLATE 15. CHINESE MAP OF THE EASTERN HEMISPHERE, 1790. (See pp. 195–96.) This map was drawn at the beginning of a scroll devoted mainly to a coastal map (p. 62 and plate 4). Diameter of the original: ca. 25 cm (?). By permission of the Staatsbibliothek zu Berlin—Preussischer Kulturbesitz, Kartenabteilung (E 530).

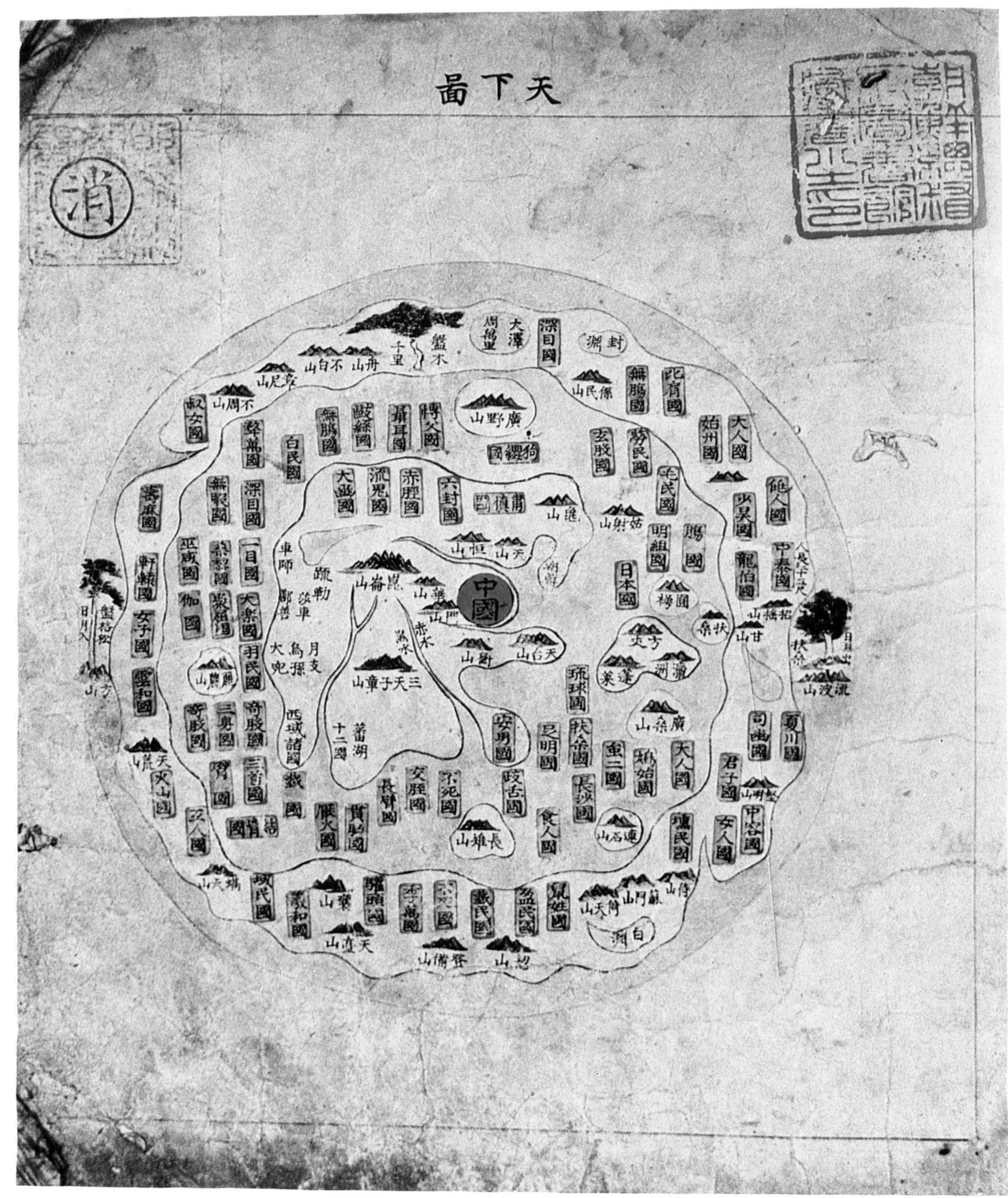

PLATE 16. *CH'ŎNHADO* (MAP OF THE WORLD). (See p. 259.) From a hand-copied atlas of perhaps the mid-eighteenth century. Although this map differs from other *ch'ŏnhado* in a few minor details, it may be called a typical example of the genre. The inner continent is encircled by a sea with various fabulous countries and mountains; the sea is surrounded by an equally fictional outer land ring. Trees at the east and west mark the places where the sun and moon rise and set. Another tree decorates the outer ring in the north. The names of most of the fictional countries and mountains, and of all three trees, come from the ancient Chinese geographical work *Shanhai jing* (Classic of mountains and seas).

Size of the original: 36.5 × 33.7 cm. By permission of the National Central Library, Seoul.

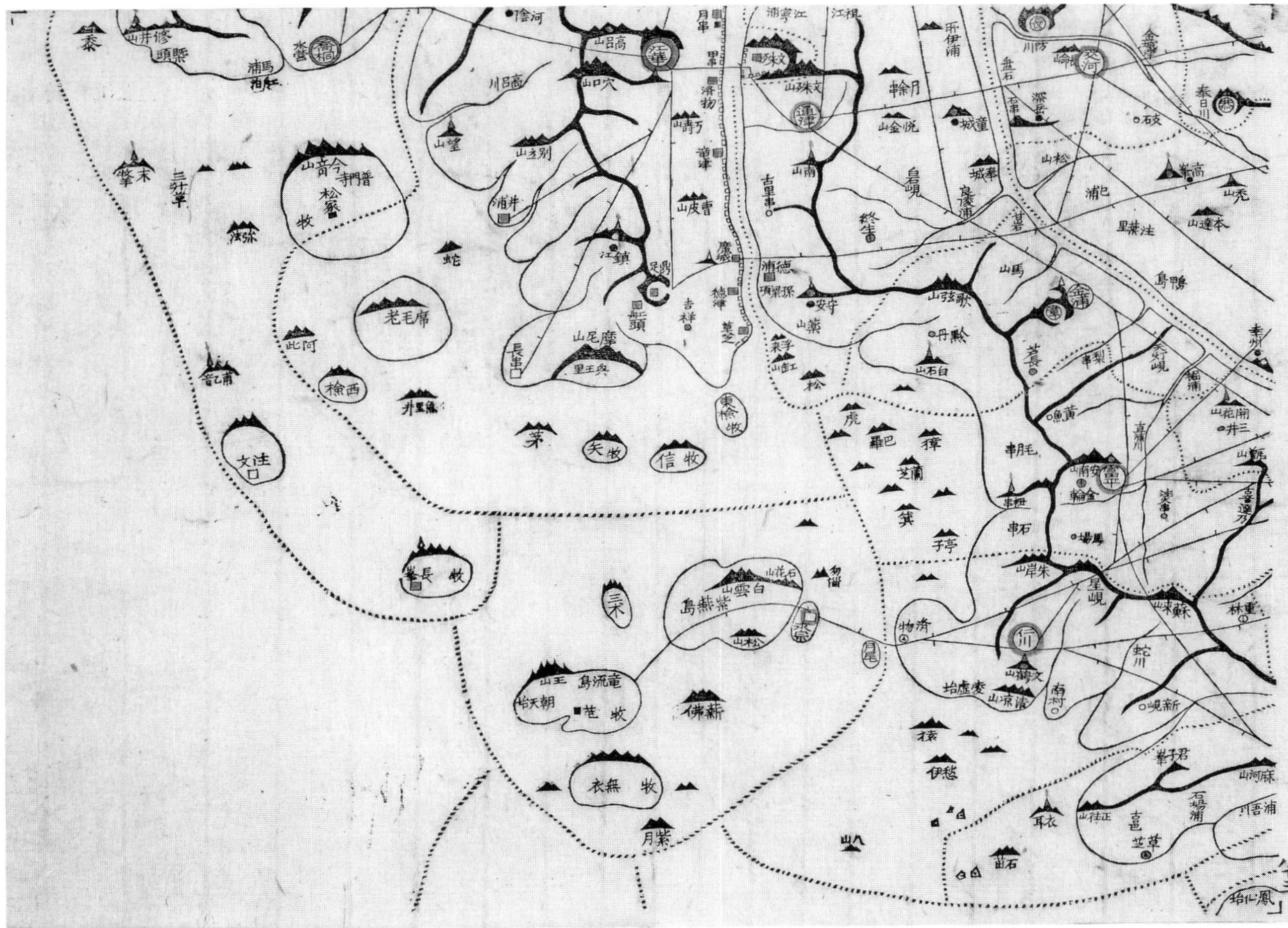

FIG. 10.43. INCH'ŎN AND SOUTHERN KANGHWA ISLAND FROM THE *TAEDONG YŎJIDO*, BY KIM CHŎNGHO, 1861. This section shows the eastern coastal section of rank 13. Compare the *Ch'ŏnggudo*'s coverage of the same area, figure 10.42. Kim has made topographical corrections on the western and southern shores of Kanghwa and in the area northwest of Inch'ŏn (circled district, lower right). Note the more abstract treatment of islands over the earlier map and the more extensive road network, extended as ferry routes to important offshore islands. Administrative jurisdiction of islands is indicated by inclusion within a district's dotted border lines (the line separating Inch'on from twelve of its islands including Chayŏn and Yongnyu is, however, an error by Kim). Circled names are district seats; double circles indicate walled towns. Squares indicate military bases, flame-shaped triangles are fire signals. In this map Kim eliminated textual notes and reduced the number of place-names but achieved much greater clarity, resulting in a net increase of information.
Size of the original: 21.5 × 28 cm. From a 1936 reprint of the *Taedong yŏjido*. Courtesy, the East Asian Library, University of California at Berkeley (Asami Library, cat. no. 20.43).

As for scale, some might think that the *Taedong yŏjido* represents a regression from the *Ch'ŏnggudo*, since the latter's margin scales on every page are abandoned. But actually the clever Kim had found a way to improve things. First he laid out a sample rectangle in his new grid, which was expanded in the new map to 80 by 120 *li*. The map itself does not show any grid lines, but this grid page could be cut out and used as a scale. But Kim's second idea probably obviated that course for most users. In his new scheme, every road became a scale bar, since every road on the map is provided with 10-*li* ticks from its beginning to its end. Road distances are in effect automatically indicated simply by counting ticks. If the user wants to gauge distances in roadless areas, he can visually borrow the closest road as a scale. In very mountainous areas the ticks are closer together, although still representing 10 *li* on the road. Thus the traveler was warned that though the road looked short, the way was circuitous and hilly.

The topographic and toponymic content of the *Ch'ŏnggudo* seems to have been completely reviewed for the *Taedong yŏjido*, which shows revisions of place-names, addition or deletion of cultural features, refinements of the coastline, and many other changes (compare figs. 10.42 and 10.43). Military sites were the object of the cartographer's special concern: close scrutiny of cor-

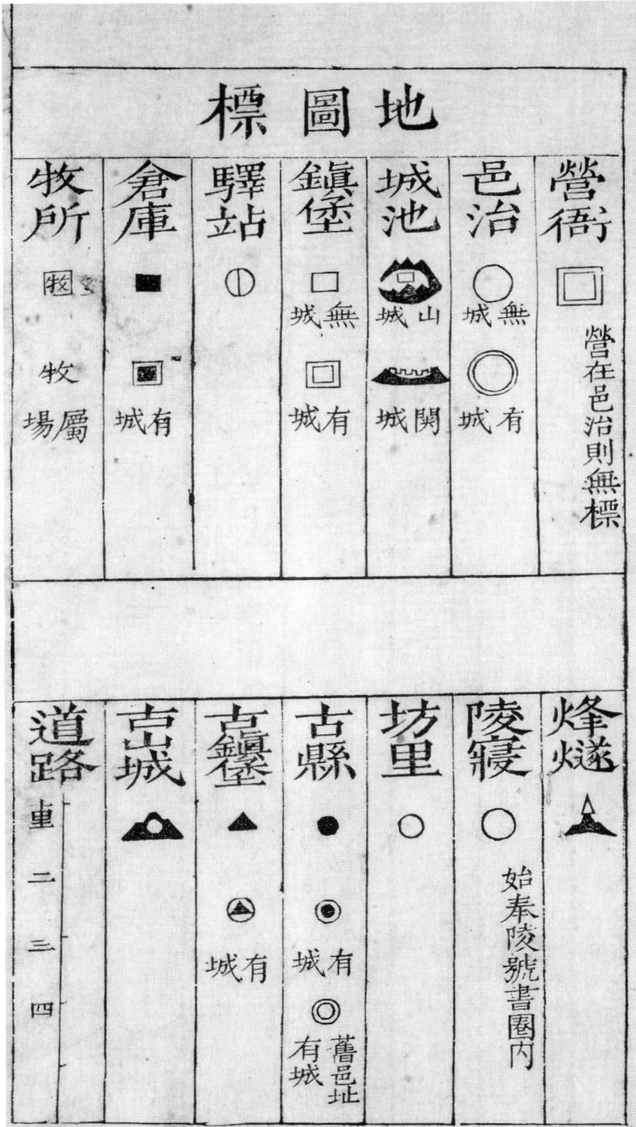

FIG. 10.44. *CHIDO P'YO* (MAP SYMBOLS). From the *Taedong yŏjido*, by Kim Chŏngho. From right to left, top: military command, district seat (name in circle), wall/moat, military station, post station, granary, pasturage. Single enclosures mark unwalled sites, double ones walled sites. Bottom: fire signal, royal tomb, small settlement, former district seat, former military station, former mountain fortress, roads (with ticks to indicate ten-*li* intervals).
Size of the original: unknown. From a 1936 reprint of the *Taedong yŏjido*. Courtesy, the East Asian Library, University of California at Berkeley (Asami Library, cat. no. 20.43).

responding areas on both maps often shows changes or additions in installations, fire signals, and so forth.

Islands are given a more abstract treatment. The basic symbol for a small island is a short line of mountain peaks, often rounded like a tiara. In the case of larger islands, this basic symbol is extended into a fine-lined circle, varying in size according to the size of the island. In general there is no effort to show the precise contours of islands, except for very large ones like Kanghwa and Cheju. Since such information was in fact not usually available, this treatment avoided implying more information than Kim actually had.

The most radical departure from the *Ch'ŏnggudo* was in the matter of mountains. With the *Taedong yŏjido*, the Korean shapes-and-forces style of depicting mountain ranges achieved its most pronounced and most complete realization. On the one hand, Kim's treatment was quite abstract: the montane arteries became solid black lines; the gentler the mountains the thinner the line, the higher the thicker, with ruggedness registered with angular teeth on one side of the line. Particularly grand heights loomed upward from the line as snowcapped peaks. On the other hand, this style well represented shapes-and-forces theory. If there is water, there must be a watershed; as stand the shapes, so move the forces. On the *Taedong yŏjido*, it is possible to start anywhere in Korea and find the single line of mountain spine, the single artery (*maek*) that will take you from your district directly to Paektusan without, in theory at least, ever crossing a stream or stepping in a puddle. In traditional times, whether or not Paektusan was within the administrative purview of the Korean court—and before 1712 it mostly was not—it was generally believed to be the source of vitality in the Korean land, and therefore the most distinctively Korean natural feature there was. On the *Taedong yŏjido*, Kim Chŏngho elaborated it with such flourish that he seriously distorted the northeastern frontier. This decorative flourish is also evident on his single-sheet version, figure 10.47 below. Today the maps have changed, but the feeling is still strong. In the Democratic People's Republic of Korea in the north, national revolutionary ideology focuses on the mountain, and pictures of it are everywhere. In the Republic of Korea in the south, Paektusan appears in the first line of the national anthem that also lauds "three thousand *li* of mountains and rivers fair." If there is a single symbol of national unity to which all Koreans respond, it is Mount Paektu. Looking at the *Taedong yŏjido*, one can see exactly where and how one stands in relation to it.

A more visible and perhaps more significant change than even the mountains in Kim's 1861 map is the general enhancement of cartographic clarity. No grid lines, horizontal or vertical, are shown. Textual indications and notes have been completely removed, banished to the realm of descriptive geography. Much more information is communicated by symbols (fig. 10.44). Subdistricts (*myŏn*) are eliminated as a category. Although some may reappear on the *Taedong yŏjido* marked with the symbol for a smaller settlement, they were very selectively chosen. The visual impact of this change can be appreciated

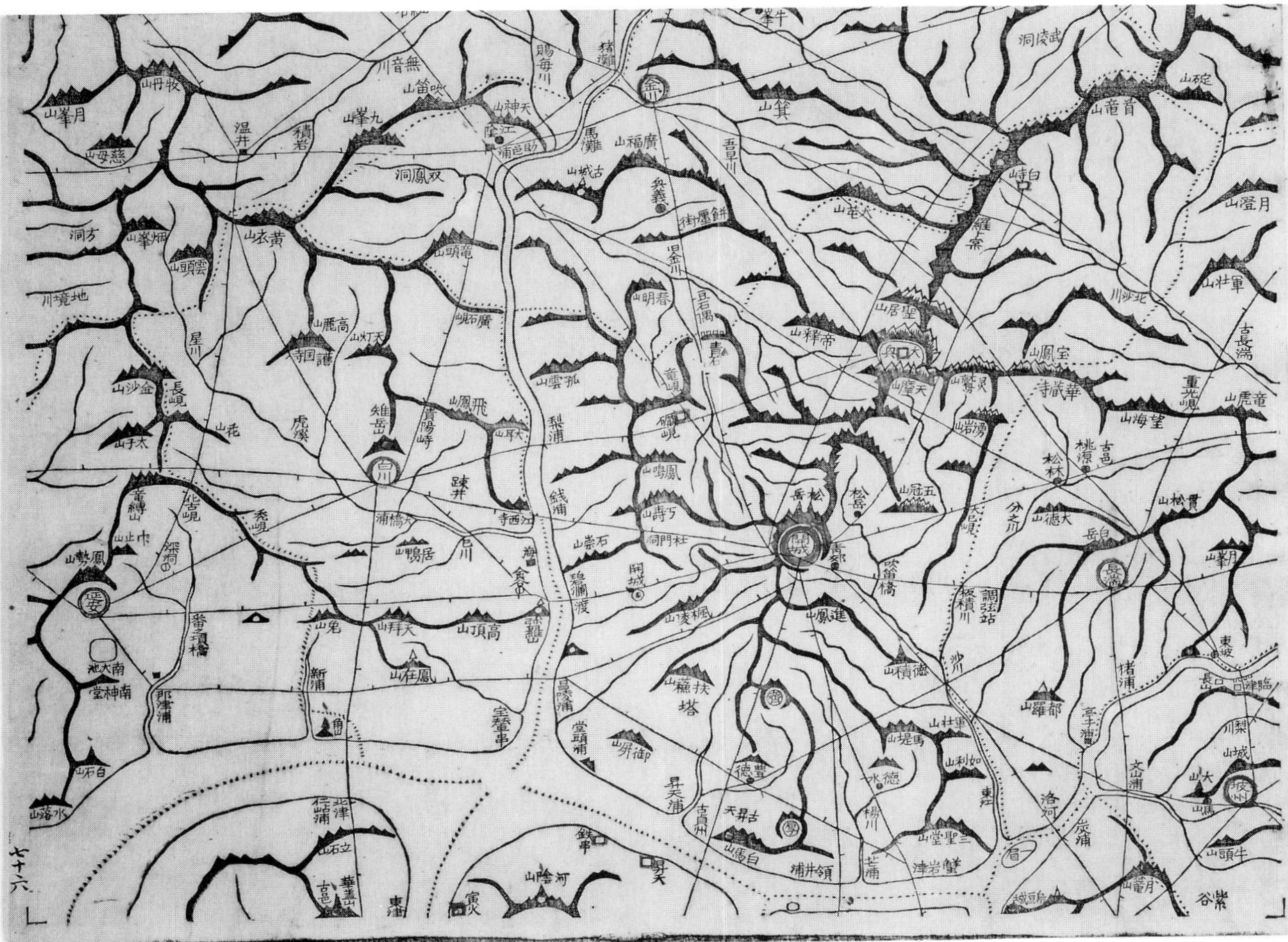

FIG. 10.45. AREA OF KAESŎNG FROM THE *TAEDONG YŎJIDO*, BY KIM CHŎNGHO. Shown is a section from rank 12. Files 13 and 14 adjoin the same files on rank 13, shown in figure 10.43. Dominant in the center is the famous chain of mountains leading to Kaesŏng, site of the Koryŏ capital (northwest of Seoul), providing its potent "artery" to Mount Paektu. Note how Kim enhances its importance with extra thickness of line and dramatic, snowy peaks. Compare the very different treatment of the same area from the *Ch'ŏnggudo* (fig. 10.35). The greater attention to mountain and road patterns, and the sacrifice of textual notes and many minor place-names, cartographically ties the region together much more clearly.
Size of the original: 21.5 × 28 cm. From a 1936 reprint of the *Taedong yŏjido*. Courtesy, the East Asian Library, University of California at Berkeley (Asami Library, cat. no. 20.43).

by comparing figures 10.35 and 10.45 or 10.42 and 10.43, which cover the same areas. Such a development constituted a radical break with tradition. While it is true that, apart from a few mapmakers like Kim Suhong (fig. 10.14) and Kim Chŏngho himself on the *Ch'ŏnggudo*, textual notes are generally not overdone on Korean maps, Kim's new treatment seems to renounce textual material in principle. Probably some users accustomed to the traditional approach missed such notes, but there is no question that their absence brings a significant new leanness and clarity to the map. "Less is more" can apply as fruitfully to cartography as to Ludwig Mies van der Rohe's architecture.

Kim Chŏngho's new format for presenting his grid sections was another radical break with the *Ch'ŏnggudo* and with the album tradition in general. As innovative and flexible as his album system had been, Kim apparently sought even more freedom of display, and with the *Taedong yŏjido* he achieved it. In the earlier arrangement it was possible to juxtapose four contiguous rectangles, or 28,000 square *li*. In the new format one could focus on as small or as large a unit as one wished, up to the entire map, which when all spread out is about seven meters high and three meters wide. The scale has been calculated at 1:160,000.[245] Because of the larger grid of 80 by 120

245. Pang, *Han'guk ŭi chido*, 181 (note 5). This is the scale accepted in Korea, where scholars make the *li* equal to 0.4 kilometer. At 0.43

li, Kim was able to achieve total coverage in twenty-two instead of twenty-nine ranks, and in nineteen rather than twenty-two files. And rather than present them on pages of an album, he sold packets of twenty-two strips folded accordion style.

The grid system is completely hidden: there are no grid lines anywhere on the map. In the ranks they are implicit, since each of the twenty-two ranks constitutes a horizontal strip of paper. But neither line nor sign nor number marks the files. The map reader must use some prominent feature, such as the coast at either end, or a major river or mountain range, to line up one rank with another. Worse, Kim provides no reference map with the ranks overlaid to help users find their place. Given his own remarks on the convenience of his grid and reference map for the *Ch'ŏnggudo* (fig. 10.33), this radical reversal on the *Taedong yŏjido* must have been purposely and carefully considered. In light of the tendency of Kim's other revisions and changes, this must have arisen from a desire to keep the map clean and clear, with nothing on it but the actual geographical and cultural features of the land, either named or symbolized. Travelers would not encounter any grid lines as they made their way on the road, nor would they find any on the map. But on such a large map, in twenty-two separate strips, users surely needed some guidance and reference. This need was admirably filled by the Japanese reprint of 1936, which numbered the files on each rank strip and appended an abbreviated but adequate reference grid map.[246]

If Kim wanted no lines, it is no surprise that he also did away with the traditional woodblock identification line that cut the page in two with the *p'ansim* (block heart), as it was called, to provide the title of the work, the chapter, and the page number. Book readers were not unduly bothered by the *p'ansim*, because it did not impede the vertical lines of text. The sheet was folded at that point into recto and verso page surfaces, and the accumulated double pages were sewn together at the matching sides to make a booklet. But in the case of maps or illustrations, this practice cut the map or picture in half and made it impossible to look at both halves together. Resourceful publishers often compensated by carving the right half of the map on the verso side of the first block, and the left half on the recto side of the second, so that the whole map could be viewed at once. This is the usual practice in the *ch'ŏnhado* albums. But that was of no help for a grid map, with its relatively long strips. The plain fact was that the *p'ansim* never did make sense in map publishing, and Kim's eliminating it was a service to cartography. His average block accommodated a carved surface of about 30 by 40 centimeters, which comprised two grid rectangles (fig. 10.46). The printed sheets were then pasted into strips for unimpeded horizontal scanning. Each strip was folded accordion style for ease of reference and portability.

After the 1861 edition, there was another in 1864. Whether this was simply a new printing or involved some revisions to the blocks themselves is unclear. Among the surviving copies of the *Taedong yŏjido* are many made by hand. Even the printed copies are often individually colored, so that there is considerable variation from copy to copy.[247]

It remains only to mention the condensed edition of the *Taedong yŏjido*, called the *Taedong yŏji chŏndo* (Complete territorial map of the Great East). This is a single-sheet version that was also published by woodblock, a distillation of the complete map set onto a single sheet of about 77 by 115 centimeters. Copies are frequently seen and were evidently popular from the moment they appeared. A short introduction printed at the side discusses the salient features of Korea's shapes-and-forces structure, gives the dimensions of the nation on all sides and as a total (10,920 *li*), and concludes with a ringing patriotic wish for the longevity of Korea (fig. 10.47).

The *Taedong yŏji chŏndo* represents the final shape of Korea according to traditional Korean cartography. In spite of its obvious filial link to the *Taedong yŏjido*, it is slightly different from the latter in its overall proportions. Table 10.9 shows the relative latitudinal distance between Seoul and the northern and southern limits of the Korean mainland on Kim Chŏngho's three major

kilometer, the value used throughout this chapter, it would be 1:172,000. An impressive permanent exhibit of the completely assembled *Taedong yŏjido* is displayed on the wall of a spacious escalator way in the Consolidated Government Office Building (Chonghap Ch'ŏngsa) in Seoul.

246. *Daitō yochizu* (*Taedong yŏjido*), together with Suematsu Yasukazu, ed., *Daitō yochizu sakuin* (Index to the *Taedong yŏjido*) (Seoul: Keijō Imperial University, College of Law, 1936). This edition is a photolithographic reprint of an 1861 printed set held by the History Compilation Society of the Japanese Government-General of Chōsen, which was reorganized as the National History Compilation Committee under the Republic of Korea. The invaluable index lists over 11,600 place-names on the map. There is a Korean reprint of the Japanese reprint: Kim Chŏngho, *Taedong yŏjido* (Seoul: Han'guk Sahakhoe, 1965), without the index. A Korean index was published separately: Pak Sŏngbong, Pang Tong'in, and Chŏng Wŏn'ok, comps., *Taedong yŏjido saegin* (Index to the *Taedong yŏjido*) (Seoul: Kyŏnghŭi University, Han'guk Chŏnt'ong Munhwa Yŏn'guso, 1976).

247. In Yi Ch'an's lists of the map holdings of nine important libraries and museums in Korea (*Han'guk ko chido*, 231–49 [note 4]), I counted eight complete and two incomplete sets of the *Taedong yŏjido*. Of the ten, five were the 1861 edition—three printed copies and two hand copies. There was only one printed copy of the 1864 edition (in Seoul National University Library). Of the remaining four undated sets (including the two incomplete ones) one was printed and three were hand copied. An undetermined number of other copies are distributed among smaller institutional collections or individual collectors in Korea and Japan.

maps, with a Chŏng Sanggi map and a modern map included for comparison. Kim's large grid maps show a slight exaggeration of the north with respect to the south, probably owing to exaggerated distance figures in the north. But the ratios show that the single-sheet *Taedong yŏji chŏndo* was no mere proportional reduction of the larger map, as has been thought. Since the ratios show constant improvement (by the standard of modern maps), they seem to demonstrate a continuing effort by Kim Chŏngho to improve his distance data. Although such north-south ratios are not uniform throughout the map because of variances in relative position between any given pair of towns in the north and the south, they are consistent in showing improvement from one map to another. In the end, although Kim had many problems still to solve—indeed to discover—with respect to east-west relationships, he came close to the north-south proportions of a very precise modern map.

Local, Regional, and Defense Maps

District Maps

Compared with the scholarly interest in world, East Asian, and national maps, relatively little attention has been paid to Korean cartographic works of smaller scope. Even provincial maps have been slighted by researchers, in spite of their prominence in *ch'ŏnhado* albums and their important structural connection with national maps in the Chŏng Sanggi style. Judging from the lists of maps at the end of Yi Ch'an's *Han'guk ko chido*, a fair number of *ŭpto* (district maps) are scattered throughout the major collections, and many fascinating examples are illustrated in the main body of his great album. Yet one cannot find in the same writer's bibliography a single monographic study of local or district maps as a general class.

At least as early as Yang Sŏngji's time, systematic attention was given to the compilation of district maps. His 1482 list (appendix 10.1) names several, in addition to the general collection implied by the title *Chiri chi nae p'alto chugundo* (Maps of the districts of the Eight Provinces from [his comprehensive geographical monograph] *Chiri chi*). The wording is ambiguous, and we cannot know whether this collection was complete, but at least we know that King Sejo's original order asked that all districts be mapped. However, neither maps nor texts from this project now exist. King Yŏngjo apparently launched a similar project in 1757; in this case many of the materials have survived but are as yet unpublished. Pang Tong'in reports that this collection of maps and written descriptions of 295 districts (out of about 335 at that time), titled *Yŏji tosŏ* (Geographical maps and texts), by Sŏ Myŏng'ŭng (d. 1787), contains district maps and maps of such installations as *kam'yŏng* (governors' compounds), military garrisons, and coastal bases. Pang includes four illustrations from this collection, but they are too tiny for study.[248] I have already discussed references to King Chŏngjo's district mapping project of 1791, in connection with which several excellent collections of maps exist. Many local maps from the royal collection (Kyujanggak) are included in Yi Ch'an's lists, but they are very late, mostly from the 1890s or 1900s, and do not seem to have yet been the object of a special study.

Judging from available illustrations, there seem to be three general types of district map. The first is what might best be called the "map-painting"; the second is a shapes-and-forces map with the district or group of districts as the unit; and the third is the grid map. I have already given illustrations of local shapes-and-forces maps (figs. 10.21 and 10.22) and grid maps of districts (figs. 10.36 to 10.38), and I now introduce some specimens of map-paintings.

As the term implies, a map-painting is not always easy to classify. But given the very broad definition of map that is operative in these volumes, there can be no hesitation in calling the *Tongnae Pusan ko chido* (Old map of Tongnae and Pusan) a map, even if it is also indubitably a painting (plate 19). As a map it contains a wide variety of cartographic information: places are indicated and named; roads and routes on land and sea are marked, with distances between various points written in; and mountains and rivers are represented with due attention to their position, orientation, and interrelationships. A person who consulted this map carefully and took notice of the surroundings could use it as a local travel guide. But the features are also painted, from a bird's-eye perspective, as a landscape: trees here and there on the mountains, a boat moving into the harbor; the walls and buildings of the district seat of Tongnae; and Pusan Harbor and the Japanese trading compound as the focal area. Faced with a choice of being a cartographer or a painter, the creator of this work chose painting: a peninsula that in fact bent away from the center was made to bend toward it, distorting the map but giving the painting cohesion and focus.

Unlike the Tongnae map-painting, which covers the entire territory of the district, the *Sin'an chido* (Map of Sin'an, or Chŏngju) focuses on the *ŭp*, or district seat, and leaves much of its other territory out of the picture (fig. 10.48).[249] Although various cartographic features are

248. Pang, *Han'guk ŭi chido*, 125–26 and pls. 5–6, 8–9 (note 5).

249. Sin'an is an alternative name for Chŏngju; it was originally the name of a post station that occupied the site before the district seat for Chŏngju was moved there. In 1812 Chŏngju was downgraded for what the national government saw as its complicity in the Hong Kyŏngnae rebellion, which was quelled in that year. The use of the name Sin'an, as well as the avoidance of the official new name (Chŏngwŏn),

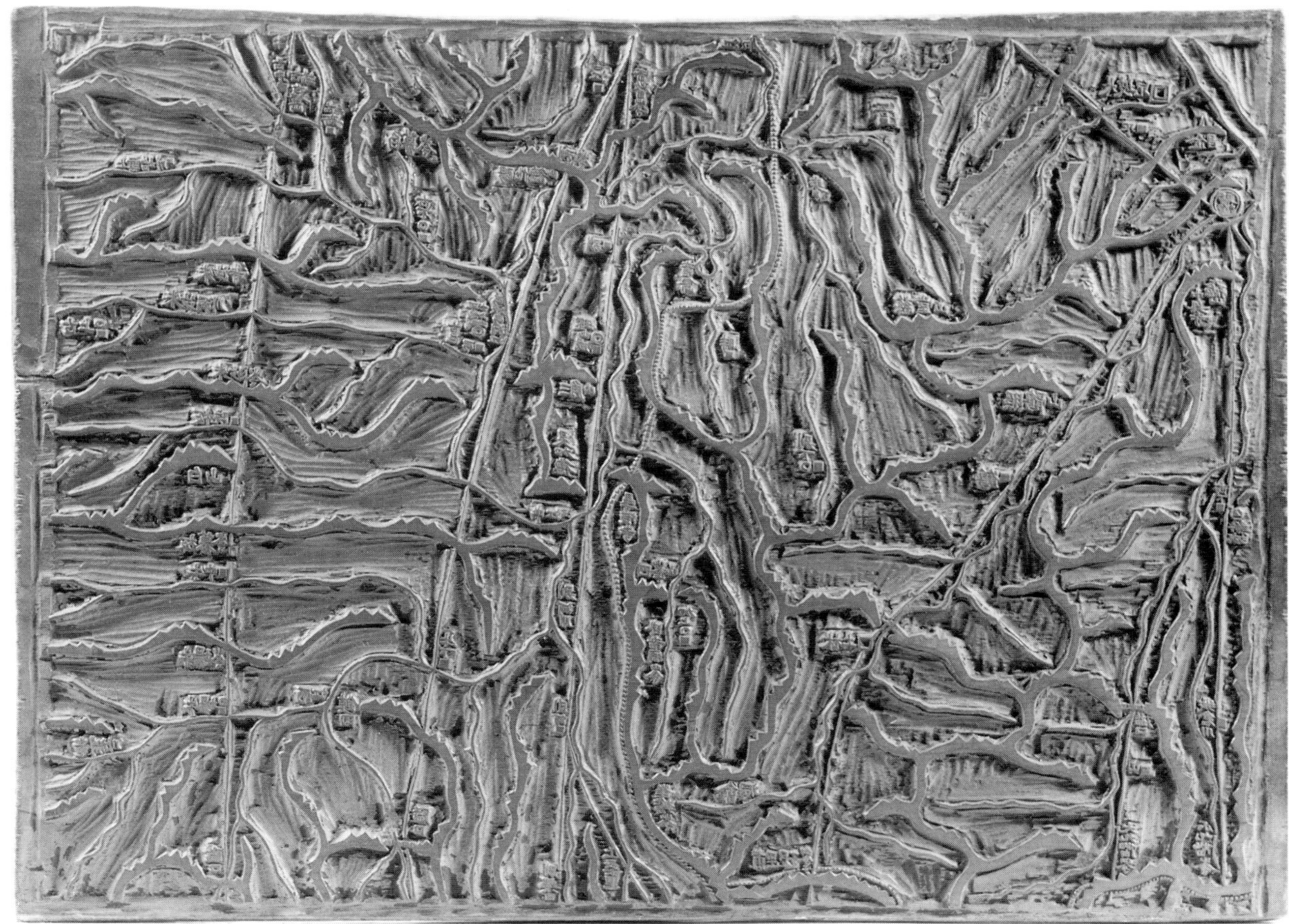

FIG. 10.46. WOODBLOCK FOR A SECTION OF THE *TAEDONG YŎJIDO*, WITH CORRESPONDING PRINTED SHEET. The block shows a section near the end of rank 6. The district seat of Pyŏktong and a bit of the Yalu River are in the upper right corner of the block (upper left on the print).

present, this work might seem to many to be more a painting than a map. The *Andong ŭpto* (Map of the district seat of Andong) is in the same tradition (fig. 10.49). Both of these map-paintings are intended to display the order and community of the district, as represented by the tight clusters of grass-roofed houses, official buildings, schools, and memorials to filial sons and faithful wives. Another feature, shared with most such works, is the prominence given to the protecting mountains and distant ranges, as if to show the district's nodal connections with the national shapes-and-forces arterial system. These evocative elements send messages that say "this is what this district is all about" in a way that a conventional map cannot. Given this purpose, basic cartographic concerns such as scale and orientation are of relatively less importance. The position and size of various features are more indicative of their psychic importance to the community than of mere spatial relationships. Although north is at the top of the picture in the three examples illustrated, orientation is quite variable and often utilizes the inside-out perspective, in which features are painted and labeled as if they were to be seen and read from inside the picture (for example, plate 21). In such cases one must sometimes guess which direction is on "top."

In the normal case, map-paintings will not fail to show mountain and river relationships, and in that sense there is an obvious overlap between them and the shapes-and-forces maps of districts. But even when the latter are more or less "painterly" in their execution, they will still be more map than painting (for example, fig. 10.21). The "north at the top" convention is invariably followed so that directional relationships, important for shapes-and-forces analysis, are clear. The genre lends itself to wide variety and a richness of individual whim. The *Kapsanbu hyŏngp'yŏndo* (Situational map of Kapsan Prefecture),

suggests that this map was done by a local person in the period following 1812.

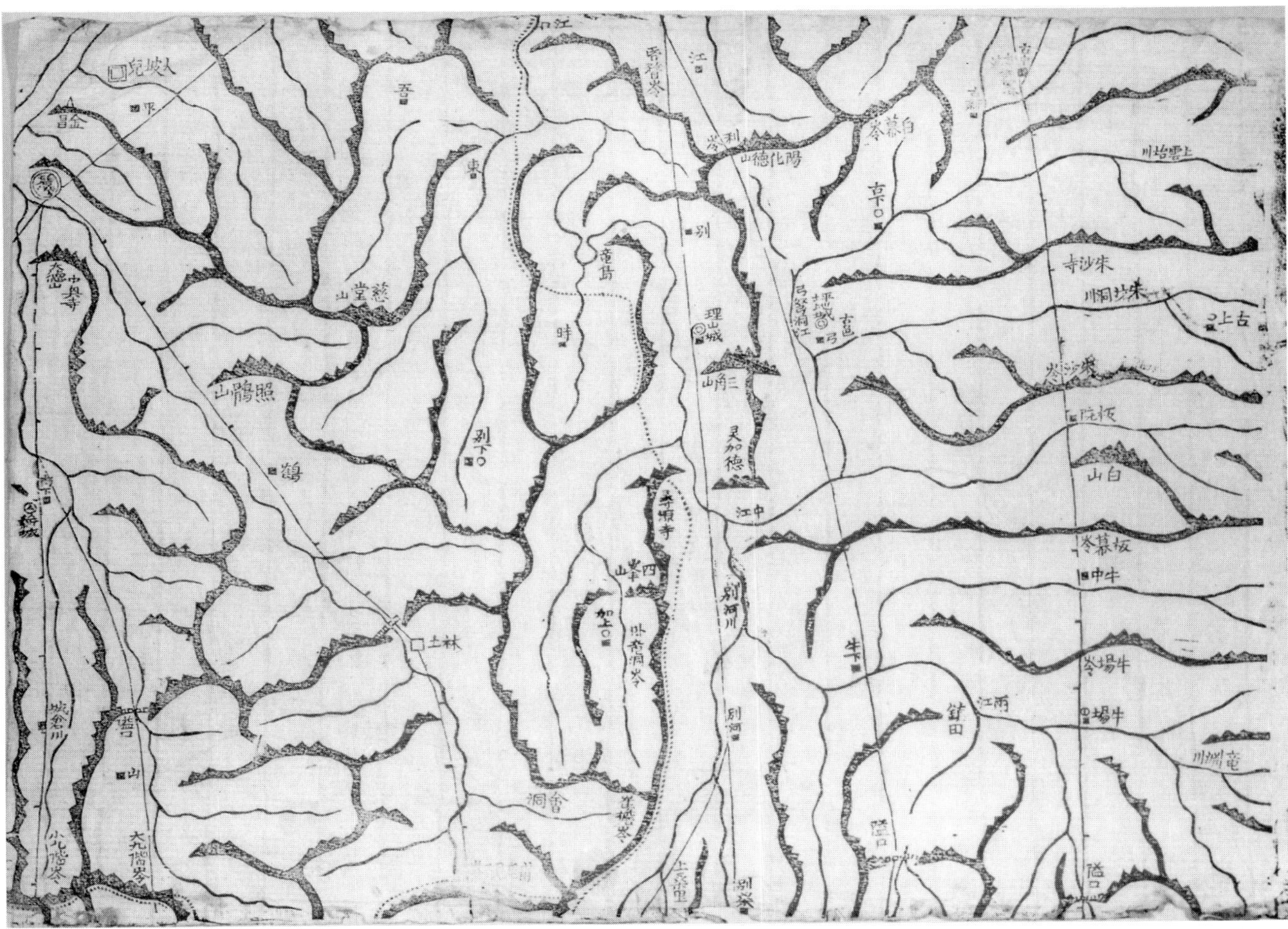

It would have taken about 120 such blocks to print the *Taedong yŏjido* and its associated texts, maps of Seoul, and diagrams. Size of the original: 30 × 40 cm. By permission of Soongsil University Museum, Seoul.

possibly of the late eighteenth or early nineteenth century, was done in a rather abstract and highly unusual style (fig. 10.22). The intent of this work is not clear, but a biological view of shapes-and-forces seems within the realm of possibility: the mountains resemble cellular tissue, the rivers are like blood-filled veins, and the Pond of Heaven at the top of Paektusan is shaped like a liver. Yet whatever the imaginative intent and however stylized the execution, the district of Kapsan is presented as a *map*. Features and places are marked and named; directional and spatial relationships are basically cartographic.

The advent of a more scientific cartography in the eighteenth century brought grids and scales to the district as well as the national map, as we saw in the district maps illustrated in figures 10.36 to 10.38. But these new and much more accurate maps did not displace the traditional map-paintings and shapes-and-forces maps, and it is not hard to see why. Compared with the latter, the district grid maps lack life and are emotionally sterile. They were not conceived in the district itself, and their purpose was not to provide an independent map of a locality, but to assemble uniform units for the construction of a national map. That map represented a noble and modern purpose, and we have seen Kim Chŏngho presenting it with patriotic flourish. His large grid maps undoubtedly impressed Koreans, especially in that important dimension of their identity that esteemed the collective national self, which was certainly experiencing an awakening at that time. But when those same Koreans thought of where they had come from, most must have preferred the traditional map-paintings, with their evocative confirmation of all the traditional verities. As for the consumer, so too for the producer: the district map remained principally the domain of the traditional landscape painter; it was probably not a medium in which a cartographer like Kim Chŏngho could have made much of a living.

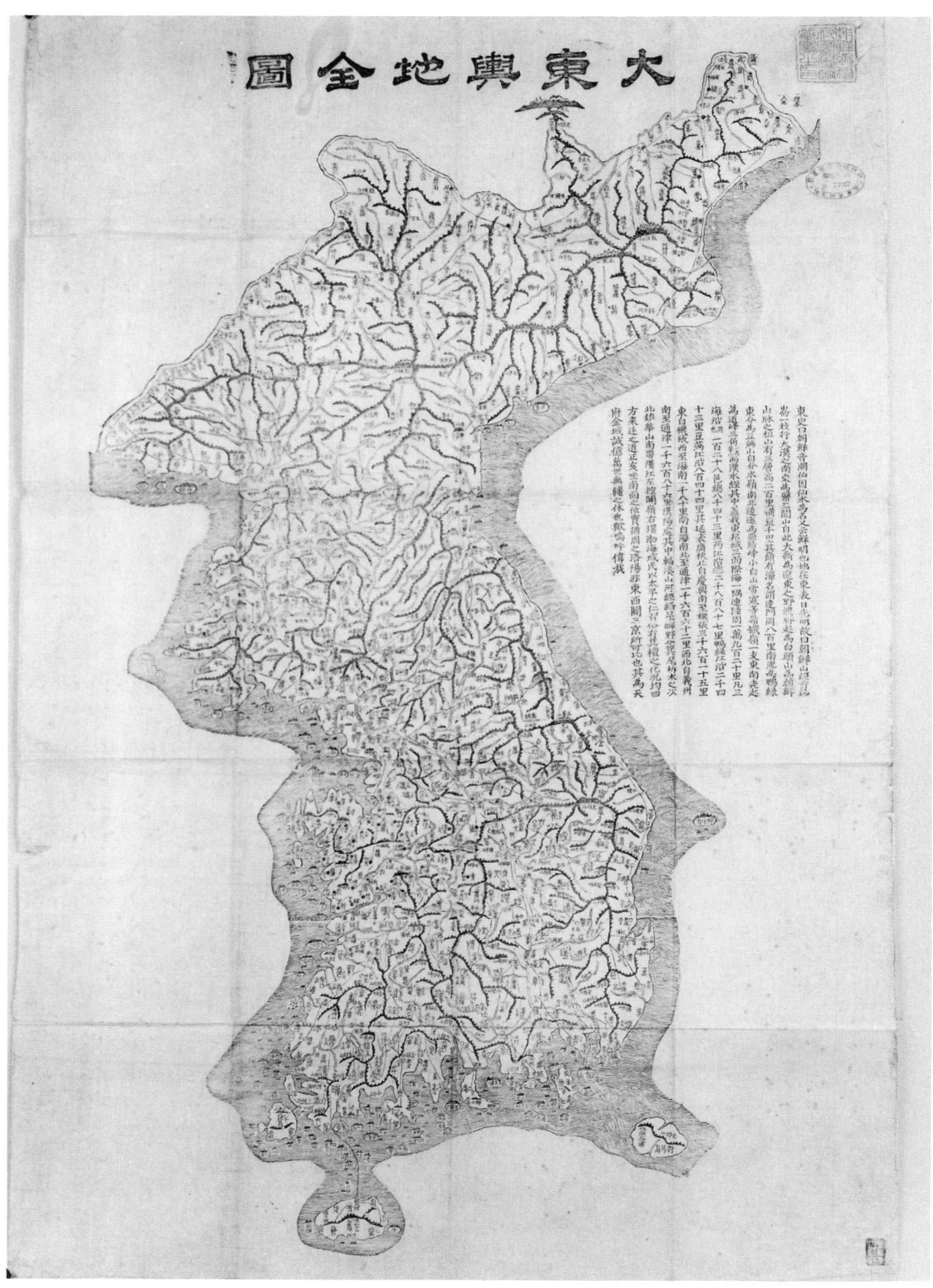

FIG. 10.47. *TAEDONG YŎJI CHŎNDO* (COMPLETE TERRITORIAL MAP OF THE GREAT EAST), BY KIM CHŎNGHO, 1861. Woodblock print. Although the northern border rivers are not as well done as on the Chŏng Sanggi maps, the overall proportions are better. This map was published at a time when Koreans saw themselves under imminent threat from the West, and it has some of the qualities of a morale poster. The text at the right begins with a discussion of the nation's shapes-and-forces and of Paektusan as "the grandfather of Korea's mountain arteries," followed by the dimensions of its seacoasts and border rivers (total perimeter, 10,920 *li*). After an evocation of the distinctive civilization bequeathed to the nation by the legendary founders Tan'gun and Kija, the essay surges to its climax: " 'Tis a storehouse of Heaven, a golden city! Truly, may it enjoy endless bliss for a hundred million myriad generations! Oh, how great it is!"
Size of the original: 115.2 × 76.4 cm. By permission of the National Central Library, Seoul (cat. no. Kojo 6-15).

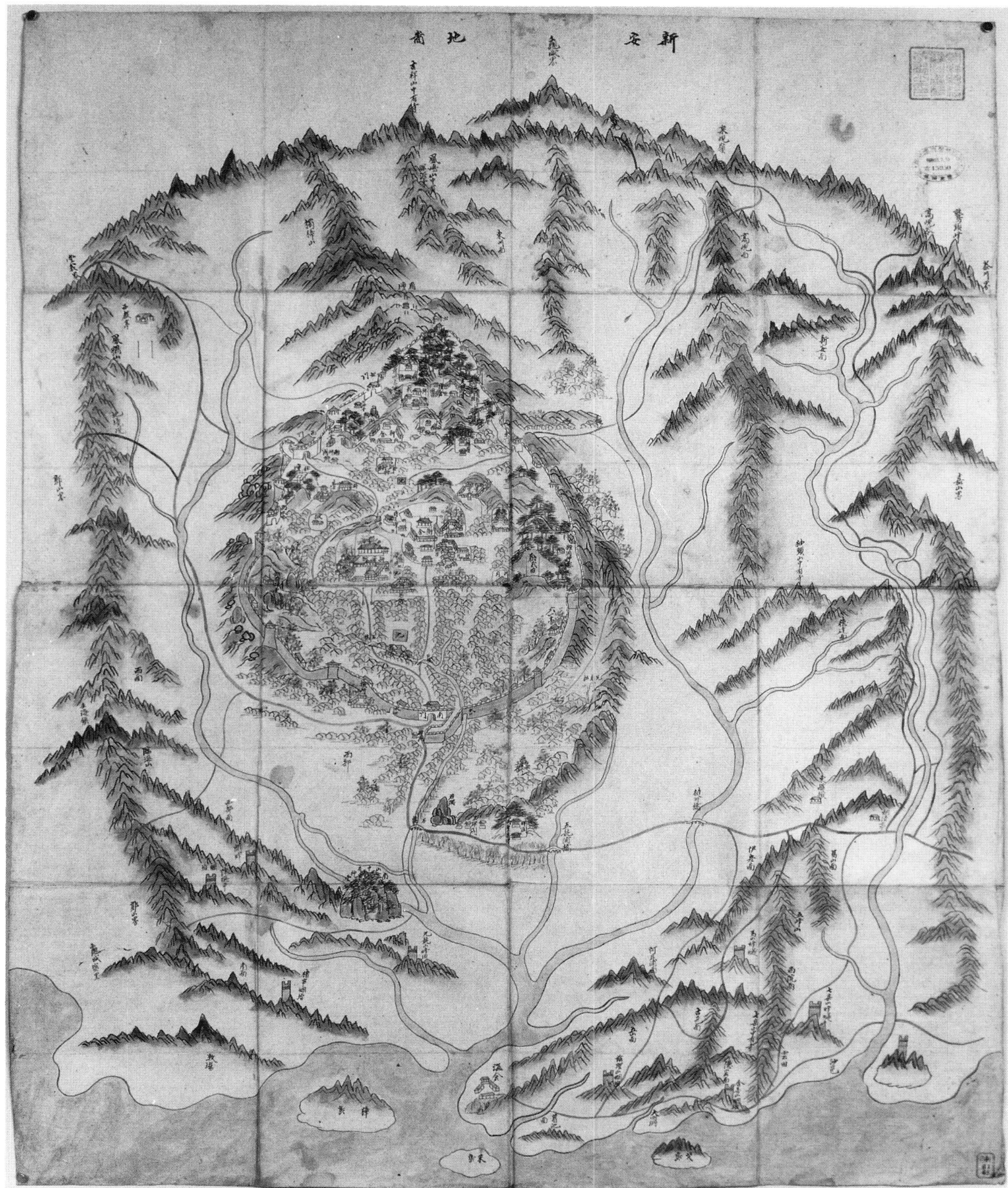

FIG. 10.48. *SIN'AN CHIDO*. Sin'an, the name of a post station that occupied the site before it became the seat of Chŏngju (northern P'yŏng'an Province) served as an alternative name for the town, especially in the years following 1812, when the district was downgraded because of what the government took to be its complicity in the Hong Kyŏngnae rebellion, which was quelled in that year. The post station and guesthouse are at center left; the magistrate's compound is just above them to the right. The local school and a restful stand of pines fill the northern end of the town. This map-painting concentrates on the district center, making no attempt to include the Chŏngju lands to the east, west, and south (front). Roads are indicated in red, and the names of some subdistricts, mountains, and bridges are entered, but in general the concerns of the painter have prevailed over those of the cartographer.
Size of the original: 115.7 × 94.8 cm. By permission of the National Central Library, Seoul (cat. no. Kojo 61-68).

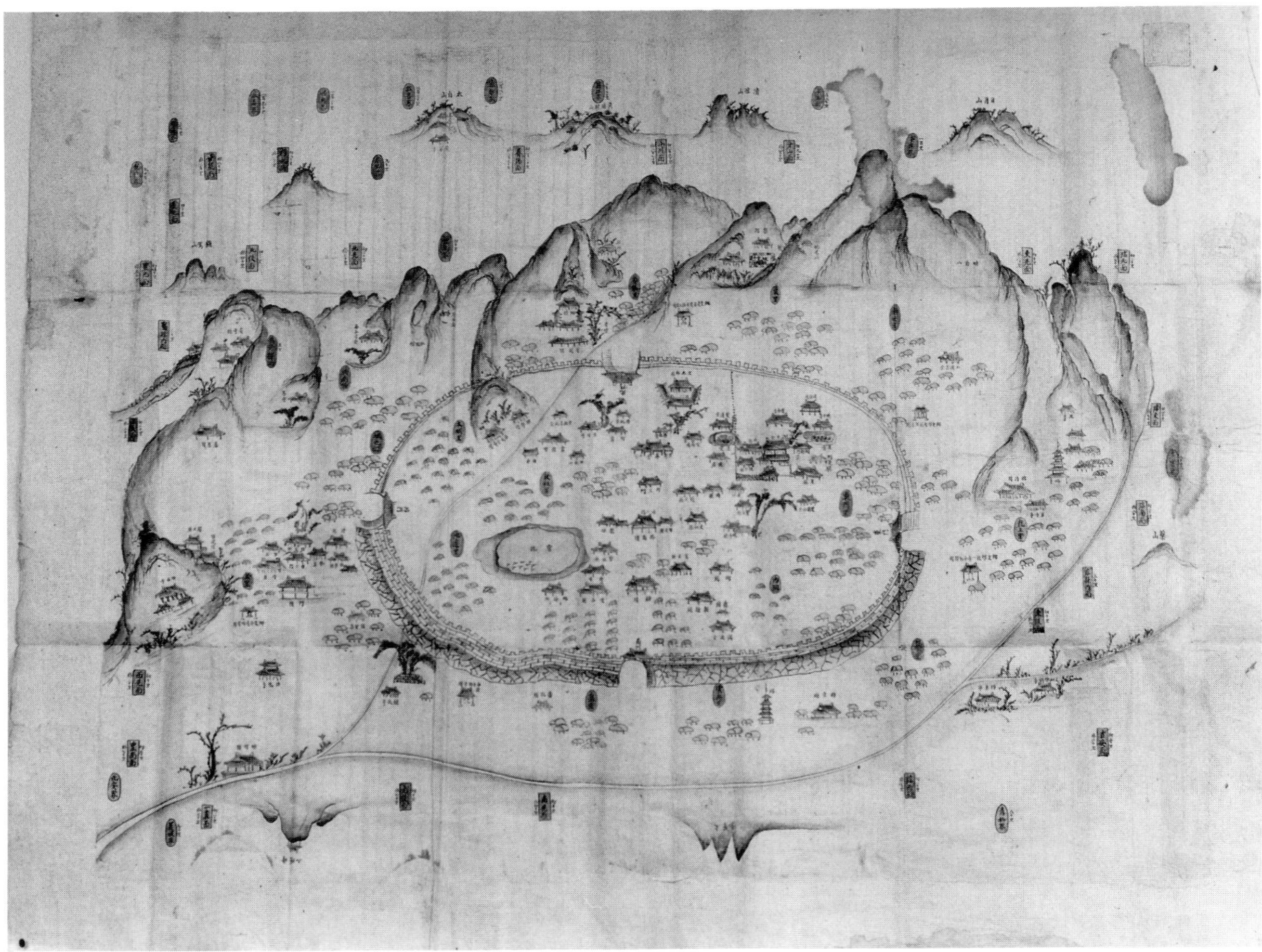

FIG. 10.49. *ANDONG ŬPTO.* Andong is a key regional center in northern Kyŏngsang Province, well known in Korea today for its conservation of much of the traditional culture. The map-painting exudes a sense of communal order, with its imposing but still restrained official buildings and pavilions looking over the orderly groupings of grass-roofed houses. The town's favorite trees have not been forgotten. Subdistricts and mountains are labeled in the background, with their distances from the town indicated. "This is indeed a town where one would want to live," Yi Chunghwan exclaimed in his eighteenth-century classic *T'aengni chi* (On selecting a village).
Size of the original: 120 × 148 cm. By permission of the National Central Library, Seoul (cat. no. Ko 0233-1).

MAPS OF SEOUL AND P'YŎNGYANG

Maps of Seoul were probably popular from the first surveys of 1393, or from Prince Suyang's mapping hikes of 1454. To a degree unrealized in China, with its vast lands and distances and strong local elites, or in Japan, with its bipolar political structure anchored both in the shogun's capitals (Kamakura, Kyōto, Edo) and in the local power bases of sixty-six daimyos, the capital of Korea monopolized and concentrated the political ambitions of the elite class. Families that came to Seoul to pursue bureaucratic careers seldom left, even if the political whirlwinds blew their and their children's ambitions aside. To be sure, it was necessary to maintain one's landowning base in the country, but the more powerful one was, the closer to Seoul that base was likely to be. Seoul not only was the center of Korean power, it was virtually its totality. There were other towns of importance—the frontier ports of Ŭiju and Tongnae (Pusan), the regional economic centers of P'yŏngyang, Chŏnju, and Taegu—and they were places where the locals concentrated on making money. But although Korea's elite disdained money less than they liked to let on, money in itself was not the basis of power. The fortunes that counted were those founded in the currency of social status, and only in Seoul, in service to the national government, could it be maintained and accumulated. A map of Seoul on a striking screen or in a well-painted album

FIG. 10.50. *TOSŎNGDO* (MAP OF THE CAPITAL). The king in his formal actions faces south, and so does this map-painting, perhaps intended for royal use. The northern mountains protecting the city and, symbolically, the nation, loom at the bottom. At the time this map was executed, in the last quarter of the eighteenth century, King Chŏngjo ruled from Ch'angdŏk palace, the large wooded area just left of center. There is a clear shapes-and-forces emphasis to this map, with mountains and streams carefully shown. The drainage system was completely reconstructed by King Yŏngjo in 1759; today it is all underground. The street plan has undergone many changes, but the general pattern of what is now only the city's inner core can already be seen in this early map. Compare figure 10.51. Size of the original scroll: 67 × 92 cm. By permission of Kyujanggak Archives, Seoul National University (cat. no. Koch'uk 4709-3).

was an emblem of success for those who lived there. It is doubtful that many of these maps were bought so that their owners could find their way around.

Maps of Seoul were as concerned with mountains and rivers, walls, and important official buildings as those of any district seat. But since these features and sites were nationally rather than merely civically important, they required from the mapmaker more dignity and weight. And unlike the average district town, which had few passageways that could properly be called streets, Seoul had a large number of relatively wide thoroughfares. The display of street networks was a particularly impressive aspect of a map of Seoul that maps of most other towns simply could not claim. Silla's capital at Kyŏngju, and to a lesser extent Later Paekche's early tenth-century capital at Chŏnju, both imitated the Tang urban grid; but though still sensible to a strolling tourist today, in general the ethos of small towns has long since compromised the original blocks and quarters. Later Korean sensibilty eschewed Chinese-style urban grids, and Kaesŏng and Seoul are distinctive among classical East Asian capitals for their oddly aligned palaces and curvy, asymmetrical urban reticulations. Of the two illustrations, the *Tosŏngdo* (Map of the capital) (fig. 10.50) is of the map-painting type, but pains were taken by its artist, and its street plan accurately reflects its time (late eighteenth century). The *Susŏn chŏndo* (Complete map of the "Source of Good" [the capital]) (fig. 10.51), its title reflecting a

FIG. 10.51. *SUSŎN CHŎNDO*. The title uses a classical Chinese phrase alluding to a national capital. This woodblock-printed map of Seoul probably dates from about the middle of the nineteenth century. Pukhan Mountain, with its well-known fortress, looms above the city to the north, which on this map is at the top. Chongno, then as now the straightest avenue in the city, divides the map into its northern and southern halves. The mountain Mongmyŏksan, now known as Namsan, dominates the city on the south. The Han River flows by what were then the southern exurbs before bending northward toward the West Sea (Yellow Sea). Some scholars attribute this map to Kim Chŏngho and date it about 1825, but evidence for such an ascription is uncertain.

Size of the original: 25.4 × 22.2 cm. By permission of the National Central Library, Seoul (cat. no. Kojo 61-47).

Han-period Chinese conception of a national capital, represents a city that has filled out somewhat and shows considerably more development outside the walls than does the *Tosŏngdo*. This woodblock production, known in many copies, is a very carefully done, "pure" map of perhaps the 1840s or 1850s. Its street plan is a bit more extended and precisely drafted than that of the *Tosŏngdo*, but it is structurally identical to the earlier map-painting and testifies to the latter's accuracy of detail.

P'yŏngyang was the capital of Koguryŏ from the fifth to the seventh century, but its roots go back at least to the late centuries B.C. It is by far Korea's oldest city. During the Koryŏ period it was the choice of many for the capital, and schemes to move the government there occasionally disrupted Koryŏ's peace. Under the Chosŏn dynasty it retained its old prestige. Dominating the northwest, it participated in the China-Korea trade economy and was the locus of one of the more powerful provincial governorships, and perhaps one of the more important day-to-day military commands in the country. It was famous for its ancient cultural attractions, its distinctive foods, and its kisaeng entertainers, who provided a more indulgent and pleasurable atmosphere than could be found in Seoul. Traditional maps of P'yŏngyang seem to be all of the map-painting type, showing the city in bird's-eye view from across the Taedong River, often with the famous Nŭngna Island in the foreground.[250] The painting shown in plate 20 is of this type, but it is also much more, a map with a great party. The busy parade of boats on the river, with officials and soldiers participating and kisaeng and others looking on from the shore, may have been painted to commemorate an official's tenure as governor of P'yŏng'an Province, a scene in which the map of P'yŏngyang itself, including the governor's own extensive compound, had a prominent and appropriate place. As a *pièce d'occasion* more than a map, this painting is not typical of P'yŏngyang maps. But the many that are known, often large screens, invariably show P'yŏngyang in the same bird's-eye view, if without the festive doings. As maps they are pictures, and as pictures they do not represent scientific cartography. But unless what has survived in the way of P'yŏngyang maps is totally misleading, these map-paintings overwhelmingly suited Korean taste.

DEFENSE MAPS

I have already mentioned Korea's fragile relations with the Manchu state following its invasions of Korea in 1627 and 1636 and its overthrow of Ming China in 1644. This situation led King Hyojong (r. 1649–59) to rebuild Korea's military and harden its economic and logistic base, in hopes that some reverse in China would provide an opening for a Korean attack on the Manchu rear, enabling Korea to help Ming back to power and wipe away the humiliation of 1637. In that year Hyojong and his two brothers, as young princes, had been marched off to a difficult seven years of confinement in Shenyang as hostages. Of course, no opportunity for Hyojong's vaunted *pukpŏl* (northern strike) ever came. But he and his successors Hyŏnjong and Sukchong, and a bureaucracy suffused with deep-seated hostility to the Manchus, kept the military strong and began a program of repairing and expanding fortifications on the frontiers and throughout the country. Policies were developed to build up manpower reserves and logistics potential. Revenue measures were pursued so that all of this could be paid for.

Defense maps (*kwanbangdo*) were already being compiled in the 1430s, by Chŏng Ch'ŏk and others, and we have seen that they constituted a significant proportion of the maps mentioned in the royal annals during the middle period of the Chosŏn dynasty (table 10.6). But with the general mobilization started by King Hyojong, they came into their heyday. Since they were usually secret and closely guarded, they did not often come to open notice; we have seen that in the face of Manchu inquiries even general maps of little military significance were guarded with wary determination. Still, a fair number of *kwanbangdo* have survived, and they constitute an interesting and important subgroup of maps within the larger compass of Korean cartography.

The *Yogye kwanbang chido* (Map of the defensive system of the Liaodong and Jizhou sector) was presented to the throne early in 1706 by Yi Imyŏng (1658–1722), then right assistant state councillor under King Sukchong (fig. 10.52).[251] According to his preface, while in Beijing on a diplomatic mission the previous year, he had purchased a book called *Chousheng bilan* (Essential reading in preparation for victory), by a late Ming scholar named Xian Kejin, which described, with maps, the Great Wall and related defense systems.[252] Upon his return to Seoul

250. An exception is the *Kisŏng chŏndo* (Complete map of Kisŏng [poetic name for P'yŏngyang]), which is painted from a very high perspective directly above and gives the feeling of an aerial photograph. See the illustration in Yi, *Han'guk ko chido*, pl. 14 (p. 26) (note 4).

251. The presentation notice and the full text of the preface are in *Sukchong sillok*, 43.3a–b (note 17); the date of the notice corresponds to 24 February 1706. The preface also appears on the painting and in Yi Imyŏng's collected works: see Yu Yŏngbak, "Yogye kwanbang chido," *Tosŏgwan* 27, no. 11 (November 1972): 32–34.

252. I have been unable to find more than a brief biographical notice of Xian Kejin, a native of the Jiangnan area who during the Wanli years served (among other positions) as an imperial censor in Shanxi Province. This would certainly have brought him into close contact with Great Wall and frontier affairs. So far I have found no listing of the *Chousheng bilan* in Chinese bibliographies. Yi Imyŏng says of this book, "Having received the order to copy it for presentation, I also took the *Ula difang tu* [Map of the Ula region], which is included in the *Shengjing tongzhi* [General history of Shengjing] compiled by a Qing man, as well as

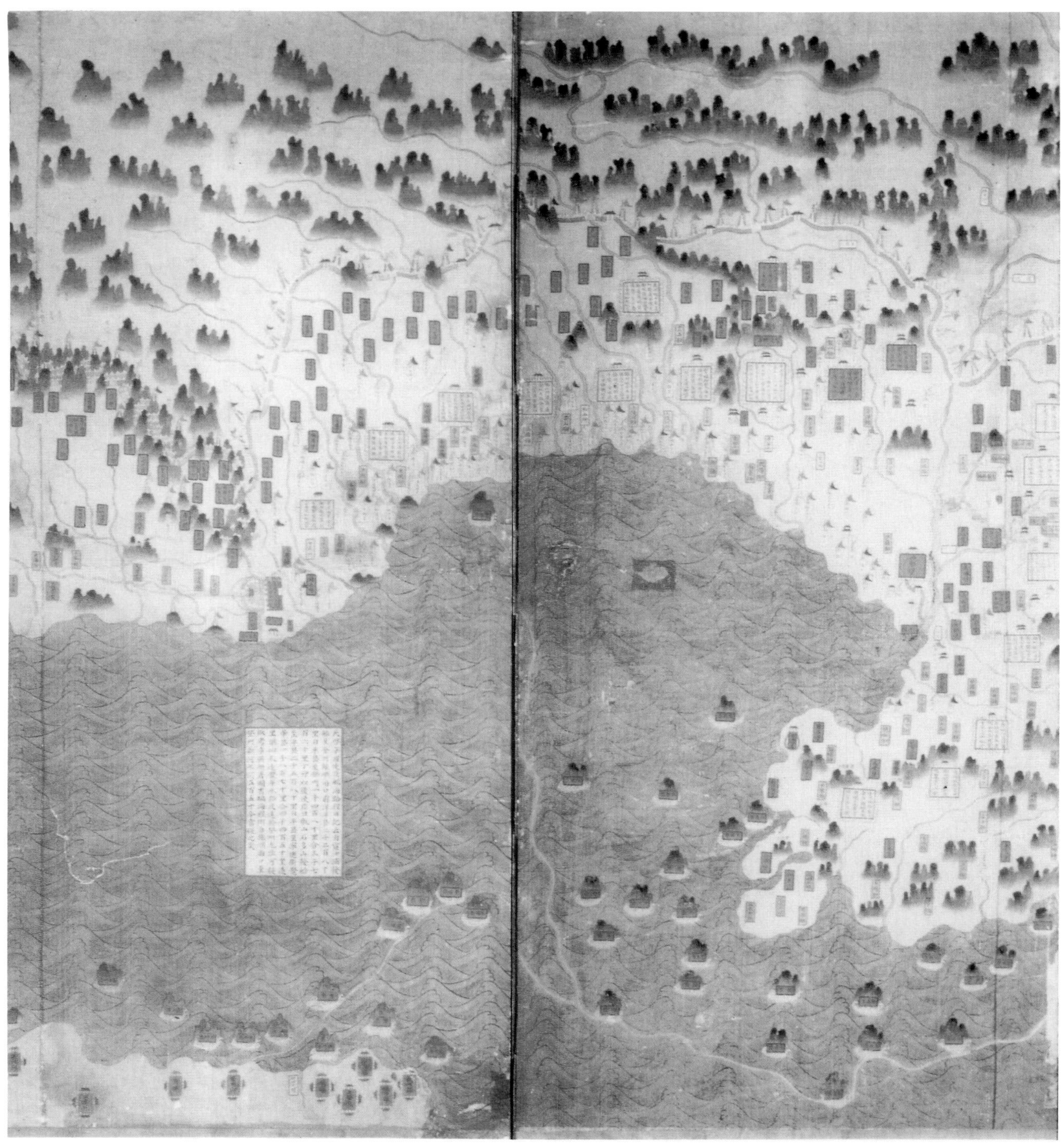

FIG. 10.52. *YOGYE KWANBANG CHIDO*, BY YI IMYŎNG, 1706. This map, known as the *Yogyedo*, in nine screen panels, covers in a single panorama the Chinese-Manchu frontier from the Pacific to the hills west of Beijing, although the section shown here does not go beyond the eastern terminus of the Great Wall. Compiled from a variety of Chinese, Manchu, and Korean sources, this map has an unusual international dimension with a distinctively Korean stamp. The Willow Palisade, branching off from the Great Wall in the west, arching up and around and then back to the Yellow Sea on the east, is clearly

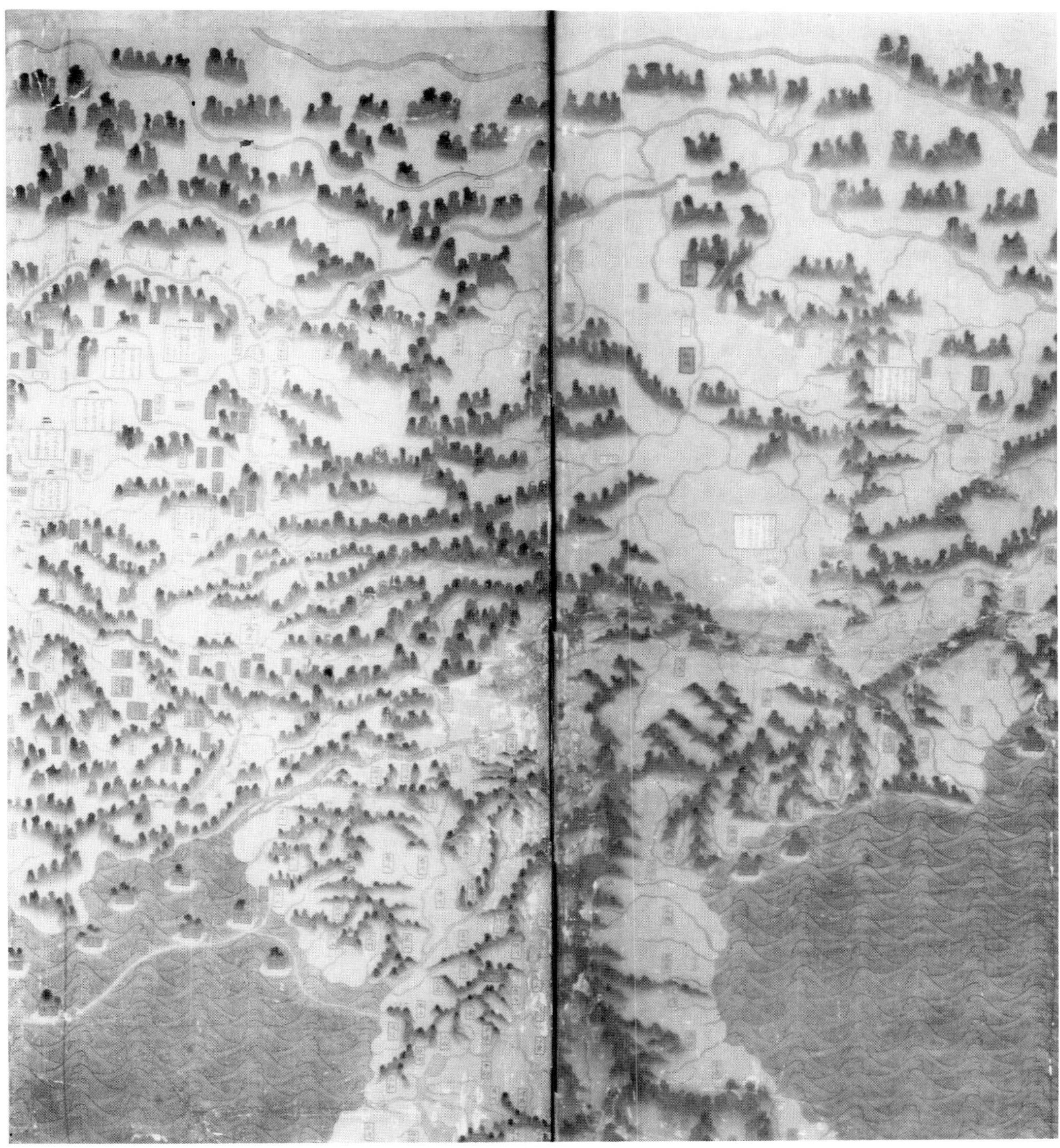

shown, although it is not as evident as on the map in figure 10.53. Koreans traveling to Shenyang and Beijing were met by Manchu officials at Fenghuang (or Bianmen), which was about fifty kilometers north of the Yalu River frontier. The major mountain ranges and rivers of Manchuria are completely shown, as is the branch of the Willow Palisade that extends northeastward to the Ula and Ninguta defense areas.
Size of the entire original: 139 × 635 cm; each panel shown here: 139 × 64 cm. By permission of Kyujanggak Archives, Seoul National University (cat. no. Kodae 4709-91).

he conceived the idea of combining the maps in Xian's book with a map of the northern Manchurian area of Ula (from an administrative history of Shenyang), plus Korean maps of Korea's two defense sectors and other materials, to compile a general map of the eastern end of the Great Wall, from Beijing to the Pacific. With the obvious aid of an unacknowledged and nameless adjutant, probably a professional painter, the project was launched and soon completed. The whole project recalls the compilation process for the *Kangnido* over three centuries earlier: a Korean diplomat obtains cartographic materials in Beijing and combines them with maps available in Seoul to create a transcending, international map beyond the scope of anything known earlier.

The *Yogyedo*, as it is known for short, is mounted on a ten-panel folding screen with total dimensions of approximately 135 by 635 centimeters. Because of its great size, published reproductions (which all feature the Korean end) emphasize the general expanse and do not permit much legibility of the many textual notes and even more numerous place-names. But the Ula-Ninguta defense line in north-central and northeastern Manchuria, and its relation to Liaodong and to the northeastern Korean areas, is clearly grasped.

The coverage of Yi Imyŏng's map was perhaps more important for its ambition and its historical reference value than for the practical requirements of Korean military men, who were not likely to get involved with the Beijing area and northeastern China proper. Still, as Yi remarked in his preface, "What we have to be concerned with is not just the Liaodong-Jizhou sector but the territories that it connects with; we must put all these [maps] together into one. Otherwise we will have no understanding of the bigger situation on the frontier, nor will we know where the cold winds are blowing from."[253] Also, the Beijing area of the frontier was a frequent topic in Chinese histories, and an accurate map that not only showed it but connected it with the Korean end was undoubtedly of great interest to Korean scholars. Two copies, essentially identical, are known; one may be the original. Further study of this important map is needed.

The more relevant eastern sector, in which Korea figured directly, understandably attracted the frequent attention of Korean defense mapmakers. A particularly striking example is the *Sŏbuk p'ia yanggye malli illam chi to* (Synoptic myriad-*li* map of the double frontier between them and us in the west and north) (fig. 10.53). Korean defenses were organized in two commands, one in the P'yŏng'an Province in the northwest, the other in Hamgyŏng in the northeast, whence the term "double frontier." Although not drawn with the large-scale éclat of the *Yogyedo*, it is cartographically superior for the area it covers. Executed sometime in the first half of the eighteenth century (but after Kangxi's "last years"—generally 1721–22—mentioned in the marginal texts), this map presented a much clearer picture of Manchuria's shapes-and-forces and the military arrangements on both the Korean and Manchu sides. Another interesting feature is the interpretation of the 1712 frontier agreement, which is somewhat different from that already shown on the *Pukkwan Changp'a chido* (fig. 10.29). This interpretation, linked to a putatively more northern source for the Tumen, figured in Chinese-Korean frontier negotiations in the 1880s and 1900s and still provides grist for perennial Korean claims concerning the Jiandao (Korean Kando) area north of the Tumen.

The early eighteenth-century frontier maps appeared too early to benefit from the more accurate delineation of the Yalu-Tumen river line on the Chŏng Sanggi maps. The treatment on these frontier maps, itself a considerable improvement over the modified Chŏng Ch'ŏk treatment common in the seventeenth century, presented the frontier in virtually a single shallow arc bearing northeast from the mouth of the Yalu, without the distinctive bends that ever since Chŏng Sanggi have instantly signaled the border.

The Manchus were the principal but not the only defense concern of Korean military strategists. There was also Japan to worry about. Our retrospective view of Japanese history gives such weight to the era of peace during the Tokugawa shogunate (1600–1868) that it requires an effort to grasp the seventeenth-century Korean view of Japan, which was realistically based in Korea's own recent and bitter experience. Japan had evacuated its troops from the peninsula in 1598, when Hideyoshi's death provided the political opportunity for what had already been mandated by logistic and strategic stalemate. But some Ming forces had remained in Korea until 1600, and for several years after that there was a high state of defensive readiness all along the southern and eastern coasts. The trade agreements of 1609, which permitted the daimyo of Tsushima to maintain a permanent trading installation in Pusan, and which reestablished a structure for Korean-Japanese relations through his intermediation, considerably stabilized the situation. But Korea never let down its guard throughout the seventeenth and eighteenth centuries. The southern coastal towns bristled with military and naval presence well into the nineteenth century.

The *YŏngHo nam yŏnhae hyŏngp'yŏndo* (Map of the natural defenses of the southern coasts of Kyŏngsang and

maritime tributary itineraries from a former [i.e., Ming] time and [maps] of the western and northern defense sectors and of the coastal and river frontiers, all our own materials, and combined them into a single map" (*Sukchong sillok*, 43.3a [note 17]). This context implies that the *Chousheng bilan* had maps, and I have assumed so.

253. *Sukchong sillok*, 43.3a (note 17).

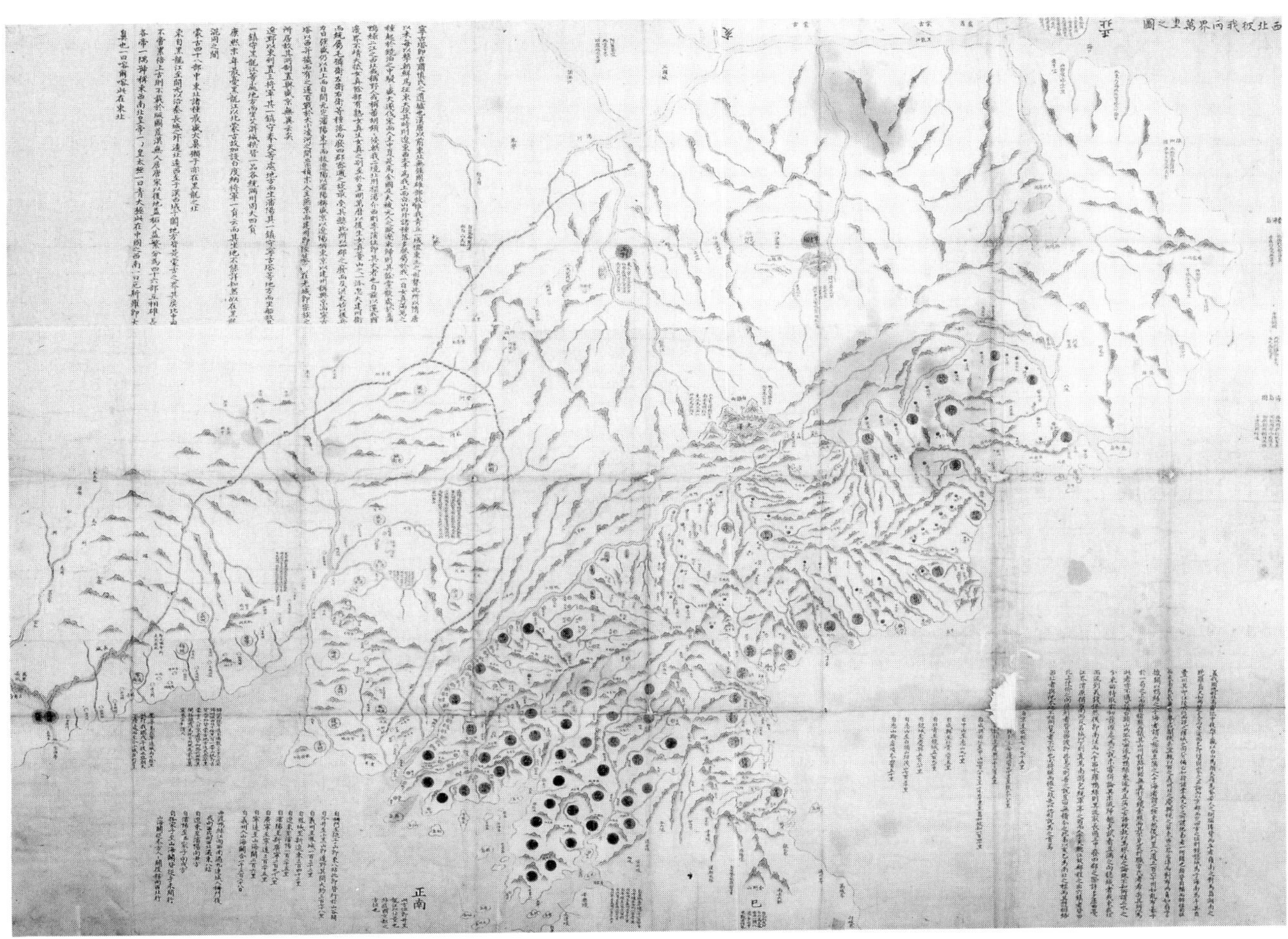

FIG. 10.53. *SŎBUK P'IA YANGGYE MALLI ILLAM CHI TO*. Of unknown authorship, this map probably dates from the mid-eighteenth century, but based on mention of Kangxi's "last years" and on the nonindication of Changjin as a district, it can be no earlier than about 1722 and no later than 1787. Its subject matter is similar to that of the parts of the *Yogyedo* shown in figure 10.52, but it is actually superior to the latter in the clarity with which it shows the topography and defense structure of Korea and Manchuria.
Size of the original: unknown. By permission of Kyujanggak Archives, Seoul National University (cat. no. Ko 4709-22).

Chŏlla provinces) is one of the relatively few surviving defense maps emphasizing this region (fig. 10.54). It is drawn on a broad strip of paper running over eight meters from right to left. Judging from style and appearance, it was probably done during the first part of the nineteenth century.[254] It spite of the impression left by the title, it covers the east coast of Kyŏngsang and the west coast of Chŏlla as well as the southern coasts of both provinces. The long, horseshoe-shaped coastline is mapped as if in a straight east-west panorama, including every town and harbor, every islet and inlet, and many other features of interest to coastal navigation and defense. A direct air flight over this length of coast would cover about 750 kilometers, but the actual length of the extremely uneven coastline, not including the thousands of islands large and small, would be several times that.

Another traditional feature of Korean defenses was the *sansŏng* (mountain fortress). Mentioned in the oldest stratum of historical records and archaeologically known for prehistoric periods, the *sansŏng* was a perennial Korean response to the problem of how to protect the population in time of war. In a situation where Korea was usually outmanned and overmatched by more powerful neighbors, it represented a realistic strategy of survival and attrition. In such places, chosen with attention to natural defenses, fortifiability, and water supply and

254. This map seems to have left no clue to its date of compilation. It does use the name Kohŭng instead of Hŭngyang for an important Chŏlla district, but attempts to use this fact for dating can prove only that the map was drawn before 1913 (not 1895 as claimed by some), which seems self-evident in any case. I put the map in the first half of the nineteenth century based on style.

FIG. 10.54. DETAIL FROM THE *YŎNGHO NAM YŎNHAE HYŎNGP'YŎNDO*. This map, probably done in the first half of the nineteenth century, shows a continuous panorama of the southeastern, southern, and southwestern coasts of Korea. The map, on paper in forty folds with an aggregate lateral length of over eight meters, shows all towns, military bases, harbors or anchorages, and such, along that coast, with distances and other notes written in. A written list of harbors and anchorages gives the docking or anchorage capacity and wind information for each site. The section illustrated here shows the area centering on T'ongyŏng (oval cartouche; now called Ch'ungmu), the general headquarters of the naval defense command. The modern headquarters of the Republic of Korea's navy is just to the northeast, at Chinhae. The large island southeast of T'ongyŏng is Kŏje. A fortified barrier/bridge links the T'ongyŏng mainland with supporting offshore bases.
Size of the entire original: 56.7 × 816.0 cm; this detail, 56.7 × 81.6 cm. By permission of the National Central Library, Seoul (cat. no. Kwi 116 Kojo 61-29).

maintained in a perpetual state of readiness and provisions, the enemy could be harried as well as avoided, and one's future might be negotiated from a position of safety. Such a strategy could cost a high price in property loss and destruction in the open areas that were evacuated, but it saved large numbers of lives and thus provided the essential conditions for rebuilding and recovery after the danger had passed. Korea's mountains were good for more than just their arteries to Paektusan.

As an example of the size and extent of some of the larger *sansŏng*, we may cite the famous Ch'ŏrong (Iron Jar fortress), already noted in a shapes-and-forces map of Yŏngbyŏn (fig. 10.21). It goes back to Koguryŏ times and probably saw frequent use during Koryŏ's various troubles. King T'aejong carried out a major refortification project in 1416, giving the Iron Jar and a large adjacent area a wall 26,815 *ch'ŏk* in circumference. Given the evidently longer *ch'ŏk* of the pre-Sejong era, this would have been considerably longer than the 5.34 kilometers indicated by the equivalence used in this chapter. When Sejong established new borders and created the district of Yŏngbyŏn in 1429, this compound became the district seat. Amid the general military strengthening of the seventeenth century, the Iron Jar proper was refortified in 1633, with its own inner perimeter wall of 550 meters. Sukchong added a northern branch fortress in 1675, then in 1685 refurbished and extended the general town fortifications, creating a walled perimeter of 13.24 kilometers, with the walls about 4 meters high. It had four main gates and enclosed three small rivers and fifty wells and springs. King Yŏngjo did some more rebuilding in 1750, and the whole complex was captured in a splendid map-painting, probably done sometime during the eighteenth century (plate 21).[255]

In general, cartography related to local and specialty maps proceeded on a separate track from national and provincial mapmaking. People of traditional mind saw no contradiction in this. Had it been conceivable to map a whole province in the style of a map-painting, or even the whole country, doubtless such maps would have found both artist-executors and buyer-appreciators. In subject areas where the map-painting style was practical and suitable, such as regional or district maps and many types of defense maps, that style seems to have been preferred.

The Historical and Social Setting of Korean Cartography

History was not kind to the record that would reveal to us the earlier stages of Korean cartography. The oldest surviving Korean map, apart from a drawing on the wall of a Koguryŏ tomb, dates only from about 1470. What came before, if not lost to the outrages of foreign invasions or internal turbulence, has yielded to the ravages of time. Even the three known copies of the 1402 *Kangnido* are today all in Japan as a result of war and pillage. And the year 1402 does not really signal a historical change in this pattern; just one other map from the fifteenth century survives, in a copy that can only be conjectured to descend from the national map of 1463, and that copy itself is also in Japan. Indeed, very few surviving maps can be said with certainty to predate the Japanese invasions of 1592, and if one subtracts from those the *Sŭngnam* maps, remaining specimens will be very few indeed. During the seventeenth century a small corpus begins to accumulate; only in the eighteenth and nineteenth centuries does that corpus acquire the variety and redundancy that promote serious study. And of these later maps, relatively few can be accurately dated or attributed to known individuals.

In spite of such obstacles, it has not been hard to identify some maps unique in the world, as well as others of outstanding originality and merit. Korea can claim the oldest record concerning Buddhist maps of the Five Indias type; the first genuine world map to have been made in East Asia; the development of an original shapes-and-forces cartography from a consciousness of its own mountainous terrain; a rare terrestrial globe of Western inspiration, uniquely set into an armillary sphere of purely Chinese mold; the cosmographical *ch'ŏnhado*, with its curious processing of ancient Chinese geographical lore; and Kim Chŏngho's giant national grid maps, innovatively designed and packaged for ease of use.

Korea's constant concern with the designs of its enemies is seen to have had a major impact on its cartographic development. The element of state security in mapmaking surfaces early and lasts until the end. It almost certainly had a negative effect on the *Sŭngnam* maps; on the other hand, it stimulated the genre of the defense map, which was both highly functional and very appealing to the eye. Security concerns were the major background factor in the Mukedeng affair, and they even entered folklore in the various legends that have grown up around Kim Chŏngho.

As a polity strongly committed to bureaucratic centralism and domination by a hereditary class of oligarchs, mapmaking understandably was centered in the government. Kwŏn Kŭn, Yang Sŏngji, and Yi Imyŏng were representatives of this dominant group. But Yi Hoe, the principal cartographer of the *Kangnido* of 1402 and of a map of Korea that was probably the standard for the first half of the fifteenth century, was by all appearances of relatively humble status. Chŏng Ch'ŏk came from a minor family of local administrators and owed such bureaucratic success as he had more to the patronage and friendship of his king than to the encouragement of the high bureaucracy. Chŏng Sanggi, though from a distinguished family, never took the civil examinations and never held government office. The social background of Kim Chŏngho is unclear, but at best he could have been only of low-ranking military background; he probably was an ordinary commoner. He lived at a time when the line between poor gentry and talented commoner could be hazy and crossable; in any case he somehow acquired a good education, and that was always the first requirement for upward mobility.

In the early periods of Chosŏn mapmaking, most of the activity is at state initiative; in the later periods, new trends in maps start outside the government and achieve either its recognition, as with the Chŏng Sanggi maps, or only its passive aquiescence (my belief) or even possible hostility (if we credit the legend), as with Kim Chŏngho. The periods when the government dominated mapmaking are also the periods for which the smallest number of maps survive: when the fires and invaders came, maps were all in a few places and were too easily lost. But when cartography acquired a broader social base from the seventeenth century on, the avenues of survival mul-

255. *Munhŏn pigo*, 30.8a (note 51). In view of the government's heavy investment in the Yŏngbyŏn-Ch'ŏrong-Puksansŏng complex during the seventeenth century and the conformity of this map-painting to the construction listed in the *Munhŏn pigo*, there seems to be no basis for attributing this painting, as some do, to the sixteenth century or earlier. The painting style seems typical of the late eighteenth century.

tiplied, and so did the maps. Involved in this were not only the mapmakers and the consumers, but painters, woodblock printers, and copyists of all kinds, all but a few of them utterly anonymous. The heyday of Korean mapmaking came when the government lost—or abandoned—its monopoly. Even the royal project of 1791, to create a national grid and organize local mapping, apparently failed to win bureaucratic support or even official historical notice, and it was redeemed only years later by the efforts of private scholars or self-chosen men of mission, such as Ch'oe Han'gi and Kim Chŏngho.

It is perhaps the lack of such broader involvement in learning and development that explains how genuinely brilliant projects, such as the *Kangnido* and the terrestrial globe, could be completed and then go nowhere. Why is it that after the first appearance of Sejong's odometer, the "*li*-counting drummobile," in 1441 or Sejo's triangulation instrument, the *kyuhyŏng*, in 1467 we never see any development of them or indeed any sign that they were ever used again? Is it that they were later constantly used and became commonplace? Or that they were tried out and found not to work so well after all? If Sejong could send observers to the extremities of the country to take measurements of polar altitude, why could no one else repeat the effort later (and this time preserve the data)? Were the scientific observational instruments that the government purchased in Beijing in 1715, or any others, ever used to determine geodetic coordinates? If there are clear sources on these problems, I, at least, have been unable to find them. They represent some questions for the future.

During the traditional period of Korean history, Korean culture borrowed heavily from Chinese civilization, and this relationship is evident in cartography. In the concrete case of Kim Chŏngho's great grid maps, there was a direct methodological connection, in that both Ch'oe Han'gi's preface to the *Ch'ŏnggudo* and Kim's own introduction to the *Taedong yŏjido* cited Pei Xiu's six principles of mapmaking with high approval. Kim clearly considered his cartographic method an application of Pei's principles. But it is interesting to observe that in spite of this cultural congruity and direct influence, Koreans did different things with maps than Chinese did, and the maps they made usually had a rather different appearance than Chinese maps. Having the same materials available from the Islamic and Western worlds, the two countries reacted differently. China produced a map of "Great Ming," while Korea added itself and Japan and produced a world map. Korea was indeed a different country, with its own sense of itself. For all the culture borrowed from China, Korea's own independent cultural tradition never lacked vitality, and in any given context Chinese culture, no matter how much it might have dominated appearances, was only part of the story. Cartography is a particularly useful medium for showing the great variety within the broader East Asian civilization, which though dominated by Chinese culture cannot be defined only in its terms.

Korea had a lively curiosity about Western observational and cartographic sciences. It imported, copied, and printed Western maps, and it made the globe for the armillary clock. But none of this seems to have had any effect on Korean maps, which continued to develop on their own as if nothing had happened. Kim Chŏngho printed a hemispherical Western map, copying everything from the graticules to the ecliptic, yet in his own maps he continued on his own path, wholly within Korean tradition. He understood the function of geodetic coordinates and made reference to them in his work. But to the extent that he used them at all, he seems to have been content with the map-derived coordinates determined by King Chŏngjo's mathematicians in 1791.

Before the nineteenth century was over, the influence of Western maps had begun to come into Korea strongly, and indeed it overwhelmed traditional cartography. But cartography was only a drop of water in a mighty wave of modern change. Two processes were involved. One was a turning away from Chinese civilization in general. The other was the introduction of Western civilization, at first through Protestant missionaries (mainly American) but more decisively through an aggressive Japanese intrusion and colonization. If a single event is to be connected with this phenomenon, it would be Japan's defeat of China in 1895. Ten years earlier, China had made Korea a civilizational cause by insisting on the continuation of the tributary system, but then it violated its cardinal tenet by blatantly interfering in Korea's internal politics during the decade of Yuan Shikai's residency. Japan's victory not only weakened China's standing as a nation, it also, for a critical balance of influential Koreans, destroyed China's credibility as a civilization. (Not that Japan did not bring plenty of pain and suffering to Korea in the process.) From that time on, the accommodation to Western ways proceeded without significant impediment. Then and only then can modern Western cartography be said to have prevailed in Korea.

APPENDIX 10.1 KOREAN MAPS LISTED BY YANG SŎNGJI IN 1482

Title	Date	Cartographer
1 *Odo yanggyedo* (Map of the five provinces and two frontier zones)[a]	Mid-Koryŏ (ca. 1150)	
2 *P'altodo* (Map of the Eight Provinces)	Early Chosŏn (ca. 1400)	Yi Hoe
3 *P'altodo* (Map of the Eight Provinces)	Sejong reign (1418–50)	Chŏng Ch'ŏk
4 *Yanggye taedo sodo* (Large and small maps of the two frontier zones)	Sejong reign (1418–50)	Chŏng Ch'ŏk
5 *P'altodo* (Map of the Eight Provinces)	Sejo reign (1455–68)	Yang Sŏngji
6 *Yŏyŏn Much'ang Unul samŭpto* (Map of the three districts Yŏyŏn, Much'ang, and Unul)[b]	Sejo reign (1455–68)	Yang Sŏngji
7 *Yŏnbyŏn sŏngjado* (Map of border fortifications)	Sejo reign (1455–68)	Yang Sŏngji
8 *Yanggye yŏnbyŏn pangsudo* (Map of border defenses in the two frontier zones)	Sejo reign (1455–68)	Yang Sŏngji
9 *Cheju Samŭpto* (Map of the three districts of Cheju Island)	Sejo reign (1455–68)	Yang Sŏngji
10 *Yŏnhae Choundo* (Map of coastal shipping and transport)	Sejo reign (1455–68)	An Ch'ölson
11 *Yŏng'ando yŏnbyŏndo* (Map of the frontier border of Yŏng'an [Hamgyŏng] Province)	Sejo reign (1455–68)	Ŏ Yuso
12 *P'yŏng'ando yŏnbyŏndo* (Map of the frontier border of P'yŏng'an Province)	Sejo reign (1455–68)	Yi Sunsuk
13 "The maps in each of the governors' headquarters in the three southern provinces"	Sejo reign (1455–68)	?
14 *Ilbon Yugu kukto* (Map of Japan and the Ryūkyūs)	Sejo reign (1455–68)	Japanese monk Tōan
15 *TaeMyŏng ch'ŏnhado* (Map of the Great Ming world [in both silk and paper scrolls])	Sejo reign (1455–68)	?
16 *Chiri chi nae p'alto chugundo* (Maps of the districts of the Eight Provinces from *Chiri chi*)	Sejo reign (1455–68)	Yang Sŏngji
17 *P'alto sanch'ŏndo* (Map of mountains and rivers of the Eight Provinces)	Sejo reign (1455–68)	Yang Sŏngji
18 *P'alto kagil yanggyedo* (Separate maps of the Eight Provinces, with the two frontier zones)[c]	Sejo reign (1455–68)	Yang Sŏngji
19 *Yodongdo* (Map of Liaodong)	Sejo reign (1455–68)	?[d]
20 *Ilbon TaeMyŏngdo* (Map of Japan and the Great Ming)	Sejo reign (1455–68)	?

Source: Sŏngjong sillok (Annals of King Sŏngjong, r. 1470–94), 138.10b, Yang Sŏngji Memorial dated Sŏngjong 13/2/13 (2 March 1482).

[a]Phrases on the order of "eight provinces," etc., may be taken as a collective name and could be rendered "Korea." During Koryŏ times and for most of the fifteenth century the "two frontier zones" were under separate military administration.

[b]Three Yalu frontier districts (in the area of modern Kanggye), established in 1416, 1443, and 1442, respectively, but abandoned as logistically impractical in 1455.

[c]In this case, separate maps of each of the eight provinces are to be understood.

[d]It is unclear whether Yang Sŏngji's authorship extends to items 19 and 20.

11 · Cartography in Japan

KAZUTAKA UNNO

INTRODUCTION: THE MAIN MAPPING TRADITIONS

Japanese cartography before the Meiji Restoration in 1867 was characterized by considerable diversity. This is a reflection not only of the practical use of maps—mainly as administrative documents and tools for wayfinding—but also of their adaptation for decoration, propaganda, and literary purposes. Practical uses were by far the more common, as we might expect in a pragmatically oriented society. Maps in this category included local maps of manors, maps of the properties of religious institutions, and maps of reclaimed land as well as maps of cities, provinces, and the country as a whole. Route maps and marine charts form another category, while in the metaphysical sphere we encounter maps from Buddhist cosmology.

Because of Japan's isolation for most of the Edo period (1603–1867), European maps of the world and of Japan adopted by the Japanese assume something of a rhetorical and ornamental character. We encounter them on folding screens used to decorate rooms, on hanging scrolls, and on smaller objects such as sword guards and plates (fig. 11.1). Although these point to the artistic applications of cartography, it is clear that map images were regarded as important in many other situations. Indeed, when Japanese culture as a whole was being crystallized in the Edo period, artists played a prominent role in mapmaking. At the same time, there is no clear-cut evidence of a European-type scientific revolution in Japanese cartography. Modern standards of accuracy were not widely disseminated until the early nineteenth century, when surveys began to be made with precise instruments and methods. But despite their technical limitations, Japanese maps were regarded as an important source of knowledge. The key to understanding them lies in the particular context of Japanese history and society.

The opening chapters of Japanese map history are placed in the Nara period (710–84), when Japan was consolidating elements of a new culture based on Chinese influences. Buddhism had made remarkable gains. Nara, the first "permanent" capital, had been constructed according to Chinese geomancy with Chang'an (modern Xi'an) as a model, while the Taihō Code of 701 affirmed commitment to the Taika Reform from 645, which were both based on Chinese legal and administrative models. It is to this period that the historian of cartography may turn for the creation of cartographic traditions in Japan. On the one hand, there are the practical paddy-field maps, a product of land reclamation associated with the endowment of Buddhist temples. On the other hand, there is the so-called Gyōki tradition, a series of maps of Japan first recorded in 805, just after the Nara period proper.

As early as 738, efforts were also made to prepare a series of provincial maps. The order for these surveys is preserved in the *Shoku Nihongi* (Chronicles of Japan, continued), an official history of Japan dating to 797, and in 796 another order, recorded in the *Nihon kōki* (Later chronicles of Japan) of 840, was given by the central government to compile maps of the provinces. The next important systematic attempt to compile such maps did not occur until after 1605, shortly after the establishment of the Tokugawa shogunate, when such maps were called *kokugunzu*; another scheme to compile provincial maps was undertaken from 1644 to about 1656. By this date Japanese cartography was adopting scientific methods, as is suggested by instructions that include a prescribed scale equivalent to 1:21,600 (six *sun* to one *ri*). This trend culminated in the surveys of Inō Tadataka and Mamiya Rinzō in the first two decades of the nineteenth century.

The Gyōki tradition of maps of Japan might have originated as early as the eighth century, but the earliest extant copies of this type date to the late thirteenth and early fourteenth century. There is no evidence that the Buddhist priest Gyōki himself compiled any such maps—indeed, the content of the known maps would date a prototype to approximately half a century at the earliest after his death—but they may have been derived from the maps following the orders of 738 and 796 for provincial surveys mentioned above. Whatever their origin, despite advances in cartographic knowledge and techniques, the Gyōki-type maps survived in modified form into the nineteenth century. They were regarded as satisfactory until the more accurate surveys of the Edo period, but even then Gyōki-type maps continued to be produced.

The compilation of provincial and national maps assumes organized surveying techniques as well as an understanding of at least the rudimentary instruments. Our knowledge in this area is incomplete. Even for the modern period there are many gaps, and the techniques and instruments of antiquity remain a matter for speculation. Apart from the few remaining artifacts in Japan, the only way forward is to draw analogies with Chinese practice, for which there is evidence in Chinese texts and their Japanese translations. From the seventeenth century onward the picture is clearer. European ideas, methods, and instruments became important, although much of the evidence is found in instruments and manuals from the eighteenth century. The Portuguese, who played a leading role in transmitting European surveying techniques and instruments to Japan, were also Japan's main European trading partners. In all likelihood they, and to some extent the Spanish and the Jesuits, initiated the process of transmission, which began when instruments such as the compass and astrolabe arrived in the early seventeenth century.

Two well-defined traditions of world maps also weave through the history of Japanese cartography. The longest lived was the Buddhist conception of the world; a later tradition originated in European knowledge from the sixteenth century onward. The first type of world map is dated after the introduction of Buddhism to Japan in the sixth century, but the exact date for the first Japanese map is not known. By the mid-seventh century, a Buddhist cosmography was accepted in high social circles. The earliest surviving world map is from 1364 (the *Gotenjiku zu* [Map of the Five Indias]), but the genre flourished throughout medieval and Tokugawa Japan until the mid-nineteenth century.

Some of the world maps based on European ideas were introduced before the period of national isolation. The so-called *Nanban* (southern barbarian) maps, introduced by the Jesuits, were derived from a variety of European originals, including the world map in Abraham Ortelius's *Theatrum orbis terrarum* (first published 1570). A map of 1602, derived ultimately from the work of Matteo Ricci, served as a model for Japanese world maps throughout the Edo period and offered an alternative to Buddhist cosmography. Like the Buddhist maps, the Ricci-type maps were published for a general audience in the later stages of the Edo period.

European influence was also felt in other areas of cartography from the early seventeenth century onward, notably in marine charts and the Jōtoku-type maps of Japan (named after the map of Japan found at Jōtoku Temple in Fukui; see below). The Jōtoku-type maps—of which four of five known examples are dated between 1592 and 1627 and a fifth to the mid-seventeenth century—were a modification of the Gyōki tradition mentioned above.

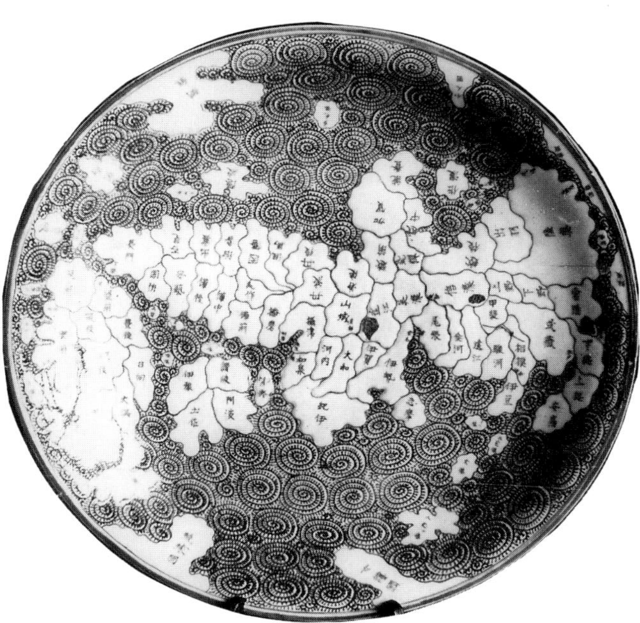

FIG. 11.1. MAP OF JAPAN ON AN IMARI PLATE, CA. 1840. In the nineteenth century, maps of Kyūshū and Japan started to be drawn on plates produced in Imari, Kyūshū, a well-known china center. Those of Japan were of the Gyōki tradition (see the discussion of Gyōki maps below).
Diameter of the original: 49 cm. Nanba Matsutarō, Nishinomiya, Hyōgo Prefecture. Photograph courtesy of Kazutaka Unno.

Improvements, such as to the outline of the coast, may have been made by the Portuguese traveler Ignacio Moreira, who lived in Japan in 1590–92. Insofar as the Jōtoku maps were used in navigation, they may be linked to the marine charts introduced by the Portuguese in the early seventeenth century. The surviving charts relate to East and Southeast Asia and to Japan. The surviving copies begin in the early seventeenth century and extend through the first two decades, but with a few late copies dating to the mid-nineteenth century. That they should have been drafted during the period of isolation is explained mainly by the fact that such charts continued to be used as evidence of a surveyor's license.

The main strength of Japanese cartography thus lies in its portrayal of the home country at different scales. This should not be surprising: Japan was geographically separate from continental Asia, it had a long history of undisturbed independence, and it was economically self-sufficient. There was little impetus to map the world beyond its own shores. It was Japan that mattered most to the Japanese, and this attitude was reinforced by the seclusion policy of the Tokugawas. The hierarchical structure of Japanese society must also be taken into account. Mapmaking had to fit into its proper place and to reflect the requirements of the governmental elite. Their concern was with Japan; the rest of the world came

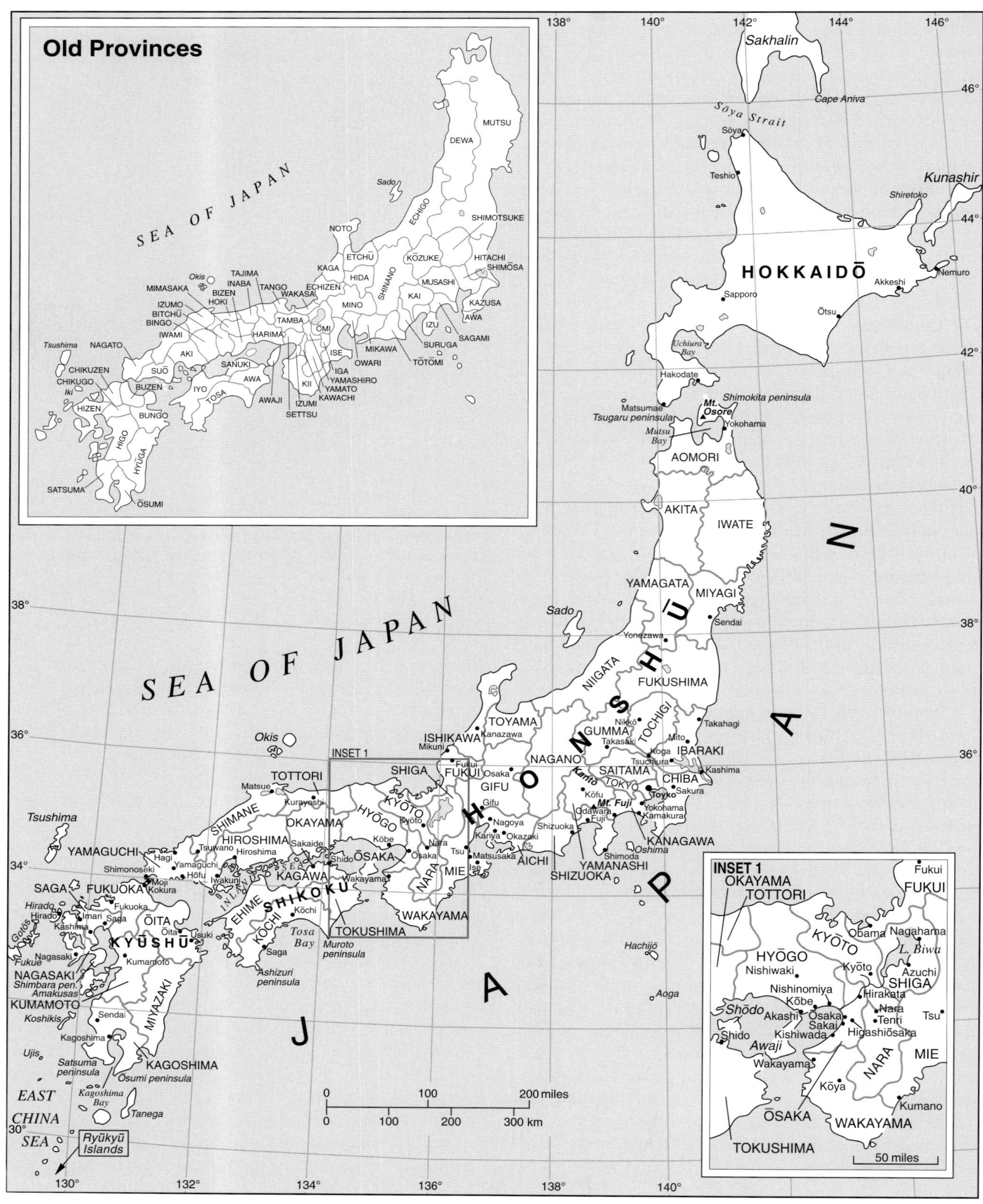

FIG. 11.2. REFERENCE MAP FOR JAPANESE CARTOGRAPHY. This map shows most of the locations mentioned in the text, including the modern prefectures and old provinces.

a distant second. (See figure 11.2 for a reference map of Japan.)

VOCABULARY, SCALE, ORIENTATION, AND MATERIALS

There are several terms for "map" and the different types of maps compiled and used in Japan in historical times. The most important root in these terms is *zu*, which might be translated as "map" or "diagram" and seems to have been in use since the eighth or ninth century. Before then, at least in the seventh-century entries in the *Nihon shoki* (Chronicles of Japan) of 720, the word *kata* 圖 was used; this was an abbreviation of *katachi* (shape). The most widely used word, though, has been *ezu* 繪圖, which translates as "pictorial diagram" and was derived from the *jōri* system of land distribution. It was used in the titles of most of the maps discussed in this chapter and even formed part of the word for "official cartographer" (*ezukata*) at the time of the official surveys under the Tokugawas.

A prefix added to *zu* or *ezu* can specify the type of map; for example, a *chizu* is a land map and a *sekaizu* is a map of the world. It appears that the term *hakuzu* (white or simple diagram) was used to denote early diagrams with a grid based on the *jōri* framework, and a *bunzu* (graphic list) gave the location and area of a parcel of land based on this system. Eighth-century maps of landownership were called *denzu* (cadastral maps), and a similar genre of *shōenzu* (manorial maps) appeared between the twelfth and sixteenth centuries. The latter was derived from the words for "villa" (*shō*) and "cultivated land" (*en*). Provincial maps such as those compiled under the Tokugawas are called *kokugunzu* or *kuniezu*, and route maps are *dōchūzu*. Whereas all of these terms are Japanese in derivation, the marine charts derive their Japanese name, *karuta*, from the Portuguese word for map, *carta*. One early eighteenth-century writer used the term *shinro hanzu* (card chart of courses), but it was not used by anyone else.[1]

Some introductory comments can also be made about scale, orientation, the qualities of the materials, and the size of Japanese maps. In the West, after the revival of Ptolemaic geography, emphasis was placed on the scientific nature of mapmaking, and this included indicating mathematical scale. In Japan, however, scale was not indicated except for special purposes. Even travel maps, which should clarify the distance between two points, were rarely drawn at a fixed scale, and the distances were usually stated literally. Points and objects considered to be important landmarks were drawn in an exaggerated fashion.

In regard to orientation, there was no established custom, which may in part be due to the way many Japanese maps were consulted. They were often large when unrolled or unfolded, but they could easily be spread out on the tatami on the floors of houses, where there was very little furniture. Users would then sit or kneel around them and rotate the maps as necessary, and so multiple points of view were preferred over a single one. In the case of the maps on the folding screens and hanging scrolls, however, it was necessary to record the information in a consistent direction, since these would not have been consulted on the floor. Nonetheless, there was no standard rule for orienting the maps, as illustrated by some of the Gyōki-type maps: west for the mid-sixteenth-century version at Tōshōdai Temple; south for the Ninna Temple version dating from 1306; east for the *Shūgaishō* (Collection of oddments) versions of 1548 and 1589; and north for the version at the Tokyo National Museum from about 1625. In town plans of Edo (now Tokyo) and Ōsaka, to give another example, orientation was contingent on the location of the city castle, which was shown toward the top of the map. North tended to be used for the general maps of Japan in the Edo period, perhaps influenced by European cartography.

The large size of most Japanese maps also may be attributed to the availability of floor space. When the maps were not being consulted they could be rolled up or folded and stored in a small area. The qualities of the paper, drawing brush, and woodblocks also contributed to the large size. The paper was thin, strong, and flexible but also rough, and the brush that was traditionally used for writing and drawing was incapable of making fine lines, although it was good for coloring. These factors naturally made it practically impossible to write or draw anything small. The same may be said in general for the woodblock process of printing, despite the vast improvements in line cutting by the middle of the nineteenth century, when some woodblock prints might have been taken for copperplate engravings.

In that it could be used more easily for multicolor maps, woodblock or relief printing had an advantage over copperplate engraving, an intaglio process. Multicolor printing in cartography appears to have been adopted at approximately the same time as the beginning of the *ukiyoe* (pictures of the floating world) color prints about 1765, although it is not known exactly when such maps were first made. Before this, maps were colored by brush or by *kappa-zuri* (stenciling), in which cut-out patterns were placed on the paper and colors applied over them. The way to tell if the *kappa* method was used is to look for uncolored places or defects caused by using stencils. Because a pattern is cut out of tanned paper, for example, a complete ring cutting cannot be colored, and in coloring a long strip, the pattern is fixed in place with string to prevent the brush from turning it over. Consequently

1. Nishikawa Joken, *Ryōgi shūsetsu*, see p. 381 and note 142 below.

a mark of stagnant paint caused by the string is left on the paper. Because more than one pattern is used to color a complicated part, one color inevitably overlaps another or some space is left uncolored. Multicolored woodblock prints are relatively free from these problems, so it is possible to tell whether the *kappa* method was used by finding places that are not colored or where there are irregularities in the color. Determining this is not always easy, however, because light colors often were used to hide these defects.

Neither the paper nor the woodblocks came in large sizes; several sheets of highly glutinous paper were patched together, and in very large maps stronger paper was pasted to the back to reinforce the patchwork. The strength of the paper made it possible and preferable to make the maps in the form of scrolls and folding books; indeed atlases—the alternative format—were rare. Hemp and vellum also were used, the former for most of the extant works from the eighth century and the latter for a few of the charts. In these cases, too, a large surface had to be created to portray information clearly.

HISTORIOGRAPHY AND MAJOR COLLECTIONS

The history of Japanese cartography is in the process of being written, and research in progress will substantially expand the existing literature. A number of volumes published in the past two decades have laid an excellent foundation for future historians, who will have the task of filling in the major gaps that still exist for topics such as medieval estate plans, river maps, and the history of ancient and medieval surveying. The history of Japanese cartography has passed its first hurdle: the general, extensive knowledge that now exists is ready to make way for the specific, intensive knowledge of present and future scholarship.

One of the pioneers of the subject was Kawada Takeshi, the author of a three-part article on early Japanese cartography written in 1895. The emphasis was on official maps in the Edo period, and ten years later Kawada published another three-part article on cartography and geographical writings through the ages.[2] Another pioneer was Takagi Kikusaburō, whose brief history of cartography and surveying was published in 1931.[3] Although this was the first separate volume to include a discussion of Japanese cartography from antiquity to the early twentieth century, the book has two shortcomings. The author, who was an employee of the Department of Land Survey, concentrates on cartography and surveying after the Meiji Restoration in 1867; moreover, the description is too fragmentary to be considered a proper history.

The first substantial history of Japanese cartography was written by Ashida Koreto and published in 1934.[4] The work treats a variety of maps ranging in scope from maps of Japan to maps of the world. Ashida's training as a historian was used effectively to give a concise but valuable description of Japanese cartographic history. In 1932 Fujita Motoharu published his history of Japanese geography, devoting most of it to a description of cartography; a revised, enlarged edition was published in 1942.[5] The latest contribution in this field is Oda Takeo's enlargement of part 2 (dealing with Japanese cartography) of a general history of maps originally published in 1973.[6]

Two works focus specifically on the history of Japanese maps of Japan. An article by Ashida is a short account of this subject, and Akioka Takejirō's book on the history of maps of Japan offers a more detailed treatment.[7] There are also volume 1 of *Nihon kochizu taisei* (1972), which provides a history of maps relating to Japan as a whole as well as of maps, plans, and charts of smaller areas within the country,[8] and Akioka's history of the making of world maps.[9]

Some literature exists in Western languages on the history of Japanese cartography. Early contributions include those by Ramming in 1937 and Akioka and Muroga in

2. Kawada Takeshi, "Honpō chizukō" (Study of the maps of our country), *Shigaku Zasshi* 6 (1895): 268–77, 349–58, and 507–18, and his "Nihon chishi gen'i o ronzu" (On the transition of geographical descriptions in Japan), *Rekishi Chiri* 7 (1905): 821–27, 916–21, and 1038–45.

3. Takagi Kikusaburō, *Nihon chizu sokuryō shōshi* (A brief history of cartography and surveying) (Tokyo: Kokon Shoin, 1931).

4. Ashida Koreto, *Honpō chizu no hattatsu* (The evolution of cartography in Japan) (Tokyo: Iwanami Shoten, 1934).

5. Fujita Motoharu, *Nihon chirigaku shi* (History of Japanese geography) (Tokyo: Tōkō Shoin, 1932), and *Kaitei zōho Nihon chirigaku shi* (Revised and enlarged history of Japanese geography) (Tokyo: Tōkō Shoin, 1942; reprinted Tokyo: Hara Shobō, 1984).

6. Oda Takeo, *Chizu no rekishi* (History of maps), 2 vols. (Tokyo: Kōdansha, 1974), vol. 2 (Japan). The earlier work was his "Nihon no chizu to sono hattatsu" (Japanese maps and their development), in *Chizu no rekishi* (History of maps) (Tokyo: Kōdansha, 1973), 211–89.

7. Ashida Koreto, "Nihon sōzu no enkaku" (History of general maps of Japan), *Kokushi Kaikokai Kiyō* 2 (1930): 17–59, and Akioka Takejirō, *Nihon chizu shi* (History of maps of Japan) (Tokyo: Kawade Shobō, 1955).

8. Unno Kazutaka, Oda Takeo, and Muroga Nobuo, eds., *Nihon kochizu taisei* (Great collection of old Japanese maps), 2 vols. (Tokyo: Kōdansha, 1972–75). The first volume of this set, with the added title *Monumenta cartographica Japonica*, contains 133 plates, most in color, followed by plate descriptions and 75 additional black-and-white figures; the second volume, titled *Nihon kochizu taisei sekaizu hen* (also on the title page: "The world in Japanese maps until the mid-19th century"), contains 138 plates, most in color, and under separate cover has plate descriptions and 127 additional black-and-white figures.

9. Akioka Takejirō, *Sekai chizu sakusei shi* (A history of making world maps) (Tokyo: Kawade Shobō Shinsha, 1988). The entire four-part set (book, twelve single-sheet color reproductions, videotape, and forty-eight slides) is titled *Sekai kochizu shūsei* (Collection of old world maps), English title, *Akioka Collection/Old World Maps/16th–19th Centuries* (Tokyo: Kawade Shobō Shinsha, 1988). It is a history of world maps mainly from the time of the European Renaissance until the end of the Edo period in Japan.

1959.[10] Muroga also wrote "The Development of Cartography in Japan" for *Old Maps in Japan*.[11] Watanabe's *Cartography in Japan: Past and Present* (1980) was published on the occasion of the Tenth International Conference of the International Cartographic Association, held in Japan.[12] An English-language description of the history of official Japanese cartography is included in the catalog of an exhibition on the subject hosted by the National Diet Library.[13] The most recent works to appear in Western languages (the present chapter excepted) are by Cortazzi and Unno.[14] There is also a parallel tradition of publishing facsimiles of early Japanese maps, but the maps they include represent only a small part of the extant corpus of Japanese maps, plans, and charts.[15]

The relevant maps are scattered among private and public collections, posing problems of access.[16] Although the major libraries and museums in Japan include maps in their published catalogs of books, there are very few catalogs of the collections themselves. Institutions with large collections of old maps that have published catalogs include the Geographical Institute in the Faculty of Letters at Kyōto University, the Saga Prefectural Library in Saga, and the Kōbe City Museum.[17] The last named has the largest collection in Japan (approximately 5,500 items). Other important collections in Japan include those of the National Archives and the National Diet Library in Tokyo.[18]

Important private collections are now in three public institutions. The collection of the map historian Akioka Takejirō has passed to the National Museum of Japanese History in Sakura, Chiba Prefecture, and the Kōbe City Museum. The National Museum contains over a thousand items relating to Japan (including surveying instruments), and the Kōbe City Museum has mainly Akioka's world maps. Although it has not been cataloged completely, a part of the collection was reproduced in Akioka's *Nihon kochizu shūsei* (1971) and in his *Sekai kochizu shūsei* (1988).[19] Nanba Matsutarō's collection is also at the Kōbe City Museum. The collection of map historian Ayusawa Shintarō, consisting of maps of the world and their related materials, was presented to the Yokohama City University Library, which has published an annotated catalog.[20]

A depository of special note because of its eighth-century maps is the Shōsōin (Treasury) at Tōdai Temple in Nara. As one of the major land reclaimers in the eighth century, the temple had many maps of manors. The sur-

10. M. Ramming, "The Evolution of Cartography in Japan," *Imago Mundi* 2 (1937): 17–22, and Takejirō Akioka and Nobuo Muroga, "The Short History of Ancient Cartography in Japan," *Proceedings of the International Geographical Union Regional Conference in Japan, 1957* (Tokyo: Organizing Committee of the IGU Regional Conference in Japan and the Science Council of Japan, 1959), 57–60. See also Takejirō Akioka and Nobuo Muroga, "The History of Cartography in Japan," in *A Catalogue of Old Maps of Japan, Exhibited at the Tenri Central Library, September, 1957*, ed. Takeo Oda (Kyōto, 1957), 1–6.

11. Nobuo Muroga, "The Development of Cartography in Japan," in *Old Maps in Japan*, ed. and comp. Nanba Matsutarō, Muroga Nobuo, and Unno Kazutaka, trans. Patricia Murray (Ōsaka: Sōgensha, 1973), 158–76. This book was originally published in Japanese: *Nihon no kochizu* (Ōsaka: Sōgensha, 1969). The color plates are identical in the two volumes, but the black-and-white figures accompanying the text vary slightly.

12. Akira Watanabe, *Cartography in Japan: Past and Present* (Tokyo: International Cartographic Information Center, 1980).

13. *Nihon no chizu: Kansen chizu no hattatsu* (Cartography in Japan: Official maps, past and present), exhibition catalog, Twenty-fourth International Geographical Congress and Tenth Conference of the International Cartographic Association (Tokyo: Kokuritsu Kokkai Toshokan, 1980).

14. Hugh Cortazzi, *Isles of Gold: Antique Maps of Japan* (Tokyo: Weatherhill, 1983), and Kazutaka Unno, "Japan" and "Japanische Kartographie," in *Lexikon zur Geschichte der Kartographie*, 2 vols., ed. Ingrid Kretschmer, Johannes Dörflinger, and Franz Wawrik (Vienna: Franz Deuticke, 1986), 1:357–61 and 1:361–66.

15. An early collection was edited by Kurita Mototsugu, *Nihon kohan chizu shūsei* (Early maps and plans printed in Japan) (Tokyo: Hakata Seishōdō, 1932). Most works, however, are more recent. Rather limited in scope is *Nihon no kochizu* (note 11). Much more ambitious are the two volumes of Unno, Oda, and Muroga, *Nihon kochizu taisei* (note 8). There are also Akioka Takejirō's *Nihon kochizu shūsei* (Collection of old maps of Japan) (Tokyo: Kajima Kenkyūjo Shuppankai, 1971), and his *Sekai kochizu shūsei* (note 9).

16. For example, the maps possessed by the feudal clans during the Edo period were handed over to prefectural and municipal libraries, and presumably not all of these pieces have come to the attention of map historians. This of course might be said of maps in private collections as well.

17. There are over five hundred items at the Kyōto depository; the catalog appeared in three installments: volumes 3, 5 (both 1934), and 9 (1937) of *Chiri Ronsō*, an academic journal edited by the Institute. The catalog of the Saga collection is *Saga Kenritsu Toshokan zō kochizu ezu roku* (Catalog of early maps and plans in the Saga Prefectural Library collection) (Saga, 1973). That for the Kōbe collection is in the cartographic sections of the *Kōbe Shiritsu Hakubutsukan kanzōhin mokuroku* (Catalog of the collections at the Kōbe City Museum), 6 vols. (Kōbe, 1984–89), Nanba Collection; included, for example, are maps of the world, Japan, provinces, cities (Edo, Kyōto, Ōsaka), the northern frontier of Japan, and foreign countries, as well as itineraries. The museum acquired a part of the Akioka Takejirō Collection (approximately 1,500 items) in 1989.

18. At the National Archives, it is the Naikaku Library that keeps early maps; it was an independent bureau until 1971, when the National Archives were established. The National Diet Library has a map room devoted to modern maps; early maps, however, are classified with books in general.

19. Akioka, *Sekai chizu sakusei shi* (note 9). A catalog of part of the collection was compiled when it was given to the National Museum of Japanese History; it was published in parts in the journal *Gekkan Kochizu Kenkyū* 7, nos. 3–11 (1976–77). For some reproductions see also Akioka, *Nihon kochizu shūsei* (note 15).

20. See note 17 on the Kōbe City Museum catalog, which has been published according to types of maps and makes an indispensable reference work. On the Ayusawa collection, see Yokohama Shiritsu Daigaku Toshokan (Yokohama City University Library), *Ayusawa Shintarō Bunko mokuroku* (Catalog of the Shintaro Ayusawa Collection) (Yokohama: Yokohama Shiritsu Daigaku Toshokan, 1990).

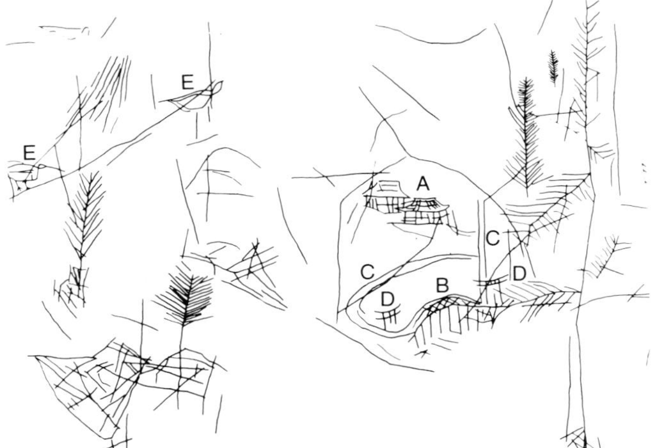

FIG. 11.3. LINE-ENGRAVED MURAL FROM TOMB 48 AT KAZUWA, KURAYOSHI, TOTTORI PREFECTURE, DISCOVERED IN 1974. The keyhole-shaped tomb is on a hill about sixty meters above sea level. The mural is on the huge stone that forms the back wall (260 cm high and 224 cm wide); it is painted red all over, and the lines were scratched into the surface with a sharp tool. The contents include houses (A), a bridge (B), roads (C), most likely Shinto archways (D), and birds (E).
Size of this part of the mural: 86 × 110 cm.

vival of twenty maps preserved at the Shōsōin may be attributed to the excellent storage conditions there, as well as to imperial restrictions on their use. The maps themselves were preserved because of their use in settling boundary disputes, but when those functions declined they were treated with less care and were damaged and even abandoned.[21]

A number of important collections are in European and North American libraries. In Vancouver, the George H. Beans Collection at the University of British Columbia Library is the largest collection outside Japan (three hundred items if books with maps are included).[22] There are also pre–Meiji Restoration maps that relate to a variety of subjects—to the world and foreign countries and to Japan and its regions, routes, and shrines—in five Munich depositories.[23] The strength of these collections is in nineteenth-century maps.

Ancient and Medieval Japanese Cartography before the Edo Period

ARCHAEOLOGICAL EVIDENCE

What may be regarded as the oldest map in Japan is a topographical drawing on the stone wall inside tomb 48 at Kazuwa in Kurayoshi, Tottori Prefecture, estimated to have been built in the sixth century A.D.[24] It is scratched on the flat surface of a huge cinnabar-coated stone, clearly depicting a landscape with houses, roads, a bridge, trees, birds, and possibly torii (gateways to Shinto shrines) (fig. 11.3). The purpose may have been to console the spirit of the dead: at least it was not intended for any practical geographical use. Similar engravings of birds, ships, fish, and trees are found in nearly forty tombs throughout the prefecture. These, like the practice of putting swords and beads in stone huts, were probably meant to provide the deceased with recognizable objects and landscapes so that they might be as active in death as in life.

There are also some colored paintings on the walls of the Takamatsuzuka burial mound (*kofun*) in Nara Prefecture that date to the end of the seventh century or the beginning of the eighth.[25] These fall into the category of celestial cartography (see chap. 14 below). A star chart is painted on the ceiling using gold foil for the stars, which are connected by straight red lines. On the north wall, opposite the entrance, is a partly damaged painting of a

21. In 1833, in order to repair the Shōsōin, everything inside, including national treasures, was moved elsewhere and examined. In the process the twenty paddy-field maps were found in a Chinese-style chest that contained discarded materials. See Kazutaka Unno, "Extant Maps of the Paddy Fields Drawn in the Eighth Century Japan," paper delivered at the Fourteenth International Conference on the History of Cartography, Uppsala, 1991, and Tōkyō Daigaku Shiryō Hensanjo (Historiographical Institute, Tokyo University), ed., *Tōdaiji kaiden zu* (Maps of paddy fields reclaimed by Tōdai Temple), 2 vols. (facsimile and explanation), in *Dainihon komonjo, Iewake* (Old documents of Great Japan, Every family), vol. 18, *Tōdaiji monjo* (Records of Tōdai Temple), pt. 4 (Tokyo: Tōkyō Daigaku Shuppankai, 1965–66; reprinted 1980).

22. George H. Beans, *A List of Japanese Maps of the Tokugawa Era* (Jenkintown, Pa.: Tall Tree Library, 1951), and supplements A, B, and C (1955, 1958, 1963), describes the collection. Since it was acquired by the University of British Columbia Library, the collection has been added to, and a catalog is now in preparation. A study of the collection undertaken in 1985 is discussed in Unno Kazutaka, "Hokubei ni okeru Edo jidai chizu no shūshū jōkyō: Bīnzu Korekushon o chūshin to shite" (Some collections of Japanese maps of the Edo period in North America: Mainly on the Beans Collection), *Jinbun Chiri* 39, no. 2 (1987): 16–41.

23. Eva Kraft, *Japanische Handschriften und Traditionelle Drucke aus der Zeit vor 1868 in München* (Stuttgart: Franz Steiner, 1986). The depositories are the Bayerische Staatsbibliothek, Deutsches Museum, Münchner Stadtmuseum-Puppentheatermuseum, Staatliches Museum für Völkerkunde, and Universitätsbibliothek.

24. See Noda Hisao, "Tottori ken no sōshoku kofun" (Decorated tombs in Tottori Prefecture), *Kyōiku Jihō* 163 (1980): 2–11.

25. This mound was excavated in 1972. The identity of the entombed person is not known, but he or she must have held a high rank in society. The rectangular stone room inside the tomb measures 1.13 by 2.6 by 1 meters (height, length, width); the walls are stuccoed, and only the floor and the southern wall where the entrance is are not filled with colored paintings. See Inokuma Kanekatsu and Watanabe Akiyoshi, *Takamatsuzuka kofun* (The Takamatsuzuka burial mound), Nihon no Bijutsu (Japanese art), no. 217 (Tokyo: Shibundō, 1984), and Kashihara Kōkogaku Kenkyūjo (Kashihara Archeological Institute), ed., *Hekiga kofun Takamatsuzuka* (Takamatsuzuka: A burial mound with mural paintings) (Nara and Asuka: Nara Ken Kyōiku Iinkai and Asuka Mura, 1972).

genbu (a turtle and snake interlocked) to represent the tutelary god of the north. The eastern and western walls also have their tutelary gods. A full explanation of the Takamatsuzuka paintings cannot be given because of the damage.[26]

A recently discovered cartographic relic is a landscape on a wooden board that was unearthed in the ruins of the ancient capital of Heijō (in present-day Nara) (fig. 11.4).[27] It is a rough sketch drawn in ink on what was probably the bottom of a tray made of Japanese cypress, measuring 62 by 10.8 by 0.8 centimeters. The picture is drawn in an oblique perspective and includes a group of buildings, a palace, walls, gates, and the like. It is not clear whether it is a sketch of a real or an imaginary place, but judging from the structure and the arrangement of buildings and the annotation *oku no in* (detached building), it is probably one of the Buddhist temples in the mountains. Because the board was unearthed with wooden tablets mentioning the date Tenpyō 8–10 (736–38), the sketch may be considered to be from the same time. Below the sketch are some Chinese characters, probably written for practice. Some of them give the name of a civil servant, Ato no Sakanushi (fl. 739–57), which also appears a few times in the archives of the Shōsōin. This sketch therefore appears to have been drawn by him.

LITERARY EVIDENCE FOR MAPS: THE OFFICIAL HISTORIES

As elsewhere in East Asia, the early cartographic record in Japan often takes the form of literary allusions rather than actual map artifacts. Japanese mythology attributes the creation of the archipelago to the male and female deities Izanagi and Izanami. The names of the islands and the order of their creation differ slightly in the A.D. 712 *Kojiki* (Records of ancient matters) and the A.D. 720 *Nihon shoki*, respectively the oldest extant history and the first official history of Japan.[28] Both works are indif-

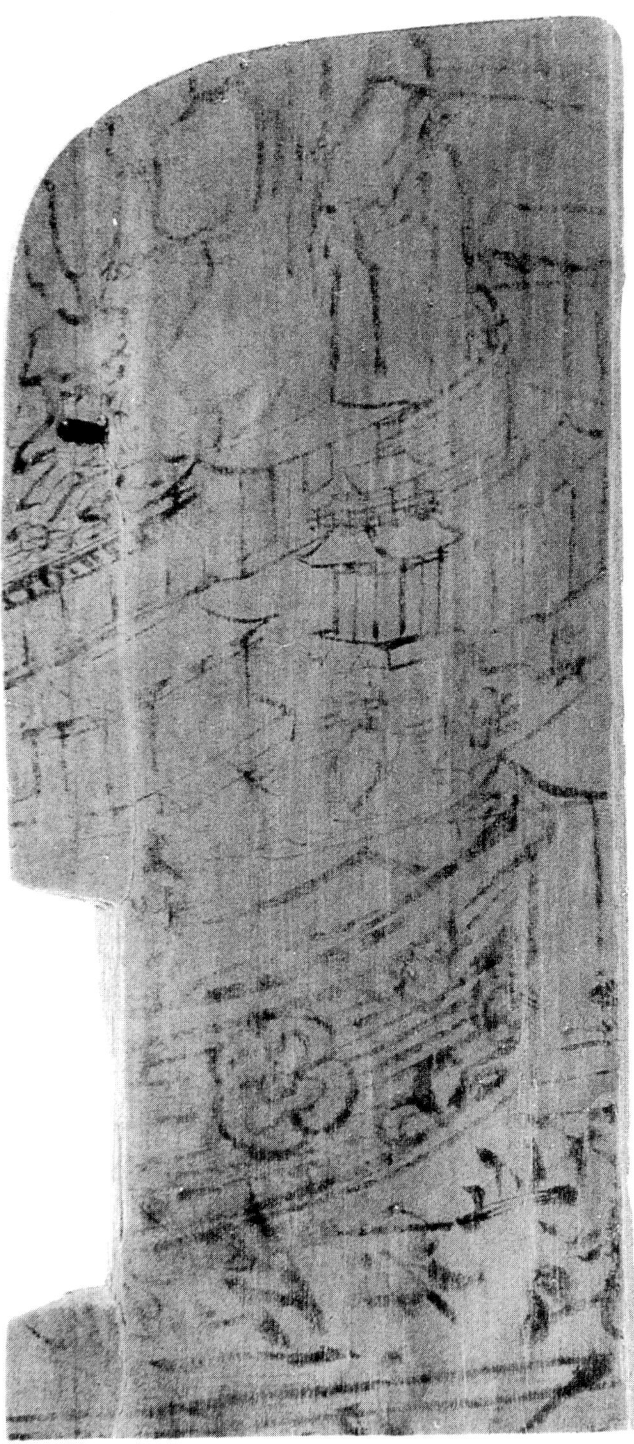

FIG. 11.4. MID-EIGHTH-CENTURY LANDSCAPE ON A WOODEN BOARD. In the foreground are some buildings that probably belonged to a Buddhist temple in the mountains; the rocky mountain in the background and the annotation *oku no in* (detached building) at center right support this idea.
Size of the original: ca. 20 × 10 cm. Nara National Cultural Properties Research Institute, Nara. Photograph courtesy of Kazutaka Unno.

26. For example, it was common in ancient Chinese murals that a three-legged crow and a toad or rabbit were painted in the sun and the moon, respectively; because of damage done by thieves, it cannot be ascertained whether these existed in the mound.

27. Discovered in 1989; see the *Yomiuri Shinbun* (Yomiuri newspaper), published by Yomiuri Shinbun Ōsaka Honsha, no. 13284, 20 October 1989, pp. 1 and 30. It was also recorded in other newspapers and is awaiting further study. Heijō translates literally as "castle of peace"; specifically it was the imperial palace, but also the city of Nara (full form Heijōkyō) during the Nara period (710–84).

28. For English translations, see Basil Hall Chamberlain, trans., *The Kojiki: Records of Ancient Matters* (1882; reprinted Tokyo: Charles E. Tuttle, 1986); and for the *Nihon shoki* see William George Aston, trans., *Nihongi: Chronicles of Japan from the Earliest Times to A.D. 697*, 2 vols. in 1 (1896; reprinted Tokyo: Charles E. Tuttle, 1985). Both contain introductions by the translators. Izanagi and Izanami are dis-

ferent to the size of the islands. This is especially so in the *Kojiki*, in which the names of small islands in the Inland Sea, the Sea of Japan, the Straits of Tsushima, and the East China Sea are recorded in order of their creation. This suggests that such islands were important to navigators, who were likely to have visited the Asian mainland even before the Japanese missions to the Chinese court in A.D. 57 and 107.[29] The geographical knowledge of the navigators is unknown, however: no extant materials provide any evidence.

The first reliable cartographic entry in the official history of Japan dates to A.D. 646, when the term *katachi* was used.[30] The *Nihon shoki*, which can be considered with some confidence when treating the seventh century, records an imperial edict of this date to the effect that the area of each province is to be reported to the central government, accompanied by a map:

> The boundaries of the provinces should be examined and a description or map [*katachi*] prepared, which should be brought here [Naniwa, the capital—in Ōsaka Prefecture today] and produced for Our inspection. The names of the provinces and districts will be settled when you come.[31]

This edict most likely was related to the Taika Reform of 645, a nationwide political reform based on Chinese models.[32]

The next cartographic entry, also in the *Nihon shoki*, concerns a map of Tanega Island that a Japanese envoy brought back with him in 681. It is reported that

> the Envoys sent to Tanegashima [Tane no shima] presented a map [*kata*] of that island [Tane no kuni]. This country is more than 5000 ri distant from the capital [that is, at the time of compilation in 720, Nara], and lies in . . . the sea south of Tsukushi [Kyūshū].[33]

Three years later, Mino no Ōkimi and others sent to Shinano (modern Nagano Prefecture) compiled a map of the province and presented it to the court. On this occasion we are told that

> on this day [in 684] Prince Mino, Tsukura [personal name], Uneme no Omi [family name], of Lower Shōkin rank, and others were sent to Shinano to inspect the conformation of the ground, perhaps with the object of having a capital there.
>
> 11th day [same year]. Prince Mino and his colleagues presented a map [*kata*] of the province of Shinano.[34]

An even more ambitious task was assigned later that year when "Prince Ise and his colleagues were sent to determine the boundaries of the provinces."[35] Subsequent official histories also show that the ancient authorities were aware of the value of maps for various administrative purposes.[36] According to the continuation of the *Nihon shoki*, the *Shoku Nihongi*, in 738 the central government ordered the provincial authorities to compile and submit a map of each province. The next official history, the *Nihon kōki*, also states that the central government ordered that maps of the provinces be composed in 796.[37] This latter entry is detailed and specifies that large and small villages, the distance between post stations, the shapes of famous mountains, and the width of large rivers should be included.

In this period, there was an infusion of Chinese meth-

---

cussed in vol. 1, secs. 2–12 of the *Kojiki* (Chamberlain, *Kojiki*, 17–52, including notes), and in vol. 1, bk. 1 of the *Nihon shoki* (Aston, *Nihongi*, 6–34). The creation of the islands is in vol. 1, secs. 3–5 of the *Kojiki* (Chamberlain, *Kojiki*, 19–27, including notes), and vol. 1, bk. 1 of the *Nihon shoki* (Aston, *Nihongi*, 10–18). In the *Nihon shoki* the "books" may also be referred to as chapters.

29. Fan Ye, *Hou Han shu* (History of the Later Han, compiled fifth century A.D.), chap. 85; see the edition in 12 vols. (Beijing: Zhonghua Shuju, 1965), 10:2821.

30. There is an account relating an earlier event that involved maps. In Chūai 9 (A.D. 391?) the empress Jingū led an expedition against Silla on the Korean peninsula, and when its king surrendered he presented her with *zuseki* (*tu ji*) of his country. The Japanese reading for the Chinese characters *tu ji* (*tu* refers to all figures including maps and *ji* means family registration) is *shirushihefumita*. *Hefumita* means family registration, so one might surmise that *shirushi* (mark) was the term used for figures, including maps, in ancient Japan. This, however, has been doubted by Iyanaga on the grounds of the questionable dating, as well as the possibility that the term *tu ji*, which had been used in China since antiquity, was simply adopted: Iyanaga Teizō, "Handen tetsuzuki to kōhandenzu" (Procedure for apportioning paddies and the maps prepared before and after), in *Shōen ezu kenkyū* (Studies on manorial maps), ed. Takeuchi Rizō (Tokyo: Tokyōdō Shuppan, 1982), 33–34. For the account of the Silla expedition, see Aston, *Nihongi*, 1:230–32 (note 28); Aston indicates that the "books of maps and registers" submitted were cadastral maps.

31. The specific citation is from book 25 of the *Nihon shoki*; see Aston, *Nihongi*, 2:225 (note 28).

32. The coup d'état was in 645, the reform edict in 646, and the series of reforms and modifications from 646 to 701. For a discussion of the Taika Reform, see George B. Sansom, *A History of Japan*, 3 vols. (Stanford: Stanford University Press, 1958–63), 1:56–60.

33. See Aston, *Nihongi*, 2:352 (note 28). Tanega Island at the time was an important port en route to China via the Ryūkyūs: see Ouyang Xiu et al., *Xin Tang shu* (New history of the Tang, compiled 1032?–60), chap. 220; see the edition in 20 vols. (Beijing: Zhonghua Shuju, 1975), 20:6209.

34. See Aston, *Nihongi*, 2:362 and 364 (note 28).

35. See Aston, *Nihongi*, 2:365 (note 28).

36. Six official histories, collectively known as the *Rikkokushi* (Six national histories), cover Japanese history until late in the ninth century, the *Nihon shoki* being the first. Following it are the *Shoku Nihongi* (covering 697 to 791), the *Nihon kōki* (792–833), the *Shoku Nihon kōki* (Later chronicles of Japan, continued, 834–50), the *Nihon Montoku Tennō jitsuroku* (Veritable records of the emperor Montoku of Japan, 851–58), and the *Nihon sandai jitsuroku* (Veritable records of three reigns of Japan, 859–88).

37. See chapter 13 of the *Shoku Nihongi* and chapter 5 of the *Nihon kōki* in volumes 2 and 3 of the *Shintei zōho kokushi taikei* (Series of histories of our country revised and enlarged), 66 vols. (Tokyo: Yoshikawa Kōbunkan, 1929–64).

ods into Japan. The sixth and seventh centuries were a critical time in Japanese history, during which elements of Chinese culture helped to define the culture of Japan even until the present. Outstanding were the arrival of Mahayana Buddhism via Korea and the application of elements of Chinese political theory. This process of using Chinese models continued throughout the Nara (710–84) and early Heian (794–1185) periods.[38]

The extant *Tōdaiji sangai shishi no zu* (Map of the premises of Tōdai Temple [Nara]) of 756,[39] for instance, contains an indication of a Chinese grid, and the official records hint at a deeper knowledge. According to the *Nihon shoki*, in 602 a Buddhist priest from Paekche named Kwallŭk presented the Japanese court with some books on astronomy and geography. The entry gives us a glimpse of the process of scientific transmission. In the tenth year under the empress Suiko (r. 593–628), we are told:

> A Pèkché priest named Kwal-leuk [Kwallŭk] arrived and presented by way of tribute books of Calendar-making, of Astronomy, and of Geography [which included geomancy], and also books of the art of invisibility and of magic [or rather the three arts of fortune-telling, weather divination, and using charms against illness]. At this time three or four pupils were selected, and made to study under Kwal-leuk. Ōchin, the ancestor of the Yako no Fumibito [or Tamafuru, Yako no Fuhito no Oya], studied the art of Calendar-making. Kōsō, Otomo [Ōtomo] no Suguri, studied Astronomy and the art of invisibility [fortune-telling]. Hinamitatsu [or Hinitachi], Yamashiro no Omi, studied magic [the arts of weather divination and using charms against illness]. They all studied so far as to perfect themselves in these arts.[40]

An official commentary on Japanese law, the *Ryō no gige* (Commentary on the codes) of 833, enumerates the titles of mathematical textbooks used at the Daigaku Ryō (Imperial University) at the former capital of Nara.[41] These included the *Kyūshō*, the *Kaitō*, and the *Shūhi*, which were respectively the ancient Chinese textbooks *Jiuzhang suanshu* (Nine chapters on mathematical art), *Haidao suan jing* (Mathematical classic for seas and islands), and *Zhoubi suan jing* (Arithmetical classic of the Zhou gnomon).[42] The first two explain how to measure the length and height of distant objects by applying the principle of right-angled triangles; the third pertains to the structure of the heavens and earth. These subjects might have been taught at the Daigaku Ryō as early as the beginning of the eighth century: the *Ryō no gige* comments on the *Yōrō ritsuryō* (Yōrō code of laws) of 718, itself a revision of the *Taihō ritsuryō* (Taihō code of laws) of 701, which also pertained to laws and regulations.[43] Another ninth-century document that catalogs books existing in Japan, Fujiwara no Sukeyo's *Nihonkoku genzaisho mokuroku* (A list of books at present in Japan) of about 891,[44] contains an entry about the *Jin shu* (History of the Jin).[45] In it is a biography of the Chinese mapmaker Pei Xiu (223–71) that suggests Pei's six principles of cartography were then available to Japanese scholars.[46]

---

38. See, for example, John Whitney Hall, *Japan: From Prehistory to Modern Times* (New York: Delacorte Press, 1970), 35–74, and Sansom, *History of Japan*, 1:45–128 (note 32). For a discussion of Chinese science—including cosmology, astrology and the occult sciences, and calendrical science—and its influence on Japan, see Shigeru Nakayama, *A History of Japanese Astronomy: Chinese Background and Western Impact* (Cambridge: Harvard University Press, 1969), 7–76.

39. This map is discussed below. Stanley-Baker uses the map to illustrate how Japanese implemented their knowledge of Chinese landscape painting, but with amendments such as the "gently rolling hillocks" and the "irregular and natural manner" of displaying the trees: see Joan Stanley-Baker, *Japanese Art* (London: Thames and Hudson, 1984), 57–58.

40. See Aston, *Nihongi*, 2:126 (note 28).

41. On the Daigaku Ryō, see appendix 3, "A Note on Higher Education, 700–1000," in Sansom, *History of Japan*, 1:474–76 (note 32). The *Ryō no gige* is in vol. 22 of the *Shintei zōho kokushi taikei* (note 37).

42. The *Jiuzhang suanshu*, a Han dynasty text, is called by Nakayama, *History of Japanese Astronomy* (note 38), the "oldest mathematical classic in China," and the *Zhoubi suan jing* (ca. 200 B.C.), the "oldest canon of Chinese scientific cosmology" (p. 273). Both works were required for students of calendar making (p. 72), and the *Zhoubi suan jing* for those studying astronomy (p. 43). The *Zhoubi suan jing* and *Haidao suan jing* (ca. 265) are mentioned in Joseph Needham, *Science and Civilisation in China* (Cambridge: Cambridge University Press, 1954–), vol. 3, with Wang Ling, *Mathematics and the Sciences of the Heavens and the Earth* (1959), 19–23 and 571–72. All three of the ancient Chinese mathematical textbooks are found in the *Siku quanshu* (Complete library from the four treasuries, compiled 1773–82) (Taipei: Taiwan Shangwu Yinshuguan, 1970–82).

43. *Ritsu* refers to penal laws and *ryō* to administrative institutions. See Hall, *Japan*, 48–61 (note 38); E. Papinot, *Historical and Geographical Dictionary of Japan* (1910; reprinted Ann Arbor: Overbeck, 1948), 616; and Sansom, *History of Japan*, 1:67–74 (note 32). On the *Yōrō ritsuryō*, see vol. 22 of the *Shintei zōho kokushi taikei*. Although the entire text of the *Taihō ritsuryō* was never included, a part of it is in the *Ryō no shūge* (Supplement of the *Ryō no gige*) of ca. 875 in vols. 23–24 of the *Shintei zōho kokushi taikei* (note 37).

44. Sansom notes the date as about 890 and says that it "records 1,579 titles and 16,790 volumes"; the record was compiled after "a fire in the palace had destroyed a great number of books in 875, so that the total number of volumes imported from China during the ninth century must have reached an impressive figure," Sansom, *History of Japan*, 1:124 (note 32). The *Nihonkoku genzaisho mokuroku* is in the *Zoku gunsho ruijū* (Classified series of various books: Continuation, 1923–28 in 71 vols.), 3d rev. ed., 67 vols. (Tokyo: Zoku Gunsho Ruijū Kanseikai, 1957–59), vol. 30, bk. 2.

45. Fang Xuanling et al., *Jin shu* (compiled 646–48), contains a treatise on astronomy; see the modern edition in 10 vols. (Beijing: Zhonghua Shuju, 1974); see also Nakayama, *History of Japanese Astronomy*, 33–40 and 272 (note 38).

46. *Jin shu*, chap. 35 (vol. 4) (note 45). For a discussion of the six principles of cartography, see above, pp. 110–13.

FIG. 11.5. THE TOMB OF THE EMPEROR NINTOKU, SAKAI, ŌSAKA PERFECTURE. The critical measurements are length, 486 meters; diameter of the circular part at the head of the "keyhole," 249 meters; height of the circular knoll, 35 meters; width at the front, 305 meters; highest point of the trapezoid, 33 meters. It is surrounded by three moats. Photograph courtesy of Kazutaka Unno.

SURVEYING INSTRUMENTS AND PROJECTS

Despite incomplete records on early surveying and cartography in Japan, there are some indications about ancient projects and instruments, some of them used until more recent times. Koreans who settled in Japan are thought to have played an important role in ancient surveying, mainly through the diffusion of Chinese methods and instruments to Japan. The construction of burial mounds in the Tomb period (ca. 300–600) attests to an early Japanese need for surveying instruments and techniques. This was at the time of the first unified state in Japanese history, the Yamato state, which was centered on the area of present-day Ōsaka and Nara. The keyhole-shaped mounds that served as mausoleums for the emperors originated in this area (fig. 11.5), and—unlike most cultural phenomena in Japan—were diffused to the west, eventually reaching Kyūshū, rather than the other way around.[47] Inasmuch as the process of building the tombs points to some degree of political authority and social structure, it links the beginnings of surveying to wider developments in Japanese society. Not only was a large, organized labor force necessary to build them, but they also required some form of intellectual preparation. Plans—although none are known to exist—were probably involved in designing the tombs, the shape of the mounds, and the surrounding moats. The sites where the tombs were to be built also had to be selected and surveyed.

We have no precise knowledge about the instruments used for compiling the plans, surveying the sites, and building the mounds. Statements from the Chinese and Japanese literature do, however, point to instruments that were used for similar purposes in ancient China. For instance, the Chinese textbook *Zhoubi suan jing* men-

47. For brief discussions of the mounds, see Hall, *Japan*, 20–23 (note 38); R. H. P. Mason and J. G. Caiger, *A History of Japan* (Melbourne: Cassell Australia, 1976), 11–14; and H. Paul Varley, *Japanese Culture*, 3d ed. (Honolulu: University of Hawaii Press, 1984), 12–14. The largest is that of the fifth-century emperor Nintoku in Sakai, Ōsaka Prefecture; it is 486 meters in length.

FIG. 11.6. ILLUSTRATION FROM THE *KASUGA GONGEN GENKI E* (PAINTINGS OF MIRACLES OF KASUGA SHRINE, 1309). Shown in use are the *mizubakari* (water level), *sumitsubo* (ink pad for carpentry), *suminawa* (ink cord for carpentry), *sumisashi* (ink stick), and *magarigane* (square). The painting, by Takashina Takakane, is an imaginary scene from the construction of the house of Fujiwara no Mitsuhiro.
Size of the original: 41.5 cm in height. By permission of the Imperial Household Agency, Tokyo.

tions the *ju* (carpenter's square), an instrument that appears to have been useful for measuring height, depth, and distance.[48] Instruments of Chinese origin are also mentioned extensively in the *Wamyō ruijū shō* (Classified glossary of Japanese terms), a Japanese encyclopedia compiled about 935 by Minamoto no Shitagō (911–83).[49] Examples include those illustrated in figure 11.6 (from another work). The Shōsōin, an annex of Tōdai Temple, has two *sumitsubo* (ink pads for carpentry) that probably date from the eighth century. One with lacquered designs appears to have been for ceremonial use, but, like the other smaller instrument, it lacks an accompanying *suminawa* (ink cord for carpentry).[50] The oldest extant *sumitsubo* of practical value is believed to date to the thirteenth or fourteenth century and was found on a beam of the southern main gate to Tōdai Temple. At the top is an iron ring seventeen millimeters in diameter for fitting a cord when the *sumitsubo* was used as a plumb line. The instrument is in the possession of the Tokyo University of the Arts. A scene in the *Matsuzaki tenjin engi emaki* (Painted scroll of the origin of the Matsuzaki Shrine) of 1311, owned by Hōfu Tenmangū Shrine, suggests that the iron ring was used in this way.[51]

48. *Zhoubi suan jing*, chap. 1. Six uses of the *ju* are enumerated: correcting the *sheng*, or ink cord, checking heights, measuring depth, calculating distance, drawing circles, and forming a rectangle by putting two *ju* together. That it could be used for calculating height, depth, and distance suggests that the *ju* was graduated.

49. See the *Wamyō ruijū shō*, 4 vols., Nihon Koten Zenshū (Comprehensive collection of Japanese classical works), 4th ser. (Tokyo: Nihon Koten Zenshū Kankōkai, 1930–32), chap. 15. Plates showing the ancient compass, square, water level, ink pad, ink cord, and ink stick are in Nihon Gakushiin (Japanese Academy), ed., *Meiji zen Nihon kenchiku gijutsu shi* (History of Japanese architectural techniques) (Tokyo: Nihon Gakujutsu Shinkōkai, 1961; reprinted 1981), 189–218, and Nakamura Yūzō, *Zusetsu Nihon mokkōgu shi* (Illustrated history of Japanese woodworking tools) (Tokyo: Shinseisha, 1968); for a simple history of these instruments, see Muramatsu Teijirō, *Daiku dōgu no rekishi* (A history of carpenters' tools) (Tokyo: Iwanami Shoten, 1973), 149–52.

50. The first of the *sumitsubo* at the Shōsōin is 11.7 by 29.6 by 9.4 centimeters (height, length, width); the other is only 2 centimeters tall and 4 centimeters long and is made of rosewood with silver designs.

51. On the *sumitsubo* see Muramatsu, *Daiku dōgu no rekishi*, 149–52 (note 49). The *Matsuzaki tenjin engi emaki* is reproduced in Komatsu Shigemi, ed., *Zoku Nihon emaki taisei* (Series of Japanese picture scrolls: Continuation), 20 vols. (Tokyo: Chūōkōronsha, 1981–85), vol. 16.

Tōdai Temple owns the oldest extant *magarigane* (square), said to have been acquired by its head priest in Edo in 1685. It is made of iron with the longer side measuring 37 centimeters and the shorter 19.8 centimeters; both are 1.46 centimeters wide and 1.2 millimeters thick on the outside and 0.6 millimeters on the inside. Only the face of the longer side is graduated, indicating that the instrument dates to a period much earlier than 1685. Modern Japanese squares are graduated much like conventional rulers, but on the reverse of the longer side there is also a graduation based on the square root of two, known as *urame* (graduations on the reverse), which Muramatsu believes to have originated at least in the eleventh or twelfth century. There is also a graduation known as *marume* (circumference graduation) on some squares: the graduations are multiplied by π so that the circumference of a circle may be measured by placing the square along the diameter.[52]

Such instruments are known to have existed in China before the Tomb period in Japan. The *Mengzi* (on the sayings and deeds of Mencius [372–289 B.C.]) mentions the *zhun* (water level), *sheng* (ink cord for carpentry), and *ju* (square); it also mentions the *gui* or compass for drawing circles.[53] In the *Huainanzi* ([Book of the] Master of Huainan, ca. 120 B.C.), reference is made to the *gui*, *zhun*, *sheng*, and *ju*, as well as to the *quan* (weight) and *heng* (steel yard). Among these instruments, the *gui* and *ju* seem to have been either the first devised or the most basic.[54] Of these items, the *sheng* or *suminawa* is the first to appear in the official Japanese histories; the *Nihon shoki* records a song composed by a carpenter on the imminent execution of a comrade about 490:

> The much to be regretted
> Carpenter of Winabe —
> The ink-cord he applied, —
> When he is no more,
> Who will apply it?
> Alas! that ink-cord![55]

This song not only won the pardon of the condemned but also contained the word *suminawa*. Because later examples of *suminawa* came with *sumitsubo*, it is likely that this tool was also available by the fifth century. By inference, the same might be said to be true of the *ju*, *zhun*, and *gui*, which are of greater importance for construction.

Recorded projects involving the possible use of such instruments for surveying and drafting include drainage and irrigation canals, a road, temples, and capital cities. The *Nihon shoki* refers to a drainage canal in the north of Naniwa no Miyako (Ōsaka) in the fifth century. This was probably constructed to help drain the low-lying, marshy Kawachi Province (now part of Ōsaka Prefecture) to the west. Three years later a road was built running approximately ten kilometers in a straight line leading south from Naniwa.[56] The rise of Buddhism in Japan also led to an influx of craftsmen and instruments. Carpenters specializing in temple construction started to arrive from Korea late in the sixth century, accompanying priests, tile makers, and painters. For example, it is recorded about 577 that

> the King of the Land of Pèkchè presented to the Emperor, through the returning Envoys Prince Oho-wake and his companions, a number of volumes of religious books, with an ascetic, a meditative monk, a nun, a reciter of mantras, a maker of Buddhist images, and a temple architect, six persons in all.[57]

Examples of their work included Hōkō Temple in Asuka, Yamato Province, started about 588, and Shitennō Temple (Temple of the Four Heavenly Kings) in Ōsaka, begun about 587 and completed in 593.[58]

Surveying skills were especially important for aligning the capital city, which until the early eighth century was relocated with each new emperor. The oldest capital to have been excavated is Naniwa, dating to the reign of the emperor Tenmu (672–86). Only the palace and government area have been unearthed, but from these an inclination of 34′35″ east of the north-south axis has been found to exist in the layout of the city.[59] The capital at Fujiwara, Yamato Province, was in 694–710 the first large-scale capital to be built on the Chinese checker-

52. About the supposition of an earlier date of the *magarigane* at Tōdai Temple and the date of origin of the *urame* system, see Muramatsu, *Daiku dōgu no rekishi*, 131–32 and 140–41 (note 49).

53. Chapter 7 of the *Mengzi*; see *Mengzi yinde* (Concordance to Mencius), Harvard-Yenching Sinological Index Series, suppl. 17 (1941; reprinted Taipei: Chengwen Chubanshe, 1966). Or, in Japanese, see *Mōshi*, 2 vols. (Tokyo: Iwanami Shoten, 1968–72).

54. See the *Huainanzi* attributed to Liu An (d. 122 B.C.), in *Huainanzi zhu* (Commentary to *Huainanzi*, third century), ed. Gao You, chap. 20; modern edition (Taipei: Shijie Shuju, 1962). Or, in Japanese, *Enanji*, trans. Togawa Yoshio et al., in *Chūgoku kotenbungaku taikei* (Series of Chinese classics), 60 vols. (Tokyo: Heibonsha, 1967–74), vol. 6.

55. See Aston, *Nihongi*, 1:361–62 (note 28) (the song is on 362). The incident took place during the reign of the emperor Yūryaku in the fifth century; Aston's chronology puts the date at 469, but it could have been about twenty years later.

56. Naniwa no miyako translates literally as "the capital of Naniwa," that of the emperor Nintoku. The drainage canal and road are mentioned in Aston, *Nihongi*, 1:280–83 (note 28).

57. Aston, *Nihongi*, 2:96 (note 28).

58. See, for instance, Aston, *Nihongi*, 2:115 (other examples are mentioned in books 20–22 [2:90–156]) (note 28). The relevant eras are those of the emperors Bitatsu (r. ca. 572–85), Yōmei (r. ca. 586–87), and Sushun (r. ca. 588–92) and the empress Suiko (r. 593–628). The *Nihon shoki* does not mention any maps as having been composed or involved with the projects in this and the previous two notes.

59. See vol. 7 of the journal *Naniwakyū Shi no Kenkyū*, issued by the Ōsaka Shi Bunkazai Kyōkai (Ōsaka City Cultural Properties Association) (1981), pt. 1, legend of pls. 1–25.

board model. Its inclination is 26′30″ to the west of true north.[60] Heijō (Nara) was also constructed on a grid network early in the eighth century. This was at a time of strong Chinese influence on Japanese culture, and the site appears to have been selected on the grounds of its important Buddhist establishments and because it satisfied the requirements of Chinese geomancy (see above, pp. 216–22). These included that it be surrounded on the western, northern, and eastern sides by mountains and that rivers and a pond be situated to the south. The city, although smaller, was modeled on the Tang Chinese capital of Chang'an, which incorporated a rectangular grid. Although excavations have uncovered only a part of the city, it has been found that the inclination of the main street is 12′40″ to the west of true north.[61]

With all these cities, the small deviations from true north indicate that an orientation of the streets and buildings took place before the actual construction. For alignment, it is highly unlikely that the magnetic compass was used, and since there was also no bright star near the celestial North Pole at that time, the gnomon may have been employed.[62] Using a gnomon for such a purpose was explained in the *Zhoubi suan jing*. The method of aligning appears to have involved marking the end of the morning and afternoon shadows of the gnomon, connecting them with a line (east to west), and drawing a perpendicular line (south to north) from the gnomon through the center of this line. No ancient gnomons have been found in Japan, but it is likely that they were imported or made domestically before the second half of the seventh century.

Only in the eighteenth century were books on surveying techniques published, reflecting in part more ancient practices.[63] None of them, however, reveals the intricacies of the techniques practiced in Japan. Apprentices learned the skills directly from their masters and submitted a pledge in their own blood that they would not disclose their knowledge to anyone. This was to protect the members of the trade at a time when patents were unknown. The custom of secrecy must have originated with the development of the techniques and the organization of its craftsmen and helps to account for the dearth of records on surveying in ancient and medieval Japan.

MAPS OF PADDY FIELDS

The earliest extant Japanese maps relate to landownership and date to the eighth century (see appendix 11.1). These maps provide tangible evidence of the state of contemporary surveying and cartography, especially on the Japanese maps of paddy fields, which exhibit a grid structure. The network employed east-to-west and north-to-south lines based on the *chō* (109.09 m), a unit of measurement used in the *jōri* system that served as a frame of reference for the administrators of land control (fig. 11.7). When a *jōri* network was used, words such as "mountain" and "sea" were entered in the squares to designate topographical features; presumably early maps of rice fields used the same method.[64] That some of the early maps have grids extending beyond the arable land into the sea or mountains suggests that such a network served more an arbitrary, theoretical function than a realistic one. Despite the use of grids, there was no fixed convention as to orientation.

The reason these maps of paddy fields were produced and have survived is to be found in the development of landownership in eighth- and ninth-century Japan. They were practical documents to record ownership, and they were useful to settle disputes in a period when aristocrats, Shinto shrines, and Buddhist temples were consolidating their private holdings. This trend had originated in the mid-eighth century, when the demand for maps of agricultural holdings would have been increased by the implementation of agrarian reforms. One attempt on the part of the authorities to alleviate the problems caused by a contemporary drift toward a decentralized feudal regime was to increase the amount of arable land through reclamation.[65] This attempt was made in 723, and to

60. See Nara Kokuritsu Bunkazai Kenkyūjo (Nara National Cultural Properties Research Institute), *Asuka Fujiwarakyū hakkutsu chōsa hōkoku* (Reports of the excavation of the site of the Fujiwara imperial palace, Asuka), vol. 6 (Nara, 1976), 21.

61. See Yamato-Kōriyama Shi Kyōiku Iinkai (Board of Education, Yamato-Kōriyama City), *Heijōkyō Rajōmon ato hakkutsu chōsa hōkoku* (Report of the excavation of the sites of the Rajō Gate, Heijōkyō) (Yamato-Kōriyama, 1972), 30. Nara was the first of the capitals not to disappear with the departure of the imperial court. See also Herbert E. Plutschow, *Historical Nara* (Tokyo: Japan Times, 1983), esp. 76–83; and refer also to Sansom, *History of Japan*, 1:82–98 (note 32), Hall, *Japan* (note 38), 48–61, and Varley, *Japanese Culture*, 30–31 (note 47).

62. See Yabuuchi Kiyoshi, "Naniwakyū sōken jidai no hōi kettei" (The determination of position at the time of constructing Naniwakyū), *Naniwakyū Shi no Kenkyū* 2 (1958): 77–82.

63. These included Hosoi Kōtaku, *Hiden chiiki zuhō daizensho* (Complete book of the secret art of surveying and mapping), a manuscript of 1717 at the National Diet Library in Tokyo; Matsumiya Toshitsugu, *Bundo yojutsu* (Techniques of protraction), a manuscript of 1728 at the National Archives in Tokyo; Murai Masahiro, *Ryōchi shinan* (Surveying guidebook, 1733); and Shimada Dōkan, *Kiku genpō chōken bengi* (Explanation of surveying, 1734). The last two are reproduced in volumes 9 and 10 of the *Edo kagaku koten sōsho* (Series of scientific classics during the Edo period), 46 vols. (Tokyo: Kōwa Shuppan, 1976–83).

64. An example of a map bearing a *jōri* grid in areas of sea is the *Settsu no kuni Yatabe gōri jōri zu* (*Jōri* map of Yatabe County, Settsu Province [in the area of modern Hyōgo Prefecture]) of 1162. It is partly extant and kept at the Kōbe City Central Library.

65. This drift toward decentralized feudalism may be traced to the Taika Reform Edict of 646, consisting of four articles that included the abolition of private landownership and a new system of taxation.

1 *ri* = 36 *tsubo*

a

6 *chō* (approximately 654.5 meters)

6 *chō*

1	12	13	24	25	36
2	11	14	23	26	35
3	10	15	22	27	34
4	9	16	21	28	33
5	8	17	20	29	32
6	7	18	19	30	31

b

6 *chō* (approximately 654.5 meters)

6 *chō*

1	7	13	19	25	31
2	8	14	20	26	32
3	9	15	21	27	33
4	10	16	22	28	34
5	11	17	23	29	35
6	12	18	24	30	36

1 *tsubo* = 10 *tan*

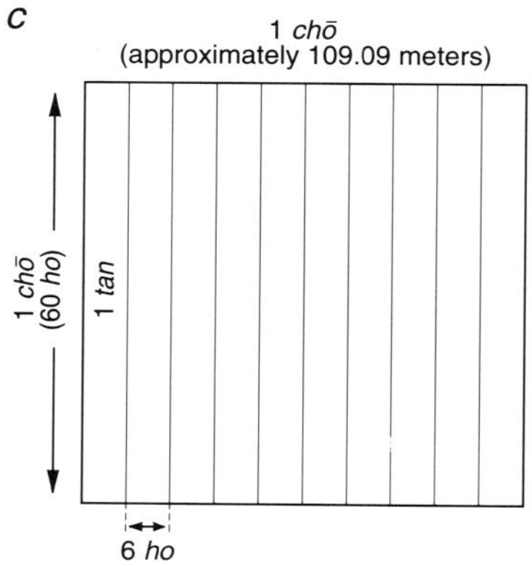

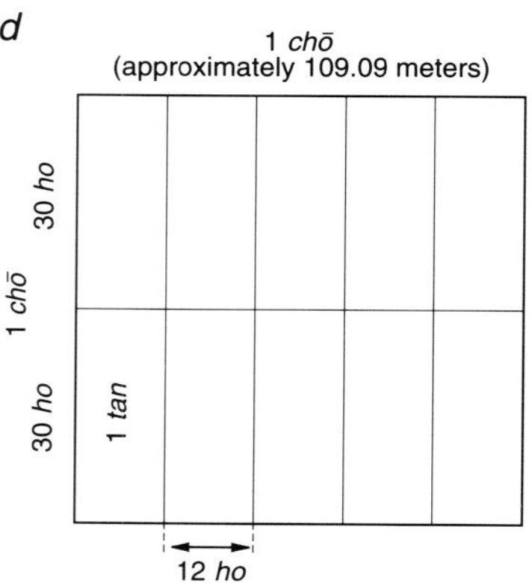

FIG. 11.7. DIAGRAM OF THE *JŌRI* SYSTEM. The main unit of length was the *chō* (109.09 m), and the main unit of area was the *tsubo* (one square *chō*). Thirty-six *tsubo* comprised one *ri*. The name *jōri* derived from the labeling of the two axes (*jō* the rows, *ri* the columns; thus "*ri*" had two usages, one to refer to the columns in the system and one to refer to the square with thirty-six *tsubo*). The numbering of the *tsubo* varied: "zig-zag" or "shuttle" numbering (*chidori shiki*) started from the top left, went down the first column and returned up the second column in boustrophedonic fashion as in *a*; "parallel" numbering (*heiko shiki*) always started at the top and went down the columns as in *b*. For area measurement, the *tsubo* was divided into ten strips known as *tan*, either 6 *ho* wide and 1 *chō* long or 12 *ho* wide and 30 *ho* long (60 *ho* [or *po*] made one *chō*). The "long-area" system (*nagachi gata*) is illustrated in *c*; the "halving" system (*haori gata*) is shown in *d*.

encourage reclamation they decreed that reclaimed land with new ponds and ditches could be owned privately down to the generation of the great-grandchild and that land with existing ponds and ditches could be handed down for one generation only. From 743, ownership of reclaimed land was extended to perpetuity, and from the middle of the eighth century, immunity to taxation on reclaimed land was awarded first to Buddhist establishments and then eventually to other religious institutions and secular landholders. When such reclamation projects were undertaken, it was common practice for the reclaimer to notify the provincial authorities and to com-

The right to allocate land and to tax was vested in the emperor, from 645 considered an absolute monarch. One of the functions of the reforms was to create a system of loyalty to the emperor, and the wide distribution of land was a significant attempt to weaken the powerful clans.

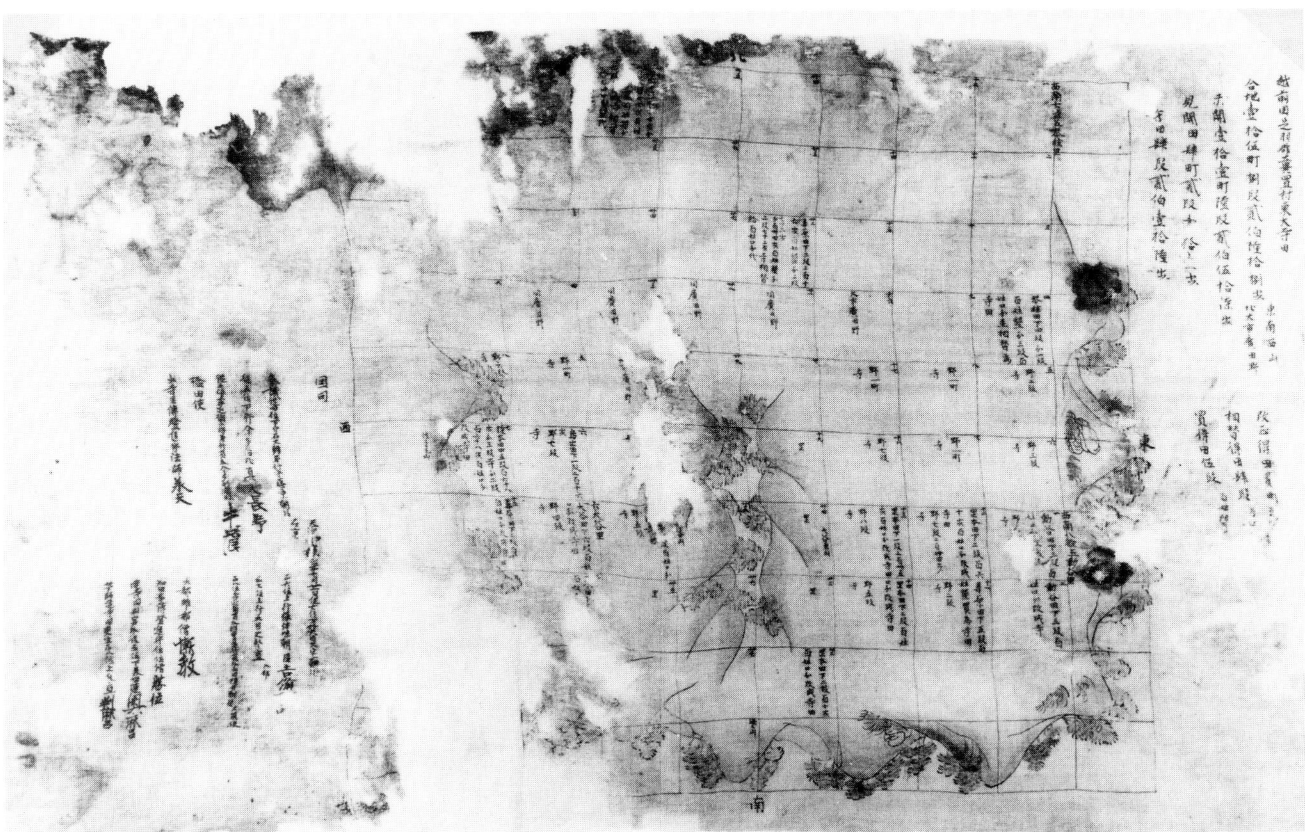

FIG. 11.8. EXAMPLE OF AN EIGHTH-CENTURY PADDY-FIELD MAP. This manuscript on hemp dating from 766 shows land belonging to Tōdai Temple (Nara) in the village of Kusooki, Asuha County, Echizen Province (today in Fukui Prefecture). Clearly depicted is the *jōri* system, each box being one square *chō* in area. The paddy field is not divided by lines, but there are written entries to show its size. Of the twelve signatures at the left of the map, one bearing the title *sanshi* (government mathematician) is found at the left of the bottom row.
Size of the original: 69 × 113 cm. Shōsōin, Tōdai Temple, Nara. Photograph courtesy of Kazutaka Unno.

pile a map as evidence in the event of future disputes. The authorities then used these documents for taxation purposes.[66]

Most of the maps of rice paddies were survey maps of land that was reclaimed in the mid-eighth century. Tōdai Temple was one such institution to absorb land in the eighth century, and the Shōsōin preserves at least twenty eighth-century maps of paddy fields that were reclaimed by the temple (fig. 11.8). The earliest of these was compiled in 751, the latest in 767. With the exception of those drawn on paper, they were on hemp (another map at the Shōsōin on hemp, but which is not a paddy-field map, is the *Tōdaiji sangai shishi no zu* of 756.)[67] Signatures of the parties involved in reclamation and, in most cases, the official provincial stamps appear on these maps to prevent alteration. Also included on approximately half of these maps are the signatures of mathematicians called *sanshi* (government mathematician), implying that students of mathematics at the Daigaku Ryō were involved in surveying and mapmaking. This seems to be borne out by the fact that the maps in the possession of the temple "appear to have been based on reasonably accurate measurements" and that "some of these estate maps included topographical particulars and points of the compass."[68] In addition to the maps at the Shōsōin, other examples of eighth-century maps exist, including one relating to agricultural land of Yamada County in Sanuki Province (now in Kagawa Prefecture) that has a grid network and dates to 736 (see no. 1, appendix 11.1).

66. Under these conditions the inequitable distribution of privately owned land led to a feudal system that lacked the strong central administration envisioned by the Taika Reform and the Taihō Code. The failure of the system of allotments was acknowledged by an edict of 902, which noted that it had fallen into disuse. After this date, it appears that no more allocations were made. For a more detailed analysis of these events, see Sansom, *History of Japan*, 1:56–59, 83–89, and 103–11 (note 32).

67. Most of these are reproduced in Tōkyō Daigaku Shiryō Hensanjo, *Tōdaiji kaiden zu* (note 21).

68. Cortazzi, *Isles of Gold*, 4 (note 14).

FIG. 11.9. MAP SHOWING THE MANORS OF KŌNO AND MAKUNI IN NAGA COUNTY, KII PROVINCE. On the reverse of the manuscript an inscription dates the map to 1143, when the boundaries of the manors were confirmed. Places where signs were put to demarcate the boundaries are shown by circles. There is no particular orientation, since the information is recorded in different directions.
Size of the original: 112.5 × 92 cm. Jingo Temple, Kyōto. Photograph courtesy of Kazutaka Unno.

MAPS OF MANORS, SHRINES, AND TEMPLES

The form of private landownership by religious institutions and the aristocracy was known as *shōen* (*shō* = villa; *en* = cultivated land). After the eighth century such ownership increased, and the tendency was to enlarge existing possessions through purchase, illegal absorption of public land, and commendation (cession of private lands to the protection of feudal lords). Half of the country was under the *shōen* system in the eleventh century, and by the thirteenth there appear to have been approximately five thousand *shōen* jurisdictions.[69] Maps of manors pertained to lands that fell under the *shōen* system, and the term *shōenzu* refers to such maps.[70]

Most of the maps that pertain to landownership, however, date to the Japanese medieval period, approximately from the beginning of the Kamakura period (1185–1333) to Oda Nobunaga's (1534–82) entrance into Kyōto in 1568.[71] Maps of manors, Shinto shrines, and Buddhist temples were part and parcel of feudalism and helped to reinforce its system of controls. An important type of map in this period was the *shīji bōji no zu* (map of boundary marks on all sides), a term that was applied to maps of both secular and religious landholdings. Rivers and roads were shown planimetrically, whereas mountainous areas were drawn pictorially and in oblique profile. The maps, which tend to contain details of roads and tracts of land, were drawn in simple fashion and did not follow any established conventions. A predecessor of this type is the previously mentioned *Tōdaiji sangai shishi no zu*, which was intended to delineate the property of Tōdai Temple.

The oldest extant manorial map is of Kōno and Makuni no Shō in Kii Province (today Wakayama Prefecture), compiled in 1143 (fig. 11.9). When the annual tax exemption was determined for this manor, its acreage and borders were to be established; the survey, however, determined only the basic directional boundaries of the property. Two other manorial maps owned by the same temple also stress the delineation of important boundary limits rather than being concerned with the details of what was within them. These were of Ashimori no Shō in Bitchū Province (now part of Okayama Prefecture) in 1169 and of Kaseda no Shō in Kii Province in 1183.[72]

---

69. For more on the *shōen* system, see *The Cambridge History of Japan*, vol. 3, *Medieval Japan*, ed. Kozo Yamamura (Cambridge: Cambridge University Press, 1990), 89–127, Hall, *Japan*, 68–72 (note 38), and Papinot, *Dictionary of Japan*, 585 (note 43).

70. See Ramming, "Evolution of Cartography," 17 (note 10). Reproductions of nearly all the extant maps of manors from antiquity and the medieval period are in Nishioka Toranosuke, ed., *Nihon shōen ezu shūsei* (Collected maps of Japanese manors), 2 vols. (Tokyo: Tokyōdō Shuppan, 1976–77). Clearer and larger reproductions exist in volume 3 of Tōkyō Daigaku Shiryō Hensanjo (Historiographical Institute, Tokyo University), ed., *Nihon shōen ezu shūei* (Collected facsimiles of maps of Japanese manors) (Tokyo: Tōkyō Daigaku Shuppankai, 1988); this will be a five-volume series when the other four are published. For reproductions of maps, plans, and views of shrines, temples, and manors that are typical of the medieval period, see Kyōto Kokuritsu Hakubutsukan (Kyōto National Museum), ed., *Koezu: Tokubetsu tenrankai zuroku* (Old picture maps: A special exhibition catalog) (Kyōto: Kyōto Kokuritsu Hakubutsukan, 1969), and Naniwada Tōru, ed., *Koezu* (Old picture maps), Nihon no Bijutsu (Japanese art), no. 72 (Tokyo: Shibundō, 1972). The old views, maps, and plans of shrines are reproduced in Miyaji Naoichi, supervisor, *Jinja kozu shū* (Collected old drawings of shrines) (Tokyo: Nippon Denpō Tsūshinsha, 1942; reprinted Rinsen Shoten, 1989), and in Fukuyama Toshio, supervisor, *Jinja kozu shū zokuhen* (Collected old drawings of shrines, continuation) (Kyōto: Rinsen Shoten, 1990).

71. The Kamakura period was followed by an interlude known as the Kenmu Restoration, which took place under the emperor Go-Daigo (r. 1318–39) in 1333–35. Its failure led to the creation of the Ashikaga shogunate, which may be said to coincide with the Muromachi period of 1336–1573.

72. The Ashimori map (157.2 × 85.4 cm) is reproduced in color in Unno, Oda, and Muroga, *Nihon kochizu taisei*, vol. 1, pl. 4 (note 8). The map of Kaseda (96 × 115.6 cm) is reproduced in Kyōto Kokuritsu Hakubutsukan, *Koezu*, pl. 64 (note 70).

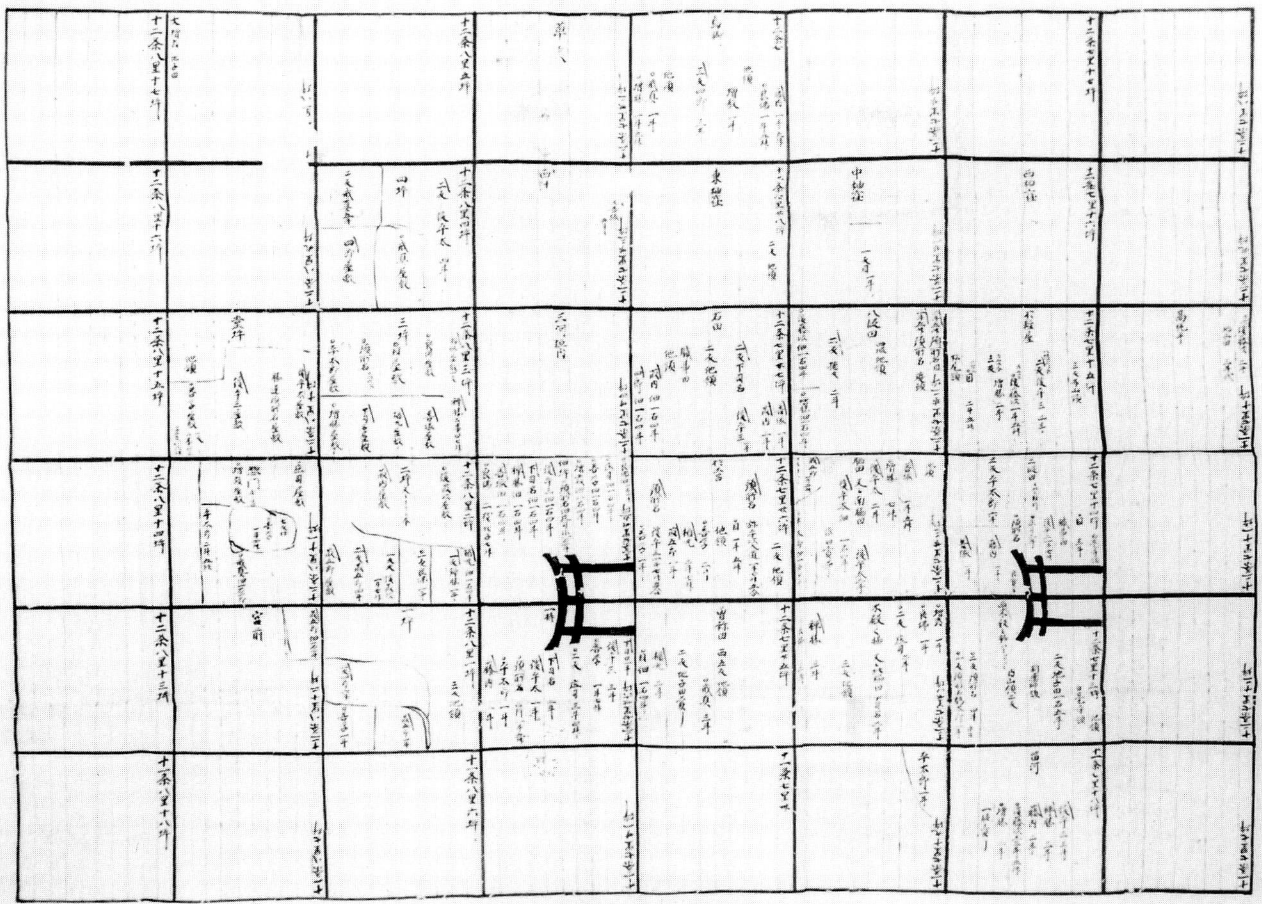

FIG. 11.10. EXAMPLE OF A *DOCHŌ:* A "LAND LEDGER" OR MAP OF THE OTOGI MANOR. The manuscript was made by Daijōin, a small branch temple of Kōfuku Temple in Nara, in 1265, so that the actual state of the manor could be shown. The *jōri*-system framework provides a clear locational reference.
Size of the original: 96.1 × 127.8 cm. National Archives, Tokyo. Photograph courtesy of Kazutaka Unno.

When the Kamakura shogunate was established in 1185, disputes concerning profits between the owners and the administrators (*jitō*) of land were common. In an attempt to resolve such disputes, the government devised a policy of dividing the manors into two equal parcels for the lord and the administrator. At this time, *shitaji chūbun no zu* (maps indicating that the land was divided into two equal parts) were compiled. Two extant maps that are representative of this process are the map of Tōgō no Shō (1258) in Hōki Province (now Tottori Prefecture), which belonged to Matsuo Shrine in Kyōto, and the map of Izaku no Shō (1324) in Satsuma Province (today Kagoshima Prefecture), which belonged to Ichijōin Temple in Nara.[73] The rivers and roads are drawn planimetrically and are relatively accurate in these works, and the division between the lord's and the administrator's lands is indicated by red lines.

The landlords needed to know the actual status of their manors, so in the thirteenth century simple maps of manors called *dochō* (literally, land ledgers) and *jikken ezu* (inspection maps) began to be compiled. In form they resembled the ancient *denzu* (cadastral maps) first compiled from surveys after the reform edict of 646. Dated examples are the *dochō* of Otogi no Shō (fig. 11.10) and the Wakatsuki no Shō of 1307, both of which were in Yamato Province (now Nara Prefecture) and belonged to Daijōin Temple in Nara.[74] The emphasis in these examples, as in others drawn for similar purposes, is on

73. The former (122.5 × 102.5 cm) is preserved in the Yanagisawa Collection in Ōsaka, and the latter (96 × 62 cm) is owned by the Historiographical Institute, Tokyo University; both are reproduced in Kyōto Kokuritsu Hakubutsukan, *Koezu*, pls. 72 and 73 (note 70).

74. The *dochō* of Wakatsuki (122.6 × 131.6 cm) is at the Ibaraki Prefectural History Hall and is reproduced in Tōkyō Daigaku Shiryō Hensanjo, *Nihon shōen ezu shūei*, vol. 3, pl. 21 (note 70). About the *denzu*, see Ramming, "Evolution of Cartography," 17 (note 10).

FIG. 11.11. A MAP SHOWING THE PROPERTY OF JINGO TEMPLE, KYŌTO, DATING FROM 1230. Signs delimiting the property were put up in eight places. The manuscript shows multiple perspectives because the inscriptions are written in different directions.
Size of the original: 199.2 × 160.8 cm. Jingo Temple, Kyōto. Photograph courtesy of Kazutaka Unno.

describing the content rather than the shape of the parcels of land. Entered into a *jōri*-system framework are placenames, rice paddies and vegetable fields, and the acreage for the lord and the administrator. Some of them contain ponds, rivers, roads, and the like, but in general they have few pictorial elements. That they were drawn in black and white also attests that they were designed to serve immediate practical purposes.

Maps were also compiled to designate the properties of shrines and temples. Two examples dating to 1230 depict the regions surrounding Jingo Temple and its branch, Kōzan Temple; these are the *Jingoji jiryō bōji ezu* (Map of the property of Jingo Temple) (fig. 11.11) and the *Kōzanji jiryō bōji ezu* (Map of the property of Kōzan Temple).[75] They were commissioned to prevent neighboring peasants from entering the grounds of the temple to cut the trees and fish in its river. The temple applied to the imperial court, which sent officials to help the head of the temple inspect the grounds, plant markers to indicate important points, and register them on the maps. These maps are typical examples of the *shīji bōji no zu* of temples. Two more examples are the *Rinsenji ryō Ōi Gō kaihan ezu* (Plan of the boundaries of Ōi Gō owned by Rinsen Temple) of 1347 and the *Ōei kinmei ezu* (Map compiled under shogunal orders in the Ōei era [1394–1428]) of 1426.[76] These differed from the *shīji bōji no zu* genre in that the roads and tracts of land were drawn with straight lines, an indication that straightedges were used. The former shows the area surrounding Rinsen Temple in Saga (now a part of metropolitan Kyōto) and notes the temple as the owner of each parcel of land. The latter contains the same area, but its scope is greater and the content more detailed: it notes over one hundred temples facing the road, but because it does not designate landownership, it was obviously made for a different purpose.

In addition to maps indicating ownership, some maps were drafted to assist in the rebuilding, repair, and restoration of shrines and temples. Detailed plans were commissioned especially at the time of rebuilding, as suggested by the *Fukōin kyūki hōkyō ezu* (Plan of the old structure of Fukōin Temple) (fig. 11.12). This was compiled when Fukōin Temple, a branch of Shōkoku Temple in Kyōto, was rebuilt and clearly marks the locations of individual pillars. Another example is the *Tsurugaoka Hachimangū shūei mokuromi ezu* (Plan for building and repairs of Tsurugaoka Hachiman Shrine) of 1591.[77] Architectural drawings of a less detailed nature were composed for repairs and restoration as well as for more general purposes. Oblique drawings were used to emphasize the features of buildings, but these were often based on original planar drawings. Examples include the *Gion oyashiro ezu* (Map of Gion Shrine) of 1331 (plate 22) and the *Usa Hachimangū ezu* (Map of Usa Hachiman Shrine) and *Shimogamo jinja ezu* (Map of Shimogamo Shrine) from about the fifteenth century.[78]

A group of landscape drawings that was characteristically medieval is the so-called mandala (Japanese

75. The latter (163.7 × 164.6 cm) is reproduced in Kyōto Kokuritsu Hakubutsukan, *Koezu*, pl. 55 (note 70), and (colored) in Unno, Oda, and Muroga, *Nihon kochizu taisei*, vol. 1, pl. 5 (note 8).

76. For the 1347 map (140 × 207 cm) see Unno, Oda, and Muroga, *Nihon kochizu taisei*, vol. 1, pl. 7 (note 8); for the 1426 map (291.2 × 241.5 cm) see Kyōto Kokuritsu Hakubutsukan, *Koezu*, pl. 61 (note 70).

77. Owned by Tsurugaoka Hachiman Shrine in Kamakura. It measures 139.2 by 105 centimeters and is reproduced in Miyaji, *Jinja kozu shū*, pl. 69, and in Kyōto Kokuritsu Hakubutsukan, *Koezu*, pl. 76 (both in note 70).

78. The first two maps (167 × 107.5 cm and 135 × 139 cm) are preserved at their respective shrines, Gion Shrine today being generally referred to as Yasaka Shrine (Kyōto) and Usa being in Ōita. The map of Shimogamo Shrine (in Kyōto), 214 by 193.5 centimeters, is owned by the Kyōto National Museum. All three are reproduced in Miyaji, *Jinja kozu shū*, the first in color in the frontispiece with a detail in pl. 27, and the others in pls. 128 and 19, and in Kyōto Kokuritsu Hakubutsukan, *Koezu*, pls. 1, 6, and 2 (both in note 70).

mandara; religious picture) type. The original use of a mandala, as a cosmic diagram, was to assist meditation in the Shingon sect of Buddhism, founded in 806 by the priest Kūkai (774–835) (see pp. 373–74). They could be drawn for individual rites by sketching on the ground, or they could be produced in a more permanent fashion as paintings and carvings. In either case, in the Heian period and subsequently, mandalas were primarily for religious purposes rather than objects of art. The term came to be applied to paintings that were not related to esoteric Buddhism but showed famous Shinto shrines and Buddhist deities.

Such are the picture maps of landscapes containing shrines and temples, used as religious objects, that are discussed next. Extant examples show that they were oblique drawings with multiple viewpoints and must have been based on plans showing the arrangement of the buildings. They substituted for personal visits and were known by the name of the depicted shrine or temple plus the Sanskrit mandala at the end.[79] They were worshiped as early as the late twelfth century. In his diary, the *Gyokuyō* (literally, Leaves of gem), the regent Kujō Kanezane (1149–1207) notes that he performed religious services in front of such a drawing that had been sent in 1184 by a Buddhist priest in Nara:

> I received a picture of Kasuga Shrine from a monk in Nara. Early in the morning after washing, I got dressed in formal wear, worshiped as if I were in front of the actual shrine, and read one thousand volumes of Buddhist scripture. It is quite a penance. I will continue to do this with my family for the next seven days.[80]

The map was a surrogate for reality, and the services were conducted in the same manner as services in person at the shrine.

The *Hanazono tennō shinki* (Autographic record of the emperor Hanazono [1297–1348]) records that in 1326 such drawings were known as mandalas. The emperor writes in February 1326, "For the last three or four years, people make offerings and perform various ceremonies before a Kasuga-mandala [picture of the shrine of that name] as they would if they were at the shrine."[81] This confirms that the term was used at least from the first quarter of the fourteenth century onward. Approximately fifteen mandala-type drawings are extant, the oldest dated example being the *Kasugamiya mandara* (Mandala of Kasuga Shrine) by the artist Kanshun (fig. 11.13).[82]

Among the mandala genre as a whole is a colorful and attractive group known as the *sankei mandara* (mandalas for visitations), dating from between the fifteenth and seventeenth centuries.[83] Their distinguishing feature is the ordinary people who appear in the scenes. Rather than being used for religious services, they were carried by shamans to believers and clients and presumably used as

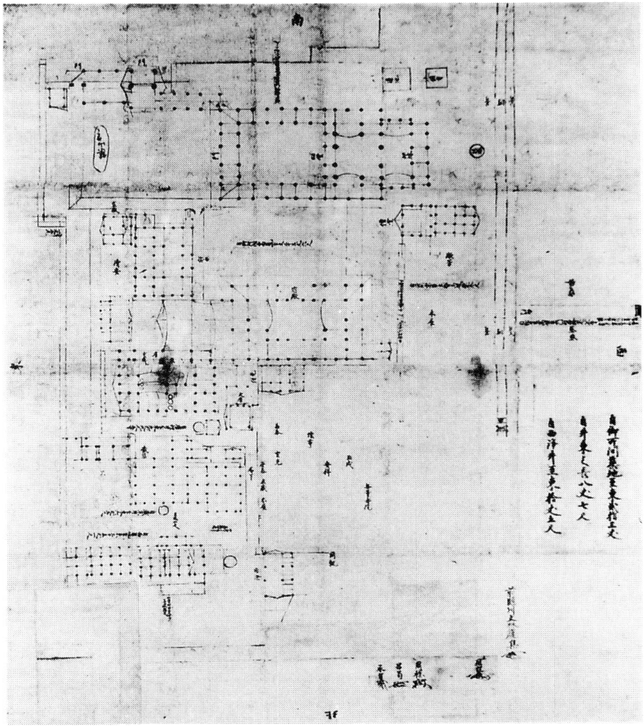

FIG. 11.12. THE *FUKŌIN KYŪKI HŌKYŌ EZU* OF 1510. This is an example of a plan used for rebuilding. Note that the locations of the pillars are clearly marked on the manuscript. (The title was given to the plan at a later date.)
Size of the original: 157.4 × 130 cm. Shōkoku Temple, Kyōto. Photograph courtesy of Kazutaka Unno.

79. See, for example, Varley, *Japanese Culture*, 49–50 (note 47), and Hugo Munsterberg, *The Arts of Japan: An Illustrated History* (Tokyo: Charles E. Tuttle, 1985), 79 and 96.

80. *Gyokuyō*, 3 vols., 66 chaps. (Tokyo: Kokusho Kankōkai, 1906–7), vol. 3, chap. 40, p. 22. Although Kasuga Shrine is Shinto and the monk and scripture are Buddhist, there is no inconsistency in Kujō's entry. Until the Meiji Restoration in 1868, when a clear division was made between shrines (Shinto) and temples (Buddhist), and also between priests of the two religions, it was not unusual for a Buddhist priest to serve as a Shinto priest as well.

81. The diary is in the *Shiryō taisei* (Series of historical materials), vols. 33–34 (Tokyo: Naigai Shoseki, 1938), quotation on 34:158.

82. Other examples include: *Ikomamiya mandara* (Mandala of Ikoma Shrine), early fourteenth century (104.9 × 41.7 cm) at the Nara National Museum; *Iwashimizu hachimangū mandara* (Mandala of Iwashimizu Hachiman Shrine), early fourteenth century (88.7 × 27.9 cm) at Rikkyokuan, Tōfuku Temple in Kyōto; and *Kakinomotomiya mandara* (Mandala of Kakinomoto Shrine), mid-fourteenth century (132.9 × 57.8 cm) at the Yamato Culture Hall. These and other extant examples are reproduced in Kyōto Kokuritsu Hakubutsukan, *Koezu*, pls. 45 (Kasuga), 43 (Ikoma), 50 (Iwashimizu), 42 (Kakinomoto), and others between 37 and 51 (note 70). Reproductions of the Ikoma and Iwashimizu mandalas are also in Miyaji, *Jinja kozu shū*, pls. 47–48 and 11, and that of Kasuga in Fukuyama, *Jinja kozu shū zokuhen*, pl. 1 (both in note 70).

83. For example, *Atsuta sangū mandara* (Mandala for pilgrimages

FIG. 11.13. THE *KASUGAMIYA MANDARA* OF 1300. In this manuscript, the oldest dated extant sample of the "mandala" landscape genre, solemnity is more important than accuracy: such maps, which served as surrogates for the actual shrines and temples, were used for religious worship. The oblique perspective with more than one viewpoint was typical of the genre. Size of the original: 108.5 × 41.5 cm. By permission of the Yuki Museum of Art, Ōsaka.

promotional material to show the splendor and prosperity of their shrines and temples.

THE GYŌKI-TYPE MAPS OF JAPAN

The official histories indicate that orders were issued for provincial maps in 646, 738, and 796.[84] Although there is no existing evidence that a map of Japan was compiled from them, there is at least indirect evidence of what a mental image of the country might have been at the time. A Chinese official history, the *Sui shu* (History of the Sui), notes that it took five months to cross the territory of Wa (Japan) from west to east and three months from north to south, according to the indigenous people.[85] Although these figures were most likely an exaggeration to impress the Chinese, they do at least suggest that the longer axis lay west to east. The *Engi shiki* (Rules pertaining to the execution of laws, edited in 927) similarly suggests that the general shape of the country had been visualized. The text refers to the boundaries of the territory as being Michinoku in the east, Tōchika in the west, Tosa in the south, and Sado in the north; respectively they are northeastern Honshū, the Gotō Islands off western Kyūshū, a province in southern Shikoku, and an island in the Sea of Japan where the northward curve of Honshū begins to be pronounced.[86] This suggests that the archipelago was thought to be elongated from east to west, and the idea that Sado forms the northern boundary, rather than some point in northern Honshū, suggests that the curve of Honshū was not yet understood. That half of Kyūshū lies farther south than southern Shikoku also appears not to have been known.

---

to Atsuta Shrine [Nagoya]), 1529 (169.7 × 144.8 cm), Tokugawa Reimeikai, Tokyo; *Fuji sangū mandara* (Mandala for pilgrimages to Mount Fuji), drawn by Kanō Motonobu (1476–1559), mid-sixteenth century (180.6 × 118.2 cm), Sengen Shrine, Fuji; and *Nachi sangū mandara* (Mandala for pilgrimages to Nachi Shrine), late sixteenth century (150 × 160 cm), Nachi Taisha Shrine, Kumano. These and other *sankei mandara* are reproduced in Kyōto Kokuritsu Hakubutsukan, *Koezu*, pls. 31 (Atsuta), 35 (Fuji), 28 (Nachi), and others between 23 and 36 (note 70). The Atsuta mandala is reproduced in Miyaji, *Jinja kozu shū*, pls. 62 and 63, and those of Nachi and Fuji in Fukuyama, *Jinja kozu shū zokuhen*, pls. 42 and 49 (both in note 70).

84. See the section on the literary evidence for maps above.

85. Wei Zheng et al., *Sui shu* (compiled 629–56), chap. 81; see the edition in 6 vols. (Beijing: Zhonghua Shuju, 1973), 6:1825. "Wa" is the term used for Japan in ancient Chinese documents. Japan, of course, was smaller then than today and extended only as far eastward as central Honshū.

86. See chapter 16 of the *Engi shiki*, in vol. 26 of the *Shintei zōho kokushi taikei* (note 37). Papinot refers to the *Engi shiki* as "a collection in 50 volumes of the regulations concerning the ceremonies of the palace, the audiences of the officials, the customs of the provinces, etc.": Papinot, *Dictionary of Japan*, 81 (note 43). These boundaries pertain to an annual ritual (Tsuina) held at the imperial court, in which prayers were offered to the Shinto gods so that they might drive evil spirits out of the country.

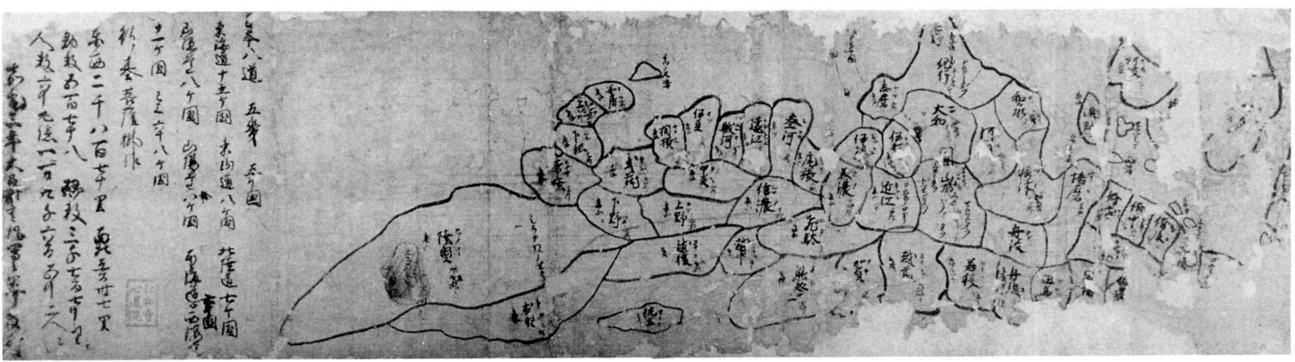

FIG. 11.14. MAP OF JAPAN OWNED BY NINNA TEMPLE, KYŌTO. This manuscript dates from 1306. It is oriented to the south, and the main routes leading from the capital province of Yamashiro (today Kyōto Prefecture) are shown in red. The part showing western Japan is damaged, so it is impossible to determine the complete image of the country at the time. Size of the original: 34.5 × 121.5 cm. Ninna Temple, Kyōto. Photograph courtesy of Kazutaka Unno.

The first general maps of Japan tended to be diagrams depicting the provinces of the country and the main routes from Yamashiro, the province where the capital of Kyōto was situated.[87] These are termed "Gyōki-type" maps after the Buddhist priest Gyōki (668–749). Gyōki played an important role in diffusing Buddhism, and he appears to have been not only a traveling monk but also a civil engineer, inasmuch as he was involved in constructing public works such as dams, canals, bridges, and roads. There is a possibility that he had something to do with the 738 order to compile provincial maps: he evidently had a great influence on the emperor Shōmu (r. 724–49) and played an important role at his court.[88]

No maps of the kind attributed to Gyōki are extant from the eighth century.[89] Such maps were known to have been compiled as early as the beginning of the ninth century, however, and particularly from the early fourteenth until the mid-nineteenth century (see appendix 11.2). The genre is typical of the conservative element in Japanese culture and established a conventional image of the country at least until the arrival of the Europeans in the sixteenth and seventeenth centuries. Cortazzi puts it thus:

> Although the Gyōgi-type maps gradually improved, they never contained much in the way of geographical information, and the shape of the Japanese islands became stereotyped, so that even when mapmakers knew better they tended to follow the old patterns. This kind of stereotyping was a common feature of many other aspects of Japanese culture from the eleventh to the nineteenth century, including poetry . . . theatre arts . . . and even the martial arts.[90]

Whether Gyōki himself actually composed any maps is not known, but the most reliable biography, the *Gyōki nenpu* (Chronological history of Gyōki) of 1175, by Izumi no Takachichi, does not mention mapmaking among his varied activities.[91] There is similarly no tangible evidence of a Gyōki-type map from the Nara period, one that would have shown the province of Yamato, where the capital of Nara was situated, as the focal point and the origin of the main routes.

How these maps came to be attributed to Gyōki is nonetheless of interest. Inscriptions on one of the oldest extant originals of a Gyōki-type map, composed in 1306 and in the possession of Ninna Temple in Kyōto, provide a clue (fig. 11.14). There is one that notes that "the author copied the map sheltered from cold winds; the map should not be shown to outsiders," and another gives the date "third year of Kagen, Tairyo (the twelfth month)" (January/February 1306). These indicate that there might have been a connection between the map and the annual ritual of Tsuina, held on the last day of the year at the imperial court to drive evil spirits beyond the boundaries of the country.[92] Gyōki is associated with

87. An exception is a manuscript map owned by Shōmyō Temple in Yokohama and kept at the Kanazawa Bunko. The extant portion covers only the western half of Japan and also refers to Mongolia (Mōkokoku). It is oriented to the south, and the shapes of the provinces are similar to those on the "Yochi zu" (see notes 98 and 99 below). See Akioka, *Nihon chizu shi*, pl. 4, pp. 19–22 (note 7). Cortazzi reproduces the map but repeats Akioka's unfounded inference of 1305 as its approximate date: see Cortazzi, *Isles of Gold*, pl. 4 and pp. 5–6 (note 14).

88. See, for instance, Papinot, *Dictionary of Japan*, 134 (note 43), and Cortazzi, *Isles of Gold*, 4 (note 14).

89. At least one source, nevertheless, gives the eighth century as the origin of the Gyōki-type maps: Akioka, *Nihon chizu shi*, 3 (note 7).

90. Cortazzi, *Isles of Gold*, 10 (note 14).

91. The *Gyōki nenpu* is in the *Zokuzoku gunsho ruijū* (Classified series of various books: Second continuation), 16 vols. (Tokyo: Kokusho Kankōkai, 1906–9; reprinted 1969–78), 3:428–37.

92. Until 1873, when Japan adopted the Gregorian calendar, a lunisolar calendar was used; new years began between the equivalents of 20 January and 19 February of the Christian calendar: see Papinot, *Dictionary of Japan*, 836 (note 43), and Nakayama, *History of Japanese Astronomy*, 65–73 (note 38).

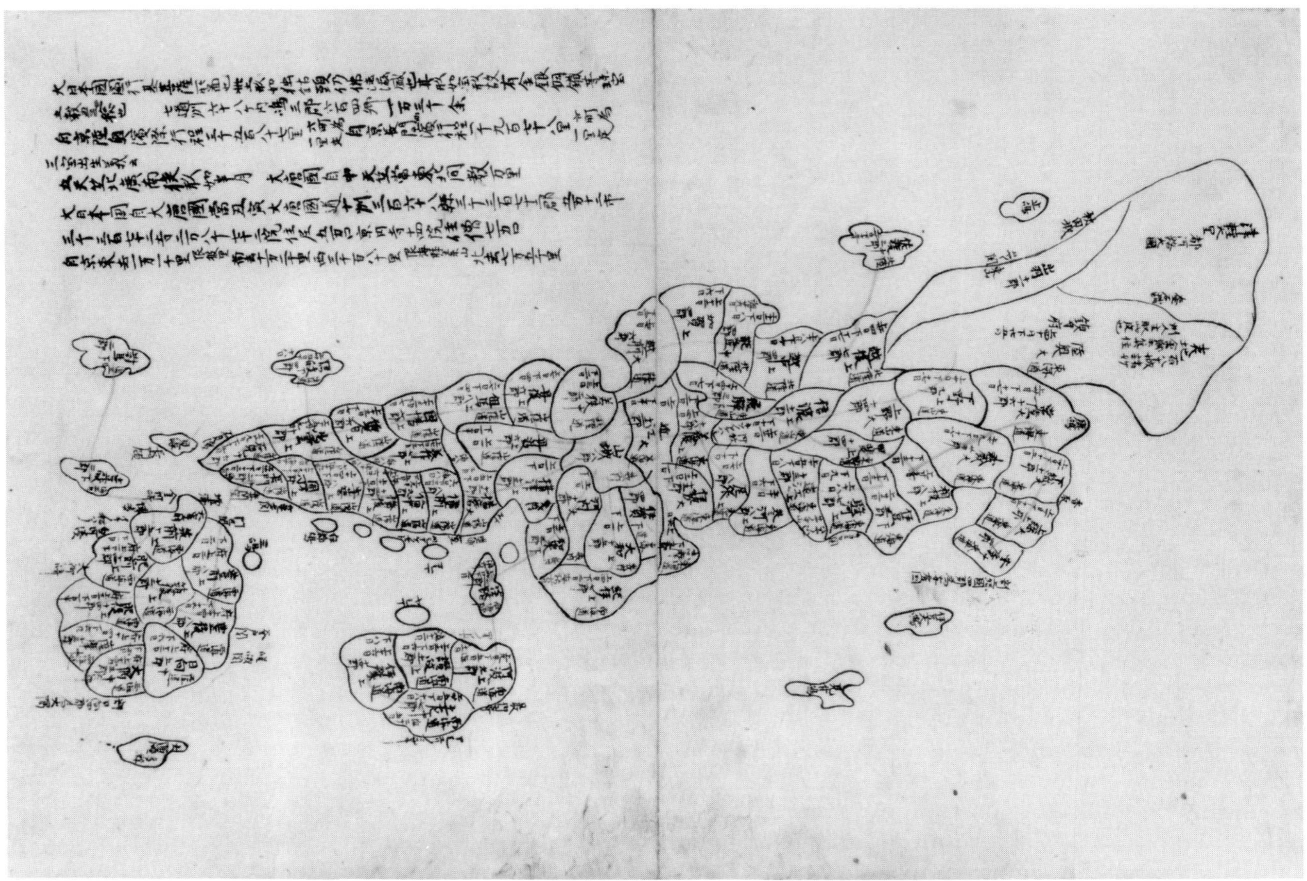

FIG. 11.15. THE "DAINIHONKOKU ZU" IN THE 1548 CODEX OF THE *SHŪGAISHŌ*. Important geographical features are the provinces and the main routes (in red), which focus on Yamashiro Province where the capital of Kyōto was situated. The explanation in the upper left corner and the information on the map are to be read from different directions, the first with the left side on top and the second with the right.
Size of the original: 26.3 × 41.3 cm. By permission of the Tenri Central Library, Tenri, Nara Prefecture.

this ritual: according to the records of Hōshaku Temple in Yamazaki, Yamashiro Province, Gyōki advised the emperor Monmu (r. 697–707) of its necessity in 706.[93] A map depicting the boundaries of the country might have been of interest to those involved in the ceremony. Originally this would have taken place at the imperial court, but later a large number of shrines and temples held their own ceremonies. The map at Ninna Temple might have been copied for this purpose. Inscriptions relating to Buddhism, such as "Buddhism prospers more and more vigorously" on the *Nansenbushū Dainihonkoku shōtō zu* (Orthodox map of Great Japan in Jambūdvīpa)[94] of about 1550 at Tōshōdai Temple in Nara, a map of Japan in the *Shūgaishō* (Collection of oddments), an encyclopedia compiled by Tōin Kinkata (1291–1360), and other maps of the Gyōki genre, strongly suggest that such maps had religious associations and praised the country. It is as such that they should be understood, rather than as practical maps designed to show geographical information in the modern sense.

Typically the Gyōki-type maps show the archipelago

93. Gyōki was the founder of this temple. See Terajima Ryōan, ed., *Wakan sansai zue* (Illustrated encyclopedia of Japanese and Chinese things related to the three powers [heaven, earth, and man], 1715), chap. 4; modern edition in 2 vols. (Tokyo: Tōkyō Bijutsu, 1982), 1:56.

94. Jambūdvīpa is the continent containing India and the surrounding territories in Buddhist cosmology (see the next section). It was common for Japanese to refer to their country as "Great Japan in Jambūdvīpa" during the long medieval period in which Buddhism heavily influenced Japanese culture (roughly the eleventh through the seventeenth centuries). The term appears to have been used to indicate that Japan was a part of the same world as India and China. Around the map are entries showing the names of administrative divisions in Japan and various statistical information. It is a manuscript map (for details see appendix 11.2); reproductions are in Akioka, *Nihon chizu shi*, pl. 12, pp. 47–53 (note 7); Unno, Oda, and Muroga, *Nihon kochizu taisei*, vol. 1, pl. 8 (note 8), and Cortazzi, *Isles of Gold*, pl. 5, with a commentary on p. 6 (note 14).

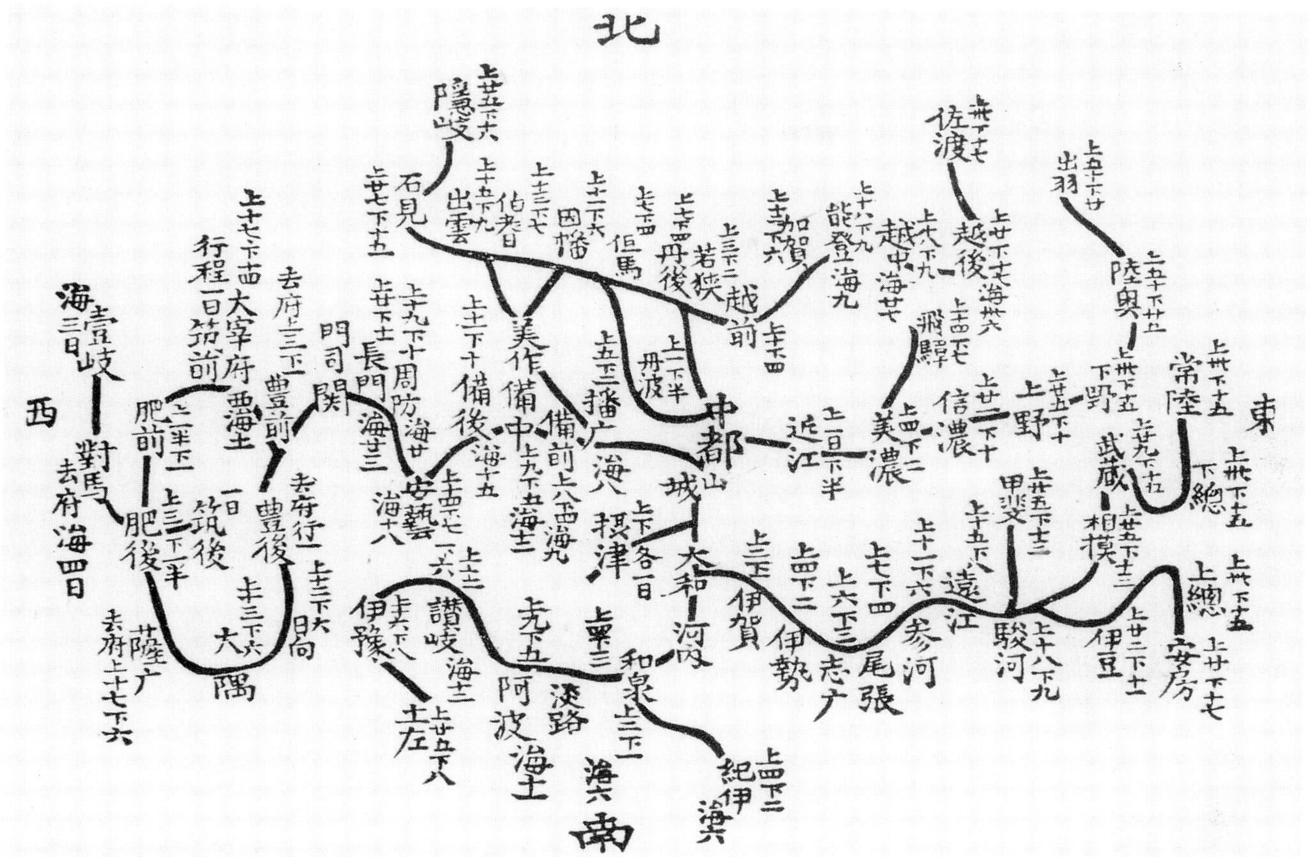

FIG. 11.16. A MAP OF JAPAN IN THE *NICHŪREKI*. This map is a schematic representation of main routes and places, and probably served as an aid for tax collection: its contents include Dazaifu (a local agency of the central government) in northern Kyūshū and Moji ga Seki on the tip of Kyūshū facing Honshū over the Strait of Shimonoseki. Tribute from the provinces west of Moji ga Seki was taken to Dazaifu, and that from the rest went to Kyōto.
Size of the original: 22.7 × 30.6 cm. From the reproduction in Kondō Heijō and Kondō Keizō, eds., *Kaitei shiseki shūran* (Revised collection of historical books), vol. 23 (Tokyo: Kondō Shuppanbu, 1901), maps p. 190.

as elongated from east to west, with a slight northward curve in the thicker eastern end of Honshū. The lack of a prominent curve may be seen on examples from as late as the latter half of the sixteenth century: the map at Tōshōdai Temple of about 1550 and copies of the "Dainihonkoku zu" (Map of Great Japan) in a 1548 and 1589 version of the *Shūgaishō* (see fig. 11.15).[95] Other characteristic features include the provinces drawn in roundish shapes, using curved lines wherever possible, thereby giving the coastline the irregular appearance of a collection of arcs. Geographical accuracy was not a particular consideration; what was important was the relative position of the provinces and a general scheme of the main roads leading from the capital province of Yamashiro. Information was recorded in marginal notes, some of which suggest a practical use for the maps. The maps in the 1548 and 1589 *Shūgaishō*, for example, contain entries about the number of days needed to carry tribute from each province to the central authorities and the names of places that were important to travelers; these entries, however, were omitted in printed editions published later in the Edo period (1603–1867).[96] The form

95. The 1548 and 1589 codices are at the Tenri Central Library, Tenri, and the Sonkeikaku Library, Tokyo. The latter was published in facsimile in 1976 (Tokyo: Kojisho Sōkan Kankōkai) and its map is reproduced in color in *Kokushi daijiten* (Large dictionary of the history of our country [Japan]) (Tokyo: Yoshikawa Kōbunkan, 1979–), vol. 11 (1990), color pages "Nihon zu" (maps of Japan), pl. 2.

96. There is one edition from the Keichō era (1596–1614) with no colophon (ca. 1607); three others with dates and publishers are 1642 by Nankyōshodō, 1642 by Nishimura Kichibee, and 1656 by Murakami Kanbee; all were published in Kyōto. There are also a few more editions without a date. The map in the Keichō edition is reproduced in Kurita, *Nihon kohan chizu shūsei*, "Kaisetsu" (explanation), folding plate (note 15); Akioka, *Nihon chizu shi*, pl. 5 (note 7), and idem, *Nihon kochizu shūsei*, pl. 7 (note 15). Cortazzi reproduces a map of Japan (17.5 × 27.5 cm) from a later edition of the *Shūgaishō* at the Tenri Central Library

and content of the entries for the number of days in the *Shūgaishō* map are similar to those found on a map in the encyclopedia *Nichūreki* (Two guides, twelfth century) (fig. 11.16).[97] Other examples of Gyōki maps include the *Yochi zu* (Land map) of 805, not extant but known from later copies,[98] and though it does not show any routes, a map showing western Japan owned by Shōmyō Temple in Yokohama. Only part of the second map has survived, and it dates to the latter part of the thirteenth century. It shows the body of either a dragon or a snake, most likely the former, since the dragon was a guardian deity of Buddhism, considered to control water, rain, and clouds, and here would be protecting Japan.[99]

The Gyōki-style image also found its way into printed Chinese and Korean maps showing Japan in the fifteenth and sixteenth centuries. Examples include a woodblock print in the Korean *Haedong cheguk ki* (Chronicle of the countries in the Eastern Sea [Japan, Ryūkyū]) of 1471 by Sin Sukchu (1417–75), and another woodblock edition printed in China in 1523, the *Ribenguo kaolüe* (Summary of Japan).[100] They were also a source for European cartographers late in the sixteenth century.[101] Two examples are known to exist in European archives: one in the Medici papers relating to the East Indian trade may date to the diplomatic mission to Europe of four young Christian Japanese nobles in 1582–90.[102] The other, in the Archivo

in Nara Prefecture; see Cortazzi, *Isles of Gold*, 8 and 71 (pl. 6) (note 14).

97. The *Nichūreki* was formed from the previous portable encyclopedias *Shōchūreki* (Hand-size guide) and *Kaichūreki* (Pocket guide), both compiled early in the twelfth century. The two maps in the *Nichūreki* are manuscript diagrams showing the main routes from Kyōto but without any geographical outline of the islands. One of them lists the number of days required to carry tribute, the other does not. The latter, according to an inscription, was copied by the student Miyoshi Yukiyasu from the first volume of the *Kaichūreki* in 1128. The first has no such inscriptions. The earliest codex of the *Nichūreki*, dating from approximately 1324–28, is owned by the Sonkeikaku Library (22.7 × 15.3 cm); a facsimile was published by the library in 1937. The *Nichūreki* is reproduced in Kondō Heijō and Kondō Keizō, eds., *Kaitei shiseki shūran* (Revised collection of historical books), vol. 23 (Tokyo: Kondō Shuppanbu, 1901), maps: p. 190. Akioka and Cortazzi reproduce the map without the information for tribute: see Akioka, *Nihon chizu shi*, pl. 2 (note 7), and Cortazzi, *Isles of Gold*, pl. 3 (note 14).

98. According to Fujii Sadamiki (1732–97), the original belonged to Shimogamo Shrine in Kyōto. Unfortunately it has not been found there or anywhere else at present. The copy is in chapter 2 of Fujii's *Shūko zu* (Illustrations of collected antiques) at the National Museum of Japanese History, Sakura, and elsewhere and contains the main routes (in red lines) from Yamashiro to the other provinces (see appendix 11.2). A copy of the map alone is at the National Archives in Tokyo. See also Akioka, *Nihon chizu shi*, pl. 1, pp. 9–13 (note 7).

99. For the map at Shōmyō Temple, see note 87 above. In the strictest sense, the Shōmyō map, like those in the *Nichūreki*, is not a Gyōki-type example; it has enough in common with others, however, to warrant consideration as such here. Whether there was a precise reason for the presumed dragon is not known, but a possible symbolic reason might be surmised from the events by and in the last third of the thirteenth century. Throughout the Kamakura period (1192–1333) a Chinese influence was felt in Japanese art, to some extent created by the influx of refugees at the time of the Mongolian conquest (1234–79). The Mongolians attempted to invade Japan in 1274 and again in 1281, both times being stymied by stormy weather, referred to as the original *kamikaze* (divine wind). It is possible to imagine the dragon as a symbol of the *kamikaze*, and the dragon around the archipelago in the Shōmyō Temple map as a type of divine savior or protector. See, for example, James Jackson Jarves, *A Glimpse at the Art of Japan* (1876; reprinted Tokyo: Charles E. Tuttle, 1984), 81–82; Basil Hall Chamberlain, *Japanese Things: Being Notes on Various Subjects Connected with Japan* (Tokyo: Charles E. Tuttle, 1985; reprinted from an edition of 1905), 443–44; Munsterberg, *Arts of Japan*, 39 and 90 (89–105 for the art of the Kamakura period) (note 79); and Noritake Tsuda, *Handbook of Japanese Art* (1941; reprinted Tokyo: Charles E. Tuttle, 1985), 221 (108–41 for the Kamakura period).

100. Akioka notes that the *Haedong cheguk ki* has the first printed individual map of Japan in the world. Both maps are mentioned: Akioka, *Nihon chizu shi*, 33–37, pl. 8–9 (note 7). The map in the *Haedong cheguk ki* was not, however, the first Gyōki-type image to appear outside Japan: one appears on a Korean world map dated 1402 (oldest surviving copy is from around 1470); this is mentioned, for example, in Unno, "Japan," 358 (note 14). On the Korean world map dated 1402 and the *Haedong cheguk ki*, see chapter 10 above. The *Haedong cheguk ki* has been reproduced in Japanese by the Iwanami Bunko, *Kaitō shokoku ki* (Chronicle of the countries in the Eastern Sea), blue series 458–1 (Tokyo: Iwanami Shoten, 1991).

101. The Gyōki-type maps were just one source of information for Western cartographers, and examples of European maps based on this style are discussed and reproduced in the works of several historians. See, for instance, Armando Cortesão, "Study of the Evolution of the Early Cartographic Representation of Some Regions of the World: Japan," in *Portugaliae monumenta cartographica*, 6 vols., by Armando Cortesão and Avelino Teixeira da Mota (Lisbon, 1960; facsimile edition, Lisbon: Imprensa Nacional–Casa da Moeda, 1987), 5:170–78 and 6:40–41 (addenda); Cortazzi, *Isles of Gold*, 17–25 and pls. 12–29, which show some European maps from 1528 to 1646–57 (note 14); Erik W. Dahlgren, *Les débuts de la cartographie du Japon* (Uppsala: K. W. Appelberg, 1911; reprinted Amsterdam: Meridian, 1977); George Kish, "The Cartography of Japan during the Middle Tokugawa Era: A Study in Cross-Cultural Influences," *Annals of the Association of American Geographers* 37 (1947): 101–19; and idem, "Some Aspects of the Missionary Cartography of Japan during the Sixteenth Century," *Imago Mundi* 6 (1949): 39–47.

102. The four young nobles represented the Christian daimyates of Bungo, Ōmura, and Arima in Kyūshū. The envoy was arranged by the Jesuit missionary Alessandro Valignani after the first of his three residencies in Japan (1579–82, 1590–92, and 1598–1603). For an account of the mission, see Otis Cary, *A History of Christianity in Japan*, 2 vols. (New York: Fleming H. Revell, 1909; reprinted 1987), vol. 1, *Roman Catholic and Greek Orthodox Missions*, 92–97. The map found in the Medici papers, titled *Iapam*, is an unattributed, undated manuscript (see appendix 11.2); for discussions on this map and reproductions see Akioka, *Nihon chizu shi*, 186–90, pl. 14 (note 7); Hiroshi Nakamura, "Les cartes du Japon qui servaient de modèle aux cartographes européens au début des relations de l'Occident avec le Japon," *Monumenta Nipponica* 2, no. 1 (1939): 100–123; Cortazzi, *Isles of Gold*, 23–24 and pl. 23 (note 14); and Kish, "Missionary Cartography," 42–46 (note 101). Kish refers to an article written by the discoverer of the map: Sebastiano Crinò, "La prima carta corografica inedita del Giappone portata in Italia nel 1585 e rinvenuta in una filza di documenti riguardanti il commercio dei Medici nelle Indie Orientali e Occidentali," *Rivista Marittima* 64 (1931): 257–84.

Histórico Nacional in Madrid, is dated 1587 and was included in the report of the 1587 mission to the Spanish viceroy of the Philippines by the feudal lord of Hirado, Hizen Province, Matsura Shigenobu (1549–1614).[103]

BUDDHIST WORLD MAPS

The indigenous Shinto mythology places the world into a vertical structure incorporating the heavenly world, the earth, and the underworld.[104] There is no horizontal structure, however, which perhaps accounts for the lack of Shinto maps. Buddhism, on the other hand, contains concrete spatial views about the universe,[105] and with its arrival in the sixth century A.D. via Korea, the foundation was laid for a whole genre of religious world maps.

At the center of the Buddhist universe, in doctrines transmitted to Japan, is a tall mountain called Sumeru, Sumi or Shumi in Japanese. It is in the middle of a flat circular earth, and around it revolve the sun and the moon. At the foot of the mountain are seven basins of water and seven mountain ranges alternating in concentric circles. Beyond them is a broad stretch of brackish ocean surrounded by yet another range of mountains. In the ocean there are four continents with different shapes, to the north, east, south, and west. The actual geographical region to contain India and the surrounding territories is represented by the southern continent, which appears as an inverted triangle and suggests the shape of the Deccan peninsula. In Sanskrit this continent is called Jambūdvīpa after a huge imaginary *jambū* tree believed to grow in the far north of India, the word *dvīpa* meaning land; its Japanese equivalents are *Enbudai* and *Senbushū*, based on the pronunciation of the Chinese translation. The other three continents originally may have been suggested by the areas surrounding India, but in Buddhism they have become purely imaginary continents.[106]

The *Nihon shoki* provides the evidence that this worldview had been accepted by the Japanese by the middle of the seventh century. It notes that a model of Mount Sumeru was constructed in the metropolitan area of Asuka, Nara Prefecture, in 657 in order to hold a welcoming party for overseas visitors. We are told: "A model of Mount Sumi was constructed to the west of the Temple of Asuka-dera. Moreover the festival of All Souls was held. In the evening the people from Tukhāra [Dvāravatī, on the lower Mae Nam River] were entertained."[107] Two other references to models of Sumeru

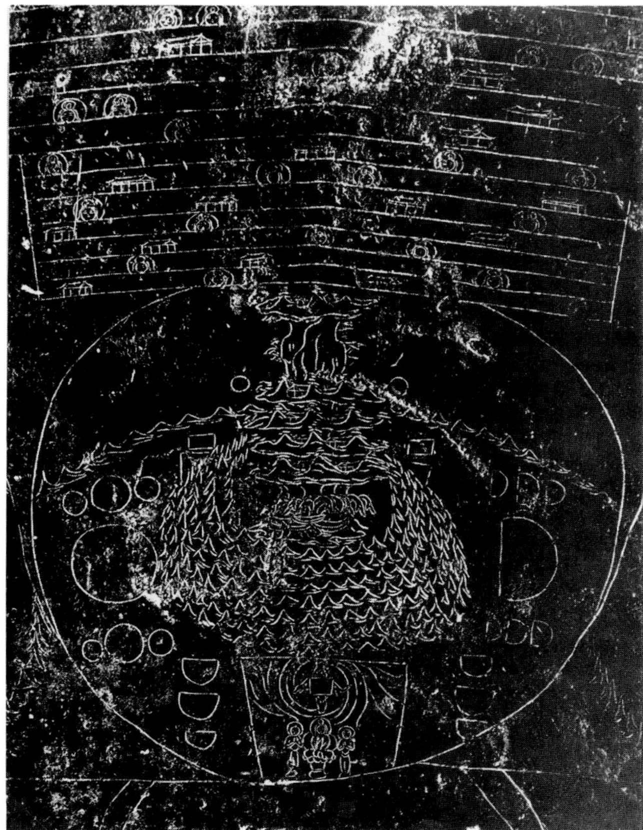

FIG. 11.17. AN OBLIQUE VIEW OF MOUNT SUMERU AND THE UNIVERSE ENGRAVED ON THE PEDESTAL OF THE GREAT STATUE OF THE BUDDHA AT TŌDAI TEMPLE. At the top are shown the twenty-five strata of the heavens, while Jambūdvīpa and its four great rivers flowing out of Lake Anotatta are shown in the lower portion. The rivers flow out of the lake through the mouths of an ox, a horse, an elephant, and a lion and make a circle around the lake. The bronze pedestal was cast in 749.
Width of the original: ca. 40 cm. Photograph courtesy of Kazutaka Unno.

103. This map, a manuscript sketch, is also mentioned by Cortazzi, *Isles of Gold*, 24 (note 14). A lengthier discussion, comparing it with the Florence map, and a copy of Nakamura's tracing of the map are found in Kish, "Missionary Cartography," 44–46 (including fig. 4) (note 101); cited is Nakamura, "Les cartes du Japon." Matsura (spelled Matsuura in this case) is mentioned in regard to trade with Spaniards based in Manila in Sansom, *History of Japan*, 2:373 (note 32).

104. These are Takamagahara (or Takama no hara, the Plain of High Heaven); Ashihara no Nakatsukuni (the Central Land of Reed Plains); and Ne no Kuni (the Land of the Roots), Yomi no Kuni (the Dark Land), or Yomotsukuni (the Land of Hades). See Chamberlain, *Kojiki*, 15, 38–43 (note 28).

105. For discussion of this in a South Asian context, see Joseph E. Schwartzberg, "Cosmographical Mapping," in *The History of Cartography*, ed. J. B. Harley and David Woodward (Chicago: University of Chicago Press, 1987–), vol 2.1 (1992), 332–87.

106. Cortazzi notes that Jambūdvīpa "represented the whole of the inhabited world" in Indian cosmology: Cortazzi, *Isles of Gold*, 9 (note 14).

107. See Aston, *Nihongi*, 2:251 (note 28). There are still some huge stones with carvings that might have been used for this reception at the site of Asuka. See Kazutaka Unno, "Japan before the Introduction of the Global Theory of the Earth: In Search of a Japanese Image of the Earth," *Memoirs of the Research Department of the Toyo Bunko* 38 (1980): 39–69, esp. 62–66.

FIG. 11.18. THE *GOTENJIKU ZU* BY JŪKAI, 1364. Lake Anotatta and the four great rivers are shown toward the top of the center. The entries in the boxes in the sea part of the manuscript are extracts from the *Da Tang xiyu ji*.
Size of the original: 177 × 166.5 cm. Hōryū Temple, Nara. Photograph courtesy of Kazutaka Unno.

date to 659 and 660.[108] The earliest extant drawing of Mount Sumeru is, however, an engraving on a lotus petal forming the pedestal of the great statue of the Buddha, enshrined at Tōdai Temple (fig. 11.17).[109] Its Jambūdvīpa clearly shows the traditional four rivers flowing out of the sacred Lake Anotatta (Manasarowar) in the north.

108. See Aston, *Nihongi*, 2:259 and 265 (note 28).

109. The great statue of the Buddha (Daibutsu) is one of the best known and most visited historical monuments in Japan. It was constructed in an attempt to unite the people of Japan after a rebellion begun in 740 by Fujiwara no Hirotsugu (d. 740). The first attempts to build a giant Buddha were made at Shigaraki (presumably Shiga in Ōmi Province, now Ōtsu in Shiga Prefecture) and Naniwa (now part of

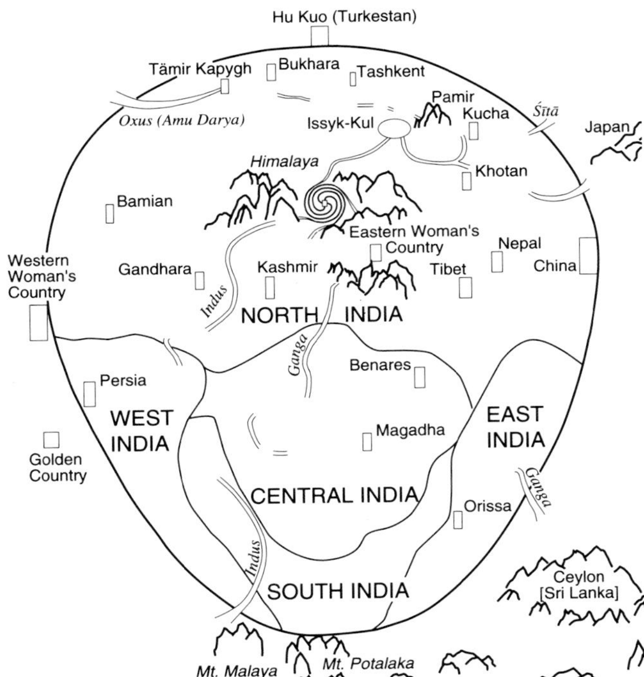

FIG. 11.19. SCHEMATIC EXPLANATION OF FIGURE 11.18.

The oldest Japanese map that may be considered a map of the world is the *Gotenjiku zu* (Map of the Five Indias), drawn by the priest Jūkai (b. 1297) (figs. 11.18 and 11.19). *Tenjiku* means India, which is divided into five geographical regions (north, east, south, west, and central), and the map's Jambūdvīpa is in the shape of an egg with the small end down. On the map there are many place-names and travel routes based on the *Da Tang xiyu ji* (Record of a journey to the western regions of Great Tang), the travel record of the Chinese priest Xuanzhuang (602–64), who visited India in 629–45. The purpose of the map, in the form of a colorful manuscript mounted as a scroll, may have been to represent the pilgrimage to India, but it also portrays the world as it was known to the peoples of East Asia. Because it depicts Xuanzhuang's pilgrimage realistically, the *Gotenjiku zu* was also an object of worship; for this reason copies are preserved by the older temples even today (see appendix 11.3).[110]

The *Gotenjiku zu* is not an original Japanese creation, but a slightly revised copy of a map made in China, where there are extant maps with exactly the same composition. Later copies are known, such as the "Nanzhanbuzhou tu" (Map of Jambūdvīpa), in the *Fajie anli tu* (Maps of the configuration of Dharmadhātu) of 1607 by Renchao, and the "Sihai Hua yi zongtu" (General map of Chinese and foreign territory within the four seas), found in the *Tushu bian* (Compilation of illustrations and writings), an illustrated encyclopedia compiled by Zhang Huang (1527–1608).[111] Although these dates of compilation are much later than the *Gotenjiku zu*, they could not have been influenced by a Japanese source because of the circumstances of cultural transmission between China and Japan, in which Japan was the recipient. This suggests an original Chinese prototype, as acknowledged by Zhang.[112]

The Japanese prototype for the surviving *Gotenjiku* maps and those that have been lost (but are known to have been associated with Shōrin Temple in Kyōto) was one owned by Tō (or Kyōōgokoku) Temple in Kyōto (table 11.1). It is no longer extant, but according to the traditional account in the *Saiiki zu sofuku nikō roku* (Two revisions on a map of the western regions) of 1737, it was brought to Japan by the priest Kūkai after a trip to China, where he had been studying.[113] A version kept at Kushuon'in Temple in Hirakata, Ōsaka Prefecture, is considered the most faithful to the original. It was probably copied by the chief priest Sōkaku (1639–1720), who was also a skillful painter involved in restoring the Ryōkai

Ōsaka) before eight attempts in Nara from 747 to 749. The original, sixteen meters in height, was completed in 752, and Plutschow notes that its casting "was the last symbolic act in the centralization of the Japanese state," (104); for a discussion on the Great Buddha and Tōdai Temple, see Plutschow, *Historical Nara*, 100–116 (note 61).

110. See Nobuo Muroga and Kazutaka Unno, "The Buddhist World Map in Japan and Its Contact with European Maps," *Imago Mundi* 16 (1962): 49–69, esp. 49 and 51. For color reproductions of the map see Nanba, Muroga, and Unno, *Nihon no kochizu/Old Maps in Japan*, pl. 1 (note 11); Unno, Oda, and Muroga, *Nihon kochizu taisei*, vol. 2, pl. 1 (note 8); Cortazzi, *Isles of Gold*, pl. 11 (note 14); and José Aguilar, ed., *Historia de la Cartografia: La tierra de papel* (Buenos Aires: Editorial Codex S.A., 1967), 181. For French and English editions of the *Da Tang xiyu ji*, see Stanislas Julien, trans. *Mémoires sur les contrées occidentales*, 2 vols. (Paris, 1857–58), and Thomas Watters, *On Yuan Chwang's Travels in India*, 2 vols. (1904–5; reprinted New York: AMS, 1970).

111. The maps in the Chinese texts are simple in form, but their model must have been a larger and more detailed map. On Renchao's map, for instance, there are some blank circles and squares where place-names seem to have been written in the model. For a discussion and reproduction of both maps, see Muroga and Unno, "Buddhist World Map," 52–57 (note 110), which includes figs. 4 and 5. The *Fajie anli tu* was reprinted in 1654 (Kyōto: Akitaya Heizaemon); an edition from 1919 was reproduced in 1977 (Taipei: Xinwenfeng). The *Tushu bian* (compiled 1562–77 and printed in 1613) was reprinted in 1971 (Taipei: Chengwen Chubanshe).

112. Zhang's map is discussed in Unno Kazutaka, "Min Shin ni okeru Mateo Ritchi kei sekaizu: Shutoshite shinshiryō no kentō" (Chinese world maps of the Ming and Qing dynasties derived from the work of Matteo Ricci: An examination of new and neglected materials), in *Shinhatsugen Chūgoku kagakushi shiryō no kenkyū: Ronkō hen* (Studies on recently discovered source materials for the history of Chinese science: Collected articles), ed. Yamada Keiji (Kyōto: Research Institute for Humanistic Studies, Kyōto University, 1985), 507–80. See also above, pp. 173, 175 (fig. 7.4), and 255–56.

113. See Muroga and Unno, "Buddhist World Map," 50 (note 110). That the original map was kept at Tō Temple is also recorded on the Kōbe City Museum (formerly Akioka) map (listed in appendix 11.3).

TABLE 11.1 Genealogy of Manuscript *Gotenjiku* Maps (those in dashed lines are not extant)

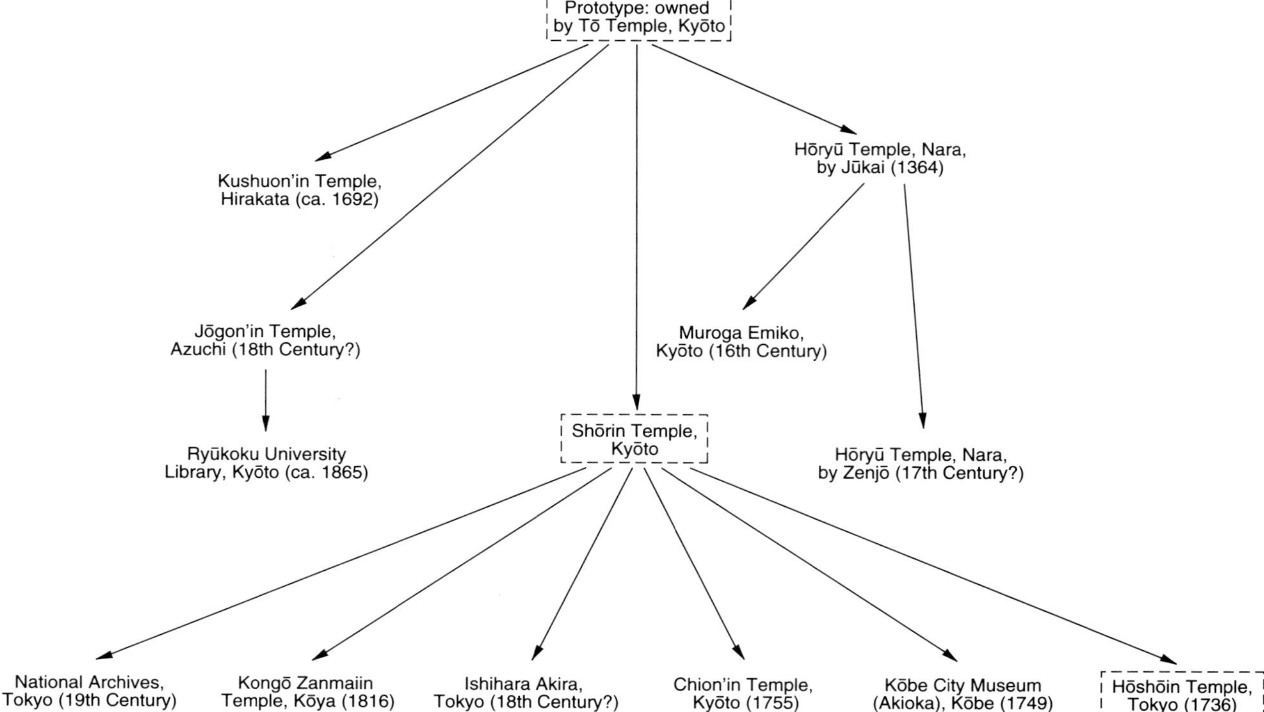

mandalas of the worlds Vajra and Garbha at Tō Temple in 1691–93 that are still in the possession of that temple. The mandalas had also been brought from China by Kūkai, and Sōkaku's work on them suggests that the copy of the map of the Five Indias dates approximately to the same time. Another copy of the Tō map, made by the monk Ekū of Shōrin Temple, came into the possession of Tsūyo, the high priest of Zōjō Temple in Edo, in 1736. This was noted in the record of the revision of it that had been kept at Hōshōin.[114] The Hōshōin, Chion'in, and Kōbe City Museum (formerly Akioka) maps record that they were derived from the Shōrin version (see appendix 11.3 and table 11.1). Extant manuscript maps of the Five Indias (including these) are notable for containing Japan but not Korea. Presumably, then, the original at Tō Temple was similar, and the evidence suggests that it was a Japanese revision rather than a Chinese original. It is doubtful that Chinese cartographers would have omitted Korea and included a large Japan. The legend of the maps' having been brought from China, therefore, may have been concocted as part of the biography of Kūkai, the founder of the temple. An alternative possibility is that it was based on a Korean map of Jambūdvīpa dating to the mid-twelfth century. An *Och'ŏnch'ukkuk to* (Map of the lands of the Five Indias) evidently based on Xuanzhuang's *Da Tang xiyu ji* was made by the scholar Yun P'o (d. 1154) and presented to the king at the time. This was probably simply a copy of a Chinese map of Jambūdvīpa with Korea added to it.[115]

Table 11.1 shows the genealogy of the *Gotenjiku* maps. The Hōryū Temple and Muroga copies share a common source in that material from the *Da Tang xiyu ji* is listed in the area of sea. This was probably the idea of Jūkai, and the sixteenth-century Muroga map was simply a copy, as was that at the temple, which is thought to date to the seventeenth century. The six maps based on the

114. This record, the *Saiiki zu sofuku nikō roku* (mentioned above), is included in the *Dainihon Bukkyō zensho* (Collected records on the Buddhism of Great Japan), 151 vols. (Tokyo, 1912–22), vol. 2 of the 4 vol. *Yūhōden sōsho* (Series of travels) (1915; reprinted Tokyo: Daiichi Shobō, 1979), 1–29. The Hōshōin map, reproduced in the frontispiece of the volume, was lost in a fire during the Second World War.

115. At the time in Korea it was common to improve the visual appeal of maps by adding Korea and Japan to old Chinese maps. Rather than having composed the map from materials in the *Da Tang xiyu ji*, as recorded in his biography, it is more likely that Yun simply attached Korea to a Chinese map of Jambūdvīpa and then colored it. See Chōsen Sōtokufu (Government-General in Korea), ed., *Chōsen kinseki sōran* (A comprehensive survey of ancient Korean inscriptions), 2 vols. (Seoul: Chōsen Sōtokufu, 1919), 1:371. Neither the original nor the copy of the *Och'ŏnch'ukkuk to* is extant.

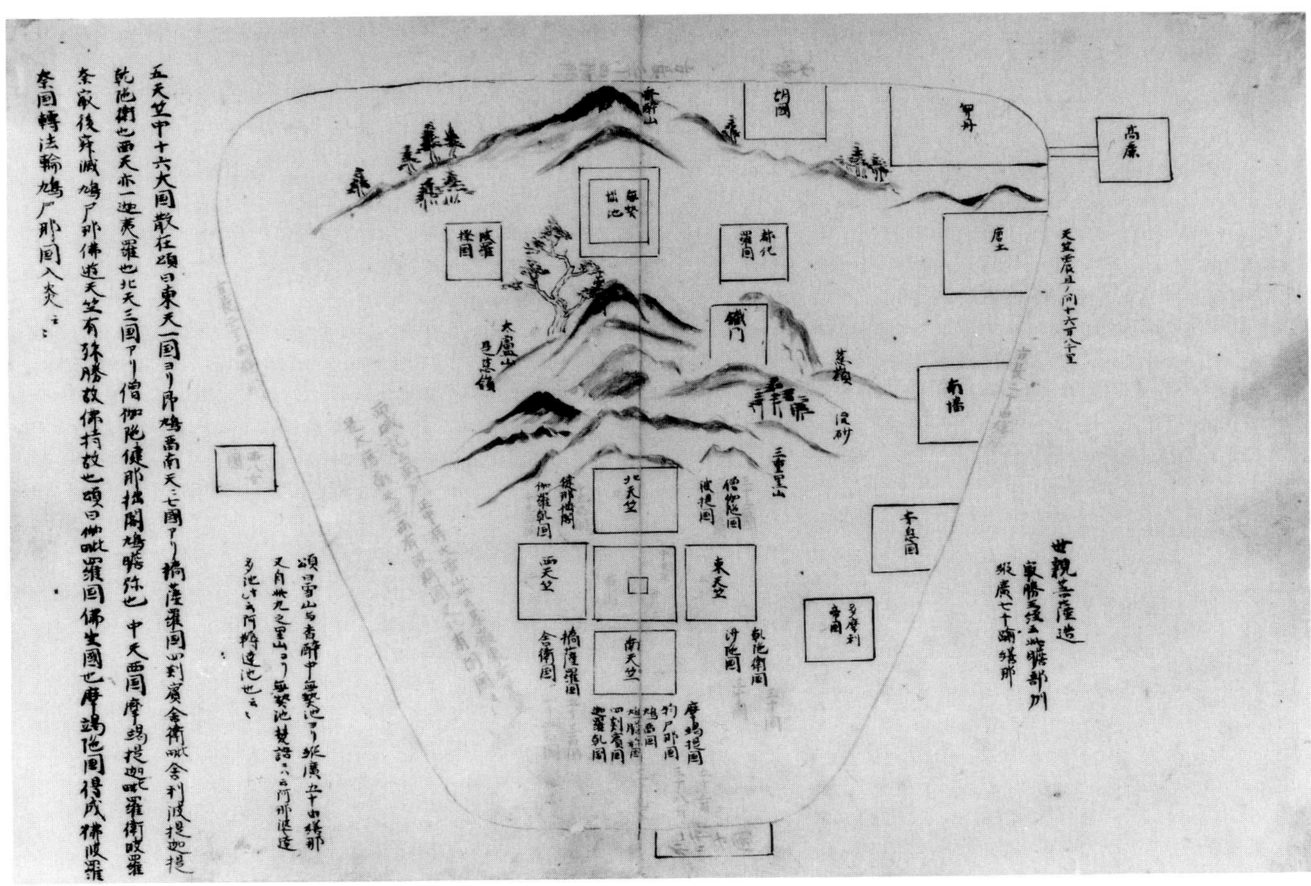

FIG. 11.20. THE "TENJIKU ZU" (MAP OF INDIA) IN THE 1548 CODEX OF THE *SHŪGAISHŌ*. Its design is simpler than that of the *Gotenjiku zu*, but it also depicts Jambūdvīpa. The Five Indias are schematized in rectangular boxes in the lower center of the manuscript.

Size of the original: 26.3 × 41.3. By permission of the Tenri Central Library, Tenri, Nara Prefecture.

Shōrin version differ from the others in that they do not contain the terms Kokoku (for Hukuo: Turkestan), Saidaijokoku (Western Women's Country in the Western Sea), and Konjikoku (Suvarna-bhūmi: Golden Country). The reason for this is that the Tō Temple original seems to have been damaged by 1736, and it was impossible to read these inscriptions when copying.[116]

Smaller and simpler maps than those listed in appendix 11.3 and table 11.1 were also made. An example is the "Tenjiku zu" (Map of India) in the *Shūgaishō*, the oldest codex with it dating to 1548 and others to 1554 and 1589 (fig. 11.20).[117] Although it is named as only a map of India, it is nonetheless similar to those of the Five Indias and may therefore be considered a Buddhist map of the world: the continent narrows toward the south, is much broader in the north, and has place-names for sites outside India proper.[118] The five Indian countries (North, East, South, West, and Central) as well as the others are disproportionately schematized using mainly rectangular frames.[119] In the case of Korea, the name is enclosed in

116. Concerning Kokoku in the other types, only the Hōryū example has both square frames and letters; that at Kushuon'in has only square frames, and neither frames nor letters exist on those at Jōgon'in and Ryūkoku University.

117. The 1554 codex is at the National Diet Library in Tokyo; it has a map of India, but its map of Japan has been lost. Two codices of the *Shūgaishō* (1548 and 1589), which are mentioned above, contain maps of both Japan and India. Other copies of the *Shūgaishō* exist, and most of them have a map of India. The first printed *Shūgaishō* dates to the Keichō period (1596–1614) but did not have a map of India. The first printed copy with a map of India was Nishimura's publication of 1642. See Muroga and Unno, "Buddhist World Map," 51–52 and fig. 2 (a reproduction of the 1554 copy) (note 110).

118. For instance, there are names for the "Western Women's Country" in a square frame in the area of sea to the west of Jambūdvīpa: on the 1554 version it is labeled Sai An Koku (Western Country Called An) and on the 1548 and 1589 codices it is Sai Hachijo Koku (Western Country of Eight Women); both erred by replacing the character 大 (*dai* = great) with 八 (*hachi* = eight). Saidaijokoku, originally referred to in chapter 11 of the *Da Tang xiyu ji*, also appears in the older maps of the Five Indias.

119. Taking into account redundancies and some mistakes, the names of the countries in India noted outside the frames are the same as those

a square frame that is connected to the continent with parallel straight lines; this was probably done to indicate that it was a peninsula. Fewer geographical mistakes were made in the 1548, 1554, and 1589 manuscripts and later printed versions than in the *Gotenjiku* maps, but nonetheless they had their share, which might be related to frequent copying of the codex. Ansoku Koku (Parthia), for example, is placed in southeastern Jambūdvīpa rather than to the west, and Harana Koku (Varanasi), which is supposed to be in the southeast, is placed in the northwest where Turkestan should be. The "Tenjiku zu" lacks rivers and, notably, the route of Xuanzhuang. This may be attributed to the geometrical schematization. The source of the four great rivers, Munetsunō Chi (Lake Anotatta), however, is shown in the north-central part of the continent. Mountainous regions in northern India and elsewhere are indicated by side views of trees to give the map a pictorial look.

Early Assimilation of European Cartography

For roughly a century, about 1543–1639, Japanese cartography was influenced by an infusion of European knowledge. Prominent in this process were European navigators—especially from Spain and Portugal—who first arrived in 1542 or 1543 and Jesuit missionaries who were active between 1549 and 1639. Their influence was felt in four important areas—marine charts, maps of Japan, maps of the world, and surveying—and their legacy can been seen in many Japanese works throughout the Edo period (1603–1867). One of the problems facing the historian, however, is the dating of works ascribed to this period of European influence: dates are lacking, so approximations must be inferred from the content of the maps.

This period of contact and cartographic assimilation was a turbulent one. Japanese cartography benefited, but we should not overestimate the contribution of Europeans, and especially of the missionaries, to Japanese society and culture as a whole. Christians played only a part in the overall events, and Japan certainly did not become Christian or Europeanized.[120] Because of the role the Jesuits played in diffusion, it is nevertheless important to understand the general picture surrounding their activity, which began in 1549 with the arrival of the Jesuit priest Francis Xavier (1506–52).[121]

At first authorities reacted to missionary work with tolerance, seeing no threat in the new religion and considering it initially as another version of Buddhism. However, one of the historical weaknesses of Christianity—its intolerance—soon made it difficult for it to coexist with Shintoism and Buddhism. Christianity became a political issue: when entire daimyates (fiefdoms) were converted, its ideology was resisted. The first expulsion of the missionaries took place in 1587, and the first martyrdoms in 1597. When the issue flared up again early in the seventeenth century, Tokugawa Ieyasu (1542–1616) promulgated his edict of 1614 to suppress Christianity.[122] A series of three edicts was also issued to diminish non-Japanese influences; they effectively banned Christianity and isolated Japan from the rest of the world. These were the Exclusion Decrees of 1633, 1635, and 1639. The first allowed only licensed Japanese ships to trade overseas; the second forbade Japanese nationals to leave or return to Japan; and the third expelled the Portuguese from Japan, essentially limiting entry to Chinese and Dutch merchants. From 1641 the Dutch were given trading quarters on the small urban island of Deshima in Nagasaki Bay. Christianity and European trade were regarded as inextricable, but since the Dutch were less involved in promoting religion, they survived.

---

of the sixteen countries of the Five Indias in the priest Gyōyo's 1446 encyclopedia *Ainōshō* (Bag of rubbish), chapter 7, paragraph 27. In it the Pamirs ("Sōrei") are marked both east (1548 and 1589 maps) and west of the central mountainous region; the eastern section and the adjacent desert ("Ryūsa") to the south were probably entered on the grounds of an entry in the *Ainōshō*: "Ryūsa and Sōrei separate India from China; to the northwest of Sōrei is Daisessen" (the Himalayas). These place-names also seem to have been entered in the process of copying this map.

120. For an in-depth survey of Christianity in Japan see Cary, *History of Christianity in Japan*, 1:13–257 (note 102). The period of Christian involvement in pre-Tokugawa Japan is discussed in most general histories. Sansom's treatise on the Western world and its interaction with Japan provides a good perspective on the subject: George B. Sansom, *The Western World and Japan: A Study in the Interaction of European and Asiatic Cultures* (New York: Alfred A. Knopf, 1962), 54–86, 105–10, 115–51, and 167–80.

121. Several studies of tangential interest have been made on the role of the missionaries in cartography pertaining to Japan: for example, Kish, "Missionary Cartography," 39–47 (note 101); Joseph F. Schütte, "Map of Japan by Father Girolamo de Angelis," *Imago Mundi* 9 (1952): 73–78; Kay Kitagawa, "The Map of Hokkaido of G. de Angelis, ca 1621," *Imago Mundi* 7 (1950): 110–14; and Chohei Kudo, "A Summary of My Studies of Girolamo de Angelis' Yezo Map," *Imago Mundi* 10 (1953): 81–86. Also relevant are studies of the Portuguese traveler Ignacio Moreira, who worked with the Jesuits without being a member of the order: Joseph F. Schütte, "Ignacio Moreira of Lisbon, Cartographer in Japan 1590–1592," *Imago Mundi* 16 (1962): 116–28, and Ryōichi Aihara, "Ignacio Moreira's Cartographical Activities in Japan (1590–2), with Special Reference to Hessel Gerritsz's Hemispheric World Map," *Memoirs of the Research Department of the Toyo Bunko* 34 (1976): 209–42.

122. This was a political decision, as the following extract from the decree indicates: "Christians have come to Japan, not only sending their merchant vessels to exchange commodities, but also longing to disseminate an evil law and to overthrow right doctrine so that they may change the government of the country and obtain possession of the land. This is the germ of great disaster and must be crushed." Cited in Cary, *History of Christianity in Japan*, 1:176–77 (note 102). The decree dates to 27 January 1614.

A number of cartographic events are usually given prominence in historical accounts describing the encounter between Japan and Europe. One such event was the introduction of the theory of a spherical earth, credited to Xavier, who resided in Japan from 1549 to 1551. In letters sent to European Jesuits from Cochin and to Father Ignatio de Loyola in Rome from Goa in 1552, Xavier noted that European astronomy and meteorology were known in Japan. It is clear from these letters that Xavier had explained the theory of a spherical earth, but it is not specified if he carried with him a globe or even a map of the world.[123] Other records suggest that the first European globes and maps of the world had appeared in Japan by 1580. In that year, according to European sources, Oda Nobunaga brought a terrestrial globe to a meeting with the Jesuits Genecchi Soldo (Soldi) Organtino (1533–1609) and Lourenço (1526–92), a Japanese convert and catechist. Oda questioned them about it as well as about Organtino's route from Europe to Japan.[124] In 1581 Oda also used a map of the world to question another Jesuit, Alessandro Valignani (1539–1606), about a route from Europe.[125] The next year Valignani directed an envoy of four young nobles representing the three Christian Kyūshū feudal lords to various places in Europe, including Rome. In Padua in 1585 they received from the German botanist Melchior Guilandini (1520–89) a copy of Ortelius's *Theatrum orbis terrarum* and the first three volumes of Georg Braun and Frans Hogenberg's *Civitates orbis terrarum* (1572, 1575, 1581). These were included in the objects they brought to Nagasaki on their return in 1590; others were maps, sea charts, an astrolabe, and a terrestrial globe.[126] We shall now see how these initial contacts led to several distinct traditions in Japanese cartography.

NANBAN WORLD MAPS

The Europeans who sailed to Japan in the sixteenth century were called *nanbanjin* (southern barbarians).[127] The term was applied mainly to the Portuguese and Spaniards who had arrived in Japan from a southerly direction. After 1639 they were prohibited from entering the country, and only the Dutch were allowed to stay, having been transferred in 1641 to Deshima, an artificial islet linked by a bridge to Nagasaki proper.[128] The term nonetheless remained in use, and among the world maps made in Japan after European models, those that appear to have been made from the late sixteenth century to approximately 1639 collectively are termed the "*Nanban* group." The maps not only are defined by their date of production, but also share similar elements in design and style, and therefore some maps produced after 1639 also fall into this classification. Examples include the three world maps listed in appendix 11.4 (equirectangular projection, type C), which presumably were made in the second half of the seventeenth century.[129]

Over thirty world maps of the *Nanban* tradition are known to exist; some of these are later copies. Appendix 11.4 classifies twenty-eight of them according to whether they are marine charts or made on an oval, equirectangular, or Mercator projection. An interesting feature of those listed as charts and those designated as equirectangular type B maps is their attempt to place Japan near the center of the world, putting the Eastern Hemisphere

123. For the letters, see Georg Schurhammer and J. Wicki, eds., *Epistolae S. Francisci Xaverii aliaque eius scripta*, 2 vols. (Rome, 1944–45), EP. 96, 110. Also see Unno Kazutaka, "Seiyō chikyūsetsu no denrai" (Introduction of the global theory to Japan), *Shizen* 34 (1979): no. 3, pp. 60–67, and no. 6, pp. 62–69.

124. Luís Fróis (d. 1597), *Historia de Japam*, pt. 2, chap. 26 (Lisbon, Arquivo Historico Ultramarino, cod. 1659); see Matsuda Kiichi and Kawasaki Momota, trans., *Furoisu Nihonshi* (History of Japan by Fróis), 15 vols. (Tokyo: Chūō Kōronsha, 1977–80), 5:29–30. Sansom considers Fróis' work (under the title *Historia do Japão*) "the best single source for an account of the Jesuit propaganda in Japan in the second half of the sixteenth century"; the period covered is 1549–78. See Sansom, *Western World and Japan*, 115 (note 120). Lourenço was a nearly blind Japanese who received this name upon baptism by Xavier in 1551; he became a lay brother of the Society in 1563 and was active in converting Japanese to Christianity. See, for example, Sansom, *Western World and Japan*, 120, and Cary, *History of Christianity in Japan*, 1:47 (note 102).

125. Alessandro Valignani, *Sumario de las cosas de Japón*, ed. José Luis Alvarez-Taladriz (Tokyo: Sophia University, 1954), 150–51. See also Matsuda Kiichi, "Nihon junsatsushi Varinyāno no shōgai" (Life of Valignani, visitor to Japan), in *Nihon junsatsu ki Varinyāno* (Valignani's summary of things Japanese), trans. Matsuda Kiichi and Sakuma Tadashi (Tokyo: Tōgensha, 1965), 100.

126. Luís Fróis, *Historia de Japam*, pt. 3, chap. 13 (Lisbon, Biblioteca da Ajuda. cod. 49-IV-57), see Matsuda and Kawasaki, *Furoisu Nihonshi*, 2:66 [note 124]), and Eduardo de Sande, *De missione legatorum Iaponensium* . . . (Macao, 1590; reprinted 1935). There is a Japanese translation of the latter: Izui Hisanosuke et al., trans., *De Sande Tenshō Ken'ō shisetsu ki* (Record of the mission to Europe in the Tenshō era [1573–91] by de Sande) (Tokyo: Yūshōdo Shoten, 1969), 548 (for the reference to Ortelius's atlas and three volumes with printed illustrations of famous cities in the world).

127. *Nan* means "south"; *ban*, "rude"; *jin*, "people."

128. There was a special term to distinguish the Dutch—*kōmōjin*, or "redheaded people." The Chinese characters for *kōmō* might also be read *Oranda* (Holland). For a discussion of the Dutch at Nagasaki, see Herbert E. Plutschow, *Historical Nagasaki* (Tokyo: Japan Times, 1983), 45–71.

129. The map in Nagahama (ca. 1652) is reproduced in color in Unno, Oda, and Muroga, *Nihon kochizu taisei*, vol. 2, pl. 38 (note 8); the map in Odawara (ca. 1652) is reproduced in Nakamura Hiroshi, "Nanban byōbu sekaizu no kenkyū" (Research on the world map on *Nanban* folding screens), *Kirishitan Kenkyū* 9 (1964): 1–273, esp. pl. 6; and the map in Nikkō (probably late seventeenth century) is reproduced in color in *Bessatsu Taiyō* (The sun, special issue), no. 8 (Tokyo: Heibonsha, 1974), folded between pp. 56 and 57. The maps of Japan paired with the three world maps are modified from the square Keichō map to fit into the rectangular space on the screens. For the Keichō map of Japan, see below.

FIG. 11.21. ANONYMOUS *NANBAN*-STYLE WORLD MAP. Important features include graduations of latitude on both sides of the map, a bar scale in the lower center, and no place-names. Areas of sea and rivers are colored navy blue, and some islands are red and green; generally, though, the map lacks color. Gold leaf is pasted all over the manuscript.
Size of the original: 154 × 352 cm, on a six-fold screen. By permission of Hosshin Temple, Obama, Fukui Prefecture.

FIG. 11.22. *TYPUS ORBIS TERRARUM*: A *NANBAN*-STYLE MAP OF THE WORLD, CA. 1625. Petrus Plancius's world map of 1592 was the source for this equirectangular projection on a six-fold screen, but the manuscript's title and illustrations come from elsewhere. As on four other known *Nanban* world maps on this type of projection, the Pacific Ocean is in the center: this gives a better geographical perspective from the point of view of Japan, the Americas being shown to the right and not as distant as on maps with the Atlantic in the center.
Size of the original: 156 × 316 cm. By permission of the Tokyo National Museum, Tokyo.

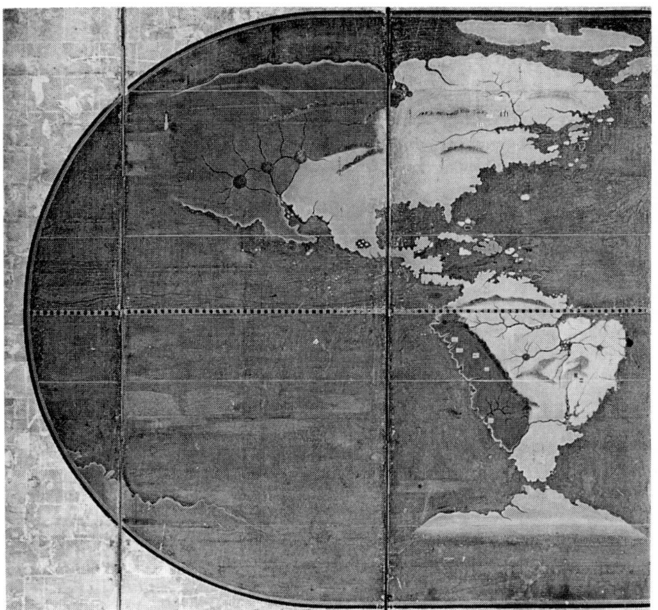

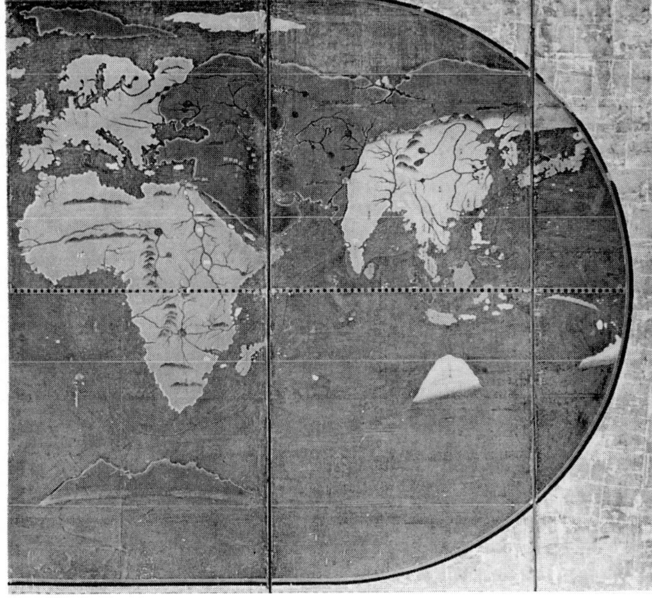

FIG. 11.23. A *NANBAN*-STYLE MAP OF THE WORLD ON AN OVAL PROJECTION, CA. 1595. *Nanban* maps were based on European sources, but the exact source for this one is not yet known. An improvement on European knowledge at the time, however, can be seen here in the image of East Asia. Other content—for example, the rivers and the islands in the Arctic Ocean—appears not to be different from that of contemporary or earlier European maps. Characteristic of this particular type of *Nanban* world map are the Atlantic Ocean in the center and the sea routes from the Iberian Peninsula to East Asia. The manuscript is on a six-fold screen and is paired with the map of Japan shown in figure 11.26.
Size of the original: 148.5 × 364 cm. By permission of Jōtoku Temple, Fukui.

to the left and the Western Hemisphere to the right (figs. 11.21 and 11.22). Most of the *Nanban* maps were executed on large folding screens that served as room dividers or decoration; the colorful embellishment on the maps and the fact that some of them do not have place-names suggest they were largely ornamental in function. Seventeen of the world maps on folding screens listed in appendix 11.4 and the Kawamori map of the Eastern Hemisphere are paired (originally made as a set) with companion illustrations on folding screens; fourteen of the companion illustrations are maps of Japan, suggesting that the Japanese at this date were keenly aware of their country as a part of a larger world.[130] Drawing maps on folding screens is in itself an unusual application of cartography, but it points to the value of maps as visual images rather than as vehicles for disseminating information.

Knowledge about the *Nanban* world maps is incomplete, however, especially in regard to important details such as dates, authorship, stylistic classifications, and their relation to European models. Some clues are nonetheless available. One in the possession of Yamamoto Hisashi and probably the earliest of the *Nanban* maps, for example, contains the tribal name Orankai. This name was first known in Japan in 1592, when Japanese forces under Katō Kiyomasa (1562–1611) sent back information on their invasion of the northeastern region beyond Korea, and therefore dates the map to 1592 at the earliest.[131] Tracing the source materials for the *Nanban* world maps is not easy, but some generalizations can be inferred. In the case of the Yamamoto, Kobayashi, Jōtoku, and Kawamura maps,[132] for instance, it appears that a Por-

130. Japanese interest in Europe continued after the edict to suppress Christianity and the Exclusion Decrees. European-style paintings and maps continued to be produced, with the support of the shogunate, on the grounds of their importance. World maps, for instance, were exempt from a 1668 law prohibiting the import of luxuries; they were considered useful. Other kinds of maps, however, were included in a list of about eighty prohibited items. See the *Nagasaki ki* (Records on Nagasaki) and *Nagasaki oboegaki* (Memorandum of Nagasaki), quoted in Kimiya Yasuhiko, *Nikka bunka Kōryūshi* (History of cultural intercourse between Japan and China) (Tokyo: Fuzanbō, 1955), 690–91. There are no records of the maps on folding screens from the early Edo period.

131. The map is reproduced in color in Okamoto Yoshitomo, *Jūroku seiki ni okeru Nihon chizu no hattatsu* (Development of the map of Japan in the sixteenth century) (Tokyo: Yagi Shoten, 1973), frontispiece 5, and in Unno, Oda, and Muroga, *Nihon kochizu taisei*, vol. 2, pl. 32 (note 8). For the invasion of Korea, refer to Sansom, *History of Japan*, 2:352–62 (note 32).

132. The maps are listed in appendix 11.4 (oval projection). The Kobayashi map is reproduced in color in Okamoto, *Jūroku seiki ni okeru Nihon chizu no hattatsu*, frontispiece 2 (note 131), and in Unno, Oda, and Muroga, *Nihon kochizu taisei*, vol. 2, pl. 33 (note 8). For the Jōtoku map, see figure 11.23. The Kawamura map is reproduced

tuguese-owned original was involved: each of these shows courses from Portugal and Spain to East Asia (fig. 11.23). Judging from the solitary protrusion along the western coast of South America just south of the equator and the fact that from that point southward the coastline was straight in a southeasterly direction, the prototype was probably Ortelius's map of the world of 1587 or a later revision.[133]

Of the maps in appendix 11.4 on an equirectangular projection, those of types A, B1, and B2 can be traced in general to the world map of 1592 of Petrus Plancius (1552–1622).[134] Those of type C were probably derived from the revision of Plancius's map made about 1598 by Hendrik Floris van Langren (ca. 1573–1648);[135] those of type D1 showing only the Eastern Hemisphere appear to have used van Langren's map as a model;[136] and those of type D2 are revisions of the D1 maps.[137] The maps on the Mercator projection were probably derived from the world map of 1609 of Pieter van den Keere (1571–ca. 1646) and other European sources (plate 23).[138] Judging from the characteristics of the other *Nanban* maps listed in appendix 11.4, it is likely that the two charts were derived from a European chart of the world.[139] In addition to those mentioned, other types of European world maps are thought to have been brought to Japan by the Jesuits, and these would have also been used in compilation.[140] Finally, although European sources were certainly important, Japanese information was also worked into the *Nanban* maps, with the result that East Asia was shown more accurately.

---

in color in Okamoto, frontispiece 4, and in Unno, Oda, and Muroga, vol. 1, pl. 15.

133. For editions of Ortelius's world maps, see Robert W. Karrow, Jr., *Mapmakers of the Sixteenth Century and Their Maps: Bio-bibliographies of the Cartographers of Abraham Ortelius, 1570* (Chicago: Speculum Orbis Press, 1993), 1–31. Ortelius's world maps of 1570, 1586, and 1587 are reproduced in Rodney W. Shirley, *The Mapping of the World: Early Printed World Maps, 1472–1700* (London: Holland Press, 1983), pls. 104, 8, and 130.

134. See Akioka Takejirō, "Momoyama jidai Edo jidai shoki no sekaizu byōbu tō no gaihō" (Outline of the world maps on folding screens of the Momoyama [ca. 1583–ca. 1602] and early Edo periods), *Hōsei Daigaku Bungakubu Kiyō* 4 (1958): 263–311; Tokita Tadamasa, "Nanban sekaizu byōbu genzu kō" (On the originals of the world maps on folding screens, 2), *Nagasaki Dansō* 57 (1975): 32–61. The Mody map and the Jingū Library map are discussed and reproduced in Unno Kazutaka, "Jingū Bunko shozō no Nanban kei sekaizu to nan'yō karuta" (A *Nanban* map of the world and a Japanese marine chart of Southeast and East Asia in the Jingū Library collection), *Nihon Yōgakushi no Kenkyū* 9 (1989): 9–36. Maps at the Tokyo National Museum, Nanban Culture Hall, and the University of California–Berkeley are reproduced in Unno, Oda, and Muroga, *Nihon kochizu taisei*, vol. 1, pl. 10, vol. 2, pl. 40, and vol. 2, fig. 37 (note 8), and Cortazzi reproduces the Nanban Culture Hall map in *Isles of Gold*, pl. 33 (note 14). The Fukushima map is reproduced in *Kokushi daijiten*, vol. 8, color pages, "Sekaizu" (maps of the world), pl. 3 (note 95), and the Nanba map is reproduced in Nanba, Muroga, and Unno, *Nihon no kochizu/Old Maps in Japan*, pl. 5 (note 11), and in Unno, Oda, and Muroga, *Nihon kochizu taisei*, vol. 2, pl. 39. For a reproduction of Plancius's map, see Frederik Caspar Wieder, *Monumenta cartographica*, 5 vols. (The Hague: Nijhoff, 1925–33), vol. 2, pls. 26–38. Other sources were also used since these maps include illustrations and, in the case of those in Tokyo and Ōsaka, the title *Typus orbis terrarum*, which were not taken from Plancius's map.

135. See Tokita, "Nanban sekaizu byōbu genzu kō," 2 (note 134). Some parts of Plancius's map revised by van Langren are reproduced in Wieder, *Monumenta cartographica*, vol. 2, pls. 39 and 40 (note 134). For each map included in type C, see note 129 above.

136. The Kawamori map is owned by Kawamori Kōji and kept at the Sakai City Museum. This map and the Myōkaku Temple map are reproduced in Unno, Oda, and Muroga, *Nihon kochizu taisei*, vol. 2, fig. 33 and pl. 35, respectively (note 8). The Usuki City Library map is in Joseph F. Schütte, ed., *Monumenta historica Japoniae* (Rome, 1975–), vol. 1, pl. 2 (facing p. 16), the Saga Prefectural Library map is in Unno Kazutaka, *Chizu no shiwa* (Map creases; or, Essays on the history of cartography) (Tokyo: Yūshōdō Press, 1985), fig. 27, and the Sōji Temple map is in Nakamura, "Nanban byōbu sekaizu no kenkyū," pl. 9 (*Nansenbu Sekaizu* [Map of the Jambūdvīpa world]) (note 129).

137. The Koga City Museum of History map copied in 1836, probably by Takami Senseki (1785–1858), from a copy of 1691 is reproduced in Nakamura, "Nanban byōbu sekaizu no kenkyū," pl. 10 (note 129), and in Unno, Oda, and Muroga, *Nihon kochizu taisei*, vol. 2, fig. 34 (note 8). The Yamakuni Shrine map is a copy dating from 1685; the map at Yokohama City University Library is titled *Yochizu* (Map of the earth), and the one at Yamaguchi University Library is titled *Bankoku sōzu* (Map of all the countries).

138. For the Imperial Household Agency map, see plate 23; the Kōbe City Museum map is reproduced in color in Nanba, Muroga, and Unno, *Nihon no kochizu/Old Maps of Japan*, pl. 3 (note 11), and the Kōsetsu Museum of Art map is reproduced in color in Akiyama Terukazu, ed., *Genshoku Nihon no bijutsu* (The fine arts of Japan in color), 30 vols. (Tokyo: Shōgakkan, 1966–72), vol. 25, *Nanban bijutsu to Yōfūga* (Nanban art and Western-style painting), pl. 5, and in Unno, Oda, and Muroga, *Nihon kochizu taisei*, vol. 2, pl. 42 (note 8). There are also color plates of the *Nanban* maps in the Yamamoto Hisashi Collection, Jōtoku Temple, Nanban Culture Hall, Shimonogō Kyōsai Library, Imperial Household Agency, and Kōsetsu Museum of Art in the frontispieces of *Tanbō daikōkai jidai no Nippon* (Japan in the age of great navigation: The inquiries), 8 vols. (Tokyo: Shōgakkan, 1978–79), vol. 5, *Nippon kara mita ikoku* (Foreign countries interpreted by the Japanese).

On van den Keere's map and its connection to Willem Janszoon Blaeu's map see Günter Schilder, "Willem Jansz. Blaeu's Wall Map of the World, on Mercator's Projection, 1606–07, and Its Influence," *Imago Mundi* 31 (1979): 36–54; idem, *Three World Maps by François van den Hoeye of 1661, Willem Janszoon (Blaeu) of 1607, Claes Janszoon Visscher of 1650* (Amsterdam: Nico Israel, 1981); and Takahashi Tadashi, "Nanban toshizu byōbu kara Kaeriusu sekaizu e" (From maps of cities on the Nanban folding screens to Kaerius's map of the world), in *Ezu no kosumorojii* (Cosmology of picture maps), ed. Katsuragawa Ezu Kenkyūkai (Katsuragawa Picture Map Research Society), vol. 1 (Kyōto: Chijin Shobō, 1988), 248–64.

139. The Hosshin Temple map and the map in the Ikenaga Hajime Collection are reproduced in Unno, Oda, and Muroga, *Nihon kochizu taisei*, vol. 2, pl. 34 and fig. 31, respectively (note 8).

140. The Jesuit Matteo Ricci, who made world maps in China at that time, also used some European maps as sources. For more on Ricci, see above, pp. 170–77, and below, pp. 404–10.

MARINE CHARTS

Marine charts were introduced by European pilots who sailed to Japan. Although the Japanese understood the differences between them and other maps, they borrowed the Portuguese word *carta* (map) to derive their own word *karuta* to designate this group of charts. This term was used in two charts of Southeast and East Asia (charts 8 and 10 of appendix 11.5) and in one chart of Japan (chart 4 of appendix 11.6). In the anonymous *Anjin no hō* (Western techniques of navigation), a collection of talks by the Nagasaki pilot Shimaya Ichizaemon Sadashige dating to 1670 and the first work to explain charts in Japan, they are referred to as *bankoku no zu* (maps of all the countries), perhaps because Shimaya might have been discussing charts showing various countries of the world.[141] Nishikawa Joken, whose *Ryōgi shūsetsu* (Collected theories of heaven and earth, 1714) contains the first systematic explanation of marine charts, translated the word *karuta* into *shinro hanzu* (card chart of courses), a term that seems to have been used only by him.[142]

Japanese charts can be divided into two groups, those of Southeast and East Asia and those of Japan alone. Compiling them was associated originally with European techniques of navigation, but exactly when they were introduced is not known. This question is likely to remain unsolved because of their low rate of survival and because they were used only by a small group of navigators, notably those who sailed the officially licensed ships for trade in the Far East between 1592 and 1636.[143] Nakamura, for instance, notes the paucity of materials, and of the "homemade" charts of Japan in particular he observes that it is astonishing that they have survived the vicissitudes of time.[144] Despite the imperfect record, we can presume that some Japanese were likely to have been exposed to charts as early as 1542 or 1543 when the Portuguese landed at Tanega Island. Our firmest evidence for the transmission, however, is found in a work on the principles of European navigation dated 1618, by Ikeda Kōun (fl. ca. 1618–36). In it he states that he learned navigation in 1616 from a European named Manoeru Gonsaru—that is, the Portuguese captain Manuel Gonzalez, who traded between Luzon and Japan at the beginning of the seventeenth century.[145] It is most probable that marine charts were included in his studies.

Charts of Southeast and East Asia

Portuguese maritime cartography was the dominant influence on Japanese marine charts. There are no surviving charts compiled in Dutch, or any that suggest a dominant Dutch influence. Possibly there was no transmission involving Dutch charts because of their policies of secrecy, but there is the possibility that any Dutch information that had been acquired might have been worked into revisions of an earlier Portuguese original. Although we can point to Dutch influence in other aspects of the sciences—especially in the eighteenth and nineteenth centuries (including a naval school at Nagasaki where the Dutch taught navigation toward the end of the Edo period)—it is important to note that during the period of

141. The *Anjin no hō* is now owned by the National Archives in Tokyo. Because of its nature, the explanations in it tend to be fragmentary. The charts are counted among the six instruments to be used at sea, the other five being the astrolabe (*isutarahi*), quadrant (*watarante*), nocturlabe (*hokkyokuzuban*), compass (*konhatsu*), and large compass (*ōkishaku* or *ōgishaku*).

142. *Ryōgi shūsetsu*, chap. 7, leaves 46–48 (following the original manuscript owned by the National Archives in Tokyo); this may be found in Nishikawa Tadasuke, ed., *Nishikawa Joken isho* (Preserved works of Nishikawa Joken), 18 vols. (Tokyo: Kyūrindō, 1898–1907), vol. 18. Earlier, in his *Kai tsūshō kō* (Trade with China and other countries, 1695) and its enlargement, *Zōho kai tsūshō kō* (Enlarged edition of *Kai tsūshō kō*, 1708), however, Nishikawa used the term *karuta*. The *Zōho kai tsūshō kō* was reprinted by Iwanami Bunko, 3384–85 (Tokyo: Iwanami Shoten, 1944), and in Ono Tadashige, ed., *Bankoku tokai nendaiki* (Chronicle of Japanese intercourse with all the countries) (Tokyo: Shōrinsha, 1942).

143. Ships licensed by the shogunate were known as *goshuinsen* (trading ships authorized by the government), and the license was called a *goshuinjō* (license bearing the "august vermilion seal"). There were evidently two reasons for this licensing, one being the protection of foreign trade and the other an effort by Toyotomi Hideyoshi (1536–98) to suppress Japanese piracy. For Japanese trade and the vermilion-seal ships, see Iwao Seiichi, *Shinpan shuinsen bōeki shi no kenkyū* (Studies on the history of trade under the vermilion-seal licenses of the Tokugawa shogunate, revised and enlarged edition) (Tokyo: Yoshikawa Kōbunkan, 1985). See also Hiroshi Nakamura, "The Japanese Portolanos of Portuguese Origin in the XVIth and XVIIth Centuries," *Imago Mundi* 18 (1964): 24–44, esp. 24–26. Tangentially, Japanese had learned from Chinese pilots since antiquity, but our knowledge in this area is not yet clear. Shipbuilding certainly was learned from the Chinese, and the *goshuinsen* for the most part were of Chinese design.

144. See Nakamura, "Japanese Portolanos," 26–27 and 35 (note 143).

145. The manuscript of Ikeda's book is at the Kyōto University Library. Although it is untitled, it is called *Genna kōkai ki* or *Genna kōkai sho* (Book of the art of navigation in the Genna era [1615–23]); reproductions are in the *Kaihyō sōsho* (Series of literature on lands overseas), 6 vols., comp. Shinmura Izuru (Kyōto: Kōseikaku, 1927–28), vol. 3; *Kaiji shiryō sōsho* (Series of materials on maritime history), 20 vols. (Tokyo: Ganshōdō Shōten, 1929–31), vol. 5; and Saigusa Hiroto, ed., *Nihon kagaku koten zensho* (Series of Japanese scientific classics) (Tokyo: Asahi Shinbun Sha, 1942–49; reprinted 1978), vol. 12. Ikeda is identified at the end of his preface as "Ikeda Yoemon nyūdō Kōun": Yoemon was his common name, Kōun was his Buddhist name, and *nyūdō* means lay priest. In the preface he says that in 1616 he was taught navigation by a European named Manoeru Gonsaru and that he sailed to Luzon with him for two years. Manuel Gonzalez is mentioned in Léon Pagès, *Histoire de la religion chrétienne au Japon, depuis 1598 jusqu'à 1651*, 2 vols. (Paris: C. Douniol, 1869–70), 1:389. It is possible that Gonzalez was a Spanish citizen, since Luzon was then a Spanish possession; the principal foreign language used in Ikeda's writings, however, is Portuguese.

Dutch monopoly on European trade with Japan (1639–1853) there was no particular Japanese need for practical charts. The Portuguese had simply arrived earlier, and what knowledge Japanese pilots had managed to acquire from them was evidently sufficient. The circumstances of contact were also different, notably in that there was no opportunity for Dutch and Japanese pilots to work together during the Age of Isolation. Interpreters at Nagasaki, furthermore, had seen some charts in the Dutch Office on Deshima and in Dutch ships, but there was no interest in copying them or bringing them into Nagasaki proper. During this time, as I will note below, marine charts were presented as graduation certificates for land surveyors; for this purpose it was not necessary to improve the content.

By comparing the contents and features—overall form, place-names, the shape of the compass roses, and the placement and embellishment of the bar scales—of the charts in appendix 11.5, it can be established that all but one (no. 16) were derived from the same Portuguese prototype, and the outstanding example seems to have been modeled on another Portuguese chart. Examples include the Portuguese place-names in charts 2, 3, 5, and 16 in appendix 11.5 and inscriptions such as "Sebastião, Afez" in chart 3;[146] there are also flags with five dots and with a cross to show, respectively, Portuguese territory and places with Christians. A common feature of the charts is that the scales—in the fashion of the Portuguese prototypes—are sometimes given in values equivalent to Spanish miles, although the units are specified as Portuguese nautical leagues (*léguas*). This suggests that copying was a mechanical process for the Japanese chartmaker, who added the scales without fully appreciating the value of the units. Some charts without scales also survive, and this suggests similar copying from secondary sources (see nos. 9, 10, 12, and 14 in appendix 11.5).

We should not be misled by such technical characteristics to think in terms of a standard Japanese chart. In geographical area, for instance, there is variation. The charts depicting the largest geographical areas (nos. 1 and 16) cover the whole of the area from Africa to Hokkaidō, with the center of the charts lying off Sri Lanka and India. Other charts are centered on the west coast of the Malay Peninsula (nos. 2, 3, and 4), and others on the west coast of Luzon (nos. 7, 8, 10, 11, 12, and 13), suggesting that they may have been abstracted from a more extensive map. Yet other charts are truncated westward at Sri Lanka (no. 5) or in the western part of the Malay Peninsula (no. 6). Areas to the west, of course, were not so important because Japanese traders did not go that far. The image of Japan itself and the adjacent coastlines also shows variation. Some charts, for instance, show a strip of the coast of mainland Asia along the northern part of the Sea of Japan (nos. 1, 4, 8, 9, 10, 11, 12, 13, 14, and 15), and in another Hokkaidō is depicted as part of the mainland (no. 7). Even the configuration of the Japanese archipelago varies; of note are charts 9 and 12, which show Japan much as it appears on the Shōhō map of Japan of about 1670, a remarkably accurate representation (see below, pp. 399 and 400 [fig. 11.34]).

The physical characteristics of the charts also vary. One, for instance, is attached to a thick wooden spindle and rolled like a scroll (no. 2), another (no. 7) is drawn on Japanese paper and stuck on the inner sides of two folding pine boards (rather like some European charts prepared for shipboard use). On at least one chart, too, we know that a waterproof lacquer had been used (no. 7). The lacquer is still transparent, so the original colors are visible. That the use of lacquer was associated with such charts is suggested in an order by the Mito clan that was commissioned in Nagasaki in 1671: forty-three *monme* (a *monme* being a unit of silver equal to 3.75 grams [.12 troy ounce]) for a chart of Southeast Asia, five *monme* for folding boards to stick the chart to, and three *monme* for lacquering it. The supplier was a certain Shimaya Ichizaemon, who had learned the arts of European navigation in Nagasaki.[147] Affixing a chart drawn on paper to a heavy pine board and then waterproofing it with lacquer was designed for convenience at sea. The hinges of the board were made of thick leather, which would last through several voyages. In the case of this chart, time has left these straps tattered.

Finally, there was a diverse group of original owners or patrons for the charts that have survived. Thus, one chart was in the possession of the Ikeda family, the leaders of the Okayama clan (no. 2);[148] an 1833 chart (no. 3)

146. The inscription means "Sebastião made this," and it is not known to whom it refers. Such an inscription is on Domingos Sanches's chart of ca. 1618: "Dominguos Sanches a fes em Lisboa anno 1618"; see Michel Mollat du Jourdin and Monique de La Roncière, *Les portulans: Cartes marines du XIIIe au XVIIe siècle* (Fribourg: Office du Livre, 1984); English edition, *Sea Charts of the Early Explorers: 13th to 17th Century*, trans. L. le R. Dethan (New York: Thames and Hudson, 1984), pl. 73 and pp. 250–51.

147. The Mito clan was a branch of the Tokugawa family based in Mito, Hitachi Province (now Ibaraki Prefecture) from 1609 until the Meiji Restoration. The head of the clan at the time in question was Tokugawa Mitsukuni (1628–1700), a patron of letters and history. From 1657 he sponsored the compilation of the *Dainihon shi* (History of Great Japan [243 vols.]), which was completed posthumously in 1906. Papinot notes that it raised "the prestige of the imperial dynasty," caused "the Tokugawa[s] to be regarded as usurpers" (p. 68), and was regarded as "the best authority on historical matters" (p. 681) at the time of Papinot's work toward the end of the Meiji era (1868–1912). Mitsukuni appears to have also taken an interest in science. See Papinot, *Dictionary of Japan*, 68 and 680–81 (note 43). For the order and the bond payable by the clan, dating to the twenty-seventh day of the sixth month of Kanbun 11 (1671), see Adachi Hiroyuki, "Kaifū sen shōkai kiji" (Navigational record of the ship *Kaifū*), *Kaiji Shi Kenkyū* 14 (1970): 120–28.

148. The Okayama in Bizen Province (now Okayama Prefecture), which was under the control of the Ikedas from 1603 to 1868. Noted

was a copy of a seventeenth-century one that belonged to Itoya Zuiemon (d. 1650) and was used for trading voyages in Southeast Asia;[149] and another chart is preserved by the descendants of the Ōsaka merchant Sueyoshi Magozaemon (1570–1617), whose representative is believed to have used it on annual voyages to Luzon or Annam (Vietnam) in the first half of the seventeenth century (no. 4). Yet another chart (no. 6) had been preserved by the house of Kadoya of Matsusaka, Mie Prefecture (fig. 11.24), and is believed to have first been the property of Kadoya Shichirobee (1610–72),[150] a merchant who lived in Kōchi (modern Hoi An). Since he had had many letters and other items sent to his brothers in Matsusaka after correspondence was allowed between Japanese inside and outside Japan from about 1660, there is a good chance that this chart was among them.[151] An interesting feature of the chart is the existence of pinholes to trace a sea route from Nagasaki to Kōchi via the Strait of Formosa. This route is divided into two lines, indicating a round-trip voyage.

Not all charts were prepared for use at sea. One is included in a report by the Nagasaki magistrate after a ship from the Bataan Islands had drifted to Japan in 1680 (no. 14). In this case, perhaps because of its nonmaritime function, there is no indication of latitude, and the distortion caused by the square shape of the chart makes the Malay Peninsula and the Sunda Islands appear very small. Another example (no. 11), probably used in legal proceedings, was a chart copied by the interpreter Ro Kōrō (1847–1923?) about 1865 in Nagasaki. The original may be one kept at the Nagasaki magistracy.[152] Another reason for compiling charts was to demonstrate mastery of drafting skills, probably the origin of the chart at the Jingū Library (no. 13). Its content is similar to that of two others (nos. 7 and 8), and it was meticulously drafted.

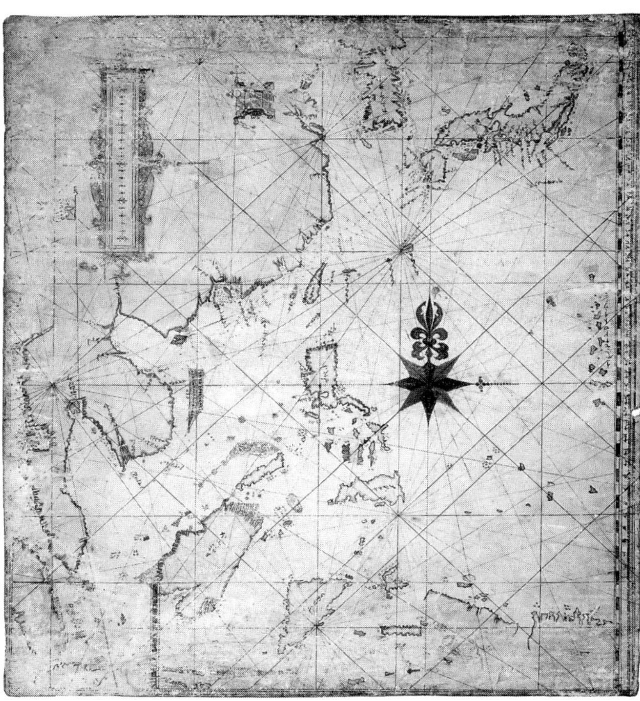

FIG. 11.24. A MARINE CHART OF SOUTHEAST AND EAST ASIA, CA. 1630. This chart, which formerly belonged to the Kadoya family, is an example of Japanese marine charts that do not show areas west of the Malay Peninsula that would have been on the prototype. This is suggested by the center of the rhumb line network, which is at the left rather than in the center of the chart as usual. Two lines of pinholes between Nagasaki and "Kōchi" (Hoi An in today's Vietnam) are evidence that the chart was used for sailing.
Size of the original: 44 × 38.8 cm, on vellum. By permission of the Jingū Historical Museum, Ise.

Charts of the Japanese Archipelago

Besides these sixteen charts of Southeast and East Asia, another group of charts relate only to the Japanese archipelago and were produced in Japan. Eight known examples are listed in appendix 11.6. For all of them the authors, dates, and circumstances of origin are uncertain. The only clue is the order for a chart, folding boards, and lacquering commissioned by the Mito clan in 1671, as noted above, in which is also written "thirty *monme* [for a] *Nihon karuta* [chart of Japan]."[153] From this entry it is possible to say that Shimaya was compiling charts of Japan by 1671. Since they had to be ordered from Nagasaki, as did those of Southeast and East Asia, it appears that chart-making skills were limited to relatively few individuals.

In contrast to the charts of Southeast and East Asia,

in Papinot, *Dictionary of Japan*, 199–200 and 479 (note 43).

149. Not much is known about Itoya except that he came originally from Kyōto, settled in Nagasaki because of the foreign trade, made twenty-four journeys overseas between 1601 and 1632, and died in 1650. There appear to have been relatives with the same surname involved with overseas trade. The copy of the chart was made by Takami Tadatsune (or Senseki, 1785–1858), a collector of old maps. See Nakamura, "Japanese Portolanos," 27–29 and table 1 (n. 8) (note 143). On Itoya's death, see Tokita Tadamasa, "Goshuinsen bōekika Itoya Zuiemon boseki ron" (On the tombstone of Itoya Zuiemon, authorized trading-ship trader), *Nagasaki Shiritsu Hakubutsukan Kanpō* 19 (1979): 1–7.

150. This chart is noted as *Gaikoku tokai no ezu* (Chart of overseas navigation) in a list of articles brought from Annam (Vietnam) in Matsumoto Dadō, *Annan ki* (Records of Annam, 1807). See Nakamura Hiroshi, *Goshuinsen kōkai zu* (Sea charts used by the authorized trading ships) (Tokyo: Nihon Gakujutsu Shinkōkai, 1965), 72–76.

151. See Kawashima Motojirō, *Shuinsen bōeki shi* (History of trade by the authorized trading ships) (Ōsaka: Kōjinsha, 1921), 449–81. Expatriates were forbidden by law to return to Japan after 1639. Nakamura suggests that the chart was used from 1631 to 1636: Nakamura, "Japanese Portolanos," 29 (note 143).

152. Ro Kōrō, *Ro Kōrō jijoden* (Ro Kōrō's autobiography), published by the author in 1922 when he was in his seventy-sixth year.

153. Adachi, "Kaifū sen shōkai kiji," 127–28 (note 147).

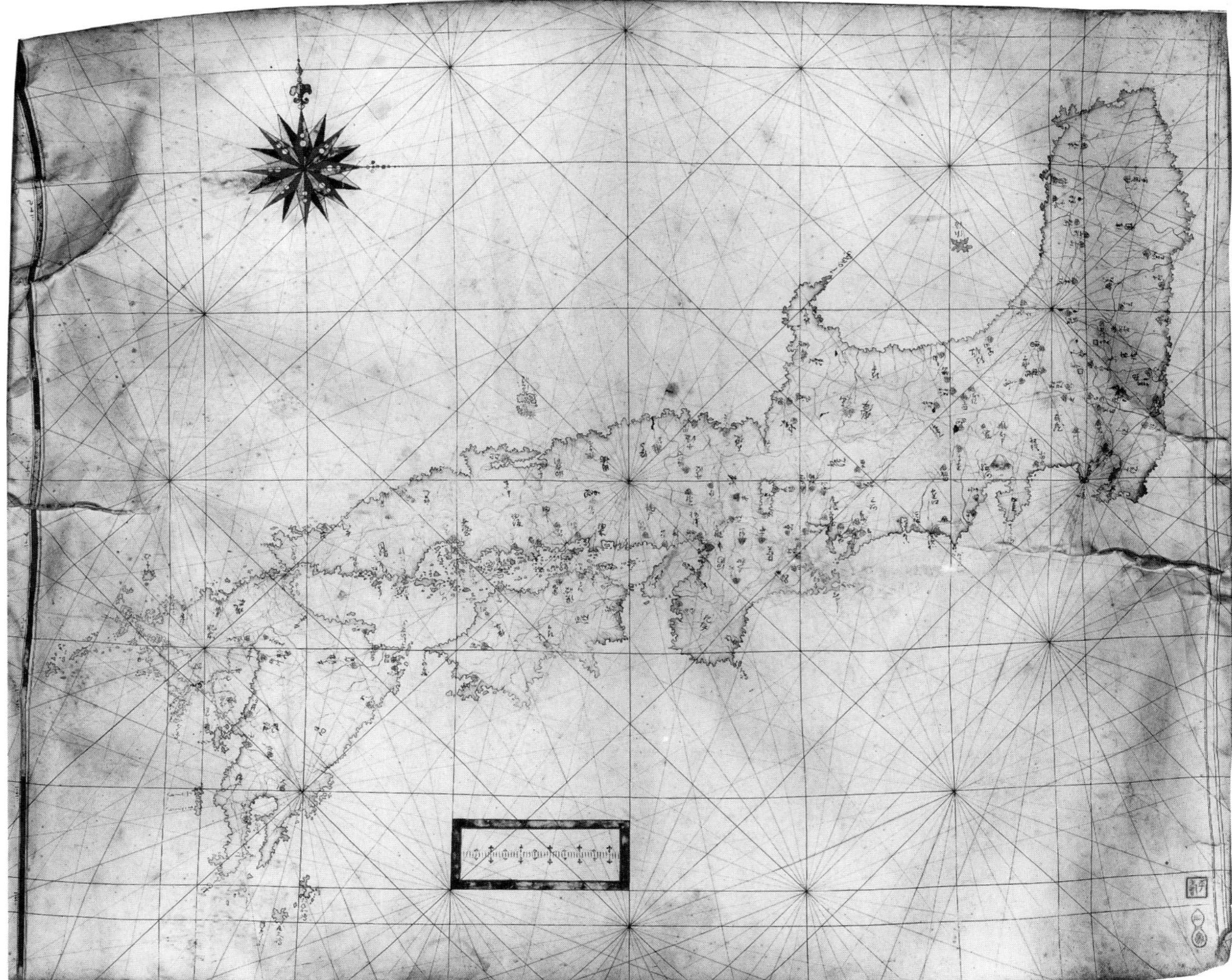

FIG. 11.25. MARINE CHART OF JAPAN, CA. 1671. Characteristic of charts of Japan is the fact that the image of Japan was not based on any of the charts of Southeast and East Asia. Rather, more recent Japanese sources were probably used: evidence here is the outline of northeastern Honshū, which is similar to that of the Keichō map. Note also that the boundaries of the provinces and main cities are included, suggesting that the origin of such a chart differed from that of charts of Southeast and East Asia. The bar scale is in Japanese *ri*.
Size of the original: 62 × 76 cm, on vellum. By permission of the Mitsui Library, Tokyo.

those of Japan were made from the outset by Japanese craftsmen, and their geographic and hydrographic details were not based on European models (fig. 11.25). They depict the Japanese archipelago more fully and correctly than do the charts of Southeast and East Asia, and they even include the names of cities and provincial boundaries that were probably taken from a terrestrial map of Japan. Three of them also have bar scales in *ri*, two of which use the western Japanese system of forty-eight *chō* to the *ri* (nos. 1 and 2 of appendix 11.6) and the other the eastern system of thirty-six *chō* to the *ri* (no. 4).[154] These suggest that there were different centers of production that used regionally recognizable units of measurement.

Once again this group of Japanese marine charts is far from a standard set. Although the eight charts may have a common prototype, they are diverse in content and form (appendix 11.6). The two earliest charts probably represent the initial stage of development (nos. 1 and 2).

154. The length of a degree of latitude in charts 1 and 2 of appendix 11.6 is between 32 and 33 *ri*; for chart 4 it is 43.75 *ri*. The value of 36 *chō* to the *ri* was used in provinces east of Kyōto, and that of 48 *chō* to the *ri* to the west; there was also another *ri* of 6 *chō* used in the Kantō area. For discussion, see Nakamura, "Japanese Portolanos," 28 and 38–42 (note 143). Also refer to Nakamura, *Goshuinsen kōkai zu*, 93–120 (note 150).

Both are drawn on vellum and contain European characteristics, notably in their system of rhumb lines indicating thirty-two directions and the decoration of the compass rose. Yet even these two charts are far from identical. In the first, the northernmost extent of Honshū is approximately 39°30′ north, whereas in the second it is approximately 41° north; to show this latter, more accurate position, itself based on a survey reading, Tōhoku (or Ōu) was elongated toward the north. The outline of northeastern Japan, the names of cities, and the provincial boundaries on the first (Mitsui) chart resemble those on the Keichō map of Japan, which appears to have been completed about 1639 (see below, p. 397 and plate 26), and suggest that it served as a model for the other charts.

The other six charts differ in various respects, and some are late copies of earlier versions (nos. 4, 7, and 8). All are drawn on rice paper, hardly a waterproof medium designed for use on board ship. The geographical content has also been tailored to individual use. Some of the charts include Korea, Hokkaidō, and the Izu Islands of Hachijō and Aoga, and northern Honshū is placed between 41°30′ and 42° north (nos. 4–7); but in other cases Korea and Hokkaidō are excluded (nos. 1–3). It is through such features and the positioning of geographical names that the lineage of the charts has been reconstructed. One of these diagnostic criteria is the number of rhumb lines used. Only one of the charts composed on rice paper has rhumb lines that indicate thirty-two directions (no. 3). Other charts (nos. 4–7) employ a system of twenty-four directions based on one of the surviving charts of Southeast and East Asia (appendix 11.5, no. 14), and another (no. 8), contains no indications of direction, although in the area of Japan it has a graduated rectangular frame and two compass roses.

Such variety makes it more difficult to establish when and for whom the first of this group of charts was made. According to tradition, one of the charts (no. 3) was compiled by the head of the Kawagoe clan in the province of Musashi (today divided into Tokyo, Saitama, and Kanagawa prefectures), Matsudaira Terutsuna (1620–71). The evidence for this link is a sheet of paper attached to the back of the chart, reading, "Tradition says that this comes from the pen of Chikōin," Chikōin being the posthumous name of Matsudaira. Although Nakamura argues for the historicity of the tradition, suggesting that Matsudaira copied the chart in 1638, the evidence as he rehearses it is tenuous.[155] Perhaps we should not entirely rule out the claims of this attribution to record the founding event of this group of charts, but it must be further noted that their documented production in Japan did not begin until 1670. This suggests that the tradition is not correct.

A more convincing theory is that these charts have an official origin. A hint lies in the way the charts differ from contemporary land maps, in particular in their directional lines and indications of latitude on both the left and right sides. This suggests that latitudinal measurements were taken at strategic points along the coast, and these measurements are so correct that they hardly differ from today's figures. That they were made for private groups is unlikely. It is more probable that such an interest in the geographical nature of the coastline originated with the central government, an idea reinforced by some documents in the collection known as Shimaya's records.[156]

According to these records, in 1669 the shogunate had a Chinese-style ship built in Nagasaki and appointed one Shimaya to be captain. He was ordered to investigate every small island between the northeast of Honshū and Nagasaki in 1670–71, starting with Edo as a point of reference. Although there is no specific mention of the chart he would have made, we know from other sources that Shimaya made charts for the Mito clan (see above) at about the same time. Moreover, when we turn to the surviving charts, their neat workmanship and detailed content suggest that the idea of compiling them originated in an order from the shogunate rather than from an individual such as Shimaya. Charts for personal use would have been less polished than any of these.

To assist his work, the shogunate probably gave Shimaya a copy of the Keichō map of Japan, a theory that is supported by the similarity of the coastline of the Sea of Japan on this map and some of the charts (nos. 1 and 2). On the other hand, the outline of the Pacific coast of Japan is based on Shimaya's observations. A minor difference between these two charts is the latitude for northern Honshū, the second being more accurate at 41° north and most likely based on better information. Judging from the latitude and shape of northeastern Honshū on the other six charts, it is likely that Shimaya produced a revision of the second chart as their model. It is possible that someone else may have been responsible, but there were few navigators skilled enough to conduct the measurements or, indeed, with the opportunity and the necessary equipment to make them. We must also remember that Shimaya was commissioned by the shogunate to undertake a voyage to the Bonin Islands, and one result was a chart depicting their location relative to Honshū.

Given the dates of Shimaya's voyage to map the islands between northeastern Honshū and Nagasaki, the charts

155. See Nakamura, "Japanese Portolanos," 35–36 (note 143).

156. Specific references are found in Hayashi Fukusai et al., *Tsūkō ichiran* (Collected documents for the history of Japanese diplomacy, ca. 1853), 8 vols. (Tokyo: Kokusho Kankōkai, 1912–13), appendix, chap. 18 (8: 508–12), and in Akioka Takejirō, "Ogasawara shotō hakken shi no kihonshiryō chizu ni tsuite" (On the fundamental documents concerning the discovery of the Bonin Islands), *Kaiji Shi Kenkyū* 9 (1967): 96–118, esp. 104–5.

FIG. 11.26. THE MAP OF JAPAN AT JŌTOKU TEMPLE, FUKUI, DATING FROM ABOUT 1595. The coastline is a revision of that on the Gyōki-type maps, which may in part reflect the activity of the Portuguese. Particularly outstanding is the delineation of Kyūshū and the Inland Sea. This manuscript, on a six-fold screen, is paired with the world map shown in figure 11.23.
Size of the original: 148.5 × 364 cm. By permission of Jōtoku Temple, Fukui.

of Japan must also date to 1670 or 1671. This conclusion is supported by the evidence of the Mito clan order. The charts, however, were not produced because of their value to Japanese navigation. Onshore landmarks were sufficient guides for ships skirting the coasts, and there was no need for charts on the open seas because of the edict prohibiting Japanese nationals from traveling abroad. They seem to have been prepared solely to reward successful students of surveying.[157] Given that there was no incentive to improve the information on them, Japanese marine charts then entered a period of stagnation.

THE JŌTOKU-TYPE MAPS OF JAPAN

The Jōtoku-type maps of Japan, also referred to as the Jōtokuji (Jōtoku Temple) type, derive their name from the map of Japan found at Jōtoku Temple in Fukui (fig. 11.26). That map is one of a pair with a map of the world (fig. 11.23 above), both of them on six-fold screens. The authorship and date of composition have not been ascertained for either map, but they certainly date to 1592 at the earliest. Some of the information on the map of Japan pertains to the invasion of Korea that began in that year, and the tribal name Orankai in a region beyond Korea on the world map was not known in Japan until then. From the evidence of the seal of the painter Kanō Eitoku (1543–90) that appears on both maps, they were undoubtedly executed by one of his apprentices.[158] Three other examples of the Jōtoku-type maps of Japan are listed as companion maps in appendix 11.4 (oval projection and equirectangular, type D1; a fifth example, at the Kōbe City Museum, will be mentioned below), and while they are partly derived from the earlier Gyōki-type maps of Japan, they are distinguished by the coastline that is shown in minute detail. Such maps were an important development during the period under discussion. They point to the influence of European ideas and knowledge from late in the sixteenth century, but at the same time they represent a synthesis of indigenous tradition and improvements in knowledge from both Japanese and European sources.

The content of the Jōtoku-type maps has an affinity with the Gyōki-type maps, but with improvements. The original Jōtoku model is thought to be a map of Japan with Kyūshū appearing as a long rectangle running north to south, rather than following the shape found in the surviving Gyōki-type maps. The map of Japan in Sin Sukchu's *Haedong cheguk ki* printed in Korea in 1471 (see p. 370) substantiates this theory. Kyūshū, patterned after a Japanese map of Japan that differed from the round mass of the Gyōki type, shows a uniform width from east to west at both the north and the south and has a slight protrusion in the southern coastline on both the

157. See below, p. 394.

158. For the invasion of Korea, see Sansom, *History of Japan*, 2:352–62 (note 32). Seals, which serve the same purpose as signatures, are usable as long as they are kept in good condition. The person who used Kanō's seal did not do anything unusual: other apprentices or followers of famous artists are known to have signed their works as if they were made by the masters.

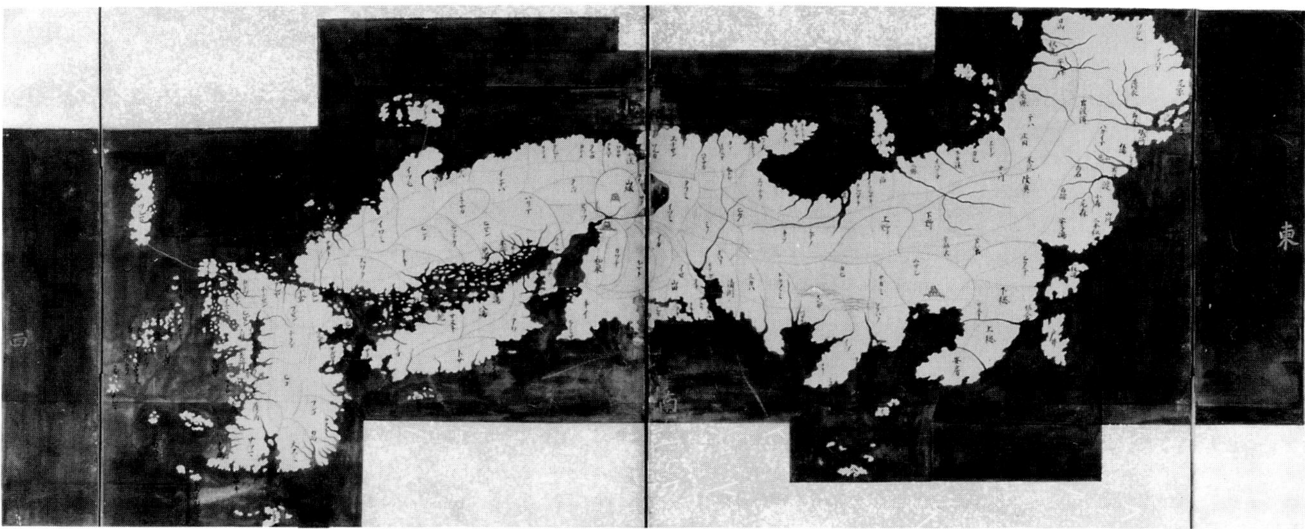

FIG. 11.27. THE *NANSENBUSHŪ DAINIHONKOKU SHŌTŌ ZU* (ORTHODOX MAP OF GREAT JAPAN IN JAMBŪDVĪPA), AN EXAMPLE OF THE JŌTOKU-TYPE MAP OF JAPAN DATING TO CA. 1627. Among the Jōtoku-type maps, this manuscript has the most place-names.

Size of the original: 48.5–56.3 × 161 cm (map), on a four-fold screen (113 × 267 cm). Kawamori Kōji, Takaishi, Ōsaka Prefecture. Photograph courtesy of Sakai City Museum and Kazutaka Unno.

east and the west. On the Jōtoku type, however, southern Kyūshū has unrealistic protrusions on the east and the west, and the tips of the Satsuma and Ōsumi peninsulas are flattened. This suggests that the model for the Jōtoku-type map included a rectangular Kyūshū like that in the *Haedong cheguk ki*.

The framework of provincial boundaries and routes to and from the national capital of Kyōto also provides evidence of continuity, but the form of the coastline exhibits a clear break with the Gyōki tradition. Although the Jōtoku maps have their imperfections, many of the irregularities of the coastline are included in a map of Japan for the first time. The shape of Kyūshū, in particular, is exceedingly accurate compared with the rest of the map: the curve of Kagoshima Bay is clearly depicted, and the peninsulas and inlets of Hizen Province (today Nagasaki and Saga prefectures) are expressed in remarkable detail. Lake Biwa in central Honshū and part of the Yodo River system that flows from it to Ōsaka Bay are depicted with relatively minor differences from their actual shapes. There are some characteristic inaccuracies: ignored, for instance, are the irregularities in the Pacific coast of Tōhoku (or Ōu) (and the rivers that flow into the Pacific), as well as in the representation of Tosa Bay and the peninsulas of Ashizuri and Muroto, rendering the Pacific coast of Shikoku completely different from reality. The simplistic rendition of this coast probably follows the Gyōki style, but it also suggests a dearth of additional information at the time. Similarly, the peninsulas of Tsugaru and Shimokita in the north of Honshū and the large Mutsu Bay between them have been excluded in favor of a slightly indented, more rounded coastline. It appears that no new materials were used to portray the archipelago on the Jōtoku-type maps. With the exception of Kyūshū and the Inland Sea, revisions were made to the coastline of the Gyōki style to produce observable changes but without regard to reality.

Besides the map at Jōtoku Temple, folding-screen examples of Jōtoku maps are in the possession of Kobayashi Ataru, Kawamura Heiemon, and Kawamori Kōji. As at Jōtoku Temple, these national maps are paired with world maps of the *Nanban* group.[159] On the Kobayashi and Jōtoku Temple examples, the two oldest, the only place-names included are the Kyūshū port cities of Hakata (Fukuoka), Nagoya (Hizen Province), and Nagasaki. Nagoya was developed in 1591 in conjunction with Hideyoshi's invasion of Korea[160] but declined rapidly with the return of the troops in 1598, suggesting that these maps were produced about or shortly after 1598. The Kawamura example also includes Nagoya and thus may be considered contemporaneous, a deduction that is supported by the style of painting on the screens. The accompanying world maps also suggest production in the last decade of the sixteenth century. Since they contain the

159. On the Kobayashi and Kawamura maps, see note 132 above; on the Kawamori map, see note 136 above.

160. Sansom notes that the construction of Hideyoshi's base at Nagoya began in 1591, and it was from there that he launched the campaign against Korea in the following year; see Sansom, *History of Japan*, 2:352 (note 32).

tribal name Orankai in the northeast beyond Korea and were compiled about the time of the invasion of Korea, they may be dated to 1592–98.

The Kawamori example (fig. 11.27) is paired with a map of the Eastern Hemisphere on a folding screen, on which there are also tables providing the distances between Japan and its trading partners, notes on the prevailing situation in those places, and a catalog of goods exported to Japan. From the information about Taiwan, we may surmise that it was composed about 1627.[161] Since these tables help date the map of the Eastern Hemisphere, the map of Japan may be similarly dated. Some of the geographical features in the map of Japan, however, are older than the supposed date of composition. Among the place-names is Nagoya in Kyūshū, as well as others that were current around the end of the sixteenth century, mainly as early feudal centers, but that had declined by 1627. Of the Jōtoku examples, only the Kawamori map includes a great number of place-names and exhibits minute detail in its execution.

One more example of the Jōtoku type, in the Nanba Collection at the Kōbe City Museum, appears at first glance to be from the sixteenth century. Judging from the values given for land productivity in each province, however, it was probably commissioned by the shogunate from the mid-seventeenth century onward.[162] All but four of the names on this map belong to the provinces, the outstanding ones being a town and islands off the western coast of Kyūshū: Arima in the Shimabara peninsula, the Gotōs, the Amakusas, and the Koshikis. Arima was included on some maps after 1580, and until 1614 the Jesuits had a seminary there. It appears that the original of the map in the Nanba Collection was composed sometime between these dates.

On the Kawamori map there are about 160 place-names other than those of the provinces. Some of these point to contact with Europeans; in particular, there are the names of two islands off the western coast of Kyūshū—Hanerasu and Santakarara—that are derived from the European place-names Pannellas and Santa Clara. The Dutch cartographer Jan Huygen van Linschoten (1563–1611) indicated that there was an island called Pannellas immediately northeast of another island, Meaxuma (that is Meshima [Island of Woman]); Pannellas, then, was the Japanese Oshima (Island of Man). Santakarara, also found in Linschoten's work as Santa Clara, is included on one of the Japanese charts of Southeast and East Asia (appendix 11.5, no. 6), and it was probably another name for the Uji Islands.[163] This is supported by the Kawamori map, on which four islands are shown almost midway between Meshima and Iō Island (Kikai Island), suggesting that they also must be the Uji Islands. The Japanese compiler of the Kawamori map evidently either was unaware that these were the same islands with different European and Japanese names or lacked the proper information to distinguish them. The two names could have been written side by side had he wanted to use the Japanese term Ujishima. A similar case exists with Hanerasu: on the Kawamori map it is indicated without the Japanese name Oshima but is placed exactly in the location of Oshima between Meshima and Ōchika Island (today Fukue Island).

The use of European place-names on the Kawamori map thus indicates that its source material included European maps. Additional evidence of this is provided by an undated and anonymous Italian manuscript map of Japan at the Tenri Central Library (fig. 11.28).[164] This resembles the Kawamori map and includes the names of Pannellas and Santa Clara, spelled as Panelas and S. Clara. Moreover, three European works included maps of Japan in the same form: the *Saverio orientale* of 1641 by Berardin Ginnaro (1577–1644), the *Fasciculus e Iapponicis floribus* of 1646 by Antonio Francisco Cardim (1596–1659), and the first volume of the *Arcano del mare* (1646–47) by Robert Dudley (1574–1649).[165] On them Santa Clara

161. See Iwao Seiichi, "Ishibashi hakushi shozō sekaizu nendai kō" (On the date of the world map in the collection of Dr. Ishibashi), *Rekishi Chiri* 61 (1933): 511–22. One of the notes on the map states that the Dutch were living in the south of Taiwan (Formosa) and the Europeans of Luzon in Tansui (Tanshui); respectively these were occupied by the Dutch in 1624 and the Spanish in 1626.

162. The Nanba map (on a two-fold screen, 56.8 × 124 cm) is reproduced in color in Nanba, Muroga, and Unno, *Nihon no kochizu/Old Maps in Japan*, pl. 21 (note 11). See also Nakamura Hiroshi, "Sengoku jidai no Nihonzu" (Maps of Japan at the time of the civil wars [1467–1568]), *Yokohama Shiritsu Daigaku Kiyō* 58 (1957): 1–98, esp. 24–27.

163. For information on these names and islands, see Jan Huygen van Linschoten, *Itinerario, voyage ofte schipvaert*, 5 vols. (The Hague: Nijhoff, 1910–39), pt. 2, *Reys-Gheschrift vande navigatien der Portugaloysers* (1595), chap. 39 (Werken uitgegeven door de Linschoten-Vereeniging 43 [1939]: 235 and 251–52). The map of the East Indies by Arnold Floris van Langren (ca. 1571–ca. 1644) attached to the book shows these islands as "Meaxuma" and "Santa Clara": see Pál Teleki, *Atlas zur Geschichte der Kartographie der japanischen Inseln* (Budapest, 1909; reprinted 1966), 24; and Cortazzi, *Isles of Gold*, 21–22 and pl. 22 (note 14).

164. This map was purchased in Rome before the Second World War by the Japanese diplomat Yoshiura Morisumi. It has been kept at Tenri Central Library since the end of the war. The paper and the frame both are distorted rectangles; see Nakamura, "Sengoku jidai no Nihonzu," 33–40 (note 162).

165. Berardin Ginnaro, *Saverio orientale; ò, Vero istorie de' Cristiani illustri dell'Oriente* . . . (Naples: Francesco Savio, 1641) (there are three parts, the map of Japan being in part 1); Antonio Francisco Cardim, *Fasciculus e Iapponicis floribus, suo adhuc madentibus sanguine* (Rome: Typis Heredum Corbelletti, 1646); Robert Dudley, *Dell'arcano del mare*, 3 vols. (Florence, 1646–47; 2d ed. 1661). See also Joseph F. Schütte, "Japanese Cartography at the Court of Florence: Robert Dudley's Maps of Japan, 1606–1636," *Imago Mundi* 23 (1969): 29–58, esp. 31 and 46, for these citations; two reproductions of Dudley's printed maps from the 1661 edition are on pp. 33–34. The maps by Ginnaro, Cardim, and Dudley are mentioned by Cortazzi, *Isles of Gold*, 44–45; he reproduces a copy of Cardim's map, *Iapponiae nova & accurata*

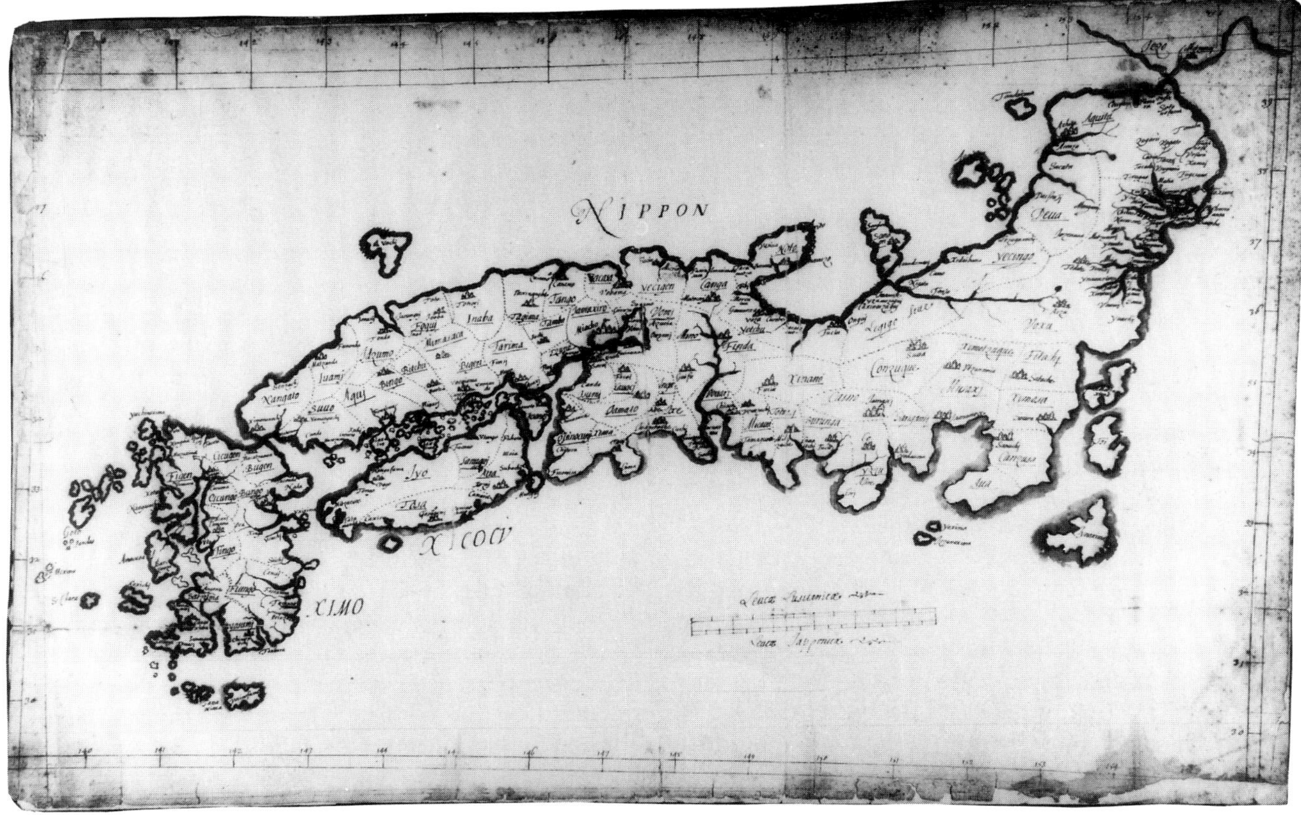

FIG. 11.28. AN ITALIAN MANUSCRIPT MAP OF JAPAN DATING TO THE LATE SIXTEENTH CENTURY. This is possibly an example of source material used by the compiler of the Kawamori map. The spelling of place-names is Portuguese on an otherwise Italian-language copy: the original might have been composed by Ignacio Moreira, who was in Japan from 1590 to 1592. The bar scales are in Lusitanian and Japanese leagues (*leucae Lusitanicae* and *leucae Japponicae*).
Size of the original: 46.5 × 72.4 cm; 40 × 67.3 cm (inside frame). By permission of the Tenri Central Library, Tenri, Nara Prefecture.

is written with minor spelling alterations. This is not the case with Pannellas: the islands are referred to on Ginnaro's map as "Osima I." and on Dudley's as "I. Oscuma." These names are versions of Oshima, suggesting that this name was already in use on Japanese maps being sent abroad. Of the three maps published in Europe, that of Cardim is the most widely known, and he has lent his name to this group of four maps, which on the grounds of their similarity probably share a common source. Schütte's research on the three Cardim-type maps in Europe points to their association with the Portuguese Ignacio Moreira (b. 1538 or 1539), who resided in Japan between 1590 and 1592.[166] According to Valignani, one of his contemporaries, Moreira compiled maps, and the content of his *Declaração da descripção de Japão* was essentially the same as that of the three European Cardim-type maps. A Latin translation entitled *Iaponicae tabulae explicatio* is also extant.[167]

Schütte, however, was not aware of the Tenri map,

descriptio (Rome, 1646; in the British Library, London), and two of Dudley's, *Asia carta diciaset[t]e piu moderna* (1661, in the British Library) and *Carta particolare della grande isola del' Giapone è di iezo con il regno di Corai et altre isole in torno* (1661, in Cortazzi's personal collection) (Cortazzi, *Isles of Gold*, pls. 64–66 [note 14]). As mentioned subsequently, a map compiled by Ignacio Moreira from information acquired in 1590–92 appears to have been the prototype of these maps.

166. See Schütte, "Ignacio Moreira," 126–27 n. 108 (note 121). Schütte here also refers to a map by Philippe Briet (1601–68) that appeared in Nicolas Sanson d'Abbeville, *L'Asie en plusieurs cartes nouvelles et exactes* (Paris, 1652); the title was *Description des isles de Iapon en sept principales parties*. An earlier map by Briet, *Royaume du Iapon* (Paris: Mariette, 1650), is reproduced in Teleki, *Atlas zur Geschichte der Kartographie*, pl. IX-1 (note 163), and Cortazzi, *Isles of Gold*, pl. 67 and (about Briet), 45–46 (note 14). See also Schütte, "Japanese Cartography at the Court of Florence," which includes two of Dudley's manuscript maps showing the coastal areas of northern Honshū and southern Hokkaidō (figs. 3–4, pp. 33–36). The discussion on "models and sources" is on 45–58 (note 165).

167. Archivum Romanum S.J., Jap. Sin. 22, fol. 300r–v; the text is included as an "annex" to Schütte, "Ignacio Moreira," 127–28 (note 121).

which was researched by Okamoto and Takahashi. Of the four maps, they concluded that the Tenri manuscript is the closest in form to Moreira's; the place-names on the manuscript map were limited to those in use while Moreira was in Japan, and Portuguese spelling was retained on an Italian-language copy.[168] The exact date is still unknown.

It is probable that the Kawamori map was modeled on one of Moreira's or one copied by a missionary in Japan. Since the Kawamori map was completed about 1627, the Moreira group of maps might have served as source material. It could therefore be said that the Cardim type of map had already been developed before 1592–98 when the early *Nanban* group of world maps was completed.

From the evidence provided by the place-names, we may therefore conclude that the Jōtoku maps were influenced by Europeans who had been in Japan. The standard copies used for subsequent revisions, however, were produced by Japanese. This conclusion has been reached by observing that the outline of Japan, with the exception of Kyūshū, differed little from that on the Gyōki-type maps. It is the many place-names along the coast that were updated. European pilots who received the Gyōki-type maps amended and revised them. It was European navigators, for example, who provided evidence of several uninhabited islands off the coast of western Kyūshū shown on the Jōtoku-type maps. The coast of Kyūshū itself was depicted in the greatest detail, reflecting the island's importance in sixteenth-century trade that from 1545 was conducted mainly with the Portuguese. The chartlike characteristics of these maps might be attributed to the requirements and efforts of the pilots. On the Japanese marine charts of Southeast and East Asia, for example, Japan appeared as it did on the Jōtoku (Cardim) type.

The Jōtoku maps therefore combined Japanese mapping with the ideas and knowledge of European pilots and missionaries. On the one hand, European navigators sought available maps of areas into which they sailed, and the revision of such charts was essential for the success of subsequent voyages. On the other hand, Japanese navigators also recorded the names of important islands that lay along their sailing routes. Some of the islands shown on the map in the *Haedong cheguk ki* might seem too small to be of importance, but they were probably included because of their positions on trading routes between Japan and the Ryūkyūs and Korea. This suggests that the outline of the Japanese coast, especially western Kyūshū and the adjacent islands, was mapped by Japanese before the arrival of Europeans. The map of Japan in Tōshōdai Temple also shows a large number of small islands off western Kyūshū, reflecting the pattern of sea travel between Japan and the continent. Modifying the Gyōki-type maps into a chartlike format is thus an important development of the mid-sixteenth century. The arrival of Europeans speeded up the process, and the Jōtoku maps were eventually the outcome of both Japanese and European efforts.

THE INTRODUCTION AND MANUFACTURE OF TERRESTRIAL GLOBES

I have already mentioned a European terrestrial globe that had been introduced to Japan by 1580. In that year, for example, in a meeting with the Jesuit Organtino, Oda Nobunaga discussed using a European globe.[169] Similarly, in 1591 the mission of three Kyūshū lords, on returning from Europe and when staying at Murotsu, Harima Province (now part of Hyōgo Prefecture), showed European globes, maps, and charts to a number of daimyos who were passing through the port town.[170] The globes were not always of European origin. In 1592 the Dominican Fray Juan Cobo, envoy of the governor-general of Spanish Manila, met Toyotomi Hideyoshi at Nagoya, Hizen Province, and presented him with a terrestrial globe with all its place-names written in Chinese characters.[171] Thereafter, European celestial and terrestrial globes continued to be imported into Japan by Christian missionaries and employees of the Dutch East India Company, presumably either as a way to proselytize or as diplomatic gifts to the shogunate. References to such globes and their use are found in both Japanese and European sources. In 1596 the Japanese Christian João Sotão accompanied his wife to a church in Kyōto, where he showed her a map of the world and a terrestrial globe.[172] In 1606 there is also a literary reference to the use of globes when the Confucianist Hayashi Razan visited the Japanese monk (*irmão*) Fabian Fukansai at a Christian church in Kyōto, where he examined a terrestrial globe and criticized its theory of the round earth.[173] From the

168. Okamoto, *Jūroku seiki ni okeru Nihon chizu no hattatsu*, 103–9 and 163–207 (note 131); Takahashi Tadashi, "Seizen seru shoki Nihon chizu ni tsuite: I. Moreira kei chizu o chūshin to shite" (On the early maps of Japan by Europeans, emphasizing the I. Moreira-type maps), *Nihongakuhō* 4 (1985): 1–33; idem, "Jūshichi seiki Nihon chizu ni okeru Teisheira gata to Moreira gata: N. Sanson to R. Daddoree no Baai" (About the Moreira- and Teixeira-type maps of seventeenth-century Japan: The atlases of N. Sanson and R. Dudley), *Nihongakuhō* 6 (1987): 111–35.

169. See note 124 above.

170. See note 126 above.

171. Emma H. Blair and James A. Robertson, eds., *The Philippine Islands, 1493–1898*, 55 vols. (Cleveland: Arthur H. Clark, 1903–9), 9:45.

172. Luís Fróis, Annual Report, 3 December 1596 (Archivum Romanum S.J., 352), 179–230v; see "1596 nendo Iezusu Kai nenpō" (Annual report of the Society of Jesus, 1596), trans. Sakuma Tadashi, *Kirishitan Kenkyū* 20 (1980): 261–410.

173. Hayashi Razan, "Hai Yaso" (Denouncing Christianity), in *Razan bunshū* (An anthology of Razan's prose, 1662), 2 pts., ed. Hayashi Gahō.

middle of the next century, on five occasions it is recorded that officials of the Dutch Office in Japan made gifts of globes to the shogunate. Thus in 1642 the head of the office, Jan van Elzerack, presented a terrestrial globe to Inoue Chikugo no Kami (Masashige, 1585–1661), a high-ranking official in the shogunate.[174] Then, in 1647, a large terrestrial globe was given by Willem Verstegen to Inoue;[175] in 1652 Adriaen van der Burgh presented a terrestrial globe and a map to Inoue;[176] and in 1657 Zacharias Wagenaar's presentation of a terrestrial and a celestial globe to the shogunate is recorded, although they were destroyed by fire shortly afterward.[177] Finally, in 1659 Wagenaar presented a second pair of globes to the shogunate.[178] Globes were also acquired by the Japanese aristocracy, as in 1661 when Hendrick Indijck, the head of the Dutch Office, hand delivered to a secretary of the late Inoue Masashige at Edo a terrestrial and a celestial globe that his lord had ordered.[179]

The earliest recorded Japanese venture in the manufacture of terrestrial globes was in 1605, when the emperor asked his "ordinary" craftsman to make a globe.[180] Later the shogunate paid attention to the maintenance and repair of globes, as suggested by the repair of the *Tenchi no zu* (Figures of heaven and earth; possibly celestial and terrestrial globes) by shogunal officials with the assistance of others such as the former Christian Okamoto San'emon (Giuseppe Chiara [1602–85]) from 1677,[181] but the dates of the globes are not known. On the other hand, a terrestrial and a celestial globe repaired by Uma Michiyoshi from 1791 to 1794 by order of the shogunate were Willem Jansz. Blaeu's globes of about 1640.[182]

The oldest extant Japanese-made terrestrial globe might be that which accompanies a padre doll on a toy dating most likely from early in the Edo period. The globe measures about 3.8 centimeters in diameter, and its geography, including a worldwide sea route beginning from Portugal, is derived from a *Nanban* map on an oval projection.[183] It was not until 1690 that the astronomer Shibukawa Harumi (see chapter 14 below) made the first Japanese terrestrial globe for practical use (plate 24), based on Ricci's world map of 1602. Seven years later Shibukawa made another globe, and his works continued to influence globe making in the eighteenth century (see appendix 11.7). Besides Ricci's world map, other European sources for Japanese globes in the eighteenth century included a globe dating to 1700 by Gerard Valck (1652–1726) and Leonard Valck (1675–1746) (see below). Throughout the nineteenth century almost all Japanese terrestrial globes were based on the world maps compiled by the Japanese *Rangaku* (Dutch studies) scholars, from various Dutch maps (fig. 11.29).

Besides the globes derived from European models, there were also some based on Buddhist ideas. The earliest extant example was made by the priest Sōkaku (1639–1720) about 1702. On this globe, at the top of the earth's axis, fixed roughly perpendicular to the base, is a columnar object made from rock crystal and shaped like Mount Sumeru; the geography on the sphere itself is also drawn according to the Buddhist image of the world.[184] Accompanying the globe is a flat model titled *Shumisengi*

Modern editions of "Hai Yaso" appear in *Razan sensei zenshū* (Collected work of the teacher Razan) (Kyōto: Heian Kōkogakkai, 1918), and in *Kirishitan sho Haiyasho* (Books on Christianity and writings denouncing Christianity), ed. Ebisawa Arimichi et al., Nihon Shisō Taikei (Series of Japanese thought), vol. 25 (Tokyo: Iwanami Shoten, 1970), 413–17.

174. In 1643 Inoue took this globe to use in the investigation of a Dutch ship that had washed up on the northeastern shore of Japan (its captain was H. C. Shaep): *Journael ofte dachregister gehouden bij den schipper Hendricq Cornelisz. Schaep . . .*, Algemeen Rijksarchief, The Hague, Overgekomen Brieven, jaar 1645, Book 2, GGG2. Kolonial Archief no. 1055 (see the translation of the journal: *Nanbu hyōchakuki* [An account of a shipwreck cast ashore at (the daimyate of) Nanbu (now in Iwate Prefecture)], trans. Nagazumi Yōko [Tokyo: Kirishitan Bunka Kenkyūkai, 1974], 61).

175. *Nagasaki Oranda shōkan no nikki* (Diary of the head of the Dutch Office in Japan, 1641–54), 3 vols., trans. Murakami Naojirō (Tokyo: Iwanami Shoten, 1956–58), 2:178.

176. Reported by Willman; see Nils Matson Kjöping, *Een Kort Beskriffning Vppå Trenne Reesor och Peregrinationer sampt Konungarijket Japan . . . III. Beskrifwes een Reesa till Ost Indien, China och Japan . . . aff Oloff Erickson Willman* (Wisingsborgh, 1667); in Japanese, *Nihon taizaiki* (An account of a sojourn in Japan), trans. Ozaki Tadashi, Shin Ikoku Sōsho (New series on foreign countries) (Tokyo: Yūshōdō Shoten, 1970), 38.

177. *Tsūkō ichiran*, chap. 242 (6:212) (note 156); Arnoldus Montanus, *Gedenkwaerdige gesantschappen der Oost-Indische maatschappy in 't vereenigde Nederland, aan de kaisaren van Japan* (Amsterdam: J. Meurs, 1669), 370–71, 386.

178. *Tsūkō ichiran*, chap. 242 (6:212) (note 156); Montanus, *Gesantschappen aan de kaisaren van Japan*, 399 (note 177).

179. Montanus, *Gesantschappen aan de Kaisaren van Japan*, 414 (note 177).

180. Pagès, *Histoire de la religion chrétienne au Japon*, 1:125 (note 145).

181. *Saken yoroku* (Diary of the Office of Christians, 1672–91) in *Zokuzoku gunsho ruijū*, 12:607 (note 91).

182. Uma Michiyoshi (Kitayama Shin'yō), *Oranda tenchi ryōkyū shūho seizō ki* (Records of the repair of the Dutch celestial and terrestrial globes, 1795); owned by the National Diet Library, Tokyo. See also Peter van der Krogt, *Old Globes in the Netherlands: A Catalogue of Terrestrial and Celestial Globes Made prior to 1850 and Preserved in Dutch Collections*, trans. Willie ten Haken (Utrecht: HES, 1984), 70.

183. The owner of the device, Kayahara Hiroshi, dates it to the Kyōhō era (1716–35), but evidence such as Toyotomi Hideyoshi's seal and the mechanical simplicity of the rotating doll suggests that it was made much earlier. The doll is twenty-two centimeters tall and resembles a Jesuit missionary carrying a whip; a handle, before it was broken, turned both the doll and the globe. See Unno, "Chikyūgi tsuki no bateren ningyō" (A terrestrial globe with a padre doll), in *Chizu no shiwa*, 248–50 (note 136).

184. Sōkaku's globe is discussed in detail in Unno Kazutaka, "Sōkaku no chikyūgi to sono sekaizō" (Sōkaku's globe and his image of the world), *Kagakushi Kenkyū* 117 (1976): 8–16.

FIG. 11.29. NUMAJIRI BOKUSEN'S TERRESTRIAL GLOBE *DAI YOCHI KYŪGI* (LARGE GLOBE OF THE EARTH) OF 1855. The cartographic image, figures, and place-names are the same as those on Shibata Shūzō's *Shintei kon'yo ryakuzenzu* (Newly revised map of the earth, 1852), its source. The globe was printed from a woodblock onto paper gores and colored by hand, and it has twelve bamboo ribs that fold like traditional Japanese umbrellas. Numajiri Bokusen (1774–1856) was a geographer who had a private school at Tsuchiura, Hitachi Province (now Ibaraki Prefecture).
Size of the original: 23 cm. Honma Takeo, Tsuchiura. Photograph courtesy of Kazutaka Unno.

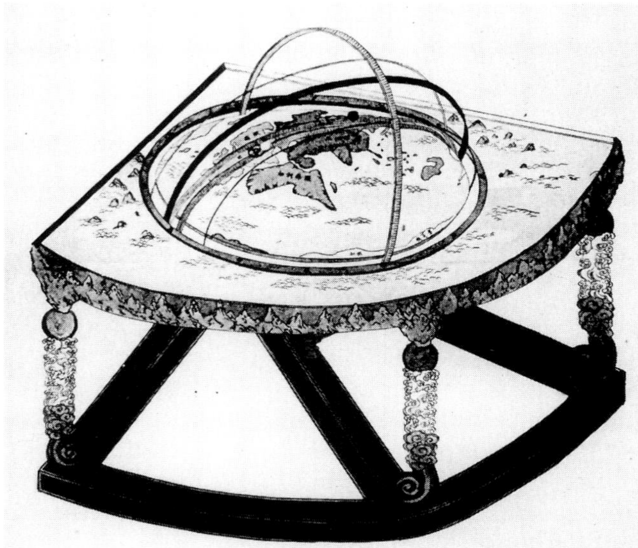

FIG. 11.30. ENZŪ'S *SHUKUSHŌGI ZU* (SKETCH OF AN INSTRUMENT OF THE BUDDHIST IMAGE OF THE FLAT EARTH) OF 1814. Enzū devised astronomical models to demonstrate the Buddhist view of a flat earth, one such model being the *Shukushōgi*, which was sketched in order to be printed from a woodblock. Although the *Shukushōgi* itself has been lost, its general structure has been determined from the prints that were made, and it includes arcs showing the orbits of the sun and the moon at the solstices and equinoxes. The image of the flat earth was adopted directly from the portrait of the Eastern Hemisphere on European maps. Not included here is Enzū's preface, which was attached to the *Shukushōgi zu*.
Size of the original: 60 cm in width; total size with the preface: 130 × 60 cm. Ryūkoku University Library, Kyōto. Photograph courtesy of Kazutaka Unno.

(Instrument of Mount Sumeru), to explain the justification of a Buddhist flat-earth theory. Later, about 1751, a simple instrument using geographical data derived from European sources was made by order of another Buddhist priest, Kakushū (d. 1756).[185]

The trend of incorporating European knowledge into the Buddhist image of the heavens and earth became stronger in the nineteenth century, an example being the *Shukushōgi* (Instrument of the Buddhist image of the flat earth, or the Buddhist orrery) of the Buddhist priest Enzū (1754–1834), who was active in propagating Buddhist astronomy (fig. 11.30). Enzū's instrument was a model to explain the passage of the seasons based on ideas from the design of the European orrery together with the *Shumisengi*, which contained a clockwork mechanism to help explain the Buddhist view of the universe. About 1848 Enzū's disciple Kanchūzenki (fl. 1834–48) and a later disciple Kōgon (d. 1871) made plans for more exquisite and refined astronomical clockwork models and had them made by the watchmaker Tanaka Hisashige (1799–1881). These instruments improved the clockwork mechanism in Enzū's *Shukushōgi* and the *Shumisengi*. About 1855 another disciple of Kanchūzenki, Sada Kaiseki (1818–82), developed another clockwork model called *Shijitsu tōshōgi* (Model showing the equality of visual and substantial objects). This demonstrated the Ptolemaic system and the theory of a flat earth; copies of this instrument were also made by Tanaka.[186]

185. Tōkai Sanjin, *Fugyō shinmon zōhyō* (Research on heaven and earth, with comments, 1751); owned by Ōtani University Library.

186. On Buddhist cosmology in Japan, see Unno Kazutaka, "Nihonjin to Shumisen" (The Japanese and Mount Sumeru), in *Ajia no uchūkan* (Cosmology in Asia), ed. Iwata Keiji and Sugiura Kōhei (Tokyo: Kōdansha, 1989), 349–71. Enzū's *Shumisengi zu* (Sketch of an instrument of Mount Sumeru) and *Shukushōgi zu* (Sketch of an instrument of the Buddhist image of the flat earth), Kanchūzenki's works of the same name, and Sada's *Shijitsu tōshōgi zu* (Sketch of a model showing the equality of visual and substantial objects) are reproduced in *Shumisen zufu* (Collection of pictures of Mount Sumeru), ed. Tokushi Yūshō (Kyōto: Ryūkoku Daigaku Shuppanbu, 1925).

SURVEYING INSTRUMENTS AND TECHNIQUES

The transfer and reception of surveying methods and instruments was also an important aspect of the transmission of European cartographic ideas into Japan in the encounter period. Although traditional models and ways of mapping survived, the availability of European instruments and techniques permitted revision of Japanese maps, noted above, to occur more effectively. At first European concepts and instruments were applied to navigational astronomy, but they were also incorporated into topographical surveying after the closure of the country. The main center for European-style surveying was Nagasaki, a thriving international port.

The legendary account of the transmission of European surveying methods to Japan places it early in the Edo period. The story alleges that a Dutchman named "Kasuparu" came to Japan in 1641 and taught the art of surveying to Higuchi Kentei (1601–84).[187] About this date, the only known Dutchman with such a name was the surgeon Caspar Schamburger, who arrived in Japan in 1647 or 1648.[188] Even if this is the same person, we cannot be sure that it was he who taught surveying. There was subsequently no strong Dutch influence on Japanese surveying, and we must conclude that the Dutch were not involved in disseminating such knowledge.

A much stronger case can be made for Portuguese involvement. According to Ro Senri, Higuchi studied astronomy and geography under a teacher named Hayashi Sensei (or Kichiemon), a Christian resident of Nagasaki who was put to death for his beliefs in 1646. Higuchi was later implicated with his teacher, and he was subsequently imprisoned for twenty-one years.[189] One of his works, the *Nigi ryakusetsu* (Brief explanation of the heavens and earth), was based on the first section of the *Compendium Catholicae veritatis* (ca. 1593), a work composed by the Jesuit Pedro Gomez (1535–1600) for his Japanese students.[190] Further evidence of Portuguese influence is that both Hayashi and Higuchi were educated not by the Dutch, but by the Jesuits and Portuguese pilots.

Evidence of assimilation is provided by the foreign terminology used in Hosoi Kōtaku's (1658–1735) *Hiden chiiki zuhō daizensho* (Complete book of the secret art of surveying and mapping), a manuscript dating to 1717. Hosoi states that surveying is the art of *pirōto* (*piloto*, pilot) as taught to Japanese by the Dutch and explains that *pirōto* is "a foreign term meaning computation or calculation." From this it is obvious that the origin and the meaning of the word had been forgotten. The Dutch

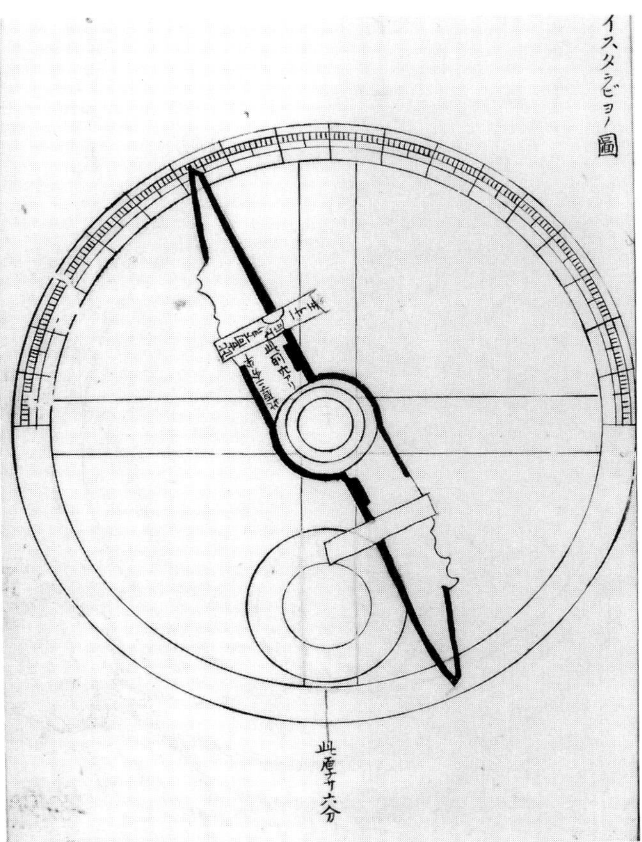

FIG. 11.31. ILLUSTRATION OF AN ASTROLABE (*ISUTARABIYO*) IN HOSOI KŌTAKU'S *HIDEN CHIIKI ZUHŌ DAIZENSHO*.
Diameter of the original: 17.2 cm. National Diet Library, Tokyo. Photograph courtesy of Kazutaka Unno.

187. Also known as Kobayashi Yoshinobu, an astronomer at Nagasaki. The date of his death is given as 9 February 1684; see Ro Senri, [*Nagasaki*] *senminden* (Biographies of the pioneers in Nagasaki, 1731) (Edo: Keigendō, 1819), chap. 1, s.v. "Kobayashi Yoshinobu."

188. Or Schambergen; he was one of the first medical doctors from Holland to arrive in Japan and founded a school of medicine known as Kasuparu ryū. Refer to Plutschow, *Historical Nagasaki*, 97 (note 128).

189. See Ro, *Senminden*, chap. 1, s.v. "Kobayashi Yoshinobu" (note 187).

190. See Ohara Satoru, "Kirishitan jidai no kagaku shisō" (Scientific thought in the Christian period), and "Pedoro Gomesu cho 'Tenkyūron' no kenkyū" (A study of "De sphaera" by Pedro Gomez), *Kirishitan Kenkyū* 10 (1965): 101–78 and 179–273; and Hirose Hideo, "Kyū Nagasaki tengakuha no gakutō seiritsu ni tsuite: 'Nigi ryakusetsu' ni kanshite" (On the formation of the old Nagasaki school of astronomy: Concerning *Nigi ryakusetsu*), *Rangaku Shiryō Kenkyūkai Kenkyū Hōkoku* 184 (1966): 3–14. For a discussion of the *Nigi ryakusetsu*, see Nakayama, *History of Japanese Astronomy*, 98–100 (note 38). The only extant copy is in the Naikaku Library of the National Archives in Tokyo; it has been published in *Kinsei kagakushisō* (Scientific thought in the modern ages), 2 vols., ed. Furushima Toshio et al., Nihon Shisō Taikei (Series of Japanese thought), vols. 62–63 (Tokyo: Iwanami Shoten, 1971–72), vol. 2. The *Compendium Catholicae veritatis* is in Rome, Biblioteca Apostolica Vaticana, Regina Lat. 426; see Joseph F. Schütte, "Drei Unterrichtsbücher für japanische Jesuitenprediger aus dem XVI. Jahrhundert," *Archivum Historicum Societatis Iesu* 8 (1939): 223–56.

word for pilot, *loods*, has no linguistic resemblance to *pirōto*. Hosoi's book contains illustrations of various surveying instruments, among them the *watarante* (quadrant), *kuhadarantei* (quadrant), *konpansu* (compass), *isutarabiyo* (astrolabe), and *asutarabiyo* (astrolabe) (fig. 11.31), all of which show a direct link to the Portuguese and Spanish languages. The terms *watarante*, *konpasu*, and *isutarahi*[191] are also included in Matsumiya Toshitsugu's (1686–1780) treatise on surveying, *Bundo yojutsu* (Techniques of protraction) of 1728.[192] Since Matsumiya was not a student of Hosoi's, there were probably different surveying schools and factions using European instruments and referring to them by their European names. Some of these instruments from the Edo period have survived.[193]

Not only are the instruments referred to by their European names in these works, but so are the months of the year because of their importance for declination tables. Examples from Hosoi's manuscript include *shanero*, *hebereiro*, *setenboro*, *nobenboro*, and *desenboro*; although Matsumiya is not as strict as Hosoi with voiced sounds, his phonetic representations of the months are the same. Both works also list the months in Dutch, as *yanwari*, *befuriwari*, and *maruto* testify. Because of their importance in European navigation, the names of the months appear to have been taught to Japanese sailors by Europeans. The *Genna kōkai sho* (Book of the art of navigation in the Genna era [1615–23]) of 1618, written by Ikeda, records the Portuguese names of the months,[194] and the surveying methods that Hosoi and Matsumiya describe were related to the techniques of navigation that Ikeda had learned from the European Manuel Gonzalez (see note 145 above). Hosoi, as mentioned previously, considered the art of *pirōto* to be the basis for surveying. Matsumiya also acknowledged that the surveying techniques he had mastered had their roots in European navigation.

The mistaken idea that it was the Dutch rather than the Portuguese who had introduced such knowledge to the Japanese was also perpetuated by Matsumiya. It was prevalent in works on surveying during the Edo period, probably as a result of the existence of Dutch trading privileges and, notably, the tendency to avoid things associated with the Portuguese after the ban on Christianity. The Portuguese were known for their evangelistic zealotry, and thus anything associated with them was avoided in a way that did not apply to the more mercantilistic Dutch.

Two important early surveyors, according to Matsumiya's work, were Higuchi and Shimaya. The latter was an accomplished pilot skilled in European navigation, but it appears that Higuchi, who had studied astronomy with European navigators as well as learning from Hayashi,[195] was his superior because of his knowledge and expertise in sailing. Before Shimaya made his exploratory voyage to the Bonin Islands in 1675, the shogunate had asked Higuchi to make the trip, but he had declined on the grounds of age, and command of the expedition was then given to Shimaya.[196]

That European navigational science was influential for Japanese surveying practices is also attested by the maps that, according to Hosoi, were presented to all successful surveying apprentices upon graduation. Two of these—the *Jagatara kaijō bundo zu* (Chart of Southeast Asia) and the *Nihon seizu* (Chart of Japan)—were marine charts; the third, the *Bankoku sōzu* (Map of all the countries), was also based on European knowledge. Any surveyor who did not possess them, Hosoi remarks, could not be considered genuine.[197] There was also a table of declinations, the *Nanban goyomi* (Western almanac), which served as a similar mark of proficiency during the Edo period.[198]

The State and Cartography

The state has exerted an important influence on the development of Japanese cartography. We have already seen, for example, how an entry in the *Nihon kōki* notes that provincial maps were ordered to be made in 796, and

191. *Watarante* or *kuhadarantei* (also *kuhatarantei*) is a corruption of the Portuguese *quadrante* or Spanish *cuadrante*; *konpansu* and *konpasu* of the Portuguese *compasso* or Spanish *compás*; *isutarabiyo*, *asutarabiyo*, or *isutarahi* of the Portuguese *astrolábio* or Spanish *astrolabio*. The reason for the different Japanese renditions is that the same non-Japanese word might be heard, and thus written down, differently. There is also the possibility that such words, although originally Portuguese, were transmitted to the Japanese by the Dutch, since Portuguese was the lingua franca for Dutch-Japanese trade as late as the 1660s. Since, however, Portuguese were involved in educating Hayashi and Higuchi, it is more likely that there was a direct Portuguese influence rather than an indirect one.

192. National Archives in Tokyo; Matsumiya notes that the *konpasu* was called *passuru* (*passer*) by the Dutch.

193. Japanese astrolabes and quadrants are in the Akioka Collection of the National Museum of Japanese History, Sakura, Chiba Prefecture. Compasses are in the Kunōzan Tōshōgū Museum, Shizuoka (allegedly, Tokugawa Ieyasu's belongings), and in the Matsuura Historical Museum, Hirado, Nagasaki Prefecture.

194. See note 145 above.

195. This is mentioned in Ro Sōsetsu's letter in the *Sokuryō higen* (Secrets of surveying), edited by Hosoi Kōtaku, 1728, a collection of statements by people involved in the study of European navigational astronomy in Nagasaki after the closure of the country. The manuscript is preserved at Tōhoku University Library, Sendai.

196. Ro, *Senminden*, chap. 2, s.v. "Shimaya Kenryū" (note 187).

197. Hosoi called the maps *karuta*; see the *Hiden chiiki zuhō daizensho* (note 63). The *Bankoku sōzu* is discussed below.

198. See Hidaka Jikichi, "Hyūga sadowara hanshi Hidaka Shigemasa no Nanban ryū chōkenjutsu sonota" (Hidaka Shigemasa's studies on the surveying of the Occidental school, etc.), *Kagakushi Kenkyū* 44 (1957): 17–24.

FIG. 11.32. MAP OF SEBA COUNTY, ECHIGO PROVINCE (TODAY NIIGATA PREFECTURE), CA. 1597. Rivers, roads, towns, villages, temples, shrines, a castle, and a fortress are shown, as are paddy fields, other cultivated land, and waste land. The standard land productivity of each village is also indicated on the manuscript. This map was probably a product of the cadastral survey undertaken by the Toyotomi regime between 1582 and 1603.
Size of the original: 243 × 693 cm. Uesugi family, Yonezawa, Yamagata Prefecture. Photograph courtesy of Kazutaka Unno.

thereafter there are fragmentary records pertaining to the compilation of regional maps. An example is in the *Azuma kagami* (Mirror of the eastern lands), a chronicle of the Kamakura shogunate from 1180 to 1266, in which it is recorded that in 1188 Minamoto no Yoritomo (1147–99) ordered that a navigational chart be compiled of Kikai Island, possibly as part of his attempt to annihilate the remnants of the Heishi army and bring the sea south of Kyūshū under his control.[199] According to the same chronicle, when he had gained control of northeastern Honshū in the following year, Yoritomo received from the inhabitants maps of the provinces of Mutsu and Dewa. The originals were probably prepared by the Fujiwara family, the former rulers of the region, since they contained detailed information on the location of mountains, rivers, seas, plains, villages, and fields.[200] Despite such references, there are no known records about the compilation of national maps until 1591, when the government of Toyotomi Hideyoshi embarked on such a project. We can only presume the reason was that compiling national maps was commonplace and that each administration kept, used, and possibly ordered maps of the country for official purposes.

The main purpose of the survey was to assess the amount of land owned privately so that taxes could be levied. Standard land productivity was to be entered into registers called *kenchi chō* (land assessment books) or *gozen chō* (books presented to the emperor) by the authorities of the feudal lord of each province and was to be accompanied by maps. The *Tamon'in nikki* (Diary of Tamon'in), for example, refers to such maps in an entry under 1591: "I hear that orders have been issued that maps of all the counties across the nation should be submitted with all paddy fields entered together with the sea, mountains, rivers, hamlets, temples, shrines, and areas of paddy fields, and that they are to be kept at the royal court."[201] It was thus the responsibility of the feudal lords to enter the location of the main topographic features within their domains, emphasizing productive land. Two examples of maps from the survey, dating to 1597 at the latest, are those of Seba County (fig. 11.32) and Kubiki County in Echigo Province (now Niigata Prefecture). Both maps are oriented to the southeast, presumably so that the coastline would be at the bottom. In addition to containing topographic detail, they are of high artistic quality.[202]

199. See the *Azuma kagami*, which is in vols. 32 and 33 of the *Shintei zōho kokushi taikei*, vol. 32, chap. 8 (note 37). The first half of the *Azuma kagami* was compiled during the latter half of the thirteenth century, the second at the beginning of the fourteenth.

200. See the *Azuma kagami* in *Shintei zōho kokushi taikei*, vol. 32, chap. 9 (note 37). According to it, Yoritomo was at a loss when he heard that the administrative maps of Mutsu and Dewa had been destroyed in the fire at Hiraizumi castle. Two brothers who were familiar with the provinces, Buzen no Suke Sanetoshi and Tachibana no Tōgo Sanemasa, then presented him with these detailed maps.

201. The entry cited is under the twenty-ninth day of the seventh month, in the nineteenth year of Tenshō (1591, the era being 1573–91): see Tsuji Zennosuke, ed., *Tamon'in nikki* (Diary of the Tamon'in), 5 vols. (Tokyo: Sankyōshoin, 1935–39), 4:306. The diary was kept from 1478 to 1617 and is preserved at Kōfuku Temple, on the grounds of which is the Tamon'in (a small branch temple), in Nara.

202. Facsimiles of the maps are in Tōkyō Daigaku Shiryō Hensanjo (Historiographical Institute, Tokyo University), ed., *Echigo no kuni gun* (or *kōri*) *ezu* (Maps of counties in Echigo Province), vol. 1 (the Kubiki map, 340 × 586 cm) and vol. 2 (the Seba map) (Tokyo: Tōkyō Daigaku, 1983, 1985, and 1987). The book published in 1987 is an explanation and index containing the inscriptions on these maps. For a discussion

The yield of the productive land was calculated according to the amount of unhulled rice harvested per square *shaku* (ca. 30 cm), with allowances being made for variables such as soil type. The assessment of tax was to be based on this fixed amount, with similar allowances for factors such as soil type, difficulty of cultivation, upkeep of irrigation channels, and distance of transportation. The main classification of wet (paddy) fields was as follows: those producing 1.5 *koku* (7.5 bushels) per square *shaku* were first-class fields; 1.3 *koku*, second-class fields; and 1.1 *koku*, third-class fields. The final register defined the value of the cultivated land according to its yield in *koku*. Land transactions thereafter were made in *koku* rather than according to area.[203]

There is a strong possibility that Hideyoshi's land survey was less complete or less useful than the original order had demanded. Although by the time it was finished every province may have been surveyed, the project fell short of its objective. This was also true despite its thoroughgoing nature and the "ferocity," as Sansom terms it, with which the edicts pertaining to it were framed. Resistance by the peasantry, experiencing a time of prosperity, was intense: compliance meant revealing to the authorities the correct area of the land owned privately as well as the amount of previously evaded taxes. Threats of execution, including crucifixion, were issued by Hideyoshi in an attempt to crush resistance so that the survey would be thorough. Sansom's remark that the provinces were "not completely" surveyed suggests that the gaps were caused by resistance and deceit on the part of landholders.[204] Such a shortfall is confirmed by contemporary documents. Registers and maps submitted by the beginning of 1593 were entered on a list sent to Maeda Gen'i (1539–1602), a high-ranking vassal of Hideyoshi, by Komai Shigekatsu (1558–1635), an assistant to Hideyoshi's nephew Toyotomi Hidetsugu (1568–95). This list notes that twenty-nine of the sixty-six provinces submitted only registers, whereas thirteen submitted both; these thirteen were Kazusa, Shimōsa, Musashi, Sagami, Shima, Iga, Wakasa, Yamashiro, Inaba, and Hōki in Honshū; Tosa in Shikoku; and Bungo and Hizen in Kyūshū.[205] It may be related to this situation that in 1605 the Tokugawa shogunate issued a further order calling for newly compiled provincial maps.[206] Whether this was to be a continuation of the Hideyoshi survey or a completely different one still has not been determined.

We do know, according to evidence presented by Brown, that there were considerable errors in the results of surveys during the early Tokugawa period.[207] Measurements were routinely biased downward because surveyors rounded down, but not up, to the next *ken* (about 1.82 m) or half a *ken*. Hemp ropes—subject to stretching or contraction with varying moisture conditions—were routinely used, despite widespread knowledge that they were a significant source of error. Further, the principle of area calculation, which relied on the gridlike crossed-rope technique, was adequate for square or rectangular areas but could not satisfactorily accommodate irregular or curved boundaries. This lack of surveying accuracy resulted in an interesting paradox. Although Japanese rulers would clearly have benefited from surveys that did not underestimate land area, and though the technology was certainly available for more accurate measurements, routine land surveying did not develop as might have been expected. Brown proposes a number of reasons for this, including the deprecation of practical sciences by the samurai class (from which most surveyors came), the lack of mathematical education, the secrecy of surveying techniques, and most of all the confinement of the demand for surveying within the public sector, stifling competition and the incentive for improvement.[208]

In total, five large projects to compile provincial maps were undertaken during the course of the Tokugawa shogunate (appendix 11.8), although no permanent organizations resulted at either the national or the provincial level. Official cartographers (*ezukata*) were appointed for each project by the shogunate and the clans, but usually only a few of them did the actual work. Also involved were painters, calligraphers, and handymen when required, and presumably also some surveying and drafting specialists. The best documented of the projects is the third, which was conducted between 1644 and the mid-1650s.[209] Detailed instructions were given for com-

---

of the maps, refer to Itō Tasaburō, "Echigo Uesugi shi ryōgoku kenkyū no nishiryō" (Two historical materials for studying the domains of the Uesugi family in Echigo), *Nihon Rekishi* 138 (1959): 2–14; and also to Kazutaka Unno, "Government Cartography in Sixteenth Century Japan," *Imago Mundi* 43 (1991): 86–91.

203. See Sansom, *History of Japan*, 2:316–19 (note 32).

204. See Sansom, *History of Japan*, 2:316–19 (note 32).

205. See Unno, "Government Cartography" (note 202). See also *Shokoku gozenchō* (Books of the standard land productivity of the provinces) by Komai Shigekatsu, in *Mikikigusa* (Collection of various writings, 1830–ca. 1865), comp. Miyazaki Shigemi, vol. 6, no. 7; this is a manuscript at the National Archives in Tokyo, and the date given for the document corresponds to 6 February 1593. This material was first mentioned academically in Kuroda Hideo, "Edo bakufu kuniezu gōchō kanken" (A personal view of provincial maps and books of standard land productivity prepared by the order of the Tokugawa shogunate), *Rekishi Chiri* 93, no. 2 (1977): 19–42.

206. See the *Tokugawa jikki* (Authentic records of the Tokugawa shogunate [1849]), bk. 1, in vol. 38 of the *Shintei zōho kokushi taikei* (note 37), and the *Kansei chōshū shokafu* (Genealogies of families revised during the Kansei era), 26 vols. (1812; printed by Zoku Gunsho Ruijū Kanseikai, 1964–67), chap. 494, s.v. "Tsuda Hidemasa."

207. Philip C. Brown, "Never the Twain Shall Meet: European Land Survey Techniques in Tokugawa Japan," *Chinese Science* 9 (1989): 53–79.

208. Brown, "Land Survey Techniques," 78–79 (note 207).

209. For details of these projects, refer to Kawamura Hirotada, *Edo bakufu sen kuniezu no kenkyū* (A study of the provincial maps com-

piling the maps from the field surveys that were undertaken by the most powerful clan in the province. We are told that the scale was to be six *sun* to one *ri* (1:21,600) and that bold red lines were to be used for arterial roads, with marks at every one *ri*, and thinner lines for smaller roads. In the case of rivers without bridges, it was to be noted whether ferries were available or a traveler had to wade across. With seashores, the maps had to record whether they were rocky or had sandy beaches, and also if ships could be moored there (plate 25).[210] Identical instructions and the same scale were later stipulated for the fourth and fifth projects.

Owing to lack of information, it is impossible to define the areas covered in the first two projects. The maps from the third project include the area from Sakhalin (Karafuto) and the Kurile Islands (Chishima) in the north to the Ryūkyū Islands in the south. The map of Sakhalin, the Kuriles, and Hokkaidō by the Matsumae clan was compiled on a scale smaller than 1:21,600, and the outlines of the territories are greatly distorted.[211] That the shogunal authorities accepted this map and did not order the area resurveyed reflects their lack of interest in this region. The maps were designed, like those of Hideyoshi's survey, to express crop productivity for purposes of taxation; the northern frontier was notorious for poor crops because of its adverse climate.

The detailed instructions for the third project do not seem to have given sufficient thought to the problems involved in joining together the provincial surveys to form a national map. This task would have been practically impossible from the maps produced because the mountainous areas along the map borders were not accurately surveyed. The peripheral mountains were depicted pictorially only as they would have been seen from inside each province. In the fourth project, therefore, an order was introduced to draw the mountainous borders in the same way as the rest of the province, thereby facilitating the construction of a national map.[212]

Among the provincial maps compiled from the five projects and submitted to the shogunate, only the Tenpō provincial maps from the fifth project are preserved as a complete set.[213] In addition to these, there are maps of six provinces on eight sheets from the Genroku provincial maps (fourth project) preserved in the National Archives in Tokyo.[214] Duplicates and drafts made by the feudal lords exist in libraries and museums throughout Japan, along with later reproductions of the originals.[215] Many of the maps are undated, with dates and names of the lords in charge of map compilation being a feature added only in the fourth and fifth projects. Difficulties in studying these maps are compounded because none of the maps can be dated with any certainty to the first and second projects except a few examples such as the map of Settsu Province (see fig. 11.33).

As we have seen, national maps based on the provincial ones were part of the government's plans. At present two prototypes of these national maps are known, one based on the third project and the other on the fourth. Also extant are two national maps likely to have been based on the first and second projects, although it is unknown which belongs to which project. These latter are the two large manuscript maps at the National Diet Library, Tokyo, and the Saga Prefectural Library, Saga. The map at the National Diet Library is traditionally called the Keichō map of Japan, after the era when the order was given (1605 being the tenth year of the Keichō era [1596–1614]) (plate 26). The date of completion is not specified but appears to have been about 1639: strips of paper attached to the map give the names of daimyos for 1639 and 1653, the latter date being surmised as that of a revision. The symbols for the seats of the clans and the configuration of the country as a whole suggest that the map was not composed from scratch in 1639 but was based on earlier information. Characteristic of this configuration is a compressed northern Honshū, the shallow

---

piled by the Tokugawa shogunate) (Tokyo: Kokon Shoin, 1984); and to Hirotada Kawamura, "*Kuni-ezu* (Provincial Maps) Compiled by the Tokugawa Shogunate in Japan," *Imago Mundi* 41 (1989): 70–75.

210. See the documents pertaining to the old and new provincial maps that were submitted to the shogunate by Kondō Morishige in 1817; they are referred to in Kawada, "Honpō chizukō" (note 2). They are also reproduced in the *Kondō Seisai zenshū* (Collection of Kondō Seisai's [Morishige] works), 3 vols. (Tokyo: Kokusho Kankōkai, 1905–6), vol. 3.

211. The map by the Matsumae clan is not extant, but its reduced image is on the *Kōkoku michinori zu* (see below, pp. 399 and 400). The maps of the Ryūkyū Islands were completed by the Satsuma clan in 1649, and they are preserved at the Historiographical Institute at Tokyo University, *Shimazu ke monjo* (Documents of the Shimazu family), 76–2–4, 5, and 6. The maps are reproduced in color in *Ryūkyū Kuniezu Shiryōshū* (Collected historical materials of provincial maps of Ryūkyū), no. 1 (Naha: Okinawa ken Kyōiku Iinkai, 1992).

212. The order was issued in 1696; see *Genroku nenroku* (Diary of the Genroku years, 1688–1703 in 64 vols.), in the *Ryūei Hinamiki* (Diary of the shogunate, 1656–1856, manuscript in 734 vols.), owned by the National Archives in Tokyo. See also Fukui Tamotsu, *Naikaku Bunko shoshi no kenkyū* (Studies on the bibliography of the Naikaku Library) (Tokyo: Seishōdō, 1980), 365.

213. The Tenpō era was 1830–43. Eighty-three sheets of the original Tenpō provincial maps, thirty-six spare sheets, and the cases for keeping the maps are preserved at the National Archives in Tokyo. See Fukui, *Naikaku Bunko shoshi no kenkyū*, 355–60 (note 212).

214. The eight sheets of the original at the National Archives cover the provinces of Hitachi, Shimōsa, Hyūga, Ōsumi, Satsuma, and (on three sheets) Ryūkyū.

215. On the duplicates, drafts, and later reproductions, see "Kagaku Kenkyūhi ni yoru Kenkyū no Hōkoku" (Reports on the research depending on scientific research expenses), "Genson Kochizu no Rekishi Chirigakuteki Kenkyū (Ippan Kenkyū A) (Historical geographical research on extant old maps [general study A])," *Tōkyō Daigaku Shiryō Hensanjo Hō* 16 (1981): 25–40, esp. 31–33.

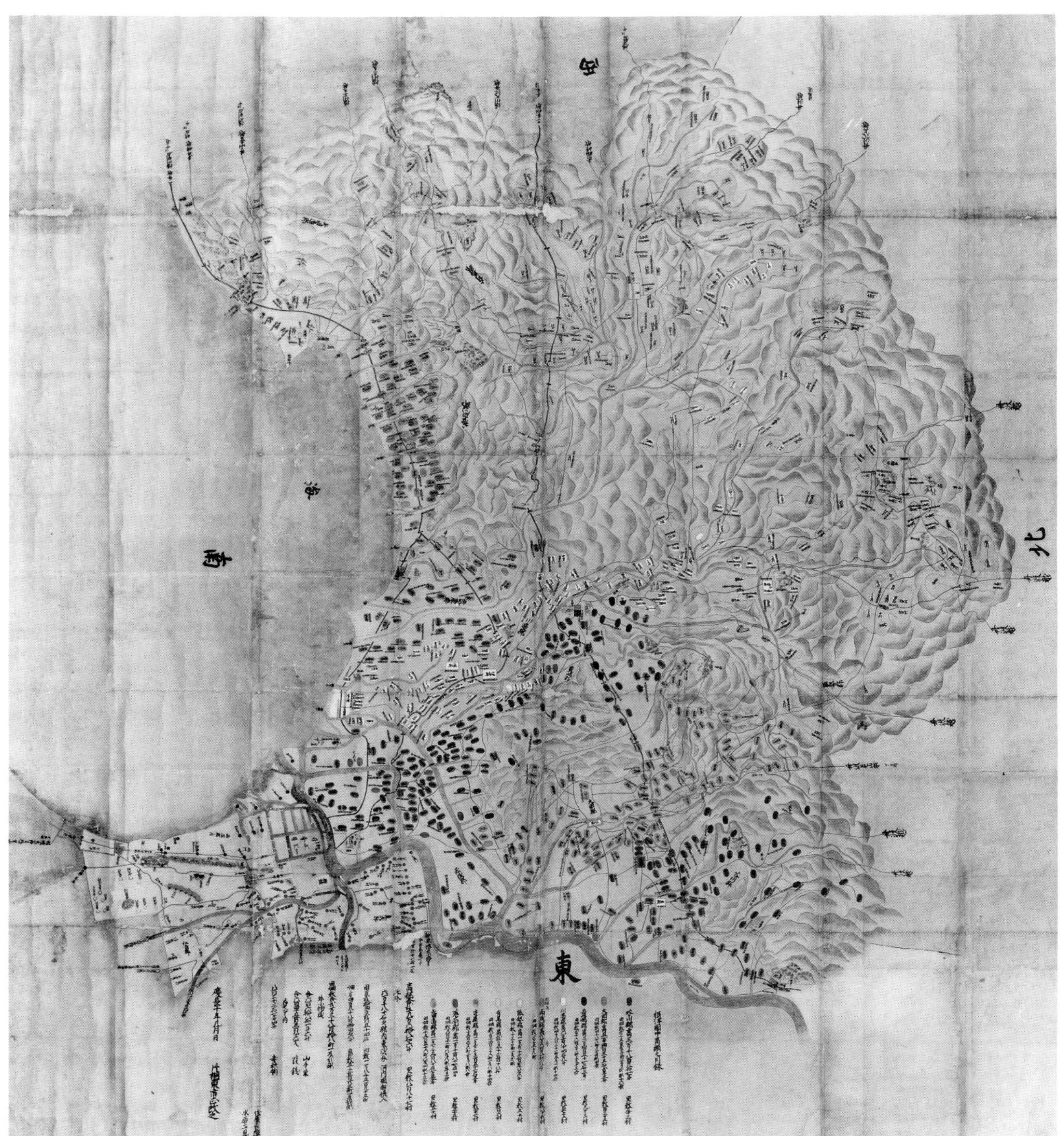

FIG. 11.33. AN EXAMPLE OF A KEICHŌ PROVINCIAL MAP: SETTSU PROVINCE (NOW SPLIT INTO PARTS OF HYŌGO AND ŌSAKA PREFECTURES). This manuscript has an inscription dating it to the ninth month of the tenth year of Keichō (1605) and noting that its execution was supervised by Katagiri Ichinokami (Katsumoto), the governor of Settsu, Kawachi, and Izumi provinces. The color of the ovals containing the names of villages differs by county. The marks along the main roads are spaced one *ri* apart. There is no particular orientation: information is recorded in different directions.

Size of the original: 249 × 225 cm. Nishinomiya City Office, Nishinomiya, Hyōgo Prefecture. Photograph courtesy of Kazutaka Unno.

curve of Mutsu Bay into the land area, and a Kyūshū elongated from north to south.

The map at the Saga Prefectural Library consists of three parts that together measure 622 by 674 centimeters.[216] There is no indication of when it was completed, even though it has the standard land productivity of each province in the margins. Northern Honshū and Kyūshū are closer to reality than on the Keichō map, but there is no difference between them with regard to the shape of Shikoku, in particular to the indistinctly drawn curve of Tosa Bay and Muroto and Ashizuri peninsulas.

The third project produced a markedly improved general map, called the Shōhō map of Japan. Although it is not known if the original draft still exists, the copy thought to be the closest to it is the *Kōkoku michinori zu* (Map of the distances of Japan), compiled from the Shōhō provincial maps by the military engineer and surveyor Hōjō Ujinaga (1608–70) (fig. 11.34).[217] On it the main part of the archipelago appears almost as accurately as on today's maps; the Ryūkyūs are excluded. The reason for the high degree of accuracy is still not known. Hōjō was provided only with the distances along the main roads between villages and towns, and since such information alone is not sufficient to make a good map, it is thought that he might have taken measurements for latitude. Errors were made in Hokkaidō, Sakhalin, and the Kuriles because for this area the erroneous map made by the Matsumae clan was used. Despite these errors, this is considered to be the earliest map preserved in the world that includes a large number of place-names in Sakhalin and the Kuriles.

The "Genroku map of Japan" is in turn based on the provincial maps from the fourth project.[218] It covers the area from Sakhalin and the Kuriles to the Ryūkyūs and Yonakuni Island in the Yaeyama group, as well as the southern part of the Korean peninsula. The content suggests that it was composed primarily to show coastal routes: not much information is given about the areas inland. The greatest errors appear in the northern tip of Honshū and in Shikoku: the first has a very small Shimokita peninsula and the second is shown as slanting more to the southwest than it should be. The authorities therefore commissioned Hōjō's son Ujisuke (1666–1727) to revise the map in 1717, but the results were not satisfactory. Two years later the shogun Yoshimune (1684–1751) made the mathematician Takebe Katahiro (1664–1739) responsible for mapmaking and personally gave him instructions for revising the map of Japan.[219] According to these instructions Takebe selected mountains to obtain a view of the peaks to be drawn on the map. His information was supplemented by that of the clans, which gave the angles at which the peaks were observed from their territories. When combined, this intersecting system of angles formed the network for Takebe's map.[220] Takebe completed a general map in 1723 and finished its revision in 1728. This is called the Kyōhō map of Japan, named after the contemporary era (1716–35).[221] Its scale was 1:216,000 (six *bu* to one *ri*), and though it corrected many of the mistakes on the Genroku map, including the slant of Shikoku, it was still rather inaccurate and indeed inferior to the Shōhō map. Takebe learned from his experience and noted that, in future, reliable maps would have to be based on observations of latitude and longitude. He concluded that the method of surveying then in use (intersection) was inadequate if precision was the objective. In other words, traversing and computing angles of high points were not enough.[222]

In addition to provincial and national maps, the Tokugawa shogunate commissioned other types of maps, plans, and charts. The third project included an order for each clan to compile and to submit plans showing the locations of their capitals; about 160 were collected by the government over sixteen years.[223] Sixty-three of these large-scale "Shōhō castle plans" (*shiro ezu*) are extant

216. The map is reproduced in Akioka Takejirō, *Nihon chizu sakusei shi* (A history of the making of Japanese maps) (Tokyo: Kajima Kenkyūjo Shuppankai, 1971), fig. 57.

217. A color reproduction is in Unno, Oda, and Muroga, *Nihon kochizu taisei*, vol. 1, pl. 19 (note 8), and *Kokushi daijiten*, vol. 11, color pages "Nihon zu" (Maps of Japan), pl. 5 (note 95). There are many other copies of the Shōhō map under different titles.

218. Maps preserved at the Meiji University Library, Tokyo (two sheets, each 309 × 222 cm at 1:324,000) and the Shizuoka Prefectural Central Library, Shizuoka (*Kōkoku enkai ritei zenzu* [Map of the sea routes along the coasts of Japan], 355 × 446 cm) are the only known samples extant today. It appears that they number fewer than those from the Shōhō project. The Meiji University map is reproduced in color in Unno, Oda, and Muroga, *Nihon kochizu taisei*, vol. 1, pl. 20 (note 8), and *Kokushi daijiten*, vol. 11, color pages "Nihon zu" (Maps of Japan), pl. 6 (note 95). The map in Shizuoka is reproduced in color in *Nihon no chizu: Kansen chizu no hattatsu* (note 13).

219. For the cartographical works of Hōjō Ujisuke and Takebe Katahiro, see Kawamura, *Edo bakufu sen kuniezu no kenkyū*, 320–49 (note 209). For Yoshimune's instructions, see Takebe Katahiro, *Nihon ezu shitate sōrō ikken* (The process of compiling a map of Japan [ca. 1723]), in vol. 3 of the *Kondō Seisai zenshū* (note 210).

220. See Kawamura, *Edo bakufu sen kuniezu no kenkyū*, 320–49 (note 209).

221. A copy of the Kyōhō map composed in 1793 (four sheets, from west to east: 147 × 188 cm, 173 × 203 cm, 172 × 207 cm, and 149 × 208 cm) was kept at the Rikuchi Sokuryō Bu (Army's Department of Land Survey) until it was lost in World War II. It is reproduced in Unno, Oda, and Muroga, *Nihon kochizu taisei*, vol. 1, fig. 20 (note 8).

222. See Takebe, *Nihon ezu shitate sōrō ikken* (note 219); chap. 37 of the *Kōsho koji* (Historical allusion of worthy books, 1826), in *Kondō Seisai zenshū*, vol. 3 (note 210); and Ōta Nanpo, *Chikkyō yohitsu besshū* (Superfluous writings at Bamboo Bridge, Edo Castle, extra volume, ca. 1803), chap. 12 (modern edition [Tokyo: Kondō Shuppansha, 1985], 332–33). Triangulation in Japan began in 1872.

223. See chap. 28 of the *Kōsho koji*, in *Kondō Seisai zenshū*, vol. 3 (note 210); and Kawamura, *Edo bakufu sen kuniezu no kenkyū*, 121–23 (note 209).

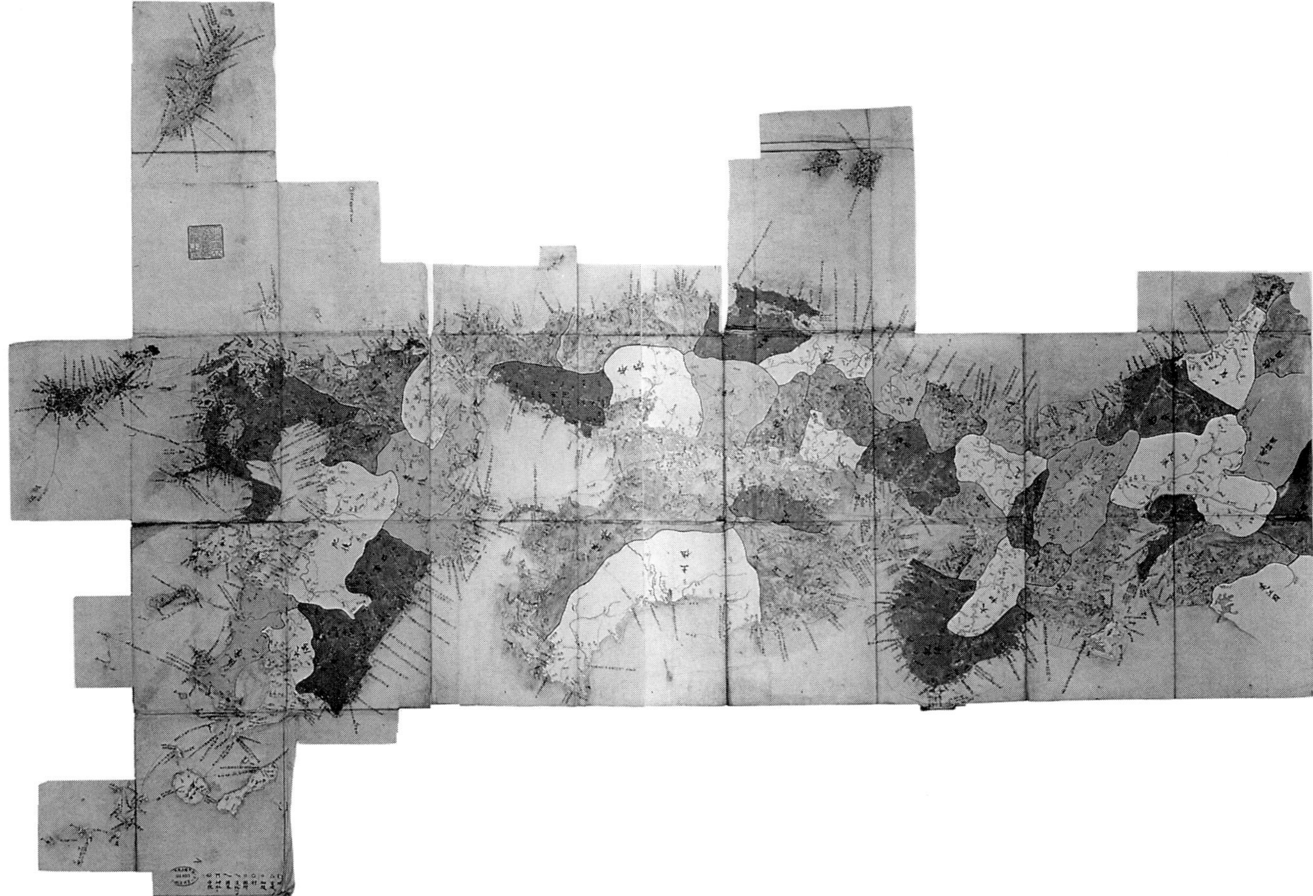

FIG. 11.34. THE *KŌKOKU MICHINORI ZU*, CA. 1670. This manuscript is compiled from the provincial maps of the third Tokugawa survey and generally known as the Shōhō map of Japan after the era when the survey began (1644–47). Whereas most of the archipelago is portrayed rather accurately, the islands to the north of Honshū (Hokkaidō, Sakhalin [Karafuto], and the Kuriles [Chishima]) are notably poor. This is because of the erroneous map compiled by the regional daimyate (Matsumae), which evidently did not follow the detailed instructions given by the shogunate. The scale is approximately 1:432,000. Size of the original: 129 × 178 cm (western part) and 162 × 83 cm (eastern part). By permission of the Ōsaka Prefectural Nakanoshima Library, Ōsaka.

and preserved at the National Archives in Tokyo.[224] They are so named because of the emphasis placed on the areas where the castles were situated (fig. 11.35). Characteristically, the moats, stone walls, and gates of the castles as well as the widths of the streets are all shown accurately on these plans.

Town plans, notable for their accuracy and minute detail, were also compiled of Edo, Kyōto, and Ōsaka, cities under the direct control of the shogunate.[225] Government experts did the surveying and drew plans that were to exert a strong influence on subsequent plans of these cities. For instance, the task of compiling the first accurate plan of Edo after the fire of 1657 (fig. 11.36) fell to the elder Hōjō.[226] Plans of Kyōto dating to 1637 and about 1642 by the Nakai family have also survived. That of 1637 is titled *Rakuchū ezu* (Plan of Kyōto) and

224. These plans have been reproduced by the National Archives since 1976 in annual publications under the title *Shōhō shiro ezu* (Shōhō castle plans). The project is still in progress, and fifty-five items had been issued as of the end of December 1991. The plans of Kokura (Buzen Province), Hiroshima (Aki Province), Matsue (Izumo Province), Kasama (Hitachi Province), and Sendai (Mutsu Province) are reproduced in color in Unno, Oda, and Muroga, *Nihon kochizu taisei*, vol. 1, pls. 96–100 (note 8).

225. A general history on city plans is Yamori Kazuhiko, *Toshizu no rekishi* (History of city maps), 2 vols. (Tokyo: Kōdansha, 1974–75), vol. 1. Another that concentrates on Edo is Iida Ryūichi and Tawara Motoaki, *Edozu no rekishi* (History of the maps of Edo), 2 vols. (Tokyo: Tsukiji Shokan, 1988). The subject is touched upon, although briefly, by Cortazzi, *Isles of Gold*, 39, 50, 54, 56–58 (note 14). Some early plans are reproduced in Harada Tomohiko and Nishikawa Kōji, eds., *Nihon no shigai kozu* (Old Japanese plans), 2 vols., *Nishi Nihon hen* (Western part of Japan), and *Higashi Nihon hen* (Eastern part of Japan) (Tokyo: Kajima Shuppankai, 1972–73).

226. It is generally accepted that Hōjō's work is associated with the

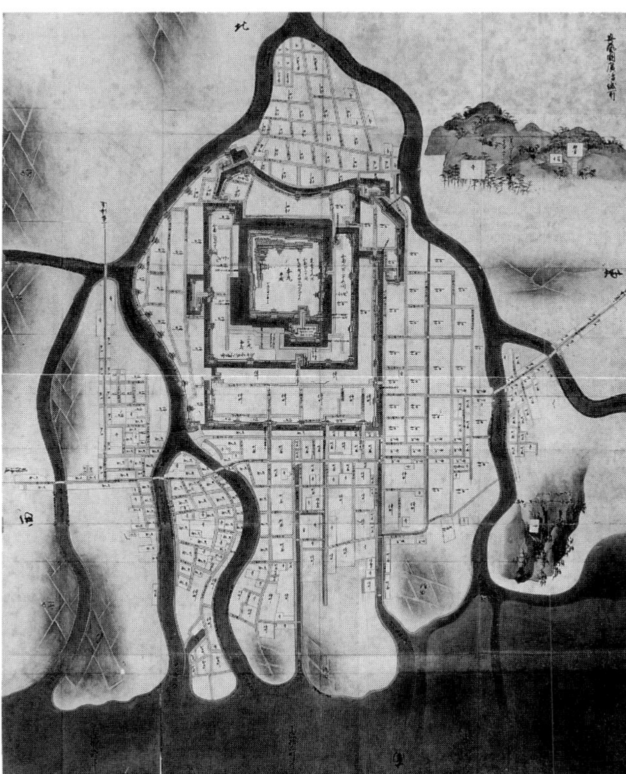

FIG. 11.35. AN EXAMPLE OF A SHŌHŌ CASTLE PLAN: HIROSHIMA, CA. 1645. Although the term *shiro ezu* (castle plan) was used for these enterprises, their content was not limited to the castles, as is demonstrated in this manuscript.
Size of the original: 242 × 193 cm. By permission of the National Archives, Tokyo.

is at a scale of 1:1,500. The revision of about 1642 is on a scale of approximately 1:1,263 (fig. 11.37).[227] It is also recorded that the Nakai family was ordered to map Ōsaka in 1613, but the oldest surviving plan of that city dates to about 1655.[228]

Maps showing sea and land routes were also commissioned by the shogunate.[229] Route maps of the Tōkaidō (Tōkai road) from Kyōto to Edo were ordered to be made in 1634, 1646, and 1651.[230] Such land itineraries were

manuscript plan of Edo in the possession of the Mitsui Library in Tokyo, which is reproduced in *Kanbun gomai zu* (Five-sheet plans of Edo published during the Kanbun era [1661–72]) (Tokyo: Haga Shoten, 1970); in *Nihon no chizu: Kansen chizu no hattasu* (note 13); and in Iida and Tawara, *Edozu no rekishi* (note 225).

227. The first (505 × 236 cm) is preserved at the Archives and Mausoleums Department of the Imperial Household Agency. The revision of ca. 1642 (636 × 283 cm and 262 × 30 cm) is kept at the Kyōto University Library. Reproductions are the *Kunaichō Shoryōbu shozō Rakuchū ezu* (Plan of Kyōto owned by the Imperial Household Agency) (Tokyo: Yoshikawa Kōbunkan, 1969) and the *Rakuchū ezu: Kan'ei Manji zen* (Plan of Kyōto: Kan'ei era [1624–43], before the Manji era [1658–60]) (Kyōto: Rinsen Shoten, 1979); the first is of the 1637 manuscript, the second of the ca. 1642 revision.

228. It is a manuscript entitled *Ōsaka sangō machi ezu* (Map of the three districts of Ōsaka) (214 × 236 cm), kept at the Ōsaka City Museum. A color reproduction is in Unno, Oda, and Muroga, *Nihon kochizu taisei*, vol. 1, pl. 87 (note 8). The Nakai family were hereditary *daiku gashira* (general carpenters) appointed by the shogunate in Kyōto.

229. Early examples are the *Kisoji Nakasendō Tōkaidō ezu* (Map of the Kiso/Nakasen road and the Tōkai road, 120 × 1,920 cm) and the *Saigokusuji kairiku ezu* (Map of the sea and land routes of the western regions, 124.8 × 732.6 cm), both of 1668 and owned by the National Diet Library. Parts of them are reproduced in color in Unno, Oda, and Muroga, *Nihon kochizu taisei*, vol. 1, pls. 114 and 115 (note 8).

230. In 1634 Miyagi Kazunami and Akiyama Masashige were ordered to investigate the roads and lodgings from Edo to Kyōto in preparation for the visit of the shogun Iemitsu (1604–51, r. 1623–51) to the imperial court at Kyōto; they went back to Edo just over a month later and

FIG. 11.36. PART OF THE PLAN OF EDO BY HŌJŌ UJINAGA, CA. 1658. This plan was compiled from the results of a survey ordered by the shogunate after the fire of 1657. There is no particular orientation, but the detail shown here has east at the top; Edo castle is in the center. Compass roses with twelve or twenty-four radiating lines begin at some of the gates of the castle, the lines being colored red or green. The scale of the manuscript is one *bu* to four *ken*, giving a value of between 1:2,400 (in which 1 *ken* = 6 *shaku*) and 1:2,600 (in which 1 *ken* = 6.5 *shaku*).
Size of the original: 318 × 418 cm. By permission of the Mitsui Library, Tokyo.

characteristically in linear form with no regard for distances and directions, but they did take into account the aesthetics of portraying landscapes (fig. 11.38). For the sea routes, a survey was ordered in 1667 of coastal Honshū from Edo to the west as well as of coastal Shikoku and Kyūshū. The charts from this survey differed from the marine charts discussed above in that the coastlines were represented as a long line with fewer indentations than had the interior been expressed correctly.[231] Shimaya's survey of 1670–71 was also sponsored by the government, and his marine charts, which were compiled

submitted the map. See the *Tokugawa jikki*, bk. 2, in vol. 39 of the *Shintei zōho kokushi taikei* (note 37). In 1647 Matsuda Sadahei and Iikawa Naonobu were ordered to investigate and make a map of the roads, post towns, and bridges from Edo to Ōsaka. See the *Tokugawa jikki*, bk. 3, in vol. 40 of the *Shintei zōho kokushi taikei*. Refer to Ashida Ijin (or Koreto), "Chizu to kōtsū bunka" (Maps and transportation culture), *Kōtsū bunka* 3–5 (1938–39): 282–90, 358–64, and 445–54.

231. These charts are referred to on the 1680 manuscript *Kaihin shūkō zu* (Chart of the sea routes along the coasts) by Ebi Gensui (or Gaishi), a participant in the survey. There are three copies of the chart, preserved at the Geographical Institute at Kyōto University (approximately 1:64,800, three folding books), at the Kōbe City Central Library

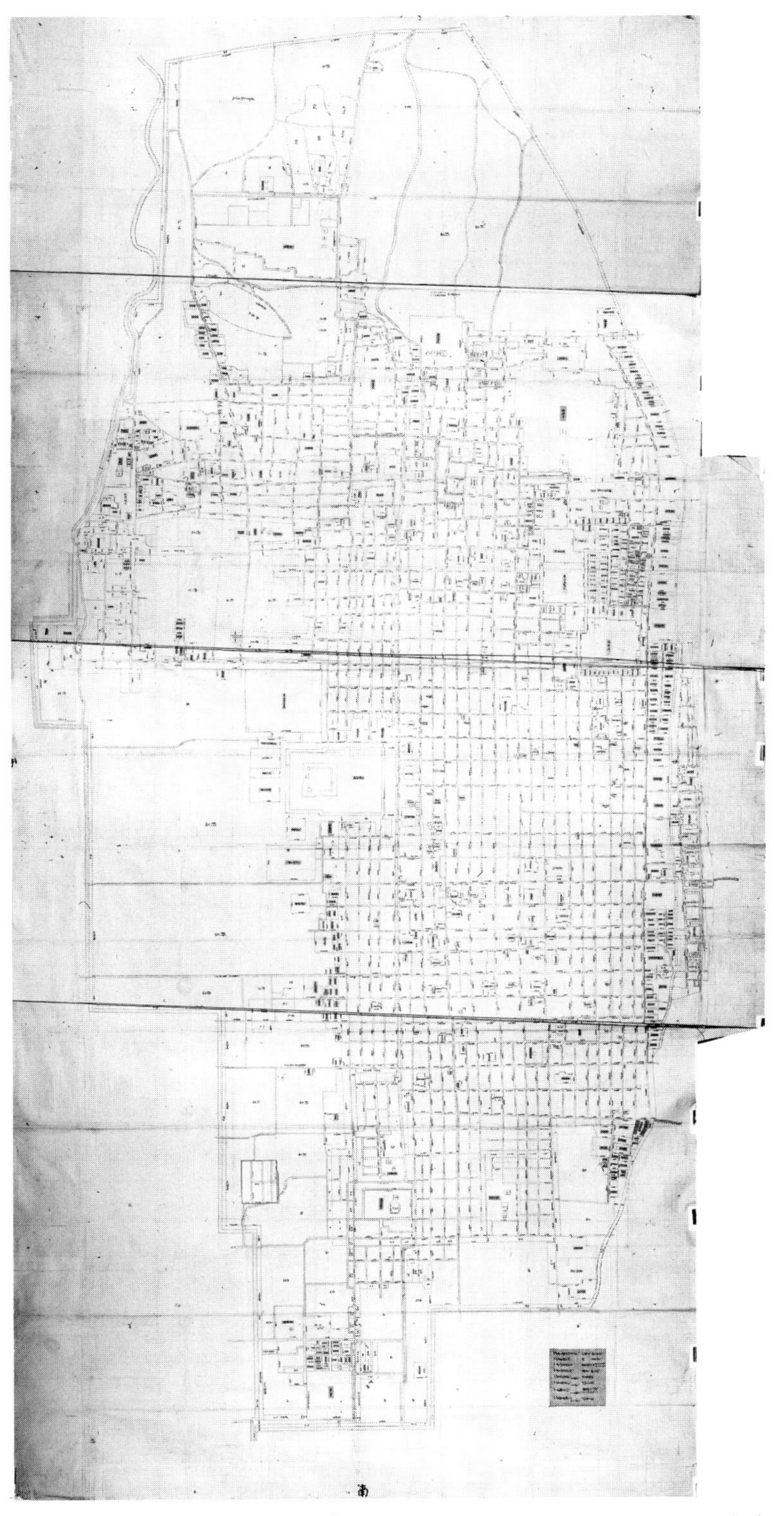

FIG. 11.37. PART OF THE REVISION OF THE *RAKUCHŪ EZU*, CA. 1642, BY SOME MEMBERS OF THE NAKAI FAMILY. Although there is no particular orientation, north is at the top of this detail, which shows the northern half of Kyōto. The manuscript was drawn on paper. The lines were made by pressing with a tracing spatula. The scale is about 1:1,263 at the ratio of one *ken* (1.81 m) to six *shaku* (1 *shaku* = 30.3 cm); each grid is 4.75 square *bu* (1 *bu* = 3 mm), which represents an area of ten square *ken*.
Size of the entire original: 636 × 283 cm and 262 × 30 cm. By permission of the Kyōto University Library, Kyōto.

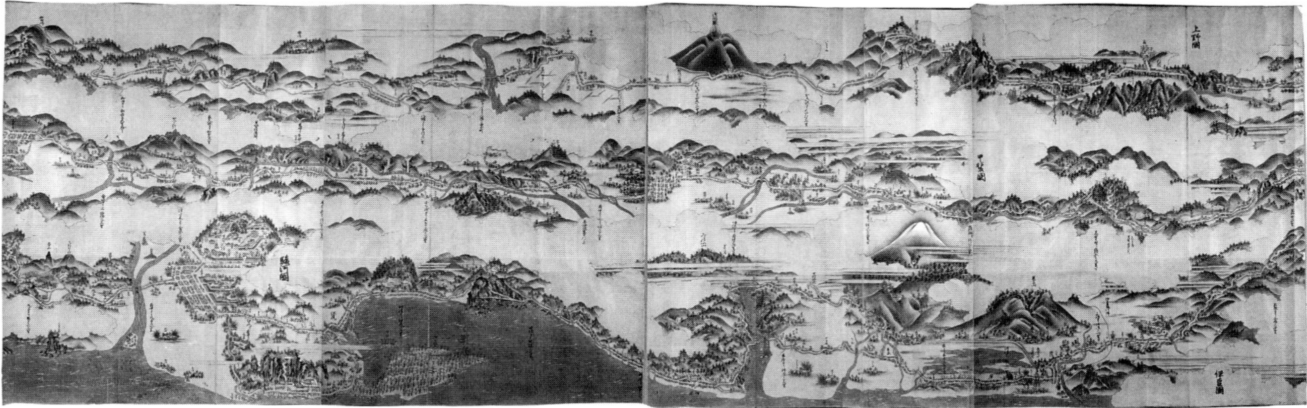

FIG. 11.38. PANORAMIC ROUTE MAP: PART OF THE *KISOJI NAKASENDŌ TŌKAIDŌ EZU* (MAP OF THE KISO/NAKASEN ROAD AND THE TŌKAI ROAD), 1668. Note that the folding-book manuscript shows the castles from a realistic bird's-eye perspective, presumably because the wooden model castles presented to the government might have been made available to the painters. Both routes connected Edo to Kyōto, the Tōkai road following the Pacific coast and the Nakasen road being inland.
Size of the entire original: 120 × 1,920 cm. By permission of the National Diet Library, Tokyo.

from it, were scientific works giving degrees of latitude.[232] Besides the types of maps mentioned so far, the shogunate also ordered its experts and the daimyos to compile other maps, plans, and charts when necessary and for official purposes.

Development of the Printed Map Trade

World Maps Derived from Matteo Ricci

The Jesuit Matteo Ricci (1552–1610) is best known for his role in the transmission of European ideas to China and as an agent by which knowledge of Chinese geography was sent back to Europe (above, pp. 171–77). In Japan, however, his place in cartographic history is somewhat different. The map of the world he compiled in China found its way to Japan, where it was printed in several versions. It becomes a fitting document with which to introduce the development of a Japanese map trade after the period of initial contact with Europeans.

In his memoirs, Ricci notes that his maps of the world not only were popular throughout China, where he was working, but were also sent to Macao and Japan.[233] The exact date when they first appeared in Japan is not known, but from 1605 copies were used for instruction in geography and astronomy at the Jesuit academy in Kyōto.[234] Among Ricci's world maps, however, the 1602 edition exerted the greatest influence in Japan, and almost all surviving examples in Japan were based on this model.[235] An important reason for the success of Ricci's map was that it was written in Chinese, and thus it was easy for the Japanese to understand. Some place-names on the

(five folding books and including the coast of eastern Japan), and by Nanba Matsutarō (four folding books plus four books of the surveying diary). Part of the Kyōto chart is reproduced in Unno, Oda, and Muroga, *Nihon kochizu taisei*, vol. 1, pl. 116 (note 8); part of the Nanba chart is reproduced in Nanba, Muroga, and Unno, *Nihon no kochizu/Old Maps in Japan*, pl. 34 (note 11).

232. See above, p. 385.

233. Matteo Ricci, *Storia dell'introduzione del Cristianismo in Cina*, 3 vols., ed. Pasquale M. d'Elia, Fonti Ricciane: Documenti Originali concernenti Matteo Ricci e la Storia delle Prime Relazioni tra L'Europa e la Cina (1579–1615) (Rome: Libreria dello Stato, 1942–49), 2:60.

234. Henri Bernard, *Matteo Ricci's Scientific Contribution to China*, trans. Edward Chalmers Werner (Beijing: Henri Vetch, 1935), 70.

235. An intact copy of the 1602 map is preserved at the Miyagi Prefectural Library in Sendai; it is on six hanging scrolls that together measure 171 by 361 centimeters. Two others are known to exist in a less than complete state. On the one at the Kyōto University Library, on six hanging scrolls (166.5 × 366 cm), the crests of the Society of Jesus have been cut out. The other at the National Archives, Tokyo (height 170.4 cm) is missing the articles around the main map, the supplementary maps, and the illustrations on astronomy; this copy is also divided and attached to the reverse of an album-type manuscript atlas of China. The topic of Ricci's world map and its influence throughout the Edo period is treated briefly by Shintaro Ayusawa, "The Types of World Map Made in Japan's Age of National Isolation," *Imago Mundi* 10 (1967): 123–27 (attached to this article is M. Ramming, "Remarks on the Reproduced Japanese Maps," 128); Ayusawa refers to Kurita Mototsugu, "Edo jidai no sekai chizu gaisetsu" (Outline of the world maps of the Edo period), *Shigaku Kenkyū* (a quarterly issued by Hiroshima University), vol. 10, no. 1 (1938): 73–80. See also Ayusawa Shintarō, "Mateo Ritchi no sekaizu ni kansuru shiteki kenkyū: Kinsei Nippon ni okeru sekai chiri chishiki no shuryū" (Historical research on Matteo Ricci's world map: On the main current of the knowledge of world geography during the Tokugawa age), *Yokohama Shiritsu Daigaku Kiyō* 18 (1953). Ricci is discussed in a broader scientific framework by Nakayama, *History of Japanese Astronomy*, 79–86 (note 38). The 1602 map at the Miyagi Prefectural Library was reproduced by the library in 1981 and is in Unno, Oda, and Muroga, *Nihon kochizu taisei*, vol. 2, pl. 58 (note 8). The version at Kyōto University was reproduced

Ricci maps, however, are entered in the Japanese katakana syllabary, which points to Jesuit involvement in transmitting knowledge. Although Japanese experts at the time could read the Chinese characters, they could not yet transliterate place-names from the roman alphabet to the Japanese syllabary and thus would have required the help of informed Jesuits.[236] The syllabary was of course valuable not only for approximating the correct pronunciation of Western place-names, but also for transcribing other names written in Chinese characters. Place-names from the 1602 original edition were used on the Japanese Ricci maps, the only changes being the addition of Kinshima and Ginshima (Islands of Gold and Silver) and the correction of one of Ricci's errors by labeling the island north of Honshū as Ezo.

Changes started to take place in 1645, when place-names were introduced from European maps other than Ricci's and placed in the general structure of Ricci's outlines of the world's landmasses and his projection. One such amended version is the 1645 *Bankoku sōzu* (Map of all the countries) (fig. 11.39), made in the form of a scroll and paired with an illustration showing people of the world. Although its authorship is not known, it was printed in Nagasaki, and it has the distinction of being the first European-style map to be printed in Japan. Despite the closure of the country, European missionaries or Japanese who understood European languages appear to have been involved in its compilation. The translations for the Tropics of Cancer and Capricorn differ from those on the Ricci original, and the accompanying illustrations of people reveal a strong influence from *Nanban* world maps based chiefly on Mercator's world map, which has similar illustrations.[237] New place-names in katakana included Ribiainderiyoru ("Libya Interior") in northern Africa and Kabotebowaesuperanshiya ("Cabo de Boa Esperança") for the Cape of Good Hope. Place-names of large areas appear in a woodblock rendition of the cursive hiragana syllabary; others, however, are written with a brush in the square katakana syllabary. This suggests that the map was incomplete at the time of printing and may have been completed by adding more place-names in manuscript and coloring by hand.[238]

The *Bankoku sōzu*-type maps were oriented with east at the top.[239] As a result, the Americas appeared at the top of the sheet, Europe and Africa were at the bottom, and a massive southern continent formed a quarter of the map on the right side. Like the accompanying illustration of people, the *Bankoku sōzu*-type maps were designed to be hung in the alcoves (*tokonoma*) of houses. The mapmakers accordingly stressed their decorative qualities: Japanese and non-Japanese sailing boats were drawn in the otherwise blank spaces on the scroll.

In origin, however, these maps appear to have served a different function. There is evidence that they were

in China by Yu Gong Xue Hui in 1936 (reprinted Tokyo: Daian, 1967). That at the National Archives is reproduced in Funakoshi Akio, "Kon'yo bankoku zenzu to sakoku Nippon" (Ricci's world maps and Japan in the age of national isolation), *Tōhō Gakuhō* (Kyōto) 41 (1970): 595–710, esp. pl. 2. At the Miyagi Prefectural Library there is also an early seventeenth-century copy of the 1602 original; in this version are the Islands of Gold and Silver (Kinshima and Ginshima) in the sea to the east of Japan. It is reproduced in color in Unno, Oda, and Muroga, *Nihon kochizu taisei*, vol. 2, pl. 57.

236. An example of the Jesuit influence on toponymy involved "Castilia del Oro" in the northern part of South America: the place-name was entered in Chinese characters reading Jin-jia-xi-la (in their Japanese reading, Kin-ka-sai-rō) and in katakana Kasuteradouno (in the case of the Miyagi Prefectural Library version mentioned above), which resembles neither the Chinese nor the Japanese reading; *jin* (or *kin*) means gold (*oro*).

237. See appendix 11.4. The type B1 map at the Nanban Culture Hall in Ōsaka and the type C map at the Idemitsu Museum of Arts in Tokyo, formerly the Matsumi Tatsuo Collection (a pair of six-fold screens, each 166 × 363 cm [map only: 166 × 484 cm]), are each accompanied by an illustration showing forty types of people from throughout the world.

238. Generally, hiragana is used for writing words and names that appear in the Japanese language per se, and katakana for loan words or names—that is, those derived from other languages—as well as for highlighting words or names that could appear in hiragana; this latter use of katakana is similar to the use of italics in several European languages. On the *Bankoku sōzu*, however, it appears that the use of the two types of script was stylistic. The only extant copy of the *Bankoku sōzu* of 1645 is at the Shimonoseki City Chōfu Museum. Originally it was made in the form of a scroll, but it is now spread out and framed. For colored reproductions of the manuscript *Bankoku sōzu* and the woodblock illustration of people (map, 134 × 57.6 cm; illustration of people, 136 × 59.5 cm) at the Kōbe City Museum, see Unno, Oda, and Muroga, *Nihon kochizu taisei*, vol. 2, pl. 60 (note 8), and Cortazzi, *Isles of Gold*, 37–38 and 112–14 (pls. 42–43) (note 14). There is also a colored reproduction of the map at the Shimonoseki City Chōfu Museum (map, 132.4 × 57.9 cm; illustration, 132 × 57.6 cm) in vol. 8 of the *Kokushi daijiten*, color pages "Sekai zu" (Maps of the world), pl. 9 (note 95). That the *Bankoku sōzu* and the illustration of people were printed in Nagasaki is known by the inscription "Hishū Sonoki gōri Nagasaki no tsu ni oite kaihan" (Published in Nagasaki, Sonoki County, Hizen Province) at the top of the illustration of people. For the *Bankoku sōzu*, see Unno Kazutaka, "Shōhō kan 'Bankoku sōzu' no seiritsu to rufu" (The *Bankoku sōzu* [Map of all the countries] published in 1645 and its popularization), *Nihon Yōgakushi no Kenkyū* 10 (1991): 9–75, and idem, " 'Bankoku sekai igyō zu' ni tsuite" (On the map of all countries and picture of the strange people in the world), *Biburia* 99 (1992): 20–33.

239. Six other versions are known to exist at present: a 1652 *Bankoku sōzu* (65.5 × 41 cm) paired with a *Sekai ninkeizu* (Illustration of the people in the world; 65 × 41.5 cm) at the Kōbe City Museum; a 1671 *Bankoku sōzu* (40 × 56 cm, one sheet with the map and illustration of people, published by Hayashi Jizaemon in Kyōto) at the British Library, London, and the National Diet Library, Tokyo; an undated map at Saidai Temple in Nara (128 × 56.3 cm, published by Eya Shōbee in Kyōto, map only); an undated map (110.5 × 57.4 cm, map only) owned by Sakaguchi Shigeru in Tsu, Mie Prefecture; an undated map at the Kōbe City Museum, Ikenaga Collection (61.5 × 39.4 cm, map only); and a map dated Teiyū (Hinoto Tori) of Shōhō (1651; the era, however, included only the years 1644–47) (paired with an illustration of people, each 137 × 59 cm) at the Kōbe City Museum and the British Library. The last is a post-Shōhō imitation with some errors; the original *Bankoku sōzu* has "Shōhō Tori" only. For colored reproductions of the Saidai version and that of 1652, see Unno, Oda, and Muroga, *Nihon*

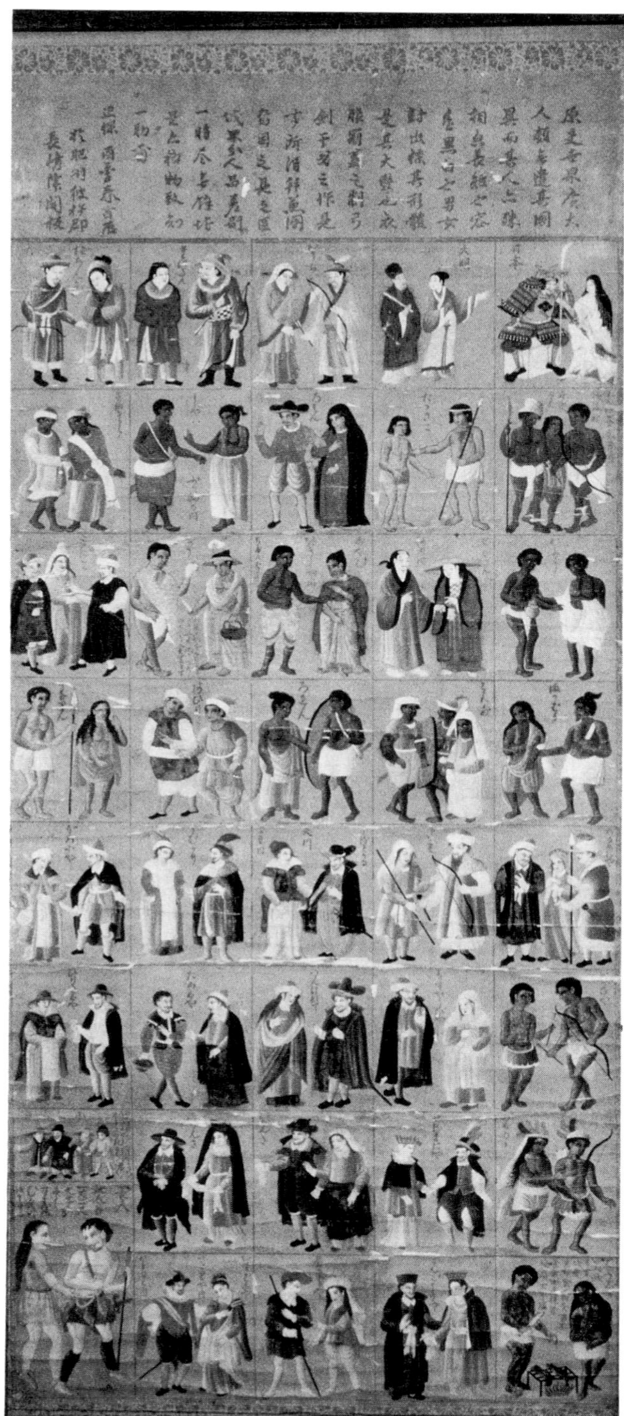

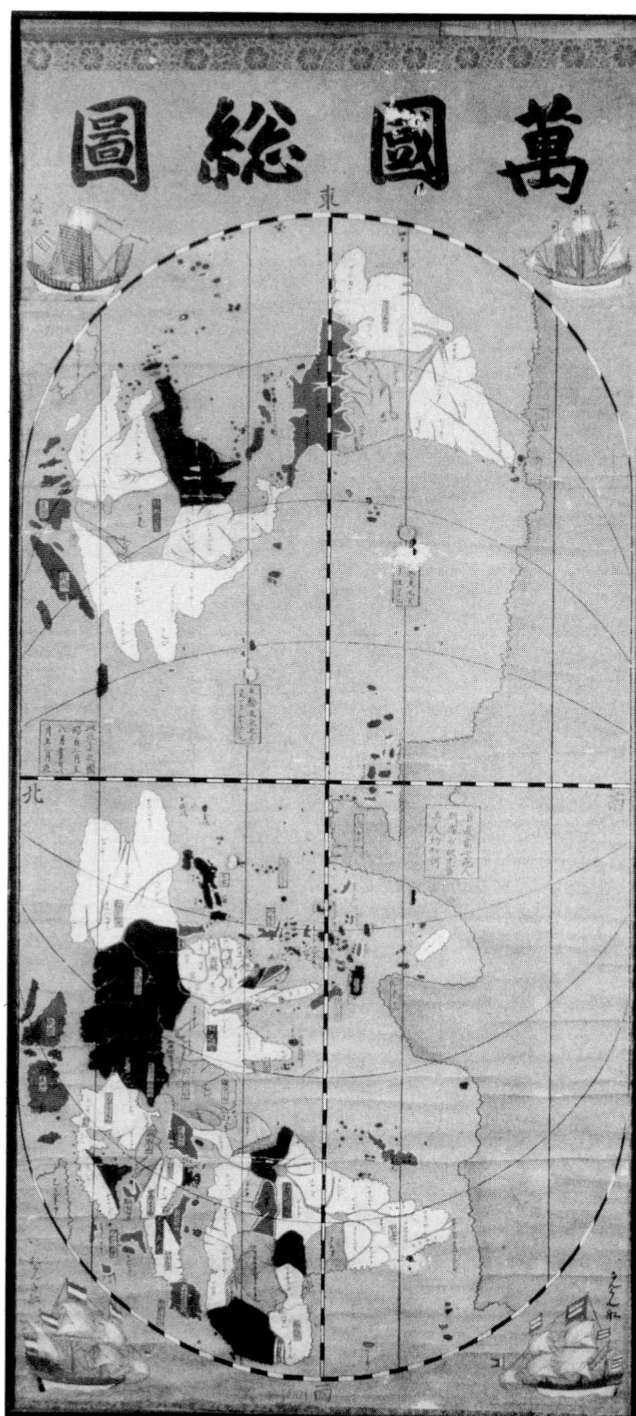

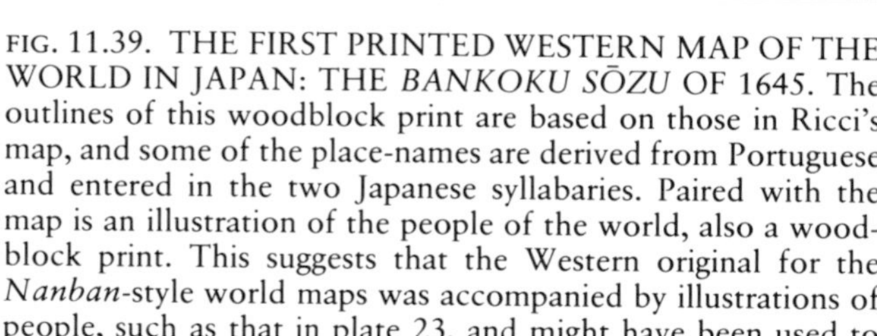

FIG. 11.39. THE FIRST PRINTED WESTERN MAP OF THE WORLD IN JAPAN: THE *BANKOKU SŌZU* OF 1645. The outlines of this woodblock print are based on those in Ricci's map, and some of the place-names are derived from Portuguese and entered in the two Japanese syllabaries. Paired with the map is an illustration of the people of the world, also a woodblock print. This suggests that the Western original for the *Nanban*-style world maps was accompanied by illustrations of people, such as that in plate 23, and might have been used to revise Ricci's information: all three of the known Mercator *Nanban*-style world maps have illustrations of the people of the world. The four ornamental ships outside the border of the map are Chinese and Japanese at the top and European at the bottom. The map is oriented to the east.

Size of the original: 132 × 57.6 cm (illustration of people) and 132.4 × 57.9 cm (map). Shimonoseki City Chōfu Museum, Shimonoseki. Photograph courtesy of Kazutaka Unno.

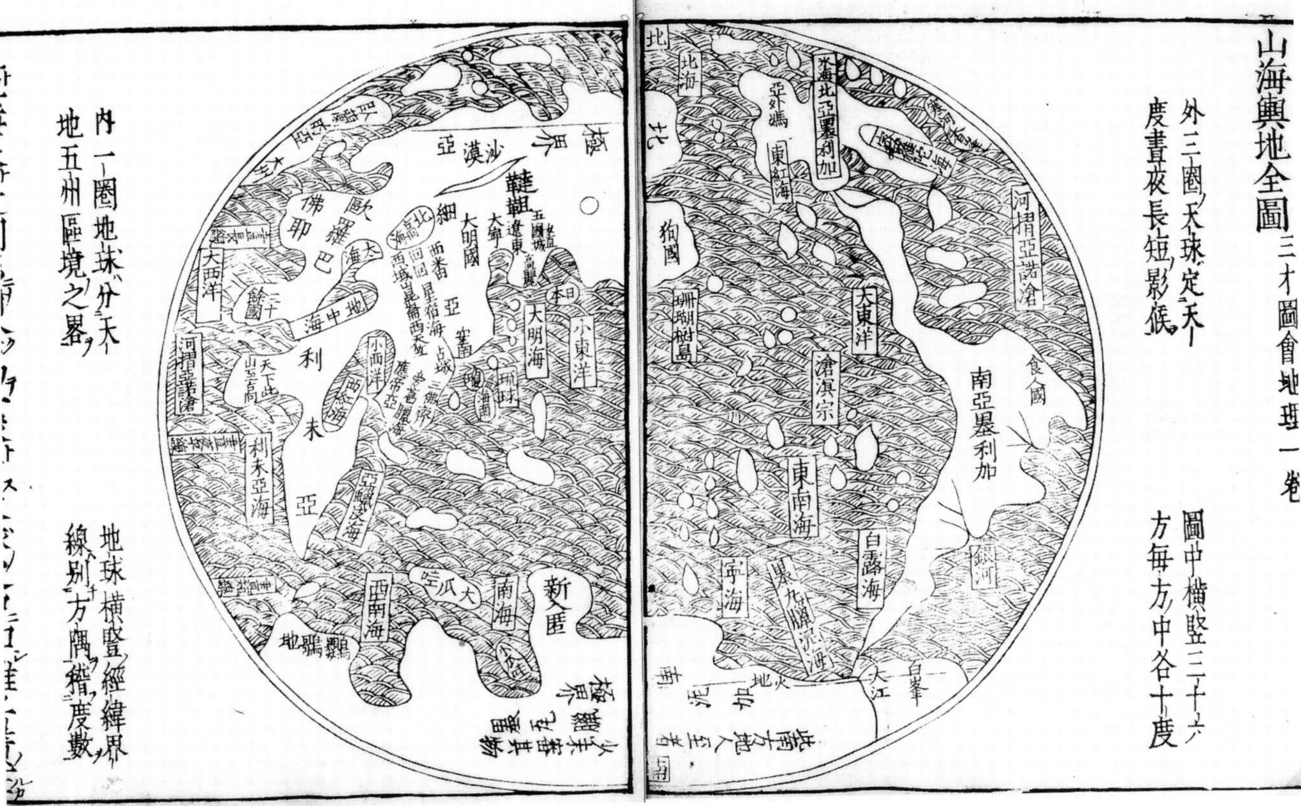

FIG. 11.40. THE "SANKAI YOCHI ZENZU" (MAP OF THE LANDS AND SEAS OF THE EARTH) IN MATSUSHITA KENRIN'S *RON'Ō BENSHŌ*, 1665. The map was reproduced from that of the same title in Chinese, "Shanhai yudi quantu" (Complete geographic map of the mountains and seas), in Wang Qi's *Sancai tuhui*. Wang's map was a reproduction of the map of the same title in Feng Yingjing's *Yueling guangyi* (Enlarged annotation of Monthly observances, 1602), a variation of the Nanjing edition of Ricci's map.
Size of the original: 19.4 × 33 cm. Collection of Kazutaka Unno.

presented as certificates to apprentices who had mastered surveying. In his manuscript book of 1717, for instance, Hosoi notes that he received a *Bankoku sōzu* as a certificate.[240] This practice was probably begun by Higuchi Kentei, a pioneer of surveying and navigation in Nagasaki. It was only after 1646, when Higuchi was imprisoned, that map publishers discovered the commercial potential of the image and produced versions of the *Bankoku sōzu*. These small copies with illustrations of people began to appear in seventeenth-century books and encyclopedias for the general public,[241] the most literate of whom were the upper classes and the urban population. According to Passin, "By the Genroku Period (1688–1704) a surprisingly modern publishing industry had developed." This included, besides professional writers and book illustrators, large publishing houses producing editions of more than ten thousand copies "to satisfy the audiences created by the spread of literacy and the cultural efflorescence of the cities." He estimates that from the middle of the eighteenth century, 40 to 50 percent of the male population was literate.[242] The geographical quality of these copies, however, tended to deteriorate. The same can be said for the only revision of the *Bankoku sōzu*,

kochizu taisei, vol. 2, pls. 59 and 61 (note 8). The 1671 map is reproduced in N. H. N. Mody, *A Collection of Nagasaki Colour Prints and Paintings* (1939; reprinted Tokyo: Charles E. Tuttle, 1969), pl. 24; and Helen Wallis, "The Influence of Father Ricci on Far Eastern Cartography," *Imago Mundi* 19 (1965): 38–45, esp. fig. 7. The Sakaguchi map is reproduced in Unno, Oda, and Muroga, *Nihon kochizu taisei*, vol. 2, fig. 55; the Teiyū map is in Nanba, Muroga, and Unno, *Nihon no kochizu/Old Maps in Japan*, pl. 7 (note 11); Mody, *Nagasaki Colour Prints*, pl. 23; and Wallis, "Father Ricci," figs. 5–6.

240. See above, p. 394.

241. Examples include the "Sekai bankoku sōzu" (General map of all the countries in the world) and the illustration of people in the *Tōsho zōho setsuyōshū taizen* (Enlarged dictionary) published by Yabuta in 1693 and in similar dictionaries dating to 1695, 1696, and 1699; and the "Bankoku no zu" (Map of all the countries) and illustration of people in the *Nendaiki eiri* (Illustrated chronicle, 1706) and in similar chronicles dating to 1710, 1711, and 1713. A reproduction of the 1711 map and illustration of people is in Unno, Oda, and Muroga, *Nihon kochizu taisei*, vol. 2, fig. 57 (note 8).

242. See Herbert Passin, *Society and Education in Japan* (New York: Teachers College Press, Columbia University, 1965), 11–12, and 47–49.

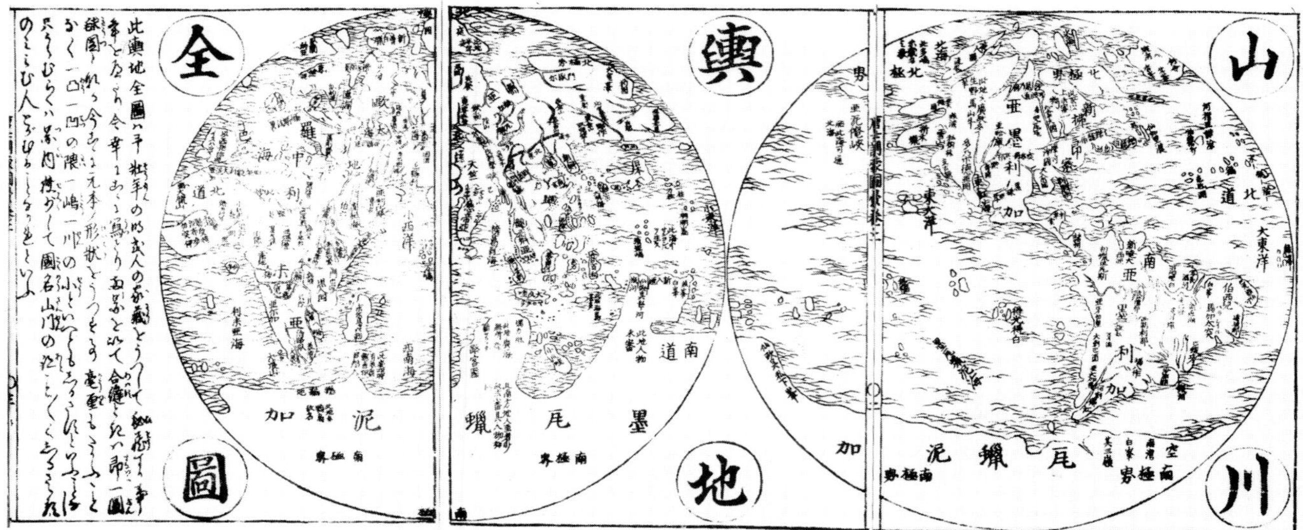

FIG. 11.41. "SANSEN YOCHI ZENZU" (MAP OF MOUNTAINS AND RIVERS ON THE EARTH) FROM HIRAZUMI SEN'AN'S *MOROKOSHI KINMŌ ZUI*, 1719. The map occupies three pages in the encyclopedia. Although meridians and parallels are not included, the following latitudes are inscribed: Arctic Circle (Hokkyokuken), Tropic of Cancer (Hokudō), Tropic of Capricorn (Nandō), and Antarctic Circle (Nankyokuken). Judging from the configurations and place-names, the source material must have been the map of the Eastern and Western hemispheres in the *Fangyu shenglüe* (Compendium of geography) by Cheng Boer et al., published in China in 1612. Size of the original: 18.3 × 42 cm. Collection of Kazutaka Unno.

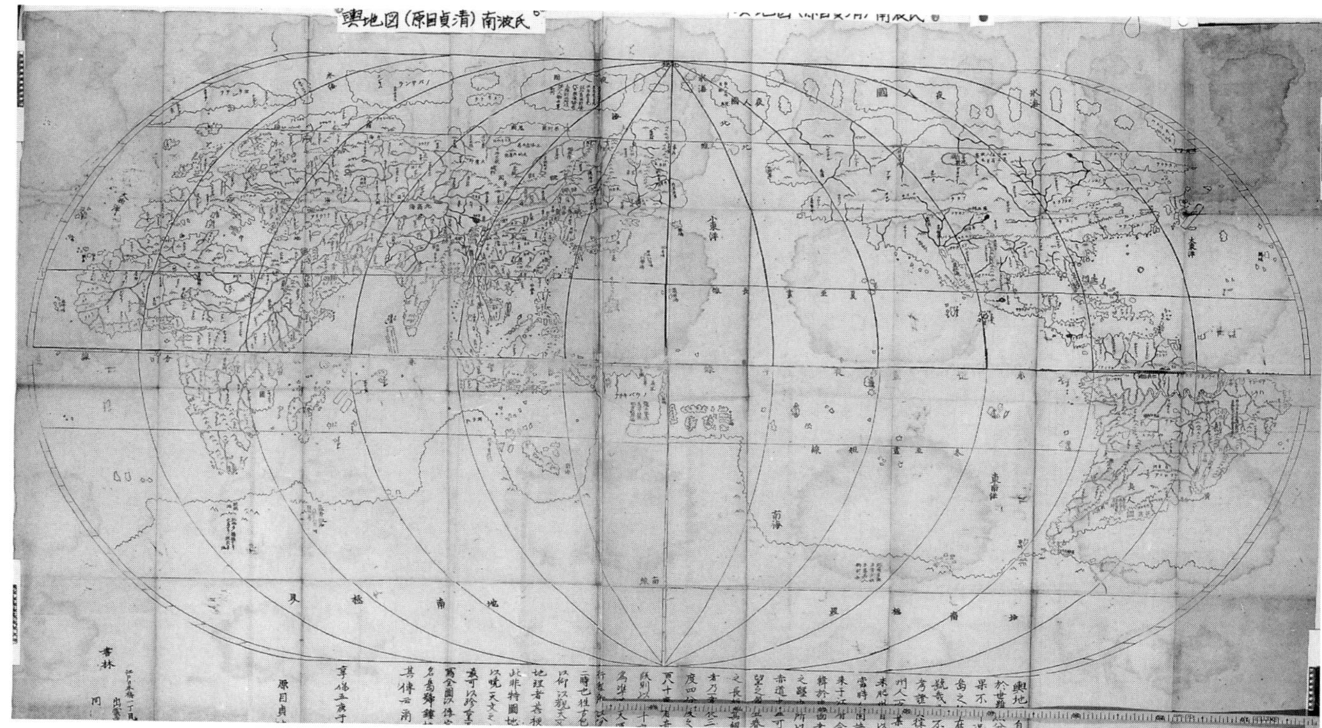

FIG. 11.42. HARAME SADAKIYO'S *YOCHI ZU* OF 1720. Unlike on Ricci's map, east of Japan in the ocean are the "Island of Gold" and the "Island of Silver." It appears that Harame's map is a reproduction of a revision of Ricci's map: various place-names throughout the world entered in katakana are the same as those on some of the revisions of Ricci's map in Japan. Size of the original: 91.5 × 154 cm. By permission of the Kōbe City Museum, Kōbe.

Ishikawa Ryūsen's (fl. 1686–1713) *Bankoku sōkaizu* (General world map) of 1688: although large areas of Asia were revised on it, the rest of the geographical content was actually worse than on the *Bankoku sōzu*.[243]

Ricci's maps were also published as book illustrations in China, and many were reprinted in Japan. An early example of a Ricci map in Japan was of the Eastern and Western hemispheres in Maezono Sobu's *Meisei tōki* (Account of the Ming-Qing war, 1661).[244] The map was reproduced from Pan Guangzu's *Huiji yutu beikao quanshu* (Reference work of the maps of China), published in China in 1633. The map had been copied from the *Fangyu shenglüe* (Compendium of geography) by Cheng Boer et al., published in China in 1612.[245] A greatly simplified version appeared in *Ron'ō benshō* (Argumentation of Unki Ron'ō, 1665) by Matsushita Kenrin (1637–1703) (fig. 11.40), and this had been reprinted from the Chinese *Sancai tuhui* (Illustrated compendium of the three powers [heaven, earth, man]), a pictorial encyclopedia compiled by Wang Qi dating to 1609.[246] In 1719 Hirazumi Sen'an also included a map of the Eastern and Western hemispheres in his *Morokoshi kinmō zui* (Illustrated encyclopedia of China) (fig. 11.41),[247] but this owed more to Ricci's map in the 1612 *Fangyu shenglüe*.

Sheet maps also continued to appear. In 1708 Inagaki Kōrō republished a copy of Ricci's 1602 map of the Northern and Southern hemispheres as the *Sekai bankoku chikyū zu* (Map of the world),[248] including on it sections of the illustration of people of the world and non-Japanese sailing boats. This was not merely a copy of Ricci's work, however, because it incorporated features from the *Bankoku sōzu*.

More innovative was the first independent version of Ricci's maps on an oval projection, published in 1720. The map, by Harame Sadakiyo, was titled *Yochi zu* (World map) (fig. 11.42).[249] Parts of Southeast Asia were revised, and most of the place-names were recorded in katakana. The *Yochi zu* seems to have stimulated further publishing ventures. Such was the anonymous *Bankoku zu* (Map of all the countries), a small and simple map issued by Hon'ya Hikoemon and appearing in 1744.[250] Harame's map appears to have been the model, but the continents of Eurasia and Magellanica (the unexplored southern landmass) had been revised: part of Eurasia was adopted from the *Nan'enbudai shokoku shūran no zu* (Outline map of the countries of Jambūdvīpa), a Buddhist map of the world also published by Hon'ya in 1744,[251] and Magellanica, elongated from east to west on Ricci's map, was here limited to the lower center and lower right of the *Bankoku zu*. The author of the Buddhist map was Kabō Hyōzō, who also composed the undated *Dainihonkoku no zu* (Map of Great Japan) similarly published by Hon'ya.[252] It is therefore suspected that he might be the author of the *Bankoku zu*. In 1783 a large Ricci-type oval map was introduced by Nakane Genran under the pen name Mihashi Chōkaku, titled *Chikyū ichiranzu* (Map of the world).[253] This was based on Harame's *Yochi*

243. Ishikawa's *Bankoku sōkaizu* (127 × 57.5 cm) was published in Edo by Sagamiya Tahee and is kept at the Kōbe City Museum; it is reproduced in color in Unno, Oda, and Muroga, *Nihon kochizu taisei*, vol. 2, pl. 62 (note 8). A second edition was issued in 1708, and like the first it did not have an accompanying illustration of people. Copies of the 1708 edition are at the Kōbe City Museum, the Yokohama City University Library, the Tōyō Library in Tokyo, the Beans Collection at the University of British Columbia Library, and elsewhere.

244. The map, titled "Tendozu" (*Chan du tu*; graduated map) appears on four pages of the book; it is reproduced in Unno, Oda, and Muroga, *Nihon kochizu taisei*, vol. 2, fig. 52 (note 8).

245. The map in the *Fangyu shenglüe* was a copy of Ricci's work published by Feng Yingjing (1555–1606) about 1604. Feng was a government official and a friend of Ricci's. For further information on the editions of Ricci's maps made in China and the various works produced under the influence of these maps, see Unno, "Min Shin ni okeru Mateo Ritchi kei sekaizu" (note 112). See also above, pp. 170–77.

246. Matsushita, a Confucian, was a physician; *unki* means luck, and *Ron'ō benshō* deals with astronomy.

247. Published in Ōsaka by Ōnogi Ichibee and in Edo by Suhara Mohee; reissued in Kyōto by Ogawa Tazaemon (1796) and in Ōsaka by Kawachiya Kichibee and three publishers (1802). The *Morokoshi kinmō zui* is at the National Archives in Tokyo and many other libraries in Japan; see Ayusawa, "Mateo Ritchi no sekaizu ni kansuru shiteki kenkyū," 205–7 (note 235).

248. The map (127.5 × 42.5 cm) was published in Ōsaka by Ikedaya Shinshirō and Iseya Heizaemon. It is at the Kōbe City Museum and is reproduced in Unno, Oda, and Muroga, *Nihon kochizu taisei*, vol. 2, pl. 63 (note 8).

249. Harame's 1720 map was published by Izumoji Izuminojō and Izumoji Sashichirō in Edo.

250. Hon'ya's map (52.8 × 71.8 cm) is at the Kōbe City Museum. A reproduction is in Muroga Nobuo and Unno Kazutaka, "Edo jidai kōki ni okeru Bukkyō kei sekaizu" (Buddhist world maps in the Late Edo period), *Chirigakushi Kenkyū* 2 (1962): 135–229, esp. pl. 6; reprinted in *Chirigakushi kenkyū* (Researches in the history of geography), 2 vols. ed. Chirigakushi Kenkyūkai (Society for Research in Historical Geography) (Kyōto: Rinsen Shoten, 1979), 2:135–229.

251. The *Nan'enbudai shokoku shūran no zu* (53 × 73 cm) is at the Kōbe City Museum and is reproduced in Unno, Oda, and Muroga, *Nihon kochizu taisei*, vol. 2, pl. 7 (note 8). This map was based on the Buddhist world map of Rōkashi (or Hōtan) published in 1710 as the *Nansenbushū bankoku shōka no zu* (Visualized map of all the countries in Jambūdvīpa). Rōkashi's maps became the prototype of the Buddhist genre in the eighteenth and nineteenth centuries; Ayusawa, "Types of World Map," reproduces it as fig. 2, to which Ramming has added some comments, p. 128 (note 235). On Rōkashi's map, see below.

252. The map of Japan has the name Kabō Sen'ichi with the title of ear doctor. On the Buddhist map of the world, shops with medicines for deafness are mentioned. The combination of this evidence suggests that Sen'ichi and Hyōzō are the same person. The *Dainihonkoku no zu* (48.8 × 69 cm) is owned by Tanaka Ryōzō in Kyōto and is reproduced in vol. 2 of *Chikusendō kosho tenkan mokuroku* (Chikusendō's catalog of an exhibition of antique books) (Kyōto: Chikusendō, 1974).

253. The map is preserved at the Waseda University Library in Tokyo and elsewhere. Among the versions, only for the one at Waseda is there a bag bearing Nakane's name, the title *Chikyū ichiran no zu* (Map of the world), and the name of the publisher, Asai Yūhidō [of] Kyōto. The cover title is *Chikyū ichiranzu*.

zu, but the figure and description of Magellanica are similar to those of the *Bankoku zu*. It is probable, then, that Mihashi had worked from a map with this earlier version of Magellanica. Several later editions of Mihashi's map are extant, each with a different title and publisher.[254] This was a common practice, not only with maps of the world but also with maps of Japan and cities: copyrights as well as printing blocks were sold.

The Ricci-type oval map that had the greatest impact was that of the Confucian scholar Nagakubo Sekisui (1717–1801), whose map of the world appeared about 1788. Originally it was titled *Chikyū bankoku sankai yochi zenzusetsu* (Map with an account of all the countries, lands, and seas in the world), but later it was called the *Kaisei chikyū bankoku zenzu* (Revised map of all the countries in the world).[255] Evidently based on Harame's *Yochi zu*, it revised the information on the northern frontier of Japan and added newly known place-names from elsewhere in the world. Like others of this type, Nagakubo's map underwent several editions and led to the publication of many compact, simple versions. Such miniature copies of Ricci's world maps were published repeatedly until the end of the Edo period and contributed to the dissemination of geographical knowledge, even while Japan was closed to outside influence.[256]

A notable example of a Ricci map from the last third of the Edo period is the *Kon'yo zenzu* (Map of the earth) by Inagaki Shisen (1764–1836). Dating to 1802, it is the printed map most faithful to Ricci's copy of 1602. It was reduced to 54.5 by 114 centimeters, but whatever information was omitted from the map was included in the accompanying book.[257] By the date when the *Kon'yo zenzu* was published, however, competing maps of the world with newer information from Dutch originals were also being compiled and sold. In spite of this, Inagaki's work was published with information that was two hundred years out of date. The reason for this might be found in the fact that other than the scholars of Dutch-learning, Japanese considered China to be a culturally advanced country; they, Inagaki included, placed a value on Ricci's maps because they thought he was Chinese, since his name appeared on them as "Li Madou" in Chinese characters.

POPULAR MAPS IN SEPARATE SHEET FORMAT

Books had been printed in Japan from the latter half of the eleventh century, but independent sheet maps were not printed until the seventeenth century. By this period there was a growing demand for separate maps not only among intellectuals but also from the wider public.[258] It is therefore appropriate to write of the "popularization" of cartography from this time on. This expanded market was served by a map trade of growing complexity that led to changes in the form and content of Japanese maps.

Maps of Japan

The earliest Japanese map of Japan to be printed on an independent sheet with a known date of publication is the anonymous *Dainihonkoku jishin no zu* (Earthquake map of Great Japan) of 1624 (fig. 11.43).[259] It was composed as a charm to protect against earthquakes and contains superstitious annotations about them. Depending on which month an earthquake occurred, it was believed to be an omen of various events: for June, for example,

254. The first to be issued was a version with two publishers, Ōnogi Ichibee in Ōsaka and Asai Shōemon in Kyōto, mentioned on the sheet (81.5 × 153.7 cm); it is reproduced in Kurita, *Nihon kohan chizu shūsei*, pl. 4 (note 15). A later edition, including a version titled *Bankoku chikyū saiken zenzu* (Detailed map of all the countries on the globe), has the names of these three publishers: Ōnogi Ichibee in Ōsaka, Umemura Saburobee in Kyōto, and Suharaya Ichibee in Edo; it is reproduced in color in Unno, Oda, and Muroga, *Nihon kochizu taisei*, vol. 2, pl. 67 (note 8).

255. The map is 103.5 × 155 centimeters, and no publisher is mentioned on the first edition. The *Kaisei chikyū bankoku zenzu* was published by Yamazaki Kinbee in Edo and Asano Yahee in Ōsaka and has an anonymous preface (by Katsuragawa Hoshū?).

256. Almost none of the compact, simple versions name the reviser or give a publishing date. Two with dates are *Shinsei bankoku yochi zenzu* (Newly made map of all the countries in the world, 1844) by Den Ken (32.5 × 91.3 cm; the map is 25.7 × 39.5 cm), and *Chikyū bankoku sankai yochi zenzusetsu* (Map with an account of all the countries, lands, and seas in the world, 1850) by Yamazaki Yoshinari (40.5 × 60.5 cm). Both are reproduced in Unno, Oda, and Muroga, *Nihon kochizu taisei*, vol. 2, fig. 100, and pl. 110 (note 8).

257. *Kon'yo zenzusetsu* (Explanation of the *Kon'yo zenzu*). The geographical explanations of Ricci's map are translated into Japanese, and the place-names given in Chinese characters are transliterated into the katakana syllabary; see Ayusawa, "Mateo Ritchi no sekaizu ni kansuru shiteki kenkyū," 182–88 (note 235).

258. For the early Japanese printed maps, see Kurita, *Nihon kohan chizu shūsei*, "Sōsetsu" (Introduction), 1–3 (note 15). Chamberlain and Sansom note that there were printed books in tenth-century Japan and that woodblock printing, via China, was known at least as early as the eighth century. In addition, Hillier reports that woodblock printing was "used for pictorial work . . . as early as the 10th century, probably a lot earlier": see Jack R. Hillier, *The Japanese Print: A New Approach* (London: G. Bell and Sons, 1960), 14; Chamberlain, *Japanese Things*, 396 and 509 (note 99); and George B. Sansom, *Japan: A Short Cultural History*, 2d rev. ed. (New York: Appleton-Century-Crofts, 1962), 434. On the woodblock prints and artists, strongly connected with the *ukiyoe* (pictures of the floating world) school, see Hillier, *Japanese Print*, 9–17; Stanley-Baker, *Japanese Art*, 184–90 (note 39); and Tsuda, *Handbook of Japanese Art*, 230–39 (note 99). See also Hugo Munsterberg, *The Japanese Print: A Historical Guide* (Tokyo: Weatherhill, 1982), esp. 11–136 (from the introduction of woodblock prints in the Nara period to the end of the Edo period).

259. See Saburō Noma, "Earthquake Map of Japan, 1624," *Geographical Reports of Tokyo Metropolitan University* 9 (1974): 97–106. An earlier printed map of Japan to appear in a book was in the Keichō edition of the *Shūgaishō*, ca. 1607 (see above, p. 369 and note 96).

the map notes that an earthquake foreshadows disease, drought, death of cows and horses, and some pleasure. Such inscriptions suggest that the map was designed for a general audience. The image of Japan follows in the Gyōki tradition, as was the case with many other maps of the country in calendars and books on fortune-telling throughout the Edo period. Some of these maps included a dragon around the archipelago, as does the *Dainihonkoku jishin no zu*.[260] In Buddhist belief the dragon is thought to cause earthquakes, and on the map its head is being crushed with the *kanameishi*, a stone kept in the sacred area of Kashima Shrine in Kashima, Ibaraki Prefecture. There is also a *waka* poem (a traditional thirty-one syllable verse form) on the map.[261]

The shape of Japan on maps printed from woodblocks changed for the first time in 1662, when the *Shinkai Nihon ōezu* (Newly revised map of Japan) was published.[262] It maintained the Gyōki tradition insofar as it still showed the routes from Yamashiro to the other provinces in the same way and included such legendary lands as the Rasetsukoku (Land of Women). The shape of the country, however, was much improved: the coastlines, including the major peninsulas and bays, were portrayed far more accurately. Its model appears to have been the modified Keichō type of map paired with the type C *Nanban*-style world maps on an equirectangular projection (appendix 11.4). The coasts on both maps have accentuated curves, perhaps to enhance their pictorial value, and representations of ships serve the same purpose. Demand for the *Shinkai Nihon ōezu* led to its reprinting in 1666.[263] This was also the year when the first atlas of Japan printed in Japan, the *Nihon bunkei*

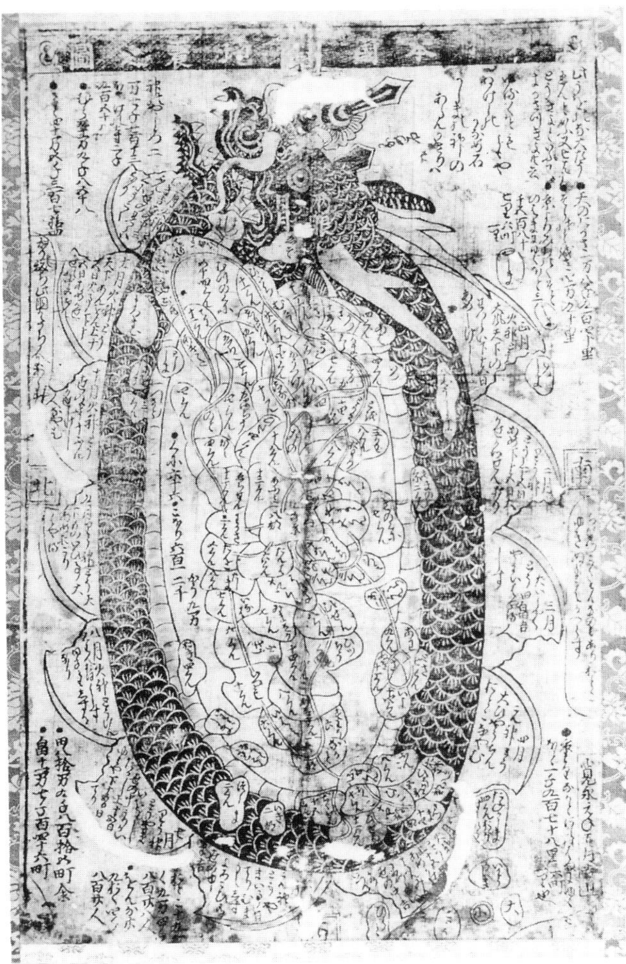

FIG. 11.43. THE ANONYMOUS *DAINIHONKOKU JISHIN NO ZU* OF 1624. On the twelve spines of the dragon surrounding the archipelago are written monthly fortunes associated with earthquakes. The portrait of Japan is based on the Gyōki-type representation, and the map shows Kamakura—the center of the shogunate from 1192 to 1333 and, in the fourteenth and fifteenth centuries, the seat of the Muromachi shogunal office in charge of Kantō—as a province.
Size of the original: 44 × 26.7 cm. By permission of the Harada Masaaki Collection, Yanagida, Ishikawa Prefecture.

260. Examples of maps with a dragon include the Ise-Koyomi (Ise Province almanac or calendar) of 1673, 1675, 1676, 1680, and 1682, and those in the anonymous *Ōzassho* (Large miscellany) of 1693, 1715, 1774, 1816, 1846, and 1852. Four examples of Gyōki-type maps of Japan not to include a dragon are the ca. 1640 *Nansenbushū Dainihonkoku shōtō zu* (Orthodox map of Great Japan in Jambūdvīpa; 70.5 × 185 cm) at the Tokyo University Library, the 1651 *Nihonkoku no zu* (Map of Japan; 109.5 × 51 cm) at the National Museum of Japanese History in Sakura, the ca. 1651 *Gyōki Bosatsu setsu Dainihonkoku zu* (Map of Great Japan according to Bodhisattva Gyōki; 80.4 × 42.3 cm) at the Kōbe City Museum, and the 1654 *Nihonkoku no zu* (Map of Japan; 121 × 53 cm) in the Kurita Kenji Collection in Nagoya. For reproductions see Unno, Oda, and Muroga, *Nihon kochizu taisei*, vol. 1, pls. 12–13 and fig. 43 (the ca. 1640, 1651, and 1654 maps) (note 8); Akioka, *Nihon kochizu shūsei*, pl. 20 (the 1651 map) (note 15); and idem, *Nihon chizu shi*, folding plate (the ca. 1651 map) (note 7).

261. Although the creature on the *Dainihonkoku jishin no zu* is said to be a dragon, there is a similar creature, Namazu, that is popularly considered to cause earthquakes by its movements inside the earth. One description of Namazu reads: "—an eel-like creature, but thicker and flat-headed and supplied with mustachios,—which dwells somewhere in the bowls [*sic*] of the earth, and whose occasional wrigglings are the cause of earthquakes" (see Chamberlain, *Japanese Things*, 444 [note 99]). When Namazu took the place of the dragon is not certain, but toward the end of the Edo period it started to appear on a lot of maps.

262. This is the cover title. Printed on the map (59 × 88 cm) is *Fusōkoku no zu* (Map of the country Fusō); Fusō was a legendary land in ancient China and was placed in the eastern sea. The map is preserved at the Kōbe City Museum (formerly in the Nanba Collection), the National Museum of Japanese History, Sakura (formerly Akioka Collection), the Geographical Institute at Kyōto University, and the Kurita Kenji Collection, Nagoya; the last three, however, have lost the cover title. The Kōbe City Museum map is reproduced in Unno, Oda, and Muroga, *Nihon kochizu taisei*, vol. 1, pl. 25 (note 8). The Kurita map is reproduced in Kurita, *Nihon kohan chizu shūsei*, pl. 16 (note 15).

263. This map is reproduced in Nanba, Muroga, and Unno, *Nihon no kochizu/Old Maps in Japan*, pl. 26 (*Fusōkoku no zu*) (note 11), in Akioka, *Nihon chizu shi*, folding plate (note 7), in idem, *Nihon kochizu*

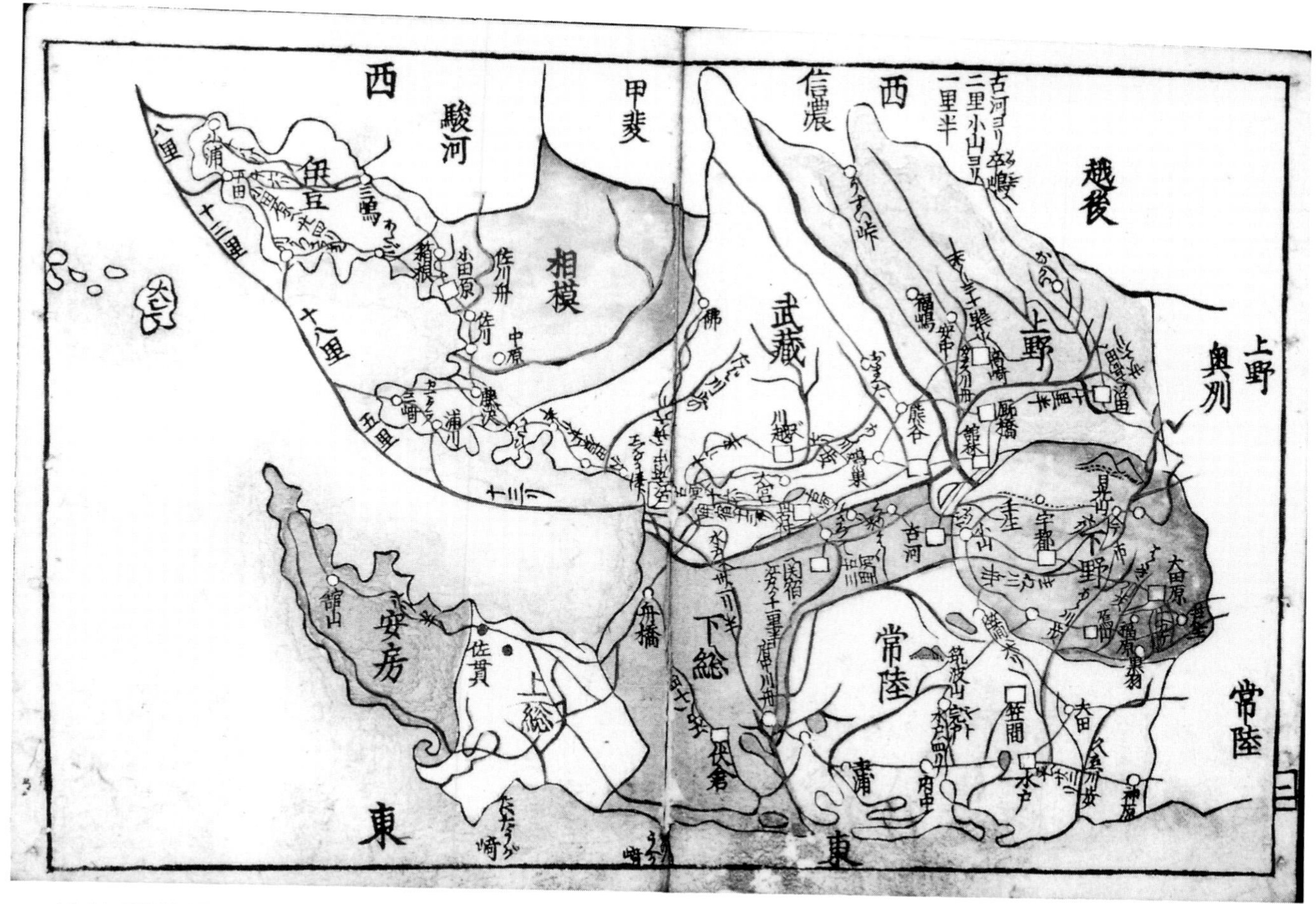

FIG. 11.44. THE FIRST AND SECOND MAPS, SHOWING THE ŌU AND KANTŌ REGIONS, OF THE ATLAS *NIHON BUNKEI ZU*, 1666. The album-style atlas was produced by dividing the Keichō map of Japan (plate 26) into sixteen parts. (Ōu is the generic name of the two provinces of Mutsu and Dewa, or the whole of northern Honshū.)
Size of each page: 18.7 × 13.5 cm. By permission of the Kōbe City Museum, Kōbe, Nanba Collection.

zu (Separate maps of Japan), was published (fig. 11.44).[264] This was produced by subdividing the Keichō map and keeping the same scale to preserve the accuracy of the original. This, however, was not as popular as the more decorative works like the *Shinkai Nihon ōezu*.

Because of the popularity of the *Shinkai Nihon ōezu*, a more decorative version was published in 1687 by Ishikawa Ryūsen, an *ukiyoe* artist. Titled *Honchō zukan kōmoku* (Outline map of Japan), it was the first of many maps published by Ishikawa, and it established a model for woodblock maps throughout most of the eighteenth century (plate 27).[265] Works based on Ishikawa's original version, and published mainly in the area of Edo, are referred to as Ryūsen-type maps of Japan.[266] Ishikawa's maps were both decorative and practical, and they served as a combined Who's Who and travel map. Useful infor-

shūsei, pl. 24 (note 15), and in Cortazzi, *Isles of Gold*, pl. 38 (note 14). Cortazzi translates the title as "Map of the land of the rising sun."

264. There are two versions of the atlas, one at the Kōbe City Museum, Nanba Collection, and the other at the Meiji University Library, Tokyo (binding 19.5 × 13.8 cm). One leaf of the Kōbe version is reproduced in color in Nanba, Muroga, and Unno, *Nihon no kochizu/Old Maps in Japan*, pl. 28 (note 11); six maps of the Meiji atlas are reproduced in color in Unno, Oda, and Muroga, *Nihon kochizu taisei*, vol. 1, pl. 21 (note 8). The whole of the atlas is reproduced with the *Shinkan jinkoku ki* (Newly published notes on the provinces and their inhabitants, 1701) in *Kinsei bungaku shiryō ruijū, kohan chishi hen* (Classified series of materials of modern literature, early printed geographical descriptions), 22 vols. (Tokyo: Benseisha, 1975–81), vol. 22 (see also note 269).

265. On *ukiyoe* refer to note 258 above. Three versions exist: in the Nanba Collection, Kōbe City Museum (60.5 × 132 cm); in the National Archives, Tokyo (58 × 127.7 cm); and in the Akioka Collection, National Museum of Japanese History, Sakura (60.5 × 130 cm). All have been reproduced in color, the first in Nanba, Muroga, and Unno, *Nihon no kochizu/Old Maps in Japan*, pl. 27 (note 11) and Cortazzi, *Isles of Gold*, pl. 44 (note 14); the second in Unno, Oda, and Muroga, *Nihon kochizu taisei*, vol. 1, pl. 27 (note 8) as well as color plate 27 (this volume); and the third in Akioka, *Nihon kochizu shūsei*, pl. 30 (note 15).

266. It was the custom of artists who belonged to the same family or school to use their given names.

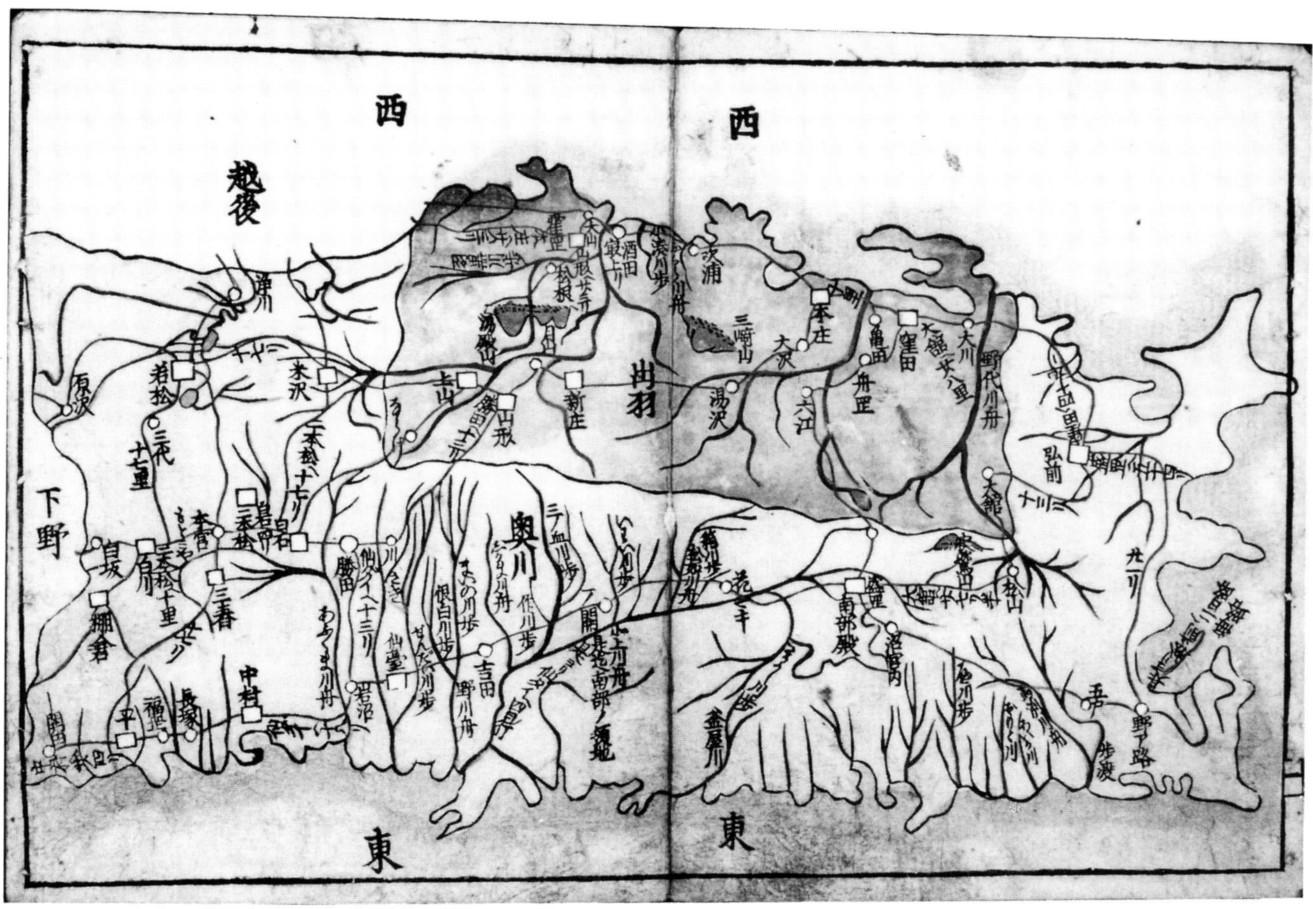

mation to administrators, travelers, and the general public included the names of feudal lords, the standard productivity of the land in *koku* of rice, and important and scenic places along the routes. Each new edition tended to expand both the informative and ornamental aspects of the work.[267]

Nor were the Ryūsen maps without their competitors. In Ōsaka about 1703 there began the publication of rival maps claiming to be more accurate. In this venture, also printed by the woodblock process, Mabuchi Jikōan collaborated with Okada Keishi to publish the *Kōsei Dainihon enbizu* (Corrected perfect map of Great Japan). This went through different titles and a reduced edition over a period of at least thirty years.[268] Its sources may

267. Examples of later maps by Ishikawa are the *Nihon kaisan chōriku zu* (Map of the seas and lands of Japan, 1691) (82.1 × 171 cm) and the *Nihon sankai zudō taizen* (Map of the mountains and seas of Japan, 1703) (98.5 × 171.5 cm). For a reproduction of the first, see Unno, Oda, and Muroga, *Nihon kochizu taisei*, vol. 1, pl. 31 (note 8); for the second, see Akioka, *Nihon kochizu shūsei*, pl. 34 (note 15).

268. Examples include maps in the Kurita Kenji Collection, Nagoya (75.5 × 121.5 cm), and at the National Museum of Japanese History, Sakura (79 × 123 cm), which are reproduced in Kurita, *Nihon kohan chizu shūsei*, pl. 21 (note 15), and in *Akioka Korekushon Nihon no kochizu* (Old maps of Japan in the Akioka Collection), exhibition catalog (Sakura: Rekishi Minzoku Hakubutsukan Shinkōkai, 1988), pl. D-2. Others include the *Kaisei Dainihon bizu* (Revised satisfactory map of Great Japan; 78.7 × 122.5 cm) and the *Kaisei Dainihon zenzu* (Revised general map of Great Japan; 81.5 × 126.5 cm), which were probably printed with the woodblocks of the *Kōsei Dainihon enbizu* (which was most likely the first issue of this group), and only their cover titles differ from it. They are reproduced in color in Unno, Oda, and Muroga, *Nihon kochizu taisei*, vol. 1, pl. 28 (note 8), and Cortazzi, *Isles of Gold*, pl. 40 (note 14). They are preserved at the National Museum of Japanese History, and copies of the *Kōsei Dainihon enbizu* are at the Meiji University Library (76.4 × 121 cm) and the Kōbe City Museum (78.8 × 126.5 cm). The *Dainihonkoku zenbizu* (Complete map of Great Japan; 65.2 × 111.3 cm), National Museum of Japanese History, with no author's name, colophon, or tables of provinces, was published in Kyōto by Uemura Yaemon in 1735, according to Higuchi Hideo and Asakura Haruhiko, revisers, *Kyōhō igo edo shuppan shomoku* (Bibliography of books printed from the Kyōhō era) (Toyohashi: Mikan Kokubun Shiryō Kankōkai, 1962), 38.

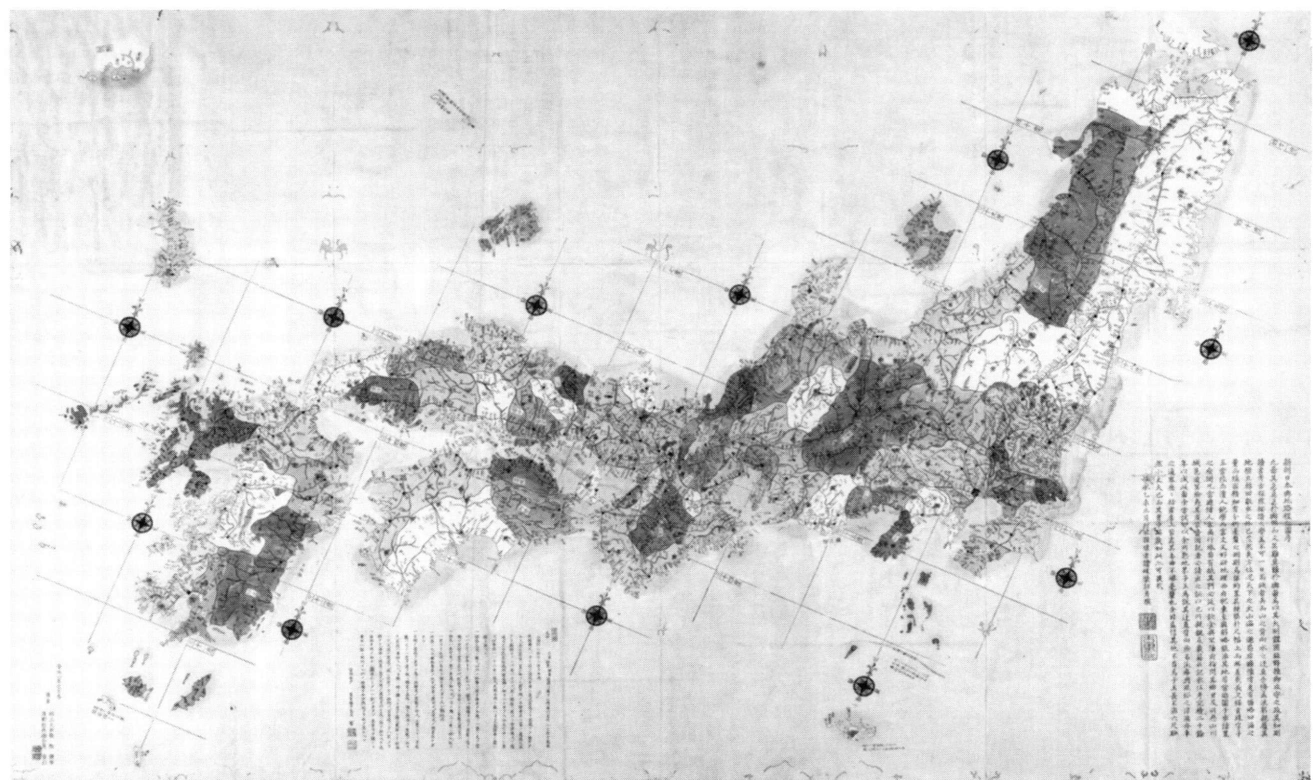

FIG. 11.45. NAGAKUBO SEKISUI'S *KAISEI NIHON YOCHI ROTEI ZENZU*, 1779. This was the first printed map of Japan with a graticule, which was superimposed from an existing official map. The copy shown here was the first issue of the first edition, attested to by the sickle-shaped Shimokita peninsula in the north of Honshū and by the lack of Mount Osore in the peninsula. Another issue and later editions show the peninsula as ax shaped as well as including Mount Osore. Size of the original: 84 × 136 cm. By permission of the Beans Collection, University of British Columbia Library, Vancouver.

have included charts of Japan, so that some coastal sections in the west appear more precisely than on the Shōhō map.[269] Although the *Kōsei Dainihon enbizu* shows improvements in northern Honshū and Tosa Bay, it is generally inferior to Ishikawa's *Honchō zukan kōmoku* in both information and decorative appeal.

Nagakubo Sekisui's *Kaisei Nihon yochi rotei zenzu* (Revised route map of Japan, 1779) (fig. 11.45) marked the end of the dominance of the Ryūsen type. Compiled at a scale of one *sun* to ten *ri*, or 1:1,296,000, it established a new model for the map trade that also lasted about a century. The map was revised and republished in 1791, and also in 1811, 1833, 1840, and 1844, after Nagakubo's death; a large number of other copies were also made.[270] In 1783, Nagakubo introduced a compact version (52 × 49.6 cm) titled *Jūsen Nihon yochi zenzu* (Reengraved map of Japan).[271] His map of 1779 is noted as the first printed map of Japan with parallels expressing degrees of latitude and meridians with no degrees for longitude. The grid of parallels and meridians was superimposed on an existing official map, probably the Shōhō map, rather than compiled from original survey data. The idea of using such a grid evidently came from the manu-

269. The Shōhō map itself was first printed in 1701 in the *Shinkan jinkoku ki* by Seki Sokō; it was divided into provinces and simplified. (The *Shinkan jinkoku ki* has been reproduced by Iwanami Bunko, blue series, no. 28-1 [Tokyo: Iwanami Shoten, 1978]). Five years previously Seki compiled the *Nihon bun'iki shishō zu* (Quickly understandable atlas of the regions of Japan, 1696), an atlas version of the Shōhō map that was never published. The general map of Japan in the *Nihon bun'iki shishō zu* is reproduced in Unno, Oda, and Muroga, *Nihon kochizu taisei*, vol. 1, fig. 23 (note 8), and in Unno, "Kinsei kankō no Nihonzu" (Maps of Japan printed in [early] modern times), in *Chizu no shiwa*, 126–38, esp. 133 (note 136).

270. As in the case of Ishikawa Ryūsen, the personal name was preferred for professional reasons, and the map became known as the "Sekisui map." Two versions of the first map issued are found in the Meiji University Library (82.2 × 132.8 cm) and in the Beans Collection (fig. 11.45); on these versions the shape of the Shimokita peninsula is different than on later editions. The first is reproduced in color in Unno, Oda, and Muroga, *Nihon kochizu taisei*, vol. 1, pl. 33 (note 8). For the second issue of the first edition (83 × 135.5 cm), see Nanba, Muroga, and Unno, *Nihon no kochizu/Old Maps in Japan*, pl. 29 (note 11). The revised editions were the same size as the 1779 copy. The 1811 reprinted version, 83 × 134.5 cm and in Cortazzi's personal collection, is reproduced in Cortazzi, *Isles of Gold*, pl. 41 (note 14).

271. The *Jūsen Nihon yochi zenzu* was reduced by Sotani Ōsei, and the meridians and parallels run parallel with the edges of the paper. It is reproduced in Unno, Oda, and Muroga, *Nihon kochizu taisei*, vol. 1, fig. 50 (note 8).

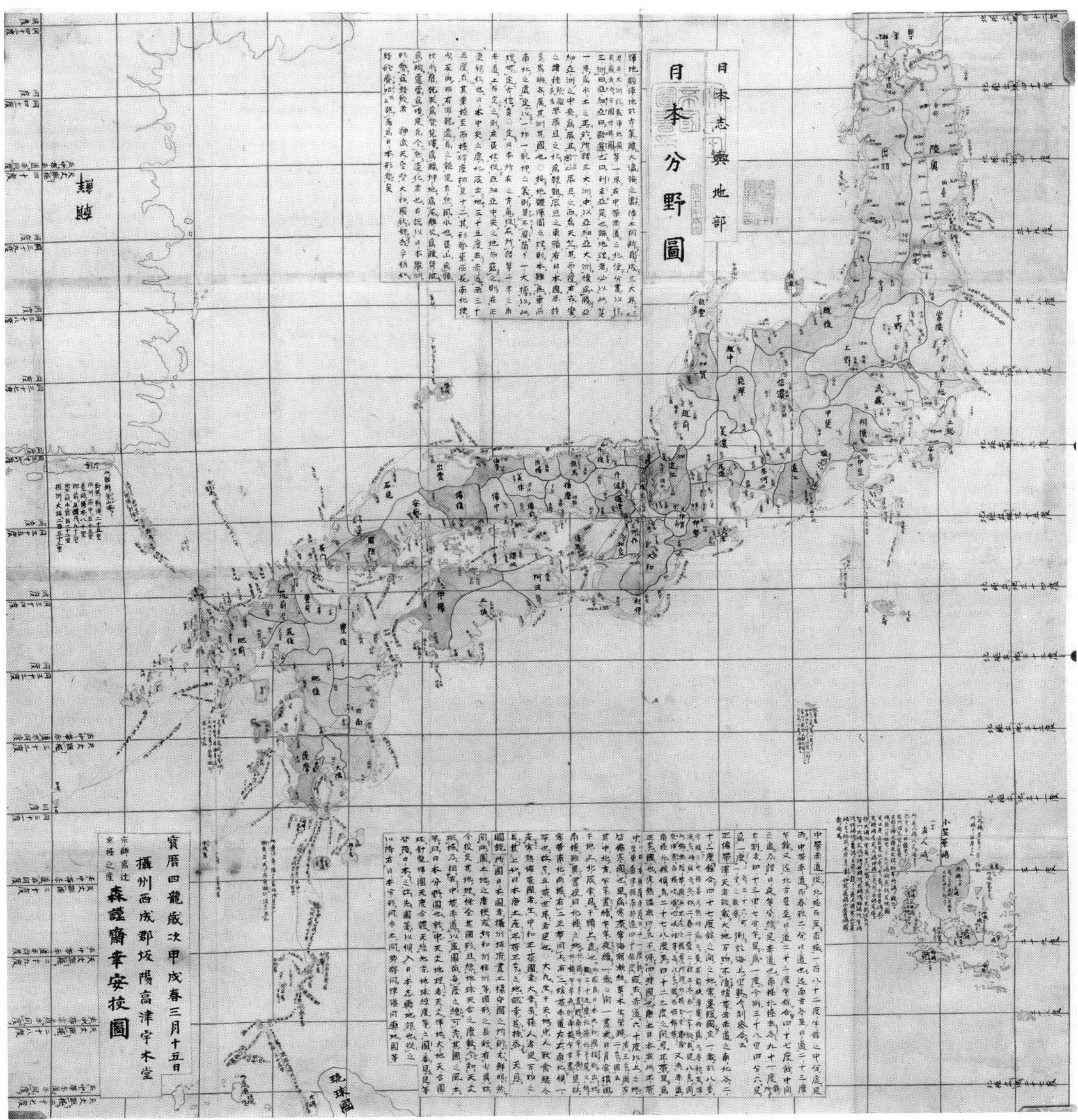

FIG. 11.46. MORI KŌAN'S *NIHON BUN'YA ZU*, 1754. This map, a copy of which was found in Nagakubo's belongings after his death, most likely gave him the idea of using a grid of parallels and meridians. As on Nagakubo's map of Japan, the manuscript gives values for latitude only. Mori's interest in latitude and longitude probably came from copying marine charts of South and East Asia.
Size of the original: 102.5 × 95 cm. By permission of the National Archives, Tokyo.

script *Nihon bun'ya zu* (Astronomical map of Japan, 1754) (fig. 11.46) by Mori Kōan (1692?–1757?).[272] On both maps Kyōto appears to be the point of origin for the longitude lines, but on neither are numerical values given.

272. Reproduced in color in Unno, Oda, and Muroga, *Nihon kochizu taisei*, vol. 1, pl. 29 (note 8).

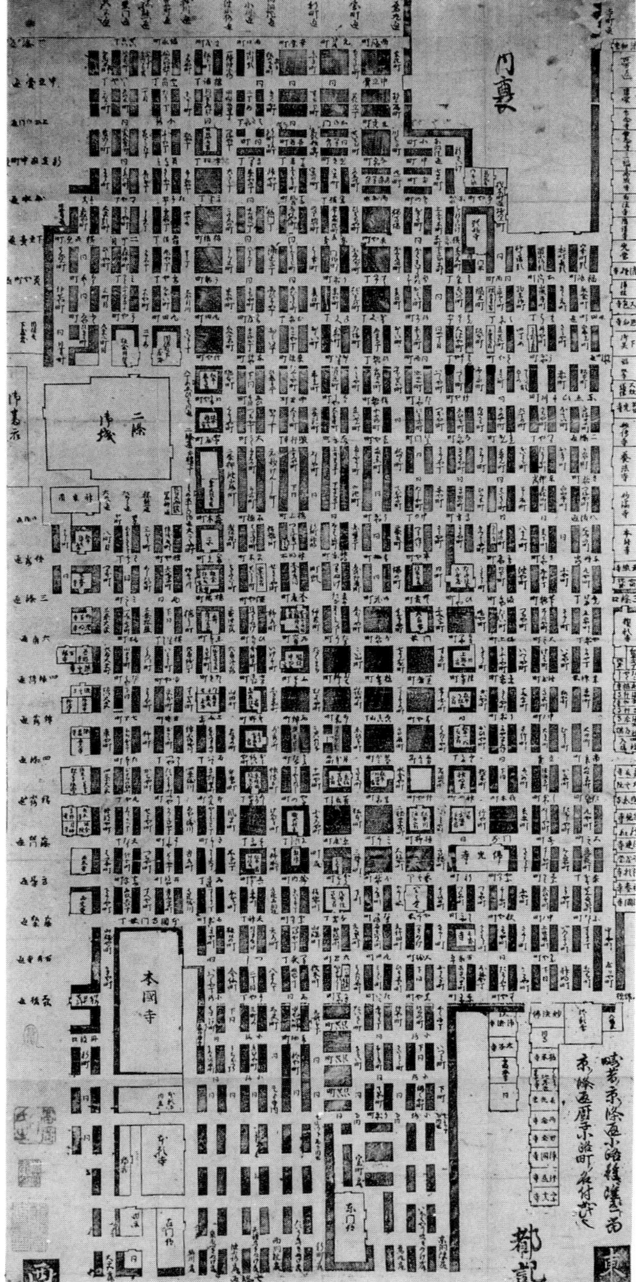

FIG. 11.47. PLAN OF KYŌTO, BEFORE 1641: A REVISION OF A TRADITIONAL DIAGRAMMATIC PLAN OF THE CAPITAL. At the upper right is the imperial palace and at the lower center are the "licensed quarters," which were moved to the western suburbs (off the plan) in 1641 and therefore date the map before this time.
Size of the original: 116.6 × 54 cm. Ōtsuka Takashi, Kyōto (formerly in the Moriya Yoshitaka Collection). Photograph courtesy of Kazutaka Unno.

Another genre of maps of Japan is exemplified in plate 28. These maps could be considered a hybrid between landscape drawings/paintings and maps in the last quarter of the Edo period, but there are very few examples of this genre. The map in plate 28 dates from about 1804.

Provincial Maps and Town Plans

Whereas the maps from the official Tokugawa provincial surveys gradually filtered into general circulation because they were not treated as state secrets, there were also provincial maps produced specifically for the popular market.[273] Beginning in 1709 with the *Kawachi no kuni ezu* (Map of Kawachi Province) by Hayashi Jōho, provincial maps of the Go Kinai region[274] were rapidly published. The dramatic increase in provincial maps at the beginning of the nineteenth century was probably a result of a popular interest in local geography: maps of thirty-six of the sixty-six provinces were published during the Edo period. Many popular encyclopedic geographies were also published. These were related to tourism, and examples include the *Miyako meisho zue* (Illustrated description of notable places in Kyōto, 6 vols.), the *Yamato meisho zue* (Illustrated description of notable places in Yamato Province, 7 vols.) of 1780 and 1791, and the *Izumi meisho zue* (Illustrated description of notable places in Izumi Province) of 1796, all by Akisato Ritō (fl. 1776–1830), and the *Edo meisho zue* (Illustrated description of notable places in Edo, 20 vols.) of 1836 by Saitō Chōshū (d. 1799) and others.[275]

Numerous maps and plans of towns were also pub-

273. For provincial maps printed in the Edo period, see Kurita Mototsugu, "Edo jidai kankō no kokugunzu" (Printed provincial maps of the Edo period), *Rekishi Chiri* 84, no. 2 (1953): 1–16, and Miyoshi Tadayoshi, "Nanba Korekushon chū no kankō shokokuzu ni tsuite" (On the printed provincial maps in the Nanba Collection), *Kōbe Shiritsu Hakubutsukan Kenkyū Kiyō* 4 (1987): 27–52.

274. The Go Kinai region includes the provinces of Yamashiro, Yamato, Kawachi, Settsu, and Izumi, and it incorporates the modern prefectures of Ōsaka and Nara and part of the prefectures of Kyōto and Hyōgo. The 1709 map (52 × 124.9 cm) is reproduced in Unno, Oda, and Muroga, *Nihon kochizu taisei*, vol. 1, pl. 50 (note 8); and in Nanba, Muroga, and Unno, *Nihon no kochizu/Old Maps in Japan*, pl. 52 (note 11). Other examples of provincial maps of Go Kinai are *Yamashiro meishōshi zu sōzu* (General map of the atlas describing scenic spots in Yamashiro Province, 1711) by Ōshima Takeyoshi (42.4 × 61.7 cm); *Yamato no kuni saiken ezu* (Detailed map of Yamato Province, 1734) by Ishikawa Shun'ei; a map of Izumi Province, 1736; and a map of Settsu Province, 1739.

275. In addition to these, there are seven others that belong to Akisato's *Meisho zue* group: *Shūi Miyako* (Kyōto, addendum, 1787); *Settsu*, 1796–98; *Tōkaidō*, 1797; *Ise sangū* (Visit to Ise Shrine, 1797); *Kawachi*, 1801; *Kisoji* (Kiso road, 1805); and *Ōmi*, 1814. See Miyoshi Manabu, "Meisho zue kaisetsu" (Explanation of the *Meisho zu*), in *Iwanami kōza chirigaku* (Iwanami lectures on geography), 76 vols. (Tokyo: Iwanami Shoten, 1931–34), Bekkō (Supplement) (1932): 1–22.

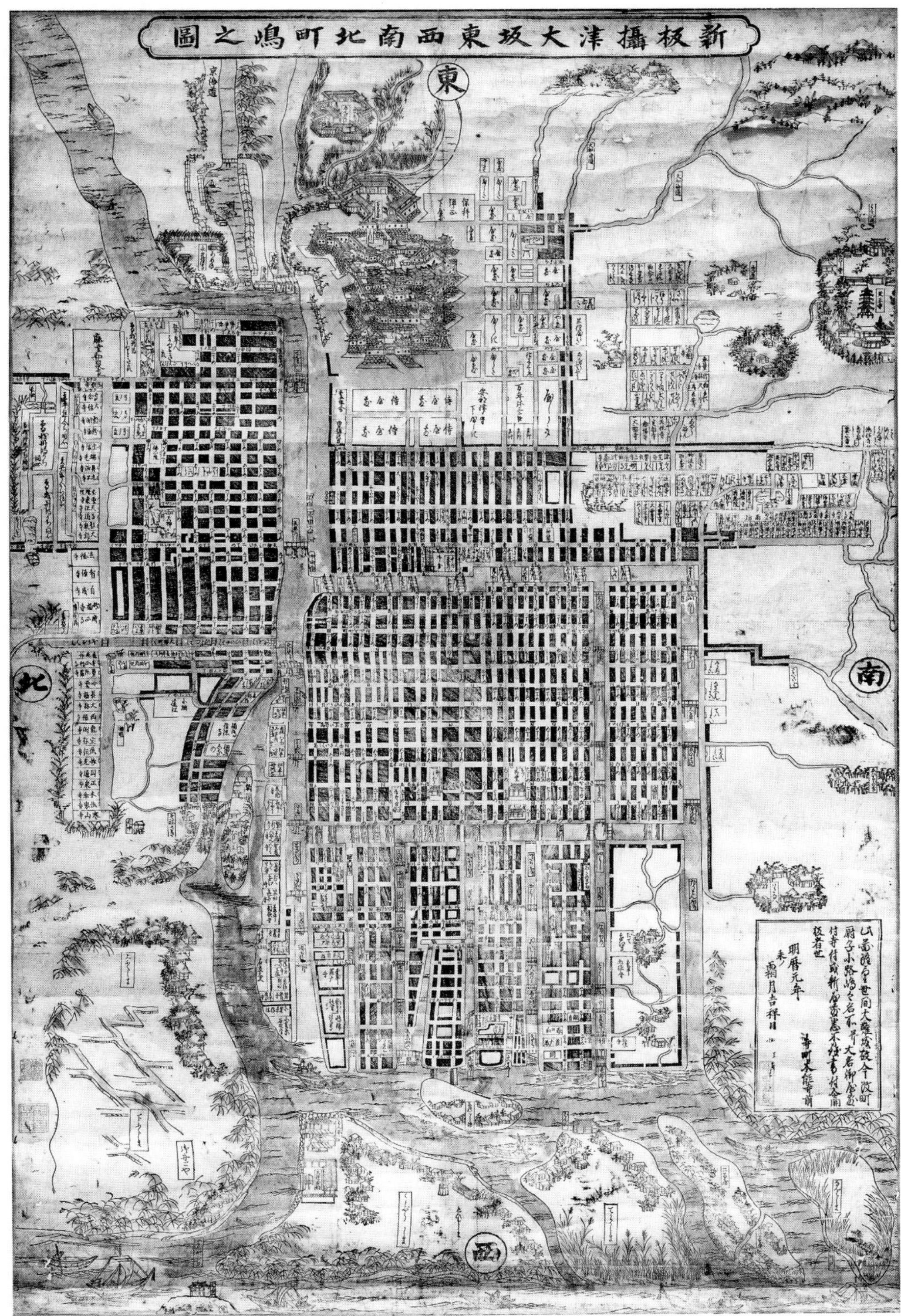

FIG. 11.48. *SHINPAN SETTSU ŌSAKA TŌZAINANBOKU MACHI SHIMA NO ZU*, 1655. This is the oldest known extant printed plan of Ōsaka. It is oriented to the east so that it would follow the convention of plans of Ōsaka by placing the castle at the top. The description is purely diagrammatic and could not have been a result of surveying.
Size of the original: 119.4 × 77.5 cm. By permission of the Beans Collection, University of British Columbia Library, Vancouver (1655.1).

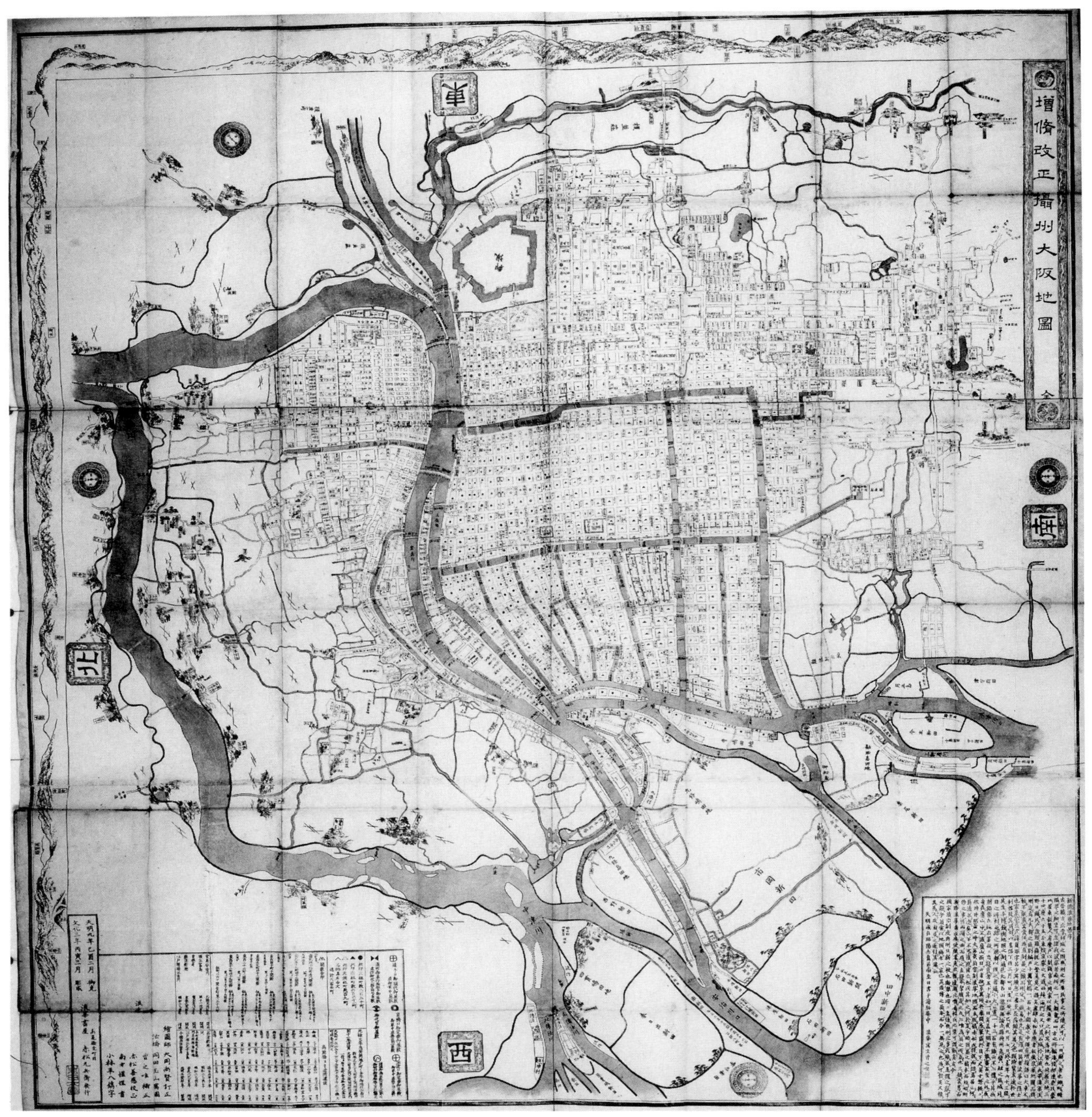

FIG. 11.49. *ZŌSHŪ KAISEI SESSHŪ ŌSAKA CHIZU* (ENLARGED AND REVISED PLAN OF ŌSAKA, SETTSU PROVINCE, 1806) BY ŌOKA SHŌKEN ET AL. With regard to accuracy in printed plans of Ōsaka, this is a milestone; revisions were published in 1844 and 1872. According to the preface by Sotani Ōsei, it was based on an unfinished plan by the cartographer Sawada Kazunori (1717–79).
Size of the original: 152 × 141 cm. Iwata Chinami Collection, Tokyo. Photograph courtesy of Kazutaka Unno.

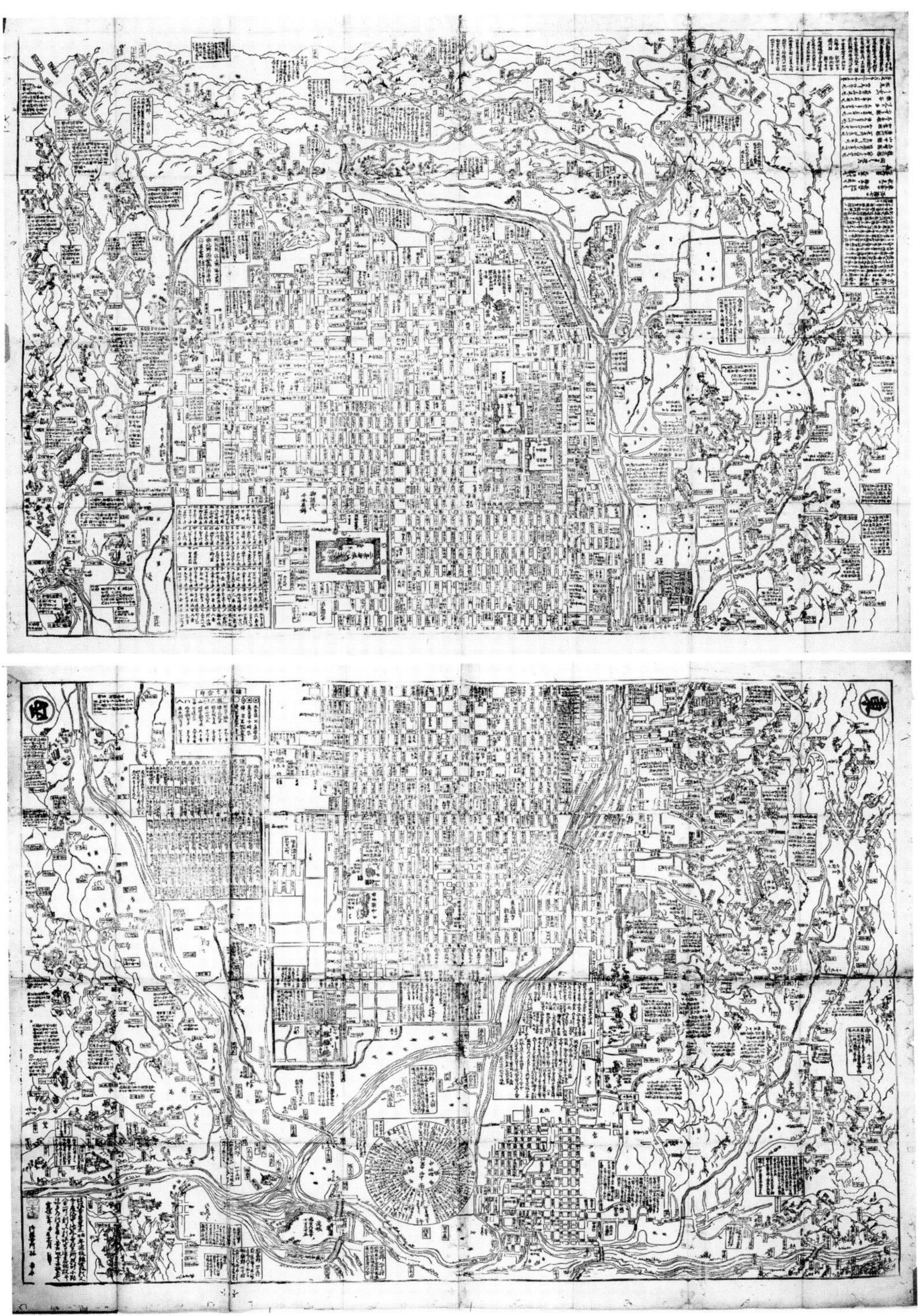

FIG. 11.50. *ZŌHO SAIHAN KYŌ ŌEZU* (LARGE PLAN OF KYŌTO, ENLARGED, SECOND EDITION, 1741), PUBLISHED BY HAYASHI YOSHINAGA. Compared with figure 11.47, the map of Kyōto seems less forceful and less artificial: shrines, temples, and notable places in the suburbs, for instance, have been added. The top sheet is northern Kyōto, the bottom southern, with Sanjō (Third Line [Street]) being the divider; the scale is eight *bu* (i.e., 2.4 cm) to one *chō* (109.09 m) or 1:4,500. Hayashi was a well-known map publisher in Kyōto in the seventeenth and eighteenth centuries.

Size of the originals: 87 × 121.5 cm (northern part), 86.5 × 120.5 cm (southern part). Iwata Chinami Collection, Tokyo. Photograph courtesy of Kazutaka Unno.

FIG. 11.51. *SHINPAN EDO ŌEZU* (NEWLY ISSUED PLAN OF EDO, 1671) BY OCHIKOCHI DŌIN. Modeled on a ca. 1658 plan of Edo (fig. 11.36), this is by far the most accurate printed plan of Edo, or rather the center of Edo. Four more sheets were published up until 1673 to show the surroundings of Edo, under the title *Shinpan Edo soto ezu* (Newly issued plan of the areas outside Edo). Ochikochi Dōin was the pseudonym of the surveyor Fujii Hanchi. One *bu* equals five *ken* (1:3,250).
Size of the original: 153.5 × 162.3 cm. Iwata Chinami Collection, Tokyo. Photograph courtesy of Kazutaka Unno.

lished throughout the Edo period (appendix 11.9).[276] These were not for administrative use but for ordinary citizens, and they emphasize notable places such as shrines, temples, and historical sites. Two of the oldest surviving examples are of Edo—the *Bushū Toshima gōri Edo no shō zu* (Plan of Edo, Toshima County, Musashi Province)—and Kyōto (fig. 11.47), thought to have been published, respectively, about 1632 and before 1641. The oldest extant plan of Ōsaka—the *Shinpan Settsu Ōsaka*

276. Town plans printed in the Edo period are described in detail in Kurita Mototsugu, "Nihon ni okeru kokan toshizu" (Old printed maps of cities in Japan), *Nagoya Daigaku Bungakubu Kenkyū Ronshū* 2 (1952): 1–13. For reproduced examples, see appendix 11.9 and Nanba, Muroga, and Unno, *Nihon no kochizu/Old Maps in Japan*, pls. 60, 64, 65, 67, 68, 70, 73, 75, and 76 (note 11); Unno, Oda, and Muroga, *Nihon kochizu taisei*, vol. 1, pls. 71–76, 78, 81, 83–86, 88, 90–92, 104–5, and 107–9 (note 8); Kurita, *Nihon kohan chizu shūsei*, pls. 41–70 (note 15); and Cortazzi, *Isles of Gold*, pp. 122, 126, pls. 50, 54, 56–58 (note 14). Despite such publications and reproductions, the study of

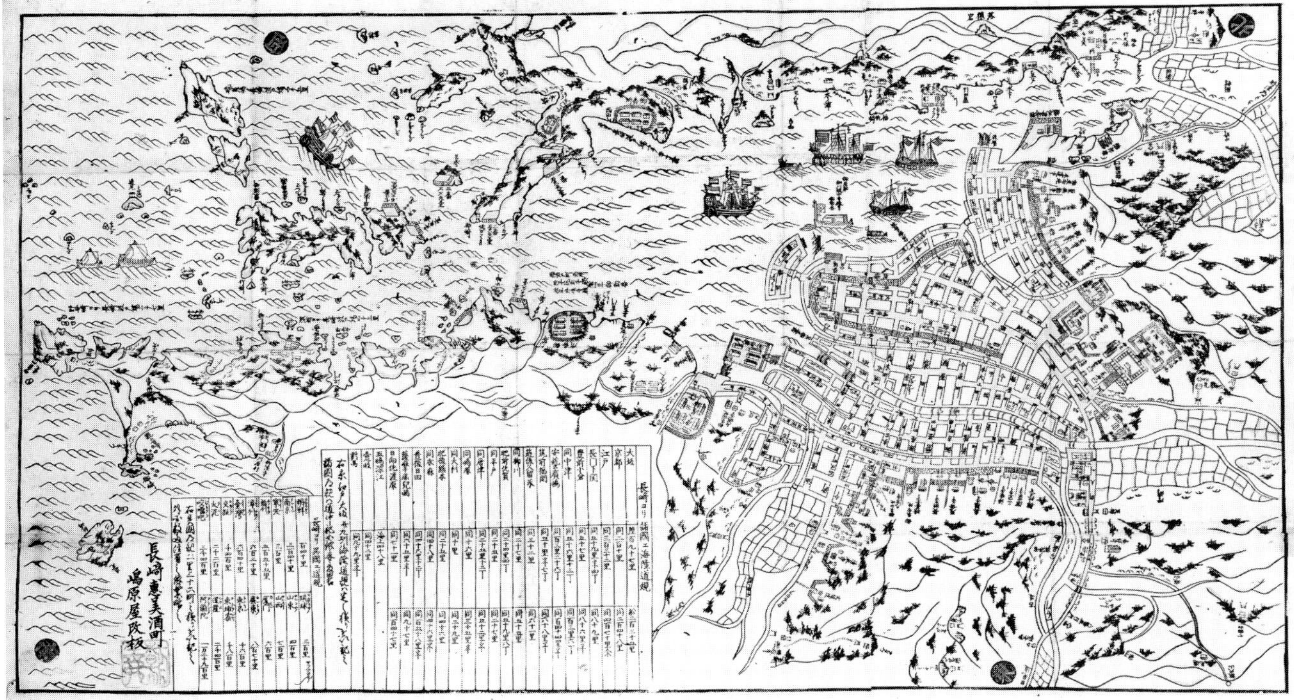

FIG. 11.52. *SHINPAN NAGASAKI ŌEZU* (NEWLY ISSUED PLAN OF NAGASAKI, CA. 1760) PUBLISHED BY SHIMABARAYA. Early printed plans of Nagasaki included Nagasaki peninsula, as does this one. Judging from the buildings and geographical features, the information may be dated to 1741–65.
Size of the original: 57 × 101.8 cm. By permission of the Kurita Kenji Collection, Nagoya.

tōzainanboku machi shima no zu (Newly issued plan of Ōsaka with the east-west and north-south streets and islands, Settsu Province)—dates from 1655 (fig. 11.48).[277] Plans of Kyōto were compiled in the same way as they had been in medieval times, merely depicting the simple grid pattern of the city. Those of Ōsaka (for example, fig. 11.49) and Edo, however, are presumed to have been compiled from ground surveys, albeit not to a high standard of accuracy. On the plans of Kyōto (for example, fig. 11.50) and Ōsaka, which had been issued by several publishers in Kyōto, the residential blocks were printed in black until about 1687, but thereafter they were left blank as on the plans of Edo (fig. 11.51). In terms of the number of plans published, Nagasaki ranks fourth after Edo, Kyōto, and Ōsaka. Its oldest city plan known to survive—the *Nagasaki ōezu* (Large plan of Nagasaki)—dates to approximately 1681. Plans published before the 1760s tended to cover the area of Nagasaki Bay (fig. 11.52); thereafter, plans focused on the port and its interior became more popular, such as the *Hishū Nagasaki no zu* (Plan of Nagasaki, Hizen Province).[278]

Other than these four cities, plans until the end of the Edo period concentrated on towns with famous shrines, temples, historical sites, and scenic spots. Examples include plans of Nara in 1666 and Kamakura about 1670, respectively the *Washū Nanto no zu* (Plan of the southern metropolis [Nara], Yamato Province) and the *Sōshū Kamakura no moto ezu* (Standard plan of Kamakura, Sagami Province).[279] Toward the end of the Edo period, the process of opening Japan to the world led to plans' being composed of the treaty ports of Shimoda, Hakodate, and Yokohama. Plans of the port of Shimoda and

town plans is incomplete. We still do not know, for example, exactly how many plans were published.

277. The plans of Edo and Ōsaka are listed in appendix 11.9; the plan of Kyōto (116.6 × 54 cm) is owned by Ōtsuka Takashi of Kyōto and was owned previously by Moriya Yoshitaka. The plans of Edo and Kyōto are reproduced in Unno, Oda, and Muroga, *Nihon kochizu taisei*, vol. 1, pls. 72 and 80 (note 8). For the early printed plans of Edo, see Nagasawa Kikuya, "Edo no hanzu ni tsuite" (On the printed plans of Edo), *Shoshigaku*, n.s., 2 (1965): 31–51; Iida and Tawara, *Edozu no rekishi* (note 225); and Iwata Toyoki, *Edozu sōmokuroku* (General catalog of plans of Edo) (Tokyo: Seishōdō Shoten, 1980). For the early printed plans of Kyōto, see Fujita Motoharu, *Toshi kenkyū Heiankyō hensenshi, tsuketari kochizu shū* (History of the Kyōto region, accompanied by collected old plans) (Kyōto: Suzukake Shuppanbu, 1930; reprinted Nihon Shiryō Kankōkai, 1976), and Ōtsuka Takashi, *Kyōtozu sōmokuroku* (General catalog of plans of Kyōto) (Tokyo: Seishōdō Shoten, 1981).

278. For details see appendix 11.9.

279. On the first, see appendix 11.9; the second measures 70.9 by 103.3 centimeters. See Kurita, "Nihon ni okeru kokan toshizu" (note 276).

Hakodate Bay, for instance, were published in 1855; these are the *Zushū Shimoda minato no zu* (Chart of Shimoda harbor, Izu Province) by Shizunoya and the *Hakodate zenzu* (General chart of Hakodate) published by Shunjudō.[280] Those of Yokohama began to be published in 1859, two examples being the *Tōkaidō Kanagawa onbōeki ba* (Map of the Kanagawa trading port, Tōkai road) and Takashima Hōdō's *Yokohama meisaizu* (Detailed plan of Yokohama).[281] Plans of the capitals of feudal domains were also being printed late in the Edo period; they were used mainly for visiting temples and shrines and for tourism. Examples include plans of Okazaki in about 1840, Sunpu (Shizuoka) in 1842 and 1868, Kōfu in 1849, Hiroshima in 1865, and two undated mid-century plans of Kanazawa.[282] In all, printed plans of over thirty-one cities and towns, including pleasure and hot-spring resorts, were made during the Edo period.

FIG. 11.53. *TŌKAIDŌ MICHIYUKI NO ZU*, CA. 1654. Oblivious to direction, the road winds its way between Kyōto at the top and Edo at the bottom right. The names of the daimyos mentioned on the map date it to 1652–54. It is the oldest known extant Japanese printed itinerary map of Japan.
Size of the original: 130.7 × 57.7 cm. Photograph courtesy of Nakao Shōsendō, Ōsaka.

Itineraries

The history of pictorial itineraries can be traced back to at least the seventeenth century. The oldest surviving printed itinerary map is the *Tōkaidō michiyuki no zu* (Itinerary map of the Tōkai road; fig. 11.53), thought to be published in 1654, a date derived from the common date (1652–54) of service for the feudal lords mentioned on the map.[283] The road curves freely without regard to

280. The chart of Shimoda (74.5 × 51.5 cm) is reproduced in color in Nanba, Muroga, and Unno, *Nihon no kochizu/Old Maps in Japan*, pl. 75 (note 11), and Kurita, *Nihon kohan chizu shūsei*, pl. 60 (note 15). The chart of Hakodate measures 72.7 by 77.3 centimeters. Shizunoya is the specialist in Dutch studies Ōtsuka Hachirō (1795–1855).

281. The former (23.5 × 60 cm) is reproduced in Unno, Oda, and Muroga, *Nihon kochizu taisei*, vol. 1, fig. 69 (note 8), and the latter (36.9 × 46.3 cm and 15.6 × 21.8 cm) in Kurita, *Nihon kohan chizu shūsei*, pl. 62 (note 15).

282. These are *Taihei Okazaki ezu* (Plan of Okazaki at peace), published by Okadaya Ichibee in Edo and Hon'ya Bunkichi in Okazaki (43.4 × 84 cm); *Sunpu hitori annai* (Guide to Sunpu for visiting alone), published by Nishinoya in Sunpu (now Shizuoka) (33.3 × 48 cm); *Sunpu meishō ichiran zu* (Visualized map of the scenic spots of Sunpu) (68.9 × 92.4 cm); *Kaihō Kōfu ezu* (Portable plan of Kōfu), published by Murataya Kōtarō in Kōfu (51.9 × 54.8 cm); *Hiroshima machimachi michishirube* (Guide to the towns of Hiroshima) (28.8 × 45.7 cm); untitled plan of Kanazawa (34.5 × 41.5 cm); and untitled plan of Kanazawa (70.9 × 97.4 cm). Reproductions are in Kurita, *Nihon kohan chizu shūsei* (note 15): the *Sunpu meishō ichiran zu* (pl. 59); the maps of Kōfu (pl. 59) and Hiroshima (pl. 67); and both maps of Kanazawa (pls. 65 and 66). See also Kurita, "Nihon ni okeru kokan toshizu," 11–12 (note 276). The *Sunpu meishō ichiran zu* and the smaller map of Kanazawa are in Nanba, Muroga, and Unno, *Nihon no kochizu/Old Maps in Japan*, pls. 73 and 70 (note 11).

283. Copies are in the possession of Nakao Shōsendō (fig 11.53) and the Kikkawa family in Iwakuni, Yamaguchi Prefecture. The second edition, with slipshod pictures and explanations, is kept at the Kōbe City Museum, reproduced in *Kōbe Shiritsu Hakubutsukan kanzō meihin zuroku* (Masterpieces of the Kōbe City Museum) (Kōbe: Kōbe Shi Supōtsu Kyōiko Kosha, 1985), fig. 17 (131 × 58 cm), and at the Beans Collection of the University of British Columbia Library, reproduced in Beans, *Japanese Maps of the Tokugawa Era*, facing p. 14 (131 × 59 cm) (note 22). See Unno Kazutaka, "Mukanki Tōkaidō michiyuki

FIG. 11.54. EXTRACTS FROM THE *TŌZAI KAIRIKU NO ZU* PUBLISHED BY NISHIDA KATSUBEE IN 1672. The top part includes Kyōto, Ōsaka, and Awaji Island, the bottom showing the western extremity of the map and including Nagasaki. Source material evidently included the *Kisoji Nakasendō Tōkaidō ezu* (fig. 11.38) and the *Saigokusuji kairiku ezu*, both manuscripts commissioned by the shogunate.
Size of the original: 33.7 × 1,530 cm. By permission of the Mitsui Library, Tokyo.

measured distance and direction, although the map does contain post towns and indicates the distance between them. Rich ornamentation is characteristic of this version, but pocket-sized editions of 1666 and 1667 were published with the more utilitarian geographic needs of travelers in mind.[284] Accuracy was again not a major concern of the mapmaker, but decorative features symbolic of the journeys were included as well as practical information such as lists of fares for travelers using horses.

The first itinerary map for travelers thought to be based on official sources was the *Tōzai kairiku no zu* (Map of the east-west sea and land routes) of 1672 (fig. 11.54).[285] It was published in Kyōto by Nishida Katsubee and was modeled on the official manuscript maps *Kisoji Nakasendō Tōkaidō ezu* (Map of the Kiso/Nakasen road and the Tōkai road) and *Saigokusuji kairiku ezu* (Map of the sea and land routes in the western regions). It would have been beyond the resources of private publishers to compile maps of such large areas of the country, so there are grounds for believing that these maps were produced under the auspices of the Tokugawa authorities. In 1690 Ochikochi Dōin (fl. 1670–96) published his *Tōkaidō bungen ezu* (Surveyed route map of the Tōkai road), on which were entered compass bearings in squares to assist in reading accurate directions (fig. 11.55).[286] The map was based on the results of the survey undertaken by Hōjō Ujinaga on behalf of the shogunate. Ochikochi himself

no zu no ihan" (Two undated editions of the *Tōkaido michiyuki no zu*, an itinerary map of the Tōkai road), *Gekkan Kochizu Kenkyū* 22, no. 6 (1991): 2–5.

284. The 1666 map, published by Fushimiya in Kyōto, measures 56 by 41 centimeters and is preserved in the Österreichische Nationalbibliothek, Vienna. It is mentioned and reproduced in Kawamura Hirotada, "Ōsutoria Kokuritsu Toshokan shūzō no Edo jidai Nihonsei chizu" (On the maps made by Japanese in the Edo period, owned by the Austrian National Library), *Gekkan Kochizu Kenkyū* 18, no. 7 (1987): 2–6. The 1667 map, published in Kyōto by Shijō Nakamachi (56.5 × 40 cm), is found in the Beans Collection at the University of British Columbia Library; it is reproduced in Unno, "Hokubei ni okeru Edo jidai chizu no shūshū jōkyō," fig. 6 (note 22). The woodblock for the 1667 map was probably the same as that for the 1666 map.

285. Two copies are known, one at the National Diet Library (35.5 × 1560 cm) and the other at the Mitsui Library, both in Tokyo (fig. 11.54). For a part of the first, see *Nihon no chizu: Kansen chizu no hattatsu*, fig. 22 (note 13); parts of the second are reproduced in Unno, Oda, and Muroga, *Nihon kochizu taisei*, vol. 1, pl. 117 (note 8). The *Tōzai kairiku zu* (Map of the east-west sea and land routes), reproduced in Kurita, *Nihon kohan chizu shūsei*, pl. 71 (note 15), is another version.

286. Ochikochi Dōin is the pseudonym of the surveyor Fujii Hanchi (or Hisane); he is discussed in detail in Fukai Jinzō, *Zuō Ochikochi Dōin* (Zuō's [Fujii's] Ochikochi Dōin) (Toyama: Katsura Shobō, 1990). Parts of the map are reproduced in Kurita, *Nihon kohan chizu shūsei*, pl. 72 (note 15); Nanba, Muroga, and Unno, *Nihon no kochizu/Old Maps in Japan*, pl. 36 (note 11); and Unno, Oda, and Muroga, *Nihon kochizu taisei*, vol. 1, pl. 118 (note 8). For reproductions of the entire map see *Tōkaidō meisho no ki; Tōkaidō bungen ezu* (Description of the famous places on Tōkai road; Surveyed route map of the Tōkai road), Nihon koten zenshū (Comprehensive collection of Japanese classical works), 4th ser. (Tokyo: Nihon Koten Zenshū Kankōkai, 1931);

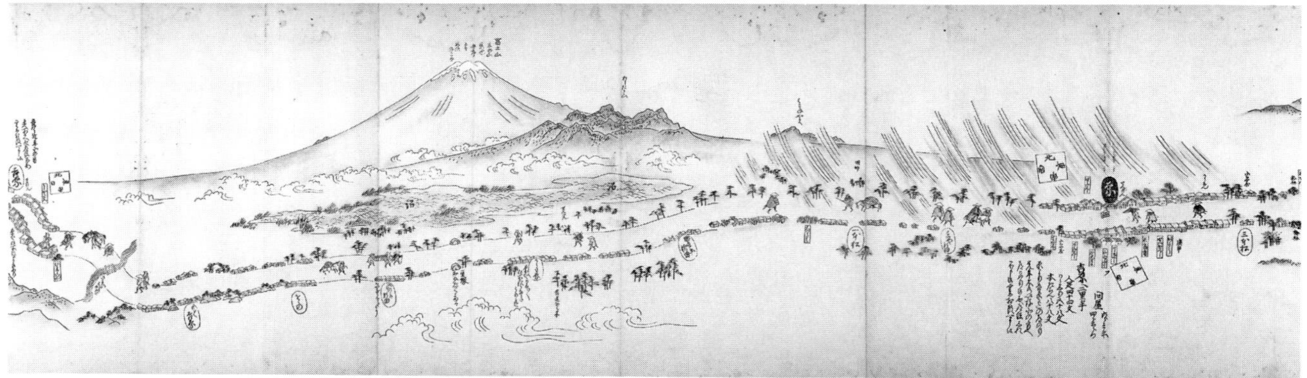

FIG. 11.55. EXTRACT FROM THE *TŌKAIDŌ BUNGEN EZU* BY OCHIKOCHI DŌIN, 1690. The vicinity of the post towns of Hara and Yoshiwara lying to the south of Mount Fuji are shown. Ochikochi, a surveyor, edited the map from a previous route map of the Tōkai road completed after a government survey in 1651, and the scenes and people were painted by the artist Hishikawa Moronobu (1618–94). This map includes a scale of three *bu* to one *chō* (1:12,000) and square signs at each place to show directions. This is the first issue of the first edition giving fares between stations next to each post town.
Size of the original folding book: 26.7 × 14.9 cm (total length: 3,610 cm). By permission of the Tokyo National Museum, Tokyo.

participated in this survey, which was to result in a map divided into five folding books, drawn on rectangular paper 28 by 3,610 centimeters, and at the scale of three *bu* to one *chō* (1:12,000). Accuracy alone, however, was not sufficient to make the map marketable: it was therefore illustrated with pictures of travelers and scenes along the roads by Hishikawa Moronobu (1618–94), an *ukiyoe* artist who was the master of Ishikawa Ryūsen.[287] The map went through several editions. One was a pocket edition revised by Sōyō in 1752 under the same title and published as a folding book that was handy to carry.[288]

Pocket-sized itinerary maps depicting the roads and sea routes of all Japan came to be published in large numbers from the first half of the eighteenth century onward, and they were very popular. Five main categories, all drawn on rectangular sheets, were produced during the Edo period: picture scrolls, mandalas, labyrinths, diagrams with straight parallel lines, and "conformal" maps to minimize distortion.

The picture scrolls, such as Nishida's *Tōzai kairiku no zu* and Ochikochi's *Tōkaidō bungen ezu*, were originally large maps for ornamental use. Later, miniaturized versions, including the 1752 pocket edition of Ochikochi's work and the *Kisoji anken ezu* (Simple map of the Kiso road) of 1756 edited by Sōyō (11 × 16 cm), served as handy itinerary maps. Many of these works are in the form of rectangular bound books with the routes laid out horizontally and with illustrative views along both sides of the routes.

Itinerary maps from the mandala[289] category showed the routes as curved lines and included roadside scenes as if they were viewed from above. Not many maps of this type were produced. They were, in fact, more like pictures than maps and were designed mainly for ornamental use. Examples include the *Tōkaidō michiyuki no zu* of about 1654 (fig. 11.53 above); two works by Katsushika Hokusai, *Tōkaidō meisho ichiran* (Panoramic view of famous places on the Tōkai road) of 1818 (43 × 58 cm) and *Kisoji meisho ichiran* (Panoramic view of famous places on the Kiso road) of 1819 (42 × 56 cm); and the *Shinkoku kaisei Tōkaidō saiken ōezu* (Detailed large map of the Tōkai road, newly revised) (70 × 142 cm), edited by Shōtei Kinsui (Nakamura Yasusada, 1797–1862) and illustrated by the painter Kuwagata Shōi in the middle of the nineteenth century.[290]

Kohan Edozu shūsei (Collection of early printed plans of Edo), Bekkan (supplement, separate volume) (Tokyo: Chūō Kōron Bijutsu Shuppan, 1960); and the *Kohan chishi sōsho* (Series of early printed geographical descriptions), vol. 12 (Tokyo: Geirinsha, 1971). The *Tōkaidō bungen ezu* is a five-volume folding book, and it is reproduced in these modern editions by relief printing. Ochikochi and the *Tōkaidō bungen ezu* are compared with John Ogilby's *Britannia* (1675) in Koji Hasegawa, "Road Atlases in Early Modern Japan and Britain," in *Geographical Studies and Japan*, ed. John Sargent and Richard Wiltshire (Folkestone, Eng.: Japan Library, 1993), 15–24.

287. Hishikawa is noted as the "true founder of the *ukiyo-e* school . . . who brought about the change from painting to woodcut printing"; see Munsterberg, *Arts of Japan*, 154 (note 79). For an introduction to the Hishikawa school, see Munsterberg, *Japanese Print*, 16–22 (note 258).

288. It measured 15.8 by 9.2 centimeters folded (total length was 1,220 cm) and was published by Yorozuya Seibee in Edo. Part of the map is reproduced in Unno, Oda, and Muroga, *Nihon kochizu taisei*, vol. 1, fig. 72 (note 8). The personal history of the reviser Sōyō is not known.

289. Since medieval times in Japan the term has been used for the precincts of shrines and temples. The term is applied to itinerary maps that resemble landscape drawings, similar to those of the medieval period mentioned previously, pp. 364–66.

290. The *Tōkaidō meisho ichiran* is reproduced in Cortazzi, *Isles of*

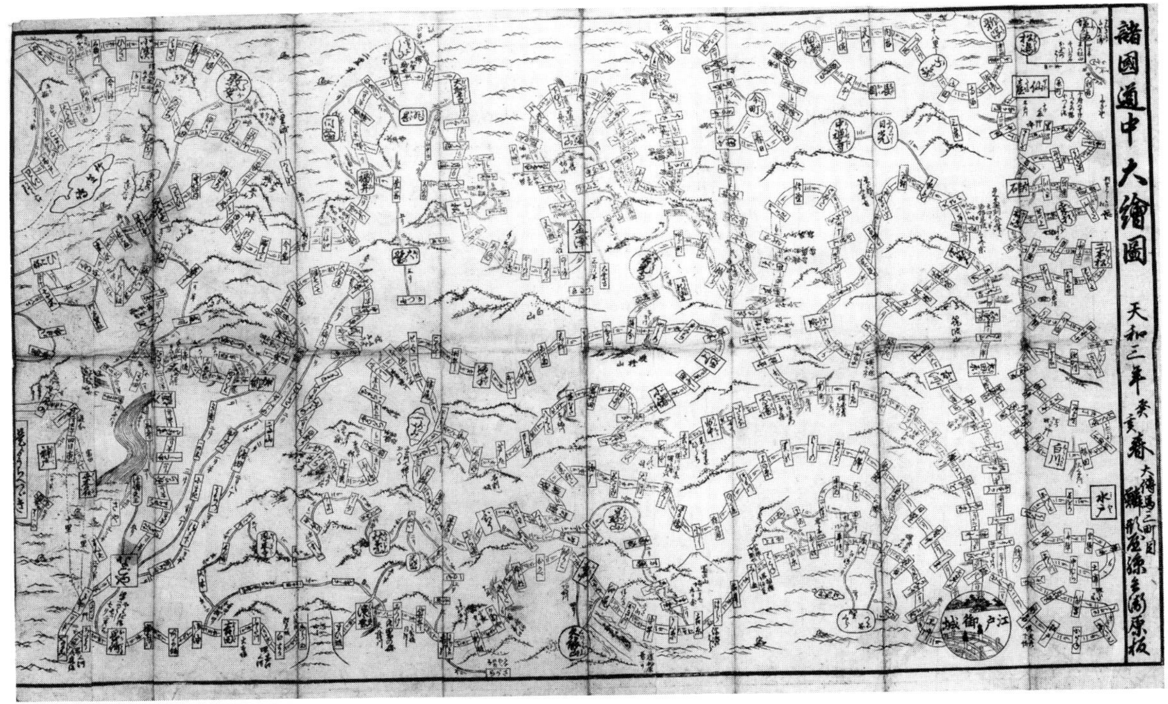

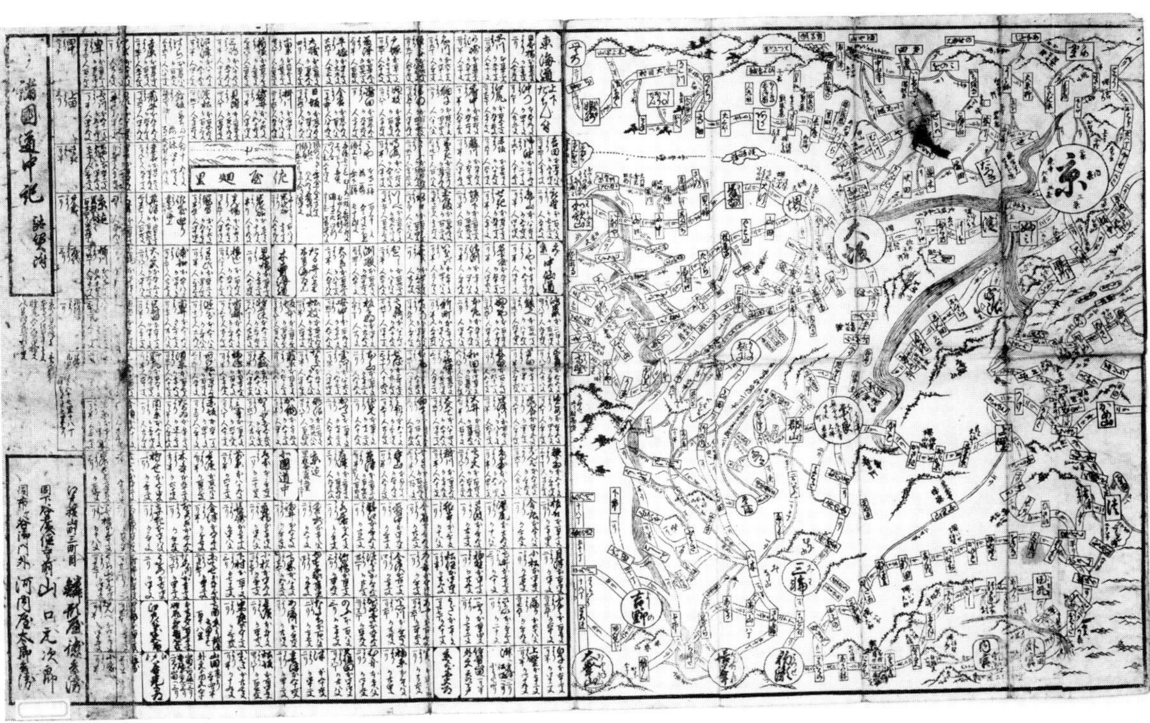

FIG. 11.56. AN EXAMPLE OF A LABYRINTH ITINERARY MAP, *SHOKOKU DŌCHŪ ŌEZU*. Originally published in 1683 by Urokogataya Magobee, the map shows the main roads and post stations in Honshū with no attention paid to distance and direction. Edo is represented by the circle in the lower right corner of the upper map, Kyōto by the circle in the upper right of the lower map. The left half of the lower map contains a table of fares between stations. They are printed on both sides of a single sheet.
Size of the original: 38.3 × 63 cm. By permission of the Kōbe City Museum, Kōbe, Nanba Collection.

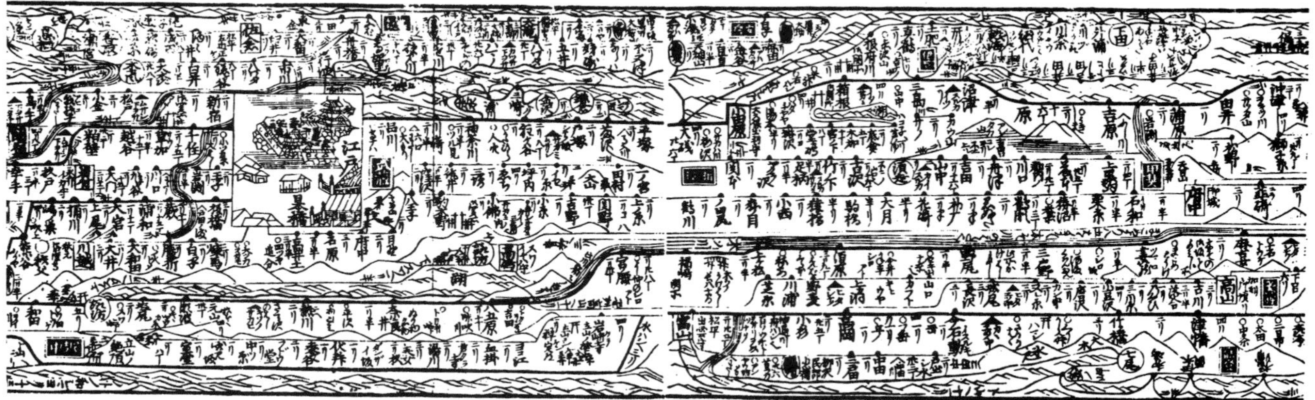

FIG. 11.57. AN EXAMPLE OF A DIAGRAMMATIC ITINERARY MAP, SHOWING THE PART CENTERED ON EDO IN KOKURYŪSAI'S *DAIZŌHO NIHON DŌCHŪ KŌTEI KI*, 1744. South is at the top: the large square toward the left is Edo, the upper part of the map is the Pacific Ocean, and the lower part is the Sea of Japan. Roads are shown by parallel straight lines.
Size of the original folding book: 16.5 × 7.3 cm (total length, 505 cm). Collection of Kazutaka Unno.

Disregarding distance, direction, and the shape of the land, "labyrinth" itinerary maps—often printed on both sides of the paper—were characterized by the scattering of roads and post towns throughout the map. As on the mandala maps, the roads were shown as curving lines. The main difference was that they were of greater practical than ornamental value. They include figure 11.56, the *Shokoku dōchū ōezu* (Large itinerary map of all the provinces) of 1683 and the *Dōchū hitori annai zu* (Map for traveling alone) of 1788 (29.9 × 77 cm; also printed on both sides).[291]

The diagrammatic maps with parallel lines include the *Kairiku Nihon dōchū hitori annai* (Guide to the sea and land routes of Japan for traveling alone) of 1722 (14.5 × 380 cm) and the *Daizōho Nihon dōchū kōtei ki* (Widely enlarged itinerary of Japan) of 1744 (fig. 11.57).[292] If it was shown at all, the shape of the land was distorted dramatically on such maps by generalizing routes and coastlines to straight parallel lines.

Finally, on maps of the conformal category ("conformal" is not used here in the sense of a map projection) there was an attempt to minimize distortion. Two examples are the *Dainihon dōchū hayabiki saiken zu* (Detailed and quickly discernible itinerary map of Great Japan) of 1830 (37.5 × 120 cm) by Akisato Ritō and the *Dainihon hayakuri dōchū ki* (Quickly discernible itinerary of Great Japan) of 1844 (39.5 × 91.5 cm; printed on both sides).[293] These versions were published late in the Edo period and were folded several times to make a portable book.

Not all of the itineraries can be categorized in these five groups. One style, known as *sugoroku* and so named after a Japanese dice game similar to backgammon, was a hybrid between the travel guidebook and a list of post towns. Examples include the *Tenmei kaisei shokoku dōchū ki taisei* (Complete itinerary of every province revised in the Tenmei era [1781–88]) of about 1785 and the *Toshidama ryōmen dōchū ki* (Itinerary printed on both sides, a New Year's gift) (fig. 11.58) dating from the mid-eighteenth century.[294] Another minor genre combined the picture scroll and diagrammatic categories; an example is the *Dainihon kairiku shokoku dōchū zukan* (Itinerary map of the sea and land routes of Great Japan) of 1864.[295]

Maps of the World and of China

As we have seen with the maps derived from Matteo Ricci, the Japanese map trade also issued a variety of printed editions of world maps during this period of popularization.[296] Until the publication of Nagakubo's

Gold, pl. 59 (note 14), and the last two are in Nanba, Muroga, and Unno, *Nihon no kochizu/Old Maps in Japan*, pls. 37 and 38 (note 11).

291. The latter was published by Kikuya Kihee, Kyōto.

292. Both were folding books published in Ōsaka; the first by Kemaya Hachirōemon and the second (edited by Kokuryūsai) by Torikai Ichibee.

293. The 1830 map is reproduced in Unno, Oda, and Muroga, *Nihon kochizu taisei*, vol. 1, pl. 121 (note 8); the 1844 map was published by Akitaya Taemon at Ōsaka and five other publishers. On the cover of the 1830 map is another title: *Nihon kairiku hayabiki dōchū ki* (Quickly discernible itinerary of the sea and land routes of Japan); a folio edition was also made because even when folded up and doubled over to form a book, the map was still too long to be practical.

294. Both are printed on both sides; the former measures 30.4 by 39.3 centimeters.

295. A bound book (8.5 × 18 cm), published by Sakaiya Naoshichi and eight others.

296. World maps from the period of isolation (1639–1854) are described in Ayusawa, "Types of World Map," 123–27 (with Ramming's comments on 128) (note 235), and Ayusawa Shintarō, "Sekai chiri no bu" (Section of world geography), in *Sakoku jidai Nihonjin*

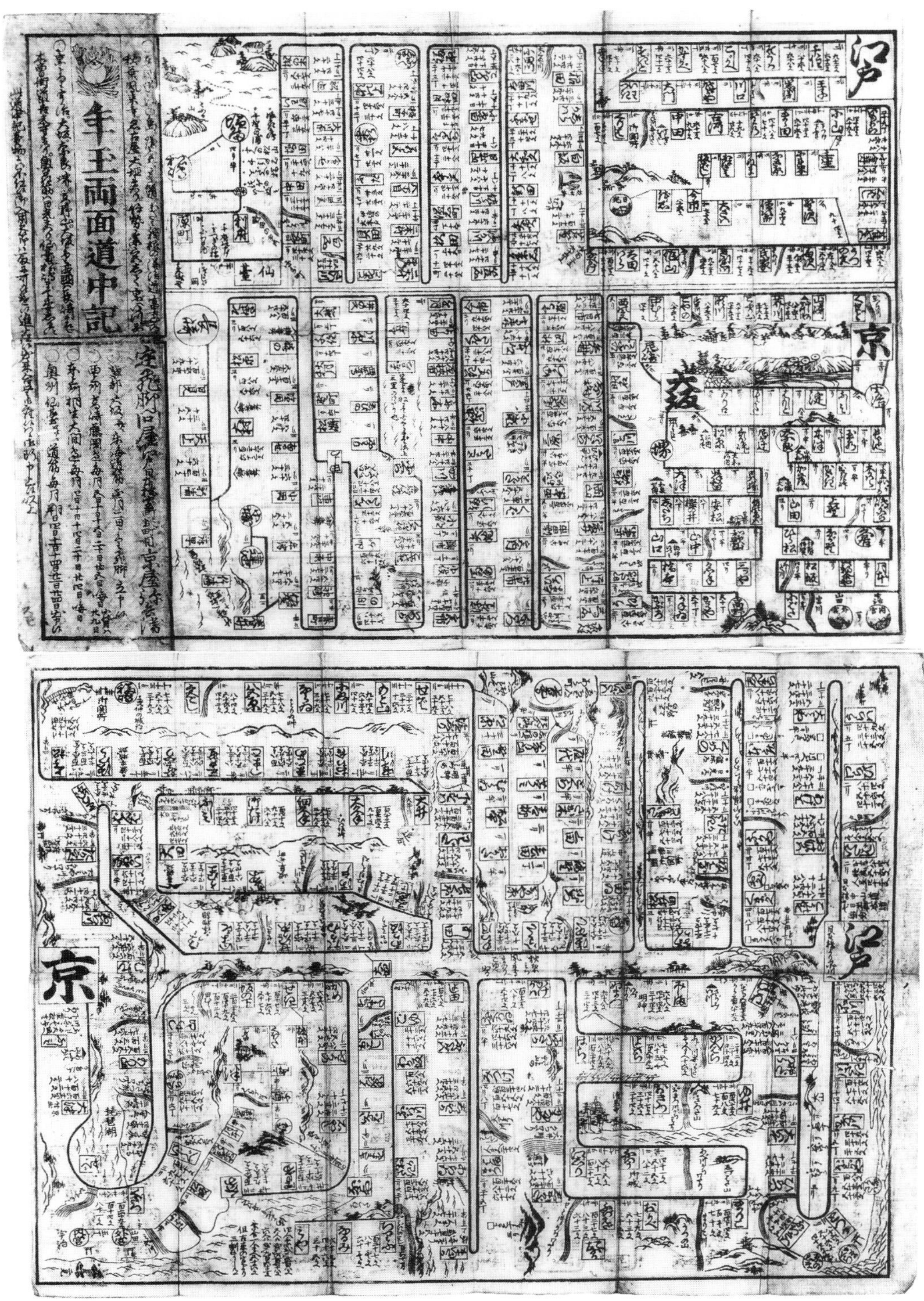

FIG. 11.58. AN EXAMPLE OF AN ITINERARY MAP OF THE *SUGOROKU* VARIETY: *TOSHIDAMA RYŌMEN DŌCHŪ KI*. This variety differs from the labyrinth maps in that the starting and finishing points of each road are clearly indicated, and the roads themselves are straight lines with curves for continuation. *Sugoroku* is a game played with dice in which the object is to move from a starting point to a finishing point, thus the association with maps that could be used for travel. This map was published by Kyōya Yahee in the mid-eighteenth century.
Size of the original: 30 × 39.5 cm. Collection of Kazutaka Unno.

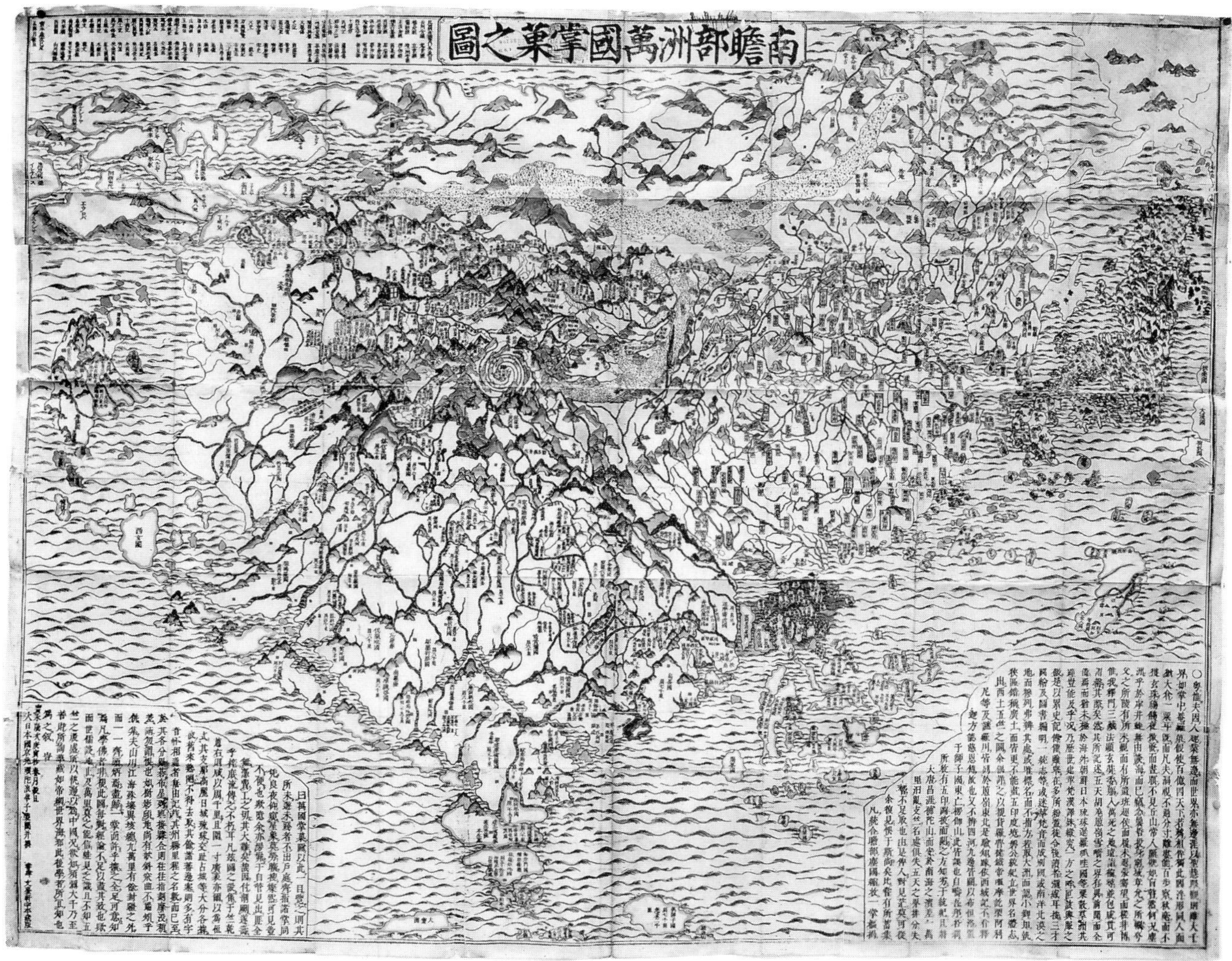

FIG. 11.59. *NANSENBUSHŪ BANKOKU SHŌKA NO ZU* BY RŌKASHI, 1710. Modeled directly on the ca. 1709 map of Jambūdvīpa at the Kōbe City Museum (plate 29), this is the first Buddhist world map printed with European geographical knowledge. In the upper left corner Europe is described as a group of islands, and in the ocean south of Japan is South America as an island. Changes from the ca. 1709 original include the omission of part of the continental outline and some unrealistic islands.
Size of the original: 113.5 × 144 cm. Geographical Institute, Faculty of Letters, Kyōto University. Photograph courtesy of Kazutaka Unno.

Chikyū bankoku sankai yochi zenzusetsu about 1788, the mainstream of printed world maps had been based on the *Bankoku sōzu* type (above). They tended to be published as illustrations in books and were probably related to an interest in the geography of the world. This interest was, however, probably superficial: the distortion was often great, and the intention was more often to arouse a sense of exoticism than to disseminate correct geographical information.

Nagakubo's map gained in popularity early in the nineteenth century, when many copies appeared on the market. Although these were small and simplified versions, often lacking the cartographer's name and date of publication, we are able to trace the models employed. These were Den Ken's *Shinsei bankoku yochi zenzu* (Newly made map of all the countries in the world, 1844) and Yamazaki Yoshinari's (1796–1856) *Chikyū bankoku sankai yochi zenzusetsu* (Map with an account of all the countries, lands, and seas in the world, 1850).[297]

no kaigai chishiki (Japanese knowledge of overseas during the age of national isolation), ed. Kaikoku Hyakunen Kinen Bunka Jigyō Kai (Society of Cultural Projects to Commemorate the One Hundredth Anniversary of the Opening of the Country) (Tokyo: Kengensha, 1953), 3–367.

297. See note 256 for information on Den Ken's and Yamazaki's maps.

The world maps based on the *Bankoku sōzu* and the Nagakubo map were influenced by European cartography, but popular Buddhist world maps continued to be published throughout the Edo period.[298] The reason for their success rests not in religious conviction or in a belief that the Buddhist image of the world was correct, but in their traditional image of Asia and in the place-names, especially in the interior of China and India, that were missing from the maps derived from European models. Some priests, nonetheless, tried to combine the conventional Buddhist image of the world with information from European geographical knowledge. A notable attempt was made by Sōkaku (1639–1720), the head priest at Kushuon'in Temple in Hirakata, Ōsaka Prefecture. Among his surviving works are probably a manuscript version of the Tō Temple map titled *Gotenjikukoku no zu* (Map of the countries of the Five Indias, ca. 1692), the manuscript *Daimin sei zu* (Map of the provinces of Ming China, 1691), and a Buddhist terrestrial globe of his own invention dating to about 1703.[299] On the evidence of the outlines and place-names on the globe, two anonymous and untitled manuscript maps of Jambūdvīpa were attributed to Sōkaku; they are not dated, but they appear to have been made about 1698 and 1709. On the first, in the collection of Muroga Emiko, the northern part of Jambūdvīpa (shaped like a radish) is left blank, and Europe is not shown. The second, at the Kōbe City Museum, shows a complete Jambūdvīpa in the shape of a fan and has Europe in the northwest (plate 29).[300] Sōkaku's image of the world was later improved by incorporating elements from the more realistic *Nansenbushū bankoku shōka no zu* (Visualized map of all the countries in Jambūdvīpa; fig. 11.59), published in 1710 by the priest Rōkashi (Hōtan, 1654–1738).[301] With demand for it enhanced by its inclusion of traditional Asian place-names, it was reprinted in the same year and frequently republished with the same date until about 1815.[302] A miniaturized version was published by Kabō Hyōzō in 1744;[303] there were several copies of this map, and new editions were published without dates late into the Edo period.[304]

China was also the subject of popular printed maps in the Edo period. Here we can trace the links with the older practice of making maps of Japan's cultural neighbors, Korea and China. In the early Edo period the maps of China were reproductions of those made in China. From the middle of the eighteenth century, however, they were edited in Japan and published either as sheet maps or as illustrations in books (fig. 11.60). The first map of China printed in Japan based on a European model appeared in the latter half of the nineteenth century.[305]

298. The development of Buddhist maps of the world in the Edo period is treated in Muroga and Unno, "Buddhist World Map," 58–68 (note 110); Muroga Nobuo and Unno Kazutaka, "Nihon ni okonowareta Bukkyō kei sekaizu ni tsuite" (On Buddhist world maps in Japan), *Chirigakushi Kenkyū* 1 (1957): 67–141, reprinted in *Chirigakushi kenkyū*, 1:67–141 (note 250); and Muroga and Unno, "Edo jidai kōki ni okeru Bukkyō kei sekaizu," 135–229 (note 250).

299. Respectively 168 by 172 centimeters, 382 by 181.5 centimeters, and 381 by 179 centimeters (two sheets), and 20 centimeters in diameter. All are reproduced in Unno, Oda, and Muroga, *Nihon kochizu taisei*, vol. 2, pls. 2, 21, and 5 (note 8).

300. On the Muroga map (138.5 × 154.5 cm), see Unno, "Sōkaku no chikyūgi to sono sekaizō" (note 184). For reproductions of both, see Muroga and Unno, "Buddhist World Map," figs. 6–7 (note 110); and Unno, Oda, and Muroga, *Nihon kochizu taisei*, vol. 2, pls. 3–4 (note 8).

301. Two editions of Rōkashi's map were published in 1710, both in Kyōto, by Bundaiken Uhei and Nagata Chōbee. The woodcut map published by Bundaiken is reproduced and discussed, for example, in Ayusawa, "Types of World Map," 124, 128 (Ramming's comments), and fig. 2 (121 × 144 cm) (note 235); in Muroga and Unno, "Buddhist World Map," 62–63 and fig. 9 (note 110); in Nanba, Muroga, and Unno, *Nihon no kochizu/Old Maps in Japan*, pl. 8 (note 11); in Unno, Oda, and Muroga, *Nihon kochizu taisei*, vol. 2, pl. 6 (note 8) (fig. 11.59); and in Cortazzi, *Isles of Gold*, pl. 48 (note 14) (118 × 145.2 cm, at the Kōbe City Museum). A reproduction of an early nineteenth-century manuscript (127.5 × 152.2 cm) at the Kōbe City Museum, Nanba Collection, modeled on Rōkashi's map, is in Muroga and Unno, "Buddhist World Map," 64–65 and fig. 11, and Cortazzi, *Isles of Gold*, 38 and pl. 49.

302. A book published by Nagata Chōbee carries an advertisement that mentions the map: see Tōkōji Sōryō, *Sessō yawa* (Night talk beside a window commanding a view of snow) (Kyōto, 1815) (a list of publications that were finally bound and for which Nagata held the copyright). Nagata also dealt with the first edition of the map, and the advertisement gives "Hōtan" as the author, meaning that Rōkashi and Hōtan are the same person. I doubted this in my "Sōkaku no chikyūgi to sono sekaizō" (note 184), and the view there should be corrected according to this evidence. The map published by Nagata is reproduced in Beans, *Japanese Maps of the Tokugawa Era*, facing p. 21 (note 22).

303. See notes 251 and 252 above.

304. Examples are the *Nan'enbudai shokoku shūran no zu* (Outline map of the countries of Jambūdvīpa) by Kabō Hyōzō, published by Mikuniya Ryūsuke in the early nineteenth century in Edo (56.5 × 86 cm), and the *Bankoku shūran zu* (cover title) or *Bankoku shōka no zu* (map title) (Visualized map of all the countries; 47 × 65 cm). Both are reproduced in Unno, Oda, and Muroga, *Nihon kochizu taisei*, vol. 2, fig. 7 and pl. 8 (note 8).

305. Chronologically, the first reprinted Chinese map of China was the *Kō Min yochi no zu*, ca. 1659 (of the *Huang Ming yudi zhi tu*, Map of Ming China, 1631) published by Rinsendō in Kyōto (124 × 57 cm); a large map of China compiled by a Japanese was Nagakubo Sekisui's *Dai Shin kōyozu* (Enlarged map of Great Qing China, 1785), published by Suharaya Ichibee and Suharaya Ihachi in Edo (fig. 11.60); an atlas of China was Tōjō Shinkō's *Shin nikei jūhassei yochi zenzu* (Atlas of the two capitals and eighteen provinces in Qing China, 1850), published by Suharaya Ihachi in Edo and others (34.4 × 23.7 cm); and the first printed map of China based on a European model was Shibata Shūzō's *Dai Shin ittō zu* (General map of Great Qing China, ca. 1865) (45.5 × 66 cm). All are reproduced in Unno, Oda, and Muroga, *Nihon kochizu taisei*, vol. 2, fig. 16 and pls. 26, 28, and 30 (note 8).

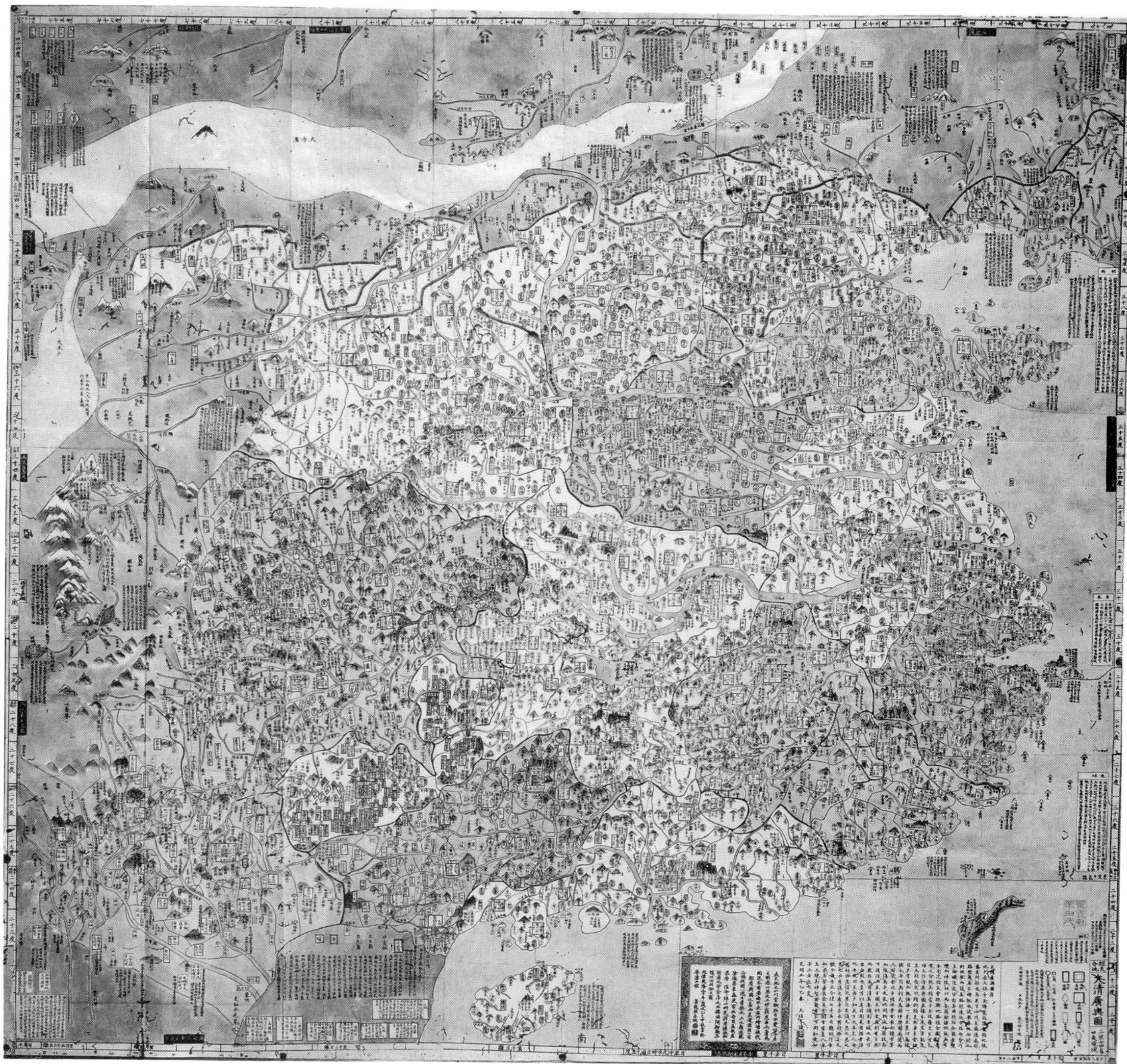

FIG. 11.60. *DAI SHIN KŌYOZU* (ENLARGED MAP OF GREAT QING CHINA) BY NAGAKUBO SEKISUI, 1785. This was modeled on the "Yushu jingtian hedi zhi tu" (Map of China with latitude, longitude, and scale clarified) in You Yi, *Tianjing huowen* (Questions and answers on astronomy), published in China in 1672 and reproduced in Japan in 1730. On Sekisui's map is written "*Keiten gatchi*" (i.e., "*Jingtian hedi*": Clarification of latitude, longitude, and scale) at the beginning of its sheet title. The information about the provinces was taken from the provincial maps in the *Da Qing yitong zhi* (Comprehensive gazetteer of the Great Qing realm, 1746).
Size of the original: 182 × 188 cm. Geographical Institute, Faculty of Letters, Kyōto University. Photograph courtesy of Kazutaka Unno.

Maps of Fictional Places

Finally, among the commercial productions there were maps of fictional places compiled late in the Edo period.[306] An early example is the "Daigepponkoku no zu" (Map of Great Geppon) in Dōjarō Maa's *Shōhi chiriki* (Geographical description of geisha girls, 1777). The title is a parody of *Dainippon* (Great Japan): the character

306. See Unno, "Tawamure no chizu" (Amusing cartographic works) and "Zoku tawamure no chizu" (Amusing cartographic works, continuation), in *Chizu no shiwa*, 5–7 and 8–17 (note 136).

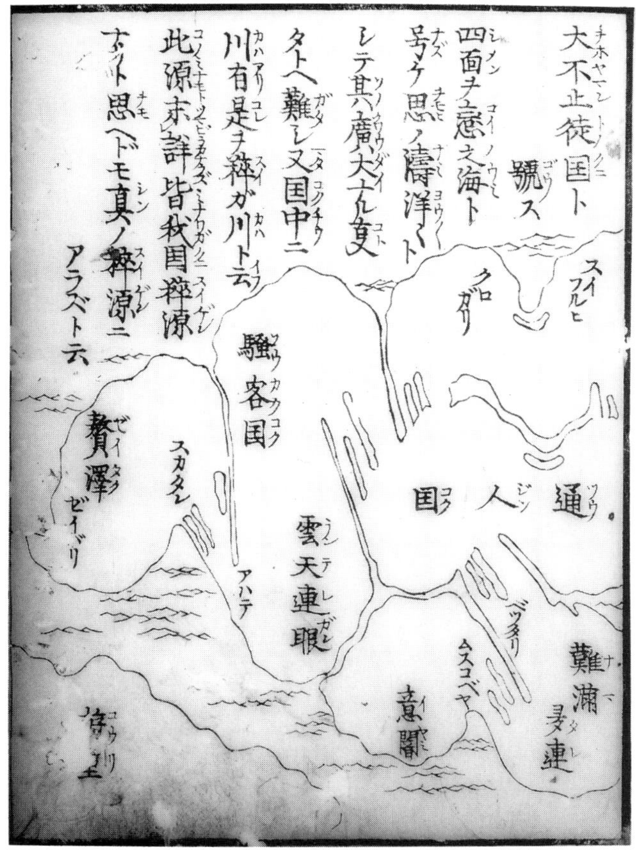

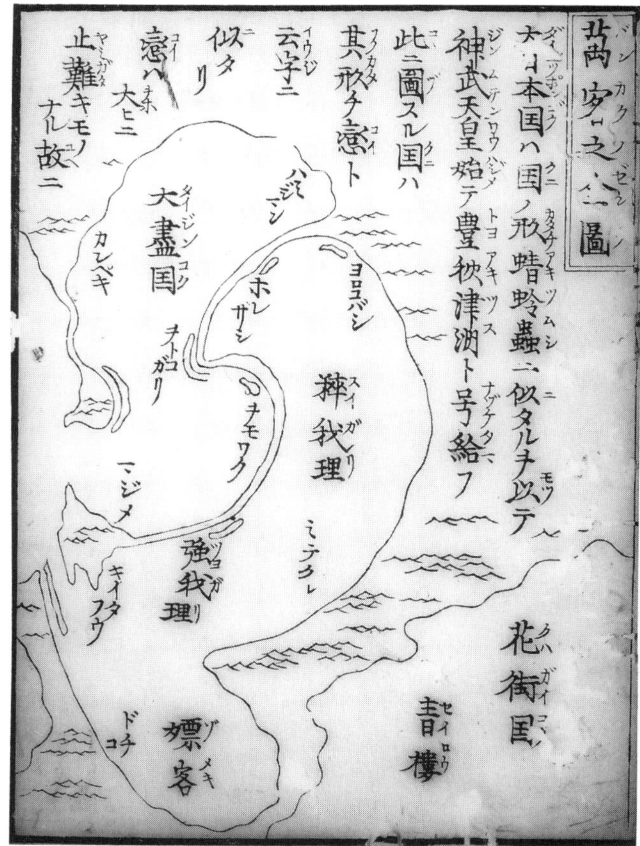

FIG. 11.61. "BANKAKU NO ZENZU," 1822. When the left side is placed at the top, the outline of the landforms is a cursive *koi*, "romance." The place-names refer to words pertaining to entertainment districts such as Yoshiwara in Edo.
Size of the original: 18.7 × 27.5 cm. Collection of Kazutaka Unno.

日, meaning sun may be read *ni*; in the title it is replaced by 月 the character for moon, and read *ge*.[307] This map likened the entertainment district of Edo, Yoshiwara, to a group of small islands.[308] Two further examples displayed similar districts in Kyōto and Ōsaka in the form of the world. These were the "Ajina Myōjū bankoku sōzu" (Map of all the countries in Ajina Myōjū [Asia]), included in the *Zatto ichiran* (Handbook of "Zatto") of Suisai in 1820, and the "Bankaku no zenzu" (Map of many guests; fig. 11.61) in *Akan sanzai zue* (Encyclopedia of the insatiable spending of money, 1822) by Akatsuki no Kanenari (1793–1860).[309] On the latter the land that resembles Japan is portrayed as a cursive 戀, the Chinese character for koi (romance). The name of the country on the map, Ōyamanto no Kuni, is a parody of Ōyamato no Kuni (The Country of Great Japan) and means "people are not stopping very much"; the reason given for using 戀 is that it is very difficult to stop loving. A similar method using the hiragana script was used on the "Godō meisho no zenzu" (Map of obstacles to spiritual awakening; fig. 11.62) in the *Zen'aku meisho zue* (Illustrated book of noted places of good and evil), published in 1846 and drawn by Ippitsuan Eisen (1790–1848).[310] The land was depicted as さとるべし and まようふ, *satorubeshi* and

307. 日本 can be read *Nihon* and *Nippon*, among other possibilities. 日 is used for *nichi* but read *ni* in the name of the country; 月 or *getsu* has here been reduced likewise to the first syllable *ge*. The sounds *shōhi chiriki* are the same as the musical instruments *shō* and *hichiriki*. Part of the compiler's pen name, Dōjarō, was taken from the jocular name of his study (room) and means "what shall I do?" or "what nonsense!" Maa is an exclamation. The map appears on two pages of the book, published by Kōshodō in Edo. A reproduction of the map is in Unno, "Tawamure no chizu," 6 (note 306).

308. On Yoshiwara, the amusement and brothel district of Edo, see Stephen Longstreet and Ethel Longstreet, *Yoshiwara: The Pleasure Quarters of Old Tokyo* (Tokyo: Yenbooks, 1988).

309. *Bankoku* and *bankaku* refer to "world"; *Ajina* is a parody of "Asia." The *Zatto ichiran*, which also means "looking at roughly," was published by Yoshinoya Jinbee in Kyōto; the *Akan sanzai zue*—this title is a pun on *Wakan sansai zue* (see note 93)—was published in Ōsaka. Both maps are reproduced in Unno, respectively, "Zoku tawamure no chizu," 9, and "Tawamure no chizu," 6 (note 306).

310. The second edition of the *Zen'aku meisho zue* was published by Chōondō, Hon'ya Matasuke in Edo in 1858.

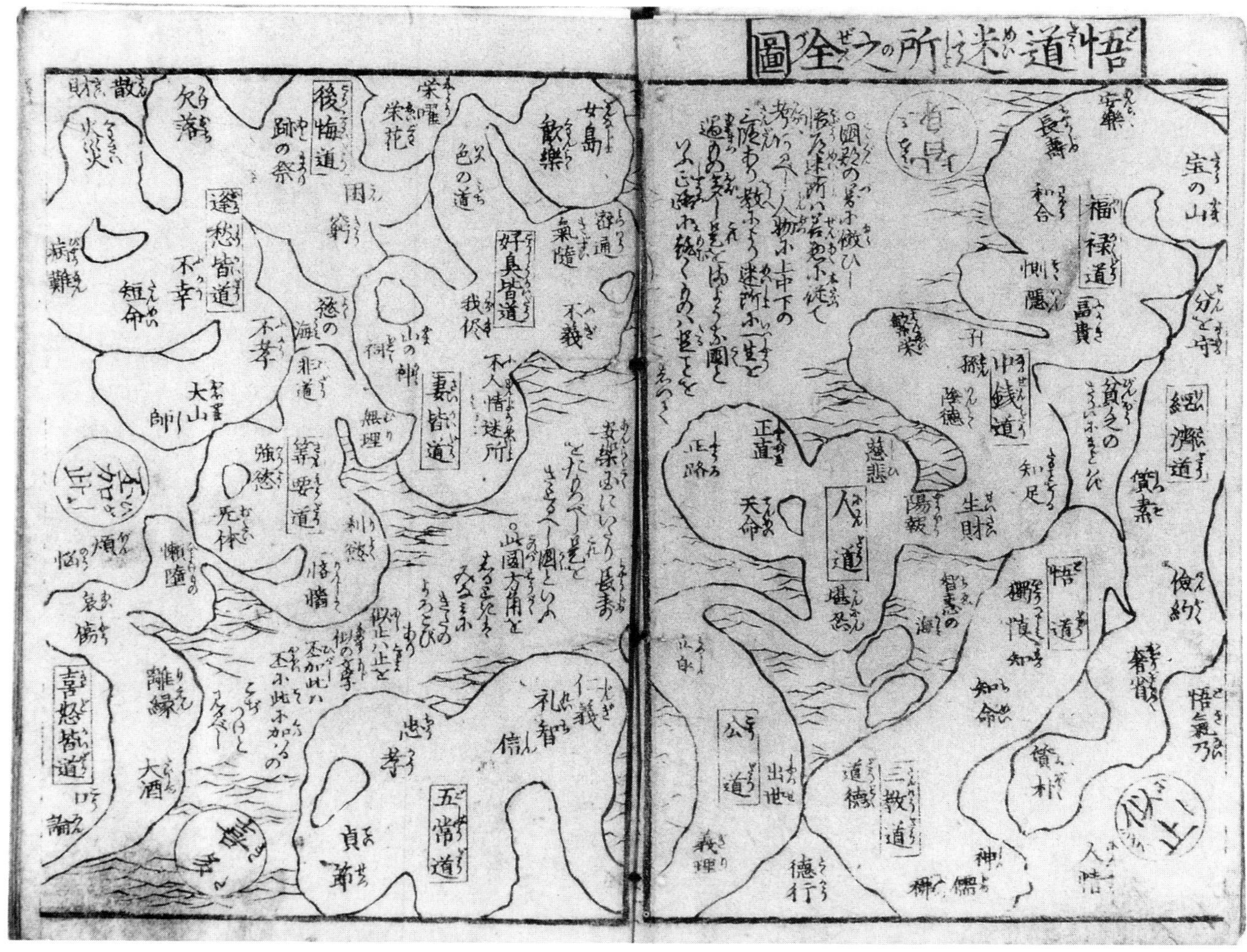

FIG. 11.62. "GODŌ MEISHO NO ZENZU," 1846. The expressions *satorubeshi* ("you should be spiritually awakened") and *mayouna* ("don't go astray") may be seen in their hiragana forms: the first by turning the map so that the left side is on top and reading on a slant from the new left side; the second by turning the map so that the right side is on top and likewise reading from the new left side. Also, some of the Chinese-character combinations are puns: different meanings have been created by reading (pronouncing) the characters differently from the way they are read (pronounced) in the place-names.
Size of the original: 17.4 × 21 cm. Collection of Kazutaka Unno.

mayouna (respectively, "you should be spiritually awakened" and "don't go astray").

Japanese Cartography and "Dutch Learning"

Like the adoption of marine charts in an earlier age, "Dutch learning" or "Dutch studies" (*Rangaku*), literally Japanese learning via Dutch-language materials, began to exert a significant influence on cartography from the mid-eighteenth century.[311] Especially while Tokugawa Yoshimune (1684–1751) was in office as shogun (from 1716 to 1745), the adoption of the new knowledge coincided with a transitional period in Japanese society. Yoshimune, the eighth of the Tokugawa shoguns, was a notable reformer.[312] Economic development increased the influence of the merchants in society, despite their low rank,[313] and nurtured a sense of freedom in the cities. Confucianism began to develop a positivist outlook at the expense of its traditional idealistic position. This was typified by the use of the inductive method for studying

311. *Rangaku* (Dutch learning) is discussed in, for example, Nakayama, *History of Japanese Astronomy*, 165–69 (note 38); Sansom, *History of Japan*, vol. 3 (1615–1867), 188–89 (note 32); and Plutschow, *Historical Nagasaki*, 95–109 (note 128).

312. His regime is discussed in Sansom, *History of Japan*, 3:154–72 (on his interest in science specifically, see 168–70) (note 32).

313. The four strata below the shogun were the daimyos and their samurais, the peasants, the artisans, and then the merchants. Below them were the outcastes.

Chinese classics, adopted by the *kogaku* (ancient learning) school under the leadership of Itō Jinsai (1627–1705) and Ogiu (Ogyū) Sorai (1666–1728). The *kogaku* school was a reaction to Neo-Confucianism—which had promoted the concept of harmony between human nature and the physical world—and it challenged the assumption that the natural order was reflected in the social hierarchy. The *kogaku* school distinguished human affairs from those of the heavens, noting that the latter were an object not of rational inquiry but of worship only.[314]

In such circumstances, academic research in agriculture and mining was promoted as a part of industrial policy, and Yoshimune undertook to reform the Japanese calendar, which was closely related to agriculture. The astronomer he had appointed for the project, Nakane Genkei (1662–1733), noted that the reform was not possible without referring to Chinese books on astronomy and calendars composed by the Jesuits in China. After consulting a Chinese book, itself only an extract from a Chinese version of a European work, Nakane told Yoshimune that "no progress could be made so long as Chinese translations of Western books were kept out of Japan for such absurd reasons as a mere mention in the text of something related to Christianity or Christians."[315] This led Yoshimune to lift an existing ban on importing books in 1720, provided they did not relate to Christianity. The shogun's interest in European science and technology also led him to take advantage of every opportunity to speak with the director and staff of the Dutch factory in Nagasaki when they paid courtesy calls to Edo. From them he also ordered books, telescopes, and other items of interest.

About 1740, a number of Japanese scholars began to study Dutch. There were at the time Japanese who were linguistically proficient for diplomatic and mercantile affairs, especially the official translators in Nagasaki, but their ability to interpret accurately the content of academic books was limited. Some of the official translators in Nagasaki nonetheless attempted to translate technical material, notably maps and books on geography.

The first result of such efforts was a translation of the terrestrial and celestial globes of the Valcks dating from 1700. This was undertaken by the astronomer Kitajima Kenshin (fl. 1719–37) and an unidentified official translator thought to be Nishi Zenzaburō (ca. 1716–68). Together they converted the globe into plane maps. The map of the celestial sphere has been lost, but the map of the world, a manuscript in the form of a scroll, is extant. The title is *Oranda shintei chikyū zu* (World map based on a Dutch source), and though neither a date nor an author is mentioned, a comparison with the description in Kitajima's booklet noted below suggests that the map is either his own work or a faithful copy of it (fig. 11.63). It is drawn on a globular projection with the Eastern and Western hemispheres halved into north and south. It is not known for certain where Kitajima obtained his knowledge of projections, but it is likely that he acquired it from Ro Sōsetsu (1675–1729), a successor of Hayashi Sensei. The booklet that Kitajima wrote in 1737 describes the task and the translation of the names on the globes and provides a brief explanation.[316] The project was undertaken on the order of the authorities, probably the Nagasaki magistrate.

Official translators were involved in several subsequent projects. Among them, special mention must be made of Motoki Ryōei (1735–94) and Matsumura Mototsuna (or Genkō, fl. 1771–92), who help reveal the place of cartography in the wider context of the history of Japanese science. It has been said of Motoki that his translations "are significant not only as the first Japanese sources on the Copernican heliocentric system, but also as a landmark in the advancement of the study of Western languages in Japan."[317] Their cartographic work, executed either jointly or singly, appeared for the most part in two periods: the first half of the 1770s and from 1790 to 1793. In the first period they concentrated on older materials, and in the second on translating more recent works.

In 1772 Motoki translated the section on map use in the 1722 Dutch edition of *Kort begryp der oude en nieuwe geographie* by Johann Hübner (1668–1731) and compiled a booklet titled *Oranda chizu ryakusetsu* (Outline of Dutch cartography).[318] The next year he completed the *Oranda chikyū zusetsu* (An explanation in a Dutch atlas of the world), which is preserved in a manuscript at Nagasaki City Museum. This is a translation of

314. See Nakayama, *History of Japanese Astronomy*, 108 and 156–58 (note 38). He refers to Ogiu Sorai, *Gakusoku furoku* (Appendix to the principles of learning, 1727) in *Nihon jurin sōsho* (Collection of Confucian writings in Japan), 6 vols. (Tokyo: Tōyō Tosho Kankōkai, 1927–29), vol. 4; and to Maruyama Masao, *Nihon seiji shisō shi kenkyū* (A study of the history of political thought in Japan) (Tokyo, 1952), 52–54, 80–82, and 210.

315. Quotation from Sansom, *History of Japan*, 3:169 (note 32); see also Nakayama, *History of Japanese Astronomy*, 166 (note 38). On Nakane Genkei see Watanabe Toshio, *Kinsei Nihon tenmongaku shi* (History of modern Japanese astronomy), 2 vols. (Tokyo: Kōseisha Kōseikaku, 1986–87), 1:91–94.

316. The map of the world is preserved at the Ōsaka Prefectural Nakanoshima Library, Ōsaka. Details are given in Unno Kazutaka, "Faruku chikyūgi denrai no hamon" (The influence of the Valcks' globe on Japanese maps and globes), *Nihon Yōgakushi no Kenkyū* 8 (1987): 9–34. Kitajima's booklet, *Kōmō tenchi nizu zeisetsu* (Explanation of Dutch celestial and terrestrial globes, 1737), is preserved at the Tokyo University Library (Nanki Library Collection), Kyōto University Library, and Tōhoku University Library, Sendai. A facsimile of the Kyōto University Library version was published by Chinsho Dōkōkai (Tokyo, 1916).

317. Nakayama, *History of Japanese Astronomy*, 173 (note 38); for a discussion of Motoki's work, see 173–79.

318. Motoki's manuscript (1772) is preserved at the Seikadō Library, Tokyo.

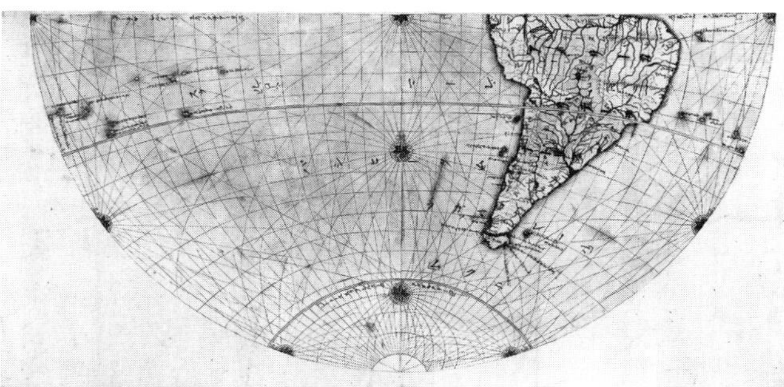

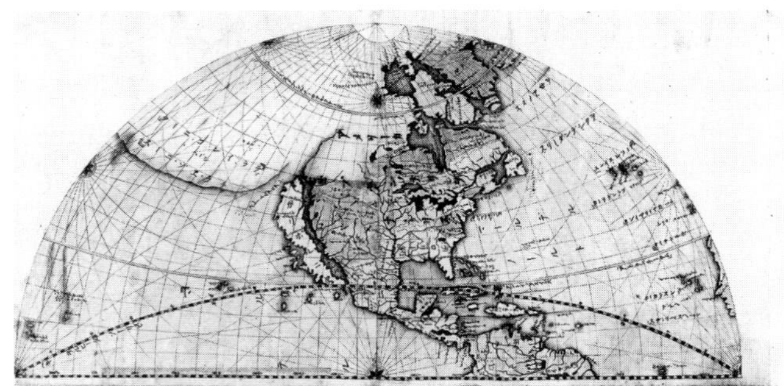

FIG. 11.63. *ORANDA SHINTEI CHIKYŪ ZU*, CA. 1737. Although he is not mentioned on the map, this manuscript scroll may be ascribed to the Nagasaki astronomer Kitajima Kenshin, who was involved in converting Gerard and Leonard Valck's terrestrial and celestial globes into flat maps.
Size of the original: 25.5 × 247 cm. By permission of the Ōsaka Prefectural Nakanoshima Library, Ōsaka.

the introduction to the *Atlas van zeevaart* of 1745.[319] Finally, in this period Motoki translated (in 1774) the *Tenchi nikyū yōhō* (The use of celestial and terrestrial globes) from Willem Jansz. Blaeu's *Tweevoudigh onderwiis van de hemelsche en aerdsche globen* (Amsterdam, 1666 [first edition, 1634]).[320]

In 1790, at the request of the high-ranking official Matsudaira Sadanobu (1758–1829),[321] Motoki translated an edition of about 1785 of the *Nieuwe atlas* (1730) of Johannes Covens and Cornelis Mortier as *Oranda zensekai chizusho yaku* (Translation of a Dutch atlas of the world). The original *Nieuwe atlas* used for the translation is in the Shizuoka Prefectural Central Library, Shizuoka; on the maps are pasted gold and silver slips of paper with the Japanese translation of the main cities on each map. In addition to these works, there was the *Seijutsu hongen taiyō kyūri ryōkai shinsei tenchi nikyū yōhōki* (The ground of astronomy, newly edited and illustrated; on the use of celestial and terrestrial globes according to the heliocentric system) in seven volumes, 1792–93. The Dutch original in this case was *Gronden der sterrenkunde* (1770), itself a translation of *A Treatise Describing and Explaining the Construction and Use of New Celestial and Terrestrial Globes* (London, 1766) by George Adams the elder (ca. 1704–73).[322] Finally, there are two translated collections of place-names that appear to have been by-products from the copies that Motoki made of Dutch atlases of the world.[323]

Matsumura appears to have been a close friend of Motoki and was mentioned as a collaborator in most of the translations, including the *Oranda chizu ryakusetsu* and *Oranda chikyū zusetsu*. Two works compiled by Matsumura alone are also known. These are his *Shinzō bankoku chimei kō* (Newly enlarged list of geographical names in the world) of 1779 and a map showing the Eastern and Western hemispheres that contains the place-names from the other work.[324]

In contrast to that at Nagasaki, Dutch learning in Edo developed by emphasizing medicine. Physicians who had gained knowledge of European medicine were, however, later encouraged to acquire more general knowledge, notably in geography and astronomy. The earliest work of a cartographic nature by a Dutch scholar in Edo was the *Shinsei chikyū bankoku zusetsu* (Explanation of the new map of all the countries in the world) of 1786. The translation was undertaken by Katsuragawa Hoshū (1751–1809) from the topographical explanation attached to Joan Blaeu's (1598–1673) world map of 1648, the *Nova totius terrarum orbis tabula*. Katsuragawa was a physician in the employ of the shogunate, which owned the Blaeu map, and had doubtless acquired it from the Dutch factory at Deshima. The map had already been

319. The *Atlas van zeevaart en koophandel door de geheele weereldt* (Amsterdam, 1745) is itself a translation of Louis Renard's *Atlas de la navigation et du commerce qui se fait dans toutes les parties du monde* (Amsterdam, 1715). See Nakayama, *History of Japanese Astronomy*, 174 (note 38).

320. Motoki's manuscript is preserved at the Nagasaki City Museum. See Nakayama, *History of Japanese Astronomy*, 175 (note 38).

321. Matsudaira rose to prominence when he was appointed president of the Council of Elders to determine national policy after riots in 1787 stemming from a succession of poor harvests. He was also appointed adviser to the shogun (Ienari [1773–1841, r. 1787–1837], the eleventh of the Tokugawas). The changes he presided over are known as the Kansei Reform, after the era name of 1789–1800. See Sansom, *History of Japan*, 3:193–206 (note 32).

322. The full Dutch title was *Gronden der starrenkunde, gelegd in het zonnestelzel, bevatlyk gemaakt; in eene beschrijving van 't maaksel en gebruik der nieuwe hemel- en aarrd-globen* (Amsterdam, 1770). See Nakayama, *History of Japanese Astronomy*, 177 and 285 (note 38).

323. One is the *Yochi kokumei yaku* (Translation of the names of countries on the earth, ca. 1777), and the other is the *Tenchi nikyū yōhō kokumei* (The use of celestial and terrestrial globes, names of countries, ca. 1794); both are manuscripts. For these two works see Unno Kazutaka, "*'Tenchi nikyū yōhō kokumei'* kō" (On the *Tenchi nikyū yōhō kokumei*), *Nihon Yōgakushi no Kenkyū* 3 (1974): 113–37.

324. The 1779 work is at the Tenri Central Library. On Matsumura's works, see Unno, "*'Tenchi nikyū yōhō kokumei'* kō," 113–37 (note 323).

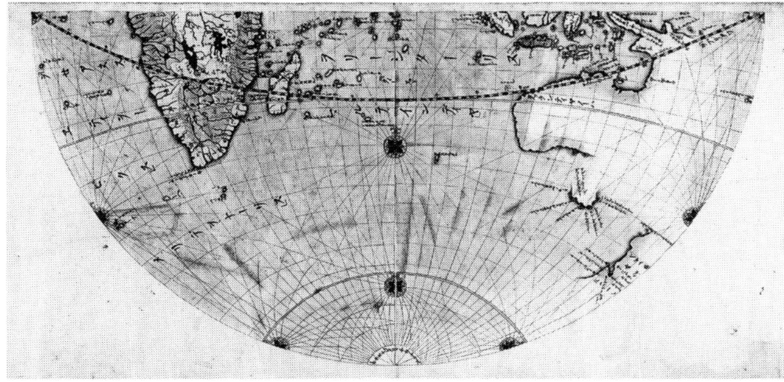

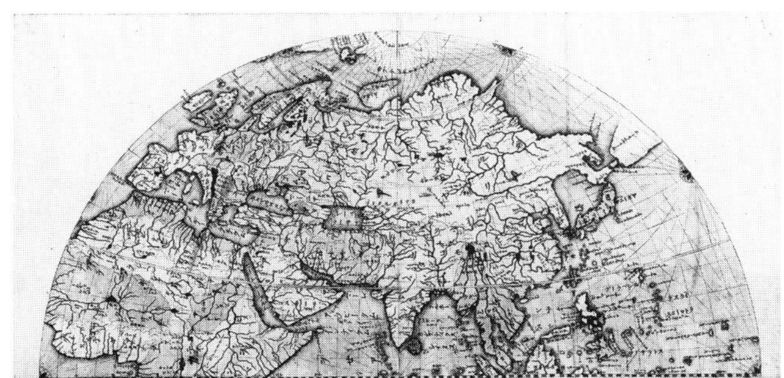

used in 1709 by Arai Hakuseki (1657–1725) to obtain geographical information from the Italian Jesuit Giovanni Battista Sidotti (1668–1715), so that it was well known by the time of the translation. An appendix to the translation, made in 1791, contains a colored miniature of the original map, copies of its various illustrations, and translated names of the main places.[325]

The first printed Japanese map of the world to be influenced by *Rangaku* is a copperplate engraving of 1792, the *Yochi zenzu* (Map of the earth) by Shiba Kōkan (1747–1818). This was also the first map to be engraved on copperplates in Japan. The *Yochi zenzu* is a translation of Alexis Hubert Jaillot's revision of Guillaume Sanson's map of the Eastern and Western hemispheres issued by Covens and Mortier (Amsterdam, ca. 1730), with revisions of Japan based on the latest information.[326] It was reprinted at least three times with the same copperplate and colophon, supplemented with place-names or illustrations added around the map. From the second edition onward, the title became *Chikyū zu* (Map of the terrestrial globe) (fig. 11.64). Explanatory booklets were published for each edition, the *Yochi ryakusetsu* (Brief explanation of the earth) for the first edition being the simplest.[327]

Dutch maps were the earliest to be translated, but by the late eighteenth century Japanese scholars were also working with maps in other foreign languages. Thus, Katsuragawa was the translator not only of a more up-to-date Dutch map of the world, but also of Russian maps that were presented to the shogunate by Adam K. Laxman (1766–96?) in 1792.[328] The first put him in competition with the publisher Shiba, whose *Yochi zenzu* was issued while his work was in progress. A mutual friend, Ōtsuki Gentaku (1757–1827), who had provided Shiba with the Jaillot map, attempted to dissuade Shiba from publishing the translation on the grounds that Katsuragawa's work was more advanced. Katsuragawa, however, never finished his work: only the Western Hemisphere

325. Katsuragawa notes that the *Shinsei chikyū bankoku zu* is a translation of the original title, *Nova totius terrarum orbis tabula.* Blaeu's map is now in the Tokyo National Museum: see Minako Debergh, "A Comparative Study of Two Dutch Maps Preserved in the Tokyo National Museum—Joan Blaeu's Wall Map of the World in Two Hemispheres, 1648 and Its Revision ca. 1678 by N. Visscher," *Imago Mundi* 35 (1983): 20–36. Katsuragawa's translation is reproduced in Ono Tadashige, ed., *Kōmō* [or *Oranda*] *Zatsuwa* (Miscellaneous notes of Holland) (Tokyo: Sōrinsha, 1943), 1–34. Codices including the appendix of 1791 are in the Ashida Collection at the Meiji University Library, Tokyo, the Hōsa Library in Nagoya, and the National Archives in Tokyo. Those in the Hōsa Library and the Ashida Collection are the most complete.

326. Shiba's 1792 map (50.5 × 92.8 cm) is reproduced in Kurita, *Nihon kohan chizu shūsei*, pls. 5 and 6 (note 15). The Covens and Mortier issue, ca. 1730, of Jaillot's map was formerly preserved at the Miyagi Prefectural Library in Sendai but was burned during a bombing in the Second World War. For a reproduction, see Ayusawa Shintarō, *Chirigakushi no kenkyū* (Studies on the history of geography) (Tokyo: Aijitsu Shoin, 1948; reprinted Hara Shobō, 1980), frontispiece, and Unno, Oda, and Muroga, *Nihon kochizu taisei*, vol. 2, fig. 69 (note 8). Shiba, a student of European culture, painted under the influence of Dutch models, studied European astronomy, geography, and history, and was the first Japanese to engrave using copperplates. A visual synthesis of the arts and sciences seems to have marked his work; a brief summary notes that "he was particularly interested in the scientific rendering of space through perspective" (Munsterberg, *Arts of Japan*, 159 [note 79]). That Shiba would have chosen to use Dutch names is no surprise; he acknowledged the superiority of Europe in such areas as the sciences and of course learned from the scholars of Dutch studies at Edo and Nagasaki. In the realm of science, his greatest achievement was probably the popularization of Copernican theory through three books: *Chikyū zenzu ryakusetsu* (Brief explanation of the world map, 1793), *Oranda tensetsu* (Dutch astronomy, 1796), and *Kopperu tenmon zukai* (Copernican astronomy illustrated, 1808); see Nakayama, *History of Japanese Astronomy*, 187 (note 38).

327. On the maps and books, see Sugano Yō, *Nihon dōhanga no kenkyū: Kinsei* (Studies on Japanese copperplate prints: The modern age) (Tokyo: Bijutsu Shuppansha, 1974), 358–75.

328. Two objectives of Laxman's voyage were to repatriate some Japanese who had been shipwrecked on the coast of Russian Asia and to negotiate for trade. One of the castaways was a certain Kōdayū, who had taught Japanese in Irkutsk under the protection of Eric Laxman, the father of the mariner and a Swede by birth. Under official order and with the assistance of Kōdayū, Katsuragawa wrote a book about

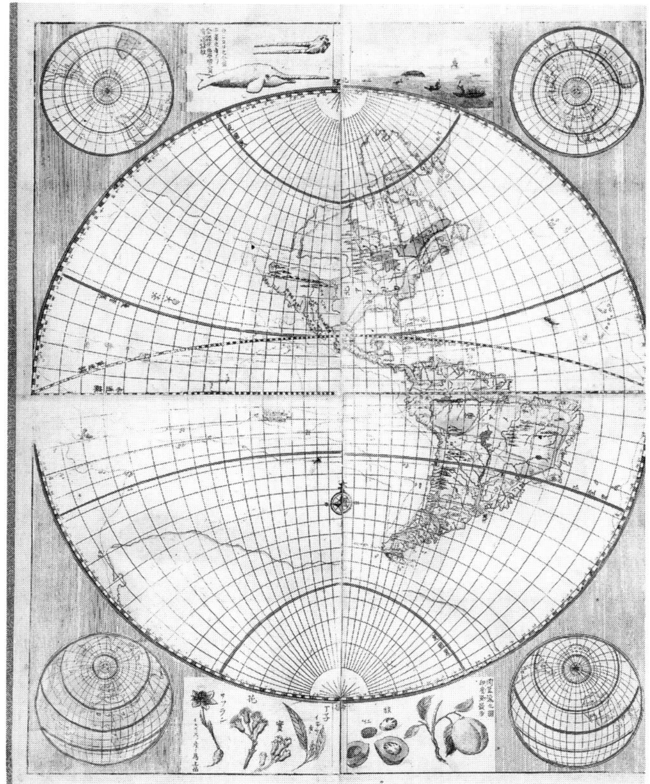

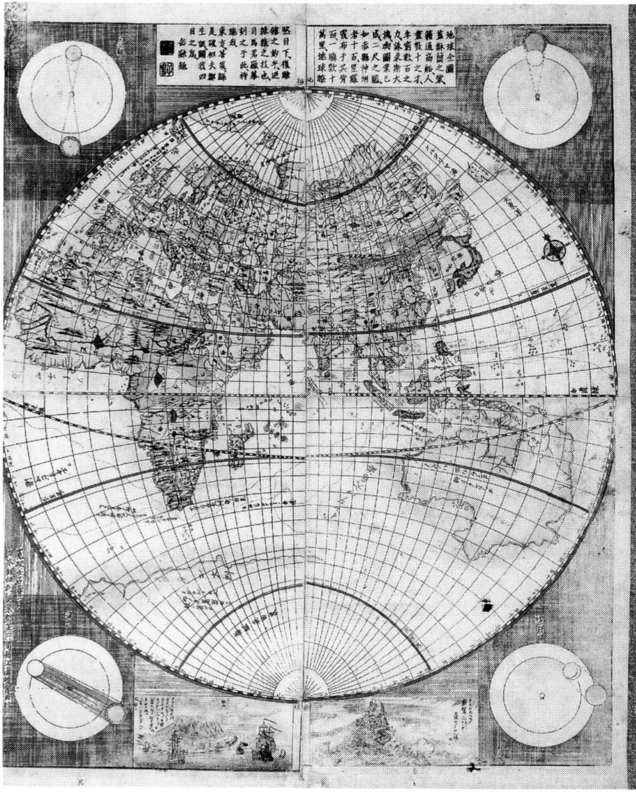

FIG. 11.64. SHIBA KŌKAN'S *CHIKYŪ ZU*, CA. 1795. This is a reprint of Shiba's *Yochi zenzu* of 1792. The Covens and Mortier issue, ca. 1730, of Jaillot's world map, owned at the time by the well-known specialist in Dutch studies Ōtsuki Gentaku, was the translated source of the copperplate, notable for its portrayal of the Japanese northern frontier based on the first shogunal expedition.
Size of the original: 55 × 86 cm. By permission of the Kayahara Hiroshi Collection, Tsu, Mie Prefecture.

was printed from a woodblock, and the manuscript of both hemispheres is lost.[329] In 1794 he did complete maps of the world, the Americas, Europe, Africa, Asia, and other places based on the Laxman maps. These contained better information than the Dutch maps, and this may explain why Katsuragawa abandoned the other project in favor of the map of the world from Russian sources. The maps translated from the Russian became appendixes to the *Hokusa bunryaku* (Story of a driftage to the north, 1794). They were kept in the shogunal library and never published.[330]

The works of Shiba and Katsuragawa in Edo influenced map publishers in Ōsaka. Most notably, Shiba's world map inspired the *Oranda shin'yaku chikyū zenzu* (Map

the castaways titled *Hokusa bunryaku* (Story of a driftage to the north, 1794). Laxman's attempt to open trade was rebuffed: he was ordered to go to Nagasaki but instead returned to Russia, dissatisfied. See Sansom, *History of Japan*, 3:202 (note 32), and Philipp Franz von Siebold, *Manners and Customs of the Japanese in the Nineteenth Century* (1841; reprinted Tokyo: Charles E. Tuttle, 1985), 193. Ten kinds of maps are attached to the *Hokusa bunryaku*; of them, the maps of the world, Asia, Africa, America, and Europe were probably presents from the Russian government. Copies of two maps of Europe and America with the Russian language are preserved at the Yokohama City University Library (Ayusawa Collection). For a reproduction of the map of America (50 × 63 cm), see Unno, Oda, and Muroga, *Nihon kochizu taisei*, vol. 2, fig. 83 (note 8).

329. The printed map does not have the name of the author or publisher or the date of publication. For details on the translation of the Dutch map of the world, see Unno Kazutaka, "Katsuragawa Hoshū no sekaizu ni tsuite" (On Katsuragawa Hoshū's map of the world), *Jinbun Chiri* 20, no. 4 (1968): 1–12. The cover title for the map of the Western Hemisphere (79.5 × 88.5 cm) is *Bankoku chikyū zenzu* (Map of all the countries on the globe, ca. 1792). The trial print is preserved by Nakao Ken'ichirō of Ōsaka. On the paper casing, Kimura Kenkadō wrote "Getchi Katsuragawa shi chikyū zu" (Getchi Katsuragawa's map of the terrestrial globe); Kimura was a well-known collector in Ōsaka, and Getchi was Katsuragawa's pseudonym. For reproductions of the map and paper casing, see Unno, Oda, and Muroga, *Nihon kochizu taisei*, vol. 2, pl. 78 and fig. 68 (note 8).

330. They are now preserved at the National Archives in Tokyo. Reproductions are in Unno, Oda, and Muroga, *Nihon kochizu taisei*, vol. 2, pl. 80 (world map of the Eastern and Western hemispheres, each 56.6 cm in diameter), pl. 81 (map of Asia, 46.8 × 58.7 cm), and fig. 72 (map of the northern Pacific, 46 × 62 cm) (note 8). See note 328 for the *Hokusa bunryaku*.

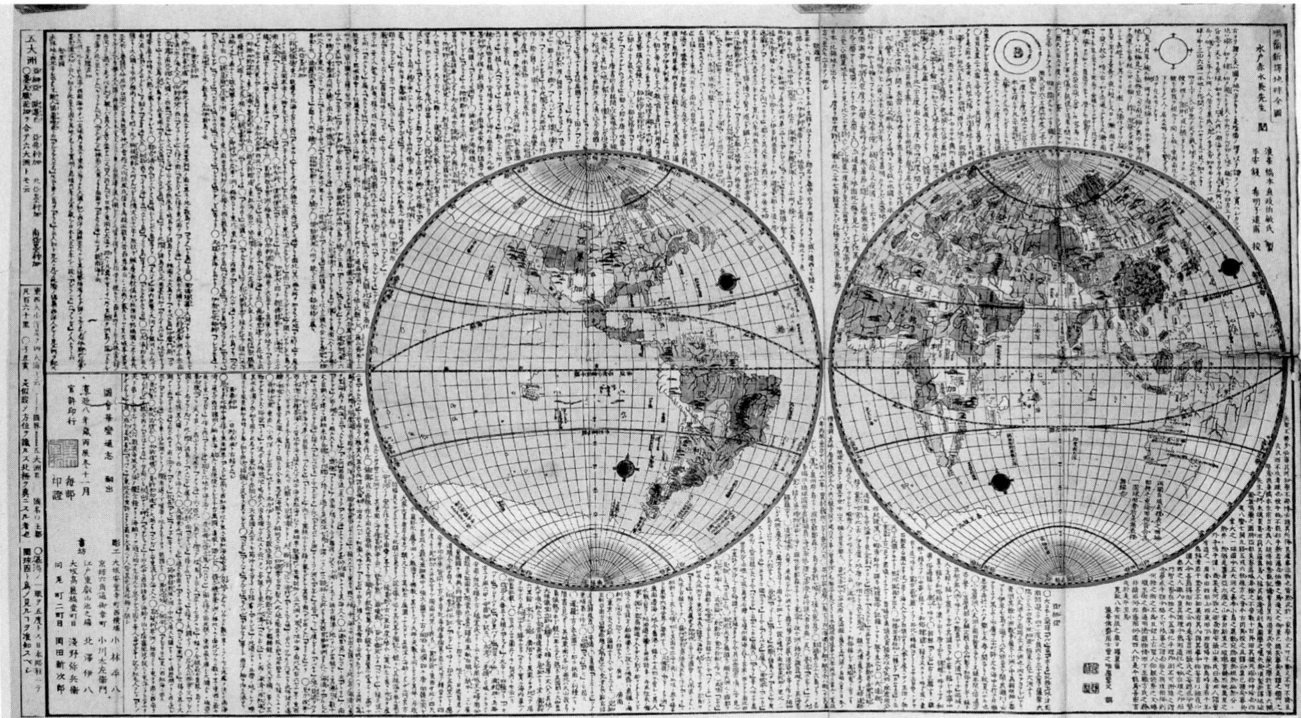

FIG. 11.65. *ORANDA SHIN'YAKU CHIKYŪ ZENZU.* The date is 1796 according to the map, or 1797 according to what is printed on the paper cover attached to the first issue. The content of the map suggests that the Dutch original dated from the first half of the eighteenth century, but not much else is known. Interesting is the rare globular projection.
Size of the original: 55.5 × 93.7 cm. Yokohama City University Library, Yokohama, Ayusawa Collection. Photograph courtesy of Kazutaka Unno.

of the world, newly translated from a Dutch edition), printed from a woodblock in 1797 (fig. 11.65). Allegedly translated by the Ōsaka *Rangaku* scholar Hashimoto Sōkichi (1763–1836), it was probably done by Sotani Ōsei (1738–97), a scholar of Chinese who wrote the preface to the map as well as composing other maps.[331] The map, drawn on a globular projection, followed Shiba's map in its depiction of northern Japan. At least four editions bearing the same year of publication are known. They do, however, have different combinations of copublishers, suggesting that the map was published repeatedly over time. Imitations were also made, and the map was even copied onto a globe.[332]

In 1804 the Russian envoy Nikolai Rezanov (1764–1807) repatriated some Japanese who had been ship-

331. The reason for doubting Hashimoto as its author lies in the explanation around the map: it gave no more information than previous books published in Japan and suggests that no Dutch literature was consulted as a primary source. It is likely that Hashimoto's name was used to lend an air of authority to the map. Ōsaka was known not to have many scholars of Dutch studies. Sotani is therefore the likely compiler, but if Hashimoto actually was involved, it may have been to read and explain the geographical names and notes on a Dutch map. This is discussed in Unno Kazutaka, "Oranda shin'yaku chikyū zenzu ni okeru sanshōshiryō" (Reference materials in a Dutch map of the world newly translated), *Nihon Yōgakushi no Kenkyū* 7 (1985): 65–102. The first edition by Sotani Rinzō and three other publishers is reproduced in Beans, *Japanese Maps of the Tokugawa Era*, facing p. 29 (note 22); an edition by Okada Shinjirō and three other publishers (measuring 51 × 92 cm) is in Kurita, *Nihon kohan chizu shūsei*, pl. 8 (note 15), and Unno, Oda, and Muroga, *Nihon kochizu taisei*, vol. 2, pl. 82 (note 8). Other maps by Sotani Ōsei include the *Jūsen Nihon yochi zenzu* of 1783 (see above, p. 414 and note 271), the *Kaisei Settsu Ōsaka zu* (Revised plan of Ōsaka, Settsu Province, 1789), and the *Zōshū kaisei Sesshū Ōsaka chizu* (Enlarged and revised plan of Ōsaka, Settsu Province, 1806) (fig. 11.49) with Sotani's preface of 1781 (Sotani corrected this plan; it is reproduced in Unno, Oda, and Muroga, *Nihon kochizu taisei*, vol. 1, pl. 88 [note 8]).

332. Publishers for four editions of Hashimoto's map of the world are Ogawa Tazaemon, Kitazawa Ihachi, Asano Yahee, and Sotani Rinzō; Ogawa, Kitazawa, Asano, and Okada Shinjirō; Ogawa, Asano, and Okada; and Ogawa, Yanagihara Kihee, and Okada. The globe, approximately thirty centimeters in diameter, is preserved at the Hagi Local Museum, Yamaguchi Prefecture. Two imitations were another *Oranda shin'yaku chikyū zenzu*, 32.8 by 20.4 centimeters, with no colophon and in an album style; and Tajima Ryūkei's *Oranda chikyū zenzu* (Complete Dutch map of the world, 1840; 45 × 69.5 cm). A reproduction of Tajima's map is in Unno, Oda, and Muroga, *Nihon kochizu taisei*, vol. 2, fig. 75 (note 8). On the imitations, see Unno, "Hashimoto Sōkichi sekaizu no ihan gihan mohōban" (Some unusual fake and imitative editions of Hashimoto Sōkichi's world map), in *Chizu no shiwa*, 305–18 (note 136).

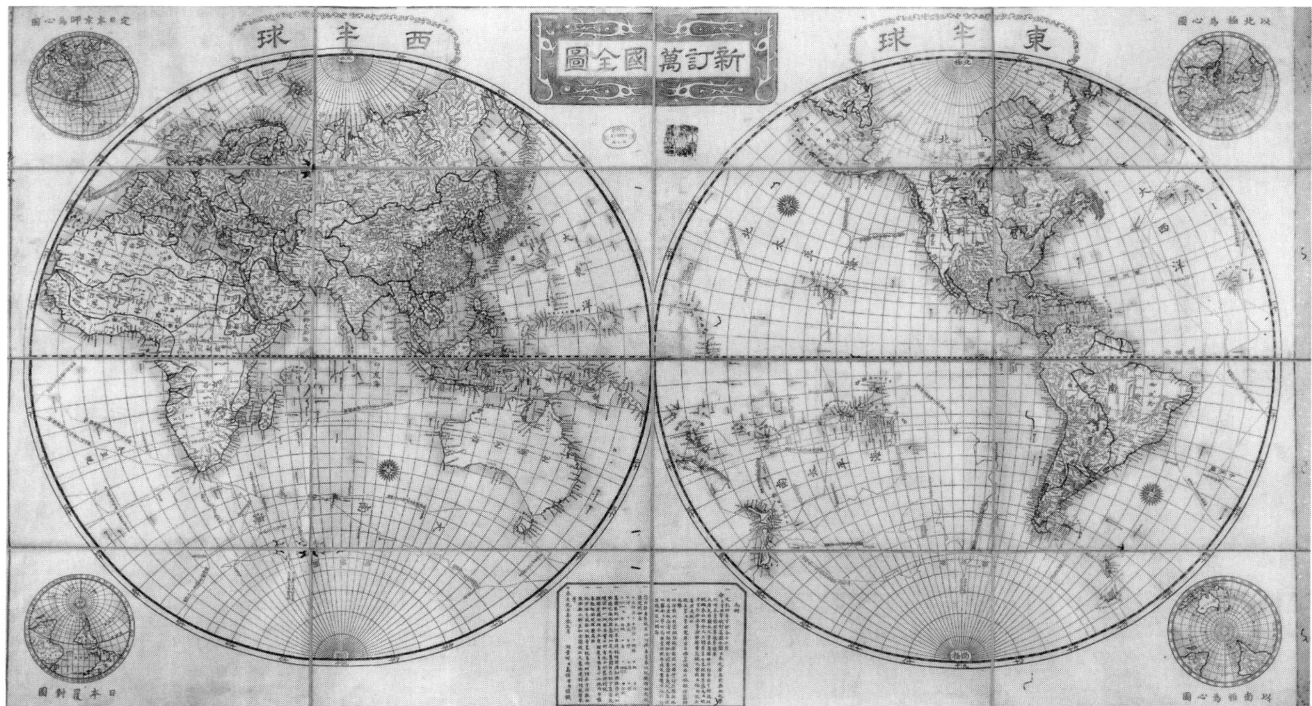

FIG. 11.66. *SHINTEI BANKOKU ZENZU* BY TAKAHASHI KAGEYASU ET AL., CA. 1816. In order to show Japan near the center of this copperplate map, the conventional Eastern Hemisphere has been placed on the left and labeled *nishi hankyū* (Western Hemisphere); the Western Hemisphere has been placed on the right and labeled *higashi hankyū* (Eastern Hemisphere). Of the four small hemispheres in the corners, the one in the upper left is centered on Kyōto.
Size of the original: 114 × 198 cm. Geographical Institute, Faculty of Letters, Kyōto University. Photograph courtesy of Kazutaka Unno.

wrecked in Russia.[333] Rezanov presented some maps to the authorities, and these were translated along with others that the shipwrecked Japanese had obtained in Russia. One of the maps the Japanese brought back was translated by the *Rangaku* scholar Matsubara Uchū (fl. 1789–1808) and titled *Bankoku yochi zenzu* (Map of all the countries in the world). It was a copperplate print of the Eastern and Western hemispheres and has an exotic look because of the Russian words in the Cyrillic alphabet copied from the original.[334] The map does not give the name of the cartographer or the date of publication, but it can be traced to Matsubara and dated to about 1808 on the grounds of a book on world geography, the *Honkoku yochi zenzu ryakusetsu* (Brief explanation of the reproduced map of the world). This book, corresponding to the map, was published in 1808 and was written by Matsubara Josui (Uchū).[335]

Japan at the time was receiving pressure from Russia and other countries to establish commercial relations.[336] The shogunate therefore decided in 1807 to commission the astronomical observatory at Asakusa in Edo to con-

333. For accounts of Rezanov's mission in 1804–5, again an unsuccessful attempt to open Japan to Russian trade, the subsequent retaliatory raids by Khvostov and Davidov in Sakhalin and Hokkaidō in 1806, and the Japanese captivity of Vasily Golovnin in 1811–13, see Sansom, *History of Japan*, 3:202–4 (note 32), and von Siebold, *Manners and Customs*, 193–203 (note 328).

334. Folding map (37 × 16.5 cm; each hemisphere measures 33.5 cm in diameter). Copies are preserved at the Tōhoku University Library in Sendai, at the Waseda University Library in Tokyo, and by Kayahara Hiroshi in Tsu, Mie Prefecture. For a reproduction, see Unno, Oda, and Muroga, *Nihon kochizu taisei*, vol. 2, pl. 85 (note 8).

335. Josui was a pseudonym; the map is discussed in Okamura Chibiki, "Wasurerareta dōban gaka Matsubara Uchū" (A forgotten copperplate artist, Matsubara Uchū), in *Kōmō bunka shiwa* (Historical essays on Dutch culture), by Okamura Chibiki (Tokyo: Sōgensha, 1953), 198–206; and in Unno Kazutaka, "Hyōryūmin Tsudayūra no kikoku to chizu no denrai" (Introduction of European cartography when Tsudayū [one of the repatriated castaways] and others returned to Japan), *Nihon Yōgakushi no Kenkyū* 4 (1977): 101–22. According to the records of the voyage, *Kankaiibun* (Novel news from a trip around the world, completed in 1807) by Ōtsuki Gentaku, the Russian map was bought by the castaways in Russia. The *Kankaiibun* has been published (Tokyo: Yasaka Shobō, 1986) and is in *Hokumon sōsho* (Northern gateway series), 6 vols., ed. Ōtomo Kisaku (Tokyo: Hokkō Shobō, 1943–44; reprinted Tokyo: Kokusho Kankōkai, 1972), vol. 4. It has been translated into modern Japanese by Ikeda Akira (Tokyo: Yūshōdō Shuppan, 1989). Matsubara's *Honkoku yochi zenzu ryakusetsu* was discovered in 1985.

336. Mainly the United States and Britain. See von Siebold, *Manners and Customs*, 188–93 (note 328).

struct a new world map that could be used for diplomatic purposes. The operations were placed under the direction of Takahashi Kageyasu (or Sakuzaemon the younger, 1785–1829), the shogunal astronomer. Others involved with the production of the map were the astronomer Hazama Shigetomi (1756–1816), the two official interpreters, Baba Sajūrō (1787–1822) and Motoki Seiei (1767–1822), and Aōdō Denzen (Nagata Zenkichi, 1748–1822), the last of whom engraved it in copper. This team collected materials from Japan, China, and Europe and in 1810 completed a manuscript version showing the Eastern and Western hemispheres. The map was titled *Shintei bankoku zenzu* (Newly revised map of all the countries). An interesting feature is that in an attempt to show Japan near the center of the world, the "Eastern" and "Western" hemispheres are transposed from their conventional positions. The Japanese labeling has been changed so that the Americas are in the "Eastern Hemisphere" (*higashi hankyū*).[337] In the margin were drawn supplementary hemispheric maps, one centered on Kyōto and the other being its opposite.

The greatest difficulty the compilers experienced was in depicting the area around Sakhalin. It was still being explored at the time and was represented in conflicting ways on different European maps. After the map was completed, Takahashi and his colleagues sought to correct it with more reliable information on the west coast of Sakhalin and around the mouth of the Amur River, such as that obtained by Mamiya Rinzō (1780–1844). To improve the image of East Asia, they used the *Huangyu quanlan tu* (Map of a complete view of imperial territory; known as the Kangxi Jesuit atlas), completed in China in 1718.[338] A revision of the 1810 *Shintei bankoku zenzu* (fig. 11.66) was issued about 1816 as a government publication, engraved on copperplates by Nagata Zenkichi but without his name on the map. Lines of longitude were entered on the map, but no numerical values were given. Overall, the map compares favorably with European maps of the same period, and it is the first in the world to show the Mamiya (Tatar) Strait.[339]

While the *Shintei bankoku zenzu* was still being compiled, Takahashi ordered a small trial print run. This was done in 1809 by Nagata, evidently to confirm his skill at engraving in copper, and titled *Shinsen sōkai zenzu* (Newly printed map of the whole world; 23.3 × 34 cm). Differences from European maps included the arrangement and names of the Eastern and Western hemispheres. The world map was paired on a scroll with the *Nihon henkai ryakuzu* (Simple map of the frontiers of Japan), the first in Japan to be executed on a predetermined equidistant conic projection with the prime meridian running through Kyōto (fig. 11.67).[340] It was, moreover, distinguished by its highly accurate delineation of Japan, derived by using the results of the official coastal survey then in progress under Inō Tadataka (or Chūkei, 1745–1818). It was this map that was translated and used by Philipp Franz von Siebold (1796–1866) in his *Nippon* of 1832 (fig. 11.68).[341]

The story of von Siebold's maps illustrates the significance Japanese authorities attached to maps. It becomes clear that although foreign maps were being sought by Japanese officials, copies of recent maps of Japan were regarded as "classified" documents. Von Siebold, who was born in southern Germany and had been the personal physician of Wilhelm I of Holland, arrived in Deshima as a physician to the Dutch East India Company in 1823. He practiced and taught in the city of Nagasaki as well as opening the first Japanese medical academy. In 1829, when he was starting his voyage back to Europe, Japanese officials discovered in his belongings a recently made map of Japan based on Inō's maps; possessing it constituted an illegal act both for von Siebold and for Takahashi, who had given it to him. Von Siebold was banished from Japan, but Takahashi was sentenced to death (or at least

337. The manuscript map (106.5 × 188 cm) is now at the National Archives in Tokyo. A color reproduction is in Unno, Oda, and Muroga, *Nihon kochizu taisei*, vol. 2, pl. 87 (note 8), and it is discussed in Akabane Sōzō, "Takahashi Kageyasu no Shintei bankoku zenzu ni tsuite" (On the *Shintei bankoku zenzu* by Takahashi Kageyasu), *Nihon Rekishi* 131–32 (1959): 78–95, 51–56.

338. On the exploration of the northern frontier, including Sakhalin, see below. The Kangxi Jesuit atlas used was in the collection of Kimura Kenkadō and was copied at the shogunal observatory about 1808. There are three copies: at the National Archives in Tokyo, at the National Diet Library in Tokyo, and in my personal collection. The two scrolls of *Jūrokusei zu* (Maps of the sixteen provinces [of China]) and *Kyūhen zu* (Maps of nine frontiers) include thirty-two maps in all. What is probably a later copy is in the Takami Collection, Koga, Ibaraki Prefecture.

339. The map (114 × 198 cm) was referred to by Ōtsuki Gentaku in *Ran'yaku teikō* (Account of my difficulties in translating Dutch books, ca. 1816); Ōtsuki's work can be found in *Yōgaku* (Western studies), 2 vols., ed. Numata Jirō et al., Nihon Shisō Taikei (Series of Japanese thought), vols. 64–65 (Tokyo: Iwanami Shoten, 1972–76), esp. 1:379, giving the date of about 1816. A reproduction is in Kurita, *Nihon kohan chizu shūsei*, pls. 10 and 11 (note 15), and in Nanba, Muroga, and Unno, *Nihon no kochizu/Old Maps in Japan*, pl. 16 (note 11). For the copperplate map, see Ayusawa, "Types of World Map," 124, 126, and fig. 1; see also Ramming's comments, 128 (note 235).

340. The *Shinsen sōkai zenzu* is reproduced in Unno, Oda, and Muroga, *Nihon kochizu taisei*, vol. 2, fig. 82 (note 8).

341. Philipp Franz von Siebold, *Nippon, Archiv zur Beschreibung von Japan und dessen Neben- und Schutzländern*, 4 vols. (Leiden, 1832–[54?]), vol. 1, pl. I ("Japan mit seinen Neben und Schutzländern"). He entered the place-names on the map in two ways: on some editions, such as the one at the Kinki University Library in Higashiōsaka (22.5 × 34.1 cm), the place-names are all written in the same direction, whereas on other editions, such as the one at the Kyūshū University Library in Fukuoka (21.9 × 34.1 cm), the place-names in western Japan are written in various directions. Von Siebold's map is described in detail in Unno Kazutaka, "Shīburoto to 'Nihon henkai ryakuzu' " (Siebold and his small map of Japan), *Nihon Yōgakushi no Kenkyū* 5 (1979): 101–28. On Inō Tadataka, see below.

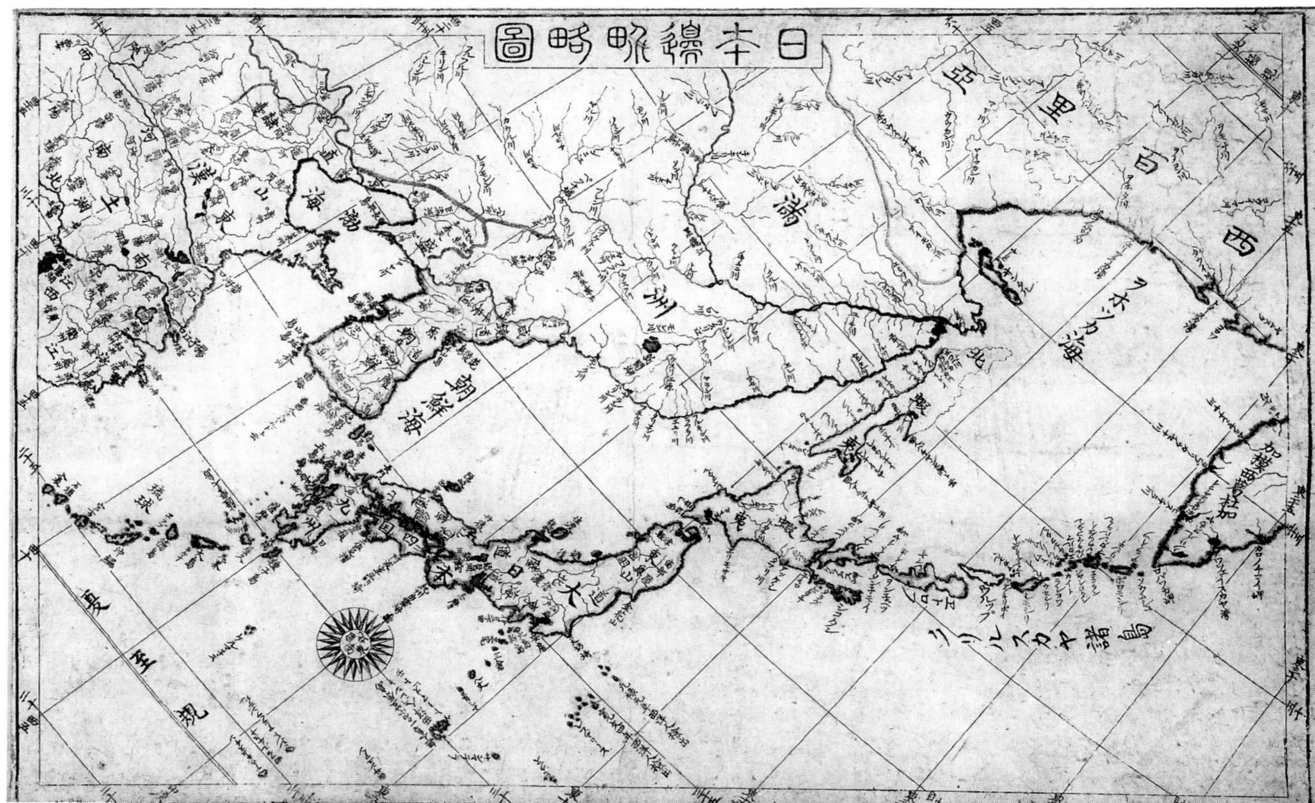

FIG. 11.67. TAKAHASHI KAGEYASU'S *NIHON HENKAI RYAKUZU*, 1809. This map is paired on a scroll with Takahashi's *Shinsen sōkai zenzu*, a small trial print of the *Shintei bankoku zenzu*. This copperplate was the first map of Japan made in Japan to be executed on a predetermined projection, conical with equidistant parallels; the prime meridian runs through Kyōto. The portrait of Sakhalin (Karafuto in Japanese) was based on surveying reports of Mamiya Rinzō before 1808. Size of the original: 21.5 × 34.5 cm. Photograph courtesy of Kazutaka Unno.

to prison, where he died), and many of von Siebold's students were also imprisoned. The map was confiscated, but von Siebold was able to make a copy and publish it.[342] In addition to the attempt to smuggle a map out of the country, von Siebold had also engaged in further cartographic espionage. It is known that he secretly made topographic measurements during a journey to Edo, undertook a hydrographic survey of the Strait of Shimonoseki, studied Japanese maps of territories north of Honshū in the shogun's library, and consulted Japanese geographical works including a map by Mogami Tokunai (1754–1836), a diary by Mamiya Rinzō, and a work by Takahashi, presumably a map of Hokkaidō and Sakhalin.[343]

As far as Japanese cartography was concerned, the exchange worked in the other direction. Takahashi's map stimulated the compilation and publication of maps of the world and foreign countries, based mainly on recently acquired European models. Examples include the *Shinsei yochi zenzu* (Newly made map of the world, 1844) by Mitsukuri (or Mizukuri) Shōgo (1821–47), *Dōhan bankoku yochi hōzu* (cover title, Copperplate square map of all the countries in the world, 1846) by Nagai Seigai (or Soku, d. 1854) (fig. 11.69), *Shintei kon'yo ryakuzenzu* (Newly revised map of the earth, 1852) by Shibata Shūzō (1820–59), *Chōtei bankoku zenzu* (Repeatedly revised map of all the countries, 1855) by Yamaji Akitsune (fl. 1835–60) and Shibata Shūzō, *Yochi kōkaizu* (Chart of

342. The map in von Siebold, *Nippon* (1840) (note 341) measures 91.8 by 67.9 centimeters (only Honshū, Shikoku, and Kyūshū). Von Siebold himself mentions the affair in *Manners and Customs*, 169–70 (note 328). For a brief discussion of von Siebold and Japan, see Cortazzi, *Isles of Gold*, 51–53 (note 14), and Plutschow, *Historical Nagasaki*, 106–9 (note 128).

343. See Cortazzi, *Isles of Gold*, 51 (note 14); one of von Siebold's maps is reproduced (pl. 83): *Karte vom japanischen Reiche*, dated 1840 (39.5 × 55 cm) and kept at the British Library, London. In regard to Mogami's work, von Siebold states that on 16 April 1826 he was given maps of Ezo, the Kuriles, and Sakhalin by Mogami and promised not to publish them for twenty-five years; he adhered to this and published them in 1852. See von Siebold in J. C. Coen, *Reize van Maarten Gerritsz. Vries in 1643 naar het noorden en oosten van Japan* . . . (Amsterdam, 1858), 336. On Mogami, see below.

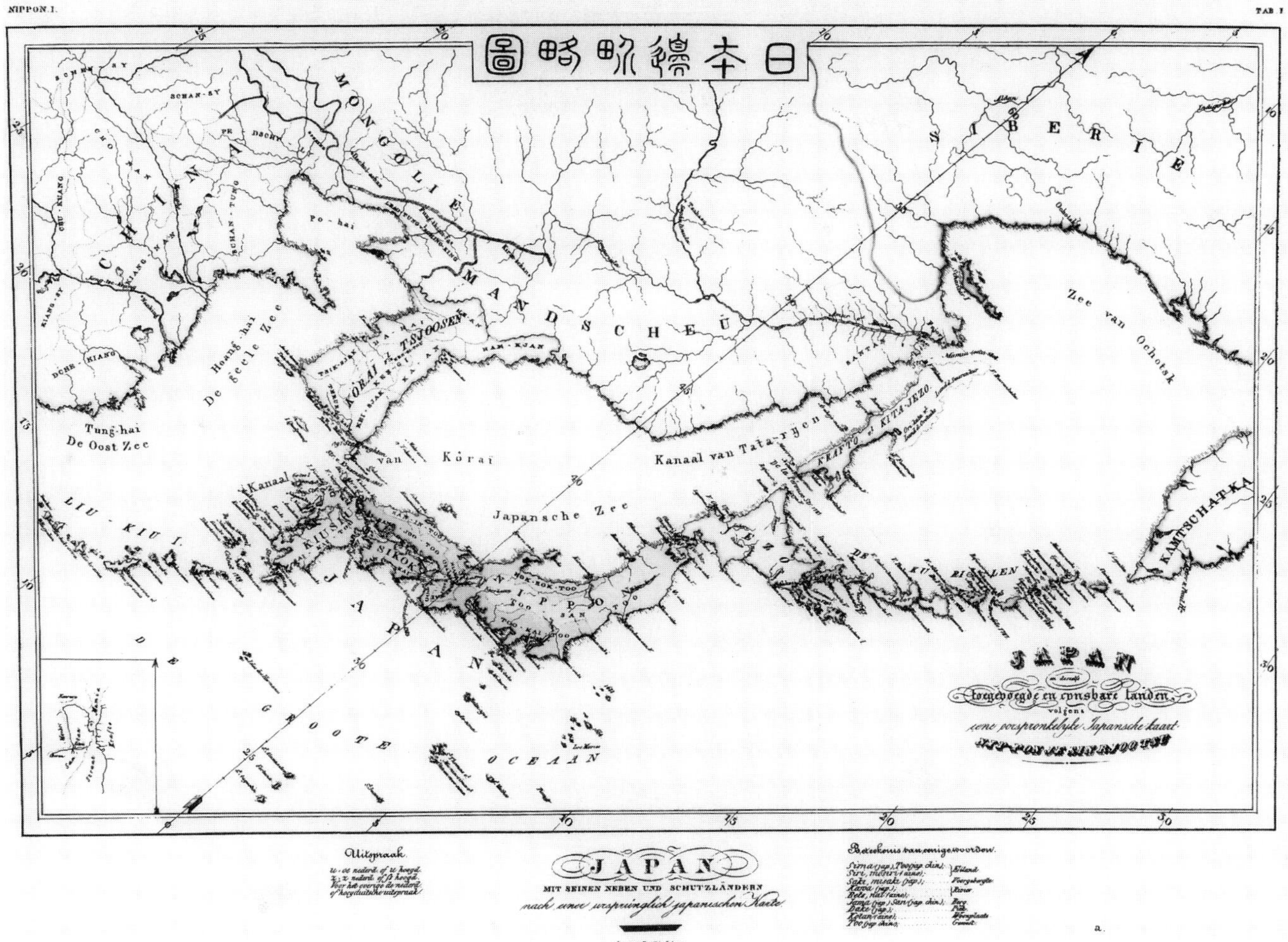

FIG. 11.68. *JAPAN MIT SEINEN NEBEN UND SCHUTZLÄNDERN* BY PHILIPP FRANZ VON SIEBOLD, 1832. The Chinese characters at the top are the same as on Takahashi's *Nihon henkai ryakuzu* (fig. 11.67), of which this map is a translation with some additional information such as "Str. Mamia" and "De la Pérouse Str." (respectively the Tatar Strait and the La Pérouse or Sōya Strait).
Size of the original: 22.5 × 34.1 cm (frame). By permission of the Kinki University Library, Higashiōsaka, Ōsaka Prefecture.

the world, 1858) by Takeda Kango (d. 1859), and *Shinkan yochi zenzu* (Newly published map of the world, 1861) by Satō Masayasu (1821–77).[344]

Each of these maps exhibited some notable feature or revision that indicates the breadth of geographical sources and knowledge available to the *Rangaku* scholars by the late Edo period. Thus Mitsukuri's map contains a revision of the southeast coast of Australia, the rest of the content being based on the *Shintei bankoku zenzu* and, evidently, partly on a French map dating from 1835. One of its characteristics is the representation of colonies so that it is possible to determine at a glance whose rule they lay under. Nagai's map claims to have been based on a British map of 1839. This appears to be true because of its use of the Mercator projection and the outlines of the land, but there are many similarities to the place-names on Mitsukuri's map, suggesting that at least one other original source was employed.[345] While appearing

344. Those of 1844, 1846, 1855, and 1858 are mentioned in Ayusawa, "Types of World Map," 125–27, and Ramming's comments, 128 (note 235). Reproductions of the six examples cited here are in Unno, Oda, and Muroga, *Nihon kochizu taisei*, vol. 2, pl. 89 (1844 map, 33.5 × 59 cm, on a scroll measuring 35 × 120 cm), pl. 91 (1846 map, 32 × 36 cm, on a scroll measuring 33.5 × 109.5 cm), pl. 95 (1852 map, 40 × 72.5 cm, on a sheet measuring 49 × 107 cm), pl. 99 (1855 map, 104 × 185 cm), pl. 100 (1858 map, 88.5 × 156.5 cm), and fig. 94 (1861 map, 136 × 133.5 cm) (note 8). A copy of the 1846 map is also in Nanba, Muroga, and Unno, *Nihon no kochizu/Old Maps in Japan*, pl. 17 (note 11).

345. Yasuda Raishū, an artist, was in charge of the copperplate printing. The first Japanese printed map to use the Mercator projection was Shiba Kōkan's *Hinkai zu* (Map of coastal regions, 1805), which includes the Indian Ocean and East Asia. It is preserved in the Beans Collection at the University of British Columbia Library (38 × 53 cm) and at the

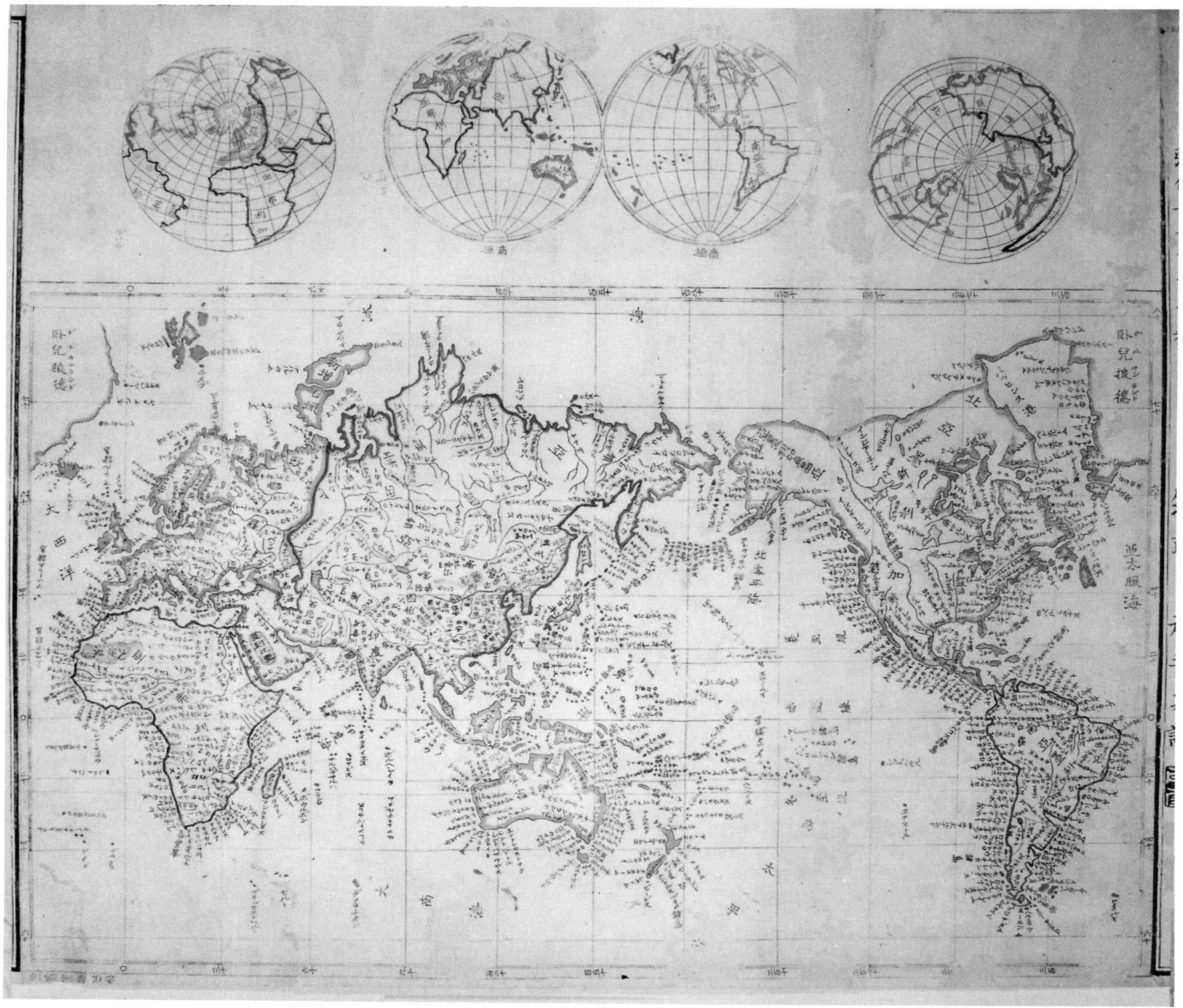

FIG. 11.69. *DŌHAN BANKOKU YOCHI HŌZU* BY NAGAI SEIGAI, 1846. Sources for this scroll included the *Shinsei yochi zenzu* and a British map dating from 1839. Size of the entire scroll: 33.5 × 109.5 cm (map only: 32 × 36 cm). By permission of the Kayahara Hiroshi Collection, Tsu, Mie Prefecture.

on an old-fashioned oval projection, Shibata's map makes a contribution through its new content.[346] Moreover, to designate mountainous areas, it employs hachures that had been introduced to Japanese printed maps in 1850 on the *Shin'yaku Orandakoku zenzu* (Newly translated map of Holland) by Takami Senseki (1785–1858).[347] The map by Yamaji and Shibata revised southern Australia and the northern part of North America according to information from Karl Sohr and Friedrich H. Handtke's *Vollständiger Universal-handatlas der neueren Erdbeschreibung über alle Theile der Erde* of 1846.[348] Though

Mitsui Library in Tokyo; for a reproduction, see Mody, *Nagasaki Colour Prints*, pl. 195 (note 239). Which British map was involved in Nagai's map is not known.

346. According to Shibata's explanation, he used an oval projection after taking into account the deficiencies of the stereographic and Mercator projections. Shibata says that a map that consists of two circles (stereographic projection) makes the central part of the circles too small, and a square map (Mercator projection) makes the parts close to the poles too large.

347. The map is 57 by 86 centimeters, and on the case is printed a statement to the effect of "permission to publish, the first month of the third year of Kaei" (1850). For a reproduction, see Unno, Oda, and Muroga, *Nihon kochizu taisei*, vol. 2, pl. 90 (note 8). A facsimile was published by Noma Kagaku Igaku Kenkyūshiryō Kan (Tokyo, 1981).

348. The *Chōtei bankoku zenzu* (also *Jūtei bankoku zenzu*) was an official publication like the *Shintei bankoku zenzu*, of which it was a revision. The *Chōtei* was compiled under the direction of Yamaji Yukitaka (or Kaiko). Ayusawa notes that the information came from the world map in Sohr and Handtke's atlas. See Ayusawa, "Types of World

other sources were also consulted, Takeda's map is a translation of John Purdy's chart of the world (1845), brought to Japan by Russian sailors in 1854.[349] Both this map and Satō's work are large and based on the Mercator projection. Satō's map was derived mainly from an unidentified Dutch map issued in 1857.

The impact of *Rangaku* on Japanese cartography can thus hardly be overestimated. Not only were new basic maps of the country compiled and published, but many of these accurate and detailed maps were also in turn popularized by the publication of reduced and simplified versions.

Japanese Mapping of Their Northern Frontier and Coastlines

Important surveying work was undertaken in the northern frontier region late in the eighteenth century and in the first quarter of the nineteenth. Although overlapping to some extent, there were two principal objectives. One brought to a conclusion a protracted episode in Japanese cartography involving the exploration and mapping of the islands to the north of Honshū.[350] In a sense we can refer to this region as the "northern frontier," although only Hokkaidō today is uncontestedly Japanese.[351] The other objective was an accurate portrait of the archipelago as a whole, which was completed in 1821 and based on the remarkable survey of the coastlines by the former sake maker Inō Tadataka.

Throughout the ancient and medieval periods the political unification of the archipelago excluded the northern part of Honshū and the islands lying to the north of it. Since antiquity the terms used for these islands were "Ezochi" and "Ezo,"[352] and although the origin is not known for certain, Ezo is thought to have been derived ultimately from the Ainu language as a corruption of "Emichiw" (man). Two corruptions of Emichiw—"Emishi" (hairy people) and "Ebisu" (barbarians)—were used before the Nara period to designate the non-Japanese to the east of the area of Japanese settlement on Honshū. As the Japanese expanded eastward and northward, the term moved northward to denote the unconquered, unassimilated peoples. From the beginning of the Edo period, the coastal areas along the Pacific Ocean and the Sea of Japan, to which ships sailed from the provincial seat of Matsumae at the southern tip of Hokkaidō, were known as East and West Ezochi. In the case of West Ezochi most of the ships headed directly north, the end

Map," 126, and Ramming's comments, 128 (note 235). A copy of Sohr and Handtke's atlas is preserved at the National Diet Library, Tokyo (formerly the Mitsukuri Shōichi Collection).

349. Ayusawa remarks that Takeda's map went through several editions before 1900: Ayusawa, "Types of World Map," 126 (note 235). Purdy's chart was brought to Japan in 1854 by the envoy Evfimy Putyatin, who concluded a treaty to open Japan to Russian trade in 1855, following in the footsteps of the United States and Britain, which had negotiated treaties in 1854; it was on board the frigate *Diana*, which was damaged by tsunamis following an earthquake while anchored at Shimoda harbor (now in Shizuoka Prefecture). For an account of the Russian mission's sojourn in Japan, see Howard F. Van Zandt, *Pioneer American Merchants in Japan* (Tokyo: Lotus, 1980), 77–142. For more information on Takeda's map, see Ayusawa Shintarō, "Takeda Kango no Yochi kōkaizu no keitō" (Genealogy of Takeda Kango's *Yochi kōkaizu*), in his *Sakoku jidai no sekaichirigaku* (World geography in the age of national isolation) (Tokyo: Nichidaidō Shoten, 1943; reprinted Hara Shobō, 1980), 331–49.

350. On Japanese exploration to the north of Honshū, see Nobuo Muroga, "Geographical Exploration by the Japanese," in *The Pacific Basin: A History of Its Geographical Exploration*, ed. Herman R. Friis (New York: American Geographical Society, 1967), 96–105. On the history of cartography of the northern frontier, refer to three articles by Takakura Shin'ichirō and Shibata Sadakichi and a fourth (the supplement) by Takakura alone: "Wagakuni ni okeru Karafuto chizu sakuseishi" (History of the development of the cartography of Sakhalin in Japan); "Wagakuni ni okeru Chishima chizu sakuseishi" (History of the development of the cartography of the Kuriles in Japan); "Wagakuni ni okeru Hokkaidō hontō chizu no hensen" (Development of the cartography of Hokkaidō in Japan), 1 and 2; and "Hokkaidō chizu no hensen hoi" (Development of the cartography of Hokkaidō: Supplement)—all in *Hokkaidō [Teikoku] Daigaku Hoppō Bunka Kenkyū Hōkoku*, nos. 2 (1939): 1–48, 3 (1940): 1–75, 6 (1942): 1–80 and 7 (1952): 97–166, and 11 (1956): 49–73, respectively. Important maps of the northern frontier are reproduced in Hoppō Ryōdo Mondai Chōsakai (Japan Society for Research on the Northern Territories), ed., *Hoppō Ryōdo: Kochizu to rekishi* (The northern territories of Japan: Old maps and history) (Tokyo: Chūōsha, 1971), and in Narita Shūichi, ed., *Ezo chizu shō* (Extracted maps of Ezo) (Tokyo: Sara Shobō, 1989). Important maps of Hokkaidō are reproduced in Takakura Shin'ichirō, ed., *Hokkaidō kochizu shūsei* (Collection of historical maps of Hokkaidō and the adjacent regions) (Sapporo: Hokkaidō Shuppan Kikaku Sentā, 1987). See also Teleki, *Atlas zur Geschichte der Kartographie* (note 163); Cortazzi, *Isles of Gold*, 54–61 (note 14); John A. Harrison, "Notes on the Discovery of Yezo," *Annals of the Association of American Geographers* 40 (1950): 254–66; and Koreto Ashida, "Old Maps of Hokkaido," in *Dainippon* (Great Japan), ed. Bunmei Kyōkai (Tokyo: Bunmei Kyōkai, 1936), 127–37.

351. From the Japanese point of view, the Kurile Islands (Chishima) lying to the south of Urup (in Japanese, Uruppu) were occupied illegally by the Soviet Union at the end of the Second World War and are thus still considered Japanese. The islands in question are, from north to south: Iturup, Kunashir, Shikotan, and the Malaya Kuril'skaya Gryada (Small Kurile Ridge); their Japanese names are Etorofu, Kunashiri, Shikotan, and the Habomai group. The rest of the Kuriles are not disputed and have been a part of Russia since 1945. Before then, the entire chain was Japanese from 1875, when an agreement was reached giving Russia Sakhalin (Karafuto in Japanese) and Japan the Kuriles. Sakhalin itself had been under joint rule from 1867 until 1875, and in 1905 the southern half of it was annexed by Japan after the 1904–5 Russo-Japanese War. During the Russian civil war of 1917–22, northern Sakhalin became a part of the Russian Far Eastern Republic of 1920–22, but it was incorporated into the Soviet Union in 1922. In 1945 the Soviet Union annexed the southern part of the island; it is not disputed and is now part of Russia.

352. Ezo or Yezo was evidently divided into three parts: Matsumae and Ezochi (Hokkaidō), Kita or Oku (northern or distant: Sakhalin), and Ezo ga Chishima or Chishima (thousand islands of Ezo or thousand islands: the Kuriles).

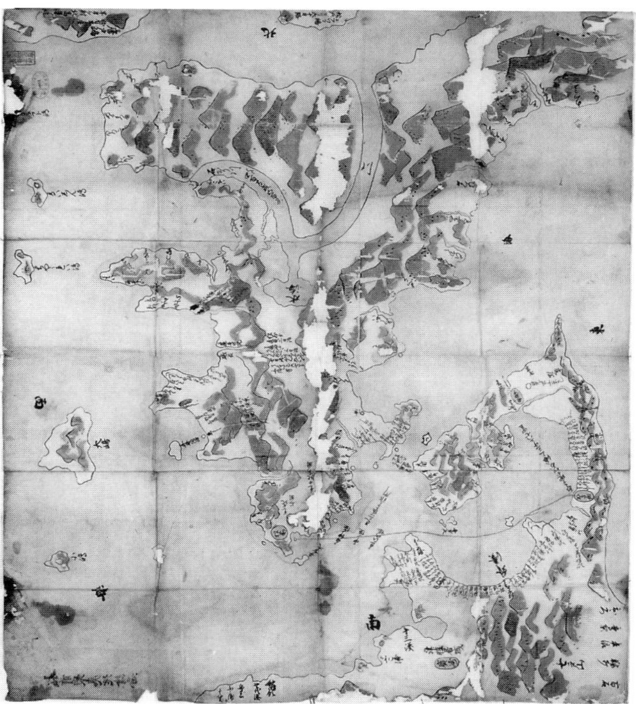

FIG. 11.70. AN EXAMPLE OF A MAP SHOWING HOKKAIDŌ AS A PENINSULA: THE SEVENTEENTH-CENTURY MANUSCRIPT *MATSUMAE EZOCHI EZU* (MAP OF MATSUMAE AND EZOCHI). In the lower right corner is the northern extremity of Honshū, with Hokkaidō shown as a peninsula stretching from the upper right corner to the center of the map. A delineation of Sakhalin (Karafuto) in the upper left is noted to be near northern Korea.
Size of the original: 113 × 96 cm. By permission of the Hokkaidō University Library, Sapporo.

of the shipping line being known as "Teshio-furo," a corruption of the Ainu term *Teshio-kuru* (people of Teshio, a place in northwestern Hokkaidō). The end of the East Ezochi line was known as "Menashi-furo," a corruption of the Ainu *menashi-kuru* or "eastern people" (today's Nemuro region).[353]

Until this time, the geography of what is now known as Hokkaidō (so called since 1869) was a question mark. Only the southwestern part of the island had been governed by the Japanese since the fourteenth century, and when Tokugawa Ieyasu recognized the Matsumae clan as overlords in 1604, it was still uncertain whether Hokkaidō was part of continental Asia or a separate island. The first recorded surveying expedition to the northern frontier dates to 1633, when the daimyo of Matsumae ordered a vassal, Takahashi Giemon, to determine the distance between East and West Ezochi. The next expedition appears to have been conducted in 1635 by another vassal, Murakami Hiroyoshi. He was ordered to circumnavigate and map Hokkaidō, but it is not known whether he completed this voyage or what kind of map, if any, he compiled. The first map to be identified from the Matsumae clan was one submitted to the shogunate in connection with the 1644 project to complete the Shōhō provincial maps. This map has been lost, but the miniature and inaccurate copy incorporated in the Shōhō map of Japan has survived. In 1661 another circumnavigation of Hokkaidō took place; this was entrusted to the vassal Yoshida Sakubee, who sailed along the eastern coast to the north and then around to Matsumae that summer. In 1700, for the fourth national project to compile provincial maps, the clan submitted only another small map with the same detail as that included in the Shōhō map.[354]

Despite the lack of original survey maps, there are other maps of the northern frontier. These were compiled from those in the possession of the Matsumae clan before the Ezochi expedition under the auspices of the shogunate in 1785–86. These maps may be classified according to their depiction of Hokkaidō as an archipelago with exaggerated rivers, as a long island stretching north to south, or as a peninsula curving from the east to the southwest (figs. 11.70 and 11.71).[355]

353. Teshio-furo and Menashi-furo appeared, for instance, as Texxoy and Menaxi on the 1621 map accompanying the report of the Italian Jesuit Girolamo de Angelis (1567–1623), the first European to visit and to compose a map of what is now Hokkaidō. It was depicted as an elongated island stretching from west to east, with these names at either end. Angelis's reports of 1618 (London, British Museum, Add MS. 9860, fols. 239–42) and of 1621, accompanied by a map (39 × 53 cm; Archivum Romanum S.J., Epistolae Martyrum, Jap. Sin. 34, fols. 49–54v), are reproduced in Hubert Cieslik, ed., *Hoppō Tanken Ki* [Record of an exploration of the northern region]: *Foreigners' Reports on Ezo in the Genna Period* (Tokyo: Yoshikawa Kōbunkan, 1962). Studies pertaining to the Angelis map include Kitagawa, "Map of Hokkaido"; Schütte, "Map of Japan"; and Kudo, "De Angelis' Yezo Map" (all in note 121). Kitagawa also reproduces the map, as does Cortazzi, *Isles of Gold*, pl. 84 (note 14).

354. For a brief history of the cartography undertaken by the Matsumae clan, see Matsumae Hironaga, *Matsumae shi* (History of Matsumae, 1781), chap. 2, 122–34, in the *Hokumon sōsho*, vol. 2 (note 335). Also see Matsumae Hironaga, ed., *Fukuyama hifu* (Important records of Fukuyama [Matsumae], 1776), *Nenrekibu* (Chronicle), in *Shinsen Hokkaidō shi* (Newly compiled history of Hokkaidō), 7 vols., ed. Hokkaidō Chō (Hokkaidō Office) (Sapporo, 1936–37), vol. 5. See above for information on the third and fourth projects to compile provincial maps and the resulting Shōhō and Genroku national maps. The Genroku map of Matsumae and Ezo was preserved at the Tokyo University Library but was lost in the fire following the Great Kantō Earthquake of 1923; a reduced copy (83 × 65 cm), however, is kept at the Hokkaidō University Library, Sapporo, and reproduced in Takakura, *Hokkaidō kochizu shūsei*, pl. 10 (note 350).

355. An example of a map showing Hokkaidō as an archipelago is the early eighteenth-century *Ezo zu* (Map of Ezo), a manuscript at the Hokkaidō University Library, Sapporo (102 × 101.5 cm). Two with Hokkaidō as an island elongated from north to south are "Ezo no zu" (Map of Ezo) in the *Wakan sansai zue* (1715) (see note 93), and Hayashi Shihei's *Ezo no kuni zenzu* (fig. 11.71). An example with Hokkaidō as a peninsula is the seventeenth-century manuscript *Matsumae Ezochi*

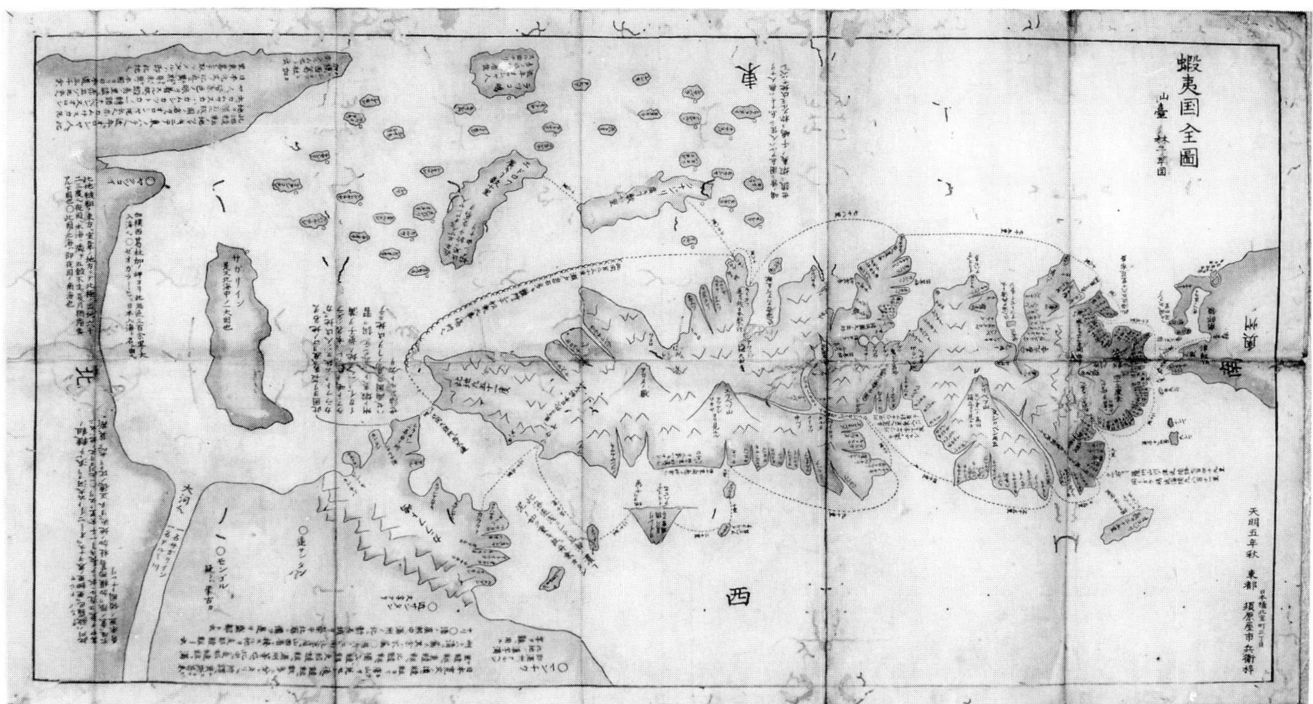

FIG. 11.71. *EZO NO KUNI ZENZU* (MAP OF THE EZO REGION) BY HAYASHI SHIHEI, 1785. Oriented to the east, this is an example of Hokkaidō's being elongated from north to south. The islands toward the top are the Kuriles (Chishima), and Sakhalin is shown as an island to the left. An interesting point is that Karafuto, later determined to be the same as Sakhalin, is shown as part of the Asian continent to the northwest of Hokkaidō and the southwest of Sakhalin. The source of information was a map submitted to the shogunate by the Matsumae clan before the official Ezochi expedition in 1785–86. Size of the original: 50 × 92 cm. By permission of the Kōbe City Museum, Kōbe, Nanba Collection.

The 1785–86 exploration of the northern frontier by the shogunate was a new departure prompted by the advance of the Russians in northeastern Asia.[356] This move had created official anxiety, and geographical information on the northern frontier was essential on the grounds of both diplomacy and national defense. Ten shogunal officers including Yamaguchi Tetsugorō were appointed as formal members of the expedition. Mogami Tokunai, who later gained fame as an explorer and surveyor, was among them as an assistant to Aoshima Shunzō, and together with Yamaguchi and two others they were in charge of East Ezochi. Among the five to explore West Ezochi were Satō Genrokurō and Ihara Yaroku. In 1785 the eastern contingent explored along the Pacific coast of Hokkaidō and eventually reached the island of Kunashir in the Kuriles before returning to Hokkaidō. After reaching Sōya, three of the western group, including Ihara, crossed to Sakhalin, where they went to Tarantomari (at approximately 47°10′N) on the west coast, and then to Shiretoko on Hokkaidō before returning to Sōya. In 1786 Yamaguchi and Mogami went to Iturup and Urup, islands in the Kurile chain; Ōishi Ippei went north along the west coast of Sakhalin to Kushunnai (Il'inskiy, at approximately 48°N) and then turned back.[357] Surveying was conducted en route as well as at

ezu (fig. 11.70). All are reproduced in Unno, Oda, and Muroga, *Nihon kochizu taisei*, vol. 1, pls. 63 and 64 and figs. 30 and 58 (note 8).

356. On Russian expansion in relation to Japan see Sansom, *History of Japan*, 3:201–5 (note 32), and Harrison, "Discovery of Yezo," 259–62 (note 350). Several treatises have been compiled about Russian exploration, which included exceptional activity in the Far East and northern Pacific during the eighteenth century. Examples include D. M. Lebedev, *Ocherki po Istorii Geografii v Rossii XVIII v. (1725–1800 gg.)* (Essays on the history of geography in Russia in the eighteenth century [1725–1800]) (Moscow: Izdatel'stvo Akademii Nauk SSSR, 1957); L. S. Berg, *Otkrytie Kamchatki i ekspeditsii Beringa, 1725–1742* (The discovery of Kamchatka and the expeditions of Bering, 1725–1742) (Moscow: Izdatel'stvo Akademii Nauk SSSR, 1946); G. P. Müller, *Voyages et découvertes faites par les Russes le long des côtes de la Mer Glaciale & sur l'Océan Oriental, tant vers le Japon que vers l'Amérique*, 2 vols. (Amsterdam: Marc-Michel Rey, 1766); and Innokenty Gerasimov, ed., *A Short History of Geographical Science in the Soviet Union* (Moscow: Progress, 1976). For reproductions of some of the maps from this period, see A. V. Efimova (Yefimov), *Atlas geograficheskikh otkrytiy v Sibiri i v severo-zapadnoy Amerike XVII–XVIII vv.* (Atlas of geographical discoveries in Siberia and northwestern America, seventeenth to eighteenth century) (Moscow: Nauka, 1964).

357. Reports of the expedition are in Satō Genrokurō, *Ezo shūi* (Supplement of Ezo, 1786) in the *Hokumon sōsho*, vol. 1 (note 335); a translation into modern Japanese is in Inoue Takaaki, trans., *Akaezo fusetsu kō* (Research on the rumors about the Red Ezo [Russians])

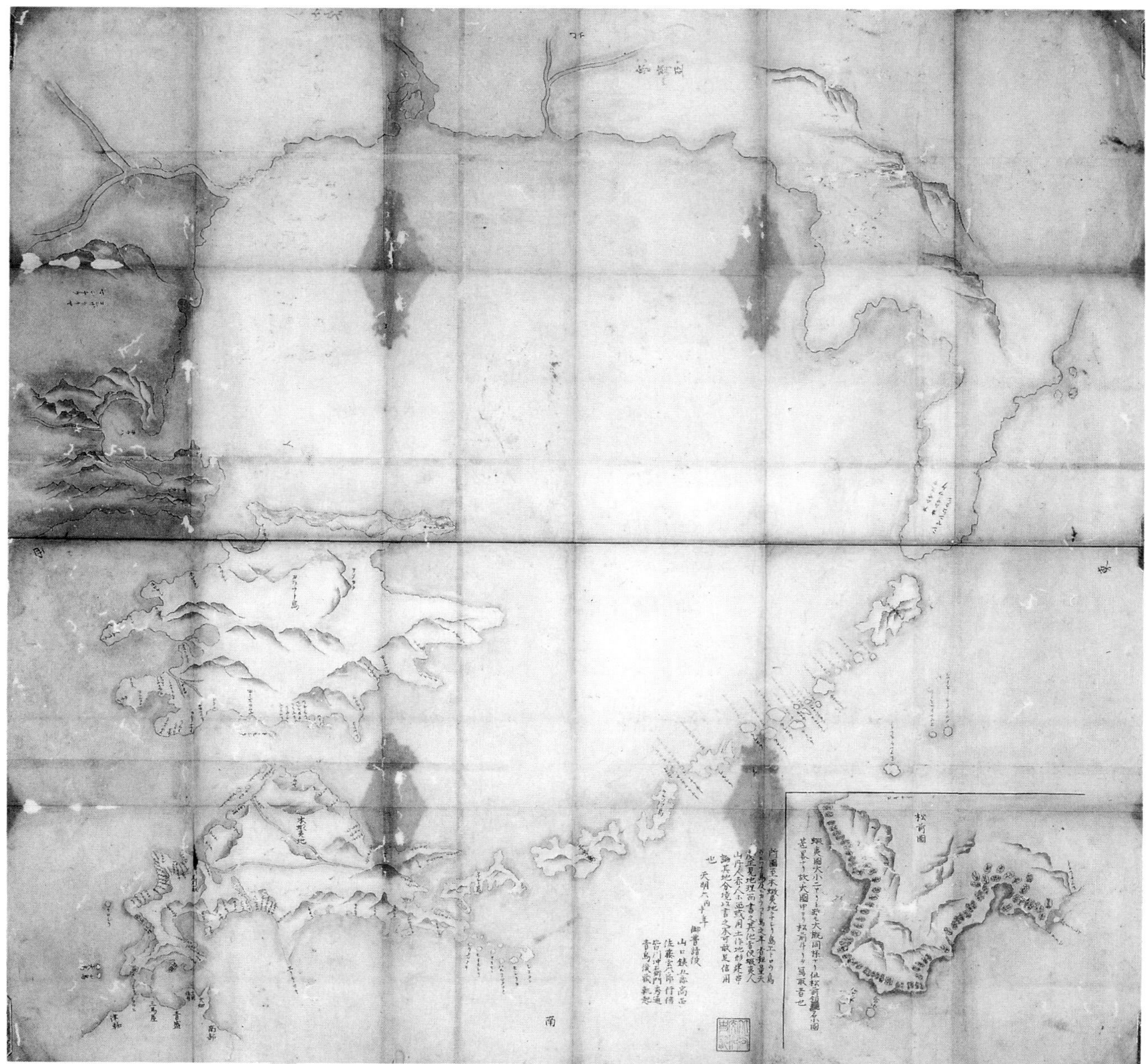

FIG. 11.72. MAP OF EZO (1786) FROM THE 1785–86 EXPEDITION COMMISSIONED BY THE SHOGUNATE. Possibly an interim report of the suspended project, the manuscript shows Sakhalin larger than it should be, but the representation of Hokkaidō and the Kuriles is reasonable. This version has the signatures of Yamaguchi Tetsugorō and three others. Size of the original: 96.5 × 100.5 cm. Muroga Emiko Collection, Kyōto. Photograph courtesy of Kazutaka Unno.

the final destinations. The resulting map, compared with its largely conjectural predecessors, showed a remarkable improvement in the representation of the territory (fig. 11.72).

The map nevertheless had its defects. Hokkaidō is shown shorter from north to south than it actually is, while Sakhalin is drawn as larger than Hokkaidō but with a similar appearance.[358] This first map may be regarded as an interim report of the expedition: the project was suspended when only half completed because of the

(Tokyo: Kyōikusha, 1979), 95–211. A work with a complete account of the 1785–86 expedition is Terui Sōsuke, *Tenmei Ezo tanken shimatsu ki* (The circumstances of the exploration of Ezo during the Tenmei era) (Tokyo: Yaedake Shobō, 1974).

358. Copies are preserved in the Muroga Emiko Collection, Kyōto, the National Archives in Tokyo, the Tenri Central Library (Matsudaira Sadanobu Collection), and the Yamaguchi Prefectural Archives (Mōri Family Collection) in Yamaguchi.

political downfall of its promoter, the chief of the cabinet Tanuma Okitsugu (1719–88). Yet even in its incomplete state it marked a new stage in knowledge of the northern frontier, and it was incorporated into Nagakubo's map of the world of about 1788. The same cartographer published the *Ezo Matsumae zu* (Map of Ezo and Matsumae), based on the shogunal expedition's map of about 1790.[359]

While the shogunate's project was still under suspension, the Matsumae clan conducted an independent survey of Sakhalin in 1790. The expedition reached Kotantoru at approximately 48°40′ north on the west coast and Cape Aniva (Nakashiretoko) at 46° north on the east coast, and the results of the survey were recorded in the manuscript map of about 1793, *Matsumae chizu* (Map of Matsumae) by Katō Kengo. This, however, perpetuated the error of the previous expedition and overextended the island from east to west. Only the image of southern Sakhalin around Aniva Bay showed an improvement.[360]

In 1791 the shogunate resumed its interest in the territory and dispatched a second party to the north, headed by Mogami Tokunai, who had served as a highly praised assistant on the first expedition. Mogami began by resurveying Iturup and Urup, and in 1792 he extended the work to Sakhalin. Here he explored Tōfutsu at 46°30′ north on the eastern coast of Aniva Bay and Kushunnai. The information he obtained was incorporated into improved maps of these areas.[361]

In the same year that Mogami was surveying on Sakhalin, Laxman sailed into Nemuro, and in 1796–97 British ships anchored in Uchiura (or Funka [Eruption]) Bay.[362] These events forced the shogunate to intensify its interest in the northern frontier. The result was an expedition launched in 1798 on an unprecedented scale, with 182 members in all, including Mogami. The expedition was really a pretext for establishing a shogunal administration in the north.[363] After examining the results of this expedition, the shogunate decided in 1799 to rule eastern Hokkaidō and the southern islands in the Kurile chain directly rather than allowing them to remain under the control of the Matsumae fiefdom. Subsequent exploration was undertaken under the aegis of the shogunate and led to the collection of more data to improve the map of Hokkaidō.[364]

The compilation of a manuscript chart—*Kōto (Edo) yori Tōkai Ezochi ni itaru shinro no zu* (Chart of the course from Edo to Ezochi)—to ensure safe travel by sea from Edo to the Pacific coast of Hokkaidō was also completed in 1799. This was undertaken by Hotta Nisuke (1745–1829), an employee of the astronomical observatory, on the orders of the shogunate.[365] In 1800 the shogunate permitted Inō Tadataka to survey the Pacific coast of Hokkaidō, and the remainder of the coastline had been completed by 1817 by Mamiya. Before the completion of this survey, however, the shapes of Hokkaidō, the southern Kuriles, and southern Sakhalin were already well delineated. The proof of these improvements is to be found in Kondō Morishige's (1771–1829) *Ezochi zushiki* (Map of Ezo) of 1802 (fig. 11.73).[366] Kondō had surveyed in the north regularly since

359. The *Ezo Matsumae zu* (33.3 × 45.5 cm) is reproduced in Nanba, Muroga, and Unno, *Nihon no kochizu/Old Maps in Japan*, pl. 47 (dated ca. 1795) (note 11). It is at the Kōbe City Museum (Nanba Collection).

360. Takahashi Hiromitsu, a vassal of the clan, was in charge of the survey, and Katō Kengo was a physician of the Matsumae lord; see Takakura and Shibata, "Wagakuni ni okeru Karafuto chizu sakuseishi," 14–18 (note 350).

361. There were six people, two of whom were in charge of the expedition, in 1791; in 1792 there were twelve all together. Those who went to Sakhalin were Mogami, Wada Hyōdayū, and Kobayashi Gennosuke. See Minagawa Shinsaku, *Mogami Tokunai* (Tokyo: Dentsū Shuppanbu, 1943), 98–129.

362. Presumably associated with the discoveries of Broughton in 1795–97. See William Robert Broughton, *A Voyage of Discovery to the North Pacific Ocean* (London: T. Cadell and W. Davies, 1804; reprinted Amsterdam: Nico Israel, and New York: Da Capo Press, 1967).

363. In the summer of 1798 Mogami went to Iturup with Kondō Morishige; see Minagawa, *Mogami Tokunai*, 140–202 (note 361). For Mogami's reports see his *Ezo sōshi* (Draft of Ezo, 1790) and his *Ezo sōshi kōhen* (Draft of Ezo: Sequel, 1800), both in the *Hokumon sōsho*, vols. 1 and 3 (note 335).

364. Besides Mogami, other important explorers and surveyors who were active in Ezo in and after 1798 included Kondō Morishige, Hata Ahakimaro (or Murakami Shimanojō, 1760–1808), and Mamiya Rinzō. Kondō went to Ezochi in 1798 with Watanabe Kyūzō, a high-level shogunal officer, and also reached Iturup that year and the next; in 1798 it was Mogami who guided them to Iturup. Kondō explored Ezochi five times in all until he was released from the office in 1808. Hata went with Kondō in 1798 and investigated eastern Hokkaidō and Kunashir; from 1799 to 1807 he often went to Hokkaidō to explore. In 1801 he and another high-ranking shogunal official, Matsudaira Tadaakira, went on an expedition around Hokkaidō, which led to a great improvement of its shape on maps. Hata is not known only for his work in the north: he was an eminent surveyor, as his maps of the Kantō provinces (Awa, Izu, and Kazusa) of 1789–93 attest. See Minagawa Shinsaku, "Murakami Shimanojō no Ezochi kinmu" (Murakami Shimanojō's service in Ezochi), *Denki* 7, nos. 4–6 (1940): 10–15, 19–24, 17–24. Mamiya went to Ezochi for the first time in 1799 as an assistant to Hata, and the next year he was appointed to a lower-level position in the shogunate. He was involved in surveying and exploring Ezochi until the shogunate abolished its direct control in 1822. He might have learned his surveying skills from Hata. See Akaba Eiichi, *Mamiya Rinzō* (Tokyo: Shimizu Shoin, 1974); and Hora Tomio, *Mamiya Rinzō*, rev. ed. (Tokyo: Yoshikawa Kōbunkan, 1987).

365. The chart (116.4 × 270.3 cm) is preserved at the Kyōdokan Museum in Tsuwano, Shimane Prefecture. It is reproduced in Takakura, *Hokkaidō kochizu shūsei*, pls. 26 and 28 (note 350).

366. A manuscript on two sheets: *Ken* map (Hokkaidō and Sakhalin) (fig. 11.73), and *Kon* map (of the Chupuka Shotō [the Kuriles], 45 × 74.5 cm). In addition to the copy preserved at the Hakodate City Library, copies are also preserved at the Tenri Central Library (Matsudaira Sadanobu Collection) and at the Monbushō Shiryōkan (Library of Historical Materials at the Ministry of Education) (Tsugaru Family Collection) in Tokyo.

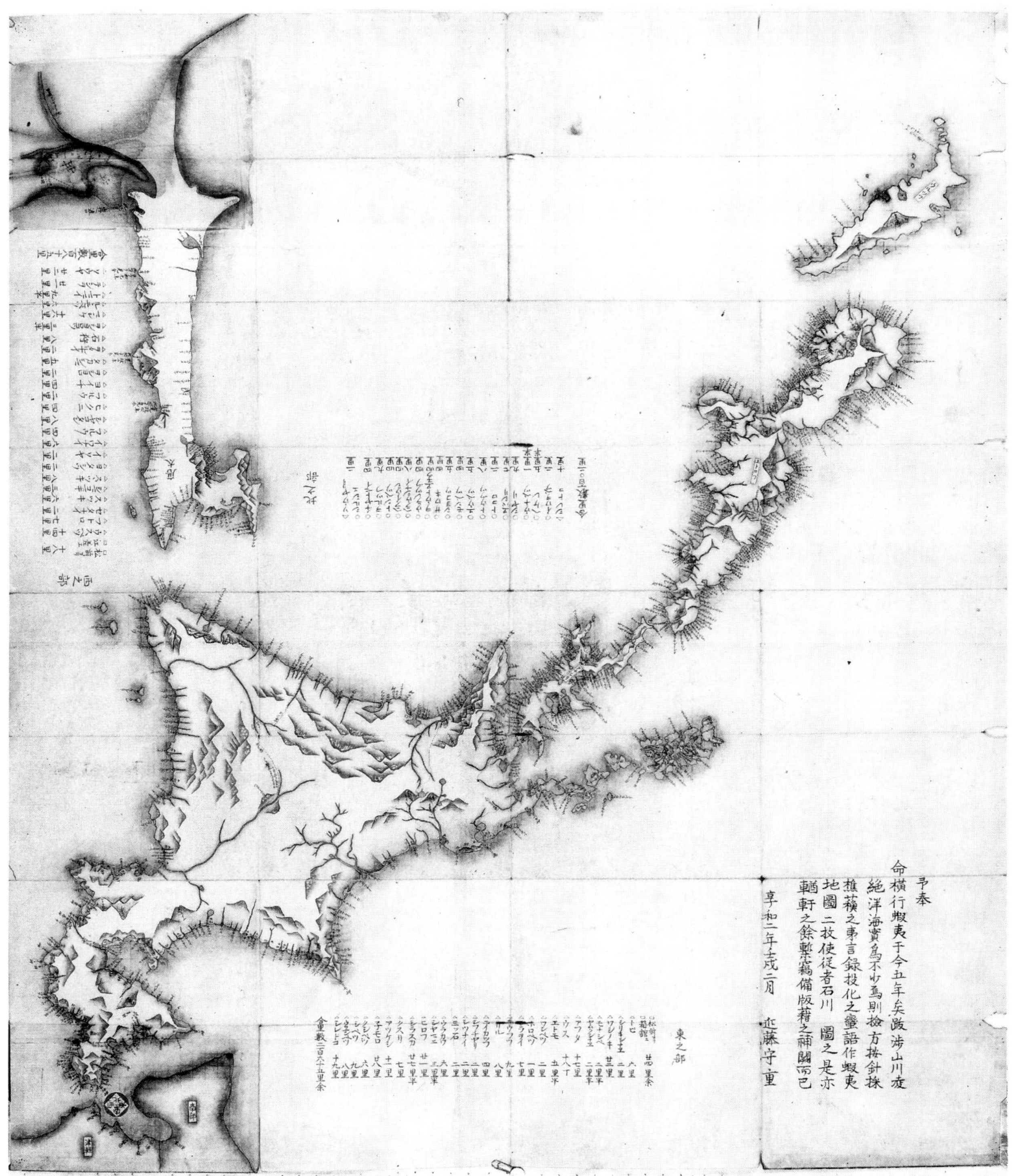

FIG. 11.73. ONE OF THE TWO SHEETS OF KONDŌ MORISHIGE'S *EZOCHI ZUSHIKI* OF 1802. Based on information from shogunal expeditions to the north of Honshū, in which Kondō himself participated, this map represents an important stage in worldwide knowledge of Hokkaidō, Sakhalin (Karafuto), and the Kuriles (Chishima). The geographical status of Sakhalin was of interest to Russians and Japanese alike at this time. On this manuscript, Kondō leaves the question open by showing two possibilities, one whereby Sakhalin is an island and the other whereby it is a peninsula.

Size of the original: 89.5 × 74.5 cm. By permission of the Hakodate City Library, Hakodate.

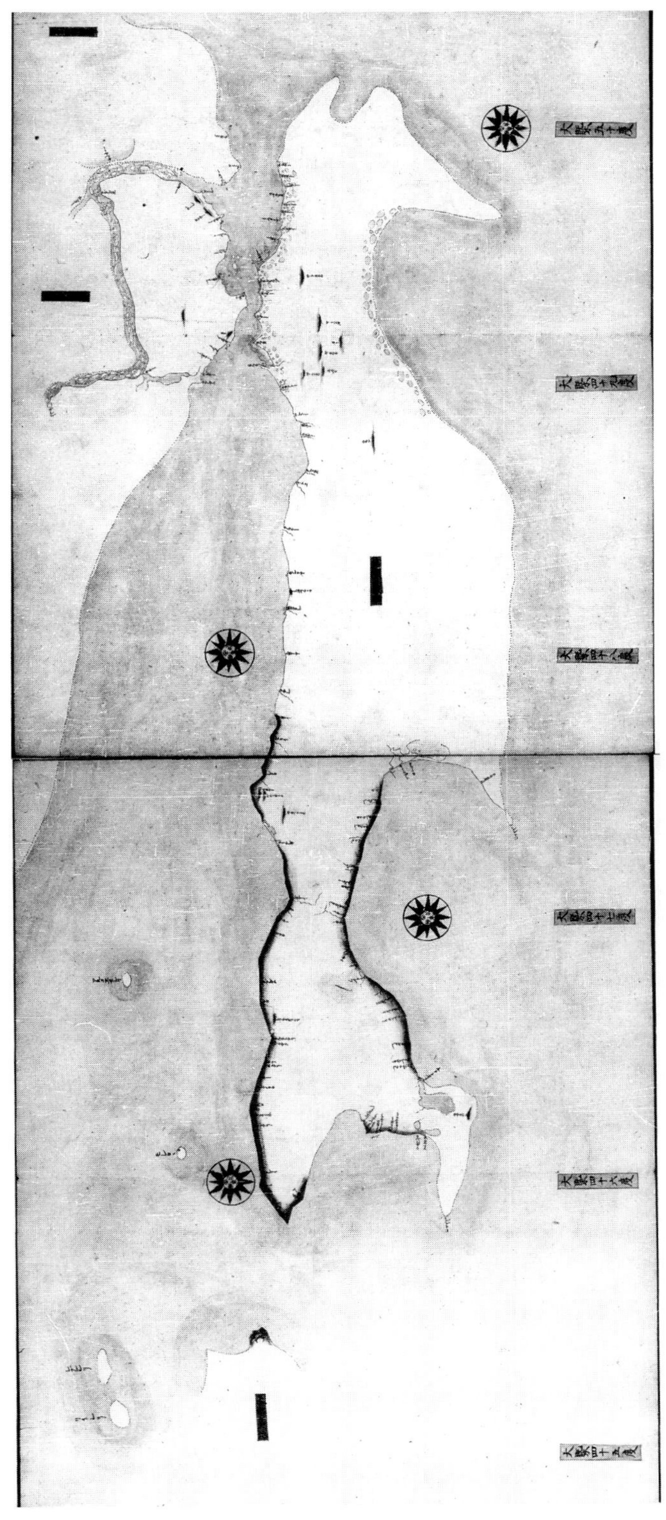

FIG. 11.74. "KITA EZOCHI" IN THE FIRST VOLUME OF THE *HOKUI BUNKAI YOWA* (MISCELLANEOUS RECORDS OF THE NORTHERN EZO REGION, 1811) BY MAMIYA RINZŌ. Mamiya's reports, comprising ten volumes in all, were submitted to the shogunate to present the findings of his 1808–9 expedition to Sakhalin. A prominent weakness of the manuscript map is its portrait of northeastern Sakhalin, where Japanese had yet to explore.
Size of the original: 72.8 × 29.6 cm. By permission of the National Archives, Tokyo.

1798, the year of his participation in the shogunal expedition. The information for his map was derived from various previous expeditions, however. In the case of Sakhalin, for example, the information came from a shogunal expedition in 1801, which reached Shōya at approximately 49°20′ north on the west coast and Naibutsu at approximately 47°20′ north on the east coast. Since proof was still lacking on whether Sakhalin was connected to continental Asia, the expedition compiled a map of Sakhalin and attached to it another map showing a possible link to the mainland, thereby juxtaposing the two theories.[367] Kondō became particularly interested in this question and studied not only the Japanese evidence but also Chinese and European maps of northeastern Asia. In 1804 his study was completed as a book titled *Hen'yō bunkai zukō* (Cartographical study of the important frontiers of Japan), and in it he concluded that "Karafuto" was a peninsula separated by a river and was therefore different from the land Europeans called Sakhalin.[368]

The task of completing the survey of the coastline of Sakhalin fell to Matsuda Denjūrō (1769–1843) and Mamiya, who were dispatched by the shogunate in 1808 on the grounds of national defense.[369] The two acted independently. Matsuda traveled along the west coast and Mamiya went from the eastern part of Aniva Bay along the east coast. Mamiya reached Cape Terpeniya (Kitashiretoko) but proceeded no farther on the advice of his local guides, who argued that navigation farther to the north would be dangerous. Rather, he turned back and crossed the island to catch up with Matsuda, who was by then at Cape Rakka at approximately 52° north on the west coast and was convinced that Sakhalin was an island. The coast to the north of the cape was difficult to reach by ship or on foot, so together they decided to return to Hokkaidō along the west coast. Mamiya regretted not having been able to continue to the north of Sakhalin, so he sailed there again later in the year. In 1809 he reached Naniō, at approximately 53°15′ north near the northern end of the strait separating Sakhalin from the mainland, crossed the strait in a local ship, and extended his journey to Delen in the lower reaches of the Amur.[370] The result was that he was convinced that Karafuto was indeed the Sakhalin depicted on European maps. A manuscript map that presented the results of his expedition to the shogunate, the *Kitaezotō chizu* (Map of northern Ezo Island) of 1810, depicted the west coast of Sakhalin and the lower reaches of the Amur in detail on a scale of three *sun* and six *bu* to one *ri* (1:36,000). The whole of Sakhalin was included on the manuscript "Kita Ezochi [zu]" ([Map of] northern Ezo) (fig. 11.74) in the itemized reports of the expedition, which were completed in 1811.[371] The unsurveyed northeastern coastline was shown as a dotted line, and the eastward bend at the north is an example of its errors.

Related in part to the geographical questions of the northern frontier was the surveying work of Inō Tadataka. In 1800, under the guidance of Takahashi Yoshitoki (or Sakuzaemon the elder, 1764–1804), an official astronomer whose calendar reform was adopted in 1797, he surveyed the land route from Edo to the coast of southeastern Hokkaidō.[372] This was designed to determine the length of a degree of latitude, in conjunction with improving the accuracy and the amount of astronomical data for further calendric study. Inō calculated the length

367. The explorers were Nakamura Koichirō (to Naibutsu) and Takahashi Jidayū (to Shōya). Their map is preserved at the Tokyo University Library (Nanki Library Collection) and at the Hokkaidō University Library in Sapporo; the latter, a manuscript (107.5 × 38.8 cm), is reproduced in Hoppō Ryōdo Mondai Chōsakai, *Hoppō Ryōdo*, pl. 30, and Narita, *Ezo chizu shō*, pl. 78 (both in note 350).

368. The *Hen'yō bunkai zukō* is preserved at the National Archives in Tokyo, the Daitōkyū Kinen Library in Tokyo, and elsewhere. The works of Kondō Morishige (common name, Jūzō; pseudonym, Seisai) have been published under the title *Kondō Seisai zenshū* (note 210); the *Hen'yō bunkai zukō* is in vol. 1. Karafuto and Sakhalin, of course, have since been proved to be the same island.

369. For a report on the exploration of Sakhalin see Matsuda Denjūrō, *Hokui dan* (Story of northern Ezo, ca. 1823), chap. 3, which is reproduced in the *Hokumon sōsho*, vol. 5 (note 335), and in the *Nihon shomin seikatsu shiryō shūsei* (Collected historical records about the lives of the Japanese people), 20 vols. (Tokyo: San'ichi Shobō, 1968–72), vol. 4. See also Mamiya Rinzō, *Kitaezo zusetsu* (Illustrated exposition on Kitaezo [Sakhalin], 1811) (Edo: Harimaya Katsugorō, 1855); it is reproduced in vol. 5 of the *Hokumon sōsho*.

370. See Mamiya Rinzō, *Tōdatsu chihō kikō* (Voyage to eastern Tartary, 1811); reproduced in the *Nihon shomin seikatsu shiryō shūsei*, vol. 4 (note 369), and also *Tōdatsu chihō kikō*, ed. Hora Tomio and Tanisawa Shōichi, Tōyō Bunko (Eastern library series), no. 484 (Tokyo: Heibonsha, 1988), 115–65. Delen was a Chinese settlement.

371. The maps, reports, and records of Mamiya's expedition are preserved at the National Archives in Tokyo. The reports, entitled *Hokui bunkai yowa* (Miscellaneous records of the northern Ezo region), were dictated by Mamiya, although his name is not mentioned; a reproduction of this book is in Hora and Tanisawa, *Tōdatsu chihō kikō*, 3–113 (note 370). The *Kitaezotō chizu* is a manuscript on seven sheets, each sheet being 306.5 by 121 centimeters; a sheet showing the Mamiya (Tatar) Strait is reproduced in Unno, Oda, and Muroga, *Nihon kochizu taisei*, vol. 1, pl. 67 (note 8).

372. Yoshitoki was the father of Kageyasu; both were also known as Sakuzaemon: see Papinot, *Dictionary of Japan*, 629 (note 43). The revision of the Japanese calendar (or rather, an ephemeris that included information useful for constructing the official calendar) was conducted jointly by Takahashi the elder and Hazama Shigetomi (1756–1816), students of the independent astronomer Asada Gōryū (1734–99) at Ōsaka. It was known as the Kansei Calendar Reform after the contemporary era (1789–1800) and was "significant because in it the Japanese for the first time successfully adopted Western measurements in an official reform." See Nakayama, *History of Japanese Astronomy*, 194–95 (note 38). On Inō, who became a surveyor and cartographer late in life, see Ryōkichi Ōtani, *Tadataka Inō, the Japanese Land-Surveyor*, trans. Kazue Sugimura (Tokyo: Iwanami Shoten, 1932); Cortazzi, *Isles of Gold*, 35–37 (note 14); E. B. Knobel, "Inō Chūkei and the First Survey of Japan," *Geographical Journal* 42 (1913): 246–50; and Norman Pye and W. G. Beasley, "An Undescribed Manuscript Copy of Inō Chūkei's Map of Japan," *Geographical Journal* 117 (1951): 178–87.

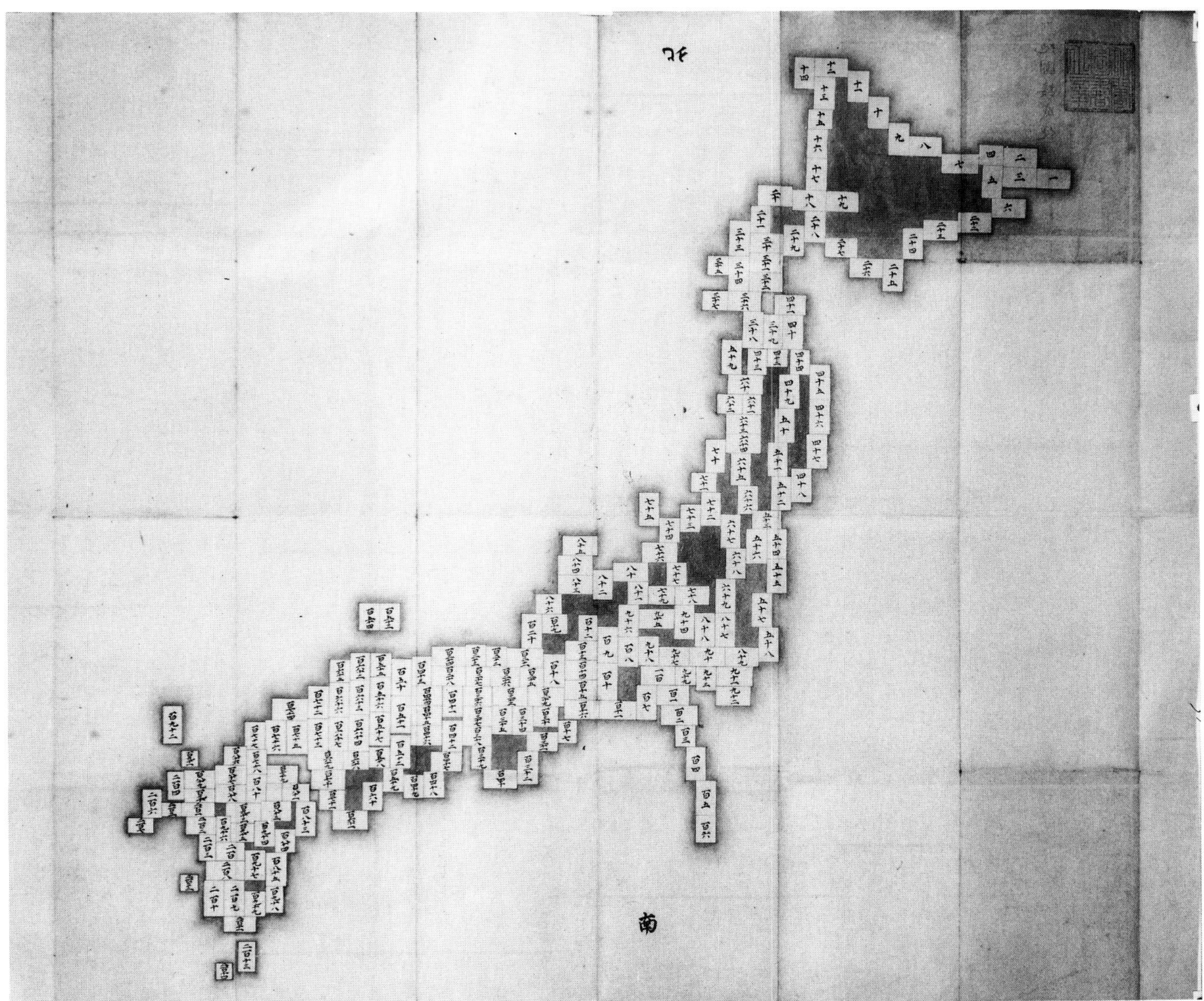

FIG. 11.75. *CHIZU SESSEI BENRAN* (INDEX TO THE DISTRIBUTION OF MAP SHEETS), 1821. This was an appendix to *Yochi jissoku roku* (Collection of land-survey data), the notes from the survey of Japan undertaken by Inō Tadataka and completed by Takahashi Kageyasu between 1800 and 1821. The index numbers and shows the correct distribution of the 214 manuscript maps at 1:36,000 that were derived from the survey. Size of the original: 107 × 121 cm. By permission of the National Archives, Tokyo.

of a degree of latitude to be 28.2 *ri* (110.85 km), departing about 130 meters from the mean for the modern value, between 35 and 41 degrees. This information was useful for predicting solar and lunar eclipses.

In the following year Inō started his survey of the coastlines of Japan, beginning on the northeast of Honshū, and continued until 1815 when the archipelago was completed. Maps were compiled from the results of these surveys, beginning in 1804 with one of the northeastern coast of Honshū that was presented to the shogunate. This led to Inō's appointment as a low-level official and to an order for the observatory to support the surveying project. Other maps were submitted periodically as the survey proceeded, including a plan of Edo, which he surveyed in 1816. The task of completing the maps of the entire coastline was finished by the observatory in 1821 under the direction of Takahashi Kageyasu.

As a group, these maps were titled *Dainihon enkai yochi zenzu* (Maps of the coastlines of Great Japan) and consist of 225 sheet maps on three different scales. The largest scale, three *sun* and six *bu* to one *ri* (1:36,000), includes 214 maps (fig. 11.75); there are eight maps on a medium scale of six *bu* to one *ri* (1:216,000); and there are three maps at the smallest scale of three *bu* to one

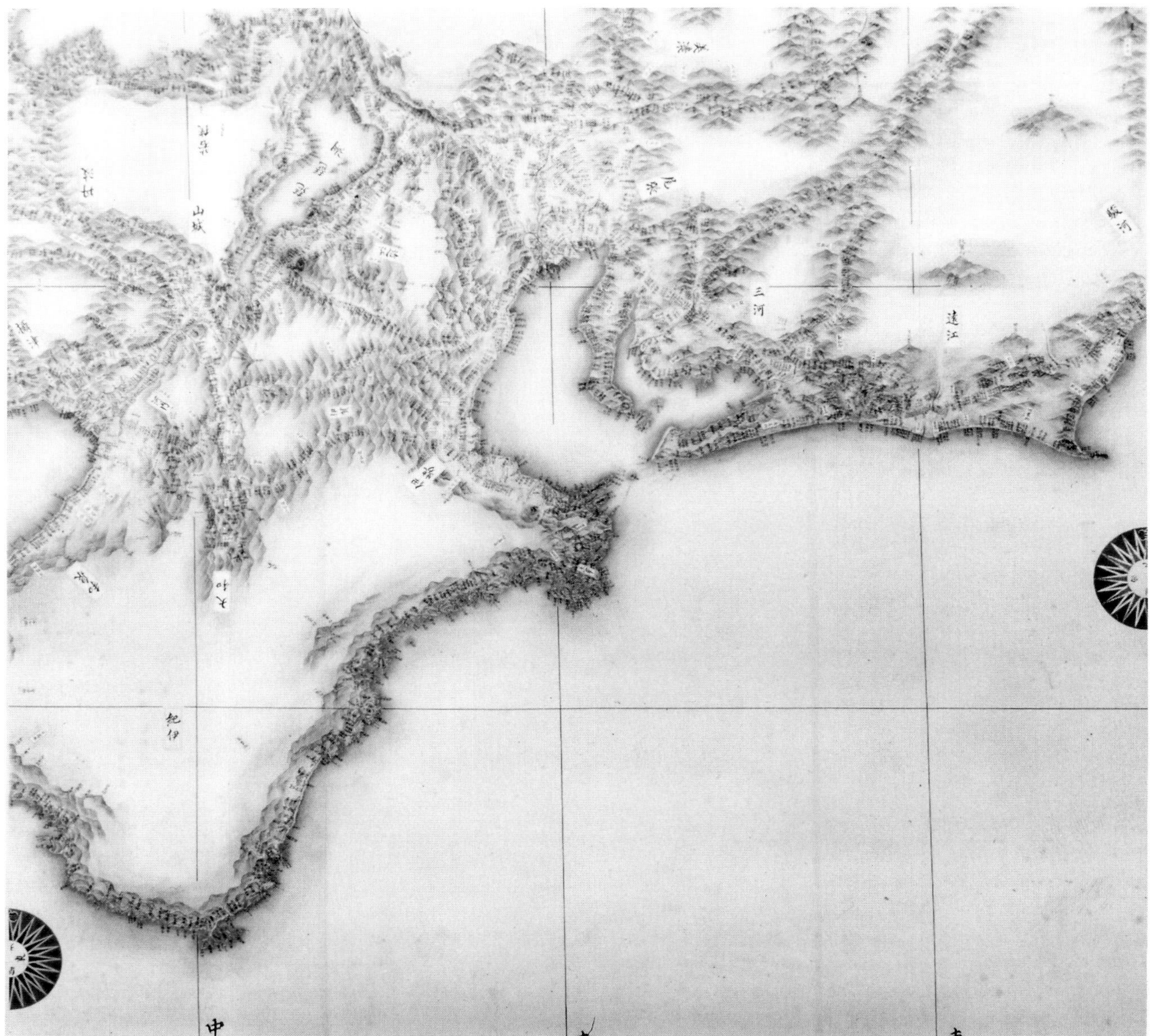

FIG. 11.76. A SHEET FROM THE *DAINIHON ENKAI YOCHI ZENZU* DATING FROM 1821. This is one of the eight sheets to show Japan at a medium scale of 1:216,000 (six *bu* to one *ri*). Although some of the maps were prepared before Inō's death in 1818, the task of completing the series was undertaken by the shogunal observatory until 1821. Since the emphasis of the map is on the coastlines, the interior is left blank except for the places where the surveyors went. The manuscript sheet shown here is centered on Nagoya.
Size of the original: 241 × 131.8 cm. By permission of the Tokyo National Museum.

ri (1:432,000).[373] The projection is trapezoidal, with the prime meridian running through Kyōto.[374] In addition to the maps, the notes from the survey were also submitted to the government.[375] Only the maps at 1:432,000 were printed about 1867, after the interior was filled in and Sakhalin added with the assistance of the Kaiseijo (Institute for Western Studies). These printed versions are entitled *Kanpan jissoku Nihon chizu* (Maps of Japan from surveys published by the government).[376]

Two manuscript sets of the 225 maps were made, one for the shogunate and the other for the Inō family. The one in the possession of the government was destroyed in the 1873 fire at Edo castle (the imperial palace). Inō's heirs were then forced to submit their set to the Meiji government, which handed it over to Tokyo University Library; unfortunately, this set was also lost in the Great Kantō Earthquake of 1923. Although the originals no longer exist, there are some copies that were made as extras from the small- and medium-scale maps while the work was being completed (fig. 11.76), as well as later copies. None of these have titles, and they are therefore known simply as "Inō's maps."

These maps of the coastlines, with only surveying courses in the interior, were highly accurate. This was recognized in 1861 by the British navy, which began to survey the coasts but settled for a copy of the three sheets at 1:432,000 and started to supplement them. The maps were presented to the British minister plenipotentiary by the Japanese government and are now at the National Maritime Museum in Greenwich.[377]

Conclusion

The functions of traditional Japanese maps described in this chapter are similar to those of China and Korea, with some significant exceptions arising from Japan's special relationships with the outside world at various times in its history and its strong Buddhist tradition. In addition to the more obvious uses of maps for political and fiscal administration in a strongly hierarchical society and for wayfinding on land and sea, maps were also used to a remarkable degree for rhetorical and symbolic purposes.

Several literary references to maps (*kata* and *katachi*) from the seventh century point to their value in recording boundaries and summarizing provincial information for the central government. In the eighth and ninth centuries, following the Taika or Great Reform of A.D. 646 (the main purpose of which was to centralize government and introduce a new system of land tenure, local government, and taxation), maps of Buddhist temple lands reclaimed for paddy fields were produced. No other civilization has preserved as many original eighth-century map documents. Similarly, maps were used to establish acreage for the annual tax exemption of manors in the *shōen* system of private landownership by religious institutions and the aristocracy covering more than half of the country by the eleventh century. Surveying for siting drainage and irrigation canals, roads, temples, and capital cities is also recorded in this early period.

It is in the sixteenth century that coordinated efforts were made to survey and map the provinces, beginning in 1591 with the government of Toyotomi Hideyoshi. During the period of Tokugawa administrations that followed, although no systematic effort was made to establish cartographic organizations, official cartographers were appointed ad hoc for various projects. Large-scale maps of 1:21,600 that covered the country from northernmost Honshū to the Ryūkyūs were completed in the middle of the seventeenth century. Despite underestimated acreages—caused by rounding down but never up and by systematic errors in surveying instruments—a national map of Japan and the islands to the north thus emerged under the Togukawa shoguns. There is evidence that such official cartography was jealously secretive, as is indicated by von Siebold's experience in the nineteenth century described above.

373. The medium-scale maps are preserved at the Tokyo National Museum, and two of the three small-scale ones are at the Kōbe City Museum. A complete set of the large-scale maps, however, is not known to exist. Inō's survey and the maps are discussed in detail in Ōtani, *Tadataka Inō* (note 372); and Hoyanagi Mutsumi, ed., *Inō Tadataka no kagakuteki gyōseki: Nihon chizu sakusei no kindaika eno michi* (A new appreciation of the scientific achievement of Inō Tadataka) (Tokyo: Kokon Shoin, 1974, rev. ed. 1980). For reproductions of Inō's maps, see Nanba, Muroga, and Unno, *Nihon no kochizu/Old Maps in Japan*, pl. 31 (one of the small-scale maps in Kōbe) and pl. 30 (Shōdo Island from the large-scale maps) (note 11); Unno, Oda, and Muroga, *Nihon kochizu taisei*, vol. 1, pl. 35 (two of the medium-scale maps), pl. 36 (the road from Takasaki to Mikuni from the large-scale maps), pl. 37 (map of Ōshima, Izu Islands), and pl. 39 (Plan of Edo) (note 8); and Cortazzi, *Isles of Gold*, pl. 45 (one of the small-scale maps in Kōbe; 203.5 × 162.1 cm) (note 14). The Tokyo National Museum maps are reproduced in facsimile (Tokyo: Buyōdō, 1993).

374. The prime meridian ran through Nishisanjōdai (now Nishigekkōchō, Nakagyō ku) in Kyōto, where the calendric office of the shogunate was located. See Watanabe, *Kinsei Nihon tenmongaku shi*, 2:469–74 (note 315). The projection was trapezoidal, not the sinusoidal Sanson-Flamsteed as incorrectly judged by Ōtani Ryōkichi and repeated by Cortazzi (*Isles of Gold*, 36–37 [note 14]); see Hoyanagi, *Inō Tadataka no kagakuteki gyōseki*, 22–24 (note 373).

375. The notes, *Yochi jissoku roku* (Collection of land-survey data), in fourteen books, are accompanied by the manuscript *Chizu sessei benran* (Index to the distribution of map sheets) of 1821 (one sheet, 107 × 121 cm) (fig. 11.75); they are preserved at the National Archives in Tokyo. The *Chizu sessei benran* shows the disposition of the 214 numbered maps and is reproduced in Unno, Oda, and Muroga, *Nihon kochizu taisei*, vol. 1, fig. 52 (note 8).

376. The *Kanpan jissoku Nihon chizu* was a woodcut on four sheets published by the Kaiseijo in Edo; a revision was issued in 1870 by the Daigaku Nankō (University Southern School, later Tokyo University) with the *Dainihon enkai jissoku roku* (Survey notes of the coastlines of Great Japan), in fourteen books.

377. Pye and Beasley, "Copy of Inō Chūkei's Map" (note 372).

During the Edo period—Japan's long history of undisturbed independence and isolation from the rest of the world—it should not be thought that Japan was immune to outside influence. As in the case of China, there was no wholesale transmission of European scientific mapping until the nineteenth and twentieth centuries. Japan did not become Christian or Europeanized despite its contacts with the Spanish and Portuguese (the "southern barbarians") in the sixteenth century. Nevertheless, there are strong influences in surveying and mapping practice from the seventeenth century. The Portuguese may be said to have introduced surveying and navigation instruments into Japan, and this influence survived despite a ban on contacts with the Portuguese in the 1630s. The Dutch, who had special dispensation from the exclusion edicts (although through very limited contact, carefully segregated from mainland Japan), influenced a trend to more empirical studies (*Rangaku*) and a separation of the once indissoluble heavenly and earthly concerns of traditional Japanese society. From the mid-eighteenth century, the influx of Dutch books, globes, and maps, the introduction of the heliocentric Copernican system, and the translation of Dutch atlases of the world and treatises on globe making into Japanese enlarged the Japanese scope of the world through Dutch eyes. Its importance can hardly be overestimated both on the widening of the Japanese horizon and also for compiling more technically accurate maps of the country and its surrounding seas.

Furthermore, an important aspect of the Edo period—domestic and international peace—might be said to have heightened popular interest in culture and travel, which in turn helped to stimulate demand for maps. The rhetorical and ornamental character of maps thus becomes important, and Western maps were frequently used for this purpose, as in the largely ornamental *Nanban* world maps on Western projections that served as large folding screens. Similarly, the Jōtoku-type maps, also on folding screens, were largely decorative, portraying a conventional national image. These were related to the Gyōki-type maps, which were largely symbolic in character and may have originally been associated with the annual purification ritual of Tsuina.

The Jesuit Matteo Ricci's role in providing models for world maps was somewhat different in Japan than in China, where only some intellectuals took an interest in his maps. His world map was printed in several versions and became the basis of a printed map trade for maps designed to be hung in houses or published in books and encyclopedias when cartography was popularized during the Edo period. Our knowledge of the sources for these maps and the process of their transmission is incomplete and needs further study. The popularization of maps is also reflected in their frequent display on personal items for daily use, such as sword guards, fans, mirrors, *inrō* (portable medicine pouches), netsuke, combs, and plates. Maps of fictional places were also popular in the late Edo period.

When the Dutch monopolized European trade with Japan, there was apparently no particular need for improved sea charts. The Portuguese information on surveying and navigation gleaned earlier was evidently sufficient, and Japanese were in any case prohibited from traveling abroad. With some interesting exceptions, therefore, charts were not updated for use at sea but fulfilled an honorary function, often being presented as graduation certificates for surveyors. The eclectic sources and widely varying characteristics of these charts thus make it impossible to conceive of a "Japanese tradition" of charting in the European sense.

A quite different tradition, that of the Buddhist world map (map of the Five Indias), was apparently current in the seventh century (although the earliest surviving map is from the fourteenth), and survived until the mid-nineteenth century. The maps, rich in Buddhist cosmology, were objects of worship, and several survive both in large painted versions and as small woodcut book illustrations. Also associated with Buddhist beliefs are a series of models of the Buddhist universe and terrestrial globes replete with Buddhist cosmological worldviews. Indeed, although celestial globes were common in China, there appears to have been far more interest in terrestrial globes in Japan than in China, and this issue needs to be explored further. There are many reports of terrestrial globes being brought by the Dutch as gifts in the seventeenth century, and records survive of Japanese globe making and of the repair of existing Dutch examples.

This chapter, which provides the most detailed account of traditional Japanese cartography in English, should serve as a springboard for further analytical studies of the maps and concepts presented. Generalizations need to be drawn that will enhance our understanding of Japanese cartography in relation to the main historical forces of the country. More comparative studies between the cartographies of Japan, China, and Korea need to be undertaken, particularly concerning the transmission of Buddhist ideas of cosmographical mapping. Detailed studies by historical geographers who can use maps as evidence for reconstructing past geographies of Japan at the local level need to be pursued. More studies are needed that trace the European sources of maps in the *Nanban* and other genres. In short, we are now at a stage where connections can be drawn and sound generalizations may emerge.

APPENDIX 11.1 EXTANT ANCIENT MAPS OF PADDY FIELDS

Content of Map					
Place	Province	Paddy-Field Owner	Year of Completion	Size (cm) (h × w)	Material
1 Yamada County	Sanuki	Gufuku Temple (in Yamato Province [Nara Prefecture])	736	28 × 127.5	Paper
2 Minuma, Inugami County 3 Heru, Inugami County	Ōmi	Tōdai Temple	751	68 × 252	Hemp
4 Minase, Shimanokami County	Settsu	Tōdai Temple	757	28.5 × 69	Paper
5 Daizudokoro, Nakata County	Awa	Tōdai Temple	758	28.6 × 51.9	Paper
6 Niijima, Nakata County	Awa	Tōdai Temple	758	57 × 103	Paper
7 Ikarugi, Tonami County	Etchū	Tōdai Temple	759	82 × 99	Hemp
8 Suka, Imizu County	Etchū	Tōdai Temple	759	81 × 108	Hemp
9 Kubota, Imizu County	Etchū	Tōdai Temple	759	79 × 126	Hemp
10 Ōyabu, Niikawa County	Etchū	Tōdai Temple	759	79 × 141	Hemp
11 Hasetsukabe, Niikawa County	Etchū	Tōdai Temple	759	79 × 115	Hemp
12 Kusooki, Asuha County	Echizen	Tōdai Temple	759	78 × 109	Hemp
13 Isawa, Tonami County	Etchū	Tōdai Temple	759	56.3 × 110	Paper
14 Naruto, Imizu County	Etchū	Tōdai Temple	759	77 × 141	Hemp
15 Nukada Temple and surroundings	Yamato	Nukada Temple	ca. 760	113.7 × 72.5	Hemp
16 Kusooki, Asuha County	Echizen	Tōdai Temple	766	69 × 113	Hemp
17 Chimori, Asuha County	Echizen	Tōdai Temple	766	144 × 194	Hemp
18 Takakushi, Sakanoi County	Echizen	Tōdai Temple	766	55.8 × 114	Paper
19 Iyama, Tonami County 20 Ikarugi, Tonami County 21 Kinahiru, Tonami County 22 Suka, Imizu County 23 Naruto, Imizu County 24 Shikata, Imizu County 25 Ōyabu, Niikawa County	Etchū	Tōdai Temple	767	68 × 623	Hemp
26 Naruto, Imizu County	Etchū	Tōdai Temple	ca. 767	62.6 × 57.7	Paper
27 Shikata, Imizu County	Etchū	Tōdai Temple	ca. 767	37 × 57.9	Paper

(IN CHRONOLOGICAL ORDER AND DATING TO THE NARA PERIOD, 710–84)

Orientation	Provincial Seal	Signature of Surveyor	Owner of Map	Remarks
South	No	No	Tawa Library, Shido, Kagawa Prefecture	Original date: 15, 12th month, 7th year of Tenpyō; possibly copied at end of eleventh century
North	Yes	Not known (damaged)	Shōsōin	Two maps, nos. 2 and 3, are connected and make scroll
North	Yes	No	Shōsōin	Original date: 16, 12th month, 8th year of Tenpyōshōhō
West	No	No	Shōsōin	Manuscript for composing a formal map
West (?)	No	No	Shōsōin	As above; actually a map of Hirakata in same county
East	Yes	Yes	Shōsōin	
East	Yes	Yes	Shōsōin	
South	Yes	Yes	Shōsōin	
East	Yes	Yes	Shōsōin	
South	Yes	Yes	Shōsōin	
North	Yes	Yes	Shōsōin	
East	Yes	Not known (damaged)	Nara National Museum, Nara	This might have been owned previously by Shōsōin
East	Yes	Yes	Fukui Seikō, Kyōto	This might have been owned previously by Shōsōin
North	Yes	No	National Museum of Japanese History, Sakura	Some parts are missing
North	No	Yes	Shōsōin	See figure 11.8
North	No	Yes	Shōsōin	Although part with date is missing, 766 is almost certain
North	No	Yes	Nara National Museum, Nara	Possibly owned previously by Shōsōin
South	Yes	No	Shōsōin	Seven maps, nos. 19–25, are connected to form a scroll
South	No	No	Nara National Museum, Nara	Some parts are missing; copy of no. 23
South	No	No	Nara National Museum, Nara	Some parts are missing; copy of no. 24

APPENDIX 11.2 EXTANT EARLY MANUSCRIPTS OF THE GYŌKI-TYPE MAP OF JAPAN, INCLUDING THE SEMI-GYŌKI TYPE (IN CHRONOLOGICAL ORDER)

Owner(s) of Map	Title	Date	Size (cm) (h × w)	Orientation	Attributed to Gyōki	Remarks
1 Shōmyō Temple, Yokohama	Not known	Second half of 13th century	34.1 × 52.2	South	Not known	Eastern half is missing; no entries for routes; kept at Kanazawa Bunko, Yokohama
2 Ninna Temple, Kyōto	None	1306	34.5 × 121.5	South	Yes	Part with Kyūshū is damaged; see figure 11.14
3 Sonkeikaku Library, Tokyo	None	ca. 1324–28	22.7 × 30.6	North	No	Included in *Nichūreki* (Two guides, late 12th century); routes only; inscription says copied from *Kaichūreki* (Pocket guide) 1128; see fig. 11.16
4 Sonkeikaku Library, Tokyo	None	ca. 1324–28	22.7 × 30.6	North	No	Included in *Nichūreki*; entries about number of days required for transporting tribute from each province to government; no inscription as in 3
5 Tenri Central Library, Tenri, Nara	"Dainihonkoku zu" (Map of Great Japan)	1548	26.3 × 41.3	East	Yes	Included in *Shūgaishō* (Collection of oddments) of 1548; see fig. 11.15
6 Tōshōdai Temple, Nara	*Nansenbushū Dainihonkoku shōtō zu* (Orthodox map of Great Japan in Jambūdvīpa)	ca. 1550	168 × 85.4	West	Yes	
7 Zhongshan University, Guangzhou (Canton); and others	*Riben xingji tu* (Gyōki's map of Japan)	ca. 1564	Not certain	South	Yes	Included in *Riben yi jian* (Outline of Japan) by Zheng Shungong, reproduced 1939
8 Archivio di Stato, Florence	*Iapam*	ca. 1585	28 × 60	South	No	All entries in Latin
9 Sonkeikaku Library, Tokyo	"Dainihonkoku zu" (Map of Great Japan)	1589	26 × 36.5	East	Yes	Included in *Shūgaishō* of 1589
10 Mutō Kinta, Kamakura	None	ca. 1595	Maximum length, 51	North	No	Map of eastern Asia drawn on fan owned by Toyotomi Hideyoshi
11 Kitano Shrine, Kyōto	None	ca. 1600	98 in diameter	North	No	Relief on back of bronze mirror by Kise Jōami; no entries of routes

APPENDIX 11.2 (*continued*)

Owner(s) of Map	Title	Date	Size (cm) (h × w)	Orientation	Attributed to Gyōki	Remarks
12 Tokyo National Museum, Tokyo	*Nansenbushū Dainihonkoku shōtō zu* (Orthodox map of Great Japan in Jambūdvīpa)	ca. 1625	156 × 315 (map only, 57.5 × 108)	North	Yes	Folding screen paired with map of world
13 Okazawa Sagenta, Nishiwaki	*Nansenbushū Dainihonkoku shōtō zu* (Orthodox map of Great Japan in Jambūdvīpa)	ca. 1640	103 × 273	South	Yes	Folding screen
14 Hosshin Temple, Obama	None	Early 17th century	154 × 352	South	No	Folding screen paired with map of world
15 Ishikawa Prefectural Gallery, Kanazawa	None	Early 17th century	155 × 364	North	No	Folding screen paired with two plans of Kyōto and area of government offices; no entries of routes
16 Formerly N. H. N. Mody, Kōbe	None	Mid-17th century	204 × 447	North	No	Folding screen paired with map of world
17 Fukushima Kitarō, Obama	*Nansenbushū Dainihonkoku shōtō zu* (Orthodox map of Great Japan in Jambūdvīpa)	Mid-17th century	96.5 × 249	South	Yes	Folding screen paired with map of world
18 National Museum of Japanese History, Sakura; and others	"Yochi zu" (Land map)	Second half of 18th century	27.5 × 85	West	No	Included in *Shūko zu* (Illustrations of collected antiques) by Fujii Sadamiki; refers to an 805 original, now lost

Appendix 11.3 List of Manuscript Maps of the Five Indias (in Chronological Order)

Owner	Title	Date	Size (cm) (h × w)	Remarks
1 Hōryū Temple, Nara	*Gotenjiku zu* (Map of the Five Indias)	1364	177 × 166.5	Drawn by priest Jūkai; see figure 11.18
2 Muroga Emiko, Kyōto (former Ayusawa Collection)	*Tenjiku ezu* (Map of India)	16th century	119.4 × 128	Upper part is missing; title may not be original one
3 Kushuon'in Temple, Hirakata	*Gotenjikukoku no zu* (Map of the Five Indias)	ca. 1692	168 × 172	Copied by priest Sōkaku
4 Hōryū Temple, Nara	*Gotenjiku zu* (Map of the Five Indias)	17th century?	167 × 175	Copy of 1364 map; drawn by priest Zenjō
5 Hōshōin Temple, Tokyo (no longer extant)	*Saiiki zu* (Map of the western regions)	1736	Not known	Lost in fire during Second World War; recorded in *Saiiki zu sofuku nikō roku* (Two revisions on a map of the western regions), which is reproduced in second volume of *Yūhōden sōsho* (Series of travels, 1915) in *Dainihon Bukkyō zensho* (Collected records on the Buddhism of Great Japan); map reproduced as frontispiece
6 Kōbe City Museum, Kōbe (formerly Akioka Takejirō, Tokyo)	*Tenjiku no zu* (Map of India)	1749	167.5 × 134.8	Copy from a copy of lost version at Shōrin Temple, Kyōto
7 Chion'in Temple, Kyōto	*Tenjiku zu* (Map of India)	1755	156.5 × 130	Copy of lost version at Shōrin Temple
8 Jōgon'in Temple, Azuchi	*Gotenjiku zu* (Map of the Five Indias)	18th century?	159.2 × 133.8	
9 Ishihara Akira, Tokyo (deceased)	*Gotenjiku zu* (Map of the Five Indias)	18th century?	Not known	Not yet researched
10 Kongō Zanmaiin Temple, Kōya	*Gotenjiku zue* (Map of the Five Indias)	1816	152 × 130.7	
11 Ryūkoku University Library, Kyōto	*Gotenjiku no zu* (Map of the Five Indias)	ca. 1865	173 × 128.7	
12 National Archives, Tokyo	*Tō Genjō Sanzō Gotenjiku zu* (Xuanzhuang, Map of the Five Indias)	19th century	164 × 133	Tō Genjō Sanzō in title means Xuanzhuang, a priest who lived during time of Tang dynasty

APPENDIX 11.4
CLASSIFICATION OF THE *NANBAN*-STYLE WORLD MAPS (MAPS SHOWING ONLY THE OLD WORLD [OR THE EASTERN HEMISPHERE] ARE INCLUDED)

MARINE CHART

Owner	Companion	Format
Hosshin Temple, Obama	Gyōki-type map of Japan	One of a pair of six-fold screens; 154 × 352 cm; see fig. 11.21
Ikenaga Hajime, Kōbe (formerly)	Scene of arrival of "southern barbarians"	One of a pair of six-fold screens; 158 × 347 cm

Characteristics:
1. Pacific Ocean in center of map
2. Graduation of latitude on both left and right sides
3. West coast of South America with protrusions near Tropic of Capricorn
4. No government areas marked (Hosshin Temple version has evidence that many slips of paper with place-names were pasted on it)
5. Bar scale at bottom in middle

OVAL PROJECTION

Owner	Companion	Format
Yamamoto Hisashi, Sakai	None	One of a pair of six-fold screens; 135.5 × 269.5 cm
Kobayashi Ataru, Tokyo	Jōtoku-type map of Japan	One of a pair of six-fold screens; 158 × 368 cm
Jōtoku Temple, Fukui	Jōtoku-type map of Japan	One of a pair of six-fold screens; 148.5 × 364 cm; see figure 11.23
Kawamura Heiemon, Obama	Jōtoku-type map of Japan	One of a pair of eight-fold screens; 117 × 375 cm

Characteristics:
1. Atlantic Ocean in center of map
2. Courses from Portugal and Spain to East Asia noted
3. West coast of South America runs straight southeast from equator

EQUIRECTANGULAR PROJECTION (TYPE A)

Owner	Companion	Format
Formerly N. H. N. Mody, Kōbe	Gyōki-type map of Japan	One of a pair of six-fold screens; 204 × 447 cm
Jingū Library, Ise	None	Folding map; 85.3 × 156.8 cm

Characteristics:
1. Atlantic Ocean in center of map
2. Supplementary maps of Northern and Southern hemispheres on Mody version
3. More place-names than other styles

EQUIRECTANGULAR PROJECTION (TYPE B1)

Owner	Companion	Format
Tokyo National Museum, Tokyo	Gyōki-type map of Japan	One of a pair of six-fold screens; 156 × 316 cm; see figure 11.22
Nanban Culture Hall, Ōsaka	Revised Jōtoku-type map of Japan	One of a pair of six-fold screens; 155 × 356.2 cm
University of California–Berkeley	Revised Jōtoku-type map of Japan	One of a pair of six-fold screens; 68 × 226.5 cm

Characteristics:
1. Pacific Ocean in center of map
2. Tierra del Fuego, Nova Guinea, and Terra Australis indicated as separate landmasses
3. With maps of Northern and Southern hemispheres and illustration of Ptolemaic theory
4. *Typus orbis terrarum* for title (except on version at California–Berkeley)

EQUIRECTANGULAR PROJECTION (TYPE B2)

Owner	Companion	Format
Fukushima Kitarō, Obama	Gyōki-type map of Japan	One of a pair of four-fold screens; 96.5 × 247 cm
Nanba Matsutarō, Nishinomiya	Revised Keichō-type map of Japan	One of a pair of six-fold screens; 97 × 273 cm

Characteristics:
1. Pacific Ocean in center of map
2. Tierra del Fuego, Nova Guinea, and Terra Australis indicated as separate landmasses
3. Does not include characteristics 3 and 4 of type B1 maps

APPENDIX 11.4 (*continued*)

EQUIRECTANGULAR PROJECTION (TYPE C)

Owner	Companion	Format
Shimonogō Kyōsai Library, Nagahama	Modified Keichō-type map of Japan	One of a pair of six-fold screens; 105 × 262 cm
Masuda Tarō, Odawara	Modified Keichō-type map of Japan	One of a pair of six-fold screens; 105 × 266 cm
Gokōin Temple, Nikkō	Modified Keichō-type map of Japan	One of a pair of six-fold screens; 86 × 239 cm; second panel missing
Idemitsu Museum of Arts, Tokyo (formerly owned by Matsumi Tatsuo)	Forty types of people from all over the world on either side	Pair of six-fold screens; each 166 × 363 cm; map of world split into two parts, each occupying two-thirds of a screen; when screens are placed together, map measures 166 × 484 cm

Characteristics:
1. Atlantic Ocean in center of map (the Idemitsu Museum version places Europe and Africa in center of map)
2. Supplementary maps of Northern and Southern hemispheres
3. Includes Novaya Zemlya, which was explored in 1596 (the Idemitsu Museum version omits parts of both top and bottom of map)

EQUIRECTANGULAR PROJECTION (TYPE D1)

Owner	Companion	Format
Kawamori Kōji, Sakai, Ōsaka Prefecture	Jōtoku-type map of Japan	One of a pair of four-fold screens; 109.5 × 273 cm (map only, 90 × 152.4 cm)
Myōkaku Temple, Okayama Prefecture	None	Six-fold screen; 97 × 272.5 cm
Usuki City Library, Ōita Prefecture	None	Folding map; 117 × 137 cm
Saga Prefectural Library, Saga	None	Folding map; 87 × 160 cm

EQUIRECTANGULAR PROJECTION (*continued*)

Owner	Companion	Format
Sōji Temple, Yokohama	None	Hanging scroll; 130 × 140 cm

Characteristics:
1. Shows only Old World or Eastern Hemisphere
2. No entry of equator, Tropics of Capricorn and Cancer, or polar circles
3. Accompanied by table about countries and regions exporting goods to Japan about 1627

EQUIRECTANGULAR PROJECTION (TYPE D2)

Owner	Companion	Format
Koga City Museum of History, Koga (formerly Takami Yasujirō, Koga)	None	Folding map; 118.5 × 117.5 cm
Yamakuni Shrine, Kyōto Prefecture	None	Folding map (?); 118.8 × 120.5 cm
Yokohama City University Library, Yokohama	None	Folding map; 116.5 × 121.5 cm
Yamaguchi University Library, Yamaguchi	None	Folding map; 114 × 120 cm

Characteristics:
1. Shows only Old World or Eastern Hemisphere
2. Has date 8th month of Kan'ei 14 (1637)
3. Enlarges accounts of table in type D1
4. Includes equator and pictures of Western sailing vessels
5. Political division different from that of type D1
6. Representation of rivers simpler than in type D1

MERCATOR PROJECTION

Owner	Companion	Format
Imperial Household Agency, Kyōto	Plans or views of twenty-eight cities	One of a pair of eight-fold screens; map 177 × 483 cm; see plate 23
Kōbe City Museum, Kōbe (Ikenaga Collection)	Views of four cities	One of a pair of eight-fold screens; map 159 × 478 cm

APPENDIX 11.4 (*continued*)

MERCATOR PROJECTION (*continued*)		
Owner	Companion	Format
Kōsetsu Museum of Art, Kōbe	Picture of Battle of Lepanto	One of a pair of six-fold screens; map 153.5 × 370 cm

Characteristics:
1. Atlantic Ocean or Europe in center of map
2. Part of Terra Australis facing South America protrudes
3. Supplementary maps of northern and southern polar regions (not on version at Kōsetsu Museum of Art)
4. Includes illustrations of people of world

APPENDIX 11.5 LIST AND GENEALOGY OF JAPANESE MARINE CHARTS OF SOUTHEAST AND EAST ASIA

Number and Owner	Title	Author or Copier	Material	Westernmost Area	Nakamura[a] (table 1)	Remarks[b]
1 Tokyo National Museum, Tokyo	[*Tōyō shokoku kōkai zu* (Chart of the eastern countries)]	Anonymous	Vellum	Madagascar	4	ca. 1615; title given later
2 Okayama Museum of Art, Okayama	None	Anonymous	Vellum	Arabian Sea	1	Latter half of 16th century
3 Koga City Museum of History, Koga (formerly Takami Yasujirō, Koga)	None	Takami Senseki (copier)	Paper	Arabian Sea	2	Copy of chart in Itoya Zuiemon's (d. 1650) belongings, 1833
4 Sueyoshi Kanshirō, Ōsaka	None	Anonymous	Vellum	Arabian Sea	5	ca. 1610; labels with Chinese characters, recorded to have been attached in 1787, indicate Tropic of Cancer, foreign place-names, and bar scale

[a]Hiroshi Nakamura, "The Japanese Portolanos of Portuguese Origin in the XVIth and XVIIth Centuries," *Imago Mundi* 18 (1964): 24–44, table 1, between 26 and 27.

[b]Numbers 1, 2, 6, 7, 8, 11, 14, and 16 are reproduced in color in Unno Kazutaka, Oda Takeo, and Muroga Nobuo, eds., *Nihon kochizu taisei* (Great collection of old Japanese maps), 2 vols. (Tokyo: Kōdansha, 1972–75), vol. 2, *Nihon kochizu taisei sekaizu hen*, plate nos. 46, 45, 47, 48, 50, 51, 52, and 49. Number 5 (*Kōmōi Kairo zu*) is reproduced in color in Okamoto Yoshitomo, *Jūroku seiki ni okeru Nihon chizu no hattatsu* (Development of the map of Japan in the sixteenth century) (Tokyo: Yagi Shoten, 1973), frontispiece 7. Number 1 is reproduced in color in Michel Mollat du Jourdin and Monique de La Roncière, *Les portulans: Cartes marines du XIIIe au XVIIe siècle* (Fribourg: Office du Livre, 1984), and the English translation, *Sea Charts of the Early Explorers: 13th to 17th Century*, trans. L. de R. Dethan (New York: Thames and Hudson, 1984), pl. 72.

APPENDIX 11.5 (*continued*)

Number and Owner	Title	Author or Copier	Material	Westernmost Area	Nakamura[a] (table 1)	Remarks[b]
5 Okamoto Michiko, Tokyo	*Kōmōi Kairo zu* (Dutch chart)	Uchiyama Hachisaburō (copier)	Paper	Sri Lanka	3	Copy dating to 1845
6 Jingū Historical Museum, Ise	None	Anonymous	Vellum	Malay Peninsula	7	ca. 1630; formerly belonged to the Kadoya family; see figure 11.24
7 Shimizu Takao, Kyōto	None	Anonymous	Paper	Malay Peninsula	—	On two wooden boards attached with hinges. Bonin Islands (then known as Tatsumi Islands) are included. Location of Bonins at 27°N was confirmed in 1675 by Shimaya, who led expedition under order of shogun to determine their location.
8 Tōhoku University Library, Sendai	*Ko karuta* (Small chart)	Anonymous	Paper	Malay Peninsula	8	
9 National Archives, Tokyo	*Kon'en tendo gattai zu* (Chart corresponding to astronomical degrees)	Mori Kōan (copier)	Paper	Malay Peninsula	10	Copy dating to 1752
10 Nagakubo Atsushi, Takahagi	*Kōmō karuta zu* (Dutch chart of sea routes)	Nagakubo Sekisui (copier)	Paper	Malay Peninsula	—	
11 Nagasaki Prefectural Library, Nagasaki	None	Ro Kōrō (copier)	Paper	Malay Peninsula	11	Copy dating to ca. 1865
12 National Archives, Tokyo	*Tensen chihō no zu* (Planisphere with astronomical lines)	Mori Kōan (copier)	Paper	Malay Peninsula	9	Copy dating to 1752; chart has several revisions, exemplified by fact that area up to 28°S has been widened so parallels at intervals of one degree fill whole area
13 Jingū Library, Ise	None	Anonymous	Paper	Malay Peninsula	—	
14 Nagasaki Prefectural Library, Nagasaki	None	Anonymous	Paper	Malay Peninsula	—	In manuscript *Batanjin emaki* (Painted scroll showing Bataan Islanders, 1680); chart 15 of Nakamura's table 1 is copy of this chart

APPENDIX 11.5 (*continued*)

Number and Owner	Title	Author or Copier	Material	Westernmost Area	Nakamura[a] (table 1)	Remarks[b]
15 (Published)	*Bankokuzu kawa shōzu* (Sketch from a map of various countries, drawn on vellum)	Anonymous	Paper	Malay Peninsula	12	In Inaba Tsūryū (Shin'emon), *Sōken kishō* (Sword ornaments, 1781)
16 Nanba Matsutarō Nishinomiya	None	Anonymous	Paper	Gulf of Guinea	6	

GENEALOGY

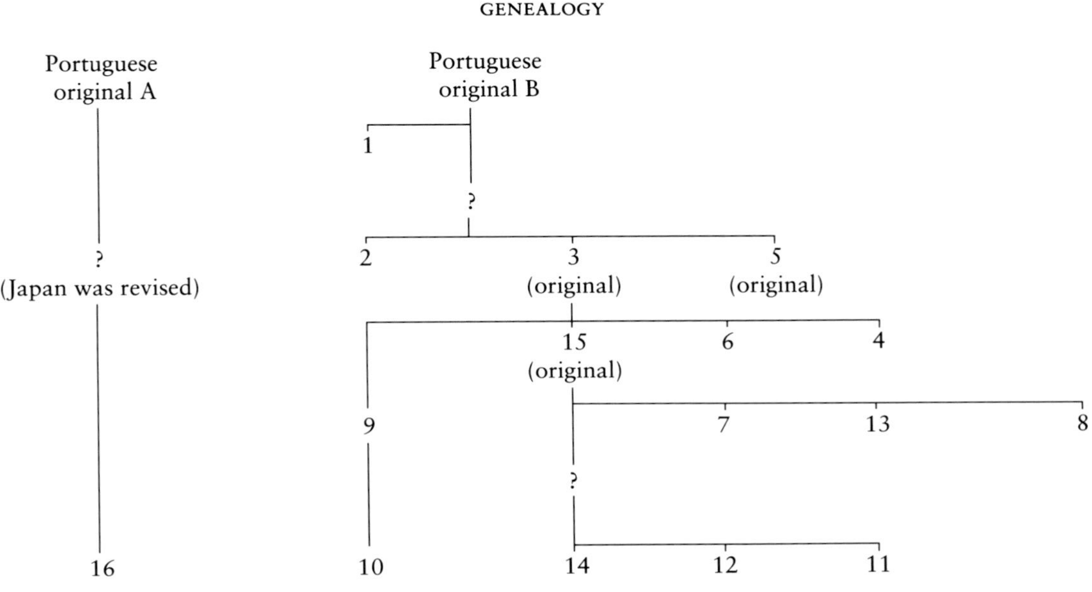

APPENDIX 11.6 LIST AND GENEALOGY OF JAPANESE MARINE CHARTS OF JAPAN

Number and Owner	Author or Copier	Number of Compass Points	Material	Nakamura[a] (table 2)	Remarks[b]
1 Mitsui Library, Tokyo	Anonymous	32	Vellum	19	Wooden stick attached to both right and left sides; see figure 11.25
2 Tokyo National Museum, Tokyo	Anonymous	32	Vellum	18	

[a]Hiroshi Nakamura, "The Japanese Portolanos of Portuguese Origin in the XVIth and XVIIth Centuries," *Imago Mundi* 18 (1964): 24–44, table 2 on p. 28.

[b]Numbers 1, 2, and 3 are reproduced in Unno Kazutaka, Oda Takeo, and Muroga Nobuo, eds., *Nihon kochizu taisei* (Great collection of old Japanese maps), 2 vols. (Tokyo: Kōdansha, 1972–75), vol. 1, pls. 18 and 17 (color), and fig. 17; number 2 is in color in Michel Mollat du Jourdin and Monique de La Roncière, *Les portulans: Cartes marines du XIIIe au XVIIe siècle* (Fribourg: Office du Livre, 1984), and the English translation, *Sea Charts of the Early Explorers: 13th to 17th Century*, trans. L. de R. Dethan (New York: Thames and Hudson, 1984), pl. 77; number 3 is in Akioka Takejirō, *Nihon kochizu shūsei* (Collection of old maps of Japan) (Tokyo: Kajima Kenkyūjo Shuppankai, 1971), pl. 16; number 5 is in Kazutaka Unno, *Chizu no shiwa* (Map creases; or, Essays on the history of cartography) (Tokyo: Yūshōdō Press, 1985), fig. 28; number 6 is in Nanba Matsutarō, Muroga Nobuo, and Unno Kazutaka, eds. and comps., *Nihon no kochizu* (Old maps in Japan) (Ōsaka: Sōgensha, 1969), English translation, *Old Maps in Japan*, trans. Patricia Murray (Ōsaka: Sōgensha, 1973), pl. 22 (color), and Akioka, *Nihon kochizu shūsei*, pl. 15.

APPENDIX 11.6 (*continued*)

Number and Owner	Author or Copier	Number of Compass Points	Material	Nakamura[a] (table 2)	Remarks[b]
3 Ōkōchi Masatoshi, Tokyo	[Matsudaira Terutsuna?]	32	Paper	17	
4 Koga City Museum of History, Koga (formerly Takami Yasujirō, Koga)	Takami Senseki (copier)	24	Paper	16	Title: *Pirōto no hō karuta* (Chart used by pilots); copy dating to 1811
5 Saga Prefectural Library, Saga	Anonymous	24	Paper	—	
6 National Museum of Japanese History, Sakura (formerly Akioka Takejirō)	Anonymous	24	Paper	20	
7 Nagakubo Atsushi, Takahagi	Nagakubo Sekisui (copier)	24	Paper	—	Two copies; one is revision of northern Honshū
8 Nagasaki City Museum, Nagasaki	Fujishima Chōzō (copier)	24	Paper	—	Copy dating to 1920; together with map of China

APPENDIX 11.7 EARLY JAPANESE TERRESTRIAL GLOBES

Owner	Author(s)/Maker(s)	Date	Diameter (cm)	MS or Printed	Material
1 Jingū Historical Museum, Ise	Shibukawa Harumi	1690	24	MS	Paper
2 National Museum of Science, Tokyo	Shibukawa Harumi	1697	33	MS	Paper
3 Kayahara Hiroshi, Tsu		17th century	ca. 3.8	MS	Wood
4 Nanban Culture Hall, Ōsaka		17th century	25.2	MS	Wood? lacquered

[a]Akioka Takejirō, *Nihon chizu sakusei shi* (A history of the making of Japanese maps) (Tokyo: Kajima Kenkyūjo Shuppankai, 1971); Akioka Takejirō, *Sekai chizu sakusei shi* (A history of making world maps) (Tokyo: Kawade Shobō Shinsha, 1988); Fujita Motoharu, *Kaitei zōho Nihon chirigaku shi* (Revised and enlarged history of Japanese geography) (Tokyo: Tōkō Shoin, 1942, reprinted Tokyo: Hara Shobō, 1984); Hirotada Kawamura, Kazutaka Unno, and Kazuhiko Miyajima, "List of Old Globes in Japan," *Der Globusfreund* 38–39 (1990–91): 173–75; Kōbe Shiritsu Hakubutsukan (Kōbe City Museum), *Kochizu ni miru sekai to Nippon* (The world and Japan as seen in old maps) (Kōbe, 1983); idem, *Akioka Kochizu Korekushon meihin ten* (A collection of masterpieces: The Akioka collection of old maps) (Kōbe, 1989); Sugano Yō, "Eisei Bunko shozō Shiba Kōkan sei chikyūgi" (Shiba Kōkan's terrestrial globe in the Eisei Library collection), *Nihon Yōgaku-*

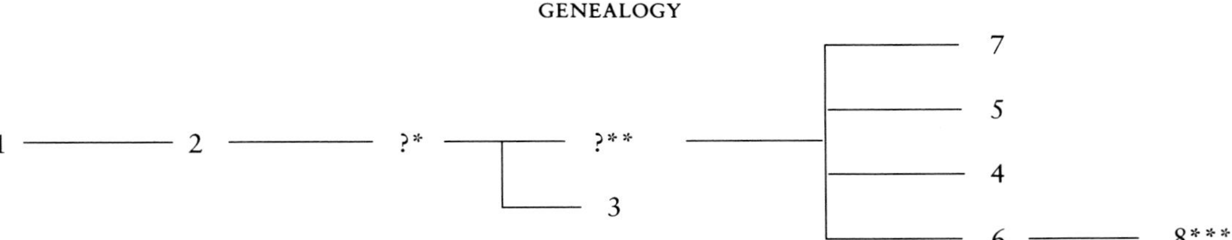

* Correction of northern Honshū
** Incorporates Hokkaidō and Korea
*** Together with a map of China

(IN CHRONOLOGICAL ORDER)

Origin of Cartographic Image	Source(s) with Reproductions[a]	Paired Celestial Globe	Remarks
Matteo Ricci's map	Unno, Oda, and Muroga, *Nihon kochizu taisei*, vol. 2, pl. 56	33 cm, paper	
Matteo Ricci's map	Akioka, *Sekai chizu sakusei shi*, 190	36 cm, paper	Formerly in Tani Kanjō Collection
Nanban style: oval projection	Unno, *Chizu no shiwa*, 249		Possible to spin both globe and attached doll
Japanese copy of Matteo Ricci's map	Kōbe Shiritsu Hakubutsukan, *Kochizu ni miru sekai to Nippon*, pl. 26	25 cm, wood?, lacquered	

shi no Kenkyū 7 (1985): 47–64; Takagi Takeo, *Tenmon kyōgu* (Tools for teaching astronomy) (Tokyo, 1973); Unno Kazutaka, Oda Takeo, and Muroga Nobuo, eds., *Nihon kochizu taisei* (Great collection of old Japanese maps), 2 vols. (Tokyo: Kōdansha, 1972–75), vol. 2, *Nihon kochizu taisei sekaizu hen*; Unno Kazutaka, *Chizu no shiwa* (Map creases; or, Essays in the history of cartography) (Tokyo: Yūshōdō Press, 1985); Unno Kazutaka, "Akashi Shiritsu Tenmonkagakukan shozō kochikyūgi ni tsuite" (On the early terrestrial globe in the Akashi Planetarium collection), *Kagakushi Kenkyū* 124 (1977): 235–36; Unno Kazutaka, "Faruku chikyūgi denrai no hamon" (The influence of the Valcks' globe on Japanese maps and globes), *Nihon Yōgakushi no Kenkyū* 8 (1987): 9–34; Yamamoto Takeshi, ed., *Kōchi ken no rekishi* (History of Kōchi Prefecture) (Tokyo: Yamakawa Shuppansha, 1970).

Owner	Author(s)/Maker(s)	Date	Diameter (cm)	MS or Printed	Material
5 Kushuon'in Temple, Hirakata	Sōkaku	ca. 1702	20	MS	Paper
6 National Museum of Japanese History, Sakura (Akioka Collection)	Irie Shūkei?	ca. 1750	20	MS	Paper
7 Yamanouchi Shrine, Kōchi	Kawatani Keizan	1762	?	MS	Wood? lacquered
8 Kayahara Hiroshi, Tsu		18th century?	21	MS	Paper
9 Muroga Emiko, Kyōto		18th century?	?	MS	Paper
10 Geographical Institute, Kyōto University		18th century	32	MS	Paper
11 Sue Fumito, Iwadeyama, Miyagi Prefecture		18th century?	ca. 28	MS	Paper
12 Kōbe City Museum, Kōbe (Ikenaga Collection)		ca. 1805	25.3	MS	Paper
13 Taikodani Inari Shrine, Tsuwano	Hotta Nisuke	1809	36	MS	Wood
14 Eisei Library, Tokyo	Shiba Kōkan	1810	45.2	MS	Wood, lacquered
15 Kamata Kyōsaikai Museum, Sakaide	Chūjō Sumitomo	1838	28.7	MS	Paper
16 Iizuka Jūzō, Himeji	Akashika Yoshisada	1843	15.5 × 16.4	MS	Clay
17 Akashi Planetarium, Akashi	Fujimura Tanjō	1847	37.6	MS	Paper
18 Hagi Local Museum, Hagi		Early 19th century	28	MS	Paper
19 Kumamoto City Museum, Kumamoto		First half of 19th century	ca. 20	MS	Paper

(*continued*)

Origin of Cartographic Image	Source(s) with Reproductions[a]	Paired Celestial Globe	Remarks
Buddhist world map	Unno, Oda, and Muroga, *Nihon kochizu taisei*, vol. 2, pl. 5	52 cm, copper copy of Shunkai's globe	
Gerard and Leonard Valck's globe, 1700	Akioka, *Nihon chizu sakusei shi*, 15; Akioka, *Sekai chizu sakusei shi*, 186; Kōbe Shiritsu Hakubutsukan, *Kochizu ni miru sekai to Nippon*, pl. 27; idem, *Akioka Kochizu Korekushon meihin ten*, pl. 37	35 cm, paper	
Shibukawa's globe	Takagi, *Tenmon kyōgu*, 122; Yamamoto, *Kōchi ken no rekishi*, frontispiece		Painted by Ikegawa Sōkurō; original date: 12th month, 11th year of Hōreki
Shibukawa's globe		23 cm, plaster, Shunkai globe	Formerly owned by Date Aki family
Matteo Ricci's map			Stand is lost
Gerard and Leonard Valck's globe, 1700	Fujita, *Kaitei zōho Nihon chirigaku shi*, 425, 428; Unno, "Faruku chikyūgi denrai no hamon"		
Matteo Ricci's map		ca. 28 cm, paper	Exhibited in Yūbikan, Iwadeyama
Kon'yo zenzu (Map of the earth) and *Kon'yo zenzusetsu* (Explanation of the *Kon'yo zenzu*), by Inagaki Shisen, 1802	Kōbe Shiritsu Hakubutsukan, *Kochizu ni miru sekai to Nippon*, pl. 24		
Katsuragawa Hoshū's globe, 1794	Unno, Oda, and Muroga, *Nihon kochizu taisei*, vol. 2, pl. 86	37 cm, wood	Katsuragawa's globe was kept at Shōkōkan Library, Mito; original date: 12th month, 5th year of Bunka
Jaillot's map, ca. 1730	Sugano, "Eisei Bunko shozō Shiba Kōkan sei chikyūgi," figs. 1, 2, 8, 10, 13		
Shintei bankoku zenzu (Newly revised map of all the countries), by Takahashi Kageyasu, 1816			The stand is lost
			Jujube type, weight: 1.78 kg
Shinsei yochi zenzu (Newly made map of the world), by Mitsukuri Shōgo, 1844	Unno, "Akashi Shiritsu Tenmonkagakukan shozō kochikyūgi ni tsuite"		
Hashimoto Sōkichi's map, 1797	Kawamura, Unno, and Miyajima, "List of Old Globes," pl. 38		The earth's axis is level
Katsuragawa globe	Akioka, *Sekai chizu sakusei shi*, 193	21 cm, paper, Reimeikan Museum Collection, Kagoshima	

Owner	Author(s)/Maker(s)	Date	Diameter (cm)	MS or Printed	Material
20 Kamata Kyōsaikai Museum, Sakaide	Kume Michikata	First half of 19th century	30.3	MS	Paper gores over plaster
21 Shimonoseki City Art Museum, Shimonoseki		Mid-19th century	30.6	MS	Paper
22 Kōbe City Museum, Kōbe (Ikenaga Collection)		Mid-19th century	31	MS	Paper
23 Shōryū Temple, Wakayama	Nakatani Sōnan	Mid-19th century	21.4	MS	Paper
24 Imperial Household Agency, Tokyo	Suzuki Shigetoki	1852	119	MS	Paper
25 Honma Takao, Tsuchiura	Numajiri Bokusen	1855	23	Printed	Paper and bamboo
26 Kōbe City Museum, Kōbe (Akioka Collection)	Numajiri Bokusen	1855	23	Printed	Paper and bamboo
27 Mōri Museum, Hōfu	Numajiri Bokusen	1855	23	Printed	Paper and bamboo
28 Kōbe City Museum, Kōbe (Nanba Collection)	Horiuchi Naotada	1855	31.7	MS	Paper
29 Shōko Shūsei Museum, Kagoshima	Takagi Hidetoyo and Miki Ikkōsai	1856	19.5	Printed	Paper
30 National Museum of Japanese History, Sakura (Akioka Collection)	Takagi Hidetoyo and Miki Ikkōsai	1856	19.5	Printed	Paper
31 Abe Masamichi, Tokyo	Tokyo Prefectural Secondary School	1871	41 (axis)	Printed	Cloth and steel
32 Kayahara Hiroshi, Tsu	Kajiki Genjirō	1873	16	Printed	Paper
33 Kayahara Hiroshi, Tsu	Ōya Gaikō	1873	21	Printed	Paper

(*continued*)

Origin of Cartographic Image	Source(s) with Reproductions[a]	Paired Celestial Globe	Remarks
	Takagi, *Tenmon kyōgu*, 122	28 cm, paper over plaster	Badly damaged
Shinsei bankoku yochi zenzu (Newly made map of all the countries in the world), by Den Ken, 1844			Formerly in Kazuki Yasuo Collection; the earth's axis is level
Revision of the globe at the Shimonoseki City Art Museum, above.	Kōbe Shiritsu Hakubutsukan, *Kochizu ni miru sekai to Nippon*, pl. 25		Earth's axis is level and form of stand is same as that of Shimonoseki globe
Hashimoto Sōkichi's map, 1797	Takagi, *Tenmon kyōgu*, 122	22 cm, paper	
	Unno, Oda, and Muroga, *Nihon kochizu taisei*, vol. 2, pl. 93		
Shintei kon'yo ryakuzenzu (Newly made revised map of the earth), by Shibata Shūzō, 1852	Unno, Oda, and Muroga, *Nihon kochizu taisei*, vol. 2, pl. 103		Twelve folding ribs; see figure 11.29
Shintei kon'yo ryakuzenzu, by Shibata Shūzō, 1852	Kōbe Shiritsu Hakubutsukan, *Akioka Kochizu Korekushon meihin ten*, pl. 32		Twelve folding ribs
Shintei kon'yo ryakuzenzu, by Shibata Shūzō, 1852	Kawamura, Unno, and Miyajima, "List of Old Globes," pl. 37		Twelve folding ribs
Shinsei yochi zenzu, by Mitsukuri Shōgo, 1844	Kōbe Shiritsu Hakubutsukan, *Kochizu ni miru sekai to Nippon*, pl. 28		
Shintei kon'yo ryakuzenzu, by Shibata Shūzō, 1852	Unno, Oda, and Muroga, *Nihon kochizu taisei*, vol. 2, pl. 102		
Shintei kon'yo ryakuzenzu, by Shibata Shūzō, 1852	Kōbe Shiritsu Hakubutsukan, *Kochizu ni miru sekai to Nippon*, pl. 29; Akioka, *Sekai chizu sakusei shi*, 193		
J. Betts's *New Portable Globe*, ca. 1860			Folding
	Fujita, *Kaitei zōho Nihon chirigaku shi*, 525		Balloon title, *Bankoku fukikyū* (Rich and joyful globe of all the countries)

APPENDIX 11.8 PROJECTS TO COMPILE PROVINCIAL MAPS UNDER THE TOKUGAWA SHOGUNATE

Enterprise	Year of Official Announcement	Year of Completion	Scale	Total Number of Sheets	General Map of Japan Based on Provincial Maps	Remarks
First	1605	Unknown	Unknown	Unknown	Years of completion not known, but works that seem identical exist	
Second	1633?	Unknown	Unknown	Unknown	Years of completion not known, but works that seem identical exist	
Third	1644	ca. 1656	6 *sun* to 1 *ri* (1:21,600)	76	Completed ca. 1670	Plans of cities where clan offices were situated and models of castles along Tōkai road were also made
Fourth	1697	1702	6 *sun* to 1 *ri* (1:21,600)	83	Completed 1702	
Fifth	1835	1838	6 *sun* to 1 *ri* (1:21,600)	83	Not made	

APPENDIX 11.9 EARLY PRINTED PLANS OF SIX MAIN CITIES

Title	Date	Author	Publisher(s)	Owner(s)	Reproduction or Remarks[a]
EDO					
1 *Bushū Toshima gōri Edo no shō zu* (Plan of Edo, Toshima County, Musashi Province)	ca. 1632	None	None	National Diet Library, Tokyo	97 × 128.5 cm; Unno, Oda, and Muroga, *Nihon kochizu taisei*, vol. 1, pl. 72
2 *Shinpan Bushū Edo no zu* (Newly printed plan of Edo, Musashi Province)	1661	None	Kyōto: Kawano Michikiyo	Mitsui Library, Tokyo	84 × 121 cm; Kurita, *Nihon kohan chizu shūsei*, pl. 42
3 *Shinpan Bushū Edo no zu* (Newly printed plan of Edo, Musashi Province)	1662	None	None	Tōyō Bunko (the Oriental Library), Tokyo	Type of Kawano's plan mentioned above

[a]Akioka Takejirō, *Nihon chizu shi* (History of maps of Japan) (Tokyo: Kawade Shobō, 1955); idem, *Nihon kochizu shūsei* (Collection of old maps of Japan) (Tokyo: Kajima Kenkyūjo Shuppankai, 1971); George H. Beans, *A List of Japanese Maps of the Tokugawa Era* (Jenkintown, Pa.: Tall Tree Library, 1951), supplements A, B, and C (1955, 1958, 1963); idem, "Some Notes from the Tall Tree Library," *Imago Mundi* 11 (1954): 146–47; Hugh Cortazzi, *Isles of Gold: Antique Maps of Japan* (Tokyo: Weatherhill, 1983); Fujita Motoharu, *Toshi kenkyū Heiankyō hensenshi, tsuketari kochizu shū* (History of the Kyōto region, accompanied by collected old plans) (Kyōto: Suzukake Shuppanbu, 1930; reprinted Nihon Shiryō Kankōkai, 1976); Kurita Mototsugu, ed., *Nihon kohan chizu shūsei* (Early maps and plans printed in Japan) (Tokyo: Hakata Seishōdō, 1928); Kyōto Koten Dōkōkai (Kyōto Classical Studies Group), comp., *Kohan Nagasaki chizushū* (Early printed plans of Nagasaki) (Kyōto: Kyōto Koten Dōkōkai, 1977); *Kyōto shi shi, chizu hen* (History of Kyōto City, section of plans) (Kyōto, 1947); N. H. N. Mody, *A Collection of Nagasaki Colour Prints and Paintings* (1939; reprinted Tokyo: Charles E. Tuttle, 1969); Nanba Matsutarō, Muroga Nobuo, and Unno Kazutaka, eds. and comps., *Nihon no kochizu* (Old maps in Japan) (Ōsaka: Sōgensha, 1969); English edition, *Old Maps in Japan*, trans. Patricia Murray (Ōsaka: Sōgensha, 1973); Unno Kazutaka, Oda Takeo, and Muroga Nobuo, eds., *Nihon kochizu taisei* (Great collection of old Japanese maps), 2 vols. (Tokyo: Kōdansha, 1972–75).

APPENDIX 11.9 (*continued*)

Title	Date	Author	Publisher(s)	Owner(s)	Reproduction or Remarks[a]
EDO (*continued*)					
4 *Shinpan Bushū Edo no zu* (Newly printed plan of Edo, Musashi Province)	1664	None	Kyōto: Kawano Michikiyo	Kōbe City Museum, Kōbe (Nanba Collection)	
5 *Shinpan Bushū Edo no zu* (Newly printed plan of Edo, Musashi Province)	1664	None	None	University of British Columbia Library, Vancouver (Beans Collection)	Beans, *Japanese Maps*, facing p. 13; type of Kawano's plan mentioned above
6 *Shinpan Bushū Edo no zu* (Newly printed plan of Edo, Musashi Province)	1666	None	Kyōto: Kawano Kakunojō	Daitōkyū Kinen Library, Tokyo	95.1 × 120.6 cm
7 None	1666	None	Edo: Daikyōji Kahee	Tokyo University Library, Tokyo	47.5 × 70.2 cm
8 *Shinpan zōho Edo zu* (Newly printed, enlarged plan of Edo)	1666	None	Kyōto: Kawano Kakunojō	Nanba Matsutarō Collection, Nishinomiya	51.4 × 71.8 cm; Nanba, Muroga, and Unno, *Nihon no kochizu/Old Maps in Japan*, pl. 76
9 *Shinpan Edo ōezu* (Newly issued plan of Edo)	1671	Ochikochi Dōin	Edo: Kyōjiya Kahee	National Diet Library, Tokyo; National Archives, Tokyo; and others	Unno, Oda, and Muroga, *Nihon kochizu taisei*, vol. 1, pl. 73
KYŌTO					
10 None	1624–41	None	None	Ōtsuka Takashi Collection, Kyōto (formerly Moriya Collection)	Kurita, *Nihon kohan chizu shūsei*, pl. 48; *Kyōto shi shi, chizu hen*, pl. 14; Unno, Oda, and Muroga, *Nihon kochizu taisei*, vol. 1, pl. 80; see figure 11.47
11 *Heianjō Motodachiuri yori Kujō made machinami no zu* (City plan of Heianjō [Kyōto] from Motodachiuri to Kujō)	1624–41	None	None	National Museum of Japanese History, Sakura (Akioka Collection)	Including eastern suburbs; Kurita, *Nihon kohan chizu shūsei*, pl. 49; Akioka, *Nihon chizu shi*, pl. 46; Akioka, *Nihon kochizu shūsei*, pl. 93
12 *Heianjō tōzainanboku machinami no zu* (East-west, south-north city plan of Heianjō)	1624–41	None	None	Mitsui Library, Tokyo; Kurita Kenji Collection, Nagoya	Including eastern and western suburbs; Fujita, *Toshi kenkyū Heiankyō hensenshi tsuketari kochizu shū*, pl. 2; *Kyōto shi shi, chizu hen*, pl. 15
13 *Heianjō tōzainanboku machinami no zu* (East-west, south-north city plan of Heianjō)	1641–52	None	None	National Museum of Japanese History, Sakura (Akioka Collection)	Including red-light district of Shimabara; Akioka, *Nihon chizu shi*, pl. 47; Akioka, *Nihon kochizu shūsei*, pl. 92

APPENDIX 11.9 (*continued*)

Title	Date	Author	Publisher(s)	Owner(s)	Reproduction or Remarks[a]
KYŌTO (*continued*)					
14 *Heianjō tōzainanboku machinami no zu* (East-west, south-north city plan of Heianjō)	1652	None	Yamamoto Gohee	Institute of Japanese History, Kyōto University	Fujita, *Toshi kenkyū Heiankyō hensenshi tsuketari kochizu shū*, pl. 3; *Kyōto shi shi, chizu hen*, pl. 16; Unno, Oda, and Muroga, *Nihon kochizu taisei*, vol. 1, pl. 81
15 *Shinkai Rakuyō narabini rakugai no zu* (Newly revised plan of Rakuyō [Kyōto] and its surroundings)	1654	None	None	?	Including Kamo and Ōi rivers; *Kyōto shi shi, chizu hen*, pl. 17; original date: 12th month, 2d year of Shōō
16 *Shinpan Heianjō tōzainanboku machinami rakugai no zu* (Newly printed east-west, south-north plan of the city and surroundings of Heianjō)	1654	None	Kitayama Shūgakuji mura (Kyōto): Muan	University of British Columbia Library, Vancouver (Beans Collection)	Fujita, *Toshi kenkyū Heiankyō hensenshi tsuketari kochizu shū*, pl. 4; Beans, "Tall Tree Library," 147
17 *Shinpan Heianjō tōzainanboku machinami rakugai no zu* (Newly printed east-west, south-north plan of the city and surroundings of Heianjō)	1657	None	Kyōto: Maruya	Satō Collection, Kariya, Aichi Prefecture	Probably later issue of Muan edition of 1654
18 *Shinkai Rakuyō narabini rakugai no zu* (Newly revised plan of Rakuyō [Kyōto] and its surroundings)	1657	None	None	Kōbe City Museum, Kōbe (Nanba Collection)	Probably later issue of 1654 edition with same title
ŌSAKA					
19 *Shinpan Settsu Ōsaka tōzainanboku machi shima no zu* (Newly issued plan of Ōsaka with the east-west and north-south streets and islands, Settsu Province)	1655	None	Kyōto: anonymous	University of British Columbia Library, Vancouver (Beans Collection), formerly owned by Kanda Kiichirō; Kidō Library Collection, Kishiwada, Ōsaka Prefecture; Hōsa Library, Nagoya	119.4 × 77.5 cm; see figure 11.48
20 *Shinpan Ōsaka no zu* (Newly printed plan of Ōsaka)	1657	None	Kyōto: Kawano Michikiyo	Hōsa Library, Nagoya; Sako Collection, Ōsaka	Kurita, *Nihon kohan chizu shūsei*, pl. 52; Unno, Oda, and Muroga, *Nihon kochizu taisei*, vol. 1, pl. 86
21 ?	1661	None	Maruya Shōzaemon	Kōbe City Museum, Kōbe (Nanba Collection)	

APPENDIX 11.9 (*continued*)

Title	Date	Author	Publisher(s)	Owner(s)	Reproduction or Remarks[a]
ŌSAKA (*continued*)					
22 *Zōho Ōsaka no zu* (Enlarged plan of Ōsaka)	ca. 1670	None	None	Sako Collection, Ōsaka	
23 *Shinpan Ōsaka no zu* (Newly printed plan of Ōsaka)	1671	None	Kyōto: Fushimiya	Kidō Library Collection, Kishiwada, Ōsaka Prefecture	
24 *Shinpan Ōsaka no zu* (Newly printed plan of Ōsaka)	1678	None	Kyōto: Fushimiya	Ōsaka Prefectural Nakanoshima Library, Ōsaka	
25 *Shinsen zōho Ōsaka ōezu* (Newly compiled, enlarged large plan of Ōsaka)	1686	None	Kyōto: Hayashi Yoshinaga	University of British Columbia Library, Vancouver (Beans Collection); Waseda University Library, Tokyo; and others	
26 *Shinsen zōho Ōsaka ōezu* (Newly compiled, enlarged large plan of Ōsaka)	1687	none	Kyōto: Hayashi Yoshinaga	Mitsui Library, Tokyo; Ōsaka Prefectural Nakanoshima Library, Ōsaka; Sako Collection, Ōsaka	Kurita, *Nihon kohan chizu shūsei*, pl. 53; Unno, Oda, and Muroga, *Nihon kochizu taisei*, vol. 1, pl. 91
27 *Shinpan Ōsaka no zu* (Newly printed plan of Ōsaka)	1687	None	Kyōto: Hayashi Yoshinaga	Kidō Library Collection, Kishiwada, Ōsaka Prefecture	
NAGASAKI					
28 *Nagasaki ōezu* (Large plan of Nagasaki)	ca. 1681	None	None	Tenri Central Library, Tenri; Kōbe City Museum, Kōbe (Ikenaga Collection); British Library, London	63.4 × 143 cm (Tenri copy); Kyōto Koten Dōkōkai, *Kohan Nagasaki chizushū*, pl. 1; Cortazzi, *Isles of Gold*, pl. 50
29 *Karafune raichō zu Nagasaki zu* (Plan of Nagasaki with illustrations of Chinese ships coming to Japan)	ca. 1690	None	Edo: Matsue (Murata Sanshirō)	Tenri Central Library, Tenri; Kurita Kenji Collection, Nagoya; Kōbe City Museum, Kōbe (Ikenaga Collection); and others	Kurita, *Nihon kohan chizu shūsei*, pl. 68; Unno, Oda, and Muroga, *Nihon kochizu taisei*, vol. 1, pl. 108; Kyōto Koten Dōkōkai, *Kohan Nagasaki chizushū*, pl. 2
30 [*Nagasaki ōezu*] [Large plan of Nagasaki]	ca. 1730	None	Nagasaki: Nakamura Sanzō (Chikujuken)	Kōbe City Museum, Kōbe (Ikenaga Collection)	

APPENDIX 11.9 (*continued*)

Title	Date	Author	Publisher(s)	Owner(s)	Reproduction or Remarks[a]
NAGASAKI (*continued*)					
31 *Kaisei Nagasaki zu* (Revised plan of Nagasaki)	1745	Hassendō Shujin	Kyōto: Hayashi Jizaemon	Kurita Kenji Collection, Nagoya; University of British Columbia Library, Vancouver (Beans Collection); Kōbe City Museum, Kōbe (Ikenaga and Nanba Collections)	Issued in 1808 and 1830; cover title of later issues changed to *Nagasaki saikenzu* (Detailed plan of Nagasaki)
32 *Shinkan Nagasaki ōezu* (New version [of the] large plan of Nagasaki)	1752	None	Nagasaki: Nakamura Sōzaburō (Chikujuken)	Formerly N. H. N. Mody Collection, Kōbe	Mody, *Collection of Nagasaki Colour Prints and Paintings*, pl. 29
33 *Shinpan Nagasaki ōezu* (Newly issued plan of Nagasaki)	ca. 1760	None	Nagasaki: Shimabaraya	Kurita Kenji Collection, Nagoya; Kōbe City Museum, Kōbe (Ikenaga Collection)	57 × 101.8; Unno, Oda, and Muroga, *Nihon kochizu taisei*, vol. 1, pl. 107; see figure 11.52
34 *Hishū Nagasaki no zu* (Plan of Nagasaki, Hizen Province)	1764	None	Nagasaki: Ōhata Bunjiemon	University of British Columbia Library, Vancouver (Beans Collection); Kōbe City Museum, Kōbe (Ikenaga and Nanba Collections)	61 × 88.5 cm (Beans Collection); Beans, *Japanese Maps*, facing p. 23; Kyōto Koten Dōkōkai, *Kohan Nagasaki chizushū*, pl. 5
35 *Hishū Nagasaki zu* (Plan of Nagasaki, Hizen Province)	1778	None	Nagasaki: Ōhata Bunjiemon	University of British Columbia Library, Vancouver (Beans Collection); Kōbe City Museum, Kōbe (Ikenaga and Nanba Collections)	Kurita, *Nihon kohan chizu shūsei*, pl. 69; Unno, Oda, and Muroga, *Nihon kochizu taisei*, vol. 1, pl. 109; Kyōto Koten Dōkōkai, *Kohan Nagasaki chizushū*, pl. 6
NARA					
36 *Washū Nanto no zu* (Plan of the southern metropolis [Nara], Yamato Province)	1666	None	Nara: Ozaki San'emon	Tōhoku University Library, Sendai; Nara Prefectural Library, Nara	99.3 × 62.4 cm
37 *Washū Nanto no zu* (Plan of Nara, Yamato Province)	1709	None	Nara: Yamamura Juzaburō	Kōbe City Museum, Kōbe (Nanba Collection)	Kurita, *Nihon kohan chizu shūsei*, pl. 55
38 *Washū Nanto ezu* (Plan of Nanto, Yamato Province)	1778	None	Ōsaka: Shibukawa Seiemon and Yanagihara Kihee	Kōbe City Museum, Kōbe (Nanba Collection)	Nanba, Muroga, and Unno, *Nihon no kochizu/Old Maps in Japan*, pl. 67

APPENDIX 11.9 *(continued)*

Title	Date	Author	Publisher(s)	Owner(s)	Reproduction or Remarks[a]
SAKAI					
39 *Senshū Sakai no zu* (Plan of Sakai, Izumi Province)	1704	None	Sakai: Takaishi Kimei	Kōbe City Museum, Kōbe (Nanba Collection)	Unno, Oda, and Muroga, *Nihon kochizu taisei*, vol. 1, pl. 104
40 *Sakai ōezu kaisei kōmoku* (Large plan of Sakai: Revised outline)	1735	Kawai Morikiyo	Ōsaka: Murakami Ihee	Kurita Kenji Collection, Nagoya; National Diet Library, Tokyo; Kōbe City Museum, Kōbe (Nanba Collection); and others	Kurita, *Nihon kohan chizu shūsei*, pl. 56; cover title is *Kaisei Sakai ezu kōmoku* (Revised outline plan of Sakai)
41 *Sakai saiken ezu* (Detailed plan of Sakai)	1798	None	Sakai: Kitamura Sahee; Ōsaka: Kashiharaya Kahee	Kōbe City Museum, Kōbe (Nanba Collection)	Unno, Oda, and Muroga, *Nihon kochizu taisei*, vol. 1, pl. 105

12 · Cartography in Vietnam

John K. Whitmore

The study of maps made in Vietnam over the past five centuries has barely begun. Any person attempting to examine the Vietnamese cartographic tradition must cast a wide net to gather the scattering of detail in various forms that helps us understand how this tradition came to be. Although scholars of the Ecole Française d'Extrême-Orient (such as Henri Maspero, Leonard Aurousseau, and Emile Gaspardone) laid the bibliographic foundation for the study of Vietnamese cartography, the colonial period saw almost no work on historical maps. The only major exception was Gustave Dumoutier's 1896 study of an early itinerary to the south (see below).

The main work available to us is a collection of maps called the *Hồng-đức bản đồ* (Maps of the Hồng-đức period [1471–97]), which was published three decades ago by the Viện Khảo-Cổ (Institute of Historical Research) in Saigon (now Ho Chi Minh City).[1] Besides reproductions of the maps, the work presents a good index to the names recorded on the maps and a brief introductory study of the maps (in Vietnamese and, in a shorter version, French) by Trương Bửu Lâm. This introduction forms the core of our knowledge about pre-1800 Vietnamese maps. It is supplemented by several articles published in Hanoi over the past decade by Bùi Thiết that look at indigenous maps of the old capital (now Hanoi).[2] Bùi covers the same time period as the previous study but extends it into the nineteenth century. Thái Văn Kiểm also produced several pieces dealing with mapping during the Nguyễn dynasty (1802–1945).[3] Yet no synthesis exists for the half millennium of premodern Vietnamese mapping.

To attempt such a study in North America is difficult. The major collection of old Vietnamese manuscripts, including maps, is that developed by the Ecole Française d'Extrême-Orient in Hanoi and taken over by the Democratic Republic of Vietnam in 1954–55. This collection is now held at the Viện Nghiên Cứu Hán Nôm (Sino-Nom Institute) in Hanoi, though its documents are still identified by the French system (A. plus a number).[4] The imperial library in Hue was divided in the late 1950s, with the Nguyễn dynasty archives going to Da Lat (now in Ho Chi Minh City?) and the pre-1800 material deposited at the Viện Khảo-Cổ. Little serious damage appears to have been done to any of these collections during the war. Upon the French withdrawal, microfilm copies were made of the most important documents in Hanoi, and collections of the microfilms were deposited in two locations, the above-mentioned Viện Khảo-Cổ and the Ecole Française d'Extrême-Orient in Paris. Other manuscript collections exist at the Société Asiatique in Paris (the Henri Maspero Collection, designated by HM plus a number) and the Tōyō Bunko (Oriental Library) in Tokyo.

The only historical Vietnamese maps available in the United States are black-and-white reproductions (and sometimes reproductions of reproductions). The maps in the published *Hồng-đức bản đồ* are photographic reproductions from the negatives of a Japanese microfilm, and the white-on-black illustrations are often none too clear.[5] In general, the American scholar of Vietnamese maps is dependent on collections of the Ecole Française d'Ex-

1. Trương Bửu Lâm, ed., *Hồng-đức bản đồ* (Saigon: Bộ Quốc-gia Giáo-dục, 1962). Although A.2499 (reel 141, no. 253) is the main copy of this work, the scholars of the Viện Khảo-Cổ used an almost identical copy (on microfilm) from the Tōyō Bunko in Tokyo (number 100.891); *Hồng-đức bản đồ*, XVI–XVII, XXVIII–XXIX.

2. See Bùi Thiết, "Sắp xếp thế hệ các bản đồ hiện biết thành Thăng Long thời Lê" (Establishing the generations of known maps of the city of Thăng Long from the Lê dynasty, 1428–1787), *Khảo Cổ Học* (Archaeology) 52, no. 4 (1984): 48–55, esp. 49–50.

3. Thái Văn Kiểm, "Lời nói đầu" (Introduction), in *Lục tỉnh Nam-Việt* (*Đại-Nam nhất-thống chí*) (The six provinces of southern Vietnam [Record of the unity of Đại Nam]) (Saigon: Phủ Quốc-vụ-Khanh Đặc-Trách Văn-Hóa, 1973), Tập Thượng, V–XIII; idem, "Interprétation d'une carte ancienne de Saigon," *Bulletin de la Société des Etudes Indochinoises*, n.s., 37, no. 4 (1962): 409–31; idem, *Cố đô Huế* (The old capital of Huế) (Saigon: Nha Văn-hóa Bộ Quốc-gia Giáo-dục, 1960).

4. For the holdings of this collection, see its catalog as microfilmed in 1954–55 (reel 3, nos. 8–9), containing cards for over 3,600 documents. A document with a map contained in this microfilm collection is noted in this chapter by the document's reel and number(s). A list of the nearly six hundred microfilmed documents may be found in G. Raymond Nunn, ed., *Asia and Oceania: A Guide to Archival and Manuscript Sources in the United States*, 5 vols. (New York: Mansell, 1985), 3:1054–60. All original documents designated by A. are in Hanoi; only microfilm copies of them exist elsewhere. For a list of early maps, including most in this chapter, see also Trần Nghĩa, "Bản đồ cổ Việt Nam" (Old maps of Vietnam), *Tạp Chí Hán Nôm* (Hán Nôm review) 2, no. 9 (1990): 3–10.

5. To avoid confusion, note that the grids of lines drawn on the maps in the published *Hồng-đức bản đồ* were not on the originals and were intended only as an index grid for the transcription of the Chinese characters.

trême-Orient microfilms now at Cornell University and the University of Hawaii, Honolulu. I have consequently found it difficult to obtain a sense of the physical composition of the maps discussed below. Since the maps are nearly all in manuscript, not printed, and bound in the Chinese fashion, questions of format, medium, size, scale, material, and exact style of illustration must be left to someone who is able to undertake a hands-on study of existing maps, preferably in Hanoi. Where such description appears below, it depends on Joseph E. Schwartzberg's examination of manuscripts held in France. He has graciously allowed me access to his notes. Any problems in interpretation are naturally my own.

In this chapter I attempt to gather what data I can on premodern Vietnamese maps, given the restrictions above, and to provide a historical outline for the further study of Vietnamese cartography. As we shall see, no maps exist for the country of Đại Việt (Great Viet) from the earliest major dynasties, the Lý (1010–1225) and the Trần (1225–1400). Countrywide mapping appears only in the first century of the Lê dynasty (1428–1527). The following Mạc dynasty (1528–92) seems to have done some mapping, but the earliest existing body of maps was produced for the two families ruling under the Lê after the latter's restoration (1592–1787), the Trịnh in the capital and the Nguyễn on the southern border. Atlases and itineraries were compiled in the north, and an itinerary was made in the south. A final surge of mapping took place during the nineteenth century when the Nguyễn dynasty reunified and took power over the entire country, now known as Đại Nam (Great South). Unfortunately, we have little information on how any of this mapping was done. During the thousand-year period, the country we call Vietnam grew from its original center in the north around Hanoi all the way south down the east coast of the Indochinese peninsula. Most of this extension occurred in the seventeenth and eighteenth centuries, as newcomers permeated the lowlands occupied by Chams and Khmers and came in contact with other Southeast Asian peoples around the Gulf of Siam. Vietnamese cartography from 1600 to 1900 reflects this growth.

The Vietnamese style of mapping is essentially a Chinese one. Its development parallels the formation of the Chinese model in Vietnam over the period from the fifteenth to the nineteenth century. The Vietnamese intelligentsia, being literate in Chinese, could read texts from China, but the question of how far they were exposed to Sinic cartographic precepts must remain open. Vietnamese terms for "map" are variants of the Chinese word *tu* (Sino-Vietnamese *đồ*), meaning illustration, drawing, plan, and by extension map—*bantu* (*bản-đồ*), *ditu* (*địa-đồ*), *yutu* (*dư-đồ*), *quantu* (*tòan-đồ*).

COSMOGRAPHY

Non nước is one term the Vietnamese use for their country. It means quite simply "mountain and water" in Vietnamese (not Sino-Vietnamese), and this concept seems to underlie the Vietnamese visual approach to spatial representation. Rolf Stein has demonstrated the importance of this concept into the twentieth century through the study of miniature gardens. Here we see, set up in the courtyards of both homes and temples, containers of water with rocks placed in them. On these rocks are grown or set live miniature plants, with ceramic models of buildings, humans, and animals. The symbolic mountain and water form the landscape in which the integration of natural and supernatural is shown. In a number of the famous temples of Vietnam whose origins go back almost a millennium, we find magical representations of the universe (heaven, earth, and water) in the form of the miniature mountains and ponds (fig. 12.1).[6] Here we have what seems to be the original form of Vietnamese cosmography, a graphic portrayal of their universe. This portrayal is simple, and for that very reason the Vietnamese appear to have retained it as their image of the world around them, of the magical and fertile forces at large in the natural world.

Linked to this image is the construction of artificial mountains by Vietnamese kings for their court rituals during the tenth and eleventh centuries. In 985, 1021, and 1028, first Lê Hoan, then the Lý kings Thái-tổ and Thái-tông, celebrated the royal birthdays by having mountains built of bamboo, called Mountain of the South, Nam Sơn, and using them as the centerpieces of the accompanying rituals. The construction of 985 was built on a boat in the middle of a river and was associated with boat races (perhaps the dragon boat festival). In the celebration of 1028 there were five peaks, one in the center (Mount Meru?) surrounded by four peripheral ones (possibly indicating the cardinal directions). Among these peaks was coiled a dragon (or water spirit).[7] Here too we have the imagery of mountain and water serving as nearly as we can tell to represent the cosmos and its powers in a smaller form and paralleling the temple structures of Angkor and Pagan in contemporary Cambodia and Burma.[8]

6. Rolf A. Stein, *The World in Miniature: Container Gardens and Dwellings in Far Eastern Religious Thought*, trans. Phyllis Brooks (Stanford: Stanford University Press, 1990), 13–21, 36–37, 52, 58, 77, 83, 89–91, 103, 104, 109.

7. Stein, *World in Miniature*, 39–40 (note 6); Ngô Sĩ Liên (fifteenth century), *Đại-Việt sử-ký toàn-thư*, 3 vols., ed. Ch'en Ching-ho (Tokyo, 1984–86), 1:190, 214, 219(14).

8. On Angkor and Pagan, see Eleanor Mannikka, "Angkor Wat: Meaning through Measurement" (Ph.D. diss., University of Michigan, 1985), and Michael Aung-Thwin, *Pagan: The Origins of Modern Burma* (Honolulu: University of Hawaii Press, 1985).

FIG. 12.1. VIETNAMESE COSMOGRAPHY. This drawing of a miniature garden in the temple of Trấn Vũ (Hanoi) in the early 1940s shows the pattern of mountains and water in Vietnamese cosmography.

Size of the original: 12.7 × 19.2 cm. From Rolf A. Stein, "Jardins en miniature d'Extrême-Orient," *Bulletin de l'Ecole Française d'Extrême-Orient* 42 (1943): 1–104, esp. pl. III.

Although there are no sources to say whether this form of royal ritual continued beyond 1028, we may easily transfer the imagery to the Buddhist structures of the following three centuries. Just north of the capital of Thăng-long (now Hanoi) lay the temple of Vạn-phúc, which we know has existed since at least 1057. It is a 140-foot brick tower built on ascending terraces and flanked by two sacred ponds. Though the temple is built in Chinese fashion, the existence of the ponds was a distinctively Vietnamese trait and continued the conjunction of mountain and water as cosmic symbolism.[9] This style of Buddhist temple, the tower, continued from the eleventh century into the fourteenth, the years of the Lý and Trấn dynasties. The Vietnamese state of these centuries was more akin to its contemporaries in Southeast Asia than to the Song state in China. The Vietnamese administration depended on personal ties, not bureaucratic ones, and its cosmology was Hindu-Buddhist.

An attempt to strengthen central power occurred in the mid-thirteenth century, but the Mongol threat and subsequent invasions in the second half of the century deflected it. Royal power primarily covered the area immediately around the capital, while outer areas were controlled indirectly through local intermediaries or powerful figures (royal or otherwise) designated by the throne. In either case, the capital had no direct control over these outer areas. The king gained access to their resources only so far as the area remained loyal to the center. No maps of Đại Việt exist from these centuries, though the Vietnamese history *Đại-Việt sử-ký toàn-thư* (Complete book of the historical records of Great Viet, 1479) mentions two that dealt with the outer reaches of the kingdom, one from the late eleventh century, the other from a century later. The first was drawn up by the famous minister Lý Thường Kiệt in 1075 and covered the then southern frontier with Champa (south of Nghệ-an Province and the Col des Nuages, the old

9. Louis Bezacier, *L'art viêtnamien* (Paris: Editions de l'Union Française, 1954), 135 ff.; idem, *Relevés de monuments anciens du Nord Việt-nam* (Paris: École Française d'Extrême-Orient, 1959), pls. 14–23; Stein, *World in Miniature*, 14–15 (note 6).

"southern border," Nam-giới).[10] The map of the 1170s, called the *Nam-bắc phiên-dới địa-đồ* (Atlas of the boundaries to the south and north), resulted, we are told, from a royal inspection tour of the coasts and the frontiers.[11] The *Đại-Việt sử-ký toàn-thư* specifically noted "the mountains and the streams" (*sơn xuyên* in Sino-Vietnamese) as being the focal points for each effort.

In general, however, there is little evidence that the Vietnamese before the fifteenth century had either much inclination to map or the centralized control needed to bring data together for an atlas of the country. The cosmography, which in its simplest form required just the symbols of a mountain and a pond, continued through these centuries. Linked to it were the spirit cults, based as they were in specific localities across the country. This spiritual geography, though unmapped, provided a sense of place for the Vietnamese. It also reflected the influence of the Chinese science of geomancy. As Ungar has noted, "One envisions a spiritual map of sacred places: nodes of potent configurations of mountains and waters, connected by subterranean 'veins' (*mạch*) through which geomantic energy flowed." During these centuries, the Vietnamese were growing ever more conscious of their cultural territory and its boundaries. Rather than a vague conception of their own territory fading into the distance, they began to develop a sense of where it ended and of what cultural configurations lay on the other side. This territory had originally included the Hundred Yue (in Sino-Vietnamese Việt) and stretched from the Yangtze River in the north to Champa in the south.[12] Across the Vietnamese northern border lay China and its encroaching power, while to the south and west were Southeast Asian peoples and kingdoms, the Chams, the Thais/Laos, and others.

By the late fourteenth century, the Vietnamese were explicitly setting themselves apart from their neighbors. In the 1370s the Vietnamese royal court banned "northern" (Chinese) clothing and the Cham and Lao languages. Over half a century later (in the 1430s), the first Vietnamese geography, *Dư địa chí* (Geographical record), modeled on the classic Chinese text the "Yu gong" (Tribute of Yu), continued this effort to draw a cultural line between the land of Đại Việt and the peoples beyond its borders.[13] Yet the Vietnamese maintained their nonvisual approach to their land. Even though both Vietnamese and Chinese texts of the late fourteenth and early fifteenth centuries were rich in detail on the riverine nature of Vietnam,[14] no maps exist to show us the view from these times of either the intricate pattern of water communications or the territorial separation of Vietnamese and non-Vietnamese.

Maps of Đại Việt

Maps became important for the Vietnamese when the government of Đại Việt adopted the Sinic bureaucratic model in the second half of the fifteenth century. Following half a century of crisis from the 1370s to the 1420s (Cham invasions, 1371–90, the Hồ regime, 1400–1407, and the Ming occupation, 1407–27), the liberator of Vietnam, Lê Lợi, established a new dynasty that continued many of the elements of the old world. The new king was, however, very interested in maintaining a stronger resource base for the central government and acted to establish a system of public land that would ensure it. At the same time, the Lê court was ideologically open, and a new strand of modernist Neo-Confucian thought modeled on that of Ming China appeared, advocated by younger scholars.

After three decades of intermittent strife, the modernist group of scholars, with a young king supporting them, overcame the resistance of the conservative military oligarchy that had helped found the dynasty. The young king, Lê Thánh-tông (1460–97), promptly took the opportunity to change the orientation of the state. During the 1460s, he initiated triennial Neo-Confucian examinations, used the successful scholars in his government, and established a centralized bureaucratic administration. The government, staffed by literati-officials, now penetrated all the way to the village and was able both to preach the new moral precepts and to gain a stronger control of local resources. Almost immediately, a large amount of information began to flow from the provinces into the capital of Thăng-long, including population figures in 1465. Local officials were expected to travel throughout their jurisdictions and to know them at first hand. Within a hundred days of arriving at their posts,

10. *Đại-Việt sử-ký toàn-thư*, 1:248 (note 7). Specifically, the mapping covered the three territories of Bồ-chính, Địa-lý (Lâm-bình), and Ma-linh (Minh-linh); see *Hồng-đức bản đồ*, 16–17, 46–48, 193 (note 1).

11. *Đại-Việt sử-ký toàn-thư*, 1:299 (note 7); Emile Gaspardone, "Bibliographie annamite," *Bulletin de l'Ecole Française d'Extrême-Orient* 34 (1934): 1–173, esp. 45–46 (#21).

12. Esta S. Ungar, "From Myth to History: Imagined Polities in 14th Century Vietnam," in *Southeast Asia in the 9th to 14th Centuries*, ed. David G. Marr and A. C. Milner (Singapore: Institute of Southeast Asian Studies, 1986), 177–86, quotation on 179; on Vietnamese geomancy, see Pierre Huard and Maurice Durand, *Connaissance du Việt-Nam* (Hanoi: Ecole Française d'Extrême-Orient, 1954), 70–71.

13. O. W. Wolters, *Two Essays on Đại-Việt in the Fourteenth Century* (New Haven: Council on Southeast Asia Studies, Yale Center for International and Area Studies, 1988), 31, 32, 41; John K. Whitmore, *Vietnam, Hồ Quý Ly, and the Ming (1371–1421)* (New Haven: Yale Center for International and Area Studies, 1985), 16; Nguyễn Trãi (1380–1442), "Dư địa chí" (Geographical record), in *Nguyễn Trãi Toàn Tập* (Complete collection of the works of Nguyễn Trãi) (Hanoi, 1969), 186–227, esp. 222–23.

14. Wolters, *Two Essays*, xvii, xxxviii n. 21 (note 13).

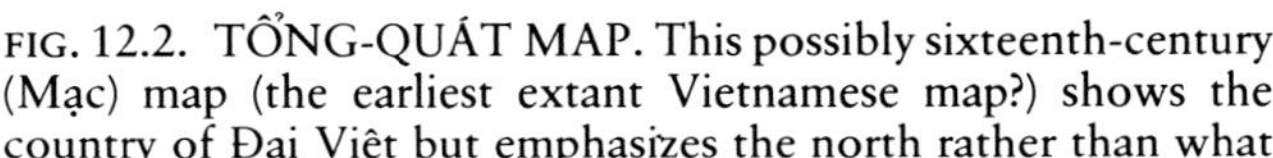

FIG. 12.2. TỔNG-QUÁT MAP. This possibly sixteenth-century (Mạc) map (the earliest extant Vietnamese map?) shows the country of Đại Việt but emphasizes the north rather than what was then the south (home of the Lê). North is at the top. Photograph courtesy of Hamilton Library, University of Hawaii at Manoa, Honolulu (microfilm collection, A.2499).

the officials had to submit a detailed report on their jurisdictions.[15]

In 1467 the king sent the twelve provinces an order to map the country. In it he commanded the officials to draw careful maps of their provinces' topography, illustrating the mountains and streams and pointing out the strategic areas, the communication routes, and historical and contemporary features. These maps were then to be sent to the capital where, in 1469, they were combined with data on the numbers and different types of local communities to establish the official maps for the twelve provinces. Twenty-one years later, in 1490, Thánh-tông accepted the *Thiên-hạ bản-đồ* (Maps of all under heaven [the empire]) as the official atlas for his realm, now expanded to thirteen provinces with the conquest of Champa in 1479 and the annexation of its northern territories.[16]

The Hanoi scholar Bùi Thiết has stated that there appear to be no existing maps from this first period of Vietnamese cartography.[17] As we shall see below, all the maps known to us that can be linked to these original maps are seventeenth century or later. Yet there is one map, unstudied as far as I know, that offers the possibility of being from before 1600. It is unlike what comes to be the standard Lê dynasty map and has much information on it. It will require a detailed examination for full understanding. Here I can only introduce it and make some initial comments.

This map, which I shall call the Tổng-quát map, is fixed as a sort of appendix to the Lê atlas in a section called

15. John K. Whitmore, *Transforming Đại Việt: Politics and Confucianism in the Fifteenth Century* (forthcoming), chap. 5.

16. *Đại-Việt sử-ký toàn-thư*, 2:665(62), 676(11), 736(8) (note 7); Gaspardone, "Bibliographie," 46 (#22) (note 11).

17. Bùi, "Bản đồ," 49–50 (note 2).

"Bản-quốc bản-đồ tổng-quát mục-lục" (Index and overview of the maps of our country).[18] It is unrelated to the written text attached to it and is totally concerned with Đại Việt, making almost no mention of any neighboring country or people (a local prefecture in southwestern China being the lone exception). Bearing a northern orientation, this map is a sketch drawn in a simple style (fig. 12.2). It apparently covers two sides of a sheet and consists of a line drawing showing the rivers and the land between them. No pattern was used to show water in the rivers, and only a few scattered mountains are drawn in, employing the standard Chinese three-ridge style.[19] Location is shown by the numerous written names, with no signs. The result is a map that gives a sense of crowded places and of water where there is no land. Hence, stylistically, it could easily represent an early Vietnamese attempt to render a picture of the country.

The information provided by the names on the map seems rather mixed, at least from my initial examination of it. There are over three hundred names, fifteen of them circled for emphasis. These fifteen place-names include the capital city, called here An-nam Long-biên Thành (a reference to the days of Tang dynasty control in the region), the West Lake outside the capital, the site of the cult to the legendary Hùng kings, and twelve provinces. The term for province, used only for the five central (Red River Delta) provinces (Kinh-bắc, Hải-dương, Sơn-tây, Sơn-nam, and An-bang), is *thừa-chính* (seat of the provincial government), derived from the term applied in the 1460s. The other seven provinces, in the northern and western mountains and the then-southern plains, are identified by name only. The curious point here is that one of these "provinces" is Cao-bằng, which did not become a separate province until the late seventeenth century. On the other hand, the normal twelfth province of 1469, the southernmost one of Thuận-hóa, is here listed uncircled in its two old parts of Thuận-châu and Hoa-châu. The new thirteenth province, listed in 1490, Quảng-nam, south of Thuận-hóa, also appears uncircled but is shown as an estuary. Overall, this map shows a decided lack of interest in the southern area. There is no emphasis on Thanh-hoa and Nghệ-an provinces, which formed the home base of the Lê dynasty, and the Western Capital (Tây-kinh), the original home of the Lê, does not appear at all.

The most important evidence for an early date for this map is that An-bang Province is not called An-quảng, as it was from the late sixteenth century on, and Thái-nguyên is not called Ninh-sóc, as it was between 1469 and 1490.[20] The capital is also known as Phụng-thiên, a term Lê Thánh-tông used for it. Although this set of terms might have come from the late fifteenth century, my inclination is to see the map as a product of the Mạc dynasty in the following century. This new dynasty ignored the sites significant to its predecessor, did not control the south—hence its lack of focus in that direction—and had a decided interest in the northern mountains (especially Cao-bằng, where it would take refuge on its defeat). In addition, the Mạc had a strong reason, as I shall discuss below, to maintain the institutions of Lê Thánh-tông's Hồng-đức period (1471–97). Until a more detailed study of the terms on this map can be made, we may see it as the only extant pre-seventeenth-century Vietnamese map.

The standard Lê dynasty atlas derives from the cartographic activity of the second half of the fifteenth century, but all extant copies show signs of having been redrawn in the seventeenth century or later. First of all, An-quảng is the contemporary name for An-bang Province, and this change occurred after the Lê Restoration of 1592. In addition, the map of the capital shows the location of the Trịnh *chúa*'s (lord's) residence, the palace of the prince (*vương phủ*). This is another indication of a post-restoration date, since the Trịnh held power in Thăng-long only after they had placed the Lê back on the throne.[21]

The Lê atlas consists of fifteen maps: the entire country, the capital, and the thirteen provinces (fig. 12.3). The maps generally have a western orientation[22] and are more sophisticated than the Tổng-quát map discussed above. In particular, water is shown in the rivers (the flow of the current) and the sea (roiling waves), thus providing a greater sense of the solidity of the land. The mountains are drawn in the same three-ridge style, but now they fill out the landscape more and provide a better sense of Vietnam's topography. Human constructions (walls, temples, and palaces), when drawn, are shown in frontal elevation on these maps. All other features are noted by written terms. The main purpose of these maps is administrative, and consequently they record the locations of the various jurisdictions (province, prefecture, and district), generally in rectangular boxes with no hierarchy present. Other names record either mountains or riverine features, with an occasional human construction noted, not drawn.

The map of Đại Việt as a whole includes borderlands of the countries to the north and south, China and

18. *Hồng-đức bản đồ*, 50–53 (note 1).

19. For the style of mountains in Chinese cartography, see David Woo, "The Evolution of Mountain Symbols in Traditional Chinese Cartography," paper presented at the 1989 annual meeting of the Association of American Geographers.

20. *Hồng-đức bản đồ*, 189, 196, 198 (note 1).

21. *Hồng-đức bản đồ*, XI, XIV–XV, XXV (note 1); see also Bùi, "Bản đồ," 50–52 (note 2).

22. *Hồng-đức bản đồ*, 2–49 (note 1). The maps of Kinh-bắc and Thái-nguyên have an eastern orientation and those of Hải-dương and Lạng-sơn a northern one. The Lê atlas also came to be known as *Hồng-đức bản đồ*, but I will refer to it in this chapter as the Lê atlas to avoid confusion with the larger collection and published volume.

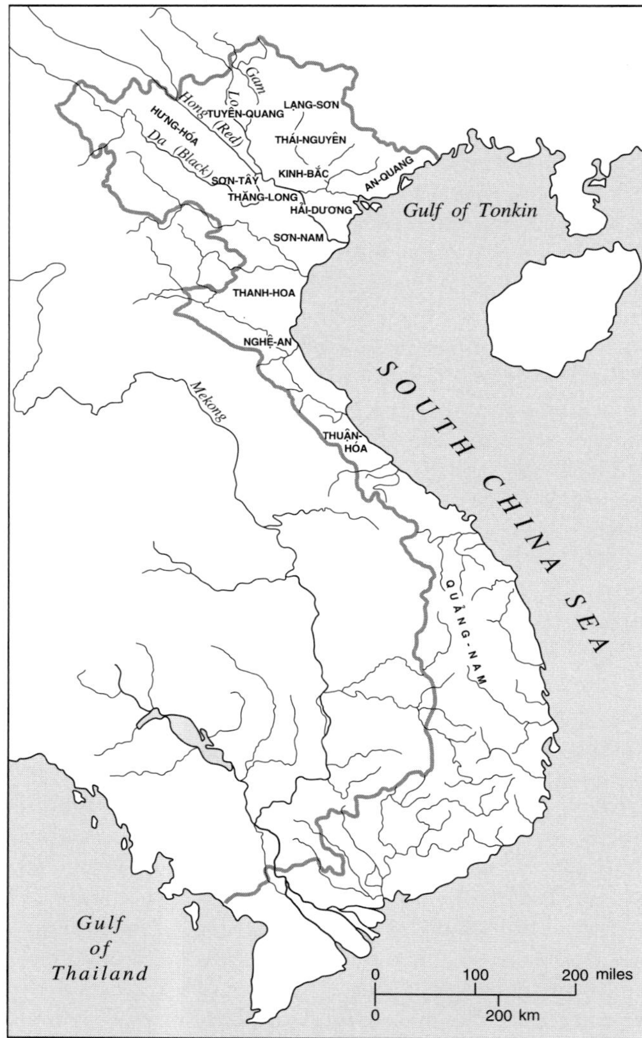

FIG. 12.3. REFERENCE MAP FOR THE STUDY OF ĐẠI VIỆT CARTOGRAPHY.
After Nguyễn Khắc Viện, *Vietnam: A Long History* (Hanoi: Foreign Languages Publishing House, 1987), 99.

Champa, and shows what demarcates Vietnam from these two states: walls on the north and a stone inscription on the south (fig. 12.4).[23] To the northwest is the southwestern Chinese province of Yunnan, and to the northeast are the southeastern Chinese provinces of Guangxi and Guangdong, marked by fortresses and noted as being respectively "location of the Hundred Yue" and "location of the Yue, capital of Chao To" (Triệu Đà, king of Nam Việt in the second century B.C.). These latter phrases are the result of the Vietnamese cultural boundary construction of the fourteenth century. The two Vietnamese capitals are located: Trung-đô (Thăng-long [Hanoi]) in the central Red River Delta and Tây-kinh (the Western Capital) in the upriver part of Thanh-hoa Province. The only temples on this map reflect a ritual configuration. At the center, just outside the capital, lies the temple to the spirit cult of Lý Ông Trọng, a mythic hero believed to have helped the great Chinese emperor Qin Shihuang (third century B.C.) fight the barbarians and to have protected the Vietnamese capital region thereafter.[24] North, east, south, and west of the capital in the delta are four Buddhist temples (Phả-lại, Quỳnh-lâm, Phổ-minh, and Thiên-phúc on Mount Phật-tích), standing guard over the Vietnamese domain.

The map of the capital, Thăng-long, is the major city map we have for Vietnam before 1800 (fig. 12.5). It is strictly for government activities and court rituals, showing little of the thriving everyday and commercial life.[25] The shape of the outer citadel is irregular, fitted as it was within the watercourses surrounding the city. The imperial city within the citadel is, however, decidedly regular, with its proper Sinic north-south orientation. Outside the walls of the citadel are shown the Bảo-thiên Buddhist tower, a central part of Vietnamese Hindu-Buddhist cosmography from the middle of the eleventh century, and the fifteenth-century site of the Nam-giao ceremony, the Confucian sacrifice to heaven. There are also the White Horse Temple (Bạch-mã từ) and the temple of Trấn Vũ, sites of ancient cults, as well as the National College (Quốc-tử-giám). Within the walls of the citadel are various government buildings and palaces, together with the grounds where the Confucian examinations were held.

The maps of the delta provinces (Kinh-bắc, Sơn-tây, Hải-dương, An-quảng, and especially Sơn-nam; fig. 12.6) reflect the riverine nature of their environment.[26] In them we find such features as the intertwining of streams meandering through the delta (*ngã-ba*), canals (*kinh*), bridges (*cầu*), channels (*cửa*), and estuaries (*môn*). In human terms, we see Buddhist temples (*tự*) and spirit cults (*miếu*), as well as some guard stations (*tuần*) and the occasional major historical site (such as the old capital at Cổ-loa) or tomb (for example, that of the old Tang governor Gao Pian). In the northern and western provinces (Thái-nguyên, Tuyên-quang, Hưng-hóa, and Lạng-sơn), mountains are the dominant feature, both drawn and noted. Passes (*ải*) are indicated, as well as a scattering

23. *Hồng-đức bản đồ*, 4–5 (note 1).

24. Keith W. Taylor, "Notes on the *Việt Điện U Linh Tập*," *Vietnam Forum* 8 (1986): 26–59, esp. 38.

25. *Hồng-đức bản đồ*, 8–9 (note 1); Hoàng Đạo Thúy, *Thăng Long, Đông Đô, Hà Nội* (Hanoi, 1971), plate between 46 and 47; Nguyen Thanh-nha, *Tableau économique du Viet Nam aux XVII*^{e} *et XVIII*^{e} *siècles* (Paris: Editions Cujas, 1970), 111–17.

26. On this environment, see Pierre Gourou, *Les paysans du delta tonkinois: Etudes de géographie humaine* (1936; Paris: Mouton, 1965), 17–108 ("Le milieu physique").

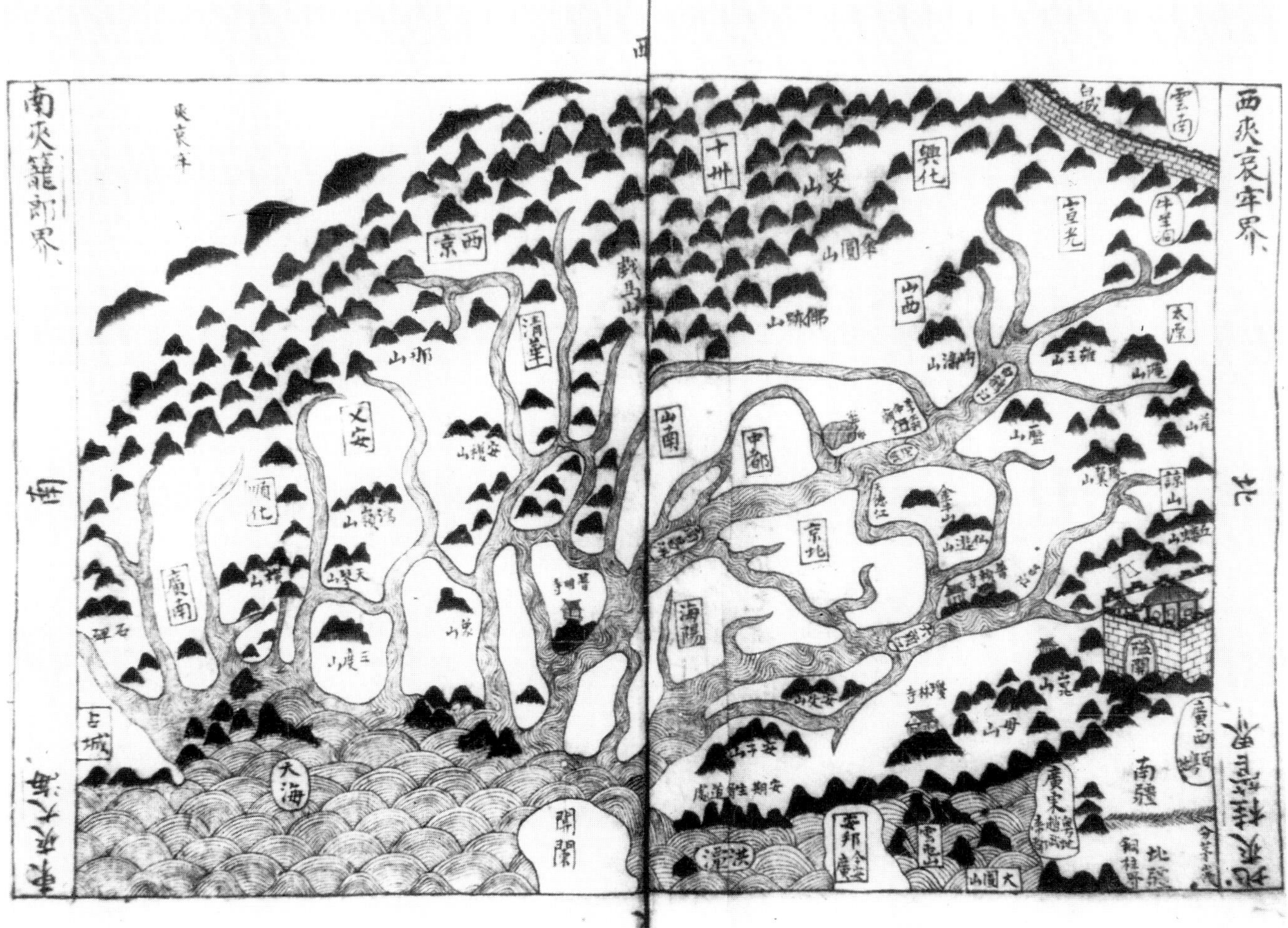

FIG. 12.4. ĐẠI VIỆT FROM THE LÊ ATLAS. This seventeenth-century (Trịnh) copy of the late fifteenth-century map shows the country in its classic premodern format (west is at the top).

Photograph courtesy of Hamilton Library, University of Hawaii at Manoa, Honolulu (microfilm collection, A.2499).

of other natural features. The main human features are military camps (*doanh*), guard posts (*tuần*), and walled cities (*thành*). Lạng-sơn Province shows a large fortress with a gate leading into the Chinese province of Guangxi. Because the Lê dynasty base was originally in the two southern provinces of Thanh-hoa and Nghệ-an, these two maps come first in the atlas. But they, together with the newer provinces of Thuận-hóa and Quảng-nam, share the same features. They are part of what is now central Vietnam, and they have narrow lowlands lying between the mountains and the sea, with short streams running parallel to each other, usually from west to east. The prominent feature is consequently the estuary (*môn*).

The political situation of the sixteenth, seventeenth, and eighteenth centuries ensured that the cartographic pattern established by the Lê king Thánh-tông in the late fifteenth century would continue. When the dynasty fell into chaos following the death of Thánh-tông's son Hiến-tông in 1504, later rulers made various efforts to restore what Thánh-tông had established in his great Hồng-đức period. This was achieved only when the powerful minister Mạc Đăng Dung seized the throne and began his own dynasty in 1528. His legitimacy and that of his family lay in rebuilding central power along the bureaucratic lines Thánh-tông had first set up sixty years earlier.[27] As far as we can tell from the surviving evidence, the Mạc retained and developed the fifteenth-century institutions. If the Tổng-quát map is a Mạc production, they continued the Hồng-đức provincial organization, at least in the territory they controlled.

27. John K. Whitmore, "*Chung-hsing* and *Ch'eng-t'ung* in Đại Việt: Historiography in and of the Sixteenth Century," in *Textual Studies on the Vietnamese Past*, ed. Keith W. Taylor (forthcoming).

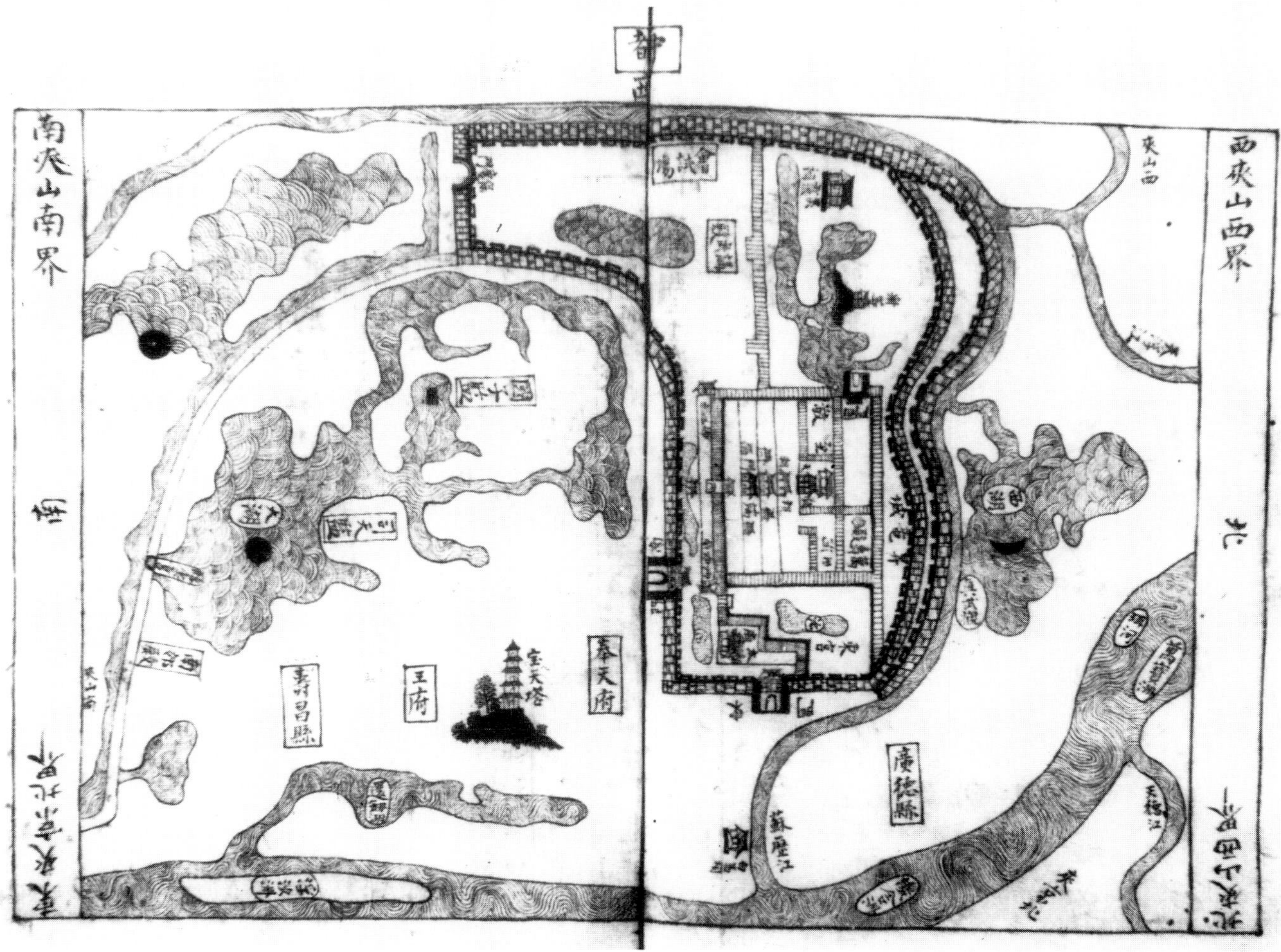

FIG. 12.5. THE CAPITAL FROM THE LÊ ATLAS. This seventeenth-century (Trịnh) copy of the late fifteenth-century map shows the Vietnamese capital of Thăng-long (now Hanoi), fitted within the waters of the Red River and its delta (west is at the top).

Photograph courtesy of Hamilton Library, University of Hawaii at Manoa, Honolulu (microfilm collection, A.2499).

The Trịnh military regime, which brought the Lê back to the throne, picked up the Hồng-đức institutions preserved by the Mạc. The Trịnh did not, however, emphasize the civilian elements as they contested the Lê legacy with their erstwhile allies, the Nguyễn. The latter had established their base on the southern frontier in the newest province of Quảng-nam during the sixteenth century and now claimed that the Trịnh were usurpers. Since these two families remained at loggerheads for almost two centuries, the cartographic tradition stayed frozen: both sides considered Thánh-tông's thirteen provinces to be the established pattern.

In the middle of the seventeenth century, the collection that we now call the *Hồng-đức bản đồ* began to come together. Lê/Trịnh literati evidently redid the Lê atlas, continuing the Hồng-đức work with minor changes. One scholar, Đỗ Bá, compiled the above-mentioned "Bản-quốc bản-đồ tổng-quát mục-lục," joined it with the (possibly) Mạc map, and placed it at the end of the redrawn atlas late in the century. He also added the set of four itineraries that I shall discuss in the following section.[28] To all appearances, these remained the official maps of Đại Việt for the following century to the end of the Lê dynasty in 1787. Despite population growth, little change occurred in the organization of the kingdom.[29]

The major geographical change for the Trịnh zone of Đại Việt came later in the seventeenth century when their

28. *Hồng-đức bản đồ*, XV–XVI, XXVII–XXVIII, 52–53, 68–69 (note 1).

29. Đặng Phương-nghi, *Les institutions publiques du Viêt-Nam au XVIIIe siècle* (Paris: Ecole Française d'Extrême-Orient, 1969), 77–79; Gaspardone, "Bibliographie," 47 (note 11), noted a report of a 1723 "newly established map" that maintained the old Hồng-đức system.

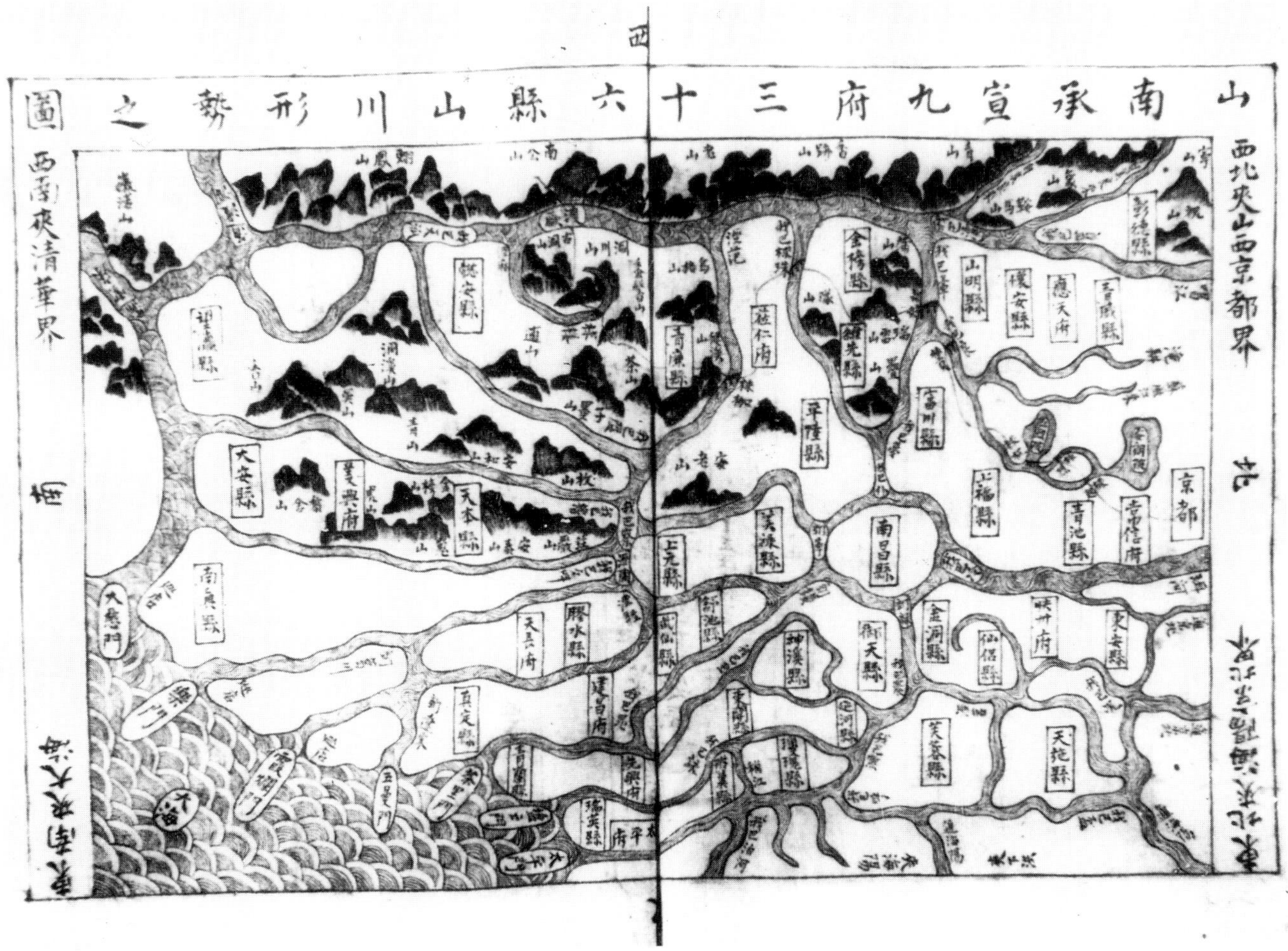

FIG. 12.6. A PROVINCE FROM THE LÊ ATLAS. This seventeenth-century (Trịnh) copy of the late fifteenth-century map shows Sơn-nam Province in the southeastern portion of the Red River Delta and its watery domain (west is at the top). Photograph courtesy of Hamilton Library, University of Hawaii at Manoa, Honolulu (microfilm collection, A.2499).

forces were finally able to drive the Mạc out of the latter's refuge in the northern mountain prefecture (*phủ*) of Cao-bằng. The Mạc remnants had fled there following their defeat in 1592 and had received diplomatic protection from the Ming dynasty in China. The Ming themselves were defeated by the Manchus in 1644, but for the next two decades the Lê/Trịnh court was, first, uncertain about the Qing dynasty's intentions and, second, totally involved in its attempt to smash its recalcitrant Nguyễn rivals in the south. Then in 1667 the Trịnh seized Cao-bằng, but the Chinese forced them to give it up through diplomatic pressure two years later. Finally, in 1677 the Trịnh managed, this time by diplomacy, to regain the territory.[30] There are three surviving maps of Cao-bằng within the *Hồng-đức bản đồ*, and they were probably drawn in the two years between the original conquest and the Chinese-induced surrender, that is, 1667–69. These maps show the prefecture as a whole, the fortified camp of Mục-mã, and the region of Cao-bằng city, known as Phục-hòa.[31] The style of the first map is once more simplistic, similar to that of the Tổng-quát map. Again the rivers are merely outlined, while the pattern of mountains in this highland area remains more similar to that of the Lê atlas, thus bestowing a better topographical sense. A different element, like the itineraries below, is the indication of roads and trails throughout the territory. Administrative units are shown by their

30. Keith W. Taylor, "The Literati Revival in Seventeenth-Century Vietnam," *Journal of Southeast Asian Studies* 18 (1987): 1–22, esp. 17; *Hồng-đức bản đồ*, XIV, XXVII (note 1).

31. *Hồng-đức bản đồ*, 174–85 (note 1).

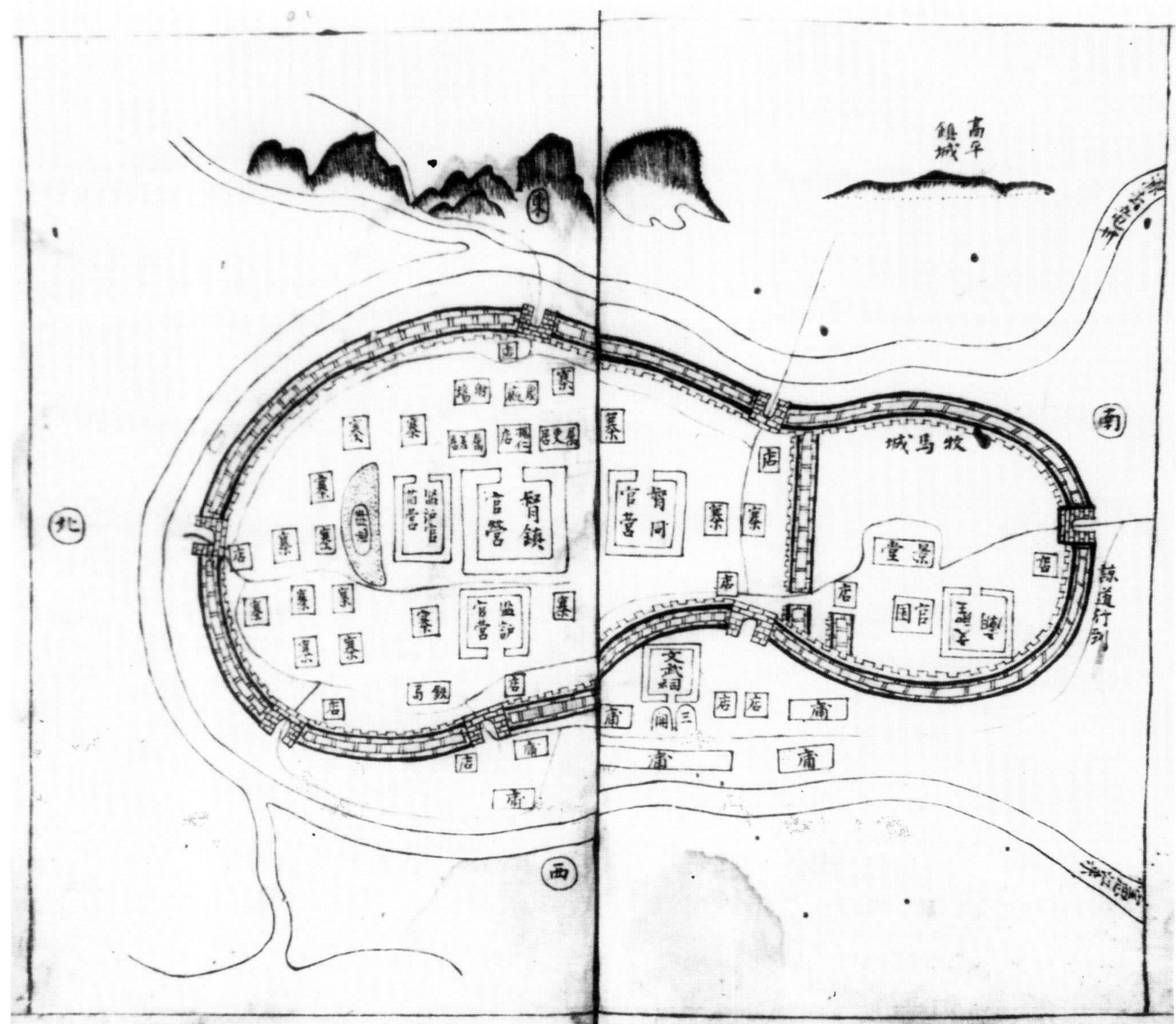

FIG. 12.7. MAP OF THE MỤC-MÃ FORTIFICATION. This seventeenth-century (Trịnh) map of the Mục-mã fortress, Cao-bằng Prefecture, in the northern mountains near the Chinese border, shows the military and civil settlement of the Mạc stronghold (east is at the top).

Photograph courtesy of Hamilton Library, University of Hawaii at Manoa, Honolulu (microfilm collection, A.2499).

names in ovals. Besides the mountain terrain and its settlements (*tổng*), the major elements of the prefectural map are military: eleven camps (*đồn*), the main fortification of Mục-mã, and the walled city of Phục-hòa. The second map is a detailed study of the Mục-mã fortification (fig. 12.7), locating the walls, twelve strong points (*điếm*), sixteen camps (*trại*) within the walls, and five settlements (*phố*) just west of the fortress. Compared with the detail of this map, the third map—of the walled city and its immediate environs—is merely sketched in. These maps, undoubtedly deriving from the 1667 military campaign and its aftermath, supplemented the standard provincial map of Thái-nguyên (to which Cao-bằng Prefecture then belonged) in the Lê atlas.

When the Nguyễn dynasty was establishing itself after 1802, an unknown scholar brought these seventeenth-century northern maps—those collected by Đỗ Bá (the Lê atlas, the Tổng-quát map, and the set of four itineraries discussed in the next section) and the three Cao-bằng maps—together with the *Bình-nam đồ*, the major

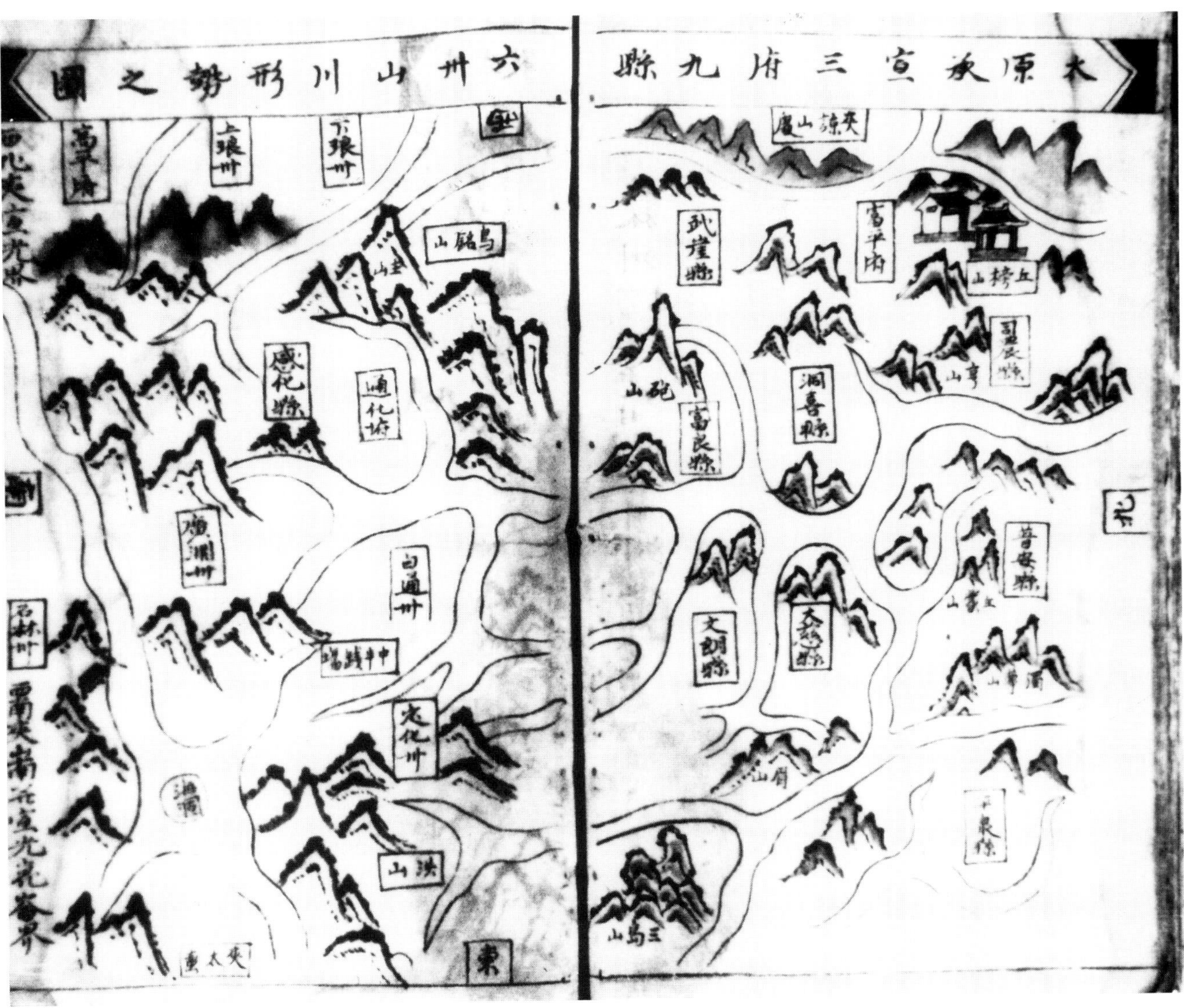

FIG. 12.8. A PROVINCE FROM THE *AN-NAM HÌNH-THẮNG ĐỒ*. This possibly mid-eighteenth-century (Trịnh) copy of the late fifteenth-century map shows the northern highland province of Thái-nguyên and the new vertical style of illustrating the mountains (west is at the top).

Photograph courtesy of Hamilton Library, University of Hawaii at Manoa, Honolulu (microfilm collection, A.3034).

map made in the south, and the *Đại-Man quốc-đồ*, a later one from the north (both discussed below), probably to help set up the administration of the then newly unified country.[32] This came to form the present collection called the *Hồng-đức bản đồ*. Although this is the most important collection, others are known to exist, evidently copies (at least in part) of it.

The chief example of a later version is the *An-nam hình-thắng đồ* (Illustrated maps of An-nan).[33] The versions of the Lê atlas maps found in the *An-nam hình-thắng đồ* are basically the same as in the *Hồng-đức bản đồ*; even the orientation of each map is the same (Kinh-bắc an eastern one, Hải-dương northern). Yet the style of the drawing differs greatly (fig. 12.8). Although again only lines are used to delineate water and land, the mountains have moved from the earlier three-ridge style toward the naturalistic landscape mode with its vertical exag-

32. *Hồng-đức bản đồ*, XIV, XVII, XXVII, XXIX (note 1).
33. A.3034 (reel 114, no. 171).

geration, just as in Chinese cartography.[34] In this way the Vietnamese imitated the karst formations of their northern border. The mountains are no longer mere outlines but are well drawn, with specks of vegetation on them. Temples and walls are given a more Chinese quality as well. One map even shows waves lapping the shore. Somewhat arbitrarily, I believe that the redrawing of the maps now in the *An-nam hình-thắng đồ* took place in the eighteenth century. The one substantive change that leaps to the eye is in the center of the map of the country. No longer is the temple to the cult of Lý Ông Trọng there (though it is on the map of Sơn-tây Province); rather, the Bảo-thiên tower now stands at the capital. The emphasis on this Buddhist protective element at the center of Đại Việt may have been a result of the Buddhist resurgence in the seventeenth and eighteenth centuries.

In general, we may say that the tradition of the Lê atlas remained a strong one in Vietnamese history. It appears in a number of different versions and artistic styles through the eighteenth and nineteenth centuries and into the twentieth. Yet the information contained on the maps varied little. As Joseph Schwartzberg has noted in viewing these collections, "There appears to be a variety of individuals drawing the maps, which would help to account for the wide variety of styles in the depiction of particular types of features. Here and there are some idiosyncratic touches."[35] A good example of how far this atlas tradition continued and of how little it actually changed may be seen in an artistically redrawn version from the early twentieth century (see below). This version almost exactly maintains (with a few errors) the information contained in the Lê atlas of at least two centuries earlier.[36]

Itineraries of Đại Việt

The other major type of Lê dynasty map is the itinerary, generally tracing the route from the Vietnamese capital to a location on or beyond either the southern or the northern border. This type of cartography seems to have begun with the great military expedition led by Lê Thánh-tông as he crushed the southern state of Champa in 1471. The *Đại-Việt sử-ký toàn-thư* noted that on this campaign the king consulted a map of Champa as his armies moved south and began to change the names linked to the mountains and rivers there. With the Vietnamese marching into Champa, Thánh-tông became apprehensive about the difficulties of knowing the terrain and had a local chief help draw up another, more detailed map of the area, focusing on strategic points and on good routes through the mountains and across the streams.[37] These efforts undoubtedly laid the foundation for the first of four sets of maps, the route south to Champa, in the *Thiên-nam tứ-chí lộ-đồ thư* (Book of maps of the major routes of Thiên-nam). Again, when the maps were originally drawn is unknown. The term Thiên-nam (South of Heaven) implies the fifteenth century, since Thánh-tông was the first to use it.

Once more, we know the text only in its seventeenth-century form. As I noted earlier, the scholar Đỗ Bá brought it together with the Lê atlas and the "Bản-quốc bản-đồ tổng-quát mục-lục" late in the seventeenth century at a time when road maintenance had taken on added importance.[38] The four routes move from the Vietnamese capital of Thăng-long: south to the Cham capital (thirty-one pages of maps); northeast to the two regions of Khâm and Niệm along the coast near the Chinese border (ten pages of maps); northwest to the southwestern Chinese province of Yunnan (ten pages of maps); and north to the great gateway into the southeastern Chinese province of Guangxi (ten pages of maps).[39] The style of these maps also returns to that of the Tổng-quát map, with only lines separating land and water. Again, however, as in the Lê atlas, the topography is shown in the form of three-ridge mountains and the administrative units are indicated by names within boxes. Some human construction is sketched in, particularly the southern fortifications. Being itineraries, the maps show the paths to be taken through the landscape (fig. 12.9). Although the text of each section seems to speak of routes by land, water, and sea, what they mean is that the maps show the distinctive features of all three environments, differing from route to route.

The route to the southern border and Champa[40] begins by noting the features for the different environments: land—inns and bridges; water—rivers, canals, and harbors; sea—estuaries, currents, shallows, and deeps. Compared with the Lê atlas, the maps for these itineraries show much more detail of the common and commercial life in Vietnam. As we move across the southern provinces, we encounter villages, markets, inns, temples, guard stations, and other local institutions lying close by the road. Bridges, ferries, creeks, confluences of streams, rapids, estuaries, and dangerous places are noted. The route

34. Woo, "Evolution" (note 19).

35. Personal communication (July 1991).

36. Tạ Trọng Hiệp, "Les fonds de livres en Hán Nôm hors du Vietnam: Eléments d'inventaires," *Bulletin de l'Ecole Française d'Extrême-Orient* 75 (1986): 267–93, esp. 285–86. See also Bùi, "Bản đồ," 51–52 (note 2).

37. *Đại-Việt sử-ký toàn-thư*, 2:682(27) and 683(1) (note 7); Gustave Dumoutier, "Etude sur un portulan annamite du XVe siècle," *Bulletin de Géographie Historique et Descriptive* 11 (1896): 141–204, esp. 141–42.

38. *Hồng-đức bản đồ*, XII–XIII, XXV–XXVI (note 1); Henri Maspero, "Le protectorat général d'Annam sous les T'ang (I): Essai de géographie historique," *Bulletin de l'Ecole Française d'Extrême-Orient* 10 (1910): 539–84, esp. 542; Nguyen, *Tableau*, 177–81 (note 25).

39. *Hồng-đức bản đồ*, 66–137 (note 1). Curiously, the routes west into Laos are ignored; see Nguyen, *Tableau*, 195 (note 25).

40. *Hồng-đức bản đồ*, 70–103 (note 1).

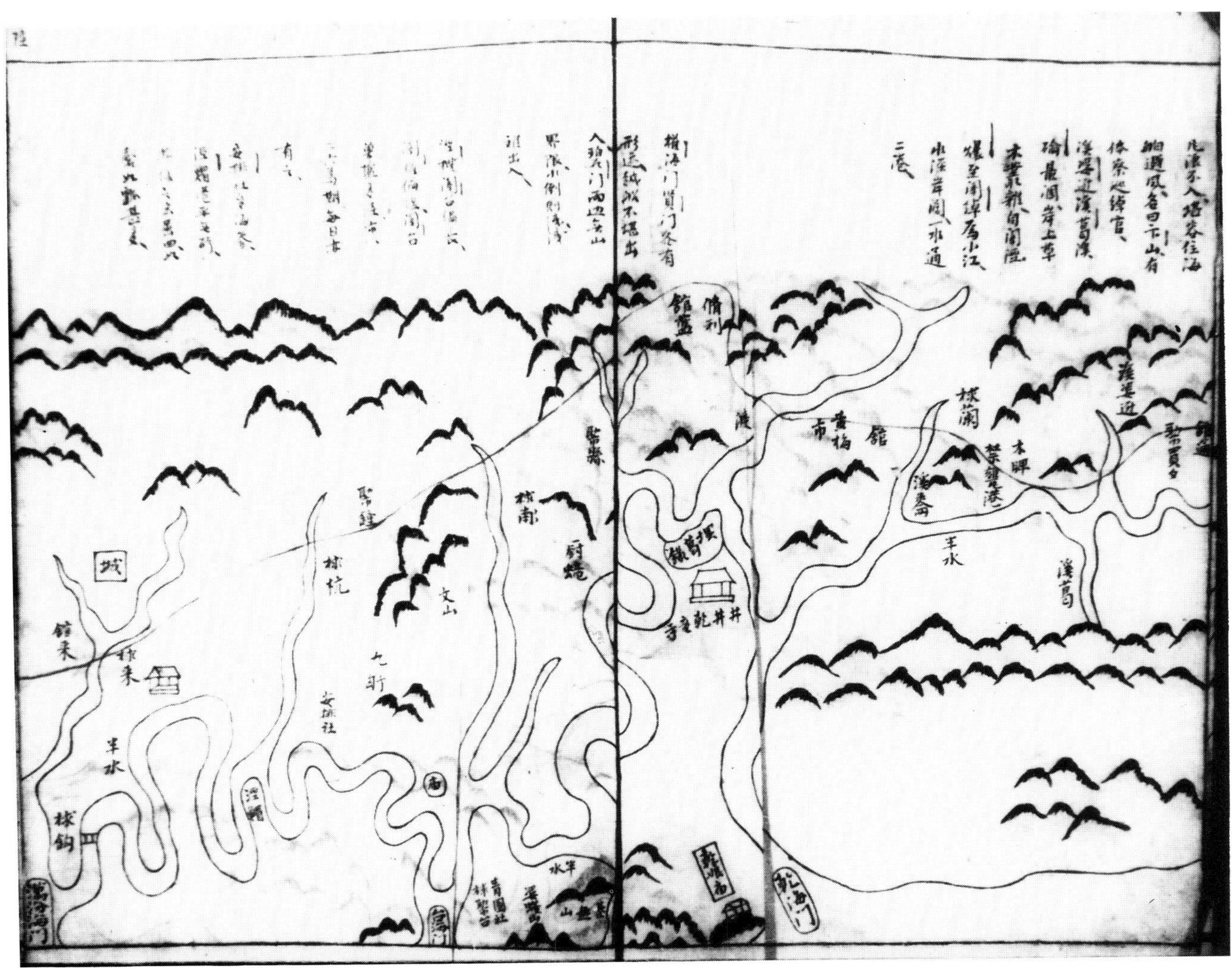

FIG. 12.9. *THIÊN-NAM TỨ-CHÍ LỘ-ĐỒ THƯ*, PORTION OF THE ITINERARY TO THE SOUTH. This portion of the seventeenth-century (Trịnh) itinerary shows the road through the mountainous area of southern Thanh-hoa Province and northern Nghệ-an Province in north-central Vietnam (west is at the top).
Photograph courtesy of Hamilton Library, University of Hawaii at Manoa, Honolulu (microfilm collection, A.2499).

moves out of the Red River Delta, across Thanh-hoa and Nghệ-an, past the old southern border (Nam-giới), and into the militarized zone. Here the Nguyễn, controlling the south, built their defensive walls, which the Trịnh assaulted sporadically without success for half a century (from the 1620s to the 1670s). Marked out on the map are the series of walls, known generally under the name Đồng-hới, and the military support accompanying them: arms depots, camps, entrenched and artillery positions, granaries, and elephant stables. Interspersed among these features are the villages and markets of everyday life. This zone ends on the map at about Đà-nẵng. After that the detail lessens somewhat—not surprisingly, since it was an itinerary to the south made in the north. Going past the major port of Hội-an, villages, lagoons, estuaries, islands, and more mountains are the major features along the coast as it stretches south. Beyond Nha-trang and Cam-ranh Bay, the capital of Champa is reached. Here distances are distorted as the northern sense of location became vague, trying to blend what mapmakers heard of the distant reality with the detail of their old maps.[41]

The other three routes in the collection all concern the north, going from the Vietnamese capital to some point on or near the Chinese border. The first moves to the northeast, going through the delta to the coast. Besides the usual inns, bridges, and streams, there is an emphasis on the sea, with its rocks and waves. The main local human institution noted is the market, perhaps indi-

41. *Hồng-đức bản đồ*, XIII, XXVI (note 1).

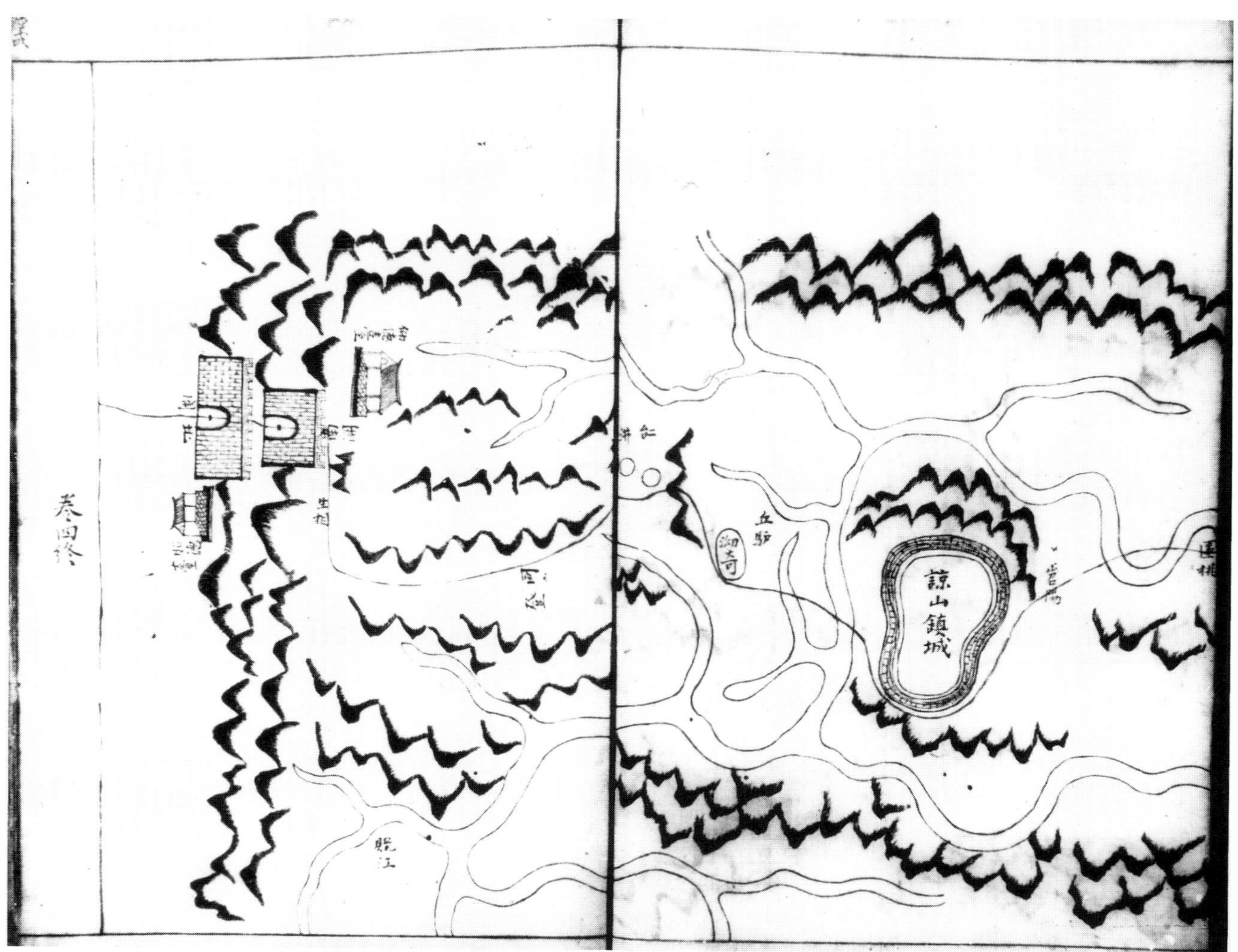

FIG. 12.10. *THIÊN-NAM TỨ-CHÍ LỘ-ĐỒ THƯ*, PORTION OF THE ITINERARY TO GUANGXI PROVINCE. This is the last portion of the itinerary to the north through Lạng-sơn Province (and its citadel) to the gateway leading into Guangxi Province, the route generally taken by Vietnamese embassies to China.
Photograph courtesy of Hamilton Library, University of Hawaii at Manoa, Honolulu (microfilm collection, A.2499).

cating a commercial route. The second route goes to the northwest, through the mountains to Yunnan. Here the emphasis is on the mountains and on the rivers passing through them. Military camps and posts are indicated, but the main human feature, noted only by name, is the *tổng* or upland community. The route moves upstream, through the headwaters, to Guangnan Prefecture in Yunnan. The final route was the official one for Vietnamese embassies traveling to the Chinese capital of Beijing. Going north from the capital, the map shows the usual features, natural and human. Markets, posts, and camps appear before the route moves upriver through various defiles past the walled city of Lạng-sơn to the great gate on the Chinese border leading into Guangxi Province (fig. 12.10). These maps do not show the mountain communities as those of the previous route do.

The *Thiên-nam tứ-chí lộ-đồ thư* is thus a guide to a good part of the kingdom of Đại Việt. The maps show us the "beaten paths," the main communication routes throughout the kingdom, and much of the natural and human detail lying along these routes. They also show us the international routes of communication by land, ignoring the important sea connections. Yet, as I have noted, the maps cannot show us all of Vietnam at the time. The Lê/Trịnh cartographers were blocked by the Nguyễn and their defensive walls from penetrating the southern lands. Here Vietnamese society was expanding southward down the coastal lowlands through Cham territory into the Khmer domain, building roads as it went. For information on this area, we need to look at the *Bình-nam đồ* (Maps of the pacification of the south), covering "from [the walls of] Đồng-hới to the border of

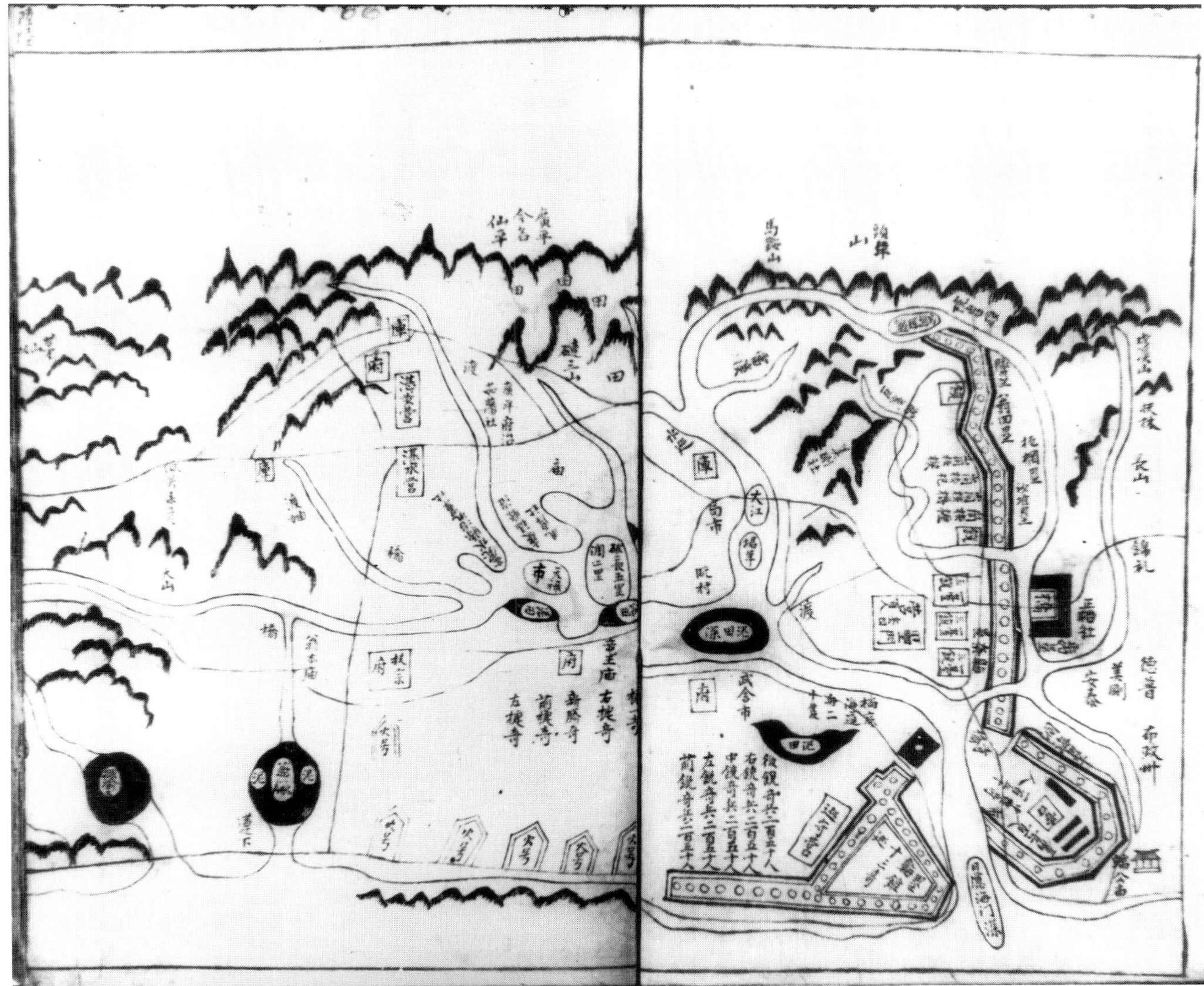

FIG. 12.11. *BÌNH-NAM ĐỒ*, NORTHERN END OF THE ITINERARY. This section of the seventeenth-century southern (Nguyễn) itinerary shows the fortifications on their northern border in Quảng-binh (formerly Thuận-hóa) Province (west is at the top).

Photograph courtesy of Hamilton Library, University of Hawaii at Manoa, Honolulu (microfilm collection, A.2499).

Cambodia" (twenty-eight pages of maps). It appears to date from 1654, according to the cyclic characters on the title page and information from the text.[42] These Nguyễn maps begin with much detail on their fortifications and the militarized zone (fig. 12.11). Their style differs little from that of the northern maps just discussed, though they are somewhat more artistic, with elephants pictured in the mountains and turtles in the sea. As we move south, we gain information not provided by the northern maps—on rice fields, residences, temples, troop strengths, and the depths of the channels in the estuaries. We even see a Buddhist tower adjacent to a lake in the old cosmographic style. We pass the old Cham city, then Đà-nẵng, the international port of Hội-an, and Quy-nhơn, go beyond Lê Thánh-tông's conquest of 1471 and Nha-trang, through the remains of Champa, until we reach Angkor in Cambodia (fig. 12.12). The detail lessens as we go farther south and gets quite thin beyond the boundaries of the fifteenth-century territory. Yet while the geographic relationship of the Mekong Delta and

42. *Hồng-đức bản đồ*, XIII–XIV, XXVI–XXVII, 138–67 (note 1); Nguyen, *Tableau*, 179–80 (note 25). Tạ Trọng Hiệp believes the date should be 1774 at the earliest; notes from Joseph E. Schwartzberg.

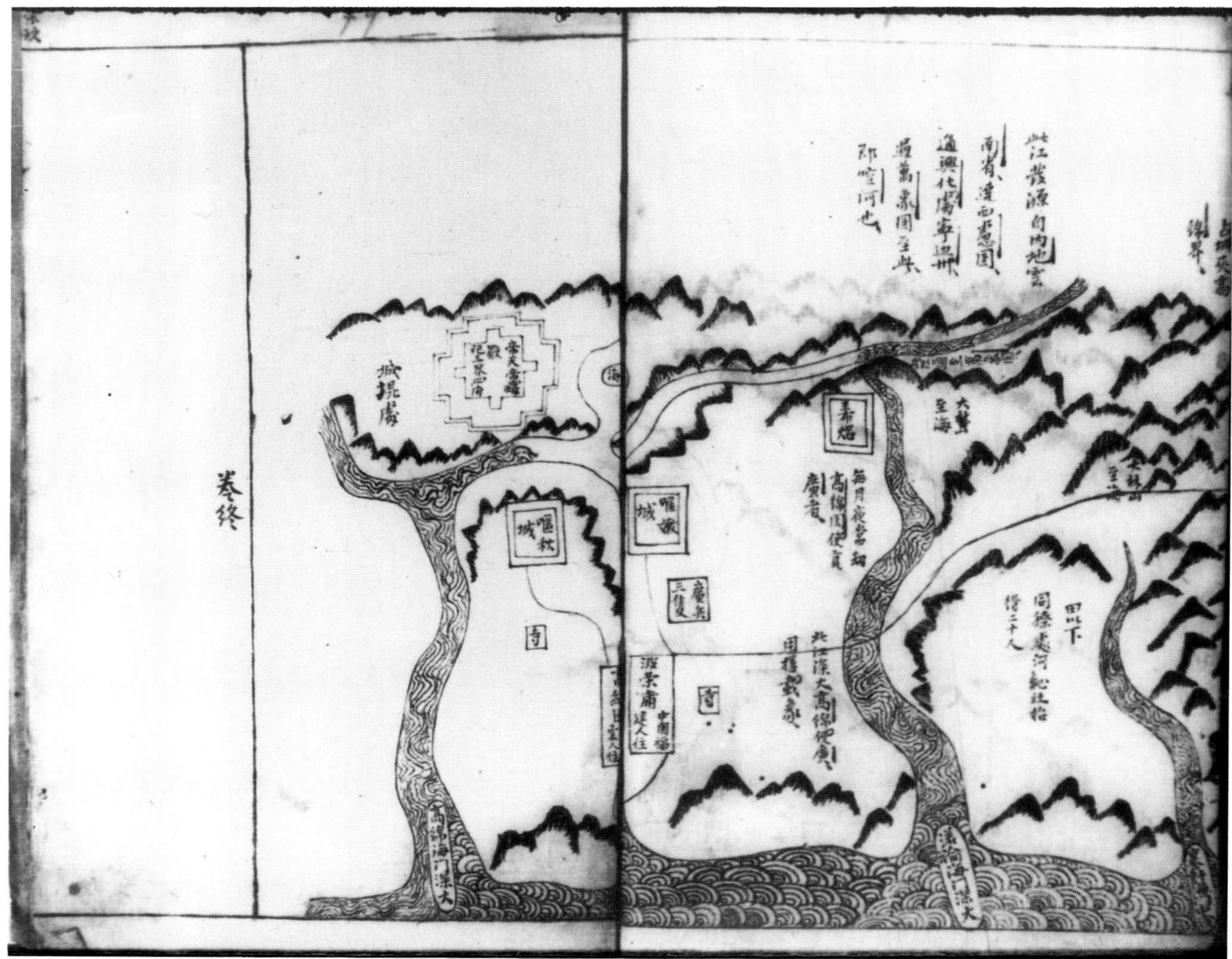

FIG. 12.12. *BÌNH-NAM ĐỒ*, SOUTHERN END OF THE ITINERARY. This section from the same map depicted in figure 12.11 shows Cambodia, giving a planimetric view of Angkor and the contemporary conception of the lower Mekong River (west is at the top).

Photograph courtesy of Hamilton Library, University of Hawaii at Manoa, Honolulu (microfilm collection, A.2499).

Angkor was beyond the ken of the Nguyễn cartographers, they did pick up some interesting detail from Cambodia. Angkor itself is shown in a planimetric view with an interesting configuration. The South China Sea commercial connection is noted, as Cambodia is stated to have a Cantonese settlement and a Fujianese entrepôt to the east (Ba-vinh Phố; whether Hokkien or Teochiu speakers cannot be determined). The Mekong connection to Laos and Yunnan is recognized, as is the King of Fire, a highland chieftain, and his connection with Cambodia. These maps explore the *Nam-tiến*, the Vietnamese push southward during the seventeenth and eighteenth centuries (fig. 12.13), and provide interesting data on the social expansion and commercial interests of the Nguyễn domain.

Despite some problems, all the itineraries described so far, both northern and southern, appear to date from the latter half of the seventeenth century. This brings us again to the *An-nam hình-thắng đồ*, the collection of maps paralleling the *Hồng-đức bản đồ*. Although the latter has been published in detail, the *An-nam hình-thắng đồ* still exists only in manuscript and has yet to be closely examined. As noted above, parts of this collection are much better illustrated than the *Hồng-đức bản đồ*. On a very general impression, I am placing it in the eighteenth century, but the question needs to be addressed in detail. It is a northern production and reproduces, with two additional sections, the Lê atlas (as I noted above) and the *Thiên-nam tứ-chí lộ-đồ thư* of the *Hồng-đức bản đồ* (table 12.1). Yet the four itineraries are here untitled, somewhat longer, and treated as internal routes, not links to the borders and foreign parts. They do not start explicitly at the Vietnamese capital, and they do not end on foreign soil. The route to the south (well illustrated with mountains and temples) goes beyond the Cham capital, as Vietnamese society itself did; the route to the northwest ends at the border of Yunnan; and we see only the great gate as the culmination of the route to the north. Instead, these routes are seen as going *through* the Vietnamese provinces.

A very interesting addition is the untitled second section of the *An-nam hình-thắng đồ* (fifty pages of maps). It is a detailed look at the northwestern mountains, their rivers, and the human settlements along the banks of these streams. Unlike the Lê atlas and the route to the south in the *An-nam hình-thắng đồ*, this set of maps is very simply sketched, even more so than the Tổng-quát map. It consists only of lines indicating either the riverbanks or the route being illustrated. Boxed-in names show communities, and names without boxes mark other features. Despite the region, no mountains are shown. The value of this section lies in the multitude of villages and other institutions shown lying along the streams.

The *An-nam hình-thắng đồ* is thus possibly an eighteenth-century collection from the northern state of Đại Việt. Other collections also exist in different styles from about this time and place, combining the Lê atlas and the itineraries or including the latter alone. We see the *Toản-tập Thiên-nam lộ-đồ thư* (Compilation of route maps of Thiên-nam), the *Thiên-nam lộ-đồ* (Route maps of Thiên-nam), and the *Kiền-khôn nhất-lãm* (An overall view of the cosmos)[43] as eighteenth-century variants of both the atlas and the itinerary traditions. Eighteenth-century variants of Nguyễn cartography in the south are the *Bình-nam chỉ-chưởng nhật-trình-đồ* (Map of the daily stages on the route to the south)[44] and again the *Kiền-khôn nhất-lãm*, whose author, Phạm Đình Hổ, probably gained access to the Nguyễn mapping after the Trịnh forces took the Nguyễn capital in the wake of the Tây-sơn revolt. Two of these manuscripts, the *Toản-tập Thiên-nam lộ-đồ thư* and the *Bình-nam chỉ-chưởng nhật-trình-đồ* (fig. 12.14), represent the northern and the southern itinerary styles, and Schwartzberg has examined them in Paris. Apparently woodblock prints, they are illustrated in black ink, with red used to highlight roads and buildings. The first also used gray shading in the mountains. The two are probably typical of the pre-1800 cartographic style.[45]

The fourth section of the *An-nam hình-thắng đồ*, however, is a strikingly illustrated trip through southern China

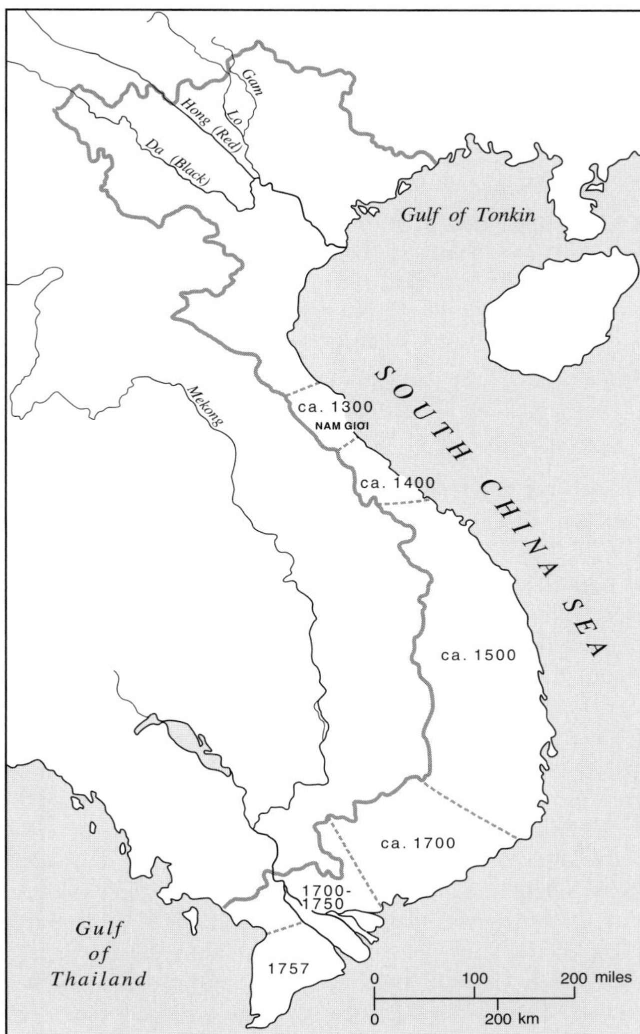

FIG. 12.13. REFERENCE MAP OF VIETNAMESE SOUTHERN EXPANSION. A very simplified map of the expansion. After *Vietnam: A Country Study*, ed. Ronald J. Cima, Federal Research Division, Library of Congress (Washington, D.C.: Federal Research Division, 1989), 22; based on information from Joseph Buttinger, *Vietnam: A Political History* (New York: Praeger, [1968]), 50.

by water (one hundred pages of maps). The drawings include mountains, trees, and settlements. The trip is called an embassy and possibly took place in 1729–30, though it ends before reaching the Chinese capital in northern China. The route goes through the mountains

43. Société Asiatique, Paris, HM2241 (all original documents designated HM are at the Société Asiatique), A.1081, and A.414 (reel 21, no. 58), respectively.

44. HM2207.

45. Gaspardone, "Bibliographie," 46–47 (note 11); Bùi, "Bản đồ," 52 (note 2); Schwartzberg notes.

TABLE 12.1 Comparison of Maps in the *Hồng-đức bản đồ* and the *An-nam hình-thắng đồ*

Hồng-đức bản đồ	*An-nam hình-thắng đồ*
1 Lê atlas, 15 maps (2 pages each for 9 maps and 1 page each for 6 maps)	1 Lê atlas, 15 maps (2 pages each for 9 maps and 1 page each for 6 maps)
2 "Bản-quốc bản-đồ tổng-quát mục-lục," 1 map (2 pages)	2 Untitled (northwest mountains), 50 pages of maps
3 *Thiên-nam tứ-chí lộ-đồ thư* Route to the south, 31 pages of maps Route to the northeast, 10 pages of maps Route to the northwest, 10 pages of maps Route to the north, 10 pages of maps	3 Untitled (*Thiên-nam tứ-chí lộ-đồ thư*) Route to the south, 38 pages of maps Route to the northeast, 12 pages of maps Route to the northwest, 11 pages of maps Route to the north, 12 pages of maps
4 *Bình-nam đồ*, 28 pages of maps	4 *Sứ trình thủy hàng* (An embassy [to China] by water), 100 pages of maps
5 *Đại-Man quốc-đồ*, 1 map (2 pages)	
6 *Cao-bằng phủ toàn đồ* (The complete map of the prefecture of Cao-bằng), 3 maps (2–3 pages each)	

of northern Vietnam, the province of Lạng-sơn, and the great gate on the Guangxi border before picking up the Chinese river system and passing through Siming, Taiping, Nanning, and Guizhou. The final map is of Huaian Prefecture in Jiangsu Province, east-central China.

Two other Vietnamese sets of maps are probably also from the eighteenth century and are similarly rich in artistry. Both go all the way through China to the capital. In the first manuscript, for example, the rivers are in terra cotta, the roads red, the mountains gray, blue, or violet, walls (single and double) of the towns and cities in red, gray, or blue, and village residences in black and white (plate 30).[46] These sets of maps provide well-illustrated Vietnamese views of southern China and of life along its rivers. The view is from the river as both sides stretch away from it—bridges, walls, banners, dwellings, temples, and mountains. The temples are in frontal elevation with considerable detail, and the mountains are in the naturalistic style with vertical emphasis.

The only other Vietnamese map we have of a foreign country before the nineteenth century is also the only map we possess from the Tây-sơn dynasty (1788–1802). This is the *Đại-Man quốc-đồ* (Map of the country of the Great Man [Southern Barbarians]), drawn in the western Vietnamese mountain province of Hưng-hóa at the end of the eighteenth century (1798, with a preface of 1800) (fig. 12.15).[47] The "country" concerned is an overview of the Thai world to the west of Vietnam, stretching from the mountains in the north to the sea in the south. It is full of *mang* (in Thai *muang*) and *trình* (*chieng* or *xieng*), local polities that ruled the wet rice villages. At the center of the map is the Southern Barbarian country capital (apparently Ayudhya rather than Bangkok). Formerly it had been part of Xiêm-la (Siam), the map tells us. The depiction of the coast and its estuaries is inaccurate, as befits a map drawn at an inland site. The map has a northern orientation, from the seacoast to the mountains. The rivers run from north to south with the Mae Nam Yom and Chao Phraya in the center. To the east is the Mekong River system and to the west another river, possibly the Salween. On the eastern side lie Xieng Khouang, the Lao-long capital (Vientiane?), the former Khmer capital (Angkor), and Champa. Like the mapmakers in the earlier *Bình-nam đồ*, these Vietnamese cartographers had difficulty with the Mekong and its delta. Neither portrays the great lake in Cambodia, and the entrances to the sea are confused. This map also shows a reliance on Chinese texts in the terms used for Siam and Cambodia. Such a mixture of prior text and contemporary report probably confused some of the political reality to the south and west. In general, though, the map appears to reflect what was then known, and it was meant to serve as a guide, listing the number of days' travel between locations. Note that all these itineraries are by land or river, not by sea. The style is standard for most of the pre-1800 Vietnamese maps. Lines separate water and land, though as in Burmese maps many of the rivers are fringed by mountains that set off the streams and help define the land. The mountains are the usual three-ridge type and form a decoration on the eastern, northern, and western land fringes of the map, while also filling much of its surface. Lines connecting locations (boxed-in names) show the communication routes. Only the Thai capital is drawn with a double set of boxes and a gate.

46. HM2182 (untitled) and HM2196 (titled *Sứ trình đồ họa* [Illustrated map of an embassy (to China)]); for a description of the different routes taken by Vietnamese embassies north to Beijing, see Trần Văn Giáp, "Relation d'une ambassade annamite en Chine au XVIIIe siècle," *Bulletin de la Société des Etudes Indochinoises*, n.s., 16, no. 3 (1941): 55–81, esp. 55–58.

47. *Hồng-đức bản đồ*, XIV, XXVII, 168–73 (note 1).

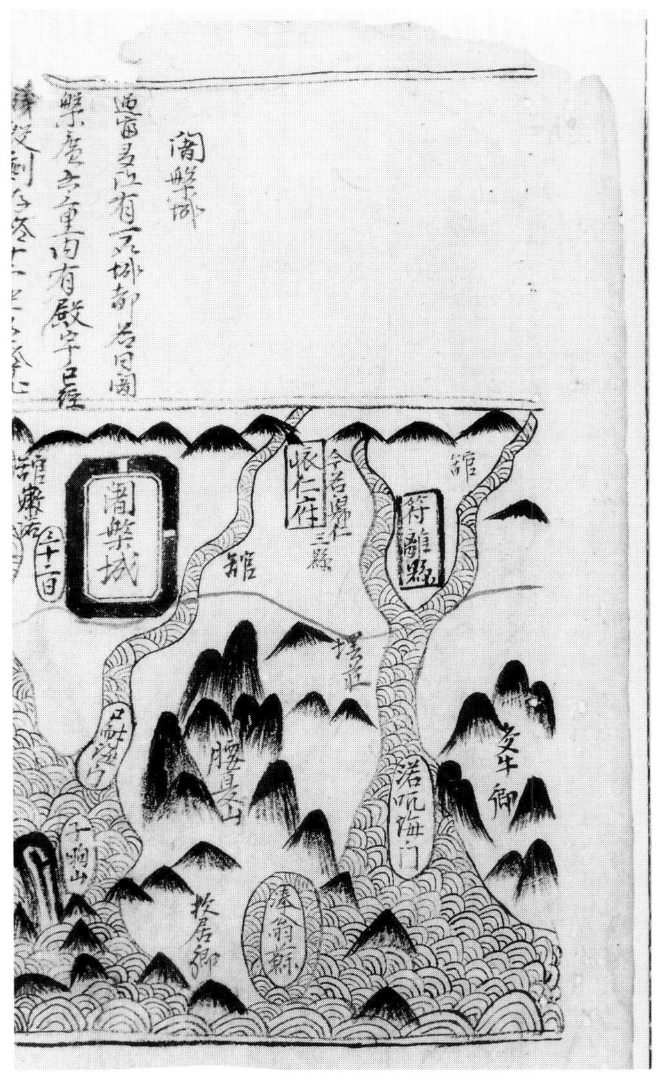

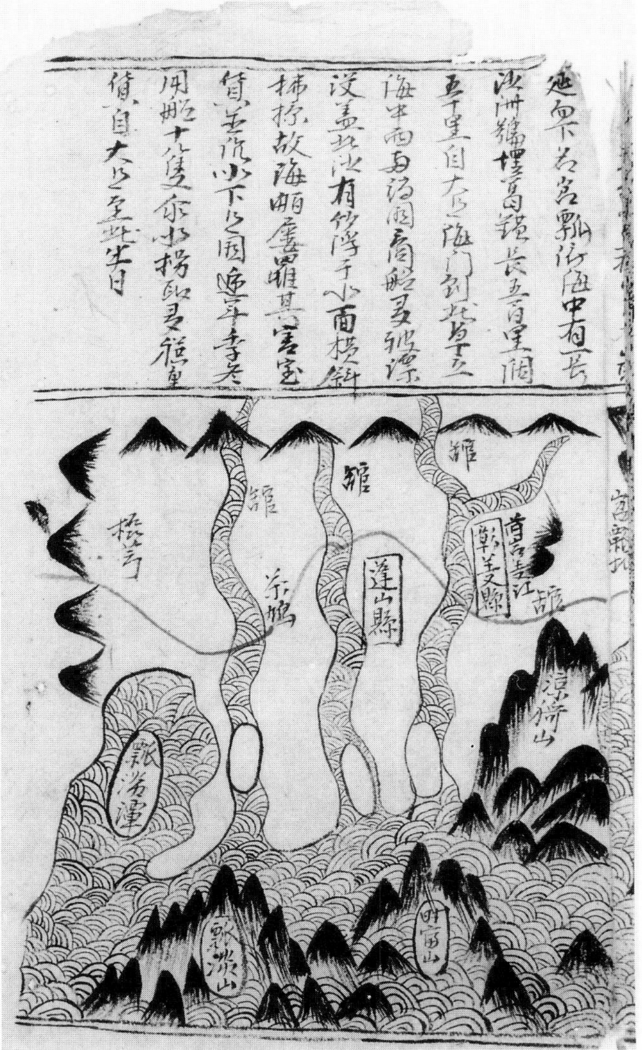

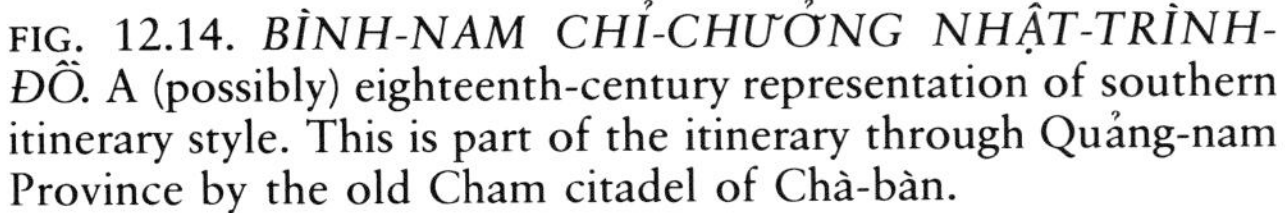

FIG. 12.14. *BÌNH-NAM CHỈ-CHƯỞNG NHẬT-TRÌNH-ĐỒ.* A (possibly) eighteenth-century representation of southern itinerary style. This is part of the itinerary through Quảng-nam Province by the old Cham citadel of Chà-bàn. Size of each folio: ca. 25.5 × 14.5 cm. By permission of the Société Asiatique, Paris (HM2207, fols. 18b–19a).

Maps of Đại Nam

Vietnamese cartography took a sharply different turn in the nineteenth century, becoming both more Western and more Chinese. The simplicity and sketchiness of the Lê style of maps were superseded by these international influences. While the Gia-long emperor (1802–19), founder of the new Nguyễn dynasty, harked back to the great Hồng-đức era of Lê Thánh-tông over three hundred years earlier, he made a conscious effort to bring Chinese influences more strongly into Vietnam. The Western influence came from the small group of Frenchmen who formed part of the international mélange the Nguyễn brought together to conquer the kingdom. This influence is particularly to be seen in the Vauban style of fortification adopted by the Nguyễn and blended with the Chinese ideological influence in building their new type of citadel.[48]

When the Nguyễn first gained power in 1802, one necessary initial task was to gather maps from all parts of the country and all prior regimes in order to begin the cartographic integration of Vietnam, now unified from China to Cambodia. The Nguyễn officials had to go beyond the frozen tradition of the Lê, based as it was

48. Alexander Barton Woodside, *Vietnam and the Chinese Model: A Comparative Study of Nguyễn and Ch'ing Civil Government in the First Half of the Nineteenth Century* (Cambridge: Harvard University Press, 1971), 16.

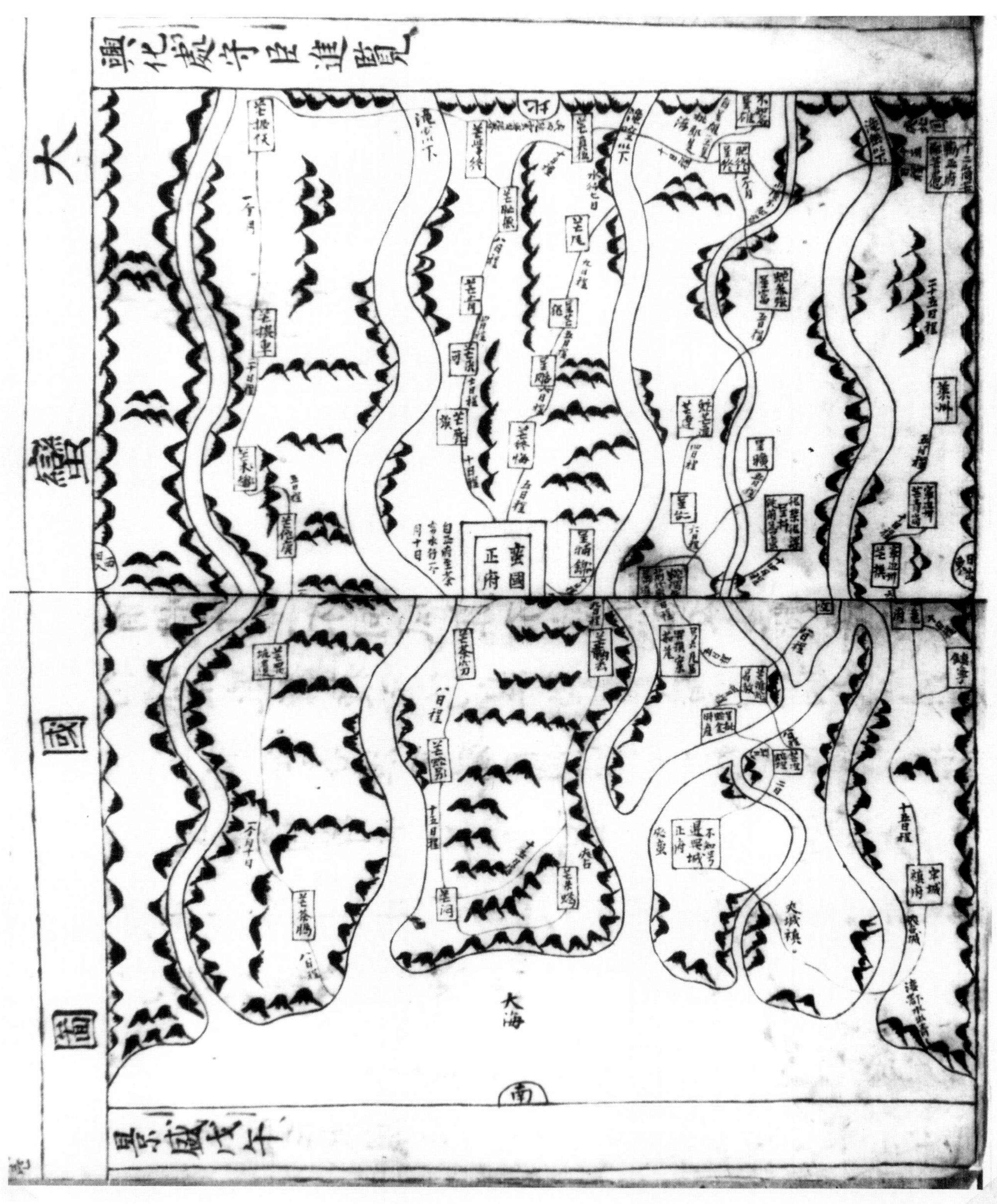

FIG. 12.15. *ĐẠI-MAN QUỐC-ĐỒ* MAP. This 1798 (Tây-sơn) map of the Tai principalities to the west of Vietnam was drawn in the northwestern mountain province of Hưng-hóa and reflects its view of the Mekong, Chao Phraya, and possibly Salween river systems (north is at the top [the book is rotated 90° counterclockwise]).
Photograph courtesy of Hamilton Library, University of Hawaii at Manoa, Honolulu (microfilm collection, A.2499).

on the fifteenth-century work, and add the territory that Vietnamese society had encompassed under its own regime in the south during the seventeenth and eighteenth centuries. In addition, there was the broader Vietnamese knowledge of the international scene with which the Nguyễn had become so familiar during their years out of power. One result of this initial effort was the *Hồng-đức bản đồ* collection that I have used for the Lê period. Though it actually originated in the seventeenth century, the memory of its derivation from the greatness of the fifteenth century remained.

In addition to the international influences, two other elements need to be considered if we are to understand the cartographic development under the Nguyễn, one political, the other bibliographic. Although the Gia-long emperor reigned over all of Vietnam, now called Đại Nam (Great South), he chose not to rule directly in all parts of it. The Nguyễn court in Huế controlled central Vietnam (the old south and the base of its power for over two centuries) but decided to place viceroys in charge of the north and the far south. These viceroys, Nguyễn Văn Thành and Lê Văn Duyệt, respectively, thus stood between the throne and the people they administered.[49] In effect, the two were warlords. In the north, capital of the Lê dynasty and home of its officials, this indirect rule was meant to soothe injured feelings. In the south, the least developed section of the domain, it was merely convenient. In any case, this tripartite division hindered bureaucratic development throughout the country and hampered the accumulation of local data. The bureaucratic and cartographic development took place gradually over three decades. First, Thành in the north was removed in 1816 following the crisis over the newly appointed heir apparent, the future Minh-mạng emperor (1820–40). The latter, when he took the throne on his father's death in 1819, worked to extend the bureaucracy, but not in the south, where Duyệt held sway until he died in 1831. Only then did the throne move in, setting off a massive rebellion in 1833–34, which was crushed. Thus an integrated bureaucratic entity came to exist in Vietnam only in the mid-1830s, and the cartographic achievements followed suit.

The bibliographic development that occurred was a rejection of the atlas form used by the Lê and the adoption of the geography as it evolved under the Ming and Qing dynasties in China. This form was initially used by the Ming government in 1461 under the name *Da Ming yitong zhi* (Comprehensive gazetteer of the Great Ming). The phrase *yitong zhi* (Vietnamese *nhất thống chí*) would continue to be used for this form, and it appeared in Vietnam in the nineteenth century. The geography was organized by province and, within each section, dealt with a standard list of topics. The first such new geography to appear in Vietnam (though without maps) had been the *Ô-châu cận-lục* (Modern record of Ô-châu), a study of the southern Thuận-an area from the sixteenth century during the Mạc dynasty. Now, in the surge of government borrowing from China, the Nguyễn chose to develop this form of writing, and the official country, capital, and provincial maps would eventually appear within it.

In 1806, probably not long after the compilation of the *Hồng-đức bản đồ* collection, Lê Quang Định completed the *Nhất-thống dư-địa chí* (Unified geography) for the Gia-long emperor. Even though it was without maps, it was similar to the work compiled by Đỗ Bá some 150 years earlier. Of the ten chapters, the first four cover routes south and north from the capital, now in the center at Huế, to the borders, and the final six look at the provinces without any particular topical organization.[50] As noted, the Nguyễn court in Huế during this first reign focused mainly on central Vietnam.

In the north, scholars carried on the Lê cartographic tradition as maps of the region appeared in various collections. In 1810 Đám Nghĩa Am compiled the *Thiên tải nhàn đàm* (Concerning ideas of a thousand years), which reproduced the Lê atlas, following its basic pattern (such as the three-ridge mountains) while developing it stylistically. The map of the entire country provided a northerner's view of the southern expansion (fig. 12.16).[51] Just beyond Champa, Am put in Gia-định (the Saigon area), and in a new map of the "southwest" he followed it with Cambodia (Cao-miên) and Siam (Xiêm-la). In the process, he skipped over the southern coast lying between Gia-định and the old Cham capital. Other maps in the text, however, came from the southern *Bình-nam đồ* and cover this missing territory. Artistic touches that were added include tigers romping in the mountains, a crab on the shore, and fish in the sea. The old capital is no longer called the capital, just Thăng-long, and certain details were added to its map that appear to reflect contemporary reality. Somewhat later (1830), another such volume appeared, this time called the *Giao-châu dư-địa đồ* (Geographic maps of Giao-châu [an old Chinese name for northern Vietnam]), with three sections of maps.[52] Drawn in black ink, the maps have an overpainting in

49. Woodside, *Chinese Model*, 102–3, 136, 141–42, 220, 284–85 (note 48); Ralph B. Smith, "Politics and Society in Viêt-Nam during the Early Nguyễn Period (1802–62)," *Journal of the Royal Asiatic Society of Great Britain and Ireland*, 1974, 153–69.

50. A.1829; A.2667; Maspero, "Essai," 543 (note 38); Leonard Aurousseau, Review of Charles B. Maybon, *Histoire moderne du pays d'Annam (1592–1820)*, in *Bulletin de l'Ecole Française d'Extrême-Orient* 20, no. 4 (1920): 73–121, esp. 83 n; Ralph B. Smith, "Sino-Vietnamese Sources for the Nguyễn Period: An Introduction," *Bulletin of the School of Oriental and African Studies* 30 (1967): 600–621, esp. 609; Thái, "Lời nói đầu," x (note 3).

51. HM2125; Bùi, "Bản đồ," 52 (note 2); Schwartzberg notes.

52. HM2240; A.2716 (reel 143, no. 267a); Schwartzberg notes.

FIG. 12.16. *THIÊN TẢI NHÀN ĐÀM* MAP OF THE COUNTRY. This 1810 (Nguyễn) copy of the late fifteenth-century map of Đại Việt (or Thiên-nam) shows the traditional northern (Lê) view of the country and its limited concern for the Vietnamese extension to the south (Gia-định) (west is at the top). Compare figure 12.3.
Size of each folio: ca. 25.3 × 18.7 cm. By permission of the Société Asiatique, Paris (HM2125).

gray for water and shading in the mountains and red on the roads. The illustrations of the mountains go from the three-ridge style to a style representing more naturalistically the strongly vertical karst landscape, though in places more rounded than before. Interestingly, the islands off the coast, instead of being drawn merely as mountains jutting out of the water, are shown in perspective, with shoreline surrounding the peaks (fig. 12.17). The result is a better topographic sense and the beginning of a more realistic approach being taken by the Vietnamese cartographers. Even so, artistic touches appear here as well, including elephants and monkeys. The work also reproduces the map of the Great Man country—the Thai world—in a clearer format while leaving out some detail (such as a reference to Yunnan). A few years after Đàm Nghĩa Am's text, there appeared the *Bắc-thành địa-dư chí* (Geographic record of the north), an examination of the northern provinces, again without maps. Each of the eleven provinces and the old capital (Thăng-long) had its own chapter.[53]

The south, operating almost independently under the viceroy Lê Văn Duyệt, had its own mapping efforts. In 1816, Trần Văn Học completed his map of Gia-định Province (the Saigon region), providing a planimetric view and strongly showing Western influence (fig. 12.18). Học,

53. A.1565 (reel 17, no. 45, and reel 143, no. 264a); A.81 (reel 188, no. 435; French copy); Maspero, "Essai," 543 (note 38); Aurousseau, Review of Maybon, 83–84 n (note 50); Smith, "Sino-Vietnamese Sources," 608 (note 50); Thái, "Lời nói đầu," x (note 3); Woodside, *Chinese Model*, 142 (note 48).

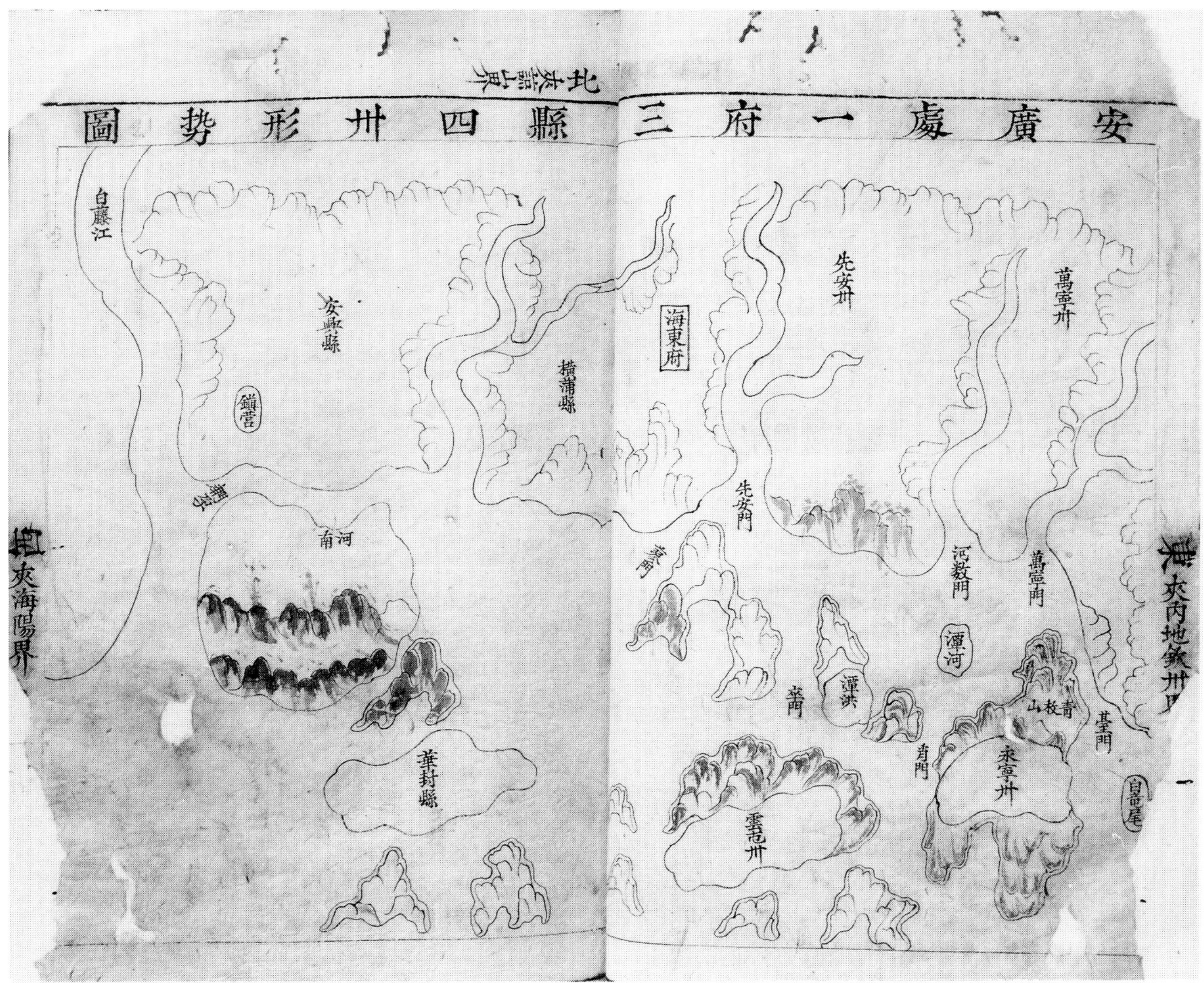

FIG. 12.17. *GIAO-CHÂU DƯ-ĐỊA ĐỒ*, A PROVINCE. This 1830 (Nguyễn) map of An-quảng Province in northeastern Vietnam was a northern production but was drawn with a European-style perspective of the islands (west is at the top). Size of each folio: ca. 25.3 × 17 cm. By permission of the Société Asiatique, Paris (HM2240).

long a companion of the Gia-long emperor, had had contact with the French in India and during the long campaign to conquer the country in the 1790s. He was involved in building the first of the Nguyễn's Vauban-style citadels, that of Saigon, in 1790 and became the main architect for later construction. The map well reflects Học's abilities in this direction. No longer do we have the sketched-in, impressionistic drawings of the Lê. The rivers give a sense of more precise measurement, and the citadel shows its Vauban origins. The roads, paths, and walls seem accurate, with buildings and ponds outlined along them. Unlike the earlier maps of Thăng-long in the north, this map provides a sense of the ordinary and commercial life going on within the city. However, no hills or changes in elevation are shown. Notations provide the locations of various significant natural and human features, but unlike the Lê maps, there are no drawings of temples or other buildings. All is strictly orthogonal, with no frontal views (as of temples, gates, walls, and such).[54] Then, in the 1820s, the former governor of the south, Trịnh Hoài Đức, compiled the *Gia-định thành thông-chí* (Geography of Gia-định) covering the five provinces of the southern region, also without maps.[55] Within a decade, there appeared the *Nam-kỳ hội-đồ* (Collected maps of the south) with maps of the

54. Thái, "Interprétation" (note 3).

55. A.1561 (reel 100, no. 154); Smith, "Sino-Vietnamese Sources," 609–10 (note 50); Thái, "Lời nói đầu," x (note 3).

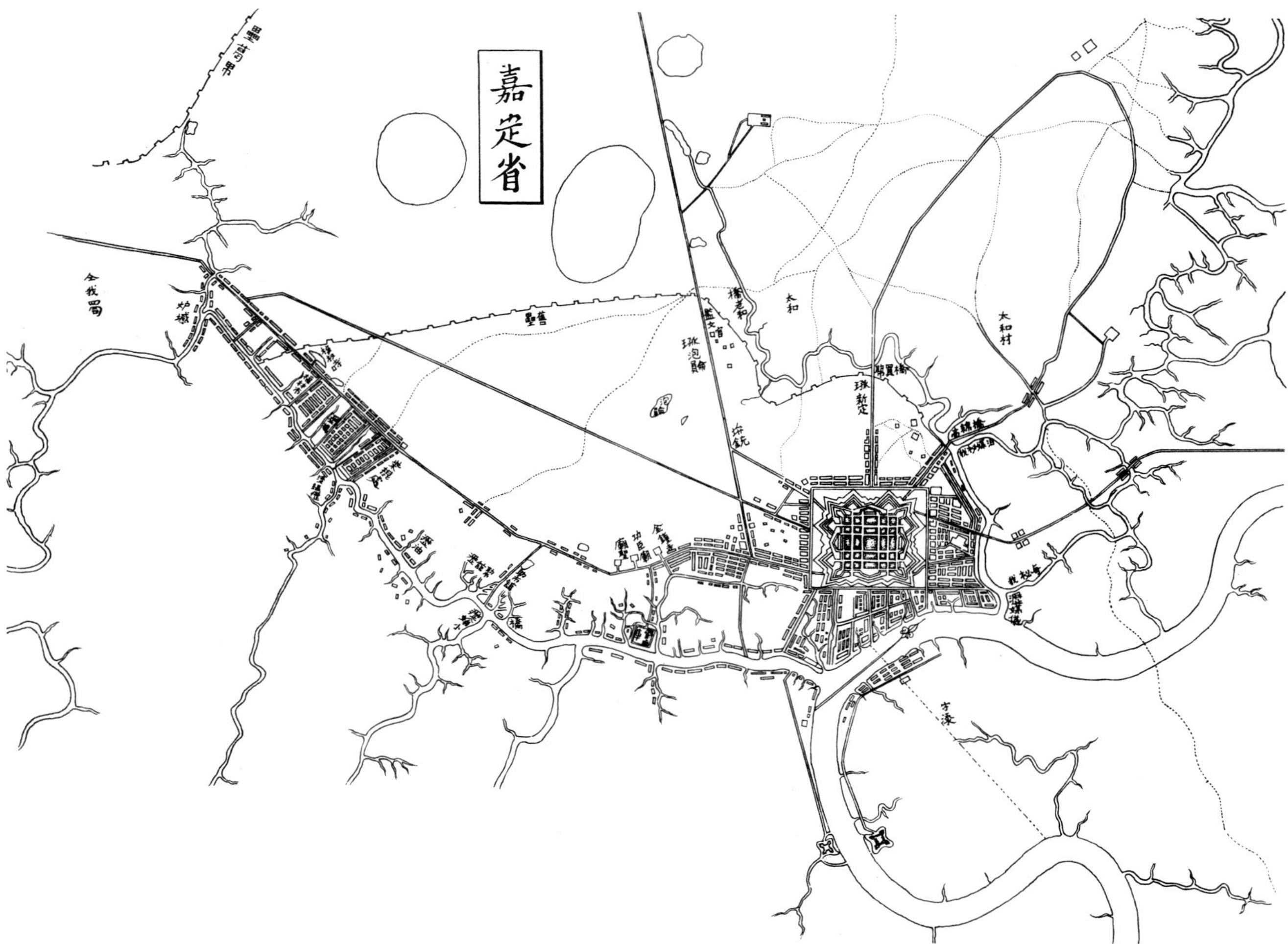

FIG. 12.18. *GIA-ĐỊNH TỈNH*, MAP OF SAIGON. This redrawing of the 1816 (Nguyễn) map depicts Saigon and its environs in a planimetric view using a European style, as drawn by the southerner Trần Văn Học (north is at the top). Size of the redrawing: ca. 27.3 × 38 cm. From Thái Văn Kiểm, "Interprétation d'une carte ancienne de Saigon," *Bulletin de la Société des Etudes Indochinoises*, n.s., 37, no. 4 (1962): 409–31, esp. fig. 29.

entire country and of each of the now six provinces. These maps were also European in style.[56]

As I noted previously, the bureaucratic integration of the country occurred slowly through the 1820s and into the 1830s. In 1833 the Minh-mạng emperor had the *Hoàng-Việt địa-dư chí* (Geography of imperial Vietnam) compiled. It was quite brief, covering the country in only two chapters, focusing on the center and the north, still with no maps. A map of Hà-nội (the new name for Thăng-long) in the early 1830s now shows the same precision and design as Trần Văn Học's southern map fifteen years earlier.[57] Only in the late 1830s was Đại Nam sufficiently integrated to have equal information from all the provinces. While the detailed results would not be seen for several decades, we begin to get an effort to show the whole of Vietnam as it had come to be under the Minh-mạng emperor. The apparent first attempt in this direction, the *Đại-Nam* (or *Nam-Việt*) *bản-đồ* (Maps of Đại Nam), reflects the old tradition. It was a new edition, redrawn in a somewhat different style, of the *Hồng-đức bản đồ* collection from over three decades earlier. It contained the Lê atlas, the four northern itineraries (*Thiên-nam tứ-chí lộ-đồ thư*), the Cao-bằng maps, the southern itinerary (*Bình-nam đồ*), and the map of the Thai world (*Đại-Man quốc-đồ*). What points to this decade as the collection's time of origin is that Cambodia is noted as Cao-miên Phủ (prefecture) and that it no longer contains

56. A.95 (reel 13, no. 29); Woodside, *Chinese Model*, 142 (note 48).

57. A.71 (reel 12, no. 25); A.1074 (reel 135, no. 220); Maspero, "Essai," 544 (note 38); Aurousseau, Review of Maybon, 83 n (note 50); Smith, "Sino-Vietnamese Sources," 609 (note 50); Hoàng, *Thăng Long*, plate between 54 and 55 (note 25).

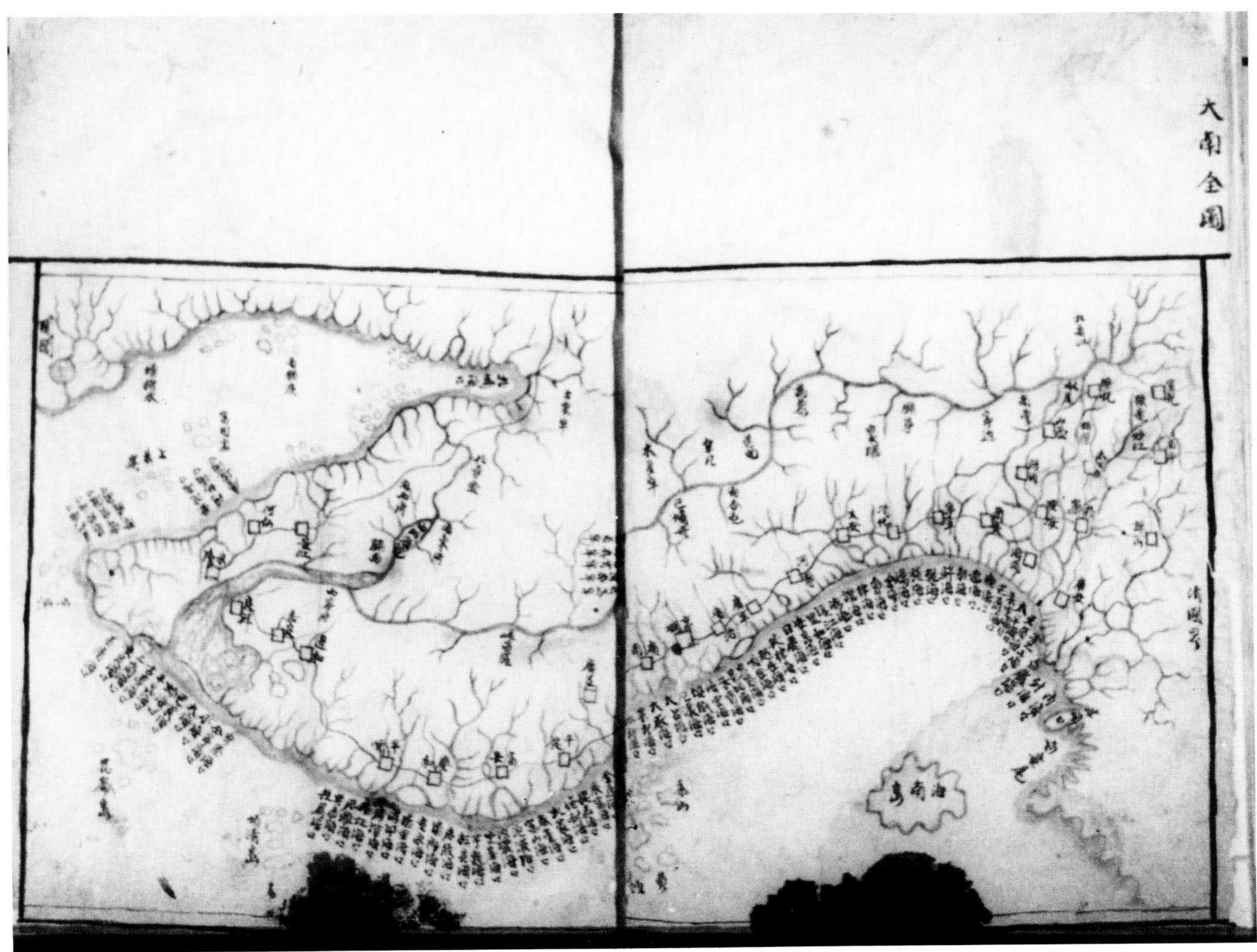

FIG. 12.19. MAP OF THE COUNTRY IN THE *ĐẠI-NAM TOÀN-ĐỒ*. The 1839 (Nguyễn) map of the entire country of Đại Nam portrays a better sense of the Mekong River system and the great lake in Cambodia (west is at the top). It is drawn in a European style.

Photograph courtesy of Hamilton Library, University of Hawaii at Manoa, Honolulu (microfilm collection, A.2559).

the drawing of Angkor.[58] Thus Cambodia had lost its sovereignty, as symbolized by Angkor, and had become in Vietnamese eyes a part of their own administrative apparatus. This was the result of Vietnam's temporary conquest of Cambodia in the mid-1830s.

At the same time, the Vietnamese court had become relatively open to technological and military advances from the West.[59] Thus encouraged, other Vietnamese cartographers had begun to use Western techniques in drawing a new set of maps for the now integrated country, which stretched all the way to the Gulf of Thailand. In 1839 there appeared the *Đại-Nam toàn-đồ* (Complete maps of Đại Nam).[60] The map of the entire country shows thirty-two provinces (including Cambodia) and eighty-two estuaries extending along the entire coastline of nineteenth-century Vietnam (fig. 12.19). This map has taken on the more realistic Western form, showing with greater accuracy the shape of the coast and the hydrographic complexities, even of the Mekong River system (as the great lake of Cambodia). The provincial maps are individualized, are European in outline, and concentrate more on the natural features. They show the river systems in a dendritic manner, and though the mountains are only jotted in, using a simple three-ridge style, many of them are named. Large forests are also shown. The maps seem to be newly drawn, breaking with the previous tradition. Yet about this time there appeared two maps now contained in a copy of the northern set of itineraries, called

58. A.1603 (reel 139, no. 245); Schwartzberg notes; *Hồng-đức bản đồ*, x (n. 4), xxiv (n. 5) (note 1).

59. Woodside, *Chinese Model*, 281–84 (note 48).

60. A.2559 (reel 18, no. 43).

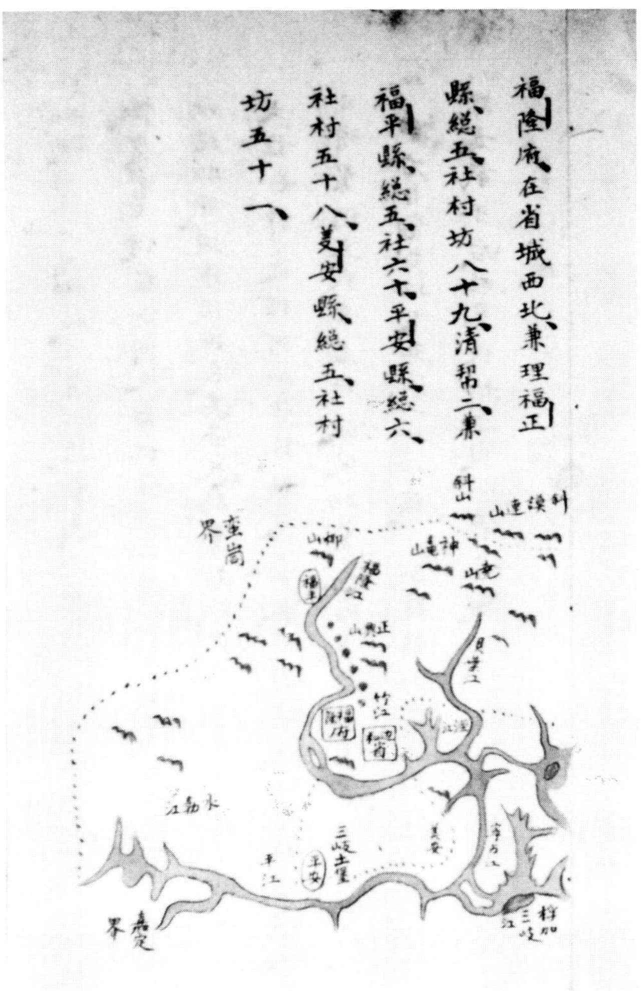

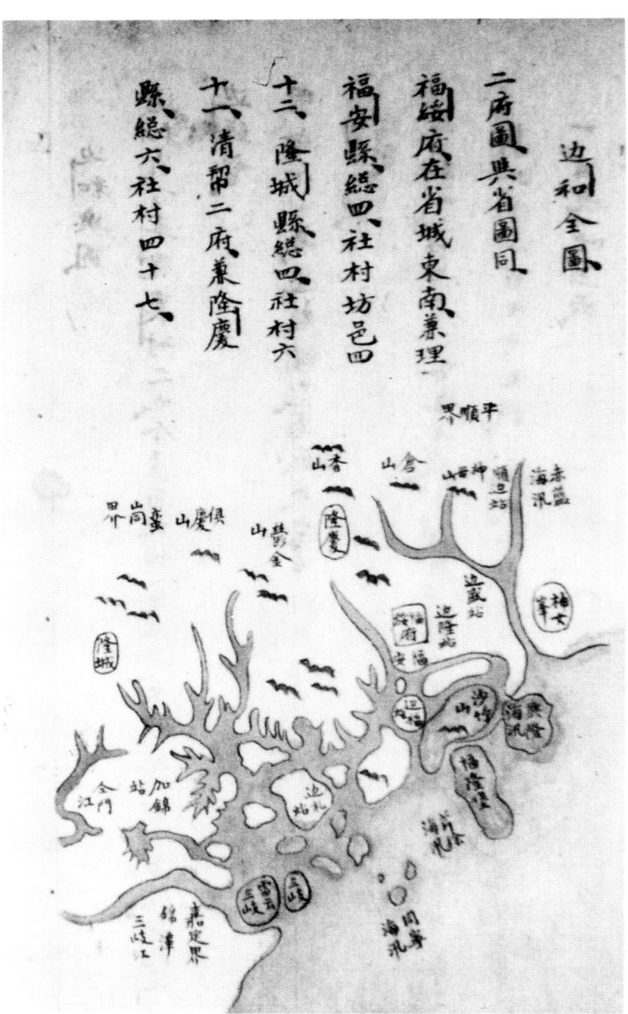

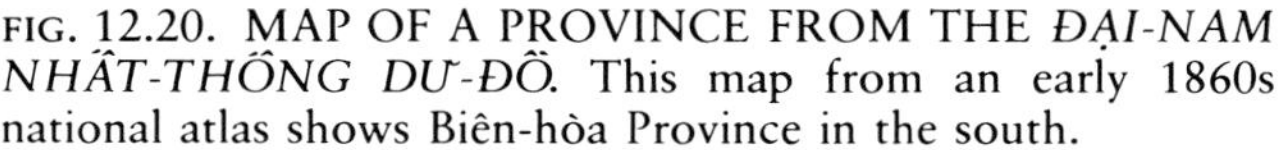
FIG. 12.20. MAP OF A PROVINCE FROM THE *ĐẠI-NAM NHẤT-THỐNG DƯ-ĐỒ*. This map from an early 1860s national atlas shows Biên-hòa Province in the south.

Photograph courtesy of Hamilton Library, University of Hawaii at Manoa, Honolulu (microfilm collection, A.68).

Thiên-nam tứ-chí lộ-đồ thư dẫn (Introduction to the book of maps of the major routes of Thiên-nam), that are simpler, more traditional versions of the map of the entire country in the *Đại-Nam toàn-đồ*.[61] All three maps show Indochina, its estuaries, and the coastline around to Cambodia (as well as the latter's great lake) from a similar perspective. The latter two maps, one simpler than the other, thus share much with the 1839 map but are less precise.

Also in the *Thư dẫn* manuscript is a sky chart, showing only patterns of stars. No identification is offered for any star or cluster of stars, and no title exists for the celestial map itself. It is merely attached to the manuscript. The only other sky chart I have seen in the Vietnamese documents is a clearly Chinese one included in a later edition of the *Thiên-hạ bản-đồ* (Maps of all under heaven [the empire]), the original name of the Lê atlas. Here the stars and star clusters are named and linked to the zodiac. In addition, at the center of the sky chart and oriented to it is an oversimplified, unscientific map of China and East Asia. Although it shows Korea, Japan, and the Ryūkyūs, the view of Vietnam (An-nan) and Southeast Asia gives no indication of Vietnamese knowledge. It is the only map of East Asia I have seen in any Vietnamese manuscript.[62]

Although the Thiệu-trị emperor (1841–47) also put together a short geography of his country (the *Đại-Nam thông-chí* [Record of Đại Nam]), it was the following Tự-đức emperor (1848–84) who was able to produce the

61. A.73 (reels 9–11, no. 22); A.588 (reel 156, no. 305); Woodside, *Chinese Model*, 145 (note 48); Schwartzberg notes.

62. A.1362 (reel 21, no. 55); Archives d'Outre-Mer, Aix-en-Provence, France, B.439–40.

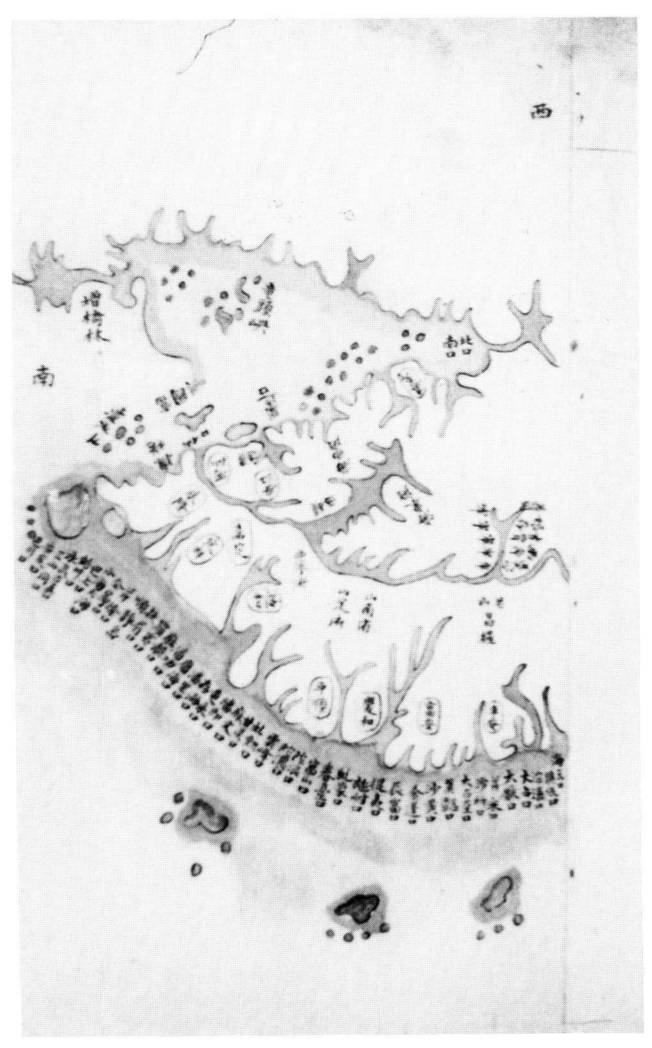

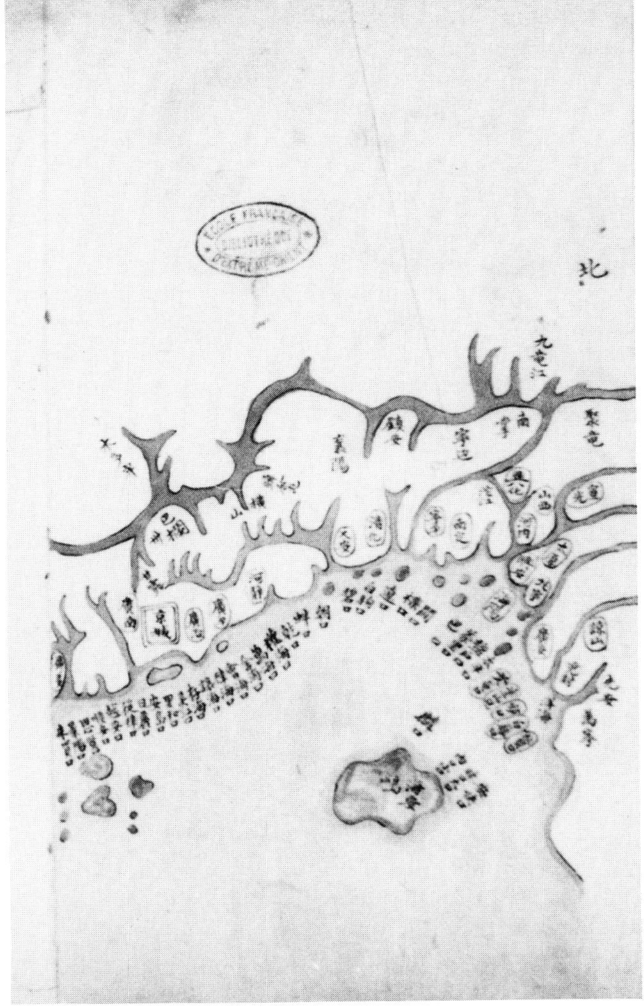

FIG. 12.21. MAP OF THE COUNTRY FROM THE *ĐẠI-NAM NHẤT-THỐNG DƯ-ĐỒ*. From the same manuscript as figure 12.20, this map shows all of Đại Nam as well as bordering countries.
Photograph courtesy of Hamilton Library, University of Hawaii at Manoa, Honolulu (microfilm collection, A.68).

major geographic work of his dynasty, finally integrating the geographical text with the maps. First, in the early 1860s, his court produced the *Đại-Nam nhất-thống dư-đồ* (Geographic maps of the unity of Đại Nam). Here we have maps of the provinces (and even of some prefectures) encompassing all of Vietnam to the southernmost province of An-giang.[63] These maps show some Western influence in their boundaries, river systems, and coastlines. The style in general, however, has the standard three ridge mountains scattered about and other physical features similar to the Lê pattern (fig. 12.20). The map of the country has become more international, showing more of the surrounding countries than earlier maps had done. The Mekong River system, for example, is better understood (fig. 12.21).

This work forms the bridge between the *Đại-Nam toàn-đồ* of the 1830s and the major project to follow. It continued and brought forward the new realistic emphasis from Europe, which the Vietnamese joined to the Chinese geographic tradition. The new work, the *Đại-Nam nhất-thống chí* (Record of the unity of Đại Nam), based on the *Da Qing yitong zhi* (Comprehensive gazetteer of the Great Qing realm, completed 1746) in China, was begun in 1865 and completed in 1882. For the capital of Huế, the capital region (Thừa-thiên), and each of the twenty-nine provinces, there is a map and a description following the Chinese pattern: the province

63. A.3142 (reel 138, no. 240); A.490; A.1307 (reel 19, no. 52); A.68 (reel 13, no. 28); A.1600 (reel 137, no. 230).

as a whole, its constituent parts, physical nature, climate, customs, cities, schools, taxes, mountains, rivers, temples, biographies, and products.[64]

The Tự-đức geography was the only one in the nineteenth century to cover the entire country. This was so even though the six southern provinces were lost by the end of 1867. The emperor never gave up hope of reacquiring them, and the cartography represented the wish to reintegrate the land. Immediately after the French conquest of northern and central Vietnam in 1884–85, the new Đồng-khánh emperor (1885–89) brought out his own geography, the *Đồng-khánh địa-dư chí-lược* (Geographic summary of the Đồng-khánh era), more Europeanized still and covering just central and northern Vietnam. There also appeared during this reign a collection of maps called the *Đại-Nam quốc cương-giới vị-biên* [or *vung-biên*] (Boundaries of Đại Nam), by Hòang Hữu Xứng.[65] There is one map for the entire country, one for the capital, and one for each of the thirty-one provinces. The country map is about halfway between the traditional style and that of the West. The Indochinese peninsula is generally recognizable, but the form of the coastline is like the old style. The Mekong system resembles that of the *Đại-Nam nhất-thống dư-đồ* map. The provincial maps are similar, partway between the realistic and the traditional styles.

Then, two decades afterward (1909) under the Duy-tân emperor (1907–16), a reedition of the *Đại-Nam nhất-thống chí* came out, though only for central Vietnam. This edition has seventeen chapters, with a map for each, covering eleven provinces, the capital, and the capital region, plus maps for the country, for central Vietnam, and for the forbidden city within the capital. These maps are even more Western in style, with the four points of the compass shown and a legend defining the signs used on the map (including railroads). The rivers and coastlines are quite realistic, but mountains are not shown.[66]

By this time, Vietnamese scholars under French colonial control had begun to develop a new style, one less realistic and more Sinic. We see this in two map collections. The first, the *Nam bắc kỳ hoa đồ* (Illustrated maps of the northern and southern regions), is a copy of the early 1860s *Đại-Nam nhất-thống dư-đồ*. It shows the developing realistic influence of French cartography. But it also has decided Chinese features—a roiling sea and the vertical, karst-style mountains. The second collection is a curious pastiche of earlier works. Called the *Tiên-Lê Nam-Việt bản-đồ mô-bản* (Copy of the atlas of Vietnam of the former Lê dynasty), it is an artistic, colorful rendering (with a few errors) of the Lê atlas.[67] Tạ Trọng Hiệp believes it comes from the early twentieth century, produced by scholars working for the Ecole Française d'Extrême-Orient in Hanoi.[68] The one addition to the Lê atlas is the last map, a new drawing of the country map from the *Đại-Nam tòan-đồ* of 1839. This map shows the coastline and river systems of Indochina. Yet its rendering is a step away from the realistic Western style of the nineteenth century and a move toward the impressionistic Sinic style. In effect, these two works seem to represent a choice by Vietnamese literati in the French colonial system to reject scientific detail for artistic Chinese style.

Other maps from the middle to late nineteenth century are of the citadels adapted by the Nguyễn from the French Vauban style. The Archive d'Outre-Mer has a number of such maps, those of Hà-nội, Sơn-tây, Tuyên-quang, Nam-định, and Nha-trang.[69] Each is drawn somewhat differently, but all in Western style, reflecting variety therein as well as in the designs of the walls themselves. Yet despite the differences, all these maps reflect the continuity from the original Vauban design as well as from Trần Văn Học's 1816 Gia-định map. These maps, however, are not the urban maps that Học's was. In this sense they are more like the maps of Thăng-long from the Lê period, showing only the walls and certain government and court ritual buildings, not the common, commercial life.

The French, in their sweep across northern Vietnam in the 1880s, seem to have availed themselves of local maps to aid their efforts. The Bibliothèque Nationale in Paris and the Bayerische Staatsbibliothek in Munich hold maps that the French apparently picked up and used in the campaign.[70] They may even have copied or commissioned some of them in order to gain local knowledge of the terrain. The existence of romanized script (*quốc ngữ*) on the maps indicates their adaptation to French needs. Three maps are of Hà-nội, two others of Ninh-bình and Nam-định, provinces in the southern portion of the Red River Delta, two of Sơn-tây Province, and the other six of mountain areas to the north and north-

64. A.1448 (reel 20, no. 54); A.69 (reel 191, no. 457); Maspero, "Essai," 544–45 (note 38); Aurousseau, Review of Maybon, 83 n (note 50); Smith, "Sino-Vietnamese Sources," 609 (note 50); Thái, "Lời nói đầu," x–xi (note 3).

65. HM2133; A.1342 (reel 179, no. 395); A.748 (reel 145[2], no. 2586).

66. A.537 (reels 9–11, no. 17); HM2133; Aurousseau, Review of Maybon, 83 n (note 50); Smith, "Sino-Vietnamese Sources," 609 (note 50); Thái, "Lời nói đầu," xi (note 3); Tạ, "Fonds," 284–85 (note 36); Schwartzberg notes.

67. Ecole Française d'Extrême-Orient, Paris, Viet.A.Geo.4; Schwartzberg notes.

68. A.95 (reel 13, no. 29); Tạ, "Fonds," 285–86, pls. 5–6 (note 36).

69. Archives d'Outre-Mer, Aix-en-Provence, France, Est. A.80–81, 83, 85, and B.331; Schwartzberg notes.

70. Département des Cartes et Plans, Res. Ge. A.394–96, 15298, D.9069–71, 9148, F.9443–44; Bayerische Staatsbibliothek, Cod. Sin. 82–84; Schwartzberg notes.

west (Lào-cai, Thái-nguyên, and Lạng-sơn). The maps continue the earlier mode, being generally drawn in black ink with other colors used to illustrate different features (roads in red, rivers in blue, and mountains in brown or gray). Citadels (Hà-nội and in the provinces) are drawn to show their Vauban characteristics. These thirteen maps provide some indication of the cartography available locally in the north during the second half of the nineteenth century.

Other local maps also exist from the Nguyễn period. The card catalog from the collection in Hanoi lists a number of village maps, but I have seen no example of them.[71] Another type of local map was most significant for the central government, the land record (*địa bộ*). These records, village by village, show each piece of land in outline, along with its dimensions and type. The government undertook the surveys somewhat piecemeal, though we as yet know little of how the officials proceeded (some of the records have been lost). In dealing with land and land taxes from 1836 to 1875, the Nguyễn government divided its territory into three zones similar to, but not exactly the same as, its earlier political jurisdictions. In the old Đại Việt of earlier centuries, from Hà-tĩnh to the Chinese border, the taxes were heavy on communal lands and light on private lands. In central Vietnam, from Quảng-bình to Khánh-hòa, taxes were moderate and equal on both communal and private lands. In southern Vietnam, from Bình-thuận through the Mekong Delta, taxes were light in general. The development of the land records followed these same zones. The oldest records were undertaken in the first years of the dynasty (1805–6) in the nine northernmost provinces of the first zone. Then, from 1810 to 1818, the surveys took place in the central zone. No original surveys exist from the Gia-long period for the four provinces in the southern section of the first zone or in the third zone to the south. About 1830, the northern and central surveys already done were confirmed, then the four provinces not yet done in the first zone held their surveys in the early 1830s. Finally, the provinces of the southern zone held theirs in 1836 following the defeat of Lê Văn Khôi's rebellion in the mid-1830s. Right after that, in 1837–40, four marginal northern provinces were resurveyed, and evidence exists for mapping of newly developed lands in the south.[72]

The final set of Vietnamese maps that concern us are quite different ones, and they bring us back to where we began—the pattern of mountain and water. These maps are diagrams of the Nguyễn imperial tombs (fig. 12.22). Differing greatly from Western diagrams of the tombs, they feature both mountains and water and reflect the increasingly strong Sinic influence of the Nguyễn period. Stein has pointed out the strong parallel between tombs and the miniature gardens he has studied.[73] Even as the Vietnamese became more Sinic through the nineteenth century, these maps show that they retained the basic magical elements of mountains and water that underlay their cartography into the modern era.

Conclusion

The development of Vietnamese cartography followed efforts of the governments to centralize and extend their control over the country. Mapping was linked to the bureaucracy, and the atlases were meant to show the locations of the different jurisdictions, often recording as well the number and types of villages in each district. The Vietnamese government sought to gain access to the resources of the realm, human and material. These resources came out of the villages, and the maps were meant to show their spread. Thus premodern Vietnamese maps as they exist today were tied strongly to the government's adoption of the Chinese model, and with it the Chinese mode of bureaucracy. That the maps were generally Sinic in style followed therefrom. The atlases served as bureaucratic tools in the efforts of the imperial court to manage the villages. In the earlier centuries, during the Lý and Trần dynasties when indirect control was the norm, the Vietnamese throne had had neither the control nor the inclination to map the country. During the Lê and Nguyễn dynasties the bureaucracy provided the means, and its purpose—resource control—the aim for doing so. Thus the major efforts at mapping (or reproducing earlier maps) took place at times of bureaucratic strength: the last third of the fifteenth century, the second half of the seventeenth century, the 1830s, and the 1860s.

Vietnamese mapping was internal, not external. The itineraries mainly led to the borders, north and south, and no farther. The southern route first went to the capital of Champa, but this territory soon became Vietnam-

71. A.1844, 1895–96, 2964 from the microfilm of the Ecole Française d'Extrême-Orient catalog list (reel 3, nos. 8–9).

72. Smith, "Sino-Vietnamese Sources," 616–17 (note 50); Nguyễn Thế Anh, "La réforme de l'impôt foncier de 1875 au Viet-Nam," *Bulletin de l'Ecole Française d'Extrême-Orient* 78 (1991): 287–96, esp. 288–89. No illustration exists of the *địa bộ*. My description derives from personal observation of the documents in Da Lat during my visit there with Ralph Smith in March 1966. This material, not yet studied to my knowledge, apparently is now held in Ho Chi Minh City. On the mapping of newly developed land in the south, 1838–40, see Paul J. Bennett, "Two Southeast Asian Ministers and Reactions to European Conquest: The Kinwun Mingyi and Phan-thanh-Gian," in *Conference under the Tamarind Tree: Three Essays in Burmese History* (New Haven: Yale University Southeast Asia Studies, 1971), 103–42, esp. 110, and Louis Malleret and Georges Taboulet, eds., "Foire Exposition de Saigon, Pavillon de l'Histoire, la Cochinchine dans le passé," *Bulletin de la Société des Etudes Indochinoises*, n.s., 17, no. 3 (1942): 1–133, esp. 40–42.

73. Stein, *World in Miniature*, 104, 111–12 (note 6); Thái, *Cố đô Huế*, pls. 29, 36 (note 3).

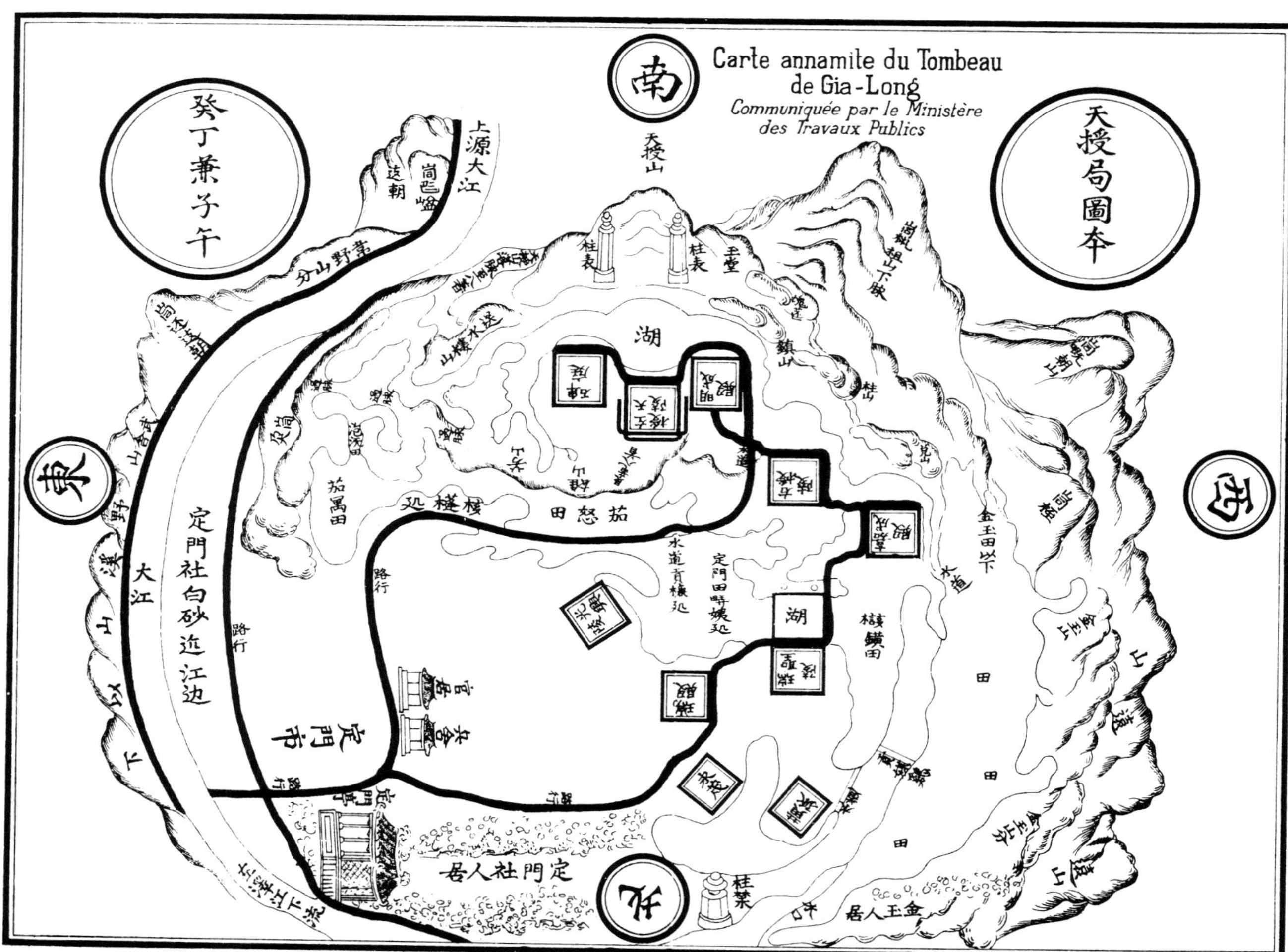

FIG. 12.22. THE NGUYỄN TOMBS, HUẾ. This redrawing of the Vietnamese diagram of the Gia-long emperor's tomb with the rare (imperial) southern orientation shows again the cosmological mix of water and mountains.

From Charles Patris and L. Cadière, *Les Tombeaux de Hué: Gia-Long* (Hanoi: Imprimerie d'Extrême-Orient, 1923), pl. XXI.

ese. It eventually went to Angkor, which in turn became (temporarily) part of Vietnam. Only the routes into China and Siam were non-Vietnamese subjects of Vietnamese cartography, and these were by land. Sea routes, as illustrated in charts, are not to be found. However, the international contacts of the Vietnamese in the late eighteenth and early nineteenth centuries did lead to inclusion of more of mainland Southeast Asia in maps of the country. Still, East Asian, general Asian, and world maps are essentially unknown in the Vietnamese collections. The itineraries supply interesting descriptive detail for Vietnam itself and remain within the cultural framework of the pre-1800 style. They do not seem to have made the change into the realistic style of the mid-nineteenth century. The new style seems to have been reserved for the government atlases.

This chapter marks a beginning, and many questions inevitably remain. Much more work is still to be done on Vietnamese maps, using both those documents on microfilm and the manuscripts held in Hanoi. Such work will undoubtedly bring forth significant cartographic materials, the interpretation of which will change the synthesis supplied here. In particular, we must examine the documentation of the nineteenth century more closely in order to gain a better sense of the cartographic efforts of the Nguyễn government. We need to understand better both the Western and the Chinese elements available to the Vietnamese and their use of them. Overall, we must use the cartographic materials to further our understanding of Vietnamese society and its development in the early modern world.

Celestial Mapping in East Asia

13 · Chinese and Korean Star Maps and Catalogs

F. Richard Stephenson

Charting the stars is an undertaking entirely different from delineating terrestrial features such as continents and islands. Since the stars appear as scattered points of light, any attempt to divide them into groups must necessarily be subjective. The arbitrariness of such an exercise is increased by the wide range in brightness among the roughly six thousand stars that are estimated to be visible to the average unaided eye over the whole of the celestial sphere. It is thus remarkable that throughout history only two distinct schemes of astral cartography have enjoyed widespread usage. These are of Babylonian-Greek and of Chinese origin, and the latter system of astrography is the subject of this chapter.[1] Interaction between the various Eurasian cultures through the centuries has been largely responsible for restricting the development of other schemes of mapping the night sky. Nevertheless, the great differences that exist between the two major systems reflect their artificiality.

Although Chinese astrography has now passed into history (except at the popular level), in recent years there has been renewed interest in its scientific value. This development has largely resulted from growing appreciation of the importance of ancient and medieval observations of temporary stars (novas and supernovas) and comets—notably Halley's comet—recorded in Chinese history. The positions of these objects on the celestial sphere are often carefully described relative to specific star groups. With only isolated exceptions, Chinese observations of such phenomena have proved to be without equal anywhere else in the world before the European Renaissance.[2]

Several star maps and catalogs identifying the constituents of Oriental star groups in terms of their Western equivalents (for example, Bayer Greek letters or Flamsteed numbers) have been produced in China in recent years.[3] These works have largely superseded the well-known concordances of Williams, Schlegel, Wylie, Tsutsihashi and Chevalier, and latterly Ho.[4]

No detailed investigation of East Asian uranography has been published in a European language since the appearance of the voluminous work by Needham more than thirty years ago, which is still of considerable use today.[5] Two Chinese books on the subject that have appeared in the past decade or so are also frequently referenced in this chapter. The *Zhongguo gudai tianwen wenwu tuji* (Album of ancient Chinese astronomical relics) exhibits many fine photographs (some in color) of celestial charts and globes (as well as other astronomical artifacts).[6] It also provides brief but useful notes. The

I am grateful to K. K. C. Yau, Jet Propulsion Laboratory, Pasadena, California, for much help and advice.

1. The system of Babylonian-Greek origin is discussed in other volumes of this *History*. The term "astrography" is a synonym for astral cartography. Another variant used elsewhere in this chapter is "uranography."

2. A detailed discussion of historical observations of supernovas, with special emphasis on Chinese records, is given by David H. Clark and F. Richard Stephenson, *The Historical Supernovae* (Oxford: Pergamon Press, 1977). For investigations of the history of Halley's comet see, for example, T. Kiang, "The Past Orbit of Halley's Comet," *Memoirs of the Royal Astronomical Society* 76 (1972): 27–66, and F. Richard Stephenson and Kevin K. C. Yau, "Far Eastern Observations of Halley's Comet, 240 BC to AD 1368," *Journal of the British Interplanetary Society* 38 (1985): 195–216.

3. For example, Chen Zungui, *Zhongguo tianwenxue shi* (History of Chinese astronomy) (Taipei: Mingwen Shuju, 1984–), vol. 2, and Yi Shitong, *Quantian xingtu: 2000.0* (All-sky star atlas for epoch 2000.0) (Beijing, 1984).

4. John Williams, *Observations of Comets from B.C. 611 to A.D. 1640* (London: Strangeways and Walden, 1871); Gustave Schlegel, *Uranographie chinoise; ou, Preuves directes que l'astronomie primitive est originaire de la Chine, et qu'elle a été empruntée par les anciens peuples occidentaux à la sphère chinoise*, 2 vols. (Leiden: E. J. Brill, 1875; reprinted Taipei: Chengwen Chubanshe, 1967); Alexander Wylie, *Chinese Researches* (Shanghai, 1897), pt. 3 (scientific), 110–39; P. Tsutsihashi and Stanislas Chevalier, "Catalogue d'étoiles observées à Pé-kin sous l'empereur K'ien-long (XVIIIe siècle)," *Annales de l'Observatoire Astronomique de Zô-sè (Chine)* 7 (1911): I-D105; and Ho Peng-yoke, trans. and annotator, *The Astronomical Chapters of the Chin Shu* (Paris: Mouton, 1966), 67 ff. and 263 ff.

5. Joseph Needham, *Science and Civilisation in China* (Cambridge: Cambridge University Press, 1954–), vol. 3, with Wang Ling, *Mathematics and the Sciences of the Heavens and the Earth* (1959). For more recent but briefer discussions, with an emphasis on Korean celestial cartography, see Sang-woon Jeon (Chŏn Sang'un), *Science and Technology in Korea: Traditional Instruments and Techniques* (Cambridge: MIT Press, 1974), esp. 22–33, and Joseph Needham et al., *The Hall of Heavenly Records: Korean Astronomical Instruments and Clocks, 1380–1780* (Cambridge: Cambridge University Press, 1986), esp. chap. 5.

6. Zhongguo Shehui Kexueyuan Kaogu Yanjiusuo (Archaeological

Zhongguo hengxing guance shi (History of stellar observations in China) by Pan Nai contains an extensive discussion of the history of Chinese uranography.[7] This work also contains numerous illustrations, regrettably of mediocre quality: these are mainly of Chinese star maps and globes, but Japanese and Korean charts are also included. Both of these recent compilations are indispensable aids to the study of East Asian celestial cartography.

Independent Developments in Chinese Celestial Cartography

Early documentary evidence suggests that astronomy developed in ancient China independently from other civilizations. Preserved texts make no positive allusion to any interaction between China and other cultures until as late as about 130 B.C.[8] This lack is hardly surprising in view of the relative isolation of China. Suggestions of cultural exchanges between China and the West in remote antiquity have been made from time to time, but these are largely based on inference rather than on authentic records.[9]

No contacts of astronomical significance between China and India are recorded until the Tang dynasty (618–907).[10] By this period, however, the complex system of political astrology that had evolved in China through the centuries had long since been codified.[11] Despite the presence of Indian astronomers at the Tang court, there were no fundamental changes in the course of Chinese astronomy, and astrography in particular. Major star charts from China in the medieval period reveal negligible traces of Western innovations. Foreign influence was felt only in unofficial culture—for instance, in the Tang vogue of horoscope astrology.[12]

Arab astronomers were active in China from the Yuan dynasty (1279–1368) onward, and at the very beginning of the Ming (1368) an Islamic astronomical bureau was set up in the capital. However, only with the arrival of the Jesuit astronomers in China from the late sixteenth century onward do we find important improvements of foreign origin in mapping the night sky. Several Jesuits obtained high positions at the Chinese court, including the directorship of the imperial observatory itself. The Jesuit astronomers charted the stars with an accuracy hitherto unrivaled in China and, furthermore, introduced detailed knowledge of the far southern constellations, but there is little evidence that they tried to supplant the traditional Chinese representation of the night sky by the Occidental constellations. Only in the twentieth century did the Western tradition of astral cartography finally gain supremacy in China.

During the latter half of the first millennium, Chinese methods of astronomy and astrology—along with other aspects of culture—spread to Korea and Japan and afterward to Vietnam. In each of these countries celestial observations began to be made in the traditional Chinese style, and this pattern continued down to relatively modern times.[13] It is thus not surprising that the extant star maps from Korea and Japan (there appear to be no significant survivals from Vietnam, but see p. 504) clearly display Chinese influence and in general reveal little that is original. Korean star maps will be discussed toward the end of this chapter; star maps from Japan are the subject of chapter 14.

Research Institute, Chinese Academy of Social Science [Academia Sinica]), *Zhongguo gudai tianwen wenwu tuji* (Beijing: Wenwu Chubanshe, 1980).

7. Pan Nai, *Zhongguo hengxing guance shi* (Shanghai, 1989).

8. The Chinese envoy Zhang Qian reached as far west as Bactria in 128 B.C. A detailed account of his travels is given by Sima Qian in chapter 123 of the *Shi ji* (Records of the grand historian, completed ca. 91 B.C.). For an English translation of this narrative, see Friedrich Hirth, "The Story of Chang K'iên, China's Pioneer in Western Asia: Text and Translation of Chapter 123 of Ssï-ma Ts'iên's Shï-ki," *Journal of the American Oriental Society* 37 (1917): 89–152.

9. See, for example, Joseph Needham, *Science and Civilisation in China* (Cambridge: Cambridge University Press, 1954–), vol. 1, with Wang Ling, *Introductory Orientations* (1954), 150 ff.; Edwin G. Pulleyblank, "Chinese and Indo-Europeans," *Journal of the Royal Asiatic Society of Great Britain and Ireland*, 1966, 9–39. A case for the cultural isolation of ancient China is made by Michael A. N. Loewe in his introduction to A. F. P. Hulsewé, *China in Central Asia: The Early Stage, 125 B.C.–A.D. 23* (Leiden: E. J. Brill, 1979), 39 ff.

10. Yabuuchi Kiyoshi (Yabuuti Kiyosi), "Researches on the *Chiu-chih Li*—Indian Astronomy under the T'ang Dynasty," *Acta Asiatica* 36 (1979): 7–48, is useful on this topic.

11. The practice of astrology in China from at least the second century B.C. involved court astronomers' maintaining a regular watch of the day and night sky for portents such as eclipses, comets, and lunar and planetary conjunctions. These omens were interpreted according to a carefully prescribed set of rules, based largely on the star group in or near which the phenomenon was seen. For extensive examples of ancient Chinese portentology, see Ho, *Astronomical Chapters* (note 4).

12. A few rather crude medieval star maps of Chinese origin—for example, a chart painted on the ceiling of a twelfth-century tomb—are preserved that portray both the signs of the Western zodiac and East Asian star groups. Examples of horoscope astrology during the Tang dynasty are given by Edward H. Schafer, *Pacing the Void: T'ang Approaches to the Stars* (Berkeley and Los Angeles: University of California Press, 1977), 58 ff.

13. As in the case of China, the records of many of these observations are still preserved today. See for example, Seong-rae Park, "Portents and Neo-Confucian Politics in Korea, 1392–1519," *Journal of Social Sciences and Humanities* 49 (1979): 53–117; Kanda Shigeru, *Nihon tenmon shiryo* (Japanese astronomical records) (Tokyo, 1935); Ho Peng-yoke, "Natural Phenomena Recorded in the *Đai-Viêt su'-ky toan-thu'*, an Early Annamese Historical Source," *Journal of the American Oriental Society* 84 (1964): 127–49.

The Beginnings of Celestial Cartography in China

The origins of uranography in China are lost in antiquity. In attempting to trace the evolution of celestial mapping in ancient China, we are seriously hampered by the low survival rate of suitable documents. No star maps or catalogs earlier than the Former Han dynasty (206 B.C.–A.D. 8) are now extant. In fact, scarcely any Chinese texts on astronomy that were written before this period are currently known to exist; as a rule, only fragmentary quotations are preserved in later works. Hence, for the most ancient—and probably the most crucial—period of the development of astronomy in China, the researcher must rely almost entirely on whatever information can be gleaned from such diverse sources as oracle texts, almanacs, chronicles, and even poetic works. Although several star groups are mentioned in inscriptions dating from before 1000 B.C., the fact remains that texts originating earlier than about 100 B.C. preserve the names of no more than about thirty separate asterisms. Yet later writings assert that by roughly 300 B.C. the night sky had already been divided into nearly three hundred constellations.[14]

In the absence of surviving astral charts, extant catalogs can give a sound indication of the level mapping had attained at the period in question. The relation between cataloging and mapping stars is not well defined, however. An accurate star map will normally be based on a catalog, but sketch maps of constellations (as found in many tombs) can be made quite independently. Also, constellation lists that give only brief indications of relative positions of stars are often constructed for purely astrological purposes.

According to tradition, preserved in texts written many centuries after the events they purport to describe, astronomy was practiced in China almost from the dawn of civilization. Thus, writing about 91 B.C., the great historian and astronomer Sima Qian (ca. 145–ca. 85 B.C.) stated in his *Shi ji* (Records of the grand historian) that to his knowledge there had never been a time when the rulers of China failed to encourage observation of the heavens.[15] Sima Qian also listed some of the principal astronomers from earliest (legendary) times down to his own era. In the seventh century A.D., Li Chunfeng (602–70), author of the astronomical treatise incorporated in the *Jin shu* (History of the Jin), followed much the same tradition, emphasizing the importance assigned to astronomy in the time of the sage-rulers of old.[16]

The semilegendary *Shu jing* (Book of documents), containing some sections written during the Zhou dynasty (ca. 1027–256 B.C.), cites an interesting account in its "Yao dian" (Canon of Yao). This relates how in remote antiquity the sage-emperor Yao commanded the Xi and He brothers "in reverent accordance with the august heavens, to compute and delineate the sun, moon and stars . . . and so deliver respectfully the seasons to be observed by the people."[17] The "Yao dian" further relates that four stars (Niao, Huo, Xu, and Mao) were to be observed by astronomers situated at the extreme limits of the Chinese dominions in order to determine the seasons. Yet the whole account is so idealized that it appears to be no more than an imaginative tale with little or no factual basis.[18]

In practice, no direct attestation of astronomical activity in China is preserved before the Shang (ca. sixteenth to eleventh century B.C.),[19] and even this evidence is sparse. It is from this period (also known as the Yin dynasty) that the earliest written records survive. Before the Shang, Chinese history gives way to legend. For instance, the existence of the Xia dynasty, which according to traditional Chinese history immediately preceded the Shang, has yet to be adequately substantiated.[20] Astronomical data that purport to originate in the Xia, but that likely are much more recent, are contained in the *Xia xiaozheng* (Lesser calendar of the Xia). This work, which is essentially a farmers' calendar, notes how certain months were marked by the visibility of particular star groups.[21]

14. See, for example, Fang Xuanling et al., *Jin shu* (History of the Jin, compiled 646–48), chap. 11; modern edition in 10 vols. (Beijing: Zhonghua Shuju, 1974). Further details are given below. "Asterism," a term of Greek origin (*asterismos*) implying a minor constellation, is often used to describe Chinese star groups.

15. This statement is recorded in the astronomical treatise (chap. 27) of the *Shi ji*. For a translation of Sima Qian's treatise, see Edouard Chavannes, trans., *Les mémoires historiques de Se-Ma Ts'ien*, 5 vols. (Paris: Leroux, 1895–1905), 3:339–412.

16. See the translation from chap. 11 of the *Jin shu* by Ho Peng-yoke, *Li, Qi and Shu: An Introduction to Science and Civilization in China* (Hong Kong: Hong Kong University Press, 1985), 115–16.

17. Translated by Ho, *Li, Qi and Shu*, 116–17 (note 16).

18. Much this same view is taken by Yabuuchi Kiyoshi, *Chūgoku no tenmon rekihō* (The history of astronomy and calendrical science in China) (Tokyo: Heibonsha, 1969; rev. ed. 1990), 267.

19. A chronological scheme showing the inclusive dates of the various Chinese dynasties is given above, table 2.1, p. 25. Most of these dates are precisely known, but the beginning and end of the Shang are still a matter for conjecture.

20. For comments regarding the status of the Xia, see, for example, Charles Patrick Fitzgerald, *China: A Short Cultural History*, 4th rev. ed. ([London]: Barrie and Jenkins, 1976), 26–28. Recently, various contributors to *Early China* 15 (1990): 87–133 have taken a rather conservative view of the Xia.

21. A translation of the *Xia xiaozheng* was made by William Edward Soothill, *The Hall of Light: A Study of Early Chinese Kingship* (London: Lutterworth Press, 1951), 237–42. Herbert Chatley, "The Date of the Hsia Calendar *Hsia Hsiao Chêng*," *Journal of the Royal Asiatic Society of Great Britain and Ireland*, 1938, 523–33, summarized the astronomical references in this work and investigated the visibility of the various asterisms it contains. He concluded that all the data in the

FIG. 13.1. SHANG ORACLE BONE WITH REFERENCE TO THE FIRE STAR (ANTARES). This inscribed ox bone, found near Anyang and dating from ca. 1300 B.C., contains one of the earliest allusions to a star in Chinese history. The star, named at the end of the central column of characters, is called Huo. (This is a negative of the rubbing.)
Size of the original: 14 × 4.5 cm. Photograph courtesy of F. Richard Stephenson.

Shang chronology is still by no means securely established; few datable texts are preserved from this early period. Original Shang records consist almost entirely of "oracle bones," divination texts inscribed on turtle shells and animal bones using a primitive form of Chinese script.[22] So far it has been possible to obtain only tantalizing glimpses of Shang astronomy. Celestial observations recorded on the oracle bones are rare, as might be expected of texts of this nature, and the names of only a few constellations are mentioned.

The oracle bone inscriptions have long been known to contain allusions to eclipses, and in the past few decades there have been several attempts to deduce the dates of these observations using astronomical calculations.[23] Recently, extensive searches have been made for other astronomical records on the Shang oracle bones.[24] Apart from eclipses, occasional references to the planet Jupiter, comets (all undatable), and certain star groups were noted. No planet other than Jupiter has yet been identified on Shang inscriptions. Constellations are almost exclusively mentioned in the context of sacrifice; evidently it was the custom to make regular offerings to them (and also to Jupiter), as the following example illustrates: "There was a divination on day *jiyou*. In the night of day *geng (-xu)*, a sacrifice was made to Dou (the Dipper). There was a divination on day *geng (-xu)*. In the night of *xin (-hai)*, a sacrifice was (again) made to Dou."[25]

Xia xiaozheng are quite consistent with a date of compilation about 350 B.C. This is more than a thousand years after the Xia is supposed to have come to an end.

22. Valuable discussions of oracle bone inscriptions and Shang divination are given by David N. Keightley, *Sources of Shang History: The Oracle-Bone Inscriptions of Bronze Age China* (Berkeley and Los Angeles: University of California Press, 1978), 3–27, and Hung-hsiang Chou, "Chinese Oracle Bones," *Scientific American* 240 (April 1979): 134–49. The relics, first discovered near Anyang at the end of the nineteenth century, probably date from the latter part of the Shang, between about 1350 and 1050 B.C. So far about 160,000 texts—mostly no more than fragments incised with a few characters—have been cataloged.

23. See, for example, Dong Zuobin (Tung Tso-pin), *Yin lipu* (On the calendar of the Yin dynasty) (Lizhuang, Szechuan: Academia Sinica, 1945), pt. 2, 1–37; Homer H. Dubs, "The Date of the Shang Period," *T'oung Pao* 40 (1951): 322–35; Zhang Peiyu, Xu Zhentao, and Lu Yang, "Zhongguo zui zao qi di rishi jilu he gongyuanqian shisi zhi gongyuanqian shiyi shiji rishi biao" (China's earliest records of solar eclipses and a solar eclipse table for the fourteenth to the eleventh century B.C.), *Nanjing Daxue Xuebao* (1982): 371–409, esp. 381–84.

24. The only detailed account in a Western language is by Xu Zhentao, Kevin K. C. Yau, and F. Richard Stephenson, "Astronomical Records on the Shang Dynasty Oracle Bones," *Archaeoastronomy* 14, suppl. to *Journal for the History of Astronomy* 20 (1989): S61–S72. A prior investigation, published in Chinese, was made by Wen Shaofeng and Yuan Tingdeng, *Yinxu buci yanjiu: Kexue jishu pian* (Studies on Yin oracle bone writings: Science and technology volume) (Chengdu: Sichuan Shehui Kexue Chubanshe, 1983), 1–66.

25. Xu, Yau, and Stephenson, "Astronomical Records," 568 (note 24); many of my subsequent remarks on Shang constellations are also derived from this source.

In the example above, *jiyou*, *gengxu*, and *xinhai* were consecutive days, the forty-sixth, forty-seventh, and forty-eighth days of a sixty-day cycle. This same cycle has continued in use, probably without interruption, until the present.

Several further references to the star group Dou are preserved on other oracle bone fragments. This asterism was probably identical with the Big Dipper or Plow in Ursa Major, whose shape is so characteristic that it probably was one of the earliest constellations to be recognized. Dou is also mentioned in texts of the subsequent Zhou dynasty, and here its identity with the Big Dipper is readily apparent (see below). Two additional stars or star groups that are referred to on the oracle bones are Huo (Fire) (see fig. 13.1) and Xin (Heart). These names, which probably relate respectively to the bright red star Antares and a group of three stars centered on Antares, remain in use in later Chinese history. Other Shang allusions to asterisms occur relatively rarely, and the identity of the constituent stars is more problematic; in particular, none of the inscriptions give any indication of celestial location.

Pending further investigation of Shang texts, it is impossible to determine whether the people of the time took notice only of the more prominent features of the night sky or whether they already recognized a variety of constellations. Although very few asterisms have so far been identified on Shang inscriptions, these records are of some significance in the history of celestial cartography; they contain the earliest known Chinese references to the grouping of stars into constellations that are in any way reliable. It is also worth emphasizing that the allusions to asterisms found on the oracle bones—as well as Shang references to the planet Jupiter—are among the oldest that survive from any civilization.

The Constellations as Envisaged during the Western and Eastern Zhou Dynasties and the Chunqiu Period (ca. 1027–468 b.c.)

Compared with those from the Shang, relatively few original documents survive from its successor, the Zhou dynasty. Nevertheless, late copies of several Zhou texts are available, some having been printed and reprinted many times. Important astronomical references are found in a few writings that either originated at this time or contain much contemporary material. These works include the *Chunqiu*, *Zuozhuan*, and *Shi jing*.

The *Chunqiu* (Spring and autumn [annals]) is a chronicle of Lu, one of the early feudal states of China, during the period from 722 to 480 B.C.[26] This chronicle, which according to tradition was compiled by Confucius, records many eclipses and also a few comets and meteors.[27] By the Chunqiu period, the rulers of individual states employed astronomers to keep a watch for celestial omens and also to maintain a reliable calendar. The names of some of these astronomers are still preserved.[28] Nevertheless, only a single star group is noted in the *Chunqiu*. In 613 B.C. it is recorded that a comet entered the asterism Beidou (Northern Dipper),[29] which is identifiable with the Big Dipper.

In 525 B.C. another comet was reported, this time at Dachen. Later known as Dahuo, Dachen was one of the *ci*, or "Jupiter stations." Here we have one of the earliest references in Chinese history to these twelve equal divisions of the sky (and later of the celestial equator) based on the motion of Jupiter, or rather its supposed invisible counterrotating correlative planet Taisu.[30] Since Jupiter completes a full circuit of the sky in almost twelve years, the sun in its annual course would spend a month in each division. Apart from the number of stations, the *ci* (which were still important in Chinese astrology in relatively recent times) had nothing in common with the signs of the Western zodiac. The latter divisions are based on the ecliptic rather than the celestial equator. In Chinese astronomy and astrology the zodiac has never held a special place except in popular thinking.

In the *Zuozhuan* (Zuo's tradition [of interpreting the *Chunqiu*]), an ancient enlargement of the spring and autumn annals, a number of scattered references to star groups occur, notably Huo (the Fire Star or Antares).[31]

---

26. Similar chronicles were kept in other feudal states, but most were presumably destroyed at the Burning of the Books in the Qin dynasty (221–207 B.C.). See Burton Watson, *Early Chinese Literature* (New York: Columbia University Press, 1962), 37.

27. That it was compiled by Confucius is asserted by Mencius, see *Mengzi*, bk. 3, pt. 2, chap. 9; modern edition with translation by James Legge, ed. and trans., *The Four Books* (1923; reprinted New York: Paragon, 1966), 676–77. For a recent discussion of the astronomical records in the *Chunqui*, see F. Richard Stephenson and Kevin K. C. Yau, "Astronomical Records in the *Ch'un-ch'iu* Chronicle," *Journal for the History of Astronomy* 23 (1992): 31–51.

28. For example, in chap. 27 of the *Shi ji* (see the modern edition in 10 vols. [Beijing: Zhonghua Shuju, 1977]) and chap. 11 of the *Jin shu* (note 14). Ho, *Astronomical Chapters*, 46–48 (note 4), gives some useful comments on individual astronomers of the period based on biographical details in the *Shi ji*.

29. At this early period, the more common name is simply Dou. Later texts also frequently allude to Nandou (Southern Dipper), a group of six stars in Sagittarius resembling Beidou.

30. For further details see, for example, Liu Tan, *Zhongguo gudai zhi xingsui jinian* (Ancient Chinese Jupiter-cycle calendar) (Beijing: Kexue Chubanshe, 1957), and Needham, *Science and Civilisation*, 3:402–4 (note 5).

31. Translations of the *Chunqiu* together with the *Zuozhuan* have been made by James Legge, *The Chinese Classics*, 5 vols. (Hong Kong: Hong Kong University Press, 1960 [reprint of last editions, 1893–95]), vol. 5, and by Séraphin Couvreur, trans., *Tch'ouen Ts'ou et Tso Tchouan*, 3 vols. (Hochienfu: Mission Press, 1914).

Several of the *ci* are also alluded to. The *Zuozhuan* covers much the same period as the *Chunqiu*, but its date of compilation has been much disputed. It was anciently attributed to Zuo Qiuming, a contemporary of Confucius,[32] but there is now general agreement that the bulk of the book was written about 300 B.C., with some later additions.[33] The *Zuozhuan* is remarkable for the large amount of extended narrative material it contains; this contrasts with the terse style of the *Chunqiu* itself. Some of this material may be legendary, but much is probably factual, based on sources that have long since disappeared. Certain of the reports in the *Zuozhuan* may be corroborated not only by other written sources but also by archaeological evidence.[34]

For dates before about 600 B.C., stars are usually referred to in the *Zuozhuan* as seasonal markers, but later we find evidence of the development of astrological prediction, based largely on previously recorded coincidences between celestial and terrestrial events. Such an example occurred in 532 B.C.:

> This spring, in the king's first month, a (strange) star appeared in (the constellation) Wu-nu. Pei Zhao of Qing said to Zichan, "In the 7th month, on [day] *wuzi*, the ruler of Jin will die. . . . It was on [day] *wuzi* that duke Feng (anciently) ascended on high, when a (strange) star appeared in this same place. Thus it is that I make this observation."
>
> On [day] *wuzi*, duke Ping of Jin died.[35]

In the various early folk songs assembled in the *Shi jing* (Book of odes) there are several references to star groups.[36] The *Shi jing* was probably compiled about 600 B.C., although many of the odes may date from several centuries earlier.[37] Only about ten separate asterisms are alluded to, but since these are mentioned almost at random, it is plausible that many more star groups were recognized at this early period. Most of the constellation names are identical with those used in later times, although there are also one or two archaic renderings.

The *Shi jing* contains the earliest recorded allusions in Chinese history to the Milky Way (Tianhanhe, or Celestial Han River). An extract from one of these odes seems worth quoting here, since it mentions both the Milky Way and several discrete asterisms:

> In the heavens there is the (celestial) Han (the Milky Way), it looks down and is bright. . . . Brilliant is the Draught Ox, but one does not yoke it to any carriage; in the east there is the Opener of Light (Lucifer); in the west there is the Long Continuer (Hesperus). . . . In the south there is the Winnowing Basket, but one cannot winnow with it, in the north there is the Ladle, but one cannot ladle wine or congee with it . . . in the north there is the Ladle, it raises its western handle.[38]

Here the Draught Ox is the asterism Niu in Capricorn, and the Winnowing Basket is the star group Ji in Sagittarius. As noted above, the Ladle (Dou) refers to the Big Dipper. The apparent rotation of the "handle" of this constellation around the north celestial pole formed a convenient hourly and seasonal marker to many early civilizations—for example, Sumerian, Indian, and Egyptian.[39] In ancient times the Big Dipper was much closer to the celestial pole than at present owing to the precession of the equinoxes.[40] Hence its rotation about the pole would be particularly obvious.

The concept of a pole star seems to date at least to the Chunqiu age. Thus in the *Lun yu* (Analects [of Confucius]), a work that dates from the fifth or fourth century B.C., it is recorded that "the Master said, 'He who exercises government by means of his virtues may be compared to the north polar star, which keeps its place and all the stars turn towards it.' "[41] At the time, the nearest bright star to the north celestial pole would be β UMi (Kochab), some seven degrees away.[42]

It is noteworthy that of the asterisms mentioned in the *Shi jing*, most (including Niu and Ji cited in the quotation above) are identifiable with what in more recent texts are

---

32. See, for example, the *Shi ji*, chap. 14 (note 28).

33. Valuable historical comments are provided by Watson, *Early Chinese Literature*, 40–66 (note 26), and Timoteus Pokora, "Pre-Han Literature," in *Essays on the Sources for Chinese History*, ed. Donald D. Leslie, Colin Mackerras, and Wang Gungwu (Canberra: Australian National University Press, 1973), 23–35, who also give important surveys of other pre-Han literature.

34. See, for example, Roland Felber, "Neue Möglichkeiten und Kriterien für die Bestimmung der Authentizität des Zuo-Zhuan," *Archiv Orientální* 34 (1966): 80–91.

35. Translated by Legge, *Chinese Classics*, 5:628–29 (note 31). In citing this and other translations and quotations throughout this chapter, I have substituted pinyin romanization.

36. There are fine English translations of the poems in the *Shi jing* by Arthur Waley, trans., *The Book of Songs* (London: Allen and Unwin, 1937), and Bernhard Karlgren, ed. and trans., *The Book of Odes* (Stockholm: Museum of Far Eastern Antiquities, 1950; reprinted 1974).

37. Useful comments on the date of the *Shi jing* are given by Watson, *Early Chinese Literature*, 202–30 (note 26).

38. This translation, part of Ode 203, is by Karlgren, *Book of Odes*, 155 (note 36).

39. A discussion of the importance of the Dipper to early civilizations is given by Donald J. Harper, "The Han Cosmic Board (*Shih* 式)," *Early China* 4 (1978): 1–10.

40. Precession is largely produced by solar and lunar torques acting on the rotating earth. These torques cause the earth's axis of rotation to describe a circle of approximately twenty-four degrees radius over a period of some 26,000 years. A gradual displacement of the celestial poles relative to the stars results.

41. *Lun yu*, bk. 2, chap. 1, translated by Legge, *Chinese Classics*, 1:145 (note 31).

42. This star appears to have remained the pole marker until the Later Han, by which time, owing to the effect of precession, its distance from the celestial pole had increased to nearly ten degrees (see below). About 500 B.C., Polaris (α UMi) was as much as fifteen degrees from the north celestial pole.

TABLE 13.1 The Twenty-eight Lunar Lodges

Number	Name (Translation)	Determinative Star	Computed Angular Extent (Han Dynasty)
1	Jue (Horn)	α Vir	12°
2	Kang (Neck)	κ Vir	9°
3	Di (Base)	α Lib	15°
4	Fang (Chamber)	π Sco	5°
5	Xin (Heart)	σ Sco	5°
6	Wei (Tail)	μ Sco	19°
7	Ji (Basket)	γ Sgr	11°
8	Nandou (Southern Dipper)	φ Sgr	27°
9	Niu (Ox)	β Cap	8°
10	Xunü (Maid)	ε Aqr	12°
11	Xu (Emptiness)	β Aqr	10°
12	Wei (Rooftop)	α Aqr	17°
13	Yingshi (Encampment)	α Peg	17°
14	Dongbi (Eastern Wall)	γ Peg	9°
15	Kui (Stride)	ζ And	16°
16	Lou (Harvester)	β Ari	11°
17	Wei (Stomach)	35 Ari	15°
18	Mao (Mane?)	17 Tau	11°
19	Bi (Net)	ε Tau	18°
20	Zuixi (Turtle Beak)	φ′ Ori	1°
21	Shen (Triad)	δ Ori	8°
22	Dongjing (Eastern Well)	μ Gem	33°
23	Yugui (Ghost Vehicle)	θ Cnc	4°
24	Liu (Willow)	δ Hya	15°
25	Qixing (Seven Stars)	α Hya	7°
26	Zhang (Extended Net)	υ Hya	17°
27	Yi (Wings)	α Crt	18°
28	Zhen (Axletree)	γ Crv	17°

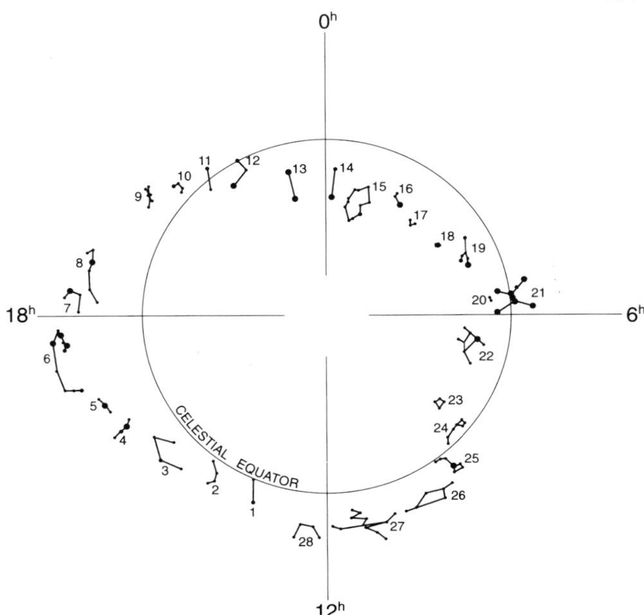

FIG. 13.2. THE UNEVEN DISTRIBUTION OF THE LUNAR LODGES. The diagram clearly shows this concept and also indicates stellar magnitude by size of symbol. Although the *xiu* were in general closer to the celestial equator in ancient times, the correspondence was never good. See also table 13.1, which lists the names, determinative stars, and computed angular extent of the lunar lodges.

termed *xiu* (lunar lodges). These are a series of twenty-eight constellations encircling the sky in the vicinity of the celestial equator. (The expression *xiu* is a derivative of *su*, "to stay the night.") The names of several lunar lodges have also been noted on certain bronze vessels that date from much the same period as many of the folk songs in the *Shi jing*.[43] This suggests that by fairly early in the first millennium B.C. at least some of the *xiu* had already acquired special significance. Nevertheless, the notion of the lunar lodges as a discrete entity cannot be firmly established by documentary evidence until as late as the fifth century B.C.[44]

From at least the Former Han dynasty onward, the lunar lodges were of prime importance in uranography. They defined the celestial coordinates of other star groups—as well as specifying the locations of the sun, moon, and planets—and were also regarded as of great astrological significance. Owing to the sparseness of the available evidence, however, the evolution of the concept of the *xiu* from rough celestial markers to the basis of

43. See Needham, *Science and Civilisation*, 3:248 note d (note 5). In this chapter the word "lodge" has been used (rather than "mansion") in translating *xiu*; lodge has been substituted for mansion in quotations as well.

44. A chest bearing the names of all twenty-eight lunar lodges in order dating from about 433 B.C. was discovered in 1978; see pp. 519–20 for further details.

a developed coordinate system cannot be traced in any detail.

Table 13.1 gives a standardized list of the twenty-eight lunar lodges as found in texts from the third century B.C. onward. Subsequently there were only minor variations in the designations of individual lodges. In this table the *xiu* are numbered in their traditional order, commencing with Jue; translations of individual names are appended. A few names among the first seven *xiu* relate to various features of the celestial dragon (see below), but the other designations seem little more than a random assemblage and in general are much more mundane than those of the Western zodiacal signs.

Astronomical calculations involving the precession of the equinoxes yield an extremely ancient date for the origin of the lunar lodges. The *xiu* lie fairly close to a great circle that today is by no means coincident with the celestial equator (fig. 13.2). By applying precession, one can readily show that in ancient times the circle of the lodges was a better approximation to the celestial equator than at present. Modern studies based on this and similar ideas have suggested the middle of the third millennium B.C. as the date when the *xiu* were first conceived,[45] but there is no evidence of an advanced culture in China at such a remote period—at least a millennium before the earliest written records from this part of the world.

At present it is not possible to ascertain whether only a few relatively well-defined asterisms were recognized in the early Zhou and Chunqiu periods or whether—as seems more likely—the stars had already been grouped into many of the patterns familiar in more recent times. Certainly there is no hint of the existence of star maps in the Zhou dynasty. The astronomers of the period seem to have possessed few instruments. For example, only the gnomon is alluded to in the *Zuozhuan*.[46] Not until the Zhanguo (Warring States) period do we find the first suggestions of systematic division of the night sky, and even then information is accessible mainly in secondary sources of questionable reliability.

Celestial Cartography in the Zhanguo Period (403–221 B.C.)

The Zhanguo period, which followed the Chunqiu era, was a time of marked philosophical speculation.[47] It is thus disappointing that little definite information can be established with regard to the development of astronomy during the Zhanguo; in particular, no star maps that date from these times are known to survive. Texts written many centuries later trace the origins of serious mapping of the night sky in China to this period.[48] There is a persistent tradition that Zhanguo mapping had an important influence on future uranography. For example, chapter 11 of the *Jin shu*, compiled about A.D. 635, contains the following statement in a section titled "Tianwen jingxing" (The stars of the heavens):

> During the time of Emperor Wudi (reign [A.D.] 265–90), Chen Zhuo the Astronomer Royal combined together the astronomical charts made by the three (ancient) schools of Gan (De), Shi (Shen) and Wu Xian, giving a total number of 283 star groups and 1,464 stars in (his) records. A general outline of the more significant ones is now given to complete the (following) section on the stars.[49]

The *Jin shu* proceeds to give a descriptive list of 240 constellations in which measurements are lacking; the main emphasis is astrological. The account above is elaborated in the astronomical treatise (chap. 19) of the *Sui shu* (History of the Sui, compiled 629–56 by Wei Zheng et al.)—much the same time as the *Jin shu*—while additional details are found in an astrological manuscript of the early seventh century A.D. that Paul Pelliot recovered from Dunhuang in 1908.[50] Further reference to these sources will be made below.

It was long supposed that a portion of a star catalog compiled by Shi Shen, who was an astronomer of the Wei state during the Zhanguo, was preserved in the much more recent *Kaiyuan zhanjing* (Kaiyuan treatise on astrology), compiled about 730 by the Indian astronomer Gautama Siddhārtha (Qutan Xida).[51] In a section titled *Xingjing* (Star classic), the *Kaiyuan zhanjing* lists the north polar distances and other details for the key stars of the

45. See, for example, Chu K'o-chên, "The Origin of the Twenty-eight Lunar Mansions," *Actes du VIIIe Congrès International d'Histoire des Sciences (1956)* (Florence: Gruppo Italiano di Storia delle Scienze, 1958), 1:364–72; David S. Nivison, "The Origin of the Chinese Lunar Lodge System," in *World Archaeoastronomy*, ed. A. F. Aveni (Cambridge: Cambridge University Press, 1989), 203–18.

46. A solstitial observation made in 655 B.C. is recorded in the *Zuozhuan*; see Legge, *Chinese Classics*, 5:142 and 144 (note 31).

47. For example, the great philosophers Mencius (372–289 B.C.), Mo Zi (470–391 B.C.?), and Han Feizi (d. 233 B.C.) all lived during the Zhanguo.

48. There are no such assertions in preserved writings dating from the centuries immediately following the Zhanguo.

49. Translated by Ho, *Astronomical Chapters*, 67 (note 4). A few quotations ascribed to Shi Shen and Gan De are contained in the astronomical treatise of the *Han shu* (History of the Former Han), compiled by Ban Gu about the first century, but the main source of alleged extracts from the works of Gan De, Shi Shen, and Wu Xian is a late treatise, the *Kaiyuan zhanjing*. From a study of these fragments, Maspero concluded that all lived sometime between about 350 and 250 B.C.; see Henri Maspero, "L'astronomie chinoise avant les Han," *T'oung Pao* 26 (1929): 267–356, esp. 269–70. In particular, Wu Xian was probably a pseudonym; the original Wu Xian was reputed to be an astronomer of the Shang dynasty (*Shi ji*, chap. 27 [note 28]).

50. Maspero, "L'astronomie chinoise," 272 and 319 (note 49).

51. See, for example, Needham, *Science and Civilisation*, 3:197 and 266–68 (note 5). (Gautama held the position of astronomer royal in China.)

twenty-eight lunar lodges and also for reference stars in ninety-two other groups—about half of the visible sky.[52] Early histories such as the *Shi ji* and the *Hou Han shu* (History of the Later Han, compiled fifth century A.D. by Fan Ye) attribute a work bearing the title *Xingjing* to Shi Shen.[53]

Several years ago, Maeyama made a detailed investigation of the stellar positional measurements in the *Xingjing* section of the *Kaiyuan zhanjing*.[54] Applying precession, he was able to deduce a date within a few decades either side of 70 B.C., and thus long after the Zhanguo. Independent research by Yabuuchi reached much the same conclusion.[55] This date is well into the Former Han and about half a century after Hipparchus produced his famous star catalog. Hence the *Xingjing* as preserved today cannot be the same as the work of that name attributed to Shi Shen. Maeyama emphatically remarks, "Thus, a widely accepted assumption, that the first systematic measurements in equatorial coordinates covering 120 constellations should originally be ascribed to Shi Shen (350 B.C.), has now turned out to be a mere fiction."[56]

It is, of course, arguable that the measurements reported in the existing *Xingjing* represent revisions of more ancient observations that no longer survive. Even if this were the case, however, the form the original data took cannot be established. Discussion of the content of the *Xingjing* will be continued below in the section devoted to the Qin and Han dynasties. Because of the lack of suitable historical records, we cannot adequately assess the contribution to astronomy and, in particular, celestial cartography made during the Zhanguo. Nevertheless, we may conclude that evidence for detailed mapping of the night sky by Shi Shen and his presumed contemporaries Gan De and Wu Xian rests on late traditions rather than on more tangible evidence.

It has long been known that an almost complete list of the names of the lunar lodges is preserved in the *Yueling* (Monthly observances), an almanac that may have been composed during the Zhanguo.[57] In this work, which is concerned with agriculture, court ceremony, and so forth, the passing months are marked by the sun's position in various constellations and the culmination of certain star groups at dawn and dusk.

In 1978, archaeologists working in Hubei Province made a discovery that shed new light on the history of the lunar lodges. This was a lacquer chest, unearthed from the tomb of Yi, a marquess of the state of Zeng, who died about 433 B.C.[58] On the lid of the chest are inscribed in a roughly circular pattern the names of twenty-eight constellations (fig. 13.3). With only a few exceptions, these ideographs—which are artistically written in a style typical of the time—are identifiable in terms of the names of the lunar lodges as found in later texts and are cited in the usual order.[59] The various characters surround a much larger ideograph denoting Dou (the Dipper) and are inserted between representations of a tiger and a dragon. This inscription provides the earliest documentary evidence for the existence of twenty-eight lunar lodges. Before its discovery, the oldest known *complete* list of the *xiu* dated from approximately two centuries later.

The asymmetrical configuration of the characters denoting the lunar lodges on the lid of the chest is intriguing. It seems likely that the illustrations of the dragon and tiger and the large character Dou were executed first and the names of the *xiu* were inserted in the remaining space afterward.[60] There does not appear to be any deliberate attempt to portray the irregular spatial distribution of the *xiu* (fig. 13.2 above). Presumably the animals were intended to represent the Azure Dragon and White Tiger—two of the five palaces (*gong*) into which the night sky is known to have been divided at least from the

52. *Kaiyuan zhanjing*, chaps. 60–68 (there is an edition published in Beijing, 1786); the *Xingjing* is also copied in the Daozang (Daoist canon), in sec. 284 titled *Tongzhan daxiangli xingjing* (The great firmament star manual common to astrology).

53. For example, *Shi ji*, chap. 27 (note 28), and *Hou Han shu*, chap. 12; see the modern edition in 12 vols. (Beijing: Zhonghua Shuju, 1965–73).

54. Yasukatsu Maeyama, "The Oldest Star Catalogue of China, Shih Shen's Hsing Ching," in *Prismata: Naturwissenschaftsgeschichtliche Studien*, ed. Yasukatsu Maeyama and W. G. Salzer (Wiesbaden: Franz Steiner, 1979), 211–45.

55. Yabuuchi Kiyoshi, "*Sekishi Seikyo* no kansoku nendai" (The observational date of the *Shi Shen Xingjing*), in *Explorations in the History of Science and Technology in China*, ed. Li Guohao et al. (Shanghai: Shanghai Chinese Classics Publishing House, 1982), 133–41.

56. Maeyama, "Oldest Star Catalogue," 212 (note 54).

57. The *Yueling* has been translated by Séraphin Couvreur, ed. and trans., *Li Ki; ou, Mémoires sur les bienséances et les cérémonies*, 2d ed., 2 vols. (Paris: Cathasia, 1913), vol. 1, chap. 4, 330–410. The work was attributed to Lü Buwei (d. 235 B.C.) by several Han critics, but the precise date of composition has been much disputed. Although in the first half of the present century some scholars favored compilation of the *Yueling* during the Chunqiu period, such an early date is not in keeping with the highly formalized and structured style of the almanac. For details, see Chūryō Nōda, *An Inquiry concerning the Astronomical Writings Contained in the Li-chi Yüeh-ling* (Kyōto: Kyōto Institute, Academy of Oriental Culture, 1938), 2. Based on astronomical computations, Nōda derived a date for the *Yueling* within about a century of 620 B.C., though the reliability of this date depends on the validity of his interpretation.

58. This chest is now in the Hubei Provincial Museum, Wuhan. For details see Wang Jianmin, Liang Zhu, and Wang Shengli, "Zeng Houyi mu chutu di ershiba xiu qinglong baihu tuxiang" (The twenty-eight lunar lodges and paintings of the Green Dragon and the White Tiger, from the tomb of Zeng Houyi), *Wenwu*, 1979, no. 7:40–45.

59. For several of the names, phonetic equivalents are used.

60. This was first suggested by Wang, Liang, and Wang, "Zeng Houyi" (note 58).

FIG. 13.3. LID OF A CHEST GIVING NAMES OF TWENTY-EIGHT LUNAR LODGES FROM A 433 B.C. TOMB. Names of individual lunar lodges, most of them identifiable with present-day names, are inscribed in a ring around the large character Dou (the Dipper). This chest, discovered in 1978 during excavations of the tomb of the marquess Yi, provides the earliest known list of all twenty-eight lunar lodges.
Size of the original: 82.8 × 47.0 × 19.8 cm. By permission of Hubei Provincial Museum, Wuhan.

Former Han dynasty.[61] As depicted on the chest lid, however, the orientation of the lunar lodges relative to the two animals is highly erroneous (roughly 180 degrees out of phase).

The Qin and Han Dynasties (221 B.C.–A.D. 220)

Present knowledge of Qin astronomy is still very incomplete, as is that for the Zhanguo and earlier periods. Nevertheless, the importance attached to stargazing in this short-lived dynasty (221–207 B.C.) can be judged from the large number of astronomers (more than three hundred) said to be in the service of the ruler.[62] Future excavations of the tomb of the first emperor Qin Shihuang—situated at Lintong near Xi'an—may possibly yield valuable information on Qin celestial cartography. To date, only an annex of the mausoleum has been examined, revealing the now world famous "terra-cotta army." The sepulcher itself has still to be excavated. Sima Qian gives the following account of the interior of this mausoleum, which was sealed in 210 B.C.:

> Liquid mercury was used to simulate the flow of the many rivers, (Chiang) Jiang, (Huang) He and the great sea. Machines were used to circulate (the mercury) and make it flow. Above, astronomical charts (tianwen) were drawn; below geographical maps were depicted.[63]

These "astronomical charts" presumably included representations of the constellations. As recent discoveries have emphasized (see below), the practice of painting star maps on the ceilings of Chinese tombs seems to have been fairly common from the Han dynasty onward.

A compendium of natural philosophy compiled in the state of Qin not long before China was unified, the *Lüshi*

61. The other three celestial palaces were named Red Bird, Dark Warrior, and Forbidden Purple (see below).

62. *Shi ji*, chap. 6 (note 28).

63. *Shi ji*, chap. 6; translation from F. Richard Stephenson and C. B. F. Walker, eds., *Halley's Comet in History* (London: British Museum Publications, 1985), 45.

Chunqiu (Master Lü's Spring and autumn [annals]), cites all twenty-eight lunar lodges in order. Until the discovery of the fifth century B.C. chest in the tomb of the marquess Yi, this was the earliest known complete list of the *xiu*. The *Lüshi Chunqiu* was compiled about the middle of the third century B.C. by a team of scholars gathered together by the prime minister Lü Buwei.[64] Names of the lodges, which are cited in a philosophical context, are essentially identical with those in table 13.1.

Archaeological excavations at Mawangdui (Hunan Province) in 1973 provided new information on astronomy during the Qin and also the early Han.[65] In that year many manuscripts were recovered from a tomb, along with an inscription giving the precise date of interment, which corresponds to 168 B.C. One of these texts, written on silk, gives the earliest reliable indication of the use of the lunar lodges to mark the positions of the planets.[66] This manuscript, now known as the *Wuxingzhan* (Prognostications from the five planets), details, among other matters, the various *xiu* in which the planet Venus rose and set between 246 and 177 B.C.[67] Such information had previously been found only in texts dating from about 100 B.C. onward—for example, in the astronomical treatise of the *Shi ji*—though the *Kaiyuan zhanjing* attributes similar data to Shi Shen and Gan De.

Han records reveal a firm belief in the correspondence between celestial and terrestrial events. An astronomer royal (*taishi ling*, literally prefect of the grand clerks) was appointed to take charge of astronomy, astrology, and the daily records of the empire. His duties were to record "any anomalous happenings in nature . . . , [which] were construed as signs of warnings by heaven toward the misbehavior or misgovernment of the ruler of man."[68] The practice of astronomy was largely centralized at the capital in the Han and all later dynasties, and the various instruments used by the official observers were closely guarded to prevent inspection by the general public. As a result, practically all of the major star maps and catalogs throughout Chinese history were produced by the imperial astronomers.

Although the oldest extant astral charts in China date from the Han, survivals are fairly rare, and only a few constellations are depicted. However, there is ample evidence that Han uranography reached a high level of attainment. Histories of the period give brief descriptions of several star maps and also a celestial globe, and two contemporary stellar catalogs have come down to us. By the Former Han we find the first definite evidence of a system of celestial coordinates in China. In this scheme, which continued in use until the present century, the lunar lodges played a key role. The choice of spherical coordinates is consistent with the development of the *huntian* (enveloping heaven) theory, the concept of a spherical heaven, by at least the Former Han; otherwise this theory is not encountered until the Later Han (25–220).[69]

Star charts appear to have been common during the Han, especially in the later phase of this dynasty. Needham has drawn attention to several contemporary references to such artifacts. For example, the bibliographical section of the *Han shu* lists a work entitled *Yueling botu* (Silken map of the path of the moon), which Geng Shouchang presented to the emperor in 52 B.C., and the biography of Wang Mang (45 B.C.–A.D. 23), in chapter 99 of the *Han shu*, mentions a *zigetu* (chart of the Purple Palace).[70] Nothing else is known about these maps. An interesting discussion among astronomers in A.D. 92 is reported in *Xu Han shu* (Supplement to the Han history). Here it was said that star maps always have methods of graduation (i.e., coordinates),[71] suggesting that by this period such maps were fairly numerous.

The numbers of asterisms and stars noted in Han texts bear no relation to the figures that were later attributed to Zhanguo astronomers such as Shi Shen. Thus, in the astrological treatise ("Tianwen zhi") that forms chapter 26 of the *Han shu*, composed toward the end of the first century A.D. by Ma Xu, the following details are given:

> In the astronomical charts there can be found 118 groups of stars that can be identified inside (i.e. north of) and outside (i.e. south of) the (equatorial belt of

64. Lü Buwei was also credited by Han critics with authorship of the *Yueling* (note 57 above). The first twelve chapters of the *Lüshi Chunqui* are practically identical with the *Yueling*.

65. For brief details of the Mawangdui discoveries, see Michael A. N. Loewe, "Manuscripts Found Recently in China: A Preliminary Survey," *T'oung Pao* 63 (1977): 99–136, and idem, *Ways to Paradise: The Chinese Quest for Immortality* (London: George Allen and Unwin, 1979), 12 ff.

66. Another silk manuscript discovered in this tomb has attracted widespread attention because of the variety of cometary tails it depicts. See Xi Zezong, "The Cometary Atlas in the Silk Book of the Han Tomb at Mawangdui," *Chinese Astronomy and Astrophysics* 8 (1984): 1–7, and Michael A. N. Loewe, "The Han View of Comets," *Bulletin of the Museum of Far Eastern Antiquities* 52 (1980): 1–31.

67. For the text of this manuscript, see Mawangdui Han Mu Boshu Zhengli Xiaozu (Study Group on the Han Silk Manuscripts from Mawangdui), " 'Wuxing zhan' fubiao shiwen" (Explanatory table for "Prognostication from the Five Planets"), *Wenwu*, 1974, no. 11:37–39. The planet Venus repeats its pattern of visibility at eight-year intervals.

68. This quotation is taken from Wang Yü-ch'üan, "An Outline of the Central Government of the Former Han Dynasty," *Harvard Journal of Asiatic Studies* 12 (1949): 134–87, esp. 165. For a recent detailed study of Han portents, see Hans Bielenstein, "Han Portents and Prognostications," *Bulletin of the Museum of Far Eastern Antiquities* 56 (1984): 97–112.

69. Maeyama Yasukatsu, "On the Astronomical Data of Ancient China (ca. −100 +200): A Numerical Analysis (Part 1)," *Archives Internationales d'Histoire des Sciences* 25 (1975): 247–76, esp. 248.

70. Needham, *Science and Civilisation*, 3:276 (note 5).

71. Cited by Needham, *Science and Civilisation*, 3:276 note d (note 5).

> the) lunar lodges. The total number of stars in these group[s] is 783. All of them are connected with particular prefectures and kingdoms, and with officials, palace (affairs) and all kinds of things.[72]

The great astronomer and mathematician Zhang Heng (78–139),[73] a contemporary of Ma Xu, gave independent estimates of stellar numbers. Appointed astronomer royal in 116, he made major contributions to Han celestial cartography. In his "Lingxian" (Spiritual constitution of the universe, ca. 118), of which only fragments are preserved today,[74] Zhang Heng wrote as follows:

> North and south of the equator there are 124 [star] groups which are always brightly shining. 320 stars can be named (individually). There are in all 2500, not including those which the sailors observe. Of the very small stars there are 11,520. All have their influences on fate.[75]

Enumeration of 2,500 stars would be an arduous but not necessarily impossible task at this early period. The remarkably precise figure of 11,520 seems incredible, however; this is roughly twice the number of stars visible to the average unaided eye over the entire celestial sphere. It is unfortunate that the quotation above is so brief. Further details might have given valuable insight into Han astrography. Regarding the stars "which the sailors observe," Needham points out that already by the Later Han seafarers were making regular voyages to Southeast Asia, where they would see constellations invisible in China.[76] Even though there are more definite records of the observation of southern stars from near the equator in later dynasties, no illustrations of the south circumpolar asterisms are preserved until the Ming (see below).

Regrettably, the astral charts Zhang Heng produced did not survive for long. In the astronomical treatise of the *Sui shu* it is related that they "got lost in the disturbances (at the end of the Han) and the names and details which they showed were not preserved."[77]

In an extant fragment of the writings of Zhang Heng is found one of the earliest clear descriptions of the *huntian* theory of the universe, a rival of the *gaitian* (covering heaven, i.e., hemispherical dome) and *xuanye* (infinite empty space) theories:

> The heavens are like a hen's egg and as round as a crossbow bullet; the earth is like the yolk of the egg, and lies alone in the centre. Heaven is large and earth small. . . .
>
> The circumference of the heavens is divided into 365¼° [*du*]; hence half of it, 182⅝° [*du*], is above the earth and the other half is below. This is why, of the 28 *xiu* . . . , only half are visible at one time.[78]

In 117 Zhang Heng built a rotating celestial globe that appears to have accurately represented the constellations:

> Zhang Heng . . . constructed a celestial sphere [*huntian*], which included the inner and outer circles, the south and north (celestial) poles, the ecliptic and the equator . . . the stars within (i.e. north of) and without (i.e. south of) the twenty-eight lunar lodges, and the paths of the sun, the moon, and the Five Planets. The sphere was rotated by a water (clock), and was placed (in a chamber) above a (palace) hall. The transits, rising and setting of heavenly bodies (shown on the sphere) in the chamber reflected those in the (actual) heavens.[79]

This sphere, which had a circumference of 14.61 *chi* (3.4 m), was carefully preserved for two centuries after its manufacture, but it eventually met an ignominious fate. In 316, the northern half of China fell into the hands of nomadic invaders. Nothing was heard of the celestial globe until a century later, when the city of Chang'an was recaptured. When the instruments of Zhang Heng were examined, however, "although the forms [of the instruments] were still recognisable, the marks of graduation had all gone, and nothing was left of the representation of stars, sun, moon and planets."[80]

It is unfortunate that no original charts having any pretensions to accuracy are known to exist from the Han dynasty—or indeed until well into medieval times. A number of Han representations of constellation patterns are still extant, some having been unearthed in recent years,[81] but these tend to be decorative rather than func-

72. Translated by Ho, *Astronomical Chapters*, 66 (note 4), from the *Han shu* account as copied in chap. 11 of the *Jin shu*.

73. For a brief biography of Zhang Heng, see Yu-che Chang, "Chang-Hen, a Chinese Contemporary of Ptolemy," *Popular Astronomy* 53 (1945): 122–26.

74. The main source of quotations from the "Lingxian" is the *Yang Hui suanfa* (Yang Hui's methods of computation), a collection of Yang Hui's work and other books, written in 1275.

75. Translated by Needham, *Science and Civilisation*, 3:265 (note 5). (Throughout this chapter, the parenthetical notes in translations from Needham have sometimes been omitted.)

76. Needham, *Science and Civilisation*, 3:265 note d (note 5).

77. *Sui shu*, chap. 19, translated by Needham, *Science and Civilisation*, 3:264 (note 5).

78. Translated by Needham, *Science and Civilisation*, 3:217 (note 5). The adopted units, *du* (degrees of 365.25 to a circle), were chosen because the average daily motion of the sun relative to the background stars is one *du* (i.e., 0.986 degree). For most purposes it is convenient to translate *du* as degrees. For more on the *huntian* and *gaitian* theories, see above, pp. 117–24.

79. Translated by Ho, *Astronomical Chapters*, 59 (note 4).

80. Chapter 23 of the *Song shu* (History of the [Liu] Song), compiled 492–93 by Shen Yue, translated by Joseph Needham, Wang Ling, and Derek J. de Solla Price, *Heavenly Clockwork: The Great Astronomical Clocks of Medieval China*, 2d ed. (Cambridge: Cambridge University Press, 1986), 95–96. Most of the parenthetical notes in translations from *Heavenly Clockwork* are omitted here.

81. For illustrations of several of these artifacts, see *Zhongguo gudai tianwen wenwu tuji* (note 6).

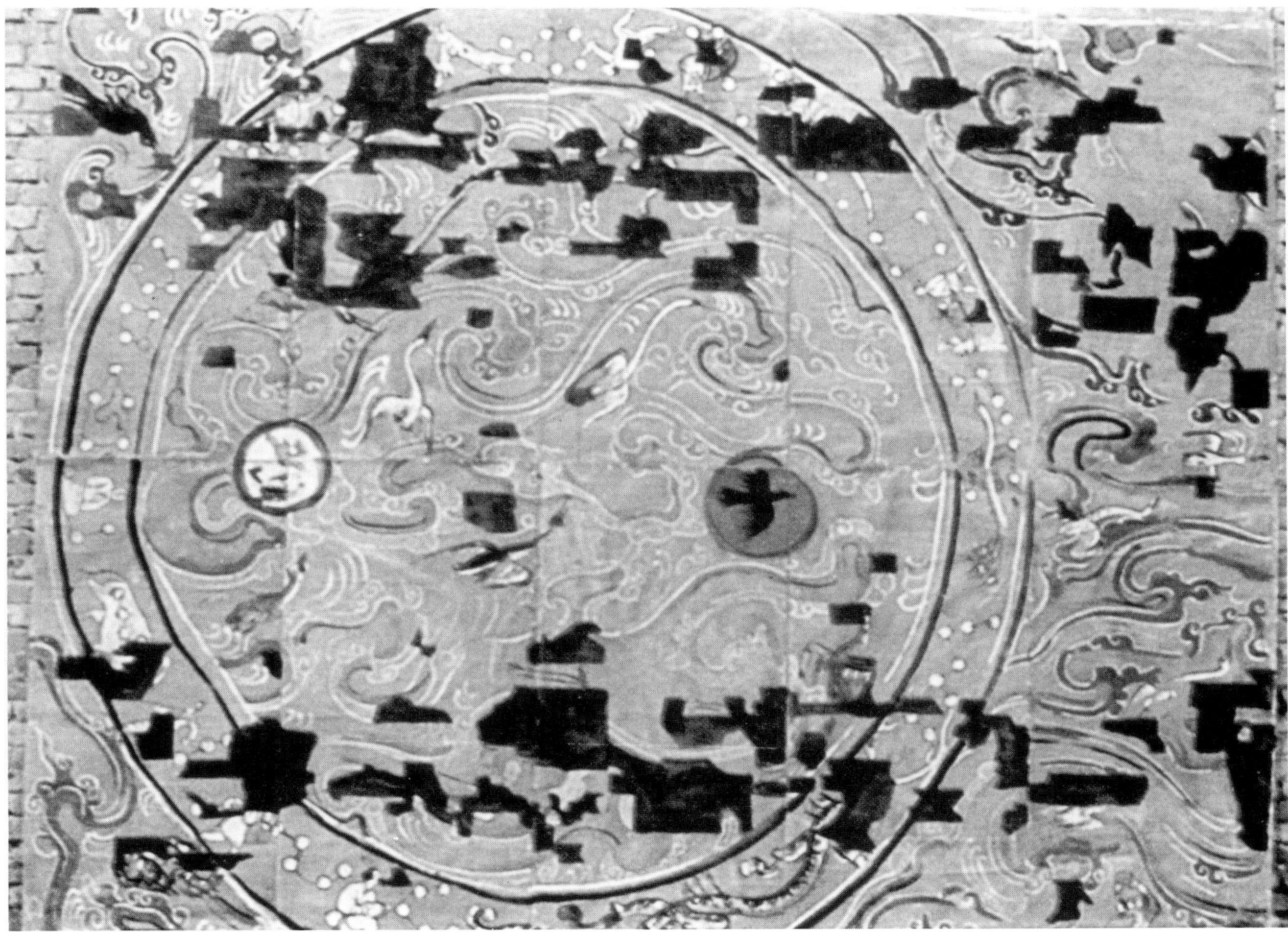

FIG. 13.4. COPY OF A PAINTING OF LUNAR LODGE ASTERISMS FOUND ON THE CEILING OF A TOMB AT XI'AN. A recent careful copy at full scale of a painting discovered on the ceiling of a Former Han tomb. In the ring at the edge of the picture are depicted the twenty-eight lunar lodges in order (commencing at the lower right and moving counterclockwise).
Diameter of the inner circle: ca. 2.5 m. Jiaotong University, Xi'an. By permission of Zhong Wanmai.

tional and typically include only a few inaccurately delineated star groups.

In 1987 the earliest known illustration of all the lunar lodge asterisms was discovered at Xi'an, near the site of the Former Han capital of Chang'an (fig. 13.4).[82] It is painted on the arched ceiling of a tomb that came to light at Jiaotong University, and its estimated date is toward the end of the Former Han, as indicated by coins and other objects found within tomb. The twenty-eight *xiu* asterisms are roughly sketched in a narrow band (some 25 cm wide) around the edge of a circle with an approximate diameter of 2.5 meters. Within this circle are depicted typically Daoist representations of the sun and moon, and also clouds and cranes. Predominant colors are blue, turquoise, red, white, and black.

The painting appears to have been deliberately defaced by grave robbers in antiquity. Although its general state of preservation is poor, several of the constellation patterns are still intact. Stars are denoted by circles of approximately equal size; there seems to be no attempt to display relative brightness. Individual stars are joined into groups by short straight lines. Both of these characteristics are typical of star charts throughout later Chinese history. The circular band portraying the lunar

82. See Shaanxi Sheng Kaogu Yanjiusuo (Shaanxi Archaeological Institute) and Xi'an Jiaotong Daxue (Xi'an Jiaotong University), "Xi'an Jiaotong daxue Xi Han bihua mu fajue jianbao" (Preliminary report on the excavation of the Western Han tomb with murals in Xi'an Jiaotong University), *Kaogu yu Wenwu*, 1990, no. 4:57–63; also Luo Qikun, "Xi'an Jiaotong daxue Xi Han muzang bihua ershiba xiu xingtu kaoshi" (On the star map showing the twenty-eight *xiu* painted on the wall of a Western Han tomb in the campus construction site of Xi'an Jiaotong University in Shaanxi), *Ziran Kexue Shi Yanjiu* 10 (1991): 236–45. The latter article is accompanied by color photographs showing several sections of this astral map.

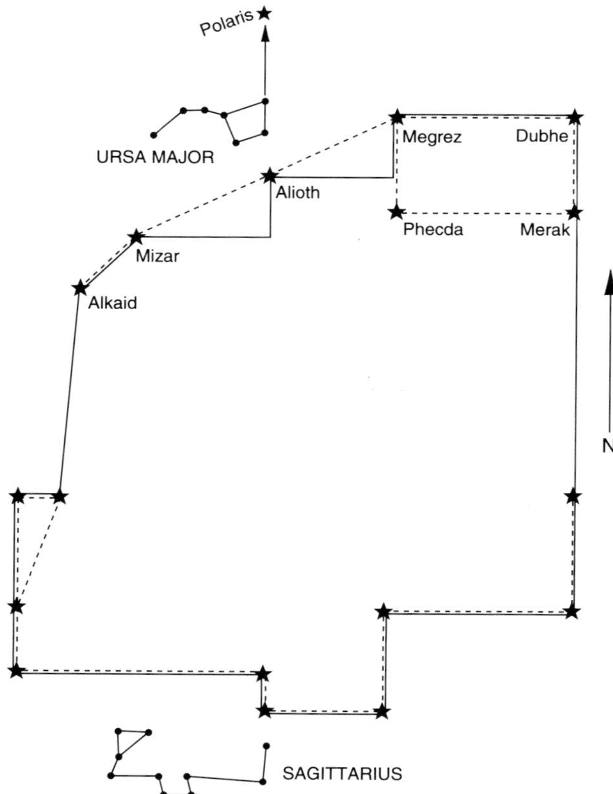

FIG. 13.5. DIAGRAM OF THE HAN CHANG'AN WALLS. Since ancient times, it has been stressed that the northern wall of Han Chang'an (built between 194 and 190 B.C.) took the shape of Beidou (the Northern Dipper, in Ursa Major) while the southern wall resembled Nandou (the Southern Dipper, in Sagittarius). Studies by Hotaling confirm this.
After Stephen James Hotaling, "The City Walls of Ch'ang-an," *T'oung Pao* 64 (1978): 1–46, fig. 22.

lodges also contains illustrations of men, animals, and such, in several cases associated with the constellations themselves.[83] This is an unusual feature; throughout East Asian history, the use of pictorial symbols for the *xiu* is very rare. Normally the star configurations—sometimes highly idealized—are depicted instead. Here we find a marked contrast with Western representations of the signs of the zodiac, which are usually purely symbolic.

What may be a medieval Korean reproduction of a detailed and fairly accurate Han chart is still preserved in a Seoul museum. This planisphere, engraved on marble in 1395, has been damaged over the centuries, but a careful copy to the same scale made in 1687 is in a fine state of preservation (several other late copies are also well preserved). These charts will be discussed in detail below, but let me make a few preliminary remarks at this point.

The Korean map is drawn on a polar (equidistant) projection. Its accompanying inscription relates that it is based on a rubbing of a previous stone chart that had been a gift from China many centuries before. During a war in the Korean peninsula in 670, the original stele had been lost. Calculations based on the positions of the equinoxes as shown on the extant copies of the planisphere indicate an original date toward the end of the first century B.C.[84]

Unique representations of two constellations on an enormous scale appear to have been produced early in the Former Han dynasty. The *Sanfu huangtu* (Yellow plans of the three capital commanderies), probably written sometime between the third and sixth centuries A.D.,[85] gives a detailed account of the building of the walls of the Han metropolis of Chang'an between 194 and 190 B.C. In particular, the text asserts: "The south of the city wall constituted the shape of the Southern Dipper (Nandou = Sagittarius), the north constituted the shape of the Northern Dipper (Beidou = Ursa Major). That until now people refer to the city wall of the Han capital as the 'dou wall' is because of this."[86]

Hotaling, who made a detailed investigation of the city walls of Han Chang'an, remarked that "it is the distinctive shape of the north wall which sets Han Changan apart from the hundreds of other walled cities in China." By the use of scale diagrams, he concluded that the two constellations "really fit the shape of the city wall."[87] Since the length of the north wall was some seven kilometers and that of the south wall only about one kilometer less, these would be the largest representations of star groups ever constructed by any civilization (fig. 13.5).

Roofing tiles from the ruins of Chang'an, depicting the four mythical creatures denoting the nonpolar celestial palaces, were discovered between 1956 and 1958.[88] As in later illustrations (see below), the Dark Warrior (a symbol of the Northern Palace) is shown as a turtle entwined with a snake, though the Azure Dragon, White Tiger, and Red Bird are depicted in more recognizable fashion.

83. For example, the lunar lodge Niu (Ox) is superimposed on a picture of an ox. Unfortunately, many other illustrations are extensively damaged. For a nontechnical account in English, with color photographs, see F. Richard Stephenson, "Stargazers of the Orient," *New Scientist* 137, no. 1854 (1993): 32–34.

84. The derivation of this date is outlined below, pp. 563–64. The term planisphere is used in this chapter to describe a circular map—often on a polar projection—of one or both hemispheres of the night sky.

85. For this suggested date of the *Sanfu huangtu*, see Dubs in Ban Gu, *The History of the Former Han Dynasty*, 3 vols., trans. Homer H. Dubs (Baltimore: Waverly Press, 1938–55), 1:125 n.

86. Translated by Stephen James Hotaling, "The City Walls of Ch'ang-an," *T'oung Pao* 64 (1978): 1–46, quotation on 6.

87. Hotaling, "City Walls," 29, 39 (note 86).

88. Pan, *Zhongguo hengxing guance shi*, pl. 3 (note 7). Samples of these tiles are exhibited in the National Museum of Shaanxi History in Xi'an.

Stellar coordinates have been briefly alluded to above. Unlike the ecliptic framework adopted in the West until recent centuries, Chinese celestial coordinates were equatorial and corresponded closely to the modern system of declination and right ascension (RA).[89] The coordinate in place of declination was known as *qiujidu* (degrees from the pole). This was equivalent to the modern north polar distance (NPD) and was measured from the celestial pole of the time. Positions were expressed in *du*, closely equivalent to degrees (see above).

Writing in the first century A.D., Cai Yong in his *Yueling zhangzhu* (Notes to the Monthly observances) provides important details regarding the use of declination circles on astral charts. He states that three concentric circles were depicted on star maps. The smallest of these was the circle of constant visibility, whose radius was equal to the latitude of the place for which it was constructed. The celestial equator was represented by a circle of intermediate radius. Finally, the outer circle of constant invisibility was set as the limit beyond which no star ever rose above the horizon.[90] The distances between the inner and outer circles and the equator were equal. These same three circles are shown on the Korean chart already mentioned and also on several maps that survive from the Five Dynasties (907–60) and subsequent periods. Clearly, this early tradition became well established in later East Asian history.

In specifying RA, the astronomers of China did not employ a single coordinate origin (such as the vernal equinox). Instead, they measured the positions of celestial bodies eastward from a series of twenty-eight unequally spaced local meridians. These meridians were defined by selected determinative stars (*juxing*), one in each of the lunar lodges; coordinates measured relative to them were termed *ruxiudu* (degrees within a lodge). The term *xiu* came to imply both the asterism itself and the zone of RA it covered. As in the case of north polar distance, RA was expressed in *du*. The equatorial extension of a particular *xiu* (the angular separation between the standard meridian of that lodge and the adjacent reference meridian of the next *xiu* to the east) could range from as small as one or two degrees to some thirty-three degrees.

The determinative stars of the twenty-eight lunar lodges, along with their computed equatorial angular extensions (to the nearest degree) during the Han dynasty are listed in table 13.1. Column 3 of this table gives the modern constellation reference (Bayer Greek letter or Flamsteed number) of each determinant star according to Maeyama,[91] while column 4 specifies the width of each *xiu* in degrees.

As Biot noted more than a century ago, there is a marked correlation between the widths of *xiu* differing in number by fourteen, as exemplified by the two widest lunar lodges, Nandou (no. 8) and Dongjing (no. 22).[92] This feature has yet to be satisfactorily explained. A hint as to why the *xiu* have such uneven spacing is found in the astronomical treatise of the *Shi ji*. This work notes the correspondence in RA between the stars of Beidou and certain of the lunar lodges (Jue, Nandou, and Shen). Biot demonstrated that the determinative stars of a variety of lunar lodges were originally chosen on account of the agreement in RA between these and circumpolar stars.[93] We might thus expect that down through the centuries the choice of certain *xiu* determinatives would be altered to allow for differential precession between the circumpolar stars and the much more southerly lunar lodges. Tradition appears to have played an important role, however. Needham quotes evidence that by medieval times any links between the *xiu* and circumpolar stars had been forgotten.[94] Not until the Ming dynasty do any changes appear to have been made in the determinant stars of the *xiu* (see below).

Numerous solar eclipses are recorded in chapter 27 of the *Han shu*, and for most of these an estimate of the RA of the sun is quoted, expressed to the nearest *du*.[95] For example, in 181 B.C., the totally eclipsed sun was said to be nine degrees in Yingshi. Since the RA of the determinative star of Yingshi (α Peg) was 319 degrees at the time, this corresponds to an RA of 328 degrees. Although the Former Han data prove to be rather crude (average error some five degrees), by the Later Han the corresponding results were of fair precision (mean error two degrees).[96]

89. Declination is a celestial coordinate equivalent to the latitude of the earth's surface. Right ascension (RA) corresponds closely to terrestrial longitude. In modern astronomical practice RA is measured parallel to the celestial equator from the vernal equinox through 360 degrees or 24 hours.

90. This summary of the text by Cai Zhu is given by Bo Shuren, "Astrometry and Astrometric Instruments," in *Ancient China's Technology and Science*, comp. Institute of the History of Natural Sciences, Chinese Academy of Sciences (Beijing: Foreign Languages Press, 1983), 15–32, esp. 18.

91. Maeyama, "Oldest Star Catalogue" (note 54).

92. J. B. Biot, review of *Ueber die Zeitrechnung der Chinesen* by Ludwig Ideler, *Journal des Savants*, 1839, 721–30, and 1840, 27–41.

93. However, several *xiu* determinatives cannot be keyed to circumpolar stars. See T. Kiang, "Notes on Traditional Chinese Astronomy," *Observatory* 104 (1984): 19–23.

94. Needham, *Science and Civilisation*, 3:239 (note 5). Needham notes that in the *Mengxi bitan* (Brush talks from Dream Brook), composed by Shen Kuo about 1088, the administrator of the imperial observatory had asked Shen why the equatorial extensions of the *xiu* were so unequal. He had replied that it was because of the convenience of having them in whole numbers of degrees. He was thus unaware of the true explanation.

95. For a compilation and translation of Former Han records of solar eclipses, see Ban, *Former Han Dynasty*, esp. 3:544–59 (note 85).

96. These figures are based on my own unpublished analysis. See also N. Foley, "A Statistical Study of the Solar Eclipses Recorded in Chinese

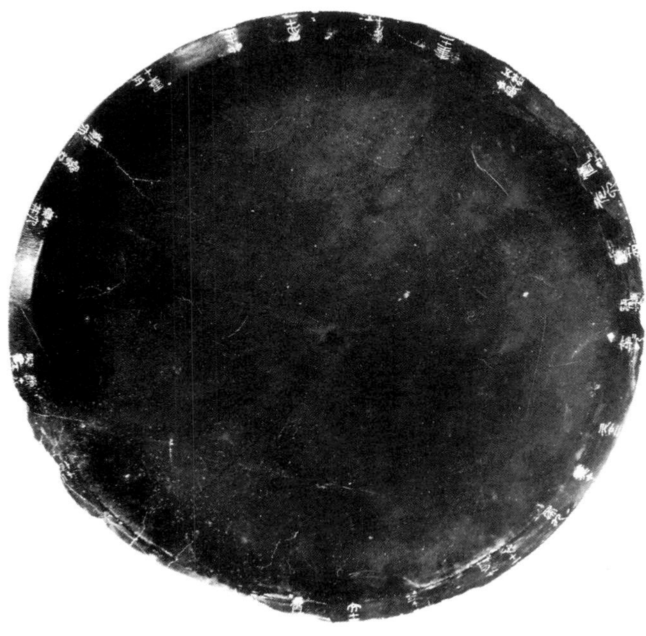

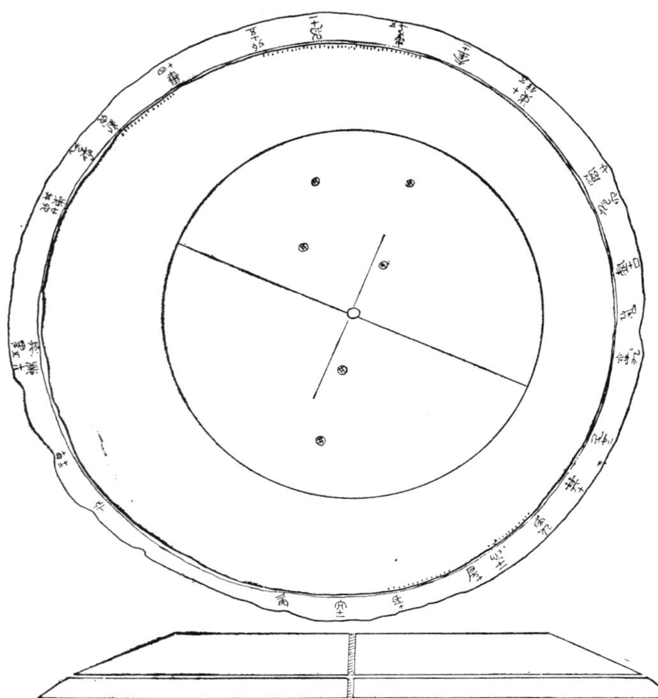

FIG. 13.6. INSCRIBED HAN LACQUER DISK SHOWING IRREGULAR SPACING OF *XIU*. Found in the tomb of a Han nobleman who died in 169 B.C., this instrument is the earliest to show the uneven angular extent of the lunar lodges. It may have been a prototype for a cosmic board, or *shi*. On the right is a reconstruction of the two parts; on the left is a photograph of the rim with the names of the lunar lodges.

Diameter of the original: 25.6 cm. By permission of Anhui Province Fuyang Regional Museum (left). Reconstruction (right) from Zhongguo Shehui Kexueyuan Kaogu Yanjiusuo (Archaeological Research Institute, Chinese Academy of Social Science [Academia Sinica]), *Zhongguo gudai tianwen wenwu tuji* (Album of ancient Chinese astronomical relics) (Beijing: Wenwu Chubanshe, 1980), 115.

The practice of using the *xiu* to specify RA is analogous to the Ptolemaic system of expressing celestial longitude relative to one of the twelve zodiacal signs, a method that was in common use in Western astronomy until recent centuries. However, apart from the choice of equatorial instead of ecliptic coordinates and the greater number of divisions, there are other fundamental differences between the lunar lodge system and the Occidental zodiac, notably the nonuniform extent of individual *xiu*.[97]

It is well known that the *xiu* have several features in common with the Indian *nakṣatras*—star groups "in the path of the moon." Thus Vedic lists from about 1000 B.C. cite either twenty-seven or twenty-eight *nakṣatras*. Yet a recent investigation by Pingree and Morrissey argues strongly against a common origin or even association.[98] From some unknown period, the *nakṣatras*—in common with the *xiu*—also had determinative stars (*yogatārās*), but measurements of the positions of these stars cannot be traced until the fifth century A.D. (in such works as the *Paitāmahasiddhānta*). From these relatively recent measurements it is evident that about one-quarter of the *yogatārās* coincide with determinative stars of the *xiu*, though there are major differences between the choices of many of the remaining determinatives. Whatever the explanation for similarities between the two systems, by the fifth century A.D. the *xiu* had already been firmly established in Chinese astronomy for many centuries. On present evidence, it would be difficult to argue a convincing case for the development of the *xiu* from the *yogatārā*.

In 1977 an instrument showing the approximate extent of each of the *xiu* was unearthed in Anhui Province. This device, consisting of an inscribed lacquer disk, was found in the mausoleum of a nobleman who died in 169 B.C.

and Korean History during the Pre-telescopic Era" (M.Sc. diss., University of Durham, 1989).

97. This is, of course, at variance with the primitive concept of the lunar lodges as representing the nightly resting places of the moon.

98. David Pingree and Patrick Morrissey, "On the Identification of the *Yogatārās* of the Indian *Nakṣatras*," *Journal for the History of Astronomy* 20 (1989): 99–119. For more on the Indian *nakṣatras* and their relation to cartography, see Joseph E. Schwartzberg, "Cosmographical Mapping," in *The History of Cartography*, ed. J. B. Harley and David Woodward (Chicago: University of Chicago Press, 1987–), vol. 2.1 (1992), 332–87.

(fig. 13.6). It may already have been of considerable age when it was entombed with the body of its owner.[99] The rim, which is graduated in *du*, is marked with the names of the lunar lodges. The instrument provides the earliest direct evidence of the uneven extent of the *xiu*.

In this same burial chamber was found a well-preserved "cosmic board" (*shi*), the oldest example so far discovered.[100] During the 1970s, a number of similar Han devices, made of either lacquer or bronze, were unearthed from other tombs.[101] These were found to have a fairly standard pattern: basically, a circular disk known as the "heaven plate" (*tianpan*) is mounted on a larger square board or "earth plate" (*dipan*). The two plates are connected by a central pin and are free to rotate relative to one another. Near the rim of the disk are engraved the names of the twenty-eight lunar lodges and also the twelve months of the year, while the pattern of the seven stars of Beidou, the Big Dipper, is marked across the center of the disk. The lower plate is inscribed with twelve compass directions and also the names of the *xiu*. Such cosmic boards show the Big Dipper as seen from the outer surface of an imaginary dome of the sky, that is, as portrayed on the surface of a celestial globe. Cosmic boards were used for divination from Han times onward. In particular, they illustrate the importance assigned to the Big Dipper in ancient China as a symbol of celestial power.[102]

The oldest complete listing of the equatorial extensions of the *xiu* is to be found in the *Huainanzi* ([Book of the] Master of Huainan, ca. 120 B.C.), a Daoist treatise. This was presented to the emperor Wu of the Former Han in 139 B.C. by his uncle, Liu An (d. 122 B.C.), the prince of Huainan (in modern Anhui Province). At the time, it was said to have been recently completed. The emperor was so pleased that he had the book placed in his private library. An unpublished analysis that I have made of the widths of the lunar lodges recorded in chapter 3 of the *Huainanzi* indicates a mean error as small as 0.5 degree.

Both Han official histories often give descriptions of the apparent paths of comets across the sky.[103] Such records are particularly interesting from the uranographic point of view, since, unlike the moon and planets, comets are not confined to the ecliptic zone. For example, the account in chapter 27 of the *Han shu* of the motion of Halley's comet in 12 B.C. notes the passage of the comet through or near more than ten separate asterisms.[104] Such records provide a valuable test of the accuracy of modern theories of the motion of this famous comet. Cometary records before the middle of the second century B.C. seldom mention constellations apart from the *xiu*, but afterward many star groups to the north and south of the lunar lodges are referred to as well. Although this feature may be significant, let me emphasize that only summaries of the original observations now exist, possibly with the loss of key information.

A report in the *Han shu* of a comet in 138 B.C. is especially interesting, since it makes one of the first known mentions of two of the three *yuan* (enclosures), major groupings of constellations bounded by fairly well defined chains of stars.[105] These are the Taiwei (Supreme Subtlety) and Ziwei (Forbidden Purple) enclosures, situated respectively in the Leo/Virgo region and the north circumpolar zone. (The other *yuan* is the Tianshi [Celestial Market] enclosure, situated largely in Hercules and Ophiuchus.) In astrology the Taiwei *yuan* was particularly important, since it was crossed by the ecliptic and hence lay in the path of the moon and planets. The Ziwei *yuan* formed the approximate boundary of the region within which stars were permanently above the horizon as seen from north-central China.

Occasional entries in the *Han shu* (after about 50 B.C.) indicate that by this period uranography had progressed to the extent that individual stars in at least some asterisms were assigned reference numbers. Thus, in recording a possible nova that appeared during the spring of 48 B.C., the *Han shu* states that a guest star (*ke xing*) was "about 4 *cun* [roughly 0.4 degree] east of the second star of Nandou."[106] Use of stellar reference numbers is relatively rare in Han records, but from the Jin dynasty (265–420) onward it was commonplace (see below).

A valuable indication of how far the night sky had been mapped by the Former Han is provided by the two major star catalogs that are preserved from this period.

99. A drawing of this instrument is provided by Yin Difei, "Xi Han Ruyinhou mu chutu di zhanpan he tianwen yiqi" (Divination board and astronomical instruments from the Western Han tomb of the marquis of Ruyin), *Kaogu*, 1978, 338–43.

100. Yin, "Xi Han Ruyinhou mu chutu di zhanpan he tianwen yiqi" (note 99), provides a photograph of this instrument.

101. Brief details are supplied by Harper, "Han Cosmic Board" (note 39), who also cites primary references in Chinese journals. Much of the subsequent description is based on that given by Harper.

102. For an interesting Han account of the use of divination boards, see chap. 99 of the *Han shu*; modern edition in 7 vols. (Beijing: Zhonghua Shuju, 1970). A translation is given by Dubs in Ban, *Former Han Dynasty*, 3:463–64 (note 85).

103. Translations of these records are given by Ho Peng-yoke, "Ancient and Mediaeval Observations of Comets and Novae in Chinese Sources," *Vistas in Astronomy* 5 (1962): 127–225, 143 ff. Chapter 26 of the *Han shu* is titled "Tianwen zhi" (Treatise on astrology). Chapter 27, "Wuxing zhi" (Treatise on the five phases), devoted mainly to meteorological phenomena, also contains many records of such celestial phenomena as comets and eclipses.

104. A translation and discussion of this record is given by Stephenson and Yau, "Far Eastern Observations," 201–2 (note 2).

105. *Han shu*, chap. 26 (note 102), see also Ho, "Ancient and Mediaeval Observations," 144 (note 103).

106. *Han shu*, chap. 26 (note 102), see also Ho, "Ancient and Mediaeval Observations," 147 (note 103).

These are contained respectively in the *Shi ji* and the *Xingjing*; the compilation in the latter work has already been briefly referred to.

The constellation list in chapter 27 of the *Shi ji* is the earliest to cover the whole sky visible from China. Although the star inventory in the *Xingjing* (*Kaiyuan zhanjing*, chaps. 60–63) appears to be roughly contemporaneous with it, the versions of this latter work that are preserved today are incomplete. In the *Shi ji*, Sima Qian gives brief descriptions of approximately one hundred asterisms (including the *xiu*). These are divided into the five celestial palaces of unknown antiquity: the Central Palace (Zhonggong), Eastern Palace (Donggong), Southern Palace (Nangong), Western Palace (Xigong), and Northern Palace (Beigong). Of these five regions, the Central Palace—also known as the Forbidden Purple Palace (Ziweigong)—was circular; it was the domain of the north circumpolar stars, which never set from the latitudes of northern China. The remaining four palaces took the form of truncated sectors, extending from the circle of constant visibility to the circle of constant invisibility. Among these latter divisions, the Eastern Palace was symbolized by the Azure Dragon (Canglong), the Southern Palace by the Red Bird (Zhuniao), the Western Palace by the White Tiger (Baihu), and the Northern Palace by the Dark Warrior (Xuanwu). Sima Qian's account is more or less repeated in chapter 26 of the *Han shu*, written some two centuries later. Not until the seventh century do we find lists of the constellations that are appreciably more extensive.

Commencing with the circumpolar stars—in the Central Palace—the *Shi ji* gives a qualitative description of each asterism. Star groups in the Eastern Palace, which covers the approximate range of RA from 12 hours to 18 hours, include the first seven lunar lodges; those in the Northern Palace (18 hours to 24 hours) the second seven *xiu*; and so on for the Western Palace and Southern Palace.[107] Positional measurements are absent, and the relative locations of individual asterisms are only vaguely described. As is clear from the frequent allusions to augury, one of the principal motives for compiling this catalog was astrological.[108]

In describing the various constellations, the *Shi ji* is by no means consistent. For some important asterisms, such as Beidou, the constituent stars are individually named. Often the number of stars in a particular group is specified, but this is not always the case, even for the lunar lodges. When cited, the number of stars in an asterism can vary from only one—as in the case of Lang (Sirius) or Laoren (Canopus)—to more than ten. In general, Chinese asterisms were much smaller in extent than the Western constellations.

Most of the names of star groups in the *Shi ji* are identical with those found in later writings. These designations tend to be prosaic compared with those of Western constellations. Instead of gods and goddesses, we find a reflection of the Chinese empire, for example: the emperor and his family; ministers and generals; domestic animals; buildings such as palaces, markets, prisons, and stables. When a change in the appearance of an asterism occurred or a celestial body entered it, this was believed to presage an event involving the terrestrial equivalent.

Reference to the *Shi ji* and later star lists shows that correspondence between Chinese and Babylonian-Greek names for constellations is rare, emphasizing their independent origins. Apart from Beidou, whose ladle shape is so obvious, little more than Wei, another well-defined star group, and Lang can be cited. Wei represents the tail of the Dragon in Chinese uranography and the tail of the Scorpion in the Occident. Lang (Wolf) is equivalent to the brilliant Sirius, the Dog Star. Whereas Sirius is a member of the constellation Canis Major, however, Lang was regarded as an isolated star.[109]

Among the entries in the *Shi ji* catalog, eleven stars are described as "large." Bo has noted that practically all of these are among the brightest stars observed today, but his efforts to suggest that Sima Qian recognized as many as five grades of brightness are far from successful.[110] Whereas Ptolemy grouped the stars into six classes (1 = brightest; 6 = faintest, the foundation of the modern system of stellar magnitudes), no comparable scheme ever found favor in China in the pre-Jesuit period. Provided even a faint star was recognized as an established member of a constellation, its astrological importance seems to have been no less than that of the brighter constituents of the same group.

The *Xingjing*, dated by both Maeyama and Yabuuchi

107. The actual order of the four noncentral palaces in chap. 27 of the *Shi ji* is East, South, West, North. Owing to the irregular widths of the *xiu*, the true angular extent of these palaces ranges from about 75 to 110 degrees.

108. For example, an entry in the *Shi ji*, chap. 27 (note 28), describes the lunar lodge Xin, which consists of the three stars in Scorpio (σ Sco, Antares, and τ Sco): "Xin (Heart) represents the 'Hall of Brilliance.' The large star is the Heavenly King; the front and rear stars represent his sons. It is an unfavorable omen when (the stars) are in a straight line. When they are in a straight line, the Heavenly King will err in judgement" (my translation). Even in later times, Chinese astronomers believed that the relative positions of the stars in a constellation could change to some extent, a curious notion that suggests lack of careful measurement of any but the principal star in each group. See chap. 11 of the *Jin shu* (note 14) and other Chinese treatises on astrology.

109. For remarks on the few similarities between pictorial representations of the constellations as found in China and the Occident, see below.

110. Bo Shuren, "Sima Qian—The Great Astronomer of Ancient China," *Chinese Astronomy and Astrophysics* 9 (1985): 261–67. Bo was able to find very few references to the brightness of stars other than for the small number described as "large."

to within about thirty years of 70 B.C.,[111] contains accurate positional information for 120 star groups. In addition to the twenty-eight lunar lodges, sixty-two asterisms to the north of the *xiu* and thirty asterisms to the south are cataloged. Apart from the lunar lodges, the present list cites only star groups in the Central Palace, Eastern Palace, and Northern Palace. The text as found in the *Kaiyuan zhanjing*—and also in the Daozang, under the title *Tongzhan daxiangli xingjing* (The great firmament star manual common to astrology)—is accompanied by rough sketches of each asterism, though these are not necessarily of early date.

Although the star catalog in the *Xingjing* is incomplete, the original version presumably included many similar measurements of position for asterisms in the Western Palace and Southern Palace. The preserved portions of the catalog appear to be representative of a systematic survey of the whole of the night sky as visible from north-central China.

From a careful investigation, Maeyama was able to confidently identify virtually all of the principal stars in the 120 asterisms listed in the *Xingjing*.[112] After correcting a few obvious copyist's errors, he found that the real accuracy of positional measurements was typically to within about one degree. Measurements of such precision were probably made with the aid of an armillary sphere. The construction of such a device, consisting of a sighting tube attached to a single polar-mounted declination ring, is attributed in later Chinese history to Luoxia Hong about 104 B.C. We know, however, that at least by 52 B.C. a second ring (at right angles to the first) had been added.[113]

The *Xingjing* determinations of celestial latitude, measured north and south of the ecliptic (the Huangdao, or Yellow Road)[114] are particularly interesting; they reveal that the apparent annual path of the sun through the constellations was already clearly defined. The earliest known estimates of the ecliptic extensions of the lunar lodges date from the Later Han; these are contained in chapter 13 of the *Hou Han shu*.[115] But it should be stressed that throughout Chinese history much more emphasis was placed on equatorial coordinates.

The Three Kingdoms to the Sui Dynasty (220–618)

The troubled centuries following the end of the Han were characterized by extensive astronomical activity—often independently in northern and southern China. Although we know that several major astral charts and celestial globes were produced during this period, none are extant. Instead, only a few crude maps of the stars are preserved.

Histories written centuries afterward mention the production of both star maps and celestial globes during the Three Kingdoms period (220–65), but the available information is tantalizingly brief. A stellar chart compiled by the Wu astronomer royal Chen Zhuo (fl. late third century A.D.) has already been mentioned in passing. This was produced sometime between 265 and 280. According to the *Jin shu*, the map constructed by Chen Zhuo depicted 1,464 stars in 283 groups.[116] A few additional details are found in the *Sui shu*. After commenting on the loss of Han star maps in the disturbances that accompanied the downfall of the Later Han dynasty, the *Sui shu* continues as follows:

> But then Chen Zhuo Astronomer-Royal of the Wu State (in the Three Kingdoms period) first constructed and made a map of the stars and constellations according to the three schools of astronomers, Master Gan, Master Shi, and Wu Xian, adding an explanation with an astrological commentary. There were 254 constellations, 1283 stars, and 28 *xiu*, with 182 additional stars, making in all 283 constellations and 1565 stars.[117]

There is an obvious error in the text above, for the total number of stars should be 1,464 rather than 1,565. Here we find a recurrence of the tradition of detailed uranography developed during the Zhanguo. The number of "additional stars" (in the twenty-eight lunar lodge asterisms) is interesting. Han compilations (including the extant versions of the *Xingjing*) assign only 164 stars to the *xiu*, but lists from the early Tang dynasty (618–907) onward include additional stars in six of the lunar lodges, making a total of 182,[118] identical with the figure attributed to Chen Zhuo. Possibly he himself was responsible for revising the number of stars in the *xiu*; there appear to have been no major amendments after his time. Regrettably, apart from the numbers of asterisms and stars depicted, nothing is known concerning the map Chen Zhuo made. He is also known to have written several books on astronomy and astrology, including

111. Maeyama, "Oldest Star Catalogue" (note 54), and Yabuuchi, "*Sekishi Seikyo* no kansoku nendai" (note 55).

112. Maeyama, "Oldest Star Catalogue" (note 54).

113. See, for example, Ho, *Li, Qi and Shu*, 124–25 (note 16).

114. The celestial equator was known as the Chidao, or Red Road.

115. See Maeyama, "Astronomical Data of Ancient China," 269 ff. (note 69).

116. See p. 518. According to chap. 11 of the *Jin shu* (note 14), Chen Zhuo produced his astral chart when the first Jin emperor Wudi (265–90) was on the throne. Since the Wu state was annexed by the Jin empire in 280, the date range reduces to between 265 and 280.

117. From the astronomical treatise of the *Sui shu*, chap. 19, translated by Needham, *Science and Civilisation*, 3:264 (note 5).

118. See, for example, chap. 11 of the *Jin shu* (note 14).

Xingshu (Description of the stars), which survived until at least the twelfth century.[119]

The *Jin shu* attributes some kind of celestial globe (*huntian*) to Lu Ji (fl. 220–45), who—like Chen Zhuo—was an astronomer of Wu State.[120] According to the *Sui shu*, a third astronomer of Wu, named Ge Heng (fl. 250), constructed a device "to show the earth fixed at the center of the heavens; these were made to revolve by a mechanism while the earth remained stationary."[121] In both cases the details are fragmentary. The Wu astronomer and mathematician Wang Fan (219–57) criticized earlier celestial globes as being either so small that the stars were overcrowded or so large that turning them was difficult. He remarked: "I have therefore re-designed the celestial sphere by taking a scale of 3 *fen* to each degree (*du*). The whole of the heavens are thus represented by a sphere with a circumference of 1 *zhang* 1 *chi* 9 *cun* and 5¾ *fen* (approx. 2.52 metres)."[122] Unfortunately, nothing is known regarding the representation of the stars on this globe. According to the *Jin shu*, the various instruments Lu Ji and Wang Fen produced all disappeared after nomadic invaders overran northern China early in the fourth century.[123]

Information on the production of star maps and celestial globes during the Jin dynasty (265–420) is negligible. Nevertheless, the extensive astronomical observations from this period recorded in the *Jin shu* (and also in the *Song shu*) seem to imply the existence of good astral charts. Numerous observations of passages of the moon, planets, comets, and meteors through or close to asterisms are preserved, and many constellations are noted in addition to the twenty-eight *xiu*.

The assignment of a reference number to individual stars within asterisms during the Han has already been mentioned. Extant Jin records contain many examples of the following form: "the moon concealed the second star of Xuanyuan [an asterism in Leo]" or "Venus invaded the second star from the south of Fang [in Scorpio]." By computing the position of the moon or planet on the appropriate date, it becomes possible to identify certain individual stars within asterisms near the ecliptic.[124] Although such numbering schemes continued in use until relatively modern times, it has yet to be established whether significant variations occurred through the centuries.

The Jin dynasty was succeeded by the Liu Song dynasty, 420–79 (one of the Six Dynasties). Not long afterward, a bronze celestial globe was constructed at Nanjing by the astronomer royal Qian Luozhi (who also produced a bronze armillary sphere). On this device, a wide variety of constellations was depicted using colored pearls, though there is disagreement regarding the individual colors. The earliest account, in the *Song shu*, runs as follows:

> In the seventeenth year [of the Yuanjia reign period, i.e., 440] a small astronomical instrument [*huntian*, a celestial globe] was also made, of diameter 2.2 *chi* (0.54 m) and circumference 6.6 *chi* (1.6 m), with (two) tenths of a *cun* (5 mm) to a degree. The twenty-eight lunar lodges were fixed on, and pearls of three colours, white, black and yellow, represented the stars of the three schools of astronomers. The sun, moon and five planets were again attached to the ecliptic [as in the case of the armillary sphere produced four years earlier by Qian Luozhi].[125]

The *Sui shu* differs on the colors used to represent the various stars. It gives two slightly different accounts in separate sections of its astronomical treatise (both in chap. 19). One of these gives a description similar to that in the *Song shu* but adds, "The twenty-eight [lunar lodges] and all the constellations both north and south of the equator were indicated by pearls of three colours, white, green and yellow, according to the three schools of astronomers."[126] The other account asserts that the stars were denoted by red, black, and white, adding that the total numbers of stars of each color agreed with those enumerated by Chen Zhuo. In the various preserved texts there is possibly some confusion between the features of the armillary sphere (*hunyi*) and celestial globe (*huntian*) manufactured by Qian Luozhi.

This practice of marking the stars on celestial globes and charts in three colors continued in China during the medieval period.[127] It also spread to Korea, where it was still in vogue in the eighteenth century (see below). Whether it actually reflected traditions originating in the Zhanguo is still unanswered. Although no information

119. See Needham, *Science and Civilisation*, 3:207 and 264 note e (note 5).

120. *Jin shu*, chap. 11 (note 14).

121. Wei Zheng et al., *Sui shu*, chap. 19; see the modern edition in 6 vols. (Beijing: Zhonghua Shuju, 1973).

122. Translated from chap. 11 of the *Jin shu* by Ho, *Astronomical Chapters*, 66 (note 4).

123. *Jin shu*, chap. 11 (note 14).

124. For a discussion of the identity of fifty zodiacal stars mentioned in Chinese texts of the fourth to sixth century A.D., see Liu Ciyuan, "You yueliang yanfan jilu dedao di wushike huangdao xing di dong Jin Nanbei chao shiqi xing ming" (Names of fifty stars on the ecliptic during the Eastern Jin and Northern and Southern dynasties, obtained from records of close lunar conjunctions), *Tianwen Xuebao* 27 (1986): 276–78 (English abstract on 278).

125. From the astronomical treatise of the *Song shu*, chap. 23, translated by Needham, Wang, and Price, *Heavenly Clockwork*, 96 (with slight amendments) (note 80).

126. Translation from Needham, Wang, and Price, *Heavenly Clockwork*, 97 (note 80).

127. For example, Needham, *Science and Civilisation*, 3:264 (note 5), remarks: "That star-maps showing the traditional colors still existed in A.D. 1220 we know from a story about an unfortunate examination candidate Xu Ziyi."

is available on the accuracy with which the stars were displayed on the celestial globe of Qian Luozhi, it was apparently still in use when China was unified under the Sui dynasty in 581. Thus it was reported that the armillary sphere and globe made by Qian were both taken to Chang'an in that year, and sixteen years later they were moved to the astronomical observatory at Luoyang.[128] Nothing further is heard of these instruments after 605.

During the Liu Song, the first extant measurements were made of the NPD of the Pole Star. For many centuries the bright star known as Diwang (Emperor) or Dadi (Great Emperor)—β UMi—had acted as a prominent North Pole marker, though it was never less than seven degrees from the true pole. However, by the Later Han its distance from the celestial pole had increased significantly owing to precession, and the astronomers of the time adopted Niuxing (Pivot star) instead.[129] This latter object, a constituent of the asterism Beiji (North Pole), may be identified with a rather faint star in the constellation of Camelopardalis (32 H Cam). Listed as 2102 in the Smithsonian Astrophysical Catalog (SAO), it is one of the few stars visible to the unaided eye near the path of the celestial pole at this period.

As recorded in the *Sui shu*, the Song astronomer royal Zu Gengzhi (429–500) measured the NPD of Niuxing and found its distance from the "place of nonmovement" (*bu dong chu*) to be rather more than one *du*. This result compares with the calculated distance for SAO 2102 about A.D. 460 of 1.9 degrees.[130]

The star SAO 2102 remained the choice as pole marker for several centuries, eventually being replaced by the present pole star Tianhuang Dadi (Great Celestial Emperor)—α UMi or Polaris.

About 550, a large celestial globe was constructed at the capital of Jiankang (Nanjing). This is described as follows:

> It was made of wood, as round as a ball, several armspans in circumference, and pivoted on the south and north poles, while round the body of it were shown the twenty-eight *xiu*, as also the stars of (each of) the Three Masters, the ecliptic, the equator, the milky way, etc. There was also an external horizontal circle surrounding it, at a height which could be adjusted, to represent the earth.... When the globe rotated from east to west, the stars which made their meridian transits morning and evening corresponded exactly with their degrees ... there was absolutely no difference from the heavens.[131]

The fate of this device is not known. When the first Sui emperor conquered the Chen dynasty of southern China (589), it is recorded that

> he captured their astronomical expert Zhou Fen and the instruments which had been handed down from the (Liu) Song time. Whereupon he ordered Yu Jicai and others to check for size and accuracy the old (star-) maps, both private and official, dating from the (Northern) Zhou, Qi, Liang, and Chen dynasties, and formerly in the keeping of Zu Gengzhi, Sun Senghua, and others. The object of this was the construction of hemispherical maps (*gaitu*) following the positions of the stars of the Three Schools.[132]

The various "old (star-) maps" that were consulted had been produced during the previous hundred years.

Despite the marked interest in uranography throughout the period from the Three Kingdoms to the Sui, only a few stellar charts of this era are preserved. Two of these artifacts merit some comment here. During excavations at Luoyang in 1973, a star map was discovered on the ceiling of a Northern Wei tomb.[133] Dating from 526, this chart—which is some three meters in diameter—portrays the stars as red circles of roughly equal size on a buff background (plate 31). Some stars are linked into groups, but most are unconnected. The whole appearance is rather sketchy, and few asterisms are readily recognizable apart from Beidou. An unusual feature is the importance assigned to the Milky Way, which is shown in blue, bisecting the night sky. This is probably the earliest known pictorial representation of the River of Heaven from China. As noted above, however, allusions to the Milky Way dating from more than a thousand years previously are found in the *Shi jing* poems.

Several years ago, a painting on silk depicting the mythical sage-rulers Fuxi and Nuwa, dating from sometime between 500 and 640, was unearthed from a tomb at Gaochang (modern Turpan, in Xinjiang Province).[134] This is one of several finds of uranographic significance from the extreme northwest of China (on or near the old Silk Road). The painting—which measures about 2.25 meters by 1 meter—displays rough sketches of about thirty star groups in white on a buff background. Part of the Milky

128. *Sui shu*, chap. 19; see Needham, Wang, and Price, *Heavenly Clockwork*, 98 (note 80).

129. Pan, *Zhongguo hengxing guance shi*, 166–69 (note 7), gives a valuable investigation of the history of stars identified by the Chinese as pole markers. This largely supersedes the discussion by Needham, *Science and Civilisation*, 3:259–62 (note 5).

130. *Sui shu*, chap. 19 (note 121).

131. *Sui shu*, chap. 19, translated by Needham, *Science and Civilisation*, 3:384 (note 5).

132. *Sui shu*, chap. 19, translated by Needham, *Science and Civilisation*, 3:264 (note 5).

133. A discussion of this star map, including a color photograph, was published by Wang Che and Chen Xu, "Luoyang Bei-Wei Yuan Yi mu di xingxiangtu" (The celestial map from the Northern Wei tomb of Yuan Yi at Luoyang), *Wenwu*, 1974, no. 12:56–60 and pl. 1. These authors have identified some thirty star groups depicted on the chart. See also *Zhongguo gudai tianwen wenwu tuji*, 8 (note 6).

134. *Zhongguo gudai tianwen wenwu tuji*, 9 and 120 (note 6).

Way is also shown. A very similar painting on silk, dating from 897, was found in a tomb at Dunhuang (Gansu Province) in 1908; it seems that such artifacts were not uncommon in tombs, typically being fixed to the ceiling.[135]

Two intriguing constellation lists that may date from the pre-Tang era deserve special comment, even though neither contains any measurements. Both are poems. The best known of these is attributed to the Sui poet Wang Ximing, whose pen name was Dan Yuanzi (fl. ca. 590). Known as the *Butian ge* (Song of the sky pacer), this work gives a brief description of almost three hundred asterisms and also enumerates the stars in each. Needham suggests that Tan "might perhaps be termed the Aratus or the Manilius of China, though so much later than they."[136] So highly regarded was the *Butian ge* that in later centuries it could be said that "all who have discussed the constellations have taken the *Butian ge* as their standard."[137] In the twelfth century, chanting portions of the poem on clear nights was recommended as a way to gain familiarity with the constellations.[138]

Later texts assert that the poem describes 283 constellations containing a total of 1,464 stars.[139] These figures are identical with the numbers of asterisms and stars said to be represented on the third-century star map of Chen Zhuo.[140]

Two manuscript versions of another constellation poem that may be about as old as, or even somewhat earlier than, the *Butian ge* were discovered at Dunhuang in 1908 and are now in the Bibliothèque Nationale in Paris. Unlike the composition by Wang Ximing, this poem, entitled *Xuan xiang shi* (Poem of the image of the heavens), seems to have had only a limited circulation. The existing renderings (which differ to some degree) were among the numerous manuscripts Pelliot found in the Caves of the Thousand Buddhas at Dunhuang.[141] One of the two manuscripts containing the poem (P. 2512) bears a date that corresponds to 621—at the very beginning of the Tang—but this may be a copy of an earlier original. The date of the other text (P. 3589) is not preserved. An extensive study was made recently by Deng Wenkuan, who was of the opinion that the poem was composed sometime between the Three Kingdoms and Sui eras.[142]

The *Xuan xiang shi* is less detailed than the *Butian ge*, and there is a definite astrological bias that is absent in the poem attributed to Wang Ximing. In addition, the *Xuan xiang shi* (which is of unknown authorship), divides the constellations into three groups according to their association with the astronomers of antiquity Shi Shen, Gan De, and Wu Xian. This is one reason Deng judges the work to be older than the *Butian ge* (which groups the constellations in the various palaces). Deng also notes that elsewhere in manuscript P. 3589 there is an abstract from an astrological work by Chen Zhuo, though neither of these features necessarily indicates an early date.

Remarks on the contents of an unspecified star chart on which the stars were said to be marked in three colors (recalling the celestial globe of Qian Luozhi) are also found in a separate section of manuscript P. 2512.[143] This portion of the text begins with a list of the twenty-eight lunar lodges. For each lunar lodge, the number of constituent stars is stated; in addition, the equatorial extent (in *du*) is given, together with the angular extent of the determinative star from the north celestial pole. These figures are nearly all identical to Han values; they are not the result of contemporaneous measurements. The number of stars in all twenty-eight *xiu* amounts to 182, including 17 stars adjoining six of the lunar lodges. This total is identical to that said to have been included by Chen Zhuo in his chart of the constellations.

The text continues with brief descriptions of 256 additional asterisms to the north and south of the lunar lodges. The first 94 of these star groups are said to be taken from Shi Shen, the next 118 from Gan De, and the remaining 44 from Wu Xian. There is no overlap among the three lists; each contains a separate set of constellations. For every asterism in a set, the number of constituent stars is given, together with a brief qualitative description of the position of the asterism relative to neighboring star groups; there are no measurements.

The final totals of asterisms and stars (284 and 1,464) are virtually identical to the figures cited in the *Jin shu* and *Sui shu* for the chart constructed by Chen Zhuo and the celestial globe of Qian Luozhi. As we will see below, these numbers remained almost canonical even in relatively recent centuries.

135. Pan, *Zhongguo hengxing guance shi*, pl. 25 (note 7).

136. Needham, *Science and Civilisation*, 3:201 (note 5).

137. This remark forms part of the text of an eighteenth-century Korean astronomical screen. See Joseph Needham and Gwei-djen Lu, "A Korean Astronomical Screen of the Mid-Eighteenth Century from the Royal Palace of the Yi Dynasty (Chosŏn Kingdom, 1392–1910)," *Physis* 8 (1966): 137–62, esp. 148.

138. Needham, *Science and Civilisation*, 3:281 (note 5).

139. Needham and Lu, "Korean Astronomical Screen" (note 137).

140. Soothill, *Hall of Light*, 244–51, translates and discusses this poem (note 21).

141. Mark Aurel Stein was the first European to explore the vast archives at Dunhuang. In 1907 he purchased numerous manuscripts for the British Museum. A year later, Paul Pelliot acquired many of the remaining texts for the Bibliothèque Nationale. Finally, the Chinese government removed what was left.

142. Deng Wenkuan, "Bi 'Butian ge' geng gulao di tongshu shixing zuopin—'Xuanxiang shi' " (A popular work for star recognition older than the "Butian ge"—"Xuanxiang shi"), *Wenwu*, 1990, no. 3:61–65.

143. See Maspero, "L'astronomie chinoise," 272 and 319 ff. (note 49). I have based the following details on a microfilm copy of the manuscript text supplied by the Bibliothèque Nationale.

The signs of the Western zodiac were first introduced into China by way of India, during the Sui dynasty. The earliest references in Chinese to the Western zodiac are found in the *Da zang jing* (Great storehouse of sutras), the Chinese translation of the Buddhist Tripiṭaka, in a sutra known as the *Dafangdeng daji jing* (Sutra of the great assembly of bodhisattvas).[144] This text, which forms section 397 of the *Da zang jing*, was translated from the Sanskrit during the Sui. Chapter 42 of the *Dafangdeng daji jing* contains a list of the zodiacal signs governing each of the twelve lunar months. The names may be translated as follows: month 1, Ram; 2, Bull; 3, Pair of Birds; 4, Crab; 5, Lion; 6, Celestial Woman; 7, Steelyard; 8, Scorpion; 9, Archer (literally "to shoot with a bow"); 10, Sea Monster (*mojie*); 11, Water Vessel; 12, Celestial Fish.[145] Most of these names are readily recognizable in terms of the familiar names of the zodiacal constellations, obvious exceptions occurring in months 6 and 10. In particular, *mojie* is a transliteration of the Sanskrit word *makara* (sea monster), the equivalent of Capricorn. It is not until the Tang that we find the first pictorial representations from China of at least portions of the zodiac.

The Tang Dynasty and Five Dynasties Period (618–960)

The Tang (618–907) was a period of great cultural attainment, not least in astronomy. Both of the official Tang histories preserve many careful observations of the motions of the sun, moon, and planets—as well as comets—relative to the constellations,[146] suggesting that accurate representations of the night sky were available to the imperial astronomers of the time. In general, however, survivals of star maps from this dynasty are scarcely better than from earlier periods. Little is known even about to what extent celestial charts and globes were produced during the Tang.

During the seventh and eighth centuries, several Indian astronomers held office at the imperial observatory in Chang'an, the Tang metropolis. Some of these—including Gautama Siddhārtha (Qutan Xida), who compiled the *Kaiyuan zhanjing* in 730—attained the position of astronomer royal,[147] but there is no evidence that they had any significant influence on the development of Chinese uranography. The Indian astronomers were especially concerned with mathematical astronomy, based on Greek methods, for predicting celestial events such as eclipses.

A native Chinese, Yixing (682–727), who was a Buddhist priest as well as a leading astronomer, constructed several important instruments at Chang'an. One of these appears to have been a celestial globe. According to the astronomical treatises of the two official Tang histories, this was "made in the image of the round heavens and on it were shown the lunar lodges in their order, the equator and the degrees of the heavenly circumference."[148] This device was rotated once every twenty-four hours by a water clock. Unfortunately, no further uranographic details are available.

Little information is available on other stellar charts produced during the Tang. Yixing is known to have initiated an expedition to Annam (modern Vietnam), which—among other objectives—observed the southern constellations invisible from China.[149] This took place in 724 and was led by the astronomer royal Nangong Yue, accompanied by Da Xiang and Yuan Tai. Although how fully the southern stars were charted is not recorded, the account in chapter 35 of the *Jiu Tang shu* makes interesting reading:

> Da Xiang and Yuan Tai say that at Jiaju [modern Hanoi, latitude 21°N] if one observes the pole it is elevated above the Earth's surface only a little more than 20°. Looking South in the 8th month from out at sea Laoren (Canopus) is remarkably high in the sky. The stars in the heavens below it are very brilliant and there are many large and bright ones which are not . . . known. In general all the stars which are more than 20° from the southern pole are all visible. Indeed it is the part of the sky which the ancient *Huntian* (Celestial Sphere) school of astronomers regarded as permanently below the horizon and therefore not to be seen.[150]

Chinese navigational charts depicting a few southern asterisms are preserved from the Ming (1368–1644), but not until the time of the Jesuits is there any evidence for extensive mapping of the south circumpolar constellations in China (see below).

144. The *Da zang jing* has been published under the title *Taishō shinshū Daizōkyō* (The Tripiṭaka in Chinese revised by Taishō University), ed. Takakusu Junjirō and Watanabe Kaigyoku, 85 vols. (Tokyo: Taishō Issaikyō Kankōkai, 1924–32). See W. Eberhard, "Untersuchungen an astronomischen Texten des chinesischen Tripitaka," *Monumenta Serica* 5 (1940): 208–62, esp. 232 ff.

145. I am grateful to A. C. Barnes, formerly of Durham University, for valuable comments and advice on the identification of the signs of the Western zodiac as found in the Tripiṭaka.

146. Liu Xu et al., *Jiu Tang shu* (Old history of the Tang, compiled 940–45), chap. 35; see the modern edition in 16 vols. (Beijing: Zhonghua Shuju, 1975). Ouyang Xiu et al., *Xin Tang shu* (New history of the Tang, compiled 1032?–60), chap. 31; see the modern edition in 20 vols. (Beijing: Zhonghua Shuju, 1975).

147. For details, see Yabuuchi, "Researches on the *Chiu-chih Li*" (note 10).

148. Translated by Needham, Wang, and Price, *Heavenly Clockwork*, 77 (note 80).

149. For details, see Arthur Beer et al., "An 8th-Century Meridian Line: I-Hsing's Chain of Gnomons and the Pre-history of the Metric System," *Vistas in Astronomy* 4 (1961): 3–28.

150. Translated by Beer et al., "Meridian Line," 10 (note 149).

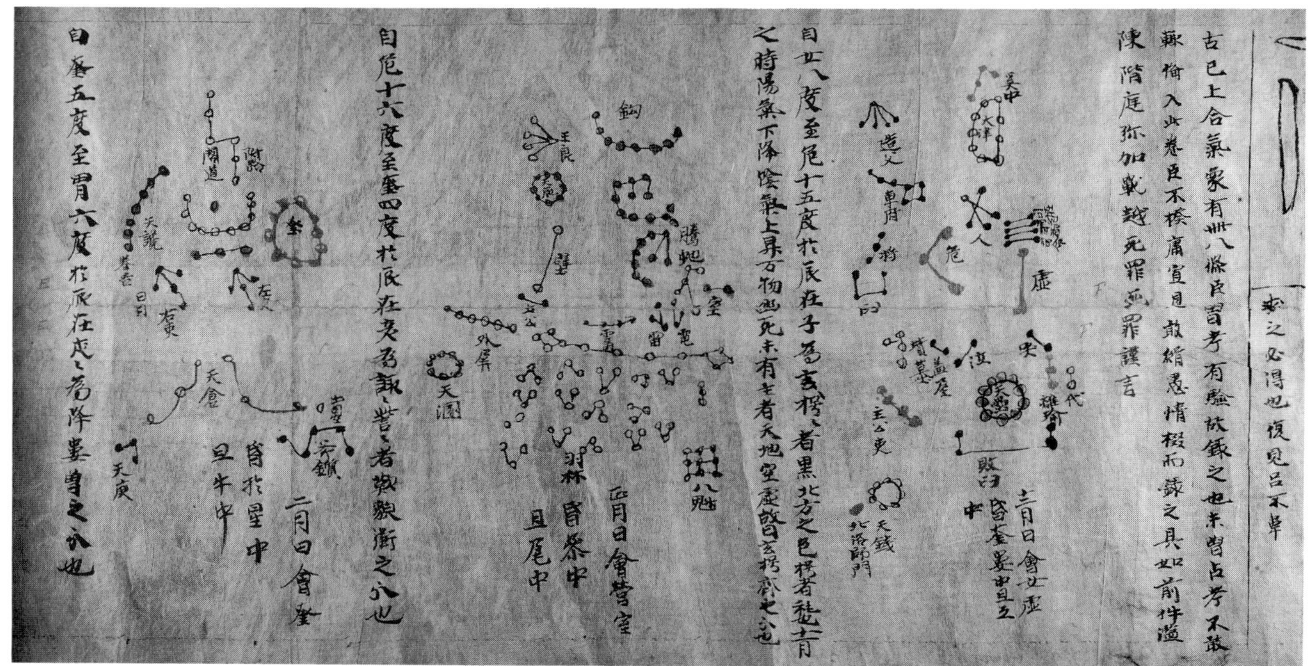

FIG. 13.7. DUNHUANG STAR MAP IN THE BRITISH LIBRARY. The map is illustrated here in four overlapping sections (see also figs. 13.8–13.10); each section reads right to left. Constellations outside the north circumpolar region are depicted. Following the depiction of the region of constant visibility, each vertical strip covers one of the twelve Jupiter stations. There is no attempt at a projection on this rather crude chart.
Size of the entire scroll: 24.5 × 340 cm. By permission of the Oriental and India Office Collections, British Library, London (Stein no. 3326).

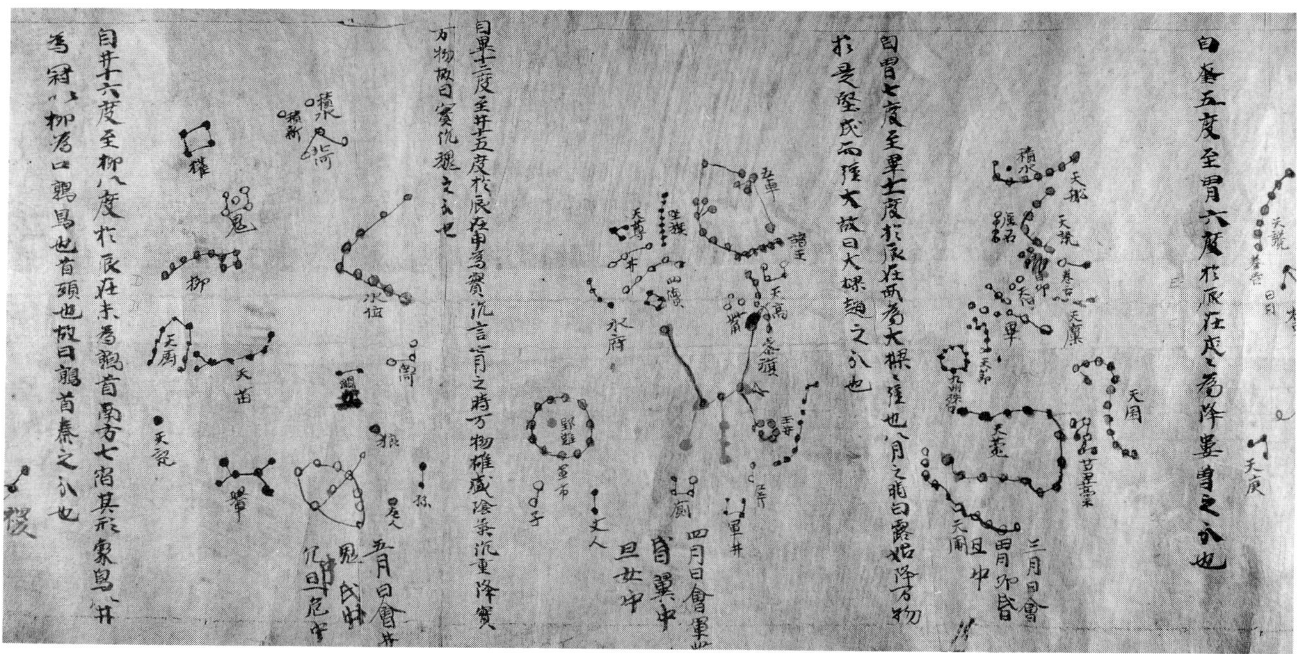

FIG. 13.8. CONTINUATION OF FIGURE 13.7. This continuation toward the east covers about ninety degrees in right ascension, apart from overlap. The band of stars at the right includes both Mao (the Pleiades) and Bi (the Hyades) in Taurus, while the central zone includes Shen (the principal stars of Orion). Note the "bow and arrow" formation in the left band pointing at Lang—the Celestial Wolf (bright star Sirius).
By permission of the Oriental and India Office Collections, British Library, London (Stein no. 3326).

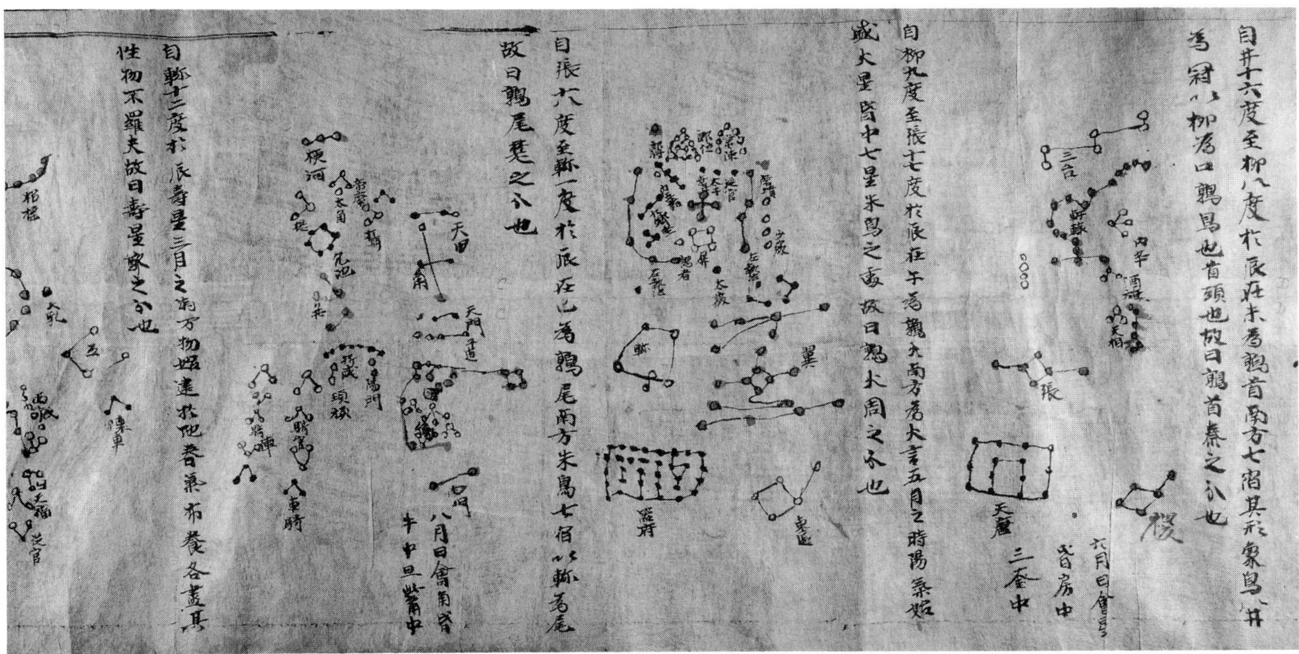

FIG. 13.9. CONTINUATION OF FIGURE 13.8. Continuing toward the east, this section also covers about ninety degrees in right ascension, apart from overlap. The vertical text to the right in this and the other sections gives, among other details, the range of RA (relative to the lunar lodges) within which the stars depicted lie.
By permission of the Oriental and India Office Collections, British Library, London (Stein no. 3326).

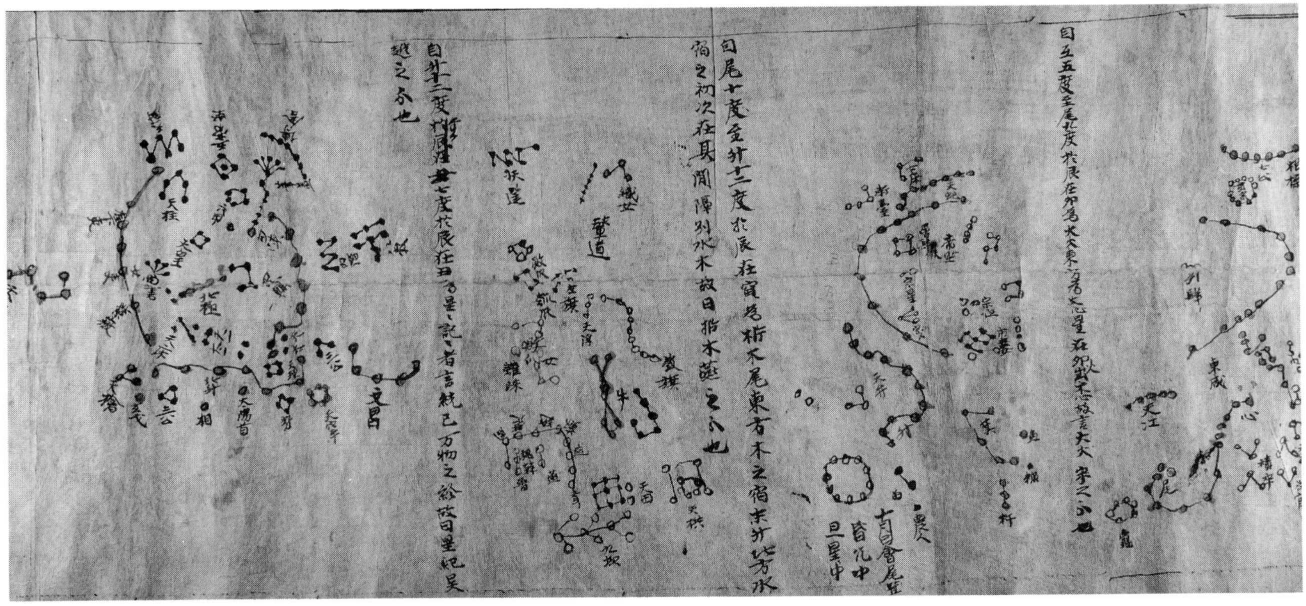

FIG. 13.10. CONTINUATION OF FIGURE 13.9. The three zones toward the right and center are a continuation of figure 13.9 toward the east, thus completing the full circuit of the sky apart from the circumpolar region. The left zone shows the north circumpolar region with the seven stars of Beidou (the Dipper) prominent at the lower edge.
By permission of the Oriental and India Office Collections, British Library, London (Stein no. 3326).

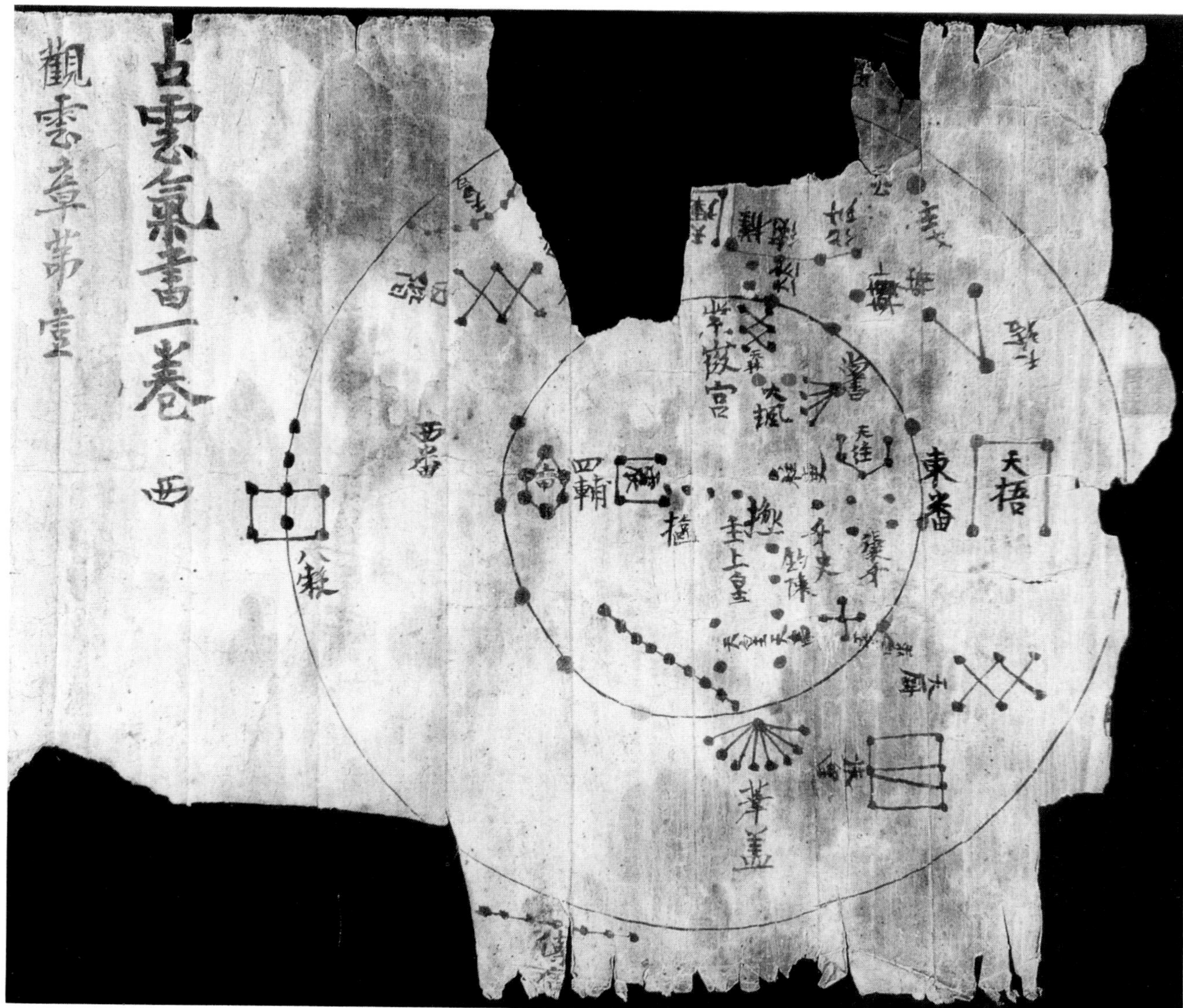

FIG. 13.11. DUNHUANG STAR MAP IN DUNHUANG (NORTH POLAR REGION). Constellations in the north circumpolar region are depicted on this fragmentary paper star map. The two circles are very approximately at declinations +50° and +70°.

Size of the original: unknown. Photograph from Zhongguo Shehui Kexueyuan Kaogu Yanjiusuo (Archaeological Research Institute, Chinese Academy of Social Science [Academia Sinica]), *Zhongguo gudai tianwen wenwu tuji* (Album of ancient Chinese astronomical relics) (Beijing: Wenwu Chubanshe, 1980), 12.

Two fairly detailed star maps that probably date from either the Tang dynasty or the Five Dynasties period (907–60) were discovered in the grottoes at Dunhuang early in the present century. The more substantial of these was among the manuscripts Stein acquired for the British Museum in 1907. A second manuscript astral chart was recovered by Chinese government personnel a few years later (after the European expeditions). It is now preserved in Dunhuang.

The map Stein acquired for the British Museum was first discussed by Needham, who also published photographs of portions of the chart.[151] Needham suggested a date of approximately 940, but he gave no justification for this assumption. However, Chinese scholars tend to prefer a date about two centuries earlier.[152] What little evidence there is seems inconclusive. Of the manuscripts Stein recovered from Dunhuang that bear a date, the earliest was written in 405, and the latest was produced in 995.[153]

151. Stein no. 3326 (this manuscript is now in the British Library); Needham, *Science and Civilisation*, 3:264 and pls. 24 and 25 (note 5).

152. Xi Zezong, "Chinese Studies in the History of Astronomy, 1949–1979," *Isis* 72 (1981): 456–70, esp. 464; Pan, *Zhongguo hengxing guance shi*, 156 (note 7).

153. Lionel Giles, *Descriptive Catalogue of the Chinese Manuscripts*

This crude but colorful chart is depicted on a scroll of buff paper 24.5 centimeters wide (figs. 13.7 to 13.10). In the British Museum catalog, the manuscript is described as of "mediocre" quality.[154] Only the last third of the scroll, whose full length is some 3.4 meters, is devoted to the stellar map. In the earlier portion, various forms of "celestial vapors" are illustrated. The star chart is in thirteen sections. One of these shows the region of constant visibility (north of about declination +55°), and the other twelve depict the remaining portion of the sky visible from China. Each of these latter rectangular strips, approximately thirty degrees wide, covers one of the twelve *ci* (Jupiter stations). There is no attempt at a projection as such; the horizontal (RA) scale is significantly exaggerated relative to the vertical (declination) scale.

Stars are represented on this map by circles of three colors—red, black, and yellow—all constituents of any one asterism being shown by a single color. (These are the same colors mentioned on the Pelliot manuscript from Dunhuang, dated 621.) The various constellations are only roughly sketched, with little or no measurement, and the boundaries of the lunar lodges, celestial equator, and so forth, are not shown. But the name of each star group is marked, and the shapes of the various asterisms resemble those on later maps. As yet, no count appears to have been made of the numbers of constellations and stars that the Dunhuang chart displays. One of the main attractions of the map is its age. It is probably the oldest original representation of the whole of the visible night sky that is still extant from any civilization.[155]

The celestial map that still remains in Dunhuang is in fragmentary condition and covers only the polar constellations (fig. 13.11). The stars are shown in two colors—black and red—on a rather stained buff manuscript. Stellar positions are more carefully marked than on the British Museum map, and two circles of declination are shown. Pan is of the opinion that this is a late copy of a seventh-century map.[156]

A Tang illustration of the twenty-eight *xiu* star groups, painted on the ceiling of a tomb at Asitana in Turpan, was discovered in 1964 (fig. 13.12).[157] The constellation patterns, which are highly idealized, are arranged in a square formation, seven to a side; each edge of the square corresponds to one of the four nonpolar palaces (*gong*). There is no attempt to indicate the irregular spacing of the lunar lodges. Also of Tang origin is a bronze mirror in the collection of the American Museum of Natural History, New York. On the back of this mirror constellation diagrams of the twenty-eight lunar lodges are shown in a circular pattern.[158] The main importance of these artifacts is that they are the oldest well preserved depictions of the *xiu* star configurations as an entity; earlier representations (e.g., of Han origin) have suffered much with the passage of time.

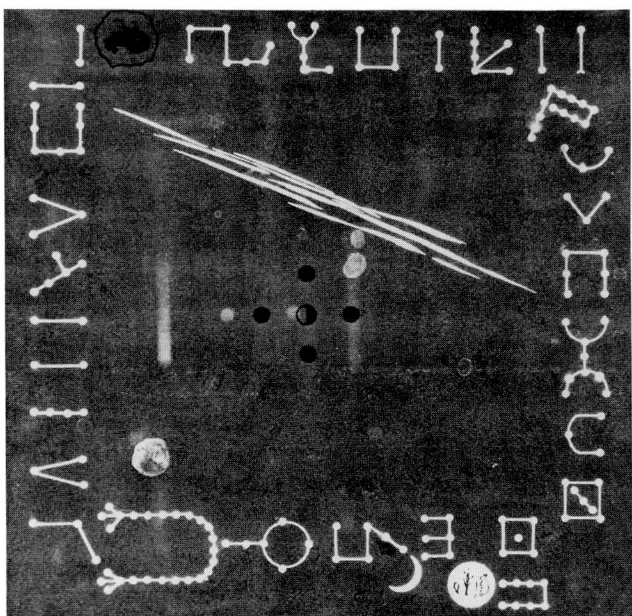

FIG. 13.12. LUNAR LODGE PATTERNS PAINTED ON THE CEILING OF A TANG DYNASTY TOMB AT ASITANA, TURPAN (XINJIANG PROVINCE). The lunar lodges are shown highly idealized, seven to each side of a square. Each group of seven corresponds to one of the four nonpolar palaces (*gong*) associated with the four cardinal directions.
Size of the original: unknown. Xinjiang Uygur Autonomous Regional Museum, Ürümqi. Photograph from Zhongguo Shehui Kexueyuan Kaogu Yanjiusuo (Archaeological Research Institute, Chinese Academy of Social Science [Academia Sinica]), *Zhongguo gudai tianwen wenwu tuji* (Album of ancient Chinese astronomical relics) (Beijing: Wenwu Chubanshe, 1980), 69.

Two catalogs of the constellations are preserved from the early Tang (ca. 635), but these are devoid of positional measurements; they consist merely of qualitative descriptions of some 250 asterisms. The catalogs, which resemble one another closely, are contained in the astronomical

from Tunhuang in the British Museum (London: British Museum, 1957), xi.

154. Giles, *Descriptive Catalogue*, 225 (note 153).

155. Xi Zezong, "Dunhuang xingtu" (A star map from Dunhuang), *Wenwu*, 1966, no. 3:27–38 provides a complete illustration and detailed discussion of the map.

156. Pan, *Zhongguo hengxing guance shi*, 156 (note 7). An excellent color illustration of the Dunhuang map has been published in *Zhongguo gudai tianwen wenwu tuji*, 12 (note 6).

157. See Xinjiang Weiwuer Zizhiqu Bowuguan (Museum of the Xinjiang Uygur Autonomous Region), "Tulufan xian Asitana-Halahezhuo gu muqun fajue jianbao" (Preliminary report on the excavation of ancient tombs at Asitana and Halahezhuo, Turpan County), *Wenwu*, 1973, no. 10:7–27, esp. 18–19; also Schafer, *Pacing the Void*, 79–81 (note 12).

158. For a discussion and photograph, see Edouard Chavannes, "Le cycle turc des douze animaux," *T'oung Pao*, 2d ser., 7 (1906): 51–122.

treatises of the *Jin shu* (chap. 11) and *Sui shu* (chaps. 19–20).[159] Both were compiled by the Tang astronomer royal Li Chunfeng (602–70) as supplements to the records of celestial events in the dynasties covered by the histories. A count indicates 240 asterisms containing 1,298 stars in the *Jin shu* list.[160] The *Sui shu* catalog includes a few asterisms not found in the *Jin shu*.

The *Jin shu* and *Sui shu* star catalogs specify the number of stars in each asterism they cite, but the descriptions of relative position of the various constellations are vague. For instance, they would be of less help than the *Butian ge* poem (see above) as a guide to the night sky. Their main use is as astrological manuals, and in this regard they are extremely comprehensive.

An interesting description of the Milky Way is given at the end of the Jin and Sui catalogs. This detailed account, which traces the path of the River of Heaven across the sky from Scorpio, northward to Cassiopeia, and then southward to Vela, is worth quoting in full:

> The River of Heaven rises up in the east and passes between Wei (6th lunar lodge) and Ji (7th lunar lodge) known by the name Hanjin ("Ford of the Heavenly River"). There the River divides into two branches (which follow different routes).
>
> The southern (route) passes (the constellations) Fuyue, Yu, Tianyo, Tianbian and Hegu. The northern (route) passes below (the stars of) Gui, penetrates beneath Ji (7th lunar lodge) and connects together the head of Nandou (8th lunar lodge) and Zuoqi. After passing below Tianjin, it rejoins the southern branch, and the two travel together in a southwesterly direction.
>
> (The River of Heaven) then encloses Hugua, and joins together (the constellations) Ren, Chu, Caofu, Tengshe, Wangliang, Fulu, the northern tip of Gedao, Tailing, Tianchuan and Juanshe. From there it travels southwards and encloses Wuche. Then it passes through the south of Beihi, enters Dongjing (22nd lunar lodge) and Shuiwei, and takes a south-easterly direction. After connecting (the constellations) Nanhe, Quejiu, Tiangou, Tianji ("Celestial Cycle") and Tianji ("Celestial Pannicled Millet"), it finally declines in the heavens south of Qixing (25th lunar lodge).[161]

This is probably the earliest detailed account from any part of the world of the circuit of the Milky Way.

Yixing and his colleagues made a number of accurate determinations of star positions, including a series of measurements of the NPDs of the determinative stars of the lunar lodges in 725. These latter data are preserved in the astronomical treatises of both the *Jiu Tang shu* and the *Xin Tang shu*, where they are compared with Han results. Most of the measurements Yixing made are expressed to the nearest degree, although a few are given to the nearest half degree. Analysis of these data indicates a typical error of rather less than a degree.[162] Yixing also made a number of determinations of the ecliptic latitudes of certain stars, noting that they appeared to have changed since Han times. His supposed variations are much too large to be explained by proper motion, however; they must merely result from errors of measurement and poor definition of the ecliptic.[163]

One of the most interesting—and valuable—of all Tang astronomical records is a careful account of the path of Halley's comet through the constellations in the spring of 837. At that time the comet was making its closest known approach to the earth (some twelve times the lunar distance) in the entire historical period. Since the motion of the comet was considerably perturbed by its near encounter with our planet, the remarkably precise Chinese observations have proved of great value to modern astronomers in investigating the past orbit.[164] The text in the astronomical treatise of the *Jiu Tang shu* is too long to translate here,[165] but on approximately ten separate nights the comet's RA (expressed relative to the lunar lodges) was measured to the nearest degree and sometimes half degree. At about its closest approach to the earth, it was traversing the sky at more than forty degrees daily. The Tang record is so precise that it enables the date and time of perihelion (when the comet came nearest to the sun) to be deduced to within an hour or so. This account of Halley's comet is by far the most extensive and accurate record of the motion of a comet from any part of the world before the European Renaissance.

As I noted in the previous section, a Sui translation of an Indian Buddhist sutra contains the oldest known Chinese references to the twelve signs of the Western zodiac. Tang translations of further Buddhist scriptures, including the sutra known as *Xiu yao jing* (Lunar lodges and planets), dating from 760, make further references

159. The *Jin shu* star catalog has been translated by Ho, *Astronomical Chapters*, 67–112 (note 4), as part of his translation of the entire astronomical treatise of the *Jin shu*. This excellent translation has been referred to several times in this chapter. No other astronomical treatise in the standard dynastic histories from the Han to the Ming has so far been translated in extenso into a Western language.

160. Ho, *Astronomical Chapters*, 19 (note 4).

161. Translated by Ho, *Astronomical Chapters*, 112–13 (note 4).

162. This result is based on my unpublished analysis.

163. Xi Zezong, "Seng Yixing guance hengxing weizhi di gongzuo" (On the observations of star positions by the priest Yixing [683–729]), *Tianwen Xuebao* 4 (1956): 212–18, esp. 212; Ang Tian Se, "I-Hsing (683–727 A.D.): His Life and Scientific Work" (Ph.D. diss., University of Malaya, Kuala Lumpur, 1979), 378–85.

164. See, for example, Donald K. Yeomans and Tao Kiang, "The Long-Term Motion of Comet Halley," *Monthly Notices of the Royal Astronomical Society* 197 (1981): 633–46.

165. A complete translation, with commentary, is given by Stephenson and Yau, "Far Eastern Observations," 206–7 (note 2).

to the zodiac.[166] Although by the Tang most of the Chinese names of the zodiacal signs had remained unchanged, Gemini was now typically identified as a man and a woman, while Virgo had become two women. Capricorn remained the sea monster, under names such as Ma-giat or Ma-kiat—further attempts to transcribe the Sanskrit *makara*. These are the usual equivalents at subsequent periods in Chinese history.

The earliest known pictorial representations of the signs of the zodiac that are of Chinese origin also date from the Tang. These sketches are among the manuscripts recovered from Turpan in 1975.[167] Although now somewhat fragmentary, they depict the symbols for a few of the zodiacal signs. In one example, the symbols for Virgo (two women) and Libra (a balance) are shown alongside drawings of adjacent *xiu* star groups.[168] The contrast between the representations of the Western signs of the zodiac and the Chinese lunar lodges is interesting; the latter are almost invariably indicated by star patterns, scarcely ever pictorial symbols (fig. 13.13). During the latter half of the Tang, horoscope astrology based on the Western zodiac arrived in China. This became fairly widespread at the popular level, especially among Buddhist adherents, but had negligible impact on the official astrology practiced at court.[169]

Little is known regarding celestial cartography during the turbulent Wudai or Five Dynasties period, which lasted from 907 to 960. The brief astronomical treatises in the official dynastic histories—the *Jiu Wudai shi* (Old history of the Five Dynasties) by Xue Juzheng (912–81), chapter 139, and the *Xin Wudai shi* (New history of the Five Dynasties) by Ouyang Xiu (1007–72), chapter 59—make no mention of the production of star maps at this time. Although a significant number of celestial observations (eclipses, lunar and planetary movements, comets, and such) are recorded in these works, positional measurements are rare. Nevertheless, some interesting representations of the lunar lodges are preserved from the time of the Five Dynasties.

Several tombstones of the Southern Tang dynasty (937–60) in Jiangsu Province are engraved with idealized outlines of the twenty-eight *xiu* arranged in sevens along the sides of a square.[170] But the most important finds from the Five Dynasties period show much more careful representations of the lunar lodges. Two maps were discovered at Hangzhou during excavations of the tombs of Qian Yuanguan (who died in 941) and his wife Wu Hanyue (d. 952). Each chart was carved on the stone ceiling of the mausoleum. Qian Yuanguan was ruler of the small state of Wuyue, which existed during the whole of the Five Dynasties period. The first chart was recovered in 1958 from the mausoleum of Wu Hanyue.[171] Further excavations in 1975 led to the recovery of the second chart from the tomb of her husband.[172] The two plani-

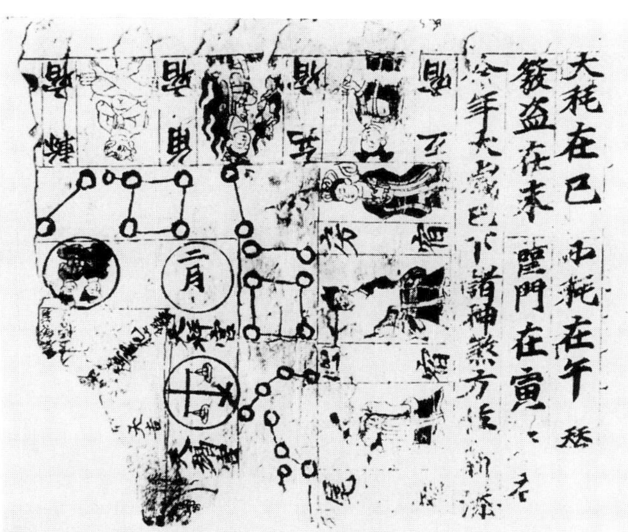

FIG. 13.13. FRAGMENTARY TANG SKETCH SHOWING THE WESTERN ZODIACAL SYMBOLS. Symbols of the zodiacal signs Virgo (characteristically shown as two women) and Libra are depicted, as well as the outlines of Chinese lunar lodges on this Buddhist manuscript unearthed at Turpan, Xinjiang Province.
Size of the original: unknown. Photograph from Zhongguo Shehui Kexueyuan Kaogu Yanjiusuo, *Zhongguo gudai tianwen wenwu tuji*, 70.

166. *Da zang jing*, section 1299; see Eberhard, "Untersuchungen an astronomischen Texten," 232 ff. (note 144).

167. For details, see Xia Nai, "Cong Xuanhua Liao mu di xingtu lun ershiba xiu he huangdao shier gong" (Discussion of twenty-eight lodges and the twelve palaces on the ecliptic based on a star map from a Liao tomb at Xuanhua), *Kaogu Xuebao*, 1976, no. 2:35–58, esp. 49 ff. Reprinted in *Kaoguxue he keji shi* (Archaeology and the history of technology), by Xia Nai (Beijing: Kexue Chubanshe, 1979), 29–50, esp. 46–47.

168. Photographs are published by Xia, "Cong Xuanhua Liao mu di xingtu lun ershiba xiu he huangdao shier gong," pl. 13 (top) (note 167), and in *Zhongguo gudai tianwen wenwu tuji*, 70 (note 6).

169. For discussions of Tang astrology, see Schafer, *Pacing the Void*, 58–119 (note 12), and Shigeru Nakayama, "Characteristics of Chinese Astrology," *Isis* 57 (1966): 442–54, esp. 450.

170. Photographs are published in *Zhongguo gudai tianwen wenwu tuji*, 75–76 (note 6).

171. For details, see Zhejiangsheng Wenwu Guanli Weiyuanhui (Committee for the Management of Cultural Relics, Zhejiang Province), "Hangzhou Lin'an wudai muzhong di tianwen tu he mise ci" (Astronomical maps and specially glazed porcelains found in the Five Dynasties tombs at Hangzhou and Lin'an), *Kaogu*, 1975, no. 3:186–94, esp. 190–91; a scale drawing is provided (the stele is no longer extant).

172. Yi Shitong, "Zuigu di shike xingtu—Hangzhou Wuyue mu shike xingtu pingjia" (The oldest star map engraved in stone—An assessment of a star map engraved on stone from the Wuyue tomb at Hangzhou), *Kaogu*, 1975, no. 3:153–57. Both a photograph of a rubbing and a scale drawing are shown in this article. The stele is damaged, but its surface is in fair condition. It measures about 4.2 by 2.7 meters and is now housed in a huge glass showcase at Hangzhou Beilin (Hangzhou Forest of Steles).

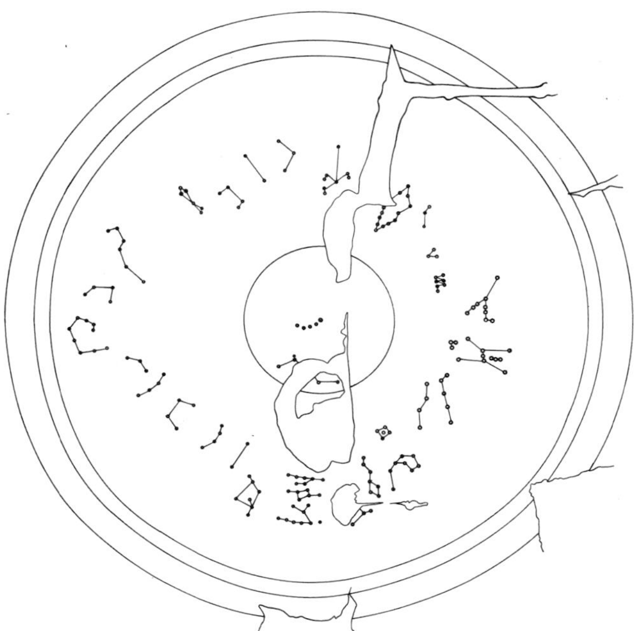

FIG. 13.14. DRAWING OF A WUYUE STAR MAP SHOWING LUNAR LODGES. The stele from which this drawing was made closely resembles that in figure 13.15. It was found in 1958 in the tomb of Wu Hanyue but is said to have been destroyed during the Cultural Revolution. Fortunately, this accurate scale drawing survives.
Diameter of the original: 180 cm. Redrawing from Zhongguo Shehui Kexueyuan Kaogu Yanjiusuo, *Zhongguo gudai tianwen wenwu tuji*, 73.

FIG. 13.15. RUBBING OF A WUYUE STAR MAP ALSO SHOWING LUNAR LODGES. Although badly damaged, the stele from which this rubbing was made shows the lunar lodges and a few polar asterisms with fair precision. It was recovered in 1975 from the tomb of Qian Yuanguan.
Diameter of the original: 190 cm. Photograph from Zhongguo Shehui Kexueyuan Kaogu Yanjiusuo, *Zhongguo gudai tianwen wenwu tuji*, 72.

spheres closely resemble one another (figs. 13.14 and 13.15).[173] Both depict only a few constellations: the twenty-eight *xiu* and a few circumpolar asterisms, notably Beidou. The projection is polar (equidistant); in each case the circle of constant visibility is shown, while the boundary of each chart represents the circle of constant invisibility. The celestial equator is also marked on the map that was found in the grave of Qian Yuanguan. Measurements on this chart indicate a circle of constant visibility of radius 37°, while the edge of the chart extends to 38° south of the equator. These correspond better to northern China; the latitude of Hangzhou is close to 30°.

The outlines of the various asterisms on the Wuyue planispheres approximate the true configurations fairly well; typical positional errors are about three degrees. Hence, though only a small number of constellations are displayed, the Qian Yuanguan stele is the oldest known Chinese star map of any precision. An interesting minor feature of both charts is the inclusion of eight stars in Beidou: the seven well-known constituents together with 80 UMa (Alcor), which is situated only eleven arc minutes from the much brighter star η UMa (Mizar). Here we have possibly the earliest known pictorial representation of this rather close binary, regarded by the Arabs as something of a test of sight.

The Song and Contemporary Dynasties (960–1279)

Many celestial charts and globes are known to have been produced during the Song dynasty, and two important products of this period still exist. These are probably the most detailed and accurate maps of the night sky that have come down to us from the whole of the pre-Jesuit period in China. Unlike the star charts that survive from previous dynasties, both Song artifacts are mentioned in the extant literature of their own time, and their history can be traced in detail. One of the charts, engraved on stone in 1247, is still in almost pristine condition, but the other map, though originally produced much earlier (1094), now exists only in late copies. The original version

173. The two charts are illustrated and discussed in *Zhongguo gudai tianwen wenwu tuji*, 72–73 and 122 (note 6).

of this latter work was the first known star chart from any part of the world to be printed. Both Song maps will be discussed in detail later in this section.

Other maps and globes of Song origin that are mentioned in history have long since disappeared. Thus in the late tenth century the *Taiping yulan* (Imperial encyclopedia of the Taiping reign period, compiled 983) notes the existence of a work titled *Liexing tu* (Map of the principal stars) but does not mention the compiler. Later, Ma Yongqing mentions in his *Lanzhen zi* (Book of the truth-through-indolence master) that about 1115 he often discussed astronomy with certain monks who possessed maps of the stars. Not long afterward (ca. 1150), Zheng Qiao (1108–66) in his *Tongzhi* (Comprehensive treatises) complained that printed star charts were generally not to be relied upon and furthermore were hard to correct.[174] The problems of cutting the necessary printing blocks (complete with names of constellations) with adequate precision and producing good-quality impressions from them must have been considerable. Zheng Qiao's remarks suggest that printed maps of the night sky were fairly common by the middle of the twelfth century—several centuries before they first appeared in Europe.

According to the official Song history, the *Song shi*, an instrument that was probably a celestial globe was designed by a student in the official bureau of astronomy at Bian (modern Kaifeng), the Song capital, in 976. This device, which was driven by a water clock, was said to depict "the Purple Palace, the lunar lodges in order, Beidou, and the celestial equator and ecliptic."[175] No further information is given concerning the number of asterisms or stars represented. It is not known how long the globe remained in use.

In chapter 80 of the *Song shi*, under a year corresponding to 1102, we find the following account by a government minister named Wang Fu:

> I chanced to meet a wandering unworldly scholar at the capital, who told me his family name was Wang and gave me a Daoist book which discussed the construction of astronomical instruments in detail. So afterwards I asked the emperor to order the Supply Department to make some models to test what it said. This they did in the space of two months. The instrument is round like a ball, graduated in 365¼ degrees, and shows the south and north poles . . . the ecliptic and equator, . . . the twenty-eight lunar lodges, the three walled regions of the heavens, and the stars of the whole heavenly round.[176]

In the quotation above, the "three walled regions" are the Ziwei *yuan*, Taiwei *yuan*, and Tianshi *yuan*. It is regrettable that information on the representation of the constellations on this celestial globe is so brief. The fate of this artifact is not known.

Not long before—in 1092—the great astronomer Su Song (1020–1101) had, on imperial command, constructed an armillary sphere and a celestial globe. These instruments were installed in a tower at Bian and were driven by a water clock employing an escapement device.[177] They were said to accurately replicate the apparent motion of the heavens. In his *Xinyi xiang fayao* (New design for an armillary [sphere] and [celestial] globe), printed in 1094, Su Song gives the following description of the celestial globe:

> The body of the celestial globe is spherical like a ball and has a diameter of 4.565 [*chi*] [some 1.7 m]. On its surface, the circumference is marked with 365 and a fraction degrees [*du*], and the constellations and stars both north and south of the equator are marked; there being 246 names (of constellations) and a total of 1281 stars. The [Purple Palace] is situated at the northern part with thirty-seven names of groups and a total of 183 stars. The sum total is therefore 1464 stars. It (the globe) is circumscribed by the ecliptic and equator. The twenty-eight lunar lodges are shown in their succession, and also the path where the sun, moon and five planets move.[178]

The total number of both constellations (283) and stars (1,464) said to be depicted follows the traditional figures of antiquity.[179] Sadly, the celestial globe and other instruments constructed by Su Song did not remain in operation long. In 1126 the emperor and his court abandoned Bian to the invading Jin armies, later establishing a new capital at Lin'an (modern Hangzhou) in the south. Chapter 22 of the *Jin shi* (History of the Jin, 1345) relates that "all the astronomical instruments were carried away [by the Jins] in carts to Yen [near modern Beijing]. The [various gear wheels], the celestial globe, the bells, . . . all broke or wore out after some years. Only the bronze armillary sphere remained in the observatory of the (Jin) Bureau of Astronomy and Calendar."[180]

Fortunately, star maps that may closely follow the

174. These references to Song star charts that are no longer extant are from Needham, *Science and Civilisation*, 3:281 (note 5).

175. Tuotuo et al., *Song shi* (History of the Song, 1346), chap. 38; see the modern edition in 40 vols. (Beijing: Zhonghua Shuju, 1977). See also Needham, Wang, and Price, *Heavenly Clockwork*, 71–72 (note 80).

176. Translated by Needham, Wang, and Price, *Heavenly Clockwork*, 119–20 (note 80).

177. A detailed description of the construction of the clock tower and its instruments is given by Needham, Wang, and Price, *Heavenly Clockwork*, chaps. 4–6 (note 80).

178. Translated by Needham, Wang, and Price, *Heavenly Clockwork*, 46 (note 80).

179. For example, the star map produced by Chen Zhuo during the third century A.D. was said to depict precisely these numbers of asterisms and stars—see above.

180. Translated by Needham, Wang, and Price, *Heavenly Clockwork*, 132 (note 80).

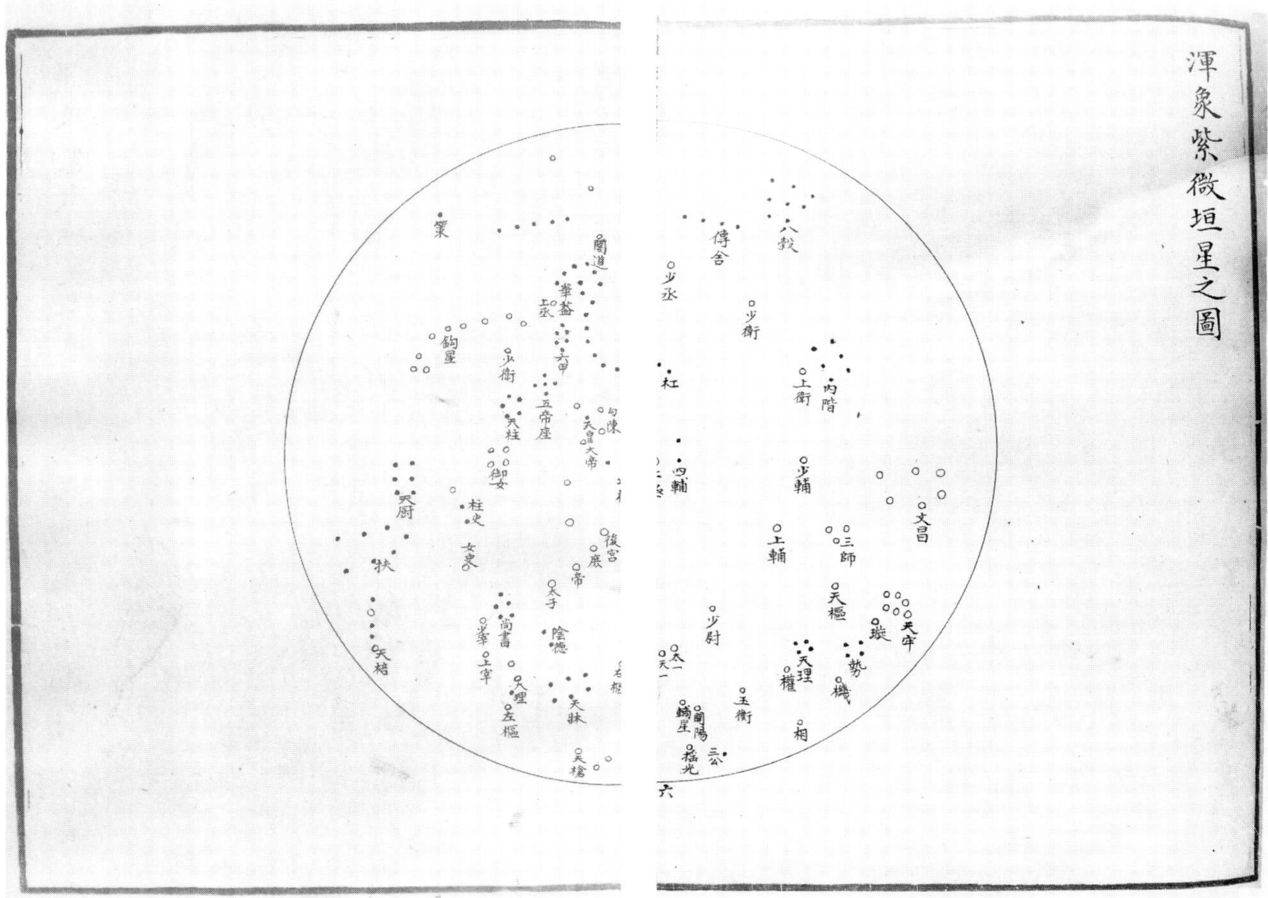

FIG. 13.16. THE SU SONG STAR MAP OF THE NORTH POLAR REGION. Thirty-seven star groups containing 174 stars are depicted in this chart of the north circumpolar region (north of declination +58°). The chart shown here and those in figures 13.16 to 13.20 originally date from 1094 and were the earliest printed star maps. The originals have long since been lost, and the figures here are from the 1781 edition of *Xinyi xiang fayao* said to be based on an 1172 manuscript.
Size of each page: 30 × 22 cm. By permission of the National Library of China, Beijing.

uranography on Su Song's celestial globe are still extant in copies of his *Xinyi xiang fayao*.[181] Today the earliest extant version dates from 1781 and is preserved in the National Library, Beijing. It is said to be based on a careful manuscript copy of an 1172 printing made about 1670 by the bibliophile Qian Zeng, which took several months to complete. The following details are recorded in the *Siku quanshu* (Complete library from the four treasures), a catalog of rare books compiled 1773–82:

> After the Southern Song (period), there were only very few copies of this book remaining. The edition we now have follows the text of that in the possession of Qian Zeng of the Ming dynasty. At the back of this book there were the two following lines, "Edition of . . . the 8th year of the Qiandao reign-period (A.D. 1172)." This shows that the present edition is a true copy of the text of the Song edition. . . . the copying of Qian Zeng was extremely skilful. . . . He himself said that "all the illustrations with their lines and details (followed the original copy) without a hair's-breadth of difference. . . . The result was in no way inferior to the Song edition itself."[182]

In the *Xinyi xiang fayao* the night sky is divided into five sections as follows: (1) the north circumpolar region (north of about declination +58°) (fig. 13.16); (2) stars between RA 12 hours and 24 hours (from the autumn to the spring equinoxes) and in the approximate declination range −58° to +58° (fig. 13.17); (3) stars between RA 0 hours and 12 hours and in the same declination range as (2) (fig. 13.18); (4) the entire Northern Hemi-

181. See *Zhongguo gudai tianwen wenwu tuji*, 77–81 and 122–23 (note 6).

182. Translated by Needham, Wang, and Price, *Heavenly Clockwork*, 12 (note 80).

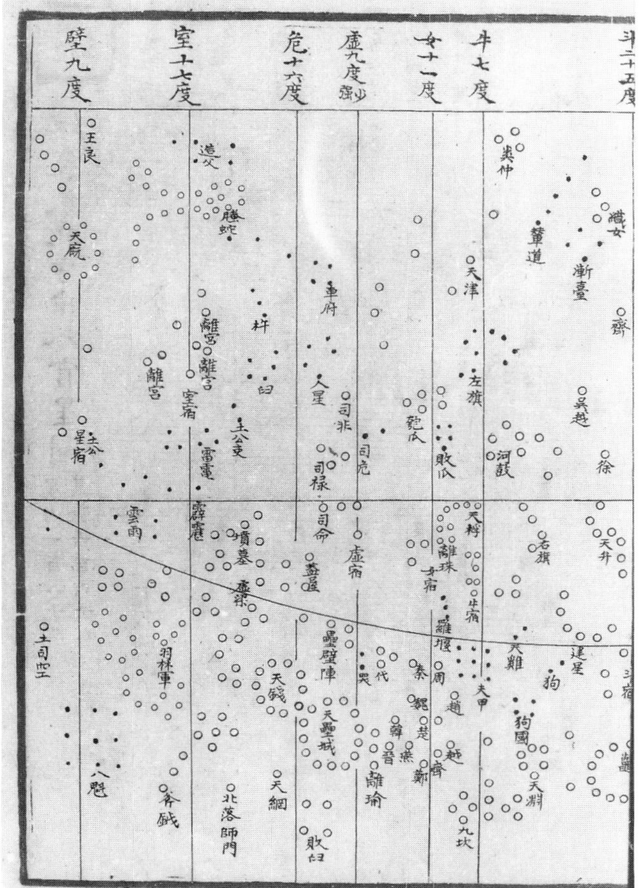

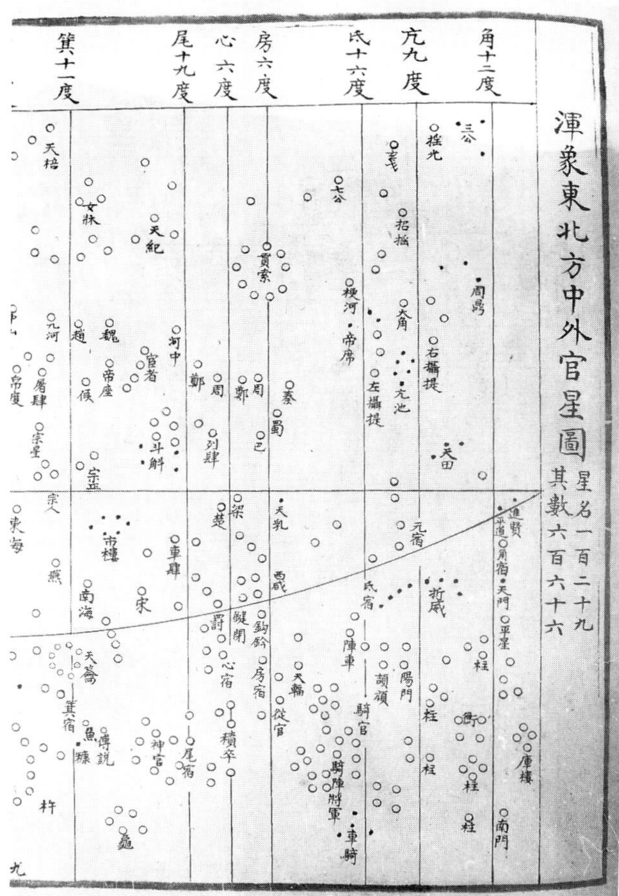

FIG. 13.17. THE SU SONG STAR MAP FROM 12 HOURS TO 24 HOURS RA. A total of 666 stars and 129 asterisms are said to be shown in this chart of the nonpolar regions between RA 12 hours and 24 hours. Boundaries of the lunar lodges are represented by vertical straight lines; the equator and ecliptic are also shown.
Size of each page: 30 × 22 cm. By permission of the National Library of China, Beijing.

sphere (fig. 13.19); and (5) the Southern Hemisphere down to the circle of constant invisibility (approximate declination −58°) (fig. 13.20). For comparison, the colatitude of Kaifeng is close to 55°.[183]

Charts 1, 4, and 5 (figs. 13.16, 13.19, and 13.20), which are circular, are on a polar (equidistant) projection, map 5 having a central void representing the region of sky permanently below the horizon. The remaining maps, which are rectangular, do not use a true projection but employ roughly equal scales for RA and declination. Needham's allusion to these charts as being on "Mercator's" projection has thus led several unwary authors astray.[184] On maps 2 and 3 (figs. 13.17 and 13.18) the boundaries of the lunar lodges are indicated by parallel lines, while on charts 4 and 5 the *xiu* limits are depicted by radial lines extending from the celestial equator to either the circle of constant visibility or the edge of the zone of constant invisibility. Both the ecliptic and the equator are indicated on charts 2 and 3, but the Milky Way is omitted. Constellations and certain stars are individually named on each chart.

The caption to chart 2 states that a total of 117 asterisms and 615 stars are represented thereon, while the corresponding figures for chart 3 are 129 asterisms and 666 stars. The totals of 246 constellations and 1,281 stars are precisely the numbers said to have been represented on the celestial globe of Su Song. A count I have made on chart 1 indicates 37 star groups and 174 stars in the north circumpolar region; the number of constellations

183. The declinations of the circles of constant visibility and constant invisibility for any given station are—ignoring refraction—numerically equal to the colatitude (90° minus the latitude) of that place.

184. Needham, *Science and Civilisation*, 3:278 and fig. 104 (note 5).

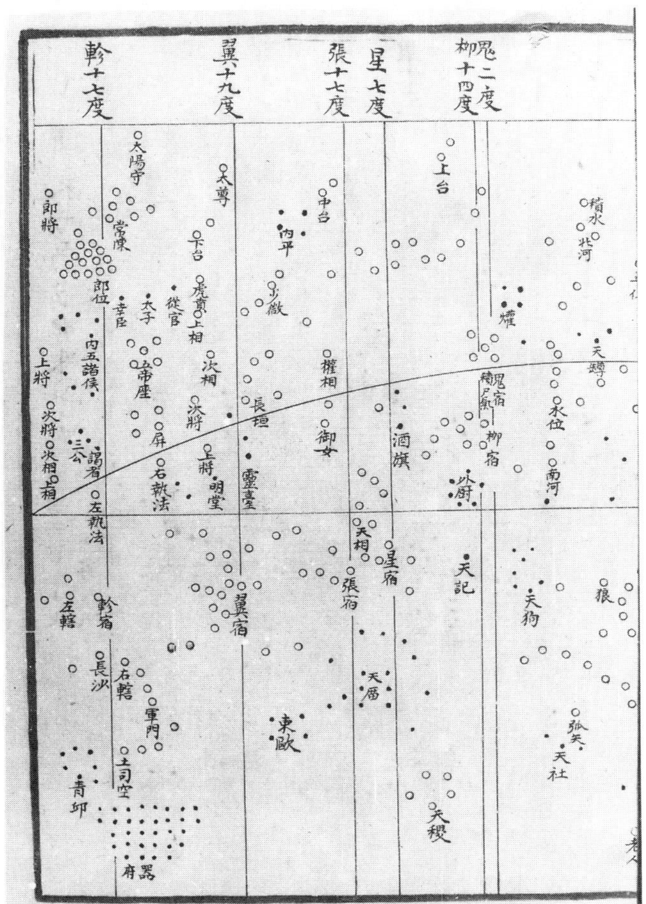

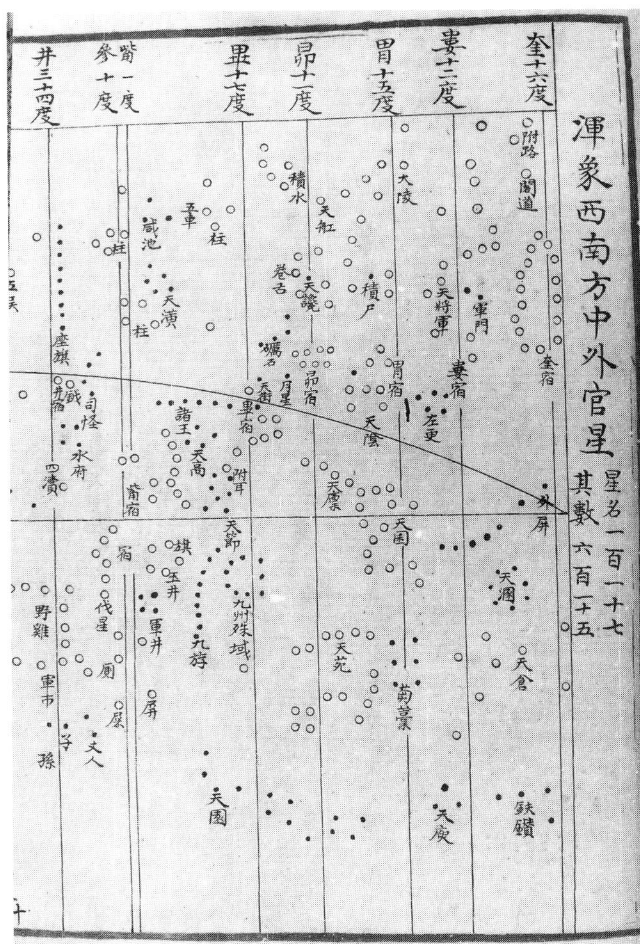

FIG. 13.18. THE SU SONG STAR MAP FROM 0 HOURS TO 12 HOURS RA. A total of 615 stars in 117 asterisms are said to be shown in this chart of the non-polar regions between RA 0 hours and 12 hours. *Xiu* boundaries, equator, and ecliptic are also depicted.

Size of each page: 30 × 22 cm. By permission of the National Library of China, Beijing.

(283) thus agrees with that on the celestial globe for the same region of sky, although there is a slight discrepancy in the star numbers (1,455, compared with the customary figure of 1,464).

On the Su Song charts, all the stars in any particular constellation are denoted by either open circles or black circles. No attempt is made to distinguish between stars of different brightness, which is essentially true of all pre-Jesuit maps of the night sky. Stars are not shown joined into groups, although their grouping is obvious. (On later versions stars are joined into groups by straight lines.) Each constellation is named, as are certain important individual stars. The distribution of asterisms whose constituents are represented by open circles agrees well with that of the star groups depicted in red or yellow on the somewhat earlier Dunhuang star charts; similarly, the grouping of black circles on the maps in the *Xinyi xiang fayao* corresponds closely to the pattern of black circles on the Dunhuang artifact. It seems highly unlikely that Su Song and his colleagues ever saw or were influenced by the Dunhuang charts. Hence we have an independent recurrence of the ancient tradition of sets of constellation groupings.

On the available printings of the Su Song astral maps, individual asterisms are neatly depicted, and on first impression the charts appear to be skillfully executed. Although the equatorial extensions of individual *xiu* are fairly accurately specified (to the nearest degree) at the upper edges of maps 2 and 3, measurements on the charts themselves indicate several errors of two or three degrees in these quantities. Many asterisms are depicted with an idealized form; circular and oval figures and other sym-

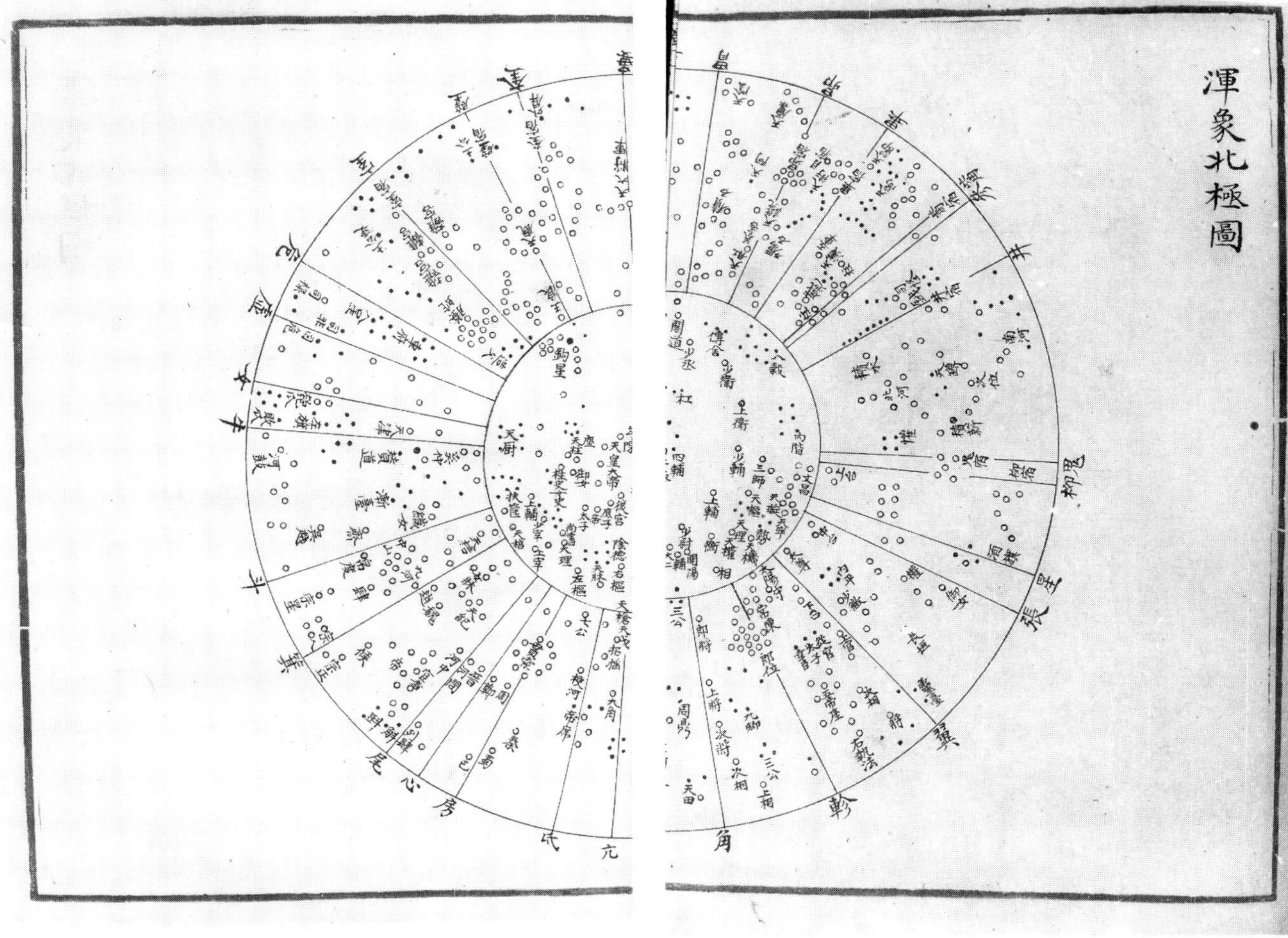

FIG. 13.19. THE SU SONG STAR MAP OF THE NORTHERN HEMISPHERE. Constellations in the entire Northern Hemisphere are depicted. The circle of constant visibility and celestial equator are shown; *xiu* boundaries are represented by radial lines. Size of each page: 30 × 22 cm. By permission of the National Library of China, Beijing.

metrical features are fairly common. An analysis (unpublished) I have made of the locations of a sample of twenty bright stars shows that the mean error in declination is as high as four degrees, some discrepancies exceeding ten degrees. These results recall the complaint of Zheng Qiao that in his time (mid-twelfth century), printed star charts were generally not to be relied on.

A considerably superior star map of Song origin, dating from 1247, is preserved in the collection of the Suzhou (Soochow) Museum in Jiangsu Province, not far from the Southern Song capital of Hangzhou. This planisphere, some 1.05 meters in diameter, is engraved on a stone block measuring approximately 2.2 by 1.1 meters (fig. 13.21). The stele is extremely well preserved, and direct rubbings of its surface can still be purchased in China. The *Tianwen tu* (Astronomical chart), as it is titled, is accompanied by a description summarizing the basic cosmological and astrological knowledge of the time.[185]

Nearly half a century ago, the Suzhou planisphere was extensively studied by Rufus and Tien.[186] A more recent investigation is by Pan.[187] Both the planisphere and the accompanying explanatory text are believed to be the work of Huang Shang, a Confucian scholar, in 1193. At

185. Since the chart was engraved during the Chunyou reign period (1241–52), this map is sometimes referred to by titles such as the *Chunyou tianwen tu* (Astronomical chart of the Chunyou reign period).

186. W. Carl Rufus and Hsing-chih Tien, *The Soochow Astronomical Chart* (Ann Arbor: University of Michigan Press, 1945).

187. Pan Nai, "Suzhou Nan Song tianwentu bei di kaoshi yu pipan" (Examination and critique of a Southern Song astronomical chart on a stone stele at Suzhou), *Kaogu Xuebao*, 1976, no. 1:47–61. See also *Zhongguo gudai tianwen wenwu tuji*, 84–85 and 123 (note 6).

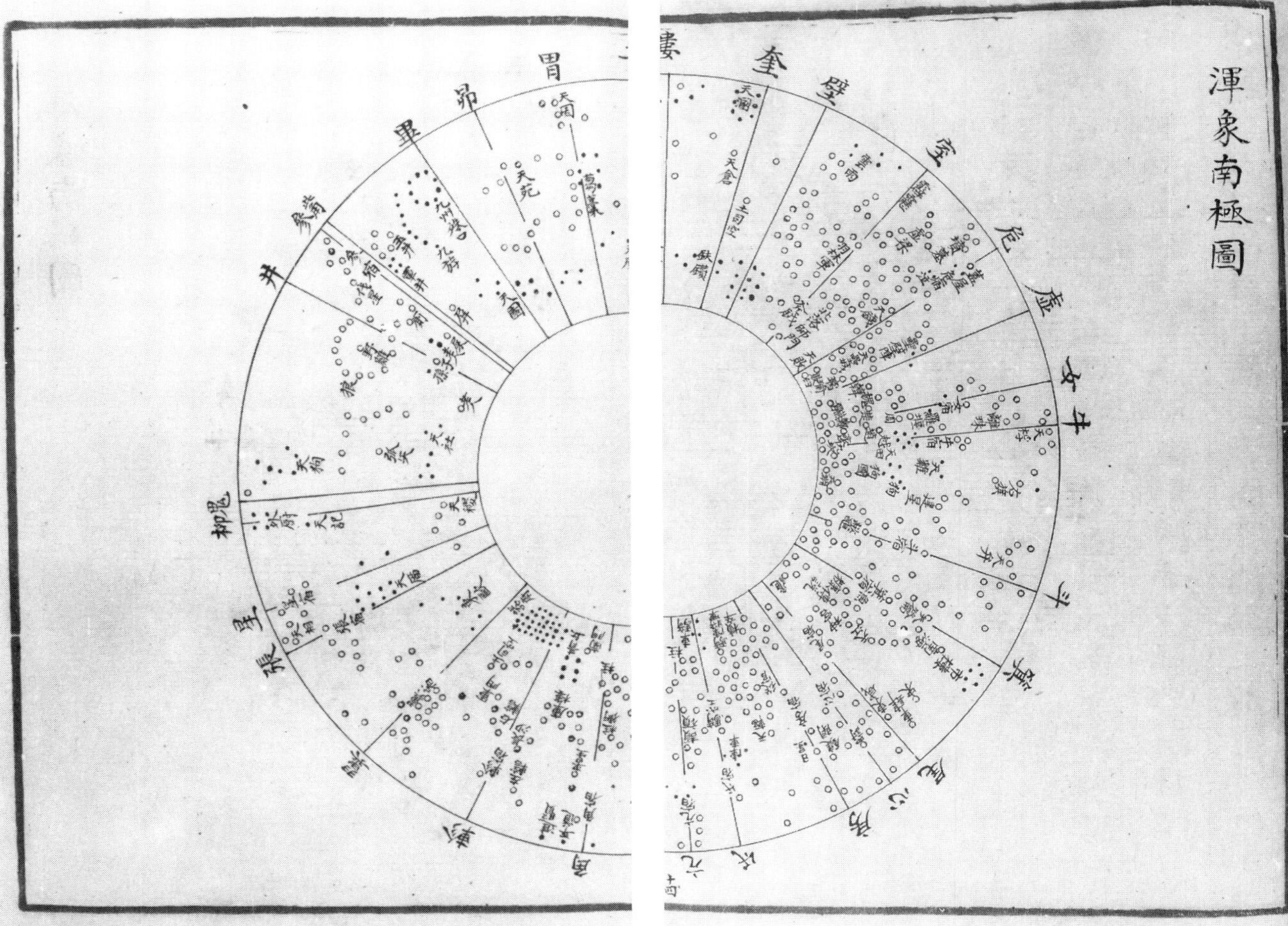

FIG. 13.20. THE SU SONG STAR MAP OF THE SOUTHERN HEMISPHERE. Constellations south of the celestial equator that are visible from China are shown. Between the circle of constant invisibility (declination −58°) and the South Pole, stars are uncharted.

Size of each page: 30 × 22 cm. By permission of the National Library of China, Beijing.

the time, Huang was tutor to Prince Jia, who soon afterward became the emperor Ningzong (r. 1194–1224). Although Huang's work was highly valued, it was not engraved on stone until 1247, more than twenty years after Ningzong's death.

The Suzhou planisphere portrays the whole of the sky visible from central China on a polar (equidistant) projection. The circle of constant visibility (declination +56°), celestial equator, ecliptic (incorrectly represented as a circle), boundaries of the twenty-eight lunar lodges, and outline of the Milky Way are all shown. *Xiu* boundaries are denoted by radial lines extending outward from the circle of constant visibility to the edge of the chart, which represents the circle of constant invisibility (declination −57°). Stars are indicated by small dots, joined into groups by straight lines. A few very bright stars such as Sirius and Canopus are represented by fairly large dots, but no systematic attempt is made to distinguish between stars of different brightness or to divide asterisms into the three ancient sets. Each constellation is named, as are the individual constituents of several important groups. Around the edge of the chart are marked the names and widths of the twenty-eight lunar lodges in *du* (inner circle) and the names of the twelve Jupiter stations, the twelve "terrestrial branches" (*zhi*)—as direction indicators—and other details (outer circle). A count by Rufus and Tien indicated 1,440 stars, and Pan enumerated 1,434. Both totals are rather less than the traditional figure of 1,464. Rufus and Tien noted as many as 313 asterisms—far more than the standard number of 283.[188]

Some idealization in the shapes of the asterisms

188. Rufus and Tien, *Soochow Astronomical Chart* (note 186); Pan, "Suzhou Nan Song tianwentu" (note 187).

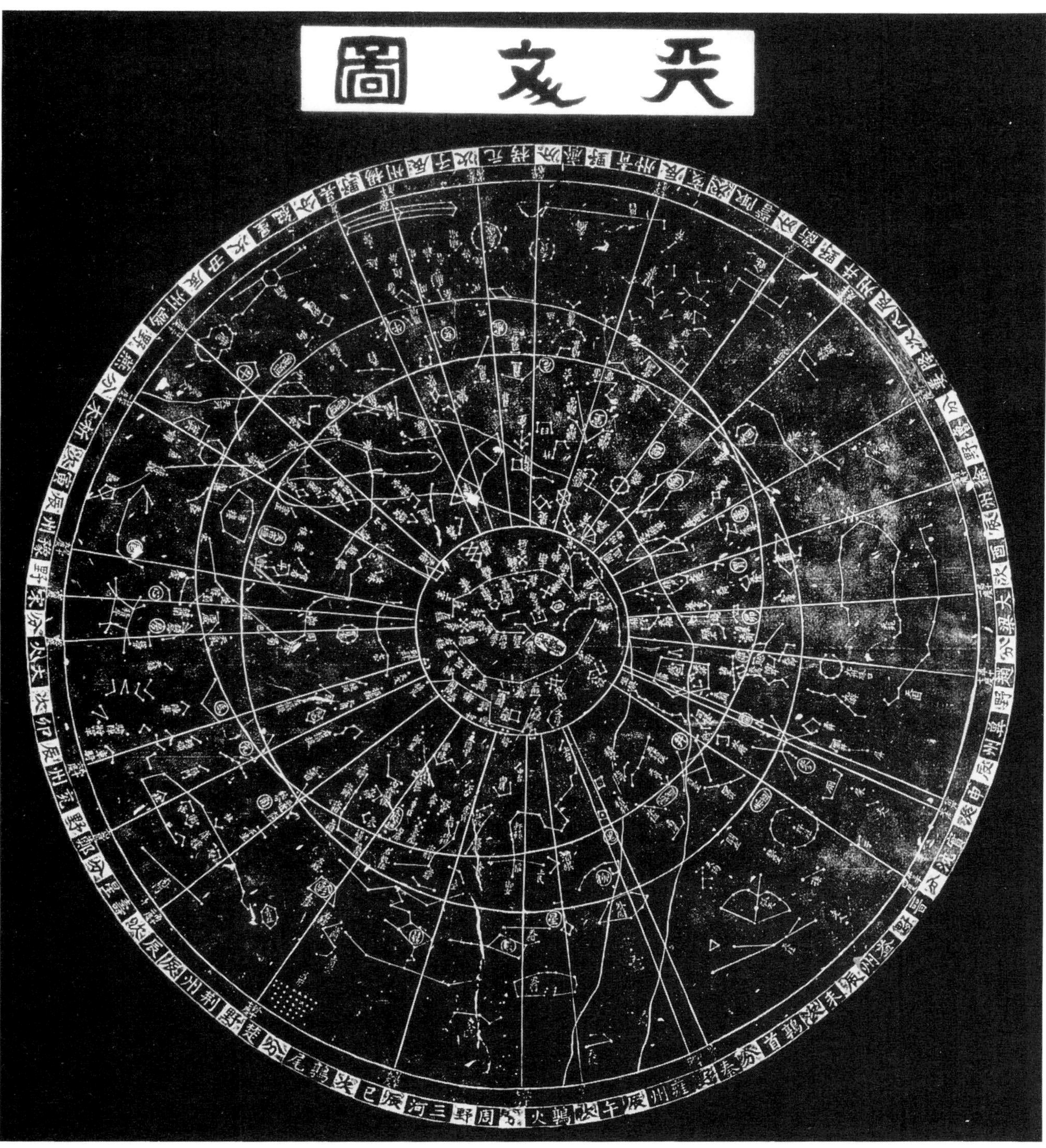

FIG. 13.21. RUBBING OF THE SUZHOU STAR MAP. This detailed star map is estimated to display some 1,440 stars. The projection is polar (equidistant). *Xiu* boundaries are represented by radial lines, and the circle of constant visibility, equator, and ecliptic are shown. The edge of the chart is at −57° declination. Diameter of the original: 105 cm. Suzhou Museum, Jiangsu Province. Photograph from W. Carl Rufus and Tien Hsing-chih, *The Soochow Astronomical Chart* (Ann Arbor: University of Michigan Press, 1945), pl. 1A.

FIG. 13.22. LIAO DYNASTY STAR MAP, 1116. This star map, painted on the ceiling of the tomb of a Buddhist official and excavated at Xuanhua near Beijing, shows all twelve zodiacal symbols (Taurus has since become defaced) as well as the twenty-eight *xiu* constellations.
Diameter of the original: ca. 220 cm. Photograph from Zhongguo Shehui Kexueyuan Kaogu Yanjiusuo, *Zhongguo gudai tianwen wenwu tuji*, 13.

depicted on the Suzhou star map can be noticed, but this is less evident than on the charts in the *Xinyi xiang fayao*. A check on the declinations of twenty selected bright stars (as undertaken above for the Su Song maps) reveals a tolerably small average error of less than two degrees. In the sample under discussion, the only really large error was for α Centauri (7.5°), but this may possibly be explained by its proximity to the edge of the map. The widths of the individual *xiu* are accurate to the nearest degree. Judging from this map, celestial cartography had reached a fairly high degree of maturity by the late Song dynasty.

This conclusion is confirmed by a recent analysis of Song determinations of the equatorial extensions of the *xiu* by Guo Shengchi.[189] These widths were measured to the nearest one-quarter *du* by the Northern Song astronomer Yao Shunfu in 1102 and are contained in the calendar treatise of the *Song shi*. In this investigation, Guo showed that typical errors of measurement are as small as about one-quarter degree, superior to the accuracy achieved at any previous period in Chinese history.

After an investigation made by Yabuuchi, Pan and Wang determined the positions of a large number of stars as represented on both the Suzhou planisphere and the *Xinyi xiang fayao* astral charts.[190] They also included a number of observations of star position recorded in the *Wenxian tongkao* (General study of literary remains)—a historical encyclopedia compiled by Ma Duanlin about 1280—and other Song works. As a result, these authors identified as many as 360 stars in terms of their Western equivalents. Before the Song dynasty it is possible to produce a list of concordances in comparable detail only by using the *Xingjing* star list of the Former Han. It is unfortunate that Pan and Wang gave no indication of the confidence (based on an assessment of errors of measurement) with which their identifications could be accepted.

As is evident from the Su Song and Suzhou star charts, very few Chinese asterisms resemble the Occidental constellations. Among larger groups we might cite little more than prominent features of the night sky such as Beidou (already frequently mentioned), Shen (Triad—equivalent to the four principal stars of Orion together with the belt and sword), and Wei (Tail—the tail of the Dragon in Chinese uranography and equivalent to the tail of the Scorpion in the West). On the contrary, the well-known W formation of the bright stars of Cassiopeia is shown divided into two distinct groups on Chinese star maps: Wangliang (named for a famed charioteer of the Zhanguo period) and Gedao (Hanging Gallery). Not surprisingly, the well-defined Pleiades and Hyades clusters in Taurus were recognized as discrete entities in China (under the names Mao and Bi), but it would be difficult to suggest many more examples of correspondence in shape between East Asian and Western asterisms.

Extensive measurements of star positions (not merely confined to the lunar lodges) were carried out several times during the Song dynasty. These surveys took place in the following reign periods: Jingyou (1034–38), Huangyou (1049–54), Xining (1068–77), and Yuanfeng (1078–85). A discussion of the relative precision of some of these measurements is given by Pan.[191] In particular, both the Su Song and Suzhou star maps were based on the results of the Yuanfeng survey.

Not long after the Jin dynasty (1115–1234) was established in northern China, official astronomers began to make observations in the traditional Chinese style. Judging from the quality of many of these observations—as recorded in the astronomical treatise (chap. 20) of the *Jin shi*—it is apparent that the Jin astronomers, like their Song counterparts, were equipped with effective star

189. Guo Shengchi, "Bei Song Heng xing guance jingdu chuyi" (On the accuracy of observations of the North Star during the Northern Song), *Tianwen Xuebao* 30 (1989): 208–16, English abstract on 216.

190. Yabuuchi Kiyoshi, "Sōdai no seishuku" (Description of the constellations in the Song dynasty), *Tōhō Gakuhō* (Kyoto) 7 (1936): 42–90, and Pan Nai and Wang De-chang, "The Huang-You Star of the Song Dynasty—A Chinese Star List of the Early Medieval Period," *Chinese Astronomy and Astrophysics* 5 (1981): 441–48.

191. Pan, *Zhongguo hengxing guance shi*, 169–75 (note 7).

maps. Nothing of this nature appears to have survived, however.

In 1971 a fairly well preserved star map from the Liao dynasty (916–1125) was brought to light during excavations at Xuanhua in Hebei Province, not far from Beijing.[192] This map is painted on the ceiling of the tomb of Zhang Shiqing, a palace official and devout Buddhist who died in 1116 (fig. 13.22). The map has no pretensions to astronomical accuracy. Its main interest lies in the juxtaposition of a complete set of both the lunar lodges and the signs of the Western zodiac. Apart from fragmentary Tang illustrations (see above), it is the earliest extant example of such joint representation.

The circular Liao map is about 2.2 meters in diameter. At its center is a bronze disk, surrounded by replicas of red lotus petals. Immediately beyond is an illustration of the asterism Beidou, also in red, and several disks that presumably signify the sun, moon, and planets. Enclosing these figures are the twenty-eight *xiu* constellation patterns, shown in red and arranged in a circle. Finally, in an outer ring are depicted the twelve signs of the Western zodiac in a variety of colors. The symbol for Taurus has since become obliterated, but the remaining symbols are well preserved. The representations of Aries, Cancer, Leo, Libra, Scorpio, and Pisces are readily recognizable, but in keeping with the names found in Buddhist sutras, Gemini is portrayed by a man and woman, Virgo by two women, Sagittarius by a man with a horse, Capricorn by a sea-monster, and Aquarius by an ornate vessel.

Two other Sinified representations of the Western zodiacal signs survive from this period. Like the Liao star chart, both are closely associated with Buddhism. Among the decorative patterns on the sides of a large bell at Xingtai in Hebei Province are attractive and well-preserved illustrations of the twelve zodiacal signs.[193] This bell, more than 2 meters high and with a base circumference of 7.2 meters, was cast in 1174 during the Jin dynasty. Originally hung in Kaiyuan Buddhist Temple, it was recently moved to a nearby park. A series of rather well executed wall paintings of the zodiacal symbols were discovered early in the present century in the Caves of the Thousand Buddhas at Dunhuang (fig. 13.23).[194] These originate from sometime during the Xi Xia period (1032–1227).

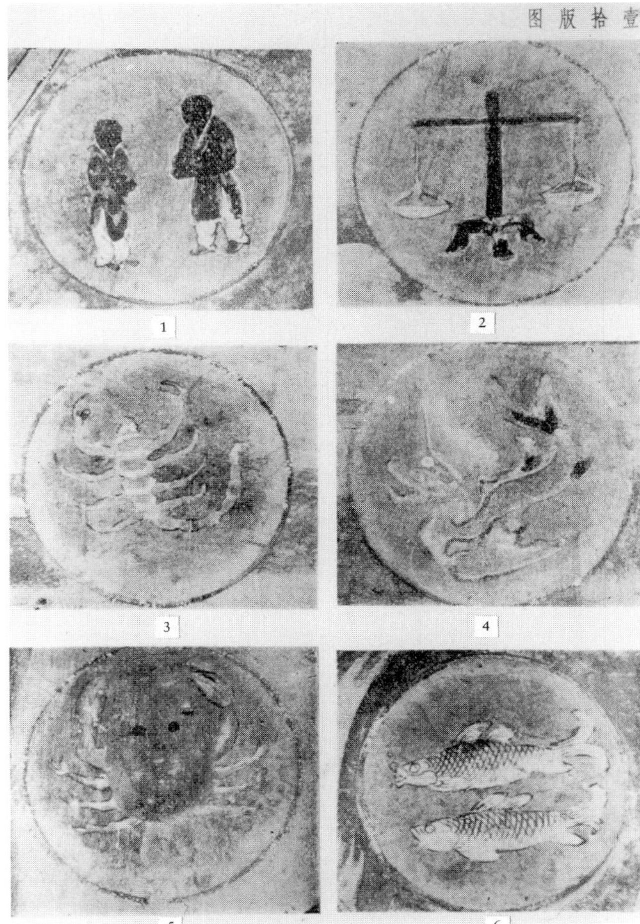

FIG. 13.23. XI XIA PERIOD ZODIACAL SYMBOLS. The Buddhist paintings shown here are part of a full series depicting the zodiacal symbols. In order are (1) Gemini, (2) Libra, (3) Scorpio, (4) Capricorn, (5) Cancer, (6) Pisces.
Size of each original: unknown. From Xia Nai, "Cong Xuanhua Liao mu di xingtu lun ershiba su he huangdao shier gong" (Discussion of the twenty-eight lunar lodges and twelve palaces of the ecliptic based on the star map from the Liao tomb at Xuanhua), reprinted in *Kaoguxue he keji shi* (Archaeology and the history of technology), by Xia Nai (Beijing: Kexue Chubanshe, 1979), 29–50.

The Yuan and Ming Dynasties (1279–1644)

The Yuan dynasty was a time of considerable astronomical activity, as is evident from the extensive celestial observations recorded in the official history of the time by Song Lian et al., *Yuan shi* (History of the Yuan, compiled 1369–70). Along with the Song, the Yuan has been

192. Hebeisheng Wenwu Guanlichu, Hebeisheng Bowuguan (Hebei Province Cultural Relic Agency, Hebei Provincial Museum), "Liaodai caihui xingtu shi woguo tianwenshishang di zhongyao faxian" (The Liao period star map, an important discovery in the history of Chinese astronomy), *Wenwu*, 1975, no. 8:40–44; Edward H. Schafer, "An Ancient Chinese Star Map," *Journal of the British Astronomical Association* 87 (1977): 162; Yi Shitong, "Hebei Xuanhua Liao Jin mu tianwen tu jianxi-jianji xingtai tiezhong huangdao shier gong tu xiang" (A brief investigation of the star map from the Liao-Jin tombs at Xuanhua in Hebei—Also the twelve zodiacal signs as found on the Jintai iron bell), *Wenwu*, 1990, no. 10:20–24.

193. Yi, "Hebei Xuanhua Liao Jin mu tianwen tu jianxi-jianji xingtai tiezhong huangdao shier gong tu xiang" (note 192).

194. Xia, "Cong Xuanhua Liao mu di xingtu lun ershiba xiu he huangdao shier gong," esp. 47 and pls. 11–12 (note 167).

aptly described as the "heyday of Chinese astronomy."[195] It is therefore unfortunate that no astral globe or major star chart of the period is now extant. A celestial globe that was constructed at Beijing by the great astronomer and mathematician Guo Shoujing (1231–1316) survived until the eighteenth century, when it was melted down. The only extant star map that is probably of Yuan origin gives no hint of the level of attainment reached by Guo Shoujing and his contemporaries. This sectional chart is devoid of coordinates, and many of the constellation patterns have idealized forms. An outline of the history of the Yuan globe is given below, followed by a brief discussion of the sectional star map.

Between 1276 and 1279, Guo Shoujing equipped the imperial observatory at Dadu (Beijing) with a variety of new instruments, including a large celestial globe (*huntian xiang*). According to chapter 48 of the *Yuan shi*, this globe was six *chi* (some 1.7 m) in diameter, with graduations in both RA and declination. Both the celestial equator and the ecliptic were delineated, the later being "elevated above and depressed below the equator by 24 degrees [*du*] and a small fraction in each case." It was further stated that the globe was "placed upon a square box, the south and north poles being below and above the surface by 40 degrees and a large fraction, half of the globe being visible and half concealed. Within the box there are hidden toothed wheels set in motion by machinery for turning the globe."[196] Unfortunately, the representation of the constellations is not described. The value used for the obliquity of the ecliptic (a little more than twenty-four *du*) agrees well with the true result of 23.5 degrees. The figure of forty *du* "and a large fraction" for the altitude of the north and south poles also corresponds closely to the latitude of Beijing (39.9°).

The celestial globe and other instruments constructed by Guo Shoujing continued in use at Beijing throughout the remainder of the Yuan dynasty. About 1370 the first Ming emperor had them moved to his capital of Nanjing, where they continued to be used, despite the great difference in latitude between the two cities (almost seven degrees). Although the third Ming emperor restored Beijing as the capital in 1421, the Yuan instruments remained at Nanjing. Eventually, in 1437 the astronomer royal requested that wooden duplicates of the instruments at Nanjing be sent to Beijing.[197] Bronze replicas were then cast to equip the new Beijing observatory, which was established in 1442.

Little more is heard of either set of instruments for more than 150 years, but their subsequent history is far from happy. In 1599 (toward the end of the Ming dynasty), the celestial globe and other instruments manufactured by Guo Shoujing were inspected at Nanjing by the great Jesuit scholar Matteo Ricci (1552–1610), known to the Chinese as Li Madou. Although Ricci was unaware of either their exact date or their origin, he wrote that "it seems certain that they were molded when the Tartars were in power in China," but he was misguided in his rather biased assertion that their designer "had some knowledge of European astronomical science."[198] The following further extract from his journal provides some interesting details:

> They had installed here [at Nanjing] certain astronomical instruments or machines, made of cast metal which, in size and in elegance of design, surpassed anything of the kind as yet ever seen or read about in Europe. These instruments had stood the test of rain and snow and change of weather for nearly two hundred and fifty years [*sic*], with no detriment to their original splendor. There were four of the larger kind. . . .
>
> The first was a large globe. Three men with outstretched arms could scarcely encircle it. It was marked with meridians and parallels according to degrees, and it stood on an axis, set into a huge bronze cube in which there was a small door, for entrance, to turn the sphere. There was nothing engraved on the surface of this globe, neither stars nor zones. Hence it appeared to be an unfinished work. . . .
>
> Later on, Father Matthew (i.e., Ricci) saw similar instruments at Beijing, or rather duplicates of these, and undoubtedly cast by the same artisan.[199]

As Needham has emphasized, the effects of weathering on the celestial globe at Nanjing over more than three centuries may have been more severe than Ricci imagined.[200] The device survived a further century or so after Ricci viewed it. In 1670 it was taken to Beijing where, along with the other Yuan instruments, it was relocated at the imperial observatory. Three years later the Belgian Jesuit Ferdinand Verbiest, considering the Yuan artifacts useless, had them placed in storage to make way for new instruments that he had constructed. In 1688 both the Yuan and Ming instruments were viewed, gathering dust, by Louis Le Comte, another Jesuit missionary. Le Comte also noticed a further celestial globe about one meter in

195. Ho, *Li, Qi and Shu*, 164 (note 16).

196. Translated by Needham, Wang, and Price, *Heavenly Clockwork*, 137 (note 80).

197. For details, see Yu Jie and Yi Shitong, "Beijing gu guanxiangtai" (An ancient observatory in Beijing), in *Zhongguo gudai tianwen wenwu lunji* (Collected essays on ancient Chinese astronomical relics), ed. Zhongguo Shehui Kexueyuan Kaogu Yanjiusuo (Archaeological Research Institute, Chinese Academy of Social Science [Academica Sinica] (Beijing: Wenwu Chubanshe, 1989), 409–14.

198. *China in the Sixteenth Century: The Journals of Matthew Ricci, 1583–1610*, trans. Louis J. Gallagher from the Latin version of Nicolas Trigault (New York: Random House, 1953), 331.

199. *China in the Sixteenth Century*, 329–31 (note 198).

200. Needham, *Science and Civilisation*, 3:368 (note 5).

diameter, which was of relatively coarse construction.[201] Nothing further is known regarding this device. In 1715 all the Yuan instruments and all but four of the Ming copies were melted down, with imperial sanction, by the Jesuit astronomer royal of China, Bernard Stumpf, who needed the bronze to manufacture new quadrants.[202] The four Ming replicas that escaped destruction included the celestial globe and three other devices: an armillary sphere, a simplified armillary, and a gnomon. Today the celestial globe seems to have disappeared, possibly as a result of the Boxer Rebellion in 1900, but the other three instruments can still be seen at Purple Mountain Observatory, Nanjing, having been transferred from Beijing in 1931.

Pan Nai has reproduced copies of a series of small sectional constellation maps (some seventy-five in all) that he attributes to Guo Shoujing. These depict up to about four asterisms. Star groups, and in many cases their constituent stars, are named, but the various diagrams contain no reference coordinates or even indications of the relative locations of the individual sections. The constellation patterns are in general highly idealized, and among them many symmetrical forms (especially circles) are evident. Pan found these maps along with other material that he attributes to Guo Shoujing in the National Library, Beijing, in a Ming compilation titled *Tianwen huichao* (A collection of manuscripts on astronomy).[203]

Guo Shoujing, who held the post of astronomer royal under Kubilay Khān, is regarded as the inventor of the equatorial mounting.[204] Among his various writings is the calendar treatise in chapters 52–57 of the *Yuan shi*. He is known to have produced two catalogs of stars soon after he constructed the equipment for the Beijing observatory. These were titled: *Xin ce ershiba shezazuo zhuxing ruxiu quji* (Newly measured positions of the twenty-eight lunar lodges and the various known asterisms) and *Xin ce wuming zhuxing* (Newly measured positions of those stars without names).[205] The latter catalog evidently contained details for previously uncharted stars. Although both catalogs were thought to have been lost long ago, Pan recently uncovered what appears to be a partial copy of the first in the same Ming collection of manuscripts in which the sectional star maps were found.[206] He made a careful examination of the text of this catalog and compared the details with the briefer information listed in the *Yuan shi* calendar treatise. As a result, he asserted that he had found a copy of Guo Shoujing's first catalog containing many original positional measurements. In this work the coordinates (RA and NPD) of 741 stars—all expressed to the nearest 0.1 *du*—are still preserved. Pan gives full details, including an annotated version of the catalog. By comparing the measured coordinates, corrected for precession, with those in a modern catalog, he was able to identify nearly all the stars in terms of their Western equivalents.

In the calendar treatise of the *Yuan shi*, Guo cites measurements of the widths of the *xiu* and the NPDs of the determinant stars he made in 1280. These values are also quoted to the nearest 0.1 *du*. Comparison with the corresponding values computed by Pan[207] reveals a mean error in the widths of the lunar lodges of about four arc minutes. They are thus of considerably higher precision than the corresponding Song determinations. Although the mean error in the Yuan measurements of NPD is relatively large (nineteen arc minutes), it is apparent from inspection of the results that the discrepancies result partly from a faulty latitude setting of the instrument.

The Yuan measurements reveal that by 1280 the width of the narrowest lunar lodge (Zuixi) had decreased to almost zero owing to relative precession between its determinative star (φ' Ori) and that of the adjacent *xiu* Shen (δ Ori) (see table 13.1). Guo's result for the equatorial extent of Zuixi (0.05 *du*), corresponds to only about three arc minutes, but actually the width had already become negative by approximately this amount, so that Zuixi had effectively vanished! The lunar lodge was tacitly assumed to exist with negligible width until its redefinition early in the Qing dynasty by the Jesuit astronomer Johann Adam Schall von Bell (see below).

Early in the Yuan dynasty (1267), the Persian astronomer Jamāl al-Dīn, of Maragheh observatory, brought a number of astronomical instruments to Beijing as a gift from Hūlāgū Khān (or his successor) to Kubilay.[208] These devices, which included a celestial globe and an astrolabe, are described in chapter 48 of the *Yuan shi*.[209] Because they were designed for ecliptic measurements (rather than equatorial) and were graduated into 360 degrees, they attracted little attention among Chinese astronomers such as Guo Shoujing. Throughout much of the Yuan dynasty, there were Arab astronomers at the court of Beijing. Afterward, when the Ming dynasty was established (1368), an Islamic astronomical bureau, known as

201. Louis Henry Le Comte, *Nouveaux mémoires sur l'état présent de la Chine*, 2 vols. (Paris: Anisson, 1696), 1:138–48.

202. Aloys Pfister, *Notices biographiques et bibliographiques sur les Jésuites de l'ancienne mission de Chine, 1552–1773*, 2 vols. (Shanghai: Mission Press, 1932–34), 2:645.

203. Pan, *Zhongguo hengxing guance shi*, fig. 40 and p. 276 (note 7).

204. Needham, *Science and Civilisation*, 3:377–82 (note 5).

205. Pan, *Zhongguo hengxing guance shi*, 276 (note 7).

206. For details, see Pan, *Zhongguo hengxing guance shi*, 276 ff. (note 7).

207. Pan, *Zhongguo hengxing guance shi*, 272–73 (note 7).

208. For details, see Needham, *Science and Civilisation*, 3:372–75 (note 5).

209. The astrolabe was unfamiliar to the Chinese, who could not even decide on a name for it. There is no evidence that astrolabes were ever used by Chinese astronomers.

the Huihui Sitianjian, was set up in parallel with the traditional bureau. The Arab astronomers were particularly concerned with calendar problems and mathematical astronomy; their impact on Chinese uranography appears to have been insignificant. Even when the Jesuits arrived at Beijing in the early seventeenth century, the Islamic bureau was still active.

After the high point astronomy reached during the Song and Yuan dynasties, the Ming was a period of decline. There were no significant advances in positional measurement, and there is no evidence that any major astral charts or star catalogs were produced.[210] In his valuable article on the astronomical bureau in Ming China, Ho cites several examples of official incompetence, including inadequate understanding of the setting of astronomical instruments, calendar discrepancies, and serious errors in predicting eclipses.[211] In keeping with this unhappy situation, surviving Ming astral charts prove to be of very mediocre quality, although one can argue that they are not necessarily representative of the best of Ming astral cartography. No Ming celestial globes are now known to exist.

Charts of the night sky in the Chinese style produced either by or in association with Jesuits during the late Ming (and also Qing [1644–1911]) dynasties will be discussed below. The rest of this section is confined to indigenous Ming artifacts. Extant Ming star maps are of two basic types: circular charts displaying the whole of the night sky as visible from northern or central China, and sets of sectional star maps.

Three circular astral charts that survive from the Ming, each depicting the night sky as seen from China, take the following forms: (1) a painting formerly on the ceiling of a Buddhist temple at Longfu, near Beijing (dated 1453); (2) a stone engraving at Changshu in Jiangsu Province (dated 1506); and (3) a paper diagram preserved at a former Daoist temple in Putian, Fujian Province (date sometime after 1572).[212] Before discussing details peculiar to each individual map, I shall summarize the principal features common to all three charts.

All the planispheres bear a general resemblance to the great Suzhou chart of the Song dynasty. Each is on a polar (equidistant) projection. Both the circle of constant visibility and the celestial equator are displayed, while the edge of each chart forms the circle of constant invisibility. The boundaries of the lunar lodges are represented by radial lines extending from the circle of constant visibility to the periphery of the map. Stars are depicted by circles that are joined into groups by straight lines. On any one map, symbols denoting the stars are of roughly equal size. As is typical of most pre-Jesuit charts, no attempt is made to distinguish between stars of different brightness. At the periphery of both the Longfu and Changshu charts, details such as the names of the *xiu*, *ci* (Jupiter stations), and directions are marked as on the Suzhou map. This information is absent in the Putian artifact.

The oldest surviving Ming star map is from the Longfu Temple near Beijing (fig. 13.24). This painting, on cloth mounted on an octagonal wooden board (diameter about 1.8 m), was discovered in 1977 when the temple, which had been built in 1453, was being demolished. In 1901 the temple had been extensively damaged by a fire, but fortunately the central hall remained intact. The chart of the night sky, which is 1.6 meters in diameter, displays 1,420 stars in gold on a blue background.[213] It has since been moved to the Beijing Ancient Observatory. Very few constellation names are marked, and neither the ecliptic nor the Milky Way is shown. If the equator is accurately positioned, the circle of constant visibility is at a declination of +60°, while the boundary of the map is at −62°. These declinations are more appropriate for southern China than for the neighborhood of Beijing. Measurements I have made on the scale drawing published by the Archaeological Research Institute, Academia Sinica[214] indicate that although the extent of each of the lunar lodges is fairly accurately depicted (to the nearest degree or so), stars are in general crudely positioned, and the form of several asterisms is highly idealized. Standard errors in the declinations of twenty selected bright stars are as large as seven or eight degrees, making identification difficult in several instances.

A somewhat more recent artifact is the Changshu planisphere, engraved on stone in 1506 (fig. 13.25). This stele is now preserved in the Office of Cultural Relics at Changshu, Jiangsu Province, not far from Suzhou. Like the Suzhou star map, it is titled *Tianwen tu* (Astronomical chart), and it appears to be modeled on the Song artifact. The Changshu stele measures 2.0 by 1.0 meters and is 24 centimeters thick; the astral chart itself is 70 centimeters in diameter.[215] All together, 1,466 stars are depicted in 284 constellations, essentially the traditional figures of antiquity. Both the Milky Way and the ecliptic are shown. The circle of constant visibility is at a declination of +52°, while the map extends to −53°; both values are

210. Other sciences also lost ground during the Ming. For example, in mathematics there was hardly any work of value between the beginning of the dynasty and 1500. See Ho, *Li, Qi and Shu*, 106 (note 16).

211. Ho Peng-yoke, "The Astronomical Bureau in Ming China," *Journal of Asian History* 3 (1969): 137–57.

212. Photographs or drawings of all three are in *Zhongguo gudai tianwen wenwu tuji*, 96–99 (note 6); the first and third are in Pan, *Zhongguo hengxing guance shi*, figs. 64 and 70 (note 7).

213. *Zhongguo gudai tianwen wenwu tuji*, 125 (note 6); Pan, *Zhongguo hengxing guance shi*, 309–10 (note 7). When I inspected the map in 1992 it was in poor condition, but restoration was planned.

214. *Zhongguo gudai tianwen wenwu tuji*, 96 (note 6).

215. For details, see *Zhongguo gudai tianwen wenwu tuji*, 125 (note 6), and Pan, *Zhongguo hengxing guance shi*, 316–18 (note 7).

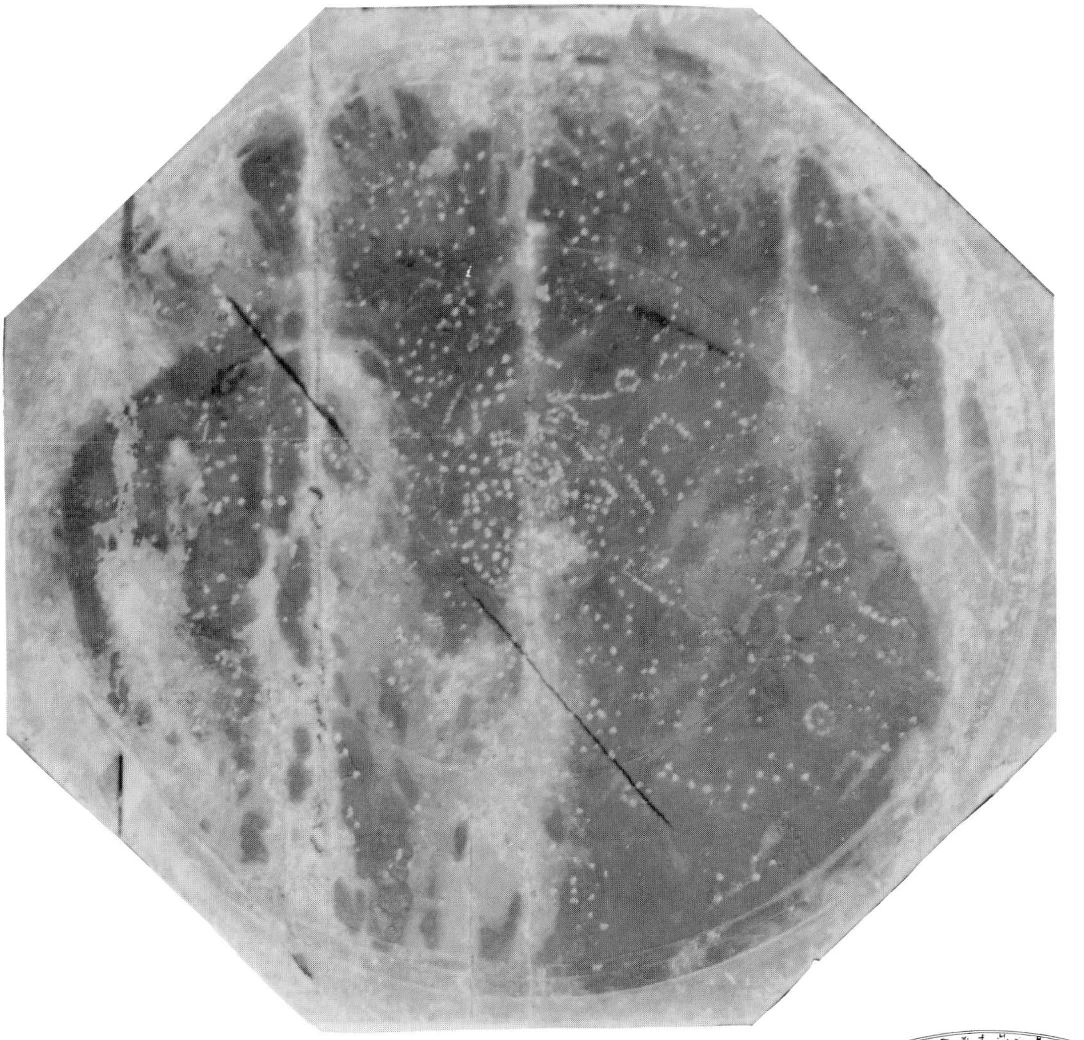

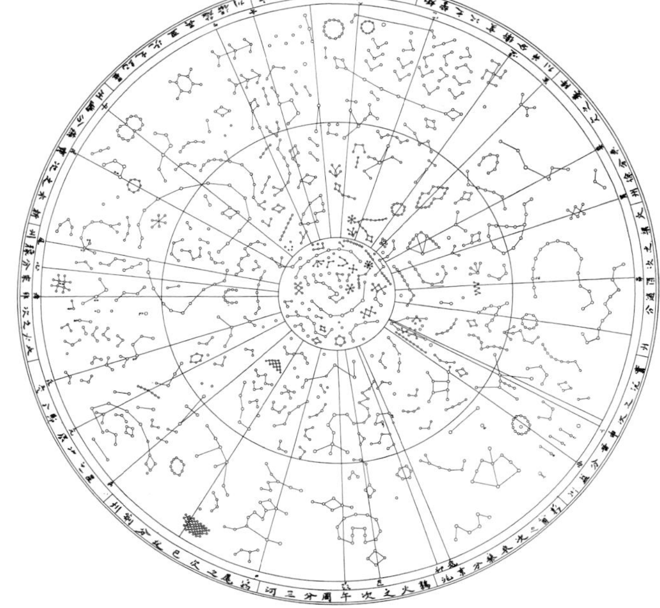

FIG. 13.24. PHOTOGRAPH AND DRAWING OF THE STAR MAP FROM LONGFU TEMPLE, 1453. The painting is on cloth and displays 1,420 stars in gold on a blue background. There are many similarities with the Suzhou star chart (fig. 13.21), although the precision is much less.
Diameter of the original: 160 cm. Photograph courtesy of Ancient Observatory, Beijing. Redrawing from Zhongguo Shehui Kexueyuan Kaogu Yanjiusuo, *Zhongguo gudai tianwen wenwu tuji*, 96.

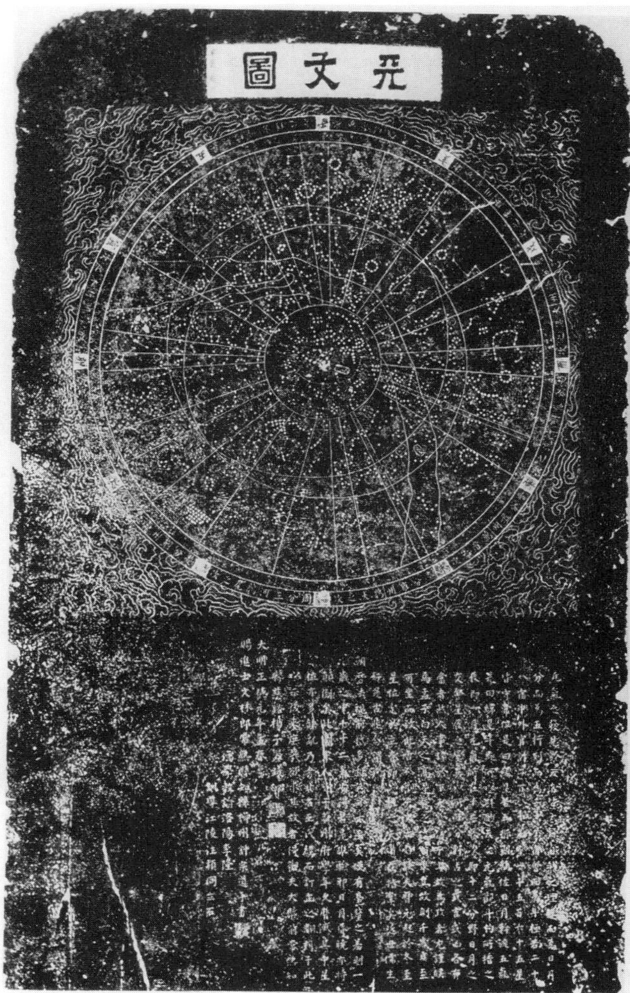

FIG. 13.25. RUBBING OF THE CHANGSHU STONE PLANISPHERE OF 1506. This chart closely resembles the Suzhou stele, but the constellations are depicted with relatively low accuracy. The equator and ecliptic are depicted in such a way that they intersect at exactly 180 degrees apart; an unusual feature at this epoch.
Diameter of the original: 70 cm. Office of Cultural Relics, Changshu, Jiangsu Province. Photograph from Zhongguo Shehui Kexueyuan Kaogu Yanjiusuo, *Zhongguo gudai tianwen wenwu tuji*, 97.

more suitable for northern China than Jiangsu. Although the equinoctial points are 180 degrees apart (a result that seems to be deliberately obtained by allowing the radii of the equator and ecliptic to vary slightly), the obliquity is only 19 degrees. Constellations are individually named. As with the Longfu planisphere, the widths of individual *xiu* are fairly accurately represented, but the stars are only approximately positioned, and some of the constellation patterns show idealized form. Although the Changshu map bears only general similarity to the Longfu painting, errors in the positioning of individual stars are of comparable magnitude.

The third Ming planisphere, printed on paper sometime after 1572, is now in poor condition. Fortunately, Pan has produced a careful scale diagram (fig. 13.26).[216] The full size of the sheet of paper is 150 by 90 centimeters; the astral chart is 60 centimeters in diameter. In all, 1,400 stars—grouped into 288 asterisms—are displayed on the chart. The Milky Way is not represented. An indication of the date when the chart was produced is given by the positioning of the 1572 guest star (a brilliant supernova that appeared in Cassiopeia) but not that of the one in 1604 (the next bright supernova, occurring in Ophiuchus).[217] Although the circle of constant visibility and the rim of the chart are concentric with the focal point of the boundaries of the lunar lodges (the north celestial pole), both the celestial equator and the ecliptic are badly misplaced. The latter two circles intersect 180 degrees apart, but their centers are both 10 degrees from—and on opposite sides of—the center of the chart. Tracing the paths of the equator and ecliptic in relation to the nearby constellations reveals fair accord with the true relative positions. The implication is that the constellations are misplaced and distorted to accommodate the highly erroneous equatorial and ecliptic circles. This curious artifice seems to be unique in Chinese history, but a late eighteenth-century Korean celestial map exhibits close parallels (probably independently; see below).

An extensive series of astral diagrams, most of them devoted to small sections of the night sky, is printed in the *Sancai tuhui* (Illustrated compendium of the three powers [heaven, earth, man], compiled by Wang Qi and printed in 1609) (fig. 13.27). The various diagrams were evidently produced with minimal measurement. The principal map in this collection depicts the night sky down to about declination $-55°$ on a polar (equidistant) projection. This is accompanied by a large number of sectional maps: for example, the region between the zones of constant visibility and constant invisibility (between about $+60°$ and $-55°$ declination) is shown in twenty-eight rectangular strips each centered on a separate *xiu*. Additional charts are devoted to the north circumpolar region. The outline of the Milky Way is delineated on those sections that cover the appropriate part of the sky. Once again, there is no attempt to discriminate between stars of different brightness; stars are represented by cir-

216. Pan, *Zhongguo hengxing guance shi*, fig. 70 (note 7).

217. *Zhongguo gudai tianwen wenwu tuji*, 99 and 125 (note 6). A detailed discussion of the supernovas of 1572 and 1604 is given by Clark and Stephenson, *Historical Supernovae*, chaps. 10 and 11 (note 2).

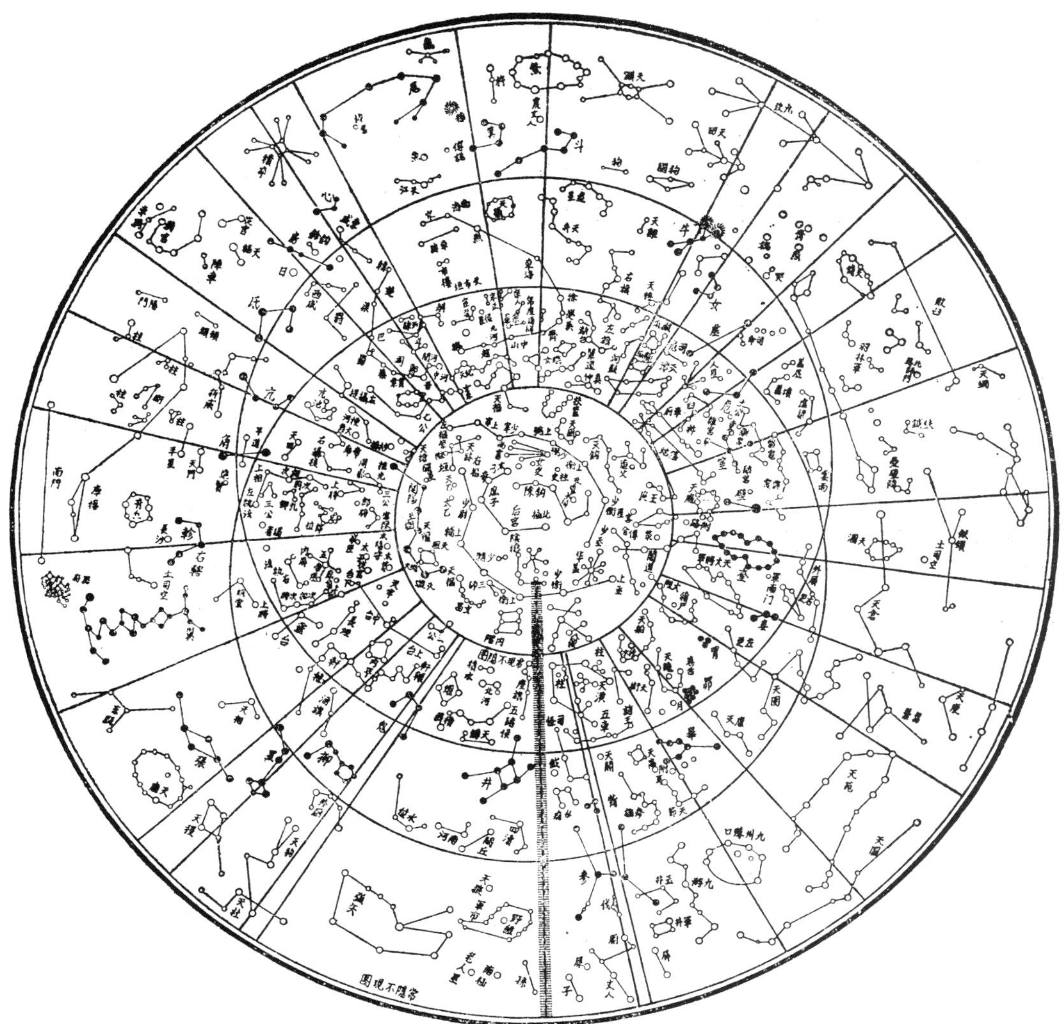

FIG. 13.26. SKETCH OF A MING PAPER PLANISPHERE. Although the precise date of this planisphere is not known, it can be deduced as sometime between 1572 and 1604, since the position of the supernova in the former year is marked but not that of the one in 1604. The original is at the Tianhougong Temple, Putian. Note that both the equator and the ecliptic are offset, an unusual device.
Diameter of the original: 60 cm. From Pan Nai, *Zhongguo hengxing guance shi* (Shanghai, 1989), fig. 70.

cles of equal size, joined into constellations by straight lines. Each constellation is individually named.

Charts showing sketches of a few selected constellations are preserved in Mao Yuanyi's *Wubei zhi* (Treatise on military preparations, compiled ca. 1621).[218] This work contains four such maps designed to assist navigation in the Indian Ocean. The diagrams, although roughly executed, are of particular interest since they depict several far southern asterisms, including Denglonggu (the Frame of the Lantern [the Southern Cross]) (fig. 13.28).[219]

Celestial Cartography in Korea

Because of its proximity to China, the history of Korea has been strongly influenced by its more powerful neighbor since ancient times, and this is true of Korean astronomy and astrology as well. The earliest relics of astronomical significance that have been uncovered in

218. These charts are discussed by George Phillips, "The Seaports of India and Ceylon, Described by Chinese Voyagers of the Fifteenth Century, Together with an Account of Chinese Navigation," *Journal of the Royal Asiatic Society, North China Branch* 20 (1885): 209–26, esp. 216–18. See also Joseph Needham, *Science and Civilisation in China* (Cambridge: Cambridge University Press, 1954–), vol. 4, pt. 3, with Wang Ling and Lu Gwei-djen, *Physics and Physical Technology: Civil Engineering and Nautics* (1971), 564–67.

219. These diagrams are reproduced and discussed in *Zhongguo gudai tianwen wenwu tuji*, 94–95 and 124–25 (note 6).

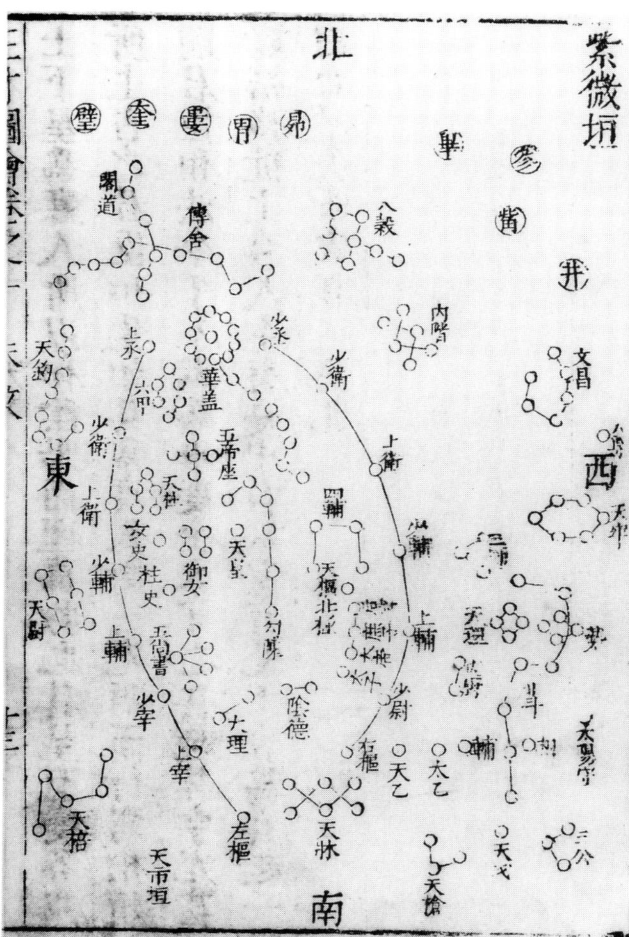

FIG. 13.27. PRINT OF A SECTION OF A STAR MAP IN THE *SANCAI TUHUI* SHOWING CONSTELLATIONS IN THE PURPLE PALACE (CONSTANT VISIBILITY) REGION. This is one of a series of crude star charts, first printed in 1609. Size of the original: 21 × 14 cm. From Wang Qi, comp., *Sancai tuhui* (1609 edition), *tianwen* 1.13a. Photograph courtesy of the Harvard-Yenching Library, Harvard University, Cambridge.

Korea are of Chinese origin. These are two cosmic boards depicting the Big Dipper and the twenty-eight lunar lodges (for details on the construction and use of cosmic boards, see above). Both instruments date from the first century B.C. and were discovered in the tombs of Chinese officials near modern P'yŏngyang.[220] At the time, the Han dynasty had established several commanderies in the northern part of present-day Korea.

No specifically Korean astronomical artifacts or reliable written records of celestial phenomena appear to have survived from before about 500. By this time, the three independent kingdoms of Koguryŏ, Paekche, and Silla had long since been established in the Korean peninsula. From this Samguk (Three Kingdoms) period, star maps have so far been found only in Koguryŏ. Several charts, consisting of representations of a few selected constellations in the Chinese style, have been uncovered in tombs dating from about 500. A scale diagram of one of these maps on the ceiling of a tomb near the Yalu River—along the present-day border between China and Korea—has been published by Jeon.[221] The chart illustrates seven of the twenty-eight lunar lodges with fair precision, Wei (the Tail of the Dragon) being particularly prominent. Similar examples occur in other Koguryŏ sepulchers, and these also clearly reveal Chinese influence.

Paintings of the four creatures, which in Chinese mythology represent the quarters of the sky, are found on the interior walls of several tombs near P'yŏngyang (the site of the Koguryŏ capital). The beautifully colored illustrations in one of these mausoleums dating from about 550 have been judged among the best paintings in the Orient that survive from this period.[222] On the east wall of the main chamber is depicted the Azure Dragon, on the south wall the Red Bird (fig. 13.29), on the west wall the White Tiger and on the north wall opposite the entrance the Dark Warrior (represented as in China by a turtle entwined with a snake) (fig. 13.30).

It is by way of Koguryŏ that the oldest example of detailed celestial cartography from anywhere in East Asia has come down to us, even though only late copies of the original star map are extant. The earliest of these reproductions, now in a Seoul museum, was engraved on stone in 1395. This chart is inscribed with brief details of its history that are essentially confirmed by Korean historical works.[223] It is related that at some unknown date a star map incised on stone was presented to the king of Koguryŏ by a Chinese emperor, an event that is unfortunately not recorded in Chinese history. This artifact was carefully preserved at P'yŏngyang until 670, when it suffered an ignominious fate; it was submerged in the nearby Taedong River when Koguryŏ was conquered by the Silla army. Although the stele was never recovered, many centuries later (1392) a rubbing from it was presented to the founder of the Yi dynasty. The king was so impressed with his gift that not long afterward he

220. For further information, see W. Carl Rufus, "Astronomy in Korea," *Transactions of the Korea Branch of the Royal Asiatic Society* 26 (1936): 4–48, esp. 4–6.

221. See Jeon, *Science and Technology in Korea*, fig. 1.1 (note 5).

222. Good-quality reproductions of these paintings have been published; see, for example, Editorial Staff of Picture Albums, ed., *Korean Central Historical Museum* (P'yŏngyang: Korean Central Historical Museum, 1979), 56–59.

223. The history of this star chart is outlined in Hongmun Kwan (Royal Library), *Chŭngbo Munhŏn pigo* (Documentary reference encyclopedia, expanded and supplemented) (Seoul: Empire of Korea, 1908)—see W. Carl Rufus, "The Celestial Planisphere of King Yi Tai-jo," *Transactions of the Korea Branch of the Royal Asiatic Society* 4, pt. 3 (1913): 23–72, esp. 37–38.

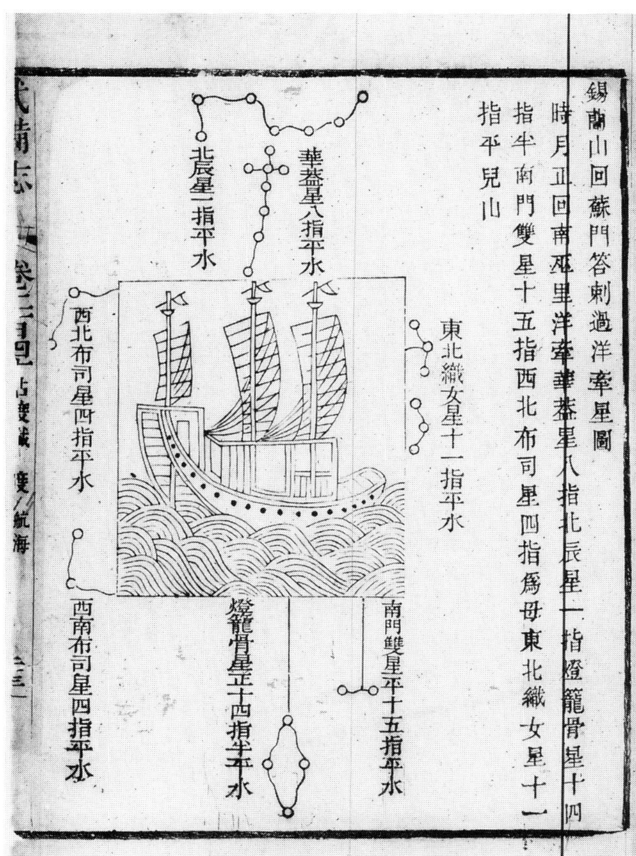

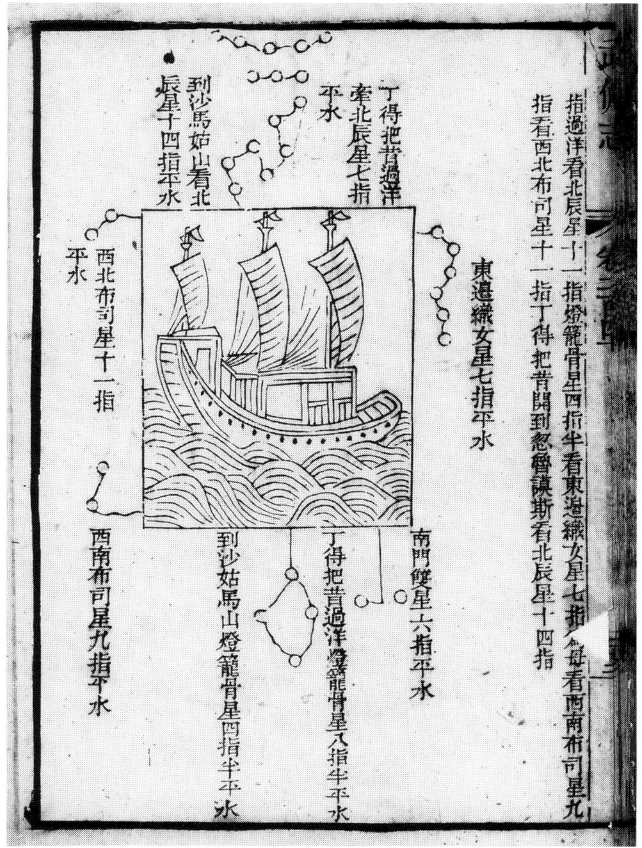

FIG. 13.28. MING NAVIGATIONAL CHARTS IN THE *WUBEI ZHI*. These charts were produced as aids to navigation in the Indian Ocean. The two far southern constellations are Nanmen (Southern Gate [α and β Centauri]) at the right and Denglonggu (The Frame of the Lantern [Southern Cross]) to the left of Nanmen.
Size of the original: unknown. Courtesy of the Geography and Map Division, Library of Congress, Washington D.C.

had a new engraving made on a block of marble. Since this stele dates from long after the Samguk period, I will consider it in detail later in this section.

What may well have been an observatory is still standing at Kyŏngju, the site of the Silla capital (fig. 13.31). This bottle-shaped structure, known as Ch'ŏmsŏngdae (Star Observing Tower), was built in 647, the sixteenth year of Queen Sŏndŏk.[224] It is some nine meters tall, and varies in diameter from about five meters at the base to three meters at the top. Although there has been dispute whether the tower was actually used as an observatory or was merely symbolic of Silla's astronomy, it implies a developed astronomy at this period.[225]

After its conquest of Koguryŏ and Paekche about 670, Silla ruled the peninsula for another 250 years, a period roughly contemporaneous with the Tang dynasty in China. The *Samguk sagi* (History of the Three Kingdoms, compiled by Kim Pusik, 1145) relates that in 692 a Buddhist monk named Tojŭng brought a star chart from China.[226] Unfortunately, nothing is known about its construction, and there appears to be no parallel account in Chinese history. Also from the *Samguk sagi*, we learn that in 749 Silla appointed a "director in charge of astrology."[227]

More than fifty separate observations of celestial phenomena, notably eclipses, comets, meteors, and lunar and planetary movements, are recorded in the *Samguk sagi* during the period of Silla ascendancy.[228] From these records it is apparent that the Silla sky watchers followed much the same style of observation as their Chinese counterparts and in particular adopted the Chinese constel-

224. For a detailed description of Ch'ŏmsŏngdae, see Kim Yongwoon, "Structure of Ch'ŏmsŏngdae in the Light of the Choupei Suanchin," *Korea Journal* 14, no. 9 (1974): 4–11.

225. Jeon, *Science and Technology in Korea*, 33–35 (note 5), favors the identification of this structure as an observatory; Kim, "Structure of Ch'ŏmsŏngdae" (note 224), is more cautious.

226. Kim Pusik, comp., *Samguk sagi*, chap. 8; see the edition in 9 vols. (Kyŏngju, 1512; reprinted Seoul, 1931).

227. *Samguk sagi*, chap. 9 (note 226).

228. These observations are scattered throughout chaps. 6–12 of the *Samguk sagi* (note 226).

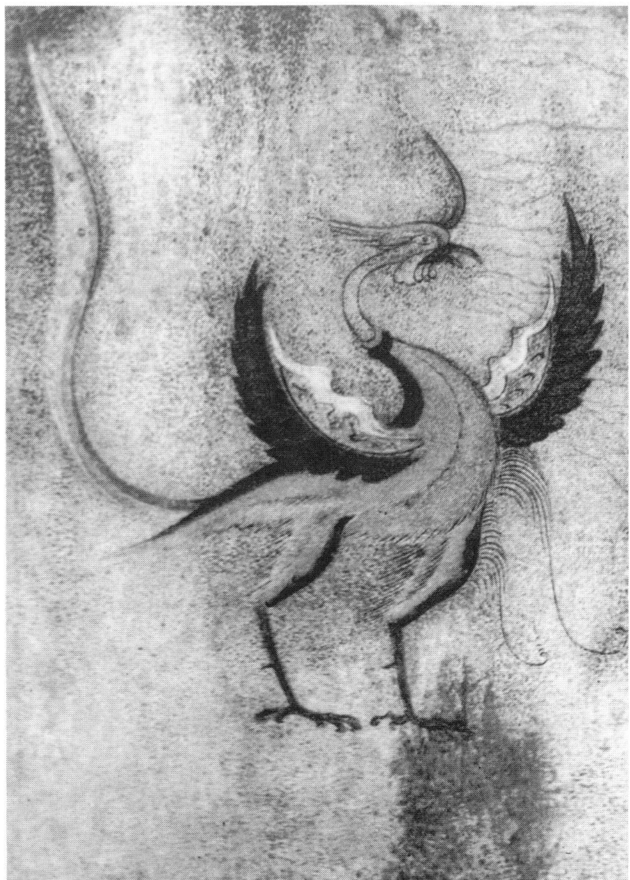

FIG. 13.29. KOGURYŎ TOMB ILLUSTRATION OF RED BIRD. Appropriately, this is painted on the south wall of a tomb. This and the other illustrations on the walls are particularly exquisite (see also fig. 13.30).
Size of the original: unknown. From Editorial Staff of Picture Albums, ed., *Korean Central Historical Museum* (P'yŏngyang: Korean Central Historical Museum, 1979), 57.

lation patterns. Thus in the *Samguk sagi* the positions of about twenty comets and meteors seen after 670 are specified in relation to Chinese star groups, including several of the lunar lodges. Observations such as these suggest that the Silla astronomers possessed star maps, presumably of Chinese origin, of at least tolerable utility. Since the *Samguk sagi* does not give any actual measurements of position (in terms of degrees), however, we have no way of knowing how accurate such charts might have been.

During the whole of the Koryŏ dynasty, founded by Wang Kŏn in 918 and lasting until 1392, virtually no direct information is available either on the acquisition of Chinese star maps and catalogs by Koreans or on the production of such artifacts by native astronomers. A brief note in the *Koryŏ sa* (History of Koryŏ, 1451), the official history of the period, mentions that a star chart was made by O Yun bu, who died in 1305. This short biographical sketch asserts that O Yun bu was an assiduous observer who watched all night long despite heat or cold. His chart is said to have "harmonized all the doctrines," but unfortunately no further description of it survives.[229]

Despite the lack of information on star maps from Korea during the Koryŏ dynasty, there is ample evidence that this was a period of great astronomical activity. The astronomical records in the *Koryŏ sa* (chaps. 47–49) include numerous reports of comets, lunar and planetary movements, and meteors. These records reveal that, as in China, the principal motive for celestial observation was astrological. Although the Chinese division of the night sky into constellations was systematically adopted by the Korean astronomers, however, they interpreted the celestial phenomena they witnessed as omens affecting their own ruler and country.[230] Some of the cometary records in the *Koryŏ sa* are particularly detailed, and these show that the sky watchers of the Korean court possessed an extensive knowledge of the constellations. For instance, the motion of a comet through the circumpolar region in 1110 is described relative to as many as nine separate asterisms.[231] No positional measurements are preserved in the *Koryŏ sa*, but one can scarcely doubt that good-quality star maps were available to the official astronomers at the capital of Songdo (Kaesŏng).

Accounts of the movements of comets, and of the moon, planets, and meteors in the *Koryŏ sa*, mention virtually all the star groups mapped by the Chinese astronomers, including the lunar lodges. Many of these asterisms are referred to time after time in separate Korean records. It is apparent that in Korea there was a negligible independent tradition of mapping the constellations. The numerous accurately dated references in the *Koryŏ sa* to the passage of the moon or five bright planets through or near asterisms are of special interest. By computing the celestial coordinates of the moon and planets on the stated dates (reduced to the Julian calendar), it is possible to approximately delineate the outlines of the star groups in the zodiacal region.[232] On this basis there do not appear to be any marked discrepancies between the constellation patterns as recognized by the Korean and Chinese astronomers at this period, though there seems to be ample

229. Chŏng Inji et al., comps and eds., *Koryŏ sa*, chap. 122; see the edition in 3 vols. (Seoul: Asea Munhwasa, 1972). The *Koryŏ sa* follows the format of a traditional Chinese dynastic history, with annals, monographs, and biographies.

230. See, for example, Park, "Portents and Neo-Confucian Politics" (note 13).

231. *Koryŏ sa*, chap. 47 (note 229).

232. Pan and Wang, "Huang-You Star," 441 (note 190). I have made similar investigations, as yet unpublished.

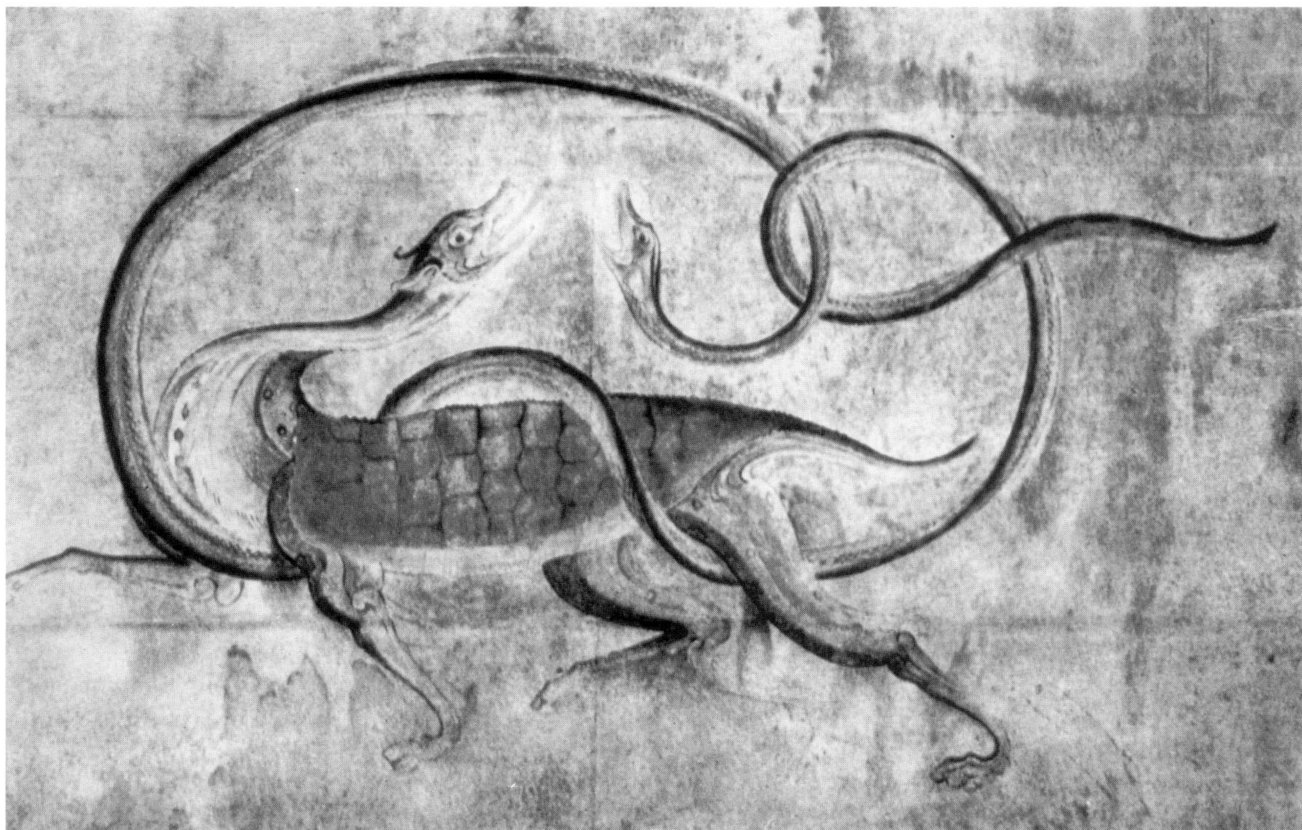

FIG. 13.30. KOGURYŎ TOMB ILLUSTRATION OF BLACK TURTLE. Painted on the northern wall of the same tomb as figure 13.29, this illustration shows a turtle entwined with a snake—a fairly common representation of the Dark Warrior of the North.

Size of the original: unknown. From Editorial Staff of Picture Albums, ed., *Korean Central Historical Museum* (P'yŏngyang: Korean Central Historical Museum, 1979), 58.

opportunity for further research in this aspect of the history of celestial cartography.

About 1200, the astronomical treatise of the *Koryŏ sa* begins to cite occasional records in which stars are numbered within their respective constellations; for example, an observation in 1223 in which the planet Venus "invaded the fifth star of Nandou."[233] Numbering of individual stars within asterisms had begun in China by at least the first century B.C. (see above), but until about 1200 there is little evidence of this practice in Korea. By this latter date Korean astrography had apparently progressed so far that such refinement was considered desirable. It is unfortunate, however, that instances of star numbering in the *Koryŏ sa* are so sporadic that systematic comparison with the contemporary Chinese numbering scheme—by calculating lunar and planetary movements—seems scarcely viable.

In 1389 the Koryŏ government was overthrown by Yi Sŏnggye, and as King T'aejo (r. 1392–98) he established a new dynasty named Yi. This was to last for more than five hundred years—until the Japanese annexation in 1910. In 1394 King T'aejo moved his capital to Hanyang (Seoul), where it was to remain throughout the dynasty. Finding the official astronomers of the fallen dynasty incompetent, he organized a new astronomical board, the *sŏun'gwan*. At the same time, new books pertaining to astronomy and astrology were compiled.[234] The observations made by the *sŏun'gwan*—in both this and later periods—are summarized in the *Chosŏn wangjo sillok* (Royal annals [*sillok*] of Chosŏn), a series of extensive chronicles of events occurring during the reigns of the first twenty-five kings of the Chosŏn dynasty.[235] These observations follow much the same style as those

233. *Koryŏ sa*, chap. 49 (note 229).

234. Rufus, "Astronomy in Korea," 22 (note 220).

235. These cover the period from 1392 to 1863. (Korea was known as Chosŏn during these years [also called the Yi dynasty].) The *Koryŏ sa* was probably largely compiled from similar material that has long since disappeared.

FIG. 13.31. SILLA OBSERVATORY (CH'ŎMSŎNGDAE), KYŎNGJU, BUILT IN 647. Opinions have differed over the exact nature of this structure. If it is an observatory (as its name suggests), it is the oldest surviving building of this nature anywhere in the world.
Photograph courtesy of F. Richard Stephenson.

reported in the *Koryŏ sa*, although they are often more detailed.

A major astrographical achievement during the reign of King T'aejo was the engraving in 1395 of a star map based on a preserved rubbing of an ancient stele. This stele had been lost when Koguryŏ fell in 670. Brief details regarding the history of this chart have already been given above. Rufus translated part of the inscription found on extant copies of the 1395 chart as follows:

> Many years having passed since it was lost, existing rubbings of the original were also [believed to be] out of stock.
>
> However, when His Majesty [King T'aejo] began to reign, a man having one of the originals tendered it to him. His Majesty prized it very highly and ordered the court astronomers to engrave it anew on a stone model. The astronomers replied that the chart was very old and the degrees of the stars were already antiquated; so it was necessary to revise it by determining the present midpoints of the four seasons and the culminations at dark and dawn and to engrave an entire new chart designed for the future.
>
> His Majesty responded, "Let it be so!"[236]

Preparations for the new celestial planisphere were begun not long after King T'aejo ascended the throne. This work was carried out by a team of astronomers under the supervision of Kwŏn Kŭn and other senior members of the board of astronomy.[237] A preliminary manuscript version was prepared in the summer of 1395. Rufus was able to inspect this chart, and he noted that the central star map was inverted compared with the final version.[238] Unfortunately, this manuscript can no longer be traced. By December 1395, the circular astral chart was engraved on a huge block of black marble bearing the caption *Ch'ŏnsang yŏlch'a punyajido* (Chart of the constellations and the regions they govern). The accompanying inscription, as well as providing a historical summary, contains a variety of astronomical tables and other information.[239]

The principal motive of King T'aejo in having a reproduction made of the old star map may well have been "to acquire new star charts as symbols of the royal authority of the new dynasty."[240] The result is the survival of what may well be a copy of a very early Chinese star map. In the *Chŭngbo Munhŏn pigo* (an eighteenth-century historical compendium), it is asserted that although the inscription on the Koguryŏ chart had been updated, "the astrography according to the old chart (itself) . . . was engraved directly on stone."[241]

Today it is not possible to assess the evidence that convinced the astronomers of King T'aejo that the rubbing presented to their ruler was indeed taken from the ancient Koguryŏ star map rather than some other early astral chart. However, although the surface of the stele, which was engraved in 1395, is now damaged, several high-quality reproductions are preserved, and these reveal that the astrography they are based on is very archaic.

The 1395 stele, which weighs about a ton, has approximate dimensions of 2.1 meters high by 1.2 meters wide by 12 centimeters thick; the circular star map itself is about 90 centimeters in diameter.[242] Damage has occurred on several occasions owing to fire, water ero-

236. Rufus, "Celestial Planisphere," 31–32 (note 223).
237. Jeon, *Science and Technology in Korea*, 26 (note 5).
238. Rufus, "Astronomy in Korea," 23 (note 220).
239. A full translation of the inscription is provided by Rufus, "Celestial Planisphere," 29 ff. (note 223).
240. Jeon, *Science and Technology in Korea*, 25 (note 5).
241. *Chŭngbo Munhŏn pigo*, chap. 2 (note 223).
242. See Na Ilsŏng, "Chosŏn sidae in chŏn'mun ŭigi yŏn'gu" (Study

sion, and transport—for example, during the Japanese invasion in 1592 when the building it was housed in was destroyed.[243] The chart is now on exhibit at the Royal Museum in Toksugung Palace in Seoul.

On a visit to Seoul in October 1993 I was able to inspect and photograph the stele. Except for a small section (covering about 10 percent of the planisphere) that is badly worn, all asterisms can be clearly discerned, as well as the Milky Way, coordinate circles, and boundaries of the lunar lodges. It is on record that 120 rubbings of the stone were made in 1571,[244] suggesting that the whole surface was then in sound condition. Presumably all of these rubbings have long since disappeared.

Fortunately, an accurate seventeenth-century replica of the planisphere of King T'aejo is still in a good state of preservation. This reproduction was engraved on a block of white marble in 1687 at the command of King Sukchong (r. 1674–1720) (figs. 13.32 and 13.33).[245] Its plane dimensions are almost the same as those of the original stele, although the thickness (30 cm) is considerably greater.[246] The newer engraving is stated to be a faithful copy of the older stele, except that the title has been moved to the top.[247] It is currently exhibited at the King Sejong Memorial Museum in Seoul, where a framed rubbing is also on view. The following description is based on an examination of photographs of rubbings made available to me.[248]

The central astral chart is circular, approximately ninety centimeters in diameter, and is on a polar (equidistant) projection centered on the north celestial pole. On this chart, stars are denoted by dots, nearly all of similar size. Canopus, Sirius, and one or two other bright stars are represented by unusually large dots, but as on Chinese astral maps of the pre-Jesuit era there is no systematic attempt to indicate brightness. Stars are joined into groups by straight lines. Two concentric circles represent the northern circumpolar boundary and the celestial equator. Assuming that the equator is accurately positioned, the declination of the north circumpolar circle is +52°. The chart is bounded by the circle of constant invisibility, whose declination is −55°. Both of these declinations would be adequate for use in central and northern China as well as Korea. As is usual in East Asian star charts produced before the Jesuit period, the ecliptic is incorrectly shown as a circle. The boundaries of the twenty-eight lunar lodges are represented by radial lines extending from the north circumpolar circle to the rim of the chart.

Although the Milky Way is portrayed with fair precision, the outlines of the various asterisms are crudely depicted. Each asterism is individually named, but the configurations of stars sometimes differ widely from the forms represented on medieval Chinese maps. A count by Rufus and Chao indicated 1,464 stars—agreeing with the canonical number according to the "ancient schools,"[249] but the number of asterisms (totaling 306) is quite different from the Zhanguo tradition (283).

The circular edge of the chart is graduated in *du*, or solar degrees. Immediately surrounding this is a narrow band divided into twelve equal arcs. Each of these sections is labeled in three separate ways: (1) with the Chinese equivalent of one of the signs of the Western zodiac; (2) with one of the twelve "terrestrial branches," as direction indicators; and (3) with the name of the archaic Chinese state that from ancient times was believed to be governed by the stars in this sector. The implied links between 1 and 3 are curious; there is, of course, no direct correspondence between the signs of the zodiac—which are based on the ecliptic—and any Oriental divisions of the sky, including the twelve Jupiter stations.[250] Additionally, since the chart is equatorial (the customary practice for pre-Jesuit artifacts), the regular spacing of the zodiacal signs represents only a very crude approximation to reality.

Most of the names of the signs of the zodiac have close parallels with the names found in Chinese translations from the Sanskrit of Buddhist sutras made from the sixth century onward (see above). In order we have: White Sheep (Aries), Golden Bull (Taurus), Male and Female (i.e., yin-yang: Gemini), Great Crab (Cancer), Lion (Leo), Two Women (Virgo), Celestial Balance (Libra), Celestial Scorpion (Scorpio), Man and Horse (Sagittarius), Sea Monster (Capricorn), Precious Water Bottle (Aquarius), and Two Fish (Pisces).[251] It seems likely that these

of astronomical instruments in the Chosŏn period), *Tongbang hakchi* 42 (1984): 205–37, esp. 209.

243. *Chŭngbo Munhŏn pigo*, chap. 3 (note 223).

244. *Sŏnjo sillok* (Annals of King Sŏnjo, r. 1567–1608), chap. 5; see Kuksa P'yŏnch'an Wiwŏnhoe (National History Compilation Committee of the Republic of Korea), ed., *Chosŏn wangjo sillok* (Royal annals of Chosŏn), 48 vols. (Seoul: Kuksa P'yŏnch'an Wiwŏnhoe, 1955–58).

245. Historical details are given by Rufus, "Celestial Planisphere," 27 (note 223).

246. See Na, "Chosŏn sidae," 212 (note 242).

247. Jeon, *Science and Technology in Korea*, 28 (note 5).

248. K. L. Pratt of the University of Durham, who on a visit to the King Sejong Memorial Museum photographed both the stele and the framed rubbing, kindly supplied me with copies of his photographs. I am also grateful to Na Ilsŏng of Yonsei University, Seoul, for a further photograph of a rubbing in his possession.

249. See, for example, Rufus, "Astronomy in Korea," fig. 24 (note 220); W. Carl Rufus, "Korea's Cherished Astronomical Chart," *Popular Astronomy* 23 (1915): 193–98, esp. pl. X; W. Carl Rufus and Celia Chao, "A Korean Star Map," *Isis* 35 (1944): 316–26, esp. 316–17 (showing a negative print of a rubbing); and Jeon, *Science and Technology in Korea*, fig 1.3 (note 5).

250. Rufus and Chao, "Korean Star Map," 326 (note 249).

251. The various translations of the Korean names of the twelve zodiacal signs are by Rufus, "Astronomy in Korea," table 3 (note 220).

FIG. 13.32. RUBBING OF A 1687 COPY OF THE 1395 STAR MAP. The stele from which this rubbing was made is reputed to be an accurate copy of a star map engraved on stone in 1395. The 1395 map was based on a rubbing of a Chinese stone chart lost in 670. Calculations yield a date of about 30 B.C.. Diameter of the original: ca. 90 cm. By permission of the King Sejong Memorial Museum, Seoul.

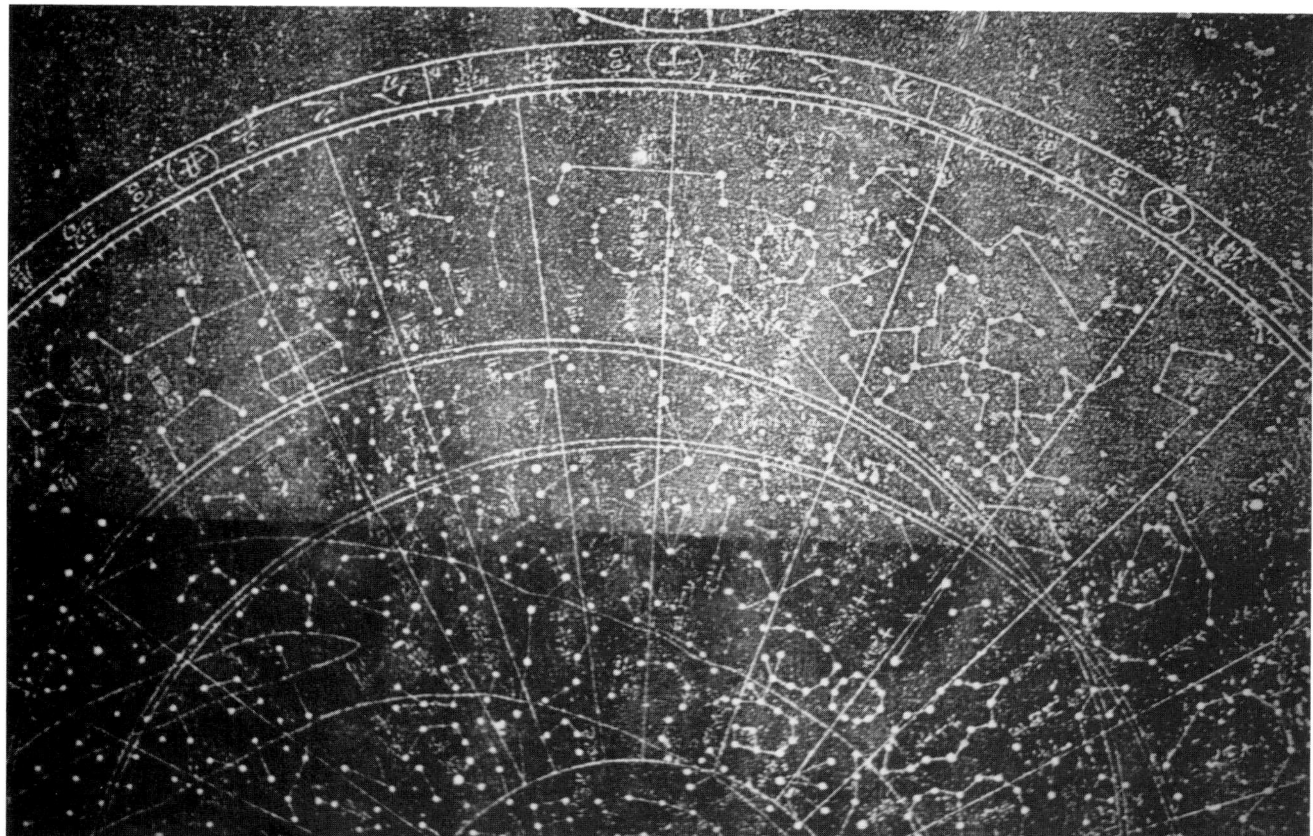

FIG. 13.33. DETAIL OF FIGURE 13.32. Section of chart shown in figure 13.32 with details of the ecliptic, equator, Milky Way, circle of constant visibility, and lunar lodge boundaries, as well as numerous constellations.

King Sejong Memorial Museum, Seoul. Photograph courtesy of K. L. Pratt, Durham University.

names were already inscribed on the 1395 stele. Incorporation of the signs of the zodiac in an officially produced star map is rather surprising. In China we find no parallel on any star map produced by official astronomers, and in any event the Yi rulers did not embrace Buddhist doctrines but were overtly Confucian in their philosophy.

The inscription accompanying the Korean star chart gives several indirect indications of the date of production of the lost original. It is stated that the equinoxes were "in the east, a little preceding the fifth degree of Jue (the first lunar lodge) and in the west a little beyond the fourteenth degree of Kui (the fifteenth *xiu*)." These positions indicate a date between 40 and 20 B.C.[252] Measuring of the location of the autumnal equinox on the star map itself yields a similar date. (The vernal equinox is shown several degrees in error, but this discrepancy arises merely because the ecliptic is depicted as a circle.) In general, the positions of the stars are inaccurately marked. On the assumption of a date about 30 B.C., analysis of the NPDs of twenty selected bright stars measured on the chart indicates a standard error of plus-or-minus five degrees.[253] Although of low precision, these NPDs are definitely more compatible with an ancient than a medieval origin.

An early date for the original chart is also indicated by the table of lunar lodges inscribed immediately below the star map in existing copies. This table lists the number of stars in each *xiu*, followed by the equatorial extent and the NPD of the determinative star. Coordinates are expressed to the nearest *du*. The widths of the various *xiu* represented on the chart itself agree fairly accurately (to within about a degree) with those in the accompanying table. The NPDs of the determinative stars are less carefully plotted, however, with typical errors amounting to several degrees. Rufus implied that King T'aejo's astronomers had included new determinations of these coordinates,[254] but this statement proves incorrect. All of

252. These are my computations, as yet unpublished.

253. These computations are also mine.

254. Rufus, "Astronomy in Korea," 24 (note 220).

FIG. 13.34. PLANISPHERE FOUND IN A JAPANESE JUNK AND NOW IN EDINBURGH. A navigational device that is an accurate copy of a Japanese bronze planisphere produced in 1668 that was in turn a careful copy of the 1395 Korean planisphere.
Diameter of the original: 34.5 cm. ©The Trustees of the National Museums of Scotland 1993 (NMS T1878.37).

the tabular *xiu* widths are in exact accord with those cited in the ancient *Xingjing* (see above). In the case of the NPDs for determinative stars, there is also good general agreement with the *Xingjing*, although the significant number of differences suggests some independence here. Comparison of the recorded NPDs in the table with computed values indicates a date within about a century of the birth of Christ,[255] a result that adequately supports the date deduced from the equinox locations. Although the inscription on the 1395 stele (as found in extant copies) alleges that it "was designed for the future," there seems to be little direct evidence to support this assertion.

A further seventeenth-century reproduction—in bronze—of the 1395 planisphere is preserved in Japan (see the discussion and illustration below, chapter 14). This artifact, known as a *Bundo no kiku*, was produced by the Japanese astronomer Fukushima Kunitaka in 1668. An exact copy of this replica, also in bronze, is now in the Royal Scottish Museum, Edinburgh. Its date of construction is unknown, but it was apparently recovered from a Japanese junk wrecked on an island off the coast of Japan sometime during the previous century (fig.

255. Computations, as yet unpublished, are mine.

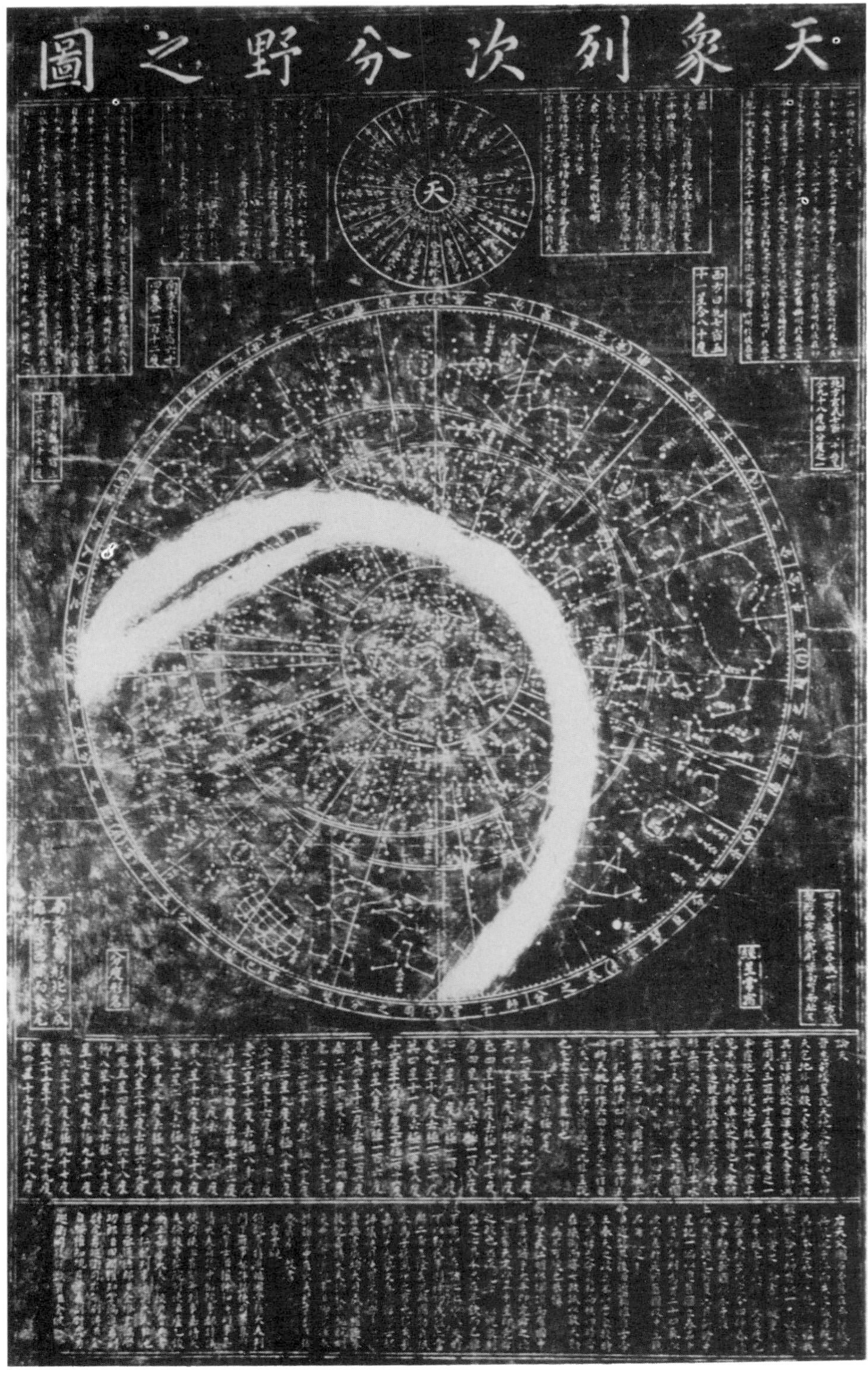

FIG. 13.35. BLOCK PRINT COPY OF THE 1395 STAR MAP. Most prints of this star map show the Milky Way very prominently, as here. The accord with the 1687 stele shown in figure 13.32 is excellent. The text contains a history and description of the chart, as on the stele itself. Size of the copy: unknown. Photograph courtesy of Na Ilsŏng, Yonsei University Observatory, Seoul.

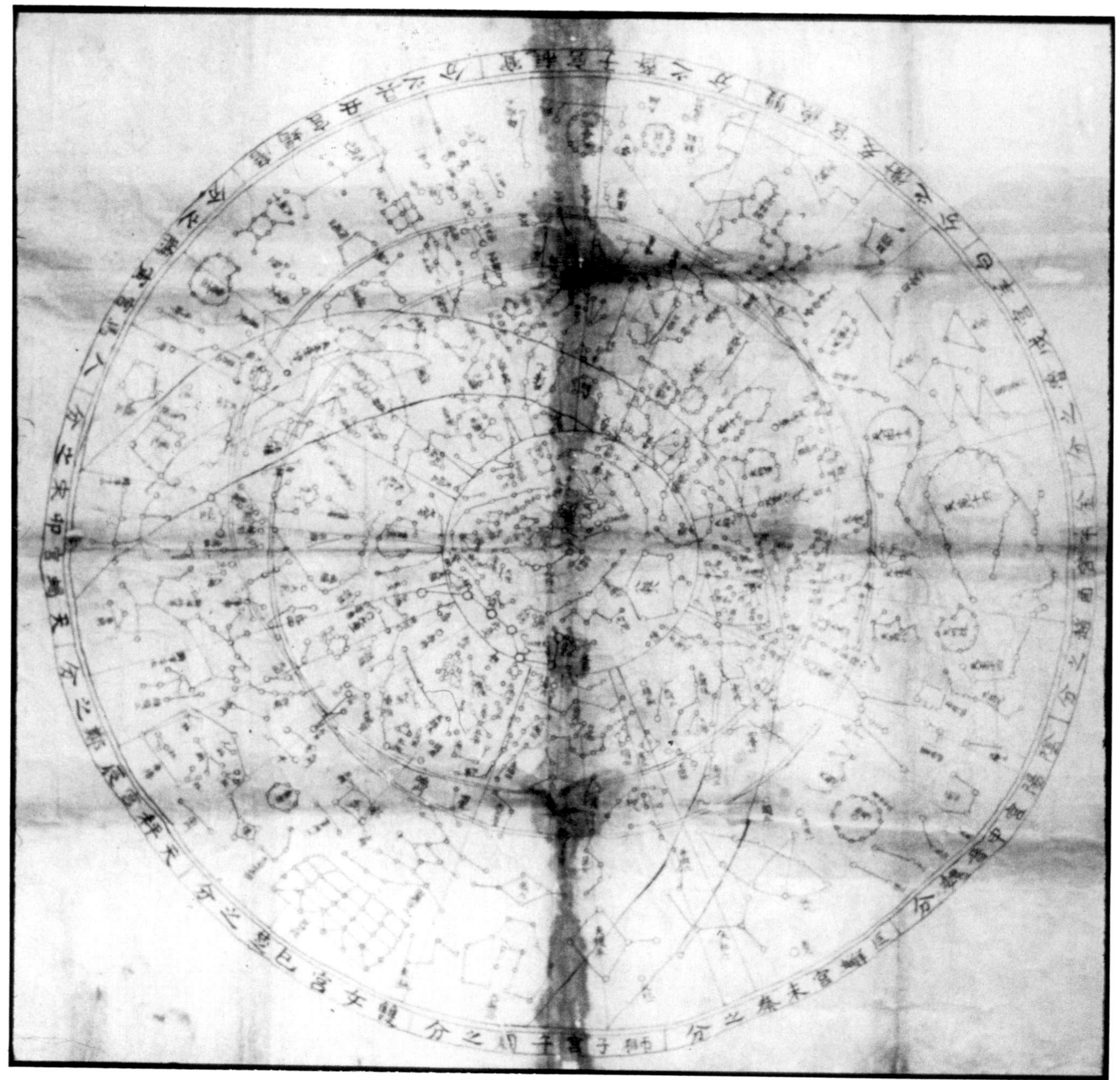

FIG. 13.36. MANUSCRIPT COPY OF THE 1395 STAR MAP. Although the date of this copy is unknown, it is probably late (perhaps nineteenth century). Size of the copy: unknown. Photograph courtesy of Na Ilsŏng, Yonsei University Observatory, Seoul.

13.34).[256] This navigational instrument has two small compasses inset near its edges. Rufus and Chao seem to have been the first to recognize the identity of its astrography with that on the Korean star map.[257] The device is in a fine state of preservation.[258]

The *Bundo no kiku* in the Edinburgh collection has an overall diameter of about thirty-four centimeters; the star chart itself is twenty-four centimeters across. Stars are represented by raised dots, joined into groups.

256. E. B. Knobel, "On a Chinese Planisphere," *Monthly Notices of the Royal Astronomical Society* 69 (1909): 435–45.

257. Rufus and Chao, "Korean Star Map," 317 (note 249).

258. Knobel, "Chinese Planisphere," pls. 17 and 18 (note 256), and F. Richard Stephenson, "Mappe celesti nell'antico Oriente," *L'Astronomia*, no. 98 (1990): 18–27, esp. 22. See also Clark and Stephenson, *Historical Supernovae*, pl. 5 (note 2).

FIG. 13.37. KOREAN FOLDING SCREEN. This colorful screen depicts both the 1395 planisphere (on the right) and a double hemisphere chart produced in China by the Jesuit astronomer Ignatius Kögler in 1723. Size of the original: 230 × 440 cm. By permission of the Whipple Museum of Science, Cambridge.

Although constellation names are absent, the uranography (including circles and *xiu* boundaries) is otherwise an accurate copy of that on the 1395 Korean map.

Several early copies of the 1687 planisphere are still preserved, in a variety of forms. These include: block prints (fig. 13.35); manuscript copies (fig. 13.36); and silk screens. At least six block prints, probably dating from the eighteenth century, are known to be preserved in museum and library collections. These are about the same size as the stele itself and depict the stars as white dots on a black background. The Milky Way is boldly represented—as a white band—but otherwise the prints closely resemble rubbings from the stone. Many hand copies of the stele, some no more than a century old, are preserved.[259]

One example of a silk screen is in the Whipple Museum of Science in Cambridge, England, to which it was presented by a Korean collector. This reproduction, dating from 1755–60 (fig. 13.37), is painted on an ornate eight-panel folding screen, approximately 4.4 by 2.3 meters, that also depicts two star charts of Jesuit origin (see below). It was discussed in detail first by Needham and Lu and more recently by Needham et al.[260] Stars are represented either by red or black dots or by open circles on a buff background and are joined by straight lines into named constellations. Although the astrography differs considerably from that on the Dunhuang charts, there are certain striking similarities. In particular, the distribution of stars marked in each color (red, black, and yellow) on the two maps agrees well, once again recalling the ancient groupings. However, much further historical research would be necessary in order to adequately explain these equivalences.

On the Korean screen, the ecliptic is depicted in yellow, as befits the "Yellow Road," and correspondingly the celestial equator is marked in red. The edges of the lunar lodges are also clearly marked. Close examination shows that this star map is a faithful reproduction of the chart of 1687—it is by no means purely decorative.

259. Information in this paragraph concerning copies 1 and 2 is from Na Ilsŏng (personal communication).

260. Needham and Lu, "Korean Astronomical Screen" (note 137), and Needham et al., *Heavenly Records*, 153–79 (note 5).

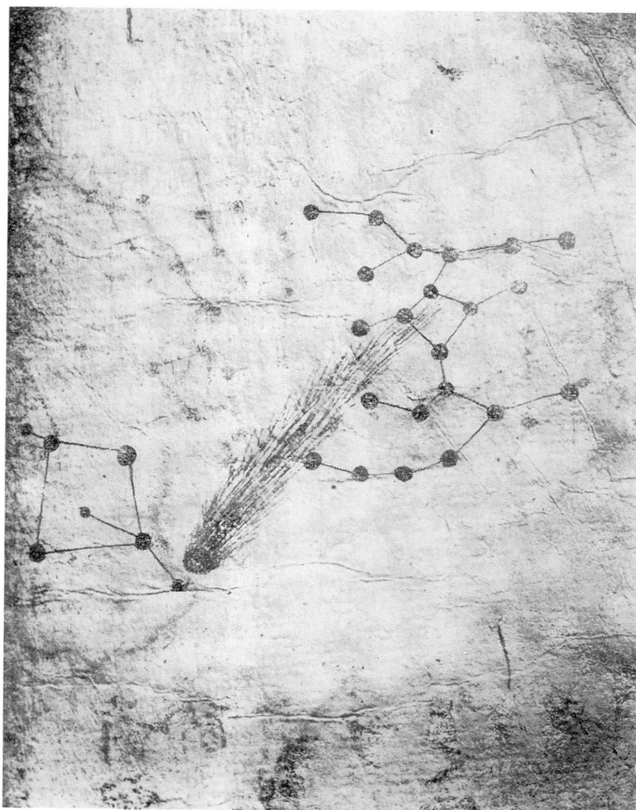

FIG. 13.38. KOREAN SKETCH OF POSITION OF COMET OF 1664. This sketch by the court astronomers is probably typical of many that were produced in Korea and probably China. Alas, virtually all of these have disappeared, including the original of the drawing shown here.
From W. Carl Rufus, "Astronomy in Korea," *Transactions of the Korea Branch of the Royal Asiatic Society* 26 (1936): 4–48, esp. fig. 27.

The existence of such a variety of copies shows that the celestial planisphere of King T'aejo was highly prized for several centuries, but little is known about how the chart was regarded by the official astronomers of the later Yi dynasty or about its influence on Korean celestial cartography. Possibly the stele was valued mainly as a historical relic; it is difficult to imagine that such an archaic representation of the night sky could have fulfilled a serious role in positional astronomy. Apart from errors in the positions of the stars, the visibility of the constellations would have been significantly affected by precession. For example, some constellations marked on the chart would no longer be visible in the latitude of Korea, while others that were not depicted would have come into view. Unfortunately, only minimal information is preserved regarding indigenous uranography during the Yi dynasty. Apart from occasional sketches of individual constellations by the court astronomers—for example, to indicate the position of comets (fig. 13.38)—no other pre-Jesuit star maps or celestial globes are known to be preserved.

The *Chŭngbo Munhŏn pigo* relates that a planisphere was carved in stone in 1433, but no details are provided. Soon afterward, in 1437, King Sejong had a celestial globe installed at his newly constructed royal observatory. At his death in 1450 this observatory is said to have "possessed one of the finest and most complete sets of astronomical instruments in the world."[261] The following description of the celestial globe is found in the official chronicle of the time, the *Sejong sillok* (Annals of King Sejong, r. 1418–50):

> The celestial globe was made of lacquered cloth (on a framework), round as a crossbow-bullet, having a circumference of 10.86 feet [equivalent diameter about 75 cm]. Coordinates were marked on it in celestial degrees [*du*]; the equator was in the middle with the ecliptic crossing it at an angle of a little under 24 degrees. All over the cloth surface were marked the constellations north and south of the equator.[262]

Unfortunately, the record is silent on whether the astrography on the globe was based on new measurements or it was at least partially dependent on the 1395 chart. The celestial globe apparently remained in use for fully a century; after possibly undergoing repairs in 1526, it was replaced by a replica in 1549. However, this new version was destroyed during the Hideyoshi invasion in 1592, when the great stele of King T'aejo was also damaged.[263] In 1601 a further celestial globe was fashioned,[264] but no information is available on its construction or duration of use. Although Korea was a vassal of China during the Yi dynasty, as it had been at earlier periods, there does not appear to be any direct evidence that Chinese star maps or globes reached Yi Korea before the era of the Jesuit astronomers.

The Jesuit Contribution

Celestial cartography in China during the last years of the Ming dynasty and much of the subsequent Qing owed a great deal to the influence of missionaries of the Society of Jesus.[265] Many of these men were skilled astronomers, and they used their knowledge "to arouse the intellectual curiosity of the Chinese and to interest them in the doc-

261. Needham et al., *Heavenly Records*, 94 (note 5).

262. *Sejong sillok*, chap. 77, translated by Needham et al., *Heavenly Records*, 74–75 (note 5).

263. Jeon, *Science and Technology in Korea*, 67–68 (note 5).

264. *Sejong sillok*, chap. 77, translated by Needham et al., *Heavenly Records*, 100 (note 5).

265. This remark, of course, also applies to other branches of astronomy, and to science in general.

trines of the West."[266] Indeed, several Jesuits attained the office of astronomer royal at the Qing court. So great was their impact on celestial cartography in both China and Korea that no significant star chart produced in either country after 1600 (until the spread of modern knowledge in the twentieth century) is free of Jesuit influence. Although no member of the Society of Jesus reached Korea before the twentieth century, Korean ambassadors to China made contact with Western science as transmitted by the Jesuits, and copies of a number of star maps showing European influence found their way to the "Hermit Kingdom" (Korea).

In 1583 the Italian scholar Matteo Ricci became the first member of the Society of Jesus to enter the Chinese mainland. Ricci eventually (in 1601) settled at Beijing, the capital, dying there in 1610. Although not specifically an astronomer, Li Madou (as Ricci became known) profoundly impressed the Chinese with his knowledge of Western astronomy, for example, in eclipse prediction and calendrical science. The era of direct Jesuit influence on the course of Chinese astronomy lasted from the pioneering efforts of Ricci until 1773, when the Society of Jesus was temporarily disbanded by Pope Clement XIV.[267] Later Roman Catholic missionaries held the office of astronomer royal until 1826, but they never matched the achievements of their Jesuit predecessors. By this time, however, Chinese astronomy was irrevocably opened up to Western ideas.

In the field of uranography, the Jesuit astronomers made several important advances over contemporary Chinese methods of mapping the night sky. As well as measuring stellar coordinates with considerably higher precision than had ever been achieved previously in China, the missionaries introduced the first detailed knowledge of stars in the south circumpolar region.[268] They also established the Western system of grouping stars into six classes of brightness (or magnitude), which had its origin in ancient Greece; hitherto Chinese astronomers had shown little concern with the marked range of brightness among the stars visible to the unaided eye. Despite these and other European innovations (e.g., ecliptic coordinates), the Jesuits did not attempt to replace the traditional Chinese asterisms with Western constellations, although they charted many additional stars. The first telescope constructed in China was made by Jesuits in 1631, and several similar instruments were brought from Europe soon afterward,[269] but the telescope never found favor among traditional Chinese and Korean official astronomers, and—like Johannes Hevelius in Europe—the Jesuits themselves preferred a sighting tube rather than a telescope for measuring star coordinates.

Matteo Ricci is known to have constructed a number of astronomical spheres and globes made of copper and iron,[270] but neither the instruments themselves nor descriptions of them appear to have survived. Ricci several times sent messages to Rome asking that astronomers be sent to China, particularly to help reform the calendar, which had last been revised by Guo Shoujing as long ago as 1280. It was not until 1630 (twenty years after Ricci's death) that his hope was realized when the German Johann Adam Schall von Bell (1592–1666, Chinese name Tang Ruowang) and the Italian Giacomo Rho (1593–1638, Chinese name Luo Yagu), two Jesuits skilled in astronomy, reached Beijing.

Not long afterward, the Christian convert Xu Guangqi (1562–1633), who was director of calendar reform in the Ming government, published several small star maps and presented them to the emperor Sizong (1628–45). Xu Guangqi (also known as Paul Xu), a distinguished scholar, had been a close friend of Ricci's. The charts he produced incorporated Western innovations, but—as Xu himself appreciated—they were too small to represent the stars accurately;[271] the largest was only about fifty centimeters in diameter. Original prints of two of these charts, on paper, are preserved in the Vatican Library, titled *Jianjie zong xingtu* (General map of the visible stars) and *Huangdao zong xingtu* (Two general maps of the stars relative to the ecliptic).[272] A third chart by Xu bore the caption "Chidao liang zong xingtu" (Two general maps of the visible stars relative to the equator). This is printed in the encyclopedia *Chongzhen lishu* (Calendar treatise of the Chongzhen reign period, 1635), produced by Schall von Bell and his Jesuit colleagues. A fourth sectional map is titled *Huangdao ershifen xingtu* (Map of the stars relative to the ecliptic in twenty parts) and is preserved at the Palace Museum in Beijing. Brief descriptions of the first three charts follow.[273]

The largest chart, *Jianjie zong xingtu*, has an external diameter of fifty-seven centimeters and an internal diameter of fifty-four centimeters.[274] It depicts the whole of the visible sky on a polar (stereographic) projection down

266. This apposite quotation is from Pasquale M. d'Elia, "The Double Stellar Hemisphere of Johann Schall von Bell S.J.," *Monumenta Serica* 18 (1959): 328–59, quotation on 328.

267. The order was reestablished by Pope Pius VII in 1814.

268. Previously, only a few southern constellations had been mapped by the Chinese (see above).

269. Pasquale M. d'Elia, *Galileo in China: Relations through the Roman College between Galileo and the Jesuit Scientist-Missionaries (1610–1640)*, trans. Rufus Suter and Matthew Sciascia (Cambridge: Harvard University Press, 1960), 41.

270. *China in the Sixteenth Century*, 169 (note 198).

271. D'Elia, "Double Stellar Hemisphere," 347 (note 266).

272. Biblioteca Apostolica Vaticana, MS. Barberini, Orient. 151/1c, 151/1d (a copy of 1c), and 151/1e. A print of the first of these charts is also in the Bibliothèque Nationale, Paris.

273. See also Pan, *Zhongguo hengxing guance shi*, pls. 58–60 (note 7).

274. D'Elia, "Double Stellar Hemisphere," 338 (note 266).

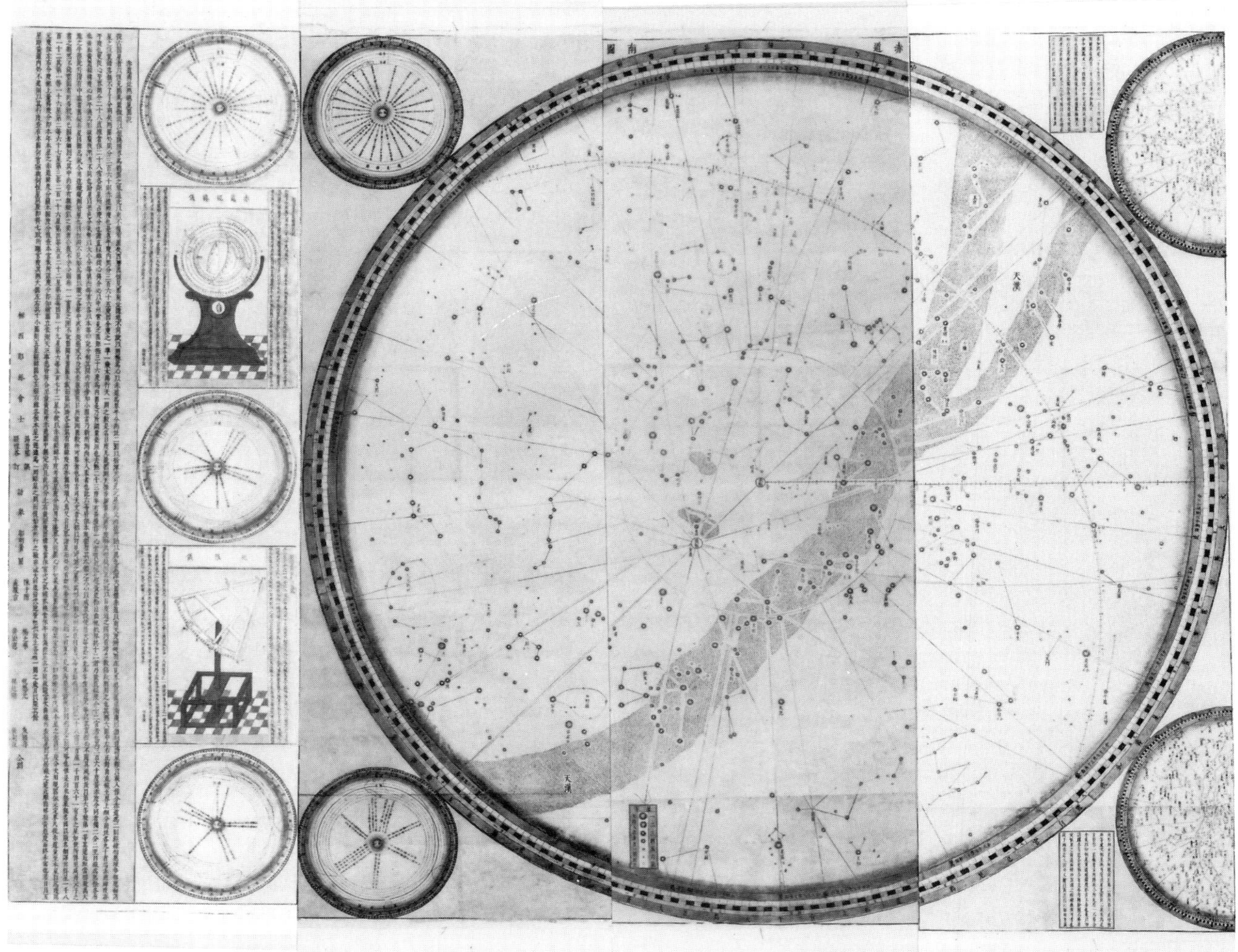

FIG. 13.39. SECTION OF SCHALL VON BELL STAR MAP, 1634, SHOWING STARS SOUTH OF THE CELESTIAL EQUATOR. This is part of a print similar to that shown in figure 13.40. Stars are grouped into six magnitudes on this equatorial chart, which is on a polar (stereographic) projection. Note the ecliptic pole and the boundaries of the zodiacal signs (shown as arcs).

to about 50° south declination and closely resembles a traditional Chinese chart. Stars are grouped into the characteristic patterns of antiquity, and there is little or no attempt to denote magnitude. The Milky Way is clearly marked. Constellations are individually named, but information on the chart is so crowded—especially toward the center—that the whole effort is of little practical use. The celestial equator, ecliptic, and circle of constant visibility are shown, along with the boundaries of the lunar lodges (denoted by radial straight lines extending from the circle of constant visibility to the edge of the chart). There appears to be no information on the number of stars depicted, but it is at least comparable with the standard tally of approximately 1,460.

The two smaller charts each depict the whole of the night sky in two separate hemispheres, bounded either by the ecliptic or by the celestial equator. Precise dimensions are available for the Vatican print: each hemisphere has an external diameter of twenty-nine centimeters and internal diameter of twenty-two centimeters.[275] The maps of the southern sky are probably the earliest surviving examples from China that delineate the full course of the Milky Way (as well as showing the Magellanic Clouds) and that represent more than a handful of constellations in the south circumpolar region; some twenty asterisms are depicted in this zone. The lunar lodge boundaries are not marked. Each hemisphere is drawn on a polar (stereographic) projection, a refinement that seems scarcely necessary on such a small scale.

The hemispheres on the *Huangdao zong xingtu* are

275. D'Elia, "Double Stellar Hemisphere," 338 (note 266).

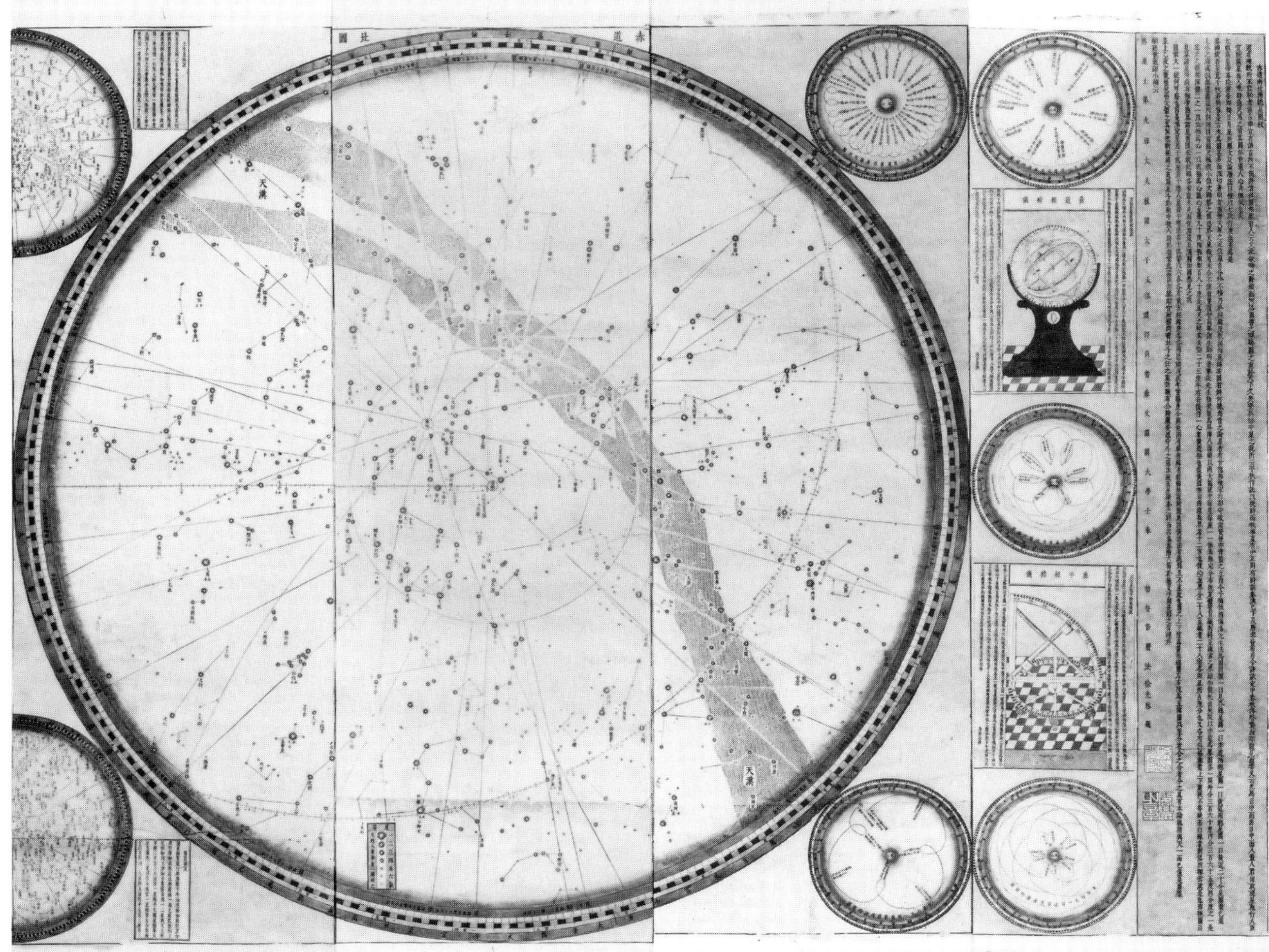

Size of the entire original: ca. 170 × 450 cm. By permission of the Biblioteca Apostolica Vaticana, Rome (MS. Barberini, Orient. 149). Photograph courtesy of Pan Nai.

centered on the appropriate (north or south) ecliptic pole, and each extends to the ecliptic. The celestial equator is not shown. Although stars are grouped into the traditional asterisms, they are represented by symbols of six different sizes to indicate magnitude. Certain nebulas (*qi*) are also marked. The limits of the twelve zodiacal signs are delineated as equally spaced radial lines extending from either pole to the ecliptic.

The astrography on the "Chidao liang zong xingtu" closely resembles that on the ecliptic maps. These maps are centered on the appropriate (north or south) celestial pole, and each extends to the celestial equator. The ecliptic is shown, and the boundaries of the zodiacal signs are depicted as arcs radiating from the ecliptic pole to the edge of each chart. In addition, each chart is divided into twelve equal sectors by radial lines extending from either celestial pole to the equator. Two of these lines pass through the equinoctial points.

In 1628, Xu Guangqi revised the determinative stars of three lunar lodges: Zuixi, Kui, and Mao. In each case he selected brighter neighboring stars. The reference star of Zuixi was changed from φ' Ori to λ Ori, that of Kui from ζ And to η And, and that of Mao from 17 Tau to η Tau. No similar alteration appears to have occurred at any previous time in Chinese history. However, Xu did nothing to solve the difficulties arising from the disappearance of the lodge Zuixi. Always the narrowest lodge, this had gradually narrowed owing to precession and had reached zero width during the Yuan. Soon after Xu's revision the situation was rectified by Schall von Bell,[276]

276. Pan, *Zhongguo hengxing guance shi*, 348 (note 7).

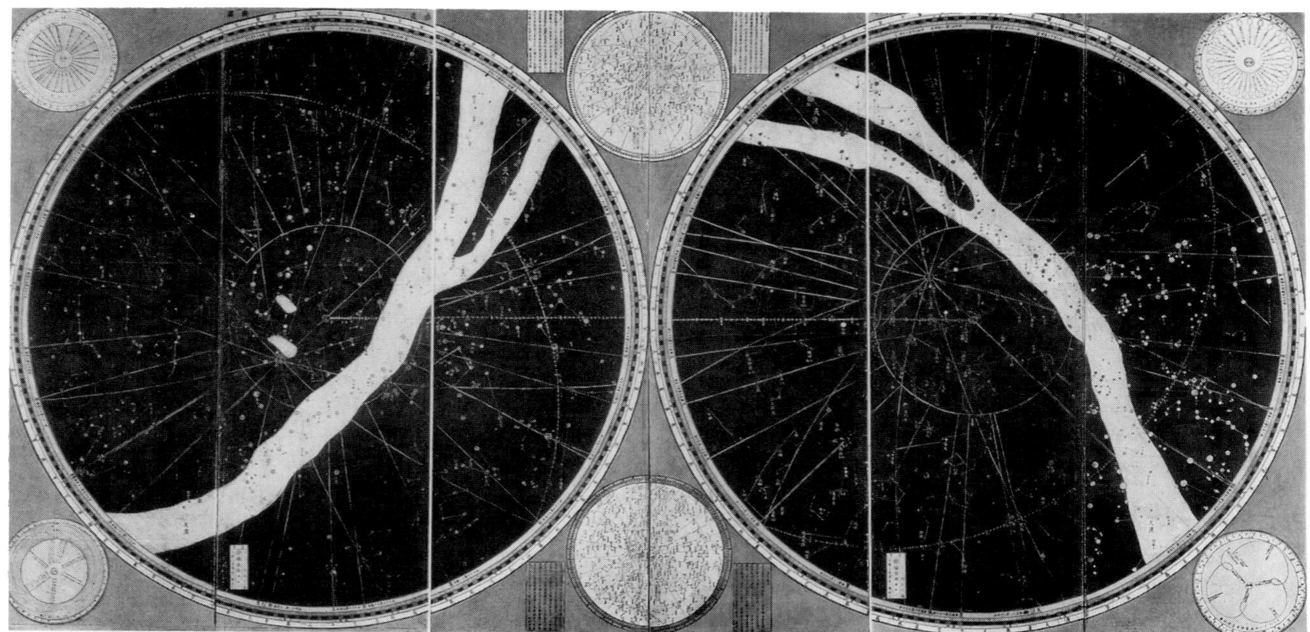

FIG. 13.40. 1634 STAR MAP BY SCHALL VON BELL, NOW IN BEIJING. This colorful print (gold stars on a blue background) is said to have been presented to the last Ming emperor. In all, 1,812 stars are depicted, as well as the Milky Way and Magellanic Clouds.

Size of the original: 1.6 × 4.2 m. First Historical Archives of China, Palace Museum, Beijing. Photograph from Zhongguo Shehui Kexueyuan Kaogu Yanjiusuo, *Zhongguo gudai tianwen wenwu tuji*, 16.

who took the bold step of reversing the order of the lodges Zuixi and Shen so that Zuixi now became the twentieth *xiu* and Shen the twenty-first. As a result, the width of Shen was reduced from eleven degrees, forty-four minutes to only twenty-four minutes, while Zuixi acquired an angular extent of eleven degrees, twenty-four minutes. The long-term effect of precession was to slowly increase the width of Shen so that there was no risk of its disappearing.

Not long before his death in 1633, Xu Guangqi initiated the production of a larger-scale map of the night sky, showing both hemispheres. This extensive project, which involved redetermining the coordinates of a large number of stars, was undertaken by Schall von Bell, Rho, and several Chinese scholars, including Xu himself. It was completed the following year. The chart was engraved on wooden blocks in eight sections, each measuring 1.6 by 0.5 meters (fig. 13.39). Composite prints, 4.2 by 1.6 meters, were suitable for screen or wall mounting. In an accompanying preface, the title of the work is given as *Huangdao nanbei liang zong xingtu* (Two general maps of the stars south and north of the equator). A print on paper in eight separate sections is preserved in Beijing (fig. 13.40). This is reported to have been presented to the emperor Sizong (the last Ming ruler) by Schall von Bell.[277] Other copies of similar dimensions, printed on paper, are in the Vatican Library (two examples), at the Bibliothèque Nationale, Paris (two examples), and at the Consiglio Nazionale di Ricerche in Bologna.[278] The Beijing print depicts the background sky as dark blue and the Milky Way as white stippled with black; individual stars are gilded. One of the Vatican prints is attractively colored: the sky is shown in pale blue, and the Milky Way and stars are gilded. This copy was originally presented to the Grand Secretariat of the last Ming emperor. Both of these presentation copies are in a fine state of preservation. The other surviving prints in Rome, Paris, and Bologna are more basic; they are devoid of color and are on ordinary Chinese paper. The Paris and Bologna prints are in good condition, but the monochrome Vatican version is much faded.[279]

277. However, Pan, *Zhongguo hengxing guance shi*, 354–55 (note 7), asserts that the print now in the Palace Museum is of early Qing origin (ca. 1650).

278. D'Elia, "Double Stellar Hemisphere," 337 (note 266), Pan, *Zhongguo hengxing guance shi*, 354 (note 7), and Pan Nai, "Shiqi shiji chu shijie shouqu yizhi di heng xingtu" (A unique star map of the early seventeenth century), *Kexue* 42 (1990): 275–80.

279. On the Beijing version and the presentation copy in the Vatican, see *Zhongguo gudai tianwen wenwu tuji*, 16 and 101 (note 6), and d'Elia, "Double Stellar Hemisphere," pls. I and II (note 266). Pan is of the opinion that the Vatican presentation copy is the oldest surviving print. For example, it contains the names of all ten compilers, whereas the Beijing version carries only the name of Schall von Bell. Pan also bases his argument on coloring and other details, see Pan, "Shiqi shiji

In general, the various prints that are now in Beijing, Rome, Paris, and Bologna differ mainly in medium and color. On each example, the inner six sections are mainly taken up by two circular maps of the constellations, each 1.55 meters in diameter; one of these covers the Northern Hemisphere and the other the Southern Hemisphere. Explanatory prefaces, written by Xu Guangqi and Schall von Bell, are contained in the two outer sections. Additional small diagrams occupy much of the remaining space. Most of these illustrations depict planetary movements and certain astronomical instruments, but two small planispheres—each forty-three centimeters in diameter—chart the stars visible from northern China. One of these maps is centered on the celestial pole, the other on the ecliptic pole.

The two large hemispheres extend from either the north or the south celestial pole to the equator, so that the whole of the sky is represented. In his accompanying preface, Xu Guangqi explains the reason for including south polar constellations:

> In the southern hemisphere beyond the visible stars there are the stars in the zone of invisibility near the pole. These stars do not figure on the old Maps. But, though they are not directly visible from our various provinces, they are all visible from the coast down to Malacca. These parts belong to the sphere of sovereignty of our country; how can the stars visible there be excluded?[280]

Each chart is accurately drawn on a polar (stereographic) projection for the epoch 1628. The basic form resembles the twin equatorial maps of Xu Guangqi. Thus the stars are grouped into traditional Chinese asterisms, and the full circuit of the Milky Way is shown (together with the Magellanic Clouds). The ecliptic is marked, at declination close to 23.5°, and the boundaries of the twelve signs of the zodiac are represented by arcs extending from the ecliptic pole to the edge of each chart. Stars are classified into six magnitudes, according to the size of the symbol. In addition, the circles of constant visibility and invisibility (at 36° north and south declination) are depicted, and the boundaries of the lunar lodges are shown as radial lines extending to the edge of the chart. The periphery of each hemisphere is graduated into both ordinary degrees and *du*. A number of telescopic nebulas are also represented.

In all, 1,812 naked-eye stars are depicted, considerably more than the customary 1,460 or so shown on indigenous Chinese charts. In his accompanying explanation Schall von Bell states that of the total number of stars represented, 16 are of the first magnitude, 67 of the second, 216 of the third, 522 of the fourth, 419 of the fifth, and 572 of the sixth. Most of the additional stars are included in the region of sky visible from northern China, and he emphasizes that earlier Chinese star charts were far from complete for this zone. The remaining excess stars (126) are grouped in twenty-three asterisms in the south circumpolar region. Schall von Bell remarked that "because hitherto [the stars] were not combined into figures, they bore no name; therefore words transliterated from their original [Western] names have been used here."[281] In practice, some of the Chinese names of these groups are direct translations of their Western equivalents: for example, Fire Bird (Phoenix) and Triangular Shape (Triangulum). There are several obvious differences, however.

Unpublished measurements I have made on the two charts reveal that the stars are accurately positioned—typically to within a small fraction of a degree. The whole work was a remarkable pioneering effort; it was undoubtedly the most complete and accurate representation of the night sky produced in China up to that date. Later charts, compiled during the Qing dynasty, display even more stars, but the compilation by Schall von Bell is truly a landmark in Chinese astral cartography.

When the Manchus conquered China in 1644, Schall von Bell became the first astronomer royal of the new Qing dynasty. He was deposed twenty years later, and a Chinese astronomer was appointed in his place. Schall von Bell died in 1666. Soon afterward it became apparent that his successor lacked competence, and in 1667 the Belgian Jesuit Ferdinand Verbiest (1623–88, Chinese name Nan Huairen) became the new astronomer royal, a position he held until his death in 1688. Roman Catholic missionaries continued to serve as astronomer royal until 1826, when they were expelled by the emperor Xuan Zong (r. 1821–50) as part of a general suppression of Christianity in China.

By the Qing dynasty, the number of surviving star maps and celestial globes escalated to such a degree that it would take a separate essay to describe them in any detail. However, among major Qing artifacts I might mention maps and globes produced as the result of (a) a revision of Schall von Bell's catalog in 1672–73 by Verbiest; (b) a detailed sky survey in 1744–52 by Ignatius Kögler (1680–1746, Chinese name Dai Jinxian) and his successors; and (c) a further survey in 1842–45 by native Chinese astronomers.

Verbiest made revised measurements of star positions and also added a small number of previously uncharted faint stars. His revised catalog listed 1,876 stars visible

chu shijie shouqu yizhi di heng xingtu" (note 278). I am grateful to the Consiglio Nazionale di Ricerche for providing large-scale photographs of the print that is on display at their Bologna office.

280. Translated by d'Elia, "Double Stellar Hemisphere," 348 (note 266).

281. Translated by d'Elia, "Double Stellar Hemisphere," 356 (note 266).

FIG. 13.41. FERDINAND VERBIEST WITH HIS CELESTIAL GLOBE. Ferdinand Verbiest, who was astronomer royal in China from 1667 to 1688, is shown dressed as a Chinese official in this mid-nineteenth-century Japanese print. Also shown are his sextant and celestial globe.
Size of the original: 37.5 × 26 cm. By permission of the British Museum, London.

to the unaided eye. An investigation of the accuracy of the coordinates of the determinant stars of the twenty-eight *xiu* indicates remarkably high precision. Positional errors seldom exceeded one arc minute.[282] The results of Verbiest's work, including detailed maps of the whole sky, were published in 1674 in his *Yixiangzhi* (Description of astronomical instruments). In the previous year, he had cast in bronze a large celestial globe displaying these stars. This globe, "six *chi*" (roughly 1.5 m) in diameter, was one of the many new instruments with which Verbiest equipped the imperial observatory at Beijing. All these instruments were constructed for unaided-eye observations using sighting tubes. After the Boxer Rebellion in 1900, the celestial globe and four other instruments were transported to Berlin, where they were placed in the royal garden at Potsdam Palace. They were not returned to China until 1921. They are still in almost pristine condition after more than three centuries in the open air and can be viewed at their original site—the Ancient Observatory in Beijing (figs. 13.41 and 13.42).[283]

The extensive survey begun in 1744 under Kögler's direction took eight years. During that time, as many as 3,083 stars in three hundred constellations were charted. At the commencement of the project, Kögler was assisted by another Jesuit, August von Hallerstein (1703–74). After Kögler's death in 1746, von Hallerstein succeeded him as Qing astronomer royal, and two other Jesuits, Anton Gogeisl (1701–71) and Felix da Rocha (1713–81), helped him complete the work. The star catalog and associated astral charts were published in 1757 in the *Yixiang kaocheng* (Treatise on astronomical instruments). These accurately drawn equatorial charts are on a polar (equidistant) projection. Rather surprisingly, there is no attempt to represent magnitude; all stars are denoted by dots of equal size. Excellent replicas of these maps were published near the turn of the present century (figs. 13.43 and 13.44).[284] In 1723 Kögler had produced an ecliptic star map showing both hemispheres. Several copies of this work are known to exist.[285]

A new survey begun in 1842, although undertaken by Chinese astronomers, still made use of the old Jesuit instruments. This task, under the direction of Jing Zheng, lasted two and a half years (until 1845), and all together 3,240 stars were charted. It was published as the *Yixiang kaocheng xupian* (Sequel to the Treatise on astronomical instruments) and contains detailed equatorial star maps for both hemispheres drawn on a polar stereographic projection.[286]

The last significant example of Qing celestial cartography dates from 1903, only a few years before the downfall of the dynasty. This is in the form of a large bronze celestial globe (.96 meter in diameter) that displays 1449 stars in the traditional constellations (fig. 13.45). It was built to replace the globe constructed by Verbiest that had been transported to Germany in 1900. The Qing globe is still in excellent condition and can be viewed at Purple Mountain Observatory, Nanjing. On the foundation of the republic, only eight years after the globe was installed, the way was clear for the introduction of astral charts depicting only the Western constellations.

There seem to have been few significant astronomical

282. Pan, *Zhongguo hengxing guance shi*, 381 (note 7).

283. An excellent photograph is published in *Zhongguo gudai tianwen wenwu tuji*, 105 (note 6).

284. Tsutsihashi and Chevalier, "Catalogue d'étoiles," D1-D16 (note 4).

285. For example, Needham et al. have published a useful photograph of an engraving in a private London collection; see *Heavenly Records*, fig. 5.6 (note 5).

286. The *Yixiang kaocheng xupian* was published in Beijing, ca. 1845.

FIG. 13.42. VERBIEST'S CELESTIAL GLOBE IN BEIJING. One of the many instruments cast by Verbiest that can still be seen at the Ancient Observatory, Beijing, this celestial globe is still in almost pristine condition (above and right).
Diameter of the original: ca. 1.5 m. Ancient Observatory, Beijing. Photographs from Zhongguo Shehui Kexueyuan Kaogu Yanjiusuo, *Zhongguo gudai tianwen wenwu tuji*, 104 and 105.

contacts between the Jesuit missionaries and Koreans during the Ming dynasty. In 1631, however, not long before the end of the Ming, the Yi ambassador Chŏng Tuwŏn returned to Korea from the Ming court with a number of books on astronomy and several scientific instruments. These acquisitions included a telescope, presented by the Portuguese Jesuit João Rodrigues, and an astronomical chart.[287] Shortly after the demise of the Ming dynasty in 1644, Crown Prince Sohyŏn of Korea, who had been held hostage at the Ming court, returned to his homeland bearing a number of gifts from Schall von Bell, including a celestial globe.[288] Not long afterward, in 1648, another Korean named Song Inyong stud-

287. Donald L. Baker, "Jesuit Science through Korean Eyes," *Journal of Korean Studies* 4 (1982–83): 207–39, esp. 219–20.

288. Needham et al., *Heavenly Records*, 178 (note 5).

赤道北恆星圖

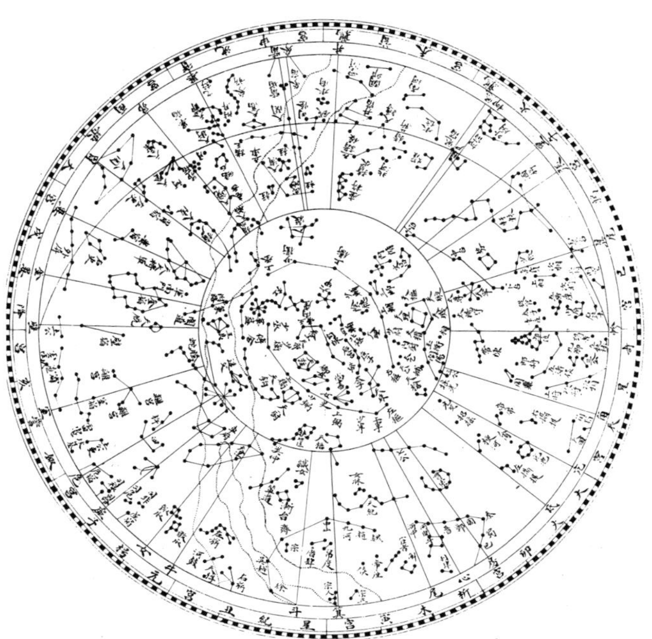

赤道南恆星圖

FIG. 13.43. REPLICA OF KÖGLER/VON HALLERSTEIN STAR MAP, 1757 (NORTHERN HEMISPHERE). This print forms a unit with figure 13.44 and depicts the night sky north of the celestial equator according to von Hallerstein. Both charts are produced for latitude 40°N, the latitude of Beijing. From P. Tsutsihashi and Stanislas Chevalier, "Catalogue d'étoiles observées à Pé-kin sous l'empereur K'ien-long (XVIIIe siècle)," *Annales de l'Observatoire Astronomique de Zô-sè (Chine)* 7 (1911): I-D105, plates between IV and V.

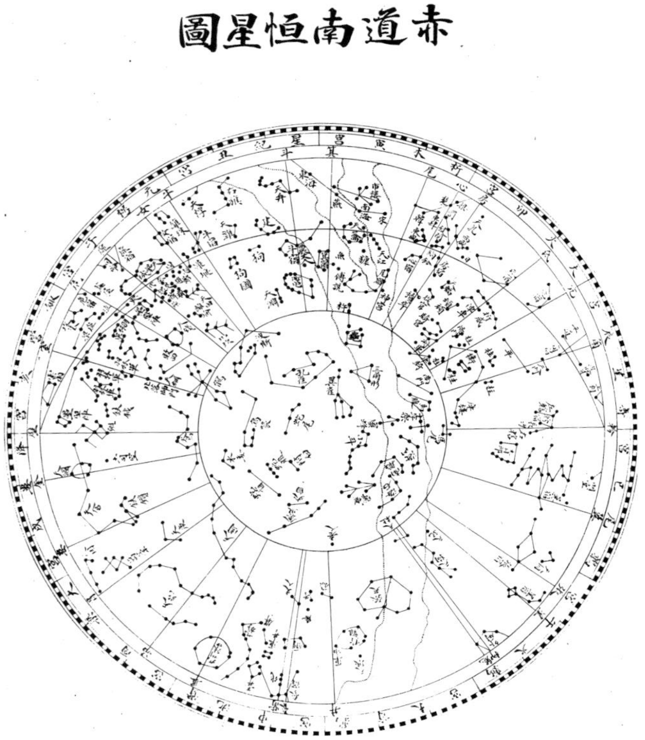

FIG. 13.44. REPLICA OF KÖGLER/VON HALLERSTEIN STAR MAP, 1752 (SOUTHERN HEMISPHERE). This is a print of one of two charts produced in 1757 by Jesuits including August von Hallerstein (see also figure 13.43). Unlike other Jesuit charts, there is no attempt to represent stellar magnitude. This equatorial chart is on a polar (equidistant) projection. From P. Tsutsihashi and Stanislas Chevalier, "Catalogue d'étoiles observées à Pé-kin sous l'empereur K'ien-long (XVIIIe siècle)," *Annales de l'Observatoire Astronomique de Zô-sè (Chine)* 7 (1911): I-D105, plates between IV and V.

ied under Schall von Bell, and he brought back a large astronomical chart.[289] Unfortunately, nothing is known about the construction of either the globe or the charts acquired by various Korean travelers. Many years later, in 1708, the Yi Bureau of Astronomy produced a replica of the 1634 chart of Schall von Bell and presented it to King Sukchong. Like its Chinese equivalent, it is known to have displayed 1,812 stars.[290] Regrettably, no copy can be traced today.

A block print of a late eighteenth-century planisphere of unusual form is preserved in Seoul. This chart, titled *Honch'ŏn chŏndo* (Complete map of the celestial sphere), depicts the night sky visible from Korea on a polar (equidistant) projection.[291] Although the accompanying text states that there are 1,449 stars in 336 constellations, whereas the number of constantly invisible stars around the South Pole amounts to 121 in 33 constellations, the planisphere itself shows little evidence of Jesuit influence. For example, the south circumpolar region is not displayed (despite the remarks in the text), and there is little or no attempt to discriminate between stars of different brightness. The whole chart is divided into twelve equal sectors separated by radial lines that extend from the center of the chart to its edge. At the periphery these sectors are labeled as *ci*, but the positions are only approximate; two of these lines are incorrectly shown as passing through the equinoctial points.

An unusual feature—also found on a late sixteenth-century Ming astral map—is the representation of both the ecliptic and the celestial equator as offset circles. The centers of each of these circles are thirteen degrees from the center of the chart (which also corresponds to the center of the circle of constant visibility), yet the circles intersect at two points exactly 180 degrees apart (i.e., at

289. Needham et al., *Heavenly Records*, 178 (note 5).

290. *Chŭngbo Munhŏn pigo*, chap. 3 (note 223).

291. For a photograph, see Jeon, *Science and Technology in Korea*, fig. 1.4 (note 5).

FIG. 13.45. QING CELESTIAL GLOBE, 1903, AT NANJING. The last important example of Qing uranography, this large celestial globe was cast in bronze only a few years before the dynasty came to an end. It displays 1,449 stars in the traditional Chinese constellations. Diameter of the original: ca. 100 cm. Purple Mountain Observatory, Nanjing. Photograph courtesy of F. Richard Stephenson.

the equinoxes). As in the case of the Ming planisphere there is clear evidence of distortion of the constellations to accommodate this device.

Several Korean copies of Kögler's 1723 star map are preserved, two of them on screens.[292] One of these screens, dating from 1755–60, is now in Cambridge, England. This illustrates, in attractive colors, both the Kögler artifact and the 1395 planisphere of King T'aejo.[293] A block print of Kögler's map was made as late as 1834.[294] By this date, although a certain amount of nostalgia still surrounded the medieval planisphere of King T'aejo, serious Korean astrography was thoroughly influenced by Western innovations.

Concluding Remarks

Although it can be established that several constellations were recognized in China by the late second millennium B.C., knowledge of the early development of Chinese uranography is still fragmentary. It is convenient for us to divide the history of celestial mapping into four discrete eras. In the earliest of these periods, which extends from about 1300 B.C. to 100 B.C., there are no surviving star maps or catalogs. No more than about thirty constellation names are extant from this long interval: little more than the lunar lodges and the Dipper. In the second period, between about 100 B.C. and A.D. 700, there is evidence of extensive astral cartography, but in general existing star charts from this period portray only a few constellations. A number of important celestial maps have survived from the third period, between about 700 and 1600, especially from the later half of this interval. Finally, from about 1600 onward, all star charts of any significance reveal Western influence. The Jesuit astronomers introduced European techniques of mapping the sky, although they did not attempt to supplant the traditional Chinese asterisms with Western constellations.

In the ancient period, perhaps the most important recent discovery has been a list of all twenty-eight *xiu* inscribed on a chest dating from about 433 B.C. This is the earliest date at which the existence of the twenty-eight lodges can be established as an entity. Although arguments based on precession tend to suggest a much earlier date for the origin of the *xiu* (the third millennium B.C.), they remain unsupported by documentary evidence. It may well be that further archaeological excavations will shed new light on this or other problems. Archaeological discoveries of astronomical importance have been very haphazard, however, and this pattern seems likely to continue, at least in the near future.

Although few star maps survive from the period between about 100 B.C. and A.D. 700, historical records assert that many charts and celestial globes were produced at this time, and their loss is to be lamented. A number of contemporary star catalogs (notably that preserved in the *Xingjing*) and constellation lists are extant, and these indicate a high level of attainment reached by astral cartography. Replicas (several times removed from the original) of a Chinese star chart of the first century B.C. appear to be preserved in Korea. The configurations of some of the constellations on these copies (the earliest dating from 1395) differ considerably from those on medieval Chinese charts. A detailed investigation seems to be a matter of some urgency. The 1395 stele was held in such regard in Korea that all surviving Korean star maps from the pre-Jesuit period are replicas of it.

The Suzhou planisphere (engraved on stone in 1247) provides direct evidence of the considerable achievements of Song celestial cartography. Other notable star maps of the period, although now existing only in late copies, were originally produced by Su Song in 1094. These were the earliest known star maps to be printed in any part of the world. A colored star chart found at Dunhuang, possibly dating from the eighth century, although crude, is the sole survivor of its era. This chart displays the stars in three colors, recalling the groupings of antiquity. The groupings on the Dunhuang and Su Song star maps have much in common, and careful comparison between them may well shed new light on the whole question of the traditions ascribed to the "ancient schools." Unfortunately, a Ming copy of a celestial globe cast by the great Yuan astronomer Guo Shoujing disappeared early in the present century. If its whereabouts can be traced, it could provide an important missing link in the study of medieval Chinese celestial cartography.

Several of the star maps produced by Jesuit astronomers in China have been extensively studied; there is ample evidence that these were produced with consummate skill. As yet, no detailed investigations of several important Qing artifacts have appeared. These items include the bronze celestial globe cast by Ferdinand Verbiest in 1673 (now at the Ancient Observatory in Beijing), several colorful astral charts (preserved at the First Historical Archive of China in in the Palace Museum, Beijing), and the bronze globe of the stars cast as recently as 1903 (displayed at Purple Mountain Observatory). Evidently, much can still be written on Qing uranography.

292. Needham et al., *Heavenly Records*, 159–69 and 175 (note 5).

293. Good illustrations are in Needham et al., *Heavenly Records*, figs. 5.1, 5.3, and 5.5 (note 5).

294. Jeon, *Science and Technology in Korea*, 31 (note 5).

14 · Japanese Celestial Cartography before the Meiji Period

KAZUHIKO MIYAJIMA

CELESTIAL MAPS IN ANTIQUITY AND THE MIDDLE AGES

Although there are few materials to tell us what kind of celestial maps there were in ancient and medieval Japan, we may assume that most were of Chinese derivation, either brought to Japan or copied there in manuscript form.[1] We have two types of evidence on which to base this claim. One is the archaeological evidence at Takamatsuzuka, a tomb in Nara Prefecture, and the other includes more standard maps and documents.

THE STAR CHART ON THE CEILING OF TAKAMATSUZUKA

The map of the heavens on the ceiling of the late seventh- or early eighth-century Takamatsuzuka is valuable in that it is the earliest known example of Japanese celestial cartography.[2] Evidence of Chinese influence is found in the twenty-eight lunar lodges, which were important in Chinese and, later, Japanese astronomy. The twenty-eight lunar lodges form a square, in the middle of which are the constellations Shiho (Four supports) and Hokkyoku (Stars of the North Pole) (fig. 14.1).[3]

In addition to the celestial ceiling, the walls of the tomb are covered with colorful paintings. They show the four sacred animals (*shi shinjū*) that are the tutelary gods of the four directions in Chinese tradition: *seiryū* (blue dragon) on the eastern wall, presumably *suzaku* (red bird) on the southern wall (almost completely ruined when the tomb was entered in the fourteenth century), *byakko* (white tiger) on the western wall, and *genbu* (turtle and snake interlocked) on the northern wall across from the entrance. Paintings of the sun on the eastern wall and the moon on the western wall use gold and silver foil, and beneath them a number of horizontal, parallel red lines represent clouds or mist. Sporadically placed along these lines are blue or green mountain signs, most likely an ancient way to show that the sun and the moon are seen beyond the mountains and clouds.

Rather than calling it a *seizu*—a star map that faithfully represents the relative positions of specific constellations—we might better refer to the ceiling of Takamatsuzuka as a *seishōzu*, a schematic picture of certain stars and constellations. The diagrammatic arrangement of the work neglects their relative locations.

Other examples of tombs displaying the twenty-eight lunar lodges are found in China and Korea (see above, esp. pp. 523–24, 537, 548–49). The lunar lodges on the ceiling along with portraits of the four sacred animals and human beings create a cosmos inside the tomb. A later Japanese example of a *seishōzu* with the lunar lodges is on the ceiling of the Naginataboko, a great, elaborately decorated, medieval carriage. Replicas of the Naginataboko are still used in the annual Gion festival in Kyōto (plate 32).

OTHER CELESTIAL MAPS AND PERTINENT DOCUMENTS BEFORE THE EDO PERIOD (TO A.D. 1600)

Other celestial maps, books, and documents attest to the

1. For a concise, accurate history of Japanese astronomy in English, see Shigeru Nakayama, *A History of Japanese Astronomy: Chinese Background and Western Impact* (Cambridge: Harvard University Press, 1969); it does not, however, discuss celestial cartography. For this latter topic, the following Japanese sources are recommended: Imoto Susumu, "Honchō seizu ryakkō" (Summary of researches on celestial maps made in Japan), pts. 1 and 2, *Tenmon Geppō* 35 (1942): 39–41 and 51–57; idem, "Zoku honchō seizu ryakkō" (Summary of researches on celestial maps made in Japan, continuation), *Tenmon Geppō* 35 (1942): 67–69; idem, "Maboroshi no seishuku zu" (A lost celestial map), *Tenmon Geppō* 65, no. 11 (1972): 290–92; Yabuuchi Kiyoshi (Yabuuti Kiyosi), "Chūgoku, Chōsen, Nihon, Indo no seiza" (Chinese, Korean, Japanese, and Indian constellations), in *Seiza* (Constellations) Shin Tenmongaku Kōza (New lecture series on astronomy), vol. 1, ed. Nojiri Hōei (Tokyo: Kōseisha, 1957), 123–56; and Watanabe Toshio, *Kinsei Nihon tenmongaku shi* (History of modern Japanese astronomy), 2 vols. (Tokyo: Kōseisha Kōseikaku, 1986–87), vol. 2, esp. 737–846.

2. See Takamatsuzuka Kofun Sōgō Gakujutsu Chōsakai (Joint Committee for the Scientific Investigation of Takamatsuzuka Burial Mound), *Takamatsuzuka kofun hekiga chōsa hōkokusho* (Report on the investigation of the Takamatsuzuka fresco by the Agency for Cultural Affairs) (Kyōto: Benrido, 1974). The internal dimensions of the tomb are 2.6 meters from north to south, 1.0 meter from east to west, and 1.13 meters in height.

3. In Shiho only three of the four stars remain intact, and in Hokkyoku four out of five. The ceiling is damaged, and some stars have either fallen off or been peeled off by thieves. Takamatsuzuka is also discussed above, see pp. 352–53.

FIG. 14.1. MAP OF THE HEAVENS ON THE CEILING OF TAKAMATSUZUKA (CA. 700). Each star is indicated by thin gold leaf nine millimeters in diameter. The stars are connected by red lines drawn using a straightedge to form constellations. Drawn as if viewed from below, the lunar lodges form four groups of seven (*shihō shuku*), situated in each of the cardinal directions (north is at the top). The number of stars in each

Chinese influence on Japanese celestial cartography. The families given hereditary charge of calendar making and astrology at the imperial court, for instance, compiled their celestial maps from Chinese originals and copies of them. Titles of Chinese celestial maps and books can be found in an ancient catalog at the Shōsōin (treasury) at Tōdai Temple and in Fujiwara no Sukeyo's *Nihonkoku genzaisho mokuroku* (A list of books at present in Japan, ca. 891).[4] Two examples of titles are *Seki shi seikei bosan* (Star catalog and text on the constellations and the practice of astrology of Master Shi) and *Bosan* (Star catalog) that were taken from the Chinese titles *Xingjing* (Star classic [on astrological divination]) and *Bu zan* (Descriptive catalog on the stars and constellations), works credited to the Wei astronomer Shi Shen (Japanese Seki Shin).

Further documentary evidence suggests that celestial maps might have existed in Japan by the first third of the twelfth century. The *Chūyūki* (Diary of the Nakamikado) of 1131 notes that when a fire broke out at the Onmyō no Tsukasa (the Yin-Yang Board, an imperial office in charge of calendar making, timekeeping, astrology, and

4. The *Nihonkoku genzaisho mokuroku* is in the *Zoku gunsho ruijū* (Classified series of various books: Continuation, 1923–28, in 71 vols.), 3d rev. ed., 67 vols. (Tokyo: Zoku Gunsho Ruijū, 1957–59), vol. 30, bk. 2.

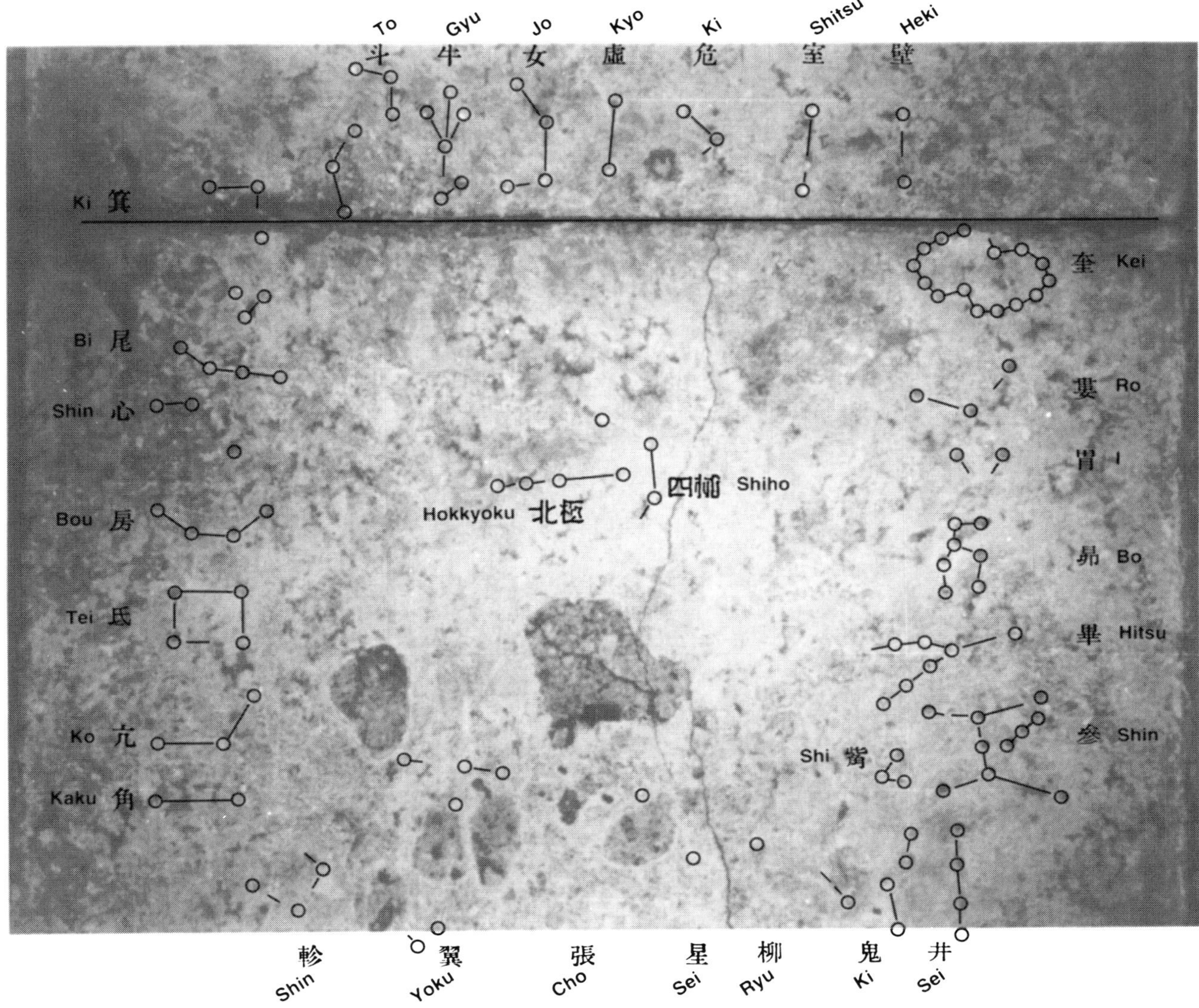

lunar lodge is typical of ancient Chinese star maps. On the right, we show the ceiling with an overlay containing the names and characters for the lunar lodges (they do not appear on the original). The horizontal line toward the top is a joint in the rocks. Size of the square formed by the constellations: 80 × 80 cm. Photograph courtesy of Yabuuchi Kiyoshi, Kyōto.

divination) in 1127, all the instruments were destroyed except a *rōkoku* (clepsydra, or water clock) and a *kontenzu* (a map or model of the celestial sphere). The latter instrument might have been a celestial globe, but we cannot be sure, because the syllable *zu* was also the word for map.[5]

Such references give us insights into the early history of Japanese celestial cartography. Further insights are provided by the manuscript *Shi Shi bu zan* (Star catalog according to Master Shi) that was handed down in the Wakasugi family, descendants of the steward of the Tsuchimikado (formerly called Abe) clan, hereditary holders of the Onmyō no Tsukasa. This is one of two scrolls that appear to be separate parts of one larger document. Its sources include Chen Zhuo's fourth-century *Bu zan* and a tripartite *Bu zan* attributed to Shi Shen, Gan De (or Gan Wenqing), and Wu Xian. These sources are mentioned in bibliographies and so are known to have been

5. Nakatsukasa Munetada, the author of this diary, is also known as Fujiwara Munetada. The relevant section of the *Chūyūki* is found in the *Shiryō taisei* (Series of historical materials), 43 vols. (Tokyo: Naigai Shoseki, 1934–43), 12:286–87. It is also included in the *Hōgibu* (Volume on technical specialists, 1909) section of the *Koji ruien* (Historical encyclopedia of Japan, 1896–1914), 51 vols. (Tokyo: Yoshikawa Kōbunkan, 1982). See also Watanabe, *Kinsei Nihon tenmongaku shi*, 2:463–64 (note 1).

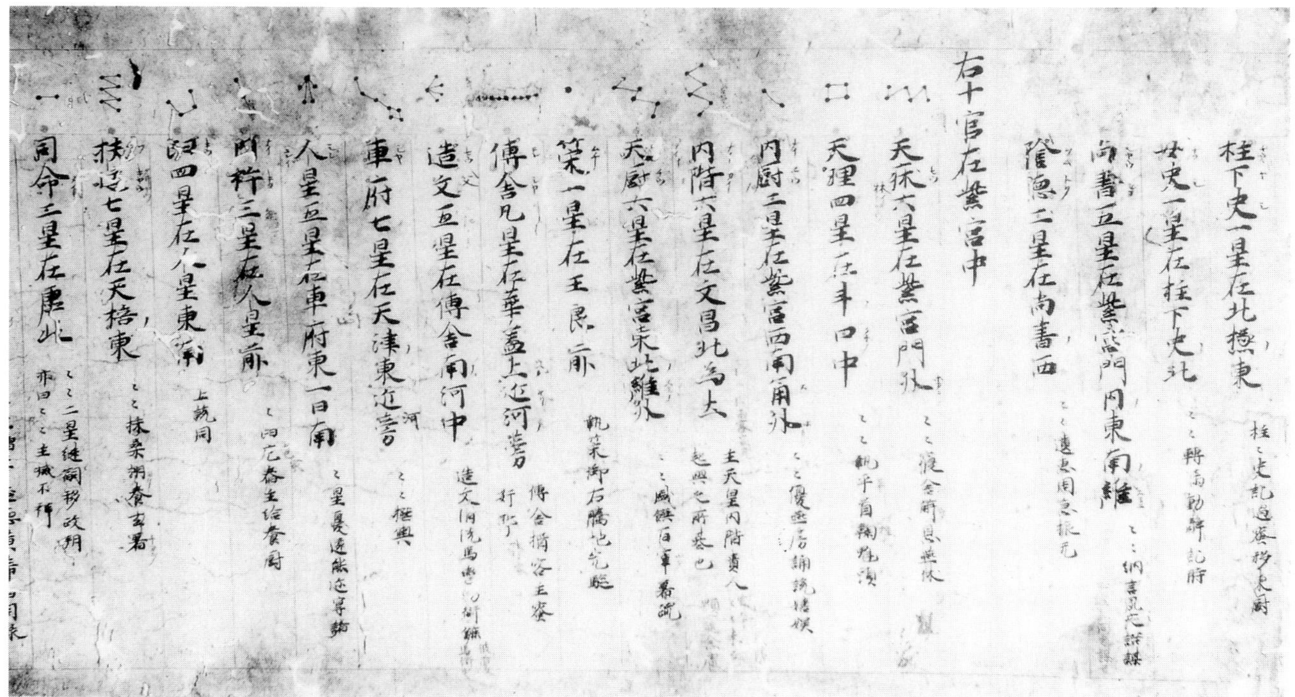

FIG. 14.2. PART OF THE *SHI SHI BU ZAN* MANUSCRIPT. This star catalog in scroll form, handed down in the Tsuchimikado clan, describes and illustrates the configuration of individual constellations according to ancient Chinese sources. Above the description of each constellation is a drawing in red of the configuration of stars. Size of the original: unknown. Kyōto Prefectural Museum. Photograph from Murayama Shūichi, ed., *Onmyōdō kiso shiryō shūsei* (Compilation of basic material on the techniques of divination) (Tokyo: Tōkyō Bijutsu, 1987), 193.

in Japan.[6] The scrolls contain brief descriptions of constellations, apparently based on the Chinese stellar records of Shi, Gan, and Wu attributed to the late Zhanguo (Warring States) period (403–221 B.C.).

It is not known how many times the original(s) brought from China were copied. We do know that about 1215 Abe no Yasutoshi made a copy and wrote remarks. Some of the constellations in Abe no Yasutoshi's version evidently were copied from the original (or a copy) and others copied from another source—the manuscript *Yoru no tsuki no susumu o tadasu no zu* (Star map for confirmation of the lunar motion at night).[7] Whereas the version used by Abe no Yasutoshi contained drawings, another source used a century later did not: according to Abe no Yasuyo in 1314, the original (or copy) that he used had no drawings, so he had to rely on books in the possession of the family in order to draw the constellations.[8] These remarks indicate that at least two different versions were used. The works of Abe no Yasutoshi and Abe no Yasuyo were later put together by an unknown copyist and annotated by Abe no Ariyo (1327–1402) (fig. 14.2).

Although these scrolls showed configurations of individual constellations, they contained no map of the heavens as a whole. The *Yoru no tsuki no susumu o tadasu no zu* (fig. 14.3), however, was a Japanese attempt to compile a celestial map that could be used to observe such phenomena as occultations of stars by the moon or close approaches between the moon and individual stars or constellations. Compiled by a member of the Abe clan at an unknown date, it was later copied by Abe no Yasuyo. Until it was destroyed in the Second World War,

6. These remarks are based on the reproductions of the two scrolls and bibliographical information in Murayama Shūichi, ed., *Onmyōdō kiso shiryō shūsei* (Compilation of basic material on the techniques of divination) (Tokyo: Tōkyō Bijutsu, 1987), 187–203 and 368–81. The scrolls are labeled A (*Seki shi bosan*) and B (*Zakka hō* [Various ways to tell fortunes]) in Murayama's book. Bibliographies such as *Nihonkoku genzaisho mokuroku* and other documents mention only the title of Chen's *Bu zan*, so we have no information on its contents. For more on Shi Shen, Gan De, and Wu Xian, see chapter 13 above.

7. In an epilogue dating from 1215, Abe states that "although there were some drawings in the original, I copied some from the *Yoru no tsuki no susumu o tadasu no zu* to replace those that were unconvincing in the original."

8. Watanabe, *Kinsei Nihon tenmongaku shi*, 2:763 (note 1), and Murayama, *Onmyōdō kiso shiryō shūsei*, 372 (note 6).

it was the oldest extant Japanese celestial map. A recent reconstruction in the form of a scroll was made from surviving pictures by Sasaki Eiji. The reconstruction captures the original's basic character, albeit with some uncertainties.[9] It consists of two maps of stars, one circular and the other rectangular. According to Watanabe Toshio, the boundaries for the lunar lodges were drawn based on values similar to those derived from observations by Yixing (682–727) during the Tang dynasty.[10] The Chinese constellations, the ecliptic,[11] and the Milky Way are also shown.

The books on Chinese constellations and the celestial maps mentioned above were all used for astrology, which in Japan—as in the rest of East Asia—was associated with governance and activities at the court. Commoners were excluded from its secrets, and the office of astrologer was made hereditary so that the practice could be controlled. According to the *Ryō no gige* (Commentary on the codes) of 833, private ownership of manuals on magic, books on astronomy, maps of stars, and devices such as armillary spheres was prohibited by law.[12] In ancient and medieval Japan circulation of celestial maps was therefore very limited. The only exception was during the period of civil wars (1467–1568), when the policy of secrecy was not heeded.

Two celestial maps from the period of civil wars are now the oldest existing Japanese star maps. Both are circular and show continuity with the Chinese tradition. The older map, *Ten no zu* (Map of the heavens) (fig. 14.4), is a hanging scroll that was probably made before 1547 (it seems to have been donated to Asakura Takakage, the lord of Echizen [now Fukui Prefecture], by Tanino Ippaku, a priest from Nara who was at Takedan Temple in 1547); the scroll is now designated an important national cultural property. Its distinctive feature is that within a narrow band representing the circle of constant visibility are entered the names of the twelve traditional *ji* (Chinese *ci* [Jupiter stations; literally, stations of the fictitious planet counter-Jupiter]). Although the Chinese divided the heavens into twelve equal *ji*, on this map they are unequal, and the degrees for each *ji* are noted. Another notable feature is the 366 meridians of right ascension radiating to the outermost boundary—the circle of constant invisibility—outside which lies the part of the celestial sphere constantly below the horizon; this is based on the Chinese degree (*du*, Japanese *do*), of which there were roughly 365¼ in a circle. Also included on the map are the celestial equator and Chinese constellations, but the ecliptic is not drawn in.[13] The second map, formerly in the possession of Imoto Susumu, was reportedly compiled during the Tenbun era (1532–55).[14] Its contents include Chinese constellations, the Milky Way, radially drawn straight lines to show the boundaries of the twenty-eight lunar lodges, the ecliptic, and three concentric circles that mark the circle of constant visibility, the celestial equator, and the circle of constant invisibility.

Another extant work from before the Edo period is a drawing of the stars by the emperor Go-Yōzei (r. 1586–1611). The map is very simple and includes Hokuto (the Big Dipper), Hokushin (or Hokkyoku-sei, the north star), and other stars in the north polar region that are recorded with the name Tengaisei (heavenly umbrella).[15]

9. The original was lost in an air raid on 25 May 1945. Sasaki Eiji, Fukui Prefecture, based his reconstruction on the photographs in Imoto, "Maboroshi no seishuku zu" (note 1) and also referred to the *Ten no zu* (Map of the heavens) at Takedan Temple (see below). Although in the original there were finely crosshatched lines, they are not drawn in the reproduction; there are also several errors in the names of the constellations. There is no record of the size of the original; the reproduction measures twenty-seven by seventy-eight centimeters (rectangular portion), twenty-seven by twenty-two centimeters with a radius of nine centimeters (circular portion), and twenty-seven by fifty-four centimeters for the section of annotations. Sasaki added an explanation of his reconstruction at the end of the reproduction.

10. Yixing, born Zhang Sui, was a patriarch of esoteric Buddhism and a leading astronomer. He had almost completed the Da yan calendar at his death in 727. This highly influential calendar was introduced to Japan in 735 and used from 763 to 861. See Watanabe, *Kinsei Nihon tenmongaku shi*, 2:760–65, esp. 762 (note 1), and Ōsaki Syōji, *Chūgoku no seiza no rekishi* (History of the Chinese constellations) (Tokyo: Yūzankaku, 1978). For more on Yixing, see above, esp. pp. 123, 533, and 538.

11. Ōsaki, *Chūgoku no seiza no rekishi* (note 10), says it is the path of the moon, not the ecliptic, but ecliptic seems correct.

12. The *Ryō no gige* is in the *Shintei zōho kokushi taikei* (Series of histories of our country revised and enlarged), 66 vols. (Tokyo: Yoshikawa Kōbunkan, 1929–64), vol. 22, chap. 10. See also the *Koji ruien, Hōgibu*, 284 (note 5).

13. The names and boundaries of the twenty-eight lunar lodges are shown around the periphery; individual ranges in right ascension are the same as those in *Rekirin mondōshū* (Questions and answers about the calendar, 1414) by Kamono Arikata.

14. According to Imoto, "Honchō seizu ryakkō" (note 1), the map was made into a hanging scroll. Around the edges of the map are quotations from the *Huntian yi* (Armillary sphere) by Zhang Heng (78–139) of the Later Han dynasty, the *Yueling zheng yi* (Commentary to the Monthly observances) by Kong Yingda (574–648), and other such works. The constellations of Beidou (Japanese Hokuto, the Big Dipper) and Beiji (Japanese Hokkyoku, the Stars of the North Pole), the twenty-eight lunar lodges, and the constellations of the three enclosures (*yuan*): Ziwei (Forbidden Purple, Japanese Shibi), Taiwei (Supreme Subtlety, Japanese Taibi), and Tianshi (Celestial Market, Japanese Tenshi) are represented by filled-in red circles. The other stars are shown as filled-in black circles. The Milky Way is painted with white wash, and the equator and the ecliptic are denoted by red and yellow circles, respectively.

15. Tengaisei in Japan may also refer to a silk umbrella placed above statues of the Buddha. The map is preserved at the library of the Imperial Household Agency. It was made into a hanging scroll and is significant only in that it was drawn by the emperor himself. No detailed study of the map has yet been made.

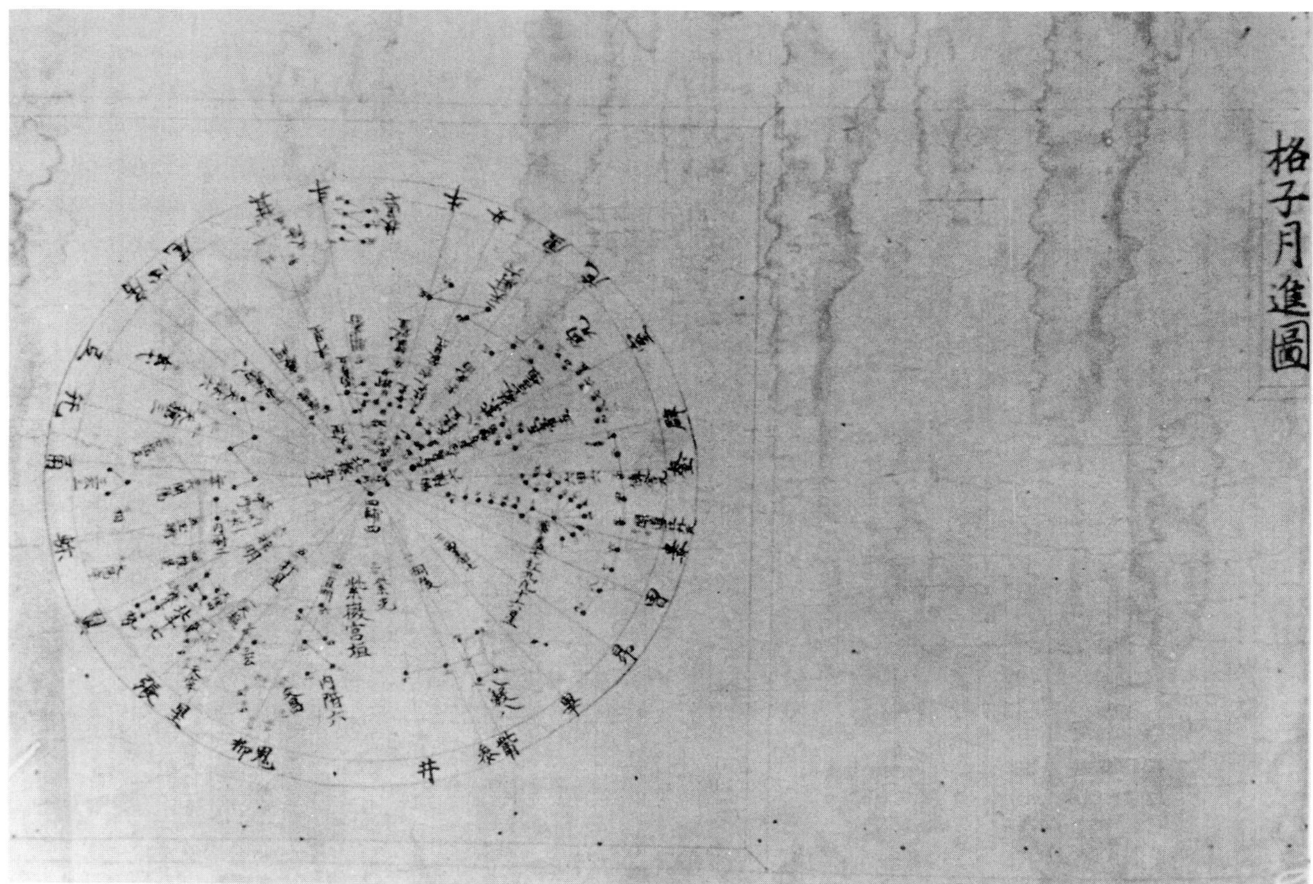

FIG. 14.3. THE *YORU NO TSUKI NO SUSUMU O TADASU NO ZU*. These two details show the circular star map, centered on the North Pole, and the equatorial zone of the rectangular map (from the vernal equinox [RA 0h] to the autumnal equinox [RA 12h]). On the latter chart the celestial equator is represented by a straight horizontal line across the middle. The irregularly

Celestial Maps in the Edo Period

Influence of Chinese and Korean Celestial Maps

Early in the Edo period (1600–1868) Chinese books containing celestial maps were brought to Japan. Most influential were Chen Yuanjing's *Shilin guang ji* (Records of many things, ca. 1250), Wang Qi's *Sancai tuhui* (Illustrated compendium of the three powers [heaven, earth, and man], completed 1607, printed 1609), and You Yi's *Tianjing huowen* (Questions and answers on astronomy, 1672). The *Shilin guan ji*, a popular encyclopedia, contains entries about the life of commoners and was written for practical use. Two versions from the Yuan dynasty, three from the Ming, and one in Japanese still exist. The *Sancai tuhui* is illustrated throughout. You's *Tianjing huowen* attempted to reconcile ancient and recent Chinese theories with Western astronomy, but his understanding of both Chinese and European ideas was inadequate. Many of the quotations from books on Western astronomy written by Jesuits and those on Chinese astronomy were erroneous.[16] From the time of their arrival, books on astronomy published in Japan included celestial maps based on these works.

The Japanese reprint of the *Shilin guang ji*, which allegedly dates from 1699, is based on the Chinese edition of 1325 but contains both a rectangular and a circular celestial map that are lacking in the extant Chinese publication.[17] These two maps are valuable not only from the Japanese perspective but also from that of Chinese celestial cartography during the Song dynasty (960–1279). This is because they differ from maps in the *Xinyi xiang fayao* (New design for an armillary [sphere] and [celestial] globe), printed in 1094 by Su Song, and the famous

16. The *Tenkei wakumon chūkai zukan* (Annotations to illustrations contained in the *Tianjing huowen*, 1750) by Irie Osamu (see note 34 below) pointed out some of these errors. Some Buddhist priests were offended by You's work, since it did not agree with the Buddhist conception of the universe.

17. 1325 is the first Chinese printed edition. Later, inferior Chinese copies survive.

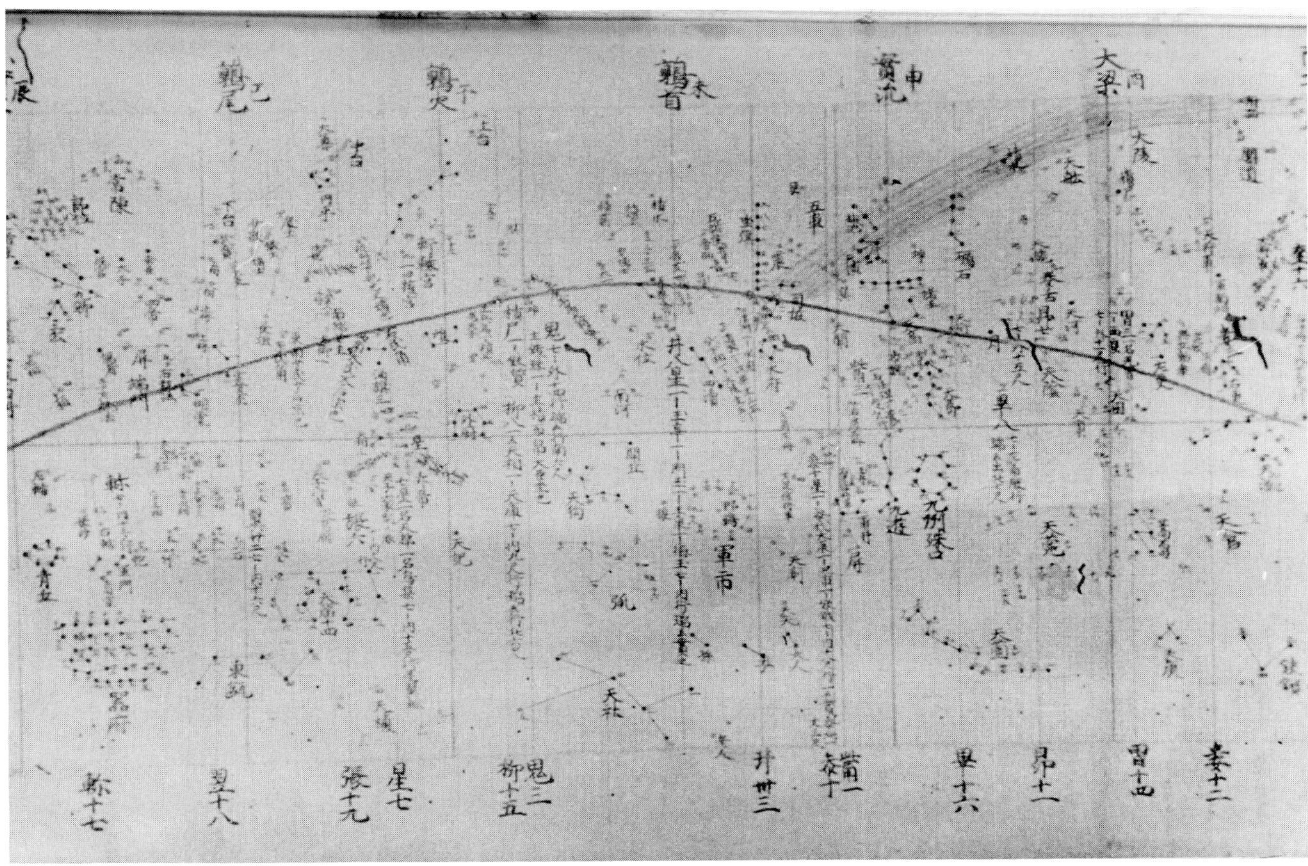

spaced vertical straight lines are lines of right ascension, marked by determinative stars (*kyo sei*), that serve as boundaries for the twenty-eight lunar lodges.

Size of the original: unknown. From Imoto Susumu, "Maboroshi no seishuku zu" (A lost celestial map), *Tenmon Geppō* 65, no. 11 (1972).

Suzhou planisphere engraved on a stele in 1247 but based on a much earlier map.[18]

The *Tianjing huowen*, brought to Japan about 1672–79, combined ancient Chinese theories, the theories of the natural philosopher Zhu Xi (1130–1200), and the recent philosophical opinions of Fang Yizhi (1611–71) with knowledge that had been obtained from the Jesuits. Although the *Tianjing huowen* did not receive much attention in China, it was widely available in Japan and inspired many explanatory and critical books, especially after Nishikawa Masayasu (1693–1756) published the first Japanese version in 1730. Of particular importance was its illustration of the stars around the South Pole, which had not been shown on previous celestial maps; the book therefore provided the Japanese with their first knowledge of such stars. The section on astronomy and astrology in Terajima Ryōan's *Wakan sansai zue* (Illustrated encyclopedia of Japanese and Chinese things related to the three powers [heaven, earth, and man], 1715) took information from the *Tianjing huowen* and *Sancai tuhui*.[19] Both Chinese works also influenced later Japanese celestial maps.

Another work that influenced Japanese celestial cartography is the *Ch'ŏnsang yŏlch'a punyajido* (Chart of the constellations and the regions they govern), a star map

18. For the *Xinyi xiang fayao* see pp. 541–45 above and also Joseph Needham, Wang Ling, and Derek J. de Solla Price, *Heavenly Clockwork: The Great Astronomical Clocks of Medieval China*, 2d ed. (Cambridge: Cambridge University Press, 1986). I studied the Chinese text while reconstructing Su Song's water-powered armillary sphere and celestial globe tower (*shui yun yixiang tai*) for Seikō, the watchmaking company; the results of the study have not yet been published. The extant edition of the *Xinyi xiang fayao* is not the original. For a discussion of the Suzhou planisphere see pp. 545–48 and Yabuuchi Kiyoshi, *Chūgoku no tenmon rekihō* (The history of astronomy and calendrical science in China) (Tokyo: Heibonsha, 1969; rev. ed. 1990). A large number of rubbings of the Suzhou planisphere are found in Japan; some of these apparently were brought to Japan before the Meiji era and influenced Japanese celestial cartography.

19. Terajima Ryōan, ed., *Wakan sansai zue* (reprinted Tokyo: Tōkyō Bijutsu, 1982).

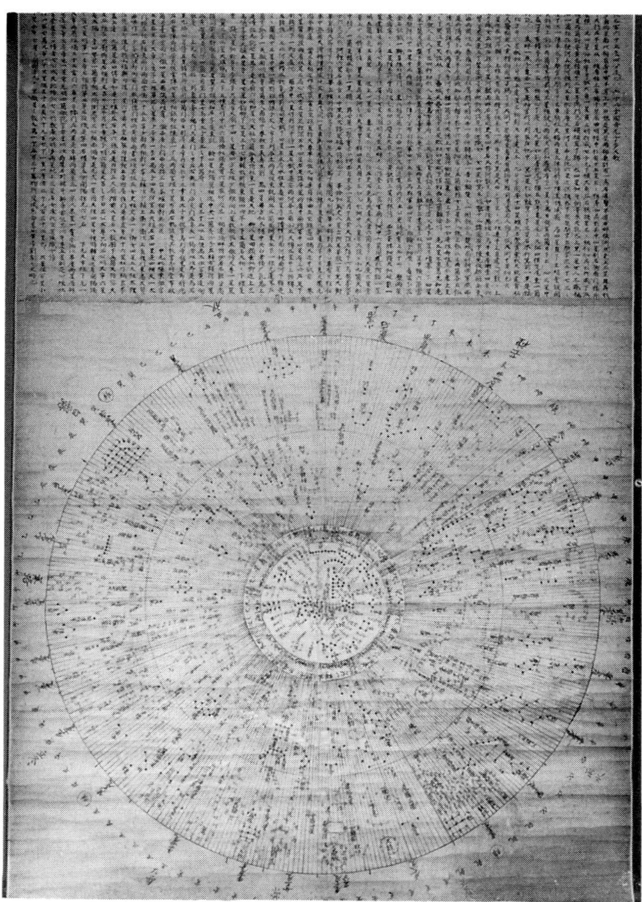

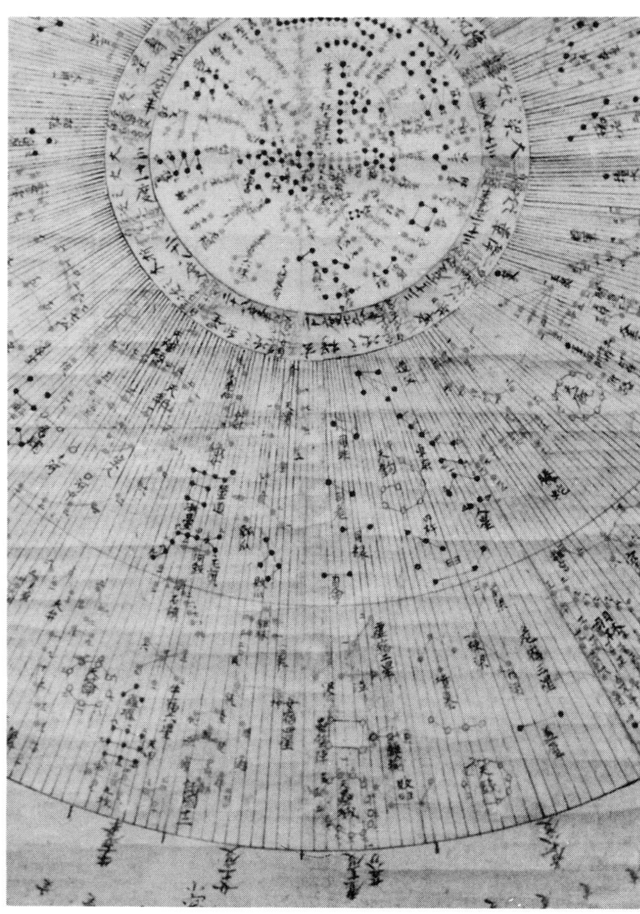

FIG. 14.4. THE *TEN NO ZU* AT TAKEDAN TEMPLE, FUKUI PREFECTURE, AND DETAIL. Shown are the circle of constant visibility, names and extents of the twelve *ji*, and the meridians of right ascension based on the Chinese degree. Above the map is written some constellation poetry known as *Hoten ka* (see below, p. 598). Surrounding the names of the lunar lodges one hundred directions are indicated by twenty characters (eight of the twelve earthly branches [*shi*], eight of the ten celestial stems, and four of the eight signs of divination), each used five times. The stars that came from Wu Xian are represented by open black circles and those of Shi Shen and Gan De are shown as red and black circles, respectively.
Size of the original: 144.2 × 156.5 cm. Photograph courtesy of Sasaki Eiji, Asahi-cho, Fukui Prefecture.

engraved on a stone in Korea in 1395.[20] Although no rubbings or prints from the 1395 stele survive in Japan, its influence is apparent. Rubbings and woodblock prints from a Korean stone copy of 1687 (identical to the 1395 stele except that the title was moved to the top) are extant. There are prints with the letters and figures in white on a black background and others with black letters and figures on white, and some copies have a colorless Milky Way on a pale blue background. The *Ch'ŏnsang yŏlch'a punyajido* provided the information for Fukushima Kunitaka's *Bundo no kiku* (literally, "standard for degree mensuration"), an instrument made in 1668, and Shibukawa Harumi's *Tenshō retsuji no zu* (Map of the arrangement of stars and constellations, 1670) and *Tenmon bun'ya no zu* (Map showing divisions of the heaven and regions they govern, 1677).[21]

20. See pp. 560–61 above and Sang-woon Jeon (Chŏn Sang'un), *Science and Technology in Korea: Traditional Instruments and Techniques* (Cambridge: MIT Press, 1974), 26–28. (The Japanese version of Jeon's book is *Kankoku kagaku gijutsu shi* [Tokyo: Koma-Shorin, 1978].)

21. Shibukawa's works are all discussed below; the two mentioned here are black-and-white woodblock prints, some of which had colored paint added. The *Ch'ŏnsang yŏlch'a punyajido* and the celestial maps in the *Sancai tuhui* also might have been used as source materials for the first printed Japanese celestial map. This map is in the *Shisho inmō ryaku zukai* (An outline illustrated with maps for correcting mistakes in the four Confucian classics), edited by Ōhara Takekiyo in 1653. Because of differences in the arrangement of the stars and shapes of

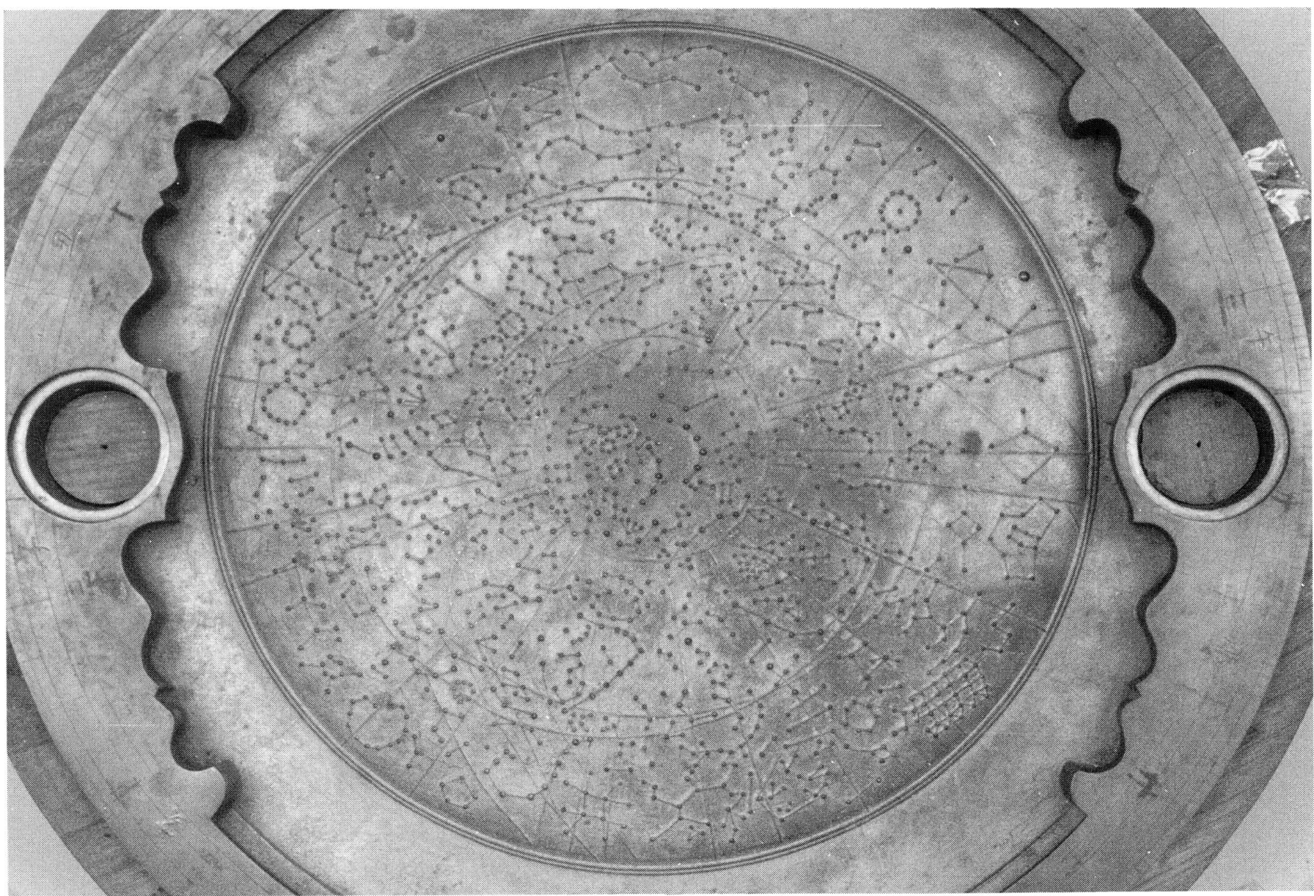

FIG. 14.5. FUKUSHIMA KUNITAKA'S *BUNDO NO KIKU*, 1683. This pan-shaped bronze instrument features a celestial map based on the *Ch'ŏnsang yŏlch'a punyajido* engraved in its concave center. On the edge are two small depressions for magnetic compasses. The original was made in 1668 on the orders of the military engineer Hōjō Ujinaga. According to Fukushima's engraving on the reverse, in 1683 the feudal lord of Hasuike, Nabeshima Naoyuki, ordered a smith named Chōken to copy the *Bundo no kiku*.
Size of the original: ca. 34 cm in diameter, celestial map ca. 24 cm in diameter. Saga Prefectural Library. Photograph courtesy of Kazuhiko Miyajima, Ōsaka.

Fukushima's *Bundo no kiku* is a pan-shaped bronze instrument featuring a celestial map based on the *Ch'ŏnsang yŏlch'a punyajido* engraved in the concave central part (fig. 14.5). The instrument discussed in chapter 13 is an accurate copy of the *Bundo no kiku* illustrated here. That copy is thought to have been taken from a sailing ship, and Needham speculates that it was used by navigators.[22] Fukushima's *Bundo no kiku* was described through illustrations and identified as a "*taien bundo*" (large circular protractor) in *Bundo yojutsu* (Techniques of protraction, 1728) by Matsumiya Toshitsugu (or Kanzan).[23] Matsumiya notes the star map, but he does not relate it to the instrument's function. The instrument's main uses were as a level, by filling the central depression

the constellations, however, we cannot firmly establish this relationship. The map is illustrated in Watanabe, *Kinsei Nihon tenmongaku shi* (note 1), and discussed by Imoto, "Honchō seizu ryakkō," pt. 2 (note 1). It is circular and spread out over two pages, with the North Pole in the center. Three concentric circles represent the circle of constant visibility, the equator, and the circle of constant invisibility; the outline of the Milky Way is drawn in. Other than the lunar lodges, there are few stars on the map, and they are depicted by black circles either left blank or filled in.

22. Joseph Needham, *Science and Civilisation in China* (Cambridge: Cambridge University Press, 1954–), vol. 3, with Wang Ling, *Mathematics and the Sciences of the Heavens and the Earth* (1959), 279 and 282. See also E. B. Knobel, "On a Chinese Planisphere," *Monthly Notices of the Royal Astronomical Society* 69 (1909): 435–45, esp. 436.

23. The *Bundo yojutsu*, a manuscript book on surveying techniques, is at the National Archives in Tokyo.

with water, and for measuring azimuths with reference to the graduations on the circumference.[24]

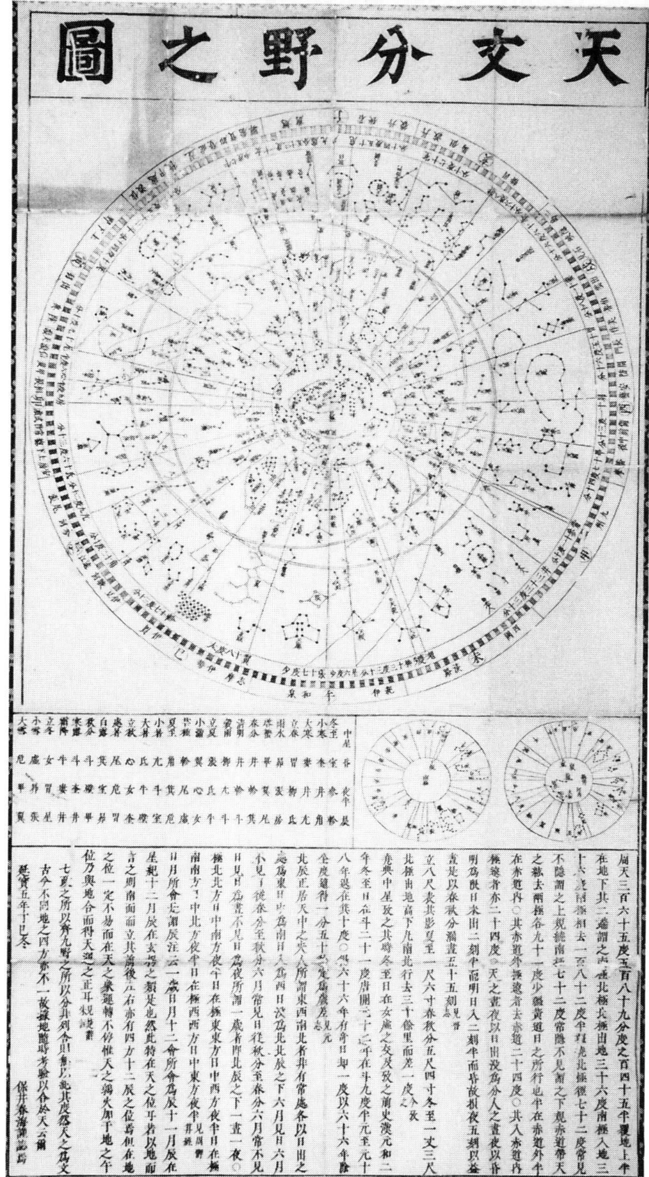

FIG. 14.6. *TENMON BUN'YA NO ZU* (MAP SHOWING DIVISIONS OF THE HEAVEN AND REGIONS THEY GOVERN), 1677, BY SHIBUKAWA HARUMI. This circular star map was based on the *Ch'ŏnsang yŏlch'a punyajido*. Shibukawa added a Japanese adaptation of Chinese "field allocation" astrology, in which terrestrial regions were associated with various celestial divisions.
Size of the original: 108 × 55.5 cm. Photograph courtesy of Kazuhiko Miyajima, Ōsaka.

SHIBUKAWA HARUMI'S STAR MAPS AND NEW CONSTELLATIONS

One of the greatest astronomers to have lived in Japan was Shibukawa Harumi (1639–1716), the son of Yasui Santetsu (1590–1652), a master of the game of go at the shogunate.[25] Shibukawa's infant name was Yasui Rokuzō. He changed his given name to Santetsu after his father died. Harumi is his pen name. He changed his surname to Yasui (written with a different first character) by the time his *Tenmon bun'ya no zu* was published in 1677. In 1702 he changed his surname again, this time to Shibukawa, which was the former name of the Yasui family.

Shibukawa is particularly known for his calendrical system, adopted in 1684 and named the Jōkyō calendar after the reign title (1684–88). Replacing the Senmyō (Chinese Xuanming) calendar, the Jōkyō calendar is notable because, unlike previous Japanese calendars that relied completely on Chinese theory, it was compiled from Shibukawa's own systematic astronomical observations. It was the first in Japan to be so produced, and it was widely adopted.[26] The calendar reform earned Shibukawa an appointment as official astronomer (*tenmonkata*) to the shogunate, a post that thereafter became hereditary. From this time onward Japan ceased to adopt Chinese calendars and made its own.

Before his calendar reform, Shibukawa compiled the *Tenshō retsuji no zu*, based partly on the *Ch'ŏnsang yŏlch'a punyajido*. However, he adopted the values for extents in right ascension of the twenty-eight lunar lodges from the Chinese Guo Shoujing's Shoushi calendar (season-granting system) of 1279.[27] The declination values of the determinative stars (*kyo sei*) that demarcated the lunar lodges were taken from the *Song shi* (History of the Song, 1346) as quoted in the *Tianwen da cheng guan kui jiyao* (Essentials of astronomy, 1653) edited by Huang

24. Kazutaka Unno, "A Surveying Instrument Designed by Hōjō Ujinaga [1609–70])," paper presented at the Seventh International Conference on the History of Science in East Asia, Kyōto, Japan, August 1993.

25. Go is played with black and white stones on a board marked with nineteen vertical and nineteen horizontal lines, the goal being to surround the larger portion of the board and capture the opponent's stones.

26. Nakayama, *History of Japanese Astronomy*, 120 (note 1).

27. The Shoushi calendar astronomical system survives in Song Lian et al., *Yuan shi* (History of the Yuan, comp. 1369–70); see the edition in 15 vols. (Beijing: Zhonghua Shuju, 1976), chaps. 52–55. The Shoushi calendar was not significantly influenced by Islamic astronomy. According to Yabuuchi, *Chūgoku no tenmon rekihō*, 145 (note 18), no Islamic influence is evident in computational procedures; it essentially followed Chinese tradition, except for some Islamic influence on the new instruments used to establish the system.

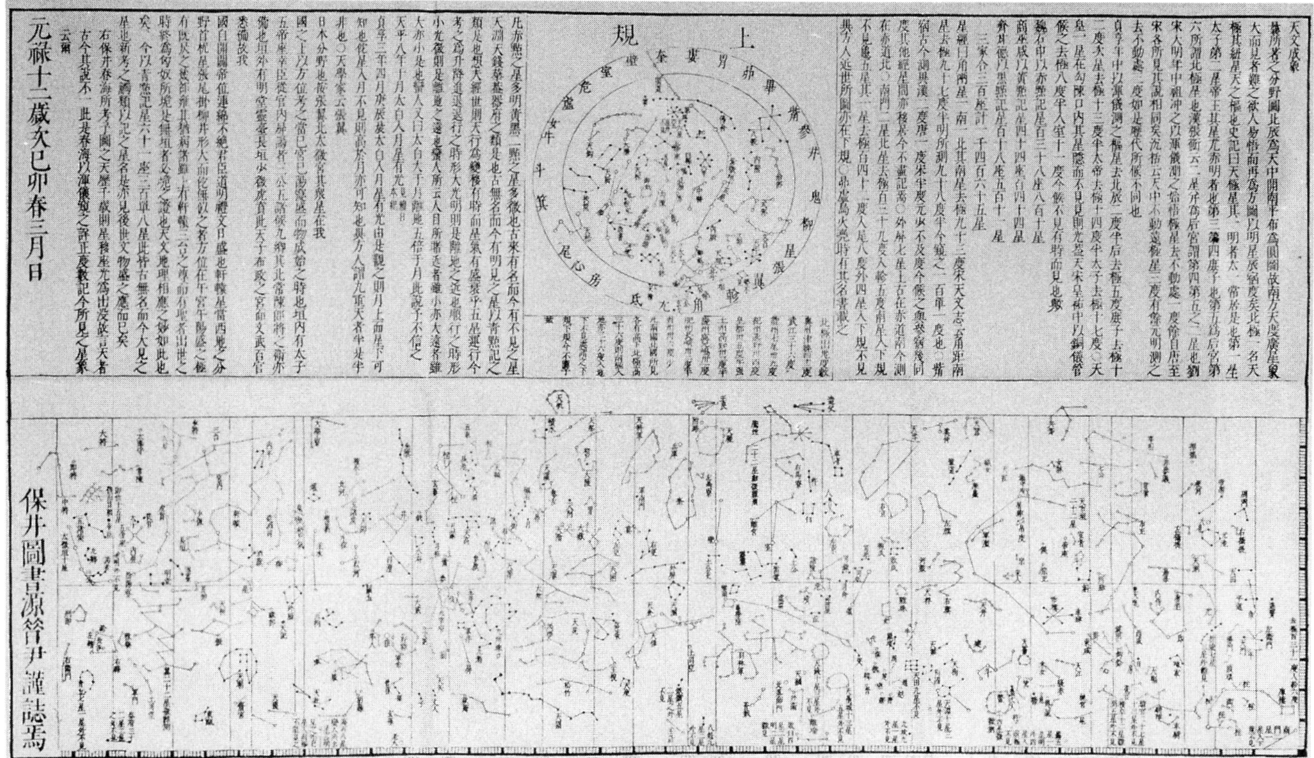

FIG. 14.7. *TENMON SEISHŌ ZU*, 1699, BY SHIBUKAWA HARUMI. The map was published under the name of Shibukawa's son Hisatada. Stars are shown as small black circles (Gan De's constellations) and as small open black circles painted yellow, red, and blue. The yellow are Wu Xian's constellations, those in red are Shi Shen's, and those in blue are Shibukawa's newly added constellations.
Size of the original: 49.9 × 82.5 cm. Photograph courtesy of Kazuhiko Miyajima, Ōsaka.

Ding.[28] To fix the position of constellations, the position of one star was measured and entered on the map; the other stars were then added without measurement as they were seen by the eye. Yamamoto Kakuan hints at the defects caused by such a method in his *Seimei kō* (Thoughts on the names of stars, 1744): "Even when observing the constellations with the help of traditional star maps, they cannot be recognized."[29]

In Chinese "field allocation" astrology, the twelve Jupiter stations were correlated with twelve ancient states of China and were used to interpret celestial omens, as were nine groups of the twenty-eight lunar lodges that were correlated with nine archaic provinces (see pp. 208–10 above). This correspondence was applied by Shibukawa to the regions of Japan in his *Tenmon bun'ya no zu* (fig. 14.6). Although this new correspondence is a major difference between the *Tenmon bun'ya no zu* and the *Tenshō retsuji no zu*, it is not the only one.[30] The location of Canopus (the star Nankyoku Rōjin [the old man in the southern polar region]; Chinese Laoren) also suggests that the *Tenmon bun'ya no zu* was revised to some extent. In both the Korean *Ch'ŏnsang yŏlch'a punyajido* and the *Tenshō retsuji no zu* it is placed to the southeast of Sirius (the star Rō [wolf]; Chinese Lang), far from its actual position to the southwest. The *Tenmon bun'ya no zu* shows it due south of Sirius, closer to its correct location. Shibukawa further revised the location of Canopus in his *Tenmon seishō zu* (Map of the arrangement of stars and constellations, 1699), in which it is almost in the right place.

The *Tenmon seishō zu* is divided into two parts, a circular celestial map above and a rectangular one below (fig. 14.7). In the top part, whose center is the North Pole, constellations in the north circumpolar region are

28. The declination values of the determinative stars in the *Tianwen da cheng guan kui jiyao* are said to be from the treatise on musical scales and astronomy in the *Song shi*, but they in fact differ slightly, and Shibukawa's values match those in the *Tianwen*. The *Tianwen*, 80 volumes including illustrations, was compiled from 143 books on astronomy and astrology. Weather phenomena and divination using animals and plants were also mentioned.

29. This quotation appears at the end of the *Seimei kō*; see Watanabe, *Kinsei Nihon tenmongaku shi*, 2:766–67 (note 1). There are no modern editions or reprints of the *Seimei kō*.

30. Watanabe, *Kinsei Nihon tenmongaku shi*, 2:733 (note 1), claimed that except for this innovation the maps were identical even in their mistakes.

drawn. Names of the lunar lodges are written around the outside, and latitudes of various parts of Japan are written below the map. In the columns on both sides of the circular map is a brief commentary on the constellations associated with the three archaic Chinese systems of Gan De, Wu Xian, and Shi Shen. Here it is mentioned that Shibukawa established new constellations and that they made this star map in accordance with positions of stars observed using an armillary sphere.

On the rectangular map, three horizontal lines are drawn: the uppermost represents the circle of constant visibility, the middle line the equator, and the lowest the circle of constant invisibility. A few star groups are also shown a little above the upper line. A scale of right ascension is marked at the bottom at both one-degree and ten-degree intervals. This follows the Chinese and Japanese tradition and marks 365 degrees. The vertical lines are the boundaries between the lunar lodges. At the right, graduations of celestial declination are marked at intervals of one and ten degrees, ranging over 108 degrees from top to bottom.

Besides the constellations that mark the twenty-eight lunar lodges, with their origins in antiquity, China also had a number of constellations that corresponded to sections of its bureaucracy. Sometimes if something unusual happened in one of these constellations, an astrologer would claim that the corresponding government department was responsible. This system dates from the Jin dynasty, when the director of the National Observatory, Chen Zhuo, integrated the constellations used by the three astronomical and astrological schools. To identify their origins, the constellations were classified under the names used by a Zhanguo astrologer. Wu Xian's set included 44 constellations of 144 stars; Shi Shen's had 138 constellations, 810 stars, and the twenty-eight lunar lodges; Gan De's contained 118 constellations of 511 stars. The Japanese government continued to use the Chinese constellations during Shibukawa's lifetime. Some stars visible to the naked eye were still not included in any established constellation. Shibukawa therefore used them to make 61 new constellations out of 308 stars, which he associated with the Japanese bureaucracy. He published this system in 1698 in his book *Tenmon keitō* (Astronomical encyclopedia). It appears in 1699 on the *Tenmon seishō zu*.[31] Whereas the three sets of Chinese constellations were colored in yellow, red, and black, as was customary on celestial maps, Shibukawa's uniquely Japanese constellations were colored in blue. The *Tenmon keitō*, furthermore, also included celestial maps based on original data obtained by Shibukawa, unlike the *Tenmon bun'ya no zu* and the *Tenshō retsuji no zu*. Both the *Tenmon keitō* and the *Tenmon seishō zu* show an improvement in the accuracy of the position of the stars.

Shibukawa's *Tenmonzu byōbu* (Drawings of the celestial sphere on a folding screen) is a six-fold gilded screen made sometime between 1697 and 1715 (fig. 14.8).[32] On the three and part of the fourth panels to the right is a large circular celestial map. The three panels on the left have two small circular celestial maps and a rectangular celestial map. The circular maps on the left are centered on the North and South Poles, the former, along with the rectangular celestial map, having come from the *Tenmon seishō zu*, and the latter probably copied from the *Tianjing huowen*. These are rather conventional, but the large map on the right is unusual in that, although it is centered on the North Pole and the circle of constant invisibility is the outermost circle, it is a mirror image of the stars, produced as if the celestial sphere were viewed from the outside.

SHIBUKAWA HARUMI'S INFLUENCE

Shibukawa's celestial maps had a great influence on contemporary and later celestial maps and globes. For example, the celestial maps in Iguchi Tsunenori's (or Jōhan) *Tenmon zukai* (Astronomy illustrated, 1689) and Namura Jōhaku's *Koreki binran bikō* (Handbook of old calendars, 1692), as well as Iguchi's *Tenshō hokusei no zu* (Celestial map of the Northern Hemisphere, 1698), are all copies of the *Tenshō retsuji no zu*.[33]

The *Tenmon zukai* is the first astronomical book published in Japan. Although all five volumes mention the study of mathematical astronomy, the first volume has circular star maps and the second has figures of the lunar lodges. Most of the stars are shown as black circles, but those of the lunar lodges and other notable stars in the circular star map are shown as white circles, as are the determinative stars in the figures of the lunar lodges. In the section on the lunar lodges, the starting points of the various lunar lodges are identical to those in the *Tenshō retsuji no zu*, including some that were in error. Namura's

31. The *Tenmon keitō* is an eight-volume astronomical manuscript, given to the shogunate family and the Ise Jingū (Ise Shrine). For matters related to astrology, the author followed the *Tianwen da cheng guan kui jiyao*; however, the contents are considerably simplified. As for astronomy, the author was greatly influenced by the *Tianjing huowen*, but he also criticizes it. Shibukawa mentions his observations and instruments for observation in detail. The *Tenmon seishō zu* was made according to these data.

32. Shibukawa's *Tenmonzu byōbu* was deposited in the Ōsaka City Museum by Nanba Matsutarō and is discussed in Akioka Takejirō, "*Kon'yo bankoku zenzu byōbu* sōsetsu, Shibukawa Harumi byō narabini Tō Kōsekishi byō no sekaizu tenmonzu byōbu" (General remarks on the *Kon'yo bankoku zenzu byōbu* [a folding screen on which a world map is drawn], and some remarks on folding screens with world maps and star maps by Shibukawa Harumi and by Tō Kōsekishi), *Hōsei Daigaku Bungakubu Kiyō* 8 (1962): 1–28.

33. Watanabe, *Kinsei Nihon tenmongaku shi*, 2:827–30 (note 1).

Koreki binran bikō contains a star map identical to the one in Iguchi's *Tenmon zukai.*

Iguchi's *Tenshō hokusei no zu*, showing the Northern Hemisphere, was published in 1698 with his *Tenshō nansei no zu* (Celestial map of the Southern Hemisphere). They are 52 centimeters wide and 125 centimeters long. Copies are rare, and the maps have not been studied in depth. In contrast to the star maps in the *Tenmon zukai*, no meridians separate the lunar lodges. The stars of the lunar lodges and notable stars in other constellations are shown as white circles.

Many other contemporary maps were influenced by Shibukawa's work, notably the revised star maps in Irie Osamu's *Tenkei wakumon chūkai zukan* (Annotations to illustrations contained in the *Tianjing huowen*, 1750),[34] which was the second volume of Irie's explanation and revision of You Yi's *Tianjing huowen*. In the beginning of the second volume Irie writes: "Though Yasui Harumi achieved excellent results in the study of constellations, the circular star map he originated did not show the shapes of constellations correctly. His later rectangular star map finally showed the real shapes. I copy the arrangement of stars from the latter using a compass and a ruler, then make revised maps and show them after the originals." The *Tianjing huowen* has eight types of circular star maps. Irie reprinted the eight original maps and printed revisions of all except the star map of the region of constant invisibility. If the original illustration or explanation was incorrect, he showed a corrected one (or one he thought was correct), basing his revisions on Shibukawa's *Tenmon seishō zu*. The map of the region of constant invisibility, "Nankyoku shoseien kenkai seizu," was not revised because "this region cannot be seen from Japan and some other countries, so it was not illustrated in old celestial globes. Western astronomy was introduced to China, and Westerners who traveled across the sea in the south observed and made this. We cannot see it from Japan, so it is impossible to revise." Of the revised map of the region of constant visibility, "Hokkyoku shibien kenkai kaisei zu," Irie wrote: "I have copied three kinds of star maps of the whole heaven; each has good points and bad points compared with the originals. Yasui Harumi's *Tenmon seishō hōzu* [*Tenmon seishō zu*] is almost real, though it is not perfect, and it is well known, so I revised this a little and made each revised map. I would like people who see these to try to revise them [further]." He then printed copies from the *Tenmon seishō zu* below the revised map of the Milky Way and the constellations along it in the southern sky ("Nankyoku kakansei kenkai kaisei zu").

There is also a celestial globe dating from 1701 at Dōshisha University that was based on Shibukawa's *Tenmon bun'ya no zu*. On the papier-mâché globe are the stars of Wu Xian, Shi Shen, and Gan De, distinguished by gold, red, and black. Gold foil is pasted on the Milky Way. Although it was made after Shibukawa's new constellations had been established, they are not on the globe. It has the same errors as on the *Tenmon bun'ya no zu*, and the connections between the stars are the same as on the *Tenmon bun'ya no zu*, so it seems that the globe was based on this map. The *kontenzu* that escaped the fire of the Yin-Yang Board in 1127 is thought to have been a celestial globe, but aside from this no celestial globe is mentioned before Shibukawa's time. The earliest extant Japanese celestial globes were made by Shibukawa and his contemporaries. These were large (about 50 cm diameter) and made of copper, but later ones were smaller (about 30 cm) and were made of papier-mâché and of paper pasted on wood or plaster. There are also examples lacquered in black.[35]

A later work apparently modeled on the *Tenmon seishō zu* is the celestial map in Hara Nagatsune's *Tenmon keii mondō wakai shō* (Questions and answers on astronomy, 1779).[36] The positions and sizes of the dots representing the stars and the boundary of the twenty-eight lunar lodges differ slightly from Shibukawa's map, and Hara's map does not distinguish between the constellations of three systems and the stars of the constellations established by Shibukawa. However, the annotations on the *Tenmon seishō zu* concerning which stars constitute each constellation are reproduced verbatim. Hara's map is noteworthy because it is one of the few star charts with the north and south transposed (the sphere appears as viewed from the outside) and with the names of the constellations in Chinese to which katakana (a phonetic syllabary) are attached to indicate pronunciation.

34. Irie Osamu (1699–1773) published *Tenkei wakumon chūkai* (Annotations to the *Tianjing huowen*) and *Tenkei wakumon chūkai zukan* as explanations for the *Tianjing huowen* (Japanese *Tenkei wakumon*). The former contains commentary on the introduction to the *Tianjing huowen*, by You Yi's friends, that explains the title of the book and technical terms. There are no illustrations. The *Tenkei wakumon chūkai zukan* comments on illustrations and explanations in the beginning of the text of the *Tianjing huowen*.

35. See Miyajima Kazuhiko, "Dōshisha Daigaku shozō Genroku 14 nen sei tenkyūgi no ichizuke" (The position of the celestial globe made in 1701 and owned by Dōshisha University), *Dōshisha Daigaku Rikōgaku Kenkyū Hōkoku* 21 (1981): 279–300. There is abundant literature introducing and reporting on individual celestial globes, but few works deal with them collectively: see Hirose Hideo, "Tenkyūgi oboegaki" (Memorandum on celestial globes), *Gotō Puranetaryumi Gakugeihō* 6 (1978), and Miyajima Kazuhiko, "Mukashi no tenmon giki" (Astronomical instruments of old days), in *Tenmongaku shi* (History of astronomy), a series of lectures on modern astronomy, vol. 15 (1983). For a list of globes in Japan based on my investigations (celestial globes) and those of Kazutaka Unno (terrestrial globes), see Hirotada Kawamura, Kazutaka Unno, and Kazuhiko Miyajima, "List of Old Globes in Japan," *Der Globusfreund* 38–39 (1990): 173–77.

36. I have not seen Hara's map, but it is reproduced and discussed in Watanabe, *Kinsei Nihon tenmongaku shi*, 2:833–34 (note 1).

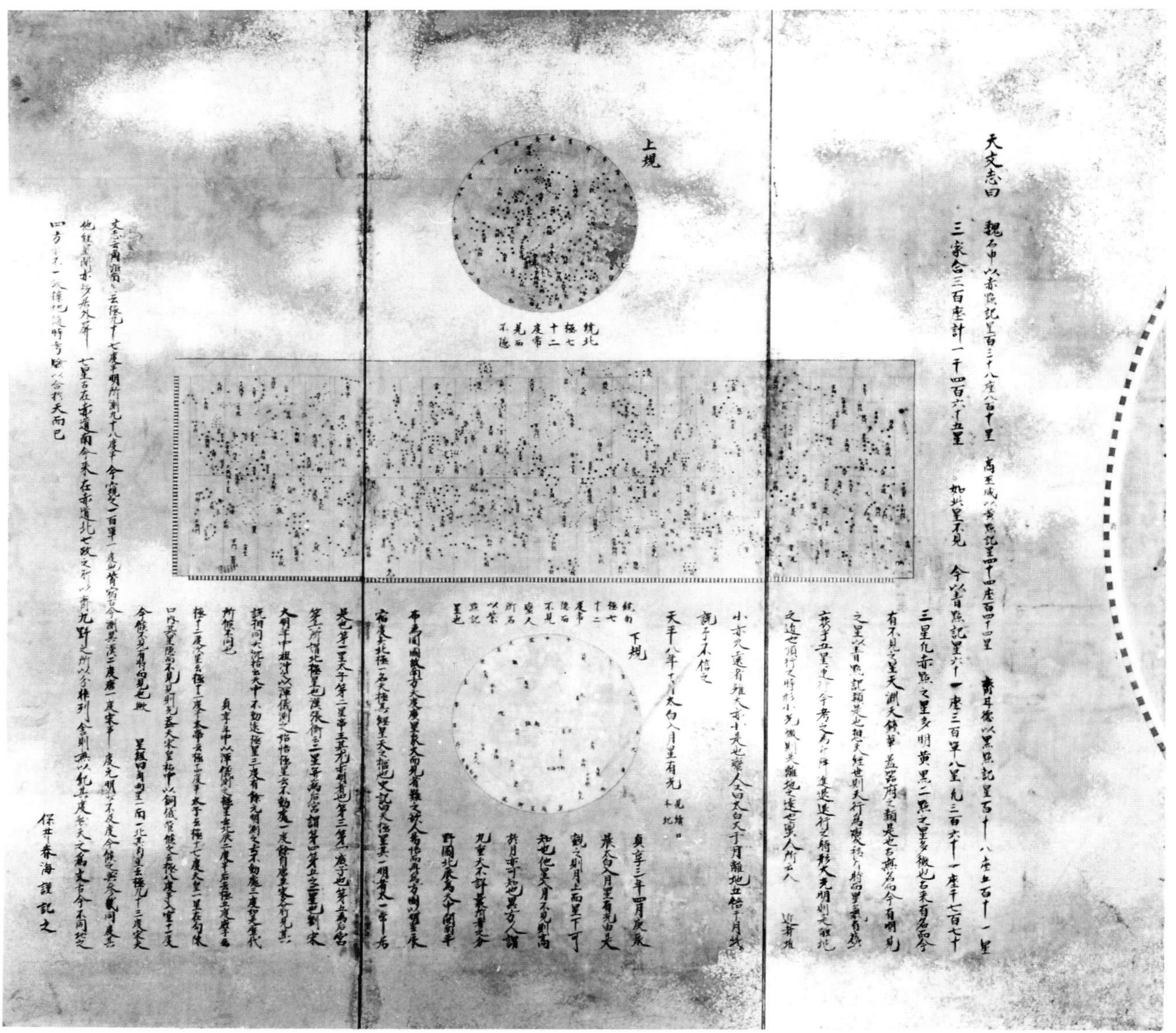

FIG. 14.8. *TENMONZU BYŌBU* BY SHIBUKAWA HARUMI, a six-fold screen with four celestial maps, three circular and one rectangular. The large circular map shows the stars of the Northern Hemisphere as though viewed from outside. Size of the entire screen: 184 × 377.5 cm. Ōsaka City Museum. Photograph courtesy of Nanba Matsutarō, Nishinomiya.

Iwahashi Zenbei (1756–1811), who produced a refracting telescope in Japan, invented a rotating diagrammical board called *heiten gi* (literally "planisphere"). Several colored disks with different radii one on top of the other rotate around the center. They represent (from the inside) the earth (with the North Pole at the center; radius ca. 4.5 cm), the tides with the moon's position, the day of the month with the sun's position determined by the lunar calendar, the twenty-eight lunar lodges, and last, the time of day (radius is approximately 12.5 cm). Basic astronomical knowledge is described in the accompanying text, *Heiten gi zukai* (Illustrated explanations of the *heiten gi*), primarily based on the *Tianjing huowen*. The maps included in this book are a rectangular map called "Kōsei no zu" (Map of the fixed stars), six pages, and two circular maps (two pages each)—one showing the northern polar region and the other the southern polar region. The rectangular map and the circular map centered on the North Pole are based on the *Tenmon seishō zu*. The other circular map centered on the South Pole is based on the *Tianjing huowen*.

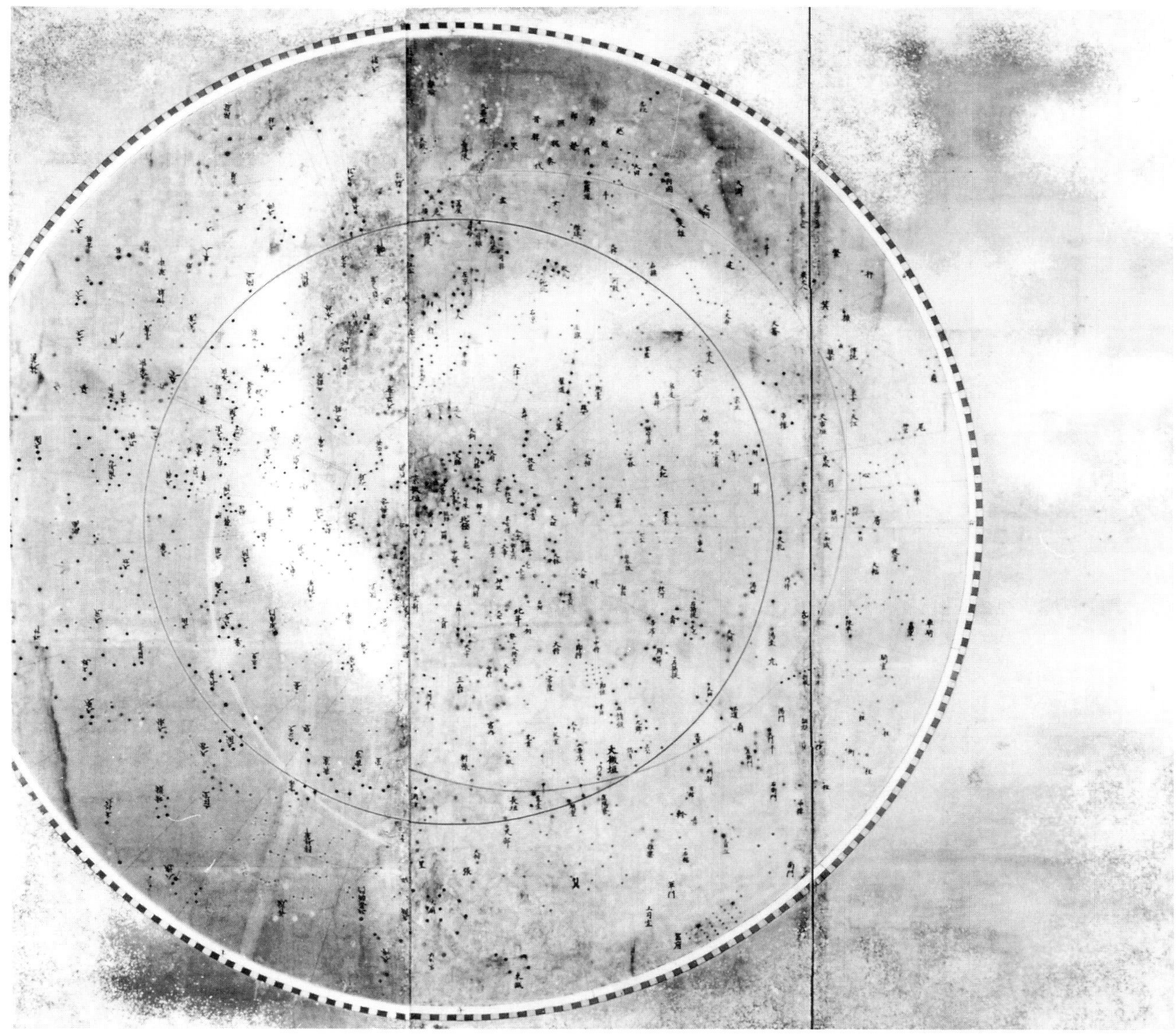

CELESTIAL MAPS BY NAGAKUBO SEKISUI

A work by the Mito geographer Nagakubo Sekisui, *Tenshō kanki shō* (Brief explanation of astronomical phenomena, 1774), contains a rare rotating planisphere, or volvelle, with the Chinese constellations.[37] The text describes why celestial chart boards were made and how to use them.

Tenshō kanki shō is a small pamphlet, and the circular map it contains is correspondingly small and fastened to one of the pages by a thread at its center (the North Pole) so it can rotate (fig. 14.9). The twenty-eight lunar lodges are depicted, but most constellations are omitted. The Milky Way is white on a blue background, the stars are either black or white circles, the equator is red, and the ecliptic is yellow. The position of the sun along the ecliptic is marked at twelve equal intervals including the winter

37. According to Watanabe, *Kinsei Nihon tenmongaku shi*, 2:832 (note 1), an 1824 work, *Tenmon seishō zukai* (Explanation of mapmaking in regard to the arrangement of the stars and astronomical phenomena), has the same content as the *Tenshō kanki shō*, including the rotating planisphere.

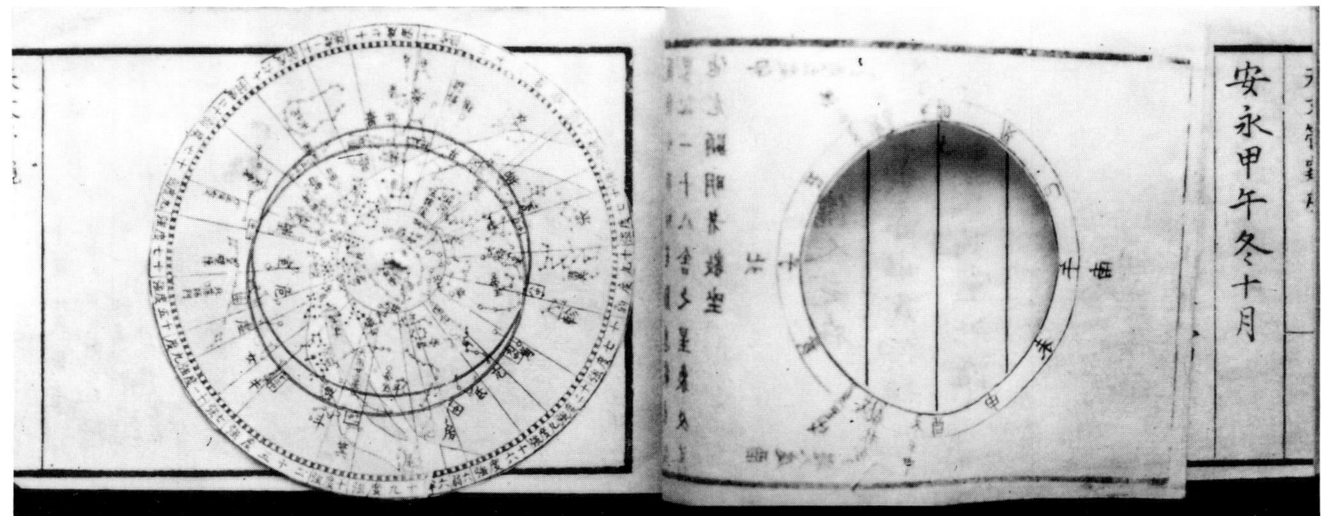

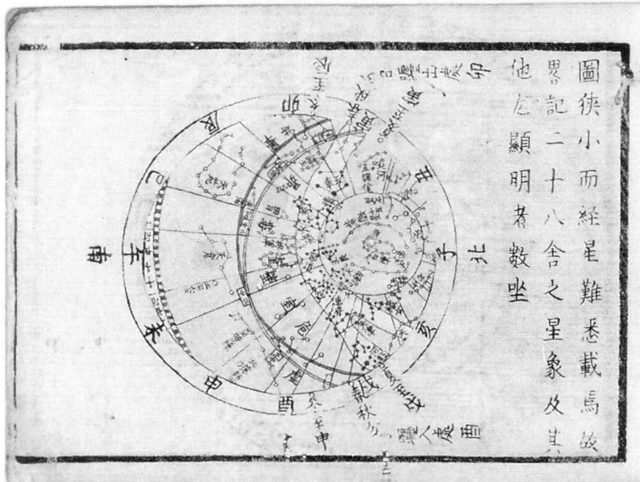

FIG. 14.9. ROTATING STAR MAP IN THE *TENSHŌ KANKI SHŌ* BY NAGAKUBO SEKISUI, 1774. In the first view is shown a rotating star map (left) over which the preceding page (right), with its cut-out circle representing the horizon, may be flipped, as in the second view. The star map is centered on the North Pole, and the equator (concentric circle) and ecliptic (eccentric circle) are also depicted.
Diameter of the rotatable star map: 10.8 cm; horizon circle: 7.7 cm. Owned by Kazu Tsuguto. Photograph courtesy of Kazuhiko Miyajima, Ōsaka.

solstice and the vernal equinox. In addition, the circle of constant visibility and the circle of constant invisibility are shown. Around the circumference, a scale in degrees and the extent of each of the twenty-eight lunar lodges are marked. Covering the map is a page with a circular window—the circle representing the horizon.

Nagakubo is also thought to be the author of an anonymous, undated *Tenmon seishō zu* (Map of the arrangement of stars and constellations) (fig. 14.10), which is larger and shows more stars than the planisphere in the *Tenmon seishō zukai*; the reason for this presumption is that the two works are similar in appearance and bear similar names.[38]

A scroll containing a rectangular celestial map, formerly in the possession of Kobayashi Yoshio (d. 1991), dates from 1796 and carries the inscription "I, Sekisui, drew this" (fig. 14.11). The arrangement of the stars and the notes on the map are duplicated precisely from Shibukawa's *Tenmon seishō zu*, except for the Milky Way, ecliptic, and thirty-five degree north declination line (corresponding to the latitude of Kyōto), lacking on Shibukawa's map.[39]

THE INFLUENCE OF JESUIT CELESTIAL MAPS AND STAR CATALOGS

When the Jesuits arrived in China toward the end of the Ming dynasty, they compiled star catalogs and maps from observations and tried to identify the stars in the Chinese constellations with those of the West. These Chinese

38. *Tenmon seishō zu* is homonymous with Shibukawa's *Tenmon seishō zu*, although there is a difference in the character *sei*. The former *Tenmon seishō zu* is illustrated by Watanabe, *Kinsei Nihon tenmongaku shi*, 2:833 (note 1), who believes Nagakubo produced the map. I was shown a celestial map by the late Kayahara Motoichirō that is identical to the *Tenmon seishō zu* reproduced in Watanabe. The map I saw was folded, and to its reddish brown cover was attached a piece of white paper that said "*Tenmon seishō zukai*, written by Master Chō Sekisui," which is Nagakubo Sekisui's pen name. I have also seen an identical celestial map owned by the University of Kyōto, and to its blue cover is attached a piece of white paper that says "*Tenmon seishō zu kan* [all of the *Tenmon seishō zu*]." Considering this, despite their slightly different titles, both the *Tenmon seishō zu* and the *Tenmon seishō zukai* must have been produced by Nagakubo.

39. About 1937, when he was studying at Daisan National High School, the astronomer Kobayashi was given the scroll, which had been found in the house of his English teacher, a Mr. Perkins.

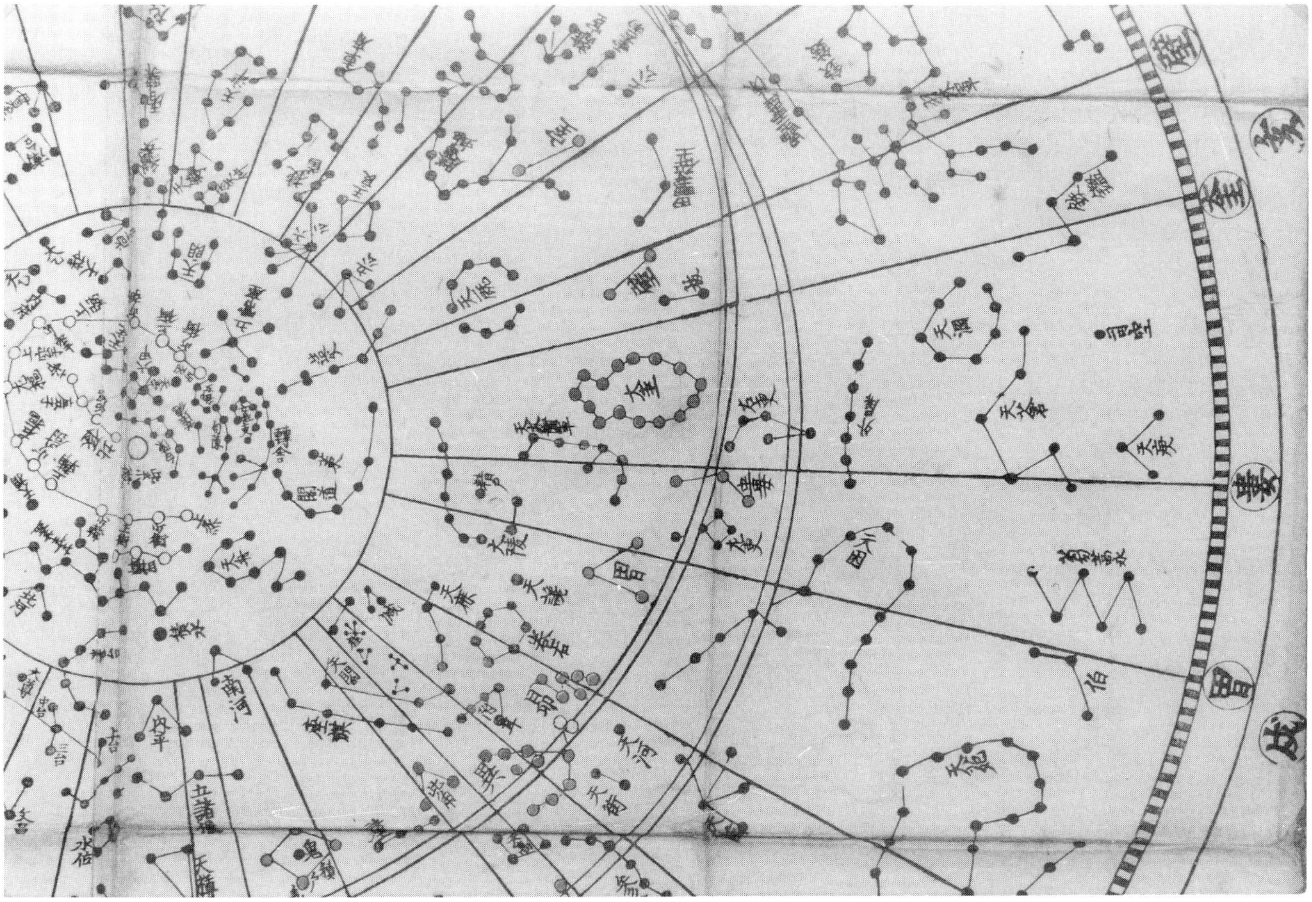

FIG. 14.10. ANONYMOUS, UNDATED *TENMON SEISHŌ ZU* AND DETAIL. On this celestial map, centered on the North Pole, the circles of the equator and ecliptic are shown as double black lines, the stars of the twenty-eight lunar lodges are painted red, and the rest of the stars are black circles or black dots. Size of the original: ca. 72 × 72 cm. Kyōto University. Photograph courtesy of Kazuhiko Miyajima, Ōsaka.

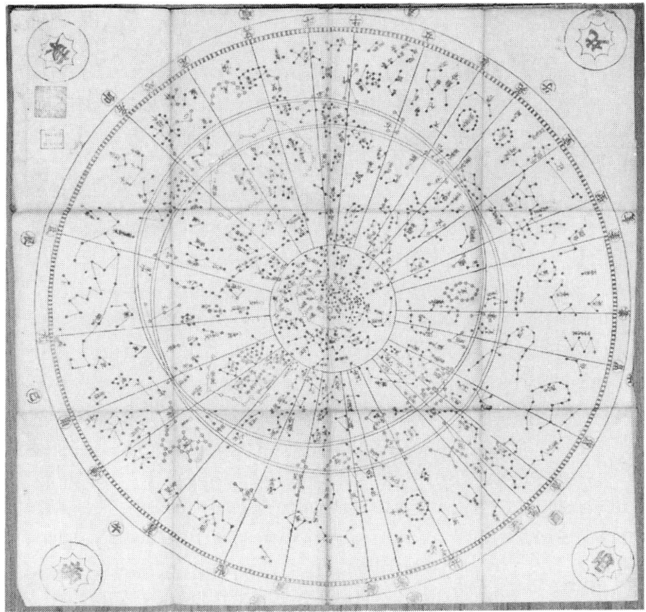

texts of the Jesuits and the aspects of European astronomy that they introduced—most notably the classification of stars by magnitude—had an influence on Japanese celestial cartography in the Edo period. We should be cautious about overrating their importance, however, as Nakayama suggests:

> Scholars who have studied the reports of such missions to China as Matteo Ricci and Johann Adam Schall von Bell tend to project their picture of seventeenth-century Chinese science onto that of Japan in the corresponding period and often conjecture that Japanese science also was substantially affected by the early contributions of the Jesuits. However, circumstances in the two countries differed greatly.
>
> The Japanese government in the beginning of the

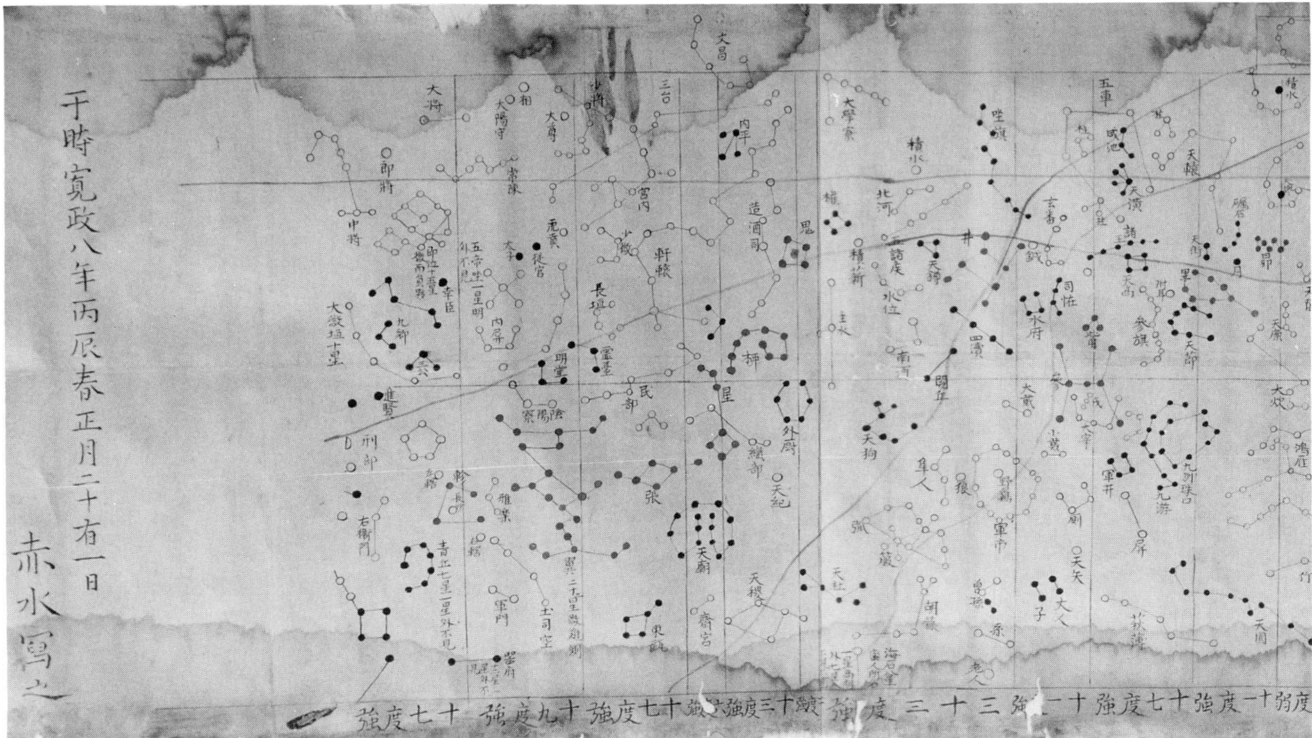

FIG. 14.11. SCROLL CONTAINING A RECTANGULAR CELESTIAL MAP DATING FROM 1796. This is based on the *Tenmon seishō zu* of Shibukawa. The field on which the constellations are drawn is surrounded by a black line, although there are a few constellations drawn in the top margin (north). The lower margin has the angular extent of each of the lunar lodges in degrees. The surrounding line, the equator, and the boundaries of the lunar lodges are drawn in black using a

> seventeenth century . . . strictly prohibited the diffusion of the Christian religion and Western learning in general. . . . The teachings of the missionaries in Japan thus were nearly eradicated.[40]

We can nonetheless see Jesuit influence in a few Japanese works from the eighteenth century and the first half of the nineteenth.

Jesuit scholarship had an influence through the *Yixiang kaocheng* (Treatise of astronomical instruments), originally compiled by Ignatius Kögler and others in 1744. In its revised version, 1755, it included the Qianlong star catalog (1752).[41] This star catalog provided star magnitudes and was probably used by the Japanese philosopher Miura Baien (1723–89) for his circular celestial maps of the Northern and Southern hemispheres (mounted as hanging scrolls), which classify the stars according to six magnitudes.[42] Traditionally, Chinese and Japanese drew stars larger on celestial maps to show they were important, but not necessarily brighter. More often than not, stars were shown as small circles of the same size without distinguishing brightness. Classification according to magnitude was introduced to China by the Jesuits. On Miura's maps, stars of the first and second magnitudes are shown as six- and five-pointed stars, respectively, with rays to highlight them. Stars of the third and fourth magnitudes are depicted as six- and five-pointed stars without rays, and stars of the fifth and sixth magnitudes are small circles with five or six rays.

The European-influenced custom of classification by magnitude may also be observed in maps by Takahashi Kageyasu (1785–1829) and Ishizaka Jōken (1783–1844). In his celestial map, *Seiza no zu* (Illustration of constellations, 1802), which he compiled after correcting the star catalog in the *Yixiang kaocheng* by taking precession into account, Takahashi classified the stars according to five magnitudes.[43] Ishizaka's *Hōen seizu* (Rectangular and circular star maps, 1826) has six classes of magnitude. The stars of the first magnitude are represented with five rays; other magnitudes are represented by the size of the circle representing the star.

40. Nakayama, *History of Japanese Astronomy*, 79–80 (note 1).

41. See Joseph Needham et al., *The Hall of Heavenly Records: Korean Astronomical Instruments and Clocks, 1380–1780* (Cambridge: Cambridge University Press, 1986), 171.

42. The positions of and connections between stars are based on older materials. These maps are kept at the house where Miura lived; there is also a celestial globe there made by Miura.

43. See Watanabe, *Kinsei Nihon tenmongaku shi*, 2:836, and Imoto, "Honchō seizu ryakkō" (both in note 1), where the *Seiza no zu* is referred to as *Tenmon sokuryō zu* (Astronomical survey map). It is

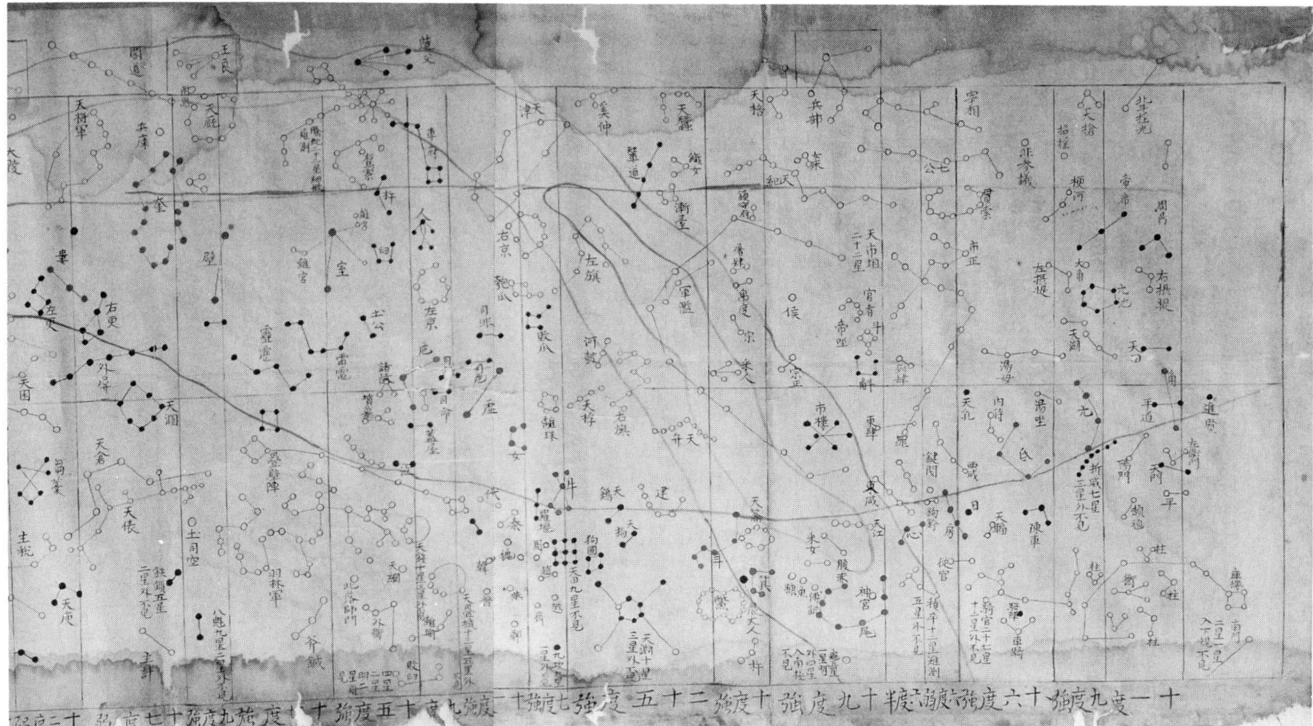

straightedge. The outline of the Milky Way, the ecliptic, and the line of declination near 35° (the latitude of Kyōto) are drawn freehand in red. The stars of Shi Shen's constellations are shown in red, those of Gan De in black, and those of Wu Xian and Shibukawa as small open circles.

Size of the original scroll: 51.5 × 178 cm. By permission of Kobayashi Tsuruko. Photograph courtesy of Kazuhiko Miyajima, Ōsaka.

One other work to be mentioned here is the *Kaisei nijū hasshuku zu* (Revised map of the twenty-eight lunar lodges), formerly in the possession of Kayahara Motoichirō and of uncertain date and authorship, but probably by Shiba Kōkan (1747–1818).[44] This map shows the equatorial zone with the lunar lodges and their surrounding stars. The equator is drawn as a straight horizontal line, and the boundaries of the twenty-eight lunar lodges are drawn as vertical straight lines, although not from top to bottom but projecting barely beyond the constellations. The Jesuit influence is apparent in that the stars are distinguished by magnitude. The determinative stars of the lunar lodges are painted in red. Some of the stars in the lunar lodges and other constellations are connected to each other by straight lines forming individual constellations. An unusual characteristic is that the extent of the lunar lodges and the values for right ascension of the determinative stars are given to a precision that far exceeds observational accuracy.[45] The widths of the twenty-eight lunar lodges (the distances of right ascensions between the standard stars) coincide with those obtained about the year 1200. There were fairly accurate observational values at this time, but inexplicably these values were not used.

CELESTIAL CARTOGRAPHY TOWARD THE END OF THE EDO PERIOD

Late in the Edo period there were maps that exhibited European influences and maps with dominant traditional elements. Of the former, two should be mentioned. One

illustrated in *Hoshi no bijutsuten: Tōzai no kichōna koseizu o atsumete* (Exhibition of stellar arts: A collection of rare old star charts of East and West), exhibition catalog, ed. Chiba Shiritsu Kyōdo Hakubutsukan (Chiba City Local Museum) (Chiba, 1989), no. 12, where it is stated that the stars are divided into six classes of brightness. Watanabe incorrectly says that only five magnitudes are distinguished. The illustration in the *Hoshi no bijutsuten* is a rectangular map, and it appears that the stars are drawn with black lines, with and without rays on six- and five-pointed stars. The boundaries of the lunar lodges are drawn vertically, as lines of right ascension. Lines of declination are drawn horizontally at ten-degree intervals (reflecting the Western division of a circle into 360 degrees, although Chinese tradition uses a division into 365¼ degrees). For Takahashi's involvement in the von Siebold incident, see above, pp. 439–40.

44. The map is thirty-two by ninety-four centimeters, and on the left margin is written "Tōto [Edo] Shinsenza, edition of Shunharō," Shunharō being Shiba Kōkan's pen name.

45. For example, the first lodge (Kaku) is measured as eleven *do* (degrees), eighty-one *fun* (minutes), twenty-one *byō* (seconds), and fifty-seven parts, and the next lodge (Kō) is measured as twenty *do*, ninety-six *fun*, thirty-one *byō*, and twenty-three parts.

is Umetani Tsunenori's *Tenshō sōsei no zu* (Map of all the stars in the heavens, 1814).[46] The map is rectangular, with the equator represented by a straight horizontal line in the center. Straight horizontal lines at the top and bottom edges represent the circles of constant visibility and invisibility. The ecliptic is a freehand curved line. Some constellations, such as the Big Dipper, are drawn above the line of constant visibility. Shibukawa's new constellations are also shown. The boundaries of the twenty-eight lunar lodges are drawn as vertical lines. The arrangement of these boundaries and the stars in the constellations are quite different from the old system used in celestial maps such as the *Tenmon seishō zu* and the system of the *Yixiang kaocheng* and *Hōen seizu*. It is similar to the old system, however, in that the constellations are relatively few and there are many large errors in declination.

The second example is Satō Tsunesada's *Shinsei tenkyū seishō zu* (New star map of the celestial sphere, 1815), on which lines of right ascension and declination are drawn every ten degrees in imitation of the equatorial stereographic projection used for maps of the world.[47] Although there are some errors, the application of this projection to a celestial map is interesting: not only is it a rarity, but it also reflects European influence via maps of the world, which were being introduced to Japan by Dutch traders. The four sheets form a map showing two circular hemispheres. One has its lines of right ascension ranging from 0° to 180°, and the other from 180° to 360°, so that the appearance of the map is similar to a European world map divided into circular Western and Eastern hemispheres. The circumference is a double circle graduated alternately in black and yellow at one-degree intervals of declination. The equator, drawn as a horizontal straight line in the center of the map, is graduated in black and red alternately at intervals of two degrees of right ascension. The northern and southern tropics are drawn in yellow. The circles of constant visibility and invisibility are drawn in red, and all other lines of right ascension and declination are drawn in black. The stars are red and black, with constellations connected by black lines. The Milky Way and large and small Magellanic clouds are drawn in off-white. A trace of an amendment in the shape of the large Magellanic cloud is visible. The stars around the South Pole are also drawn.

Some maps show a continuity with tradition. One is an individual, remarkably accurate map of Chinese constellations; this is the *Tenshō kaisei no shin zu* (Revised star map), of uncertain date.[48] It is a pamphlet in which some constellations are drawn on every page. The brightness of stars is distinguished by size. The angular distance between stars is written in feet (*shaku*) and inches (*sun*), originally used to denote the length of objects on earth in China and Japan. Although equivalents differed by country, periods, and use, one *shaku* was approximately 22.5 to 33.3 centimeters, and ten *sun* equaled one *shaku*. Used to represent angular distance in this context, one *sun* equals about one degree.

An interesting map that was composed by the Tsuchimikado family in 1824 is the *Seizu hoten ka* (Star map poetry). Evidently it was based on the *Butian ge* (Song of the sky pacer)—the Chinese pronunciation of *Hoten ka*—a rhymed mnemonic used to teach the constellations written by Wang Ximing (pen name Dan Yuanzi, fl. ca. 590).[49] The map, included in a pamphlet and folded, measures twenty by ten centimeters. Abe no Haruchika's preface on the second page explains that it was published for beginners. The circular celestial map of the region of constant visibility appears on the third and fourth pages. Two types of stars are distinguished by black and white circles. In each constellation the stars are linked by straight lines and the names of the twenty-eight lunar lodges are written in the space between the second and third concentric circles. Shibukawa's constellations are not drawn. There is no poem on the circular celestial map, but there are verses from the *Hoten ka* in the book, probably within the rectangular celestial map.[50] As I mentioned above, the Tsuchimikado (Abe) family was in charge of astronomy and astrology at the imperial court. Because their work was secret, Watanabe points out that it is noteworthy that this celestial map was published.[51]

Tangentially, several copies survive of the Chinese work *Huntian yitong xingxian quantu* (Map indicating the arrangement of the stars throughout the entire heavens), with versions from 1822 and 1826 subsequently brought to Japan. The celestial map consists of eight sheets joined together. Since the map was engraved (intaglio) on wood, the prints are white on indigo (fig. 14.12).[52]

46. The *Tenshō sōsei no zu* is kept at the Chiba City Local Museum and illustrated in the *Hoshi no bijutsuten*, 22–23, no. 14 (note 43). On the left of the illustration is written "Umetani Tsunenori has produced this based on the Master Hokusui's celestial map that Umetani reduced to one-sixteenth of the original size" (the original of Master Hokusui was 266 × 809 cm). Master Hokusui is probably Asano Hokusui, the disciple of the writer, script writer, naturalist, and technician Hiraga Gennai (1728–79). Asano Hokusui gave lectures on ancient traditional astronomy, wrote a popular book on astronomy, and was active as an artist (Watanabe, *Kinsei Nihon tenmongaku shi*, 1:423 [note 1]).

47. The *Shinsei tenkyū seishō zu* is held at the Chidō Museum, Tsuruoka City, Yamagata Prefecture, and illustrated in *Hoshi no bijutsuten*, 10–11, no. 1, with explanation on 44 (note 43). Satō Tsunesada was a clansman of the Shōnai (northwestern part of what is now Yamagata Prefecture) clan.

48. Owned by Ogi Sadami, Nagoya, the map is reproduced in the *Hoshi no bijutsuten*, 22, no. 15, explanation on 45 (note 43).

49. See chapter 13 above, esp. p. 532.

50. Watanabe Toshio, personal correspondence.

51. Watanabe, *Kinsei Nihon tenmongaku shi*, 2:834–36 (note 1), discusses and reproduces the first four pages of this work.

52. I have introduced this here because neither of two important

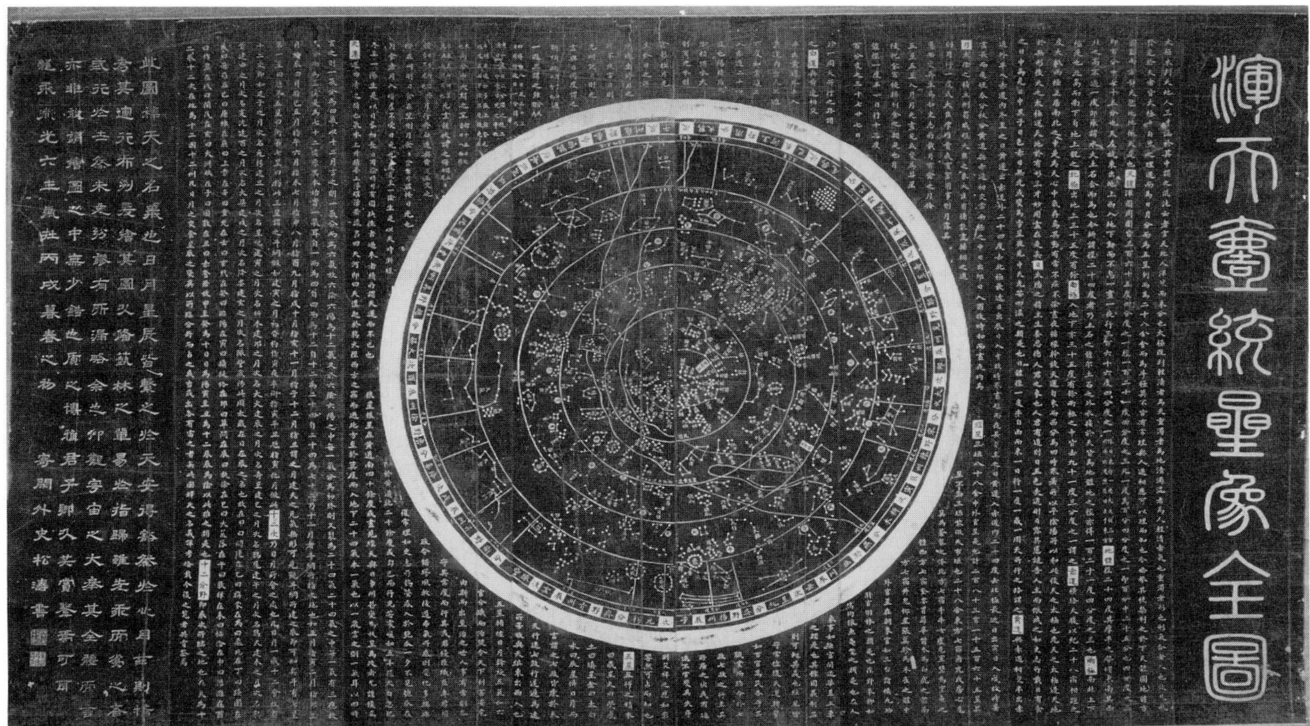

FIG. 14.12. *HUNTIAN YITONG XINGXIAN QUANTU*, 1826. On this Chinese map centered on the North Pole, the outline of the Milky Way is drawn, as well as the concentric circles of constant visibility, the Tropic of Cancer, the equator, the Tropic of Capricorn, and constant invisibility. Between the circle of constant invisibility and the outermost circle twelve *shi* and twelve Jupiter stations are marked.
Size of the original: 125 × 219.5 cm. Tsuyama Local Museum. Photograph courtesy of Kazuhiko Miyajima, Ōsaka.

CELESTIAL MAPS SHOWING WESTERN CONSTELLATIONS

Because of the Edo policy of national isolation until 1720, a few Chinese books such as the *Tianjing huowen* provided the only means for the Japanese to learn about European astronomy until the ban on importing foreign books was partially lifted. This led to the appearance of a great number of Chinese and Dutch books in Japan and to the rise of *Rangaku*, the study of European science through Dutch-language sources. After Shimoda was opened to American commerce in 1854, the study of Western science became known as *Yōgaku*, or "Western learning." *Rangaku* and *Yōgaku* laid the foundation for celestial maps based on Western astronomy to predominate in the Meiji period (1868–1912) and after.

One map from the Edo period by Shiba Kōkan, a polymath known for having popularized Copernican theory in Japan,[53] established this tendency toward European-style celestial maps. The copperplate print *Tenkyū zu* (Illustration of the spheres) (fig. 14.13) is the first Japanese celestial map to be published with Western constellations. It includes the Northern and Southern hemispheres, each centered on the ecliptic pole. From the center of the two circular celestial maps, straight lines of ecliptic longitude are drawn radially at twenty-degree intervals. On the chart of the Northern Hemisphere, eccentric circles indicate the equator and the Tropic of Cancer. In the Southern Hemisphere similar circles represent the equator and the Tropic of Capricorn. Both have circles of declination at 66°30′ representing the Arctic and Antarctic circles. The constellations are drawn in various colors.

The *Tenkyū zu* was made from a tracing of Frederick de Wit's *Planisphærium Cœleste* of about 1660–80, with

sources on Chinese celestial mapping, the Zhongguo Shehui Kexueyuan Kaogu Yanjiosuo (Archaeological Research Institute, Chinese Academy of Social Science [Academica Sinica]), *Zhongguo gudai tianwen wenwu tuji* (Album of ancient Chinese astronomical relics) (Beijing: Wenwu Chubanshe, 1980), and Pan Nai's *Zhongguo hengxing guance shi* (History of stellar observations in China) (Shanghai, 1989), mentions it.

53. See Sugano Yō, "Shiba Kōkan no chosho *Shutō dempō* to dōhan *Tenkyū zu* ni tsuite" (On Shiba Kōkan's book *Shutō dempō* [The introduction of vaccine] and the copperplate print *Tenkyū zu*), *Nihon Yōgakushi no Kenkyū* 5 (1979): 65–100. For outstanding research on Shiba, see Kuroda Genji, *Shiba Kōkan* (Tokyo: Tōkyō Bijutsu, 1972).

FIG. 14.13. COPPERPLATE PRINT OF SHIBA KŌKAN'S *TENKYŪ ZU*. The circular celestial map at the right is centered on the ecliptic North Pole and has the ecliptic as an outer circumference. In the top right corner Saturn is drawn, and at top left is Jupiter. On the left, the map centered on the South Pole has a waxing crescent moon in the upper corners, and surveying instruments are shown in the lower corners. To the left of the Northern Hemisphere are explanations, and to the

the names of Chinese constellations and stars attached and the two hemispheres transposed. The *Tenkyū zu* is interesting because it illustrates Western constellations. Traditional Chinese and Japanese celestial maps used only lines to connect the stars and did not have such figures drawn in.[54]

Hoshi Mandaras

Among the mandalas that are used for esoteric Buddhism are the so-called *hoshi mandaras* (star mandalas), both circular and rectangular (figs. 14.14 and 14.15).[55] Schematically, at the center is Mount Sumeru, the center of the Buddhist universe where the Buddha resides, and surrounding it are the figures of the Big Dipper god (the seven traditional stars and Alcor = 80 UMa), the god of the nine stars (sun, moon, five planets, and the imaginary celestial bodies Ketu and Rāhu),[56] the twelve signs of the zodiac, and the god of the twenty-eight lunar lodges.

54. The *Planisphærium Cœleste* was published in Amsterdam. It is said that de Wit's celestial map is based on the celestial map drawn on the top corner of Willem Jansz. Blaeu's *Nova totius terrarum orbis tabula*; see Imai Itaru, "Edo Jidai kagakushi no naka no Blaeu" (Blaeu in the history of science during the Edo period), *Rangaku Shiryō Kenkyūkai Kenkyū Hōkoku* 136 (1963), and Hirose Hideo, "Oranda tensetsu" (European astronomical theory), in *Yōgaku* (Western studies), 2 vols., ed. Numata Jirō et al., Nihon Shisō Taikei (Series of Japanese Thought), vols. 64–65 (Tokyo: Iwanami Shoten, 1972–76), vol. 1.

55. On Japanese mandalas and Buddhist cartography, see pp. 364–66.

56. These are the Sanskrit names. They usually indicate the lunar nodes.

left of the Southern Hemisphere are the names of Shiba Kōkan and Honda Saburōemon (the reviser) and the date of publication. The circumference of both hemispheres is graduated in one-degree increments. The names of Chinese constellations and stars are written on the map.
Size of the original: 40.7 × 90.1 cm. Owned by Miyamoto Masayuki. Photograph courtesy of Kazuhiko Miyajima, Ōsaka.

Aboriginal Celestial Cartography

Although many aboriginal Ainus lived in northern Japan (Tōhoku and Hokkaidō) as well as in the Kuriles and on Sakhalin, only a few remain today. Their original constellations have been handed down, but we have no evidence that star maps were ever compiled.

To the southwest of Japan proper are the Nansei or Ryūkyū Islands, comprising the Amamis, Okinawas, and Yaeyamas. The Ryūkyū kingdom in the Okinawa and Yaeyama Islands paid tribute to Japan from about the seventh century and to China from about the fourteenth, even during the Edo period, when it was controlled by the Satsuma clan based in Kyūshū. After the Sino-Japanese War of 1894–95 the islands came completely under Japanese control. Culturally, both China and Japan have exerted influence on the Ryūkyūs since antiquity, and this joint influence may still be seen. In regard to the indigenous view of the heavens, however, constellations that are not related to those of either China or Japan have been passed down to us.

An example of these constellations is found on a celestial map that has recently come to my attention (fig. 14.16).[57] It covers two pages in a book titled *Seizu* (Star map) that, according to its cover, was copied in 1827.

57. The map was first discussed by Yaeyama Museum curator Hanaki Yasuo in "Seizu ni tsuite" (On a star chart), *Ishigaki Shi Shi no Hiroba* (Ishigaki Municipal History Forum) 11 (1987): 1, 3–7. Although it has been known locally since that time, it has not come to the attention of many scholars in the main archipelago. I had the opportunity to examine it in 1992.

FIG. 14.14. A CIRCULAR *HOSHI MANDARA*. Mount Sumeru is at the center, surrounded by representations of celestial deities. The mandala dates from the end of the Heian period (794–1185).
Size of the original: 117 × 83 cm. Horyū Temple, Nara. Photograph courtesy of Yabuuchi Kiyoshi, Kyōto.

FIG. 14.15. A RECTANGULAR *HOSHI MANDARA*. The layout is similar to figure 14.14.
Size of the original: unknown. Sanukibō Temple. Photograph courtesy of Yabuuchi Kiyoshi, Kyōto.

The book contains information about the arrangement of the stars, the times and places of their appearance, the conditions of the winds and waves at sea, and farming. Most of the constellations on the map have been identified with modern stars and constellations, but some have not because they were used only locally or because additional data (such as direction, dates of compilation, times of appearance) are not in the book. Presumably the map depicts the night sky at a particular date and time, but this warrants further investigation.

Conclusion

One of the oldest extant star maps is that on the ceiling at the burial mound of Takamatsuzuka. Although it does not cover the entire heavens and is not accurate, including incorrectly drawn constellations, some parts are rather realistic. Chinese influence may be seen in all extant Japanese star maps, including that at Takamatsuzuka, until European astronomy was introduced. The boundaries of the twenty-eight lunar lodges at irregular intervals take the place of lines of right ascension or celestial longitude on European star maps. Only a few circles, notably the celestial equator and the ecliptic, correspond to Western circles of declination or ecliptic latitude. Seldom were

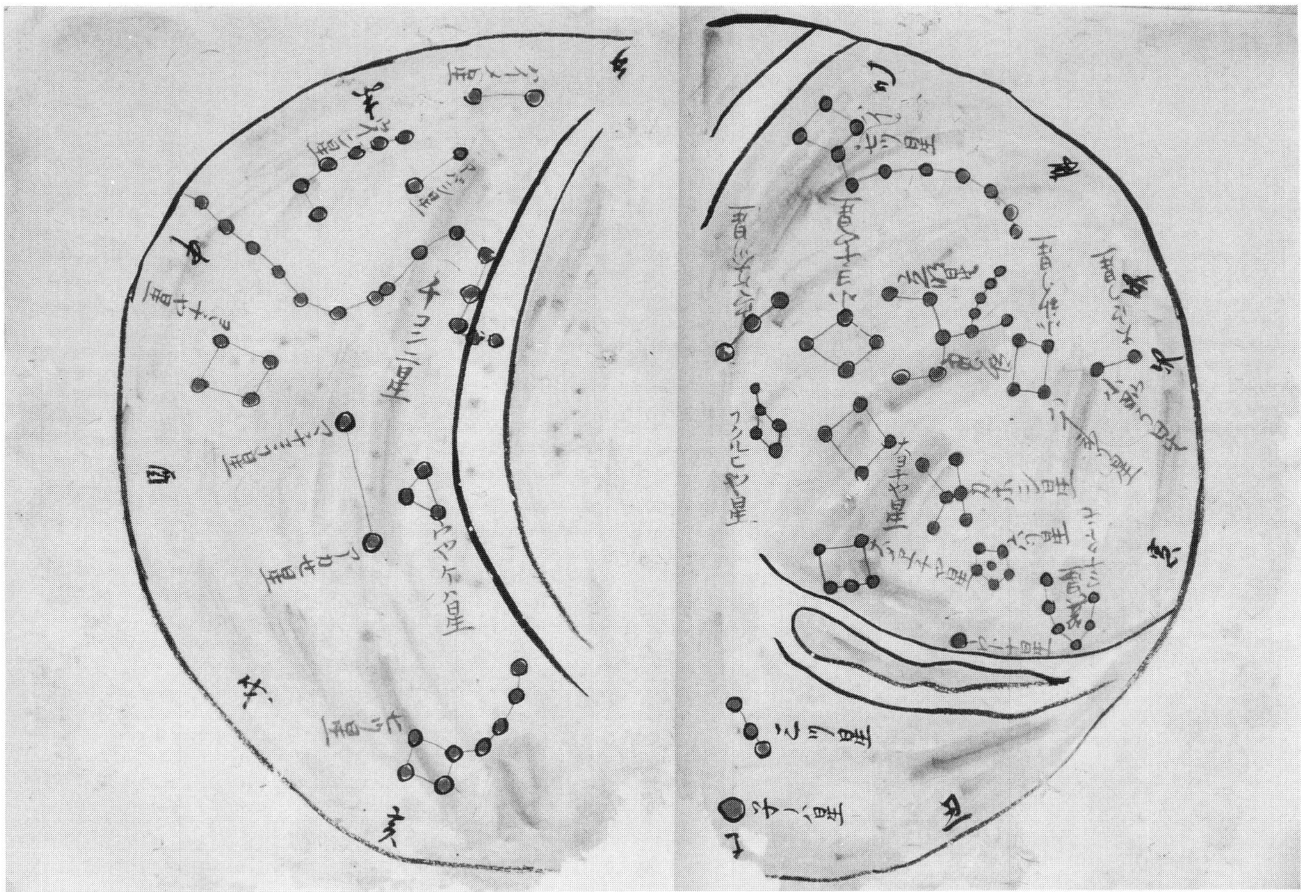

FIG. 14.16. MAP FROM THE RYŪKYŪ ISLANDS. The map is encircled by a black line around which are the names of the twelve *shi*, which represent the points of the compass. The outline of the Milky Way is also shown in black, and the stars are depicted by open black circles painted red and connected to other stars by red lines. The names of the constellations are entered in black or red, and prominently drawn at the bottom center is Ninoha-bushi, the Pole Star.
Size of the original: 25 × 21 cm. Yaeyama Museum, Ishigaki Island. Photograph courtesy of Kazuhiko Miyajima, Ōsaka.

stars represented according to brightness, nor were illustrations of the constellations drawn; the dots or small circles used to represent the stars were linked only by straight lines; and the constellations were Chinese, with the exception of the Japanese ones added in the seventeenth century by Shibukawa Harumi.

In the Edo period European astronomy was introduced originally through Chinese literature and then through Dutch works and other books. Gradually European elements in Japanese star maps increased, and the Japanese started to draw more constellations around the celestial South Pole. From the Meiji era to the present, Japanese maps of the heavens are mainly based on European astronomy.

Farmers and fishermen also had their own constellations, but it appears that they were hardly ever drawn. The indigenous peoples of northern Japan and Okinawa used other constellations. They are depicted in books from the island of Ishigaki in Okinawa Prefecture dating to the Edo period.

Cartography in Greater Tibet and Mongolia

15 · Maps of Greater Tibet

Joseph E. Schwartzberg

Introduction

Despite its limited population, the area referred to in this chapter as Greater Tibet has given rise to a remarkably rich and varied cartographic tradition. Moreover, this tradition remains vibrant and resistant to the cultural influence of far more populous and powerful neighbors. For this history, Greater Tibet is taken to correspond roughly to the area between India and China proper in which a distinctive form of Buddhism, widely known as Lamaism, is the dominant faith. But since Lamaism is a term that Tibetans themselves avoid, the preferred form, Tibetan Buddhism, will be followed throughout this work. The area of Tibetan Buddhism that I shall consider includes the whole of Tibet and Bhutan as now constituted as well as adjacent territories in India (particularly Ladakh and Sikkim), Nepal, and China (virtually all of Qinghai and parts of Xinjiang, Gansu, Sichuan, and Yunnan) in which Tibetan culture either predominates or is significantly represented.

In areas where Tibetan Buddhism blends with other cultural traditions, the omission of maps from these other traditions in this chapter should not be interpreted to mean that such works do not exist. Some of those maps—for example, several from the culturally eclectic region of Nepal—have already been discussed in volume 2, book 1 of *The History of Cartography*, along with others that fall unquestionably within the Hindu tradition. However, other Nepali maps associated with Hindu patrons are discussed here if they follow a style more consistent with Tibet than with India. Apart from the well-known faiths of Buddhism and Hinduism, over much of Greater Tibet there is continued adherence to a set of beliefs and practices associated with Bön, a pre-Buddhist cult, that has, as we shall see, relevance for the history of cartography. Over the centuries Bön and Buddhist practices became inextricably intertwined, and it will not always be possible to differentiate between the two in discussing the sources of various aspects of Tibetan cartography. Figure 15.1 shows the locations of most of the places to be discussed in this chapter.

I shall first consider the state of knowledge about the indigenous cartography of Greater Tibet, note where Tibetan maps are known to exist or are likely to be discovered, discuss the sources and general nature of the highly diversified cartographic corpus, point out some of the canons of cartography followed by monks, scholars, and other artists charged with making maps, and consider the nature of what one might call "folk cartography." I shall then consider cosmographic mapping, which has a long and rich history in the region and remains an important aspect of its culture. This will be followed by a discussion of geographic maps, ranging in coverage from the entire world as known to Tibetans, through maps of regions of widely varying territorial extent, to plans of towns and individual shrines. The conclusion will assess the nature of the entire cartographic corpus of the region and seek to relate it to the traditions of other cultures.

The State of Our Knowledge

As was true of the cartography of India and Southeast Asia, maps from Greater Tibet have not previously been given due recognition in standard histories of the subject. Bagrow, for example, ignores the area completely.[1] Santarém, citing Francis Wilford, calls attention to an eighteenth-century three-dimensional map of the kingdom of Nepal that was presented to Warren Hastings, but that no longer extant work does not appear to have belonged within the Buddhist tradition.[2] Similarly, Adler, citing Pullé, discusses a Nepalese map that clearly belongs more

1. Leo Bagrow, *History of Cartography*, rev. and enl. R. A. Skelton, trans. D. L. Paisey (Cambridge: Harvard University Press; London: C. A. Watts, 1964; reprinted and enlarged, Chicago: Precedent Publishing, 1985). Bagrow does refer, on 207, to "a Buddhist map showing the world as a floating lotus-blossom," but what he is describing is actually a Hindu cosmographic conception. See Joseph E. Schwartzberg, "Introduction to South Asian Cartography," in *The History of Cartography*, ed. J. B. Harley and David Woodward (Chicago: University of Chicago Press, 1987–), vol. 2.1 (1992), 295–331, esp. 300.

2. Manuel Francisco de Barros e Sousa, Viscount of Santarém, *Essai sur l'histoire de la cosmographie et de la cartographie pendant la Moyen-Age et sur les progrès de la géographie après les grandes découvertes du XVe siècle*, 3 vols. (Paris: Maulde et Renou, 1849–52), 1:364. Wilford's description of the map in question is quoted at length in Schwartzberg, "South Asian Cartography," 326 (note 1).

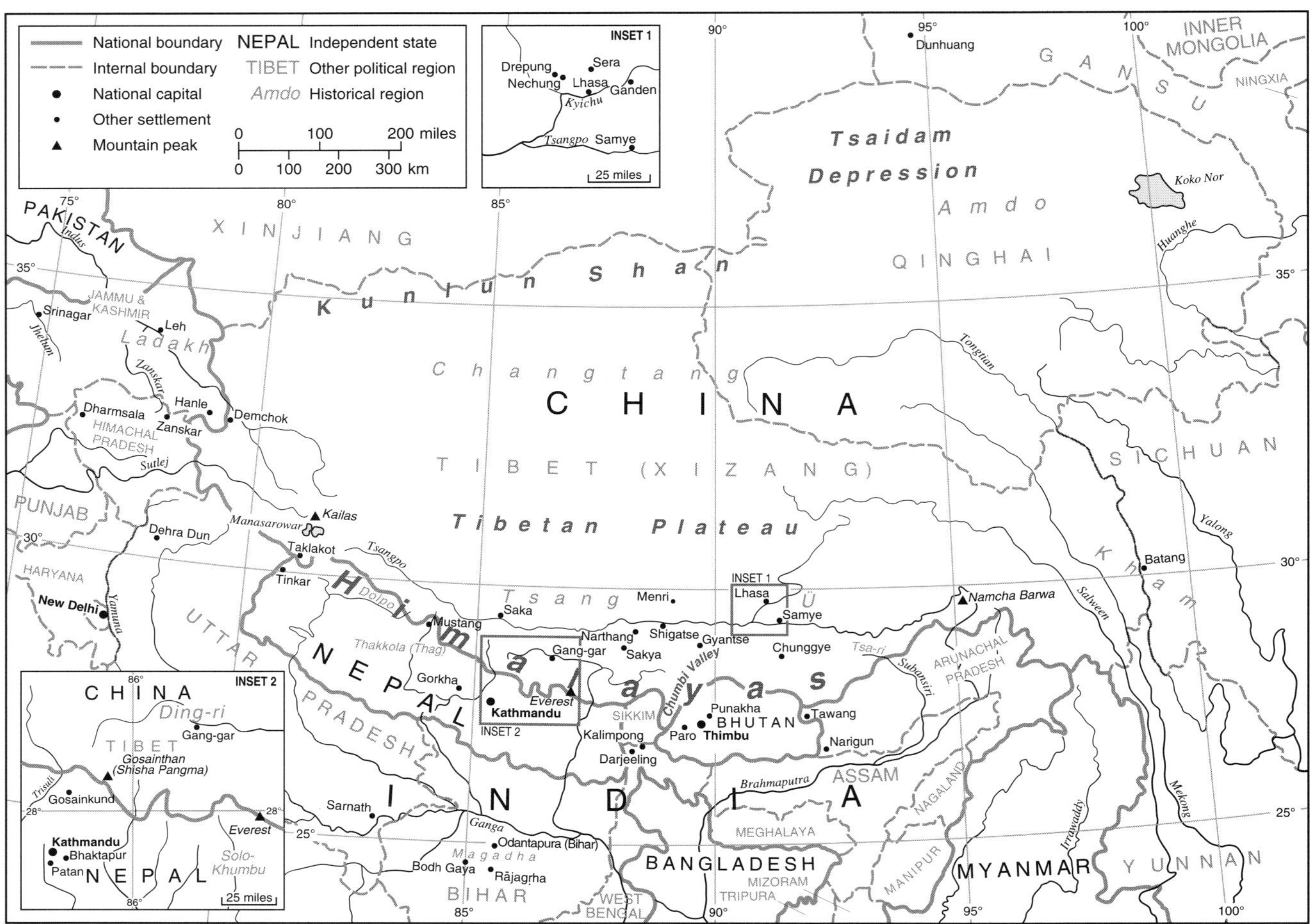

FIG. 15.1. REFERENCE MAP FOR GREATER TIBET.

to the Hindu than to the Buddhist tradition.[3] Other well-known histories totally fail to note maps from Greater Tibet, but there is little point in providing a detailed list of them.[4]

Perhaps the earliest outside scholarly reconstruction of any form of indigenous cartography from this region is that of the Augustinian missionary Antonio Agostino Giorgi (1711–97), who in 1762 published an engraved Tibetan cosmography that was almost certainly copied from a local source.[5] Among relatively modern scholars, the task of explicating and illustrating Tibetan cosmography was resumed by Waddell, whose exposition of Tibetan Buddhism first appeared in 1895.[6] Many others have since contributed to our understanding of this subject. One lavishly illustrated work, *Ajia no kosumosu + mandara*, is particularly worthy of mention in that it puts the cosmography of Greater Tibet into a sweeping pan-Asian context.[7] The importance of cosmography in the

3. Bruno F. Adler, "Karty pervobytnykh narodov" (Maps of primitive peoples), *Isvestiya Imperatorskogo Obshchestva Lyubiteley Yestestvoznaniya, Antropologii i Etnografii: Trudy Geograficheskogo Otdeleniya* 119, no. 2 (1910), esp. 231; Francesco L. Pullé, *La cartografia antica dell'India*, Studi Italiani di Filologia Indo-Iranica, Anno IV, vol. 4 (Florence: Tipografia G. Carnesecchi e Figli, 1901).

4. The notes relative to this point in Schwartzberg, "South Asian Cartography," 296 and 298 (note 1), are as applicable to Greater Tibet as they are to India.

5. Antonio Agostino Giorgi, *Alphabetum Tibetanum missionum apostolicarum* (Rome: Typis Sacrae Congregationis de Propaganda Fide, 1762). The original manuscript is in the Bibliotheca Apostolica Vaticana, Rome. Giorgi's detailed illustration, Cosmografia Buddhistica Tibetana, facing page 472 of the published version, has an explanation keyed to the original on 470–86. His diagram has been copied in several subsequent works including those by Laurence Austine Waddell, *The Buddhism of Tibet, or Lamaism, with Its Mystic Cults, Symbolism, and Mythology, and Its Relation to Indian Buddhism*, 2d ed. (Cambridge: W. Heffer, 1935), 79 (originally published in 1895); Pullé, *La cartografia antica dell'India*, 23 (note 3); and Sugiura Kōhei, ed., *Ajia no kosumosu + mandara* (The Asian cosmos), catalog of exhibition, "Ajia no Uchūkan Ten," held at Rafōre Myūjiamu in November and December 1982 (Tokyo: Kōdansha, 1982), 18.

6. Waddell, *Buddhism of Tibet*, esp. 77–111 (note 5), which include numerous cosmographic diagrams.

7. Sugiura, *Ajia no kosumosu + mandara* (note 5).

lives of Tibetan Buddhists cannot be overemphasized. Cosmographies form objects of contemplation in daily worship and serve a variety of other, largely didactic purposes.

With respect to geographic mapping, I noted in my discussion of South Asian cartography that a number of early British travelers in the Himalayan region—beginning with Francis Hamilton (formerly Buchanan) who spent fourteen months in Nepal in 1802–3—commented on having received "native maps," which they subsequently used in preparing maps of their own.[8] These maps came from diverse ethnic groups—not always identifiable—and probably varied in style accordingly. Some undoubtedly were from Tibetans or closely related peoples. Regrettably, none of the earliest maps Hamilton alluded to survives, and I know of no published account that describes them.[9] Not until 1863, when the results of the Tibetan expedition of the Schlagintweit brothers were published, do we have any careful reproduction of a Tibetan geographic map (fig. 15.42 below).[10] Two further examples of Tibetan cartography, both relating to Lhasa and its great Potala palace, were included in the work by Waddell, published several decades later.[11] Thereafter, a variety of additional maps appeared with increasing frequency in diverse publications; but as a rule the authors emphasized subjects other than cartography. Not until 1975–78, when several works by the anthropologist Barbara Nimri Aziz appeared, did any published work compare Tibetan maps and discuss their intrinsic qualities and the cognitive processes involved in their construction.[12] Shortly thereafter another anthropologist, Mary Shepherd Slusser, published several insightful works on traditional maps from the Vale of Kathmandu in Nepal.[13] Since then a considerable number of publications treating maps have appeared, mostly by art historians, students of religion, and others with little specialized knowledge of cartography. Nevertheless, some of these studies are very informative and are cited in this chapter. Although there is as yet no comprehensive work on the indigenous mapping of Greater Tibet, such a volume could be rather substantial. A trilogy of well-researched articles by Norwick reviews the geographic mapping of Tibet as a whole and discusses the ways that region has appeared on East Asian, Arabic, and Western maps, as well as on a few Tibetan works, since classical antiquity.[14] For the much more limited area of Nepal, Gurung has brought out a small, well-illustrated volume, *Maps of Nepal*.[15] Almost all the maps treated within that work, however, are either European or essentially modern maps based on surveys conducted since the mid-nineteenth century, and the style of the few maps that might be regarded as essentially indigenous is more Indian than Tibetan. Such works were discussed, therefore, in relation to South Asia.[16] There is also a published bibliography of Nepali maps, but it contains no reference to an indigenous work.[17]

In addition to these wide-ranging surveys, there are a few published works by historians of cartography that deal with specific issues relating to Tibetan maps. The oldest is a 1947 study by Nakamura of "Chinese" world maps preserved in Korea, which discusses what was in fact originally a Tibetan map of indeterminate age (fig. 15.30 below), a copy of which was brought from China to Japan during the ninth century.[18] The map was dis-

8. Joseph E. Schwartzberg, "Geographical Mapping," in *The History of Cartography*, ed. J. B. Harley and David Woodward (Chicago: University of Chicago Press, 1987–), vol. 2.1 (1992), 388–493, esp. 430–31.

9. The extensive journals of Hamilton in the Oriental and India Office Collections, British Library, London, may throw some light on the matter, however, since he had a very keen interest in cartography and discussed at great length maps he had acquired in other areas. See, for example, discussions and reproductions of Burmese maps in chapter 18 and table 18.1 below.

10. Hermann von Schlagintweit-Sakünlünski, Adolphe von Schlagintweit, and Robert von Schlagintweit, *Results of a Scientific Mission to India and High Asia*, vol. 4 (Leipzig: Brockhaus; London: Trübner, 1863), geographical map 3.

11. Waddell, *Buddhism of Tibet*, illustrations on 40 and facing 287 (note 5).

12. Barbara Nimri Aziz, "Tibetan Manuscript Maps of Dingri Valley," *Canadian Cartographer* 12 (1975): 28–38; idem, *Tibetan Frontier Families: Reflections of Three Generations from D'ing-ri* (Durham, N.C.: Carolina Academic Press, 1978), esp. 10–22; and idem, "Maps and the Mind," *Human Nature* 1, no. 8 (1978): 50–59.

13. Mary Shepherd Slusser, "Serpents, Sages, and Sorcerers in Cleveland," *Bulletin of the Cleveland Museum of Art* 66, no. 2 (1979): 67–82; idem, *Nepal Mandala: A Cultural Study of the Kathmandu Valley*, 2 vols. (Princeton: Princeton University Press, 1982); idem, "On a Sixteenth-Century Pictorial Pilgrim's Guide from Nepal," *Archives of Asian Art* 38 (1985): 6–36; idem, "The Cultural Aspects of Newar Painting," in *Heritage of the Kathmandu Valley: Proceedings of an International Conference in Lübeck, June 1985*, ed. Niels Gutschow and Axel Michaels (Saint Augustin: VGH Wissenschaftsverlag, 1987), 13–27.

14. Braham Norwick, "Locating Tibet—The Maps," in *Tibetan Studies: Proceedings of the 4th Seminar of the International Association for Tibetan Studies, Schloss Hohenkammer, Munich, 1985*, ed. Helga Uebach and Jampa L. Panglung (Munich: Kommission für Zentralasiatische Studien, Bayerische Akademie der Wissenschaften, 1988), 301–20; idem, "Why Tibet Disappeared from 'Scientific' 16th–17th Century European Maps," in *Tibetan Studies: Proceedings of the 5th Seminar of the International Association for Tibetan Studies, Narita, 1989*, ed. Ihara Shōren and Yamaguchi Zuihō (Narita: Naritasan Shinshoji, 1992), 633–44; and idem, "Modern Mapping of Tibet: A Cautionary Tale," presented at the 6th IATS Seminar, August 1992, forthcoming, read by me in draft form. Norwick's bibliography in the first of these three papers promises to be particularly useful for future historical research.

15. Harka Gurung, *Maps of Nepal: Inventory and Evaluation* (Bangkok: White Orchid Press, 1983).

16. Schwartzberg, "Geographical Mapping," 429–35 and 455–57 (note 8).

17. *Bibliographie du Népal*, vol. 3, *Sciences naturelles*, bk. 1, *Cartes du Népal dans les bibliothèques de Paris et de Londres*, by L. Boulnois (Paris: Editions du Centre National de la Recherche Scientifique, 1973).

18. Hiroshi Nakamura, "Old Chinese World Maps Preserved by the

FIG. 15.2. SCROLL ON ANNUAL PUBLIC DISPLAY IN NEPAL. This cotton scroll, dating from the seventeenth or eighteenth century, is shown hanging in the courtyard of the Guita-bahil monastery in Patan, in the Vale of Kathmandu. It is one among a number of sacred relics that the monastery exhibits to the public for several days each year. The many panels of the upper and lower registers of the scroll constitute a series of vignettes depicting successive scenes in a narrative. These are usually separated by clumps of trees or similar conventional devices. Thus one could construe the ensemble as a kind of narrative route map. Scrolls of this type have been recorded in the Vale of Kathmandu since about the mid-fifth century A.D.
Size of the original: unknown. Photograph courtesy of Mary Shepherd Slusser.

cussed in a wider context in other works by Nakamura.[19] In a similar vein, Unno discusses numerous Chinese, Korean, and Japanese maps embodying concepts originating in India and Tibet or transferred from India to Tibet and thence transmitted farther east.[20] Finally, a work originally written in Russian and later translated into English deals with a modern recension of what is perhaps the most ancient Tibetan map to survive in any form, allegedly a world map (fig. 15.27 below), said to be of an "Iranian-Tibetan" type, as well as a cosmography of the much better known "Indian-Tibetan" tradition.[21]

In contrast to the environmental conditions in India and Southeast Asia, where humidity, vermin, and other factors contributed to the decay and outright destruction of indigenous maps, the prevailingly cold, dry climate of Greater Tibet is conducive to their preservation. Hence, in the numerous palaces, monasteries, temples, and other religious edifices that dot the landscape of the region one finds mural paintings and manuscripts that, even after the passage of centuries, remain quite legible. Moreover, Buddhist monasteries, like Jain monasteries in India, assumed responsibility for preserving documents, and the tasks of copying documents and adorning monasteries were important aspects of monastic life. Not only in monasteries, however, but wherever Tibetan Buddhists assemble and in numerous private homes as well, one finds various types of cosmographies. These include both abstract mandalas and other types on which the various parts of the cosmos are identifiable, as well as maplike depictions of holy places and topographical paintings of places important in the lives of the Buddha, of numerous bodhisattvas, and of other holy personages.

Figures 15.2 and 15.3 show two of the many genres of sacred maps. Figure 15.2 depicts a small part of a long painted scroll map hanging in a Nepali monastery courtyard as part of an annual display of sacred relics. Other works are much more ephemeral. Figure 15.3 shows the preparation of a cosmogram in the form of an elaborate mandala (Tibetan *dkyil'khor*) prepared with powdered sand of various colors. Although such works may take a week or more to prepare, they are used only briefly in a religious service at which, following certain offerings, the mandala is swept away, reminding the faithful of the transitory nature of all existence.[22]

Although nature has been kind to the cultural legacy of Greater Tibet, non-Tibetans have been less sparing.

Koreans," *Imago Mundi* 4 (1947): 3–22. See also relevant notes in A. L. Mackay, "Kim Su-hong and the Korean Cartographic Tradition," *Imago Mundi* 27 (1975): 27–38.

19. See especially Hiroshi Nakamura, *East Asia in Old Maps* (Tokyo: Centre for East Asian Cultural Studies, 1962), 8 and fig. 1.

20. Unno Kazutaka, *Chizu no shiwa* (Map creases; or, Essays on the history of cartography) (Tokyo: Yūshōdō Press, 1985); and idem, "The Asian Lake Chiamay in the Early European Cartography," in *Imago et mensura mundi: Atti del IX Congresso Internazionale di Storia della Cartografia*, 3 vols., ed. Carla Clivio Marzoli (Rome: Istituto della Enciclopedia Italiana, [1985]), 2:287–96.

21. L. N. Gumilev and B. I. Kuznetsov, "Dve traditsii drevnetibetskoy kartografii (landschaf i etnos, VIII)," *Vestnik Leningradskogo Universiteta* 24 (1969): 88–101, translated as "Two Traditions of Ancient Tibetan Cartography (Landscape and Ethnos, VIII)," *Soviet Geography: Review and Translation* 11 (1970): 565–79.

22. At least two examples of colored sand mandalas that have been preserved, notwithstanding the rules calling for their destruction, are a rather small one, kept under a glass dome in the Jacques Marchais Center of Tibetan Art in Staten Island, New York, and a much larger work constructed by visiting lamas at the Minneapolis Institute of Arts in 1992. The latter work and its preparation are illustrated and described in *Arts: The Magazine for Members of the Minneapolis Institute of Arts*, June 1992, front and back cover and 6–8.

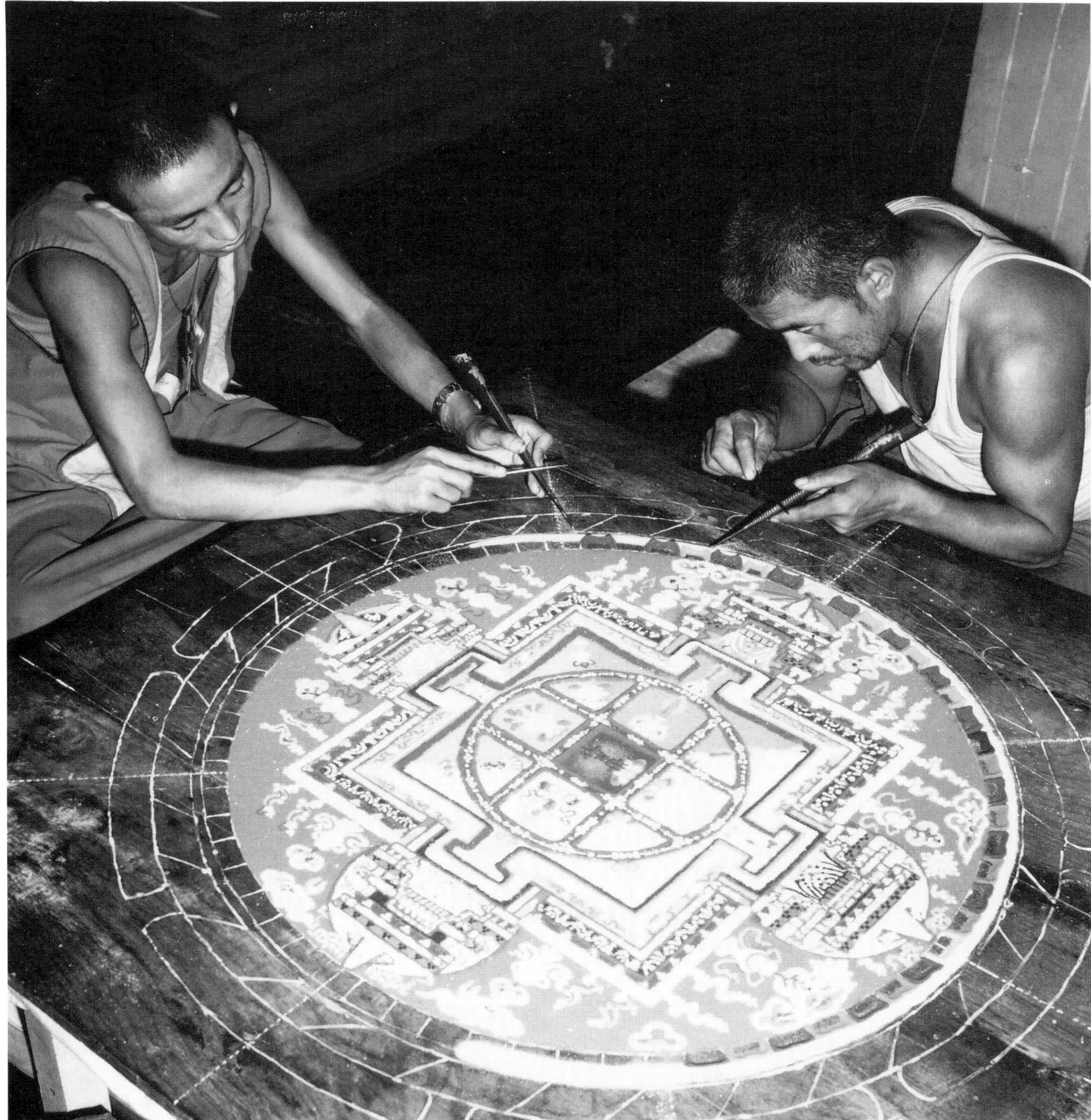

FIG. 15.3. CONSTRUCTING A SAND MANDALA. Monks may work for a week or more to construct an elaborate cosmographic mandala of powdered sand that, when finished, will provide the focal object of a religious ceremony. Once certain prescribed ceremonial offerings have been made, they will sweep away the sand with no visible regret. The work in progress in this diagram is being made at Ganden (dGa'ldan) monastery, which serves a Tibetan refugee community in southern India. By permission of Hakusuisha Publishing Co., Ltd., Tokyo.

Since the takeover of Tibet proper by the Communist regime of China, literally thousands of monasteries have been razed or badly damaged, with incalculable loss of their treasures. Fortunately, some of the greatest shrines, in particular the Potala, the traditional residence of the Dalai Lama, have been somewhat less affected, largely because of their appeal to tourists. Moreover, in surviving Buddhist temples and municipal and provincial museums throughout China, Tibetan objects are held in high regard. China's recent fostering of the tourist industry in the once-forbidden territory of Tibet now provides a significant incentive for cultural preservation. It is also one among many factors—including the mystique Tibet holds for many Westerners and the growing religious appeal of Buddhism—that have sparked widespread interest in Tibetica and a commensurate burgeoning of relevant scholarship, much of it of a very high standard.

Cosmographies and other maps from greater Tibet may now be found in museums, libraries, and private collections throughout the world, including many that do not specialize in Asian art. There are at least several dozen important public collections of Tibetica outside Tibet and other areas of China. Even in relatively small museums, one may find significant specialized collections—for example, in respect to Bhutanese art, at the Musée d'Ethnographie de Neuchâtel in Switzerland. The subject is still very much in its infancy, and this chapter can do no more than offer a sample of what exists in major collections worldwide.

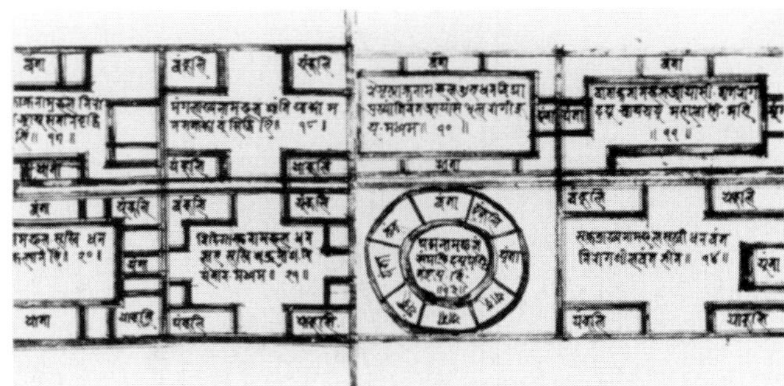

FIG. 15.4. SECTIONS OF A NEWARI TEXT ON ARCHITECTURE. This is part of a series of drawings from an architectural manual still in use among Newaris, the dominant ethnic group of the Vale of Kathmandu and early recipients of cultural influences from India. The drawings stipulate types of floor plans appropriate for a wide range of situations, depending on the orientation of the house and on the situation of the immediate neighborhood and quarter of the town, and they indicate blessings and curses that may be expected from building in certain ways.
Size of the original: unknown. From Gerhard Auer and Niels Gutschow, *Bhaktapur: Gestalt, Funktionen, und religiöse Symbolik einer nepalischen Stadt im vorindustriellen Entwicklungsstadium* (Darmstadt: Technische Hochschule, 1974), 36.

THE DEVELOPMENT AND NATURE OF TIBETAN MAPPING

Maps from Greater Tibet have evolved over many centuries and assume many forms. Innumerable mandalas, objects of Buddhist meditation, may be regarded as cosmographic maps. Many of these are highly abstract microcosmic representations of the entire universe. Others, styled as *bhavacakras* (Tibetan *srid-pa'i-'khor-lo*, wheels of life) are more obviously representational and depict various realms and states of existence in time and space. Still others envisage the universe as a geometrically ordered assembly of its deities. Various views represent the plane of terrestrial existence and the set of continents therein, centered on the sacred Mount Meru, the axis not only of the earth, but of the universe as a whole. Many focus on specific portions of the universe, such as its various heavens or hells. A popular cosmographic game focuses on the paths by which salvation may be reached. Much-used charts, which I shall consider only briefly, are those that relate to astrology and other forms of divination. And finally, paintings incorporating maplike components commonly depict the places associated with the lives of various deities, saints, or epic heroes. Among geographic maps, most relate to relatively small localities, especially towns where important monasteries are located or the monasteries themselves. But maps of larger regions, even including at least two believed to be world maps, have also been produced over the centuries, as have route maps intended especially for pilgrims.

Copies of Indian Buddhist cosmographies entered Tibet relatively early. The *bhavacakra*, a type of cosmographic diagram found today throughout Greater Tibet (e.g., fig. 15.18 below), is derived from an Indian prototype, arguably from one painted in either the fifth or the sixth century on the wall at the entrance to cave 17 of the great cluster of cave temples at Ajanta; but it seems likely that both that and the Tibetan versions were derived from a still earlier source.[23] Waddell notes that a copy of a *bhavacakra* is alleged, "with reason," to have been brought to the country by an Indian monk in the eighth century.[24] Whether or not this is true, it seems likely that some version of that cosmogram was transmitted to Tibet before the long period of decline of Indian Buddhism, which set in about the ninth century. A more clearly datable transmission, early in the thirteenth century, was that of a true-scale stone model of the renowned Mahābodhi Temple at the Indian town of Bodh Gaya, where the Buddha attained enlightenment. This relic and another construction copied from it were seen at Narthang (sNar thang) monastery in Tibet in 1936, and both may still be extant.[25]

23. John Huntington, an art historian specializing in Tibet, believes that both the Ajanta and Tibetan representations are based on some unknown pre-Ajanta prototype (personal communication, 12 May 1992).

24. Waddell, *Buddhism of Tibet*, 108 (note 5).

25. For relevant details see Rāhula Sāṅkṛityāyana, "Second Search of

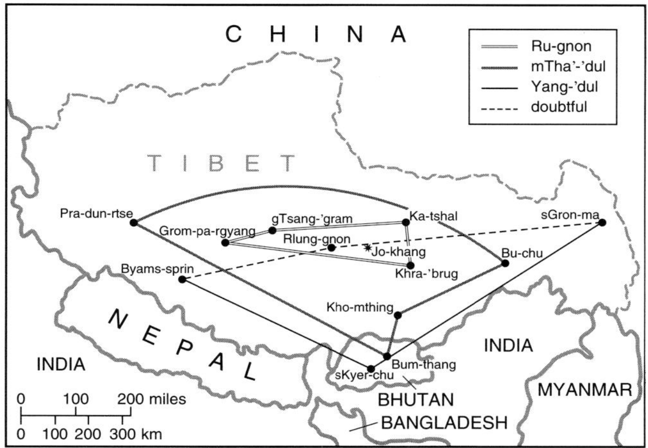

FIG. 15.5. DISPOSITION OF MAJOR TIBETAN TEMPLES. This figure indicates the geographic distribution of twelve important temples in three groups of four each, arranged in a rough approximation of what ideally ought to have been successively larger quadrilaterals, with the Jo-khang Temple at Lhasa at the center. The situation of Rlung-gnon Temple is especially problematic according to this schema. (The spellings here follow Aris.)
After Michael Aris, *Bhutan: The Early History of a Himalayan Kingdom* (Warminster, Eng.: Aris and Phillips, 1979), 16.

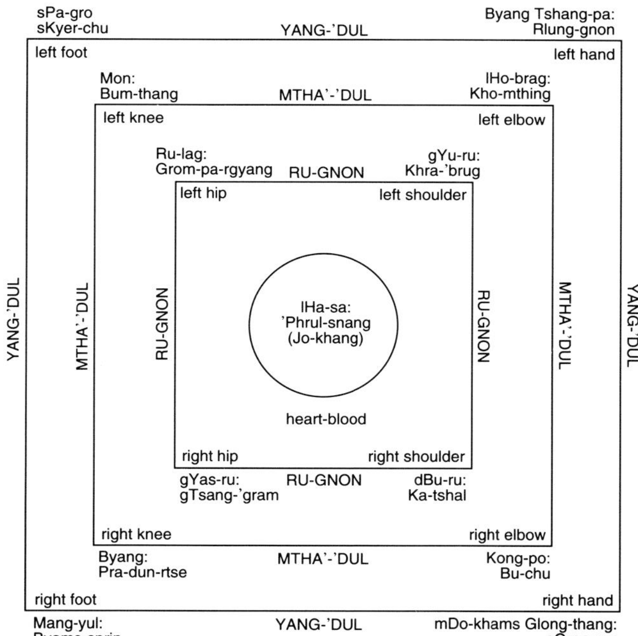

FIG. 15.6. IDEALIZED ARRANGEMENT OF MAJOR TIBETAN TEMPLES. This figure is a closer approximation to the idealized arrangement, according to the Tibetan derivative of an originally Chinese model. The names in capital letters are regional designations in three zones of diminishing civilization outward from Lhasa; those in upper- and lowercase letters are the temple designations; and the text in all lowercase letters tells the parts of the body of the underlying cosmic demoness to which each of the temples corresponds. Although shown here as a square, the Tibetan model actually is intended to be aligned along an east-west axis, in contrast to the Chinese model. (The spellings here follow Aris.)
After Michael Aris, *Bhutan: The Early History of a Himalayan Kingdom* (Warminster, Eng.: Aris and Phillips, 1979), 16.

Other evidence of cultural transmissions from India relates to architectural and town planning. A number of Nepali and Tibetan manuscripts, for example, include diagrams that set forth rules for laying out houses and whole communities.[26] Figure 15.4 provides an example from one such text. Other diagrams suggest that Chinese architectural models were also borrowed.[27] Influences in respect to architectural plans were also transmitted widely between Tibet and other Buddhist lands.[28]

Sanskrit Palm-Leaf MSS. in Tibet," *Journal of the Bihar and Orissa Research Society* 23 (1937): 1–57; and Charles E. A. W. Oldham, "Some Remarks on the Models of the Bodh Gaya Temple Found at Narthang," *Journal of the Bihar and Orissa Research Society* 23 (1937): 418–28.

26. For similar South Asian examples, see Schwartzberg, "South Asian Cartography," 318–21 (note 1). The period from the eighth to the twelfth century in particular was one in which there was a massive diffusion of artistic techniques and motifs from northeastern India, the last major bastion of Buddhism on the Indian subcontinent, to Nepal, Tibet, and Southeast Asia. This is abundantly documented in Susan L. Huntington and John C. Huntington, *Leaves from the Bodhi Tree: The Art of Pala India (8th–12th Centuries) and Its International Legacy* (Seattle: Dayton Art Institute in Association with the University of Washington Press, 1990).

27. The most comprehensive scholarly discussion of Tibetan architecture is Paola Mortari Vergara and Gilles Béguin, eds., *Dimore umane, santuari divini: Origini, sviluppo e diffusione dell'architettura tibetana/Demeures des hommes, sanctuaires des dieux: Sources, développement et rayonnement de l'architecture tibétaine* (Rome: Università di Roma "La Sapienza," 1987). Apart from its numerous illustrations, this volume is noteworthy for its exhaustive bibliography of primary and secondary sources (569–79). Other illustrations from architectural manuals appear in Pratapaditya Pal, *Art of Nepal: A Catalogue of the Los Angeles County Museum of Art Collection* (Berkeley: Los Angeles County Museum of Art in association with University of California Press, 1985), 176–77, diagrams for the construction of stupas, among others; Slusser, *Nepal Mandala*, vol. 2, fig. 89 (note 13), "depicting the optimal city plan as a mandala of eighty-one squares upon which is superimposed a circular plan (*alavau*) 'such as a serpent would make by bringing head and tail together' "; and Gerhard Auer and Niels Gutschow, *Bhaktapur: Gestalt, Funktionen, und religiöse Symbolik einer nepalischen Stadt im vorindustriellen Entwicklungsstadium* (Darmstadt: Technische Hochschule, 1974), 39 and 87 (respectively, a plan for a filter system for an underground well in the shape of a swastika and a set of illustrations for house construction, including those shown in fig. 15.4).

28. See, for example, Anne Chayet, "The Jehol Temples and Their Tibetan Models," in *Soundings in Tibetan Civilization*, ed. Barbara Nimri Aziz and Matthew Kapstein (New Delhi: Manohar, 1985), 65–72. Of particular interest in this article is the way the mixed axonometric and divergent perspectives of Tibetan paintings (see fig. 15.50) were distorted by Chinese architects who used only the former and were unfamiliar with the underlying conventions of the originals.

Not only were the layout and external appearance of temples of concern to the Buddhists of Greater Tibet, but so was placing them in accordance with systems of divination and geomancy. These systems were a "complex amalgam of Indian, Chinese, and Tibetan forms"[29] but were essentially Chinese in origin. It is believed that they were introduced by a Chinese princess, Kong-jo, who in about A.D. 640 was sent to marry the great Tibetan monarch Songtsen Gampo (Srong-brtsan-sGam-po) and brought with her, as a gift from her father, "a divination chart in 300 sections executed according to the Chinese divinatory sciences."[30] One important consequence of the belief in geomancy relates to the positioning of important Tibetan (including Bhutanese) temples. Although the origins of these locations are partially shrouded in myth, no fewer than ten texts arrange them (though with less than perfect agreement as to their names and locations) in three sets of four each, illustrated in figure 15.5. Each of these texts presumes the existence in the redactor's mind of a mental map, which is the Tibetan analogue, shown in figure 15.6, of the prior Chinese scheme conveyed by figure 4.1 above (p. 76) (with five rather than three concentric zones out from the capital).[31]

The origins of geographic maps of Tibetan provenance are obscure. If Gumilev and Kuznetsov's interpretation of the world map discussed below (pp. 639–42) is to be credited, that work should be regarded as an "Iranian-Tibetan" creation reflecting Persian geographic knowledge from the third and second centuries B.C.[32] How old the Sino-Tibetan world map brought to light by Teramoto might be is also open to question, though it cannot be later than the ninth century when, as noted above, a copy of it was transmitted to Japan. Very different from either of the foregoing is the oldest known locality map, which was found on a fragment of a Dunhuang manuscript thought to date from the Tibetan imperial era in the ninth century (fig. 15.7).[33] Although there would probably have been even earlier opportunities for Tibetans to observe and learn from the mapping activities of their Chinese and other neighbors, the earliest I can document relates to an expedition sent by the Mongol emperor Kubilay Khān (r. 1260–94) to explore and map the source region of the Huanghe (Yellow River), which lies in what is now Qinghai, a culturally Tibetan province. The oldest surviving regional map based on that expedition (not counting Mongol maps of the whole of China) is in a text dated 1366.[34] Subsequent Chinese mapping ventures in Tibet are well documented and are discussed elsewhere in this volume. Particularly noteworthy is the Jesuits' training two lamas to be surveyors in Beijing. These surveyors submitted a map of Tibet to those fathers in 1717; it was regarded as the most accurate to that date.[35] Obviously, further transmission of mapmaking skills among Tibetans would have been relatively easy. Some such transmissions will be noted below in the discussion of topographic mapping.

FIG. 15.7. FRAGMENT OF AN OLD MAP OF THE DUNHUANG REGION OF GANSU. This map fragment, painted in black ink on paper, is thought to date from the ninth century and is the oldest known Tibetan locality map as well as the oldest work to show any form of architectural plan, in this case of an unspecified Buddhist monastery. The locality depicted is believed to be Dunhuang in the present province of Gansu. The mixture of perspectives—oblique, frontal, and planimetric—is noteworthy. Though the text is in Tibetan, the style is essentially Chinese.
Size of the original: 30 × 48 cm. By permission of the Bibliothèque Nationale, Paris (Pelliot tibétain 933).

The peoples of Greater Tibet had thus developed a keen sensitivity to place, position, and relative location, both geographical and cosmographic. They were receptive to new ideas on such matters from a diversity of

29. Michael Aris, *Bhutan: The Early History of a Himalayan Kingdom* (Warminster, Eng.: Aris and Phillips, 1979), 12.

30. Quoted from a Tibetan legend in Aris, *Bhutan*, 12 (note 29).

31. Aris, *Bhutan*, 3–33 (note 29).

32. Gumilev and Kuznetsov, "Two Traditions," 568 (note 21).

33. This work is illustrated and discussed in Vergara and Béguin, *Dimore umane, santuari divini*, 89–91 (note 27), as well as in *Dieux et démons de l'Himâlaya: Art du Bouddhisme lamaïque*, catalog of an exhibition at the Grand Palais, 25 March to 27 June 1977 (Paris: Secrétariat d'Etat à la Culture, 1977), 66–67; and in Per Kvaerne, "Tibet: The Rise and Fall of a Monastic Tradition," in *The World of Buddhism: Buddhist Monks and Nuns in Society and Culture*, ed. Heinz Bechert and Richard Gombrich (London: Thames and Hudson, 1984), 253–70, esp. 258.

34. This map is illustrated and discussed at length in Herbert Franke, "Die Erforschung der Quellgebiete des Gelben Flusses in Nordosttibet unter dem Mongolenkaiser Qubilai," in *Der Weg zum Dach der Welt*, ed. Claudius C. Müller and Walter Raunig (Innsbruck: Pinguin-Verlag, [1982]), 59–61.

35. This is discussed in Clements R. Markham, ed., *Narratives of the Mission of George Bogle to Tibet and of the Journey of Thomas Manning to Lhasa* (1876; New Delhi: Mañjuśrī Publishing House, 1971), lxi–lxii.

sources, sometimes as a consequence of Tibetan conquest of neighboring lands and often resulting from travels by pilgrims between Tibet and other lands to which Buddhism had been disseminated. They devised numerous ways to express cartographically their view of the universe and its constituent parts. Their appreciation of maps also resulted in their using cartography for a variety of didactic purposes, largely teaching cosmographic ideas, but also, for example, in drawing anatomical and medical diagrams, preparing architectural models, and conceptualizing specific intellectual problems.[36] In this chapter I shall examine a small sample of Tibetan maps, but before doing so, it is in order to demonstrate the existence of certain canons by which they were drawn.

CANONS OF CARTOGRAPHY

One is struck by the amount of conventionalization that characterizes so many Tibetan maps, though not to the extent of denying them their originality. Much has been written by art historians about the stylistic and iconographic canons of Tibetan painting, sculpture, and architecture; and it is clear that the rules that have evolved (and continue to evolve) are also largely applicable to maps, both cosmographic and geographic, assuming that those maps are made by monks and others whose training requires learning the conventions.

Starting with the most abstract of cosmographic maps, a mandala, we see in figure 15.8 a monk demonstrating the laying out of the basic structure of what will ultimately become a rather complicated design (compare fig. 15.3). Constructing a mandala is an important exercise in the initiation of many novice monks, and learning the requisite discipline takes much practice. Long poems, comprising a series of rhyming verses, provide mnemonics to assist in this endeavor. Brief excerpts from one such poem, in translation, follow:

> Having thus completed these preliminary rites,
> one must construct the place for the temple [for which the mandala is a metonym] . . .
> Take light-coloured soil which is viscous, pliable and soft, . . .
> Sprinkle it and make it completely smooth with pure consecrated water and sweet-smelling medicinal incense.
> Smear cotton threads with white and red colouring and consecrate them as Method and Wisdom possessed of no duality.
> Cover the sphere of the void (viz. the space for the *maṇḍala*) with rays (viz. lines) of white and red,
> (the four) bordering lines (of the square),
> (the four) crossing lines (two diagonal and two straight across),
> the encircling line (inside the square).[37]

FIG. 15.8. PREPARATION OF A MANDALA AS A PART OF THE INITIATION RITUAL FOR A TIBETAN MONK. Here a novice monk at Chimre monastery is seen laying out the basic plan of what will probably become an elaborate mandala following well-established conventions for the proportions of the elements. Other conventions will guide him in the use of color, the choice of signs, and so forth.
From Manfred Gerner, *Architekturen im Himalaja* (Stuttgart: Deutsche Verlags-Anstalt, 1987), 48.

And so on for several more pages.

Though usually prepared in two dimensions, mandalas are always perceived as three-dimensional in the act of meditation. Among actual three-dimensional structures that may also be seen as mandalas (though their principal use is as reliquaries) are *chortens*, the term by which stupas are known in Greater Tibet. Every structural element of a *chorten* has a specific cosmographic meaning.[38]

36. Diagrams of various types of systems in the human body and of the method of applying a medical treatment known as moxibustion appear in *Trésors du Tibet: Région autonome du Tibet, Chine* (Paris: Muséum National d'Histoire Naturelle, 1987), fig. 89. Other anatomical tables are illustrated in Liu Lizhong, *Buddhist Art of the Tibetan Plateau*, ed. and trans. Ralph Kiggell (Hong Kong: Joint Publishing, 1988), 314–15; Günter Schüttler, *Die letzten tibetischen Orakelpriester: Psychiatrisch-neurologische Aspekte* (Wiesbaden: Franz Steiner, 1971), facing 65. Architectural models have been discussed above. A well-illustrated discussion of what may be construed as a road map used for didactic purposes appears in Charles Genoud, *Buddhist Wall-Painting of Ladakh*, trans. Tom Tillemans (Geneva: Edition Olizane, 1982), painting no. 1 on unnumbered page in section on Spituk, and 103–6. A smaller photograph of the same map and a briefer explanation appear in Richard Gombrich, "The Buddhist Way," in *The World of Buddhism: Buddhist Monks and Nuns in Society and Culture*, ed. Heinz Bechert and Richard Gombrich (London: Thames and Hudson, 1984), 9–40, esp. 28, pl. 22.

37. David L. Snellgrove, ed. and trans., *The Nine Ways of Bon: Excerpts from "gZi-brjid"* (London: Oxford University Press, 1967), 199.

38. A well-illustrated, though very speculative, study of the subject for South, Southeast, and East Asia is Adrian Snodgrass, *The Symbolism of the Stupa* (Ithaca, N.Y.: Southeast Asia Program, Cornell University,

FIG. 15.9. SYNTHETIC LANDSCAPE EMBODYING A DIVERSITY OF CONVENTIONAL ELEMENTS EMPLOYED IN TIBETAN MAPS AND OTHER FORMS OF PAINTING. From David P. Jackson and Janice A. Jackson, *Tibetan Thangka Painting: Methods and Materials* (London: Serindia Publications, 1984), 154, drawn by Robert Beer.

FIG. 15.10. THE POTALA OF LHASA. This detail from a nineteenth-century native drawing of Lhasa demonstrates the divergent perspective so common on Tibetan paintings and drawings, which permits the viewer to take in much more of the object being presented than would the convergent perspective that characterizes Western art. Of particular note is the depiction of the complete path around the Potala, which, as the residence of the revered Dalai Lama, is a shrine worthy of circumambulation by devout Buddhist pilgrims. In actuality, no more than half of this path would be visible at any one time. Several pilgrims, one prostrate, are shown along the path.
Size of the original: unknown. From Laurence Austine Waddell, *The Buddhism of Tibet, or Lamaism, with Its Mystic Cults, Symbolism, and Mythology, and Its Relation to Indian Buddhism*, 2d ed. (Cambridge: W. Heffer, 1935), facing 287.

Though *chortens* are architectural features and hence, strictly speaking, outside our purview, plans for constructing them that appear in various manuals may be considered cartographic. Such paradigmatic drawings have been reproduced in a number of published works.[39] The rules for construction are set forth by Olschak and Thupten Wangyal, who introduce the subject by noting, "The perfect proportions of the Buddha's body correspond to the design of religious monuments."[40] At a more elaborate level of detail are the rules for drawing *bhavacakras* (compare fig. 15.18). Although these rules will not be discussed, I note them here simply to demonstrate that well-established conventions apply to drawings of considerable complexity.[41] As with all such rules, however, exceptions developed that in time themselves became conventions.

Finally we come to conventions followed in many geographic maps as well as in various types of hagiographic and narrative painting that commonly embody maplike depictions of landscape. Figure 15.9 is a synthetic diagram incorporating a large number of the conventional landscape signs used in such works. It is taken from a book that discusses the theory and practice of painting various forms of *thankas* (*thang-kas*, painted cloth hangings), in respect to composition, iconometry, color, shading, outlining, and so forth, specifically from an appendix that deals with motifs and signs. Within that appendix we find numerous (as many as several dozen) acceptable ways to draw trees, rocks, clouds, "cloud thrones," and other

1985). A more complete and reliable study is *The Stūpa: Its Religious, Historical and Architectural Significance*, ed. Anna Libera Dallapiccola in collaboration with Stephanie Zingel-Avé Lallemant (Wiesbaden: Franz Steiner, 1980).

39. For example, Pal, *Art of Nepal*, 176–77 (note 27); and Blanche Christine Olschak and Geshe Thupten Wangyal, *Mystic Art of Ancient Tibet* (1973; Boston: Shambhala, 1987), 18.

40. Olschak and Thupten Wangyal, *Mystic Art*, 18 (note 39).

41. For further details, see Geshe Sopa, "The Tibetan 'Wheel of Life': Iconography and Doxography," *Journal of the International Association of Buddhist Studies* 7, no. 1 (1984): 125–45.

features that characterize Tibetan maps and other representational paintings.[42]

Distinctive conventionalized ways of depicting settlements and individual works of architecture have also evolved in Greater Tibet. Some of these relate to colors used for particular kinds of edifices (e.g., red is often selected to represent religious shrines and white for ordinary residences), while others have to do with visual perspective. Figure 15.10, for example, depicting the famous Potala palace in Lhasa, illustrates what has been called "divergent perspective," which is almost opposite to the optically "correct" Western perspective, with one or more vanishing points. Here we see several sides of the Potala represented simultaneously, which one could not normally see from a single location on the ground. But to Tibetans familiar with the edifice, this approach makes eminent sense, since it conveys more of the information they want than would a Western view.

FOLK CARTOGRAPHY

Although most maps from Greater Tibet that have come to light were made by specially trained individuals who learned to follow the cartographic canons just noted, others are "naive" productions that nevertheless demonstrate impressive cartographic skills. Many, perhaps most, such maps were made at the request of foreign travelers and scholars. I have already noted that a number of European travelers in the Himalayas benefited from maps made by local informants; but I can recall no statement by a Himalayan traveler who had tried to obtain a map locally and failed. The map obtained by the Schlagintweit brothers during their expedition of 1854–58 (fig. 15.42 below) is a good example of what may be regarded as folk cartography. Made by a Bhotia chief who had evidently traveled much and seen many other works employing the conventions discussed above, this map does embody some of them in simplified form; but overall it has a spontaneous, untutored appearance. The same may be said of a map painted by a Sherpa a century or so later.[43] This charming work depicts the artist's home village at the foot of a mountain bearing rock formations and surmounted by clouds strikingly like some of those in figure 15.9. Within and around the village are men and women, represented as taller than their houses, going about their day-to-day activities (one of which is praying together with a lama), as well as animals drawn in a similarly exaggerated size.

A third example of the partial absorption of Tibetan cartographic conventions by folk artists appears in a painting of a townscape brought to light by Aziz.[44] This remarkably detailed map, painted in 1970 by an elderly artist of Ding-ri Gang-gar, the sole town of the Ding-ri region, north of Mount Everest, also adheres to some widespread Tibetan conventions. Among these are the use of a divergent (rather than a convergent) perspective, the inclusion of more features than the eye could possibly take in at a single viewing, the essentially pyramidal arrangement of the composition (compare fig. 15.48), and the use of off-white throughout for residences and of red for temples.

But not all Tibetan folk cartography reflects the seeming familiarity with more schooled cartographic creations found in the three examples previously discussed. A more naive, but nevertheless very informative, map of Ding-ri by a trader from that region has also been reproduced and discussed by Aziz.[45] This work (fig. 15.11) is more idiosyncratic than the others. It was drawn spontaneously at Aziz's request, and the artist, then about forty years old, had no familiarity with modern mapping. He had, however, traveled widely in Tibet and had probably seen various traditional maps from that region. Although his map does incorporate certain common premodern conventions, such as placing the artist in the center of the area portrayed and arranging various features and text as if seen from that vantage point, the sorts of specifically Tibetan iconographic conventions displayed in figure 15.9 are nowhere evident. Compared with maps of individual settlements, generally drawn from an oblique perspective, essentially planimetric maps of regions are relatively rare in Tibet. Nevertheless, if one compares figure 15.11 with figure 15.12, one can readily discern many features on the trader's map that reflect the topography shown on a modern map. It would probably be difficult to obtain a comparably detailed and topographically reliable rendering of the home region by an untrained layperson in most parts of the world.

Whether living in an environment of high mountains or largely barren plateau such as characterizes most of Greater Tibet somehow enhances cartographic ability or conduces to a particular approach to mapmaking is a question worthy of scholarly investigation. Some relevant experiments, admittedly on a very modest scale, have in

42. See David P. Jackson and Janice A. Jackson, *Tibetan Thangka Painting: Methods and Materials* (London: Serindia Publications, 1984), 150–72. The conventions illustrated are even in contexts that have nothing to do with the landscape of Greater Tibet. On a visit to the Institut für Tibetologie und Buddhismuskunde in Vienna in 1983, I was given a rather detailed map of that city prepared for a conference on Tibet by a visiting Tibetan scholar in a style that was immediately and unmistakably identifiable as Tibetan.

43. Herbert Tichy, *Himalaya* (Vienna: Anton Schroll, 1968), frontispiece and relevant text.

44. Aziz, "Tibetan Manuscript Maps," 35, with explanatory text on 34 and 36–37 (note 12); also reproduced in Aziz, *Tibetan Frontier Families*, among unnumbered plates between 96 and 97 (note 12).

45. Aziz, "Tibetan Manuscript Maps," 32, text on 34 and 36–37 (note 12). Some of the details relating to this map were conveyed to me by Aziz on 12 November 1992.

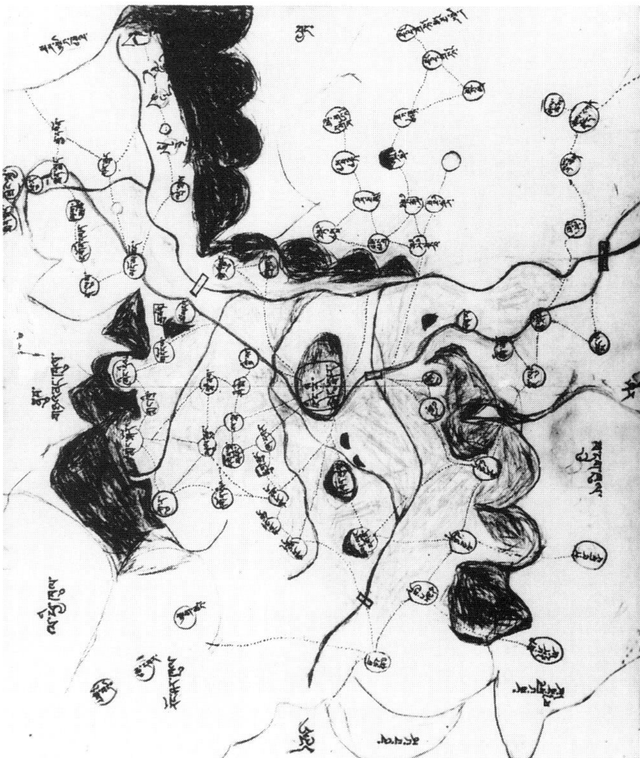

FIG. 15.11. INDIGENOUS MAP OF DING-RI VALLEY, TIBET. This is a map of the region of Ding-ri drawn from memory in 1970 by Dinggang Nima Woser, a local Tibetan trader, who had never seen a modern surveyed map. Drawn in ink and various pencils on paper. The orientation of map features and text is outward from the center, where the town of Ding-ri Gang-gar is situated. The meanings of the signs employed by the artist are, by and large, self-evident. The map is oriented here with north at the top. Mount Everest is in the lower right corner, and Rongphun is the circle directly above it. Langkor is shown as the central circle in the range that runs north-south on the left side of the map. Yoldon is the circle southeast of the northernmost circle and is connected to it by a dashed line.
Size of the original: ca. 48 × 30 cm. Photograph courtesy of Barbara Nimri Aziz.

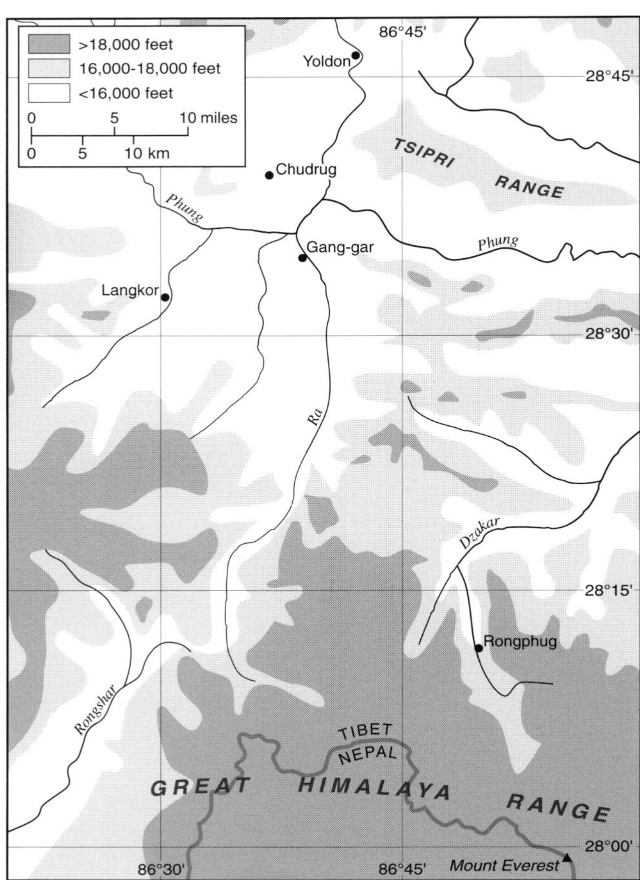

FIG. 15.12. MODERN MAP OF DING-RI VALLEY, TIBET. This map covers the same area as figure 15.11 and is provided for comparison.
After Barbara Nimri Aziz, "Tibetan Manuscript Map of Dingri Valley," *Canadian Cartographer* 12 (1975): 28–38, esp. fig. 4.

fact been carried out. In Nepal the cognitive skills of schoolchildren from various ethnic groups have been evaluated by testing their ability to draw a map of the route between home and school. Similar tests have been conducted among American students. Tests of Sherpa children among others were carried out and analyzed by an American anthropologist, James F. Fisher, who has had much experience working in the Himalayas. They yielded a total sample of twenty-eight maps from pupils in grades four through seven from a single school in the Solo-Khumbu region of Nepal, virtually in the shadow of Mount Everest. Of this sample Fisher reports:

> The maps of all but the youngest Sherpa children . . . are generally more sophisticated than similar maps drawn by Newar, Limbu, Chhetri, and Gurung children. . . .
>
> The maps of seventh-grade children [three out of four of which Fisher illustrates] attempt to symbolize spatial relationships abstractly and clearly. All three maps not only show the house and school as recognizable buildings but keep them relatively small, much more nearly proportional to the distances involved. The number of landmarks increases, and everything is labeled so that a stranger looking at the map will understand its symbols. Where trail intersections

> might cause confusion, these older students are careful to put in the other trail. That correct choices must be made to arrive at the school, not somewhere else, is understood.[46]

Noting that the maps of American children may seem to be more accurate (Fisher again uses the term "sophisticated") in that they are closer to the model Western planimetric ideal, Fisher observes that the differences

> can be explained by cultural and environmental differences between the two groups. Sherpas are used to an environment of steep slopes with relatively few flat places and must therefore handle spatial relations in three, rather than only two, dimensions. . . . To represent height on a two-dimensional surface is a challenging topographic problem. In a bird's-eye view of a steep slope . . . , two points may look close together although they are actually far apart—but on a vertical, not a horizontal, plane. The problem that Sherpa and most Nepalese children (and adults, for that matter) face is that faced by any mountain people explaining how far away a place is. A destination may be only half a mile away but up a steep and difficult slope. Therefore, as any trekker in Nepal knows, distance is measured in time, not linear units.
>
> The Sherpa children thus tend to construct their maps to show the relation of higher and lower, sacrificing that of depth and width, so that the map represents a vertical cross section rather than a bird's-eye view.[47]

The probable validity of the foregoing insights will, I believe, be demonstrated repeatedly in the analysis of a variety of maps from Greater Tibet in the rest of this chapter.

Cosmographic Maps

In chapters 8 and 17 of this book, as well as in the chapter on cosmographic mapping in South Asia in volume 2, book 1 of *The History of Cartography*, cosmographic ideas associated with various forms of Buddhism practiced outside Greater Tibet have been illustrated and discussed. Space limitations preclude comparably detailed coverage of the exceedingly elaborate cosmography of Greater Tibet, but in brief we may note that the region shares with other Buddhist areas the following general characteristics (most of them also applicable to Hinduism and Jainism).

1. Our universe is but one among many millions existing in infinite space and in time measured in many millions of years.
2. Each universe is vertically ordered and centered on an *axis mundi*, Mount Meru (Sumeru or Ri-rab).
3. The essential structure of the universe is tripartite, consisting of a lowermost realm, the Kāmadhātu, or Realm of Desire; a middle realm, the Rūpadhātu, or Realm of Form; and a uppermost realm, the Ārupyadhātu, or Realm of Nonform.[48]
4. The horizontal terrestrial plane, within the Kāmadhātu, constitutes only an insignificant part of the cosmos as a whole.
5. Around Mount Meru on earth are symmetrically arranged continents and seas, differently shaped in each of the four cardinal directions.
6. Each continent is of stupendous extent and is characterized by a unique set of denizens, some divine and others not.
7. Neither the cosmos nor the terrestrial plane is anthropocentric, since the southern continent, Jambūdvīpa, is the only one where human beings live.
8. Above and below the terrestrial plane within the Kāmadhātu are other planes of heavens and hells, where still other beings live.
9. The universe, like all animate creatures within it, goes through cycles of emergence, decline, dissolution, and rebirth.
10. The most exalted cosmic state is nirvana, and attaining it frees souls from the painful process of rebirth; but this state lies beyond the three levels of existence and is theoretically without form and therefore dimensionless.

Much of the reality just described is represented graphically, and also in three dimensions, in a great variety of ways, some of them quite abstract and others rather obviously representational. In this section I shall consider several of those ways, beginning with relatively abstract mandalas and proceeding to genres, also largely construed as mandalas, in which the component elements are increasingly discernible to laymen and even to uninitiated non-Buddhists. I shall also briefly illustrate how maplike cosmographic images are incorporated into paintings of deities, saints, and epic heroes.

Abstract Mandalas

Although there are many varieties of mandalas, all have in common the idea that they are a means of recognizing, through meditation, a divine palace, the abode of oneself conceived as a Buddha. Abstractly a mandala may denote

---

46. James F. Fisher, *Sherpas: Reflections on Change in Himalayan Nepal* (Berkeley and Los Angeles: University of California Press, 1990), 82–89, quotations from 83–84.

47. Fisher, *Sherpas*, 84–85 (note 46).

48. See table 17.1, p. 718, relating to the vertical ordering of the Hinayana Buddhist cosmos and the placement therein of the major realms and subrealms. Although there are differences between this schema and that of Tibetan Buddhism, there is an overall similarity of structure.

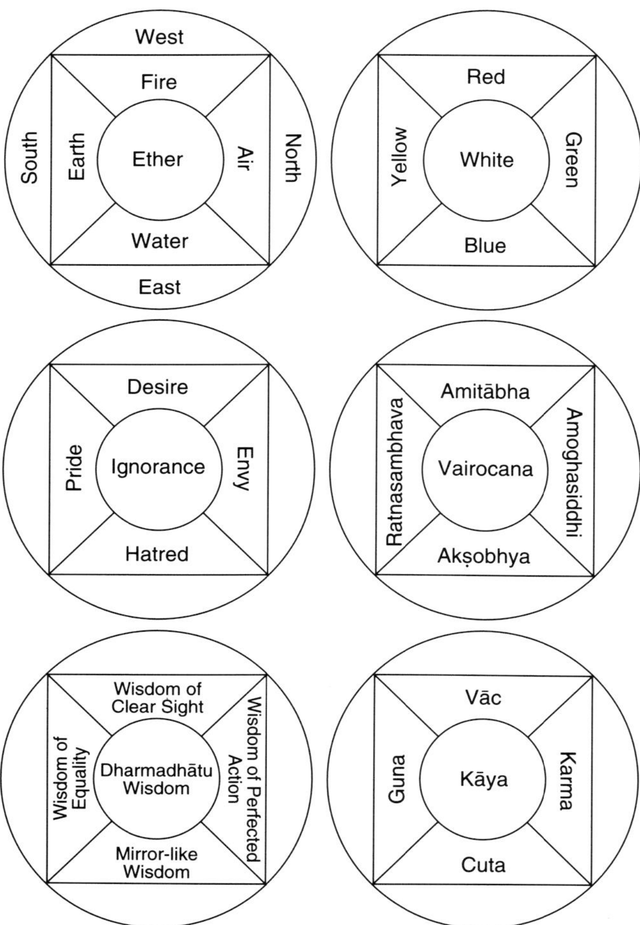

FIG. 15.13. BASIC SCHEMA OF THE *TATHĀGATAMAṆḌALA*. Six essential and inherently inseparable aspects of this particular mandala, associated with the process of rebirth and the attainment of nirvana, are shown in this diagram. It illustrates the essential plan of numerous other mandalas that are also used as objects of meditation by Tibetan Buddhists, though there are numerous differences of detail.
After Detlef Ingo Lauf, *Secret Doctrines of the Tibetan Books of the Dead*, trans. Graham Parkes (Boulder, Colo.: Shambhala, 1977), 111.

the divinization of our environment. This is true irrespective of the scale or form in which the mandala is perceived. Tucci defines a mandala as "a psychocosmogram."[49] Figure 15.13 shows a form of mandala associated with the Tibetan *Bar-do thos-grol* (Book of the dead), the so-called *tathāgatamaṇḍala*, that of five "transcendent Buddhas . . . , who, together with their female counterparts, the Buddha-Dākinīs, open up a realm of symbolism rich in interrelationships."[50] The sets of related meanings assigned to the various components of the mandala are elucidated by the six variations of the basic diagram. It is noteworthy that at the elemental level (view 1), showing water, earth, fire, and air, there is a remarkable accord between the Tibetan schema and the alchemical diagrams of the elements found in Western medieval and Renaissance astrological texts. The principal difference is that the latter assigns a central position to the human world, whereas the ineffable element "ether" occupies that position in the Tibetan view. In fact, the "human" world is not even an element in the type of Tibetan view presented here. The connections between the two call for additional research.[51] Only in respect to the basic elemental structure are the spatial and directional aspects of the mandala obvious; but it would be a mistake to adopt a purely materialist perspective and suppose that the remaining views are aspatial. Rather, one should see each view as suffusing all the others in a complex, multivalent cosmic whole. But even at the elemental level the parts are regarded as being related in the ways they are successively subsumed into one another until only the psychic center representing awareness remains. Lauf illustrates this point in a sequence of diagrams that begins with the "Unity of All Elements" and then proceeds successively with the dissolution of earth as it sinks into water, the dissolution of water as it sinks into fire, the dissolution of fire as it sinks into air, and at the very end, the dissolution of air as it sinks into awareness, the ultimate reality. This is what is said to occur in the "process of death," which initiates each stage of transmigration toward or away from the ultimate goal of attaining nirvana.[52]

The earliest surviving Nepali mandala, dating from the thirteenth century, is one of only three known examples of Nepali paintings predating the fourteenth century (plate 33). The structure of this work is fundamentally the same as that of the model presented in figure 15.13 (or the one being created in fig. 15.3).

Although most of the hundreds of mandalas that one

49. Giuseppe Tucci, *The Theory and Practice of the Maṇḍala, with Special Reference to the Modern Psychology of the Subconscious*, trans. Alan Houghton Brodrick (New York: Samuel Weiser, 1970; first published by Rider, 1961), 25. This work provides an extensive, though often controversial, discussion of the many forms and meanings of Tibetan mandalas and both supplements and complements the better known, but also controversial, work by Carl Gustav Jung, *Mandala Symbolism*, trans. R. F. C. Hull (Princeton: Princeton University Press, 1972). Many individual mandalas are discussed by Tucci in his magnum opus, *Tibetan Painted Scrolls*, 3 vols. (Rome: Libreria dello Stato, 1949). Also noteworthy is José Argüelles and Miriam Argüelles, *Mandala* (Berkeley, Calif.: Shambhala, 1972), which examines figures that may be construed as mandalas from a cross-cultural perspective. For the general interpretation of mandalas provided in this text I am largely indebted to Roger Jackson, Carleton College, Northfield, Minnesota.

50. Detlef Ingo Lauf, *Secret Doctrines of the Tibetan Books of the Dead*, trans. Graham Parkes (Boulder, Colo.: Shambhala, 1977), 64.

51. For a discussion of similar and even earlier ideas from medieval Europe, see John E. Murdoch, *Antiquity and the Middle Ages*, Album of Science (New York: Charles Scribner's Sons, 1984), 360–68.

52. Lauf, *Secret Doctrines*, 90–91, esp. fig. 15 (note 50).

can see in published works on Tibetan art are paintings, they also exist in other forms. Many are produced as relatively simple black-and-white block prints. I have already noted that elaborate mandalas are made out of colored sand for use—and destruction—in religious ceremonies. Others, comparably ephemeral, are sculpted by lamas out of butter mixed with various colored powders and worshiped by ordinary people in nightlong ceremonies.[53] Though most, as previously noted, are essentially two-dimensional, many forms of three-dimensional mandalas may also be found in monasteries, museums, and elsewhere. Among these are certain stupas, each part of which embodies multiple meanings. For example, the five vertically arrayed components of a Tibetan stupa correspond, in ascending order, to earth, water, fire, air, and ether, and each of those elements has corresponding properties such as color, *tathāgata* (form of a transcendant Buddha), and so forth, analogous to those indicated in figure 15.13.[54]

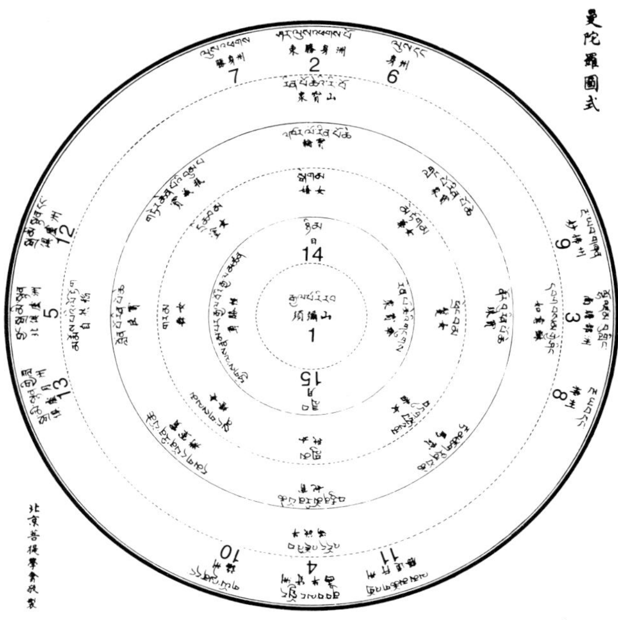

1. Sumeru; 2. Pūrvavideha; 3. Jambūdvīpa; 4. Aparagodānīya; 5. Uttarakuru; 6. Deha; 7. Videha; 8. Cāmara; 9. Aparacāmara; 10. Śāthā; 11. Uttaramantriṇa; 12. Kurava; 13. Kaurava; 14. Sun; 15. Moon.

FIG. 15.14. THE GENERAL SCHEME OF THE MERU MANDALA. This representation of the mandala, with text in both Tibetan and Chinese, was produced, presumably for academic purposes, at the Peking Buddhist Institute, probably in the late 1930s. The size and media of the original are not known. A description of the diagram appears in a compilation of minor Tibetan works in a Western format (but bearing no entry regarding the date or place of publication) that was acquired by Alex Wayman at the Tibetan Press in Dharmsala, Himachal Pradesh, India, in 1970. An analogous diagram appears in figure 15.15. Some of the more obvious geographical and astronomical features have been identified by the numbers 1–15.
After Alex Wayman, *The Buddhist Tantras: Light on Indo-Tibetan Esotericism* (New York: Samuel Weiser, 1973), pl. 8 (which contains a more complete translation than I have provided).

VIEWS FOCUSING ON THE TERRESTRIAL PLANE

As I noted in the brief introductory outline of Tibetan cosmography, within the terrestrial plane of the Kāmadhātu a set of primary and satellite continents are symmetrically arranged around Mount Meru. So too are a number of additional features, including the sun and the moon, which may be seen in figure 15.14, representing the so-called Meru mandala.[55] The work depicted appears to have been prepared as part of an academic undertaking in Beijing rather than as an object for worship. Although some of its components are more obviously geographic than others, each has its properly assigned place within the cosmos as a whole. Whether examples exist that show every such feature is not known, though some include a variety of features not indicated there.

Depictions of the Meru system are very numerous and assume a multitude of forms. Many appear as block prints, some quite simple and others relatively complex.[56] Still others are on painted, embroidered, and appliquéd *thankas*.[57] A particularly powerful and expressive assemblage of at least five large cosmographic mandalas, all focusing on the terrestrial plain, is to be found in central Bhutan. All are well preserved, but undated, fresco paintings on the walls of Bhutanese *dzongs* (castles), three in Paro and two in Punakha. Another considerably more abstract, but nevertheless striking, cosmographic mural is at a nearby *dzong* in Simtokha.[58]

53. Antoinette K. Gordon, *Tibetan Religious Art* (New York: Columbia University Press, 1952), 92–93.

54. Detlef Ingo Lauf, *Tibetan Sacred Art: The Heritage of Tantra*, trans. Ewald Osers (Berkeley, Calif.: Shambhala, 1976), 139.

55. See Alex Wayman, *The Buddhist Tantras: Light on Indo-Tibetan Esotericism* (New York: Samuel Weiser, 1973), 101–3.

56. Examples of such prints may be found in Nik Douglas, *Tantric Charms and Amulets* (New York: Dover, 1978), figs. 70–74, with accompanying text; Peter Gold, *Tibetan Reflections: Life in a Tibetan Refugee Community* (London: Wisdom Publications, 1984), 28; and Jackson and Jackson, *Tibetan Thangka Painting*, 38 (note 42).

57. A very clear example appears in Louis P. Van der Wee, "Rirab Lhunpo and a Tibetan Narrative of Creation," *Ethnologische Zeitschrift*, 1976, no. 2, 67–80, illustration on 77. The original work is in the Teo Sørensen Collection of the Ethnographical Museum of the University of Oslo. The cosmography drawn by Giorgi, *Alphabetum Tibetanum* (note 5), is also of this type.

58. Four such works from Bhutan are well illustrated, in color, in Sugiura, *Ajia no kosumosu + mandara*, 39–41 (note 5). The same source also illustrates a painting from nineteenth-century Tibet in a very different style (21). Other sources depicting these and other Bhutanese paintings include Manfred Gerner, *Architekturen im Himalaja* (Stutt-

FIG. 15.15. MERU MANDALA, PAINTED ON A WALL OF THE PARO *DZONG*, BHUTAN. This view, of unknown date, is probably based on Vasubandu's *Abhidharmakośa*. Like figure 15.14, it indicates the terrestrial continents situated in the four cardinal compass directions from Mount Meru, each with its distinctive shape and flanked by two smaller tributary continents. Meru, the home of the gods, rises from the center of the painting and is surrounded by seven ranges of golden mountains (depicted as a set of nested squares). It obscures the northern continent, Uttarakuru, whose presence is implied by the two square tributary continents that are visible. Unlike figure 15.14, which is oriented with east at the top, this view is oriented toward the north, thus giving greater prominence to the southern continent, Jambūdvīpa, where humans live. Numerous icons appear within the various continents, in the intervening oceans, and on Meru itself. Although these have not been individually identified, one may assume they correspond in the main to individual elements indicated in figure 15.14.
Size of the original: unknown. From Sugiura Kōhei, ed., *Ajia no kosumosu + mandara* (The Asian cosmos), catalog of exhibition, "Ajia no Uchūkan Ten," held at Rafōre Myūjiamu in November and December 1982 (Tokyo: Kōdansha, 1982), 41. ©Kōdansha Ltd. 1982.

Although the works at Paro and Punakha are based on two different texts, the *Abhidharmakośa* (a fourth- or fifth-century work by Vasubandu, an Indian Buddhist scholar) and a tenth-century Tibetan recension of the originally Sanskrit *Kālacakra Tantra* (text of the wheel of time), Olschak and Thupten Wangyal discuss them collectively as a sequentially ordered graphic testament to the earth's evolution.[59] The earliest phase of the cosmogony, depicted on one of the Paro murals, is based on the second of these texts. The painting gives much prominence to the elements out of which our world is aggregated. On it the cosmos is seen from above in the form of four concentric rings, whose varying colors symbolize the four elements, beginning with the outermost band, representing air (yellow) and progressing inward through fire (red), water (light blue), and earth (dark blue). On this last ring the main and satellite continents are ranged about Meru in the four cardinal directions, while inside the ring are eighteen narrow, concentric, brightly colored rings representing six successive triads of continent-mountain-ocean, a set of features reminiscent of an Indian Puranic view.[60] The most central part of the painting is Meru itself, but a Meru not yet fully formed, marked by a spiral wind out of which the elements are condensed. Additionally, the painting includes twelve narrow intersecting circles, all of the same size but with centers spaced at thirty-degree intervals with reference to Meru, that stand for the months of the year and a brick red ellipse intersecting all the monthly circles that represents the trajectory of the sun. All of these features, in turn, overlap the two wide bands symbolizing water and earth.[61]

---

gart: Deutsche Verlags-Anstalt, 1987), 46; Chantal Massonaud, "Le Bhoutan," in *Les royaumes de l'Himâlaya: Histoire et civilisation* (Paris: Imprimerie Nationale, 1982), 67–116, esp. 113; Blanche Christine Olschak, *Ancient Bhutan: A Study on Early Buddhism in the Himâlayas* (Zurich: Swiss Foundation for Alpine Research, 1979), 42; idem, *The Dragon Kingdom: Images of Bhutan* (Boston: Shambhala, 1988), front and back inside covers; Olschak and Thupten Wangyal, *Mystic Art*, 108–9 (note 39); and Françoise Pommaret-Imaeda and Yoshiro Imaeda, *Bhutan: A Kingdom of the Eastern Himalayas*, trans. Ian Noble (Boston: Shambhala, 1985), pls. 23 and 26.

59. Olschak and Thupten Wangyal, *Mystic Art*, 108–9 (note 39).

60. Joseph E. Schwartzberg, "Cosmographical Mapping," in *The History of Cartography*, ed. J. B. Harley and David Woodward (Chicago: University of Chicago Press, 1987–), vol. 2.1 (1992), 332–87, esp. fig. 16.2 (text on 336–37) and fig. 16.9 (text on 345).

61. The foregoing description is based mainly on Pommaret-Imaeda and Imaeda, *Bhutan*, 112 (note 58), and in part on Olschak and Thupten Wangyal, *Mystic Art*, 108–9 (note 39). For a somewhat different interpretation of the same works, see Massonaud, "Le Bhoutan," 112–14 (note 58), which also differs from the quoted description in stating that, rather than representing the months of the year, the intersecting circles actually stand for "douze vents qui sont autant de chemins" (twelve winds that are, at the same time, roads) (112); but she does not provide the basis or significance of this interpretation. A very large *thanka* in the Potala palace in Lhasa that appears to incorporate much of the same content as the Paro mural painting, and probably additional astronomic information as well, is illustrated, though with minimal explanatory text, in *Trésors du Tibet*, fig. 84 (note 36).

The second and third paintings in the sequence, from Punakha and Paro, respectively, also appear to be based on the *Kālacakra Tantra*. The second shows the primordial world system just described as it begins to assume a more differentiated form, with Meru rising out of the earth and into the clouds and with the abodes of the gods depicted on its various levels. On the third painting plants and flowers begin to appear around the sacred mountain. Both views depict Meru in frontal perspective in the upper half of the painting and half the earth, in planimetric perspective, on the lower half. The fourth and fifth paintings, again from Punakha and from Paro, are based on the *Abhidharmakośa*. On the fourth the world continents are shown in greater detail than previously, and on the fifth (fig. 15.15), the fully formed world system, they are even more prominently depicted. Both paintings combine several perspectives: frontal for Meru itself (thereby obscuring Uttarakuru, the northern continent, which lies to its rear), oblique for the platform out of which Meru rises, and planimetric for the rest of the system.[62]

Apart from representations of the terrestrial plane of the type just discussed, there are others on which the surrounding continents are either reduced to insignificant size or omitted altogether, as in figure 15.16.[63] Other representations vary widely in appearance and are presented in a number of published works.[64] There are also some striking three-dimensional representations of Meru that appear to conform to the Tibetan Buddhist conception. A particularly large example, in the Yonghegong Temple in Beijing, provides a remarkably good match to the view presented in figure 15.16, though without the surrounding continents.[65] Another, also from northern China, is in the form of an elaborate gilt brass mandala with a diameter of 35.1 centimeters and a height of 37.5 centimeters. In this model the surrounding primary and satellite continents are presented as four sets of three

FIG. 15.16. MOUNT MERU AND ASSOCIATED FEATURES ON A TIBETAN TEMPLE BANNER. Painted on a cloth *thanka*. Date and provenance within Tibet not known. At the base of this painting are what appear to be eight horizontal rows of mountains (though only seven should properly be shown) with intervening ocean. Above these are the four stages of Meru itself. Within the uppermost stage are the sun and moon, below which appear the Big Dipper and the Pleiades, respectively. On Meru's summit stands the palace of the gods. Ranged about Meru are signs for the primary and secondary continents, Jambūdvīpa in the south (bottom center, said to be in the shape of the "shoulder blade of a sheep"), Pūrvavideha in the east, Aparagodānīya in the west, and Uttarakuru, recognizable only by the two square satellites flanking the mountain (the primary continent being hidden behind it), in the north. Also shown are the "Seven Jewels of a Universal Monarch," "Eight Buddhist Symbols" (identified with specific ladies), and wish-fulfilling trees.

Size of the original: unknown. Courtesy of the Department of Library Services, American Museum of Natural History, New York (neg. no. 319427).

62. In this paragraph I have followed Olschak and Thupten Wangyal, *Mystic Art* (note 39).

63. This work is illustrated in Wayman, *Buddhist Tantras*, 104–5, with explanation on 106–9 (note 55); Gordon, *Tibetan Religious Art*, 47, with explanation on 44 and 49 (note 53); and Sugiura, *Ajia no kosumosu + mandara*, 19 (note 5).

64. Two undated views of painted *thankas* are presented in Van der Wee, "Rirab Lhunpo" (note 57) (apart from the planimetric view already indicated therein), one of unknown provenance in Van der Wee's personal collection (fig. 1) and the other, from Ladakh, held by the Völkerkundemuseum der Universität Zürich (fig. 6). The first is also depicted in Armand Neven, *Etudes d'art lamaïque et de l'Himalaya* (Brussels: Oyez, 1978), 28 (fig. 12); and the latter in Sugiura, *Ajia no kosumosu + mandara*, 20 (note 5), and Unno, *Chizu no shiwa*, 192 (fig. 89) (note 20). A very clear oblique Sikkimese version, with an accompanying key, appears in Mark Tatz and Jody Kent, *Rebirth: The Tibetan Game of Liberation* (Garden City, N.Y.: Anchor Books, 1977), 38–39. Van der Wee, "Rirab Lhunpo" (note 57), also reproduces a number of paintings in which Meru figures only as a subsidiary element (e.g., as being held in the hand of the bodhisattva Vajrapāṇi), on which I will make no further comment. A very large fresco painting of Meru is reproduced in Müller and Raunig, *Der Weg zum Dach der Welt*, 293 (note 34).

65. Illustrated in Gerner, *Architekturen im Himalaja*, 46 (note 58).

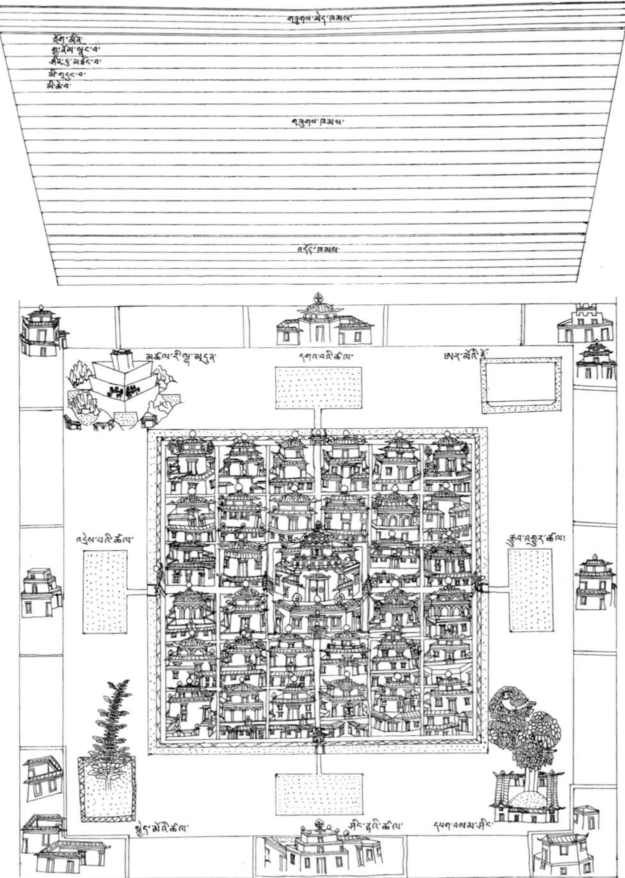

FIG. 15.17. BÖN DRAWING OF THE *LHA SUM-CU-SA-GSUM GYI GŻAL-YAS-KHAṄ* (THE PALACES OF THE THIRTY-THREE GODS). This pen-and-ink drawing was probably made in England by a Tibetan refugee, Tenzin Namdak (bsTan-'dzin rnam-dag), a learned lama of the Bön cult. It illustrates a portion of the text outlining the major doctrines of his faith, specifically some verses from "The Way of the Shen of the Visual World." In addition to the palaces, the map depicts a sacred bird known as Khyung (the Garuda of Indian mythology) in a Tree of Paradise at the lower right, a number of parks, and, it appears, at least four lakes (the stippled rectangular areas), presumably in the four cardinal directions from the central Palace of Victory. At the top of the diagram Mount Meru, becoming ever broader from base to summit, is unmistakable.
Size of the original: unknown. From David L. Snellgrove, ed. and trans., *The Nine Ways of Bon: Excerpts from "gZi-brjid"* (London: Oxford University Press, 1967), pl. 21. By permission of Oxford University Press.

pavilions each.[66] Commenting on these differences, Van der Wee cites a cosmogonic narrative related by Thubten Jigme Norbu, who stated:

> If I think of Rirab Lhunpo [one of the many names for Meru], it doesn't matter whether it be round or square. . . . What is important is what my faith signifies for me. Perhaps it is square; but if I believe it to be round, for me it is round. Even if it doesn't exist at all, for me it will always remain round.[67]

Meru is sacred not only to Tibetan Buddhists, but also to Bön-pos, as the followers of the pre-Buddhist Bön cult are called. Over the centuries the two faiths have borrowed so much from one another that it is no longer possible to untangle the lineage of particular practices and concepts, including those relating to cosmography. But it is clear that Bön cosmography, while incorporating much that is unquestionably within the Buddhist tradition, also contains much that is idiosyncratic. Until recently Bön had received little intensive study outside Tibet itself; but in a pioneering work, mainly of translation, Snellgrove has set forth the more important portions of Bön doctrine as given in a twelve-volume work, whose short title is *gZi-brjid* (The glorious).[68] Within that work, a section titled "The Way of the Shen [a religious officiant] of the Visual World" relates largely to cosmology. The *gZi-brjid* seems to have attained its present form in the late fourteenth or early fifteenth century, and the specific text Snellgrove worked from is believed to be about four hundred years old. Several maps drawn by a refugee Bön lama, Tenzin Namdak (bsTan-'dzin rnam-dag), who collaborated with Snellgrove in preparing the translation, are included in an appendix. One of these is the Bön version of the Meru mandala, which in most important respects conforms to that depicted in figure 15.14.[69] Another diagram from the same work, also relating to Meru, but more to its immediate precincts, is reproduced here as figure 15.17.[70] What is left to conjecture is the period during which maps of this type began to be drawn in Tibet. My surmise, based on the known antecedents of other works noted in this and the preceding section and on the likelihood (to be discussed below) that Bön cartography predates the arrival of Buddhism in Tibet, would be that cosmological portions of the Bön

66. Illustrated in Marilyn M. Rhie and Robert A. F. Thurman, *Wisdom and Compassion: The Sacred Art of Tibet* (San Francisco: Asian Art Museum, 1991), 382.

67. Van der Wee, "Rirab Lhunpo," 72 (note 57), my translation from the French original.

68. Snellgrove, *Nine Ways of Bon* (note 37).

69. Snellgrove, *Nine Ways of Bon*, fig. XX (note 37). This and all the other drawings accompanying Snellgrove's translation relate to items "which bonpos take for granted as the normal possessions of their high dignitaries and scholars" (264).

70. For more on this map, see Snellgrove, *Nine Ways of Bon*, 91 (Tibetan text on 90), from "The Way of the Shen of the Visual World," in relation to fig. XXI (note 37). Since Bön is a little-studied minority cult (though Bön practices permeate Tibetan Buddhism, and vice versa) and since Bön lamas are not numerous outside Tibet, in which country, were it more accessible, older Bön maps could surely be found, it has been necessary to use a modern drawing here. Given the source, however, I believe it may be considered as adhering rather closely to traditional models.

texts, like those of Tibetan Buddhists, would quickly have given rise to various forms of cartographic expression.

COSMOGRAPHIES FOCUSING LARGELY ON THE TEMPORAL DIMENSION

In contrast to the present-oriented and materialistic secular worldview of most inhabitants of industrialized states, that of Tibetan Buddhists is largely concerned with the hereafter and the rewards or punishments that one will reap for the cumulative merit earned in past and present lives. This attitude regarding *saṃsāra* (Tibetan *'khor ba*, the cycle of transmigration/rebirth) is reflected in numerous cosmographic diagrams in which the temporal dimension figures prominently, either implicitly or explicitly or both. Such diagrams form the subject of the following section.

Bhavacakras (wheels of life/existence/becoming) are important cosmographic means for conveying to the masses the essentially soteriological belief central to Tibetan Buddhism. Such works, often painted at the entrances to temples and monasteries, remind viewers of the realms of existence souls pass through in successive incarnations and warn them of the awful fate awaiting those who cling to worldly desires. In keeping with their largely didactic purpose, such diagrams are conventionally simplified depictions of a portion of a complex cosmography relating to the Kāmadhātu, or Realm of Desire. The *bhavacakra* is discussed in considerable detail by Waddell.[71] It typically illustrates six subrealms within the Kāmadhātu, each as a segment of a wheel, and by implication includes the Rūpadhātu and Ārupyadhātu as well, since the wheel is believed to contain all of *saṃsāra*. The lowermost segment generally includes a multiplicity of hells. Waddell designates it "the Great Judgment" and shows it as compounded of images of eight hot hells and eight cold hells, as well as representations of the sinner, the weigher (of good and evil past deeds), the judge, and angels of good and evil. The segments to the left and right of the Great Judgment represent respectively the world of tantalized spirits (*pretas*) and the animal world. The uppermost segment, predictably, represents heaven(s), while the segments to its left and right signify the human world and the world of titans (*asuras*). Like the Great Judgment, each of the other five segments can be subdivided into a number of subrealms.[72] Figure 15.18, showing the upper portion of a *bhavacakra*, is not identical in its arrangement to the work described by Waddell, though it is close in spirit.[73] Alternative views show a five-segment *bhavacakra*, with hell at the bottom, animal and *preta* realms to its left and right, respectively, and, at the top, a realm of both gods and titans to the left and of humans to the right.[74]

Detailed analysis of this and many other Tibetan paintings reveals what are in effect maps within maps, since some of the scenes depicted in the several portions of the work would individually qualify as maplike views of townscapes, landscapes, or combinations of the two, usually drawn in a characteristically Tibetan oblique perspective. This tendency in Tibetan cosmographic art to create a collage of diverse elements compounded of views of mythic and nonmythic places is even more evident in figure 15.25 below.

A type of Tibetan cosmography, little known or understood by non-Tibetans, is the *tshogs-zhing*, or field of assembly (fig. 15.19). Like *bhavacakras*, such works are largely concerned with cosmic time as well as with cosmic space. Their principal function is to help worshipers visualize "the objects to which one goes for refuge," here conceived as a virtually numberless host of Buddhas and other spiritually advanced beings (*aryas*) whose existence may be apprehended in both time and space.[75] The image presented appears largely in the form of a tree atop which sits a guru (teacher), depicted as either Śākyamuni Buddha or another major religious figure, in this case Tsong-khapa (1357–1419), the founder of the Gelukpa (bGe-lugs pa, Yellow Hat) monastic order. The literal meaning of *tshogs-zhing*, however, is "merit sphere or object," and it may be interpreted unambiguously as referring to a field, not a tree as some commentators have suggested. Further, although the field is presented on the two-dimensional surface of a *thanka*,

> just as in the visualization of a *mandala*—what is two-dimensional becomes three dimensional when visualized. Therefore, although . . . all the deities below the guru appear to be on the front of the tree, they are actually arrayed around the tree. For example, . . . the four world protectors . . . all appear along the lower front of the tree, while in a visualization each would be found in his appropriate direction. . . . The guru who is the focus of the *tshogs-zhing* sits, both

71. Waddell, *Buddhism of Tibet*, 77–122 (note 5).

72. Waddell, *Buddhism of Tibet*, illustrations on 102 and 109 (note 5).

73. See Eleanor Olson, "The Wheel of Existence," *Oriental Art*, n.s., 9 (1963): 204–9. Olson also provides an illustration of the work as a whole.

74. See, for example, the work illustrated by Sopa, "Tibetan 'Wheel of Life,' " 131 (note 41). Other examples of *bhavacakras* may be seen in numerous museums and in illustrations in many publications, including several cited in this chapter.

75. Roger Jackson, "The Tibetan Tshogs Zhing (Field of Assembly): General Notes on Its Function, Structure and Contents," *Asian Philosophy* 2, no. 2 (1992): 157–72; quotation on 159. Jackson provides a list of fourteen published representations of fields of assembly (170 n. 3), in addition to the one his own account relates to (157, pl. 1). The similarities from one view to another in the examples I have seen are striking.

FIG. 15.18. A TIBETAN *BHAVACAKRA*. This picture from an eighteenth-century (?) *thanka*, painted on cotton cloth, is representative of many such works depicting the various realms, or domains of existence, within the Kāmadhātu, the lowermost major component of the vertically structured tripartite universe. Within the wheel, the three favorable upper domains are, from left to right, those of *asuras* (titans or demigods), gods, and humans, while the unfavorable domains below, again from left to right, are those of *pretas* (hungry ghosts), hells, and animals. Within the circular area at the center of the *bhavacakra* is a representation of the principal factors responsible for reincarnation in particular realms, each symbolized by an animal: a pig for ignorance, a snake for anger, and a cock for desire. On the periphery of the wheel there are symbols denoting each link on a twelvefold chain of causation. The entire wheel is in the grip of Shinje, a powerful monster, believed to be the wrathful counterpart of the compassionate Avalokiteśvara, the bodhisattva regarded as the protector of Tibet. Shinje's spiritual insight is indicated by his third eye.

Size of the original: ca. 109 × 86 cm. By permission of the Newark Museum, Newark, N.J. (acc. no. 36.535).

FIG. 15.19. THE TIBETAN *TSHOGS-ZHING*, OR FIELD OF ASSEMBLY. The assemblage of divinities in this diagram represents to Tibetan worshipers a field of merit and a place of spiritual refuge defined in both temporal and spatial terms. The bottom half of the diagram, occupied by a hierarchy of eleven discrete lineages of gurus arrayed on a tree, is conceived synchronically and seen as representing, in effect, a sphere within which their power is manifested. The three lineages of deities in the upper half are conceived as representing a temporal sphere in which power is manifested diachronically. The two spheres converge at the center in the body of a major guru, seen here as the focus of the two spheres. Maitreya, the future Buddha, in his palace in the Tuṣita heaven, at the upper left, and Amitābha, the heavenly Buddha, in his palace in Sukhāvatī, the western paradise, at the upper right, seemingly preside over the entire assembly.
Size of the original: unknown. By permission of the Gerd-Wolfgang Essen Tibetica Collection, Hamburg.

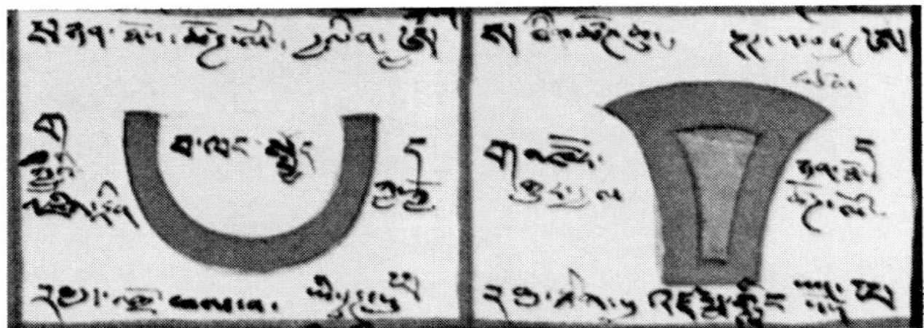

FIG. 15.20. TWO SPACES FROM THE TIBETAN GAME OF REBIRTH. These two spaces, 17 (*right*) and 18 (*left*), represent Jambūdvīpa and Aparagodānīya, the continents lying to the south and west of Mount Meru. The former is an easily recognizable shape, said to resemble the shoulder blade of a sheep. The same shape may be seen for Jambūdvīpa in figure 15.15, for example. The shape of Aparagodānīya, however, departs slightly from its customary circular form. The board as a whole may be construed as an aggregation of maps composing, in effect, a larger cosmic constellation by which the sacred geography of Buddhism may be learned through play.
From Mark Tatz and Jody Kent, *Rebirth: The Tibetan Game of Liberation* (Garden City, N.Y.: Anchor Books, 1977).

> literally and figuratively, at the intersection of the diachronic spiritual power of the guru-lineage [in the sky above] and the synchronic spiritual power of the pantheon [arranged on the tree below]. It is into spheres corresponding to these two types of power that the *tshogs-zhing* basically is organized.[76]

The place referents in the diagram are numerous. For example, the bodhisattvas Maitreya and Amitābha are seated in their palaces in the Tuṣita heaven and the western paradise, Sukhāvatī (depicted in the upper-left and upper-right corners), and the world protectors stand guard over the four cardinal directions at the base of the guru lineage.[77]

Concern with the cosmography relating to rebirth is not merely associated with worship among Tibetans, but also finds expression in play. One vehicle by which such concern is channeled is the game of "Rebirth" (full correct title, "Determination of the Ascension of Stages"), allegedly invented in the thirteenth century by Kunga Gyaltsen (Kun-dga'-rgyal-mtshan), a scholar of the Sakya (Sa-skya) sect.[78] This game, played by Tibetans of all ages, is an important educational device, "inculcating in children the Buddhist map of the world and an understanding of the workings of karma."[79] Since then, many regional variants of the game have spread throughout the world of northern (Mahayana) Buddhism as well as in India, from whence it was adapted by the English in its present popular form as "Snakes and Ladders."[80] In a remarkably thorough and abundantly illustrated study of the game, Tatz and Kent discuss its origins, development, subsequent diffusion, and philosophic basis in respect to the Buddhist belief in karma and rebirth; the nature of the game board, which may be construed as a map; and the rules of play.

The current Tibetan version of the game board consists of a gridded field of 104 rectangular spaces, in thirteen rows and eight columns, each of which represents a specific place in the Buddhist cosmos, whether it be an actual realm in space (e.g., Jambūdvīpa, Meru, Shambhala, or "The Black Rope and Crushing Hells") or a state of mind or being (e.g., "Wisdom Holder among the Gods of Sense Desire").[81] To Westerners the distinctions may seem obvious, but to Buddhists they are not. The players' object is to advance through the field, according to throws of a die, with the goal of reaching nirvana. These places and states are distributed on the board in an ordered manner, with hells occupying much of the bottom two rows and increasingly exalted realms or states occurring as one moves upward. Each space contains a symbol indicating its general and more specific nature (e.g., all hells are indicated by circles, while individual hells are differentiated by color or other added graphic devices), as well as statements specifying the consequences of particular throws of the die that cause one to be reborn in a higher or lower state. Figure 15.20 illustrates two adjacent spaces, those relating to the southern and western continents of the terrestrial realm. Overall, the scheme constitutes a complex "combination of several overlapping systems," which are not apparent from mere visual inspection of the game board, including, for example, a path through the realm of sense desire within the Mount Meru world system and a path and system relating to the various Buddhist heavens.[82] "The

76. Jackson, "Tibetan Tshogs Zhing," 162 (note 75).

77. Jackson, "Tibetan Tshogs Zhing," 164 (note 75).

78. Tatz and Kent, *Rebirth*, 1 and 19 (note 64).

79. Tatz and Kent, *Rebirth*, 1 (note 64).

80. F. E. Pargiter, "An Indian Game: Heaven or Hell," *Journal of the Royal Asiatic Society of Great Britain and Ireland*, 1916, 539–42. Pargiter ascribes the game to the Vaishnavite form of Hinduism but offers no probable date for its origin. Despite the alleged Tibetan invention, one cannot rule out the diffusion of some prototype of the game from India to Tibet.

81. Tatz and Kent, *Rebirth*, 62–63 (note 64). A complete folding game board, drawn by a Tibetan artist, is provided in the end-cover pocket of the book. Photographs of other versions of the game, with different numbers and arrangements of spaces, are also provided, including a Tibetan block print of the original nine-by-nine-square field, a modern Bhutanese version, and a nineteenth-century version; a Korean version is also described, but not illustrated (11–15). Waddell, *Buddhism of Tibet*, 471–73 (note 5), illustrates and discusses a version of the game that he acquired, probably in the late nineteenth century, with a small playing field of only eight by seven squares. Slusser, *Nepal Mandala*, vol. 2, pl. 331 (note 13), illustrates a Hindu version of the game from the Vale of Kathmandu. Loden Sherap Dagyab notes that Tibetan game boards he has seen may be as large as two meters high and one and a half meters wide (*Tibetan Religious Art*, 2 vols. [Wiesbaden: Otto Harrassowitz, 1977], 1:41).

82. Tatz and Kent, *Rebirth*, 32–47 (note 64).

various paths may be taken to represent the cultural diversity of humankind. . . . These may also be interpreted as differing attitudes toward religion."[83] Though most squares, paths, and systems relate to the world of Tibetan Buddhism, there are also several paths by which non-Buddhists—Hindus and Muslims—can rise to the level of "wisdom holder," and another by which followers of the indigenous Bön cult of Tibet can attain that same happy state, yet far short of nirvana.[84]

MAPS OF SPECIFIC PORTIONS OF THE COSMOS

Apart from features associated with the Mount Meru system, many other parts of the Tibetan spiritual universe—heavens, paradises presided over by particular deities, other sacred places, and hells—all find places in the cosmographic art of the region. Here I note only a few examples, beginning with an especially popular subject for religious paintings, the western paradise, Sukhāvatī (literally, the Happy Land), chiefly identified with the Buddha in the form of Amitābha (Immeasurable Glory). In figure 15.21 Sukhāvatī is shown on a *thanka* that forms the focal object of worship at a Buddhist altar. Detailed descriptions of this peaceful realm appear in numerous ancient Buddhist texts, going back perhaps as far as the third century A.D. It is said to loom from a dark sea as a mountain of gleaming copper. From the summit a bejeweled palace rises into a heaven populated by a host of cloud-mounted gods and goddesses. Those fortunate enough to be reborn in Sukhāvatī enjoy continuous festivity, marked by music, dancing, bright banners, and other delights. Surrounding the palace is a zone of silence marked off by a high embankment built of skulls. Nevertheless a winding, narrow road, suggestive of a navel cord, does cross the sea and ascends from the earthly domain to a golden gate in the wall, thus providing access to the palace. Through appropriate meditation, the worshiper aspires to traverse this difficult spiritual path. The sea that must be crossed represents the state of *bardo*, the period between lives, when the soul, in purgatory, is judged in terms of past good and bad deeds to determine the state to which it will be consigned in its next birth.[85]

Paintings of the northern paradise of Shambhala (fig. 15.22), in contrast to those of Sukhāvatī, generally employ an essentially planimetric perspective, thereby emphasizing its wheel-like appearance. Because Shambhala is identified with the deity Kālacakra, a name signifying "wheel of time," the depiction of his domain as a mandala (literally, circle) is especially appropriate. Within the wheel, however, details are shown in an oblique perspective. Curiously, despite the prominence of Shambhala in the popular religion of Tibet and the existence of several Tibetan and Sanskrit guidebooks describing the way by which it can be reached (in increasingly vague terms as the journey progresses), it is not commonly depicted in paintings, perhaps because it is regarded as attainable only by accomplished yogis.[86] Nevertheless,

> storytellers used to wander Tibet with paintings they would unroll before audiences in order to illustrate subjects such as the journey to Shambhala. In one such performance, witnessed . . . in Lhasa, the storyteller indicated on his painting how the traveler to Shambhala must climb a stairway to the top of a mountain, where his body will become light as an insect, enabling him to walk to the kingdom on clouds.[87]

The Tibetan version of the earlier mentioned *Kālacakra Tantra* derives from one brought from India, probably in the tenth or eleventh century, though a still earlier version is traditionally believed to have been brought to India from Shambhala itself. The text describes in great detail that utopian realm, founded by an Indian king, Sucandra, in a blessed, mountain-girt land in a location revealed to him by the Buddha during his lifetime on earth (800 B.C. in the Tibetan tradition). Thereafter Shambhala was ruled by a long line of monarchs, each reigning for a hundred years and each an incarnation of a particular bodhisattva. Tibetans continue to believe in the earthly existence of Shambhala and in the prophecy of the *Kālacakra Tantra* that when the world reaches a certain state of moral decay, a savior-king will emerge from Shambhala, defeat the forces of evil, and bring about a new order of world peace.

The Tibetan recension of the *Kālacakra Tantra*

---

83. Tatz and Kent, *Rebirth*, 41 (note 64).

84. Tatz and Kent, *Rebirth*, 33 (note 64), present a "Map of the Game Board" that is in fact a deconstruction of the five systems and several subsystems incorporated within it.

85. This description was abstracted from a much fuller discussion in Gerd-Wolfgang Essen and Tsering Tashi Thingo, *Die Götter des Himalaya: Buddhistische Kunst Tibets*, 2 vols. (Munich: Prestel-Verlag, 1989), 1:201–2, including a full-page view of the *thanka*. The authors refer to Sukhāvatī as the southwestern paradise rather than the western paradise, but most texts use the latter designation. Other representations, quite different from the one illustrated here, appear in various publications: one that appears to be a woodblock print, with an accompanying detailed Tibetan text, is presented by Waddell, *Buddhism of Tibet*, 140 (note 5); a painted essentially planimetric view in Detlef Ingo Lauf, *Verborgene Botschaft tibetische Thangkas/Secret Revelation of Tibetan Thangkas* (Freiburg im Breisgau: Aurum, 1976), 49, with text on 48; and a fresco painting three and a half stories high on the *chorten* of Dumtse lhakang at Paro in Bhutan, of what I take to be Sukhāvatī, in Massonaud, "Le Bhoutan," 110 (fig. 39) (note 58).

86. An exhaustive and well-illustrated treatment is provided by Edwin Bernbaum, *The Way to Shambhala* (Garden City, N.Y.: Anchor Press, 1980). Also useful, though without indigenous maps, is Peter Bishop, *The Myth of Shangri-La: Tibet, Travel Writing and the Western Creation of Sacred Landscape* (London: Athlone Press, 1989).

87. Edwin Bernbaum, "The Hidden Kingdom of Shambhala," *Natural History* 92, no. 4 (1983): 54–63, esp. 59.

FIG. 15.21. ALTAR WITH *THANKA* (SCROLL HANGING) SHOWING SUKHĀVATĪ, THE WESTERN PARADISE IN THE COSMOGRAPHY OF TIBETAN BUDDHISM. From Tibet, nineteenth century. This painted scroll occupies a central place in the altar and is one among many sacred images that may form focal points in the meditation that is characteristic of the worship of Tibetan Buddhists.
Size of the original: 58 × 39 cm. By permission of the Gerd-Wolfgang Essen Tibetica Collection, Hamburg.

FIG. 15.22. SHAMBHALA, THE NORTHERN PARADISE OF TIBETAN BUDDHISM. From Tibet, late eighteenth century, a painted cloth *thanka*. To reach the hidden idyllic kingdom of Shambhala, travelers must undertake an epic journey across deserts and mountains and overcome all manner of intervening natural obstacles. Those who achieve their goal will encounter a country of beautiful cities and parks divided into the shape of a lotus blossom with eight petals; in the center, in a magnificent palace, lives the king, an incarnate bodhisattva. The lotus here represents the emergence of purity and enlightenment amid the depravity and delusion of the surrounding world. In this picture mountain ranges not only set off Shambhala from the outside world, but also separate the capital from the eight outer portions of the realm and divide those regions from one another. Other paintings show streams, rather than mountains, between neighboring outer regions and depict their shapes with less rigid regularity than in this view.

Size of the original: 65 × 46 cm. By permission of the Gerd-Wolfgang Essen Tibetica Collection, Hamburg.

"reflects the influence of various non-Buddhist religions that were found in Central Asia . . . most notably Nestorian Christianity, Manichaeism, and Islam," suggesting to some Western scholars that there once really was a kingdom of Shambhala somewhere within that general region.[88] Helmut Hoffmann actually sought to find it, using the guidebooks in much the same way as Heinrich Schliemann drew on the *Iliad* and the *Odyssey* to locate Troy, and he concluded that it lay in the Pamir Mountains to the east of Samarkand. Others have suggested the Tarim basin or the Turfan Depression as likely regions.[89] Based on the ring of mountains shown on the available cosmographic maps, a case can be made for any of those areas, especially if one recognizes that they were once much better watered and therefore more felicitous environments than they are at present.

Within Shambhala, according to followers of the Bön cult, there exists a "Nine-Stage Swastika Mountain." A very elaborate drawing of this mountain, together with a lengthy caption in Tibetan, accompanies Snellgrove's translation of Bön doctrines. The much briefer English caption simply states that the mountain "representing the Nine Ways of *Bon* . . . [is] surrounded by its eight royal palaces in the country known variously as *sTag-gzigs*, *'Ol-mo-luṅ-riṅ*, *Śambhala*, etc.," but it fails to identify the other features shown in that intriguing diagram, which in some respects resembles Jain renditions of Jambūdvīpa and surrounding islands.[90]

Apart from the examples noted above, there are numerous additional maps of heavens, hells, and other portions of the Buddhist cosmos. Many form parts of detailed hagiographic compositions (to be examined below), while others focus on a single place or a group of closely related places.[91]

ASTROLOGICAL AND DIVINATORY DIAGRAMS

As previously noted, astrology and other forms of divination play an important role in the culture of Greater Tibet.[92] Manuals devoted to the subject abound in graphic devices that may be consulted to determine outcomes preordained by one's time of birth or by the apparition of various signs believed to be diagnostic guides to the future. Although much could be written on the subject of divination, it lies at the margin of the concerns of this volume. Hence I shall do no more than direct attention to a few works in which relevant diagrams appear and comment briefly on them, taking note of several genres that relate to Indian cosmography.

The Tibetan system of astrology, an amalgam of Indian, Chinese, and indigenous concepts, is discussed and illustrated in a number of nineteenth-century scholarly publications.[93] The diagrams included in almanacs consulted by practitioners of astrology can be quite complex and esoteric. A particularly rich assemblage of astrological drawings from a nineteenth-century Tibetan almanac appears in *Die Götter des Himalaya*, together with a detailed key to its contents. The almanac is of special interest because charts based on the different original systems appear alongside one another on the same large page.[94]

A genre of astrological divination that can be found in related forms in China (its apparent place of origin), in Tibet, and with considerable modification, in India is of the type illustrated in figure 15.23.[95] This artifact happens to be in bronze; but more commonly works of this type, the making of which was traditionally an important source of income for astrologers, would be painted on cloth or paper, especially those kept on family altars. The principal component of diagrams of this type is the body of a tortoise (rather abstractly depicted in this instance) or a frog. In the Chinese and Tibetan versions, the body of whichever creature is used is divided into nine parts in a magic square containing nine smaller squares, in three rows and three columns, each with a number from 1 to

88. Bernbaum, "Hidden Kingdom," 56 (note 87).

89. Bernbaum, "Hidden Kingdom," 56 (note 87). Among the extant maps of Shambhala is a striking nineteenth-century Mongolian *thanka*, remarkable in that it shows Shambhala only on the right side of the painting. This work is held by the Musée Guimet, Paris, and illustrated in Rhie and Thurman, *Wisdom and Compassion*, 378–79 (note 66).

90. Snellgrove, *Nine Ways of Bon*, fig. XXII (note 37); see Schwartzberg, "Cosmographical Mapping," 367–72 (note 60), for Jain examples.

91. For example, a frighteningly vivid Tibetan depiction of a series of hells arranged in eight concentric rings below Mount Meru (presumably the "Eight Hot Hells" sometimes shown in the lowermost segment of the *bhavacakra*), painted on a *thanka* is depicted in Sugiura, *Ajia no kosumosu + mandara*, 22–23 (note 5). An exquisitely painted *thanka* in the Potala palace in Lhasa shows not only the same set of concentrically ordered hells, but also a heavenly field above into which rises a many-layered Meru. This remarkable composition is illustrated in *Trésors du Tibet*, 83 (note 36). A comprehensive view of the hot and cold hells is provided in a woodblock print reproduced in Lauf, *Secret Doctrines*, 132, with a key diagram on 133 and explanatory text on 130–31 and 134–37 (note 50). Though it is similar in subject to the previously noted work, the two are completely different in appearance, a fact only partly accounted for by the different media employed.

92. Réne de Nebesky-Wojkowitz, *Oracles and Demons of Tibet: The Cult and Iconography of the Tibetan Protective Deities* (The Hague: Mouton, 1956), 291–98 and 455–66, discusses more than a dozen methods of divination used in Tibet in addition to resort to oracles. Few of these, however, appear to rely heavily, if at all, on graphic aids.

93. Especially useful are Emil Schlagintweit, *Buddhism in Tibet, Illustrated by Literary Documents and Objects of Religious Worship, with an Account of the Buddhist Systems Preceding It in India* (1863; London: Susil Gupta, 1968), 290–328; and Waddell, *Buddhism of Tibet*, 450–74 (note 5).

94. Essen and Thingo, *Die Götter des Himalaya*, 2:229–30 and 1:256–57 (note 85).

95. Along with one other bronze example (in a private collection), it is illustrated and discussed in Siegbert Hummel, "Kosmische Strukturpläne der Tibeter," *Geographica Helvetica* 9 (1964): 34–42.

FIG. 15.23. TIBETAN BRONZE ASTROLOGICAL TABLE. The central component in this assemblage (age not known) is the rather abstract shape of a tortoise, whose head, legs, tail, and flanks represent the four cardinal and intermediate directions. On the carapace is a magic square. Other parts of the diagram, primarily with calendrical associations, are also consulted, together with its central portion, so that—based on one's date of birth and the current arrangement of zodiacal constellations—one's future can be foretold. At the top of the diagram appear three particularly revered manifestations of the Buddha: Avalokiteśvara, Mañjuśrī, and Vajrapāṇi.
Size of the original: 35.4 × 28.7 cm. By permission of the Völkerkundemuseum der Universität Zürich (cat. no. 12664).

9 so arranged that the sum of the numbers in each column and row adds up to 15. The circular band surrounding the magic square contains eight segments, here shown as the petals of a lotus, representing the eight quarters and their respective elements. In sequence, north signifies fire, northeast earth, east iron, southeast heaven, south water, southwest mountain, west tree, and northwest air. The center (the carapace of the tortoise) stands for the nadir of a vertical cosmic axis. On painted versions of such diagrams, each direction would have its corresponding color. It is likely that this system is related to the Indian *kūrmavibhāga*.[96]

A variant of the type of divination diagram just discussed, also with apparent Indian analogues, takes the form of a wheel, normally with eight radii dividing it into eight segments, with the same directional associations with elements as noted above, and with eight concentric rings.[97] Thus there is a total of sixty-four compartments, each representing a combination of two elements. But the order of elements, while regular in regard to the sequence of segments, varies within each segment, no two being identical. Obviously this leads to a more complicated set of divinatory formulas. Some Tibetan divination diagrams of this type appear to be even more complicated than the basic sixty-four-compartment type just noted, containing sixteen or more spokes and more than eight concentric rings. What I take to be one such diagram is painted on the wall of Nechung (gNas-chung) monastery near Lhasa, the former seat of the Tibetan state oracle.[98]

What appear to be Indian analogues of the types of divinatory diagrams just discussed are for the most part square rather than circular, but they are also composed of a set of directionally identified spokes and, usually, of nested squares rather than concentric rings. Moreover, the two genres differ in that a particular city of reference takes the place of the cosmic nadir at the center of the map, and individual cities (seemingly randomly ordered rather than arranged according to their actual azimuths or distances from the center) take the place of the two-element combinations of each map segment in the Tibetan case. Yet the structure and functions of both genres seem to be similar and tied to a set of astrological determinants. The cultural transmission of the underlying organizational principles from India to Tibet, or in the opposite direction (not excluding the possibility of a Chinese link), warrants investigation.[99]

CARTOGRAPHIC ELEMENTS IN HAGIOGRAPHY AND MYTHOLOGY

Exceedingly common types of Tibetan painting are those relating to past, present, and even future lives of the Buddha, of bodhisattvas, and of lamas and other saintly mortals, as well as of mythic heroes, especially those of the national Gesar epic. Such paintings, which have been produced at least since the fourteenth century, abound in maplike scenes of places important in the lives of the personages they honor. Some such scenes are painted with a great sense of verisimilitude, others are rendered fancifully. The paintings are constructed in many ways, often with great ingenuity, and reflect schools associated with particular monastic sects. Most place the protagonist in the central or most visually prominent portion of the composition. Others relegate that individual to a marginal position, giving greater prominence to a particular place or set of places within the Buddhist cosmos or to historically identifiable sacred places. Here I can do no more than illustrate a few such paintings and direct attention to useful sources in which others can be studied.

Figure 15.24 illustrates one panel of a rather large late eighteenth-century triptych dedicated to the deity Kālacakra. The two side panels of this work display most prominently the realm of Shambhala, the left one as it had been conceived in a bygone era and the right one (our illustration) as it was believed to be when the painting was made. Above and below Shambhala on both panels are tutelary figures and other locales associated with it. The central panel reverses this pattern. There Kālacakra, a giant and fearsome deity, occupies the central position while associated places, largely unidentified, are consigned to marginal positions. Places and personages depicted in the upper portions of the painting are celestial, while those at lower levels appear to be terres-

96. The account of the Tibetan diagrams is based on Gordon, *Tibetan Religious Art*, 24, 27, and 29–30 (note 53); and Schlagintweit, *Buddhism in Tibet*, 304–11 (note 93). The figure that Gordon presents is a cloth *thanka* and is much more complex than the one illustrated in figure 15.23. On page 26 she also presents photographs of illustrations, mainly astrological, from two Tibetan divination manuals. For the Indian *kūrmavibhāga*, see Schwartzberg, "Cosmographical Mapping," 337–39 (note 60).

97. Discussed in Schlagintweit, *Buddhism in Tibet*, 311–13 (note 93).

98. Letter from Toni Huber of the Department of Philosophy and Religious Studies, University of Canterbury, 10 December 1989. Huber, who enclosed a picture of this painting, was unable to determine the system for using the diagram, but there is little doubt that it serves an astrological purpose. An accompanying diagram of another such wall painting, seen in the inner courtyard of Zhwa-lu monastery to the south of Shigatse city, shows a circle with sixteen spokes and nine rings. Huber writes that his map was designed by the famous fourteenth-century Tibetan scholar Buton Rinchen Drub (Bu-ston Rin-chen-grub) and that "it is intended to correlate the transit of various heavenly bodies with the Tibetan calendrical cycle (which is a lunar sexagenary one). If you read the inscriptions its sophistication becomes apparent." Photographs of both paintings were taken by Huber in 1987.

99. Two such maps, one square and one circular, are illustrated and discussed in Schwartzberg, "Cosmographical Mapping," 348–51 (note 60). One of those and three others (all square) are illustrated and briefly discussed in Susan Gole, *Indian Maps and Plans: From Earliest Times to the Advent of European Surveys* (New Delhi: Manohar, 1989), 23–24 and 50–53.

FIG. 15.24. PANEL FROM A TIBETAN TRIPTYCH DEDICATED TO THE DEITY KĀLACAKRA. This right panel is the largest of an exceptionally large and remarkably well preserved work dated about 1780. Dominating the work is the realm of Shambhala, described in the sacred *Kālacakra Tantra*, with Kalapa, the royal capital, at its center. In the heavens above are, at left, identifiable representations of the four great tutelary deities of the Gelukpa monastic order, one of whom is Kālacakra, while in the corner to the right are Tsong-kha-pa, the fifteenth-century founder of the order, and two other abbots. In the lower left are the tombs of the Gelukpa abbots, including that of the Panchen Lama (upper quadrangle), author of a mystical work on Shambhala. To their right appears the great monastery of Tashilunpo, a chief Gelukpa monastic center, where this painting was probably made, and the surrounding town. Many individual structures therein can be identified. Especially prominent is the tall, windowless "Tower of Silk Paintings" from which giant *thankas* were unrolled for public exhibition on major festive occasions. One wonders if any of these were maps similar to those displayed in Patan, as shown in figure 15.2.

Size of this panel: 136 × 86.5 cm. From Armand Neven, *Etudes d'art lamaïque et de l'Himalaya* (Brussels: Oyez, 1978).

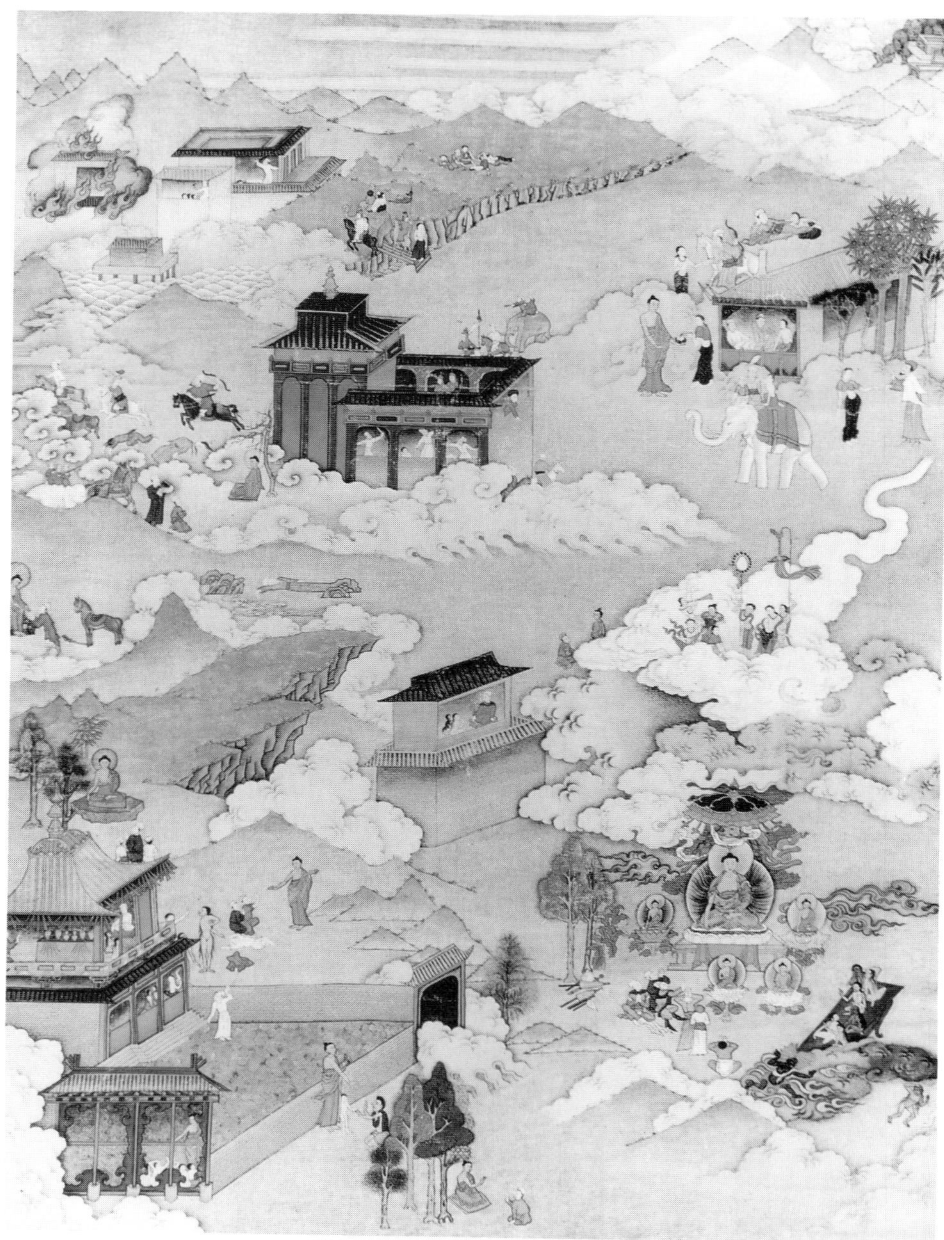

FIG. 15.25. MORAL LANDSCAPE OF THREE JATAKA TALES. This nineteenth-century painting from eastern Tibet provides a visual backdrop for the narration of three Jatakas (12, 13, and 14), all primarily set in what is now the Indian state of Bihar and the adjacent region of Nepal. Jataka 12 occupies roughly the upper half of the map. The action begins in the upper left corner and proceeds downward in a zigzag line, but this progression is not rigidly followed. For example, below the battle scene involving the Śākya tribe (into which the Buddha was born), we see a warrior laying his sword down before the Buddha and rendering homage to him, both depicted near the left margin; yet the next event relates to a stupa that warrior built to honor the Buddha, as shown near the upper-right corner. Jataka 13 occupies the lower-left quarter of the painting and relates to events involving the historical personage of Bimbisāra, king of Magadha, in southern Bihar, whose palace is the most prominent feature depicted. The lower-right quarter of the painting relates to Jataka 14, which takes place around Rājagṛha, a locality in Magadha, which in the previous story is indicated simply as a grove (here rendered as a small clump of trees) under which the Buddha sits. Except for the possibility that the line of mountains, conventionally placed near the top of the painting, might be intended as the Himalayas, there is no particular geographic logic to this composite moral landscape, but the flow of action generally suggests a temporal logic, implying a kind of a route map through time. The ingenious way of combining conventionally painted landscape elements in this painting, both as sites for the action depicted and as temporal dividers separating one story or domain from the next, is a common device in Tibetan religious art.
Size of the original: 75 × 56 cm. By permission of the Gerd-Wolfgang Essen Tibetica Collection, Hamburg.

trial and are, in part, geographically recognizable. This indicates a tendency to have the organizational structure of such paintings reflect that of the universe itself. The most prominent terrestrial component of figure 15.24 is the monastery of Tashilunpo (bKra-shis lhun-po), founded by the first Dalai Lama and subsequently the residence of the Panchen Lama and a major center of the Gelukpa monastic order. Adjoining it is the town of Shigatse (gZhis-ka-rtse). Because the third Panchen Lama (1737–80) was the author of a mystical doctrinal work, *The Way of Shambhala*, Tashilunpo is closely asociated with that sacred realm. This monastery is painted in great detail and with fidelity to its former appearance. Within it many individual structures, some no longer surviving, can be recognized from late eighteenth-century descriptions by the British ambassador George Bogle (1746–81) and lieutenant (later captain) Samuel Turner (1749?–1802). Nearby, as in reality, flows the river Tsangpo, while to the left are the enclosures of the tombs of the Panchen Lama and other great abbots.[100]

In marked contrast to figure 15.24 is the composite moral landscape shown in figure 15.25, providing the field for a pictorial accompaniment to the narration of three successive Jatakas (stories of the past lives of the Buddha). The stories in question all took place in or near what is now the Indian state of Bihar and, like the painting itself, combine mythical and historical events in both mythical and identifiable geographic settings. Though it is impossible to sort out one from the other by visual inspection of the painting, to the devout believer all are equally real. Unlike much of figure 15.24, the manner of depicting the features shown in figure 15.25 depends primarily on the artist's imagination, constrained only by adherence to established iconographic conventions. There is, for example, no way of knowing what the palace of the Magadhan king Bimbisāra (d. ca. 490 B.C.) looked like, yet it is depicted in some detail in the lower-left corner of the painting. On the whole, the spatial logic of the painting is narrative, with places shown in proximity according to the sequence of events transpiring there. This generally applies both to the sequence of stories and to the action within a particular Jataka. But this formula is not rigidly followed; at least one historical place, Rājagṛha, appears twice, being first symbolized by a grove of trees under which the Buddha sits near the left margin of the painting and forming the general locale for the whole of Jataka 14.[101]

A final illustration (fig. 15.26) relates to the Gesar epic, known throughout Tibet and Mongolia. Put into its present form in the late fourteenth or early fifteenth century, probably in the Kham region in eastern Tibet, but incorporating portions of older stories, the epic recounts the career of Gesar (a cognate of Caesar), a warrior-king sometimes associated with Shambhala and sometimes thought to be an incarnation of the third Panchen Lama. In brief, after an early life lacking in virtue and a long period of banishment and wandering, Gesar returns purified to his native land, leads an army against the evil forces of the world, and ultimately becomes a Buddha. The number of known paintings relating to the Gesar epic is not great, but those I have seen are rich in place detail that, as with many hagiographic works, combines cosmographic and geographic images.[102] The images seem to lack the spatiotemporal order noted for figure 15.25, and it therefore is not easy to follow the narrative they depict.[103] Nevertheless, the scenes are arranged in vivid maplike assemblages that make the epic come alive to viewers familiar with its content. Bards often convey the story:

> Paintings of these types are used in conjunction with a sung or chanted narrative, the storyteller using a stick to point out the scenes on the painting as he goes along. Before circulating in Tibet, this technique of illustrated recitation had been employed by monks in India, China and Japan. As in those countries, the illustrations could be painted in fresco or on portable scrolls. . . . The "literary" or narrative character of these paintings is [often] emphasized by the presence of captions, sometimes quite long, written under each scene, which serve to identify the figures and episodes and often reproduce the text of a corresponding manual, with a reference number.[104]

100. This description is based on Neven, *Etudes d'art lamaïque*, 45 (note 64). The entire triptych is illustrated and discussed on 40–45, while a larger-scale view of Tashilunpo appears on 11. Another, less detailed view of the monastery appears on a quite different hagiographic painting on which the first Dalai Lama occupies the central position. This undated work is one of many such paintings illustrated in Hiroki Fujita, *Tibetan Buddhist Art* (Tokyo: Hakusuisha, 1984), text on 189–90.

For the descriptions of Bogle and Turner, see Markham, *Mission of George Bogle to Tibet* (note 35), and Samuel Turner, *An Account of an Embassy to the Court of the Teshoo Lama in Tibet* (London, 1800; reprinted New Delhi: Mañjuśrī Publishing House, 1971).

101. The painting is one among numerous hagiographic works depicted and discussed in Essen and Thingo, *Die Götter des Himalaya*, 1:39–40 (note 85).

102. Figure 15.25 is analyzed in Alexander W. Macdonald and Pema Tsering, "A Note on Five Tibetan Thaṅ-kas of the Ge-sar Epic," in *Die Mongolischen Epen: Bezüge, Sinndeutung und Überlieferung (Ein Symposium)*, ed. Walther Heissig, Asiatische Forschungen, vol. 68 (Wiesbaden: Otto Harrassowitz, 1979), 150–57. For a brief account of the origin and nature of the epic, see Rolf A. Stein, *Tibetan Civilization*, trans. J. E. Stapleton Driver (London: Faber and Faber, 1972), 278–80.

103. Rolf A. Stein, "Peintures tibétaines de la vie de Gesar," *Arts Asiatiques* 5 (1958): 243–71, esp. 244. Stein notes that the customary order of presenting scenes on other biographical paintings, for example, those relating to the Bön saint Milarepa (Mi-la-ras-pa), is from the bottom up, usually beginning in the lower-left corner. Conceivably this provides an intentional means of distinguishing Bön from Tibetan Buddhist hagiography.

104. Stein, *Tibetan Civilization*, 285 (note 102).

FIG. 15.26. TIBETAN *THANKA* FROM A SERIES ON THE GESAR EPIC. This undated, but apparently recent (late nineteenth or early twentieth century), painting is one of a set of five, not necessarily complete, relating to the Gesar epic. The set was acquired by a Swiss citizen in Kalimpong, Sikkim, in 1949. The painting depicts numerous events described in the epic, occurring in recognizable places of both heaven and earth. For example, in the upper-right corner six divine personages are shown in Lha-ling, the land of the gods, while the bottom of the painting represents the (mythic?) land of Ma. The logic by which the assemblage of scenes is organized, however, is not obvious. The four undulating striated bands leading into the heavens either may be paths by which earthly and celestial characters in the epic can reach one another's domains or may refer to the attainment by great adepts of a "rainbow body" at the time of death.
Size of the original: ca. 71 × 53 cm. By permission of Alexander W. Macdonald.

I am not aware of any attempt to study the rich corpus of biographical/hagiographical paintings from Greater Tibet with a view to determining its underlying canons for representing and ordering geographic and cosmographic places. Such an undertaking would present a frontier for research and promises to offer important insights into the cartographic thinking of a highly distinctive culture.[105]

Geographical Maps

Although the number of known traditional geographic maps from Greater Tibet is substantial, it appears small in comparison with the vast body of cosmographic materials from the same region. This contrast may be attributed in part to the lesser interest that geographic maps hold for art historians, students of religion, and other scholars specializing on the region and in part to the lesser value Tibetans attach to nonreligious works. Accordingly, relatively little effort has been expended by outsiders to obtain geographic maps, and relatively little care seems to have been taken, by either Tibetans or outsiders, to preserve whatever maps may once have existed. Hence, while one need not doubt that the disparity between the number of surviving cosmographic and geographic maps is real, one cannot draw firm conclusions about the relative size of the two groups of maps in the past.

Here I shall examine the only two known Tibetan maps that are believed to be attempts to represent the world, a not very large number of regional maps, a variety of pilgrimage guides and other route maps, and, the most common of all Tibetan geographic maps, those that depict towns, monasteries, and other specific localities.

World Maps

Figures 15.27, 15.28, and 15.29 present a modern recension, a key, and a reference map for what appears to be the oldest known map of Tibetan provenance. In its original form, the map was undoubtedly associated with the Bön faith, which predated Tibet's acceptance of Buddhism by many centuries. This map, in Tibetan, was published in *Sgra yi don sdeb snaṅ gsal sgron me bžugs so* (Tibetan Zhang-zhung dictionary), a 1965 edition of a seventeenth-century collection of Bön texts with Tibetan translations and commentary.[106] Zhang-zhung (Shang-shung) is the early language, possibly Indo-European, of western Tibet, to which area the Bön faith seems to have been disseminated from an original source area in Iran.[107]

105. The following include an abundance of hagiographical paintings incorporating substantial maplike components: Gordon, *Tibetan Religious Art* (note 53); Per Kvaerne, "Peintures tibétaines de la vie de sTon-pa-gçen-rab," *Arts Asiatiques* 41 (1986): 36–81, for an important Bön figure with particularly detailed analyses; Lauf, *Verborgene Botschaft* (note 85); Liu, *Buddhist Art* (note 36); Olschak and Thupten Wangyal, *Mystic Art* (note 39); Julien L. Tondriau, *20 rouleaux peints tibétains et népalais* (Brussels: Musées Royaux d'Art et d'Histoire, [1964–65?]); Chögyam Trungpa, *Visual Dharma: The Buddhist Art of Tibet* (Berkeley, Calif.: Shambhala, 1975); and Waddell, *Buddhism of Tibet* (note 5).

106. Ñi-ma-grags-pa (1616–70), *Sgra yi don sdeb snaṅ gsal sgron me bžugs so* (reprinted Delhi, 1965).

107. For succinct discussions of the origins of Bön and Zhang-zhung, see Stein, *Tibetan Civilization*, 35–37 and 230–35 (note 102).

I am not aware of any premodern version of the map or of any copy held by a museum, library, or other public agency outside Tibet. It is likely, however, that manuscript maps, copied from older sources, are held in Tibetan monasteries or by refugee Tibetans in India or elsewhere. The published Delhi version of the Zhang-zhung map was studied by two Russian scholars, L. N. Gumilev, a historical geographer, and B. I. Kuznetsov, a Tibetan phililogist. The following remarks are based largely on the translated account of their thorough analysis.[108]

Gumilev and Kuznetsov's assumption—with which I concur—is that the extent of interaction between Tibet and lands to its west, especially Persia, was considerably more extensive in the early historical period than most historians recognize and that considerable information about distant lands reached Tibetan geographers, at either first or second hand, and was incorporated into the original precursor of the Zhang-zhung map. A key to their analysis is their interpretation of what lies at the center of the map, a locale named Bar-po-so-brgyad, which they identify as Parsogard (Greek Pasargadae), capital of the Persian Empire from 550 to 522 B.C., under the emperors Cyrus the Great and Cambyses. Within that central rectangle is a crudely drawn ten-story edifice that, it is said, represents the tomb of Cyrus. Although various Greek historians left contradictory accounts of that tomb, Aristobulus said that it had the form of a small tower, and Onesicritus, who accompanied Alexander on his Persian campaign, stated that it was ten stories high. A Tibetan inscription on the map itself, "swastika hill, nine stories high," apparently refers to the central figure and also notes "crystal columns with inscriptions," "the garden of the swastika," "the garden of the wheel," "the lotus garden," and "the precious garden," almost all of which can be reconciled with Greek descriptions.[109] Gumilev and Kuznetsov explain the disparity between the references to a nine- and a ten-story structure by saying that in one instance "the top story was treated separately."[110] For all but seven of the sixty-two other places named or annotated on the map (one other is outlined but not named), Gumilev and Kuznetsov are able to offer a translation or commentary or both; and in more than forty cases they suggest, with varying degrees of persuasiveness, a specific place or broad region with which the original map feature can be identified. The identifications sometimes overlap and are sometimes repetitious, however. Some are exceedingly terse and vague while other comments, comparably terse, are quite precise, for example: "Ne-khri-'bum-thang—the 'Nekribum' plateau"; "Grong-khyer-lang-ling—Jerusalem"; "Ne-seng-dra-ba'i-gromg-khyer—'City of Nesendra,' i.e., Alexandria."[111] (Of course there were many ancient Alexandrias; but the proximity of this item to a locality identified as Egypt indicates that that still-surviving city was intended.) Other arguments run into lengthy paragraphs. The conclusion is that the map, which obviously postdates Alexander's conquest of most of the area covered, also predates the Roman campaigns against Parthia and, based on numerous pieces of internal evidence, can best be ascribed to the second century B.C.

Although Gumilev and Kuznetsov's arguments are informed, carefully reasoned, and (in the absence of contrary evidence) plausible, some are less than convincing, such as the following:

> Ma-thang-bsgral-gling—the transcription of a local place name. Judging from its location in the southern part of the world ocean, south of the Nicobars [the Nicobar and Andaman Islands are earlier described as places where (human) flesh is eaten], and from its sound, this may be Madagascar. The name Madagascar, which is not used by its residents, was first reported by Marco Polo, and the first description of the island in European geography was given in *Periplus of the Erythrean Sea*, i.e., later than the time of our map and without a name. Consequently, this old Malagasy word must have reached Tibet through India. The Malagasy settled Madagascar about the 3rd century B.C. from Indonesia and thus came to the attention of the Indians who were navigating in the Indian Ocean. In other words, the Tibetan cartographer, in addition to Iranian sources, also relied on Indian sources, so that our map is not the product of plagiarism, but an original work reflecting the level of geographic knowledge in Tibet in the 2d century B.C.[112]

Apart from the identifications in the Indian Ocean basin and eastern Africa, most relate to the Near and Middle East and to Central Asia as far north as the general region of Lake Balkhash. The westernmost place named is Ionia. Curiously, however, there is no place that falls clearly within what is now Tibet. This raises the prospect (not suggested by Gumilev and Kuznetsov) that the prototype map was brought to Tibet from Iran and then "naturalized" in that area, with considerable fidelity, and transmitted by frequent copying over the subsequent centuries. Though some alterations in the original may have occurred, the arguments of Gumilev and Kuznetsov suggest that they were not great. This leads to the sup-

108. Gumilev and Kuznetsov, "Two Traditions" (note 21). The two traditions referred to in the title are called "the Irano-Tibetan tradition," which is the one that concerns us here, and "the Indian-Tibetan tradition," which refers to the Buddhist cosmographic tradition. Nothing is said in the article about the latter that will add to what has already been stated in this volume.

109. Gumilev and Kuznetsov, "Two Traditions," 570–71 (note 21).

110. Gumilev and Kuznetsov, "Two Traditions," 571 (note 21).

111. Gumilev and Kuznetsov, "Two Traditions," 575 (note 21).

112. Gumilev and Kuznetsov, "Two Traditions," 574 (note 21).

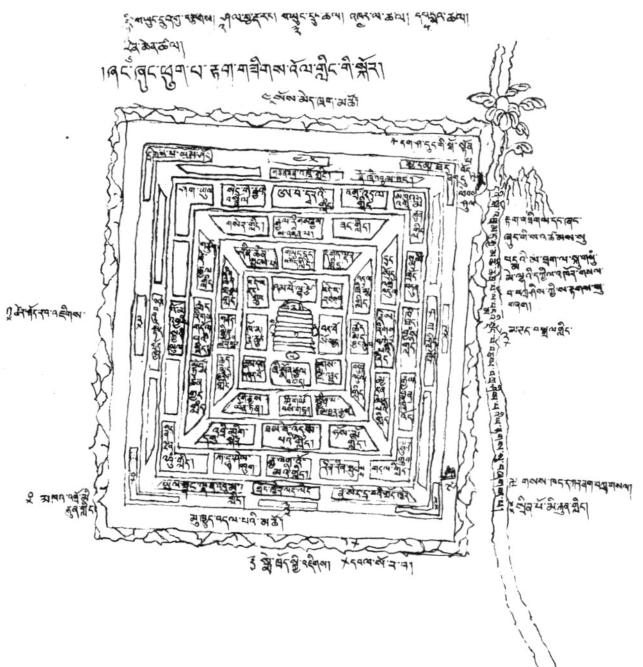

FIG. 15.27. MODERN RECENSION OF AN ANCIENT TIBETAN VIEW OF THE WORLD. The information on this map appears to represent a Tibetan view of the world as of the second century B.C. It was made to fit into the form of a mandala centering on the old Persian capital of Parsogard (Pasargadae), within which the tomb of Cyrus the Great (d. 529 B.C.), said to be ten stories high, provides the focus of the map. No old version of this map is known to survive.
From Ñi-ma-grags-pa (1616–70), *Sgra yi don sdeb snaṅ gsal sgron me bžugs so* (reprinted Delhi, 1965).

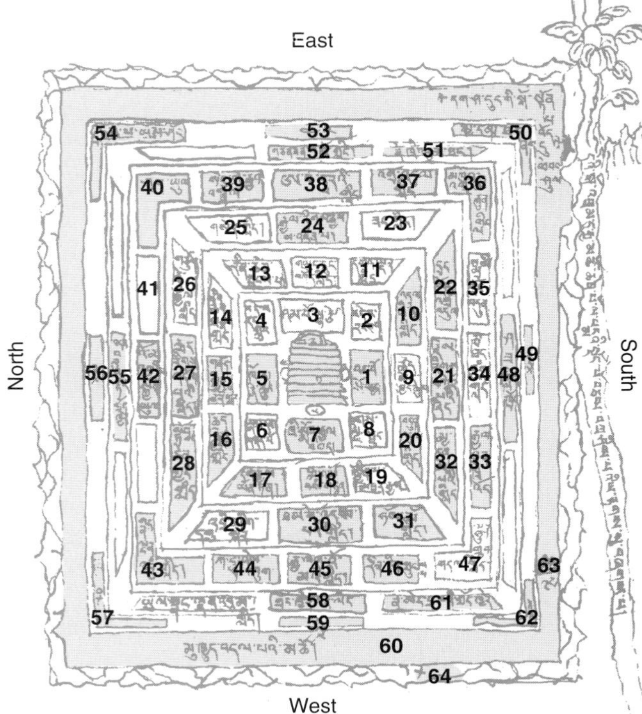

FIG. 15.28. PARTIAL KEY TO THE CONTENTS OF FIGURE 15.27. The key numbers are those assigned by Gumilev and Kuznetsov for their analysis of the map text. Numbers within shaded fields are keyed to figure 15.29, showing on a modern map the ancient regions, cities, and other geographical features indicated on the Zhang-zhung map. The analyses of the text for unshaded numbered areas are insufficient to establish any equivalent features or do not go beyond the transliteration of an unidentifiable name.
After L. N. Gumilev and B. I. Kuznetsov, "Two Traditions of Ancient Tibetan Cartography (Landscape and Ethos, VIII)," *Soviet Geography: Review and Translation* 11 (1970): 565–79, esp. fig. 3 (which contains a more complete translation than I have provided).

position that the map was accorded a sacred status and thus its integrity had to be maintained.

The Zhang-zhung map bears a remarkable resemblance to another Bön work, which has already been briefly noted under the heading of cosmographic maps. That map, identified by Snellgrove as "The Nine-Stage Swastika Mountain (representing the Nine Ways of *Bon*) . . . in the country known . . . as . . . *Śambhala*," was drawn (probably in London, and quite possibly from memory) by a refugee Bön lama about 1967.[113] Gumilev and Kuznetsov recognized the unmistakable affinity between the Zhang-zhung map, also said to represent the land of Shambhala, and the work brought to light by Snellgrove, but they observe that the latter omits several of the names on the Zhang-zhung map and was "compiled somewhat later and independently of the first version," incorporating evidence of the expansion in the region of the Kuṣāṇa dynasty in the first century A.D.[114] The map of the Nine-Stage Swastika Mountain, unfortunately, has yet to be translated, and the legends on the published version are too small to permit an independent translation. Whether it should best be regarded as a cosmographic work, as Snellgrove's title suggests, or a geographic map is moot. The place of the Kuṣāṇas on the later map, however, is noteworthy in two respects that might relate to the myth of Shambhala. First, the Kuṣāṇa Empire, a major trading state, was known for its prosperity, which could have given rise to the notion of Shambhala as an idyllic realm. Additionally, the original regional hearth of the Kuṣāṇas in Central Asia was considerably farther north than the original core area of Persia, and some of the areas under their early control lay in the Tarim basin due north of western Tibet. Hence the current Tibetan notion that Shambhala is the northern paradise, although finding no

113. Snellgrove, *Nine Ways of Bon*, v and fig. XXII (note 37).
114. Gumilev and Kusnetsov, "Two Traditions," 576 (note 21).

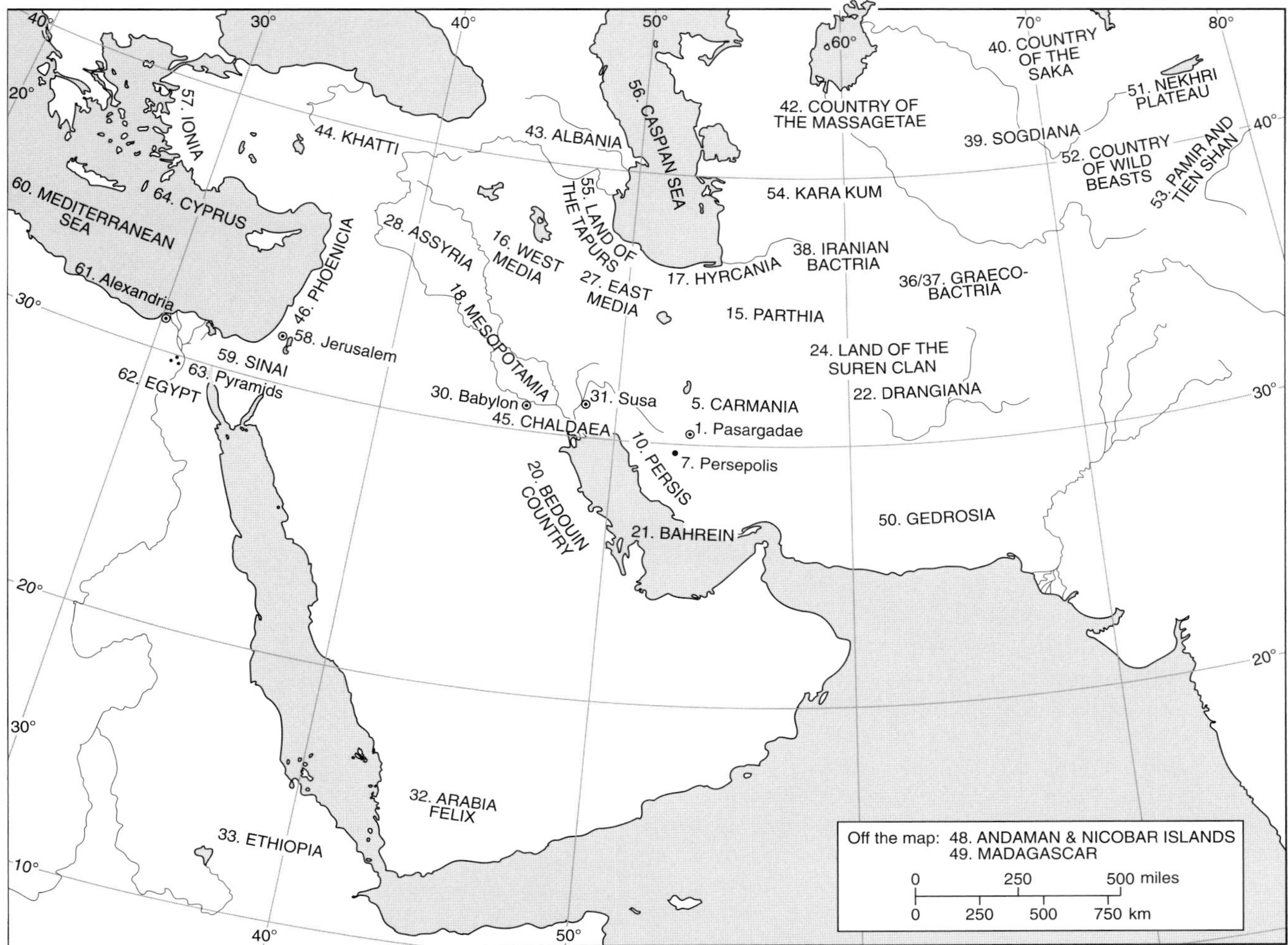

FIG. 15.29. IDENTIFIABLE ANCIENT GEOGRAPHICAL FEATURES OF THE ZHANG-ZHUNG MAP. The numbers indicated on this map are keyed to those of figure 15.28. In addition to the locales in Southwest and Central Asia and in northeastern Africa that are shown on this map, there are two others at more remote locations, the Andaman and Nicobar Islands and Madagascar, whose inclusion on the Zhang-zhung map is suggested by Gumilev and Kuznetsov.

support in the Zhang-zhung map, might be more or less appropriate for the one illustrated by Snellgrove, especially if the Kuṣāṇa domains came to occupy in the minds of ancient Tibetans the place of Persia as the archetype of Shambhala and if some great and benevolent monarch, such as the Kuṣāṇa emperor Kaniṣka, supplanted Cyrus as the archetypal divine monarch.[115] A question that remains in comparing the two maps is why the layout remained virtually identical while the content shifted. Very likely, that particular form of mandala acquired—even earlier than the time of the Zhang-zhung map—a procrustean sanctity that required fitting the content to the form and not the opposite.

A. L. Mackay dismisses Gumilev and Kuznetsov's dating of the Zhang-zhung map in a single brief paragraph. He observes that the map's

> topographic information . . . was pressed . . . into a mandala type of framework without much regard for actual geography. The map itself . . . can hardly be of the epoch represented by the place-names on it since the earliest Tibetan inscription dates only from A.D. 767. The religious cosmological view is comparable with the T-O maps of European culture exemplified by that of Cosmas Indicopleustes (*fl.* A.D. 540).[116]

But this criticism misses the point. The authenticity of the original map in no way depends on the date when Tibetan inscriptions first became known. The ancient

115. Numerous problems exist in respect to the extent and chronology of the Kuṣāṇa Empire. These are dealt with in Joseph E. Schwartzberg, ed., *A Historical Atlas of South Asia* (Chicago: University of Chicago Press, 1978), xxix–xxx, xxxiii–xxxiv, 21, and 174–76.

116. Mackay, "Kim Su-hong," 27–38; quotation on 31–32 (note 18).

Zhang-zhung language and later Tibetan were not the same. But no matter what language the original map may have been in, there is no reason its script might not have been transliterated many centuries later into Tibetan, while retaining some approximation of the original phonetic attributes of the original toponyms.

The second and only other known Tibetan world map was brought to light in 1931 by Teramoto.[117] This enigmatic map (fig. 15.30), with script in Tibetan and Chinese, composed only a small part of a long scroll document relating to Buddhist iconography and was not deciphered before the 1931 study. The original map must have been drawn by the ninth century at latest, since a copy of it was brought from the Blue Dragon (T'sing-loung-tseu) Temple at Chang'an (modern Xi'an), then the capital of China, to Japan by a priest named En-tin who died in A.D. 891. Where the map was originally made is not known. The Japanese copy of the map was kept in the Onzyōzi Temple, and over the following few centuries several additional copies were made, including one in 1194 by a priest named Zenkaku, attached to Onzyōzi Temple. Some modern copies were also made, the most recent in 1890. Zenkaku's copy was lost in the great fire in Tokyo following the earthquake of 1923, but an album of photographs previously made of the entire scroll was preserved. From 1194 to 1220 there were at least seven unsuccessful attempts to decipher the map. One apparent reason for the failures was that the map's Tibetan text was mistaken for Sanskrit. In 1893 a copy of the map was presented for study to the Japanese Bureau for Enquiry into the National Treasures; but the committee that examined it was "lukewarm about its historical and geographical value."[118] Not until the studies of Teramoto, begun in 1921, could anyone claim to have deciphered the map.

The content of the map is sparse, with only twenty-one names identified. As with the Zhang-zhung map, the identifications vary in plausibility. The territory covered appears to be vast, even greater than in the Zhang-zhung map, extending from the Byzantine Empire in the west to Korea in the east and from Turkic Central Asia in the northwest to Cambodia (Chen-la) in the southeast. Tibet itself is designated once as "Bo" and, implicitly, a second time as "Pa-man" (Eight Barbarian Lands, which were generally understood in China to comprise Korea, Chen-la, Persia, Tibet, Chien-pi, the land of the Turks, Khitan, and Mo-ho).[119] Nakamura does not suggest a principle by which the names are arranged; and in my own examination of his abstract of the map I could discern none. The map certainly does not have the symmetry of a mandala. Neither distance nor direction of places from one another is indicated consistently, and it is therefore impossible to specify the direction toward which the map is oriented. Nor does importance appear to be a criterion for decisions on what places to include and how much space to assign them. Thus, three of Teramoto's identifications are of Indian cities and the rest are of regions or even vast empires. Tibet occupies a more or less central position in the middle panel of the map; see figure 15.31 for the identification of other parts of the map.[120]

It is also difficult to put forward a plausible rationale for making the map. Paraphrasing Teramoto, Nakamura states:

> The names on this map, giving a general view of the communications between India, Persia, Tibet and Central Asia, seem to be based on Chinese and translated [into Tibetan] merely to make more easy their understanding by the Tibetans . . . [at a time when] Tibet's national prestige was at its height. . . . It was therefore most probable during this period that the map was made for political and administrative purposes.[121]

Although I find this most unconvincing, I can offer no more likely raison d'être for the work. The map does not, for example, have any apparent religious purpose. Although two of the cities named in India are important in the early history of Buddhism, the most obvious candidate for inclusion, Bodh Gaya, seems not to be noted.

The significance of the work, however, seems clearer—assuming that most of Teramoto's identifications of the far-flung regions depicted are valid—in that it proves, as previous maps do not, that Chinese knowledge of the rest of the Asia was more extensive than what could be gleaned from the travel accounts of a number of Buddhist pilgrims, among whom Xuanzhuang (602–64) was the

117. Teramoto Enga, "Waga kokushi to Toban to no Kankei" (The relation between our [Japanese] history and Tibet), *Ōtani Gakuhō* 12, no. 4 (1931): 44–83. It was also briefly discussed in Nakamura, "Old Chinese World Maps," 19–22 (note 18). Including this map (known copies of which exist only in Japan) in an article on Chinese maps preserved in Korea relates to its relevance for ascertaining the geographic knowledge of the Chinese during the Tang dynasty.

118. Nakamura, "Old Chinese World Maps," 20 (note 18). Nakamura's study of the map is based on the photographs in the album mentioned above.

119. Nakamura, "Old Chinese World Maps," 20 n. 49 (note 18). I do not know what Chien-pi refers to.

120. Nakamura, "Old Chinese World Maps," 20 n.49 and fig. 8 (note 18). The identification of Mangalore, which had little importance as far back as the ninth century, strikes me as particularly problematic. The Tibetan "Mon," which according to Stein "covers all sorts of aboriginal tribes of the wooded Himalayan Hills . . . and is possibly related to the word 'Man' used in literary Chinese for all southern 'barbarians' " (*Tibetan Civilization*, 34–35 [note 102]), suggests a more plausible association with the original "Mom."

121. Nakamura, "Old Chinese World Maps," 22 n. 51 (note 18). Nakamura does not say how proficient Teramoto was in Tibetan. If Teramoto was much more proficient in Chinese than in Tibetan, it would hardly be surprising for him to see more associations on the map with essentially Chinese toponyms than with Tibetan ones.

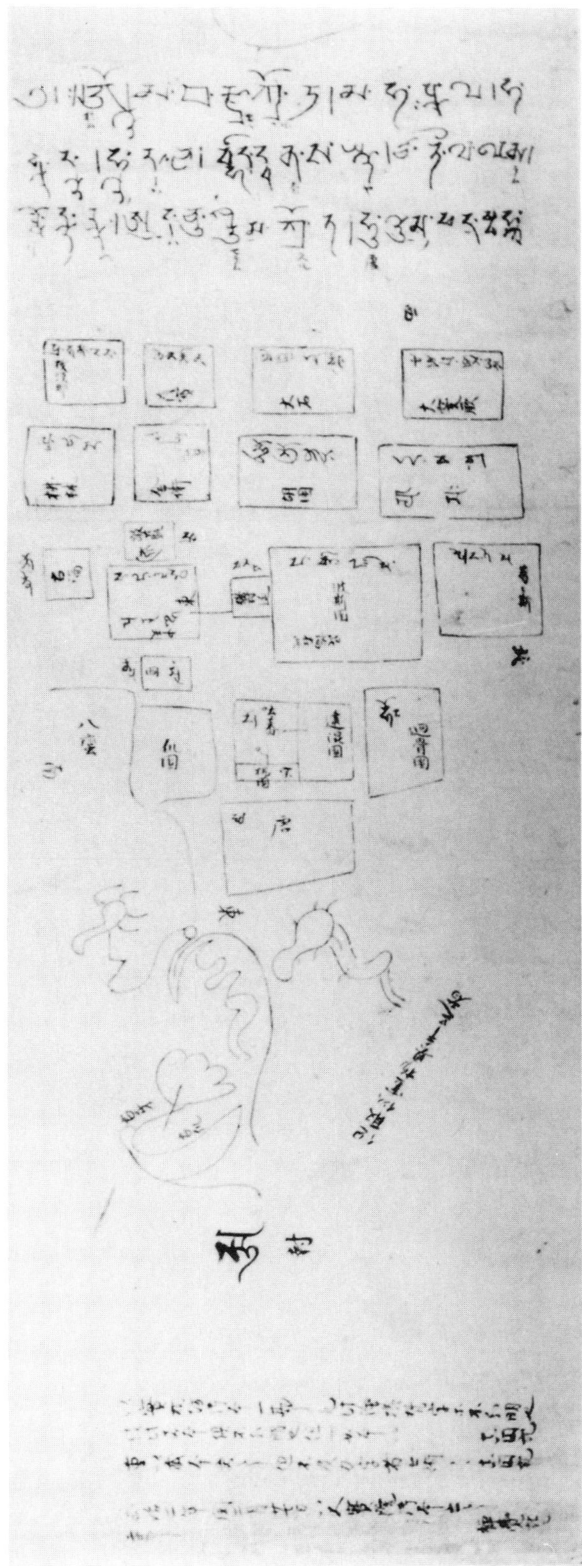

FIG. 15.30. FACSIMILE OF SINO-TIBETAN WORLD MAP. This is one of several facsimiles of the Sino-Tibetan world map. Photograph courtesy of Kazutaka Unno.

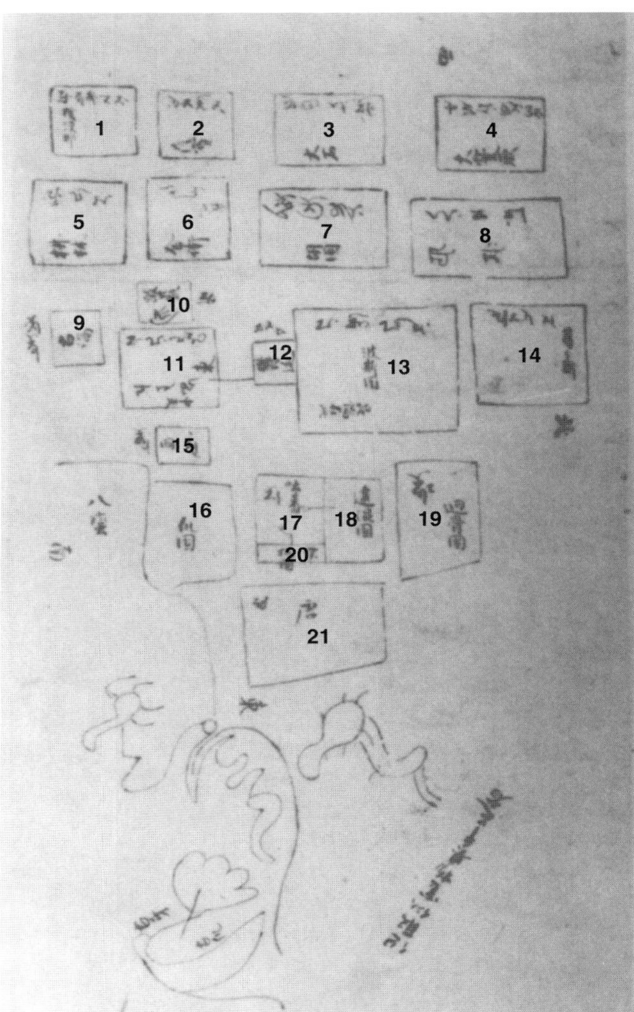

1. Pahanyan, Pahan-na? (Fergana); 2. Kin-hin (Kashmir); 3. Tha-ku-sha-si (Takṣaśilā/Taxila); 4. Taha Thor-kus (Turkestan); 5. Pu-lin (Byzantine Empire); 6. Sha-he (Śrāvastī/Set Mahet); 7. Kuo-kuo (Tokhara); 8. Ya-mah, ? Ghah (Tokhara); 9. Koron (Kurana); 10. Pran (Polan); 11. Pa-la-lor (Bolor, in Gilgit region of modern Kashmir); 12. Thahan (Tashkent); 13. Pa-sin-go (Persia); 14. Ped-she-la, ? Pitasela (Pītaśila/Khuzdar in Baluchistan); 15. Mom, Mangali, ? Mangale (Mangalore); 16. Koua-kouo (Touen-houang, ? Dunhuang in modern Gansu); 17. Po (Bed/Tibet); 18. Kin-ku-ko (Kyrghiz); 19. Hor (Uighur); 20. Than (China); 21. Pa-man (Eight Barbarian Lands, generally understood to comprise Korea, Chen-la [Cambodia], Persia, Tibet, Chien-pi [unknown], Kitan, and Mo-ho (Manchuria). Where two names are given within parentheses, the modern one comes second.

FIG. 15.31. KEY TO FIGURE 15.30. The key to the map indicates the spellings of various toponyms, as provided in a 1931 publication by Teramoto. The organization appears rather anarchic, and no obvious purpose for the map emerges from study of it. Although this may be due in part to geographic shortcomings in the original, it is also likely that errors introduced through repeated copying, together with errors of interpretation, account for the confusing picture the map presents. After Hiroshi Nakamura, "Old Chinese World Maps Preserved by the Koreans," *Imago Mundi* 4 (1947): 3–22, esp. 21.

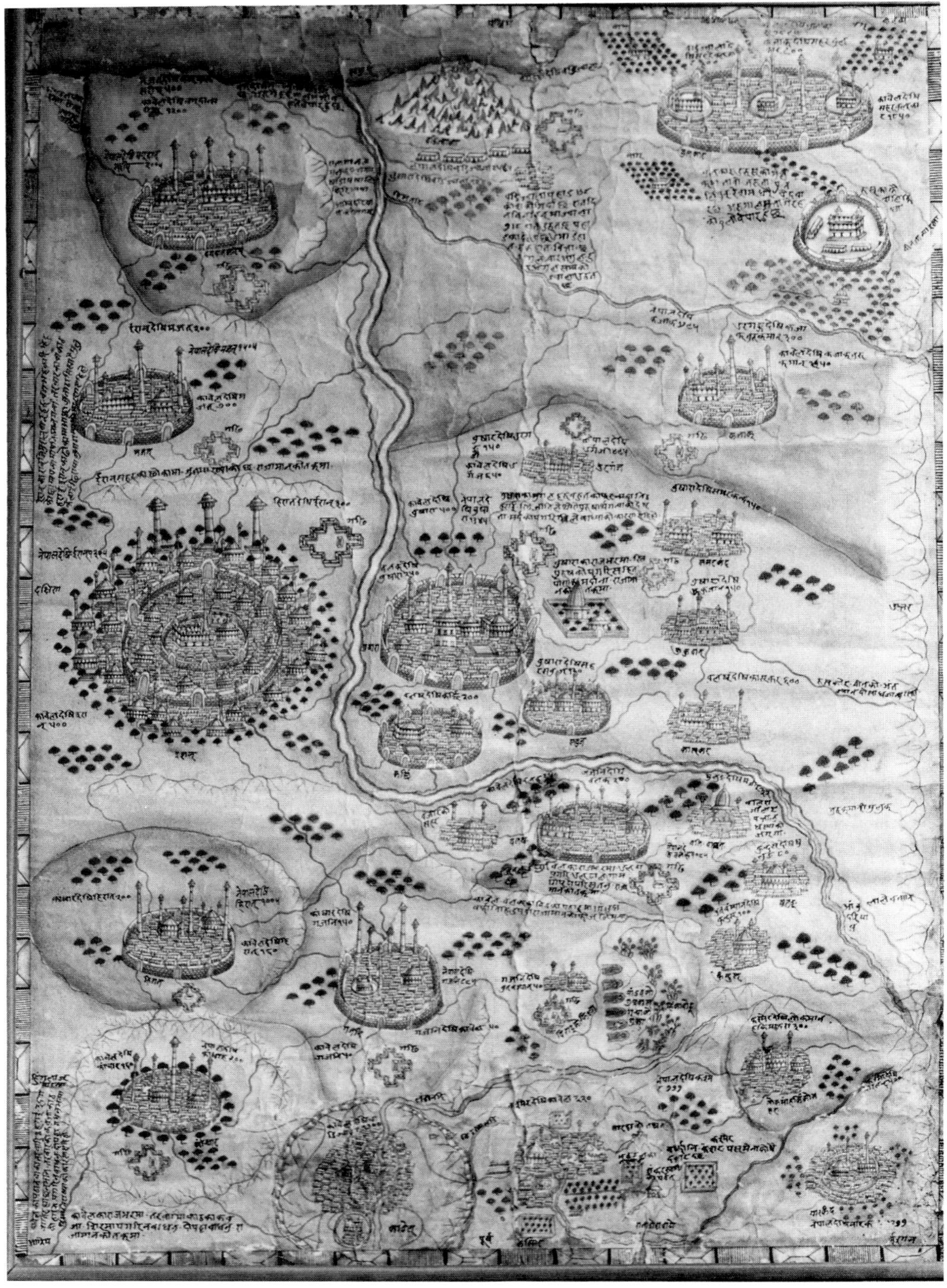

FIG. 15.32. NEPALI MAP OF CENTRAL ASIA. In Nepali, written in Devanagari script, probably latter half of the nineteenth century, painted paper mounted on cloth. Most of the details on this map are easily identified with known geographic features, while a few are mythological. Overall, the level of accuracy is not high. The large body of water at the top of the map, which is oriented toward the west, is a conflation of the Aral Sea—into which the largest river on the map, the Amu Darya, actually flows—and the Caspian Sea. The adjacent burning mountain, which should signify the vicinity of Baku, is incorrectly shown on the eastern, not the western, shore of that sea. Near the left margin the positions of Mashhad, second city from the top, and "Iran" (Tehran), the city below it, are inverted. European influences appear in the ruled, graduated border, in the naturalistic manner of depicting mountains, and in the use of color in a not particularly successful attempt to signify political jurisdictions.
Size of the original: 76 × 54 cm. Private collection, New Delhi. Photograph courtesy of Susan Gole, London.

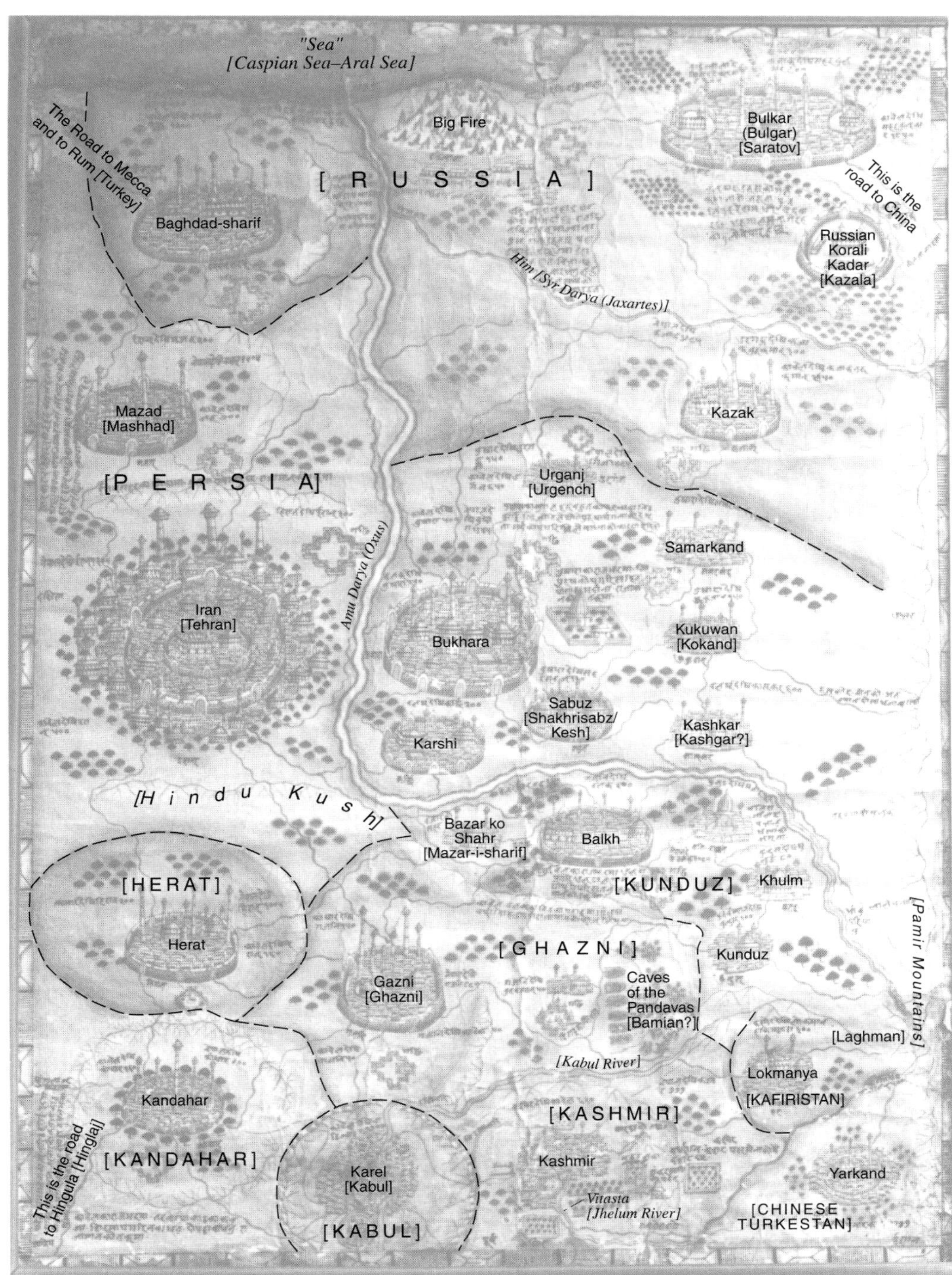

FIG. 15.33. KEY TO NEPALI MAP OF CENTRAL ASIA (FIG. 15.32). Names appearing in neither parentheses nor brackets are transliterations from the original text. Names in parentheses are modern equivalents. Names in brackets are my inferred identifications. Names in uppercase letters are inferred political jurisdictions. Dashed lines are presumed boundaries between political jurisdictions shown by color on the original.

most prominent. Conceivably a part of this knowledge, especially that relating to western Asia, was relayed eastward via the Bön Zhang-zhung tradition.

REGIONAL MAPS

A Nepali Map of Central Asia

Of the known regional maps from Greater Tibet, the one covering by far the greatest area is that illustrated in figure 15.32. Although the map is in Nepali and undoubtedly was drawn by a Hindu, it is discussed here rather than (as in the case of other works of Nepali provenance) with maps of South Asia because its style is so much more akin to Tibetan than to Indian cartography.[122]

The area covered by the map, including most of Central Asia and even extending into southern Russia, is indicated by a map key (fig. 15.33) and by its modern counterpart (fig. 15.34) and may be defined by the cities shown nearest the four corners of the map: Baghdad in the upper left, Bulgar (modern Saratov) in the upper right, Yarkand (in Chinese Turkestan) in the lower right, and Kandahar (in southern Afghanistan) in the lower left. Curiously, no part of either Nepal or Tibet is included. The orientation is toward the west, but the accuracy of the map tends to decrease in that direction. Several prominent physical features appear. The large body of water in the upper-left corner appears to be a conflation of the Aral and Caspian seas, and the large river running into it is the Amu Darya (Oxus). Several other rivers are indicated: the Him, probably intended to represent the Syr Darya (Jaxartes), even though it does not run into the Amu Darya as shown; and what appear to be the Kabul and Jhelum rivers, running through the city of Kabul and through Kashmir, respectively. Several mountain ranges are also shown. The one skirting the great bend in the Amu Darya appears to represent the Hindu Kush; but on the whole the representation of orography, including most of the large mountain chain running along the right edge of the map, appears fanciful. Less prominent, but unquestionably meaningful, are the indications of mountains around several cities to show that they are wholly or partially hemmed in by highlands. Vegetation signs scattered throughout the map, seem to connote the presence of orchards or wooded areas but probably are largely decorative, especially when not in the immediate vicinity of a city.

Though most of the places shown are fairly well known and were important in the nineteenth century when the map was probably drawn, there are also several features that are mythological, such as the caves of the Pandavas, one of the two contending forces in the Indian epic *Mahābhārata*, which are shown by the six horizontal oblongs above (west of) Kabul. Some places that appear mythological probably have a basis in fact, such as the burning mountain near the top of the map, which probably relates to an ancient Hindu fire temple near Baku that was maintained by priests from Punjab. (Oil seepages in this area, occasionally ignited, have been known for millennia.) Near ten towns on the map there is a conventional sign, along with the word *garhi* (little fort), suggesting their military importance. What appear to be Hindu temples are drawn at several places, possibly indicating resident communities of Hindu traders. Much more common, however, are minarets (present in every town shown except for Russian Korali Kadar) signifying the dominance of Islam in the region. Also near the towns are notes—often derogatory—about the nature of the local population or about matters of commercial interest, as well as figures giving distances of the towns from either Kabul or Nepal (i.e., Kathmandu), presumably in *khos* (a unit of about two miles). Roads appear as thin red lines. Notes along the edges of the map point the way to several places not shown, including China, Mecca, Rum (Turkey), and Hinglaj, a place of pilgrimage for both Hindus and Muslims on the coast of Baluchistan.

The degree of detail and the specificity of what is shown tend to be somewhat greater in the eastern portion of the map than farther west, especially in and around Kabul and Kashmir, where Gole was able to identify a number of important landmarks. For example, even at the greatly reduced scale of figure 15.32 one can easily recognize the famous Shalimar Gardens adjacent to the latter locale. This is one of a number of reasons to suppose that the artist either had firsthand knowledge of those areas or worked from relatively reliable secondhand information. Elsewhere, however, egregious errors abound, such as the already noted garbled physical geography or the inversion of the relative positions of Iran and Mazad (Tehran and Mashhad).

Although its style is essentially Tibetan, especially the way cities are depicted (compare plate 34 discussed below), the map also embodies certain European traits. One of these is the ruled, graduated border. Another, arguably, is the manner of depicting mountains. Especially significant is the use of color to suggest varying political control over particular areas, for example, salmon for Iran, yellow for the Russian empire, beige for Bukhara, and so forth; but if that is indeed the intention, the execution is not particularly clear and will not match

122. This map was brought to my attention by Susan Gole, who has had most of the text translated and has studied the map extensively. The text discussion is based mainly on her findings. See Gole, *Indian Maps and Plans*, 142–43 (a full-page color photograph and two additional black-and-white enlarged excerpts, along with a brief discussion) (note 99). A much more penetrating analysis, with numerous photographs, is Susan Gole, "A Nepali Map of Central Asia," *South Asian Studies* 8 (1992): 81–89.

the actual configuration of territorial control at any single historical date. The area I take to be Russia appropriately envelops most of the conflated Caspian-Aral Sea in the upper left, but Baghdad's placement within this region makes no sense. Speculating on the date and purpose of the map, Gole wrote in her initial effort at interpreting the map:

> It is possible that the map was drawn in the early years of the 19th century, when attempts were being made from Nepal to unite the neighbouring countries and throw out the foreign European invaders, before they became too powerful. The map may have accompanied a travel journal kept by one of the ambassadors sent to negotiate with the courts of far off countries, figured in the map. Attempts to learn more about its provenance in Kathmandu have been unsuccessful.[123]

Although the seriousness of the errors over much of the map argues against its having been made in the course of, or as a result of, the travels of any ambassador or other single individual, the idea that a map so elaborate in execution would have served a diplomatic purpose strikes me as eminently plausible. In Gole's later and deeper study of the map, however, she backed away from that position, which appeared no longer tenable in light of her discovery in Russian Korali Kadar (upper right) of a "square building [which] appears to be the fort built by the Russians . . . in 1855, known as Kazala."[124] Assuming the correctness of the 1855 date, that would rule out the possibility that the map was made in the period of rapid territorial expansion by the Nepali House of Gorkha, which was terminated abruptly by the Anglo-Nepali War of 1814–16.

Nevertheless, there are grounds to pursue Gole's original hypothesis in a modified form. Not too long after its decisive defeat by the British, Nepal, thereafter a British protectorate, was witness to the great convulsion of the so-called Sepoy Mutiny of 1857–59, which for a brief moment in history appeared as if it might signal the end of British rule in the subcontinent. Many north Indian princes and landlords then joined forces with the mutineers, and though Nepal officially remained loyal, the temptation among antigovernment factions in Nepal—a country remarkably prone to political intrigue—to pursue an opposite course must have been considerable. To any such faction, forging a grand alliance of Asian powers would be an understandable objective. It is even conceivable that Nepal's close contacts with Tibet inclined people toward the idea, embodied in the Gesar epic, that a savior would come out of the west, or else from the mythic realm of Shambhala. In any event, it would have seemed reasonable for Nepali conspirators to seek out whatever intelligence they could obtain that might bear on the feasibility of their grandiose aspirations. One way

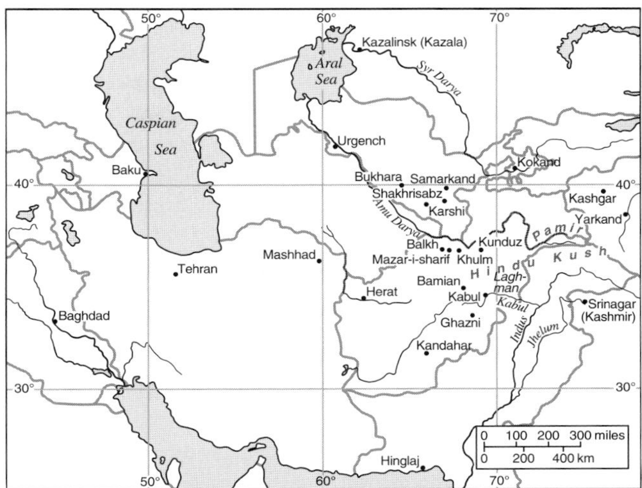

FIG. 15.34. REFERENCE MAP OF CENTRAL ASIA FOR FIGURES 15.32 AND 15.33. Only relevant place-names and physical features are indicated. Contemporary political boundaries are shown for reference purposes only. No specific referent for Kazak can be identified. Laghman is included as a regional designation. Saratov is off the map to the north.

of doing this would have been to commission a map, with relevant notes on the character of the regions and peoples in question and on the distances that would have to be traversed if their forces were to act in unison. The potential key ally in the undertaking would have been Afghanistan, which had dealt Britain one of its most humiliating defeats in the Afghan War of 1839–43 and which was strategically best situated to strike a new blow against the raj. This would explain why many of the road distances on the map were to Kabul rather than to Nepal itself.

To obtain the desired intelligence the hypothesized plotters would need to engage a person believed to possess considerable knowledge of the world beyond Nepal as well as familiarity with mapmaking. The evidence of the map itself, specifically its combining of the features of both British and Tibetan styles of cartography and the richness of the textual information provided, suggests that such an individual was indeed commissioned for the task. That much of what was presented in the map was factually incorrect is beside the point, for there was probably no one among the conspirators who could challenge the map's authenticity. The mapmaker, of course, need not have been sympathetic to the aims of those he was working for or have believed their goals were obtainable. In fact the evidence suggests quite the opposite—that he sought, rather, to dissuade the plotters from embarking

123. Gole, *Indian Maps and Plans*, 143 (note 99).

124. Gole, "Nepali Map," 82 (note 122).

on a venture doomed to failure and one that might cost him his head were he subsequently to be seen as supporting it.[125] That would explain the numerous, largely counterfactual, disparaging remarks about the peoples in the areas portrayed; for example: "The Pathans of Kabul are cowards, it has been determined, they are also deceitful." "Throughout the kingdom of Kabul, there are screws of wood in the swords." "Having determined that the mughals of Bokhara are absolute cowards, the womenfolk at least did some brave deeds." "Iranian Muslims are very arrogant. They take a bath if even the shadow of a Hindu falls on them."[126]

The colors on the map, seemingly used—albeit ineptly—to indicate political control, also provide grounds for supposing that the work dates from the mid-nineteenth century, even if they do not enable us to point to a more specific date near 1857. For example, the distinctive rose color of the area around Herat suggests that the mapmaker viewed it as an independent state, which in fact it was during 1839–57 or, arguably, until its annexation by Kabul in 1862. Similarly, the beige coloration of what seems to be the emirate of Bukhara suggests that it too was regarded as independent, as was indeed the case from 1753 until its subjugation by Russia in 1868. Although these and previously noted arguments are inconclusive, they are certainly plausible.

Maps Emphasizing Sacred Places in the Regions around Lhasa and Kathmandu

Perhaps the most sumptuous of Tibetan regional maps available for public display outside Tibet itself are a group of large and similar works depicting Lhasa and other important monastic towns in central Tibet. I know of five such works, though others almost surely exist. Two are at the Musée Guimet in Paris: one (plate 34) said to be of the eighteenth century (but possibly the nineteenth), and a very similar work dating from the nineteenth century.[127] Two are in Belgium: one believed to date from the second half of the nineteenth century, at the Ethnografisch Museum, Antwerp, and a similar but somewhat smaller, undated work in the Musées Royaux d'Art et d'Histoire in Brussels.[128] Finally, the largest of the group, thought to be from either the Kham region of eastern Tibet or Mongolia and to date from either the eighteenth century or the first half of the nineteenth, is in the Royal Ontario Museum in Toronto.[129] This work is so strikingly similar to the one in Brussels that it seems safe to assert that one is a copy of the other or that both were copied from some unknown third model. Before the Chinese takeover in Tibet, paintings of this kind were displayed for pilgrims in monasteries and at important points of passage to provide them with a visual guide to the places to be visited. The monasteries also edited textual guides providing topographical information and serving as commentaries on the paintings. As Lauf notes, "The precision of the details . . . allows us however to assume that the artist must have seen these buildings with his own eyes."[130] Yet the desideratum of verisimilitude does not prevent the composition from being ordered "like a mandala of sacred centers of Buddhist doctrine in and near Lhasa."[131]

The style of depicting landscape on these paintings is said to descend from that of works produced during the Qianlong period in China (1736–96).[132] Taken as a whole, the presentation is exalted, being "treated as a vision of a Pure Land here on earth, which is what Tibetans believe it to be. There are rolling hills, winding streams, jeweled trees, and pastel clouds, on a broad and tilted picture plane."[133] The manner of composition of all five maps is not greatly different from that of figure 15.32, being marked in particular by the conventionalized representation in exaggerated size of the major urban places within a walled circular field, and the consequent contraction of the much larger nonsacred spaces between them. Also, because of the seeming reluctance to leave any large

125. For an account of some of the antigovernment conspiracies in Nepal immediately before the Sepoy Mutiny, including a potential revolt by Gurung troops within the Nepalese army, see Ikbal Ali Shah, *Nepal: The Home of the Gods* (London: Sampson Low, Marston, [1938]), 84–97.

126. Gole, "Nepali Map," 86–87 (note 122).

127. The second of the two maps (inv. no. M.G. 21248) is illustrated in *Dieux et démons de l'Himâlaya*, 241, text on 238 (note 33); and in Vergara and Béguin, *Dimore umane, santuari divini*, cover illustration with caption on inside cover page (note 27). A full-scale reproduction of the map proper (61 x 45 cm) was available for sale by the Musée Guimet in the form of a jigsaw puzzle.

128. The Antwerp map (inv. AE 73.25) is extensively discussed in Detlef Ingo Lauf, *Lhasa: De heilige stad van Tibet en haar omgeving* (Antwerp: Etnografisch Museum van de Stad, 1974), in Dutch with summaries in French, German, and English. Readers using this generally exemplary booklet should be advised that the captions therein generally do not match the numerous photographs next to which they appear. A separately published three-page set of errata corrects this problem. The Brussels map (Collection Léon Verbert 349) is illustrated with a brief note in Pia Van der Wee, Louis P. Van der Wee, and Janine Schotsmans, *Symbolisme de l'art lamaïque* (Brussels: Musées Royaux d'Art et d'Histoire, 1988), 32–33; and also in *Dieux et démons de l'Himâlaya*, 240 (note 33). It also appears in four museum catalogs listed in the former.

129. This map, from the George Crofts Collection of the Royal Ontario Museum (cat. no. 2193), is published and discussed in Rhie and Thurman, *Wisdom and Compassion*, 374–75 (note 66). The museum description (personal communication, September 1989) suggests an eastern Tibetan provenance, and Rhie and Thurman suggest either that region or, for reasons not stated, Mongolia, which strikes me as unlikely; the museum assigns the earlier date and Rhie and Thurman the later.

130. Lauf, *Lhasa*, 21 (note 128).

131. Rhie and Thurman, *Wisdom and Compassion*, 374 (note 66).

132. Rhie and Thurman, *Wisdom and Compassion*, 374 (note 66).

133. Rhie and Thurman, *Wisdom and Compassion*, 374 (note 66).

empty space, many landscape elements are incorporated into the painting, though some of them may have no more than a purely decorative function. Of the five maps, those in Antwerp, Brussels, and Toronto are so compressed that what is depicted might pass for a single conurbation; yet the area included on those works and on the two in Paris would extend over several thousand square kilometers. The Paris maps show several rivers and mountain ranges, none of which I have identified with certainty. These are probably intended to serve largely as regional dividers, much as mountains do on the hagiographic paintings discussed in the preceding section, and it is possible that most of the mountains, if not the rivers, are not meant to represent actual physical features. The rivers and mountains shown on the maps in Belgium and Toronto are more sublimated than on those on the maps in Paris and are more naturalistic in appearance. In general, the oblique perspective of the former maps is that of an observer viewing the region from a much lower position than in the case of the Paris examples. Hence the format of the Paris maps has a long vertical dimension, whereas the other three examples (Antwerp, Brussels, and Toronto) greatly compress the vertical dimension.[134]

The published guide to the Antwerp map identifies all the major settlements and many individual structures within them. Although no comparable descriptions of the other works are available, comparison with the Antwerp example should let one identify most places. For example, on all the maps the identity of the centrally situated Potala in Lhasa is unmistakable, and the distinctive appearance of the monastery in the lower-left corner of the works at Paris and Antwerp lets one immediately recognize it as Samye (bSam yas) (compare fig. 15.47 below), nearly forty kilometers to the southeast, though in plate 34 it appears to be only a stone's throw away. The relationship between those two settlements suggests a general westward orientation to the map; but there does not appear to be a concerted effort by any of the artists to maintain consistent directional relations among the places shown. Monasteries that could be intended as Samye show up near the lower-right corner of the the Brussels and Toronto examples. I have not established all the correspondences among the five works, since it would be exceedingly difficult using only small-scale photographs, but the task would be worthy of carrying out from direct inspection of the paintings. Also worthy of investigation is the consistency of orientation with respect to the individual centers depicted. My impression is that the directional perspective from which particular towns are shown reflects the way they are most likely to be seen when approached from Lhasa, without reference to actual compass bearings.

There is another large scroll map (*paubha*) of the Vale of Kathmandu (fig. 15.35) that is conceptually something of a hybrid between the biographical (actually hagiographical) maps and the types of regional maps discussed above. This beautiful, richly colored work is painted in a distinctly Newari style. It is unusual in that it bears a precise dedicatory date, a specific day in the year Newar Samvat 923 (A.D. 1802) and in that among its abundant text is an inscription stating that it was commissioned by Sri Cikhidi of the Śākya clan, whose genealogy is then provided. This map is the subject of a detailed analysis by Banerjee, who states that "such maps are often personalized" but does not indicate the grounds for implying, as he does, that maps of the type we are considering are a common genre. The map, he says, was not only a souvenir of the pilgrimage but in itself "an object of reverence."[135]

The content of Cikhidi's map provides a vivid impression of the domestic architecture of the Vale of Kathmandu, of the appearance of a number of its important Buddhist and Brahmanical monuments, of the iconography of its religious sculpture and mural painting, and of the former mode of dress of its Newari inhabitants (whose religion is itself a Buddhist-Hindu hybrid faith). The perspective of the artist is mixed, usually frontal for single edifices and oblique for large architectural complexes, with individual components of them in a frontal perspective. As in Tibetan maps, people are shown in exaggerated size. Almost all the available map space is given over to interesting cultural and physical detail. The road network and drainage pattern are more completely and realistically presented than on any of the previous regional maps, though the courses of individual roads and rivers have been distorted to fit the demands of the composition. The mountain range at the top of the map is obviously the Himalayas, lying just north of the Vale of Kathmandu, including a distant range of snowcapped peaks and a nearer snow-free crest. But Banerjee errs, I believe, when he states: "A typical pictorial map is oriented . . . towards the north, in the style of modern maps, and is, therefore, scientific in its outlook."[136] The northern orientation in this case merely enables a composition in which the Himalayas provide an appropriate skyline away from which rivers flow toward the bottom of the

134. The Antwerp map measures 90.5 x 148 cm, Brussels 88.5 x 120 cm, and Toronto 135.4 x 184.6 cm.

135. N. R. Banerjee, "A Painted Nepalese *Paubhā* in the Collection of the National Museum, New Delhi," in *Buddhist Iconography* (New Delhi: Tibet House, 1989), 154–63 and figs. 27–37; quotations on 155 and 159. An abridged version of this article, but with larger and clearer photographs, is N. R. Banerjee and O. P. Sharma, "A Note on a Painted Map of the Kathmandu Valley at the National Museum, New Delhi," *Marg* 38, no. 3 [1986]: 77–80. The work is also illustrated and discussed by Gole, *Indian Maps and Plans*, 68 (note 99).

136. Banerjee, "Painted Nepalese *Paubhā*," 155 (note 135).

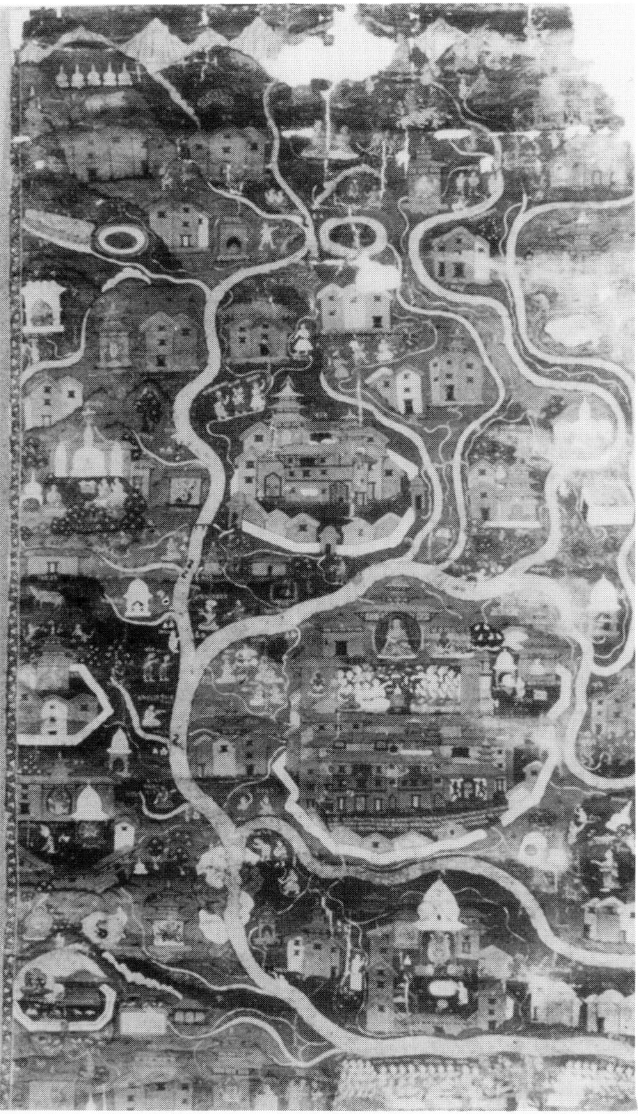

FIG. 15.35. NEPALI MAP OF THE VALE OF KATHMANDU. Newari style, text in Newari and Sanskrit, Newar Samvat 923 (A.D. 1802). Painted on cloth in black, red, light and dark green, yellow, gold, and white. This map provides a remarkably vivid and accurate impression of many aspects of the cultural landscape of the area around Kathmandu in the early nineteenth century, especially in respect to places of religious significance. The patron who commissioned the work is shown, along with his family and other members of his retinue, at many pilgrimage sites they jointly visited in that region. The abundant text of the map aids in the identification of all the major places depicted. The orientation is toward the north, where the snowcapped Himalayas form the skyline.
Size of the original: 152 × 82 cm. National Museum, New Delhi (acc. no. 56.1168). From Susan Gole, *Indian Maps and Plans: From Earliest Times to the Advent of European Surveys* (New Delhi: Manohar, 1989), 68.

map. A similarly rendered skyline appears clearly at the top of one of the two regional maps at the Musée Guimet, and one is also evident, though sometimes more muted, on the other maps of the Lhasa region (compare plate 34). I am also inclined to discount Banerjee's implication that the artist might have tried to adhere to any particular scale, "between half an inch to a mile and an inch to a mile," in painting the map.[137] Finally, I would reject his suggestion that the artist who painted the map for Cikhidi might have received training from the British (although he thought it unlikely), given that comparably detailed maps were made in Tibet well before the time of significant British intercourse with Nepal.[138] It seems more probable that the stylistic traditions of Tibet and Nepal influenced one another more than either tradition derived from exposure to British influence, at least until after Nepal's military defeat by the British in 1814–16.

Tibetan Maps in the Wise Collection

Perhaps the most comprehensive set of relatively large-scale Tibetan regional maps is the Wise Collection, comprising three large folios of maps and drawings held by the Oriental and India Office Collections, in the British Library, London. Most of the maps in the collection are assembled from multiple sheets of European paper spliced together to form extensive and very detailed cartographic documents. The constituent map sheets vary in size, but most approximate forty-eight by sixty centimeters. Figure 15.36, for example, illustrates a portion of one such map (less than a third of its total surface), showing all of two sheets and parts of two others that compose it. Although the maps in the Wise Collection were at some unknown period the object of considerable study, as evinced by the hundreds of English annotations and key numbers they bear, the circumstances of their compilation and history before their acquisition by the India Office are shrouded in mystery. The maps and drawings, all in black ink and watercolor, are attributed in the undated, anonymous catalog description to "a Tibetan artist, probably a lama, who had had contact with Europeans and developed a semi-European style of drawing."[139] In my view, however, the map style remains essentially indigenous. The maps, which could justifiably be categorized under the traditional Tibetan genre of *sa-khra* (picture map), are described as "similar to pilgrim maps showing villages and buildings in pictorial form," while the "25 drawings of places, monasteries, people and ceremonies in Tibet . . . appear to have been com-

137. Banerjee, "Painted Nepalese *Paubhā*," 155 (note 135).
138. Banerjee, "Painted Nepalese *Paubhā*," 159 (note 135).
139. Oriental and India Office Collections, British Library, London, Add. Or. 3013–43, unpublished catalog description, nine pages, p. 1.

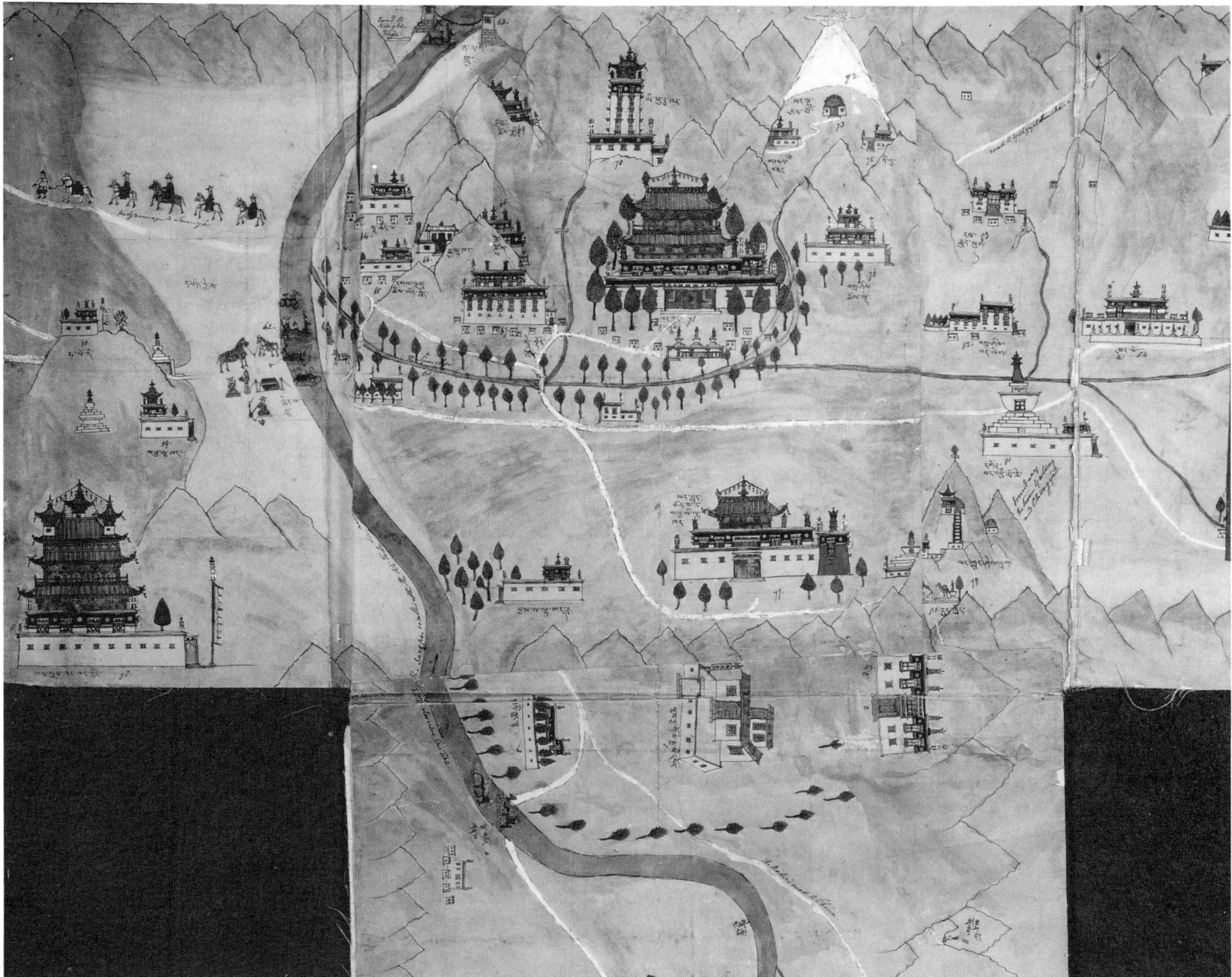

FIG. 15.36. PART OF A LARGE MAP OF A PORTION OF CENTRAL AND EASTERN TIBET. This illustration covers somewhat less than a third of an irregularly shaped painted map that forms one of seven large "picture maps" in the Wise Collection of Tibetan maps. The portion shown in the photograph comprises all or parts of five pages, out of a total of six, that were pasted together to form the complete map. The principal center shown in this illustration is the famous monastery at Samye, the oldest in Tibet. Included on the map are numerous explanatory inscriptions in Tibetan for which—in contrast to other portions of the collection—no translation is currently available. The map is believed to have been made by a Tibetan who was specially trained by a British mentor (possibly Wise, about whom nothing is known) to carry out mapping and other intelligence gathering sometime between 1844 and 1862.
Size of the detail: ca. 70 × 85 cm. By permission of the Oriental and India Office Collections, British Library, London (Add. Or. 3017).

missioned by the writer of the explanatory text."[140] Just who that writer was is not known.

> Although the bindings (c. 1890–1900) are lettered Wise, no record of anyone of that name connected with Tibet or adjacent areas can be found in the records. The water-marks [1849–56] of the explanatory text as well as internal evidence . . . suggest a date for the drawings of between 1844 and 1862. The notes suggest that "Wise" was acquainted with the western side of Tibet and had travelled in Ladakh.[141]

140. Unpublished catalog description (note 139).

141. Unpublished catalog description (note 139). I would dismiss as incorrect the handwritten suggestion added to the prefatory note for the catalog description that the Wise in question could have been James F. N. Wise (1834–85), author of *Notes on the Races, Castes, and Tribes of Eastern Bengal* (London: Harrison, 1883). I have checked the latter's career via his several contributions to the *Journal of the Asiatic Society of Bengal*, beginning in 1873, and, in addition to the fact that he would have been too young to be involved with our anonymous lama as early as 1844, I find no grounds to link him in any way with Tibet.

In light of the British concern with military intelligence beyond the frontiers of India at the time the maps were made, a period straddling the establishment in 1849 of a British protectorate over Jammu and Kashmir, of which Ladakh was a part, it is reasonable to suggest that the "lama" (or lamas) who made the maps in the Wise collection was (were) secretly recruited for that task and that the name Wise was simply a pseudonym for some now forgotten British official who saw to his (their) recruitment and training and who, for diplomatic reasons, could not be correctly identified. Much, of course, has been written about the later "pundits" who worked for the Survey of India. They were brilliant individuals selected from among the peoples of several faiths living on India's borders, who were specially trained to carry out transfrontier intelligence and mapping missions, using assumed names and disguised as pilgrims and traders, during the period 1863–93. But such missions had precursors as early as the Central Asian expedition of Mir Izzet Ullah in 1812–13, and it seems likely that the maps in the Wise Collection were the outcome of a number of such precursor expeditions.[142] Several facts support this line of reasoning. First, the maps were on European paper. Second, they were pieced together in such a way—some pages taking off at right angles from those that had previously been assembled (as in fig. 15.36)—as to suggest that the mapmaker could not fully anticipate what and how much mappable territory lay before him as he made his way into unknown country. No other Tibetan maps I have seen have this characteristic. Finally, one must consider the map content—for example, the great detail in which bridges were portrayed, the annotations relative to routes and the distances to places beyond the limits of the maps themselves, and so forth.[143] The maps and appended notes also contain much ethnographic and other detail that, though of only marginal interest for military purposes, would be highly relevant for the conduct of diplomacy.[144] Inclusion of certain more esoteric religious details might be explained on two grounds: first, as being of personal interest to the map compiler, who, though working for the British, was probably nonetheless a devout Buddhist; and second, as a means of seeming to legitimize an activity that would otherwise raise unwanted suspicions by Tibetan officials curious to know what the mapmakers were up to.

Appendix 15.1 provides a synopsis of some of the important particulars relative to the Wise maps, including several that do not properly qualify as "regional." To the details provided there, I can add a few more general notes. First, what appears to be most characteristically Tibetan in the style of the maps is their attention to architectural detail. In general, monasteries, *chortens* (stupas), palaces, and other major cultural features are rendered in conventionalized colors and, along with people, animals and plants, are drawn at scales vastly greater than those employed for physiographic features. The actual map size seems to be mainly a function of the amount of detail the artist feels it necessary to depict. But whereas architectural features are often shown in an oblique perspective in other Tibetan maps, in the Wise maps they are, with relatively few exceptions, shown in frontal perspective, presumably from the most likely direction of approach. In respect to map orientation, no general rule appears to be followed; though each page of a multisheet map does appear to have one predominant orientation, the orientation may shift as one goes from one sheet to another, as one can see by contrasting the lower portion of figure 15.36 with the other visible portions of the map. Further, whereas in a purely Tibetan style mountain summits are usually portrayed as pointing away from an observer on the ground (e.g., in the opposite direction as seen from a road following a river valley), that convention is frequently violated in the Wise maps. Another seeming concession to Western cartographic style is the depiction of linear features such as roads and many rivers as narrower than they would be on most Tibetan maps. The common Tibetan device of having smaller streams appear from and then disappear behind mountains is largely retained, as might be expected from a cartographer unable to follow every stream to its source and trying to map only what he actually saw as he made his way through a topographically complex countryside. Finally, in landscapes largely devoid of forests, lines of trees and village groves are prominently depicted wherever they occur.

Whether the maps in the Wise Collection were made by a single artist, as the catalog description suggests, or by two or more individuals is open to question. Although most of the maps in the collection are anything but crude, there are exceptions (items a and b of appendix 15.1), and I am inclined to think that at least two individuals were at work. But, it is possible that these exceptions are relatively crude because some circumstance (e.g., the need

142. A very readable and generally authoritative account is provided by Derek Waller, *The Pundits: British Exploration of Tibet and Central Asia* (Lexington: University Press of Kentucky, 1990), 14 and 22 for Mir Izzet Ullah in particular.

143. Examples of annotations that would have been militarily useful include the following (all taken from Add. Or. 3014): "Another bridge across the Indus. The river frozen in winter and passable." "Kera—a village; road not passable here in summer when the river is very large." "Hanle is eleven marches from Lé [Leh]."

144. Examples from Add. Or. 3014 include: "Pituk . . . a gunpo [monastery] of the Gelukpa . . . sect who came from Tashi-lunpo." "Tok, the jaghier [personal estate] of the present raja of Ladak, given to him by Maharaja Golab Singh, after he conquered the country, . . . Masho, residence of the brother of the Ladak Raja, who is a lama." "Liktse, a village & where a tax gatherer of the Ladak govt. stays."

FIG. 15.37. PORTION OF A TIBETAN MAP SHOWING THE AREA IN THE GREAT BEND OF THE TSANGPO/ BRAHMAPUTRA RIVER. The map illustrated here was drawn on non-European paper in black ink and yellow and brown watercolor. It probably dates from the late nineteenth or early twentieth century and was obtained by Heinrich Harrer in Tibet, about 1950. The artist employed many conventional symbols of Tibetan cartography in preparing this vivid depiction of the difficult topography through which the Tsangpo flows in eastern Tibet. Particularly striking is the drawing of summits pointing away from the main axis of travel along the river itself and their apparent convergence within the river bend. As the map text has yet to be translated, a full interpretation of its content is not yet possible.
Size of the original: 52 × 171 cm. By permission of the Völkerkundemuseum der Universität Zürich (cat. no. 14495).

for security) forced the mapmaker to prepare them with more haste than was customary.

Although a number of published works have drawn attention to the maps in the Wise Collection, no one so far has analyzed them in any detail.[145] Such analysis—not only of the maps, but of the accompanying drawings and text—by a scholar with the requisite linguistic skills and knowledge of Tibetan culture and history is sorely needed and bound to be richly rewarding.

Tibetan Maps in the Harrer Collection

Another noteworthy collection of mainly regional maps from Tibet is in the Völkerkundemuseum der Universität Zürich. These constitute a portion of the Tibetan collection of Heinrich Harrer, acquired by the museum in 1972 and still awaiting detailed study. The maps I have seen are on non-Western paper and rendered in ink and various watercolors. One of the maps in the collection is illustrated in figure 15.37. Appendix 15.2 provides details of those maps and some of the others. The appendix is based on a brief firsthand inspection in 1987, study of several published photographs, and notes in a museum catalog relating to an exhibit of Harrer's materials.[146]

145. Among the works that cite the Wise Collection are Aziz, "Tibetan Manuscript Maps," in which fig. 1, labeled "folk drawing," is excerpted from one of the Wise maps (which one and the precise locale depicted are not stated); and idem, "Maps and the Mind," in which the same excerpt appears on p. 55, labeled "pilgrim's map" (both in note 12); Martin Brauen, *Feste in Ladakh* (Graz: Akademische Druck- u. Verlagsanstalt, 1980), 14 (fig. 1); and Müller and Raunig, *Der Weg zum Dach der Welt* (note 34), in which there are map excerpts showing the monastery complexes at Shigatse (p. 112), Gyantse (rGyal-rtse) (p. 263), and an unspecified locale (p. 357).

146. The photographs are published in Martin Brauen, *Heinrich Harrers Impressionen aus Tibet* (Innsbruck: Pinguin-Verlag, 1974), pls. 9 and 110 in color, and fig. 7 (p. 104) in black and white. The book includes a fifteen-page catalog of the exhibition at the museum that opened in December 1974. The same map excerpt that appears in pl. 9 of Brauen is reproduced on p. 88 of Müller and Raunig, *Der Weg zum Dach der Welt* (note 34), where it is labeled "Landschaft am Knie des Tsangpo-Flusses," which would only barely place it, as Brauen does, in the Tibetan province of Kham. Another excellent photograph, which appears on stylistic grounds to be remarkably similar to most in the

The age of the maps is uncertain, but I would judge that none predates the late nineteenth century or post-dates the first quarter of the twentieth. Since Harrer left Tibet early in 1951, after a sojourn of seven years, that year represents an absolute terminus ad quem for the entire collection.[147] How and why Harrer acquired these specific maps is at present not known. Since several of the works in his collection appear to relate to a military campaign, however, as is evident from the depiction of weapons, military encampments, a field with a few corpses, and what appear to be battle lines, we may suppose that he hoped to use at least some of the maps to put forward a Tibetan view of the several Anglo-Tibetan skirmishes resulting from the British Younghusband expedition to Lhasa in 1904.[148]

Although it is virtually certain, on stylistic grounds, that the Harrer maps were made by several different hands, most show broad similarities, and all use conventions of the type depicted in figure 15.9 to show landscape elements, especially various types of mountains. Other conventions for depicting religious edifices and settlement also tend to be broadly similar and comparable to those used in the earlier maps of the Wise Collection.

A singular exception, however, is one uncataloged map, a portion of which—judging from the house types and rope suspension bridges shown on it—relates to an area of transition from Tibetan culture to the tribal culture of what is now the northeast Indian state of Arunachal Pradesh. That area was of considerable interest to Anglo-Tibetan and Anglo-Chinese diplomacy during the period leading up the Simla Conference of 1914, when Britain and Tibet agreed on the McMahon Line as their common border in that region, a border which China stoutly rejects. The area was first surveyed for Britain by Henry T. Morshead and Frederick M. Bailey in 1912–13, but it was visited in 1884 by Kintup ("KP"), the illiterate Sikkimese servant of a Chinese "lama" employed by the Survey of India to determine whether the Tsangpo and the Brahmaputra were the same river, as had been suspected.[149]

Other Regional Maps

A small section of a large and elaborate cotton scroll map is illustrated in figure 15.38. This map has been studied in great detail by Huber.[150] It was given to Hugh Richardson in Lhasa about 1944, when he was the representative of the British government in Tibet. The work, which Richardson described as "well used" at the time he acquired it, is not dated, but it is ascribed by the British Museum to the nineteenth or twentieth century and designated a "pilgrim map." The map was noted and partially reproduced in several publications by Aziz, who describes it as a pilgrimage map of Tsa-ri (rTsa-ri) rong skor.[151] But, as Huber pointed out, although "one end of the map does indeed depict *some* [my emphasis] of the southwestern section of the long Tsa-ri *rong-skor* pilgrimage route, . . . the majority of it covers areas well to the west

Harrer Collection, appears in Tichy, *Himalaya*, 133 (note 43). This very detailed map, labeled "Alte tibetische Landkarte," is not described in the work cited and is not attributed in any way to Harrer. Rather, it is stated that the photograph came from Ella Goldschmidt of Vienna (who may well have obtained it from or through Harrer). My attempts to establish contact with her proved futile.

147. Harrer, an Austrian mountaineer, was interned in India during World War II and in 1944 escaped from detention in the Indian town of Dehra Dun, subsequently making his way into Tibet. After a long and difficult journey to Lhasa, he was allowed to reside in that city, and in time he became a trusted employee of the Tibetan government and a tutor of the young Dalai Lama. He did not leave Tibet until March 1951, some months after the Chinese occupation of that country. He describes his Tibetan sojourn in *Seven Years in Tibet*, trans. Richard Graves (London: Rupert Hart-Davis, 1953), but he does not mention in that work that Tibetan maps were among the substantial body of Tibetan materials he took with him to India.

148. I am, however, aware of no attempt by Harrer to use the maps in the way suggested. Of greater concern to Harrer was the cause of Tibetan independence. To have called attention to Tibet's past humiliation by Britain, a nation generally inclined to support Tibet's interests vis-à-vis China, would have to be a diversionary exercise that would in no way further the aims of Tibet in the mid-twentieth century. Harrer makes no mention of the maps in his work *Return to Tibet*, trans. Ewald Osers (New York: Schocken Books, 1984).

149. Escaping twice from Tibetan captivity, Kintup managed to carry on with the mission (1880–84) after the defalcation of his Chinese master, who sold him into slavery in May 1881. His account of the routes he explored was largely discounted, and Kintup was all but forgotten until 1914, when his achievements were finally recognized. The stories of Kintup, Morshead and Bailey, and others concerned with making known the region in question are recounted in Waller, *Pundits*, 214–47 (note 142). It is conceivable that information obtained from Kintup while he was a captive in Tibet was in some way related to the decision by the Tibetans themselves to prepare the map and possibly others that may yet come to light. It is also conceivable, though less likely, that the map was made using information the Tibetans extracted from Kintup, whose powers of recollection were said to be extraordinary. In either event, Harrer's being entrusted with the map when he left Tibet for India in 1951 may have related to some now unfathomable Tibetan diplomatic purpose.

150. Although I have had an opportunity to study the map at first hand, my discussion of it is based on work by Huber, who in conjunction with a Tibetan colleague, Tashi Tsering, resident in Dharmsala, India, has translated all of the abundant text on the map and prepared glosses, some of them rather lengthy, keyed to most of its 140 inscriptions; Toni Huber, "A Tibetan Map of lHo-Kha in the South-eastern Himalayan Borderlands of Tibet," *Imago Mundi* 44 (1992): 9–23. Huber's analysis is by far the most extensive and thorough scholarly commentary on a Tibetan map I have seen. His kindness in making this work available to me in draft form is acknowledged with gratitude.

151. Aziz, "Tibetan Manuscript Maps," 29, with a partial illustration on page 30, and idem, "Maps and the Mind," 54–55, with a different illustration overlapping those two pages (both in note 12). A small portion of the map has been printed in color on the dust jacket of Aziz and Kapstein, *Soundings in Tibetan Civilization* (note 28), but this will probably have been removed by most libraries.

FIG. 15.38. SMALL PORTION OF A LARGE TIBETAN MAP OF THE TSA-RI REGION AND AN ADJACENT PORTION OF ARUNACHAL PRADESH, INDIA. Painted on a cotton scroll, this undated Tibetan map shows an area along the upper Subansiri River straddling Tibet's border with the present Indian state of Arunachal Pradesh. Long thought to be a pilgrimage map, the work appears instead to have been prepared, very likely at the behest of the Tibetan government, as a cartographic record of physical, administrative, ethnographic, and other needed intelligence relating to an area of conflicting political interests in the early part of this century. The map is exceedingly rich in place-specific information, as is evident not only from its visual detail, but also from its numerous inscriptions.
Size of the entire original: 63.5 × 351 cm. By permission of the British Museum, London (1986.5-26.01).

of Tsa-ri proper."[152] It is also clear from the map text that its author had little or no concern with religion; but rather, the map was related to travel in general, questions of political and fiscal administration, and topographic and ethnographic intelligence.[153]

152. Huber, "Tibetan Map," 9 (note 150). Aziz appears also to have erred in stating that Richardson acquired the map "before World War II" and in placing Tsa-ri in northern Tibet and in a valley in a "legendary area accessible only by passage through dangerous lands inhabited by brigands and wild forest people," rather than near Tibet's southern border ("Maps and the Mind," 55 [note 12]). Although the area covered is indeed difficult to reach, estimates of the number of pilgrims from all parts of Tibet making the Tsa-ri pilgrimage (held every twelve years) range from as few as 10,000 to 15,000 to as many as 100,000 (Huber, "Tibetan Map," 17 [note 150]).

153. Consider, for example, the following translated map inscriptions (using Huber's numbering system):

4. "Crossing the Gla-'khol pass from bKra-shis monastery it is one day's journey as far as lHun-rtse (rDzong) [castle]."

35. "A forty-five-step ladder [clearly shown on map] to the upper valley."

49. "Peaks on the eastern bDag-ri range."

53. "The precipice which is the extremity of the (bDag-ri [referring to another note]) mountains."

55. "From this resthouse above the Bya-chu [a river] it is five days' journey as far as the nine passes and nine valleys."

58. "All (this area is) the continuous frontier of the country with the lDing-klo [a tribal group of Arunachal Pradesh] in those (places) behind, to the south of here."

67. "The narrow footpath to Klung."

75. "There is only Klo(-pa) [tribal] country behind the passes on the entire southern flank of these snow mountains (i.e., the Grand Himalayan range)."

From Huber, "Tibetan Map," 4–18, passim (note 150).

Those who wish to compare the content of the map with that of the Tibetan guidebook discussing a number of the features the map

The map content is largely conveyed by inscriptions and, even more, by its highly varied graphic elements/signs, using both shape and color to convey meaning. It is thus a particularly good example of the *sa-khra* (picture map) genre. Several different signs are used to show mountains, some suggesting glacial terrain, others karstic highlands, and still others lower, rounded summits. As in many other Tibetan maps the summits point away from the observer's route in a valley, in this case of the Subansiri River, that forms the main axis of the map. Although it might at first appear that the mapmaker had little concern for what lay away from this axis, the notes on the map—many of them referring to the number of days' march to other locales and the passes to be crossed to get there—prove otherwise. Other map notes point out areas of special interest such as grassy flats and plains. Many regions and areas occupied by particular tribes and clans are explicitly identified. Dozens of settlements—not all of them identifiable—are depicted, as are more than a dozen monasteries, a comparable number of estates, smaller numbers of temples and stupas, several forts, and at least one trading post. Footpaths are frequently labeled as such, and at least one "cattle path" is noted.

> The western portions of the map . . . show the barren, rolling landscape of the Tibetan plateau marked regularly by small villages (in general, only 1 or 2 buildings are used to represent a village), religious establishments and plots of land under cultivation. Moving into the eastern portions the settlements become less frequent and the Tibetan style dwellings begin to give way to the thatched huts used by the pre-literate Arunachal tribals who visited the upper Subansiri valleys on the Tibetan border in order to trade. The scene painted here is marked by dramatic cliffs, waterfalls, and a variety of vegetation . . . including bamboo, flowering herbs and different tree species, all reflecting the high rainfall that this area is well known for.[154]

Huber does not suggest a date for the map or a responsible agency, but I believe it is similar in date, origin, and purpose to the uncataloged map in the Harrer Collection discussed above. The circumstances surrounding the drawing of the Harrer map would have applied equally for the map obtained by Harrer's contemporary, Richardson, since both works refer to a troubled and politically sensitive frontier region.

We can only speculate on how many and what sorts of additional regional maps survive in Tibet. Todd Lewis, an anthropologist specializing in the Himalayan region and Tibet, reports having seen "very extensive frescoes that depict major regions of Tibet" in the Potala and Norbulinka (Nor-bu-ling-ka, Summer Palace) in Lhasa, but he was not able to obtain photographs of such works.[155] The Potala is vast, and its treasures have yet to be made fully known beyond Tibet. Its mural paintings, begun in the year 1648 and repeatedly painted over during the following three centuries, are said to adorn "every hall, every chapel, and every corridor," often being drawn with a "vertical view and diversified perspective" suggesting maplike images.[156] Further, it is reasonable to suppose that what was true of the Potala was also true, though obviously to a lesser degree, for other major monasteries. Regrettably, however, relatively few Tibetan monasteries have survived the cultural vandalism and officially sanctioned destruction that Tibet has suffered since 1950, especially during the Chinese Cultural Revolution of 1966–69.[157]

Mural and scroll maps are also known to exist in Buddhist monasteries outside Tibet. For example, a seventeenth-century map of Bhutan may be seen on a wall of the Punakha *dzong* (castle). And as noted in the introduction, long cloth map scrolls (*paubhas*) are periodically put on public display in the courtyards of various Nepali monasteries. No inventory of mural and scroll maps has yet been attempted, but it is highly probable that more will come to light in the near future. It is not always easy to classify the few known examples as regional, route, or locality maps. One such painting appears in figure 15.2 and an excerpt from another, an enormous work of the eighteenth or nineteenth century, in figure 15.46. Although the latter could be regarded as a regional map, it seems preferable to discuss it below under locality maps.

PILGRIMAGE AND OTHER ROUTE MAPS

Travel and visibility in Tibet and neighboring areas are largely constrained by difficult mountain topography marked by long and deep orographic troughs (such as those along the upper Indus, Sutlej, and Tsangpo rivers). Consequently, what might be intended as maps of major regions may take the form of rather elongated strips (as do several in the Wise Collection), making the distinction

---

depicts may consult Elena Rossi Filibeck, "A Guide-Book to Tsa-ri," in *Reflections on Tibetan Culture: Essays in Memory of Turrell V. Wylie*, ed. Lawrence Epstein and Richard F. Sherburne (Lewiston, N.Y.: Edwin Mellen Press, 1990), 1–10.

154. Huber, "Tibetan Map," 10 (note 150).

155. Letters from Todd Lewis to me dated 1 September 1987, 12 February 1988, and 28 March 1988.

156. *The Potala Palace of Tibet*, comp. Cultural Relics Administration, Tibet Autonomous Region (Shanghai: People's Art Publishing House, 1982), 52. This work contains numerous photographs with glimpses of mural paintings, but none are good enough to permit a meaningful assessment of their cartographic quality.

157. Since the opening of Tibet to foreign tourists beginning in 1982, the Chinese authorities have restored portions of some of the more important monasteries. What effect, if any, this may have had on the state of their cartographic murals is not known. Photography within the monasteries is largely forbidden.

FIG. 15.39. MAP PORTION OF A TIBETAN SCROLL SHOWING PILGRIMAGE PLACES IN THE VALE OF KATHMANDU. This map occupies the upper half of a large fragment of a two-part Nepali cloth scroll completed in Newar Samvat 755 (A.D. 1634). Referred to as a "banner painting," the entire work was executed in opaque watercolors (various greens, blues, salmon, red, black, white, and gilt) on cotton. The lower register of the work (not shown) is a narrative painting relating to a traditional Buddhist history of the deity Svayambhūnāth. The portion shown here is designated a *tīrtha mahātmya*, which signifies that its function was to glorify *tīrthas* (pilgrimage places), in this case in the Vale of Kathmandu, and guide pilgrims to them. The *tīrthas* are symbolized by icons representing their presiding deities, occupy distinctive fields of color, and are identified by inscriptions in the Newari script. Only a few of the identified *tīrthas* shown, however, correspond clearly to currently frequented sites. The rivers on the map reflect the actual network of the Vale (but with the topological distortion required by the scroll format) and group those sites by subregions. We may assume that this work, like other banner paintings, was hung out for periodic display in the courtyard of some Nepali *vihāra* (monastery), and it is likely that exposure to the elements is the reason only a fragment of the original remains. Size of the entire original: 39 × 130.2 cm. By permission of the Cleveland Museum of Art, gift of Mrs. Albert S. Ingalls (54.788).

between regional and route maps more arbitrary and problematic than in other parts of the world. Nevertheless, I have selected here a number of works whose principal objectives appear to be to illustrate routes or places with relatively little regard for the nature of the terrain away from those routes. Some of these have the elongated shape characteristic of many route maps, while others do not.

Oldest among the known works of the latter type is a fragment of a Newari scroll painting (fig. 15.39) that illustrates an ancient text, the *Svayambhū Purāṇa*.[158] In its original form, the entire scroll probably was of a type similar to the one illustrated in figure 15.2. The fragment shown here comprises two horizontal components, a lower narrative register of scenes and an upper register that has been designated as a *tīrtha mahātmya* (i.e., a record of "eminent places of pilgrimage"). The map includes fifteen or so named pilgrimage places in the Vale of Kathmandu, of which most can be identified iconographically, and a few additional scenes of worship. These are regionally grouped into elongated fields (shapes necessitated by the scroll format), set off from one another by the course of the Bagmati River and several identifiable tributaries.[159] In its use of rivers as dividers, this scroll map is similar in appearance to many cosmographic scenes in the lengthy scroll presentations of the Thai Buddhist *Trai phum*, discussed in chapter 17 of this volume (compare fig. 17.23). This raises the possibility of the existence of an intermediate Indian form, which might have spread to both Nepal and Siam from a sacred site such as Bodh Gaya, frequented by pilgrims from both countries.

Totally different in form from the *tīrtha mahātmya*, though similar in function, is an undated Tibetan work (fig. 15.40) that relates to the sacred Mount Kailas (Kailāsa) and other places of pilgrimage in its vicinity. The work in question is a bas-relief bronze tablet that includes a rather detailed representation of the physically visible, as well as mythological, features on and in the vicinity of Kailas (Ti-se to Tibetans). This is taken by Buddhists, Bön-pos, Hindus, and Jains to be the earthly manifestation of the sacred Mount Meru, the axis of the universe.[160] Despite its absence of text, the most important features shown on the tablet (see fig. 15.41) may be readily recognized on medium- and large-scale maps of the region it depicts, in particular Mount Kailas, which rises 6,700 meters above sea level; the lesser mountain, Dorje Tijung to its west (the tablet being oriented with the viewer looking to the north); and Lakes Manasarowar

158. This map is discussed at length in Slusser, "Serpents, Sages, and Sorcerers" (note 13).

159. The entire fragment measures 39 by 130.2 centimeters, of which the upper half is the map portion. Slusser, "Serpents, Sages, and Sorcerers," 68 (note 13), provides a key diagram on which the several map elements are noted and, wherever possible, identified. Each is further discussed in the accompanying text. The work is also illustrated in Slusser, *Nepal Mandala*, vol. 2, fig. 569 (note 13). A photograph of a more modern, though undated, painting, somewhat similar to the work discussed by Slusser and relating to the same region—but with a distinctly more Indian appearance—was given to me in Brussels in 1987 by Armand Neven, a historian of South Asian art. At the time the work was in the London gallery of Jean-Claude Ciancimino, but I do not know its present location. As in the earlier work, the various *tīrthas* are grouped into fields by a network of rivers and are recognizable both by the images of their tutelary deities and by names written on the painting itself. A line of mountain peaks near the top of the painting presumably represents the Himalayas.

160. See the expert analysis by Eva Stoll, "Ti-se, der heilige Berg in Tibet," *Geographica Helvetica* 21 (1966): 162–67. The tablet is also illustrated and more briefly discussed, in the broad context of Buddhist beliefs about Mount Sumeru, by Van der Wee, "Rirab Lhunpo," 71–72 and fig. 7 (note 57).

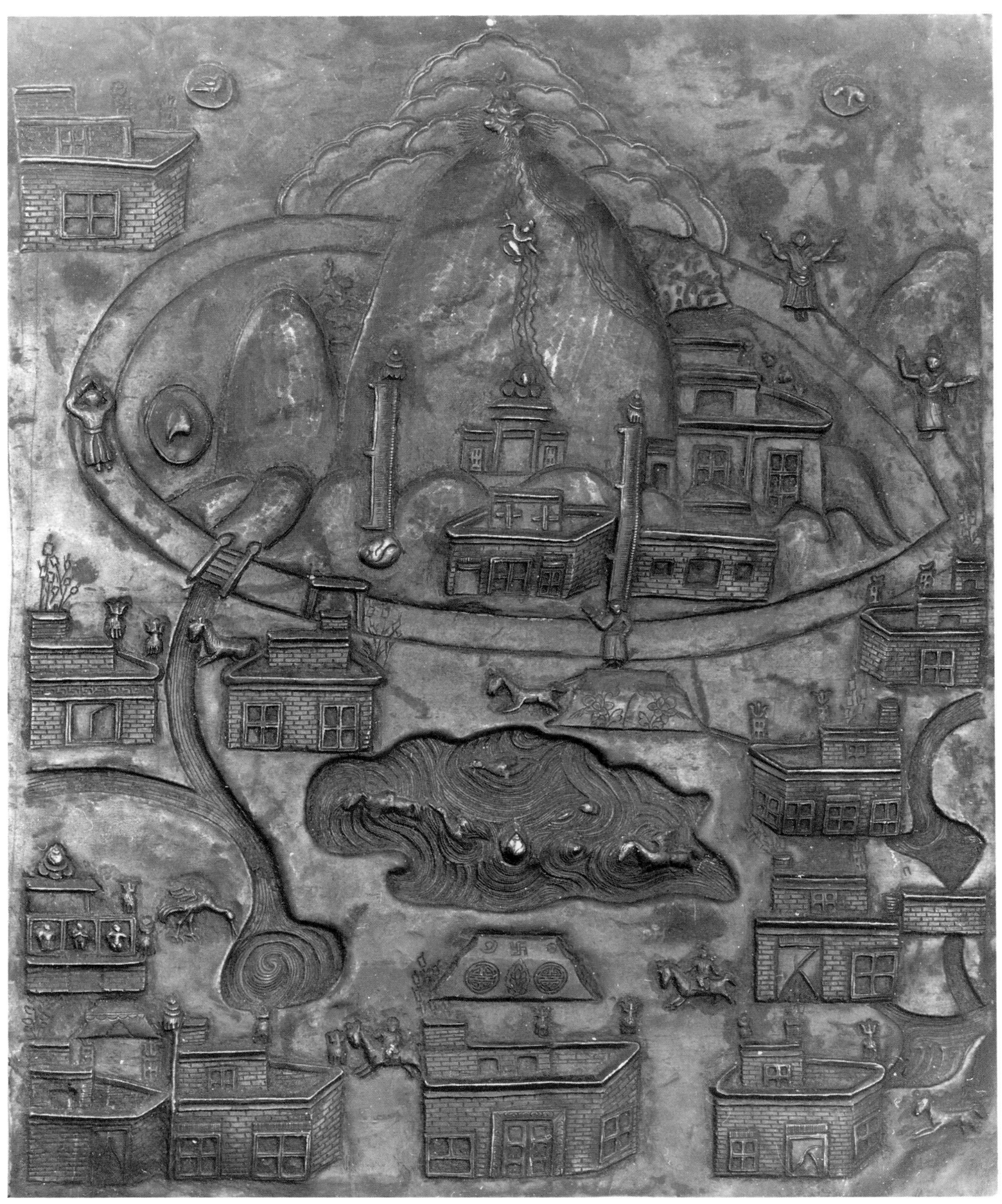

FIG. 15.40. TIBETAN BRONZE TABLET SHOWING PLACES OF PILGRIMAGE IN THE AREA OF MOUNT TI-SE (KAILAS). This undated bas-relief tablet shows not only many places and monasteries that pilgrims are enjoined to visit, but also unreachable features that are purely mythological. Of particular note are the pilgrims circumambulating the mountain clockwise.
Size of the original: 34.2 × 28 cm. By permission of the Völkerkundemuseum der Universität Zürich (cat. no. 12665).

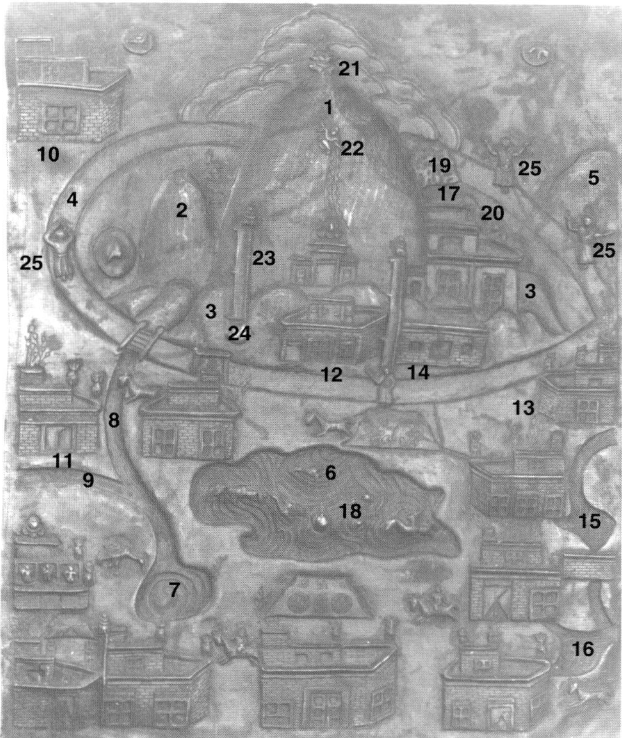

1. Ti-se (Mount Kailas); 2. Mount Dorje Tijung; 3. other unspecified neighboring mountains; 4. pilgrim route around Ti-se; 5. peak signifying the Tsering-Changa group; 6. Lake Manasarowar; 7. Langag, Lake of the Demons; 8 and 9. Tarchen-C'u and Lha-C'u, tributaries to Langag; 10 to 14. five *gompas* (monasteries) of which four lie in the cardinal directions around Ti-se (10. Tint'ipu, north; 11. C'oku, west; 13. Zupral, east; 14. Gyantrag, south; and the fifth, 12. Silung); 15 to 17. Lakes Kurkyal C'ungo, Ding Ts'o, and T'uki Zingbo; 18. lotus bud in Manasarowar; 19. traces of the struggle between Saint Mila Rêpa and the Bon-po; 20. Mila Rêpa's hand and footprints; 21. Mila Rêpa on Ti-se; 22. Mila Rêpa circumambulating Ti-se; 23. flagpole; 24. the Buddha's footprint; 25. pilgrims of whom the one in the high hat is a Bon-po.

FIG. 15.41. KEY TO FEATURES SHOWN ON FIGURE 15.40. After Eva Stoll, "Ti-se, der heilige Berg in Tibet," *Geographica Helvetica* 21 (1966): 162–67, esp. fig. 1.

(Ma pham) and Langag (La-ng) to the south. Some idea of the scale of the area shown may be obtained from the knowledge that circumambulation of Manasarowar, the larger lake, normally takes several days, the circuit extending for seventy kilometers, or roughly one hundred kilometers if one visits all eight of the surrounding monasteries that are also depicted on the tablet. A shorter circuit is that of the mountain itself, but that the more arduous trip is also enjoined on the devout is made evident from the clearly marked path on the tablet along which two pilgrims and a monk are shown making their clockwise circumambulation. Regrettably, the published accounts of the tablet say nothing about the circumstances of its acquisition, or about its date or provenance. What is reasonably certain, however, is that the work conforms closely to published Tibetan pilgrimage guides for the region, and that it would have served well in orienting pilgrims toward realizing the several objects of their journey.

Included in Stoll's article are photographs of several paintings of Ti-se and its surroundings. These are among a number of pilgrimage maps that are remarkably similar in content to the Zurich tablet: (*a*) a work by a Tibetan lama, Nav-Kushok, taken from a book on the region by the Indian explorer Swami Pranavānanda, who spent years in the vicinity of Kailas; (*b*) a wall painting, rendered in a naive folk style, found in a Nepali village near the Tibetan border; (*c*) a *thanka* painted by Tulku Tsewang, remarkably similar to the earlier-mentioned wall map, found in a monastery in Dolpo, Nepal; and (*d*) a painting on paper of unknown date and provenance, but with accompanying Tibetan text, that is rather different in style and composition from all of the foregoing, though generally similar in content.[161]

Well to the southeast of Kailas and Manasarowar is another sacred mountain-lake dyad, the peak of Gosainthan (Shisha Pangma) and the lake of Gosainkund, with respect to which one pilgrimage map is known to exist. This work, painted on cloth and measuring 125 by 79 centimeters, is held by the Department of Oriental Antiquities, British Museum, London (registration number 1928.0707.1). Found in Kathmandu, the map is believed to date from the early nineteenth century. Although I know it only from a black-and-white photograph, I would judge that it is the product of a Shaivite Hindu artist, despite some apparent influences from Tibetan models, notably in its embossed border, similar to those found on many Buddhist *thankas*, and its composition, in some ways reminiscent of the cosmography of Sukhāvatī depicted in figure 15.21. The map is richly detailed and seems to show many Shaivite temples on either side

161. Only works a and b are illustrated by Stoll, "Ti-se" (note 160). Work a also appears in Swami Pranavānanda, *Kailās-Mānasrōvar* (Calcutta: S. P. League, 1949), fig. 93, with numbers keyed to an accompanying list of forty-four specific features it depicts, largely duplicating those of the bronze tablet in Zurich. Pranavānanda provides even more detailed explanations of these features (pp. 10, 14, and 123) than does Stoll and appears to be the source Stoll largely relied on. Stoll's source for work b is Olschak and Thupten Wangyal, *Mystic Art*, 6 (note 39), which presents the work at a larger scale. The map was discovered by the Swiss geologist Augusto Gansser en route to Mount Kailas, on the wall of a small monastery in Tinkar, southwest of the Tibetan town of Taklakot and near the junction of the borders of India, Nepal, and Tibet. Both it and work c are illustrated in Blanche Christine Olschak, Augusto Gansser, and Andreas Gruschke, *Himalayas* (New York: Facts on File, 1987), 192 and 81 respectively. Work d is illustrated in Louis P. Van der Wee, "A 'Cloister-City'—Ṭanka," *Journal of the Indian Society of Oriental Art*, n.s., 4 (1971–72): 108–20; relevant note on 113 and illustration in fig. 5. The *thanka*, in the Rijksmuseum voor Volkenkunde, Leiden, is incompletely and somewhat misleadingly identified in the relevant museum note that Van der Wee cites. I have referred here only to works I have seen photographs of; others, however, are known to exist.

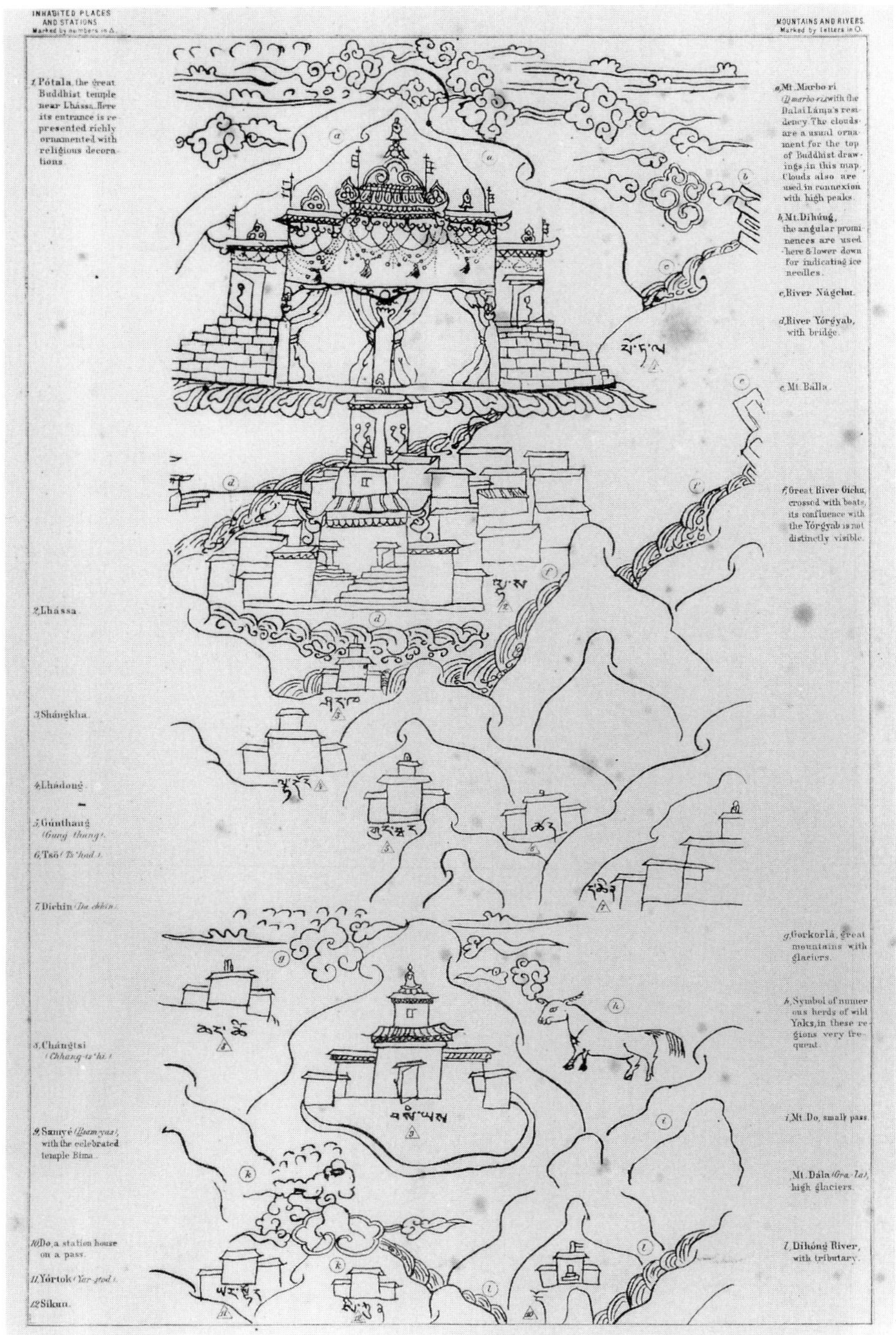

FIG. 15.42. TIBETAN MAP OF THE COMMERCIAL ROUTE FROM LHASA TO ASSAM. This seemingly naive map makes use of numerous conventional signs in Tibetan cartography, evident from the annotations along the margins. It is the northern half of the map taken from Hermann von Schlagintweit-Sakünlünski, Adolphe von Schlagintweit, and Robert von Schlagintweit, *Results of a Scientific Mission to India and High Asia* (1861). The map embodies far more specific detail than the uninitiated non-Tibetan viewer might suppose. The route depicted begins at the Potala, shown at the top of the map, continues through Lhasa itself, and thence through Samye. The second half of the map (not shown) continues through Tawang, now in India, and south to a point near the plains of Assam. The scale is notably larger in the better-known northern part of the map than in its southern portion.
Size of this portion: ca. 30 × 20 cm. By permission of the Bodleian Library, Oxford (Maps 206a.2).

of the Trisuli River along the pilgrimage route to Gosainkund and beyond to Gosainthan, as well as lines of pilgrims wending their difficult way toward those high Himalayan destinations whose sanctity for Nepali Hindus rivals that of the more renowned region around Kailas.

Despite the recent inroads of modern technology and Western education in the Himalayan region, it appears that pilgrimage maps embodying elements of traditional cartography are still being made. Snellgrove illustrates one such map, taken from a modern printed Tibetan pilgrimage guide to the holy places in the region of Thakkola (Thag)—now included in the Nepali region of Mustang, just south of the Tibetan border—and he also provides a key to its contents, drawn on a modern geographic base.[162] The guide the map is taken from is one of many such works that have been produced in the areas of Tibetan Buddhism over a period of centuries. A systematic search through such guidebooks may well uncover many more pilgrimage maps.

The final route map that I shall consider is among the earliest works of Tibetan cartography to be brought to the attention of Western scholars. Drawn on a scroll for Hermann von Schlagintweit-Sakünlünski in the monastery town of Narigun in January 1856, it was published in a black-and-white facsimile, at two-thirds the original size, in the atlas volume that formed part of the massive report of the scientific expedition of the Bavarian Schlagintweit brothers to India and "High Asia."[163] The map illustrates a route followed by Bhotia traders between Lhasa and Assam, via Samye and Tawang. Figure 15.42 reproduces its northern (upper) half. A note on the map, printed in English, indicates that it was drawn by Dávang Dórje, a Bhotia chief, with the assistance of some lamas. His signature appears in a panel near the bottom of the map, a rare instance of attributing authorship on a Tibetan map. All the original map text is in Tibetan, but along both sides of the facsimile appear notes in English keyed to its contents: thirty-two, along the left margin, relating to "inhabited places and stations," and twenty-three on the right referring to mountains, rivers, and other physical and biological features.

Notwithstanding the circumstances attending its preparation, the map is in an unadulterated Tibetan style, employing many common conventions. For example, map annotations relating to clouds indicate not only that those behind the Potala are the "usual ornaments for the top of Buddhist drawings," but also that others "are used in connection with high peaks." Elsewhere, in relation to the lone yak depicted not far from the lower-right corner of the illustration, the note reads, "Symbol of numerous herds of wild Yaks, in these regions very frequent." Thus, just as a single house, on this and other Tibetan maps, may signify a village, one yak may indicate an entire herd or herds.

The route this map deals with runs generally north-south at right angles to the predominant grain of the terrain. Hence the mountains do not point away from the observer following a route in a river valley, as they conspicuously do so on several other maps we have examined. Rather, they point away from the southerly locale of Narigun, from which the author's dominant view is toward the north. As with most Tibetan maps, the scale is not consistent from one part of the map to another; in this case the scale for the northern portion is greater than for the south. However, that nearly half of the vertical dimension of the northern half is taken up by the Potala, the most prominent feature shown, and the city of Lhasa would in itself account for most of this difference.

PLANS OF TOWNS, MONASTERIES, AND OTHER SMALL LOCALITIES

Apart from cosmographic maps, the most common type of cartographic artifacts from Greater Tibet are those that relate to specific localities, especially major religious centers or particular monasteries or other religious edifices within those centers. Many of these depictions, as we have seen, are appended to much larger, essentially hagiographical works, being designed to indicate the sanctified places associated with a particular religious figure, whether mortal or divine. Here I shall consider only paintings in which a particular locality forms the dominant focus of the composition. Only a small sample of such maps—chosen to demonstrate the diversity of their modes of representation—will be individually discussed. Brief notes relating to the remainder of those known to me—probably a reasonably representative sample of a much larger corpus—are provided in appendix 15.3.[164]

Possibly the most intriguing, best-studied, and oldest among locality maps is one focusing on Svayambhūnāth stupa, on the outskirts of Kathmandu (fig. 15.43).[165]

162. David L. Snellgrove, trans. and ed., "Places of Pilgrimage in Thag (Thakkhola)," *Kailash* 7 (1979): 72–132; Tibetan text 133–70, maps on 72 and 132. The date of the text the map relates to may be 1607, 1667, or 1727, depending on the specific sixty-year Tibetan calendric cycle the stated date refers to. The map itself, however, has clearly been made to conform broadly to recent geographic knowledge of the region depicted.

163. Schlagintweit-Sakünlünski, Schlagintweit, and Schlagintweit, *Results of a Scientific Mission*, vol. 4, map 3 (note 10).

164. Appendix 15.3 includes a few maps that I have seen personally and a much larger number that I know only through photographs, either published—with or without explanatory text—or sent to me by various sources. Obviously any attempt at analysis based solely on small-scale photographs has serious limitations. The citations of published works in the final column of the appendix are limited to those I have personally examined.

165. The first attempt at scholarly analysis of this map was that of Theodore Riccardi, Jr., "Some Preliminary Remarks on a Newari Paint-

FIG. 15.43. NEWARI MAP OF THE PRECINCTS OF SVAYAMBHŪNĀTH STUPA AND THE VALE OF KATHMANDU. This unusual painted cloth map was prepared in connection with the rededication of the ancient Svayambhūnāth stupa in A.D. 1565. The work has two relatively discrete components: an upper portion, in which the stupa and the structures in its immediate vicinity are depicted at a very large scale against a field of cinnabar, and a lower portion in which several dozen places in the Vale of Kathmandu are shown within a field of muted gray. Prominent features of the map include the mountain rim of the Vale, shown in a highly stylized manner in the arch that largely envelops the stupa and also at the bottom of the map, and the Bagmati River system. The placement of specific towns and religious sites within this system illustrates the basic orderliness of the presentation. The abundant text on the map, in both Newari and Sanskrit, enables the positive identification of most of the features shown.

Size of the original: 101 × 85 cm. Photograph by Mary Shepherd Slusser. By permission of Princeton University Press.

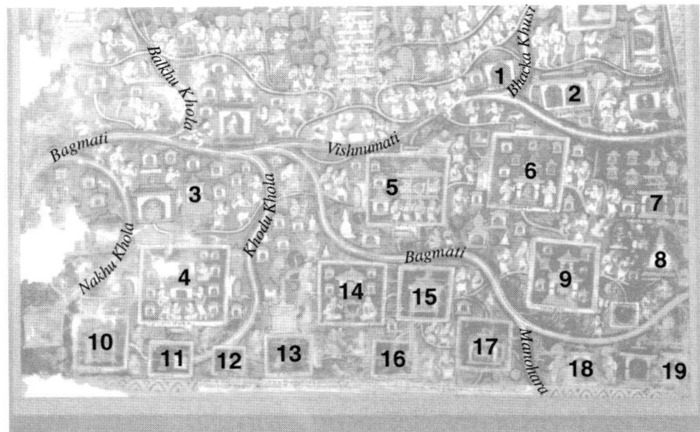

1. Vidyāśvarī; 2. Sobhā Bhagavatī; 3. Pulchok; 4. Patan; 5. Yaṅgal (southern Kathmandu); 6. Yaṃbu (northern Kathmandu); 7. Thaṃ Bahī; 8. Bodhnāth; 9. Deopatan; 10. Kumbheśvara (?); 11. Theeva; 12. Sanagaon; 13. Harasiddhi; 14. Pharping (?); 15. Thimi; 16. Lubhu; 17. Bhaktapur; 18. Sūrya Vināyaka; 19. an unnamed *vihāra*.

FIG. 15.44. KEY TO PART OF FIGURE 15.43. The numbered sites are those that bear identifiable labels on figure 15.43. With two exceptions (10 and 19), these are named and similarly numbered on figure 15.45.
After Mary Shepherd Slusser, "The Cultural Aspects of Newar Painting," in *Heritage of the Kathmandu Valley: Proceedings of an International Conference in Lübeck, June 1985*, ed. Niels Gutschow and Axel Michaels (Saint Augustin: VGH Wissenschaftsverlag, 1987), 13–27, esp. fig. 2.

FIG. 15.45. REFERENCE MAP FOR FIGURE 15.43. This is a modern map of the Vale of Kathmandu. Key numbers on this map are identical to those of figure 15.44. Sites 5 and 6 represent southern and northern Kathmandu respectively. Sites 10 and 19 on figure 15.44 cannot be located on a modern map.
After Mary Shepherd Slusser, "The Cultural Aspects of Newar Painting," in *Heritage of the Kathmandu Valley: Proceedings of an International Conference in Lübeck, June 1985*, ed. Niels Gutschow and Axel Michaels (Saint Augustin: VGH Wissenschaftsverlag, 1987), 13–27, esp. fig. 4.

Although the area of this map includes most of the Vale of Kathmandu and is similar to that of the regional map illustrated in figure 15.35, I have chosen to consider it here as a locality map because of the extraordinary emphasis given to a single dominant religious complex. The date of the map is Newar Samvat 685 (A.D. 1565), when it was, according to its inscription, commissioned in a monastery in the Nepali town of Patan in connection with the reconsecration of the nearby renowned ancient stupa that is its central object. Long forgotten and seriously deteriorated over the following four centuries, the map resurfaced in 1967 and passed into the hands of a private collector, who saw to its restoration.

> The painting is arranged in two sections, the upper one devoted to the glory of Svayambhū, the lower to the towns and sacred sites—Buddhist and Hindu—that bathe in its refulgence. In both sections the arrangement is by no means haphazard but represents on the one hand an exact site plan of the Svayambhū compound and on the other a correct map of the Kathmandu Valley. Indeed, the painting . . . serves as a pilgrim's guide to the sacred geography of the Kathmandu Valley. In short, it is a map.[166]

Arching above the stupa and to both sides is a highly stylized representation of the mountain wall that forms the western, northern, and southern rims of the Vale, a pattern one sees also at the bottom of the map, which represents the Vale's eastern margin. Between the two runs the Bagmati River, exiting the Vale to the left (south). Over much of the map, captions in Newari, varying in legibility, identify many of the places depicted. Between the places so marked, we can identify the major components of the drainage system by interpolation. Figure 15.44 shows the major features of the map apart from the stupa itself and the buildings in its immediate vicinity; these may then be compared with the corresponding features on a modern map as presented in figure 15.45.[167]

Also from Patan is another map relating to the city

ing of Svayambhūnāth," *Journal of the American Oriental Society* 93 (1973): 335–40. A more complete study is that of Slusser, "Cultural Aspects," and Slusser, *Nepal Mandala*, vol. 1, 299, and vol. 2, fig. 495 (both in note 13); and Pratapaditya Pal, *The Arts of Nepal*, pt. 2, *Painting* (Leiden: E. J. Brill, 1978), 80–81, 152, and fig. 108.

166. Slusser, "Cultural Aspects," 20 (note 13).

167. Slusser, "Cultural Aspects," figs. 7 and 8 (note 13); Slusser's fig. 6 is a modern plan of the Svayambhūnāth compound to which one can also relate the features in the appropriate portion of the original map.

FIG. 15.46. NEPALI SCROLL MAP OF THE VALE OF KATHMANDU. This is a small portion of an enormous cloth banner painting that is periodically displayed in the courtyard of the Kwa Bahal monastery in the town of Patan in the Vale of Kathmandu. It depicts numerous sacred places and several towns of the region, but it has not been analyzed sufficiently to determine the full extent of the area it relates to. Many of the features depicted have been labeled, presumably in Newari, while other features are distinctive enough in their representation to be readily identifiable. Despite the restrictive format of the painting, there is an unmistakable topographical logic to the organization of the features, including various meandering streams, rolling hills, and—marking the horizon—cloud-capped mountains. In many locales flora, fauna, and people enliven the work.
Size of the entire original: ca. 90 × 1,800 cm. Photograph courtesy of Mary Shepherd Slusser.

and neighboring places in the Vale of Kathmandu. This work, believed to date from the mid-nineteenth century, appears on an enormous painted cloth scroll that is hung from time to time in the courtyard of the Kwa Bahal monastery. A small part of the map is illustrated in figure 15.46. Neither Slusser nor Pal, who discuss it in several works, provides an estimate of its overall dimensions, but based on color slides and black-and-white prints of the entire painting, I estimate its length as approximately eighteen meters and its height as about ninety centimeters.[168] On this map

> all the principal sacred places and a number of towns are graphically illustrated, some further identified with written labels. They occupy a landscape of cloud-capped mountains, rolling hills, and meandering streams enlivened with diverse flora, fauna, and people, the latter largely occupied with religious affairs. Given the limitations of the long narrow format, the shrines are so well organized topographically that, labeled or not, most can be easily identified.[169]

Regrettably, this is the extent of Slusser's description. One might also wish to know how far beyond the urban core of the Vale the map extends and also the point of view of the artist. Is the map a panoramic view, as if seen from a particular point, or does some other principle guide its composition? These questions also apply to the other scroll map from Patan noted in appendix 15.3.

In striking contrast to the Nepali map focusing on Svayambhūnāth are a number of Tibetan representatations of the renowned Tibetan monastery of Samye, founded in A.D. 775 and restored several times since. Of these, the best known is undoubtedly the one reproduced in figure 15.47, which has been dated as early as the sixteenth century but more likely dates from the seventeenth or eighteenth. Samye is of particular interest because it is the first and oldest of all Tibetan monasteries and was also consciously designed, largely following the model of the monastery of Odantapura, in what is now the Indian state of Bihar, to symbolize the Buddhist universe.

> The central square temple represents Mount Meru. . . . Its three stories are each in a different architectural

168. This work is illustrated in several works by Slusser: *Nepal Mandala*, vol. 2, fig. 98; "Pilgrim's Guide from Nepal," 30–31; and "Cultural Aspects," 26 (all in note 13); and also by Pal, *Arts of Nepal*, fig. 164, plus note on 132 (note 165). I am indebted to Todd Lewis for his sending me sets of black-and-white photographs and color slides covering the entire painting.

169. Slusser, "Pilgrim's Guide from Nepal," 30 (note 13).

FIG. 15.47. TIBETAN MAP OF SAMYE MONASTERY. This exquisitely painted *thanka* faithfully depicts the eclectic nature of its central temple, which is compounded of Tibetan, Indian, and Chinese styles, one for each of its three stories, and makes evident the attempt to re-create, within the complex as a whole, a model of the universe as conceived by Tibetan Buddhists.
Size of the original: 53 × 38 cm. By permission of the Newark Museum, Newark, N.J., Shelton Tibetan Collection (acc. no. 20.271).

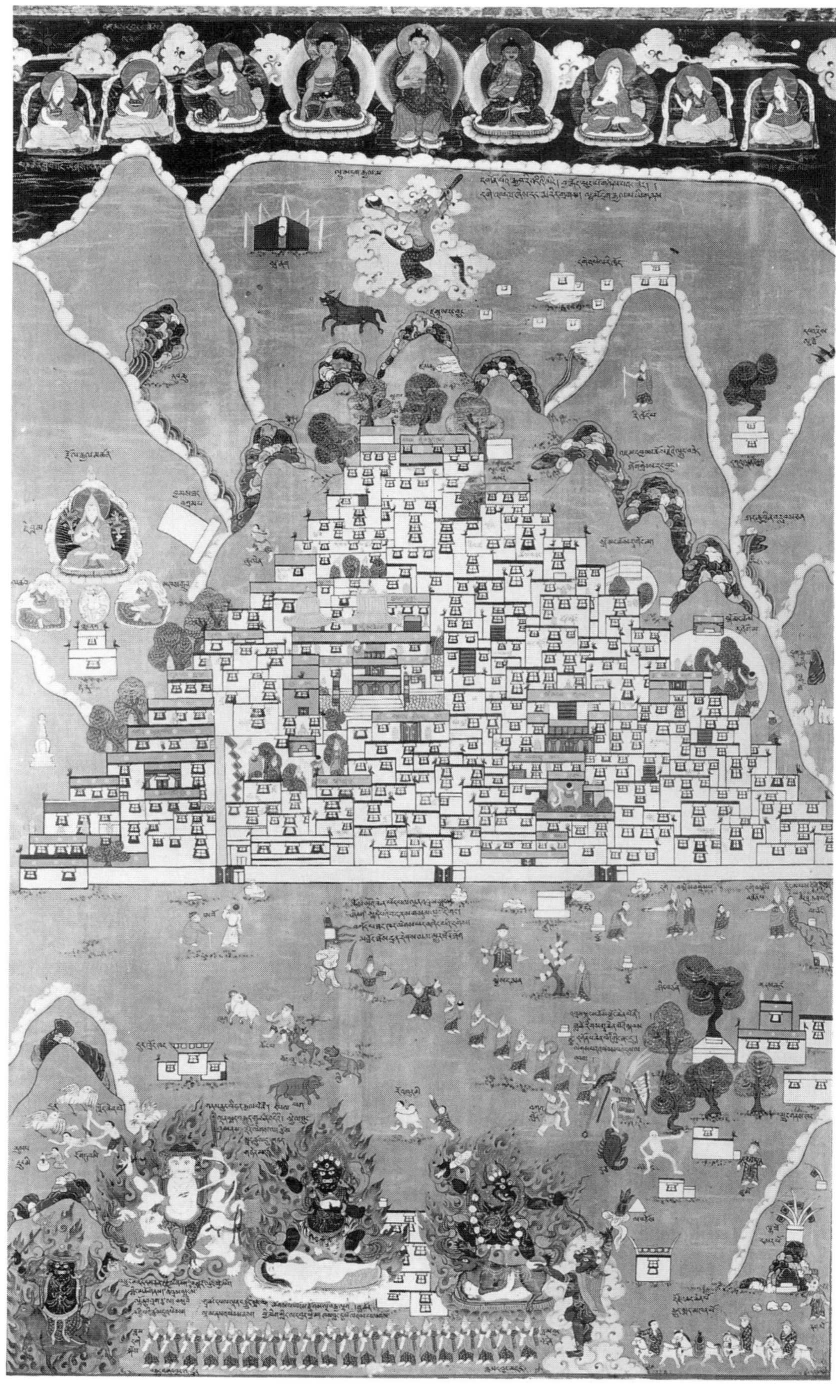

FIG. 15.48. MAP OF DREPUNG MONASTERY. This stylistically distinctive mid-eighteenth century *thanka* of the place where the Dalai Lama resided before the construction of the Potala is believed to have been made by a pilgrim from the Amdo region of northeastern Tibet. Though not evident at the scale of the photograph, the painting abounds in descriptive text, which has enabled it to be studied more intensively, perhaps, than any other Tibetan locality map. The text identifies not only scores of individualistically portrayed buildings within the city, most of them serving religious functions (including a number associated with a Tantric college with which the pilgrim was presumably affiliated), but also numerous features of the natural landscape. The seeming disregard for perspective—note in particular the vertical bar representing a street in the lower left portion of the city—is deceptive; generally speaking, the higher the feature within the image, the greater its distance from the observer. An inscription on the map indicates that its sanctity is such that individuals apprehending it, by sight or touch, will be rewarded with a speedy attainment of liberation (nirvana).
Size of the original: 115 × 68 cm. By permission of the Musées Royaux d'Art et d'Histoire, Brussels (Collection Léon Verbert 350).

FIG. 15.49. MAP DEPICTING A PROCESSION OF MONKS ENCIRCLING THE POTALA IN LHASA. This large map, of uncertain date and provenance, appears to combine Tibetan and Chinese stylistic elements and may well have been produced by a follower of Tibetan Buddhism in China. It was acquired in Beijing in 1931. The work has yet to receive careful study, and it not certain whether the small areas of settlement surrounding the Potala are intended to be villages in the vicinity of Lhasa or sacred centers in their own right. Possible Sinic elements in the map include the rings of trees that surround these peripheral areas and the wave pattern in the stream running along its lower edge.
Size of the original: unknown. ©The National Museum of Ethnography, Stockholm (H.2882). Photograph by Bo Gabrielsson.

> style: Tibetan, Indian, and Chinese. Four temples to the north, south, east and west represent the four worlds [i.e., *dvīpas* or island continents]; smaller temples represent the [tributary] islands. . . . Two further temples symbolize the sun and the moon.[170]

Most of the elements just noted can be readily discerned on the painting. For example, the three cusp-shaped structures in the foreground presumably represent the continent of Jambūdvīpa and its two island tributaries, while the temples in the upper-right and upper-left corners probably signify the sun and moon. Additional noteworthy features are the two zigzag walls enclosing the compound, said to represent the ring of mountains surrounding the terrestrial plane of the universe.[171] The fidelity of the map is noted by Reynolds: "Except for the fanciful hill and stream landscape, this painting faithfully represents the main elements of Sam-ye as it looks in photographs taken in the mid-20th century."[172] A similar concern for fidelity seems to infuse the three other paintings of Samye of which I have seen photographs (all noted in appendix 15.3), though none of these is quite as detailed as the Newark example.

About eight kilometers west of Lhasa and some seventy kilometers northwest of Samye lies Drepung ('Bras-spung), another important monastery town (formerly inhabited by 7,000 to 10,000 monks); a painted map is illustrated in figure 15.48. The contrast between the maps of Drepung and Samye is so striking that one would hardly suppose that both fell within the same broad Tibetan Buddhist tradition. Although it is not apparent at the scale of our photograph, the Drepung map is

170. *Catalogue of the Tibetan Collection and Other Lamaist Material in the Newark Museum*, 5 vols. (Newark, N.J.: Newark Museum, 1951–71), 3:65.

171. Trungpa, *Visual Dharma*, 130 (note 105).

172. Valrae Reynolds, *Tibet: A Lost World*, exhibition catalog for the Newark Museum Collection of Tibetan Art and Ethnography (New York: American Federation of Arts, 1978), 118.

replete with text identifying more than one hundred buildings, natural features, icons, and other elements. Hence the map has lent itself to what is perhaps the most thorough scholarly analysis ever made of any Tibetan locality map.[173] Based on both inscriptional and stylistic considerations, it has been suggested that the map was made in the mid-eighteenth century, probably by a pilgrim from the region of Amdo in northeastern Tibet. Of particular concern to the artist were the institutions associated with a Tantric college in Drepung, and notes on them are especially numerous. A particularly important legend, a "formula of consecration" inscribed just below the town's lower wall, reads: "Through the merit of [i.e., conveyed by] the perfect realization of this image of the glorious monastery of Drepung, well known in the three worlds, one will obtain speedy liberation by seeing it, by understanding it through touching it, and by remembering it."[174] The composition of the painting has a deceptively naive appearance. Although there appears to be a general lack of concern for perspective, the distance of buildings from the hypothetical observer increases with their apparent altitude on the prominence that dominates the composition, which in fact may not be particularly prominent. The buildings are drawn in considerable detail, mostly painted white, while temples, abbots' residences, and other religious edifices are surmounted by an ocher band.

The view of Lhasa illustrated in figure 15.49 resembles the maps of Samye and Drepung in its strong focus on a major monastic complex, and it also resembles that of Svayambhūnāth in placing its dominant object within a much more inclusive spatial and religious context. This painting has not previously been published, so far as I am aware, and I know it only from several photographs and the tantalizingly brief descriptive note provided to me by the museum, which reads, "Big painting depicting a procession encircling the Potala in Lhasa—acquired in Peking 1931."[175] Neither the provenance nor the date of the painting is known. The style appears more or less Tibetan, but the work could be by other followers of Tibetan Buddhism, possibly resident in China's capital. The map does not appear to have any inscriptions. A characteristic it shares with several other paintings I have noted (see especially plate 34) is that it shows Lhasa surrounded by a cluster of smaller centers, none of which I can identify positively. In this case, however, the striking degree to which those lesser centers are subordinate to the Potala justifies classifying this work as essentially a locality map. It is not clear whether the map's peripheral settlements are meant to be distinct from Lhasa and at a distance from it or merely quarters of what one might consider "greater Lhasa." On other maps the lesser centers are generally shown surrounded by walls, but here they seem to be ringed by circles of broadleaf trees. This is contrary to what one would actually find in the prevailingly bleak environment of the Tibetan Plateau and would support the hypothesis of a non-Tibetan origin. What appears to be a more Chinese than Tibetan feature of the map is the wave pattern within the stream that runs along the bottom edge.

Shigatse, the locale of Tashilunpo, the monastic seat of the Panchen Lama and the second largest city of Tibet, is the subject of the final illustration, figure 15.50. This seemingly unfinished and undated painting on canvas is on display at the American Museum of Natural History, New York. It is of interest for a number of reasons. First, it effectively combines a remarkable diversity of perspectives: planimetric, oblique, and frontal. Second, the features portrayed are rendered with great fidelity to what actually existed in Shigatse at the time the painting appears to relate to, roughly the third quarter of the nineteenth century. Third, it is one of at least two maps that appear to have been copied from the whole or a part of some unknown prototype. The upper-left portion of the painting, depicting Tashilunpo, bears a resemblance to another work (item kk in appendix 15.3) in a private collection, which is believed to be based on the same prototype.[176] Finally, there exists for this map a very detailed key (though not as complete as for the map of Drepung), made to accompany a faithful copy of it reproduced in a work by Sarat Chandra Das, a celebrated Bengali student of the religion, geography, ethnography, and history of Tibet, who between 1879 and 1883 made several extensive journeys to that country and to China proper as an agent of the British Indian government.[177]

173. The most extensive discussion, together with a detailed map key, is that of Philippe van Heurck, "Description de la *thaṅ-ka* représentant le monastère de Drepung," *Bulletin des Musées Royaux d'Art et d'Histoire* 57, no. 2 (1986): 5–29. Briefer notes with illustrations appear in *Dieux et démons de l'Himâlaya*, 58, 237–38 (note 33); Lauf, *Tibetan Sacred Art*, 12, 16 (color photograph) (note 54); and Vergara and Béguin, *Dimore umane, santuari divini*, 80 (note 27), on which striking similarities in the compositions of the Drepung and Samye paintings are demonstrated, notwithstanding their huge stylistic differences.

174. Heurck, "Description de la *thaṅ-ka*," 22 (note 173).

175. Letter from Claes Hallgren, 23 October 1989.

176. The privately owned map of Tashilunpo is illustrated and described in Van der Wee, " 'Cloister-City'—Ṭanka," 109, 114–20, and figs. 1 and 7 (note 161). It appears at first glance to be only a fragment of a larger work covering the same area as figure 15.50, but Van der Wee observed, on removing the map from its frame, that the right border, which would have been painted were that the case, was a blank area in which the artist tried his colors before working on the rest of the composition. Thus that map could not have been the one from which figure 15.50, or any other showing the whole of Shigatse, was copied.

177. Sarat Chandra Das, *Journey to Lhasa and Central Tibet* (1902; reprinted New Delhi: Mañjuśrī Publishing House, 1970), Shigatse map facing 45. The caption for figure 15.50 is based primarily on Das's keyed notations for many of the major features depicted. Das's work includes,

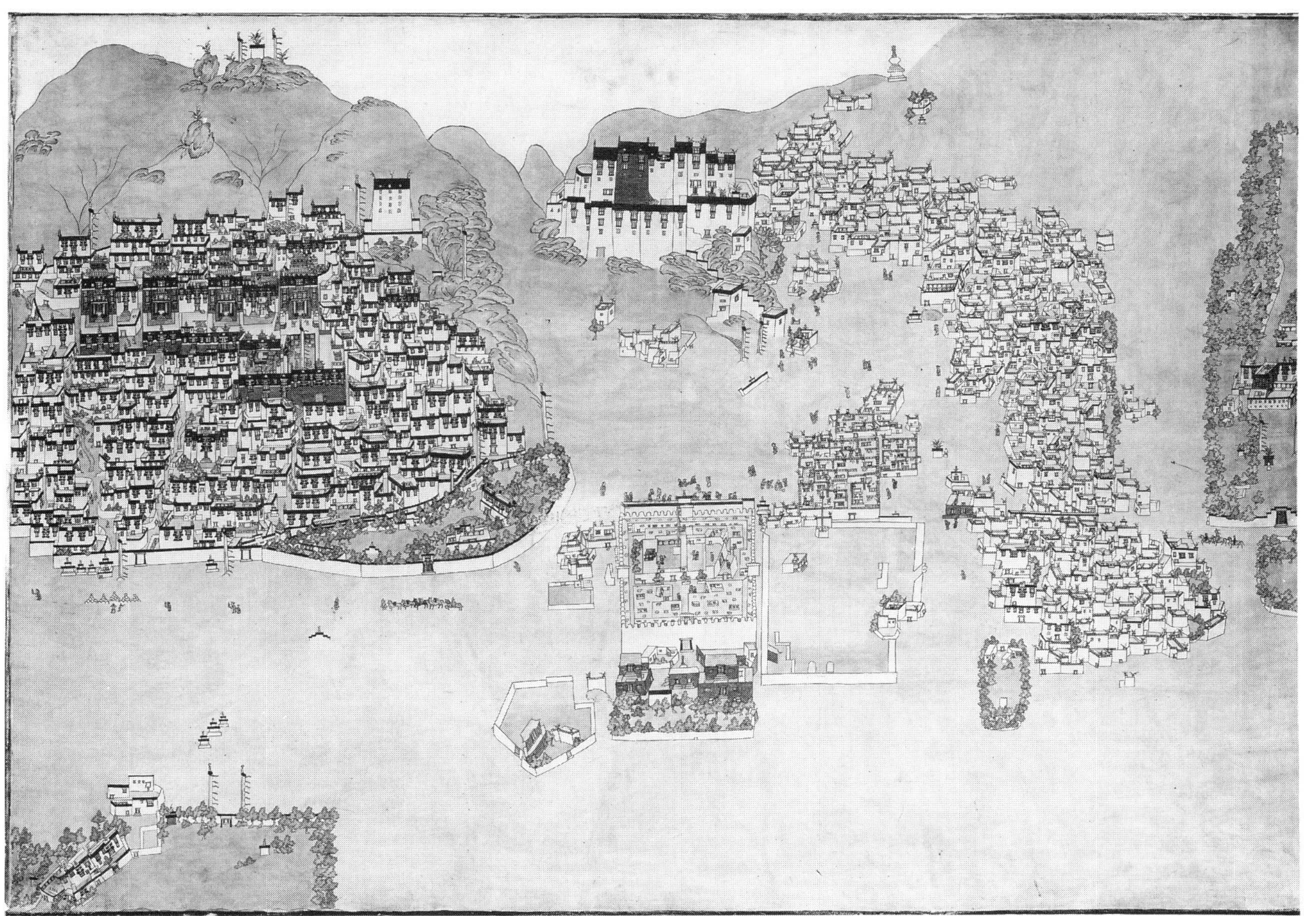

FIG. 15.50. TIBETAN PAINTING OF THE TOWN OF SHIGATSE, INCLUDING THE TASHILUNPO MONASTIC COMPLEX. This seemingly incomplete yet remarkably detailed painting on canvas appears to have been copied from a larger original. Although this map is undated, the work it is based on appears to have been painted in the latter half of the nineteenth century. Where that original may be is not known; but it is possible that a painted cloth map, restricted to the area of Tashilunpo (shown here at the left), now in a private collection, may be a fragment of that work. The map is marked by a remarkably wide range of perspectives, generally looking north: planimetric, oblique (from various angles and occasionally divergent), and frontal. The artist seemingly chose whichever type of view would best enable him to convey a visual impression of particular types of features. Despite this inconsistency, or perhaps because of it, structures are rendered with great fidelity, and dozens of them have been positively identified. Among the main components of the map, in addition to Tashilunpo, are Shigatse *dzong* (fort), in the top center; the town of Shigatse proper, in a descending arc to the right of the *dzong*; the compound of an important nobleman, shown by the oblong at the lower right; a "camp for exercise," the nearly square feature to the left of the nobleman's compound; another low-lying *dzong*, marked by a wall with battlements, a bit farther to the left; the temple of the war god, Gesar, directly below the *dzong*; a market area, in the empty space above the *dzong*; an important stupa (not in the usual form), between the market area and Shigatse *dzong*; and the summer palace of the Grand (Panchen?) Lama, lower left. Among the more conspicuous features in Tashilunpo itself are the mausoleums of the first four Panchen Lamas, marked by their Chinese-style roofs and shown in a straight line upper left in Tashilunpo (but not the mausoleum of the fifth Panchen Lama, who died in 1882); a tall stone structure (said to be more than thirty-five meters high) above them and to the right, from which great banners were hung at certain festivals every year; and "the Park of Happiness," the walled garden attached to the monastery at the lower right, where the Grand Lama's parents reside.
Size of the original: ca. 66 × 94 cm. Courtesy of the Department of Library Services, American Museum of Natural History, New York (cat. no. 70.2-187; neg. no. 335058).

No record exists of how the museum obtained the Shigatse map, but it is not unlikely that it was a bequest from Das's friend and fellow student of Tibet, W. W. Rockhill (1854–1914), an American diplomat, who may have received it from Das himself. In his 1899 introduc-

in addition to the copy of the map of Shigatse, two illustrations that appear to be faithful copies of Tibetan locality maps of the Potala and the Jo-khang palaces and provides a few notes, in English, on their major features (facing 154 and 160). I have not been able to determine the originals from which they were taken and have not noted the works in appendix 15.3.

tion to Das's account of his Tibetan travels, Rockhill acknowledges his "lasting debt of gratitude for the valuable information which he gave me while in Peking." Rockhill, who edited Das's travel narrative, worked for the Smithsonian Institution for several years; that he may also have had a connection with the American Museum of Natural History in New York City is suggested by the fact that he was residing in Block Island, New York, when he wrote his introduction.[178] As to the date of the unknown original, one may safely infer that it was painted between 1854 and 1882, the years when the fourth and fifth Panchen Lamas died, since the painting prominently shows, in a single line, the mausoleums of the first four such dignitaries but fails to show the fifth.[179] This would accord with the period of Das's visits to Shigatse.

Though not illustrated here, one additional locality map worthy of brief consideration is a mural painting of Chang'an, the capital of China under the Tang dynasty (618–907). It is one of a large number of paintings of localities occupying the corridors of the Potala in Lhasa. That undated map is illustrated in a Chinese work on Buddhist art from Tibet, but unfortunately it appears without any explanatory text apart from the caption, "This is an interesting depiction of the famous grid layout of the Tang capital."[180] The painting's combination of an overall planimetric frame for the city and the set of vertical elevations for the buildings within its sixteen compartments and horizontal perspectives to depict the landscape above and below provides a starker duality of perspectives than in any other of the Tibetan maps I have considered. Obviously the painting is no more than an idealized and relatively abstract view of what Chang'an was supposed to look like, and it lacks the sense of fidelity, or at least personal familiarity with the subject, that characterizes most other Tibetan locality maps. It warrants notice, however, in that its manner of execution reinforces the proposition that Tibetan painters were not locked into a uniform artistic style as they prepared the maps I have sought to describe in this chapter.

Conclusions

For Greater Tibet, no less than for India and Southeast Asia, the standard histories of cartography to this point lead one to suppose there was virtually no indigenous tradition of cartography. As this history has demonstrated, such a conclusion is clearly untenable for all three regions. The cartographic legacy of Greater Tibet in particular displays a richness, variety, and vigor that are remarkable given the region's meager population and its relative isolation, both physical and political, from the rest of the world.

Although several scholars, mainly art historians and anthropologists, have written exemplary articles relating to individual maps and provided valuable insights into the mental processes underlying them, none has tried to assess the entire corpus. Only Aziz and Slusser have sought to make generalizations based on the study of a number of maps, and in both cases that number was small.[181] We still await a comprehensive study by a scholar with the requisite linguistic skills, cultural knowledge, and sensitivity. What is provided in this chapter should be considered no more than an overview of the many types of maps that call for further investigation.

The most important point to be made about indigenous cartography from Greater Tibet is that its impetus was overwhelmingly religious. While exceedingly informative maps (e.g., figs. 15.32 and 15.38) for essentially secular purposes were certainly made (though few premodern examples survive), the vast majority of the maps that have come to light served some religious end. Among these were cosmographies—often including a temporal dimension—of widely varying complexity, intended to aid in religious education and meditation; maps forming components of hagiographical and biographical compositions, used to promote the veneration of important religious figures; maps painted as adjuncts to illuminated religious texts, such as the Jataka stories; *tīrtha mahātmyas*, glorifying places of pilgrimage; other types of pilgrimage guides; and cartographic records of pilgrimages successfully undertaken (e.g., fig. 15.35). Additionally, though they are only cursorily considered in this work, there are innumerable religiously sanctioned astrological charts and other graphic aids (e.g., fig. 15.23) that still guide Tibetans in their day-to-day lives and are regarded as indispensable resources in making major decisions.

Although for organizational purposes a basic distinction has been made in this history between cosmographic and geographic mapping, one must note that such a distinction might appear arbitrary to most people from the region that concerns us. The road to Shambhala, for example, wherever that land may be, certainly began, in the minds of the faithful, in locales that some of them experienced at first hand, as is evident from the scroll maps showing the route that were once carried about by itinerant Tibetan storytellers. Conversely, pilgrimage maps, such as those of the sacred region around Mount Kailas, included numerous wholly mythological places.

The scale of maps from Greater Tibet varies enormously. Leaving aside cosmographies, the range is from

178. Rockhill's introduction to Das, *Journey to Lhasa*, xv (note 177).

179. Van der Wee, " 'Cloister-City'—Ṭanka," 115–16 (note 161).

180. Liu, *Buddhist Art*, 187 (note 36).

181. Aziz, "Tibetan Manuscript Maps," *Tibetan Frontier Families*, and "Maps and the Mind" (all in note 12); Slusser, "Serpents, Sages, and Sorcerors," *Nepal Mandala*, "Pilgrim's Guide from Nepal," and "Cultural Aspects" (all in note 13).

what were probably intended to be world maps (e.g., figs. 15.27 and 15.30) to plans of an individual edifice. Within this broad spectrum, maps of individual localities appear to be most common (see appendix 15.3). Furthermore, some maps of considerably larger regions, especially of the area around Lhasa, are drawn with so much compression of sacrally "empty" space between major religious centers that the uninitiated observer gets the impression that a single, though rather expansive, locality is being depicted.

Within individual maps uniformity of scale seldom appears to be a desideratum. The use of varying scales for different map elements seems to be a general feature of Tibetan cartography. Thus it is not uncommon to show human beings and occasionally animals at a much larger scale than the buildings they move among. What rules govern such decisions by mapmakers is not entirely clear, but there are several probable reasons for depicting certain map elements at an exceptionally large scale. First, the need to draw features large enough so the artist can show all their essential attributes (e.g., within a city, the main gates, towers, important religious edifices, and so forth or, within a building, the number of stories and the general nature of the construction). Second, the desire to underscore the religious or political importance of particular places. Third, the need to make visible small but significant features that might otherwise be overlooked (frequently to aid a teacher or storyteller using a map in a didactic way). Finally, the need to highlight the existence of a common, but physically small, element in a particular locale by depicting only one or a few such elements at a large scale, assuming that the map reader will recognize it/them as signifying a general class. Conversely, scale compression might have been regarded as appropriate when none of the foregoing considerations applied.

There appears to be no general rule with respect to the "correct" orientation of maps in Greater Tibet. It is not especially significant that a number of maps from Nepal appear to be oriented toward the north. Since it is a common convention for mapmakers within the region to show a line of mountain summits at the top of a map, the choice of the Himalayas for the purpose in Nepal appears to be the most appropriate; on the other hand, the frequent use of other real or aesthetically contrived crests in Tibet proper, without particular regard to their direction vis-à-vis the map viewer, suggests that the Nepali use of north is merely fortuitous.

It is common for mapmakers of Greater Tibet to orient features so that they point away from the map reader. This is particularly true of route maps, especially when the routes depicted lie within long river valleys. In such cases mountain peaks typically point away from the valley bottom, in opposite directions on its two flanks, as might also trees, houses, and other features. A more common practice, however, is to adopt an oblique perspective, as if from a perch in space. Even more common is the use of multiple perspectives (e.g., fig. 15.50), showing some features, such as buildings, mountains, and trees, from a horizontal perspective (in frontal elevation); others from one or more oblique perspectives (oblique frontal also being very common for buildings); and still others, such as lakes and large compounds, from a vertical (planimetric) perspective. The use of a divergent perspective (e.g., fig. 15.10), the very opposite of that conventionally used in Western drawing, is also common. On maps showing a number of towns, the perspective chosen for each would presumably be the one that normally obtains for travelers approaching it along the most traveled routes. Thus Lhasa would generally be shown as if one were looking toward the north, while on the same set of maps another town, say Shigatse, might be seen as if one were looking at it toward the south.

The foregoing generalizations on perspective apply as much to cosmographic as to geographic maps. In the former, when—as is often the case—the emphasis is on the vertical dimension of the universe, the obvious choice is to employ a frontal perspective such that the dominant orientation of the map is at a right angle to the horizontal plane, which is generally assumed to be the norm in modern cartography. Similarly, in showing the cities or celestial palaces in which specific divinities are believed to reside, there is usually one widely accepted correct view based on specific religious texts or precedents drawn from earlier cosmographies. But in composite depictions (e.g., fig. 15.15), multiple perspectives are also common.

If there is any general principle at work here, it is that cartographers consciously choose the perspective or combination of points of view that will best allow communication with their intended audience. They feel in no way hamstrung by a scientistic need for consistency in how they accomplish that purpose, just as they feel no obligation to maintain consistency of scale. Moreover, mapmakers seem to give their audience credit for being able to make the mental adjustments necessary to derive the intended meaning from the map, even when the task is relatively difficult—for example, in seeing the "field of assembly" (fig. 15.19) as constituting, in essence, two tangent spheres rather than a mere two-dimensional representation of a tree.

The rules that govern the composition of maps in Greater Tibet appear to be far from uniform, and it is not at all clear, especially outside the realm of cosmographic maps, in what ways and to what extent general principles apply. For cosmographies, of course, especially for mandalas (see figs. 15.3 and 15.8), rules tend to be relatively rigid, and learning them is an important aspect of the training of many monks. Also, in Tibetan painting in general, there are well-developed canons of composition and principles of aesthetics that guide the artist. In geographic maps, on the other hand, individual cartographers appear to have enjoyed considerable personal discretion, judging from the idiosyncratic nature of much of the available corpus. This question calls for further study. It is certainly reasonable to assume that part of the difficulty in interpreting the Zhang-zhung world map (fig. 15.27) derives from the fact that its author felt constrained by the necessity of placing all the places shown in such a way as to constitute a mandala. But what of more modern maps? It would be interesting, for example, to compare all the regional maps of Lhasa and its environs (i.e., plate 34 and maps analogous to it) to determine how far the selection, placement, and directional relationships of and among the various places depicted were kept more or less uniform and were guided by aesthetic, hierarchic, and other nongeographic considerations. Unfortunately, such an exercise is virtually impossible to undertake using only small photographs and would require firsthand study in the various museums that hold these works. Further, on the same set of maps, as well as on many others, one might wish to ascertain whether the mountains and rivers that set off one part of the map from another had any function other than to define sections of the map for the reader to scan in turn. That is, do those features on the map even exist in nature, rather than being mere aesthetic or didactic devices used by the cartographer? And if they do exist, how much liberty may the cartographer take in literally bending them to a particular purpose, as obviously happened in the depiction of rivers on the map focused on Svayambhūnāth (e.g., fig. 15.43)? Likewise, in the case of purely cosmographic maps, despite the existence of widely shared views, one should not rule out the role of an individual's faith in determining the image, as evinced by the passage describing the diversity of ways Mount Meru is depicted (p. 624).

That there are widely used graphic conventions in Tibetan mapping is beyond question. Many of these conventions may be seen in figure 15.9. The signs used there apply not only to geographic maps but also to many cosmographies and are commonly employed in works of art that have no specific cartographic quality. Hence their meanings are, on the whole, self-evident and easy for laymen and even non-Tibetans to comprehend. Other conventions relate to color; for example, red is used to indicate religious buildings and white to show ordinary residences or to identify specific cosmic elements. A category of sign that calls for special mention is the religious icon. Such signs are of particular importance in cosmographic maps but are used in others as well. The icon (be it a particular deity, a bejeweled palace, or a particular type of tree) is often taken to represent a specific place and may be used with no other sign (such as a circle or square) to designate that place. Tibetan Buddhists, Bön-pos, and Nepali Hindus are likely to have no difficulty in making the necessary associations, and what laymen might fail to see, their preceptors will make clear to them; but followers of alien faiths will encounter much difficulty in interpreting, or even recognizing, the signs that many maps employ. Even more difficult for the uninitiated will be the interpretation of the many religious icons that are metaphoric or allegorical—for example, the narrow, winding road, suggestive of a navel cord, that leads across the black sea of *bardo* (an intermediate state of existence that prefigures one's next rebirth) to Sukhāvatī, the "happy land" of the west (shown in fig. 15.21).

To this point we have been considering what is shown on maps from Greater Tibet and how. But it is also necessary to ask what is omitted. Silences are often pregnant with meaning, yet easy to overlook. One wonders, for example, if Tibetan Buddhist maps made a point of omitting holy places sacred only to Bön-pos and vice versa. Similarly, within the former tradition, did maps by followers of the dominant Gelukpa (Yellow Hat) sect fail to show establishments associated primarily with other sects? I have noted that on primarily secular maps, even those believed to have been drawn primarily for intelligence purposes at the behest of the British, places of religious importance figured prominently. But was the opposite true? What place, if any, did sites with little religious importance find on explicitly religious maps?

The roots of cartography in Greater Tibet extend far back in time and probably first took hold outside the region itself. Among the earliest surviving objects from Tibet that we may designate as maps are works such as the fragment, possibly of the ninth century, of a depiction of the Dunhuang region of Gansu (fig. 15.7) and two detailed models of the Mahābodhi Temple thought to have been carried away from Bodh Gaya early in the thirteenth century (noted on p. 612). Even earlier, and certainly not later than the tenth century, some sort of model must have been taken to Tibet from India to enable the construction, dated 971, of Samye monastery in the form of the one at Odantapura in what is now Bihar. Also borrowed from India were cosmographic schemata such as the *bhavacakra*, or wheel of life, which would have been transmitted at a still earlier date. The diffusion of cosmological ideas back and forth between Tibet and China also appears to be indisputable, and one may reasonably suppose that with them went cosmographic artifacts. Turning again toward the west, if we accept the views of Gumilev and Kuznetsov relative to the Zhang-zhung world map, we would have to believe that substantial geographic knowledge from Persia and lands even farther west made its way to Tibet long before the establishment of Buddhism in that region and was incorporated in the earliest of all Tibetan maps. Other seeming relationships between Tibetan and Western mapping relate to cosmography. How, for example, do we explain the remarkable similarity between the basic schema of the *tathāgatamaṇḍala* (fig. 15.13) and that of many Western cosmographic schemata? At a much more recent date, what accounts for the similarity in the style of the Nepali map of western Asia (fig. 15.32), obviously the work of a Hindu, and various regional maps showing the Buddhist holy cities centered on Lhasa (of which plate 34 is an excellent example)? Can we establish a connection, as was suggested, between the kindred Tibetan beliefs in a blessed land to the west, Sukhāvatī, and in the virtuous warrior-king, the protagonist of the Gesar epic, who comes out of Shambhala to rid the world of evil forces (p. 637) and the notion that this Nepali map may have been commissioned, perhaps during the time of the Indian Mutiny of 1857–59, with a view to forging a grand alliance to drive the British out of Asia?

Although the specific events, routes of transmission, and particular ideas and artifacts transmitted in all directions to and from Greater Tibet are at present only dimly discernible, it seems clear that the cartography of Greater Tibet, especially in its formative phase, owes much to foreign cultural influences. As historical research advances, it should shed fresh light on the relevant diffusion processes. This is clearly a topic on which new thought and research are needed. In this regard the intriguing, though partially tendentious, arguments of Gumilev and Kuznetsov and of Teramoto (as put forward by Nakamura) with respect to the two supposed Tibetan world maps (pp. 639–43) call for reexamination and independent confirmation or refutation.

Beyond Greater Tibet, particularly in Mongolia, there are other areas where virtually the same form of Buddhism is practiced and for which cosmographic and other cartographic artifacts are known to exist. I have not had an opportunity to study those works in detail. Many of the topographic maps in Mongolian script have been studies by Walther Heissig, especially for their toponymy, but the corpus has not been analyzed from the viewpoint of the history of cartography. A brief introduction, with basic bibliographic references and representative illustrations, is found in appendix 15.4 and in figures 15.51 and 15.52.

Within Greater Tibet itself there was a progressive differentiation of ideas relative to the style and content of cosmographic and geographic maps. Again, research is needed to establish the periods, routes, and nature of transmissions and to discover when and where significant endogenous developments occurred. It is probably too early to state whether identifiable "schools" of cartography arose, but distinctive regional styles, identifiable by art historians (e.g., that of the map of Drepung, illustrated in fig. 15.48), certainly came into being. Whether these styles were different in respect to cartography than in respect to painting in general needs to be investigated. To what extent, one should ask, were they a product of competition among various faiths (Hindu vs. Buddhist in Nepal, Buddhist vs. Bön in Tibet) or of religious orders within the community of monks? What did they owe to the personal influence of particularly charismatic lamas or of scholarly pilgrims? What role was played by military conquest, from which not even Tibet was immune? How was mapping affected by the introduction of paper? What other factors were at work?

The areas of provenance of known Tibetan maps suggest that only a few centers account for a very large proportion of the total corpus: central Tibet (especially Lhasa), the Vale of Kathmandu, and arguably Ladakh. It is not clear, however, whether other centers also existed, since the interests of the outside world, especially of the British, focused on only a handful of places in Greater Tibet. In particular, the portions of the Tibetan cultural sphere within the Chinese provinces of Qinghai and Sichuan are poorly represented among the artifacts I have considered, even though the number of ethnic Tibetans living in those areas is considerably greater than in Tibet proper (the Autonomous Region of Tibet). It is therefore doubtful that my sample of artifacts is representative.

With respect to the few known centers of mapmaking, one is struck by, and must try to explain, the profound diversity among surviving maps. This is especially noteworthy in the case of the Vale of Kathmandu, long the hearth of the particularly vibrant Newari culture. Compare, for example, the works illustrated in figures 15.35, 15.39, 15.43, and 15.46, to which might be added other strikingly different maps that I have not illustrated. Of course the works in question were not contemporaneous; but it is doubtful if that factor alone explains the differences.

Finally, one cannot ignore the possible influence of the physical environment on the development of cartographic sensibility and on the propensity to use and understand maps. Over most of Greater Tibet, there exist high vantage points from which largely barren expanses of land stretch before the observer, who would see them through the clear mountain air as if they were living maps. Can viewing such sweeping vistas and, in the case of habitual travelers, comparing one such view with many others possibly be without effect in the development of mapping skills? Might not habituation to such views largely account for the popularity of high oblique perspectives in Tibetan mapping? Further cognitive studies such as those carried out by Fisher and others of the mapmaking skills of Nepali schoolchildren (pp. 618–19), and analogous studies of Tibetans and control groups including adult populations, appear warranted.

To conclude, I quote an observation by Slusser that,

though written about Nepali maps, has much more general applicability within the region that concerns us here: "We know little about the Newar penchant for map making, perhaps because we have not thought about it." After noting some of the known, though admittedly remote, antecedents of Nepali mapping going as far back as an Indian map sent to China in A.D. 648, Slusser suggests that "it may well be that in Nepal as well, map making is a venerable science."[182] In my judgment that suggestion is likely to prove valid for most, if not all, of Greater Tibet. Surely the time has come for historians of cartography to take a fresh and much more penetrating look at the remarkable evolution of mapping in that fascinating remote corner of the world.

182. Slusser, "Cultural Aspects," 26 (note 13).

APPENDIX 15.1 TIBETAN MAPS IN THE WISE COLLECTION, ORIENTAL AND INDIA OFFICE COLLECTIONS, BRITISH LIBRARY, LONDON

Volume and Catalog Number	Number and Arrangement of Sheets	Approximate Dimensions[a] (h × w) (cm)	Area Covered	Remarks
a. Vol. 1, Add. Or. 3013	Three, in a line	62 × 153	Lhasa and environs, especially the Potala and Jo-khang	More a locality than a regional map. Relatively crudely rendered.
b. Vol. 1, Add. Or. 3014	Four, in a line; one end sheet at a right angle to others	70 × 283	Ladakh, from Leh to Hanle, about 200 km to the southeast	Abundant annotations are keyed to numbers on the map. Even more crudely rendered than item a.
c. Vol. 1, Add. Or. 3015	Seven, in a line; one irregular end sheet at a right angle to others	74 × 335	From west of Demchok in Ladakh, past Mount Kailas and Lake Manasarowar to Pekut Lake in Tsang Province of central Tibet, i.e., from approximately 79° to 86°E	Seen as if the imaginary viewer is looking south (actually south-southwest). A prominent river on the map, labeled "Makchakabad," flowing out of Lake Manasarowar and into the Tsangpo, cannot be identified on modern survey maps. Parts of five sheets left blank.
d. Vol. 1, Add. Or. 3016	Seven, in a line	48 × 300	Central Tibet, middle sheet includes Tashilunpo monastery in Shigatse (ca. 89°E)	Sheets 6 and 7 of the map are said to relate to "Namtsho?" the location of which has not been determined. More than 150 key numbers on map, but no accompanying text.
e. Vol. 2, Add. Or. 3017(a)	Six, some joined to others at right angles	160 × 185	Central and eastern Tibet, including Samye and Chunggye	Includes numerous Tibetan captions, but without accompanying explanatory text. Partially illustrated in figure 15.36. See also item f.
f. Vol. 2, Add. Or. 3017(b)	One, affixed to right of item e	48 × 62	Chunggye	Scale much larger than that of item e, to which this is affixed. Fifty-seven edifices, taken to be the tombs of Tibetan kings, are named in Tibetan. Text to match Arabic key numbers is not available.
g. Vol. 2, Add. Or. 3018	Five, in a line	33 × 194	Zanskar valley in Ladakh and adjacent areas	Accompanied by abundant explanatory text

[a]Because most of the maps do not form perfect rectangles, their actual areas may be significantly smaller than the maximum vertical and horizontal dimensions cited suggest.

APPENDIX 15.2 TIBETAN MAPS IN THE HARRER COLLECTION, VÖLKERKUNDEMUSEUM DER UNIVERSITÄT ZÜRICH

Catalog Number	Approximate Dimensions (h × w) (cm)	Medium	Area Depicted	Description/Remarks	Reference in Brauen, *Impressionen aus Tibet*[a]
a. 14481	87 × 97	Black ink; red and gray wash	Southeastern Tibet	Area of relatively dense settlement. Seems to relate to a military campaign, probably the Younghusband expedition of 1904.	104 and catalog, p. 3
b. 14482, side A	52 × 67	Black, red, blue, and yellow watercolor	Presumably similar to item a	Undoubtedly related to a military campaign; seems to be by the same artist as item a	
c. 14482, side B	52 × 67	Black ink	A single locality, presumably in southeastern Tibet	Very large-scale depiction. Compounds shown planimetrically, but individual houses, of several types, shown in frontal elevation. Much detail relating to religious structures. Association with item b not clear.	
d. 14485	32 × 95	Painted	Not known	Not seen by me. Described as "*Landkarte.*"	catalog, p. 3
e. 14488	31 × 190	Painted	Southern Tibet	As for item d	catalog, p. 3
f. 14493	32 × 94	Painted	Not known	As for item d	catalog, p. 3
g. 14495	52 × 171	Black ink; yellow and brown watercolor	Area of eastern bend of Tsangpo/Brahmaputra	See figure 15.37 and text	pl. 110
h. ?	52 × 62	Same as for item b	Kham Province in eastern Tibet	Style similar to item b but richer in detail, especially in regard to vegetation. Appears also to have a military purpose. One building flies British flag. Ferries prominently shown.	pl. 9
i. ?	62 × 95 (L-shaped, two sheets of paper)	Black ink; brown, yellow, green, and mauve wash	Presumably on or near Tibet's border with Arunachal Pradesh in India	Bears note: "Restauriert Winter 1982"	
j. ?	26 × 46	Black ink and red watercolor	A small locality	Style similar to item b. Monastery shown in oblique perspective, with several hamlets.	
k. ?	32 × 106	Black ink; blue, red, and yellow watercolor	One main river valley (Chumbi?) with several tributaries	Filed in same drawer as items b and c. Also seems to relate to military campaign. Flags flying in several crudely drawn settlements.	

[a]Martin Brauen, *Heinrich Harrers Impressionen aus Tibet* (Innsbruck: Pinguin-Verlag, 1974), including a fifteen-page catalog of the exhibition at the museum that opened in December 1974.

APPENDIX 15.3

Locality[a]	Place Where Map Is Held	Provenance and Date	Dimensions (h × w) (cm)	Language/Text[b]
a. Chang'an, China	Potala, Lhasa	Lhasa; date unknown	Not known	None evident
b. Chokpuri Hill, Lhasa	Potala, Lhasa	Lhasa; date unknown	Not known	None evident
c. Chokpuri Hill, Lhasa	Potala, Lhasa	Lhasa; date unknown	Not known	None evident
d. Drepung, near Lhasa	Musées Royaux d'Art et d'Histoire, Brussels, Collection Léon Verbert 350	Probably around Amdo in northeastern Tibet; eighteenth century	115 × 68	Numerous inscriptions in cursive Tibetan
e. Dza-rung, Nepal (?)	Not known	Gi-long Nor-bu (artist); 1975	Not known	None evident
f/g. Gang-gar, Ding-ri, Tibet	Private collection of Barbara Nimri Aziz?	Pän-tän P'a-la (artist), former inhabitant of town; ca. 1975	Not known	None evident
h. Gorkha, Nepal	Museum, Bhaktapur, Nepal	Nepal (Newari style); early nineteenth century	Not known	None
i. Gyantse, Tibet	Palchor monastery, Gyantse, Tibet	Gyantse, Tibet, date unknown	Not known, but quite large	None evident
j. Jo-khang Palace, Lhasa	Potala, Lhasa	Potala, Lhasa; date unknown	?	None evident
k. Lhasa	Not known	Probably Darjeeling, India; late nineteenth century	?	None
l. Lhasa	Gerd-Wolfgang Essen Tibetica Collection, Hamburg	Tibet; ca. 1900	93 × 160	None

[a]Chokpuri Hill, Drepung, Jo-khang, and the Potala are all within the orbit of Lhasa; Tashilunpo is similarly within the orbit of Shigatse.

[b]The note "none evident" signifies that text cannot be discerned on available photographs and is not mentioned in relevant references.

[c]The citations in this column are: Barbara Nimri Aziz, *Tibetan Frontier Families: Reflections of Three Generations from D'ing-ri* (Durham, N.C.: Carolina Academic Press, 1978); *Catalogue of the Tibetan Collection and Other Lamaist Material in the Newark Museum*, 5 vols. (Newark, N.J.: Newark Museum, 1951–71), vol. 3; Sarat Chandra Das, *Indian Pandits in the Land of Snow* (1893; reprinted Delhi: Delhi Printers Prakashan, 1978); Gerd-Wolfgang Essen and Tsering Tashi Thingo, *Die Götter des Himalaya: Buddhistische Kunst Tibets*, 2 vols. (Munich: Prestel-Verlag, 1989); Werner Forman and Bedrich Forman, *Art of Far Lands*, ed. Lubor Hájek, trans. W. Cungh and H. Watney ([London]: Spring Books, [1958?]); Manfred Gerner, *Architekturen im Himalaja* (Stuttgart: Deutsche Verlags-Anstalt, 1987); Philippe van Heurck, "Description de la *thaṅ-ka* représentant le monastère de Drepung," *Bulletin des Musées Royaux d'Art et d'Histoire* 57, no. 2 (1986): 5–29; Lumír Jisl, *Tibetan Art*, trans. Ilse Gottheiner (London: Spring Books, 1957); Detlef Ingo Lauf, *Tibetan Sacred Art: The Heritage of Tantra*, trans. Ewald Osers (Berkeley, Calif.: Shambhala, 1976); Lobsang P. Lhalungpa, *Tibet, the Sacred Realm: Photographs, 1880–1950*, exhibition catalog, Philadelphia Museum of Art, 20 March–22 May 1983 ([Millerton, N.Y.]: Aperture, 1983); Li Jicheng, *The Realm of Tibetan Buddhism* (New Delhi: UBS Publishers' Distributors, 1986); Liu Lizhong, *Buddhist Art of the Tibetan Plateau*, ed. and trans. Ralph Kiggell (Hong Kong: Joint Publishing, 1988); Alexander W. Macdonald and Anne Vergati Stahl, *Newar Art: Nepalese Art during the Malla Period* (Warminster, Eng.: Aris and Phillips, 1979); Claudius C. Müller and Walter Raunig, eds., *Der Weg zum Dach der Welt* (Innsbruck: Pinguin-Verlag [1982]); Ngapo Ngawang Jigmei et al., *Tibet* (New York: McGraw-Hill, 1981); Blanche Christine Olschak, Augusto Gansser, and Andreas Gruschke, *Himalayas* (New York: Facts on File, 1987); Blanche Christine Olschak and Geshe Thupten Wangyal, *Mystic Art of*

Greater Tibetan Locality Maps

Medium	Description	Where Published[c]
Mural painting	See text, p. 670	Liu, *Buddhist Art*, 187 (fig. 343)
Mural painting	Various unidentified buildings in oblique frontal perspective ranged about hill; rugged hills in background	Li, *Realm of Tibetan Buddhism*, 166–67 (mislabeled)
Mural painting	As for item b, but gentler terrain in background	Liu, *Buddhist Art*, 187 (fig. 344)
Painted *thanka*	See text, pp. 666–68, and figure 15.48	Heurck, "Description de la *thaṅ-ka*," 5–29; Lauf, *Tibetan Sacred Art*, pl. 5; Lhalungpa, *Tibet*, 22; *Tibetische Kunst*, fig. 96 and pp. 72–73; Van der Wee, Van der Wee, and Schotsmans, *Symbolisme*, 34–35; Vergara and Béguin, *Dimore umane*, 80
Ink and paint on paper	Town, within sight of Mount Everest; oblique perspective; background naturalistically rendered	Aziz, *Tibetan Frontier Families*, unnumbered plate between 96 and 97
Painted on paper	Oblique perspective of mountainside town	Aziz, *Tibetan Frontier Families*, unnumbered plate between 96 and 97
Wall painting	Palace and neighboring buildings set in middle ground in circuit of trees; forests to rear and cultivated fields and river in foreground, each marked by characteristic animals drawn larger than many buildings; very diverse treatment of vegetation	Vergati, "Les royaumes de la vallée de Katmandou," 202–3
Mural painting	Several clusters of temples and monasteries in oblique frontal perspective, separated by areas of relatively verdant terrain	Liu, *Buddhist Art*, fig. 302
Mural painting	Festival scene with large gathering of monks; mixed oblique and frontal perspectives	Liu, *Buddhist Art*, 184 and fig. 339
Ink on paper	Multiple, partially divergent perspectives	Waddell, *Buddhism of Tibet*, facing 287
Painted *thanka* with brocade border	Painted as guide map for pilgrims; varying oblique perspectives; much open terrain between city proper and Potala	Essen and Thingo, *Die Götter des Himalaya*, 1:245–47 and 2:221–22

Ancient Tibet (1973; Boston: Shambhala, 1987); Valrae Reynolds, *Tibet: A Lost World*, exhibition catalog for the Newark Museum Collection of Tibetan Art and Ethnography (New York: American Federation of Arts, 1978); Theodore Riccardi, Jr., "Some Preliminary Remarks on a Newari Painting of Svayambhūnāth," *Journal of the American Oriental Society* 93 (1973): 335–40; Mary Shepherd Slusser, *Nepal Mandala: A Cultural Study of the Kathmandu Valley*, 2 vols. (Princeton: Princeton University Press, 1982); idem, "On a Sixteenth-Century Pictorial Pilgrim's Guide from Nepal," *Archives of Asian Art* 38 (1985): 6–36; idem, "The Cultural Aspects of Newar Painting," in *Heritage of the Kathmandu Valley: Proceedings of an International Conference in Lübeck, June 1985*, ed. Niels Gutschow and Axel Michaels (Saint Augustin: VGH Wissenschaftsverlag, 1987), 13–27; David L. Snellgrove and Hugh Richardson, *A Cultural History of Tibet* (London: Weidenfeld and Nicolson, 1968); *Tibetische Kunst: Katalog zu Ausstellung, 8.–30. März, 1969, Helmshaus, Zürich, 17 Apr. bis 11 Mai 1969, Gesellschaftshaus zu Schützen, Luzern* (Bern: TIBETA, 1969); Herbert Tichy, *Himalaya* (Vienna: Anton Schroll, 1968); Chögyam Trungpa, *Visual Dharma: The Buddhist Art of Tibet* (Berkeley, Calif.: Shambhala, 1975); Louis P. Van der Wee, "A 'Cloister-City'—Ṭanka," *Journal of the Indian Society of Oriental Art*, n.s., 4 (1971–72): 108–20; Pia Van der Wee, Louis P. Van der Wee, and Janine Schotsmans, *Symbolisme de l'art lamaïque* (Brussels: Musées Royaux d'Art et d'Histoire, 1988); Paola Mortari Vergara and Gilles Béguin, eds., *Dimore umane, santuari divini: Origini, sviluppo e diffusione dell'architettura tibetana/ Demeures des hommes, sanctuaires des dieux: Sources, développement et rayonnement de l'architecture tibétaine* (Rome: Università di Roma "La Sapienza," 1987); Anne Vergati, "Les royaumes de la vallée de Katmandou," in *Les royaume de l'Himâlaya: Histoire et civilisation* (Paris: Imprimerie Nationale, 1982), 164–208; Laurence Austine Waddell, *The Buddhism of Tibet, or Lamaism, with Its Mystic Cults, Symbolism, and Mythology, and Its Relation to Indian Buddhism*, 2d ed. (Cambridge: W. Heffer, 1935 [first published 1895]); Siddiq Wahid, *Ladakh: Between Earth and Sky* (New York: Norton, 1981).

APPENDIX 15.3

Locality[a]	Place Where Map Is Held	Provenance and Date	Dimensions (h × w) (cm)	Language/Text[b]
m. Lhasa	Private collection of David Tremayne, London	Tibet (?); probably twentieth century	?	None evident
n. Lhasa and environs	National Museum of Ethnography, Stockholm, cat. no. H 2882	Provenance and date unknown	Very large	None evident
o. Menri monastery, Tsang Province, Tibet	Oriental and India Office Collections, British Library, London	Menri (?), Tibet; ca. 1900	?	None
p. Nepal /Tibet border? Unknown stupa and surrounding holy places	Ashmolean Museum, Oxford University	Tibet or Nepal	?	None evident
q. Patan, Nepal	Guita-bahil monastery, Patan, Nepal	Patan, Nepal; seventeenth to eighteenth century	Very large, more than 10 m long	Newari
r. Patan and environs in Vale of Kathmandu, Nepal	Kwa Bahal monastery, Patan, Nepal	Patan, Nepal; ca. 1850	Est. 90 × 1,800	Newari
s. Potala, Lhasa	Private collection of Blanche Christine Olschak	Tibetan monastery in northern Nepal; seventeenth century (?)	?	Devanagari script
t. Potala, Lhasa	Private collection	Tibet; seventeenth-eighteenth century	81 × 56	None evident
u. Potala, Lhasa	Probably Darjeeling, India; in late nineteenth century	Tibet; probably late nineteenth century	?	None evident
v. Potala, Lhasa	Potala, Lhasa	Potala, Lhasa; date unknown	Height ca. 3 m; width not known	Abundant text, presumably Tibetan
w. Potala, Lhasa	Potala, Lhasa	Potala, Lhasa; date unknown	As for item t	Tibetan
x. Potala, Lhasa	Potala, Lhasa	Potala, Lhasa; date unknown	?	As for item t
y. Potala, Lhasa	Potala, Lhasa	Potala, Lhasa; date unknown	Very large	None evident
z. Potala, Lhasa	Potala, Lhasa	Potala, Lhasa; date unknown	?	None
aa. Potala, Lhasa (future site of)	Potala, Lhasa	Potala, Lhasa; date unknown	?	None evident
bb. Samye, Tibet	Newark Museum, Newark, N.J.; acc. no. 20.271	Kham region, eastern Tibet, seventeenth-eighteenth century	53 × 38	Tibetan text in red and gold
cc. Samye, Tibet	Samye (?), Tibet	Samye (?), Tibet; date unknown	?	Cursive Tibetan

(*continued*)

Medium	Description	Where Published[c]
Painted on paper	Oblique perspective, terrain presented in relatively naturalistic style	
Painted on cloth	See text, p. 668, and figure 15.49	
Paint and ink on paper	Shows a major monastery of the Bön religion; buildings in frontal and terrain in oblique perspective, considerable attention to vegetation	Müller and Raunig, *Der Weg zum Dach der Welt*, 373, 375
Temple banner painted on cloth	Stupa at a very large scale and surrounding holy places at much smaller scales, all in frontal perspective within mountainous terrain presumably near Nepal/Tibet border; work illustrates a pilgrimage	Snellgrove and Richardson, *Cultural History*, appended photographic plate
Painted cloth scroll	Detailed townscape in panoramic form, oblique perspective. See also figure 15.2.	Slusser, *Nepal Mandala*, vol. 2, fig. 97; idem, "Pilgrim's Guide," 30
Painted cloth scroll	See text, pp. 663–64, and figure 15.46	Slusser, *Nepal Mandala*, vol. 2, fig. 98; idem, "Pilgrim's Guide," 31 and 32; idem, "Cultural Aspects," fig. 11
Painted *thanka*	Does not show buildings constructed in reign of fifth Dalai Lama (1617–82); oblique perspective	Olschak and Thupten Wangyul, *Mystic Art*, 78
Paint and ink on cloth	Oblique perspective, structures very individualistically portrayed	Forman and Forman, *Art of Far Lands*, 204
Ink on paper	Multiple, partially divergent perspectives	Waddell, *Buddhism of Tibet*, 40
Mural painting	Shows Potala being extended; oblique frontal perspective	Li, *Realm of Tibetan Buddhism*, 166; Jisl, *Tibetan Art*, 29 and fig. 33
Mural painting	As for item t; further stage in construction	Jisl, *Tibetan Art*, 29 and fig. 34
Mural painting	Shows a near terminal phase of construction; frontal perspective	Gerner, *Architekturen im Himalaja*, 98
Mural painting	Festive scene showing unfurling of giant *thanka* on south wall; oblique frontal perspective	Liu, *Buddhist Art*, 184 and fig. 338
Mural painting	Shows Potala before expansion; mixed frontal and oblique perspectives	Liu, *Buddhist Art*, 187 and fig. 342
Mural painting	Shows original temple and fortifications where Potala was to be; oblique perspective	Jisl, *Tibetan Art*, fig. 32
Painted and gilt *thanka* on Chinese silk	See text and figure 15.47	*Catalog of the Tibetan Collection*, 3:64 and 91; Reynolds, *Tibet*, 118 and 126; Trungpa, *Visual Dharma*, 37; Vergara and Béguin, *Dimore umane*, 79–80
Fresco painting	Extraordinary composition; monastery complex shown in varying oblique perspectives, within circular wall, set against variegated landscape depicted at much smaller scale, also in mixed perspectives	Ngapo et al., *Tibet*, 247–49

APPENDIX 15.3

Locality[a]	Place Where Map Is Held	Provenance and Date	Dimensions (h × w) (cm)	Language/Text[b]
dd. Samye, Tibet	Samye monastery	Samye, Tibet; date unknown	?	None evident
ee. Samye, Tibet	Main temple in Gangtok, Sikkim	Sikkim, date unknown	?	None evident
ff. Sherpa village, Nepal	Not known	Kalden, a twentieth-century Sherpa artist	?	None
gg. Shigatse, Tibet	American Museum of Natural History, Hall of Asian Peoples	Tibet; late nineteenth or twentieth century	ca. 66 × 94	None evident
hh. Svayambhūnāth, Nepal	Collection of A. Peter Burleigh	Patan, Nepal; 1565	101 × 85	Newari and Sanskrit
ii. Svayambhūnāth, Nepal	Private collection, Paris	Central Nepal; late seventeenth century	90 × 70	None evident
jj. Svayambhūnāth, Nepal	Private collection, Brussels	Nepal; late eighteenth century (?)	82 × 55	Devanagari script, language not known
kk. Tashilunpo, Shigatse, Tibet	Shigatse, Tibet, owner not known	Shigatse, Tibet; late eighteenth century	?	None evident
ll. Tashilunpo, Shigatse, Tibet	Private collection	Central Tibet; between 1854 and 1882	?	None evident
mm. Tashilunpo, Shigatse, Tibet	Library at Alchi monastery, Ladakh	Ladakh (?); date unknown	?	None evident
nn. Tashilunpo, Shigatse, Tibet	Traktok monastery, Ladakh	Traktok monastery, date unknown	?	None evident
oo. Unknown city	Somewhere in Tibet?	Tibet?	Very large	Considerable text, presumably Tibetan
pp. Unknown monastery complex	National Museum of Ethnography, Stockholm	Northern Tibet or Qinghai (?); date unknown	138 × 186	None evident

(continued)

Medium	Description	Where Published[c]
Fresco painting	Oblique view of monastic complex in its original form	Gerner, *Architekturen im Himalaja*, 53
Fresco painting	Approximation of Western one-point perspective	Olschak, Gansser, and Gruschke, *Himalayas*, 175
Painted, on paper (?)	Village painted against background of vividly depicted Mount Khumbila. People and animals shown larger than houses. Oblique frontal perspective.	Tichy, *Himalaya*, frontispiece
Paint and ink on paper	See text, pp. 668–70, and figure 15.50	
Painted on cloth	See text, pp. 661–63, and figure 15.43	Riccardi, "Preliminary Remarks"; Slusser, *Nepal Mandala*, vol. 2, pl. 495; idem, "Pilgrim's Guide"; idem, "Cultural Aspects," 20–27
Painted on cloth	Frontal perspective for stupa, oblique perspective for its surroundings	Macdonald and Vergati Stahl, *Newar Art*, frontispiece
Painted on cloth	Frontal perspective for stupa, oblique perspective for some nearby structures	
Painted *thanka*	Mixed frontal and oblique perspectives; terrain presented in relatively naturalistic style	Liu, *Buddhist Art*, 228 (fig. 435)
Painted on cloth, formerly in silk frame	Oriented toward north, many structures can be individually identified; wall in foreground, mountains in background; oblique frontal perspective	Van der Wee, "Ṭanka," passim
Painted *thanka*	Oblique frontal perspective	Wahid, *Ladakh*, 88–89
Painted, on paper (?)	Essentially a frontal perspective	Gerner, *Architekturen im Himalaja*, 85
Appears to be mural painting	Very detailed representation of a townscape with many large edifices (monasteries?) and places of assembly in a rather verdant setting. Mixed oblique and frontal perspectives. Conventional crest of mountains in background.	Das, *Indian Pandits*, frontispiece
Painted on cloth	Monastery complex mountain girt on three sides; foreground shows laymen with horses, camels, wagons, and tents; varying oblique perspectives	

Appendix 15.4 Mongolian Cartography

G. Henrik Herb

Although the Mongolian people of the Central Asian steppes have had enormous historical influence on China, the Middle East, and even Europe, an authoritative account of a Mongolian cartographic tradition is still lacking. Indigenous sources for the reconstruction of this tradition either have not survived or have yet to come to light.[1] The only Mongolian geographical maps that have been preserved were made after the middle of the eighteenth century—that is, after close contact with well-established foreign cartographic traditions.[2] As a result, it is extremely difficult to isolate indigenous Mongolian elements from the existing maps. Apart from geographical maps, there is the question of cosmographical artifacts. It would be surprising if the traditional use of such works among Tibetan Buddhists were not mirrored to some extent by their Mongol coreligionists. However, I have not had the opportunity to investigate that issue.

Even the two maps customarily cited as the oldest Mongolian artifacts—which have been termed Renat 1 and Renat 2 in the literature—thus appear to have been made with little Mongolian contribution.[3] These maps were brought to Europe in 1734 by a Swedish officer, Johan Gustav Renat, after a seventeen-year captivity by the West Mongolian Kalmyks. Both Renat 1, which covers the area from about 70° to 90° east longitude (the western terminus is Samarkand), and Renat 2, which covers an even larger region stretching from 79° to 107° east longitude (the eastern terminus is Ulan Bator), are oriented to the south and use Oirat, a modified Mongolian script.[4] Beyond the script, there is little that indicates a Mongolian tradition; they are believed to be copies of Chinese originals.[5]

The remaining Mongolian maps—about 565 artifacts are known—differ considerably from the Renat maps and form a unified corpus. They are regional manuscript maps that depict Mongolian tribal and pasture territories. Their most striking features, which are clearly visible in most of them, are annotations extending beyond the boundaries in different directions (see figs. 15.51 and 15.52). A few of the maps date from the middle of the eighteenth century, but most were made between the middle of the nineteenth century and the early twentieth century.

The maps are held in several depositories: the largest collection, 335 maps, is part of the Mongolian State Archive in Ulan Bator, followed by the Staatsbibliothek zu Berlin—Preussischer Kulturbesitz, with 182 maps and the Tenri Central Library in Tenri, Japan, with 44.[6] Additional maps are preserved at the Royal Library in Copenhagen, the Archive of the Polish Academy of Sciences in Krakow, and in private collections.[7]

The German scholar Walther Heissig—who was the first to describe these maps in detail and on whose work this brief account is based—believes that the incentive for their production came from the outside.[8] He links their production to the division of Mongols into "banners" that took place in 1649, shortly after submission to Manchu rule. The creation of banners—that is, the delimitation of enduring territorial boundaries for the pasture areas of Mongolian tribal units—was intended to prevent border disputes and to limit the groups' movements. By restricting pastoral nomadism, it fundamentally changed the

1. "Mongolian" in this context is clearly not restricted to the present Mongolian state, since this political unit does not encompass all Mongolian people. Large numbers of Mongols inhabit Inner Mongolia in China, Chinese Turkestan (now Xinjiang Uygur Autonomous Region), parts of Manchuria, and the Buryat-Mongol lands of Siberia.

2. This does not mean there was no indigenous Mongolian cartographic tradition. It is possible that surviving maps might not have come to light and that others might have been drawn directly on the ground, or on materials such as animal hides and wooden boards, as well as paper or cloth, that did not survive the rigors of time.

3. The maps are preserved at the University Library, Uppsala. They are reproduced in John F. Baddeley, *Russia, Mongolia, China; Being Some Record of the Relations between Them from the Beginning of the XVIIth Century to the Death of the Tsar Alexei Mikhailovich, A.D. 1602–1676*, 2 vols. (London, 1919; reprinted New York: B. Franklin, 1964), and Nicholas Poppe, "Renat's Kalmuck Maps," *Imago Mundi* 12 (1955): 157–59.

4. The Oirat script was introduced in 1648 by Zaya Pandita, a Buddhist monk and scholar; see Poppe, "Kalmuck Maps," 157–58 (note 3).

5. There is considerable uncertainty about the authors of the maps. Heissig stated that Renat 1 was drawn by the Kalmyk ruler Galdan Tseren (1727–1845), whereas Poppe mentioned that the map was given to Renat by Galdan Tseren. In the case of Renat 2, Heissig believed that the map was made by Oirats (*Ölöten*), who made improvements to a Chinese original, whereas Poppe claimed that Renat had obtained the map from Chinese troops who had attacked the Kalmyks. See Walther Heissig, "Über Mongolische Landkarten," *Monumenta Serica* 9 (1944): 123–73, esp. 124 and 127, and Poppe, "Kalmuck Maps," 157 (note 3).

6. The holdings in Ulan Bator are mentioned in Walther Heissig, ed., *Mongolische Ortsnamen*, 3 vols. (Wiesbaden: Franz Steiner, 1966–81), 1:VIII, and see the Mongolian reference cited there. The maps in Berlin, which were collected by Herrmann Consten and Walther Heissig, are cataloged in Walther Heissig, *Mongolische Handschriften, Blockdrucke, Landkarten* (Wiesbaden: Franz Steiner, 1961). Twenty-six of the maps were reproduced in 1966 (2 in color and 24 in black and white), and 118 of them were reproduced in 1978 (black-and-white facsimiles); see Heissig's *Mongolische Ortsnamen*, vols. 1 and 2. The maps in this collection were made between 1890 and 1920; most date from 1907 (Heissig, *Mongolische Handschriften*, 338; this source also contains a list of the maps in Tenri, 493–94, with references).

7. The map in Copenhagen, MS. Mong. 562, is of the Setsen Khan district (Khalkha), 1913. Heissig mentioned that he saw a map depicting the Khalkha district in the collection of L. Kotwicz in Krakow (*Mongolische Ortsnamen*, 2:XII [note 6]). Heissig has mentioned that there are three maps in private collections in addition to the 182 maps kept at the Staatsbibliothek zu Berlin, referring to his 1944 article for further information. However, only two maps could be identified from Heissig's sources. (1) Map of the Khangghin Banner of the Ordos region from 1909, owned by the German professor Walter Fuchs (described, including a list of 189 toponyms, and reproduced in "Über Mongolische Landkarten," 126, 136–47, pl. XIII [note 5]). A similar map is part of the Berlin collection (Hs. Or. 108); see *Mongolische Handschriften*, 337 (note 6). (2) Map of the Dzungghar Banner of the Ordos region, undated, which is owned by Heissig (described, including a list of 210 toponyms, and reproduced in "Über Mongolische Landkarten," 126, 160–70, and pl. XV). A similar map is part of the Berlin collection (Hs. Or. 836); see *Mongolische Handschriften*, 337).

8. Heissig, "Über Mongolische Landkarten," 128 (note 5).

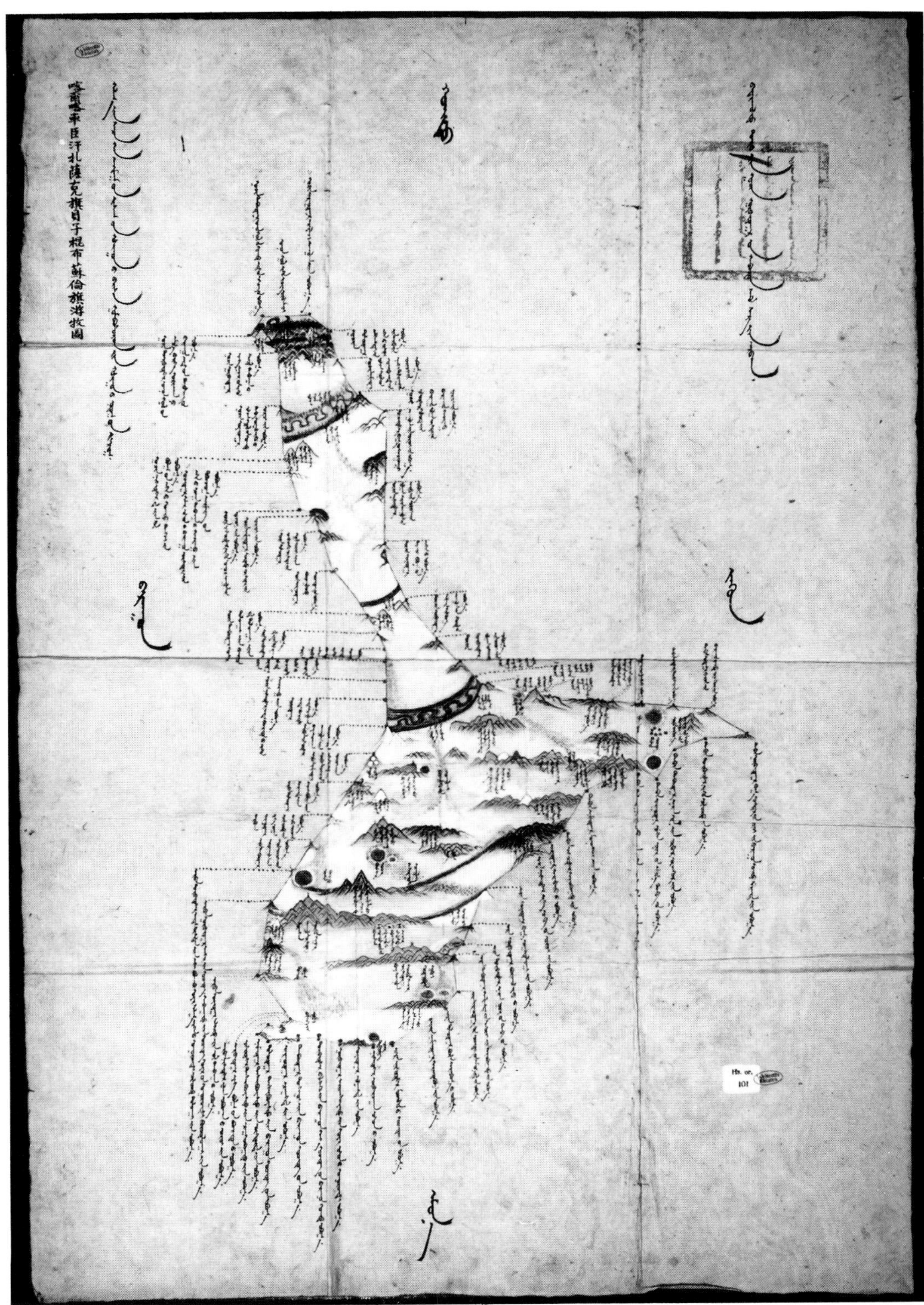

FIG. 15.51. MAP OF THE ΓOMBOSURUN BANNER IN THE SETSEN KHAN DISTRICT (KHALKHA). Mongolian manuscript map from 1907, depicting sixty-nine boundary markers (*oboγa*) extending beyond the pasture territory. In addition to a variety of toponyms, it also shows six caravan routes that traverse the banner from west to east, leading to Küriyen (Urga).

Size of the original: 96.3 × 64.3 cm. By permission of the Staatsbibliothek zu Berlin—Preussischer Kulturbesitz, Orientabteilung (Hs. Or. 101).

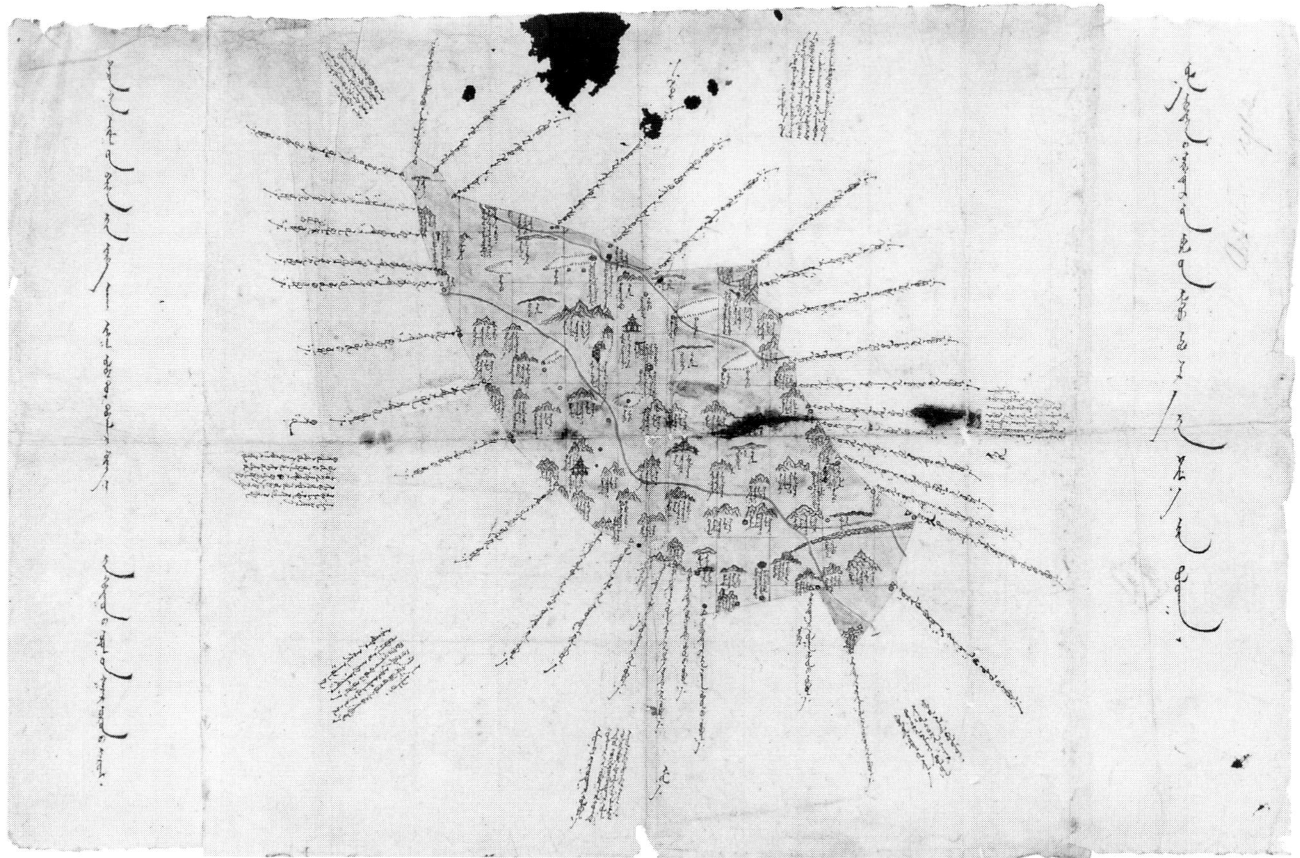

FIG. 15.52. MAP OF THE TUSALAΓČI GÜNG BANNER IN THE SETSEN KHAN DISTRICT (KHALKHA). This Mongolian manuscript map from 1913 depicts a tribal pasture territory. Thirty-five annotations extending radially from the boundary identify the directional position of boundary markers, called *oboγa*. It also depicts the names of fifty-five mountains and mountain ridges, sand dunes, the monastery Yerügel-i daγaγaγsan sümee, and the site of a religious school as well as a homestead. Seven annotations in the margins define the distances between the border markers.
Size of the original: 49.5 × 75 cm. By permission of the Oriental Department, Royal Library, Copenhagen (Mong. 562).

tribal social structures. The banner boundary markers, called *oboγa*, were not permanent. They consisted mostly of piles of stones, sand, or earth, and the imperial government ordered that they be checked and repaired periodically.[9]

The production of maps is not mentioned in the context of the first Manchu orders to fix banner boundaries, but it is documented for later years. A Mongolian treatise from 1802 mentions that maps were made in 1686 at the order of an imperial envoy to document newly established banner boundaries. A letter from the colonial office in Beijing to the ruler of the Tümet Banner in Köke Khota in the summer of 1690, written in Mongolian, asked for detailed geographical descriptions about the individual Mongolian banners.[10]

The information requested, which was to be used in preparing the *Da Qing yitong zhi* (Comprehensive gazetteer of the Great Qing realm, completed 1746), included "the pasture areas of each banner and also the name of the banner and its territory; the extent of each banner territory in miles in all cardinal directions; and also the numbers of paths and, furthermore, from which direction the banner borders meet with the borders of another country." Also requested were the "mountains, rock formations and ruins, temples and monasteries, bridges, ravines and mountain passes that exist in each banner; moreover, the products of the region and the submitted taxes, persons of rank and name, laws and customs, the number of settlements, and the size of the families."[11]

A printed map of the Dolonor (Dolon Nor) region, which has been dated to about 1911, also provides corroborative evi-

9. Documents indicate that yearly checks were requested in 1832, 1855, and 1889; see Heissig, "Über Mongolische Landkarten," 130–31 (note 5).

10. Heissig, "Über Mongolische Landkarten," 128–30 (note 5), and idem, *Mongolische Ortsnamen*, 1:VII (note 6). I have followed Heissig's spellings for Mongolian place-names throughout this appendix; Heissig also provides maps of some of these areas.

11. From the German in Heissig's "Über Mongolische Landkarten," 128–30 (note 5), where he used the word *aufschreiben* for the recording of this geographical information, which could also mean preparing a list. However, he later used the word *aufzeichnen*, which implies drawing a map (*Mongolische Ortsnamen*, 1:VII–VIII [note 6]).

dence. Its title states that it was based on a manuscript map submitted at the request of the imperial government in 1742.[12]

Imperial involvement in the production of maps of Mongolian banners is documented for yet another case. Mostaert reports that on 16 December 1739, a meeting took place that included the imperial emissary Liou Bayar, the head of the Mongolian confederation in the Dolonor region, and the rulers of the seven Dolonor banners.[13] The purpose of the meeting was to delimit the boundaries for the seven banners because a dispute had arisen. A map was drawn, and each of the seven rulers affixed his seal to state his approval. The map was sent to the colonial office in Beijing.[14]

Heissig claims that these individual requests by the imperial government for maps of Mongolian territories were routine procedure: all Mongolian districts, confederations, and banners were required to submit maps every ten years to the colonial office in Beijing, where they were registered and translated.[15] The maps in the collection of the Staatsbibliothek zu Berlin give testimony to five stages in the process from production to use by the colonial office: (1) initial Mongolian drafts; (2) completed Mongolian maps, not colored; (3) finished Mongolian maps; (4) finished maps with small glued-on pieces of paper bearing Chinese transcriptions and translations; (5) maps with Chinese lettering that were based on Mongolian originals.[16]

Despite foreign incentives for the production of these maps, Mongolian scholars claim they are part of a Mongolian tradition. According to Heissig, B. Goncigdorž traced the maps and their construction to the Khitans, whose cartographic activities apparently have been documented for as early as 1179.[17] Chagdarsurung (Shagdarsurung in other transcriptions) pointed out that the maps of Mongolian banners and districts were based on what he called the "compass rose system of Mongolian cartography," which used twenty-four directional points to define the locations of the boundary markers, or *oboγa*.[18] These directional points were made up of the twelve elements of the Chinese zodiac (which included the four cardinal directions), in combination with eight colors and with four points related to elements of divination.[19] In addition, he identified a set of intermediary points that enabled a further division into forty-eight directions. To support his argument he gave a translation of the Mongolian lettering on a map of the Čeringwangduyibabudorǰi Banner in the Setsen Khan district, Khalkha. On this map, thirty-two annotations extended vertically from each boundary marker beyond the boundary line, similar to the representation in figure 15.51. Each of these annotations identified the location of that marker by giving reference to the directional points and also its distance to the next boundary marker clockwise.[20]

Apart from these tentative findings by Mongolian scholars, little attention has been given in the general literature to the precursors and designs of regional Mongolian pasture area maps. Most efforts have been directed at their geographical content, because the maps are unique resources for place-name research. Virtually all authors who discuss them include long lists of toponyms in their works. As comparisons between maps from different time periods have shown, the place-names mentioned in the maps have changed little over the last two to three hundred years.[21] To make this resource more accessible, Walther Heissig initiated an ambitious project of collecting, alphabetizing, and cross-referencing the place-names on many of the maps in the Staatsbibliothek zu Berlin. This resulted in the most comprehensive and authentic collection of Mongolian place-names to date and provides a solid basis for etymological research.[22]

Although the scarcity and late date of the extant sources makes it difficult to trace the origins of Mongolian cartography, further synthesis is still feasible and desirable. A first step would be a translation from Mongolian and a critical examination of the work by Goncigdorž mentioned by Heissig. Goncigdorž's findings appear to be very promising, particularly his claim that the maps of the Khitans were precursors to Mongolian maps. In order to free the study of Mongolian cartography from its linguistic isolation, and to prevent the introduction of nationalist bias, it will be necessary to enlist the help of scholars from different disciplines and with different linguistic competence.

---

12. The scripts on the map were Chinese and Mongolian (Heissig, "Über Mongolische Landkarten," 130–31 [note 5]). Note that Heissig never consulted the original printed map, but had only seen a reproduction of the printed map in a journal.

13. See Mostaert's introduction in Saγang Sečen (Ssanang Ssetsen), *Erdeni-yin Tobči: Mongolian Chronicle*, 4 vols., ed. Antoine Mostaert (Cambridge: Harvard University Press, 1956), 1:85–86.

14. The 1739 map has not been preserved, but Mostaert showed that it was the archetype of a 1740 map he illustrated. Mostaert reproduced a 1903 photo of the original 1740 map as well as a 1927 copy; Saγang Sečen (Ssanang Ssetsen), *Erdeni-yin Tobči*, pl. 1 and additional plate (note 13). Joseph Klér, "A propos de cartographie mongole," *Bulletin de la Société Royale Belge de Géographie* 24, pts. 1 and 2 (1956): 26–51, also reproduced the map. However, as Mostaert pointed out, Klér reproduced not the original map, as he claimed, but the 1927 copy. According to Mostaert, the 1740 map is preserved at the "*ia men* de la bannière d'Otoγ" (126 and additional note at the end of the book).

15. Heissig, *Mongolische Handschriften*, 338 (note 6).

16. Heissig, *Mongolische Handschriften*, 338 (note 6). In addition to works conforming to these five stages, Heissig described two maps that were unique: Hs. Or. 253, which he considered "so modern in design that it is certainly purely Chinese," and Hs. Or. 33, which has Manchu script (p. 337).

17. Heissig, *Mongolische Ortsnamen*, 2:XII (note 6). I have been unable to locate the Goncigdorž work cited by Heissig.

18. Ts. Chagdarsurung, "La connaissance géographique et la carte des Mongols," *Studia Mongolica*, vol. 3 (2) (1975): 345–70, esp. 347.

19. The twelve elements are, clockwise: rat (north), bull, tiger, hare (east), dragon, snake, horse (south), sheep, monkey, rooster (west), dog, and pig; and the eight colors are two shades each for black, blue, red, and white. See Chagdarsurung, "La connaissance géographique," 347–50 (note 18), and Heissig, *Mongolische Ortsnamen*, 2:XII–XIII (note 6).

20. Chagdarsurung, "La connaissance géographique," 355–58 (note 18); Heissig notes that the descriptions of the 32 boundary markers correspond exactly with Hs. Or. 146 in the Berlin collection (*Mongolische Ortsnamen*, 2:XV and pl. 65 [note 6]).

21. Heissig, *Mongolische Ortsnamen*, 1:IX (note 6).

22. Magadbürin Haltod collected 13,644 Mongolian place-names from the 182 maps in the Staatsbibliothek zu Berlin—Preussischer Kulturbesitz, and they were published as an alphabetical list in volume 1 of *Mongolische Ortsnamen* (note 6). Volume 2 contains facsimiles of 118 of the 182 maps. In volume 3 the place-names have been cross-referenced by Š. Rasidondug, H.-R. Kämpfe, and V. Veit. Other lists of place-names were often based on inaccurate Chinese translations; Heissig, *Mongolische Ortsnamen*, 1:IX–XI (note 6).

Cartography in Southeast Asia

16 • Introduction to Southeast Asian Cartography

JOSEPH E. SCHWARTZBERG

For this history, Southeast Asia is defined as the portion of mainland Asia to the south of China and between India and Vietnam, together with that portion of insular Asia in which Malay peoples predominate (fig. 16.1). Hence it comprises the whole of Burma (Myanmar), Thailand, Laos, and Cambodia, an area within which Hinayana (Theravada) Buddhism is the dominant faith, and the Malay world—Malaysia, most of Indonesia, Brunei, and the Philippines—within which Islam and Christianity have come to be the leading religions. Vietnam is excluded because of its cultural affinity to China, an affinity that is strongly reflected in its rich cartographic heritage (see chap. 12).

Southeast Asia, as here defined, shows relatively little unity with regard to the surviving corpus of its premodern indigenous maps. To treat the entire region together in this history is, to a considerable extent, an editorial convenience that reflects the division of labor among the contributors to this work. There are, to be sure, certain commonalities in the widespread cosmographic artifacts (architectural and cartographic) that reflect the related Hindu and Buddhist worldviews. There are also similarities among the many terrestrial maps from Burma and the few from Thailand. There is, however, relatively little similarity between the maps of those two countries and the handful of surviving maps from the Malay world. Even more remarkably, within the latter area there is almost no similarity from one map to another. Finally, one must note that for certain countries, most notably the Philippines, but also Laos and Cambodia, not a single indigenous, premodern noncosmographic map is known to have survived to the present day.

Apart from certain cosmographic aspects of architecture and some maplike bas-reliefs on temple friezes, there is no surviving map from Southeast Asia dating from before the sixteenth century. Hence, despite a few references in the literature to earlier works, all of the existing corpus dates from the period after the advent of the Portuguese. Almost all the maps we know of date from the eighteenth century or later. Nevertheless, in most of what I shall discuss in this and subsequent chapters, there is little, apart from maritime maps, that points to the diffusion of cartographic ideas from the West. This contrasts markedly with the situation respecting influences from China, which appear to be present in many maps from Burma and Thailand. Regrettably, however, I am aware of nothing in the literature that makes it clear when and how Chinese cartographic concepts were transmitted.

In the rest of this chapter I shall discuss first the state of knowledge with respect to the indigenous cartography of Southeast Asia and then the nature of the surviving corpus. The next chapter relates to cosmography, and there I will treat both the dominant cosmographic ideas and their expression in two-dimensional maps and in other forms, including works of architecture and the layout of cities. Chapter 18 will be devoted mainly to terrestrial maps: topographic maps, route maps, town plans, and a few examples of other genres. The few surviving maritime maps and their largely Western antecedents are discussed in chapter 19. A brief concluding chapter will evaluate the surviving corpus and suggest lines for future research.

THE STATE OF OUR KNOWLEDGE

Many of the observations made concerning the meagerness of the relevant literature on South Asian cartography (which was discussed in volume 2, book 1 of *The History of Cartography*) apply with equal, if not greater, force to maps of Southeast Asia. This region too is virtually ignored by most cartographic historians. Bagrow, for example, devotes only a single paragraph to Southeast Asia, which I quote in full:

> Maps from Siam, Cambodia and the Malay Archipelago are also unknown in Europe. Such maps must have existed, however: Affonso d'Albuquerque, governor of India, wrote to the King of Portugal in 1512 that he was sending him a copy of a large map made by a Javanese pilot, showing the Indian Ocean from the Cape of Good Hope, with the Red Sea, the Persian Gulf, the Moluccas, and the sea-routes to China and Formosa, as well as land routes in the interior. The Javanese were experienced sailors: in 1513 the King of Djapara alone had a fleet of 80 warships. Unfortunately, nothing is known of their maps, and there

> is little hope that anything will ever be discovered, because the palm-leaves from which they were made are not very durable.[1]

Why Bagrow (writing sometime before 1943) failed to note Burma, Laos, and what is now Vietnam under the heading "Southeast Asia" is not clear. Conceivably he still considered Burma a part of India, despite its having been made a separate crown colony in 1937. Vietnam he might have considered, as we do in this volume, a cultural extension of China. Similarly, Laos might have been regarded merely as a cultural appendage of Siam. But since no maps of those areas are mentioned in his discussions of either China or India, we may assume that he was totally ignorant of any surviving maps from Southeast Asia, as we have defined the region, even though a number of relevant publications had appeared in European journals by the time his history was compiled. Additionally, there already existed in European libraries maps from Southeast Asia that had not yet been the objects of scholarly study.

Bagrow erred also in stating that "there is little hope that anything will ever be discovered" because Southeast Asian maps were made from palm leaves and hence were fragile. (Where he obtained such faulty information is not indicated.) Although some maps were indeed made on palm leaves, we now know that most were not. What makes Bagrow's concluding observation unfortunate is not merely its factual inaccuracy but, more important, the dampening effect it presumably exerted on the search for indigenous Southeast Asian maps by other scholars. This criticism may also be applicable, though in lesser degree, to the author of the article on cartography in the Dutch *Encyclopaedie van Nederlandsch-Indië*, who observed, probably erroneously, that "there are no [surviving] Javanese or indigenous maps from the pre-Dutch period."[2]

Whether later historians of cartography, such as Brown, Crone, and Kish, took their cues from Bagrow is not clear; but it is noteworthy that none mentions Southeast Asian indigenous maps.[3] Unno, however, has brought to my attention maps from Burma, Thailand, Vietnam, and Indonesia.[4] The Indonesian examples include two no-longer-extant Javanese maps: the one noted by Bagrow, which is described in Albuquerque's missive of about 1512 to the king of Portugal, and another presented by a Javanese prince to a Mongol military commander in 1293 to signify his submission to Mongol rule. There is also a surviving Javanese map, allegedly dating from the sixteenth century, that was brought to light by an Indonesian geologist, Rachmat Kusmiadi, and subsequently illustrated in Harvey's *History of Topographical Maps*.[5] But it appears that neither Kusmiadi, Unno, nor Harvey was aware that the map was discovered as early as 1858 and that a Dutch philologist, K. F. Holle, had translated its detailed Sundanese text and published a paper on it in 1877, a full century before Kusmiadi's presentation.[6]

Long before Holle's article, however, a series of descriptions and engraved copies of Burmese maps were presented by Francis Hamilton (formerly Buchanan; 1762–1829) in the *Edinburgh Philosophical Journal* (from 1821 to 1824) and then in the *Edinburgh Journal of Science* (1824).[7] Hamilton had collected several dozen such works during a sojourn in Burma in 1795 and was later to collect additional indigenous maps in India and Nepal and to gain distinction as the pioneer compiler of district gazetteers in India. He was, so far as I am aware, the first European to provide scholarly descriptions of indigenous maps from Southeast Asia.

After Hamilton, several other British collectors, most notably Henry Burney (1792–1845) and James George Scott (1851–1935), also acquired substantial numbers of indigenous maps in Burma.[8] Burney, an excellent scholar of Eastern languages and cultures, was the first permanent British resident at the Burmese court (1830–38) and also served earlier in Thailand and the Malay States. Scott, who first went to Burma as a war correspondent in 1884, became the British resident in the Northern and Southern Shan States and served continuously in Burma and Thai-

1. Leo Bagrow, *History of Cartography*, rev. and enl. R. A. Skelton, trans. D. L. Paisey (Cambridge: Harvard University Press; London: C. A. Watts, 1964; reprinted and enlarged Chicago: Precedent Publishing, 1985), 208.

2. Frederik Caspar Wieder, "Oude Kaartbeschrijving" section in "Kaartbeschrijving," in *Encyclopaedie van Nederlandsch-Indië*, 8 vols. (The Hague: Martinus Nijhoff, 1917–40), 2:227–36; quotation on 229.

3. Lloyd A. Brown, *The Story of Maps* (Boston: Little, Brown, 1949; reprinted New York: Dover, 1979); Gerald Crone, *Maps and Their Makers: An Introduction to the History of Cartography* (London: Hutchinson University Library, 1953, and four subsequent editions up to 1978); and George Kish, *La carte: Image des civilisations* (Paris: Seuil, 1980).

4. I would like to thank Kazutaka Unno for communicating this information.

5. Rachmat Kusmiadi, "A Brief History of Cartography in Indonesia" (paper presented at the Seventh International Conference on the History of Cartography, Washington, D.C., 7–11 August 1977), 1–3; P. D. A. Harvey, *The History of Topographical Maps: Symbols, Pictures and Surveys* (London: Thames and Hudson, 1980), 114.

6. K. F. Holle, "De Kaart van Tjiëla of Timbangantĕn," *Tijdschrift voor Indische Taal-, Land- en Volkenkunde* 24 (1877): 168–76, and endcover folding map.

7. The maps appeared in thirteen separate articles; full citations are provided in chapter 18 and in the Bibliographical Index.

8. For biographical details, see Thaung Blackmore, *Catalogue of the Burney Parabaiks in the India Office Library* (London: British Library, 1985); and Andrew Dalby and Sao Saimöng Mangrai, "Shan and Burmese Manuscript Maps in the Scott Collection, Cambridge University Library" (unpublished manuscript, n.d. [ca. 1984]), 15 pages and catalog (47 pp.) with 39 figs. and 8 pls.

FIG. 16.1. REFERENCE MAP FOR THE STUDY OF INDIGENOUS SOUTHEAST ASIAN CARTOGRAPHY.

land from 1891 until his retirement in 1910. Many of the maps he collected were Shan maps, made at his behest. Though these maps were from an area of Burma, they relate to a culture that is more Thai than Burmese. Neither Burney nor Scott is known to have written about the maps he collected (though both wrote extensively about other matters), and it was not until about 1979–85 that any sort of public record of their coverage was made. This was thanks to the efforts of Patricia Herbert, who cataloged the Burney papers at the Library of the Royal Commonwealth Society, and to Andrew Dalby and Shan historian Sao Saimöng Mangrai, who performed a similar service for the Scott Collection at Cambridge University.[9]

Following Burmese independence, historians within that country also demonstrated an interest in its cartographic heritage. Many maps that might otherwise have been consigned to oblivion were collected and preserved by members of the Burmese Historical Commission, most notably U Maung Maung Tin and Than Tun. But little of what they found, as far as I know, has been discussed in published accounts. A noteworthy exception is the lucid exposition of King Mindon Min's plan of about 1850 for the development of a new Burmese capital at Mandalay.[10] Whether other such discussions exist in Burmese is not known.

Premodern indigenous Thai maps are much fewer than for either Burma or Vietnam. Nevertheless, several valuable works on that limited corpus have appeared: a brief study by Wenk on the illustrated geographic portion of the *Traibhūmikathā* (Story of three worlds), a Thai cosmological text (Thai *Trai phum*); an exceptionally thorough study by Kennedy of an early nineteenth-century military map of northeastern Thailand; and an outline history by Freitag of Thai cartography of both the premodern and modern periods.[11] This last work is the only study of any country of Southeast Asia by a historian of cartography that has any claim to comprehensiveness, apart from Quirino's study of Philippine cartography, which, in the absence of old indigenous maps, relates entirely to foreign (either Chinese or European) and modern works.[12]

Surviving maps of the Malay world are very few, and the literature relating to them is meager. Not counting discussions of cosmographic works, to be considered below, only a few maps have been the subject of scholarly articles. Of these, I have already noted the commentaries of Holle and Kusmiadi on a supposed sixteenth-century Javanese map. One other undated Javanese, or possibly Balinese, map that has been the object of some study appears on a batik shawl that is described in a 1933 catalog.[13] From the aboriginal Sakai tribe of peninsular Malaysia come a number of bamboo artifacts inscribed with diverse and enigmatic drawings that have been interpreted as locality maps and explained in considerable detail by the ethnographer Hrolf Vaughan Stevens.[14] Apart from these works, there are published commentaries on maps that are either copied from or strongly influenced by European prototypes. The former include several early nineteenth-century Buginese sea charts of the greater part of Southeast Asia that are discussed at considerable length in an exemplary study by Le Roux. The latter, a supposed early eighteenth-century map of the Malay Peninsula, was described by Phillimore in *Imago Mundi*.[15]

This list of sources, although not exhaustive, does cover the principal known published references to surviving indigenous Southeast Asian maps that are not essentially cosmographic in nature. Asian cosmography, however, is a subject to which historians of cartography have as yet given scant attention. For commentaries on

9. Parts of Herbert's catalog appear as an appendix to the work by Blackmore, *Burney Parabaiks* (note 8); but this appendix mentions no maps, in contrast to an unpublished portion of the catalog available at the Royal Commonwealth Society in London. The paper by Dalby and Saimöng, "Shan and Burmese Manuscript Maps" (note 8), provides descriptions of the maps in the Scott Collection. I extend my gratitude to both Herbert and Dalby for their exceptional kindness in helping me locate and understand Burmese and Shan maps. The personal introductions Herbert arranged were of invaluable assistance.

10. U Maung Maung Tin and Thomas Owen Morris, "Mindon Min's Development Plan for the Mandalay Area," *Journal of the Burma Research Society* 49, no. 1 (1966): 29–34.

11. Klaus Wenk, "Zu einer 'Landkarte' Sued- und Ostasiens," in *Felicitation Volumes of Southeast-Asian Studies Presented to His Highness Prince Dhaninivat Kromamun Bidyalabh Bridhyakorn . . . on the Occasion of His Eightieth Birthday*, 2 vols. (Bangkok: Siam Society, 1965), 1:119–22 with 1 pl.; Victor Kennedy, "An Indigenous Early Nineteenth Century Map of Central and Northeast Thailand," in *In Memoriam Phya Anuman Rajadhon: Contributions in Memory of the Late President of the Siam Society*, ed. Tej Bunnag and Michael Smithies (Bangkok: Siam Society, 1970), 315–48 and 11 appended maps; and Ulrich Freitag, "Geschichte der Kartographie von Thailand," in *Forschungsbeiträge zur Landeskunde Süd- und Südostasiens*, Festschrift für Harald Uhlig zu seinem 60. Geburtstag, vol. 1, ed. E. Meynen and E. Plewe (Wiesbaden: Franz Steiner, 1982), 213–32; see also Klaus Wenk, *Thailändische Miniaturmalereien nach einer Handschrift der indischen Kunstabteilung der Staatlichen Museen Berlin* (Wiesbaden: Franz Steiner, 1965), 64 and pl. XI.

12. Carlos Quirino, *Philippine Cartography (1320–1899)*, 2d rev. ed. (Amsterdam: Nico Israel, 1963).

13. Koninklijk Instituut voor de Tropen, *Aanwinsten op ethnografisch en anthropologisch gebied van de Afdeeling Volkenkunde van het Koloniaal Instituut over 1933*, Afdeeling Volkenkunde 6 (Amsterdam, 1934), 24–26.

14. Hrolf Vaughan Stevens, "Die Zaubermuster der Ôrang hûtan," pt. 2, "Die 'Toon-tong'-Ceremonie," *Zeitschrift für Ethnologie* 26 (1894): 141–88 and pls. IX and X.

15. C. C. F. M. Le Roux, "Boegineesche zeekaarten van den Indischen Archipel," *Tijdschrift van het Koninklijk Nederlandsch Aardrijkskundig Genootschap*, 2d ser., 52 (1935): 687–714 and folding map, and Reginald Henry Phillimore, "An Early Map of the Malay Peninsula," *Imago Mundi* 13 (1956): 175–79.

representations of the cosmos, beyond the realm of mortals, we must rely mainly on ethnographers, students of religion, and art historians. In the case of Southeast Asia, where all the major religions are exogenous, much of the basic cosmological research has been done mainly in and on other areas, especially India, the hearth of Hinduism and Buddhism, which jointly held sway in the region for most of its recorded history.[16]

So far as I am aware, the earliest published images of the Buddhist cosmos are the facsimiles of Burmese and Laotian representations of the universe that appear in Bastian's *Ideale Welten* (1892).[17] A set of more generalized drawings by Bastian attempts to elucidate the major components of the Hinayana Buddhist cosmos.[18] At about the same time, Gerini provided a detailed drawing of the Cakravāla, the central portion of the cosmos, as conceived by the Thais and reconstructed (three-dimensionally and on a grand scale) for the performance, over the period of one week (25 December 1892 to 1 January 1893), of the sacred tonsure ceremony.[19] Yet another seminal work of this period was Temple's *Thirty-seven Nats*, which shows the interplay of Burmese spirit worship and popular Buddhism and includes one cosmographic world map.[20] Additional illustrated discussions of Southeast Asian Buddhist cosmographies do not appear, as far as I know, until after World War II. Some of these will be introduced in their appropriate contexts in the following chapter.

Attempts to understand the cosmological systems of the numerous tribal peoples of Southeast Asia are relatively recent. Although many anthropological studies of the postcolonial period address the subject, and some even make it a dominant concern, reproductions of visual representations of major portions of the cosmos, as conceived by animistic tribal people, are not abundant. Particularly noteworthy in this regard, however, is the pioneering work of the Swiss missionary-ethnographer Hans Schärer, who first brought to light maps associated with the mortuary cult of the Ngaju Dayaks of interior Borneo.[21] Other scholars have since shed additional light on similar practices among other Dayak groups; but for no other part of the Malay Archipelago have cosmographic artifacts emerged that one would clearly identify as maps. Anthropologists have, however, imputed cosmographic significance to the spatial arrangements of houses, villages, and even petty states in various localities of Indonesia.[22]

The Nature of the Southeast Asian Cartographic Corpus

We do not know when maps were first made in Southeast Asia. No early cave paintings or other prehistoric archaeological remains that include maplike elements have come to light in the region. Nor do we know of any ancient historical texts that make explicit references to what one might consider a map, even though many Hindu and Buddhist texts rich in cosmographic and geographic detail—the Indian epics, the Puranas, the Jataka tales, and so forth—have become a part of the literary heritage of the Indianized portions of Southeast Asia. Similarly, I am unaware of any wholly indigenous literature of the period before the advent of Europeans that makes reference to maps. Although it might not seem unreasonable to assume that early Indian migrants to Southeast Asia, dating as far back as the late first millennium B.C., brought with them some of the limited cartographic competence that then existed in their home countries, there is neither tangible nor textual evidence to support such a supposition. Nor do we have firm evidence of the transfer of cartographic skills from China to the lands to its south, not even during the Mongol expansion into those two areas in the thirteenth century. It may also seem reasonable to suppose that Indonesians, who settled Madagascar about the beginning of the Christian Era and remained in maritime contact with that island until the fifteenth century, must have developed some sort of nautical charts as navigational aids; but again no supporting evidence survives.

One must turn then to architecture for the first unequivocal visual representations of cosmographic ideas in Southeast Asia. Royal Hindu temples, essential symbols of political power in the many Indianized states of the region, were centered on shrines representing Mount Meru, the cosmic *axis mundi*. The earliest such temples

16. For research on South Asian cosmology, see Joseph E. Schwartzberg, "Cosmographical Mapping," in *The History of Cartography*, ed. J. B. Harley and David Woodward (Chicago: University of Chicago Press, 1987–), vol. 2.1 (1992), 332–87.

17. Adolf Bastian, *Ideale Welten nach uranographischen Provinzen in Wort und Bild: Ethnologische Zeit- und Streitfragen, nach Gesichtspunkten der indischen Völkerkunde*, 3 vols. (Berlin: Emil Felber, 1892), vols. 1 and 3.

18. Adolf Bastian, "Graphische Darstellung des buddhistischen Weltsystems," *Verhandlungen der Berliner Gesellschaft für Anthropologie, Ethnologie und Urgeschichte*, 1894, 203–15 and pls. 3–7, in *Zeitschrift für Ethnologie*, vol. 26.

19. Gerolamo E. Gerini, *Chūḷākantamaṅgala; or, The Tonsure Ceremony as Performed in Siam* (Bangkok: Siam Society, 1976; first published in 1895), diagram 1 (facing p. 136).

20. Richard C. Temple, *The Thirty-seven Nats: A Phase of Spirit-Worship Prevailing in Burma* (London: W. Griggs, 1906); world map facing 8. There is also a new edition with an essay and bibliography by Patricia M. Herbert (London: P. Strachan, 1991).

21. Hans Schärer, *Die Gottesidee der Ngadju Dajak in Süd-Borneo* (Leiden: E. J. Brill, 1946), English translation, *Ngaju Religion: The Conception of God among a South Borneo People*, trans. Rodney Needham (The Hague: Martinus Nijhoff, 1963); and idem, "Die Vorstellungen der Ober- und Unterwelt bei den Ngadju Dajak von Süd-Borneo," *Cultureel Indië* 4 (1942): 73–81.

22. See the discussion below, pp. 739–40.

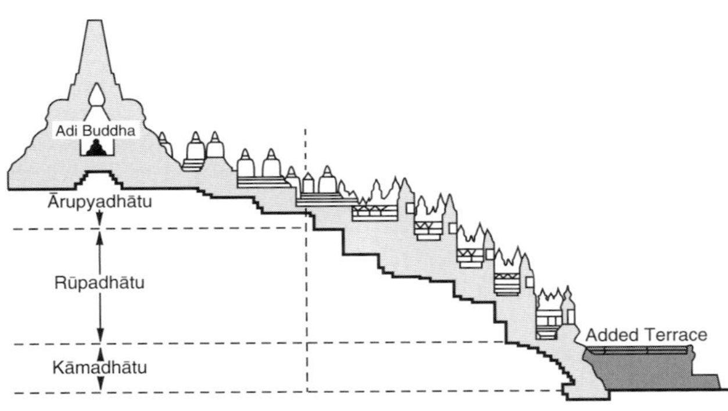

FIG. 16.2. BOROBUDUR. This great Mahayana Buddhist "temple mountain" was constructed in central Java. It comprises five more or less square terraces of gradually diminishing size over a large square plinth, surmounted in turn by three diminishing circular terraces surrounding a large circular stupa. The entire edifice symbolizes the unity of the cosmos, with the Buddhist transition from the low states of reality toward the base, through various increasingly exalted psychological states, to the blissful condition of enlightenment at the summit. Carvings in stone on the walls of each terrace depict the acts that contribute to or detract from merit at various stages of existence. At the lower levels the scenes represent mundane subjects, and at higher levels the subject matter is increasingly metaphysical. Of a design inspired by Indian stupas, Borobudur was the likely source of inspiration for the numerous temple mountains built at Angkor.

Photograph courtesy of Robert Harding Picture Library Limited, London. Drawing after Philip Rawson, *The Art of Southeast Asia* (London: Thames and Hudson, 1967), 224–25.

FIG. 16.3. ANGKOR WAT. An enormous structure (1,550 m long × 1,400 m wide), this essentially Vaishnavite Hindu "temple mountain," built in the first half of the twelfth century, was one among many constructed by Khmer monarchs to symbolize their divinity and universal dominion. Its central tower represents Mount Meru and its surrounding moat the encompassing ocean. More than a mile of sculpted bas-relief adorns the friezes at various levels of the temple and illustrates a wide range of mundane and mythological themes, prominent among which is that of the god-king as the giver of water and source of fertility. The Hindu cosmogonic myth of serpents wrapped around Mount Meru and used as ropes to churn the world ocean is also rendered in massive sculpture.
From George Groslier, *Angkor*, 2d ed. (Paris: Librairie Renouard, 1931), fig. 80.

surviving are in Java and date from the seventh century. Buddhist temples, incorporating similar cosmological principles, go back at least as far as the late eighth century. Borobudur, the greatest of all Javanese Buddhist temples (fig. 16.2), was constructed about 800, and the Lara Janggrang Temple, the most renowned of Javanese Hindu edifices, was built at Prambanan about 900–930. On the mainland, shortly thereafter, the foundations were laid for the magnificent temple complex at Angkor. There each successive Khmer monarch built one or more of his own temple mountains, culminating in the construction of the Vaishnavite Angkor Wat (fig. 16.3) in the first half of the twelfth century and the even more extensive Mahayana Buddhist Angkor Thom and Bayon assemblages (ca. 1200). Thai conquest of the Khmers in the late thirteenth century, the area's conversion to Hinayana Buddhism, already dominant in lands farther to the west, and other factors brought an end to such grandiose projects. But whether Hindu, Mahayana Buddhist, or Hinayana Buddhist (the traditions were often blended), Southeast Asian religious architecture retained a strong cosmographic symbolism. Among Hinayana Buddhists, the dominant form became the *chetiya*, a variant of the South Asian stupa, also symbolizing Mount Meru. Since literature on the subject is voluminous and readily accessible, there is little point in trying to summarize here the development of Southeast Asian architecture, even in regard to its strictly cosmographic aspects.

Some architectural assemblages were intended as reconstructions of the sacred geography of the group of sites in and around Bodh Gaya in northeastern India, where the Buddha attained enlightenment.[23] At least four attempts were made to recreate, more or less faithfully,

23. See, for example, Donald M. Stadtner, "King Dhammaceti's Pegu," *Orientations* 37, no. 2 (1990): 53–60.

FIG. 16.4. BAS-RELIEF PLAN OF A JAVANESE TEMPLE COMPOUND. This plan, sculpted in stone, appears on the north wall of Jawi Temple, constructed in central Java in the late thirteenth century.
Size of the bas-relief: ca. 30 × 70 cm. From Jacques Dumarçay, "Notes d'architecture Javanaise et Khmère," *Bulletin de l'Ecole Française d'Extrême-Orient* 71 (1982): 87–147, esp. pl. 11.

the great temple at Bodh Gaya that the Mauryan emperor Aśoka is believed to have had built in the third century B.C. These include temples built at Pagan and Pegu, in Burma, in the thirteenth and fifteenth centuries, respectively, and two in the northern Thai towns of Chiang Mai and Chiang Rai, also built in the fifteenth century.

> While the chronicles in Thailand are silent about where Tiloka [the king responsible for building the two Thai reconstructions] obtained his temple plans, those in Burma that mention King Dhammacetiya's copy—built in Pegu at almost the same time as Tiloka's—specifically state that he sent a large group of artisans to Bodhgaya to make plans and models.[24]

Apart from their architectonic aspect, temples were interesting for their sculptural details and their mural painting. Much of this painting in relatively recent temples, as I shall note in the next chapter, is itself cosmographic; and it seems virtually certain that a tradition of such painting goes back many centuries. But because painting seldom lasts more than a few hundred years, I can point to no good examples of very old mural cosmographies.[25] Sculptural detail on Southeast Asian temples ranged over a wide variety of themes, both secular and religious. Some bas-reliefs appear to present bird's-eye views of scenes including temples, courtyards, villages, cultivated fields, and occasionally streams and other natural features. In the absence of a systematic inventory of Southeast Asian temple sculpture, I cannot venture an opinion on the frequency of such maplike images, though they do not appear to have been common. Figures 16.4 and 16.5 present examples from late thirteenth- and fourteenth-century Java.

The pervasive Buddhist concern with cosmology found expression in the illustrations that accompanied literary texts. One such text, the *Traibhūmikathā* or *Trai phum*, allegedly adapted from the Sri Lankan Pali canon by the Thai king Phra Ruang in 1345, describes in great detail the structure of the three worlds and thirty-one realms composing the universe. Some of the surviving recensions (no complete copy predates the eighteenth century) are profusely illustrated. In fact the illustrations may take up far more space than the accompanying text. Whether the original and other early recensions were similarly adorned is not known, but it is reasonable to suppose that some were.[26]

Cosmographies were also prepared in other forms: in paintings, on palm-leaf manuscripts, as three-dimensional sculptures, as decorations on furniture, and presumably, if contemporary practice in South Asia may be taken as a guide, in various types of folk art. Gates and other temple structures, together with temple compounds in Bali, are characteristically carved to represent Mount Meru and its several flanking ranges. In Powun-daung (Powin Taung), on the lower Chindwin River of Upper Burma, there is a group of cave temples dating from the ninth to the twentieth century; on the exterior of one of them (date unspecified), a cross-sectional representation of Mount Meru and its flanking ranges is sculpted in an ensemble eight meters high.[27]

Finally, we must note that cosmography is not an exclusive concern of the great religious traditions but is also relevant for animistic religions. Much more detailed field research is needed, however, to uncover the visual manifestations of the cosmological conceptions of Southeast Asia's numerous tribal peoples. Yet one can safely assert that they too assume a variety of forms and that

24. Robert L. Brown, "Bodhgaya and South-east Asia," in *Bodhgaya: The Site of Enlightenment*, ed. Janice Leoshko (Bombay: Marg Publications, 1988), 101–24; quotation on 110. For a note on even earlier Tibetan models of Bodh Gaya, see chapter 15.

25. A representation of Mount Meru surrounded by seven concentric mountain chains was painted on the wall of Ānanda monastery in Pagan, in central Burma, about 1776. This work is illustrated in Jane Terry Bailey, "Some Burmese Paintings of the Seventeenth Century and Later, Part II: The Return to Pagan," *Artibus Asiae* 40 (1978): 41–61; illustration facing 59.

26. The first full translation of the text, with extensive critical commentary, is that of George Coedès and C. Archaimbault, *Les trois mondes (Traibhūmi Braḥ R'vaṅ)* (Paris: Ecole Française d'Extrême-Orient, 1973). A handsomely illustrated English translation, also with extensive commentary, is Frank E. Reynolds and Mani B. Reynolds, *Three Worlds according to King Ruang: A Thai Buddhist Cosmology* (Berkeley, Calif.: Asian Humanities Press, 1982). For an account of the background of the work, the dating of its various recensions, and the uses it was put to, see Craig J. Reynolds, "Buddhist Cosmography in Thai History, with Special Reference to Nineteenth-Century Culture Change," *Journal of Asian Studies* 35 (1976): 203–20.

27. Robert Heine-Geldern, "Weltbild und Bauform in Südostasien," *Wiener Beiträge zur Kunst- und Kulturgeschichte Asiens* 4 (1928–29): 28–78, esp. picture facing 70 and text on 71–72.

they have been influenced, in varying degrees, by exposure to the belief systems and material culture of neighboring nontribal communities, Buddhist, Muslim, and Christian. A few examples will be considered in the following chapter.

The first written notice we have of a Southeast Asian map (in a relatively narrow sense of the term) is from the *Yuan shi* (History of the Yuan), an official Chinese history compiled in 1369–70 by Song Lian and others. This work refers to a Yuan military invasion of Java in 1292–93 during which Raden Vijaya, one of the leaders of the then state of Kediri, presented the invaders with a map and census record of the country as an act of surrender. This appears to indicate the existence of an official administrative map of the country. Since the map was a symbol of territorial possession, its submission to the enemy was, in Unno's view, tantamount to relinquishing the territory.[28] This interpretation is consistent with the stated purpose of a textually rich surviving Javanese map (from the western Sundanese region), probably dating from the late sixteenth century, that I shall discuss in a later chapter. It suggests the possibility of an official Javanese mapping tradition that has long since faded from memory.

Apart from the two items just noted, the Malay world provides us with very few additional maps. I have already mentioned the so-called Rodrigues map (the one noted by Bagrow). This chart was copied from a Javanese nautical chart about 1513 but was lost in a shipwreck en route to Portugal.[29] And the Venetian traveler Ludovic Varthema mentions that a chart was consulted by a native pilot on a voyage he took from Borneo to Java in 1505.[30] Much later there were nautical charts copied, in varying degrees, from European prototypes. One was an early eighteenth-century work depicting the Malay Peninsula and the Gulf of Thailand.[31] The others, presumably all adapted from one or more European originals of the early nineteenth century, depicted virtually the whole of Southeast Asia. Of the four such maps known to have been drawn, I could find only two.[32] These are large, exceedingly detailed charts with abundant text in the Bugi script. Bugis, whose home region is on the island of Sulawesi (Celebes), have been in recent centuries the leading indigenous seamen of Indonesia, and there is evidence that they revised their original map source(s) based on their own nautical experience and attached considerable value to their work.

Only three other Indonesian maps have come to my attention, none of them in any way resembling the others discussed: the undated batik map, from eastern Java or Bali;[33] a presumably nineteenth-century map from central Java, probably used for some administrative purpose; and a map of the domains of the sultan of Pontianak in Kalimantan (Borneo), tentatively dated 1826 and, though written in Malay, at least partly European in inspiration.

FIG. 16.5. JAVANESE LANDSCAPE IN BAS-RELIEF. This sculpted landscape includes a temple, some nearby streams, two bridges, boulders along the water's edge, paddy fields, and a village (*upper left*). The sculpture, which originally came from a fourteenth-century temple in the central Javanese village of Trawulan (Trowulan), was badly damaged by fire while on loan for the Paris Exhibition of 1931.
Height of the original: 67 cm; total width unknown. From August Johan Bernet Kempers, *Ancient Indonesian Art* (Amsterdam: C. P. J. van der Peet, 1959), pl. 288.

Finally, there are the earlier mentioned works by Sakai aborigines from the interior of West Malaysia, which, though recent, might well represent a tradition that is centuries old.

Burma also appears to have had officially sanctioned maps serving a variety of purposes—military, engineering, cadastral, and so forth—in the late eighteenth and nineteenth centuries. The record is unclear, however, on whether any such cartography preceded the period just specified. Several well-educated Europeans lived in Burma for protracted times during the late eighteenth and early nineteenth centuries, had extensive friendly interaction with the Burmese, and might well have influenced the subsequent development of Burmese cartography. Among these was an Italian Barnabite missionary, Gaetano Mantegazza, whose sojourn in Burma lasted from

28. I thank Kazutaka Unno for this information on the notice in the *Yuan shi*, 49:13.

29. Discussed in Heinrich Winter, "Francisco Rodrigues' Atlas of ca. 1513," *Imago Mundi* 6 (1949): 20–26.

30. Ludovic Varthema, *The Travels of Ludovico di Varthema in Egypt, Syria, Arabia Deserta, and Arabia Felix, in Persia, India, and Ethiopia, A.D. 1503 to 1508*, trans. John Winter Jones, ed. George Percy Badger (London: Printed for the Hakluyt Society, 1863), 249.

31. Phillimore, "Map of the Malay Peninsula" (note 15).

32. For the known present and alleged former locations of all five copies and a detailed discussion of several of them, see Le Roux, "Boegineesche zeekaarten" (note 15).

33. See note 13 above.

1772 to 1784. Mantegazza prepared what, for that time, was a rather good map of the country, a feat he could hardly have achieved without significant assistance from indigenous informants.[34] Another Barnabite missionary, Vincenzo Sangermano, who lived in Burma from 1783 until 1807 or 1808, established a college in Rangoon. There "fifty students ... were instructed in several branches of learning and science; so that besides some ecclesiastics, it ... produced skilful engineers, physicians and even pilots." Sangermano must have had considerable competence as a cartographer. He accepted an English commission "to make a chart of the port of Rangoon, which he executed with so much ability, as to receive [from the British government] a pension for life." Among Sangermano's principal concerns were the cosmography of Burmese Buddhism and the geography of the Buddhist empire, both of which he described at length in a posthumously published manuscript.[35]

Francis Hamilton, to whom I have already drawn attention, collected a number of maps of Burma during an eight-month visit in 1795. Most, if not all, of these were drawn at his behest and with some obvious coaching on his part. Since he does not mention seeing any indigenous maps, one might suppose that few if any then existed. On the other hand, the sophistication of the maps he collected and the non-European style of many of their cartographic signs suggests that those who drew them were not cartographic neophytes. In any event, a very large, detailed Burmese map illustrating the country traversed during the Burmese invasion of Manipur in the late 1750s (plate 37 below) and another military map relating to the Burmese campaign against the Thai capital at Ayutthaya in 1767 (fig. 18.38 below) indicate that mapmaking could not have been a wholly new endeavor for the Burmese.

Burmese appreciation for maps in regard to the needs of the state is further illustrated in that the Burmese king engaged Eurasian shipowner and captain William Gibson, who had long resided in Burma, to make "a map of the Burman dominions, together with the adjacent countries of Hindustan, Siam and Cochin-China." How long he was employed in this task, on which he was reportedly working in 1819, is not known. Nor do we known whether it was ever completed. But it is noteworthy that Gibson was asked in 1822 to accompany a Burmese diplomatic mission to Vietnam with the object of forging an anti-Siamese alliance. We may reasonably assume that additional mapping was among his responsibilities on that journey.[36]

Siam, though never submitting to colonial rule, also saw its share of European sojourners—missionaries, merchants, diplomats, and soldiers of fortune—beginning early in the sixteenth century. But an abortive French bid for dominance in 1688 led to the expulsion of most foreigners, and it was not until the nineteenth century that relations with the West were resumed.[37] Before that period we have no direct evidence of Western influence on Siamese cartography, though European ships and crews and geographic knowledge are illustrated in the geographic portion of an eighteenth-century recension of the cosmographic text the *Traibhūmikathā*. Given that Siam was the most powerful state in Southeast Asia during most of the eighteenth and nineteenth centuries, as well as one of the most culturally advanced, it may seem odd that so few premodern maps have survived from that country. One probable cause is the destruction of the royal library during the Burmese sack of Ayutthaya in 1767. An additional explanation may lie in a Siamese bureaucratic institution known as *chamra*, which, according to Wilson, is a periodic purging of official documents so that they represent only the currently recognized and acceptable truth.[38] Thus, with the advent of

34. Gaetano Maria Mantegazza, *La Birmania: Relazione Inedita del 1784 del Missionario Barnabita G. M. Mantegazza* (Rome: Ed. A. S., 1950); text is in the original French, preface and editorial notes are in Italian.

35. Vincenzo Sangermano, *A Description of the Burmese Empire Compiled Chiefly from Native Documents*, trans. and ed. William Tandy (Rome: Oriental Translation Fund of Great Britain and Ireland, 1833), preface by N. Wiseman, iii–iv.

36. B. R. Pearn, "The Burmese Embassy to Vietnam, 1823–24," *Journal of the Burma Research Society* 47, no. 1 (1964): 149–57, esp. 150, citing an account by an American missionary who knew Gibson well. Gibson's journal recounting the mission (which proved unsuccessful) was turned over shortly before his death in 1824 to John Crawfurd, who published it in London in 1828. I have not had an opportunity to study that document.

37. The nature and extent of these contacts are discussed by Reynolds, "Buddhist Cosmography," 211–13 (note 26). Reynolds notes the particularly important roles played by Crown Prince Mongkut during his years as a monk (1824–51), when he assiduously cultivated contacts with well-educated foreigners. Among his tutors were the American missionary Jesse Casswell, who perhaps provided Mongkut's principal intellectual window to the West, and the French bishop Jean Baptiste Pallegoix, from whom Mongkut learned astronomy.

38. Constance Wilson, "Cultural Values and Record Keeping in Thailand," *CORMOSEA* [Committee on Research Materials on Southeast Asia] *Bulletin* 10, no. 2 (1982): 2–17, esp. 4–6: "A ... concept important in understanding archival management in Thailand is that of *chamra* (cleansing, purification). *Chamra* is an integral and fully acceptable process in Thai culture, one which has a long history behind it. *Chamra* may refer only to minor cleansing or editing of a document: correcting grammatical errors, modernizing the spelling, or updating the titles of individuals. *Chamra* can also refer to a more drastic cleansing process: the filling in of blank areas in manuscripts, the rewriting of materials, or the removal of documents from circulation. *Chamra* is a continuous process. As ideas change so does the need for cleansing. Records which were cleansed for one set of reasons in the early twentieth century can be recleansed for a different set of reasons today. As a consequence of *chamra* we cannot look upon the physical collections of manuscripts which exist in Thailand as a stable fixed entity but instead must look at them as being in a constant state of flux as manuscripts are added, removed, recataloged, and shifted around. The process of *chamra* is

modern survey methods and the demonstrable "inferiority" of early maps, *chamra* would very likely have required their destruction.

Apart from cosmographies and a few unimpressive maps held by libraries in the West, the only currently noteworthy premodern indigenous maps to have surfaced in Thailand are a very long, multifolio folding map (probably dating from the late seventeenth or early eighteenth century), largely relating to the religious establishments on the Sathing Phra peninsula, a spit of land on the east coast of the Malay Peninsula, and a large and rather detailed military map, supposedly made in the early nineteenth century for the pacification of the ethnically Lao region of the Korat Plateau in what is now the northeastern part of the country.[39] But evidence of Thai mapmaking capability was provided in a letter dated 3 April 1824 by Lieutenant James Low, who then transmitted to the secretary to the government of Prince of Wales Island (Penang) a large "Map of Siam, Camboja, and Laos" that, in his words, he "first completed . . . in the year 1822 from a large mass of original, and I firmly believe, authentic Native Plans, & Itineraries, and oral information . . . which I have since that period amended and enlarged." Low's letter also referred to a forthcoming "Geographical Memoir to which the Map will afterwards be an accompaniment."[40] Whether Low's sources included maps drawn by ethnic groups other than Thais is uncertain, though his letter of transmittal refers to the "constant resort to this Island [Penang] of Siamese . . . and of Natives of other Indo Chinese Countries" from whom his informants were drawn.[41] Finally, it is not clear whether the indigenous maps referred to were prepared at Low's request—like the Burmese maps collected by Hamilton—or whether some existed before his inquiries. In 1830 Low published a second map during a sojourn in Calcutta. Despite its expansive title, "A Map of Siam, North Laos, Martaban, Tenasserim and Part of the Malayan Peninsula," this work was essentially an updated version of his initial effort of 1822. Among its sources Low notes, but does not itemize, "Numerous Burman and Siamese original Charts & Itineraries collected during the last ten years."[42]

Among peoples ethnically related to the Thais are the Shans, mainly living in eastern Burma, and the Laos. Numerous Shan maps are to be found in various British libraries, especially in the remarkable and now well-documented Scott Collection at Cambridge.[43] These maps are strikingly rich in detail and, despite certain commonalities, display considerable personal idiosyncrasy in style. Although most were made, as noted, at the behest of Scott, a British political agent in the Shan States, one senses, as with the maps collected by Hamilton and Burney, that many of the productions are not those of neophyte cartographers.

Although I am aware of no surviving Lao maps, we do have a brief description by a French explorer in Laos of how they were customarily made.[44] For the neighboring, but culturally distinctive, region of Cambodia, I know of neither surviving maps (two cosmographies excepted) nor, apart from the possible inclusion of Cambodian sources in Low's maps, textual references to maps. Cambodia was in a state of marked decline during the seventeenth to the nineteenth century and was periodically tributary to and ravaged by its more powerful Thai and Vietnamese neighbors. That mapmaking might not have been a serious Khmer concern in these circumstances is understandable.

It should be obvious from this account that the maps on which we can shed some light in this volume can in no way be regarded as a representative sample of Southeast Asia's total cartographic corpus over the indeterminate period when maps were made. I can say very little about the relative distribution of traditional Southeast Asian maps by provenance, period, or genre. Many factors contribute to this conclusion. Most important, perhaps, is that the rate of loss of maps will vary greatly by genre and because of differing environmental and historical circumstances. As was the case in South Asia, a great many maps can be presumed to have been destroyed by the ravages of climate, rot, vermin, fire, other accidents, and war. Intentional destruction of maps that had gone out of date or had lost ecclesiastical or political favor must also have played a role. The Thai institution of *chamra* is relevant in this context. Many maps were presumably of only ephemeral interest to begin with and

---

never finished. Many of the separate classifications of documents have been recataloged [by the National Library] several times since the manuscripts were first brought together."

39. On the former, see pp. 784–85; on the latter, see Kennedy, "Nineteenth Century Map" (note 11).

40. Low's letter is in the Oriental and India Office Collections, British Library, London (Straits Settlement Factory Records, Prince of Wales Island Public Consultations, vol. G/34/94, 442–45); the map is kept separately (Maps VII, 51). The letter is reprinted in Larry Sternstein, " 'Low' Maps of Siam," *Journal of the Siam Society* 73 (1985): 132–57, quotations on 132 and 133.

41. Quoted in Sternstein, " 'Low' Maps of Siam," 132 (note 40).

42. Quoted in Sternstein, " 'Low' Maps of Siam," 153 (note 40). The map is in the Oriental and India Office Collections (Maps VII, 52). Regrettably, for our purposes, Sternstein's aim in the article is to offer a critical review of Low's own cartographic efforts rather than of Low's sources, which were both indigenous and European. Hence he does not discuss or even speculate about the quality of whatever indigenous maps Low might have consulted. Nor does he indicate whether any such maps are known to have survived.

43. Dalby and Saimöng, "Shan and Burmese Manuscript Maps" (note 8).

44. "Im Innern von Hinterindien (nach dem Französischen des Dr. Harmand)," *Globus* 38, no. 14 (1880): 209–15.

were made with no thought to their preservation beyond the period for which they had some utility. Obviously, too, some maps are more prone to decay than others, given the fragile nature of their materials and the exposure of the surfaces on which they were drawn (especially in the case of cosmographies painted on temple walls).

In comparison with the known South Asian corpus, that of Southeast Asia is considerably better in respect to topographic maps but substantially less rich in plans of cities, forts, and temples. The relative lack of city maps, less marked for Burma than for other areas, may of course be partially attributed to the fact that until recently Southeast Asia was much less urbanized than the Indian subcontinent; but the disparity between the two regions in respect to the role of temples and other religious edifices, for which we have few Southeast Asian plans, was less pronounced. Also scarce in the Southeast Asian corpus are maps relating to pilgrimage, which in South Asia were relatively more important. This is not too surprising given that pilgrimage plays a substantially less important role in Southeast Asia than in South Asia. Moreover, for the Muslim population of Southeast Asia, roughly half the total, the principal places of pilgrimage are situated in the remote Middle East. Route maps, engineering maps, cadastral maps, and military maps each constitute a relatively small portion of the Southeast Asian corpus, in the case of military maps considerably less than for India. As in South Asia, celestial and nautical charts also form only a small part of the surviving works. Cosmographic maps, on the other hand, are abundant for both regions, though probably proportionally underrepresented among the works brought to light in this history. Finally, we have almost totally excluded from our purview astrological charts and other diagrams related to various forms of divination, which await the interpretation of scholars specializing in these subjects.

17 · Cosmography in Southeast Asia

JOSEPH E. SCHWARTZBERG

INTRODUCTION

The diversity of cosmographic conceptions encountered in Southeast Asia reflects the ethnic and religious diversity of the region's inhabitants. Alongside Southeast Asia's principal religious traditions of Buddhism and Islam and its locally dominant Christian and Hindu traditions, one encounters a bewildering variety of animistic beliefs. Each tradition, of course, has its own way of envisaging the cosmos, and some have given graphic expression to their cosmographic views in strikingly distinctive forms, while others seemingly have not. Of the myriad tribal and other ethnic groups who inhabit Southeast Asia, a great many have received only scant scholarly attention. Even so, the corpus of anthropological and related literature that might provide visual evidence of the indigenous cosmographies of the region is exceedingly large, and we have only scratched its surface. Among the many groups who have embraced Islam and Christianity, I have come across very little visual evidence of specifically Christian or Islamic components in the composite cosmographies that have sometimes emerged as a result of such conversions. For the Buddhist and Hindu population, however, and for certain animistic groups in Indonesia, we have discovered numerous interesting examples of ways of portraying the cosmos or specific aspects of it or of projecting ideas about the cosmos onto the terrestrial plane. These form the substance of this chapter.

I begin with tribal cosmographies, with particular reference to mortuary cults, divination, astronomy, and horology. I then take up the cosmographies of Buddhism, especially in its dominant Hinayana (Theravada) form, and the related cosmographies of Hinduism. (Separating the two in Southeast Asia is not always possible.) In dealing with those two great traditions, both ultimately of Indian origin, I first consider the underlying cosmographic conceptions of the two faiths. I then proceed to an exposition of the ways the universe as a whole and specific portions of it have actually been portrayed, in two and three dimensions. Next I consider astronomy and graphic examples of aids to geomancy and divination. Finally, I note the relation of cosmographic conceptions to the ordering of earthly human constructions, architectural, ecclesiastical, and political. Although I do not suggest that my presentation provides a representative, much less a complete, exposition of the wealth of surviving graphic cosmographies, I do dare to hope it conveys a sense of the remarkable range of relevant artifacts and suggests lines along which future attempts might be made to obtain a more comprehensive view of the subject.[1]

The distinction between tribal and nontribal societies in Southeast Asia is often far from clear. In general, scholars tend to consider as tribal all those groups whose religious beliefs might be described as animistic, as well as other groups who have only recently converted to one of the major world faiths: groups whose environments are only weakly linked with those of the dominant ethnic groups of the states they inhabit and who speak languages different from that of the politically dominant majority; groups whose populations are small (sometimes only in the hundreds and but rarely in excess of a million); and groups who depend mainly on shifting cultivation for their livelihood. But many tribal groups showing some of these attributes do not exhibit others, especially with respect to religion. A clear majority of the large group of Shans in Burma (Myanmar) and Thailand, for example, have long since become Buddhists, while most of the Minangkabaus of Sumatra, also a rather numerous people, are said to have become devout Muslims. Conversely, among many regionally differentiated segments of the

1. The only work, so far as I am aware, that tries to survey the many dimensions of cosmogony, cosmology, and cosmography longitudinally over the entire area that concerns us in this chapter is Horace Geoffrey Quaritch Wales, *The Universe around Them: Cosmology and Cosmic Renewal in Indianized South-east Asia* (London: Arthur Probsthain, 1977). This scholarly tour de force seeks to outline the origins and evolution of cosmological concepts among the region's tribal, Hindu, and Buddhist populations from prehistoric times to the modern era and traces many ideas to sources as remote as ancient Sumeria, the shamanic traditions of Siberia, and the prehistoric, megalithic Dong-son culture of the northern Southeast Asian mainland. The implicit lesson of such a study for the historian of cartography is the need to keep in mind that the origins of an idea that is being portrayed and the origins of the graphic or symbolic form through which it is conveyed at a particular period in history, though possibly related, are often two quite separate questions. The gap between the two will often be measured in centuries or even in millennia.

dominant national groups, who may profess the Islamic, Buddhist, or Christian faith, animistic beliefs and practices still form vital components of their worldview and cultural behavior.

Religious syncretism is, in fact, so common in Southeast Asia that only rarely does a people practice one of the great world faiths to the total exclusion of one or more earlier traditions. Moreover, not only have the allegedly "more advanced" groups retained animistic practices, but the clearly tribal groups appear to have borrowed, over the millennia since Indians first penetrated Southeast Asia, many aspects of Hinduism and Buddhism and adapted them in accordance with their own cultural proclivities. In time these groups reacted similarly to the intrusion of other alien faiths. This has unquestionably affected the cosmological systems of many tribal and quasi-tribal groups and shaped their mental maps of the cosmos, if not also the physical expressions of those mental maps. It follows that any search for the Ur-cosmography of a particular Southeast Asian people, even among groups who appear to be relatively isolated and primitive, will probably be in vain. Whatever tangible artifacts survive in the humid tropical conditions that characterize the region are likely to incorporate diverse cultural influences.

In reviewing creation myths of ostensibly Muslim villagers from what is now peninsular Malaysia, for example, Skeat refers to the following: the "Mountains of Caucasus . . . called . . . Bukit Kof," forming a wall to the earth protecting it from "excessive winds and beasts of prey"; the "people called Yajuj and Majuj (Gog and Magog)," who are boring through the wall and who, when successful, will bring an end to all things; "a great central mountain called Mahameru (Saguntang Maha Biru)." From another account, found in a Malay charm-book, Skeat describes "the Kaʿbah, which is the Navel of the Earth, whose growth is comparable to a Tree, . . . whose branches are four in number, and . . . extend unto the north, south, east, and west, where they are called the Four Corners of the World"; Gabriel, whom God asks to kill an anthropomorphic serpent, Sakatimuna, very probably "a reminiscence of the Indian 'Naga' "; the "Earth of the width of a tray" and "Heaven of the width of an umbrella." There is also reference to the earth's being "carried by a colossal buffalo upon the tip of its horns" and numerous other phenomena. The origins of these ideas can be traced to ancient Greek, Christian, Arabic, Iranian, Hindu-Buddhist, and Indo-Muslim sources.[2] Although I am not aware of any visual cosmographies from the Malay world in which these elements are evident, I would not be surprised to find them, given the richness of the plastic and graphic arts that characterize so much of the region and the lack of any strong taboo—despite the area's allegiance to Islam—with respect to the iconic representation of people, animals, and mythological beings.

Tribal Cosmographies

Cosmographies in Mortuary Cults

Pan-Indonesian concepts that are incorporated in many tribal cosmologies are those based on binary oppositions: life/death, humans/animals, village/forest, metal/cloth, male/female, warfare/farming, and so forth.[3] Also common are the ideas of the opposition of an upper world and a lower world, with earth between the two and the individual tribal territory, village, or house as a microcosmic analogue of the macrocosm.[4] Many animistic societies posit not only a multitude of spirits investing animate and inanimate objects in the world of humans, but also special spirit worlds separate from our own yet interacting with it. These spirit worlds, over various parts of Indonesia, include realms, which may be either upper worlds or lower worlds or both, to which the dead are ritually transported after appropriate intervals following their initial interment. Among several interior and coastal tribes on the island of Borneo (or Kalimantan, as Indonesians prefer to call it) such mortuary cults are associated with an elaborate cultural complex incorporating all of the attributes of tribal cosmology noted earlier. Constituting an important part of that complex are the elaborate secondary burial ceremonies during a season specially set aside for the purpose.

The largest and best-studied Bornean tribe associated with mortuary cults is the Ngaju Dayaks, who inhabit the regions along the Barito, Kapuas, and Kahayan rivers in the southern part of the island. Their cult has given rise to some of the most striking maps to emerge from the Malay world. Some of these are illustrated and discussed in a classic study that formed the doctoral thesis of a Swiss Protestant missionary, Hans Schärer, who lived among the Ngaju Dayaks and studied their religion from 1932 to 1939 and again from 1946 until his death in

2. Walter William Skeat, *Malay Magic: Being an Introduction to the Folklore and Popular Religion of the Malay Peninsula* (1900; London: Macmillan, 1960), 1–5. Most of the referents will be obvious to readers who have read volume 2, book 1, of this history. The idea of the world's being held up on the horns of a buffalo, for example, is clearly a variant of the Indo-Islamic view that the earth was supported on the horns of a bull, which in turn has an origin in Iran. See Joseph E. Schwartzberg, "Cosmographical Mapping," in *The History of Cartography*, ed. J. B. Harley and David Woodward (Chicago: University of Chicago Press, 1987–), vol. 2.1 (1992), 332–87, esp. 378.

3. Susan Rodgers, "Batak Religion," in *The Encyclopedia of Religion*, 16 vols., ed. Mircea Eliade (New York: Macmillan, 1987), 2:81–83, esp. 82.

4. Horace Geoffrey Quaritch Wales, "The Cosmological Aspect of Indonesian Religion," *Journal of the Royal Asiatic Society of Great Britain and Ireland*, 1959, 100–139.

1947.[5] Schärer's study includes four elaborate cosmographic maps, three of the Upperworld and one of the Underworld, together with more limited views of particular portions of the cosmos (e.g., a village of the dead) and of associated objects (e.g., the "soul boat" used to transport the dead to the lower world, other boats used by the supreme deities to make their voyage to earth, the "Tree of Life," an individual house within the Upperworld). Such drawings, says Schärer, were of great importance ritually. They were made by Ngaju Dayak priests "in the exercise of their functions" and laid "before them during the recitation of the chants in which the journey to these regions [the Upperworld and the Underworld] is described."[6] Such maps provide priests with a set of visual mnemonics as they chant, often over the course of several nights, the sacred hymns that will guide the spirits of dead males, accompanied by the *sangiang* (a class of demigods), to their final resting places.[7] Males go to the Upperworld and females to the Underworld. All the maps are said to "relate to the sacred events of early time, . . . [and] represent that which is handed down in the songs and priestly chants as divine revelation."[8] The examples Schärer referred to, however, were all allegedly made within fifty years before the publication of his book in 1946. No specific dates are given, and Schärer does not say when the custom of preparing such cosmographies began; but Stöhr advances reasons, based on subsequent writings by Schärer, for believing that all the maps he illustrates were made after the turn of the twentieth century.[9] In any event it seems clear that, despite their unique appearance, they were influenced to some degree by contacts with various outsiders, including the Dutch, who first penetrated the territory of the Ngaju Dayaks in 1836. Schärer does not specify the medium or the size of the maps, but those he illustrates were apparently drawn on paper with pencils and crayons of Western provenance. On some of the maps made known by him and others there is considerable text, much of it in Ngaju Dayak (which was not a written language before the advent of European missionaries), in cursive roman script. Since other maps nave no text at all, one may assume that the text was inserted either by missionaries or by Dayak informants at their behest to explicate the extraordinarily detailed map content. Two of the four maps Schärer illustrates, both annotated, are attributed to "Joh. Salilah," quite likely a convert to Christianity; the others, not annotated, are simply said to be by priests.[10]

A number of additional cosmographies are known to exist. Schärer himself refers to a renowned Ngaju Dayak priest and teacher, Massaid Singkoh, who lived on the middle Kahayan River and who, about 1900, first used maps to instruct his students. "These originally very simple drawings were later copied and amplified, and we now own numerous such works, some of which display much artistic feeling."[11] It is not known where these numerous examples are to be found at present, but a likely repository is the Baseler Mission in Basel, Switzerland, with which Schärer was affiliated until his death. I briefly studied one undated map of the Upperworld in a display on the Ngaju Dayak mortuary cult at the Völkerkundemuseum in Basel. Also in Switzerland, at the Völkerkundemuseum of the University of Zurich, is another undated map of the Upperworld. Of all the maps known, this one (to be discussed below) has been most thoroughly studied.[12]

At the Rautenstrauch-Joest-Museum für Völkerkunde in Cologne are two large cosmographic maps. These were bequeathed to the museum by the family of the German missionary Philipp Zimmermann, who worked in Borneo, mainly among the Ngaju Dayaks, from 1903 to 1914 and from 1920 to 1928.[13] The works were presumably

5. Hans Schärer, *Die Gottesidee der Ngadju Dajak in Süd-Borneo* (Leiden: E. J. Brill, 1946). That work has been translated into English as *Ngaju Religion: The Conception of God among a South Borneo People*, trans. Rodney Needham (The Hague: Martinus Nijhoff, 1963). A more succinct discussion by Schärer appears in "Die Vorstellungen der Ober- und Unterwelt bei den Ngadju Dajak von Süd-Borneo," *Cultureel Indië* 4 (1942): 73–81. Yet another work by Schärer is the posthumously published *Der Totenkult der Ngadju Dajak in Süd-Borneo*, 2 vols., Verhandelingen van het Koninklijk Instituut voor Taal-, Land- en Volkenkunde, vol. 51, pts. 1–2 (The Hague: Martinus Nijhoff, 1966). Other discussions of the Ngaju Dayak mortuary cult include Philipp Zimmermann, "Studien zur Religion der Ngadju-Dajak in Südborneo," *Ethnologica*, n.s., 4 (1968): 314–93 (published posthumously from a 1911 manuscript); two works by Waldemar Stöhr, "Über einige Kultzeichnungen der Ngadju-Dajak," *Ethnologica*, n.s., 4 (1968): 394–419 and 12 pls. (a commentary on the cosmographical drawings collected by Zimmermann), and "Das Totenritual der Dajak," *Ethnologica*, n.s., 1 (1959): 1–245 (a comparative analysis of the mortuary cults among all the Bornean tribes in which such cults are found); Verena Münzer, *Tod, Seelenreise und Jenseits bei den Ngadju Dajak in Kalimantan*, Lizentiatsarbeit, Universität Zürich, Philosophische Fakultät I, Abt.: Ethnologie (Zurich: Published by the author, 1976); and Wales, "Cosmological Aspect" (note 4). Works relating to mortuary cults among other Bornean tribes include Charles Hose and William McDougall, *The Pagan Tribes of Borneo: A Description of Their Physical, Moral and Intellectual Condition with Some Discussion of Their Ethnic Relations*, 2 vols. (London: Macmillan, 1912); and Peter Metcalf, *A Borneo Journey into Death: Berawan Eschatology from Its Rituals* (Philadelphia: University of Pennsylvania Press, 1982).

6. Schärer, *Ngaju Religion*, 11 (note 5).

7. Schärer, *Der Totenkult*, 441–42 (note 5).

8. Schärer, *Ngaju Religion*, 11 (note 5).

9. Stöhr, "Über einige Kultzeichnungen," 415 (note 5).

10. Schärer, *Ngaju Religion*, pls. 3 (Upperworld) and 4 (Underworld) (note 5).

11. Schärer, *Der Totenkult*, 441 (note 5). The date for Massaid Singkoh comes from Münzer, *Tod, Seelenreise und Jenseits*, 105 (note 5).

12. The Basel Völkerkundemuseum map is approximately 108 by 75 centimeters; no catalog number was noted. The map at the University of Zurich (catalog no. 15650) measures 98 by 43 centimeters.

13. Stöhr's foreword to Zimmermann, "Studien zur Religion der

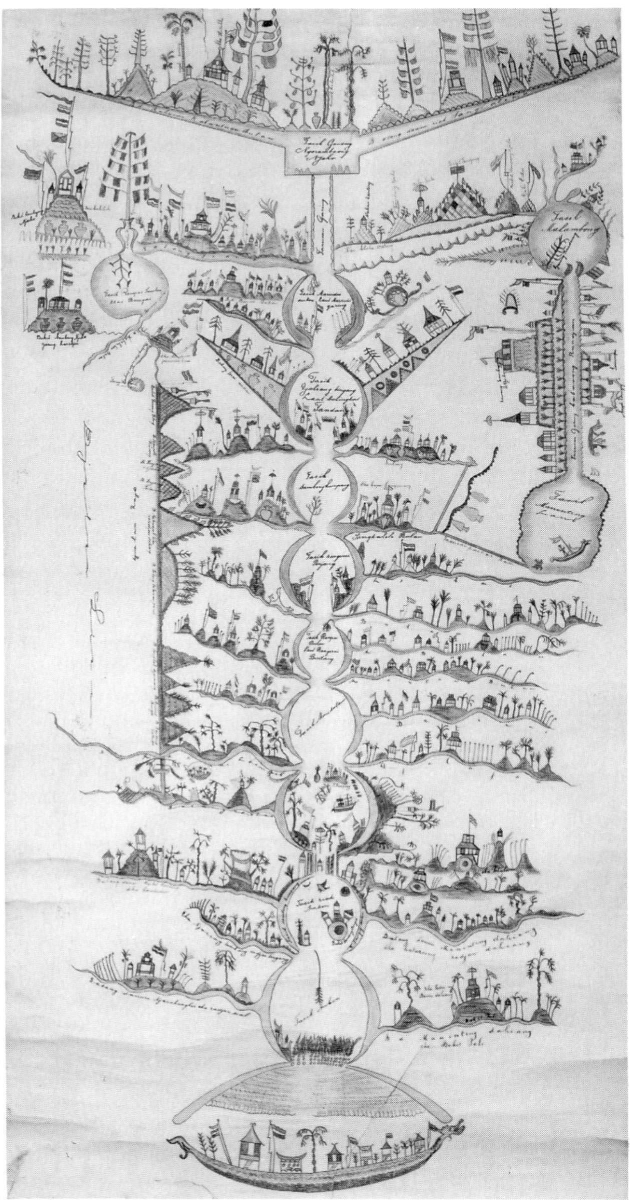

FIG. 17.1. NGAJU DAYAK MAP OF THE UPPERWORLD. From southern Kalimantan (Borneo), early twentieth century (ca. 1905); black ink and wax crayon on European paper. The map bears the notation *Djalan liau* (Route of the souls of the dead). At its base one sees the earth, or Middleworld, which the Ngaju Dayaks believe is carried on the back of a *naga* (world snake). Rising immediately above the earth is the thirty-one-layered firmament traversed by a route through foggy seas. Beyond lie the three lowest seas, whose surrounding territory, including that along the rivers flowing into them, is associated with six malevolent *rajas* who bring misfortune to the earth. Between this area and the uppermost part of the map lies a rather complicated assemblage of features within the domain inhabited by the *sangiangs* and other demigods. Here one finds rivers by which the soul ship must pass, a Route of the Rajas (the long vertical line toward the left margin), enchanted mountains, restorative mountains, a whirlpool of fire (*upper left*), and so forth. Near the top of the map is a rectangular sea into which two mighty streams flow and out of which flows a canal providing the only link to the seas below. Above these features lies the domain of Mahatala, whose magnificent castle is on a high hill in the upper left, as well as the houses of other gods among Trees of Life and, near the castle, an ivory coconut palm.
Size of the original: 138 × 69.5 cm. Courtesy of the Rautenstrauch-Joest-Museum für Völkerkunde, Cologne (cat. no. 51 288).

obtained by Zimmermann during the earlier of his two sojourns in Borneo. Stöhr infers, on stylistic grounds, that the maps predate those illustrated by Schärer, but beyond that he makes no attempt to date them.[14]

Also at the Rautenstrauch-Joest-Museum is a photograph of a very detailed but rather tattered map of the Upperworld. The provenance of the map is uncertain. The photograph has been in the museum since about 1967, but the present location of the map is unknown, and no information is now available as to how it was acquired. It is noteworthy on several counts: first, it appears to be on indigenous bark cloth rather than European paper; second, it bears an abundance of text in what appears to be the Arabic script, in which the Malay lingua franca used to be written (this is present both on the map proper and on two attached side flaps, which may have been used to wrap the map after it was folded); third, a number of its features are stylistically distinctive; and finally, it includes abundant depiction of anthropomorphic figures, presumably to represent deities rather than animal totems, which are the virtually universal norm on other Dayak maps.

Although they are arguably less aesthetically pleasing and less interesting, as well as simpler, than Schärer's maps, I have chosen to illustrate the aforementioned pair (figs. 17.1 and 17.2). Figure 17.1, depicting the Upperworld, shows the domain associated with the supreme deity, Mahatala, who is totemically represented by a bird, the hornbill. The following précis by Schärer conveys the flavor of Ngaju Dayak thought in respect to this portion of the cosmos, even though some particulars of what is described do not accord fully with what is indicated on the map caption:

> Mahatala lives on the primeval mountain in the Upperworld, which is raised above the world inhabited by

Ngadju-Dajak," 314; and Stöhr, "Über einige Kultzeichnungen," 415 (note 5).

14. Stöhr, "Über einige Kultzeichnungen," 414–15 (note 5). The similarity of the Zurich example to the one in figure 17.1 suggests that it too predates those of Schärer.

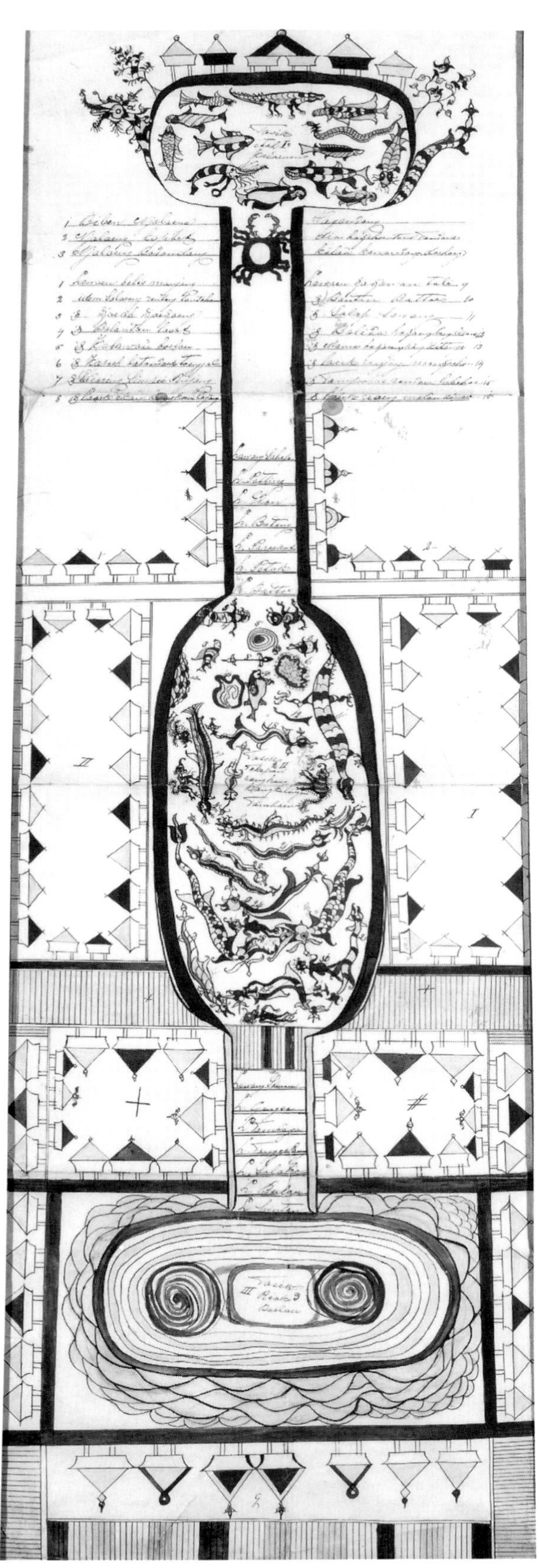

FIG. 17.2. NGAJU DAYAK MAP OF THE UNDERWORLD. Provenance and date are the same as for figure 17.1; black ink, watercolor, and wax crayon on European paper. The textual glosses on the map, largely illegible, are thought to have been written by the German missionary Philipp Zimmermann. In contrast to figure 17.1, this map is to be read from the top down. As in figure 17.1, earth, or the Middleworld, provides the place of departure. It is represented by a seaside village borne up by world snakes. The sea itself is enclosed by the head of a snake, signifying the identity of the two. Just below the entrance to the Underworld proper a giant crab stands watch. Since much of the writing left and right of the downward canal is illegible, its meaning cannot be discerned. Farther down are seven gates of successively harder and more precious materials, beginning with fallen leaves and ending with stone. Below these lies what could be the primeval sea in the shadow of the water snake, within which swim many distinctive water creatures, dependents of Jata. Seven more gates, successively of iron, yellow copper, red copper, tin, silver, gold, and diamonds, lead to a sea of golden waves surrounded by a village, the proper realm of Jata. Size of the original: 115 × 36 cm. Courtesy of the Rautenstrauch-Joest-Museum für Völkerkunde, Cologne (cat. no. 51 989).

> mankind. Its entrance, in the shape of a wide river, is reached by ascending through forty-two layers of cloud, each with its own name. The Upperworld is a faithful image of this world, but everything there is richer and more beautiful. On the many rivers and lakes there live the *sangiang*, descendants of two of the three brothers of the first human couple. These *sangiang* come to the aid of the third brother, who was left below on earth, on the occasion of all important religious ceremonies. . . . At the headwaters of the river of the Upperworld live the higher spirits, and on the primeval mountain, from which all rivers originate, Mahatala is enthroned.[15]

A more explicit, though far from complete, description based on Stöhr appears in the legend of figure 17.1.[16]

Of the Underworld, presided over by the goddess Jata, whose totem is a snake, Schärer has this to say:

> Jata lives in the Underworld (or primeval waters) lying beneath the world of men. It is approached in the vicinity of a village where a tributary joins the main stream, where the water is exceptionally deep. . . . By this way one enters the extensive villages under the

15. Schärer, *Ngaju Religion*, 16 (note 5). There is an unexplained disagreement between the "forty-two layers of cloud" in this description and the thirty-one-layered firmament shown in figure 17.1. The latter view could have been derived from some Hinayana Buddhist source region, since the cosmos as conceived by Hinayana Buddhism comprises just thirty-one stratified realms (discussed below in this chapter). In either case the striving for exactitude runs counter to what one might stereotypically expect of a "primitive" people. In fact, the elaborateness of the entire mortuary culture complex rivals the rituals of virtually any of the world's major religions.

16. Stöhr, "Über einige Kultzeichnungen," 415–16 (note 5).

> river where the subjects of Jata live and also Jata herself. The subjects of Jata are crocodiles. . . .
>
> Jata's village lies on the river of the Underworld called Basuhun Bulau, Saramai Rabia, i.e. the river of heaped-up gold, of washed-up gold dust.[17]

For a more complete picture, we must turn again to Stöhr, whose description provides the basis for the legend of figure 17.2.[18] In considering both figures, bear in mind that we are essentially analyzing their visual content, which, to the uninitiated, does not do justice to their underlying meaning. Although the latter merits deeper analysis, the issue is too complex to be adequately treated within this brief survey.

The most complete content analysis by far of any Ngaju Dayak cosmography is that provided by Münzer for the abundantly annotated map of the Upperworld in Zurich.[19] Münzer treats the map in six sections, providing a photograph for each to bring out the details. She discusses these under eighty-two numbered subheads, each keyed by number to the map. The actual number of features indicated is much larger, however. For example, rivers alone number 160. Despite her prodigious efforts, Münzer admits that many of her translations and interpretations of glosses on the map are conjectural. Of particular interest in her discussion are the various ways the map shows the Tree of Life, a very important symbol throughout much of Indonesia, which may be represented by a more or less naturalistic tree, by an umbrella, by a spear, or by a golden head covering, adorned with precious stones, for the god Mahatala.[20] Also noteworthy, and commented on by several authors, are the scores of flags that fly over many features of the map. Many of these resemble the Dutch tricolor, though other colors are used as well. Schärer warns us against assuming, as one writer did, that this was a Dayak acknowledgment of being under the protection of the Dutch. He notes that the Dayaks had pennants before the establishment of Dutch rule in Borneo and that their colors were associated with specific deities who were only aspects of the total godhead. Thus red, together with yellow or gold, probably signified the Underworld and Jata; white the Upperworld and Mahatala; and black the evil aspects of the godhead as a whole. Each village had a flag whose colors represented its own associated deity, and the flag and flagstaff together symbolized the Tree of Life.[21]

There is a great deal more to Ngaju Dayak symbolism, whether on maps or in other forms of sacred art. Much of it relates to the opposed male Upperworld and female Underworld principles and to their ritual union in numerous contexts on earth and in transporting souls to their final resting place. The soul boat, for example, has a prow shaped like the head of the hornbill representing Mahatala and a stern like the snake that signifies Jata.

Dayak belief systems were, it seems, remarkably plastic. Hence places were found in their maps of the Upperworld for separate rivers for Hindus, Arabs, Chinese, and Europeans and other foreigners as well as for the tribal villages of other strangers they came in contact with.[22] The rapid evolution of styles over the past century, very likely accelerated by access to new graphic tools and media, shows that the Dayak propensity for innovation is strong. Can we then hope to uncover some cosmography in which European influence has little or no trace?

Three undoubtedly cosmographic artifacts, foreshadowing those I have just discussed, are decorated bamboo containers on which are portrayed the Upperworld, the Underworld, and, between them, the earth. Two of these belong to the family of the missionary, Philipp Zimmermann. Of those, one is of Ngaju Dayak provenance (fig. 17.3) and the other from the nearby and related Ot Danum tribe. The third container, held in Leiden, bears so striking a similarity to that in figure 17.3 that Stöhr supposes both may have been made by the same individual, even though that the Leiden example was acquired in 1893 (how soon after its completion is not known) whereas that in figure 17.3, according to Zimmermann's notes, was made by a talented artist whom he obviously came to know between the time of his arrival in Borneo and 1911 (when the notes were written) and who died shortly after completing the work.[23] A full interpretation

17. Schärer, *Ngaju Religion*, 16–17 (note 5).

18. Stöhr, "Über einige Kultzeichnungen," 417–18 (note 5).

19. Münzer, *Tod, Seelenreise und Jenseits*, 105–27, with six photographic plates (note 5). The entire map is shown in an exhibition catalog by Elisabeth Biasio and Verena Münzer, *Übergänge im menschlichen Leben: Geburt, Initiation, Hochzeit und Tod in aussereuropäischen Gesellschaften* (Zurich: Völkerkundemuseum der Universität Zürich, 1980), fig. 134, with note on p. 185.

20. The issue of the association between a cosmic mountain as *axis mundi* and a Tree of Life serving a similar purpose has been much discussed and debated by cultural historians. Perhaps the most vigorous exponent of the primacy of the latter is Frederick Bosch, whose concern is with areas touched by Indian civilization. He sees the Meru-centered macrocosm as—at least from a Western perspective—"mechanically governed and built up from inanimate material. In contrast with this," he continues, "I take the view that this macrocosmic system in its turn is rooted in an even older and deeper soil; that it is the expression of a system not built up of lifeless matter but inspired within by Life itself. It is the organism of the Cosmic Tree of Life which set the example to all creation and left its mark on the conception of the greatest and most sublime of all things created, the macrocosmos." Frederick David Kan Bosch, *The Golden Germ: An Introduction to Indian Symbolism* (The Hague: Mouton, 1960), 231; originally published in Dutch as *De gouden kiem: Inleiding in de Indische symboliek* (Amsterdam: Elsevier, 1948). The point is made early in this chapter because, as the reader will observe, sacred trees and sacred mountains are so frequently conjoined in Southeast Asian cosmographies.

21. Schärer, *Ngaju Religion*, 25 (note 5).

22. Schärer, *Ngaju Religion*, 12–15 (note 5).

23. Stöhr, "Über einige Kultzeichnungen," 394–95 (note 5); a discussion of all three works appears on 394–99. Only brief remarks relate

FIG. 17.3. COSMOGRAPHIC ENGRAVING ON A BORNEAN BAMBOO CONTAINER. Ngaju Dayak, ca. 1905. The three registers depict the Upperworld, Middleworld (earth), and Underworld from top to bottom, not counting the purely decorative lower border. The creatures inhabiting the Upperworld and Underworld are largely known from the chants that accompany the mortuary ceremonies that form a prominent part of the religion of most Bornean peoples. Not all, however, can be identified here. Among the objects depicted are Trees of Life, the hornbill (the bird totem of the supreme deity, Mahatala), the spirit ship that transports the dead from earth to the Upperworld, and the *sangiangs* (demigods who guide the souls of the dead on their cosmic journey). The activities taking place on earth relate to the preparations for a feast that will inaugurate the days of ceremonial activity during which the sacred mortuary rites will be performed.

Size of the original: height 27.5 cm; computed circumference 15.1 cm. By permission of the Rautenstrauch-Joest-Museum für Volkerkunde, Cologne.

can be offered for none of the works, but all three show the major divisions of the tripartite Dayak cosmos in relation to one another and represent phenomena associated with the mortuary cult. Both the container in figure 17.3 and the Leiden container are incised, whereas the scenes on the Ot Danum container are executed in basrelief. The tools and processes used in making the three objects are not indicated. The two containers illustrated by Stöhr were derived from Zimmermann's own rolled-out impressions of the originals. They contain many of the same elements, from an iconographic perspective, but are stylistically rather different, the Ot Danum work being considerably less detailed and less meticulously executed than the Ngaju Dayak piece.

The meanings of many of the elements portrayed have not been ascertained, but most are clearly appropriate for the portions of the cosmos where they are placed. There are, however, a few elements in the Middleworld of the figure 17.3 cosmos that are shown—more properly in Stöhr's view—in the Upperworld in the Leiden example. Stöhr suggests the reason might be that in the former work the artist ran out of space. One such example is the palm tree surmounted by a lance (one of the many forms of the Tree of Life) on which is perched a huge fabulous bird (as big as a house in the local mythology) near the left margin of the middle register. Immediately to the right of this tree is a small structure identified as an ossuary. Farther to the right, a house on stilts extends from the bottom of the middle register into the register signifying the Upperworld. The proffered explanation in this case—as in many Indonesian cultures—is that the house in itself represents a microcosm uniting all three worlds. The attic thus partakes of the Upperworld even when the house itself is on the earth. Hanging in the attic are several vessels of the type Dayaks use to store sacred objects. In the right portion of the attic one sees an act of "sacral prostitution," a rite that is supposed to take place on the night before the commencement of the chanting that recites the journey of dead souls to the afterworld. Elsewhere in the middle register one sees other preparations for a ritual feast. Some of the figures in the picture are recognizable as priests and priestesses. Among the more cryptic elements is the tree near the right margin (seen again near the left margin because of the overlap in the illustration) that seems to be sprouting rifles and other weapons. This is among the very few evidences of European influence in the work.

In the Upperworld one is struck by the seemingly mundane nature of most of the elements portrayed. The explanation is that most of what we desire on earth is also to be found in the Upperworld, but in larger and more splendid form. Moreover, although hunting and fishing—presumably pleasurable activities—still take place there, the pursuit of the quarry is easier, which may be why one of the figures, at the far left (and again at the far right), is armed with a rifle. Although many of the animals and birds depicted can be identified, at least generically, others are fabulous. The more or less anthropomorphic figures are *sangiangs*, the demigods who assist in transporting spirits from earth to the Upperworld. Prominent elements of the upper register are the two giant Trees of Life, with their lower, fruit-bearing limbs intertwined; the large hornbill, the symbol of the supreme deity, Mahatala, perched between the lower and upper limbs of the tree on the right; and the hornbill-prowed spirit boat that transports the dead, seen in the triangle on the left.

The lowest register in figure 17.3 (not counting the purely decorative border), representing the Underworld, is in some ways the hardest to interpret. Its aquatic denizens are, of course, appropriate for the domain of the goddess Jata, whose own form is that of a water snake. That some of them are swimming into a weirlike fish trap is tentatively explained in that fish form the principal food for those who live in the Upperworld, and they are here offering themselves as such for a specific ceremonial occasion.

Stöhr illustrates one additional Ngaju Dayak bamboo container from Zimmermann's collection that depicts a scene of strife between *sangiangs* and other mythological figures.[24] The style resembles that in figure 17.3. Although the work does contain some cosmological elements, the arrangement is less orderly than in the work just discussed, and no case could be made for calling it a map.

Were it not for the scholarship of individuals such as Schärer, Zimmermann, and Stöhr, one might reasonably infer that maps as detailed as those of the Ngaju Dayaks must have evolved over a rather long period. Yet the contrary seems to be true. That does not mean, however, that cartography was alien to Borneo before the advent of Europeans. In what remains of my discussion of that region, I shall present evidence suggesting that mapping probably did form a part of the tribal tradition and that the thinking of Dayaks incorporated a strong element of

to the third container, held by the Rijksmuseum voor Volkenkunde, Leiden (inventory no. 942132), and it is not illustrated by Stöhr. Notices and illustrations are provided in several other works, however: Hendrik Herman Juynboll, *Borneo*, 2 vols. (Leiden: E. J. Brill, 1909–10), 2:376–77; J. A. Loebèr, "Merkwaardige kokersversieringen uit de zuider- en oosterafdeeling van Borneo," *Bijdragen tot de Taal-, Land- en Volkenkunde van Nederlandsch-Indie* 65 (1911): 40–52; and B. A. G. Vroklage, "Das Schiff in den Megalithkulturen Südasien und der Südsee," *Anthropos* 31 (1936): 712–57, esp. fig. 13. The similarity to figure 17.3 is, as Stöhr has indicated, remarkable. The relevant notes, however, are exceedingly meager and do not relate to cosmography.

24. Stöhr, "Über einige Kultzeichnungen," pl. XX, description on 399 (note 5).

cartographic imagery—that is, that they possessed a keenly developed sense for making mental maps.

Figure 17.4 reproduces a map made by a chief of the Madang tribe of Sarawak, presumably early in the present century, to portray "the land of the shades [dead souls] and of the country traversed by the ghost on its journey thither." According to Hose and McDougall,

> This was done in the way maps of their own country are always made by the Borneans, namely, he laid upon the floor bits of stick and other small objects to represent the principal topographical features and relations. We tested the trustworthiness of his account by asking him to repeat it on a subsequent occasion; when he did so without any noteworthy departure from the former description.[25]

In considering the Madang map in the larger context of Bornean mortuary cults, Hose and McDougall observe that it conforms to the essential general view that "the land of shades is the basin of a river divided by a mountain ridge from that which the ghost departs."[26] Noteworthy in the first quotation are the use of the word "always" and the absence of any suggestion that the chief's accomplishment was at all exceptional.

In his discussion of the mortuary practices of the very small (1,600 or so souls) Berawan tribe of Sarawak, Metcalf makes clear the geographic specificity of the chants by which the souls of the dead are guided on their upstream journey toward the land of the dead. The initial part of this journey leads to a high mountain divide, analogous to the mountain range shown in figure 17.4, beyond which the route becomes mythical. The Berawan rite is recitative. The lead singer repeatedly asks the assembled community, "What place is this, you live ones?" to which the community replies with the names of successive sidestreams upriver from the longhouse ("The place of *long* [name] . . . , you dead one"). The *longs* (river mouths or confluences) so named along the route may be as little as a few hundred yards from one another. "So," says Metcalf,

> the soul's journey continues, verse by verse, rivermouth by rivermouth. . . . For the Berawan, the river provides their principal highway. They use it to get to their farms, and to travel to other villages. Consequently, they know all its twists and turns and sidestreams intimately. This journey was one that everyone present could readily imagine. . . .
>
> But this is not merely a journey in space, it is also a journey in time. Every time the site of a previous longhouse is passed that place is referred to . . . just as [it was when it first became known]. Consequently, the song reiterates the migration route of the ancestors of the community.[27]

One might say, then, that the song gives expression to a consensual mental map of the historical geography of the Berawans, a part of which lies in the real world and another part, more vaguely limned, in the world of myth. One might further hypothesize that the type of ephemeral map the Madang chief created for Hose could have been provided by members of any number of other Bornean tribes among whom chants relating to the journey of the dead form part of the local mortuary cult.

Finally, note that, although the Berawan example indicates a mental map that is essentially linear (i.e., a route map), there are reasons to suppose that most groups share more multidimensional mental maps, as did the Madang chief. In discussing the "sacred land," that is, the territory collectively occupied by the Ngaju Dayaks, Schärer makes it clear that they are sensitive to both its mythical and its real-world boundaries. Lying between the Upperworld and the Underworld and resting on the back of

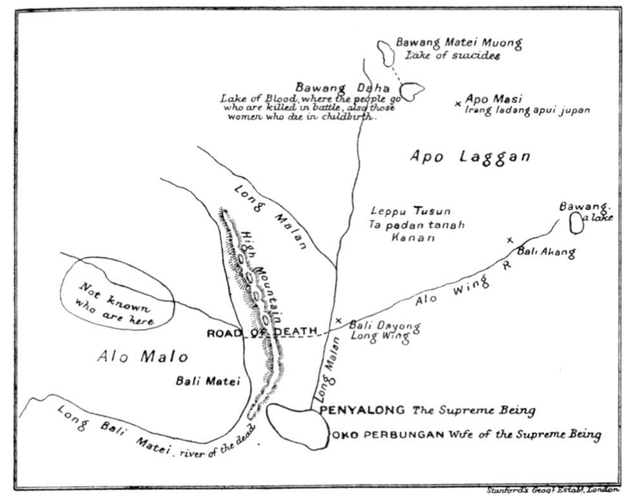

FIG. 17.4. COPY OF EPHEMERAL MADANG MAP OF THE "LAND OF SHADES" AND OF THE WAY THERE. The original of this map was made in the early part of the twentieth century from bits of stick and other small objects and was copied by a colonial administrator at whose behest it was made. Although the details differ notably from those of more or less contemporaneous Ngaju Dayak maps, the basic ideas in this map—that there is a land, or lands, for dead souls and a difficult and perilous way to get there—are in general agreement with the belief systems of tribes throughout Borneo whose religion centers on an elaborate mortuary cult. The administrator for whom the map was made found the work in no way extraordinary.
From Charles Hose and William McDougall, *The Pagan Tribes of Borneo: A Description of Their Physical, Moral and Intellectual Condition with Some Discussion of Their Ethnic Relations*, 2 vols. (London: Macmillan 1912), vol. 2, fig. 78.

25. Hose and McDougall, *Pagan Tribes*, 2:43–44 (note 5).
26. Hose and McDougall, *Pagan Tribes*, 2:44 (note 5).
27. Metcalf, *Borneo Journey into Death*, 216–17 (note 5).

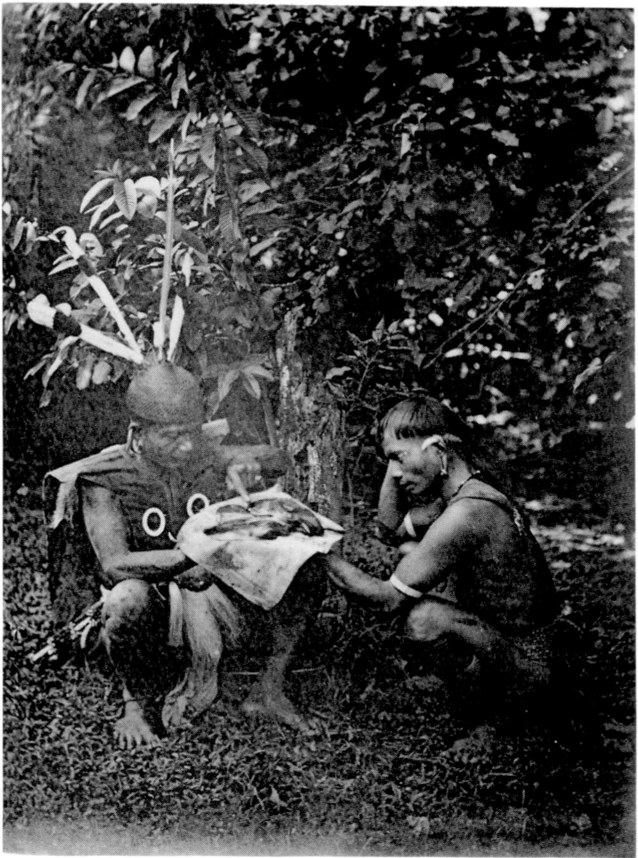

FIG. 17.5. KAYAN TRIBESMEN OF SARAWAK EXAMINING A PIG'S LIVER FOR PURPOSES OF DIVINATION. Early twentieth century. This practice is (or was) associated with tribal groups throughout Borneo. Among some groups, the various lobes of the ventral side of the liver were believed to represent specific geographic regions or spirit worlds with respect to which knowledge of the future was sought. See also figure 17.6.
From Charles Hose and William McDougall, *The Pagan Tribes of Borneo: A Description of Their Physical, Moral and Intellectual Condition with Some Discussion of Their Ethnic Relations*, 2 vols. (London: Macmillan, 1912), vol. 2, pl. 159.

Jata, the water snake (see fig. 17.2), the sacred land is "bounded by the raised tail and head of the deity of the Underworld."[28] This land, the primeval village Batu Nindan Tarong, becomes conflated with the Ngaju Dayaks' own terrestrial home villages, each regarded by its residents as "the largest and most beautiful place in the whole world," where alone "peace, safety, happiness, and the good life are to be found."[29] As Ngaju Dayaks move through the space beyond the physically marked boundaries of their sacred land, including the worlds of the Indians, Chinese, Arabs, Europeans, and other friendly Dayak groups, all of whom have been assigned visible places in their divine cosmographies, they exhibit forms of ritual behavior appropriate to the meanings attached to such places.[30]

Malay mortuary cults extend as far west as Madagascar, and research on that island may reveal cosmographies related to those discussed for Borneo. For a less distant aboriginal Malay culture, that of the Semelais in the interior of the West Malaysian states of Perak and Pahang, Gianno has been able to draw a rather elaborate map of the Underworld and describe the accompanying beliefs with respect to the means by which the dead reach that domain. Regrettably, however, she does not indicate whether her map is copied from an original Semelai artifact or is a reconstruction based on a wholly oral account.[31]

GRAPHIC AIDS TO DIVINATION

Many methods of divination are practiced in tribal Southeast Asia, where the belief that the future can be foretold is widespread, if not universal. Here I can do no more than point out briefly a few of the methods that embody a cartographic component.

To begin, one need look no further than some of the Bornean tribes considered above in connection with mortuary cults. Among these tribes the markings on the liver of a ritually sacrificed pig are commonly consulted to provide information on the future course of events. Indeed, Hose and McDougall aver that this is a pig's "most important function."[32] Figures 17.5 and 17.6 illustrate how various parts of the liver were taken to represent particular regions and peoples in a given divination rite. "The various lobes and lobules are taken to represent the various districts concerned in the question on which light is desired, and according to the strength and intimacy of the connections between these lobes, the people of the districts represented are held to be bound in more or less lasting friendship."[33]

There is remarkable similarity between the divination scheme just described and that surmised for the "Bronze Liver of Piacenza," an Etruscan object dating from the third century B.C. (though the features shown on the latter appear to be mainly extraterrestrial), and possibly also that for an even older Chaldean terra-cotta liver in the British Museum.[34] Since divination from animal entrails

28. Schärer, *Ngaju Religion*, 60 (note 5).

29. Schärer, *Ngaju Religion*, 61 (note 5).

30. Schärer, *Ngaju Religion*, 64–65 (note 5).

31. Rosemary Gianno, *Semelai Culture and Resin Technology*, Memoirs of the Connecticut Academy of Arts and Sciences, vol. 22 (New Haven, 1990), map 4, with key on p. xi and discussion on 46–48.

32. Hose and McDougall, *Pagan Tribes*, 2:61 (note 5).

33. Hose and McDougall, *Pagan Tribes*, 2:62 (note 5).

34. Both of these are discussed, and the Piacenzo liver illustrated, in O. A. W. Dilke, "Maps in the Service of the State: Roman Cartography

is widespread, it would not be surprising to find comparable objects that would help fill the enormous temporal and spatial gap between the ancient Western and modern Bornean examples.

To the southwest of the Kayans, among the Ibans, sometimes referred to as Sea Dayaks, similar practices apply, but here the territories denoted are not all of this earth. The gallbladder is of particular importance in that it divides the liver's left central lobe, representing the territory of men, from the right central and right lateral lobes, associated with the domain of the spirits. At the base of the gallbladder is a ligament that appears to function as a bridge between the lobes at either side, between men and benevolent spirits, thereby providing the latter with access to the afflicted in times of sickness.[35]

Several types of divination totally different from the one just described are found among the Bataks of northern Sumatra. The Bataks are a congeries of tribal groups whose cosmology in many ways resembles that of the Dayaks, but who have had considerably more contact with nontribal peoples over a very long period and who have developed—as few other tribal groups have done—their own written language and literature. Modes of Batak divination are discussed in considerable detail in a doctoral thesis in theology by Parkin, who notes that similar, though simpler, systems are followed in Javanese and Malay societies. The Batak system, of indisputable Hindu derivation, is both spatial and horological.[36] I shall consider only its visual aspect.

Parkin's thesis unfortunately provides no photograph of the designs drawn by Bataks in their divination practices, though it does include some simple black-and-white diagrams of them. The key element in most of these is an eight-armed figure, the *desa na ualu*, showing the eight principal points of the compass. In these diagrams north and south are equated with "upstream" and "downstream," respectively, while east and west signify "rising" and "setting," as with the sun. These directions describe a cosmic field through which move not only people but also the *naga* (snake), "which controls space and time" and "whose movements determine the course of history in the middleworld." (For some groups a dragon, scorpion, or turtle takes the place of the snake.) It is believed that the *naga* "begins his annual circumlocution of space in the East and spends three months in each of the cardinal points in turn."[37] The *bindu*, or point from which the eight arms radiate, is the place where cosmic power is concentrated.[38]

The drawing of the *desa na ualu* often accompanies an animal sacrifice, the design being laid out at the base of the sacrificial pole to which the animal to be slaughtered is tethered. Events or ceremonies entailing a spatial risk or hazard (e.g., establishing new villages, planning new irrigation systems, or preparing for battle or for legal actions) can be carried out only after drawing the requisite diagram. This cosmogram

> is not a direction-finding compass, it is not merely an instrument to ascertain required information, it is, rather, the magico-religious re-enactment of space "in illo tempore." . . . It re-establishes, in visible and local form, the primeval focal point of totality power. "Salvation," or success, depends upon bringing the local event into harmony with that special spatial kratophany.[39]

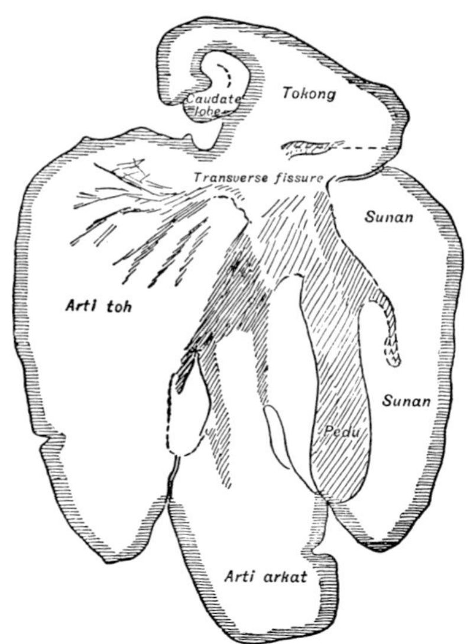

FIG. 17.6. VENTRAL SIDE OF PIG'S LIVER FROM WHICH OMENS ARE READ. The several parts, the local Kayan terms for which are provided here, are taken as signifying the territories about which information is sought. The future relations between the inhabitants of these territories are foretold by examining the ligaments, fissures, and various markings connecting or separating the parts in respect to both their distribution and their state. In one augury discussed by Hose and McDougall (*Pagan Tribes*, 2:63), three districts and one specific village were recognized. The short gallbladder (*pedu*) separating two of the districts identified was seen as a good omen, as was the large development of the caudal lobe and the strength of the ligaments. See also figure 17.5.
From Hose and McDougall, *Pagan Tribes*, vol. 2, fig. 79.

to the End of the Augustan Era," in *The History of Cartography*, ed. J. B. Harley and David Woodward (Chicago: University of Chicago Press, 1987–), 1:201–11, esp. 202–4 and figs. 12.2 and 12.3.

35. Erik Jensen, *The Iban and Their Religion* (Oxford: Clarendon Press, 1974), 138–39.

36. Harry Parkin, *Batak Fruit of Hindu Thought* (Calcutta: Christian Literature Society, 1978), esp. chap. 8, "Magic and Divination," 199–217. A brief account of the fundamentals of the Javanese system of divination appears in Clifford Geertz, *The Religion of Java* (Glencoe, Ill.: Free Press, 1960), 32–35, and Skeat's *Malay Magic*, 532–80 (note 2), deals with a wide variety of Malay divinatory practices.

37. Parkin, *Batak Fruit of Hindu Thought*; quotations on 199–200 (note 36).

38. Parkin, *Batak Fruit of Hindu Thought*, 207 (note 36).

39. Parkin, *Batak Fruit of Hindu Thought*, 201 (note 36). Another

FIG. 17.7. KENYAH DAYAK TRIBESMEN OF BORNEO MEASURING THE LENGTH OF THE SHADOW OF A GNOMON AT NOON TO DETERMINE THE TIME FOR SOWING PADDY.
From Charles Hose and William McDougall, *The Pagan Tribes of Borneo: A Description of Their Physical, Moral and Intellectual Condition with Some Discussion of Their Ethnic Relations*, 2 vols. (London: Macmillan, 1912), vol. 2, pl. 60.

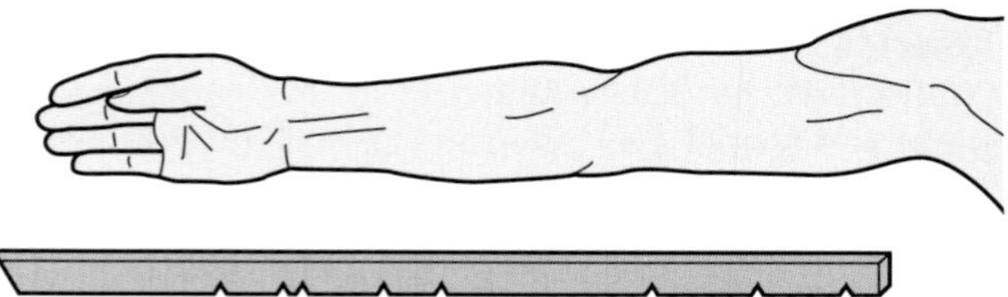

FIG. 17.8. BASE OF GNOMON EMPLOYED BY KENYAH DAYAKS. Notches indicate times for conducting various agricultural operations. The periods when the midday sun's shadow falls between successive notches are those that are appropriate for specific tasks, based on prior experience of the seasons. These notches are calibrated against specified lengths of portions of the arm of the professional village weather clerk.
After Gene Ammarell, "Sky Calendars of the Indo-Malay Archipelago: Regional Diversity/Local Knowledge," *Indonesia* 45 (1988): 84–104, esp. 87.

What follow in Parkin's account are notes on the relation of the Batak system of divination to others practiced in the Malay world and in India;[40] a list of some of the rules according to which movements in specific directions are prescribed or proscribed at particular times; and copies of diagrams that are variations on the basic *desa na ualu* motif, with appropriate commentary.

Another type of diagram, several variations of which Parkin discusses, is an elaboration of the *bindu* from a simple point to a *bindu matoga*, "the authoritative point of power," represented as a double-square diagram (a diamond placed within a square so as to bisect all four of its sides, a symbol signifying the Middleworld). Such diagrams are drawn in red, black, and white, colors that represent the cosmic trinity, and they form the field around which dances of cosmic renewal are performed. Added to the diagram are cosmic symbols such as the *naga*, Trees of Life, an egg, axes and adzes, and others. One such drawing is described as having sides one meter long and being drawn on the ground outside the house of the person for whom the rite is performed. A particular side had to be parallel to the front of the house and a particular corner had to be toward the east.[41] I note these matters, without further elaboration, simply to demonstrate the sorts of considerations that enter into the cosmographic thinking of supposedly simple tribal societies and that inform the expression of such thought in a host of cosmographic diagrams of which we at present have only rudimentary and fragmentary knowledge.

ASTRONOMY AND HOROLOGY

Some knowledge of astronomy, however rudimentary, may reasonably be assumed among virtually all peoples whose lives are regulated by the seasons and who look to the heavens for indicators of their progression. Not sufficiently understood, however, are the ways various nonliterate societies express and transmit graphically the astronomical knowledge they possess and the specific forms in which such expressions are rendered. Given the

account, differing in minor respects from that of Parkin, is that of Anicetus B. Sinaga, *The Toba-Batak High God: Transcendence and Immanence* (Saint Augustin, West Germany: Anthropos Institute, 1981), 127–32. This account, for example, notes that a tortoise, not a snake (though still called *naga*), is buried beneath the sacrificial pole to represent the Underworld, that the Middleworld is symbolized by an octagon, and that the hornbill represents (as among the Dayaks) the Upperworld. Both accounts agree on the necessity to draw the cosmic design on the ground at the base of the slaughtering pole, which represents the Tree of Life.

40. Indian divination charts are discussed in Schwartzberg, "Cosmographical Mapping," 343–51 (note 2).

41. Parkin, *Batak Fruit of Hindu Thought*, 207–9 (note 36).

hundreds of tribal groups inhabiting Southeast Asia, one cannot hope to do justice to that subject in a work not devoted exclusively to astronomical lore, or even to sketch adequately the limited knowledge about it that modern scholars have thus far been able to acquire. Suffice it to say that it has been established that numerous tribal groups in Southeast Asia do recognize and portray some of the more prominent constellations and do understand and chart, after a fashion, regularities in the motions of the sun and moon.

The first attempt to summarize the state of astronomical knowledge among peoples of the Malay Archipelago was that of Maass in 1924.[42] Maass's study made no special distinction between tribal and nontribal or literate and nonliterate peoples and, not surprisingly, was devoted mainly to detailing the knowledge and beliefs of the former groups, a large part of which is actually astrological and not central to our present concerns.

A much briefer attempt at synthesis, concentrating on what might be considered tribal peoples, was put forward by Ammarell in 1988. Ammarell observes that the Malay Archipelago reveals a "tradition that is, at once, unique to this cultural area and richly diverse in its local variation." He classifies Malay observations of celestial phenomena under three broad rubrics: those that depend on measuring the altitude of the sun at noon, those that relate to annual changes in the apparitions of familiar groups of stars; and those that trace cyclical changes in the phases of the moon.[43]

An illustration of a method falling under the first of Ammarell's three general rubrics is provided by figures 17.7 and 17.8. This method, derived from the previously cited study of Bornean tribes by Hose and McDougall, is but one of several Bornean ways of determining the time for sowing paddy. The tribe in question, the Kenyah Dayaks, relies on observing the changing lengths of shadows cast by a gnomon specially constructed for the purpose. Another Bornean group, the Kayans, relies on noting the position of a midday sunbeam (*kleput doh*, "the blowpipe of the spirit") shining through a hole in the roof into the chamber of the tribal weather prophet in the communal longhouse. Still other groups employ altogether different techniques based on stellar observations.[44]

Among the various types of gnomons that Ammarell reports and illustrates, a particularly sophisticated and accurate type, called a *bencet*, was in use in Java from about 1600 to 1855. Derived from the *bencet* was a unique Javanese calendar of twelve unequal months, ranging in length from twenty-three to forty-three days.[45]

Stellar observational techniques, according to Ammarell, are of two types, one based on fixing the heliacal apparitions of individual stars or of constellations, the other based on observing their culminations. Heliacal apparitions refer to stellar positions with respect to the horizon when they are first or last visible in the twilight of dawn or of dusk, while heliacal culminations refer to the times when particular stars transit the meridian (are at their highest altitude) at either dawn or dusk. Observations of heliacal culminations for calendrical purposes appear to be confined to the Malay cultural realm.[46]

An example of the way two constellations, equivalent to Scorpio in Western astronomy, are perceived by the seafaring Bugis of the island of Sulawesi (Celebes) is given in figure 17.9. Maass provides comparable illustrations for other Malay peoples. Far more complete than either of these two expositions, however, is the depiction in figure 17.10 of the entire celestial vault as seen from the island of Palawan in the Philippines, comparing the Palawan and Western systems.[47] It is not known when, why, or even whether members of various Malay ethnic groups had occasion to draw diagrams of constellations without being asked by outsiders to do so; but where there is agreement from one group to another in the manner of their perception, as often appears to be the case, it suggests that constellations must occasionally have been graphically rendered, since it is difficult to imagine how knowledge of them could otherwise be transmitted from one nonliterate society to another speaking a different language. Similarly, transmission from one generation to another within a given society also argues for constellations' being drawn in some simple form (e.g., scratched on a leaf, outlined in the sand).

42. Alfred Maass, "Sternkunde und Sterndeuterei im Malaiische Archipel," *Tijdschrift voor Indische Taal-, Land- en Volkenkunde* 64 (1924): 1–172, 347–460, and appendix in 66 (1926): 618–70.

43. Gene Ammarell, "Sky Calendars of the Indo-Malay Archipelago: Regional Diversity/Local Knowledge," *Indonesia* 45 (1988): 84–104; quotation on 86. Curiously, Ammarell does not cite Maass among his many sources, most of which postdate Maass.

44. Ammarell, "Sky Calendars," 90 (note 43); and Hose and McDougall, *Pagan Tribes*, 1:105–9 (note 5).

45. Ammarell, "Sky Calendars," 90 (note 43).

46. Ammarell, "Sky Calendars," 91, 95, and illustrations and captions on 92–93 (note 43). Not all the ways described entail graphic methods of recording time. I have confined the discussion to those that utilize a graphic solution.

47. The system of Palawan astronomy is discussed in considerable detail in Nicole Revel, *Fleurs de paroles: Histoire naturelle Palawan*, 3 vols. (Paris: Editions Peeters, 1990), vol. 2, *La maîtrise d'un savoir et l'art d'une relation*, 189–242. The work includes numerous illustrations by Anna Fer. In addition to mapping the entire night sky as seen from Palawan, this remarkably thorough work includes the configurations of numerous individual constellations and their figurative equivalents in the Palawan view and also provides a map of the regions of the Palawan sky that correspond to five motifs of Palawan mythology.

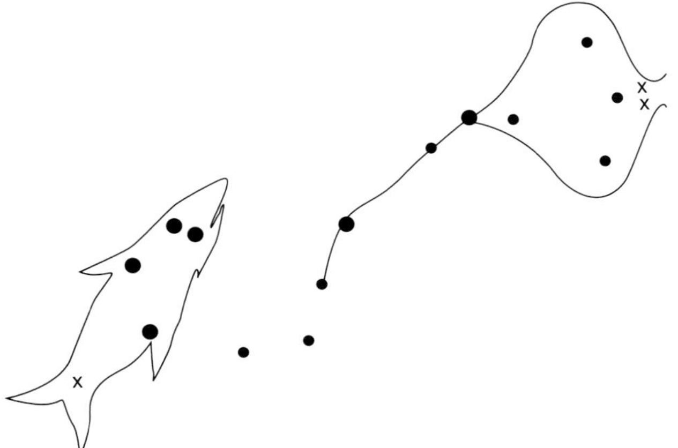
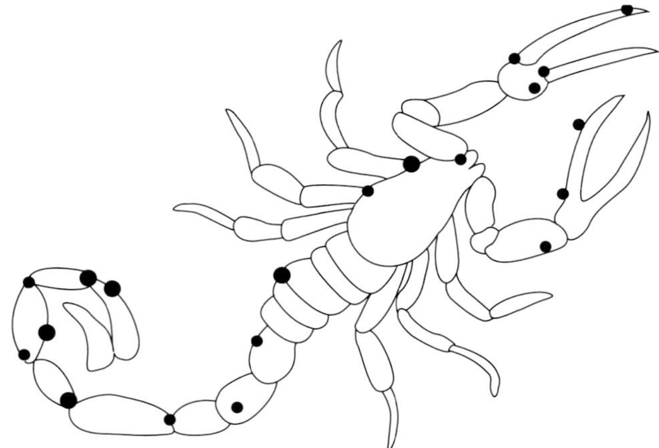

FIG. 17.9. BUGINESE PERCEPTION OF THE STARS CONSTITUTING THE INTERNATIONALLY DESIGNATED CONSTELLATION SCORPIO. Though most of the same stars are included in each of the two views, those that appear to Western observers as a single cluster, internationally designated Scorpio (*right*), are seen by Bugi mariners as two separate asterisms, one perceived as a ray, the other as a shark (Xs indicate stars drawn by informants but not pointed out in the sky). After Gene Ammarell, "Navigation Practices of the Bugis Seafarers" (paper presented at the Forty-first annual meeting of the Association for Asian Studies, Washington, D.C., 18 March 1989).

Buddhist and Hindu Cosmographies

Underlying Conceptions

In an earlier volume of this history, I outlined a number of attributes of certain cosmological conceptions that were common to the Hindu, Buddhist, and Jain traditions.[48] These traditions flourished and coexisted in India for more than one and a half millennia. Two of them, the Hindu and the Buddhist, also took root in Southeast Asia by the beginning of the Christian Era and subsequently diffused over much of the region. Thus the basic cosmological perspectives associated with India were embraced first by the elite and ultimately by the masses in the lands to India's east. Hence I need not repeat here in detail earlier discussion of the development and nature of Indian cosmology in general. Rather, I shall provide only an initial synoptic overview and then discuss certain particularities of the specifically Hinayana Buddhist views that now predominate over most of mainland Southeast Asia.[49] Of the formerly important Mahayana cosmology, which still informs Vietnamese Buddhism, I shall say virtually nothing. In brief, Indian cosmology, whether in its home region or in Southeast Asia, can be characterized as follows:

1. The cosmos is vast, exceedingly complex, and composed of a very large number of discrete constituent units whose dimensions and shapes are precisely specifiable.

2. There is a multiplicity of universes. (In Hinayana Buddhist cosmography their number is infinite.)

3. Each universe is centered on its own gigantic Mount Meru (Sumeru in the Pali canon of Hinayana tradition), which, as *axis mundi*, runs through the center of the earth and unites it to the realms above and below. Meru itself comprises several vertically stratified layers that, like the surrounding universe, have their characteristic degrees of merit, denizens, plants, and so forth. Ranged about Meru are buttress ranges of mountains, terrestrial continents lying in the four cardinal directions, concentric ring continents, and intervening ring oceans.

4. Each universe within the cosmos is vertically structured. (Consequently, visual representations of a universe are often rendered on a vertical plane rather than on the horizontal plane that characterizes most terrestrial maps.)

5. The fundamental division of the universe is tripartite. The arrangement of the major divisions, however, and the number of subdivisions within each basic division, varies from one system and region to another. (Variations also apply to other aspects of the cosmos.)

6. The motions of the sun, moon, and stars within the universe and their spatial relation to one another and to other parts of the universe are specified.

48. Schwartzberg, "Cosmographical Mapping," 332–43 (note 2).

49. The most comprehensive secondary source on Buddhist cosmography is Willibald Kirfel, *Die Kosmographie der Inder nach Quellen dargestellt* (Bonn: Kurt Schroeder, 1920; reprinted Hildesheim: Georg Olms, 1967; Darmstadt: Wissenschaftliche Buchgesellschaft, 1967). Also exceedingly useful is W. Randolph Kloetzli, *Buddhist Cosmology, from Single World System to Pure Land: Science and Theology in the Images of Motion and Light* (Delhi: Motilal Banarsidass, 1983), which, apart from its own original analysis, contains a valuable guide to the primary and secondary literature.

A brief overview will be found in W. Randolph Kloetzli, "Buddhist Cosmology," in *The Encyclopedia of Religion*, 16 vols., ed. Mircea Eliade (New York: Macmillan, 1987), 4:113–19. For a comparative historical and cultural perspective, Wales, *Universe around Them* (note 1), is recommended.

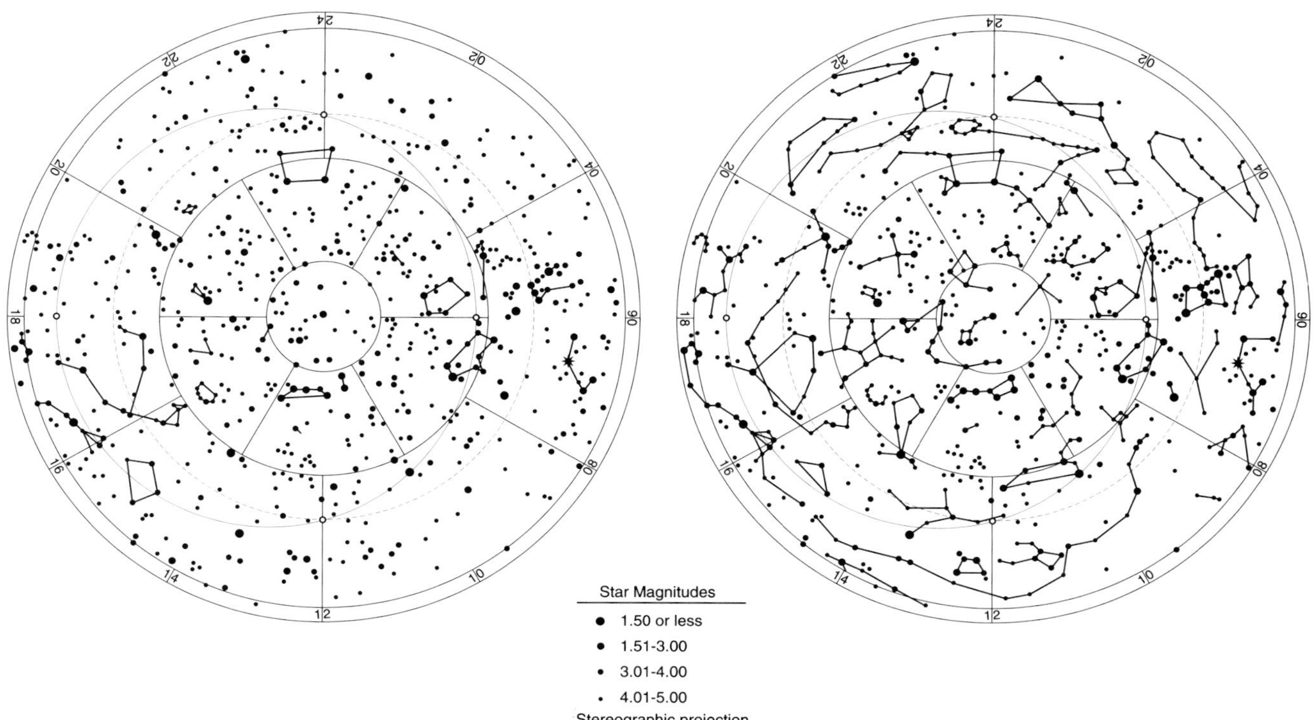

FIG. 17.10. CONSTELLATIONS OF THE NORTHERN HEMISPHERE AS SEEN FROM PALAWAN. The constellations perceived by the Palawan Islanders (*left*) are much less numerous than those known to Western astronomy (*right*). Although they tend to incorporate many of the same brighter stars, they have substantially different configurations. The Southern Cross, known to Palawanese as Büntal, does not appear on the Palawan Islanders' map.
Simplified from Nicole Revel, *Fleurs de paroles: Histoire naturelle Palawan*, 3 vols. (Paris: Editions Peeters, 1990), vol. 2, *La maîtrise d'un savoir et l'art d'une relation*, 226–27.

7. The cosmos has been ethicized; that is, its constituent units are perceived as having greater or lesser merit, as are the various periods of time. As a general rule, higher components of the universe connote greater merit than lower components. Moksa (Hindu) or nirvana (Buddhist; often romanized as nibbana from Southeast Asian languages) is the highest domain.

8. The religiophilosophical preoccupation with cosmology is essentially eschatological and is rooted in the axiomatic belief in reincarnation and the quest for escape from the cycle of rebirth through attainment of moksa or nirvana.

9. Like humans and other sentient beings within the cosmos, universes themselves are reborn and go through cycles of emanation, degeneration, and annihilation.

10. The various parts of any universe are populated by a multitude of denizens in addition to human beings: various types of spirits, titans, deities of various degrees of potency, bodhisattvas (in Buddhist views), and so forth. The parts are also characterized by numerous real and fabulous plants and beasts whose phenomenal life spans, vast sizes, and even shapes may be related to the part of the universe they inhabit. (Individually or in combination, these entities may serve in cosmographies as icons to represent particular portions of the universe.)

11. In all systems, the realm of humans is far from being the most meritorious, and none of the systems is anthropocentric. Similarly, the current age, rather than being a golden era, is one in which the universe is debased and approaching dissolution.

With specific reference to Buddhism in general or the Buddhism of Southeast Asia in particular, there is no single system of cosmology. But the oldest and still the most common view is the "singleworld" or Cakravāla system. The broad outline of this system is provided both in the Pali canonical texts and in Buddhist Sanskrit texts, in which the designation Cakravāla (Pali Cakkavāla) refers to the iron mountain ring that is believed to encompass the world and provide the outermost limit within which heavenly bodies move around the axis of Meru.

The following account from Kloetzli is based on the *Abhidharmakośa* of Vasubandu, an Indian text of the fourth or fifth century. It relates to the Cakravāla in a restricted sense of the word, namely the portion of universe where the earth is situated. It should be considered

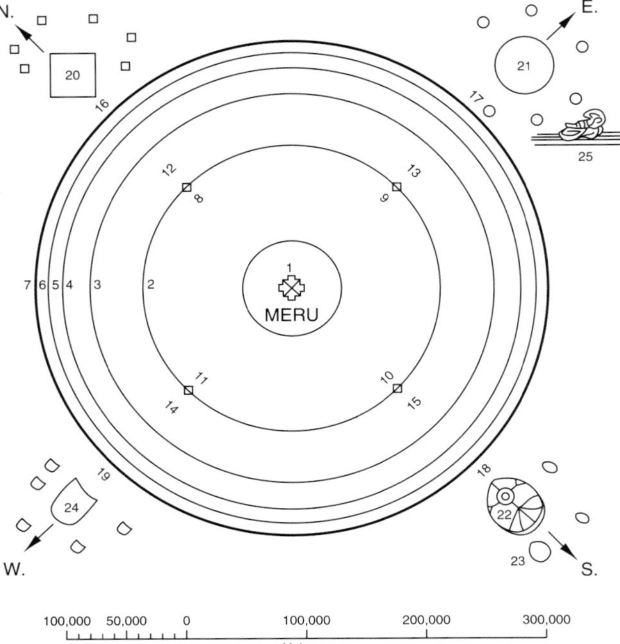

1. Indra's palace atop Mount Meru, sphere of the Tāvatiṃsadevaloka (Sanskrit Trayāstriṃśadevaloka), thirty-three gods.

2-7. The Kulācalas or concentric ranges of mountains.

8-11. Mansions of the Cātummahārājika (Sanskrit Caturmahārājakāyika), four great kings.

12-19. Oceans: 12, golden; 13, silvery; 14, rosy; 15, blue; 16, yellow; 17, milky; 18, salt; and 19, crystalline.

20-24. Continents and islands: 2, Uttarakuru; 21, Pubbavideha (Sanskrit Pūrvavideha); 22, Jambudīpa (Sanskrit Jambūdvīpa); 23, Laṅkā (Sri Lanka); and 24, Aparagoyāna (Sanskrit Aparagodānīya).

25. Vishnu floating on the sea of milk and reclining on the coiled body of the serpent Śeṣa as a couch.

FIG. 17.11. PLANIMETRIC VIEW OF THE CAKRAVĀLA ACCORDING TO THE *TRAI PHUM*. This drawing was prepared by Gerini to help explain an important rite of passage for a crown prince of Thailand incorporating the cosmic symbolism derived from the *Trai phum*, Thailand's principal cosmographic text. The diagram is only approximately to scale. Thus Meru, which ought to have a diameter of 10,000 *yojanas* (1 *yojana* = 15+ km), has been presented at a much larger size to render it more distinguishable. The outer wall of the Cakravāla is not shown, since it lies very far out and would have occupied too much space. These departures from true scale, however, are compensated for by the more correct view presented in figure 17.12.
After Gerolamo E. Gerini, *Chūḷākantamaṅgala, or, The Tonsure Ceremony as Performed in Siam* (Bangkok: Siam Society, 1976; first published in 1895), diagram no. 1.

along with figures 17.11 and 17.12, which for the most part provide correct scale representations of the Cakravāla from a planimetric perspective and in cross section, respectively:

> The *cakravāla* is represented as a disk ringed with a series of seven circular, golden mountain ranges, arranged concentrically with Mount Meru at the center and the *cakravāla* wall of iron at the perimeter. Proceeding outward from the center, the mountains are known as Meru, Yugandhara, Īṣadhāra, Khadirika, Sudarśana, Aśvakarṇa, Vinataka, Nimindhara, and Cakravāla. Mount Meru has a height of eighty thousand *yojanas* and penetrates the waters in equal measure; each of the mountain ranges is half the height and depth of the preceding range. The waters of various seas (sītā) fill the regions between the mountain ranges.[50]

Between the Nimindhara and Cakravāla mountain ranges lies a great ocean (*mahāsamudra*) within which are four islands (*dvīpas*) situated in the four cardinal directions from Meru: Pūrvavideha to the east, Jambūdvīpa to the south, Aparagodānīya to the west, and Uttarakuru to the north. The names suggest not only spatial but also theological directions. For example, Jambūdvīpa, named for the *jambū* tree that is found there, suggests the fruit of the path of Buddhism.

> All these entities rest on a layer of golden earth (*kāñcanamayībhūmi*), and all of the mountains except the *cakravāla* are composed of excrescences of this golden earth. . . . The golden earth of the *cakravāla* rests on a circle of water (*ābmaṇḍala*); a layer of wind (*vāyumaṇḍala*) supports the water and in turn rests on empty space (*ākāśa*).[51]

The four islands in the *mahāsamudra* differ in size and shape. Uttarakuru, for example, is square and measures two thousand *yojanas* on a side. The islands differ also in the duration of life for their inhabitants, a thousand years in the case of Uttarakuru. Symmetrically flanking each mentioned island are two similarly shaped islands that are only one-tenth as large. The faces of the inhabitants of both the main and the satellite islands are said to conform in shape to that of the islands themselves.

The vertical plan of the universe (table 17.1) is summarized by Kloetzli as follows:

> A series of heavens is arrayed above the *cakravāla* in three great divisions: (1) those heavens in the "realm of desire" (*kāmadhātu*) corresponding to the six classes of the "gods of desire" (*kāmadeva*); (2) the seventeen heavens belonging to the "realm of form" (*rūpadhātu*), grouped into four classes of "meditation realms" (*dhyāna*); and (3) the four "infinities" of the "realm of nonform" (*ārūpyadhātu*). . . . Several of the heavens have characteristics worth noting. The ruler of the Trāyastriṃśa is Indra, or Śakra, whose abode rests atop Mount Meru. The Tuṣita is distinguished by the fact that it is here that the *bodhisattva* is born

50. Kloetzli, "Buddhist Cosmology," 114 (note 49).
51. Kloetzli, "Buddhist Cosmology," 114 (note 49).

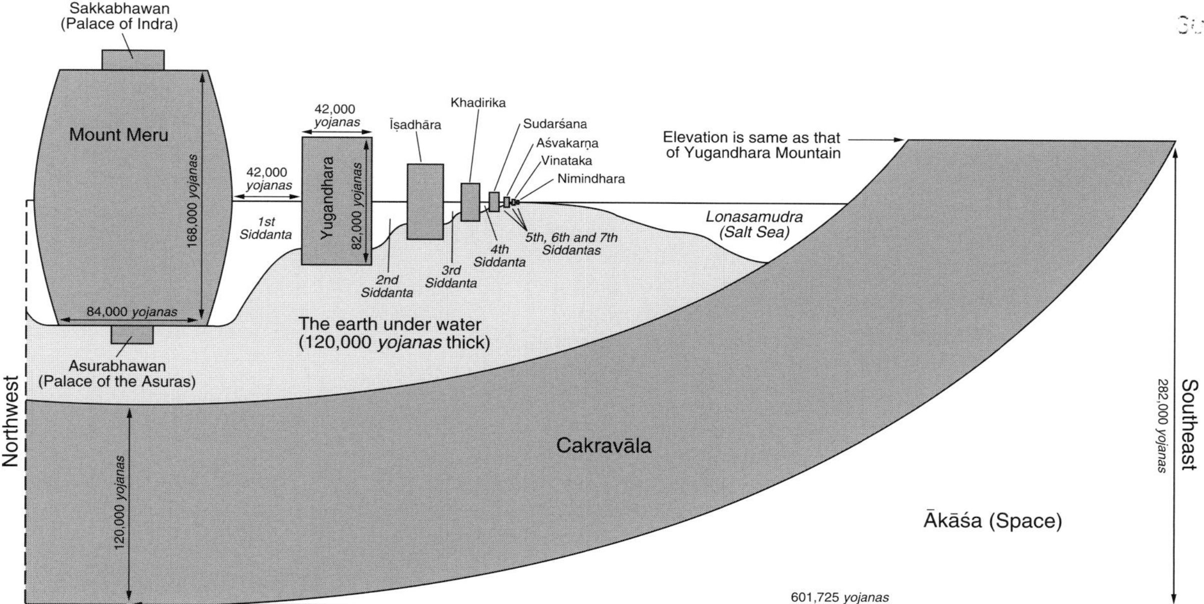

FIG. 17.12. PORTION OF A CROSS SECTION THROUGH THE CAKRAVĀLA. This cross section is adapted from one of a series of drawings, all to scale, that collectively set forth the horizontal and vertical dimensions of all the major components of the Buddhist cosmos as given in the Pali canon. Those drawings were based on another set copied from a monastery in Sri Lanka to which the originals had been brought from Burma at some unspecified date. Because of the symmetry of the Cakravāla about Mount Meru, there is no need to show its full horizontal extent. The relation of the portions of this presentation to what is shown in figure 17.11 is self-evident. The height of the Cakravāla wall is the same as that of the Yugandhara Mountain. In the annular space between those two features the sun, moon, and stars revolve about Meru. The differential distance of the sun from Yugandhara and its changing elevation in the sky will cause seasonal and diurnal variations in the length of Yugandhara's shadow, and hence in the length of a day and of shadows during the day on the several island continents. Because this profile extends from northwest to southeast rather than along one of the two cardinal axes, none of the four island continents shown in figure 17.11 appear as they otherwise would within the Lonasamudra. Eight levels of hell, situated within the earth underwater and centered under each of the four continents, are for the same reason omitted from this diagram.
Based on Adolf Bastian, "Graphische Darstellung des buddhistischen Weltsystems," *Verhandlungen der Berliner Gesellschaft für Anthropologie, Ethnologie und Urgeschichte*, 1894, 203–15, in *Zeitschrift für Ethnologie*, vol. 26, esp. pl. 5 and key on 210–13.

> immediately prior to being born as a Buddha in Jambudvīpa. The duration of life in the Tuṣita corresponds to the ages in which a Buddha appears. The uppermost heaven is the Akaniṣṭha; the fourth infinity is designated *bhavāgra* ("pinnacle of being").[52]

In my discussion of South Asian cosmography, I noted a Hindu tradition associated primarily with divination, which had little relation to the Meru-centered systems that predominate in religions of Indian origin.[53] The *kūrma* (tortoise), one among many icons of this tradition, also appears, as we shall see, as a symbol for the earth in contemporary cosmography of the still Hindu island of Bali—often, as in India, in association with a *naga* (snake)—and figures also in the cosmographies of other parts of Indonesia.

Southeast Asian cosmology incorporates a strong belief in astrology, utilizing a system of Indian origin. Many published and manuscript manuals and charts abound in visual representations of the heavenly bodies and celestial forces said to be at work is determining the destiny of humans. The subject, however, is too extensive and specialized to be treated adequately in this work.

REPRESENTATIVE COSMOGRAPHIC ARTIFACTS

Although my search for Southeast Asian cosmographies may best be described as an introductory reconnaissance, it has revealed a remarkable variety—if not a particularly large number—of artifacts on paper, palm leaf, stone, and other media that reflect concepts of Indian origin in both the Hinayana Buddhist and Hindu traditions. In this section I make no attempt to document all the cosmographic

52. Kloetzli, "Buddhist Cosmology," 114 (note 49). A different and much simpler model of the vertical cosmos, as envisaged by the Burmese, appears in John B. Ferguson, "The Symbolic Dimensions of the Burmese Sangha" (Ph.D. diss., Cornell University, 1975), table 1. One reason for the different Burmese perspective was the necessity for them to find places in their cosmos for the thirty-seven *nats*, pre-Buddhist spirits who continue to be propitiated by Burmese and accommodated within Buddhism as practiced in Burma.

53. Schwartzberg, "Cosmographical Mapping," 337–38 (note 2).

TABLE 17.1 The Vertical Aspect of the Buddhist Universe in the Cakravāla World System

Realm	Division	Abode	Vertical extent	Group	Worlds	Loka
III. ĀRUPYADHĀTU (REALM OF NON-FORM)		31. Naivasañjñānāsañjña (Neither Consciousness nor Not Consciousness)		Caturārūpya Brahmaloka (Fourfold Formless World of Brahmas)		
		30. Ākincanya (Realm of Nothingness)				
		29. Vijñañānantya (Infinity of Intellect)				
		28. Ākāśānantya (Infinity of Space)				
II. RŪPADHĀTU (REALM OF FORM)	4th Dhyāna (4th Meditation)	27. Akaniṣtha (Not Youngest)	$2^{22} \times 40$ Ky	Suddhāvāsika (Abodes of the Pure Ones)	1,000 × 1,000 × 1,000 Worlds	Brahmaloka (World of Brahmas)
		26. Sudarśana (Well Seeing)	$2^{21} \times 40$ Ky			
		25. Sudṛśa (Beautiful)	$2^{20} \times 40$ Ky			
		24. Atapa (No Heat)	$2^{19} \times 40$ Ky			
		23. Avṛha (Effortless)	$2^{18} \times 40$ Ky			
		22. Bṛhatphala (Abundant Fruit)	$2^{17} \times 40$ Ky			
		21. Punyapravasa (Merit Born or Merit Begetting)	$2^{16} \times 40$ Ky			
		20. Anabhraka (Cloudless)	$2^{15} \times 40$ Ky			
	3d Dhyāna (3d Meditation)	19. Śubhakṛtsna (Complete Beauty)	$2^{14} \times 40$ Ky		1,000 × 1,000 Worlds	
		18. Apramāṇaśubha (Immeasurable Beauty)	$2^{13} \times 40$ Ky			
		17. Parittaśubha (Limited Beauty)	$2^{12} \times 40$ Ky			
	2d Dhyāna (2d Meditation)	16. Ābhāsvara (Radiant Gods)	$2^{11} \times 40$ Ky		1,000 Worlds	
		15. Apramanābha (Immeasurable Splendor)	$2^{10} \times 40$ Ky			
		14. Parittābha (Limited Splendor)	$2^{9} \times 40$ Ky			
	1st Dhyāna (1st Meditation)	13. Mahābrahma (Great Brahmas)	$2^{8} \times 40$ Ky		Four-Island Universe	
		12. Brahmapurohita (Brahma Priests)	$2^{7} \times 40$ Ky			
		11. Brahmakāyika (Retinue of Brahma)	$2^{6} \times 40$ Ky			
I. KĀMADHĀTU (REALM OF DESIRE)	Kāmadeva (Gods of Desire)	10. Paranirimitavaśavartin (Rulers over Things Created by Others)	$2^{5} \times 40$ Ky			Devaloka (World of Gods)
		9. Nirmāṇarati (Those Who Have Pleasure in Creation)	$2^{4} \times 40$ Ky			
		8. Tuṣita (Blissful Gods)	$2^{3} \times 40$ Ky			
		7. Yāma (Yāma)	$2^{2} \times 40$ Ky			
		6. Trāyastriṃśa (Meru) (Gods Who Have the Thirty-three at Their Head)	2×40 Ky			
	Parīṣaṇḍā (Beings on the Four Steps of Meru)	5. Caturmahārājakāyika (Meru) (Four Great Kings)	40 Ky			
		Saddāmattas (Meru) (Always Drunk)				
		Mālādharas (Meru) (Bearing Garlands)				
		Karoṭapāṇis (Meru) (Bowl in Hand)				
		4. Manuṣya (Mankind)				
		3. Preta (Ghosts) (Abode is Yama, below Meru, not to be confused with Yāma)				
		2. Tiryagyoni (Animals) (Special abode is the Exterior Ocean)				
		1. Naraka (Damned) (Abodes are the Hells)				
Layers upon which the Cakravāla rests		Kāñcanamayībhūmi (Golden Earth)	320 Ky	1,120 Ky		
		Ābmaṇḍala (Circle of Water)	800 Ky			
		Vayumaṇḍala (Circle of Wind)	1,600 Ky			
		Ākāśa (Space)				

Note: Sanskrit forms are the designations of various portions of the universe or of their denizens. These are followed in parentheses by English translations. Figures preceding the abbreviation "40Ky" (for 40,000 *yojanas*; one *yojana* = 15+ km) indicate the vertical extent of specific portions of the universe, where indicated in the *Abhidharmakośa*, the text on which this figure is based.

This figure is adapted from W. Randolph Kloetzli, *Buddhist Cosmology, from Single World System to Pure Land: Science and Theology in the Images of Motion and Light* (Delhi: Motilal Banarsidass, 1983), figs. 2, 3, and 8, supplemented by personal communication, 4 September 1991. Not indicated in this figure, but provided by Kloetzli, are the horizontal dimensions of various portions of the cosmos (in two systems, his figs. 4 and 5), the duration of life of the denizens of each portion of the cosmos (fig. 6), and the heights of these respective denizens (fig. 7).

works encountered, but I will discuss and illustrate a fairly representative set of objects. Beginning with views that relate to the universe as a whole, I next narrow my concern to Manuṣya, the realm of humankind; then I focus further on Mount Meru, the *axis mundi*, and its associated surrounding features; and finally I proceed to consideration of a small sample of other specific, still more circumscribed portions of the cosmos. The subsequent section deals mainly with less tangible constructions considered in relation to the cosmic forces believed by Southeast Asians to govern the universe. Here I touch briefly on astronomy, the indispensable handmaiden of astrology, on geomancy, and on divination. Finally, I take note of the mental cosmographic maps that have exerted an important influence on the layout of architectural monuments, the distribution of temples and pilgrimage routes, and the conduct of statecraft.

Views Showing the Structure of the Universe

The vertical axis of the universe and its essentially tripartite division are, as we have seen, givens in the several Indian views of the cosmos and in their Southeast Asian derivatives. Figure 17.13, an early nineteenth-century representation of the many levels of the universe, illustrates this perception. With its multiplicity of heavens and hells and the several layers of the Kāmadhātu (the material world of desire), including the level of the earth itself, it reflects fairly closely the cosmological schema presented in table 17.1.[54] I know of no other Thai cosmography that encapsulates so much of the universe in a single image. For obvious reasons, given the enormous range in the dimensions of particular parts of the cosmos, no attempt was made to draw this—or any other known traditional Southeast Asian cosmography—to scale. But that does not mean that Southeast Asian Buddhists were unmindful of the stupendous dimensions of their universe. In 1882, for example, a Burmese scholar, U Kalyāṇa, probably a monk, prepared a text and chart titled *Hbôn sin* (A chart of the universe according to the Buddhist cosmography), in which the dimensions of each portion of the cosmos are specified. A large printed version of the chart, with the relevant dimensions (but not

FIG. 17.13. THE VERTICALLY ARRAYED UNIVERSE. Commissioned painting on European paper by the Chinese-Thai artist Bun Khong, in the Thai album of James Low, Thailand, about 1825. The uppermost column of small pavilions, each containing its own divinity, represents the many heavenly levels rising above the seven-tiered Mount Meru. Meru is flanked by the chariots of the sun and the moon, the stars, and two unidentified mythological figures. At the base of the mountain are magical forests inhabited by wild animals, and below that, ranged around Meru and shown in planimetric perspective, the four terrestrial island continents in an ocean ringed by a great double *naga* (snake). Still lower are various levels of hell, each characterized by a particular form of punishment for those consigned there for their past misdeeds.
Size of the original: 146 × 27 cm. By permission of the Oriental and India Office Collections, British Library, London (Add. MS. 27370, fol. 5).

54. Although I have no grounds to contest the authenticity of this representation, which is included in at least two reputable works on Thai painting, it must be noted that the work was commissioned by a British Lieutenant, James Low, while he was on temporary duty in Ligor, in southern Siam, and was executed by Bun Khong, an ethnic Chinese artist resident in that part of the country. Portions of this work are illustrated, in color, in Wladimir Zwalf, ed., *Buddhism: Art and Faith* (London: British Museum Publications, 1985), 179, caption on 184; and in Henry Ginsburg, *Thai Manuscript Painting* (Honolulu: University of Hawaii Press, 1989), 15. Notes on Low were provided to me by the Library of the Royal Asiatic Society, 21 December 1983.

the associated text), is in the British Library.[55] Dimensions are also noted, as we shall see, on various older manuscript maps.

A broadly similar, but stylistically different, Burmese view of the entire universe was published by the German ethnologist Adolf Bastian.[56] That view, meticulously drawn in black ink, was prepared from a copy by a Sinhalese monk of a Burmese drawing, accompanied by abundant text explaining the various parts of the universe. The Sinhalese copyist tried to adhere to the original Burmese form of the Pali text but in doing so made many errors. Hence, in presenting the work Bastian did not follow the often incomprehensible glosses that came with it but rendered all the names in their Sanskrit form, using a language he was more familiar with. The dimensions of each part of the cosmos are also provided by Bastian but, regrettably, not those of the drawing or of the copy made of it.

The most important and best-known text on Southeast Asian cosmography is the *Trai phum* (Story of three worlds) (*Traibhūmikathā* in Sanskrit).[57] Compiled from more than thirty Buddhist sources, this work, the oldest of Thai literature, is attributed to the then crown prince of the kingdom of Sukhothai, Phya Lithai, who is supposed to have composed it in A.D. 1345, primarily for the religious edification of his mother, but also to spread the message of the *dhamma* (Buddhist faith) to his future subjects. It has appeared in subsequent centuries in a number of different recensions, and beautifully illustrated copies may be found in several museums and libraries in Europe and the United States and, of course, in Thailand itself. None of the surviving examples, however, predate the sixteenth century.[58] Of the copies believed to have existed in the former Thai capital of Ayutthaya, few are known to have survived the city's sack by Burmese invaders in 1767.[59]

55. The work is in the Oriental and India Office Collections, with the pressmark OP 218 (32). The chart (60 × 48 cm) bears an abundance of text, entirely in Burmese except for the notation "Bengalee Job Printing Press.—Ranoon [*sic*]." At each of the many levels of the cosmos depicted there is a modicum of embellishment by a repetitive, stylized design, but this does not appear to convey any distinctive meaning. I am indebted to Patricia Herbert for tracking down this document, which had been misplaced, and transmitting a copy of it to me.

56. Adolf Bastian, *Ideale Welten nach uranographischen Provinzen in Wort und Bild: Ethnologische Zeit- und Streitfragen, nach Gesichtspunkten der indischen Völkerkunde*, 3 vols. (Berlin: Emil Felber, 1892), vol. 1, pls. II and III, with detailed key on 280–82. The principal one of those two illustrations also appears in Robert Heine-Geldern, "Weltbild und Bauform in Südostasien," *Wiener Beiträge zur Kunst- und Kulturgeschichte Asiens* 4 (1920–29): 28–78, fig. 22, opposite p. 65.

57. Complete translations of the *Trai phum*, with extensive commentary, appear in George Coedès and C. Archaimbault, *Les trois mondes (Traibhūmi Braḥ R'vaṅ)* (Paris: Ecole Française d'Extrême-Orient, 1973); and Frank E. Reynolds and Mani B. Reynolds, *Three Worlds according to King Ruang: A Thai Buddhist Cosmology* (Berkeley, Calif.: Asian Humanities Press, 1982) (which contains twelve color plates of the Berlin manuscript cited below and three additional reproductions of cosmographic paintings, in a modernistic style, by a contemporary Thai artist). Craig J. Reynolds has written an excellent historiographic essay, "Buddhist Cosmography in Thai History, with Special Reference to Nineteenth-Century Culture Change," *Journal of Asian Studies* 35 (1976): 203–20. A very useful account of the history and nature of the *Trai phum* is provided in Klaus Wenk, *Thailändische Miniaturmalereien nach einer Handschrift der indischen Kunstabteilung der Staatlichen Museen Berlin* (Wiesbaden: Franz Steiner, 1965), 14–22, which includes twenty-four superb color illustrations (with gold leaf laid on by hand) from the Berlin manuscript. Another beautifully illustrated work, with fifty-three quarto-sized pages of photographs in color, is *Samutphāp traiphūm burān chabap Krung Thon Burī/Buddhist Cosmology Thonburi Version* (Bangkok: Khana Kammakān Phičhāranā læ Čhatphim ʿĒkkasān thāng Prawattisāt, Samnak Nāyok Ratthamontrī, 1982). The text of this work, except for the title page and very brief photo captions, is entirely in Thai.

A set of views different from those in the Bangkok Thonburi manuscript is illustrated in Sugiura Kōhei, ed., *Ajia no kosumosu + mandara* (The Asian cosmos), catalog of exhibition, "Ajia no Uchūkan Ten," held at Rafōre Myūjiamu in November and December 1982 (Tokyo: Kōdansha, 1982), 28–33. A work by an art historian that includes numerous illustrations, largely in color, from both the earliest (Ayutthaya) and Thonburi manuscripts in Bangkok is Jean Boisselier, *Thai Painting*, trans. Janet Seligman (Tokyo: Kodansha International, 1976). Other useful notes and illustrations appear in Ginsburg, *Thai Manuscript Painting*, 13–18, with four illustrations, three from the Berlin manuscript and one from the manuscript at Harvard (note 54); and a review of Wenk, *Thailändische Miniaturmalereien*, by Elizabeth Lyons in *Artibus Asiae* 29 (1967): 104–6. Finally, an easy-to-read summary of the work is provided by George Coedès, "The Traibhūmikathā Buddhist Cosmology and Treaty on Ethics," *East and West* (Rome) 7 (1957): 349–52.

58. Three incomplete illuminated manuscript copies of the *Trai phum* are held by the National Library in Bangkok. Of these the shortest, dating from the first half of the sixteenth century or from the early seventeenth century (respectively, Wenk, *Thailändische Miniaturmalereien*, 20, and Boisselier, *Thai Painting*, 89 [note 57]), is the oldest surviving example. A second, stylistically similar to the one just noted, is from the first half of the seventeenth century. A third, called the Thonburi version, is precisely dated, being from the year 2319 of the Buddhist Era (A.D. 1776). (The Thonburi period of Thai history was from 1767 to 1782, when, after the Burmese sack of the previous capital of Ayutthaya, the seat of government was at that site across the Mekong from modern Bangkok.) Libraries outside Thailand known to possess manuscript copies include the Museum für Indische Kunst, Berlin (MIK II 650, acquired by Adolf Bastian in 1893), a more complete manuscript than the Thonburi manuscript in Bangkok of the same date; the New York Public Library, with three examples, Thai Manuscripts 1 (twentieth century) and 25 and 26 (both nineteenth century); and the Harvard University Art Museum, Hofer Collection, 517 (1984), which, despite its recent date, has been prepared in the traditional accordion style. Portions of another *Trai phum* manuscript, consisting of sixty-one folios written in Lanna Thai script, were unearthed during the excavation of a stupa in the vicinity of Chiang Mai, in northern Thailand; see Sommāi Prēmčhit, Kamon Sīwichainan, and Surasingsamrūam Chimphanao, *Phračhēdī nai Lānnā Thai* (Stupas in Lanna Thai) (Chiang Mai: Khrongkan Suksā Wichai Sinlapa Sathapattayakam Lānnā, Mahāwitthayālai Chīang Mai, 1981). This as yet uncataloged manuscript, which has not been translated or studied by any Western scholar, is in the Echols Southeast Asian Collection of the Cornell University Library, which acquired it about 1965.

59. Reynolds and Reynolds, *Three Worlds*, 37–38 (note 57).

Most of the illuminated manuscripts contain relatively little text. Of three manuscripts in Bangkok, for example, only one has extensive text, and even that covers no more than a tenth of the entire work, which in the published English translation comes to more than three hundred pages. That the *Trai phum* continues to retain a strong hold on the minds of modern Thais is evident in that between 1912 and 1972 eight edited versions of the full text were published in Thailand, largely for inclusion in university curricula, and a number of abridged and simplified versions have also appeared.[60] Not known, however, is how far these recent works incorporate the visual cosmographic elements we are principally concerned with here.

In subject matter, successive recensions of the *Trai phum* are not necessarily consistent but reflect modifications of previously presented views in light of the beliefs of the day. This will become apparent from the discussion in the chapter on terrestrial cartography of some of the geographic details the work portrays. As one would expect, there are also modifications over time in the artistic style of the illustrations—for example, in the adoption by the late nineteenth century of an essentially Western perspective.

Of the known illustrated manuscripts containing portions of the *Trai phum*, one in Berlin has received the most scholarly attention. Most of its illustrations relate to the cosmography of the three worlds (the three *dhātus* shown in table 17.1) and to a rather mythicized geography of Southeast Asia and the Indian Ocean. Other portions focus on the earthly life of the Buddha and various Jatakas (stories of the Buddha's previous lives, often set in mythical locales).[61] The manuscript copies of the *Trai phum* known to me are lengthy documents, each including scores of illustrated, primarily cosmographic folios. They are made on thick indigenous paper and pasted together accordion-style so that the work can be extended to provide as lengthy a view of particular portions as the reader might wish. Hence the work is said to contain "the world's longest cosmic picture."[62] Books prepared in this manner are known as *samud khoi* manuscripts (after the tree, sometimes described as mulberry, from whose bark a type of unbleached paper is made). They were bound between wooden boards and were designed to be read continuously from panel to panel on one side and then turned over and read in the same manner on the verso side. The same system was common among manuscripts from Burma, where such works were known as *parabaiks*. (The Burmese, however, often wrote in white ink or steatite on black paper.)

As mentioned above, the *Trai phum* manuscripts are lengthy documents—one manuscript, according to Boisselier, extends to 34.72 meters.[63] The comparable, but more complete, Berlin manuscript, 272 leaves in all, is 50.90 meters long. Individual folios have an average width of 51.5 centimeters and a height of 23.9 centimeters. Thus the four longest sixteen-folio illustrations just noted might have covered 8 meters of continuous illustration, within which single scenes often appear to extend over numerous continuous folios. Not surprisingly, therefore, there is no easily reproducible section that will provide a succinct view of the entire cosmos such as that shown in figure 17.13.[64]

The introductory text to the Thonburi manuscripts in Berlin and Bangkok informs the reader that the work was executed in 1776 at the command of King Phrya Taksin under the guidance of the supreme patriarch of Wat (Temple) Rakang. The names of four illustrators and four copyists are given, and the trained art historian can distinguish unevenness and personal idiosyncrasies in the quality of the work. Thai manuscript artists employed a broad range of colors. The palette for the Thonburi manuscripts comprised "white, yellow, vermilion, lapis-lazuli blue, malachite green and black, as well as all the tones that can be obtained by mixing these colours and fine gold."[65] Illustrations in the *Trai phum* relate to each of the thirty-one realms that compose the three worlds of Hinayana Buddhism. Subjects covered for each such realm are the mode of birth (instantaneous, from moisture, from an egg, or from a womb), the quality of existence within the realm, the character of beings and events to be encountered, and the ways of death and subsequent fates of the realm's inhabitants. The five lower realms, those of hell beings, of animals, of suffering ghosts, of *asuras* (semidivine and semidemonic creatures), and of humans, are treated in particular detail. Following this exposition, a chapter is devoted to

> entities that have no minds and indistinct material factors—that is, the mountains, rivers, trees, etc.—that provide the "natural" environment for the beings who live in the lower realms of the world of desire. In this chapter Phya Lithai includes an account of Mt. Sumeru and the surrounding seas, mountain ranges, and continents that form the *cákkavāḷa* or cosmological unit within which we live. He includes an obviously archaic but nevertheless quite logical system of astronomy that describes the way in which the sun,

60. Reynolds and Reynolds, *Three Worlds*, 24 (note 57).

61. Wenk, *Thailändische Miniaturmalereien*, 16 (note 57).

62. *Ajia no kosumosu*, 28 (note 57); translation by Amy Weeks.

63. Illustrated in *Ajia no kosumosu* are four continuous sections of the text (shown side-by-side on two facing pages), each consisting of sixteen continuous folios; two other sections of seven folios each, at a larger scale; and a section three and a half folios in length at a still larger scale, 28–33; see also Boisselier, *Thai Painting*, 90–94 (note 57).

64. The set of ten continuous folios illustrated in *Ajia no kosumosu*, 32–33 (note 57), right column, do, however, appear to depict a series of heavens very much like those illustrated in figure 17.13.

65. Boisselier, *Thai Painting*, 90, 92, and quotation on 94 (note 57).

> the moon, the stars, and the planets revolve around the peak of the central mountain. He also incorporates a highly mythical geography of northern India and the surrounding areas, which he identifies as the southern continent in the *cakkavāḷa* and calls Jambudīpa.[66]

From a stylistic point of view both Boisselier and Wenk see various external influences both in the Thonburi manuscripts and in earlier recensions. "In the landscapes, trees, rocks and water continue on occasion to reflect the Chinese influence . . . , while as regards figures, the attire of Muslims, Chinese, and Europeans of the seventeenth and eighteenth centuries perpetuates the classic note of exoticism."[67] One reviewer, however, takes issue with Wenk and, by implication, with Boisselier, on the extent of Chinese influence in Thai painting.[68] It seems reasonable to suggest the plausibility of some Chinese and Western influence, whatever the amount, and to raise the related question of the extent of such influence on other aspects of cartography, especially those we shall consider in the chapter on terrestrial mapping.

Within a number of Thai temples one finds mural paintings illustrating some of the same cosmographic themes treated in the *Trai phum*, though many of these are falling to the ravages of time.[69] When, in the reign of King Rama I (1782–1809), Bangkok's Wat Phrachettuphon was being restored and expanded, paintings from the *Trai phum* were added to the walls of the northern *vihāra* (monastery), which was one of several so adorned. An American visitor of that period wrote that the walls of one monastery

> were completely covered with representations of heaven, earth, hell, and one of the stars of which their books speak. There were angels, men, and monkeys, foreigners, or caricatures of white men, and dignified natives—scenes of gaiety and sadness—by land and sea—of war and peace—temples and brothels, with almost every sketch which could be framed from their sacred books, or conceived by their versatile limners. . . . My informant, the prince, remarked that the object of these paintings was to instruct the illiterate, through the medium of their senses.[70]

The *Trai phum*, it appears, still influences contemporary Thais, as do cosmological considerations in general. Although not nearly as popular as the *Trai phum*, other old cosmographic texts are also still read. One, a fifteenth-century northern Thai text, the *Phra malai sutta*, "deals only with the realms of woe and heavenly reward and focuses soteriological interest on the popular hope concerning the future advent of the Buddha Metteya [Maitreya]," the next Buddha who will descend to earth.[71]

As evidence of the popular appeal of cosmographic themes, Reynolds and Reynolds cite two remarkable recent constructions. At Wat Phairongwua in Suphan Buri Province, for example, "there is a gigantic exhibit

66. Reynolds and Reynolds, *Three Worlds*, 35 (note 57).

67. Boisselier, *Thai Painting*, 92 (note 57).

68. Lyons, review of *Thailändische Miniaturmalereien*, 106 (note 57).

69. Reynolds and Reynolds, *Three Worlds*, 21–22; Wenk, *Thailändische Miniaturmalereien*, 7; and Boisselier, *Thai Painting*, esp. 10 (all in note 57).

70. David Abeel, *Journal of a Residence in China, and the Neighboring Countries, from 1829 to 1833* (New York: Leavitt, Lord, 1834), 258. Reynolds, "Buddhist Cosmography in Thai History," adds that "paintings of the Hindu-Buddhist heavens also covered walls within the Grand Palace compound," 211 (note 57). A map of major sites in Thailand where monasteries with mural paintings, often cosmographic, are to be found is provided in Elizabeth Wray, Clare Rosenfield, and Dorothy Bailey, *Ten Lives of the Buddha: Siamese Temple Paintings and Jataka Tales* (New York: Weatherhill, 1972), 118. Figure 2 (p. 124) of that work is a photograph showing one such vast mural on the south wall of the ordination hall of Wat Suwannaram at Thonburi. Regrettably, the mural is in the shade and partly obscured by a large statue of the Buddha placed before it, making it difficult to appreciate this work. Of murals depicting the Jatakas, Wray, Rosenfield, and Bailey (133–34) note that "scenes are not placed in chronological sequence; rather, they are arranged according to where they occur. For example, even though Sama [a character in a story] being wounded and Sama being mourned are two incidents separated in time by several events, they are painted in the same locale, the forest." Although such depictions might not quite qualify as cartographic, they do indicate a special sensitivity to spatial and topographic concerns and a mode of organizing narrative fundamentally different from that taken for granted in Western cultures. (The same organizational system is to be found in the bas-relief sculptures of the Jatakas on the railing medallions around the Bharhut stupa [in what is now the Indian state of Madhya Pradesh], dating from the mid-second century B.C., and also in relatively modern Tibetan and Nepalese paintings depicting events in the lives of lamas.)

Mural paintings of cosmographic themes are also found in Burmese temples, although I do not know how common those that might be described as cosmographies may be. One such illustration appears in Jane Terry Bailey, "Some Burmese Paintings of the Seventeenth Century and Later, Part II: The Return to Pagan," *Artibus Asiae* 40 (1978): 41–61. Figure 24 of that article, facing p. 59, shows a painting in the Ānanda monastery at Pagan that depicts Mount Meru and its seven surrounding concentric mountain chains. The date is about 1776, and the style is much like that of a similar scene painted on the doors of an eighteenth-century Siamese cabinet in the National Museum of Ayutthaya (see fig. 17.21 below). The similarity could well be explained by Bailey's suggestion (p. 61) that "Siamese painters, taken as prisoners of war after the capture of Ayudhyā in 1767, may have been set to work in Burma." I will point out below at least one Burmese cosmographic manuscript that suggests a Thai antecedent and that could be accounted for in like manner.

71. Reynolds and Reynolds, *Three Worlds*, 21 (note 57). Ginsburg, *Thai Manuscript Painting*, 72–73 (note 54), notes that illustrated *Phra malai* manuscripts are known only from the late eighteenth and early nineteenth centuries and states that their illustrations "include visions of heaven, hell, and of daily life on earth." Although the ten reproductions he has selected lack any quality that one might call cartographic, others in Boisselier, *Thai Painting* (note 57), come close to the tone set by the *Trai phum*. It would be worthwhile to examine views from others among the large number of manuscripts in museums in the United States and Europe for which Ginsburg provides the locations.

spread over several acres of land within which one can observe vivid, three-dimensional depictions of the pleasures of the heavenly realms reserved for those who make merit and even more striking representations of the realms of woe to which those who do various kinds of evil are condemned." No less remarkable, in a temple building in the Buddhist center Wat Phutudom in Pathum Thani Province one finds

> representations of each of the three worlds and the thirty-one realms. In the dark recesses of the basement visitors pass through and observe the three lowest realms of woe; on the main floor where a great Buddha image is seated, they pass through and observe the realms of the animals and of men; and as they climb through a series of nine upper rooms, built on top of one another, and reach the temple roof, they pass through and observe the six heavenly realms of the *devatā*, the sixteen realms of the world with only a remnant of material factors, and the four realms of the world without material factors. Moreover, a fascinating twist is given to the symbolic significance of the tour of the three worlds by the fact that the heavenly realms are depicted in nine sets . . . each of which is associated with an event in the Buddha's life, with a particular planet, and with a particular day of the week. Thus as visitors climb upward through the twenty-six realms above the realm of men, they also pass by and observe the crucial events in the life of the Buddha, the nine major astronomical segments of the celestial regions, and, in addition, they pass through the full temporal cycle represented by a calendrical week.[72]

Though our discussion to this point has focused on what is now Thailand, the cosmographic notions embodied in the *Trai phum* were not confined to that area.

> All Theravāda Buddhist peoples shared the same cosmography as it was inherent in the the Pāli scriptural tradition. . . . Fragments of the Pāli canon and commentaries also reached Burmese, Lao, Mon, and Khmer courts; and through the centuries, the cosmography penetrated deeper and deeper into mainland Buddhist belief. Even in non-Theravāda Southeast Asia, artists portrayed aspects of the cosmography in Buddhist monuments. The bas reliefs on the Mahāyāna Buddhist Borobudur temples in central Java depicting sinners being punished in the hells have their origins in a Nepalese text. The cosmography was not, in fact, exclusively Theravāda but was part of a larger Hindu-Buddhist tradition.[73]

Although Hinduism is virtually extinct in Southeast Asia (leaving aside its following among recent immigrants from India), Hindu concepts inform the worldviews of many Southeast Asian peoples and, as we have seen, have had some influence on a number of tribal religious systems. The only area in which Hinduism remains the dominant faith is the Indonesian island of Bali and a part of the adjacent island of Lombok. Cosmography in this area, perhaps even more than in most of the rest of Southeast Asia, is an important aspect of popular culture, and cosmographic symbolism plays an important role in architecture, cremation rites, other religious practices, modes of governance, and some of the dances for which Bali is famous. Every Balinese village, for example, has its temple, and every temple has one or more multitiered Meru pagodas. Perhaps the most characteristic as well as spectacular Balinese representations of the cosmos are the great towers in which corpses are borne to the cremation ground.[74]

Though I lack space to discuss other aspects of Balinese cosmography, I should point out that a variety of cosmographic paintings also exist. Bosch provides a striking illustration of the multilayered, vertical axiality of the cosmos through an undated painting of Shiva, the dominant god in the Balinese pantheon, "in the shape of the primeval liṅga . . . [who], rising from the underworld and cleaving the sky, penetrates into the highest heaven."[75]

The Four-Continent Earth and the Mount Sumeru System

As previously noted, in the cosmography of Buddhism the earth is perceived, as it is in Hinduism and Jainism, as a horizontal disk centered on Mount Sumeru (Sanskrit Meru; in this section I shall adhere to the Pali forms of proper names). In each of the four cardinal directions outward from Sumeru lies an island continent, distinguishable by its shape. Plate 35 and figure 17.14 show Burmese examples of this conception. Of the two illustrations, the former, though much simpler, seems in some respects more faithful to the canonical descriptions, especially in regard to the clearly differentiated shapes of the four continents. The wedge-shaped Jambudīpa (Sanskrit Jambūdvīpa) is the southern continent, inhabited by

72. Reynolds and Reynolds, *Three Worlds*, 25–26 (note 57).

73. Reynolds, "Buddhist Cosmography in Thai History," 206–7 (note 57). Two footnotes to the passage quoted refer to a number of works dealing with various parts of Southeast Asia beyond Thailand that I have not yet had an opportunity to explore. It would not be surprising if those works were to lead to further graphic representations of the cosmos. Later in this chapter I will refer to several Burmese cosmographic paintings that, but for their distinctive style, could be right out of the *Trai phum*.

74. See Miguel Covarrubias, *Island of Bali* (1936; New York: Alfred A. Knopf, 1956), 371–72, for a detailed description and explanation.

75. Bosch, *Golden Germ*, pl. 62.b and text on 165 (note 20). For additional paintings see Covarrubias, *Island of Bali*, 6–7, illustrating the turtle-snake motif with explanatory text (note 74); and *Ajia no kosumosu*, 110 (same painting just noted), 111 (another version of the same theme), and 124–25 and 126 (modern paintings of cremation ceremonies illustrating the cremation towers) (note 57).

FIG. 17.14. BURMESE PALM-LEAF COSMOGRAPHY. These seven leaves are part of a manuscript of 113 folios (plus several unused leaves), written on one side only, encased in painted wooden endpieces. Etched with a stylus, the design is subsequently fixed with lampblack. Purchased in 1949, the manuscript is of unknown age but thought to date from the nineteenth century. The subjects of the leaves depicted are generally similar to those shown in plate 35, but the Buddha, for reasons unknown, here is sitting in what is taken to be the northern continent.
Size of each leaf: ca. 6 × 25 cm. By permission of the British Library, London (Add. MS. 17699 A).

humans like ourselves. There the Buddha sits under the *jambu* tree from which the continent takes its name. The shapes of two of the continents shown on this cosmography differ from those of Thai and Tibetan cosmographies of similar derivation. Here the northern continent, Uttarakuru, is round and the western continent, Aparagoyāna (Sanskrit Aparagodānīya), is square. In the non-Burmese cases the reverse is true. That the arrangement of shapes is correct here, however (at least for Burma), is suggested in that their disposition with respect to one another is identical to that of the cosmography presented in another version on palm leaf.[76] One would suppose that the latter, a published palm-leaf manuscript, would be exemplary. Although we do not know the date of that illustration, whose the map portion covers nine leaves, its appearance and style suggest that it is significantly older

76. This palm-leaf cosmography is published in a Burmese encyclopedia, *Myanma swezoun kyan*, 2d ed., 15 vols. (Rangoon, 1968), 10:295.

than the one depicted in figure 17.14 (thought to date from the nineteenth century) and possibly, on that count, closer to the ancient tradition.

Since the orientation of the writing in plate 35 varies from continent to continent, one cannot definitively say which part of the map is "up," but the manner of binding of all three manuscripts suggests that east is at the top. Despite their overall similarities, several significant points of difference exist among the three maps. Particularly striking are the variations in the number and shapes of tributary continents adjacent to the four principal continents. On both plate 35 and the palm-leaf manuscript not illustrated here, the shapes of main and tributary continents are identical; in figure 17.14 they vary in the west. The two palm-leaf manuscripts, however, are alike in having four symmetrically distributed tributary continents, two on each side of the principal continents, whereas in plate 35 each main continent has a different number (as is true in some Pali texts where, however, the numbers given are in the hundreds). Reading clockwise from Jambudīpa, which has four tributaries, we come to Aparagoyāna with six, Uttarakuru with five, and Pubbavideha (Sanskrit Pūrvavideha), the eastern continent, with only three. (Differences from one continent to another also exist in Tibetan Buddhist cosmographies, though the numbers there do not accord with those in plate 35. This seems to be a matter in which a certain degree of artistic license is in order.) A final point of similarity between plate 35 and the undepicted palm-leaf cosmography is that both show the sun and the moon by large circles symmetrically disposed with respect to Pubbavideha, whereas neither sun nor moon is evident in figure 17.14.

Since palm-leaf manuscripts do not lend themselves to the use of color, the lack of it in figure 17.14 does not constitute an important difference from plate 35, which was painted with a multihued pastel palette. Although particular colors do figure prominently in Buddhist and Hindu cosmographies—signifying, among other things, particular directions—the use of color in the map shown in plate 35 seems in some respects rather arbitrary. Although, as one might expect, tributary continents and main continents are always in the same hue, the four main continents, which one would expect to be of distinctive colors, appear in two cases to be alike. (Possibly the pigment of one has faded with time.) Around Sumeru, each of the ring seas is colored light green, while Sumeru, the seven ring mountain ranges around it, and the outermost rim, the Cakkavāla, are all compounded of bands of several hues. The gold core of sacred Sumeru itself is attested in the canon.

Figure 17.15 presents the Thai equivalent of the Cakkavāla cosmography presented in plate 35 and figure 17.14. This illustration, from the Berlin manuscript of

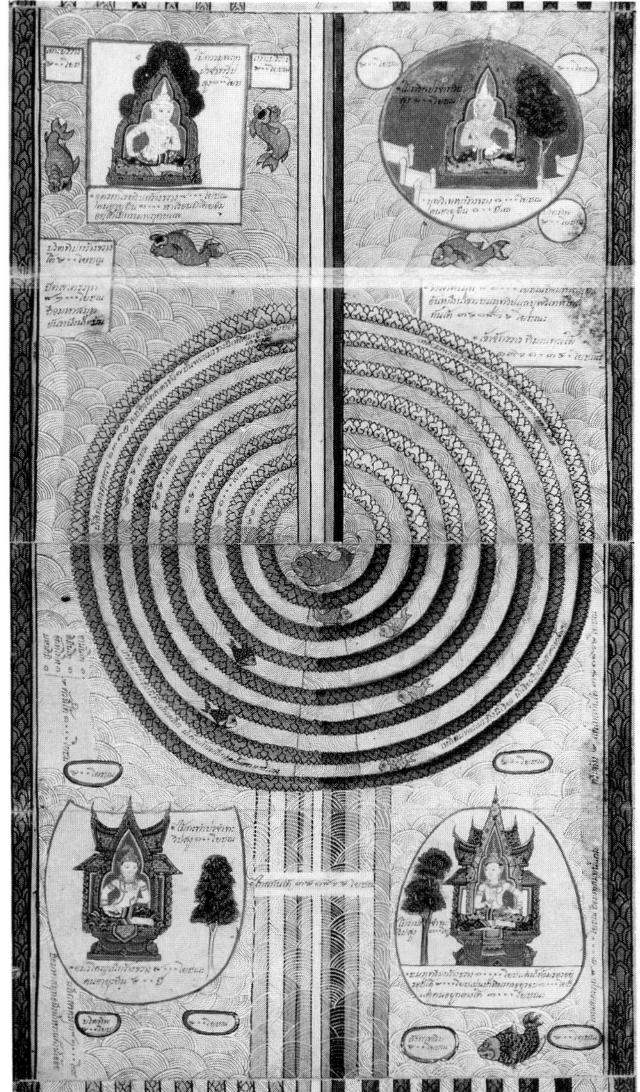

FIG. 17.15. THE FOUR CONTINENTS AROUND SUMERU AS SHOWN IN THE *TRAI PHUM*. Painted, bark-paper (*samud khoi*) manuscript, bound acccordion-style, Thailand, 1776 (the year 2319 of the Buddhist Era). Four successive folios, out of a total of the 272 composing the Berlin manuscript of the *Trai phum*, are depicted here. The work was made on the bidding of the then king Taksin of Siam, and the names of the four artists and four scribes are specified in the manuscript. This illustration shows the vertical shaft of Sumeru, the seven surrounding mountain ranges and seas, the four principal continents inhabited by humans, each with its distinctive shape, and the multiple tributary continents nearby (here arbitrarily shown as only three in number). Within each continent a tutelary deity sits within his palace, near or behind which appears a tree, one of the characteristic features of each continent.

Size of the entire original: 51.8 × 3,195 cm; each panel: 51.8 × 23 cm. By permission of the Museum für Indische Kunst, Staatliche Museen zu Berlin—Preussischer Kulturbesitz (MIK II 650/RF 10–16).

the previously discussed *Trai phum* cosmography, in most particulars bears a strong resemblance to the comparable representation in the Bangkok manuscript of the same date. In figure 17.15, judging from their shapes, Jambudīpa is in the lower left corner, with the western, northern, and eastern continents ranged clockwise from it, each with three satellite continents of more or less the same shape. Inexplicably, however, the southern and western continents and the northern and eastern ones are reversed from left to right in the Bangkok case, as compared with the Berlin manuscript. Other differences of note relate to the scenes depicted in each of the four continents. In the Berlin manuscript, for example, each contains what appears to be a Buddha within a palace with a characteristic tree alongside or to the rear, whereas in the Bangkok manuscript the scenes are more complex, the palaces being set within white-walled compounds outside two of which (Jambudīpa and Aparagoyāna) a man and a woman are depicted. The background color of the fields where the compounds are situated in the Bangkok manuscript is distinctive, and the color of each set of tributary continents is appropriately matched. This touch, following specifications in the text itself, is absent in the Berlin manuscript. In both, however, the colors of the four quadrants of the seven ring mountain ranges are distinct, and in the Bangkok case that color is also keyed to the color of the continents.

In both manuscripts the dominant perspective is planimetric, while Sumeru appears as a vertical shaft running through the ensemble and linking it to other portions of the universe above and below. That Sumeru goes beneath the level of the seas is indicated in both cases by a wave pattern overpainted on the portion of the mountain in the lower part of the field and the absence of any such overpainting in the upper half.

Stylistically, with respect to the manner of rendering mountains, seas, trees, human figures, fish, and animals and in the use of color, the *Trai phum* cosmography differs markedly from the Burmese examples. But most of these differences are to be expected given the different artistic traditions of the two countries and the varying dates of the works in question, and they have little cartographic significance.

Recently excavated from a stupa in northern Thailand is a *samud khoi* manuscript, written in the Lanna Thai

FIG. 17.16. SUMERU AND SEVEN SUCCESSIVE RANGES OF RING MOUNTAINS. This east-facing cross-sectional representation forms part of a very extensive assemblage of monastic caves carved into the live rock at Powun-daung. The height from the ground to the top of the quatrefoil ornaments below the two uppermost niches is approximately 8.2 meters, and the width at the base is 4.6 meters. The arched entrance to the cave is roughly 2.4 meters broad and 2.1 meters high. Atop Sumeru are two niches, one representing Indra's nine-tiered palace in the heaven of the thirty-three *devatās*, and the other formed by the interlocking limbs of two of the wish-granting Pāricchattaka trees by which that heavenly realm (line 6 in table 17.1) is characterized. Below this level are two horizontal rows of four niches each. The images that presumably once occupied the upper of those two rows are no longer present; those in the lower row are said to have a goblinesque appearance (not easily discerned in this photograph). It seems likely that the no longer extant former set represented the Four Great Kings and the latter, either humans on the four continents of Manuṣya or demigods (*asuras*) living below the world of humans. Although the placement of the latter within Sumeru would not be in accordance with the letter of the canon, the need to accommodate the various elements of the design to the position of the cave entrance could have played a decisive role in the composition. The caves and associated monuments at Powun-daung were used for meditation and prayer and not as residences for the monastic community living nearby. They are believed to have been commenced in the ninth or tenth century, probably by the now extinct Pyu (a Tibeto-Burman people who occupied Upper Burma before the advent of the Burmans proper), and have been continually expanded up to the twentieth century. The date of this particular part is not known, but it is probably not among the more recent parts of the assemblage.
Size of the original: height greater than 10 m. From Charles Duroiselle, "The Rock-Cut Temples of Powun-daung," in *Archaeological Survey of India Annual Report, 1914–15* (Calcutta, 1920), 42–55, esp. pl. XXX.

script, that includes a cosmographic map of Sumeru and the four surrounding continents. This map, executed in black ink, was supposed to establish the cosmic link between the stupa and Sumeru. It is in a nonpictorial style altogether different from the southern Thai examples and considerably simpler, showing only a single ring mountain range around Sumeru, out from which is a circular band divided into four quadrants, each bearing the name of the continent in its particular cardinal direction from the cosmic axis and a statement of the width of the continent in *yojanas*.[77]

For sheer grandeur, no essentially two-dimensional Buddhist cosmography I know of can match the one shown in figure 17.16, which presents a cross-sectional view of Sumeru and its surrounding mountain ranges hewn out of a sandstone cliff face at Powun-daung (Powin Taung) in Upper Burma. More than ten meters in height, this as-yet-undatable work forms part of an extensive monastic cave assemblage that was begun in the ninth or tenth century and continually expanded up to the present century. As far as I know, it has not been carefully studied by any modern scholar familiar with the cosmology it is based on, and some of the suggestions offered in the legend must therefore be viewed as tentative.[78]

The complex of Sumeru and the surrounding mountain ranges seen in cross section appears in Southeast Asia in a variety of architectural forms: as a design element in the finials above doors and windows of religious edifices and other important buildings and as an accent on the gates of religious compounds. The styles are quite varied. In his work on the art of Laos, Parmentier illustrates a number of movable carved wooden altars, which he calls *portes-luminaires*, in which the Sumeru complex figures prominently. Attached to the altars were curved iron rods that held votive candles whose light was reflected from the altar to provide the desired aura of sanctity. A particularly graceful undated work of this genre is found in a temple in the northern Lao village of Ban Mang. This lyre-shaped object has a maximum height and width of approximately 2.35 and 1.42 meters, respectively. Each mountain range on the altar is a narrow, tiered spire in the characteristic form of the Lao pagoda. These rise regularly in height inward toward the central spire above Sumeru, which then rises well above the others. Prominent features of this intricately carved composition are pairs of protective *nagas* entwined about Sumeru and one another at two different levels, with additional entwined *nagas* at Sumeru's base and still others helping to frame the composition. Parmentier does not discuss the iconographic significance of this serpentine emphasis.[79]

I know of no Sumeru cosmographies from Cambodia comparable to those I have just discussed. But since Cambodia's history as a Hinayana Buddhist nation, following the late thirteenth-century Thai conquest of the Mahayana Khmer empire, has much in common with that of Thailand and Burma, the likelihood of finding similar works in that country appears high.[80]

Apart from the essentially two-dimensional representations of the Cakkavāla system and the Mount

77. The map occupies two folios of the manuscript's total of sixty-one. The manuscript (which I shall refer to again in the discussion of terrestrial route maps) is held in the Echols Southeast Asian Collection of the Cornell University Library. According to Phramaha Wan, who examined a microfilm print of it for me, it includes selections from the *Trai phum*. In addition to the brief text in each of the continents, there are eight lines of text within the circle of Sumeru itself, rather brief notes outside the circular band containing the continents, and ornamental projections from the band, each with a few words of text, leading to the four corners of the map.

Yet another Cakkavāla cosmography appears in an illuminated folding manuscript that, though attributed to Burma, shows many points of similarity to the *Trai phum* excerpt in figure 17.15. This work is illustrated in *Ajia no kosumosu*, 35 (note 57). The illustration, on six consecutive folios, presents a continuous, but incomplete, picture depicting Mount Sumeru, with the palace of Indra on its summit; the seven surrounding ranges of progressively decreasing altitude, each surmounted by a palace of its own; a red sun bearing the image of a cock to the right of Sumeru; a white moon with the image of a hare to its left; and additional iconic details on the shaft of the central mountain. Unfortunately the manuscript is in a serious state of decay and very likely incomplete (portions of ten or so additional folios are visible beneath those photographed). No information is available on the date, physical attributes, authorship, or context of this richly and meticulously painted work. The catalog credits Takao Inoue with providing the photographs but does not indicate the manuscript's present whereabouts.

78. The assemblage is illustrated and discussed at length by Charles Duroiselle, "The Rock-Cut Temples of Powun-daung," in *Archaeological Survey of India Annual Report, 1914–15* (Calcutta, 1920), 42–55, esp. 49–51, and more briefly noted by Heine-Geldern, "Weltbild und Bauform," fig. 26 and pp. 71–72 (note 56).

79. Henri Parmentier, *L'art du Laos*, 2 vols. (Paris: Imprimerie Nationale and Hanoi: Ecole Française d'Extrême-Orient, 1954), 1:45–47 and 262–66, vol. 2, pl. XIV.

80. A pair of cosmographic paintings, allegedly from Cambodia, is illustrated in *Ajia no kosumosu*, 93 (note 57). Apart from their attribution to Cambodia, their joint designation, "Kumeru Mandara" (Sumeru mandalas), and the fact that they were owned by a London art dealer, Jean-Claude Ciancimino, no further information about the paintings (size, date, and so forth) was provided. I have seen one of the pair in London and take it to be fairly recent, possibly even of the twentieth century. The paintings bear a striking similarity to a very large pair of paintings in Mīnākṣī Temple in the south Indian city of Madurai, one titled *Bhūgolam* (Globe/Geography) and the other *Khagolam* (Celestial dome). The first of the allegedly Cambodian pair is clearly a painting of the terrestrial disk as portrayed in Hindu cosmography, essentially as shown in Schwartzberg, "Cosmographical Mapping," fig. 16.9 and a portion of 16.10 (note 2). The other, no less clearly an astronomical work, resembles the Madurai painting illustrated as figure 16.19 of the same essay. The Hindu, rather than Buddhist, content of the pair argues against a Cambodian provenance. Since the Cambodian script bears some resemblance to the Tamil script, both being derived from an ancient Pali prototype, a mistaken attribution would be understandable.

Sumeru complex, both are symbolically reproduced in the architecture of Southeast Asia. This was noted briefly in chapter 16 and illustrated with respect to the great "temple mountain" at Borobudur in central Java (fig. 16.2). Since the literature on Southeast Asian architecture is extensive—if not always definitive—in its treatment of the associated cosmographic concepts, little purpose would be served by attempting to recapitulate here the detailed exposition and sophisticated analyses of art historians who are specialists in the field.[81]

Even ephemeral constructions, when created for ritual purposes, may portray cosmography in remarkable detail and on a rather grand scale. This is particularly true of works associated with important rites of passage, such as coronations, consecrations, cremations, and the first tonsure ceremony, especially for persons of royal lineage. Although not as grandiose as in past generations, such works still serve as cosmic legitimizers of the Thai monarchy and were characteristic also of royal ritual in Burma and Cambodia.[82] Not only the rituals were cosmicized, but also the very person of the monarch himself, as is evident from official documents that identify him as the axis of the universe and describe various parts of his body, as well as his regalia and associated paraphernalia, as representing specific portions of the cosmos.[83] Finally, we may note that ephemeral cosmographic structures were also employed in popular religious practices. Among the Burmese, for example, light bamboo constructions of Sumeru are erected for the celebration of the Tawadeintha festival, which celebrates the Buddha's ascent to the Tāvatiṃsa heaven to preach to his parents.[84] It seems probable that a thorough study of religious practices in mainland Southeast Asia would reveal other examples of the use of cosmography in popular ritual.

Reviewing the importance of Sumeru in both Hindu and Buddhist belief systems, Mabbett summarizes the issue as follows (and simultaneously adumbrates certain matters I have yet to touch on):

> The ambiguities and multivalences of symbolism which are exhibited by temple and *stūpa* architecture, *maṇḍalas* and tantric ritual, make clear to us a duality in Mount Meru's nature that could never emerge from the contemplation of it merely as a point on a two-dimensional map, even a cosmic map. For the cosmic map represents only space "out there"—even if the data of cosmography, with its elephant-sized jambu fruits and circumambulating rivers, are invisible to the fleshly eye (so are molecules and magnetic fields), still they are regarded as concrete and physical in a very real sense. The principles of Meru-centric cosmography are continuous with what its authors regarded as hard objective science.[85]

Views of Limited Portions of the Cosmos

Given the essentially didactic purpose of the *Trai phum* and other major cosmographic texts of Hinayana Buddhism in Southeast Asia, it is only natural that illuminated manuscripts, mural paintings in temples, and other forms of pictorial art should portray views of those portions of the cosmos that would present viewers with evidence of the delights or punishments awaiting them as the karmic consequences of good or bad behavior in the present life or in lives yet to come. Scenes of various hells are particularly abundant, which is hardly surprising when one considers that there was a specific type of hell for each category of sinful conduct. Thus, for each of the eight major hells ruled over by Lord Yama there are sixteen associated hells, whose topological relation to one another finds its logic in the moral domain. Figure 17.17, taken from the Bangkok manuscript of the Thonburi recension of the *Trai phum*, provides the sort of awe-inspiring picture—in this case of Roruva, the screaming hell—that would give the viewer an incentive to adhere to the straight and narrow path. But even worse hells, beginning with Maha Roruva, the great screaming hell, and ending with Maha Avici, the great hell of suffering without respite, lie at successively lower levels. The following translated passage will convey some sense of how faithful the picture presented is to the text:

81. Among the major works that deal with the issue of cosmic symbolism in Southeast Asian architecture are Bosch, *Golden Germ* (note 20); Heine-Geldern, "Weltbild und Bauform" (note 56); Paul Mus, *Barabudur: Esquisse d'une histoire du Bouddhisme fondée sur la critique archéologique des textes*, 2 vols. (Hanoi: Imprimerie d'Extrême-Orient, 1935); and Wales, *Universe around Them* (note 1).

82. Among published works that deal with this subject are Wales, *Universe around Them* (note 1); Horace Geoffrey Quaritch Wales, *Siamese State Ceremonies: Their History and Function* (London: Bernard Quaritch, 1931); Dhani Nivat, "The Gilt Lacquer Screen in the Audience Hall of Dusit," *Artibus Asiae* 24 (1961): 275–82; Gerolamo E. Gerini, *Chūḷākantamaṅgala; or, The Tonsure Ceremony as Performed in Siam* (Bangkok: Siam Society, 1976; first published in 1895); Robert Heine-Geldern, *Conceptions of State and Kingship in Southeast Asia* (Ithaca, N.Y.: Cornell University, 1956), a revised version of an article of the same title published in *Far Eastern Quarterly* 2 (1942): 15–30, which in turn recapitulates many of the ideas in Heine-Geldern "Weltbild und Bauform" (note 56); and *Ajia no kosumosu* (note 57).

83. For example, Heine-Geldern, "Conceptions of State and Kingship," 21–22 (note 82), notes that at the coronation of King Sisowath of Cambodia in 1906 an official document pronounced that "the king is identified with Mount Meru itself, his right eye representing the sun, his left eye the moon, his arms and legs the four cardinal points, the six-tiered umbrella above his head the six lower heavens, his pointed crown the spire of Indra's palace on the summit of the Meru and his slippers the earth. This means that the king is identified with the axis of the universe." Or to put it another way, the king had himself become a map.

84. Heine-Geldern "Weltbild und Bauform," 72 (note 56).

85. I. W. Mabbett, "The Symbolism of Mount Meru," *History of Religions* 23 (1983): 64–83, esp. 79.

FIG. 17.17. RORUVA HELL IN THE THAI *TRAI PHUM* COSMOGRAPHY. This hell, full of flame-spouting lotus blossoms, is the place to which are consigned those who give false testimony, malign others, commit thefts, or plunder. Creatures so damned are reborn in the blossoms and are unable to push themselves completely out of them. Also seen in this illustration are the innermost four of sixteen ancillary hells that extend in lines of four from each of the cardinal directions outward from the main hell.
Size of the original: unknown. From *Samutphāp traiphūm burān chabap Krung Thon Burī/Buddhist Cosmology Thonburi Version* (Bangkok: Khana Kammakān Phičhāranā læ Čhatphim ʿĒkkasān thāng Prawattisāt, Samnak Nāyok Ratthamontrī, 1982), pl. 36.

> These eight large hells each have four corners with gates at the four cardinal points. The floors that form the bottom of these hells are made of fiery red iron, and the ceilings that cover the top are also formed of fiery red iron. These hells are square and they measure 100 *yojana* [about 1,500 km]. . . . The four sides, the floor, and the ceilings are each nine *yojana* thick, and there are no empty places in these hells; they are full of hell beings who are pressed tightly together and fill the entire hell. . . .
>
> Surrounding each of these eight large hells are sixteen small auxiliary hells, which are their satellites, four on each side. Each of these . . . is surrounded by countless numbers of smaller hells, like the villages surrounding a city in our human world; and each of the auxiliary hells is ten *yojana* wide. These auxiliary hells, plus the large hells, are 136 in number.[86]

There is, of course, much more to the description, including frightening details as to the duration of suffering in each hell and the duration of the hell itself until the dissolution of the universe at the end of a *kappa* (Sanskrit *kalpa*), the termination of one cycle of creation and destruction. Such information reminds us that each portion of the cosmos has not only its spatial dimensions, which are relatively easy to portray graphically, but its temporal dimensions as well. The latter are no less important an aspect of their reality in the mind of Buddhists.

Depictions of heavens and other blessed cosmic spaces, including nibbana, are also common in painted Buddhist cosmographies. As a rule these are portrayed in frontal elevation or in a frontal oblique perspective as palaces

86. Reynolds and Reynolds, *Three Worlds*, 67–68 (note 57).

FIG. 17.18. THE GREAT CITY OF NIBBANA, AS SHOWN IN THE ILLUMINATED *TRAI PHUM* TEXT. Painted on paper, two folios of a manuscript of the Thonburi recension, bound in the traditional Siamese accordion style. The literal rendering of a portion of the text accompanying this illustration reads:

> Its ground laid with sands of crystal,
> A lake filled to the brim with clear cool water, blooming with lotuses.
> Bees are busy fondling the stamens;
> Melodious songs are heard from peacocks, cranes, and wild ducks,
> and white and red hansas [swans].

This is a far cry from the canonical idea of a nibbana characterized by the absence of form or consciousness, although some suggestion of formlessness for one attaining that goal is conveyed by the divan with no reclining figure. The idea of a city is conveyed by the walls, the gates therein, the paved paths, and the fact that there are buildings shown in addition to just a single palace.
Size of the original: unknown. From *Samutphāp traiphūm burān chabap Krung Thon Burī/Buddhist Cosmology Thonburi Version* (Bangkok: Khana Kammakān Phičhāranā læ Čhatphim 'Ēkkasān thāng Prawattisāt, Samnak Nāyok Ratthamontrī, 1982), pl. 1.

within which is shown a locally reigning deity or grouping of divine creatures. Adjacent features, such as gardens or groves of trees bearing sumptuous fruit, may or may not be depicted. The cartographic nature of such scenes, viewed in isolation, is not always apparent; but when one recognizes that they have a logically correct place, based on their position within a much larger cosmic ensemble, it becomes evident that they form an integral element of a cosmic map. Higher realms that appear to be particularly common subjects for cosmographic paintings and three-dimensional constructions include the Tāvatiṃsa heaven, at the summit of Sumeru, where Indra reigns over thirty-three lesser deities and nibbana itself. The popularity of the former realm may be explained largely in terms of its close association with Sumeru and partly because it represents a proximate goal whose attainment—in contrast to more distant heavens—is not beyond the aspirations of ordinary, imperfect mortals.

Nibbana, shown in figure 17.18, on the other hand, is popular despite the remote likelihood of soon being born again at that level. Rather, it is appealing because it represents the ultimate goal toward which all good Buddhists aspire. Yet there is a paradox in its being visibly portrayed in cosmographies such as the *Trai phum*, in that it represents the highest level of Arūpadhātu (Sanskrit Ārupyadhātu), the "Realm of Nonform," even above the level of "Nothingness." Although, as table 17.1 shows, this highest level is theoretically one of "Neither Consciousness nor Not Consciousness," the *Trai phum*, written for the instruction of laypersons, for whom formlessness was probably too esoteric a concept, resolves the problem with the following passage:

> As for Nibbāna, there are two kinds that are gained by those who have completely rid themselves of the multitude of defilements. . . . one kind is called the Nibbāna with some substratum of life remaining, and the other kind is called the Nibbāna in which there is no substratum of life remaining. That which is attained as the fruit of fully perfected sainthood is said to reach the first treasure. . . . That which is attained when the five aggregates are left behind is said to reach the Nibbāna in which there is no substratum of life remaining.[87]

It seems, therefore, that what is being depicted is merely the first level of nibbana. (The five aggregates referred to in the quoted passage are material factors, feelings, perceptions, mental formations, and mental processes.)[88]

Figure 17.19 presents a Burmese view of one of the heavenly abodes. Though not dated, this view is roughly contemporaneous with the depiction of nibbana that I have just discussed and is similar in spirit, if not in manner of execution. The twenty-nine-folio palm-leaf manuscript from which the figure is extracted shows a series of heavenly mansions in their appropriate order on one side of the work and a series of hells on the other.

Although the description of the geography of the Cak-

87. Reynolds and Reynolds, *Three Worlds*, 329–30 (note 57).
88. Reynolds and Reynolds, *Three Worlds*, 336–37 (note 57).

FIG. 17.19. ŚAKA IN HIS HEAVENLY MANSION, TĀVATIṂSA, IN BURMESE PALM-LEAF MANUSCRIPT. Shown are five leaves of a lavishly illustrated twenty-nine-leaf Burmese cosmographic manuscript. On one side of the set of leaves are arrayed a series of increasingly exalted heavenly mansions and on the other a series of increasingly debased hells. Many of the leaves in this work individually depict specific heavenly or infernal locales, whereas elsewhere, as in this excerpt, a number of leaves collectively portray such locales. Though undated this work is thought to be from the eighteenth century.
Size of the original: each leaf ca. 7.0 × 49.5 cm. By permission of the Oriental and India Office Collections, British Library, London (Or. 12168, fols. 3–7).

kavāla, and of Jambudīpa in particular, occupies only a portion of one of the eleven chapters of the *Trai phum* (the rest of that chapter being given over mainly to astronomy), the illustrations relating to that predominantly mythic geography account for approximately half of the manuscript of the Thonburi recension, that is, for one entire side of the many continuously joined folios of illustration and text. In much of what is shown one sees from one to five rivers flowing through fabulous landscapes, often mountainous and forested and populated by a diversity of humans and mythic anthropomorphic creatures, not to mention wild animals, fish, and birds. Although the text—at least the recension translated by Reynolds and Reynolds—bears no evidence of Western influences, this is not entirely true of the illustrations. For example, in one scene in the Berlin manuscript there are two hunters, one carrying a rifle. Seemingly, then, the illustrators of the Thonburi recension were given considerable license to insert into their work much that was either only hinted at in the text or, as in the case of hunters armed with rifles, not mentioned at all. (The next chapter will illustrate how new knowledge of the geography of lands remote from Siam found its way into the illustrations.)

The origin of the five major rivers of Jambudīpa is a favorite theme in Hinayana Buddhist cosmography. Its illustration in figure 17.20 is taken not from the *Trai phum* (though there are a number of similar presentations in that work), but from the lavishly painted Burmese manuscript from which plate 35 was excerpted.[89] Since no translation of the accompanying Burmese text is avail-

89. For more on this manuscript, see Zwalf, *Buddhism: Art and Faith*, 173–74 (note 54).

FIG. 17.20. LAKE ANOTATTA AND THE ORIGIN OF THE WORLD'S RIVERS, FROM AN UPPER BURMESE COSMOGRAPHIC MANUSCRIPT. Two folios of a brightly painted and partially gilded paper book in fifty-nine folds with Burmese text, nineteenth century. This scene depicts the mountainous terrain of Himavanta (the Himalayas), within which lies the circular, mountain-girt Lake Anotatta. Through rocky openings in the mountain rim shaped like the heads of an ox, horse, lion, and elephant, rivers flow to the south, east, north, and west, respectively. Though various texts specify that each river initially flows three times around Anotatta, here that idea is suggested by swirling lines in the ring of water that surrounds the lake, rather than by three spiraling watercourses as on other paintings. The river flowing to the right, the proto-Ganga, dashes against a mountain, spurts high into the air, falls back to earth, follows an underground course, and eventually reemerges to form five rivers, whose names can be traced to north Indian rivers associated with the hearth area of Buddhism.
Size of each page: 20.5 × 54 cm. By permission of the British Library, London (Or. 14004, fol. 33).

able, however, I quote here some apposite passages from the *Trai phum*:

> In the Himavanta [Himalayas] there are seven large bodies of water: one is called the Anotatta lake. . . .
>
> . . . Anotatta lake is surrounded by five mountain ranges. One mountain range is called the Sudassana range. . . .
>
> . . . The mountain range called Sudassana is replete with gold and surrounds the Anotatta lake like a wall. . . .
>
> There are four outlets on the sides of the Anotatta lake, one in each of the cardinal directions. The outlet in one of these directions resembles the face of a lion; that in another direction resembles the face of an elephant; that in another direction resembles the face of a horse; and that in another direction resembles the face of an ox. . . .
>
> The water that flows from the side of the head of the bed [i.e., from the south, since people in Jambudīpa were supposed to sleep with their heads pointed away from Sumeru] goes in a circle three times around the Anotatta lake. It is called Avaṭṭagangā. . . . It then flows in the direction of the head of the bed . . . and is called Kanhagaṅgā; it then dashes against a mountain, shoots upward for 60 *yojana*, . . . and is called Ākāsangaṅgā [identified with the Milky Way]; it then falls on a rock called Tiyaggala rock . . . and goes

> through the rocks . . . and is called Bahalagaṅgā; the water then drops beneath the rocks . . . and is called Ummagga-gaṅgā; it then flows against the mountain called the Vijjhanatiracchāna mountain, gushes up above the ground, and becomes . . . the five great rivers. One of these is called the Gaṅgā; another is called the Yamunā, another is called the Aciravatī, another is called the Mahī; and another is called the Sarabhū. These rivers [the modern Indian Ganga, its right-bank tributary, the Yamuna, and its left-bank tributaries, the Rapti, Buri Gandak, and Gogra, respectively] flow into the country where the people live [India], and then flow out to the ocean.[90]

This passage does not fit the illustration in every particular, and the elided portions mask discordances beyond those evident from what I have quoted; nevertheless, the correspondences are for the most part obvious. But in any case, we would find here that the world of myth begins to merge into the known geography of South and Southeast Asia. This, we observed, was also true of the Hindu and Jain cosmographies of South Asia.[91]

In his sumptuous and extensive study of Thai painting, Boisselier presents several photographs, both in color and in black and white, of additional representations of the the mythic landscape in the environs of Lake Anotatta. Perhaps the most decorative of the many such creations appears on an eighteenth-century gold-and-black lacquer library cupboard. This exuberant composition depicts the entire Sumeru complex, including the palaces atop each of its mountain ranges, among which that of the city of Indra is the most splendid of all (fig. 17.21). Above the Himavanta, but below the summit of Sumeru, are the sun, the moon, and the palaces of the planets and constellations, with crowds of gods hovering in the celestial interstices between them, while below the zigzag crest of the Himavanta the earth teems with animals, luxuriant forests, and lotus-filled lakes.[92]

Dating from many centuries earlier than the paintings just described is a three-dimensional Khmer representation of Lake Anotatta and the rivers issuing from it. This intriguing architectural assemblage (the Neak Pean complex) was discovered among the vast remains of the former imperial capital at Angkor. Erected in 1191, during the reign of Jayavarman VII (1181 to ca. 1219), it forms the dominant element of what appears to have been an important cult center at Angkor. Since a full and well-illustrated account of it is provided by Boisselier, and since further critical analysis is offered by Wales, among others, I shall not describe it further here.[93]

The significance of the Neak Pean complex lies not only in its conception but in its antiquity. It seems reasonable to suggest that the attraction of the portion of the Buddhist cosmography relating to Kelāsa (Sanskrit Kailāsa) and Anotatta, while perhaps not quite as powerful as that relating to Sumeru, was such that it would have resulted in many more cosmographic representations than have survived to the present. Less durable cosmographies would not have withstood the ravages of time, as Boisselier notes about painting among various civilizations dating back to the sixth century in the area that is now Thailand.[94] And if my suggestion about lost representations of Anotatta is correct, it should be no less true of painted, woven, carved, and sculptural representations of other components of the exceedingly vast and complex Buddhist cosmos.

In his study of popular Burmese religious practices, Temple includes several cosmographic paintings, at least two of which show signs of Western influence. Figure 17.22 is particularly interesting. Although Temple's caption, "A Burmese Map of the World, showing traces of Mediaeval European Map-making," makes little sense, much of his commentary on the map warrants quotation in extenso:

> The native map shows Zambúdék [Jambudīpa] with the *zambú* tree at its top or northern end. . . . [It] is shown as being surrounded by the Sakyáwála [Cakkavāla] Range and cut across by the Himawun (Himavanta or Himálaya). The Himalayas form the Fairyland of the Burmans, on and beyond which to the north everything is marvellous. Here are the seven great lakes, including Anawdat (Anótatta) in the centre, where grows the lotus and whence spring all the great rivers, after forming concentric rings around it. Here also is the jewelled Myinmódaung (Mt. Méru) [curiously off center] with the seven rings of mountains round it.
>
> Below the Himalayan line lies the world inhabited by all the human beings of whom the Burmese have any experience, with the sacred Bó Tree and the sites in the Buddhist Holy Land in the centre. These are quaintly shown by small red squares and circular patches, about and south of the holy trees.
>
> To the south of everything lies the Ocean, Thamóddayá (Sanskrit Samudrá), studded with the 500 lesser islands attached to Zambúdék, on which dwell

90. Reynolds and Reynolds, *Three Worlds*, 292–96 (note 57). The identification of the rivers with their modern Indian names is based on Joseph E. Schwartzberg, ed., *A Historical Atlas of South Asia* (Chicago: University of Chicago Press, 1978).

91. Schwartzberg, "Cosmographical Mapping" (note 2).

92. Boisselier, *Thai Painting*, fig. 109, and caption on facing page (140); see also figs. 26, 115, and 116 (note 57). The lacquer cupboard is also depicted at a significantly larger scale in *Ajia no kosumosu*, 25, with a still larger detail of the region around Lake Anotatta on 26–27 (note 57).

93. Jean Boisselier, "Pouvoir royale et symbolisme architectural: Neak Pean et son importance pour la royauté angkorienne," *Arts Asiatiques* 21 (1970): 91–108, and Wales, *Universe around Them*, 128–33 (note 1).

94. Boisselier, *Thai Painting*, 13–19 (note 57).

FIG. 17.21. THAI LIBRARY CUPBOARD DEPICTING MYTHIC LANDSCAPE IN THE ENVIRONS OF LAKE ANOTATTA. Black and gold lacquer on wood. Mount Sumeru forms the central axis of this sumptuous composition and is flanked by the seven ranges of surrounding mountains, each, like Sumeru itself, surmounted by a palace inhabited by one or more tutelary deities. In the spaces between the successive ranges one sees a conventional water pattern indicating the intervening oceans. The entire complex is seen as if in vertical cross section. The cosmic elephants, which here support the plateau at the summit of Meru, are not noted in the *Trai phum* text, which this depiction generally follows. On the plateau atop Sumeru is the city of Tāvatiṃsa (Thirty-three Gods) and the abode of the god Indra (Śakra) to which the Buddha ascended to visit his deceased mother. Somewhat lower, in a line with the palaces above Yugandhara, the highest of the seven flanking mountains, are the sun and moon. The jagged mountains at the foot of Sumeru are the Trikūṭas, the buttress mountains that support the cosmic axis. Within them may be seen the city of the *asuras* (demons). Flanking the Trikūṭas are the Himavanta (Himalayas), whose equally jagged crest serves as a divider between the heavenly realm and Jambudīpa, the land of humans and also of all sorts of mythical creatures and exuberant vegetation. Near the lower left is Lake Anotatta, from which issue the same four rivers shown in figure 17.20. One of these, flowing toward the lower right, is the proto-Ganga, which, again as in figure 17.20, strikes against a mountain, rises high into the air, and then dashes into the earth.
Size of the original: 196 × 122 × 87 cm. Courtesy of the Chao Sam Phraya National Museum, Ayutthaya.

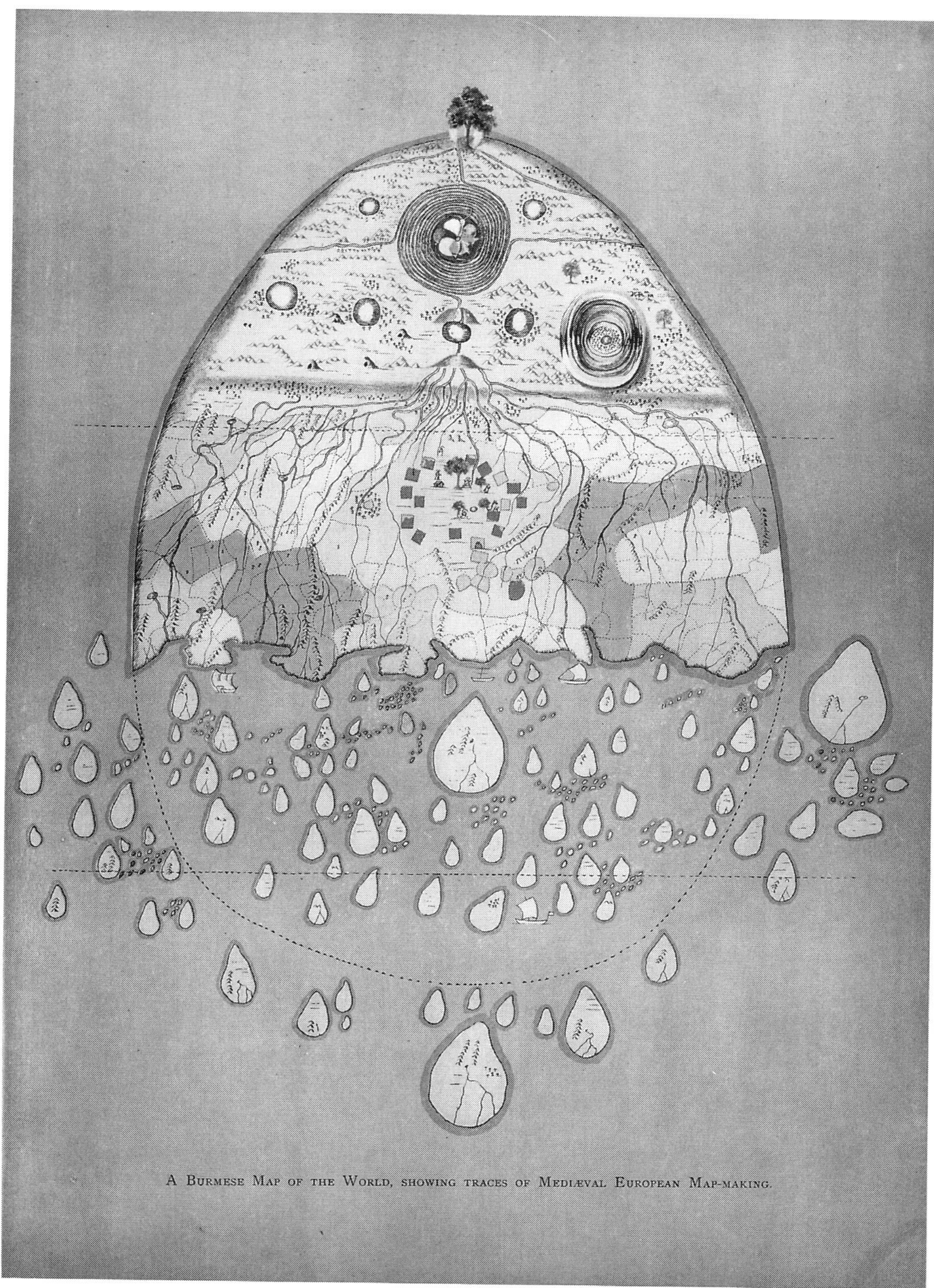

A Burmese Map of the World, showing traces of Mediæval European Map-making.

FIG. 17.22. BURMESE MAP OF JAMBUDĪPA, THE SOUTHERN CONTINENT. Ink and watercolor on paper. The locale, date (probably nineteenth century), author, size, and present whereabouts of this cosmographic map are not known. The map portrays many of the key features of Jambudīpa as noted in Pali Buddhist sources: the shape of the continent, its many auxiliary islands, the great *jambu* tree, the northern Himavanta region, Lake Anotatta and the four great rivers issuing from and swirling around it, Sumeru and its surrounding ranges (to the southeast of the great lake), the river-laced plains of India, and the Cakkavāla mountain wall surrounding the great island. The traditional spatial relationships, however, are in large measure garbled. There is also substantial evidence of European influence in the cartographic signs used, the bounded spaces colored as if they were discrete political entities, and the imposition of what appear to be two parallels of latitude.
From Richard C. Temple, *The Thirty-seven Nats: A Phase of Spirit-Worship Prevailing in Burma* (London: W. Griggs, 1906), figure facing p. 8.

the inferior peoples that come from across the sea. . . .

It is interesting to note that all this is "natural" geography. All the seas known to the Burmans are to the South, and the land of great mountains is chiefly to the North and also to the East and West of them. And beyond these dwell people who are marvellous to them, at any rate by hearsay.

As to the form of the map. It is an attempt to copy a coloured European map of the 17th century, as is shown by the method of indicating mountains and rivers especially, and also of depicting men, animals and ships. Perhaps the most interesting point in the copying is the use of the colours for dividing off the different countries and the dotted lines for their apparent subdivisional boundaries, while in fact no countries are named or meant. The compiler has even gone so far as to represent the Equator and the Tropic of Capricorn [equator and Tropic of Cancer would seem to make more sense] without having any idea as to the meaning of either expression.[95]

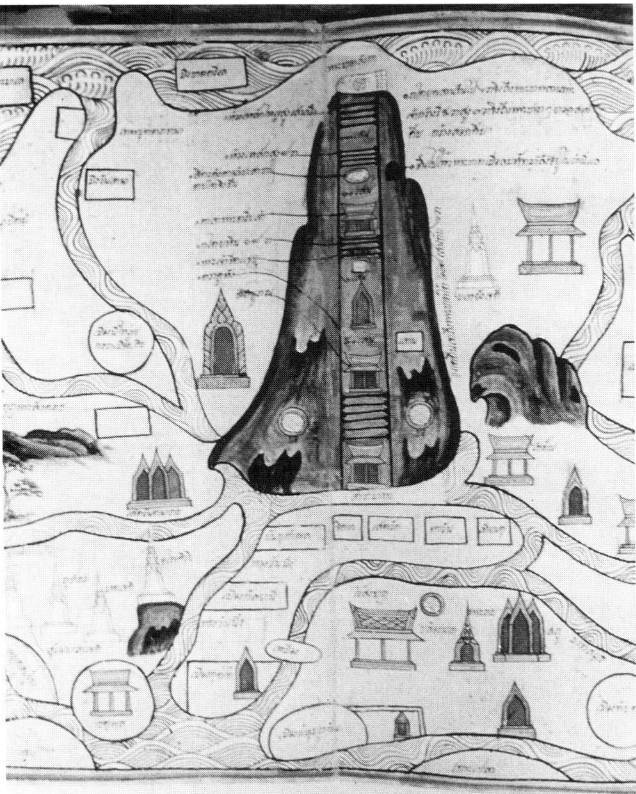

FIG. 17.23. ADAM'S PEAK IN SRI LANKA AS SHOWN IN THE *TRAI PHUM*. At the summit of this peak, the most visually prominent in Sri Lanka, is a large indentation, approximately 1.5 meters long, which Buddhists take to be the footprint (*sripada*) of the Buddha. (Hindus and Muslims, respectively, ascribe the footprint to Śiva and to Adam; hence the peak is a major place of pilgrimage for all three faiths). The height of the peak is given as either 428 or 528 *sens* (figure is not wholly legible) and 6 *wās*, equivalent to approximately 8,572 or 10,572 meters, as against the surveyed altitude of 2,238 meters. Halfway up the peak is a feature identified as the Nila cave. The other named features of the peak, as well as those elsewhere on the island and most of those on other islands, have yet to be determined. Although the shape of Sri Lanka is not recognizable here, the steep, asymmetrical aspect of the peak itself, which has been described as resembling the Swiss Matterhorn, does suggest an attempt to represent the visual impression it conveys to pilgrims. However, the relationship of other identified places on the map to Sri Lanka bears little resemblance to geographic reality. For example, the large island near the lower-right corner bears the toponyms Phuket, Thalang, and Songkhla, which denote an island off the west coast of the Malay Peninsula, a town on Phuket Island, and a town on the east coast of the peninsula.
Size of the original: unknown. From the Bangkok manuscript (Thonburi recension) of the *Trai phum*. Photography courtesy of Joseph E. Schwartzberg.

It is a pity that Temple does not say more about how and why this fascinating map came to be drawn, but it seems not to have to have been made at his behest. It would also be interesting to know on what grounds, other than inference, he could assert that the compiler had knowledge of some unnamed seventeenth-century European map that allegedly provided his model.

For a much more detailed representation of the religious topography of Jambudīpa than that of the map described by Temple, we may return to the *Trai phum*, a very large part of which is devoted to what appear to be terrestrial locales traversed by the rivers flowing south from Lake Anotatta. More than two hundred places are situated along these rivers and on the bodies of water they flow into. Most are towns or countries, with names included in rectangles or, less commonly, in ovals or circles. Others are religious sites, shown by drawings of the events or personages they are associated with. Many rectangles are devoid of names or other text. The places identified include many in India and Nepal that are historically linked with the life of the Buddha as well as many others within South and Southeast Asia whose association with the Buddha is wholly mythological. The topological arrangement of such places cannot be reconciled with that revealed by modern maps. The course of the rivers themselves is for the most part mythological. Individual streams and groups of streams repeatedly divide and reconverge as they wend their often parallel ways over successive folios of the several manuscripts. Based on my recollection of the manuscripts I have personally inspected, the pattern is more or less consistent,

95. Richard C. Temple, *The Thirty-seven Nats: A Phase of Spirit-Worship Prevailing in Burma* (London: W. Griggs, 1906), illustration following p. 8 and explanation on verso side of same, both pages unnumbered (there is also a new edition with an essay and bibliography by Patricia M. Herbert [London: P. Strachan, 1991]).

suggesting that the artists tried to adhere faithfully to the original textual description.

Twelve consecutive panels of the Bangkok manuscript of the *Trai phum* were reproduced in 1984 on a calendar that was distributed as a complimentary New Year's greeting by the Thai magazine *Sinlapa Wattanatham* (Art and culture). Within the first four and a half panels of this portion of the manuscript much of the terrain traversed by the rivers is associated with northern India; but two rivers flowing out of that region lead directly into and through various countries of Southeast Asia, with no indication of the intervening mountain barriers shown on modern maps. Thus the function of the rivers in certain areas may be to provide a pictorial guide to the flow of the narrative rather than to convey a credible sense of geography. The arbitrariness of the sequencing of places is evident in that Takṣaśilā (Taxila), in what is now Pakistan, appears in panel 4, just short of the transition in panel 5 to northern Siam. Panel 6 includes various places in Burma and northern Siam. Panel 7 covers a vast area and includes the former Burmese capital at Pagan, the centrally positioned former Siamese capital at Ayutthaya, places in the Korat Plateau in northeastern Siam, and two small squares at the top of the map labeled Yuan (Vietnam) and Champa, a former Hindu kingdom in southern Vietnam that was conquered by the Vietnamese in the fifteenth century. Tavoy and Martaban on the Tenasserim coast of southern Burma, a number of towns in central Siam and on the Siamese portion of the Malay Peninsula, and islands in the Gulf of Siam are found on panel 8. The positioning of places on panels 7 and 8 indicates a general orientation toward the east. Although most of panel 8 appears to relate to known surviving places, localities such as Bean Island, Banana Island, and Squash Island are more problematic. Also depicted on panel 8 is Hanuman, the monkey hero of the Indian *Rāmāyaṇa* epic, advancing along a causeway to Hanuman's Island, which appears in panel 9, along with one additional Siamese town on the Malay Peninsula. Most of panel 9 consists of water, however, and within it are two large ships, one seemingly Chinese and the other Siamese. A third ship, presumably European, appears in panel 10, which is also mainly water, along with one unidentified circular island and parts of several others including Sri Lanka. Panel 11, illustrated in figure 17.23, includes a confusing cluster of islands, the largest of which is Sri Lanka, as is apparent from the depiction there of Adam's Peak, on whose the summit the Buddha is supposed to have left his footprint. Finally, panel 12 shows the continuation of several of the previously noted islands; a sugarloaf-shaped island bearing the caption "Naga Island, where naked people live"; and a diverse assortment of fish, both realistic and fabulous, including one with the trunk and tusks of an elephant. Nearby a large rectangular island (or perhaps simply an informational box) bears the caption, "Place where the big fish come together, the biggest of which, Anon, is the head of them all; when Anon shakes his body there is an earthquake."[96]

Astronomy, Astrology, Geomancy, and Mental Maps in Relation to Fields of Cosmic Force

Throughout recorded history, most Southeast Asians—animists, Hindus, Buddhists, and Muslims, whatever their estate—have believed that their lives are largely controlled by unseen forces emanating from various parts of the cosmos and, in matters of moment, have ordered their behavior accordingly. As put by Heine-Geldern, among others, a central concern in this regard is

> the parallelism between Macrocosmos and Microcosmos, between the universe and the world of men. According to this belief humanity is constantly under the influence of forces emanating from the directions of the compass and from stars and planets. These forces may produce welfare and prosperity or work havoc, according to whether or not individuals and social groups, above all the state, succeed in bringing their lives and activities in harmony with the universe. Individuals may attain such harmony by following the indications offered by astrology, the lore of lucky and unlucky days and many other minor rules. Harmony between the empire and the universe is achieved by organizing the former as an image of the latter, as a universe on a smaller scale.[97]

Citing as his evidence the designs on Dong-son Bronze Age drums, Wales asserts that the inhabitants of a large part of mainland and insular Southeast Asia "had by the fourth or third century B.C. become aware of the planetary cosmology."[98] But whether or not that inference is correct, it is certain that the subsequent Indianization of much of the region entailed the diffusion of Hindu astronomical knowledge and its associated astrological lore and iconography. This knowledge served a wide range of purposes: calendrical, horological, astrological, as aids in navigation and the regulation of agriculture,

96. I am grateful to Lorraine Gesick, professor of history at the University of Nebraska, Omaha, for sending me copies of the twelve panels of the calendar and for translating selected portions of the text, and also to Chanthaburi Thawut of Rambhai Barni Teachers' College, Chanthaburi, Thailand, who was particularly helpful in the interpretation of panels 11 and 12.

97. Heine-Geldern, "Conceptions of State and Kingship," 15 (note 82), and see also his "Weltbild und Bauform" (note 56). Heine-Geldern's examples and arguments are also largely repeated, frequently amplified, and occasionally challenged in the more extended monograph by Wales, *Universe around Them* (note 1).

98. Wales, *Universe around Them*, 5 (note 1).

and as guides to architecture, town planning, and the spatial organization of state polities.

A diversity of calendars and astrological almanacs based on knowledge of the movements of the sun and moon are currently in use in Southeast Asia. Many, perhaps most, were illustrated. Some were adopted with little change from Hindu models, and others were developed in Southeast Asia itself. Both the Indian and the derivative Southeast Asian systems necessitated the recognition of the appropriate zodiacal constellations. Manuscripts containing horoscopic diagrams—in effect, maps of the positions of the principal heavenly bodies at a particular moment in time—are common, and such diagrams are sometimes seen painted or portrayed in bas-relief on the walls of temples and other important edifices, presumably to commemorate the auspicious day when they were formally consecrated.

Approximately half of chapter 9 of the *Trai phum* is devoted to a carefully reasoned exposition of astronomy. It commences as follows:

> The space in the middle between the region of the Cakkavāḷa mountain range and the Yugandhara mountain range is the area where there are paths for the sun, the moon, the planets, and the multitude of stars, and where they travel back and forth in an orderly fashion in paths that enables us to know the years and the months, the days and the nights, and to know the events, both good and bad.[99]

The text also provides guides for recognizing the constellations; for example: "The group of stars called *assayuja* has five gem abodes in a line. The lunar mansion called *bharaṇī* has three gem abodes close together like the three stones that hold a cooking pot."[100] Diagrams in the illustrated manuscripts exhibit some of these heavenly relationships and aid in the recognition of particular zodiacal configurations.

Despite the importance of astrology, Southeast Asia has never had, it appears, any form of astronomical observatory—comparable, say, to the observatories established by Sawai Jai Singh in India—to take careful measurements of the positions and paths of various heavenly bodies over the course of years.[101] Nor were planispheric or spherical astrolabes or any other comparably sophisticated astronomical instruments in use in that region. Nevertheless, Eade reports that a number of dated inscriptions from Pagan, the Burmese capital from A.D. 1044 to 1287, give planetary positions in degrees and minutes, which he found accurate to with two or three minutes, and in a few cases to the very minute. How such accuracy was attained is not indicated.[102]

Evidence from many parts of Southeast Asia attests to the sensitivity of ordinary people to their orientation in cosmic space, usually with particular reference to compass directions, but sometimes with respect to other dyadic referents, such as sky/earth (simultaneously symbolizing male/female), Upperworld/Underworld (also male/female), mountain (inland)/sea, upstream/downstream, interior/exterior (with respect to houses), toward the capital/away from the capital, and so forth. To know that one is positioned correctly in a specific situation entails carrying in one's head an ever-changing contextually specific mental map; for what is a horoscopically correct pattern of movement on one day of the week or month may be altogether incorrect on another. To treat this issue in detail would take us beyond the scope of this study, but it is appropriate to direct attention to it, since it forms a particularly promising area for future research that may yield valuable insights on the circumstances in which it was deemed necessary to move from the realm of mental map to tangible graphic artifacts.[103]

Apart from astrology and other forms of divination, beliefs in Southeast Asia that incorporate considerations of directionally determined cosmic forces include those based on geomantic principles. I do not know to what degree the Chinese system of *feng shui* (wind and water) has found acceptance among various Southeast Asian ethnic groups. Certainly the drawings that constitute figure 17.24, extracted from a Burmese text giving the rules for siting halls for the ordination of Buddhist monks, do appear to conform to a similar set of principles. But the work in question, published in 1967, is allegedly a translation of a part of the Tipiṭaka (Three baskets; Sanskrit Tripiṭaka) of the Pali Buddhist canon that was originally codified in the third century B.C. in India, a country that lacks a well-articulated geomantic tradition. (The specific portion of the Tipiṭaka from which the text is said to have been taken is the *Vinayapiṭaka*, the portion of the law prescribing rules for monks.) The explanation of this seeming paradox may be that the recent Burmese recension of the Pali text (one of a number prepared by various sects over time) deviated from the original in much the same way as does the noncanonical *Trai phum*. This

99. Reynolds and Reynolds, *Three Worlds*, 277–89; quotation on 277 (note 57). See also the unnumbered color plate from the Berlin manuscript showing "the astral paths" and the diagrams of the astronomical system by Reynolds and Reynolds on 360–61.

100. Reynolds and Reynolds, *Three Worlds*, 280 (note 57).

101. On Jai Singh, see Schwartzberg, "Cosmographical Mapping," 361–67 (note 2).

102. J. C. Eade, *Southeast Asian Ephemeris: Solar and Planetary Positions, A.D. 638–2000* (Ithaca, N.Y.: Cornell Southeast Asia Program, 1989), 7.

103. Among works that discuss the subject are Geertz, *Religion of Java* (note 36); H. L. Shorto, "The Planets, the Days of the Week and the Points of the Compass: Orientation Symbolism in 'Burma,' " in *Natural Symbols in South East Asia*, ed. G. B. Milner (London: School of Oriental and African Studies, 1978), 152–64; and Wales, "Cosmological Aspect" (note 4).

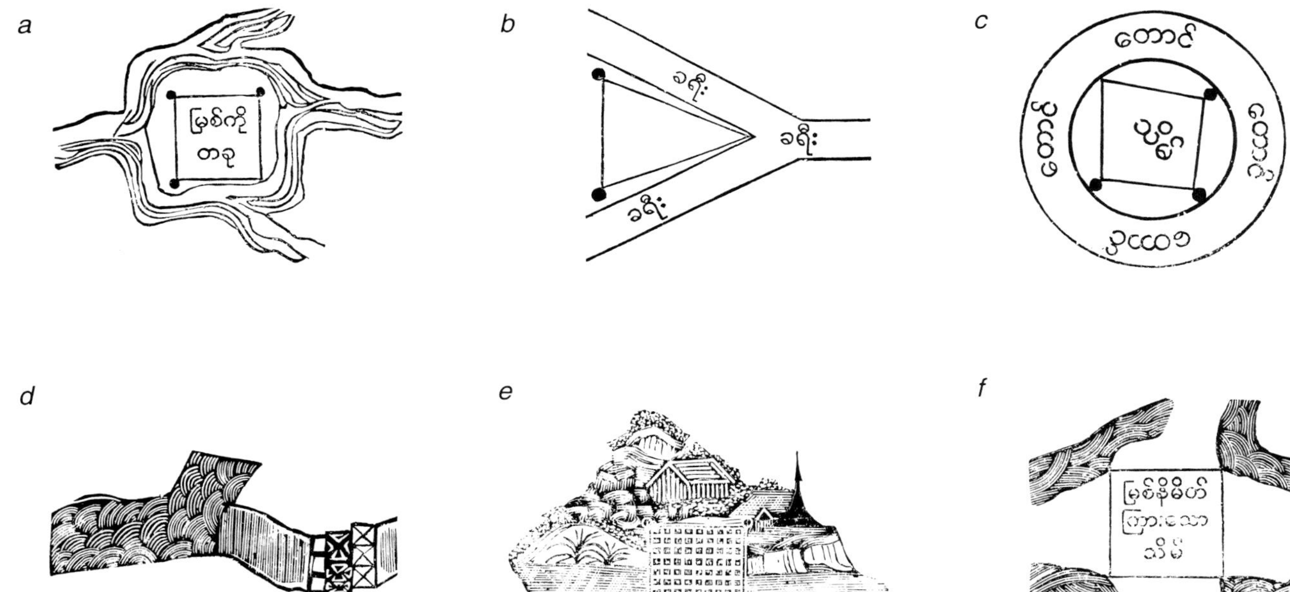

FIG. 17.24. DIAGRAMS PRESCRIBING RULES FOR THE SITING OF ORDINATION HALLS FOR BUDDHIST MONKS IN BURMA. These are six out of a total of eighty-seven paradigmatic illustrations, each accompanied by a brief explanatory text that prescribes what is permissible and what is proscribed in the siting and manner of construction (e.g., with respect to height) of ordination halls so as to remain in cosmic harmony with the physical characteristics of the locality where the hall is to be placed. The rules for the hall in the island (*a*), for example, state that the hall should have four corners (the number of dots plus one), with only one corner touching a river, while for *b* the rule is that a hall built in the fork of a road should have three corners and (presumably) that one corner should extend to the fork. The rule relating to the hall within a double circle (*c*) states that termite mounds may be used (but how and why?) for locating the boundaries of the hall. Judging solely from appearances, the rules for *d–f* and other diagrams seem to relate to construction on various types of terrain (smooth, rocky, cultivated, hilly, forested, etc.), on streams with different flow characteristics, in proximity to sacred trees, in areas that are fenced or artificially bounded in various ways, and so forth. The book in which these diagrams appear is a 1967 edition of a compilation of four works forming part of an 1863 recension of the ancient Pali Buddhist canon, the Tipiṭaka. It is printed in the Burmese script but written in the Pali language.
From *Wī thi bhuṃ cañ‘ chan‘" puṃ sim‘ puṃ* (Rangoon: Ū" Po‘ Rañ‘-Do‘ Co Rañ‘, 1967), 162, 165, 168, 169.

seems likely in that the published book the diagrams were taken from was a new edition of an 1863 compilation of four texts that came out of the Fifth Buddhist Council, convened in that year by the then reigning Burmese king Mindon. By this time Chinese geomantic ideas may have taken strong enough root in Burma to have been incorporated into the monastic rules.[104]

Although perhaps not geomantic in the Chinese sense of the term, the principles underlying temple construction in Southeast Asia also embody ideas of harmony with cosmic forces and their directional orientation. Moreover, they did commonly lead to the graphic depiction on the ground of the cosmic giant Vāstupuruṣa, represented by an eighty-one-square mandala, following the Hindu rites of construction that have been previously described and illustrated. Since the chief gods of the Hindu pantheon are embodied in the mandala, its emplacement in the temple foundation establishes the desired bond between the temple as microcosm and the universe as macrocosm.[105]

Similarly, in laying out traditional state capitals Southeast Asian monarchs and their counselors were acutely sensitive to the need for harmony with forces operating within the macrocosm and for displaying through cosmic symbolism the legitimacy of their rule. The architectural

104. The collection of texts in question was compiled by a Burmese monk, Maingkhaing Sayadaw. The book they appeared in is titled *Wī thi bhuṃ cañ‘ chan‘" puṃ sim‘ puṃ* (The way to nibbana: Its stages, how wonderful, how gentle). The 1967 edition (in the Pali language, but printed in the Burmese script) was published in Rangoon by Ū" Po‘ Rañ‘-Do‘ Co Rañ‘ for the Ratana'wādi Piṭakat‘ Bookshop. The four works contained therein were "Wi htē pon" (Mental processes), "Bon-sin" (Planes of existence; obviously, judging from the illustrations, a work of cosmography and a part of the *Abhidhammapitaka* [Basket of metaphysics]), "S'an-pon" (Poetics), and "Sein pon sima" (Ordination hall construction). The selected diagrams were excerpted from the last of these four sections (pp. 161–88). I am indebted to Reverend Kelatha of the Burmese Buddhist monastery of Montgomery County, Maryland, and to Mr. Maung of Washington, D.C., for assistance with the interpretation of this book.

105. Schwartzberg, "Cosmographical Mapping," 378–79 (note 2), and Wales, *Universe around Them*, 40–41 (note 1).

plans, which were undoubtedly drawn up as a guide to the layout of some, if not all, of those capitals, are as far as I know not extant (with one known exception, that for Mandalay, discussed in the following chapter). However, the guiding plan is evident in archaeological remains, which are in effect giant cosmographies in stone. The layout of the states themselves also commonly followed cosmographic principles, even though in this respect no graphic representations are known to have existed. Writing of Southeast Asia as a whole, Heine-Geldern observes that

> there is overwhelming evidence of the cosmological basis of state and kingship in this area. This evidence is found in numerous passages in literature and inscriptions, in the titles of kings, queens and officials, in the "cosmic" numbers of queens, ministers, court priests, provinces, etc., in rites and customs, in works of art, in the lay-out and structure of capital cities, palaces and temples. One need only put these various items together to obtain a relatively clear picture. This picture will be more complete in continental Southeast Asia, where the old forms of Buddhist state and kingship survived into very recent times. It will be hazier in the Archipelago as a result of Mohammedan and European influences.
>
> In Southeast Asia, even more than in Europe, the capital stood for the whole country. . . . the capital city could be shaped architecturally as a much more "realistic" image of the universe, a smaller microcosmos within that microcosmos, the empire. The remains of some of the ancient cities clearly testify to the cosmological ideas which pervaded the whole system of government.[106]

Thus, in Angkor and elsewhere moats and artificial basins came to represent cosmic rivers, lakes, and oceans; and sculptural compositions, often on a monumental scale, were employed to recreate cosmogonic myths. Considerations of space and focus preclude a more extended discussion of this topic, but interested readers will find it amply treated in numerous works by art historians and other scholars.[107]

Finally, one must note the connections of the state not only with Buddhist cosmography, but also with Hindu astronomy. As Heine-Gildern notes, from a passage in the *Xin Tang shu* (New history of the Tang, compiled 1032?–60),

> the kingdom of Java in the 9th century was divided into twenty-eight provinces, their governors together with the four ministers again having numbered thirty-two high officials. This may have been a somewhat older form of the same system, in which the provinces corresponded to constellations, the twenty-eight "Houses of the Moon," and the four ministers to the guardian gods of the cardinal points. It is clear that in all these cases the empire was conceived as an image of the heavenly world of stars and gods.[108]

106. Heine-Geldern, "Conceptions of State and Kingship," 16–17 (note 82).

107. Among those relating mainly to Angkor, in chronological order, are Henri Marchal, *Guide archéologique aux temples d'Angkor, Angkor Vat, Angkor Thom, et les monuments du petit et du grand circuit* (Paris: G. Van Oest, 1928), particularly useful for its many site plans; George Groslier, *Angkor*, 2d ed. (Paris: Librairie Renouard, 1931), a pioneering descriptive work; George Coedès, *Angkor: An Introduction*, ed. and trans. Emily Floyd Gardiner (Hong Kong: Oxford University Press, 1963), originally published as *Pour mieux comprendre Angkor* (Hanoi, 1943), an interpretative analysis; and Bernard Philippe Groslier and Jacques Arthaud, *Angkor: Art and Civilization*, rev. ed., trans. Eric Ernshaw Smith (New York: Frederick A. Praeger, 1966), with magnificent photographs.

108. Heine-Geldern, "Conceptions of State and Kingship," 21 (note 82).

Postscript

Since this chapter was set in type, an important Cambodian cosmographic manuscript has come to light. It is an undated (probably late eighteenth or early nineteenth century) richly illuminated multicolor painted adaptation of the Thai *Trai phum*, prepared in the same accordion-style format on indigenous paper. The work is held in the National Library in Bangkok and is the subject of a well-illustrated seventy-nine-page monograph by the National Library, Fine Arts Department, *Traiphum chabap phasa khamen* (The Cambodian manuscript of the *Trai phum* cosmography), translated into Thai by Amphai Khamtho (Bangkok: Ammarin Printing Group, 1987).

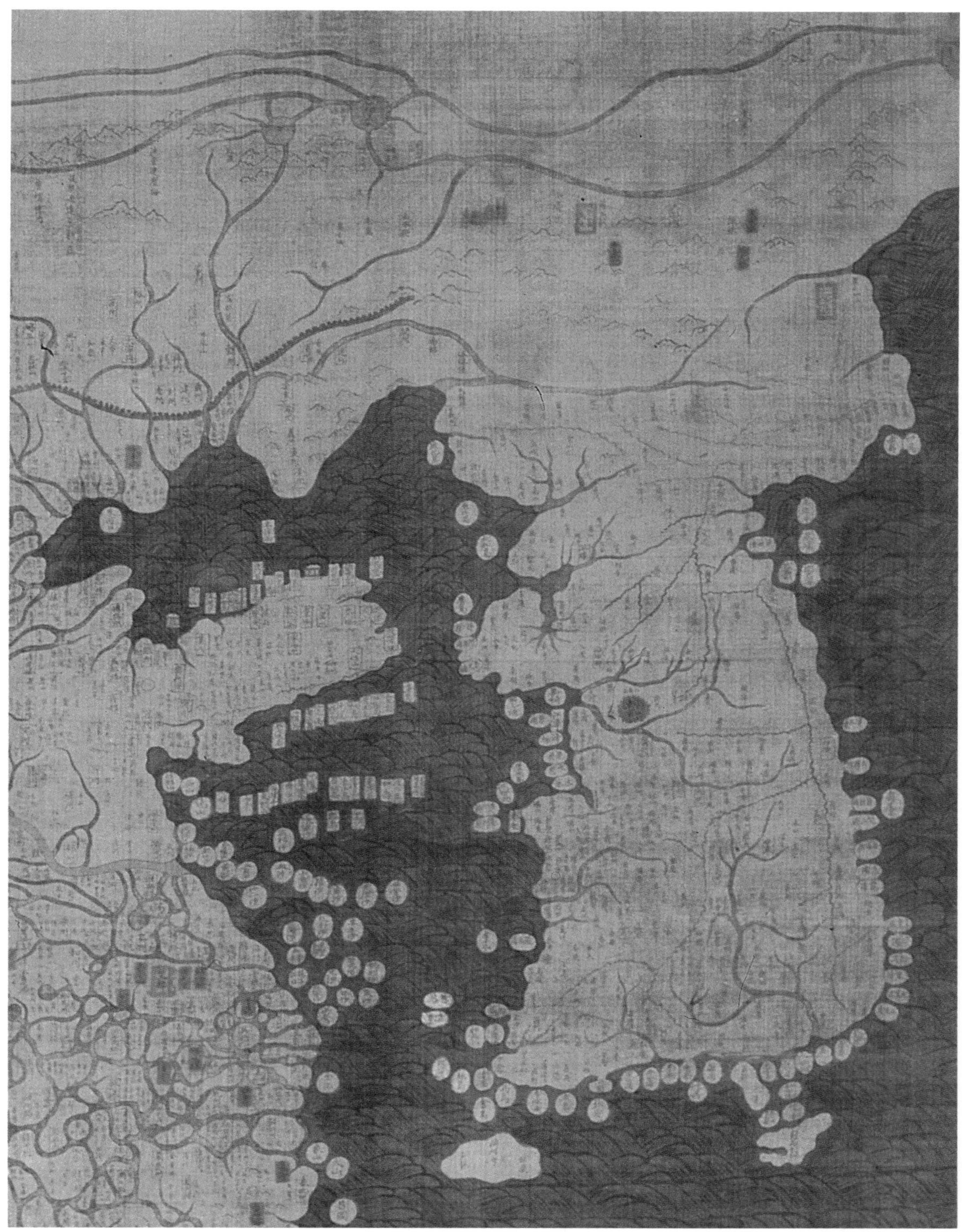

PLATE 17. THE KOREAN PENINSULA FROM THE *KANGNIDO*, BY YI HOE AND KWŎN KŬN. (See pp. 283 and 289.) This is the oldest Korean map of Korea to survive, with an amazingly accurate coastline for its time. The northern frontier has a flattened appearance, as on many later maps down to the seventeenth century. The map has an abbreviated shapes-and-forces character, with the main mountain "artery" (*maek*) along the east coast and several more extending westward. One of them reaches to the area of Seoul, marked with a crenellated circle. Many inlets and harbors are indicated with elliptical cartouches. Presence and absence of certain northern districts shows that this is a copy of about 1470.
Size of the detail: ca. 80 × 60 cm. By permission of Ryūkoku University Library, Kyōto, Japan.

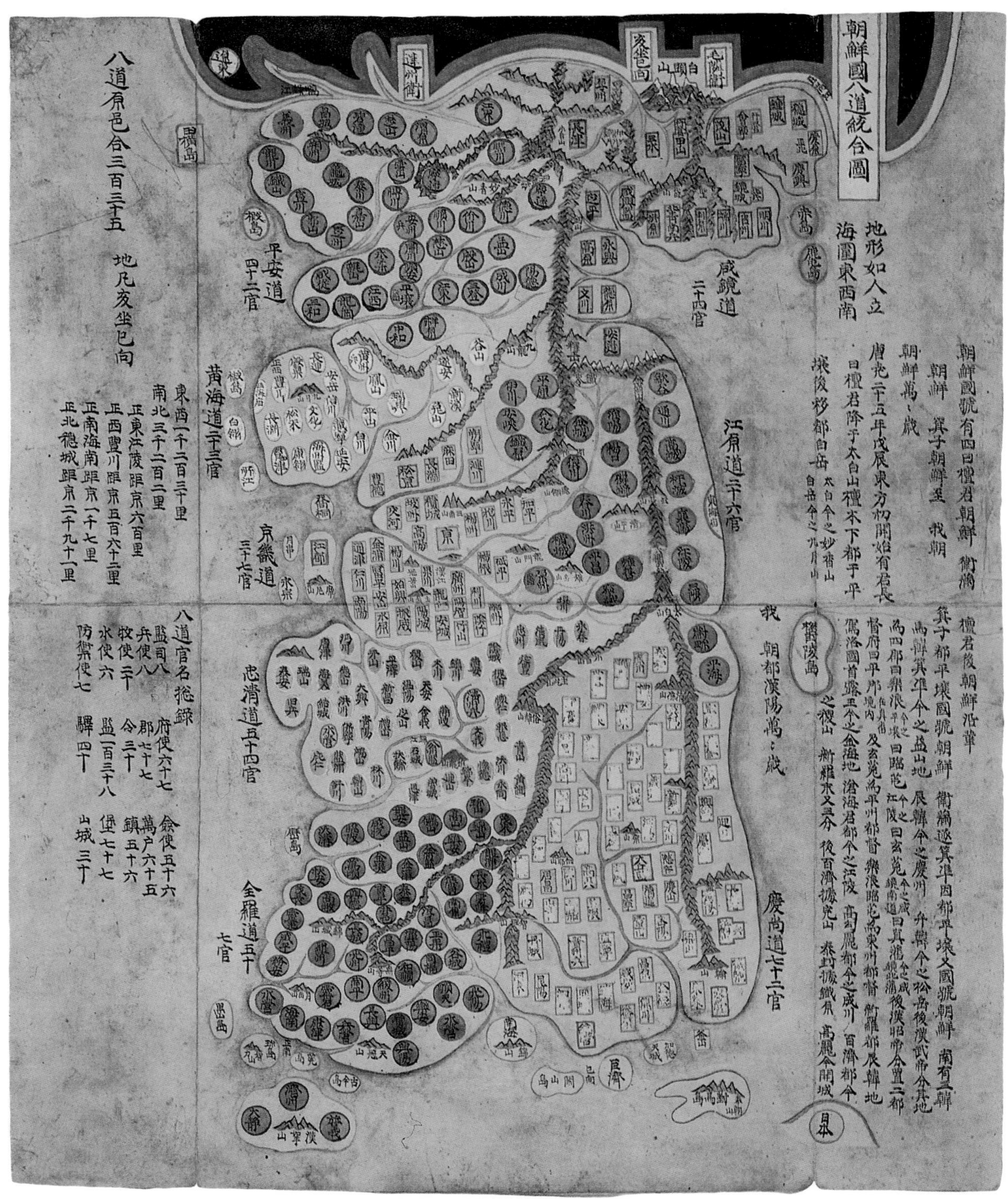

PLATE 18. *CHOSŎN'GUK P'ALTO T'ONGHAPTO* (CONSOLIDATED MAP OF THE EIGHT PROVINCES OF CHOSŎN). (See p. 292.) The flat northern frontier and the distinction of district name cartouches by provincial colors are characteristics of Chŏng Ch'ŏk–style maps, but the indication of the major mountain ranges evokes the treatment of the Korean portion of the *Kangnido*, by Yi Hoe and Kwŏn Kŭn. But though typologically early, the distinctive folk-art style marks it as an eighteenth- or nineteenth-century product. The left margin has information on the overall dimensions of the country and the number of its administrative districts. On the right are notes on the origins of Korea, including the slogan "Chosŏn for myriad and myriad years!"
Size of the original: 50.3 × 40.8 cm. By permission of Yi Ch'an, Seoul.

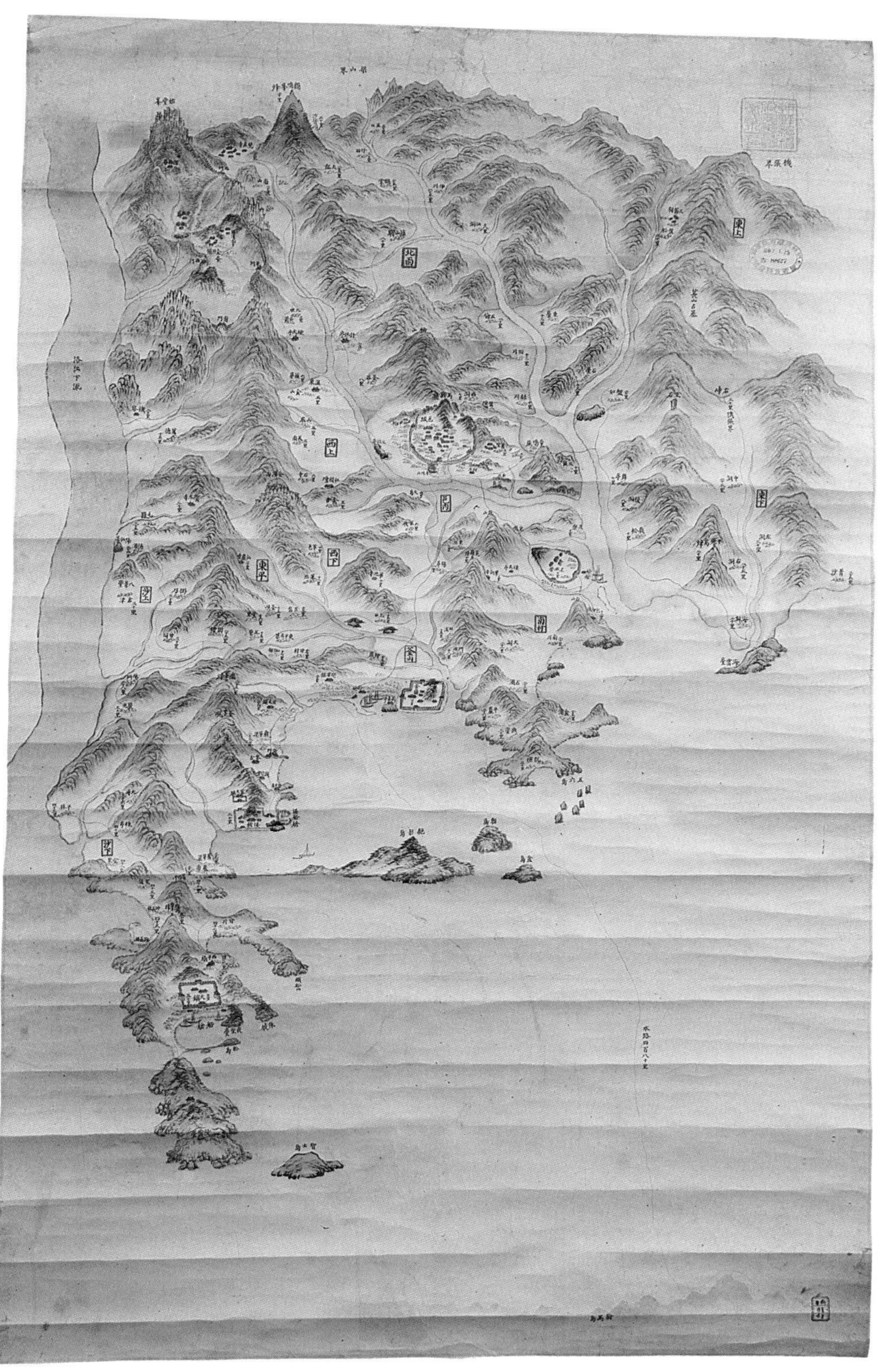

PLATE 19. *TONGNAE PUSAN KO CHIDO* (OLD MAP OF TONGNAE AND PUSAN). (See p. 329.) This representation of the district of Tongnae, from a scroll of unknown date, is a typical example of the map-painting, which was generally favored for local maps and coexisted with the more scientific cartography of the provincial and national levels. It filled many of the functions of a map, and all the lands of the district are included. Distances are indicated by notes. In the center is the walled seat of Tongnae district, with a road leading south about twenty *li* to the "Japan House" (Waegwan) at Pusan Harbor. A major Korean naval base lies across the hills to the northeast; other military installations watch the Japanese from various positions around the harbor. The large river at the left is the mouth of the Naktong.
Size of the scroll: 133.4 × 82.7 cm. By permission of the National Central Library, Seoul (cat. no. Kwi 112 Kojo 61–41).

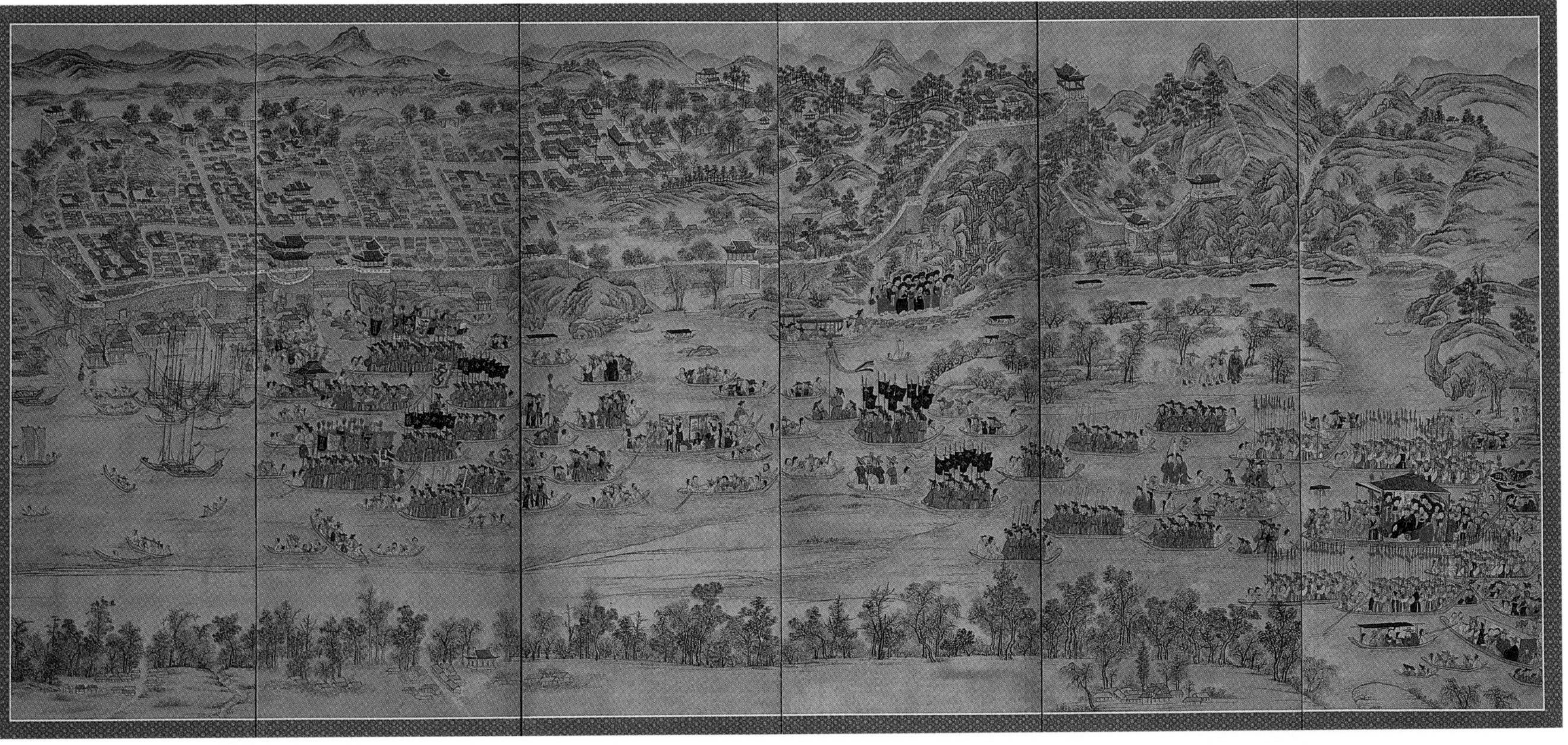

PLATE 20. UNTITLED VIEW OF P'YŎNGYANG, WITH PARTICIPANTS IN A FESTIVE OCCASION ON THE RIVER IN THE FOREGROUND. (See p. 337.) This large screen shows a procession of boats escorting a high official (far right), with the city of P'yŏngyang, capital of P'yŏng'an (or Kwansŏ) Province, in the background. A boat in the center of the procession has a pennon reading "Commander of All Military Forces in Kwansŏ," which would describe one of the powers of the provincial governor. Many of the boats in the procession carry soldiers. The painting possibly commemorates the arrival of a new P'yŏng'an governor. The walls and gates of P'yŏngyang, the governor's compound, and the streets and residential areas of the city can be seen in the background. Although the festive and celebratory character of this painting transcends its character as a map of the city, map-paintings of P'yŏngyang, without the fanfare and fun, frequently feature such bird's-eye views eastward across the Taedong River and Nŭngna Island, as here. The map of P'yŏngyang adds an important symbolic element to this highlight from a prominent official's career.

Size of the original: 125.5 × 286.6 cm. Photograph courtesy of Christie's, New York.

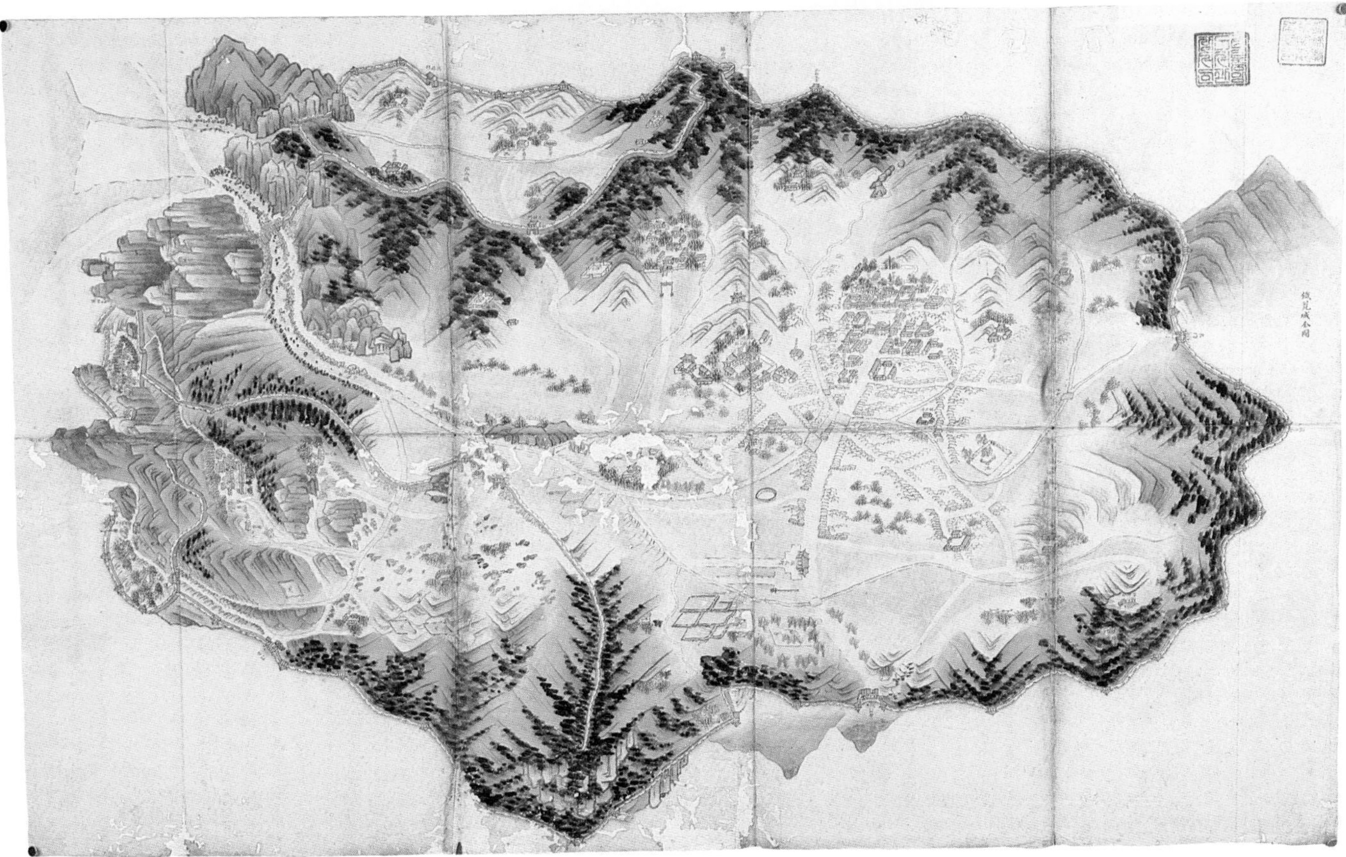

PLATE 21. *CH'ŎRONGSŎNG CHŎNDO* (COMPLETE MAP OF IRON JAR FORTRESS). (See pp. 330 and 343.) This map-painting, which on art-historical grounds was done during the eighteenth century, shows the fortress of Ch'ŏrong proper on the peak known as Yaksan, at the extreme left; it has its own separate wall. Spreading out from left to right is the adjacent district seat of Yŏngbyŏn, headquarters of a major military command; its wall was over thirteen kilometers in perimeter. At the upper left is an entrance into Puk Sansŏng (North Fortress), which also had independent fortifications. The buildings of the military command and the district government are at center right, with the grass-roofed homes of the townspeople clustered to the south. The Chosŏn kings invested heavily in this complex during the seventeenth century (see text); the results are evident here. Yaksan is famous in Korea for its azaleas. The area at left, already overwhelming in its scenic grandeur, must have been ravishing in late March.
Size of the original: 78.7 × 120.3. By permission of the National Central Library, Seoul (cat. no. Ko 2702–20).

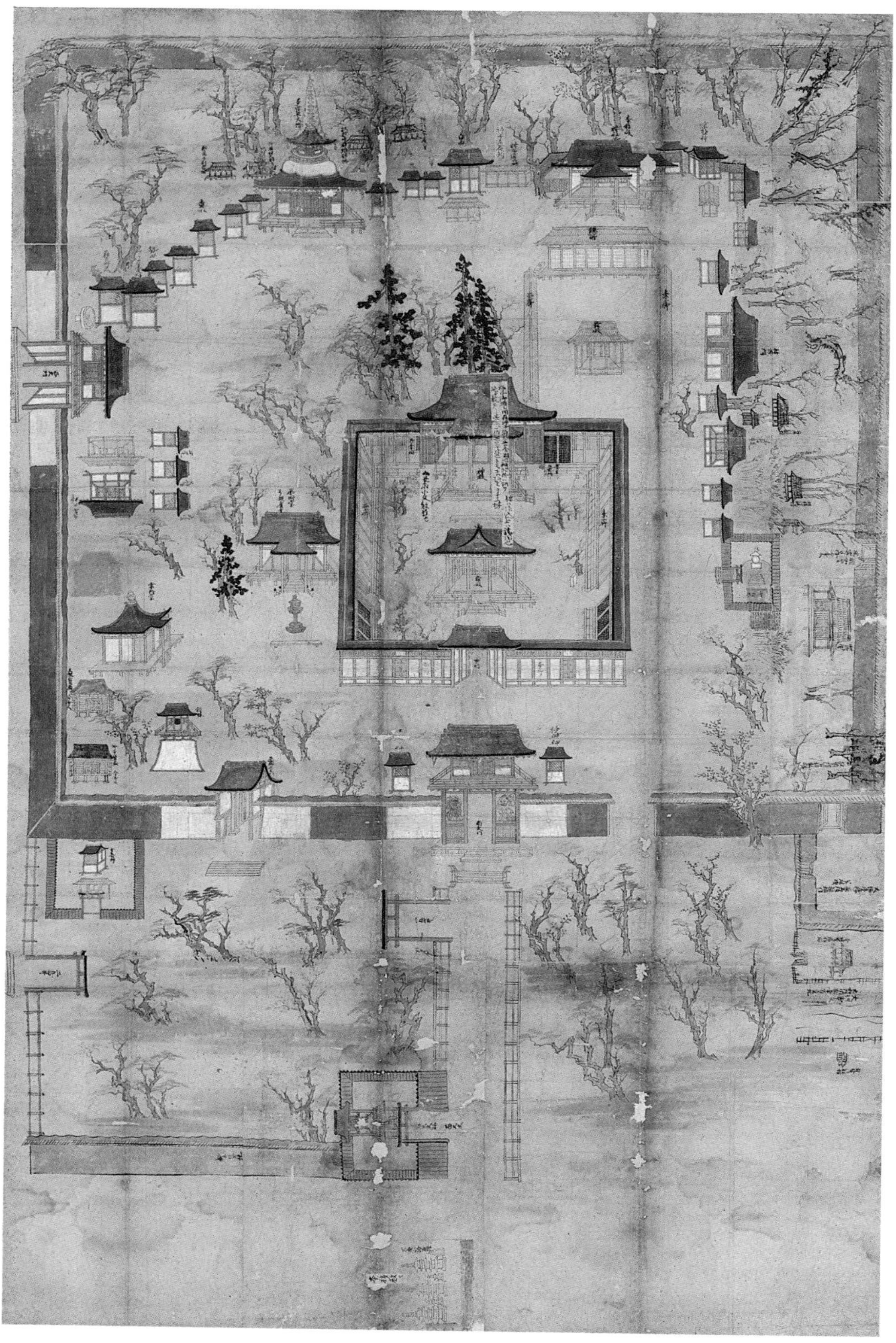

PLATE 22. THE *GION OYASHIRO EZU* (MAP OF GION SHRINE) OF 1331. (See p. 364.) Emphasis in the manuscript is placed on the buildings, which for the most part are two-dimensional from a head-on perspective. The map is oriented so as to be viewed from two directions, as above and with the left side at the bottom.
Size of the original: 167 × 107.5 cm. By permission of Yasaka Shrine, Kyōto.

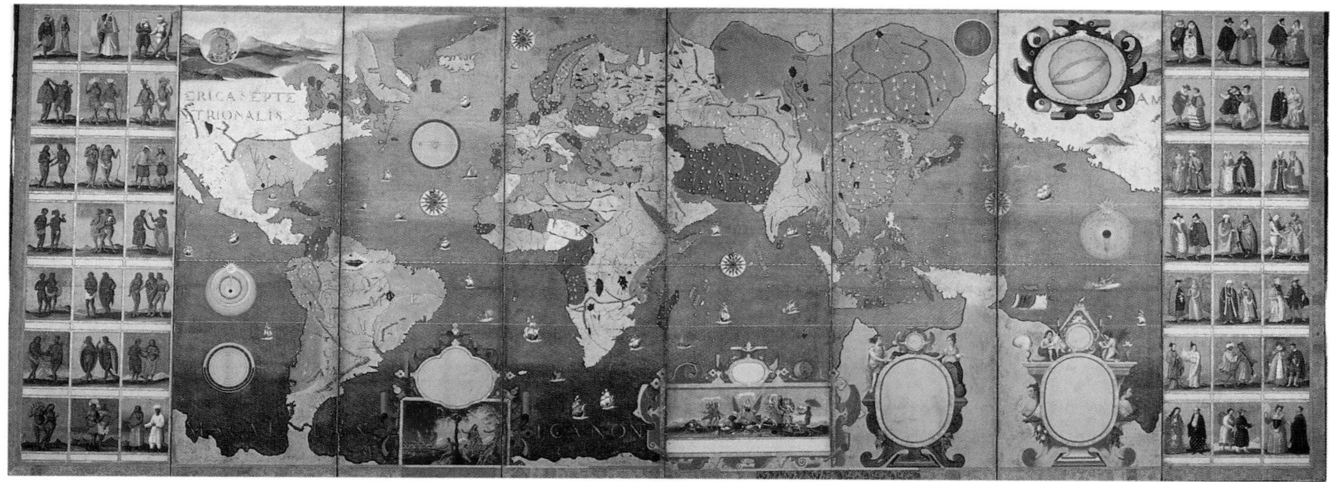

PLATE 23. THE EARLY SEVENTEENTH-CENTURY *BANKOKU EZU* (MAP OF ALL THE COUNTRIES) WITH ITS COMPANION SCREEN SHOWING VIEWS OF TWENTY-EIGHT CITIES. (See p. 380.) These are on a pair of eight-fold screens. The world map is one of the last *Nanban*-style and is presumed to have been modeled on van den Keere's 1609 map on the Mercator projection. Although the van den Keere map has not been found, this presumption is based on the 1619 revision of it and on Willem Jansz. Blaeu's world map of 1606–7 that served as a model for van den Keere. The illustrations of costumed people and the views of the cities are purely European in style and are also thought to have been copied from van den Keere's work.

Size of the original: 177 × 483 cm (map) and 178 × 465 cm (views). By permission of the Imperial Household Agency, Tokyo.

PLATE 24. SHIBUKAWA HARUMI'S TERRESTRIAL GLOBE OF 1690. (See p. 391.) The earliest of the terrestrial globes made by Shibukawa, it was dedicated to the Grand Shrine at Ise along with his celestial globe made the following year. On it may be seen the influence of Matteo Ricci's world map, especially Magellanica and the place-name "Nohakinea" (Nova Guinea) written in katakana. Each country is painted in a different color, Japan being in gold.

Diameter of the original: 24 cm (manuscript on paper). By permission of the Jingū Historical Museum, Ise.

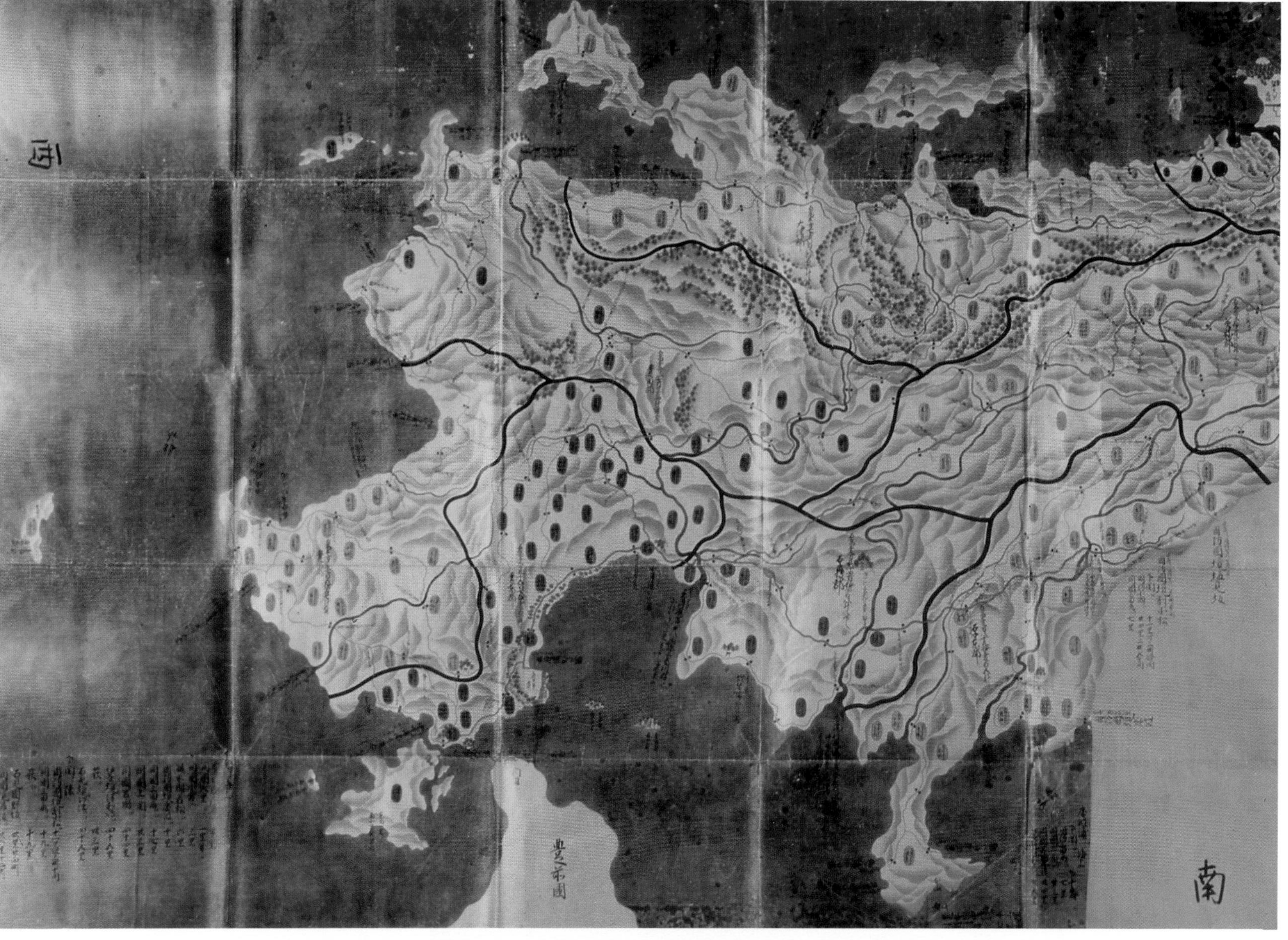

PLATE 25. AN EXTRACT FROM A SHŌHŌ PROVINCIAL MAP: NAGATO PROVINCE (NOW PART OF YAMAGUCHI PREFECTURE) OF 1649. (See p. 397.) The Shōhō enterprise was the third of five projects to compile provincial maps undertaken by the Tokugawa shogunate, and the first to include detailed instructions for compiling the maps from field surveys. On the manuscript, there are marks one *ri* apart along the main roads, and the color of the ovals symbolizing towns and villages differs by county.

Size of the entire original: 334 × 480 cm. By permission of the Yamaguchi Prefectural Archives, Yamaguchi.

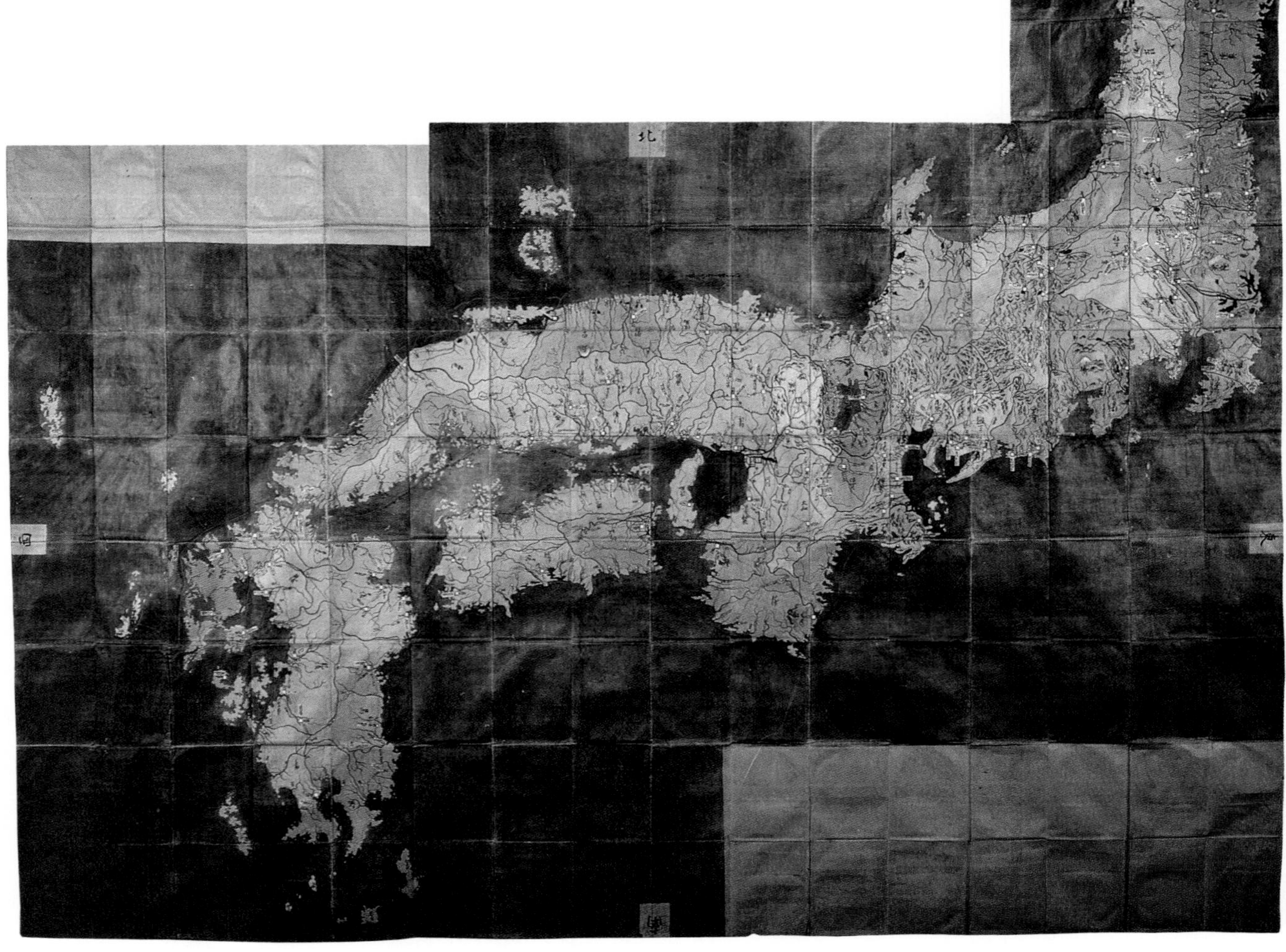

PLATE 26. KEICHŌ MAP OF JAPAN, CA. 1653. (See p. 397.) Dating and identifying this map with the Keichō project, which was inaugurated in 1605, is based on three important sets of information: the seats of the daimyates as they were before 1633, the names of the daimyos in 1639, and the names of the daimyos in 1653. Although it cannot be said with absolute certainty, the geographical information most likely came from the Keichō project (the Kan'ei project, the second of five, began about 1633), and the original map of the country was compiled about 1639. This manuscript, then, would have been a revision. Size of the original: 370 × 434 cm. By permission of the National Diet Library, Tokyo.

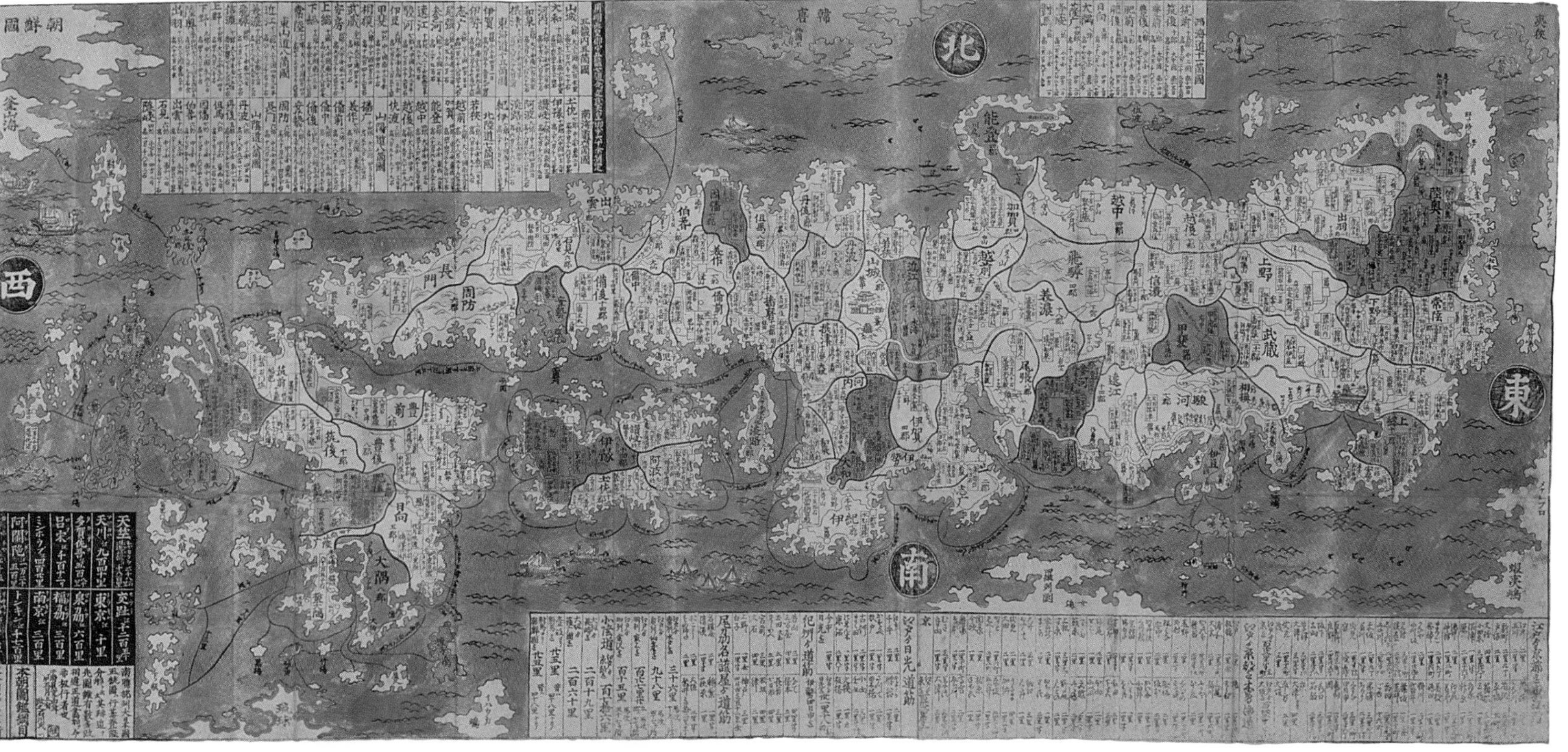

PLATE 27. THE *HONCHŌ ZUKAN KŌMOKU* (OUTLINE MAP OF JAPAN, 1687) BY THE *UKIYOE* ARTIST ISHIKAWA RYŪSEN. (See p. 412.) Based on the modified Keichō-type map, it was made more useful and decorative by adding such things as post towns, distances between these towns, waves, and ships. Revisions of this woodblock print were published for nearly a century, and works based on this original are known collectively as Ryūsen-type maps of Japan. This is the first edition of a map of Japan to have the name Ishikawa Ryūsen.

Size of the original: 58 × 127.7 cm. By permission of the National Archives, Tokyo.

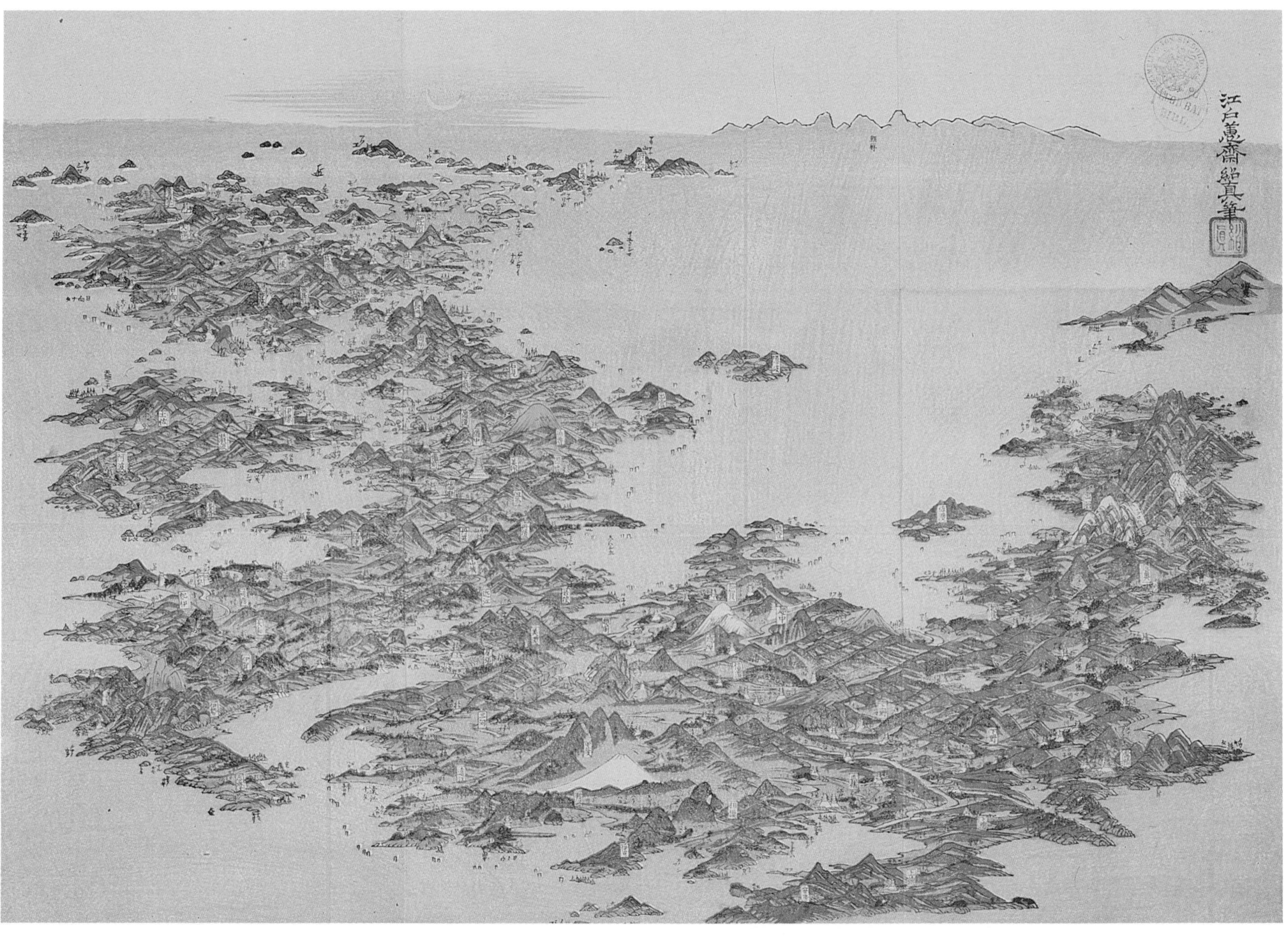

PLATE 28. THE *NIHON MEISHO NO E* (PANORAMIC VIEW OF THE NOTED PLACES OF JAPAN, CA. 1804). (See p. 416.) The oblique aerial composition shown in this woodblock print was done by the *ukiyoe* artist Kuwagata Keisai. Kuwagata was the first painter to portray Japan as if seen from above.

Size of the original: 42 × 59 cm. By permission of the Bibliotheek der Rijksuniversiteit, Leiden.

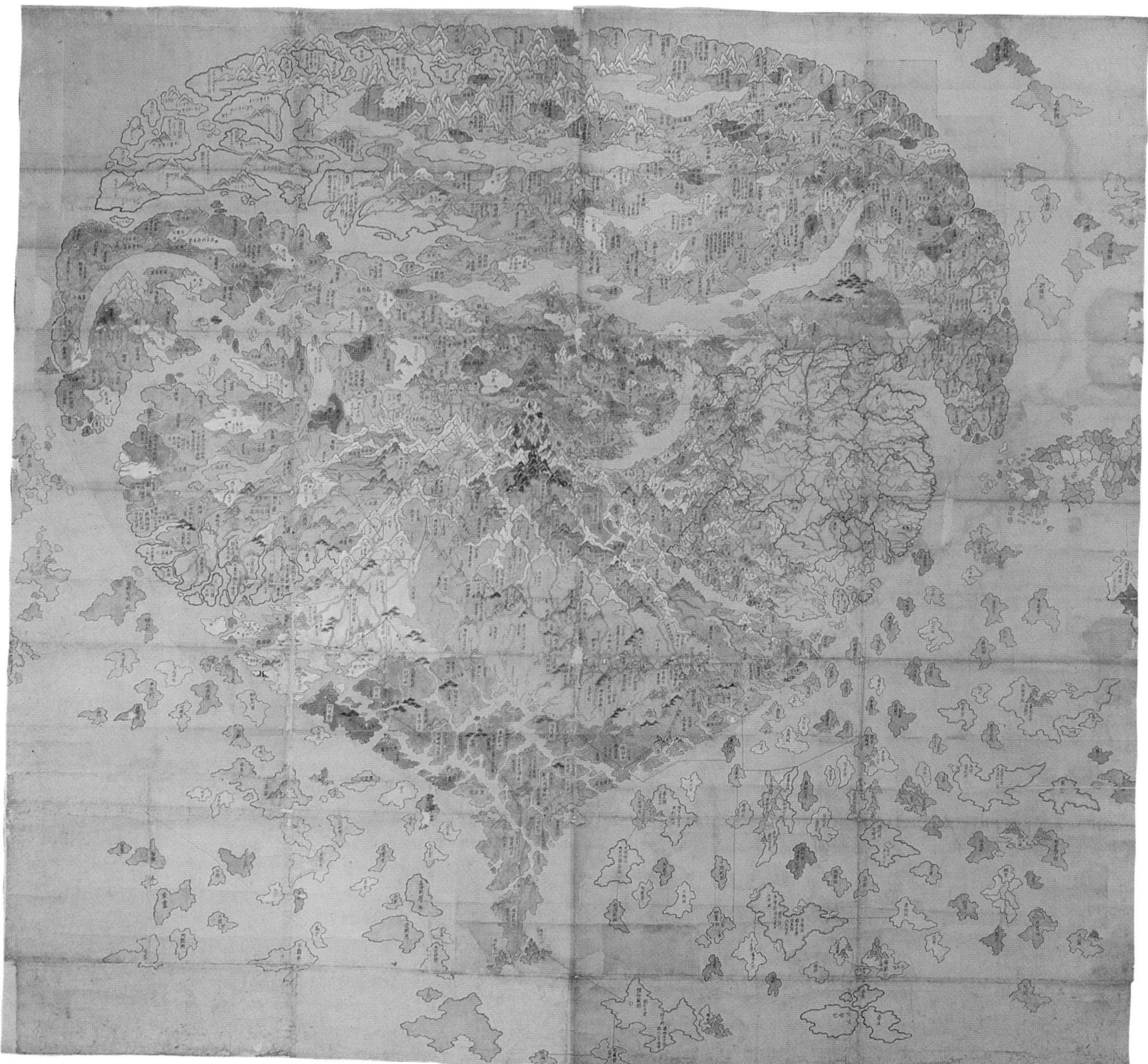

PLATE 29. A MAP OF JAMBŪDVĪPA, CA. 1709. (See p. 429.) This example of later Buddhist world maps has incorporated within it European geographical knowledge, including Europe itself in the upper left corner. Judging from the content and basic structure, it was most likely composed by the monk-painter Sōkaku. The manuscript is an intermediate stage between Sōkaku's ca. 1698 and Rōkashi's 1710 maps of Jambūdvīpa (see fig. 11.59).

Size of the original: 152 × 156 cm. By permission of the Kōbe City Museum, Kōbe, Nanba Collection.

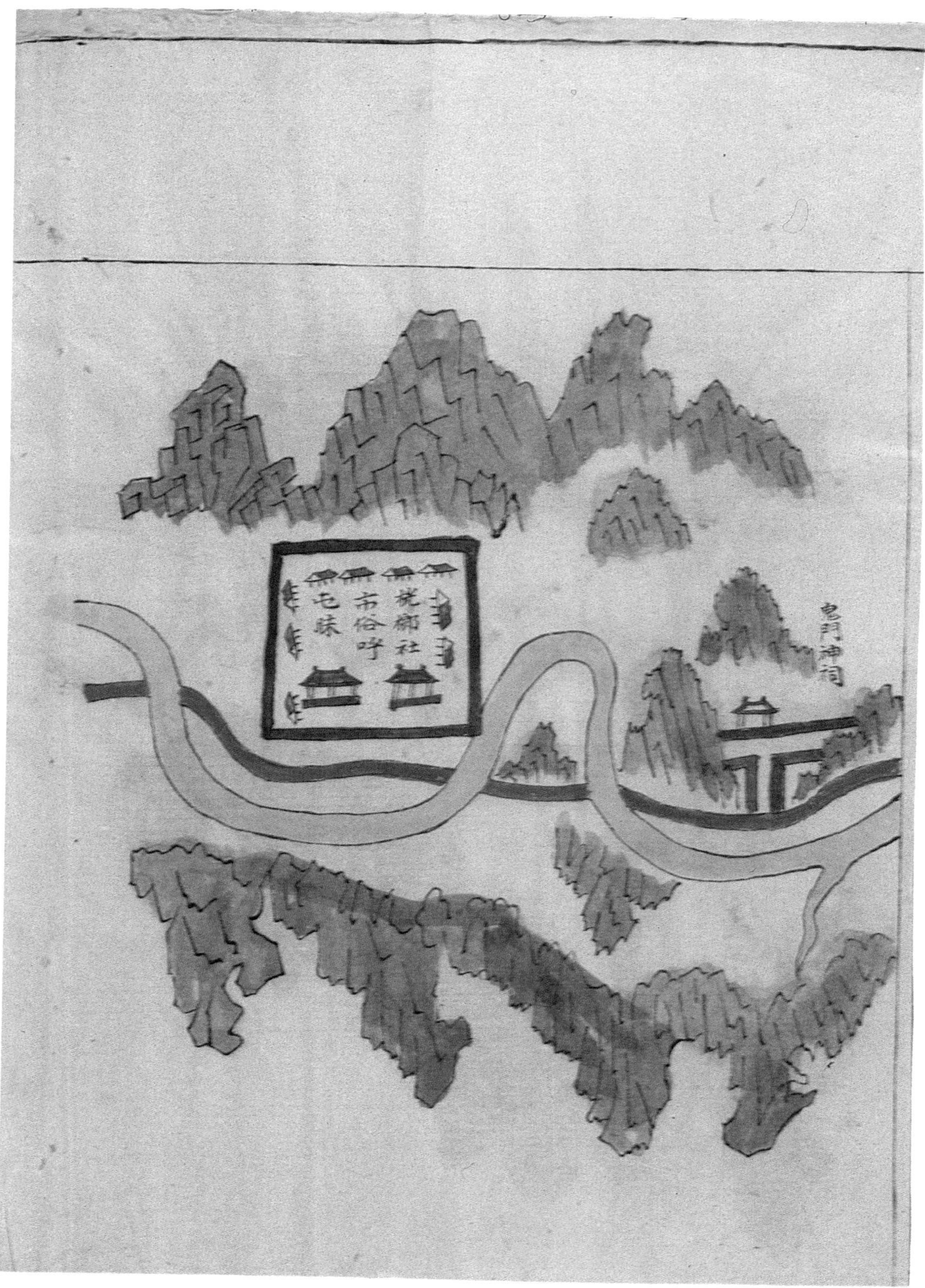

PLATE 30. EMBASSY TO CHINA. (See p. 496.) This portion of a probably eighteenth-century (Trịnh) map shows part of the route of a Vietnamese embassy to the Chinese capital of Beijing and portrays the view in both directions from the river (east is at the top). Size of the original: ca. 24.5 × 20.5 cm. By permission of the Société Asiatique, Paris (HM2182).

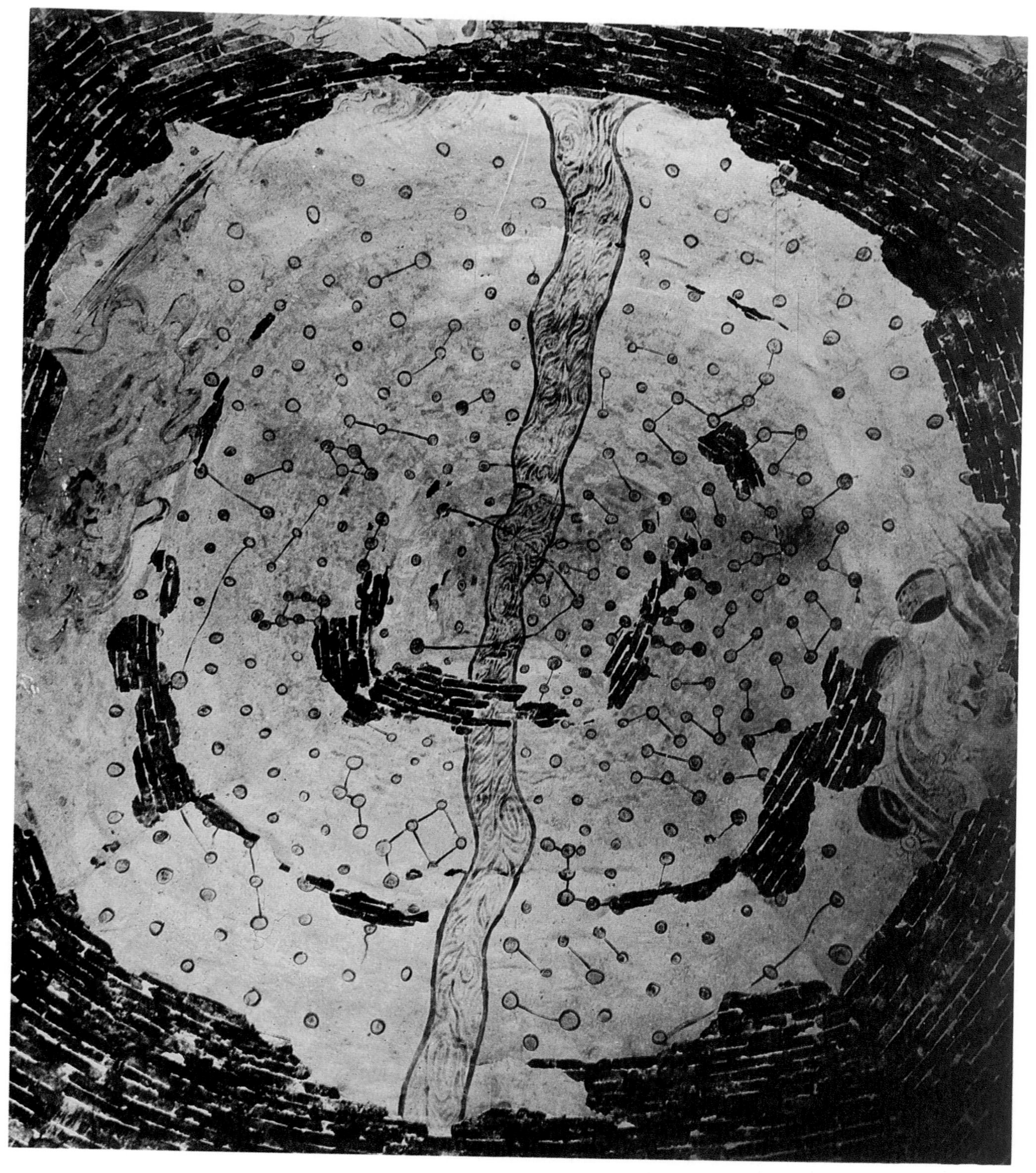

PLATE 31. PAINTING OF CONSTELLATIONS AND THE MILKY WAY ON THE CEILING OF A NORTHERN WEI TOMB. (See p. 531.) This painting is dominated by the Milky Way (the River of Heaven), shown in blue. There is little effort to depict the true configuration of the stars, but this is the earliest surviving attempt to portray the whole of the visible sky.

Diameter of the original: ca. 3 m. Photograph from Zhongguo Shehui Kexueyuan Kaogu Yanjiusuo (Archaeological Research Institute, Chinese Academy of Social Science [Academia Sinica]), *Zhongguo gudai tianwen wenwu tuji* (Album of ancient Chinese astronomical relics) (Beijing: Wenwu Chubanshe, 1980), 8 (pl. 6).

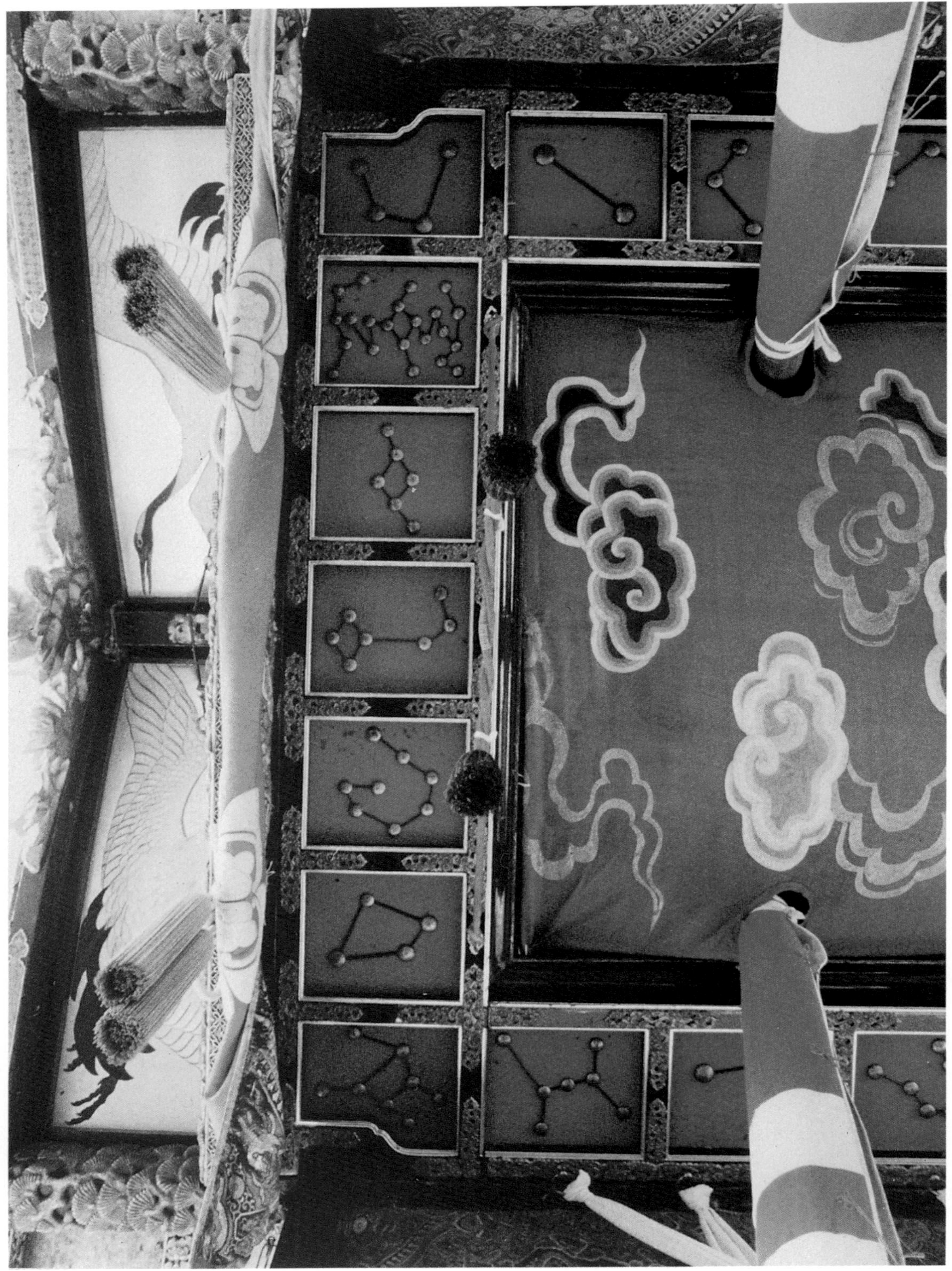

PLATE 32. PART OF THE NAGINATABOKO *SEISHŌZU* (See p. 579.) This shows part or all of thirteen of the twenty-eight lunar lodges on the ceiling of the Naginataboko (Naginata carriage). The carriage was thought to have been made about 1441, although it is not known if the *seishōzu* was there originally. In this photograph, the front of the carriage is to the left.
Size of the entire *seishōzu*: 350 × 263 cm. Photograph courtesy of Kazuhiko Miyajima, Ōsaka.

PLATE 33. MANDALA OF AN ESOTERIC FORM OF THE BODHISATTVA MAÑJUŚRĪ AND HIS CONSORT. (See p. 620). Mañjuśrī is among the more frequently depicted bodhisattvas, aspirants to Buddhahood who voluntarily postpone attaining that state to work for the welfare of mortals. Mañjuśrī's special function is to stimulate understanding. Although the profusion of anthropomorphic icons on this mandala tends to blind uninitiated viewers to its underlying cosmographic properties, they nevertheless remain an essential element of the work.

Size of the original: unknown. The current location of this mandala is not known.

PLATE 34. THE POTALA AND OTHER PRINCIPAL HOLY PLACES OF CENTRAL TIBET. (See pp. 648–50 and 668.) In Tibetan, painted on cloth, probably late eighteenth century. The size of the area covered (several thousand square kilometers) belies what the view might suggest in that the major features on the map are painted at an exaggerated scale, whereas the less sacred space between them is commensurately compressed. The subregions of central Tibet that the various towns depicted fall within are marked off from one another by streams and conventionally rendered mountain ranges. The orientation of the map is generally toward the west. A peculiar feature of the map is its depiction, in front of the Potala, of two elephants, gifts of the king of Nepal to the Dalai Lama. They are said to have died not long after their arrival on the high plateau of Tibet. Also noteworthy are the many pilgrims shown in various parts of the painting, at a scale much larger than that of the adjacent buildings.

Size of the original: 100 × 72 cm. Musée National des Arts Asiatiques—Guimet, Paris. By permission of the Réunion des Musées Nationaux, Paris (inv. no. M.A. 1043).

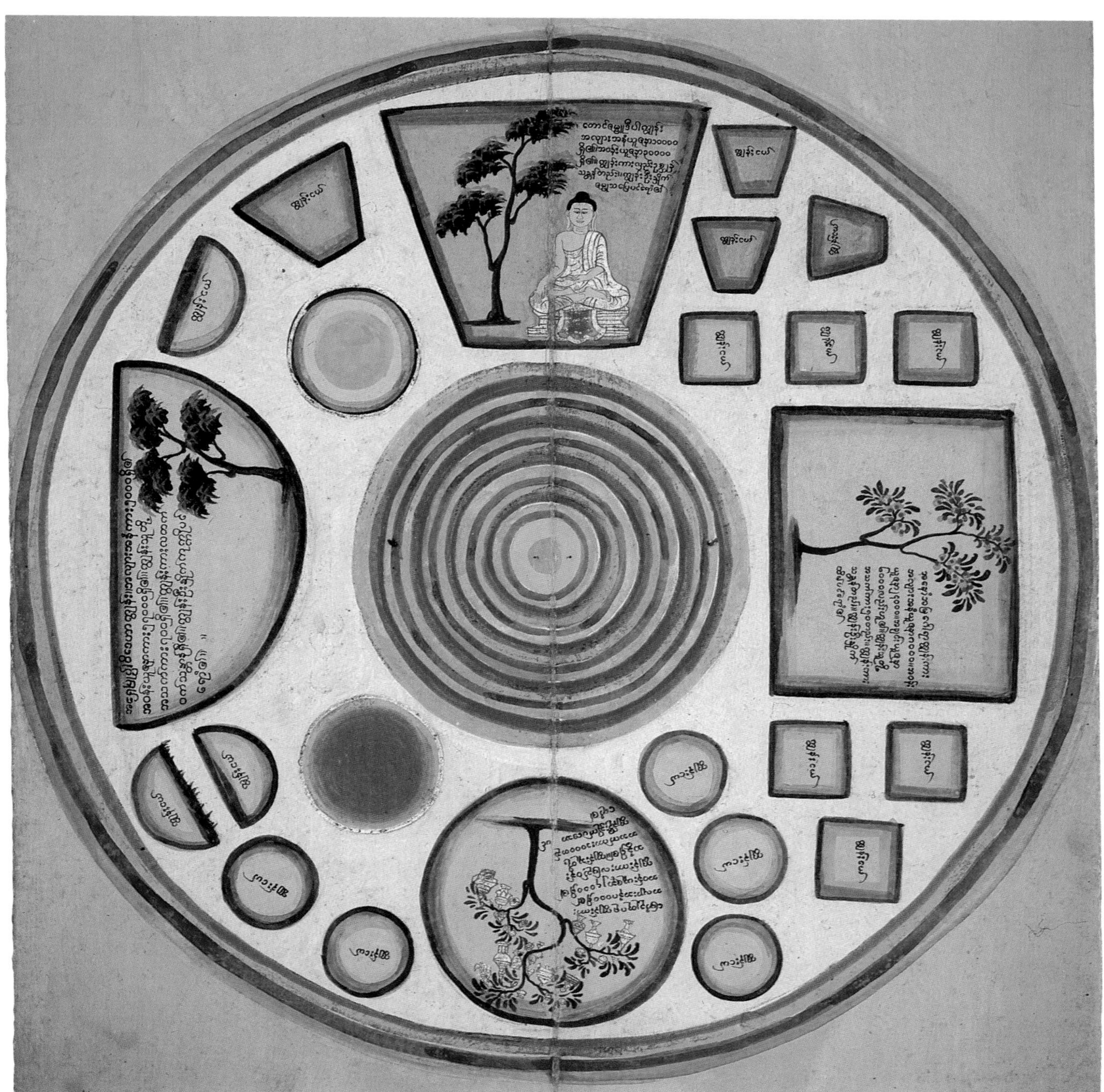

PLATE 35. BURMESE PAINTING OF THE CAKKAVĀLA. (See pp. 723–25 and 731.) Painted on heavy mulberry bark paper, part of an illustrated late nineteenth-century Burmese cosmological manuscript, folded accordion-style. Surrounding the central Mount Sumeru are seven ring mountain ranges, each one compounded of narrower bands of red, orange, pink, and dark green, separated by light green ring seas. In the four cardinal directions outward from these central mountains are the four continents on which humans of different forms live, each type having faces in the shape of the continent they inhabit. Those of our own type inhabit the wedge-shaped southern continent, Jambudīpa, colored in a brownish violet, in which the Buddha sits beside the eponymous *jambu* tree. The square western continent, colored tan, is Aparagoyana; the circular and semicircular northern and eastern continents, both yellow, are Uttarakuru and Pubbavideha, respectively. Each continent has from three to six tributary continents of the same shape and color. Both the continents and their tributaries are bounded by green borders. The two large circles near Pubbavideha are presumed to be the sun (orange) and the moon (yellow). Encompassing the entire world is the rim of the the universe, the iron Cakkavāla, composed of narrow red, pink, dark green, and light green bands. The field these features appear on was left unpainted. This is from a manuscript titled "The Bliss of the Country of the Nats, Buddhist Religion," acquired in Mandalay in 1886. On the reverse side of this manuscript is another, "The Horrors of Hell according to the Buddhist Religion," also illustrated and also dealing with cosmographic themes.
Diameter of the original: ca. 41.5 cm. By permission of the British Library, London (Or. 14004, fol. 27).

PLATE 36. MAP OF ASIA FROM THE ARABIAN SEA TO KOREA AND JAPAN. (See pp. 741–42 and 745.) This work covers six panels of the Berlin manuscript of the Thonburi recension of the *Trai phum* (Story of three worlds), a long Thai cosmographic treatise. This folding manuscript, dated 1776, is of the type known as *samud khoi* and is painted on indigenous paper. The names of four artists responsible for this work are noted in its colophon. The map is oriented to the south on the right and to the east on the left. The Indian peninsula appears near the right border and the Indo-Pacific peninsula (with no Malay peninsular extension) is seen to its left, while China dominates the left half of the map. Paralleling the upper margin over two-thirds of the map is a phalanx of islands that collectively represent the archipelagoes from Japan to what is now Indonesia.

Size of the entire original: 51.8 × 3,195 cm; each panel: 51.8 × 23 cm. By permission of the Museum für Indische Kunst, Staatliche Museen zu Berlin, Preussischer Kulturbesitz (MIK II 650/RF 4–9).

PLATE 37. MAP OF THE VALE OF MANIPUR SHOWING THE ROUTE OF KING ALAUNGPAYA'S INVASION OF 1758–59. (See pp. 752–54.) This large map, painted on two pieces of cloth, relates to one of several Burmese invasions of Manipur, now in northeastern India. The date when it was drawn is not known but could be several generations later than the events to which it relates. It is not known whether the map was compiled from field notes taken at the time of the invasion or from subsequently gathered intelligence. The work has no consistent orientation, but we have chosen to show it with east at the top. Most highland features and man-made structures shown in vertical elevation point toward a hypothetical observer within the Vale (in contrast to the use noted in figure 18.6). The second river from the eastern (top) edge of the map, as we have presented it, is the Nan-twee, whose actual existence was a point of dispute in later diplomatic exchanges between Great Britain and Burma. Within the Vale the main features depicted can be readily identified on modern topographic maps. An important feature of the map is its consistent use of color to distinguish between the settlements and battle arrays of the Manipuris.

Size of the original: 203 × 284 cm (map area 201 × 264 cm). By permission of the Map Room, Royal Geographical Society, London (Burma S.59).

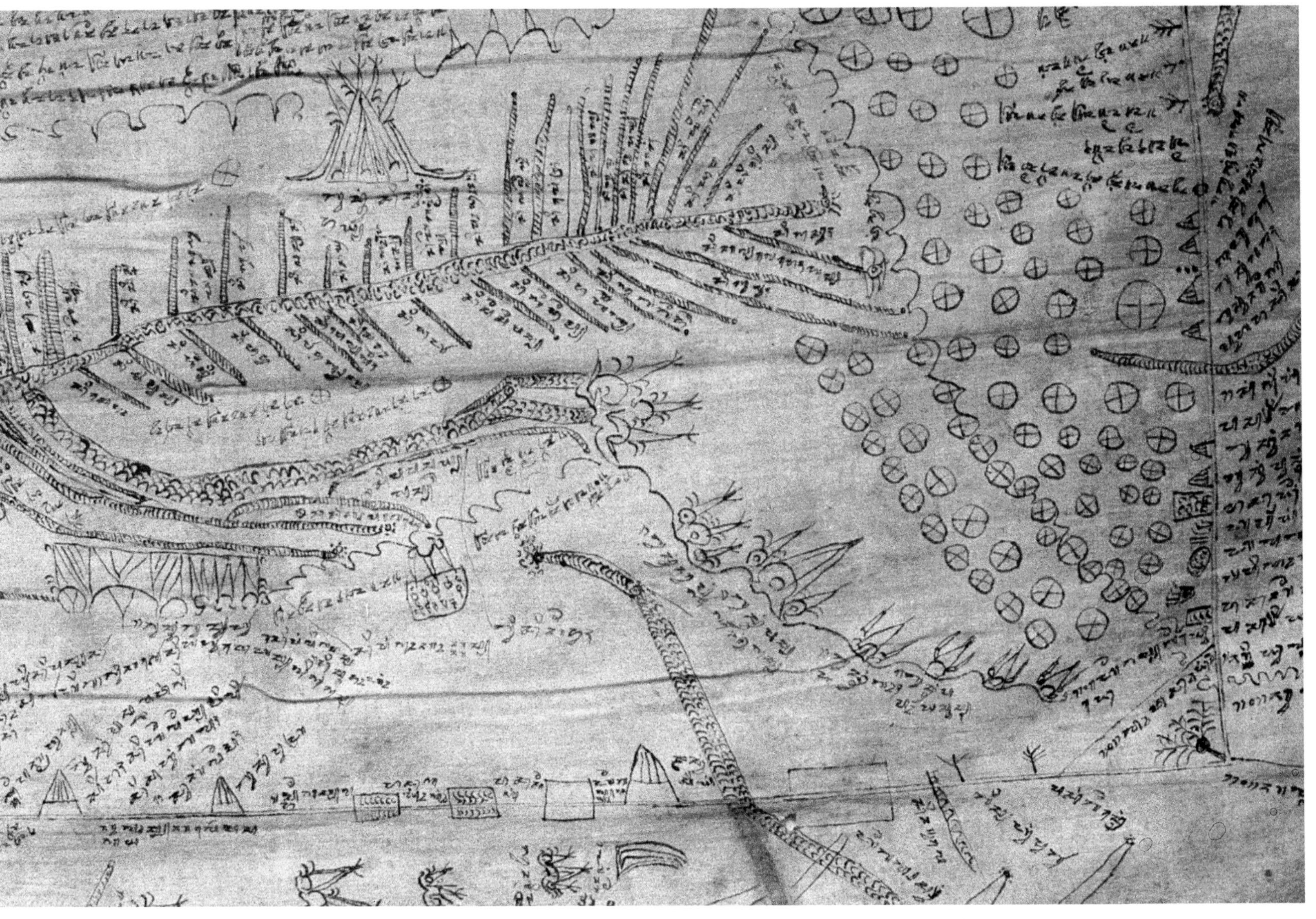

PLATE 38. CENTRAL PORTION OF THE SACRED MAP OF TIMBANGANTEN. (See pp. 767–69.) This excerpt from the map depicted in figure 18.18 provides a clearer view of the variety of signs used, not all of which can be unambiguously identified. Particularly prominent here is Cikuray volcano, in the upper-left quarter of the illustration, below which lies the village of Ciela. Along the right vertical and lower horizontal lines that signify the territorial limits of Timbanganten are various peaks, trees, and rectangular features (reservoirs?) that appear to be boundary markers. The circles with plus signs reportedly represent broadleaf forest.
Photograph courtesy of Joseph E. Schwartzberg.

PLATE 39. PRESUMED ADMINISTRATIVE MAP OF A SMALL PART OF CENTRAL JAVA. (See pp. 773–75.) This relatively modern map (probably of the mid-nineteenth century) is rendered in ink and seven tones of watercolor on European paper. Despite its southerly orientation and its Javanese text, it is likely that the map was made at the behest of local Dutch authorities, possibly as an aid in collecting revenue and exacted labor from the approximately 230 *kampongs* (settlements) shown, within variously colored areas that presumably indicate low-order administrative subdivisions. The two dark circular areas near the right margin of the map are the volcanoes Merbabu (the larger of the two) and Telomoyo, whose summits are sixteen kilometers apart.

Size of the original: 37.9 × 53.6 cm. By permission of the Bibliothèque Nationale, Paris (acc. no. Rés. Ge. D 7776).

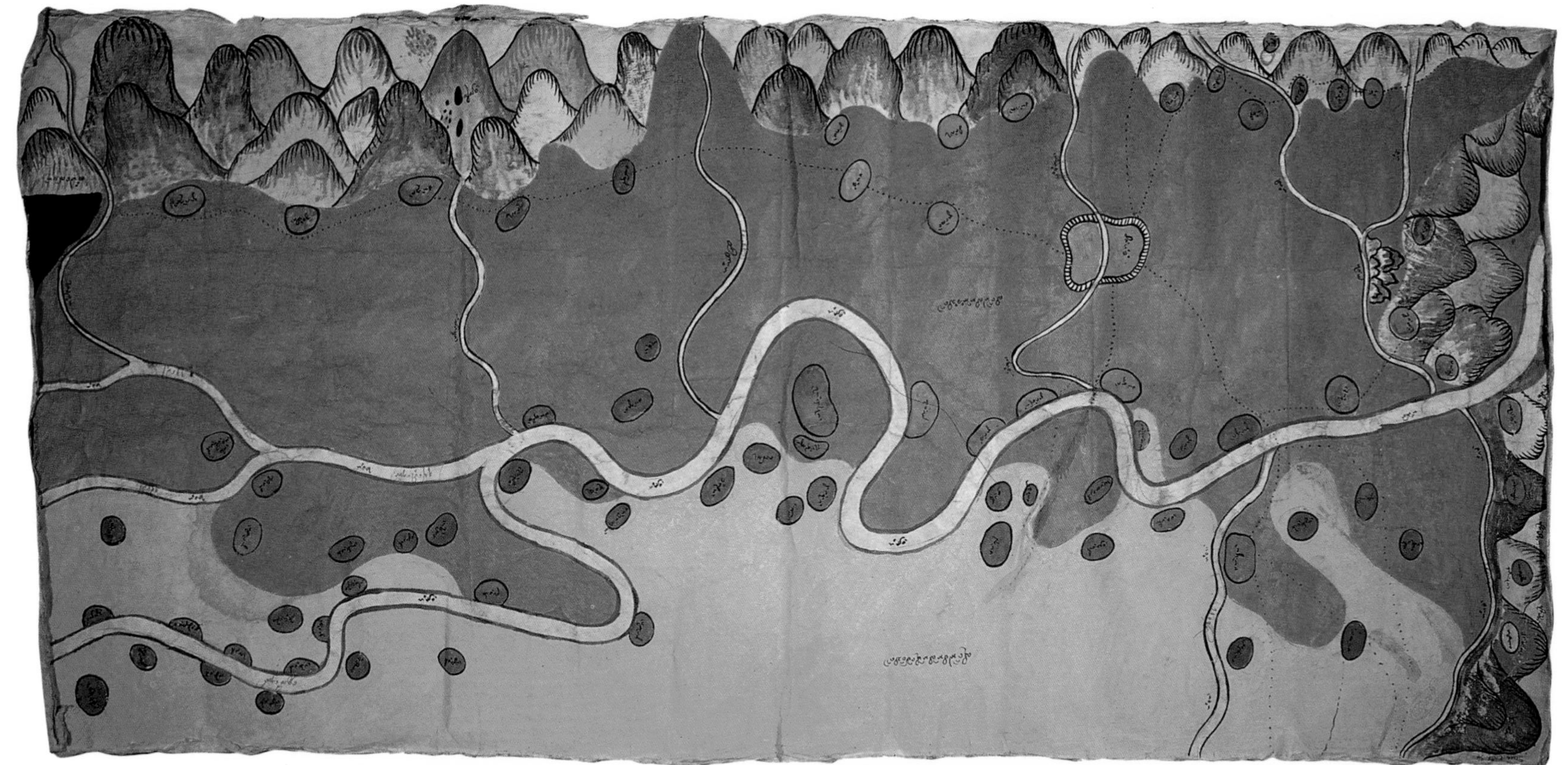

PLATE 40. SHAN MAP RELATING TO A BORDER DISPUTE BETWEEN (BRITISH) BURMA AND CHINA ALONG THE NAM MAO RIVER. (See pp. 794–95.) Covering an area of perhaps twenty-five kilometers, this large-scale and remarkably accurate map was probably drawn in 1889. It was painted on Shan paper, with text in Chinese Shan and Burmese notes subsequently added in pencil. Bright tempera colors are here used to differentiate Chinese territory (in yellow on the upper part of the map) and two Burmese states tributary to the British (the larger in red and the smaller in black).
Size of the original: 75 × 163 cm. By permission of the Syndics of Cambridge University Library (Scott LR. 13.34).

18 · Southeast Asian Geographical Maps

JOSEPH E. SCHWARTZBERG

INTRODUCTION

Surviving noncosmographic maps from Southeast Asia assume a wide variety of forms, ranging from cryptic, preliterate representations of very localized microenvironments by West Malaysian aborigines to small-scale Burmese maps covering areas of over a million square kilometers and a less detailed Thai map of the greater part of Asia. No extant map, so far as I am aware, dates from before the sixteenth century. The distribution of known maps by areas of provenance is very uneven. Burma (Myanmar) is the source of by far the largest corpus, though many of the maps from that country were drawn by individuals who were ethnically Shans rather than Burmans. (Vietnam may account for even more surviving maps than Burma, but for purposes of this history it is treated as part of the Sinic cultural realm rather than of Southeast Asia; see chapter 12.) Maps from Thailand and the Malay world are, by contrast, very few. Within mainland Southeast Asia, I know of not a single surviving noncosmographic map from Cambodia or Laos. Similarly, not a single surviving traditional map has been reported in the scholarship on Philippine cartography.[1] Elsewhere in the Malay world, only West Malaysia, Java, and Borneo (Kalimantan) have yielded any cartographic artifacts. Sulawesi (Celebes) may be added to the list if we also consider nautical charts, but they are the subject of the following chapter.

The materials considered here do not lend themselves to any neat mode of analysis or compellingly logical succession of topics. My general organizational plan is to proceed from maps of large areas through those of ever-decreasing coverage. I shall first consider the single Thai map (though known from more than one manuscript) of near continental coverage. Next I shall examine regional maps relating to rather extensive areas at the scale of an entire country or a major region, dealing first with mainland and then with insular Southeast Asia. I shall then turn successively to route maps that also relate to extensive areas, to large-scale maps of relatively small rural localities (those where all parts might be reached within a day or so from some central location), large-scale maps of cities, and finally, architectural plans at a still larger scale.

A MAP OF THE GREATER PART OF ASIA

As noted in chapter 17 on Southeast Asian cosmography, the places depicted in the richly illustrated Thai *Trai phum* (Story of three worlds), while mostly belonging to the realm of sacred myth, do merge gradually into the physical world as known to Thais when the various recensions of that sacred text were written. Thus we find rivers flowing out of a rather mythologized India into what is recognizably the north of the area that is now Thailand and thence, as Thai rivers actually do, south to the Gulf of Thailand. The several intervening mountain ranges, of which the Thais could not have been wholly ignorant, are simply forgotten in creating such a view. The artists' seeming willingness to ignore geographic facts shows up in other respects, notably in the *Trai phum*'s depicting at least one river flowing south past the northern head of the gulf down the Malay Peninsula to the area of Pattani, near the border of modern Malaysia. This suggests a convention of using rivers as visual linking devices in telling a story about places. If such was in fact the case, the apparent map errors could not be properly be regarded as a sign of geographic ignorance or cartographic ineptitude; and it seems likely that persons viewing the maps in the more geographic portions of the illuminated manuscript would have known enough not to take them too literally.

With the information now at my disposal, I am unable to say whether any of the recensions of the *Trai phum* that preceded the two Thonburi manuscripts of 1776 contained geographical maps. Nor can I specify the precise number of folios with such maps in the Thonburi manuscripts in Bangkok and Berlin and in later examples in other places.[2] Of those folios, as the previous paragraph indicates, some relate mainly to the area that is now Thailand, while others are more sweeping in their coverage. The best example of the latter is illustrated in plate 36.[3] The area depicted extends from the Arabian

1. Carlos Quirino, *Philippine Cartography (1320–1899)*, 2d rev. ed. (Amsterdam: Nico Israel, 1963). None of the pre-European maps cited in this work are of Filipino provenance.

2. The dates and locations of the several known manuscripts of the *Trai phum* are provided in note 58 of chapter 17.

3. Klaus Wenk illustrates the entire work in black and white in "Zu einer 'Landkarte' Sued- und Ostasiens," in *Felicitation Volumes of*

Sea (on the right) past India (the first large peninsula), the Bay of Bengal, the Indochinese peninsula (with the Gulf of Martaban recognizable but not, curiously, the narrow projection of the Malay Peninsula), and eastern China to Korea (on the farthest left of the mainland). Off the coast on the left half of the map are a multitude of islands—mainly nebulously shaped—that collectively suggest the Malay Archipelago (including the Philippines) and Japan. In the first panel on the right the large, roughly triangular island is undoubtedly Sri Lanka, and the two islands along the right border presumably represent the Maldives and Laccadives. Also near the right border is a large peninsula that is almost certainly Kathiawar, in the present Indian state of Gujarat. The map, broadly conceived, may be said to be oriented toward the south on the right half and toward the east on the left.

Named along the several blue-green rimmed coasts are numerous cities and several provinces. Eleven cities in Thailand are specifically identified, as are the mouths of the Mae Klong and Tha Chin rivers. Oddly, the contemporary capital, Thonburi, is not named, but the recently destroyed previous capital of Ayutthaya is. Few of the many islands named could be identified by Wenk (on whose description this account is partially based), but among those few are Sri Lanka, Java, and the four islands composing Japan. Almost all the islands on the left half of the map are virtually identical in size, shape, and north-south orientation, but one prominent exception, possibly Java, shows a deep embayment on its northern shore. All the islands are colored yellow, recalling the ancient Indian characterization of Southeast Asia as Suvarṇadvīpa (Islands of Gold, which for Ptolemy became the Golden Chersonese). A number of the cities and islands are linked by fine ochre lines along which are given distances from place to place, expressed in *yojanas*. Since Wenk was unable to identify several of the places named, his efforts to test the accuracy of the measurements proved inconclusive.[4] One might suppose from the presentation of so many distances that the map was intended as an aid to navigation. But given the lack of a strong Thai maritime tradition, such an inference appears unwarranted. Moreover, the map bears little resemblance to any of the nautical maps from various parts of Asia that are described in this work or in volume 2, book 1 of *The History of Cartography*. Further, the inclusion of the map within a cosmographic manuscript, the already-discussed *Trai phum*, does not support a belief that it was meant as an aid to navigators. Finally, the map's indication of distances to places (mainly islands) that appear to be mythical, a practice not uncommon on cosmographic maps, also supports the conclusion that it was not made for sailors.

Prominently drawn on the right and middle folios of the map are mermaids and other fabulous sea creatures, a European sailing ship (with a crew member in the bow looking through a telescope), and in the Yellow Sea, a Chinese junk. These features underscore the fact that contacts with both Europe and China were instrumental in altering the Thai view of the world during the period when the map was drawn. Further, they, along with other considerations, raise the question whether either a European or a Chinese map might have provided a model for the Thai painter(s) of this map. Since there is no evidence of independently derived Thai knowledge of most of the areas depicted, and since there is no apparent attempt to make the configuration of the coast conform to what is shown in much greater detail on other folios of the *Trai phum*, an exogenous source appears almost certain. Since very few features away from the coasts are shown, the most likely model for the map would have been a sea chart. European sea charts of the area were already numerous by 1776, when the map was painted, and some might well have been available to the artist(s), though probably not during the period after 1688 when virtually all Westerners were expelled from the country, an event noted in chapter 16. Western presence was not to be significantly felt again until the nineteenth century. For works of the pre-1688 period, I know of no chart that configures the coasts of Asia in the manner of this map. The large number of islands in a continuous chain over the center and left portions of the work, however, does recall a similar depiction on the easternmost of the twelve sheets of the so-called Catalan atlas of about 1375, attributed to Abraham Cresques.[5] Wenk, who also considers

Southeast-Asian Studies Presented to His Highness Prince Dhaninivat Kromamun Bidyalabh Bridhyakorn . . . on the Occasion of His Eightieth Birthday, 2 vols. (Bangkok: Siam Society, 1965), 1:119–22, with one plate. He also provides a large two-page color illustration of the same area as plate 36 in *Thailändische Miniaturmalereien nach einer Handschrift der indischen Kunstabteilung der Staatlichen Museen Berlin* (Wiesbaden: Franz Steiner, 1965), pl. XI and text on 64. In neither of these accounts does Wenk cite any example of maps before those of the Thonburi-period manuscripts, and he does not recognize as "maps" the numerous "geographische, kosmographische, und mythologische Vorstellungen [representations]" of the manuscript that accompany the view that I have illustrated. Nor are any such maps acknowledged in two articles on the cartography of Thailand by the geographer Ulrich Freitag: "Zur Periodisierung der Geschichte der Kartographie Thailands," in *Kartenhistorisches Colloquium Bayreuth '82, 18.–20. März 1982: Vorträge und Berichte*, ed. Wolfgang Scharfe, Hans Vollet, and Erwin Herrmann (Berlin: Dietrich Reimer, 1983), 213–27; and "Geschichte der Kartographie von Thailand," in *Forschungsbeiträge zur Landeskunde Süd- und Südostasiens*, Festschrift für Harald Uhlig zu seinem 60. Geburtstag, vol. 1, ed. E. Meynen and E. Plewe (Wiesbaden: Franz Steiner, 1982), 213–32. Whether Freitag was personally aware of their existence is not clear from the articles cited.

4. Wenk, "Zu einer 'Landkarte' Sued- und Ostasiens," 121 (note 3).

5. Bibliothèque Nationale, Paris (MS. Esp. 30); this map is illustrated, among others, in Kenneth Nebenzahl, *Atlas of Columbus and the Great Discoveries* (Chicago: Rand McNally, 1990), 6–7, with detail on the eastern islands on 8.

the map's likely source, is of the opinion that it was most likely a Chinese prototype, possibly similar to one of the *Wubei zhi* charts.[6]

Several prominent features on the left half of the map may be cited in support of a Chinese prototype. First, there are a number of small, distinctively drawn rocky islands surrounding the junk in the South China Sea that could represent Hainan and the shoals and reefs of the Paracel Islands and Spratly Islands (compare the two dark islands to the left of Sri Lanka, which could be the Andaman Islands and Nicobar Islands). Second, of all the rectangles on the map that signify ports, the largest, depicted within a unique rectangular embayment, is situated where the Indochinese peninsula gives way to the eastward bulge of southern China proper. This could represent the then prominent port of Guangzhou (Canton), not so distant from this bend in the Asian coast. Third, there is a prominent (though unidentified) inland mountain feature to the left of that port for which there is no parallel on the rest of the map. Most persuasive, perhaps, is the very large river (with two interior towns on its banks) that I judge to be the Yangtze, virtually bisecting China proper, again with no parallel feature elsewhere on the map. No river in India, Thailand, or other parts of the Indochinese peninsula is explicitly shown inland from its mouth. Finally, the conventionalized style in which the ocean is rendered also characterizes a number of Chinese maps. As for the seemingly conflicting evidence that points to the Catalan atlas or a similar Western work as a model, we must recall that an important source for that atlas was the account of Marco Polo, who reported that a large chain of islands lay off the eastern coast of Asia. Since Polo would only have been repeating received Chinese knowledge, there is, in fact, no conflict between the two conceptions.[7]

Although the foregoing description relates to the map of Asia in the Berlin manuscript of the Thonburi version of the *Trai phum*, it also broadly applies to the manuscript in the National Library at Bangkok, the left half of which is illustrated by Coedès.[8] Numerous differences of detail may be cited, however: the size of the Chinese ship in the South China Sea (more modest than in the Berlin manuscript), the greater number and variety of features shown in the seas to the east of China (providing further support for believing in a Chinese progenitor for the map), the number and alignment of islands (though not their general shape), the shape of the mountain feature in the interior of China, and the number of ports and other coastal features depicted.

Maps of Countries and Regions

Maps of Burmese Provenance

Thanks to the efforts of four British officials, Henry Burney (1792–1845), Francis Hamilton (1762–1829), Arthur Purves Phayre (1812–85), and James George Scott (1851–1935) (the activities of Burney, Hamilton, and Scott in Burma were noted in chapter 16), we are fortunate in having available to us a substantial corpus of Burmese regional maps dating from the late eighteenth and the nineteenth centuries. To these are added several others that have been preserved through the efforts of U Maung Maung Tin and Than Tun of the Burmese Historical Commission.[9] Since it is not practicable to deal with all of these in detail, I have selected a representative sample to illustrate and discuss and will treat the others summarily in appendixes 18.1 and 18.2. In discussing Burmese maps I will need to allude frequently to places that are not generally familiar to nonspecialists on Southeast Asia. The locations of the more important ones are provided in figure 18.1.

Maps Collected by Francis Hamilton

Figures 18.2 to 18.6 are copies of maps from among the dozens that were drawn for Hamilton (see also appendix 18.1) during his eight-month sojourn in Burma in 1795, serving as an aide to Captain Michael Symes, the first British ambassador to that country. Although Hamilton viewed these maps as "very deficient in accuracy," he also stated that those who drew them were "wonderfully quick in comprehending the nature of our maps; and some of them, to whom I could render the occupation advantageous, very soon improved their plans, and produced drawings, which have tended to throw much light on the geography of what . . . is called the *Farther Peninsula of India*."[10] Although no earlier precisely datable geographical maps are known, one may argue that the Burmese engaged in mapmaking before the Symes mission. If they were as adept in learning from European

6. Wenk, "Zu einer 'Landkarte' Sued- und Ostasiens," 122 (note 3); on the *Wubei zhi* charts, see pp. 52–55.

7. The numerous islands to the south of Jambudīpa on the Burmese cosmographic map illustrated in figure 17.22 above (from Richard C. Temple, *The Thirty-seven Nats: A Phase of Spirit-Worship Prevailing in Burma* [London: W. Griggs, 1906], facing p. 8) suggest a similar conceptual derivation.

8. George Coedès, *The Vajirañāṇa National Library of Siam* (Bangkok: Bangkok Times Press, 1924), pl. XIX.

9. I have not been able to ascertain from Than Tun which specific maps are or were in his possession.

10. Francis Hamilton (formerly Francis Buchanan), "An Account of a Map of the Countries Subject to the King of Ava, Drawn by a Slave of the King's Eldest Son," *Edinburgh Philosophical Journal* 2 (1820): 89–95, 262–71, and pl. X; quotation on 90.

FIG. 18.1. REFERENCE MAP OF MAINLAND SOUTHEAST ASIA. This map shows the locations of most of the places, physical features, and ethnic groups mentioned in this chapter. Those not shown here may be found on figure 16.1 or on more narrowly focused regional maps later in this chapter.

maps as Hamilton avers, they should have been no less adept at learning from examples they might have encountered in their extensive intercourse with China, which long antedated their dealings with Europeans. Moreover, the extent of detail and the cartographic idiosyncrasies of the maps drawn for Hamilton could in no way be attributed to whatever guidance he may have provided.

Hamilton published engraved copies of fourteen of the maps he collected in the *Edinburgh Philosophical Journal*, in twelve articles over the period 1820–24, and its successor, the *Edinburgh Journal of Science*, in two articles in 1824. Anticipating this series of publications, Hamilton stated, "I think it may be interesting to publish some of the original maps, exactly as drawn by the natives, but reduced to as small a scale as can be done consistent with exactness."[11] The discussion below of the maps in question is based primarily on the author's commentaries on these works. Scholars, however, will be able to pursue inquiries on Burmese cartography in much greater detail by referring to the many unpublished volumes of Hamilton's journals that are in the Oriental and India Office Collections, British Library, London.

So far as I am aware, the original maps are not to be found with the journals, and it is not known whether any survive. Several reasonably faithful copies appear to have been made, however, with the original Burmese text translated or transliterated into English. On some of these copies annotations have subsequently been made, in ink or pencil, that help one interpret what is being portrayed. One set of copies was sent to Europe, while another, communicated to the then governor-general of India, was used by Alexander Dalrymple as source material for some of his own maps of Southeast Asia. I do not know whether these latter copies are the same as the copies held in the Hamilton Collection at the National Archives of India, New Delhi. Nor do I know whether the maps at the National Archives are identical in size to the Burmese originals or where any copy still exists of several of the maps discussed in the publications by Hamilton that are not held by the National Archives. Thus, without access to his journals, I cannot provide the exact number of maps he collected over his remarkably active eight-month sojourn in Burma. Apart from those discussed in the text that follows, some particulars of the others I have some definite knowledge about are provided in the appendixes to this chapter, especially appendix 18.1.

Figure 18.2 is the published version of a map, ostensibly of the "Dominions of the King of Ava," as Burma was often formerly designated; but in fact it also includes large, vaguely limned areas to the east. This work is one of a number procured by Hamilton

> from a man in great poverty, who said, that . . . he was a slave of . . . the heir-apparent of the kingdom, and who probably had been reduced to the servile state by debt, as his intelligence and manners denoted a person who had held considerable rank, and received a good education. Before he succeeded so far as he has done, he made several attempts, with less success; and the nature of our maps, together with the manner of laying down places by bearing and distance, had been repeatedly explained to him.[12]

Hamilton does not report the slave's ethnic identity, but I am inclined to think he was a Shan, since several of the other maps he drew relate to areas then tributary to the court of Ava, which are at present in the areas occupied by Shans and kindred Tai populations in what now comprises Myanmar, Thailand, and China. He seemed to know these areas in some detail, whereas he expressed little confidence in a map that he drew of the former kingdom of Pegu in southern Burma, relatively remote from the main area of Shan occupance.

Hamilton's general characterization of the map is admirably succinct:

> In this map, the then boundary of the empire of Ava is marked by interrupted lines, while the country properly belonging to each subjected nation is distinguished by dotted lines. Mountains are represented by a straight line, like the plain on which they stand, while a waved line represents their summits. This map is chiefly useful, as showing the connection and relative position of the different nations occupying the peninsula; for the outline has many great defects. The extent from east to west is much too great in proportion to the length from north to south; and the peninsulas projecting to the south especially, are very much curtailed, the Gulf of Siam and the Malaya peninsula being altogether omitted. [The virtual omission of the Malay Peninsula occurs repeatedly on Burmese maps as well as, more curiously, on the Thai map illustrated in plate 36.] The former omission especially occasions a most enormous distortion, making the course of the Cambodia [Mekong] river shorter than that of Siam [Mae Nam]. The [Shan] countries, again, to the north of the capital, are too much extended. . . .
>
> Cities in this map are denoted by squares, those which have been the seat of empire having within a smaller square or dot, and those which are the seat of tributary princes having within a cross.[13]

Hamilton makes no mention of having suggested to the slave any of the conventions cited in the quoted passages. With respect to cities, he notes that space limitations in the reduced-scale published version precluded

11. Hamilton, "Map of the Countries Subject to the King of Ava," 90 (note 10).

12. Hamilton, "Map of the Countries Subject to the King of Ava," 93–94 (note 10).

13. Hamilton, "Map of the Countries Subject to the King of Ava," 94 (note 10).

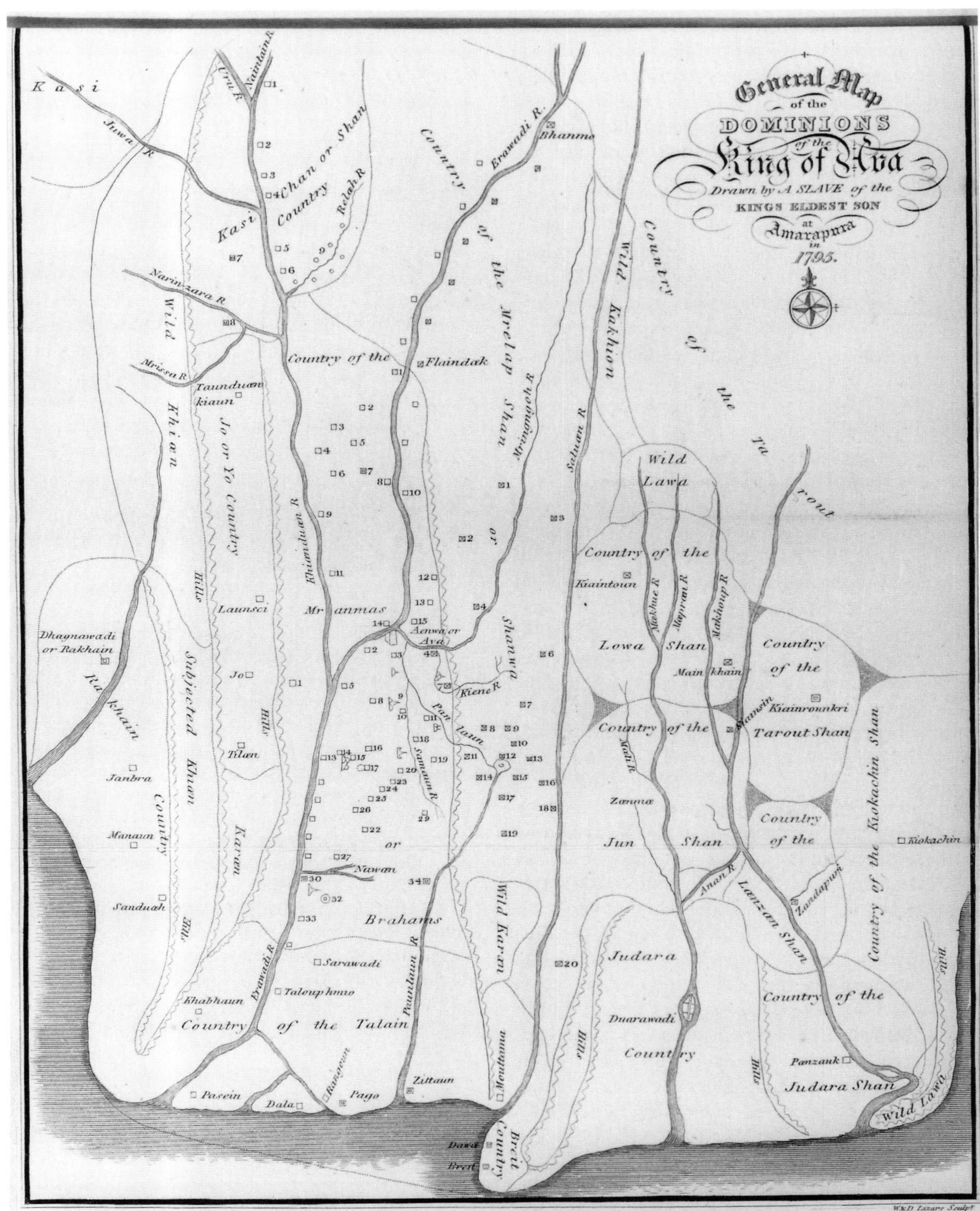

FIG. 18.2. MAP OF THE KINGDOM OF AVA AND LANDS TO THE EAST. The Salween River running north-south for virtually the entire length of the map is a rough approximation of the eastern limit of Ava except in the far south, where "Breit Country" signifies the southern extension of Burma down the Tenasserim coast. As in plate 36, the Malay Peninsula is effectively excluded. The area to the east of the Salween, including the Mae Nam and Mekong rivers, provides a sketchy view of the non-Burmese portion of the Indo-Pacific peninsula.

Size of the published map: 25.6 × 19.8 cm. From Francis Hamilton, "An Account of a Map of the Countries Subject to the King of Ava, Drawn by a Slave of the King's Eldest Son," *Edinburgh Philosophical Journal* 2 (1820), pl. X.

inserting the names; but he provides lists keyed to the seventy-eight numbers on the map, nine for the country of the Kasi Chan or Shan (the Kachin tribal region), fifteen for the Country of the Mranmas (Burman) territory north of Ava (actually, the then nearby capital of Amarapura, sometimes referred to as New Ava), thirty-four for the Burman territory south of Ava, and twenty in the Country of the Mrelap Shan or Shanwa (more or less equivalent to the present Shan States).[14]

The eastern half of the map includes the Country of the Tarout (China), various areas dominated by mountain tribes (Wild Kakhion and Wild Lawa), and other territories of a number of Tai peoples. The term "Shan" appears to be used here as a generic designation for a host of non-Burman peoples, including the Kasi Shan (west of whom lie the Kasis or Manipuris), Lowa Shan, Tarout Shan, Jun Shan (with their capital at Zaenmae, or Chiang Mai), Judara (Ayutthaya) Country (Siam proper), and Lanzaen Shan (Laos). East of these are the Country of the Judara Shan (Cambodia, which was then tributary to Siam) and the Country of the Kiokachin Shan (Cochin China), within which the Wild Lawa Hills, a perplexing toponym, suggest the Annamite Cordillera. Beyond lies the South China Sea. In the southwest, Rakhain Country represents Arakan on the Bay of Bengal, while toward the northwest Jo or Yo Country and Wild Khiaen signify Chin tribal regions between Burma proper and India. In the south the Malay Peninsula, as noted, is virtually absent, but a suggestion of it is provided by the Breit Country in which two towns, Dawae and Breit, represent the towns of Tavoy and Mergui, respectively. The northernmost point signified on the map is the Kasi Shan town of Main Ghain (no. 1 on the map), which I take to be the present-day town of Maingkwan.

If all these identifications are correct, the overall extent of the map would be from about 12° to 27° north latitude and from about 92° to 108° east longitude. That any one individual could have knowledge of so vast an area in 1795, rudimentary as that knowledge undoubtedly was for much of the area depicted, indicates a Shan or Burman tradition in which geographic knowledge was remarkably well developed and, most likely, the existence of an accompanying capacity for, if not quite an active tradition of, geographical cartography of which only meager evidence survives for the period before 1795. It is noteworthy in this context that the slave's map, showing the Kasi Shan territory as tributary to Burma, was taken by the British, on the recommendation of Henry Burney, British ambassador to the court of Ava, as sufficiently authoritative to support the restoration to Burma in 1832 of certain territory annexed to Manipur by the Treaty of Yandabo (1826) concluding the first Anglo-Burmese War.[15]

Figure 18.3, Hamilton's reduced-scale copy of a map of northern Burma drawn for him at Amarapura by "a native of Taunu" (Toungoo) who was related to a highly placed servant of the Burmese government, gives further evidence of an indigenous cartographic tradition in that the manner of depicting hill ranges by wavy lines and vegetation by diversely rendered arboreal forms and grass signs is so totally unlike anything that might have come from either a European or a Chinese source.[16] It does, however, find echoes, as we shall see, in many later maps of Burmese provenance. That Toungoo, the mapmaker's home city, is situated on the Sittang River in Lower Burma, hundreds of miles south of the large areas depicted, attests additionally to the gathering and internalization of geographic knowledge so it could be depicted on a map accurately enough to be immediately recognized by one familiar with the geography of the area portrayed.

In Hamilton's view, the mapmaker "was not so intelligent and quick as the slave who gave me the general map" (fig. 18.2). That the mapmaker was not himself an experienced cartographer is suggested by the description of the way the map was drawn:

> This person's first attempts, as might be expected, were very rude. He began at a given place, . . . and, going on in a certain direction, he laid down the places occurring, until his paper afforded no more room. He then twisted round his line, until he completed the route with which he had commenced. Then he returned to the first point, and commencing with a second route, proceeded in the same manner, and continued so on until he traced the whole of what he intended. The remote parts were thus distorted in a most extraordinary degree. After some pains, however, he improved much, and produced the map, now published.[17]

The general frame to which most of the map features can be related is that provided by the river system, various components of which can be discerned on a modern map. The most prominent of these are, of course, the Irrawaddy itself and its principal right-bank tributary, the Chindwin ("Khiaenduaen"). The former has been bent toward the northeast (instead of continuing north) to obtain space for introducing the town of Khandi (upper left), whose modern equivalent is not evident—perhaps Myitkyina. Other prominent features of the map include the two lakes to the west of the Irrawaddy, greatly exag-

14. Hamilton, "Map of the Countries Subject to the King of Ava," 95 (note 10).

15. W. S. Desai, "A Map of Burma (1795) by a Burmese Slave," *Journal of the Burma Research Society* 26, no. 3 (1936): 147–52.

16. The published account of this map and its origins is in Francis Hamilton, "Account of a Map of the Country North from Ava," *Edinburgh Philosophical Journal* 4 (1820–21): 76–87 and pl. II, esp. 76.

17. Hamilton, "Map of the Country North from Ava," 76 (note 16).

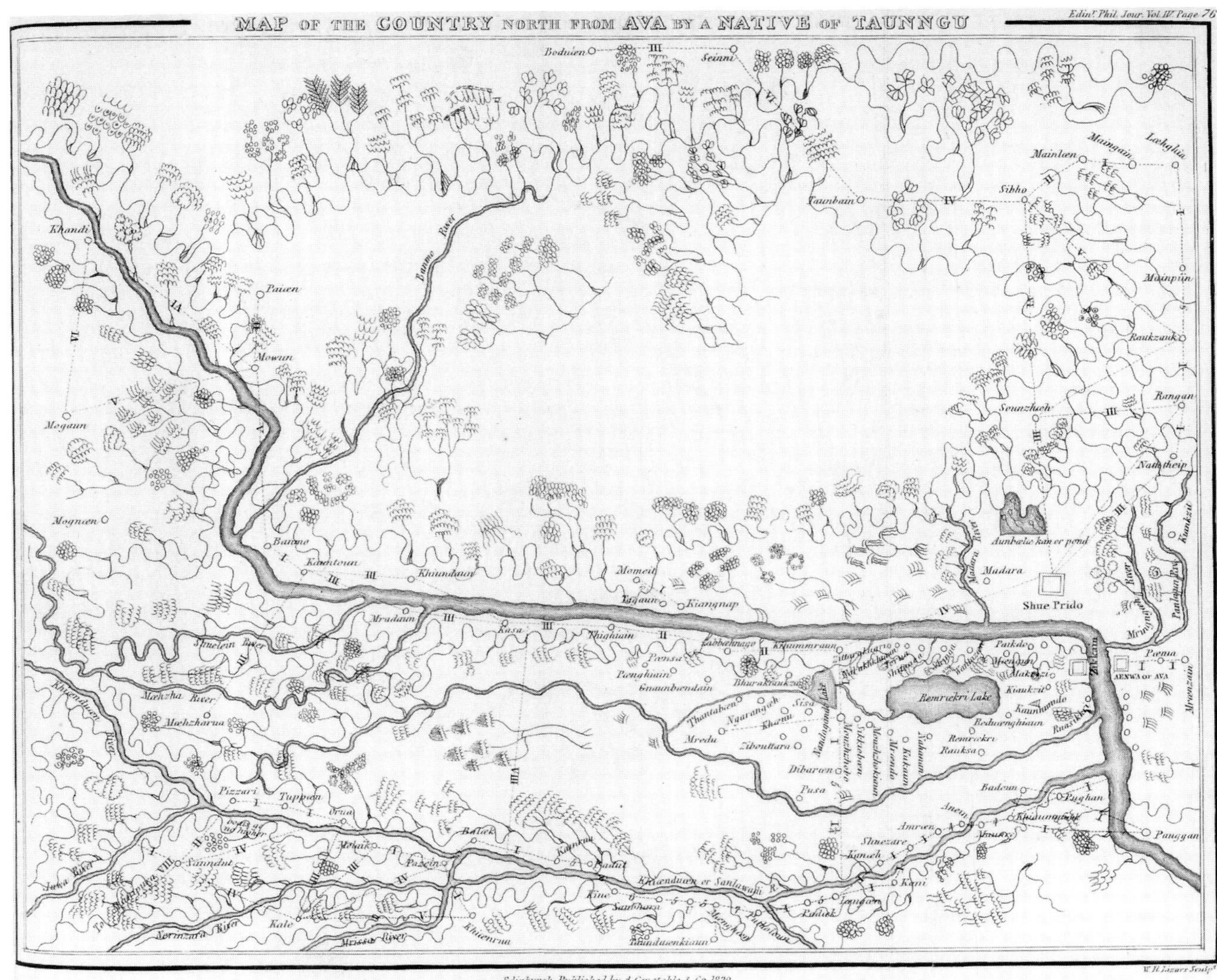

FIG. 18.3. MAP OF THE COUNTRY NORTH FROM AVA. East is at the top of the map, and the then capital, (New) Ava, actually Amarapura, not far from what is now Mandalay, is near the bend in the Irrawaddy close to the map's right margin. The most characteristic feature of this map, as of so many others of Burmese provenance, is the lavish detail with which vegetation is depicted.
Size of the published map: 19.3 × 23.8 cm. From Francis Hamilton, "Account of a Map of the Country North from Ava," *Edinburgh Philosophical Journal* 4 (1820–21), pl. II.

gerated in size but easily found on modern large-scale maps; Aunbælæ Kun, a pond in which lotus plants are depicted, to the east of Shue (Shwe) Prido (Amarapura); and a cluster of present and former capitals, each marked by a double square, including Shue Prido, Ava, and Zikkain (Sagaing). All other settlements are marked by small circles. The interpretation of the vegetation signs is problematic. Hamilton says he has "reason to think, that . . . [the mapmaker] has given to each kind a sort of appropriate form, although the resemblance is by no means striking to a botanist. . . . Some forms were, however, pointed out, especially one representing the tea-tree, which grows spontaneously in many parts of the peninsula."[18] (It is not clear to me, however, which of the several types of vegetation shown represents the tea tree.)

An important feature of the map, as with most of the other maps Hamilton collected, is its inclusion of distances between named places, expressed either in days of travel (indicated by roman numerals) or, for shorter distances, in Burmese leagues (*dain*), each the equivalent of 2.2 British miles, indicated by arabic numerals.[19] By comparing linear distances on the map with travel times,

18. Hamilton, "Map of the Country North from Ava," 76–77 (note 16).

19. Hamilton, "Map of the Country North from Ava," 77 (note 16).

one can derive some idea of the varying degree of scale distortion from one part of the map to another. In general, the scale tends to decrease with distance from Amarapura. No attempt is made to trace the actual routes between places; rather, the connections are shown by ruled dashed lines, which are obviously the work not of the mapmaker, but of Hamilton himself. This raises the question of how else and how frequently he may have intervened in the mapmaking and underscores the need for caution in interpreting all the maps that reach us because of his efforts.

It is clear that Hamilton's influence was brought to bear significantly when the author of the map just discussed rendered for him a map of Burma to the south of Amarapura, complementing the one just discussed. The map of the South, says Hamilton, differs from that of the North "in style chiefly, by the compiler having omitted the imitations of trees, which were mentioned to him as rather an encumbrance than an ornament."[20] (We may therefore perhaps infer that the absence or paucity of vegetation on maps subsequently made for Hamilton runs counter to the propensity of Burmese map compilers to include that important element of the landscape and reflects instead the British collector's eagerness to obtain a useful map in what to him appeared the most expeditious way possible.) Although this map will not be illustrated or discussed in detail, let me point out one statement in Hamilton's description of it that reflects what appears to be a fairly common European misconception in regard to the drainage courses of many Southeast Asian rivers:

> One of the most remarkable features of the country represented in this Map is, that although it is far from level, consisting chiefly of swelling grounds, many of which contain rock; and although in many places these rise even to hills disposed in ridges of considerable length, though of no great elevation, yet the rivers anastomose almost as much as in the low lands of Bengal, where there is not the slightest trace of rock, stone, or eminence.[21]

Hamilton is correct in stating that the map he has copied and published, like others he has presented, does show anastomosis to a remarkable degree. He errs, however, in supposing that the apparent drainage pattern of the map, as presented, is reasonably reliable. Although in the Irrawaddy delta, of which he had some firsthand information, anastomosis is indeed common, he is badly mistaken in giving credence to the numerous links shown on the map between the Irrawaddy and the Sittang to its east and between both the middle Irrawaddy and the Sittang and the Salween, still farther east and beyond an intervening mountain range. It appears that the map compiler, in depicting the largely riverine routes between places in two different river basins, maintained the double line for a river in showing the route beyond a stretch near its headwaters, carrying it over a divide and then down to another stretch of the route near the headwaters of another river on the far side of that divide. This seems to be the identical process employed by native Americans in making certain of their maps, as described by Lewis.[22] I shall refer to this misunderstood Burmese mapping convention again in reference to several other regional maps, including some made late in the nineteenth century.

How far Hamilton could influence the final appearance of a Burmese or, more properly in this case, a Shan map may be seen by comparing figures 18.4 and 18.5, both of which are meant to portray the country of a people called the Jun Shan, focusing on Chiang Mai in what is now northern Thailand and adjacent areas of the Burmese Shan States and Laos, an area at the time tributary to Burma and estimated by Hamilton as being some 46,000 square miles in extent. The topological correspondence among many elements of the two is immediately obvious, though the relative distances and areas involved vary considerably. The first of the two maps under review was obtained soon after the receipt of the general map; but "as this contained no distances, he [the slave], at my request, made out the second map, in which these are given; and the manner of delineating the country is altered."[23]

Most of the alterations, including the reorientation of the map from east to north, the omission of all depiction of vegetation, and the suppression of the three prominently drawn temples (including the famous Shue Daun [Temple of Seven Pagodas]), are too obvious to require much comment; but it is worth noting that numerous towns that appear on the first map (toward the Salween River in the west, toward the frontiers of Siam and Kiainroungri [modern equivalent unknown], and east of the Mekong ["Maekhaun"]) are omitted on the second because the compiler could not recall the distances to or between them. Although an obvious attempt has been made in the second map to maintain a greater consistency of scale, the compiler's success was not remarkable. Thus Sinhoun (Lamphun), which is allegedly only six leagues (roughly 20 km) from Chiang Mai, is placed farther from

20. Francis Hamilton, "Account of a Map Constructed by a Native of Taunu, of the Country South of Ava," *Edinburgh Philosophical Journal* 5 (1821): 75–84 and pl. V; quotation on 75.

21. Hamilton, "Map Constructed by a Native of Taunu," 75 (note 20).

22. G. Malcolm Lewis, "Indian Maps," in *Old Trails and New Directions: Papers of the Third North American Fur Trade Conference*, ed. Carol M. Judd and Arthur J. Ray (Toronto: University of Toronto Press, 1980), 9–23, esp. 19.

23. Francis Hamilton, "Account of Two Maps of Zaenmae or Yangoma," *Edinburgh Philosophical Journal* 10 (1823–24): 59–67 and pl. III; quotation on 59.

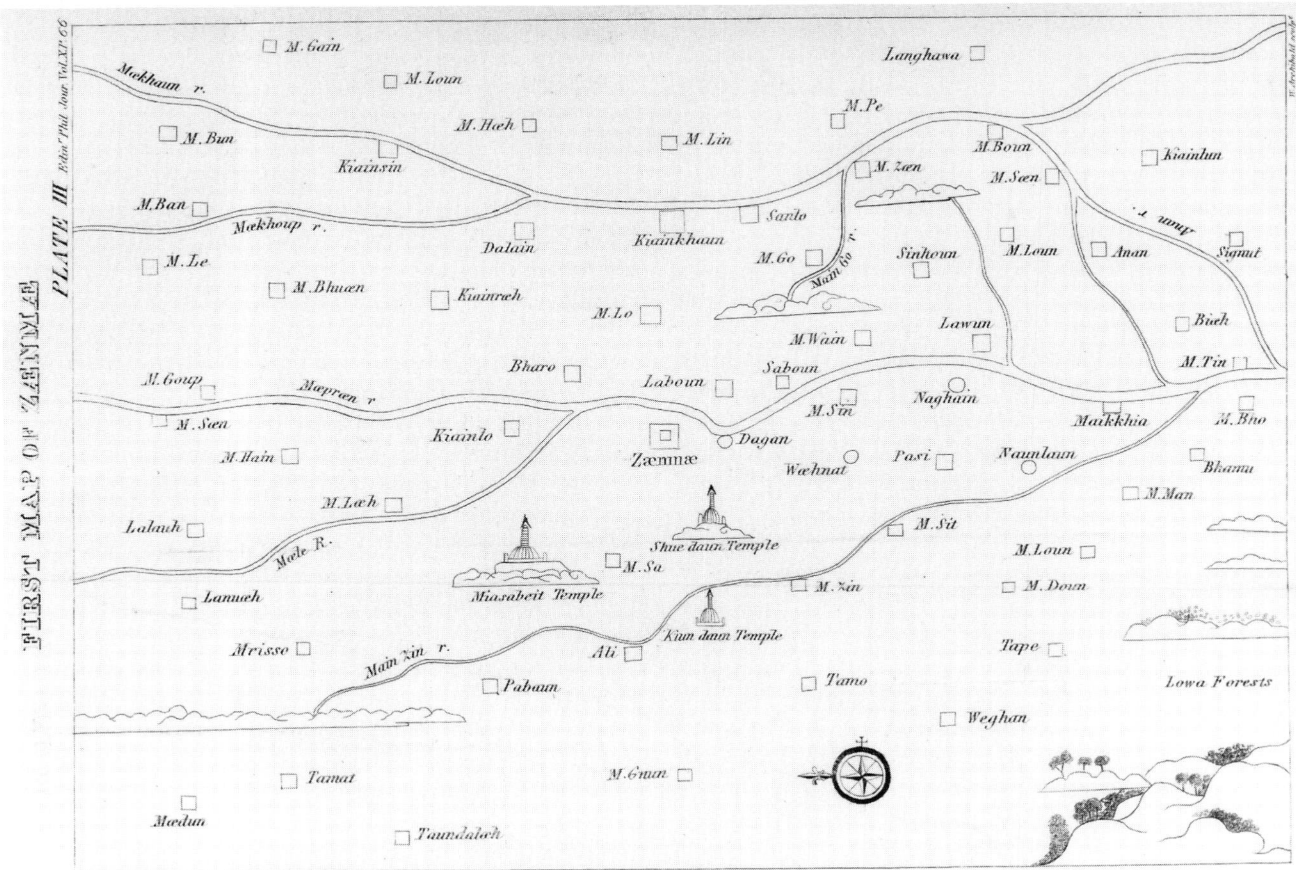

FIG. 18.4. MAP OF A LARGE AREA CENTERED ON CHIANG MAI [ZAENMAE] IN WHAT IS NOW NORTHERN THAILAND. This map was drawn by the same slave who was responsible for those in figures 18.2 and 18.5. It is a copy of the map as originally drawn. Compare figure 18.5.
Size of the published map: 12.2 × 16.8 cm. From Francis Hamilton, "Account of Two Maps of Zaenmae or Yangoma," *Edinburgh Philosophical Journal* 10 (1823–24), pl. III.

that city than from Anan (modern equivalent unknown), which is said to be three days distant.[24] Similarly, although a modest effort has gone into showing rivers as flowing in their true directions rather than in the essentially north-south direction of the first map, the second map does not come close to portraying the bewildering complexity of the drainage pattern in the area it relates to, which is marked by numerous captured streams and zigzag river courses.

Of the rivers shown, the Salween, in the far west of the second map and said to mark the left margin of the first; the Mae Nam (or Chao Phraya), in the center of the map; and the Mekong, in the east, are all well known. The Anan (which I take to be the modern Mae Nam Nan)—said on the second map to be navigable—connects the Mae Nam with the Mekong, which no river actually does. This seems to be another case of extending the map sign for a river beyond the actual extent of that river and over a divide when a route along it continues on to join another riverine route.

On the whole, neither of the maps portrayed in figures 18.4 and 18.5 is easy for a nonspecialist to interpret, since the toponyms for places lying almost entirely outside Burma have been rendered in an outmoded anglicized form of the Burmese script used in the late eighteenth century. There appears to be little correspondence between the sound of the names as given by Hamilton and the essentially Thai forms shown on modern maps.

The final regional map I shall examine from among the large number Hamilton collected was drawn for him in Amarapura by a native of the town of "Dawae" (Tavoy) on the Tenasserim coast (fig. 18.6). It covers a north-south distance of about 350 miles from "Mouttama" (Martaban) to "Breit" (Mergui) but is somewhat less complete in the north beyond the area of the then Dawae Province, whose boundary is indicated by a dashed line, than in Dawae itself. Perhaps the most striking feature of this map is its varying perspective. Although all the names are

24. Hamilton, "Two Maps of Zaenmae or Yangoma," 61 (note 23).

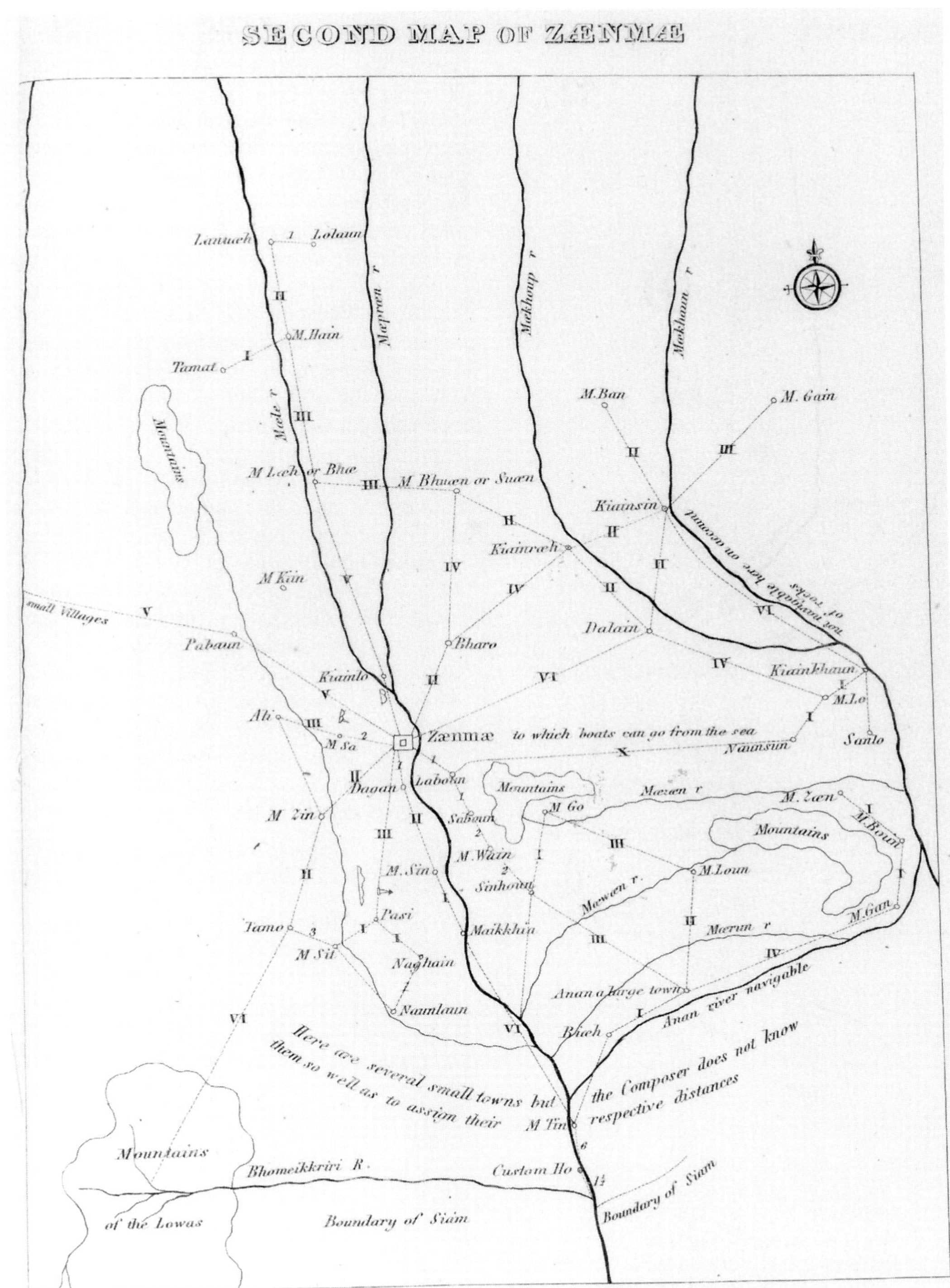

FIG. 18.5. REVISION OF THE MAP CENTERED ON CHIANG MAI [ZAENMAE]. This is the map shown in figure 18:4 as it was redrawn after critical review of the initial effort. Added to this version are lines along which are indicated the traveling time, in days, between pairs of named localities. The "Mountains of the Lowas" (Lawas), a mountain tribe, are shown in the southwest corner of both maps. The Mekong runs more or less parallel to the other rivers and appears to be linked to the Maepraen (Mae Nam Ping, a northwestern extension of the Mae Nam) by the Anan (Mae Nam Nan), which is not actually the case.

Size of the published map: 16.6 × 12.2 cm. From Francis Hamilton, "Account of Two Maps of Zaenmae or Yangoma," *Edinburgh Philosophical Journal* 10 (1823–24), pl. III.

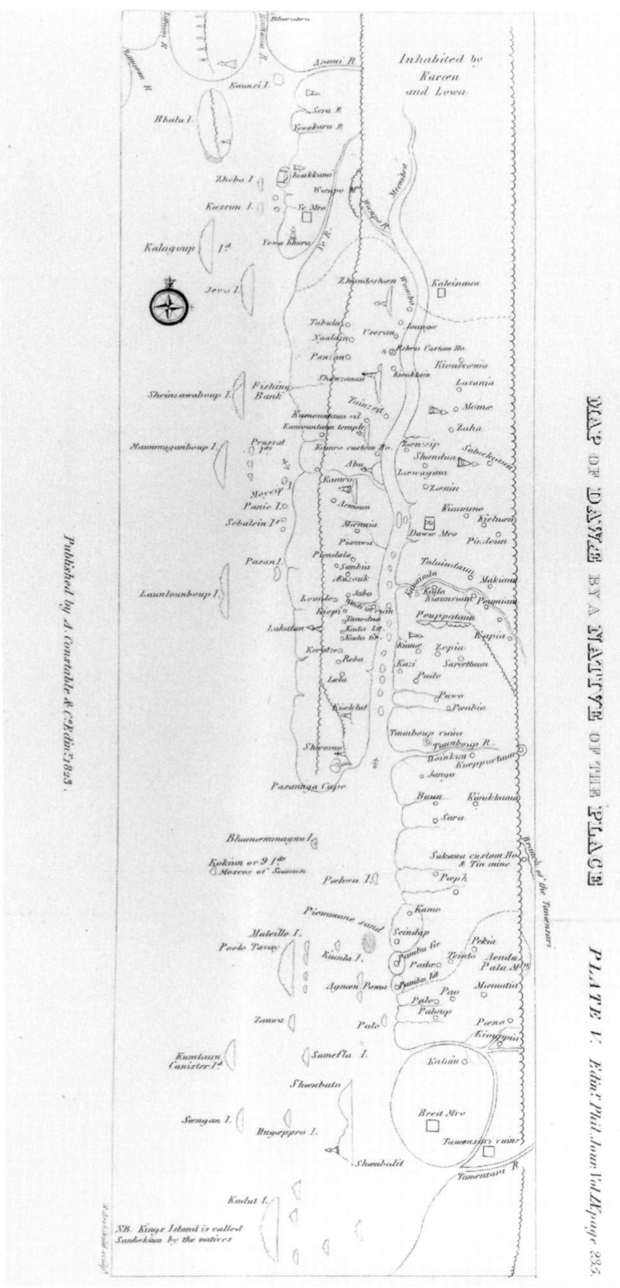

FIG. 18.6. MAP OF A PORTION OF THE TENASSERIM COAST OF AVA. Originally drawn by a native of Dawae (Tavoy), with a steatite pencil on indigenous paper, for Francis Hamilton. The north-south extent of the area depicted, from the Mouttama (Martaban) in the north to Breit (Mergui) and King Island in the south, is not quite 475 kilometers. A characteristic feature of this map, shared by many others from Burma, is the depiction of elevated features, such as mountains and pagodas, pointing away from the viewer as they would most usually be seen.
Size of the published map: 24.3 × 7.6 cm. From Francis Hamilton, "Account of a Map Drawn by a Native of Dawae or Tavay," *Edinburgh Philosophical Journal* 9 (1823), pl. V.

written as if the map was meant to have an orientation toward the north, that tells us nothing about the mapmaker's intention because the original draft was transferred by "a Mahommedan painter" from the black Burmese paper, on which it was initially drawn with a steatite pencil, to European paper. The names were added in roman script by Hamilton himself "as the man read them from his black book."[25] Names aside, many features are shown in simplified frontal elevation, pointing away from the direction from which they are most likely to be seen by a resident of the region. Thus the summits of the long hill range near the eastern edge of the map that forms the spine of Tenasserim and the frontier with Siam point toward the east, while the summits of the hills running through the large peninsula in the northern two-thirds of the map point westward, away from the town of Dawae, a bit north of center, as do the islands seen in profile as from the mainland looking toward the Gulf of Martaban. Pagodas—often, though not necessarily, built on hilltops—also point toward the east or west away from the central axis of the map. Considerable detail in regard to settlement is provided and, as on many other Burmese maps, towns, shown by squares, are differentiated from villages, shown by circles. Rounding out the content of the map are a sketchy indication of the drainage pattern and miscellaneous bits of useful intelligence, such as the location of a customs house and a tin mine on the Siamese border, indications of the locales of non-Burmese ethnic groups, and a few details of coastal hydrography.

Burmese and Shan Regional Maps Predating the British Annexation of Upper Burma

Of all the traditional Burmese maps known to exist, the largest, and conceivably the oldest, relates to one of several military campaigns that King Alaungpaya launched against Manipur during his turbulent reign from 1752 to 1760—almost certainly the campaign of 1758–59, which the king personally commanded. This detailed map covers the whole of the Vale of Manipur, together with the surrounding region of northeastern India and northern Burma (plate 37). It is on two pieces of cloth (probably Indian muslin) sewn together and was given to the Royal Geographical Society, London, in 1928 by G. U. Yule of St. John's College of Cambridge University. He discovered it among some family possessions and supposed it could have been obtained by his uncle, Sir Henry Yule, when he accompanied the British mission to the court

25. Francis Hamilton, "Account of a Map Drawn by a Native of Dawae or Tavay," *Edinburgh Philosophical Journal* 9 (1823): 228–36 and pl. V; quotation on 228.

of Ava in 1855.[26] At G. U. Yule's request, the map was examined by A. G. Cooke, a lecturer in Burmese (presumably at Cambridge), who sought, with mixed success, to establish correspondences between the places and rivers named on the map and those shown on maps by the Survey of India. Cooke suggested that the person who drew the map was probably not on the expedition it depicted, since he would then have known that the border town of Tammu (modern Tamu), from which the actual penetration of Manipur would have begun, lay on the Yu Yabwe River, not below the junction of that river with the Chindwin, as in Cooke's view it is shown on the map. He concludes that the artist "probably worked by hearsay and trusted for the rest to intuition."[27] The Chindwin is nowhere named, however, and Cooke very likely is wrong in equating it with the Nan-twee (variously spelled). Cooke's expertise may also be called into question in that he states that the "route is that actually taken by the king [Alaungpaya] . . . (1750 circa)," a date that is two years before Alaungpaya gained the throne and eight years before the expedition he led.[28]

A second map of the same area, in the same Burmese style and with the same matter relating to the Burmese invasion but with map text entirely in English, forms part of the Burney Collection of the Royal Commonwealth Society, London (a detail is shown in fig. 18.7). (In the rest of this discussion the maps will be referred to as the RGS and the RCS maps, respectively.) In 1984 the content of the RCS map had not been properly determined, and an appended note erroneously suggested that it might be of an area on the Salween River, which in fact lies far to the east of Manipur. The RCS map, acquired with the rest of Burney's Burma collection in 1921, is drawn on paper and is far smaller than its RGS counterpart, from which it was presumably copied. Since Burney could have obtained it only during his sojourn in Burma as British ambassador, from 1829 to 1837, we have the dates 1759 and 1837 as the terminus a quo and terminus ad quem for the RGS map and, most likely, 1829 and 1837 for the RCS map.

It is not known whether there was an earlier map, possibly made in the field, that might have served as a prototype for the maps under review; but given the long-standing relationship between Burma and Manipur, its former tributary, and the existence before the first Anglo-Burmese War of some form of Burmese military maps (discussed below), that is entirely possible. Moreover, the Burmese had a long history of systematically gathering geographic intelligence, in connection with revenue surveys and census operations, from which mapmaking would have been relatively easy. For example, shortly after his accession in 1781, King Bodawpaya ordered a general revenue inquest whose results have been characterized as the "Burmese Domesday Book."[29]

With specific reference to their disputed northwestern frontier, it is noteworthy that between 1829, when Burney first arrived in Ava (restored as Burma's capital in 1823 and remaining as such until 1837), and 1832 there was extensive diplomatic correspondence on the subject.[30] Within this correspondence the Burmese made many references to maps, revenue surveys, and other documents supporting their position. These records went as far back as the reign of the Burmese king Minkhaung II (1481–1502), when a Shan chronicle was allegedly compiled fixing the boundaries and, in considerable detail, the territorial constituents of Manipur in nine directions, by which was probably meant "4 cardinal points and 4 half-cardinal points of the compass plus the central point."[31] Regrettably, however, none of the maps mentioned in the Burmese texts appears to have been dated or described in sufficient detail to permit their being unambiguously related to the two under consideration. Of note in this context are Burmese protestations about the "overconfidence" the British placed in their own map of the border region (as of 1829); their assertion that "it is our practice to draw big bold lines in our maps to represent rivers, and thinner lines to indicate streams"; their reference to the powers long since delegated by the Burmese to *myetaings* (surveyors of land for revenue assessments) in the border regions; their recognition of the "unfortunate" fact that a map the Burmese sent to Calcutta in support of their claims "fails to show the course of [a disputed river] with exactitude, due to the fault of our cartographer"; their reference to revenue surveys carried out in the border region in the years 1126 and 1145 B.E. (A.D. 1764 and 1783); their inclusion in one missive of a rough map of the route between Burma and India over the An Pass; and the existence of a disputed stream, the Nantwee, which appeared on Burmese maps but which the British believed to be fictitious.[32] Among maps that so showed the "Nantwee" are the RGS and RCS maps; and, whatever the present name for that stream might be, that river is only one among several between the Chindwin and the modern border with

26. Letter from G. U. Yule to the Secretary, Royal Geographical Society, 30 January 1928.

27. Handwritten note by A. G. Cooke, dated 29.1.28.

28. Cooke, handwritten note. Cooke could have been influenced in rendering his judgment by knowledge of an Anglo-Burmese diplomatic dispute relative to river courses in the area. I shall refer to that dispute below.

29. Daniel George Edward Hall, *A History of South-east Asia*, 3d ed. (New York: St. Martin's Press, 1968), 585.

30. Much of the Burmese correspondence is translated and discussed by Thaung Blackmore, *Catalogue of the Burney Parabaiks in the India Office Library* (London: British Library, 1985).

31. Blackmore, *Burney Parabaiks*, 69 n. 3 (note 30).

32. Blackmore, *Burney Parabaiks*, 61, 62, 68, 82, 85, and 89 (note 30).

Manipur. One such map (probably not the one illustrated in plate 37), brought to a British post on the Chindwin River in March 1828, showed "a large river called the Ningtee running to the west of the Kabaw Valley and this, they claimed, was the true boundary between Burma and Manipur"; but a Lieutenant Pemberton, the resident British officer, who had himself surveyed the region, pronounced the Burmese map a fake, asserting that the Ningtee was the local name for the Chindwin.[33] In any event, in 1832, at Burney's urging, the British restored the Kabaw Valley to Burma from Manipur. On balance, in light of the intense interest in the area from 1826 to 1832, it seems most likely that that was the period when the RGS map of Manipur was drawn and that Burney's (the RCS) map was probably made between 1829 and 1832, though possibly as late as 1837.

The style of the two maps is distinctly non-European. Though the RCS map was almost surely copied from the RGS map (unless both were taken from a third source), it underwent some slight simplification in the copying. Mountain and hill ranges in the RGS map are rather naturalistically rendered, in black and mauve. On the RCS version they are painted in a blue wash and several of them, all minor, have been omitted, probably for want of space. On both maps summits generally point westward, away from Burma, but there is no consistency in this matter. Virtually all the ranges are capped with tree signs, suggesting forests. These are depicted in pale gray on the RGS map and by much bolder stylized green forms on the RCS copy. Both maps emphasize lakes and streams. These are shown in blue with a basket-weave pattern, but the somewhat simplified shapes of lakes in the RCS map do not faithfully reflect those of the RGS version. On the larger RGS map, space permits naming rivers and lakes within oval cartouches, often several times per river. A peculiarity of the rivers is that none is shown draining out of the Vale of Manipur (the western two-thirds of the map). Moreover, what must be the Manipur River, running from the mountains of the north to those of the south, is not linked with the large lake "Loup Tait" (Laktak), as shown on modern maps, and its actual southern egress from the Vale is, at best, hinted rather than clearly shown. In the north-central part of the map are two nebulous patches, identified as areas of rice cultivation; these are rendered in tan on the RGS map and in aqua on the RCS map.

Settlement is similarly indicated on both maps. It is clear that a hierarchy of places is intended. All settlements (and many other features as well) are outlined in red ink on the RCS map. Two important places near the border (one, the town of Tammu, is labeled a stockade) are shown by double squares; eight others, all *myo* (towns, six in Burma and two in Manipur) are rendered by squares with openings on four sides suggesting gates; and all other settlements (122 in Manipur and 47 in Burma on the RCS map) are shown by small squares or rectangles. On both maps Burmese settlements are shown in yellow, whereas those in Manipur are shown in either lavender or tan on the RGS map and in red only on the RCS map. Although on the former the two colors in Manipur are partially interspersed, the places shown in lavender are possibly villages or towns inhabited by Hinduized Manipuris, and those in tan may be occupied by the not yet converted Naga tribes, who are also locally prominent. Within Burma, more than two dozen pagodas are shown in frontal elevation, all with spires pointing west. This is in striking contrast with Manipur, where only a single Hindu temple, at the capital, Imphal (here called "Munipura Myo"), is depicted.

Some of the signs for pagodas in Burma are accompanied by a staff topped by a waving pennant whose symbolic significance is not evident. On later Burmese maps this is the sign for a military post, but the association with pagodas suggests a different meaning here. On the other hand, both pagodas and military posts would likely be found in settlements of consequence.

The focal aspect of both maps is Alaungpaya's military campaign, the key portion of which is shown in figure 18.7. The invasion route is marked by a dotted yellow line. Not far within Manipur, the battle alignments of the two opposing forces are shown by straight parallel lines. Those of Burma are, like all other Burmese features, in yellow and are labeled, on the RCS map, "Alaungphra's use of great mud defences." Those of Manipur, lavender on the RGS map and red on the RCS map, also consistent with the color-coding of settlement, are labeled "Kelzein Munipore mud defences." The invasion route continues past the latter defense line and terminates at the Manipuri capital, suggesting not only success in battle but subsequent military occupation of the enemy's territory. All of these map features, it appears, were intended to legitimize Burmese claims to Manipur.

I cannot say whether the map under consideration is an attempt to render a reasonably objective picture of the area and events depicted or whether parts of it were concocted to vindicate the Burmese position in diplomatic dealings with the British. Like Cooke in 1928, I have tried with mixed success to relate the content of the map to what is shown on modern maps of the area, concentrating perforce on elements of topography. In broad terms, the map conveys a good sense of the region: the depiction of Manipur as a mountain-girt vale; the forested mountain crests; the presence within the Vale of numerous lakes, with Laktak to the southwest and

33. Daniel George Edward Hall, *Henry Burney: A Political Biography* (London: Oxford University Press, 1974), 186; additional particulars on 214.

several smaller lakes to the northeast; and a rather complex drainage pattern with a dominant roughly north-south alignment are all consistent with reality. But in trying to trace individual river courses one quickly runs into difficulty, and the inability to match the Burmese or Shan toponyms of the map with the Manipuri, Naga, and Chin names found on modern maps poses insurmountable problems for the nonspecialist. It is easy to see why a British observer in the early nineteenth century might have labeled such a map a fake, even if it was made totally in good faith. A similar harsh judgment might have been made about many of the maps prepared for Hamilton, which clearly were (in most cases) good-faith efforts. In any event, as my description makes evident, there is much about the Manipur map, especially in regard to cartographic signs including the use of color, that is relatively sophisticated in comparison to the earlier military map illustrated in figure 18.38 below. Hence it may be viewed as paradigmatic in the study of Burmese cartography. Accordingly, this discussion largely reduces the need for equally detailed analysis of many other similar cartographic productions of Burmese provenance collected during the course of the nineteenth century.

Although it is clear that Henry Burney took a keen interest in Burmese maps, we have no definitive record of all that he saw and used. Even before his posting as ambassador to the court of Ava, he drew up a "Large Map of the Empires of Ava, Siam and Cochin-China compiled by collating the personal knowledge of the Principal Members of a Burman Embassy to the King of Cochin China, with the best European authorities." That map, which almost certainly incorporated material from the various works prepared for Hamilton, among other sources, was dated Calcutta, 22 October 1824. Burney also prepared a sketch map of the routes between Martaban, Tavoy, and Bangkok, to which he appended a note, "This sketch is taken from several sketches & descriptions furnished by native Christian Burmese & Siamese travellers." Additionally, there are in the library of the Royal Commonwealth Society in London two other maps, clearly Burmese in style, that Burney copied from originals, presumably no longer extant, he received at Ava.[34] For these maps, as well as the others discussed in this section, see also appendix 18.2.

Among surviving Burmese regional maps, there is one bearing a date, 1183 B.E. (A.D. 1821), that precedes Burney's arrival at Ava. This painted cloth map is the earliest of three maps of Shan territories that were acquired by the Oriental and India Office Collections. The letter of transmittal, by L. A. Goss, dated 26 February 1907, states that all three were drawn for the Burmese government, and that they were acquired by the British before their occupation of Upper Burma in 1885.[35] The 1821 map relates to two of the many tributary Shan *sawbwas* (chiefdoms), Maing Tsait (Möng Sit) and Maing Pone (Möng Pawn), over which Burma was at the time seeking to maintain control. The map was rich in detail relating to

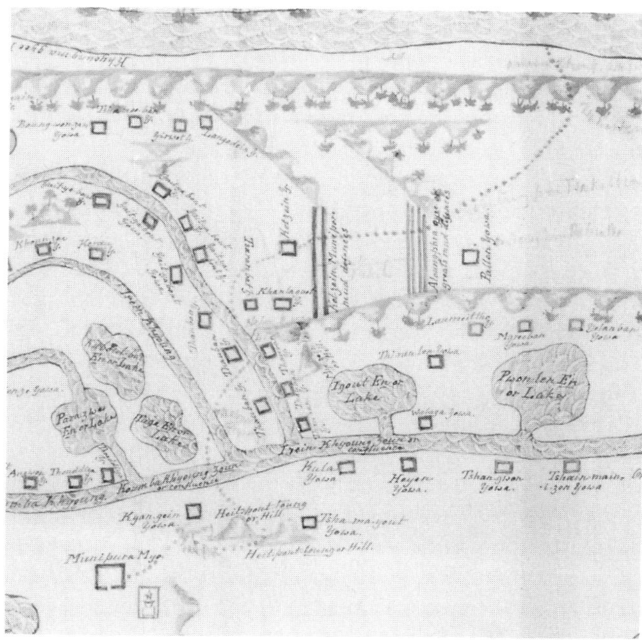

FIG. 18.7. THE FIELD OF BATTLE BETWEEN THE BURMESE AND THE MANIPURIS AND THE SUBSEQUENT BURMESE ROUTE OF INVASION TO IMPHAL IN 1759. The area shown here is a small part of a map that appears to have been copied from the one illustrated in plate 37, but painted on paper and at a much smaller scale. The copy, which maintains a traditional Burmese style but has a text entirely in English, presumably dates from the period 1829–37 and was very likely made for intelligence purposes at the behest of Henry Burney, the then British ambassador to Burma. On the whole, the copy is remarkably faithful to the original, though slightly less detailed. We have here retained the eastern orientation employed for plate 37. Like its prototype, this map maintains a consistent distinction, by color, of features associated with the Burmese (in yellow) and those associated with Manipur (in red). The invasion route of the Burmese is shown by a dotted yellow line, which has no counterpart in respect to the movement of the forces opposing them; the battle lines of the two armies, however, are shown in the appropriate colors.
Size of the entire original: 34 × 47 cm; this detail: ca. 15 × 14 cm. By permission of the Burney Collection, Royal Commonwealth Society, London (box XV, fol. 9, map C).

34. All four maps are listed in the appendix of Burney papers at the Royal Commonwealth Society, by Evans Lewis, Patricia Herbert, and D. K. Wyatt, in Blackmore, *Burney Parabaiks*, 101–18, esp. 104 and 117 (note 30).

35. I am greatly indebted to Patricia Herbert of the Oriental and India Office Collections of the British Library for assisting me with the interpretation of these three maps, and also with those of the Burney Collection, noted above, by reading for me enough map text to let me place them in their proper geographic contexts and render brief summaries of their content.

settlement and drainage and was in many respects similar to the map of Manipur—for example, in showing all hill ranges parallel to the four edges of the cloth it was drawn on and in the colors and types of cartographic signs it employed.

About the military significance of the two later maps in the Oriental and India Office Collections, however, there can be no doubt. One of these, datable to the 1850s or 1860s, covers a large region between the Salween and Mekong rivers in what is now eastern Myanmar and northern Thailand. It is rich in detail relating to routes (shown by dotted lines), settlements, drainage, and hill features. Many of the villages named within ovals are identified as (military) "camps," including the name of a specific *sawbwa*, while within a number of squares the word "army" appears, also with the accompanying name of a *sawbwa*. These appear to relate to Shan forces under Burmese command in an attempt to wrest the northern cities of Chiang Rai and Chiang Mai, both included on the map, from Siamese control, those areas having long been contested between the two neighboring powers. Unlike the other two maps, which have Burmese text, on this one the text is Shan, written in the Burmese script. Visually the map, rendered entirely in black ink, is the most distinctive of the three. Of particular note is its manner of depicting hills by undulating ink lines on which deep and shallow waves alternate with some regularity. Rather than drawing in vegetation, differentiated by type, as is so common on Burmese maps, the mapmaker merely suggested forests by closely spaced ciliate lines along the undulations.

The third regional map in the Oriental and India Office Collections is dated 1223 B.E. (A.D. 1861). It covers a large area astride the Salween River, largely overlapping that of the map just discussed but extending less far to the east and somewhat farther to the west. A very small part of this map is illustrated in figure 18.8. The map's most striking characteristic is the care lavished on depicting hill ranges in a highly stylized Sinic style and the even greater care in depicting forest, both on the hills and in groves separate from them. The vegetation differs so much from one part of this large map to another (not evident in the area excerpted) that one wonders whether it was all drawn by a single artist. An equally plausible explanation, of course, is that the variations reflect fundamentally different vegetation types. Notes on the map in various otherwise blank areas read, in Burmese, "woodcutting forest" and "woodmaking forest," signifying stands that are mature enough to cut and others that are not. Elsewhere, where forest signs are painted, notes indicate that the forest is to be cut by persons coming from such-and-such a town. On both this map and the one previously discussed, the method of aligning hills differs from that of the 1821 map in that they do not have simple north-south or east-west trends parallel to the edges of the map itself, but presumably follow their actual orientations. In this and other ways there appears to have been a significant development in cartographic expertise. As on many Burmese maps, large rivers are here generically differentiated from smaller streams. Distinctions are also made in giving distances along roads, using two different units of measurement.

Not dated, but presumably of roughly the same period as the two maps of the Shan territories just discussed, are three smaller and simpler Burmese works in the British Library. These works, none of them of particular interest for this history, form part of the collection of Arthur Purves Phayre, who served in various parts of Burma for most of the period from 1834 to 1867.

From Moulmein, which passed to British rule in 1824, come two additional maps now in the National Archives of India, New Delhi, one dated 1871 and the other 1871(?), that relate to the interest the new rulers had, in the interim, developed in the area to the northeast, which they looked on both as a source of high quality timber and as a potential trading route (never actually realized) to China via Chiang Mai. Although both maps were made by Burmese foresters at the behest of the British, there can be little doubt that the mapmakers were tapping cartographic traditions that were essentially indigenous.

The reign of the Burmese king Mindon (1853–78) was for the most part peaceful; but toward its end relations with the British had begun to deteriorate, in part because Burma developed diplomatic relations with France and other European powers whom the British viewed as potential rivals in what they had come to regard as their own sphere of influence. The regnal period of Mindon's less prudent successor, Thibaw, Burma's last monarch, was marked by a progressive deterioration of Anglo-British relations and by the looming threat, ultimately realized in 1885, of renewed hostilities. Thus it is not surprising that among the regional maps that can be assigned to this general period are several that are clearly military in nature.[36] The first of these three maps, all assumed to date from between 1870 (when Burma, with subsequent British diplomatic opposition, occupied the area of the Red Karens to which it relates in part) and 1885, covers an extensive north-south stretch of territory straddling a part of the border between British and Upper

36. Photographs of all three maps (items m, n, and o in appendix 18.2), map legends, and other relevant notes were provided by Tin Maung Oo, a former student of U Maung Maung Tin of the Burmese Historical Commission, both of Mandalay. The latter possesses a large collection of Burmese maps, originals or copies, to which I make frequent reference in appendixes 18.2 to 18.5. Hereafter, wherever I refer to works in that collection, the assistance of Tin Maung Oo in making them accessible through correspondence with me since our meeting in Mandalay in 1984 is gratefully acknowledged.

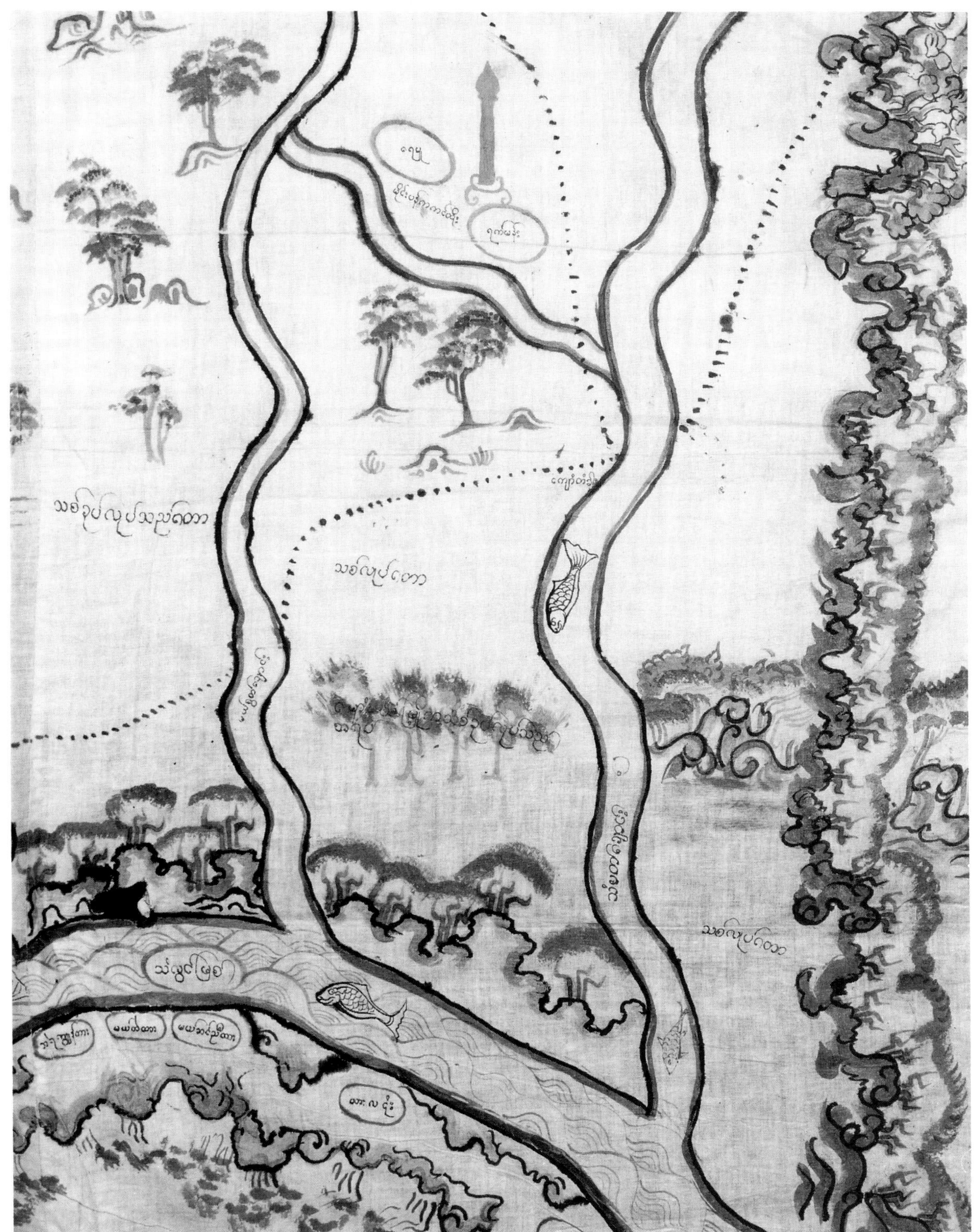

FIG. 18.8. SMALL EXCERPT FROM A LARGE BURMESE MAP OF MUCH OF THE EASTERN SHAN STATES. This large painted cloth map, dated 1861, covers an extensive area astride the Salween River in what are now Myanmar and Thailand. The specific locale covered by this excerpt has not been identified. The map provides a wealth of detail on settlements, routes, and drainage, but its most distinctive characteristic is the opulence and variety with which it depicts vegetation and, to a somewhat less marked degree, hill ranges.

Size of the entire original: 299 × 275 cm. By permission of the Oriental and India Office Collections, British Library, London.

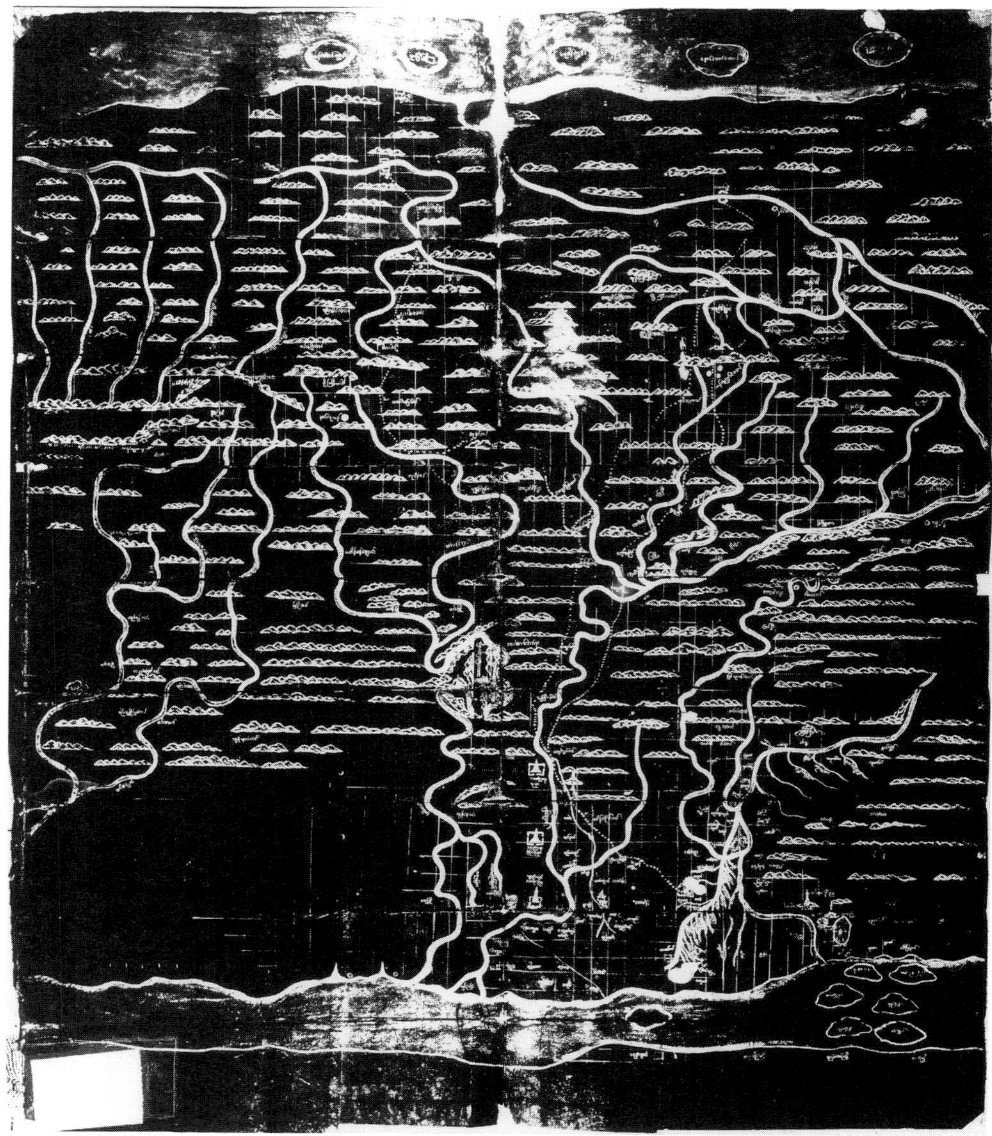

FIG. 18.9. MILITARY MAP OF AN AREA BETWEEN THE BAY OF BENGAL AND THE IRRAWADDY RIVER. This map presumably dates from the period 1870–85 but could be based in large measure on surveys carried out by the Burmese roughly one century earlier, of an area that largely passed into British hands in 1824. That area, Arakan, was one of intermittent Burmese control and was last subdued by Burma after an elaborately planned multipronged invasion in 1784. The original is a black *parabaik* (indigenous paper) map. Perhaps the most striking feature of the original map is its depiction of what appear to be literally scores of discrete hill ranges. In fact, the region has only one noteworthy mountain range, the Arakan Yoma. The multiplicity of small hill ranges, then, must be interpreted to signify simply that much of the area shown was continuously hilly. Also noteworthy are several dotted routes (which could have been those used in the 1784 invasion and intended for a reconquest a century later).

Of particular note is the depiction of what appears to be an all-water route between the Bay of Bengal and the Irrawaddy, across the Arakan Yoma. In fact, no such route exists; rather, the land route through An Pass was shown no differently from the largely riverine route of which it formed an important segment (a usage similar to that along the Anan route, noted on fig. 18.5), one of two traditional overland routes between central Burma and Arakan. Many topographic eminences on this map, presumably landmarks, are singled out for special notice (not apparent at the scale of this reproduction). These are shown by variously colored circles whose meaning is not clear. Color-coding is also clearly evident, however, in the depiction of military posts, those of the British and the Burmese being shown by red and yellow flags, respectively. Towns are represented by squares and villages by smaller circles. Pagodas are also abundantly shown. A final noteworthy map feature is its rectangular grid (ruled in yellow). Such grids are common on Burmese *parabaik* maps. Such a feature should not be regarded, however, as a modern military grid. Rather, it helped in copying maps from preliminary sketches, usually with an increase in scale.
Size of the original: 131 × 109 cm. Collection of U Maung Maung Tin, Mandalay.

FIG. 18.10. THE PRINCIPAL FEATURES SHOWN IN FIGURE 18.9. This is adapted from an annotated sketch made of the map by Tin Maung Oo in 1984. Names in brackets are not on the map.

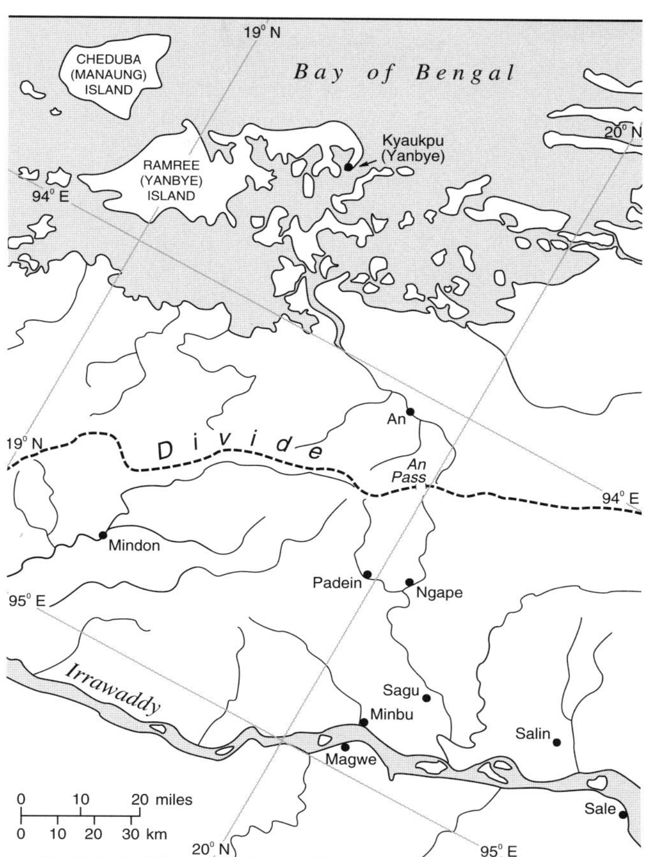

FIG. 18.11. MODERN MAP OF THE REGION SHOWN IN FIGURES 18.9 AND 18.10.

Burma. Although both British and Burmese military posts and garrisons (each in its distinctive color), as well as border posts, figure prominently in this very detailed map, it is noteworthy that, as in most other Burmese military maps, religious features such as pagodas, reliquaries, and spirit houses also are abundantly depicted. Partially overlapping the area of the previously noted map is another, extending east-west along the entire central portion of the then Anglo-Burmese frontier. This much simpler work was described as a "map for putting [military] outposts between the . . . Western Yoma [Arakan range] and the Red Karen Area." In depicting its military features the same set of signs was used as for the foregoing production.

The third military map (fig. 18.9), tentatively assigned to 1870–85 despite a suggestion that it belonged to the period before the first Anglo-Burmese War in 1824, depicts an area from Arakan, on the Bay of Bengal, and a portion of the central Irrawaddy basin.[37] The probable reason for suggesting a pre-1824 date is the map's inclusion of a large area, Arakan, that passed from Burma to the British in that year, although most of the territory included, as on the two previously noted works, was in Burma. My own suggested date is based on the similarity of this map, in both purpose and style, to the first of those described in the previous paragraph. Stylistically, both seem to have evolved considerably from maps that can unambiguously be assigned to a period before the second Anglo-Burmese War (1855). Although Burma was not in a position to map Arakan directly from the field after 1824, it did prepare a map of the region noted by Burney, and it amassed a great deal of information on it between 1784, when it was annexed following a carefully planned, multipronged invasion, and the hostilities with Britain that led to its loss. Moreover, since no fewer than twenty thousand Arakanese were deported to Burma proper in 1785, there would have been no dearth of sources for the intelligence on the region needed to prepare the subject map.[38] Figures 18.10 and 18.11 highlight the more important features shown in figure 18.9.

Post-1885 Maps of the Cambridge Scott Collection

After annexing Upper Burma in 1886, the British assumed

37. The suggestion was made in the notes on the map by Tin Maung Oo.

38. Hall, *History of South-east Asia*, 585 (note 29).

that they thereby became the legal inheritors of Burmese suzerainty over the roughly three dozen Shan states lying to the east. Making that suzerainty effective, however, engaged the British continuously until 1890, and the task of doing so with a minimum of bloodshed fell largely to James George Scott, whose career in Burma had begun in 1879. Scott was appointed resident for the Northern Shan States in 1891 and spent most of the rest of his service in the region, retiring in 1910 as superintendent for all of the Shan States. In the interim he served as a member of the three boundary commissions, which delimited Burma's borders with Siam, French Indochina, and China during 1889–1900. In his various official capacities he collected a wealth of manuscripts on Burma, which were passed on to his brother, Robert Forsyth Scott, Master of St. John's College, Cambridge, and bequeathed to the university by Sir Robert's widow in 1933.[39]

Among Scott's manuscripts was a remarkable collection of forty-seven maps, most obtained in the Shan States and the rest in Upper Burma. These maps have been cataloged by Andrew Dalby of the Cambridge University Library with the assistance of Sao Saimöng Mangrai, a Shan historian who was invited to Cambridge in 1984 expressly to help with the ordering and analysis of the Scott Collection. Dalby notes that most if not all of the maps in the collection "were drawn at the orders of Scott himself or of one of his colleagues in the British administration." Of the seven maps of Upper Burma, all seem to have been drawn by Burmese clerks, often recruited from the previous Burmese administration, who accompanied the newly appointed British district officers as they familiarized themselves with their districts by personal inspection. Of the maps of the Shan States, at least seven are characterized as general orientation maps of extensive areas. The language of these maps is Burmese, "indicating that they date from the early years of the annexation when communication with British officers was only possible in the presence both of a Burmese-speaking Shan and an English-speaking Burmese." On these maps the representation of the extensive areas to the east of the Salween River, an area little known even to the Burmese, was judged very inaccurate. Another seven maps are of individual states. These maps emphasize internal communications, state borders, or both, even though the concept of a linear territorial boundary was alien in the ethnically mixed tribal territories dominated by the Shans (who constituted only about half the total population in the region that bears their name). Of these maps four were in Burmese and one each in Shan, Khün, and Lü (all tribal languages). The last three maps, in Dalby's view, "show less European influence than any of the others." Of the remaining twenty-six maps, seven are said to relate to British expeditions, three to interstate disputes, eleven to external frontiers, one to a road transport scheme, and one to "an application for the post of village headman." The purpose of the remaining three maps is unclear. In all, there are twelve maps in Shan, one "mostly in Chinese Shan," one in Khün, one in Lü, and thirty-two in Burmese.[40]

Physically, the maps vary substantially. The more elaborate ones, including most of the general topographic ones we are principally concerned with here (brief notes on other types of maps will be provided in subsequent sections of this chapter), were drawn on white calico. Most of the others were on "full-size sheets of the strong, durable, cream-coloured Shan paper, locally made from shredded bark," and one was on the thicker paper used for *parabaiks*, which are the Burmese equivalents of the *samud khoi* documents previously described for Siam. In size, the maps range from as little as 21.8 by 30.8 centimeters (actual map area) to as much as 172 by 255 centimeters and 164 by 269 centimeters.

Stylistically, too, there is considerable variation. The maps of Upper Burma were rendered in black ink and several bright painted colors. "Clearly they were drawn in an office at leisure and display the almost obsessive neatness for which Burmese manuscripts of all types are noted." The less elaborate maps were always on paper, drawn with a European pencil and sometimes one or two added colors of ink. On some of the maps scribbled pencil notes, often transliterated names and occasionally titles (not always accurate), were added by Scott himself. On the whole, the Shan maps (with one major exception, to be discussed below) are of a simpler style and, in Dalby's view, less aesthetic. Of maps that he judges to be of a "pure Shan cartographic style," he observes:

> In their depiction of mountains, rivers and most noticeably towns—large, sometimes decorated circles, with names and other details written inside the symbol—these maps resemble one another much more than they resemble any European map that might have been shown to the artist as a model. Yet having been acquired on different missions, quite widely separated in time and space, it is difficult to believe that they influenced one another. Indeed, to judge from the condition of the maps in the collection, none of them . . . [with one cited exception] can have remained long enough in the Shan States to make it likely that there

39. *Dictionary of National Biography*, suppl. 1931–40 (London: Oxford University Press, 1949), 797–99; and Andrew Dalby and Sao Saimöng Mangrai, "Shan and Burmese Manuscript Maps in the Scott Collection, Cambridge University Library," unpublished manuscript, n.d. (ca. 1984), 15 pages, followed by a 47-page illustrated catalog (39 figs. and 8 pls.).

40. Dalby and Saimöng, "Shan and Burmese Manuscript Maps," introduction (pages not numbered) (note 39). All subsequent quotations relating to maps in the Scott Collection, except where otherwise noted, are from this six-page introduction.

> was much stylistic interchange among them. . . . This suggests to us that there was a pre-European Shan-Khün-Lü style of cartography, from which each artist who has contributed to the collection deviates to his own degree under the influence of the Europeans for whom these maps were drawn.

When Dalby wrote the remarks above he had not, I gather, seen any of the pre-1885 Shan maps that are discussed here and listed in appendix 18.2. His observations, however, confirm my own views on the existence of what might be considered a distinctive Shan cartographic style and, I would hypothesize, a Shan tradition of cartography that predated Hamilton's 1795 visit to India and enabled the remarkable works prepared for him by the (probably Shan) slave of the Burmese crown prince.

Dalby notes that the maps in the Scott Collection are generally oriented toward the east, although some lacked any consistent orientation in respect to lettering and pictographic symbols. The Burmese words for the four cardinal directions were often written along the appropriate map edges. Vegetation is depicted abundantly and in considerable variety. Towns are generally shown by squares on Burmese maps and by circles on Shan maps. Pagodas frequently serve as landmarks. Roads are usually rendered by dotted lines, often in red. Other signs relate to weirs, irrigation channels, bunds, tanks, and bridges. Boundaries were only rarely shown, and even more rare—on three maps only—was the use of a color wash to differentiate territories subject to two different political jurisdictions. Textual notes referring to administrative jurisdiction, on the other hand, are common. Most map scales appear to be large, though inconsistent. Many maps have notations relating to stages ("posts") between settlements. A few note that a certain number of posts equal one "thumb" (ca. one inch) on the map and state the number of English miles to the thumb. On at least one map, distances between places are given as so many "nights" spent en route, the number being as high as fifteen in at least one instance. Pencil grids, such as were noted on many pre-1885 Burmese maps, are common.[41]

I have singled out for illustration only two regional maps from the Scott Collection. (Several other maps of localities will be discussed later in this chapter.) Figure 18.12 is an excerpt from a relatively simple map of a frontier area that appears to be the oldest of the group, since it bears a note stating that it was received at Cambridge on 17 August 1890. The text is in Burmese. Much more elaborate than figure 18.12, and possibly the latest in the collection, is the map of Kengtung (fig. 18.13), the largest (approximately 31,000 square km) of all the Shan states. The area depicted extends from the Salween in the west (top of the map) to the Mekong in the east. Both of those rivers are painted in a singular fashion that largely makes them appear to flow just above, rather than on, the neighboring land surface. Smaller rivers appear as thin single lines. Filling most of the area of the map are more or less naturalistically depicted mountains, painted mainly in various shades of gray. The area of what today is mainly Laos is left blank, and very little detail (and no settlement) is depicted in areas of Shan states other than Kengtung, the areas being rendered in a blue or purple wash.[42]

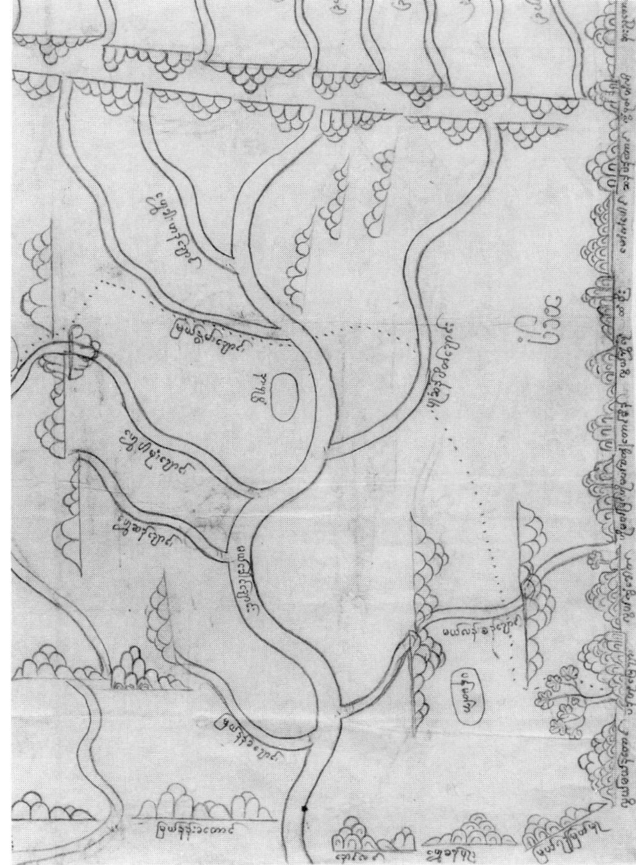

FIG. 18.12. EXCERPT FROM A BURMESE MAP OF MÖNG MÄU AND MÈ HSA KUN, TWO TRANS-SALWEEN SHAN STATES. This view represents about a fourth of the total map, which probably dates from the investigation of that area by a boundary commission in 1888–89. The entire map, drawn in pencil on rice paper, is judged to cover an area of approximately 40 by 40 kilometers mainly lying just within the present border with Thailand in the southernmost part of what is now the Shan state of Myanmar. The map text is in Burmese.
Size of the entire map: 52 × 52 cm. By permission of the Syndics of Cambridge University Library (Scott LL. 9.92).

41. The observations in this paragraph are largely my own, and the terms quoted are from my own notes of 1984.

42. For a fuller description see Dalby and Saimöng, "Shan and Burmese Manuscript Maps" (note 39).

FIG. 18.13. MAP OF KENGTUNG. This map, painted in six colors and black ink on a large pencil-gridded sheet of calico, dates from sometime after 1896. The Burmese text suggests that the artist was a Burmese commissioned by the *sawbwa* (Shan chief) to execute the work, perhaps for Scott, to whom it was presented. Kengtung State extended from the Salween River in the west (top of the map) to the Mekong in the east. The significance of the various shapes and colors used to depict mountains on this map has not been determined. Nor is it known why different signs are used for the twenty-six settled places that are shown, though either an ethnic or an administrative meaning seems plausible. Trails are shown by red dotted lines.

Size of the original: 90 × 91 cm. By permission of the Syndics of Cambridge University Library (Scott LL. 9.101).

SIAMESE MAPS

In contrast to the substantial number of Burmese regional maps, the dearth of such maps from what was historically included within the state of Siam[43] is remarkable, especially when one considers the extent of intercourse—admittedly often hostile—between the two countries and the fact that a significant proportion of the maps I have included among the group from Burma were actually drawn by Shans, an ethnic group closely affiliated with the Siamese. The probable explanation for this anomaly in terms of the Siamese institution of *chamra*, the periodic bureaucratic purging of outdated documents, was provided above (pp. 698–99). In any event, not counting the small portion of the essentially cosmographic *Trai phum* that one may consider geographic, we can point to only a handful of Siamese maps, of which only two can be regarded as regional.

Of these regional maps, the older, dubbed sometime after 1910 "A Logistics Map from the Reign of Phra Bat Somdet Phra Ramathibodi I," is by far the more detailed and accurate. It is also, fortunately, the subject of an exceptionally thorough and well-informed study by Kennedy. The map he discusses, however, is not the original, allegedly dating from the reign of Rama I (A.D. 1782–1809), but rather a twentieth-century reproduction of a map that, despite the assertion in the title, could be either a revision of the original, assignable by internal evidence to the period between 1809 and 1834,[44] or more likely an essentially new work (fig. 18.14). If the latter, it was "compiled from intelligence reports taken during the campaign of 1827 [against Laotian forces in the Korat Plateau] and the form it takes [was] shaped by the quality of those reports and the routes the armies took." In either case, the latest date of 1834 would still apply. Regrettably, we know the present location of neither the assumed late eighteenth-century original nor the nineteenth-century version. Older maps do, however, appear to have been consulted in making the map, and corrections of the information they presented, especially in regard to river systems, seem to have been incorporated in it.[45]

By and large, the map compares favorably with the Burmese regional maps I have examined. This provides strong support for the notion that a now forgotten Thai tradition of cartography may have existed, comparable in age and development to that of Burma. On the whole, the map works well as a guide to the routes between the places it depicts and within any single region of Siam (see figs. 18.15 and 18.16). The scale and directional relationships among places are usually fairly consistent. However, Kennedy observes,

> Links between neighbouring regions are often poorly made. Towns on either side of a mountain range, such as the one that separates the Pa Sak Valley proper from the hilly northwest Khorat plateau, are not in correct mutual relationship, and the same can be said for any pair of adjacent regions that are separated by a barrier that is not frequently crossed. Under such circumstances, scales change suddenly and true directions of travel cannot be read from the map.[46]

Two regions on the map are shown with less than usual internal consistency. The first, surprisingly, is the central riverine portion of Siam, "presumably for the simple reason that it was too well known to require being drawn as carefully as the other regions."[47] The other, a relatively remote region illustrated in figure 18.17, is on the Mekong River near Vientiane. There the "local scale varies from around 1:8,000,000 in the north-south direction to 1:2,500,000 in the other." In the same general area, directions and other details become badly distorted, so that a geographic grid Kennedy fitted to the map as an aid to its analysis shows that "between Phan Na and Vientiane the latitude grid line becomes parallel to longitude lines a little further west. [Furthermore,] a route from Phan Na to Nong Han is drawn through a range of mountains . . . when what the road crosses is in fact paddy fields."[48]

Despite its lack of accuracy, the image in the area illustrated in figure 18.17 does convey some sense of the highly pictographic symbolization employed on the map. Cartographic signs are for the most part self-explanatory and in many respects (e.g., the manner of depicting rivers) similar to those employed on many Burmese and Chinese maps. Oddly, none of the main rivers are named, though

43. I use the words "Siam" and "Siamese" here in preference to "Thailand" and "Thai." As a political designation, the latter would be anachronistic for the time period under consideration, and as an ethnolinguistic designation, Thai or Tai would be confusing in that the cartography of the Shans, a Thai people, has already been discussed along with that of other peoples resident in or politically subject to Burma.

44. Victor Kennedy, "An Indigenous Early Nineteenth Century Map of Central and Northeast Thailand," in *In Memoriam Phya Anuman Rajadhon: Contributions in Memory of the Late President of the Siam Society*, ed. Tej Bunnag and Michael Smithies (Bangkok: Siam Society, 1970), 315–48 and 11 appended map plates; see esp. 315–16 and 322–23.

45. Kennedy, "Nineteenth Century Map," esp. 348 (note 44).

46. Kennedy, "Nineteenth Century Map," 322 (note 44). Although I have not checked any of the Burmese regional maps as diligently as Kennedy has studied the map under review, it is my impression that the criticism conveyed by the quoted passage would apply equally to them.

47. Kennedy, "Nineteenth Century Map," 322 (note 44). For analysis, Kennedy divides the area covered by the map into sixteen naturally defined regions and devotes the greater part of his paper (pp. 324–43) to discussing the content and accuracy of the map in respect to each of them.

48. Kennedy, "Nineteenth Century Map," 343 n. 45 (note 44).

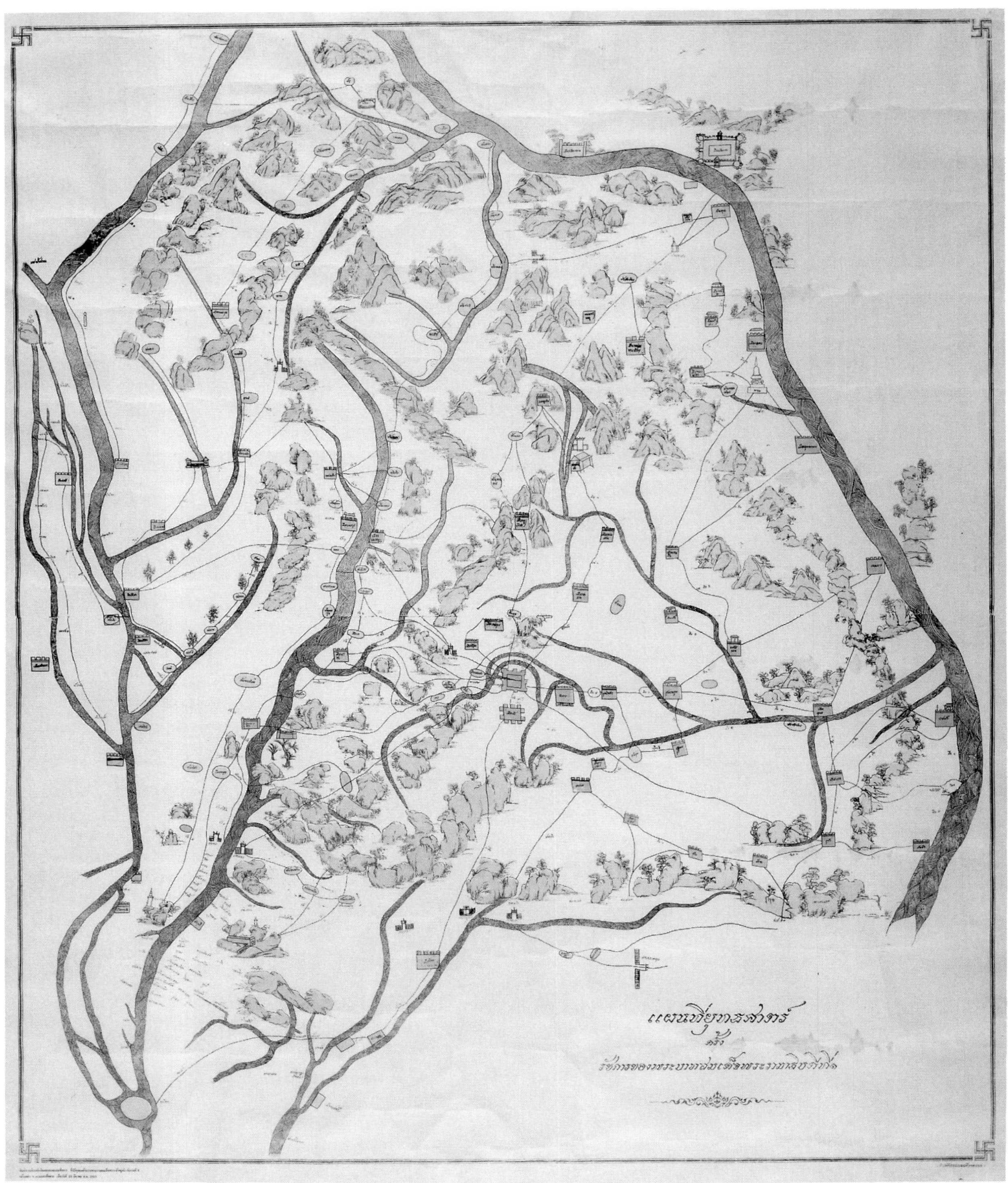

FIG. 18.14. MILITARY MAP OF THE KORAT PLATEAU AND ADJOINING AREAS OF SIAM. This figure depicts a twentieth-century reproduction of a map allegedly dating from the reign of the Siamese monarch Rama I (1782–1809), but apparently either revised or completely redrawn sometime before 1834. In its present form the map seems to incorporate intelligence gathered during a military campaign in the region in 1827. This copy is drawn in black ink on paper. The size, media, and location of the original from which this copy was made are not known. See also figures 18.15 to 18.17.
Size of this facsimile: 101.5 × 84.5 cm. By permission of the Royal Thai Survey, Bangkok.

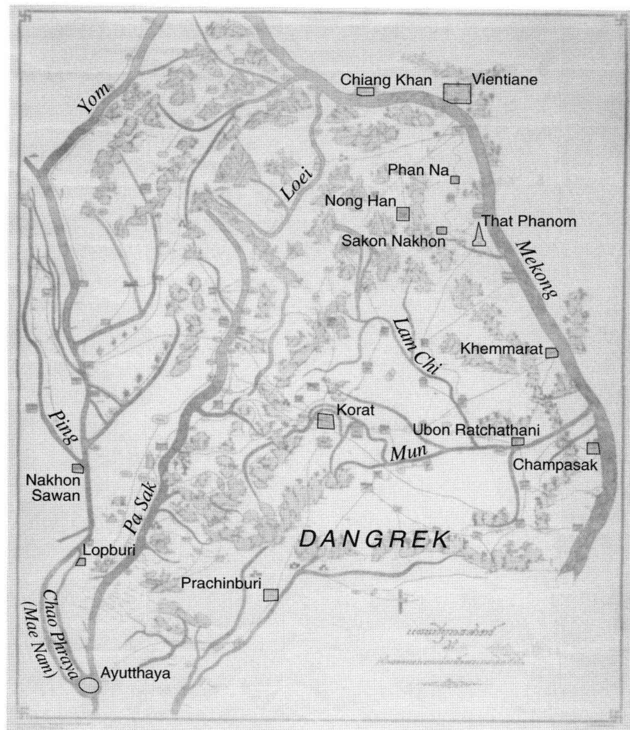

FIG. 18.15. MILITARY MAP OF THE KORAT PLATEAU WITH TRANSLITERATION OF NAMES OF SELECTED FEATURES. Compare figure 18.16.

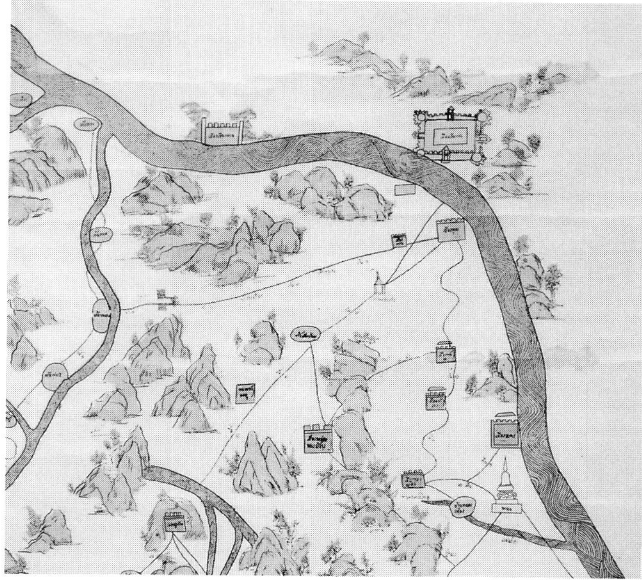

FIG. 18.17. EXCERPT FROM A MILITARY MAP OF THE KORAT PLATEAU. This view is only a small portion of the facsimile depicted in figure 18.14, including an area in the far northeast of what is now Thailand and an adjacent area across the Mekong in Laos. Although there is no way to determine the fact by casual visual inspection, the area shown is probably the least accurately portrayed portion of the entire map. For example, the route traversing a range of hills in the lower central portion actually runs through a lowland area of paddy fields. Nevertheless, on the whole the map obviously embodies a great deal of detailed and apparently reliable data about settlement, roads, and terrain.
Size of the detail: ca. 30 × 30 cm. By permission of the Royal Thai Survey, Bangkok.

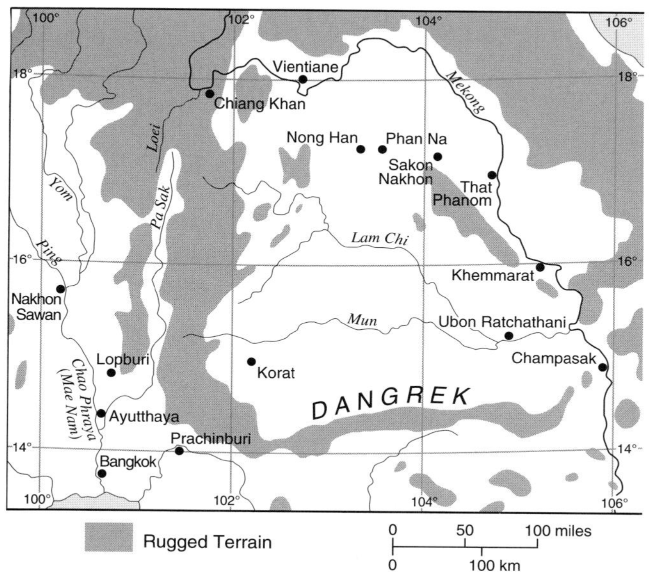

FIG. 18.16. MODERN MAP OF THE KORAT PLATEAU AND ADJOINING AREAS OF SIAM. On this map are identified most of the features shown in figure 18.15.

some of the lesser streams are.[49] Also noteworthy is the existence of a settlement hierarchy (as in many Burmese maps), with towns indicated by squares, here almost always surmounted by battlements, and selected villages shown by ovals. A striking feature of the map is the great attention given to areas of rugged terrain, shown in a variety of forms and with considerable exaggeration of vertical scale. Vegetation is also shown in considerable variety, though with less prominence than on most Burmese maps. To what extent the variation in the manner of depicting landforms and vegetation represents an attempt at symbolic differentiation of recognized types actually existing in nature is not known.

According to Constance Wilson, a historian who has done extensive archival research in Thailand, among the ten to twenty thousand *samud khoi* manuscripts in the National Library of Thailand only five or six include maps.[50] Among those maps, the nature of only one is

49. Kennedy, "Nineteenth Century Map," 316 (note 44).

50. Conversation with Constance Wilson, Northern Illinois University, DeKalb, March 1984.

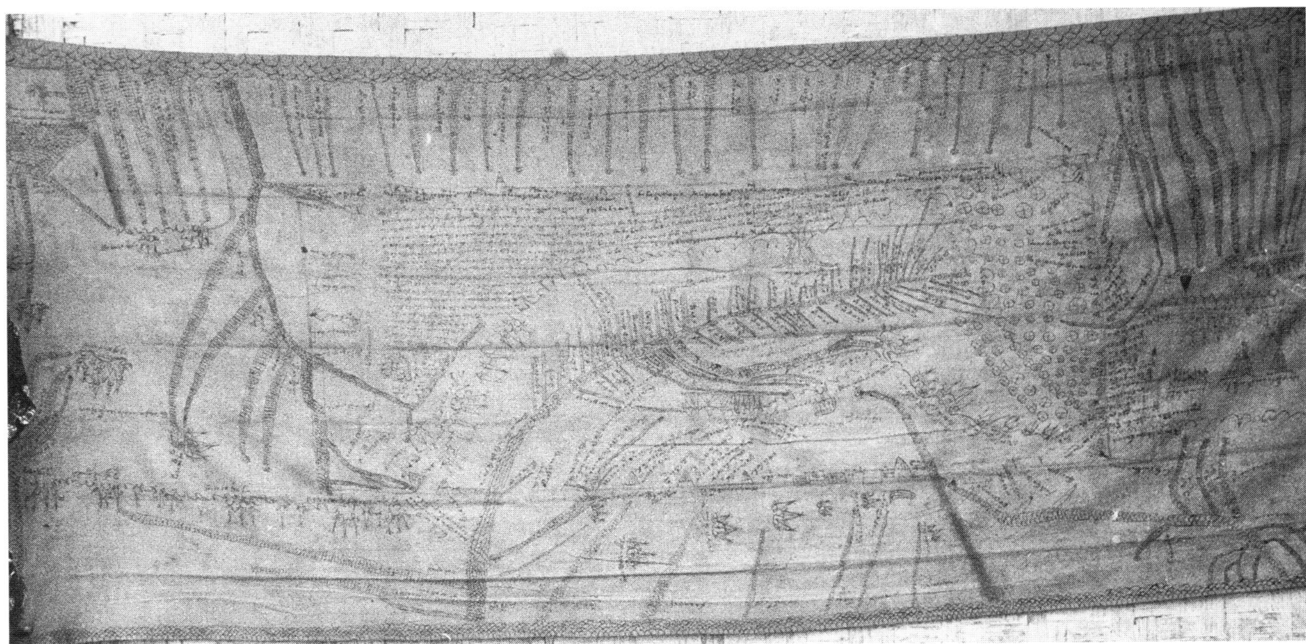

FIG. 18.18. SACRED MAP OF THE SUNDANESE CHIEFDOM OF TIMBANGANTEN, LATE SIXTEENTH CENTURY. This large and exceedingly detailed cloth map is drawn in ink, oriented with south at the top, and covers virtually all of the western third of Java. The large rectangle in the middle of the map is the former chiefdom of Timbanganten, depicted at a greatly exaggerated scale, within which lay the village of Ciela, where the map is still held and venerated as a sacred relic. Within the rectangle is a lengthy inscription in Sundanese indicating when, why, by whom, and for whom the map was drawn (though the era cannot be stated with certainty), noting that its purpose was to identify the lands that belonged to the *sunan* (prince) who commissioned it and who proclaims himself a good Muslim. The map is especially noteworthy for the originality of its cartographic signs, incorporating as it does some signs that are notably different from any seen on other maps from Southeast Asia.
Size of the original: 91 × 223 cm. Photograph courtesy of Joseph E. Schwartzberg.

known to me. That unprepossessing work forms the second of the known Siamese regional maps. The map is drawn in white chalk on heavy charcoal-coated bark paper in eight folding panels and is dated 1867. The orientation is roughly toward the south. The area covered is bounded on the south mainly by the Mae Nam Mun River and on the east by the Mekong, into which the Mun flows. The scale appears to be considerably compressed in the west so as to extend the map to the town of Korat (Nakhon Ratchasima), one of only a few named places lying to the south of the Mun. In all, then, the mapped area extends approximately 400 kilometers east-west. The northern margin is indeterminate but probably does not extend more than 150 kilometers north of the Mun. There is also some north-south compression to accommodate the stretch of the Mekong to the north of the Mun confluence; and given the strip format of the map, it was necessary to bend the Mekong's course counterclockwise, making the two rivers appear to flow almost parallel to one another rather than roughly at right angles as they actually do. The map appears to have been hastily drawn. In addition to showing five rivers (by double lines with a crude wave pattern in between), it identifies seventeen locales. Ten are shown within rectangles (nine *muangs*, or towns, and one *ban*, or village), five within ovals (the town of Korat, two places cryptically designated "Royal Field," and two *bans*), and two more by circles (one a *ban* and one not identified). All but seven of these places are shown connected by roads, drawn as thin white lines, and all but one (a *ban*) lie on either a road or a river. Two administrative districts, Ubon (modern Ubon Ratchathani) and Suwannaphum, are named, but neither is bounded. I assume that the map relates to local administration.[51]

MALAY MAPS

The few known surviving regional maps from the Malay

51. The manuscript in which the map appears is at the National Library of Thailand, Bangkok, C.S. 1229/161 (ca. 24 × 61 cm) and is titled *Chotmaihet Ratchakan thi 4*. Wilson was kind enough to send me a photocopy of the map, to translate almost all the text on it, and to provide some useful explanatory notes (correspondence 10 April 1984).

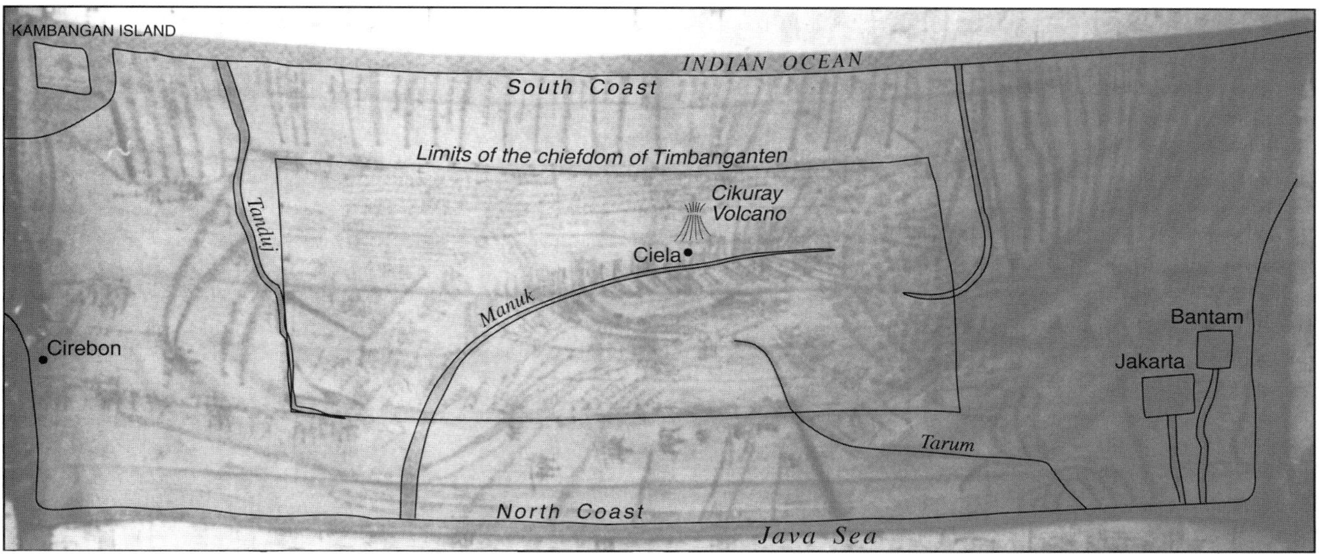

FIG. 18.19. THE PRINCIPAL FEATURES ON FIGURE 18.18.

world form a remarkably disparate group. In all, there are one map of western Javanese provenance, presumably of the late sixteenth century; another enigmatic and undatable batik map from either eastern Java or Bali; a third map from east-central Java, probably drawn in the mid-nineteenth century; and finally, a map from western Borneo tentatively dated 1826. There is no indication that any two of these were informed by a common cartographic tradition. In addition, there are a number of works that may best be discussed as nautical charts. I shall here consider only the four regional maps, in the order given above.

Figure 18.18 and plate 38 present, respectively, a comprehensive view and a detail from the western Javanese map, a large and very detailed work on cloth. The map, which is held to be a *pusaka* (sacred relic), is carefully preserved in the village of Ciela in the Garut District in the Sundanese region of Java, where it is periodically worshiped. Ciela is about sixty-five kilometers southeast of Bandung at the foot of Cikuray volcano, which the map shows very prominently. There, in 1976, the map was discovered (actually, as we shall soon see, rediscovered) by Rachmat Kusmiadi, an Indonesian geologist on duty with the country's Geological Survey. Kusmiadi brought the map to light the following year in a paper presented at the Seventh International Conference on the History of Cartography.[52] It was subsequently illustrated and briefly discussed by Harvey, who stated that it "has the appearance of being, like the Early Han maps from China, on the borderline between symbol-map and picture-map."[53] I was able to inspect the map personally in Java in 1984. Only then did I learn that it had been previously discovered in 1858 by a Dutch official, J. C. Lammers van Toorenburg (aided by a Javanese informant), and that after some fruitless attempts by others to interpret it the task was assigned to a philologist, K. F. Holle, who studied it at length in 1862, made two copies, and published an article on it, including a detailed reduced-scale drawing, in 1877.[54] In 1990 the map was promi-

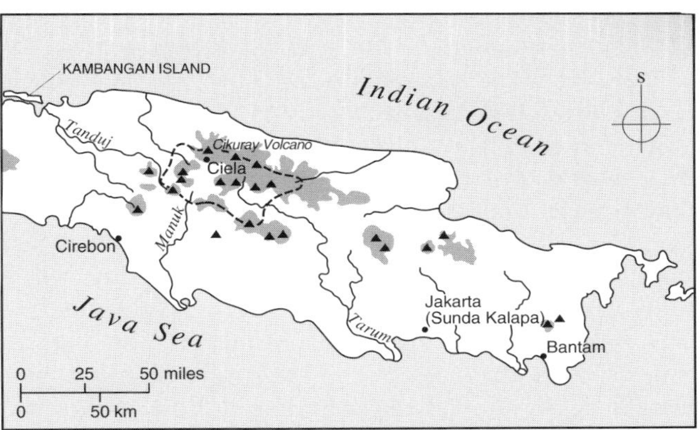

FIG. 18.20. MODERN MAP OF THE REGION DEPICTED IN FIGURES 18.18 AND 18.19. South is at the top.

52. Rachmat Kusmiadi, "A Brief History of Cartography in Indonesia," paper presented at the Seventh International Conference on the History of Cartography, Washington D.C., 7–11 August 1977.

53. P. D. A. Harvey, *The History of Topographical Maps: Symbols, Pictures and Surveys* (London: Thames and Hudson, 1980), 114.

54. K. F. Holle, "De kaart van Tjiëla of Timbanganten," *Tijdschrift voor Indische Taal-, Land- en Volkenkunde* 24 (1877): 168–76 and endcover folding map (this article was translated for me by L. Ruyter

nently featured in the television documentary "The Shape of the World."[55]

The village of Ciela was at one time the seat of a minor chiefdom called Timbanganten, one of many petty states formed on the conquest and breakup in the late 1570s of the interior Sundanese Hindu kingdom of Pajajaran (whose capital was at Pakuan, near modern Bogor) by the sultan of Bantam, which had accepted Islam only half a century earlier. The local magnates were then forcibly converted to Islam.[56] The chief of Timbanganten, named on the map and identified as a Muslim, was possibly among those converts. But whether it was that chief or one of his forebears who adopted the new faith, Kusmiadi's assertion that the map dates from the fifteenth century is clearly untenable, since no part of Java had accepted Islam before 1525.[57] The Indonesian museum label for one of Holle's copies, now on display in the Jakarta City Museum, states that "it is estimated that the age [of the map] is more than 300 years."[58] In any event, in the long inscription within the map, said to be in "corrupt old Javanese," the mapmaker identifies himself as Masjaya and states that his *sunan* (prince) Lawas Jaya (otherwise unknown to history), a Muslim, ordered the map to be made to establish the limits of his domain. That Lawas Jaya's original name was Was Jaya Cacandran suggests that it was indeed he who converted and that he changed his name at that time.[59] The map text lists, based on a census, seventy-eight *kampongs* (villages) included within the chiefdom and states that Lawas Jaya "divided the region [realm] into three sections, which were in turn subdivided" and that this was done on "Friday, the 14th, the month of Muharam, the year of Alip."[60] Alip, the first letter of the Arabic alphabet, here would signify the year 1, indicating that the *sunan* was inaugurating a new era in his realm. Prominently shown on the map are a number of what appear to be boundary markers along what I take to be the rectangular boundary of the entire domain. Although internal divisions are not evident, Holle states, in seeming contradiction to the museum caption, that Timbanganten was partitioned in two parts by the river Manuk, which runs in an arcuate course to the north of Cikuray volcano.[61]

Both the museum copy and the published version of the map carry very detailed (and, it appears, identical) sets of numerical and alphabetical keys to the map. Portions of the text are identified by roman numerals I to III, and at least 250 map features are identified by Arabic numerals, upper- and lowercase letters, and upper- and lowercase letters followed by prime signs. No mention of the key, presumably made by Holle, appears in his article, however, and my attempts to find the key at the Jakarta City Museum proved fruitless. Nor does Holle dwell at length in his article on the geographic content of the map.

Without the map key, attempts by nonspecialists to make sense of the map will necessarily be speculative. A few important matters, however, are certain. First, the orientation of the map, though not uniform, is, as with most maps of Islamic provenance, dominantly toward the south. Second, the area covered (see figs. 18.19 and 18.20) is virtually the whole of Java to the west of Cirebon (not itself shown), except for the far western tip of the island,

in June 1984). The published map is unusually large, 29.7 by 78.7 centimeters, and will reward detailed study, especially by scholars with the requisite linguistic competence. The example I procured in the Netherlands, however, was on rather brittle paper and in poor condition. On one corner of the original map (not shown on the reproduction) Holle inscribed the sentence "Den 1 Daag 1862 een copy genommen door K. F. Holle." This appears not to have been noticed by Kusmiadi. Whether Lammers was the first European to see the map is itself open to doubt, since a Mr. Netscher reported that an anonymous German scholar already had examined the map some time earlier. Netscher (as cited by Holle) observed that the scholar's research received no general publicity (p. 172), by which one may assume that it never was published, and nothing more is known of him.

55. The map was discussed in the introductory and final episodes of the six-part series produced by Granada Television Limited, England.

56. Hall, *History of South-east Asia*, 215 (note 29).

57. Kusmiadi, "Cartography in Indonesia," 1–2 (note 52). Kusmiadi was unaware that the map had been studied by Holle a century earlier and did not know of the museum copy in Jakarta. Harvey, *Topographical Maps*, 114 (note 53), following Kusmiadi, also stated that the map was from the fifteenth century.

58. The label was translated for me in full by Alan Feinstein in Jakarta in March 1984; his assistance is gratefully acknowledged. Another part of the label states that the age was "300," not "more than 300," years. The date when the label was written, however, is not known. It could be an Indonesian translation of an initial inscription written in Dutch by Holle or a colleague as much as a century before I copied it, with no alteration of the original text. If the date when the map was made immediately followed the conversion of the ruler in the 1570s, and if the label (not dated) was written about 1877 when Holle wrote his article on the map, either statement would be reasonable. Holle, "De kaart," 174 (note 54), stated that the map was just under three hundred years old. One bit of internal evidence in support of a date in the late sixteenth century is the map's designation of Jacatra (Jakarta) by its old name, Sunda Kalapa. Although the name had been changed on its conquest by Bantam about 1525, the new designation might have taken some time to be recognized. But the probability of designating the place as Jacatra would have steadily increased with time.

59. The introductory portion of the map inscription, translated in the museum label, is somewhat perplexing. As translated for me by Feinstein, it reads: "This is the assignment, or affair, of the ruler in the region of Timbanganten, that is Susunan Cantayam. But all of this [land?] is the inheritance, or estate of Maharaja Tunggal, who held the title Maharaja Sukma, and who was the son-in-law of Ratu Tunggal, who was called Tuwinis." Feinstein calls attention to the ambiguity of the relative pronoun "who" in the previous sentence. Also ambiguous in my mind is the relationship of Susunan Cantayam and Maharaja Tunggal to Lawas Jaya. It seems likely that we are dealing here with several levels of political authority in a more or less feudal system. I surmise that a *susunan*, to cite the title of Cantayam, is higher than a *sunan*, the title enjoyed by Lawas Jaya, but lower than a maharaja. But why maharaja, rather than sultan, was used after conversion is not clear.

60. Holle, "De kaart," 174–75 (note 54).

61. Holle, "De kaart," 175 (note 54).

an area of roughly 40,000 square kilometers, including all of the traditionally Sundanese territories. Third, within that area the chiefdom of Timbanganten, represented by the rectangle in the center of the map, is shown at a greatly exaggerated scale, occupying slightly more than a fourth of the entire map surface and nearly three-tenths of its land surface, when in actuality it would probably have accounted for a rather small though indeterminate fraction of that amount. Fourth, despite this exaggeration and the absence of a key, one can relate the drainage features and topography on the map, as well as those few features that Holle's article identifies, to the known contemporary geography of Java well enough to recognize the work as a remarkable cartographic achievement.

The cartographic signs used on the Timbanganten map are highly distinctive. Among the most expressively rendered features is Cikuray volcano (see plate 38). Other mountains assume a variety of forms, but it is not clear in any instance which of them are volcanic. Well over a hundred rivers and tributary streams are shown, all characteristically vermicular in appearance. Many issue from what appear to be springs, shown by a roseate line pattern, especially on the south side of Timbanganten, that suggests the karstic topography of that region.[62] The principal river systems are those of the Manuk, which virtually bisects the chiefdom, and the Tarum, with its large delta near the northwest corner of the map (lower right in fig. 18.18). It appears that the mapmaker recognized, while drawing the map, that the course of the latter river was not being carried nearly far enough to the west and felt it necessary to correct this error. One can see (on the original, but not on Holle's published reproduction) the darkening out of the original more easterly route and the crossing of the tributaries of the new route with those of the old (in each case three to the south and one to the north), which could not be adequately effaced. A third river system, which I cannot identify positively, runs in part along the eastern edge of the chiefdom, and it seems clear that the chiefdom's northeastern corner, and perhaps the southeastern as well, is defined in terms of that river. Along the northern border of the chiefdom appear four rectangles. A wavy pattern in two of the rectangles suggests that they are lakes or artificial tanks; the others are empty. Elsewhere along the border are nine triangular features, presumably distinctive mountain peaks (six in the north and three in the south) that also appear to serve as boundary markers. Around most of the map perimeter a scalloped pattern represents the sea. In Ciela it was suggested to me that the circles with plus signs inside, clustered near the source of the Manuk, represented broadleaf forest. Vegetation signs elsewhere are generally small, scattered, and not easy to distinguish at the scale of our illustrations. The most prominent is a palm tree on the square island in Penanjung Bay in the far southeast of the map; others appear to be boundary markers along the rectangular perimeter of the chiefdom, especially at its northwest and southwest corners.

Conspicuously absent on the map is any distinctive sign for a town or village. One reason may be that rural settlement in this part of Java was not and still is not highly nucleated. But the abundance of map text makes it highly likely that all seventy-eight *kampongs* within Timbanganten, and possibly others as well, were named in their appropriate places. The two large squares near the northwest corner of the map have been identified by Holle as Sunda Kalapa, an old name for Jakarta (as mentioned above, an anachronistic usage), and Bantam (Banten)—or more precisely, given their connections to the sea, their harbors. Within the area of Bantam harbor are a number of houses, probably representing the coastal dwellings on piles that are common in the region. (The considerable distance of both places from the sea is puzzling.) A third town, Indramayu, where the Manuk River enters the Java Sea, is indicated by a crudely drawn square enclosure. There is no sign at any of these coastal locations, or for that matter anywhere else on the map, of European presence, which commenced in a modest way in western Java in 1522, though no factory (trading post governed by a factor) was established in the region until 1611 or 1612.

As I noted at the outset, the map of Timbanganten is regarded as a sacred relic. Along with other *pusakas* in Ciela, it is kept in the care of a *kuncen*, a hereditary officeholder in the village who commands great respect. When Holle visited Ciela, the *kuncen* was an elderly woman; during my own visit it was a middle-aged man. The other *pusakas* included a kris (a Javanese dagger), a javelin, the barrel of a small brass cannon, a ring stone, and a cloth pennant bearing an image of a kris with two separate blades.[63] The last was presumably the emblem of the state and possibly represented its division north and south of the Manuk River. All these objects are kept in two *karpeks* (black wooden reliquaries). Among them, the map appears to possess particular sanctity. Holle writes:

> That this map is greatly revered can be seen on Friday eve. Then, the *kuncen* unwraps it so that the population can view it. Then, a brazier is lighted with

62. See the geological map of Java in Charles A. Fisher, *South-east Asia: A Social, Economic and Political Geography* (London: Methuen, 1964), 226–27 and relevant text on 228–29.

63. Keeping *pusakas*, especially krises, as protective relics is a characteristic feature of popular religion in Java, notwithstanding the island's universal adherence to Islam, in which the veneration of such objects is theoretically anathema. The people of Ciela nevertheless thought of themselves as good Muslims and proudly showed me their village mosque and the adjoining *madrassah* (religious school). Of the objects named, based in part on Kusmiadi's description, I did not personally see the javelin or the ring stone.

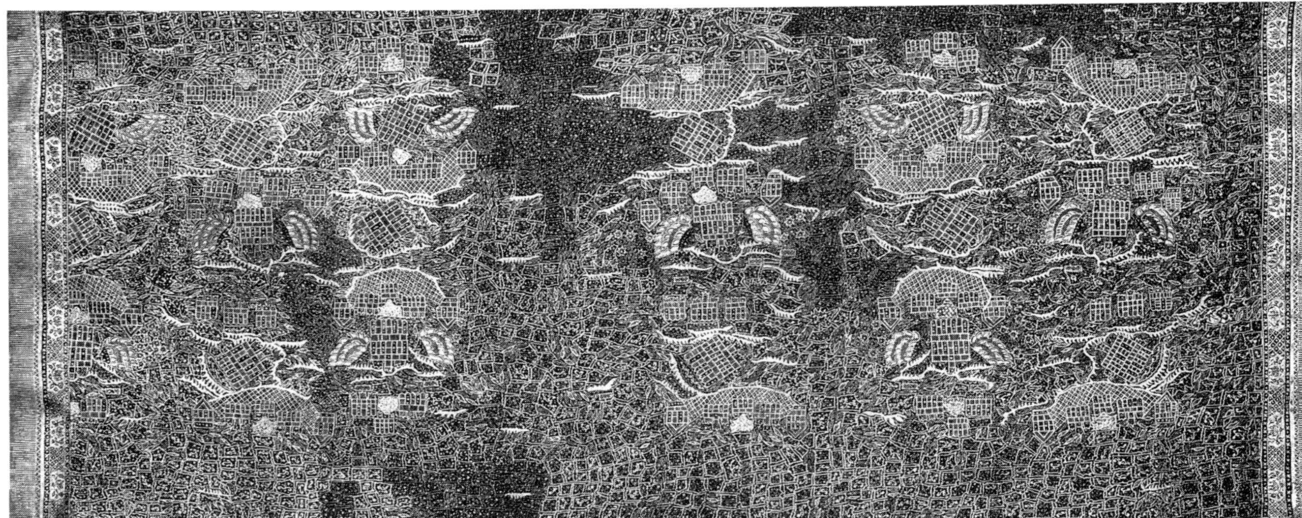

FIG. 18.21. BATIK MAP OF AN UNKNOWN LOCALITY FROM EITHER EASTERN JAVA OR BALI. This exquisite work was executed at an unknown date on a large *slèndang* (shawl). The basic batik pattern has been printed in pale blue and orange against a dark indigo field, while *prada* work (application of gold foil) highlights particular features. The total absence of map text makes interpretation of the work difficult; and it is not certain whether it refers to a real or imagined area, or perhaps combines geographic and cosmographic elements. There can be no doubt, however, that it incorporates various signs suggestive of the intensively cultivated landscape of either eastern Java or Bali. See also figure 18.22.
Size of the original: 93 × 233 cm. By permission of the Collection Royal Tropical Institute Tropenmuseum, Amsterdam.

> incense, and the *kuncen* mutters an Arabic prayer, which only she is allowed to pray, and which she repeats continuously. The population lingers around the house, taking in eagerly the sounds of prayer, although it is inaudible to them, and watch with equal respect the sacred cloth which is equally obscure to them. The thought that it once belonged to one of their former rulers, and that it even was made by a powerful *dalem* [ruler], leaves them in their worshipful mood.[64]

Previously, however, the local population held back their curiosity to see the *pusakas*, since the *kuncen* threatened them with all manner of dire consequences if they did. On the occasion of a visit by Lammers, however, several prominent villagers found the courage to enter the *kuncen*'s house in his company to view the relics.[65] During my own visit the relics were revealed to me by the *kuncen* only after appropriate ceremony, including several offerings of fruit and coffee to the spirits of the relics and a prayer by the *kuncen*, who faced south toward Cikuray volcano rather than west toward Mecca as he recited his decidedly un-Islamic invocation.[66] Unfortunately, I did not ascertain how frequently the relics are now viewed by the villagers.

Although the second (possibly) Javanese map I shall consider is no less traditional than the one of Timbanganten, the contrast between the two is dramatic. The latter (figs. 18.21 and 18.22), acquired by the Koninklijk Koloniaal Instituut in Amsterdam in 1933, is in the form of a large and elaborate batik *slèndang* (shawl). It is ascribed to either eastern Java or Bali, but no informed opinion as to its age or the specific area it depicts has yet been offered. Interpretation is rendered particularly difficult because the map bears not a single word of text. One cannot even state with certainty that it refers to a specific area as opposed to one or more generic landscape types. There seems to be little doubt, however, that this very well preserved map was made for persons retaining some aspects of Hindu culture, because of the number of motifs from that religion that it embodies. This alone would tell us little about the date if the map was from the still Hindu area of Bali; but if the area of provenance was eastern Java, that would suggest a date not much later than 1800, and more likely earlier, since Mataram, the region's (and Java's) last Hindu state, did not accept Islam until the late eighteenth century.[67] A detailed description of the map with a tentative attempt at analysis was published in 1934 and forms the basis for most of the discussion that follows.[68]

64. Holle, "De kaart," 171–72 (note 54).

65. Holle, "De kaart," 172 (note 54).

66. The ceremonial aspect of viewing the map may be seen on the Granada Television series (note 55).

67. Hall, *History of South-east Asia*, 215 (note 29).

68. Koninklijk Instituut voor de Tropen, *Aanwinsten op ethnografisch en anthropologisch gebied van de Afdeeling Volkenkunde van het Koloniaal Instituut over 1933*, Afdeeling Volkenkunde 6 (Amsterdam, 1934), 24–26. I am greatly indebted to Tjeerd R. Tichelaar, formerly

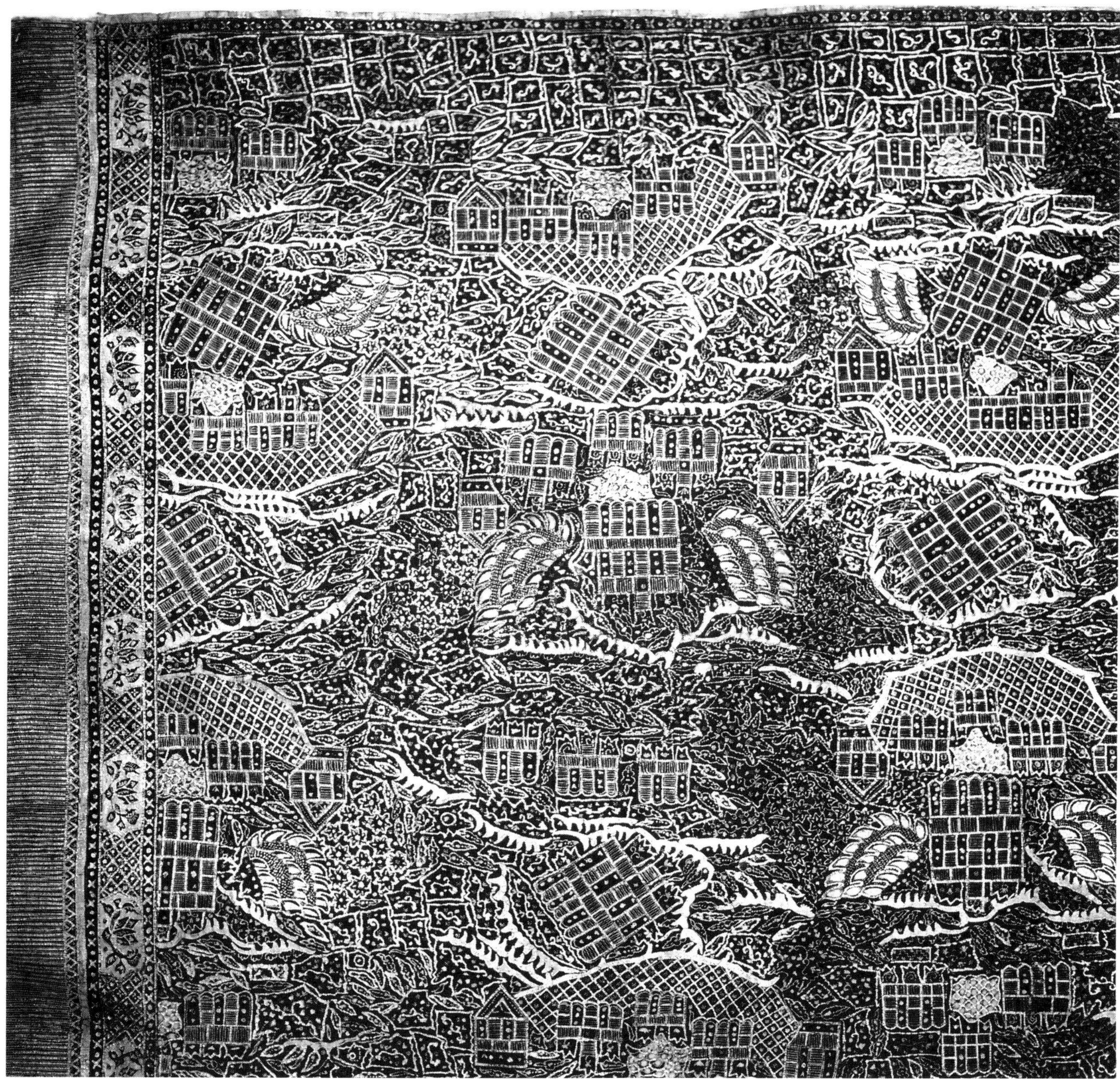

FIG. 18.22. DETAIL OF BATIK MAP FROM EITHER EASTERN JAVA OR BALI. Although the total absence of text makes it impossible to assign a precise time, place, or purpose of the map, which is here illustrated in part (see fig. 18.21 for the entire work), the lavish execution and the motifs included clearly indicate that it was made for persons of substantial means retaining some aspect of Hindu culture. The most prominent among such motifs are the so-called Puri gates, the winged ornamental entrances symbolizing the portals to heaven. These were characteristic features of Javanese Hindu temples. If, as believed, the map's provenance is eastern Java, which did not embrace Islam until the late eighteenth century, its likely date would be not much later than 1800.
By permission of the Collection Royal Tropical Institute Tropenmuseum, Amsterdam.

The sumptuous batik pattern of the map is printed in pale blue and orange, with *prada* work (gold foil) highlights, set against a dark indigo field covering the entire surface of the shawl except its border. The border is

of the Geografisch Instituut, Rijksuniversiteit Utrecht, for translating this description and analysis for me. A photograph of a portion of the shawl and a brief description appear in H. Paulides, "Oude en nieuwe kunst op Bali, tegen den achtergrond van het Westen," *Cultureel Indië* 2 (1940): 169–85, esp. 174 and 180.

relatively simple and narrow (perhaps 2 cm wide) along the length of the map and wider (10.5 cm) and much more ornate at both ends. The map pattern itself is exceedingly dense; thus there appears to be no place where one could lay a coin seven millimeters in diameter without obscuring a bit of the design. The map comprises a left complex and a right complex, covering respectively roughly one-third and two-thirds of the total area. Within each, certain basic design motifs are repeated in a somewhat, but far from perfectly, symmetrical manner.

The dominant motif in both complexes is a square area divided into rectangular compartments crowned with three-pointed roofs and flanked on both sides by large wings. There are four of these arranged in the form of a diamond on the right and five, arranged as on the points and center of a large X, on the left. No two are quite alike, however, and some face the top of the map (as presented) while others face the bottom. This key motif is taken to be the traditional so-called Puri gates of Javanese Hindu temples, which symbolize the portals of heaven on the sacred mountain (*gunung*, i.e., Meru); but the rectangles within each square are thought to signify *kampongs*. Just below each square are pentagonal areas marked by wave patterns that are thought to symbolize either the world ocean or the lake surrounding Mount Meru. Adjacent to each such gatelike assemblage are smaller houses with single gabled roofs and areas of a small diagonal checkerboard pattern believed to represent gardens.

A second major motif is a square, much like that of the dominant motif, but without the flanking wings and with the other adjacent features slightly differently disposed in respect to it. These are interspersed with the first motif like a large X in the right complex and like a diamond in the left complex (the opposite pattern from the first motif).

Most of the map not occupied by these motifs is filled with seemingly randomly arranged, more or less square shapes that are taken to represent *sawah* (wet rice) fields. Also scattered about are leafy and branching forms thought to indicate woods and brushy vegetation and starry floral shapes that probably connote *tamans* (pleasure gardens), as well as several other patterns. Conspicuously distributed throughout the map are horizontally disposed serrated lines of varying length, thought to represent riverbanks, ravines, and artificial embankments. Many of these suggest terraces, a characteristic feature of areas of intensive agriculture in Java. Most of the serrations point toward the bottom of the map (as shown in the illustration), implying a general, though not uniform, slope of the land in that direction; and the concentration of *sawah* lands in the lower part of the map, especially the lower right, is consistent with the implied general flow of water that way.

Although the purpose for which the batik map was drawn is not known, it has been suggested that the ensemble was intended to depict "twin cities," each consisting of a group of *kampongs* arranged in a traditional Hindu-Javanese pattern known as *Moncå-pat*.[69] Writing of this pattern in 1918, van Ossenbruggen observed:

> *Moncå-pat* [*Montjå-pat* in Dutch orthography] in Middle Java points to a unit of a *desa* [locality] with its four neighboring *desas*, which one should think of as being arranged according to the four cardinal points. This unity of *desas* extends further, including even more distant areas, and is connected with the old Javanese system of unity of which, even today, traces can be found in the law of the princely states.[70]

The continuation into the twentieth century of adherence to this system appears to have been limited to the responsibilities of various villages in dealing with crimes in a particular area and their obligation to assist one another in maintaining order and peace and apprehending criminals. In preceding centuries the system would have had much wider applicability. Within it, the bonds of one's village with those in the cardinal directions from it were taken to be stronger than those with villages in the intermediate directions.[71] Hence it may well have been regarded as necessary to assign to all settlements outward from a specific point of reference, say a capital, an unambiguous directional designation so as to know what the mutual obligations were. If this supposition is correct, one can understand why the maker of the batik map might have imposed a greater regularity in the spacing of signs representing settlement than in respect to other features of the map to which the schema did not apply, such as *sawah* lands, garden plots, ravines, and terraces, all of which appear to be more naturalistically disposed over the map surface. This, then, would account for the previously noted symmetrical regularity of the two dominant map motifs.

As to the places depicted—assuming they are not mythical—I am inclined to believe that, rather than Bali, the map relates to the neighboring principalities of Yogyakarta and Surakarta in south-central Java (within the eastern half of the island). I suggest further, counter to the

69. *Aanwinsten op ethnografisch en anthropologisch gebied*, 25–26 (note 68).

70. F. D. E. van Ossenbruggen, "De oorsprong van het Javaansche begrip Montjå-pat, in verband met primitieve classificaties" (The origins of the Javanese concept of *Moncå-pat* in connection with primitive classifications), *Verslagen en Mededeelingen der Koninklijke Akademie van Wetenschappen, Afdeeling Letterkunde*, 5th ser., pt. 3 (1918): 6–44, esp. 6. I am indebted to Marcus Vink, a doctoral student in history at the University of Minnesota, for translating key portions of this important text.

71. Van Ossenbruggen, "De oorsprong van het Javaansche begrip Montjå-pat," 7 (note 70).

idea put forward above that the map relates mainly to "twin cities," that two associated "states" were meant to be shown, including a network of places associated with their respective capitals of the same name. Since the *Moncå-pat* relationship among places was based on spatial considerations rather than size of settlement, the absence on the map of any apparent hierarchy among the places shown would not be unreasonable. Yogyakarta and Surakarta, the successors to the kingdom of Mataram after its dissolution in 1755, were the only two states to survive under nominal indigenous rule after 1830, by which time the Dutch had assumed power over the rest of the island. Both areas had vigorous traditions of batik manufacture and would have been able to provide the type of royal patronage needed to support a work as splendid as the one under consideration. The winged Puri gates that appear so prominently on the map are a conspicuous architectural feature of the region; and these, along with the inferred *Moncå-pat* pattern of *desas*, also argue more for a Javanese than a Balinese provenance. On the other hand, the liberal use of *prada* work is more characteristic of Balinese textiles.[72]

An examination of the pattern of terrain on the map also argues for situating it in Java. One important visual aspect is the lack of any feature suggesting a volcano. For Java as a whole (no less than for Bali) volcanoes dominate the landscape, and one would expect to find them shown somehow (as they are on the other Javanese maps examined in this chapter) if any were present in the region depicted. In the area of predominately *sawah* landscape directly between Yogyakarta and Surakarta, however, a distance of approximately sixty kilometers, there are no volcanoes, although there is a lobe of talus slope on the southeastern flank of Mount Merapi that might be signified by the area with little *sawah* land in the upper middle portion of the map as shown in figure 18.21. Further, the more or less parallel southwest-northeast trend of ravines on the map, roughly in line with its long axis, is reasonably consistent with the pattern of drainage in the region suggested. On Bali the drainage pattern is more or less radial, and there is no area without volcanoes that is comparable in size to that just noted for Java.[73] But if the suggestion that the map relates to Yogyakarta and Surakarta is correct, it would appear to relate not to the whole of those two principalities, since both reached the south coast of Java and there is no indication of the sea on the map. Rather, the abundance of *sawah* land depicted suggests that the area covered would be the fertile corridor between and around those two cities, in which that landscape predominates.

The third of the three known Javanese regional maps (plate 39) is cataloged at the Bibliothèque Nationale as "Carte javanaise ms. en couleurs à identifier (S.l., n.d.) [without place or date]." The map was part of a lot of thirty-six documents given to the library in 1878. Its language and script are Javanese. (The Javanese script is no longer in use, and few Javanese can still read it.) With the assistance of a Javanese journalist I have been able to identify the region the map relates to, an area of central Java mainly to the east of the two features most prominently shown on it, the volcanoes Merbabu (3,142 m) and Telomoyo (1,894 m), and to determine that the orientation is to the south. The map has a relatively modern look, particularly in its depiction of roads (one main artery and two secondary routes) and drainage, and was probably prepared not long before its acquisition by the Bibliothèque Nationale. The routes indicated can be found on a modern large-scale map of Java, but I have not determined the date of their construction. The main route shown is the one that links the port of Semarang to the north with the old princely state capital of Surakarta to the southeast. The area of coverage may be estimated as approximately thirty-five kilometers north-south by fifty kilometers east-west.

I have not determined at whose behest the map was made, or the purpose it was intended to serve. One striking feature of the map, which covers a portion of one of the most productive and densely settled regions of Java, is its depiction of settlement. Approximately 230 *kampongs* are shown. Their names appear within more or less ovoid forms, which often run into one another. These forms are bounded by rings of dots of varying colors, generally a darker hue of the color of the surrounding field. Those background colors—yellow, orange, red, blue, light and dark green, and light and dark gray—form fourteen blocks of territory set off from one another by thin boundary lines, by streams, or, over one short stretch, by the main road. I take these blocks to be low-order administrative subdivisions or areas within which revenue assessments were levied. The blocks vary substantially in size and include anywhere from one to fifty-one *kampongs* each (in the portion of their total territory within the map; it is conceivable that the territory with only one *kampong* is an exclave of another of the same color nearby). Only the lone area in light gray, stretching between the two volcanoes but lying mainly to the south of Mount Merbabu, contains no settlement. Settlement is sparse in the southern portion (top) of the map, especially in the southeast corner. There is no obvious reason this should be so, but it is conceivable that the southeast corner fell wholly or partially within the jurisdiction of the sultan of Surakarta. That

72. *Aanwinsten op ethnografisch en anthropologisch gebied*, 26 (note 68).

73. For relevant maps see *Atlas van Tropisch Nederland* (Batavia: Koninklijk Nederlandsch Aardrijkskundig Genootschap, 1938), maps 16.a and b, 17.a, 19.d, 21, and 22.

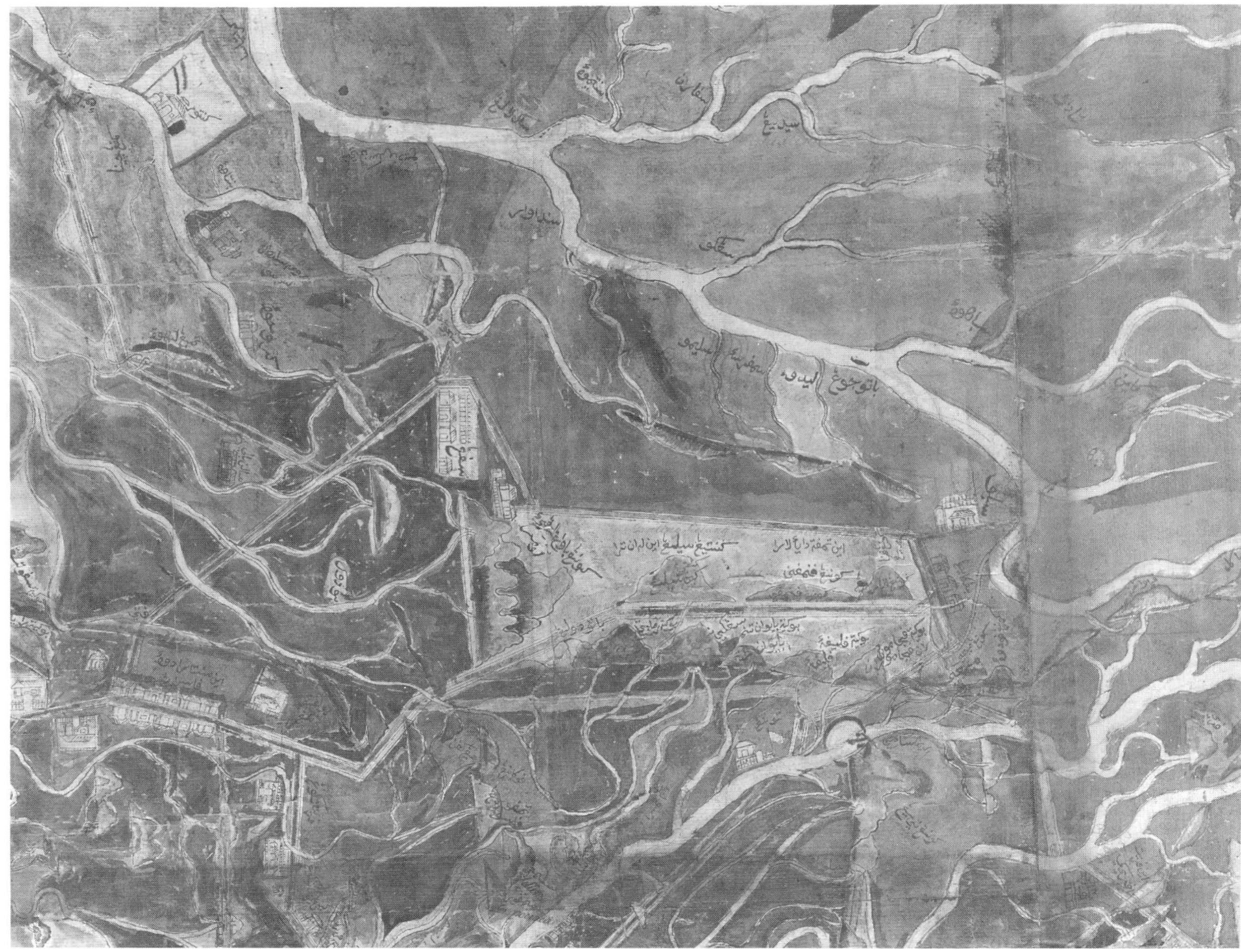

FIG. 18.23. DETAIL FROM A LARGE MAP OF THE FORMER SULTANATE OF PONTIANAK IN KALIMANTAN (BORNEO). This map is drawn in black ink and yellow, brown, blue, and red watercolor on European paper. It is believed to have been made in 1826. The orientation is to the south, and it depicts a portion of the deep interior of Kalimantan, of which the Dutch presumably then had no firsthand knowledge. It is noteworthy for the degree of topographic detail it incorporates and the depiction of streams issuing from what appear to be springs in the mountain wall in the lower portion of the photograph.
Size of the entire original: 92.7 × 83.1 cm. By permission of the Kaartenverzameling, Geografisch Instituut, Rijksuniversiteit te Utrecht (acc. no. VIII.C.d.1).

the areas of color extend to the very edge of the map, and that four *kampongs* lie directly on the map border, suggests that the map formed part of a series, all drawn for the same purpose.

Given the prominence of streams on the map and the local importance of irrigation, one might suppose that the map was tied to water management. However, the lack of any logical disposition of the territories mapped with respect to watercourses makes that unlikely. Both streams and roads are shown in white and outlined in black. On the main road are seven bridges depicted in two different styles and one gate, conceivably a checkpoint or a toll station. The two volcanoes are shown by circles and colored in the same tone of dark gray as much of the territory adjoining them. The absence of streams flowing from their western flanks and to the south of Merbabu and the dearth of *kampongs* in that portion of the map suggest that part of the area covered were essentially beyond the mapmaker's purview.

Though relatively modern in appearance, the map certainly does not look Dutch, even after we make allowances for the Javanese text. Nevertheless, despite the map's orientation to the south, I surmise that it was made at the behest of the Dutch as they sought to regularize their administrative control over the area. I surmise further that, lacking sufficient Dutch surveyors at the time,

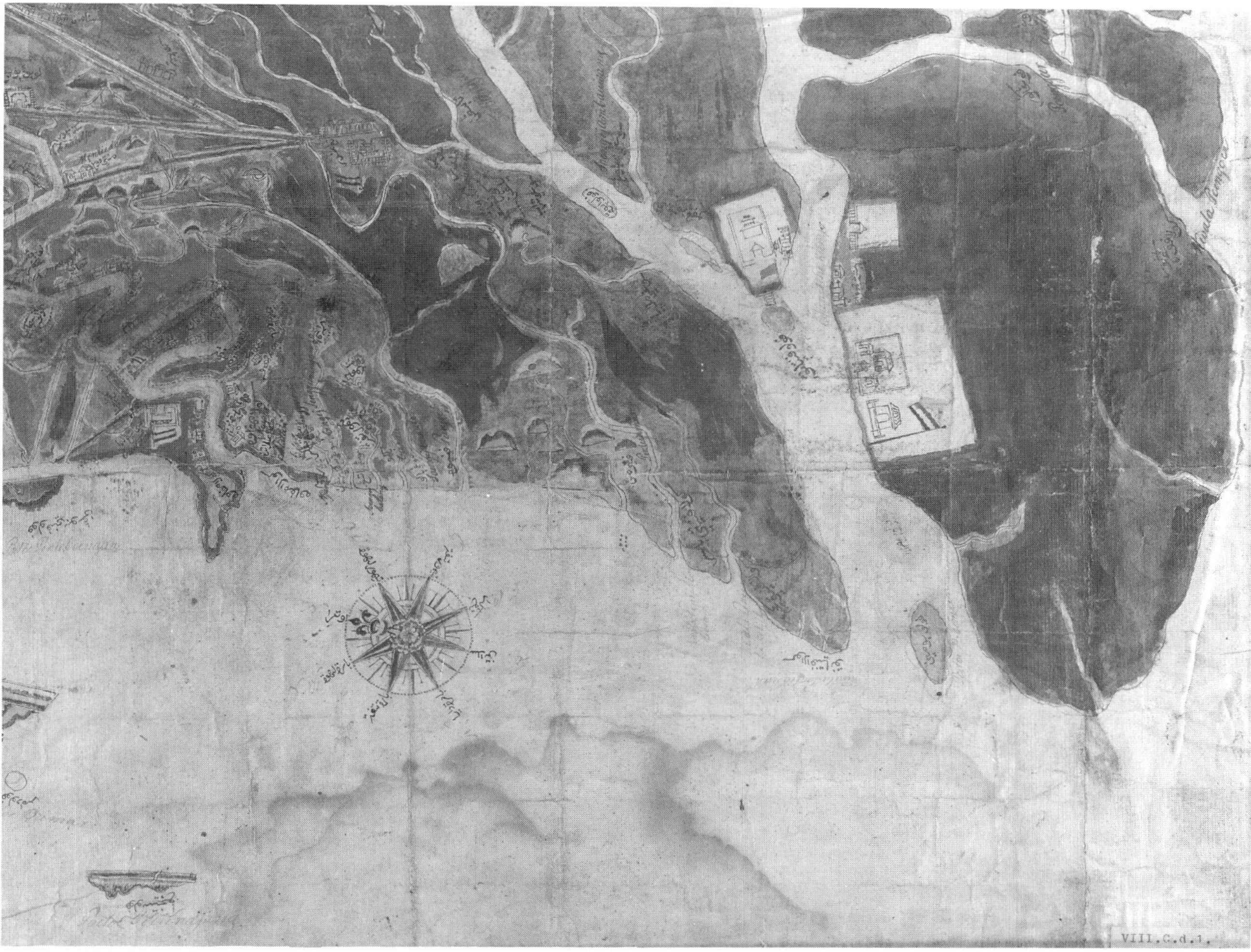

FIG. 18.24. ANOTHER DETAIL OF THE MAP OF PONTIANAK IN FIGURE 18.23. This detail shows what is unmistakably a delta at the mouth of the Kapuas River and includes the sultan's *kraton* (palace) and the nearby "Residentie," the seat of the Dutch resident administrator. European influence is apparent in the inclusion here of a compass rose and also (outside the area depicted) an ornate cartouche that seems to represent the sultan's coat of arms.
Size of the entire original: 92.7 × 83.1 cm. By permission of the Kaartenverzameling, Geografisch Instituut, Rijksuniversiteit te Utrecht (acc. no. VIII.C.d.1).

they fell back on locally recruited staff, utilizing wherever possible those known or presumed to have the requisite cartographic aptitude, more or less as the British appear to have done in similar circumstances in Burma. Further research on this question is obviously needed.

The final regional map (figs. 18.23 and 18.24) from what is now Indonesia focuses on the area of Pontianak in the west of Borneo (Kalimantan). This map is assigned the date "1826?" The language used to show several hundred toponyms and for related notes is Malay, written in Arabic script. Many later additions of both toponyms and notes (running to several lines) were made in the Roman script in pencil, but these are faded and now illegible. A few additional names and notes appear in ink. Accompanying the map is a modern pen-and-ink sketch of the area covered, which extends from the large delta of the Kapuas River in the south more than three hundred kilometers northward, approximately to what is now the border with Sarawak. The map appears to be greatly compressed in its east-west dimension, which measures less than that from north to south, but it may in fact extend all the way to the Müller Mountains, more than five hundred kilometers from the coast.

If the suggested date of 1826 is correct, the map was made only a short time after the Dutch established their control over Pontianak, in 1822. Their knowledge of the interior of Borneo in that period would have been negligible at best. Thus, practically all of the details away from the coast could have come only from indigenous informants; and it seems likely that such individuals actually drew the map, though with some supervision from the Dutch. Although the map bears a compass rose with thirty-two marked and eight named directions, of which north is the most prominent, the map text, including that on a fairly elaborate cartouche, probably bearing the local sultan's coat of arms, indicates a primary orientation toward the south, as was true of both of the maps from Java.

Apart from the abundance of named settlements, the map abounds in details relating to topography. Hills and mountains are drawn rather naturalistically in frontal elevation, usually in blue, with some shading along the edges. Although they are variously oriented, there is some tendency for them to be shown trending parallel to the edges of the map. What I assume to be the Müller range of central Borneo, the eastern limit of the Kapuas drainage basin, is shown in profile against the sky near the eastern edge of the map, with no land beyond. Rivers of varying thickness are shown in white, also fairly naturalistically. The Kapuas Delta is unmistakably shown as such (fig. 18.24). Some streams seem to be emerging over mountain crests, between two peaks. Many other streams are shown issuing from a hole drawn in the front of a mountain, well above its base, reminiscent of the special attention given to springs on the map of Timbanganten.

A distinctive feature about the text for various places is that it takes the form of complete sentences, for example, "This mountain is Pandang" or "This place is called Daya [place of the people of] Lava."[74] This longhand didactic form suggests that the map was intended for the elucidation of strangers to the region, namely the Dutch.

Settlement is generally shown by white houses drawn in frontal elevation, also variously oriented, but here and there one finds planimetric views of enclosures. Among these are the *kraton* (palace) of the sultan, shown within a red wall, and a nearby yellow rectangle, labeled "Residentie," indicating the seat of the presiding Dutch resident. In this and two other yellow enclosures are Dutch flags. It is not clear whether some red lines outside these several enclosures represent walls or roads. Background color for the map is predominantly a light brown wash, presumably indicating the sultan's domains, while some small coastal areas, shown by a yellow wash, appear to be the then limited holdings of the Dutch. In the northern part of the map what appears to be the word "London" (in Malay) is written on a mountain, which may signify that the area beyond was controlled by the English. The area in question could have been the sultanate of Sambas, just south of Sarawak, an area that came under British control in 1813 and was taken over by the Dutch sometime between 1824 and 1830, or possibly Sarawak itself, ceded by the sultan of Brunei to Rupert Brooke in 1841. In the latter case, the proposed map date of 1826 would not be tenable.[75]

A clue to how and when the Pontianak map came to be drawn is provided in an article by Le Roux that deals primarily with Malay nautical charts. He notes that early in the nineteenth century "a sketch map of South Celebes was produced by someone by the name Daeng Mamangung . . . at the request of the Dutch government." In recognition of the quality of that map, the author was awarded an official certificate in Dutch and Buginese, dated 25 August 1824, stating that the map "deserves the highest praise" and encouraging the mapmaker "to continue through all times, with renewed zest and diligence on this path that he has started so honorably."[76] If in the 1820s the Dutch would have deemed it expedient to induce a learned Bugi to make a map of his home, they would have had no less a reason to do the same in Pontianak and, for that matter, in other areas into which they had projected their power. In the case of Pontianak, which was in an area the British had also shown an interest in, the case for locally grounded intelligence would in fact have been even more compelling. Le Roux expressed the opinion that the Celebes map might be found in one of the archives in Bogor, Jakarta (Batavia), or Bandung, and that he intended to explore the matter further; but no more information about the matter is available to me.

Route Maps

Some of the maps designated route maps for purposes of analysis will not, at first glance, be recognizably very different from those classified as regional maps, especially some of those brought to light by Francis Hamilton. The criterion used in setting aside a certain group as route maps is that their emphasis is primarily on delineating the course of some type of feature, whether roads, pilgrimage paths, rivers, or even in one instance a telegraph line, and according little or no attention to other features, except perhaps in the immediate vicinity of the routes of principal concern. Some route maps have a strip form, which is particularly well suited to their purpose, but most do not. The discussion that follows is divided into two parts, the first relating to maps that are believed to relate to

74. These and a few other readings from the text on not particularly clear slides of portions of the map were made for me by Mohammed Radzi Haji Othman, a student from Malaysia at the University of Minnesota. I extend my thanks for his assistance.

75. In Othman's reading "London," it was not made clear to me where the mountain stood in relation to the Sambas River, which would have suggested (assuming his reading was correct) which of the two boundary hypotheses was the more likely.

76. C. C. F. M. Le Roux, "Boegineesche zeekaarten van den Indischen Archipel," *Tijdschrift van het Koninklijk Nederlandsch Aardrijkskundig Genootschap*, 2d ser., 52 (1935): 687–714 and folding map; quotations on 701. This article was translated for me in its entirety by L. Ruyter, whom I thank.

Buddhist pilgrimage and the second to maps of a secular nature.

PUTATIVE ROUTE MAPS RELATING TO THE SACRED PLACES OF BUDDHISM

In each of two nineteenth-century Burmese cosmographic manuscripts in the British Library is a chart depicting the Buddha in the center of an array of sixteen panels, arranged along twelve spokes, each panel indicating a specific place associated with his life on earth, its general direction from Bodh Gaya (the place in northeastern India where he attained enlightenment), and the time required to reach those places from Bodh Gaya.[77] Though one chart is on palm leaf and the other on paper, they are otherwise very similar. Each indicates the cardinal and intermediate directions along the spokes emanating from Bodh Gaya, the sacred center of Buddhism. The sequence seemingly begins at the top with Mithilā, which is said to lie to the east, though its true azimuth from Bodh Gaya is only twenty-five degrees. The directions of other places are also improperly given, often with comparable or even larger errors. The distances cited range from five days to a month and are also far from reliable.

To call such diagrams "route maps" would perhaps be stretching a point; but it is appropriate to discuss them here because of their apparent relation to figure 18.25, to be taken up shortly, which is without question a route map. Despite their geographic inaccuracy, the layouts are clearly not arbitrary but rather appear to follow some untranslated prescription in the manuscripts of which each forms a part.[78] A possible use for the diagrams, simple though they are, was as initial guides to pilgrims who wished to visit the most sacred shrines of Buddhism in India; but if so, they would serve that purpose poorly. More likely, they had some didactic function in instructing the faithful about the life and partially mythologized travels of the Buddha.

Also dealing largely with the places sacred to Buddhism are two diagrams that appear in a forty-eight-page *samud khoi* manuscript (twenty-four folios, with text on a total of thirty-nine pages), mainly in the Lanna Thai script, from somewhere in the vicinity of Chiang Mai in northern Thailand. Given its relatively well preserved condition, the undated manuscript probably dates, in the opinion of Wyatt, from the early twentieth century, but it could have been copied in whole or in part from a much older work. The manuscript was studied in 1978 by a Thai scholar, Sommāi Prēmčhit, and discussed by him in a work in Thai on the stupas of northern Thailand.[79] It tells, among other things, how to go about building a *chedi* (stupa) in the style of the Mahābodhi Temple at Bodh Gaya in India.

One of the diagrams contained in this manuscript is of a type that indicates the main holy places in and around the stupa at Bodh Gaya. This diagram, a *yantra* (sacred visual charm) on folio 21 of the manuscript, was allegedly devised by a powerful monk, Upaguttathera (Upagupta in Sanskrit, Phra Uppakhuṭ in Thai), who was supposed to have lived in northern India in the time of Aśoka (third century B.C.).[80] It was to be inscribed on a silver plate and then, with appropriate ceremonies, placed within the pinnacle of the stupa to protect it from subsequent harm. Which *chedi* its Lanna Thai version was originally prepared for is not known and there is no reason why the diagram could not have been used for more than one stupa. There are at least four *chedis* in Burma and northern Thailand modeled on the prototype at Bodh Gaya.[81]

The diagram consists of a square made up of nine smaller squares (three by three). Of the nine squares, the middle one, represented as the "Buddha's throne" (at Bodh Gaya), is said to be the "centre of all cities in the

77. The two diagrams form parts of the manuscripts illustrated in figures 17.14, 17.20, and plate 35 above.

78. For translating the text of one of the two charts discussed here, I am indebted to Patricia Herbert of the British Library. That chart is illustrated in Heinz Bechert, " 'To Be a Burmese Is to Be a Buddhist': Buddhism in Burma," in *The World of Buddhism: Buddhist Monks and Nuns in Society and Culture*, ed. Heinz Bechert and Richard Gombrich (London: Thames and Hudson, 1984), 147–58, esp. 155. Several possibilities suggest themselves in regard to there being twelve spokes. They may indicate months when a visit to the places named along a particular spoke would be particularly auspicious or the sequence of months in which visits to particular sites ought ideally to be made. Alternatively, the spokes may relate to the twelve-year cycle of the Chinese calendar and indicate places of pilgrimage to be visited by persons born in a particular year of that cycle. For a discussion of a prescribed list of pilgrimage places for northern Thai Buddhists based on year of birth, see Charles F. Keyes, "Buddhist Pilgrimage Centers and the Twelve-Year Cycle: Northern Thai Moral Orders in Space and Time," *History of Religions* 15 (1975): 71–89.

79. Sommāi Prēmčhit, Kamon Sīwichainan, and Surasingsamrūam Chimphaneo, *Phračhēdī nai Lānnā Thai* (Stupas in Lanna Thai) (Chiang Mai: Khrongkan Suksā Wichai Sinlapa Sathapattayakam Lānnā, Mahāwitthayālai Chīang Mai, 1981), 89–90 and 104–5 (in Thai).

80. Despite the widespread belief of his association with Aśoka, Upaguttathera's historical existence may have been anytime between the third century B.C. and the first century A.D. A very popular cult venerating him has arisen in mainland Southeast Asia. It is explored in John S. Strong, *The Legend and Cult of Upagupta: Sanskrit Buddhism in North India and Southeast Asia* (Princeton: Princeton University Press, 1992). A thorough study of this work may shed light on the maps discussed in this section.

81. The first of these was built in Pagan in Burma in the thirteenth century; three others, one in Pegu (Burma) and the two Thai examples, one each in Chiang Mai and Chiang Rai, date from the fifteenth century. Both of the Thai structures go by the name Wat Jed Yod (Temple of Seven Stupas). Details on the history of these edifices and on contacts between Southeast Asia and India that led to their construction are provided by Robert L. Brown, "Bodhgaya and South-east Asia," in *Bodhgaya: The Site of Enlightenment*, ed. Janice Leoshko (Bombay: Marg Publications, 1988), 101–24.

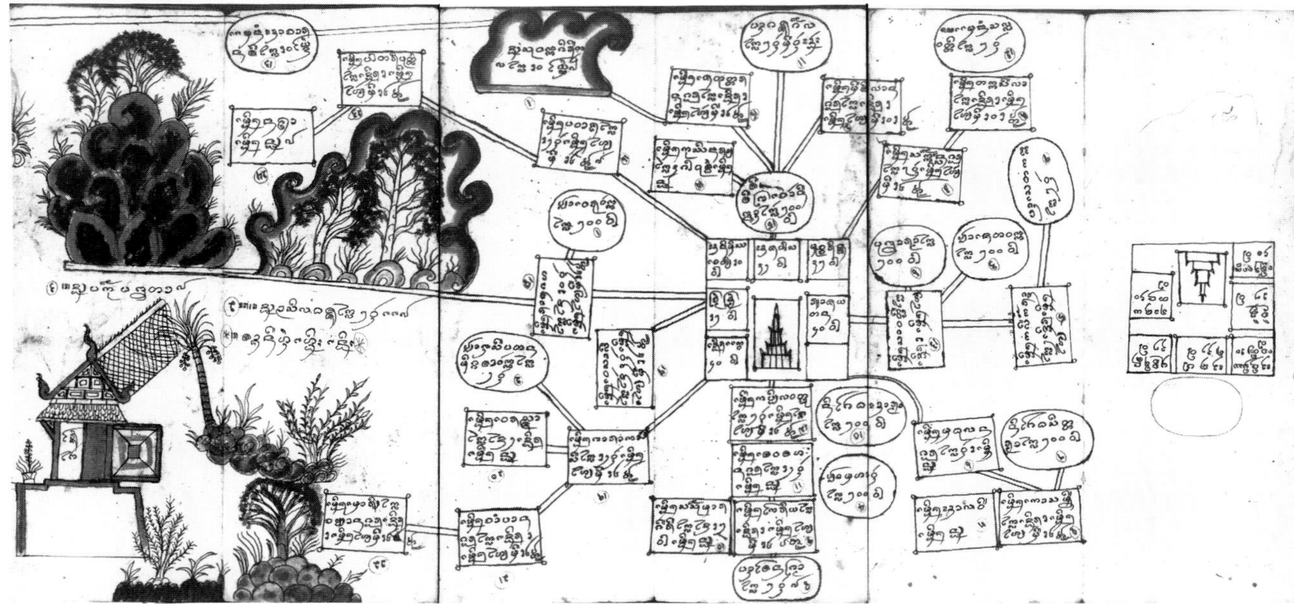

FIG. 18.25. PORTION OF A LANNA THAI MAP BELIEVED TO RELATE TO PLACES OF PILGRIMAGE VISITED BY A LONG-RESIDENT THAI PILGRIM IN INDIA. This enigmatic northern Thai map, of unspecified date, forms part of a forty-eight-page *samud khoi* manuscript, essentially cosmographic, said to have been placed in a stupa—possibly near the town of Chiang Rai—to protect the structure from harm. The manuscript was acquired for the Echols Collection on Southeast Asia sometime before 1981 by Donna Markham (then a graduate student there) and has yet to be cataloged or fully translated. The map shows a number of cities, villages, sacred places, and natural features in some sort of relation to Bodh Gaya and its precincts. Of forty-one additional places shown, those that could be identified all relate to sites significant in the lifetime of the Buddha or known to have been important in the centuries following his death.
Size of the entire map: 36.5 × 277 cm. By permission of Cornell University Libraries, Ithaca, New York, Echols Collection (the map portion shown here comprises all or parts of fols. 23–26).

TABLE 18.1 Key to Places Shown in Figure 18.25

Key Number/ Letter (folio)[a]	Text in Translation[b]	Remarks
0 (24)	No text (illustration shows temple of Bodh Gaya)	Place where the Buddha attained enlightenment
0.a (24)	Ratana-ghara (Ruenkaeo), 40 *wās*	Sites 0.a (24) to 0.f (24) are in the immediate vicinity of Bodh Gaya and are associated with specific events in the Buddha's quest for enlightenment
0.b (24)	Cankama (Conkrem), 15 *wās*	
0.c (24)	Animissa Cetiya, 10 *wās*	
0.d (24)	Ajapāla, 32 *wās*	
0.e (24)	Muccalinda, 35 *wās*	
0.f (24)	Rajāyatana, 40 *wās*	
1.1.a (24)	Rājagaha, 10 days distant, big city with 16 gates	Site of first Buddhist Council, early fifth century B.C.
1.1.b (24)	Veluvana, forest, 500 *wās*	Close to 1.1.a
1.2 (25)	Usila-dhajja, mountain, 2 days distant; this way is to the north	Not identified

[a]See figure 18.26 for key to locations.

[b]Courtesy of Phramaha Wan Surote and David Wyatt; words in square brackets were added by me.

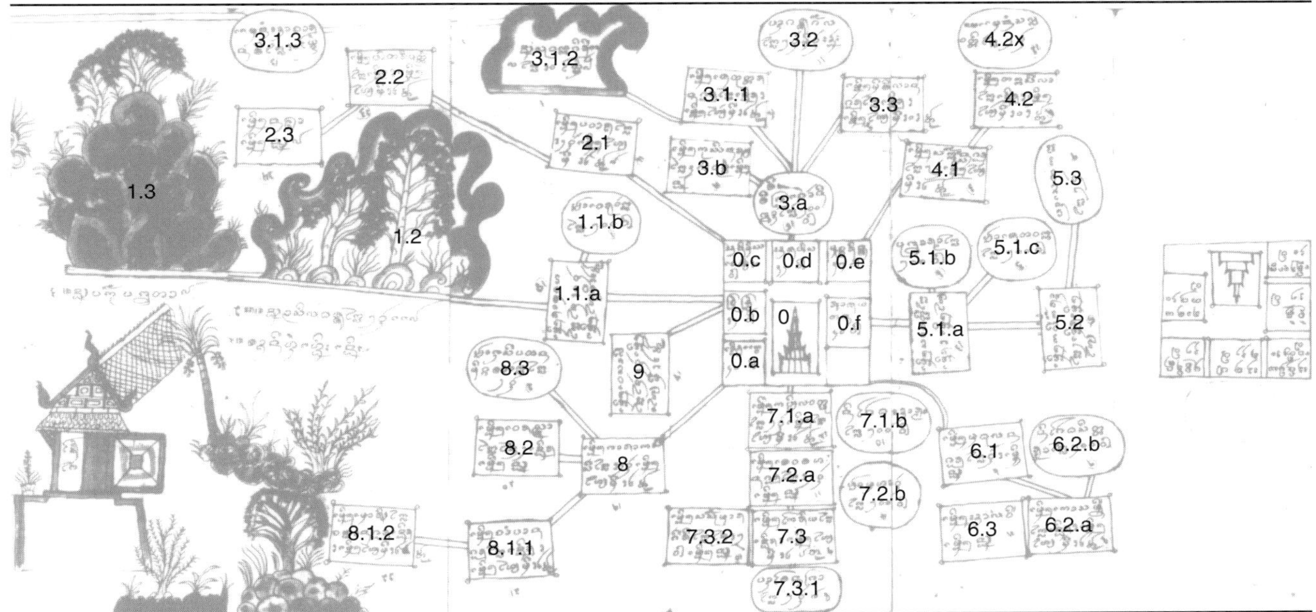

FIG. 18.26. KEY TO PLACES SHOWN ON FIGURE 18.25. All toponyms are indicated by key numbers and letters that are ordered clockwise and listed (first digits 1–9) according to the point of departure from Bodh Gaya of the map path on which they lie. Along any given path, numbers (second or third digits) ascend away from the node of origin. Letter *a*, *b*, or *c* following a given number signifies that all places with that number lie close to one another (within 2 km). The letter *x* following a number signifies that the place in question is somehow connected to another place with the same number, but at an indeterminate distance, and is not linked to it by a map path. The translated map text is given in table 18.1, as provided to me, but with minor editing for consistency.

TABLE 18.1 (*continued*)

Key Number/ Letter (folio)[a]	Text in Translation[b]	Remarks
1.3 (25)	Panku-paeenta, mountain(s)	Not identified
2.1 (24)	Padmāvatī (Pavāra), 12 days distant; big city with 16 gates	Modern Pawaya
2.2 (24)	Pātaliputta, 1 mountain distant; big city with 16 gates	Capital of Mauryan empire, modern Patna
2.3 (24)	Nāḷa, small city	Nālandā, site of major Buddhist university
3.a (24)	Where the lord attained nirvana, 500 *wās* distant	Close to 3.b
3.b (24)	Kusīnārā, 3 *gāvuddhas* distant; small city	See 3.a; *gāvuddha* is a unit of distance
3.1.1 (24)	Jetuttanagara, 1 month distant; big city with 16 gates	Also called Madhyamikā
3.1.2 (24)	Suvannagiritala Mountain(s), 10 *yojanas* distant	Name signifies "golden mountain"; adjoins modern south Indian city of Gooty
3.1.3 (25)	Ācāra-nadi [river], 10 *yojanas*	Not identified
3.1.4 (26)	Cetarabba, city otherwise called Maddava, 10 *yojanas*	Not identified, off figure 18.25 to left

TABLE 18.1 (*continued*)

Key Number/ Letter (folio)[a]	Text in Translation[b]	Remarks
3.1.4.x (26)	Palileyyaka, very big elephant forest	Not identified, off figure 18.25 to left
3.1.5 (26)	Ketumati-nadi [river], 10 *yojanas* distant	Not identified, off figure 18.25 to left
3.2 (24)	Gajjangala, village, 5 days distant to the east	Presumably Kajaṅgala
3.3 (23, 24)	Mithilanagara, 1 month distant; big city with 101 gates	Mithilā; site of second and third Buddhist Councils in early fourth and mid-third centuries B.C., respectively
4.1 (23)	Sankassanagara, 7 days distant; big city with 16 gates	Saṅkissa
4.2 (23)	Takkasilā, 1 month distant; big city with 101 gates	Takṣaśilā, site of major Buddhist university and stupa, in northwest Pakistan
4.2.x (23)	Sallavatti, river, 5 days distant	Salalavatī River; location problematic
5.1.a (23)	Sāvatthi, 1 month distant; big city	Modern Set Mahet
5.1.b (23)	Pubbarama [monastery], 500 *wās* distant	Close to 5.1.a
5.1.c (23)	Jetavana forest, 500 *wās* distant	Close to 5.1.a. Grove serving as frequent retreat of Buddha
5.2 (23)	Kalingaraja, 2 months distant; big city	Assumed to be Kaliṅganagara, capital of an ancient state
5.3 (23)	Setakanna village, 5 days distant	Not identified
6.1 (23)	Madhulanagara, 1 day distant; small city	Not identified
6.2.a (23)	Kosambi, 1 month distant; big city with 16 gates	Site of monastery and stupa
6.2.b (23)	Nigodhasitārāim monastery, 700 [400?] *wās* distant	Close to 6.2.a
6.3 (24)	Ālovī, small city	Not identified
7.1.a (23)	Kapilavātthu, 5 days distant; big city with 16 gates	Place of the Buddha's renunciation
7.1.b (23, 24)	Nigodhārām monastery, 500 *wās* distant	Close to 7.1.a
7.2.a (24)	Devadahanagara, 12 days distant; small city	Devapaṭṭana, visited by Mauryan emperor Aśoka
7.2.b (23, 24)	Mahāvana forest, 500 *wās* distant	Close to 7.2.a; means "great grove"
7.3 (24)	Koliya, 1 month distant; big city with 16 gates	Koliya was a republican city-state in time of Buddha; the location of its capital, Rāmagama, is problematic
7.3.1 (24)	Donabrahma village, 5 days distant	Not identified
7.3.2 (24)	Sumsumāragiri, 15 *wās* distant; small city	Location is problematic; 15 *wās* is assumed to be a mistake (of either mapmaker or copyist); 12 days may be intended
8 (24)	Bārānasī, 12 days distant; big city with 16 gates	Varanasi/Benares
8.1.1 (24)	Campanagara, 1 month distant; big city with 16 gates	Campā
8.1.2 (25)	Moriya, 1 month distant from Campanagara; big city with 16 gates	Republican city-state in the time of Buddha; capital was Pipphalivana
8.2 (24)	Verañjā, 2 months distant; small city	Also called Adarañjiya
8.3 (24)	Isipatana-migadayavana, 2 days distant	Sarnath, place of Buddha's first sermon; close to Verañjā
9 (24)	Vesāli, 3 days distant; big city with 16 gates	Vaisāll, capital of ancient Licchavi confederacy of city-states

universe," while six others (all but the lower-center and lower-right squares, which show nothing but "sacred words") include the names, within triangles, of the other principal sacred places in Champu Dipa (Jambūdvīpa, here roughly India) associated with the Buddha in their (spatial?) relationship to Bodh Gaya.[82] Though the few details just provided differ in some respects from those that relate to the Burmese twelve-spoked charts discussed above, the similarities are sufficient to suggest a genre of Buddhist religious mapping that warrants further investigation.

More interesting than the *yantra* however, is a much more complex diagram extending across folios 23–27, a portion of which is reproduced as figure 18.25. A partial guide to its contents is presented in figure 18.26 and table 18.1. The focal feature is a square (in folio 24) including a stylized representation of the Mahābodhi Temple at Bodh Gaya and six named holy sites in its immediate vicinity. The arrangement of these sites reflects that of the *yantra* described above, though with an obvious difference in the scale of the area of reference. From the square emanate what look like nine routes, some with two or more branches, that connect or closely approach twenty-four rectangles, each bearing the name of a city; fourteen ovals, each containing the name of a village, monastery, forest (grove), river, or other holy place; and three pictorially rendered mountains. Brief descriptive comments follow the place-names. Although all forty-one names along these presumed routes were transliterated for me, I have thus far been able to identify only twenty-seven of the toponyms.[83] In each case the name relates to a place in India that was important in the lifetime of the Buddha or in the few centuries immediately following. A representative entry (in the rectangle numbered 7.1.a on fig. 18.26) reads as follows: "Kapilavātthu, five days distant, big city with sixteen gates." The place referred to (Sanskrit Kapilavastu) is where the Buddha renounced the pursuit of worldly goals.

Of the twenty-four cities named, numbers 3.3 and 4.2, Mithilanagara and Takkasilā (ancient Takṣaśilā; modern Taxila), are described as big cities with 101 gates; another twelve are said to be big cities with 16 gates; two big cities, with no specification of gates; seven small cities; and one (3.1.4) merely as a city, with no size specification. (The rationale behind the assigning of city size is often not clear, nor is that relating to the numbers of gates, 16 or 101, which are obviously conventional.) Of the fourteen toponyms within ovals, three refer to villages, three to monasteries, one to a sacred site adjacent to a city, three to rivers, and four to *vanas* (forests or sacred groves). Apart from the three pictorially rendered mountains along the routes from Bodh Gaya, there are six other pictorially rendered features on folios 25–27 (to the left of what is shown on fig. 18.25): a building simply labeled "kitchen"; Acutta, a "hermit's cottage"; "the hunter's son's residence"; Palileyyaka, an elephant forest; Muccalinda Pond (where a blind serpent sheltered the Buddha under its hood during a week of rain); and Gandhamadana Mountain (a name formerly applied to the eastern Himalayas).

The topological logic of the map is far from clear. With respect to distances from Bodh Gaya, there is no correlation between the straight-line or route distances of the places named and the straight-line or inferred route distances of the identifiable places as measured on a modern map of India. One point of consistency, however, is the close spatial association of place-names in ovals with the nearest place-names in rectangles. Otherwise the sequence of places along particular routes seems, in a number of instances, to make little sense. The route from Bodh Gaya to cities 8 (Bārānasī/Varanasi), 8.1.1 (Campanagara/Campā), and 8.1.2 (Moriya/Pipphalivana), which we can examine by comparing figures 18.26 and 18.27, will illustrate the point. The first stage to Bārānasī, said to be twelve days from Bodh Gaya, takes one 190 kilometers to the west-northwest. The second, said to be one month distant, presumably from Bārānasī, takes one to a town on the Ganga 205 kilometers to the east-northeast of Bodh Gaya. The third and final stage, another month distant, takes one to the capital of an ancient republican city-state 290 kilometers northwest of Bodh Gaya.[84] In light of these locations, it becomes obvious

82. The foregoing remarks are based on readings of Sommāi Prēmčhit's text by Thong-chai Winichakul, University of Wisconsin, and David Wyatt, Cornell University. The latter provided me with a written translation of one key paragraph. Their assistance is acknowledged with gratitude.

83. The transliteration of names and translation of ancillary text were provided for me by Phramaha Wan Surote, a northern Thai monk, and an anonymous Thai layman at the Thai Buddhist monastery in Silver Spring, Maryland. They were subsequently confirmed and slightly augmented by David Wyatt. I am most indebted to all of them for their assistance. All but one of the places identified appear on maps or photographs in Joseph E. Schwartzberg, ed., *A Historical Atlas of South Asia* (Chicago: University of Chicago Press, 1978), 16, 19, 21, and 23, and may be found by referring to the atlas index. Devadahanagara, the one exception, was identified in *The Geographical Encyclopaedia of Ancient and Medieval India*, ed. K. D. Bajpai (Varanasi: Indic Academy, 1967–), 1:107, s.v. "Devadaha."

84. It is noteworthy that the only place on the entire map where it is specified that the distance is from the previously named place along a given route is between Campanagara and Moriya. I do not know whether this is significant. The routes shown in figure 18.27 generally follow those known to have existed in Mughal times (seventeenth century) as shown in Irfan Habib, *An Atlas of the Mughal Empire: Political and Economic Maps with Detailed Notes, Bibliography and Index* (Delhi: Oxford University Press, 1982). Although no reference period for the Lanna Thai map is available, I assume that the routes I have depicted would provide a reasonably good approximation of those most likely followed by the presumed original pilgrim to whose travels the map relates.

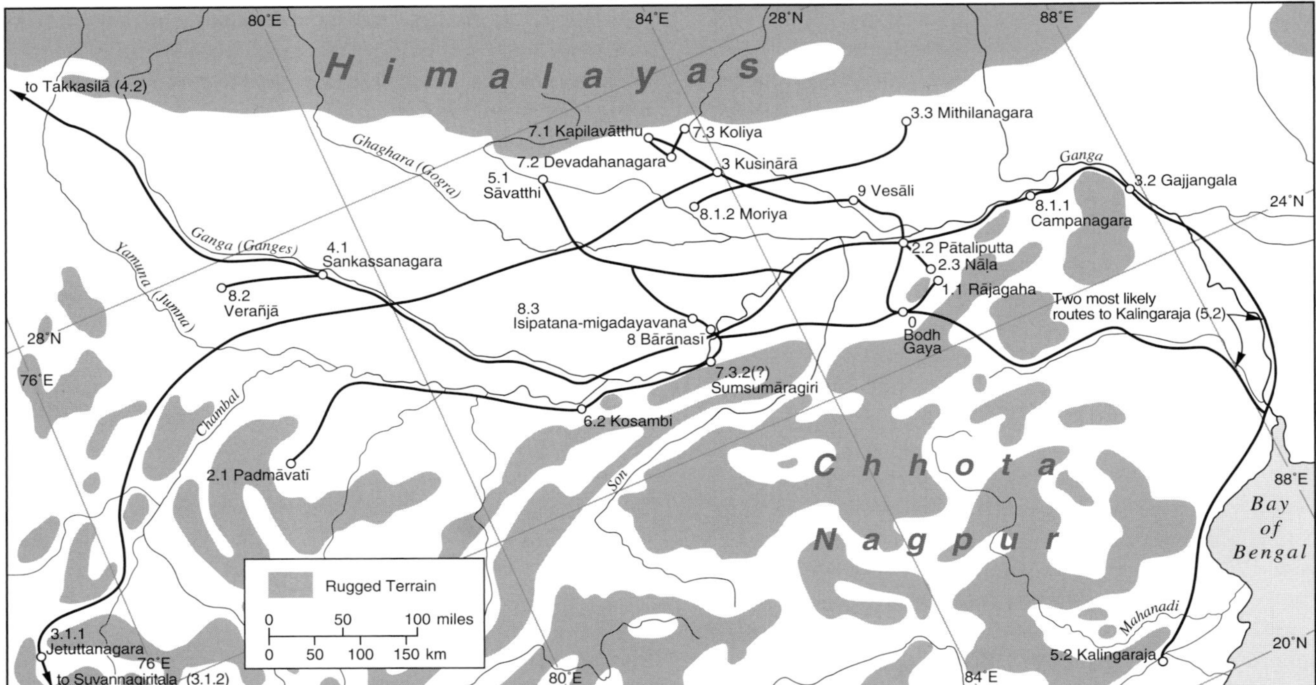

FIG. 18.27. HYPOTHETICAL RECONSTRUCTION OF SELECTED ROUTES SHOWN IN FIGURE 18.25. The routes shown are those along paths between places that could be identified. The key numbers (without attached letters) and names of such places match those of figure 18.26 and table 18.1. Not shown directly on this map are Takkasilā, which lies well to the northwest (near the modern Pakistani city of Rawalpindi), and Suvannagiritala (modern Gooty in the south Indian state of Andhra Pradesh). A sojourn in India of at least several years would have been required for a single premodern pilgrim to cover all the routes (including the implied routes to the two most distant locales just noted).

that the orientation of the original map cannot be specified. Nor can an approximate scale be given. Nor is there is any apparent connection between the actual geographic directions of identified places either from Bodh Gaya or from one another and the directions shown on the map.

The verbally indicated distances that accompany the toponyms on the map do, however, provide a more meaningful set of relationships. Distances are specified in either linear or temporal units. Generally these appear to refer to the previous place on a particular route rather than the cumulative time or distance from Bodh Gaya, the place from which all routes originate, though in the case of the route from Bodh Gaya to Kapilavātthu, Devadahanagara, and Koliya (7.1, 7.2, and 7.3) the latter possibility would make more sense. Long distances are normally specified in months (two months in the case of the route from Sāvatthi to Kalingaraja [5.1 to 5.2] and from Bārānasī to Verañjā [8 to 8.2], and one month in the case of connections to nine other cities) or *yojanas*. (Presumably a "month" in the context of this map might mean anything from fifteen to forty-five days.) A slightly more common specification, accounting for twelve cases, is in days (ranging from two to twelve days). Linear distances are given in *yojanas* (a variable measure, the often-cited specification for which—about sixteen kilometers—is here clearly much too short); *gāvuddhas* (in only one instance; equivalency also not determined), and *wās* (approximately 2 m). The only route in which *yojanas* figure is from 3.1.1 to 3.1.5, each place being said to be ten *yojanas* from the place preceding it). *Wās* figure in seven cases (five of five hundred *wās*, one of seven hundred *wās*, and one—almost certainly a misinterpretation—of only 15 *wās*). Distances to places named in ovals are invariably given either in days or in *wās* and are always less than those given for the toponyms in the rectangles they are adjacent to. This suggests that the places named (a number of which have yet to be identified) were somehow associated with the nearby city. In some cases an association is historically verifiable.

For the eighteen pairs of connected cities for which verbal time-distances are provided, I have checked the times stipulated against measured geographic route distances to test their plausibility. In fifteen cases a person could cover the distance via the hypothetical routes indicated on figure 18.27 within the time indicated at speeds of less than forty kilometers a day or, in eleven cases, less than twenty-five kilometers a day. In three cases, however, the stated travel time is not credible. In the most extreme such case, the route from Bodh Gaya to Sankassanagara, for which "seven days distant" is indicated, one would have to travel more than 85 kilometers a day to cover the distance. Here, and perhaps elsewhere, there may have been an error either in the original map, in copying from one manuscript to another, or in the translation of the map text. Of the times given, several are far more than required for a particular geographic distance; for example, ten days for the route from Bodh Gaya to Rājagaha (locale no. 1), which is only about seventy kilometers distant. But if the time specified is taken not as the minimum required to make a particular transit but the time actually taken by a devout pilgrim making many stops en route (and possibly also considering the time at a given site in addition to the travel time), this seeming difficulty in interpreting the map disappears.

The original author of the map was almost certainly from northern Thailand. He was probably either a monk recording his own pilgrimage to the holy places of Buddhism in India, during a sojourn that would necessarily have lasted several years, or the transcriber of an account of the earlier travels of some other individual or group. I am inclined to believe that the map relates to the travels of an individual monk. (I have called attention elsewhere to other cases of pilgrims' recording their travels in cartographic form.)[85] If this hypothesis is correct, each of the nine routes radiating out from Bodh Gaya might represent a single period of travel (perhaps that of a specific year) from the monk's base in that most holy place, without any indication of closure on returning to Bodh Gaya. At first glance, the sequence of identified places along any given route may appear rather arbitrary. But when one considers that convenience and optimizing on time spent in travel are not major concerns for devout pilgrims, and that any route would be in part contingent on weather, on the astrological auspiciousness of traveling in particular directions during particular months, and on the interpretation of various unpredictable omens, seemingly quixotic routes may become understandable. (By way of example, the reconstructed paths of some of the early Chinese Buddhist pilgrims in India were often rather tortuous.)[86]

The map provides no clear clues to when it was originally drawn. The absence of any features suggestive of Islamic or British rule in the area covered, the former dating from the late twelfth century, proves nothing, since the author's concerns were exclusively with places sacred to Buddhists. A likely time for his Indian sojourn would be the fifteenth century during the reign of the devout Lanna Thai monarch Tiloka, who is presumed to have sent monks to Bodh Gaya (as did a contemporary Burmese monarch) to make plans of the temple there to enable copies of it to be rebuilt in Thailand (see above).[87] Assuming, then, that the map was recopied repeatedly from that date, copyists' errors could easily explain some of the puzzling features I have noted.

If, as supposed, each of the nine routes emanating from Bodh Gaya represents a single trip away from that center, the sequence of those trips requires consideration, since it could explain the logic of the map. Although it is possible that the arrangement is wholly arbitrary, I am inclined to think they follow a temporal order clockwise from the route to Rājagaha (shown as site 1 on fig. 18.26). The sanctity of the latter place and its proximity to Bodh Gaya would have made it a likely candidate for a first sortie away from that center. Although I have not succeeded in identifying them, two large pictorially depicted mountains to the left of Rājagaha may be among the five hills surrounding that town that are identified in the *Mahābhārata* and in the Pali annals.[88] Other pictorial details of folios 25–27 may also be among the many sacred sites associated with the region. The prominence of the features on the left portion of the map seems to give it a position of primacy in the author's view of the religious geography of India.[89] The reason for hypothesizing a clockwise sequence is that in circumambulating (*padakkina*) a sacred site, Buddhists always proceed to the right. What the map might then be intended to convey is the author's circuit of pilgrimages from Bodh Gaya, irrespective of their true direction, as if they formed a visual *padakkina*, the temporal dimension of which was given priority over the spatial.

Several alternative hypotheses may also be suggested. First, the map could have been made as a guide for prospective pilgrims but, being based on hearsay, might represent little more than a garbled version of the actual facts. The idiosyncratic nature of the routes, however, is

85. See, for example, the Jain pilgrimage map discussed in Joseph E. Schwartzberg, "Geographical Mapping," in *The History of Cartography*, ed. J. B. Harley and David Woodward (Chicago: University of Chicago Press, 1987–), vol. 2.1 (1992), 388–493, esp. 440–42; or the Nepali map commissioned by a certain Cikhidi described on pp. 649–50 above.

86. These are mapped in Schwartzberg, *Historical Atlas of South Asia*, 28 (note 83).

87. The Chiang Rai Temple was probably destroyed during the Burmese occupation of the area dating from the mid-sixteenth century. It has been partially restored since 1844. See Brown, "Bodhgaya and South-east Asia," 111 (note 81).

88. The many holy places near Rājagaha are described in Bimala Churn Law, *Rājagṛiha in Ancient Literature*, Memoirs of the Archaeological Survey of India, no. 58 (Delhi: Manager of Publications, 1938).

89. Lending support to the supposition that the map area to the left of Bodh Gaya is meant to be given special prominence, the artist initially drew the focal square of Bodh Gaya on folio 23, upside down with respect to its position on folio 24, so that there was minimal space to the left of it before running into the text of folio 22. Hence the prominent features drawn to the left of Rājagaha could not be accommodated, and a new start appears to have been made.

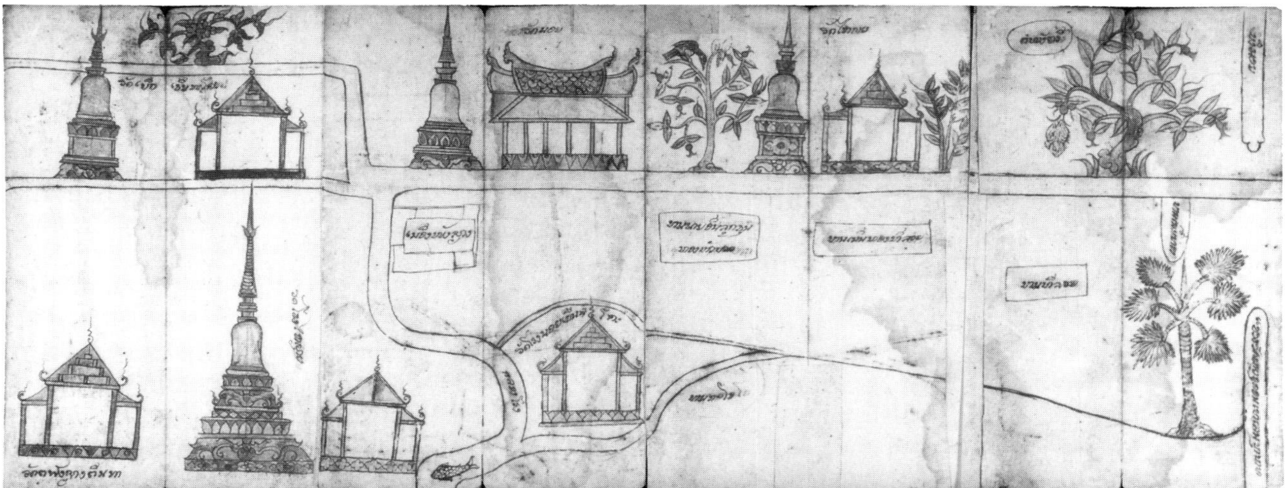

FIG. 18.28. SMALL PORTION OF A VERY LONG THAI ROUTE MAP OF AN AREA BETWEEN NAKHON SI THAMMARAT AND SONGKHLA ON THE MALAY PENINSULA. This accordion-style *samud khoi* map, painted on indigenous Thai paper, dates from the late seventeenth or early eighteenth century and is the oldest known geographical map from mainland Southeast Asia. It forms an addition to an older (1615) manuscript that deals with the lands tributary to a certain temple, Wat Phra Kho, and its tributary temples, as well as those falling within the civil domain. The most distinctive attributes of this map are the prominence with which it depicts religious edifices, the concomitant subordination of the signs for settlement in general, and the richness of detail relating to plants and animals.
Size of the entire original: perhaps 40 × 1,200 cm. National Library of Thailand, Bangkok. Photograph courtesy of Cornell University Libraries, Ithaca, New York, Echols Collection (Wason film 4309).

such that it is hard to imagine that, even in garbled form, they would recommend a prescription for pilgrims in general. Second, the map may be a largely mythologized representation of the travels of the monk Upaguttathera. Finally, it may be a mythologized itinerary of the Buddha himself.[90] At present, none of the suggested hypotheses is susceptible of proof.[91]

SECULAR ROUTE MAPS

A short segment of the oldest Southeast Asian secular route map I have any knowledge of is illustrated in figure 18.28. That work, misleadingly identified as "Map of Nakhon Si Thammarat" (the name of a town near, but not actually within, the area of the map), dates from the late seventeenth or early eighteenth century.[92] It is part of a historical text relating to an area of southern Thailand on the Malay Peninsula. The manuscript is of the accordion-style *samud khoi* variety, and the map portion, when photographed, consisted of forty folios. The orientation of the map is toward the east, that is, toward the Gulf of Thailand. There almost surely was more to the map at one time, since there is clear evidence that the manuscript, which is badly damaged in places, had been torn into several parts and that in putting it back together a considerable lacuna was left between what are now folios 33 and 34 (numbers assigned by the National Library). A lesser gap exists between folios 35 and 36. Although folio 39 does appear to be the true southern end of the map, it has not been established that folio 1

90. Although a number of the places shown could not possibly have been visited by the Buddha in his historically verified existence, myths that arose after his death have entailed visits by him to many places in India and elsewhere for which there is no historical evidence. His leaving his footprint on Adam's Peak in Sri Lanka (see fig. 17.23) is a case in point.

91. Sommāi Prēmčhit observes that in his study of Lanna Thai manuscripts beginning in 1973 he has never come across any map comparable to the one I have discussed here (Sommāi, Kamon, and Surasingsamrūam, *Phračhēdī nai Lānnā Thai*, 105 [note 79]). Hence there is no other known model against which to test the ideas I have advanced in this chapter.

92. The work is at present in the manuscript collection of the National Library of Thailand, Bangkok, where it has been dated Čhulasakarat 977 (1615). A microfilm copy (Wason film 4309) made by David Wyatt, who owns the negatives, is in the Cornell University Library, Ithaca, New York. This map was brought to my attention in 1984 by Lorraine Gesick, at the time a visiting scholar at Cornell University and now of the University of Nebraska, Omaha. On 15 April 1985 she sent me a microfilm printout of the entire work. Much of the account here is based on relevant information she provided, for which I am most grateful.

A monograph on the work, including a nearly complete facsimile (one and half folios are missing) at a reduced scale, has since been published: Suthiwong Phongphaibūn, *Phutthasātsanā Thǣp Lum Thalēsāp Songkhlā Fang Tawan'ǭk samai Krung Sī 'Ayutthayā: Rāingān kānwičhăi* (Report on the research on the Buddhist religion around the Thale Sap basin on the eastern shore in the Ayutthaya period) (Songkhla, 1980).

represents the original northern end. The extant part of the map depicts the Sathing Phra peninsula (essentially a coastal sandbar, approximately 70 km from north to south and nowhere more than 10 km wide). Within the area shown is a combined land and water route, part road and part coastal backwater, leading from Songkhla northward toward Nakhon Si Thammarat. The dimensions of the original map are not available, but what remains is thought to be more than twelve meters long and about thirty-five to forty centimeters or so in width. Accepting the figures above as reasonable, I derive an average north-south scale for the map of roughly 1:6,000.

The initial portion of the manuscript includes a copy of a royal decree dated to a year equivalent to 1610 as well as a local history, immediately following folio 39, that ends about 1700, which is thought to be the approximate date of the map. The map focuses on a prominent temple, Wat Phra Kho (shown in folio 26), and designates numerous other wats as *khyn* (subordinate to) that temple. An important purpose of the map was to show which rice fields were tributary to the temple and which fell within the civil domain. Though not to a uniform scale, the map is said to be "otherwise very accurate, [presumably] indicating local authorship," as does the southern Thai style of orthography.[93] The most prominent among the roughly 250 features shown on the map appear to be temples, *chedis*, and other (religious?) edifices. But throughout the map, more modestly presented, are scores of rectangles and ovals, with names written in them, that presumably represent large and small settlements, respectively, or perhaps the fields belonging to those settlements. Most of these lie on or close to what seem to be roads. But many, possibly most, of those apparent roads, especially in the northern portion of the map, are in fact canals, backwaters, and streams, as is evinced by the occasional fish, crocodile, or crayfish depicted. Such features, singly or in parallel paths, extend the entire length of the map, which, along with the strip format, provides the rationale for classifying the work as a route map. It appears that, as on some Burmese maps, no clear differentiation is made between the water and the land stretches of certain routes. Thus, without being able to read the text or having local knowledge of the area depicted, one frequently cannot ascertain which is which.

A puzzling feature of the map is the attention it lavishes on fauna and flora. As on so many Burmese maps, vegetation figures very conspicuously. A dozen or so types of trees are depicted (most of them numerous times). Almost all are shown individually, rather than as groves or forests. These trees are often laden with fruit, and in some are found birds, monkeys, and other animals. Only on the several rocky hills, which seem to anchor the southern tip of the sandbar forming the Sathing Phra peninsula, and across the narrow strait toward Songkhla does one see what looks like a forest. Within the forest, what appears to be a tiger chases a deer. The strait between the peninsula and Songkhla is indicated by a narrow, sweeping arc within which a wave pattern is drawn. Such a pattern also appears on the margins of other portions of the map, reflecting their littoral location. Within these bands of sea appear numerous forms of marine life, the most curious being a swimming elephant (similar to the creature depicted in the assembly of fish described in chapter 17 in respect to one of the panels of the *Trai phum*).

Appendix 18.3 provides basic data on nine maps of Burmese provenance that I have chosen to designate as route maps. The oldest three of these are among the large group of maps that Francis Hamilton procured during his sojourn in Burma (1795). Three others relate to the courses of rivers: one of the lower Irrawaddy, presumably made before the British annexed that area in 1852, and two, drawn sometime before 1867, of which one shows a stretch of the middle Irrawaddy south of Pagan and the other a tract along the Shweli, which flows out of China and joins the Irrawaddy in Upper Burma. The remaining three, all drawn between the second and third Anglo-Burmese Wars, include a map drawn by a forester at the behest of British interested in exploiting trading opportunities with northern Siam; a large map showing the main routes between Siam and Cochin China; and a long *parabaik* showing the route of the telegraph line linking the then British border in Lower Burma with Mandalay. Several of these maps warrant discussion in addition to the brief notes provided in the appendix.

Of the three maps procured by Hamilton, only one calls for special comment. Despite its simplicity, this map is of more than passing historical interest in that it relates to a tribute mission, sent in some unspecified year, by the Burmese king to the Chinese emperor. The original work (which may no longer survive) was probably drawn by the *zabua* (governor) of the frontier city of Bhamo at the request of Hamilton, who says it was

> one of the rudest, which I procured; but . . . important as tending to settle, by high and perfectly informed authority, many most interesting points respecting the rivers which enter the farther peninsula of India [i.e., the Indo-Pacific peninsula] from Thibet and China, and thus enabling us to decide with more confidence on the relative situations of different places [in China] in other maps.[94]

The map covers the route all the way from the then

93. Lorraine Gesick correspondence (1985).

94. Francis Hamilton, "Account of a Map of the Route between Tartary and Amarapura, by an Ambassador from the Court of Ava to the Emperor of China," *Edinburgh Philosophical Journal* 3 (1820): 32–42 and pl. I; quotation on 32.

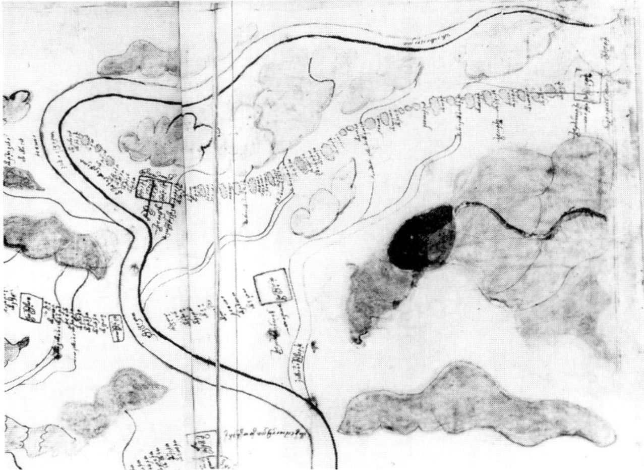

FIG. 18.29. DETAIL FROM A BURMESE MAP OF THE "ROUTES BY SHANS FROM COCHIN CHINA YAHME." This map covers a very large part of mainland Southeast Asia. Although it is precisely dated, to its very day in 1871, and although the name of the artist, U Yit, is specified, much about this large, crude, yet detailed work is cryptic, including the cited descriptive title assigned to it at the archives where it is now held. The map was drawn in ink and watercolor on European paper. The portion shown here indicates a grossly distorted course of the Mekong River, parts of the Annamite Cordillera (the dark, undulating shapes in the right portion of the photograph), and the initial parts of two of three routes presented on the map—the upper route, commencing at what may be the city of Hue (the square in the upper-right corner), and a middle route, commencing at what is almost certainly Saigon (the square near the center of the map). See figures 18.30 and 18.31. Size of the entire original: 64.5 × 122.5 cm. By permission of the National Archives of India, New Delhi (Historical map fol. 91, no. 14).

capital at Amarapura (called "Shue Prido," or Golden City) not only to Beijing ("Udhin Pri"), the Chinese capital, but on to Jehol ("Taraek Pri"; Chengde) in Manchuria, the seat of the emperor's hunting lodge, to which the mission was obliged to proceed. About three-fifths of the map's length was devoted to showing in some detail the portion of the itinerary in Burma, which took the mission only 14 days to negotiate, and the remaining two-fifths covered the portion in China, traversed in 121 days. The number of travel days required for each stage of the journey is indicated between the named towns and cities in much the same way as on other maps prepared for Hamilton.

Within China the route is shown by a double line along part of which appears a note, "ten days by canal," and then, farther north, "carriage road"; but otherwise no symbolic differentiation enables us to distinguish one such part of the route from another. Few details not directly relating to travel appear on the map, especially within its Chinese portion.

Another Burmese route map focusing on routes beyond that nation's frontiers covers the area between Siam in the west and Cochin China in the east. A small portion of this map is illustrated in figure 18.29. Its author is identified as U Yit, and its date is precisely given as the "8th day of waxing of the eleventh [lunar] month, 1232 B.E.," equivalent to 28 January 1871.[95] This large and detailed map was rather crudely executed in brown ink and brown, blue, and yellow watercolors on four pieces of paper pasted together. Much about the work is very puzzling. But before attempting to unravel its several mysteries, it is necessary to specify as much about its content as is at present feasible.

Basically the map relates to three routes between the Mae Nam Ping Valley of Siam proper and what appear to be Phnom Penh, Saigon, and another city to the north of Saigon, very likely Hue. Although the routes converge in the west at a point in the Ping Valley that does not appear to have any special importance (see figs. 18.30 and 18.31, which provide a provisional abstract of the entire map and the location of the features shown there on a modern map), the three eastern or southeastern termini are not directly linked. Along the northernmost of the three routes appear approximately 105 place-names (not counting the names of eleven rivers crossed), six of which appear within squares, presumably signifying towns; on the middle route approximately seventy more place-names (plus five river crossings), of which four are written in squares); and on the south approximately eighty names (plus thirteen river crossings), of which three are in squares. Additionally, four place-names within circles that lie to the south of the middle route are joined to the route by thin lines that seem to be roads. The places within squares and circles are bounded in brown, and some are emphasized by an additional yellow outline or yellow wash.

The map shows numerous rivers, some by variously spaced double brown lines and others by single lines.

95. I have seen no published reference to this map. My initial attempts to identify places on it were greatly assisted by Patricia Herbert of the British Library, for which I am most grateful. Translating from my own handwritten copies of the date and what I took to be the more important place-names, Herbert was able to positively identify some features and provide tentative identifications of several others. From that base I was able to identify tentatively a few additional features. Full-scale, but not particularly legible, photocopies of my own photocopy of the map (in six parts spliced together) were then sent to Michael Aung-Thwin on 13 January 1992, with the request that he transliterate a number of strategically selected toponyms. His reply, dated 30 January, provided the requested transliterations where the map was sufficiently legible, but few of these Burmese names could be matched unambiguously with their modern Thai equivalents. Additionally, Aung-Thwin's letter and a long telephone conversation on 4 February confirmed some of my own thinking about the map. For any shortcomings that remain in the balance of this analysis, the responsibility is mine.

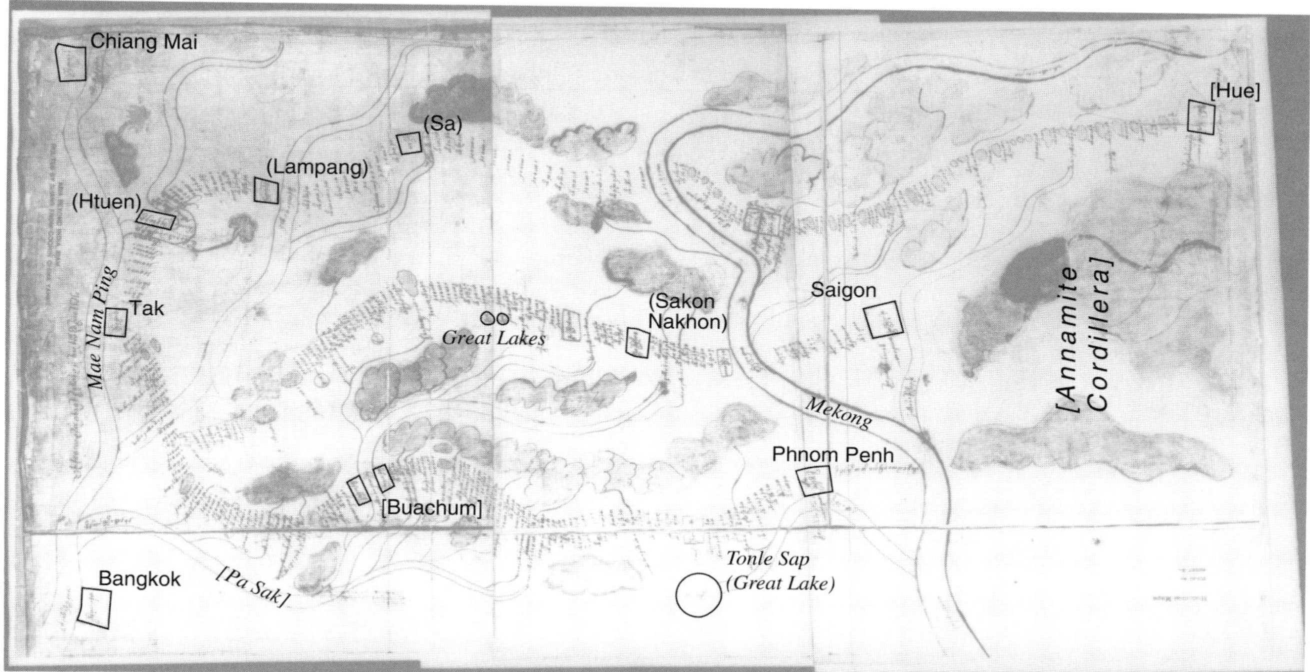

FIG. 18.30. BURMESE MAP OF THE "ROUTES BY SHANS FROM COCHIN CHINA YAHME." This is a reproduction of the entire map shown in figure 18.29, with tentative identifications of various features (see legend to fig. 18.31).

Not infrequently, a single line becomes a double line downstream from the point where a stream crosses one of the three routes. The principal rivers on the map have been identified as the Mekong and the Ping, the former reinforced by a band of brown wash. The Bassac, leading from the Tonle Sap, the great lake of Cambodia, may also be considered positively identified; but the identification of all other streams is conjectural, and one can place no trust in the apparent direction of their flow. Conspicuously absent is the South China Sea, which ought to appear both adjacent to Hue (in the upper right) and in the large blank space to the southeast of Saigon (lower right). Mountains and hills are crudely represented in blue wash, outlined with brown, occasionally reinforced by an additional band of yellow. Most such areas include amorphous shapes, some of them quite large, appearing either singly or in linear clusters. One gets the impression that the cartographer made little attempt to portray their proper orientation, but rather tried only to show that they lay between other map features or to the

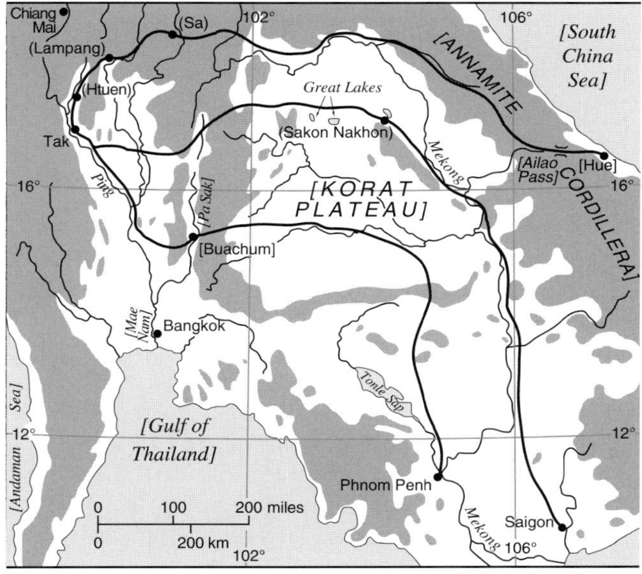

FIG. 18.31. A HYPOTHETICAL RECONSTRUCTION ON A MODERN BASE MAP OF THE THREE ROUTES DEPICTED IN FIGURE 18.30. The supposition underlying this reconstruction is that the routes taken formed parts of a coordinated diplomatic mission of Vietnamese and Cambodians (not "Shans," as misleadingly stated) that was aimed at seeking Burmese assistance against the takeover of Indochina by the French, which was then progressing rapidly. Since the routes of the three groups of envoys converged in western Thailand and they appear never to have actually reached their presumed destination in Burma, we may infer that the envoys were intercepted and advised not to proceed with their mission.

left or right of one of the three routes on the map. Along all three routes, but mainly the northern one, are many features that resemble small billowy clouds inserted between successively named settlements. I judge that each such sign specifies a crest to be crossed. There are twenty-nine of these on the northern route between the Mekong River and the route's eastern terminus, suggesting that the route followed a rather long path through the Annamite Cordillera, which extends to within a few miles of the South China Sea at Hue, the presumed northern terminus. Elsewhere on the map there are only a half dozen such features on the route itself, apart from four other places where the route is shown running between a pair of such signs, presumably to suggest a pass.

The title of the map has been translated (it is not known by whom) as "Asia beyond India, Siam, Routes by Shans from Cochin China Yahme." The title itself raises several questions. The word Yahme, for example, is one to which I and several specialists I consulted can attach no meaning. As for the name Cochin China, bear in mind that that term used to have a much more inclusive connotation than when it referred to one of the five constituent units of French Indochina. Previously it included the whole of Annam.[96] The word "Shan," referring to the travelers whose routes the map depicts, should not I believe be taken to refer to any ethnic group that would today be recognized as Shan, or even to any other ethnically Thai people. If the Shans in question were such a group, the title would, from a Burmese perspective, logically have been worded "to Cochin China Yahme," not "from" that place. In my examination of several of the maps prepared for Hamilton I noted that a number of groups who were not ethnic Burmans were designated there as Shan, with one or another prefixed modifier. On the map of Ava and nearby countries (fig. 18.2), for example, the "Country of the Judara Shan" and the "Country of the Kiokachin Shan" obviously can only refer to Cambodia and to Cochin China, respectively, neither being an area of Thai peoples. This suggests that in certain contexts the word "Shan" formerly had, for Burmans, the generic connotation of foreigner. If this reasoning is correct (and assuming that the map title is accurately translated), the map we are dealing with may show routes taken by certain non-Burmans from three separate places in what was loosely designated Cochin China to the central Ping Valley of Siam. Supporting the idea that the origins of the three routes were in the east rather than in Siam, the eastern half of the map seems to be drawn at a considerably larger scale than the western half, with the Mekong and the Annamite Cordillera given particular prominence. It is noteworthy that neither Bangkok nor Chiang Mai is reached by any of the three routes shown. The appearance of both in the corners of the map suggests that they were intended as important locational referents to help place the terminal portion of the routes. It is not known whether the routes depicted were traversed only once or repeatedly. The former seems more likely, since the routes are highly idiosyncratic and would hardly have served as regular arteries of trade.

It appears, then, that neither Bangkok nor Chiang Mai, the two most important places in the western portion of the map, was the object of the travelers along the three routes shown. What, then, might have been the point of three such long and arduous journeys? The motive of both the travelers and the mapmaker was, I believe, political or military, or both. Yet no Siamese destination of obvious political importance appears along the route. One possibility is that the travelers' missions were never completed, and that they had left their respective starting places with the intention of proceeding to the Ping Valley and then meeting in that general region at whatever points the various groups would reach, given their differing rates of march. The political context of the time in Cambodia and Annam suggests a plausible motive. Saigon had been occupied by the French in 1859; Cambodia had become a French protectorate in 1863; Cochin China (in the narrow sense of southern Annam) had been annexed by the French in stages between 1862 and 1867; and the French were poised to assume control over the rest of the country. Given that threat, and assuming latent opposition to French rule in the areas already taken over, delegations from Saigon, Phnom Penh, and Hue to Siam and Burma to solicit support to stop the spread of French rule, or even to throw it off completely, would have made sense. So too would the departure of multiple embassies, since the likelihood that one or more would be intercepted by the French would surely have to be considered.

If one accepts this admittedly speculative hypothesis, one must still explain why the map is in Burmese, not Siamese, Khmer, or Vietnamese, even though there is no indication that any of the three groups ever reached Burma. One plausible explanation is that the Burmese (not necessarily to the exclusion of the Thais) had somehow been given advance notice that the embassies would be dispatched as of a certain date (or dates). If the desirability of coming to the aid of the Annamese and Cambodians was discussed in Mandalay and a negative decision reached, a Burmese, possibly the map author, U Yit, might have been sent to Siam to intercept the embassies, dissuade them from continuing to Mandalay (lest the French be needlessly offended), and obtain from the travelers as much information as they could recall about the

96. On Cochin China, see Henry Yule and A. C. Burnell, *Hobson Jobson: A Glossary of Colloquial Anglo-Indian Words and Phrases, and of Kindred Terms, Etymological, Historical, Geographical and Discursive*, 2d ed., ed. William Crooke (1903; Delhi: Munshiram Manoharlal, 1968), 226–27.

routes they had taken, for use in case the Burmese decision on intervention might be changed at some later date.[97] That the map was probably made hastily in the field, at a place where artist's supplies were not abundantly available, is suggested by the uncharacteristic crudeness of the map's execution and the limited three-color palette, in contrast to the half dozen or more colors in which many of the Burmese maps we have examined were rendered.

If, then, as supposed, the map was made as a Burmese intelligence document, it would have been carefully stored at the Burmese capital, Mandalay. That it was subsequently acquired by the British is obvious from its now being held by the National Archives of India, New Delhi, which presumably received it from its most likely former storage place in the Survey of India Archives in Calcutta. Finally, it seems likely that whatever notes the English may have made on the map, apart from providing an English title, were lost during one of its several transfers.[98]

In striking contrast to the enigmatic map I have just discussed is a map showing the plan for the route of the telegraph line built from Nyaungu on the border of British Burma to Mandalay, built sometime between 1860 and 1880 (fig. 18.32). This map is labeled "Parabaik no. 191."[99] It is drawn in white ink and red chalk on a folding black *parabaik*. The map is sufficiently self-explanatory to require little comment. Nevertheless, it is noteworthy that it includes abundant detail that seems to have no direct connection to the matter of telegraphy, including, as do so many Burmese maps, pagodas and other prominent buildings.

Before closing this discussion of route maps, let me call attention to the single unambiguous indication I have come across that Laotians, like other peoples of mainland Southeast Asia, also made maps. My authority for this assumption is the German adaptation of an account by a French traveler, "Dr. Harmand," of a journey he made from Khemmerat on the Mekong River to the Annamese city of Hue over the period from February to August 1877.[100] This serialized, more or less popularly written account, presented in the general form of a log but in the third person, indicates that on 19 June a local official in a small Lao village drew Harmand a map of the course of the Se Bang Hieng, along which his party was traveling. Of this route map, a black-and-white copy of which accompanies the published narrative, Harmand observed:

> [It] was drawn with white chalk on a black lacquered tablet, which one used there for the drawing up of documents. This same sketch was later etched with a stylus onto a palm leaf and then the characters were rubbed in with a mixture of oil and lamp black, so that they emerged distinctly and indelibly in black.[101]

The map (assuming the fidelity of the copy to the original) was rather simple. It showed the Se Bang Hieng emerging from an area of squiggly lines that presumably represented the Annamite Cordillera and flowing in a single sweeping curve to the Mekong. On each of its banks the mouths of four tributaries are shown. Although the tributaries on each bank appear in their correct sequence, their order is not correct if one considers both banks together. For example, it is said that the Se Pahom (modern equivalent not identified), a left-bank tributary, actually enters the Se Bang Hieng to the east of the Se Tamouk (modern Se Thamouk), a right-bank tributary, but it is shown entering considerably to the west of the latter stream. Obviously one cannot say how representative this single, hastily made example might be of Laotian cartography.

Maps of Primarily Rural Localities

Known relatively large-scale maps of Southeast Asia are mainly of Burmese provenance. Notes on thirteen pre-1885 Burmese examples are provided in appendix 18.4,

97. At the time the map was made, Mindon, the reigning Burmese monarch, was purposefully cultivating cordial relations with several European powers, most notably France and Italy, as a counterpoise to the British, who were the principal threat facing his own kingdom. This issue is discussed in Hall, *History of South-east Asia*, 626 and 628 (note 29). It is noteworthy that the hypothesized embassies to Burma had something of a reverse counterpart when, in 1823–24, a diplomatic mission consisting of two Burmese officers and an Anglo-Indian named Gibson was sent to Cochin China "bearing presents and a royal letter in which it was suggested that the two countries co-operate in conquering Thailand and partitioning that country" (B. R. Pearn, "The Burmese Embassy to Vietnam, 1823–24," *Journal of the Burma Research Society* 47, no. 1 [1964]: 149–57, esp. 149). Obviously, given the nature of the mission, it was sent by sea, rather than overland.

For an excellent discussion of the type of diplomacy practiced by both Burma and Vietnam in the final years of their struggles to remain independent, see Paul J. Bennett, "Two Southeast Asian Ministers and Reactions to European Conquest: The Kinwun Mingyi and Phan-thanh-Gian," in *Conference under the Tamarind Tree: Three Essays in Burmese History* (New Haven: Yale University Southeast Asia Studies, 1971), 103–42. I am indebted to Aung-Thwin for this reference.

98. The loss and distortion of information over time may be seen in that Sri Nandan Prasad, ed., *Catalogue of the Historical Maps of the Survey of India (1700–1900)* (New Delhi: National Archives of India, ca. 1975), 220, states that this map is not dated, despite its being precisely dated, and says that the title refers to routes from "Cochin China Yahune," not "Cochin China Yahme."

99. Photographs of this map and various relevant particulars were provided to me by Tin Maung Oo, a student of U Maung Maung Tin (who now has the map), in 1985 (undated dispatch). His authority for dating the map between 1860 and 1880 was not stated, but the map could not be later than 1885, when the British annexed Upper Burma.

100. "Im Innern von Hinterindien (nach dem Französischen des Dr. Harmand)," *Globus* 38, no. 14 (1880): 209–15.

101. "Im Innern von Hinterindien," 213; the copy of the map appears on 212 (note 100).

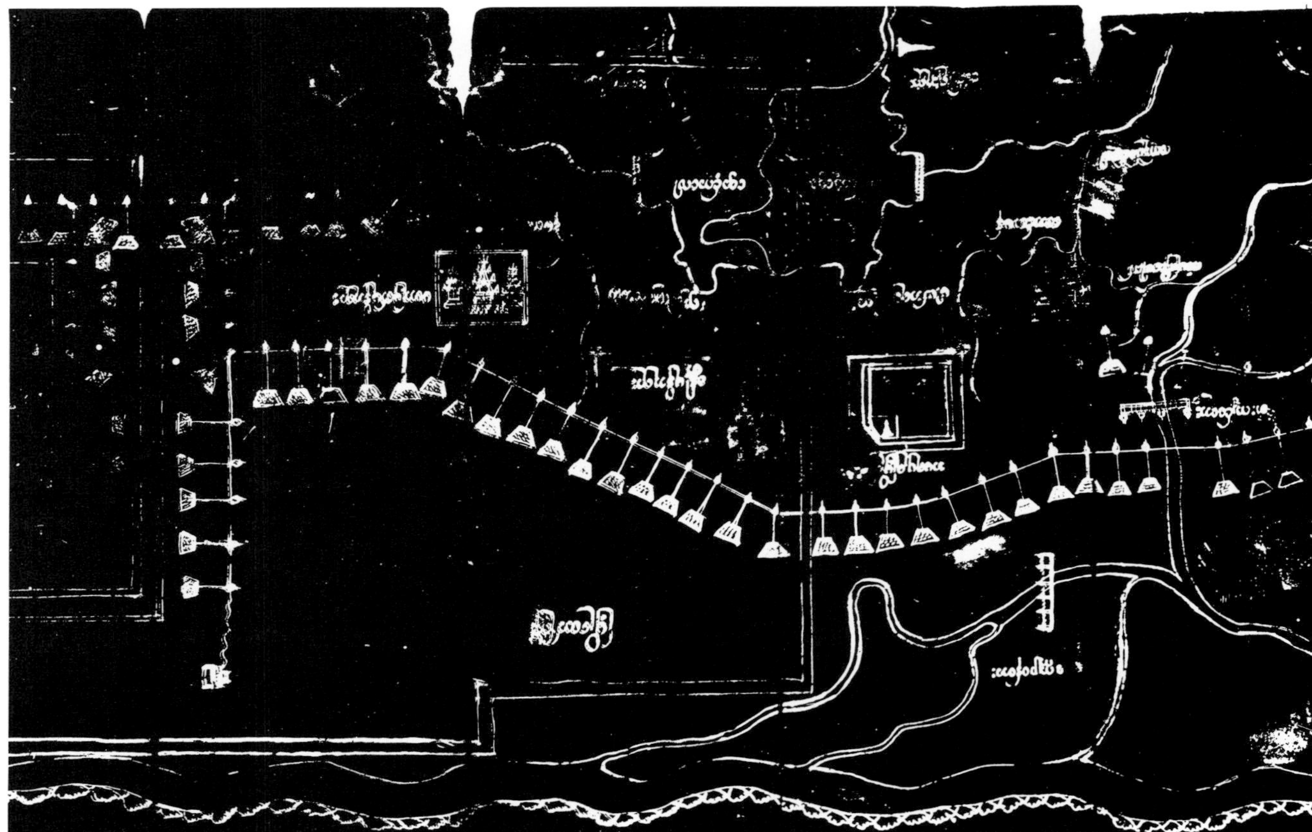

FIG. 18.32. SMALL PORTION OF A MAP OF THE TELEGRAPH LINE FROM THE THEN BORDER OF BRITISH BURMA TO MANDALAY. This map was drawn between 1860 and 1880 in white ink and red chalk on a folding black *parabaik* of at least thirty-one panels. Although most of the information relates to telegraphy, considerable ancillary detail is also provided.
Size of each panel: 37.5 × 12.8 cm. Collection of U Maung Maung Tin, Mandalay.

where the maps are ordered by known or presumed date. Excluded from that listing are a certain number of later works that form part of the Scott Collection at Cambridge University, for which a comprehensive but unpublished catalog was prepared about 1985.[102] Apart from the Burmese works, I am aware of only one locality map of Siamese provenance, and a remarkable group from the aboriginal Sakai tribe of West Malaysia.[103] Discussed below are, first, a small sample of the Burmese maps and then, in the order indicated, the Siamese and Malay maps.

Figure 18.33 is the first of four essentially cadastral maps that relate to lands in several villages inherited by a royal minister, Mahamingyaw Raza, who presumably ordered the survey by which the maps were made.[104] I have been able to determine neither the location of the area depicted nor the date of the maps, but I judge that they relate to a part of Upper Burma not very distant from Mandalay and to a time shortly before 1885. Since the fourth map of the group is clearly not finished and the *parabaik* the maps appear in has six empty pleats, one might infer from the ratio of utilized to empty space in the manuscript that the survey was less than three-fourths complete. Notations on the map state which parcels of land belonged to the minister and which to other persons, including the minister's steward and grandmother. Numbers written in most of the fields indicate how many bundles of rice could be transplanted there, while other fields are identified as seedling nurseries. Fallow rice fields have no number. Uncultivated fields are shown in brown. Field boundaries are prominently depicted in green, but notes in some of the larger fields state that individual divisions are not shown. Fields

102. Dalby and Saimöng, "Shan and Burmese Manuscript Maps" (note 39).

103. Omitted from the following discussion is any analysis of the several maplike depictions of rural localities in bas-relief on stone that appear on the friezes of old Hindu temples in central Bali. These were noted in chapter 16, but particulars relative to the places they represent are not available.

104. For assistance in interpreting this map I am indebted to U Nyunt Maung, director of the Manuscripts Section, and May Kyi Win, assistant librarian, University of Rangoon Library.

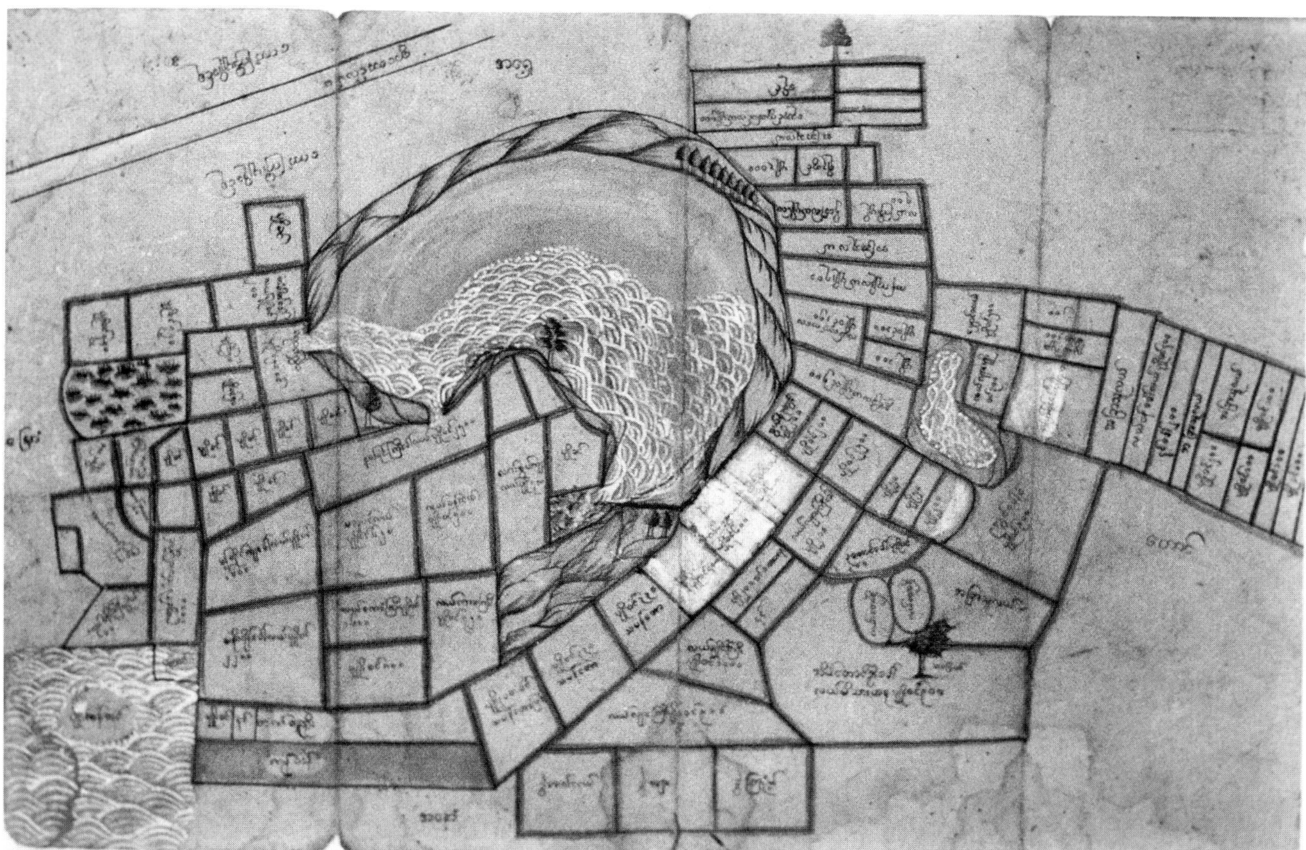

FIG. 18.33. CADASTRAL MAP, PRESUMABLY OF AN AREA IN CENTRAL BURMA. This mid- or late nineteenth-century map is one of a set of four that relate to lands in several villages inherited by a royal minister who presumably ordered the survey from which the maps were made. Notations on the map state which of the parcels of land depicted belonged to the minister and which to other persons, while numbers specify how many bundles of rice could be transplanted in various fields. Details are also provided about fields set aside for more specialized uses.
Size of the original: 52 × 81.6 cm. Rangoon University Library (MS. 9108). Photograph courtesy of Joseph E. Schwartzberg.

devoted to crops other than rice are noted. Palms and various broadleaf trees are individually and naturalistically depicted, as are clumps of high herbaceous vegetation, all in various shades of green and blue. Distinctive signs exist for earth embankments, tanks and ponds (marked by a scalloped wave pattern), roads, and so forth. On maps other than the one illustrated, built-up areas of villages are indicated by a few houses, which are drawn more or less realistically in an oblique view, as if either from ground level or from an elevated perspective. On the northern margin of the third map a hill range is shown naturalistically in shades of blue and brown. The cardinal directions are noted on the edges of all four maps. Three of the maps are oriented toward the east. The lone exception, with a northern orientation, may be explained by the fact that the oblong shape of the village did not lend itself to the usual orientation within the *parabaik* mode of presentation.

The rest of the Burmese maps I shall consider are from the Scott Collection. The most painstakingly and sensitively drawn map of these (and perhaps of the collection as a whole) is one of the environs, mainly to the south, of the town of Meiktila, situated in the Irrawaddy Plain about 115 kilometers south-southwest of Mandalay.[105] Figure 18.34 presents a partial view of this work. The focus of this excerpt is the town itself (1891 population 4,155) and its adjacent lake. The town, states the *Imperial Gazetteer*, "stands on the margin of a large artificial lake, with an irregular indented margin. The lake is practically divided into two bodies of water, the north and the south lake. Over the strip of water uniting the two . . . [runs] a narrow wooden bridge," which is visible on the map.[106] The map brims with local detail. Much of it relates to

105. Dalby and Saimöng, "Shan and Burmese Manuscript Maps" (note 39).

106. *Imperial Gazetteer of India*, new ed., 26 vols. (Oxford: Clarendon Press, 1907–9), 17:287.

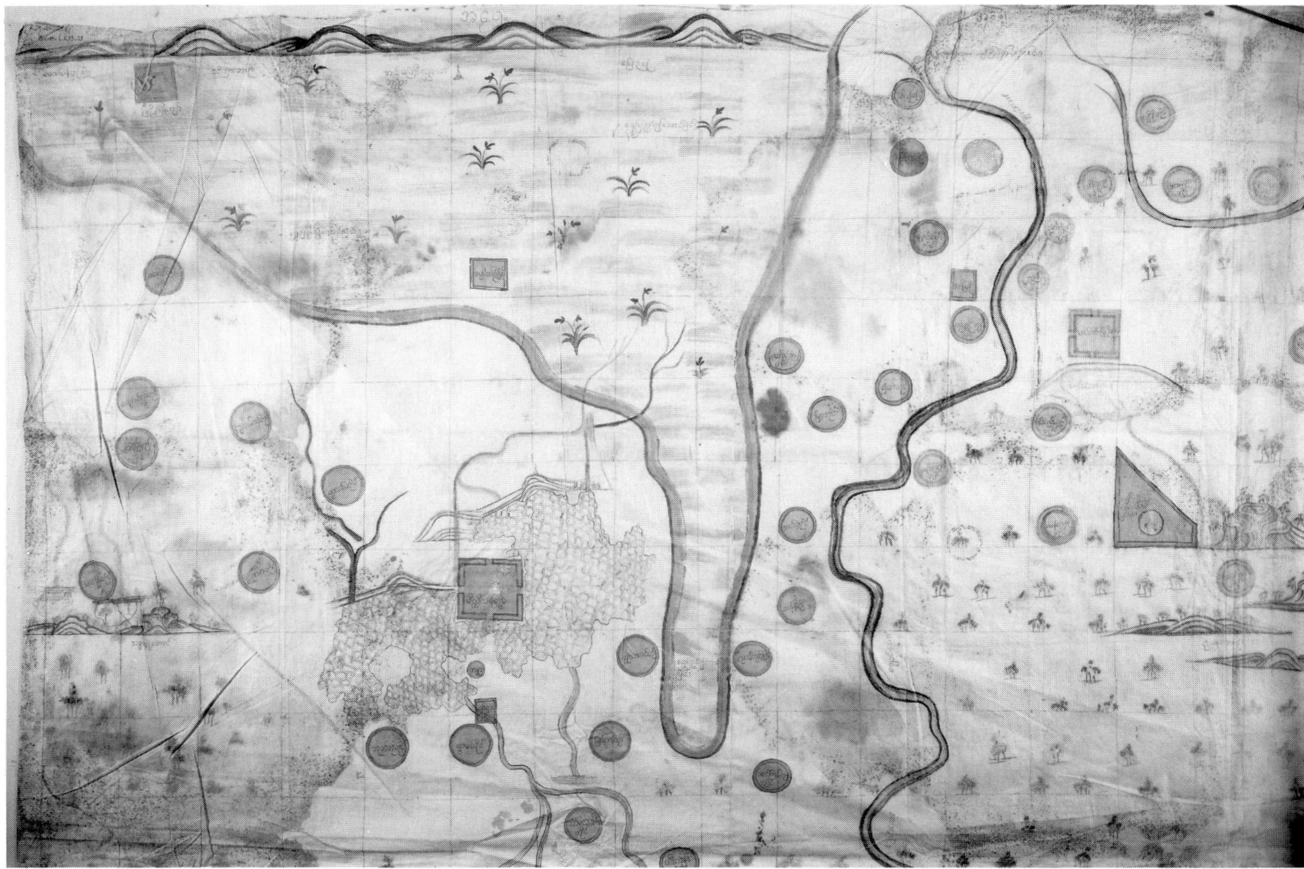

FIG. 18.34. DETAIL FROM A LARGE-SCALE LAND-USE MAP OF THE ENVIRONS OF MEIKTILA IN CENTRAL BURMA. This late nineteenth-century map is painted in nine colors and black ink on two pencil-gridded calico sheets sewn together. It focuses on the town of Meiktila, indicated by the square on the margin of a large artificial reservoir, and shows, seemingly with considerable accuracy, an abundance of detail relating to irrigation, land use, and settlement but, curiously, not to roads.
Size of the entire original: 255 × 178 cm (sewn together). By permission of the Syndics of Cambridge University Library (Scott LR. 13.25).

the irrigation for which Meiktila was well known. Channels feed the lake from the west, and embankments are clearly depicted on the lake's eastern margins. Judging from the colors, two types of distributaries flow eastward beyond these embankments. Some of them feed into an irregularly shaped area outlined by a thick band of color containing the sign of a plant (presumably irrigated rice). Several other plant signs, some apparently representing other types of cultivation and some groves of trees, are scattered over the map. A pair of tree-topped hilly eminences are shown to the north of the town. Settlements are shown by yellow squares and circles, rimmed by various colors to indicate the political divisions they belong to. Surprisingly, no roads are shown. The map is seriously stained and damaged by mildew, which might be mistaken for an intentional stippled pattern.

There is yet another map in the Scott Collection relating to Meiktila District, in this case to the single township of Taungbo, which is rendered at a very large scale (fig. 18.35).[107] This map is also painted in many colors on cloth and relates to many of the same features as the map just discussed.

Figure 18.36 presents a rather simple map of the environs of Kang Hung (modern Jinghong), just west of the Mekong in what is now the Chinese province of Yunnan.[108] The style may be characterized as pure Shan. Text is in the Khün dialect, a variant form of Shan, with Burmese translations subsequently added. This map, presumably drawn during Scott's visit to the area in March 1891, reached Cambridge on 18 January 1892. It obviously relates mainly to settlement, shown by the forty-three

107. Dalby and Saimöng, "Shan and Burmese Manuscript Maps" (note 39).

108. Dalby and Saimöng, "Shan and Burmese Manuscript Maps" (note 39).

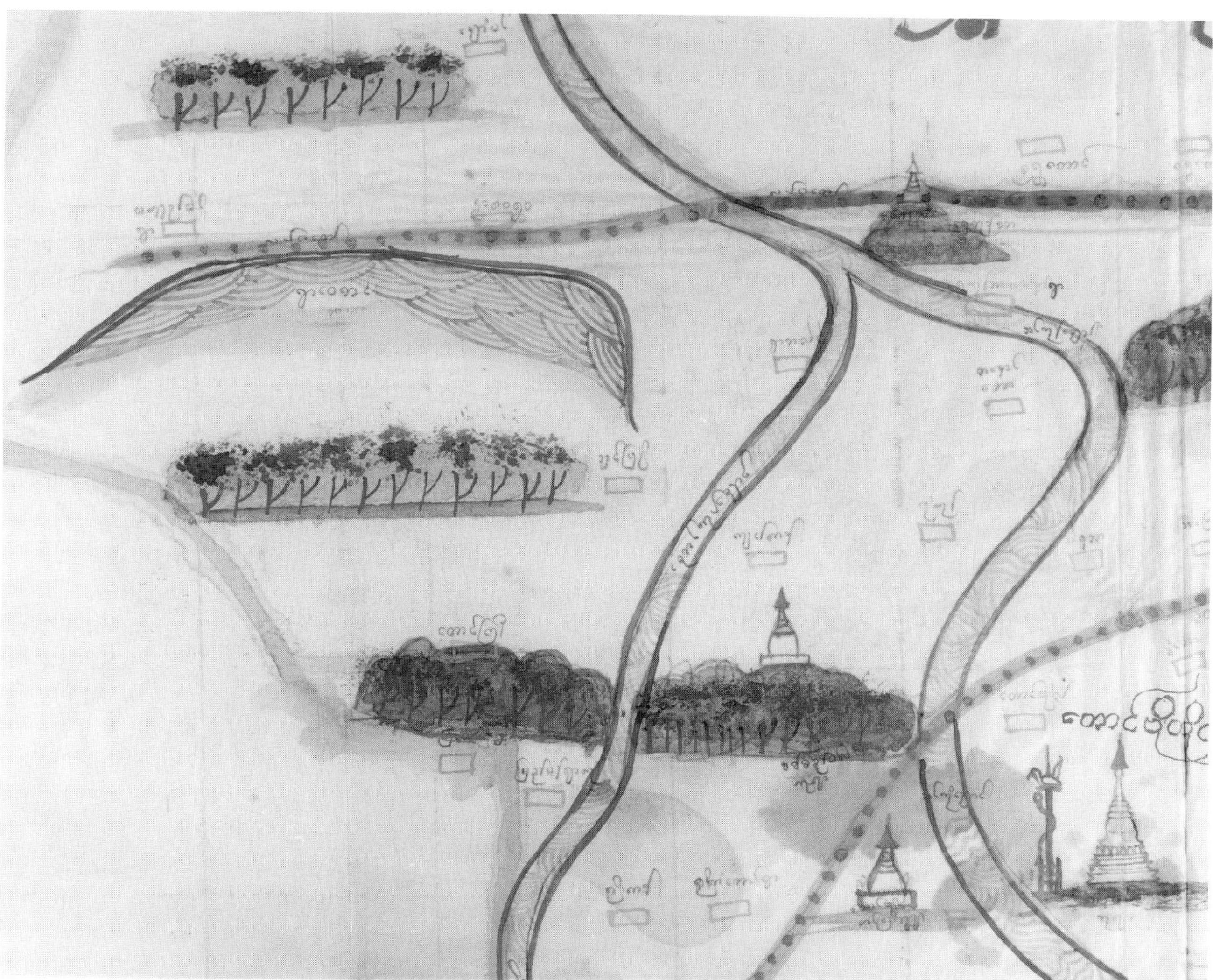

FIG. 18.35. DETAIL FROM A LARGE-SCALE MAP OF TAUNGBO TOWNSHIP IN MEIKTILA DISTRICT OF CENTRAL BURMA. The principal difference between this painted cloth map and that illustrated by figure 18.34 is that here many features are individually and distinctively drawn in frontal perspective. For example, no two pagodas look quite the same. Roads are also shown.
Size of the original: unknown. By permission of the Syndics of Cambridge University Library (Scott LR. 13.32).

circular signs. These circles, however, appear to be consciously differentiated; the circumference of some is a single line, of others a double line, and of still others a scalloped band with the scallops sometimes facing out and sometimes facing in. To a certain extent they appear to be regionally clustered. Since the area is one where numerous tribal groups live close to one another, one might surmise that the three or four types of circles connoted different ethnic groups. The circles are also differentiated by size, perhaps to make them roughly proportional to population. If so, we have here a unique Southeast Asian case of a graduated circle map. It is not clear what is indicated by the lone doubly bounded rectangle, but if my surmise about ethnic affiliations is correct, it might be a settlement exercising administrative oversight over the set of associated villages shown by simple double circles. These are certainly matters in which Scott would have taken an interest.

The prominent depiction of rivers on the map does not call for comment, but the faint indications of river crossings are noteworthy, as is the equally faint rendering of certain trails. Finally, there is an exceptionally stylized and rather faded rendering of what seem to be forested hills running around virtually the entire map, enclosing all but one of the villages depicted. This suggests that the map relates to a small settled vale. One such vale,

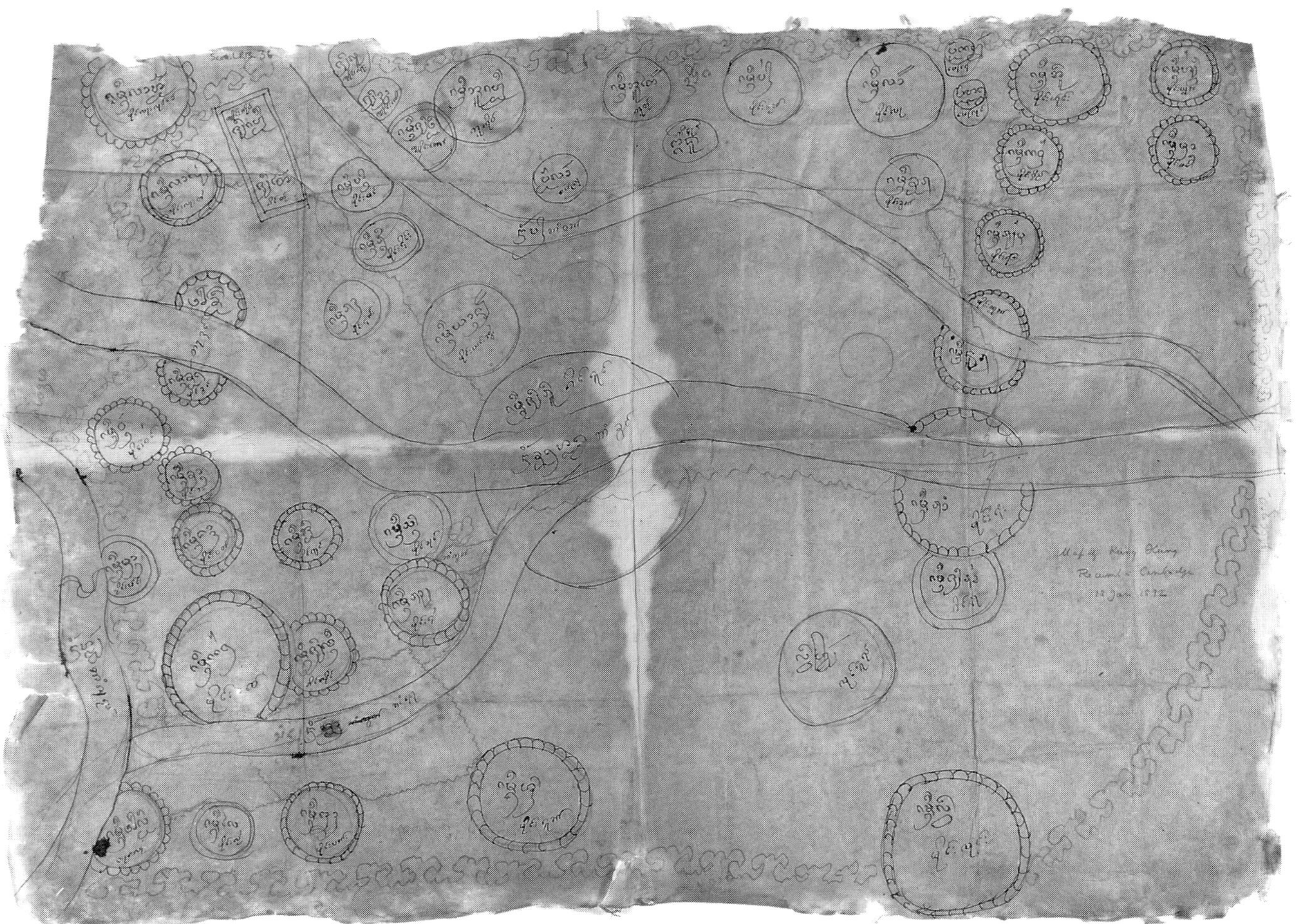

FIG. 18.36. SHAN MAP OF THE ENVIRONS OF KANG HUNG (MODERN JINGHONG). Drawn in black ink and pencil on paper, this map relates mainly to settlement and was probably drawn for James George Scott when he visited in 1891. Since the area depicted is one of considerable tribal admixture, it seems plausible that the differentiation of the patterns along the circumferences of the forty-three circles shown was intended to signify villages dominated by different tribal groups, and that the variations in circle size were meant to convey some idea of their populations. If so, this use of graduated circles would represent a rather sophisticated cartographic achievement by a supposedly "primitive" people.
Size of the original: 79 × 57 cm. By permission of the Syndics of Cambridge University Library (Scott LR. 13.36).

including Kang Hung, can be discerned on some modern maps of India.[109]

Another rather sophisticated work from the Scott Collection is shown in plate 40.[110] This map covers an area extending perhaps fifteen miles in a west-southwest to east-northeast direction on each side of the Nam Mao River (Burmese Shweli), along the then disputed Sino-Burmese border. The text is in Chinese Shan, though Burmese notes have been added in pencil. The northern part of the map (on the side of the river without the parallel mountains), situated in China, is colored yellow and labeled in Shan, "All this is Möng Mäo territory." The southern part, in Burma, bears the label, "The red is all Namkham territory." The bright tempera colors used to denote territorial possession were, of course, added at the behest of the concerned British authorities. A black enclave in the east is identified as Sèlan (modern equivalent not identified). The level of accuracy of the map is remarkably high. One can easily pick out the individual meanders of the Nam Mao on a modern map. In addition to showing the river system, the map names and depicts more than eighty villages, the town of Namkhan (the irregularly bounded shape toward the western portion of the Burmese side), and, in various shades of blue and purple, stylized mountains.

Dalby quotes from the diary of H. Daly, the superin-

109. See, for example, the hachured 1:1,000,000 sheet 102 of the Survey of India's *Imperial Atlas of India* (surveyed in 1904 and updated in 1910).

110. Dalby and Saimöng, "Shan and Burmese Manuscript Maps" (note 39).

tendent for the Northern Shan States, a passage that very likely relates to this map (suggesting that it is one of the few—possibly the only one—among the Scott Collection for which Scott was not personally responsible):

> 15th [of May, 1889], Selan. The whole valley is full of villages and hamlets, but those belonging to Meungmow [Möng Mäo in China] and those subject to Theinni [Hsinwi? in Burma] are much intermixed, in explanation of which it is stated that the course of the river has changed in recent years. I have had maps prepared by local men showing the villages which belong to each state and circle.[111]

The sole rural locality map of Siamese provenance is from the Burney Collection. Like figure 18.33, it appears to relate primarily to land use, though not necessarily to revenue assessment. Because many of the features shown run to the edge of the page the map was drawn on, it may be regarded as a fragment of a larger whole for which no identifying information is now available. The style of writing suggests that the map dates from the early nineteenth century, and since Henry Burney's sojourn in Thailand lasted from 1825 to 1827, the map can scarcely date from a later period.[112] It is conceivable that the map was made at Burney's request, but since it has no annotation in English and deals with no subject or place of obvious importance, it is hard to imagine why he would have had it made. The map is obviously of a rather small area, but none of the three place-names on it are recognizable. It is completely planimetric and totally devoid of pictographic symbols. The map shows all or parts of forty-seven bounded areas, varying considerably in size and separated by sinuous dividers. These appear in some instances to be canals or streams and in others to be roads or, perhaps, roads running alongside watercourses. Which is which is not visually apparent and must be inferred largely by the map context. Only one roadway is explicitly identified as such. Almost all the field separations are shown by double black ink lines, reinforced with a yellow wash. Every field and outlined area but one is labeled as to its nature or use, in some cases with more than one designation. The translated designations include eleven parcels, all fairly small, labeled "island"; two swamps, one creek, and one curiously shaped area called "canal mouth"; various types of fields (six for rice, one "open," one "grassy," two for keeping water buffalo, and one for elephants); areas designated as various types of forest or simply as "forests"; one hill and one "mound hill"; several hamlets, not all named; two pagodas; two temples in disrepair; and a cremation ground, as well as a separate "burial ground for spirits."

The only artifacts from the Malay world that appear to fit under the rubric I have designated as maps of rural localities are a set of images on bamboo that form part of a group of about 1,500 late nineteenth-century artifacts obtained by Hrolf Vaughan Stevens from several Negrito tribes in what is now West Malaysia. A large number of the artifacts in this collection were engraved designs on bamboo, perhaps ten of which may be interpreted as incorporating maplike components (one example is illustrated in fig. 18.37).[113]

All the designs of interest to us came from a single tribal group, the Sakais, who lived mainly by hunting, fishing, and gathering, with some supplemental slash-and-burn cultivation. The designs were executed as "magical designs" on musical instruments called *tuang-tuang*, employed in ceremonies intended to ward off specific types of harm that Sakais were likely to encounter or, less frequently, to help realize some practical goal such as inducing rain or finding suitable house-building materials. Two such instruments were generally used together, one in each hand, and were struck on the ground to produce musical notes to achieve the desired end.[114] The cartographic elements within the engraved designs are often highly abstract and far from obvious in meaning to an outsider. They were meant to indicate important features of specific localities within the territory where the Sakais lived. Among the features shown are nipa swamps, agricultural clearings, areas planted in specific crops, areas with fish traps, anthills, places infested with deadly snakes, house sites, and so forth. The legend for the example illustrated here points out the symbolism

111. Quoted in Dalby and Saimöng, "Shan and Burmese Manuscript Maps" (note 39).

112. Royal Commonwealth Society, London (ser. E, box III, no. VIII), measuring 29.8 by 32.3 centimeters. I am indebted to Henry Ginsburg of the Oriental and India Office Collections, British Library, for his painstaking translation of the map text.

113. The collection was presented to and remains with the Berlin Museum für Volkerkünde. The works of particular interest here are discussed in Hrolf Vaughan Stevens, "Die Zaubermuster der Ôrang hûtan," pt. 2, "Die 'Toon-tong'-Ceremonie," *Zeitschrift für Ethnologie* 26 (1894): 141–88 and pls. IX and X. Stevens's theories in regard to the interpretation of the art of the several Negrito tribes were subjected to severe criticism by Walter William Skeat and Charles Otto Blagden, *Pagan Races of the Malay Peninsula*, 2 vols. (London: Macmillan, 1906), 1:395–401. Much of the criticism hinged on the fact that Stevens's research was carried out through the medium of Malay, a language foreign to both him and his informants, which led to certain fundamental misunderstandings. Nevertheless, Skeat and Blagden acknowledged the pioneering nature of the work, which they then presented (401–92) in almost as much detail as the German original, quoting large portions of the original text in translation yet leaving out the "obvious mistakes" (401). With respect to those artifacts that most concern us, Skeat and Blagden have raised no basic objections. The account here relies primarily on their recapitulation of Stevens's interpretations. The illustrations in Stevens's article are completely duplicated in Skeat and Blagden, including all with maplike elements. Their discussions are also fuller than those I can provide here.

114. Skeat and Blagden, *Pagan Races*, 1:471–72 (note 113).

FIG. 18.37. ONE EXAMPLE OF INCISED BAMBOO *TUANG-TUANG* OF THE SAKAI TRIBE OF WEST MALAYSIA INCORPORATING CARTOGRAPHIC ELEMENTS. This late nineteenth-century charm (and others like it) bears drawings that are intended to protect the owner from specific types of harm or to help bring about some needed event. To make the charm effective the *tuang-tuang* had to be beaten on the ground to produce a musical tone. This example is a charm to protect growing crops near a house from damage by animals. The patterns in the lowest third of the picture represent a house, with a ladderlike pattern at its upper-right corner to show the steps leading up to it, and a field planted with sweet potatoes, part of which, signified by the more compact cluster at the bottom, is on a hill. Within the central division various crops are depicted between dead trees, shown by simple vertical lines. From right to left these include: maize and yams, tapioca, sugarcane (three stalks), maize again, tapioca, banana (lower plant), and another type of yam (above). The dots around these plants denote grass. A diverse group of animals against which the charm is directed is shown in the upper portion of the drawing. This is a photograph of an exact copy—made in situ—of an original that the owner was unwilling to part with.
Size of the original: height 24 cm. By permission of the Museum für Völkerkunde, Staatliche Museen zu Berlin—Preussischer Kulturbesitz.

employed and its relation to the purpose for which the *tuang-tuang* were made.

I am unaware of preliterate tribal groups in Southeast Asia, other than the Sakais, who produce (or formerly produced) comparable maps relating to real-world localities. But many small, relatively isolated, and scarcely studied groups live in the region, and there is no reason to assume that the Sakai case is unique. This issue calls for further research.[115]

Maps of Primarily Urban Localities

Although cities figure prominently on a number of the regional maps I have considered and are in some instances shown with an attempt at differentiation of their internal structure, the known corpus of maps that relate primarily or wholly to urban places is meager and limited almost entirely to Burma.[116] There is no reason, however, to suppose that further research will not uncover additional examples, not only from that country but from others as well. Sixteen known maps from Burma are noted and briefly described in appendix 18.5. These maps focus exclusively on former capitals, not only of Burma, but also of neighboring countries. Mandalay is represented by eleven examples, of which seven cover only part of the city. The adjacent and previous capital at Amarapura is the focus of two maps, but it is also shown, along with other nearby former capitals at Ava and Sagaing, on some of the maps of Mandalay. A single map relates to the former Thai capital of Ayutthaya, which the Burmese sacked in 1767. Finally, "Zimmay" (Chiang Mai), capital of the northern Thai kingdom of Lanna, which was intermittently subject to Burma, is depicted on a very simple sketch that appears to have been drawn by a Burmese

115. It is appropriate to call attention to a rather sophisticated anthropological study of the cognitive mapping process among the nearly extinct Onge tribe, an exceedingly primitive group in terms of material culture, as elicited through fieldwork conducted on the island of Little Andaman in 1983–84. This work is detailed in Vishvajit Pandya, "Movement and Space: Andamanese Cartography," *American Ethnologist* 17 (1990): 775–97. Although the Andaman Islands, which belong to India, lie just outside Southeast Asia as defined for this volume, they are culturally more akin to Southeast than to South Asia. Nevertheless, because the Andamanese maps illustrated in Pandya's article were produced under rather artificial experimental conditions and are not indicative of what sort of maps, if any, the Onges might have drawn in the absence of foreign investigators, I have chosen not to illustrate them in this chapter.

116. The only example I can cite of a traditional non-Burmese map of a town or city from Southeast Asia is a fairly simple drawing (55 × 75 cm) in pencil and ink on European paper of the southern Thai town of Ligor (modern Nakhon Si Thammarat) drawn by a local artist, an acculturated ethnic Chinese named Bun Khong, for Captain James Low about 1825, when Low was on a diplomatic mission to that city. This diagram is one of a series of drawings made for Low that is held by the Royal Asiatic Society of London (RAS 340).

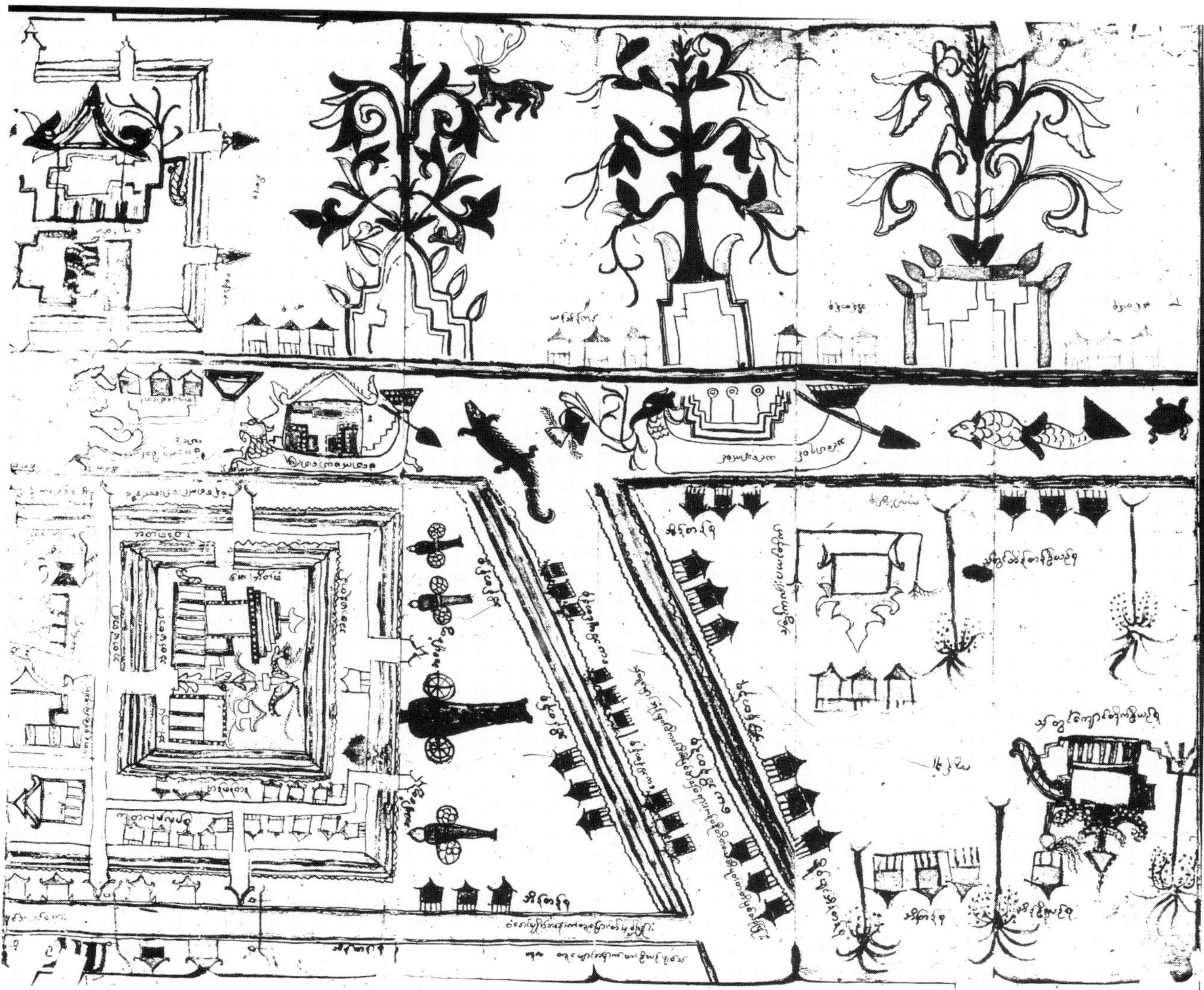

FIG. 18.38. DETAIL FROM A LARGE BURMESE MILITARY INTELLIGENCE MAP. The map, of which the portion shown here forms only a very small fraction, relates to the successful campaign by the Burmese against the then Siamese capital of Ayutthaya in 1767. It appears to have been based on intelligence reports gathered before the campaign, since the city was razed during the conflict. The size of the entire map is not known, and parts of it may be lost. The map was painted on at least forty-eight folding panels of white *parabaik*. A striking feature of the map is its combining details of obvious military significance—such as artillery emplacements—with others relating to fauna, flora, and religious edifices that presumably are irrelevant militarily.
Size of each panel: 87.6 × 17.8 cm. National Archives, Rangoon. Photograph courtesy of Joseph E. Schwartzberg.

Shan. I shall here describe only three of the maps just cited (that of Ayutthaya, one of the two of Amarapura, and one of Mandalay) and shall illustrate the first two. Several essentially architectural plans will also be discussed.

The Burmese map of Ayutthaya (a small part of which is shown in fig. 18.38) is said to have been made for military intelligence just before and in anticipation of the campaign directed toward that city in 1767. Not counting cosmographies, this makes it the oldest of all known surviving Burmese maps. The map, on white *parabaik*, appears to have been torn in a number of places, and the panels cannot be put together to form a continuous image. At a minimum, one can say that the map area was approximately 7.5 square meters.[117] Whether the *para-*

117. I have not personally seen this important work and know it only from one 19 by 24 centimeter color photograph, showing many of the panels laid out (often discontinuously) in four rows, and three black-and-white larger-scale photographs of a much smaller portion of the whole that were sent to me along with a page of relevant notes. The color photograph was supplied by Andrew Dalby, then on the staff of the library of Cambridge University, and the other materials by Tin

baik was intended to be read continuously as one long strip presentation or as several parallel strips laid side by side is not clear. The latter would better suit the purpose of the map, to provide military intelligence, given the nonlinear area it related to; but I know of no other case of a *parabaik*—as opposed to a map on cloth or a large piece of paper, or several pieces pasted together in a nonaccordion fashion—that was so used.

How the map was prepared is not known. Since it related to the capital city of what was at the time Burma's principal enemy, I assume it was made by a spy. In that event it would probably have had to be drawn from notes taken as the spy roamed the streets and canals of Ayutthaya, presumably being put together in its present form only after he returned to the safety of his own country. Figure 18.38 depicts one of numerous compounds, almost twenty of which may be dimly discerned. Although the map shows considerable detail about the interiors of these compounds, which have the look of enclosing important structures such as government offices, temples, and perhaps the residences of high officials, the interior of the largest compound, which I take to be the royal palace (at the terminus of the canal shown in the figure), is relatively empty.[118] This suggests that the presumed spy was more successful in obtaining information about, and possibly even direct access to, the other important places depicted. Annotations in black ink, generally rather terse and probably running into the hundreds, appear on all parts of the map. A note sent to me from Burma states that the map "describes the location of houses, wooden stockades, granaries, cannons, monasteries, and guards."[119] In the illustration one sees one large and three small cannons, their wheels splayed sideways so that there is no doubt what is depicted.

The map is remarkably ornate in style, and its palette is extraordinarily vivid. Houses are painted in blue, blue-green, green, yellow, red, black, and white; trees in blue-green, yellow, red, and brown; roads in black and five other colors; barges in four colors; and so forth. As on so many Burmese maps, vegetation is shown very prominently and in many different, though perhaps conventionalized, forms. Animals too loom large on the map. The canals teem with fish, crocodiles, and an occasional crane or tortoise. Elsewhere appear at least one elephant, deer, monkey, pig, and bird. All of this exuberance may seem counter to the alleged purpose of the map; but taking into consideration the likelihood that it was to be presented directly to the king and that time would not have been of the essence in making it ready for so august a personage, the seeming contradiction is readily explained.

Compared with the map of Ayutthaya, that of Amarapura and its environs, made about 1850, has a much more modern look. The original map was a four-panel *parabaik* whose dimensions are unknown. The description that follows is based on what was said to be a faithful copy, made in the 1970s on European paper by six monks at the Taung Lay None monastery at Amarapura in the 1970s (fig. 18.39).[120] Despite the alleged fidelity of the copy, it is questionable whether its rather intense colors (applied using modern felt-tipped pens) closely resemble those of the original work; but in general the style of the map conforms well to other Burmese works of the mid-nineteenth century. Among the features shown on the map are the old royal walled city, its internal street pattern, the wards and hamlets lying outside the royal enclosure, monasteries, pagodas, canals, ditches, periodically inundated areas, two wooden causeways over such areas, various forms of vegetation (rendered more or less naturalistically), and across the Irrawaddy, the forested range of the Sagaing Hills, shown in frontal elevation near the top of the map with the sky beyond punctuated by conventionally rendered clouds somewhat similar to those on certain Tibetan maps. A feature that distinguishes this map from most others of Burmese provenance is its orientation toward the west.

Of Mandalay, as noted, there are at least eleven surviving maps that predate the British occupation of that city. All of these maps are believed to date from the early 1850s to 1885, the date of the British occupation. The earliest is the map King Mindon ordered to guide the construction of his proposed new capital. This work, now very faded and worn, was found in 1954 in the Shwenandaw monastery in Mandalay, where it had been kept at least since 1886. It is at present on display in the museum of the Burmese Department of the University of Mandalay. The work has been described as "probably the best surviving example of Burmese surveying practise of more than a century ago," and it has been suggested that the plan, motivated by the king's desire to shift the capital away from the unsanitary and crowded site at Amarapura, was commenced in 1853 and completed in

Maung Oo. All the photographs were taken at the home of U Maung Maung Tin in Mandalay. The color photograph is itself of a set of color photographs taken before the map was turned over to the National Archives in Rangoon. I was not able to see the map or to determine its accession number at that institution. I extend my deep appreciation for the help rendered by all three. Much of the discussion of the map is based on the notes provided by Tin Maung Oo.

118. The layout conforms to my own recollections of the remains of Ayutthaya from a visit there in 1980.

119. Notes by Tin Maung Oo, undated, in dispatch sent in 1985.

120. I saw and photographed this map at the monastery on 4 March 1984. I am grateful to U Ba Khet of Mandalay for guiding me to the place on the outskirts of Amarapura and for interpreting for me with the resident monks. I am also grateful to U Pyin Nya Zaw Ta, the head monk, for permitting me to study the map, the original of which was in the possession of Than Tun of the Burmese Historical Commission, then resident in Japan.

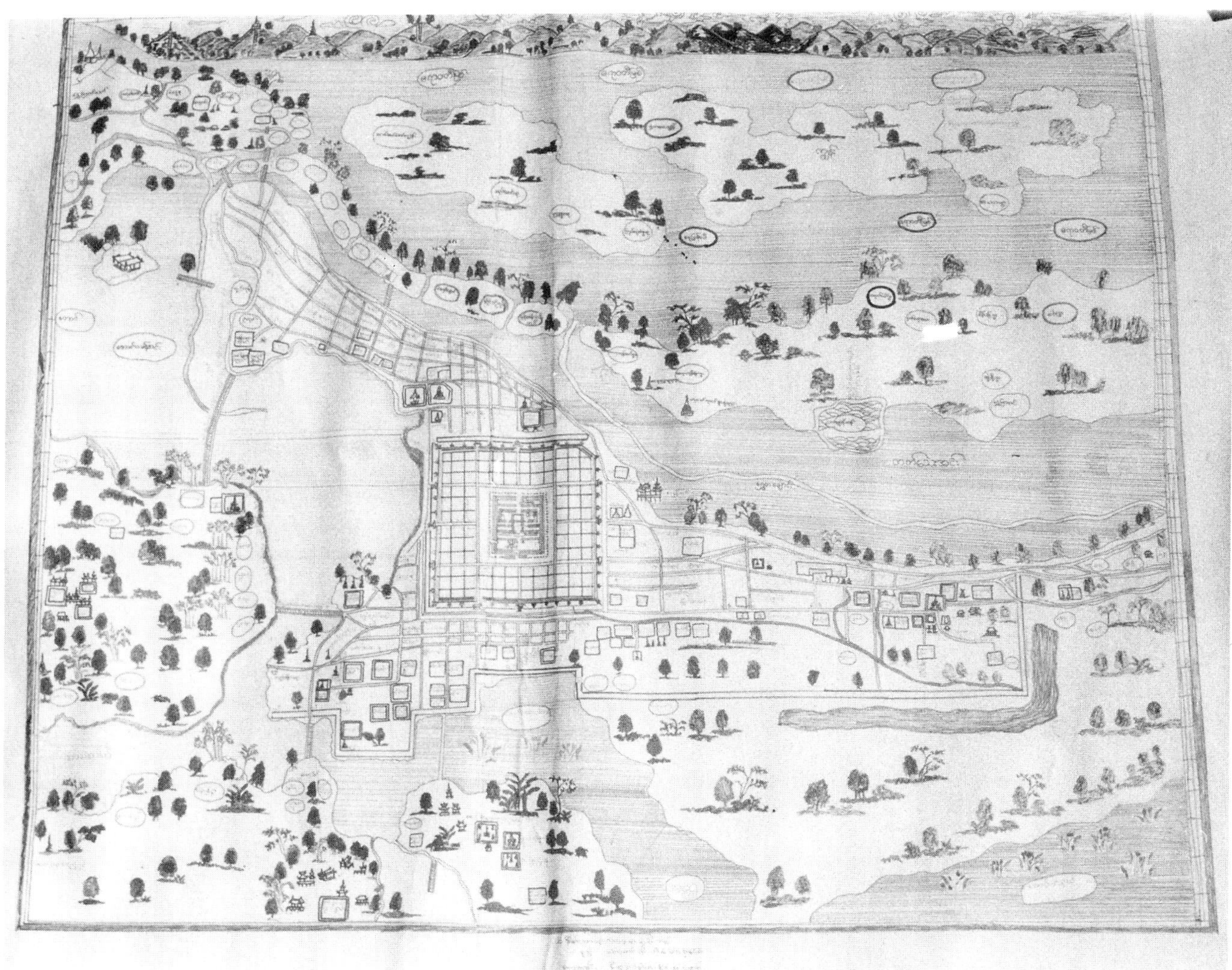

FIG. 18.39. MAP OF AMARAPURA, CAPITAL OF BURMA BEFORE THE FOUNDING OF MANDALAY. This is a modern copy, executed by Burmese monks at the Taung Lay None monastery in Amarapura, where it is now held, of a mid-nineteenth-century original. Although drawn with felt-tipped pens, the map was said to adhere closely to the style of the original four-panel *parabaik*.
Size of this copy: 93.7 × 105.7 cm. Photograph courtesy of Joseph E. Schwartzberg.

1855.[121] The surveyors and engineers charged with laying out the framework for constructing the new capital apparently did not include the mapmaker, though the names of none of those individuals is recorded on the map itself. There is no evidence that any European was directly involved in devising the plan, but those who did so "could not have been entirely ignorant of the effects which were being produced by Captain Fraser's town planning in Rangoon."[122] It is noteworthy that the plan does not tally completely with what exists on the ground. The construction of Mandalay did not begin until 1857, and departures from the initial plan were sanctioned for a variety of reasons. For example, it was originally intended to make the eastern and western walls of the royal compound somewhat longer than those on the north and south; but ultimately a perfect square was decided on, each side being 600 *tas* (a Burmese yard) in length, so that the total number of *tas* in the perimeter would be 2,400 (roughly 2.06 km), equal to the number of years elapsed since the Buddha attained nirvana. Other changes were made to ensure an adequate water supply

121. U Maung Maung Tin and Thomas Owen Morris, "Mindon Min's Development Plan for the Mandalay Area," *Journal of the Burma Research Society* 49, no. 1 (1966): 29–34 and two maps; quotation on 29.

122. Maung Maung Tin and Morris, "Development Plan," 30 (note 121).

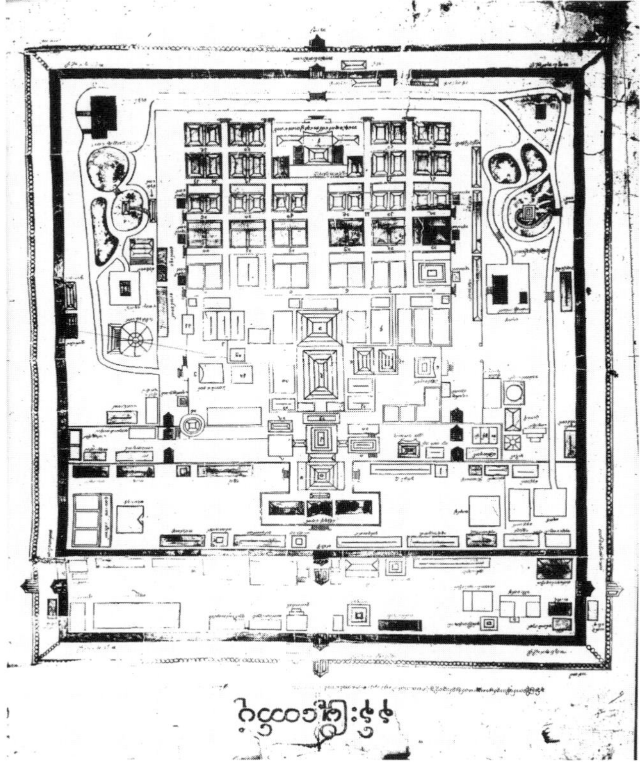

FIG. 18.40. PLAN OF THE KING'S APARTMENTS, ROYAL PALACE, MANDALAY, CA. 1870. This faithful modern copy of the original white *parabaik* plan was made by U Maung Maung Tin and appears to be drawn in black ink on European paper. The buildings within the palace enclosure are rendered in great detail and, it seems, with considerable fidelity in both scale and internal differentiation.
Size of this copy: ca. 76 × 56 cm. Photograph courtesy of Joseph E. Schwartzberg.

to the new city and to stiffen its defenses.[123]

The plan, drawn in five colors on a large sheet of gridded black *parabaik* approximately 229 by 145 centimeters, covers the Mandalay Plain from the mountains to the east to the banks of the Irrawaddy on the west and extends over a somewhat greater distance from north to south. Each grid square on the map measures 3.18 centimeters on a side, which was calculated to represent approximately 893 meters. Curiously, the point of origin of the grid is not within the proposed palace, but at a particular spot on an embankment some distance away. "The orientation arrow [not a traditional feature of Burmese maps], however, is drawn through the exact centre of the area designed for the Shwemyodaw [royal compound]."[124] The alignments for the boundaries of the Shwemyodaw and the city's defensive ditches and ramparts were all demarcated parallel to the grid lines and were subsequently so constructed. The coordinates, however, rather than running exactly along cardinal directions, are actually about three and a half degrees in error. Although no scale appears on the plan, it is clear from comparing actual distances in Mandalay against map distances that the map, at least in its central portion, was drawn to a scale that works out to be 2.253 inches to the mile (1:28,123).[125] Away from the city itself, however, one notes appreciable differences in the location of actual constructions and their implied places on the plan, though these are said to be "not unreasonably large when one considers the rudimentary methods and materials available to the surveyors and to the draftsman." The method of surveying is explained in the following passage:

> A certain amount of rough triangulation must have been employed to fix the positions of the islands and the villages in the Irrawaddy, and of its right bank. Otherwise all the control seems to have been by measurements along and offsets from the lines of the grid. If all the grid lines shown on the plan were completely traversed to fix the positions of the villages, canals, rivers, lakes, and the mountain foot, it would mean that considerably more than 1600 miles of line were measured, recorded, and plotted during the survey.[126]

From the grand scale of the proposed new city it seems clear that it was envisaged as a long-term capital for the kingdom of Burma. The associated water supply, flood protection, and drainage systems that appear on the plan (including numerous new and enlarged embankments and canals and an artificial lake approximately 15 km long and 4 km wide) indicate a degree of engineering sophistication remarkable for its day. How much additional progress the Burmese might have made in surveying, mapping, and engineering had the third and final Anglo-Burmese War not set the country on a completely new course is a question we shall never be able to answer.

At least five of the remaining maps of Mandalay, one of the entire city and four relating to particular portions of it, appear to have been drawn to aid planning and engineering operations. One map of the entire city, whose purpose is not clear, provides a detailed layout of streets noting the residences of the local dignitaries, while another, including extensive surrounding areas, obviously serves a cadastral function. Three maps of relatively small areas are particularly rich in architectural detail. It is not always clear whether these maps were prepared before or after the structures shown on them were built (for

123. Maung Maung Tin and Morris, "Development Plan," 30–31 (note 121).

124. Maung Maung Tin and Morris, "Development Plan," 31 (note 121).

125. Maung Maung Tin and Morris, "Development Plan," 31 (note 121).

126. Maung Maung Tin and Morris, "Development Plan," 31–32 (note 121).

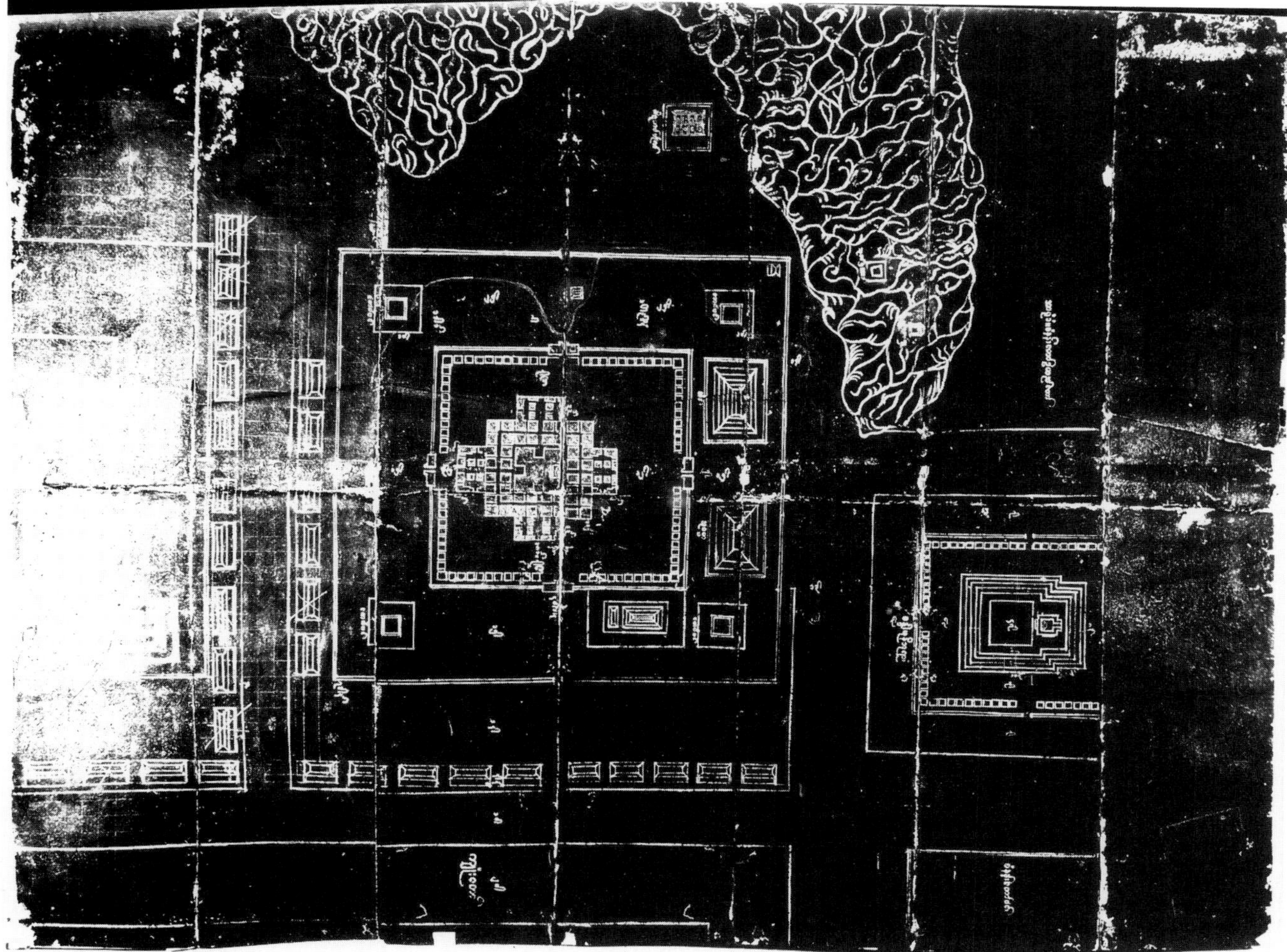

FIG. 18.41. ROYAL TEMPLE COMPLEX AT THE FOOT OF MANDALAY HILL, CA. 1875. This view includes a substantial portion of a rather detailed large-scale architectural plan of a cluster of temples and associated edifices (ordination hall, rest houses, library, and such) built sometime after 1857 but destroyed about 1890. Drawn with white chalk on black *parabaik*, with six pleats and one longitudinal fold, the entire plan covers an area of about 1.3 by 0.5 kilometers and is oriented to the east.
Size of the original: ca. 114 × 61 cm. Collection of U Maung Maung Tin, Mandalay.

more on these maps, see appendix 18.5, items g, h, i, o, and p).

Figure 18.40 depicts the king's apartments (bounded, I would judge, by the thin black line that forms the nearly perfect 2,400 *ta* square noted in the discussion of the Mandalay planning map) situated within the wooden stockade inside the Mandalay fort. The nature of the detail depicted suggests that this was not a planning map but was made as a historical record. In addition to the buildings identified as the royal apartments, the plan shows apartments for officials, a mint, a place for storing presents, a clock tower, servants' quarters, a kitchen, gardens, a swimming bath, and a "water palace." Religious structures include a monastery, a rest house for *nats* (Burmese demigods), and a tower for keeping the relic of the Buddha's tooth. Military features include the wooden stockade (the plan's outermost perimeter), guard posts, barracks, an armory, and sheds for carriages and elephants.[127]

My final example of a map of a small urban locality (fig. 18.41) depicts a plan of the royal temple complex at the base of Mandalay Hill. This plan may well have preceded the actual construction of the buildings shown, which include the Sandamuni Pagoda, at the top of the

127. The original is in the possession of U Maung Maung Kyaw (whom I am unable to identify). The description given here is based on information provided by Tin Maung Oo. Regrettably, he did not indicate which features on the plan corresponded to each of the items noted in my description, though some of them are obvious.

plan, between an area named as the royal garden and the temporary palace grounds; the larger Kyauktawgyi Pagoda, in the center of the plan; a library (the small building to the left); and a number of rest houses, shown by a set of similar small rectangular structures. The squares at each of the four corners of the larger pagoda are identified as the sites of sacred bodhi trees (*Ficus religiosa*). The structure near the upper left corner of the pagoda compound is the Pathana Sima, an ordination hall. The very large bounded area that is barely visible at the base of the map has not been identified.[128]

128. Description based primarily on information supplied by Tin Maung Oo, with some additional historical details from Andrew Dalby (letter dated 10 December 1984).

APPENDIX 18.1 REGIONAL MAPS OF VARIOUS PARTS OF SOUTHEAST ASIA

Area Covered	Place Where Map Is Held, Accession Number	Provenance and Date	Dimensions (cm) (h × w)
a. The whole of Burma as of 1795 and extensive, though indeterminate surrounding areas, especially to the east	National Archives of India, New Delhi, Historical Map Folio[1] 157, no. 13	Originally drawn in Amarapura in 1795 by a slave of the crown prince of Burma	60 × 47.1[2]
b1 and b2. Northern Burma, from Amarapura (near present Mandalay) to vicinity of Mogaung, approximately 400 km to north	NAI, HMF 157, nos. 14 and 16	Amarapura, 1795; drawn by a native of Toungoo, which is on the Sittang River in Lower Burma	Two maps, each 47.5 × 60, of almost the identical area
c. Southern Burma, from Amarapura south to the Gulf of Martaban, a north-south distance of about 700 km (a southern continuation, in effect, of b)	Was presumably held by the National Archives of India but not seen during visit to same	Amarapura, 1795; drawn by same native noted for b above	Not available; published version measures 24.5 × 19.8. Original presumably same as b.
d. The region of Upper Burma between the Irrawaddy and the Chindwin ("Khiaenduaen") rivers from their confluence north to the Chinese frontier	Location of original not known; copy may be at the National Archives of India	Originally drawn at Amarapura in 1795 by the same slave as noted for a	Original presumably 39.6 × 23.8, based on statement by Hamilton that the printed version is at half the scale of the original

1. Hereafter abbreviated NAI, HMF.

2. Unless otherwise specified, dimensions refer to copies at the National Archives of India, New Delhi.

Drawn for Francis Hamilton during His Sojourn in Burma in 1795

Orientation	Medium	Description (unless otherwise noted, maps are translated from Burmese into English)	Published Accounts
North	Black ink and gray and orange wash on European paper, now cloth backed and laminated	See text and figure 18.2	Francis Hamilton, "An Account of a Map of the Countries Subject to the King of Ava, Drawn by a Slave of the King's Eldest Son," *Edinburgh Philosophical Journal*[3] 2 (1820): 89–95, 262–71, and pl. X; W. S. Desai, "A Map of Burma (1795) by a Burmese Slave," *Journal of the Burma Research Society* 26, no. 3 (1936): 147–52.
East	Black ink with added pencil notes, now cloth backed and laminated	See text and figure 18.3. One map (sheet no. 14, map b1) is much richer in detail than the other (sheet no. 16, map b2), presumably because Hamilton sought in the latter a less cluttered version, as in the case of c (below).	Hamilton, "Account of a Map of the Country North from Ava," *EPJ* 4 (1820–21): 76–87 and pl. II.
North	Original presumably same as b	Similar to b, but without the depiction of vegetation. For note on anastomosis of rivers, see text.	Hamilton, "Account of a Map Constructed by a Native of Taunu, of the Country South from Ava," *EPJ* 5 (1821): 75–84 and pl. V.
North	Presumably ink on European paper	Emphasis is on showing travel times, in days or Burmese leagues (2.2 miles), in a network of twenty-six named places, all indicated by circles. Rivers shown, according to importance, by double or single lines. Yemyet Lake, north of Amarapura, is greatly exaggerated in size. Several hill ranges are suggested by wavy lines. Scale toward the northwest is greatly contracted.	Hamilton, "Account of a Map of the Country between the Erawadi and Khiaenduaen Rivers," *EPJ* 6 (1821–22): 107–11 and pl. IV.

3. Hereafter abbreviated *EPJ*.

Area Covered	Place Where Map Is Held, Accession Number	Provenance and Date	Dimensions (cm) (h × w)
e. An area to the east of the Irrawaddy from Amarapura south to Prome in the west to just beyond Toungoo in the east	As for d	As for d	Not known for original; 19.8 × 11.6 for published version
f. An indeterminate area in the south-central portion of what is now Yunnan Province, northwest Tonkin, and northern Laos	NAI, HMF 157, no. 38	Same as for a	27 × 38
g. An area of central Burma focusing on Pagan	NAI, HMF 157, no. 25	Original drawn by the town clerk of Pagan in 1795 (after having been shown a number of European maps by Hamilton)	39 × 31
h. The Tenasserim coast from the vicinity of Tenasserim in the south to that of Martaban in the north; a distance of roughly 525 km	Location of original or of copy not known	Amarapura, 1795, by a native of Tavoy (in Tenasserim)	Not known for original; 24.3 × 7.6 in published version
i. The north of what is now Thailand, known then to the Burmese as the country of the Jun Shan, and an adjacent area of Laos	As for h	As for a	12.2 × 16.8 in published form
j. Essentially the same as i	NAI, HMF 157, no. 22	As for a	52 × 38
k. A large area of Shan territory between the Irrawaddy and the Salween River from about 19°N to 25°N	As for h	As for a	17.3 × 7.6 in published form

(continued)

Orientation	Medium	Description (unless otherwise noted, maps are translated from Burmese into English)	Published Accounts
North	As for d	Essentially the same as for d. However, map distortion is greater toward the south, and the western edge of the Shan Plateau is naturalistically depicted as a hill range running north-south for the entire length of the map and forming, in effect, its eastern limit.	Hamilton, "Account of a Map by a Slave to the Heir-Apparent of Ava," *EPJ* 6 (1821–22): 270–73 and pl. IX.
North	Black ink on European paper	A simple map with the Mekong River flowing near the western edge and the Mainmain Kiaung (apparently the Song-koi or Red River flowing into Tonkin), with two tributaries, flowing through the center of the map. Boundaries of several tribal areas with one another and with China, Tonkin, and Burma shown by dashed lines. Twelve radiating dashed lines connect the central town of Kiaunroungye (not identified) and fifteen surrounding towns with travel time to same indicated along each line (ranging from three to twelve days).	Hamilton, "Account of a Map of the Tarout Shan Territory," *EPJ* 7 (1822): 71–75 and pl. III.
Northwest	Ink on European paper	A rather simple map of an area within several days' travel time from Pagan. Pagan shown by rectangle and seventeen other towns by circles. Irrawaddy River and several tributary streams delineated. Two hill ranges and two isolated hills pictorially shown, as are six temples at varying distances from Pagan.	Hamilton, "An Account of a Map of the Vicinity of Paukgan, or Pagan," *EPJ* 7 (1822): 230–39 and pl. IV.
Varies	Originally drawn on black (Burmese) paper with a steatite pencil, then copied onto European paper by a "Mahommedan painter"	See text and figure 18.6	Hamilton, "Account of a Map Drawn by a Native of Dawae or Tavay," *EPJ* 9 (1823): 228–36 and pl. V.
East	As for a	See text and figure 18.4	Hamilton, "Account of Two Maps of Zaenmae or Yangoma," *EPJ* 10 (1823–24): 59–67 and pl. III.
North	As for a	See text and figure 18.5	Hamilton, "Account of Two Maps of Zaenmae or Yangoma."
North	As for a	Shows the two cited rivers by single lines, a line of hills marking the western escarpment of the Shan Plateau, Amarapura (near western edge of map) by a double circle, and twenty-five other towns by circles. Distances between places are shown as in b.	Hamilton, "An Account of a Map of Koshanpri," *EPJ* 10 (1823–24): 246–50 and pl. VIII.

Area Covered	Place Where Map Is Held, Accession Number	Provenance and Date	Dimensions (cm) (h × w)
l. An area focusing on Kengtung in what is now the eastern part of the Shan States plus a small part of Laos	NAI, HMF 157, no. 27	As for a	27 × 38
m. Much of southern Burma, comprising the former kingdom of Pegu	Location of original not known	As for a	Not known for original; published map is 10.1 × 12.3
n. A very large part of Burma from the Gulf of Martaban to somewhat north of Bhamo	NAI, HMF 157, no. 12	Burma, 1795	120 × 49
o. A portion of Lower Burma from Toungoo in the north to Tavoy (on the Tenasserim coast) in the south	NAI, HMF 157, no. 15	Burma, 1795, apparently by the same native of Taungoo as did b and c	59.8 × 47.5
p. The coast of Burma from Arakan to the Gulf of Martaban	NAI, HMF 157, no. 18	Burma, 1795, by a native of Tavoy on the Tenasserim coast	62.4 × 53.2

(*continued*)

Orientation	Medium	Description (unless otherwise noted, maps are translated from Burmese into English)	Published Accounts
East	As for a	"Kiaintoun" (Kengtung) town, shown by double square, twelve other towns by single squares, and twelve other settlements by circles; all but four places are linked to Kiaintoun by fifteen straight dashed lines radiating from that place, with travel times to same (up to twelve days) indicated as on b. About ten ranges of hills are more or less naturalistically depicted in frontal elevation.	Hamilton, "Account of a Map of Upper Laos, or the Territory of the Lowa Shan," *Edinburgh Journal of Science* 1 (1824): 71–73 and pl. II.
North	As for a	Irrawaddy and deltaic distributaries and associated lakes shown in some detail; Salween shown near eastern border of map; two sandy coastal forelands indicated; several hill ranges depicted in manner of map a. Fourteen named settlements shown by small circles, eight other named places by an *x*, and seven named pagodas by drawings in frontal elevation. Hamilton comments that the map author's reliability for this area is much less than for areas to the north.	Hamilton, "Account of a Map of the Kingdom of Pegu," *Edinburgh Journal of Science* 1 (1824): 267–74 and pl. X.
North	Black ink on European paper	Very detailed map. Emphasis is on travel time (in days) or distance (in leagues) between settlements, shown in a three-tier hierarchy (double squares, single squares, and circles). Pagodas, rivers, and hill and mountain ranges are also depicted.	None known
West (?)	Black ink on paper, with added pencil notes. Now backed by cloth and laminated.	Very similar in conception to b1, described in text. Depiction of vegetation is especially rich, and in three places the words "teak woods" are added.	None; but discussed in Hamilton's journal (note on map reads "no. 22, Journal p. 173").[4]
North	Black ink on paper	Details suggest an attempt to provide the sort of information needed for a hydrographic chart. Features recognizable by special signs include rocks (coasts?), shoals, an anchorage, a customs house, and a few pagodas. Estuarine indentations are shown in exaggerated size, and places are identified by the creeks on which they are situated. The eastern half of a crude compass rose is shown along the left margin of the map. Rivers are shown by wavy double lines and the Arakan range, parallel to the coast of the same name, by a thin zigzag line. Settlements are sometimes shown by circles and sometimes with no sign next to name. Relatively few inland features are plotted. Map is translated from Burmese into English, with six Burmese words in the upper right corner.	None, but discussed in Hamilton's journal (note on map reads "no. 27. See Journal p. 257").

4. The many volumes of Hamilton's journals are held at the Oriental and India Office Collections, British Library, London. Those that relate to his time in Burma form two volumes.

Area Covered	Place Where Map Is Held, Accession Number	Provenance and Date	Dimensions (cm) (h × w)
q. Coastal regions of Burma to north and northeast of the Gulf of Martaban	NAI, HMF 157, no. 19	Burma, 1795	60 × 48
r. An area of Burma with Amarapura at the north, Prome in the southwest, and Taungoo in the southeast	NAI, HMF 157, no. 20	Burma, 1795	75 × 27.5
s. "Lawa Yain or Wild Lawa," an area of the Eastern Shan States and/or northern Thailand	NAI, HMF 157, no. 23	Burma, 1795	38 × 27
t. An area of Burma focusing on, but mainly to the northeast of, Amarapura	NAI, HMF 157, no. 26	Burma, 1795	46.5 × 34.5
u. An area of the Shan States to the east of Amarapura	NAI, HMF 157, no. 28	Burma, 1795	37 × 27
v. Area of Burma and Thailand from Cape Negrais in the southwest to "Saymmay" (Chiang Mai) in the northeast	NAI, HMF 157, no. 29	Burma, 1795	38 × 54.5
w. Area around Gulf of Martaban, including whole of Irrawaddy delta and Tenasserim coast south to "Breit" (Margui)	NAI, HMF 157, no. 30	Burma, 1795	54.5 × 38
x. Most of what is now Thailand and some adjacent areas, seemingly including much of Laos and Cambodia	NAI, HMF 157, no. 35	Burma, 1795, with annotations, most likely by Hamilton	37 × 27
y. An area centering on the Mekong River in Cambodia, Laos, and Thailand, and seemingly extending southeast to Saigon ("Sagun")	NAI, HMF 157, no. 36	Burma, 1795	37 × 27

(*continued*)

Orientation	Medium	Description (unless otherwise noted, maps are translated from Burmese into English)	Published Accounts
North	Black ink on European paper	Similar to p but with more detail on inland villages and some variations in style and cartographic signs.	None known
North	Black ink on European paper	Shows mainly settlement, arranged in a four-tier hierarchy from "Royal City" (Amarapura), to small villages, with travel times (in days) or distances (in leagues) along dashed lines connecting settlements	None known
North	Black ink on European paper	Sixty-nine settlements and/or pagodas are depicted and named. Settlements are shown by squares and circles, pagodas are distinctively drawn in frontal elevation. Several north-south trending mountain ranges are more or less naturalistically depicted, but in a style that is neither obviously Burman nor European.	None known
North	Black ink on European paper	Very similar to j	None known
North	Black ink on European paper	Very similar to t, but at a larger scale	None known
North	Black ink on European paper	Similar to j but very sketchily and crudely drawn, especially in its eastern third. Malay Peninsula omitted.	None known
North	Black ink and gray wash on paper	Emphasis is on waterways, those that are presumed to be navigable being emphasized by a gray wash. Approximately fifty towns, six pagodas, and several mountain ranges are also shown.	None known
North	Black ink on European paper	Map covers a very large area and provides names of numerous countries, regions (e.g., Korat), and towns. Although the Mekong is not shown, what might be the Annamite Cordillera is. Many travel times (in days) between named places are given. An added note reads, "A boat from Sammay [Chiang Mai] to Siam [Ayutthaya] 7 days."	None known, but described in Hamilton's journal (note on map reads "no. 114, Journal p. 169").
North	Black ink on European paper	Shows Mekong and some tributary streams including the Bassac, providing a connection to Tonle Sap. Towns are differentiated by size, the largest being Zandapure (in the center of the map, probably Phnom Penh), Sagun (presumably Saigon), and Mainlaung (in the north, not identified). Numerous notations of travel time (in days) between principal places. Zandapure is said to be thirty-three days from Saymmay (Chiang Mai) in the northwest. A vaguely depicted boundary of Zandapure (Cambodia) is indicated.	None known

APPENDIX 18.1

Area Covered	Place Where Map Is Held, Accession Number	Provenance and Date	Dimensions (cm) (h × w)
z. Cambodia and southern Vietnam	NAI, HMF 157, no. 37	Burma, 1795, with annotations by Hamilton	37 × 27

APPENDIX 18.2 BURMESE AND SHAN REGIONAL

Area Covered	Place Where Map Is Held, Accession Number	Provenance and Date	Dimensions (cm) (h × w)
a. Manipur, India, and an adjacent area of Burma	Royal Geographical Society, London, Burma S. 59	Burma, between 1759 and 1837	Approximately 203 × 284
b. "Maing Tsait and Maing Pone"; Maing Pone appears to be Möng Pawn, on the Pawn River in the Southern Shan States	Oriental and India Office Collections,[1] British Library, London, acquired in 1907 (with f and g below)	Shan States, 1183 B.E. (A.D. 1821)	173 × 133
c. Almost identical to a	Royal Commonwealth Society Library,[2] London, box XV, fol. 9, map C	Burma, presumably copied from map a between 1829 and 1837	34 × 47
d. Eastern Shan States (?)	RCSL, box XV, fol. 9, map iv	Burma, probably early nineteenth century (for original); copied between 1829 and 1832	Original "12 feet by 7 feet"; Burney's copy 74.9 × 50.2
e. "Territory to East of Karenni"	RCSL, box XV, fol. 9, map v	Ava or northern Thailand, for original, presumably early nineteenth century; copied between 1831 and 1837 at Ava	Dimensions of original not known; copy is 38 × 49

1. Hereafter abbreviated OIOC.
2. Hereafter abbreviated RCSL.

(*continued*)

Orientation	Medium	Description (unless otherwise noted, maps are translated from Burmese into English)	Published Accounts
North	Black ink on European paper	Among the features shown are the South China Sea ("Kio Bain"), the Annamite Cordillera (graphically rendered along the Vietnamese coast, but not named), a hill range in the southwest (Elephant range?), and what appear to be Saigon ("Sankaung") and Phnom Penh ("Pyayn Zouk"). Notations relative to travel as in y. Travel time from "Siam or Yoodnya" (Ayutthaya) to Phnom Penh totals nineteen days.	None known

Maps of the Period up to 1885

Orientation	Medium	Language and Script, with Description	Published Accounts
Varies	Painted in numerous colors on cloth	Burmese with some Shan toponyms. See text and plate 37.	None known
Varies	Ink and watercolor (more than five colors) on cloth	Shan in Burmese script. Similar in style to f. Area apparently covers two Shan chiefdoms. Two towns, approximately ninety villages, and seven pagodas are shown. All hill ranges are shown parallel to edge of map as in a.	None known
Varies	Black and red ink, plus wash in several colors, on European paper. Original painted on cloth.	Burmese and Shan translated into English. See text and figure 18.7.	Thaung Blackmore, *Catalogue of the Burney Parabaiks in the India Office Library* (London: British Library, 1985), 117.
Not known	Copy is in ink on European paper; original presumably on cloth	In English, copied from Burmese or Shan original. Described as a large sketch map, the map bears a note that it was "copied from a large map secretly brought to Lt. Col. Burney from the Palace of the King of Ava."	Blackmore, *Burney Parabaiks*, 117.
Not known	Ink on European paper	Shan (presumably), translated into English. Map bears note, "From a Map belonging to a Widow of King Mrudaragyee, a native of Zenmay [Chiang Mai] said to be 42 years of age. Ava, 30 Jan 1831." No other details are available.	Blackmore, *Burney Parabaiks*, 117.

Area Covered	Place Where Map Is Held, Accession Number	Provenance and Date	Dimensions (cm) (h × w)
f. Area between Salween and Mekong rivers in Eastern Shan States and northern Thailand, from about 18°30′ to 21°N and from about 98°30′ to 100°15′E	OIOC, Map Division, R & L 196/07, acquired in 1907 (with b and g)	Shan States, 1850s or 1860s; acquired by British before 1887 and transmitted to OIOC in 1907	292 × 172
g. Large area astride Salween River in Shan States of Burma and northern Thailand from about 18°90′ to 20°20′N and from about 98° to 99°45′E	OIOC, acquired in 1907 (with b and f above)	Shan States, 1223 B.E. (A.D. 1861)	299 × 275
h. Area of Burma to east of Mandalay, between Irrawaddy and Salween Rivers; north-south extent not determined	British Library, London (Or. T.C.I.d), Or. 3478, no. 1, part of Phayre Collection	Burma, pre-1867	147 × 165
i. Roughly the same area as g	British Library (Or. T.C.I.d), Or. 3478, no. 2, part of Phayre Collection	Burma, pre-1867	95 × 91
j. Roughly the same area as g	British Library (Or. T.C.I.d), Or. 3478, no. 6, part of Phayre Collection	Burma, pre-1867	93 × 63
k. Country between the Moulmein on the Tenasserim coast of Burma and Chiang Mai in northwestern Thailand	National Archives of India, New Delhi, Historical Map Folio[3] 90, no. 20	Burma, probably Moulmein, 1871. Authors are identified as Tsayafa and Ko Shong Kho.	112.5 × 71.5
l. Country between Moulmein or Tenasserim coast of Burma and "Zinmay" (Chiang Mai) in northwest Thailand	NAI, HMF 90, no. 19	Moulmein, Burma, "1871?"	40.8 × 34.3

3. Hereafter abbreviated NAI, HMF.

(continued)

Orientation	Medium	Language and Script, with Description	Published Accounts
Varies	Black ink on white paper	Shan, written in Burmese script. See text.	None known
Varies	Painted in various watercolors on cloth	Burmese. See text and figure 18.8.	None known
East	Painted in numerous colors on European paper	Burmese. Similar to f and g but more crudely executed. Mountains shown mainly in parallel rows, rivers by double wavy lines. Two-level settlement hierarchy, forts, pagodas, and roads depicted.	None known
East	Painted in red, green, and black on European paper	Burmese. Similar to h, but with a somewhat more European appearance. Settlement hierarchy in three levels	None known
East	Black ink on European paper	Burmese. A simple sketch map. Rivers shown by single lines, roads by dotted lines, settlements by small circles, mountains by rows of commas (sometimes joined by a line)	None known
North	Red, brown, and blue ink and blue, brown, and yellow watercolor on paper	Burmese with English translations added. English title added to map reads, "Map composed by Tsaya Pai & Ko Shong Kho of the District between Moulmein and Zimmay (Original)." Rivers outlined in blue ink, with blue watercolor between banks. Mountains outlined with caterpillar hatch marks. Very detailed settlement hierarchy, with gates and walls of towns shown, all in brown ink. Villages shown by circles with the word *ywa* (village) therein and no name. Roads are in heavy solid brown lines. Although this map was obviously made at the behest of the British, it retains enough of a Burmese flavor to warrant notice.	None known
East (?)	Red ink and black pencil on paper	Burmese with English translations added in pencil. Though this map appears to have been made at the behest of British interested in prospects for forestry, the style is Burmese. Rivers are shown by double lines, trails by dotted lines, about a dozen settlements by squares, all with the word *ywa* (village) therein, and what are probably meant to signify forests by wispy lines in pencil. This may be a much simplified adaptation of k.	None known

APPENDIX 18.2

Area Covered	Place Where Map Is Held, Accession Number	Provenance and Date	Dimensions (cm) (h × w)
m. Toungoo and Yamethin districts of Burma, an area approximately 130 km north-south × 65 km east-west, astride the border established following the second Anglo-Burmese War	Collection of U Maung Maung Tin, Mandalay	Burma, nineteenth century, assumed to be between 1870 and 1885	Approximately 75 × 160
n. An east-west strip across Burma between the Arakan range and the Red Karen tribal area, an area approximately 122 km east-west × 25 km north-south	Collection of U Maung Maung Tin, Mandalay; obtained from Yethaphan monastery	Burma, between 1857 and 1885	Approximately 140 × 55.5
o. Burma from the Bay of Bengal to the Irrawaddy River and from Mindon town (about 19°20′N) to Sale town (about 70°50′N)	Transferred in 1978 from Yethaphan monastery to the collection of U Maung Maung Tin, Mandalay	Burma, allegedly before 1824 (before first Anglo-Burmese War), but more likely between 1870 and 1885 because of similarity to m	Approximately 131 × 109
p. Pakokku District, on west of Irrawaddy in Upper Burma	University of Rangoon Library, P/26144	Relatively recent copy of a late nineteenth-century original	101.5 × 120.5
q. Kyauk Ye District astride Chindwin River in northern part of Upper Burma, between 22° and 26°N	Collection of U Maung Maung Tin, Mandalay	Burma, late (?) nineteenth century	Approximately 37.5 × 154.5
r. Auntgyi Kin Chaung Forest; location not known	Copied from a map in the collection of Than Tun, Tokyo; copy in collection of U Maung Maung Tin, Mandalay	Burma, date not known	Approximately 14 × 119

(continued)

Orientation	Medium	Language and Script, with Description	Published Accounts
East	Painted in red, green, white, and yellow on black gridded *parabaik* and stored between wooden staves	Burmese. An exceptionally detailed map very similar to o and very likely prepared for military intelligence in anticipation of further hostilities between the second and third Anglo-Burmese Wars (1852–85). Relevant signs identified include "outpost for a town," "outpost for a village," "British forces at a town (Garrison)," "British force [other]," "border pillar," "path" (in two categories). Additional details on settlement, drainage, hills, forests, etc., are also abundant, as are religious signs (for pagodas, tumuli, spirit houses, and rest houses).	None known
East	Painted in black, red, green, and yellow on white gridded *parabaik* in seven panels	Burmese. Described as "Map for posting [military] outposts between the bank of the Western Yoma [Arakan Range] and Red Karen Area. . . ." Though less detailed than m, this map uses the same set of signs for features with military references.	None known
West	Green, red, and yellow on black *parabaik*, bearing a yellow square grid (with intervals of approximately 1.9 cm)	Burmese. See figure 18.9 and text.	None known
West	Red and black ink and green watercolor on cloth	Burmese. A general map with naturalistic signs (rivers with wavy lines therein, forested hills in frontal elevation, etc.). Villages named within oblong cartouches. Miscellaneous notes (e.g., "Many crops can be grown here") in several parts of map.	None known
East	White talc on black *parabaik* of nine panels	Burmese. Rather simple sketch map with several rivers, two parallel mountain chains, and four towns depicted.	None known
Not known	Original was a black *parabaik* in fifteen panels with details in white talc and yellow and red paint	Burmese. Forest paths and boundaries shown, as are villages, pagodas, streams, and hills, all in characteristic late nineteenth-century fashion. Area depicted appears to be rather small.	None known

Area Covered	Place Where Map Is Held, Accession Number	Provenance and Date	Dimensions (cm) (h × w)
a. The route between Amarapura and Taraek ("Tartary"), more specifically to the Chinese emperor's hunting seat at Jehol (Chengde), beyond Beijing	Location of original not known	Given to Captain Symes at Amarapura in 1795 by the Zabua (prince) of Bhamo, near the Chinese border northeast of Ava	Long, narrow map. Original dimensions not known; 30.4 × 9.0 map area in printed form.
b. Routes from Bassein ("Pathein") to Bangkok ("Banteouk")	National Archives of India, New Delhi, Historical Map Folio[1] 157, no. 21	Burma, 1795	33 × 48
c. Much of Indochinese peninsula, with Chiang Mai, Toungoo, and Martaban in the northwest and Phnom Penh (?) and Saigon (?) in the southeast	NAI, HMF 157, no. 34	Burma, 1795, with annotations, presumably by Hamilton	27 × 37
d. Lower Irrawaddy River	In possession of U Maung Maung Tin, Mandalay	Burma, uncertain date, presumably predates British occupation of 1852	Not available; *parabaik* (Burmese paper) of many folds
e. Area along the Shweli River in the Shan States astride the Burma-China border	British Library (Or. T.C.I.d), Or. 3478, no. 5, part of the Phayre Collection	Burma, pre-1867	180 × 49

1. Hereafter abbreviated NAI, HMF.

Orientation	Medium	Language and Script, with Description	Published Accounts
Roughly north for most of the map and northeast for the portion relating to China	Presumably on European paper	Translated from Burmese to English. The original probably drawn by an officer of the Zabua of Bhamo at the request of Francis Hamilton. Towns, including twenty-five in Burma and only ten in China, are shown by rectangles (with double outlines for important places), and rivers are of varying widths, some being only noted (e.g., "4 small rivers") and not drawn. Traveling times between towns are given in days. Three-fifths of the length of the map is taken up for the distance from "Shue Prido" (Amarapura) to Bhamo, a distance with stages adding up to 14 days of travel and the remaining two-fifths almost entirely in China, for a distance requiring 121 days of travel. The route from Bhamo to Tarek Pri (Jehol) is shown by a double line along part of which a note "10 days by camel" appears and, farther north, "carriage road." Between Udhin Pri (Beijing) and Jehol is a wavy line and the note "mountains," signifying "those along which the great wall is built." No other physical features are depicted.	Francis Hamilton, "Account of the Route between Tartary and Amarapura, by an Ambassador from the Court of Ava to the Emperor of China," *Edinburgh Philosophical Journal* 3 (1820): 32–42 and pl. I.
North	Presumably on European paper	Translated from Burmese to English. A single route, via Rangoon, is shown between Bassein and Martaban ("Monttama"), but thence two routes go to Bangkok, one more directly (presumably by Three Pagodas Pass and the Kwai River valley), and the other via Tavoy ("Davoy") and Mergui ("Byeit") on the Tenasserim coast. Distances along various stages are given either in days of travel or in leagues. What appears to be the Burmese-Siamese border is indicated by a dashed line. The Malay Peninsula below Mergui is omitted as on map a in appendix 18.1.	None known
Northeast	Presumably on European paper	Translated from Burmese into English, with pencil notations added. Map focuses on Ayutthaya ("Ayoitaya"). The easternmost city, thirty-five days distant from Ayutthaya, is "Tyeinseen," next to which "prob. Saigon" appears in pencil. The map is very crude and provides little information other than travel times between a dozen important travel nodes.	None known
Varies; features point away from river	White *parabaik* bleached with chalk; details in black, green, and red	Burmese. On gridded paper. Rivers very naturalistically rendered. Main streams colored in green wash, tributaries (shown only for short distances) not colored. Other features outlined in red or black. Text in black.	None known
Not determined	Painted on European paper	Language not determined, either Shan or Burmese. Highly pictographic map at what seems to be a very large scale. Few toponyms. Fish in all the rivers, boat with two fishermen in the Shweli River.	None known

APPENDIX 18.3

Area Covered	Place Where Map Is Held, Accession Number	Provenance and Date	Dimensions (cm) (h × w)
f. The Irrawaddy River below Pagan	British Library (Or. T.C.I.d), Or. 3478, no. 7, part of the Phayre Collection	Burma, pre-1867	189 × 31
g. The route from Papun ("Paphoon") in Burma to Chiang Mai ("Zimmay") in northwest Thailand	NAI, HMF 90, no. 12	Burma, 1870	40.4 × 35.8
h. Area from Bangkok and Chiang Mai east to Phnom Penh and Saigon	NAI, HMF 91, no. 14	Burma, 1232 B.E./A.D. 1871, author identified as U Yit	64.5 × 122.5
i. Telegraph line from Nyaungu or the Irrawaddy (near the then border of British Burma) to Mandalay	In possession of U Maung Maung Tin, Mandalay, "Parabaik no. 191"	Burma, between 1860 and 1880	Thirty-one panels, each panel approximately 37.5 × 12.8; total length is 3.97 m

APPENDIX 18.4 BURMESE MAPS OF

Area Covered	Place Where Map Is Held, Accession Number	Provenance and Date	Dimensions (cm) (h × w)
a. Amarapura and its environs, an area estimated as 80 × 65 km	National Archives of India, New Delhi, Historical Map Folio[1] 157, no. 17	Burma, 1795	60 × 48
b. Area focusing on a large lake (Yemyet?) and another smaller lake to the north of Amarapura, covering an area estimated as 75 × 45 km	NAI, HMF 157, no. 31	Burma, 1795	42 × 27
c. Approximately the same as b	NAI, HMF 157, no. 32	Burma, 1795	47.5 × 30
d. An area to the southeast of the great bend of the Irrawaddy River at Amarapura, estimated as 10 × 50 km	NAI, HMF 157, no. 33	Burma, 1795	47.5 × 30
e. "Mhineloonghee Forest," north of an east-west stretch of the Salween River, perhaps in the Shan States	Royal Geographical Society, Burma S 29, no. 34. Also NAI, HMF 90, no. 11.	Copied in Calcutta in June 1871 from a "native map in the possession of Messrs. Todd Findlay & Co., Moulmein"	82.9 × 68.2

1. Hereafter abbreviated NAI, HMF.

(*continued*)

Orientation	Medium	Language and Script, with Description	Published Accounts
North (?)	Ink (?) on European paper	Burmese. A simple sketch map. Streams are shown by double lines, settlements (almost exclusively along the river) by small circles, district boundaries by dotted lines; boundary (stakes?) are indicated.	None known
West (?)	Black ink on European paper, two sheets pasted together	Burmese text with English translation added later. Titled "Rough sketch made by a forester during a discussion of some of the other maps forwarded with this. It shows the route from Paphoon to Zimmay (original)." ("Other maps" are presumably items j and k of appendix 18.2.) Rivers, three towns (in squares), two other localities (in ovals), and a trail (dashed line) are depicted.	None known
East	Brown ink and brown, blue, and yellow watercolor on paper	See text and figure 18.29	None known
Varies; features point away from the Irrawaddy	White ink and red chalk on folding black *parabaik*	Burmese. See text and figure 18.32. Abundant detail in addition to telegraph line itself, such as pagodas and other prominent buildings along the route, drainage levees along the Irrawaddy, and hills distant from Irrawaddy.	None known

Primarily Rural Localities

Orientation	Medium	Language and Script, with Description	Published Accounts
North	Ink on European paper	Burmese translated into English. Emphasis is on settlement. Towns are shown by squares and named; a great many villages are shown by circles and not named. Pagodas, rivers, lakes (including two large ones in the northwest), and hill ranges are depicted.	None known
North	Ink on European paper	Burmese translated into English. Emphasis is on travel times between settlements.	None known
North	Ink on European paper	Burmese translated into English. Emphasis is on travel times between settlements.	None known
North	Ink on European paper	Burmese translated into English. Emphasis is on drainage pattern. What appear to be some remarkable examples of stream capture are depicted. Western escarpment of Shan Plateau is shown as a range of hills.	None known
North	Black ink on European paper	Burmese. Individual portions of the forest are named. Rivers very prominently shown by double lines up to 4.5 cm apart. Rocky barrier in river pictographically shown. Mountains individually depicted (no two alike). One settlement has individual houses on piles shown in a ring in frontal elevation.	None known

Area Covered	Place Where Map Is Held, Accession Number	Provenance and Date	Dimensions (cm) (h × w)
f (i–iv). Several villages, location not determined	Rangoon University Library, MS. 9108	Upper Burma, late nineteenth century, presumably pre-1885	(i) 52 × 81.6; (ii) 26.5 × 52; (iii) 52 × 141.5; (iv) 52 × 61. Each pleat of the *parabaik* measures 52 × 20.5.
g. A portion of Shwebo District in Upper Burma	Not known	Burma, 1881; map was executed on the order of King Thitand	Scale is four times that of reproduction in source cited
h. A portion of Kyaukse District in Upper Burma	Not known	Burma, presumably ca. 1881 by association with g	Not known

(continued)

Orientation	Medium	Language and Script, with Description	Published Accounts
East on i, ii, iv; North on iii. Cardinal directions are marked on the edges of each map.	Red, two greens, brown, blue, yellow, and white watercolor, plus ink, on *parabaik*	Burmese. See figure 18.33 and text. These are essentially cadastral maps relating to the lands in several villages that were inherited by a minister, Mahamingyaw Raza, who presumably ordered the survey from which the maps were made. Since map iv is clearly not finished, one cannot be sure about the total area over which the survey was intended to extend; but six empty pleats in the *parabaik* suggests that the survey was less than three-fourths complete. Notations on maps state which parcels of land belong to the minister and which to other persons. Numbers written in most fields indicate how many bundles of rice could be planted there, while other fields are identified as seedling nurseries. Field boundaries are prominently depicted in green, but in some large fields a note states that individual divisions are not shown. Fields devoted to crops other than rice are noted. Distinctive signs exist for earth embankments, tanks and ponds, roads, etc. Built-up areas of villages are depicted by a few houses, more or less realistically drawn in oblique perspective. A great variety of vegetation, including palms, broadleaf trees, tall herbaceous plants, etc., is shown over many parts of the maps. A naturalistically rendered hill range forms the northern margin of map iii.	None known
Not known	*Parabaik*; media otherwise not known	Burmese. "It shows the irrigation system of the Shwebo Myinnè [modern equivalent unknown] the Ma embankment and the Mahananda Tank." The original gives the names of some eighty pipes. It also shows villages and village jurisdictions.	A. Williamson, comp., *Burma Gazetteer, Shwebo District*, vol. A (Rangoon: Superintendent, Government Printing and Stationery, 1929), reference on 54 and copy of map in cover pocket.
Not known	*Parabaik*; media otherwise not known	Burmese. Presumably similar to g.	Ralph Neild, H. F. Searle, and J. A. Steward, *Burma Gazetteer, Kyaukse District*, vol. A (Rangoon: Government Printing and Stationery, 1925), reference to irrigation technology (pre-British) on 72–74.

APPENDIX 18.4

Area Covered	Place Where Map Is Held, Accession Number	Provenance and Date	Dimensions (cm) (h × w)
i. A small rural area, presumably in Upper Burma	Collection of U Maung Maung Tin, Mandalay	Burma, pre-1885	Not available
j. A part of Yesagyo District of Upper Burma astride the Chindwin River not far from its confluence with the Irrawaddy	Collection of U Maung Maung Tin, Mandalay	Burma, probably latter half of the nineteenth century	Approximately 93 × 105

APPENDIX 18.5 BURMESE MAPS OF

Area Covered	Place Where Map Is Held, Accession Number	Provenance and Date	Dimensions (cm) (h × w)
a. The former Thai capital of Ayutthaya	National Archives, Rangoon	Burma, ca. 1767	Approximately 87.6 × 356; map consists of twenty panels, each measuring 87.6 × 17.8 cm
b. Environs of Pagan	British Library, Oriental Manuscripts and Printed Books, Add. MS. 18069	Burma, pre-1850, since it was acquired by the British Library in February of that year	One dimension is 44 cm, the other is not known
c. Amarapura and environs, including Sagaing and outskirts of Mandalay	Original held by Than Tun, Tokyo University. Copy is at Taung Lay None monastery, Amarapura.	Amarapura, ca. 1850 for original; copy from 1970s, hand drawn and colored by six monks	93.7 × 105.7 for copy (and presumably also for original)
d. Amarapura and environs, about 9 × 6 km	Original is said to be in London, but its present location is not known. A duplicate was sent from London by U Tet Htut to U Maung Maung Tin in Mandalay in 1962.	Allegedly eighteenth century, but more likely mid-nineteenth century (Amarapura was the capital of Burma from 1783 to 1823 and from 1837 to 1857)	51.4 × 48.9

(*continued*)

Orientation	Medium	Language and Script, with Description	Published Accounts
Not known	White and various colors, including red, on black *parabaik*	Burmese. Map bears inscription, "Notes of the Surveyor Nga Thein," in Burmese. It is essentially cadastral and shows lands distributed to the royal prince, princess, and officials; notes grain and vegetable crops grown on same; and indicates fields growing two or more crops per year. Field boundaries are shown by white lines punctuated with red dots. What appear to be a reservoir, canals, roads, and a range of hills are also shown.	None known
East	Painted black, red, and blue-green on cloth	Burmese. Rather large scale. Several rivers shown, in varying widths, in blue-green. Same color used for some isolated, forested hills in northeast corner of the map. Nearly sixty villages are named within red rectangles. Script in black.	None known

Primarily Urban Localities

Orientation	Medium	Language and Script, with Description	Published Accounts
Not known	Painted in numerous colors and black ink on white *parabaik*	Burmese. See figure 18.38 and text. The ornateness of the illustration on this map belies its essentially military intelligence function.	None known
Not known	White steatite pencil on black *parabaik* of forty folds, each measuring 44 × 17 cm	Burmese; captions in English added to one of the three maps that appear on this *parabaik* manuscript. The largest is of Pagan, showing the city's many temples and the Irrawaddy River for some distance north and south of the city. There are two other small sketch maps of Pagan, but their content has not been ascertained.	None known
West	Original was on *parabaik* with four panels, presumably executed in ink and paint. Copy is on two pieces of European paper pasted together, executed in black, red, blue, violet, brown, and green felt-tipped pens.	Burmese. Features relating to urban settlement in red. These include the old royal city, its walls, palace, and streets; woods and hamlets outside the city, with hamlet names in black or blue oval cartouches; monasteries and pagodas in brown; rivers, canals, ditches, areas of periodic inundation, and two wooden causeways in blue; vegetation, in various forms, naturalistically drawn in frontal elevation in green and violet; text in black. Forested range of Sagaing Hills, shown in frontal elevation near top (western) edge of map, with clouds in sky just beyond (rendered in a Tibetan style). Colors are said to reflect those of the original.	None known
West	Painted in red, blue, two greens, yellow, and pink, plus ink, on white *parabaik*	Burmese. The description for c is broadly applicable to this map. Among the differences are that here some hamlets are named within oval cartouches, while others are given irregular, though more or less ovoid, shapes; the Sagaing Hills are shown with less forest cover and often with pagodas on the summit, and a side-wheeler steamboat appears in the Irrawaddy River (if this last feature appeared in the original, that would rule out any date before the mid-nineteenth century).	None known

Area Covered	Place Where Map Is Held, Accession Number	Provenance and Date	Dimensions (cm) (h × w)
e. Mandalay	Burmese Department, University of Mandalay	Amarapura or Mandalay, ca. 1853	Not available
f. Mandalay and its environs, an area about 43 × 27 km	Collection of U Maung Maung Tin, Mandalay; previously held in Sagaing Sin Nin monastery (in the town of Sagaing?)	Mandalay, ca. 1856	Approximately 108 × 54
g. Mandalay and small area to west and south, approximately 18 × 21 km	Pagan Atwin Wan monastery in Mandalay; notes based on a copy held by U Maung Maung Tin, Mandalay	Mandalay, between 1857 and 1866	Approximately 71 × 102
h. Mandalay and its environs, including Amarapura and Ava	Photostat negative at the Library of Congress, Burma, n.d., 1: --. Location of original not known.	Presumably in reign of King Mindon, 1853–78, after he established the capital at Mandalay in 1857	43 × 56
i. A portion of Mandalay said to measure about 890 × 830 m, including the king's apartments within the wooden stockade inside the fort	Location of original not known (said to be in possession of U Maung Maung Kyaw); copy was made by U Maung Maung Tin, Mandalay	Burma; between 1857 and 1885, probably in latter half of that period	Approximately 76 × 56

(continued)

Orientation	Medium	Language and Script, with Description	Published Accounts
Not available	Not available	Burmese. Said to be a development plan.	U Maung Maung Tin and Thomas Owen Morris, "Mindon Min's Development Plan for the Mandalay Area," *Journal of the Burma Research Society* 49, no. 1 (1966): 29–34 with two maps.
East	Painted in white, green, yellow, and red on black *parabaik*	Burmese. Map titled "Map of the Boundaries of the Royal City." Seems to have been intended in part for planning. Few details of the urbanized area are provided, but names relative to the royal gardens are given. There are numerous details of hill features (all shown in frontal elevation) and natural and artificial drainage. About fifteen pagodas are shown, all in frontal elevation, and also several monasteries.	Than Tun, "Mandalay Maps," *Papers of the Upper Burma Writers' Society*, 1966; I have been unable to locate this reference.
East	*Parabaik*, details painted in blue, red, and yellow, text and outlines in ink	Burmese. Emphasis is on the street plan and different quarters of the city; quarters of distinguished people are named. Various channels of the Irrawaddy are shown very naturalistically and in considerable detail. The Sagaing Hills to the west are shown in frontal elevation.	None known
East	Based on the translation of the map legend, the original was in two yellows, red, and blue (and possibly other colors), probably a mixture of ink and paint on paper	Burmese. Very detailed map, essentially cadastral in nature. Drawn on graticule of squares 7.5 mm on a side. General layout of city and palace compound therein with wide surrounding areas, within which are hundreds of uniformly small circles, each with an adjacent name, presumably the owner of the field within which the circle lies. The legend states that one type of yellow indicates fields owned by the king, another yellow field, owned by civil servants, red "inherited" fields, and blue "private" fields. Roads are indicated by dotted lines; drainage is shown in detail, with streams of varying width; hill ranges are depicted naturalistically, but all in uniform height, as if seen in frontal elevation. Other map features include areas of clustered rural settlement, numerous pagodas (in frontal elevation), what might be monasteries (in oblique pictorial perspective), groves of trees (in frontal elevation), irrigation tanks, etc.	None known
Not known	Copy appears to be in black ink on European paper. Media of original, drawn on white *parabaik*, not known.	Burmese. See text and figure 18.40. Features shown include wooden stockade, guard posts, barracks, armory, elephant shed, carriage shed, official apartments, water palace, apartment for storing presents (?), servants' house, kitchen, monastery, rest house for *nats* (Burmese demigods), tower for relic of Buddha's tooth, clock tower, mint, swimming bath, and gardens.	None known

Area Covered	Place Where Map Is Held, Accession Number	Provenance and Date	Dimensions (cm) (h × w)
j. Eastern and southern sides of Mandalay; covers an area of about 6.5 × 5 km	Collection of U Maung Maung Tin, Mandalay; obtained from Lone Taw monastery in the western part of the city	Mandalay, between 1859 and 1885	Approximately 71.5 × 48.5
k. A section of northeastern Mandalay, about 9.5 × 3.5 km	National Library and Museum, Mandalay, acc. no. 143, 25.5–60	Presumably same as j	Approximately 72.5 × 125
l. Plan of Sedawgyi Dam at Lone Taw in western Mandalay	Collection of U Maung Maung Tin, Mandalay; obtained from Lone Taw monastery	Presumably same as j	Approximately 43 × 95
m. Chiang Mai ("Zimmay") and its approaches	National Archives of India, New Delhi, Historical Map fol. 90, no. 17. Copy at the Royal Geographical Society, Thailand S/S2 no. 39.	Northern Thailand or Shan States, 1870, by Sa-ya-pay	27.5 × 20.8
n. A small area east of Mandalay near Yankin Hill	Collection of U Maung Maung Tin, Mandalay	Upper Burma, 1876 (?), by the surveyor Nga Thein	Approximately 42.5 × 60
o. An area on the northern outskirts of Mandalay, an area of about 1.3 × 0.5 km	Collection of U Maung Maung Tin, Mandalay	Mandalay, late nineteenth century	Approximately 114 × 61
p. Golden Temple and environs, Mandalay, including a portion of Mandalay Hill	Not known	Presumably Mandalay, late nineteenth century	Not known

(*continued*)

Orientation	Medium	Language and Script, with Description	Published Accounts
Not known	Painted in yellow, red, and white on black *parabaik* of four panels	Burmese. Described as "Drainage Plan of Eastern and Southern Side of the Royal City." Features relate largely to what the description implies: streams, channels, drainage ditches, moat, and bridges. Other features include the outline of a fort, gateways, streets, and gardens.	None known
East	Painted in red, green, white, and yellow on black *parabaik*	Burmese. Appears to be related to engineering works in respect to drainage. Cartographic signs by and large duplicate those of j. One sign indicates "rod used for surveying." A unique sign on this map is "structure for keeping inscription."	None known
East	White talc and yellow paint on gridded black *parabaik* in six panels	Burmese. Very large-scale representation of the precinct of a dam showing adjacent physical features, partly in planimetric perspective (for drainage), partly in frontal elevation (hills and forest), and partly in oblique perspective (a single house).	None known
North	Blue pencil on European paper, with later addition in ink of English translation of text	Presumably Shan, with English translations subsequently added. A simple outline map of the town, its major urban roads, the palace, six exterior and two interior city gates, and three approach roads.	None known
East	Painted in white, red, green, and brown on gridded black *parabaik*	Burmese. Described as "Note of the Surveyor Nga Thein." Very likely a map of some royal estate, since it includes the lands surrounding a palace. Other features include villages, religious edifices, roads, a canal, streams, and hills.	None known
East	White chalk on black *parabaik*	Burmese. See text and figure 18.41. Includes carefully rendered architectural plans of Kyauktawgyi Pagoda, Sandamuni Pagoda, Pathana Sima (ordination hall), and other religious and royal edifices near the southern base of Mandalay Hill.	None known
Varies	Not known; appears to be painted and drawn in ink on white *parabaik*; number of panels not known	Burmese. Only about one-fourth of the palace is shown in the published illustration cited. A mixture of a planimetric perspective to show general layout of temple and frontal perspectives to show gates, towers, specific architectural features, and Mandalay Hill.	Barbara Nimri Aziz, "Maps and the Mind," *Human Nature* 1, no. 8 (1987): 50–59, illustration on 58–59.

19 · Southeast Asian Nautical Maps

JOSEPH E. SCHWARTZBERG

The diffusion of Austronesian cultures in late prehistoric time extended over a vast, primarily oceanic domain covering more than 208 degrees of longitude, from Madagascar in the west to Easter Island in the east. So wide a distribution presupposes an early and substantial development of maritime technology and a concomitant accumulation of knowledge of the sea and the stars. Some of this knowledge resulted in celestial mapping, examples of which have been noted above (pp. 712–13), and one cannot help but speculate on how far it also gave rise to various forms of nautical charts. The stick charts of Marshall Islanders, of course, have been amply documented; but what, if anything, might have been their cartographic antecedents? Was there a process of transmission from the nearby Malay world? Did Malay seafarers learn to use maps, as Ferrand suggests, in their repeated voyages not only to Madagascar, but at least as far as the Cape of Good Hope, beginning about the first century A.D.?[1] Unfortunately, firm evidence for answering such questions is meager. It is uncontestable, however, that the Portuguese did make use of Javanese charts early in the sixteenth century. A letter dated 1 April 1512, sent by their Indian viceroy Afonso de Albuquerque from the port of Cochin to King Manuel, refers to an accompanying piece of a map made by Francisco Rodrigues, a cartographer and the pilot-major of the armada that discovered the Molucca Islands. That map fragment was described as being from

> a large map of a Javanese pilot, containing the Cape of Good Hope, Portugal and the land of Brazil, the Red Sea and the Sea of Persia, the Clove Islands, the navigation of the Chinese and the *Gores* [inhabitants of Formosa], with their rhumbs and direct routes followed by the ships, and the hinterland, and how the kingdoms border on each other. It seems to me, Sir, that this was the best thing I have ever seen, and Your Highness will be very pleased to see it; it had the names in Javanese writing, but I had with me a Javanese who could read and write. I send this piece to Your Highness, which Francisco Rodrigues traced from the other, in which Your Highness can truly see where the Chinese and *Gores* come from, and the course your ships must take to the Clove Islands, and where the gold mines lie, and the islands of Java and Banda, of nutmeg and maces, and the land of the king of Siam, and also the end of the navigation of the Chinese, the direction it takes, and how they do not navigate farther. The main map was lost in *Frol de la Mar*. With the pilot and Pero de Alpoim I discussed the meaning of this map, in order that they could explain it to Your Highness; you can take this piece of map as a very accurate and ascertained thing, because it is the real navigation, whence they come and whither they return.[2]

Though Ferrand appears to accept the notion, seemingly implicit in Albuquerque's letter, that the Javanese actually knew of places such as Portugal and Brazil, that proposition strikes me as untenable and based on too literal a reading of what was probably a carelessly written original statement. What Albuquerque probably meant to say was that the map in question, essentially a map of the then known world, was based *in part* on a Javanese map. But even that more modest interpretation demonstrates that the Javanese did indeed make maps to aid navigators.

Apart from this map, which went down with the *Frol de la Mar*, Rodrigues drew many others. Twenty-six such maps, based both on direct observation and on a variety of secondhand sources—including, it appears, the lost original Javanese map—were assembled, along with numerous other illustrations, in the form of an atlas titled *O livro de Francisco Rodrigues*, published in 1513, a copy of which is at present held in the Bibliothèque de l'Assemblée Nationale in Paris.[3] The original manuscript

1. Gabriel Ferrand, "A propos d'une carte javanaise du XVe siècle," *Journal Asiatique*, 11th ser., 12 (1918): 158–70. For the history of Malay maritime connections with Madagascar, see Gabriel Ferrand, "Les voyages des Javanais à Madagascar," *Journal Asiatique*, 10th ser., 15 (1910): 281–330.

2. Quoted from the translation of the Portuguese original by Armando Cortesão in *The Suma Oriental of Tomé Pires . . . and The Book of Francisco Rodrigues . . .*, 2 vols. (London: Hakluyt Society, 1944), 1:lxxviii–lxxix.

3. Cortesão briefly describes each of these maps in appendix 2 of *Suma Oriental*, 2:519–26 (note 2); he also discusses them in his *Cartografia e cartógrafos portugueses dos séculos XV e XVI*, 2 vols. (Lisbon, 1935), 2:122–30. Facsimiles of all the maps at the Bibliothèque de l'Assemblée Nationale are found in Manuel Francisco de Barros e

was allegedly held by a well-known Portuguese theologian, philosopher, and historian, Jerónimo Osório (1506–80), and the Paris copy is said to date from sometime after 1520.[4] The Rodrigues atlas is the subject of an analysis by Winter, who discusses the remarkable difference in style between those maps that were based on firsthand observations and those that were not.[5] Figures 19.1 and 19.2 illustrate this difference. Since much has been written about Rodrigues's opus, I shall not discuss it further.

Figure 19.3 presents a map of the Malay Peninsula that was brought to light and analyzed in detail by Phillimore, with the aid of two Malay scholars.[6] As of 1956 the map was held by S. T. C. Parsons-Smith, having been passed down through his mother's family, the Cracrofts, one of whom probably obtained it while taking a sea voyage from India, where a number of them had been in service. The work is thought to date from the early eighteenth century. Many natural features on the map are highlighted with an indigo tint. Also highlighted are the hulls of the several fully rigged European sailing ships, whose style helps date the work. The remarkably clear script used on the map suggests that its author was an educated Malay.

The obvious contortion in the alignment of the peninsula is somewhat surprising, given that the cardinal and secondary directions are all noted on the map by straight lines and in words as well. Of this Phillimore observes:

> The southern half of the peninsula has, indeed, been so twisted round to the east . . . that the general direction of the east coast is made to run but slightly north of west, till it narrows up to the Isthmus of Kra. From Kra it then sweeps away to the east to suggest the confines of the Gulf of Siam. It is as if the map-maker found that his paper was not long enough from north to south.
>
> The disregard of more than four degrees of latitude between Singapore and Penang is hardly consistent with the map-maker being a professional sailor, skilled in the art of the astrolabe, and yet, on the other hand, he seems entirely taken up with coastal details, of interest to a navigator, and to be particularly interested in the safe harbour of Patani. . . . Most of the names that appear on the map are those of river mouths (*kuala*), promontories (*tanjong*), islands (*pulau*), or hills (*bukit*) that might be valuable aids to navigation.[7]

A particularly noteworthy map feature is Bukit Pattani, the prominently drawn hill from which the eight directional lines radiate. Also identified are near-coastal shoals and a number of features that have no direct bearing on navigation, such as three areas of rice cultivation, areas near Johor and Pattani said to be plains, the boundary of Pattani (noted by text, but not drawn), and an island off the "frontier" of Pattani. The frequent mention of places associated with Pattani, the fact that no other boundaries or frontiers are noted, and the exaggerated size of the Pattani Point, harbor, and region suggest (though Phillimore was silent on this point) that the mapmaker was a resident of that place. Until its conquest by Siam in 1785, Pattani was the capital of an independent Malay sultanate that carried on a vigorous trade with both the Dutch and the British, and before them with the Portuguese, as well as with ships from China and Japan. I assume further that the mapmaker, or perhaps his employer, was himself engaged in frequent trading contacts with Europeans; first, because the only ships shown on his chart are European, and second, because he inserts into the map the eight directional lines, which are essentially a pointless Western affectation that adds little or nothing to the utility of the work. Finally, I hypothesize, for reasons to be developed below, that the presumed merchant used Pattani as an entrepôt, gathering spices, rice, and other goods in small consignments, carried mainly by small native craft plying coastal routes from many of the other ports shown on the map, and then reselling those goods to the Europeans and other nonlocal traders.

However inaccurately plotted the details on the map may seem, they are all useful and correctly disposed in their sequence along the coast or, in the case of offshore features, in respect to proximate peninsular features. The map provides little information, however, on areas away from the coast. Of this Phillimore observes:

> A remarkable disregard, or ignorance, of the internal geography is displayed by the joining up of the Pahang and Perak rivers to continue right through the heart of the peninsula from coast to coast, as if the map-maker had no conception of the formidable mountain mass of the Cameron Highlands that rise to peaks of over 2000 metres. There are no indications of national or political boundaries [not quite correct, as I have demonstrated with respect to Pattani] or subdivisions. No town sites or forts, and no land communications.[8]

A question that Phillimore fails to address, however, is why the mapmaker would have chosen to indicate this

---

Sousa, Viscount of Santarém, *Atlas composé de mappemondes, de portulans et de cartes hydrographiques et historiques depuis le VIe jusqu'au XVIIe siècle*, 3 vols. (Paris, 1849); facsimile ed., *Atlas de Santarem*, with explanatory text by Helen Wallis and A. H. Sijmons (Amsterdam: R. Muller, 1985).

4. Hiroshi Nakamura, *East Asia in Old Maps* (Tokyo: Centre for East Asian Cultural Studies, 1962), 28–35 and fig. 8. Nakamura ventures no opinion as to the present whereabouts of the original.

5. Heinrich Winter, "Francisco Rodrigues' Atlas of ca. 1513," *Imago Mundi* 6 (1949): 20–26.

6. Reginald Henry Phillimore, "An Early Map of the Malay Peninsula," *Imago Mundi* 13 (1956): 174–79.

7. Phillimore, "Early Map," 178 (note 6).

8. Phillimore, "Early Map," 178 (note 6).

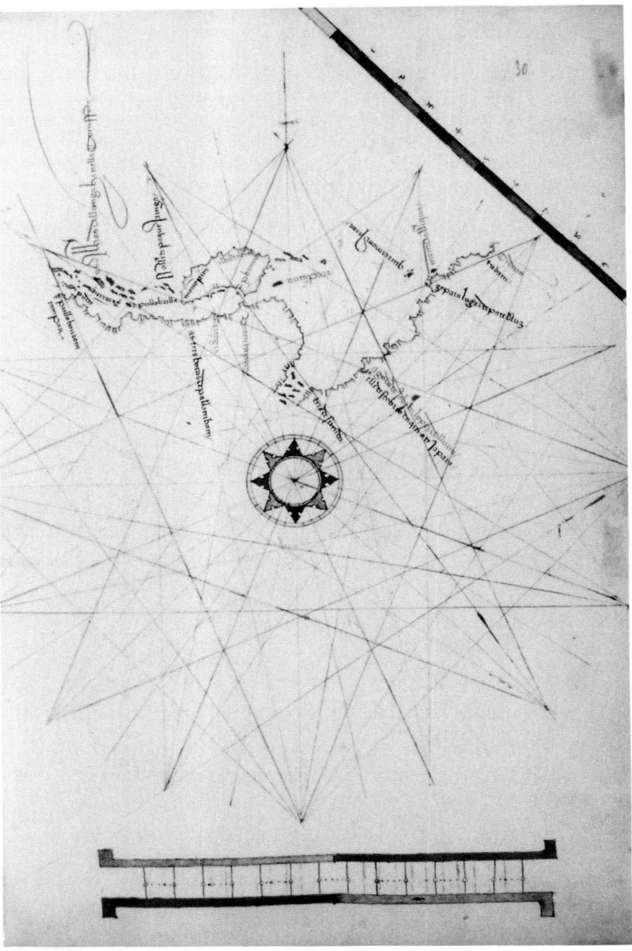

FIG. 19.1. ISLANDS IN THE WESTERN PORTION OF THE MALAY ARCHIPELAGO FROM THE ATLAS OF FRANCISCO RODRIGUES. Figures 19.1 and 19.2 are judged to be from a copy, believed to date from sometime after 1520, of Francisco Rodrigues's original atlas of 1513. The manner of depicting islands on these two folios indicates that the maps were based on entirely different sources. This figure shows a portion of the northeast coast of Sumatra, the adjacent Lingga Islands, the island of Bangka, and the northwest coast of Java. It was undoubtedly drawn from personal observation and rendered in the style of other Portuguese maps of the period.
Size of the original: 39 × 27 cm. By permission of Bibliothèque de l'Assemblée Nationale, Paris (*Journal du Pilote portugais Francisco Roïs*, MS. 1248, fol. 30).

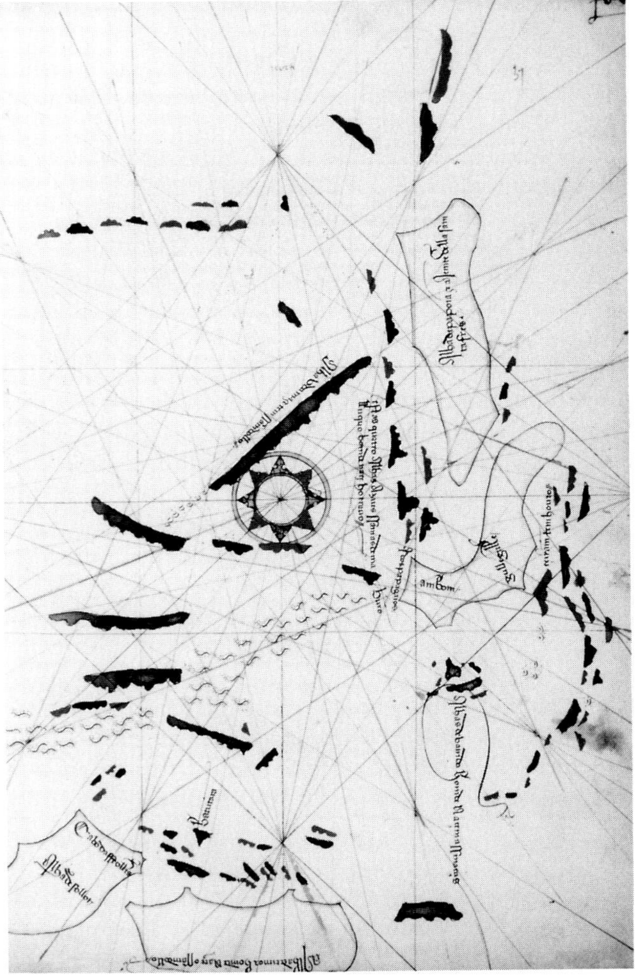

FIG. 19.2. ISLANDS IN THE EASTERN PORTION OF THE MALAY ARCHIPELAGO FROM THE ATLAS OF FRANCISCO RODRIGUES. This map displays a completely different cartographic style from that shown in figure 19.1. Here, Solor and the eastern portion of Flores are merged into a single island. Other islands shown include Timor, Banda, the Moluccas, Ambon (Amboina), and Ceram. The combination of sketchy planimetric views and coastal profiles seems to be a distinctive Javanese map trait. The stark contrast between the conventions of this map and those of the map shown in figure 19.1 underlines the fact that Rodrigues himself failed to reach this area and had to depend on indigenous sources for his information.
Size of the original: 39 × 27 cm. By permission of Bibliothèque de l'Assemblée Nationale, Paris (*Journal du Pilote portugais Francisco Roïs*, MS. 1248, fol. 37).

one transpeninsular connection and no other. Rather than being a pointless whim, I suggest that the apparent all-river connection on this map, like many we have noted on a number of Burmese and Siamese maps (including also cosmographies), was intended to signify not an all-water route, but merely a continuous route that was negotiated partly via rivers and partly via overland portages across an intervening hill barrier. If, as suggested, our merchant was concerned primarily with trade and with gathering goods to sell from many widely scattered localities, it would have mattered little to him whether his goods came by sea or overland. Although sea transport would normally have been easier and cheaper than overland shipment and therefore preferred, there were

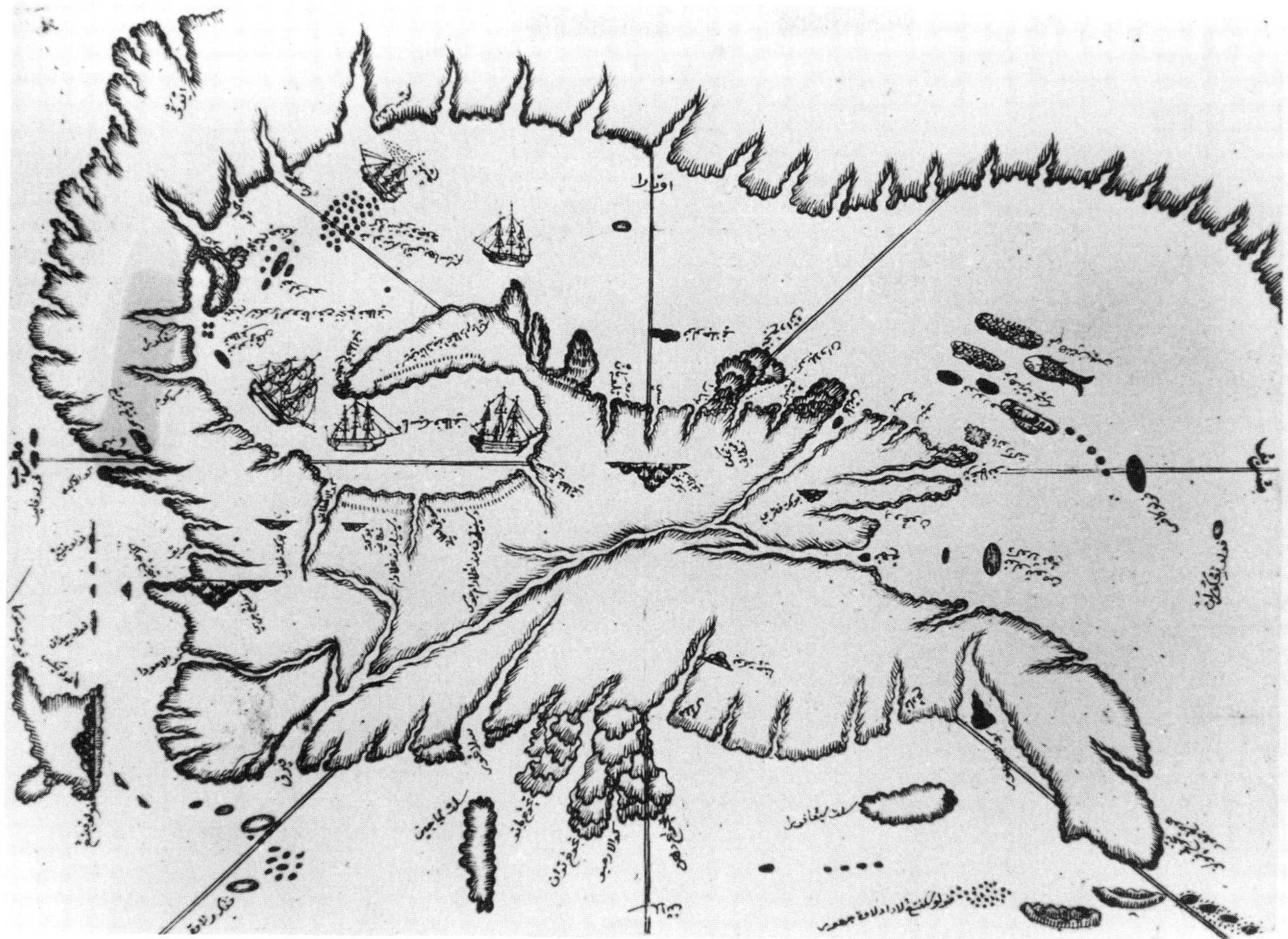

FIG. 19.3. MALAY CHART OF THE MALAY PENINSULA AND THE GULF OF SIAM. This privately held early eighteenth-century work is drawn in black ink on tough locally made paper. The text is in Malay, written in the Arabic script. The map is a curious admixture of indigenous and European elements. Among the latter are the directional lines radiating outward from a hill, Bukit Pattani, a landmark near the port and Bay of Pattani to its left. The map, despite its grossly inaccurate representations of shape, distance, and direction, is believed to have had some utility as an aid to coastal trade focused on Pattani. The puzzling transpeninsular channel conveys the impression that there was a navigable route across what is now West Malaysia; but analogues of such nonexistent waterways seen on maps from Burma and Thailand suggest a different interpretation of what they were actually intended to signify. Size of the original: roughly 30 × 40 cm. From Reginald Henry Phillimore, "An Early Map of the Malay Peninsula," *Imago Mundi* 13 (1956): 174–79.

situations, as the history of the region repeatedly demonstrates, in which land transport would have made sense. Perhaps the most common such situation was when pirates—for centuries active in the waters shown on the map (especially the Straits of Malacca)—become so dangerous that ships had to avoid the routes they frequented. The second circumstance that would lead to overland shipments was when traders at one entrepôt, say Penang or Malacca, acting in collusion or as monopolists, fixed the prices they were willing to pay for certain commodities so low that it became worthwhile for middlemen to bear the extra expense of finding a different, though distant, outlet. This was the case, for example, when the Dutch tried to monopolize the spice trade of India's Malabar Coast, only to find that the commodity they most sought was rerouted over the Palghat Pass to the eastern Coromandel coast and sold to a different set of traders.[9] Thus it seems plausible that the route depicted was the one actually used, as occasion warranted, by the merchant's suppliers or, more generally perhaps, by small traders carrying goods from Malaysia's west coast to Pattani.

9. The Dutch experience in India is well documented in Marcus Vink, "The Dutch East India Company and the Pepper Trade between Kerala and Tamilnad, 1663–1795: A Geohistorical Analysis," unpublished paper, University of Minnesota, December 1990.

There were, to be sure, various other routes that might have been selected. Two of these are illustrated and discussed by Wheatley in a paper dealing with the persistent depiction, on European maps over the period 1519 to 1623, of nonexistent transpeninsular rivers.[10] Wheatley also notes that these routes were on occasion used for trade. He does not go so far as to suggest that the initial placement of the rivers on the European maps stems from some European's having seen them on an indigenous map and, in the absence of firsthand knowledge of the area, copied them on faith. But that possibility seems worthy of further investigation. The opposite situation is of course also possible—that the Malay mapmaker copied his transpeninsular river from a European map. However, since there are no known European maps showing such a feature after 1623 (assuming Wheatley's list is complete), a century or so before the assumed date of the Malay map, that does not appear especially likely.

A final question worth considering is why drawing the map under review was deemed necessary. The text would probably have been meaningless to many, if not most, pilots of small coastal ships, who one may assume were illiterate. And the visual differentiation on the map of one harbor or stretch of coast from another, in contrast to the maps we shall next examine, is subtle at best and probably unintentional. In sailing the coastal routes for which the map might seem relevant, local ships would rarely be far from land and would largely sail within sight of the coast. Furthermore, unlike South Asian sailing maps, this one includes no specification of sailing times or directional guides based on the sighting of asterisms and shows little that a local navigator could not have committed to memory.[11] Writing of navigation, as currently practiced by seamen from Trengganu, a Malay state not far south of Pattani, Gosling observes:

> Navigation is simple, based primarily on piloting from known landmarks. The use of a simple and usually inaccurate hand compass for extended voyages out of sight of land is common. No charts are used and successful piloting depends a great deal on experience and judgment: "One must have a map in the heart." Whether in the heart or the head, the "mental" maps of the Trengganu navigators are impressive. They can accurately identify navigation routes by approximate bearing, and the intervening ports by name, for the entire Gulf of Thailand. They are able to sketch charts with remarkable accuracy, distorted only by the use of time as a variable instead of distance. In addition to these "heart" maps, complex sailing directions are passed on in the oral tradition.[12]

These observations lead me to suggest that the map was made to ease trading transactions between our hypothetical Pattani merchant and his foreign clients. More specifically, it would inform those individuals of areas from which the merchant might be expected to obtain the goods they sought and, incidentally, would let them know that an overland route was available for obtaining or disposing of goods if its use became necessary.

To conclude my discussion, I take note of a group of three remarkably similar nautical charts from a century or so later than the Malay map. Each of these covers the greater part of Southeast Asia. All three are exhaustively analyzed by Le Roux in a scholarly tour de force on which most of the following discussion is based.[13] All include abundant toponymic detail in Bugi, as well as hundreds of depth soundings noted in arabic numerals written in the Western style. Though all employ the language and script of the Bugis, who are to this day the foremost among the indigenous navigators of the Malay world, they appear to have been adapted from one or more European prototypes. One of the maps is shown in full in figure 19.4. This map was reported, as of 1935, to be part the map collection of the Koninklijk Bataviaasch Genootschap van Kunsten en Wetenschappen, in what is now Jakarta; but I do not know what successor organization came into existence after Indonesia obtained its independence or whether the map is still there. An excerpt from another nautical chart appears in figure 19.5. The latter work is in the map library of the Geografisch Instituut van de Rijksuniversiteit, Utrecht. A third chart is in the Museo Naval in Madrid. In the rest of this account I shall refer to these maps as the Batavia, Utrecht, and Madrid maps.[14]

10. Paul Wheatley, "A Curious Feature on Early Maps of Malaya," *Imago Mundi* 11 (1954): 67–72. Wheatley lists thirty-two such European maps. Most show only one transpeninsular river, but some show two. All the routes shown lie to the south of the one on the map we are now considering (fig. 2, p. 68), which is appropriate in light of Pattani's relatively northerly location, but a number do have the same eastern terminus in the Pahang River.

11. On South Asian sailing maps, see Joseph E. Schwartzberg, "Nautical Maps," in *The History of Cartography*, ed. J. B. Harley and David Woodward (Chicago: University of Chicago Press, 1987–), vol. 2.1 (1992), 494–503.

12. L. A. Peter Gosling, "Contemporary Malay Traders in the Gulf of Thailand," in *Economic Exchange and Social Interaction in Southeast Asia: Perspectives from Prehistory, History, and Ethnography*, ed. Karl L. Hutterer (Ann Arbor: Center for South and Southeast Asian Studies, University of Michigan, 1977), 73–95; quotation on 85.

13. C. C. F. M. Le Roux, "Boegineesche zeekaarten van den Indischen Archipel," *Tijdschrift van het Koninklijk Nederlandsch Aardrijkskundig Genootschap*, 2d ser., 52 (1935): 687–714. This article was translated for me in its entirety by L. Ruyter, whom I hereby thank.

14. The Batavia map, the only one of the three I have not personally seen, was registered in the catalog of maps in the year 1913 as number 1410. The Utrecht map has been thoroughly studied, in respect to the island of Sulawesi, by Edouard Cornelius Abendanon, *Midden-Celebes-expeditie: Geologische en geographische doorkruisingen van Midden-Celebes (1909–1910)*, 4 vols. (Leiden: E. J. Brill, 1915–18), 4:1868–71, particularly 1870–71 and pl. 183. The Madrid map, according to Le Roux, had the catalog number "R. 151," but it had none, I was told,

FIG. 19.4. REDRAWING OF A BUGI NAUTICAL CHART COMPILED FROM EUROPEAN SOURCES, CA. 1830. The original (current location unknown) is painted in several colors on cowhide, and is one of three similar works adapted from various Dutch cartographic sources and used by Bugi navigators over large parts of the Malay Archipelago. It was obtained from Bugi pirates on an island off the coast of Sumatra. In addition to the abundant toponymic detail and numerous depth soundings, the map abounds in details relating to coastal configurations, shore profiles (as seen from the sea), and the locations of reefs and shoals and of harbors, and it even shows the flags of European powers occupying some of the key maritime positions within the region.
Size of the original: unknown. Redrawing from C. C. F. M. Le Roux, "Boegineesche zeekarten van den Indischen Archipel," *Tijdschrift van het Koninklijk Nederlandsch Aardrijkskundig Genootschap*, 2d ser., 52 (1935), map IV.

The manner of acquisition of two of the three maps is of considerable interest. The Madrid map was found in a bamboo tube on a Philippine "Moro" pirate ship captured near Jolo, in the Sulu archipelago, and was subsequently given to a Spanish naval officer, who donated it to the Naval Museum in 1847. Given the language of the map, one may safely infer that it was acquired—whether by force or otherwise—from a Bugi source. A note accompanying the Batavia map indicates that it was found in 1859 by J. H. G. Jordens, a Dutch naval officer, off the coast of Sumatra "in the pirate *kampong* of Santhel, in the Bay of Sekana, on the Island of Singkep."[15] There is unfortunately no record of how the Utrecht map reached its present location.

All three of these charts are drafted in ink on cowhide, with highlights in various watercolors. They vary slightly in size and in area of coverage. The Utrecht map, the best-preserved example, measures approximately 76 by 105 centimeters; that in Batavia 75 by 105 centimeters; and that in Madrid 72 by 90 centimeters. The northern limit in all three cases is at about 17° or 18°N latitude, so that they include much of mainland Southeast Asia and almost the whole of the Philippines. To the south,

when I examined it in September 1984. It was then referred to simply as "Carta indígena filipina [*sic*] en dialecto BUGI MAKASSAR"; its inventory number was 90. It was, at the time, undergoing restoration. The map had been studied in detail two months previously by Jean-Paul G. Potet of Clichy, France, but I am not aware that any publication has come from his study.

15. Le Roux, "Boegineesche zeekarten," 687 (note 13).

all include Timor, and the Utrecht map also shows a very small portion of the coast of Australia. That map also extends farthest west, including the Andaman and Nicobar islands, while the others stop just west of the northern tip of Sumatra. To the east, the Utrecht and Madrid maps extend to the eastern tip of the island of Ceram in the Moluccas, and the Batavia map also encompasses several additional islands as far as Aru, lying to the south of New Guinea, which is not indicated.

Almost all the information on the maps relates to features on or near the coast. Along many stretches of coast, highlands are drawn in frontal elevation, as they would be seen from the sea, and descriptive notes appear along some of them. Within the seas, features such as shoals, shallows, marine banks, and sea depths are shown in considerable detail. The number of depth soundings on each map runs into many hundreds. The Paracel Islands, in the South China Sea, are particularly prominent. Estuaries and embayments are greatly exaggerated in size, especially in respect to their depth, more so than one would expect of Dutch charts of the period. Three major rivers are also indicated as deep embayments, the Kapuas in western Kalimantan (Borneo); the Mekong, as far inland, it appears, as Phnom Penh; and the Mae Nam (Chao Phraya), at least as far as Ayutthaya. In virtually all these respects there are variations from map to map, suggesting that the mapmakers' local intelligence played some role in what was included. For example, the large Gulf of Tomini in northern Sulawesi (Celebes) is only suggested on the Batavia and Utrecht maps, on both of which the line of the coast is interrupted, whereas it appears with greater certainty on the Madrid map.

Distinctive colors are systematically used to differentiate particular types of features. Of particular note is that, at least on the well-preserved Utrecht map, where the colors are most vivid, the traditional nests of pirates occur among the islands that are drawn with a red border. In this connection Le Roux cites an 1873 Dutch report on piracy in the East Indies mentioning that "pirates indicated, by a smoking chimney *on their own charts*, the locations where they would run into steamships. . . . This served to warn them not to show up there."[16]

At many points flags—mainly Dutch, but in a few cases British—indicate the presence of European powers. These too differ substantially from map to map and are important features in dating each of them. For example, on none of the maps does a British flag appear at Singapore, which was established as a city in 1824, five years after the island was ceded to Britain. Curiously, on the Utrecht map there is a Dutch flag at Manila.

Another key feature used to date the maps is the sidewheeler steamboat, with auxiliary sails, drawn near the lower-left corner of the Batavia map. Since it was not until 1825 that the *Van der Cappelen* (the first small private steamboat to be built in the East Indies) was launched, no earlier date can be assigned to the map. But the boat shown appears to be of a substantial size and would more likely represent either the first steamboat to arrive from the Netherlands—in 1836 (only to be wrecked the following year)—or the large *Konigin der Nederlanden*, purchased from an English company in 1840, or some English ship. The flag of the steamship is Dutch; but the English firm of Maclaine Watson had already become established in Batavia by 1840, and the red jacket of the ship's captain suggests to Le Roux that he was English.[17] Additionally, all three maps are inscribed with dates in the Hijra (Muslim) Era, which tells when they were initially completed, though not necessarily the date of the information they contain, which either could be based on knowledge of a somewhat earlier date or could reflect updating with supplemental information of a later time. The Utrecht map is inscribed as A.H. 1231 (A.D. 1816) and the Batavia map as A.H. 1244 (A.D. 1828). The date on the Madrid map, seen only on a photograph, could not be read by A. A. Cense, the Dutch linguist who worked with Le Roux in interpreting all three charts.[18]

All three maps have one or more Western-style bar scales with regular distances marked off in French or German nautical miles, or both, and numbered in arabic numerals. The approximate ratio scale works out to be 1:4,500,000. All three maps are oriented toward the north and include compass roses (one on the Madrid map, two on the Utrecht map, and three on the Batavia example) and regularly spaced rhumb lines. Those running in the cardinal and secondary directions are drawn as thick black lines, and the sixteen intermediate directions are shown by lighter dotted or pencil lines.

Since Singkep, where the Batavia map was found, lies well over a thousand miles west of Jolo, where the Madrid map was obtained, it appears that the area of use of nautical charts copied by Bugis from European sources was extensive. One wonders, naturally, how many more copies may have been made and modified in addition to those we can document. Cense, to whom the Batavia map was lent, showed it to some Bugi sailors in Makassar, presumably in the early 1930s, and was told that only a few persons were left who undertook long sea voyages; but one of his informants stated that as a youth he had seen his grandfather consult a chart drawn on cowhide that might well have been of the type we are discussing.[19]

A puzzling feature of the maps is their maintaining a European system of rhumb lines. The position of the lines

16. Le Roux, "Boegineesche zeekaarten," 692; quotation on 690 n. 1 (note 13).

17. Le Roux, "Boegineesche zeekaarten," 693–94 (note 13).

18. Le Roux, "Boegineesche zeekaarten," 694–95 (note 13).

19. Le Roux, "Boegineesche zeekaarten," 694 (note 13).

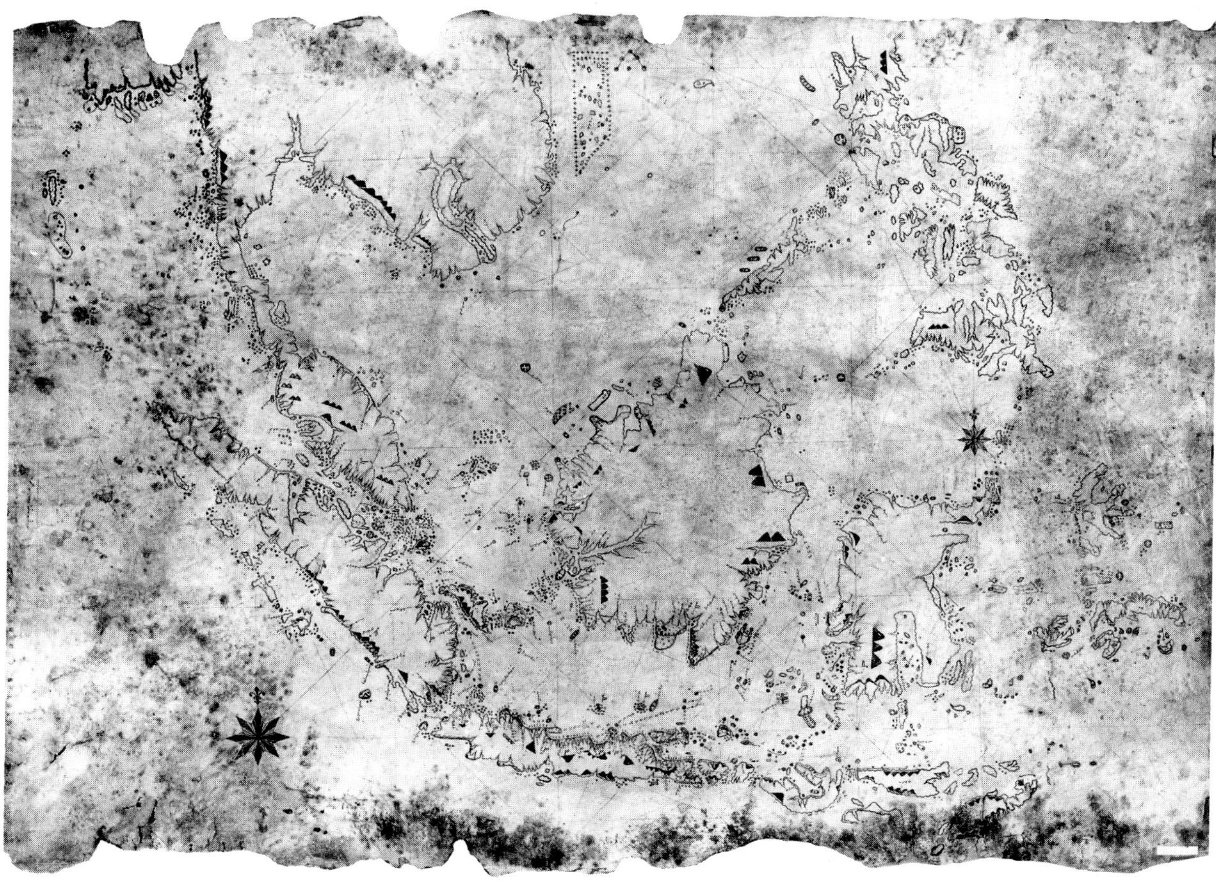

FIG. 19.5. JAVA AND NEIGHBORING ISLANDS, AS DEPICTED ON A BUGI NAUTICAL CHART. The chart from which this view is excerpted is the largest, best preserved, and territorially most extensive of the three such works that are known. Like the others, it is on cowhide. This small portion of the map conveys some idea of the remarkable detail that characterizes all three.
Size of the entire original: 76 × 105 cm. By permission of the Kaartenverzameling, Geografisch Instituut, Rijksuniversiteit te Utrecht (Sign. VIII, C.a.2.).

varies slightly from one map to another, but the set of main and subordinate directions is the same throughout. The use of rhumb lines runs counter to what I stated above, quoting Gosling, about the contemporary Malay method of navigation. But we must be remember that Bugi navigators traversed greater distances than did other maritime groups in Southeast Asia and might, even by the early nineteenth century, have altered their navigational practices as a result of contacts with Europeans. Alternatively, they may have copied the rhumb lines out of some sense of respect for the originals that led them to adhere faithfully to what was portrayed there, whether or not they perceived it as useful.

But, specifically, what map or maps provided the model for the remarkably detailed Bugi charts? To answer that question Le Roux compared the charts with a large number of European maps, mainly Dutch, that might have found their way into Bugi hands. He concludes that, in general, the configuration of coasts and other features of the Bugi charts corresponds to those of parchment manuscript charts of the second half of the seventeenth century and first half of the eighteenth, and he cites five such works in particular that are in the Dutch state archives in The Hague. Here and there, however, specific details appear to have been taken from later works, including, among others that are named, *Die nieuwe groote lichtende Zee-Fakkel* (The new great illuminating sea torch), published by Johannes Van Keulen II in 1753, especially in respect to its sixth section dealing with the Dutch East Indies; the Gerrit De Haan manuscript atlas of 1760–61, *Ligtende zee fakkel off de geheele Oost Indische waterweereldt* (Illuminating sea torch of the entire water world of the East Indies); and for much of the Moluccas and the Philippines, various works by François Valentyn.[20] Hence one must conclude that the Bugi mapmakers acquired a number of European maps and had the ability to compile from them an original work at a more or less uniform scale, choosing from their diverse sources those features they had faith in while rejecting other portions.

We are quite certain of one of the Bugis' sources of European maps. A British navigator, Thomas Forrest (1729?–1802?), spent many years in the service of the East India Company, had a good command of the Malay language (the lingua franca of the East Indies) and, in a small ship (a ten-ton galley of local design) with a predominantly Malay crew, charted much of the coast of New Guinea, the Moluccas, and Borneo and visited Mindanao as well, during 1774–76. Of the Bugis, Forrest observed:

> They are fond of sea charts, I have given many to certain *Noquedas* (commanders of Prows) for which they were very grateful, and often wrote names of places in their own language, which I read to them on the charts; and they were always very inquisitive about Europe, and *Neegree Telinga* (Indostan). . . . Long before the passage round the Cape [of Good Hope] was discovered; and in those days, before Dutch oppression, the Buggesses certainly traded largely to most of the eastern islands . . . and held many of them in subjection.

Elsewhere Forrest writes,

> I shall now describe the great gulf (*Sewa*) [Gulf of Bone in southern Sulawesi] from the information of *Noquedah Inankee*. . . . I presented the *Noquedah* with a set of the charts (*Pata*) and views of land (*Toolisan*) of my *New Guinea* voyage; on each of which he wrote name and explanation in the Buggess language, and was much gratified with the present.[21]

Forrest's generosity extended not only to the Bugis, but to other groups he came in contact with. One such group were the Illanos of the region around Illana Bay in southwestern Mindanao. He writes that during a prolonged layover during the time of the southwest monsoon, when he was the guest of the Illano sultan, whose hospitality he wished to repay, he

> constructed upon two thick planks, well pinned together, a map of the world; it was 8½ feet by 4½, allowing a margin, and when finished, by cutting a strong outline to mark both continents and islands (taken from a small plain chart), it was hung up in Rajah Moodo's hall, where, unless destroyed by fire, it is likely long to remain: whilst paper maps, had I such to present him, would, it is most likely, be lost, tore, or neglected.[22]

This world map is illustrated in figure 19.6. He also reports having given several mariner's compasses and a planisphere as gifts to various Illanos and Sulu Islanders.[23]

Of course, the maps that Forrest would willingly have parted with most probably were the most out of date among those in his possession. If we assume that those maps formed the models for the Bugi maps we are now considering, that would account for the coastal configurations of the late seventeenth or early eighteenth century as described by Le Roux.

But Forrest was hardly the first European to make maps available to curious individuals within the Malay world. Forrest also noted that Francisco Domingo Fernández Navarrete, who visited Makassar in 1650, observed even then that he was shown some European maps and books (as well as Chinese books) that formed part of the famous library of his host's father.[24] Le Roux suggests that the bibliophile in question was a renowned scholar named Karaëng Pattingalloang, who was appointed by the Dutch as ruler of Makassar. Pattingalloang was "a great lover of geography" and conversant with several languages, including Latin. For his services to the Dutch East India Company he was rewarded with a giant copper globe made by Joan Blaeu.[25]

It appears, then, that we can take it as established that Bugi navigators appreciated maps and had little difficulty understanding those of foreign origin. Le Roux states his conviction that apart from maps received from Forrest, the Bugis bought maps from employees of the Dutch East India Company, although these likely were poorly printed maps from atlases and not patented company maps.[26] But when they and neighboring seafaring peoples of South-

20. Le Roux, "Boegineesche zeekaarten," 696–97 (note 13).

21. Thomas Forrest, *Voyage from Calcutta to the Mergui Archipelago, Lying on the East Side of the Bay of Bengal* (London: J. Robson, 1792), 82 and 87. A very revealing account of Forrest appears in Leslie Stephen and Sidney Lee, eds., *The Dictionary of National Biography: From Earliest Times to 1900*, 22 vols. (first published in 66 vols., 1885–1901; reprinted London: Oxford University Press, 1937–38), 7:443–44.

22. Forrest, *Voyage from Calcutta*, 139 (note 21). The context in which the quoted passage appears is of some interest. It forms part of a several-page disquisition titled "Idea of Making a Map of the World," in which Forrest asks, "Why does nobody turn a level verdant plain of a very few acres into a map of the world?" He then proposes the means of executing such a project, which he suggests "would be pleasant and healthful to young folks, . . . and make very young persons expert in simple geography, far beyond what they get from books and maps even at a more advanced age" (139). The point of citing this passage is that it demonstrates, as do others in Forrest's narrative, that he was a teacher by inclination. To what extent, over his many years of working amicably among Malay peoples, he might have imparted useful cartographic knowledge to his indigenous shipmates and to the chiefs he befriended is worthy of conjecture.

23. Le Roux, "Boegineesche zeekaarten," 699 (note 13).

24. Forrest, *Voyage from Calcutta*, 81 (note 21).

25. Le Roux, "Boegineesche zeekaarten," 699–700 (note 13).

26. Le Roux, "Boegineesche zeekaarten," 701 (note 13).

FIG. 19.6. WORLD MAP MADE BY THOMAS FORREST ON THE WALL OF THE ROYAL CHAMBERS OF THE SULTAN OF MAGHINDANO (MINDANAO), 1774. Forrest, a British navigator and hydrographer who was obliged by circumstances to spend some months as the guest of the sultan, drew on wood an illustrated large map of the world, which he then presented to his host in return for his hospitality. The romanticized portrayal of a nuptial scene has nothing to do with the essential point that this was only one of numerous examples of the two-way transmission of cartographic information between Europeans and various seafaring Malay peoples, of whom the Bugis displayed the most avid curiosity about all matters relating to geography and navigation.
From Thomas Forrest, *Voyage from Calcutta to the Mergui Archipelago, Lying on the East Side of the Bay of Bengal* (London: J. Robson, 1792).

east Asia began to make maps on their own is a question that calls for additional comment. Leaving aside the already discussed question of the alleged fifteenth-century Javanese map of eastern Indonesia that was imparted to Rodriguez in 1511, there is other unmistakable evidence of Indonesian maps.

Although he did give maps to Malays, a hasty perusal of two of Forrest's books on his voyages in Southeast Asia reveals no instance of his having received maps from them. Notices in an 1832 catalog of Malay manuscripts, however, provide evidence to that effect. Among the Bugi manuscripts in the library of the renowned Malay scholar William Marsden (1764–1838) were noted maps with place-names written in Bugi characters, presented by Forrest to Marsden. The number of maps is not stated. A footnote to the notice says that Marsden gave the originals to the "Bibliothèque de la Société asiatique de Londres."[27] Le Roux's search for the maps proved unsuccessful. The catalog also notes that a Prince Fakkymoulana, brother of King (i.e., sultan or "rajah") Pahareddin of Maghindano (an old name for Mindanao) presented Captain Forrest with several navigation maps written in "Maghindano."[28] These were very likely in exchange for the wooden map of the world that Forrest made for the sultan. One of these maps, said by Le Roux to be in the London Museum (British Museum?), is mentioned in Forrest's account of his trip to Mindanao. I do not know whether that or any of the associated maps survives. Le Roux appears not to have been aware of the existence of any but the published example.[29] He does, however, note a statement in the catalog of the Bibliotheca Marsdeniana in London that says "a map of the Dutch East Indies is in the collection of the Nederlandsch Bijbelgenootschap [Dutch Bible Society], on which the names are written intermittently in Makassar [Bugi] script." Le Roux's attempt to find that map was also fruitless.[30]

Forrest's more illustrious contemporary, Alexander Dalrymple, was also the recipient of Malay maps. He noted, for example, that in 1764 while he was in the Philippines a servant, Pedro Manuel from Ilocos on the island of Luzon, prepared a plan on which "the bearings of several places ... [were] found to agree with their positions as I [Dalrymple] had determined them."[31] There are additional indications of Dalrymple's reliance on local sources, including maps, to supplement his own hydrographic charting. His "Map of Part of Borneo and the Sooloo Archipelago: Laid down Chiefly from Observations Made in 1761, 2, 3, and 4," published on 30 November 1770, bears a note just below the title that reads, "The Coast of Borneo from Unsang [east of Sandakan in what is now Sabah] Southward is not confirmed by any exact Observations but is laid down from a Sketch of Dato Saraphodin and from a Chart of Noquedah Koplo who came up that Coast in 1761." The stretch of coast in question is several hundred miles long and shown with a degree of detail not remarkably different from that of the rest of the map. Another note on the map indicates a number of places "well determined" from Dalrymple's own observations, as well as others based on bearings by C. Alves, and then adds, "The other Places are from Sketches I received from the Sooloos [a Malay people of the Sulu archipelago], but chiefly from the information of Bahatol an intelligent old Pilot [ethnic identity not given, but surely of some Malay group]." A Dalrymple chart at a larger scale, "The Sooloo Archipelago Laid down Chiefly from Observations in 1761, 1762, 1763, and 1764," published on 30 November 1770, bears a note that certain observations marked "F" were laid down "by the Falmouth, Man of War not very correctly determined," while others marked "S," were "from the Report of the Sooloos." I counted nine places of the latter type. Five of these were individually delineated banks, the largest of which covered an area of about fifteen by three kilometers, and four were points indicated as "no ground," presumably signifying soundings that failed to determine the depth of the sea floor. Thus, in the absence of adequate surveys by European hydrographers, Dalrymple had enough confidence in the observational and sketching capabilities of local Malay seamen to put their findings on his own charts.[32] There are, however, no firm grounds for supposing that they were then in the habit of preparing charts or sketches of any kind except at the request of Europeans, despite one or more possible exceptions cited above.

Until now no one besides Le Roux, so far as I am aware, has made a sustained and systematic effort to uncover evidence of Malay nautical maps, and no one at all has tried to find such maps from mainland Southeast Asia—which, if they existed, would not likely be numerous. But further searches within the libraries and private archives in that region and in the relevant archives of the former colonial powers definitely seem warranted.[33]

27. "Cartes de l'archipel oriental, avec les noms des lieux écrits en caractères boughis (reçu du C. Th. Forrest)," in E. Jacquet, "Mélanges malays, javanais et polynésiens," *Nouveau Journal Asiatique*, 2d ser., 9 (1832): 97–132 and 222–67, esp. 262–63.

28. Jacquet, "Mélanges Malays, Javanais et Polynésiens," 263 (note 27).

29. Le Roux, "Boegineesche zeekaarten," 699 (note 13); Thomas Forrest, *A Voyage to New Guinea, and the Moluccas, from Balambangan, Including an Account of Magindano, Sooloo, and Other Islands ... during the Years 1774, 1775, and 1776*, 2d ed. (London: G. Scott, 1880; reprinted Kuala Lumpur: Oxford University Press, 1969), pl. 18 (in two parts).

30. Le Roux, "Boegineesche zeekaarten," 689 (note 13).

31. Alexander Dalrymple, *A Collection of Charts and Memoirs* (London, 1772), viii. I am indebted to Andrew Cook, Oriental and India Office Collections, British Library, for this reference. Though described by Dalrymple as a "native," Pedro Manuel may have been, in Cook's opinion, partially of Spanish extraction (correspondence dated 17 March 1993).

32. I have consulted only a few of the numerous charts by Dalrymple at the James Ford Bell Library, University of Minnesota, and think it likely that a more extended search would yield many examples of the use of indigenous sources in addition to those I have cited.

33. I was told, for example, by the librarian of the Centre for Southeast Asian Studies in Singapore that there were indigenous maps in the Sarawak Archives in Kuching; but my letter to that agency was not answered. A personal visit would probably have yielded the information I sought.

20 · Conclusion to Southeast Asian Cartography

JOSEPH E. SCHWARTZBERG

NATURE AND DISTRIBUTION OF THE SURVIVING CORPUS

As we have seen to be true of South Asia, the received notion that Southeast Asia had virtually no tradition of cartography turns out to be invalid. A dearth of relevant scholarship, rather than an absence of surviving maps, is at the root of the dismissive opinions about Southeast Asian cartography that have been voiced by Leo Bagrow, R. A. Skelton, and other historians of the field. It is true that there are no known Southeast Asian globes or noncosmographic world maps and that only one map, from Thailand (plate 36), is even near continental in scope. And there are certain countries for which the known corpus of premodern maps is indeed meager or, as in the case of Cambodia, Laos, and the Philippines, virtually nonexistent. But there are others, most notably Burma (Myanmar), that have provided a wide diversity of maps. Moreover, one sees within Burma a continuous development of cartography, which becomes increasingly sophisticated over the period beginning in the latter half of the eighteenth century. Although that development, and less easily documented development elsewhere, undoubtedly owes much to contacts with both the West and China, and possibly also to Southwest Asian Islamic influences, the influence of foreign models on mapmaking was never sufficient, before Southeast Asia's near total absorption by European colonial powers, to eradicate the distinctive and varied flavors of indigenous maps.

One of the intriguing questions about the surviving corpus is the remarkable difference between what remains from the similar cultures of Burma and Thailand. The number of known surviving maps from Thailand—if we leave aside essentially cosmographic works—is meager, though the oldest known map, of the Sathing Phra peninsula (fig. 18.28), is perhaps as much as a hundred years older than the earliest Burmese map, relating to the sack of Ayutthaya in 1767. Of the Thai maps that survive, several are fairly sophisticated, leaving the impression that there must have been many developmental links. On the nature of those links we can at present do no more than speculate. For Burma, my relatively brief searches have revealed close to 150 traditional maps, ranging from detailed cadastral maps and large-scale plans of architectural complexes and individual cities to maps covering more than a million square kilometers. These maps include, however, not only works by ethnic Burmans, but also a substantial number of maps made by Shans, from areas within Burma. The hill-dwelling Shans, as has been noted, are a Thai ethnic group, less advanced technologically than either the lowland Thais or the Burmans, but intermittently within the political orbits of both. Several reasons account for the striking differences between what survives from Burma and from Thailand. First, in the case of Burma, British diplomatic and administrative personnel, such as Francis Hamilton, Henry Burney, and James George Scott, appreciated the value of indigenous maps and saw to it that they were made, copied, or preserved. Second, Burmese themselves valued maps, as is evident in that a number of them were preserved in monasteries and also in that two members of the Burmese Historical Commission, U Maung Maung Tin and Than Tun, made a point of tracking down and acquiring maps they felt were worthy of preservation. Third, in what is now Thailand, the institution of *chamra*, the periodic purging of manuscripts deemed out of date, resulted in

Before 1981 I was not aware of a single indigenous Southeast Asian map and was inclined to believe, based on the existing histories, that there were few to be found. Nevertheless I agreed—since no other candidate for the job emerged—to search for maps of Southeast Asia concurrently with my investigations of the history of South Asian cartography, which was my principal interest in respect to the *History of Cartography* project. Whatever I have learned about Southeast Asian maps in the years since then stems from widespread correspondence with Southeast Asian specialists—most of them experts on only a single country—in a variety of fields, especially history, art history, and religion; from library research at the University of Minnesota; and from brief visits to museums and archives elsewhere in the United States, Europe, India, and Southeast Asia. My time looking for maps in Southeast Asia was spent in Yangon, Pagan, Rangoon, Bangkok, and Java. It is important to make these points because they show how much more awaits discovery by motivated and qualified area experts. What appears in the preceding pages does not, however, exhaust the material that has come to my attention; it relates only to matters I have had a chance to study in the time available. Yet there is no doubt in my mind that future scholarship will demonstrate that these chapters are no more than a much overdue beginning to a systematic analysis of the hitherto unrecognized richness of traditional Southeast Asian cartography.

the destruction of innumerable documents of which a certain number, we can safely assume, would have been maps. But it must be noted that my search for Thai maps, based on visits to the National Library in Bangkok and to a handful of museums and libraries outside Thailand, and on correspondence with specialists on that country, was substantially more limited than my search for those of Burma. What the numerous Thai monasteries might be holding in the way of maps, in addition to the cosmographic paintings known to adorn the walls of some of them, is yet to be investigated.

If the scarcity of known Thai maps may be explained in part by limitations in the search for them, it follows that the near total absence of surviving geographical maps from Laos and Cambodia is also understandable, since neither country has been accessible to Western scholars since the 1970s. On the other hand, both countries interacted frequently with their neighbor and sometimes suzerain Vietnam, which had a cartographic tradition even more vigorous than that of Burma. Why we have no evidence of diffusion of cartographic knowledge from one area to the other warrants further investigation.

The surviving cartographic works from the Malay world could scarcely be more varied, given the limited number of known artifacts. These range from the enigmatic engraved bamboo *tuang-tuang* (fig. 18.37), made by the aboriginal forest-dwelling Sakais of West Malaysia as charms to ward off a variety of potential evils, to large, remarkably detailed nautical charts (figs. 19.4 and 19.5) drawn on cowhide, used by Bugi and Illano (Mindanao and Sulu Island) pirates and, one supposes, by other seamen of the seafaring Bugi community of southern Sulawesi. Although the European antecedents of the nautical charts are evident, that detracts little from the ingenuity that went into their construction. As a rule, however, we know exceedingly little about the missing links that presumably led up to the works discussed in this history, and we must recognize that speculation is likely to go wide of the mark. So fragmentary is the cartographic record that nowhere in the Malay realm can one detect clear evidence of an enduring cartographic style. Consider, for example, the extraordinary differences among three highly detailed maps from Java (or possibly from Bali in one case): the probable late sixteenth-century map focusing on the chiefdom of Timbanganten (fig. 18.18 and plate 38), full of text and identifiable places and physical features; the undatable, but undoubtedly later, cryptic (and conceivably mythic) batik map (figs. 18.21 and 18.22), with not a single word of text or one provable place referent; and the relatively modern map from central Java (plate 39), presumably made for some administrative purpose.

Among cosmographies, there is also an incredible diversity of objects that may be considered maps, ranging from pigs' livers used as instruments for divination (figs. 17.5 and 17.6) to such architectural wonders as Borobudur and Angkor Wat (figs. 16.2 and 16.3). Between those extremes one finds, of course, many portrayals on paper, cloth, palm leaf, and other essentially two-dimensional surfaces that come much closer to the conventional view of what a map should look like. Among these are works such as the Thai *Trai phum* (Story of three worlds; figs. 17.15, 17.17, 17.18, and 17.23), rooted in the rich and widespread cosmographic conceptions of the Hinayana Buddhist canon, and highly localized conceptions such as those associated with the mortuary cult of the tribe of Ngaju Dayaks of Kalimantan (figs. 17.1 and 17.2), which from an aesthetic standpoint are among the most appealing maps to emerge from Southeast Asia.

The chapter on cosmography does not begin to do justice to the vast architectural corpus, many works of which are believed to be representations of the cosmos as a whole, while others represent specific locales or features of particular cosmic significance, such as the Mount Meru complex (inter alia fig. 17.16) or the four great rivers flowing from Lake Anotatta (for example, Neak Pean, discussed but not illustrated above). Art historians and specialists in religion have of course dealt amply, if not yet exhaustively, with these themes. I have attempted little more than to demonstrate that in all major regions of traditional Southeast Asia maplike cosmographic symbols had, and often retain, great cultural significance and to suggest that it makes little difference to those who have faith whether the cultural construct that encapsulates the whole or even a portion of the cosmos is carved in stone, painted on paper, or rendered in the diverse media that go into the making of temporary structures built for the performance of major rites of passage. All cosmographies are, in effect, maps of sacred space.

Physical Attributes of Southeast Asian Maps

Southeast Asian two-dimensional maps are drawn in a variety of media, are made of many different materials, and assume diverse forms and sizes. Important traditional materials include palm leaf, used almost exclusively for cosmographic maps; thick indigenous paper, either black, brown, or bleached white, made from the bark of the mulberry and other trees; cloth; and cowhide. Sheets of indigenous paper were often pasted together to form long accordion-style folding books (called *parabaiks* in Burma and *samud khoi* manuscripts in Thailand), and in many of them the map was integrated with an accompanying manuscript text. But maps with six or so accordion pleats and a single longitudinal fold were fairly common in

Burma. European paper was used as early as the late eighteenth century and became increasingly common in the nineteenth. Not a few of the maps on such paper were made at the behest of Europeans themselves, like the maps drawn for Hamilton and Scott.

The surface a map was to be drawn on partially dictated the medium. Works on palm leaf had to be etched with a stylus, and the grooves were subsequently made more visible by applying lampblack. Maps on black paper were commonly drawn in white talc or chalk. As a rule, a combination of paints and inks was preferred, creating a variegated image. Map text was usually written in ink. The time available to the artist was also a factor. When a map had to made in haste, a single medium was used, most frequently black ink but occasionally pencil or chalk. A substantial number of the maps considered in this history, including virtually all of those associated with Hamilton, are copies of originals whose locations, if they survive at all, are not known.

Maps varied greatly in size. Almost none were as small as the page of a typical octavo book. It was not uncommon to paste together several sheets of paper to provide a large enough surface for a given map, but most of the very large maps were drawn on cloth, occasionally on two pieces sewn together. Many were more than a square meter in area. The largest cloth map I have seen, the Burmese map relating to their invasion of Manipur in 1758–59, measures 2.03 by 2.84 meters. Accordion-style maps also vary greatly in length. The longest of all, the Berlin manuscript of the Thai *Trai phum* cosmography, is 50.9 meters long, and the map of the Sathing Phra peninsula on the Malay Peninsula is said to extend to about 40 meters.

Cartographic Attributes of Southeast Asian Maps

In the following paragraphs I shall confine my remarks to two-dimensional geographical (noncosmographic) maps and consider the ways they differ among themselves and from conventional modern cartography, as well as whatever commonalities they may exhibit. I shall list the items noted. For ease of reference, where no spatial referent is given, the whole of Southeast Asia is implied.

1. Perhaps the most obvious general difference between Southeast Asian maps and modern maps is that almost none of the former are drawn to a uniform scale. The only known exceptions are a few large-scale maps from Burma during the period from about 1850 to 1885, when that country lost its independence, and the Bugi nautical charts, the only maps on which scales are drawn.

2. No map is drawn on a recognizable projection. Again, the Bugi charts may be seen as an exception insofar as the projection of one of the European prototype maps may be considered as having been retained, though more inadvertently than consciously so far as the Bugis were concerned.

3. Without exception, maps lack a geographic grid of longitude and latitude. Many Burmese maps, however, do have regularly ruled rectangular grids, probably following Chinese models, which would aid in copying the map from one scale to another (usually going from a small sketch to the larger final size).

4. Almost no map bears a legend key to the cartographic signs employed, even though on a large proportion of such maps it is possible to deduce what such a legend would be. Over the course of the nineteenth century, there was a definite trend toward standardizing the signs on Burmese maps, and there were also certain tendencies toward uniformity between Burmese and Siamese maps. For example, in depicting settlement it was customary to show major cities and towns, especially those with administrative functions, by squares or rectangles and lesser places by ovals or circles.

5. Certain map signs, such as for settlements, rivers, coastlines, and lakes, are almost always drawn planimetrically, whereas others, especially for mountain and hill ranges, vegetation, and prominent edifices (stupas, temples, monasteries, etc.), are generally drawn in frontal perspective. Mountains and vegetation tend to be drawn in a more or less naturalistic style. Vegetation, especially on Burmese, Shan, and Thai maps, is prominently shown in respect to both the size of map signs and their variety and richness. Fish and other aquatic creatures commonly appear in rivers and other bodies of water and, less frequently, land animals are also depicted. Stylized wave patterns often fill the space of large bodies of water and sometimes rivers as well.

6. Multicolor maps, especially from Burma, are characterized by conventionalized use of particular colors, not very different from many modern topographic maps. Settlement is typically shown in red or in yellow outlined in red; roads in black or red; mountains in mauve or, less commonly, brown, or both; vegetation in various shades of green, often with details in other colors as well; water in blue or occasionally green; and so forth.

7. Orientation varies by region. The most common orientation for Burmese maps, following Indian practice, is toward the east, though exceptions are numerous. Nontribal maps from the Malay world, following Arabic practice, are oriented toward the south. The number of Siamese examples is too small to warrant a generalization. Even where there is one dominant direction, however, the convenience of the artist may cause certain features—especially those, such as hill ranges, that are drawn in frontal perspective—to be oriented in some other direc-

tion. There is an increasing frequency of maps with orientation toward the north over the course of the nineteenth century.

8. Maps are rarely dated. Those that are may be assumed to have been made for some political purpose or at the behest of some European.

9. The names of the artists or surveyors responsible for maps are almost never given. Notable exceptions are the Javanese map of Timbanganten (fig. 18.18) and the map of the routes of "Shans" from Cochin China westward (fig. 18.29), as well as those maps made for Europeans, where the names are inserted by the individual who requested they be drawn.

10. Almost no map has a neat line. There are a few Burmese exceptions to this rule. There are also some instances (several from Burma, one from Siam, and one from Java) where the map content extends right to the edge of the page or cloth sheet on which the map was drawn, suggesting that the work was intended as part of a larger series. But this cannot be proved.

11. There are no atlases of geographic maps. One can, however, consider the Thai *Trai phum* a cosmographic atlas, since the maps it contains do not form a continuous image, as do maps in other long *samud khoi* manuscripts.

12. Apart from the Bugi nautical charts and perhaps the map of most of Asia in the Thai *Trai phum* (plate 36), there is no clear evidence of maps' having been compiled by selecting data from a variety of preexisting maps.

Because of their apparent clarity, there is a danger of assuming that certain "obvious" signs have the same meaning on Southeast Asian maps as on maps that modern scholars are more familiar with. The most obvious case is interpreting certain "rivers" as if they were continuous waterways rather than routes that were largely riverine but partly overland between connecting rivers. Thus Hamilton expressed wonder at the remarkable anastomosing quality of the rivers on one map of southern Burma—where, in fact, no riverine connections existed—and reproduced without comment a number of Burmese maps with similar fictive riverine connections in even less likely regional contexts such as the Shan Plateau and adjacent areas to the east. Similarly, Phillimore attributed the all-water route across what is now West Malaysia on a Malay nautical chart to the ignorance of the cartographer. Other examples of this type could be cited. A second source of danger in interpreting Southeast Asian maps is the scale of pictorially rendered features. Although it is obvious that the vegetation shown cannot be many miles high, as the apparent scale might suggest, it is not at all obvious that what appears to be a major mountain barrier might actually be a rather insignificant range of hills.

Future Tasks

The study of Southeast Asian cartography has barely begun. Most of the maps from that region have not been the object of careful and well-informed scholarly investigation of the type Victor Kennedy gave to the Thai military map of the Korat Plateau or that C. C. F. M. Le Roux gave to the Bugi nautical charts. A sine qua non for any comparable investigation of other maps (unless done by a native speaker of the language in which the map text is written) is to obtain a full translation or transliteration of the map text and then to identify as many places as possible on modern maps of the area depicted. Without this, even the purpose for which a map was drawn will not always be obvious.

Relating the maps to their historical contexts is also necessary; but given the fragmentary nature of many of the archival sources or, in certain cases, the total absence of relevant sources, this will often not be an easy task, especially for maps that cannot be precisely dated.

Learning more about the administrative organization within which officially sanctioned mapping was conducted is also necessary. It can be established that, at least in Burma, there were government-sponsored surveys and some standardization of surveying and mapping efforts; but we cannot infer that similar activities were nonexistent in other parts of Southeast Asia.

With respect to cosmographic maps, it will be necessary for historians of cartography to learn enough about the cultures in which they were created to interpret them on their own terms. Many of these maps, we may safely assume, will be found in monasteries and temples, where they are still most highly valued; and it may often prove necessary to enlist the support of the local monks and priests to interpret them satisfactorily.

Finally and, I would argue, most important, it is necessary to launch a sustained investigation to uncover what remains of the traditional map corpus. No one, in principle, can do this better than Southeast Asians themselves; but so far as I am aware, there is no serious student of the history of cartography in the whole of that region. In the foregoing chapters I have repeatedly noted the disappearance of maps known to have existed. The number of such disappearances, whether from neglect, accidental loss, or wanton destruction, will inevitably increase with time. If, then, we are to do justice to the elucidation of the much underestimated and little recognized mapping achievements of the peoples of Southeast Asia, the time to take up that task has come. I hope that this exposition will help pave the way for the requisite future scholarship.

21 · Concluding Remarks

David Woodward, Cordell D. K. Yee,

and Joseph E. Schwartzberg

The chapters assembled here, together with those in the first book of this volume, present what we believe is the most extensive survey of traditional Asian cartography ever attempted. In this conclusion we draw together some of the themes that have emerged in our consideration of the cartographies of China, Japan, Korea, Greater Tibet, Vietnam, and the rest of Southeast Asia. As such, it complements the conclusion to volume 2, book 1, which covered Islamic and South Asian cartography. The work here represents a major effort toward righting an imbalance in previous accounts of the history of cartography, an imbalance usually tilted toward the achievements of Mediterranean and western Europe. When they were dealt with in the past, which was not often, the cartographies surveyed here tended to be relegated to short chapters in more comprehensive histories of European cartography, either as afterthoughts at the end of such works or at the beginning where they could somehow be identified with prehistoric cartography as forming the "primitive" origins of an ever improving scientific map. Their artifacts have thus often been treated as curiosities and exotica, to be included in general histories of cartography only as antiquarian diversions. Indeed, from a Western perspective all Asian maps are by definition "exotic."

There is also no shortage of curiosities: divinatory maps on pig livers, maps carved as charms on bamboo poles, a relief model with streams flowing with mercury in a tomb guarded by crossbow booby traps. But beyond such artifacts, the chapters in this book have shown that the cultures outside the European sphere of influence have rich cartographic traditions of their own. In a number of cases, the non-Western histories of cartography are as long as, and perhaps even longer than, those of Europe. Of the East and Southeast Asian cartographies, those of China and Japan have been the most studied. The scarcity of literature on the traditional cartographies of Southeast Asia and Tibet has fostered the impression that the Chinese tradition was dominant. But as has been amply illustrated here, the notion that non-Chinese cartography in East and Southeast Asia was generally linked to that of the "Middle Kingdom" needs to be qualified. Chinese culture may have been a powerful influence in the region, but the surrounding cultures were not passive recipients. The cultures on the receiving end sometimes disagreed with the Chinese on what was significant in Chinese culture. For example, the *Shanhai jing* (Classic of mountains and seas) exerted greater influence on Korean mapmaking and geographic consciousness than it did on the Chinese tradition, at least what survives. A similar situation occurs with Islamic cartography. It seems to have been preserved in China only long enough to be transmitted to Korea, where it exerted a greater influence. Likewise, Buddhist cosmography had a more profound effect on the Japanese mapmaking tradition than on the Chinese, though Buddhism made its way to Japan through China and Korea.

Although the cartographic histories explored here rival those of the West in length and variety, they have not been presented as fully as the history of cartography in the West was in volume 1 and will be in volumes 3–6. One reason for this—as several authors in this book have noted—is that the study of non-European cartography is still only in its formative phase, even though it has gone on for more than half a century. For some time spans, such as the Tang in China, material is scant; for others, such as the Qing, the material is so vast that it has yet to be cataloged and assessed. Where artifacts are few, it might be possible to learn about maps from textual sources; but in a number of instances understanding is impeded by the lack of a word that corresponds to "map." Thus, as happened in other parts of this project, discontinuities in the record hamper attempts to construct a history.

In addition, as has been emphasized elsewhere in this work, the social contexts of mapping and mapmaking are just beginning to be understood. As in Europe, the Islamic world, and South Asia, those who made maps in the cultures surveyed here were generally members of intellectual elites. With the possible exception of Burma in the eighteenth and nineteenth centuries, there were no professional or specialist mapmakers. Those who made maps were often scholars striving for breadth of knowledge. They can be called cartographers, geographers, or astronomers only if one remembers that their activities crossed the disciplinary boundaries implied by those terms in their modern usages. Mapmaking impinged upon art, literature, science, religion, divination, magic, philosophy, and politics. The form and content of map images varied with their varying purposes. Not until relatively late in history did mapmakers in these cultures feel an allegiance to the canon of rationality and mathematical accuracy that characterizes the enterprise today.

European and Asian Cartographies Compared

This summary of East and Southeast Asian cartographic histories has so far offered a number of parallels with the history of the map in Europe. If these histories do parallel those of Europe, one might well ask whether their separation from the European history told in other volumes is justified: Why even make a distinction between the cartographic histories of Asia and Europe? It is possible—but not, in our view, justified—to conceive of the cartographies of the West and East as converging toward the goal of ever increasing verisimilitude and accuracy. The argument goes that nonspecialist forms of cartography were displaced by a specialist form deriving from the techniques of applied geometry advocated most influentially by Claudius Ptolemy. This was one of the results of contact with Europeans. In Europe too one might claim under the "convergent view" that the nonmathematical cartographies practiced during the Middle Ages were also displaced after the rediscovery of Ptolemy during the Renaissance. And, as during the Middle Ages, one could point out the development of mathematical techniques in China, Japan, and Korea that could serve as a foundation for adoption of Ptolemaic techniques. Even before the Europeans introduced Ptolemaic cartography into Asia, Chinese, Japanese, and Korean mapmakers had produced maps impressive for their mathematical accuracy.

The result of this convergent history is an approach to a worldwide practice of cartography that was specialized in two senses: executed by practitioners with training in techniques specific to mapmaking, and reflecting a sharper division between maps and other forms of representing space graphically. It makes less sense to speak of cultural or national styles of mapmaking today than to speak of such styles with regard to painting and calligraphy. Map readers of different cultural backgrounds can, without difficulty, recognize maps made today as such no matter the country of origin, though legends in different scripts may still offer resistance to reading. This may not have been true in the past, if the slow progress of cartographic Westernization is any indication.

The convergent view of cartographic history makes sense only if one accepts two propositions: first, that cartographic practice across most cultures was striving for today's mathematical cartography as its end; second, that the Western introduction of Ptolemaic techniques into Asia only accelerated that development. In opposition to this view, however, the research undertaken for this project has gathered compelling evidence that mapmakers in East and Southeast Asian cultures were not aiming to establish a mathematical art. This is particularly true of mapmakers in Southeast Asia and Greater Tibet, and generally true of mapmakers in China, Japan, and Korea. East Asian mapmakers in many cases had available to them some early statements on the usefulness of measured mapping and measurements from topographic surveys. They possessed the means to produce remarkable maps of large areas like the *Yu ji tu* (Map of the tracks of Yu, 1136). They also had developed the astronomical instruments that would have allowed them to make maps using projections and coordinate systems. A need for these technical features of Ptolemaic cartography never arose, however, because none of the cultures studied here conceived of the earth as spherical or, more precisely, found it necessary to do so. Looking at the mathematical techniques and instrumentation alone, one might be tempted to conclude that East Asian mapmakers had independently arrived one step away from modern cartography. But despite the availability of a variety of mathematical and mensurational technology, they often chose not to represent the measurements in the form of scale maps. Because such mapmakers were often members of well-educated elites, this inconsistency cannot be attributed to an imperfect understanding of mapmaking, much less to a lack of skill or to backwardness. It resulted from dedication to the written text as the primary authority for communication.

With the possible exception of East Asian astronomical mapping, what the examples of the Asian cartographies explored here suggest instead is that the history of European cartography does not necessarily provide the ideal model of the map. It does not seem proper to speak of a "normal" pattern of cartographic development in terms of a movement from pictures to hybrid picture maps to the modern mathematical map. Even if we recognize that certain classes of maps—such as topographic maps, often thought to represent the pinnacle of objectivity—have become more alike across the world in the twentieth century, it is unwise to assume that homogeneity has resulted. Casual inspection of today's topographic maps from various nations reveals wide differences in styles, criteria for inclusiveness of content, and approach that cannot be ignored. Indeed, a promising line of research might center on how traditional cartographies in the regions we have discussed have affected the modern "official" maps of national governments. Equally promising might be a comparison of the effect of European cartography in Asian countries that succumbed to colonial rule (including most states of Southeast Asia) with its effect in countries that did not (in particular, China, Korea, and Japan). In the former, the indigenous cartographies were effectively suppressed, creating a disjunction that was not apparent in East Asia.

The displacement of traditional mapmaking practices by European mathematical cartography did not necessarily represent "progress." In some respects it may have been a loss. What has been lost is an explicit and prominent human element in the cartographic image. Modern topographic maps tend to homogenize the landscape, adopting conventional signs to represent features on the earth's surface. As a result, they often lack humanity because they are drawn not from the perspective of a single observer but from a multiplicity of viewpoints: a view from everywhere is also a view from nowhere. Moreover, the uniform planimetric perspective of modern topographic maps often masks essential aspects of the features being portrayed, whereas the use of oblique perspectives or the combination of diverse perspectives manages to highlight those same essential qualities. Typically, then, the modern map experientially distances the map reader from the world, treating it as a mathematical object. This generalizing and abstracting power is one of the strengths of cartography, but it may come at the price of dehumanizing the representation.

Maps representing large areas that produced an abstract, distancing effect were made in traditional East and Southeast Asian societies, as noted in the case of the *Yu ji tu*, and these maps have drawn much scholarly attention in the past. Far more prevalent than those maps, however, were large-scale maps of small local areas that involved the reader in the landscape. One means of fostering that involvement was variable or reverse perspective, which often required a map reader to imagine turning bodily or being in several different places in order to make the map representations conform to the way they are normally viewed. By this means the mapmaker attempted to place the reader in the landscape, where one would have to turn one's head or body in order to view the terrain in all directions or to move about an object depicted, such as a monastery, so as to view its several sides. This sense of geographic illusion was enhanced by the traditional mapmaker's preference for pictorial over abstract representations of topographic features. On many maps, to be sure, the pictorial representations appear stylized and follow conventions of their own, but they impart at least something of the look of the landscape.

The response of the reader was often a major concern of traditional East and Southeast Asian mapmakers. This is yet another reason that the division between map and picture is often difficult to maintain. A map not only helped store geographic information, but also evoked aesthetic and religious responses. In China, for example, there survive texts of poems inscribed on maps that record the emotions of map readers when confronted by a cartographic image. In composite Tibetan paintings, representations of saints and deities often complement maps of the sacred places they are associated with, sometimes in a dominant position and sometimes subordinate to the cartographic component.

Map and Text

Poetic inscriptions on maps bring out another significant aspect of much traditional Asian cartography: the textuality of geographic representation. In cartography this textuality is not limited to the typography or calligraphy of the labels on maps or the presence of cartouches. It frequently extends to a complementary relationship between word and image. Most Asian societies attached great importance to the written word. Accordingly, members of the literate elite in these societies customarily preserved quantitative data on the locations of specific places and the distances between them in written narratives. Map images served primarily to reflect the appearance of the area depicted and to show spatial relationships among topographic features or hierarchical relationships based on the relative importance or sanctity of the features depicted. In these circumstances, drawing maps to a systematic geometric scale was not as important as previous accounts have made it appear. More important for the practice of aesthetic and religious cartography was that a mapmaker have the freedom to vary scale for rhetorical emphasis and emotional impact. What are now regarded as functions of the cartographic image—the preservation of shape and distance—were shared by text and image. What would now be thought of as an atlas might in certain earlier contexts have been conceived as an illustrated book, such as the geographic compendiums described in the preceding chapters. For this reason, one can hardly evaluate the accuracy of a culture's geographic knowledge solely on the evidence of the cartographic image. To properly assess a culture's knowledge of geography, one must often look to a diversity of texts.

Among the East Asian literati, geographic information was often transmitted through printed media—books—rather than by direct human contact. The literate elites in Japan, Korea, and Vietnam, for example, made it a point to import and read Chinese books, and they often used Chinese maps and other geographic materials in making their own maps. In compiling the *Honil kangni yŏktae kukto chi to* (Map of integrated lands and regions of historical countries and capitals, 1402), Korean mapmakers relied on what were considered the best Chinese, Korean, and Japanese maps available. Similarly, when incorporating Korea into their "maps of all under heaven," Chinese mapmakers seem to have relied on

Korean-made images. As these examples and others in the preceding chapters suggest, the mapmaker's task in East Asia was as much a bibliographical endeavor as an empirical one. Comparable examples from Southeast Asia and Greater Tibet are not known. Nevertheless, the making of maps based on the accumulation of knowledge from diverse sources was evident, as exemplified by the Nepali map of Central Asia or the map of the greater part of Asia in some versions of the Thai *Trai Phum.* Thus mapmaking largely, though not exclusively, involved reprocessing and reinterpreting source materials. It would not be an exaggeration to say that for a geographer, particularly in East Asia, to study a place meant above all to read the relevant graphic materials, not to journey there for a personal inspection.

Representing the Physical and Metaphysical World

Understanding a place could be a matter of mensuration, but it also could be a matter of inspiration. This suggests another aspect of the geographic illusion created with multiple viewpoints. A map was made not only to convey information in the sense of facts, but also to communicate the mapmaker's experience of the land—intellectual, perceptual, and affective. From a traditional perspective, modern practice seems somewhat impoverished in these aspects.

There is yet at least one other way traditional East and Southeast Asian mapmaking exhibits a pronounced human dimension. As in European medieval mapping, cosmological inquiry often fostered cartographic development. The difference was that in Asia mapmaking did not sever its relationship with this sphere of human activity. The preceding chapters have shown how closely related were the religious and political pursuits that fostered map production; in some cases it is hard to distinguish the two. Political centers often coincided with religious centers, and these were also usually centers of map production. Maps were useful for exerting political control, as is often pointed out above. But besides serving the interests of ruling elites by providing them with information about their domains, maps had less secular purposes. They not only represented the observable world, but also modeled what was unseen: the entire cosmos including realms of spiritual beings, heavens and netherworlds, different realms of existence, and the configurations of invisible natural forces—as, for example, in siting and divination. Mapping was important not only for journeying through geographic space but also for spiritual wayfinding. As suggested in a number of chapters here and in previous volumes, the different kinds of space were often not sharply delineated. The secular was often also sacred. Political space was often simultaneously spiritual space. Spiritual space often overlapped into architectural space, particularly in temples, tombs, and reliquaries, which themselves were often three-dimensional cosmological models. The distinction between spiritual and physical was rendered invalid.

Maps purporting to represent unobserved macrocosms—unseen levels of existence—often seem to have an abstract, geometric quality, understandably so since the awareness they are based on supposedly transcends sensory perception. Such maps—for example, Tibetan mandalas and Southeast Asian temples—are often supposed to represent purer forms of reality, less subject to the irregularity found in the material world and reflected on traditional geographic maps in East and Southeast Asia. Thus, in those areas increasing abstraction in map images did not correspond with increased knowledge of the physical world. Frequently abstraction, simplification, and regularity were qualities desirable in the representation of the immaterial, the nonworldly. Transcendent experiences, after all, were valued as ways of escaping the contingencies, the disorder, and the unpredictability of the material world. To be accurate, the representations inspired by such experiences often needed to evoke the harmony, the constancy, and the tranquillity of nonearthly realms.

In contrast, accuracy in representing the terrestrial world often meant irregularity and concreteness. True, abstract schemes like the nonary square and grid pattern do appear on images of the terrestrial world, often on maps with political purposes such as imposing or imaging order and stability. In such cases the map image may be expressing a political desire to unify the terrestrial and the celestial, the secular and the religious, just as the Chinese emperor was supposed to do. Despite this rhetorical use of maps, it seems that observation of the terrain led to different conclusions about the order of the terrestrial landscape. It was composed of heterogeneous elements, as the difficulties of maintaining territory repeatedly drove home to rulers of all cultures. This realization of and respect for the particularity of place perhaps provides another explanation for why pictorial modes of mapmaking persisted so long after the development of quantitative techniques applicable to cartography. Perhaps this attitude also helps account for why the Westernization of Asian cartography did not take place as swiftly and totally as formerly believed. The modern European geographic map was perceived as too abstract, too homogeneous, and therefore implied insufficient materiality to be a valid geographic representation. Depending on what one expects from a cartographic image, a pictorial map may be more realistic than a modern mathematical map, and vice versa. For some purposes a heterogeneous mapping space may be more useful than a homogeneous mapping space, and vice versa. The homogeneous mapping space of modern cartography is not intrinsically better.

This last statement is not intended to slight modern cartography, but is meant as a reminder that modern cartography can still learn from what it has left behind. As a result of their human dimensions, traditional East and Southeast Asian maps leave one with an impression of the world that contrasts with that left by their modern mathematical counterparts: the world of the traditional artifacts is a vital place, one that interacts with human beings, not an inert conglomeration of physical forms. One may thus view with a sense of loss the triumph of the modern. There is little doubt that it was and continues to be a triumph in many ways. Modern cartography dazzles: it is more highly technologized, it is faster, it is more efficient, it is more accurate than premodern cartography; and it is constantly being refined. But it does not capture as broad a range of human culture and experience as traditional Asian mapping. There is something, a certain expressiveness, in the hand-drawn or hand-carved line that is inimitable by modern mechanical means of graphic reproduction. After looking at the artifacts reproduced here, one begins to understand more fully the need for an aesthetics, as well as a science, of mapmaking.

The word "begins" in the preceding sentence should be emphasized. As Laozi says in the *Dao de jing* (Classic of the way and its power), the further one goes, the less one knows. Similarly, the further this project goes, the less manageable it seems to have become, particularly in respect to the range of issues that call out for analysis and the kinds of questions that need to be answered. Thus the more we understand, the clearer it becomes that the days when one person could hope to write a comprehensive history of cartography have long since passed. This is particularly true in an age when scholars have specific disciplines, for a map is best understood and appreciated as an intersection of what are often regarded as diverse disciplines. The conclusions presented here are meant to serve merely as preliminary findings—or often, rather, as provisional hypotheses—and to suggest the importance of continuing inquiry into traditional Asian mapmaking.

The Asian stage of the History of Cartography Project may be concluding, but work in this area has to continue. Several tasks of basic scholarship remain. Collections of Asian maps need to be identified; catalogs of those collections need to be compiled; and the maps themselves need to have their contents analyzed. As these tasks are accomplished, it will become increasingly possible to draw reliable conclusions about map types and styles, subjects on which there is room for considerable refinement. Also in need of development is the basic matter of consistent descriptive terminology.

A Tentative Typology

Some regularities in dominant map function emerge from the range of artifacts described in this book. It is possible to construct a tentative typology of map functions related to the scales of human experience at which they are represented. Such a task is fraught with difficulties and pitfalls, however, for different map functions are not exclusively represented at particular scales, nor is the dominant function always clear. Rather than identify a number of map types or genres, therefore, we are offering broad categories of function as a basis for discussion. There would be great risk in putting this typology in a diagram, because it might be taken too literally and because its apparent authority might mask its limitations. Nevertheless, we provide a tentative list of functions with scales of human experience that could form the components of such a matrix as table 21.1.

The typology is made more complex by the distinction between representations of the physical and metaphysical worlds. Clearly both are equally "real" to the makers and users of these representations, so that the issue is not simply one of gradations of verisimilitude. Moreover, in many maps there is a quite natural merging from one domain to another (for example, from the cosmographical to the physical geographic) within the same representation.

Some generalizations may be made from this list. Although the cosmos is entered at the smallest scale, there is a sense in which cosmographical representations of the upper worlds or netherworlds are scaleless, even though in Buddhist cosmographies the dimensions of heavens and hells are often quite explicit.

Maps whose main function is wayfinding appear largely in the middle scales (such as provincial or regional), as do intelligence maps showing topographic information such as mountains and river systems and strategic sites such as forts and defenses. Inventory is a pervasive category that appears to be called for at many scales (shrines, landed property, towns and regional resources, countries). Divination functions, on the other hand, are often appropriate at more local scales such as in siting graves, altars, ordination halls, houses, temples, and cities. When applied to directions of auspicious or inauspicious travel or military planning the scale is more regional.

Better descriptive terminology will develop when the origins and context of the maps become clearer. Dating of artifacts is uncertain in a number of cases, since we do not understand the processes by which they were made and what conventions governed copying. An

TABLE 21.1 Tentative List of Map Functions with Scales of Human Experience

DOMINANT MAP FUNCTION
(physical and metaphysical)

- Inventory
  - Places
  - Natural resources
  - Landed property
- Wayfinding
  - Navigation
  - Pilgrimage
  - Commerce
- Intelligence
  - Topographic information
  - Strategic installations
- Education
- Planning and Engineering
  - Architectural
  - City
  - Hydrological works
  - Gardens
- Administration
  - Political
  - Fiscal
- Commemoration
  - Political influence
  - Events
- Divination
  - Astrology
  - Siting
- Worship
  - Relics
  - Meditation
  - Mnemonic

SCALE OF HUMAN EXPERIENCE
(small to large)

- Room
- Building
- Neighborhood
- Town
- Local administrative area (e.g., county)
- Regional administrative area (e.g., province)
- Country
- Continent
- World
- Cosmos

improved chronology of artifacts would help us better understand the relation between specific artifacts and map types. When the artifacts are properly classified and dated, then it may be possible—in terms of aesthetic, religious, technical, or historical significance—to establish criteria for informed assessment. Writers from within the tradition, such as Shen Kuo, made a start, but no one really followed up on their work in a serious and sustained manner.

FUTURE NEEDS

The study of East and Southeast Asian cartography still awaits scholars with the necessary cultural sensitivity and knowledge to undertake the foundational work. Those interpreting traditional Asian maps in the future will have to broaden their knowledge in order to deepen it for a better understanding of the functions maps served. Future historians of Asian cartography will need to study not only science and technology, but also art, mythology, and religion. The origins of astronomy in religious needs have long been understood; less well understood and appreciated are the origins of cartography in religion. To a great extent, mapmakers in Europe broke off this relationship during the Renaissance. Those in Asia maintained it well into the nineteenth century, and in some areas, especially Greater Tibet, the nexus remains strong. Nevertheless, maps in the Buddhist tradition, to name but one Asian religion, are still little studied, particularly those produced in China and Korea. Not much is known about what artifacts survive, and the same holds for the range of uses these maps served and the conventions regarding their manufacture.

Another promising research frontier is the technological aspects of map production. The vast majority of East Asian maps were disseminated through the woodcut medium, but the processes of block making and taking impressions were quite different from their Western equivalents. In a culture that valued fineness and subtlety of expression in the graphic arts, we need to ask why line engraving on metal plates was not used for map printing in China when it was quite clearly present for ornamental metalwork, and why the comparative crudity of the woodcut was deemed acceptable for maps. We need to know far more about the processes of copying and publication and the means of dissemination, selling, and ownership in those cultures where multiple copies of maps were made. Since printed maps almost invariably appeared in books, these questions will also have to be asked of students of book production in China, since they do not seem to have been answered previously.

Another obvious desideratum is obtaining a clearer and more complete picture of what cartographic impulses were transmitted at particular periods from one part of Asia to another and between the several major regions of Asia and other parts of the world. Specific routes, dates, and agents of transmission as well as the reasons for transmission have to be ascertained. In general, determining such facts will require a much deeper search of relevant documents than has been possible so far, as well as the accumulation of a body of datable artifacts sufficiently complete to reconstruct a reasonably continuous record of the types of maps produced in different periods of history for each of the regions that concern us. The meagerness of our existing knowledge in respect to the diffusion of cartographic knowledge and artifacts is perhaps the greatest single lacuna confronting historians of Asian cartography. Some previous attempts have been made to suggest lines of transmission of certain cartographic ideas, such as the grid. Joseph Needham published a diagram in *Science and Civilisation in China* (vol. 3, table 40) showing the possible transmission of the idea of parallels and meridians from Ptolemy to a tradition of grids for which he claims Pei Xiu was famous. On the same diagram he traces a possible reciprocal route of the idea from the grid maps of the Song dynasty through al-Qazwīnī and Marino Sanudo and so to the European Renaissance. But as we have seen, the idea of the Chinese grid (seen as of great value for imposing order on towns and the landscape) was very different from the idea of a global graticule of parallels and meridians, for which the Chinese had little use or interest. Studies of the transmission of cartographic ideas must therefore start with a precise understanding of the ideas themselves, and of what they really meant for the cultures involved.

The scope of the study of traditional Asian cartography could be broadened in other ways as well. Our knowledge so far has been based largely on investigations focused on the social elites. For a complete understanding of map use and production in the cultures studied here, we have to determine whether and how the map consciousness of elites differed from that of commoners. Maps from nonliterate, tribal societies present a different set of problems, and it remains to be demonstrated whether they are in some ways fundamentally different from those made in societies where maps often complement or supplement written texts.

Toward a New Cartographic Historicism

This book appears at a time of renewed interest in historical studies. The work presented here accords with the interdisciplinary tendencies of much recent work in the field. It also validates the current reaction against totalizing schemes or models as a means of understanding history across cultures. We have tried to be as empirical and inductive as possible, to treat cultures on their own terms, to let the data lead us to conclusions, and not to let modern presuppositions skew our perceptions.

Our approach has much in common with the "new historicism," according to which cultural artifacts should be interpreted in light of historical evidence in context. The objections to this approach, namely the incompleteness of the historical record and the loss of standards that cultural relativism entails, do not seem to seriously invalidate the contribution to knowledge of which our method is capable. On the first score, as noted above, some materials may be irretrievably lost, but that does not preclude constructing narratives of the past. The result may be a plurality of narratives, differing in how they reconstruct the gaps in the record. On the second point, contextualism—which, along with the plurality of narratives, is often associated with a relativist view of truth (that there is no certain ground for knowledge)—need not result in the loss of standards or values. The result can be a reformulation or a refounding of values. In this history, for example, by looking at other cartographies, we have learned to see modern Western cartography in a new light. Western cartography developed out of practices that recognized, at least to some extent, the human dimensions discussed above. The traditional practices of East and Southeast Asia represent paths that Western cartography could have followed but did not, at some cost.

Nor need a plurality of narratives lead to confusion or even despair, as those who wish for a coherent central narrative might claim. It is still possible to separate plausible from implausible, convincing from unconvincing, and good from bad narratives. The narratives presented here are the ones that seem to make the most sense based on available materials. We have tried to explain why we believe these accounts are an improvement over past narratives. We look forward to hearing future retellings and to learning from other storytellers.

Editors

J. B. Harley (1932–91) was professor of geography, University of Wisconsin–Milwaukee.

David Woodward is professor of geography, University of Wisconsin–Madison.

Associate editor Joseph E. Schwartzberg is professor of geography and South Asian studies, University of Minnesota, Minneapolis.

Assistant editor Cordell D. K. Yee teaches in the Great Books Program, St. John's College, Annapolis.

Authors

Catherine Delano Smith is research fellow, Institute of Historical Research, University of London.

John B. Henderson is professor of history, Louisiana State University, Baton Rouge.

G. Henrik Herb is visiting assistant professor of geography, Middlebury College, Vermont.

Gari Ledyard is professor of Korean history, Columbia University, New York.

Kazuhiko Miyajima is professor of astronomy and history of science, Dōshisha University, Kyōto.

Nathan Sivin is professor of Chinese culture and of the history of science, University of Pennsylvania, Philadelphia.

F. Richard Stephenson is senior research fellow in physics, University of Durham.

Kazutaka Unno is professor emeritus of geography, Ōsaka University.

John K. Whitmore works at the Graduate Library, University of Michigan, Ann Arbor.

Project Staff

Judith Leimer, managing editor
Susan MacKerer, project administrator
Kevin Kaufman, research associate
Christina Dando, project assistant
Barbara Whalen, project assistant
Charles Dean, program assistant

Bibliographical Index

Bibliographical Access to This Volume

Two modes of access to bibliographical information are used in this volume: the footnotes and the Bibliographical Index.

The footnotes provide the full form of a reference the first time it is cited in each chapter, with short titles in subsequent citations. In each of the short-title references, the note number where the fully cited work can be found is given in parentheses.

The Bibliographical Index constitutes a complete list of all works cited in the footnotes, tables, appendixes, and figure and plate legends. Numbers in bold type indicate the pages on which the references are cited. This index is divided into five parts. The first and second parts provide characters and translations for Chinese, Korean, and Japanese collections of works and journals. The third and fourth parts provide characters and translations and also page citations for all Chinese, Korean, and Japanese sources (pre-1900 alphabetized by title and post-1900 alphabetized by author). The fifth part indexes sources in all other languages, alphabetized by author.

Chinese, Korean, and Japanese Sources

COLLECTIONS OF WORKS

Baibu congshu jicheng 百部叢書集成 (Complete collection of collectanea from one hundred classifications). Taipei: Yiwen Yinshuguan, 1965–71.

Chosŏn wangjo sillok 朝鮮王朝實錄 (Royal annals of Chosŏn). 48 vols. Ed. Kuksa P'yŏnch'an Wiwŏnhoe 國史編纂委員会 (National History Compilation Committee of the Republic of Korea). Seoul: Kuksa P'yŏnch'an Wiwŏnhoe, 1955–58.

Hokumon sōsho 北門叢書 (Northern gateway series). 6 vols. Ed. Ōtomo Kisaku 大友喜作. Tokyo: Hokkō Shobō, 1943–44. Reprinted Tokyo: Kokusho Kankōkai, 1972.

Lidai tianwen lüli deng zhi huibian 歷代天文律曆等志彙編 (Collected treatises on astrology, astronomy, and harmonics in the standard histories). 9 vols. Beijing: Zhonghua Shuju, 1976.

Nihon shomin seikatsu shiryō shūsei 日本庶民生活史料集成 (Collected historical records about the lives of the Japanese people). 20 vols. Tokyo: San'ichi Shobō, 1968–72.

Shintei zōho kokushi taikei 新訂續國史大系 (Series of histories of our country revised and enlarged). 66 vols. Tokyo: Yoshikawa Kōbunkan, 1929–64.

Sibu beiyao 四部備要 (Essential collection [of books] from the four classifications). Shanghai: Zhonghua Shuju, 1927–35.

Sibu congkan 四部叢刊 (Collection [of books] from the four classifications). Shanghai: Shangwu Yinshuguan, 1920–36.

Siku quanshu 四庫全書 (Complete library from the four treasuries, comp. 1773–82). Taipei: Taiwan Shangwu Yinshuguan, 1970–82.

Zoku gunsho ruijū 續群書類從 (Classified series of various books: Continuation, 1923–28 in 71 vols.). 3d rev. ed. 67 vols. Tokyo: Zoku Gunsho Ruijū Kanseikai, 1957–59.

Zokuzoku gunsho ruijū 續續群書類從 (Classified series of various books: Second continuation). 16 vols. Tokyo: Kokusho Kankōkai, 1906–9; reprinted 1969–78.

JOURNALS

Bessatsu Taiyō 別册太陽 (The sun, special issue)

Biburia ビブリア (Biblia [Bulletin of the Tenri Central Library])

Chirigakushi Kenkyū 地理學史研究 (Research in the history of geography)

Chiri Ronsō 地理論叢 (Collected articles in geography)

Chōsen Gakuhō 朝鮮學報 (Journal of the Academic Association of Koreanology in Japan)

Denki 傳記 (Biography)

Dixue Zazhi 地学杂志 (Geographical journal)

Doshisha Daigaku Rikōgaku Kenkyū Hōkoku 同志社大學理工學研究報告 (The science and engineering review of Doshisha University)

Fudan Xuebao 复旦学报 (Journal of Fudan University in the social sciences)

Gekkan Kochizu Kenkyū 月刊古地圖研究 (Antique maps)

Gotō Puranetaryumi Gakugeihō 五島プラネタリウム學藝報 (Gotō Planetarium literature report)

Han'guk Hakpo 韓國學報 (Journal of Korean studies)

Han'guk Munhwa 韓國文化 (Korean culture)

Han'guk sa Yŏn'gu 韓國史研究 (Korean historical studies)

Hokkaidō [Teikoku] Daigaku Hoppō Bunka Kenkyū Hōkoku 北海道 [帝國] 大學北方文化研究報告 (Studies from the Research Institute for Northern Culture, Hokkaido [Imperial] University)

Hōsei Daigaku Bungakubu Kiyō 法政大學文學部紀要 (Journal of the Faculty of Letters, Hōsei University)

Ishigaki Shi Shi no Hiroba 石垣市史のひろば (Ishigaki Municipal History Forum)

Jinbun Chiri 人文地理 (Human geography)

Kagakushi Kenkyū 科學史研究 (Research in the history of science)

Kaiji Shi Kenkyū 海事史研究 (Journal of maritime history)

Kaogu 考古 (Archaeology)

Kaogu Xuebao 考古學報 (Journal of archaeology [Acta Archaeologia Sinica])

Kaogu yu Wenwu 考古与文物 (Archaeology and cultural relics)

Keji Shi Wenji 科技史文集 (Collected works on the history of science and technology)

Kexue 科學 (Science)

Kexue Shi Jikan 科学史集刊 (History of science)

Kirishitan Kenkyū キリシタン研究 (Christian research)

Kōbe Shiritsu Hakubutsukan Kenkyū Kiyō 神戸市立博物館研究紀要 (Bulletin of the Kōbe City Museum)

Kōkogaku Zasshi 考古學雜誌 (Journal of the Archaeological Society of Nippon)

Kokushi Kaikokai Kiyō 國史回顧會紀要 (Bulletin of the Society for Recollecting Japanese History)

Kōtsū bunka 交通文化 (Traffic culture; Cultural intercourse)
Kyōiku Jihō 教育時報 (Journal of Education)
Nagasaki Dansō 長崎談叢 (Memoirs of Nagasaki)
Nagasaki Shiritsu Hakubutsukan Kanpō 長崎市立博物館館報 (Journal of the Nagasaki City Museum)
Nagoya Daigaku Bungakubu Kenkyū Ronshū 名古屋大學文學部研究論集 (Journal of the faculty of literature, Nagoya University)
Naniwakyū Shi no Kenkyū 難波宮址の研究 (Reports of the historical investigation of the forbidden city of Naniwa)
Nanjing Daxue Xuebao 南京大学学报 (Journal of Nanjing University, Natural Science)
Nihongakuhō 日本學報 (Japanese studies [Ōsaka University])
Nihon Rekishi 日本歷史 (Japanese history)
Nihon Yōgakushi no Kenkyū 日本洋學史の研究 (Studies on the history of Western learning in Japan)
Ōsaka Gakugei Daigaku Kiyō 大阪學藝大學紀要 (Research papers of Ōsaka Gakugei University)
Ōtani Gakuhō 大谷學報 (Journal of Ōtani University)
Rangaku Shiryō Kenkyūkai Kenkyū Hōkoku 蘭學資料研究會研究報告 (Reports of the Society of Dutch Sources in Japan)
Rekishi Chiri 歷史地理 (Historical geography)
Ryūkoku Daigaku Ronshū 龍谷大學論集 (Research papers of Ryūkoku University)
Seikyū Gakusō 青丘學叢 (Blue Hills journal)
Shigaku Kenkyū 史學研究 (Review of historical studies)
Shigaku Zasshi 史學雜誌 (Journal of the Historical Society of Japan)
Shizen 自然 (Nature)
Shoshigaku 書誌學 (Bibliography)
Tenmon Geppō 天文月報 (Astronomical monthly/Astronomical herald)
Tianwen Xuebao 天文學報 (Journal of astronomy [Acta Astronomica Sinica])
Tōhō Gakuhō 東方學報 (Journal of Oriental studies)
Tōhō Shūkyō 東方宗敎 (Eastern religions)
Tōkyō Daigaku Shiryō Hensanjo Hō 東京大學史料編纂所報 (Journal of the Historiographical Institute, University of Tokyo)
Tongbang hakchi 東方學志 (Journal of Far Eastern studies)
Tosŏgwan 圖書館 (Bulletin of the Central National Library of Korea)
Wenwu 文物 (Cultural relics)
Wenxian 文獻 (Documents)
Yokohama Shiritsu Daigaku Kiyō 橫濱市立大學紀要 (Journal of the Yokohama City University)
Yomiuri Shinbun 讀賣新聞 (Yomiuri newspaper)
Yu Gong Banyuekan 禹贡半月刊 ("Tribute of Yu" semimonthly [Chinese historical geography])
Zhongyang Yanjiuyuan Jindaishi Yanjiusuo Jikan 中央研究院近代史研究所季刊 (Bulletin of the Institute of Modern History, Academia Sinica)
Ziran Kexue Shi Yanjiu 自然科学史研究 (Studies in the history of natural science)

PRE-1900

"1596 nendo Iezusu Kai nenpō" 一五九六年度イエズス會年報 (Annual report of the Society of Jesus, 1596). Luís Fróis. In *Kirishitan Kenkyū* 20 (1980): 261–410. Trans. Sakuma Tadashi 佐久間正. **390**
Ainōshō 壒嚢鈔 (Bag of rubbish, 1446). Gyōyo 行譽. **376**
Akan sanzai zue 無飽三財圖會 (Encyclopedia of the insatiable spending of money, 1822). Akatsuki no Kanenari 曉鐘成. **431**
Azuma kagami 吾妻鏡 (Mirror of the eastern lands, compiled late thirteenth to early fourteenth century). **395**
Baopuzi 抱朴子 ([Book of] the master who embraces simplicity, ca. fourth century). Ge Hong 葛洪. **226**
Beitang shuchao 北堂書鈔 (Transcriptions from the Northern Hall, compiled ca. 630). Yu Shinan 虞世南. **112**
Bundo yojutsu 分度餘術 (Techniques of protraction, manuscript of 1728). Matsumiya Toshitsugu 松宮俊仍. **359, 587**

Changjiang tushuo 長江圖說 (Illustrated account of the Yangtze, 1871). Ma Zhenglin 馬徵麟. **103**
Changxing ji 長興集 (Collected works of [the viscount of] Changxing). Shen Kuo 沈括 (1031–95). In *Shen shi san xiansheng wen ji* 沈氏三先生文集 (Collected works of the three masters of the Shen clan, compiled 1718). **114**
Chaoshi congzai 朝市叢載 (Collected notes for going to market, 1833). Yang Jingting 楊靜亭. **188**
Chikkyō yohitsu besshū 竹橋餘筆別集 (Superfluous writings at Bamboo Bridge, Edo Castle, extra volume, ca. 1803). Ōta Nanpo 大田南畝. **399**
Ch'ŏnggudo 青邱圖 (Map of the Blue Hills, 1834). Kim Chŏngho 金正浩. Modern edition ed. Yi Pyŏngdo 李丙燾. **314, 316, 321, 323**
Chŏngjong sillok 定宗實錄 (Annals of King Chŏngjong, r. 1398–1400). **245, 247**
Chŏngjo sillok 正祖實錄 (Annals of King Chŏngjo, r. 1776–1800). **310, 311, 320**
Chōsen nichinichi ki 朝鮮日日記 (Record of days in Korea [Keinen's diary], 1597–98). Keinen 慶念. **292**
Chŭngbo Munhŏn pigo 增補文獻備考 (Documentary reference encyclopedia, expanded and supplemented, 1903–8, revising first edition of 1770 and unpublished revision, 1790). **252, 253, 275, 285, 286, 288, 298, 305, 310, 311, 343, 556, 560, 561, 576**
Chungjong sillok 中宗實錄 (Annals of King Chungjong, r. 1506–44). **249, 267**
Chunqiu Guliang zhuan 春秋穀梁傳 (Spring and autumn [annals] with the Guliang commentary, fourth century). Comp. Fan Ning 范甯. **206**
Chūyūki 中右記 (Diary of the Nakamikado, 1131). Nakatsukasa Munetada 中務宗忠.
In vol. 12 of *Shiryō taisei* 史料大成 (Series of historical materials). 43 vols. Tokyo: Naigai Shoseki, 1934–43. **581**
In *Koji ruien* 古事類苑 (Historical encyclopedia of Japan, 1896–1914). 51 vols. Tokyo: Hyōgensha, 1927–30). *Hōgibu* 方技部 (Volume on technical specialists). **581**
Cunxing bian 存性編 (Treatise on preserving the nature, completed 1669, first printed 1705). Yan Yuan 顏元. In *Sicun bian* 四存編 (Treatises on the four preservations). Taipei: Shijie Shuju, 1966. **225**

Da Jin diaofa lu 大金弔伐錄 (Record of the Great Jin's consolation [of the people] and punishment [of the guilty], compiled ca. twelfth century). **83**
Da Ming yitong zhi 大明一統志 (Comprehensive gazetteer of the Great Ming, 1461). Li Xian 李賢 et al. **57, 60, 294, 296**
Da Qing huidian 大清会典 (Qing administrative code, 1732 edition). **305, 312**
Da Qing Shengzu Ren (Kangxi) huangdi shilu 大清聖祖仁（康熙）皇帝實錄 (Veritable records of Shengzu, emperor Ren [Kangxi], of the Great Qing, compiled ca. 1739). **82–83, 92, 180**
Da Qing Shizu Zhang (Shunzhi) huangdi shilu 大清世祖章（順治）皇帝實錄 (Veritable records of Shizu, emperor Zhang [Shunzhi], of the Great Qing, compiled ca. 1672). **177, 178, 180**
Da Qing yitong yutu 大清一統輿圖 (Comprehensive geographic map of the Great Qing, 1863). Comp. Hu Linyi 胡林翼 et al. **198, 202**

Nihonkoku genzaisho mokuroku 日本國見在書目錄 (A list of books at present in Japan, ca. 891). Fujiwara no Sukeyo 藤原佐世. **355, 580, 582**

Nihon taizaiki 日本滯在記 (An account of a sojourn in Japan [1667]). Oloff Erickson Willman. Trans. Ozaki Tadashi 尾崎義. Shin Ikoku Sōsho 新異國叢書 (New series on foreign countries). Tokyo: Yūshōdō Shoten, 1970. **391**

Oju yŏnmun changjŏn san'go 五洲衍文長箋散稿 (Oju's extended essays and lengthy notes, undated manuscript of ca. 1840s). Yi Kyugyŏng 李圭景. **254, 314**

Oranda tenchi ryōkyū shūho seizō ki 阿蘭陀天地兩球修補製造記 (Records of the repair of the Dutch celestial and terrestrial globes, 1795). Uma Michiyoshi 馬道良 (Kitayama Shin'yō 北山晉陽). **391**

Poxie lun 破邪論 (Discussion confuting heterodoxy, seventeenth century). Huang Zongxi 黃宗羲. In *Lizhou yizhu huikan* 黎州遺著彙刊 (A collection of [Huang] Lizhou's extant works, 1910). Taipei: Long Yan Chubanshe, 1969. **223**

Qinding Da Qing huidian 欽定大清會典 (Imperially commissioned collected statutes of the Great Qing, 1899). **82, 198, 200, 201**

Qingdai yitong ditu 清代一統地圖 (Comprehensive map of the Qing period, printed 1760). **186**

Rakuchū ezu: Kan'ei Manji zen 洛中繪圖: 寬永萬治前 (Plan of Kyōto: Kan'ei era [1624–43], before the Manji era [1658–60], ca. 1642). **401**

Ran'yaku teikō 蘭譯梯航 (Account of my difficulties in translating Dutch books, ca. 1816). Ōtsuki Gentaku 大槻玄澤. In *Yōgaku* 洋學 (Western studies). 2 vols. Ed. Numata Jirō 沼田次郎 et al. Nihon Shisō Taikei 日本思想大系 (Series of Japanese thought), vols. 64–65. Tokyo: Iwanami Shoten, 1972–76. **439**

Rekirin mondōshū 曆林問答集 (Questions and answers about the calendar, 1414). Kamono Arikata 賀茂在方. **583**

Rizhi lu 日知錄 (Record of daily knowledge, ca. 1670). Gu Yanwu 顧炎武. In *Yuanchaoben Rizhi lu* 原抄本日知錄 (The original manuscript version of the Record of daily knowledge, 1958). Taipei: Minglun Chubanshe, 1970. **227**

Ryōchi shinan 量地指南 (Surveying guidebook, 1733). Murai Masahiro 村井昌弘. In vol. 9 of *Edo kagaku koten sōsho* 江戸科學古典叢書 (Series of scientific classics during the Edo period). 46 vols. Tokyo: Kōwa Shuppan, 1976–83. **359**

Ryōgi shūsetsu 兩儀集說 (Collected theories of heaven and earth, 1714). Nishikawa Joken 西川如見. **349**

In vol. 18 of *Nishikawa Joken isho* 西川如見遺書 (Preserved works of Nishikawa Joken). 18 vols. Ed. Nishikawa Tadasuke 西川忠亮. Tokyo: Kyūrindō, 1898–1907. **381**

Ryō no gige 令義解 (Commentary on the codes, 833). **355, 583**

Saiiki zu sofuku nikō roku 西域圖纛覆二校錄 (Two revisions on a map of the western regions, 1737). In *Dainihon Bukkyō zensho* 大日本佛敎全書 (Collected records on the Buddhism of Great Japan). 151 vols. Tokyo, 1912–22. Vol. 2 of *Yūhōden sōsho* 遊方傳叢書 (Series of travels), 1–29. 4 vols. 1915. Reprinted Tokyo: Daiichi Shobō, 1979. **374**

Saken yoroku 查祆餘錄 (Diary of the Office of Christians, 1672–91). **391**

Samguk sagi 三國史記 (History of the Three Kingdoms, 1145). Comp. Kim Pusik 金富軾. **238–39, 273, 557**

"Samgukto husŏ" 三國圖後序 (Postface to a map of the Three Kingdoms). Yi Ch'ŏm 李詹. In *Tongmun sŏn* (below). **242, 279**

Samguk yusa 三國遺事 (Legends of the Three Kingdoms, compiled ca. 1285). Iryŏn 一然. Modern edition ed. and trans. Yi Pyŏngdo 李丙燾. **239**

Sancai tuhui 三才圖會 (Illustrated compendium of the three powers [heaven, earth, man], completed 1607, printed 1609). Comp. Wang Qi 王圻. **122, 123, 154, 176, 205, 206, 208, 211, 215, 556**

Sanyutang riji 三魚堂日記 (Diary from Three Fish Hall, written 1659–92). Lu Longqi 陸隴其. **223, 224**

Sejong sillok 世宗實錄 (Annals of King Sejong, r. 1418–50). **247, 274, 279, 284, 285, 287, 288**

Sejo sillok 世祖實錄 (Annals of King Sejo, r. 1455–68). **283, 288, 291**

Sessō yawa 雪窓夜話 (Night talk beside a window commanding a view of snow, published 1815). Tōkōji Sōryō 東光寺桑梁. **429**

Shaanxi tongzhi 陝西通志 (Comprehensive gazetteer of Shaanxi Province, 1735). **192**

Shangqing lingbao dafa 上清靈寶大法 (Great Praxis of the supernal magic treasure). In *Zhengtong Daozang* (below). **262**

Shang shu guangting lu 尙書廣聽錄 (Record of a broad understanding of the documents, seventeenth century). Mao Qiling 毛奇齡. **224**

Shang shu guwen shuzheng 尙書古文疏證 (Inquiry into the authenticity of the ancient text version of the documents, first printed 1745). Yan Ruoju 閻若璩. **223**

Shang shu Kongzhuan 尙書孔傳 (Documents of antiquity [Book of documents] with Kong's commentary). Kong Anguo 孔安國 (fl. ca. 100 B.C). **214**

Shanhai jing 山海經 (Classic of mountains and seas, compiled ca. second century B.C.). In *Shanhai jing jiaozhu* 山海經校注 (Edited and annotated *Shanhai jing*). Ed. Yuan Ke 袁珂. Shanghai: Shanghai Guji Chubanshe, 1980. **133, 261, 262, 263, 266, 314**

Shengzu Renhuangdi shengxun 聖祖仁皇帝聖訓 (Sacred instructions of Shengzu, the emperor Ren [Kangxi], presented 1732). In *Da Qing shichao shengxun* 大清十朝聖訓 (Sacred instructions of the ten reigns of the Great Qing). Taipei: Wenhai Chubanshe, 1965. **91**

Shenji zhidi Taibai yin jing 神機制敵太白陰經 (Secret classic of the Grand White on wondrous contrivances for subduing an enemy). See *Taibai yin jing* (below).

Shenyan 愼言 (Prudent words, 1533). Wang Tingxiang 王廷相. In *Wang Tingxiang zhexue xuanji* 王廷相哲學選集 (Selected philosophical works of Wang Tingxiang, 1965). Taipei: He-Lo Tushu Chubanshe, 1974. **227**

Shenyin yu 呻吟語 (Groaning words, 1593). Lü Kun 呂坤. **226**

Shi ji 史記 (Records of the grand historian, completed ca. 91 B.C.). Sima Qian 司馬遷. **73, 75, 76, 77, 78, 88, 97, 98, 128, 129, 133, 265, 515, 516, 518, 519, 520, 528**

In *Xinjiao Shi ji sanjia zhu* 新校史記三家注 (Newly collated *Shi ji* with three principal commentaries). 5 vols. Taipei: Shijie Shuju, 1972. **204**

Shi jing 詩經 (Book of odes, ca. twelfth to seventh century B.C.). **72**

Shinkan jinkoku ki 新刊人國記 (Newly published notes on the provinces and their inhabitants, 1701). Seki Sokō 關祖衡. **414**

In vol. 22 of *Kinsei bungaku shiryō ruijū, kohan chishi hen* 近世文學資料類從: 古板地誌編 (Classified series of materials of modern literature, early printed geographical descriptions). 22 vols. Tokyo: Benseisha, 1975–81. **412**

Shinsei chikyū bankoku zusetsu 新製地球萬國圖說 (Explanation of the new map of all the countries in the world, 1786). Trans. Katsuragawa Hoshū 桂川甫周. In *Kōmō* [or *Oranda*] *Zatsuwa* 紅毛雜話 (Miscellaneous notes of Holland). Ed. Ono Tadashige 小野忠重. Tokyo: Sōrinsha, 1943. **435**

Tongzhou zhi 通州志 (Gazetteer of Tongzhou [in modern Hebei Province, near Beijing], 1879). **90, 163, 196**

Tsūkō ichiran 通航一覽 (Collected documents for the history of Japanese diplomacy, ca. 1853). Hayashi Fukusai 林復齋 et al. **385, 391**

Tuhua jianwen zhi 圖畫見聞誌 (Record of experiences in painting, 1074). Guo Ruoxu 郭若虛. Seventeenth-century edition reprinted in *Songren huaxue lunzhu* 宋人畫學論著 (Song treatises on painting). Ed. Yang Jialuo 楊家駱. Taipei: Shijie Shuju, 1962. **135, 136**

Tushu bian 圖書編 (Compilation of illustrations and writings, compiled 1562–77). Comp. Zhang Huang 章潢. **121, 122, 166, 172, 175, 373**

Wakan sansai zue 和漢三才圖會 (Illustrated encyclopedia of Japanese and Chinese things related to the three powers [heaven, earth, and man], 1715). Ed. Terajima Ryōan 寺島良安. **368, 444, 585**

Wamyō ruijū shō 倭名類聚抄 (Classified glossary of Japanese terms, compiled ca. 935). Minamoto no Shitagō 源順. In Nihon Koten Zenshū 日本古典全集 (Comprehensive collection of Japanese classical works), 4th ser. Tokyo: Nihon Koten Zenshū Kankōkai, 1930–32. **357**

"Wen fu" 文賦 (Rhapsody on literature, third century). Lu Ji 陸機. In *Wen xuan* (below). **134**

Wenxian tongkao 文獻通考 (General study of literary remains, completed ca. 1280). Ma Duanlin 馬端臨. **222**

Wenxin diaolong 文心雕龍 (The literary mind and the carving of dragons, ca. 500). Liu Xie 劉勰. **128**
In *Wenxin diaolong yizhu* 文心雕龙译注 (*Wenxin diaolong* interpreted and annotated). 2 vols. Ed. Lu Kanru 陆侃如 and Mou Shijin 牟世金. Jinan: Qilu Shushe, 1981–82. **133, 134, 135**

Wen xuan 文選 (Literary selections, completed ca. 526–31). Comp. Xiao Tong 蕭統. 1809 edition ed. Hu Kejia 胡克家. **77, 132**

Wujing zongyao 武經總要 (Conspectus of essential military techniques, 1044). Zeng Gongliang 曾公亮. **116, 117**

Wushang xuanyuan santian yutang dafa 無上玄元三天玉堂大法 (Great rituals of the jade hall from the three heavens of the supreme, sublime, and primal, compiled 1158). Lu Shizhong 路時中. In *Zhengtong Daozang* (below). **167**

Wu shi xianying zhi 吳氏先塋志 (Record of the Wu family's ancestral graves, 1635). **85**

Wuxing dayi 五行大義 (Main principles of the five phases [or elements: wood, fire, earth, metal, water], ca. 600). Xiao Ji 蕭吉. **229**

Xiangshan xian zhi 香山縣志 (Gazetteer of Xiangshan County [in modern Guangdong Province], 1750). **82**

Xiaoan xinfa 曉庵新法 ([Wang] Xiaoan's new system, 1663). Wang Xichan 王錫闡.
In *Xiaoan yishu* 曉菴遺書 (Surviving works of [Wang] Xiaoan, compiled ca. 1682). **226**
In *Muxi xuan congshu* 木犀軒叢書 (A collection of books from Muci Studio, 1883–91). Comp. Li Shengduo 李盛鐸. **226**

"Xici zhuan" 繫辭傳 (Commentary on the attached verbalizations, ca. third century B.C.). In *Zhou yi yinde* 周易引得 (Concordance to the *Zhou yi* [*Yi jing*]). 1935; reprinted Taipei: Chengwen Chubanshe, 1966. **130, 131**

Xingshui jinjian 行水金鑑 (Golden mirror of flowing waters, 1725). Fu Zehong 傅澤洪. **104**

Xin Tang shu 新唐書 (New history of the Tang, compiled 1032?–60). Ouyang Xiu 歐陽修 et al. **82, 83, 88, 210, 238, 354, 533**

Xinzeng xiangji beiyao tongshu 新增象吉備要通書 (Newly amplified almanac of auspicious images, 1721). **209**

Xunzi 荀子 (ca. third century B.C.). Attributed to Xun Qing 荀卿. In *Xunzi yinde* 荀子引得 (A concordance to *Xunzi*). 1950; reprinted Taipei: Chengwen Chubanshe, 1966. **74**

Yangch'on chip 陽村集 (Collected writings of Kwŏn Kŭn [pen name Yangch'on, 1352–1409]). Kwŏn Kŭn 權近. **245**

Yejong sillok 睿宗實錄 (Annals of King Yejong, r. 1468–69). **267**

Yi jing Laizhu tujie 易經來注圖解 (Explanations of the diagrams [associated with] the Book of changes with Lai's commentary, 1598). Lai Zhide 來知德. **214**

Yinghuan zhilüe 瀛環志略 (Short account of the maritime circuit, completed 1848). Xu Jiyu 徐繼畬. **107, 108, 109**

"Yi tu lun" 易圖論 (Discussion of the diagrams associated with the [Book of] changes). In Gui Youguang 歸有光, *Zhenchuan xiansheng ji* 震川先生集 (Collected writings of [Gui] Zhenchuan). **224**

Yi tu mingbian 易圖明辨 (A clarifying critique of the diagrams [associated with the Book of] changes, 1706). Hu Wei 胡渭. **212**

Yi tu tiaobian 易圖條辨 (Systematic critique of the diagrams associated with the [Book of] changes, ca. 1800). Zhang Huiyan 張惠言. In *Zhang Huiyan yixue shishu* 張惠言易學十書 (Ten books by Zhang Huiyan on the study of the [Book of] changes). 2 vols. Taipei: Guangwen Shuju, 1970. **214, 225**

Yiwen leiju 藝文類聚 (Classified compendium of arts and letters). Comp. Ouyang Xun 歐陽詢 (557–641). **110**

Yixue bianhuo 易學辨惑 (Exposé of delusions regarding the study of the [Book of] changes, seventeenth century). Huang Zongyan 黃宗炎. In *Zhaodai congshu* 昭代叢書 (A collection of books from our illustrious era, 1876). Ed. Zhang Chao 張潮. **224**

Yixue qimeng 易學啓蒙 (Primer on the study of the [Book of] changes, 1186). Zhu Xi 朱熹. **214**

Yixue xiangshu lun 易學象數論 (Discussion of the images and numerologies associated with the study of the [Book of] changes, written ca. 1661). Huang Zongxi 黃宗羲. **224**

Yŏngjo sillok 英祖實錄 (Annals of King Yŏngjo, r. 1724–76). **307**

Yongle dadian 永樂大典 (Grand encyclopedia of the Yongle reign period [1403–24], presented 1409). **144, 145**

Yongping fu zhi 永平府志 (Gazetteer of Yongping Prefecture [in modern Hebei Province], 1879). **160, 195**

Yōrō ritsuryō 養老律令 (Yōrō code of laws, 718). **355**

Yuanchaoben Rizhi lu. See *Rizhi lu* (above).

Yuanhe junxian tuzhi 元和郡縣圖志 (Illustrated record of the prefectures and counties of the Yuanhe reign period [806–20], written 814). Li Jifu 李吉甫. **90**

Yuan shi 元史 (History of the Yuan, compiled 1369–70). Song Lian 宋濂 et al. **53, 119, 123, 588, 697**

Yuding lidai tihua shi lei 御定歷代題畫詩類 (Imperially commissioned, classified [collection] of poems inscribed on paintings through the dynasties, commissioned 1707). Comp. Chen Bangyan 陳邦彥 et al. **153, 160, 163, 164**

Yu gong huijian 禹貢會箋 (Collected commentaries on the "Yu gong," 1753). Xu Wenjing 徐文靖. **93**

"Yu gong wufu tu shuo" 禹貢五服圖說 (Explanation of the diagrams of the five domains of the Tribute of Yu). Kong Yingda 孔穎達 (574–648). In *Tushu bian* (above). **207**

Yu gong zhuizhi 禹貢錐指 (Using an awl to gauge the depths of the *Yu gong*, written 1694–97). Hu Wei 胡渭. **207**
In *Huang Qing jing jie* 皇清經解 (Explications of the classics from the Imperial Qing, compiled 1825–29). Ed. Ruan Juan 阮元 and Yan Jie 嚴杰. **27**

Yuhan shanfang ji yishu 玉函山房輯佚書 (Fragments of lost writings collected in the Jade Box Mountain Studio, compiled 1853). Comp. Ma Guohan 馬國翰. **264**

Yunlu manchao 雲麓漫抄 (Random jottings at Yunlu, 1206). Zhao Yanwei 趙彥衛. **138**

Yuzhou zhi 禹州志 (Gazetteer of Yuzhou [in present-day Henan Province], 1835). **153**

Zatto ichiran 襍土一覽 (Handbook of "Zatto," 1820). Suisai 醉齋子. **431**

Zen'aku meisho zue 善惡迷所圖會 (Illustrated book of noted places of good and evil, 1846). **431**

Zengguang Haiguo tuzhi 增廣海國圖志 (Expanded Illustrated record of maritime kingdoms, 1847 and later eds.). Wei Yuan 魏源. **199, 200**

Zhangde fu zhi 彰德府志 (Gazetteer of Zhangde Prefecture [in modern Henan Province], 1787). **90**

Zhanguo ce 戰國策 (Intrigues of the Warring States, perhaps third century B.C.). **73, 74**

Zhangzi zhengmeng zhu 張子正蒙注 (Master Zhang's correcting youthful ignorance with commentary). Zhang Zai 張載 (1020–77). **225**

Zhejiang tongzhi 浙江通志 (Comprehensive gazetteer of Zhejiang, 1736). **156**

Zhengding fu zhi 正定府志 (Gazetteer of Zhengding Prefecture [in modern Hebei Province], 1762). **90**

Zhengtong Daozang 正統道藏 (Daoist canon of the Zhengtong reign period [1436–49]). **167, 262**

Zhenjiang fu zhi 鎮江府志 (Gazetteer of Zhenjiang Prefecture [in present-day Jiangsu Province], no date, presumably the 1596 edition). **100, 101**

Zhihe fanglüe 治河方略 (Summary of river-control methods, 1767). Jin Fu 靳輔. Ed. Cui Yingjie 崔應階. **100**

Zhoubi suan jing 周髀算經 (Arithmetical classic of the Zhou gnomon, ca. 200 B.C.). **42, 118, 355, 357**

Zhou li 周禮 (Ritual forms of Zhou, compiled during the Han). **75, 76, 77, 88**

Zhou li Zhengzhu 周禮鄭注 (Ritual forms of Zhou with Zheng's commentary, second century). Zheng Xuan 鄭玄. **210**

Zhouxian tigang 州縣提綱 (Essentials of prefectural and county [government], eleventh century). Chen Xiang 陳襄. **91**

Zhuzi yulei 朱子語類 (Classified conversations of Master Zhu, 1270). Zhu Xi 朱熹. Comp. Li Jingde 黎靖德. **119**

Zōho kai tsūshō kō 增補華夷通商考 (Enlarged edition of *Kai tsūshō kō* [Trade with China and other countries, 1695], 1708). Nishikawa Joken 西川如見. In *Bankoku tokai nendaiki* 萬國渡海年代記 (Chronicle of Japanese intercourse with all the countries). Ed. Ono Tadashige 小野忠重. Tokyo: Shōrinsha, 1942. **381**

Zuozhuan 左傳 (Zuo's tradition [of interpreting the *Chunqiu*], ca. 300 B.C.). In *Chunqiu jingzhuan yinde* 春秋經傳引得 (Concordance to the *Chunqiu* [Spring and autumn annals] and its commentaries). 4 vols. 1937. Reprinted Taipei: Chengwen Chubanshe, 1966. **71–72, 99, 129**

POST-1900

Adachi Hiroyuki 安達裕之. "Kaifū sen shōkai kiji" 快風船渉海紀事 (Navigational record of the ship *Kaifū*). *Kaiji Shi Kenkyū* 14 (1970): 120–28. **382, 383**

Akaba Eiichi 赤羽榮一. *Mamiya Rinzō* 間宮林藏. Tokyo: Shimizu Shoin, 1974. **447**

Akabane Sōzō 赤羽壯造. "Takahashi Kageyasu no Shintei bankoku zenzu ni tsuite" 高橋景保の新訂萬國全圖について (On the *Shintei bankoku zenzu* by Takahashi Kageyasu). *Nihon Rekishi* 131–32 (1959): 78–95, 51–56. **439**

Akioka Korekushon Nihon no kochizu 秋岡コレクション日本の古地圖 (Old maps of Japan in the Akioka Collection). Exhibition catalog. Sakura: Rekishi Minzoku Hakubutsukan Shinkōkai, 1988. **413**

Akioka Takejirō 秋岡武次郎. *Nihon chizu shi* 日本地圖史 (History of maps of Japan). Tokyo: Kawade Shobō, 1955. **247, 249, 350, 367, 368, 369, 370, 411, 473**

———. "Momoyama jidai Edo jidai shoki no sekaizu byōbu tō no gaihō" 桃山時代, 江戸時代初期の世界圖屏風等の概報 (Outline of the world maps on folding screens of the Momoyama [ca. 1583 to ca. 1602] and early Edo periods). *Hōsei Daigaku Bungakubu Kiyō* 4 (1958): 263–311. **380**

———. "*Kon'yo bankoku zenzu byōbu* sōsetsu, Shibukawa Harumi byō narabini Tō Kōsekishi byō no sekaizu tenmonzu byōbu" 坤輿萬國全圖屏風總說, 澁川春海描並に藤黄赤子描の世界圖天文圖屏風 (General remarks on the *Kon'yo bankoku zenzu byōbu* [a folding screen on which a world map is drawn], and some remarks on folding screens with world maps and star maps by Shibukawa Harumi and by Tō Kōsekishi). *Hōsei Daigaku Bungakubu Kiyō* 8 (1962): 1–28. **590**

———. "Ogasawara shotō hakken shi no kihonshiryō chizu ni tsuite" 小笠原諸島發見史の基本資料・地圖について (On the fundamental documents concerning the discovery of the Bonin Islands). *Kaiji Shi Kenkyū* 9 (1967): 96–118. **385**

———. *Nihon chizu sakusei shi* 日本地圖作成史 (A history of the making of Japanese maps). Tokyo: Kajima Kenkyūjo Shuppankai, 1971. **399, 469**

———. *Nihon kochizu shūsei* 日本古地圖集成 (Collection of old maps of Japan). Tokyo: Kajima Kenkyūjo Shuppankai, 1971. **351, 369, 411, 412, 413, 465, 473**

———. *Sekai kochizu shūsei* 世界古地圖集成 (Collection of old world maps). English title, *Akioka Collection/Old World Maps/16th-19th Centuries*. Tokyo: Kawade Shobō Shinsha, 1988. Including *Sekai chizu sakusei shi* 世界地圖作成史 (A history of making world maps). **350, 351, 467, 469, 471**

Akiyama Terukazu 秋山光和, ed. *Genshoku Nihon no bijutsu* 原色日本の美術 (The fine arts of Japan in color). 30 vols. Tokyo: Shōgakkan, 1966–72. Vol. 25, *Nanban bijutsu to Yōfūga* 南蠻美術と洋風畫 (Nanban art and Western-style painting). **380**

Aoyama Sadao 青山定雄. "Gendai no chizu ni tsuite" 元代の地圖について (On maps of the Yuan dynasty). *Tōhō Gakuhō* (Tokyo) 8 (1938): 103–52. **245, 246, 248, 289**

———. "Richō ni okeru nisan no Chōsen zenzu ni tsuite" 李朝に於ける二三の朝鮮全圖について (On several Yi [Chosŏn] dynasty maps of Korea). *Tōhō Gakuhō* (Tokyo) 9 (1939): 143–71. **291, 292**

Ashida Koreto 蘆田伊人. "Nihon sōzu no enkaku" 日本總圖の沿革 (History of general maps of Japan). *Kokushi Kaikokai Kiyō* 2 (1930): 17–59. **350**

———. *Honpō chizu no hattatsu* 本邦地圖の發達 (The evolution of cartography in Japan). Tokyo: Iwanami Shoten, 1934. **350**

———. "Chizu to kōtsū bunka" 地圖と交通文化 (Maps and transportation culture). *Kōtsū bunka* 3–5 (1938–39): 282–90, 358–64, 445–54. **402**

Ayusawa Shintarō 鮎澤信太郎. "Mateo Ritchi no sekaizu ni kansuru shiteki kenkyū: Kinsei Nippon ni okeru sekai chiri chishiki no shuryū" マテオ・リッチの世界圖に關する史的研究: 近世日本における世界地理知識の主流 (Historical research on Matteo Ricci's world map: On the main current of the knowledge of world geography during the Tokugawa age). *Yokohama Shiritsu Daigaku Kiyō* 18 (1953). **404, 409, 410**

———. "Sekai chiri no bu" 世界地理の部 (Section of world geography). In *Sakoku jidai Nihonjin no kaigai chishiki* 鎖國時代日本人の海外知識 (Japanese knowledge of overseas during the age of national isolation), ed. Kaikoku Hyakunen Kinen Bunka Jigyō Kai 開國百年記念文化事業會 (Society of Cultural Projects to Commemorate the One Hundredth Anniversary of the Opening of the Country), 3–367. Tokyo: Kengensha, 1953. **426–28**

———. *Chirigakushi no kenkyū* 地理學史の研究 (Studies on the history of geography). Tokyo: Aijitsu Shoin, 1948; reprinted Hara Shobō, 1980. **435**

———. "Takeda Kango no Yochi kōkaizu no keitō" 武田簡吾の輿地航海圖の系統 (Genealogy of Takeda Kango's *Yochi kōkaizu*). In *Sakoku jidai no sekaichirigaku* 鎖國時代の世界地理學 (World geography in the age of national isolation), by Ayusawa Shintarō, 331–49. Tokyo: Nichidaidō Shoten, 1943; reprinted Hara Shobō, 1980. **443**

Bessatsu Taiyō (The sun, special issue). No. 8. Tokyo: Heibonsha, 1974. **377**

Cao Wanru 曹婉如. "Lun Shen Kuo zai dituxue fangmian di gongxian" 论沈括在地图学方面的贡献 (On Shen Kuo's contributions to cartography). *Keji Shi Wenji* 3 (1980): 81–84. **50, 114**

———. "Youguan Tianshui Fangmatan Qin mu chutu ditu di jige wenti" 有关天水放马滩秦墓出土地图的几个问题 (Several problems concerning the maps excavated from the Qin tomb at Fangmatan in Tianshui). *Wenwu*, 1989, no. 12:78–85. **39**

Cao Wanru et al., eds. *Zhongguo gudai ditu ji* 中國古代地圖集 (Atlas of ancient Chinese maps). Beijing: Wenwu Chubanshe, 1990–. Vol. 1, *Zhanguo-Yuan* 战国 —— 元 (Warring States to the Yuan dynasty). **36, 38, 39, 40, 41, 47, 55, 64, 69**

Chang Yong 常勇 and Li Tong 李同. "Qin Shihuang lingzhong maicang gong di chubu yanjiu" 秦始皇陵中埋藏汞的初步研究 (Preliminary study of the mercury interred in Qin Shihuang's tomb). *Kaogu*, 1983, no. 7:659–63, 671. **79**

Chen Feiya 陈菲亚 et al., eds. *Zhongguo gudai dilixue shi* 中国古代地理学史 (History of ancient Chinese geography). Beijing: Kexue Chubanshe, 1984. **46, 62, 164, 170**

Chen Guansheng 陳觀勝 (Kenneth Ch'en). "*Fangyu shenglüe* zhong geguo dufen biao zhi jiaoding" 方輿勝略中各國度分表之校訂 (Edited table of geographic coordinates for various countries in the *Fangyu shenglüe*). *Yu Gong Banyuekan* 5, nos. 3–4 (1936): 165–94. **175**

———. "Li Madou dui Zhongguo dilixue zhi gongxian ji qi yingxiang" 利瑪竇對中國地理學之貢獻及其影響 (Matteo Ricci's contributions to and influence on Chinese geography). *Yu Gong Banyuekan* 5, nos. 3–4 (1936): 51–72. **173**

Chen Zhengxiang 陳正祥 (Chen Cheng-siang). *Zhongguo dituxue shi* 中國地圖學史 (History of Chinese cartography). Hong Kong: Shangwu Yinshuguan, 1979. **170**

Chen Zungui 陳遵嬀. *Zhongguo tianwenxue shi* 中國天文學史 (History of Chinese astronomy). Taipei: Mingwen Shuju, 1984–. **511**

Chikusendō kosho tenkan mokuroku 竹僊堂古書展觀目錄 (Chikusendō's catalog of an exhibition of antique books). Kyōto: Chikusendō, 1974. **409**

Ch'oe Ch'angjo 崔昌祚. "Chosŏn hugi sirhakchadŭl ŭi p'ungsu sasang" 朝鮮後期實學者들의風水思想 (The geomantic thought of "practical learning" scholars in the late Chosŏn dynasty). *Han'guk Munhwa* 11 (1990): 469–504. **275**

Ch'oe Pyŏnghŏn 崔柄憲. "Tosŏn ŭi saeng'ae wa Namal Yŏch'o ŭi p'ungsu chiri sŏl" 道詵의 生涯와 羅末麗初의 風水地理說 (Tosŏn's career and geomantic theory in late Silla and early Koryŏ). *Han'guk sa Yŏn'gu* 11 (1975): 102–46. **277**

Chŏn Sang'un 全相運 (Sang-woon Jeon). *Kankoku kagaku gijutsu shi* 韓國科學技術史 (Science and technology in Korea). Tokyo: Koma-Shorin, 1978. **586**

Chōsen Sōtokufu 朝鮮總督府 ([Japanese] Government-General in Korea), ed. *Chōsen kinseki sōran* 朝鮮金石總覽 (A comprehensive survey of ancient Korean inscriptions). 2 vols. Seoul: Chōsen Sōtokufu, 1919. **240, 255, 374**

———. *Chōsen shi* 朝鮮史 (History of Korea). Six series comprising 37 vols. Seoul: Chōsen Sōtokufu, 1932–37. **250, 254**

Deng Wenkuan 邓文宽. "Bi 'Butian ge' geng gulao di tongshu shixing zuopin—'Xuanxiang shi' " 比《步天歌》更古老的通俗识星作品 —— 《玄象诗》(A popular work for star recognition older than the "Butian ge"—"Xuanxiang shi"). *Wenwu*, 1990, no. 3:61–65. **532**

Ding Fubao, ed. *Shuowen jiezi gulin*. See *Shuowen jiezi* (above).

Dong Zuobin 董作賓 (Tung Tso-pin). *Yin lipu* 殷曆譜 (On the calendar of the Yin dynasty). Lizhuang, Szechuan: Academia Sinica, 1945. **514**

Fu Xinian 傅熹年. "Zhanguo Zhongshan wang Cuo mu chutu di 'zhaoyu tu' ji qi lingyuan guizhi di yanjiu" 战国中山王嚳墓出土的《兆域图》及其陵园规制的研究 (A study of the mausoleum map unearthed from the tomb of King Cuo of the Zhanguo period's Zhongshan kingdom and the planning of the mausoleum). *Kaogu Xuebao*, 1980, no. 1:97–118. **37**

Fuchs, Walter. "Pekin no Mindai sekaizu ni tsuite" 北京の明代世界圖について (On the Ming-period world map in Beijing). *Chirigakushi Kenkyū* 2 (1962): 3–4, with 2 pls. Reprinted in *Chirigakushi kenkyū* 地理學史研究 (Researches in the history of geography), 2 vols., ed. Chirigakushi Kenkyūkai 地理學史研究會 (Society for Research in Historical Geography), 2:3–4 and pls. 1–2. Kyōto: Rinsen Shoten, 1979. **246**

Fujita Motoharu 藤田元春. *Toshi kenkyū Heiankyō hensenshi, tsuketari kochizu shū* 都市研究平安京變遷史附古地圖集 (History of the Kyōto region, accompanied by collected old plans). Kyōto: Suzukake Shuppanbu, 1930; reprinted Nihon Shiryō Kankōkai, 1976. **421, 473, 474**

———. *Nihon chirigaku shi* 日本地理學史 (History of Japanese geography). Tokyo: Tōkō Shoin, 1932. **350**

———. *Kaitei zōho Nihon chirigaku shi* 改訂增補日本地理學史 (Revised and enlarged history of Japanese geography). Tokyo: Tōkō Shoin, 1942; reprinted Tokyo: Hara Shobō, 1984. **350, 469, 471**

Fukai Jinzō 深井甚三. *Zuō Ochikochi Dōin* 圖翁遠近道印 (Zuō's [Fujii's] Ochikochi Dōin). Toyama: Katsura Shobō, 1990. **423**

Fukui Tamotsu 福井保. *Naikaku Bunko shoshi no kenkyū* 內閣文庫書誌の研究 (Studies on the bibliography of the Naikaku Library). Tokyo: Seishōdō, 1980. **397**

Fukuyama Toshio 福山敏男, supervisor. *Jinja kozu shū zokuhen* 神社古圖集續編 (Collected old drawings of shrines, continuation). Kyōto: Rinsen Shoten, 1990. **362, 365, 366**

Funakoshi Akio 船越昭生. "Kon'yo bankoku zenzu to sakoku Nippon" 『坤輿萬國全圖』と鎖國日本 (Ricci's world maps and Japan in the age of national isolation). *Tōhō Gakuhō* (Kyōto) 41 (1970): 595–710. **405**

Gansusheng Bowuguan 甘肃省博物馆 (Gansu Provincial Museum). "Wuwei Mozuizi sanzuo Hanmu fajue jianbao" 武威磨咀子三座汉墓发掘简报 (Brief report on the excavations of the three Han tombs at Mozuizi in Wuwei County). *Wenwu*, 1972, no. 12:9–21. **120**

Gansusheng Wenwu Kaogu Yanjiusuo 甘肃省文物考古研究所 and Tianshui Beidaoqu Wenhuaguan 天水市北道区文化馆 (Institute of Archaeology, Gansu Province, and Cultural Center of Beidao District of Tianshui). "Gansu Tianshui Fangmatan Zhanguo Qin Han muqun di fajue" 甘肃天水放马滩战国秦汉墓群的发掘 (Excavation of the tombs from the Qin state of the Warring States period and from the Han dynasty). *Wenwu*, 1989, no. 2:1–11. **40**

Gongzhongdang Qianlong chao zouzhe 宮中檔乾隆朝奏摺 (Palace memorials from the Qianlong reign period in the palace archives). 69 vols. Taipei: Guoli Gugong Bowuyuan, 1982–88. **102**

Gu Jiegang 顧頡剛 et al., eds. *Gushi bian* 古史辨 (Essays on ancient history). 7 vols. 1926–41; reprinted Hong Kong: Taiping Shuju, 1962. **76**

Gu ditu lunwenji 古地图论文集 (Essays on ancient maps). Beijing: Wenwu Chubanshe, 1977. **43, 45, 52**

Guo Shengchi 郭盛炽 . "Bei Song Heng xing guance jingdu chuyi" 北宋恒星观测精度刍议 (On the accuracy of observations of the North Star during the Northern Song). *Tianwen Xuebao* 30 (1989): 208–16. **548**

Han Ugŭn 韓㳓劤 et al., eds. *Yŏkchu Kyŏngguk taejŏn: Chusŏk pyon* 譯註經國大典：註釋篇 (The annotated *Kyŏngguk taejŏn*: Notes and commentary, translated [from Chinese into Korean] and annotated). Seoul: Han'guk Chŏngsin Munhwa Yŏng'guwŏn, 1986. **285, 295**

Han Zhongmin 韩仲民 . "Guanyu Mawangdui boshu gu ditu di zhengli yu yanjiu" 关于马王堆帛书古地图的整理与研究 (Concerning the restoration and study of the ancient silk maps from Mawangdui). In *Zhongguo gudai ditu ji* 中國古代地圖集 , ed. Cao Wanru 曹婉如 et al., 1:12–17. Beijing: Wenwu Chubanshe, 1990–. **54, 64**

Hanaki Yasuo 玻名城泰雄 . "Seizu ni tsuite" 『星圖』について (On a star chart). *Ishigaki Shi Shi no Hiroba* (Ishigaki Municipal History Forum) 11 (1987): 1, 3–7. **601**

Harada Tomohiko 原田伴彥 and Nishikawa Kōji 西川幸治 , eds. *Nihon no shigai kozu* 日本の市街古圖 (Old Japanese plans). 2 vols. *Nishi Nihon hen* 西日本編 (Western part of Japan) and *Higashi Nihon hen* 東日本編 (Eastern part of Japan). Tokyo: Kajima Shuppankai, 1972–73. **400**

Hashimoto Masukichi 橋本增吉 . *Shina kōdai rekiho shi kenkyū* 支那古代曆法史研究 (Studies on the history of ancient Chinese calendrical astronomy). Tokyo: Tōyō Bunko, 1943. **208**

He Shuangquan 何双全 . "Tianshui Fangmatan Qin mu chutu ditu chutan" 天水放马滩秦墓出土地图初探 (Preliminary study of the maps excavated from the Qin tomb at Fangmatan in Tianshui). *Wenwu*, 1989, no. 2:12–22. **38, 39**

———. "Tianshui Fangmatan Qin jian zongshu" 天水放马滩秦简综述 (Comprehensive account of the Qin bamboo slips from Fangmatan in Tianshui). *Wenwu*, 1989, no. 2:23–31. **38**

Hebeisheng Wenwu Guanlichu 河北省文物管理处 (Hebei Province Cultural Relic Agency). "Hebeisheng Pingshan xian Zhanguo shiqi Zhongshanguo muzang fajue jianbao" 河北省平山县战国时期中山国墓葬发掘简报 (Excavation of the tombs of the Zhongshan kingdom of the Zhanguo period at Pingshan County, Hebei Province). *Wenwu*, 1979, no. 1:1–31. **37**

Hebeisheng Wenwu Guanlichu, Hebeisheng Bowuguan 河北省文物管理处，河北省博物馆 (Hebei Province Cultural Relic Agency, Hebei Provincial Museum). "Liaodai caihui xingtu shi woguo tianwenshishang di zhongyao faxian" 辽代彩绘星图是我国天文史上的重要发现 (The Liao period star map, an important discovery in the history of Chinese astronomy). *Wenwu*, 1975, no. 8:40–44. **549**

Hidaka Jikichi 日高次吉 . "Hyūga sadowara hanshi Hidaka Shigemasa no Nanban ryū chōkenjutsu sonota" 日向佐土原藩士日高重昌の南蠻流町見術その他 (Hidaka Shigemasa's studies on the surveying of the Occidental school, etc.). *Kagakushi Kenkyū* 44 (1957): 17–24. **394**

Higuchi Hideo 樋口秀雄 and Asakura Haruhiko 朝倉治彥 , revisers. *Kyōhō igo edo shuppan shomoku* 享保以後江戸出版書目 (Bibliography of books printed from the Kyōhō era). Toyohashi: Mikan Kokubun Shiryō Kankōkai, 1962. **413**

Hiraoka Takeo 平岡武夫 . *Chōan to Rakuyō: Chizu* 長安と洛陽地圖 (Chang'an and Luoyang: Maps). T'ang Civilization Reference Series, no. 7. Kyōto: Jinbunkagaku Kenkyūsho, Kyōto University, 1956. **140**

Hirose Hideo 廣瀨秀雄 . "Kyū Nagasaki tengakuha no gakutō seiritsu ni tsuite: 'Nigi ryakusetsu' ni kanshite" 舊長崎天學派の學統成立について —— 二儀略說に關して (On the formation of the old Nagasaki school of astronomy: Concerning *Nigi ryakusetsu*). *Rangaku Shiryō Kenkyūkai Kenkyū Hōkoku* 184 (1966): 3–14. **393**

———. "Oranda tensetsu" 和蘭天說 (European astronomical theory). In *Yōgaku* 洋學 (Western studies). 2 vols. Ed. Numata Jirō 沼田次郎 et al. Nihon Shisō Taikei 日本思想大系 (Series of Japanese thought), vols. 64–65. Tokyo: Iwanami Shoten, 1972–76. **600**

———. "Tenkyūgi oboegaki" 天球儀覺え書き (Memorandum on celestial globes). *Gotō Puranetaryumi Gakugeihō* 6 (1978). **591**

Hoppō Ryōdo Mondai Chōsakai 北方領土問題調查會 (Japan Society for Research on the Northern Territories), ed. *Hoppō Ryōdo: Kochizu to rekishi* 北方領土，古地圖と歷史 (The northern territories of Japan: Old maps and history). Tokyo: Chūōsha, 1971. **443, 450**

Hora Tomio 洞富雄 . *Mamiya Rinzō* 間宮林藏 . Rev. ed. Tokyo: Yoshikawa Kōbunkan, 1987. **447**

Hoshi no bijutsuten: Tōzai no kichōna koseizu o atsumete 星の美術展 —— 東西の貴重な古星圖を集めて (Exhibition of stellar arts: A collection of rare old star charts of East and West). Exhibition catalog, ed. Chiba Shiritsu Kyōdo Hakubutsukan 千葉市立鄉土博物館 (Chiba City Local Museum). Chiba, 1989, no. 12. **597, 598**

Hoyanagi Mutsumi 保柳睦美 , ed. *Inō Tadataka no kagakuteki gyōseki: Nihon chizu sakusei no kindaika eno michi* 伊能忠敬の科學的業績 —— 日本地圖作製の近代化への道 (A new appreciation of the scientific achievement of Inō Tadataka). Tokyo: Kokon Shoin, 1974; rev. ed. 1980. **453**

Hu Daojing, ed. *Mengxi bitan jiaozheng*. See *Mengxi bitan* (above).

———. *Xin jiaozheng Mengxi bitan*. See *Mengxi bitan* (above).

Hunansheng Bowuguan 湖南省博物馆 and Zhongguo Kexueyuan Kaogu Yanjiusuo 中国科学院考古研究所 (Hunan Provincial Museum and Institute of Archaeology, Academia Sinica). "Changsha Mawangdui er, sanhao Han mu fajue jianbao" 长沙马王堆二、三号汉墓发掘简报 (Preliminary excavation report on Han tombs 2 and 3 at Mawangdui, Changsha). *Wenwu*, 1974, no. 7:39–48 and 63. **150–51**

Iida Ryūichi 飯田龍一 and Tawara Motoaki 俵元昭 . *Edozu no rekishi* 江戸圖の歷史 (History of the maps of Edo). 2 vols. Tokyo: Tsukiji Shokan, 1988. **400, 401, 421**

Imai Itaru 今井湊. "Edo Jidai kagakushi no naka no Blaeu" 江戸時代科學史の中の Blaeu (Blaeu in the history of science during the Edo period). *Rangaku Shiryō Kenkyūkai Kenkyū Hōkoku* 136 (1963). **600**

Imoto Susumu 井本進. "Honchō seizu ryakkō" 本朝星圖略考 (Summary of researches on celestial maps made in Japan), pts. 1 and 2. *Tenmon Geppō* 35 (1942): 39–41 and 51–57. **579, 583, 587, 596**

———. "Zoku honchō seizu ryakkō" 續本朝星圖略考 (Summary of researches on celestial maps made in Japan, continuation). *Tenmon Geppō* 35 (1942): 67–69. **579**

———. "Maboroshi no seishuku zu" まぼろしの星宿圖 (A lost celestial map). *Tenmon Geppō* 65, no. 11 (1972): 290–92. **579, 583, 585**

Inokuma Kanekatsu 猪熊兼勝 and Watanabe Akiyoshi 渡邊明義. *Takamatsuzuka kofun* 高松塚古墳 (The Takamatsuzuka burial mound). Nihon no Bijutsu 日本の美術 (Japanese art), no. 217. Tokyo: Shibundō, 1984. **352**

Itō Tasaburō 伊東多三郎. "Echigo Uesugi shi ryōgoku kenkyū no nishiryō" 越後上杉氏領國研究の二史料 (Two historical materials for studying the domains of the Uesugi family in Echigo). *Nihon Rekishi* 138 (1959): 2–14. **396**

Iwao Seiichi 岩生成一. "Ishibashi hakushi shozō sekaizu nendai kō" 石橋博士所藏世界圖年代考 (On the date of the world map in the collection of Dr. Ishibashi). *Rekishi Chiri* 61 (1933): 511–22. **388**

———. *Shinpan shuinsen bōeki shi no kenkyū* 新版朱印船貿易史の研究 (Studies on the history of trade under the vermilion-seal licenses of the Tokugawa shogunate, revised and enlarged edition). Tokyo: Yoshikawa Kōbunkan, 1985. **381**

Iwata Toyoki 岩田豊樹. *Edozu sōmokuroku* 江戸圖總目錄 (General catalog of plans of Edo). Tokyo: Seishōdō Shoten, 1980. **421**

Iyanaga Teizō 彌永貞三. "Handen tetsuzuki to kōhandenzu" 班田手續と校班田圖(Procedure for apportioning paddies and the maps prepared before and after). In *Shōen ezu kenkyū* 莊園繪圖研究 (Studies on manorial maps), ed. Takeuchi Rizō 竹内理三, 33–34. Tokyo: Tokyōdō Shuppan, 1982. **354**

"Kagaku Kenkyūhi ni yoru Kenkyū no Hōkoku" 科學研究費によ る研究の報告 (Reports on the research depending on scientific research expenses), "Genson Kochizu no Rekishi Chirigakuteki Kenkyū (Ippan Kenkyū A)" 現存古地圖の歷史地理學的研究 (一般研究 A) (Historical geographical research on extant old maps [general study A]). *Tōkyō Daigaku Shiryō Hensanjo Hō* 16 (1981): 25–40. **397**

Kanda Shigeru 神田茂. *Nihon tenmon shiryo* 日本天文史料 (Japanese astronomical records). Tokyo, 1935. **512**

Kashihara Kōkogaku Kenkyūjo 橿原考古學研究所 (Kashihara Archeological Institute), ed. *Hekiga kofun Takamatsuzuka* 壁畫古墳高松塚 (Takamatsuzuka: A burial mound with mural paintings). Nara and Asuka: Nara Ken Kyōiku Iinkai and Asuka Mura, 1972. **352**

Kawada Takeshi 河田羆. "Honpō chizukō" 本邦地圖考 (Study of the map of our country). *Shigaku Zasshi* 6 (1895): 268–77, 349–58, and 507–18. **350, 397**

———. "Nihon chishi gen'i o ronzu" 日本地誌源委を論ず (On the transition of geographical descriptions in Japan). *Rekishi Chiri* 7 (1905): 821–27, 916–21, 1038–45. **350**

Kawamura Hirotada 川村博忠. *Edo bakufu sen kuniezu no kenkyū* 江戸幕府撰國繪圖の研究 (A study of the provincial maps compiled by the Tokugawa shogunate). Tokyo: Kokon Shoin, 1984. **396–97, 399**

———. "Ōsutoria Kokuritsu Toshokan shūzō no Edo jidai Nihonsei chizu" オーストリア國立圖書館收藏の江戸時代日本製地圖 (On the maps made by Japanese in the Edo period, owned by the Austrian National Library). *Gekkan Kochizu Kenkyū* 18, no. 7 (1987): 2–6. **423**

Kawashima Motojirō 川島元次郎. *Shuinsen bōeki shi* 朱印船貿易史 (History of trade by the authorized trading ships). Ōsaka: Kōjinsha, 1921. **383**

Kikutake Jun'ichi 菊竹純一 and Yoshida Hiroshi 吉田宏, eds. *Kōrai butsuga* 高麗佛畫 (Korean Buddhist paintings of the Koryŏ dynasty). Exhibition catalog. Nara: Yamato Bunkakan, 1978. **256**

Kim Yangsŏn 金良善 (pen name Maesan). *Maesan kukhak san'go* 梅山國學散稿 (Selected writings in Korean studies by Maesan). Seoul: Sungjŏn Taehakkyo Pangmulgwan, 1972. **237, 249, 250, 253, 254, 260, 264, 267, 289, 301**

Kimiya Yasuhiko 木宮泰彥. *Nikka bunka Kōryūshi* 日華文化交流史 (History of cultural intercourse between Japan and China). Tokyo: Fuzanbō, 1955. **379**

Kōbe Shiritsu Hakubutsukan 神戸市立博物館 (Kōbe City Museum). *Kochizu ni miru sekai to Nippon* 古地圖にみる世界と日本 (The world and Japan as seen in old maps). Kōbe, 1983. **467, 469, 471**

———. *Akioka Kochizu Korekushon meihin ten* 秋岡古地圖コレクション名品展 (A collection of masterpieces: The Akioka collection of old maps). Kōbe, 1989. **469**

Kōbe Shiritsu Hakubutsukan kanzōhin mokuroku 神戸市立博物館藏品目錄 (Catalog of the collections at the Kōbe City Museum). 6 vols. Kōbe, 1984–89. **351**

Kōbe Shiritsu Hakubutsukan kanzō meihin zuroku 神戸市立博物館藏名品圖錄 (Masterpieces of the Kōbe City Museum). Kōbe: Kōbe Shi Supōtsu Kyōiko Kosha, 1985. **422**

Kohan chishi sōsho 古版地誌叢書 (Series of early printed geographical descriptions). Vol. 12. Tokyo: Geirinsha, 1971. **424**

Kohan Edozu shūsei 古版江戸圖集成 (Collection of early printed plans of Edo). Bekkan 別卷 (supplement, separate volume). Tokyo: Chūō Kōron Bijutsu Shuppan, 1960. **424**

Kokushi daijiten 國史大辭典 (Large dictionary of the history of our country [Japan]). Tokyo: Yoshikawa Kōbunkan, 1979-. **369, 380, 399, 405**

Komatsu Shigemi 小松茂美, ed. *Zoku Nihon emaki taisei* 續日本繪卷大成 (Series of Japanese picture scrolls: Continuation). 20 vols. Tokyo: Chūōkōronsha, 1981–85. **357**

Kurita Mototsugu 栗田元次. "Edo jidai no sekai chizu gaisetsu" 江戸時代の世界地圖概說 (Outline of the world maps of the Edo period). *Shigaku Kenkyū* 10, no. 1 (1938): 73–80. **404**

———. "Nihon ni okeru kokan toshizu" 日本に於ける古刊都市圖 (Old printed maps of cities in Japan). *Nagoya Daigaku Bungakubu Kenkyū Ronshū* 2 (1952): 1–13. **420, 421, 422**

———. "Edo jidai kankō no kokugunzu" 江戸時代刊行の國郡圖 (Printed provincial maps of the Edo period). *Rekishi Chiri* 84, no. 2 (1953): 1–16. **416**

Kurita Mototsugu, ed. *Nihon kohan chizu shūsei* 日本古版地圖集成 (Early maps and plans printed in Japan). Tokyo: Hakata Seishōdō, 1932. **351, 369, 410, 411, 413, 420, 422, 423, 435, 437, 439, 472, 473, 474, 475, 476, 477**

Kuroda Genji 黑田源次. *Shiba Kōkan* 司馬江漢. Tokyo: Tōkyō Bijutsu, 1972. **599**

Kuroda Hideo 黑田日出男. "Edo bakufu kuniezu gōchō kanken" 江戸幕府國繪圖鄉帳管見 (A personal view of provincial maps and books of standard land productivity prepared by the order of the Tokugawa shogunate). *Rekishi Chiri* 93, no. 2 (1977): 19–42. **396**

Kwŏn Sangno 權相老. *Han'guk chimyŏng yŏnhyŏk ko* 韓國地名沿革考 (A study of historical changes in Korean place-names). Seoul: Tongguk Munhwa Sa, 1961. **311**

Kyōto Kokuritsu Hakubutsukan 京都國立博物館 (Kyōto National Museum), ed. *Koezu: Tokubetsu tenrankai zuroku* 古繪圖——特別展覽會圖錄 (Old picture maps: A special exhibition catalog). Kyōto: Kyōto Kokuritsu Hakubutsukan, 1969. **362, 363, 364, 365, 366**

Kyōto Koten Dōkōkai 京都古典同好會 (Kyōto Classical Studies Group), comp. *Kohan Nagasaki chizushū* 古版長崎地圖集 (Early printed plans of Nagasaki). Kyōto: Kyōto Koten Dōkōkai, 1977. **475, 476**

Kyōto shi shi, chizu hen 京都市史，地圖編 (History of Kyōto City, section of plans). Kyōto, 1947. **473, 474**

Li Daoping 李道平, ed. *Zhou yi jijie zuanshu* 周易集解纂疏 (Collected commentaries and annotations on the Zhou change [Book of changes]). Taipei, 1967. **214**

Liang Fangzhong 梁方仲. *Zhongguo lidai hukou, tiandi, tianfu tongji* 中国历代户口、田地、田赋统计 (Population, field acreage, and land tax statistics for China through the dynasties). Shanghai: Renmin Chubanshe, 1980. **71**

Liu Ciyuan 刘次沅. "You yueliang yanfan jilu dedao di wushike huangdao xing di dong Jin Nanbei chao shiqi xing ming" 由月亮掩犯记录得到的五十颗黄道星的东晋南北朝时期星名 (Names of fifty stars on the ecliptic during the Eastern Jin and Northern and Southern dynasties, obtained from records of close lunar conjunctions). *Tianwen Xuebao* 27 (1986): 276–78. **530**

Liu Laicheng 刘来成 and Li Xiaodong 李晓东. "Shi tan Zhanguo shiqi Zhongshanguo lishishang di jige wenti" 试谈战国时期中山国历史上的几个问题 (Tentative discussion of certain problems in the history of the Zhanguo period's Zhongshan kingdom). *Wenwu*, 1979, no. 1:32–36. **37**

Liu Tan 劉坦. *Zhongguo gudai zhi xingsui jinian* 中國古代之星歲紀年 (Ancient Chinese Jupiter-cycle calendar). Beijing: Kexue Chubanshe, 1957. **515**

Lu Liangzhi 卢良志. *Zhongguo dituxue shi* 中国地图学史 (History of Chinese cartography). Beijing: Cehui Chubanshe, 1984. **28, 47, 57, 62, 164, 170, 186, 193**

Luo Qikun 雒启坤. "Xi'an Jiaotong daxue Xi Han muzang bihua ershiba xiu xingtu kaoshi" 西安交通大学西汉墓葬壁画二十八宿星图考释 (On the star map showing the twenty-eight *xiu* painted on the wall of a Western Han tomb in the campus construction site of Xi'an Jiaotong University in Shaanxi). *Ziran Kexue Shi Yan Jiu* 10 (1991): 236–45. **523**

Maruyama Masao 丸山眞男. *Nihon seiji shisō shi kenkyū* 日本政治思想史研究 (A study of the history of political thought in Japan). Tokyo, 1952. **433**

Matsuda Kiichi 松田毅一. "Nihon junsatsushi Varinyāno no shōgai" 日本巡察師ヴァリニャーノの生涯 (Life of Valignani, visitor to Japan). In *Nihon junsatsu ki Varinyāno* 日本巡察記ヴァリニャーノ (Valignani's summary of things Japanese). Trans. Matsuda Kiichi 松田毅一 and Sakuma Tadashi 佐久間正. Tokyo: Tōgensha, 1965. **377**

Mawangdui Han Mu Boshu Zhengli Xiaozu 马王堆汉墓帛书整理小组 (Study Group on the Han Silk Manuscripts from Mawangdui). " 'Wuxing zhan' fubiao shiwen" 《五星占》附表释文 (Explanatory table for "Prognostication from the Five Planets"). *Wenwu*, 1974, no. 11:37–39. **521**

———. "Changsha Mawangdui sanhao Han mu chutu ditu di zhengli" 长沙马王堆三号汉墓出土地图的整理 (Restoration of the maps excavated from Han tomb 3 at Mawangdui, Changsha). *Wenwu*, 1975, no. 2:35–42. **41**

———. "Mawangdui sanhao Han mu chutu zhujun tu zhengli jianbao" 马王堆三号汉墓出土驻军图整理简报 (Preliminary restoration report on the military map found in Han tomb 3 at Mawangdui). *Wenwu*, 1976, no. 1:18–23. **41, 148**

Minagawa Shinsaku 皆川新作. "Murakami Shimanojō no Ezochi kinmu" 村上島之允の蝦夷地勤務 (Murakami Shimanojō's service in Ezochi). *Denki* 7, nos. 4–6 (1940): 10–15, 19–24, 17–24. **447**

———. *Mogami Tokunai* 最上德內. Tokyo: Dentsū Shuppanbu, 1943. **447**

Miyaji Naoichi 宮地直一, supervisor. *Jinja kozu shū* 神社古圖集 (Collected old drawings of shrines). Tokyo: Nippon Denpō Tsūshinsha, 1942; reprinted Rinsen Shoten, 1989. **362, 364, 365, 366**

Miyajima Kazuhiko 宮島一彥. "Dōshisha Daigaku shozō Genroku 14 nen sei tenkyūgi no ichizuke" 同志社大學所藏元祿 14 年製天球儀の位置づけ (The position of the celestial globe made in 1701 and owned by Dōshisha University). *Dōshisha Daigaku Rikōgaku Kenkyū Hōkoku* 21 (1981): 279–300. **591**

———. "Mukashi no tenmon giki" 昔の天文儀器 (Astronomical instruments of old days). In vol. 15 of *Tenmongaku shi* 天文學史 (History of astronomy). 1983. **591**

Miyoshi Manabu 三好學. "Meisho zue kaisetsu" 名所圖會解說 (Explanation of the *Meisho zu*). In *Iwanami kōza chirigaku* 岩波講座地理學 (Iwanami lectures on geography). 76 vols. Bekkō 別項 (supplement), 1932, 1–22. Tokyo: Iwanami Shoten, 1931–34. **416**

Miyoshi Tadayoshi 三好唯義. "Nanba Korekushon chū no kankō shokokuzu ni tsuite" 南波コレクション中の刊行諸國圖について (On the printed provincial maps in the Nanba Collection). *Kōbe Shiritsu Hakubutsukan Kenkyū Kiyō* 4 (1987): 27–52. **416**

Mok Yŏngman 목용만. *Chido iyagi* 지도 이야기 (Map conversations). P'yŏngyang: Kunjung Munhwa Ch'ulp'ansa, 1965. **236**

Mun'gyobu 文敎部 (Ministry of Education, Republic of Korea), comp. *Kugŏ 5–2* 國語 5-2 (Fifth-grade Korean reader). Seoul: Ministry of Education, 1987. **314**

Muramatsu Teijirō 村松貞次郎. *Daiku dōgu no rekishi* 大工道具の歷史 (A history of carpenters' tools). Tokyo: Iwanami Shoten, 1973. **357, 358**

Murayama Shūichi 村山修一, ed. *Onmyōdō kiso shiryō shūsei* 陰陽道基礎史料集成 (Compilation of basic material on the techniques of divination). Tokyo: Tōkyō Bijutsu, 1987. **582**

Muroga Nobuo 室賀信夫 and Unno Kazutaka 海野一隆. "Nihon ni okonowareta Bukkyō kei sekaizu ni tsuite" 日本に行われた佛敎系世界圖について (On Buddhist world maps in Japan). *Chirigakushi Kenkyū* 1 (1957): 67–141. Reprinted in *Chirigakushi kenkyū* 地理學史研究 (Researches in the history of geography), 2 vols., ed. Chirigakushi Kenkyūkai 地理學史研究會 (Society for Research in Historical Geography), 1:67–141. Kyōto: Rinsen Shoten, 1979. **225, 256, 429**

———. "Edo jidai kōki ni okeru Bukkyō kei sekaizu" 江戸時代後期における佛敎系世界圖 (Buddhist world maps in the Late Edo period). *Chirigakushi Kenkyū* 2 (1962): 135–229. Reprinted in *Chirigakushi kenkyū* 地理學史研究 (Researches in the history of geography), 2 vols., ed. Chirigakushi Kenkyūkai 地理學史研究會 (Society for Research in Historical Geography), 2:135–229. Kyōto: Rinsen Shoten, 1979. **409, 429**

Na Ilsŏng 羅逸星. "Chosŏn sidae in chŏn'mun ŭigi yŏn'gu" 朝鮮時代의 天文儀器 硏究 (Study of astronomical instruments in the Chosŏn period). *Tongbang hakchi* 42 (1984): 205–37. **560, 561**

Nagasawa Kikuya 長澤規矩也. "Edo no hanzu ni tsuite" 江戸の版圖について (On the printed plans of Edo). *Shoshigaku*, n.s., 2 (1965): 31–51. **421**

Nakamura Hiroshi 中村拓. "Sengoku jidai no Nihonzu" 戰國時代の日本圖 (Maps of Japan at the time of the civil wars [1467–1568]). *Yokohama Shiritsu Daigaku Kiyō* 58 (1957): 1–98. **388**

———. "Nanban byōbu sekaizu no kenkyū" 南蠻屏風世界圖の研究 (Research on the world map on *Nanban* folding screens). *Kirishitan Kenkyū* 9 (1964): 1–273. **377, 380**

———. *Goshuinsen kōkai zu* 御朱印船航海圖 (Sea charts used by the authorized trading ships). Tokyo: Nihon Gakujutsu Shinkōkai, 1965. **383, 384**

———. "Chōsen ni tsutawaru furuki Shina sekai chizu 朝鮮に傳わる古きシナ世界地圖 (Mappemondes antiques chinoises conservées chez les Coréens)." *Chōsen Gakuhō* 39–40 (1966): 1–73. **255, 259, 260, 261, 262, 263**

Nakamura Yūzō 中村雄三. *Zusetsu Nihon mokkōgu shi* 圖說日本木工具史 (Illustrated history of Japanese woodworking tools). Tokyo: Shinseisha, 1968. **357**

Nanba Matsutarō 南波松太郎, Muroga Nobuo 室賀信夫, and Unno Kazutaka 海野一隆, eds. and comps. *Nihon no kochizu* 日本の古地圖 (Old maps in Japan). Ōsaka: Sōgensha, 1969. **351, 373, 380, 388, 404, 407, 411, 412, 414, 416, 420, 422, 423, 426, 429, 439, 441, 447, 453, 465, 473, 476**

Naniwada Tōru 難波田徹, ed. *Koezu* 古繪圖 (Old picture maps). Nihon no Bijutsu 日本の美術 (Japanese art), no. 72. Tokyo: Shibundō, 1972. **362**

Naniwakyū Shi no Kenkyū 難波宮址の研究 (Reports of the historical investigation of the forbidden city of Naniwa). Vol. 7, 1981. Issued by the Ōsaka Shi Bunkazai Kyōkai 大阪市文化財協會 (Ōsaka City Cultural Properties Association). **358**

Nanjing Bowuyuan 南京博物院 (Nanjing Museum). *Nan Tang erling fajue baogao* 南唐二陵發掘報告 (Report on the excavation of two Southern Tang mausoleums). Ed. Zeng Zhaoyue 曾昭燏. Beijing: Wenwu Chubanshe, 1957. **80**

Nara Kokuritsu Bunkazai Kenkyūjo 奈良國立文化財研究所 (Nara National Cultural Properties Research Institute). *Asuka Fujiwarakyū hakkutsu chōsa hōkoku* 飛鳥藤原宮發掘調查報告 (Reports of the excavation of the site of the Fujiwara imperial palace, Asuka). Vol. 6. Nara, 1976. **359**

Narita Shūichi 成田修一, ed. *Ezo chizu shō* 蝦夷地圖抄 (Extracted maps of Ezo). Tokyo: Sara Shobō, 1989. **443, 450**

Nihon Gakushiin 日本學士院 (Japanese Academy), ed. *Meiji zen Nihon kenchiku gijutsu shi* 明治前日本建築技術史 (History of Japanese architectural techniques before the Meiji era). Tokyo: Nihon Gakujutsu Shinkōkai, 1961; reprinted 1981. **357**

Nihon no chizu: Kansen chizu no hattatsu 日本の地圖 —— 官撰地圖の發達 (Cartography in Japan: Official maps, past and present). Exhibition catalog, National Diet Library, Twenty-fourth International Geographical Congress and Tenth Conference of the International Cartographic Association. Tokyo: Kokuritsu Kokkai Toshokan, 1980. **351, 399, 401, 423**

Niida Noboru 仁井田陞. "Shina no tochi daichō 'gorinsetsu' no shiteki kenkyū" 支那の土地臺帳「魚鱗圖册」の史的研究 (Historical study of Chinese land register "fish-scale" maps). *Tōhō Gakuhō* (Tokyo) 6 (1936): 157–204. **86**

Nishioka Toranosuke 西岡虎之助, ed. *Nihon shōen ezu shūsei* 日本莊園繪圖集成 (Collected maps of Japanese manors). 2 vols. Tokyo: Tokyōdō Shuppan, 1976–77. **362**

Niu Zhongxun 钮仲勋. " 'Zheng He hanghai tu' di chubu yanjiu" 《郑和航海图》的初步研究 (Preliminary study of Zheng He's nautical chart). In *Zheng He xia Xiyang lunwenji* 鄭和下西洋論文集 (Collected essays on Zheng He's expedition to the Western Ocean), ed. Zhongguo Hanghai Shi Yanjiuhui 中国航海史研究会 (Research Association for the History of Chinese Navigation), 1:238–48. Beijing: Renmin Jiaotong Chubanshe, 1985. **61**

Noda Hisao 野田久男. "Tottori ken no sōshoku kofun" 鳥取縣の裝飾古墳 (Decorated tombs in Tottori Prefecture). *Kyōiku Jihō* 163 (1980): 2–11. **352**

Oda Takeo 織田武雄. "Nihon no chizu to sono hattatsu" 日本の地圖とその發達 (Japanese maps and their development). In *Chizu no rekishi* 地圖の歷史 (History of maps), 211–89. Tokyo: Kōdansha, 1973. **350**

———. *Chizu no rekishi* 地圖の歷史 (History of maps). 2 vols. Tokyo: Kōdansha, 1974. **350**

Ogawa Takuji 小川琢治. *Shina rekishi chiri kenkyū* 支那歷史地理研究 (Studies in Chinese historical geography). 2 vols. Tokyo: Kobundō Shobō, 1928–29. **245**

Ohara Satoru 尾原悟. "Kirishitan jidai no kagaku shisō" キリシタン時代の科學思想 (Scientific thought in the Christian period). *Kirishitan Kenkyū* 10 (1965): 101–78. **393**

———. "Pedoro Gomesu cho 'Tenkyūron' no kenkyū" ペドロ・ゴメス著「天球論」の研究 (A study of "De sphaera" by Pedro Gomez). *Kirishitan Kenkyū* 10 (1965): 179–273. **393**

Okamoto Yoshitomo 岡本良知. *Jūroku seiki ni okeru Nihon chizu no hattatsu* 十六世紀における日本地圖の發達 (Development of the map of Japan in the sixteenth century). Tokyo: Yagi Shoten, 1973. **379, 380, 390, 463**

Okamura Chibiki 岡村千曳. "Wasurerareta dōban gaka Matsubara Uchū" 忘れられた銅版畫家松原右仲 (A forgotten copperplate artist, Matsubara Uchū). In *Kōmō bunka shiwa* 紅毛文化史話 (Historical essays on Dutch culture), by Okamura Chibiki, 198–206. Tokyo: Sōgensha, 1953. **438**

Ōsaki Shōji 大崎正次. *Chūgoku no seiza no rekishi* 中國の星座の歷史 (History of Chinese constellations). Tokyo: Yūzankaku, 1987. **583**

Ōtsuka Takashi 大塚隆. *Kyōtozu sōmokuroku* 京都圖總目錄 (General catalog of plans of Kyōto). Tokyo: Seishōdō Shoten, 1981. **421**

Pak Sŏngbong 朴性鳳, Pang Tong'in 方東仁, and Chŏng Wŏn'ok 鄭元玉, comps. *Taedong yŏjido saegin* 大東輿地圖索引 (Index to the *Taedong yŏjido*). Seoul: Kyŏnghŭi University, Han'guk Chŏnt'ong Munhwa Yŏn'guso, 1976. **328**

Pan Nai 潘鼐. "Suzhou Nan Song tianwentu bei di kaoshi yu pipan" 苏州南宋天文图碑的考释与批判 (Examination and critique of a Southern Song astronomical chart on a stone stele at Suzhou). *Kaogu Xuebao*, 1976, no. 1:47–61. **545**

———. *Zhongguo hengxing guance shi* 中国恒星观测史 (History of stellar observations in China). Shanghai, 1989. **512, 524, 531, 532, 536, 537, 548, 551, 552, 554, 555, 569, 571, 572, 574, 599**

———. "Shiqi shiji chu shijie shouqu yizhi di heng xingtu" 十七世纪初世界首屈一指的恒星图 (A unique star map of the early seventeenth century). *Kexue* 42 (1990): 275–80. **572, 573**

Pang Tong'in 방동인. *Han'guk ŭi chido* 한국의 지도 (Korean maps). Seoul: Sejong Taewang Kinyŏm Saŏphoe, 1976. **237, 266, 284, 287, 295, 297, 301, 305, 307, 314, 316, 318, 324, 327, 329**

Pi Xirui 皮錫瑞. *Jingxue lishi* 經學歷史 (History of classical studies, printed 1907). Annotated Zhou Datong 周大同. Taipei: Yiwen Yinshuguan, 1966. **224**

———. *Jingxue tonglun* 經學通論 (Comprehensive discussions of classical studies, printed 1907). 4 vols. Taipei: He-Lo Tushu Chubanshe, 1974. **224**

Qing shi 清史 (History of the Qing). 8 vols. Taipei: Guofang Yanjiuyuan, 1961. **181**

Qing shi gao jiaozhu 清史稿校註 (Edited and annotated draft history of the Qing, original draft completed 1927). 15 vols. Taipei: Guoshiguan, 1986–. **181, 183, 187**

Rekishi ni okeru minshū to bunka: Sakai Tadao Sensei koki shukuga kinen ronshu 歴史における民衆と文化 —— 酒井忠夫先生古稀祝賀記念論集 (Peoples and cultures in Asiatic history: Collected essays in honor of Professor Tadao Sakai on his seventieth birthday). See Strickman, Michel (below).

Ren Jincheng 任金城. "Xibanya cang Ming ke *Gujin xingsheng zhi tu*" 西班牙藏明刻《古今形胜之图》(The *Gujin xingsheng zhi tu* printed during the Ming and preserved in Spain). *Wenxian* 17 (1983): 213–21. **59**

Ro Kōrō 盧高朗. *Ro Kōrō jijoden* 盧高朗自敘傳 (Ro Kōrō's autobiography). Published by the author, 1922. **383**

Ryūkyū Kuniezu Shiryōshū 琉球國繪圖史料集 (Collected historical materials of provincial maps of Ryūkū). No. 1. Naha: Okinawa ken Kyōiku Iinkai, 1992. **397**

Saga Kenritsu Toshokan zō kochizu ezu roku 佐賀縣立圖書館藏古地圖繪圖錄 (Catalog of early maps and plans in the Saga Prefectural Library collection). Saga, 1973. **351**

Shaanxi Sheng Kaogu Yanjiusuo 陕西省考古研究所 (Shaanxi Archaeological Institute) and Xi'an Jiaotong Daxue 西安交通大学 (Xi'an Jiaotong University). "Xi'an Jiaotong daxue Xi Han bihua mu fajue jianbao" 西安交通大学西汉壁画墓发掘简报 (Preliminary report on the excavation of the Western Han tomb with murals in Xi'an Jiaotong University). *Kaogu yu Wenwu*, 1990, no. 4:57–63. **523**

Sŏul Kungnip Taehakkyo Tosŏgwan 서울國立大學校圖書館 (Seoul National University Library), comp. *Han'guk ko chido haeje* 韓國古地圖解題 (Bibliographical notices of old Korean maps). Seoul: Seoul National University, 1971. **250**

Suematsu Yasukazu 末松保和, ed. *Daitō yochizu sakuin* 大東輿地圖索引 (Index to the *Taedong yŏjido*). Seoul: Keijō Imperial University, College of Law, 1936. **328**

Sugano Yō 菅野陽. *Nihon dōhanga no kenkyū: Kinsei* 日本銅版畫の研究 —— 近世 (Studies on Japanese copperplate prints: The modern age). Tokyo: Bijutsu Shuppansha, 1974. **435**

———. "Shiba Kōkan no chosho *Shutō dempō* to dōhan *Tenkyū zu* ni tsuite" 司馬江漢の著書『種痘傳法』と銅版「天球圖」について (On Shiba Kōkan's book *Shutō dempō* [The introduction of vaccine] and the copperplate print *Tenkyū zu*). *Nihon Yōgakushi no Kenkyū* 5 (1979): 65–100. **599**

———. "Eisei Bunko shozō Shiba Kōkan sei chikyūgi" 永青文庫所藏司馬江漢製地球儀 (Shiba Kōkan's terrestrial globe in the Eisei Library collection). *Nihon Yōgakushi no Kenkyū* 7 (1985): 47–64. **469**

Sugiura Kōhei 杉浦康平, ed. *Ajia no kosumosu + mandara* アジアのコスモス + マンダラ (The Asian cosmos). Catalog of exhibition, "Ajia no Uchūkan Ten," held at Rafōre Myūjiamu in November and December 1982. Tokyo: Kōdansha, 1982. **608, 621, 622, 623, 632, 720, 721, 723, 727, 728, 733**

Sun Wenqing 孫文青. *Zhang Heng nianpu* 張衡年譜 (Chronological biography of Zhang Heng). Rev. ed. Shanghai: Shangwu Yinshuguan, 1956. **132**

Takagi Kikusaburō 高木菊三郎. *Nihon chizu sokuryō shōshi* 日本地圖測量小史 (A brief history of cartography and surveying). Tokyo: Kokon Shoin, 1931. **350**

Takagi Takeo 高城武夫. *Tenmon kyōgu* 天文教具 (Tools for teaching astronomy). Tokyo, 1973. **469, 471**

Takahashi Tadashi 高橋正. "Tōzen seru chūsei isurāmu sekaizu" 東漸せる中世イスラーム世界圖 (Eastward diffusion of Islamic world maps in the medieval era). *Ryūkoku Daigaku Ronshū* 374 (1963): 86–94. **247, 248**

———. "Seizen seru shoki Nihon chizu ni tsuite: I. Moreira kei chizu o chūshin to shite" 西漸せる初期日本地圖について —— I. Moreira 系地圖を中心として (On the early maps of Japan by Europeans, emphasizing the I. Moreira-type maps). *Nihongakuhō* 4 (1985): 1–33. **390**

———. "Jūshichi seiki Nihon chizu ni okeru Teisheira gata to Moreira gata: N. Sanson to R. Daddoree no Baai" 17 世紀日本地圖におけるテイシェイラ型とモレイラ型 —— N. サンソンと R. ダッドレーの場合 (About the Moreira- and Teixeira-type maps of seventeenth-century Japan: The atlases of N. Sanson and R. Dudley). *Nihongakuhō* 6 (1987): 111–35. **390**

———. "Nanban toshizu byōbu kara Kaeriusu sekaizu e" 南蠻都市圖屏風からカエリウス世界圖へ (From maps of cities on the Nanban folding screens to Kaerius's map of the world). In *Ezu no kosumorojii* 繪圖のコスモロジー (Cosmology of picture maps), ed. Katsuragawa Ezu Kenkyūkai 葛川繪圖研究會 (Katsuragawa Picture Map Research Society), 1:248–64. Kyōto: Chijin Shobō, 1988. **380**

Takakura Shin'ichirō 高倉新一郎. "Hokkaidō chizu no hensen hoi" 北海道地圖の変遷，補遺 (Development of the cartography of Hokkaidō: Supplement). *Hokkaidō [Teikoku] Daigaku Hoppō Bunka Kenkyū Hōkoku* 11 (1956): 49–73. **443**

———, ed. *Hokkaidō kochizu shūsei* 北海道古地圖集成 (Collection of historical maps of Hokkaidō and the adjacent regions). Sapporo: Hokkaidō Shuppan Kikaku Sentā, 1987. **443, 444, 447**

Takakura Shin'ichirō 高倉新一郎 and Shibata Sadakichi 柴田定吉. "Wagakuni ni okeru Karafuto chizu sakuseishi" 我國に於ける樺太地圖作製史 (History of the development of the cartography of Sakhalin in Japan). *Hokkaidō [Teikoku] Daigaku Hoppō Bunka Kenkyū Hōkoku* 2 (1939): 1–48. **443, 447**

———. "Wagakuni ni okeru Chishima chizu sakuseishi" 我國に於ける千島地圖作製史 (History of the development of the cartography of the Kuriles in Japan). *Hokkaidō [Teikoku] Daigaku Hoppō Bunka Kenkyū Hōkoku* 3 (1940): 1–75. **443**

———. "Wagakuni ni okeru Hokkaidō hontō chizu no hensen" 我國に於ける北海道本島地圖の變遷 (Development of the cartography of Hokkaidō in Japan), 1 and 2. *Hokkaidō [Teikoku] Daigaku Hoppō Bunka Kenkyū Hōkoku* 6 (1942): 1–80 and 7 (1952): 97–166. **443**

Takamatsuzuka Kofun Sōgō Gakujutsu Chōsakai 高松塚古墳總合學術調查會 (Joint Committee for the Scientific Investigation of Takamatsuzuka Burial Mound). *Takamatsuzuka kofun hekiga chōsa hōkokusho* 高松塚古墳壁畫調查報告書 (Report on the investigation of the Takamatsuzuka fresco by the Agency for Cultural Affairs). Kyōto: Benrido, 1974. **579**

Tan Qixiang 譚其驤. "Erqian yibaiduo nian qian di yifu ditu" 二千一百多年前的一幅地图 (A map from more than 2,100 years ago). *Wenwu*, 1975, no. 2:43–48. **41, 51**

Tanbō daikōkai jidai no Nippon 探訪大航海時代の日本 (Japan in the age of great navigation: The inquiries). 8 vols. Tokyo: Shōgakkan, 1978–79. Vol. 5, *Nippon kara mita ikoku* 日本からみた異國 (Foreign countries interpreted by the Japanese). **380**

Tang Ruchuan 唐如川. "Zhang Heng deng huntianjia di tian yuan di ping shuo" 張衡等渾天家的天圓地平說 (On the theory of Zhang Heng and other uranosphere school cosmologists that the sky is spherical and the earth flat). *Kexue Shi Jikan*, 1962, no. 4:47–58. **118**

Tao Maoli 陶懋立. "Zhongguo dituxue faming zhi yuanshi ji gailiang jinbu zhi cixu" 中國地圖學發明之原始及改良進步之次序 (The origins of cartographic invention and steps toward reform and progress in China). *Dixue Zazhi* 2 (1911): no. 11, 1–9, and no. 13, 1–9. **27–28**

Teramoto Enga 寺本婉雅. "Waga kokushi to Toban to no Kankei" 我が國史と吐蕃との關係 (The relation between our [Japanese] history and Tibet). *Ōtani Gakuhō* 12, no. 4 (1931): 44–83. **642**

Terui Sōsuke 照井壯助. *Tenmei Ezo tanken shimatsu ki* 天明蝦夷探檢始末記 (The circumstances of the exploration of Ezo during the Tenmei era). Tokyo: Yaedake Shobō, 1974. **446**

Tokita Tadamasa 鴇田忠正. "Nanban sekaizu byōbu genzu kō" 「南蠻世界圖屛風」原圖考 (On the originals of the world maps on folding screens, 2). *Nagasaki Dansō* 57 (1975): 32–61. **380**

———. "Goshuinsen bōekika Itoya Zuiemon boseki ron" 御朱印船貿易家糸屋隨右衛門墓石論 (On the tombstone of Itoya Zuiemon, authorized trading-ship trader). *Nagasaki Shiritsu Hakubutsukan Kanpō* 19 (1979): 1–7. **383**

Tokushi Yūshō 禿氏祐祥, ed. *Shumisen zufu* 須彌山圖譜 (Collection of pictures of Mount Sumeru). Kyōto: Ryūkoku Daigaku Shuppanbu, 1925. **392**

Tōkyō Daigaku Shiryō Hensanjo 東京大學史料編纂所 (Historiographical Institute, Tokyo University), ed. *Tōdaiji kaiden zu* 東大寺開田圖 (Maps of paddy fields reclaimed by Tōdai Temple). 2 vols. (facsimile and explanation). In *Dainihon komonjo, Iewake* 大日本古文書家わけ (Old documents of Great Japan, Every family), vol. 18, *Tōdaiji monjo* 東大寺文書 (Records of Tōdai Temple), pt. 4. Tokyo: Tōkyō Daigaku Shuppankai, 1965–66; reprinted 1980. **352, 361**

———. *Echigo no kuni gun* (or *kori*) *ezu* 越後國郡繪圖 (Maps of counties in Echigo Province). Tokyo: Tōkyō Daigaku, 1983, 1985, 1987. **395**

———. *Nihon shōen ezu shūei* 日本莊園繪圖聚影 (Collected facsimiles of maps of Japanese manors). Tokyo: Tōkyō Daigaku Shuppankai, 1988. **362, 363**

Unno Kazutaka 海野一隆. "Tenri toshokan shozō DaiMin kokuzu ni tsuite" 天理圖書館所藏大明國圖について (On the 'Map of Ming' held by the Tenri Library). *Ōsaka Gakugei Daigaku Kiyō* 6 (1958): 60–67, with 2 pls. **249**

———. "Katsuragawa Hoshū no sekaizu ni tsuite" 桂川甫周の世界圖について (On Katsuragawa Hoshū's map of the world). *Jinbun Chiri* 20, no. 4 (1968): 1–12. **436**

———. "Kodai Chūgokujin no chiriteki sekaikan" 古代中國人の地理的世界觀 (The ancient Chinese people's geographical conception of the world). *Tōhō Shūkyō* 42 (1973): 35–51. **120**

———. " '*Tenchi nikyū yōhō kokumei*' kō" 「天地二球用法國名」考 (On the *Tenchi nikyū yōhō kokumei*). *Nihon Yōgakushi no Kenkyū* 3 (1974): 113–37. **434**

———. "Sōkaku no chikyūgi to sono sekaizō" 宗覺の地球儀とその世界像 (Sōkaku's globe and his image of the world). *Kagakushi Kenkyū* 117 (1976): 8–16. **391, 429**

———. "Akashi Shiritsu Tenmonkagakukan shozō kochikyūgi ni tsuite" 明石市立天文科學館所藏古地球儀について (On the early terrestrial globe in the Akashi Planetarium collection). *Kagakushi Kenkyū* 124 (1977): 235–36. **469**

———. "Hyōryūmin Tsudayūra no kikoku to chizu no denrai" 漂流民津太夫らの歸國と地圖の傳來 (Introduction of European cartography when Tsudayū [one of the repatriated castaways] and others returned to Japan). *Nihon Yōgakushi no Kenkyū* 4 (1977): 101–22. **438**

———. "Seiyō chikyūsetsu no denrai" 西洋地球說の傳來 (Introduction of the global theory to Japan). *Shizen* 34, no. 3 (1979): 60–67, and 34, no. 6 (1979): 62–69. **377**

———. "Shīburoto to 'Nihon henkai ryakuzu' " シーボルトと『日本邊界略圖』(Siebold and his small map of Japan). *Nihon Yōgakushi no Kenkyū* 5 (1979): 101–28. **439**

———. *Chizu no shiwa* ちずのしわ (Map creases; or, Essays on the history of cartography). Tokyo: Yūshōdō Press, 1985. **380, 465, 467, 610, 623**

"Chikyūgi tsuki no bateren ningyō" 地球儀付きのバテレン人形 (A terrestrial globe with a padre doll), 248–50. **391**

"Hashimoto Sōkichi sekaizu no ihan gihan mohōban" 橋本宗吉世界圖の異版・僞版・模倣版 (Some unusual fake and imitative editions of Hashimoto Sōkichi's world map), 305–18. **437**

"Kinsei kankō no Nihonzu" 近世刊行の日本圖 (Maps of Japan printed in [early] modern times), 126–38. **414**

"Tawamure no chizu" たわむれの地圖 (Amusing cartographic works) and "Zoku tawamure no chizu" 續たわむれの地圖 (Amusing cartographic works, continuation), 5–7, 8–17. **430, 431**

———. "Min Shin ni okeru Mateo Ritchi kei sekaizu: Shutoshite shinshiryō no kentō" 明清におけるマテオ・リッチ系世界圖——主として新史料の檢討 (Chinese world maps of the Ming and Qing dynasties derived from the work of Matteo Ricci: An examination of new and neglected materials). In *Shinhatsugen Chūgoku kagakushi shiryō no kenkyū: Ronkō hen* 新發現中國科學史資料の研究，論考篇 (Studies on recently discovered source materials for the history of Chinese science: Collected articles), ed. Yamada Keiji 山田慶兒, 507–80. Kyōto: Research Institute for Humanistic Studies, Kyōto University, 1985. **373, 409**

———. "Oranda shin'yaku chikyū zenzu ni okeru sanshōshiryō" 喎蘭新譯地球全圖における參照資料 (Reference materials in a Dutch map of the world newly translated). *Nihon Yōgakushi no Kenkyū* 7 (1985): 65–102. **437**

———. "Faruku chikyūgi denrai no hamon" ファルク地球儀傳來の波紋 (The influence of the Valcks' globe on Japanese maps and globes). *Nihon Yōgakushi no Kenkyū* 8 (1987): 9–34. **433, 469**

———. "Hokubei ni okeru Edo jidai chizu no shūshū jōkyō: Bīnzu Korekushon o chūshin to shite" 北米における江戸時代地圖の收集狀況——ビーンズ・コレクションを中心として (Some collections of Japanese maps of the Edo period in North America: Mainly on the Beans Collection). *Jinbun Chiri* 39, no. 2 (1987): 16–41. **352, 423**

———. "Jingū Bunko shozō no Nanban kei sekaizu to nan'yō karuta" 神宮文庫所藏の南蠻系世界圖と南洋カルタ (A *Nanban* map of the world and a Japanese marine chart of Southeast and East Asia in the Jingū Library collection). *Nihon Yōgakushi no Kenkyū* 9 (1989): 9–36. **380**

———. "Nihonjin to Shumisen" 日本人と須彌山 (The Japanese and Mount Sumeru). In *Ajia no uchūkan* アジアの宇宙觀 (Cosmology in Asia), ed. Iwata Keiji 岩田慶治 and Sugiura Kōhei 杉浦康平, 349–71. Tokyo: Kōdansha, 1989. **392**

———. "Mukanki Tōkaidō michiyuki no zu no ihan" 無刊記東海道路行之圖の異版 (Two undated editions of the *Tōkaido michiyuki no zu*, an itinerary map of the Tōkai road). *Gekkan Kochizu Kenkyū* 22, no. 6 (1991): 2–5. **422–23**

———. "Shōhō kan 'Bankoku sōzu' no seiritsu to rufu" 正保刊「萬國總圖」の成立と流布 (The *Bankoku sōzu* [Map of all the countries] published in 1645 and its popularization). *Nihon Yōgakushi no Kenkyū* 10 (1991): 9–75. **405**

———. " 'Bankoku sekai igyō zu' ni tsuite" 『萬國世界異形圖』について (On the map of all countries and picture of the strange people in the world). *Biburia* 99 (1992): 20–33. **405**

Unno Kazutaka 海野一隆, Oda Takeo 織田武雄, and Muroga Nobuo 室賀信夫, eds. *Nihon kochizu taisei* 日本古地圖大成 (Great collection of old Japanese maps). 2 vols. (Vol. 1, added title *Monumenta cartographica Japonica*. Vol. 2, *Nihon kochizu taisei sekaizu hen* 日本古地圖大成世界圖編 [Great collection of old Japanese maps, volume of world maps].) Tokyo: Kōdansha, 1972–75. **350, 351, 362, 364, 368, 373, 377, 379, 380, 399, 400, 401, 404, 405, 407, 409, 410, 411, 412, 413, 414, 415, 416, 420, 421, 422, 423, 424, 426, 429, 435, 436, 437, 438, 439, 441, 442, 445, 450, 453, 463, 465, 467, 469, 471, 472, 473, 474, 475, 476, 477**

Wang Che 王车 and Chen Xu 陈徐. "Luoyang Bei-Wei Yuan Yi mu di xingxiangtu" 洛阳北魏元义墓的星象图 (The celestial map from the Northern Wei tomb of Yuan Yi at Luoyang). *Wenwu*, 1974, no. 12:56–60 and pl. 1. **531**

Wang Guowei 王國維. *Renjian cihua* 人間詞話 (Poetic remarks in the human world, ca. 1910). Hong Kong: Zhonghua Shuju, 1961. **128**

Wang Jianmin 王健民, Liang Zhu 梁柱, and Wang Shengli 王胜利. "Zeng Houyi mu chutu di ershiba xiu qinglong baihu tuxiang" 曾侯乙墓出土的二十八宿青龙白虎图象 (The twenty-eight lunar lodges and paintings of the Green Dragon and the White Tiger, from the tomb of Zeng Houyi). *Wenwu*, 1979, no. 7:40–45. **519**

Wang Meng'ou 王夢鷗. "Gu mingtang tu kao" 古明堂圖考 (An investigation of the plan of the ancient luminous hall). In *Sanli yanjiu lunji* 三禮研究論集 (A collection of articles on the three ritual classics), Li Yuegang 李曰剛 et al., 289–300. Taipei: Liming Wenhua Shiye, 1981. **212**

Wang Ningsheng 汪宁生. *Yunnan Cangyuan bihua di faxian yu yanjiu* 云南沧源崖画的发现与研究 (The rock paintings of Cangyuan County, Yunnan: Their discovery and research). Beijing: Wenwu Chubanshe, 1985. **4, 5, 19, 21**

Wang Yong 王庸. *Zhongguo dilixue shi* 中國地理學史 (History of geography in China). 1938; reprinted Taipei: Shangwu Yinshuguan, 1974. **28, 35, 64, 125, 128, 170, 208**

———. *Zhongguo dili tuji congkao* 中國地理圖籍叢考 (Collected studies on Chinese geographic maps and documents, 1st ed. 1947). Rev. ed. Shanghai: Shangwu Yinshuguan, 1956. **176**

———. *Zhongguo ditu shi gang* 中國地圖史綱 (Brief history of Chinese cartography). Beijing: Sanlian Shudian, 1958. **28, 51, 64, 125, 128, 133, 139, 170**

Watanabe Toshio 渡邊敏夫. *Kinsei Nihon tenmongaku shi* 近世日本天文學史 (History of modern Japanese astronomy). 2 vols. Tokyo: Kōseisha Kōseikaku, 1986–87. **433, 453, 579, 581, 582, 583, 587, 589, 590, 591, 593, 594, 596, 598**

Wei Qingyuan 韋慶遠. *Mingdai huangce zhidu* 明代黃冊制度 (Yellow book system of the Ming period). Beijing: Zhonghua Shuju, 1961. **101**

Wen Shaofeng 溫少峰 and Yuan Tingdeng 袁庭栋. *Yinxu buci yanjiu: Kexue jishu pian* 殷墟卜辞研究 —— 科学技术篇 (Studies on Yin oracle bone writings: Science and technology volume). Chengdu: Sichuan Shehui Kexue Chubanshe, 1983. **514**

Wu Chengluo 吳承洛. *Zhongguo duliangheng shi* 中國度量衡史 (History of Chinese weights and measures). Shanghai: Shangwu Yinshuguan, 1937. **286, 312**

Xi Zezong 席澤宗. "Seng Yixing guance hengxing weizhi di gongzuo" 僧一行觀測恆星位置的工作 (On the observations of star positions by the priest Yixing [683–729]). *Tianwen Xuebao* 4 (1956): 212–18. **538**

———. "Dunhuang xingtu" 敦煌星图 (A star map from Dunhuang). *Wenwu*, 1966, no. 3:27–38. **537**

Xia Nai 夏鼐. "Cong Xuanhua Liao mu di xingtu lun ershiba xiu he huangdao shier gong" 从宣化辽墓的星图论二十八宿和黄道十二宫 (Discussion of twenty-eight lodges and the twelve palaces on the ecliptic based on a star map from a Liao tomb at Xuanhua). *Kaogu Xuebao*, 1976, no. 2:35–58. Reprinted in *Kaoguxue he keji shi* 考古学和科技史 (Archaeology and the history of technology), by Xia Nai, 29–50 and pls. 11–12. Beijing: Kexue Chubanshe, 1979. **539, 549**

Xinjiang Weiwuer Zizhiqu Bowuguan 新疆维吾尔自治区博物馆 (Museum of the Xinjiang Uygur Autonomous Region). "Tulufan xian Asitana-Halahezhuo gu muqun fajue jianbao" 吐鲁番县阿斯塔那 —— 哈拉和卓古墓群发掘简报 (Preliminary report on the excavation of ancient tombs at Asitana and Halahezhuo, Turpan County). *Wenwu*, 1973, no. 10:7–27. **537**

Xu Yuhu (Hsü Yü-hu) 徐玉虎. *Mingdai Zheng He hanghai tu zhi yanjiu* 明代鄭和航海圖之研究 (Study of Zheng He's nautical chart from the Ming period). Taipei: Xuesheng Shuju, 1976. **53**

Yabuuchi Kiyoshi (Yabuuti Kiyosi) 藪內清. "Sōdai no seishuku" 宋代の星宿 (Description of the constellations in the Song dynasty). *Tōhō Gakuhō* (Kyoto) 7 (1936): 42–90. **548**

———. "Chūgoku, Chōsen, Nihon, Indo no seiza" 中國・朝鮮・日本・印度の星座 (Chinese, Korean, Japanese, and Indian constellations). In *Seiza* 星座 (Constellations), Shin Tenmongaku Kōza 新天文學講座 (New lecture series on astronomy), vol. 1, ed. Nojiri Hōei 野尻抱影, 123–56. Tokyo: Kōseisha, 1957. **579**

———. "Naniwakyū sōken jidai no hōi kettei" 難波宮創建時代の方位決定 (The determination of position at the time of constructing Naniwakyū). *Naniwakyū Shi no Kenkyū* 2 (1958): 77–82. **359**

———. *Chūgoku no tenmon rekihō* 中國の天文曆法 (The history of astronomy and calendrical science in China). Tokyo: Heibonsha, 1969; rev. ed. 1990. **513, 585, 588**

———. "*Sekishi Seikyo* no kansoku nendai" 「石氏星經」の觀測年代 (The observational date of the *Shi Shen Xingjing*). In *Explorations in the History of Science and Technology in China*, ed. Li Guohao 李國豪 et al., 133–41. Shanghai: Shanghai Chinese Classics Publishing House, 1982. **519, 529**

Yamaguchi Masayuki 山口正之. "Shōken seishi to Tō Jakubō" 昭顯世子と湯若望 (Prince Sohyŏn and Tang Ruowang [Adam Schall]). *Seikyū Gakusō* 5 (1931): 101–17. **249–50**

Yamamoto Takeshi 山本大, ed. *Kōchi ken no rekishi* 高知縣の歷史 (History of Kōchi Prefecture). Tokyo: Yamakawa Shuppansha, 1970. **469**

Yamato-Kōriyama Shi Kyōiku Iinkai 大和郡山市教育委員會 (Board of Education, Yamato-Kōriyama City). *Heijōkyō Rajōmon ato hakkutsu chōsa hōkoku* 平城京羅城門跡發掘調查報告 (Report of the excavation of the sites of the Rajō Gate, Heijōkyō). Yamato-Kōriyama, 1972. **359**

Yamori Kazuhiko 矢守一彥. *Toshizu no rekishi* 都市圖の歷史 (History of city maps). 2 vols. Tokyo: Kōdansha, 1974–75. **400**

Yan Dunjie 严敦杰. *Zhongguo gudai kejishi lunwen suoyin 1900–1982* 中国古代科技史论文索引一九〇〇 —— 一九八二 (Index of essays on the history of ancient Chinese science and technology, 1900–1982). Nanjing: Jiangsu Kexue Jishu Chubanshe, 1986. **28**

Yang Hongxun 杨鸿勋. "Zhanguo Zhongshan wang ling ji zhaoyu tu yanjiu" 战国中山王陵及兆域图研究 (A study of the mausoleum of the king of the Zhanguo period's Zhongshan kingdom and the mausoleum map). *Kaogu Xuebao*, 1980, no. 1:119–38. **37**

Yang Wenheng 杨文衡. "Shilun Changsha Mawangdui sanhao Han muzhong chutu ditu di shuli jichu" 试论长沙马王堆三号汉墓中出土地图的数理基础 (On the mathematical foundation of the maps excavated from Han tomb 3 at Mawangdui). *Keji Shi Wenji* 3 (1980): 85–92. **52**

Yi Ch'an 李燦 (Chan Lee). "Han'guk ŭi ko segye chido" 韓國의 古世界地圖 (Old Korean world maps). *Han'guk Hakpo* 2 (1976): 47–66 with 9 pls. **259, 263**

———. *Han'guk ko chido* 韓國古地圖 (Old Korean maps). Map commentaries by Che Honggyu 諸洪圭 . Seoul: Han'guk Tosŏgwanhak Yŏn'guhoe, 1977. **237, 249, 254, 259, 267, 289, 291, 292, 294, 295, 303, 305, 307, 309, 312, 316, 317, 318, 319, 320, 321, 324, 328, 337**

———. *Han'guk ŭi ko chido* 韓國의 古地圖 (Old maps of Korea). Seoul: Pŭm'usa, 1991. **237**

Yi Chinhŭi 李進熙 . "Kaihō go Chōsen kōkogaku no hatten: Kōkuri hekiga kofun no kenkyū" 解放後朝鮮考古學の發展 —— 高句麗壁畫古墳研究 (The development of postwar Korean archaeology: Studies of Koguryŏ wall-painted tombs). *Kōkogaku Zasshi* 45, no. 3 (1959): 43–64. **238**

Yi Hongjik 李弘稙 , comp. *Kuksa taesajŏn* 國史大事典 (Encyclopedia of Korean history). 4th ed. Seoul: Samyŏng Ch'ulp'ansa, 1984. **305**

Yi Nŭnghwa 李能和 . *Chosŏn kidokkyo kŭp oegyo sa* 朝鮮基督敎及外交史 (History of Korean Christianity and foreign relations). Seoul: Chosŏn Kidokkyo Changmun Sa, 1928. **249**

Yi Pyŏngdo 李丙燾 . *Koryŏ sidae ŭi yŏn'gu* 高麗時代의 研究 (Study of the Koryŏ period). Seoul: Ŭryu Munhwasa, 1954. **278**

———. "*Ch'ŏnggudo* haeje" 青邱圖解題 (Biographical note to the *Ch'ŏnggudo*). In *Ch'ŏnggudo* 青邱圖 (Map of the Blue Hills, 1834), by Kim Chŏngho 金正浩 , 2 vols., vol. 1, introductory pp. 1–6. Seoul: Minjok Munwa Ch'iyinhoe, 1971. **314, 318**

———. "*Taedong chiji* haeje" 大東地志解題 (Bibliographical note to the *Taedong chiji*). In *Taedong chiji* 大東地志 (Administrative geography of the Great East [Korea], 1864), by Kim Chŏngho 金正浩 , 8 unnumbered pages following p. 840. Seoul: Hanyang Taehakkyo Kukkak Yŏn'guwŏn, 1974. **314**

Yi Shitong 伊世同 . "Zuigu di shike xingtu—Hangzhou Wuyue mu shike xingtu pingjia" 最古的石刻星图 ——杭州吴越墓石刻星图评介 (The oldest star map engraved in stone—An assessment of a star map engraved on stone from the Wuyue tomb at Hangzhou). *Kaogu*, 1975, no. 3:153–57. **539**

———. "Liangtian chi kao" 量天尺考 (Study of a sky-measuring scale). *Wenwu*, 1978, no. 2:10–17. **286**

———. *Quantian xingtu: 2000.0* 全天星圖 2000.0 (All-sky star atlas for epoch 2000.0). Beijing, 1984. **511**

———. "Hebei Xuanhua Liao Jin mu tianwen tu jianxi—jianji xingtai tiezhong huangdao shier gong tu xiang" 河北宣化辽金墓天文图简析 兼及邢台铁钟黄道十二宫图象 (A brief investigation of the star map from the Liao-Jin tombs at Xuanhua in Hebei—Also the twelve zodiacal signs as found on the Jintai iron bell). *Wenwu*, 1990, no. 10:20–24. **549**

Yin Difei 殷涤非 . "Xi Han Ruyinhou mu chutu di zhanpan he tianwen yiqi" 西汉汝阴侯墓出土的占盘和天文仪器 (Divination board and astronomical instruments from the Western Han tomb of the marquis of Ruyin). *Kaogu*, 1978, 338–43. **527**

Yokohama Shiritsu Daigaku Toshokan 横濱市立大學圖書館 (Yokohama City University Library). *Ayusawa Shintarō Bunko mokuroku* 鮎澤信太郎文庫目錄 (Catalog of the Shintaro Ayusawa Collection). Yokohama: Yokohama Shiritsu Daigaku Toshokan, 1990. **351**

Yomiuri Shinbun (Yomiuri newspaper). No. 13284, 20 October 1989. **353**

Yu Jie 于杰 and Yi Shitong 君世同 . "Beijing gu guanxiangtai" 北京古观象台 (An ancient observatory in Beijing). In *Zhongguo gudai tianwen wenwu lunji* 中国古代天文文物论集 (Collected essays on ancient Chinese astronomical relics), ed. Zhongguo Shehui Kexueyuan Kaogu Yanjiusuo 中国社会科学院考古研究所 (Archaeological Research Institute, Chinese Academy of Social Science [Academica Sinica]), 409–14. Beijing: Wenwu Chubanshe, 1989. **550**

Yu Yingshi 余英時 (Ying-shih Yü). "Qingdai sixiang shi di yige xin jieshi" 清代思想史的一個新解釋 (New interpretation of Qing intellectual history). In *Lishi yu sixiang* 歷史與思想 (History and thought), by Yu Yingshi, 121–56. Taipei: Lianjing Chuban Gongsi, 1976. **104**

Yu Yŏngbak 류영박. "Yogye kwanbang chido" 遼薊關防地圖 (The *Yogye kwanbang chido*). *Tosŏgwan* 27, no. 11 (November 1972): 32–34. **337**

Yuan Ke, ed. *Shanhai jing jiaozhu* (Edited and annotated *Shanhai jing*). See *Shanhai jing* (above).

Yunnan Sheng Lishi Yanjiusuo Diaochazu 云南省历史研究所调查组 (Investigative team of the Yunnan Historical Research Institute). "Yunnan Cangyuan yahua" 云南滄源崖画 (Cliff paintings of Cangyuan, Yunnan). *Wenwu*, 1966, no. 2:7–16 and 38. **4**

Zeng Zhaoyue 曾昭燏 et al., eds. *Nan Tang er ling fajue baogao* 南唐二陵發掘報告 (Report on the excavation of two Southern Tang tombs). Beijing: Wenwu Chubanshe, 1957. **81**

Zhan Libo 詹立波 . "Mawangdui Han mu chutu di shoubei tu tantao" 马王堆汉墓出土的守备图探讨 (Investigation of the garrison map excavated from the Han tomb at Mawangdui). *Wenwu*, 1976, no. 1:24–27. **41**

Zhang Cunwu 張存武 (Chang Ts'un-wu). "Qingdai Zhong-Han bianwu wenti tanyuan" 清代中韓邊務問題探源 (An inquiry into the Sino-Korean border question during the Qing dynasty). *Zhongyang Yanjiuyuan Jindaishi Yanjiusuo Jikan* 2 (1971): 463–503. **194, 298, 299, 301**

Zhang Guogan 張國淦 . *Zhongguo gu fangzhi kao* 中國古方志考 (Study of ancient local gazetteers in China). Beijing: Zhonghua Shuju, 1962. **89**

Zhang Peiyu 张培瑜 , Xu Zhentao 徐振韬 , and Lu Yang 卢央 . "Zhongguo zui zao qi di rishi jilu he gongyuanqian shisi zhi gongyuanqian shiyi shiji rishi biao" 中国最早期的日食记录和公元前十四至公元前十一世纪日食表 (China's earliest records of solar eclipses and a solar eclipse table for the fourteenth to the eleventh century B.C.). *Nanjing Daxue Xuebao* (1982): 371–409. **514**

Zhang Shunhui, ed. *Shuowen jiezi yuezhu*. See *Shuowen jiezi* (above).

Zhang Weihua 張維華 . *Ming shi Ouzhou si guo zhuan zhushi* 明史歐州四國傳注釋 (Commentary on the chapters on four European countries in the *History of the Ming*). 1934; reprinted Shanghai: Shanghai Guji Chubanshe, 1982. **171, 172**

Zhang Xincheng 張心澂 . *Weishu tongkao* 僞書通考 (Comprehensive study of forged books). 1939; reprinted Taipei: Hongye Shuju, 1975. **133**

Zhang Xiugui 张修桂 . "Tianshui 'Fangmatan ditu' di huizhi niandai" 天水《放马滩地图》的绘制年代 (Date of the maps from Fangmatan, Tianshui). *Fudan Xuebao*, 1991, no. 1:44–48. **38**

Zhang Xuan 張瑄 (Chang Hsüan). *Zhongwen changyong sanqian zixing yishi/The Etymologies of 3000 Chinese Characters in Common Usage* 中文常用三千字形義釋 . Hong Kong: Hong Kong University Press, 1968. **26**

Zhang Yi 張怡 . "Zhongguo gudai ditu zhi bijiao" 中國古代地圖之比較 (A comparison of ancient Chinese maps). *Dixue Zazhi* 2, no. 5 (1911): 1–8. **27**

Zhang Zhidong 張之洞 . *Zhang Wenxiang gong quanji* 張文襄公全集 (Complete works of the honorable Zhang Wenxiang [Zhidong], 1928). 1937; reprinted Taipei: Wenhai Chubanshe, 1970. **195**

Zhao Gang (Chao Kang) 趙岡 . "Ming-Qing diji yanjiu" 明清地籍研究 (Study of Ming and Qing land records). *Zhongyang Yanjiuyuan Jindaishi Yanjiusuo Jikan* 9 (1980): 37–59. **86**

Zhao Gang (Chao Kang) 趙岡 and Chen Zhongyi (Ch'en Chung-i) 陳鍾毅. *Zhongguo tudi zhidu shi* 中國土地制度史 (History of Chinese land-tenure systems). Taipei: Lianjing Chuban Shiye Gongsi, 1982. **100–101**

Zhejiangsheng Wenwu Guanli Weiyuanhui 浙江省文物管理委员会 (Committee for the Management of Cultural Relics, Zhejiang Province). "Hangzhou Lin'an wudai muzhong di tianwen tu he *mise* ci" 杭州、临安五代墓中的天文图和秘色瓷 (Astronomical maps and specially glazed porcelains found in the Five Dynasties tombs at Hangzhou and Lin'an). *Kaogu*, 1975, no. 3:186–94. **539**

Zheng Xihuang 郑锡煌. "*Jiu yu shouling tu* yanjiu" 九域守令图研究 (A study of the *Jiu yu shouling tu*). In *Zhongguo gudai ditu ji* 中國古代地圖集, ed. Cao Wanru 曹婉如 et al., 1:35–40. Beijing: Wenwu Chubanshe, 1990–. **46**

Zhongguo Hanghai Shi Yanjiuhui 中国航海史研究会 (Research Association for the History of Chinese Navigation), ed. *Zheng He yanjiu ziliao xuanbian* 鄭和研究資料选編 (Selected research materials on Zheng He). Beijing: Renmin Jiaotong Chubanshe, 1985. **53**

Zhongguo Kexueyuan Beijing Tianwentai 中國科學院北京天文臺 (Beijing Observatory, Chinese Academy of Sciences), ed. *Zhongguo difangzhi lianhe mulu* 中國地方志聯合目錄 (Union catalog of Chinese gazetteers). Beijing: Zhonghua Shuju, 1985. **89**

Zhongguo Shehui Kexueyuan Kaogu Yanjiusuo 中国社会科学院考古研究所 (Archaeological Research Institute, Chinese Academy of Social Science [Academia Sinica]). *Zhongguo gudai tianwen wenwu tuji* 中国古代天文文物图集 (Album of ancient Chinese astronomical relics). Beijing: Wenwu Chubanshe, 1980. **56, 81, 129, 130, 511–12, 522, 526, 531, 536, 537, 539, 540, 542, 545, 548, 552, 553, 554, 555, 572, 574, 575, 599, plate 31**

Zhu Jianqiu 朱鉴秋. " 'Zheng He hanghai tu' zai woguo haitu fazhanshizhong di diwei he zuoyong" 《郑和航海图》在我国海图发展史中的地位和作用 (The place and role of Zheng He's nautical chart in the history of the development of our country's nautical charts). In *Zheng He xia Xiyang lunwenji* 鄭和下西洋論文集 (Collected essays on Zheng He's expedition to the Western Ocean), ed. Zhongguo Hanghai Shi Yanjiuhui 中国航海史研究会 (Research Association for the History of Chinese Navigation), 1:229–37. Beijing: Renmin Jiaotong Chubanshe, 1985. **64**

Sources in Other Languages

Abeel, David. *Journal of a Residence in China, and the Neighboring Countries, from 1829 to 1833.* New York: Leavitt, Lord, 1834. **722**

Abendanon, Edouard Cornelius. *Midden-Celebes-expeditie: Geologische en geographische doorkruisingen van Midden-Celebes (1909–1910).* 4 vols. Leiden: E. J. Brill, 1915–18. **832**

Acker, William Reynolds Beal, trans. and annotator. *Some T'ang and Pre-T'ang Texts on Chinese Painting.* 2 vols. Leiden: E. J. Brill, 1954–74. **139**

Adler, Bruno F. "Karty pervobytnykh narodov" (Maps of primitive peoples). *Isvestiya Imperatorskogo Obshchestva Lyubiteley Yestestvoznaniya, Antropologii i Etnografii: Trudy Geograficheskogo Otdeleniya* 119, no. 2 (1910). **608**

Aguilar, José, ed. *Historia de la Cartografia: La tierra de papel.* Buenos Aires: Editorial Codex S.A., 1967. **373**

Aihara, Ryōichi. "Ignacio Moreira's Cartographical Activities in Japan (1590–2), with Special Reference to Hessel Gerritsz's Hemispheric World Map." *Memoirs of the Research Department of the Toyo Bunko* 34 (1976): 209–42. **376**

Akioka, Takejirō, and Nobuo Muroga. "The History of Cartography in Japan." In *A Catalogue of Old Maps of Japan, Exhibited at the Tenri Central Library, September, 1957*, ed. Takeo Oda, 1–6. Kyōto, 1957. **351**

———. "The Short History of Ancient Cartography in Japan." In *Proceedings of the International Geographical Union Regional Conference in Japan, 1957*, 57–60. Tokyo: Organizing Committee of the IGU Regional Conference in Japan and the Science Council of Japan, 1959. **351**

Alexander, Jonathan J. G. Review of volume 1 of *The History of Cartography*, "Mapping the Medieval World." *Journal of Historical Geography* 16 (1990): 230–33. **19**

Ammarell, Gene. "Sky Calendars of the Indo-Malay Archipelago: Regional Diversity/Local Knowledge." *Indonesia* 45 (1988): 84–104. **713**

———. "Navigation Practices of the Bugis Seafarers." Paper presented at the forty-first annual meeting of the Association for Asian Studies, Washington, D.C., 19 March 1989. **714**

Andersson, Johan Gunnar. *Children of the Yellow Earth: Studies in Prehistoric China.* Trans. from the Swedish by E. Classen. New York: Macmillan, 1934. **12, 13**

Ang, Tian Se. "I-Hsing (683–727 A.D.): His Life and Scientific Work." Ph.D. diss., University of Malaya, Kuala Lumpur, 1979. **538**

Anville, Jean Baptiste Bourguignon d'. *Nouvel atlas de la Chine, de la Tartarie chinoise et du Thibet.* The Hague: H. Scheurleer, 1737. **183, 299**

Argüelles, José, and Miriam Argüelles. *Mandala.* Berkeley, Calif.: Shambhala, 1972. **620**

Aris, Michael. *Bhutan: The Early History of a Himalayan Kingdom.* Warminster, Eng.: Aris and Phillips, 1979. **613, 614**

Arts: The Magazine for Members of the Minneapolis Institute of Arts. June 1992. **610**

Ashida, Koreto. "Old Maps of Hokkaido." In *Dainippon* (Great Japan), ed. Bunmei Kyōkai, 127–37. Tokyo: Bunmei Kyōkai, 1936. **443**

Aston, William George, trans. *Nihongi: Chronicles of Japan from the Earliest Times to* A.D. *697.* 2 vols. in 1. 1896; reprinted Tokyo: Charles E. Tuttle, 1985. **353, 354, 355, 358, 371, 372**

Atlas van Tropisch Nederland. Batavia: Koninklijk Nederlandsch Aardrijkskundig Genootschap, 1938. **773**

Auer, Gerhard, and Niels Gutschow. *Bhaktapur: Gestalt, Funktionen, und religiöse Symbolik einer nepalischen Stadt im vorindustriellen Entwicklungsstadium.* Darmstadt: Technische Hochschule, 1974. **612, 613**

Aujac, Germaine, and the editors. "The Growth of an Empirical Cartography in Hellenistic Greece." In *The History of Cartography*, ed. J. B. Harley and David Woodward, 1:148–60. Chicago: University of Chicago Press, 1987–. **98**

Aung-Thwin, Michael. *Pagan: The Origins of Modern Burma.* Honolulu: University of Hawaii Press, 1985. **479**

Aurousseau, Leonard. Review of Charles B. Maybon, *Histoire moderne du pays d'Annam (1592–1820).* In *Bulletin de l'Ecole Française d'Extrême-Orient* 20, no. 4 (1920): 73–121. **499, 500, 502, 506**

Ayusawa, Shintaro. "The Types of World Map Made in Japan's Age of National Isolation." *Imago Mundi* 10 (1967): 123–27. **404, 409, 426, 429, 439, 441, 442, 443**

Aziz, Barbara Nimri. "Tibetan Manuscript Maps of Dingri Valley." *Canadian Cartographer* 12 (1975): 28–38. **609, 617, 618, 653, 654, 670**

———. "Maps and the Mind." *Human Nature* 1, no. 8 (1978): 50–59. **609, 653, 654, 655, 670, 827**

———. *Tibetan Frontier Families: Reflections of Three Generations from D'ing-ri.* Durham, N.C.: Carolina Academic Press, 1978. **609, 617, 670, 677**

Aziz, Barbara Nimri, and Matthew Kapstein, eds. *Soundings in Tibetan Civilization.* New Delhi: Manohar, 1985. **654**

Bacon, Francis. *The New Organon.* In *The New Organon and Related Writings,* ed. Fulton H. Anderson. New York: Liberal Arts Press, 1960. **223**

Baddeley, John F. "Father Matteo Ricci's Chinese World-Maps." *Geographical Journal* 50 (1917): 254–70. **174**

———. *Russia, Mongolia, China; Being Some Record of the Relations between Them from the Beginning of the XVIIth Century to the Death of the Tsar Alexei Mikhailovich,* A.D. 1602–1676. 2 vols. London, 1919; reprinted New York: B. Franklin, 1964. **682**

Bagrow, Leo. *History of Cartography.* Rev. and enl. R. A. Skelton. Trans. D. L. Paisey. Cambridge: Harvard University Press; London: C. A. Watts, 1964. Reprinted and enlarged, Chicago: Precedent Publishing, 1985. **xxiii, 607, 690**

Bailey, Jane Terry. "Some Burmese Paintings of the Seventeenth Century and Later, Part II: The Return to Pagan." *Artibus Asiae* 40 (1978): 41–61. **696, 722**

Bajpai, K. D., ed. *The Geographical Encyclopaedia of Ancient and Medieval India.* Varanasi: Indic Academy, 1967–. **781**

Baker, Donald L. "Jesuit Science through Korean Eyes." *Journal of Korean Studies* 4 (1982–83): 207–39. **575**

Balazs, Etienne. *Chinese Civilization and Bureaucracy: Variations on a Theme.* Trans. H. M. Wright. Ed. Arthur F. Wright. New Haven: Yale University Press, 1964. **71, 87**

Ban Gu. *The History of the Former Han Dynasty.* 3 vols. Trans. Homer H. Dubs. Baltimore: Waverly Press, 1938–55. **71, 77, 524, 525, 527**

Banerjee, N. R. "A Painted Nepalese *Paubhā* in the Collection of the National Museum, New Delhi." In *Buddhist Iconography,* 154–63. New Delhi: Tibet House, 1989. **649, 650**

Banerjee, N. R., and O. P. Sharma. "A Note on a Painted Map of the Kathmandu Valley at the National Museum, New Delhi." *Marg* 38, no. 3 [1986]: 77–80. **649**

Bary, William Theodore de, and Irene Bloom, eds. *Principle and Practicality: Essays in Neo-Confucianism and Practical Learning.* New York: Columbia University Press, 1979. **24**

Bary, William Theodore de, Wing-tsit Chan, and Burton Watson, comps. *Sources of Chinese Tradition.* New York: Columbia University Press, 1960. **207, 222, 226**

Bastian, Adolf. *Ideale Welten nach uranographischen Provinzen in Wort und Bild: Ethnologische Zeit- und Streitfragen, nach Gesichtspunkten der indischen Völkerkunde.* 3 vols. Berlin: Emil Felber, 1892. **693, 720**

———. "Graphische Darstellung des buddhistischen Weltsystems." *Verhandlungen der Berliner Gesellschaft für Anthropologie, Ethnologie und Urgeschichte,* 1894, 203–15. In *Zeitschrift für Ethnologie,* vol. 26. **693, 717**

Baxter, William H. *A Handbook of Old Chinese Phonology.* Berlin: Mouton de Gruyter, 1992. **26**

Beans, George H. "Some Notes from the Tall Tree Library." *Imago Mundi* 11 (1954): 146–47. **474**

———. *A List of Japanese Maps of the Tokugawa Era.* Jenkintown, Pa.: Tall Tree Library, 1951. Supplements A, B, and C (1955, 1958, 1963). **352, 422, 429, 437, 473, 476**

Bechert, Heinz. " 'To Be a Burmese Is to Be a Buddhist': Buddhism in Burma." In *The World of Buddhism: Buddhist Monks and Nuns in Society and Culture,* ed. Heinz Bechert and Richard Gombrich, 147–58. London: Thames and Hudson, 1984. **777**

Beer, Arthur, et al. "An 8th-Century Meridian Line: I-Hsing's Chain of Gnomons and the Pre-history of the Metric System." *Vistas in Astronomy* 4 (1961): 3–28. **53, 123, 533**

Bennett, Paul J. "Two Southeast Asian Ministers and Reactions to European Conquest: The Kinwun Mingyi and Phan-thanh-Gian." In *Conference under the Tamarind Tree: Three Essays in Burmese History,* 103–42. New Haven: Yale University Southeast Asia Studies, 1971. **507, 789**

Bennett, Steven J. "Patterns of the Sky and Earth: A Chinese Science of Applied Cosmology." *Chinese Science* 3 (1978): 1–26. **154, 216, 217, 219, 220, 222, 276–77, 278**

Berg, L. S. *Otkrytie Kamchatki i ekspeditsii Beringa, 1725–1742* (The discovery of Kamchatka and the expeditions of Bering, 1725–42). Moscow: Izdatel'stvo Akademii Nauk SSSR, 1946. **445**

Bergman, Folke. "Travels and Archaeological Field-work in Mongolia and Sinkiang—A Diary of the Years 1927–1934." In *History of the Expedition in Asia, 1927–1935,* 4 vols., by Sven Anders Hedin, 4:1–192. Stockholm: [Göteborg, Elanders Boktryckeri Aktiebolag], 1943–45. **6**

Bernard, Henri. *Matteo Ricci's Scientific Contribution to China.* Trans. Edward Chalmers Werner. Beijing: Henri Vetch, 1935. **404**

Bernbaum, Edwin. *The Way to Shambhala.* Garden City, N.Y.: Anchor Press, 1980. **629**

———. "The Hidden Kingdom of Shambhala." *Natural History* 92, no. 4 (1983): 54–63. **629, 632**

Berthelot, René. *La pensée de l'Asie et l'astrobiologie.* Paris: Payot, 1949. **8**

Bezacier, Louis. *L'art viêtnamien.* Paris: Editions de l'Union Française, 1954. **480**

———. *Relevés de monuments anciens du Nord Việt-nam.* Paris: Ecole Française d'Extrême-Orient, 1959. **480**

Biasio, Elisabeth, and Verena Münzer. *Übergänge im menschlichen Leben: Geburt, Initiation, Hochzeit und Tod in aussereuropäischen Gesellschaften.* Zurich: Völkerkundemuseum der Universität Zürich, 1980. **706**

Bielenstein, Hans. *The Bureaucracy of Han Times.* Cambridge: Cambridge University Press, 1980. **88**

———. "Han Portents and Prognostications." *Bulletin of the Museum of Far Eastern Antiquities* 56 (1984): 97–112. **521**

Biot, Edouard. *Le Tcheou-li; ou, Rites des Tcheou.* 3 vols. 1851; reprinted Taipei: Chengwen Chubanshe, 1969. **75**

Biot, J. B. Review of *Ueber die Zeitrechnung der Chinesen* by Ludwig Ideler. In *Journal des Savants,* 1839, 721–30, and 1840, 27–41. **525**

Bishop, Peter. *The Myth of Shangri-La: Tibet, Travel Writing and the Western Creation of Sacred Landscape.* London: Athlone Press, 1989. **629**

Blackmore, Thaung. *Catalogue of the Burney Parabaiks in the India Office Library.* London: British Library, 1985. **690, 692, 753, 755, 811**

Blair, Emma H., and James A. Robertson, eds. *The Philippine Islands, 1493–1898.* 55 vols. Cleveland: Arthur H. Clark, 1903–9. **390**

Bo, Shuren. "Astrometry and Astrometric Instruments." In *Ancient China's Technology and Science,* comp. Institute of the History of Natural Sciences, Chinese Academy of Sciences, 15–32. Beijing: Foreign Languages Press, 1983. **525**

———. "Sima Qian—The Great Astronomer of Ancient China." *Chinese Astronomy and Astrophysics* 9 (1985): 261–67. **528**

Boisselier, Jean. "Pouvoir royale et symbolisme architectural: Neak Pean et son importance pour la royauté angkorienne." *Arts Asiatiques* 21 (1970): 91–108. **733**

———. *Thai Painting.* Trans. Janet Seligman. Tokyo: Kodansha International, 1976. **720, 721, 722, 733**

Boltz, Judith M. *A Survey of Taoist Literature: Tenth to Seventeenth Centuries.* Berkeley, Calif.: Institute of East Asian Studies, 1987. **230**

———. "Cartography in the Taoist Canon." *Asia Major,* forthcoming. **230**

Bosch, Frederick David Kan. *De gouden kiem: Inleiding in de Indische symboliek.* Amsterdam: Elsevier, 1948. English translation, *The Golden Germ: An Introduction to Indian Symbolism.* The Hague: Mouton, 1960. **706, 723, 728**

Boulding, Kenneth Ewart. *The Image.* Ann Arbor: University of Michigan Press, 1956. **20**

Boulnois, L. *Bibliographie du Népal,* vol. 3, *Sciences naturelles,* bk. 1, *Cartes du Népal dans les bibliothèques de Paris et de Londres.* Paris: Editions du Centre National de la Recherche Scientifique, 1973. **609**

Brauen, Martin. *Heinrich Harrers Impressionen aus Tibet.* Innsbruck: Pinguin-Verlag, 1974. **653, 675**

———. *Feste in Ladakh.* Graz: Akademische Druck- u. Verlagsanstalt, 1980. **653**

Bray, Francesca. See Needham, Joseph. *Science and Civilisation in China.*

Brook, Timothy. *Geographical Sources of Ming-Qing History.* Ann Arbor: Center for Chinese Studies, University of Michigan, 1988. **188–89**

Brooks, Robert R. R., and Vishnu S. Wakankar. *Stone Age Painting in India.* New Haven: Yale University Press, 1976. **18**

Broughton, William Robert. *A Voyage of Discovery to the North Pacific Ocean.* London: T. Cadell and W. Davies, 1804; reprinted Amsterdam: Nico Israel and New York: Da Capo Press, 1967. **447**

Brown, Lloyd A. *The Story of Maps.* Boston: Little, Brown, 1949; reprinted New York: Dover, 1979. **690**

Brown, Philip C. "Never the Twain Shall Meet: European Land Survey Techniques in Tokugawa Japan." *Chinese Science* 9 (1989): 53–79. **396**

Brown, Robert L. "Bodhgaya and South-east Asia." In *Bodhgaya: The Site of Enlightenment,* ed. Janice Leoshko, 101–24. Bombay: Marg Publications, 1988. **696, 777, 783**

Buchanan, Francis. See Hamilton, Francis.

Bùi Thiết. "Sắp xếp thế hệ các bản đồ hiện biết thành Thăng Long thời Lê" (Establishing the generations of known maps of the city of Thăng Long from the Lê dynasty, 1428–1787). *Khảo Cổ Học* (Archaeology) 52, no. 4 (1984): 48–55. **478, 482, 483, 490, 495, 499**

Bulling, A. Gutkind. "Ancient Chinese Maps: Two Maps Discovered in a Han Dynasty Tomb from the Second Century B.C." *Expedition* 20, no. 2 (1978): 16–25. **41**

Bush, Susan. *The Chinese Literati on Painting: Su Shih (1037–1101) to Tung Ch'i-ch'ang (1555–1636).* Cambridge: Harvard University Press, 1971. **136**

Bush, Susan, and Hsio-yen Shih, comps. and eds. *Early Chinese Texts on Painting.* Cambridge: Harvard University Press, 1985. **143**

Buttinger, Joseph. *Vietnam: A Political History.* New York: Praeger, [1968]. **495**

Cahill, James F. *Chinese Painting.* Geneva: Editions d'Art Albert Skira, 1960; reprinted New York: Rizzoli International Publications, 1977. **225**

———. "Confucian Elements in the Theory of Painting." In *The Confucian Persuasion,* ed. Arthur F. Wright, 115–40. Stanford: Stanford University Press, 1960. **136**

———. "Style as Idea in Ming-Ch'ing Painting." In *The Mozartian Historian: Essays on the Works of Joseph R. Levenson,* ed. Maurice Meisner and Rhoads Murphey, 137–56 Berkeley and Los Angeles: University of California Press, 1976. **225**

———. *The Compelling Image: Nature and Style in Seventeenth-Century Chinese Painting.* Cambridge: Harvard University Press, 1982. **153**

The Cambridge History of Japan. Vol. 3, *Medieval Japan.* Ed. Kozo Yamamura. Cambridge: Cambridge University Press, 1990. **362**

Cammann, Schuyler. "The 'TLV' Pattern on Cosmic Mirrors of the Han Dynasty." *Journal of the American Oriental Society* 68 (1948): 159–67. **215**

———. "The Magic Square of Three in Old Chinese Philosophy and Religion." *History of Religions* 1 (summer 1961): 37–79. **204, 214**

Cao, Wanru. "Maps 2,000 Years Ago and Ancient Cartographical Rules." In *Ancient China's Technology and Science,* comp. Institute of the History of Natural Sciences, Chinese Academy of Sciences, 250–57. Beijing: Foreign Languages Press, 1983. **41**

———. "Ancient Maps Unearthed from Qin Tomb of Fangmatan and Han Tomb of Mawangdui: A Comparative Research." *Journal of Chinese Geography* 3, no. 2 (1992): 39–50. **39**

Cardim, Antonio Francisco. *Fasciculus e Iapponicis floribus, suo adhuc madentibus sanguine.* Rome: Typis Heredum Corbelletti, 1646. **388**

Cary, Otis. *A History of Christianity in Japan.* 2 vols. New York: Fleming H. Revell, 1909; reprinted 1987. **370, 376, 377**

Catalogue of the Tibetan Collection and Other Lamaist Material in the Newark Museum. 5 vols. Newark, N.J.: Newark Museum, 1951–71. **667, 679**

Chaffee, John W. *The Thorny Gates of Learning in Sung China: A Social History of Examinations.* Cambridge: Cambridge University Press, 1985. **71**

Chagdarsurung, Ts. "La connaissance géographique et la carte des Mongols." *Studia Mongolica,* vol. 3 (2) (1975): 345–70. **685**

Chamberlain, Basil Hall. *Japanese Things: Being Notes on Various Subjects Connected with Japan.* Tokyo: Charles E. Tuttle, 1985; reprinted from an edition of 1905. **370, 410, 411**

———, trans. *The Kojiki: Records of Ancient Matters.* 1882; reprinted Tokyo: Charles E. Tuttle, 1986. **353, 354**

Chang, Kang-i Sun. "Description of Landscape in Early Six Dynasties Poetry." In *The Vitality of the Lyric Voice: Shih Poetry from the Late Han to the T'ang,* ed. Shuen-fu Lin and Stephen Owen, 105–29. Princeton: Princeton University Press, 1986. **134**

Chang, Kuei-sheng. "The Han Maps: New Light on Cartography in Classical China." *Imago Mundi* 31 (1979): 9–17. **41, 51**

Chang, Kwang-chih. *The Archaeology of Ancient China.* Rev. and enl. New Haven: Yale University Press, 1977. **204**

———. *Shang Civilization.* New Haven: Yale University Press, 1980. **72**

———. *The Archaeology of Ancient China.* 4th ed. New Haven: Yale University Press, 1986. **130**

Chang, Sen-dou. "Manuscript Maps in Late Imperial China." *Canadian Cartographer* 11 (1974): 1–14. **191**

Chang, Yu-che. "Chang-Hen, a Chinese Contemporary of Ptolemy." *Popular Astronomy* 53 (1945): 122–26. **522**

Chatley, Herbert. "The Date of the Hsia Calendar *Hsia Hsiao Chêng.*" *Journal of the Royal Asiatic Society of Great Britain and Ireland,* 1938, 523–33. **513**

Chavannes, Edouard. "Les deux plus anciens spécimens de la cartographie chinoise." *Bulletin de l'Ecole Française d'Extrême-Orient* 3 (1903): 214–47. **27, 35, 37, 113, 164**

———. "Le cycle turc des douze animaux." *T'oung Pao,* 2d ser., 7 (1906): 51–122. **537**

———. "L'instruction d'un futur empereur de Chine en l'an 1193." *Mémoires concernant l'Asie Orientale* 1 (1913): 19–64. **86**

———, trans. *Les mémoires historiques de Se-Ma Ts'ien.* See Sima Qian.

Chayet, Anne. "The Jehol Temples and Their Tibetan Models." In *Soundings in Tibetan Civilization*, ed. Barbara Nimri Aziz and Matthew Kapstein, 65–72. New Delhi: Manohar, 1985. **613**

Chen, Cheng-siang (Chen Zhengxiang). "The Historical Development of Cartography in China." *Progress in Human Geography* 2 (1978): 101–20. **113, 125**

Ch'en, Kenneth (Chen Guansheng). "Matteo Ricci's Contribution to, and Influence on, Geographical Knowledge in China." *Journal of the American Oriental Society* 59 (1939): 325–59. **173, 174, 176, 177**

Chen, Qi-xin, and Li Xing Guo. "The Unearthed Paperlike Objects Are Not Paper Produced before Tsai-Lun's Invention." *Yearbook of Paper History* 8 (1990): 7–22. **40**

Chen, Zhao Fu. "Ancient Rock Art in China." *Bollettino del Centro Camuno di Studi Preistorici* 23 (1986): 91–98. **4, 18**

———. *Cina: L'arte rupestre preistorica.* Italian trans. Giuliana Aldi Pompili. Milan: Jaca Books, 1988. **2, 4, 8, 18, 21**

Cheng, Yen-tsu. "Cosmological Theories in Ancient China." *Scientia Sinica* 19 (1976): 291–309. **117**

Cheung, Kwong-yue. "Recent Archaeological Evidence relating to the Origin of Chinese Characters." Trans. Noel Barnard. In *The Origins of Chinese Civilization*, ed. David N. Keightley, 323–91. Berkeley and Los Angeles: University of California Press, 1983. **130**

Chinese Cultural Art Treasures: National Palace Museum Illustrated Handbook. 3d ed. Taipei: National Palace Museum, 1967. **27**

Chou, Hung-hsiang. "Chinese Oracle Bones." *Scientific American* 240 (April 1979): 134–49. **514**

Christie, A. H. "The Megalithic Problem in South East Asia." In *Early South East Asia: Essays in Archaeology, History and Historical Geography*, ed. R. B. Smith and W. Watson, 242–52. New York: Oxford University Press, 1979. **9**

Chu, K'o-chên. "The Origin of the Twenty-eight Lunar Mansions." *Actes du VIIIe Congrès International d'Histoire des Sciences (1956)*, 1:364–72. Florence: Gruppo Italiano di Storia delle Scienze, 1958. **518**

Cieslik, Hubert, ed. *Hoppō Tanken Ki* [Record of an exploration of the northern region]: *Foreigners' Reports on Ezo in the Genna Period.* Tokyo: Yoshikawa Kōbunkan, 1962. **444**

Cima, Ronald J., ed. *Vietnam: A Country Study.* Washington, D.C.: Federal Research Division, 1989. **495**

Clark, David H., and F. Richard Stephenson. *The Historical Supernovae.* Oxford: Pergamon Press, 1977. **511, 554, 566**

Clément, Sophie, Pierre Clément, and Shin Yong Hak. *Architecture du paysage en Asie orientale.* Paris: Ecole Nationale Supérieure des Beaux Arts, 1982. **276, 278, 283**

Cockburn, John. "On the Recent Existence of *Rhinoceros indicus* in the North Western Provinces, and a Description of a Tracing of an Archaic Rock Painting from Mirzapore Representing the Hunting of This Animal." *Journal of the Asiatic Society of Bengal* 52, pt. 2 (1883): 56–64. **17**

———. "Cave Drawings in the Kaimūr Range, North-West Provinces." *Journal of the Royal Asiatic Society of Great Britain and Ireland*, 1899, 89–97. **17**

Coedès, George. *The Vajirañāṇa National Library of Siam.* Bangkok: Bangkok Times Press, 1924. **743**

———. *Pour mieux comprendre Angkor.* Hanoi, 1943. English edition, *Angkor: An Introduction.* Ed. and trans. Emily Floyd Gardiner. Hong Kong: Oxford University Press, 1963. **740**

———. "The Traibhūmikathā Buddhist Cosmology and Treaty on Ethics." *East and West* (Rome) 7 (1957): 349–52. **720**

Coedès, George, and C. Archaimbault. *Les trois mondes (Traibhūmi Braḥ R'vañ).* Paris: Ecole Française d'Extrême-Orient, 1973. **696, 720**

Coen, J. C. *Reize van Maarten Gerritsz. Vries in 1643 naar het noorden en oosten van Japan. . . .* Amsterdam, 1858. **440**

Coleridge, Samuel Taylor. *Coleridge on the Seventeenth Century.* Ed. Roberta Florence Brinkley. Durham, N.C.: Duke University Press, 1955. **203**

Cordier, Henri. *Description d'un atlas sino-coréen manuscrit du British Museum.* Recueil de voyages et de documents pour servir à l'histoire de la géographie depuis le XIIIe jusqu'à la fin du XVIe siècle, section cartographique. Paris: Ernest Leroux, 1896. **259**

Cortazzi, Hugh. *Isles of Gold: Antique Maps of Japan.* Tokyo: Weatherhill, 1983. **351, 361, 367, 368, 370, 371, 373, 380, 388, 389, 400, 405, 412, 413, 414, 420, 424–26, 429, 440, 443, 444, 450, 453, 475**

Cortesão, Armando. *Cartografia e cartógrafos portugueses dos séculos XV e XVI.* 2 vols. Lisbon, 1935. English translation, *History of Portuguese Cartography.* 2 vols. Coimbra: Junta de Investigações do Ultramar-Lisboa, 1969–71. **828**

———. "Study of the Evolution of the Early Cartographic Representation of Some Regions of the World: Japan." In *Portugaliae monumenta cartographica*, 6 vols., by Armando Cortesão and Avelino Teixeira da Mota, 5:170–78 and 6:40–41 (addenda). Lisbon, 1960. Facsimile edition, Lisbon: Imprensa Nacional-Casa da Moeda, 1987. **370**

Cosgrove, Denis E. *Social Formation and Symbolic Landscape.* London and Sydney: Croom Helm, 1984. **10**

Courant, Maurice. *Bibliographie coréenne.* 3 vols. Paris: Ernest Leroux, 1894–96. **257, 259**

Couvreur, Séraphin, ed. and trans. *Li Ki; ou, Mémoires sur les bienséances et les cérémonies.* 2d ed. 2 vols. Paris: Cathasia, 1913. **519**

———, trans. *Tch'ouen Ts'ou et Tso Tchouan*, 3 vols. Hochienfu: Mission Press, 1914. **515**

Covarrubias, Miguel. *Island of Bali.* 1936. New York: Alfred A. Knopf, 1956. **723**

Creel, Herrlee Glessner. *Studies in Early Chinese Culture, First Series.* Baltimore: Waverly Press, 1937. **76**

Crinò, Sebastiano. "La prima carta corografica inedita del Giappone portata in Italia nel 1585 e rinvenuta in una filza di documenti riguardanti il commercio dei Medici nelle Indie Orientali e Occidentali." *Rivista Marittima* 64 (1931): 257–84. **370**

Crone, Gerald. *Maps and Their Makers: An Introduction to the History of Cartography.* London: Hutchinson University Library, 1953, and four subsequent editions up to 1978. **690**

Crump, J. I., Jr., trans. *Chan-kuo Ts'e.* Oxford: Clarendon Press, 1970. **73, 74**

Cullen, Christopher. "A Chinese Eratosthenes of the Flat Earth: A Study of a Fragment of Cosmology in *Huai Nan Tzu.*" *Bulletin of the School of Oriental and African Studies* 39 (1976): 106–27. **119**

Dagyab, Loden Sherap. *Tibetan Religious Art.* 2 vols. Wiesbaden: Otto Harrassowitz, 1977. **628**

Dahlgren, Erik W. *Les débuts de la cartographie du Japon.* Uppsala: K. W. Appelberg; 1911, reprinted Amsterdam: Meridian, 1977. **370**

Dalby, Andrew, and Sao Saimöng Mangrai. "Shan and Burmese Manuscript Maps in the Scott Collection, Cambridge University Library." Unpublished manuscript, [ca. 1984]. **690, 692, 699, 760, 761, 790, 791, 792, 794, 795**

Dallapiccola, Anna Libera, ed., in collaboration with Stephanie Zingel-Avé Lallemant. *The Stūpa: Its Religious, Historical and Architectural Significance.* Wiesbaden: Franz Steiner, 1980. **616**

Dalrymple, Alexander. *A Collection of Charts and Memoirs.* London, 1772. **838**

Đặng Phu'o'ng-nghi. *Les institutions publiques du Viêt-Nam au XVIIIe siècle.* Paris: Ecole Française d'Extrême-Orient, 1969. **486**

Dardess, John W. *Confucianism and Autocracy: Professional Elites in the Founding of the Ming Dynasty.* Berkeley and Los Angeles: University of California Press, 1983. **205**

Das, Sarat Chandra. *Indian Pandits in the Land of Snow.* 1893; reprinted Delhi: Delhi Printers Prakashan, 1978. **681**

———. *Journey to Lhasa and Central Tibet.* 1902; reprinted New Delhi: Mañjuśrī Publishing House, 1970. **668, 670**

Debergh, Minako. "A Comparative Study of Two Dutch Maps Preserved in the Tokyo National Museum—Joan Blaeu's Wall Map of the World in Two Hemispheres, 1648 and Its Revision ca. 1678 by N. Visscher." *Imago Mundi* 35 (1983): 20–36. **435**

Delano Smith, Catherine. In *The History of Cartography,* ed. J. B. Harley and David Woodward. Chicago: University of Chicago Press, 1987–.

"Cartography in the Prehistoric Period in the Old World: Europe, the Middle East, and North Africa," 1:54–101. **1, 2, 3, 6, 9, 13, 14, 15, 18**

"Prehistoric Maps and the History of Cartography: An Introduction," 1:45–49. **1, 20**

Desai, W. S. "A Map of Burma (1795) by a Burmese Slave." *Journal of the Burma Research Society* 26, no. 3 (1936): 147–52. **747, 803**

Destombes, Marcel. "Wang P'an, Liang Chou et Matteo Ricci: Essai sur la cartographie chinoise de 1593 à 1603." In *Actes du Troisième Colloque International de Sinologie: Appréciation par l'Europe de la tradition chinoise à partir du dix-septième siècle,* 47–65. Paris: Belles Lettres, 1983. **229**

Devlet, M. A. *Bol'shaya Boyarskaya pisanitsa/Rock Engravings in the Middle Yenisei Basin.* Moscow: Nauka, 1976. **3, 4, 5**

———. *Petroglify Ulug-Khema.* (Petroglyphs of Ulug-Khem). Moscow: Nauka, 1976. **6, 7, 21, 22**

———. *Petroglify Mugur-Sargola* (Petroglyphs of Mugur-Sargol). Moscow: Nauka, 1980. **6, 7, 21, 22**

Dictionary of National Biography: From Earliest Times to 1900. 22 vols. Ed. Leslie Stephen and Sidney Lee. First published in 66 vols., 1885–1901. Reprinted London: Oxford University Press, 1937–38. **836**

Dictionary of National Biography. Suppl. 1931–40. London: Oxford University Press, 1949. **760**

Dien, Albert E. "Chinese Beliefs in the Afterworld." In *The Quest for Eternity: Chinese Ceramic Sculptures from the People's Republic of China,* ed. Susan L. Caroselli, 1–15. Los Angeles: Los Angeles County Museum of Art, 1987. **80**

Dieux et démons de l'Himâlaya: Art du Bouddhisme lamaïque. Catalog of an exhibition at the Grand Palais, 25 March to 27 June 1977. Paris: Secrétariat d'Etat à la Culture, 1977. **614, 648, 668**

Dilke, O. A. W. In *The History of Cartography,* ed. J. B. Harley and David Woodward. Chicago: University of Chicago Press, 1987–.

"The Culmination of Greek Cartography in Ptolemy," 1:177–200. **114, 124**

"Maps in the Service of the State: Roman Cartography to the End of the Augustan Era," 1:201–11. **710–11**

Diringer, David. *Writing.* New York: Frederick A. Praeger, 1962. **17**

Dorn, Ronald I., Margaret Nobbs, and Tom A. Cahill. "Cation-Ratio Dating of Rock-Engravings from the Olary Province of Arid South Australia." *Antiquity* 62 (1988): 681–89. **15**

Dorzh, D. "Rock 'Art Galleries' of Mongolia." *Canada Mongolia Review* 1, no. 2 (1975): 49–55. **7**

Douglas, Nik. *Tantric Charms and Amulets.* New York: Dover, 1978. **621**

Drake, Fred W. *China Charts the World: Hsu Chi-yü and His Geography of 1848.* Cambridge: East Asian Research Center, Harvard University, 1975. **108**

Dubs, Homer H. "The Date of the Shang Period." *T'oung Pao* 40 (1951): 322–35. **514**

———, trans. *The History of the Former Han Dynasty.* See Ban Gu.

———. *The Works of Hsüntze.* See Xun Qing.

Dudley, Robert. *Dell'arcano del mare.* 3 vols. Florence, 1646–47. 2d ed. 1661. **388**

Du Halde, Jean Baptiste. *Description géographique, historique, chronologique, politique, et physique de l'empire de la Chine et de la Tartarie chinoise.* 4 vols. Paris: Lemercier, 1735. **181, 183, 299**

———. *A Description of the Empire of China and Chinese-Tartary, Together with the Kingdoms of Korea, and Tibet.* 2 vols. London: Edward Cave, 1738–41. **181, 185**

———, ed. *Lettres édifiantes et curieuses, écrites des missions étrangères par quelques missionnaires de la Compagnie de Jésus.* 27 vols. Paris: Nicolas le Clerc, 1707–49. **180**

Dumarçay, Jacques. "Notes d'architecture Javanaise et Khmère." *Bulletin de l'Ecole Française d'Extrême-Orient* 71 (1982): 87–147. **696**

Dumoutier, Gustave. "Etude sur un portulan annamite du XVe siècle." *Bulletin de Géographie Historique et Descriptive* 11 (1896): 141–204. **490**

Durand, John D. "The Population Statistics of China, A.D. 2–1953." *Population Studies* 13 (1960): 209–56. **71**

Duroiselle, Charles. "The Rock-Cut Temples of Powun-daung." In *Archaeological Survey of India Annual Report, 1914–15,* 42–55. Calcutta, 1920. **726, 727**

Eade, J. C. *Southeast Asian Ephemeris: Solar and Planetary Positions,* A.D. 638–2000. Ithaca, N.Y.: Cornell Southeast Asia Program, 1989. **738**

Eberhard, W. "Untersuchungen an astronomischen Texten des chinesischen Tripitaka." *Monumenta Serica* 5 (1940): 208–62. **533, 539**

Eberhard, Wolfram. *Conquerors and Rulers: Social Forces in Medieval China.* 2d rev. ed. Leiden: E. J. Brill, 1965. **206**

Ecsedy, Ildikó. "Far Eastern Sources on the History of the Steppe Region." *Bulletin de l'Ecole Française d'Extrême-Orient* 69 (1981): 263–76. **11**

Edgerton, Samuel Y., Jr. *The Renaissance Rediscovery of Linear Perspective.* 1975; reprinted New York: Harper and Row, 1976. **145, 146**

Editorial Staff of Picture Albums, ed. *Korean Central Historical Museum.* P'yŏngyang: Korean Central Historical Museum, 1979. **556, 558, 559**

Edney, Matthew H. "John Brian Harley (1932–1991): Questioning Maps, Questioning Cartography, Questioning Cartographers." *Cartography and Geographic Information Systems* 19 (1992): 175–78. **xxiii**

Efimova (Yefimov), A. V. *Atlas geograficheskikh otkrytiy v Sibiri i v severo-zapadnoy Amerike XVII-XVIII vv.* (Atlas of geographical discoveries in Siberia and northwestern America, seventeenth to eighteenth century). Moscow: Nauka, 1964. **445**

Elia, Pasquale M. d'. "The Double Stellar Hemisphere of Johann Schall von Bell S.J." *Monumenta Serica* 18 (1959): 328–59. **569, 570, 572, 573**

———. *Galileo in China: Relations through the Roman College between Galileo and the Jesuit Scientist-Missionaries (1610–1640).* Trans. Rufus Suter and Matthew Sciascia. Cambridge: Harvard University Press, 1960. **569**

———. "Recent Discoveries and New Studies (1938–1960) of the World Map in Chinese of Father Matteo Ricci SJ." *Monumenta Serica* 20 (1961): 82–164. **171**

Eliade, Mircea. *The Sacred and the Profane: The Nature of Religion.* Trans. Willard R. Trask. New York: Harcourt, Brace and World, 1959. **19**

Elman, Benjamin A. "Geographical Research in the Ming-Ch'ing Period." *Monumenta Serica* 35 (1981–83): 1–18. **224**

———. *From Philosophy to Philology: Intellectual and Social Aspects of Change in Late Imperial China.* Cambridge: Council on East Asian Studies, Harvard University, 1984. **104**

Elwin, Verrier. *The Tribal Art of Middle India: A Personal Record.* Bombay: Geoffrey Cumberlege, Oxford University Press, 1951. **19**

Essen, Gerd-Wolfgang, and Tsering Tashi Thingo. *Die Götter des Himalaya: Buddhistische Kunst Tibets.* 2 vols. Munich: Prestel-Verlag, 1989. **629, 632, 637, 677**

Fang, Chaoying, and Else Glahn. "Mao K'un." In *Dictionary of Ming Biography, 1368–1644,* 2 vols., ed. Luther Carrington Goodrich and Chaoying Fang, 2:1042–47. New York: Columbia University Press, 1976. **82**

Fang Xuanling. *The Astronomical Chapters of the Chin Shu.* Trans. and annotated Ho Peng-yoke. Paris: Mouton, 1966. **210, 511, 512, 515, 518, 522, 530, 538**

Fawcett, F. "Notes on the Rock Carvings in the Edakal Cave, Wynaad." *Indian Antiquary* 30 (1901): 409–21. **12, 17**

Felber, Roland. "Neue Möglichkeiten und Kriterien für die Bestimmung der Authentizität des Zuo-Zhuan." *Archiv Orientální* 34 (1966): 80–91. **516**

Ferguson, John B. "The Symbolic Dimensions of the Burmese Sangha." Ph.D. diss., Cornell University, 1975. **717**

Ferrand, Gabriel. "Les voyages des Javanais à Madagascar." *Journal Asiatique,* 10th ser., 15 (1910): 281–330. **828**

———. "A propos d'une carte javanaise du XVe siècle." *Journal Asiatique,* 11th ser., 12 (1918): 158–70. **828**

Feuchtwang, Stephan D. R. *An Anthropological Analysis of Chinese Geomancy.* Vientiane, Laos: Editions Vithagna, 1974. **216, 217, 219, 220, 221**

Filibeck, Elena Rossi. "A Guide-Book to Tsa-ri." In *Reflections on Tibetan Culture: Essays in Memory of Turrell V. Wylie,* ed. Lawrence Epstein and Richard F. Sherburne, 1–10. Lewiston, N.Y.: Edwin Mellen Press, 1990. **656**

Fisher, Charles A. *South-east Asia: A Social, Economic and Political Geography.* London: Methuen, 1964. **769**

Fisher, James F. *Sherpas: Reflections on Change in Himalayan Nepal.* Berkeley and Los Angeles: University of California Press, 1990. **619**

Fitzgerald, Charles Patrick. *China: A Short Cultural History.* 4th rev. ed. [London]: Barrie and Jenkins, 1976. **513**

Foley, N. "A Statistical Study of the Solar Eclipses Recorded in Chinese and Korean History during the Pre-telescopic Era." M.Sc. diss., University of Durham, 1989. **525–26**

Fong, Wen C. "Archaism as a 'Primitive' Style." In *Artists and Traditions: Uses of the Past in Chinese Culture,* ed. Christian F. Murck, 89–109. Princeton: Art Museum, Princeton University, 1976. **135**

———. "The Study of Chinese Bronze Age Arts: Methods and Approaches." In *The Great Bronze Age of China: An Exhibition from the People's Republic of China,* ed. Wen Fong, 20–34. New York: Metropolitan Museum of Art, 1980. **130**

Fontein, Jan, and Wu Tung. *Han and T'ang Murals Discovered in Tombs in the People's Republic of China and Copied by Contemporary Chinese Painters.* Boston: Museum of Fine Arts, 1976. **147**

Forman, Werner, and Bedrich Forman. *Art of Far Lands.* Ed. Lubor Hájek. Trans. W. Cungh and H. Watney. [London]: Spring Books, [1958?]. **679**

Forrest, Thomas. *Voyage from Calcutta to the Mergui Archipelago, Lying on the East Side of the Bay of Bengal.* London: J. Robson, 1792. **836, 837**

———. *A Voyage to New Guinea, and the Moluccas, from Balambangan, Including an Account of Magindano, Sooloo, and Other Islands . . . during the Years 1774, 1775, and 1776.* 2d ed. London: G. Scott, 1880; reprinted Kuala Lumpur: Oxford University Press, 1969. **838**

Foss, Theodore N. "A Western Interpretation of China: Jesuit Cartography." In *East Meets West: The Jesuits in China, 1582–1773,* ed. Charles E. Ronan and Bonnie B. C. Oh, 209–51. Chicago: Loyola University Press, 1988. **180, 181, 185, 186, 299**

Franke, Herbert. "Die Erforschung der Quellgebiete des Gelben Flusses in Nordosttibet unter dem Mongolenkaiser Qubilai." In *Der Weg zum Dach der Welt,* ed. Claudius C. Müller and Walter Raunig, 59–61. Innsbruck: Pinguin-Verlag, [1982]. **614**

———. "Sung Embassies: Some General Observations." In *China among Equals: The Middle Kingdom and Its Neighbors, 10th-14th Centuries,* ed. Morris Rossabi, 116–48. Berkeley and Los Angeles: University of California Press, 1983. **83**

Fraser, Douglas. *Village Planning in the Primitive World.* New York: George Braziller, 1968. **10**

Freedman, Maurice. *Chinese Lineage and Society: Fukien and Kwang-tung.* London: Athlone Press, 1971. **218**

Freeman, Michael D. "From Adept to Worthy: The Philosophical Career of Shao Yong." *Journal of the American Oriental Society* 102 (1982): 477–91. **225**

Freitag, Ulrich. "Geschichte der Kartographie von Thailand." In *Forschungsbeiträge zur Landeskunde Süd- und Südostasiens,* Festschrift für Harald Uhlig zu seinem 60. Geburtstag, vol. 1, ed. E. Meynen and E. Plewe, 213–32. Wiesbaden: Franz Steiner, 1982. **692, 742**

———. "Zur Periodisierung der Geschichte der Kartographie Thailands." In *Kartenhistorisches Colloquium Bayreuth '82, 18.-20. März 1982: Vorträge und Berichte,* ed. Wolfgang Scharfe, Hans Vollet, and Erwin Herrmann, 213–27. Berlin: Dietrich Reimer, 1983. **742**

Fróis, Luís. *Historia de Japam.* [ca. 1597]. **377**

Frumkin, Grégoire. *Archaeology in Soviet Central Asia.* Leiden: E. J. Brill, 1970. **16**

Fu, Lo-shu, comp., trans., and annotator. *A Documentary Chronicle of Sino-Western Relations (1644–1820).* 2 vols. Tucson: University of Arizona Press, 1966. **83, 84, 85**

Fu, Shen C. Y., et al. *Traces of the Brush: Studies in Chinese Calligraphy.* New Haven: Yale University Press, 1977. **158**

Fuchs, Walter. "Materialien zur Kartographie der Mandju-Zeit." *Monumenta Serica* 1 (1936): 386–427. **185**

———. *Der Jesuiten-Atlas der Kanghsi-Zeit.* 2 vols. Beijing: Fu Jen [Furen] University, 1943. **184, 185, 186**

———. *The "Mongol Atlas" of China by Chu Ssu-pen and the Kuang-yü-t'u.* Monumenta Serica Monograph 8. Beijing: Fu Jen [Furen] University, 1946. **60**

Fujita, Hiroki. *Tibetan Buddhist Art.* Tokyo: Hakusuisha, 1984. **637**

Galileo. *The Starry Messenger.* In *Discoveries and Opinions of Galileo,* trans. and annotated Stillman Drake. Garden City, N.Y.: Doubleday, 1957. **223**

Gaspardone, Emile. "Bibliographie annamite." *Bulletin de l'Ecole Française d'Extrême-Orient* 34 (1934): 1–173. **481, 482, 486, 495**

Gaubil, Antoine. *Correspondance de Pékin, 1722–1759*. Geneva: Librairie Droz, 1970. **180, 181**

Geertz, Clifford. *The Religion of Java*. Glencoe, Ill.: Free Press, 1960. **711, 738**

———. *Local Knowledge: Further Essays in Interpretive Anthropology*. New York: Basic Books, 1983. **20**

Genoud, Charles. *Buddhist Wall-Painting of Ladakh*. Trans. Tom Tillemans. Geneva: Edition Olizane, 1982. **615**

Gerasimov, Innokenty, ed. *A Short History of Geographical Science in the Soviet Union*. Moscow: Progress, 1976. **445**

Gerini, Gerolamo E. *Chūḷākantamaṅgala; or, The Tonsure Ceremony as Performed in Siam*. Bangkok: Siam Society, 1976 [first published 1895]. **693, 716, 728**

Gerner, Manfred. *Architekturen im Himalaja*. Stuttgart: Deutsche Verlags-Anstalt, 1987. **615, 621–22, 623, 679, 681**

Gernet, Jacques. *China and the Christian Impact: A Conflict of Cultures*. Trans. Janet Lloyd. Cambridge: Cambridge University Press, 1985. **170**

Ghosh, Rai Sahib Manoranjan. *Rock-Paintings and Other Antiquities of Prehistoric and Later Times*. Memoirs of the Archaeological Survey of India, no. 24. Calcutta: Government of India, Central Publication Branch, 1932; reprinted Patna: I. B. Corporation, 1982. **2**

Gianno, Rosemary. *Semelai Culture and Resin Technology*. Memoirs of the Connecticut Academy of Arts and Sciences, vol. 22. New Haven, 1990. **710**

Giles, Lionel. *Descriptive Catalogue of the Chinese Manuscripts from Tunhuang in the British Museum*. London: British Museum, 1957. **536, 537**

Ginnaro, Berardin. *Saverio orientale; ò, Vero istorie de' Cristiani illustri dell'Oriente. . . .* Naples: Francesco Savio, 1641. **388**

Ginsburg, Henry. *Thai Manuscript Painting*. Honolulu: University of Hawaii Press, 1989. **719, 720, 722**

Giorgi, Antonio Agostino. *Alphabetum Tibetanum missionum apostolicarum*. Rome: Typis Sacrae Congregationis de Propaganda Fide, 1762. **608, 621**

Gleick, James. *Chaos: Making a New Science*. New York: Viking, 1987. **227**

Glover, I. C., B. Bronson, and D. T. Bayard. "Comment on 'Megaliths' in South East Asia." In *Early South East Asia: Essays in Archaeology, History and Historical Geography*, ed. R. B. Smith and W. Watson, 253–54. New York: Oxford University Press, 1979. **9**

Gold, Peter. *Tibetan Reflections: Life in a Tibetan Refugee Community*. London: Wisdom Publications, 1984. **621**

Gole, Susan. *Indian Maps and Plans: From Earliest Times to the Advent of European Surveys*. New Delhi: Manohar, 1989. **634, 646, 647, 649, 650**

———. "A Nepali Map of Central Asia." *South Asian Studies* 8 (1992): 81–89. **646, 647, 648**

Goloubew, Victor. "L'Age du Bronze au Tonkin et dans le Nord-Annam." *Bulletin de l'Ecole Française d'Extrême-Orient* 29 (1929): 1–46. **13**

Gombrich, Richard. "The Buddhist Way." In *The World of Buddhism: Buddhist Monks and Nuns in Society and Culture*, ed. Heinz Bechert and Richard Gombrich, 9–40. London: Thames and Hudson, 1984. **615**

Gordon, Antoinette K. *Tibetan Religious Art*. New York: Columbia University Press, 1952. **621, 623, 634, 638**

Gordon, Douglas Hamilton. "The Rock Engravings of Kupgallu Hill, Bellary, Madras." *Man* 51 (1951): 117–19. **6**

———. *The Pre-historic Background of Indian Culture*. Bombay: N. M. Tripathi, 1958. **1, 18**

Gordon, M. E., and Douglas Hamilton Gordon. "The Artistic Sequence of the Rock Paintings of the Mahadeo Hills." *Science and Culture* 5 (1939–40): 322–27, 387–92. **18**

Gosling, L. A. Peter. "Contemporary Malay Traders in the Gulf of Thailand." In *Economic Exchange and Social Interaction in Southeast Asia: Perspectives from Prehistory, History, and Ethnography*, ed. Karl L. Hutterer, 73–95. Ann Arbor: Center for South and Southeast Asian Studies, University of Michigan, 1977. **832**

Gourou, Pierre. *Les paysans du delta tonkinois: Etudes de géographie humaine*. 1936; reprinted Paris: Mouton, 1965. **484**

Graham, A. C. *Later Mohist Logic, Ethics and Science*. Hong Kong: Chinese University Press; London: School of Oriental and African Studies, University of London, 1978. **97**

Groot, J. J. M. de. *The Religious System of China: Its Ancient Forms, Evolution, History and Present Aspect*. 6 vols. Leiden: E. J. Brill, 1892–1910. Vol. 3, bk. 1, *Disposal of the Dead*, pt. 3, *The Grave*. Reprinted Taipei: Chengwen Chubanshe, 1972. **216, 217**

Groslier, Bernard Philippe, and Jacques Arthaud. *Angkor: Art and Civilization*. Rev. ed. Trans. Eric Ernshaw Smith. New York: Frederick A. Praeger, 1966. **740**

Groslier, George. *Angkor*. 2d ed. Paris: Librairie Renouard, 1931. **695, 740**

Gumilev, L. N., and B. I. Kuznetsov. "Dve traditsii drevnetibetskoy kartografii (landschaf i etnos, VIII." *Vestnik Leningradskogo Universiteta* 24 (1969): 88–101. Translated as "Two Traditions of Ancient Tibetan Cartography (Landscape and Ethnos, VIII)." *Soviet Geography: Review and Translation* 11 (1970): 565–79. **610, 614, 639, 640**

Gurung, Harka. *Maps of Nepal: Inventory and Evaluation*. Bangkok: White Orchid Press, 1983. **609**

Guy, R. Kent. *The Emperor's Four Treasuries: Scholars and the State in the Late Ch'ien-lung Era*. Cambridge: Council on East Asian Studies, Harvard University, 1987. **74, 104**

Habib, Irfan. *An Atlas of the Mughal Empire: Political and Economic Maps with Detailed Notes, Bibliography and Index*. Delhi: Oxford University Press, 1982. **781**

Hagstrum, Jean H. *The Sister Arts: The Tradition of Literary Pictorialism and English Poetry from Dryden to Gray*. Chicago: University of Chicago Press, 1958. **134**

Halén, Harry. *Mirrors of the Void: Buddhist Art in the National Museum of Finland*. Helsinki: Museovirasto, 1987. **plate 14**

Hall, Daniel George Edward. *A History of South-east Asia*. 3d ed. New York: St. Martin's Press, 1968. **753, 759, 768, 770, 789**

———. *Henry Burney: A Political Biography*. London: Oxford University Press, 1974. **754**

Hall, John Whitney. *Japan: From Prehistory to Modern Times*. New York: Delacorte Press, 1970. **205, 207, 355, 356, 359, 362**

Hamilton, Francis. "An Account of a Map of the Countries Subject to the King of Ava, Drawn by a Slave of the King's Eldest Son." *Edinburgh Philosophical Journal* 2 (1820): 89–95, 262–71. **743, 745, 746, 747, 803**

———. "Account of a Map of the Route between Tartary and Amarapura, by an Ambassador from the Court of Ava to the Emperor of China." *Edinburgh Philosophical Journal* 3 (1820): 32–42. **785, 817**

———. "Account of a Map of the Country North from Ava." *Edinburgh Philosophical Journal* 4 (1820–21): 76–87. **747, 748, 803**

———. "Account of a Map Constructed by a Native of Taunu, of the Country South from Ava." *Edinburgh Philosophical Journal* 5 (1821): 75–84. **749, 803**

———. "Account of a Map by a Slave to the Heir-Apparent of Ava." *Edinburgh Philosophical Journal* 6 (1821–22): 270–73. **805**

———. "Account of a Map of the Country between the Erawadi and Khiaenduaen Rivers." *Edinburgh Philosophical Journal* 6 (1821–22): 107–11. **803**

———. "Account of a Map of the Tarout Shan Territory." *Edinburgh Philosophical Journal* 7 (1822): 71–75. **805**

———. "An Account of a Map of the Vicinity of Paukgan, or Pagan." *Edinburgh Philosophical Journal* 7 (1822): 230–39. **805**

———. "Account of a Map Drawn by a Native of Dawae or Tavay." *Edinburgh Philosophical Journal* 9 (1823): 228–36. **752, 805**

———. "An Account of a Map of Koshanpri." *Edinburgh Philosophical Journal* 10 (1823–24): 246–50. **805**

———. "Account of Two Maps of Zaenmae or Yangoma." *Edinburgh Philosophical Journal* 10 (1823–24): 59–67. **749, 750, 751, 805**

———. "Account of a Map of the Kingdom of Pegu." *Edinburgh Journal of Science* 1 (1824): 267–74. **807**

———. "Account of a Map of Upper Laos, or the Territory of the Lowa Shan." *Edinburgh Journal of Science* 1 (1824): 71–73. **807**

Han Feizi. *The Complete Works of Han Fei Tzŭ.* 2 vols. Trans. W. K. Liao. London: Arthur Probsthain, 1939–59. **73, 74**

Harley, J. B., and David Woodward, eds. *The History of Cartography.* Chicago: University of Chicago Press, 1987–. **xxiii, 27, 54, 279** *See also entries under individual authors.*

[Harmand]. "Im Innern von Hinterindien (nach dem Französischen des Dr. Harmand)." *Globus* 38, no. 14 (1880): 209–15. **699, 789**

Harper, Donald J. "The Han Cosmic Board (*Shih* 式)." *Early China* 4 (1978): 1–10. **516, 527**

Harrer, Heinrich. *Seven Years in Tibet.* Trans. Richard Graves. London: Rupert Hart-Davis, 1953. **654**

———. *Return to Tibet.* Trans. Ewald Osers. New York: Schocken Books, 1984. **654**

Harrison, John A. "Notes on the Discovery of Yezo." *Annals of the Association of American Geographers* 40 (1950): 254–66. **443, 445**

Harvey, P. D. A. *The History of Topographical Maps: Symbols, Pictures and Surveys.* London: Thames and Hudson, 1980. **4, 126, 152, 690, 767, 768**

Hasegawa, Koji. "Road Atlases in Early Modern Japan and Britain." In *Geographical Studies and Japan*, ed. John Sargent and Richard Wiltshire, 15–24. Folkestone, Eng.: Japan Library, 1993. **424**

Hearn, Maxwell K. "The Terracotta Army of the First Emperor of Qin (221–206 B.C.)." In *The Great Bronze Age of China: An Exhibition from the People's Republic of China*, ed. Wen Fong, 353–68. New York: Metropolitan Museum of Art, 1980. **79**

Hedin, Sven Anders. *History of the Expedition in Asia, 1927–1935.* 4 vols. Stockholm: [Göteborg, Elanders Boktryckeri Aktiebolag], 1943–45. **19**

Heine-Geldern, Robert. "Weltbild und Bauform in Südostasien." *Wiener Beiträge zur Kunst- und Kulturgeschichte Asiens* 4 (1928–29): 28–78. **696, 720, 727, 728, 737**

———. "Conceptions of State and Kingship in Southeast Asia." *Far Eastern Quarterly* 2 (1942): 15–30. **728, 737, 740**

———. *Conceptions of State and Kingship in Southeast Asia.* Ithaca, N.Y.: Cornell University, 1956. **728**

Heissig, Walther. "Über Mongolische Landkarten." *Monumenta Serica* 9 (1944): 123–73. **682, 684, 685**

———. *Mongolische Handschriften, Blockdrucke, Landkarten.* Wiesbaden: Franz Steiner, 1961. **682, 685**

———, ed. *Mongolische Ortsnamen.* 3 vols. Wiesbaden: Franz Steiner, 1966–81. **682, 684, 685**

Henderson, John B. *The Development and Decline of Chinese Cosmology.* New York: Columbia University Press, 1984. **225**

———. "Ch'ing Scholars' Views of Western Astronomy." *Harvard Journal of Asiatic Studies* 46 (1986): 121–48. **105**

Heurck, Philippe van. "Description de la *thaṅ-ka* représentant le monastère de Drepung." *Bulletin des Musées Royaux d'Art et d'Histoire* 57, no. 2 (1986): 5–29. **668, 677**

Hillier, Jack R. *The Japanese Print: A New Approach.* London: G. Bell and Sons, 1960. **410**

Hirth, Friedrich. "The Story of Chang K'iên, China's Pioneer in Western Asia: Text and Translation of Chapter 123 of Ssï-ma Ts'iên's Shï-ki." *Journal of the American Oriental Society* 37 (1917): 89–152. **512**

Ho, Peng-yoke. "Ancient and Mediaeval Observations of Comets and Novae in Chinese Sources." *Vistas in Astronomy* 5 (1962): 127–225. **527**

———. "Natural Phenomena Recorded in the *Đai-Viêt su'-ky toan-thu'*, an Early Annamese Historical Source." *Journal of the American Oriental Society* 84 (1964): 127–49. **512**

———. "The Astronomical Bureau in Ming China." *Journal of Asian History* 3 (1969): 137–57. **210, 552**

———. *Li, Qi and Shu: An Introduction to Science and Civilization in China.* Hong Kong: Hong Kong University Press, 1985. **513, 529, 550, 552**

———, trans. and annotator. *The Astronomical Chapters of the Chin Shu.* See Fang Xuanling.

Ho, Ping-ti. *Studies on the Population of China, 1368–1953.* Cambridge: Harvard University Press, 1959. **71, 86**

———. *The Ladder of Success in Imperial China: Aspects of Social Mobility, 1368–1911.* New York: Columbia University Press, 1962. **71**

Hoàng Đạo Thúy. *Thăng Long, Đông Đô, Hà Nội.* Hanoi, 1971. **484, 502**

Holle, K. F. "De Kaart van Tjiëla of Timbangantĕn." *Tijdschrift voor Indische Taal-, Land- en Volkenkunde* 24 (1877): 168–76. **690, 767, 768, 770**

Hopkins, Gerard Manley. *Gerard Manley Hopkins.* Ed. Catherine Phillips. Oxford: Oxford University Press, 1986. **165**

Hose, Charles. "Various Methods of Computing the Time for Planting among the Races of Borneo." *Journal of the Straits Branch of the Royal Asiatic Society*, no. 42 (1905): 1–5. **19**

Hose, Charles, and William McDougall. *The Pagan Tribes of Borneo: A Description of Their Physical, Moral and Intellectual Condition with Some Discussion of Their Ethnic Relations.* 2 vols. London: Macmillan, 1912. **703, 709, 710, 711, 712, 713**

Hotaling, Stephen James. "The City Walls of Ch'ang-an." *T'oung Pao* 64 (1978): 1–46. **524**

Hsu, Cho-yun. *Ancient China in Transition: An Analysis of Social Mobility, 722–222 B.C.* Stanford: Stanford University Press, 1965. **206**

Hsu, Francis L. K. *Under the Ancestors' Shadow: Kinship, Personality, and Social Mobility in Village China.* 1948; reprinted Garden City, N.Y.: Anchor Books, 1967. **215**

Hsu, Mei-ling. "The Han Maps and Early Chinese Cartography." *Annals of the Association of American Geographers* 68 (1978): 45–60. **41, 46, 51, 52, 56, 64**

Huard, Pierre, and Maurice Durand. *Connaissance du Việt-Nam.* Hanoi: Ecole Française d'Extrême-Orient, 1954. **481**

Huber, Toni. "A Tibetan Map of lHo-Kha in the South-eastern Himalayan Borderlands of Tibet." *Imago Mundi* 44 (1992): 9–23. **654, 655, 656**

Hulbert, Homer B. "An Ancient Map of the World." *Bulletin of the American Geographical Society of New York* 36 (1904): 600–605. Reprinted in *Acta Cartographica* 13 (1972): 172–78. **259**

Hulsewé, A. F. P. *China in Central Asia: The Early Stage, 125 B.C.-A.D. 23*. Introduction by Michael A. N. Loewe. Leiden: E. J. Brill, 1979. **512**

Hummel, Arthur. "Atlases of Kwangtung Province." In *Annual Report of the Librarian of Congress for the Fiscal Year Ended June 30, 1938*, 229–31. Washington, D.C.: United States Government Printing Office, 1939. **190**

———, ed. *Eminent Chinese of the Ch'ing Period, 1644–1912*. 2 vols. Washington, D.C.: United States Government Printing Office, 1943–44. **305**

Hummel, Siegbert. "Kosmische Strukturpläne der Tibeter." *Geographica Helvetica* 9 (1964): 34–42. **632**

Huntington, Susan L., and John C. Huntington. *Leaves from the Bodhi Tree: The Art of Pala India (8th-12th Centuries) and Its International Legacy*. Seattle: Dayton Art Institute in Association with the University of Washington Press, 1990. **613**

Huttman, William. "On Chinese and European Maps of China." *Journal of the Royal Geographical Society* 14 (1844): 117–27. **27**

Imperial Atlas of India. Survey of India, 1904; updated 1910. **794**

Imperial Gazetteer of India. New ed. 26 vols. Oxford: Clarendon Press, 1907–9. **791**

Jackson, David P., and Janice A. Jackson. *Tibetan Thangka Painting: Methods and Materials*. London: Serindia Publications, 1984. **616, 617, 621**

Jackson, Roger. "The Tibetan Tshogs Zhing (Field of Assembly): General Notes on Its Function, Structure and Contents." *Asian Philosophy* 2, no. 2 (1992): 157–72. **625, 628**

Jacobson, Esther. "Siberian Roots of the Scythian Stag Image." *Journal of Asian History* 17 (1983): 68–120. **7**

Jacquet, E. "Mélanges malays, javanais et polynésiens." *Nouveau Journal Asiatique*, 2d ser., 9 (1832): 97–132, 222–67. **838**

Jain, Jyotindra. *Painted Myths of Creation: Art and Ritual of an Indian Tribe*. New Delhi: Lalit Kala Akademi, 1984. **19**

Jarves, James Jackson. *A Glimpse at the Art of Japan*. 1876; reprinted Tokyo: Charles E. Tuttle, 1984. **370**

Jensen, Erik. *The Iban and Their Religion*. Oxford: Clarendon Press, 1974. **711**

Jeon, Sang-woon (Chŏn Sang'un). *Science and Technology in Korea: Traditional Instruments and Techniques*. Cambridge: MIT Press, 1974. **221, 252, 279, 284, 286, 288, 310, 511, 556, 557, 560, 561, 568, 576, 578, 586**

Jisl, Lumír. *Tibetan Art*. Trans. Ilse Gottheiner. London: Spring Books, 1957. **679**

"John Brian Harley, 1932–1991." *Cartographica* 28, no. 4 (1991): 92–93. **xxiii**

Jung, Carl Gustav. *Mandala Symbolism*. Trans. R. F. C. Hull. Princeton: Princeton University Press, 1972. **620**

Juynboll, Hendrik Herman. *Borneo*. 2 vols. Leiden: E. J. Brill, 1909–10. **708**

Kalinowski, Marc. "Les instruments astro-calendériques des Han et la méthode *liu ren*." *Bulletin de l'Ecole Française d'Extrême-Orient* 72 (1983): 311–419. **115**

Karakhanyan, Grigor Hovhannesi, and Pavel Geworgi Safyan. *Syownik'i zhayrhapatkernerě* (Rock carvings of Syunik). Yerevan, 1970. **2, 3**

Karamustafa, Ahmet T. "Introduction to Islamic Maps." In *The History of Cartography*, ed. J. B. Harley and David Woodward, vol. 2.1 (1992), 3–11. Chicago: University of Chicago Press, 1987–. **27**

Karlgren, Bernhard. "Glosses on the *Ta Ya* and Sung Odes." *Bulletin of the Museum of Far Eastern Antiquities* 18 (1946): 1–198. Reprinted in Karlgren, *Glosses on the Book of Odes*. **72**

———. "Legends and Cults in Ancient China." *Bulletin of the Museum of Far Eastern Antiquities*, no. 18 (1946): 199–365. **213**

———. "Glosses on the *Book of Documents*." *Bulletin of the Museum of Far Eastern Antiquities* 20 (1948): 39–315. **76**

———. *Grammata Serica Recensa*. Stockholm, 1957. Reprinted from the *Bulletin of the Museum of Far Eastern Antiquities* 29 (1957): 1–332. **26, 27**

———. *Glosses on the Book of Odes*. Stockholm: Museum of Far Eastern Antiquities, 1964. **72**

———, ed. and trans. "The *Book of Documents*." *Bulletin of the Museum of Far Eastern Antiquities* 22 (1950): 1–81. **72, 76**

———. *The Book of Odes*. Stockholm: Museum of Far Eastern Antiquities, 1950; reprinted 1974. **72, 516**

Karrow, Robert W., Jr. *Mapmakers of the Sixteenth Century and Their Maps: Bio-bibliographies of the Cartographers of Abraham Ortelius, 1570*. Chicago: Speculum Orbis Press, 1993. **380**

Kashina, T. I. "Semantika ornamentatsii neoliticheskoy keramiki Kitaya" (Semantics of ornamentation of China's Neolithic pottery). In *U istokov tvorchestva* (At the sources of art), 183–202. Novosibirsk: "Nauka," 1978. **11**

Kawamura, Hirotada. "*Kuni-ezu* (Provincial Maps) Compiled by the Tokugawa Shogunate in Japan." *Imago Mundi* 41 (1989): 70–75. **397, 471**

Kawamura, Hirotada, Kazutaka Unno, and Kazuhiko Miyajima. "List of Old Globes in Japan." *Der Globusfreund* 38–39 (1990–91): 173–75. **469, 591**

Keates, John. Review of David Woodward, ed., *Art and Cartography: Six Historical Essays*. *Cartographic Journal* 25 (1988): 179–80. **96**

Keightley, David N. "The Religious Commitment: Shang Theology and the Genesis of Chinese Political Culture." *History of Religions* 17 (February-May 1978): 211–25. **204**

———. *Sources of Shang History: The Oracle-Bone Inscriptions of Bronze Age China*. Berkeley and Los Angeles: University of California Press, 1978. **72, 514**

———. "The Late Shang State: When, Where, and What?" In *The Origins of Chinese Civilization*, ed. David N. Keightley, 523–64. Berkeley and Los Angeles: University of California Press, 1983. **72**

Kempers, August Johan Bernet. *Ancient Indonesian Art*. Amsterdam: C. P. J. van der Peet, 1959. **697**

Kennedy, Victor. "An Indigenous Early Nineteenth Century Map of Central and Northeast Thailand." In *In Memoriam Phya Anuman Rajadhon: Contributions in Memory of the Late President of the Siam Society*, ed. Tej Bunnag and Michael Smithies, 315–48. Bangkok: Siam Society, 1970. **692, 699, 763, 765**

Keyes, Charles F. "Buddhist Pilgrimage Centers and the Twelve-Year Cycle: Northern Thai Moral Orders in Space and Time." *History of Religions* 15 (1975): 71–89. **777**

Kiang, T. "The Past Orbit of Halley's Comet." *Memoirs of the Royal Astronomical Society* 76 (1972): 27–66. **511**

———. "Notes on Traditional Chinese Astronomy." *Observatory* 104 (1984): 19–23. **525**

Kim, Yong-woon. "Structure of Ch'ŏmsŏngdae in the Light of the Choupei Suanchin." *Korea Journal* 14, no. 9 (1974): 4–11. **557**

Kirfel, Willibald. *Die Kosmographie der Inder nach Quellen dargestellt*. Bonn: Kurt Schroeder, 1920; reprinted Hildesheim:

Georg Olms, 1967; Darmstadt: Wissenschaftliche Buchgesellschaft, 1967. **714**

Kish, George. "The Cartography of Japan during the Middle Tokugawa Era: A Study in Cross-Cultural Influences." *Annals of the Association of American Geographers* 37 (1947): 101–19. **370**

———. "Some Aspects of the Missionary Cartography of Japan during the Sixteenth Century." *Imago Mundi* 6 (1949): 39–47. **370, 371, 376**

———. *La carte: Image des civilisations.* Paris: Seuil, 1980. **690**

Kitagawa, Kay. "The Map of Hokkaido of G. de Angelis, ca 1621." *Imago Mundi* 7 (1950): 110–14. **376, 444**

Kiyota, Minoru. *Shingon Buddhism: Theory and Practice.* Los Angeles: Buddhist Books International, 1978. **215**

Kjöping, Nils Matson. *Een Kort Beskriffning Vppå Trenne Reesor och Peregrinationer sampt Konungarijket Japan . . . III. Beskrifwes een Reesa till Ost Indien, China och Japan . . . aff Oloff Erickson Willman.* Wisingsborgh, 1667. **391**

Klér, Joseph. "A propos de cartographie mongole." *Bulletin de la Société Royale Belge de Géographie* 24, pts. 1–2 (1956): 26–51. **685**

Kloetzli, W. Randolph. *Buddhist Cosmology, from Single World System to Pure Land: Science and Theology in the Images of Motion and Light.* Delhi: Motilal Banarsidass, 1983. **714, 718**

———. "Buddhist Cosmology." In *The Encyclopedia of Religion*, 16 vols., ed. Mircea Eliade, 4:113–19. New York: Macmillan, 1987. **714, 716, 717**

Knechtges, David R., trans. and annotator. *Wen Xuan; or, Selections of Refined Literature.* See Xiao Tong.

Knobel, E. B. "On a Chinese Planisphere." *Monthly Notices of the Royal Astronomical Society* 69 (1909): 435–45. **566, 587**

———. "Inō Chūkei and the First Survey of Japan." *Geographical Journal* 42 (1913): 246–50. **450**

Koeman, Cornelis. "Die Darstellungsmethoden von Bauten auf alten Karten." *Wolfenbütteler Forschungen* 7 (1980): 147–92. **4**

Kohn, Livia. *Early Chinese Mysticism: Philosophy and Soteriology in the Taoist Tradition.* Princeton: Princeton University Press, 1992. **215**

Koninklijk Instituut voor de Tropen. *Aanwinsten op ethnografisch en anthropologisch gebied van de Afdeeling Volkenkunde van het Koloniaal Instituut over 1933.* Afdeeling Volkenkunde 6. Amsterdam, 1934. **692, 770, 772, 773**

Koppar, D. H. *Tribal Art of Dangs.* Baroda: Department of Museums, 1971. **19**

Kosasih, Engkos A. "Rock Art in Indonesia." In *Rock Art and Prehistory: Papers Presented to Symposium G of the AURA Congress, Darwin, 1988*, ed. Paul Bahn and Andrée Rosenfeld, 65–77. Oxford: Oxbow Books, 1991. **13**

Kraft, Eva. *Japanische Handschriften und Traditionelle Drucke aus der Zeit vor 1868 in München.* Stuttgart: Franz Steiner, 1986. **352**

Krogt, Peter van der. *Old Globes in the Netherlands: A Catalogue of Terrestrial and Celestial Globes Made prior to 1850 and Preserved in Dutch Collections.* Trans. Willie ten Haken. Utrecht: HES, 1984. **391**

Kšica, Miroslav. *Umění staré Eurasie: Skalní obrazy v SSSR* (The art of ancient Eurasia: Rock pictures in the Soviet Union). Brno: Dům Umění, [1974]. **11, 13, 16, 18**

Kudo, Chohei. "A Summary of My Studies of Girolamo de Angelis' Yezo Map." *Imago Mundi* 10 (1953): 81–86. **376, 444**

Kusch, Heinrich. "Rock Art Discoveries in Southeast Asia: A Historical Summary." *Bollettino del Centro Camuno di Studi Preistorici* 23 (1986): 99–108. **18**

Kusmiadi, Rachmat. "A Brief History of Cartography in Indonesia." Paper presented at the Seventh International Conference on the History of Cartography, Washington, D.C., 7–11 August 1977. **690, 767, 768**

Kvaerne, Per. "Tibet: The Rise and Fall of a Monastic Tradition." In *The World of Buddhism: Buddhist Monks and Nuns in Society and Culture*, ed. Heinz Bechert and Richard Gombrich, 253–70. London: Thames and Hudson, 1984. **614**

———. "Peintures tibétaines de la vie de sTon-pa-gçen-rab." *Arts Asiatiques* 41 (1986): 36–81. **638**

Kyakshto, N. B. "Pisanitsa Shaman-Kamnya" (The cliff drawings of Shaman-Kamnya). *Soobshcheniya Gosudarstvennoy Akademii Istorii Materialnoy Kul'tury* (GAIMK: Report of the State Academy for the History of Material Culture), July 1931, 29–30. **20**

Lang, David Marshall. *Armenia: Cradle of Civilization.* 3d corrected ed. London: George Allen and Unwin, 1980. **12**

Lau, D. C., trans. *Mencius.* See Mencius.

Lauf, Detlef Ingo. *Lhasa: De heilige stad van Tibet en haar omgeving.* Antwerp: Etnografisch Museum van de Stad, 1974. **648**

———. *Tibetan Sacred Art: The Heritage of Tantra.* Trans. Ewald Osers. Berkeley, Calif.: Shambhala, 1976. **621, 668, 677**

———. *Verborgene Botschaft tibetische Thangkas/Secret Revelation of Tibetan Thangkas.* Freiburg im Breisgau: Aurum, 1976. **629, 638**

———. *Secret Doctrines of the Tibetan Books of the Dead.* Trans. Graham Parkes. Boulder, Colo.: Shambhala, 1977. **620, 632**

Laufer, Berthold. "The Wang Ch'uan T'u, a Landscape of Wang Wei." *Ostasiatische Zeitschrift* 1, no. 1 (1912): 28–55. **152**

Law, Bimala Churn. *Rājagṛiha in Ancient Literature.* Memoirs of the Archaeological Survey of India, no. 58. Delhi: Manager of Publications, 1938. **783**

Lebedev, D. M. *Ocherki po Istorii Geografii v Rossii XVIII v. (1725–1800 gg.)* (Essays on the history of geography in Russia in the eighteenth century [1725–1800]). Moscow: Izdatel'stvo Akademii Nauk SSSR, 1957. **445**

Le Comte, Louis Henry. *Nouveaux mémoires sur l'état présent de la Chine.* 2 vols. Paris: Anisson, 1696. **551**

Ledderose, Lothar. "Some Taoist Elements in the Calligraphy of the Six Dynasties." *T'oung Pao* 70 (1984): 246–78. **136**

Ledyard, Gari. "Yamatai." In *Kodansha Encyclopedia of Japan*, 9 vols., 8:305–7. Tokyo: Kodansha International, 1983. **272**

Lee, Sang Hae. "Feng-Shui: Its Context and Meaning." Ph.D. diss., Cornell University, 1986. **216, 217, 218, 219**

Lee, Sherman E. *Chinese Landscape Painting.* Cleveland: Cleveland Museum of Art, 1954. **151**

Leeming, Frank. "Official Landscapes in Traditional China." *Journal of the Economic and Social History of the Orient* 23 (1980): 153–204. **125**

Legge, James. *The Chinese Classics.* 5 vols. 1893–95 editions; reprinted Hong Kong: Hong Kong University Press, 1960. **99, 515, 516, 518**

———, ed. and trans. *The Four Books.* 1923; reprinted New York: Paragon, 1966. **515**

Leonard, Jane Kate. *Wei Yuan and China's Rediscovery of the Maritime World.* Cambridge: Council on East Asian Studies, Harvard University, 1984. **192**

Le Roux, C. C. F. M. "Boegineesche zeekaarten van den Indischen Archipel." *Tijdschrift van het Koninklijk Nederlandsch Aardrijkskundig Genootschap*, 2d ser., 52 (1935): 687–714. **692, 697, 776, 832, 833, 834, 836, 838**

Leslie, D. D., and K. H. J. Gardiner. "Chinese Knowledge of Western Asia during the Han." *T'oung Pao* 58 (1982): 254–308. **231**

Lewis, G. Malcolm. "Indian Maps." In *Old Trails and New Directions: Papers of the Third North American Fur Trade Conference*, ed. Carol M. Judd and Arthur J. Ray, 9–23. Toronto: University of Toronto Press, 1980. **749**

———. "The Origins of Cartography." In *The History of Cartography*, ed. J. B. Harley and David Woodward, 1:50–53. Chicago: University of Chicago Press, 1987–. **1**

Lhalungpa, Lobsang P. *Tibet, the Sacred Realm: Photographs, 1880–1950.* Exhibition catalog, Philadelphia Museum of Art, 20 March-22 May 1983. [Millerton, N.Y.]: Aperture, 1983. **677**

Li, Jicheng. *The Realm of Tibetan Buddhism.* New Delhi: UBS Publishers' Distributors, 1986. **677, 679**

Li, Xueqin. *Eastern Zhou and Qin Civilizations.* Trans. Kwang-chih Chang. New Haven: Yale University Press, 1985. **78, 79**

Liang, Ch'i-ch'ao (Liang Qichao). *Intellectual Trends in the Ch'ing Period.* Trans. with introduction and notes by Immanuel C. Y. Hsü. Cambridge: Harvard University Press, 1959. **104**

Liao, W. K., trans. *The Complete Works of Han Fei Tzŭ.* See Han Feizi.

Libbrecht, Ulrich. *Chinese Mathematics in the Thirteenth Century: The Shu-shu Chiu-chang of Ch'in Chiu-shao.* Cambridge: MIT Press, 1973. **118**

Linschoten, Jan Huygen van. *Itinerario, voyage ofte schipvaert. . . .* 5 vols. The Hague: Nijhoff, 1910–39. Pt. 2, *Reys-Gheschrift vande navigatien der Portugaloysers*, 1595. **388**

Liu, Lizhong. *Buddhist Art of the Tibetan Plateau.* Ed. and trans. Ralph Kiggell. Hong Kong: Joint Publishing, 1988. **615, 638, 670, 677, 679, 681**

Liu Xie. *The Literary Mind and the Carving of Dragons.* Trans. Vincent Yu-chung Shih. Hong Kong: Chinese University of Hong Kong, 1983. **133**

Livingstone, David N. "Science, Magic and Religion: A Contextual Reassessment of Geography in the Sixteenth and Seventeenth Centuries." *History of Science* 26 (1988): 269–94. **65**

Loebér, J. A. "Merkwaardige kokersversieringen uit de zuider- en oosterafdeeling van Borneo." *Bijdragen tot de Taal-, Land- en Volkenkunde van Nederlandsch-Indie* 65 (1911): 40–52. **708**

Loehr, Max. "Some Fundamental Issues in the History of Chinese Painting." *Journal of Asian Studies* 23 (1964): 185–93. **135, 166**

———. "The Fate of the Ornament in Chinese Art." *Archives of Asian Art* 21 (1967–68): 8–19. **129**

———. *Ritual Vessels of Bronze Age China.* New York: Asia Society, 1968. **129**

Loewe, Michael A. N. "Manuscripts Found Recently in China: A Preliminary Survey." *T'oung Pao* 63 (1977): 99–136. **521**

———. *Ways to Paradise: The Chinese Quest for Immortality.* London: George Allen and Unwin, 1979. **80, 115, 521**

———. "The Han View of Comets." *Bulletin of the Museum of Far Eastern Antiquities* 52 (1980): 1–31. **521**

Longstreet, Stephen, and Ethel Longstreet. *Yoshiwara: The Pleasure Quarters of Old Tokyo.* Tokyo: Yenbooks, 1988. **431**

Lu, Gwei-djen. See Needham, Joseph. *Science and Civilisation in China.*

Lyons, Elizabeth. Review of Klaus Wenk, *Thailändische Miniaturmalereien. Artibus Asiae* 29 (1967): 104–6. **720, 722**

Ma, Chengyuan. "The Splendor of Ancient Chinese Bronzes." In *The Great Bronze Age of China: An Exhibition from the People's Republic of China*, ed. Wen Fong, 1–19. New York: Metropolitan Museum of Art, 1980. **130**

Ma Huan. *Ying-yai Sheng-lan: "The Overall Survey of the Ocean's Shores" [1433].* Ed. and trans. J. V. G. Mills. Cambridge: Cambridge University Press, 1970. **53**

Ma, Laurence J. C. "Peking as a Cosmic City." In *Proceedings of the 30th International Congress of Human Sciences in Asia and North Africa: China 2*, ed. Graciela de la Lama, 141–64. Mexico City: El Colegio de México, 1982. **210**

Maass, Alfred. "Sternkunde und Sterndeuterei im Malaiische Archipel." *Tijdschrift voor Indische Taal-, Land- en Volkenkunde* 64 (1924): 1–172, 347–460, and appendix in 66 (1926): 618–70. **713**

Mabbett, I. W. "The Symbolism of Mount Meru." *History of Religions* 23 (1983): 64–83. **728**

McCune, Shannon. "Some Korean Maps." *Transactions of the Korean Branch of the Royal Asiatic Society* 50 (1975): 70–102. **299–301, 307**

———. "The Chonha Do—A Korean World Map." *Journal of Modern Korean Studies* 4 (1990): 1–8. **259, 263**

Macdonald, Alexander W., and Anne Vergati Stahl. *Newar Art: Nepalese Art during the Malla Period.* Warminster, Eng.: Aris and Phillips, 1979. **681**

Macdonald, Alexander W., and Pema Tsering. "A Note on Five Tibetan Thaṅ-kas of the Ge-sar Epic." In *Die Mongolischen Epen: Bezüge, Sinndeutung und Überlieferung (Ein Symposium)*, ed. Walther Heissig, Asiatische Forschungen, vol. 68, 150–57. Wiesbaden: Otto Harrassowitz, 1979. **637**

Mackay, A. L. "Kim Su-hong and the Korean Cartographic Tradition." *Imago Mundi* 27 (1975): 27–38. **263, 267, 268, 610, 641**

McMullen, David. *State and Scholars in T'ang China.* Cambridge: Cambridge University Press, 1988. **74, 120**

Maeda, Robert J. "*Chieh-hua*: Ruled-Line Painting in China." *Ars Orientalis* 10 (1975): 123–41. **142**

Maeyama, Yasukatsu. "On the Astronomical Data of Ancient China (ca. -100 ~ +200): A Numerical Analysis (Part 1)." *Archives Internationales d'Histoire des Sciences* 25 (1975): 247–76. **521, 529**

———. "The Oldest Star Catalogue of China, Shih Shen's Hsing Ching." In *Prismata: Naturwissenschaftsgeschichtliche Studien*, ed. Yasukatsu Maeyama and W. G. Salzer, 211–45. Wiesbaden: Franz Steiner, 1979. **519, 525, 529**

Mailla, Joseph-Anne-Marie de Moyriac de. *Histoire générale de la Chine ou annales de cet empire.* 13 vols. Paris: Grosier, 1777–85. **181, 185, 186**

Major, John S. "The Five Phases, Magic Squares, and Schematic Cosmography." In *Explorations in Early Chinese Cosmology*, ed. Henry Rosemont, Jr., 133–66. Chico, Calif.: Scholars Press, 1984. **110, 207**

Malleret, Louis, and Georges Taboulet, eds. "Foire Exposition de Saigon, Pavillon de l'Histoire, la Cochinchine dans le passé." *Bulletin de la Société des Etudes Indochinoises*, n.s., 17, no. 3 (1942): 1–133. **507**

Mannikka, Eleanor. "Angkor Wat: Meaning through Measurement." Ph.D. diss., University of Michigan, 1985. **479**

Mantegazza, Gaetano Maria. *La Birmania: Relazione Inedita del 1784 del Missionario Barnabita G. M. Mantegazza.* Rome: Ed. A. S., 1950. **698**

March, Benjamin. "Linear Perspective in Chinese Painting." *Eastern Art* 3 (1931): 113–39. **147**

Marchal, Henri. *Guide archéologique aux temples d'Angkor, Angkor Vat, Angkor Thom, et les monuments du petit et du grand circuit.* Paris: G. Van Oest, 1928. **740**

Marchand, Ernesta. "The Panorama of Wu-t'ai Shan as an Example of Tenth Century Cartography." *Oriental Art*, n.s., 22 (1976): 158–73. **152**

Marion, Donald J. "Partial Translation of *Chung-kuo ti-t'u shih kang* by Wang Yung: A Study of Early Chinese Cartography with Added Notes, an Introduction and a Bibliography." M.A. thesis, Graduate Library School, University of Chicago, 1971. **28**

Markham, Clements R., ed. *Narratives of the Mission of George Bogle to Tibet and of the Journey of Thomas Manning to Lhasa.* 1876; reprinted New Delhi: Mañjuśrī Publishing House, 1971. **614, 637**

Martirosyan, A. A. "Sémantique des dessins rupestres des Monts de Guégam (Arménie)." Moscow, 1971. **12**

Martirosyan, A. A., and A. R. Israelyan. *Naskal'nye izobrazheniya Gegamskikh gor* (The rock-carved pictures of the Gegamskiy Khrebet). Yerevan, 1971. **3, 12, 13, 18**

Martzloff, Jean-Claude. *Histoire des mathématiques chinoises.* Paris: Masson, 1988. **111**

Mason, R. H. P., and J. G. Caiger. *A History of Japan.* Melbourne: Cassell Australia, 1976. **356**

Maspero, Henri. "Le protectorat général d'Annam sous les T'ang (I): Essai de géographie historique." *Bulletin de l'Ecole Française d'Extrême-Orient* 10 (1910): 539–84. **490, 499, 500, 502, 506**

———. "L'astronomie chinoise avant les Han." *T'oung Pao* 26 (1929): 267–356, esp. 269–70. **518, 532**

———. "Le Ming-t'ang et la crise religieuse chinoise avant les Han." *Mélanges Chinois et Bouddhiques* 9 (1948–51): 1–70. **212**

Massonaud, Chantal. "Le Bhoutan." In *Les royaumes de l'Himâlaya: Histoire et civilisation*, 67–116. Paris: Imprimerie Nationale, 1982. **622, 629**

Mather, Richard. "The Landscape Buddhism of the Fifth-Century Poet Hsieh Ling-yün." *Journal of Asian Studies* 18 (1958): 67–79. **136**

Mathpal, Yashodhar. *Prehistoric Rock Paintings of Bhimbetka, Central India.* New Delhi: Abhinav Publications, 1984. **13, 15**

Maung Maung Tin, U, and Thomas Owen Morris. "Mindon Min's Development Plan for the Mandalay Area." *Journal of the Burma Research Society* 49, no. 1 (1966): 29–34. **692, 799, 800, 825**

Meacham, William. *Rock Carvings in Hong Kong.* Hong Kong: Christian Study Centre on Chinese Religion and Culture, 1976. **18**

Meijer, M. J. "A Map of the Great Wall of China." *Imago Mundi* 13 (1956): 110–15. **189**

Mencius. *Mencius.* Trans. D. C. Lau. Harmondsworth, Eng.: Penguin Books, 1970. **205**

Metcalf, Peter. *A Borneo Journey into Death: Berawan Eschatology from Its Rituals.* Philadelphia: University of Pennsylvania Press, 1982. **703, 709**

Meyer, Jeffrey F. *Peking as a Sacred City.* Taipei: Chinese Association for Folklore, 1976. **212**

———. "*Feng-Shui* of the Chinese City." *History of Religions* 18 (November 1978): 138–55. **217, 221**

Mills, J. V. "Chinese Coastal Maps." *Imago Mundi* 11 (1954): 151–68. **188**

———, ed. and trans. *Ying-yai Sheng-lan: "The Overall Survey of the Ocean's Shores" [1433].* See Ma Huan.

Mitchell, W. J. Thomas. *Iconology: Image, Text, Ideology.* Chicago: University of Chicago Press, 1986. **1**

Mody, N. H. N. *A Collection of Nagasaki Colour Prints and Paintings.* 1939; reprinted Tokyo: Charles E. Tuttle, 1969. **407, 442, 476**

Mollat du Jourdin, Michel, and Monique de La Roncière. *Les portulans: Cartes marines du XIIIe au XVIIe siècle.* Fribourg: Office du Livre, 1984. English edition, *Sea Charts of the Early Explorers: 13th to 17th Century.* Trans. L. le R. Dethan. New York: Thames and Hudson, 1984. **382, 463, 465**

Montanus, Arnoldus. *Gedenkwaerdige gesantschappen der Oost-Indische maatschappy in 't vereenigde Nederland, aan de kaisaren van Japan.* Amsterdam: J. Meurs, 1669. **391**

Müller, Claudius C., and Walter Raunig, eds. *Der Weg zum Dach der Welt.* Innsbruck: Pinguin-Verlag, [1982]. **623, 653, 679**

Müller, G. P. *Voyages et découvertes faites par les Russes le long des côtes de la Mer Glaciale & sur l'Océan Oriental, tant vers le Japon que vers l'Amérique.* 2 vols. Amsterdam: Marc-Michel Rey, 1766. **445**

Munakata, Kiyohiko. *Sacred Mountains in Chinese Art.* Urbana: University of Illinois Press, 1991. **63, 168**

Munro, Neil Gordon. *Prehistoric Japan.* Yokohama, 1911. **12, 13**

Munsterberg, Hugo. *The Japanese Print: A Historical Guide.* Tokyo: Weatherhill, 1982. **410, 424**

———. *The Arts of Japan: An Illustrated History.* Tokyo: Charles E. Tuttle, 1985. **365, 370, 424, 435**

Münzer, Verena. *Tod, Seelenreise und Jenseits bei den Ngadju Dajak in Kalimantan.* Lizentiatsarbeit, Universität Zürich, Philosophische Fakultät I, Abt.: Ethnologie. Zurich: published by the author, 1976. **703, 706**

Murdoch, John E. *Antiquity and the Middle Ages.* Album of Science. New York: Charles Scribner's Sons, 1984. **620**

Muroga, Nobuo. "Geographical Exploration by the Japanese." In *The Pacific Basin: A History of Its Geographical Exploration*, ed. Herman R. Friis, 96–105. New York: American Geographical Society, 1967. **443**

———. "The Development of Cartography in Japan." In *Old Maps in Japan*, ed. and comp. Nanba Matsutarō, Muroga Nobuo, and Unno Kazutaka, trans. Patricia Murray, 158–76. Ōsaka: Sōgensha, 1973. **351**

Muroga, Nobuo, and Kazutaka Unno. "The Buddhist World Map in Japan and Its Contact with European Maps." *Imago Mundi* 16 (1962): 49–69. **255, 256, 263, 373, 375, 429**

Mus, Paul. *Barabudur: Esquisse d'une histoire du Bouddhisme fondée sur la critique archéologique des textes.* 2 vols. Hanoi: Imprimerie d'Extrême-Orient, 1935. **728**

Myanma swezoun kyan. 2d ed. 15 vols. Rangoon, 1968. **724**

Nakamura, Hiroshi. "Les cartes du Japon qui servaient de modèle aux cartographes européens au début des relations de l'Occident avec le Japon." *Monumenta Nipponica* 2, no. 1 (1939): 100–123. **370, 371**

———. "Old Chinese World Maps Preserved by the Koreans." *Imago Mundi* 4 (1947): 3–22. **255, 262, 610, 642, 643**

———. *East Asia in Old Maps.* Tokyo: Centre for East Asian Cultural Studies, 1962. **610, 829**

———. "The Japanese Portolanos of Portuguese Origin in the XVIth and XVIIth Centuries." *Imago Mundi* 18 (1964): 24–44. **381, 383, 384, 385, 463, 465**

Nakayama, Shigeru. "Characteristics of Chinese Astrology." *Isis* 57 (1966): 442–54. **539**

———. *A History of Japanese Astronomy: Chinese Background and Western Impact.* Cambridge: Harvard University Press, 1969. **118, 355, 367, 393, 404, 432, 433, 434, 435, 450, 579, 588, 596**

———. *Academic and Scientific Traditions in China, Japan, and the West.* Trans. Jerry Dusenbury. Tokyo: University of Tokyo Press, 1984. **89**

Nanba Matsutarō, Muroga Nobuo, and Unno Kazutaka, eds. and comps. *Old Maps in Japan.* Trans. Patricia Murray. Ōsaka: Sōgensha, 1973. **351, 373, 380, 388, 404, 407, 411, 412, 414, 416, 420, 422, 423, 426, 429, 439, 441, 447, 453, 465, 473, 476**

Nebenzahl, Kenneth. *Atlas of Columbus and the Great Discoveries.* Chicago: Rand McNally, 1990. **742**

Nebesky-Wojkowitz, Réne de. *Oracles and Demons of Tibet: The Cult and Iconography of the Tibetan Protective Deities.* The Hague: Mouton, 1956. **632**

Needham, Joseph. *Science and Civilisation in China.* Cambridge: Cambridge University Press, 1954–.

Vol. 1, *Introductory Orientations* (1954). With Wang Ling. **512**

Vol. 2, *History of Scientific Thought* (1956). With Wang Ling. **145, 216, 220, 265**

Vol. 3, *Mathematics and the Sciences of the Heavens and the Earth* (1959). With Wang Ling. **10, 11, 28, 48, 57, 83, 110, 113, 118, 124, 125, 164, 170, 184, 186, 207, 246, 252, 253, 262, 289, 355, 511, 515, 517, 518, 521, 522, 525, 529, 530, 531, 532, 536, 541, 543, 550, 551, 587**

Vol. 4, pt. 3, *Physics and Physical Technology: Civil Engineering and Nautics* (1971). With Wang Ling and Lu Gwei-djen. **142, 555**

Vol. 5, pt. 1, *Chemistry and Chemical Technology: Paper and Printing* (1985). By Tsien Tsuen-hsuin. **137**

Vol. 6, *Biology and Biological Technology,* pt. 2, *Agriculture* (1984). By Francesca Bray. **78**

Needham, Joseph, and Gwei-djen Lu. "A Korean Astronomical Screen of the Mid-Eighteenth Century from the Royal Palace of the Yi Dynasty (Chosŏn Kingdom, 1392–1910)." *Physis* 8 (1966): 137–62. **532, 567**

Needham, Joseph, Wang Ling, and Derek J. de Solla Price. *Heavenly Clockwork: The Great Astronomical Clocks of Medieval China.* 2d ed. Cambridge: Cambridge University Press, 1986. **522, 530, 531, 533, 541, 542, 550, 585**

Needham, Joseph, et al. *The Hall of Heavenly Records: Korean Astronomical Instruments and Clocks, 1380–1780.* Cambridge: Cambridge University Press, 1986. **252, 253, 286, 288, 511, 567, 568, 574, 576, 578, 596**

Neild, Ralph, H. F. Searle, and J. A. Steward. *Burma Gazetteer, Kyaukse District.* Vol. A. Rangoon: Government Printing and Stationery, 1925. **821**

Nelson, Howard. "Maps from Old Cathay." *Geographical Magazine* 47 (1975): 702–11. **268**

Nemeth, David J. "A Cross-Cultural Cosmographic Interpretation of Some Korean Geomancy Maps." In *Introducing Cultural and Social Cartography,* comp. and ed. Robert A. Rundstrom. Monograph 44. *Cartographica* 30, no. 1 (1993): 85–97. **283**

Neumayer, Erwin. *Prehistoric Indian Rock Paintings.* Delhi: Oxford University Press, 1983. **3, 4, 14, 15, 17, 18, 21**

Neven, Armand. *Etudes d'art lamaïque et de l'Himalaya.* Brussels: Oyez, 1978. **623, 635, 637**

Ngapo Ngawang Jigmei et al. *Tibet.* New York: McGraw-Hill, 1981. **679**

Ngô Sĩ Liên (fifteenth century). *Đại-Việt sử-ký toàn-thư.* 3 vols. Ed. Ch'en Ching-ho. Tokyo, 1984–86. **479, 481, 482, 490**

Nguyễn Khắc Viện. *Vietnam: A Long History.* Hanoi: Foreign Languages Publishing House, 1987. **484**

Nguyen Thanh-nha. *Tableau économique du Viet Nam aux XVIIe et XVIIIe siècles.* Paris: Editions Cujas, 1970. **484, 490, 493**

Nguyễn Thế Anh. "La réforme de l'impôt foncier de 1875 au Viet-Nam." *Bulletin de l'Ecole Française d'Extrême-Orient* 78 (1991): 287–96. **507**

Nguyễn Trãi. "Dư địa chí" (Geographical record). In *Nguyễn Trãi Toàn Tập* (Complete collection of the works of Nguyễn Trãi), 186–227. Hanoi, 1969. **481**

N̄i-ma-grags-pa. *Sgra yi don sdeb snaṅ gsal sgron me bźugs so* (Tibetan Zhang-zhung dictionary). Reprinted Delhi, 1965. **638, 640**

Nivat, Dhani. "The Gilt Lacquer Screen in the Audience Hall of Dusit." *Artibus Asiae* 24 (1961): 275–82. **728**

Nivison, David S. "The Origin of the Chinese Lunar Lodge System." In *World Archaeoastronomy,* ed. A. F. Aveni, 203–18. Cambridge: Cambridge University Press, 1989. **518**

Nōda, Chūryō. *An Inquiry concerning the Astronomical Writings Contained in the Li-chi Yüeh-ling.* Kyōto: Kyōto Institute, Academy of Oriental Culture, 1938. **519**

Noma, Saburō. "Earthquake Map of Japan, 1624." *Geographical Reports of Tokyo Metropolitan University* 9 (1974): 97–106. **410**

Norwick, Braham. "Locating Tibet—The Maps." In *Tibetan Studies: Proceedings of the 4th Seminar of the International Association for Tibetan Studies, Schloss Hohenkammer, Munich, 1985,* ed. Helga Uebach and Jampa L. Panglung, 301–20. Munich: Kommission für Zentralasiatische Studien, Bayerische Akadamie der Wissenschaften, 1988. **609**

———. "Modern Mapping of Tibet: A Cautionary Tale." Paper presented at the 6th IATS Seminar, August 1992. **609**

———. "Why Tibet Disappeared from 'Scientific' 16th-17th Century European Maps." In *Tibetan Studies: Proceedings of the 5th Seminar of the International Association for Tibetan Studies, Narita, 1989,* ed. Ihara Shōren and Yamaguchi Zuihō, 633–44. Narita: Naritasan Shinshoji, 1992. **609**

Novgorodova, E. A. *Alte Kunst der Mongolei.* Trans. Lisa Schirmer. Leipzig: E. A. Seemann, 1980. **6, 8, 9, 10, 21**

———. *Mir petroglifov Mongolii* (The world of Mongolian petroglyphs). Moscow: Nauka, 1984. **6, 8, 21**

Nunn, G. Raymond, ed. *Asia and Oceania: A Guide to Archival and Manuscript Sources in the United States.* 5 vols. New York: Mansell, 1985. **478**

Okladnikov, A. P. *Ancient Population of Siberia and Its Cultures.* Cambridge: Peabody Museum, 1959. **7, 9, 12, 13**

———. *Olen' zolotye roga* (Deer with the golden antlers). Leningrad, 1964. **7**

———. "The Petroglyphs of Siberia." *Scientific American* 221, no. 2 (1969): 78–82. **7**

———. *Yakutia before Its Incorporation into the Russian State.* Ed. Henry N. Michael. Montreal: McGill-Queen's University Press, 1970. **20**

———. *Der Hirsch mit dem goldenen Geweih: Vorgeschichtliche Felsbilder sibiriens.* Wiesbaden: F. A. Brockhaus, 1972. **7, 9, 21**

———. *Petroglify Mongolii.* Leningrad: Nauka, 1981. **8, 21**

Okladnikov, A. P., and A. I. Martynov. *Sokrovishcha tomskikh pisanits* (Treasures of the Tomsk petroglyphs). Moscow, 1972. **20**

Okladnikov, A. P., and V. D. Zaporozhskaya. *Petroglify Zabaykal'ya.* 2 vols. Leningrad: Nauka, 1969–70. **8, 9, 21**

Oldham, Charles E. A. W. "Some Remarks on the Models of the Bodh Gaya Temple Found at Nar-thang." *Journal of the Bihar and Orissa Research Society* 23 (1937): 418–28. **613**

Olschak, Blanche Christine. *Ancient Bhutan: A Study on Early Buddhism in the Himâlayas.* Zurich: Swiss Foundation for Alpine Research, 1979. **622**

———. *The Dragon Kingdom: Images of Bhutan.* Boston: Shambhala, 1988. **622**

Olschak, Blanche Christine, Augusto Gansser, and Andreas Gruschke. *Himalayas.* New York: Facts on File, 1987. **659, 681**

Olschak, Blanche Christine, and Geshe Thupten Wangyal. *Mystic Art of Ancient Tibet.* 1973. Boston: Shambhala, 1987. **616, 622, 623, 638, 659, 679**

Olson, Eleanor. "The Wheel of Existence." *Oriental Art,* n.s., 9 (1963): 204–9. **625**

Ong, Walter J. *Orality and Literacy: The Technologizing of the Word.* London: Methuen, 1982. **134**

Ossenbruggen, F. D. E. van. "De oorsprong van het Javaansche begrip Montjå-pat, in verband met primitieve classificaties" (The origins of the Javanese concept of *Moncå-pat* in connection with primitive classifications). *Verslagen en Mededeelingen der Koninklijke Akademie van Wetenschappen, Afdeeling Letterkunde,* 5th ser., pt. 3 (1918): 6–44. **772**

Ōtani, Ryōkichi. *Tadataka Inō, the Japanese Land-Surveyor.* Trans. Kazue Sugimura. Tokyo: Iwanami Shoten, 1932. **450, 453**

Oxnam, Robert B. *Ruling from Horseback: Manchu Politics in the Oboi Regency, 1661–1669.* Chicago: University of Chicago Press, 1975. **177**

Pagès, Léon. *Histoire de la religion chrétienne au Japon, depuis 1598 jusqu'à 1651.* 2 vols. Paris: C. Douniol, 1869–70. **381, 391**

Pal, Pratapaditya. *The Arts of Nepal.* Pt. 2, *Painting.* Leiden: E. J. Brill, 1978. **663, 664**

———. *Art of Nepal: A Catalogue of the Los Angeles County Museum of Art Collection.* Berkeley: Los Angeles County Museum of Art in association with University of California Press, 1985. **613, 616**

Pan, Nai, and Wang De-chang. "The Huang-You Star of the Song Dynasty—A Chinese Star List of the Early Medieval Period." *Chinese Astronomy and Astrophysics* 5 (1981): 441–48. **548, 558**

Pandya, Vishvajit. "Movement and Space: Andamanese Cartography." *American Ethnologist* 17 (1990): 775–97. **796**

Papinot, E. *Historical and Geographical Dictionary of Japan.* 1910; reprinted Ann Arbor, Mich.: Overbeck, 1948. **355, 362, 366, 367, 382, 383, 450**

Pargiter, F. E. "An Indian Game: Heaven or Hell." *Journal of the Royal Asiatic Society of Great Britain and Ireland,* 1916, 539–42. **628**

Park, Seong-rae. "Portents and Neo-Confucian Politics in Korea, 1392–1519." *Journal of Social Sciences and Humanities* 49 (1979): 53–117. **512, 558**

Parkin, Harry. *Batak Fruit of Hindu Thought.* Calcutta: Christian Literature Society, 1978. **711, 712**

Parmentier, Henri. *L'art du Laos.* 2 vols. Paris: Imprimerie Nationale; Hanoi: Ecole Française d'Extrême-Orient, 1954. **727**

Parsamian, E. S. "Astronomical Notes from Prague." *Sky and Telescope,* November 1967, 297. **12**

Passin, Herbert. *Society and Education in Japan.* New York: Teachers College Press, Columbia University, 1965. **407**

Patris, Charles, and L. Cadière. *Les Tombeaux de Hué: Gia-Long.* Hanoi: Imprimerie d'Extrême-Orient, 1923. **508**

Paulides, H. "Oude en nieuwe kunst op Bali, tegen den achtergrond van het Westen." *Cultureel Indië* 2 (1940): 169–85. **771**

Pearn, B. R. "The Burmese Embassy to Vietnam, 1823–24." *Journal of the Burma Research Society* 47, no. 1 (1964): 149–57. **698, 789**

Pepper, Stephen C. *World Hypotheses: A Study in Evidence.* Berkeley and Los Angeles: University of California Press, 1942. **145**

Pfister, Aloys. *Notices biographiques et bibliographiques sur les Jésuites de l'ancienne mission de Chine, 1552–1773.* 2 vols. Shanghai: Mission Press, 1932–34. **254, 551**

Phillimore, Reginald Henry. "An Early Map of the Malay Peninsula." *Imago Mundi* 13 (1956): 175–79. **692, 697, 829, 831**

Phillips, George. "The Seaports of India and Ceylon, Described by Chinese Voyagers of the Fifteenth Century, Together with an Account of Chinese Navigation." *Journal of the Royal Asiatic Society, North China Branch* 20 (1885): 209–26. **555**

Pingree, David, and Patrick Morrissey. "On the Identification of the *Yogatārās* of the Indian *Nakṣatras.*" *Journal for the History of Astronomy* 20 (1989): 99–119. **526**

Pires, Thome. *The Suma Oriental of Tomé Pires . . . and The Book of Francisco Rodrigues. . . .* Ed. and trans. Armando Cortesão. 2 vols. London: Hakluyt Society, 1944. **828**

Plutschow, Herbert E. *Historical Nagasaki.* Tokyo: Japan Times, 1983. **377, 393, 432, 440**

———. *Historical Nara.* Tokyo: Japan Times, 1983. **359, 373**

Pokora, Timoteus. "Pre-Han Literature." In *Essays on the Sources for Chinese History,* ed. Donald D. Leslie, Colin Mackerras, and Wang Gungwu, 23–35. Canberra: Australian National University Press, 1973. **516**

Pommaret-Imaeda, Françoise, and Yoshiro Imaeda. *Bhutan: A Kingdom of the Eastern Himalayas.* Trans. Ian Noble. Boston: Shambhala, 1985. **622**

Poppe, Nicholas. "Renat's Kalmuck Maps." *Imago Mundi* 12 (1955): 157–59. **682**

Porkert, Manfred. *The Theoretical Foundations of Chinese Medicine: Systems of Correspondence.* Cambridge: MIT Press, 1974. **136**

The Potala Palace of Tibet. Comp. Cultural Relics Administration Committee, Tibet Autonomous Region. Shanghai: People's Art Publishing House, 1982. **656**

Potter, Jack M. "Wind, Water, Bones and Souls: The Religious World of the Cantonese Peasant." *Journal of Oriental Studies* 8 (1970): 139–53. **220**

Powell, Joseph Michael. *Mirrors of the New World: Images and Image-Makers in the Settlement Process.* Folkestone, Eng.: Dawson; Hamden, Conn.: Archon Books, 1977. **20**

Pranavānanda, Swami. *Kailās-Mānasrōvar.* Calcutta: S. P. League, 1949. **659**

Prasad, Sri Nandan, ed. *Catalogue of the Historical Maps of the Survey of India (1700–1900).* New Delhi: National Archives of India, [ca. 1975]. **789**

Prescott, John Robert Victor. *Map of Mainland Asia by Treaty.* Carlton, Victoria: Melbourne University Press, 1975. **194**

Pullè, Francesco L. *La cartografia antica dell'India.* Studi Italiani di Filologia Indo-Iranica, Anno IV, vol. 4. Florence: Tipografia G. Carnesecchi e Figli, 1901. **608**

Pulleyblank, Edwin G. "Chinese and Indo-Europeans." *Journal of the Royal Asiatic Society of Great Britain and Ireland,* 1966, 9–39. **512**

———. *Lexicon of Reconstructed Pronunciation in Early Middle Chinese, Late Middle Chinese, and Early Mandarin.* Vancouver: UBC Press, 1991. **26**

Pye, Norman, and W. G. Beasley. "An Undescribed Manuscript Copy of Inō Chūkei's Map of Japan." *Geographical Journal* 117 (1951): 178–87. **450, 453**

Quirino, Carlos. *Philippine Cartography (1320–1899).* 2d rev. ed. Amsterdam: Nico Israel, 1963. **692, 741**

Ramaswami, N. S. "Prehistoric Rock Paintings Discovered in Tamil Nadu." *Indian News,* 6 February 1984, 7. **3**

Ramming, M. "The Evolution of Cartography in Japan." *Imago Mundi* 2 (1937): 17–22. **351, 362, 363**

Ravenhill, William. "John Brian Harley." *Transactions of the Institute of British Geographers,* n.s., 17 (1992): 120–25. **xxiii**

Rawson, Philip. *The Art of Southeast Asia.* London: Thames and Hudson, 1967. **694**

Renard, Louis. *Atlas de la navigation et du commerce qui se fait dans toutes les parties du monde.* Amsterdam, 1715. **434**

Revel, Nicole. *Fleurs de paroles: Histoire naturelle Palawan.* 3 vols. Paris: Editions Peeters, 1990. Vol. 2, *La maîtrise d'un savoir et l'art d'une relation.* **713, 715**

Reynolds, Craig J. "Buddhist Cosmography in Thai History, with Special Reference to Nineteenth-Century Culture Change." *Journal of Asian Studies* 35 (1976): 203–20. **696, 698, 720, 722, 723**

Reynolds, Frank E., and Mani B. Reynolds. *Three Worlds according to King Ruang: A Thai Buddhist Cosmology.* Berkeley, Calif.: Asian Humanities Press, 1982. **696, 720, 721, 722, 723, 729, 730, 733, 738**

Reynolds, Valrae. *Tibet: A Lost World.* Exhibition catalog for the Newark Museum Collection of Tibetan Art and Ethnography. New York: American Federation of Arts, 1978. **667, 679**

Rhie, Marilyn M., and Robert A. F. Thurman. *Wisdom and Compassion: The Sacred Art of Tibet.* San Francisco: Asian Art Museum, 1991. **624, 632, 648**

Riccardi, Theodore, Jr. "Some Preliminary Remarks on a Newari Painting of Svayambhūnāth." *Journal of the American Oriental Society* 93 (1973): 335–40. **661–63, 681**

Ricci, Matteo. *Storia dell'introduzione del Cristianesimo in Cina.* 3 vols. Ed. Pasquale M. d'Elia. Fonti Ricciane: Documenti Originali concernenti Matteo Ricci e la Storia delle Prime Relazioni tra l'Europa e la Cina (1579–1615). Rome: Libreria dello Stato, 1942–49. **120, 174, 175, 404**

———. *China in the Sixteenth Century: The Journals of Matthew Ricci, 1583–1610.* Trans. Louis J. Gallagher from the Latin version of Nicolas Trigault. New York: Random House, 1953. **120, 137, 138, 171, 172, 174, 550, 569**

Rickett, W. Allyn, trans. *Guanzi: Political, Economic, and Philosophical Essays from Early China.* Princeton: Princeton University Press, 1985–. **27, 73**

Ripa, Matteo. *Memoirs of Father Ripa, during Thirteen Years' Residence at the Court of Peking in the Service of the Emperor of China.* Trans. and ed. Fortunato Prandi. London: John Murray, 1846. **181, 185, 186, 299**

Robertson, Maureen. "Periodization in the Arts and Patterns of Change in Traditional Chinese Literary History." In *Theories of the Arts in China,* ed. Susan Bush and Christian Murck, 3–26. Princeton: Princeton University Press, 1983. **135**

Robinson, Arthur H., and Barbara Bartz Petchenik. *The Nature of Maps: Essays toward Understanding Maps and Mapping.* Chicago: University of Chicago Press, 1976. **96**

Rodgers, Susan. "Batak Religion." In *The Encyclopedia of Religion,* 16 vols., ed. Mircea Eliade, 2:81–83. New York: Macmillan, 1987. **702**

Rogers, Michael C. "Sung-Koryø Relations: Some Inhibiting Factors." *Oriens* 11 (1958): 194–202. **241**

———. "Factionalism and Koryŏ Policy under the Northern Sung." *Journal of the American Oriental Society* 79 (1959): 16–25. **241**

———. "The Regularization of Koryŏ-Chin Relations (1116–1131)." *Central Asiatic Journal* 6 (1961): 51–84. **278**

———. "*P'yŏnnyŏn T'ongnok*: The Foundation Legend of the Koryŏ State." *Korean Studies* 4 (1982–83): 3–72. **277, 278**

Rossbach, Sarah. *Feng Shui: The Chinese Art of Placement.* New York: E. P. Dutton, 1983. **216, 220**

Rowley, George. *Principles of Chinese Painting.* Princeton: Princeton University Press, 1959. **147**

Rudolph, Richard C., ed. *Chinese Archaeological Abstracts.* Monumenta Archaeologica, vol. 6. Los Angeles: Institute of Archaeology, University of California, 1978. **4**

Rufus, W. Carl. "The Celestial Planisphere of King Yi Tai-jo." *Transactions of the Korea Branch of the Royal Asiatic Society* 4, pt. 3 (1913): 23–72. **556, 560, 561**

———. "Korea's Cherished Astronomical Chart." *Popular Astronomy* 23 (1915): 193–98. **561**

———. "Astronomy in Korea." *Transactions of the Korea Branch of the Royal Asiatic Society* 26 (1936): 4–48. **556, 559, 560, 561, 563, 568**

Rufus, W. Carl, and Celia Chao. "A Korean Star Map." *Isis* 35 (1944): 316–26. **561, 566**

Rufus, W. Carl, and Won-chul Lee. "Marking Time in Korea." *Popular Astronomy* 44 (1936): 252–57. **252**

Rufus, W. Carl, and Hsing-chih Tien. *The Soochow Astronomical Chart.* Ann Arbor: University of Michigan Press, 1945. **89, 124, 545, 546, 547**

Rydh, Hanna. "On Symbolism in Mortuary Ceramics." *Bulletin of the Museum of Far Eastern Antiquities* 1 (1929): 71–120 and plates. **13**

Rykwert, Joseph. *The Idea of a Town: The Anthropology of Urban Form in Rome, Italy and the Ancient World.* London: Faber and Faber, 1976. **10**

Samutphāp traiphūm burān chabap Krung Thon Burī/Buddhist Cosmology Thonburi Version. Bangkok: Khana Kammakān Phičhāranā læ Čhatphim 'Ēkkasān thāng Prawattisāt, Samnak Nāyok Ratthamontrī, 1982. **720, 729, 730**

Saγang Sečen (Ssanang Ssetsen). *Erdeni-yin Tobči: Mongolian Chronicle.* 4 vols. Ed. Antoine Mostaert. Cambridge: Harvard University Press, 1956. **685**

Sande, Eduardo de. *De missione legatorum Iaponensium . . .* Macao, 1590; reprinted 1935. **377**

Sangermano, Vincenzo. *A Description of the Burmese Empire Compiled Chiefly from Native Documents.* Trans. and ed. William Tandy. Rome: Oriental Translation Fund of Great Britain and Ireland, 1833. **698**

Sāṅkṛityāyana, Rāhula. "Second Search of Sanskrit Palm-Leaf MSS. in Tibet." *Journal of the Bihar and Orissa Research Society* 23 (1937): 1–57. **612–13**

Sansom, George B. *A History of Japan.* 3 vols. Stanford: Stanford University Press, 1958–63. **354, 355, 359, 361, 371, 379, 386, 387, 396, 432, 433, 434, 436, 438, 445**

———. *Japan: A Short Cultural History.* 2d rev. ed. New York: Appleton-Century-Crofts, 1962. **410**

———. *The Western World and Japan: A Study in the Interaction of European and Asiatic Cultures.* New York: Alfred A. Knopf, 1962. **376, 377**

Sanson d'Abbeville, Nicolas. *L'Asie en plusieurs cartes nouvelles et exactes.* Paris, 1652. **389**

Santarém, Manuel Francisco de Barros e Sousa, Viscount of. *Atlas composé de mappemondes, de portulans et de cartes hydrographiques et historiques depuis le VIe jusqu'au XVIIe siècle.* 3 vols. Paris, 1849. Facsimile ed., *Atlas de Santarem,* with explanatory text by Helen Wallis and A. H. Sijmons. Amsterdam: R. Muller, 1985. **829**

———. *Essai sur l'histoire de la cosmographie et de la cartographie pendant la Moyen-Age et sur les progrès de la géographie après les grandes découvertes du XVe siècle.* 3 vols. Paris: Maulde et Renou, 1849–52. **607**

Schafer, Edward H. "An Ancient Chinese Star Map." *Journal of the British Astronomical Association* 87 (1977): 162. **549**

———. *Pacing the Void: T'ang Approaches to the Stars.* Berkeley and Los Angeles: University of California Press, 1977. **512, 537, 539**

Schärer, Hans. "Die Vorstellungen der Ober- und Unterwelt bei den Ngadju Dajak von Süd-Borneo." *Cultureel Indië* 4 (1942): 73–81. **693, 703**

———. *Die Gottesidee der Ngadju Dajak in Süd-Borneo*. Leiden: E. J. Brill, 1946. English translation, *Ngaju Religion: The Conception of God among a South Borneo People*. Trans. Rodney Needham. The Hague: Martinus Nijhoff, 1963. **693, 703, 705, 706, 710**

———. *Der Totenkult der Ngadju Dajak in Süd-Borneo*. 2 vols. Verhandelingen van het Koninklijk Instituut voor Taal-, Land- en Volkenkunde, vol. 51, pts. 1–2. The Hague: Martinus Nijhoff, 1966. **703**

Schilder, Günter. *Australia Unveiled: The Share of the Dutch Navigators in the Discovery of Australia*. Trans. Olaf Richter. Amsterdam: Theatrum Orbis Terrarum, 1976. **253**

———. "Willem Jansz. Blaeu's Wall Map of the World, on Mercator's Projection, 1606–07, and Its Influence." *Imago Mundi* 31 (1979): 36–54. **380**

———. *Three World Maps by François van den Hoeye of 1661, Willem Janszoon (Blaeu) of 1607, Claes Janszoon Visscher of 1650*. Amsterdam: Nico Israel, 1981. **380**

Schipper, Kristofer M., and Wang Hsiu-huei. "Progressive and Regressive Time Cycles in Taoist Ritual." In *Time, Science, and Society in China and the West*, Study of Time, vol. 5, ed. J. T. Fraser, N. Lawrence, and F. C. Haber, 185–205. Amherst: University of Massachusetts Press, 1986. **215**

Schlagintweit, Emil. *Buddhism in Tibet, Illustrated by Literary Documents and Objects of Religious Worship, with an Account of the Buddhist Systems Preceding It in India*. 1863. London: Susil Gupta, 1968. **632, 634**

Schlagintweit-Sakünlünski, Hermann von, Adolphe von Schlagintweit, and Robert von Schlagintweit. *Results of a Scientific Mission to India and High Asia*. Vol. 4. Leipzig: Brockhaus; London: Trübner, 1863. **609, 660, 661**

Schlegel, Gustave. *Uranographie chinoise; ou, Preuves directes que l'astronomie primitive est originaire de la Chine, et qu'elle a été empruntée par les anciens peuples occidentaux à la sphère chinoise*. 2 vols. Leiden: E. J. Brill, 1875; reprinted Taipei: Chengwen Chubanshe, 1967. **511**

Schleiermacher, Friedrich D. E. "*The Hermeneutics*: Outline of the 1819 Lectures." Trans. Jan Wojcik and Roland Haas. *New Literary History* 10 (1978): 1–16. **127**

Schuessler, Axel. *A Dictionary of Early Zhou Chinese*. Honolulu: University of Hawaii Press, 1987. **26, 27**

Schulz, Juergen. "Jacopo de' Barbari's View of Venice: Map Making, City Views, and Moralized Geography before the Year 1500." *Art Bulletin* 60 (1978): 425–74. **91**

Schurhammer, Georg, and J. Wicki, eds. *Epistolae S. Francisci Xaverii aliaque eius scripta*. 2 vols. Rome, 1944–45. **377**

Schütte, Joseph F. "Drei Unterrichtsbücher für japanische Jesuitenprediger aus dem XVI. Jahrhundert." *Archivum Historicum Societatis Iesu* 8 (1939): 223–56. **393**

———. "Map of Japan by Father Girolamo de Angelis." *Imago Mundi* 9 (1952): 73–78. **376, 444**

———. "Ignacio Moreira of Lisbon, Cartographer in Japan 1590–1592." *Imago Mundi* 16 (1962): 116–28. **376, 389**

———. "Japanese Cartography at the Court of Florence: Robert Dudley's Maps of Japan, 1606–1636." *Imago Mundi* 23 (1969): 29–58. **388, 389**

———, ed. *Monumenta historica Japoniae*. Rome, 1975–. **380**

Schüttler, Günter. *Die letzten tibetischen Orakelpriester: Psychiatrisch-neurologische Aspekte*. Wiesbaden: Franz Steiner, 1971. **615**

Schwartzberg, Joseph E. Section on South Asian Cartography in *The History of Cartography*, ed. J. B. Harley and David Woodward, vol. 2.1 (1992). Chicago: University of Chicago Press, 1987–.

"Cosmographical Mapping," 332–87. **371, 526, 622, 632, 634, 693, 702, 712, 714, 717, 727, 733, 738, 739**

"Geographical Mapping," 338–493. **609, 783**

"Introduction to South Asian Cartography," 295–331. **607, 608, 613**

"Nautical Maps," 494–503. **832**

———, ed. *A Historical Atlas of South Asia*. Chicago: University of Chicago Press, 1978. **641, 733, 781, 783**

Sebes, Joseph. *The Jesuits and the Sino-Russian Treaty of Nerchinsk (1689): The Diary of Thomas Pereira, S.J.* Rome: Institutum Historicum S.I., 1961. **194**

Ser-Odzhav, N. *Bayanligiyn Khadny Zurag* (Rock drawings of Bayan-Lig). Ed. D. Dorj. Ulan Bator, 1987. **2, 3**

Shagdarsurung. See Chagdarsurung, Ts.

Shah, Ikbal Ali. *Nepal: The Home of the Gods*. London: Sampson Low, Marston, [1938]. **648**

Sharma, R. K., and Rahman Ali. *Archaeology of Bhopal Region*. Delhi: Agam Kala Prakashan, 1980. **13**

Sher, Ya. A. *Petroglify Sredney i Tsentral'noy Azii* (Petroglyphs of Middle and Central Asia). Moscow: Nauka, 1980. **6**

Shih, Vincent Yu-chung, trans. *The Literary Mind and the Carving of Dragons*. See Liu Xie.

Shirley, Rodney W. *The Mapping of the World: Early Printed World Maps, 1472–1700*. London: Holland Press, 1983. **380**

Shore, A. F. "Egyptian Cartography." In *The History of Cartography*, ed. J. B. Harley and David Woodward, 1:117–29. Chicago: University of Chicago Press, 1987–. **13**

Shorto, H. L. "The Planets, the Days of the Week and the Points of the Compass: Orientation Symbolism in 'Burma.' " In *Natural Symbols in South East Asia*, ed. G. B. Milner, 152–64. London: School of Oriental and African Studies, 1978. **738**

Siebold, Philipp Franz von. *Nippon, Archiv zur Beschreibung von Japan und dessen Neben- und Schutzländern*. 4 vols. Leiden, 1832-[54?]. **439, 440**

———. *Manners and Customs of the Japanese in the Nineteenth Century*. 1841; reprinted Tokyo: Charles E. Tuttle, 1985. **436, 438, 440**

Siikala, A. L. "Finnish Rock Art, Animal Ceremonialism and Shamanic Worldview." In *Shamanism in Eurasia*, 2 vols., ed. Mihály Hoppál, 1:67–84. Göttingen: Edition Herodot, 1984. **13**

Sima Qian. *Les mémoires historiques de Se-Ma Ts'ien*. 5 vols. Trans. Edouard Chavannes. Paris: Leroux, 1895–1905. **513**

———. *Records of the Grand Historian of China*. 2 vols. Trans. Burton Watson. New York: Columbia University Press, 1961. **128, 129**

Sinaga, Anicetus B. *The Toba-Batak High God: Transcendence and Immanence*. Saint Augustin, West Germany: Anthropos Institute, 1981. **712**

Sivin, Nathan. *Cosmos and Computation in Early Chinese Mathematical Astronomy*. Leiden: E. J. Brill, 1969. **88**

———. "Shen Kua." In *Dictionary of Scientific Biography*, 16 vols., ed. Charles Coulston Gillispie, 12:369–93. New York: Charles Scribner's Sons, 1970–80. **83, 87**

———. "Wang Hsi-shan." In *Dictionary of Scientific Biography*, 16 vols., ed. Charles Coulston Gillispie, 14:159–68. New York: Charles Scribner's Sons, 1970–80. **104**

———. "Copernicus in China." In *Colloquia Copernicana 2: Etudes sur l'audience de la théorie héliocentrique*, Studia Copernicana 6, 63–122. Warsaw: Zakład Narodowy im. Ossolińskich, 1973. **186, 200**

———. "Why the Scientific Revolution Did Not Take Place in China—Or Didn't It?" *Chinese Science* 5 (1982): 45–66. **122**

———. "On the Limits of Empirical Knowledge in the Traditional Chinese Sciences." In *Time, Science, and Society in China and the West*, Study of Time, vol. 5, ed. J. T. Fraser, N. Lawrence, and F.

C. Haber, 151–69. Amherst: University of Massachusetts Press, 1986. **137**

———. "Science and Medicine in Imperial China—The State of the Field." *Journal of Asian Studies* 47 (1988): 41–90. **96, 122**

Skeat, Walter William. *Malay Magic: Being an Introduction to the Folklore and Popular Religion of the Malay Peninsula.* 1900; reprinted London: Macmillan, 1960. **702, 711**

Skeat, Walter William, and Charles Otto Blagden. *Pagan Races of the Malay Peninsula* 2 vols. London: Macmillan, 1906. **795**

Skinner, G. William. "The Structure of Chinese History." *Journal of Asian Studies* 44 (1985): 271–92. **24**

Skinner, Stephen. *The Living Earth Manual of Feng-Shui: Chinese Geomancy.* London: Routledge and Kegan Paul, 1982. **215, 220, 221, 222**

Sleeswyk, André Wegener. "Reconstruction of the South-Pointing Chariots of the Northern Sung Dynasty: Escapement and Differential Gearing in 11th Century China." *Chinese Science* 2 (1977): 4–36. **115**

Slusser, Mary Shepherd. "Serpents, Sages, and Sorcerers in Cleveland." *Bulletin of the Cleveland Museum of Art* 66, no. 2 (1979): 67–82. **609, 657, 670**

———. *Nepal Mandala: A Cultural Study of the Kathmandu Valley.* 2 vols. Princeton: Princeton University Press, 1982. **609, 613, 628, 657, 663, 664, 670, 679, 681**

———. "On a Sixteenth-Century Pictorial Pilgrim's Guide from Nepal." *Archives of Asian Art* 38 (1985): 6–36. **609, 664, 670, 677, 681**

———. "The Cultural Aspects of Newar Painting." In *Heritage of the Kathmandu Valley: Proceedings of an International Conference in Lübeck, June 1985*, ed. Niels Gutschow and Axel Michaels, 13–27. Saint Augustin: VGH Wissenschaftsverlag, 1987. **609, 663, 664, 670, 674, 677, 681**

Smith, Ralph B. "Sino-Vietnamese Sources for the Nguyễn Period: An Introduction." *Bulletin of the School of Oriental and African Studies* 30 (1967): 600–621. **499, 500, 501, 502, 506, 507**

———. "Politics and Society in Viêt-Nam during the Early Nguyễn Period (1802–62)." *Journal of the Royal Asiatic Society of Great Britain and Ireland*, 1974, 153–69. **499**

Smith, Richard J. *China's Cultural Heritage: The Ch'ing Dynasty, 1644–1912.* Boulder, Colo.: Westview Press, 1983. **204**

———. *Fortune-Tellers and Philosophers: Divination in Traditional Chinese Society.* Boulder, Colo.: Westview Press, 1991. **216, 217, 218, 219, 220, 222**

Smith, Vincent A. "Pygmy Flints." *Indian Antiquary*, July 1906, 185–95. **17**

Snellgrove, David L., ed. and trans. *The Nine Ways of Bon: Excerpts from "gZi-brjid."* London: Oxford University Press, 1967. **615, 624, 632, 640**

———. "Places of Pilgrimage in Thag (Thakkhola)." *Kailash* 7 (1979): 72–132; Tibetan text 133–70. **661**

Snellgrove, David L., and Hugh Richardson. *A Cultural History of Tibet.* London: Weidenfeld and Nicolson, 1968. **679**

Snodgrass, Adrian. *The Symbolism of the Stupa.* Ithaca, N.Y.: Southeast Asia Program, Cornell University, 1985. **615–16**

Sommāi Prēmčhit, Kamon Sīwichainan, and Surasingsamrūam Chimphanao. *Phračhēdī nai Lānnā Thai* (Stupas in Lanna Thai). Chiang Mai: Khrongkan Suksā Wichai Sinlapa Sathapattayakam Lānnā, Mahāwitthayālai Chīang Mai, 1981. **720, 777, 784**

Soothill, William E. "The Two Oldest Maps of China Extant." *Geographical Journal* 69 (1927): 532–55. **113**

———. *The Hall of Light: A Study of Early Chinese Kingship.* London: Lutterworth Press, 1951. **513, 532**

Sopa, Geshe. "The Tibetan 'Wheel of Life': Iconography and Doxography." *Journal of the International Association of Buddhist Studies* 7, no. 1 (1984): 125–45. **616, 625**

Soper, Alexander C. "Early Chinese Landscape Painting." *Art Bulletin* 23 (1941): 141–64. **139**

Sotheby's. *The Library of Philip Robinson.* Pt. 2, *The Chinese Collection.* Catalog. Day of sale, 22 November 1988. **268**

Spence, Jonathan D. *Emperor of China: Self-Portrait of K'ang-hsi.* 1974; reprinted New York: Vintage-Random House, 1975. **87**

Stadtner, Donald M. "King Dhammaceti's Pegu." *Orientations* 37, no. 2 (1990): 53–60. **695**

Stanley-Baker, Joan. *Japanese Art.* London: Thames and Hudson, 1984. **355, 410**

Stein, Rolf A. "Jardins en miniature d'Extrême-Orient." *Bulletin de l'Ecole Française d'Extrême-Orient* 42 (1943): 1–104. **480**

———. "Peintures tibétaines de la vie de Gesar." *Arts Asiatiques* 5 (1958): 243–71. **637**

———. *Tibetan Civilization.* Trans. J. E. Stapleton Driver. London: Faber and Faber, 1972. **637, 638, 642**

———. *The World in Miniature: Container Gardens and Dwellings in Far Eastern Religious Thought.* Trans. Phyllis Brooks. Stanford: Stanford University Press, 1990. **479, 480, 507**

Stephenson, F. Richard. "Mappe celesti nell'antico Oriente." *L'Astronomia*, no. 98 (1990): 18–27. **566**

———. "Stargazers of the Orient." *New Scientist* 137, no. 1854 (1993): 32–34. **524**

Stephenson, F. Richard, and Kevin K. C. Yau. "Far Eastern Observations of Halley's Comet, 240 BC to AD 1368." *Journal of the British Interplanetary Society* 38 (1985): 195–216. **511, 527, 538**

———. "Astronomical Records in the *Ch'un-ch'iu* Chronicle." *Journal for the History of Astronomy* 23 (1992): 31–51. **515**

Stephenson, F. Richard, and C. B. F. Walker, eds. *Halley's Comet in History.* London: British Museum Publications, 1985. **520**

Sternstein, Larry. " 'Low' Maps of Siam." *Journal of the Siam Society* 73 (1985): 132–57. **699**

Stevens, Hrolf Vaughan. "Die Zaubermuster der Ôrang hûtan," pt. 2, "Die 'Toon-tong'-Ceremonie." *Zeitschrift für Ethnologie* 26 (1894): 141–88. **692, 795**

Stöhr, Waldemar. "Das Totenritual der Dajak." *Ethnologica*, n.s., 1 (1959): 1–245. **703**

———. "Über einige Kultzeichnungen der Ngadju-Dajak." *Ethnologica*, n.s., 4 (1968): 394–419. **703, 704, 705, 706, 708**

Stoll, Eva. "Ti-se, der heilige Berg in Tibet." *Geographica Helvetica* 21 (1966): 162–67. **657, 659**

Strickmann, Michel. "The Tao among the Yao: Taoism and the Sinification of South China." In *Rekishi ni okeru minshū to bunka: Sakai Tadao Sensei koki shukuga kinen ronshu* (Peoples and cultures in Asiatic history: Collected essays in honor of Professor Tadao Sakai on his seventieth birthday), 23–30. Tokyo: Kokusho Kankōkai, 1982. **24**

Strong, John S. *The Legend and Cult of Upagupta: Sanskrit Buddhism in North India and Southeast Asia.* Princeton: Princeton University Press, 1992. **777**

Sugimoto, Masayoshi, and David L. Swain. *Science and Culture in Traditional Japan: A.D. 600–1854.* Cambridge: MIT Press, 1978. **24**

Sullivan, Michael. *The Birth of Landscape Painting in China.* Berkeley and Los Angeles: University of California Press, 1962. **131, 139**

———. *A Short History of Chinese Art.* Berkeley and Los Angeles: University of California Press, 1967. **225**

———. *The Three Perfections: Chinese Painting, Poetry, and Calligraphy.* London: Thames and Hudson, 1974. **158**

———. *The Arts of China.* 3d ed. Berkeley and Los Angeles: University of California Press, 1984. **204**

Suthiwong Phongphaibūn. *Phutthasātsanā Thǣp Lum Thalēsāp*

Songkhlā Fang Tawan'ọ̄k samai Krung Sī 'Ayutthayā: Rāingān kānwičhai (Report on the research on the Buddhist religion around the Thale Sap basin on the eastern shore in the Ayutthaya period). Songkhla, 1980. **784**

Tạ Trọng Hiệp. "Les fonds de livres en Hán Nôm hors du Viet-nam: Eléments d'inventaires." *Bulletin de l'Ecole Française d'Extrême-Orient* 75 (1986): 267–93. **490, 506**

Tatz, Mark, and Jody Kent. *Rebirth: The Tibetan Game of Liberation.* Garden City, N.Y.: Anchor Books, 1977. **623, 628, 629**

Taylor, Keith W. "Notes on the *Việt Điện U Linh Tập.*" *Vietnam Forum* 8 (1986): 26–59. **484**

———. "The Literati Revival in Seventeenth-Century Vietnam." *Journal of Southeast Asian Studies* 18 (1987): 1–22. **487**

Taylor, Peter J. "Politics in Maps, Maps in Politics: A Tribute to Brian Harley." *Political Geography* 11 (1992): 127–29. **xxiii**

Teleki, Pál. *Atlas zur Geschichte der Kartographie der japanischen Inseln.* Budapest, 1909; reprinted 1966. **388, 389, 443**

Temple, Richard C. *The Thirty-seven Nats: A Phase of Spirit-Worship Prevailing in Burma.* London: W. Griggs, 1906. New edition with essay and bibliography by Patricia M. Herbert. London: P. Strachan, 1991. **693, 735, 736, 743**

Teng, Ssu-yü, and Knight Biggerstaff, comps. *An Annotated Bibliography of Selected Chinese Reference Works.* 3d ed. Harvard-Yenching Institute Studies 2. Cambridge: Harvard University Press, 1971. **204**

Thái Văn Kiểm. *Cố đô Huế* (The old capital of Huế). Saigon: Nha Văn-hóa Bộ Quốc-gia Giáo-dục, 1960. **478, 507**

———. "Interprétation d'une carte ancienne de Saigon." *Bulletin de la Société des Etudes Indochinoises,* n.s., 37, no. 4 (1962): 409–31. **478, 501, 502**

———. "Lời nói đầu" (Introduction). In *Lục tỉnh Nam-Việt (Đại-Nam nhất-thống chí)* (The six provinces of southern Vietnam [Record of the unity of Đại Nam]). Saigon: Phủ Quốc-vụ-Khanh Đặc-Trách Văn-Hóa, 1973. **478, 499, 500, 501, 506**

Than Tun. "Mandalay Maps." *Papers of the Upper Burma Writers' Society,* 1966. **825**

Thomassen à Thuessink van der Hoop, Abraham Nicolaas Jan. *Indonesische siermotieven.* [Batavia]: Koninklijk Bataviaasch Genootschap van Kunsten en Wetenschappen, 1949. **13, 19**

Thorp, Robert L. "Burial Practices of Bronze Age China." In *The Great Bronze Age of China: An Exhibition from the People's Republic of China,* ed. Wen Fong, 51–64. New York: Metropolitan Museum of Art, 1980. **80**

———. "An Archaeological Reconstruction of the Lishan Necropolis." In *The Great Bronze Age of China: A Symposium,* ed. George Kuwayama, 72–83. Los Angeles: Los Angeles County Museum of Art, 1983. **79**

———. "The Qin and Han Imperial Tombs and the Development of Mortuary Architecture." In *The Quest for Eternity: Chinese Ceramic Sculptures from the People's Republic of China,* ed. Susan L. Caroselli, 17–37. Los Angeles: Los Angeles County Museum of Art, 1987. **80**

———. *Son of Heaven: Imperial Arts of China.* Exhibition catalog. Seattle: Son of Heaven Press, 1988. **81, 130**

Thrower, Norman J. W., and Young Il Kim (Kim Yŏng'il). "Dong-Kook-Yu-Ji-Do: A Recently Discovered Manuscript of a Map of Korea." *Imago Mundi* 21 (1967): 30–49. **263, 307**

Tibetische Kunst: Katalog zu Ausstellung, 8.-30. März, 1969 Helmshaus, Zürich, 17 Apr. bis 11 Mai 1969, Gesellschaftshaus zu Schützen, Luzern. Bern: TIBETA, 1969. **677**

Tichy, Herbert. *Himalaya.* Vienna: Anton Schroll, 1968. **617, 654, 681**

Tondriau, Julien L. *20 rouleaux peints tibétains et népalais.* Brussels: Musées Royaux d'Art et d'Histoire, [1964–65?]. **638**

Toulmin, Stephen. *Human Understanding.* Princeton: Princeton University Press, 1972–. **96**

———. *Cosmopolis: The Hidden Agenda of Modernity.* New York: Macmillan, 1990. **65**

Traiphum chabap phasa khamen (The Cambodian manuscript of the *Trai phum* cosmography). Trans. Amphai Khamtho. Bangkok: Ammarin Printing Group, 1987. **740**

Trần Nghĩa. "Bản đồ cổ Việt Nam" (Old maps of Vietnam). *Tạp chí Hán Nôm* (Hán Nôm review) 2, no. 9 (1990): 3–10. **478**

Trần Văn Giáp. "Relation d'une ambassade annamite en Chine au XVIIIe siècle." *Bulletin de la Société des Etudes Indochinoises,* n.s., 16, no. 3 (1941): 55–81. **496**

Trésors du Tibet: Région autonome du Tibet, Chine. Paris: Muséum National d'Histoire Naturelle, 1987. **615, 622, 632**

Trungpa, Chögyam. *Visual Dharma: The Buddhist Art of Tibet.* Berkeley, Calif.: Shambhala, 1975. **638, 667, 679**

Trương Bửu Lâm, ed. *Hồng-đức bản đồ* (Maps of the Hồng-đức period). Saigon: Bộ Quốc-gia Giáo-dục, 1962. **478, 481, 483, 484, 486, 487, 489, 490, 491, 493, 496, 503**

Tsien, Tsuen-hsuin. *Written on Bamboo and Silk: The Beginnings of Chinese Books and Inscriptions.* Chicago: University of Chicago Press, 1962. **138**

———. See also Needham, Joseph. *Science and Civilisation in China.*

Tsuda, Noritake. *Handbook of Japanese Art.* 1941; reprinted Tokyo: Charles E. Tuttle, 1985. **370, 410**

Tsutsihashi, P., and Stanislas Chevalier. "Catalogue d'étoiles observées à Pé-kin sous l'empereur K'ien-long (XVIIIe siècle)." *Annales de l'Observatoire Astronomique de Zô-sè (Chine)* 7 (1911): I-D105. **511, 574, 576**

Tuan, Yi-fu. *Space and Place: The Perspective of Experience.* Minneapolis: University of Minnesota Press, 1977. **160**

Tucci, Giuseppe. *Tibetan Painted Scrolls.* 3 vols. Rome: Libreria dello Stato, 1949. **620**

———. *The Theory and Practice of the Maṇḍala, with Special Reference to the Modern Psychology of the Subconscious.* Trans. Alan Houghton Brodrick. New York: Samuel Weiser, 1970 [first published 1961]. **620**

Turner, Samuel. *An Account of an Embassy to the Court of the Teshoo Lama in Tibet.* London, 1800; reprinted New Delhi: Mañjuśrī Publishing House, 1971. **637**

Ungar, Esta S. "From Myth to History: Imagined Polities in 14th Century Vietnam." In *Southeast Asia in the 9th to 14th Centuries,* ed. David G. Marr and A. C. Milner, 177–86. Singapore: Institute of Southeast Asian Studies, 1986. **481**

Unno, Kazutaka. "Concerning a MS Map of China in the Bibliothèque Nationale, Paris, Introduced to the World by Monsieur M. Destombes." *Memoirs of the Research Department of the Toyo Bunko (the Oriental Library)* 35 (1977): 205–17. **229**

———. "Japan before the Introduction of the Global Theory of the Earth: In Search of a Japanese Image of the Earth." *Memoirs of the Research Department of the Toyo Bunko* 38 (1980): 39–69. **120, 371**

———. "The Asian Lake Chiamay in the Early European Cartography." In *Imago et mensura mundi: Atti del IX Congresso Internazionale di Storia della Cartografia,* 3 vols., ed. Carla Clivio Marzoli, 2:287–96. Rome: Istituto della Enciclopedia Italiana, [1985]. **610**

———. "Japan." In *Lexikon zur Geschichte der Kartographie,* 2 vols., ed. Ingrid Kretschmer, Johannes Dörflinger, and Franz Wawrik, 1:357–61. Vienna: Franz Deuticke, 1986. **351, 370**

———. "Japanische Kartographie." In *Lexikon zur Geschichte der Kartographie*, 2 vols., ed. Ingrid Kretschmer, Johannes Dörflinger, and Franz Wawrik, 1:361–66. Vienna: Franz Deuticke, 1986. **351**

———. "Maps as Picture: The Old Chinese Views of Maps." Paper presented at the Thirteenth International Conference on the History of Cartography, Amsterdam and The Hague, 26 June to 1 July 1989. **153**

———. "Extant Maps of the Paddy Fields Drawn in the Eighth Century Japan." Paper delivered at the Fourteenth International Conference on the History of Cartography, Uppsala, 1991. **352**

———. "Government Cartography in Sixteenth Century Japan." *Imago Mundi* 43 (1991): 86–91. **396**

———. "A Surveying Instrument Designed by Hōjō Ujinaya (1609–70)." Paper presented at the Seventh International Conference on the History of Science in East Asia, Kyōto, Japan, August 1993. **588**

Valignani, Alessandro. *Sumario de las cosas de Japón*. Ed. José Luis Alvarez-Taladriz. Tokyo: Sophia University, 1954. **377**

Vanderstappen, Harrie. "Chinese Art and the Jesuits in Peking." In *East Meets West: The Jesuits in China, 1582–1773*, ed. Charles E. Ronan and Bonnie B. C. Oh, 103–26. Chicago: Loyola University Press, 1988. **153**

Van der Wee, Louis P. "A 'Cloister-City'—Ṭanka." *Journal of the Indian Society of Oriental Art*, n.s., 4 (1971–72): 108–20. **659, 668, 670, 681**

———. "Rirab Lhunpo and a Tibetan Narrative of Creation." *Ethnologische Zeitschrift*, 1976, no. 2, 67–80. **621, 623, 624, 657**

Van der Wee, Pia, Louis P. Van der Wee, and Janine Schotsmans. *Symbolisme de l'art lamaïque*. Brussels: Musées Royaux d'Art et d'Histoire, 1988. **648, 677**

Van Zandt, Howard F. *Pioneer American Merchants in Japan*. Tokyo: Lotus, 1980. **443**

Varley, H. Paul. *Japanese Culture*. 3d ed. Honolulu: University of Hawaii Press, 1984. **356, 359, 365**

Varthema, Ludovic. *The Travels of Ludovico di Varthema in Egypt, Syria, Arabia Deserta, and Arabia Felix, in Persia, India, and Ethiopia, A.D. 1503 to 1508*. Trans. John Winter Jones. Ed. George Percy Badger. London: Printed for the Hakluyt Society, 1863. **697**

Vergara, Paola Mortari, and Gilles Béguin, eds. *Dimore umane, santuari divini: Origini, sviluppo e diffusione dell'architettura tibetana/Demeures des hommes, sanctuaires des dieux: Sources, développement et rayonnement de l'architecture tibétaine*. Rome: Università di Roma "La Sapienza," 1987. **613, 614, 648, 668, 677, 679**

Vergati, Anne. "Les royaumes de la vallée de Katmandou." In *Les royaumes de l'Himâlaya: Histoire et civilisation*, 164–208. Paris: Imprimerie Nationale, 1982. **677**

Vink, Marcus. "The Dutch East India Company and the Pepper Trade between Kerala and Tamilnad, 1663–1795: A Geohistorical Analysis." Unpublished paper, University of Minnesota, December 1990. **831**

Vitashevskiy, V., ed. "Izobrazheniya na skalkh po r. Olekme" (Drawings on the cliffs along the Olekma River). *Izvestiya Vostochno-Sibirskago Otdela Imperatorskago Russkago Geograficheskago Obshchestva* (East Siberian department of the Imperial Russian Geographical Society's News) 28, no. 4 (1897). **20**

Vroklage, B. A. G. "Das Schiff in den Megalithkulturen Südasien und der Südsee." *Anthropos* 31 (1936): 712–57. **708**

Waddell, Laurence Austine. *The Buddhism of Tibet, or Lamaism, with Its Mystic Cults, Symbolism, and Mythology, and Its Relation to Indian Buddhism*. 2d ed. Cambridge: W. Heffer, 1935 [first published 1895]. **608, 609, 612, 616, 625, 628, 629, 632, 638, 677, 679**

Wahid, Siddiq. *Ladakh: Between Earth and Sky*. New York: Norton, 1981. **681**

Wakankar, Vishnu S. "Painted Rock Shelters of India." *IPEK: Jahrbuch für Prähistorische und Ethnographische Kunst* 21 (1964–65): 78–83. **13**

———. "Bhimbetka—The Prehistoric Paradise." *Prachya Pratibha* 3, no. 2 (July 1975): 7–29. Reprinted (but without some illustration and appendix material) in *Indische Felsbilder von der Arbeitsgemeinschaft der Ge-Fe-Bi*, 72–93. Graz: Gesellschaft für Vergleichende Felsbildforschung, 1978. **2, 3**

Wakeman, Frederic, Jr. *The Great Enterprise: The Manchu Reconstruction of Imperial Order in Seventeenth-Century China*. 2 vols. Berkeley and Los Angeles: University of California Press, 1985. **177, 178**

———, ed. *Ming and Qing Historical Studies in the People's Republic of China*. Berkeley: Institute of East Asian Studies, University of California, Berkeley, Center for Chinese Studies, 1980. **100, 189**

Wales, Horace Geoffrey Quaritch. *Siamese State Ceremonies: Their History and Function*. London: Bernard Quaritch, 1931. **728**

———. *Prehistory and Religion in South-east Asia*. London: Bernard Quaritch, 1957. **13, 14, 19**

———. "The Cosmological Aspect of Indonesian Religion." *Journal of the Royal Asiatic Society of Great Britain and Ireland*, 1959, 100–139. **702, 703, 738**

———. *The Universe around Them: Cosmology and Cosmic Renewal in Indianized South-east Asia*. London: Arthur Probsthain, 1977. **701, 714, 728, 733, 737, 739**

Waley, Arthur, trans. *The Book of Songs*. London: Allen and Unwin, 1937. **516**

Waller, Derek. *The Pundits: British Exploration of Tibet and Central Asia*. Lexington: University Press of Kentucky, 1990. **652, 654**

Wallis, Helen. "The Influence of Father Ricci on Far Eastern Cartography." *Imago Mundi* 19 (1965): 38–45. **120, 407**

———. "Chinese Maps and Globes in the British Library and the Phillips Collection." In *Chinese Studies: Papers Presented at a Colloquium at the School of Oriental and African Studies, University of London, 24–26 August 1987*, ed. Frances Wood, 88–96. London: British Library, 1988. **124, 181**

Wallis, Helen M., and Arthur H. Robinson, eds. *Cartographical Innovations: An International Handbook of Mapping Terms to 1900*. Tring, Hertfordshire: Map Collector Publications in association with the International Cartographic Association, 1987. **96, 117**

Wang, Ju Hua. "The Inventor of Paper Technology—Ts'ai Lun." *Yearbook of Paper History* 8 (1990): 156–63. **40**

Wang, Ling. See Needham, Joseph. *Science and Civilisation in China*.

Wang, Sung-hsing. "Taiwanese Architecture and the Supernatural." In *Religion and Ritual in Chinese Society*, ed. Arthur P. Wolf, 183–92. Stanford: Stanford University Press, 1974. **215**

Wang, Yü-ch'üan. "An Outline of the Central Government of the Former Han Dynasty." *Harvard Journal of Asiatic Studies* 12 (1949): 134–87. **521**

Wang, Zhongshu. *Han Civilization*. Trans. Kwang-chih Chang et al. New Haven: Yale University Press, 1982. **78**

Watanabe, Akira. *Cartography in Japan: Past and Present*. Tokyo: International Cartographic Information Center, 1980. **351**

Watson, Burton. *Early Chinese Literature*. New York: Columbia University Press, 1962. **515, 516**

———, trans. *Records of the Grand Historian of China*. See Sima Qian.

Watters, Thomas. *On Yuan Chwang's Travels in India*. 2 vols. 1904–5; reprinted New York: AMS, 1970. **373**

Wayman, Alex. *The Buddhist Tantras: Light on Indo-Tibetan Esotericism*. New York: Samuel Weiser, 1973. **621, 623**

Wechsler, Howard J. *Offerings of Jade and Silk: Ritual and Symbol in the Legitimation of the T'ang Dynasty*. New Haven: Yale University Press, 1985. **212, 213**

Wenk, Klaus. *Thailändische Miniaturmalereien nach einer Handschrift der indischen Kunstabteilung der Staatlichen Museen Berlin*. Wiesbaden: Franz Steiner, 1965. **692, 720, 721, 722, 742**

———. "Zu einer 'Landkarte' Sued- und Ostasiens." In *Felicitation Volumes of Southeast-Asian Studies Presented to His Highness Prince Dhaninivat Kromamun Bidyalabh Bridhyakorn . . . on the Occasion of His Eightieth Birthday*, 2 vols., 1:119–22. Bangkok: Siam Society, 1965. **692, 741, 742, 743**

Wheatley, Paul. "A Curious Feature on Early Maps of Malaya." *Imago Mundi* 11 (1954): 67–72. **832**

———. *The Pivot of the Four Quarters: A Preliminary Enquiry into the Origins and Character of the Ancient Chinese City*. Chicago: Aldine, 1971. **10, 211**

Whitmore, John K. *Vietnam, Hồ Quý Ly, and the Ming (1371–1421)*. New Haven: Yale Center for International and Area Studies, 1985. **481**

———. "*Chung-hsing* and *Ch'eng-t'ung* in Đại Việt: Historiography in and of the Sixteenth Century." In *Textual Studies on the Vietnamese Past*, ed. Keith W. Taylor. Forthcoming. **485**

———. *Transforming Đại Việt: Politics and Confucianism in the Fifteenth Century*. Forthcoming. **482**

Wieder, Frederik Caspar. "Oude Kaartbeschrijving" section in "Kaartbeschrijving." In *Encyclopaedie van Nederlandsch-Indië*, 8 vols., 2:227–36. The Hague: Martinus Nijhoff, 1917–40. **690**

———. *Monumenta cartographica*. 5 vols. The Hague: Nijhoff, 1925–33. **380**

Wilhelm, Hellmut. *Change: Eight Lectures on the "I Ching."* Trans. Cary F. Baynes. 1960; reprinted New York: Harper and Row, 1964. **130**

Wilhelm, Richard, and Cary F. Baynes, trans. *The I Ching or Book of Changes*. 3d ed. Princeton: Princeton University Press, 1967. **130, 131**

Williams, John. *Observations of Comets from B.C. 611 to A.D. 1640*. London: Strangeways and Walden, 1871. **511**

Williamson, A., comp. *Burma Gazetteer, Shwebo District*. Vol. A. Rangoon: Superintendent Government Printing and Stationery, 1929. **821**

Willman, Oloff Erickson. See Kjöping, Nils Matson.

Wilson, Constance. "Cultural Values and Record Keeping in Thailand." *CORMOSEA* [Committee on Research Materials on Southeast Asia] *Bulletin* 10, no. 2 (1982): 2–17. **698**

Winter, Heinrich. "Francisco Rodrigues' Atlas of ca. 1513." *Imago Mundi* 6 (1949): 20–26. **697, 829**

Wise, James F. N. *Notes on the Races, Castes, and Tribes of Eastern Bengal*. London: Harrison, 1883. **651**

Wī thi bhuṃ cañ' chan'" puṃ sim' puṃ. Rangoon: Ū" Po' Raṅ'-Do' Co Raṅ', 1967. **739**

Wittfogel, Karl A. *Oriental Despotism: A Comparative Study of Total Power*. New Haven: Yale University Press, 1957. **97**

Wolters, O. W. *Two Essays on Đại-Việt in the Fourteenth Century*. New Haven: Council on Southeast Asia Studies, Yale Center for International and Area Studies, 1988. **481**

Wong, George H. C. "China's Opposition to Western Science during Late Ming and Early Ch'ing." *Isis* 54 (1963): 29–49. **120**

Woo, David. "The Evolution of Mountain Symbols in Traditional Chinese Cartography." Paper presented at the annual meeting of the Association of American Geographers, 1989. **483, 490**

Woodside, Alexander Barton. *Vietnam and the Chinese Model: A Comparative Study of Nguyễn and Ch'ing Civil Government in the First Half of the Nineteenth Century*. Cambridge: Harvard University Press, 1971. **497, 499, 500, 502, 503, 504**

Woodward, David. "Brian Harley, 1932–1991." *Map Collector* 58 (1992): 40. **xxiii**

———. "J. B. Harley: A Tribute." *Imago Mundi* 44 (1992): 120–25. **xxiii**

———. "John Brian Harley, 1932–1991." *Special Libraries Association, Geography and Map Division Bulletin* 167 (1992): 50–52. **xxiii**

Wray, Elizabeth, Clare Rosenfield, and Dorothy Bailey. *Ten Lives of the Buddha: Siamese Temple Paintings and Jataka Tales*. New York: Weatherhill, 1972. **722**

Wright, Arthur F. "The Cosmology of the Chinese City." In *The City in Late Imperial China*, ed. George William Skinner, 33–73. Stanford: Stanford University Press, 1977. **210, 213**

Wu, Silas H. L. *Communication and Imperial Control in China: Evolution of the Palace Memorial System, 1693–1735*. Cambridge: Harvard University Press, 1970. **91**

Wylie, Alexander. *Chinese Researches*. Shanghai, 1897. **511**

Xi, Zezong. "Chinese Studies in the History of Astronomy, 1949–1979." *Isis* 72 (1981): 456–70. **536**

———. "The Cometary Atlas in the Silk Book of the Han Tomb at Mawangdui." *Chinese Astronomy and Astrophysics* 8 (1984): 1–7. **521**

Xiao Tong. *Wen Xuan; or, Selections of Refined Literature*. Trans. and annotated David R. Knechtges. Princeton: Princeton University Press, 1982–. **77, 120, 132**

Xu, Zhentao, Kevin K. C. Yau, and F. Richard Stephenson. "Astronomical Records on the Shang Dynasty Oracle Bones." *Archaeoastronomy* 14, suppl. to *Journal for the History of Astronomy* 20 (1989): S61-S72. **514**

Xuanzhuang. *Mémoires sur les contrées occidentales*. 2 vols. Trans. Stanislas Julien. Paris, 1857–58. **373**

Xun Qing. *The Works of Hsüntze*. Trans. Homer H. Dubs. London: Arthur Probsthain, 1928. **74**

Yabuuchi Kiyoshi (Yabuuti Kiyosi). "Researches on the *Chiu-chih Li*—Indian Astronomy under the T'ang Dynasty." *Acta Asiatica* 36 (1979): 7–48. **512, 533**

———. "The Influence of Islamic Astronomy in China." In *From Deferent to Equant: A Volume of Studies in the History of Science in the Ancient and Medieval Near East in Honor of E. S. Kennedy*, ed. David A. King and George Saliba, 547–59. New York: New York Academy of Sciences, 1987. **119**

Yang, Ch'ing K'un. *Religion in Chinese Society: A Study of Contemporary Social Functions of Religion and Some of Their Historical Factors*. Berkeley and Los Angeles: University of California Press, 1961. **215**

Yee, Cordell D. K. "A Cartography of Introspection: Chinese Maps as Other Than European." *Asian Art* 5, no. 4 (1992): 29–47. **xxiv**

Yeomans, Donald K., and Tao Kiang. "The Long-Term Motion of Comet Halley." *Monthly Notices of the Royal Astronomical Society* 197 (1981): 633–46. **538**

Yi, Ik Seup (Yi Iksŭp). "A Map of the World." *Korean Repository* 1 (1892): 336–41. **259, 260**

Yoon, Hong Key. "The Expression of Landforms in Chinese Geomantic Maps." *Cartographic Journal* 29 (1992): 12–15. **221**

Yu, Pauline. "Formal Distinctions in Chinese Literary Theory." In *Theories of the Arts in China*, ed. Susan Bush and Christian Murck, 27–53. Princeton: Princeton University Press, 1983. **136**

Yü, Ying-shih. "Life and Immortality in the Mind of Han China." *Harvard Journal of Asiatic Studies* 25 (1964–65): 80–122. **80**

———. "New Evidence on the Early Chinese Conception of Afterlife—A Review Article." *Journal of Asian Studies* 41 (1981): 81–85. **151**

———. " 'O Soul, Come Back!' A Study in the Changing Conceptions of the Soul and Afterlife in Pre-Buddhist China." *Harvard Journal of Asiatic Studies* 47 (1987): 363–95. **80**

Yule, Henry, and A. C. Burnell. *Hobson Jobson: A Glossary of Colloquial Anglo-Indian Words and Phrases, and of Kindred Terms, Etymological, Historical, Geographical and Discursive.* 2d ed. Ed. William Crooke. 1903; reprinted Delhi: Munshiram Manoharlal, 1968. **788**

Zelin, Madeleine. *The Magistrate's Tael: Rationalizing Fiscal Reform in Eighteenth-Century Ch'ing China*. Berkeley and Los Angeles: University of California Press, 1984. **197**

Zeng, Zhaoyue, et al., eds. *Report on the Excavation of Two Southern T'ang Mausoleums: A Summary in English*. Beijing: Cultural Objects Press, 1957. **81**

Zimmermann, Philipp. "Studien zur Religion der Ngadju-Dajak in Südborneo." *Ethnologica*, n.s., 4 (1968): 314–93. **703–4**

Zwalf, Wladimir, ed. *Buddhism: Art and Faith*. London: British Museum Publications, 1985. **719, 731**

General Index

Italic page numbers indicate that the topic appears in an illustration or in its caption on the cited page. Authors are listed in this index only when their ideas or works are discussed; full listings of authors' works as cited in this volume may be found in the Bibliographical Index (along with characters for Chinese, Japanese, and Korean authors and works). Chinese, Japanese, and Korean characters are provided in this index for terms, map titles, people, and works (except for those already given in the Bibliographical Index).

Library of Congress Cataloging-in-Publication Data

The History of cartography.

Includes bibliographical references and indexes.
Contents: v. 1. Cartography in prehistoric, ancient, and medieval Europe and the Mediterranean — v. 2, bk. 1. Cartography in the traditional Islamic and South Asian societies — v. 2, bk. 2. Cartography in the traditional East and Southeast Asian societies.
1. Cartography—History. I. Harley, J. B. (John Brian), 1932–91 II. Woodward, David, 1942–
GA201.H53 1987 526′.09 86-6995
ISBN 0-226-31633-5 (v. 1; alk. paper)
ISBN 0-226-31635-1 (v. 2, bk. 1; alk. paper)
ISBN 0-226-31637-8 (v. 2, bk. 2; alk. paper)

DATE DUE

Demco, Inc. 38-293